Safety Symbols

Safety symbols in the following table are used in the lab activities to indicate possible hazards. Learn the meaning of each symbol. **It is recommended that you wear safety goggles and apron at all times in the lab. This might be required in your school district.**

Safety Symbols	Hazard	Examples	Precaution	Remedy
Disposal	Special disposal procedures need to be followed.	certain chemicals, living organisms	Do not dispose of these materials in the sink or trash can.	Dispose of wastes as directed by your teacher.
Biological	Organisms or other biological materials that might be harmful to humans	bacteria, fungi, blood, unpreserved tissues, plant materials	Avoid skin contact with these materials. Wear mask or gloves.	Notify your teacher if you suspect contact with material. Wash hands thoroughly.
Extreme Temperature	Objects that can burn skin by being too cold or too hot	boiling liquids, hot plates, dry ice, liquid nitrogen	Use proper protection when handling.	Go to your teacher for first aid.
Sharp Object	Use of tools or glassware that can easily puncture or slice skin	razor blades, pins, scalpels, pointed tools, dissecting probes, broken glass	Practice common-sense behavior and follow guidelines for use of the tool.	Go to your teacher for first aid.
Fume	Possible danger to respiratory tract from fumes	ammonia, acetone, nail polish remover, heated sulfur, moth balls	Be sure there is good ventilation. Never smell fumes directly. Wear a mask.	Leave foul area and notify your teacher immediately.
Electrical	Possible danger from electrical shock or burn	improper grounding, liquid spills, short circuits, exposed wires	Double-check setup with teacher. Check condition of wires and apparatus. Use GFI-protected outlets.	Do not attempt to fix electrical problems. Notify your teacher immediately.
Irritant	Substances that can irritate the skin or mucous membranes of the respiratory tract	pollen, moth balls, steel wool, fiberglass, potassium permanganate	Wear dust mask and gloves. Practice extra care when handling these materials.	Go to your teacher for first aid.
Chemical	Chemicals that can react with and destroy tissue and other materials	bleaches such as hydrogen peroxide; acids such as sulfuric acid, hydrochloric acid; bases such as ammonia, sodium hydroxide	Wear goggles, gloves, and an apron.	Immediately flush the affected area with water and notify your teacher.
Toxic	Substance may be poisonous if touched, inhaled, or swallowed.	mercury, many metal compounds, iodine, poinsettia plant parts	Follow your teacher's instructions.	Always wash hands thoroughly after use. Go to your teacher for first aid.
Flammable	Flammable chemicals may be ignited by open flame, spark, or exposed heat.	alcohol, kerosene, potassium permanganate	Avoid open flames and heat when using flammable chemicals.	Notify your teacher immediately. Use fire safety equipment if applicable.
Open Flame	Open flame in use, may cause fire.	hair, clothing, paper, synthetic materials	Tie back hair and loose clothing. Follow teacher's instruction on lighting and extinguishing flames.	Notify your teacher immediately. Use fire safety equipment if applicable.

Eye Safety	Clothing Protection	Animal Safety	Radioactivity	Handwashing
Proper eye protection should be worn at all times by anyone performing or observing science activities.	This symbol appears when substances could stain or burn clothing.	This symbol appears when safety of animals and students must be ensured.	This symbol appears when radioactive materials are used.	After the lab, wash hands with soap and water before removing goggles.

GLENCOE

CHEMISTRY

MATTER & CHANGE

McGraw Hill Education

COVER: ElementalImaging/Getty Images

mheducation.com/prek-12

McGraw Hill Education

Send all inquiries to:
McGraw-Hill Education
STEM Science
8787 Orion Place
Columbus, OH 43240

ISBN: 978-0-07-677461-6
MHID: 0-07-677461-9

Printed in the United States of America.

1 2 3 4 5 6 7 8 9 DOW 20 19 18 17 16

CONTENTS IN BRIEF

Magnus Hjorleifsson/Getty Images

Welcome to

GLENCOE CHEMISTRY
MATTER & CHANGE

We are your partner in learning by meeting your diverse 21st century needs. Designed for today's tech-savvy high school students, the McGraw-Hill Education's *Chemistry: Matter and Change* program offers hands-on investigations, rigorous science content, and engaging, real-world applications to make science fun, exciting, and stimulating.

Quick Start Guide
Glencoe Chemistry | Teacher Center

Login information

1. Go to **connected.mcgraw-hill.com**
2. Enter your registered Username and Password.
3. For **new users** click here to create a new account
4. Get **ConnectED Help** for creating accounts, verifying master codes, set up classes and more.

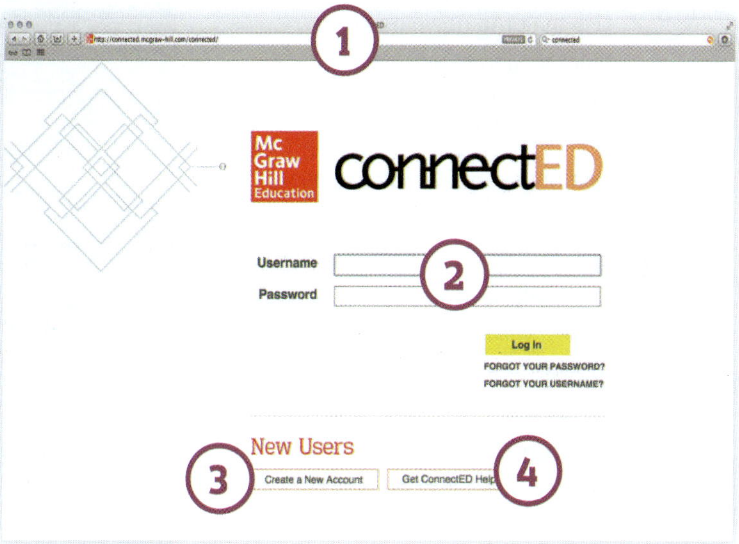

Your ConnectED Center

5. Scroll down to find the program from which you would like to work.

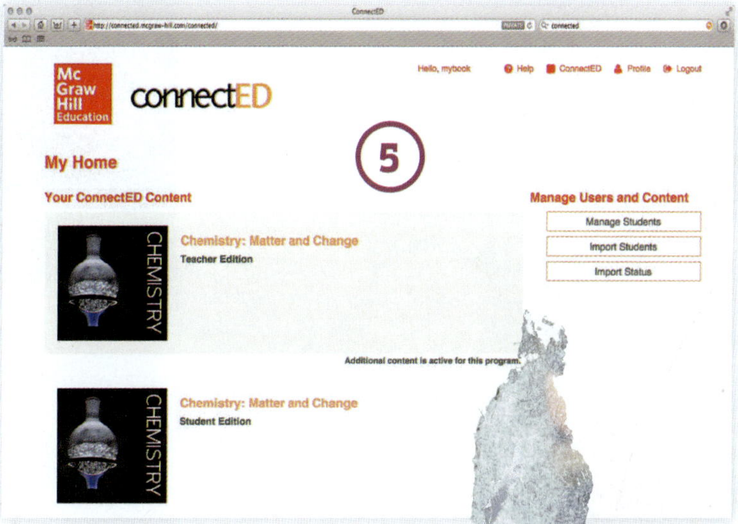

(l to r) ©Hero/Corbis/Glow Images, Rubberball/Mike Kemp/Getty Images, Hill Street Studios/Blend Images/Getty Images, ©Chris Ryan/age fotostock

Quick Start Guide
Glencoe Chemistry | Teacher Center

1. The Menu allows you to easily jump to anywhere you need to be.
2. Click the **program icon** at the top left to **return to the main page** from any screen.
3. **Select a Chapter and Lesson** Use the drop down boxes to quickly jump to any lesson in any chapter.
4. Return to your **My Home** page for all your **ConnectED** content.
5. The **Help** icon will guide you to online help. It will also allow for a quick logout.
6. **Search Bar** allows you to search content by topic or standard.
7. **Access the eBook** **Use** the **Teacher Edition** to see content and suggested answers.

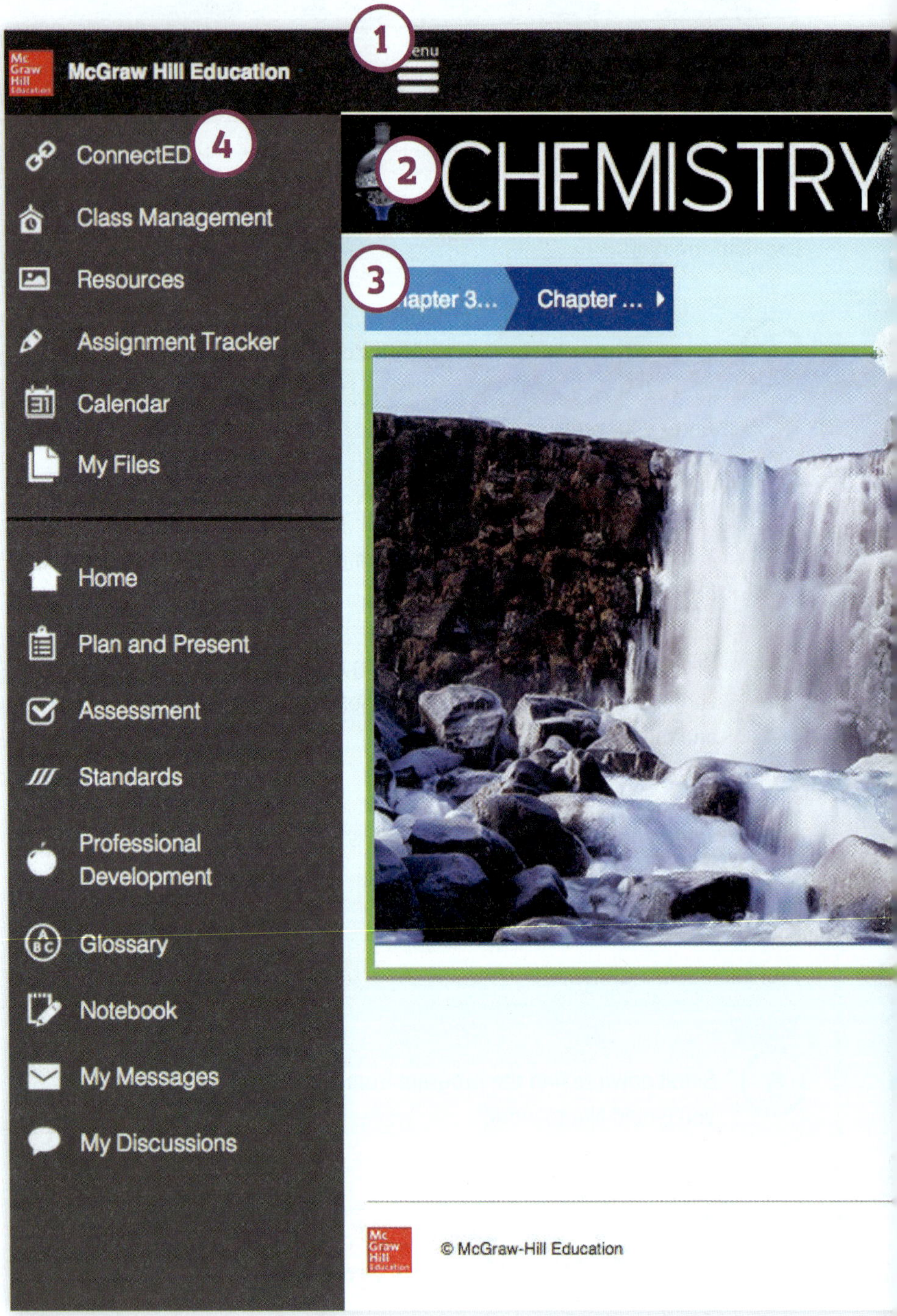

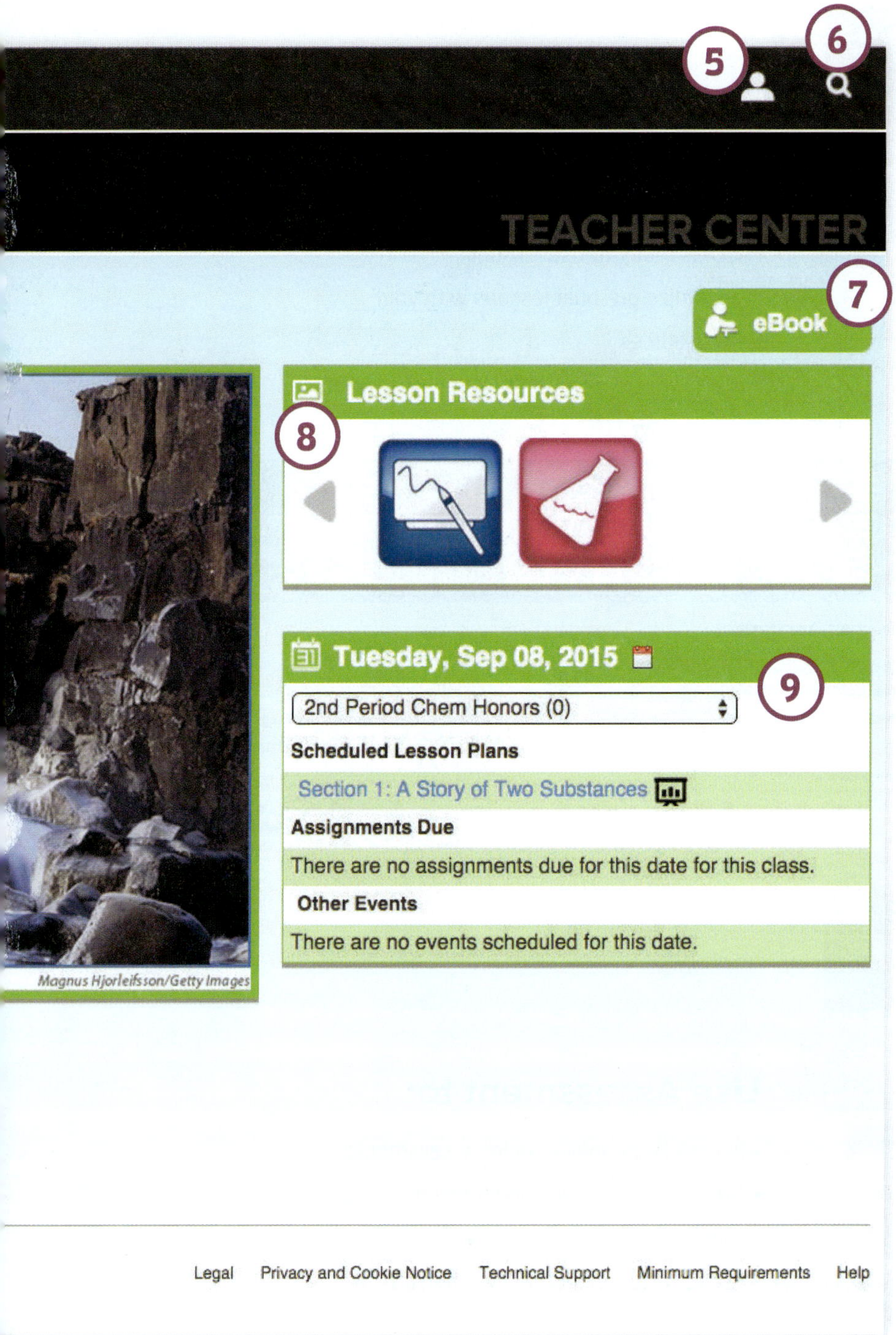

8 Quickly access helpful links to **multiple resources**, such as:

- LearnSmart
- Study Guide, WebQuest, Applying Practices
- Animations, Videos, Interactive tables
- Self-check, Quizzes, Tests
- Project-Based Learning Activities
- Lab Manuals, Safety Videos, Virtual Labs & other tools
- Vocabulary, Multilingual eGlossary, Vocab eGames, Vocab eFlashcard
- Personal Tutors
- Classroom Presentation Toolkit with Powerpoints
- Science Notebook
- Chemistry Math Handbook, Science and Engineering Practices Handbook
- Professional Development

9 **Daily Spotlight** is your preview of that day's lesson plans and assignments for each class.

Quick Start Guide
Glencoe Chemistry | Teacher Center

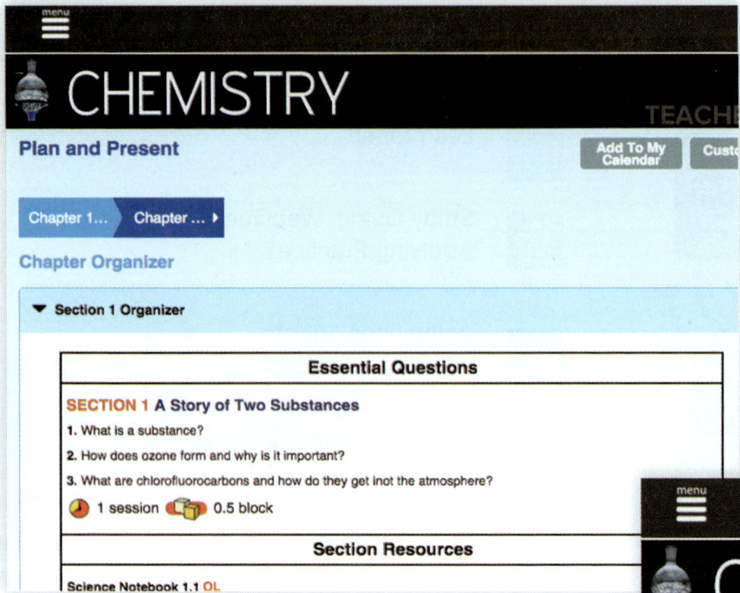

Use Plan and Present to:

- Review teacher instructional content
- Access ready-to-use lessons
- Customize pre-built lessons with your own resources

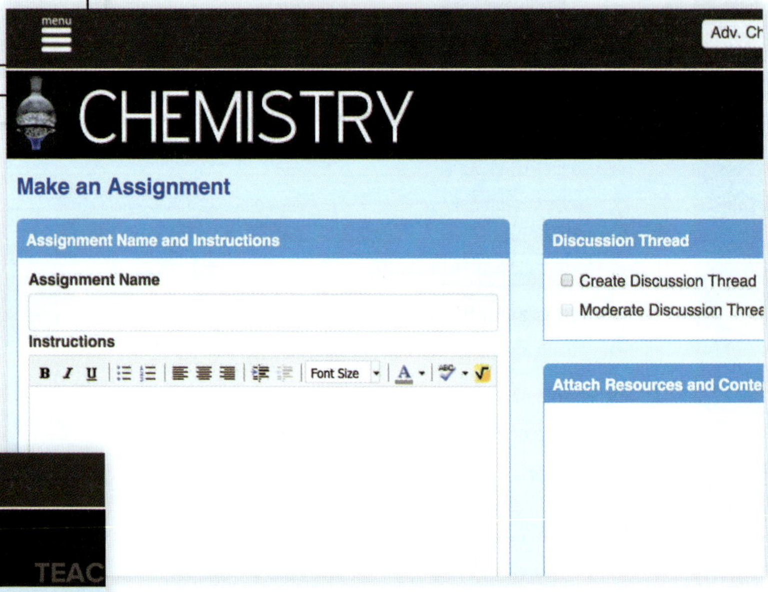

Use Assignment Tracker to:

- View recently created assignments
- Assign homework and activities
- Send and reply to student messages

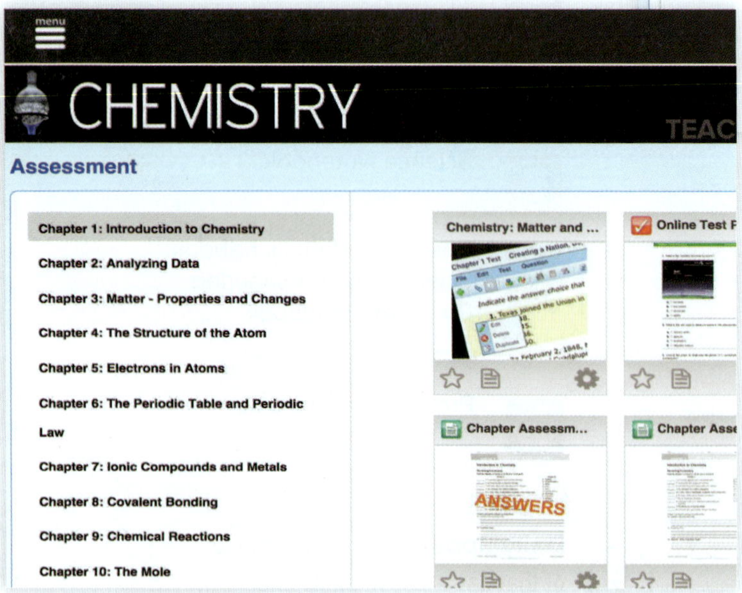

Use Assessment to:

- Access test banks to tailor assessments
- Create customized worksheets to assign online or print and distribute
- Prepare students for assessments with online testing

Mc Graw Hill Education

connected.mcgraw-hill.com

TEACHER HANDBOOK

Table of Contents

JUPITERIMAGES/Brand X/Alamy

Program Framework

Welcome to the Teacher Edition of *Chemistry: Matter and Change*. We have created this teacher edition based on input from experienced science teachers and educational consultants. Our goal is to provide you with research-based teaching strategies and activities, which are labeled for you at point-of-use.

Point-of-Use
- Strategies and activities apply directly to content.
- Concepts in Motion, Virtual Investigations, and Video icons indicate where your students can interact with tables and figures or use interactive online resources.
- Additional Practice Problems with each Example Problem provide additional student practice.

Review and Reinforcement
- Scaffolding—the introduction and reinforcement of skills and content—is incorporated throughout the lessons.
- Assessments check student understanding of key concepts and provide opportunities for reteaching at the end of each section.

Standards-Based Instruction

Differentiated Instruction
- Leveled activities and options for differentiated instruction help meet the needs of all your students, including English learners.

Hierarchical Structure
- **Level 1: Themes** The book is organized around themes—Patterns, Cause and Effect, Scale, Proportion, and Quantity, Systems and Sytem Models, Energy and Matter, Structure and Funtion, and Stability and Change.
- **Level 2: BIGIDEA** Each chapter has a Big Idea, which summarizes the content in an overarching statement.
- **Level 3: MAINIDEA** Each section of the chapter has a Main Idea that describes the focus of the section.

Assessment and Intervention
- Lessons provide standards practice.
- Assessments gauge student mastery of standards.
- Additional resources provide intervention options.

Differentiated Instruction

Activity Leveling

Teaching strategies and activities have been coded for ability-level appropriateness. A competency level is given for each activity using the following code:

AL Activities for students working above grade level

OL Activities for students working on grade level

BL Activities for students working below grade level

EL Activities for English learners

COOP LEARN Activities or strategies that promote cooperative learning

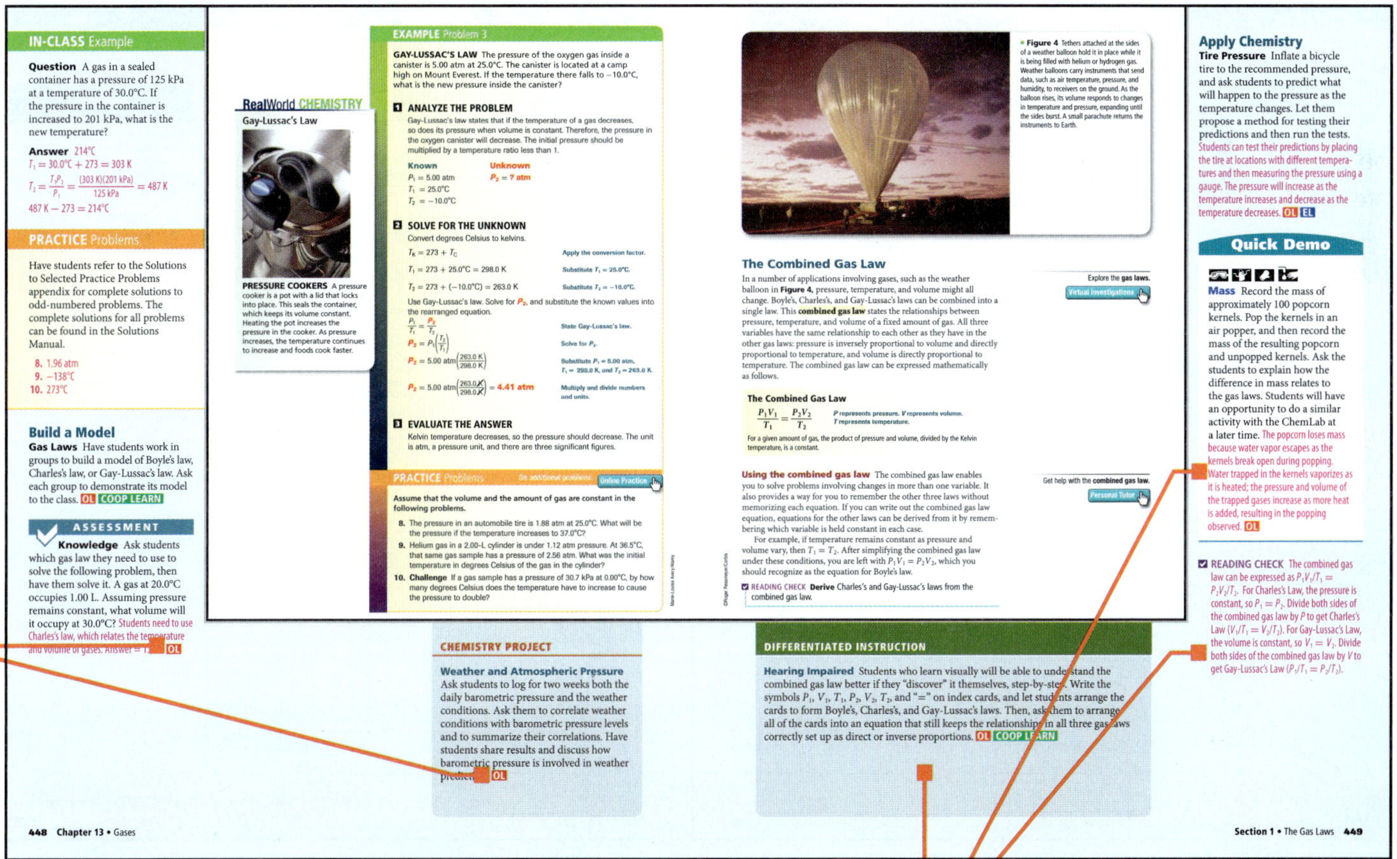

IN-CLASS Example

Question A gas in a sealed container has a pressure of 125 kPa at a temperature of 30.0°C. If the pressure in the container is increased to 201 kPa, what is the new temperature?

Answer 214°C

$T_1 = 30.0°C + 273 = 303 K$

$T_2 = \frac{T_1P_2}{P_1} = \frac{(303 K)(201 kPa)}{125 kPa} = 487 K$

$487 K - 273 = 214°C$

PRACTICE Problems

Have students refer to the Solutions to Selected Practice Problems appendix for complete solutions to odd-numbered problems. The complete solutions for all problems can be found in the Solutions Manual.

8. 1.96 atm
9. −138°C
10. 273°C

Build a Model

Gas Laws Have students work in groups to build a model of Boyle's law, Charles's law, or Gay-Lussac's law. Ask each group to demonstrate its model to the class. OL COOP LEARN

ASSESSMENT

Knowledge Ask students which gas law they need to use to solve the following problem, then have them solve it. A gas at 20.0°C occupies 1.00 L. Assuming pressure remains constant, what volume will it occupy at 30.0°C? Students need to use Charles's law, which relates the temperature and volume of gases. Answer = 1.■ OL

RealWorld CHEMISTRY

Gay-Lussac's Law

PRESSURE COOKERS A pressure cooker is a pot with a lid that locks into place. This seals the container, which keeps its volume constant. Heating the pot increases the pressure in the cooker. As pressure increases, the temperature continues to increase and foods cook faster.

EXAMPLE Problem 3

GAY-LUSSAC'S LAW The pressure of the oxygen gas inside a canister is 5.00 atm at 25.0°C. The canister is located at a camp high on Mount Everest. If the temperature there falls to −10.0°C, what is the new pressure inside the canister?

1 ANALYZE THE PROBLEM

Gay-Lussac's law states that if the temperature of a gas decreases, so does its pressure when volume is constant. Therefore, the pressure in the oxygen canister will decrease. The initial pressure should be multiplied by a temperature ratio less than 1.

Known	Unknown
$P_1 = 5.00$ atm	$P_2 = ?$ atm
$T_1 = 25.0°C$	
$T_2 = -10.0°C$	

2 SOLVE FOR THE UNKNOWN

Convert degrees Celsius to kelvins.

$T_K = 273 + T_C$ — Apply the conversion factor.

$T_1 = 273 + 25.0°C = 298.0 K$ — Substitute $T_1 = 25.0°C$.

$T_2 = 273 + (-10.0°C) = 263.0 K$ — Substitute $T_2 = -10.0°C$.

Use Gay-Lussac's law. Solve for P_2, and substitute the known values into the rearranged equation.

$\frac{P_1}{T_1} = \frac{P_2}{T_2}$ — State Gay-Lussac's law.

$P_2 = P_1\left(\frac{T_2}{T_1}\right)$ — Solve for P_2.

$P_2 = 5.00 \text{ atm}\left(\frac{263.0 K}{298.0 K}\right)$ — Substitute $P_1 = 5.00$ atm, $T_1 = 298.0$ K, and $T_2 = 263.0$ K.

$P_2 = 5.00 \text{ atm}\left(\frac{263.0 \cancel{K}}{298.0 \cancel{K}}\right) = 4.41 \text{ atm}$ — Multiply and divide numbers and units.

3 EVALUATE THE ANSWER

Kelvin temperature decreases, so the pressure should decrease. The unit is atm, a pressure unit, and there are three significant figures.

PRACTICE Problems See additional problems. Online Practice

Assume that the volume and the amount of gas are constant in the following problems.

8. The pressure in an automobile tire is 1.88 atm at 25.0°C. What will be the pressure if the temperature increases to 37.0°C?
9. Helium gas in a 2.00-L cylinder is under 1.12 atm pressure. At 36.5°C, that same gas sample has a pressure of 2.56 atm. What was the initial temperature in degrees Celsius of the gas in the cylinder?
10. **Challenge** If a gas sample has a pressure of 30.7 kPa at 0.00°C, by how many degrees Celsius does the temperature have to increase to cause the pressure to double?

Marie-Louise Avery/Alamy

Figure 4 Tethers attached at the sides of a weather balloon hold it in place while it is being filled with helium or hydrogen gas. Weather balloons carry instruments that send data, such as air temperature, pressure, and humidity, to receivers on the ground. As the balloon rises, its volume responds to changes in temperature and pressure, expanding until the sides burst. A small parachute returns the instruments to Earth.

The Combined Gas Law

In a number of applications involving gases, such as the weather balloon in **Figure 4**, pressure, temperature, and volume might all change. Boyle's, Charles's, and Gay-Lussac's laws can be combined into a single law. This **combined gas law** states the relationships between pressure, temperature, and volume of a fixed amount of gas. All three variables have the same relationship to each other as they have in the other gas laws: pressure is inversely proportional to volume and directly proportional to temperature, and volume is directly proportional to temperature. The combined gas law can be expressed mathematically as follows.

Explore the **gas laws.** Virtual Investigations

The Combined Gas Law

$$\frac{P_1V_1}{T_1} = \frac{P_2V_2}{T_2}$$

P represents pressure. *V* represents volume. *T* represents temperature.

For a given amount of gas, the product of pressure and volume, divided by the Kelvin temperature, is a constant.

Using the combined gas law The combined gas law enables you to solve problems involving changes in more than one variable. It also provides a way for you to remember the other three laws without memorizing each equation. If you can write out the combined gas law equation, equations for the other laws can be derived from it by remembering which variable is held constant in each case.

For example, if temperature remains constant as pressure and volume vary, then $T_1 = T_2$. After simplifying the combined gas law under these conditions, you are left with $P_1V_1 = P_2V_2$, which you should recognize as the equation for Boyle's law.

Get help with the **combined gas law.** Personal Tutor

READING CHECK **Derive** Charles's and Gay-Lussac's laws from the combined gas law.

©Roger Ressmeyer/Corbis

Apply Chemistry

Tire Pressure Inflate a bicycle tire to the recommended pressure, and ask students to predict what will happen to the pressure as the temperature changes. Let them propose a method for testing their predictions and then run the tests. Students can test their predictions by placing the tire at locations with different temperatures and then measuring the pressure using a gauge. The pressure will increase as the temperature increases and decrease as the temperature decreases. OL EL

Quick Demo

Mass Record the mass of approximately 100 popcorn kernels. Pop the kernels in an air popper, and then record the mass of the resulting popcorn and unpopped kernels. Ask the students to explain how the difference in mass relates to the gas laws. Students will have an opportunity to do a similar activity with the ChemLab at a later time. The popcorn loses mass because water vapor escapes as the kernels break open during popping. Water trapped in the kernels vaporizes as it is heated; the pressure and volume of the trapped gases increase as more heat is added, resulting in the popping observed. OL

READING CHECK The combined gas law can be expressed as $P_1V_1/T_1 = P_2V_2/T_2$. For Charles's Law, the pressure is constant, so $P_1 = P_2$. Divide both sides of the combined gas law by P to get Charles's Law ($V_1/T_1 = V_2/T_2$). For Gay-Lussac's Law, the volume is constant, so $V_1 = V_2$. Divide both sides of the combined gas law by V to get Gay-Lussac's Law ($P_1/T_1 = P_2/T_2$).

CHEMISTRY PROJECT

Weather and Atmospheric Pressure Ask students to log for two weeks both the daily barometric pressure and the weather conditions. Ask them to correlate weather conditions with barometric pressure levels and to summarize their correlations. Have students share results and discuss how barometric pressure is involved in weather predict■ OL

DIFFERENTIATED INSTRUCTION

Hearing Impaired Students who learn visually will be able to understand the combined gas law better if they "discover" it themselves, step-by-step. Write the symbols P_1, V_1, T_1, P_2, V_2, T_2, and "=" on index cards, and let students arrange the cards to form Boyle's, Charles's, and Gay-Lussac's laws. Then, ask them to arrange all of the cards into an equation that still keeps the relationships in all three gas laws correctly set up as direct or inverse proportions. OL COOP LEARN

448 Chapter 13 • Gases

Section 1 • The Gas Laws 449

Answers and Additional Support

- Along the side and bottom of the Teacher Edition, you will find answers to questions in the student edition,
- Demonstrations and activities that help you quickly and easily address key concepts,
- Teacher FYI and elements that provide you with additional content background information, and
- Differentiated Instruction strategies that help you meet the needs of all students.

Planning the Chapter

Planning pages appear at the beginning of each chapter.

Chapter Organizers detail the Big Idea, the Essential Questions from each section, standards covered, and resources and materials needed to teach the chapter.

Suggested Pacing provides lesson suggestions for the chapter. When used in conjunction with the Course Planning Guide starting on page 12T, you can tailor your instruction to the individual needs of your classes.

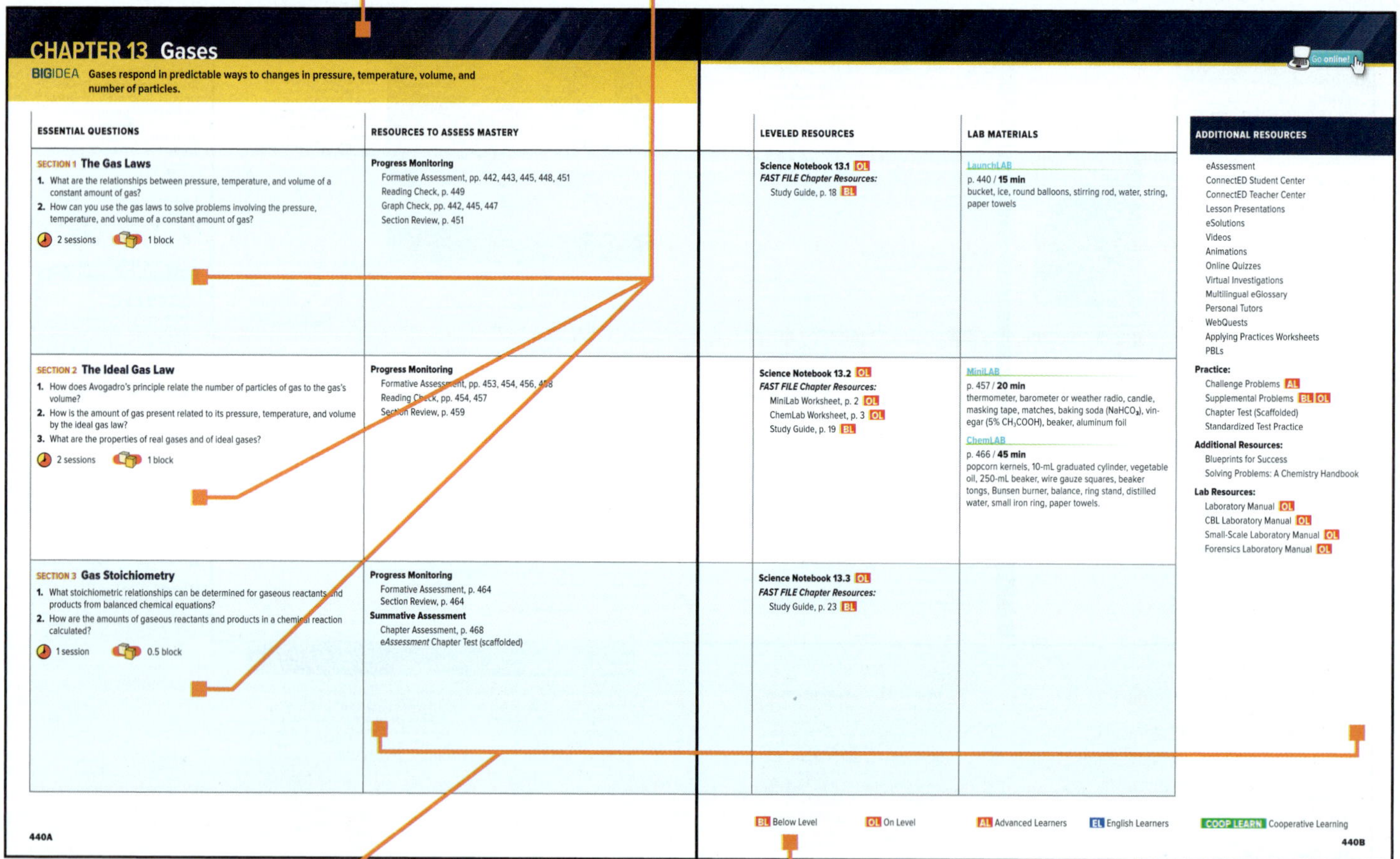

CHAPTER 13 Gases

BIGIDEA Gases respond in predictable ways to changes in pressure, temperature, volume, and number of particles.

ESSENTIAL QUESTIONS	RESOURCES TO ASSESS MASTERY	LEVELED RESOURCES	LAB MATERIALS
SECTION 1 **The Gas Laws** 1. What are the relationships between pressure, temperature, and volume of a constant amount of gas? 2. How can you use the gas laws to solve problems involving the pressure, temperature, and volume of a constant amount of gas? 2 sessions 1 block	**Progress Monitoring** Formative Assessment, pp. 442, 443, 445, 448, 451 Reading Check, p. 449 Graph Check, pp. 442, 445, 447 Section Review, p. 451	**Science Notebook 13.1** OL ***FAST FILE Chapter Resources:*** Study Guide, p. 18 BL	LaunchLAB p. 440 / **15 min** bucket, ice, round balloons, stirring rod, water, string, paper towels
SECTION 2 **The Ideal Gas Law** 1. How does Avogadro's principle relate the number of particles of gas to the gas's volume? 2. How is the amount of gas present related to its pressure, temperature, and volume by the ideal gas law? 3. What are the properties of real gases and of ideal gases? 2 sessions 1 block	**Progress Monitoring** Formative Assessment, pp. 453, 454, 456, 458 Reading Check, pp. 454, 457 Section Review, p. 459	**Science Notebook 13.2** OL ***FAST FILE Chapter Resources:*** MiniLab Worksheet, p. 2 OL ChemLab Worksheet, p. 3 OL Study Guide, p. 19 BL	MiniLAB p. 457 / **20 min** thermometer, barometer or weather radio, candle, masking tape, matches, baking soda ($NaHCO_3$), vinegar (5% CH_3COOH), beaker, aluminum foil ChemLAB p. 466 / **45 min** popcorn kernels, 10-mL graduated cylinder, vegetable oil, 250-mL beaker, wire gauze squares, beaker tongs, Bunsen burner, balance, ring stand, distilled water, small iron ring, paper towels.
SECTION 3 **Gas Stoichiometry** 1. What stoichiometric relationships can be determined for gaseous reactants and products from balanced chemical equations? 2. How are the amounts of gaseous reactants and products in a chemical reaction calculated? 1 session 0.5 block	**Progress Monitoring** Formative Assessment, p. 464 Section Review, p. 464 **Summative Assessment** Chapter Assessment, p. 468 *eAssessment* Chapter Test (scaffolded)	**Science Notebook 13.3** OL ***FAST FILE Chapter Resources:*** Study Guide, p. 23 BL	

Go online!

ADDITIONAL RESOURCES

eAssessment
ConnectED Student Center
ConnectED Teacher Center
Lesson Presentations
eSolutions
Videos
Animations
Online Quizzes
Virtual Investigations
Multilingual eGlossary
Personal Tutors
WebQuests
Applying Practices Worksheets
PBLs

Practice:
Challenge Problems AL
Supplemental Problems BL OL
Chapter Test (Scaffolded)
Standardized Test Practice

Additional Resources:
Blueprints for Success
Solving Problems: A Chemistry Handbook

Lab Resources:
Laboratory Manual OL
CBL Laboratory Manual OL
Small-Scale Laboratory Manual OL
Forensics Laboratory Manual OL

BL Below Level OL On Level AL Advanced Learners EL English Learners COOP LEARN Cooperative Learning

440A 440B

Resource Lists identify at a glance the lab materials, ancillaries, and technology resources needed to teach the chapter.

Leveling Key describes the differentiated instruction used throughout the Teacher Edition.

Teaching the Chapter

The Big Idea and Use the Photo at the beginning of the chapter help you teach the standards.

Introduce the BIGIDEA

Predictable Responses To introduce this chapter's Big Idea, lead students in a discussion about things that happen in a predictable way. Ask students what happens when you step on a car's gas pedal. The car accelerates. Ask students what happens when you change the pressure applied to the gas pedal. If you press lightly, the car accelerates slowly. If you press down harder, the car accelerates faster. Another way to demonstrate this concept would be to place a thermometer in hot and cold water and have students predict what will happen to the temperature reading on the thermometer. The temperature reading will rise in the hot water and decrease in the cold water.

Tie to Previous Knowledge Have students review the following concepts before studying this chapter.

- density
- balancing equations
- stoichiometry of reactions
- kinetic-molecular theory
- gas pressure

Use the Photo
Hot Air Balloons Ask students to predict why a hot air balloon needs a burner to stay aloft. The burner is used to heat the gas in the balloon. As temperature increases, both the volume and the pressure of the gas increase, enabling the balloon to stay aloft. In Section 1, students will read about Jacques Charles and Joseph Gay-Lussac, whose work will help students answer this question.

BIGIDEA activities help students understand the conceptual structure of the chapter—starting with the Big Idea overarching the chapter to the Main Ideas that are the focus of each section.

Use the Photo relates the chapter opener photograph to the chapter content, providing information and question prompts that will engage students.

The three-step teaching model in each section gives you direction when you want it and options to enrich the section or to meet the diverse needs of your students.

1 Focus The first step when teaching a set of concepts or skills is to focus the students' attention.

2 Teach Instruct students through a variety of active-learning strategies.

3 Assess Skill, Knowledge, and Performance assessment strategies provide a variety of assessment for all your students.

SECTION 1

1 Focus

MAINIDEA

Change in Variables Refer students to this section's Launch Lab and ask them to recall what happened to the balloon when it was placed in cold water. It decreased in size. Ask if the change in temperature affected the volume of the gas in the balloon and how they know. Yes, because the size of the balloon decreased, the volume of the gas decreased. Ask if the change in temperature affected the pressure of the gas in the balloon. Have them explain their answers. Yes. Because the sides of the balloon were less firm, the pressure on the gas decreased. BL OL AL

2 Teach

Concept Development
Gas Remind students that most of a gas is empty space. The volume of a sample of gas is much greater than the total volume of its particles. BL

ASSESSMENT
Performance Ask students to use a beaker, a graduated cylinder, a piece of rubber tubing, and water to design a procedure that can be used to measure the volume of gas in a balloon. One way students could measure the volume is by filling the cylinder with water, inverting it into a beaker full of water, releasing the air from the balloon into the cylinder using the tubing, and then recording the volume of water displaced in the cylinder. OL

GRAPH CHECK FIG. 1 4 L

SECTION 1

Essential Questions
- What are the relationships between pressure, temperature, and volume of a constant amount of gas?
- How can you use the gas laws to solve problems involving the pressure, temperature, and volume of a constant amount of gas?

Review Vocabulary
scientific law: describes a relationship in nature that is supported by many experiments

New Vocabulary
Boyle's law
absolute zero
Charles's law
Gay-Lussac's law
combined gas law

The Gas Laws

MAINIDEA For a fixed amount of gas, a change in one variable—pressure, temperature, or volume—affects the other two.

CHEM 4 YOU What might happen to the gas in a balloon if you decreased its volume by squeezing it? You would feel increasing resistance as you squeeze and might see part of the balloon bulge.

Boyle's Law

As the balloon example illustrates, the pressure of a gas and its volume are related. Robert Boyle (1627–1691), an Irish chemist, described this relationship between the pressure and the volume of a gas.

How are pressure and volume related? Boyle designed experiments like the one shown in **Figure 1.** He showed that if the temperature and the amount of gas are constant, doubling the pressure decreases the volume by one-half. On the other hand, reducing the pressure by one-half doubles the volume. A relationship in which one variable increases proportionally as the other variable decreases is known as an inversely proportional relationship.

Boyle's law states that the volume of a fixed amount of gas held at a constant temperature varies inversely with the pressure. Look at the graph in **Figure 1,** in which volume versus pressure is plotted for a gas. The plot of an inversely proportional relationship results in a downward curve.

Figure 1 As the external pressure on the cylinder's piston increases, the volume inside the cylinder decreases. The graph shows the inverse relationship between pressure and volume.

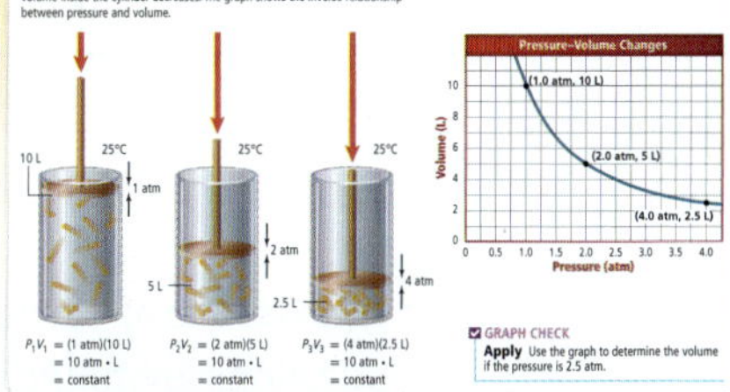

GRAPH CHECK
Apply Use the graph to determine the volume if the pressure is 2.5 atm.

Demonstration

Demonstrate Boyle's Law
Purpose
to demonstrate Boyle's law
Materials
Empty, colorless, 2-L soft drink bottle with cap; dropper; food coloring
Safety Precautions
Disposal Bottles and droppers can be reused or recycled. Rinse water and dye down a drain.

Procedure
Remove the label from an empty, colorless, 2-L soft drink bottle. Add water up to 4 cm from the top. Fill a dropper halfway with water colored with vegetable dye. Put the dropper into the 2-L bottle, and screw the cap on tightly. Tell students to observe closely.
Squeeze the sides of the bottle firmly until the dropper descends. Then, stop squeezing the bottle.

442 Chapter 13 • Gases

Assessment: Sections

Student activities and questions throughout the book provide opportunities for ongoing assessment and remediation.

Explore the gas laws with an interactive table. Concepts In Motion

Table 1 The Gas Laws

Law	Boyle's	Charles's	Gay-Lussac's	Combined
Formula	$P_1V_1 = P_2V_2$	$\frac{V_1}{T_1} = \frac{V_2}{T_2}$	$\frac{P_1}{T_1} = \frac{P_2}{T_2}$	$\frac{P_1V_1}{T_1} = \frac{P_2V_2}{T_2}$
What is constant?	amount of gas, temperature	amount of gas, pressure	amount of gas, volume	amount of gas
Graphic organizer	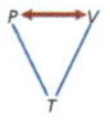	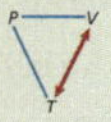	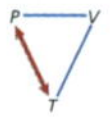	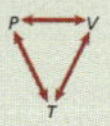

Temperature scales and the gas laws You might have noticed that the work done by Charles and Gay-Lussac preceded the development of the Kelvin scale, yet their laws require the use of temperature in kelvins. In the 1700s and early 1800s, scientists worked with several different scales. For example, a scale called the Réaumur scale was often used in France around Charles's time. On this scale—or any scale not based on absolute zero—the expression for Charles's law is more complex, requiring two constants in addition to V and T. The Kelvin scale simplified matters, resulting in the familiar gas laws presented here.

You have now seen how pressure, temperature, and volume affect a gas sample. You can use the gas laws, summarized in **Table 1**, as long as the amount of gas remains constant. But what happens if the amount of gas changes? In the next section, you will add the fourth variable, amount of gas present, to the gas laws.

SECTION 1 REVIEW

Section Self-Check

Section Summary

- Boyle's law states that the volume of a fixed amount of gas is inversely proportional to its pressure at constant temperature.
- Charles's law states that the volume of a fixed amount of gas is directly proportional to its Kelvin temperature at constant pressure.
- Gay-Lussac's law states that the pressure of a fixed amount of gas is directly proportional to its Kelvin temperature at constant volume.
- The combined gas law relates pressure, temperature, and volume in a single statement.

14. **MAINIDEA State** the relationships between pressure, temperature, and volume of a fixed amount of gas.
15. **Explain** Which of the three variables that apply to equal amounts of gases are directly proportional? Which are inversely proportional?
16. **Analyze** A weather balloon is released into the atmosphere. You know the initial volume, temperature, and air pressure. What information will you need to predict its volume when it reaches its final altitude? Which law would you use to calculate this volume?
17. **Infer** why gases such as the oxygen used at hospitals are compressed. Why must compressed gases be shielded from high temperatures? What must happen to compressed oxygen before it can be inhaled?
18. **Calculate** A rigid plastic container holds 1.00 L of methane gas at 660 torr pressure when the temperature is 22.0°C. How much pressure will the gas exert if the temperature is raised to 44.6°C?
19. **Design a concept map** that shows the relationships between pressure, volume, and temperature in Boyle's, Charles's, and Gay-Lussac's laws.

ASSESSMENT

Skill Have students determine what additional information is needed to solve this problem. If a sample of gas at 12.0°C and 1.06 atm pressure is moved to a 2.30-L container at 24.9°C, what is the final pressure of the gas? To use the combined gas law to solve this problem, students need to know the initial volume of the gas. OL

3 Assess

Check for Understanding

Have students write explanations that compare and contrast direct and inverse relationships. OL

Reteach

Ask students to explain why hot-air balloons rise. The heated air is less dense than the surrounding air and thus has less mass per unit volume. OL

Extension

Ask students to explain why the standard pressure of 1 atm is not always equal to the atmospheric pressure. Atmospheric pressure varies due to varying altitude and weather conditions. OL

SECTION 1 REVIEW

14. This relationship is given by the combined gas law: $P_1V_1/T_1 = P_2V_2/T_2$. For example: when the temperature increases, either the volume or pressure increases (or both).
15. P and V are directly proportional to T, and P and V are inversely proportional to each other.
16. You would need to know the final temperature and final pressure to calculate the final volume. Use the combined gas law.
17. A greater mass confined to a smaller volume makes transporting and storing of gases easier. Increasing temperature increases pressure, and the cylinders might explode. Before compressed oxygen can be breathed, it must be decompressed.
18. 711 torr
19. The concept map should show how P, V, and T are proportional to one another. Students should also label each pair of variables used in the gas laws. Refer to the Solutions Manual.

Section 1 • The Gas Laws 451

ASSESSMENT tasks are found throughout the section and provide opportunities to measure student understanding.

3 Assess provides an evaluation of a key concept and an activity to reteach for students struggling to meet that learning objective.

SECTION REVIEW provides students with summary statements and questions that tie to the **MAIN**IDEA and Essential Questions for that section.

Answers to all assessment questions are found in the Teacher Edition.

Assessment: Chapters

MIXED REVIEW and THINK CRITICALLY

require students to demonstrate higher-order thinking and use their writing skills.

Mastering Concepts and Mastering Problems

assess comprehension of the vocabulary and key concepts in each section.

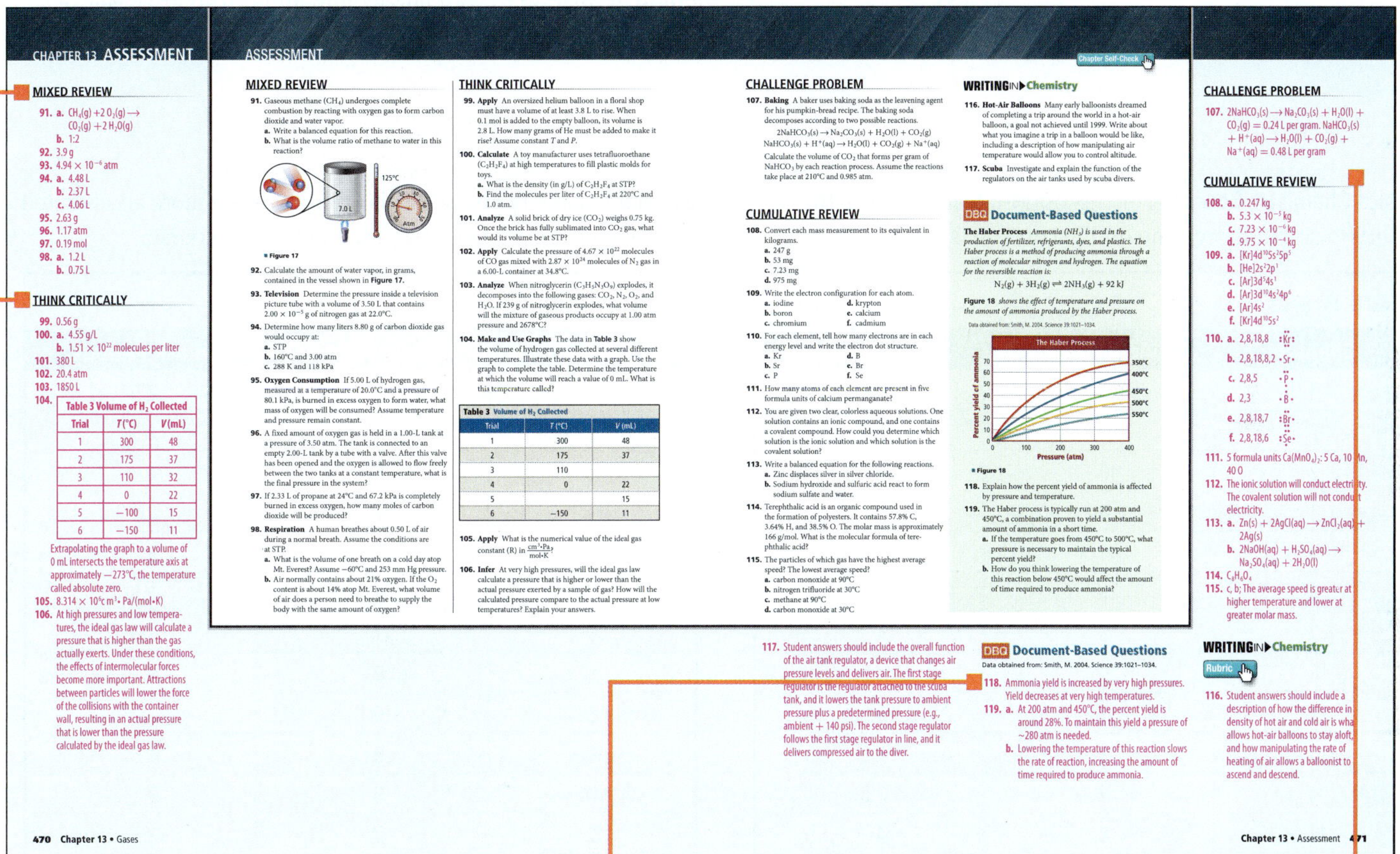

DBQ Document-Based Questions

connect students to real-world applications as they evaluate real data from current research. Students analyze graphs, charts, and other displays of data from recognized scientific journals and classic historic documents.

CUMULATIVE REVIEW

questions assess student retention of material from earlier chapters.

COURSE PLANNING GUIDE

Chemistry: Matter and Change provides a complete selection of core concepts that can be presented in a way to meet the needs of all your students. As the teacher, you are in the best position to design a chemistry course that meets the needs of individual students and classes, set the pace at which the content is covered, and determine what material should be given the most emphasis. To assist you in planning the course, the following Course Planning Guide is provided.

Chemistry: Matter and Change may be used in a full-year, two-semester program that is comprised of 160 periods of approximately 45 minutes each day. (Single Periods) Also included in the table is a half-year, one semester program that assumes the program will be taught in 80 class periods of approximately 90 minutes each day. (Block Periods)

Please remember that the planning guide is provided as an aid in planning the best course for your students. Use the guide in relation to your curriculum and the ability levels of the classes you teach.

Basics Chemistry Course—For this option, teachers have the ability to spend more time on the core areas.

Core Areas		
Chapter	**Single Periods**	**Block Periods**
1	5	2.5
2	5	2.5
3	10	5
4	8	4
5	7	3.5
6	5	2.5
7	8	4
8	9	4.5
9	7	3.5
10	9	4.5
11	7	3.5
12	8	4
13	7	3.5
14	9	4.5
15	8	4
16	8	4
17	7	3.5
18	8	4
19	5	2.5
20	7	3.5
21	8	4
22	9	4.5
23	8	4
24	8	4

General Chemistry Course—This option offers a more accel-erated pace and covers more material than the basic course.

Core Areas		
Chapter	**Single Periods**	**Block Periods**
1	5	2.5
2	5	2.5
3	8	4
4	7	3.5
5	6	3
6	5	2.5
7	7	3.5
8	8	4
9	6	3
10	9	4.5
11	6	3
12	7	3.5
13	6	3
14	7	3.5
15	8	4
16	7	3.5
17	6	3
18	7	3.5
19	5	2.5
20	6	3
21	7	3.5
22	8	4
23	7	3.5
24	7	3.5

Honors Chemistry Course—This option is designed to take students through the content at the depth and pace appropriate for an honors class. Time is built into the pacing for group and individual projects.

Core Areas

Chapter	Single Periods	Block Periods	Project
1	2	1	
2	3	1.5	
Individual Project (1)	5	2.5	**Earth's Atmospheric Ozone Levels** Have students research ozone levels in Earth's stratosphere throughout the past decade for North America and South America. Direct each student to develop a table to collect his or her data and then develop a graph to show the comparison of the data collected. Have the student note any trends or variances in ozone levels.
3	7	3.5	
4	6	3	
5	5	2.5	
6	4	2	
Group Project (1)	5	2.5	**Energy-Level Transitions** Direct the group to research an experimental method for analyzing an element's atomic emission spectrum to determine how much energy is involved in a particular energy-level transition. Work with the group to set up an apparatus that enables students to implement the method. Have the group compare its experimentally-obtained ΔE value with the accepted value. Prior to experimentation, be sure to assess any hazards and instruct students as to appropriate precautions, techniques, and protective measures.
7	6	3	
8	7	3.5	
9	5	2.5	
Individual Project (2)	5	2.5	**Solution Identification** Supply each student with a porcelain well plate, watch glasses, distilled or deionized water in a dispensing bottle, and five Beral pipettes labeled *Q1–Q5*. Each Beral pipette contains one of the following aqueous solutions, all approximately 0.1*M*; silver nitrate, hydrochloric acid, nitric acid, calcium chloride, and sodium carbonate. Through observation and experimentation, students will determine the identity of the solutions. Have each student write the chemical equations for all possible double-replacement reactions involving the five solutions and research the properties of all reactants and products in a chemistry handbook under "Physical Properties of Inorganic Compounds." Prior to investigation, assess any hazards, and instruct students about appropriate precautions, techniques, and protective measures.
10	8	4	
11	5	2.5	
12	6	3	
13	5	2.5	

Core Areas

Chapter	Single Periods	Block Periods	Project
Group Project (2)	10	5	**Quantitative Analysis** Assign each group a non-uniform mixture of three solid compounds: silver nitrate, sodium acetate, and sodium chloride. Provide 0.1*M* solutions of sodium chloride and silver nitrate. Have each student write the chemical equations for all possible double-replacement reactions involving the three compounds and two solutions. Students can research properties of all reactants and products in a chemistry handbook under "Physical Properties of Inorganic Compounds." Direct each group to devise and carry out a procedure that will enable them to determine the mass percent of each compound in the mixture. Prior to investigation, assess any hazards, and instruct students about appropriate precautions, techniques, and protective measures.
14	6	3	
15	7	3.5	
16	6	3	
17	6	3	
18	7	3.5	
19	5	2.5	
Individual Project (3)	5	2.5	**Preparation and Testing of a Buffer** Direct each student to prepare a given volume of an inorganic buffer solution of a given pH at 298K, the solution having a given total solution molarity (excepting the concentrations of hydrogen and hydroxide ions). Have the student devise and carry out a procedure to prepare the assigned buffer solution, then check the pH of the solution with a calibrated pH meter. If the pH differs significantly from the assigned value, have the student revise his or her calculations and procedure and prepare the solution correctly. Prior to experimentation, be sure to assess any hazards and instruct students as to appropriate precautions, techniques, and protective measures.
20	6	3	
21	7	3.5	
22	7	3.5	
23	7	3.5	
24	7	3.5	
Individual or Group Project	10	5	**Vitamin-C Analysis** Direct the group to research the chemical and structural formulas of vitamin C (ascorbic acid), its properties, and how the vitamin functions in the body. Have the group research the methodology for determining the vitamin-C content of various liquids by oxidation-reduction titration. Then, have students prepare the necessary solutions to carry out such an analysis and perform the analysis on liquids such as pineapple juice, orange juice, grapefruit juice, white grape juice, tomato juice, apple juice, milk, sports drinks, etc. Prior to experimentation, be sure to assess any hazards and instruct students as to appropriate precautions, techniques, and protective measures. Because all students will find the results interesting, have the group prepare and give a presentation to the entire class.

Safety in the Laboratory

The activities in *Chemistry: Matter and Change* are designed to minimize dangers in the laboratory. Careful planning and preparation as well as awareness of hazards can keep accidents to a minimum. Practice good laboratory housekeeping and management by observing these guidelines.

Personal Protection

The use of personal protection equipment is required when potentially hazardous material is present. Personal protection equipment includes eyewear, protective gloves, and laboratory aprons and coats.

Eyewear

Safety (Chemical Splash) goggles are required for science laboratory and field activities involving any hazardous chemical, which could cause damage if splashed or rubbed into the eye. Chemical Splash goggles provide eye protection from fine dusts, liquids, splashes, mists, and sprays. They also prevent splashes and sprays from body fluids or dangerous chemicals.

Safety (Chemical Splash) goggles should be large enough to protect and form a seal around the eyes. If not able to seal, goggles should contain side shields to prevent contamination to the eyes.

Eyewear should meet the ANSI Standard Z87.1-*Practice for Occupational and Educational Eye and Face Protection.* Eyewear meeting this standard will bear markings such as "Z87.1" on the frames, and the lens will be marked with the manufacturer's trademark.

Eye protection may also be provided with safety glasses. Safety glasses with side shields will not provide adequate protection from chemical splashes. They are designed primarily to protect the eyes from flying objects. Safety glasses should also have the "Z87.1" ANSI standard markings on the frames.

Protective Gloves

Gloves protect hands from heat, absorb perspiration, provide a shield from corrosive chemicals and body fluids, and prevent the transmission of microorganisms from person to person. Always check gloves to be sure that there are no tears, punctures, or holes. When removing gloves, peel the gloves off your hand, starting at the wrists and working toward the fingers. Keep the working surface of the gloves from contact with the skin during removal. (Some students and teachers are allergic to rubber or latex gloves. Alternatives must be provided in this case.)

Protective Gloves and Their Function

Glove Type	Function
Plastic	protects against light corrosives and irritants
Latex	provides protection against biological materials; should be changed as soon as they are soiled NOTE: *Some people may have an allergic reaction to latex, which can lead to serious medical problems.*
Natural Rubber	protects against electric shock and light corrosive material
Neoprene	use when working with solvents, oils, or light corrosive material
Cotton	Absorbs perspiration; wear under latex gloves.
Asbestos	insulates against heat NOTE: *Asbestos gloves are labeled with a warning about the danger of cancer. Asbestos is a known carcinogen.*

Laboratory Aprons and Coats

Laboratory aprons and coats are designed to protect clothing and skin from splashed and spilled chemicals and biological materials. They should fit the wearer properly to provide maximum protection. A laboratory coat or apron should be worn at all times in the laboratory.

Aprons are usually listed as "bib type," which are suitable for laboratory use. Aprons should be worn over clothing that covers the arms and body. Laboratory coats are usually fire retardant and made of cotton or paper. They are good for protection against flying objects, sharp or rough edges, splashes and spills, and fire. (Make sure aprons are the correct length in that they can present a trip fall hazard if too long.)

Fire Protection

Fire is one of the most frequent mishaps in the science laboratory. The first line of defense from a fire is fire prevention. Effective fire prevention centers on thorough understanding of combustion and the required ingredients. As long as air is present, oxygen will be available for combustion to take place. The areas where prevention measures are best exercised are the fuel and ignition sources.

Fires are classified by the chemical properties of the fuel. The basic classifications are grouped as follows.

- Class A – ordinary combustible (i.e., paper, wood)
- Class B – organic solvents or Flammables (i.e., acetone, alcohols, ethers)
- Class C – electrical wiring or static charges
- Class D – active metals (i.e., sodium, potassium, magnesium)

These symbols are accepted for the different classifications of fire. They are applied to fire extinguishers and extinguisher locations to indicate their suitability in extinguishing the different types of fires.

The following precautions should be taken to prevent fires from occurring in the science classroom, laboratory, storage, and preparation area.

- Be aware of ignition sources in your laboratory area (open flames, heat, and electrical equipment).
- Purchase and store flammable reagents in the smallest quantities possible.
- Do not store flammable liquids in standard refrigerators. An explosion-proof refrigerator should be used. Use appropriate signage on the door of the refrigerator— e.g.: For laboratory chemicals only, no edible foods allowed!
- Store flammable liquids in appropriate safety cabinets and/or safety cans.
- Do not store incompatible reagents together (e.g., acids with flammables).
- Do not store ethers for extended periods of time (no more than one year) as explosive peroxides can form.
- Make sure that all electrical cords are in good condition. All electrical outlets should be grounded and should accommodate a 3-pronged plug. All circuits should be GFCI protected. Do not use frayed or exposed wires missing insulation.

Each science classroom, laboratory, storage room and preparation area should have a fire blanket and an appropriate fire extinguisher.

Fire Extinguishers

In most school environments, hand-held, portable fire extinguishers are the first fire-extinguishing agent used. Therefore, a multipurpose ABC fire extinguisher must be located in each science classroom, laboratory, storage room, and preparation area. Extinguishers must be

- located in an open and highly visible area;
- inspected on a regular basis;
- used by well-trained teachers and students.

Fire extinguishers are labeled in accordance with NFPA standards. Have appropriate signage posted at levels easy to observe.

Fire Blankets

Actual fire control requires proper types of control devices such as a fire blanket. Fire blankets are made of specially treated fabric and should be located at strategic areas for all science laboratories where hazardous chemicals are stored and used. Students can use fire blankets if they are unable to reach the safety shower. Have appropriate signage posted at levels easy to observe.

Electrical Protection

Electrical safety needs must be considered for all new, old, and renovated science classrooms, laboratories, storage rooms, and preparation areas. Prevention should be the emphasis of electrical safety. Minimum considerations for electrical safety include the following.

- Ground-fault interrupters (GFI) should be installed to protect against major shock and electrical fires by preventing short circuits.
- All outlets must be grounded to prevent electrical accidents. Sufficient outlets should be provided to eliminate the need for extension cords. If floor boxes are used, they should not be located near water sources or areas where water is used.
- Surge protectors should be used to protect computers and other electronic devices from power surges.
- Emergency shut-off controls (electricity, gas, and water) should be located in an area that is easily accessible for laboratory occupants.
- Circuits should not be overloaded.
- Use only spark-free refrigeration in laboratories, storage rooms, and preparation areas for storage of flammable chemicals.
- Avoid the use of extension cords.
- Wiring should not have frayed or bare areas.

In developing the school's safety program, include provisions for handling electrical emergencies. All teachers of science should know where the master electrical cutoff switch and the control box are located, and how to operate both of these. Before an activity is conducted that requires the use of an electrical device, the teacher and students must be familiar with its operation and safety features.

Chemical Safety and Disposal

The teaching of chemistry requires the use of certain supplies and safety equipment if a safe classroom is to be maintained.

Safety Equipment

As a chemistry teacher, you should be familiar with the use and maintenance of the following equipment.

Fume Hoods

Each laboratory facility and preparation room where chemicals are used should be equipped with a fume hood. No open face fume hood with a low face velocity can provide complete safety against events that take place in the hood. However, a properly designed hood in a properly ventilated room can provide adequate protection. To provide proper protections, follow these recommended guidelines.

- Keep the interior light on so that the working area is properly illuminated.
- Check the exhaust system—air movement is inward and upward in the hood.
- Conduct all portions of the experiment that cause contaminants to form inside the hood.
- Large objects should not be placed directly on the hood's working surface. Use blocks under the large object to allow proper airflow under the object.
- Move the vertical sash to the lowest position that allows access so that manipulation is possible. The sash should protect the head and upper body in case of an explosion. Safety (Chemical Splash) goggles are required.
- Do not place your head in the hood when contaminants are being generated.
- Do not use the fume hood as a waste disposal unit except for very small amounts of volatile chemicals.
- Do not store chemicals or apparatuses in the hood.
- Keep the slots in the hood baffle (air intake) free from obstructions.
- Keep the laboratory door closed unless the manufacturer indicates otherwise.
- Remove the sash or panels only when necessary for apparatus setup. Do not operate until sash has been replaced.
- Do not place electrical receptacles or other sources that may produce a spark in the fume hood when using flammable chemicals or when gases may be present.
- The hood's sash should be marked for appropriate closure point when it is necessary to partially close the sash during an operation.
- The sash should be closed when the hood's exhaust system is not operating.
- Provide regular maintenance on the hood's exhaust system. Use static pressure gauges on the hood throat and across filters in the exhaust system to ensure proper exhaust flow.

Eyewashes

For a student or teacher who has hazardous material in the eye, the first response prior to medical treatment is flushing with tepid water to dilute chemicals, wash out debris, or irrigate the eyes. It is very important that the eyelids are held open and the eyeballs rolled so water can flow on all surfaces and in the folds surrounding the eyeballs. **Note:** Squeeze bottles are not sufficient and should not be used as eyewashes in any science laboratory.

An eyewash that can wash both eyes simultaneously should be located in every science laboratory, classroom, and preparation room where hazardous chemicals are used. To ensure that the eyewash station in your science facility will meet safety requirements, it should

- be located no more than 10 seconds from any student work station;
- comply (one eyewash) with ADA regulations on accessibility;
- be provided sufficient water pressure to operate correctly; flushing fluid not less than 1.5 L/min (0.4 gpm) for 15 minutes;
- wash both eyes simultaneously;
- supply an instant flow of recirculated tempered water continuously for at least 15 minutes;
- have a water control valve that remains on, allowing the user to use both hands;
- be clearly marked (with appropriate signage) and unobstructed for immediate use.

Note: Eyewashes should be flushed for 5 minutes once a week to remove any harmful contaminant that may form or grow in the eyewash.

Safety Showers

A safety shower should be found in any laboratory where hazardous chemicals are used. Safety showers

- must meet the standards for height, spray pattern, tepid water temperature, and water flow of 20 gallons per minute at 30 psi;
- should have a control valve that can remain on without requiring the use of the operator's hands;
- should be located no more than 10 seconds away from any student work station;
- should each be marked with a highly visible sign;
- should be large enough to accommodate the injured person and a teacher assisting with the emergency;
- should have a fixed valve handle or a chain with a large ring that can be pulled down to start the flow of water;
- must have sufficient water flow and pressure to function properly for immediate use;
- should be flushed once a week to eliminate contamination and check for proper working conditions.

Tepid water should not exceed the temperature of a person's eyeball (approximately 29°C) on combination eyewash/shower systems. Mixing valves for tepid water to eyewashes or eyewash/shower systems should have a minimum flow of 2 gallons per minute and a maximum flow of 60 gallons per minute. (Floor drains are not necessarily required. However, water on the lab floor can cause slip/fall and electrical issues. Use caution.)

Ventilation

A well-maintained ventilation system is an important contributor to a healthy environment in science classroom and laboratories. According to NFPA 45, forced ventilation at a minimum rate of more than 8 room changes of air per hour should be provided for science laboratories. Unoccupied labs and chemical storage areas require 4 room changes of air per hour. All exhausts should be vented to the outside of the building, not recirculated in the building's ventilation system.

Chemical storage rooms need systems that vent directly outside, usually to the roof, and away from fresh-air intake pipes. Stage cabinets for flammables should not be ventilated to the outside.

Every science room should be equipped with exhaust fans designed for rapid venting of smoke or bad odors created by an investigation.

Mary Kate Denny/PhotoEdit

Material Safety Data Sheets

Every chemical manufacturer is required by law to supply a recent Material Safety Data Sheet (MSDS) with each chemical it produces. These should be sent with each chemical that is purchased by a school or school district. If MSDS forms are not available, request them from the chemical manufacturer. Every teacher needs to know and understand the information on an MSDS. The following information should be listed on MSDS forms for science teachers.

- product identification
- hazardous components
- physical data
- fire and explosion hazards
- health hazards
- fire and explosion data
- spill and disposal procedures
- protective equipment
- storage and handling procedures
- transportation data and additional information

MSDS information should be reviewed with students for hazardous chemicals prior to working on the laboratory activity or experiment.

Chemical Storage

Safe chemical storage is based on the chemical properties of the substances that are to be stored. The hazards of the chemicals are closely associated with their chemical and physical properties. Proper storage of chemicals should follow these guidelines.

1. The chemicals must be properly labeled with the identity of the contents, hazards, and manufacturer's name and address.
2. The chemicals must be stored in compatible families.
3. The chemical storage room should be located so that damage and human injury will be minimal if an explosion or fire occur.
4. The chemical storeroom must have two exits and doors that lock.
5. The doors should be labeled *Authorized Personnel Only* or *Hazardous Materials.*
6. The storage room must have continuous (24/7) forced air ventilation that is vented to the outside away from air intakes.
7. Storage cabinets and shelves must be resistant to corrosion.
8. Chemicals must be stored in an upright position and no more than two or three containers deep.
9. Shelves should be equipped with a lip to prevent chemicals from being jarred off the shelf.
10. Corrosives (acids and bases) should be stored in an approved corrosive cabinet.
11. Nitric acid should be stored separate from other acids. Acids should not be stored on bare metal surface or in cabinets with flammable liquids. Store acids on plastic trays.
12. Flammables should be stored in an approved flammable cabinet.
13. Chemicals should not be stored above eye level and should never be stored on the floor.
14. Water-reactive chemicals (metals) should be stored where they will remain dry.
15. Proper safety equipment must be clearly marked inside the storeroom.
    - ABC fire extinguisher
    - Fire blanket
    - Safety goggles
    - Spill kit
    - First aid kit
16. The room must be adequately lighted.
17. Smoke detectors should be present.
18. An ongoing chemical inventory should be maintained.

Chemical Disposal

There are several important steps that should be taken prior to proceeding with chemical disposal. First, deter-mine how many chemicals are designated for disposal.
Remove

- out-of-date or contaminated chemicals;
- chemicals without legible labels;
- chemicals that are too hazardous for student use.

Then, make arrangements with an appropriate disposal company.

Options for Proper Chemical Disposal

Contact commercial chemical disposal companies in your area. Many waste disposal companies recycle chemicals and resell them. Other options may be available by contacting

- other schools in your area and combining quantities for disposal;
- industries in the area for assistance in disposal;
- institutions of higher education; they may allow you to use their system of disposal;
- the Natural Resources Conservation Commission in your state;
- local health department or water treatment plant superintendent.

Disposal of Liquid Waste

A common question from chemistry teachers is about which liquids may safely be disposed of in the sanitary drains. First, the teacher must be certain that the sewer flows to a wastewater-treatment plant and not to a stream or other natural waterway. Second, any substance from a laboratory should be flushed with at least 100 times its own volume of tap water. Third, disposal methods should be checked with local authorities because local regulations are often more stringent than federal requirements. The book *Prudent Practices in the Laboratory* published by the National Research Council, Washington, DC: National Academy Press, 2011, lists many substances that can be disposed of in the sanitary drain.

Some positive and negative ions from the *Prudent Practices* lists are given in the table shown on the next page. Note that it is important that both the positive and negative ion of a salt be listed in order for its drain disposal to be considered safe. Also note that although hydrogen and hydroxide ions are listed, acids and bases should be neutralized before disposal. A good rule of thumb is that nothing of pH less than 3 or greater than 8 should be discarded without neutralizing it first.

Prudent Practices	
Positive Ions	**Negative Ions**
aluminum	borate
ammonium	bromide
bismuth	carbonate
calcium	chloride
copper	hydrogen sulfate
hydrogen	hydroxide
iron	iodide
lithium	nitrate
magnesium	phosphate
potassium	sulfate
sodium	sulfite
strontium	tetraborate
tin	
titanium	
zinc	

Disposal of Organic Waste

Of the organic compounds most often found in high school laboratories, the following can be disposed of in the drain: methanol, ethanol, propanols, butanols, pentanols, ethylene, glycol, glycerol, sugars, formaldehyde, formic acid, acetic acid, oxalic acid, sodium and potassium salts of carboxylic acids, esters with fewer than five carbon atoms, and acetone. More extensive lists can be found in *Prudent Practices.*

Disposal of Other Waste

Other chemicals can be properly treated and disposed of in one of the following ways.

- Treat waste chemically to convert it to a form that is drain-disposable. A good example is the iodate ion, which is too strong an oxidizing agent to be disposed of untreated. However, it is readily reduced to iodide (disposable) by acidified sodium hydrogen sulfite. Many procedures for processing laboratory waste to a discardable form can be found in *Prudent Practices.*

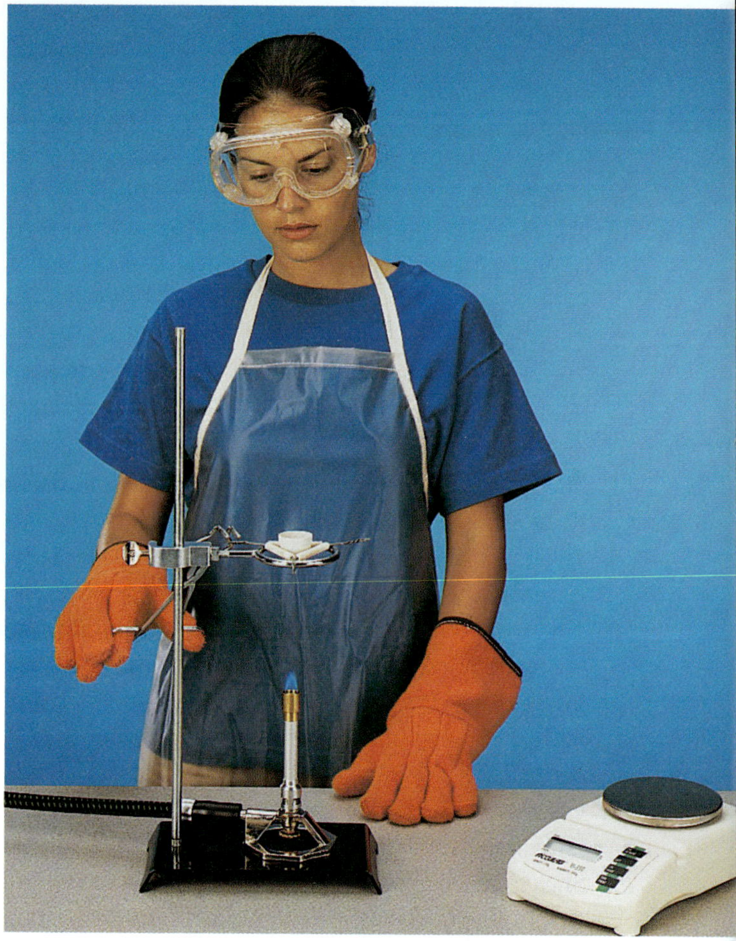

- Recycle waste. A good example is the recovery of valuable metals such as silver. Solvents can be recycled through distillation.
- If a waste cannot be recycled or processed to disposable form, then it must be packed and shipped by a Department of Transportation-approved shipper to a landfill designated to receive chemical and hazardous waste.

Matt Meadows

Microscale Chemistry

Microscale science techniques might seem out of place in a discussion of laboratory safety. However, microscale techniques offer a pedagogically sound approach to science laboratory investigations and provide simple solutions to many safety issues.

Major concepts emphasized in laboratory instruction have shifted from a simple reduction in amounts of chemicals used to a reeducation approach. The primary construct put forth by advocates of microscale chemistry is "If you don't need that much, then don't use it." The purpose is to move away from the throwaway mentality in our society. One way to express to students that the prevention of environmental pollution begins at the source is to reduce the amount of materials being wasted while teaching them basic science concepts.

Hazardous-waste management is a problem confronting all educational institutions. Schools are finding it increasingly difficult to fund the disposal of hazardous waste while attempting to maintain a hands-on chemical learning environment. Microscale experiments make it possible to promote critical-thinking skills while addressing major safety issues for a safe learning environment. Advantages of incorporating microscale/small scale chemistry into the learning environment include the

- reduction in the cost of chemicals;
- reduction of possible fire and explosion danger;
- reduction of chemical waste disposal costs;
- introduction of experiments where chemicals once thought to be too expensive or hazardous can be made available with the reduced amounts;
- recycling products for future experiments (plan ahead and use it again);
- introduction of less expensive glassware;
- reduction of exposure to toxic materials.

Hill Street Studios/Getty Images

Safety Advantages

Using microscale techniques means that dispensing liquids from plastic dropper bottles greatly reduces many spills. Accidental glass container breakage and fire hazards are greatly reduced, and air quality is improved due to smaller amounts of escaping vapor. Although some liquid is lost due to vaporization during both techniques, the microscale technique reduces hazardous organic chemicals that must be recovered and disposed of properly. In addition, this technique represents a responsible step in reducing environmental pollutants. In the science storeroom, chemicals in smaller containers (approximately 10% of the size used in macroscale) reduce the need for storage space.

Chemical Spills

If a chemical spill occurs in the laboratory or preparatory room, quick action by the teacher can reduce the possibility of injury to a student or themselves. A chemical spill such as a liter bottle of hydrochloric acid breaking in the laboratory is considered a major spill. The teacher should

- immediately evacuate all students through the exits farthest from the spill; fumes from a chemical spill can cause severe damage to the body;
- immediately assist any person splashed with the chemical to the safety shower or eyewash as appropriate;
- turn on emergency exhaust fan;
- contain the spill wearing proper protective clothing; do not allow the spill to trap you;
- call for help; the school safety plan should contain agencies or departments in your community that will assist in containment and removal of the chemical; have appropriate telephone numbers posted to call for help in case of an emergency.

Spill-Control Materials

There are many types of commercial materials that have been developed for containment and removal of chemical spills. They range from absorbent pads that quickly absorb chemicals in a liquid state, to porous bags filled with an amorphous silicate, to materials that neutralize an acid or a caustic spill.

Schools typically have a plastic, 5-gallon bucket filled with dry sand or dry clay materials. These do not neutralize an acid or base, but they do absorb the liquid or contain it in a small area. The disadvantage of using sand is that it is heavy and difficult to transport.

Once the chemical has been contained and neutralized, use clean up equipment made of plastic or polypropylene so that the equipment does not react with any of the chemical that remains. Place the contaminated material in plastic bags or containers and mark appropriately. Inform the custodial staff of the material so that it can be disposed of properly.

Disclaimer

McGraw-Hill Education makes no claims to the completeness of this discussion of laboratory safety and chemical storage. The information presented is not all-inclusive, nor does it address all of the hazards associated with the handling, storage, and disposal of chemicals, or with laboratory practices and management.

Solution Preparation

Many of the chemical reactions you will demonstrate—as well as many that your students will perform in ChemLabs, MiniLabs, and Launch Labs—take place in solution. Fortunately, you can prepare these solutions quickly and safely. **WARNINGS:** ***Read the HAZARD ALERT label on every chemical and follow all recommended safety precautions when using the chemical. When preparing solutions, wear chemical-resistant gloves, a chemical-resistant apron, and chemical (splash) safety goggles and a full-face shield if appropriate. In addition, toxic fumes given off by some chemicals, especially concentrated acids, require that you open and use them only in a fume hood or in a room with adequate (more than 8 room air changes per hour) ventilation.***

The following equipment will enable you to prepare solutions in a correct, safe, and time-efficient manner: balance; ovenproof glass beakers (1000 mL, 600 mL, 400 mL, 250 mL, 150 mL, 100 mL); glass graduated cylinders (1000 mL, 500 mL, 100 mL, 50 mL); ovenproof glass volumetric flasks (1000 mL, 500 mL, 250 mL, 100 mL); standard-stem glass funnels; polypropylene powder funnels; glass stirring rods; polyethylene wash bottle (for distilled or demineralized water). Additional equipment that can further reduce your time and effort include a magnetic stirrer, magnetic stirring bars, and a stirring-bar retriever.

Because tap water contains minerals and dissolved gases that can lead to unwanted reactions, you should prepare most solutions using only distilled or demineralized (deionized) water. Distilled water is free of inorganic materials, suspended impurities, and most organic contaminants. Demineralized water, which is less expensive than distilled water, is free of inorganic materials and most suspended contaminants and is suitable for most solutions. Many laboratory supply companies carry demineralizer cartridges that enable you to demineralize tap water in your school laboratory.

You will find that the solutions needed for the demonstrations and activities in *Chemistry: Matter and Change* are specified by amount and solute concentration. Concentration units include molarity (*M*), mass percent, and volume percent.

Molarity

Procedure for preparing a solution by dissolving a solid chemical in water

Example: Prepare 100.0 mL aqueous 0.0500*M* $CuSO_4$ solution using solid $CuSO_4{\cdot}5H_2O$.

1. Calculate moles of $CuSO_4{\cdot}5H_2O$ required.

   $M = \dfrac{\text{moles of solute}}{\text{liters of solution}}$

   moles of solute = $M \times$ liters of solution

   moles of solute = 0.0500 moles/L × 0.100 L = 0.00500 mol

2. Calculate mass of $CuSO_4{\cdot}5H_2O$ needed.

   $$0.00500 \text{ mol } CuSO_4{\cdot}5H_2O \times \frac{249.68 \text{ g } CuSO_4{\cdot}5H_2O}{1 \text{ mol } CuSO_4{\cdot}5H_2O} = 1.25 \text{ g } CuSO_4{\cdot}5H_2O$$

3. Mass out 1.25 g $CuSO_4{\cdot}5H_2O$ on a balance. Use a powder funnel to add the solid to a 100-mL volumetric flask. Add approximately 75 mL distilled or demineralized water. Swirl or stir until all the solid dissolves. If you used a magnetic stirrer, retrieve the stirring bar and use a wash bottle to rinse the bar and retriever, being sure the rinse water goes into the flask. Add water up to the 100-mL mark on the flask. If the dissolution of a solute is highly exothermic or highly endothermic, wait until the solution returns to room temperature before adding water up to the mark.

Laboratory Solutions

Example: Prepare 500.0 mL aqueous 3.0*M* H_2SO_4 by diluting concentrated (18.0*M*) H_2SO_4.

1. Let M_1 and V_1 represent the molarity and volume of the starting, more concentrated solution. Let M_2 and V_2 represent the molarity and volume of the final, more dilute solution. Substitute M_1, M_2, and V_2 into the equation $M_1V_1 = M_2V_2$ and solve for V_1.
   $(18.0M)(V_1) = (3.0M)(500.0\text{ mL})$
   $V_1 = 83\text{ mL}$
2. Fill a 500-mL volumetric flask approximately one-half full with distilled or demineralized water. Slowly and carefully add 83 mL 18*M* H_2SO_4 to the water in the flask. Swirl or stir to mix the two liquids. **WARNING:** ***Always add a concentrated acid to water, never the reverse.*** After the solution has cooled to room temperature, add water up to the 500-mL mark on the flask.

The molarities of other commonly used, concentrated acids and bases are given below.
ammonium hydroxide (NH_4OH), 14.8*M*
hydrochloric acid (HCl), 12*M*
nitric acid (HNO_3), 15.8*M*

Mass Percent

Mass percent solutions are determined by the grams of solute per 100 grams of solution.

Procedure for preparing a solution of a given mass percent solute

Example: Prepare 300.0 g of aqueous sodium chloride solution that is 10.0% NaCl by mass.

1. Calculate the mass of NaCl required.
   $$\frac{10.0\text{ g NaCl}}{100\text{ g solution}} \times 300.0\text{ g solution} = 30.0\text{ g NaCl}$$
2. Calculate the mass of water required.
   300.0 grams solution – 30.0 g NaCl = 270.0 g water
3. Prepare the solution by dissolving 30.0 grams NaCl in 270.0 g water. *Note: Because water's density equals 1.00 g/mL at 4°C, and is close to 1.00 g/mL at room temperature, you can measure 270.0 mL of water in a graduated cylinder rather than weigh 270.0 g on a balance.*

Volume Percent

Volume percent solutions are determined by milliliters of solute per 100 mL of solution.

Procedure for preparing a solution of a given volume percent solute

Example: Prepare 75.0 mL of aqueous ethanol solution that is 50.0% C_2H_5OH by volume.

1. Calculate the volume of C_2H_5OH required.
   $$\frac{50.0\text{ mL }C_2H_5OH}{100\text{ mL solution}} \times 75.0\text{ mL solution} = 37.5\text{ mL }C_2H_5OH$$
2. Prepare the solution by mixing 37.5 mL C_2H_5OH with sufficient water to make 75.0 mL of solution.

SOLUTIONS PREPARATION

Quantities are for a class of 30 students.

Chapter	Lab Type	Solution	Preparations
1	ChemLAB	hard water	To make 127 mg/L (ppm) hard water, dissolve 0.304 g of anhydrous calcium chloride in 500 mL of distilled water. Dissolve 0.139 g of magnesium chloride hexahydrate in 500 mL of distilled water. When both substances are thoroughly dissolved, combine the two solutions and stir.
1	ChemLAB	soft water	To prepare soft water (50 mg/L), add 590 mL of hard water to 910 mL of distilled water.
3	LaunchLAB	3*M* hydrochloric acid	Add 100 mL concentrated (12*M*) hydrochloric acid to 300 mL distilled water while stirring. **WARNING:** ***Do NOT add the water to the acid.***
3	ChemLAB	1*M* silver nitrate ($AgNO_3$)	Dissolve 170 g $AgNO_3$ in 600 mL distilled water. Make up to 1 L final volume with distilled water. **WARNING:** ***Silver nitrate is toxic and harmful to skin and clothing.***
6	Demonstration	phenolphthalein indicator solution	Dissolve 0.1 g phenolphthalein powder in 70 mL 95% ethanol. Make up to 100 mL final volume with distilled water.
6	ChemLAB	1.0*M* hydrochloric acid	Add 25 mL concentrated (12*M*) hydrochloric acid to 275 mL distilled water while stirring. **WARNING:** ***Do NOT add the water to the acid.***
7	Demonstration	Ammonium metavanadate	Dissolve 0.7 g ammonium metavanadate (NH_4VO_3) in 150 mL distilled water. **WARNING:** ***Solution is moderately toxic.***
9	ChemLAB	1*M* copper(II) nitrate ($Cu(NO_3)_2$)	Dissolve 242 g $Cu(NO_3)_2 \cdot 3H_2O$ in 600 mL distilled water. Make up to 1 L final volume with distilled water.
9	ChemLAB	1*M* aluminum nitrate ($Al(NO_3)_3$)	Dissolve 375 g $Al(NO_3)_3 \cdot 9H_2O$ in 600 mL distilled water. Make up to 1 L final volume with distilled water.
9	ChemLAB	1*M* magnesium nitrate ($Mg(NO_3)_2$)	Dissolve 256 g $Mg(NO_3)_2 \cdot 6H_2O$ in 600 mL distilled water. Make up to 1 L final volume with distilled water.
9	ChemLAB	1*M* zinc nitrate ($Zn(NO_3)_2$)	Dissolve 298 g $Zn(NO_3)_2 \cdot 6H_2O$ in 600 mL distilled water. Make up to 1 L final volume with distilled water.
11	LaunchLAB	0.01*M* potassium permanganate	Dissolve 0.3 g potassium permanganate ($KMnO_4$) in 200 mL distilled water.
11	LaunchLAB	0.01*M* sodium hydrogen sulfite	Dissolve 0.2 g sodium hydrogen sulfite ($NaHSO_3$) in 200 mL distilled water.
11	Demonstration	0.10*M* $CuSO_4$	Dissolve 2.50 g copper sulfate ($CuSO_4 \cdot 5H_2O$) in 50 mL distilled water. Make up to 100 mL final volume with distilled water. **WARNING:** ***$CuSO_4$ is toxic and corrosive.***
11	Demonstration	0.20*M* KOH	Dissolve 1.12 g potassium hydroxide (KOH) in 50 mL distilled water. Make up to 100 mL final volume with distilled water. **WARNING:** ***KOH is toxic and corrosive.***
11	Demonstration	0.1*M* NH_4OH	Add 1 mL of concentrated (15*M*) NH_3 (aq) to 150 mL of distilled water.

SOLUTIONS PREPARATION

Quantities are for a class of 30 students.

Chapter	Lab Type	Solution	Preparations
13	MiniLAB	5% vinegar	Add 50 mL glacial (17*M*) acetic acid to 950 mL water while stirring. **WARNING: *Do NOT add the water to the acid.***
15	Demonstration	thermite mixture	Mix 125 g iron(III) oxide (Fe_2O_3) with 50 g aluminum powder (100 mesh). *Note: For best results, ensure that the mixing process is thorough.*
15	Demonstration	thermite starter	Potassium chlorate ($KClO_3$) may be used in place of commercially prepared thermite starter. Substitute potassium chlorate in the procedure given. **WARNING: *Do NOT attempt to prepare thermite starter.***
16	LaunchLAB	3% hydrogen peroxide	Add 30 mL 30% hydrogen peroxide to 270 mL of distilled water. Store in a tightly sealed brown glass bottle. **WARNING: *Avoid skin contact with 30% hydrogen peroxide.***
16	ChemLAB	6*M* hydrochloric acid	Add 250 mL concentrated (12*M*) hydrochloric acid to 250 mL distilled water while stirring. **WARNING: *Do NOT add the water to the acid.***
17	MiniLAB	0.1*M* $CoCl_2$	Dissolve 2.4 g cobalt(II) chloride hexahydrate ($CoCl_2 \cdot 6H_2O$) in 100 mL distilled water.
17	ChemLAB	$AgNO_3$ solution	Dissolve 4 g $AgNO_3$ in 100 mL distilled water. **WARNING: *Silver nitrate is toxic and will stain skin and clothing.***
17	ChemLAB	NaCl solution	Dissolve 4 g NaCl in 100 mL distilled water.
17	ChemLAB	Na_2S solution	Dissolve 6 g sodium sulfide ($Na_2S \cdot 9H_2O$) in 50 mL distilled water.
18	LaunchLAB, ChemLAB	Phenolphthalein solution	Dissolve 0.1 g phenolphthalein powder in 70 mL 95% ethanol. Make up to 100 mL final volume with distilled water.
18	MiniLAB	6*M* acetic acid	Add 34 mL of glacial acetic acid to 66 mL of water. **WARNING: *Do NOT add the water to the acid.***
18	Demonstration	4*M* NaOH solution	Carefully add 16 g NaOH to 75 mL distilled water. Make up to 100 mL final volume with distilled water. **WARNING: *Solution will become hot.***
19	LaunchLAB	1*M* $CuSO_4$	Dissolve 25 g ($CuSO_4 \cdot 5H_2O$) in 70 mL distilled water. Make up to 100 mL final volume with distilled water.
19	Demonstration **WARNING: *Prepare fresh solutions for each demonstration. To dispose, dilute mixed solution with large quantities of water and flush immediately. The leftover mixed solution forms an explosive precipitate if allowed to stand for several hours.***		
19	Demonstration	Solution A	Dissolve 12.8 g silver nitrate ($AgNO_3$) in 150 mL distilled water. **WARNING: *This solution is toxic.***
19	Demonstration	Solution B	Dissolve 3.4 g potassium hydroxide (KOH) in 75 mL distilled water.

Quantities are for a class of 30 students.

Chapter	Lab Type	Solution	Preparations
19	Demonstration	Solution C	Dissolve 2.3 g dextrose in 50 mL distilled water.
20	Demonstration	6*M* hydrochloric acid	Add 50 mL concentrated (12*M*) hydrochloric acid to 50 mL distilled water while stirring. **WARNING:** ***Do NOT add the water to the acid.***
20	MiniLAB	saltwater solution	Dissolve 60 g sodium chloride (NaCl) in 2 L water.
20	ChemLAB	1*M* aluminum nitrate	Dissolve 37.5 g $Al(NO_3)_3{\cdot}9H_2O$ in 60 mL distilled water. Make up to 100 mL final volume with distilled water.
20	ChemLAB	1*M* copper(II) nitrate	Dissolve 24.2 g $Cu(NO_3)_2{\cdot}3H_2O$ in 60 mL distilled water. Make up to 100 mL final volume with distilled water.
20	ChemLAB	1*M* magnesium nitrate	Dissolve 14.8 g $Mg(NO_3)_2$ in 60 mL distilled water. Make up to 100 mL final volume with distilled water.
20	ChemLAB	1*M* potassium nitrate	Dissolve 10.1 g KNO_3 in 60 mL distilled water. Make up to 100 mL final volume with distilled water.
20	ChemLAB	1*M* zinc nitrate	Dissolve 29.7 g $Zn(NO_3)_2{\cdot}6H_2O$ in 60 mL distilled water. Make up to 100 mL final volume with distilled water.
22	LaunchLAB	4% polyvinyl alcohol	Slowly sprinkle 40 g polyvinyl alcohol powder into 900 mL water with constant stirring. Heat the mixture on a hot plate to 80°C while continuing to stir. Add a few drops of food coloring. Do not overheat the solution. Cool and store in a stoppered bottle.
22	LaunchLAB	4% tetraborate (borax) solution	Dissolve 12 g sodium tetraborate decahydrate ($Na_2B_4O_7{\cdot}10H_2O$) in 300 mL of warm water.
23	LaunchLAB	10% glucose solution	Dissolve 15 g of glucose in 135 mL of warm water.
23	LaunchLAB	10% starch solution	Soften 20 g starch in 50 mL distilled water. Add 100 mL boiling distilled water to dissolve. Make up to 200 mL final volume with distilled water.
23	LaunchLAB	10% gelatin solution	Soften 20 g gelatin in 50 mL distilled water. Add 100 mL boiling water to dissolve. Make up to 200 mL final volume with distilled water.
23	LaunchLAB	Benedict's solution	Dissolve 86.5 g sodium citrate and 100 g sodium carbonate decahydrate ($Na_2CO_3{\cdot}10H_2O$) in 400 mL distilled water, heating if necessary. Dissolve 8.65 g copper(II) sulfate pentahydrate ($CuSO_4{\cdot}5H_2O$) in 50 mL distilled water. Pour the copper(II) sulfate solution, with constant stirring, into the first solution. Make up to 500 mL final volume with distilled water.
23	MiniLAB	6.0*M* NaOH	Dissolve 48 g NaOH in 125 mL water, then dilute to 200 mL final volume. **WARNING:** ***Solution will become very hot.***
23	MiniLAB	Saturated NaCl solution	Slowly add 450 g NaCl to 1 L of water while stirring constantly. Some NaCl will not dissolve. Decant or filter if desired.

Equipment and Materials List

This table of equipment and inexpensive, easily accessible materials can help you prepare for your chemistry classes for the year. Refer to the Chapter Organizer in front of each chapter for a list of equipment and materials used for each laboratory activity in the chapter.

Nonconsumable

Item	LaunchLAB	MiniLAB	ChemLAB
alpha source			892
balance, centigram			230, 670, 776
balance, laboratory	2,172, 474, 514	39, 120, 301, 342, 378, 526, 571, 800	60, 126, 356, 390, 466, 550
ball bearing	400		
barometer		457	776
beta source			892
beaker, 50-mL	558		92
beaker, 100-mL	280, 366, 474, 592	726	230
beaker, 150-mL		301, 763	390
beaker, 250-mL		227, 342, 457, 526, 571, 800, 837	24, 466, 550, 670, 850
beaker, 400-mL	824	502, 683	390
beaker, 600-mL		837	
bolts	172		
bucket, 2-gallon	440		
burner, Bunsen		144, 378	92, 230, 356, 466
burner, laboratory		227	
buret, 50-mL			670
calculator, graphing			60, 126
clamp	68		164
clamp, buret			670
clay triangle		378	230, 356
clock			506, 850, 892
comb, plastic	100		
conductivity tester	204	648	196, 230
crucible		378	230, 356
cylinder, metal		526	
diffraction grating			164
dish, evaporating		837	550
dominoes	858		
dropper			24, 432, 698
flask, 275-mL polystyrene culture			164
flask, 250-mL Erlenmeyer			92, 670
flask, 500-mL Florence with rubber stopper			670

Nonconsumable			
Item	**LaunchLAB**	**MiniLAB**	**ChemLAB**
forceps, long		227, 763	698, 734
funnel		837	92
gamma source			892
Geiger counter			892
graduated cylinder, 10-mL	366	648, 800	196, 466, 584
graduated cylinder, 25-mL	30, 68, 280, 514, 824	13, 837	24, 506, 850
graduated cylinder, 50-mL	238, 474, 592, 784	39	92
graduated cylinder, 100-mL	400	301, 526	60, 390, 401, 550, 776
hammer			196
hole punch	100		
hot plate	824	242, 526, 571, 683, 800, 837	390, 850
lightbulb, 40-watt tubular with socket and power cord			164
meterstick			892
microplate, 24-well	632	648	310, 624, 698, 734
molecular-model kit	742		272
mortar and pestle			506
mystery object		39	
pennies		873	
pennies, pre- and post-1982		120	60
petri dish, plastic		800	92
petri dish with lid		13	
pipet, Beral-type		800	734, 816
pipette, disposable	204		
pipette			310
pipette, 10-mL graduated			584
pipette filler (bulb)			584
pipette, dropping	280	648	
pipette, thin-stem			624
pneumatic trough			776
radio, weather	456		
ring, small iron		378	92, 230, 356, 466, 550
ring stand	68	378	92, 164, 230, 356, 466, 550, 670
rubber stopper, test tube			192
ruler, metric	172, 318, 400	763	24, 60, 584, 850
scissors		82, 342, 422	584
screws	172		
spatula			196, 356, 506, 670
spectrum tubes (hydrogen, sodium, and neon) with power supplies			164
stirring rod, glass	280, 366, 440, 474, 784, 824	301, 342, 502, 763, 837	92, 230, 390, 506, 550, 584

Nonconsumable			
Item	**LaunchLAB**	**MiniLAB**	**ChemLAB**
stopwatch (timer, clock)	2, 400, 858	571	432, 584, 816
table of standard reduction potentials			734
test tube, large	68, 678	611	24, 196, 584, 850
test tube, small	514, 824	800	506
test-tube clamp		800	850
test-tube holder		800	
test tube rack	678		24, 196, 506, 584, 850
thermometer, nonmercury	280, 514	457, 502, 526, 571, 611	550, 776, 816, 850
tongs	824	683, 837	92, 584, 698
tongs, beaker			390, 466
tongs, crucible		378, 526	230, 356
towel, cloth			816
tube, glass	592		
tubing, burner			776
tweezer			506
voltmeter with leads	706		734
wash bottle			670
weighing bottle			670
wire cutters			310

Chemicals

Item	LaunchLAB	MiniLAB	ChemLAB
acetic acid, glacial (90 mL)		648	
acetone (30 mL)			432
alcohol (330 mL)	30		432
aluminum nitrate (270 mL)			310, 734
aluminum strip (30)			734
ammonia, 0.1*M* (20 mL)	280		
ammonia, household (30 mL)			432
ammonium chloride (300 g)	474		
ammonium nitrate (30 g)	514		
Benedict's solution (360 mL)	824		
calcium carbide (30 pea-sized pieces)		763	
calcium chloride (340 g)	474, 514	144	
cobalt(II) chloride hexahydrate (120 mL)		611	
copper strip (90)	706	726	734
copper wire			698
copper(II) nitrate (270 mL)			310, 734
copper(II) sulfate pentahydrate (390 g)	678		390, 506
ethanol, 95% (750 mL)	238	837	432, 816
glucose (30 g)	824		
glycerol (glycerin) (300 mL)	30		
hydrochloric acid, 0.1*M* (60 mL)			698
hydrochloric acid, 1*M* (900 mL)			196
hydrochloric acid, 3*M* (300 mL)	68		
hydrochloric acid, 6*M* (300 mL)			584
hydrochloric acid, concentrated (100 mL)		611	
hydrogen peroxide, 3% (1500 mL)	558		850
iron filings (20 mesh) (120 g)			390, 698
lead shot (60 g)			698
lithium chloride (30 g)		144	
litmus paper, blue, red (120 each)	632		
magnesium nitrate (15 g)			734
magnesium ribbon (1500 cm)		726	230, 310, 584
magnesium sulfate heptahydrate, (Epsom salts) (270 g)		301	356
magnesium strip (30)			734
magnesium turnings (30)			698
methanol (150 mL)		800	816
paraffin (60 g)		242	
phenolphthalein (50 mL)	632	763	670
polyvinyl alcohol solution (600 mL)	784		

LABORATORY MATERIALS

Chemicals			
Item	**LaunchLAB**	**MiniLAB**	**ChemLAB**
potassium chloride (30 g)		144	
potassium hydrogen phthalate (12 g)			670
potassium nitrate (41 g)	514		734
potassium permanganate, 0.01*M* (150 mL)	366		
2-propanol (60 mL)			816
salicylic acid (15 g)		800	
silver nitrate (200 g)			92, 624, 698
sodium chloride (table salt) (949 g)	204	144, 242, 502, 611, 683, 837	624
sodium hydrogen carbonate (baking soda) (1150 g)		378, 457, 683	
sodium hydrogen sulfite solution, 0.01*M* (150 mL)	366		
sodium hydroxide (288 g)		301, 837	670
sodium silicate solution (600 mL)	238		
sodium sulfide (6 g)			624
sodium tetraborate solution (180 mL)	784		
starch solution, 10 % (200 mL)	824		
strontium chloride (30 g)		144	
sucrose (table sugar) (180 g)	204	242	
sulfuric acid, 18*M* (10 mL)		800	
universal indicator (30 mL)	280		
unknown solution (50 mL)		144	698
vinegar, 5% (1200 mL)		457	
zinc metal (30 small pieces)	68		
zinc nitrate (270 mL)			310, 734
zinc sulfate (30 mL)			698
zinc strip (180)	706		310, 734

Other Consumables

Item	LaunchLAB	MiniLAB	ChemLAB
aluminum foil		457, 683	
aluminum pie pan, disposable, 9-inch		242	
balloon, round	440		
boiling chip	824		
box, wrapped with object inside	134		
candle	2	457	
card, index		193	
cardboard			816, 892
cheesecloth		837	
corn oil	30		
cotton ball		800	
cotton swabs		144	
cup, foam		526	
cup, paper	238		
cup, 5-oz. plastic	784		60, 432
cup, 9-oz. plastic	558	82, 873	
detergent, liquid dish		13, 763	24
effervescent antacid tablet	280	571	
emery cloth			310, 584
filter paper		82	92, 734
food coloring (red, blue, green, yellow)		13	164
food coloring, blue	30		
food coloring, red	30, 238, 592		
gelatin	824		
gum, chewing		342	
gumdrops, small	318		
hairpin		227	
honey	824		
household products, food or cleaning products	632		
ice	440		850
ice bath		611	
ice, crushed		502	
lemon	706		
liver, raw fresh pulp		242	
marker, permanent		242	

Other Consumables			
Item	**LaunchLAB**	**MiniLAB**	**ChemLAB**
marker, water-soluble black		82	
marshmallows, mini-sized	318		
matches	2, 68	457, 763	356, 550
milk, whole		13	
nails	172		
nail, iron	678	726	
object, small tarnished		683	
paper clip	318		92
paper towel	238, 440	342	432, 466, 776
paper, graph			60
paper, white	100		
pen, marking			432
pencil, glass-marking			196
pencil, grease			24, 432
pencils, colored, assorted			164
plastic			892
popcorn kernels			466
potato chip, large			550
potato, red-skin pulp			850
rubber band, small		763	816
samples, elements			196
sandpaper, fine (10 cm × 10 cm)		726	92, 734
screen, window		342	
shoebox		873	
snack bag of snack mix			126
soda bottle with cap, plastic 1-L			776
soda bottle with cap, plastic 2-L			776
splint, wooden	68, 238	763	
steel wool, pad	678	683	734
straw, plastic soda		423	
string	440		
tape, masking	100	457	432
tissue, facial			816
toothpick	558	13	
twist tie			816
vegetable oil	400	13	466
vegetable shortening, solid		837	

Other Consumables

Item	LaunchLAB	MiniLAB	ChemLAB
water	30, 400, 440, 474, 824	13, 39, 82, 227, 342, 502, 571, 611, 648, 683, 763, 800, 837	24, 60, 850
water, distilled	204, 280, 514	144, 301, 526, 726, 763, 800	24, 230, 390, 432, 466, 506, 584, 670
water, salt		726	
waxed paper			432
weighing paper		342, 800	390
wire, 12-gauge copper			310
wire, 22- or 26- gauge copper		423	
wire, aluminum			310
wire, copper			92
wire gauze squares			466, 550
yeast, active dry	558		

A^2 DEVELOPMENT PROCESS

Why A^2?

Accuracy Assurance is central to McGraw-Hill Education's commitment to high-quality, learner-oriented, real-world, and error-free products. Also at the heart of our A^2 Development Process is a commitment to make the text **Accessible** and **Approachable** for both students and teachers. A collaboration among authors, content editors,academic advisors, and classroom teachers, the A^2 Development Process provides opportunities for continual improvement through customer feedback and thorough content review.

The A^2 Development Process begins with a review of the previous edition and a look forward to state and national standards. The authors for *Chemistry: Matter and Change* combine expertise in teacher training and education with a mastery of chemistry content knowledge. As manuscript is created and edited, consultants review the accuracy of the content while our Teacher Advisory Board members examine the program from the points of view of both teacher and student. Student labs and teacher demonstrations are reviewed for both accuracy of content and safety. As design elements are applied, chapter content is again reviewed, as are photos and diagrams.

Throughout the life of the program, McGraw-Hill Education continues to troubleshoot and incorporate improvements. Our goal is to deliver to you a program that has been created, refined, tested, and validated as a successful tool for your continued **Academic Achievement.**

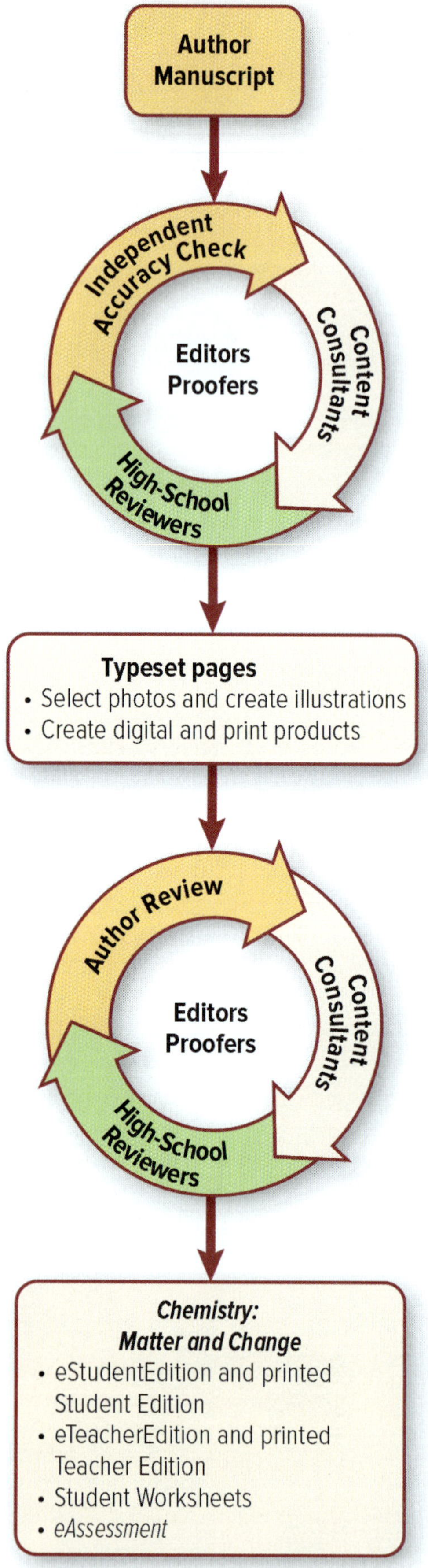

Authors

The authors of *Chemistry: Matter and Change* used their content knowledge and teaching expertise to craft manuscript that is accessible and accurate, geared toward student achievement.

Thandi Buthelezi

is Associate Professor of Chemistry at Wheaton College, Norton, MA. She earned her BA in Chemistry from Williams College, Williamstown, MA, and PhD in Experimental Physical Chemistry from the University of Florida, Gainesville, FL. Dr. Buthelezi has taught Chemistry at the undergraduate and graduate (master's) level. She is the co-founder and co-director of the Girls in Science Outreach Program at Western Kentucky University. She is a member of the American Chemical Society, the American Association for the Advancement of Science, and Sigma Xi. She has co-authored over two dozen research papers published in peer-reviewed journals.

Laurel Dingrando

is currently serving as the Secondary Science Coordinator for the Garland Independent School District. Mrs. Dingrando has a BS in Microbiology with a minor in Chemistry from Texas Tech University and an MAT in Science from the University of Texas at Dallas. She taught Chemistry for 35 years in the Garland Independent School District. She is a member of the American Chemical Society, National Science Teachers Association, Science Teachers Association of Texas, Texas Science Educators Leadership Association, and T3 (Teachers Teaching with Technology).

Nicholas Hainen

taught chemistry and physics in the Worthington City Schools, Worthington, Ohio, for 31 years. Mr. Hainen holds BS and MA degrees in Science Education from The Ohio State University, majoring in chemistry and physics. His honors and awards include: American Chemical Society Outstanding Educator in Chemical Sciences; The Ohio State University Honor Roll of Outstanding High School Teachers; Ashland Oil Company Golden Apple Award; and Who's Who Among America's Teachers. Mr. Hainen is a member of the American Chemical Society and the ACS Division of Chemical Education.

Cheryl Wistrom

is Associate Professor of Chemistry at Saint Joseph's College in Rensselaer, Indiana, where she has been honored with both the Science Division and college faculty teaching awards. She has taught chemistry, biology, and science education courses at the college level since 1990 and is also a licensed pharmacist. She earned her BS degree in biochemistry at Northern Michigan University, a BS in pharmacy at Purdue University, and her MS and PhD in biological chemistry at the University of Michigan. Dr. Wistrom is a member of the Indiana Academy of Science, the National Science Teachers Association, and the American Society of Health-System Pharmacists.

Dinah Zike

is an international curriculum consultant and inventor who has developed educational products and three-dimensional, interactive graphic organizers for over 30 years. As president and founder of Dinah-Might Adventures, L.P., Dinah is the author of more than 100 award-winning educational publications, including *The Big Book of Science*. Dinah has a BS and an MS in educational curriculum and instruction from Texas A&M University. Dinah Zike's *Foldables* are an exclusive feature of McGraw-Hill textbooks.

A² DEVELOPMENT PROCESS

Teacher Advisory Board

The Teacher Advisory Board gave the editorial staff and design team feedback on the content and design of both the Student Edition and Teacher Edition. We thank these teachers for their hard work and creative suggestions.

Ann Cooper
Science Teacher
United Local Schools
Hanoverton, OH

David L. French
Chemistry Teacher
Milford High School
Milford, OH

Richard Glink
Chemistry/Physics Teacher
Indian Lake High School
Lewistown, OH

Susan Godez
Chemistry/Physics Teacher
Grandview Heights High School
Columbus, OH

Judith Johnston
Science Teacher, Department Chair
Wilmington High School
Wilmington, OH

Christine Lewis
Science Teacher
Martins Ferry High School
Martins Ferry, OH

Jennifer L. Most
Chemistry Teacher,
Science Department Chair
West Holmes High School
Millersburg, OH

Sandra Petrie-Forgey
National Board Certified
Science Teacher
Gallia Academy High School
Gallipolis, OH

Jason J. Zaros
Chemistry/Physics Teacher
Waterford High School
Waterford, OH

Teacher Reviewers

Each teacher reviewed selected chapters of *Chemistry: Matter and Change* and provided feedback and suggestions regarding the effectiveness of the instruction.

Bridget B. Adkins
Ravenwood High School
Brentwood, TN

Deborah Bennett
Canoga Park High School
Canoga Park, CA

James Breaux
Stratford High School
Goose Creek, SC

Bob Callender
Warren Mott High School
Warren, MI

Betsy Hamrick
Crest High School
Shelby, NC

Treva Jeffries
Scott High School
Toledo, OH

Dr. Aruna Kailasa
Benjamin E. Mays High School
Atlanta, GA

Phil Lampe
Upper Arlington High School
Columbus, OH

Les McSparrin
Sharpsville Area High School
Sharpsville, PA

Delores Miller
Alden High School
Alden, NY

Leon Olivier
Union Grove High School
McDonough, GA.

Dan Reid
Central High School
Champaign, IL

Jay Wilder
Franklin County High School
Frankfort, KY

Content Consultants

Content consultants each reviewed selected chapters of *Chemistry: Matter and Change* for content accuracy and clarity.

Alton J. Banks, PhD
Professor of Chemistry
North Carolina State University
Raleigh, NC

Howard Drossman, PhD
Professor of Chemistry and Environmental Science
Colorado College
Colorado Springs, CO

Michael O. Hurst, Sr., PhD
Associate Professor of Chemistry
Georgia Southern University
Statesboro, GA

Kristen Kulinowski, PhD
Faculty Fellow, Department of Chemistry
Rice University
Houston, TX

Maria Pacheco, PhD
Associate Professor of Chemistry
Buffalo State College
Buffalo, NY

Safety Consultant

The safety consultant reviewed labs and lab materials for safety and implementation.

Kenneth R. Roy, PhD
Director of Environmental Health and Safety
Glastonbury Public Schools
Glastonbury, CT

Contributing Writers

Additional science writers added feature content, teacher materials, assessment, and laboratory investigations.

Peter Carpico
Louisville, OH

Jennifer Gonya
Galena, OH

Cindy Klevickis
Elkton, VA

Jack Minot
Columbus, OH

Richard G. Smith
Ocean Isle Beach, NC

Stephen Whitt
Columbus, OH

Jenipher Willoughby
Forest, VA

Margaret K. Zorn
Yorktown, VA

CHEMISTRY STUDY TOOLS

BE THE SCIENTIST! BE THE ENGINEER!

ConnectED is your one-stop online resource to explore real-world challenges that deepen understanding of core ideas and cross-cutting concepts!

- Chemistry ebook
- Science and Engineering Practices Handbook
- Applying Practices activities
- PBLs
- Design-your-own Labs
- Guided investigations

Get the LearnSmart© advantage! Improve your performance by using this interactive and adaptive study tool. Your personalized learning path will help you practice and master key chemistry concepts.

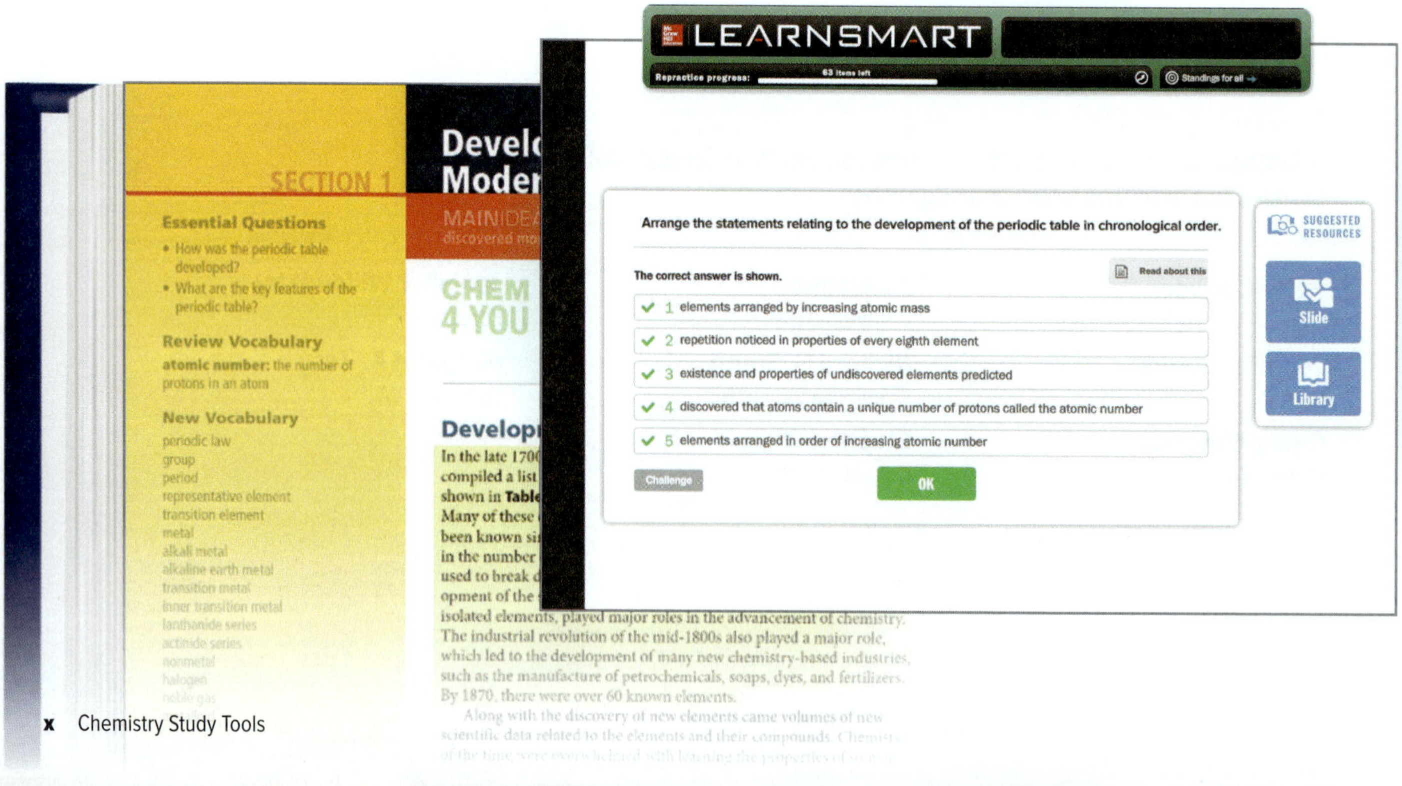

Go online!

“It’s EASY to get my assignments online and QUICK to find everything I need.”

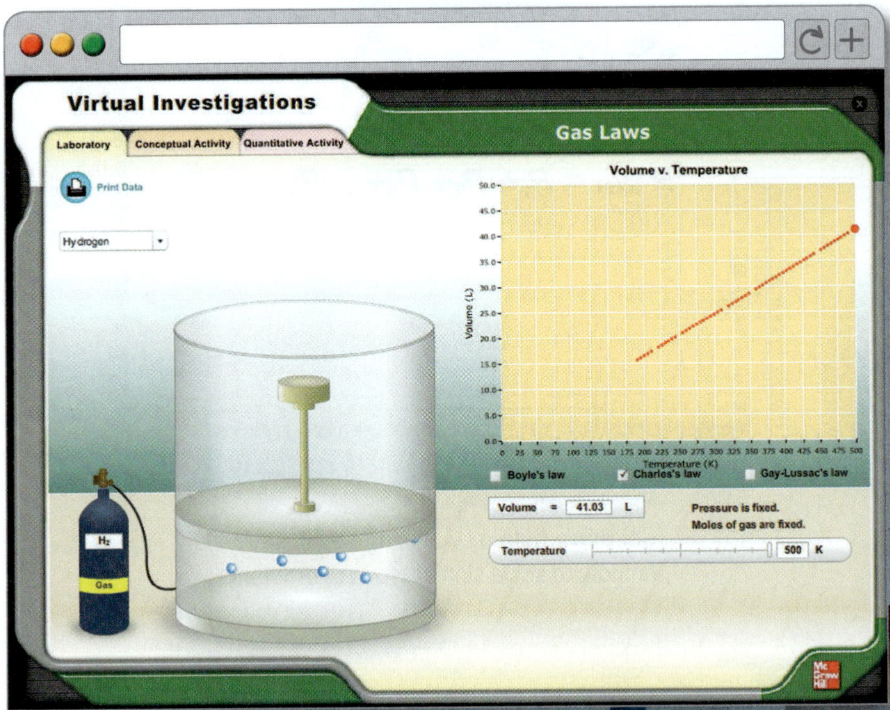

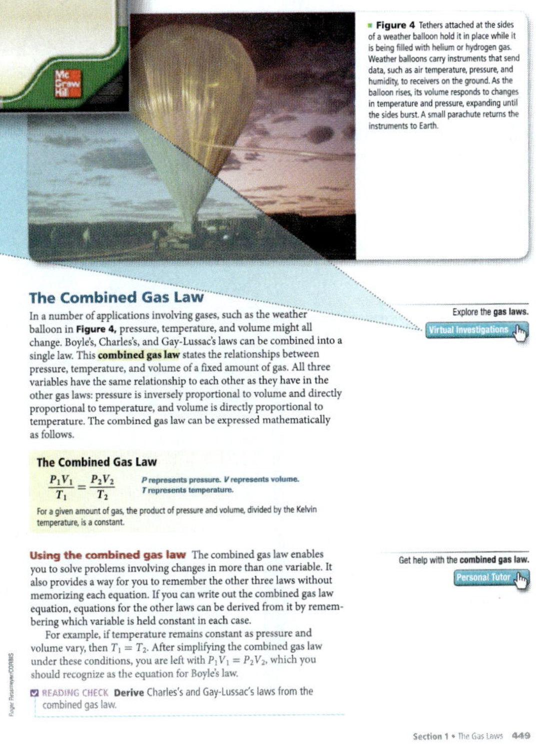
Figure 4 Tethers attached at the sides of a weather balloon hold it in place while it is being filled with helium or hydrogen gas. Weather balloons carry instruments that send data, such as air temperature, pressure, and humidity, to receivers on the ground. As the balloon rises, its volume responds to changes in temperature and pressure, expanding until the sides burst. A small parachute returns the instruments to Earth.

The Combined Gas Law

In a number of applications involving gases, such as the weather balloon in **Figure 4,** pressure, temperature, and volume might all change. Boyle’s, Charles’s, and Gay-Lussac’s laws can be combined into a single law. This **combined gas law** states the relationships between pressure, temperature, and volume of a fixed amount of gas. All three variables have the same relationship to each other as they have in the other gas laws: pressure is inversely proportional to volume and directly proportional to temperature, and volume is directly proportional to temperature. The combined gas law can be expressed mathematically as follows.

Explore the **gas laws.**
Virtual Investigations

The Combined Gas Law

$$\frac{P_1V_1}{T_1} = \frac{P_2V_2}{T_2}$$

P represents pressure. *V* represents volume. *T* represents temperature.

For a given amount of gas, the product of pressure and volume, divided by the Kelvin temperature, is a constant.

Using the combined gas law The combined gas law enables you to solve problems involving changes in more than one variable. It also provides a way for you to remember the other three laws without memorizing each equation. If you can write out the combined gas law equation, equations for the other laws can be derived from it by remembering which variable is held constant in each case.

For example, if temperature remains constant as pressure and volume vary, then $T_1 = T_2$. After simplifying the combined gas law under these conditions, you are left with $P_1V_1 = P_2V_2$, which you should recognize as the equation for Boyle’s law.

Get help with the **combined gas law.**
Personal Tutor

READING CHECK **Derive** Charles’s and Gay-Lussac’s laws from the combined gas law.

Roger Ressmeyer/CORBIS

Section 1 • The Gas Laws 449

Virtual Investigations

Enter an **online laboratory** where you can perform investigations that would be difficult or impossible to complete in the classroom. Manipulate atoms, build molecules, experiment with radioactive elements, and more!

Laura Doss/Fancy Photography/Veer

CHEMISTRY STUDY TOOLS

Example Problems throughout each chapter give you step-by-step instructions and hints on how to solve each type of problem. Each Example Problem is followed by a number of **Practice Problems** on the same topic, allowing you to practice the skill.

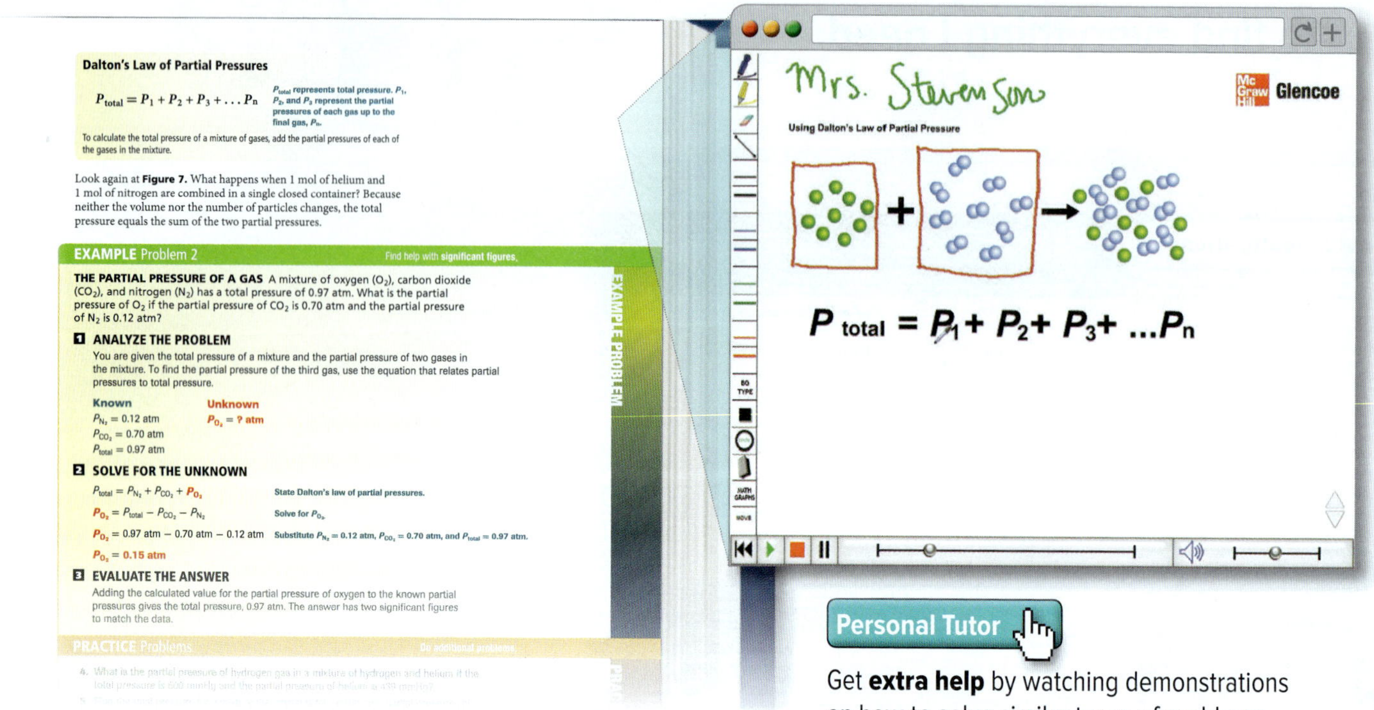

Personal Tutor

Get **extra help** by watching demonstrations on how to solve similar types of problems.

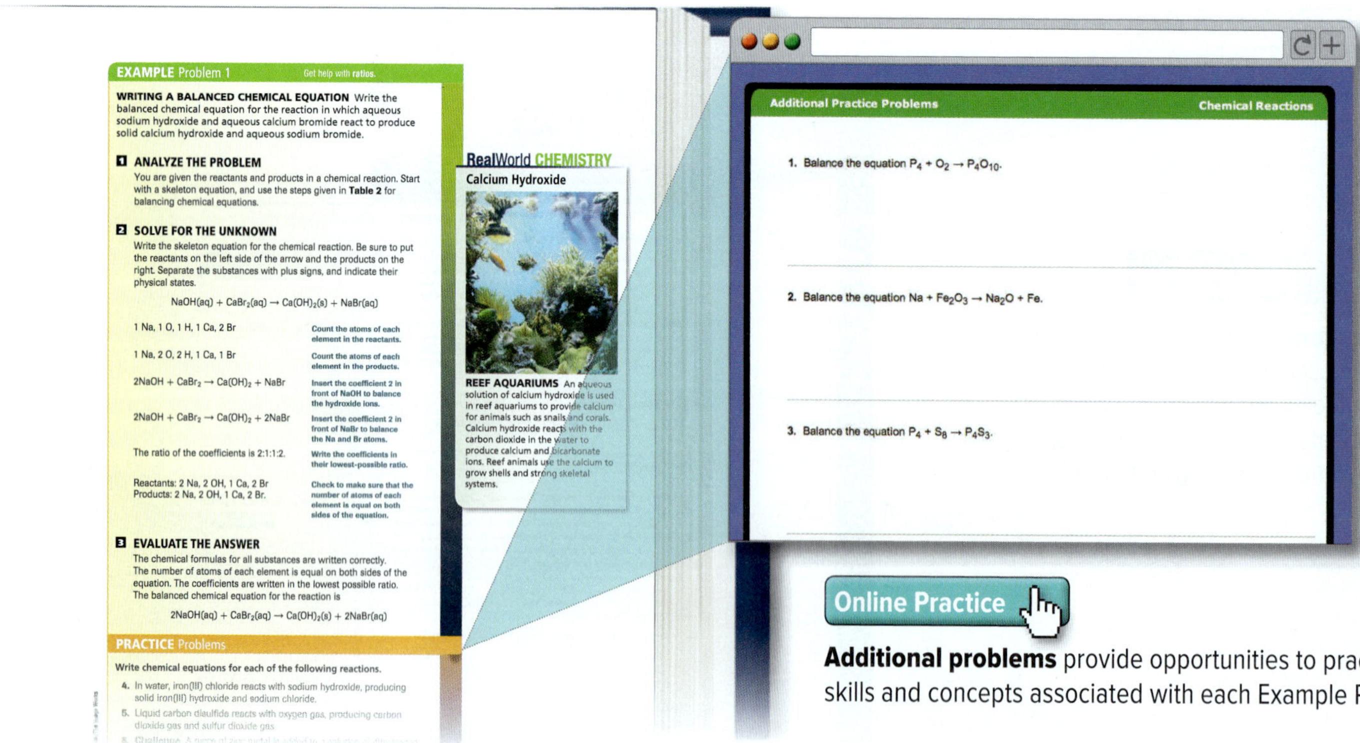

Online Practice

Additional problems provide opportunities to practice the skills and concepts associated with each Example Problem.

Go online!

"VIDEOS, ANIMATIONS, and tools help me LEARN."

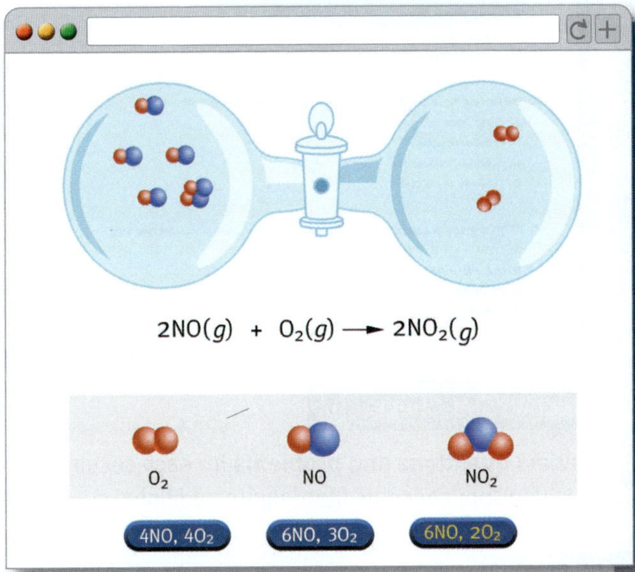

Concepts In Motion

See chemistry content come to life in **animated figures** and moving diagrams.

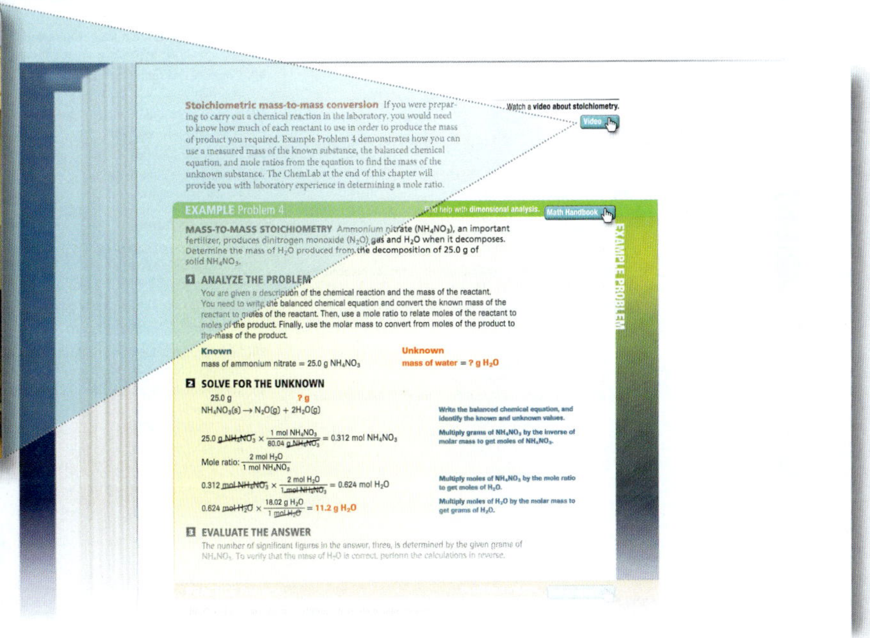

Video

Discover real-world applications of chemistry as **videos** connect to chapter content.

CHEMISTRY STUDY TOOLS

Quizzes, assessments, and study tools provide opportunities for self-assessment, review, and additional practice.

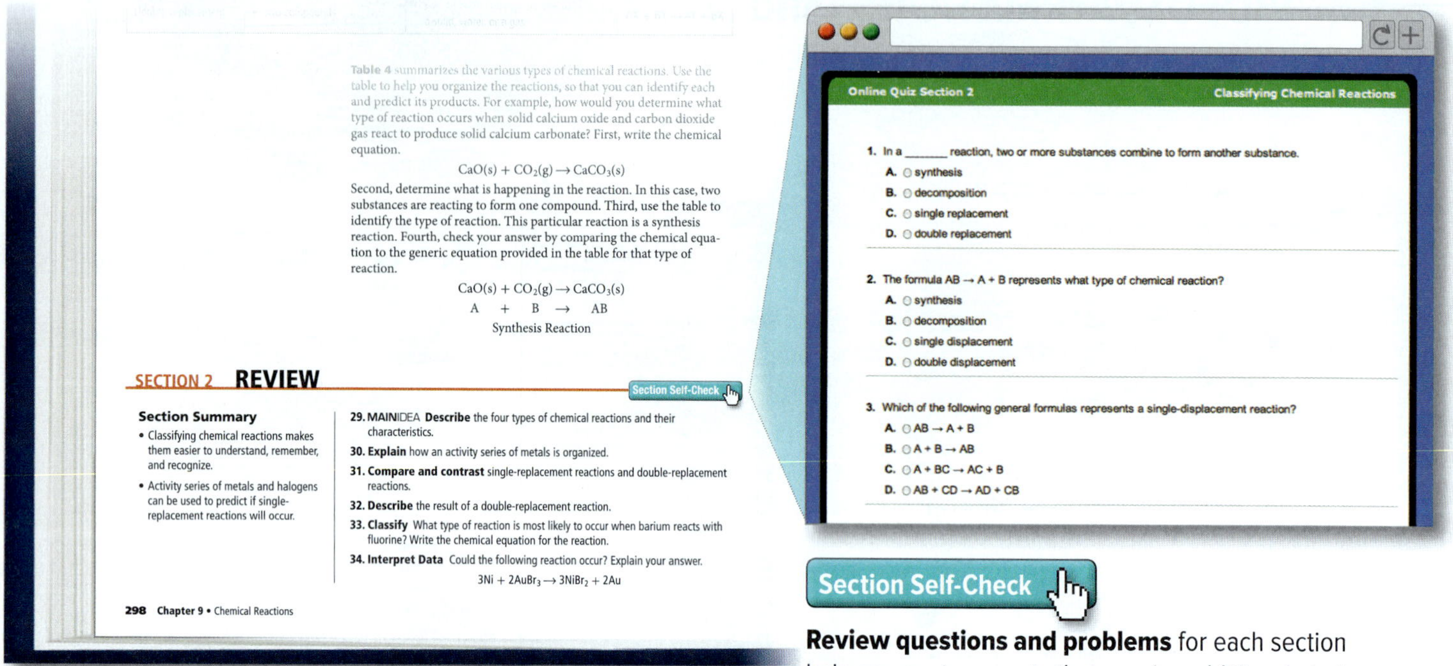

Table 4 summarizes the various types of chemical reactions. Use the table to help you organize the reactions, so that you can identify each and predict its products. For example, how would you determine what type of reaction occurs when solid calcium oxide and carbon dioxide gas react to produce solid calcium carbonate? First, write the chemical equation.

$$CaO(s) + CO_2(g) \rightarrow CaCO_3(s)$$

Second, determine what is happening in the reaction. In this case, two substances are reacting to form one compound. Third, use the table to identify the type of reaction. This particular reaction is a synthesis reaction. Fourth, check your answer by comparing the chemical equation to the generic equation provided in the table for that type of reaction.

$$CaO(s) + CO_2(g) \rightarrow CaCO_3(s)$$
$$A + B \rightarrow AB$$

Synthesis Reaction

SECTION 2 REVIEW — Section Self-Check

Section Summary
- Classifying chemical reactions makes them easier to understand, remember, and recognize.
- Activity series of metals and halogens can be used to predict if single-replacement reactions will occur.

29. MAINIDEA Describe the four types of chemical reactions and their characteristics.
30. Explain how an activity series of metals is organized.
31. Compare and contrast single-replacement reactions and double-replacement reactions.
32. Describe the result of a double-replacement reaction.
33. Classify What type of reaction is most likely to occur when barium reacts with fluorine? Write the chemical equation for the reaction.
34. Interpret Data Could the following reaction occur? Explain your answer.
$3Ni + 2AuBr_3 \rightarrow 3NiBr_2 + 2Au$

298 Chapter 9 • Chemical Reactions

Online Quiz Section 2 — Classifying Chemical Reactions

1. In a ______ reaction, two or more substances combine to form another substance.
A. synthesis
B. decomposition
C. single replacement
D. double replacement

2. The formula AB → A + B represents what type of chemical reaction?
A. synthesis
B. decomposition
C. single displacement
D. double displacement

3. Which of the following general formulas represents a single-displacement reaction?
A. AB → A + B
B. A + B → AB
C. A + BC → AC + B
D. AB + CD → AD + CB

Section Self-Check

Review questions and problems for each section help you spot concepts that require additional study.

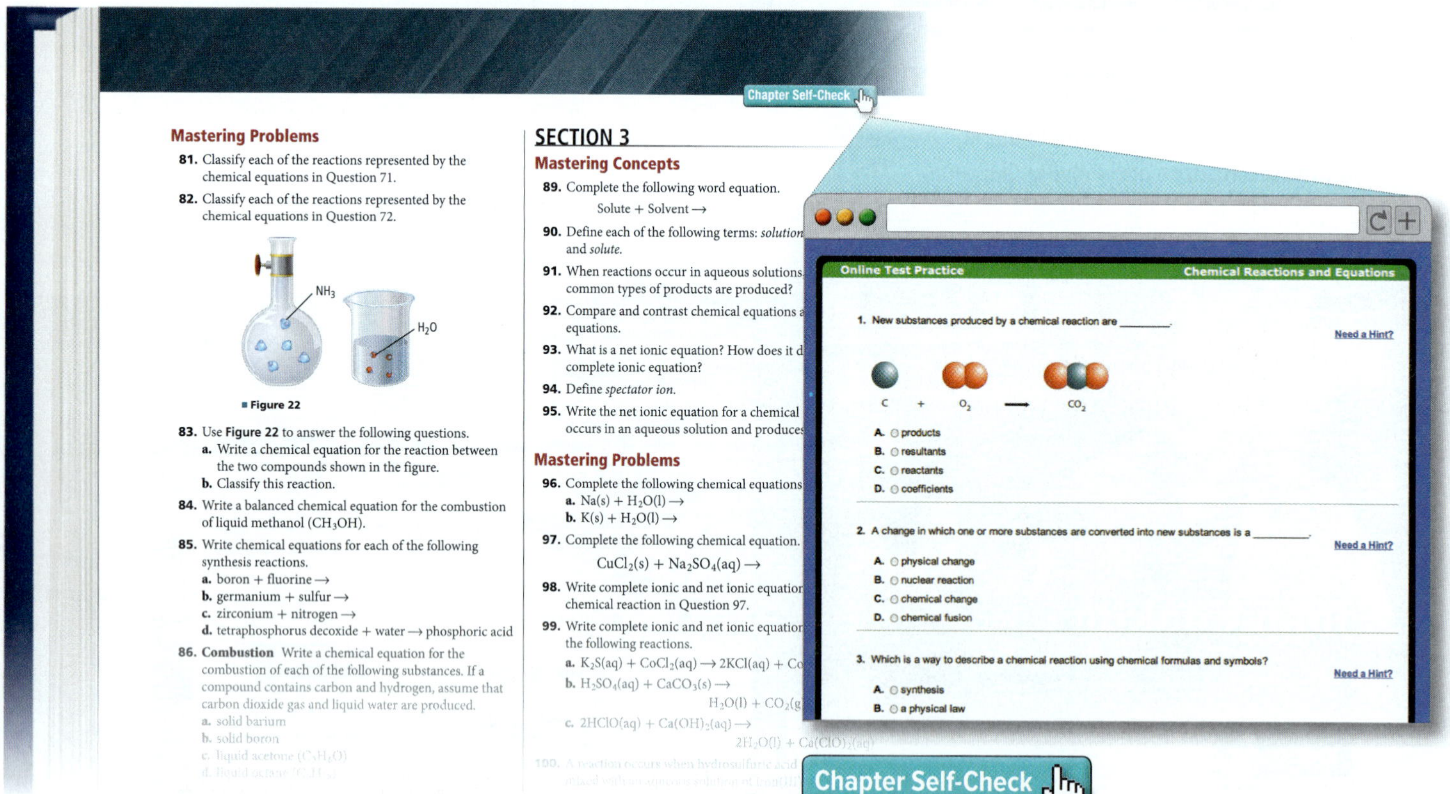

Chapter Self-Check

Mastering Problems

81. Classify each of the reactions represented by the chemical equations in Question 71.
82. Classify each of the reactions represented by the chemical equations in Question 72.

■ Figure 22

83. Use Figure 22 to answer the following questions.
a. Write a chemical equation for the reaction between the two compounds shown in the figure.
b. Classify this reaction.
84. Write a balanced chemical equation for the combustion of liquid methanol (CH_3OH).
85. Write chemical equations for each of the following synthesis reactions.
a. boron + fluorine →
b. germanium + sulfur →
c. zirconium + nitrogen →
d. tetraphosphorus decoxide + water → phosphoric acid
86. Combustion Write a chemical equation for the combustion of each of the following substances. If a compound contains carbon and hydrogen, assume that carbon dioxide gas and liquid water are produced.
a. solid barium
b. solid boron

SECTION 3

Mastering Concepts

89. Complete the following word equation.
Solute + Solvent →
90. Define each of the following terms: *solution* and *solute*.
91. When reactions occur in aqueous solutions, common types of products are produced?
92. Compare and contrast chemical equations equations.
93. What is a net ionic equation? How does it d complete ionic equation?
94. Define *spectator ion*.
95. Write the net ionic equation for a chemical occurs in an aqueous solution and produce

Mastering Problems

96. Complete the following chemical equations
a. $Na(s) + H_2O(l) \rightarrow$
b. $K(s) + H_2O(l) \rightarrow$
97. Complete the following chemical equation.
$CuCl_2(s) + Na_2SO_4(aq) \rightarrow$
98. Write complete ionic and net ionic equation chemical reaction in Question 97.
99. Write complete ionic and net ionic equation the following reactions.
a. $K_2S(aq) + CoCl_2(aq) \rightarrow 2KCl(aq) + Co$
b. $H_2SO_4(aq) + CaCO_3(s) \rightarrow H_2O(l) + CO_2(g$
c. $2HClO(aq) + Ca(OH)_2(aq) \rightarrow 2H_2O(l) + Ca(ClO)_2(aq)$

Online Test Practice — Chemical Reactions and Equations

1. New substances produced by a chemical reaction are ______. Need a Hint?

A. products
B. resultants
C. reactants
D. coefficients

2. A change in which one or more substances are converted into new substances is a ______. Need a Hint?
A. physical change
B. nuclear reaction
C. chemical change
D. chemical fusion

3. Which is a way to describe a chemical reaction using chemical formulas and symbols? Need a Hint?
A. synthesis
B. a physical law

Chapter Self-Check

Chapter-level **study tools** provide more opportunities for review and practice.

Go online!

> “I have the online **STUDY TOOLS** I need to help me **SUCCEED**.”

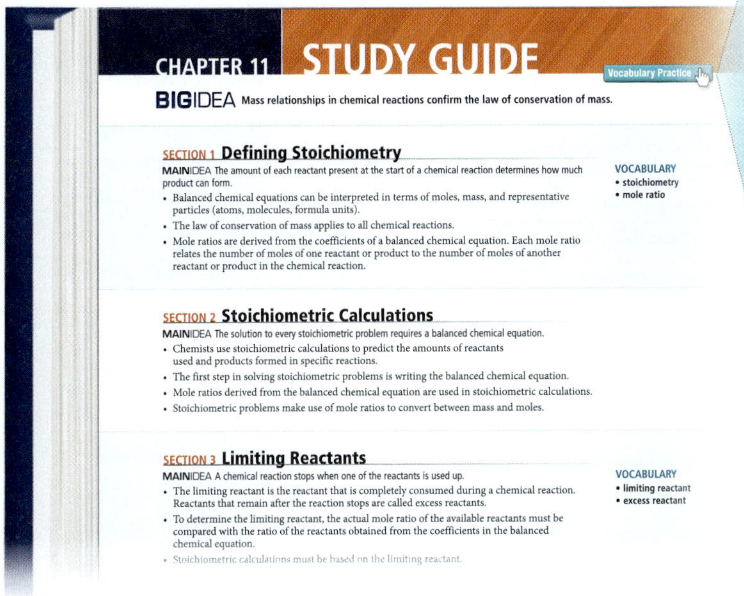
CHAPTER 11 STUDY GUIDE

Vocabulary Practice

BIGIDEA Mass relationships in chemical reactions confirm the law of conservation of mass.

SECTION 1 Defining Stoichiometry

MAINIDEA The amount of each reactant present at the start of a chemical reaction determines how much product can form.

- Balanced chemical equations can be interpreted in terms of moles, mass, and representative particles (atoms, molecules, formula units).
- The law of conservation of mass applies to all chemical reactions.
- Mole ratios are derived from the coefficients of a balanced chemical equation. Each mole ratio relates the number of moles of one reactant or product to the number of moles of another reactant or product in the chemical reaction.

VOCABULARY
- stoichiometry
- mole ratio

SECTION 2 Stoichiometric Calculations

MAINIDEA The solution to every stoichiometric problem requires a balanced chemical equation.

- Chemists use stoichiometric calculations to predict the amounts of reactants used and products formed in specific reactions.
- The first step in solving stoichiometric problems is writing the balanced chemical equation.
- Mole ratios derived from the balanced chemical equation are used in stoichiometric calculations.
- Stoichiometric problems make use of mole ratios to convert between mass and moles.

SECTION 3 Limiting Reactants

MAINIDEA A chemical reaction stops when one of the reactants is used up.

- The limiting reactant is the reactant that is completely consumed during a chemical reaction. Reactants that remain after the reaction stops are called excess reactants.
- To determine the limiting reactant, the actual mole ratio of the available reactants must be compared with the ratio of the reactants obtained from the coefficients in the balanced chemical equation.
- Stoichiometric calculations must be based on the limiting reactant.

VOCABULARY
- limiting reactant
- excess reactant

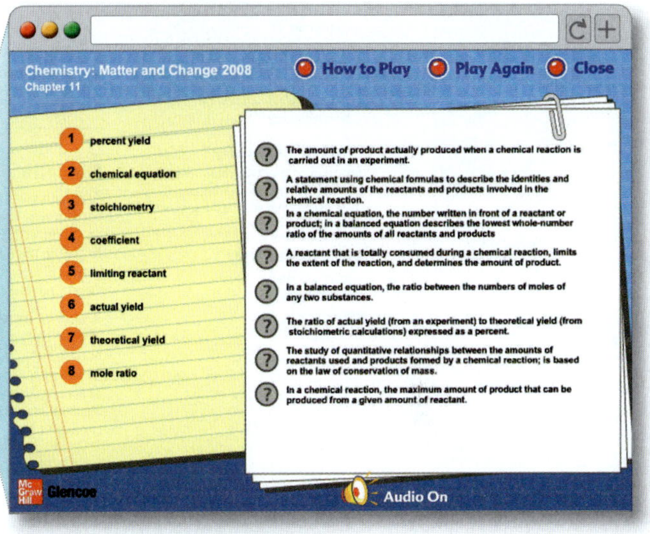

Vocabulary Practice

The **multilingual e-Glossary** and vocabulary **study tools** drive home important concepts.

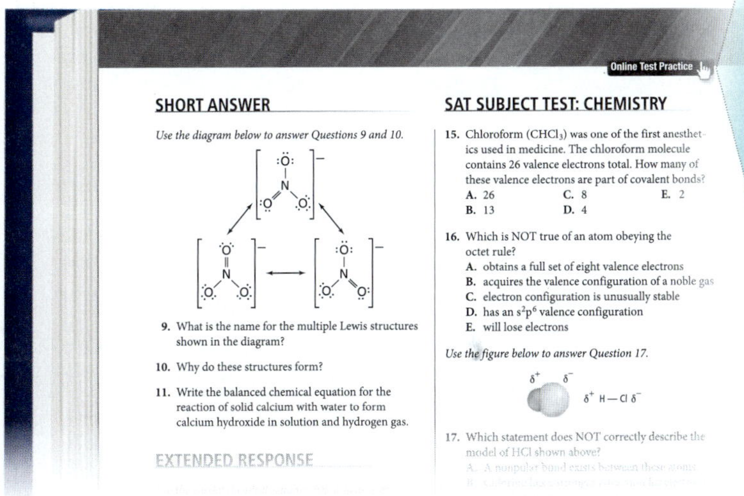
Online Test Practice

SHORT ANSWER

Use the diagram below to answer Questions 9 and 10.

9. What is the name for the multiple Lewis structures shown in the diagram?
10. Why do these structures form?
11. Write the balanced chemical equation for the reaction of solid calcium with water to form calcium hydroxide in solution and hydrogen gas.

EXTENDED RESPONSE

SAT SUBJECT TEST: CHEMISTRY

15. Chloroform ($CHCl_3$) was one of the first anesthetics used in medicine. The chloroform molecule contains 26 valence electrons total. How many of these valence electrons are part of covalent bonds?
   A. 26 C. 8 E. 2
   B. 13 D. 4
16. Which is NOT true of an atom obeying the octet rule?
   A. obtains a full set of eight valence electrons
   B. acquires the valence configuration of a noble gas
   C. electron configuration is unusually stable
   D. has an s^2p^6 valence configuration
   E. will lose electrons

Use the figure below to answer Question 17.

17. Which statement does NOT correctly describe the model of HCl shown above?

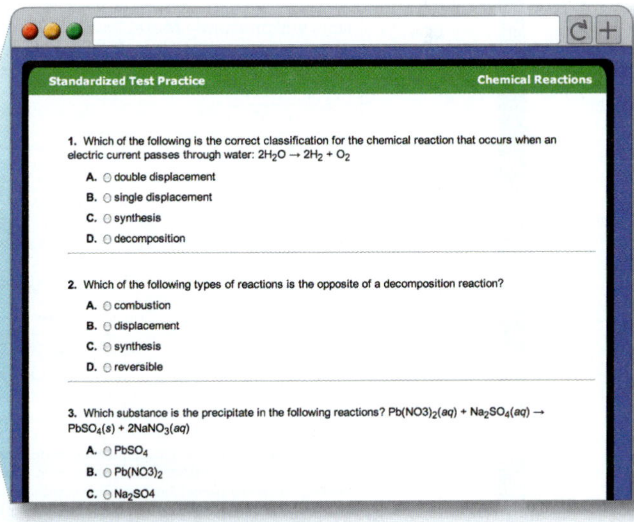

Online Test Practice

Practice taking **standardized tests** as you also review chapter content.

PhotoDisc/SuperStock

REAL-WORLD STEM

Chemistry: Matter and Change **connects chemistry to your world. Throughout the text, find personal science connections, surprising examples of chemistry in careers, and how chemists are engaged in cutting-edge science research.**

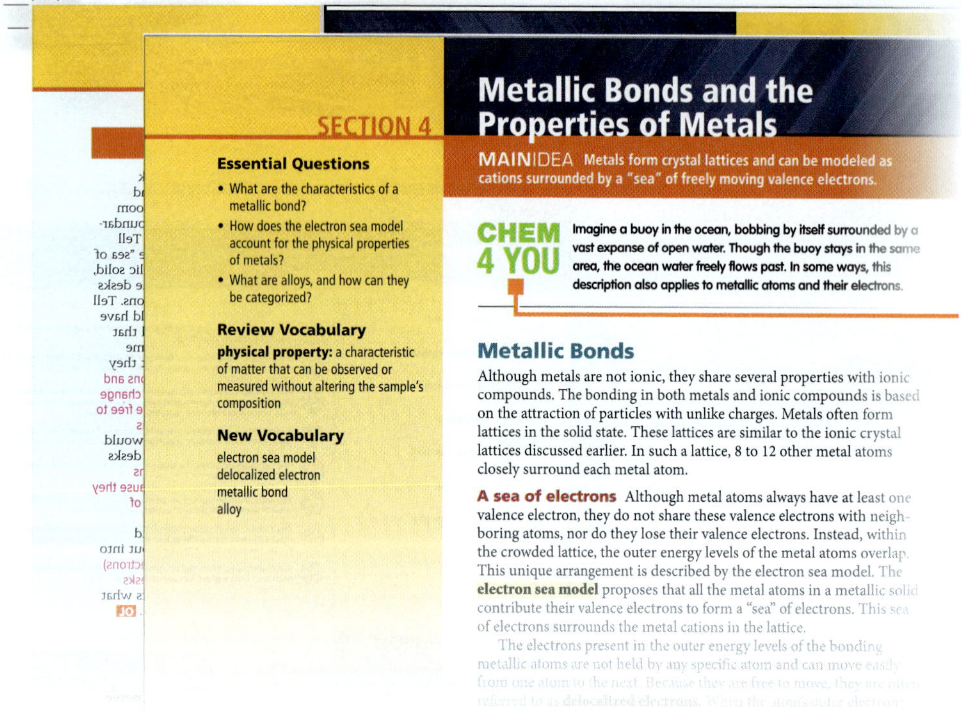

SECTION 4

Metallic Bonds and the Properties of Metals

MAIN IDEA Metals form crystal lattices and can be modeled as cations surrounded by a "sea" of freely moving valence electrons.

Essential Questions

- What are the characteristics of a metallic bond?
- How does the electron sea model account for the physical properties of metals?
- What are alloys, and how can they be categorized?

Review Vocabulary

physical property: a characteristic of matter that can be observed or measured without altering the sample's composition

New Vocabulary

electron sea model
delocalized electron
metallic bond
alloy

CHEM 4 YOU Imagine a buoy in the ocean, bobbing by itself surrounded by a vast expanse of open water. Though the buoy stays in the same area, the ocean water freely flows past. In some ways, this description also applies to metallic atoms and their electrons.

Metallic Bonds

Although metals are not ionic, they share several properties with ionic compounds. The bonding in both metals and ionic compounds is based on the attraction of particles with unlike charges. Metals often form lattices in the solid state. These lattices are similar to the ionic crystal lattices discussed earlier. In such a lattice, 8 to 12 other metal atoms closely surround each metal atom.

A sea of electrons Although metal atoms always have at least one valence electron, they do not share these valence electrons with neighboring atoms, nor do they lose their valence electrons. Instead, within the crowded lattice, the outer energy levels of the metal atoms overlap. This unique arrangement is described by the electron sea model. The **electron sea model** proposes that all the metal atoms in a metallic solid contribute their valence electrons to form a "sea" of electrons. This sea of electrons surrounds the metal cations in the lattice.

The electrons present in the outer energy levels of the bonding metallic atoms are not held by any specific atom and can move easily from one atom to the next. Because they are free to move, they are often referred to as delocalized electrons. When the atoms' outer electrons

CHEM 4 YOU at the beginning of each section connects chemistry content to your life.

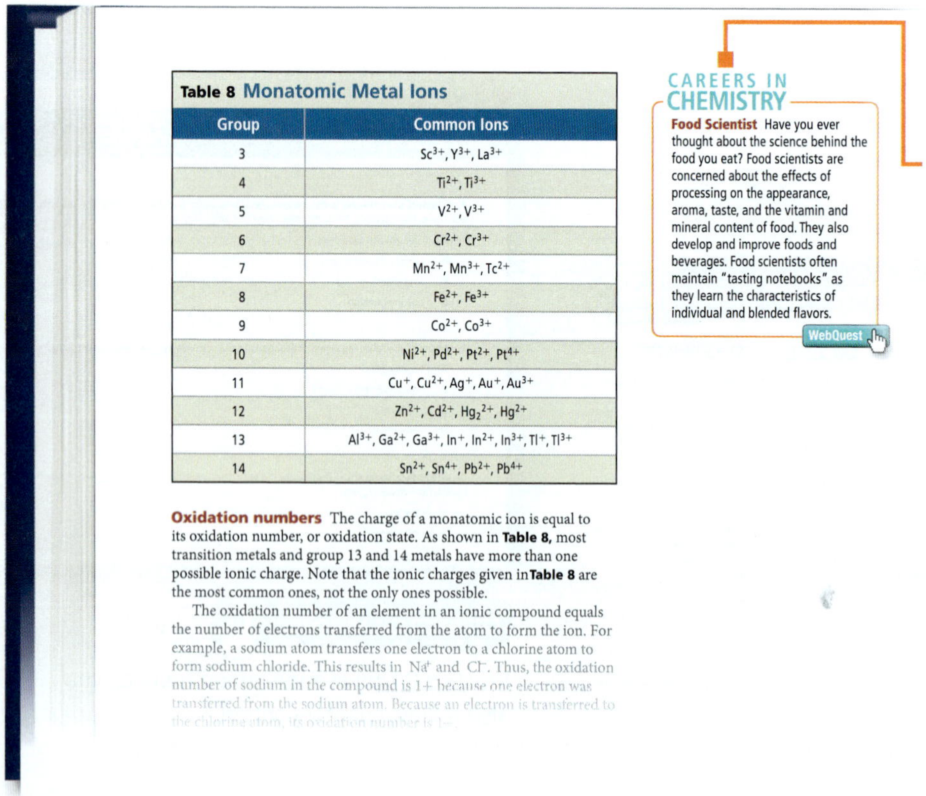

Table 8 Monatomic Metal Ions

Group	Common Ions
3	Sc^{3+}, Y^{3+}, La^{3+}
4	Ti^{2+}, Ti^{3+}
5	V^{2+}, V^{3+}
6	Cr^{2+}, Cr^{3+}
7	Mn^{2+}, Mn^{3+}, Tc^{2+}
8	Fe^{2+}, Fe^{3+}
9	Co^{2+}, Co^{3+}
10	Ni^{2+}, Pd^{2+}, Pt^{2+}, Pt^{4+}
11	Cu^{+}, Cu^{2+}, Ag^{+}, Au^{+}, Au^{3+}
12	Zn^{2+}, Cd^{2+}, Hg_2^{2+}, Hg^{2+}
13	Al^{3+}, Ga^{2+}, Ga^{3+}, In^{+}, In^{2+}, In^{3+}, Tl^{+}, Tl^{3+}
14	Sn^{2+}, Sn^{4+}, Pb^{2+}, Pb^{4+}

Oxidation numbers The charge of a monatomic ion is equal to its oxidation number, or oxidation state. As shown in **Table 8,** most transition metals and group 13 and 14 metals have more than one possible ionic charge. Note that the ionic charges given in **Table 8** are the most common ones, not the only ones possible.

The oxidation number of an element in an ionic compound equals the number of electrons transferred from the atom to form the ion. For example, a sodium atom transfers one electron to a chlorine atom to form sodium chloride. This results in Na^{+} and Cl^{-}. Thus, the oxidation number of sodium in the compound is 1+ because one electron was transferred from the sodium atom. Because an electron is transferred to the chlorine atom, its oxidation number is 1−.

CAREERS IN CHEMISTRY

Food Scientist Have you ever thought about the science behind the food you eat? Food scientists are concerned about the effects of processing on the appearance, aroma, taste, and the vitamin and mineral content of food. They also develop and improve foods and beverages. Food scientists often maintain "tasting notebooks" as they learn the characteristics of individual and blended flavors.

WebQuest

Throughout the book, **CAREERS IN CHEMISTRY** demonstrates how the chapter content applies to everyday careers.

End-of-chapter features highlight chemistry as it applies to careers, how it connects to the real world, and what chemistry means to your life.

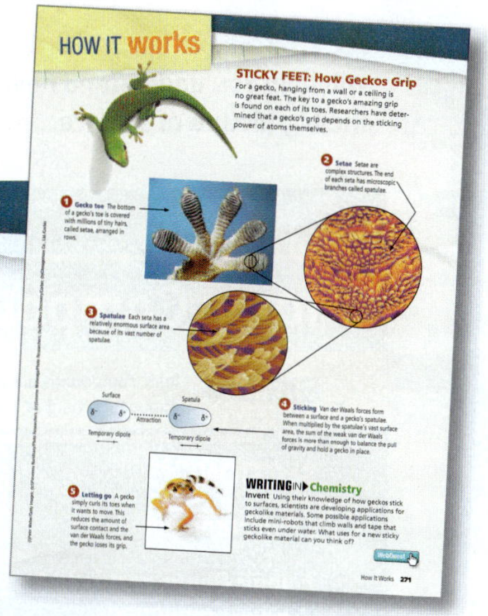
HOW IT works

STICKY FEET: How Geckos Grip

HOW IT works

Explore the chemistry involved in everyday objects or natural phenomena by discovering how they work.

everyday CHEMISTRY

Cocoa Chemistry

everyday CHEMISTRY

Did you use chemistry today? You might be surprised to discover how chemistry can be found in everyday experiences.

CHEMISTRY & health

Tiny Tweezers

CHEMISTRY & health

Did you know that chemistry happens inside your body? Investigate how chemistry and health are interrelated.

CHEMISTRY & careers

Career: Art Restorer

Painting Restoration

CHEMISTRY & careers

Put chemistry to work on the job. Explore jobs that unexpectedly rely on an understanding of chemistry.

UNDERSTANDING CHEMISTRY

At the start of each chapter, you will see the BIGIDEA that will help you understand how what you are about to investigate fits into the big picture of science.

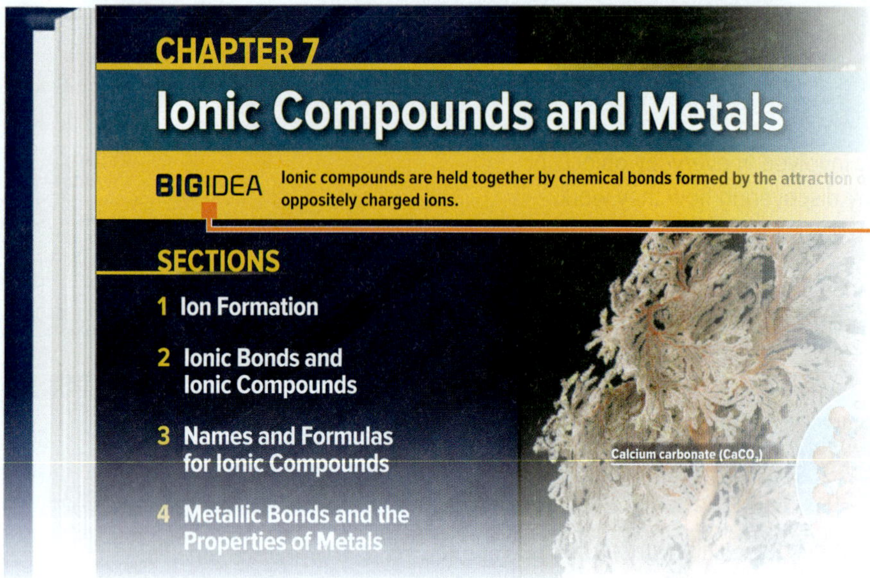
CHAPTER 7

Ionic Compounds and Metals

BIGIDEA Ionic compounds are held together by chemical bonds formed by the attraction of oppositely charged ions.

SECTIONS

1 Ion Formation

2 Ionic Bonds and Ionic Compounds

3 Names and Formulas for Ionic Compounds

4 Metallic Bonds and the Properties of Metals

The **BIGIDEA** is the focus of the chapter. The labs, text, and other chapter content will build an in-depth understanding of these major concepts.

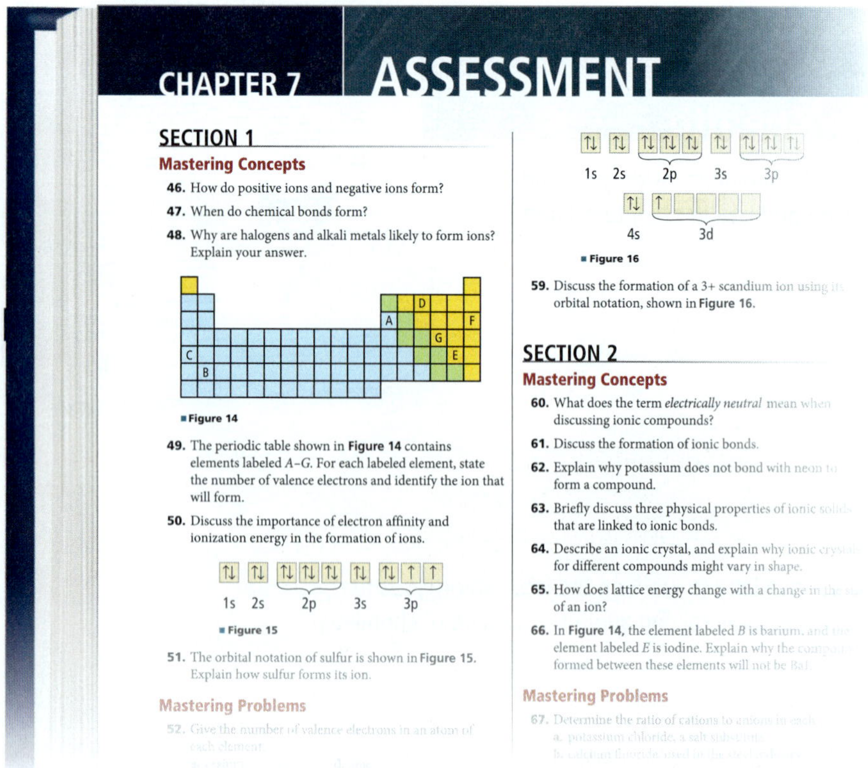
CHAPTER 7 ASSESSMENT

SECTION 1

Mastering Concepts

46. How do positive ions and negative ions form?

47. When do chemical bonds form?

48. Why are halogens and alkali metals likely to form ions? Explain your answer.

Figure 14

49. The periodic table shown in **Figure 14** contains elements labeled *A–G*. For each labeled element, state the number of valence electrons and identify the ion that will form.

50. Discuss the importance of electron affinity and ionization energy in the formation of ions.

Figure 15

51. The orbital notation of sulfur is shown in **Figure 15**. Explain how sulfur forms its ion.

Mastering Problems

Figure 16

59. Discuss the formation of a 3+ scandium ion using its orbital notation, shown in **Figure 16**.

SECTION 2

Mastering Concepts

60. What does the term *electrically neutral* mean when discussing ionic compounds?

61. Discuss the formation of ionic bonds.

62. Explain why potassium does not bond with neon to form a compound.

63. Briefly discuss three physical properties of ionic solids that are linked to ionic bonds.

64. Describe an ionic crystal, and explain why ionic crystals for different compounds might vary in shape.

65. How does lattice energy change with a change in the size of an ion?

66. In **Figure 14**, the element labeled *B* is barium, and the element labeled *E* is iodine. Explain why the compound formed between these elements will not be BaI.

Mastering Problems

The Chapter Assessment will help you evaluate your understanding of the **BIGIDEA.**

At the start of each section, you will find a reading preview that summarizes what you will learn while exploring the section.

The **MAIN**IDEA is the core concept covered in the section. Together, the Main Ideas from all the sections in the chapter support the chapter's Big Idea.

Essential Questions reflect the important goals of the section. Together, an understanding of these questions will lead toward an understanding of the section's Main Idea.

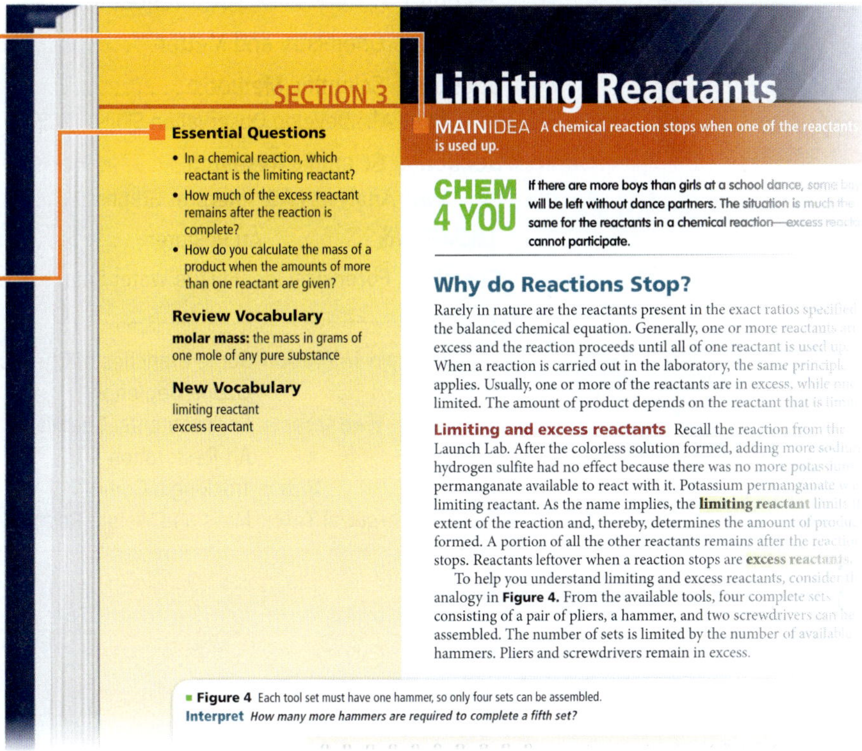

SECTION 3

Limiting Reactants

MAINIDEA A chemical reaction stops when one of the reactants is used up.

Essential Questions

- In a chemical reaction, which reactant is the limiting reactant?
- How much of the excess reactant remains after the reaction is complete?
- How do you calculate the mass of a product when the amounts of more than one reactant are given?

Review Vocabulary

molar mass: the mass in grams of one mole of any pure substance

New Vocabulary

limiting reactant
excess reactant

CHEM 4 YOU If there are more boys than girls at a school dance, some b… will be left without dance partners. The situation is much the same for the reactants in a chemical reaction—excess reac… cannot participate.

Why do Reactions Stop?

Rarely in nature are the reactants present in the exact ratios specified… the balanced chemical equation. Generally, one or more reactants… excess and the reaction proceeds until all of one reactant is used up. When a reaction is carried out in the laboratory, the same principle applies. Usually, one or more of the reactants are in excess, while one… limited. The amount of product depends on the reactant that is…

Limiting and excess reactants Recall the reaction from the Launch Lab. After the colorless solution formed, adding more sodi… hydrogen sulfite had no effect because there was no more potassium permanganate available to react with it. Potassium permanganate… limiting reactant. As the name implies, the **limiting reactant** limits… extent of the reaction and, thereby, determines the amount of prod… formed. A portion of all the other reactants remains after the reaction stops. Reactants leftover when a reaction stops are **excess reactants**.

To help you understand limiting and excess reactants, consider the analogy in **Figure 4.** From the available tools, four complete sets… consisting of a pair of pliers, a hammer, and two screwdrivers can be assembled. The number of sets is limited by the number of available hammers. Pliers and screwdrivers remain in excess.

Figure 4 Each tool set must have one hammer, so only four sets can be assembled.
Interpret *How many more hammers are required to complete a fifth set?*

In the Section Review, you will find a question that will help you to assess your understanding of the section's **MAIN**IDEA.

Essential Questions are assessed by the remaining review questions.

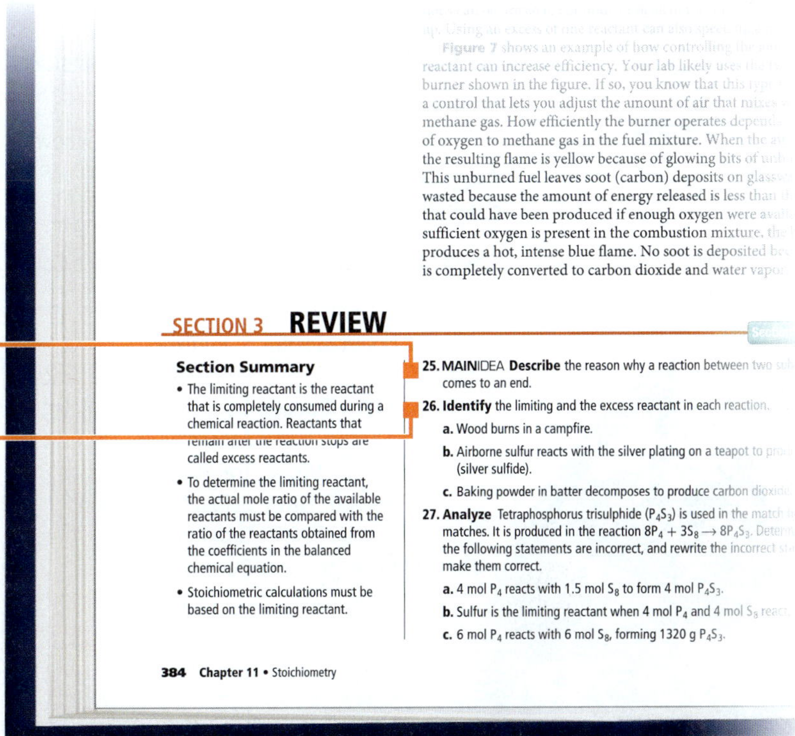

…**Figure 7** shows an example of how controlling the… reactant can increase efficiency. Your lab likely uses the… burner shown in the figure. If so, you know that this type… a control that lets you adjust the amount of air that mixes… methane gas. How efficiently the burner operates depends… of oxygen to methane gas in the fuel mixture. When the air… the resulting flame is yellow because of glowing bits of unb… This unburned fuel leaves soot (carbon) deposits on glass… wasted because the amount of energy released is less than… that could have been produced if enough oxygen were avail… sufficient oxygen is present in the combustion mixture, the… produces a hot, intense blue flame. No soot is deposited be… is completely converted to carbon dioxide and water vapor.

SECTION 3 **REVIEW**

Section Summary

- The limiting reactant is the reactant that is completely consumed during a chemical reaction. Reactants that remain after the reaction stops are called excess reactants.
- To determine the limiting reactant, the actual mole ratio of the available reactants must be compared with the ratio of the reactants obtained from the coefficients in the balanced chemical equation.
- Stoichiometric calculations must be based on the limiting reactant.

25. **MAIN**IDEA **Describe** the reason why a reaction between two s… comes to an end.
26. **Identify** the limiting and the excess reactant in each reaction.
    a. Wood burns in a campfire.
    b. Airborne sulfur reacts with the silver plating on a teapot to pro… (silver sulfide).
    c. Baking powder in batter decomposes to produce carbon dioxi…
27. **Analyze** Tetraphosphorus trisulphide (P_4S_3) is used in the match… matches. It is produced in the reaction $8P_4 + 3S_8 \rightarrow 8P_4S_3$. Deter… the following statements are incorrect, and rewrite the incorrect s… make them correct.
    a. 4 mol P_4 reacts with 1.5 mol S_8 to form 4 mol P_4S_3.
    b. Sulfur is the limiting reactant when 4 mol P_4 and 4 mol S_8 react.
    c. 6 mol P_4 reacts with 6 mol S_8, forming 1320 g P_4S_3.

384 Chapter 11 • Stoichiometry

TABLE OF CONTENTS

CHAPTER 1

Go online!

Concepts in Motion Some Branches of Chemistry
Ozone Depletion
Web Quests Environmental Chemist
Art Restoration
Video Tracking a Crime
Personal Tutor Mass and Weight Relationships

CHAPTER 2

Go online!

Concepts in Motion SI Prefixes
Precision and Accuracy
Web Quests Calibration Technician
Assessing Health Risk
Video Tracking a Crime
Personal Tutor Scientific Notation
Graphing
Virtual Investigations Density

Go online!

CHAPTER 3

CHAPTER 4

Charles D. Winters/Photo Researchers

CHAPTER 5

CHAPTER 6

©Mark Karrass/Corbis

Go online!

CHAPTER 7

Ionic Compounds and Metals 204

Go online!

CHAPTER 8

Covalent Bonding 238

Go online!

CHAPTER 9

CHAPTER 10

Mihaela Ninic/Alamy

Go online!

CHAPTER 11

CHAPTER 12

CHAPTER 13

Go online!

Concepts in Motion The Gas Laws (Animation)
The Gas Laws (Interactive Table)
Web Quests Meteorologist
Hyperbaric Oxygen
Video Gases
Leagues Under the Sea
Personal Tutor Combined Gas Law
Virtual Investigations The Gas Laws

CHAPTER 14

Go online!

Concepts in Motion Types and Examples of Solutions
Dissolution of Compounds
Strong, Weak, and Nonelectrolytes
Osmosis
Web Quests Pharmacy Technician
Environmental Chemistry
Video Water and Its Solutions
Personal Tutor Calculating Molarity and Molality
Virtual Investigations Salts and Solubility
Colligative Properties

Steve Allen/Brand X Pictures

Go online!

CHAPTER 15

Go online!

Concepts in Motions Calorimetry
Heat Flow in Endothermic and Exothermic Reactions
Reaction Spontaneity
Web Quests Heating and Cooling Specialist
Alternative Fuel Vehicles
Personal Tutor Probability
Virtual Investigations Phase Changes
Calorimetry

CHAPTER 16

Go online!

Concepts in Motion Molecular Orientation and Collision Effectiveness
Web Quests Chemical Engineer
Reaction Rate and Metabolism
Personal Tutor Solving Reaction Rates
Virtual Investigations Reaction Rates

Stephen Frisch/McGraw-Hill Education

CHAPTER 17

CHAPTER 18

Sisse Brimberg/Getty Images

Go online!

CHAPTER 19

Go online!

Concepts in Motions Redox Reaction
Summary of Redox Reactions
Web Quests Potter
Crime Scene Investigation
Personal Tutor Balancing Redox Reactions
Virtual Investigations Redox Titration

CHAPTER 20

Go online!

Concepts in Motion Voltaic Cell
Web Quests Alternative Energy Specialist
Batteries and Medical Devices
Video Hydrogen Fuel Cells
Personal Tutor Calculating Potentials
Virtual Investigations Electrochemical Cells

Terry Whittaker/Alamy

CHAPTER 21

CHAPTER 22

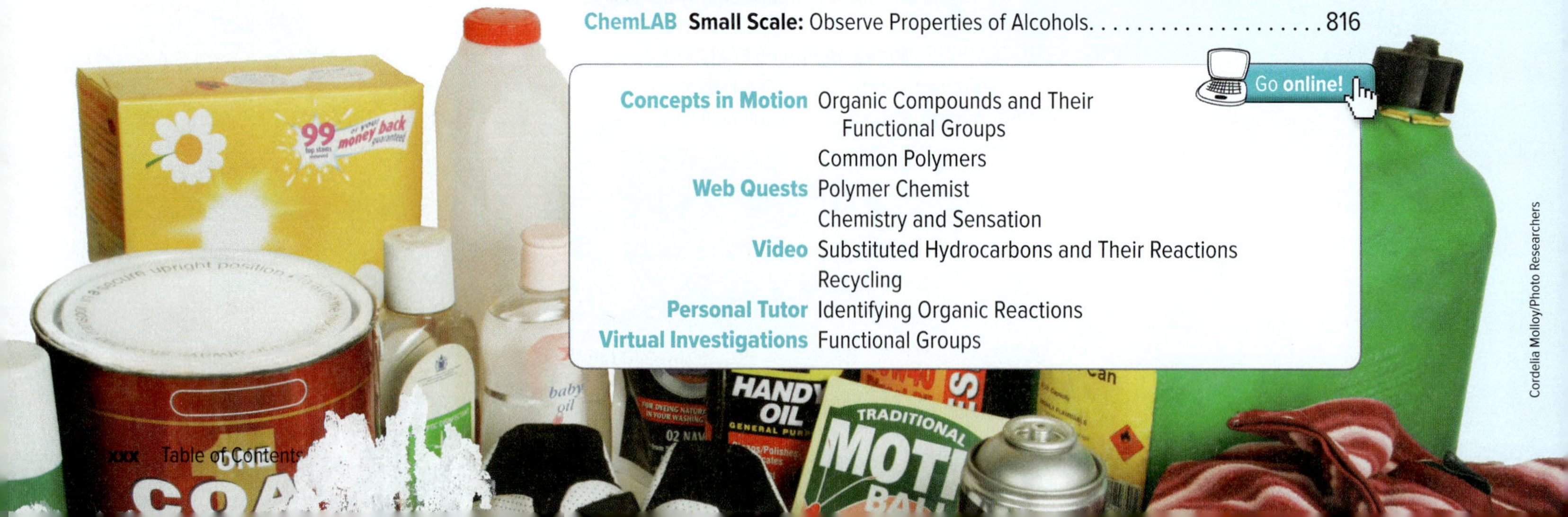

Cordelia Molloy/Photo Researchers

Go online!

CHAPTER 23

Go online!

Concepts in Motion Amino Acid Examples
Structure of DNA
Web Quests Baker
Molecular Paleontology
Video Enzymes and Surfactants
Personal Tutor Photosynthesis and Respiration

CHAPTER 24

Go online!

Concepts in Motion Summary of Radioactive Decay Processes
Nuclear Chain Reaction
Critical Mass
Nuclear Power
Web Quests Radiation Therapist
Neutron Activation Analysis
Video Radioisotopes
Personal Tutor Exponential Graphing
Virtual Investigations Half-Life

Chris Johnson/Alamy

STUDENT RESOURCES

Elements Handbook 901

Math Handbook 946

Reference Tables 968

Supplemental Practice Problems 977

Selected Solutions 993

Glossary • Glosario 1006

Index 1032

FOLDABLES® by Dinah Zike

Folding Instructions

The following pages offer step-by-step instructions to make the Foldables study guides.

Layered-Look Book

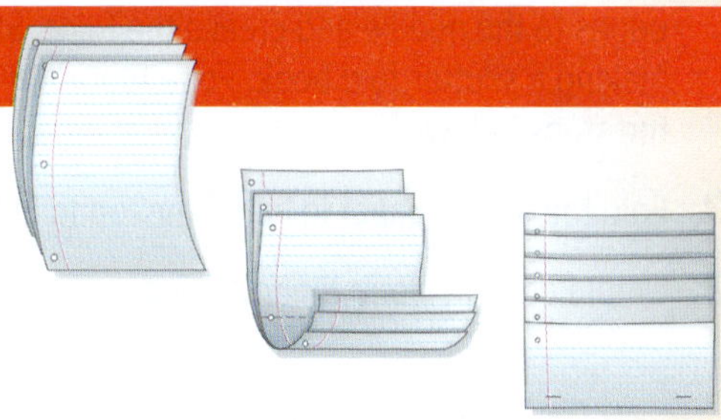

1. Collect three sheets of paper and layer them about 1 cm apart vertically. Keep the edges level.
2. Fold up the bottom edges of the paper to form 6 equal tabs.
3. Fold the papers and crease well to hold the tabs in place. Staple along the fold. Label each tab.

Trifold Book

1. Fold a vertical sheet of paper into thirds.
2. Unfold and label each row.

Three-Tab Book

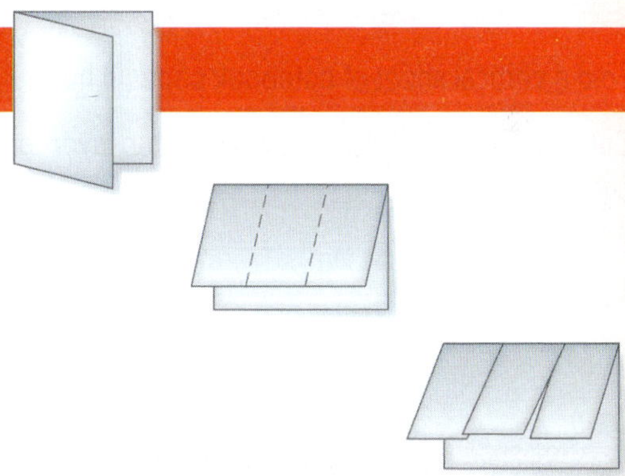

1. Fold a horizontal sheet of paper from side to side. Make the front edge about 2 cm shorter than the back edge.
2. Turn length-wise and fold into thirds.
3. Unfold and cut only the top layer along both folds to make three tabs. Label each tab.

Two- and Four-Tab Books

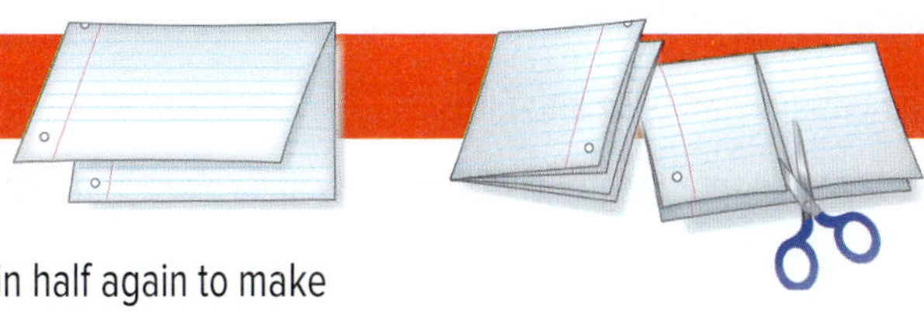

1. Fold a sheet of paper in half.
2. Fold in half again. If making a four-tab book, then fold in half again to make three folds.
3. Unfold and cut only the top layer along the folds to make two or four tabs. Label each tab.

FOLDABLES®

by Dinah Zike

Shutter-Fold and Four-Door Books

1. Find the middle of a horizontal sheet of paper. Fold both edges to the middle and crease the folds. Stop here if making a shutter-fold book. For a four-door book, complete the steps below.
2. Fold the folded paper in half, from top to bottom.
3. Unfold and cut along the fold lines to make four tabs. Label each tab.

Concept-Map Book

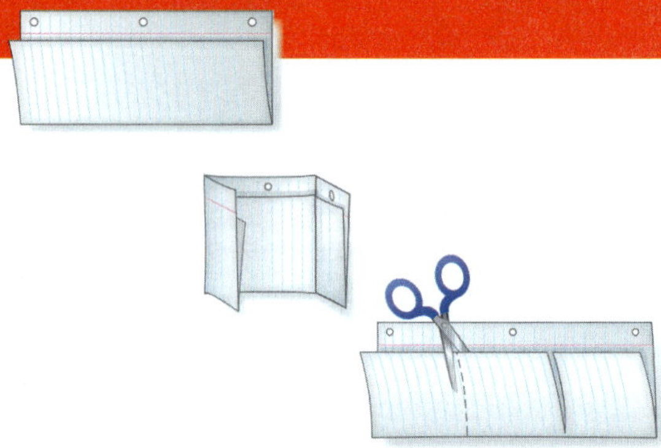

1. Fold a horizontal sheet of paper from top to bottom. Make the top edge about 2 cm shorter than the bottom edge.
2. Fold width-wise into thirds.
3. Unfold and cut only the top layer along both folds to make three tabs. Label the top and each tab.

Vocabulary Book

1. Fold a vertical sheet of notebook paper in half.
2. Cut along every third line of only the top layer to form tabs. Label each tab.

Folded Chart

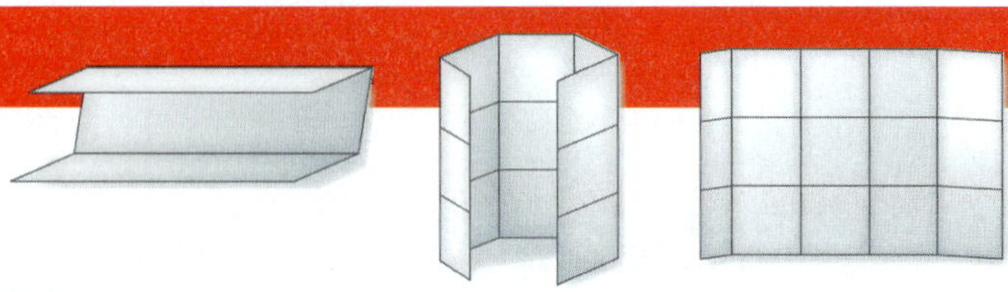

1. Fold a sheet of paper length-wise into thirds.
2. Fold the paper width-wise into fifths.
3. Unfold, lay the paper length-wise, and draw lines along the folds. Label the table.

Pocket Book

1. Fold the bottom of a horizontal sheet of paper up about 3 cm.
2. If making a two-pocket book, fold in half. If making a three-pocket book, fold in thirds.
3. Unfold once and dot with glue or staple to make pockets. Label each pocket.

Bound Book

1. Fold several sheets of paper in half to find the middle. Hold all but one sheet together and make a 3-cm cut at the fold line on each side of the paper.
2. On the final page, cut along the fold line to within 3-cm of each edge.
3. Slip the first few sheets through the cut in the final sheet to make a multi-page book.

Top-Tab Book

1. Layer multiple sheets of paper so that about 2–3 cm of each can be seen.
2. Make a 2–3-cm horizontal cut through all pages a short distance (3 cm) from the top edge of the top sheet.
3. Make a vertical cut up from the bottom to meet the horizontal cut.
4. Place the sheets on top of an uncut sheet and align the tops and sides of all sheets. Label each tab.

Accordion Book

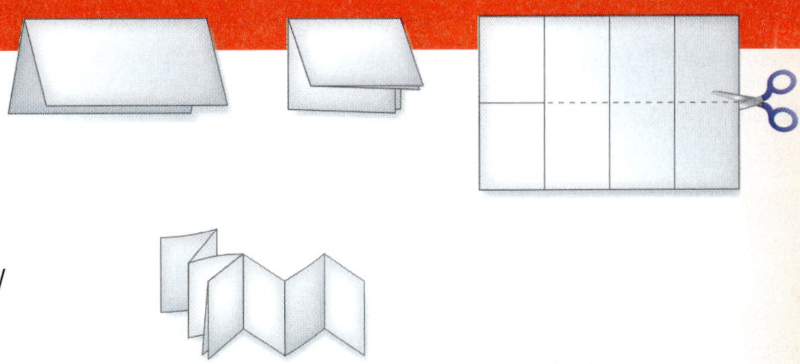

1. Fold a sheet of paper in half. Fold in half and in half again to form eight sections.
2. Cut along the long fold line, stopping before you reach the last two sections.
3. Refold the paper into an accordion book. You may want to glue the double pages together.

CHAPTER 1 Introduction to Chemistry

BIGIDEA **Chemistry is a science that is central to our lives.**

ESSENTIAL QUESTIONS	RESOURCES TO ASSESS MASTERY
SECTION 1 **A Story of Two Substances** **1.** What is a substance? **2.** How does ozone form and why is it important? **3.** What are chlorofluorocarbons and how do they get into the atmosphere? 1 session 0.5 block	**Progress Monitoring** Formative Assessment, pp. 5, 6, 7 Reading Check, pp. 5, 7 Graph Check, p. 8 Section Review, p. 8
SECTION 2 **Chemistry and Matter** **1.** How do mass and weight compare and contrast? **2.** Why are chemists interested in a submicroscopic description of matter? **3.** What defines the various branches of chemistry? 1 session 0.5 block	**Progress Monitoring** Formative Assessment, p. 11 Reading Check, p. 10 Section Review, p. 11
SECTION 3 **Scientific Methods** **1.** What are the common steps of scientific methods? **2.** What are the similarities and differences between qualitative data and quantitative data? **3.** In an experiment, which variable is the independent variable, which is the dependent variable, and which are controls? **4.** What is the difference between a theory and a scientific law? 1 session 0.5 block	**Progress Monitoring** Formative Assessment, pp. 12, 13, 16 Reading Check, pp. 13, 14 Section Review, p. 16
SECTION 4 **Scientific Research** **1.** How do pure research, applied research, and technology compare and contrast? **2.** What are some of the important rules for laboratory safety? 1 session 0.5 block	**Progress Monitoring** Formative Assessment, pp. 19, 22 Reading Check, p. 20 Graph Check, p. 20 Section Review, p. 22 **Summative Assessment** Chapter Assessment, p. 26 *eAssessment* Chapter Test (scaffolded)

Go online!

LEVELED RESOURCES	LAB MATERIALS
Science Notebook 1.1 OL ***FAST FILE Chapter Resources:*** Study Guide, p. 12 BL	LaunchLAB p. 2 / **15 min** laboratory balance, candle, match, stopwatch or clock with a second hand
Science Notebook 1.2 OL ***FAST FILE Chapter Resources:*** Study Guide, p. 13 BL	
Science Notebook 1.3 OL ***FAST FILE Chapter Resources:*** MiniLab Worksheet, p. 2 OL Study Guide, p. 15 BL	MiniLAB p. 13 / **20 min** water, two petri dishes, graduated cylinder, vegetable oil, toothpick, liquid dishwashing detergent, whole milk, four different food colorings
Science Notebook 1.4 OL ***FAST FILE Chapter Resources:*** ChemLab Worksheet, p. 3 OL Study Guide, p. 17 BL	ChemLAB p. 24 / **45 min** test tubes with stoppers, test tube rack, grease pencil, 25-mL graduated cylinder, distilled water, dropper, 250-mL beaker, Water Sample 1, Water Sample 2, dish detergent, metric ruler

ADDITIONAL RESOURCES

- eAssessment
- ConnectED Student Center
- ConnectED Teacher Center
- Lesson Presentations
- eSolutions
- Videos
- Animations
- Online Quizzes
- Virtual Investigations
- Multilingual eGlossary
- Personal Tutors
- WebQuests
- Applying Practices Worksheets
- PBLs

Practice:

- Challenge Problems AL
- Supplemental Problems BL OL
- Chapter Test (Scaffolded)
- Standardized Test Practice

Additional Resources:

- Blueprints for Success
- Solving Problems: A Chemistry Handbook

Lab Resources:

- Laboratory Manual OL
- CBL Laboratory Manual OL
- Small-Scale Laboratory Manual OL
- Forensics Laboratory Manual OL

BL Below Level   OL On Level   AL Advanced Learners   EL English Learners   COOP LEARN Cooperative Learning

CHAPTER 1

LaunchLAB

Rubric

Purpose Students will observe a chemical reaction and relate their observations to definitions of matter and chemistry.

Safety Precautions Approve lab safety forms before work begins. Warn students to use care around an open flame.

Disposal Have students place used matches in a flameproof container or in a container of water.

Teaching Strategies
- Ask students what evidence they observe that indicates a change in matter is occurring. Answers might include a change in appearance or odor produced, or gas and energy produced.
- Use the odor produced by this experiment to introduce a discussion of air pollution. This lab can be used to introduce the ozone problem discussed in this chapter.

Expected Results The candle loses mass after it is burned.

Procedure

1. Read and complete the lab safety form.
2. Use a **laboratory balance** to measure the mass of a **candle.** Record this measurement, and record detailed observations about the candle.
3. Place the candle on a burn-resistant surface, such as a lab table. Carefully strike a **match** and light the candle. Use a **stopwatch** or a **clock with a second hand** to measure the time. Allow the candle to burn for 5 min. Then, blow out the flame. Record your observations.
   WARNING: ***Do not place matches in the sink.***
4. Allow the candle to cool. Measure and record the mass of the extinguished candle.
5. Place the extinguished candle in a container designated by your instructor.

CHAPTER 1
Introduction to Chemistry

BIGIDEA Chemistry is a science that is central to our lives.

SECTIONS

1 A Story of Two Substances

2 Chemistry and Matter

3 Scientific Methods

4 Scientific Research

LaunchLAB

Where did the mass go?

When an object burns, the mass of what remains is less than the original object. What happens to the mass of the object?

FOLDABLES Study Organizer

Scientific Methods

Make a concept-map book. Label the tabs as follows: *Observation, Hypothesis, Experiments,* and *Conclusion.* Use it to summarize what you learn about scientific methods.

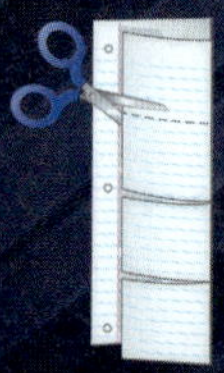

(t)TED KINSMAN/SCIENCE PHOTO LIBRARY/Photo Researchers; (b)Beateworks Inc./Alamy; (b)Daniel Sambraus/Photo Researchers; (bkgd)ping han/Alamy

Analysis

1. **Summarize** your observations of the candle as it was burning and after the flame was extinguished. Sample Summary: A small amount of smoke came from the candle and the candle got smaller as it burned. After it burned, the candle looked smaller than it did before it burned.
2. **Evaluate** Where is the matter that appears to have been lost? It changed to a gas and it was released into the room.

Inquiry Can the change in the mass of the candle vary? Plan an investigation to determine what factors might contribute to a different outcome. Yes, the amount of matter lost is a variable. Possible factors that might contribute to a different outcome are the chemical composition of the candle, the chemical composition of the wick, and the diameter of the candle.

Many of the processes that occur around you—such as the freezing of water, the burning of a log, and the rusting of a nail— are examples of chemistry in action.

Introduce the BIGIDEA

Chemistry and Life To introduce this chapter's Big Idea, lead students in a discussion of chemistry that occurs around them every day. Ask students to give an example of chemistry that is occurring in the classroom not related to lab experiments. Accept all reasonable responses. Possible answers: respiration, digestion, and various examples of heat transfer. Ask students to give examples of chemistry that occur in their homes. Accept all reasonable responses. Possible answers: cooking food, burning candles, and boiling water.

Use the Photo

Chemistry in the Home Ask students to look at the photo of the cabin. Ask students to suggest examples of additional chemical reactions that might occur inside a cabin or inside their own home. Possible answers: cooking food, burning natural gas, fuel oil, or propane for heat, or burning candles.

SECTION 1

1 Focus

MAINIDEA

Chemistry Around You Ask students to recall the last time someone baked something, such as cake, bread, or cookies, at their houses. Ask students if chemistry was involved. Yes, several chemical reactions usually occur when something is baked, such as the chemical reaction when baking powder is mixed with water and a gas is produced, or the chemical reaction when baking soda is mixed with buttermilk and a gas is produced. Ask students if chemistry is involved when an MP3 player is turned on. Yes, a chemical reaction occurs in the battery. Hold up a calculator, and ask students what type of chemistry takes place during the manufacturing of the calculator. Possible answers: Chemistry is used during the manufacturing process for the case and for the chips that perform the calculations. **OL** **EL**

2 Teach

Concept Development

Chemicals Ask students what comes to mind when they hear the term *chemical.* Often, the term has a negative connotation. Emphasize that chemicals are all around and that humans could not exist without them. Some chemicals can be harmful. Others are not only helpful, but also essential. **OL** **EL**

SECTION 1

Essential Questions

- What is a substance?
- How does ozone form and why is it important?
- What are chlorofluorocarbons and how do they get into the atmosphere?

Review Vocabulary

matter: anything that has mass and takes up space

New Vocabulary

chemistry
substance

A Story of Two Substances

MAINIDEA Chemistry is the study of everything around us.

CHEM 4 YOU Have you ever moved a piece of furniture to a new location, only to discover that the new location won't work? Sometimes, moving furniture creates a new problem, such as a door will not open all the way or an electric cord will not reach an outlet. Solving a problem only to find that the solution creates a new problem also occurs in science.

Why study chemistry?

Take a moment to observe your surroundings and **Figure 1.** Where did all the "stuff" come from? All the stuff in the universe, including everything in the photos, is made from building blocks formed in stars. Scientists call these building blocks and the "stuff" made from these building blocks *matter.*

As you begin your study of **chemistry**—the study of matter and the changes that it undergoes—you are probably asking yourself, "Why is chemistry important to me?" The answer to this question can be illustrated by real-life events that involve two discoveries. One discovery involves something that you probably use every day—refrigeration. If you go to school in an air-conditioned building or if you protect your food from spoilage by using a refrigerator, this discovery is important to you. The other discovery involves energy from the Sun. Because you eat food and spend time outdoors, this discovery is also important to you. These two seemingly unrelated discoveries became intertwined in an unexpected way—as you will soon learn.

Figure 1 Everything in the universe, including particles in space and things around you, is composed of matter.

CHEMISTRY JOURNAL

Chemistry Ask students to write a few paragraphs in their chemistry journal describing what they would like to learn in chemistry class and what they expect to learn. Refer to these paragraphs later in the year to see if their expectations were met. **BL** **OL** **AL** **EL**

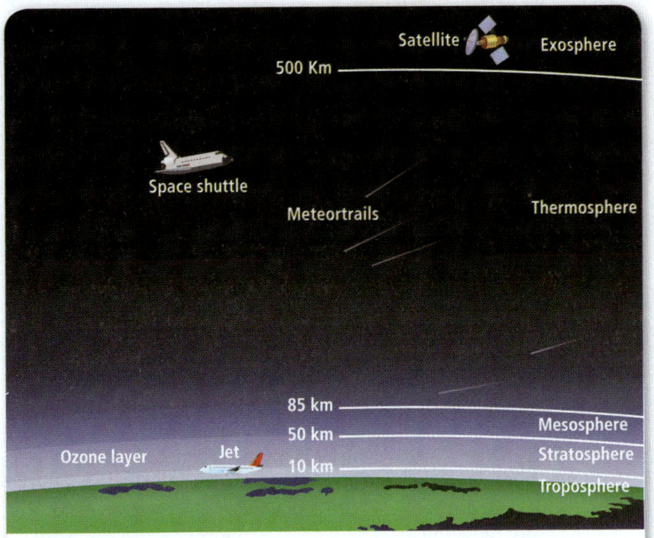

Figure 2 Earth's atmosphere consists of several layers. The protective ozone layer is located in the stratosphere.

The Ozone Layer

If you have ever had a sunburn, you have experienced the damaging effects of ultraviolet radiation from the Sun. Overexposure to ultraviolet radiation is harmful to both plants and animals. Increased levels of a type of ultraviolet radiation called UVB can cause cataracts and skin cancer in humans, lower crop yields in agriculture, and disrupted food chains in nature.

Living organisms have evolved in the presence of UVB, and cells have some ability to repair themselves when exposed to low levels of UVB. However, some scientists believe that when UVB levels reach a certain point, the cells of living organisms will no longer be able to cope, and many organisms will die.

Earth's atmosphere Living organisms on Earth exist because they are protected from high levels of UVB by ozone. Ozone, which is made up of oxygen, is a substance in the atmosphere that absorbs most harmful radiation before it reaches Earth's surface. A **substance,** which is also known as a chemical, is matter that has a definite and uniform composition.

About 90% of Earth's ozone is spread out in a layer that surrounds and protects our planet. As you can see in **Figure 2,** Earth's atmosphere consists of several layers. The lowest layer is called the troposphere and contains the air we breathe. The troposphere is where clouds occur and where airplanes fly. All of Earth's weather occurs in the troposphere. The stratosphere is the layer above the troposphere. It extends from about 10 to 50 kilometers (km) above Earth's surface. The ozone layer that protects Earth is located in the stratosphere.

READING CHECK **Explain** the benefits of ozone in the atmosphere.

RealWorld CHEMISTRY
The Ozone Layer

SUNSCREEN To offer some protection from harmful UV radiation, sunscreen can be applied to the skin. Sunscreen helps prevent sunburn and skin cancer. Health professionals recommend the use of sunscreen anytime that you are outdoors and exposed to the Sun's ultraviolet radiation.

VOCABULARY
WORD ORIGIN
Ozone
comes from the Greek word *ozōn,* which means *to smell*

Quick Demo

Substances or Chemicals
Demonstrate that chemicals are everywhere. Burn a large candle in front of the classroom. Explain that the wax in the candle is a chemical, as is the oxygen in the air needed to burn the wax. If you hold your hand too close to the flame, the flame will burn the chemicals that form your skin. UV light from the Sun can also burn your skin. Sunscreen contains chemicals that absorb UV light before it reaches your skin. Ozone is a chemical in the air that absorbs UV light before it reaches Earth's surface.

ASSESSMENT

Knowledge Provide students with two lists. One list should contain the layers of the atmosphere, listed in random order. The other list should contain one characteristic of each layer. The characteristics should relate to the chemistry of the layer. Ask students to match each layer to its characteristic. OL

READING CHECK Ozone absorbs harmful UV light from the Sun, preventing it from reaching Earth's surface where it could damage living organisms.

CHEMISTRY PROJECT

UV-B and Living Organisms Divide students into small groups, and have them research the effects of increased UV-B radiation on living organisms. Students should prepare an oral presentation with visual aids to present to the rest of the class. OL COOP LEARN

■ **Caption Question Fig. 3** Ozone and oxygen gas molecules continuously break apart and reform in the stratosphere.

ASSESSMENT

Skill Have students draw the layers of the atmosphere and indicate where ozone is formed and stored. **OL** **EL**

Quick Demo

The Equator Bring in a globe and an electric lamp. Ask students to identify the equator, where ozone is produced in the greatest quantity. Show students how the light rays strike Earth directly at the equator. Have a student volunteer confirm that more energy is felt where light rays strike directly than where they strike at an angle. Also, show how convection currents, caused by unequal heating of the atmosphere, flow ozone from the equator to the poles. **OL**

Reinforcement

Ozone Formation Ask students why the majority of ozone would be formed over the equator. Ozone formation in the stratosphere depends on UV rays from the Sun striking and breaking up oxygen. The concentration of direct rays at the equator is greater than the concentration of rays striking other parts of Earth. **OL**

■ **Figure 3** Ultraviolet radiation from the Sun causes some oxygen gas (O_2) to break into individual particles of oxygen (O). These individual particles combine with oxygen gas (O_2) to form ozone (O_3).

Explain ***why there is a balance between oxygen gas and ozone levels in the stratosphere.***

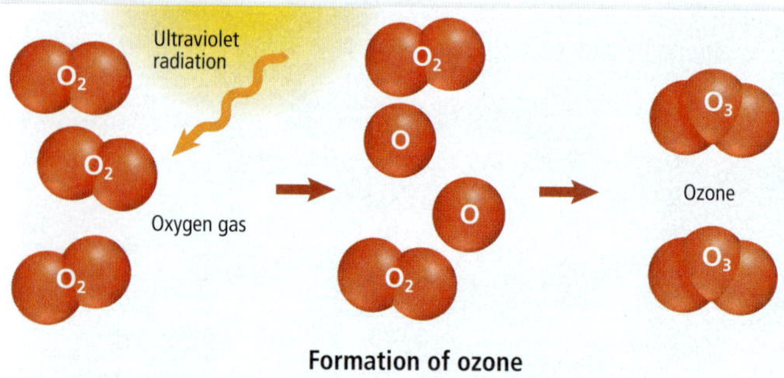

Ozone formation How does ozone enter the stratosphere? When oxygen gas (O_2) is exposed to ultraviolet radiation in the upper regions of the stratosphere, ozone (O_3) is formed. Molecules of oxygen gas are made of two smaller oxygen particles. The energy of the radiation breaks the oxygen gas into individual oxygen particles (O), which then interact with O_2 to form O_3. **Figure 3** illustrates this process. Ozone can also absorb radiation and break apart to reform oxygen gas. Thus, there tends to be a balance between oxygen gas and ozone levels in the stratosphere.

Ozone was first identified and measured in the late 1800s, so its presence has been studied for a long time. It was of interest to scientists because air currents in the stratosphere move ozone around Earth. Ozone forms over the equator, where the rays of sunlight are the strongest, and then flows toward the poles. Thus, ozone makes a convenient marker to follow the flow of air in the stratosphere.

In the 1920s, British scientist G.M.B. Dobson (1889–1976) began measuring the amount of ozone in the atmosphere. Although ozone is formed in the higher regions of the stratosphere, most of it is stored in the lower stratosphere. Ozone can be measured in the lower stratosphere by instruments on the ground or in balloons, satellites, and rockets. Dobson's measurements helped scientists determine the normal amount of ozone that should be in the stratosphere. Three hundred Dobson units (DU) is considered the normal amount of ozone in the stratosphere. Instruments, like those shown in **Figure 4,** monitor the amount of ozone present in the stratosphere today.

■ **Figure 4** Scientists use a variety of equipment, including this Brewer spectrometer, to take ozone measurements.

David Hay Jones/Science Photo Library/Photo Researchers

CULTURAL DIVERSITY

Skin Tones Darker skin tones offer more protection from harmful UV rays in sunlight. Therefore, where rays from the Sun are strong near the equator, human skin tones have evolved to be darker. UV rays are less intense in areas farther from the equator, so human skin tones are lighter there. Although the Inuit people live far north of the equator, their skin tones are darker than would be expected because snow reflects UV light. The Inuit need darker skin to protect them from the increased levels of UV light they receive from reflection.

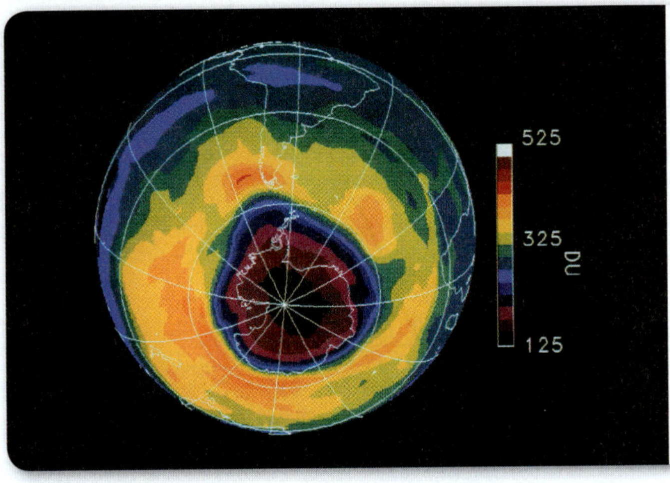

Figure 5 Satellite photos confirmed the British Antarctic Survey team's measurements that the ozone layer was thinning over Antarctica. On this satellite map, the area over Antarctica appears pink, purple, and black. The color-key on the right indicates that the ozone level ranges from 125 to about 200 Dobson Units, which is well below the normal level of 300 Dobson units.

Between 1981 and 1983, a research group from the British Antarctic Survey was monitoring the atmosphere above Antarctica. They measured surprisingly low levels of ozone—readings as low as 160 DU—especially during the Antarctic spring in October. They checked their instruments and repeated their measurements. In October 1985, they reported a confirmed decrease in the amount of ozone in the stratosphere and concluded that the ozone layer was thinning. **Figure 5** shows how the thinning ozone layer looked in October 1990.

Although the thinning of the ozone layer is often called the ozone hole, it is not a hole. The ozone is still present in the atmosphere. However, the protective layer is much thinner than normal. This fact has alarmed scientists, who never expected to find such low levels. Measurements made from balloons, high-altitude planes, and satellites have supported the measurements made from the ground. What could be causing the ozone hole?

Chlorofluorocarbons

The story of the second substance in this chapter begins in the 1920s. Large-scale production of refrigerators, which at first used toxic gases such as ammonia as coolants, was just beginning. Because ammonia fumes could escape from the refrigerator and harm the members of a household, chemists began to search for safer coolants. Thomas Midgley, Jr. synthesized the first chlorofluorocarbons in 1928. A chlorofluorocarbon (CFC) is a substance that consists of chlorine, fluorine, and carbon. Several different substances are classified as CFCs. They are all made in the laboratory and do not occur naturally. CFCs are nontoxic and stable—they do not readily react with other substances. At the time, they seemed to be ideal coolants for refrigerators. By 1935, the first self-contained home air-conditioning units and eight million new refrigerators in the United States used CFCs as coolants. In addition to their use as refrigerants, CFCs were also used in plastic foams, solvents, and as propellants in spray cans.

☑ READING CHECK **Explain** why scientists thought CFCs were safe for the environment.

CAREERS IN CHEMISTRY

Environmental Chemist An environmental chemist uses tools from chemistry and other sciences to study how chemicals interact with the physical and biological environment. This includes identifying the sources of pollutants such as ozone and their effects on living organisms.

WebQuest

Science Source/Photo Researchers

Concept Development

Coolants Have a local mechanic or air conditioning specialist speak to the class about the safeguards that currently exist to keep harmful coolants from affecting the atmosphere. Ask him or her to clarify that refrigerants shown to be harmful to the environment can be either eliminated or replaced with a coolant that is less harmful.

Visual Learning

Figure 5 Have students look at Figure 5 and describe in detail what the image shows. Discuss the image as a class to make sure that all students understand the image. OL

☑ READING CHECK CFCs do not readily react with other substances; therefore, scientists thought the molecules were stable.

ASSESSMENT

Knowledge Ask students where chemistry is used in their everyday lives. Answers might include the fuel used to power their cars or heat their homes, the clothing they wear, and the food they eat. OL EL

CAREERS IN CHEMISTRY

WebQuest

CHEMISTRY JOURNAL

Thomas Midgely Have students research Thomas Midgely, Jr. Then, have them write a short biographical summary of his life. OL

DIFFERENTIATED INSTRUCTION

Visually Impaired Have sighted students work with visually-impaired students to create a three-dimensional, tactile model of the layers of the atmosphere. Have the sighted student explain the location of the troposphere and stratosphere and the process of ozone formation and storage. OL EL COOP LEARN

☑ GRAPH CHECK The global concentration of CFC-11 in the atmosphere increased until around 1993, where it leveled out. It has been decreasing since around 1994.

3 Assess

Check for Understanding

What is the normal level of ozone in the stratosphere? 300 DU What was the lowest level that scientists found over the Antarctic in the early 1980s? 160 DU Have students explain why scientists were alarmed by these findings. OL EL

Reteach

Bring in a thin sock or item of clothing. Demonstrate that the material is still present, but it is thinner than normal and allows more light to pass through. Have students explain how this model is similar to the ozone hole. BL EL

Extension

Discuss with students how the processes used to manufacture products change over time. Include in the discussion the role chemistry has in these changes. Mention the development of refrigerants as discussed in the text. Bring to class a cardboard milk container and a plastic milk jug. Ask students to describe the advantages and disadvantages of each type of container. The cardboard will decompose over time; the plastic will not. Both can be recycled. Milk will stay fresher in the plastic. AL

■ **Figure 6** Scientists collected data on the global use of CFCs and the accumulation of CFCs over Antarctica. CFC-11 is one particular type of CFC. In the graph, the concentration of CFC-11 in the atmosphere is shown in parts per trillion (ppt).

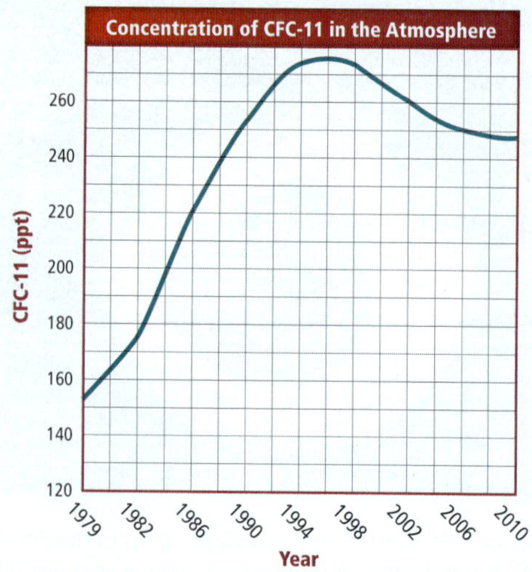

☑ GRAPH CHECK
Describe the trend in the data from 1979 through 2010.

Scientists first began to detect the presence of CFCs in the atmosphere in the 1970s. They decided to measure the amount of CFCs in the stratosphere and found that quantities in the stratosphere increased year after year. By 1990, the concentration of CFCs had reached an all-time high, as shown in **Figure 6.** However, it was widely thought that CFCs did not pose a threat to the environment because they are so stable, and consequently many scientists were not alarmed.

Scientists had noticed and measured two separate phenomena: the protective ozone layer in the atmosphere was thinning, and increasingly large quantities of CFCs were drifting into the atmosphere. Could there be a connection between the two occurrences? Before you learn the answer to this question, you need to understand some of the basic ideas of chemistry and know how chemists—and most scientists—solve scientific problems.

SECTION 1 REVIEW

Section Self-Check

Section Summary

- Chemistry is the study of matter.
- Chemicals are also known as substances.
- Ozone is a substance that forms a protective layer in Earth's atmosphere.
- CFCs are synthetic substances made of chlorine, fluorine, and carbon that were originally thought to be the ideal coolants for refrigeration.

1. MAINIDEA **Explain** why the study of chemistry should be important to everyone.
2. **Define** *substance* and give two examples of things that are substances.
3. **Describe** how the ozone layer forms and why it is important.
4. **Explain** why chlorofluorocarbons were developed and how they were used.
5. **Explain** If cells have the ability to repair themselves after exposure to UVB, why do the increasing levels of UVB in the atmosphere concern scientists?
6. **Explain** why the concentration of CFCs increased in the atmosphere.
7. **Evaluate** why it was important for Dobson's data to be confirmed by satellite photos.

SECTION 1 REVIEW

1. Chemistry is the study of matter, and everything and everyone is made of matter.
2. A substance, which is also known as a chemical, is matter that has a definite composition. Possible examples: table salt (NaCl) and table sugar (sucrose, $C_{12}H_{22}O_{11}$)
3. When oxygen gas (O_2) is exposed to ultraviolet radiation in the upper regions of the stratosphere, the molecule breaks apart. The individual oxygen particles (O) combine with other oxygen gas molecules to form ozone (O_3). Ozone is important because it forms a protective layer in the atmosphere that protects living organisms from harmful radiation.
4. Chlorofluorocarbons were developed as a safe alternative to ammonia, a common refrigerant. Chlorofluorocarbons were used as refrigerants, in foams, and as propellants in spray cans.
5. Cells have some ability to repair themselves, but some scientists believe that cells have a limit in the amount of UVB exposure that can be tolerated.
6. The use of CFCs continued to increase.
7. All scientific hypotheses, tests, experiments, and data must be independently confirmed to make them valid.

SECTION 2

Chemistry and Matter

MAINIDEA Branches of chemistry involve the study of different kinds of matter.

Essential Questions

- How do mass and weight compare and contrast?
- Why are chemists interested in a submicroscopic description of matter?
- What defines the various branches of chemistry?

Review Vocabulary

technology: a practical application of scientific information

New Vocabulary

mass
weight
model

CHEM 4 YOU Chemistry is sometimes called the central science. Research and technology, such as greener energy and cures for disease, rely on chemistry. Even when you brush your teeth or digest your breakfast, important chemical processes are at work.

Matter and its Characteristics

Matter, the stuff of the universe, has many different forms. Everything around you, like the things in **Figure 7,** is matter. Some matter occurs naturally, such as ozone, and other substances are not natural, such as CFCs, which you read about in Section 1.

You might realize that everyday objects are composed of matter, but how do you define matter? Recall that matter is anything that has mass and takes up space. Also recall that **mass** is a measurement that reflects the amount of matter. You know that your textbook has mass and takes up space, but is air matter? You cannot see it and you cannot always feel it. However, when you inflate a balloon, it expands to make room for the air. The balloon gets heavier. Thus, air must be matter. Is everything matter? The thoughts and ideas that fill your head are not matter; neither are heat, light, radio waves, nor magnetic fields. What else can you name that is not matter?

Mass and weight Have you ever used a bathroom scale to measure your weight? **Weight** is a measure not only of the amount of matter but also of the effect of Earth's gravitational pull on that matter. This force is not exactly the same everywhere on Earth and actually becomes less as you move away from Earth's surface at sea level. You might not notice a difference in your weight from one place to another, but subtle differences do exist.

Figure 7 Everything in this photo is matter and has mass and weight.
Compare and contrast *mass and weight.*

Fuse/Getty Images

SECTION 2

1 Focus

MAINIDEA

Branches of Chemistry Write the term *biochemistry* on the board. Ask students what they think a biochemist would study. the chemistry of life Write the term *environmental chemistry* on the board. Have students infer what an environmental chemist studies. chemistry and the environment Point out to students that the study of chemistry is broad and includes many areas. Many chemists specialize and focus their studies on a narrow part of chemistry.

2 Teach

- **In-Text Question** Answers might include feelings, emotions, microwaves, and sound.
- **Caption Question Fig. 7** Mass is a measure of the amount of matter. It is independent of gravity. Weight is the effect of gravity on matter.

Quick Demo

Chemistry and Matter

Light a candle used in the Launch Lab. Discuss the burning of the candle in terms of matter. Chemistry involves the study of the composition of matter, such as the wax in the candle and the oxygen in the air, and changes in matter, such as the changes that occur in the wax as it burns.

DIFFERENTIATED INSTRUCTION

Visually Impaired Have students choose several objects, such as their textbooks, and describe them. Characteristics might include that they have weight and shape. Help students understand these properties as mass and volume. Blow up a balloon. Have students touch it to "feel" the mass of the balloon and volume of air contained in the balloon. BL EL

CHEMISTRY PROJECT

Space Travel Have students research how astronauts perform common tasks, such as working with tools and eating, in the weightlessness of space. Have students prepare a short report detailing their findings. BL OL AL EL

Concept Development

Matter Bring to class cubes of different shapes, sizes, and materials. You could use wood, plastic, paper, and foam blocks. Ask students to compare and contrast the blocks. Ask students if the blocks are matter, and have them justify their answers. All are matter because they have mass and take up space. They differ in that each type of block contains a different type and amount of matter. **OL**

■ **Caption Question Fig. 8** The concept of atoms is difficult to understand because the unaided human eye cannot see them. Models help chemists "see" atoms and study them.

☑ READING CHECK Answers will vary, but might include models of automobiles, consumer products, the atmosphere, and so on.

FOLDABLES® Rubric

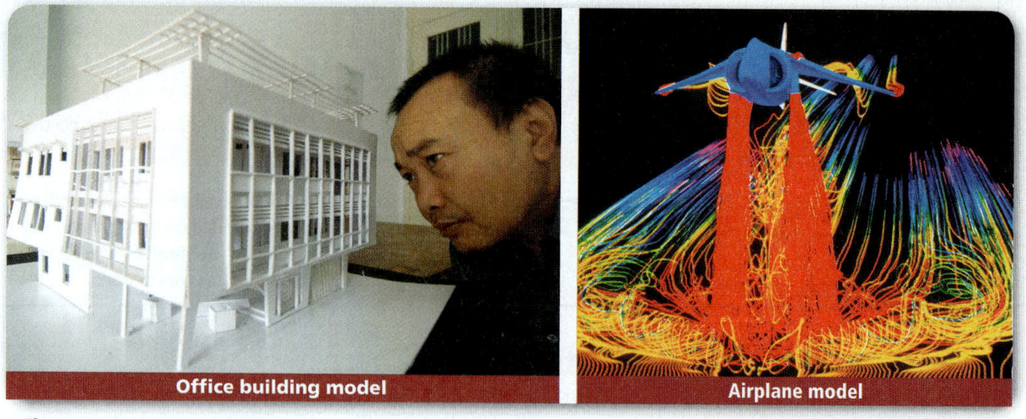

■ **Figure 8** Scientists use models to visualize complex ideas, such as the materials and structure used to build office buildings. They also use models to test a concept, such as a new airplane design, before it is mass produced.

Infer *why chemists use models to study atoms.*

It might seem more convenient for scientists to simply use weight instead of mass. Why is it so important to think of matter in terms of mass? Scientists need to be able to compare the measurements that they make in different parts of the world. They could identify the gravitational force every time they weigh something, but that would not be practical or convenient. They use mass as a way to measure matter independently of gravitational force.

Structure and observable characteristics What can you observe about the outside of your school building? You know that there is more to the building than what you can observe from the outside. Among other things, there are beams inside the walls that give the building structure, stability, and function. Consider another example. When you bend your arm at the elbow, you observe that your arm moves, but what you cannot see is that muscles under the skin contract and relax to move your arm.

Much of matter and its behavior is macroscopic; that is, you do not need a microscope to observe it. You will learn in Chapter 3 that the tremendous variety of stuff around you can be broken down into more than a hundred types of matter called elements, and that elements are made up of particles called atoms. Atoms are so tiny that they cannot be seen even with optical microscopes. Thus, atoms are submicroscopic. They are so small that over a trillion atoms could fit onto the period at the end of this sentence. The structure, composition, and behavior of all matter can be explained on a submicroscopic level—or the atomic level. All that we observe about matter depends on atoms and the changes they undergo.

Chemistry seeks to explain the submicroscopic events that lead to macroscopic observations. One way this can be done is by making a model. A **model** is a visual, verbal, or mathematical explanation of experimental data. Scientists use many types of models to represent things that are hard to visualize, such as the structure and materials used in the construction of a building and the computer model of the airplane shown in **Figure 8.** Chemists also use several different types of models to represent matter, as you will soon learn.

FOLDABLES® Incorporate information from this section into your Foldable.

VOCABULARY

SCIENCE USAGE V. COMMON USAGE

Weight

Science usage: the measure of the amount of matter in and the gravitational force exerted on an object

The weight of an object is the product of its mass and the local acceleration of gravity.

Common usage: the relative heaviness of an object

The puppy grew so quickly it doubled its weight in a matter of weeks.

☑ READING CHECK **Identify** two additional types of models that are used by scientists.

CHEMISTRY JOURNAL

Weightlessness Have students write about how they think they would feel to be in an environment without gravity. How would lack of gravity affect their weight? Would they still fit the definition of matter? Yes, they would have mass and take up space. Use this scenario to differentiate mass and weight. **OL**

Explore the **branches of chemistry** with an interactive table. Concepts In Motion

Table 1 Some Branches of Chemistry

Branch	Area of Emphasis	Examples of Emphasis
Organic chemistry	most carbon-containing chemicals	pharmaceuticals, plastics
Inorganic chemistry	in general, matter that does not contain carbon	minerals, metals and nonmetals, semiconductors
Physical chemistry	the behavior and changes of matter and the related energy changes	reaction rates, reaction mechanisms
Analytical chemistry	components and composition of substances	food nutrients, quality control
Biochemistry	matter and processes of living organisms	metabolism, fermentation
Environmental chemistry	matter and the environment	pollution, biochemical cycles
Industrial chemistry	chemical processes in industry	paints, coatings
Polymer chemistry	polymers and plastics	textiles, coatings, plastics
Theoretical chemistry	chemical interactions	many areas of emphasis
Thermochemistry	heat involved in chemical processes	heat of reaction

Chemistry: The Central Science

Recall from Section 1 that chemistry is the study of matter and the changes that it undergoes. A basic understanding of chemistry is central to all sciences—biology, physics, Earth science, ecology, and others. Because there are so many types of matter, there are many areas of study in the field of chemistry. Chemistry is traditionally broken down into branches that focus on specific areas, such as those listed in **Table 1.** Although chemistry is divided into specific areas of study, many of the areas overlap. For example, as you can see from **Table 1,** an organic chemist might study plastics, but an industrial chemist or a polymer chemist could also focus on plastics.

Get help with **mass and weight relationships.** Personal Tutor

SECTION 2 REVIEW

Section Self-Check

Section Summary

- Models are tools that scientists, including chemists, use.
- Macroscopic observations of matter reflect the actions of atoms on a submicroscopic scale.
- There are several branches of chemistry, including organic chemistry, inorganic chemistry, physical chemistry, analytical chemistry, and biochemistry

8. MAINIDEA **Explain** why there are different branches of chemistry.
9. **Explain** why scientists use mass instead of weight for their measurements.
10. **Summarize** why it is important for chemists to study changes in the world at a submicroscopic level.
11. **Infer** why chemists use models to study submicroscopic matter.
12. **Identify** three models that scientists use, and explain why each model is useful.
13. **Evaluate** How would your mass and weight differ on the Moon? The gravitational force of the Moon is one-sixth the gravitational force on Earth.
14. **Evaluate** If you put a scale in an elevator and weigh yourself as you ascend and then descend, does the scale have the same reading in both instances? Explain your answer.

ASSESSMENT

Knowledge Ask students to identify a current issue and determine which field of chemistry would most likely study the issue. Possible answers: cure for cancer or AIDS, biochemistry. OL EL BL AL

3 Assess

Check for Understanding

Ask students to define the terms *mass* and *weight.* Mass is a measurement that reflects the amount of matter. Weight is the force of Earth's gravitational pull on matter. OL EL

Reteach

Use the equation *weight* = *mass* × *acceleration due to gravity* ($W = mg$) to show students how mass and weight are related mathematically. Point out that mass must be multiplied by the acceleration due to gravity to get the numerical value of weight. OL

Extension

Ask students to point out applications, products, or processes that occur in their everyday lives that a specific branch of chemistry would include. Possible answers: The materials used to make sneakers might be studied by a polymer chemist. The biological processes in the human body might be studied by a biochemist. OL

SECTION 2 REVIEW

8. The study of chemistry is a broad field, so chemists specialize in small areas.
9. Mass is constant and is not affected by gravity. Weight varies with gravity.
10. The changes you see with your eyes begin with changes at the submicroscopic level.
11. Models enable chemists to understand difficult concepts that they cannot normally see.
12. Possible answers: Aircraft models allow scientists to test their designs before money is spent on the aircraft. Computer models of chemical processes allow chemists to test processes before manufacturing facilities are built. Car models allow scientists to test certain features, such as wind resistance, before a car is built.
13. Your mass would be the same, but your weight would be 1/6 your weight on Earth.
14. As the elevator moves up and down at constant velocity, the reading on the scale will be the same as it is when the elevator is stationary. However, during the time the elevator accelerates upward, the scale reading will be higher; during the time when the elevator accelerates downward, the scale reading will be lower.

SECTION 3

1 Focus

MAINIDEA

Scientific Methods Ask students to give examples of questions that scientists might want to answer. Possible question: How fuel efficient is a car prototype? Write a few of the questions on the board. Have students propose ways in which scientists might find the answer(s) to each question. Possible answer: build a model and test it. OL EL

2 Teach

ASSESSMENT

Skill Ask students to draw the steps of a scientific method as a flowchart. Under each heading, have them write a one-sentence description of that step. OL EL

Caption Question Fig. 10 Qualitative: one substance is blue and the other substance is green; quantitative: the flask contains 500 mL and the graduated cylinder contains 100 mL.

READING CHECK Hypotheses are not proven facts. They are educated guesses, and they are subject to change when new data or evidence is available.

SECTION 3

Scientific Methods

MAINIDEA Scientists use scientific methods to systematically pose and test solutions to questions and assess the results of the tests.

Essential Questions

- What are the common steps of scientific methods?
- What are the similarities and differences between qualitative data and quantitative data?
- In an experiment, which variable is the independent variable, which is the dependent variable, and which are controls?
- What is the difference between a theory and a scientific law?

Review Vocabulary

systematic approach: an organized method of solving a problem

New Vocabulary

scientific method
qualitative data
quantitative data
hypothesis
experiment
independent variable
dependent variable
control
conclusion
theory
scientific law

CHEM 4 YOU When packing for a long trip, how do you start? Do you throw all of your clothes into a suitcase, or do you plan what you are going to wear? Usually, it is most effective to make a plan. Similarly, scientists develop and follow a plan that helps them investigate the world.

A Systematic Approach

You might have worked with a group on an experiment in the laboratory in a previous science course. If so, you know that each person in the group probably has a different idea about how to do the lab. Having many different ideas about how to do the lab is one of the benefits of many people working together. However, communicating ideas effectively to one another and combining individual contributions to form a solution can be difficult in group work.

Scientists approach their work in a similar way. Each scientist tries to understand his or her world based on a personal point of view and individual creativity. Often, the work of many scientists is combined in order to gain new insight. It is helpful if all scientists use common procedures as they conduct their experiments.

A **scientific method** is a systematic approach used in scientific study, whether it is chemistry, biology, physics, or another science. It is an organized process used by scientists to do research, and it provides a method for scientists to verify the work of others. An overview of the typical steps of a scientific method is shown in **Figure 9.** The steps are not meant to be used as a checklist, or to be done in the same order each time. Therefore, scientists must describe their methods when they report their results. If other scientists cannot confirm the results after repeating the method, then doubt arises over the validity of the results.

Figure 9 The steps in a scientific method are repeated until a hypothesis is supported or discarded.

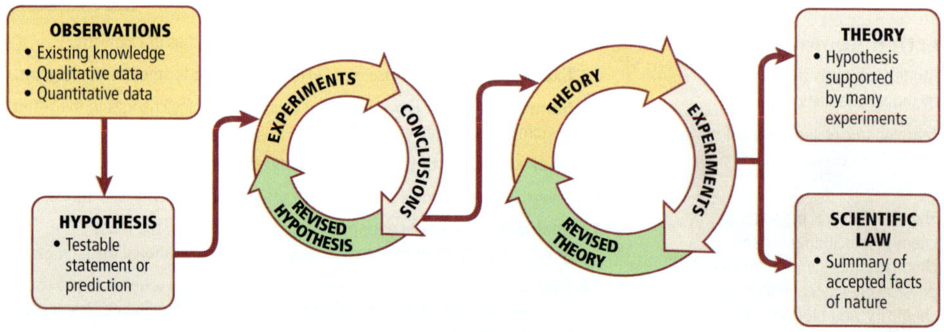

Demonstration

The Magic of Chemistry

Purpose
to demonstrate that one substance can be changed into another substance having different properties

Materials
$KMnO_4$ (0.05 g); $NaHSO_3$ (1 g); $BaCl_2{\cdot}2H_2O$ (1 g); 400-mL beakers (3); small test tubes (2)

Safety Precautions

Disposal Filter the solution through filter paper. Dispose of the solid in a landfill approved to receive chemical waste. Pour liquids down a drain.

Procedure
Before the demo, dissolve three or four small crystals of $KMnO_4$ in 250 mL of water in a beaker. In a test tube, add 1 g of $NaHSO_3$ to 1 mL of water. In a second test tube, add 1 g $BaCl_2{\cdot}2H_2O$ to 1 mL of water. **WARNING: *The solutions are toxic.*** Place the $NaHSO_3$ solution in a beaker labeled 1 and the $BaCl_2$ solution in a beaker labeled 2. To start the demonstration, show students the $KMnO_4$ solution. Pour the $KMnO_4$ solution into beaker 1. Pour the resulting solution into beaker 2.

Results
The purplish solution turns into a clear solution, and the clear solution is then changed into a milky-colored solution. The chemistry of this demo will be of

MiniLAB

Develop Observation Skills

Why are observation skills important in chemistry? Observations are often used to make inferences. An inference is an explanation or interpretation of observations.

Procedure

1. Read and complete the lab safety form.
2. Add **water** to a **petri dish** to a height of 0.5 cm. Use a **graduated cylinder** to measure 1 mL of **vegetable oil,** then add it to the petri dish.
3. Dip the end of **a toothpick** into **liquid dishwashing detergent.**
4. Touch the tip of the toothpick to the water at the center of the petri dish. Record your detailed observations.
5. Add **whole milk** to a **second petri dish** to a height of 0.5 cm.
6. Place one drop each of **four different food colorings** in four different locations on the surface of the milk. Do not put a drop of food coloring in the center.
7. Repeat Steps 3 and 4.

Analysis

1. **Describe** what you observed in Step 4.
2. **Describe** what you observed in Step 7.
3. **Infer** Oil, the fat in milk, and grease belong to a class of substances called lipids. What can you infer about the addition of detergent to dishwater?
4. **Explain** why observations skills were important in this chemistry lab.

Observation You make observations throughout your day in order to make decisions. Scientific study usually begins with simple observation. An observation is the act of gathering information. Often, the types of observations scientists make first are **qualitative data**—information that describes color, odor, shape, or some other physical characteristic. In general, anything that relates to the five senses is qualitative: how something looks, feels, sounds, tastes, or smells.

Chemists frequently gather another type of data. For example, they can measure temperature, pressure, volume, the quantity of a chemical formed, or how much of a chemical is used up in a reaction. This numerical information is called **quantitative data.** It tells how much, how little, how big, how tall, or how fast. What kind of qualitative and quantitative data can you gather from **Figure 10?**

Figure 10 Quantitative data are numerical information. Qualitative data are observations made by using the human senses.
Identify *the quantitative and qualitative data in the photo.*

Hypothesis Recall the stories of the two substances that you read about in Section 1. Even before quantitative data showed that ozone levels were decreasing in the stratosphere, scientists observed CFCs there. Chemists Mario Molina and F. Sherwood Rowland were curious about how long CFCs could exist in the atmosphere.

Molina and Rowland examined the interactions that can occur among various chemicals in the troposphere. They determined that CFCs were stable there for long periods of time, but they also knew that CFCs drift upward into the stratosphere. They formed a hypothesis that CFCs break down in the stratosphere due to interactions with ultraviolet light from the Sun. In addition, the calculations they made led them to hypothesize that chlorine produced by this interaction would break down ozone.

A **hypothesis** is a tentative, testable statement or prediction about what has been observed. Molina and Rowland's hypothesis stated what they believed to be happening, even though there was no formal evidence at that point to support the statement.

☑ READING CHECK **Infer** why a hypothesis is tentative.

MiniLAB

Rubric

Purpose Students develop a hypothesis from their observations.

Process Skills observe and infer, draw a conclusion, hypothesize, design an experiment

Safety Precautions Approve lab safety forms before work begins.

Teaching Strategies
Students can float pepper on the milk to trace movement if food coloring is not available.

Expected Result When the toothpick touches the milk, the detergent temporarily destroys the surface tension. The colors move to the outside of the dish. The detergent emulsifies any fat in the milk. Convection-like currents cause the colors to move from the outside toward the center.

Analysis

1. The oil moved away from the detergent.
2. The colors moved to the outside of the dish.
3. It helps remove grease and oil from items being washed.
4. If observations are not made carefully, there might not be enough information to explain or infer what is occurring.

ASSESSMENT

Performance Have different lab groups test milk samples that have different fat content and have students compare their observations. OL EL COOP LEARN

little value at this time. Explain that a chemist learns how to change one substance into another substance having different properties. Also, point out that a color change is a sign that a chemical reaction has occurred.

Analysis
Were your observations of these changes qualitative or quantitative data? Qualitative data; students' sense of sight was used to observe color changes.

ASSESSMENT

Knowledge Ask students how this lab illustrates why it is important not to taste anything in the lab even if it looks like a familiar food or drink. Products in the lab might look like a familiar food, but it could be a toxic substance. It is important to never taste anything in the lab. BL EL

Quick Demo

Systematic Approaches Ask groups of four students each to develop a list of the steps they would use to make a peanut butter and jelly sandwich. Then, have each group list their steps on the board. Ask the class to compare the lists. Even though each group ends up with the same product, the methods used might vary. Have students relate this analogy to the development and use of scientific methods. OL EL COOP LEARN

Extension

Solubility Using the example of dissolving salt in water, ask students to design an experiment to determine how much salt dissolves at various temperatures. Have them identify constants, such as the amount of water, and set up a control. Have them identify an independent variable, such as the water temperature, and state how they want to vary it. Then, have them identify the dependent variable, such as the amount of salt dissolved. If time is available, have students perform the experiment and analyze the data. OL

☑ READING CHECK Independent variables are the variables that are changed within the experiment. Dependent variables change in response to changes in the independent variable.

■ **Caption Question Fig. 12** By comparing the color change of the unknown solution to the controls.

Math in Chemistry

CFCs Scientists estimate that a single chlorine atom can destroy about 100,000 ozone molecules. If 7000 atoms of chlorine are released from an accidental CFC spill, how many ozone molecules will be destroyed? 7×10^8 ozone molecules OL

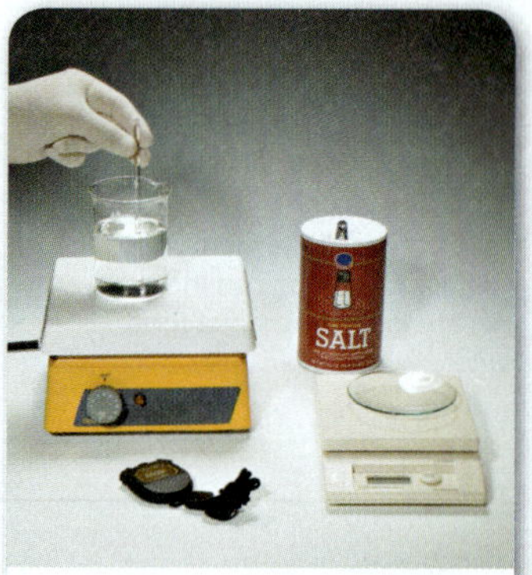

■ **Figure 11** These materials can be used to determine the effect of temperature on the rate at which table salt dissolves.

Experiments A hypothesis is meaningless unless there are data to support it. Thus, forming a hypothesis helps the scientist focus on the next step in a scientific method—the experiment. An **experiment** is a set of controlled observations that test the hypothesis. The scientist must carefully plan and set up one or more laboratory experiments in order to change and test one variable at a time. A variable is a quantity or condition that can have more than one value.

Suppose your chemistry teacher asks your class to use the materials shown in **Figure 11** to design an experiment to test the hypothesis that table salt dissolves faster in hot water than in water at room temperature (20°C). Because temperature is the variable that you plan to change, it is the **independent variable.** Your group determines that a given quantity of salt completely dissolves within 1 min at 40°C, but that the same quantity of salt dissolves after 3 min at 20°C. Thus, temperature affects the rate at which the salt dissolves. This rate is called the **dependent variable** because its value changes in response to a change in the independent variable. Although your group can determine the way in which the independent variable changes, it has no control over the way the dependent variable changes.

☑ READING CHECK **Explain** the difference between a dependent and an independent variable.

Other factors What other factors could you vary in your experiment? Would the amount of salt you try to dissolve make a difference? The amount of water you use? Would stirring the mixture affect your results? The answer to all of these questions might be yes. You must plan your experiment so that these variables are the same at each temperature, or you will not be able to tell clearly what caused your results. In a well-planned experiment, the independent variable should be the only condition that affects the experiment's outcome. A constant is a factor that is not allowed to change during the experiment. The amount of salt, water, and stirring must be constant at each temperature.

In many experiments, it is valuable to have a **control,** that is, a standard for comparison. In the above experiment, the room-temperature water is the control. **Figure 12** shows a different type of control. A chemical indicator has been added to each of three test tubes. An acidic solution is in the test tube on the left, and the indicator turns red. The test tube in the middle contains water, and the indicator is yellow. The test tube on the right contains a basic solution, and the indicator turns blue.

■ **Figure 12** Because the acidity of the solutions in these test tubes is known, these solutions can be used as controls in an experiment.

Infer *If the same chemical indicator were added to a solution of unknown acidity, how could you determine if it was acidic, neutral, or basic?*

(t)Matt Meadows, (b)Martyn F. Chillmaid/Science Photo Library/Photo Researchers

DIFFERENTIATED INSTRUCTION

Advanced Learners Have gifted students find articles in a recently published scientific journal on a research topic of interest. Have them identify each step of the scientific method used in the research described in the article. AL

CHEMISTRY JOURNAL

Qualitative and Quantitative Have students describe themselves using qualitative and quantitative data. Examples of qualitative data should use as many senses as possible: color of hair, tall or short, color of eyes, etc. Quantitative data might include their heights or lengths of hair. OL EL

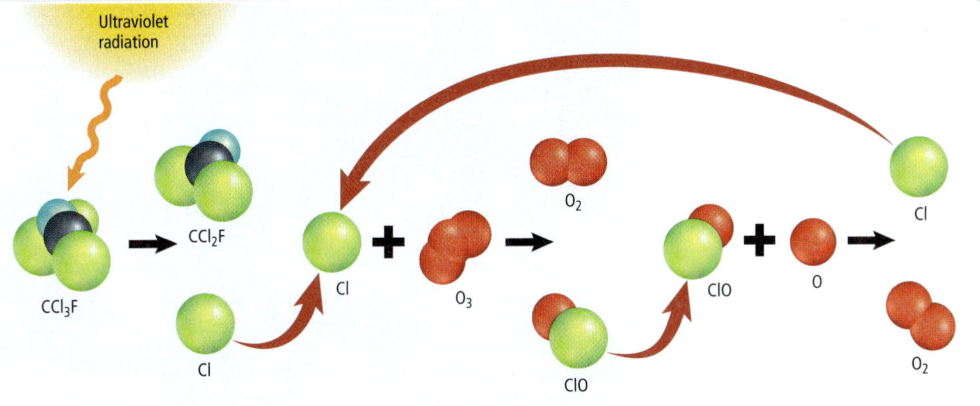

Figure 13 Molina and Rowland's model showed how CFCs could destroy ozone.

Controlling variables The interactions described between CFCs and ozone in Molina and Rowland's hypothesis take place high overhead. Many variables are involved. For example, there are several gases present in the stratosphere. Thus, it would be difficult to determine which gases, or if all gases, are causing decreasing ozone levels. Winds, variations in ultraviolet light, and other factors could change the outcome of any experiment on any given day, making comparisons difficult. Sometimes, it is easier to simulate conditions in a laboratory, where the variables can be more easily controlled.

View an **animation about ozone depletion.** Concepts In Motion

Conclusion An experiment might generate a large amount of data. Scientists take the data, analyze it, and check it against the hypothesis to form a conclusion. A **conclusion** is a judgment based on the information obtained. A hypothesis can never be proven. Therefore, when the data support a hypothesis, this only indicates that the hypothesis might be true. If further evidence does not support it, then the hypothesis must be discarded or modified. The majority of hypotheses are not supported, but the data might still yield new and useful information.

Molina and Rowland formed a hypothesis about the stability of CFCs in the stratosphere. The data that they gathered supported their hypothesis. They developed a model in which the chlorine formed by the breakdown of CFCs would react over and over again with ozone.

A model can be tested and used to make predictions. Molina and Rowland's model predicted the formation of chlorine and the depletion of ozone, as shown in **Figure 13.** Another research group found evidence of interactions between ozone and chlorine when taking measurements in the stratosphere, but they did not know the source of the chlorine. Molina and Rowland's model predicted a source of the chlorine. They came to the conclusion that ozone in the stratosphere could be destroyed by CFCs, and they had enough support for their hypothesis to publish their discovery. They won the Nobel Prize in 1995.

FOLDABLES®
Incorporate information from this section into your Foldable.

Identify Misconceptions

Students often do not understand what a scientific theory is. Many people use the term *theory* to explain something in the world around them or human behavior. What they call a theory might be a hypothesis or just an idea or prediction.

Uncover the Misconception Have students work in groups to distinguish between a fact and a prediction. Conclusions should include that a fact has been tested and found to be reliable. A prediction might be based on information, but is yet to be supported. Relate these terms to theory and hypothesis.

Demonstrate the Concept With students, develop an events-chain concept map that shows the hierarchy of the terms *theory, hypothesis, experiments,* and *observations.* Concept maps should show that a theory requires numerous hypotheses, which are supported by experiments that include observations.

Assess New Knowledge Ask students to describe a common theory. Help them see that simple statements based on observations are often hypotheses. OL EL COOP LEARN

FOLDABLES® Rubric

DIFFERENTIATED INSTRUCTION

Struggling Learners Pair struggling students with other students that understand the chemical reaction taking place in Figure 13. Have the struggling student explain the reaction to the other student, and have the other student correct any misconceptions that the struggling student might have. BL COOP LEARN

CHEMISTRY PROJECT

Biographies Divide the class into small groups. Have each group choose one of the scientists discussed in this chapter and research his or her life. Have each group prepare a short presentation for the class. BL OL AL EL COOP LEARN

3 Assess

Check for Understanding

Ask students to explain the difference between *qualitative data* and *quantitative data.* Qualitative data is observed with the senses, such as color and odor. Quantitative data is numerical information, such as 3 m or 5 mL. **OL** **EL**

Reteach

Have students clarify the difference between a theory and a scientific law. A theory is a statement that provides an explanation based on supported hypotheses. A scientific law describes something known to happen without error, such as gravity but does not explain how it happens. **OL** **EL**

Extension

Bring in a newspaper or journal article about a development in environmental chemistry. Have students identify the steps of the scientific method used as well as any controls and variables used. **OL** **EL**

ASSESSMENT

Performance Cut out large pieces of paper and write a vocabulary term from this section on each piece. Have students place these papers in the order they are used in a scientific method. Some words might be subsets of specific steps. Accept any order that students can justify. **OL** **EL**

Figure 14 It does not matter how many times skydivers leap from a plane; Newton's law of universal gravitation applies every time.

moodboard/Getty Images

Theory and Scientific Law

A **theory** is an explanation of a natural phenomenon based on many observations and investigations over time. You might have heard of Einstein's theory of relativity or the atomic theory. A theory states a broad principle of nature that has been supported over time. All theories are still subject to new experimental data and can be modified. Also, theories often lead to new conclusions. A theory is considered valid if it can be used to make predictions that are proven true.

Sometimes, many scientists come to the same conclusion about certain relationships in nature and they find no exceptions to these relationships. For example, you know that no matter how many times skydivers, like those shown in **Figure 14,** leap from a plane, they always return to Earth's surface. Sir Isaac Newton was so certain that an attractive force exists between all objects that he proposed his law of universal gravitation. Newton's law is a **scientific law**—a relationship in nature that is supported by many experiments. It is up to scientists to develop further hypotheses and experiments to explain why these relationships exist.

SECTION 3 REVIEW

Section Self-Check

Section Summary

- Scientific methods are systematic approaches to problem solving.
- Qualitative data describe an observation; quantitative data use numbers.
- Independent variables are changed in an experiment. Dependent variables change in response to the independent variable.
- A theory is a hypothesis that is supported by many experiments.

15. MAINIDEA **Explain** why scientists do not use a standard set of steps for every investigation they conduct.
16. **Differentiate** Give an example of quantitative and qualitative data.
17. **Evaluate** You are asked to study the effect of temperature on the volume of a balloon. The balloon's size increases as it is warmed. What is the independent variable? The dependent variable? What factor is held constant? How would you construct a control?
18. **Distinguish** Jacques Charles described the direct relationship between temperature and volume of all gases at constant pressure. Should this be called Charles's law or Charles's theory? Explain.
19. **Explain** Good scientific models can be tested and used to make predictions. What did Molina and Rowland's model of the interactions of CFCs and ozone in the atmosphere predict would happen to the amount of ozone in the stratosphere as the level of CFCs increased?

SECTION 3 REVIEW

15. The nature of investigations varies a great deal, and the steps needed to perform a wide array of investigations must also vary.
16. Possible answers: qualitative, silver-colored liquid; quantitative, 5 mL.
17. independent variable, the temperature; dependent variable, the size of the balloon; factor held constant, the amount of air in the balloon; control, an identical balloon kept at room temperature
18. It is called Charles's law because it describes a phenomenon that consistently takes place.
19. Their models predicted that as CFC concentrations increased, ozone levels would decrease.

SECTION 4

Scientific Research

MAINIDEA Some scientific investigations result in the development of technology that can improve our lives and the world around us.

Essential Questions

- How do pure research, applied research, and technology compare and contrast?
- What are some of the important rules for laboratory safety?

Review Vocabulary

synthetic: something that is human-made and does not necessarily occur in nature

New Vocabulary

pure research
applied research

CHEM 4 YOU Much of the information that scientists obtain through basic research is used to meet a specific need. For example, X-rays were discovered by scientists conducting basic research on electrical discharge through gases. Later, it was discovered that X-rays could be used to diagnose medical problems.

Types of Scientific Investigations

Every day in the media—through TV, newspapers, magazines, or the Internet—the public is bombarded with the results of scientific investigations. Many deal with the environment, medicine, or health. As a consumer, you are asked to evaluate the results of scientific research and development. How do scientists use qualitative and quantitative data to solve different types of scientific problems?

Scientists conduct **pure research** to gain knowledge for the sake of knowledge itself. Molina and Rowland were motivated by curiosity and, thus, conducted research on CFCs and their interactions with ozone as pure research. No environmental evidence at the time indicated that there was a correlation to their model in the stratosphere. Their research showed only that CFCs could speed the breakdown of ozone in a laboratory setting.

By the time the ozone hole was reported in 1985, scientists had made measurements of CFC levels in the stratosphere that supported the hypothesis that CFCs could be responsible for the depletion of ozone. The early pure research done only for the sake of knowledge became applied research. **Applied research** is research undertaken to solve a specific problem. Scientists continue to monitor the amount of CFCs in the atmosphere and the annual changes in the amount of ozone in the stratosphere, as shown in **Figure 15.** Applied research is also being conducted to find replacement chemicals for the CFCs that are now banned.

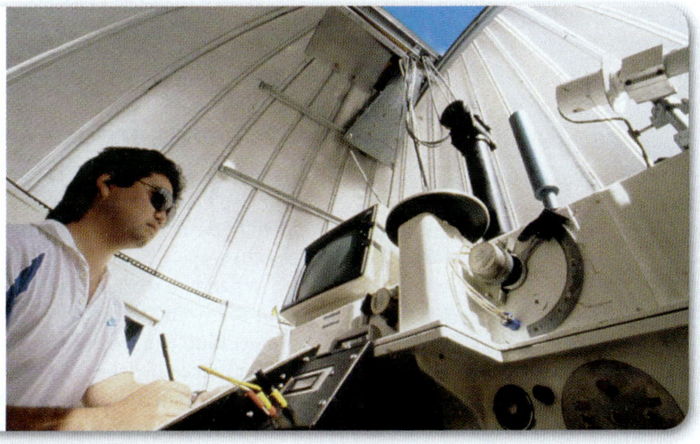

Figure 15 This UV-visible spectrometer (UV-Vis) is used to measure ozone and other stratospheric gases during the dark winter months in Antarctica.

Hank Morgan/Photo Researchers

SECTION 4

1 Focus

MAINIDEA

Technology Have students brainstorm about different types of technology and how it has affected people's lives. Ask students if all technology is beneficial to humans. Allow students to discuss this briefly because many of the topics are controversial. However, it is important for students to understand that technology can have advantages and disadvantages. OL EL

2 Teach

Quick Demo

Pure Research Bring in samples of nylon cloth and nylon cord. Nylon is a good example of a synthetic compound that has many uses. Explain that many such applications are chance discoveries that are by products of pure research.

In-Text Question Possible answer: Qualitative data might be the color change of a solution as an indication that a chemical reaction has taken place. Quantitative data might be used to determine the solution concentration that produces the highest yield of product in a chemical process.

CHEMISTRY JOURNAL

Research or Discovery Have students clip a newspaper article describing a scientific study. Ask them to discuss whether this example is pure research, applied research, or a chance discovery. OL EL

ChemLAB

The ChemLab located at the end of the chapter can be used at this point in the lesson.

Apply Chemistry

CFCs Ask students if the discovery of CFCs was pure research, applied research, or chance discovery. It was applied research to find another source of refrigerants. OL EL

Content Background

Hook-and-Loop Fasteners The design of hook-and-loop fastener tape is based on nature. After a walk in the countryside, Swiss inventor, George de Mestral, was interested in finding out why cockleburs cling to his clothing so tenaciously. With the aid of a microscope, Mestral discovered that the cocklebur was covered in tiny hooks and his clothing had tiny loops on the surface. He began searching for a material that could be used to duplicate nature's design. By accident, Mestral discovered that nylon, sewn under ultraviolet light, formed tiny hooks. The first hook-and-loop fastener tape was mass produced in France in the 1950s.

Figure 16 After its discovery, nylon was used mainly for war materials and was not available for home use until after World War II. Today it is used in a variety of products.

Strands of nylon can be pulled from the top layer of solution.

Nylon fibers are used to make hook-and-loop fastener tape.

Chance discoveries Often, a scientist conducts experiments and reaches a conclusion that is far different from what was predicted. Some truly wonderful discoveries in science have been made unexpectedly. You might be familiar with the two examples described below.

Connection to Biology Alexander Fleming is famous for making several accidental discoveries. In one accidental discovery, Fleming found that one of his plates of *Staphylococcus* bacteria had been contaminated by a greenish mold, later identified as *Penicillium.* He observed it carefully and saw a clear area around the mold where the bacteria had died. In this case, a chemical in the mold—penicillin—was responsible for killing the bacteria.

The discovery of nylon is another example of an accidental discovery. In 1930, Julian Hill, an employee of E.I. DuPont de Nemours and Company, dipped a hot glass rod in a mixture of solutions and unexpectedly pulled out long fibers similar to those shown in **Figure 16.** Hill and his colleagues pursued the development of these fibers as a synthetic silk that could withstand high temperatures. They eventually developed nylon in 1934. During World War II, nylon was used as a replacement for silk in parachutes. Today, nylon is used extensively in textiles and some kinds of plastics. It is also used to make hook-and-loop tape, as shown in **Figure 16.**

Students in the Laboratory

Watch a video about **criminal science investigation.**

Video

In your study of chemistry, you will learn many facts about matter. You will also do investigations and experiments in which you will be able to form hypotheses, test hypotheses, gather data, analyze data, and draw conclusions.

When you work in the chemistry laboratory, you are responsible for your safety and the safety of people working nearby. Often, many people are working in a small space during a lab, so it is important that everyone practice safe laboratory procedures. **Table 2** lists some safety rules that you should follow each time you enter the lab. Chemists and all other scientists use these safety rules as well.

What's CHEMISTRY Got To Do With It?

Video

Tracking a Crime

DIFFERENTIATED INSTRUCTION

English Learners Have a student explain the difference between pure and applied research and chance discovery to an English-learning student. The student can use figures in the textbook to help him or her explain the concepts. EL OL

Table 2 Safety in the Laboratory

1. Study your lab assignment before you come to the lab. If you have any questions, ask your teacher for help.	**16.** Keep combustible materials away from open flames.
2. Do not perform experiments without your teacher's permission. Never work alone in the laboratory. Know how to contact help, if necessary.	**17.** Handle toxic and combustible gases only under the direction of your teacher. Use the fume hood when such materials are present.
3. Use the table on the inside front cover of this textbook to understand the safety symbols. Read and adhere to all **WARNING** statements.	**18.** When heating a substance in a test tube, be careful not to point the mouth of the test tube at another person or yourself. Never look down into the mouth of a test tube.
4. Wear safety goggles and a laboratory apron whenever you are in the lab. Wear gloves whenever you use chemicals that cause irritations or can be absorbed through the skin. If you have long hair, you must tie it back.	**19.** Do not heat graduated cylinders, burettes, or pipettes with a laboratory burner.
5. Do not wear contact lenses in the lab, even under goggles. Lenses can absorb vapors and are difficult to remove during an emergency.	**20.** Use caution and proper equipment when handling a hot apparatus or glassware. Hot glass looks the same as cool glass.
6. Avoid wearing loose, draping clothing and dangling jewelry. Wear only closed-toe shoes in the lab.	**21.** Dispose of broken glass, unused chemicals, and products of reactions only as directed by your teacher.
7. Keep food, beverages, and chewing gum out of the lab. Never eat in the lab.	**22.** Know the correct procedure for preparing acid solutions. Always add the acid to the water slowly.
8. Know where to find and how to use the fire extinguisher, safety shower, fire blanket, first-aid kit, and gas and electrical power shutoffs.	**23.** Keep the balance area clean. Never place chemicals directly on the pan of a balance.
9. Immediately clean up spills on the floor and keep all walkways clear of objects, such as backpacks, to prevent accidental falls or tripping. Report any accident, injury, incorrect procedure, or damaged equipment to your teacher.	**24.** After completing an experiment, clean and put away your equipment. Clean your work area. Make sure the gas and water are turned off. Wash your hands with soap and water before you leave the lab.
10. If chemicals come in contact with your eyes or skin, flush the area immediately with large quantities of water. Immediately inform your teacher of the nature of the spill.	
11. Handle all chemicals carefully. Check the labels of all bottles before removing the contents. Read the label three times: before you pick up the container, when the container is in your hand, and when you put the bottle back.	
12. Do not take reagent bottles to your work area unless instructed to do so. Use test tubes, paper, or beakers to obtain your chemicals. Take only small amounts. It is easier to get more than to dispose of excess.	
13. Do not return unused chemicals to the stock bottle.	
14. Do not insert droppers into reagent bottles. Pour a small amount of the chemical into a beaker.	
15. Never taste any chemicals. Never draw any chemicals into a pipette with your mouth.	

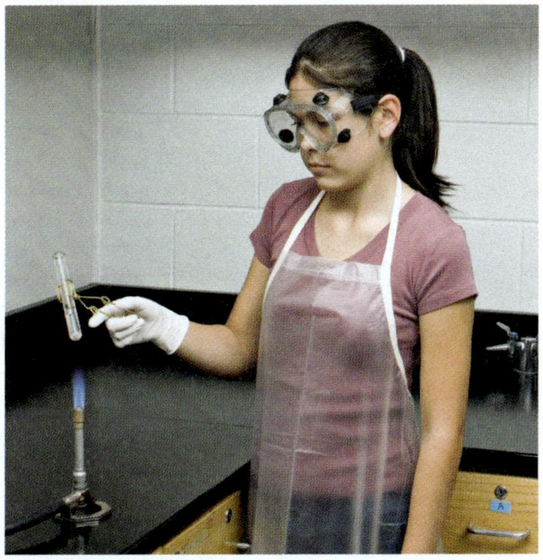

Matt Meadows

Reinforcement

Safety Divide students into groups of four. Give each group several safety tips from Table 2. Have them create a quick skit to demonstrate potential dangers and lab safety. OL COOP LEARN

Concept Development

Lab Safety Walk into the lab demonstrating poor lab safety, such as wearing loose clothing and chewing gum. Have students identify as many broken rules as they can. Have a prize, such as a special pair of lab goggles, for the student or group who identifies the greatest number of broken rules. Emphasize that although this example can be humorous, lab safety is a serious issue. OL

ASSESSMENT

Knowledge Ask students to choose a safety rule from the table. Have students read the rule aloud and then explain in their own words why the rule is important to ensure lab safety. Possible answers: Do not wear contact lenses in the lab, because the lenses can absorb vapors or the chemicals can get under the lenses, which can be hazardous to the eyes. OL EL BL AL

CHEMISTRY PROJECT

Lab Safety Have student groups create posters or a bulletin board that emphasizes safe lab procedures. Be sure the products reflect what the safe behavior is and why it should be used. OL EL

Enrichment

Time Line Divide the class into small groups. Assign each group the task of researching significant events or milestones that occurred during their assigned time period in the areas of ozone depletion, CFC accumulation in the atmosphere, and new CFCs or replacement chemicals on the market that are more environmentally friendly. Consider dividing the class into the following time periods: 1970–1975; 1976–1981; 1982–1987; 1988–2003; 2003–present. Have students write the date and a short description of the event or milestone on a sheet of paper. Use the papers to create a time line of events and display the time line on a classroom wall. OL COOP LEARN

GRAPH CHECK CFC-11: about 1992; CFC-12: about 1998

READING CHECK When temperatures drop below −78°C, stratospheric ice clouds promote the production of chemically active chlorine and bromine. When temperatures begin to warm in the spring, the chemically active chlorine and bromine react with the ozone. These chemical reactions consume much of the ozone creating a hole over Antarctica.

FOLDABLES® Rubric

■ **Figure 17** This graph shows the concentration of two common CFCs in the atmosphere over Antarctica and the global consumption of CFCs from 1980 to 2000. While CFC consumption began to decrease drastically a few years after the signing of the Montreal Protocol, the concentration of CFCs above Antarctica continued to increase for awhile before slowly leveling out.

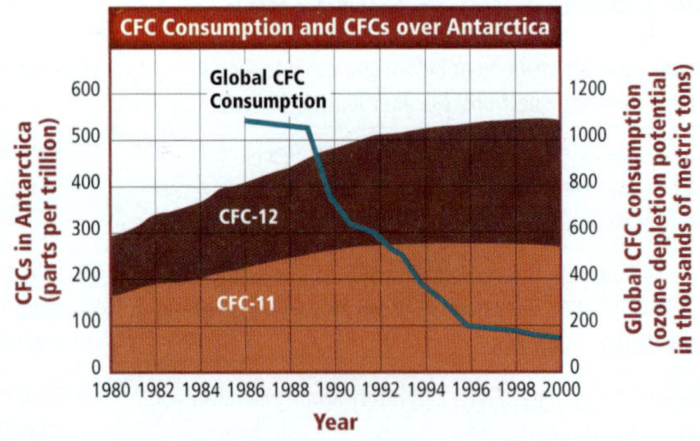

GRAPH CHECK **Identify** when CFCs in Antarctica began to level off after national leaders signed the Montreal Protocol.

The Story Continues

Now, back to the two substances that you have been reading about. A lot has happened since the 1970s, when Molina and Rowland hypothesized that CFCs broke down stratospheric ozone. The National Oceanic and Atmospheric Administration (NOAA) and many other groups are actively collecting historic and current data on CFCs in the atmosphere and ozone concentrations in the stratosphere. Through applied research, scientists determined that not only do CFCs react with ozone, but a few other substances react as well. Carbon tetrachloride and methyl chloroform are two additional substances that harm the ozone. Substances that contain bromine can also damage the ozone.

FOLDABLES® Incorporate information from this section into your Foldable.

The Montreal Protocol Because ozone depletion is an international concern, nations banded together to try to solve this problem. In 1987, leaders from many nations met in Montreal, Canada, and signed the Montreal Protocol. By signing this agreement, nations agreed to phase out the use of these compounds and place restrictions on how they should be used in the future. As you can see from **Figure 17,** the global use of CFCs began to decline after the Montreal Protocol was signed. However, the graph shows that the amount of CFCs measured over Antarctica did not decline immediately.

The ozone hole today Scientists have also learned that the ozone hole forms each year over Antarctica during the spring. Stratospheric ice clouds form over Antarctica when temperatures there drop below −78°C. These clouds produce changes that promote the production of chemically active chlorine and bromine. When temperatures begin to warm in the spring, this chemically active chlorine and bromine react with ozone, causing ozone depletion. This ozone depletion causes the ozone hole to form over Antarctica. Some ozone depletion also occurs over the Arctic, but temperatures do not remain low for as long, which means less ozone depletion in the Arctic. Upon further research, scientists have also determined that ozone thinning has occurred above every continent.

READING CHECK **Explain** what triggers the formation of the ozone hole over Antarctica.

CHEMISTRY PROJECT

Ozone Layer Divide students into small groups and have them research the current status of the ozone layer. Have students prepare an oral presentation that includes visual aids and present it to the class. OL COOP LEARN

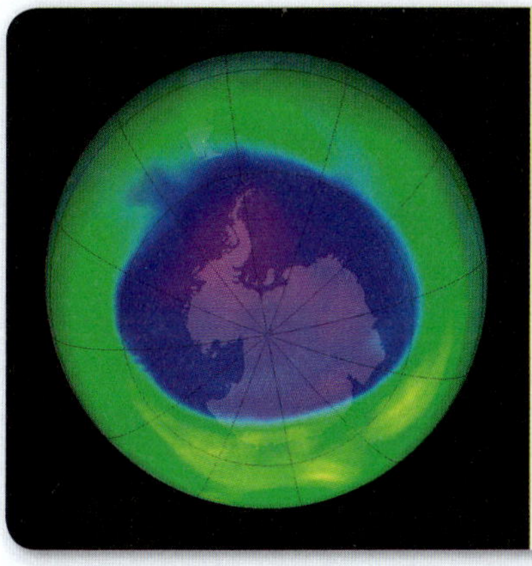

■ **Figure 18** The ozone hole over Antarctica reached its maximum level of thinning in September 2005. The color-key below shows what the colors represent in this colorized satellite image.

Compare *How do these ozone levels compare with what is considered normal?*

Figure 18 shows the ozone hole over Antarctica in September 2005. The ozone thinning over Antarctica reached its maximum for the year during this month. If you compare the color-coded key to the satellite image, you can see that the ozone level is between 110 and 200 DU. Notice the area surrounding the ozone hole. Much of this area has ozone levels around 300 DU, which is considered normal.

Scientists are not sure when the ozone layer will begin to recover. Originally, scientists predicted that it would begin to recover in 2050. However, new computer models predict that it will not begin to recover until 2068. The exact date of its recovery is not as important as the fact that it will recover given time.

VOCABULARY

WORD ORIGIN

Recover
to bring back to normal
It takes several days to recover from the flu.

Data Analysis LAB

Based on Real Data*

Interpret Graphs

How do ozone levels vary throughout the year above Antarctica? Many agencies monitor the concentration of ozone in the stratosphere over Antarctica.

Think Critically

1. **Describe** the trend in the data for 1979–2008.
2. **Evaluate** how the 2009 data compare with the data from 1979–2008.
3. **Identify** the month during which the ozone levels were the lowest in 1979–2008. In 2009?
4. **Assess** Do these data points back up what you learned in this chapter about ozone depletion? Explain your answer.

Data and Observation

This graph displays data from NASA collected over Antarctica. Data values below 220 Dobson Units are defined as the region of the ozone hole area.

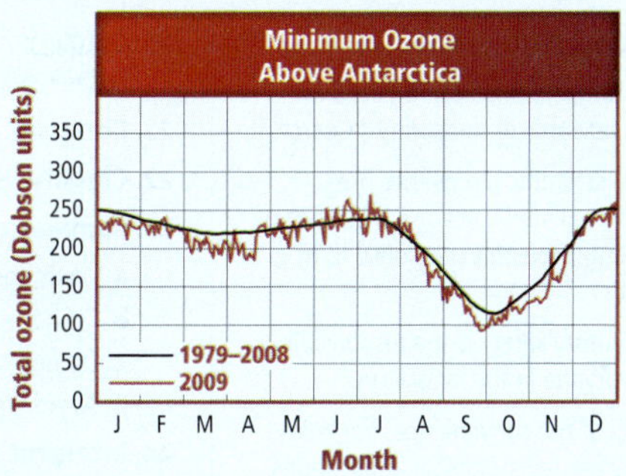

*Data obtained from Ozone Hole Watch. 2010. *National Aeronautics and Space Administration.*

NASA/Science Photo Library/Photo Researchers

Data Analysis LAB

About the Lab

- The data in this lab was collected by NASA from 1979 to 2009 using Total Ozone Mapping Spectrometers (TOMS) located on the Nimbus satellites. Since 2004, the Ozone Monitoring Instrument (OMI) located on the Aura satellite has been used.

Think Critically

1. The minimum ozone declines from 250 DU in January to around 220 DU in March. From March until mid-July, the median increases slightly. At mid-July there is a sharp decrease to around 120 DU at the beginning of October where the least amount of ozone is present. This is followed by a sharp increase to above 250 DU where it levels out in mid-December.
2. The 2009 data follows the same trend as the median value for years 1979–2008.
3. early October/late September
4. Yes, the cold temperatures allow chemically active chlorine and bromine to form. When the temperatures begin to warm, the chlorine and bromine begin to react with the ozone until the chlorine and bromine are depleted.

■ **Caption Question Fig. 18**
The normal is 300 DU, so 110 to 200 DU is below normal.

DIFFERENTIATED INSTRUCTION

Visually Impaired Have a student help a visually impaired student trace the ozone graph with their finger so that they can comprehend the trends in the graph. **COOP LEARN**

3 Assess

Check for Understanding

Ask students if research on new refrigerants that are not harmful to the environment is pure research or applied research. It is applied research because it is conducted to solve a problem. OL

Reteach

Emphasize that pure research is often a foundation for applied research. Pure research can create compounds or increase knowledge that scientists might not know how to use for years. However, when that knowledge or material is needed, it is readily available. OL

Extension

Ask students to bring in newspaper or magazine articles about scientific research and explain to the class how that research applies to their lives. Examples might include a new drug or treatment for a disease that affects someone they know or a new technology that affects the environment. OL

ASSESSMENT

Knowledge Have students summarize the articles used in the Extension feature. OL

■ **Figure 19** This car, which is powered by compressed air, and this tiny submarine, which is only 4 mm long, are examples of technologies that are made possible by the study of matter.

The Benefits of Chemistry

Chemists are an important part of the team of scientists that solve many of the problems or issues that we face today. Chemists are not only involved in resolving the ozone depletion problem. They are also involved in finding cures or vaccines for diseases, such as AIDS and influenza. Almost every situation that you can imagine involves a chemist, because everything in the universe is made of matter.

Figure 19 shows some of the advances in technology that are possible because of the study of matter. The car on the left is powered by compressed air. When the compressed air is allowed to expand, it pushes the pistons that move the car. Because the car is powered by compressed air, no pollutants are released. The photo on the right shows a tiny submarine that is made by computer-aided lasers. This submarine, which is only 4 mm long, might be used for detecting and repairing defects in the human body.

SECTION 4 REVIEW

Section Self-Check

Section Summary

- Scientific methods can be used in pure research or in applied research.
- Some scientific discoveries are accidental, and some are the result of diligent research in response to a need.
- Laboratory safety is the responsibility of everyone in the laboratory.
- Many of the conveniences we enjoy today are technological applications of chemistry.

20. MAINIDEA **Name** three technological products that have improved our lives or the world around us.
21. **Compare and contrast** pure research and applied research.
22. **Classify** Is technology a product of pure research or applied research? Explain.
23. **Summarize** the reason behind each of the following.
    - **a.** Wear goggles and an apron in the lab even if you are only an observer.
    - **b.** Do not return unused chemicals to the stock bottle.
    - **c.** Do not wear contact lenses in the laboratory.
    - **d.** Avoid wearing loose, draping clothing and dangling jewelry.
24. **Interpret Scientific Diagrams** What safety precautions should you take when the following safety symbols are listed?

SECTION 4 REVIEW

20. Possible answers: computer, internal combustion engine, and vaccinations.
21. Pure research is done for the sake of knowledge. Applied research is done to solve a specific problem.
22. Technology can be the product of either one. It can be a product of pure research when scientists realize their discovery has a practical application. It can also be a product of applied research when scientists perform research to solve a particular problem.
23. **a.** Harmful substances can get in your eyes and on your clothing if you are performing an experiment or just watching it being performed.
    b. The chemicals might be contaminated, and you do not want to contaminate the stock bottle.
    c. Contact lenses can absorb gases that can damage your eyes, and they are difficult to remove during an emergency situation.
    d. It is easy to drag these items through chemicals and across flames, which might create a hazardous situation.
24. Protect your hands from hot or cold objects; protect yourself from possible hazardous fumes; protect yourself from substances that can irritate your skin and mucous membranes of the respiratory tract; substances are flammable, do not have an open flame in the lab

CHEMISTRY&careers

Career: Art Restorer

Painting Restoration

Art does not last forever. It is damaged by events such as people sneezing on it, touching it, or by smoke during a fire. The repair of damage to artwork is the job of art restorers. Art repair is not always an easy task, because the materials used to correct the damage can also damage the artwork.

Help from above Oxygen makes up 21% of Earth's atmosphere. Near the ground, almost all the oxygen exists as oxygen gas (O_2). However, high in the atmosphere, ultraviolet light from the Sun splits oxygen gas into atomic oxygen (O). While oxygen gas is chemically reactive, atomic oxygen is even more reactive. It can damage spacecraft in orbit, which is why NASA actively studies the reactions between atomic oxygen and other substances.

Oxygen and art Atomic oxygen is especially reactive with the element carbon—the main substance found in soot from a fire. When NASA scientists treated the soot-damaged painting shown in **Figure 1** with atomic oxygen, the carbon in the soot reacted with oxygen to produce gases that floated away.

Figure 1 The photo on the left shows soot damage to an oil painting. The photo on the right shows the painting after oxygen treatment. Removal of a small amount of glossy binder was the only damage to the painting during the treatment.

NASA

Figure 2 The lipstick stain could not be removed using conventional techniques. However, atomic oxygen removed the stain without damage to the painting.

On the surface Because atomic oxygen acts only on what it touches, paint layers below the soot or other surface impurities are unaffected. If you compare the image on the left with the image on the right in **Figure 1,** you will notice that the soot was removed, but the painting was not harmed. This is in contrast to more conventional treatments, in which organic solvents are used to remove the soot. These solvents often react with the paint as well as the soot.

The kiss Another successful restoration was the Andy Warhol painting called *The Bathtub.* It was damaged when a lipstick-wearing viewer kissed the canvas. Most conventional restoration techniques would have driven the lipstick deeper into the painting, leaving a permanent pink stain. When atomic oxygen was applied to the stain using the equipment shown in **Figure 2,** the pink color vanished.

WRITING IN ▶ Chemistry

Research a recent art restoration project. Prepare a news article explaining why the object needed restoration, the challenges presented, and the chemistry used to complete the project.

WebQuest

WRITING IN ▶ Chemistry

Rubric WebQuest

Research The articles should contain information similar to what is in the feature about how atomic oxygen is used in art restoration. Student articles should also contain additional information that students find when doing their research.

Purpose

Students will learn about the reactive properties of atomic oxygen and how it can be used to repair damaged works of art.

Background

Because atomic oxygen is more reactive than molecular oxygen, it can be used to remove soot and other impurities from paintings. The oxygen can be applied using a hand-held device, or the entire painting can be placed in a chamber flooded with atomic oxygen.

Atomic oxygen does not replace other art restoration techniques. While many paint pigments will not react with atomic oxygen, some, such as lead, become discolored when treated with oxygen.

Teaching Strategies

- Develop a list of materials that can cause stains and another list of materials used to clean stains. Examples of materials that can cause stains: include dark liquids, dirt, and oily fingerprints. Have students investigate the chemistry of stains and cleaners. The chemicals in cleaners can lighten a dark stain or can make it easier for the stain to be washed away in water or nonaqueous substances.
- Discuss some of the difficulties involved in restoring works of art. Examples of difficulties include maintaining the historic value and not compromising the original artwork during restoration.
- How long might paintings be expected to last? Paintings will last for varying lengths of time depending on the media used and the conditions, such as temperature and humidity, under which the art is stored. Discuss some of the causes of damage to art works and ways this damage might be minimized. Examples of some causes and damages to art works are acid burn, foxing, tape and adhesive, water damage, creasing, abrasions, surface dirt, and pigment loss. These can be minimized by only exposing the art to the proper environmental conditions and handling properly.

ChemLAB

Rubric

Preparation

Time Allotment one class period

Process Skills acquire and analyze information, classify, compare and contrast, draw a conclusion, hypothesize, measure, observe and infer, think critically, use numbers

Safety Precautions Approve lab safety forms before work begins. Hard water containing magnesium chloride hexahydrate can irritate the eyes and respiratory tract.

Preparation of Materials Any type of liquid detergent will work for this lab.

Preparation of Solutions See the Teacher Handbook for preparation of all solutions.

Procedure

- The same size test tubes must be used.
- Make sure students rinse out the test tubes when finished.
- Make sure the magnesium chloride hexahydrate is covered; it is hygroscopic, and will absorb water from the air.

Expected Results

The distilled water is the control and would be considered soft water according to the Classification of Water Hardness table. The control should contain the most suds, while the hard water sample should contain the least amount of suds.

Analyze and Conclude

1. The distilled water produces the most suds. The hard water produces the least suds.
2. According to the background introduction, the soft water came from Community A. The hard water came from Community B.
3. 6.3 mg Mg/0.05L = 126 mg Mg/L; hard
4. independent variable, volume of water samples and amount of detergent; dependent variable, amount of suds produced; Yes, the distilled water was the control because it has a lack of minerals dissolved in it. Comparison of results will vary.
5. The volume of the liquids and the detergent could be measured with more precision.

INQUIRY EXTENSION

There are a number of products that claim to soften water. Visit a grocery store or home improvement store to find these products and design an experiment to test their claims. Student designs will vary but should include an independent variable, dependant variable, and a control.

ChemLAB

Forensics: Identify the Water Source

Background: **The contents of tap water vary from community to community. Water is classified as hard or soft based on the amount of calcium or magnesium in the water, measured in milligrams per liter (mg/L). Imagine a forensics lab has two samples of water. One sample comes from Community A, which has soft water. The other sample comes from Community B, which has hard water.**

Question: *From which community did each water sample originate?*

Materials

test tubes with stoppers (3)
test-tube rack
grease pencil
graduated cylinder (25-mL)
distilled water
dropper
beaker (250-mL)
Water sample 1
Water sample 2
dish detergent
metric ruler

Safety Precautions

Procedure

1. Read and complete the lab safety form.
2. Prepare a data table like the one shown. Then, use a grease pencil to label three large test tubes: *D* (for distilled water), *1* (for Sample 1), and *2* (for Sample 2).
3. Use a graduated cylinder to measure out 20 mL of distilled water. Pour the water into Test Tube D.
4. Place Test Tubes 1 and 2 next to Test Tube D and make a mark on each test tube that corresponds to the height of the water in Test Tube D.
5. Obtain 50 mL of Water Sample 1 in a beaker from your teacher. Slowly pour the water sample into Test Tube 1 until you reach the marked height.
6. Obtain 50 mL of Water Sample 2 in a beaker from your teacher. Slowly pour Water Sample 2 into Test Tube 2 until you reach the marked height.
7. Add one drop of dish detergent to each test tube. Stopper the tubes tightly. Then, shake each sample for 30 s to produce suds. Use a metric ruler to measure the height of the suds.
8. Use some of the soapy solutions to remove the grease marks from the test tubes.

Data Table

Sample	Height of Suds
D	
1	
2	

9. **Cleanup and Disposal** Rinse all of the liquids down the drain with tap water. Return all lab equipment to its designated location.

Analyze and Conclude

1. **Compare and Contrast** Which sample produced the most suds? Which sample produced the least amount of suds?
2. **Conclude** Soft water produces more suds than hard water. Determine from which community each water sample originated.
3. **Calculate** If the 50 mL of hard water that you obtained contained 6.3 mg of magnesium, how hard would the water be according to the table below? (50 mL = 0.05 L)

Classification of Water Hardness

Classification	mg of Calcium or Magnesium /L
Soft	0–60
Moderate	61–120
Hard	121–180
Very hard	>180

4. **Apply Scientific Methods** Identify the independent and dependent variables in this lab. Was there a control in this lab? Explain. Did all your classmates have the same results as you? Why or why not?
5. **Error Analysis** Could the procedure be changed to make the results more quantitative? Explain.

INQUIRY EXTENSION

Investigate There are a number of products that claim to soften water. Visit a grocery store or home-improvement store to find these products and design an experiment to test their claims.

CHAPTER 1 STUDY GUIDE

Vocabulary Practice

BIGIDEA Chemistry is a science that is central to our lives.

SECTION 1 A Story of Two Substances

MAINIDEA Chemistry is the study of everything around us.

- Chemistry is the study of matter.
- Chemicals are also known as substances.
- Ozone is a substance that forms a protective layer in Earth's atmosphere.
- CFCs are synthetic substances made of chlorine, fluorine, and carbon that were originally thought to be the ideal coolants for refrigeration.

VOCABULARY
- chemistry
- substance

SECTION 2 Chemistry and Matter

MAINIDEA Branches of chemistry involve the study of different kinds of matter.

- Models are tools that scientists, including chemists, use.
- Macroscopic observations of matter reflect the actions of atoms on a submicroscopic scale.
- There are several branches of chemistry, including organic chemistry, inorganic chemistry, physical chemistry, analytical chemistry, and biochemistry.

VOCABULARY
- mass
- weight
- model

SECTION 3 Scientific Methods

MAINIDEA Scientists use scientific methods to systematically pose and test solutions to questions and assess the results of the tests.

- Scientific methods are systematic approaches to problem solving.
- Qualitative data describe an observation; quantitative data use numbers.
- Independent variables are changed in an experiment. Dependent variables change in response to the independent variable.
- A theory is a hypothesis that is supported by many experiments.

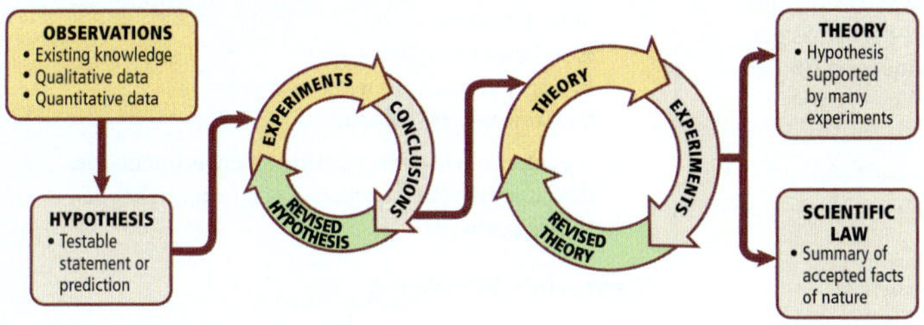

VOCABULARY
- scientific method
- qualitative data
- quantitative data
- hypothesis
- experiment
- independent variable
- dependent variable
- control
- conclusion
- theory
- scientific law

SECTION 4 Scientific Research

MAINIDEA Some scientific investigations result in the development of technology that can improve our lives and the world around us.

- Scientific methods can be used in pure research or in applied research.
- Some scientific discoveries are accidental, and some are the result of diligent research in response to a need.
- Laboratory safety is the responsibility of everyone in the laboratory.
- Many of the conveniences we enjoy today are technological applications of chemistry.

VOCABULARY
- pure research
- applied research

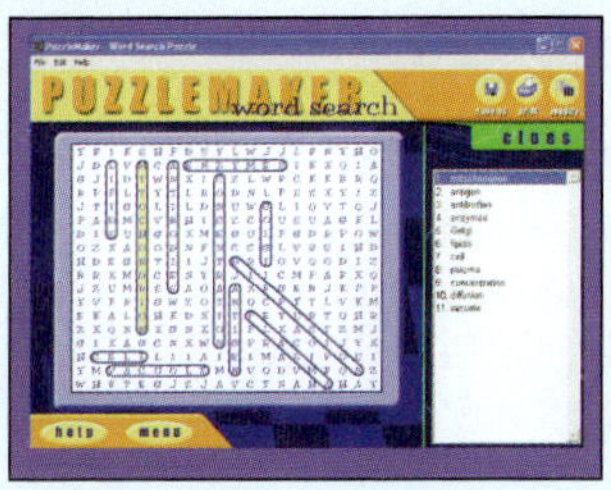

Vocabulary Practice

For additional practice with vocabulary, have students access the Vocabulary PuzzleMaker.

Use the Vocabulary

To reinforce chapter vocabulary, have students write a sentence using each term. OL EL

Review Strategies

- Have students define the terms *chemistry, chemical, matter, mass,* and *weight.* OL
- Have students summarize a possible sequence of steps in a scientific method approach, giving examples of each step. OL
- Have students explain safe practices in the laboratory. OL

Students can review with these online resources.

Vocabulary Practice

Vocabulary eGames
Vocabulary eFlashcards
Vocabulary PuzzleMaker

Section Self-Check

Chapter Self-Check

Online Test Practice

Use *eAssessment* to:

- create multiple versions of tests
- edit existing questions and add your own questions
- build tests aligned with select state curriculum standards using built-in state curriculum tags
- track students' progress

CHAPTER 1 ASSESSMENT

SECTION 1

Mastering Concepts

25. Define *substance* and *chemistry*.

26. Ozone Where is ozone located in Earth's atmosphere?

27. What three elements are found in chlorofluorocarbons?

28. CFCs What were common uses of CFCs?

29. Scientists noticed that the ozone layer was thinning. What was occurring at the same time?

■ **Figure 20**

30. Why do chemists study regions of the universe, such as the one shown in **Figure 20?**

Mastering Problems

31. If three oxygen particles are needed to form ozone, how many units of ozone could be formed from 6 oxygen particles? From 9? From 27?

32. Measuring Concentration **Figure 6** shows that the CFC level was measured at about 272 ppt (parts per thousand) in 1995. Because percent means *parts per hundred,* what percent is represented by 272 ppt?

SECTION 2

Mastering Concepts

33. Why is chemistry called the central science?

34. Which measurement depends on gravitational force—mass or weight? Explain.

35. Which branch of chemistry studies the composition of substances? The environmental impact of chemicals?

Mastering Problems

36. Predict whether your weight in the city of Denver, which has an altitude of 1.7 km above sea level, will be the same as, more than, or less than your weight in New Orleans, a city located at sea level.

37. The text tells you that 1 trillion atoms could fit onto a period at the end of this sentence. Write out the number 1 trillion using the correct number of zeros.

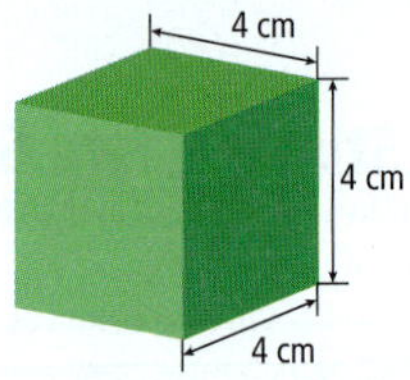

■ **Figure 21**

38. How much mass will the cube in **Figure 21** have if a 2-cm^3 cube of the same material has a mass of 4.0 g?

SECTION 3

Mastering Concepts

39. How does qualitative data differ from quantitative data? Give an example of each.

40. What is the function of a control in an experiment?

41. What is the difference between a hypothesis, a theory, and a law?

42. Laboratory Experiments You are asked to study how much table sugar can be mixed or dissolved in water at different temperatures. The amount of sugar that can dissolve in water goes up as the water's temperature goes up. What is the independent variable? Dependent variable? What factor is held constant?

43. Label each of the following pieces of data as qualitative or quantitative.
- **a.** A beaker weighs 6.6 g.
- **b.** Sugar crystals are white and shiny.
- **c.** Fireworks are colorful.

44. If evidence you collect during an experiment does not support your hypothesis, what should happen to that hypothesis?

Mastering Problems

45. One carbon (C) and one ozone (O_3) react to form one carbon monoxide (CO) and one oxygen gas (O_2) particle. How many ozone particles are needed to form 24 particles of oxygen gas (O_2)?

SECTION 4

Mastering Concepts

46. Laboratory Safety Finish each statement about laboratory safety so that it correctly states a safety rule.
- **a.** Study your lab assignment
- **b.** Keep food, beverages, and
- **c.** Know where to find and how to use the

SECTION 1

Mastering Concepts

25. Substance—any substance with a definite composition; chemistry—the study of matter and the changes it undergoes.

26. 90% in the stratosphere

27. carbon, fluorine, and chlorine

28. refrigerants, foams, propellants for spray products

29. increased usage of CFCs

30. because chemists study matter, and matter is found throughout the universe

Mastering Problems

31. 2 units; 3 units; 9 units

32. 27.2%

SECTION 2

Mastering Concepts

33. An understanding of chemistry is central to all sciences and to our everyday lives.

34. Weight; weight is calculated using the acceleration due to gravity; mass is independent of gravity.

35. Analytical chemistry studies the composition of substances; environmental chemistry studies the environmental impact of chemicals.

Mastering Problems

36. Your weight would be less in Denver because the acceleration due to gravity is less in Denver than in New Orleans.

37. 1,000,000,000,000

38. $x = 128$ g

SECTION 3

Mastering Concepts

39. Qualitative data, such as color or shape, are determined with the five senses. Quantitative data, such as mass or length, are measurements.

40. A control is a standard used for comparison.

41. A hypothesis is a tentative explanation about what has been observed. A theory is an explanation that has been supported by many experiments. A scientific law describes a relationship in nature.

42. temperature; amount of sugar dissolved; amount of water

43. **a.** quantitative
b. qualitative
c. qualitative

44. The hypothesis should be rewritten based on the new information and the new hypothesis should be tested.

Mastering Problems

45. 1 particle O_3/1 particle O_2 = x particles O_3/24 particles O_2; $x = 24$ particles O_3

SECTION 4

Mastering Concepts

46. a. before you come to the lab.
b. chewing gum out of the lab.
c. fire extinguisher, safety shower, fire blanket, and first-aid kit.

Chapter Self-Check

Mastering Problems

47. If your lab procedure instructs you to add two parts acid to each one part of water and you start with 25 mL of water, how much acid will you add, and how will you add it?

THINK CRITICALLY

48. Compare and Contrast Match each of the following research topics with the branch of chemistry that would study it: water pollution, the digestion of food in the human body, the composition of a new textile fiber, metals to make new coins, and a treatment for AIDS.

49. Interpret Scientific Diagrams Decide whether each of the diagrams shown below in **Figure 22a and b** is displaying qualitative or quantitative data.

a Types of Apples Grown in Bioscience Greenhouse

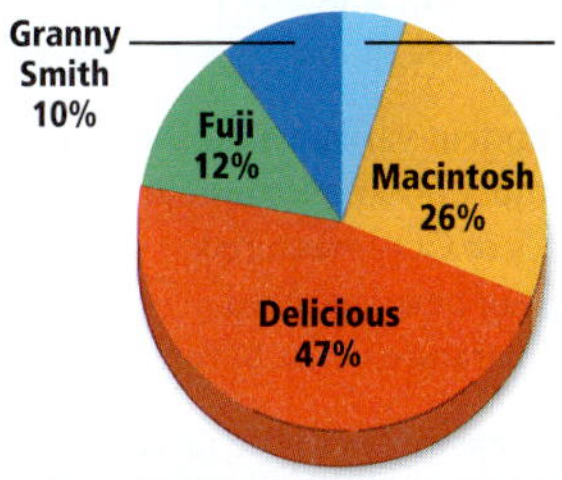

b

Data: Characteristics of Product Formed	
Color	white
Crystal Form	needles
Odor	none

■ **Figure 22**

50. Classify CFCs break down to form chemicals that react with ozone. Is this a macroscopic or a microscopic observation?

51. Infer A newscaster reports, "The air quality today is poor. Visibility is only 1.7 km. Pollutants in the air are expected to rise above 0.085 parts per million (ppm) in the next eight-hour average. Spend as little time outside today as possible if you suffer from asthma or other breathing problems." Which of these statements are qualitative and which are quantitative?

CUMULATIVE REVIEW

In Chapters 2 through 24, this heading will be followed by questions that review your understanding of previous chapters.

WRITING IN ▶ Chemistry

52. Ozone Depletion Based on your knowledge of chemistry, describe the research into depletion of the ozone layer by CFCs in a timeline.

53. CFC Reduction Research the most recent measures taken by countries around the world to reduce CFCs in the atmosphere since the Montreal Protocol. Write a short report describing the Montreal Protocol and more recent environmental measures to reduce CFCs.

54. Technology Name a technological application of chemistry that you use everyday. Prepare a booklet about its discovery and development.

DBQ Document-Based Questions

Ozone Depletion *The area of low-ozone varies over the Antarctic. NOAA collects data and monitors the low-ozone area at the poles.*

Figure 23 *shows the single-day maximum area of the ozone hole for each year from 1995 to 2009.*

Data obtained from: Southern Hemisphere Winter Summary. 2009. *National Oceanic and Atmospheric Administration.*

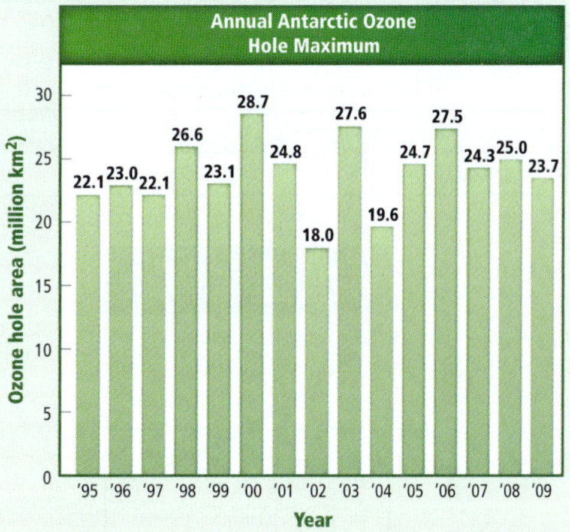

■ **Figure 23**

55. In what year did the largest maximum area of the ozone hole occur? The smallest?

56. What is the average maximum area of the ozone hole between the years 2005 and 2009? Between 2000 and 2004? Between 1995 and 1999?

Mastering Problems

47. 50 mL acid; Always add acid to the water very slowly.

THINK CRITICALLY

48. Possible answers: water pollution, environmental chemistry; the digestion of food, biochemistry; textile fibers, polymer chemistry; metal coins, inorganic chemistry; AIDS treatment, biochemistry.

49. a. quantitative data
b. qualitative data

50. microscopic observation

51. The qualitative statements are that air quality is poor and that people should spend little time outside. Quantitative statements include that visibility is only 1.7 km and that the pollutants will rise above 0.085 ppm in the next eight-hour average.

WRITING IN ▶ Chemistry

Rubric

52. Answers will vary but should include increased use of CFCs and the decrease in the ozone layer, including the effects of the depletion of life on Earth.

53. Answers will vary but should include the measures taken by the United States to limit the use and control of the disposal of CFCs. Answers should also include the measures taken by several other countries.

54. Check student booklets for accuracy. Be sure students explain clearly how the application is related to chemistry.

DBQ Document-Based Questions

Data obtained from: Southern Hemisphere Winter Summary. 2009. *National Oceanic and Atmospheric Administration.*

55. largest, 2000; smallest, 2002

56. 2005–2009: 25.0 million km^2; 2000–2004: 23.7 million km^2; 1995–1999: 23.4 million km^2

MULTIPLE CHOICE

1. B
2. C
3. A
4. D
5. D
6. A
7. A

MULTIPLE CHOICE

1. When working with chemicals in the laboratory, which is something you should NOT do?
   A. Read the label of chemical bottles before using their contents.
   B. Pour any unused chemicals back into their original bottles.
   C. Use a lot of water to wash skin that has been splashed with chemicals.
   D. Take only as much as you need of shared chemicals.

Use the table and graph below to answer Questions 2–5.

Page From a Student's Laboratory Notebook	
Step	**Notes**
Observation	Carbonated beverages taste fizzier when they are warm than when they are cold. (Carbonated beverages are fizzy because they contain dissolved carbon dioxide gas.)
Hypothesis	At higher temperatures, greater amounts of carbon dioxide gas will dissolve in a liquid. This is the same relationship between temperature and solubility seen with solids.
Experiment	Measure the mass of carbon dioxide (CO_2) in different samples of the same carbonated beverage at different temperatures.
Data analysis	See graph below.
Conclusion	

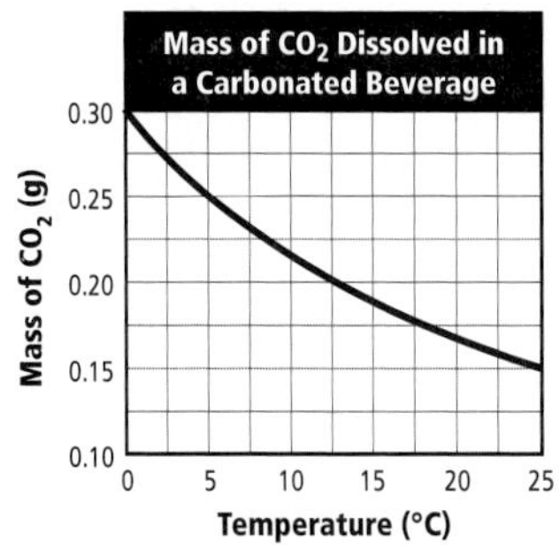

2. What must be a constant during the experiment?
   A. temperature
   B. mass of CO_2 dissolved in each sample
   C. amount of beverage in each sample
   D. independent variable

3. Assuming that all of the experimental data are correct, what is a reasonable conclusion for this experiment?
   A. Greater amounts of CO_2 dissolve in a liquid at lower temperatures.
   B. The different samples of beverage contained the same amount of CO_2 at each temperature.
   C. The relationship between temperature and solubility seen with solids is the same as the one seen with CO_2.
   D. CO_2 dissolves better at higher temperatures.

4. The scientific method used by this student showed that
   A. the hypothesis is supported by the experimental data.
   B. the observation accurately describes what occurs in nature.
   C. the experiment is poorly planned.
   D. the hypothesis should be thrown out.

5. The independent variable in this experiment is
   A. the number of samples tested.
   B. the mass of CO_2 measured.
   C. the type of beverage used.
   D. the temperature of the beverage.

6. Which is an example of pure research?
   A. creating synthetic elements to study their properties
   B. producing heat-resistant plastics for use in household ovens
   C. finding ways to slow down the rusting of iron ships
   D. searching for fuels other than gasoline to power cars

Use the table below to answer Question 7.

What is the effect of drinking soda on heart rate?		
Student	**Cans of Soda**	**Heart Rate (beats per minute)**
1	0	73
2	1	84
3	2	89
4	4	96

7. In this experiment testing the effects of soda on students' heart rates, which student serves as the control?
   A. Student 1 **C.** Student 3
   B. Student 2 **D.** Student 4

SHORT ANSWER

Use the table below to answer Questions 8 and 9.

Physical Properties of Three Elements				
Element	Symbol	Melting Point (°C)	Color	Density (g/cm^3)
Sodium	Na	897.4	Grey	0.986
Phosphorus	P	44.2	White	1.83
Copper	Cu	1085	Orange	8.92

8. Give examples of qualitative data that are true for the element sodium.

9. Give examples of quantitative data that are true for the element copper.

10. A student in class announces that he has a theory to explain why he scored poorly on a quiz. Is this a proper use of the term *theory*? Explain your answer.

EXTENDED RESPONSE

11. Explain why scientists use mass for measuring the amount of a substance instead of using weight.

Consider the following experiment as you answer Questions 12 and 13.

A chemistry student is investigating how particle size affects the rate of dissolving. In her experiment, she adds a sugar cube, sugar crystals, or crushed sugar to each of three beakers of water, stirs the mixtures for 10 seconds, and records how long it takes the sugar to dissolve in each beaker.

12. Identify the independent and dependent variables in this experiment. How can they be distinguished?

13. Identify a feature of this experiment that should be kept constant. Explain why it is important to keep this feature constant.

SAT SUBJECT TEST: CHEMISTRY

14. A scientist from which field of chemistry investigates a new form of packaging material that breaks down rapidly in the environment?
   A. biochemistry
   B. theoretical chemistry
   C. environmental chemistry
   D. inorganic chemistry
   E. physical chemistry

Use the safety symbols below to answer Questions 15–18. Some choices may be used more than once; others will not be used at all.

A. 

B. 

C. 

D. 

E. 

Select the symbol for the safety rule being described in each case.

15. Safety goggles should be worn whenever you are working in the lab.

16. Use chemicals in rooms with proper ventilation in case of strong fumes.

17. Wear proper protective clothing to prevent stains and burns.

18. Objects may be extremely hot or extremely cold; use hand protection.

19. Which statement is NOT true about mass?
   A. It has the same value everywhere on Earth.
   B. It is independent of gravitational forces.
   C. It becomes less in outer space, farther from Earth.
   D. It is a constant measure of the amount of matter.
   E. It is found in all matter.

NEED EXTRA HELP?																			
If You Missed Question . . .	1	2	3	4	5	6	7	8	9	10	11	12	13	14	15	16	17	18	19
Review Section . . .	1.4	1.3	1.3	1.3	1.3	1.4	1.3	1.3	1.3	1.3	1.2	1.3	1.3	1.2	1.4	1.4	1.4	1.4	1.2

SHORT ANSWER

8. Sodium is grey in color. It has the symbol Na. Its density is low. Its melting point is between the other values.

9. Copper has the melting point of 1085°C and a density of 8.92 g/cm^3.

10. No; a theory is an explanation of how nature behaves and is based on many repeated experiments. This student might be proposing a hypothesis.

EXTENDED RESPONSE

11. Because weight is affected by gravity, it can change depending on its location on Earth. Mass measures the amount of matter in a substance regardless of the effect of gravity on the substance, which makes it a more reliable measurement when comparing measurements made in different parts of the world.

12. The dependent variable is the amount of time required for dissolving, while the independent variable is how much the sugar is crushed before it is added. The independent variable can be identified because it is the factor that the researcher is changing, while the dependent variable changes in response to a change in the independent variable.

13. Answers will vary but can include temperature of water, volume of water, or mass of sugar added. It is important to keep these features constant in order for the different trials to be compared appropriately. If too many factors change in an experiment, the researcher cannot identify what effect each individual factor has on the outcome of the experiment.

SAT SUBJECT TEST: CHEMISTRY

14. C
15. C
16. E
17. D
18. B
19. C

CHAPTER 2 Analyzing Data

BIGIDEA **Chemists collect and analyze data to determine how matter interacts.**

ESSENTIAL QUESTIONS	RESOURCES TO ASSESS MASTERY
SECTION 1 **Units and Measurements** **1.** What are the SI base units for time, length, mass, and temperature? **2.** How does adding a prefix change a unit? **3.** How are the derived units different for volume and density? 1 session 0.5 block	**Progress Monitoring** Formative Assessment, pp. 33, 38 Reading Check, pp. 34, 37 Section Review, p. 39
SECTION 2 **Scientific Notation and Dimensional Analysis** **1.** Why use scientific notation to express numbers? **2.** How is dimensional analysis used for unit conversion? 1 session 1 block	**Progress Monitoring** Formative Assessment, pp. 41, 45 Reading Check, p. 42 Section Review, p. 46
SECTION 3 **Uncertainty in Data** **1.** How do accuracy and precision compare? **2.** How can the accuracy of experimental data be described using error and percent error? **3.** What are the rules for significant figures and how can they be used to express uncertainty in measured and calculated values? 1 session 0.5 block	**Progress Monitoring** Formative Assessment, p. 53 Reading Check, p. 49 Section Review, p. 54
SECTION 4 **Representing Data** **1.** Why are graphs created? **2.** How can graphs be interpreted? 1 session 0.5 block	**Progress Monitoring** Formative Assessment, p. 57 Reading Check, p. 57 Graph Check, pp. 55, 56, 57, 58 Section Review, p. 58 **Summative Assessment** Chapter Assessment, p. 62 *eAssessment* Chapter Test (Scaffolded)

Go online!

LEVELED RESOURCES	LAB MATERIALS
Science Notebook 2.1 OL ***FAST FILE Chapter Resources:*** MiniLab Worksheet, p. 26 OL Study Guide, p. 38 BL OL	LaunchLAB p. 30 / **15 min** graduated cylinder, alcohol, red food coloring, corn oil, water, glycerol MiniLAB p. 39 / **15 min** unknown objects, balance, water, graduated cylinder
Science Notebook 2.2 OL ***FAST FILE Chapter Resources:*** Study Guide, p. 40 BL	
Science Notebook 2.3 OL ***FAST FILE Chapter Resources:*** ChemLab Worksheet, p. 27 OL Study Guide, p. 41 BL	ChemLAB p. 60 / **45 min** water, 100-mL graduated cylinder, small plastic cup, balance, metric ruler, pencil, graphing calculator (optional), pre and post-1982 pennies, graph paper
Science Notebook 2.4 OL ***FAST FILE Chapter Resources:*** Study Guide, p. 43 BL	

ADDITIONAL RESOURCES

- eAssessment
- ConnectED Student Center
- ConnectED Teacher Center
- Lesson Presentations
- eSolutions
- Videos
- Animations
- Online Quizzes
- Virtual Investigations
- Multilingual eGlossary
- Personal Tutors
- WebQuests
- Applying Practices Worksheets
- PBLs

Practice:

- Challenge Problems AL
- Supplemental Problems BL OL
- Chapter Test (Scaffolded)
- Standardized Test Practice

FAST FILE Resources:

- Section Focus Transparency Masters
- Math Skills Transparency Masters and Worksheets
- Teaching Transparency Masters and Worksheets

Additional Resources:

- Blueprints for Success
- Solving Problems: A Chemistry Handbook

Lab Resources:

- Laboratory Manual OL
- CBL Laboratory Manual OL
- Small-Scale Laboratory Manual OL
- Forensics Laboratory Manual OL

BL Below Level OL On Level AL Advanced Learners EL English Learners COOP LEARN Cooperative Learning

CHAPTER 2

LaunchLAB

Rubric

How can you form layers of liquids?

Purpose Students will use various liquids to demonstrate the property of density.

Safety Precautions Approve lab safety forms before work begins. **WARNING: *Alcohol is flammable.*** Review MSDS with students. Remind students never to taste anything in the laboratory.

Disposal Have students add liquid soap to the contents of the cylinder, then flush them down a drain with plenty of water.

Teaching Strategies

- Emphasize to students the importance of pouring the liquids slowly down the side of the cylinder. If they are poured quickly, the layers will mix at the boundaries.
- Be sure students add the liquids in the order specified. Adding a more dense liquid to one that is less dense might cause mixing of the layers.

Expected Results The different liquids do not mix but form distinctive layers.

CHAPTER 2

Analyzing Data

BIGIDEA Chemists collect and analyze data to determine how matter interacts.

SECTIONS

1 Units and Measurements

2 Scientific Notation and Dimensional Analysis

3 Uncertainty in Data

4 Representing Data

LaunchLAB

How can you form layers of liquids?

You know that this skydiver falls through the air. And you know that ice floats in water, whereas a rock sinks. In this lab you will investigate what can happen when you pour one liquid into another.

FOLDABLES Study Organizer

Types of Graphs

Make a layered-look book. Label it as shown. Use it to organize information about types of graphs.

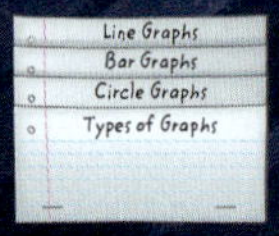

Procedure

1. Read and complete the lab safety form.
2. Observe **5-mL samples of alcohol** (dyed red), **glycerol** (dyed blue), **corn oil,** and **water.** Plan the order in which to add the liquids to a **graduated cylinder** to form four layers.
   WARNING: ***Keep alcohol away from open flames.***
3. Test your plan by adding the liquids, one at a time, to the graduated cylinder. When adding each liquid, tilt the graduated cylinder, and slowly pour the liquid so it runs down the inside. When adding the glycerol, allow it to settle before adding the next liquid.
4. Did the liquids form four distinct layers? If not, rinse out the graduated cylinder and repeat Steps 2 and 3 using a different order.

Analysis

1. **Identify** the order, from top to bottom, of the layers in the graduated cylinder. The layers from top to bottom are alcohol, corn oil, water, and glycerol.
2. **Hypothesize** what property of the liquids is responsible for the arrangement of the layers. Student hypotheses should reflect differences in densities of the liquids, although that term will probably not be used. Students might refer to differences in "heaviness."

Skydivers typically jump from an altitude of about 4000 m. They can reach speeds of 190 km/h and even greater. Collecting and analyzing data is important for a skydiver in order to maximize safety.

Inquiry What do you think would happen if small pieces of metal, plastic, and wood were added to the layers of liquids in the graduated cylinder? Student responses will likely mention that the objects will rest at some particular level relative to the liquid layers.

Introduce the BIGIDEA

Data Analysis Have students measure their heights using meter sticks and yard sticks you have placed 1 m from the floor around the room. Place a chart on the wall for students to record their heights in both centimeters and inches. Remind them to make measurements as accurately as they can.

Tie to Previous Knowledge

Have students review the following concepts before studying this chapter.

- qualitative and quantitative, variables and analysis

Use the Photo

Calibration and Accuracy Have students notice the altimeter on the skydiver. Tell them that an altimeter is an instrument used in skydiving, which uses changes in atmospheric pressure to determine the change in altitude. Have groups of students discuss the importance of the altimeter to a skydiver in free fall. The skydiver uses the altimeter to determine when to release the parachute. Altimeters must be calibrated using a known altitude before the skydiver uses it for a free fall. Skydivers can fall at a rate of 52 m/sec. Ask student groups why an accurate altimeter reading is important. If the reading is inaccurate, the skydiver might release the parachute late or too early. OL EL

SECTION 1

1 Focus

MAIN IDEA

Systems of Measurements Tell students that the United States and Liberia are the only nations that use the "English system" of measurements (yards, miles, quarts, gallons, and so on). Display several items used to measure volume in the United States: a quart measuring cup, a pint measuring cup, an 8-oz. measuring cup, a teaspoon, a gallon pitcher, and a tablespoon. Also, display items used to measure volume in the chemistry laboratory: 10-mL, 50-mL, 100-mL, and 1000-mL graduated cylinders. Ask students what each set of instruments has in common with the others. They are all used to measure volume. Ask students why graduated cylinders are used in the chemistry laboratory. Students should point out that graduated cylinders give a more-precise measurement and all use the same units of measurement. Ask students to list differences in the two systems of measurement. The English system uses a variety of quantities, whereas the metric system uses the same standard. Ask students to discuss why standard units of measurement are important to scientists. Students should indicate that by using a standard unit to measure a quantity, everyone can be assured that the measured quantity is the same. **OL**

■ **Caption Question Fig. 1** The fluid ounce is the larger unit; the larger unit is paired with the smaller numerical value.

SECTION 1

Units and Measurements

MAIN IDEA Chemists use an internationally recognized system of units to communicate their findings.

Essential Questions

- What are the SI base units for time, length, mass, and temperature?
- How does adding a prefix change a unit?
- How are the derived units different for volume and density?

Review Vocabulary

mass: a measurement that reflects the amount of matter an object contains

New Vocabulary

base unit
second
meter
kilogram
kelvin
derived unit
liter
density

CHEM 4 YOU **Have you ever noticed that a large drink varies in volume depending on where it is purchased? Wouldn't it be better if you always knew how much drink you would get when you ordered the large size? Chemists use standard units to ensure the consistent measurement of a given quantity.**

Units

You use measurements almost every day. For example, reading the bottled water label in **Figure 1** helps you decide what size bottle to buy. Notice that the label uses a number and a unit, such as 700 mL, to give the volume. The label also gives the volume as 23.7 fluid ounces. Fluid ounces, pints, and milliliters are units used to measure volume.

Système Internationale d'Unités For centuries, units of measurement were not exact. A person might measure distance by counting steps, or measure time using a sundial or an hourglass filled with sand. Such estimates worked for ordinary tasks. Because scientists need to report data that can be reproduced by other scientists, they need standard units of measurement. In 1960, an international committee of scientists met to update the existing metric system. The revised international unit system is called the Système Internationale d'Unités, which is abbreviated SI.

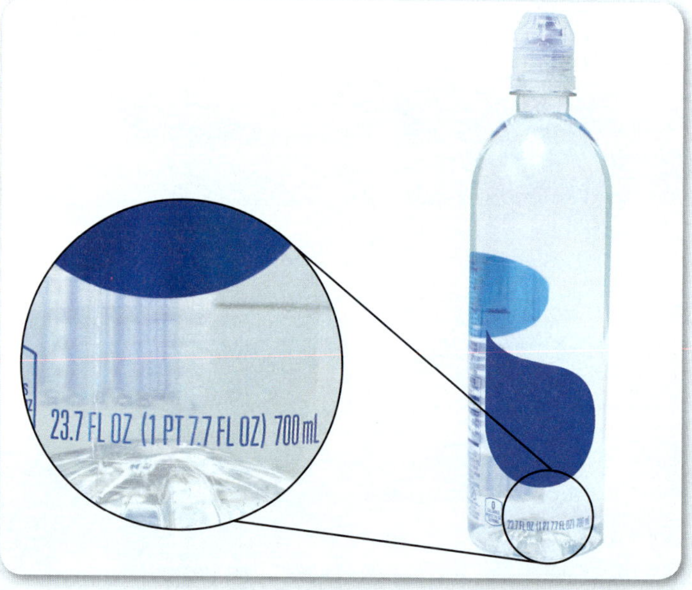

■ **Figure 1** The label gives the volume of water in the bottle in three different units: fluid ounces, pints, and milliliters. Notice that each volume includes a number and a unit.

Infer *Which is the larger unit of volume: a fluid ounce or a milliliter?*

CHEMISTRY JOURNAL

Same Meaning, Different Language Have students write about times they disagreed with someone only to realize later that they meant the same thing but used different language to say it. Ask them how SI units can solve that problem among scientists. **OL**

Base Units and SI Prefixes

There are seven base units in SI. A **base unit** is a defined unit in a system of measurement that is based on an object or event in the physical world. A base unit is independent of other units. **Table 1** lists the seven SI base units, the quantities they measure, and their abbreviations. Some familiar quantities that are expressed in base units are time, length, mass, and temperature.

To better describe the range of possible measurements, scientists add prefixes to the base units. This task is made easier because the metric system is a decimal system—a system based on units of 10. The prefixes in **Table 2** are based on factors of ten and can be used with all SI units. For example, the prefix *kilo-* means one thousand; therefore, 1 km equals 1000 m. Similarly, the prefix *milli-* means one-thousandth; therefore, 1 mm equals 0.001 m. Many mechanical pencils use lead that is 0.5 mm in diameter. How much of a meter is 0.5 mm?

Time The SI base unit for time is the **second** (s). The physical standard used to define the second is the frequency of the radiation given off by a cesium-133 atom. Cesium-based clocks are used when highly accurate timekeeping is required. For everyday tasks, a second seems like a short amount of time. In chemistry, however, many chemical reactions take place within a fraction of a second.

Length The SI base unit for length is the **meter** (m). A meter is the distance that light travels in a vacuum in 1/299,792,458 of a second. A vacuum exists where space contains no matter.

A meter is close in length to a yard and is useful for measuring the length and width of a small area, such as a room. For larger distances, such as between cities, you would use kilometers. Smaller lengths, such as the diameter of a pencil, are likely to be given in millimeters.

Table 1 SI Base Units

Quantity	Base Unit
Time	second (s)
Length	meter (m)
Mass	kilogram (kg)
Temperature	kelvin (K)
Amount of a substance	mole (mol)
Electric current	ampere (A)
Luminous intensity	candela (cd)

Vocabulary

Science usage v. Common usage

Meter

Science usage: the SI base unit of length

The metal rod was 1 m in length.

Common usage: a device used to measure

The time ran out on the parking meter.

Explore **SI prefixes** with an interactive table. Concepts In Motion

Table 2 SI Prefixes

Prefix	Symbol	Numerical Value in Base Units	Power of 10 Equivalent
Giga	G	1,000,000,000	10^9
Mega	M	1,000,000	10^6
Kilo	k	1000	10^3
–	–	1	10^0
Deci	d	0.1	10^{-1}
Centi	c	0.01	10^{-2}
Milli	m	0.001	10^{-3}
Micro	μ	0.000001	10^{-6}
Nano	n	0.000000001	10^{-9}
Pico	p	0.000000000001	10^{-12}

2 Teach

Concept Development

Scientific Measurements Have students of varying heights count the number of paces they need to walk across the classroom. Elicit from students that a pace can vary in length from person to person, and that such differences are a problem when developing a system of measurement. Scientific measurements must be exact, and two scientists must be able to compare the same measurement. This is why measurements are based on set units, such as the meter. OL EL

ASSESSMENT

Knowledge Ask students to identify items that they would measure using the base units for time, length, and mass. Examples might include the amount of time it takes to bounce a ball (s), the distance across a room (m), and the mass of a student (kg). OL

Quick Demo

Metric Masses Bring out various masses such as 1 g, 1 dg, 1 kg, and 1 mg, if available. Have students feel the weights and compare them. Remind them that prefixes on the metric scale represent a ten-fold difference in the property measured (in this case, mass).

In-Text Question

$0.5\ \text{mm} \times (1\ \text{m})/(1000\ \text{mm}) = 0.0005\ \text{m}$

CULTURAL DIVERSITY

Zero The first known use of the zero was by a seventh-century Hindu, Brahmagupta. Ancient Egyptians and Greek mathematicians computed without a zero. Babylonians used a placeholder similar to the zero. Turkish and Greek mathematicians used an inverted *h* as a symbol for zero.

DIFFERENTIATED INSTRUCTION

Struggling Learners Make sure the students handle measuring equipment and are taking measurements, rather than just reading and observing. BL

■ **In-Text Question** 1 g = 1000 mg

☑ **READING CHECK** 25°C is warmer than 25°F

Identify Misconceptions

Many students confuse accuracy and precision.

Uncover the Misconception Place a transparent container with a small target painted on the bottom on the overhead projector. Fill it three-fourths full with water. Have a student drop coins into the container, attempting to hit the target. After several coins are dropped, have students discuss the accuracy and precision the results model.

Demonstrate the Concept Choose one coin and dry it thoroughly. Carefully measure the mass of the dry coin. Divide students into groups. Give each group a balance. Have each group take several mass readings of the dry coin and record their data.

Assess New Knowledge Tell students the mass you measured of the dry coin. Have students discuss their results. They should note that, if readings were taken carefully, all of them are precise. Then discuss whether all the readings are accurate. Have students compare precision and accuracy. **OL**

■ **Figure 2** Scientists at the National Institute of Standards and Technology are experimenting with redefining the kilogram using an apparatus known as a watt balance. The watt balance uses electric current and a magnetic field to measure the force required to balance a one-kilogram mass against the force of gravity. Other scientists are counting the number of atoms in a one-kilogram mass to redefine the kilogram.

Mass Recall that mass is a measure of the amount of matter an object contains. The SI base unit for mass is the **kilogram** (kg). Currently, a platinum and iridium cylinder kept in France defines the kilogram. The cylinder is stored in a vacuum under a triple bell jar to prevent the cylinder from oxidizing. As shown in **Figure 2,** scientists are working to redefine the kilogram using basic properties of nature.

A kilogram is equal to about 2.2 pounds. Because the masses measured in most laboratories are much smaller than a kilogram, scientists often measure quantities in grams (g) or milligrams (mg). For example, a laboratory experiment might ask you to add 35 mg of an unknown substance to 350 g of water. When working with mass values, it is helpful to remember that there are 1000 g in a kilogram. How many milligrams are in a gram?

Temperature People often use qualitative descriptions, such as hot and cold, when describing the weather or the water in a swimming pool. Temperature, however, is a quantitative measurement of the average kinetic energy of the particles that make up an object. As the particle motion in an object increases, so does the temperature of the object.

Measuring temperature requires a thermometer or a temperature probe. A thermometer consists of a narrow tube that contains a liquid. The height of the liquid indicates the temperature. A change in temperature causes a change in the volume of the liquid, which results in a change in the height of the liquid in the tube. Electronic temperature probes make use of thermocouples. A thermocouple produces an electric current that can be calibrated to indicate temperature.

Several different temperature scales have been developed. Three temperature scales—Kelvin, Celsius, and Fahrenheit—are commonly used to describe how hot or cold an object is.

Fahrenheit In the United States, the Fahrenheit scale is used to measure temperature. German scientist Gabriel Daniel Fahrenheit devised the scale in 1724. On the Fahrenheit scale, water freezes at 32°F and boils at 212°F.

Celsius Another temperature scale, the Celsius scale, is used throughout much of the rest of the world. Anders Celsius, a Swedish astronomer, devised the Celsius scale. The scale is based on the freezing and boiling points of water. He defined the freezing point of water as 0 and the boiling point of water as 100. He then divided the distance between these two fixed points into 100 equal units, or degrees. To convert from degrees Celsius (°C) to degrees Fahrenheit (°F), you can use the following equation.

$$°F = 1.8(°C) + 32$$

Imagine a friend from Canada calls you and says that it is 35°C outside. What is the temperature in degrees Fahrenheit? To convert to degrees Fahrenheit, substitute 35°C into the above equation and solve.

$$1.8(35) + 32 = 95°F$$

If it is 35°F outside, what is the temperature in degrees Celsius?

$$\frac{35°F - 32}{1.8} = 1.7\ °C$$

Get help with **conversions.**

Personal Tutor

☑ **READING CHECK** **Infer** Which is warmer, 25°F or 25°C?

Robert Rathe

Demonstration

Measurement of Temperature

Purpose

to investigate the measurement of temperature with a thermometer and by sensation of temperature on the skin

Materials

1-L beakers (3); ice; hot plate; paper towels; timer; thermometers (3)

Safety Precautions

Test warm water to make sure it will not burn the students. Be careful of slippery floors if water is spilled.

Disposal Items can be cleaned and reused.

Procedure

Place three labeled beakers side by side. They should contain room-temperature water, ice water, and warm water, respectively. Measure

Kelvin The SI base unit for temperature is the **kelvin** (K). The Kelvin scale was devised by a Scottish physicist and mathematician, William Thomson, who was known as Lord Kelvin. Zero kelvin is a point where all particles are at their lowest possible energy state. On the Kelvin scale, water freezes at 273.15 K and boils at 373.15 K. Later, you will learn why scientists use the Kelvin scale to describe properties of a gas.

Figure 3 compares the Celsius and Kelvin scales. It is easy to convert between the Celsius scale and the Kelvin scale using the following equation.

Kelvin-Celsius Conversion Equation

$$K = °C + 273$$

K represents temperature in kelvins.
°C represents temperature in degrees Celsius.

Temperature in kelvins is equal to temperature in degrees Celsius plus 273.

As shown by the equation above, to convert temperatures reported in degrees Celsius to kelvins, you simply add 273. For example, consider the element mercury, which melts at −39°C. What is this temperature in kelvins?

$$-39°C + 273 = 234 \text{ K}$$

To convert from kelvins to degrees Celsius, just subtract 273. For example, consider the element bromine, which melts at 266 K. What is this temperature in degrees Celsius?

$$266 \text{ K} - 273 = -7°C$$

You will use these conversions frequently throughout chemistry, especially when you study how gases behave. The gas laws you will learn are based on kelvin temperatures.

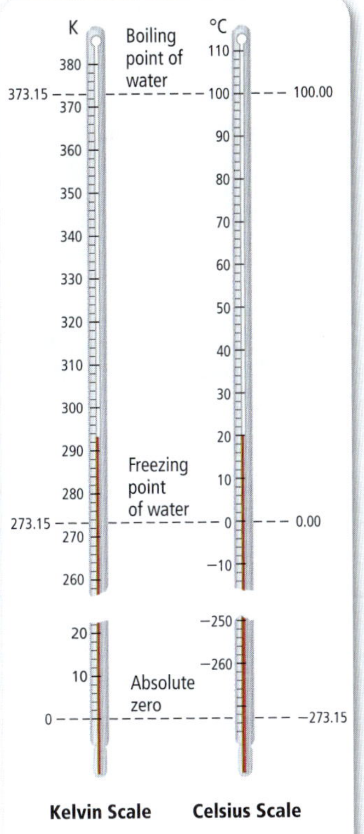

Figure 3 A change of 1 K on the Kelvin scale is equal in size to a change of 1°C on the Celsius scale. Notice also that the degree sign (°) is not used with the Kelvin scale.

Derived Units

Not all quantities can be measured with SI base units. For example, the SI unit for speed is meters per second (m/s). Notice that meters per second includes two SI base units—the meter and the second. A unit that is defined by a combination of base units is called a **derived unit.** Two other quantities that are measured in derived units are volume (cm^3) and density (g/cm^3).

Volume Volume is the space occupied by an object. The volume of an object with a cubic or rectangular shape can be determined by multiplying its length, width, and height dimensions. When each dimension is given in meters, the calculated volume has units of cubic meters (m^3). In fact, the derived SI unit for volume is the cubic meter. It is easy to visualize a cubic meter; imagine a large cube whose sides are each 1 m in length. The volume of an irregularly shaped solid can be determined using the water displacement method, a method used in the MiniLab in this section.

The cubic meter is a large volume that is difficult to work with. For everyday use, a more useful unit of volume is the liter. A **liter** (L) is equal to one cubic decimeter (dm^3), that is, 1 L equals 1 dm^3. Liters are commonly used to measure the volume of water and beverage containers. One liter has about the same volume as one quart.

Concept Development

Derived Units Ask students why volume is considered to be a derived unit. Volume is calculated as length × width × height. All are measured in units of length. OL

Enrichment

Temperature Students, in the classroom, think of temperature in Fahrenheit, but chemists think of temperature in Kelvin. Have students covert 245 K to Fahrenheit and 78°F to Kelvin.
245 K − 273 = −28°C
1.8 (−28°C) + 32 = −18°F
(78°F − 32)/1.8 = 25.6°C
25.6°C + 273 = 299 K OL EL

the temperature of the water in each beaker. Have a student place one hand in the ice water and the other hand in the warm water for two minutes. Ask the student to describe the temperature of each hand. Then, have the student simultaneously place both hands in the beaker containing room-temperature water. Have the student describe the temperature of this water as sensed by each hand.

Results

The room-temperature water will feel warm to the hand coming from ice water but cool to the hand coming from warm water.

Analysis

1. How do you classify something as hot or cold? By whether heat flows from it or to it; they are qualitative terms.
2. How does a thermometer classify something as hot or cold? It gives a numerical value to the temperature. It is quantitative.

■ **Caption Question Fig. 4** 1000 cm^3

Visual Learners

Powers of Ten After viewing a video on the powers of ten, discuss the importance of a factor of ten. Ask students why the metric system is a base ten system. Students should understand that every prefix in the system represents a factor of ten. Ask students to discuss the difference between a centimeter and a millimeter. Students should understand that a millimeter is ten times smaller than a centimeter and there are ten millimeters in one centimeter. **OL**

Content Background

History of a Meter The Metric System was first adopted in France in 1791. The units were designed to be logical and practical, neutral, and universally adopted. The definitions of the base units were designed so that any laboratory equipped with proper instruments should be able to make their own models of those base units. The historical definition of the meter defined by the French Academy of Sciences as 1/10,000,000 of the quadrant of the earth's circumference running from the North Pole, to the equator, through the city of Paris. Over time the definitions have become more exact. By the 1960s, the meter was defined in terms of an emission line of the M-86 isotope of the rare gas Kr—specifically, 1 m = 1,650,763.73 wavelengths of that orange emission line. This of course had the tremendous advantage that any well equipped lab had access to the primary standard of length. Advances in technology have developed a further definition of a meter as the distance light travels in a vacuum in 1/299,792,458th of a second.

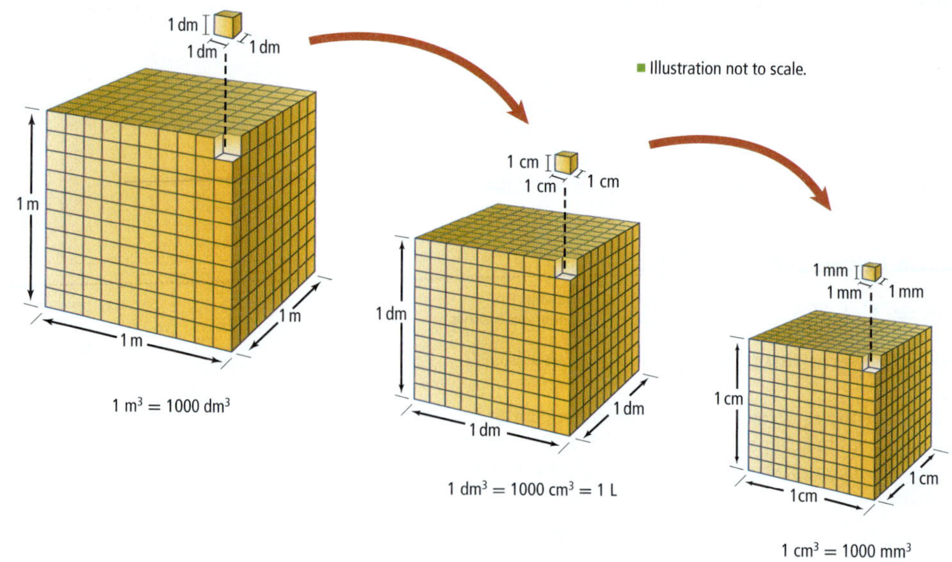

■ **Figure 4** The three cubes show volume relationships between cubic meters (m^3), cubic decimeters (dm^3), cubic centimeters (cm^3), and cubic millimeters (mm^3). As you move from left to right, the volume of each cube gets 10 × 10 × 10, or 1000 times, smaller.
Interpret *How many cubic centimeters (cm^3) are in 1 L?*

For smaller quantities of liquids in the laboratory, volume is often measured in cubic centimeters (cm^3) or milliliters (mL). A milliliter and a cubic centimeter are equal in size.

$$1 \text{ mL} = 1 \text{ cm}^3$$

Recall that the prefix *milli-* means one-thousandth. Therefore, one milliliter is equal to one-thousandth of a liter. In other words, there are 1000 mL in 1 L.

$$1 \text{ L} = 1000 \text{ mL}$$

Figure 4 shows the relationships among several different SI units of volume.

Explore **density**

Virtual Investigations

Density Why is it easier to lift a backpack filled with gym clothes than the same backpack filled with books? The answer can be thought of in terms of density—the book-filled backpack contains more mass in the same volume. **Density** is a physical property of matter and is defined as the amount of mass per unit volume. Common units of density are grams per cubic centimeter (g/cm^3) for solids and grams per milliliter (g/mL) for liquids and gases.

Consider the grape and the piece of foam in **Figure 5.** Although both have the same mass, they clearly occupy different amounts of space. Because the grape occupies less volume for the same amount of mass, its density must be greater than that of the foam.

DIFFERENTIATED INSTRUCTION

Struggling Learners Have students attach one Celsius and one Fahrenheit thermometer side by side on a piece of cardboard. Measure the temperature of several locations. Have students make a chart to record the readings on each thermometer. Compare the scales used to measure the same temperatures. Ask students to make several inferences from their observations. Possible answers: Celsius scale is based on 100 increments between boiling and freezing. The numbers indicating the temperatures go up and down with a bigger spread on the Fahrenheit scale. The Fahrenheit numbers are greater for the same temperature. **BL**

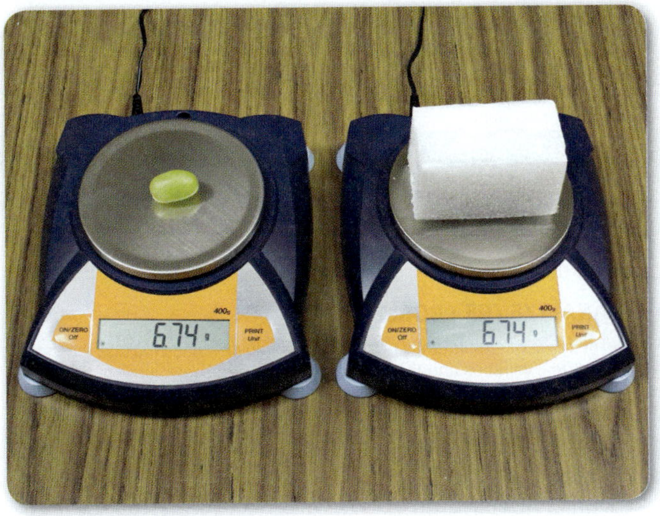

■ **Figure 5** The grape and the foam have the same mass but different volumes because the grape is more dense.

Interpret ***How would the masses compare if the volumes were equal?***

The density of a substance usually cannot be measured directly. Rather, it is calculated using mass and volume measurements. You can calculate density using the following equation.

Density Equation

$$\text{density} = \frac{\text{mass}}{\text{volume}}$$

The density of an object or a sample of matter is equal to its mass divided by its volume.

Because density is a physical property of matter, it can sometimes be used to identify an unknown element. For example, imagine you are given the following data for a piece of an unknown metallic element.

$$\text{volume} = 5.0\ \text{cm}^3$$
$$\text{mass} = 13.5\ \text{g}$$

Substituting these values into the equation for density yields:

$$\text{density} = \frac{13.5\ \text{g}}{5.0\ \text{cm}^3} = 2.7\ \text{g/cm}^3$$

Now go to **Table R-7** and scan through the given density values until you find one that closely matches the calculated value of 2.7 g/cm^3. What is the identity of the unknown element?

Connection to Earth Science As air at the equator is warmed, the particles in the air move farther apart and the air density decreases. At the poles, the air cools and its density increases as the particles move closer together. When a cooler, denser air mass sinks beneath a rising warm air mass, winds are produced. Weather patterns are created by moving air masses of different densities.

☑ READING CHECK **State** the quantities that must be known in order to calculate density.

(t)Matt Meadows, (b)Charles Falco/Science Source

RealWorld CHEMISTRY

Liquid Density Measurement

HYDROMETERS A hydrometer is a device that measures the specific gravity (the ratio of the fluid's density to that of water) of a fluid. Fluids of different densities result in different readings. Hydrometers are often used at service stations to diagnose problems with an automobile's battery.

■ **Caption Question Fig. 5** The mass of the foam would be less.

Enrichment

Ocean Currents Circulation in the ocean is density driven by differences in temperature (thermo) and salinity (haline). The thermohaline circulation in the oceans is responsible for ocean currents. Surface winds push water at the surface toward the poles from the equator. As the warmer surface water moves toward the poles, it cools and becomes denser. Evaporation of water also increases the salinity of the water, and eventually it sinks below the surface at high latitudes. The cooler, denser water flows downhill into deep ocean basins where it can remain for up to 1200 years before resurfacing. These ocean currents transfer heat and salt across the oceans and play a central role in Earth's climate.

Have students dissolve salt into water at a variety of temperatures. They should record the volume and mass of the solutions they produce and calculate the density of each solution. Have the students predict how these solutions would layer. Denser solutions would be at the bottom. **OL** **AL**

■ **In-Text Question** aluminum

☑ **READING CHECK** mass and volume

CHEMISTRY PROJECT

Density and Gravity Ask students to make predictions on how gravity will affect objects of the same volume but greater density. Have them research the effect of gravity on objects when there is no air resistance. In a vacuum, which will hit the ground first, a lead shot or a polystyrene ball of the same size? They will hit at the same time. **OL**

IN-CLASS Example

Question 116 g of sunflower oil is used in a recipe. The density of the oil is 0.925 g/mL. What is the volume of the sunflower oil in mL?

Answer

density = mass/volume
density = 0.925 g/mL
mass: 116 grams
volume = mass/density
volume = 116 g/0.925 g/mL
volume = 125 mL

PRACTICE Problems

Have students refer to the Selected Solutions appendix for complete solutions to odd-numbered problems. The complete solutions for all problems can be found in the Solutions Manual.

1. No; the density of aluminum is 2.7 g/cm^3; the density of the cube is 4 g/cm^3.
2. volume = 5 mL
3. volume = 41 mL

3 Assess

Check for Understanding

Ask students whether a liter is a base unit or a derived unit. A liter is a derived unit of volume; because volume is calculated as length × width × height. OL

Reteach

Have students hold a small piece of foam in one hand and a large piece in the other hand. Ask them if the two pieces of foam have the same mass. no Ask if they have the same volume. no Then, ask them if the densities of the two pieces of foam are the same. Have them justify their answers. Yes, they are the same material. OL EL

Your textbook includes many Example Problems, each of which is solved using a three-step process. Read Example Problem 1 and follow the steps to calculate the mass of an object using density and volume.

THE PROBLEM

1. Read the problem carefully.
2. Be sure that you understand what is being asked.

ANALYZE THE PROBLEM

1. Read the problem again.
2. Identify what you are given, and list the known data. If needed, gather information from graphs, tables, or figures.
3. Identify and list the unknowns.
4. Plan the steps you will follow to find the answer.

SOLVE FOR THE UNKNOWN

1. Determine whether you need a sketch to solve the problem.
2. If the solution is mathematical, write the equation and isolate the unknown factor.
3. Substitute the known quantities into the equation.
4. Solve the equation.
5. Continue the solution process until you solve the problem.

EVALUATE THE ANSWER

1. Reread the problem. Is the answer reasonable?
2. Check your math. Are the units and the significant figures correct? (Refer to Section 3.)

EXAMPLE Problem 1

USING DENSITY AND VOLUME TO FIND MASS When a piece of aluminum is placed in a 25-mL graduated cylinder that contains 10.5 mL of water, the water level rises to 13.5 mL. What is the mass of the aluminum?

1 ANALYZE THE PROBLEM

The mass of aluminum is unknown. The known values include the initial and final volumes and the density of aluminum. The volume of the sample equals the volume of water displaced in the graduated cylinder. According to **Table R-7,** the density of aluminum is 2.7 g/mL. Use the density equation to solve for the mass of the aluminum sample.

Known	Unknown
density = 2.7 g/mL	mass = ? g
initial volume = 10.5 mL	
final volume = 13.5 mL	

2 SOLVE FOR THE UNKNOWN

volume of sample = final volume − initial volume — State the equation for volume.

volume of sample = 13.5 mL − 10.5 mL — Substitute final volume = 13.5 mL and initial volume = 10.5 mL.

volume of sample = 3.0 mL

$$\text{density} = \frac{\text{mass}}{\text{volume}}$$ State the equation for density.

mass = volume × density — Solve the density equation for mass.

mass = 3.0 mL × 2.7 g/mL — Substitute volume = 3.0 mL and density = 2.7 g/mL.

mass = 3.0 $\cancel{\text{mL}}$ × 2.7 g/$\cancel{\text{mL}}$ = 8.1 g — Multiply, and cancel units.

3 EVALUATE THE ANSWER

Check your answer by using it to calculate the density of aluminum.

$$\text{density} = \frac{\text{mass}}{\text{volume}} = \frac{8.1\text{ g}}{3.0\text{ mL}} = 2.7\text{ g/mL}$$

Because the calculated density for aluminum is correct, the mass value must also be correct.

PRACTICE Problems

Do additional problems. Online Practice

1. Is the cube pictured at right made of pure aluminum? Explain your answer.
2. What is the volume of a sample that has a mass of 20 g and a density of 4 g/mL?
3. **Challenge** A 147-g piece of metal has a density of 7.00 g/mL. A 50-mL graduated cylinder contains 20.0 mL of water. What is the final volume after the metal is added to the graduated cylinder?

Mass = 20 g
Volume = 5 cm^3

Extension

Ask students to explain the difference between how skin and a thermometer respond to temperature. Skin responds to temperature in a qualitative way by indicating the relative warmth or coldness of an object compared to your body's temperature. A thermometer measures temperature quantitatively, against a standard. OL EL

ASSESSMENT

Skill Have students determine what 437 K is in Celsius. 164°C What is 23°C in kelvins? 296 K OL EL

MiniLAB

Determine Density

What is the density of an unknown and irregularly shaped solid? To calculate the density of an object, you need to know its mass and volume. The volume of an irregularly shaped solid can be determined by measuring the amount of water it displaces.

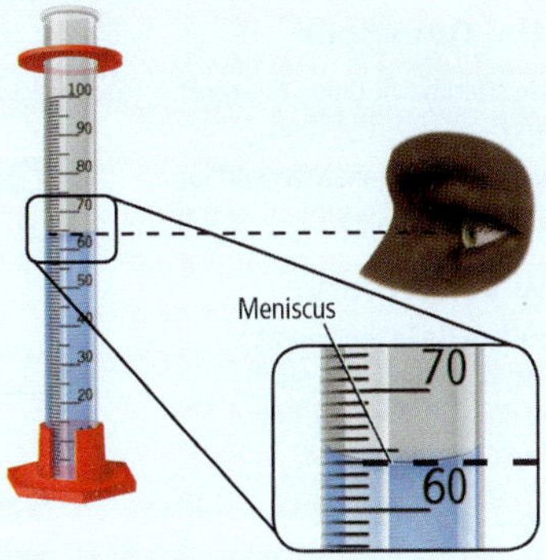

Procedure

1. Read and complete the lab safety form.
2. Obtain several **unknown objects** from your teacher. *Note: Your teacher will identify each object as A, B, C, and so on.*
3. Create a data table to record your observations.
4. Measure the mass of the object using a **balance.** Record the mass and the letter of the object in your data table.
5. Add about 15 mL of **water** to a **graduated cylinder.** Measure and record the initial volume in your data table. Because the surface of the water in the cylinder is curved, make volume readings at eye level and at the lowest point on the curve, as shown in the figure. The curved surface is called a meniscus.
6. Tilt the graduated cylinder, and carefully slide the object down the inside of the cylinder. Be sure not to cause a splash. Measure and record the final volume in your data table.

Analysis

1. **Calculate** Use the initial and final volume readings to calculate the volume of each mystery object.
2. **Calculate** Use the calculated volume and the measured mass to calculate the density of each unknown object.
3. **Explain** Why can't you use the water displacement method to find the volume of a sugar cube?
4. **Describe** how you can determine a washer's volume without using the water displacement method. Note that a washer is similar to a short cylinder with a hole through it.

MiniLAB

Rubric

Purpose Students will measure volume and mass and calculate density.

Process Skills use numbers, measure, acquire and analyze information

Safety Precautions Approve lab safety forms before work begins.

Teaching Strategies

- Because of the small volume of the washer, students might obtain a more accurate measurement by determining the average volume of several washers.
- If students use washers or other objects of known composition, have them compare their calculated densities to the accepted density values for the substance.

Expected Results Density is determined in g/mL by dividing mass by volume.

Analysis

1. $V_{object} = V_{final} - V_{initial}$
2. Answers will vary depending on object chosen. Students will use the equation, *mass = volume/density*.
3. The sugar cube would dissolve in the water.
4. Measure the outside diameter of the washer and calculate its area. Measure the diameter of the hole and calculate its area. Subtract the area of the hole from the area of the washer and multiply the answer by the thickness of the washer.

SECTION 1 REVIEW

Section Self-Check

Section Summary

- SI measurement units allow scientists to report data to other scientists.
- Adding prefixes to SI units extends the range of possible measurements.
- To convert to kelvin temperature, add 273 to the Celsius temperature.
- Volume and density have derived units. Density, which is a ratio of mass to volume, can be used to identify an unknown sample of matter.

4. MAINIDEA **Define** the SI units for length, mass, time, and temperature.
5. **Describe** how adding the prefix *mega-* to a unit affects the quantity being described.
6. **Compare** a base unit and a derived unit, and list the derived units used for density and volume.
7. **Define** the relationships among the mass, volume, and density of a material.
8. **Apply** Why does oil float on water?
9. **Calculate** Samples A, B, and C have masses of 80 g, 12 g, and 33 g, and volumes of 20 mL, 4 cm^3, and 11 mL, respectively. Which of the samples have the same density?
10. **Design** a concept map that shows the relationships among the following terms: *volume, derived unit, mass, base unit, time,* and *length.*

SECTION 1 REVIEW

4. length: meter; mass: kilogram; time: second; temperature: Kelvin
5. It multiplies the quantity by 10^6.
6. Base units are defined based on a physical object or process. Derived units are defined based on a combination of base units. The derived units for density are g/cm^3 or g/mL. The derived units for volume are cm^3 or m^3.
7. Density is the mass-to-volume ratio of a material.
8. Oil floats on top of water because the density of oil is less than that of water.
9. density of A = 80 g/20 mL = 4 g/mL; density of B = 12 g/4 cm^3 = 3 g/cm^3; density of C = 33 g/11 mL = 3 g/mL; B and C have the same density.
10. Student concepts maps will vary, but should show the following relationships: SI units are divided into base units and derived units; volume and density are derived units; and mass, time, and length are base units.

SECTION 2

1 Focus

MAINIDEA

Numbers for Science Tell students that a person with a height of 5 feet and 9 inches is 175.3 cm tall. Ask students to convert this height in centimeters to height in meters, height in kilometers, and height in millimeters. 1.753 m, 0.001753 km, 1753 mm Ask them if they are all the same quantity. Yes, they are all the same quantity expressed in different units. Ask students how the kilometer value could be written to make it easier to calculate using the calculator. 1.753×10^{-4} km OL BL

2 Teach

Quick Demo

Scientific Notation Show students a large jar of popcorn and have them guess the number of kernels. Then, show them the same number of kernels, but have the kernels divided equally into small paper cups. Tell the students approximately how many kernels are in each cup, and have them guess the number of kernels that would fill the large jar. Explain that scientific notation is similar to dividing the large number of kernels into smaller cups, making it easier to identify large or small quantities. OL EL

SECTION 2

Essential Questions

- Why use scientific notation to express numbers?
- How is dimensional analysis used for unit conversion?

Review Vocabulary

quantitative data: numerical information describing how much, how little, how big, how tall, how fast, and so on

New Vocabulary

scientific notation
dimensional analysis
conversion factor

Scientific Notation and Dimensional Analysis

MAINIDEA Scientists often express numbers in scientific notation and solve problems using dimensional analysis.

CHEM 4 YOU If you have ever had a job, one of the first things you probably did was figure out how much you would earn per week. If you make 10 dollars per hour and work 20 hours per week, how much money will you make? Performing this calculation is an example of dimensional analysis.

Scientific Notation

The Hope Diamond, which is shown in **Figure 6,** contains approximately 460,000,000,000,000,000,000,000 atoms of carbon. Each of these carbon atoms has a mass of 0.00000000000000000000002 g. If you were to use these numbers to calculate the mass of the Hope Diamond, you would find that the zeros would get in your way. Using a calculator offers no help, as it won't let you enter numbers this large or this small. Numbers such as these are best expressed in scientific notation. Scientists use this method to conveniently restate a number without changing its value.

Scientific notation can be used to express any number as a number between 1 and 10 (known as the coefficient) multiplied by 10 raised to a power (known as the exponent). When written in scientific notation, the two numbers above appear as follows.

Coefficient — Exponent

$$\text{carbon atoms in the Hope Diamond} = 4.6 \times 10^{23}$$

$$\text{mass of one carbon atom} = 2 \times 10^{-23}\ \text{g}$$

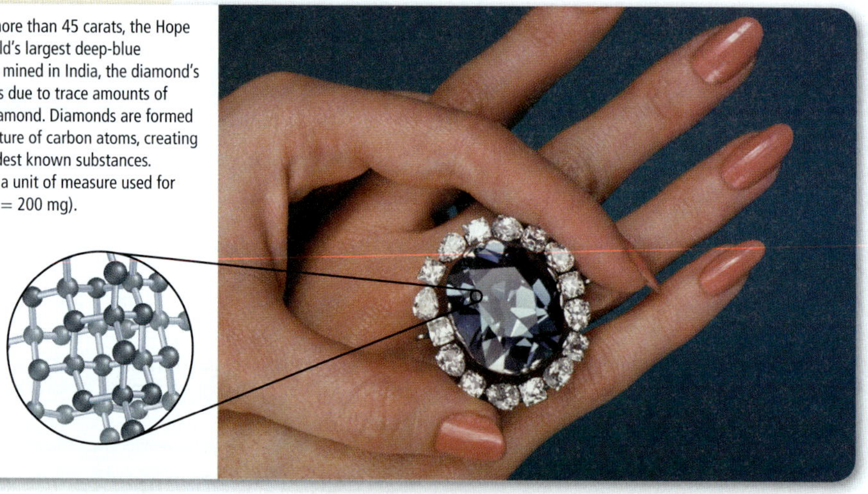

Figure 6 At more than 45 carats, the Hope Diamond is the world's largest deep-blue diamond. Originally mined in India, the diamond's brilliant blue color is due to trace amounts of boron within the diamond. Diamonds are formed from a unique structure of carbon atoms, creating one of nature's hardest known substances. Note that a carat is a unit of measure used for gemstones (1 carat = 200 mg).

DIFFERENTIATED INSTRUCTION

Struggling Learners Ask students to guess the number of beans in a pile of dried beans. Have them separate the pile into groups of ten beans and then use the number of piles to calculate the number of beans. Ask if it was easier to count the number of beans by grouping them into groups of ten. Relate the activity to scientific notation. BL EL

Let's look at these two numbers more closely. In each case, the number 10 raised to an exponent replaced the zeros that preceded or followed the nonzero numbers. For numbers greater than 1, a positive exponent is used to indicate how many times the coefficient must be multiplied by 10 in order to obtain the original number. Similarly, for numbers less than 1, a negative exponent indicates how many times the coefficient must be divided by 10 in order to obtain the original number.

Determining the exponent to use when writing a number in scientific notation is easy: simply count the number of places the decimal point must be moved to give a coefficient between 1 and 10. The number of places moved equals the value of the exponent. The exponent is positive when the decimal moves to the left and the exponent is negative when the decimal moves to the right.

Get help with **scientific notation.** Personal Tutor

$$460{,}000{,}000{,}000{,}000{,}000{,}000{,}000. \rightarrow 4.6 \times 10^{23}$$

Because the decimal point moves 23 places to the left, the exponent is 23.

$$0.00000000000000000000002 \rightarrow 2 \times 10^{-23}$$

Because the decimal point moves 23 places to the right, the exponent is −23.

EXAMPLE Problem 2

Find help with **scientific notation.** Math Handbook

SCIENTIFIC NOTATION Write the following data in scientific notation.

a. The diameter of the Sun is 1,392,000 km.

b. The density of the Sun's lower atmosphere is 0.000000028 g/cm^3.

1 ANALYZE THE PROBLEM

You are given two values, one much larger than 1 and the other much smaller than 1. In both cases, the answers will have a coefficient between 1 and 10 multiplied by a power of 10.

2 SOLVE FOR THE UNKNOWN

Move the decimal point to give a coefficient between 1 and 10. Count the number of places the decimal point moves, and note the direction.

1,392,000. — **Move the decimal point six places to the left.**

0.000000028 — **Move the decimal point eight places to the right.**

a. 1.392×10^6 km

b. 2.8×10^{-8} g/cm^3

Write the coefficients, and multiply them by 10^n where n equals the number of places moved. When the decimal point moves to the left, n is positive; when the decimal point moves to the right, n is negative. Add units to the answers.

3 EVALUATE THE ANSWER

The answers are correctly written as a coefficient between 1 and 10 multiplied by a power of 10. Because the diameter of the Sun is a number greater than 1, its exponent is positive. Because the density of the Sun's lower atmosphere is a number less than 1, its exponent is negative.

EXAMPLE PROBLEM

PRACTICE Problems

Do additional problems. Online Practice

11. Express each number in scientific notation.

a. 700 **b.** 38,000 **c.** 4,500,000 **d.** 685,000,000,000 **e.** 0.0054 **f.** 0.00000687 **g.** 0.000000076 **h.** 0.0000000008

12. Challenge Express each quantity in regular notation along with its appropriate unit.

a. 3.60×10^5 s **b.** 5.4×10^{-5} g/cm^3 **c.** 5.060×10^3 km **d.** 8.9×10^{10} Hz

DIFFERENTIATED INSTRUCTION

Visually Impaired Have sighted students use paper to make cubes that are 1 cm and 1 dm on each side, with volumes of 1 cm^3 and 1 dm^3, respectively. Have visually impaired students determine tactilely the dimensions that must be measured to calculate volume. Explain how volume is a derived unit. OL EL

IN-CLASS Example

Question Each cell in the human body contains a complete genome which is composed of base pairs. Each base pair is 0.000,000,034 m in length. There are 6,000,000,000 base pairs in each human cell. Change the above information into scientific notation.

Answer Remember that the coefficient is a number between 1 and 10. Move the decimal left or right until you have a number between 1 and 10. Then count the number of places the decimal was moved. Right gives a negative exponent and left gives a positive exponent.

a. 3.4×10^{-8} m

b. 6×10^9 base pairs

PRACTICE Problems

Have students refer to the Selected Solutions appendix for complete solutions to odd-numbered problems. The complete solutions for all problems can be found in the Solutions Manual.

11. a. 7×10^2
b. 3.8×10^4
c. 4.5×10^6
d. 6.85×10^{11}
e. 5.4×10^{-3}
f. 6.87×10^{-6}
g. 7.6×10^{-8}
h. 8×10^{-10}

12. a. 360,000 s
b. 0.000054 g/cm^3
c. 5060 km
d. 89,000,000,000 Hz

ASSESSMENT

Knowledge Have students write the following numbers in scientific notation: 4,803,000 km. 4.803×10^6 km; 0.000000342 ng 3.42×10^{-7} ng OL

Apply Chemistry

Common Measure Have students investigate the Mars Climate Orbiter disaster. Which two systems of measurements did the engineers use? One team used the English system; the other team used the metric system. What are scientists now doing to prevent this problem from recurring? There is now an extensive system of checks and balances and more formal communication among engineers to prevent this from recurring. **OL**

☑ **READING CHECK** Make sure both numbers have the same exponent, then add the coefficients.

PRACTICE Problems

Have students refer to the Selected Solutions appendix for complete solutions to odd-numbered problems. The complete solutions to all problems can be found in the Solutions Manual.

13. **a.** 7×10^{-5}
b. 3×10^{8}
c. 2×10^{2}
d. 5×10^{-12}

14. **a.** 1.51×10^{4} kg
b. 7.18×10^{-3} kg
c. 4.11×10^{5} kg
d. 4.62×10^{2} g

■ **Figure 7** The uneven heating of Earth's surface causes wind, which powers these turbines and generates electricity.

GYRO PHOTOGRAPHY/amanaimagesRF/Getty Images

VOCABULARY

ACADEMIC VOCABULARY

Sum
the whole amount; the result of adding numbers
At the checkout counter, all of the items came to a sizable sum.

Addition and subtraction In order to add or subtract numbers written in scientific notation, the exponents must be the same. Suppose you need to add 7.35×10^2 m and 2.43×10^2 m. Because the exponents are the same, you can simply add the coefficients.

$$(7.35 \times 10^2 \text{ m}) + (2.43 \times 10^2 \text{ m}) = 9.78 \times 10^2 \text{ m}$$

How do you add numbers in scientific notation when the exponents are not the same? To answer this question, consider the amounts of energy produced by renewable energy sources in the United States. Wind-powered turbines, shown in **Figure 7,** are one of several forms of renewable energy used in the United States. Other sources of renewable energy include hydroelectric, biomass, geothermal, and solar power. In 2008, the energy production amounts from renewable sources were as follows.

Hydroelectric	2.643×10^{18} J*
Biomass	4.042×10^{18} J
Geothermal	3.89×10^{17} J
Wind	5.44×10^{17} J
Solar	7.8×10^{16} J

* J stands for joules, a unit of energy.

To determine the sum of these values, they must be rewritten with the same exponent. Because the two largest values have an exponent of 10^{18}, it makes sense to convert the other numbers to values with this exponent. These other exponents must increase to become 10^{18}. As you learned earlier, each place the decimal shifts to the left increases the exponent by 1. Rewriting the values with exponents of 10^{18} and adding yields the following.

Hydroelectric	2.643×10^{18} J
Biomass	4.042×10^{18} J
Geothermal	0.389×10^{18} J
Wind	0.544×10^{18} J
Solar	0.078×10^{18} J
Total	7.696×10^{18} J

☑ **READING CHECK** **Restate** the process used to add two numbers that are expressed in scientific notation.

PRACTICE Problems

Do additional problems. Online Practice

13. Solve each problem, and express the answer in scientific notation.
a. $(5 \times 10^{-5}) + (2 \times 10^{-5})$
b. $(7 \times 10^{8}) - (4 \times 10^{8})$
c. $(9 \times 10^{2}) - (7 \times 10^{2})$
d. $(4 \times 10^{-12}) + (1 \times 10^{-12})$

14. **Challenge** Express each answer in scientific notation in the units indicated.
a. $(1.26 \times 10^{4}$ kg$) + (2.5 \times 10^{6}$ g$)$ in kg
b. $(7.06$ g$) + (1.2 \times 10^{-4}$ kg$)$ in kg
c. $(4.39 \times 10^{5}$ kg$) - (2.8 \times 10^{7}$ g$)$ in kg
d. $(5.36 \times 10^{-1}$ kg$) - (7.40 \times 10^{-2}$ kg$)$ in g

CHEMISTRY PROJECT

The Price of Gasoline Have students determine any needed conversion factors for the following problem and solve the problem. Assume that gasoline in Europe is sold at \$0.989 per liter and the current price in the United States is \$2.56/gallon. Where is gasoline more expensive? 1 L is equal to approximately 1 quart, and 4 quarts equals 1 gallon. The price of gas in Europe is \$0.989/L × 1 L/1 quart × 4 quarts/1 gallon = \$3.96/gallon. Gasoline is more expensive at the European price. **OL**

Multiplication and division Multiplying and dividing numbers in scientific notation is a two-step process, but it does not require the exponents to be the same. For multiplication, multiply the coefficients and then add the exponents. For division, divide the coefficients, then subtract the exponent of the divisor from the exponent of the dividend.

To calculate the mass of the Hope Diamond, multiply the number of carbon atoms by the mass of a single carbon atom.

$$(4.6 \times 10^{23} \text{ \cancel{atoms}})(2 \times 10^{-23} \text{ g/\cancel{atom}}) = 9.2 \times 10^{0} \text{ g} = 9.2 \text{ g}$$

Note that any number raised to a power of 0 is equal to 1; thus, 9.2×10^0 g is equal to 9.2 g.

EXAMPLE Problem 3

Find help with **scientific notation.** Math Handbook

MULTIPLYING AND DIVIDING NUMBERS IN SCIENTIFIC NOTATION Solve the following problems.

a. $(2 \times 10^3) \times (3 \times 10^2)$
b. $(9 \times 10^8) \div (3 \times 10^{-4})$

1 ANALYZE THE PROBLEM

You are given numbers written in scientific notation to multiply and divide. For the multiplication problem, multiply the coefficients and add the exponents. For the division problem, divide the coefficients and subtract the exponent of the divisor from the exponent of the dividend.

$\frac{9 \times 10^8}{3 \times 10^{-4}}$ — The exponent of the dividend is 8. The exponent of the divisor is −4.

2 SOLVE FOR THE UNKNOWN

a. $(2 \times 10^3) \times (3 \times 10^2)$	State the problem.
$2 \times 3 = 6$	Multiply the coefficients.
$3 + 2 = 5$	Add the exponents.
6×10^5	Combine the parts.
b. $(9 \times 10^8) \div (3 \times 10^{-4})$	State the problem.
$9 \div 3 = 3$	Divide the coefficients.
$8 - (-4) = 8 + 4 = 12$	Subtract the exponents.
3×10^{12}	Combine the parts.

3 EVALUATE THE ANSWER

To test the answers, write out the original data and carry out the arithmetic. For example, Problem **a** becomes $2000 \times 300 = 600{,}000$, which is the same as 6×10^5.

EXAMPLE PROBLEM

PRACTICE Problems

Do additional problems. Online Practice

15. Solve each problem, and express the answer in scientific notation.
a. $(4 \times 10^2) \times (1 \times 10^8)$ **c.** $(6 \times 10^2) \div (2 \times 10^1)$
b. $(2 \times 10^{-4}) \times (3 \times 10^2)$ **d.** $(8 \times 10^4) \div (4 \times 10^1)$

16. Challenge Calculate the areas and densities. Report the answers in the correct units.
a. the area of a rectangle with sides measuring 3×10^1 cm and 3×10^{-2} cm
b. the area of a rectangle with sides measuring 1×10^3 cm and 5×10^{-1} cm
c. the density of a substance having a mass of 9×10^5 g and a volume of 3×10^{-1} cm^3
d. the density of a substance having a mass of 4×10^{-3} g and a volume of 2×10^{-2} cm^3

PRACTICE PROBLEMS

IN-CLASS Example

Question Solve the following problems.
a. $(2.5 \times 10^{-4}) \times (2.8 \times 10^5)$
b. $(4.6 \times 10^5)/(2.3 \times 10^{-3})$

Answer
a. $2.5 \times 2.8 = 7$
$10^{-4+5=1} = 10^1$
7×10^1
b. $4.6/2.3 = 2$
$10^{5-(-3)=8} = 10^8$
2×10^8

PRACTICE Problems

Have students refer to the Selected Solutions appendix for complete solutions to odd-numbered problems. The complete solutions for all problems can be found in the Solutions Manual.

15. a. 4×10^{10}
b. 6×10^{-2}
c. 3×10^1
d. 2×10^3
16. a. area $= 9 \times 10^{-1}$ cm^2
b. area $= 5 \times 10^2$ cm^2
c. density $= 3 \times 10^6$ g/cm^3
d. density $= 2 \times 10^{-1}$ g/cm^3

CHEMISTRY JOURNAL

Scientific Notation Challenges Have students identify any difficulty they have in doing mathematical calculations with scientific notation. Have pairs of students brainstorm and identify strategies to help them overcome their difficulties. OL COOP LEARN

Build a Model

Model Dimensional Analysis

Students will use dominoes to model dimensional analysis. Tell students the dots on the dominoes represent the units that accompany measured values. The object is to change the starting pattern of dots into the desired dot pattern using the least number of dominoes. Students must match the starting upper half of the domino with the lower half of the second domino in order to cancel the unit (dot pattern). Each subsequent domino must have the lower half match the upper half of the previous domino.

Have students continue to lay out dominoes until they have the desired upper half or both upper and lower halves. If converting the lower half, students must match the upper half of the second domino with the lower half of the previous domino. When converting a double six domino to a three over two dominos students could use the following dominoes.

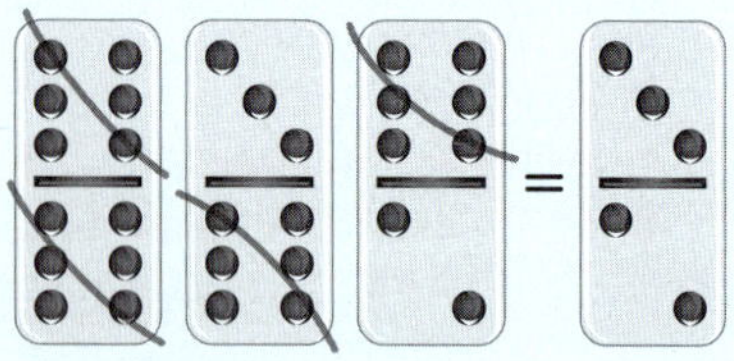

OL BL EL

■ **Figure 8** Dimensional analysis can be used to calculate the number of pizzas that must be ordered for a party. How many pizzas will you need if 32 people eat 3 slices per person and there are 8 slices in each pizza?

$$(32\ \cancel{\text{people}})\left(\frac{3\ \cancel{\text{slices}}}{\cancel{\text{person}}}\right)\left(\frac{1\ \text{pizza}}{8\ \cancel{\text{slices}}}\right) = 12\ \text{pizzas}$$

Dimensional Analysis

When planning a pizza party for a group of people, you might want to use dimensional analysis to figure out how many pizzas to order. **Dimensional analysis** is a systematic approach to problem solving that uses conversion factors to move, or convert, from one unit to another. A **conversion factor** is a ratio of equivalent values having different units.

How many pizzas do you need to order if 32 people will attend a party, each person eats 3 slices of pizza, and each pizza has 8 slices? **Figure 8** shows how conversion factors are used to calculate the number of pizzas needed for the party.

Writing conversion factors As you just read, conversion factors are ratios of equivalent values. Not surprisingly, these conversion factors are derived from equality relationships, such as 12 eggs = 1 dozen eggs, or 12 inches = 1 foot. Multiplying a quantity by a conversion factor changes the units of the quantity without changing its value.

Most conversion factors are written from relationships between units. For example, the prefixes in **Table 2** are the source of many conversion factors. From the relationship 1000 m = 1 km, the following conversion factors can be written.

$$\frac{1\ \text{km}}{1000\ \text{m}} \quad \text{and} \quad \frac{1000\ \text{m}}{1\ \text{km}}$$

A derived unit, such as a density of 2.5 g/mL, can also be used as a conversion factor. The value shows that 1 mL of the substance has a mass of 2.5 g. The following two conversion factors can be written.

$$\frac{2.5\ \text{g}}{1\ \text{mL}} \quad \text{and} \quad \frac{1\ \text{mL}}{2.5\ \text{g}}$$

Percentages can also be used as conversion factors. A percentage is a ratio; it relates the number of parts of one component to 100 total parts. For example, a fruit drink containing 10% sugar by mass contains 10 g of sugar in every 100 g of fruit drink. The conversion factors for the fruit drink are as follows.

$$\frac{10\ \text{g sugar}}{100\ \text{g fruit drink}} \quad \text{and} \quad \frac{100\ \text{g fruit drink}}{10\ \text{g sugar}}$$

DIFFERENTIATED INSTRUCTION

Struggling Learners Give students blank index cards. On each index card have students write a conversion factor and on the opposite side write the reciprocal of that conversion factor. Give students a sample problem that involves the conversion factor. Have them define the known quantity and write it on a separate card. Ask students to identify the desired end and write the unit on an index card. Have the students line up the cards from known to desired providing a connecting path from the known to the desired result.

Once students have mastered one-variable problems, they might begin two-variable problems, writing conversion cards for each conversion needed to solve for the desired units in the answer. BL

PRACTICE Problems

Do additional problems. Online Practice

17. Write two conversion factors for each of the following.

a. a 16% (by mass) salt solution

b. a density of 1.25 g/mL

c. a speed of 25 m/s

18. Challenge What conversion factors are needed to convert:

a. nanometers to meters?

b. density given in g/cm^3 to a value in kg/m^3?

Using conversion factors A conversion factor used in dimensional analysis must accomplish two things: it must cancel one unit and introduce a new one. While working through a solution, all of the units except the desired unit must cancel. Suppose you want to know how many meters there are in 48 km. The relationship between kilometers and meters is 1 km = 1000 m. The conversion factors are as follows.

$$\frac{1\ \text{km}}{1000\ \text{m}} \quad \text{and} \quad \frac{1000\ \text{m}}{1\ \text{km}}$$

Because you need to convert km to m, you should use the conversion factor that causes the km unit to cancel.

$$48\ \cancel{\text{km}} \times \frac{1000\ \text{m}}{1\ \cancel{\text{km}}} = 48{,}000\ \text{m}$$

When converting a value with a large unit, such as km, to a value with a smaller unit, such as m, the numerical value increases. For example, 48 km (a value with a large unit) converts to 48,000 m (a larger numerical value with a smaller unit). **Figure 9** illustrates the connection between the numerical value and the size of the unit for a conversion factor.

$$\frac{1\ \text{km}}{1000\ \text{m}}$$

Figure 9 The two quantities shown above are equivalent; that is, 1 km = 1000 m. Note that a smaller numerical value (1) accompanies the larger unit (km), and a larger numerical value (1000) accompanies the smaller unit (m).

Now consider this question: How many eight-packs of water would you need if the 32 people attending your party each had two bottles of water? To solve the problem, identify the given quantities and the desired result. There are 32 people and each of them drinks two bottles of water. The desired result is the number of eight-packs. Using dimensional analysis yields the following.

$$32\ \cancel{\text{people}} \times \frac{2\ \cancel{\text{bottles}}}{\cancel{\text{person}}} \times \frac{1\ \text{eight-pack}}{8\ \cancel{\text{bottles}}} = 8\ \text{eight-packs}$$

PRACTICE Problems

Do additional problems. Online Practice

Use Table 2 to solve each of the following.

19. a. Convert 360 s to ms.

b. Convert 4800 g to kg.

c. Convert 5600 dm to m.

d. Convert 72 g to mg.

e. Convert 2.45×10^2 ms to s.

f. Convert 5 μm to km.

g. Convert 6.800×10^3 cm to km.

h. Convert 2.5×10^1 kg to Mg.

20. Challenge Write the conversion factors needed to determine the number of seconds in one year.

PRACTICE Problems

Have students refer to the Selected Solutions appendix for complete solutions to odd-numbered problems. The complete solutions for all problems can be found in the Solutions Manual.

17. a. (100 g solution)/(16 g salt) (16 g salt)/(100 g solution)
b. (1.25 g)/(1 mL); (1 mL)/(1.25 g)
c. (25 m)/(1 s); (1 s)/(25 m)

18. a. (10^{-9} m)/(1 nm)
b. (1 kg)/(1000 g) and (10^6 cm^3)/(1 m^3)

19. a. 360,000 ms
b. 4.8 kg
c. 560 m
d. 72,000 mg
e. 0.245 s
f. 5×10^{-9} km
g. 0.068 km
h. 0.025 Mg

20. (365 d/1 yr)(24 h/1 d)(60 min/1 h) (60 s/1 min)

ASSESSMENT

Skill Have students identify common conversion factors and prepare a table. Have each student write one question using a conversion factor. Conversion factors can be those used in the English system, in the metric system, or factors used to convert between the English and metric systems. OL

3 Assess

Check for Understanding

Ask students which metric prefix is equal to 1×10^6. mega, M OL EL

Reteach

Write additional problems involving calculations with scientific notation on pieces of poster board. Hang the pieces around the classroom. Have student groups complete the problems, then present their work to the rest of the class. BL OL

Extension

Have students research the number of stars in our galaxy and express it in scientific notation. OL AL

IN-CLASS Example

Question The density of peanut oil is 0.92 g/mL. There are 237 mL in a measuring cup. If a recipe calls for ¼ cup of peanut oil, how many grams are required?

Answer

Known:

density = 0.920 g/mL

1 cup = 237 mL

Unknown:

? g of peanut oil

Need ¼ cup (0.250 cup):

$$0.250 \text{ cup} \times \frac{237 \text{ mL}}{\text{cup}} \times \frac{0.920 \text{ g}}{\text{mL}} = 54.5 \text{ g}$$

PRACTICE Problems

Have students refer to the Selected Solutions appendix for complete solutions to odd-numbered problems. The complete solutions for all problems can be found in the Solutions Manual.

21. 1.0×10^2 km/h

22. 86,400 s

23. mass = 9.45 g acetic acid

EXAMPLE Problem 4

EXAMPLE PROBLEM

Find help with **unit conversion.** Math Handbook

USING CONVERSION FACTORS In ancient Egypt, small distances were measured in Egyptian cubits. An Egyptian cubit was equal to 7 palms, and 1 palm was equal to 4 fingers. If 1 finger was equal to 18.75 mm, convert 6 Egyptian cubits to meters.

1 ANALYZE THE PROBLEM

A length of 6 Egyptian cubits needs to be converted to meters.

Known

length = 6 Egyptian cubits

7 palms = 1 cubit

1 palm = 4 fingers

1 finger = 18.75 mm

1 m = 1000 mm

Unknown

length = ? m

2 SOLVE FOR THE UNKNOWN

Use dimensional analysis to convert the units in the following order.

cubits → palms → fingers → millimeters → meters

$$6 \text{ cubits} \times \frac{7 \text{ palms}}{1 \text{ cubit}} \times \frac{4 \text{ fingers}}{1 \text{ palm}} \times \frac{18.75 \text{ mm}}{1 \text{ finger}} \times \frac{1 \text{ meter}}{1000 \text{ mm}} = ? \text{ m}$$

Multiply by a series of conversion factors that cancels all the units except meter, the desired unit.

$$6 \text{ cubits} \times \frac{7 \text{ palms}}{1 \text{ cubit}} \times \frac{4 \text{ fingers}}{1 \text{ palm}} \times \frac{18.75 \text{ mm}}{1 \text{ finger}} \times \frac{1 \text{ meter}}{1000 \text{ mm}} = 3.150 \text{ m}$$

Multiply and divide the numbers as indicated, and cancel the units.

3 EVALUATE THE ANSWER

Each conversion factor is a correct restatement of the original relationship, and all units except for the desired unit, meters, cancel.

PRACTICE Problems

Do additional problems. Online Practice

21. The speedometer at right displays a car's speed in miles per hour. What is the car's speed in km/h? (1 km = 0.62 mile)

22. How many seconds are in 24 h?

23. Challenge Vinegar is 5.00% acetic acid by mass and has a density of 1.02 g/mL. What mass of acetic acid, in grams, is present in 185 mL of vinegar?

SECTION 2 REVIEW

Section Self-Check

Section Summary

- A number expressed in scientific notation is written as a coefficient between 1 and 10 multiplied by 10 raised to a power.
- To add or subtract numbers in scientific notation, the numbers must have the same exponent.
- To multiply or divide numbers in scientific notation, multiply or divide the coefficients and then add or subtract the exponents, respectively.
- Dimensional analysis uses conversion factors to solve problems.

24. MAINIDEA **Describe** how scientific notation makes it easier to work with very large or very small numbers.

25. Express the numbers 0.00087 and 54,200,000 in scientific notation.

26. Write the measured distance quantities 3×10^{-4} cm and 3×10^{4} km in regular notation.

27. Write a conversion factor relating cubic centimeters and milliliters.

28. Solve How many millimeters are there in 2.5×10^2 km?

29. Explain how dimensional analysis is used to solve problems.

30. Apply Concepts A classmate converts 68 km to meters and gets 0.068 m as the answer. Explain why this answer is incorrect, and identify the likely source of the error.

31. Organize Create a flowchart that outlines when to use dimensional analysis and when to use scientific notation.

SECTION 2 REVIEW

24. When numbers are expressed in scientific notation, the space-wasting placeholder zeros are eliminated, thus making it easier to perform arithmetic on the numbers.

25. 8.7×10^{-4}; 5.42×10^{7}

26. 0.0003 cm; 30,000 km

27. 1 cm^3/1 mL

28. 2.5×10^{8} mm

29. It is a method of problem solving focusing on the units used to describe matter. A given value is multiplied by a conversion factor that relates the given unit to the desired unit.

30. Because meters are smaller than kilometers, there should be more meters than kilometers. The 68 km was divided by 1000, not multiplied by 1000.

31. Students' flowcharts should include the following yes/no decisions branches: 1) Does the given value have trailing zeros or leading zeros? (If yes, use scientific notation.); 2) Is the unit of the given value the desired unit? (If no, then use dimensional analysis.)

Essential Questions

- How do accuracy and precision compare?
- How can the accuracy of experimental data be described using error and percent error?
- What are the rules for significant figures and how can they be used to express uncertainty in measured and calculated values?

Review Vocabulary

experiment: a set of controlled observations that test a hypothesis

New Vocabulary

accuracy
precision
error
percent error
significant figure

Uncertainty in Data

MAINIDEA Measurements contain uncertainties that affect how a calculated result is presented.

CHEM 4 YOU When making cookies from a recipe, amounts are measured in cups, tablespoons, and teaspoons. Would a batch of cookies turn out well if you measured all of the ingredients using only a teaspoon? Most likely not, because measurement errors would build up.

Accuracy and Precision

Just as each teaspoon you measure in the kitchen contains some amount of error, so does every scientific measurement made in a laboratory. When scientists make measurements, they evaluate both the accuracy and the precision of the measurements. Although you might think that the terms accuracy and precision basically mean the same thing, to a scientist, they have very different meanings.

Accuracy refers to how close a measured value is to an accepted value. **Precision** refers to how close a series of measurements are to one another. The archery target in **Figure 10** illustrates the difference between accuracy and precision. For example, you measure in the lab the mass of an object three times. The arrows represent each measurement and the center of the target is the accepted value.

■ **Figure 10** An archery target illustrates the difference between accuracy and precision. An accurate shot is located near the bull's-eye; precise shots are grouped closely together.
Apply *Why is it important to measure the same data more than once?*

View an **animation about precision and accuracy.**
Concepts In Motion

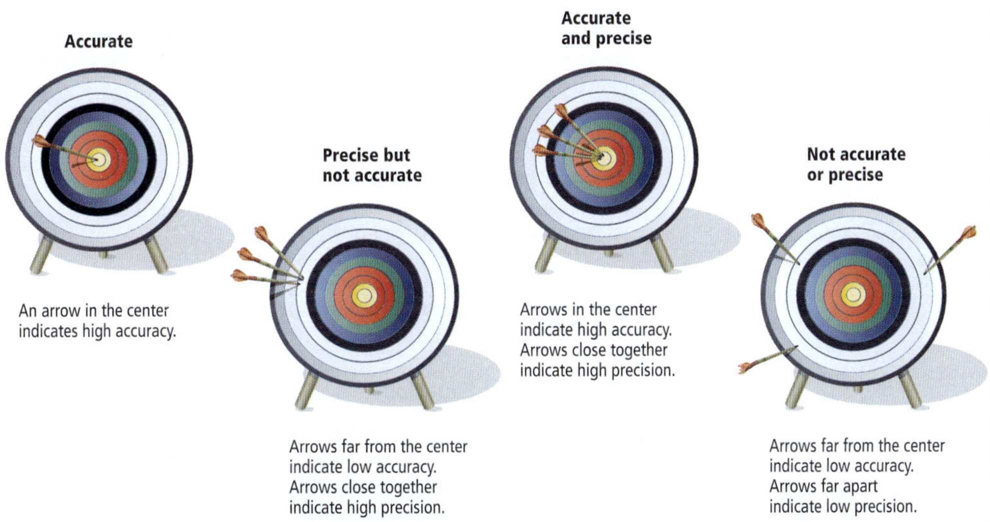

CHEMISTRY JOURNAL

Applied Accuracy and Precision Have students write about areas of their lives that require accuracy and precision. Some common examples might be sports, playing a musical instrument, a hobby, even academics. Ask students to identify the role of precision and accuracy in each example, and strategies they use to accomplish their goal. OL

SECTION 3

1 Focus

MAINIDEA

Ways of Measuring Have students look at the data they collected from the Big Idea activity. Ask them if their height data make sense. They should notice that their recorded values are not equal to their actual heights. This is because the meterstick and yardstick were both placed one meter from the floor. Ask students what they must do to make their readings accurate. They should add 100 centimeters to their centimeter reading and 39.37 inches to their inches reading. Ask students if their recorded height values are accurate. Some will say that their shoes make them taller. Others will recognize that they did not measure accurately. Have students repeat their measurements. Ask them if their recorded heights are precise. They might end up with different measurements, but will likely be fairly close. Their measurements would be reasonably precise. OL

2 Teach

Quick Demo

Accuracy and Precision
Bring in a dartboard game that uses hook and loop darts. Have students break into teams of four and play a round of darts. Emphasize that even in science, accuracy and precision require skill and repeated effort. OL EL

■ **Caption Question Fig. 10**
to evaluate the accuracy and precision of the measurements

Extension

Percent Error After calculating the percent error for each student's data in Table 3, have student groups compare the percent error to the average density obtained by each student in Table 3. Ask them if it is easier to identify which student had the most accurate data by looking at the average or the percent error in Table 3. Help student groups to see the value in the percent error; it quickly and easily identifies the accuracy of a measurement. OL COOP LEARN

Identify Misconceptions

Uncover the Misconception

Many times students assume that each measurement they make in the laboratory is both accurate and precise. They also assume that the values they derive through laboratory experimentation are accurate.

Demonstrate the Concept

Help students to recognize that experimental value is an observed value. They might need to refer to reference tables to obtain the true, accurate, or accepted value.

Access New Knowledge Give students a variety of experimental data, and have them look up the accepted value and determine the percent error.

1. Students determined the molar volume of a gas to be 21.8 L/mol. 2.7% error
2. Students determined the density of aluminum to be 2.55 g/cm³. 5.5% error
3. Students determined the specific heat of water to be 4.28 J/g•°C. 2.3% error OL

Table 3 Student Density and Error Data (Unknown was sucrose; density = 1.59 g/cm³)

	Student A		Student B		Student C	
	Density	Error (g/cm³)	Density	Error (g/cm³)	Density	Error (g/cm³)
Trial 1	1.54 g/cm³	−0.05	1.40 g/cm³	−0.19	(a) 1.70 g/cm³	+0.11
Trial 2	1.60 g/cm³	+0.01	1.68 g/cm³	+0.09	1.69 g/cm³	+0.10
Trial 3	1.57 g/cm³	−0.02	1.45 g/cm³	−0.14	1.71 g/cm³	+ 0.12
Average	(b) 1.57 g/cm³		1.51 g/cm³		1.70 g/cm³	

(a) These trial values are the most precise.
(b) This average is the most accurate.

Consider the data in **Table 3.** Students were asked to find the density of an unknown white powder. Each student measured the volume and mass of three separate samples. They reported calculated densities for each trial and an average of the three calculations. The powder, sucrose (table sugar), has a density of 1.59 g/cm³. Which student collected the most accurate data? Who collected the most precise data? Student A's measurements are the most accurate because they are closest to the accepted value of 1.59 g/cm³. Student C's measurements are the most precise because they are the closest to one another.

Recall that precise measurements might not be accurate. Looking at just the average of the densities can be misleading. Based solely on the average, Student B appears to have collected fairly reliable data. However, on closer inspection, Student B's data are neither accurate nor precise. The data are not close to the accepted value, nor are they close to one another.

Get help with **precision measurement.**

Personal Tutor

Error and percent error The density values reported in **Table 3** are experimental values, which means they are values measured during an experiment. The known density of sucrose is an accepted value, which is a value that is considered true. To evaluate the accuracy of experimental data, you can compare how close the experimental value is to the accepted value. **Error** is defined as the difference between an experimental value and an accepted value. The errors for the experimental density values are also given in **Table 3.**

VOCABULARY
WORD ORIGIN
Percent
comes from the Latin words *per,* which means *by,* and *centum,* which means *100*

Error Equation

$$\text{error} = \text{experimental value} - \text{accepted value}$$

The error associated with an experimental value is the difference between the experimental value and the accepted value.

Scientists often want to know what percent of the accepted value an error represents. **Percent error** expresses error as a percentage of the accepted value.

Percent Error Equation

$$\text{percent error} = \frac{|\text{error}|}{\text{accepted value}} \times 100$$

The percent error of an experimental value equals the absolute value of its error divided by the accepted value, multiplied by 100.

DIFFERENTIATED INSTRUCTION

Struggling Learners Have student pairs explain to one another the concepts of accuracy and precision. Have the student pairs work on Example Problem 6 and Practice Problems 35 and 36. BL COOP LEARN

Notice that the percent-error equation uses the absolute value of the error. This is because only the size of the error matters; it does not matter whether the experimental value is larger or smaller than the accepted value.

☑ READING CHECK **Summarize** why error is important.

Percent error is an important concept for the machinist who made the nut shown in **Figure 11.** The machinist must check the tolerances of the nut. Tolerances are a narrow range of allowable dimensions based on acceptable amounts of error. If the dimensions of the nut do not fall within the acceptable range—that is, the nut exceeds its tolerances—it will be retooled or possibly discarded.

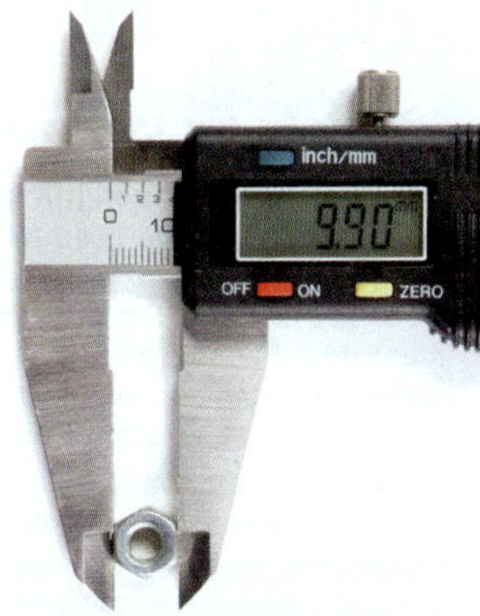

■ **Figure 11** This digital caliper is being used to check the size of a nut to one-hundredth of a millimeter (0.01 mm). Skill is required to correctly position the part in the caliper. Experienced machinists will obtain more precise and more accurate readings than inexperienced machinists.

EXAMPLE Problem 5

Find help with **percents.** Math Handbook

CALCULATING PERCENT ERROR Use Student A's density data in **Table 3** to calculate the percent error in each trial. Report your answers to two places after the decimal point.

1 ANALYZE THE PROBLEM

You are given the errors for a set of density calculations. To calculate percent error, you need to know the accepted value for density, the errors, and the equation for percent error.

Known	**Unknown**
accepted value for density = 1.59 g/cm^3	percent errors = ?
errors: −0.05 g/cm^3; 0.01 g/cm^3; −0.02 g/cm^3	

2 SOLVE FOR THE UNKNOWN

$$\text{percent error} = \frac{|\text{error}|}{\text{accepted value}} \times 100$$

State the percent error equation.

$$\text{percent error} = \frac{|-0.05\ \cancel{g/cm^3}|}{1.59\ \cancel{g/cm^3}} \times 100 = 3.14\%$$

Substitute error = −0.05 g/cm^3, and solve.

$$\text{percent error} = \frac{|0.01\ \cancel{g/cm^3}|}{1.59\ \cancel{g/cm^3}} \times 100 = 0.63\%$$

Substitute error = 0.01 g/cm^3, and solve.

$$\text{percent error} = \frac{|-0.02\ \cancel{g/cm^3}|}{1.59\ \cancel{g/cm^3}} \times 100 = 1.26\%$$

Substitute error = −0.02 g/cm^3, and solve.

3 EVALUATE THE ANSWER

The percent error is greatest for Trial 1, which had the largest error, and smallest for Trial 2, which was closest to the accepted value.

PRACTICE Problems

Do additional problems. Online Practice

Answer the following questions using data from Table 3.

32. Calculate the percent errors for Student B's trials.

33. Calculate the percent errors for Student C's trials.

34. Challenge Based on your calculations in questions 32 and 33, which student's trial was the most accurate? The least accurate?

Chris Gibson/Alamy

IN-CLASS Example

Question The melting point of paradichlorobenzene is 53.0°C. In a laboratory activity, two students try to verify this value. The first student records 51.5°C, 53.5°C, 55.0°C, 52.3°C, and 54.2°C. The second student records 52.3°C, 53.2°C, 54.0°C, 52.5°C, and 53.5°C.

a. Calculate the average value for the two students.

b. Calculate the percent error for each student.

c. Which student is most precise? Most accurate? Explain.

Answer

a. Student 1: 51.5°C, 53.5°C, 55.0°C, 52.3°C, and 54.2°C. Average Value = 53.3°C
Student 2: 52.3°C, 53.2°C, 54.0°C, 52.5°C, and 53.5°C. Average Value = 53.1°C

b. Student 1:
Percent error = (53.0 − 53.3)/(53.0 × 100) = 0.566% error
Student 2:
Percent error = (53.0 − 53.1)/(53.0 × 100) = 0.189% error

c. Student 2 is the most precise, with a range of values from 52.3 to 54.0. Student 2 is also most accurate, with 0.189% error.

☑ READING CHECK Error is important to evaluate the accuracy of experimental data.

PRACTICE Problems

Have students refer to the Selected Solutions appendix for complete solutions to odd-numbered problems. The complete solutions for all problems can be found in the Solutions Manual.

32. |1.40−1.59|/(1.59) × 100 = 11.9%
|1.68−1.59|/(1.59) × 100 = 5.66%
|1.45−1.59|/(1.59) × 100 = 8.80%

33. (0.11)/(1.59) × 100 = 6.92%
(0.10)/(1.59) × 100 = 6.29%
(0.12)/(1.59) × 100 = 7.55%

34. Most accurate: Student B, Trial 2
Least accurate: Student B, Trial 1

CHEMISTRY PROJECT

Accuracy of Measurement Tools Have students look for various measurement tools in their homes, documenting the type of tool and the accuracy of the measuring device. Have students share their findings by making a chart on the wall in the classroom. BL

■ **Caption Question Fig. 12**
Estimated digit is the final zero in a reported measurement of 5.00 cm.

Problem-Solving LAB

Purpose Students will identify an unknown from data.

Process Skills analyze and interpret data, apply concepts

Teaching Strategies

- Demonstrate calculating the density of an object by water displacement.
- Have students practice density calculations using the data from the demonstration.

Think Critically

1. Volume: Sample 1, 10.2 mL; Sample 2, 12.7 mL; Sample 3, 11.3 mL; Sample 4, 11.1 mL; Sample 5, 15.0 mL; Sample 6, 13.3 mL
   Density: Sample 1, 4.93 g/mL; Sample 2, 5.00 g/mL; Sample 3, 5.10 g/mL; Sample 4, 4.99 g/mL; Sample 5, 4.99 g/mL; Sample 6, 5.10 g/mL
   Average density = 5.02 g/mL
2. The average density of the samples was 5.02 g/mL which is very close to the accepted value of 5.01 g/cm^3 for pyrite. The samples are probably pyrite.
3. Errors: Sample 1, 0.08 g/mL; Sample 2, 0.01 g/mL; Sample 3, 0.09 g/mL; Sample 4, 0.02 g/mL; Sample 5, 0.02 g/mL; Sample 6, 0.09 g/mL
   Percent errors: Sample 1, 1.6%; Sample 2, 0.20%; Sample 3, 1.8%; Sample 0.40%; Sample 5, 0.40%; Sample 6, 1.8%
4. The student's values ranged from 0.20% error to 1.8% error. The average error was 1.03%. The data are accurate.

Problem-Solving LAB

Identify an Unknown

How can mass and volume data for an unknown sample be used to identify the unknown? A student collected several samples from a stream bed that looked like gold. She measured the mass of each sample and used water displacement to determine each sample's volume. Her data are given in the table.

Mass and Volume Data for an Unknown Sample

Sample	Mass	Initial Volume (water only)	Final Volume (water + sample)
1	50.25 g	50.1 mL	60.3 mL
2	63.56 g	49.8 mL	62.5 mL
3	57.65 g	50.2 mL	61.5 mL
4	55.35 g	45.6 mL	56.7 mL
5	74.92 g	50.3 mL	65.3 mL
6	67.78 g	47.5 mL	60.8 mL

Analysis

For a given sample, the difference in the volume measurements made with the graduated cylinder yields the volume of the sample. Thus, for each sample, the mass and volume are known, and the density can be calculated. Note that density is a property of matter that can often be used to identify an unknown sample.

Think Critically

1. **Calculate** the volume and density for each sample and the average density of the six samples. Be sure to use significant figure rules.
2. **Apply** The student hopes the samples are gold, which has a density of 19.3 g/cm^3. A local geologist suggested the samples might be pyrite, which is a mineral with a density of 5.01 g/cm^3. What is the identity of the unknown sample?
3. **Calculate** the error and percent error of each sample. Use the density value given in Question 2 as the accepted value.
4. **Conclude** Was the data collected by the student accurate? Explain your answer.

Significant Figures

Often, precision is limited by the tools available. For example, a digital clock that displays the time as 12:47 or 12:48 can record the time only to the nearest minute. With a stopwatch, however, you might record time to the nearest hundredth second. As scientists have developed better measuring devices, they have been able to make more precise measurements. Of course, for measurements to be both accurate and precise, the measuring devices must be in good working order. Additionally, accurate and precise measurements rely on the skill of the person using the instrument; the user must be trained and use proper techniques.

The precision of a measurement is indicated by the number of digits reported. A value of 3.52 g is more precise than a value of 3.5 g. The reported digits are called significant figures. **Significant figures** include all known digits plus one estimated digit. Consider the rod in **Figure 12.** The end of the rod falls between 5.2 cm and 5.3 cm. The 5 and 2 are known digits corresponding to marks on the ruler. To these known digits, an estimated digit is added. This last digit estimates the rod's location between the second and third millimeter marks. Because it is an estimate, one person might report the measurement as 5.22 cm and another as 5.23 cm. Either way, the measurement has three significant figures—two known and one estimated.

Remember that measurements reported with a lot of significant figures might be precise but not accurate. For example, some chemistry labs have balances that report mass to the nearest hundreth of a gram. If you and each of your classmates measured the same copper cylinder on the same balance, you would probably have a group of very precise measurements. But what if the balance had been previously damaged by an object that was too large for it? Your precise measurements would not be very accurate.

■ **Figure 12** The markings on the ruler represent known digits. The reported measurement includes the known digits plus the estimated digit. The measurement is 5.23 cm.
Infer *What is the estimated digit if the length of an object being measured falls exactly on the 5-cm mark?*

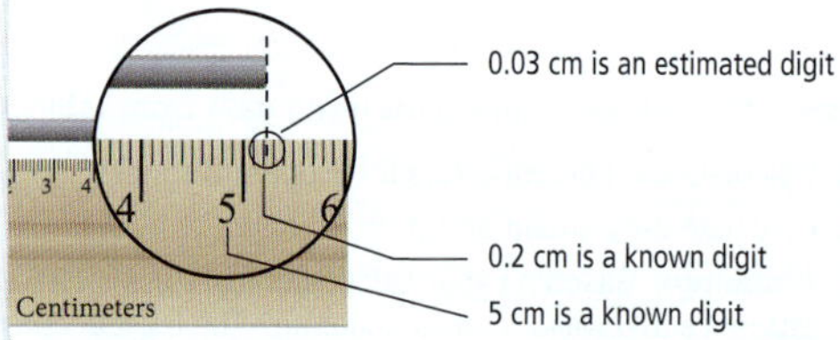

PROBLEM-SOLVING STRATEGY

Recognizing Significant Figures

Learning these five rules for recognizing significant figures will help you when solving problems. Examples of each rule are shown below. Note that each of the highlighted examples has three significant figures.

Rule 1. Nonzero numbers are always significant. — 72.3 g has three.

Rule 2. All final zeros to the right of the decimal are significant. — 6.20 g has three.

Rule 3. Any zero between significant figures is significant. — 60.5 g has three.

Rule 4. Placeholder zeroes are not significant. To remove placeholder zeros, rewrite the number in scientific notation. — 0.0253 g and 4320 g (each has three)

Rule 5. Counting numbers and defined constants have an infinite number of significant figures. — 6 molecules; 60 s = 1 min

EXAMPLE Problem 6

Find help with **significant figures.** Math Handbook

SIGNIFICANT FIGURES Determine the number of significant figures in the following masses.

a. 0.00040230 g

b. 405,000 kg

1 ANALYZE THE PROBLEM

You are given two measured mass values. Apply the appropriate rules to determine the number of significant figures in each value.

2 SOLVE FOR THE UNKNOWN

Count all nonzero numbers, zeros between nonzero numbers, and final zeros to the right of the decimal place. **(Rules 1, 2, and 3)**

Ignore zeros that act as placeholders. **(Rule 4)**

a. 0.00040230 g has five significant figures.

b. 405,000 kg has three significant figures.

3 EVALUATE THE ANSWER

One way to verify your answers is to write the values in scientific notation: 4.0230×10^{-4} g and 4.05×10^{5} kg. Without the placeholder zeros, it is clear that 0.00040230 g has five significant figures and that 405,000 kg has three significant figures.

PRACTICE Problems

Do additional problems. Online Practice

Determine the number of significant figures in each measurement.

35. a. 508.0 L **b.** 820,400.0 L **c.** 1.0200×10^{5} kg **d.** 807,000 kg

36. a. 0.049450 s **b.** 0.000482 mL **c.** 3.1587×10^{-4} g **d.** 0.0084 mL

37. Challenge Write the numbers 10, 100, and 1000 in scientific notation with two, three, and four significant figures, respectively.

CHEMISTRY PROJECT

Importance of Significant Figures Discuss with students when approximate measurements that have fewer significant figures are adequate. Have students measure a bookcase or another long object in the classroom. Have students measure the object to 1 significant digit. If the object measured 93 cm, the measurement to 1 significant digit would be 90 cm. Have students measure the same object to 2 significant digits. If the object measured 93 cm, the measurement to 2 significant digits would be 93 cm. OL

Identify Misconceptions

Uncover the Misconception

Students often do not understand why significant figures are important when using measured values.

Demonstrate the Concept

Have students discuss the precision of several measuring devices and determine the relationship between precision and significant figures. Explain to students that only those digits that are known from the scale plus the first uncertain digit should be considered.

Assess New Knowledge Have students count the students in the classroom. Ask them about significant figures in the count. This number is a counting number with unlimited significant figures. OL

IN-CLASS Example

Question Determine the number of significant figures in the following measured values.

a. 0.0546 3
b. 298.206 6
c. 102000 3
d. 0.003145 4
e. 7.847000 7

PRACTICE Problems

Have students refer to the Selected Solutions appendix for complete solutions to odd-numbered problems. The complete solutions for all problems can be found in the Solutions Manual.

35. a. 4 **b.** 7 **c.** 5 **d.** 3

36. a. 5 **b.** 3 **c.** 5 **d.** 2

37. Two significant figures: 1.0×10^{1}, 1.0×10^{2}, 1.0×10^{3}
Three significant figures: 1.00×10^{1}, 1.00×10^{2}, 1.00×10^{3}
Four significant figures: 1.000×10^{1}, 1.000×10^{2}, 1.000×10^{3}

Reinforcement

Rounding Divide students into eight groups. Give each group a large piece of poster board or paper that can be posted in the classroom and one of the numbers in Practice Problems 32 and 33. Have each group write the number to be rounded, the number of significant figures needed, the rule for rounding that applies, and the answer with the correct number of significant figures. Place the poster boards or papers around the room for students to refer to as they practice rounding and significant figures. **OL** **COOP LEARN**

Math in Chemistry

Round the Number Have students round 45.867 to 3 significant figures. 45.9 Have students round 20,856 to 2 significant figures. 21,000 Have students round the answers to the following problems to the correct number of significant figures. 2.53 + 6.0095 + 4.725 + 12.78654 26.05 25.4231 − 6.34 19.08 **BL** **EL**

Use Science Terms

Everyday Significant Figures Have students compare the definition of the term *significant* in everyday usage with usage in science. **OL** **BL**

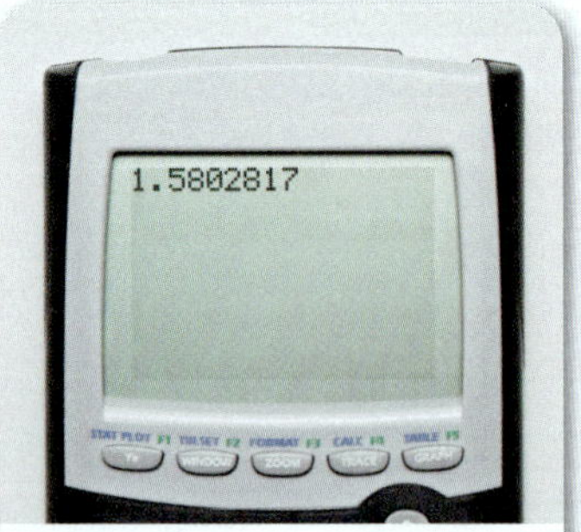

■ **Figure 13** You need to apply the rules of significant figures and rounding to report a calculated value correctly.

Rounding Numbers

Calculators perform flawless arithmetic, but they are not aware of the number of significant figures that should be reported in the answer. For example, a density calculation should not have more significant figures than the original data with the fewest significant figures. To report a value correctly, you often need to round. Consider an object with a mass of 22.44 g and volume of 14.2 cm^3. When you calculate the object's density using a calculator, the displayed answer is 1.5802817 g/cm^3, as shown in **Figure 13.** Because the measured mass had four significant figures and the measured volume had three, it is not correct to report the calculated density value with eight significant figures. Instead, the density must be rounded to three significant figures, or 1.58 g/cm^3.

Consider the value 3.515014. How would you round this number to five significant figures? To three significant figures? In each case, you need to look at the digit that follows the desired last significant figure.

To round to five digits, first identify the fifth significant figure, in this case 0, and then look at the number to its right, in this case 1.

3.515014 — **Last significant figure** (0); **Number to right of last significant figure** (1)

Do not change the last significant figure if the digit to its right is less than five. Because a 1 is to the right, the number rounds to 3.5150. If the number had been 5 or greater, you would have rounded up.

To round to three digits, identify the third significant figure, in this case 1, and then look at the number to its right, in this case 5.

3.515014 — **Last significant figure** (1); **Number to right of last significant figure** (5)

If the digits to the right of the last significant figure are a 5 followed by 0, then look at the last significant figure. If it is odd, round it up; if it is even, do not round up. Because the last significant digit is odd (1), the number rounds up to 3.52.

PROBLEM-SOLVING STRATEGY

Rounding Numbers

Learn these four rules for rounding, and use them when solving problems. Examples of each rule are shown below. Note that each example has three significant figures.

Example	Rule
2.532 → 2.53	**Rule 1.** If the digit to the right of the last significant figure is less than 5, do not change the last significant figure.
2.536 → 2.54	**Rule 2.** If the digit to the right of the last significant figure is greater than 5, round up the last significant figure.
2.5351 → 2.54	**Rule 3.** If the digits to the right of the last significant figure are a 5 followed by a nonzero digit, round up the last significant figure.
2.5350 → 2.54 2.5250 → 2.52	**Rule 4.** If the digits to the right of the last significant figure are a 5 followed by 0 or no other number at all, look at the last significant figure. If it is odd, round it up; if it is even, do not round up.

Matt Meadows

CHEMISTRY JOURNAL

Everyday Estimation Have students write examples of when they round numbers in everyday life. Examples might include estimating how much several items cost to make sure they have enough money, or estimating the mass of something to determine whether it is too heavy to lift. **OL**

PRACTICE Problems

Do additional problems. Online Practice

38. Round each number to four significant figures.

a. 84,791 kg **c.** 256.75 cm

b. 38.5432 g **d.** 4.9356 m

39. Challenge Round each number to four significant figures, and write the answer in scientific notation.

a. 0.00054818 g **c.** 308,659,000 mm

b. 136,758 kg **d.** 2.0145 mL

Addition and subtraction When you add or subtract measurements, the answer must have the same number of digits to the right of the decimal as the original value having the fewest number of digits to the right of the decimal. For example, the measurements 1.24 mL, 12.4 mL, and 124 mL have two, one, and zero digits to the right of the decimal, respectively. When adding or subtracting, arrange the values so that the decimal points align. Identify the value with the fewest places after the decimal point, and round the answer to that number of places.

Multiplication and division When you multiply or divide numbers, your answer must have the same number of significant figures as the measurement with the fewest significant figures.

EXAMPLE Problem 7

Find help with **significant figures.** Math Handbook

EXAMPLE PROBLEM

ROUNDING NUMBERS WHEN ADDING A student measured the length of his lab partners' shoes. If the lengths are 28.0 cm, 23.538 cm, and 25.68 cm, what is the total length of the shoes?

1 ANALYZE THE PROBLEM

The three measurements need to be aligned on their decimal points and added. The measurement with the fewest digits after the decimal point is 28.0 cm, with one digit. Thus, the answer must be rounded to only one digit after the decimal point.

2 SOLVE FOR THE UNKNOWN

$$\begin{array}{r} 28.0 \text{ cm} \\ 23.538 \text{ cm} \\ +\ 25.68 \text{ cm} \\ \hline 77.218 \text{ cm} \end{array}$$

Align the measurements and add the values.

The answer is **77.2 cm.** **Round to one place after the decimal; Rule 1 applies.**

3 EVALUATE THE ANSWER

The answer, 77.2 cm, has the same precision as the least-precise measurement, 28.0 cm.

PRACTICE Problems

Do additional problems. Online Practice

40. Add and subtract as indicated. Round off when necessary.

a. 43.2 cm + 51.0 cm + 48.7 cm

b. 258.3 kg + 257.11 kg + 253 kg

41. Challenge Add and subtract as indicated. Round off when necessary.

a. $(4.32 \times 10^3 \text{ cm}) - (1.6 \times 10^4 \text{ mm})$

b. $(2.12 \times 10^7 \text{ mm}) + (1.8 \times 10^3 \text{ cm})$

PRACTICE Problems

Have students refer to the Selected Solutions appendix for complete solutions to odd-numbered problems. The complete solutions for all problems can be found in the Solutions Manual.

38. a. 84,790 kg
b. 38.54 g
c. 256.8 cm
d. 4.936 m

39. a. 5.482×10^{-4} g
b. 1.368×10^5 kg
c. 3.087×10^8 mm
d. 2.014×10^0 mL, or 2.014 mL

IN-CLASS Example

Questions

a. Have students round 51.379 m to 3 significant figures.

b. Have students round 20,236 L to 2 significant figures.

Answers

a. 51.4 m
b. 2.0×10^4 L

PRACTICE Problems

Have students refer to the Selected Solutions appendix for complete solutions to odd-numbered problems. The complete solutions for all problems can be found in the Solutions Manual.

40. a. 142.9 cm
b. 768 kg

41. a. 2.7×10^3 cm
b. 2.12×10^7 cm

3 Assess

Check for Understanding

Ask students to solve the following problem. A 5.00-m board is sold to a construction crew. They measure it four times and obtain the following values: 4.98 m, 4.95 m, 5.08 m, and 5.03 m. How precise are these values? Fairly imprecise; the error ranges from −0.05 m to +0.08 m. OL

Reteach

Have students average the measurements for the board given above. Make sure they explain the number of significant figures in their answers. How accurate are the measurements? The average measurement is 5.01 m. The average measurement is very accurate. OL

Extension

Ask students to calculate the percent error for the average length of the board. Make sure they explain the number of significant figures in their answer. The percent error is 0.2%. The error, 0.01 m, has only one significant figure, so the answer has only one. OL

ASSESSMENT

Knowledge Continue the example of the construction crew by asking students what range of values count as a 5.00-m board. Have them explain their answers. Answers will vary; students might say that any value that rounds to 5.00 counts as a 5.00-m board. OL

IN-CLASS Example

Question Round the answer of the following problems to the correct number of significant figures.

a. 4,980,000 km × 0.0028 km
b. 364.5300 mm/0.00204 s

Answer

a. 14000 km^2
b. 179,000 or 1.79×10^5 mm/s

PRACTICE Problems

Have students refer to the Selected Solutions appendix for complete solutions to odd-numbered problems. The complete solutions for all problems can be found in the Solutions Manual.

42. a. 78 m^2 b. 12 m^2 c. 2.5 m^2 d. 81.1 m^2

43. a. 2.0 m/s b. 3.00 m/s c. 2.00 m/s d. 2.9 m/s

44. Divide coefficients: 1.32/2.5 = 0.528
Divide exponents: $10^3/10^2 = 10^1$
Combine parts and round:
0.528×10^1 g/cm^3; 5.3 g/cm^3

ChemLAB

The ChemLab located at the end of the chapter can be used at this point in the lesson.

EXAMPLE Problem 8

Find help with **rounding.** Math Handbook

EXAMPLE PROBLEM

ROUNDING NUMBERS WHEN MULTIPLYING Calculate the volume of a book with the following dimensions: length = 28.3 cm, width = 22.2 cm, height = 3.65 cm.

1 ANALYZE THE PROBLEM

Volume is calculated by multiplying length, width, and height. Because all of the measurements have three significant figures, the answer also will.

Known
length = 28.3 cm
width = 22.2 cm
height = 3.65 cm

Unknown
Volume = ? cm^3

2 SOLVE FOR THE UNKNOWN

Calculate the volume, and apply the rules of significant figures and rounding.

Volume = length × width × height — State the formula for the volume of a rectangle.

Volume = 28.3 cm × 22.2 cm × 3.65 cm = 2293.149 cm^3 — Substitute values and solve.

Volume = **2290 cm^3** — Round the answer to three significant figures.

3 EVALUATE THE ANSWER

To check if your answer is reasonable, round each measurement to one significant figure and recalculate the volume. Volume = 30 cm × 20 cm × 4 cm = 2400 cm^3. Because this value is close to your calculated value of 2290 cm^3, it is reasonable to conclude the answer is correct.

PRACTICE Problems

Do additional problems. Online Practice

Perform the following calculations. Round the answers.

42. a. 24 m × 3.26 m b. 120 m × 0.10 m
c. 1.23 m × 2.0 m d. 53.0 m × 1.53 m

43. a. 4.84 m ÷ 2.4 s b. 60.2 m ÷ 20.1 s
c. 102.4 m ÷ 51.2 s d. 168 m ÷ 58 s

44. Challenge $(1.32 \times 10^3$ g$) \div (2.5 \times 10^2$ $cm^3)$

SECTION 3 REVIEW

Section Self-Check

Section Summary

- An accurate measurement is close to the accepted value. A set of precise measurements shows little variation.
- The measurement device determines the degree of precision possible.
- Error is the difference between the measured value and the accepted value. Percent error gives the percent deviation from the accepted value.
- The number of significant figures reflects the precision of reported data.
- Calculations are often rounded to the correct number of significant figures.

45. MAINIDEA **State** how a measured value is reported in terms of known and estimated digits.

46. Define *accuracy* and *precision*.

47. Identify the number of significant figures in each of these measurements of an object's length: 76.48 cm, 76.47 cm, and 76.59 cm.

48. Apply The object in Question 47 has an actual length of 76.49 cm. Are the measurements in Question 47 accurate? Are they precise?

49. Calculate the error and percent error for each measurement in Question 47.

50. Apply Write an expression for the quantity 506,000 cm in which it is clear that all the zeros are significant.

51. Analyze Data Students collected mass data for a group of coins. The mass of a single coin is 5.00 g. Determine the accuracy and precision of the measurements.

Number of coins	5	10	20	30	50
Mass (g)	23.2	54.5	105.9	154.5	246.2

SECTION 3 REVIEW

45. A measured valued is reported with all of the known digits and one estimated digit.

46. Accuracy is defined as how close a value is to the accepted value. Precision is defined as how close a series of measurements are to one another.

47. They each have four significant figures.

48. Answers will vary but might include the following: They are not precise for values recorded to four significant figures. The first and second values are close enough to the accepted value to be called accurate.

49. (76.49 ~~cm~~ − 76.48 ~~cm~~)/(76.49 ~~cm~~) × 100 = 0.01307%
(76.49 ~~cm~~ − 76.47 ~~cm~~)/(76.49 ~~cm~~) × 100 = 0.02615%
(76.49 ~~cm~~ − 76.59 ~~cm~~)/(76.49 ~~cm~~) × 100 = 0.1307%

50. 5.06000×10^5 cm

51. The mass of an individual coin calculated for each trial are as follows: 5 coins, 4.6 g; 10 coins, 5.5 g; 20 coins, 5.3 g; 30 coins, 5.2 g; and 50 coins, 4.9 g. Knowing that the accepted value for the mass of the coin is 5.0 g, the data in the table are too varied to be considered precise and differ too greatly from the accepted value to be considered accurate.

SECTION 4

Essential Questions

- Why are graphs created?
- How can graphs be interpreted?

Review Vocabulary

independent variable: the variable that is changed during an experiment

New Vocabulary

graph

Representing Data

MAINIDEA Graphs visually depict data, making it easier to see patterns and trends.

CHEM 4 YOU Have you ever heard the saying, "A picture is worth a thousand words"? A graph is a "picture" of data. Scientists use graphs to present data in a form that allows them to analyze their results and communicate information about their experiments.

Graphing

When you analyze data, you might set up an equation and solve for an unknown, but this is not the only method scientists have for analyzing data. A goal of many experiments is to discover whether a pattern exists in a certain situation. Does raising the temperature change the rate of a reaction? Does a change in diet affect a rat's ability to navigate a maze? When data are listed as shown in **Table 4,** a pattern might not be obvious. However, using data to create a graph can help to reveal a pattern if one exists. A **graph** is a visual display of data.

Circle graphs Newspapers and magazines often feature circle graphs. A circle graph, like the one shown in **Figure 14,** is sometimes called a pie chart because it is divided into wedges that look like a pie. A circle graph is useful for showing parts of a fixed whole. The parts are usually labeled as percents with the whole circle representing 100%. The circle graph shown in **Figure 14** is based on the percentage data given in **Table 4.**

Figure 14 Although the percentage data presented in the table and the circle graph are basically the same, the circle graph makes it much easier to analyze.

Table 4 Sources of Chlorine in the Stratosphere

Source	Percent
Hydrogen chloride (HCl)	3
Methyl chloride (CH_3Cl)	15
Carbon tetrachloride (CCl_4)	12
Methyl chloroform ($C_2H_3Cl_3$)	10
CFC-11	23
CFC-12	28
CFC-13	6
HCFC-22	3

Chlorine in the Stratosphere

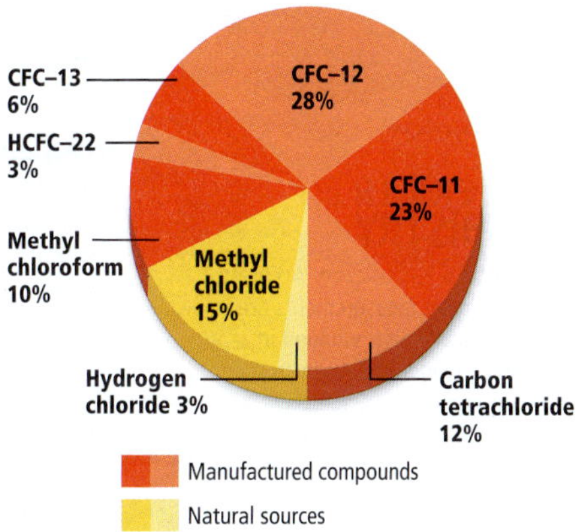

GRAPH CHECK
Analyze What percent of the chlorine sources are natural? What percent are manufactured compounds?

SECTION 4

1 Focus

MAINIDEA

Unit Conversion Using graph paper or a graphing calculator, have students graph the class data from The Big Idea activity using inches as the independent variable and centimeters as the dependent variable. Once students have graphed the results, have them calculate the slope of the line. The slope represents the conversion from centimeters to inches. The actual value should be 2.54 cm/inch. Ask students if their slope is accurate. OL

2 Teach

Concept Development

Graph the Data Have students graph the data from auto repair statistics. Number of Scheduled Maintenances performed over the useful life of the auto compared with major repair costs: 3:$5500; 2:$7500; 5:$5000; 2:$7500; 4:$4500; 1:$6500; 5:$3000; 6:$1000; 4:$4500; 3:$4500; 7:$700; 3:$6000; 1:$8000; 6:$800. Ask the students what type of graph would be most appropriate for these data. linear Ask students to draw a line that connects all the points; then ask them to draw a line that shows the trend of the graph. The line should have a negative slope. Ask students the significance of the downward slope of the line. Less money might be spent on major repairs if regularly scheduled maintenance is done. OL EL

GRAPH CHECK 18% natural sources; 82% manufactured compounds

CHEMISTRY JOURNAL

Visual Imagery Have students explain how circle graphs give both qualitative and quantitative information. A glance gives qualitative information, such as which wedge of the circle is largest. The precise percentage for each slice of the pie provides quantitative information. OL

GRAPH CHECK, FIG. 15 spinach and cashews

- **In-Text Question** independent variable value is 20.0 cm^3; dependent variable value is 54 g

Extension

Types of Graphs Have groups of students make a list from their own experiences of data that could be graphed. Have them discuss the best way to represent that data on a graph. Have groups share information, discuss the graphing methods and reasoning and come to a consensus on the best way to graph that information. OL COOP LEARN

FOLDABLES® Rubric

CAREERS IN CHEMISTRY
WebQuest

- **Figure 15** A bar graph is an effective way to present and compare data. This graph shows various dietary sources of the element magnesium. Magnesium plays an important role in the health of your muscles, nerves, and bones.

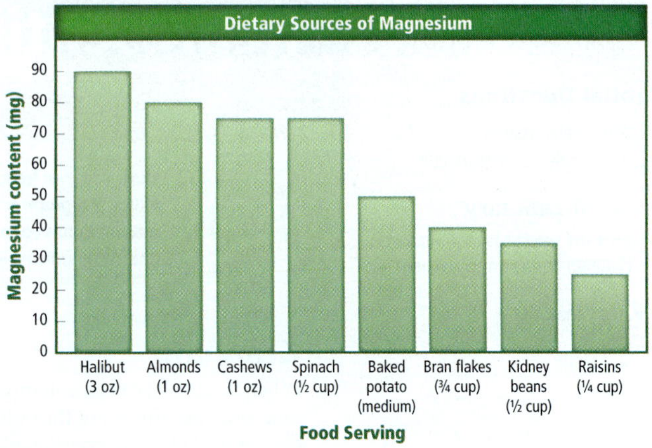

GRAPH CHECK

Interpret Which two food servings provide equal amounts of magnesium?

Get help with **graphing.**
Personal Tutor

FOLDABLES®
Incorporate information from this section into your Foldable.

Bar graphs A bar graph is often used to show how a quantity varies across categories. Examples of categories include time, location, and temperature. The quantity being measured appears on the vertical axis (*y*-axis). The independent variable appears on the horizontal axis (*x*-axis). The relative heights of the bars show how the quantity varies. A bar graph can be used to compare population figures for a single country by decade or the populations of multiple countries at the same point in time. In **Figure 15,** the quantity being measured is magnesium, and the category being varied is food servings. When examining the graph, you can quickly see how the magnesium content varies for these food servings.

Line Graphs In chemistry, most graphs that you create and interpret will be line graphs. The points on a line graph represent the intersection of data for two variables.

Independent and dependent variables The independent variable is plotted on the *x*-axis. The dependent variable is plotted on the *y*-axis. Remember that the independent variable is the variable that a scientist deliberately changes during an experiment. In **Figure 16a,** the independent variable is volume and the dependent variable is mass. What are the values for the independent variable and the dependent variable at Point B? **Figure 16b** is a graph of temperature versus elevation. Because the data points do not fit perfectly, the line cannot pass exactly through all of the points. The line must be drawn so that about as many points fall above the line as fall below it. This line is called a best-fit line.

Relationships between variables If the best-fit line for a set of data is straight, there is a linear relationship between the variables and the variables are said to be directly related. The relationship between the variables can be described further by analyzing the steepness, or slope, of the line.

CAREERS IN CHEMISTRY

Calibration Technician Accurate and repeatable measurements are essential to chemists working in research and industry. An instrument calibration technician adjusts, troubleshoots, maintains, and repairs the instruments used in laboratories and manufacturing plants. Their jobs require an understanding of the instrument's electronics and the use of computers and calibration software.

WebQuest

DIFFERENTIATED INSTRUCTION

Visually Impaired Use a magnetized board and magnets or pasta glued to paper to create different graphs. Make bar graphs and line graphs, including axes. Have the visually impaired student feel the graph and explain the information contained in the size of the bar for a bar graph or the slope of the line for a line graph. OL EL

If the best-fit line rises to the right, then the slope of the line is positive. A positive slope indicates that the dependent variable increases as the independent variable increases. If the best-fit line sinks to the right, then the slope of the line is negative. A negative slope indicates that the dependent variable decreases as the independent variable increases. In either case, the slope of the line is constant.

You can use two pairs of data points to calculate the slope of the line. The slope is the rise, or change in y, denoted as Δy, divided by the run, or change in x, denoted as Δx.

Slope Equation

$$\text{slope} = \frac{\text{rise}}{\text{run}} = \frac{\Delta y}{\Delta x} = \frac{y_2 - y_1}{x_2 - x_1}$$

y_2, y_1, x_2, and x_1 are values from data points (x_1, y_1) and (x_2, y_2).

The slope of a line is equal to the change in *y* divided by the change in *x*.

When the mass of a material is plotted against its volume, the slope of the line represents the material's density. An example of this is shown in **Figure 16a.** To calculate the slope of the line, substitute the *x* and *y* values for Points A and B in the slope equation and solve.

$$\text{slope} = \frac{54\text{ g} - 27\text{ g}}{20.0\text{ cm}^3 - 10.0\text{ cm}^3}$$
$$= \frac{27\text{ g}}{10.0\text{ cm}^3}$$
$$= 2.7\text{ g/cm}^3$$

Thus, the slope of the line, and the density, is 2.7 g/cm^3.

When the best-fit line is curved, the relationship between the variables is nonlinear. In chemistry, you will study nonlinear relationships called inverse relationships. Refer to the Math Handbook for more discussion of graphs.

Interpreting Graphs

You should use an organized approach when analyzing graphs. First, note the independent and dependent variables. Next, decide if the relationship between the variables is linear or nonlinear. If the relationship is linear, is the slope positive or negative?

Interpolation and extrapolation When points on a line graph are connected, the data are considered to be continuous. Continuous data allow you to read the value from any point that falls between the recorded data points. This process is called interpolation. For example, from **Figure 16b,** what is the temperature at an elevation of 350 m? To interpolate this value, first locate 350 m on the *x*-axis; it is located halfway between 300 m and 400 m. Project upward until you hit the plotted line, and then project that point horizontally to the left until you reach the *y*-axis. The temperature at 350 m is approximately 17.8°C.

You can also extend a line beyond the plotted points in order to estimate values for the variables. This process is called extrapolation. It is important to be very careful with extrapolation, however, as it can easily lead to errors and result in very inaccurate predictions.

☑ READING CHECK **Explain** why extrapolation might be less reliable than interpolation.

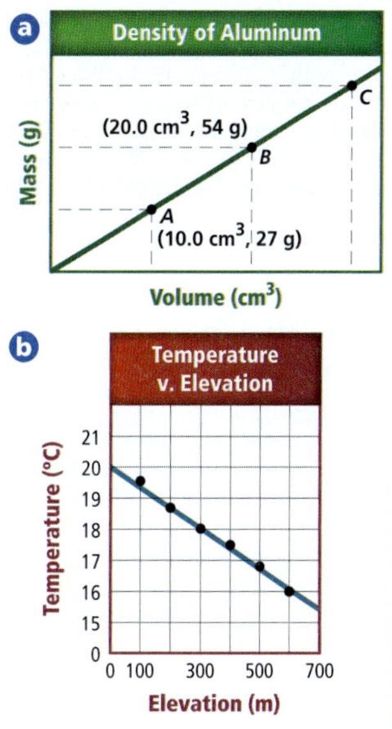

■ **Figure 16** Both of these line graphs show linear relationships. The slope of each line is defined as the ratio of rise over run.

☑ GRAPH CHECK **Identify** the graph that shows a direct relationship.

Watch a video about **tracking crime.**

Video

☑ **GRAPH CHECK** Graph A

ASSESSMENT

Performance Ask each student to draw a circle graph, a bar graph, and a line graph, and have them identify the uses for each. A circle graph demonstrates the relationship of parts of a whole. A bar graph indicates how a quantity varies with different factors, such as location or time. A line graph indicates the relationship between two variables, and data can be interpolated. OL

Extension

Graphing Calculators Many high school chemistry and algebra students use graphing calculators, but regulations for their use are not the same from country to country. Many U.S. high school mathematics teachers instruct students in the use of graphing calculators, however, only certain types of graphing calculators are allowed on AP and SAT exams. In Finland, 3D graphic calculators are not allowed on university entrance exams. In Norway, calculators with wireless communications capabilities are not allowed. Ask students what they view to be the advantages and disadvantages of using a graphing calculator. OL

☑ **READING CHECK** Extrapolation deals with values beyond the range of those measured.

CHEMISTRY PROJECT

Linear and Nonlinear Relationships Have students observe cause and effect and independent and dependent variables in their everyday lives. Have them predict whether the observed relationships are linear or nonlinear. OL

What's CHEMISTRY Got To Do With It?

Video

Tracking a Crime

GRAPH CHECK peak is ~280 DU, minimum is ~140 DU; variation = 280 DU − 140 DU = 140 DU

3 Assess

Check for Understanding

Write $y = mx + b$ on the board and ask students to identify the meaning of each symbol. Give them a data point on a line graph and the y-intercept and ask them to calculate the slope. **OL**

Reteach

Have students of varying heights stand against the board. Outline the tops of their heads, making a crude bar graph on the board. Students can stand in order of increasing height, with the tallest person in the middle, or in random order. Ask the class if the order of the students affects the image and the information they obtain from the graph. Yes, standing in different orders gives different information. **OL**

Extension

Plot two lines on a graph, one with a positive slope and one with a negative slope. For each line, ask students to explain what happens to the dependent variable when the independent variable increases. Positive: the dependent variable increases; negative: the dependent variable decreases. **OL**

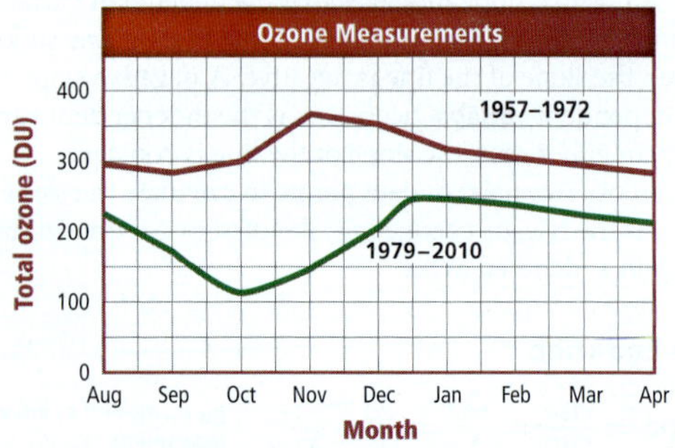

Figure 17 The two lines in this graph represent average ozone levels for two time periods, 1957–1972 and 1979–2010. The graph shows clearly that ozone levels in recent years have been lower overall than in 1957–1972. The ozone hole is generally considered to be the area where total ozone is less than 220 Dobson Units (DU).

GRAPH CHECK

Interpret By how much did the total ozone vary during the 9-month period shown for 1979–2010?

Interpreting ozone data The value of using graphs to visualize data is illustrated by **Figure 17.** These important ozone measurements were taken at the Halley Research Station in Antarctica. The graph shows how ozone levels vary from August to April. The independent and dependent variables are the month and the total ozone, respectively.

Each line on the graph represents a different period of time. The red line represents average ozone levels from 1957 to 1972, during which time ozone levels varied from about 285 DU (Dobson units) to 360 DU. The green line shows the ozone levels from 1979 to 2010. At no point during this period were the ozone levels as high as they were at corresponding times during 1957–1972.

The graph makes the ozone hole clearly evident—the dip in the green line indicates the presence of the ozone hole. Having data from two time periods on the same graph allows scientists to compare recent data with data from a time before the ozone hole existed. Graphs similar to **Figure 17** helped scientists identify a significant trend in ozone levels and verify the depletion in ozone levels over time.

SECTION 4 REVIEW

Section Self-Check

Section Summary

- Circle graphs show parts of a whole. Bar graphs show how a factor varies with time, location, or temperature.
- Independent (x-axis) variables and dependent (y-axis) variables can be related in a linear or a nonlinear manner. The slope of a straight line is defined as rise/run, or $\Delta y/\Delta x$.
- Because line-graph data are considered continuous, you can interpolate between data points or extrapolate beyond them.

52. MAINIDEA **Explain** why graphing can be an important tool for analyzing data.
53. **Infer** What type of data must be plotted on a graph for the slope of the line to represent density?
54. **Relate** If a linear graph has a negative slope, what can you say about the dependent variable?
55. **Summarize** What data are best displayed on a circle graph? On a bar graph?
56. **Construct** a circle graph for the composition of air: 78.08% N, 20.95% O_2, 0.93% Ar, and 0.04% CO_2 and other gases.
57. **Infer** from **Figure 17** how long the ozone hole lasts.
58. **Apply** Graph mass versus volume for the data given in the table. What is the slope of the line?

Volume (cm^3)	7.5	12	15	22
Mass (g)	24.1	38.5	48.0	70.1

SECTION 4 REVIEW

52. Often, trends in data are not easily seen when the data are presented in a table. Graphing the data allows patterns and trends to be more easily spotted and interpreted. Graphing provides visual information about relationships between variables, relative amounts, or parts of a whole.
53. Mass and volume data must be plotted; the y value must be mass and the x value must be volume.
54. It decreases in value as the independent variable increases.
55. circle graph: parts of a whole; bar graph: how a quantity varies with a factor such as location
56. Student graphs should have four wedge-shaped areas that are sized proportionately with the composition percentages given.
57. from September to November, which is approximately three months
58. Student graphs should show mass on the y-axis and volume on the x-axis. The slope of the line is 3.2 g/cm^3.

CHEMISTRY & health

Toxicology: Assessing Health Risk

It is likely that a closet or cupboard in your home or school contains products labeled with the symbol shown in **Figure 1.** Many cleaning, painting, and gardening products contain poisonous chemicals. Exposure to these chemicals can be dangerous. Possible effects are headaches, nausea, rashes, convulsions, coma, and even death. A toxicologist works to protect human health by studying the harmful effects of the chemicals and determining safe levels of exposure to them.

Figure 1 A skull-and-crossbones is the symbol for poison.

Jonathan Nourok/PhotoEdit

Keys to toxicity Warfarin is a drug used to prevent blood clots in people who have had a stroke or heart attack. It is also an effective rat poison. How is this possible? One key to toxicity is the dose—the amount of the chemical taken in by an organism. Exposure time can also be a factor; even low-dose exposure to some chemicals over long periods of time can be hazardous. Toxicity is also affected by the presence of other chemicals in the body, the age and gender of the individual, and the chemical's ability to be absorbed and excreted.

A dose-response curve, such as the one shown in **Figure 2,** relates the toxicity of a substance to its physical effects. This dose-response curve shows the results of an experiment in which different doses of a possible carcinogen were given to mice. The mice were checked for tumors 90 days after exposure. The graph indicates a noticeable increase in the incidence of tumors.

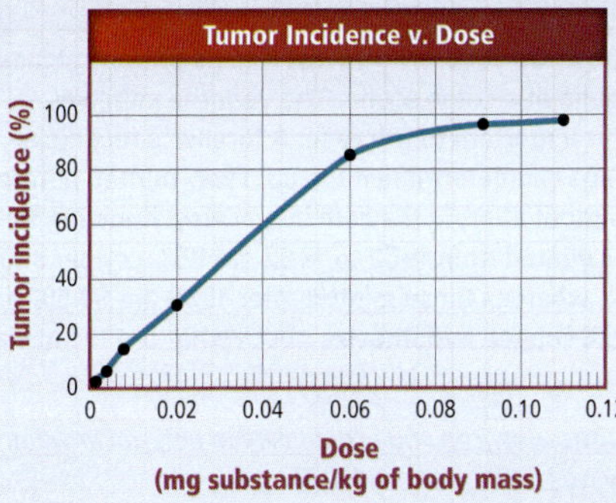

Figure 2 The seven data points correspond to seven groups of mice that were given different doses of a possible carcinogen.

Applying toxicity data How do toxicologists predict health risks to people? Toxicity data might be available from studies of routine chemical exposure in the workplace, as well as from medical records of accidental chemical contact. Toxicity testing is often carried out using bacteria and cell cultures. Toxicologists observe the effect of chemical doses on bacteria. If mutations occur, the chemical is considered potentially harmful.

MSDS Toxicologists apply mathematical models and knowledge of similar substances to toxicity data to estimate safe human exposure levels. How can you obtain this information? Every employer is required to keep Material Safety Data Sheets (MSDS) of the potentially hazardous chemicals they use in their workplace. The MSDS describe possible health effects, clothing and eye protection that should be worn, and first-aid steps to follow after exposure. You can also consult the Household Products Database, which provides health and safety information on more than 5000 commonly used products.

WRITING IN ▶ Chemistry

Research Access the MSDS for several products used at home. Compare the possible adverse health effects of exposure to the products and list the first aid requirements.

WebQuest

CHEMISTRY & health

Purpose

Students will learn about the field of toxicology, explore factors that affect chemical toxicity in organisms, and understand how data from toxicology tests can be applied in the home, school, and workplace.

Background

Selective toxicity is an important application of toxicology research. The concept is based on the fact that different species might respond differently following simultaneous exposure to a dose of a given chemical. For example, antibiotic drugs are deadly to bacteria, but far less toxic to a human host. Certain pesticides are designed to destroy insect species, while causing no adverse effects to the plants they infest.

Teaching Strategies

- Discuss examples of how the human body protects itself from exposure to toxic substances. liver and kidney cells break down toxic substances and the body excretes them, spontaneous vomiting can expel harmful substances that are swallowed
- Have students brainstorm general characteristics and information a doctor might need to know when determining potential health risks to an emergency room patient who has been exposed to a chemical. age, sex, length of time exposed, exposure route, food or liquid ingested prior to exposure

WRITING IN ▶ Chemistry

Rubric WebQuest

Research Answers will vary based on the student's product selection. Sample answer: The MSDS of an automatic dishwashing detergent indicates that it is an irritation to eyes and skin. If swallowed, it will irritate the mouth, throat, and stomach. First aid is as follows: for the eyes, flush with water for 15 min; for the skin, wash with soap and water; and if ingested give large amounts of milk or water. Do not induce vomiting. Call a doctor if ingested or the eye and skin irritation continues.

ChemLAB

Rubric

Preparation

Time Allotment one class period

Process Skills collect and organize data, interpret data, observe and infer

Safety Precautions Approve lab safety forms before work begins. Have students wear aprons and goggles. Caution students to clean up any spilled water.

Disposal Drain water from the graduated cylinder, pour pennies onto paper towel.

Procedure

- Require that the students' notebooks contain objectives, outline or flowcharts of the procedure, and data tables before they begin.
- Have paper towels available to clean up any spills.
- Have students practice reading a meniscus before beginning the activity.

Analyze and Conclude

1. Refer to the Solutions Manual.
2. Refer to the Solutions Manual.
3. The slope for pre-1982 pennies is 9.0 g/mL. The slope for post-1982 pennies is 7.1 g/mL.
4. Verifying the slopes of the lines gives you the density of the pre-1982 pennies and density of the post-1982 pennies.
5. Both pennies have similar volume but the mass is different, therefore the density is different. Mass can be used to identify both pre- and post-1982 pennies.
6. Pre-1982 pennies: $(9.0 - 8.8)/8.8 \times 100 = 2.3\%$ error
   Post-1982 pennies: $(7.1 - 7.2)/7.2 \times 100 = 1.4\%$ error

ChemLAB

Forensics: Use Density to Date a Coin

Background: **A penny that has had its date scratched off is found at a crime scene. The year the coin was minted is important to the case. A forensics technician claims she can determine if the coin was minted before 1982 without altering the coin in any way. Knowing that pennies minted from 1962 to 1982 are 95% copper and 5% zinc, whereas those minted after 1982 are 97.5% zinc and 2.5% copper, hypothesize about what the technician will do.**

Question: *How can you use density to determine whether a penny was minted before 1982?*

Materials

water
small plastic cup
pre-1982 pennies (25)
metric ruler
graph paper
100-mL graduated cylinder
balance
post-1982 pennies (25)
pencil
graphing calculator (optional)

Safety Precautions

Procedure

1. Read and complete the lab safety form.
2. Record all measurements in your data table.
3. Measure the mass of the plastic cup.
4. Pour about 50 mL of water into the graduated cylinder. Record the actual volume.
5. Add 5 pre-1982 pennies to the cup, and measure the mass again.
6. Add the 5 pennies to the graduated cylinder, and read the volume.
7. Repeat Steps 5 and 6 four times. After five trials there will be 25 pennies in the graduated cylinder.
8. **Cleanup and Disposal** Pour the water from the graduated cylinder down a drain, being careful not to lose any of the pennies. Dry the pennies with a paper towel.
9. Repeat Steps 3 through 7, using post-1982 pennies.

Data Table for the Density of a Penny

Trial	Mass of Pennies Added (g)	Total Number of Pennies	Total Mass of Pennies (g)	Total Volume of Water Displaced (mL)
1		5		
2		10		
3		15		
4		20		
5		25		

Analyze and Conclude

1. **Calculate** Complete the data table by calculating the total mass and the total volume of water displaced for each trial.
2. **Make and Use Graphs** Graph total mass versus total volume for the pre-1982 and post-1982 pennies. Plot and label two sets of points on the graph, one for pre-1982 pennies and one for post-1982 pennies.
3. **Make and Use Graphs** Draw a best-fit line through each set of points. Use two points on each line to calculate the slope.
4. **Infer** What do the slopes of the lines tell you about the two groups of pennies?
5. **Apply** Can you determine if a penny was minted before or after 1982 if you know only its mass? Explain how the relationships among volume, mass, and density support using a mass-only identification technique.
6. **Error Analysis** Determine the percent error in the density of each coin.

INQUIRY EXTENSION

Compare your results with those from the rest of the class. Are they consistent? If not, explain how you could refine your investigation to ensure more accurate results. Calculate a class average density of the pre-1982 pennies and the density of the post-1982 pennies. Determine the percent error of each average.

INQUIRY EXTENSION

The results should be consistent. More-accurate results could be achieved with a graduated cylinder that reads a more-accurate volume. Make sure the pennies are dry before they are massed.

CHAPTER 2 STUDY GUIDE

Vocabulary Practice

BIGIDEA **Chemists collect and analyze data to determine how matter interacts.**

SECTION 1 Units and Measurements

MAINIDEA Chemists use an internationally recognized system of units to communicate their findings.

- SI measurement units allow scientists to report data to other scientists.
- Adding prefixes to SI units extends the range of possible measurements.
- To convert to Kelvin temperature, add 273 to the Celsius temperature.

$$K = °C + 273$$

- Volume and density have derived units. Density, which is a ratio of mass to volume, can be used to identify an unknown sample of matter.

$$\text{density} = \frac{\text{mass}}{\text{volume}}$$

VOCABULARY
- base unit
- second
- meter
- kilogram
- kelvin
- derived unit
- liter
- density

SECTION 2 Scientific Notation and Dimensional Analysis

MAINIDEA Scientists often express numbers in scientific notation and solve problems using dimensional analysis.

- A number expressed in scientific notation is written as a coefficient between 1 and 10 multiplied by 10 raised to a power.
- To add or subtract numbers in scientific notation, the numbers must have the same exponent.
- To multiply or divide numbers in scientific notation, multiply or divide the coefficients and then add or subtract the exponents, respectively.
- Dimensional analysis uses conversion factors to solve problems.

VOCABULARY
- scientific notation
- dimensional analysis
- conversion factor

SECTION 3 Uncertainty in Data

MAINIDEA Measurements contain uncertainties that affect how a calculated result is presented.

- An accurate measurement is close to the accepted value. A set of precise measurements shows little variation.
- The measurement device determines the degree of precision possible.
- Error is the difference between the measured value and the accepted value. Percent error gives the percent deviation from the accepted value.

$$\text{error} = \text{experimental value} - \text{accepted value}$$

$$\text{percent error} = \frac{|\text{error}|}{\text{accepted value}} \times 100$$

- The number of significant figures reflects the precision of reported data.
- Calculations are often rounded to the correct number of significant figures.

VOCABULARY
- accuracy
- precision
- error
- percent error
- significant figure

SECTION 4 Representing Data

MAINIDEA Graphs visually depict data, making it easier to see patterns and trends.

- Circle graphs show parts of a whole. Bar graphs show how a factor varies with time, location, or temperature.
- Independent (x-axis) variables and dependent (y-axis) variables can be related in a linear or a nonlinear manner. The slope of a straight line is defined as rise/run, or $\Delta y/\Delta x$.

$$\text{slope} = \frac{y_2 - y_1}{x_2 - x_1} = \frac{\Delta y}{\Delta x}$$

- Because line graph data are considered continuous, you can interpolate between data points or extrapolate beyond them.

VOCABULARY
- graph

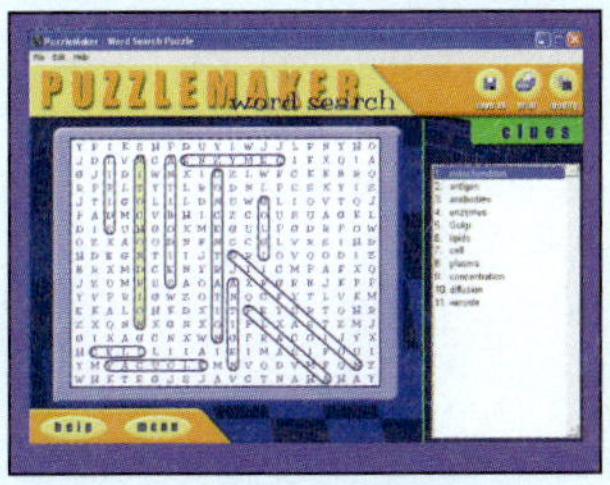

Vocabulary Practice

For additional practice with vocabulary, have students access the Vocabulary PuzzleMaker.

Use the Vocabulary

To reinforce chapter vocabulary, have students write a sentence using each term. OL EL

Review Strategies

- Have students list the SI and one other common unit for volume, pressure, and temperature. OL
- Have students list the rules for rounding significant figures and determining the correct number of significant figures following calculations. OL

Students can review with these online resources.

Vocabulary Practice

Vocabulary eGames
Vocabulary eFlashcards
Vocabulary PuzzleMaker

Section Self-Check

Chapter Self-Check

Online Test Practice

Use *eAssessment* to:

- create multiple versions of tests
- edit existing questions and add your own questions
- build tests aligned with select state curriculum standards using built-in state curriculum tags
- track students' progress

SECTION 1

Mastering Concepts

59. The number gives you the quantitative value, and the unit indicates what was measured.

60. Scientists from different countries have different languages and cultures but must be able to share and compare data.

61. Prefixes give the magnitude of the measurement.

62. 1 km = 1000 m; 1 dm = 0.1 m

63. The SI unit for volume is the cubic meter, m^3, which is equal to the product of three SI measurements of length.

64. The sizes of the units are equal; $°C + 273 = K$

65. Student drawings should show the layers in the following order from top to bottom: cork, ethyl alcohol, wood (oak), motor oil, isopropyl alcohol, vegetable oil and plastic (at the same level), water, glycerin, corn syrup, and bone.

Mastering Problems

66. density = 1 g/mL

67. volume = 3.0 mL

68. density = 1.5 g/mL

69. No, the Celsius thermometer could not be used to make this candy because the thermometer is out of the temperature range.

SECTION 2

Mastering Concepts

70. Scientific notation uses a number between 1 and 10 times a power of ten to indicate the size of very large or small numbers.

71. positive

72. a. $X.XXXX \times 10^2$
b. $X.XX \times 10^{-7}$

73. Subtract them.

74. It decreases.

75. Meters will be in the denominator so that the units will cancel when the starting value is multiplied by the conversion factor.

CHAPTER 2 ASSESSMENT

SECTION 1

Mastering Concepts

59. Why must a measurement include both a number and a unit?

60. Explain why standard units of measurement are particularly important to scientists.

61. What role do prefixes play in the metric system?

62. How many meters are in one kilometer? In one decimeter?

63. **SI Units** What is the relationship between the SI unit for volume and the SI unit for length?

64. Explain how temperatures on the Celsius and Kelvin scales are related.

65. Examine the density values for several common liquids and solids given in **Table 5.** Sketch the results of an experiment that layered each of the liquids and solids in a 1000-mL graduated cylinder.

Table 5 Density Values

Liquids (g/mL)		Solids (g/cm^3)	
Ethyl alcohol	0.789	Bone	1.85
Glycerin	1.26	Cork	0.24
Isopropyl alcohol	0.870	Plastic	0.91
Corn syrup	1.37	Wood (oak)	0.84
Motor oil	0.860		
Vegetable oil	0.910		
Water at 4°C	1.000		

Mastering Problems

66. A 5-mL sample of water has a mass of 5 g. What is the density of water?

67. The density of aluminum is 2.7 g/mL. What is the volume of 8.1 g?

68. An object with a mass of 7.5 g raises the level of water in a graduated cylinder from 25.1 mL to 30.1 mL. What is the density of the object?

69. **Candy Making** The directions in the candy recipe for pralines instruct the cook to remove the pot containing the candy mixture from the heat when the candy mixture reaches the soft-ball stage. The soft-ball stage corresponds to a temperature of 236°F. After the soft-ball stage is reached, the pecans and vanilla are added. Can a Celsius thermometer with a range of −10°C to 110°C be used to determine when the soft-ball stage is reached in the candy mixture?

SECTION 2

Mastering Concepts

70. How does scientific notation differ from ordinary notation?

71. If you move the decimal place to the left to convert a number to scientific notation, will the power of 10 be positive or negative?

72. Two undefined numbers expressed in regular notation are shown below, along with the number of places the decimal must move to express each in scientific notation. If each X represents a significant figure, write each number in scientific notation.
a. XXX.XX
b. 0.000000XXX

73. When dividing numbers in scientific notation, what must you do with the exponents?

74. When you convert from a small unit to a large unit, what happens to the number of units?

75. When converting from meters to centimeters, how do you decide which values to place in the numerator and denominator of the conversion factor?

Mastering Problems

76. Write the following numbers in scientific notation.
a. 0.0045834 mm
b. 0.03054 g
c. 438,904 s
d. 7,004,300,000 g

77. Write the following numbers in ordinary notation.
a. 8.348×10^6 km
b. 3.402×10^3 g
c. 7.6352×10^{-3} kg
d. 3.02×10^{-5} s

78. Complete the following addition and subtraction problems in scientific notation.
a. $(6.23 \times 10^6 \text{ kL}) + (5.34 \times 10^6 \text{ kL})$
b. $(3.1 \times 10^4 \text{ mm}) + (4.87 \times 10^5 \text{ mm})$
c. $(7.21 \times 10^3 \text{ mg}) + (43.8 \times 10^2 \text{ mg})$
d. $(9.15 \times 10^{-4} \text{ cm}) + (3.48 \times 10^{-4} \text{ cm})$
e. $(4.68 \times 10^{-5} \text{ cg}) + (3.5 \times 10^{-6} \text{ cg})$
f. $(3.57 \times 10^2 \text{ mL}) - (1.43 \times 10^2 \text{ mL})$
g. $(9.87 \times 10^4 \text{ g}) - (6.2 \times 10^3 \text{ g})$
h. $(7.52 \times 10^5 \text{ kg}) - (5.43 \times 10^5 \text{ kg})$
i. $(6.48 \times 10^{-3} \text{ mm}) - (2.81 \times 10^{-3} \text{ mm})$
j. $(5.72 \times 10^{-4} \text{ dg}) - (2.3 \times 10^{-5} \text{ dg})$

79. Complete the following multiplication and division problems in scientific notation.
a. $(4.8 \times 10^5 \text{ km}) \times (2.0 \times 10^3 \text{ km})$
b. $(3.33 \times 10^{-4} \text{ m}) \times (3.00 \times 10^{-5} \text{ m})$
c. $(1.2 \times 10^6 \text{ m}) \times (1.5 \times 10^{-7} \text{ m})$
d. $(8.42 \times 10^8 \text{ kL}) \div (4.21 \times 10^3 \text{ kL})$
e. $(8.4 \times 10^6 \text{ L}) \div (2.4 \times 10^{-3} \text{ L})$
f. $(3.3 \times 10^{-4} \text{ mL}) \div (1.1 \times 10^{-6} \text{ mL})$

Mastering Problems

76. a. 4.5834×10^{-3} mm
b. 3.054×10^{-2} g
c. $4.389\,04 \times 10^5$ s
d. 7.0043×10^9 g

77. a. 8,348,000 km
b. 3402 g
c. 0.0076352 kg
d. 0.0000302 s

78. a. 1.157×10^7 kL
b. 5.18×10^5 mm
c. 1.159×10^4 mg
d. 1.263×10^{-3} cm
e. 5.03×10^{-5} cg
f. 2.14×10^2 mL
g. 9.25×10^4 g
h. 2.09×10^5 kg
i. 3.67×10^{-3} mm
j. 5.49×10^{-4} dg

79. a. $9.6 \times 10^8 \text{ km}^2$
b. $9.99 \times 10^{-9} \text{ m}^2$
c. $1.8 \times 10^{-1} \text{ m}^2$
d. 2.00×10^5 kL
e. 3.5×10^9 L
f. 3.0×10^2 mL

Chapter Self-Check

80. Convert the following measurements.
- **a.** 5.70 g to milligrams
- **b.** 4.37 cm to meters
- **c.** 783 kg to grams
- **d.** 45.3 mm to meters
- **e.** 10 m to centimeters
- **f.** 37.5 g/mL to kg/L

81. Gold A troy ounce is equal to 480 grains, and 1 grain is equal to 64.8 milligrams. If the price of gold is $560 per troy ounce, what is the cost of 1 g of gold?

82. Popcorn The average mass of a kernel of popcorn is 0.125 g. If 1 pound = 16 ounces, and 1 ounce = 28.3 g, then how many kernels of popcorn are there in 0.500 pounds of popcorn?

83. Blood You have 15 g of hemoglobin in every 100 mL of your blood. 10.0 mL of your blood can carry 2.01 mL of oxygen. How many milliliters of oxygen does each gram of hemoglobin carry?

84. Nutrition The recommended calcium intake for teenagers is 1300 mg per day. A glass of milk contains 305 mg of calcium. One glass contains a volume of 8 fluid ounces. How many liters of milk should a teenager drink per day to get the recommended amount of calcium? One fluid ounce equals 29.6 mL.

SECTION 3

Mastering Concepts

85. Which zero is significant in the number 50,540? What is the other zero called?

86. Why are percent error values never negative?

87. If you report two measurements of mass, 7.42 g and 7.56 g, are the measurements accurate? How might you evaluate the precision of these measurements? Explain your answers.

88. Which produce the same number when rounded to three significant figures: 3.456, 3.450, or 3.448?

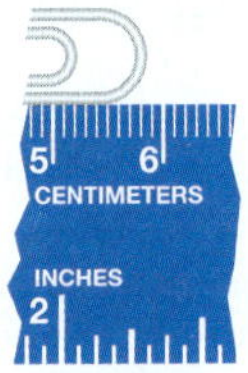

■ **Figure 18**

89. Record the measurement shown in **Figure 18** to the correct number of significant figures.

90. When subtracting 61.45 g from 242.6 g, which value determines the number of significant figures in the answer? Explain.

Mastering Problems

91. Round each number to four significant figures.
- **a.** 431,801 kg
- **b.** 10,235.0 mg
- **c.** 1.0348 m
- **d.** 0.004384010 cm
- **e.** 0.00078100 mL
- **f.** 0.0098641 cg

92. Round the answer for each problem to the correct number of significant figures.
- **a.** $(7.31 \times 10^4) + (3.23 \times 10^3)$
- **b.** $(8.54 \times 10^{-3}) - (3.41 \times 10^{-4})$
- **c.** 4.35 dm × 2.34 dm × 7.35 dm
- **d.** 4.78 cm + 3.218 cm + 5.82 cm
- **e.** 38,736 km ÷ 4784 km

93. The accepted length of a steel pipe is 5.5 m. Calculate the percent error for each of these measurements.
a. 5.2 m **b.** 5.5 m **c.** 5.7 m **d.** 5.1 m

94. The accepted density for copper is 8.96 g/mL. Calculate the percent error for each of these measurements.
- **a.** 8.86 g/mL
- **b.** 8.92 g/mL
- **c.** 9.00 g/mL
- **d.** 8.98 g/mL

SECTION 4

Mastering Concepts

95. Heating Fuels Which type of graph would you use to depict how many households heat with gas, oil, or electricity? Explain.

96. Gasoline Consumption Which type of graph would you choose to depict gasoline consumption over a 10-year period? Explain.

97. How can you find the slope of a line graph?

Mastering Problems

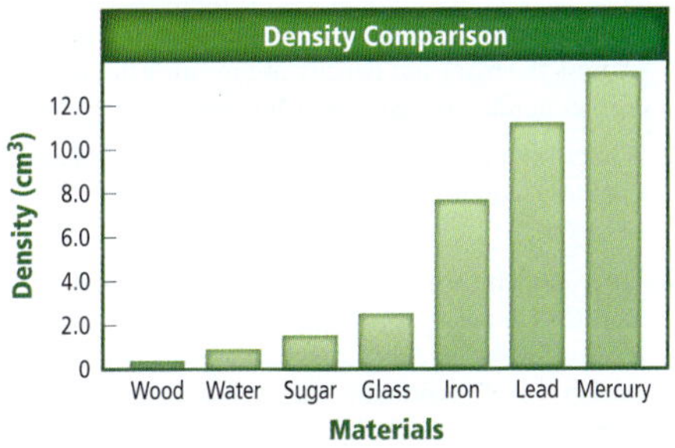

■ **Figure 19**

98. Use **Figure 19** to answer the following questions.
- **a.** Which substance has the greatest density?
- **b.** Which substance has the least density?
- **c.** Which substance has a density of 7.87 g/cm³?
- **d.** Which substance has a density of 11.4 g/cm³?

---

80. a. 5.70×10^3 mg
b. 4.37×10^{-2} m
c. 7.83×10^5 g
d. 4.53×10^{-2} m
e. 1000 cm
f. 37.5 kg/L

81. $18/g gold

82. 1810 kernels popcorn

83. 1.34 mL oxygen/g hemoglobin

84. 1 L milk

SECTION 3

Mastering Concepts

85. the first one; placeholder

86. because the percent error equation uses the absolute value of the error

87. You must know the accepted value to know if the measurements are accurate. Students might suggest checking precision by comparing the measurements to those made by others and those made on other balances.

88. 3.450 and 3.448

89. 5.85 cm

90. 242.6, is the number that has the fewest digits to the right of the decimal point; it is less precise.

Mastering Problems

91. a. 431,800 kg
b. 10,240 m
c. 1.035 m
d. 0.004384 cm
e. 0.0007810 mL
f. 0.009864 cg

92. a. 7.63×10^4
b. 8.20×10^{-3}
c. 74.8 dm³
d. 13.82 cm
e. 8.097 km

93. a. error = 5.5%
b. error = 0
c. error = 3.6%
d. error = 7.3%

94. a. percent error = 1.12%
b. percent error = 0.446%
c. percent error = 0.446%
d. percent error = 0.223%

SECTION 4

Mastering Concepts

95. A bar graph could be used with the method of heating on the *x*-axis and the number of households on the *y*-axis. If the data include all the households for a region, relative numbers could be converted to a percentage and expressed as a circle graph.

96. line or bar graph because they can show how consumption varies with time

97. Choose two points on the line. Divide the difference in the *y* values by the difference in the *x* values.

Mastering Problems

98. a. mercury
b. wood
c. iron
d. lead

MIXED REVIEW

99. **a.** $1.31 \times 10^4\ cm^2$
b. $2.73 \times 10^6\ m^2$
c. $9.26 \times 10^{-8}\ m^2$
d. 3.1×10^2
e. 2.2×10^{-5}
f. 2.00×10^1

100. **a.** 301 cg
b. 6.2 km
c. $6.24 \times 10^{-1}\ \mu g$
d. 0.2 dm^3
e. 0.00013 kcal/g
f. 0.00321 L

101. density = 6.82 g/mL
% error = 1.87%

102. No, the conversion is not correct because the units of rate should be m/min. This expression yields the units $m \cdot h/min^2$. The last conversion factor should be 60 min/1 h.

103. volume = 29 mL

104. mass = 445.20 g
density = 7.15 g/mL

105. 12.5 g lead

106. The third student (2.87 cm) is correct. A meterstick has markings to the millimeter, so a third digit should be estimated.

107. $density_{black\ hole} = 4.5273 \times 10^{10}\ g/cm^3$

108. The density of the black hole is $4.5273 \times 10^{10}\ g/cm^3$ (almost fifty billion) times greater than that of water.

109. 3.72 m determines the number of significant figures in the answer because it is the original value having the fewest number of significant figures.

110. **a.** 0.00321 g
b. 3.88 kg
c. 219,000 m
d. 25.4 L
e. 0.0876 cm
f. 0.00311 mg

111. slope = 2.7 g/mL

112. 0.24 g dextromethorphan/bottle

MIXED REVIEW

99. Complete these problems in scientific notation. Round to the correct number of significant figures.
a. $(5.31 \times 10^{-2}\ cm) \times (2.46 \times 10^5\ cm)$
b. $(3.78 \times 10^3\ m) \times (7.21 \times 10^2\ m)$
c. $(8.12 \times 10^{-3}\ m) \times (1.14 \times 10^{-5}\ m)$
d. $(9.33 \times 10^4\ mm) \div (3.0 \times 10^2\ mm)$
e. $(4.42 \times 10^{-3}\ kg) \div (2.0 \times 10^2\ kg)$
f. $(6.42 \times 10^{-2}\ g) \div (3.21 \times 10^{-3}\ g)$

100. Convert each quantity to the indicated units.
a. 3.01 g → cg
b. 6200 m → km
c. $6.24 \times 10^{-7}\ g \rightarrow \mu g$
d. 0.2 L → dm^3
e. 0.13 cal/g → kcal/g
f. 3.21 mL → L

101. Students used a balance and a graduated cylinder to collect the data shown in **Table 6.** Calculate the density of the sample. If the accepted density of this sample is 6.95 g/mL, calculate the percent error.

Table 6 Volume and Mass Data

Mass of sample	20.46 g
Volume of water	40.0 mL
Volume of water + sample	43.0 mL

102. Evaluate the following conversion. Will the answer be correct? Explain.

$$rate = \frac{75\ m}{1\ s} \times \frac{60\ s}{1\ min} \times \frac{1\ h}{60\ min}$$

103. You have a 23-g sample of ethanol with a density of 0.7893 g/mL. What volume of ethanol do you have?

104. **Zinc** Two separate masses of zinc were measured on a laboratory balance. The first zinc sample had a mass of 210.10 g, and the second zinc sample had a mass of 235.10 g. The two samples were combined. The volume of the combined sample was found to be 62.3 mL. Express the mass and density of the zinc sample in the correct number of significant figures.

105. What mass of lead (density 11.4 g/cm^3) would have a volume identical to 15.0 g of mercury (density 13.6 g/cm^3)?

106. Three students use a meterstick with millimeter markings to measure a length of wire. Their measurements are 3 cm, 3.3 cm, and 2.87 cm, respectively. Explain which answer was recorded correctly.

107. **Astronomy** The black hole in the M82 galaxy has a mass about 500 times the mass of the Sun. It has about the same volume as the Moon. What is the density of this black hole?

mass of the Sun = 1.9891×10^{30} kg
volume of the Moon = $2.1968 \times 10^{10}\ km^3$

108. The density of water is 1 g/cm^3. Use your answer from Question 107 to compare the densities of water and a black hole.

109. When multiplying 602.4 m by 3.72 m, which value determines the number of significant figures in the answer? Explain.

110. Round each figure to three significant figures.
a. 0.003210 g
b. 3.8754 kg
c. 219,034 m
d. 25.38 L
e. 0.08763 cm
f. 0.003109 mg

111. Graph the data in **Table 7,** with the volume on the *x*-axis and the mass on the *y*-axis. Then calculate the slope of the line.

Table 7 Density Data

Volume (mL)	Mass (g)
2.0	5.4
4.0	10.8
6.0	16.2
8.0	21.6
10.0	27.0

112. **Cough Syrup** A common brand of cough syrup comes in a 4-fluid ounce bottle. The active ingredient in the cough syrup is dextromethorphan. For an adult, the standard dose is 2 teaspoons, and a single dose contains 20.0 mg of dextromethorphan. Using the relationships, 1 fluid ounce = 29.6 mL and 1 teaspoon = 5.0 mL, determine how many grams of dextromethorphan are contained in the bottle.

THINK CRITICALLY

113. **Interpret** Why does it make sense for the line in **Figure 16a** to extend to (0, 0) even though this point was not measured?

114. **Infer** Which of these measurements was made with the most precise measuring device: 8.1956 m, 8.20 m, or 8.196 m? Explain your answer.

115. **Apply** When subtracting or adding two numbers in scientific notation, why do the exponents need to be the same?

116. **Compare and Contrast** What advantages do SI units have over the units commonly used in the United States? Are there any disadvantages to using SI units?

117. **Hypothesize** Why do you think the SI standard for time was based on the distance light travels through a vacuum?

THINK CRITICALLY

113. Extrapolation of measured data extends the line to this point. The graph shows that an object with no mass will have no volume.

114. 8.1956 m, because it has the greatest number of significant figures.

115. Equal place values should be added to each other.

116. Answers will vary but might include that units based on powers of ten are easy to convert from one to another. Most disadvantages involve the initial changing from another system to SI.

117. There is no chance for matter to interfere with the speed measurement in a vacuum.

Chapter Self-Check

118. Infer Why does knowing the mass of an object not help you identify what material the object is made from?

119. Conclude Why might property owners hire a surveyor to determine property boundaries rather than measure the boundaries themselves?

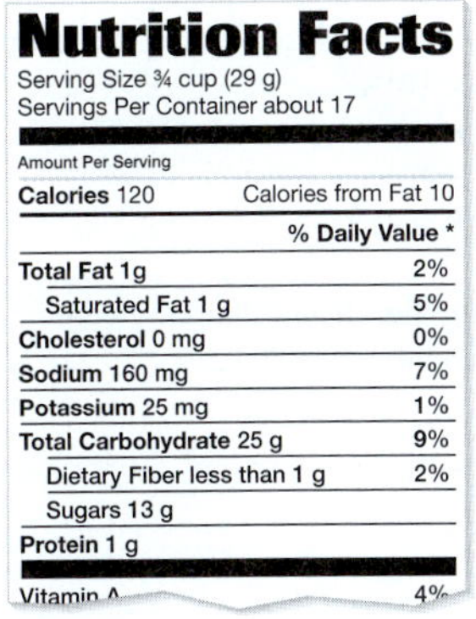

Nutrition Facts	
Serving Size ¾ cup (29 g)	
Servings Per Container about 17	
Amount Per Serving	
Calories 120	Calories from Fat 10
	% Daily Value *
Total Fat 1g	2%
Saturated Fat 1 g	5%
Cholesterol 0 mg	0%
Sodium 160 mg	7%
Potassium 25 mg	1%
Total Carbohydrate 25 g	9%
Dietary Fiber less than 1 g	2%
Sugars 13 g	
Protein 1 g	
Vitamin A	4%

■ **Figure 20**

120. Apply Dimensional Analysis Evaluate the breakfast cereal nutritional label shown in **Figure 20.** This product contains 160 mg of sodium in each serving. If you eat 2.0 cups of cereal a day, how many grams of sodium are you ingesting? What percent of your daily recommended sodium intake does this represent?

121. Predict Four graduated cylinders each contain a different liquid: A, B, C, and D.

Liquid A: mass = 18.5 g; volume = 15.0 mL
Liquid B: mass = 12.8 g; volume = 10.0 mL
Liquid C: mass = 20.5 g; volume = 12.0 mL
Liquid D: mass = 16.5 g; volume = 8.0 mL

Examine the information given for each liquid, and predict the layering of the liquids if they were carefully poured into a larger graduated cylinder.

Challenge Problem

122. Carboplatin ($C_6H_{12}N_2O_4Pt$) is a platinum-containing compound that is used to treat certain forms of cancer. This compound contains 52.5% platinum. If the price for platinum is $1047/troy ounce, what is the cost of the platinum in 2.00 g of this compound? A troy ounce is equal to 480 grains, and one grain is equal to 64.8 mg.

CUMULATIVE REVIEW

123. You record the following in your lab book: a liquid is thick and has a density of 4.58 g/mL. Which data is qualitative? Which is quantitative?

WRITING IN ▶ Chemistry

124. Kilogram Standard Although the standard kilogram is stored at constant temperature and humidity, unwanted matter can build up on its surface. Scientists have been looking for a more reliable standard for mass. Research and describe alternative standards that have been proposed. Find out why no alternative standard has been chosen.

125. Units Research and report on unusual units of measurement such as bushels, pecks, firkins, and frails.

126. Product Volume Research the range of volumes used for packaging liquids sold in supermarkets.

127. Dosing Error In hospitals, medicines are given by dose. Find out what amount of error in the administered dose is acceptable for various medicines.

DBQ Document-Based Questions

Ocean Water *The density of pure water is 1.00 g/cm³ at 4°C. Ocean water is denser because it contains salt and other dissolved substances. The graph in* **Figure 21** *shows the relationships among temperature, density, and salinity versus depth for ocean water.*

Data obtained from: *Windows to the Universe*, at the University Corporation for Atmospheric Research (UCAR).

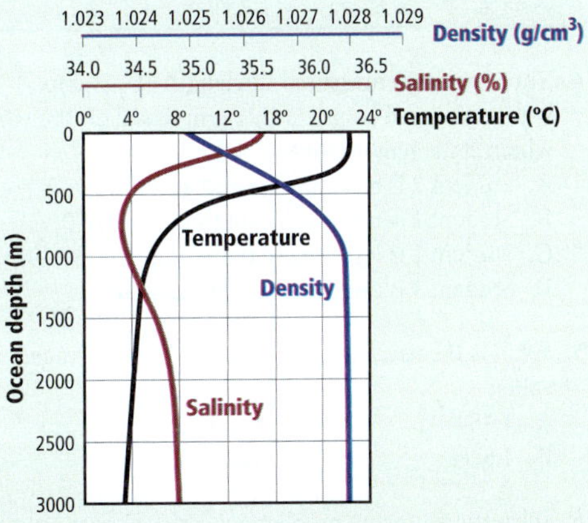

■ **Figure 21**

128. How is temperature related to the density of ocean water at depths less than 1000 m?

129. Describe the effect of depth on salinity.

130. Describe how salinity changes as the ocean water cools.

118. Mass itself has no meaning without a measurement of its volume. If the object is a pure substance and its mass and volume are known, its density can help identify it.

119. Surveyors are impartial and use equipment that is not affected by terrain or obstacles.

120. 0.43 g salt; 19%

121. Liquid A density = 1.23 g/mL
Liquid B density = 1.28 g/mL
Liquid C density = 1.71 g/mL
Liquid D density = 2.1 g/mL
From top to bottom the order of the liquids would be Liquid A, Liquid B, Liquid C, and Liquid D.

CHALLENGE PROBLEM

122. $35.30

CUMULATIVE REVIEW

123. Thick is qualitative; a density of 4.58 g/mL is quantitative.

WRITING IN ▶ Chemistry

Rubric

124. Two alternative methods of defining the standard kilogram would base the unit on the Avogadro constant, which is the number of atoms in 12 grams of pure carbon-12. One method would depend in part on X-ray measurements in silicon crystals. Another method would depend on electrical measurements that determine the ratio of the mechanical watt to the electrical watt. At this time, scientists have not gained universal acceptance for either alternative method.

125. Student answers will vary. For example, a firkin (a small wooden tub used for butter and lard) is a unit of volume equal to ¼ barrel.

126. Student answers will likely include fluid ounces, quarts, half-gallons, gallons, liters, and milliliters.

127. Student answers will vary. For definitive information on the subject, encourage students to contact the U.S. National Institute of Standards and Technology (NIST), manufacturers, pharmacists, or hospital pharmacies.

DBQ Document-Based Questions

Data obtained from Windows to the Universe, at the University Corporation for Atmospheric Research (UCAR).

128. The temperature is fairly stable through the first 200 meters, then decreases rapidly to a depth of 1000 m. As the temperature decreases, the density of the ocean increases. Below 1000 m the density of the ocean remains constant with a slight decrease in temperature.

129. The salinity rapidly decreases through the first 500 m, then increases with an increase in depth.

130. As the ocean water cools below 1000 m the salinity increases.

MULTIPLE CHOICE

1. C
2. C
3. C
4. A
5. B
6. D
7. D
8. D
9. D
10. B
11. C

MULTIPLE CHOICE

1. Which is NOT an SI base unit?
   A. second
   B. kilogram
   C. degree Celsius
   D. meter

2. Which value is NOT equivalent to the others?
   A. 500 m
   B. 0.5 km
   C. 5000 cm
   D. 5×10^{11} nm

3. What is the correct representation of 702.0 g in scientific notation?
   A. 7.02×10^3 g
   B. 70.20×10^1 g
   C. 7.020×10^2 g
   D. 70.20×10^2 g

Use the table below to answer Questions 4 and 5.

Measured Values for a Stamp's Length

	Student 1	Student 2	Student 3
Trial 1	2.60 cm	2.70 cm	2.75 cm
Trial 2	2.72 cm	2.69 cm	2.74 cm
Trial 3	2.65 cm	2.71 cm	2.64 cm
Average	2.66 cm	2.70 cm	2.71 cm

4. Three students measured the length of a stamp whose accepted length is 2.71 cm. Based on the table, which statement is true?
   A. Student 2 is both precise and accurate.
   B. Student 1 is more accurate than Student 3.
   C. Student 2 is less precise than Student 1.
   D. Student 3 is both precise and accurate.

5. What is the percent error for Student 1's averaged value?
   A. 1.48%
   B. 1.84%
   C. 3.70%
   D. 4.51%

6. Solve the following problem with the correct number of significant figures.

   $$5.31 \text{ cm} + 8.4 \text{ cm} + 7.932 \text{ cm}$$

   A. 22 cm
   B. 21.64 cm
   C. 21.642 cm
   D. 21.6 cm

7. Chemists found that a complex reaction occurred in three steps. The first step takes 2.5731×10^2 s to complete, the second step takes 3.60×10^{-1} s, and the third step takes 7.482×10^1 s. What is the total amount of time elapsed during the reaction?
   A. 3.68×10^1 s
   B. 7.78×10^1 s
   C. 1.37×10^1 s
   D. 3.3249×10^2 s

8. How many significant figures are there in a distance measurement of 20.070 km?
   A. 2
   B. 3
   C. 4
   D. 5

Use the graph below to answer Questions 9 and 10.

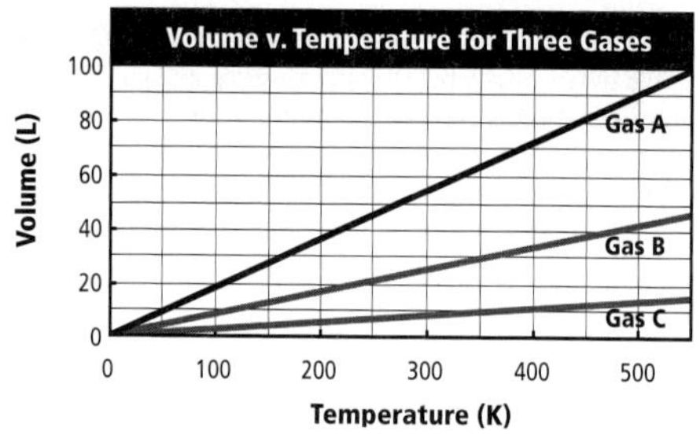

9. What volume will Gas A have at 450 K?
   A. 23 L
   B. 31 L
   C. 38 L
   D. 82 L

10. At what temperature will Gas B have a volume of 30 L?
    A. 170 K
    B. 350 K
    C. 443 K
    D. 623 K

11. Which is NOT a quantitative measurement of a pencil?
    A. length
    B. mass
    C. color
    D. diameter

Online Test Practice

SHORT ANSWER

Use the diagram below to answer Questions 12 and 13.

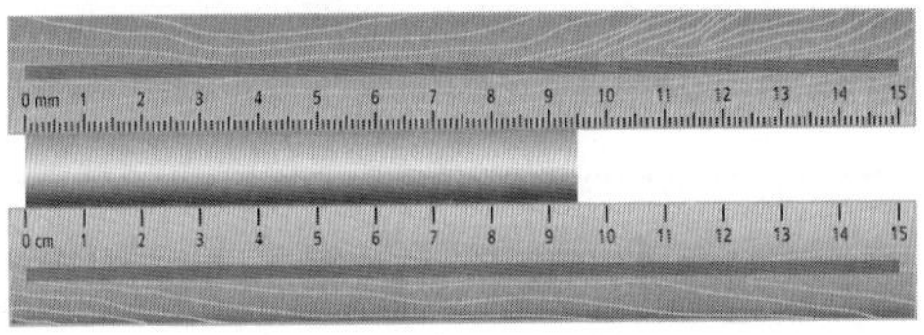

12. Explain which ruler you would use to make the more precise measurement.

13. What is the length of the rod using significant digits?

EXTENDED RESPONSE

Use the table below to answer Questions 14 to 16.

Temperature of a Solution While Heating

Time (s)	Temperature (°C)
0	22
30	35
60	48
90	61
120	74
150	87
180	100

14. A student recorded the temperature of a solution every 30 s for 3 min while the solution was heating on a Bunsen burner. Graph the data.

15. Show the setup to calculate the slope of the graph you created in Question 14.

16. Choose and explain two safety precautions the student should use with this experiment.

SAT SUBJECT TEST: CHEMISTRY

Use the graph below to answer Questions 17 to 20.

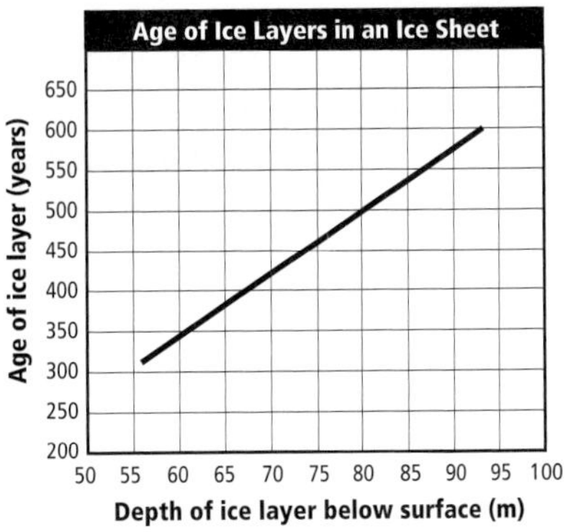

17. A student reported the age of an ice layer at 70 m as 425 years. The accepted value is 427 years. What is the percent error of the student's value?
- **A.** 0.468%
- **B.** 0.471%
- **C.** 1.00%
- **D.** 49.9%
- **E.** 99.5%

18. What is the approximate slope of the line?
- **A.** 0.00 m/y
- **B.** 0.13 m/y
- **C.** 0.13 y/m
- **D.** 7.5 m/y
- **E.** 7.5 y/m

19. What is the depth of an ice layer 450 years old?
- **A.** 74 m
- **B.** 75 m
- **C.** 76 m
- **D.** 77 m
- **E.** 78 m

20. What is the relationship between ice depth and age?
- **A.** linear, positive slope
- **B.** linear, negative slope
- **C.** linear, slope = 0
- **D.** nonlinear, positive slope
- **E.** nonlinear, negative slope

NEED EXTRA HELP?

If You Missed Question . . .	1	2	3	4	5	6	7	8	9	10	11	12	13	14	15	16	17	18	19	20
Review Section . . .	2.1	2.1	2.2	2.3	2.3	2.3	2.2	2.3	2.4	2.4	1.3	2.1	2.3	2.4	2.4	1.4	2.4	2.4	2.4	2.4

SHORT ANSWER

12. The top ruler allows more precise measurements because it has more divisions.

13. 9.50 mm (accept from 9.48 mm through 9.52 mm due to estimation)

EXTENDED RESPONSE

14. The graph should show a constant linear positive slope.

15. slope = Δ Temp/Δ Time
= (87°C − 74°C)/(150 s − 120 s)
= 0.43°C/s

16. Acceptable answers include wearing safety goggles, tying hair back, using hand protection, keeping flammable chemicals away, knowing the location of fire safety equipment.

SAT SUBJECT TEST: CHEMISTRY

17. A
18. E
19. A
20. A

CHAPTER 3 Matter—Properties and Changes

BIGIDEA **Everything is made of matter.**

ESSENTIAL QUESTIONS	RESOURCES TO ASSESS MASTERY
SECTION 1 Properties of Matter **1.** What characteristics identify a substance? **2.** What distinguishes physical properties from chemical properties? **3.** How do the properties of the physical states of matter differ? 1 session 0.5 block	**Progress Monitoring** Formative Assessment, pp. 71, 72 Reading Check, pp. 71, 72, 73, 74 Section Review, p. 75
SECTION 2 Changes in Matter **1.** What is a physical change and what are several common examples? **2.** What defines a chemical change? How can you tell that a chemical change has taken place? **3.** How does the law of conservation of mass apply to chemical reactions? 1 session 0.5 block	**Progress Monitoring** Formative Assessment, pp. 77, 78 Reading Check, p. 77 Section Review, p. 79
SECTION 3 Mixtures of Matter **1.** How do mixtures and substances differ? **2.** Why are some mixtures classified as homogeneous, while others are classified as heterogeneous. **3.** What are several techniques used to separate mixtures? 2 sessions 1 block	**Progress Monitoring** Formative Assessment, p. 80 Reading Check, p. 81 Section Review, p. 83
SECTION 4 Elements and Compounds **1.** What distinguishes elements from compounds? **2.** How is the periodic table organized? **3.** What are the laws of definite and multiple proportions and why are they important? 2 sessions 1 block	**Progress Monitoring** Formative Assessment, pp. 85, 90 Reading Check, pp. 85, 86, 87, 89 Graph Check, p. 89 Section Review, p. 90 **Summative Assessment** Chapter Assessment, p. 94 *eAssessment* Chapter Test (scaffolded)

LEVELED RESOURCES	LAB MATERIALS
Science Notebook 3.1 OL ***FAST FILE* Chapter Resources:** Study Guide, p. 68 BL	LaunchLAB p. 68 / **15 min** piece of zinc metal, large test tube, clamp, ring stand, 3*M* hydrochloric acid, graduated cylinder, wood splint, match
Science Notebook 3.2 OL ***FAST FILE* Chapter Resources:** ChemLab Worksheet, p. 53 OL Study Guide, p. 70 BL	
Science Notebook 3.3 OL ***FAST FILE* Chapter Resources:** MiniLab Worksheet, p. 52 OL Study Guide, p. 71 BL	MiniLAB p. 82 / **20 min** two 9-oz wide-mouth plastic cups, water, round filter paper, black water-soluble pen or marker, scissors, empty cup, 1/4 piece of 11-cm round filter paper
Science Notebook 3.4 OL ***FAST FILE* Chapter Resources:** ChemLab Worksheet, p. 53 OL Study Guide, p. 72 BL	ChemLAB p. 92 / **45 min** $AgNO_3$ solution, sandpaper, stirring rod, funnel, filter paper, 50-mL beaker, 50-mL graduated cylinder, 250-mL Erlenmeyer flask, small iron ring, ring stand, plastic petri dish, Bunsen burner, tongs, paper clip, copper wire

Go online!

ADDITIONAL RESOURCES

eAssessment
ConnectED Student Center
ConnectED Teacher Center
Lesson Presentations
eSolutions
Videos
Animations
Online Quizzes
Virtual Investigations
Multilingual eGlossary
Personal Tutors
WebQuests
Applying Practices Worksheets
PBLs

Practice:

Challenge Problems AL
Supplemental Problems BL OL
Chapter Test (Scaffolded)
Standardized Test Practice

Additional Resources:

Blueprints for Success
Solving Problems: A Chemistry Handbook

Lab Resources:

Laboratory Manual OL
CBL Laboratory Manual OL
Small-Scale Laboratory Manual OL
Forensics Laboratory Manual OL

BL Below Level   OL On Level   AL Advanced Learners   EL English Learners   COOP LEARN Cooperative Learning

CHAPTER 3

LaunchLAB

Rubric

How can you observe chemical change?

Purpose Students will examine several types of chemical changes.

Safety Precautions Review materials safety data sheet for HCl with students before doing the lab. Approve lab safety forms before work begins. Use gloves when handling HCl. **WARNING:** ***Use appropriate ventilation or a fume hood.***

Disposal The HCl should be neutralized with sodium bicarbonate, and then rinsed down a drain with a lot of water. The unreacted zinc can be reused. $ZnCl_2$ solution can be rinsed down a drain.

Teaching Strategies

- If you are using smaller test tubes, use about 3 to 5 mL of 6*M* HCl.

Expected Results The zinc reacts with hydrochloric acid to produce zinc chloride (which is soluble) and hydrogen gas (seen as bubbles). When a lighted splint is brought close to the mouth of the test tube, the hydrogen gas reacts with oxygen in the air to produce gaseous water. The reaction produces a pop.

CHAPTER 3

Matter—Properties and Changes

BIGIDEA Everything is made of matter.

SECTIONS

LaunchLAB

How can you observe chemical change?

Many objects in the everyday world do not change very much over time. However, when substances are mixed together, change is possible. In this lab, you will observe the changes that occur when two substances are combined.

FOLDABLES Study Organizer

Properties and Changes

Make a pocket book. Label it as shown. Use it to organize your study of chemical and physical changes and properties of matter.

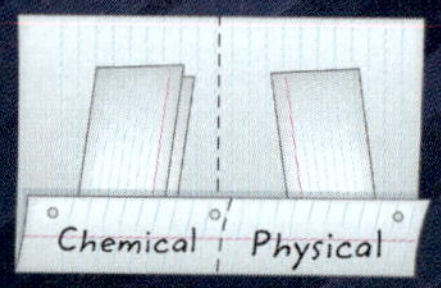

Procedure

1. Read and complete the lab safety form.
2. Place a **piece of zinc metal** in a **large test tube.**
3. Place the test tube in a **clamp** and attach the clamp to a **ring stand** so that the mouth of the test tube is pointing away from you.
4. Measure 10 mL of **3*M* hydrochloric acid** in a **graduated cylinder,** and place it on the benchtop. **WARNING:** ***HCl could cause burns and produce hazardous fumes.***
5. Light a **wood splint** with a **match.** Dispose of the match as directed by your teacher. Allow the wood to burn for 5 s, then blow out the flame to leave a glowing ember.
6. Place the glowing ember at the mouth of the test tube, then move the ember to the mouth of the graduated cylinder. Record your observations. **WARNING:** ***Be sure the test tube is facing away from you when the splint is brought near.***
7. Dispose of the ember as directed by your teacher.
8. Carefully pour the hydrochloric acid into the test tube.
9. Wait 1 min. Repeat Step 5.
10. Place the ember at the mouth of the test tube. Record your observations.

Analysis

1. **Describe** any changes you observed during the test. Bubbles were produced when the HCl was poured into the tube with zinc. Then a pop was produced when the wood splint was brought close to the tube.

About 70% of Earth's surface is covered with water. Water is the only common substance on Earth that exists naturally as a solid, a liquid, and a gas.

2. **Infer** What caused the bubbles to form when you added the hydrochloric acid to the zinc metal? The zinc reacted with the HCl, producing gas.

3. **Infer** What happened to the glowing ember in Step 10? Why did this not happen in Step 6? The pop was produced because the gas produced in the reaction reacted with the oxygen in the air. It did not happen in Step 6 because no gas was produced.

Inquiry Why did you wait before using the wood splint? Design an experiment to determine if the results vary over time. Answers will vary.

Introduce the BIGIDEA

Predictable Responses To introduce this chapter's Big Idea, lead students in a discussion about the fundamental building blocks of a pencil. Ask students to list four main parts of a pencil. Graphite, wood, eraser and a metal band. Ask students to state what each part of a pencil is made of or where it originates from. Graphite is made of carbon atoms, wood is from trees, the eraser originates from soybean oil, latex—from trees, and the metal band is either made out of aluminum (silvery color) or brass (yellow color). Everything in our universe is made of matter–the air, a desk, clothes, food, the human body, and the pencil. Because a pencil is comprised of graphite, wood, eraser and a metal band, all of these items are also made of matter. This chapter will focus on properties of matter, changes in matter, mixtures of matter, elements, and compounds.

Tie to Previous Knowledge

Have students review the following concepts before studying this chapter.

- matter, qualitative and quantitative aspects of chemistry
- significant figures

Use the Photo

States of Matter Ask students to identify phases of water that are presented in the figure. liquid and gas Ask students to relate temperature with the phase of water. At room temperature, water is a liquid and at higher temperatures—100°C or higher—water is a vapor. Ask students to give an example of another substance that exists in solid, liquid and gas phases. Example: CO_2(l)—under specific temperature and pressures— and CO_2(s)-dry ice; many other examples are acceptable. In Section 1 students will learn about the different phases of matter and their properties.

SECTION 1

1 Focus

MAINIDEA

Phase Changes Put 200 g of ice cubes in a beaker placed on a hot plate. Ask students to predict what will happen as the temperature increases from 0 to 100°C, at 1 atm. The frozen water will go through the three phases of matter. As the beaker is heated, the ice begins to melt. Once all the ice (solid) is converted to water (liquid), the temperature of the water will rise, and at 100°C and 1 atm the water will start to boil and the water vapor or steam (gas) will form. In general, density increases from gas to liquid to solid. However, water is an exception to this rule, as the solid phase is less dense than the liquid phase because of the open structure of ice. OL

2 Teach

Quick Demo

Solution or Mixture? Show students a beaker of distilled water and a beaker of unsaturated salt solution. Ask them if the beakers contain pure substances and how they would confirm their hypothesis. A visual inspection would lead students to believe both were pure substances. However, some students might realize that further tests would be necessary to confirm this. Students might suggest evaporation or the use of a conductivity apparatus to show that the second beaker is a mixture of substances. It is a solution. OL

SECTION 1

Properties of Matter

MAINIDEA Most common substances exist as solids, liquids, and gases, which have diverse physical and chemical properties.

Essential Questions

- What characteristics identify a substance?
- What distinguishes physical properties from chemical properties?
- How do the properties of the physical states of matter differ?

Review Vocabulary

density: a ratio that compares the mass of an object to its volume

New Vocabulary

states of matter
solid
liquid
gas
vapor
physical property
extensive property
intensive property
chemical property

CHEM 4 YOU Picture a glass of ice water. The ice floats, and you know the ice will eventually melt if left long enough at room temperature. When the water changes from solid to liquid, does the composition of the water also change?

Substances

As you know, matter is anything that has mass and takes up space. Everything around us is matter, including things that we cannot see, such as air and microbes. For example, table salt is a simple type of matter that you are probably familiar with. Table salt has a unique and unchanging chemical composition. Its chemical name is sodium chloride. It is always 100% sodium chloride, and its composition does not change from one sample to another. Salt harvested from the sea or extracted from a mine, as shown in **Figure 1,** always has the same composition and properties.

Recall that matter with a uniform and unchanging composition is called a substance, also known as a pure substance. Table salt is a pure substance. Another example of a pure substance is pure water. Water is always composed of hydrogen and oxygen. Seawater and tap water, on the other hand, are not pure substances because samples taken from different locations will often have different compositions. That is, the samples will contain different amounts of water, minerals, and other dissolved substances. Substances are important; much of your chemistry course will be focused on the composition of substances and how they interact with one another.

Figure 1 Whether harvested from the sea or extracted from a mine, salt always has the same composition.

Salt from the sea

Salt from a mine

(l)Luca Trovato/Getty Images, (r)Gabriela Insuratelu/iStock/Getty Images

CHEMISTRY PROJECT

Properties and Uses Have students use the *CRC Handbook of Chemistry and Physics* to research the physical and chemical properties of a particular element or compound. Then, have them write a summary relating how that particular property is useful for a product made with that element or compound. For example, copper's malleability, conductivity, and high melting point might make it ideal for cookware. OL

■ **Figure 2** A solid has a definite shape and does not take the shape of its container. Particles in a solid are tightly packed.

States of Matter

Imagine you are sitting on a bench, breathing heavily and drinking water after playing a game of soccer. You are in contact with three different forms of matter—the bench is a solid, the water is a liquid, and the air you breathe is a gas. In fact, all matter that exists naturally on Earth can be classified as one of these physical forms, which are called **states of matter.** Each of the three common states of matter can be distinguished by the way it fills a container. Scientists also recognize other states of matter. One of them is called plasma. It can occur in the form of lightning bolts and in stars.

☑ READING CHECK **Name** the common states of matter.

Solids A **solid** is a form of matter that has its own definite shape and volume. Wood, iron, paper, and sugar are all examples of solids. The particles of matter in a solid are tightly packed; when heated, a solid expands, but only slightly. Because its shape is definite, a solid might not conform to the shape of the container in which it is placed. If you place a rock into a container, the rock will not take the shape of the container, as shown in **Figure 2.** The tight packing of particles in a solid makes it incompressible; that is, it cannot be pressed into a smaller volume. It is important to understand that a solid is not defined by its rigidity or hardness. For instance, although concrete is rigid and wax is soft, they are both solids.

Liquids A **liquid** is a form of matter that flows, has constant volume, and takes the shape of its container. Common examples of liquids include water, blood, and mercury. The particles in a liquid are not rigidly held in place and are less closely packed than the particles in a solid. Liquid particles are able to move past each other. This property allows a liquid to flow and take the shape of its container, as shown in **Figure 3,** although it might not completely fill the container.

A liquid's volume is constant: regardless of the size and shape of the container in which the liquid is held, the volume of the liquid remains the same. Because of the way the particles of a liquid are packed, liquids are virtually incompressible. Like solids, however, liquids tend to expand when they are heated.

☑ READING CHECK **Compare** the properties of solids and liquids in terms of their particle arrangements.

■ **Figure 3** A liquid takes the shape of its container. Particles in a liquid are not held in place rigidly.

☑ READING CHECK The three states of matter are solid, liquid, and gas.

☑ READING CHECK The particles in solids are tightly packed. This arrangement gives solids their definite shapes and makes them incompressible. The particles in liquids are not rigidly held in place. This arrangement explains why liquids flow and take the shape of their containers.

ASSESSMENT

Knowledge Ask student to decide whether the following statements are true or false, and to support their choice with text references.

- A liquid conforms to its container. True; liquids take the shape of their container.
- A solid has definite shape but changeable volume. False; solids have definite shape and volume. OL

DIFFERENTIATED INSTRUCTION

Struggling Learners Distribute three plastic petri dishes and a collection of marbles or large-gauge metal shot. Have the students arrange the marbles or shot in the dishes to show the relative packing arrangement of solids, liquids, and gases. These can also be displayed on a projector to show the atomic motion in the three states of matter. BL EL

READING CHECK The term *gas* refers to a substance that is in the gaseous state at room temperature. The term *vapor* refers to the gaseous state of a substance that is solid or liquid at room temperature.

Problem-Solving LAB

Purpose Students will deduce the method by which the flow of gas out of a compressed tank is controlled.

Process Skills infer, predict, think critically, recognize cause and effect

Teaching Strategies

- Identify several real-world situations (outside a laboratory) where compressed gases are used.
- Identify control and safety features in these situations.

Think Critically

1. The flow of compressed gas must be controlled to control the amount and the rate at which gas is released.
2. Without the regulator device, the gas would rush out of the tank with a force powerful enough to transform it into a dangerous, uncontrolled projectile.

ASSESSMENT

Knowledge Have students research the breathing apparatus that a scuba diver uses and compare that equipment with compressed gas equipment in a laboratory. **OL**

Figure 4 Gases take the shape and volume of their containers. Particles in a gas are very far apart.

View an **animation about the three common states of matter.**

Concepts In Motion

Gases A **gas** is a form of matter that not only flows to conform to the shape of its container but also fills the entire volume of its container, as shown in **Figure 4.** If you flow gas into a container and close the container, the gas will expand to fill the container. Compared to solids and liquids, the particles of gases are far apart. Because of the significant amount of space between particles, gases are easily compressed.

You are probably familiar with the word *vapor* as it relates to the word *gas*. However, the words *gas* and *vapor*, while similar, do not mean the same thing, and should not be used interchangeably. The word *gas* refers to a substance that is naturally in the gaseous state at room temperature. The word **vapor** refers to the gaseous state of a substance that is a solid or a liquid at room temperature. For example, steam is a vapor because water exists as a liquid at room temperature.

READING CHECK **Differentiate** between gas and vapor.

Problem-Solving LAB

Recognize Cause and Effect

How is compressed gas released? Tanks of compressed gases are common in chemistry laboratories. For example, nitrogen is often flowed over a reaction in progress to displace other gases that might interfere with the experiment. Given what you know about gases, explain how the release of compressed nitrogen is controlled.

Analysis

The particles of gases are far apart, and gases tend to fill their containers—even if the container is a laboratory room. Tanks of compressed gas come from the supplier capped to prevent the gas from escaping. In the lab, a chemist or technician attaches a regulator to the tank and secures the tank to a stable fixture.

Think Critically

1. **Explain** why the flow of a compressed gas must be controlled for practical and safe use.
2. **Predict** what would happen if the valve on a full tank of compressed gas were suddenly opened all the way or if the tank were accidentally punctured.

DIFFERENTIATED INSTRUCTION

Visually Impaired Ask these students to describe physical properties that can be identified by non-visual means. Possible answers: solid—shape, size, texture, density, smell; liquid—thickness, slipperiness, smell; gas—smell **OL**

Table 1 Physical Properties of Common Substances

Substance	Color	State at 25°C	Melting Point (°C)	Boiling Point (°C)	Density (g/cm^3) at 25°C
Oxygen	colorless	gas	−219	−183	0.0013
Mercury	silver	liquid	−39	357	13.5
Water	colorless	liquid	0	100	0.997
Sucrose	white	solid	186	decomposes	1.58
Sodium chloride	white	solid	801	1465	2.17

Physical Properties of Matter

You are probably used to identifying objects by their properties—their characteristics and behavior. For example, you can easily identify a pencil in your backpack because you recognize its shape, color, weight, or some other property. These characteristics are all physical properties of the pencil. A **physical property** is a characteristic of matter that can be observed or measured without changing the sample's composition. Physical properties also describe pure substances. Because substances have uniform and unchanging compositions, they also have consistent and unchanging physical properties. Density, color, odor, hardness, melting point, and boiling point are common physical properties that scientists record as identifying characteristics of a substance. **Table 1** lists several common substances and their physical properties.

☑ READING CHECK **Define** *physical property* and provide examples.

Extensive and intensive properties Physical properties can be further described as being one of two types. **Extensive properties** are dependent on the amount of substance present. For example, mass is an extensive property. Length and volume are also extensive properties. Density, on the other hand, is an example of an intensive property of matter. **Intensive properties** are independent of the amount of substance present. For example, the density of a substance (at constant temperature and pressure) is the same no matter how much substance is present.

A substance can often be identified by its intensive properties. In some cases, a single intensive property is unique enough for identification. For instance, most of the spices shown in **Figure 5** can be identified by their scent.

RealWorld CHEMISTRY
Physical Properties

MINERALS Scientists use physical properties such as color and hardness to identify minerals. For instance, malachite is always green and relatively soft. Malachite was used as a pigment in paint and is now mainly used to make jewelry.

■ **Figure 5** Many spices can be identified by their scent, which is an intensive property.
Infer *Name an extensive property of one of the spices pictured.*

■ **Caption Question Fig. 5** The mass of the spices is an extensive property.

☑ READING CHECK A physical property is a characteristic that can be observed or measured without changing the sample's composition. Color, boiling point, and density are examples of physical properties.

Identify Misconceptions

Students might think that because extensive and intensive properties are both physical properties that they both depend on the amount of the substance being investigated.

Uncover the Misconception
Review the definitions of extensive and intensive properties. Reinforce the idea that the concepts of extensive and intensive properties are distinct.

Demonstrate the Concept
Show the students two pieces of the same type of wood that are obviously different lengths. Explain that length is an extensive physical property because it varies for the different amount of wood present. Place the pieces of wood in water and show how both float. Explain how you have proven that the density of the wood is intensive: it is the same for the long piece and the short piece because both pieces float regardless of the amount of wood in the water.

Assess New Knowledge
Have the students research physical properties of some common substances and create a table listing some extensive and some intensive properties. Ask the students to choose one property of a substance they investigated and explain why that property is extensive or intensive. Have them illustrate the concept with a demonstration. OL

FOLDABLES® Rubric

☑ READING CHECK Physical properties can be observed without changing the sample's composition. Chemical properties are not always obvious unless the substance reacts with other substances and changes its composition.

Quick Demo

Observe Properties Show students some copper turnings or foil and have them list the observable properties of these items. Roll the copper into a ball small enough to fit into a crucible. Determine the mass of the crucible with the copper in it. Then, heat the copper strongly for 5 to 10 minutes and have students make observations again. Ask them to predict what has happened to the mass, then weigh the copper oxide product. (When finished, CuO can be thrown away.) Ask students to distinguish between physical and chemical properties that they have witnessed. OL

■ **Figure 6** One of the physical properties of copper is that it can be shaped into different forms, such as the wires on circuit boards. The fact that copper turns from reddish to green when reacting with substances in the air is a chemical property.

FOLDABLES® Incorporate information from this section into your Foldable.

Chemical Properties of Matter

Some properties of a substance are not obvious unless the substance has changed composition as a result of its contact with other substances or the application of thermal or electric energy. The ability or inability of a substance to combine with or change into one or more other substances is called a **chemical property.**

Iron forming rust when combined with the oxygen in air is an example of a chemical property of iron. Similarly, the inability of a substance to change into another substance is also a chemical property. For example, when iron is placed in nitrogen gas at room temperature, no chemical change occurs.

☑ READING CHECK **Compare** physical and chemical properties.

Watch a video about the **composition of matter.** Video

Observing Properties of Matter

Every substance has its own unique set of physical and chemical properties. **Figure 6** shows physical and chemical properties of copper. Copper can be shaped into different forms, which is a physical property. When copper is in contact with air for a long time, it reacts with the substances in the air and turns green. This is a chemical property. **Table 2** lists several physical and chemical properties of copper.

Table 2 Properties of Copper

Physical Properties	Chemical Properties
• reddish brown, shiny • easily shaped into sheets (malleable) and drawn into wires (ductile) • a good conductor of heat and electricity • density = 8.96 g/cm^3 • melting point = 1085°C • boiling point = 2562°C	• forms green copper carbonate compound when in contact with moist air • reacts with nitric acid and sulfuric acid, forming new substances • one type of compound forms a deep-blue solution when in contact with ammonia

Video Lab

Video

Matter is Made up of Atoms

CHEMISTRY JOURNAL

Life on Other Planets Ask students to write a creative story about noncarbon-based life on another planet. As they research the elements that make up the life forms of their planet, they should identify whether the properties are physical or chemical. If the properties are physical, students should identify them as either extensive or intensive. OL

■ **Figure 7** Because the density of ice is lower than the density of water, icebergs float on the ocean.

British Antarctic Survey/Photo Researchers

Properties and states of matter The properties of copper listed in **Table 2** might vary depending on the conditions under which they are observed. Because the particular form, or state, of a substance is a physical property, changing the state introduces or adds another physical property to its characteristics. It is important to state the specific conditions, such as temperature and pressure, under which observations are made because both physical and chemical properties depend on these conditions. Resources that provide tables of physical and chemical properties of substances, such as the *CRC Handbook of Chemistry and Physics*, generally include the physical properties of substances in all of the states in which they can exist.

Consider the properties of water, for example. You might think of water as a liquid (physical property) which is not particularly chemically reactive (chemical property). You might also find that water has a density of 1.00 g/cm^3 (physical property). These properties, however, apply only to water at standard temperature and pressure. At temperatures greater than 100°C, water is a gas (physical property) with a density of about 0.0006 g/cm^3 (physical property) that reacts rapidly with many different substances (chemical property). Below 0°C, water is a solid (physical property) with a density of about 0.92 g/cm^3 (physical property). The lower density of ice accounts for the fact that icebergs float on the ocean, as shown in **Figure 7.** Clearly, the properties of water are dramatically different under different conditions.

VOCABULARY

ACADEMIC VOCABULARY

Resource
a source of information or expertise
The library contains resources for many different subjects.

SECTION 1 REVIEW

Section Self-Check

Section Summary

- The three common states of matter are solid, liquid, and gas.
- Physical properties can be observed without altering a substance's composition.
- Chemical properties describe a substance's ability to combine with or change into one or more new substances.
- External conditions can affect both physical and chemical properties.

1. MAINIDEA **Create** a table that describes the three common states of matter in terms of their shape, volume, and compressibility.
2. **Describe** the characteristics that identify a sample of matter as a substance.
3. **Classify** each of the following as a physical or a chemical property.
   - **a.** Iron and oxygen form rust.
   - **b.** Iron is more dense than aluminum.
   - **c.** Magnesium burns brightly when ignited.
   - **d.** Oil and water do not mix.
   - **e.** Mercury melts at −39°C.
4. **Organize** Create a chart that compares physical and chemical properties. Give two examples for each type of property.

3 Assess

Check for Understanding

Fill a 50-mL graduated cylinder with 25 mL of water. Fill another graduated cylinder with 25 mL of alcohol (methyl, ethyl, or isopropyl can be used). Ask students to predict what will occur when the alcohol is poured into the 50-mL cylinder containing the water and the liquids are mixed. To dispose of alcohols, evaporate under a hood. Students might think the volumes will add to a full 50 mL of alcohol–water mixture; however, the volume is not additive. Molecules of alcohol slip in between the molecules of water. Reinforce this by filling a beaker to the top with marbles and ask students if more material can be added. Then, add sand or water to demonstrate. OL

Reteach

Have students act out the packing of the different states of matter by "becoming" atoms themselves. Isolate a particular area of the classroom as the stage so that students can visualize the relative packing of atoms in the different states. BL EL

Extension

Have students research examples of new materials being developed by various chemical manufacturers. Ask students to relate the properties of the new materials to the arrangement of the atoms or molecules as solids, liquids, or gases. OL

SECTION 1 REVIEW

1. The table should list solid (definite volume, definite shape, incompressible); liquid (definite volume, fills container shape, virtually incompressible); gas (fills volume of container, takes shape of container, compressible).
2. The sample of matter must have a uniform and unchanging composition to be a substance.
3. **a.** chemical
   b. physical
   c. chemical
   d. physical
   e. physical
4. The chart should make clear that physical properties can be observed without changing the composition of the sample, which is not the case for chemical properties. Mass and density are examples of physical properties. Fermentation and rusting are examples of chemical properties.

SECTION 2

1 Focus

MAINIDEA

Chemical or Physical Change
Have students research mercury poisoning in humans. Indicate sources of mercury, and whether the mercury in the body undergoes a physical or a chemical change. The source of mercury poisoning might be from a broken thermometer or methyl mercury. The latter is found when a pollutant in water is eaten by fish; and the fish is eaten by humans. Because mercury bioaccumulates in the fish, humans who eat fish that have high concentrations of methyl mercury are also at high risk of mercury poisoning. In mercury poisoning, the mercury reacts with enzymes in the human body and leads to an irreversible activity of the enzymes which can be fatal. The mercury in the body undergoes a chemical reaction because it binds to enzymes. **OL**

2 Teach

Quick Demo

Changing States of Matter
WARNING: ***Perform this demo under a hood.*** Place several mothballs in a beaker. Place a half-filled flask of cold water on top to serve as a condensing "lid." Heat the mothballs until they have melted, and ask students to observe and diagram the changes. Remove the heat source and ask students to again observe the changes. With heat, students will see the change from solid to liquid, solid to vapor, and liquid to vapor. On cooling, students will see to change from liquid to solid and vapor to solid, or vapor to liquid. **OL**

SECTION 2

Essential Questions

- What is a physical change and what are several common examples?
- What defines a chemical change? How can you tell that a chemical change has taken place?
- How does the law of conservation of mass apply to chemical reactions?

Review Vocabulary

observation: orderly, direct information gathering about a phenomenon

New Vocabulary

physical change
phase change
chemical change
law of conservation of mass

Changes in Matter

MAINIDEA Matter can undergo physical and chemical changes.

CHEM 4 YOU Charcoal is a black solid. As it burns, it glows red. Eventually it ends up as ashes, carbon dioxide, and water. The dramatic changes to charcoal's appearance result from an important chemical property – the ability to burn.

Physical Changes

A substance often undergoes changes that result in a dramatically different appearance yet leave the composition of the substance unchanged. An example is the crumpling of aluminum foil. While the foil goes from a smooth, flat, mirrorlike sheet to a round, compact ball, the actual composition of the foil is unchanged—it is still aluminum. A change such as this, which alters a substance without changing its composition, is a **physical change.** Cutting a sheet of paper and breaking a crystal are other examples of physical changes in matter.

Phase change As with other physical properties, the state of matter depends on the temperature and pressure of the surroundings. As temperature and pressure change, most substances undergo a change from one state (or phase) to another. A **phase change** is a transition of matter from one state to another.

Connection to Earth Science **The water cycle** This is the case with the water cycle, which allows life to exist on Earth. At atmospheric pressure and at temperatures of 0°C and below, water is in its solid state, which is known as ice. As heat is added to the ice, it melts and becomes liquid water. This change of state is a physical change because even though ice and water have different appearances, they have the same composition. If the temperature of the water increases to 100°C, the water begins to boil and liquid water is converted to steam. Melting and formation of a gas are both physical changes and phase changes. **Figure 8** shows condensation and solidification, two common phase changes. Terms such as *boil, freeze, condense, vaporize,* or *melt* in chemistry generally refer to a phase change in matter.

■ **Figure 8** Condensation can occur when a gas is in contact with a cool surface, causing droplets to form. Solidification occurs when a liquid cools. Water dripping from the roof forms icicles as it cools.

Demonstration

Atoms in Motion

Purpose
to observe atomic motion and the formation of an alloy, brass

Materials
New pennies (3); zinc (2 g); NaOH (5 g); porcelain evaporating dish; forceps; paper towels; hot plate

Safety Precautions

Disposal Decant the cooled NaOH solution from the solid zinc. Neutralize the NaOH and rinse down a drain. Dry the zinc in an evaporating dish and reuse. Do not allow the zinc plating solution to come in contact with flammable materials.

Procedure
Dissolve 5 g NaOH in 25 mL of water in an evaporating dish. **WARNING: *NaOH can burn; avoid skin contact.*** Rinse spills with

The temperature and pressure at which a substance undergoes a phase change are important physical properties. These properties are called the melting and boiling points of the substance. Look again at **Table 1** to see this information for several common substances. Like density, the melting and boiling points are intensive physical properties that can be used to identify unknown substances. Tables of intensive properties, such as those given at the end of this textbook or in the *CRC Handbook of Chemistry and Physics,* are useful tools in identifying unknown substances from experimental data.

Watch a video about **wastewater treatment.**

Video

Chemical Changes

A process that involves one or more substances changing into new substances is called a **chemical change,** commonly referred to as a chemical reaction. The new substances formed in the reaction have different compositions and different properties from the substances present before the reaction occurred. For example, the formation of rust when iron reacts with oxygen in moist air is a chemical change. Rust, shown in **Figure 9,** is a chemical combination of iron and oxygen.

In chemical reactions, the starting substances are called reactants, and the new substances that are formed are called products. Terms such as *decompose, explode, rust, oxidize, corrode, tarnish, ferment, burn,* or *rot* generally refer to chemical reactions.

FOLDABLES® Incorporate information from this section into your Foldable.

☑ READING CHECK **Define** *chemical change.*

Evidence of a chemical reaction As **Figure 9** shows, rust is a brownish-orange powdery substance that looks very different from iron and oxygen. Rust is not attracted to a magnet, whereas iron is. The observation that the product (rust) has different properties than the reactants (iron and oxygen) is evidence that a chemical reaction has taken place. A chemical reaction always produces a change in properties. Spoiled food, such as rotten fruit and bread, is another example of chemical reactions. The properties of spoiled food, like its taste and its digestibility, differ from fresh food. Examples of food that have undergone chemical reactions are shown in **Figure 9.**

Figure 9 When iron rusts and food rots, new substances are formed due to chemical change.
Identify *the reactants and the products in the formation of rust.*

Conservation of Mass

It was only in the late eighteenth century that scientists began to use quantitative tools to study and monitor chemical changes. The analytical balance, which was capable of measuring small changes in mass, was developed at that time. By carefully measuring mass before and after many chemical reactions, it was observed that, although chemical changes occurred, the total mass involved in the reaction remained constant. Assuming this was true for all reactions, chemists summarized this observation in a scientific law. The **law of conservation of mass** states that mass is neither created nor destroyed during a chemical reaction—it is conserved. In other words, the mass of the reactants equals the mass of the products. The equation form of the law of conservation of mass is as follows.

The Law of Conservation of Mass

$$\text{mass}_{\text{reactants}} = \text{mass}_{\text{products}}$$

Mass is conserved in a chemical reaction; products have the same mass as reactants.

☑ READING CHECK A chemical change is the change of one or more substances into new substances.

Caption Question Fig. 9 Iron and oxygen are the reactants. Rust is the product.

Enrichment

Chemists in History The first true chemist is considered by many to be Robert Boyle (1627–1691), known for studying the relationship between gas pressure and volume.

The German chemist Georg Stahl (1660–1734) was also interested in gases, and particularly in combustion. He believed that when a substance burned, "phlogiston" flowed out of the material.

The English scientist Joseph Priestley (1773–1804) is credited with the discovery of the active role of oxygen in combustion. He called oxygen *dephlogisticated* air.

Antoine Lavoisier, often thought of as the father of modern chemistry, showed that mass was conserved in combustion reactions involving oxygen. It was Lavoisier who named the gas that supports combustion *oxygene,* meaning *generator of acid.* (Oxygen was originally believed to be an important part of all acids.)

Have students research and report on one of these early pioneers. OL

FOLDABLES® Rubric

What's CHEMISTRY Got To Do With It?

Video

Open the Faucet

water. Wear goggles and apron. Add 2 g zinc to the NaOH solution. Place on the hot plate and heat to just below boiling. Keep a penny as a control. Immerse two pennies in the NaOH-zinc plating bath. After one minute, turn the pennies over. Keep them immersed until silver in color. Remove the coins from the bath, rinse with cold water, and pat dry with paper towels. Remove the dish from the hot plate. Place one of the plated pennies on the hot surface. The plated coin will quickly return to its copper color, then turn golden. Remove it from the hot plate and allow it to cool. Show students the three pennies.

Results

The control penny remains copper colored. The zinc-plated penny is silver colored. The third penny is gold colored.

Analysis

Describe the brass alloy. Brass is a solid composed of the metallic elements copper and zinc.

ASSESSMENT

Knowledge Ask students why heat is needed to cause the zinc coating to diffuse into the copper. Heat increases the motion of the atoms and separates and expands the layers of atoms, thus allowing them to migrate more easily. OL

ASSESSMENT

Performance The reaction of iron with oxygen costs American industry a great deal of money. Have the students choose a particular industry (such as agriculture, transportation, or construction) and give an oral report on how the industry deals with the problem of rust. OL

IN-CLASS Example

Question In a catalytic converter, found in car exhaust systems, carbon monoxide (CO) reacts with oxygen gas (O_2) to form carbon dioxide gas (CO_2). If CO reacts with 16 g of O_2 to form 44 g of CO_2, what is the mass of CO in the catalytic converter?

Answer $mass_{reactants} = mass_{products}$
$m_{CO} + m_{O_2} = m_{CO_2}$
16 g + x = 44 g,
x = 44 g − 16 g = 28 g of CO

PRACTICE Problems

Have students refer to the Selected Solutions appendix for complete solutions to odd-numbered problems. The complete solutions for all problems can be found in the Solutions Manual.

5. 91.5 g of bromine reacted and 101.8 g of aluminum bromide were formed.
6. 89.4 g
7. 24.1 g of chlorine gas is used in the reaction. Because the sodium reacts with excess chlorine, all of the sodium (15.6 g) is used in the reaction.
8. 6.6 g
9. 156.3 g − 106.5 g = 49.8 g. Yes. Mass of reactants equals mass of products.

EXAMPLE Problem 1

Find help with **algebraic equations.** Math Handbook

EXAMPLE PROBLEM

CONSERVATION OF MASS In an experiment, 10.00 g of red mercury(II) oxide powder is placed in an open flask and heated until it is converted to liquid mercury and oxygen gas. The liquid mercury has a mass of 9.26 g. What is the mass of oxygen formed in the reaction?

1 ANALYZE THE PROBLEM

You are given the mass of a reactant and the mass of one of the products in a chemical reaction. According to the law of mass conservation, the total mass of the products must equal the total mass of the reactants.

Known
$m_{mercury(II)\ oxide} = 10.00$ g
$m_{mercury} = 9.26$ g

Unknown
$m_{oxygen} = ?$ g

Get help with **writing equations.** Personal Tutor

2 SOLVE FOR THE UNKNOWN

$Mass_{reactants} = Mass_{products}$ — State the law of conservation of mass.

$m_{mercury(II)\ oxide} = m_{mercury} + m_{oxygen}$

$m_{oxygen} = m_{mercury(II)\ oxide} - m_{mercury}$ — Solve for m_{oxygen}.

$m_{oxygen} = 10.00\ g - 9.26\ g$ — Substitute $m_{mercury(II)\ oxide} = 10.00$ g and $m_{mercury} = 9.26$ g.

$m_{oxygen} = 0.74\ g$

3 EVALUATE THE ANSWER

The sum of the masses of the two products equals the mass of the reactant, verifying that mass has been conserved. The answer is correctly expressed to the hundredths place, making the number of significant digits correct.

PRACTICE Problems

Do additional problems. Online Practice

PRACTICE PROBLEMS

5. Use the data in the table to answer the following questions.

Aluminum and Liquid Bromine Reaction		
Substance	**Before Reaction**	**After Reaction**
Aluminum	10.3 g	0.0 g
Liquid bromine	100.0 g	8.2 g
Compound	0.0 g	

How many grams of bromine reacted? How many grams of compound were formed?

6. From a laboratory process designed to separate water into hydrogen and oxygen gas, a student collected 10.0 g of hydrogen and 79.4 g of oxygen. How much water was originally involved in the process?
7. A student carefully placed 15.6 g of sodium in a reactor supplied with an excess quantity of chlorine gas. When the reaction was complete, the student obtained 39.7 g of sodium chloride. Calculate how many grams of chlorine gas reacted. How many grams of sodium reacted?
8. A 10.0-g sample of magnesium reacts with oxygen to form 16.6 g of magnesium oxide. How many grams of oxygen reacted?
9. **Challenge** 106.5 g of HCl(g) react with an unknown amount of NH_3(g) to produce 156.3 g of NH_4Cl(s). How many grams of NH_3(g) reacted? Is the law of conservation of mass observed in the reaction? Justify your answer.

CHEMISTRY PROJECT

Antoine Lavoisier Have students research Lavoisier's law of conservation of mass and the influence it had on the chemistry of his day and in the decades after his death. In particular, ask students to find out how his findings influenced the work of Joseph Proust and John Dalton, and how the combined works of these three chemists remain relevant today. Students can describe their findings in a written report or present them orally to the class. OL

CHEMISTRY JOURNAL

The Conservation of Energy The law of conservation of matter, formalized by Lavoisier, was a major achievement of the eighteenth century. However, it is really an incomplete statement without considering it with the law of conservation of energy. Have students research and write a short report on the experiments of the twentieth century that lead to the formulation of the law of conservation of energy. OL

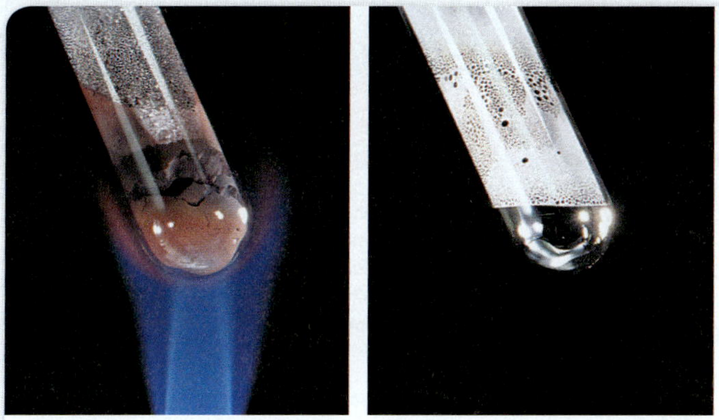

View an **animation about the conservation of mass.**

Concepts In Motion

■ **Figure 10** When mercury(II) oxide is heated, it reacts to form liquid mercury and oxygen gas. The sum of the masses of liquid mercury and oxygen gas produced during the reaction equals the mass of the mercury(II) oxide.

French scientist Antoine Lavoisier (1743–1794) was one of the first to use an analytical balance to monitor chemical reactions. He studied the thermal decomposition of mercury(II) oxide, known then as *calx of mercury*. Mercury(II) oxide, shown in **Figure 10,** is a powdery red solid. When it is heated, the red solid reacts to form silvery liquid mercury and colorless oxygen gas.

The color change and production of a gas are indicators of a chemical reaction. When the reaction occurs in a closed container, the oxygen gas cannot escape and the mass before and after the reaction can be measured. The masses will be the same. The law of conservation of mass is one of the most fundamental concepts of chemistry.

SECTION 2 REVIEW

Section Self-Check

Section Summary

- A physical change alters the physical properties of a substance without changing its composition.
- A chemical change, also known as a chemical reaction, involves a change in a substance's composition.
- In a chemical reaction, reactants form products.
- The law of conservation of mass states that mass is neither created nor destroyed during a chemical reaction; it is conserved.

10. MAINIDEA **Classify** each example as a physical change or a chemical change.
   a. crushing an aluminum can
   b. recycling used aluminum cans to make new aluminum cans
   c. aluminum combining with oxygen to form aluminum oxide

11. Describe the results of a physical change and list three examples of physical change.

12. Describe the results of a chemical change. List four indicators of chemical change.

13. Calculate Solve each of the following.
   a. In the complete reaction of 22.99 g of sodium with 35.45 g of chlorine, what mass of sodium chloride is formed?
   b. A 12.2-g sample of *X* reacts with a sample of *Y* to form 78.9 g of *XY*. What is the mass of *Y* that reacted?

14. Evaluate A friend tells you, "Because composition does not change during a physical change, the appearance of a substance does not change." Is your friend correct? Explain.

Stephen Frisch/McGraw-Hill Education

3 Assess

Check for Understanding

In a small beaker, place a scoop of baking soda and add 10 mL of vinegar. Invert a larger beaker over the top of the smaller beaker to capture the carbon dioxide–water that is produced. In a different small beaker, pour several mL of club soda and place on a hotplate. Invert a larger beaker over the top to capture the carbon dioxide–water that is produced. Ask the students how the first experiment is different from the second. The first experiment generates carbon dioxide through a chemical change, the second through a physical change. **OL**

Reteach

Set up a beaker of water and another beaker of ethanol. Place several ice cubes in each beaker. Ask students to relate their observations to the physical states of the matter involved. The ice will float on the water because the solid is less dense than the liquid; however, ice is more dense than liquid ethanol (0.7893 g/cm^3) and will sink. To dispose of ethanol, evaporate under a hood. **BL** **EL**

Extension

If you have an automatic ice-cube maker in your freezer, you may have noticed that the older ice cubes at the bottom of the tray are smaller than the newer ones at the top of the tray. Why does this happen? The process of sublimation, through which a solid becomes a gas without passing through the liquid phase, is the reason. **AL**

SECTION 2 REVIEW

10. a. physical
   b. physical
   c. chemical

11. During a physical change, a substance is altered, but its composition does not change. Examples will vary but might include changes such as melting, freezing, boiling, bending, and tearing.

12. During a chemical change, the composition of a substance is altered. Possible indicators of chemical change include a change in color, odor, or temperature, and the formation of a gas or solid from a liquid.

13. a. 58.44 g of sodium chloride
   b. 66.7 g of Y

14. The statement is incorrect. While the composition does not change, a change in appearance often accompanies a physical change.

SECTION 3

1 Focus

MAINIDEA

Chemical Composition Have students research the chemical composition of gasoline and distinguish between the octane ratings. Have students include three charts–one for each gasoline grade–showing the isooctane and heptane compositions. Typical composition of gasoline is as follows: 30–50% straight chain aliphatic hydrocarbons (such as heptane); 30–50% branched aliphatic hydrocarbons (such as isooctane); 20–30% cyclic hydrocarbons (such as cyclopentane); 20–30% aromatic hydrocarbons (such as ethyl benzene). Isooctane is assigned the octane number of 100 and the heptane is assigned the octane number of 0. Gasoline with an assigned octane number 87 (low-grade) is a mixture of 87% isooctane and 13% heptane; octane number 89 (mid-grade) is a mixture of 89% isooctane and 11% heptane; and octane number 92 (premium) is a mixture of 92% isooctane and 8% heptane. OL

2 Teach

ASSESSMENT

Performance Have the students look around the room and generate a list of items that are mixtures and items that are substances. Students might need to review the definition of *substance* from chapter 1. Have them identify the criteria they are using for classification and the rough percentages of each item in each category. Answers will vary but there are usually higher percentages of materials that are classified as mixtures. OL

SECTION 3

Mixtures of Matter

MAINIDEA Most everyday matter occurs as mixtures—combinations of two or more substances.

Essential Questions

- How do mixtures and substances differ?
- Why are some mixtures classified as homogeneous, while others are classified as heterogeneous?
- What are several techniques used to separate mixtures?

Review Vocabulary

substance: a form of matter that has a uniform and unchanging composition; also known as a pure substance

New Vocabulary

mixture
heterogeneous mixture
homogeneous mixture
solution
filtration
distillation
crystallization
sublimation
chromatography

CHEM 4 YOU That familiar hiss when you open a soft-drink bottle is the sound of gas escaping. You might have noticed that when you leave the bottle open, eventually most of bubbles are gone and the drink is flat. But the soft drink remains sweet no matter how long you leave the bottle open.

Mixtures

You have already read that a pure substance has a uniform and unchanging composition. What happens when two or more subtances are combined? A **mixture** is a combination of two or more pure substances in which each pure substance retains its individual chemical properties. The composition of mixtures is variable, and the number of mixtures that can be created by combining substances is infinite. Although much of the focus of chemistry is the behavior of substances, it is important to remember that most everyday matter occurs as mixtures. Substances tend to mix naturally; it is difficult to keep any substance pure.

Two mixtures are shown in **Figure 11.** Although you cannot distinguish between the components of the mercury-silver mixture in **Figure 11a,** you can separate them by heating the mixture. The mercury will evaporate before the silver does, and you will obtain two separate substances: mercury vapor and solid silver. The mercury and silver physically mixed to form the mixture but did not chemically react with each other. They could be separated by the physical method of boiling. When oil, seasonings, and vinegar are mixed, as shown in **Figure 11b,** the substances are in contact, but they do not react. In fact, you can still distinguish all of the substances. If the mixture remains undisturbed long enough, the oil will form a layer on top of the vinegar.

Figure 11 There are different types of mixtures. **a.** It is not possible to see the different components of some mixtures, such as this mercury-silver filling. **b.** The components of other types of mixtures are visible, as in this salad dressing.

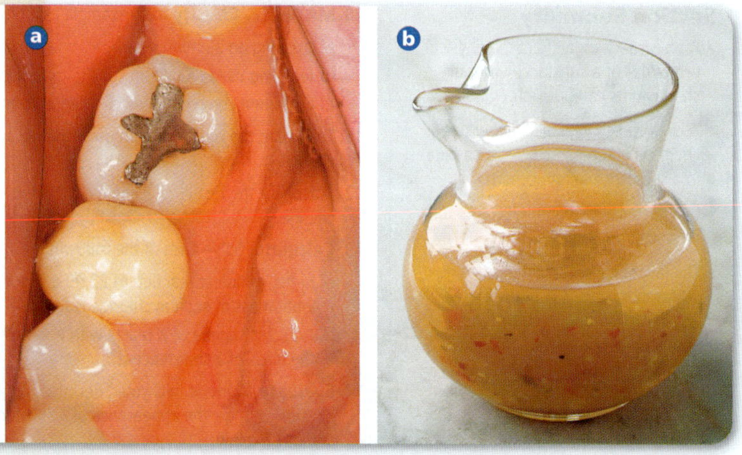

DIFFERENTIATED INSTRUCTION

Visually Impaired Set up stations with examples of heterogeneous and homogeneous mixtures that can be discerned by touch. Partner visually impaired students with sighted students and have them rotate through the stations to classify each type of mixture. BL EL COOP LEARN

Types of mixtures The combinations of pure substances shown in **Figure 11** are both mixtures, despite their obvious visual differences. Mixtures can be defined in different ways and are classified as either heterogeneous or homogeneous.

A **heterogeneous mixture** is a mixture that does not blend smoothly throughout and in which the individual substances remain distinct. The salad dressing mixture is an example of a heterogeneous mixture. Its composition is not uniform–the substances have not blended smoothly and remain distinct. In another example, fresh-squeezed orange juice is a heterogeneous mixture of juice and pulp. The pulp component floats in the juice component. We can therefore say that the existence of two or more distinct areas indicates a heterogeneous mixture.

A **homogeneous mixture** is a mixture that has constant composition throughout; it always has a single phase. If you cut two pieces out of a silver mercury amalgam, their compositions will be the same. They will contain the same relative amounts of silver and mercury, no matter the size of each piece.

☑ READING CHECK **Compare and contrast** heterogeneous and homogeneous mixtures. Give examples of each.

Homogeneous mixtures are also referred to as **solutions.** You are probably most familiar with solutions in a liquid form, such as tea and lemonade, but solutions can be solids, liquids, or gases. They can be a mixture of a solid and a gas, a solid and a liquid, a gas and a liquid, and so on. **Table 3** lists the various types of solution systems and examples. Each solution system in the table is also represented in **Figure 12.**

The solid-solid solution known as steel is called an alloy. An alloy is a homogeneous mixture of metals, or a mixture of a metal and a nonmetal in which the metal substance is the major component. For instance, steel is a mixture of iron and carbon. Adding carbon atoms increases the hardness of the metal.

Manufacturers combine the properties of various metals in an alloy to achieve greater strength and durability in their products. Jewelry is often made of alloys such as bronze, sterling silver, pewter, and 14-karat gold.

Explore **solution systems with an interactive table.** Concepts In Motion

Table 3 Types of Solution Systems

System	Example
Gas-gas	Air in a scuba tank is primarily a mixture of nitrogen, oxygen, and argon gases.
Gas-liquid	Oxygen and carbon dioxide are dissolved in seawater.
Liquid-gas	Moist air exhaled by the scuba diver contains water droplets.
Liquid-liquid	When it is raining, fresh water mixes with seawater.
Solid-liquid	Solid salts are dissolved in seawater.
Solid-solid	The air tank is made of an alloy—a mixture of two metals.

Tane Casserley NOAA/NMS Maritime Heritage Program

VOCABULARY
WORD ORIGIN
Mixture
from the Latin word *misceo,* meaning *to mix* or *blend*

CAREERS IN CHEMISTRY

Materials Scientist Materials scientists synthesize new materials and analyze their properties. They work in national laboratories, in industry, and in academia. For example, scientists at NASA developed new aluminum-silicon alloys that can be employed to build lighter and stronger engines.

WebQuest

Figure 12 All types of solution systems are represented in this photo.

☑ READING CHECK Individual substances remain distinct in a heterogeneous mixture. A homogeneous mixture has the same composition throughout and always appears as a single phase. Examples will vary but may include salad dressing (heterogeneous), water and oil mixture (heterogeneous), air (homogeneous), and tap water (homogeneous).

CAREERS IN CHEMISTRY

WebQuest

Quick Demo

The Tyndall Effect The Tyndall effect is a quick test that differentiates a solution from a colloid, a heterogeneous mixture with particles that never settle. Shine a thin beam of light (a laser is ideal) through a beaker containing a solution such as copper sulfate or sodium chloride; the beam will not be visible because of the small size of the solute particles (smaller than 10^{-6} mm). However, when you shine light through a mixture such as a weak milk colloid or clay suspension, the particles are large enough (1×10^{-6} mm to 1×10^{-3} mm and greater than 1×10^{-3} mm, respectively) to refract the beam; thus, the beam is plainly visible.

CHEMISTRY JOURNAL

Separating a Mixture Pass around a small sandwich bag containing a mixture, such as small pieces of foam, sand, copper sulfate, and iron filings. Ask students to design a branching flowchart that shows how the mixture could be separated without physically picking out the pieces. They might use the *CRC Handbook of Chemistry and Physics* to develop their separation methods. Approve each design and approve lab safety forms before work begins. After they have submitted their flowcharts, they can carry out the experiment. Have students write their results in their journals. OL

MiniLAB

Rubric

Purpose Students will separate a mixture using paper chromatography.

Process Skills apply concepts, collect and interpret data, compare and contrast, observe and infer

Safety Precautions Approve lab safety forms before work begins. Be sure students wear safety goggles and aprons.

Teaching Strategies
- Remind students to handle the filter paper as little as possible. The oil from their fingers will distort the results.
- Any size of round filter paper will work. Green and yellow dyes separate best if the paper is relatively large in diameter.
- Use several different types of black felt pens and black markers. Each brand and type has different ink combinations.

Expected Results As the water spreads out on the paper, different dyes in the ink will spread out from the center and be deposited on the filter paper at different distances from the center.

Analysis
1. Drawings should show the filter paper with the ink spot in the center and different dyes spreading out from the center.
2. Different components of the ink have varying attraction for the filter paper. Therefore, the colors that comprise the ink will be deposited at different distances from the center of the paper.
3. Answers will vary. Different makes and types of black ink have different dyes in them.

■ **Figure 13** As the mixture passes through the filter, the solids remain in the filter, while the filtrate (the remaining liquid) is collected in the beaker.

Separating Mixtures

Most matter exists naturally in the form of mixtures. To gain a thorough understanding of matter, it is important to be able to separate mixtures into their component substances. Because the substances in a mixture are physically combined, the processes used to separate a mixture are physical processes that are based on differences in the physical properties of the substances. For instance, a mixture of iron and sand can be separated into its components with a magnet because a magnet will attract iron but not sand. Numerous techniques have been developed that take advantage of different physical properties in order to separate various mixtures.

Filtration Heterogeneous mixtures composed of solids and liquids are easily separated by filtration. **Filtration** is a technique that uses a porous barrier to separate a solid from a liquid. As **Figure 13** shows, the mixture is poured through a piece of filter paper that has been folded into a cone shape. The liquid passes through, leaving the solids trapped in the filter paper.

Distillation Most homogeneous mixtures can be separated by distillation. **Distillation** is a physical separation technique that is based on differences in the boiling points of the substances involved. In distillation, a mixture is heated until the substance with the lowest boiling point boils to a vapor that can then be condensed into a liquid and collected. When precisely controlled, distillation can separate substances that have boiling points differing by only a few degrees.

MiniLAB

Observe Dye Separation

How does paper chromatography allow you to separate substances? Chromatography is an important diagnostic tool used by chemists and forensic technicians to separate and analyze substances.

Procedure

1. Read and complete the lab safety form.
2. Fill a **9-oz wide-mouth plastic cup** with **water** to about 2 cm from the top. Wipe off any water drops on the lip of the cup.
3. Place a piece of **round filter paper** on a clean, dry surface. Make a concentrated ink spot in the center of the paper by firmly pressing the tip of a **black water-soluble pen or marker** onto the paper.
4. Use **scissors** or another sharp object to create a small hole, about the diameter of a pen tip, in the center of the ink spot.
WARNING: *Sharp objects can puncture skin.*
5. Roll one quarter of an **11-cm round filter paper** into a tight cone. This will act as a wick to draw the ink. Work the pointed end of the wick into the hole in the center of the round filter paper.
6. Place the paper/wick apparatus on top of the cup of water, with the wick in the water. The water will move up the wick and outward through the round paper.
7. When the water has moved to within about 1 cm of the edge of the paper (about 20 min), carefully remove the paper from the water-filled cup and put it on a second empty **cup.**

Analysis

1. **Record** the number of distinct dyes you can identify on a drawing of the round filter paper. Label the color bands.
2. **Infer** why you see different colors at different locations on the filter paper.
3. **Compare** your chromatogram with those of your classmates. Explain any differences you might observe.

CHEMISTRY PROJECT

Colloids A mixture such as low-fat milk looks like a solution, but it is not. Low-fat milk is a type of mixture called a colloid. Have students conduct library research on types of colloids and their uses. They should write a detailed description of each type, and include information about the sources they consulted. OL

What's CHEMISTRY Got To Do With It?

Video

Hair: A Chemical Storehouse

Crystallization Making rock candy from a sugar solution is an example of separation by crystallization. **Crystallization** is a separation technique that results in the formation of pure solid particles of a substance from a solution containing the dissolved substance. When the solution contains as much dissolved substance as it can possibly hold, the addition of even a tiny amount more often causes the dissolved substance to come out of solution and collect as crystals on any available surface.

In the rock candy example, as water evaporates from the sugar-water solution, the solution becomes more concentrated. This is equivalent to adding more of the dissolved substance to the solution. As more water evaporates, the sugar forms a solid crystal on the string, as shown in **Figure 14.** Crystallization produces highly pure solids.

Sublimation Mixtures can also be separated by **sublimation,** which is the process during which a solid changes to vapor without melting, i.e. without going through the liquid phase. Sublimation can be used to separate two solids present in a mixture when one of the solids sublimates but not the other.

Chromatography **Chromatography** is a technique that separates the components of a mixture dissolved in either a gas or a liquid (called the mobile phase) based on the ability of each component to travel or to be drawn across the surface of a fixed substrate (called the stationary phase). For example, chromatography paper is a stationary phase with a solid substrate. During paper chromatography, the separation occurs because the various components of the mixture in the liquid mobile phase spread through the paper at different rates. Components with the strongest attraction for the paper travel slower.

Figure 14 As the water evaporates from the water-sugar solution, the sugar crystals form on the string.

Watch a video about **analyzing chemicals in hair.**

Video

SECTION 3 REVIEW

Section Self-Check

Section Summary

- A mixture is a physical blend of two or more pure substances in any proportion.
- Solutions are homogeneous mixtures.
- Mixtures can be separated by physical means. Common separation techniques include filtration, distillation, crystallization, sublimation, and chromatography.

15. MAINIDEA **Classify** each of the following as either a heterogeneous or a homogeneous mixture.

a. tap water **b.** air **c.** raisin muffin

16. Compare mixtures and substances.

17. Describe the separation technique that could be used to separate each of the following mixtures.

a. two colorless liquids
b. a nondissolving solid mixed with a liquid
c. red and blue marbles of the same size and mass

18. Design a concept map that summarizes the relationships among matter, elements, mixtures, compounds, pure substances, and homogeneous and heterogeneous mixtures.

3 Assess

Check for Understanding

Set out four unknowns in bottles: three of the bottles are solutions (salt water, sugar water, a very dilute milk mixture) and one is a pure substance (distilled water). Ask the students to determine which of the unknowns is a pure substance. OL

Reteach

Have teams of students develop a concept map on an acetate sheet or large sheet of paper. The map should show the relationships among the following terms: matter, substance, heterogeneous mixture, solid, liquid, gas, vapor, physical change, chemical change, homogeneous mixture, solution, alloy. OL COOP LEARN

Extension

Suspend samples of solutions, colloids, and suspensions in dialysis bags or tubing (with the ends tied) in beakers of distilled water. Ask students to classify the samples and support their classifications with diagrams or writings. Solutions have particles small enough to fit through the pores of a dialysis tube. This will be evident by the coloration of the distilled water outside the dialysis tubing (if you are using a species with color, such as copper sulfate) or by testing with a conductivity apparatus (if you are using a colorless species, such as sodium chloride). OL EL

SECTION 3 REVIEW

15. a. homogeneous
b. homogeneous
c. heterogeneous

16. Substances have a constant composition, mixtures do not. Each substance in a mixture retains its own properties, whereas the properties of a substance are different from those of the elements that comprise it.

17. a. distillation
b. filtration
c. manually separating the marbles by color

18. Concept map should show that matter is made of pure substances and mixtures. Mixtures are further divided into homogeneous and heterogeneous mixtures. Pure substances are further divided into elements and compounds. Refer to Figure 19.

SECTION 4

1 Focus

MAINIDEA

Gas in Air Have students research gases that are found in air and list the elements that are present in each compound. Air is a mixture of gases mainly nitrogen (N_2) and oxygen (O_2). Many other gases are present in smaller amounts—such as carbon dioxide (CO_2), helium (He), and argon (Ar). OL

2 Teach

Visual Learning

Periodic Table Have the students use the periodic table to determine the number of elements that are gaseous, liquid, solid, metallic, and radioactive at room temperature. OL

SECTION 4

Elements and Compounds

MAINIDEA **A compound is a combination of two or more elements.**

Essential Questions

- What distinguishes elements from compounds?
- How is the periodic table organized?
- What are the laws of definite and multiple proportions and why are they important?

Review Vocabulary

proportion: the relation of one part to another or to the whole with respect to quantity

New Vocabulary

element
periodic table
compound
law of definite proportions
percent by mass
law of multiple proportions

CHEM 4 YOU **When you eat fruit salad, you can separate your favorite fruit from the others. However, the jelly in your sandwich cannot be easily separated into its ingredients—fruit juice, sugar, and pectin. Similarly, compounds cannot be separated into their component atoms by physical processes.**

Elements

Earlier in this chapter, you considered the diversity of your surroundings in terms of matter. Although matter can take many different forms, all matter can be broken down into a relatively small number of basic building blocks called elements. An **element** is a pure substance that cannot be separated into simpler substances by physical or chemical means. On Earth, over 90 elements occur naturally. Copper, oxygen, and gold are examples of naturally occurring elements. There are also several elements that do not exist naturally but have been developed by scientists.

Each element has a unique chemical name and symbol. The chemical symbol consists of one, two, or three letters; the first letter is always capitalized, and the remaining letter(s) are always lowercase. The names and symbols of the elements are universally accepted by scientists in order to make the communication of chemical information possible.

The naturally occurring elements are not equally abundant. For example, hydrogen is estimated to make up approximately 75% of the mass of the universe. Oxygen and silicon together comprise almost 75% of the mass of Earth's crust, while oxygen, carbon, and hydrogen account for more than 90% of the human body. Francium, on the other hand, is one of the least-abundant naturally-occurring elements. There is probably less than 20 g of francium dispersed throughout Earth's crust. Elements are found in different physical states in normal conditions, as shown in **Figure 15.**

Figure 15 In normal conditions, elements exist in different states.

Copper pot—solid

Mercury switch—liquid

Helium balloon—gas

(l)Barry Mason/Alamy; (c)Tony Freeman/PhotoEdit; (r)Randy Brooke/AP Images

DIFFERENTIATED INSTRUCTION

Struggling Learners Choose several elements for students to study. Have students make flash cards with the name of the element on one side and its corresponding symbol on the other side. Provide some class time for students to quiz each other using the flash cards. BL EL

Tabelle I.

Typische Elemente							
			K = 39	Rb = 85	Cs = 133	—	—
			Ca = 40	Sr = 87	Ba = 137	—	—
			—	?Yt = 88?	?Di = 138?	Er = 178?	—
			Ti = 48?	Zr = 90	Ce = 140?	?La = 180?	Th = 231
			V = 51	Nb = 94	—	Ta = 182	—
			Cr = 52	Mo = 96	—	W = 184	U = 240
			Mn = 55	—	—	—	—
			Fe = 56	Ru = 104	—	Os = 195?	—
			Co = 59	Rh = 104	—	Ir = 197	—
			Ni = 59	Pd = 106	—	Pt = 198?	—
H = 1	Li = 7	Na = 23	Cu = 63	Ag = 108	—	Au = 199?	—
	Be = 9,4	Mg = 24	Zn = 65	Cd = 112	—	Hg = 200	—
	B = 11	Al = 27,3	—	In = 113	—	Tl = 204	—
	C = 12	Si = 28	—	Sn = 118	—	Pb = 207	—
	N = 14	P = 31	As = 75	Sb = 122	—	Bi = 208	—
	O = 16	S = 32	Se = 78	Te = 125?	—	—	—
	F = 19	Cl = 35,5	Br = 80	J = 127	—	—	—

Figure 16 Mendeleev was one of the first scientists to organize elements in a periodic manner, as shown in this chart, and to observe periodic patterns in the properties of the elements.

Science Museum/SSPL/The Image Works

A first look at the periodic table As many new elements were being discovered in the early nineteenth century, chemists began to observe and study patterns of similarities in the chemical and physical properties of particular sets of elements. In 1869, Russian chemist Dmitri Mendeleev (1834–1907) devised a chart, shown in **Figure 16,** which organized all of the elements that were known at the time. His classification was based on the similarities and masses of the elements. Mendeleev's table was the first version of what has been further developed into the periodic table of the elements. The **periodic table** organizes the elements into a grid of horizontal rows called periods and vertical columns called groups or families. Elements in the same group have similar chemical and physical properties. The table is called periodic because the pattern of similar properties repeats from period to period. The periodic table can be found at the end of this book and will be examined in great detail throughout your chemistry courses.

Compounds

Many pure substances can be classified as compounds. A **compound** is made up of two or more different elements that are combined chemically. Most matter in the universe exists in the form of compounds.

Today, there are more than 50 million known compounds, and new compounds continue to be developed and discovered at the rate of about 100,000 per year. There appears to be no limit to the number of compounds that can be made or that will be discovered. Considering this virtually limitless potential, several organizations have assumed the task of collecting data and indexing the known chemical compounds. The information is stored in databases.

READING CHECK **Define** *element* and *compound.*

The chemical symbols of the periodic table make it easy to write the formulas for chemical compounds. For example, table salt, which is called sodium chloride, is composed of one part sodium (Na) and one part chlorine (Cl), and its chemical formula is NaCl. Water is composed of two parts hydrogen (H) and one part oxygen (O), and its chemical formula is H_2O. The subscript 2 indicates that two hydrogen elements combine with one oxygen element to form water.

VOCABULARY

SCIENCE USAGE V. COMMON USAGE

Element

Science usage: a pure substance that cannot be separated into simpler substances by ordinary chemical means

Lead is one of the heaviest elements.

Common usage: the state or sphere that is natural or suited to any person or thing

In snow, huskies are in their element.

READING CHECK An element is a pure substance that cannot be separated into simpler substances by physical or chemical means. A compound is made of two or more different elements that are combined chemically.

Reinforcement

Element Symbols Reinforce students' knowledge of elemental symbols by playing "Bowling for Elements." Insert ten element flashcards in pin-like formation in pockets of a large cardboard or on the board. Divide the class into teams and have individuals take turns naming as many of the element "pins" by moving from the lead pin to any contiguous pin. Keep score as in bowling. BL EL COOP LEARN

Extension

Scientific Collaboration Dmitri Mendeleev (1834–1907) was a delegate at the First International Chemistry Congress in 1860, and from that experience he began the undertaking that would immortalize him in the history of chemistry. Have students use the Internet and other resources to learn about the meetings for chemists and scientists that take place today. OL

ASSESSMENT

Skill Have students use the *CRC Handbook of Chemistry and Physics* (section on element information) to research the following sample questions about the elements.

- How many elements are named for people? Countries? Cities or states?
- How many elements are named for mythological figures or gods?
- What town has four elements named for it?
- What elements were known to ancient humans?
- What elements were discovered before 1800? Between 1801 and 1900? Between 1901 and the present?

Ask students to choose one particular category and portray the information graphically. OL

CHEMISTRY PROJECT

Elementary Patterns Have students research the work of scientists who, like Mendeleev, worked in the area of assigning meaningful patterns among the elements. Some of these scientists include Dobereiner, Newlands, Meyer, Moseley, and Seaborg. OL

☑ READING CHECK During electrolysis, one end of a long platinum electrode is exposed to the water in the tube and the other end is attached to a power source. An electric current splits water into hydrogen gas in the compartment on the right and oxygen gas in the compartment on the left. Because water is composed of two parts hydrogen and one part oxygen, there is twice as much hydrogen gas than oxygen gas.

■ **Caption Question Fig. 17** 2:1

Quick Demo

Electrolysis of Water Using a Hoffman apparatus, show the electrolysis of water emphasizing the constant 2:1 ratio of the volumes of hydrogen to oxygen. Drain off a test-tube amount of each gas and perform a lighted splint test on the hydrogen (it barks) and a glowing splint test on the oxygen (it relights). **OL**

Enrichment

Resident Experts Have each student "adopt" an element and prepare a brochure on the properties, uses, abundance, method of mining and purifying, costs, and other pertinent information. Students can present a report per class, per week, or simply be "on-call" as the resident expert on that element when the discussion merits it. **OL**

■ **Figure 17** An electric current breaks down water into its components, oxygen and hydrogen.
Determine *What is the ratio between the amount of hydrogen and the amount of oxygen released during electrolysis?*

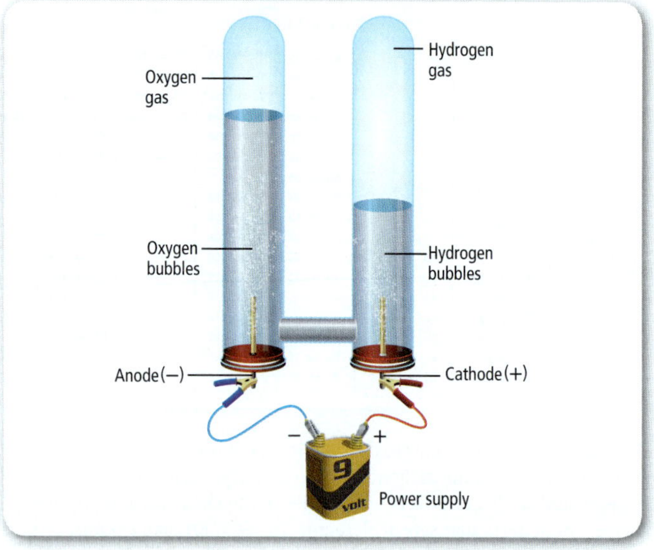

Separating compounds into components As you have read earlier in this chapter, elements can never be separated into simpler substances. However, compounds can be broken down into simpler substances by chemical means. In general, compounds that occur naturally are more stable than the individual component elements. Separating a compound into its elements often requires external energy, such as heat or electricity. **Figure 17** shows the setup used to produce the chemical change of water into its component elements—hydrogen and oxygen—through a process called electrolysis. During electrolysis, one end of a long platinum electrode is exposed to the water in a tube and the other end is attached to a power source. An electric current splits water into hydrogen gas in the compartment on the right and oxygen gas in the compartment on the left. Because water is composed of two parts hydrogen and one part oxygen, there is twice as much hydrogen gas as there is oxygen gas.

☑ READING CHECK **Explain** the process of electrolysis.

■ **Figure 18** When potassium and iodine react, they form potassium iodide, a compound with different properties.

Properties of compounds The properties of a compound are different from those of its component elements. The example of water in **Figure 17** illustrates this fact. Water is a stable compound that is liquid at room temperature. When water is broken down, its components, hydrogen and oxygen, are dramatically different than the liquid they form when combined. Oxygen and hydrogen are colorless, odorless gases that undergo vigorous chemical reactions with many elements. This difference in properties is a result of a chemical reaction between the elements. **Figure 18** shows the component elements—potassium and iodine—of the compound called potassium iodide. Note how different the properties of potassium iodide are from its component elements. Potassium is a light silver metal that reacts with water. Iodine is a black solid that changes into a purple gas at room temperature. Potassium iodide is a white salt.

CULTURAL DIVERSITY

The History of Elements Our understanding of the elements has changed drastically in the past 2500 years. The ancient Greek philosophers believed that all the elements of the universe were combinations of water, air, earth, and fire. Aristotle suggested a fifth element, aether, of which all the heavens were made. The philosopher Democritus (c. 470–380 B.C.) was the first to believe that all matter consisted of tiny particles and coined the term *atomos* meaning *indivisible*. Egyptian practices in chemistry combined with Greek philosophy and mathematics to form the science of *khemia*. The Arabs inherited these traditions and placed the *al*-before the name, forming the pseudoscience *alchemy*. Alchemists sought to find a mysterious substance that could be taken apart and recombined to form more precious elements, such as

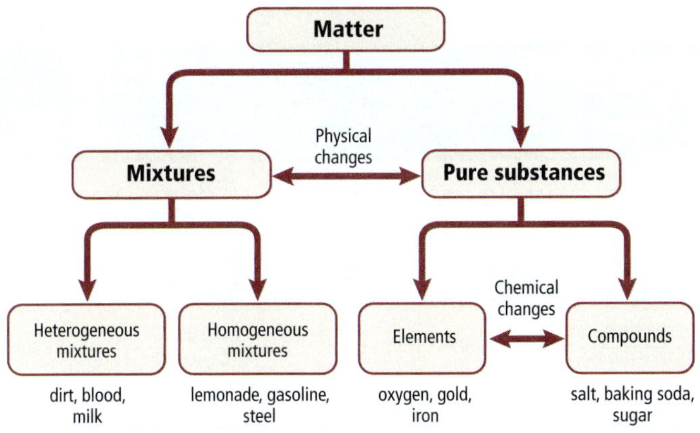

■ **Figure 19** Matter can be classified into different categories that have defined properties.

Examine ***How are mixtures and substances related? Elements and compounds?***

Recall what you have read about the organization of matter. You know that matter is classified as pure substances and mixtures. As you learned in the previous section, a mixture can be homogeneous or heterogeneous. You also know that an element is a pure substance that cannot be separated into simpler substances, whereas a compound is a chemical combination of two or more elements and can be separated into its components. Use **Figure 19** to review the classification of matter and how its components are related to each other.

☑ READING CHECK **Summarize** the different types of matter and how they are related to each other.

Law of Definite Proportions

An important characteristic of compounds is that the elements comprising them always combine in definite proportions by mass. This observation is so fundamental that it is summarized as the law of definite proportions. The **law of definite proportions** states that a compound is always composed of the same elements in the same proportion by mass, no matter how large or small the sample. The mass of the compound is equal to the sum of the masses of the elements that make up the compound.

The relative amounts of the elements in a compound can be expressed as percent by mass. The **percent by mass** is the ratio of the mass of each element to the total mass of the compound expressed as a percentage.

Percent by Mass

$$\text{percent by mass (\%)} = \frac{\text{mass of element}}{\text{mass of compound}} \times 100$$

Percent by mass is obtained by dividing the mass of the element by the mass of the compound and then by multiplying this ratio by 100 to express it as a percentage.

☑ READING CHECK **State** the law of definite proportions.

gold. Although the practice of alchemy was a dead end in itself, much useful chemical information and processes were discovered. Early Chinese philosophers believed that all materials cycled through five fundamental stages: earth, wood, fire, metal, and water. Each step of the cycle led to a state of greater order. For instance, the earth allowed a tree to grow, which in turn could be chopped down and burned to transform metals. In the heating process, condensation provided the water to renew the cycle.

Math in Chemistry

Percent by Mass

Have students practice using the percent-by-mass equation by determining the percent by mass of oxygen in water (H_2O). First, students should determine the mass of water. Using the periodic table, students should determine that each hydrogen atom has a mass of 1.00 and each oxygen atom has a mass of 16.0. (Note: the mass units used on the periodic table have not been introduced yet. You can introduce the units here or use the exercise to point out that the units will ultimately cancel.) The mass of water equals:
$(2 \text{ atoms} \times 1.00) + (1 \text{ atom} \times 16.0) = 18$
Use the formula to calculate the percent by mass of oxygen.
percent by mass (%) $= (16.0/18.0) \times 100 = 88.9\%$

Have students use the same method to calculate the percent by mass of oxygen in hydrogen peroxide (H_2O_2). 94% **OL**

☑ READING CHECK Matter is divided into mixtures and substances, which are related to each other through physical changes. Mixtures can be homogeneous or heterogeneous. Substances are divided into elements and compounds, which are related to each other through chemical changes.

☑ READING CHECK The law of definite proportions states that a compound is always composed of the same elements in the same proportion by mass, no matter how large or small the sample.

■ **Caption Question Fig. 19**
Mixtures and substances are related through physical changes—mixtures are created by physically combining substances. Elements and compounds are related through chemical changes—compounds are made of elements that are chemically bonded to one another.

Quick Demo

Percent Composition

WARNING: ***Perform this demo under a hood.*** Measure the mass of any two similar-looking organic compounds of low melting point (sucrose and lactic acid, for example) in separate crucibles, and record the mass on the board. Heat each one strongly to eliminate the hydrogen and oxygen in the form of water until all that is left is pure carbon. Reweigh the crucibles with the carbon residue, and calculate the percentage of carbon in each compound. Ask students if the two compounds could be the same. No, each has a different percentage of carbon. Ask students if the percentage of carbon had been the same in each compound, would that prove they were in fact the same compound. No, not necessarily, because the percentages of the other elements in the compound (hydrogen and oxygen in this case) could be different. **OL**

PRACTICE Problems

Have students refer to the Selected Solutions appendix for complete solutions to odd-numbered problems. The complete solutions for all problems can be found in the Solutions Manual.

19. 15.9%

20. 5.0%

21. 25% *X* and 75% *Y*

22. compound I: mass percentage hydrogen $= 15.0\text{ g} / (15.0\text{ g} + 120.0\text{ g}) = 11.1\%$; compound II: mass percentage hydrogen $= 2.0\text{ g} (2.0\text{ g} + 32.0) = 5.9\%$ Because the mass compositions of the compounds are different, the compounds themselves must be different.

23. No, you cannot be sure. Having the same mass percentage of a single element does not guarantee that the composition of each compound is the same.

Table 4 Sucrose Analysis

	20.00 g of Granulated Sugar		500.0 g of Sugarcane	
Element	**Analysis by Mass (g)**	**Percent by Mass (%)**	**Analysis by Mass (g)**	**Percent by Mass (%)**
Carbon	8.44	$\frac{8.44\text{ g C}}{20.00\text{ g sucrose}} \times 100 = 42.20\%$	211.0	$\frac{211.0\text{ g C}}{500.0\text{ g sucrose}} \times 100 = 42.20\%$
Hydrogen	1.30	$\frac{1.30\text{ g H}}{20.00\text{ g sucrose}} \times 100 = 6.50\%$	32.5	$\frac{32.50\text{ g H}}{500.0\text{ g sucrose}} \times 100 = 6.500\%$
Oxygen	10.26	$\frac{10.26\text{ g O}}{20.00\text{ g sucrose}} \times 100 = 51.30\%$	256.5	$\frac{256.5\text{ g O}}{500.0\text{ g sucrose}} \times 100 = 51.30\%$
Total	20.00	100%	500.0	100%

Get help with **percentages.**

Personal Tutor

For example, consider the compound granulated sugar (sucrose). This compound is composed of three elements—carbon, hydrogen, and oxygen. The analysis of 20.00 g of sucrose from a bag of granulated sugar is given in **Table 4.** Note that the sum of the individual masses of the elements found in the sugar equals 20.00 g, which is the amount of the granulated sugar sample that was analyzed. This demonstrates the law of conservation of mass as applied to compounds: the mass of a compound is equal to the sum of the masses of the elements that make up the compound.

Suppose you analyzed 500.0 g of sucrose from a sample of sugarcane. The analysis is shown in **Table 4.** The percent-by-mass values for the sugarcane are equal to the values obtained for the granulated sugar. According to the law of definite proportions, samples of a compound from any source must have the same mass proportions. Conversely, compounds with different mass proportions must be different compounds. Thus, you can conclude that samples of sucrose will always be composed of 42.20% carbon, 6.50% hydrogen, and 51.30% oxygen, no matter their sources.

PRACTICE Problems

Do additional problems. Online Practice

19. A 78.0-g sample of an unknown compound contains 12.4 g of hydrogen. What is the percent by mass of hydrogen in the compound?

20. 1.0 g of hydrogen reacts completely with 19.0 g of fluorine. What is the percent by mass of hydrogen in the compound that is formed?

21. If 3.5 g of element X reacts with 10.5 g of element Y to form the compound XY, what is the percent by mass of element X in the compound? The percent by mass of element Y?

22. Two unknown compounds are tested. Compound I contains 15.0 g of hydrogen and 120.0 g of oxygen. Compound II contains 2.0 g of hydrogen and 32.0 g of oxygen. Are the compounds the same? Explain your answer.

23. Challenge All you know about two unknown compounds is that they have the same percent by mass of carbon. With only this information, can you be sure the two compounds are the same? Explain.

CHEMISTRY PROJECT

Same Elements, Major Differences Have students use the *CRC Handbook of Chemistry and Physics* to look up the properties of compounds that exemplify the law of multiple proportions (for example, copper(I) chloride and copper(II) chloride; hydrogen peroxide and water). Have them write a brief paper or make an oral report discussing the distinctions between the two compounds. **OL**

Law of Multiple Proportions

Compounds composed of different elements are obviously different compounds. However, different compounds can also be composed of the same elements. This happens when those different compounds have different mass compositions. The **law of multiple proportions** states that when different compounds are formed by a combination of the same elements, different masses of one element combine with the same fixed mass of the other element in a ratio of small whole numbers. Ratios compare the relative amounts of any items or substances. The comparison can be expressed using numbers separated by a colon or as a fraction. With regard to the law of multiple proportions, ratios express the relationship of elements in a compound.

☑ READING CHECK **State** the law of multiple proportions in your own words.

Water and hydrogen peroxide The two distinct compounds water (H_2O) and hydrogen peroxide (H_2O_2) illustrate the law of multiple proportions. Each compound contains the same elements (hydrogen and oxygen). Water is composed of two parts hydrogen and one part oxygen. Hydrogen peroxide is composed of two parts hydrogen and two parts oxygen. Hydrogen peroxide differs from water in that it has twice as much oxygen. When you compare the mass of oxygen in hydrogen peroxide to the mass of oxygen in water, you get the ratio 2:1.

Compounds made of copper and chlorine In another example, copper (Cu) reacts with chlorine (Cl) under different sets of conditions to form two different compounds. **Table 5** provides an analysis of their compositions. The two copper compounds must be different because they have different percents by mass. Compound I contains 64.20% copper; Compound II contains 47.27% copper. Compound I contains 35.80% chlorine; Compound II contains 52.73% chlorine.

Using **Figure 20** and **Table 5**, compare the ratio of the mass of copper to the mass of chlorine for each compound. Notice that the mass ratio of copper to chlorine in Compound I (1.793) is exactly 2 times the mass ratio of copper to chlorine in Compound II (0.8964).

$$\frac{\text{mass ratio of Compound I}}{\text{mass ratio of Compound II}} = \frac{1.793 \text{ g Cu/g Cl}}{0.8964 \text{ g Cu/g Cl}} = 2.000$$

☑ GRAPH CHECK **Explain** why the ratio of the relative masses of copper in both compounds is 2:1.

Table 5 Analysis Data of Two Copper Compounds

Compound	% Cu	% Cl	Mass Cu (g) in 100.0 g of Compound	Mass Cl (g) in 100.0 g of Compound	Mass Ratio $\left(\frac{\text{mass Cu}}{\text{mass Cl}}\right)$
I	64.20	35.80	64.20	35.80	$\frac{1.793 \text{ g Cu}}{1 \text{ g Cl}}$
II	47.27	52.73	47.27	52.73	$\frac{0.8964 \text{ g Cu}}{1 \text{ g Cl}}$

Figure 20 Copper and chlorine can form different compounds.

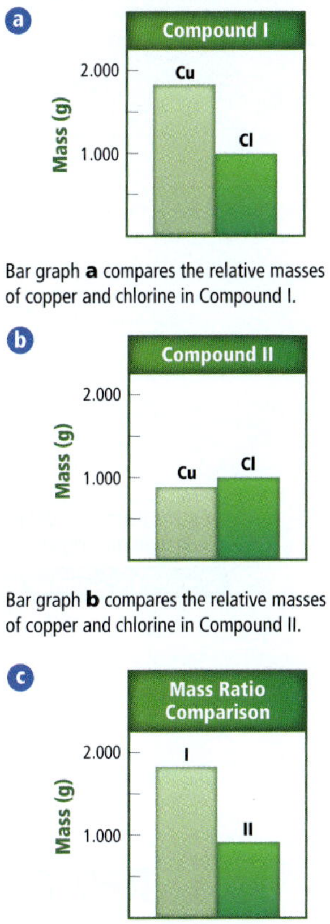

Bar graph **a** compares the relative masses of copper and chlorine in Compound I.

Bar graph **b** compares the relative masses of copper and chlorine in Compound II.

Bar graph **c** shows a comparison between the relative masses of copper in both compounds. The ratio is 2:1.

ChemLAB

The ChemLab located at the end of the chapter can be used at this point in the lesson.

☑ READING CHECK Answers will vary but should include the idea that the relative masses of the common element in two or more different compounds are in small integer ratios to each other.

Content Background

Law of Multiple Proportions

The law of multiple proportions was developed by the English chemist John Dalton, who made the observation public in 1803. One of the most fundamental principles of stoichiometry, it built on the prior work of French chemist Antoine Lavoisier, who had proposed the law of conservation of mass, and French chemist Joseph Proust, who had developed the law of definite proportions. Dalton's work, while not free of error, proved to be integral to his atomic theory. He had observed that oxygen and carbon combine to form not one, but two compounds: CO and CO_2. Further, he noted that while each compound had its own weight ratio, these ratios were proportional: the proportion of oxygen to carbon in CO was 1.33:1 and the ratio for CO_2 was 2.66:1, with the amount of oxygen in CO_2 being exactly twice that in CO.

☑ GRAPH CHECK because of the law of multiple proportions

CHEMISTRY JOURNAL

Periodic Table Ask students to draw a chart about Mendeleev's periodic table, addressing what was correct in his original table and what it was incorrect. Some questions to consider include: did Mendeleev's table have the ability to correctly estimate the positions of elements that had yet to be discovered? Was he correct to arrange the elements in order of mass? Did he arrange the periods by the energy level of the valence electrons? Did he arrange the groups by the number of valence electrons? OL

ASSESSMENT

Skill Clip photos of various mixtures from magazines or other sources and have students identify them as heterogeneous mixtures, solutions, or pure substances in various phases. OL

3 Assess

Check for Understanding

Have students draw diagrams featuring atoms and molecules that show packing arrangements in heterogeneous mixtures, solutions, and pure substances. OL

Reteach

Students might find it easier to memorize a triangular arrangement of matter from most general at the bottom of the pyramid to most specific (element, compound, solution, heterogeneous mixture) at the top and with classification becoming more ordered as you move from right to left. BL EL

Extension

Give each student an "information square" on which they describe themselves by a category such as name, address, height, hair color, eye color, interests, pets, number of siblings. Have them put a small picture of themselves in the center. Then pass out copies of the squares and ask the students to arrange them into as many meaningful patterns as possible and describe their pattern in a presentation. BL EL COOP LEARN

■ **Figure 21** Different compounds are formed when different relative masses of each element are combined. Although they are both made of copper and chlorine, Compound I has a greenish color, whereas Compound II has a bluish color.

Explore the laws of **definite and multiple proportions.**

Virtual Investigations

Figure 21 shows the two compounds formed by the combination of copper and chlorine and presented in **Table 5** and **Figure 20.** These compounds are called copper(I) chloride and copper(II) chloride. As the law of multiple proportions states, the different masses of copper that combine with a fixed mass of chlorine in the two different copper compounds can be expressed as a small whole-number ratio. In this case, the ratio is 2:1.

Considering that there is a finite number of elements that exist today and an exponentially greater number of compounds that are composed of these elements under various conditions, it becomes clear how important the law of multiple proportions is in chemistry.

SECTION 4 REVIEW

Section Self-Check

Section Summary

- Elements cannot be broken down into simpler substances.
- Elements are organized in the periodic table of the elements.
- Compounds are chemical combinations of two or more elements, and their properties differ from the properties of their component elements.
- The law of definite proportions states that a compound is always composed of the same elements in the same proportions.
- The law of multiple proportions states that if elements form more than one compound, those compounds will have compositions that are whole-number multiples of each other.

24. MAINIDEA **Compare and contrast** elements and compounds.

25. Describe the basic organizational feature of the periodic table of the elements.

26. Explain how the law of definite proportions applies to compounds.

27. State the type of compounds that are compared in the law of multiple proportions.

28. Complete the table, and then analyze the data to determine if Compounds I and II are the same compound. If the compounds are different, use the law of multiple proportions to show the relationship between them.

Analysis Data of Two Iron Compounds

Compound	Total Mass (g)	Mass Fe (g)	Mass O (g)	Mass Percent Fe	Mass Percent O
I	75.00	52.46	22.54		
II	56.00	43.53	12.47		

29. Calculate the mass percent of hydrogen in water and the mass percent of oxygen in water.

30. Graph Create a graph that illustrates the law of multiple proportions.

SECTION 4 REVIEW

24. Elements cannot be broken down into simpler substances by ordinary chemical means, whereas compounds can.

25. The periodic table of the elements is organized by rows (called "periods") and columns (called "groups" or "families"). Elements in the same group have similar physical and chemical properties. The patterns of similar properties repeat from period to period.

26. The law of definite proportions describes the mass composition of a substance.

27. The law of multiple proportions relates the compositions of two compounds composed of the same elements.

28. Compound I mass percent Fe = 69.95%; mass percent O = 30.05%. Compound II mass percent Fe = 77.73%; mass percent O = 22.27%. The compounds are not the same. The relative mass ratio of Fe in compound I and compound II is 2:3.

29. mass % of hydrogen in water = 2.0/18.0 × 100% = 11%
mass % of oxygen in water = 16.0/18.0 × 100% = 89%

30. Graph should be similar to Figure 20.

CHEMISTRY & careers

Career: Arson Investigator

Forensic Ignitable Liquid Detection

Inside a burning warehouse, havoc and destruction reign. Intense heat and smoke fill closed spaces. Leaping flames spread; walls and ceilings collapse. Was the fire accidental or the work of an arsonist?

Ignitable liquids Fire investigators analyze evidence to determine how a fire began and spread. If arson is suspected, it is likely that ignitable liquids—chemicals that speed the spread of a fire—were involved.

The properties of an ignitable liquid The properties that make ignitable liquids useful as fuels also make them dangerous in fire situations. Ignitable liquids are readily absorbed and are powerful solvents. They do not mix well with water, often floating on top and at room temperature, and they form vapors that can ignite and burn.

Evidence of an ignitable liquid What evidence indicates the presence of an ignitable liquid? One indicator is an unusual burn pattern, like that present on the floor joists in **Figure 1.** In this case, called a "rundown" burn pattern, an ignitable liquid was likely poured in this area, running down between the floorboards to the joists below.

Robert Corry

Figure 1 Ignitable liquids can cause a rundown burn pattern.

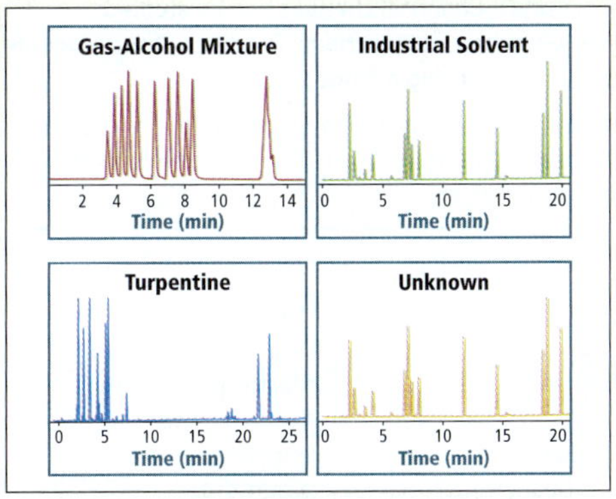

Figure 2 Chromatograms display individual components of mixtures.

Another indicator is a small slick on top of any wet material, similar to the automobile-oil slick floating on a puddle on a wet street. If investigators see such clues, they can take samples of the affected materials for testing.

Chemical analysis Investigators take any samples they collect to the lab for chemical analysis. In the lab, volatiles are extracted from a sample. The extract is separated and analyzed using a process called gas chromatography-mass spectrometry. The components of the mixture are displayed in a chromatogram, like the ones shown in **Figure 2.** By comparing the chromatogram of the unknown with those of known compounds, the identity of the ignitable liquid can be determined. Mass spectra can be obtained for each sample component for confirmation.

WRITING IN ▶ Chemistry

Think Critically Look at the chromatogram of the unknown sample and compare it to the three known samples. Can you determine which accelerant was used? Could that knowledge give you any insight into who might have committed the crime? Explain your answer.

WebQuest

Purpose

Students will describe the properties of accelerants and explain how investigators analyze chemical evidence and identify accelerants.

Background

In gas chromatography-mass spectrometry (Gc-Ms), a volatile or semi-volatile liquid sample is injected into a machine and gently heated, causing vaporization. A stream of inert gas (nitrogen or helium) carries the mixture along a long column containing a fixed substrate called a stationary phase. Mixture components that interact more with the stationary phase are slow to move through the column and emerge later than components that have less interaction with the stationary phase. As each component emerges from the column, it enters the mass spectrometer, where it is ionized. The ions are separated in the analyzer based on their mass-to-charge ratios. Each mass is individually accelerated to the detector. In Gc-Ms, the chromatogram obtained displays the individual sample components based on their retention times (how long they were retained on the column). The mass spectra obtained for each sample component provides information on the structure of each component in a sample.

Teaching Strategies

- Review the difference between physical and chemical properties of matter. Have students identify each ignitable liquid property described in the feature as physical or chemical.
- Explain the process of gas chromatography. Have students compare the procedure in the liquid chromatography MiniLab to gas chromatography.
- The definition of an accelerant is anything that hastens a process. Items such as carpet, clothing, or hand towels fall under the definition of an accelerant. However, this does not require that a fire occurrence is the result of arson if those items are present. Ignitable liquids are usually the accelerants used in arson. They fall into the following eight categories: gasoline, petroleum distillates, isoparaffinic products, aromatic products, napthenic paraffinic products, normal alkane products, oxygenated products, and others.

WRITING IN ▶ Chemistry

Rubric WebQuest

Think Critically The unknown chromatogram most closely matches that of an industrial solvent. The criminal would require access to industrial-grade chemicals, perhaps through his or her workplace or the workplace of a family member or friend.

ChemLAB

Rubric

Preparation

Time Allotment one class period

Process Skills observe and infer, classify

Safety Precautions Review materials safety data sheet for $AgNO_3$ with students before doing lab. Approve lab safety forms before work begins. Be sure students wear aprons and goggles. Remind students to tie back long hair and use caution around a flame. Paper clip will remain hot for several minutes. Do not allow solutions to contact skin or clothing.

Disposal Rinse filtrate down a drain with plenty of water. The solid wastes can be collected for recycling according to local regulations.

Procedure

- The lengths of wire can be cut ahead.
- Copper coil can be allowed to sit in $AgNO_3$ solution overnight.
- Darken room to view flame color.

Analyze and Conclude

1. A grayish solid formed on the wire. The solution turned blue-green. Yes, a solid formed and a color change occurred. The products are silver and copper nitrate.
2. Silver metal is white to gray. Copper nitrate is blue-green.
3. Experimental results should agree with blue-green light.
4. homogeneous; heterogeneous; homogeneous

INQUIRY EXTENSION

Experiments will vary. There are other possible hypotheses. The copper wire might not have been clean. The better observations will be more detailed.

ChemLAB

Identify the Products of a Chemical Reaction

Background: Chemical changes can be studied by observing chemical reactions. Products of the reaction can be identified using a flame test.

Question: *Is there a chemical reaction between copper and silver nitrate? Which elements react, and what is the compound they form?*

Materials

$AgNO_3$ solution	small iron ring
sandpaper	ring stand
stirring rod	plastic petri dish
funnel	Bunsen burner
filter paper	tongs
50-mL beaker	paper clip
50-mL graduated cylinder	copper wire
250-mL Erlenmeyer flask	

Safety Precautions

WARNING: ***Silver nitrate is highly toxic. Avoid contact with eyes and skin.***

Procedure

1. Read and complete the lab safety form.
2. Rub 8 cm of copper wire with sandpaper until it is shiny. Observe and record its physical properties.
3. Measure 25 mL $AgNO_3$ (silver nitrate) solution into a 50-mL beaker. Record its physical properties.
4. Coil the copper wire so that it fits into the beaker. Make a hook and suspend it from the stirring rod.
5. Place the stirring rod across the top of the beaker, immersing part of the coil in the $AgNO_3$ solution.
6. Make and record observations of the wire and the solution every 5 min for 20 min.
7. Set up a filtration apparatus: attach the iron ring to the ring stand, and adjust its height so the end of the funnel is inside the neck of the Erlenmeyer flask.
8. Fold the circle filter paper in half twice to form a quarter of a circle. Tear off the lower-right corner of the flap facing you. Open the folded paper into a cone, and place it into the funnel.
9. Remove the coil from the beaker, and dispose of it as directed by your teacher.

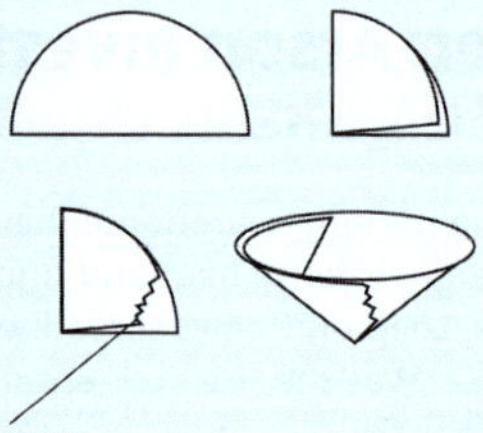

10. Slowly pour the liquid down the stirring rod into the funnel to catch the solid products in the filter paper.
11. Collect the filtrate in the Erlenmeyer flask, and transfer it to a petri dish.
12. Adjust a Bunsen burner flame until it is blue. Hold the paper clip in the flame with tongs until no additional color is observed.
13. Using tongs, dip the hot paper clip into the filtrate. Then, hold the paper clip in the flame. Record the color you observe. After removing the clip from the burner, let it cool before handling.
14. **Cleanup and Disposal** Dispose of materials as directed by your teacher. Clean and return all lab equipment to its proper place.

Analyze and Conclude

1. **Observe and Infer** Describe the changes you observed in Step 6. Is there evidence that a chemical change occurred? Predict the products formed.
2. **Compare** Use resources such as the *CRC Handbook of Chemistry and Physics* to determine the colors of silver metal and copper nitrate in water. Compare this information with your observations of the reactants and products in Step 6.
3. **Identify** Copper emits a blue-green light in flame tests. Do your observations confirm the presence of copper in the filtrate collected in Step 11?
4. **Classify** Which type of mixture is silver nitrate in water? Which type of mixture is formed in Step 6? In Step 10?

INQUIRY EXTENSION

Compare your recorded observations with those of several other lab teams. Form a hypothesis to explain any differences; design an experiment to test it.

Vocabulary Practice

BIGIDEA Everything is made of matter.

SECTION 1 Properties of Matter

MAINIDEA Most common substances exist as solids, liquids, and gases, which have diverse physical and chemical properties.

- The three common states of matter are solid, liquid, and gas.
- Physical properties can be observed without altering a substance's composition.
- Chemical properties describe a substance's ability to combine with or change into one or more new substances.
- External conditions can affect both physical and chemical properties.

VOCABULARY
- states of matter
- solid
- liquid
- gas
- vapor
- physical property
- extensive property
- intensive property
- chemical property

SECTION 2 Changes in Matter

MAINIDEA Matter can undergo physical and chemical changes.

- A physical change alters the physical properties of a substance without changing its composition.
- A chemical change, also known as a chemical reaction, involves a change in a substance's composition.
- In a chemical reaction, reactants form products.
- The law of conservation of mass states that mass is neither created nor destroyed during a chemical reaction; it is conserved.

$$\text{mass}_{\text{reactants}} = \text{mass}_{\text{products}}$$

VOCABULARY
- physical change
- phase change
- chemical change
- law of conservation of mass

SECTION 3 Mixtures of Matter

MAINIDEA Most everyday matter occurs as mixtures—combinations of two or more substances.

- A mixture is a physical blend of two or more pure substances in any proportion.
- Solutions are homogeneous mixtures.
- Mixtures can be separated by physical means. Common separation techniques include filtration, distillation, crystallization, sublimation, and chromatography.

VOCABULARY
- mixture
- heterogeneous mixture
- homogeneous mixture
- solution
- filtration
- distillation
- crystallization
- sublimation
- chromatography

SECTION 4 Elements and Compounds

MAINIDEA A compound is a combination of two or more elements.

- Elements cannot be broken down into simpler substances.
- Elements are organized in the periodic table of the elements.
- Compounds are chemical combinations of two or more elements and their properties differ from the properties of their component elements.
- The law of definite proportions states that a compound is always composed of the same elements in the same proportions.

$$\text{percent by mass} = \frac{\text{mass of the element}}{\text{mass of the compound}} \times 100$$

- The law of multiple proportions states that if elements form more than one compound, those compounds will have compositions that are whole-number multiples of each other.

VOCABULARY
- element
- periodic table
- compound
- law of definite proportions
- percent by mass
- law of multiple proportions

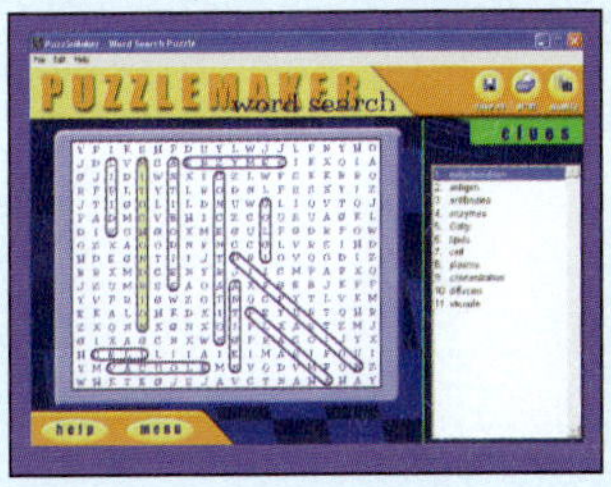

Vocabulary Practice

For additional practice with vocabulary, have students access the Vocabulary PuzzleMaker.

CHAPTER 3 STUDY GUIDE

Use the Vocabulary

To reinforce chapter vocabulary, have students write a sentence using each term. OL EL

Review Strategies

- Have students summarize the differences between physical and chemical properties and physical and chemical change and give an example of each. OL
- Have students be able to use the symbols of the most frequently used elements. BL
- Have students classify matter according to the scheme given in the chapter. OL
- Problems from the Supplemental Practice Problems appendix or the Supplemental Problems booklet can be used for review. BL

Students can review with these online resources.

Vocabulary Practice

Vocabulary eGames
Vocabulary eFlashcards
Vocabulary PuzzleMaker

Section Self-Check

Chapter Self-Check

Online Test Practice

Use *eAssessment* to:

- create multiple versions of tests
- edit existing questions and add your own questions
- build tests aligned with select state curriculum standards using built-in state curriculum tags
- track students' progress

SECTION 1

Mastering Concepts

31. Answers will vary. Water, salt, and sugar are all substances. Substances have unique and unchanging compositions.

32. pure substance; constant composition

33. Answers will vary. Water is colorless, a liquid, freezes at approximately 0°C, and boils at approximately 100°C.

34. **a.** intensive
b. extensive
c. intensive
d. extensive

35. The statement is false. Properties are affected by changes in temperature and pressure. Explanations will vary.

36. Solid, liquid, and gas. Possible examples: ice, water, and helium; Substances that are in the gas phase at room temperature are referred to as gases, whereas substances such as water vapor that are not gases at room temperature are vapors.

37. **a.** liquid **b.** gas **c.** solid **d.** gas **e.** solid **f.** solid

38. **a.** physical **b.** physical **c.** chemical **d.** physical **e.** chemical **f.** physical

39. The volume of the milk remains unchanged. Milk, which is a liquid, conforms to the shape of its container, thus the shape of the milk changes as it is poured from the carton into the bowl.

40. 100°C; 100°C; Boiling point is an intensive property because it is independent of quantity or amount.

Mastering Problems

41. Table 6 shows two compounds that are white solids, but it is sucrose that decomposes before a boiling point is determined. Therefore, the unknown substance is sucrose.

SECTION 2

Mastering Concepts

42. **a.** physical **b.** chemical

43. **a.** physical **b.** physical **c.** chemical **d.** chemical **e.** chemical

44. It is a chemical change. Green bananas have different properties compared to yellow bananas.

45. It is a physical change. The composition of the substance does not change.

CHAPTER 3 ASSESSMENT

SECTION 1

Mastering Concepts

31. List three examples of substances. Explain why each is a substance.

32. Is carbon dioxide gas a pure substance? Explain.

33. List at least three physical properties of water.

34. Identify each physical property as extensive or intensive.
a. melting point
b. mass
c. density
d. length

35. "Properties are not affected by changes in temperature and pressure." Is this statement true or false? Explain.

36. List the three states of matter, and give an example for each state. Differentiate between a gas and a vapor.

37. Classify each as either a solid, a liquid, or a gas at room temperature.
a. milk
b. air
c. copper
d. helium
e. diamond
f. candle wax

38. Classify each as a physical property or a chemical property.
a. Aluminum has a silvery color.
b. Gold has a density of 19 g/cm^3.
c. Sodium ignites when dropped in water.
d. Water boils at 100°C.
e. Silver tarnishes.
f. Mercury is a liquid at room temperature.

39. A carton of milk is poured into a bowl. Describe the changes that occur in the milk's shape and volume.

40. **Boiling Water** At what temperature would 250 mL of water boil? 1000 mL? Is the boiling point an intensive or extensive property? Explain.

Mastering Problems

41. **Chemical Analysis** A scientist wants to identify an unknown compound on the basis of its physical properties. The substance is a white solid at room temperature. Attempts to determine its boiling point were unsuccessful. Using **Table 6,** name the unknown compound.

Table 6 Physical Properties of Common Substances

Substance	Color	State at 25°C	Boiling Point (°C)
Oxygen	colorless	gas	−183
Water	colorless	liquid	100
Sucrose	white	solid	decomposes
Sodium chloride	white	solid	1465

SECTION 2

Mastering Concepts

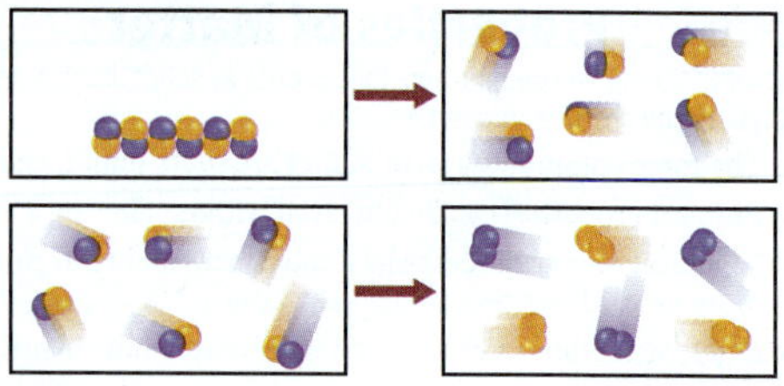

■ **Figure 22**

42. Label each set of diagrams in **Figure 22** as a physical or a chemical change.

43. Classify each as a physical change or a chemical change.
a. breaking a pencil in two
b. water freezing and forming ice
c. frying an egg
d. burning wood
e. leaves changing colors in the fall

44. **Ripening** Is the process of bananas ripening a chemical change or a physical change? Explain.

45. Is a change in phase a physical change or a chemical change? Explain.

46. List four indicators that a chemical change has probably occurred.

47. **Salt** Sodium and chlorine combine to form sodium chloride, or table salt. List the reactants and products of this reaction.

48. **Burning Candle** After burning for three hours, a candle has lost half of its mass. Explain why this example does not violate the law of conservation of mass.

49. Describe the difference between a physical change and a chemical change.

Mastering Problems

50. **Ammonia Production** A 28.0-g sample of nitrogen gas combines completely with 6.0 g of hydrogen gas to form ammonia. What is the mass of ammonia formed?

51. A 13.0-g sample of X combines with a 34.0-g sample of Y to form the compound XY_2. What is the mass of the reactants?

52. If 45.98 g of sodium combines with an excess of chlorine gas to form 116.89 g of sodium chloride, what mass of chlorine gas is used in the reaction?

53. A substance breaks down into its component elements when it is heated. If 68.0 g of the substance is present before it is heated, what is the combined mass of the component elements after heating?

46. Probable indicators of a chemical reaction include a change in color, odor, or temperature, and/or the production of a gas or a solid upon mixing.

47. reactants: sodium and chlorine; product: sodium chloride

48. The mass of the candle is conserved if you consider the gaseous products from the reaction.

49. A physical change alters a substance without changing its composition, while a chemical change involves a change in composition.

Mastering Problems

50. $Mass_{ammonia} = 34.0$ g

51. 47.0 g

52. $Mass_{chlorine} = 70.91$ g

53. 68.0 g

54. $Mass_{copper\ sulfide} = 159$ g

55. (180 g + 192 g) = (108 g + x g) Using the law of conservation of mass, x = 264 g.

Chapter Self-Check

54. Copper(I) sulfide is formed when copper and sulfur are heated together. In this reaction, 127 g of copper reacts with 41 g of sulfur. After the reaction is complete, 9 g of sulfur remains unreacted. What is the mass of copper sulfide formed?

55. When burning 180 g of glucose in the presence of 192 g of oxygen, water and carbon dioxide are produced. If 108 g of water is produced, how many grams of carbon dioxide are produced?

SECTION 3

Mastering Concepts

56. Describe the characteristics of a mixture.

■ **Figure 23**

57. Name the separation method illustrated in **Figure 23**.

58. Describe a method that could be used to separate each mixture.
a. iron filings and sand
b. sand and salt
c. the components of ink
d. helium and oxygen gases

59. "A mixture is the chemical bonding of two or more substances in any proportion." Is this statement true or false? Explain.

60. Which of the following are the same and which are different?
a. a substance and a pure substance
b. a heterogeneous mixture and a solution
c. a substance and a mixture
d. a homogeneous mixture and a solution

61. Describe how a homogeneous mixture differs from a heterogeneous mixture.

62. **Seawater** is composed of salt, sand, and water. Is seawater a heterogeneous or homogeneous mixture? Explain.

63. **Iced Tea** Use iced tea with and without ice cubes as examples to explain homogeneous and heterogeneous mixtures. If you allow all of the ice cubes to melt, what type of mixture remains?

64. **Chromatography** What is chromatography, and how does it work?

SECTION 4

Mastering Concepts

65. State the definition of element.

66. Correct the following statements.
a. An element is a combination of two or more compounds.
b. When a small amount of sugar is completely dissolved in water, a heterogeneous solution is formed.

67. Name the elements contained in the following compounds.
a. sodium chloride (NaCl)
b. ammonia (NH_3)
c. ethanol (C_2H_6O)
d. bromine (Br_2)

68. What was Dmitri Mendeleev's major contribution to the field of chemistry?

69. Is it possible to distinguish between an element and a compound? Explain.

70. How are the properties of a compound related to those of the elements that comprise it?

71. Which law states that a compound always contains the same elements in the same proportion by mass?

72. **a.** What is the percent by mass of carbon in 44 g of carbon dioxide (CO_2)?
b. What is the percent by mass of oxygen in 44 g of carbon dioxide (CO_2)?

73. Complete **Table 7** by classifying the compounds as 1:1 or 2:2, 1:2 or 2:1, and 1:3 or 3:1.

Table 7 Ratios of Elements in Compounds

Compound	Simple Whole-Number Ratios of Elements
NaCl	
CuO	
H_2O	
H_2O_2	

Mastering Problems

74. A 25.3-g sample of an unknown compound contains 0.8 g of oxygen. What is the percent by mass of oxygen in the compound?

75. Magnesium combines with oxygen to form magnesium oxide. If 10.57 g of magnesium reacts completely with 6.96 g of oxygen, what is the percent by mass of oxygen in magnesium oxide?

76. When mercury oxide is heated, it decomposes into mercury and oxygen. If 28.4 g of mercury oxide decomposes, producing 2.0 g of oxygen, what is the percent by mass of mercury in mercury oxide?

SECTION 3

Mastering Concepts

56. Mixtures are a physical blend of two or more substances in any proportion. Mixtures do not have a constant composition. Their properties are largely those of their component substances.

57. distillation

58. **a.** A magnet can be used to draw the iron filings from the sand.
b. Add water to the mixture to dissolve the salt. Filter the mixture to remove the sand, and then boil off the water so only the salt remains.
c. Paper chromatography should be used to separate the ink's components. If enough ink is available, distillation can also be used, but is more complicated than chromatography.
d. Cool the gas mixture until it condenses, then distill the condensate.

59. The statement is false because mixtures are a physical blend of substances, not a chemical bonding of substances.

60. **a.** same **c.** different
b. different **d.** same

61. Homogeneous mixtures contain a single phase. Heterogeneous mixtures can have many phases.

62. heterogeneous; composition is not uniform

63. Ice tea with ice cubes is a heterogeneous mixture. Ice tea without ice cubes is a homogeneous mixture. When all ice cubes melt, the remaining mixture is homogeneous.

64. Chromatography is a technique used to separate components of a mixture.

SECTION 4

Mastering Concepts

65. An element is a substance that cannot be broken down into simpler substances by physical or chemical means.

66. **a.** A compound is combination of two or more elements.
b. When a small amount of sugar is completely dissolved in water, a homogeneous solution is formed.

67. **a.** sodium and chlorine
b. nitrogen and hydrogen
c. carbon, hydrogen, and oxygen
d. bromine

68. Mendeleev developed the first widely accepted periodic table of the elements.

69. Yes, elements can be distinguished from compounds. Compounds can be broken down into their component elements, whereas elements cannot be broken down into simpler substances.

70. The properties of a compound are unique to that compound and different from those of its component elements.

71. the law of definite proportions

72. **a.** $(12\ g/44\ g) \times 100\% = 27\%$
b. $(32\ g/44\ g) \times 100\% = 73\%$
or $(100\% - 27\% = 73\%)$ if you use the result from part (a)

73. NaCl is 1:1; CuO is 1:1; H_2O is 2:1; H_2O_2 is 2:2.

Mastering Problems

74. $\text{mass percent}_{oxygen} = 3\%$

75. $\text{mass percent}_{oxygen} = 39.7\%$

76. $\text{mass percent}_{mercury} = 93.0\%$

77. The ratio of carbon to a fixed mass of oxygen in compound I is 0.748:1, while in compound II it is 0.375:1.

78. 64%

79. The law of multiple proportions. CO_2 will have the highest percent by mass of oxygen because it has more oxygen atoms for the same number of carbon atoms.

80.

Compound	Mass % of oxygen	Mass of second element in the compound (g)
CuO	20	Cu = 64
H_2O	89	2H = 2
H_2O_2	94	2H = 2
CO	57	12
CO_2	73	12

MIXED REVIEW

81. Gases are the most compressible state of matter, solids the least. Liquids are virtually incompressible. Compressibility is determined by the amount of space between particles in each state. Gases have the greatest amount of space between particles, solids the least.

82. **a.** homogeneous
b. heterogeneous
c. heterogeneous
d. homogeneous

83. initial $mass_{hydrogen} = 21.6$ g; $mass_{hydrogen} = 10.6$ g

84. 50 units; no; 50 particles of oxygen will remain

85. **a.** homogeneous mixture
b. heterogeneous mixture
c. homogeneous or heterogeneous mixture depending on the soil sample
d. pure substance
e. heterogeneous mixture
f. heterogeneous mixture

86. **a.** compound
b. homogenous mixture
c. element
d. heterogeneous mixture
e. homogenous mixture

87. before cooking: liquid, white (clear) and (bright) yellow; after cooking: solid/opaque, white and dull yellow; chemical change

77. Carbon reacts with oxygen to form two different compounds. Compound I contains 4.82 g of carbon for every 6.44 g of oxygen. Compound II contains 20.13 g of carbon for every 53.7 g of oxygen. What is the ratio of carbon to a fixed mass of oxygen for the two compounds?

78. A 100-g sample of an unknown salt contains 64 g of chlorine. What is the percent by mass of chlorine in the compound?

79. Which law would you use to compare CO and CO_2? Explain. Without doing any calculations, determine which of the two compounds has the highest percent by mass of oxygen in the compound.

80. Complete **Table 8.**

Table 8 Elements in Compounds

Compound	Mass of Compound (g)	Mass of Oxygen (g)	Mass % of Oxygen	Mass of Second Element in the Compound (g)
CuO	80.0	16		
H_2O	18.0	16		
H_2O_2	34.0	32		
CO	28.0	16		
CO_2	44.0	32		

MIXED REVIEW

81. Which state(s) of matter are compressible? Which state(s) of matter are not compressible? Explain.

82. Classify each mixture as homogeneous or heterogeneous.
a. brass (an alloy of zinc and copper)
b. a salad
c. blood
d. powdered drink mix dissolved in water

83. Phosphorus combines with hydrogen to form phosphine. In this reaction, 108.3 g of phosphorus combines with excess hydrogen to produce 129.9 g of phosphine. After the reaction, 11.0 g of hydrogen remains unreacted. What was the initial mass of hydrogen before the reaction? What mass of hydrogen is used in the reaction?

84. If you have 100 particles of hydrogen and 100 particles of oxygen, how many units of water can you form? Will you use all the particles of both elements? If not, what will remain?

85. Classify each substance as a pure substance, a homogeneous mixture, or a heterogeneous mixture.
a. air **c.** soil **e.** sediment
b. aerosol **d.** water **f.** muddy water

86. Identify each as a homogenous mixture, a heterogeneous mixture, a compound, or an element.
a. pure drinking water **d.** seawater
b. salty water **e.** air
c. helium

87. Cooking List physical properties of eggs before and after they are cooked. Based on your observations, does a physical change or chemical change occur when eggs are cooked? Justify your answer.

88. Ice Cream You might have noticed that while eating ice cream on a hot day, some of the ice cream begins to melt. Is the observed change in the state of the ice cream a physical or a chemical change? Justify your answer.

89. Pizza Is a pizza a homogeneous or heterogeneous mixture? Explain.

90. Sodium reacts chemically with chlorine to form sodium chloride. Is sodium chloride a mixture or a compound?

91. Is air a solution or a heterogeneous mixture? What technique can be used to separate air into its components?

92. Indicate whether combining the following elements yields a compound or a mixture.
a. $H_2(g) + O_2(g) \rightarrow$ water
b. $N_2(g) + O_2(g) \rightarrow$ air

THINK CRITICALLY

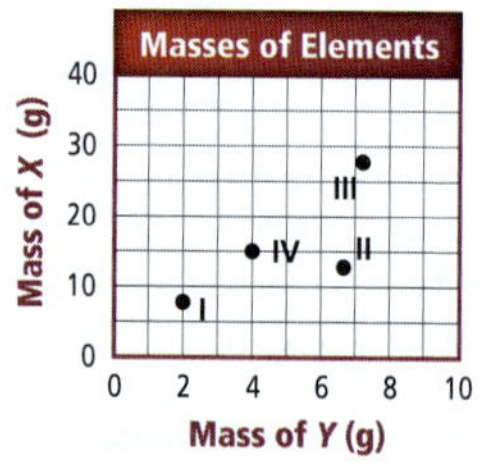

■ **Figure 24**

93. Interpret Data A compound contains the elements *X* and *Y*. Four samples with different masses were analyzed, and the masses of *X* and *Y* in each sample were plotted on a graph shown in **Figure 24.** The samples were labeled *I, II, III,* and *IV.*
a. Which samples are from the same compound? How do you know?
b. What is the approximate ratio of the mass of *X* to the mass of *Y* in the samples that are from the same compound?
c. What is the approximate ratio of the mass of *X* to the mass of *Y* in the sample(s) that are not from the same compound?

88. Ice cream melting is a physical change because the chemical composition remains constant.

89. A pizza is a heterogeneous mixture because the individual parts of the pizza—dough, sauce, cheese, toppings—remain separate.

90. Sodium chloride is a compound because a chemical reaction occurred between sodium and chlorine.

91. Air is a solution. Air can be separated into its components, oxygen, nitrogen and carbon dioxide by using gas chromatography.

92. **a.** compound
b. mixture

THINK CRITICALLY

93. **a.** Samples *I, III,* and *IV* are the same compound. A straight line can be drawn through these three plotted points. The slope of the line is equivalent to ratio $Mass_X/Mass_Y$. The fact that all three points are on the same line shows that they all have the same mass ratio of *X* to *Y* and must be the same compound.
b. Mass ratio of *X* to *Y* for samples *I, III,* and *IV* is 3.75:1.
c. Sample *II* mass ratio is 1.9:1

Chapter Self-Check

94. Apply Air is a mixture of many gases, primarily nitrogen, oxygen, and argon. Could distillation be used to separate air into its component gases? Explain.

95. Analyze Is gas escaping from an opened soft drink an example of a chemical or a physical change? Explain.

96. Apply Give examples of heterogeneous mixtures for the systems listed in **Table 9.**

Table 9 Heterogeneous Mixtures

System	Example
Liquid-liquid	
Solid-liquid	
Solid-solid	

Challenge Problem

97. Identify Lead Compounds A sample of a certain lead compound contains 6.46 g of lead for each gram of oxygen. A second sample has a mass of 68.54 g and contains 28.76 g of oxygen. Are the two samples the same? Explain.

CUMULATIVE REVIEW

98. What is chemistry?

99. What is mass? Weight?

100. Express the following numbers in scientific notation.
a. 34,500 **d.** 789
b. 2665 **e.** 75,600
c. 0.9640 **f.** 0.002189

101. Perform the following operations.
a. $10^7 \times 10^3$
b. $(1.4 \times 10^{-3}) \times (5.1 \times 10^{-5})$
c. $(2 \times 10^{-3}) \times (4 \times 10^5)$

102. Convert 65°C to kelvins.

103. Graph the data in **Table 10.** What is the slope of the line?

Table 10 Energy Released by Carbon

Mass (g)	Energy Released (kJ)
1.00	33
2.00	66
3.00	99
4.00	132

WRITING IN ▶ Chemistry

104. Synthetic Elements Select a synthetic element, and prepare a short written report on its development. Be sure to discuss recent discoveries, list major research centers that conduct this type of research, and describe the properties of the synthesized element.

DBQ Document-Based Questions

Pigments *Long before scientists understood the properties of elements and compounds, artists used chemistry to create pigments from natural materials.* **Table 11** *gives some examples of such pigments used in ancient times.*

Data obtained from: Orna, Mary Virginia. 2001. Chemistry, color, and art. *Journal of Chemical Education* 78 (10): 1305

Table 11 Common Artists' Pigments Used in Early Times

Common Name	Chemical Identity	Comments
Charcoal	elemental carbon (carbon black)	produced by dry distillation of wood in a closed vessel
Egyptian blue	calcium copper tetrasilicate, $CaCuSi_4O_{10}$	crystalline compound containing some glass impurity
Indigo	indigotin, $C_{16}H_{10}N_2O_2$	derived from different plants of the genus *Indigofera*
Iron oxide red	Fe_2O_3	in continuous use in all geographic regions and time periods
Verdigris	dibasic acetate of copper, $Cu(C_2H_3O_2)_2 \cdot 2Cu(OH)_2$	other copper compounds, including carbonate, are also called verdigris

105. a. Compare the mass percent of carbon in charcoal, indigo, and verdigris.
b. Compare the mass percent of oxygen in iron oxide and Egyptian blue.

106. List an example of an element and a compound from **Table 11.**

107. Is the production of charcoal from the dry distillation of wood a chemical or a physical change? Explain.

94. If the mixture of gases is cooled sufficiently, it will condense into a mixture of liquids. This mixture could then be distilled.

95. physical change because the composition of CO_2 gas is still the same inside and outside of the soft drink

96. Examples might include: liquid-liquid — water and oil; solid-liquid — sand and water; solid-solid — iron filings and sugar.

CHALLENGE PROBLEM

97. sample I:
$mass_{lead}/mass_{oxygen} = 6.46$
sample II:
$mass_{lead}/mass_{oxygen} = 1.383$
The two samples are not the same because the two $mass_{lead}/mass_{oxygen}$ ratios are not the same.

CUMULATIVE REVIEW

98. Chemistry is the study of matter and the changes that it undergoes.

99. Mass is the measure of the amount of matter an object contains. It is measured on a balance. The weight of an object is the amount of gravitational pull acting on that matter. It is measured on a scale.

100. a. 3.45×10^4
b. 2.665×10^3
c. 9.640×10^{-1}
d. 7.89×10^2
e. 7.56×10^4
f. 2.189×10^{-3}

101. a. 10^{10}
b. 7.1×10^{-8}
c. 8×10^2

102. 338 K

103. slope: 33 kJ/g

WRITING IN ▶ Chemistry

Rubric

104. Answers will vary.

DBQ Document-Based Questions

Data obtained from: Orna, Mary Virginia. 2001. Chemistry, Color and Art. *Journal of Chemical Education* 78 (10): 1305.

105. a. charcoal: mass percent carbon = 100%; indigo: mass percent carbon = 73%; and verdigris: mass percent carbon = 14%
b. iron oxide: mass percent oxygen = 30%; Egyptian blue: mass percent oxygen = 43%

106. element, carbon; compound, Fe_2O_3

107. chemical change, because the composition of the dry wood changes into charcoal

MULTIPLE CHOICE

1. D
2. D
3. C
4. C
5. D
6. D
7. D
8. C
9. D

MULTIPLE CHOICE

Use the table below to answer Questions 1 and 2.

Mass Analysis of Two Chlorine-Fluorine Samples				
Sample	Mass of Chlorine (g)	Mass of Fluorine (g)	% Cl	% F
I	13.022	6.978	65.11	34.89
II	5.753	9.248	?	?

1. What are the values for % Cl and % F, respectively, for Sample II?
A. 0.6220 and 61.65
B. 61.65 and 38.35
C. 38.35 and 0.6220
D. 38.35 and 61.65

2. Which statement best describes the relationship between the two samples?
A. The compound in Sample I is the same as in Sample II. Therefore, the mass ratio of Cl to F in both samples will obey the law of definite proportions.
B. The compound in Sample I is the same as in Sample II. Therefore, the mass ratio of Cl to F in both samples will obey the law of multiple proportions.
C. The compound in Sample I is not the same as in Sample II. Therefore, the mass ratio of Cl to F in both samples will obey the law of definite proportions.
D. The compound in Sample I is not the same as in Sample II. Therefore, the mass ratio of Cl to F in both samples will obey the law of multiple proportions.

3. After two elements react to completion in a closed container, the ratio of their masses in the container will be the same as before the reaction. Which law describes this principle?
A. law of definite proportions
B. law of multiple proportions
C. law of conservation of mass
D. law of conservation of energy

4. Which is NOT a physical property of table sugar?
A. forms solid crystals at room temperature
B. appears as white crystals
C. breaks down into carbon and water vapor when heated
D. tastes sweet

5. Which describes a substance that is in the solid state?
A. Its particles can flow past one another.
B. It can be compressed into a smaller volume.
C. It takes the shape of its container.
D. Its particles of matter are close together.

Use the diagram below to answer Questions 6 and 7.

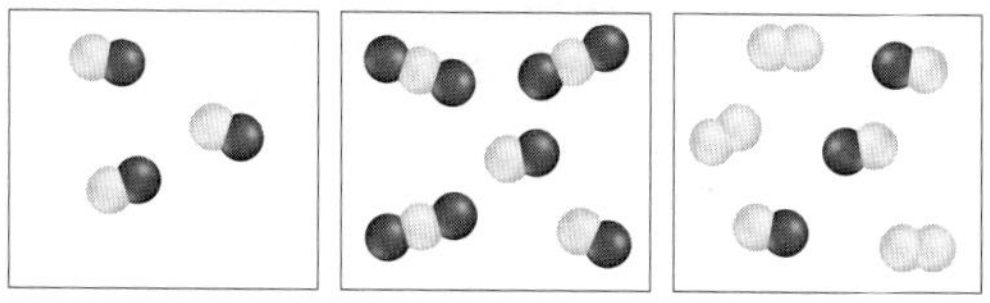

A B C

6. Which best describes Figure A?
A. element
B. mixture
C. solution
D. compound

7. Which statement is false?
A. Figure B is composed of two different compounds.
B. Figure C is composed of two different compounds.
C. Figure B represents 13 total atoms.
D. Three different types of elements are represented in Figure C.

8. Na, K, Li, and Cs all share similar chemical properties. In the periodic table of elements, they most likely belong to the same
A. row.
B. period.
C. group.
D. element.

9. Magnesium reacts explosively with oxygen to form magnesium oxide. Which is NOT true of this reaction?
A. The mass of magnesium oxide produced equals the mass of magnesium consumed plus the mass of oxygen consumed.
B. The reaction describes the formation of a new substance.
C. The product of the reaction, magnesium oxide, is a chemical compound.
D. Magnesium oxide has physical and chemical properties similar to both oxygen and magnesium.

Online Test Practice

SHORT ANSWER

10. Compare and contrast the independent variable in an experiment with the dependent variable.

11. A student reports the melting point of a gas as −295°C. Explain why his claim is unlikely to be correct.

12. Place the following metric prefixes in order from the smallest value to the largest value: deci, kilo, centi, micro, mega, milli, giga, nano.

EXTENDED RESPONSE

Use the table below to answer Questions 13 to 15.

Selected Properties of Substances in a Mixture

Item	Soluble in Water?	Soluble in Alcohol?	Density (g/cm^3)	Particle Size (mm)
Sawdust	no	no	0.21	1
Mothball flakes	no	yes	1.15	3
Table salt	yes	no	2.17	2

13. Is the mixture described in the table homogeneous or heterogeneous? Explain how you can tell.

14. Do the data describe chemical or physical properties? Explain your answer.

15. Propose a method to separate the three substances based on the properties described above.

16. Explain the difference between a chemical change and a physical change. Is the combustion of gasoline a chemical change or a physical change? Explain your answer.

SAT SUBJECT TEST: CHEMISTRY

17. Which is a correct statement about methods for separating mixtures?
    A. Distillation results in the formation of solid particles of a dissolved substance.
    B. Filtration depends on differences in sizes of particles.
    C. Separations depend on the chemical properties of the substances involved.
    D. Chromatography depends on the different boiling points of substances.
    E. Sublimation can be used to separate two gases present in a mixture.

Use the table below to answer Questions 18 and 19.

Percent by Mass of Carbon, Hydrogen, and Oxygen in Selected Compounds

Compound	% H	% C	% O
Carbonic acid (H_2CO_3)	3.2	19.4	77.4
Acetic acid (CH_3COOH)	6.7	40.0	53.3
Methanol (CH_3OH)	12.5	37.5	40.0
Methanal (H_2CO)	6.7	40.0	53.3
Isopropanol (C_3H_8O)	13.3	60.0	26.6

18. You have a 125-g sample of one of these substances. You determine that it is made of 16.7 g H, 75.0 g C, and 33.3 g O. Which compound is it?
    A. acetic acid
    B. carbonic acid
    C. methanal
    D. methanol
    E. isopropanol

19. In another experiment, you determine that a sample of acetic acid consists of 56.8% oxygen. What is your percent error?
    A. 3.50%
    B. 6.57%
    C. 1.07%
    D. 12.6%
    E. 2.06%

NEED EXTRA HELP?

If You Missed Question . . .	1	2	3	4	5	6	7	8	9	10	11	12	13	14	15	16	17	18	19
Review Section . . .	3.4	3.4	3.2	3.1	3.1	3.4	3.4	3.4	3.1	1.3	2.1	2.1	3.3	3.1	3.3	3.2	3.3	3.4	2.3

SHORT ANSWER

10. Both the independent and the dependent variables can have different values during the course of an experiment. The independent variable has specific values that are predetermined by the researcher, while the dependent variable has values that are measured as a result of the experiment and therefore cannot be determined in advance.

11. The value of −295°C is equivalent to −22 K. This answer does not make sense because 0 K is absolute zero and no temperature can be below this.

12. nano, micro, milli, centi, deci, kilo, mega, giga

EXTENDED RESPONSE

13. The mixture is heterogeneous. Based on particle size and color, the different substances can be readily distinguished.

14. These are physical properties because they depend only on the substance itself. Chemical properties depend on the behavior of substances as they react with other substances.

15. Use the solubility of mothballs in alcohol first: dissolve the mothballs and filter or decant the liquid. Use crystallization to remove the alcohol. Next, add water to the remaining substances: the sawdust will float because its density is less than that of water, and the salt will dissolve. Filter or decant to separate the two substances. Evaporate the water to obtain the salt.

16. A physical change does not change the composition of the substance, whereas a chemical change is a process in which one or more substances are changed into new substances. The combustion of gasoline is a chemical change because the gasoline is changed into other substances during combustion.

SAT SUBJECT TEST: CHEMISTRY

17. B
18. E
19. B

CHAPTER 4 The Structure of the Atom

BIGIDEA **Atoms are the fundamental building blocks of matter.**

ESSENTIAL QUESTIONS	RESOURCES TO ASSESS MASTERY
SECTION 1 **Early Ideas About Matter** **1.** What are the similarities and differences of the atomic models of Democritus, Aristotle, and Dalton? **2.** How does Dalton's theory explain the conservation of mass? 1 session 0.5 block	**Progress Monitoring** Formative Assessment, p. 105 Reading Check, pp. 103, 104 Section Review, p. 105
SECTION 2 **Defining the Atom** **1.** What is an atom? **2.** How can the subatomic particles be distinguished between in terms of relative charge and mass? **3.** Where are the locations of the subatomic particles within the structure of an atom? 2 sessions 1 block	**Progress Monitoring** Formative Assessment, pp. 107, 112, 113, 114 Reading Check, pp. 108, 109, 110, 112 Section Review, p. 114
SECTION 3 **How Atoms Differ** **1.** How is the atomic number used to determine the identity of an atom? **2.** What is an isotope? **3.** Why are atomic masses not whole numbers? **4.** Given the mass number and atomic number, how are the number of electrons, protons, and neutrons in an atom calculated? 1 session 0.5 block	**Progress Monitoring** Formative Assessment, p. 117 Reading Check, p. 120 Section Review, p. 121
SECTION 4 **Unstable Nuclei and Radioactive Decay** **1.** What is the relationship between unstable nuclei and radioactive decay? **2.** How are alpha, beta, and gamma radiation characterized in terms of mass and charge? 1 session 0.5 block	**Progress Monitoring** Formative Assessment, p. 124 Section Review, p. 124 **Summative Assessment** Chapter Assessment, p. 128 *eAssessment* Chapter Test (scaffolded)

Go online!

LEVELED RESOURCES	LAB MATERIALS
Science Notebook 4.1 OL ***FAST FILE Chapter Resources:*** Study Guide, p. 96 BL	LaunchLAB p. 100 / **15 min** paper, hole punch, plastic comb, tape
Science Notebook 4.2 OL ***FAST FILE Chapter Resources:*** Study Guide, p. 97 BL	
Science Notebook 4.3 OL ***FAST FILE Chapter Resources:*** MiniLab Worksheet, p. 82 OL ChemLab Worksheet, p. 83 OL Study Guide, p. 98 BL	MiniLAB p. 120 / **20 min** pennies, balance ChemLAB p. 126 / **45 min** balance, calculator, bag of snack mix
Science Notebook 4.4 OL ***FAST FILE Chapter Resources:*** Study Guide, p. 101 BL	

ADDITIONAL RESOURCES

- eAssessment
- ConnectED Student Center
- ConnectED Teacher Center
- Lesson Presentations
- eSolutions
- Videos
- Animations
- Online Quizzes
- Virtual Investigations
- Multilingual eGlossary
- Personal Tutors
- WebQuests
- Applying Practices Worksheets
- PBLs

Practice:

- Challenge Problems AL
- Supplemental Problems BL OL
- Chapter Test (Scaffolded)
- Standardized Test Practice

Additional Resources:

- Blueprints for Success
- Solving Problems: A Chemistry Handbook

Lab Resources:

- Laboratory Manual OL
- CBL Laboratory Manual OL
- Small-Scale Laboratory Manual OL
- Forensics Laboratory Manual OL

BL Below Level OL On Level AL Advanced Learners EL English Learners COOP LEARN Cooperative Learning

CHAPTER 4

LaunchLAB

Rubric

How can the effects of electric charges be observed?

Purpose Students will observe the behavior of charged objects.

Safety Precautions Approve lab safety forms before work begins. **WARNING:** ***Some students might be allergic to wool.***

Teaching Strategies

- Any brand of plastic tape should give the same results.
- When rubbed, plastic and hard rubber gain a negative charge; glass, fur, and wool gain a positive charge.
- Remind students that charging the objects is a separation of existing electric charges, not the creation of new electric charges.
- Make a list of materials with the same charge as the tape stuck to the table and a list of those materials with the same charge as the pieces of tape that were stuck together.

Expected Results

Step 3: The pieces of paper are attracted to the comb.
Step 4: The pieces of tape repel each other as they are brought together.
Step 5: The pieces of tape attract each other as they are brought together.

CHAPTER 4
The Structure of the Atom

BIGIDEA Atoms are the fundamental building blocks of matter.

SECTIONS

1 Early Ideas About Matter

2 Defining the Atom

3 How Atoms Differ

4 Unstable Nuclei and Radioactive Decay

LaunchLAB

How can the effects of electric charges be observed?

Electric charge plays an important role in atomic structure. In this lab, you will study the effects of like and opposite charges.

FOLDABLES® Study Organizer

The Atom

Make a concept-map book. Label it as shown. Use it to help you organize your study of the structure of the atom.

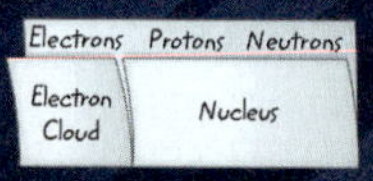

Procedure

1. Read and complete the lab safety form.
2. Cut out small round pieces of **paper** using a **hole punch,** and spread them out on a table.
3. Run a **plastic comb** through your hair. Bring the comb close to the pieces of paper. Record your observations.
4. Obtain two **10-cm pieces of tape.** Fold a 1-cm portion of each piece back on itself to form a handle. Stick both pieces of tape firmly to your desktop. Then, quickly pull both pieces off of the desktop and bring them close together so that their nonsticky sides face each other. Record your observations.
5. Stick a third piece of tape to your desktop. Stick a fourth piece of tape on top of it. Quickly pull the pieces of tape off of the desktop and pull them apart. Bring the two pieces close together so that their nonsticky sides face each other. Record your observations.

Analysis

1. **Interpret** your observations using your knowledge of electric charge. Determine which charges are similar and which ones are different. The pieces of paper and the comb, and the pieces of tape in Step 5 had opposite charges. The pieces of tape in Step 4 had the same charges.
2. **Explain** how you can tell. Like electrical charges repel each other, whereas opposite electrical charges attract each other. The

The graphite in this pencil and a diamond are made from atoms of the same element—carbon.

objects that repelled each other had similar charges. The objects that attracted each other had opposite charges.

3. **Infer** why neutral pieces of paper were attracted to the charged comb in Step 3 above. Neutral matter is made of negative and positive charges which can be attracted to other charges.

Inquiry How can you relate the different charges you have observed to the structure of matter? Answers will vary, but should contain some indication that atoms must have charged particles.

Introduce the BIGIDEA

Fundamental Building Blocks To introduce this chapter's Big Idea, lead students in a discussion about the fundamental building blocks of everyday items. Ask students to state what a red rosebush is made of at a macroscopic level. A red rosebush has red roses, green leaves, a stem, and roots. What about a car? A car has wheels, windows, doors, an engine, seats, and a steering wheel; this is not an exhaustive list. Ask students to state what water, table salt, and sugar are made of at a submicroscopic level. Water, table salt, and sugar are made of atoms.

Tie to Previous Knowledge

Have students review the following concepts before studying this chapter.

- matter, scientific laws, and theories
- elements, laws of definite and multiple proportions

Use the Photo

Atomic Structure Ask students what they think is the smallest particle of matter and whether they can see it with an unaided eye. Answers will vary. In section 2 students will learn about the building blocks of matter and how they can be observed.

SECTION 1

1 Focus

MAINIDEA

Development of the Atomic Model Refer students to Figure 10. Starting from 1860, go through the time line with students asking them to state the key discoveries of each period. OL

2 Teach

Enrichment

Observing Properties Prior to the seventeenth century, researchers were thought of as natural philosophers who relied on the method of deductive reasoning to explain the world around them. Using this method, philosophers observed events and extrapolated those observations to relate to other specific situations. The process was deductive because it went from general to specific, from universal to individual.

Sir Francis Bacon (1561-1626), an English lawyer and philosopher, argued passionately that although deductive reasoning might work in the realm of mathematics, the laws of science had to be induced to be established as generalizations drawn from a vast amount of experimental observations. Have students compare inductive and deductive reasoning, citing examples of each. OL

SECTION 1

Early Ideas About Matter

MAINIDEA The ancient Greeks tried to explain matter, but the scientific study of the atom began with John Dalton in the early 1800s.

Essential Questions

- What are the similarities and differences of the atomic models of Democritus, Aristotle, and Dalton?
- How was Dalton's theory used to explain the conservation of mass?

Review Vocabulary

theory: an explanation supported by many experiments; is still subject to new experimental data, can be modified, and is considered successful if it can be used to make predictions that are true

New Vocabulary

Dalton's atomic theory

CHEM 4 YOU A football team might practice and try different plays in order to develop the best-possible game plan. As they see the results of their plans, coaches can make adjustments to refine the team's play. Similarly, scientists over the last 200 years have hypothesized different models of the atom, refining their models as they collected new data.

The Roots of Atomic Theory

Science as we know it today did not exist several thousand years ago. No one knew what a controlled experiment was, and there were few tools for scientific exploration. In this setting, the power of the mind and intellectual thought were considered the primary avenues to the truth. Curiosity sparked the interest of scholarly thinkers known as philosophers who considered the many mysteries of life. As they speculated about the nature of matter, many of the philosophers formulated explanations based on their own life experiences.

Many of them concluded that matter was composed of things such as earth, water, air, and fire, as shown in **Figure 1.** It was also commonly accepted that matter could be endlessly divided into smaller and smaller pieces. While these early ideas were creative, there was no method available to test their validity.

Figure 1 Many Greek philosophers thought that matter was composed of four elements: earth, air, water, and fire. They also associated properties with each element. The pairing of opposite properties, such as hot and cold, and wet and dry, mirrored the symmetry they observed in nature. These early ideas were non-scientific.

CHEMISTRY JOURNAL

The Point of the Matter Show students a reproduction of any of Georges Seurat's artwork, which use the pointillistic technique. *A Sunday Afternoon on La Grande Jatte* is a good example. Show them the work at a distance, and then gradually decrease the viewing distance. Ask them to write how the artwork parallels Democritus's ideas about the composition of matter. OL

Democritus The Greek philosopher Democritus (460–370 B.C.) was the first person to propose the idea that matter was not infinitely divisible. He believed matter was made up of tiny individual particles called *atomos,* from which the English word *atom* is derived. Democritus believed that atoms could not be created, destroyed, or further divided. Democritus and a summary of his ideas are shown in **Table 1.**

While a number of Democritus's ideas do not agree with modern atomic theory, his belief in the existence of atoms was amazingly ahead of his time. However, his ideas were met with criticism from other philosophers who asked, "What holds the atoms together?" Democritus could not answer the question.

Aristotle Other criticisms came from Aristotle (384–322 B.C.), one of the most influential Greek philosophers. He rejected the notion of atoms because it did not agree with his own ideas about nature. One of Aristotle's major criticisms concerned the idea that atoms moved through empty space. He did not believe that empty space could exist. His ideas are also presented in **Table 1.** Because Aristotle was one of the most influential philosophers of his time, Democritus's atomic theory was eventually rejected.

In fairness to Democritus, it was impossible for him or anyone else of his time to determine what held the atoms together. More than two thousand years would pass before scientists would know the answer. However, it is important to realize that Democritus's ideas were just that—ideas, not science. Without the ability to conduct controlled experiments, Democritus could not test the validity of his ideas.

Unfortunately for the advancement of science, Aristotle was able to gain wide acceptance for his ideas on nature—ideas that denied the existence of atoms. Incredibly, the influence of Aristotle was so great and the development of science so primitive that his denial of the existence of atoms went largely unchallenged for two thousand years!

☑ READING CHECK **Infer** why it was hard for Democritus to defend his ideas.

VOCABULARY
WORD ORIGIN
Atom
comes from the Greek word *atomos,* meaning *indivisible*

Table 1 Ancient Greek Ideas About Matter

Philosopher		Ideas
Democritus (460–370 B.C.)		• Matter is composed of atoms, which move through empty space. • Atoms are solid, homogeneous, indestructible, and indivisible. • Different kinds of atoms have different sizes and shapes. • Size, shape, and movement of atoms determine the properties of matter.
Aristotle (384–322 B.C.)		• Empty space cannot exist. • Matter is made of earth, fire, air, and water.

(t)Science Photo Library/Photo Researchers, (b)©QEDimages/Alamy

Quick Demo

Discovery of the Atom
Build a mousetrap type of device using a marble that descends down a series of shoots, turns wheels, rings bells, and makes various noises. Many of these are available in children's board games. Shield the device with a cardboard box. You might wish to make part of the marble's descent visible through cut-out windows. Ask the students to draw what they think is going on inside the box based on their observations, and have them draw parallels to the discovery of the parts of the atom. OL

Enrichment

Subatomic Exploration Obtain the video of Ray and Charles Eames' "Powers of Ten," a short dramatization that takes students on a journey from the subatomic level to intergalactic space. The video also provides an excellent review of scientific notation.

☑ READING CHECK It was difficult for Democritus to defend his ideas because he did not perform experiments.

CHEMISTRY PROJECT

Greek Philosophers Have students research Democritus and Aristotle and their contributions to the field of chemistry. Students should find that Aristotle was not considered a chemist, whereas Democritus' ideas were ahead of his time. OL

Extension

Dalton's Atomic Theory John Dalton, a practicing Quaker, left school at age 11 and then returned to teach at age 12! Dalton's first love was meteorology, and he kept careful daily recordings of the weather for 57 years. His book *Meteorological Observations and Essays* (1793) qualified him as one of the pioneers in this area of study. Such meteorological studies led Dalton to think about the composition and properties of air. By studying the works of Democritus, Boyle, and Proust, Dalton went on to formalize his own atomic theory. This theory extended beyond gases—it encompassed all matter. Dalton advanced this theory in his book *New System of Chemical Philosophy* (1808). Have students research Dalton's original writings on his atomic theory, citing examples of where the theory has been found to be in error and where it is still held to be true. **AL**

☑ **READING CHECK** similarities: matter composed of atoms; atoms are indestructible and indivisible; changes in matter are due to changes in groupings of atoms; differences: Democritus stated that matter is composed of empty space through which atoms move, whereas Dalton made no such claims. Dalton's theory states that atoms can combine to form compounds.

Table 2 Dalton's Atomic Theory

Scientist		Ideas
Dalton (1766–1844)		• Matter is composed of extremely small particles called atoms. • Atoms are indivisible and indestructible. • Atoms of a given element are identical in size, mass, and chemical properties. • Atoms of a specific element are different from those of another element. • Different atoms combine in simple whole-number ratios to form compounds. • In a chemical reaction, atoms are separated, combined or rearranged.

John Dalton Although the concept of the atom was revived in the eighteenth century, it took another hundred years before significant progress was made. The work done in the nineteenth century by John Dalton (1766–1844), a schoolteacher in England, marks the beginning of the development of modern atomic theory. Dalton revived and revised Democritus's ideas based on the results of scientific research he conducted. In many ways, Democritus's and Dalton's ideas are similar.

Thanks to advancements in science since Democritus's day, Dalton was able to perform experiments that allowed him to refine and support his hypotheses. He studied numerous chemical reactions, making careful observations and measurements along the way. He was able to determine the mass ratios of the elements involved in those reactions. The results of his research are known as **Dalton's atomic theory,** which he proposed in 1803. The main points of his theory are summarized in **Table 2.** Dalton published his ideas in a book, an extract of which is shown in **Figure 2.**

☑ READING CHECK **Compare and contrast** Democritus' and Dalton's ideas.

Figure 2 In his book *A New System of Chemical Philosophy*, John Dalton presented his symbols for the elements known at that time and their possible combinations.

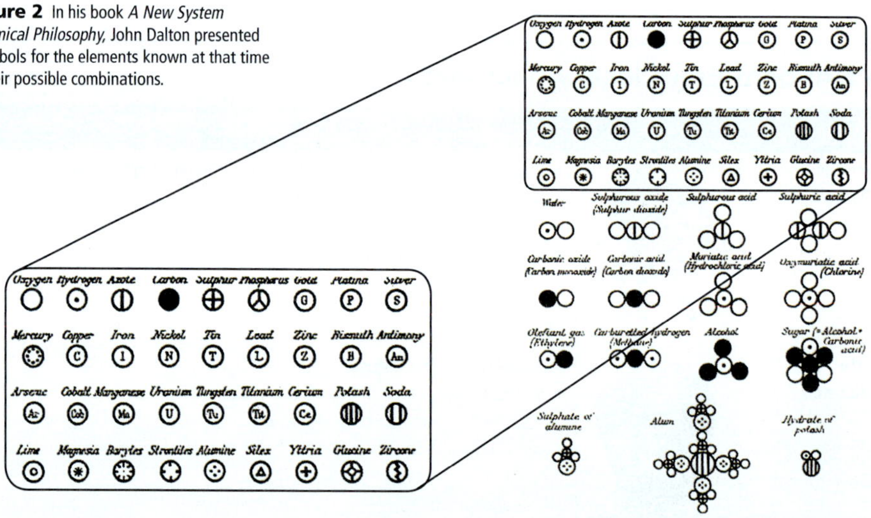

DIFFERENTIATED INSTRUCTION

Hearing Impaired John Dalton believed that chemical formulas should be encoded as pictographs, in which each element would have a particular drawing associated with it. Have students research Dalton's pictographs, and draw both the pictograph and its letter symbol counterpart. **OL**

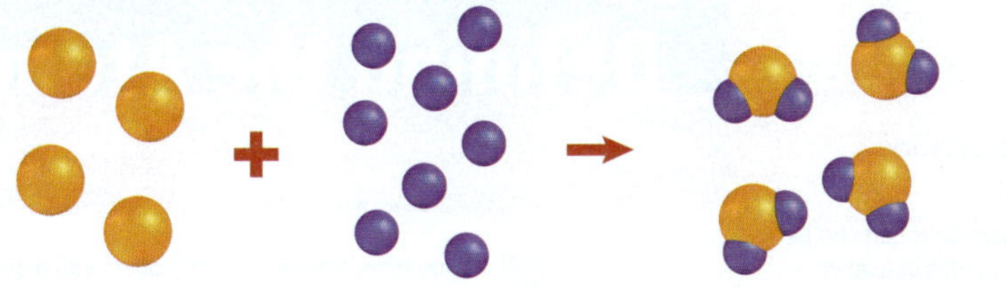

Figure 3 When atoms of two or more elements combine to form a compound, the number of atoms of each element is conserved. Thus, the mass is conserved as well.

Conservation of mass Recall that the law of conservation of mass states that mass is conserved in any process, such as a chemical reaction. Dalton's atomic theory easily explains that the conservation of mass in chemical reactions is the result of the separation, combination, or rearrangement of atoms—atoms that are not created, destroyed, or divided in the process. The formation of a compound from the combining of elements and the conservation of mass during the process are shown in **Figure 3.** The number of atoms of each type is the same before and after the reaction. Dalton's convincing experimental evidence and clear explanation of the composition of compounds, and conservation of mass led to the general acceptance of his atomic theory.

Dalton's atomic theory was a huge step toward the current atomic model of matter. However, not all of Dalton's theory was accurate. As is often the case in science, Dalton's theory had to be revised as additional information was learned that could not be explained by the theory. As you will learn in this chapter, Dalton was wrong about atoms being indivisible. Atoms are divisible into several subatomic particles. Dalton was also wrong about all atoms of a given element having identical properties. Atoms of the same element can have slightly different masses.

SECTION 1 REVIEW

Section Self-Check

Section Summary

- Democritus was the first person to propose the existence of atoms.
- According to Democritus, atoms are solid, homogeneous, and indivisible.
- Aristotle did not believe in the existence of atoms.
- John Dalton's atomic theory is based on numerous scientific experiments.

1. **MAIN IDEA Contrast** the methods used by the Greek philosophers and Dalton to study the atom.
2. **Define** *atom* using your own words.
3. **Summarize** Dalton's atomic theory.
4. **Explain** how Dalton's theory of the atom and the conservation of mass are related.
5. **Apply** Six atoms of Element A combine with eight atoms of Element B to produce six compound particles. How many atoms of Elements A and B does each particle contain? Are all of the atoms used to form compounds?
6. **Design** a concept map that compares and contrasts the atomic ideas proposed by Democritus and John Dalton.

ASSESSMENT

Performance Ask students to build a device such as the one mentioned in the Quick Demo. The contraptions can be shared in class, and students can draw or describe how they think the contraptions work. Emphasize the connection between the observation of a phenomenon and the model developed to explain it. OL EL

3 Assess

Check for Understanding

Have students look through newspaper and magazine articles for examples of opinions, laws, and theories in science, government, economics, or any other area of interest. Have them copy the articles and highlight the examples. OL

Reteach

Read aloud sections from the childhood favorite *The Cat in the Hat Comes Back* by Dr. Seuss. Ask students to write how the story parallels the development of the various atomic models discussed in this chapter. OL

Extension

Have students research information about alchemy and how the work of alchemists in the Middle Ages influenced the development of modern chemistry. OL

SECTION 1 REVIEW

1. Greek philosophers could not conduct experiments to verify their hypothesis, whereas Dalton could make careful measurements.
2. An atom is the smallest component of an element that exhibits all of the characteristic properties of that element.
3. Matter is composed of atoms that are indivisible. Atoms of an element are identical in size, mass, and chemical properties. Atoms of a specific element are different from atoms of another element. In a chemical reaction, atoms are separated, combined, or rearranged.
4. Dalton explained that atoms are neither created nor destroyed in chemical reactions but only rearranged.
5. One of each; no, two atoms of Element B are left.
6. Concept maps will vary, but should reflect the following summary. Both believed: matter composed of extremely small particles called atoms; all atoms of a given element are identical, but differ from the atoms of other elements; atoms could not be created, divided, or destroyed. Democritus also believed that matter is composed of empty space through which atoms move, that different kinds of atoms come in different sizes and shapes, and that the differing properties of atoms are due to the size, shape, and movement of the atoms. Dalton specified that different atoms combine in simple whole-number ratios to form compounds.

SECTION 2

1 Focus

MAINIDEA

Structure of the Atom Bring a fruit (peach or plum) cut in halves to class. Use the fruit to demonstrate the structure of an atom. The pit resembles the nucleus, whereas the pulp represents the regions where electrons are constantly moving. BL

2 Teach

Quick Demo

Small Sizes Place a 1-cm grid on an overhead projector. Place a clear lid or tray filled with water on top of the grid. Make an oleic acid solution (0.50% dissolved in methanol). Using a pipette and a small graduated cylinder, have students calculate the volume of a single drop of 0.50% oleic acid. Next, sprinkle the surface of the water with lycopodium or talcum powder. Place a single drop of the oleic acid in the middle of the water. The acid will spread out to the thickness of one molecule. Have students calculate the area by counting the squares. The density of oleic acid is 0.895 g/cm^3 and the molecule has a length and width one-tenth of its thickness. Have students calculate the mass and thickness of the molecule. Dispose of the oleic acid, lycopodium powder, talcum powder, and methanol by rinsing them down a drain with a large volume of water. Tell students mass = (volume)(density) and thickness = (volume/area). OL

SECTION 2

Essential Questions

- What is an atom?
- How can the subatomic particles be distinguished in terms of relative charge and mass?
- Where are the locations of the subatomic particles within the structure of the atom?

Review Vocabulary

model: a visual, verbal, and/or mathematical explanation of data collected from many experiments

New Vocabulary

atom
cathode ray
electron
nucleus
proton
neutron

Defining the Atom

MAINIDEA An atom is made of a nucleus containing protons and neutrons; electrons move around the nucleus.

CHEM 4 YOU If you have ever accidentally bitten into a peach pit, you know that your teeth pass easily through the fruit, but cannot dent the hard pit. Similarly, many particles that pass through the outer parts of an atom are deflected by the dense center of the atom.

The Atom

Many experiments since Dalton's time have proven that atoms do exist. So what exactly is the definition of an atom? To answer this question, consider a gold ring. Suppose you decide to grind the ring down into a pile of gold dust. Each fragment of gold dust still retains all of the properties of gold. If it were possible—which it is not without special equipment—you could continue to divide the gold dust particles into still smaller particles. Eventually, you would encounter a particle that could not be divided any further and still retain the properties of gold. This smallest particle of an element that retains the properties of the element is called an **atom.**

To get an idea of its size, consider the population of the world, which was about 6.5×10^9 in 2006. By comparison, a typical solid-copper penny contains 2.9×10^{22} atoms, almost five trillion times the world population! The diameter of a single copper atom is 1.28×10^{-10} m. Placing 6.5×10^9 copper atoms side by side would result in a line of copper atoms less than 1 m long. **Figure 4** illustrates another way to visualize the size of an atom. Imagine that you increase the size of an atom to be as big as an orange. To keep the proportions between the real sizes of the atom and of the orange, you would have to increase the size of the orange and make it as big as Earth. This illustrates how small atoms are.

Figure 4 Imagine that you could increase the size of an atom to make it as big as an orange. At this new scale, an orange would be as big as Earth.

(l)©Stockdisc/PunchStock; (r)European Space Agency/Photo Researchers

Demonstration

Thomson's Experiment

Purpose
to observe the characteristics of a cathode-ray tube

Materials
Crookes's tube (cathode-ray tube); high-voltage DC power supply; bar magnet

Safety Precautions

Disposal Items can be saved and reused.

Procedure
Connect the two electrodes of the power supply to the ends of the tube using wires with alligator clips on each end. Turn on the power supply. **WARNING**: ***The high-voltage power supply can cause severe electric shock.*** If the tube fails to light, increase the voltage from the power supply. If the tube still fails to light, turn off the power supply and reverse the wire connections on the tube. Have students

Connection to Biology **Looking at atoms** You might think that because atoms are so small, there would be no way to see them. However, an instrument called the scanning tunneling microscope (STM) allows individual atoms to be seen. Just as you need a microscope to study cells in biology, the STM allows you to study atoms. An STM works as follows: a fine point is moved above a sample and the interaction of the point with the superficial atoms is recorded electronically. **Figure 5** illustrates how individual atoms look when observed with an STM. Scientists are now able to move individual atoms around to form shapes, patterns, and even simple machines. This capability has led to the exciting new field of nanotechnology. The promise of nanotechnology is molecular manufacturing—the atom-by-atom building of machines the size of molecules. As you will later read, a molecule is a group of atoms that are bonded together and act as a unit.

Figure 5 This image, recorded with an STM, shows the individual atoms of a fatty acid on a graphite surface. The false colors were added later on to improve the contrast between each atom.

The Electron

Once scientists were convinced of the existence of atoms, a new set of questions emerged. What is an atom like? Is the composition of an atom uniform throughout, or is it composed of still-smaller particles? Although many scientists researched the atom in the 1800s, it was not until almost 1900 that some of these questions were answered.

The cathode-ray tube As scientists tried to unravel the atom, they began to make connections between matter and electric charge. For instance, has your hair ever clung to your comb? To explore the connection, some scientists wondered how electricity might behave in the absence of matter. With the help of the newly invented vacuum pump, they passed electricity through glass tubes from which most of the air had been removed. Such tubes are called cathode-ray tubes.

A typical cathode-ray tube used by researchers for studying the relationship between mass and charge is illustrated in **Figure 6.** Note that metal electrodes are located at opposite ends of the tube. The electrode connected to the negative terminal of the battery is called the cathode, and the electrode connected to the positive terminal is called the anode.

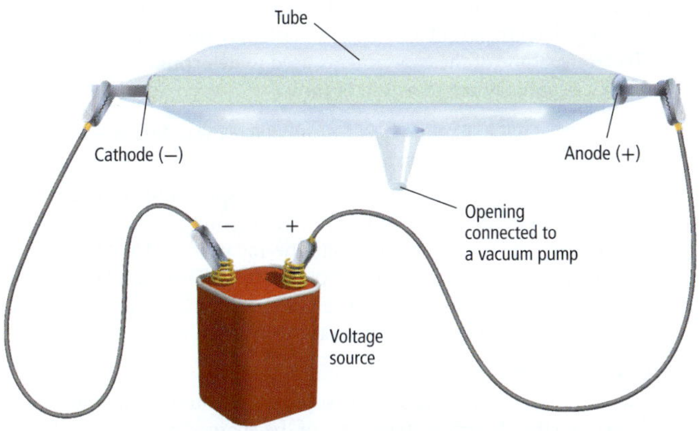

Figure 6 A cathode-ray tube is a tube with an anode at one end and a cathode at the other end. When a voltage is applied, electricity travels from the cathode to the anode.

Quick Demo

Electric Charge Use an electroscope or pith balls to show the transfer of electrons and the consequences. Rub an acrylic wand with fleece or silk (this pulls off the electrons from the material) and then touch the wand to the electroscope or pith balls. The foil leaves of the electroscope (or the pith balls themselves) will acquire a negative charge that will cause them to separate from each other. Reinforce the rule that like charges repel.

Extension

STM Have students research how scanning tunneling microscopes (STM) and scanning probe microscopes (SPM) function and why the technology is important. OL

ASSESSMENT

Skill Given the number of atoms in a pre-1982 penny, 2.9×10^{22}, have students calculate the approximate volume of a single atom of copper. Then, have them determine the number of atoms it would take to fill up any volume of interest, such as a 2-L soda bottle. OL

observe the electron beam on the fluorescent screen. Deflect the beam with the magnet. Reverse the magnet's field by flipping it around and use it to show deflection of the beam in the opposite direction.

Results

Students will see evidence of an electron beam. The phosphors on the plate glow along the path of the beam. Point out that Thomson observed deflection by means of both magnets and charged plates.

Analysis

1. What properties account for the bending of the cathode ray in a magnetic field? The ray has an electric charge.
2. What is the function of the electromagnets on a TV picture tube? They cause the electron beam to move back and forth, forming an image on the inside face of the luminescent chemical-coated screen.

ASSESSMENT

Skill Have students research an article describing how a CRT or television works. OL

Enrichment

Discoveries and Errors The discovery of the inert gases in the 1890s led Dimitri Mendeleev to speculate about ether, which scientists of the time believed was the component of empty space through which light traveled.

Because Mendeleev predicted the existence and properties of elements heavier than hydrogen, he believed that he could also predict the existence of an element lighter than hydrogen (element X). In his writings, he noted that element X was "capable of moving freely everywhere throughout the universe, with an atomic weight nearly one millionth that of hydrogen and traveling with a velocity of about 2250 kilometers per second." He placed element X at the top of the family of inert gases. Mendeleev resisted the concept of the electron because he felt that atoms had no further internal structure.

Have students investigate and write about other scientists who, while credited with one or more discoveries, were in error with some of their other theories. AL

FOLDABLES® Rubric

READING CHECK Crookes observed a flash of light within one of his tubes while he was working in a darkened laboratory.

■ **Figure 7** A tiny hole located in the center of the anode produces a thin beam of electrons. A phosphor coating allows the position of the beam to be determined as it strikes the end of the tube.

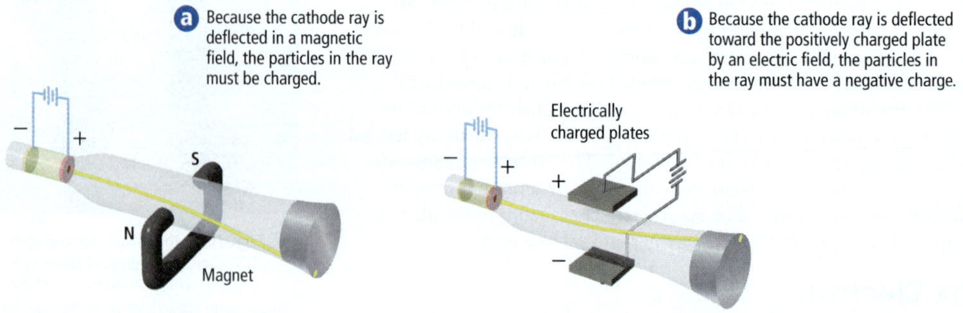

RealWorld CHEMISTRY

Cathode Ray

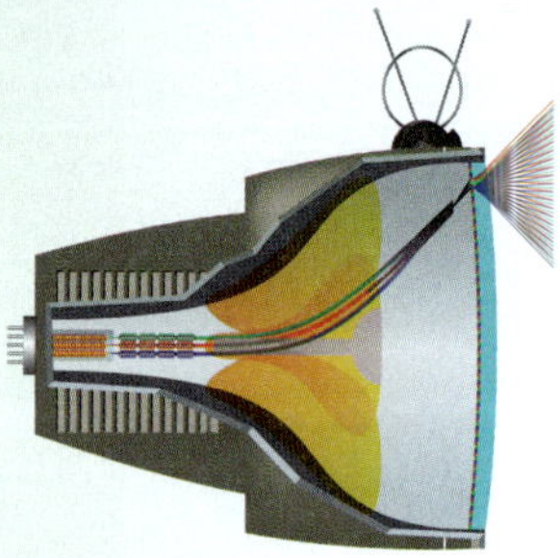

TELEVISION Television was invented in the 1920's. Conventional television images are formed as cathode rays strike light-producing chemicals that coat the back of the screen.

Sir William Crookes While working in a darkened laboratory, English physicist Sir William Crookes noticed a flash of light within one of the cathode-ray tubes. A green flash was produced by some form of radiation striking a zinc-sulfide coating that had been applied to the end of the tube. Further work showed that there was a ray (radiation) going through the tube. This ray, originating from the cathode and traveling to the anode, was called a **cathode ray.** The accidental discovery of the cathode ray led to the invention of television. A conventional television is nothing more than a cathode-ray tube.

Scientists continued their research using cathode-ray tubes, and they were fairly convinced by the end of the 1800s of the following:

- Cathode rays were a stream of charged particles.
- The particles carried a negative charge. (The exact value of the negative charge was not known.)

Because changing the metal that makes up the electrodes or varying the gas (at very low pressure) in the cathode-ray tube did not affect the cathode ray produced, researchers concluded that the ray's negative particles were found in all forms of matter. These negatively charged particles that are part of all forms of matter are now known as **electrons.** Some of the experiments used to determine the properties of the cathode ray are shown in **Figure 7.**

READING CHECK **Explain** how the cathode ray was discovered.

FOLDABLES® Incorporate information from this section into your Foldable.

Mass and charge of the electron In spite of the progress made from all of the cathode-ray tube experiments, no one succeeded in determining the mass of a single cathode-ray particle. Unable to measure the particle's mass directly, English physicist J. J. Thomson (1856–1940) began a series of cathode-ray tube experiments at Cambridge University in the late 1890s to determine the ratio of its charge to its mass.

Charge-to-mass ratio By carefully measuring the effects of both magnetic and electric fields on a cathode ray, Thomson was able to determine the charge-to-mass ratio of the charged particle. He then compared that ratio to other known ratios.

DIFFERENTIATED INSTRUCTION

Hearing Impaired Provide students with two magnets each and have them determine which sides are attracted to each other and which ones are repelled. Ask students to label the sides that are attracted to each other as positive and negative, such that each magnet has a positive and a negative end. Place one magnet on the desk. Bring two positive ends close to each other and record observations. Repeat the experiment with one positive end and one negative end, and record observations. Ask students to make a parallel with the cathode-ray tube experiment. Like ends of the magnets will repel (positive-positive or negative-negative), whereas opposite ends (positive-negative or negative-positive) will attract. This concept was used in the cathode-ray tube experiment using electrically charged plates, as depicted in Figure 7. It was concluded from observations that the cathode ray is negatively charged. OL

Thomson concluded that the mass of the charged particle was much less than that of a hydrogen atom, the lightest known atom. The conclusion was shocking because it meant there were particles smaller than the atom. In other words, Dalton had been incorrect—atoms were divisible into smaller subatomic particles. Because Dalton's atomic theory had become so widely accepted and Thomson's conclusion was so revolutionary, many other scientists found it hard to accept this new discovery. But Thomson was correct. He had identified the first subatomic particle—the electron. He received a Nobel Prize in 1906 for this discovery.

☑ READING CHECK **Summarize** how Thomson discovered the electron.

The oil-drop experiment and the charge of an electron The next significant development came in the early 1910s, when the American physicist Robert Millikan (1868–1953) determined the charge of an electron using the oil-drop apparatus shown in **Figure 8.** In this apparatus, oil is sprayed into the chamber above the two parallel charged plates. The top plate has a small hole through which the oil drops. X-rays knock out electrons from the air particles between the plates and the electrons stick to the droplets, giving them a negative charge. By varying the intensity of the electric field, Millikan could control the rate of a droplet's fall. He determined that the magnitude of the charge on each drop increased in discrete amounts and determined that the smallest common denominator was 1.602×10^{-19} coulombs. He identified this number as the charge of the electron. This charge was later equated to a single unit of negative charge noted 1−; in other words, a single electron carries a charge of 1−.

So good was Millikan's experimental setup and technique that the charge he measured almost one hundred years ago is within 1% of the currently accepted value.

Mass of an electron Knowing the electron's charge and using the known charge-to-mass ratio, Millikan calculated the mass of an electron. The equation below shows how small the mass of an electron is.

$$\text{Mass of an electron} = 9.1 \times 10^{-28}\ \text{g} = \frac{1}{1840}\ \text{the mass of a hydrogen atom}$$

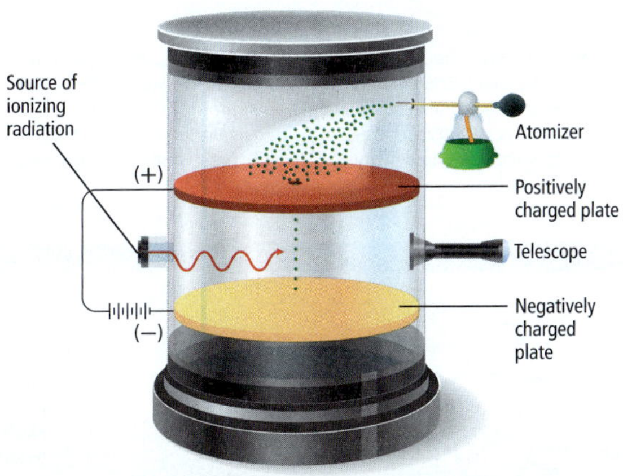

■ **Figure 8** The motion of the oil droplets within Millikan's apparatus depends on the charge of droplets and on the electric field. Millikan observed the droplets with the telescope. He could make the droplets fall more slowly, rise, or pause as he varied the strength of the electric field. From his observations, he calculated the charge on each droplet.

Apply Chemistry

Television A more sophisticated version of the cathode-ray tube can be found in many homes in the form of a television set or computer monitor. A conventional television uses magnets to direct the electron beam back and forth across the back side of its screen. The screen is coated with luminescent chemicals that produce the colors we see when it is illuminated by the electron beam.

☑ READING CHECK Thomson performed a series of experiments with a cathode-ray tube. He measured the effect of magnetic and electric fields to determine the mass-to-charge ratio of particles. He concluded that the charged particles were much lighter than the lightest known atom, which meant that atoms had a structure.

DIFFERENTIATED INSTRUCTION

Advanced Learners Interested students can research and prepare a tutorial for struggling students on the oil drop experiment performed by Robert Millikan. AL

Quick Demo

CRT Show the students a CRT with a paddlewheel or use a photograph of one (lab equipment catalogs are a good source of such photographs). Emphasize to the students that the only way for the paddlewheel to move uphill against gravity is to be bombarded by a stream of particles emanating from the cathode.

Concept Development

Lise Meitner Have students verify the date of Lise Meitner's work on the time line, then brainstorm world events affecting Europe during that period. Inform students that Meitner was born into a Jewish family in Austria when females were discouraged from pursuing higher education, and that she fled Berlin for Sweden in 1938.

Ask Students As a scientist, what obstacles did Meitner face due to the period in which she lived? How would you compare obstacles she faced to those faced by scientists working in the United States today? As a female, Meitner overcame societal objections to pursue high school, college, and post-graduate degrees. Identified as a Jew, Meitner faced potential Nazi persecution. The work of U.S. scientists today is not bound by race, religion, or gender, but can be affected by funding for higher education and research. AL OL

☑ READING CHECK Thomson's model of the atom consists of electrons distributed throughout a uniform spherical shape, like plums are distributed in a uniform pudding.

■ **Figure 9** J. J. Thomson's plum pudding model of the atom states that the atom is a uniform, positively charged sphere containing electrons.

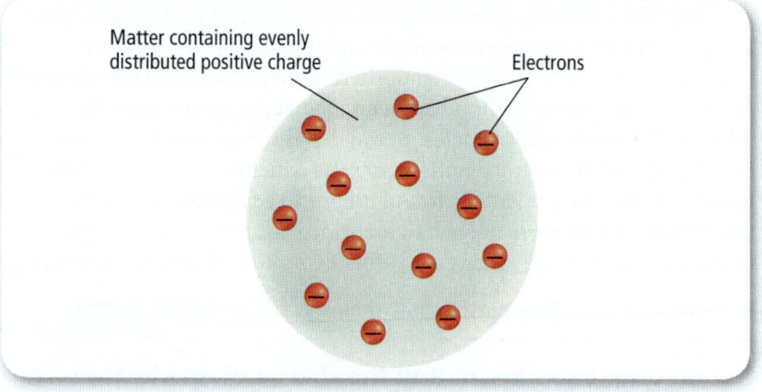

The plum pudding model The existence of the electron and the knowledge of some of its properties raised some interesting new questions about the nature of atoms. It was known that matter is neutral—it has no electric charge. You know that matter is neutral from everyday experience: you do not receive an electric shock (except under certain conditions) when you touch an object. If electrons are part of all matter and they possess a negative charge, how can all matter be neutral? Also, if the mass of an electron is so small, what accounts for the rest of the mass in a typical atom?

In an attempt to answer these questions, J. J. Thomson proposed a model of the atom that became known as the plum pudding model. As you can see in **Figure 9,** Thomson's model consisted of a spherically shaped atom composed of a uniformly distributed positive charge in which the individual negatively charged electrons resided. As you are about to read, the plum pudding model of the atom did not last for long. **Figure 10** summarizes the numerous steps in understanding the structure of the atom.

☑ READING CHECK **Explain** why Thomson's model was called the plum pudding model.

SSPL/The Image Works

■ **Figure 10**

Development of Modern Atomic Theory

Current understanding of the properties and behavior of atoms and subatomic particles is based on the work of scientists worldwide during the past two centuries.

1860 — 1885 — 1910

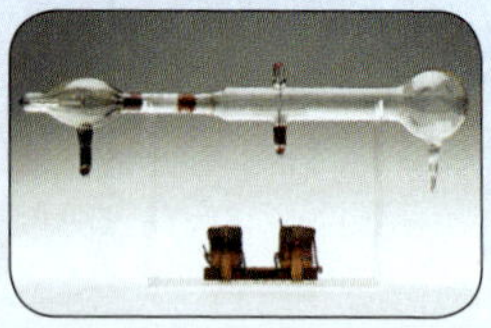

1897 Using cathode-ray tubes, J. J. Thomson identifies the electron and determines the ratio of the mass of an electron to its electric charge.

1911 With the gold foil experiment, Ernest Rutherford determines properties of the nucleus, including charge, relative size, and density.

1913 Niels Bohr publishes a theory of atomic structure relating the electron arrangement in atoms and atomic chemical properties.

1932 Scientists develop a particle accelerator to fire protons at lithium nuclei, splitting them into helium nuclei and releasing energy.

1932 James Chadwick proves the existence of neutrons.

CHEMISTRY PROJECT

Atomic Force Microscope Have students research on how atomic force microscopes (AFM) can be used to learn about atoms adsorbed on a surface. Have students create an informational brochure on their findings. OL

CHEMISTRY JOURNAL

Comparing Models Have students use their journals to sketch the structure of an atom according to the plum pudding and nuclear atomic models. Have them label the subatomic particles in each sketch and list the key points associated with each model. They should clearly label that the nuclear model is the currently accepted model of the atom. OL EL

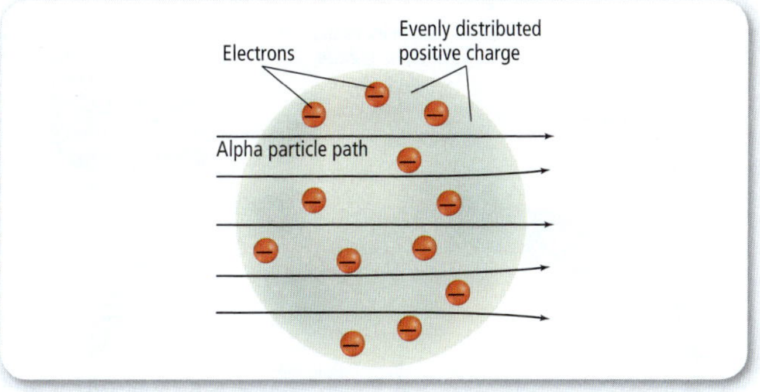

■ **Figure 11** Based on Thomson's model, Rutherford expected the light alpha particles to pass through gold atoms. He expected only a few of them to be slightly deflected.

The Nucleus

In 1911, Ernest Rutherford (1871–1937) began to study how positively charged alpha particles (radioactive particles you will read more about later in this chapter) interacted with solid matter. With a small group of scientists, Rutherford conducted an experiment to see if alpha particles would be deflected as they passed through a thin gold foil.

Rutherford's experiment In the experiment, a narrow beam of alpha particles was aimed at a thin sheet of gold foil. A zinc-sulfide-coated screen surrounding the gold foil produced a flash of light when struck by an alpha particle. By noting where the flashes occurred, the scientists could determine if the atoms in the gold foil deflected the alpha particles.

Rutherford was aware of Thomson's plum pudding model of the atom. He expected the paths of the massive and fast-moving alpha particles to be only slightly altered by a collision with an electron. And because the positive charge within the gold atoms was thought to be uniformly distributed, he thought it would not alter the paths of the alpha particles, either. **Figure 11** shows the results Rutherford expected from the experiment.

1938 Lise Meitner, Otto Hahn, and Fritz Straussman split uranium atoms in a process they called fission.

1954 CERN, the world's largest nuclear physics research center, located in Switzerland, is founded to study particle physics.

2010 The Large Hadron Collider at CERN enters a new era when it smashes beams of protons at the highest energy ever.

1960 — 1985 — 2010

1939–1945 Scientists in the United States and Germany each work on projects to develop the first atomic weapon.

1968 Scientists provide the first experimental evidence for subatomic particles known as quarks.

1995 Scientists discover the top quark, the last of the six quarks predicted to exist. The top quark mass has helped scientists to theorize about the mass of the yet discovered Higgs boson particle.

Content Background

Particle Accelerators Particle accelerators are the most commonly used tool for scientists studying the particles that comprise atoms. There are two main types of particle accelerators: linear particle accelerators (linacs) and circular accelerators. The longest linac in the world is the Stanford Linear Accelerator, located in California, which is 3 km long. The largest circular accelerator, is the Large Hadron Collider in France and Switzerland, which has a circumference of about 27 km. These facilities are used by scientists from many cooperating nations.

DIFFERENTIATED INSTRUCTION

Struggling Learners Consider creating a gold foil "shooting gallery" within a part of your classroom to conceptually recreate aspects of Rutherford's experiment. Using string, hang tennis balls from the ceiling at 60-cm intervals (horizontally and vertically) so that they form a small rectangular field. These tennis balls represent nuclei of the gold atoms in the foil. Have students take turns throwing a table tennis ball "alpha particle" through the field of tennis ball "gold atoms." Record the number of throws that go through versus those that are deflected. Ask students to compare the in-classroom setup and results with the those of Rutherford's gold foil experiment. BL EL

ASSESSMENT

Knowledge Ask students to react to the following statement "Most of matter is nothing." Have them use specific experimental evidence discussed in this section. **OL**

READING CHECK In Rutherford's model, the atom is composed of a small, dense nucleus that contains all the positive charge and almost all its mass. The electrons move through the empty space surrounding the nucleus.

Extension

Rutherford Ask students to research information on Rutherford's work, especially his contribution to the study of alpha-ray properties. Students can write an essay about his life in their chemistry journals. **AL**

Caption Question Fig. 13 The positive alpha particles are repelled by the positive nucleus.

View an **animation about the gold foil experiment.**

Concepts In Motion

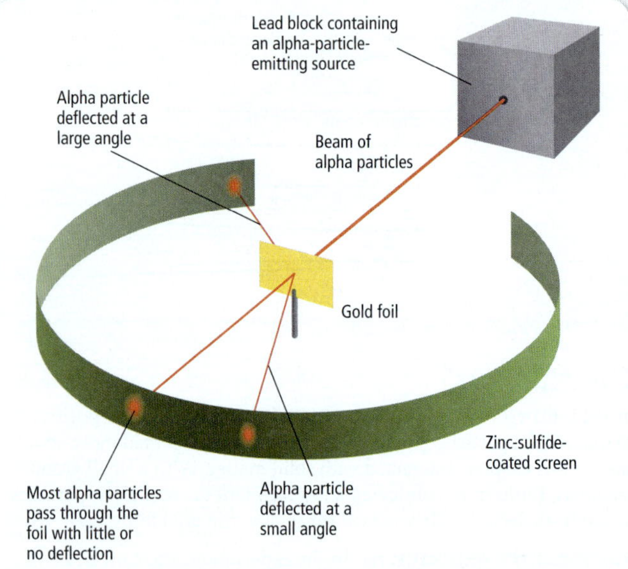

Figure 12 During Rutherford's experiment, a beam of alpha particles bombarded a thin gold foil. Most of the alpha particles went through the gold foil. However, a few of them bounced back, some at large angles.

The actual results observed by Rutherford and his colleagues are shown in **Figure 12.** A few of the alpha particles were deflected at large angles. Several particles were deflected straight back toward the source. Rutherford likened the results to firing a large artillery shell at a sheet of paper and the shell coming back at the cannon.

Rutherford's model of the atom Rutherford concluded that the plum pudding model was incorrect because it could not explain the results of the gold foil experiment. Considering the properties of the alpha particles and the electrons, and the frequency of the deflections, he calculated that an atom consisted mostly of empty space through which the electrons move. He also concluded that almost all of the atom's positive charge and almost all of its mass were contained in a tiny, dense region in the center of the atom, which he called the **nucleus.** The negatively charged electrons are held within the atom by their attraction to the positively charged nucleus. Rutherford's nuclear atomic model is shown in **Figure 13.**

Because the nucleus occupies such a small space and contains most of an atom's mass, it is incredibly dense. If a nucleus were the size of the dot in the exclamation point at the end of this sentence, its mass would be approximately as much as that of 70 automobiles! The volume of space through which the electrons move is huge compared to the volume of the nucleus. A typical atom's diameter is approximately 10,000 times the diameter of the nucleus. If an atom had a diameter of two football fields, the nucleus would be the size of a nickel.

READING CHECK **Describe** Rutherford's model of the atom.

Figure 13 In Rutherford's nuclear model, the atom is composed of a dense, positively charged nucleus that is surrounded by negative electrons. Alpha particles passing far from the nucleus are only slightly deflected. Alpha particles directly approaching the nucleus are deflected at large angles.
Infer *what force causes the deflection of alpha particles.*

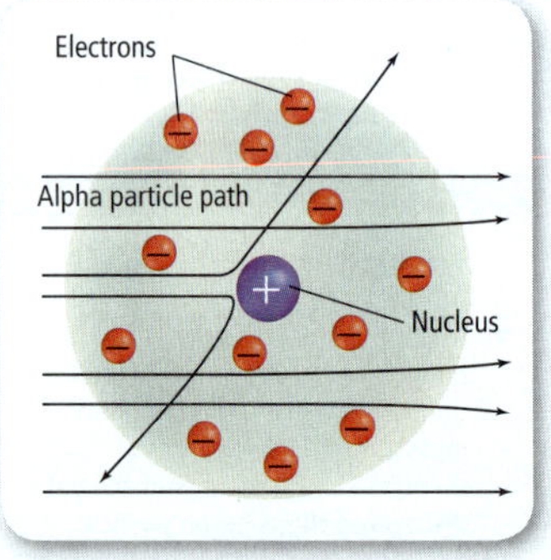

Demonstration

The Gold Foil Experiment

Purpose

to demonstrate Rutherford's Gold Foil Experiment

Materials

Laser pointer (preferably green, red will also work but not as effective), Inkjet transparency film, two empty picture frames, glue, three ring stands and clamps

Safety Precautions

Laser pointers are potential eye hazards. Instructors must be careful when performing this demonstration and be certain that none of the light paths impinge on students.

Procedure

1. Place two picture frames fixed on ring stands and a laser pointer fixed in a three-fingered clamp on the table.
2. Rutherford expected to find very little scattering based on the plum pudding model and confirm the structure of the atom.
3. Turn the laser pointer on, shine the beam through the frame containing no transparency film, and strike a screen on a wall.
4. Tell students that Rutherford's experiment did not support what he expected to see.
5. Turn the laser pointer on, shine the beam through the frame containing a transparency film and strike a screen on a wall.

The repulsive force produced between the positive nucleus and the positive alpha particles causes the deflections. **Figure 13** illustrates how Rutherford's nuclear atomic model explained the results of the gold foil experiment. The nuclear model also explains the neutral nature of matter: the positive charge of the nucleus balances the negative charge of the electrons. However, the model still could not account for all of the atom's mass.

The proton and the neutron By 1920, Rutherford had refined the concept of the nucleus and concluded that the nucleus contained positively charged particles called protons. A **proton** is a subatomic particle carrying a charge equal to but opposite that of an electron; that is, a proton has a charge of 1+. In 1932, Rutherford's coworker, English physicist James Chadwick (1891–1974), showed that the nucleus also contained another subatomic neutral particle, called the neutron. A **neutron** is a subatomic particle that has a mass nearly equal to that of a proton, but it carries no electric charge. In 1935, Chadwick received the Nobel Prize in Physics for proving the existence of neutrons.

VOCABULARY
SCIENCE USAGE V. COMMON USAGE
Neutral
Science usage: to have no electric charge
Neutrons have a charge of zero. They are neutral particles.
Common usage: not engaged in either side
Switzerland remained neutral during World War II.

Data Analysis LAB

Based on Real Data*

Interpret Scientific Illustrations

What are the apparent atomic distances of carbon atoms in a well-defined crystalline material? To visualize individual atoms, a group of scientists used a scanning tunneling microscope (STM) to test a crystalline material called highly ordered pyrolytic graphite (HOPG). An STM is an instrument used to perform surface atomic-scale imaging.

Data and Observations

The image shows all of the carbon atoms in the surface layer of the graphite material. Each hexagonal ring, indicated by the drawing in the figure, consists of three brighter spots separated by three fainter spots. These bright spots are from alternate carbon atoms in the surface layer of the graphite structure. The cross-sectional view below the photo corresponds to the line drawn in the image. It indicates the atomic periodicity and apparent atomic distances.

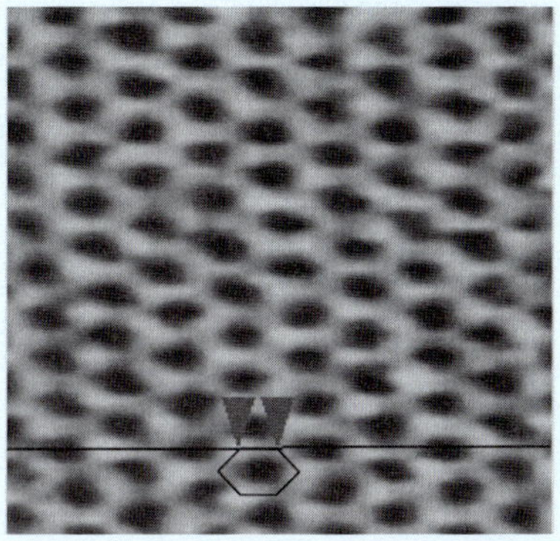

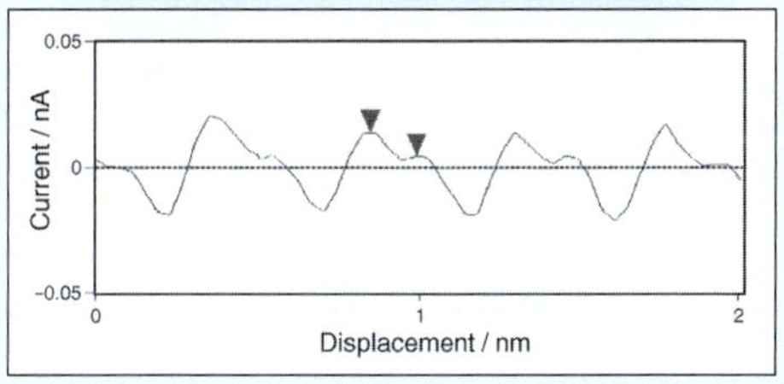

*Data obtained from: Chaun-Jian Zhong et al. 2003. Atomic scale imaging: a hands-on scanning probe microscopy laboratory for undergraduates. *Journal of Chemical Education* 80: 194–197.

Think Critically

1. **Estimate** the distance between two nearest bright spots.
2. **Estimate** the distance between two nearest neighbor spots (brighter–fainter, marked with triangles in the figure).
3. **State** What do the black spots in the image represent?
4. **Explain** How many carbon atoms are across the line drawn in the image?

Data Analysis LAB

About the Lab

- Scanning Tunneling Microscopy (STM) is a technique used to visualize individual atoms.
- An example of a highly ordered pyrolytic graphite (HOPG) material is the lead used in pencils.
- STM can be used to provide information about the structure of materials: examples—arrangement of carbon atoms in graphite or arrangement of carbon atoms in diamond.
- STM can be used to estimate the size of atoms and separation distances between adjacent atoms.
- Data obtained from: Chuan-Jian Zhong et al. 2003. Atomic Scale Imaging: A Hands-On Scanning Microscopy Laboratory for Undergraduates. *Journal of Chemical Education* 80:194–197

Think Critically

1. 0.14 nm
2. 0.25 nm
3. The black spots represent the cavity in the structure of graphite.
4. 9. Each positive peak and each shoulder in the cross-section view represents a carbon atom.

ASSESSMENT

Performance Using photos that have a feature with a commonly known or determinable length in them, have students determine the sizes of other objects in the photos. OL

Results

A beam of light, unimpeded by any object, is observed on the wall when the laser beam passes through an empty picture frame. Shining the laser beam through the frame containing a transparency film produces a similar strong beam of light. However, scattered smaller points are also observed. This suggests that some of the light was deflected out of the beam as it struck the target.

The result of shining the laser beam through an empty picture frame is analogous to what Rutherford expected to see based on the plum pudding model. Using a transparency film depicts what he observed.

Analysis

Ask these questions.

1. Which picture frame demonstrated the correct atomic structure? transparency film
2. Compare this demonstration to Rutherford's experiment. The laser beam represents the beam of alpha particles, and the picture frame represents the gold foil or target.

ASSESSMENT

Performance Ask students to create a table to compare and contrast what they observed in the demonstration and Rutherford's experiment. OL

ASSESSMENT

Skill Using the mass of an electron as the basic unit of mass, have students set up a table that relates the masses of the other subatomic particles, other atoms, and other objects of their choosing. **OL**

3 Assess

Check for Understanding

Ask students to compare functions and dimensions of the basic unit of chemistry, the atom, with the basic unit of biology, the cell. **OL**

Reteach

Have groups of students build the various atomic models using clay, construction paper, gum drops, or other appropriate materials. **BL** **EL**

Extension

Ask interested students to research the more recent discoveries about the structure of protons and neutrons, and the discovery of quarks. Students might want to prepare a short presentation on their findings for their classmates. **AL**

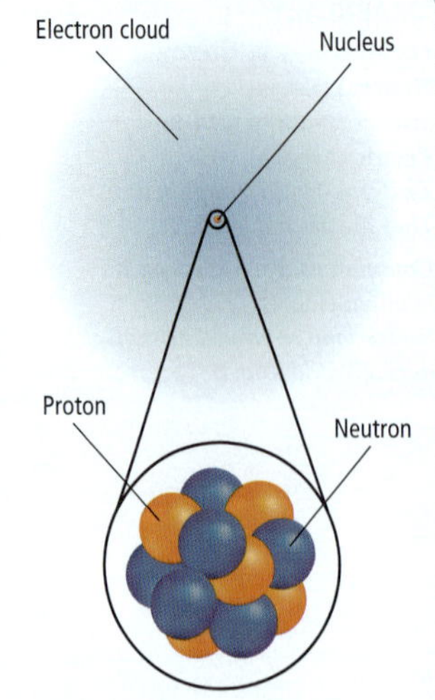

Figure 14 Atoms are composed of a nucleus containing protons and neutrons, and surrounded by a cloud of electrons.

View an **animation about the structure of the atom.** Concepts In Motion

Explore the **properties of subatomic particles with an interactive table.** Concepts In Motion

Table 3 Properties of Subatomic Particles

Particle	Symbol	Location	Relative Electric Charge	Relative Mass	Actual Mass (g)
Electron	e^-	In the space surrounding the nucleus	1−	$\frac{1}{1840}$	9.11×10^{-28}
Proton	p	In the nucleus	1+	1	1.673×10^{-24}
Neutron	n	In the nucleus	0	1	1.675×10^{-24}

Completing the model of the atom All atoms are made up of the three fundamental subatomic particles—the electron, the proton, and the neutron. Atoms are spherically shaped, with a small, dense nucleus of positive charge surrounded by one or more negatively charged electrons. Most of an atom consists of fast-moving electrons traveling through the empty space surrounding the nucleus. The electrons are held within the atom by their attraction to the positively charged nucleus. The nucleus, which is composed of neutral neutrons (hydrogen's single-proton nucleus is an exception) and positively charged protons, contains all of an atom's positive charge and more than 99.97% of its mass. It occupies only about one ten-thousandth of the volume of the atom. Because an atom is electrically neutral, the number of protons in the nucleus equals the number of electrons surrounding the nucleus. The features of a typical atom are shown in **Figure 14,** and the properties of the fundamental subatomic particles are summarized in **Table 3.**

Subatomic particle research is still a major interest to modern scientists. In fact, scientists have determined that protons and neutrons have their own structures. They are composed of subatomic particles called quarks. These particles will not be covered in this textbook because scientists do not yet understand if or how they affect chemical behavior. As you will learn in later chapters, chemical behavior can be explained by considering only an atom's electrons.

SECTION 2 REVIEW

Section Self-Check

Section Summary

- An atom is the smallest particle of an element that maintains the properties of that element.
- Electrons have a 1− charge, protons have a 1+ charge, and neutrons have no charge.
- An atom consists mostly of empty space surrounding the nucleus.

7. **MAIN IDEA Describe** the structure of a typical atom. Identify where each subatomic particle is located.
8. **Compare and contrast** Thomson's plum pudding atomic model with Rutherford's nuclear atomic model.
9. **Evaluate** the experiments that led to the conclusion that electrons are negatively charged particles found in all matter.
10. **Compare** the relative charge and mass of each of the subatomic particles.
11. **Calculate** What is the difference expressed in kilograms between the mass of a proton and the mass of an electron?

SECTION 2 REVIEW

7. A typical atom consists of a central, small, dense nucleus containing protons and neutrons. The nucleus is surrounded by a cloud of negatively charged electrons.
8. Thomson's plum pudding model describes atoms as spherical particles with uniformly distributed positive charge in which individual, negatively charged electrons are located in fixed positions. In contrast, Rutherford's model states that an atom is mostly empty space, with a small, dense, central nucleus containing all of an atom's positive charge and most of its mass. The negatively charged electrons move through the empty space and are held in the atom by their attraction to the positively charged nucleus.
9. The deflection toward positively charged plates demonstrated the negatively charged nature of electrons; the fact that changing the type of electrode or the type of gas used in the cathode-ray tube did not affect the ray produced led to the conclusion that electrons are present in all matter.
10.

Particle	Relative Charge	Relative Mass
Electron	−1	1/1840
Proton	+1	1
Neutron	0	~1

11. 1.672×10^{-27} kg

SECTION 3

How Atoms Differ

MAIN IDEA The number of protons and the mass number define the type of atom.

Essential Questions

- How is the atomic number used to determine the identity of an atom?
- What is an isotope?
- Why are atomic masses not whole numbers?
- Given the mass number and atomic number, how are the number of electrons, protons, and neutrons in an atom calculated?

Review Vocabulary

periodic table: a chart that organizes all known elements into a grid of horizontal rows (periods) and vertical columns (groups or families) arranged by increasing atomic number

New Vocabulary

atomic number
isotope
mass number
atomic mass unit (amu)
atomic mass

CHEM 4 YOU You are probably aware that numbers are used every day to identify people and objects. For example, people can be identified by their Social Security numbers and computers by their IP addresses. Atoms and nuclei are also identified by numbers.

Atomic Number

As shown in the periodic table of the elements inside the back cover of this textbook, there are more than 110 different elements. What makes an atom of one element different from an atom of another element?

Not long after Rutherford's gold foil experiment, the English scientist Henry Moseley (1887–1915) discovered that atoms of each element contain a unique positive charge in their nuclei. Thus, the number of protons in an atom identifies it as an atom of a particular element. The number of protons in an atom is referred to as the **atomic number.** The information provided by the periodic table for hydrogen is shown in **Figure 15.** The number 1 above the symbol for hydrogen (H) is the number of protons, or the atomic number. Moving across the periodic table to the right, you will next come to helium (He). It has two protons in its nucleus, and thus it has an atomic number of 2. The next row begins with lithium (Li), atomic number 3, followed by beryllium (Be), atomic number 4, and so on. The periodic table is organized left-to-right and top-to-bottom by increasing atomic number.

Because all atoms are neutral, the number of protons and electrons in an atom must be equal. Thus, once you know the atomic number of an element, you know the number of protons and the number of electrons an atom of that element contains. For example, an atom of lithium, atomic number 3, contains three protons and three electrons.

Atomic number

$$\text{atomic number} = \text{number of protons} = \text{number of electrons}$$

The atomic number of an atom equals its number of protons and its number of electrons.

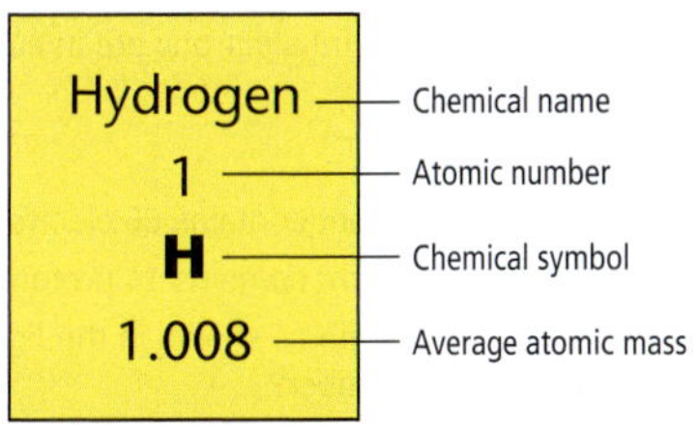

Figure 15 In the periodic table, each element is represented by its chemical name, atomic number, chemical symbol, and average atomic mass.
Determine *the number of protons and the number of electrons in an atom of gold.*

SECTION 3

1 Focus

MAIN IDEA

Defining the Atom Ask students: What information can be used for one's home address? number, street, city, state, and zip code; Each residential address is as unique as the particles that are used to define the type of atom. BL

2 Teach

Concept Development

Isotopes Pass around a sandwich bag that contains several varieties–or "isotopes"–of the same thing. Good choices are pastas, beans, and small multicolored candies. Ask students to name one similarity and one difference among the elements in the sandwich bags. similarity: same type of matter (for example, pasta); difference: mass Point out the parallel with isotopes. Isotopes of an elements are the same type of matter but have different masses. OL

Caption Question Fig. 15
79 protons and 79 electrons

DIFFERENTIATED INSTRUCTION

Struggling Learners Use three different colors of gum drops to allow students to build models of atoms that show the relationship of protons, neutrons, and electrons. Be sure to emphasize that the scaling of the model is in no way accurate; electrons are 1/1840 the size of protons and neutrons, and the diameter of an atom is about 10,000 times the size of its nucleus. BL EL

EXAMPLE Problem 1

Find help with **solving algebraic equations.** Math Handbook

EXAMPLE PROBLEM

ATOMIC NUMBER Complete the following table.

Composition of Several Elements

	Element	Atomic Number	Protons	Electrons
a.	Pb	82		
b.			8	
c.				30

1 ANALYZE THE PROBLEM

Apply the relationship among atomic number, number of protons, and number of electrons to complete most of the table. Then, use the periodic table to identify the element.

Known

a. element = Pb, atomic number = 82

b. number of protons = 8

c. number of electrons = 30

Unknown

a. number of protons **(N_p)**, number of electrons **(N_e) = ?**

b. **element,** atomic number **(Z), N_e = ?**

c. **element, Z, N_p = ?**

2 SOLVE FOR THE UNKNOWN

a. number of protons = atomic number — **Apply the atomic-number relationship.**

$N_p = 82$ — **Substitute atomic number = 82.**

number of electrons = number of protons

$N_e = 82$

The number of protons and the number of electrons is 82.

b. atomic number = number of protons — **Apply the atomic-number relationship.**

$Z = 8$ — **Substitute number of protons = 8.**

number of electrons = number of protons

$N_e = 8$

The atomic number and the number of electrons is 8.

The **element** is **oxygen (O).** — **Consult the periodic table to identify the element.**

c. number of protons = number of electrons — **Apply the atomic-number relationship.**

$N_p = 30$ — **Substitute number of electrons = 30.**

atomic number = number of protons

$Z = 30$

The atomic number and the number of protons is 30.

The **element** is **zinc (Zn).** — **Consult the periodic table to identify the element.**

3 EVALUATE THE ANSWER

The answers agree with atomic numbers and element symbols given in the periodic table.

PRACTICE Problems

Do additional problems. Online Practice

12. How many protons and electrons are in each atom?

a. radon

b. magnesium

13. An atom of an element contains 66 electrons. Which element is it?

14. An atom of an element contains 14 protons. Which element is it?

15. Challenge Do the atoms shown in the figure to the right have the same atomic number?

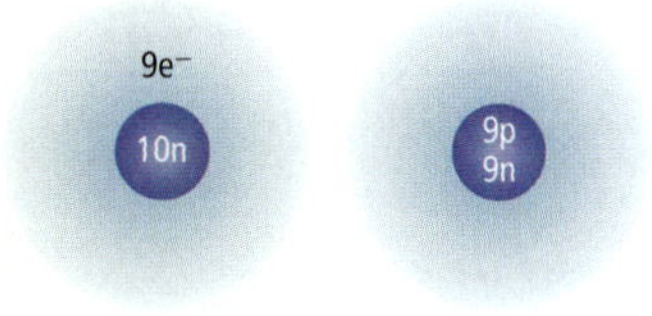

Enrichment

Sequence of Elements in the Periodic Table Of the scientists working under Rutherford at Cambridge University, Henry Moseley was the youngest. He beamed X-rays at samples of the different elements to find that the wavelength of the resultant X-rays decreased with increasing atomic weight. Moseley attributed this to the increasing number of electrons in the atom and, consequently, the increasing number of protons in the nucleus. This led to the final arrangement of Dmitri Mendeleev's periodic table—elements arranged by increasing number of protons (atomic number). This arrangement solved problems with the sequence of elements with nearly similar atomic masses and clearly identified "holes" in the table for yet-to-be discovered elements. Moseley's method was used to refute Urbain's claim of discovering a new element that he called celtium and to support Hevesey's discovery of hafnium. Have students compare Mendeleev's periodic table with Moseley's periodic table and identify discrepancies in the sequence of elements. AL

IN-CLASS Example

Question Complete the following table.

Element	Atomic Number	Protons	Electrons
K	19	19	19
Br	35	35	35
Ne	10	10	10

PRACTICE Problems

Have students refer to the Selected Solutions appendix for complete solutions to odd-numbered problems. The complete solutions for all problems can be found in the Solutions Manual.

12. a. radon, 86 protons, 86 electrons
b. magnesium, 12 protons, 12 electrons

13. dysprosium

14. silicon

15. yes, 9

CULTURAL DIVERSITY

A Woman with Designs Patsy Sherman had a temporary job at the 3M Corporation in St. Paul, Minnesota, when a lab assistant accidentally spilled some chemicals on her shoes. She noticed that the site of the spill became waterproof and stain-resistant. She decided to pursue the chemical mixture, which later was marketed as the fabric protectant Scotchgard™. The protectant consists of a rubbery molecule, one side of which is sticky, allowing it to cling to fabric, while the other is slippery enough to repel stains.

3M was impressed with Sherman's discovery and awarded her with a full-time position. Scotchgard™ became one of 3M's most profitable products. Sherman has gone on to earn 16 U.S. patents and serves on the board of the National Inventors Hall of Fame.

Isotopes and Mass Number

Dalton was incorrect about atoms being indivisible and in stating that all atoms of an element are identical. All atoms of an element have the same number of protons and electrons, but the number of neutrons might differ. For example, there are three types of potassium atoms that occur naturally. All three types contain 19 protons and 19 electrons. However, one type of potassium atom contains 20 neutrons, another 21 neutrons, and still another 22 neutrons. Atoms with the same number of protons but different numbers of neutrons are called **isotopes.**

Mass of isotopes Isotopes containing more neutrons have a greater mass. In spite of these differences, isotopes of an atom have the same chemical behavior. As you will read later in this textbook, chemical behavior is determined only by the number of electrons an atom has.

Isotope notation Each isotope of an element is identified with a number called the mass number. The **mass number** is the sum of the atomic number (or number of protons) and neutrons in the nucleus.

Mass number

$$\text{mass number} = \text{atomic number} + \text{number of neutrons}$$

The mass number of an atom is the sum of its atomic number and its number of neutrons.

For example, copper has two isotopes. The isotope with 29 protons and 34 neutrons has a mass number of 63 (29 + 34 = 63), and is called copper-63 (also written ^{63}Cu or Cu-63). The isotope with 29 protons and 36 neutrons is called copper-65. Chemists often write out isotopes using a notation involving the chemical symbol, atomic number, and mass number, as shown in **Figure 16.**

Natural abundance of isotopes In nature, most elements are found as mixtures of isotopes. Usually, no matter where a sample of an element is obtained, the relative abundance of each isotope is constant. For example, in a banana, 93.26% of the potassium atoms have 20 neutrons, 6.73% have 22 neutrons, and 0.01% have 21 neutrons. In another banana, or in a different source of potassium, the percentage composition of the potassium isotopes will still be the same. The three potassium isotopes are summarized in **Figure 17.**

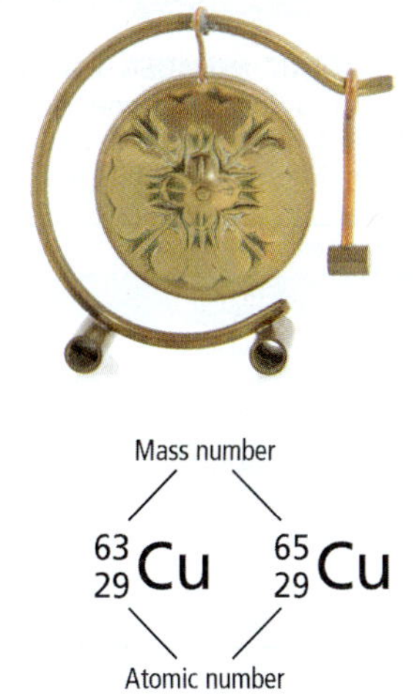

■ **Figure 16** Cu is the chemical symbol for copper. Copper, which was used to make this Chinese gong, is composed of 69.2% copper-63 and 30.8% copper-65.

	Potassium-39	Potassium-40	Potassium-41
Protons	19	19	19
Neutrons	20	21	22
Electrons	19	19	19

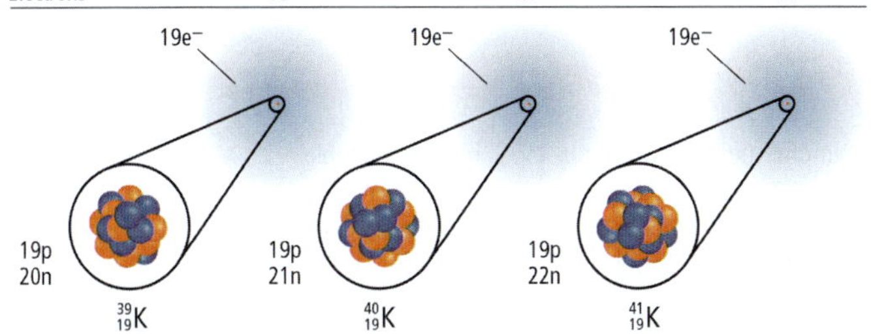

■ **Figure 17** Potassium has three naturally occurring isotopes: potassium-39, potassium-40, and potassium-41.

***List** the number of protons, neutrons, and electrons in each potassium isotope.*

Identify Misconceptions

Students might think that isotopes contain different numbers of electrons and protons.

Uncover the Misconception Ask students if and how isotopes of an atom differ from each other. Isotopes of an atom have different numbers of neutrons.

Demonstrate the Concept Explain to students that an atom's identity is defined solely by the number of protons in its nucleus. The number of neutrons can vary, resulting in the existence of different isotopes. The number of electrons in the neutral isotopes is the same. Reinforce that various ions of an element can have different numbers of electrons.

Assess New Knowledge Ask a student to show on the board how three isotopes of oxygen (oxygen-16, oxygen-17, and oxygen-18) are the same and how they are different. All isotopes have 8 protons, but the three isotopes differ by having 8, 9, and 10 neutrons respectively. BL

ASSESSMENT

Performance Have students use the overhead projector and a set of colored dots cut from acetate sheets representing protons, neutrons, and electrons to build configurations for various isotopes. BL EL

CHEMISTRY JOURNAL

Useful Isotopes Have students research the uses of isotopes in medicine and radiochemical dating. In their journals, have them describe how radiochemical dating makes use of radioactive isotopes. Also, have them list at least three isotopes used in medicine, and describe how those isotopes are used. OL

■ **Caption Question Fig. 17** potassium-39: 19 protons, 20 neutrons, 19 electrons; potassium-40: 19 protons, 21 neutrons, 19 electrons; potassium-41: 19 protons, 22 neutrons, 19 electrons

IN-CLASS Example

Question A national laboratory has analyzed the composition of isotopes of several elements. The data is displayed in the table below. Determine the number of protons, electrons, and neutrons in the given isotope of iron. Name the isotope, and give its symbol.

Element	Atomic Number	Mass Number
Boron	5	11
Chlorine-35	17	35
Chlorine-37	17	37
Iron	26	56
Magnesium-24	12	24
Magnesium-26	12	26

Answer Iron-56, $^{56}_{26}$Fe, has 26 protons, 26 electrons, and 30 neutrons.

Math in Chemistry

Atomic Mass Show students how to calculate the atomic mass of an element, X.

1. If an element does not have any isotopes, then its atomic mass is equal to its mass.
2. If an element has isotopes, then multiply each mass contribution of each isotope by its percent abundance.
3. Remember to convert the percent abundance to a fraction. For example:

   25% abundance = 25/100 = 0.25
4. Sum all of the results from Step 2 calculations and the answer is the atomic mass of the element, X. Refer to **Figure 18** for an example of how to calculate the atomic mass of chlorine, Cl.

EXAMPLE Problem 2

EXAMPLE PROBLEM

USE ATOMIC NUMBER AND MASS NUMBER A chemistry laboratory has analyzed the composition of isotopes of several elements. The composition data is given in the table below. Determine the number of protons, electrons, and neutrons in the isotope of neon. Name the isotope and give its symbol.

Isotope Composition Data

	Element	Atomic Number	Mass Number
a.	Neon	10	22
b.	Calcium	20	46
c.	Oxygen	8	17
d.	Iron	26	57
e.	Zinc	30	64
f.	Mercury	80	204

1 ANALYZE THE PROBLEM

You are given some data for neon in the table. The symbol for neon can be found on the periodic table. From the atomic number, the number of protons and electrons in the isotope are known. The number of neutrons in the isotope can be found by subtracting the atomic number from the mass number.

Known
element: neon
atomic number = 10
mass number = 22

Unknown
number of protons (N_p), electrons (N_e), and neutrons (N_n) = ?
name of isotope = ?
symbol for isotope = ?

2 SOLVE FOR THE UNKNOWN

number of protons = atomic number = **10**
number of electrons = atomic number = **10**
Apply the atomic number relationship.

number of neutrons = mass number − atomic number
Use the atomic number and the mass number to calculate the number of neutrons.

$N_n = 22 - 10 = 12$
Substitute mass number = 22 and atomic number = 10

The **name** of the isotope is **neon-22.**
Use the element name and mass number to write the isotope's name.

The **symbol** for the isotope is $^{22}_{10}$**Ne.**
Use the chemical symbol, mass number, and atomic number to write out the isotope in symbolic notation form.

3 EVALUATE THE ANSWER

The relationships among number of electrons, protons, and neutrons have been applied correctly. The isotope's name and symbol are in the correct format. Refer to the Elements Handbook to learn more about neon.

PRACTICE Problems

Do additional problems. Online Practice

16. Determine the number of protons, electrons, and neutrons for isotopes **b.–f.** in the table above. Name each isotope, and write its symbol.

17. Challenge An atom has a mass number of 55. Its number of neutrons is the sum of its atomic number and five. How many protons, neutrons, and electrons does this atom have? What is the identity of this atom?

PRACTICE Problems

Have students refer to the Selected Solutions appendix for complete solutions to odd-numbered problems. The complete solutions for all problems can be found in the Solutions Manual.

16.

	Protons and Electrons	Neutrons	Isotope	Symbol
b.	20	26	calcium-46	$^{46}_{20}$Ca
c.	8	9	oxygen-17	$^{17}_{8}$O
d.	26	31	iron-57	$^{57}_{26}$Fe
e.	30	34	zinc-64	$^{64}_{30}$Zn
f.	80	124	mercury-204	$^{204}_{80}$Hg

17. 25 protons, 25 electrons, 30 neutrons; manganese.

Table 4 Masses of Subatomic Particles

Particle	Mass (amu)
Electron	0.000549
Proton	1.007276
Neutron	1.008665

Mass of Atoms

Recall from **Table 3** that the masses of both protons and neutrons are approximately 1.67×10^{-24} g. While this is a small mass, the mass of an electron is even smaller—only about 1/1840 that of a proton or a neutron.

Atomic mass unit Because these extremely small masses expressed in scientific notation are difficult to work with, chemists have developed a method of measuring the mass of an atom relative to the mass of a specific atomic standard. That standard is the carbon-12 atom. Scientists assigned the carbon-12 atom a mass of exactly 12 atomic mass units. Thus, one **atomic mass unit (amu)** is defined as one-twelfth the mass of a carbon-12 atom. Although a mass of 1 amu is nearly equal to the mass of a single proton or a single neutron, it is important to realize that the values are slightly different. **Table 4** gives the masses of the subatomic particles in terms of amu.

Atomic mass Because an atom's mass depends mainly on the number of protons and neutrons it contains, and because protons and neutrons have masses close to 1 amu, you might expect the atomic mass of an element to always be nearly a whole number. However, this is often not the case. The explanation involves how atomic mass is defined. The **atomic mass** of an element is the weighted average mass of the isotopes of that element. Because isotopes have different masses, the weighted average is not a whole number. The calculation of the atomic mass of chlorine is illustrated in **Figure 18.**

VOCABULARY

ACADEMIC VOCABULARY

Specific
characterized by precise formulation or accurate restriction
Some diseases have specific symptoms.

Explore **decoding the periodic table.**

Virtual Investigations

■ **Figure 18** To calculate the weighted average atomic mass of chlorine, you first need to calculate the mass contribution of each isotope.

Get help with **finding an average.**

Personal Tutor

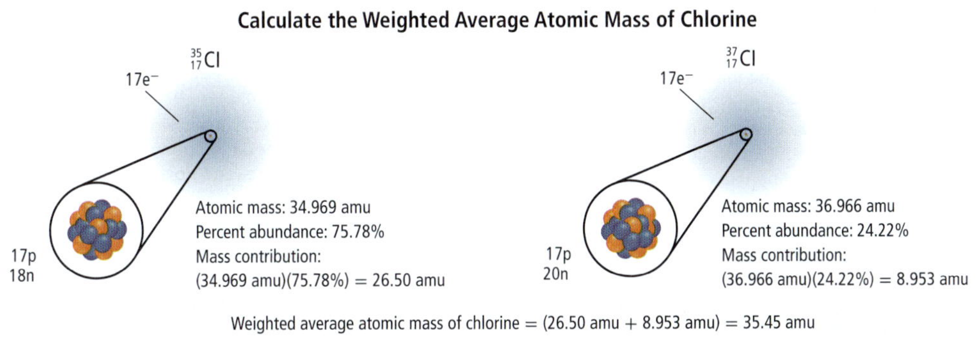

Extension

Table of Isotopes Familiarize your students with the Table of Isotopes in the *CRC Handbook of Chemistry and Physics*. At this point, they should be able to identify how many isotopes are found per element and understand how the isotope notation is listed. Check student understanding by quizzing them on facts they look up in the table. **OL**

ChemLAB

The ChemLab located at the end of the chapter can be used at this point in the lesson.

Reinforcement

Atomic Mass Point out to students that the answer to a weighted average atomic mass problem will probably be a mass that is closest to the element with the highest percent abundance.

Quick Demo

Comparing Mass Put 1 kernel of corn in a self-sealing bag. Place 1840 kernels in a second self-sealing bag. Compare the mass of the two bags. Ask the students how this compares to the mass of an electron and a mass of a proton. This demonstrates the relative mass of an electron (1 kernel) to the mass of a proton (1840 kernels). **BL**

CHEMISTRY PROJECT

Separate Isotopes Have students research how isotopes are separated for commercial, medical, and industrial purposes. For instance, in nuclear power plants, only the uranium-235 is suitable for use as a fissionable fuel and it must be separated from the more abundant uranium-238. **OL**

DIFFERENTIATED INSTRUCTION

Advanced Learners Ask students to discern which isotope of nickel is most abundant given the masses below and that the atomic mass of Ni is 58.69 amu.

	Mass	% Abundance
^{58}Ni	57.93	68.08
^{60}Ni	59.93	26.22
^{61}Ni	60.93	1.14
^{62}Ni	61.93	3.63
^{64}Ni	63.93	0.93

The isotope that has a mass closest to the atomic mass of the element is likely the most abundant isotope. For example the atomic mass of chlorine is 35.45 amu; therefore, Cl-35 is likely the most abundant isotope. The atomic mass of Ni is 58.69 amu. There is no isotope for Ni-59 listed; therefore, Ni-58 is likely the most abundant. Show students the actual percent abundances after they have completed their reasoning. **AL**

READING CHECK To calculate atomic mass, sum the products of each isotope's percent abundance multiplied by its isotopic mass.

MiniLAB

Rubric

Purpose Students will determine the atomic mass of a penny given a mixture of its pre- and post-1982 "isotopes."

Process Skills classify, measure, use numbers

Safety Precaution Approve the lab safety forms before work begins.

Teaching Strategies

- Review the reason for the change in composition of the penny in 1982 during the pre-lab discussion. cost savings
- Review the calculation of weighted averages with students.
- Do not use any pennies produced in 1982.

Expected Results Pre-1982 pennies have a greater mass than post-1982 pennies.
Mass of ten pre-1982 pennies = 31.10 g
Average mass of a pre-1982 penny = 3.11g
Mass of ten post-1982 pennies = 25.48 g
Average mass of a post-1982 penny = 2.55 g
The atomic mass depends on the mixture analyzed.

Analysis

1. The relative number of pre- and post-1982 pennies determines the percentage abundance of each group.
2. The atomic mass of a penny depends upon the mixture of pennies each student receives. Sample data is shown here.
mass contribution (pre-1982) = (55.0%)(3.11 g) = 1.71 g
mass contribution (post-1982) = (45.0%)(2.55 g) = 1.15 g
atomic mass = (1.71 g + 1.15 g) = 2.86 g
3. A different mixture would have a different relative abundance and a different atomic mass.
4. Masses of individual pennies will vary due to wear.

Figure 19 Bromine is extracted from sea water and salt lakes. The Dead Sea area in Israel is one of the major bromine production sites in the world. Applications of bromine include microbe and algae control in swimming pools. It is also used in medicines, oils, paints, pesticides, and flame-retardants.

Eitan Simanor/Alamy

Chlorine exists naturally as a mixture of about 76% chlorine-35 and 24% chlorine-37. It has an atomic mass of 35.453 amu. Because atomic mass is a weighted average, the chlorine-35 atoms, which exist in greater abundance than the chlorine-37 atoms, have a greater effect in determining the atomic mass. The atomic mass of chlorine is calculated by multiplying each isotope's percent abundance by its atomic mass and then adding the products. The process is similar to calculating an average grade. You can calculate the atomic mass of any element if you know the number of naturally occurring isotopes, their masses, and their percent abundances.

READING CHECK **Explain** how to calculate atomic mass.

Isotope abundances Analyzing an element's mass can indicate the most abundant isotope for that element. For example, fluorine (F) has an atomic mass that is extremely close to 19 amu. If fluorine had several fairly abundant isotopes, its atomic mass would not likely be so close to a whole number. Thus, you might conclude that all naturally occurring fluorine is probably in the form of fluorine-19 ($^{19}_{9}F$). Indeed, 100% of naturally occurring fluorine is in the form of fluorine-19. While this type of reasoning generally works well, it is not foolproof. Consider bromine (Br). It has an atomic mass of 79.904 amu. With a mass so close to 80 amu, it seems likely that the most common bromine isotope would be bromine-80. However, bromine's two isotopes are bromine-79 (78.918 amu, 50.69%) and bromine-81 (80.917 amu, 49.31%). There is no bromine-80 isotope. **Figure 19** shows one of the major production sites of bromine, located in the Dead Sea area. Refer to the Elements Handbook to learn more about chlorine, fluorine, and bromine.

MiniLAB

Model Isotopes

How can you calculate the atomic mass of an element using the percentage abundance of its isotopes?
Because they have different compositions, pre- and post-1982 pennies can be used to model an element with two naturally occurring isotopes. From the penny 'isotope' data, you can determine the mass of each penny isotope and the average mass of a penny.

Procedure

1. Read and complete the lab safety form.
2. Get a bag of **pennies** from your teacher, and sort the pennies by date into two groups: pre-1982 pennies and post-1982 pennies. Count and record the total number of pennies and the number in each group.
3. Using a **balance,** determine the mass of 10 pennies from each group. Record each mass to the nearest 0.01 g. Divide the total mass of each group by 10 to get the average mass of a pre- and post-1982 penny isotope.

Analysis

1. **Calculate** the percentage abundance of each group using data from Step 2. To do this, divide the number of pennies in each group by the total number of pennies.
2. **Determine** the atomic mass of a penny using the percentage abundance of each "isotope" and data from Step 3. To do this, use the following equation:
mass contribution = (% abundance)(mass)
Total the mass contributions to determine the atomic mass. Remember that the percent abundance is a percentage.
3. **Infer** whether the atomic mass would be different if you received another bag of pennies containing a different mixture of pre- and post-1982 pennies. Explain your reasoning.
4. **Explain** why the average mass of each type of penny was determined by measuring 10 pennies instead of by measuring and using the mass of a single penny from each group.

CHEMISTRY JOURNAL

Mass Spectrometer Ask students to research information on mass spectrometers. Ask them to write an essay in their journal explaining how a mass spectrometer works and what its uses are. Encourage them to include a diagram of the apparatus and to create a flowchart of its function. AL

EXAMPLE Problem 3

EXAMPLE PROBLEM

CALCULATE ATOMIC MASS Given the data in the table, calculate the atomic mass of unknown Element X. Then, identify the unknown element, which is used medically to treat some mental disorders.

1 ANALYZE THE PROBLEM

Calculate the atomic mass and use the periodic table to confirm.

Isotope Abundance for Element X		
Isotope	Mass (amu)	Percent Abundance
^{6}X	6.015	7.59%
^{7}X	7.016	92.41%

Known
^{6}X: mass = 6.015 amu
abundance = 7.59% = 0.0759
^{7}X: mass = 7.016 amu
abundance = 92.41% = 0.9241

Unknown
atomic mass of X = ? amu
element X = ?

2 SOLVE FOR THE UNKNOWN

^{6}X: mass contribution = (mass)(percent abundance) — Calculate ^{6}X's contribution.

mass contribution = (6.015 amu)(0.0759) = 0.456 amu — Substitute mass = 6.015 amu and abundance = 0.0759.

^{7}X: mass contribution = (mass)(percent abundance) — Calculate ^{7}X's contribution.

mass contribution = (7.016 amu)(0.9241) = 6.483 amu — Substitute mass = 7.016 amu and abundance = 0.9241.

atomic mass of X = (0.4565 amu + 6.483 amu) = **6.939 amu** — Total the mass contributions to find the atomic mass.

The **element** with a mass nearest 6.939 amu is **lithium (Li)**. — Identify the element using the periodic table.

3 EVALUATE THE ANSWER

The result of the calculation agrees with the atomic mass given in the periodic table. The masses of the isotopes have four significant figures, so the atomic mass is also expressed with four significant figures. Refer to the Elements Handbook to learn more about lithium.

PRACTICE Problems

Do additional problems. Online Practice

18. Boron (B) has two naturally occurring isotopes: boron-10 (abundance = 19.8%, mass = 10.013 amu) and boron-11 (abundance = 80.2%, mass = 11.009 amu). Calculate the atomic mass of boron.

19. **Challenge** Nitrogen has two naturally occurring isotopes, N-14 and N-15. Its atomic mass is 14.007. Which isotope is more abundant? Explain your answer.

SECTION 3 REVIEW

Section Self-Check

Section Summary

- The atomic number of an atom is given by its number of protons. The mass number of an atom is the sum of its neutrons and protons.
- Atoms of the same element with different numbers of neutrons are called isotopes.
- The atomic mass of an element is a weighted average of the masses of all of its naturally occurring isotopes.

20. MAINIDEA **Explain** how the type of an atom is defined.

21. **Recall** Which subatomic particle identifies an atom as that of a particular element?

22. **Explain** how the existence of isotopes is related to the fact that atomic masses are not whole numbers.

23. **Calculate** Copper has two isotopes: Cu-63 (abundance = 69.2%, mass = 62.930 amu) and Cu-65 (abundance = 30.8%, mass = 64.928 amu). Calculate the atomic mass of copper.

24. **Calculate** Three magnesium isotopes have atomic masses and relative abundances of 23.985 amu (78.99%), 24.986 amu (10.00%), and 25.982 amu (11.01%). Calculate the atomic mass of magnesium.

SECTION 3 REVIEW

20. by the atomic number
21. the proton
22. Atomic masses are not whole numbers because they represent weighted averages of the masses of the isotopes of an element.
23. 63.5 amu
24. 24.31 amu

IN-CLASS Example

Question Calculate the atomic mass of the element X and identify the element from the periodic table knowing that it has the following isotopes: ^{54}X (mass: 53.940, abundance: 5.9%), ^{56}X (mass: 55.935, abundance: 91.72%), ^{57}X (mass: 56.935, abundance: 2.1%), and ^{58}X (mass: 57.933, abundance: 0.28%).

Answer Atomic mass number
= (53.940)(0.059) + (55.935)(0.9172) + (56.935)(0.021) + (57.935)(0.0028)
= 55.9 g/mol
X = Fe (The unknown element is iron)

PRACTICE Problems

Have students refer to the Selected Solutions appendix for complete solutions to odd-numbered problems. The complete solutions for all problems can be found in the Solutions Manual.

18. 10.81 amu
19. N-14 is more abundant because the atomic mass is closer to 14 than 15.

3 Assess

Check for Understanding

Obtain a copy of a mass spectrograph from the Internet or reference texts. Have students analyze the abundance of each isotope and calculate an average atomic mass. OL

Reteach

Compare the calculation of an average atomic mass with the calculation of a grade using weighted averages. For instance, a student's grade in chemistry is weighted as follows: 50% tests; 40% lab reports; 10% homework. If the student averaged 85 on tests, 95 on lab reports, and 70 on homework, what would be his final grade? 87.5 BL

Extension

Have students do a scavenger hunt through the Table of Isotopes in the *CRC Handbook of Chemistry and Physics.* Provide a list of clues, such as element with the highest number of isotopes, three isotopes, a weighted average of 24.305, and so on. OL

SECTION 4

1 Focus

MAINIDEA

Chain Reaction Ask students what happens if several dominoes are placed in a vertical position very close to each other and one of them falls down. What about the reverse process? Suppose one of the dominoes is put back into a vertical position. What will happen? The forward process knocks all of the dominoes down. In the reverse process, nothing will happen. The one domino will be in the vertical position, while the other dominoes remain in the horizontal position. Dominoes are stable when they are placed horizontally on a surface. Dominoes that fall down gain stability. In a similar way, nuclei emit energy and gain stability. BL

2 Teach

Concept Development

Changes Emphasize to students the three major kinds of changes: physical, chemical, and nuclear. Students should be able to describe the mechanism of each change and give clear examples of each.

SECTION 4

Essential Questions

- What is the relationship between unstable nuclei and radioactive decay?
- How are alpha, beta, and gamma radiation characterized in terms of mass and charge?

Review Vocabulary

element: a pure substance that cannot be broken down into simpler substances by physical or chemical means

New Vocabulary

radioactivity
radiation
nuclear reaction
radioactive decay
alpha radiation
alpha particle
nuclear equation
beta radiation
beta particle
gamma ray

Unstable Nuclei and Radioactive Decay

MAINIDEA Unstable atoms emit radiation to gain stability.

CHEM 4 YOU Try dropping a rock from the height of your waist. The rock goes from a higher energy state at your waist to a lower energy state on the floor. A similar process happens with nuclei in an unstable state.

Radioactivity

Recall that a chemical reaction is the change of one or more substances into new substances and involves only an atom's electrons. Although atoms might be rearranged, their identity remains the same. Another type of reaction, called a nuclear reaction, can change an element into a new element.

Nuclear reactions In the late 1890s, scientists noticed that some substances spontaneously emitted radiation in a process they named **radioactivity.** The rays and particles emitted by the radioactive material were called **radiation.** Scientists discovered that radioactive atoms undergo changes that can alter their identities. A reaction that involves a change in an atom's nucleus is called a **nuclear reaction.** The discovery of these nuclear reactions was a major breakthrough, as no chemical reaction had ever resulted in the formation of new kinds of atoms.

Radioactive atoms emit radiation because their nuclei are unstable. Unstable systems, whether they are atoms or people doing handstands, as shown in **Figure 20,** gain stability by losing energy.

Radioactive decay Unstable nuclei lose energy by emitting radiation in a spontaneous process called **radioactive decay.** Unstable atoms undergo radioactive decay until they form stable atoms, often of a different element. Just as a rock loses gravitational potential energy and reaches a stable state when falling to the ground, an atom can lose energy and reach a stable state when emitting radiation.

Image Source/Getty Images

Figure 20 Being in a handstand position is an unstable state. Like unstable atoms, people doing handstands eventually return to a more stable state—standing on their feet—by losing potential energy.

CHEMISTRY JOURNAL

Nuclear News Have students scan through daily newspapers and magazines for articles dealing with nuclear phenomena. Have them keep a listing of the articles and classify them by categories, such as weaponry, power, food, medicine, and industrial. Based on their lists, have students estimate rough percentages for each category. OL

Types of Radiation

Scientists began researching radioactivity in the late 1800s. They investigated the effect of electric fields on radiation. By directing radiation from a radioactive source between two electrically charged plates, scientists were able to identify three different types of radiation based on their electric charge. As shown in **Figure 21,** radiation was deflected toward the negative plate, the positive plate, or not at all.

Alpha radiation The radiation that was deflected toward the negatively charged plate was named **alpha radiation.** It is made up of alpha particles. An **alpha particle** contains two protons and two neutrons, and thus has a 2+ charge, which explains why alpha particles are attracted to the negatively charged plate as shown in **Figure 21.** An alpha particle is equivalent to a helium-4 nucleus and is represented by $^{4}_{2}\text{He}$ or α. The alpha decay of radioactive radium-226 into radon-222 is shown below.

$$^{226}_{88}\text{Ra} \rightarrow {}^{222}_{86}\text{Rn} + \alpha$$

radium-226 → radon-222 + alpha particle

Note that a new element, radon (Rn), is created as a result of the alpha decay of the unstable radium-226 nucleus. The type of equation shown above is known as a **nuclear equation.** It shows the atomic numbers and mass numbers of the particles involved. The mass number is conserved in nuclear equations.

Beta radiation The radiation that was deflected toward the positively charged plate was named **beta radiation.** This radiation consists of fast-moving beta particles. Each **beta particle** is an electron with a 1− charge. The negative charge of the beta particle explains why it is attracted to the positively charged plate shown in **Figure 21.** Beta particles are represented by the symbol β or e^-. The beta decay of carbon-14 into nitrogen-14 is shown below. The beta decay of unstable carbon-14 results in the creation of the new atom, nitrogen (N).

$$^{14}_{6}\text{C} \rightarrow {}^{14}_{7}\text{N} + \beta$$

carbon-14 → nitrogen-14 + beta particle

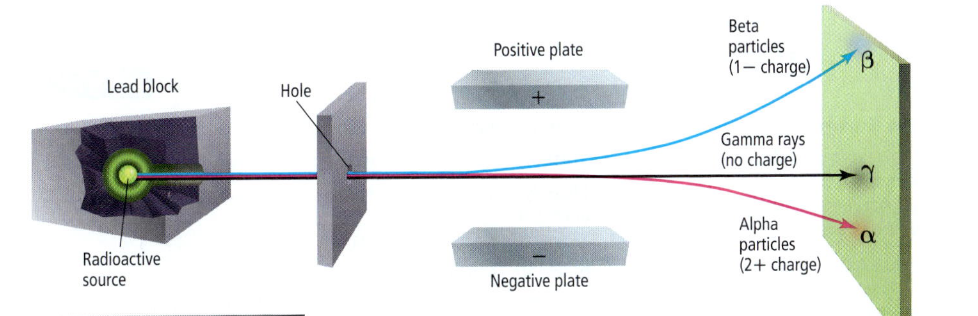

View an **animation about radiation deflection.**

Concepts In Motion

Figure 21 An electric field will deflect radiation in different directions, depending on the electric charge of the radiation.

Explain *why beta particles are deflected toward the positive plate, alpha particles are deflected toward the negative plate, and gamma rays are not deflected.*

CAREERS IN CHEMISTRY

Chemistry Teacher Chemistry teachers work in high schools and colleges. They lecture, guide discussions, conduct experiments, supervise lab work, and lead field trips. High school teachers might also be asked to monitor study halls and serve on committees. College instructors might be required to do research and publish their findings.

WebQuest

Quick Demo

Radiation To acquaint students with the concept of radiation, purchase an inexpensive radon kit from a local hardware store and test your classroom radon levels. These kits are sent through the mail for laboratory analysis. Radon, a by-product of the decay of uranium-238 in soils and building materials, is considered hazardous due to the fact that it decays into radioactive isotopes of polonium, bismuth, and lead. These heavy metal ions are not eliminated easily from the body. It is estimated that 10,000 to 20,000 lung cancer deaths are caused by radon gas exposure in the United States each year.

Caption Question Fig. 21 Beta particles are deflected toward the positive plate because they are negatively charged. Alpha particles are deflected toward the negative plate because they are positively charged. Gamma rays are not deflected because they have no charge.

CAREERS IN CHEMISTRY

WebQuest

DIFFERENTIATED INSTRUCTION

Visually Impaired Give students sets consisting of two differently shaped objects representing protons and neutrons. Use small containers, such as petri dishes, to represent the nucleus and contain the protons and neutrons. Ask students to model various isotopes and then follow up by asking them to demonstrate an alpha or beta decay by removing the appropriate numbers of protons or neutrons. BL EL

CHEMISTRY PROJECT

Radioactive Elements Have students research on the naturally occurring radioactive elements, their sources, everyday applications and present their findings to their classmates. OL

ASSESSMENT

Knowledge Have each student write two questions from each section of this chapter. Use the questions for a game show similar to a quiz tournament. OL

3 Assess

Check for Understanding

Have student groups of three use the Table of Isotopes in the *CRC Handbook of Chemistry and Physics* to find any isotope that undergoes radioactive decay. The first student in the group writes down the listed isotope, then passes the paper to a second student in the group. The second student writes down the ejected particle listed, then passes it to the third student in the group. This student should complete the nuclear equation by writing the resulting nuclide. Have students continue with this process for five minutes. Each group should check with the teacher to see whether its equations are correct. The group with the most correct nuclear equations wins. OL COOP LEARN

Reteach

Have students complete the following nuclear reactions. OL

$^{60}_{27}Co \rightarrow ^{60}_{28}Ni + ?$ β

$^{241}_{95}Am \rightarrow ^{237}_{93}Np + ?$ $^{4}_{2}He$

Extension

Have students research the use of radioactive isotopes in classical experiments in chemistry and physics. OL

Table 5 Characteristics of Radiation

	Alpha	Beta	Gamma
Symbol	$^{4}_{2}He$ or α	e^- or β	γ
Mass (amu)	4	$\frac{1}{1840}$	0
Mass (kg)	6.65×10^{-27}	9.11×10^{-31}	0
Charge	2+	1−	0

Gamma radiation The third common type of radiation is called gamma radiation, or gamma rays. A **gamma ray** is a high-energy radiation that possesses no mass and is denoted by the symbol γ. Because they are neutral, gamma rays are not deflected by electric or magnetic fields. They usually accompany alpha and beta radiation, and they account for most of the energy lost during radioactive decays. For example, gamma rays accompany the alpha decay of uranium-238.

$$^{238}_{92}U \rightarrow ^{234}_{90}Th + \alpha + 2\gamma$$

uranium-238 → thorium-234 + alpha particle + gamma rays

Because gamma rays are massless, the emission of gamma rays by themselves cannot result in the formation of a new atom. **Table 5** summarizes the basic characteristics of alpha, beta, and gamma radiation.

Nuclear stability The primary factor in determining an atom's stability is its ratio of neutrons to protons. Atoms that contain either too many or too few neutrons are unstable and lose energy through radioactive decay to form a stable nucleus. They emit alpha and beta particles and these emissions affect the neutron-to-proton ratio of the newly created nucleus. Eventually, radioactive atoms undergo enough radioactive decay to form stable, nonradioactive atoms. This topic will be covered later in the book.

SECTION 4 REVIEW

Section Self-Check

Section Summary

- Chemical reactions involve changes in the electrons surrounding an atom. Nuclear reactions involve changes in the nucleus of an atom.
- There are three types of radiation: alpha (charge of 2+), beta (charge of 1−), and gamma (no charge).
- The neutron-to-proton ratio of an atom's nucleus determines its stability.

25. MAINIDEA **Explain** how unstable atoms gain stability.
26. **State** what quantities are conserved when balancing a nuclear reaction.
27. **Classify** each of the following as a chemical reaction, a nuclear reaction, or neither.
    a. Thorium emits a beta particle.
    b. Two atoms share electrons to form a bond.
    c. A sample of pure sulfur emits heat energy as it slowly cools.
    d. A piece of iron rusts.
28. **Calculate** How much heavier is an alpha particle than an electron?
29. **Create** a table showing how each type of radiation affects the atomic number and the mass number of an atom.

SECTION 4 REVIEW

25. They undergo a series of radioactive decay reactions until they reach a stable element.
26. mass number and atomic number
27. a. nuclear b. chemical
    c. neither d. chemical
28. 6.65×10^{-27} kg

29.

Particle	Atomic Number	Mass Number
α	−2	−4
β	+1	no change
γ	no change	no change

HOW IT works

Mass Spectrometer: Chemical Detective

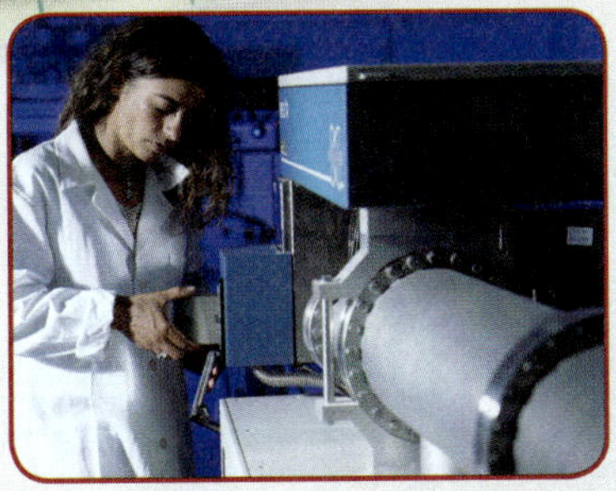

Imagine a forensic scientist needs to identify the inks used on a document to test for possible counterfeiting. The scientist can analyze the inks using a mass spectrometer, such as the one shown at left. A mass spectrometer breaks the compounds in a sample of an unknown substance into smaller fragments. The fragments are then separated according to their masses, and the exact composition of the sample can be determined. Mass spectrometry is one of the most important techniques for studying unknown substances.

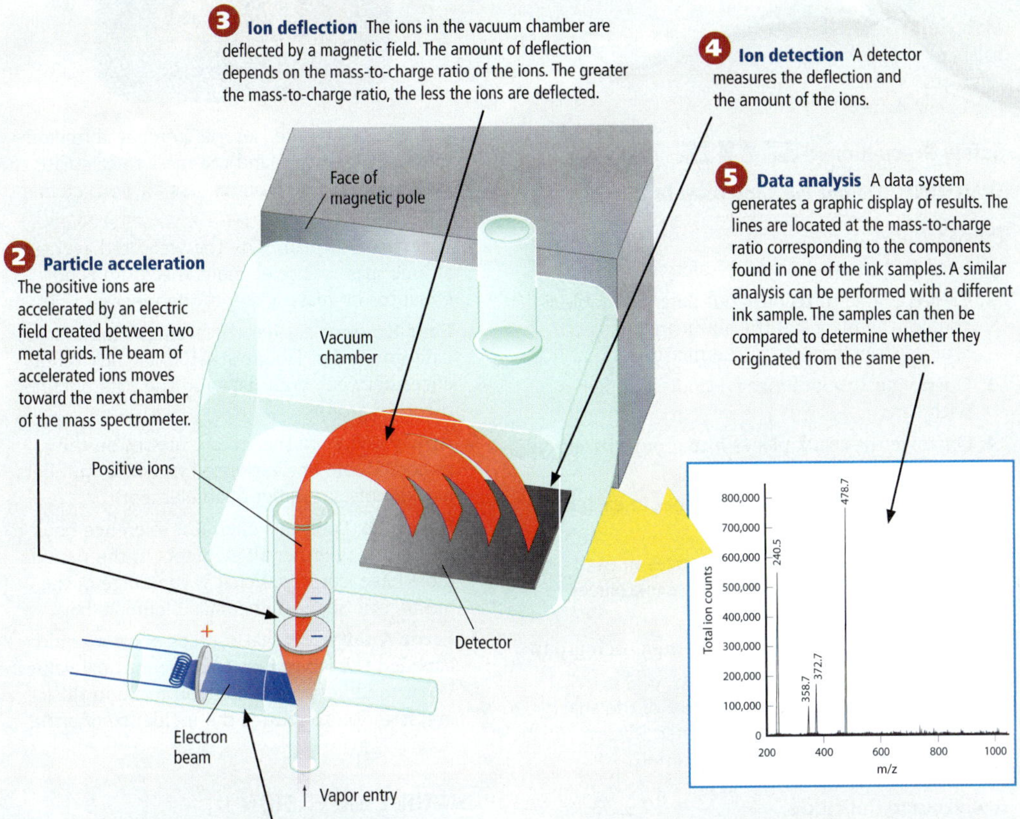

Mauro Fermariello/Photo Researchers

WRITING IN ▶ Chemistry

Summarize Research a case in which a mass spectrometer was used to distinguish between different types of ink, and write a summary of the procedure and results. WebQuest

HOW IT works

Purpose

Students will describe how a mass spectrometer functions to determine the composition of an unknown sample.

Background

Mass spectrometry is utilized across scientific fields. Applications include distinguishing dioxins in contaminated marine life, as well as detecting and identifying the presence of steroids in athletes' blood. Geologists use mass spectrometry to locate oil deposits by measuring petroleum compounds in rock strata. Mass spectrometry can be used by anesthesiologists to monitor a patient's breath during surgery, while astronomers utilize this tool to analyze molecules found in space.

Teaching Strategies

- Review key terms and concepts from found in the diagram, including *atom, electron, ion,* and *charge-to-mass ratio.*
- Have students create a chart that compares the components and functions of a cathode-ray tube and a mass spectrometer.

WRITING IN ▶ Chemistry

Rubric WebQuest

Summarize Students should be able to explain how mass spectrometers work and how they are used to distinguish between various ink samples. Their reports should clearly summarize the steps of mass spectrometry and how ink samples are compared to each other.

ChemLAB

Rubric

Preparation

Time Allotment one class period

Process Skills formulate models, error analysis, research, infer, draw conclusions, communicate, think critically, calculate, measure, collect and organize data

Safety Precautions Approve lab safety forms before work begins. Have students wear an apron and goggles. Remind students that they should not eat the snacks.

Alternate Materials Different size paper clips

Procedure

- Provide each lab group with a 1.5 oz snack bag.
- Make sure they choose snack bags that contain snacks of similar mass.

Analyze and Conclude

1. Answers will vary depending on the type of snack bags.
2. Answers will vary depending on the type of snack bags. If the snacks all have similar masses, the average atomic mass will be closest to the most abundant snack.
3. Answers will vary. The students should find that the atomic mass is closest to the most abundant piece. In the sample data, the atomic mass is closer to the cereal pieces than the pretzels or bagel chips due to their high percent abundance.
4. Answers will vary. Students should realize that the small sample size and difference in samples will lead to differences.
5. The atomic mass is different from the mass number because it is an average and therefore will not be a whole number.
6. Answers will vary depending on the reference sources used. They should find most elements have numerous natural isotopes.
7. The error in the lab is due to small sample size. The experiment could be modified by starting with a much larger sample size.

INQUIRY EXTENSION

Answers will vary. The students should be able to predict the most abundant element based on the atomic mass listed on the periodic table.

ChemLAB

Model Atomic Mass

Background: Most elements in nature occur as a mixture of isotopes. The weighted average atomic mass of an element can be determined from the atomic mass and the relative abundance of each isotope. In this activity, you will model the isotopes of the imaginary element "snackium." The measurements you make will be used to calculate a weighted average mass that represents the average atomic mass of snackium.

Question: *How are the atomic masses of the natural isotopic mixtures calculated?*

Materials

balance
calculator
bag of snack mix

Safety Precautions

WARNING: ***Do not eat food used in lab work.***

Procedure

1. Read and complete the lab safety form.
2. Create a table to record your data. The table will contain the mass and the abundance of each type of snack present in the mixture.
3. Open your snack-mix bag. Handle the pieces with care.
4. Organize the snack pieces into groups based on their types.
5. Count the number of snack pieces in each of your groups.
6. Record the number of snack pieces in each group and the total number of snack pieces in your data table.
7. Measure the mass of one piece from each group and record the mass in your data table.
8. **Cleanup and Disposal** Dispose of the snack pieces as directed by your teacher. Return all lab equipment to its designated location.

Analyze and Conclude

1. **Calculate** Find the percent abundance of the pieces by dividing the individual-piece quantity by the total number of snack pieces.
2. **Calculate** Use the isotopic percent abundance of the snack pieces and the mass to calculate the weighted average atomic mass for your element snackium.
3. **Interpret** Explain why the weighted average atomic mass of the element snackium is not equal to the mass of any of the pieces.
4. **Peer Review** Gather the average atomic mass data from other lab groups. Explain any differences between your data and the data obtained by other groups.
5. **Apply** Why are the atomic masses on the periodic table not expressed as whole numbers like the mass number of an element?
6. **Research** Look in a chemical reference book to determine whether all elements in the periodic table have isotopes. What is the range of the number of isotopes chemical elements have?
7. **Error Analysis** What sources of error could have led the lab groups to different final values? What modifications could you make in this investigation to reduce the incidence of error?

INQUIRY EXTENSION

Predict Based on your experience in this lab, look up the atomic masses of several elements on the periodic table and predict the most abundant isotope for each element.

CHAPTER 4 STUDY GUIDE

Vocabulary Practice

BIGIDEA Atoms are the fundamental building blocks of matter.

SECTION 1 Early Ideas About Matter

MAINIDEA The ancient Greeks tried to explain matter, but the scientific study of the atom began with John Dalton in the early 1800s.

- Democritus was the first person to propose the existence of atoms.
- According to Democritus, atoms are solid, homogeneous, and indivisible.
- Aristotle did not believe in the existence of atoms.
- John Dalton's atomic theory is based on numerous scientific experiments.

VOCABULARY
- Dalton's atomic theory

SECTION 2 Defining the Atom

MAINIDEA An atom is made of a nucleus containing protons and neutrons; electrons move around the nucleus.

- An atom is the smallest particle of an element that maintains the properties of that element.
- Electrons have a 1− charge, protons have a 1+ charge, and neutrons have no charge.
- An atom consists mostly of empty space surrounding the nucleus.

VOCABULARY
- atom
- cathode ray
- electron
- nucleus
- proton
- neutron

SECTION 3 How Atoms Differ

MAINIDEA The number of protons and the mass number define the type of atom.

- The atomic number of an atom is given by its number of protons. The mass number of an atom is the sum of its neutrons and protons.

atomic number = number of protons = number of electrons

mass number = atomic number + number of neutrons

- Atoms of the same element with different numbers of neutrons are called isotopes.
- The atomic mass of an element is a weighted average of the masses of all of its naturally occurring isotopes.

VOCABULARY
- atomic number
- isotope
- mass number
- atomic mass unit (amu)
- atomic mass

SECTION 4 Unstable Nuclei and Radioactive Decay

MAINIDEA Unstable atoms emit radiation to gain stability.

- Chemical reactions involve changes in the electrons surrounding an atom. Nuclear reactions involve changes in the nucleus of an atom.
- There are three types of radiation: alpha (charge of 2+), beta (charge of 1−), and gamma (no charge).
- The neutron-to-proton ratio of an atom's nucleus determines its stability.

VOCABULARY
- radioactivity
- radiation
- nuclear reaction
- radioactive decay
- alpha radiation
- alpha particle
- nuclear equation
- beta radiation
- beta particle
- gamma ray

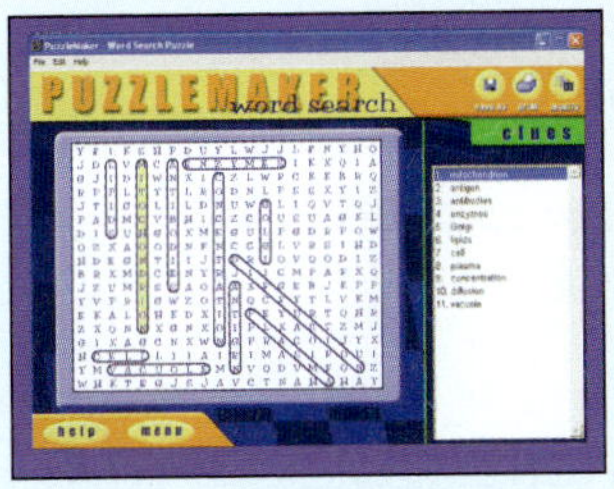

Vocabulary Practice

For additional practice with vocabulary, have students access the Vocabulary PuzzleMaker.

Use the Vocabulary

To reinforce chapter vocabulary, have students write a sentence using each term. OL EL

Review Strategies

- Have students summarize how the atomic model has changed from Dalton's concept to our present understanding. OL
- Have students list the parts of the atom and give their charge, location, and relative size. OL
- Have students determine the average atomic mass for an element given isotope masses and corresponding percent abundance. OL
- Problems from the Supplemental Practice Problems appendix or the Supplemental Problems booklet can be used for review. OL

Students can review with these online resources.

Vocabulary Practice

Vocabulary eGames
Vocabulary eFlashcards
Vocabulary PuzzleMaker

Section Self-Check

Chapter Self-Check

Online Test Practice

Use *eAssessment* to:

- create multiple versions of tests
- edit existing questions and add your own questions
- build tests aligned with select state curriculum standards using built-in state curriculum tags
- track students' progress

SECTION 1

Mastering Concepts

30. Democritus

31. John Dalton

32. Democritus's ideas: matter is composed of empty space through which atoms move; different kinds of atoms have different sizes and shapes; the movement, size, and shape of atoms result in unique properties of matter; changes in matter result from changes in the groupings of atoms. Dalton's atomic theory: matter is composed of atoms; different atoms combine in simple whole number ratios to form compounds; atoms cannot be created, destroyed, or divided.

33. Ideas. He had no experimental evidence.

34. He had no scientific instruments to research matter at the atomic level.

35. He did not believe that atoms could move through empty space.

36. Atoms are not indivisible and all atoms of an element are not identical.

37. Mass is conserved because atoms cannot be created, divided, or destroyed. Chemical reactions involve only the separation and rearrangement of atoms.

38. Matter is anything that occupies space and has mass. For example: desk, chair.

SECTION 2

Mastering Concepts

39. protons and neutrons; positive charge equal to the number of protons

40. evenly through a sphere

41. They would only be slightly deflected.

42. **a.** electron cloud
b. protons
c. neutron

43. electron $<$ proton $=$ neutron

44. The number of protons equals the number of electrons.

45. 89 +

46. protons and neutrons

47. 1836

48. the electron

49. changing the type of electrode or type of gas did not affect the ray produced.

50. A cathode ray is composed of a stream of electrons that originate at the cathode and flow through the tube to the anode.

SECTION 1

Mastering Concepts

30. Who originally proposed the concept that matter is composed of tiny, indivisible particles?

31. Whose work is credited with being the beginning of modern atomic theory?

32. Distinguish between Democritus's ideas and Dalton's atomic theory.

33. **Ideas and Scientific Methods** Was Democritus's proposal of the existence of atoms based on scientific methods or ideas? Explain.

34. Explain why Democritus was unable to experimentally verify his ideas.

35. What was Aristotle's objection to the atomic theory?

36. State the main points of Dalton's atomic theory using your own words. Which parts of Dalton's theory were later found to be erroneous? Explain why.

37. **Conservation of Mass** Explain how Dalton's atomic theory offered a convincing explanation of the observation that mass is conserved in chemical reactions.

38. Define *matter* and give two everyday examples.

SECTION 2

Mastering Concepts

39. What particles are found in the nucleus of an atom? What is the charge of the nucleus?

40. How was the overall charge distributed in the plum pudding model?

41. How did the charge distribution in the plum pudding model affect alpha particles passing through an atom?

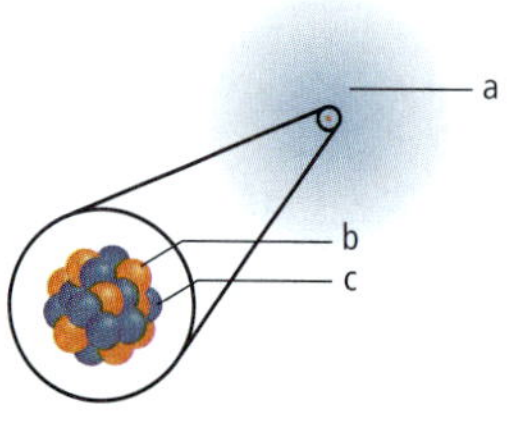

■ **Figure 22**

42. Label the subatomic particles shown in **Figure 22**.

43. Arrange the following subatomic particles in order of increasing mass: neutron, electron, and proton.

44. Explain why atoms are electrically neutral.

45. What is the charge of the nucleus of element 89?

46. Which particles account for most of an atom's mass?

47. If you had a balance that could determine the mass of a proton, how many electrons would you need to weigh on the same balance to measure the same mass as that of a single proton?

48. **Cathode-Ray Tubes** Which subatomic particle was discovered by researchers working with cathode-ray tubes?

49. What experimental results led to the conclusion that electrons were part of all forms of matter?

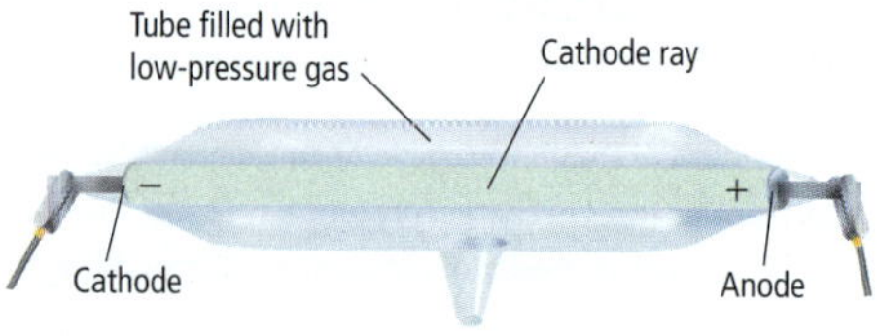

■ **Figure 23**

50. **Cathode Ray** Use the elements labeled in **Figure 23** to explain the direction of a cathode ray inside a cathode-ray tube.

51. Briefly explain how Rutherford discovered the nucleus.

52. **Particle Deflection** What caused the deflection of the alpha particles in Rutherford's gold foil experiment?

53. **Charge of Cathode Rays** How was an electric field used to determine the charge of a cathode ray?

54. Explain what keeps the electrons confined in the space surrounding the nucleus.

55. What is the approximate size of an atom?

56. **Visualizing Atoms** What technique can be used to visualize individual atoms?

57. What are the strengths and weaknesses of Rutherford's nuclear model of the atom?

SECTION 3

Mastering Concepts

58. How do isotopes of a given element differ? How are they similar?

59. How is an atom's atomic number related to its number of protons? To its number of electrons?

60. How is the mass number related to the number of protons and neutrons an atom has?

51. Rutherford aimed a beam of alpha particles at a thin foil of gold. He expected most of the alpha particles to pass through the gold atoms, confirming the plum pudding model. However, a few particles were deflected at very large angles, which led to the discovery of the positively charged nucleus.

52. The α particles were deflected by the positively charged gold nuclei.

53. The cathode ray was attracted to the positive end of the magnet implying that the cathode ray is negative.

54. attraction to the positively charged nucleus

55. It is in the range of 10^{-10} m.

56. STM

57. Strengths: explained the results of the gold-foil experiment and why an atom is electrically neutral. Weaknesses: could not account for the total mass of an atom or the arrangement of the electrons.

SECTION 3

Mastering Concepts

58. differ: number of neutrons, masses; similar: chemical properties, number of protons and electrons

59. They are all equal.

60. mass number $=$ number of p $+$ number of n

Chapter Self-Check

61. How can you determine the number of neutrons in an atom if its mass number and its atomic number are known?

62. What do the superscript and subscript in the notation $^{40}_{19}K$ represent?

63. Standard Units Define the atomic mass unit. What were the benefits of developing the atomic mass unit as a standard unit of mass?

64. Isotopes Are $^{24}_{12}Mg$, $^{25}_{12}Mg$, $^{26}_{12}Mg$ isotopes of each other? Explain.

65. Does the existence of isotopes contradict part of Dalton's original atomic theory? Explain.

Mastering Problems

66. How many protons and electrons are contained in an atom of element 44?

67. Carbon A carbon atom has a mass number of 12 and an atomic number of 6. How many neutrons does it have?

68. Mercury An isotope of mercury has 80 protons and 120 neutrons. What is the mass number of this isotope?

69. Xenon An isotope of xenon has an atomic number of 54 and contains 77 neutrons. What is the xenon isotope's mass number?

70. If an atom has 18 electrons, how many protons does it have?

71. Sulfur Determine the atomic mass of sulfur given the following naturally occurring isotopes: $^{32}_{16}S$ (31.972 amu, 94.99%), $^{33}_{16}S$ (32.971 amu, 0.75%), $^{34}_{16}S$ (33.968 amu, 4.25%), and $^{36}_{16}S$ (35.967 amu, 0.01%).

72. Fill in the blanks in **Table 6.**

Table 6 Chlorine and Zirconium

Element	Cl	Cl	Zr	Zr
Atomic number	17		40	
Mass number	35	37		92
Protons				40
Neutrons			50	
Electrons		17		

73. How many electrons, protons, and neutrons are contained in each atom?
a. $^{132}_{55}Cs$ **c.** $^{163}_{69}Tm$
b. $^{59}_{27}Co$ **d.** $^{70}_{30}Zn$

74. How many electrons, protons, and neutrons are contained in each atom?
a. gallium-69 **c.** titanium-48
b. fluorine-23 **d.** tantalum-181

75. For each chemical symbol, determine the number of protons and electrons an atom of the element contains.
a. V **c.** Ir
b. Mn **d.** S

76. Gallium, which has an atomic mass of 69.723 amu, has two naturally occurring isotopes, Ga-69 and Ga-71. Which isotope occurs in greater abundance? Explain.

77. Atomic Mass of Silver Silver has two isotopes: $^{107}_{47}Ag$, which has a mass of 106.905 amu and a percent abundance of 52.00%, and $^{109}_{47}Ag$, which has a mass of 108.905 amu and a percent abundance of 48.00%. What is the atomic mass of silver?

78. Data for chromium's four naturally occurring isotopes are provided in **Table 7.** Calculate chromium's atomic mass.

Table 7 Chromium Isotope Data

Isotope	Percent Abundance	Mass (amu)
Cr-50	4.35	49.946
Cr-52	83.79	51.941
Cr-53	9.50	52.941
Cr-54	2.36	53.939

SECTION 4

Mastering Concepts

79. What is radioactive decay?

80. Why are some atoms radioactive?

81. Discuss how radioactive atoms gain stability.

82. Define *alpha particle, beta particle,* and *gamma ray.*

83. Write the symbols used to denote alpha, beta, and gamma radiation and give their mass and charge.

84. What type of reaction involves changes in the nucleus of an atom?

85. Radioactive Emissions What change in mass number occurs when a radioactive atom emits an alpha particle? A beta particle? A gamma particle?

86. What is the primary factor that determines whether a nucleus is stable or unstable?

87. Explain how energy loss and nuclear stability are related to radioactive decay.

88. Explain what must occur before a radioactive atom stops undergoing further radioactive decay.

89. Boron-10 emits alpha particles and cesium-137 emits beta particles. Write balanced nuclear reactions for each radioactive decay.

61. number of n^0 = mass number − atomic number

62. superscript = mass number (40) and subscript = atomic number (19)

63. amu = 1/12 of the mass of a C-12 atom; It's a relative standard that is closer in size to atomic and subatomic masses.

64. Yes; isotopes are atoms of the same element with different atomic mass but the same number of protons.

65. Yes; not all atoms of an element are identical in mass.

Mastering Problems

66. 44 protons, 44 electrons

67. 6 neutrons

68. mass number is 200

69. mass number is 131

70. 18 protons

71. $S = (31.972\text{ amu})(0.9499) + (32.971\text{ amu})(0.0075) + (33.968\text{ amu})(0.0425) + (35.967\text{ amu})(0.0001) = 32.065\text{ amu}$

72.

Element	Cl	Cl	Zr	Zr
Atomic Number	17	17	40	40
Mass Number	35	37	90	92
Protons	17	17	40	40
Neutrons	18	20	50	52
Electrons	17	17	40	40

73. a. 55 electrons, 55 protons, 77 neutrons
b. 27 electrons, 27 protons, 32 neutrons
c. 69 electrons, 69 protons, 94 neutrons
d. 30 electrons, 30 protons, 40 neutrons

74. a. 31 electrons, 31 protons, 38 neutrons
b. 9 electrons, 9 protons, 14 neutrons
c. 22 electrons, 22 protons, 26 neutrons
d. 73 electrons, 73 protons, 108 neutrons

75. a. 23, 23 **c.** 77, 77
b. 25, 25 **d.** 16, 16

76. Ga-69 must be more abundant because the atomic mass of gallium is closer to the mass of Ga-69 than the mass of Ga-71.

77. 107.86 amu

78. 51.99 amu

SECTION 4

Mastering Concepts

79. Radioactive decay occurs when unstable nuclei spontaneously (without input of energy) lose energy by emitting radiation.

80. The stability of atoms depends on their neutron-to-proton ratio. When this ratio is either too large or too small the nucleus of an atom becomes unstable causing an atom to be radioactive.

81. by emitting radiation or a particle

82. alpha particle: helium atom with a 2+ charge; beta particle: electron; gamma rays: high energy radiation

83. $^{4}_{2}He$ 4 +2
β 1/1840 −1
γ 0 0

84. nuclear reaction

85. α, mass number decreases by 4; β, no change in mass number; γ, no change in mass number

86. the neutron-to-proton ratio

87. Radioactivity results when unstable nuclei emit energy in order to gain stability.

88. stable, nonradioactive atom must be formed

89. $^{10}_{5}B \rightarrow ^{6}_{3}Li + ^{4}_{2}He$

$^{137}_{55}Cs \rightarrow ^{137}_{56}Ba + ^{0}_{-1}e$

MIXED REVIEW

90. Atoms are indivisible and atoms of the same element can have different mass. An atomic structure consists of subatomic particles: electrons, protons and neutrons.

91. A cathode-ray tube has a metal electrode at each end. The electrodes are connected to a power supply. When current flows, electrons are emitted from the cathode and travel through the tube to the anode.

92. Thomson showed that the electron's mass was less than the mass of hydrogen, the lightest atom. This showed that there were smaller subatomic particles. Atoms are divisible.

93. Rutherford expected α particles to be slightly deflected when passing through a gold foil. But some particles were deflected at very large angles.

94. 12; protons are the only charged particles in the nucleus. To balance their positive charge, there must be the same number of electrons as protons.

95. There are 143 neutrons. Uranium.

96.

isotope	S-32	Ca-44	Zn-64	F-19	Na-23
Atomic number	16	20	30	9	11
Mass number	32	44	64	19	23
p^+	16	20	30	9	11
n^0	16	24	34	10	12
e^-	16	20	30	9	11

97. An atom's diameter is about 10,000 times the diameter of its nucleus. The density of the nucleus must be large.

98. The nucleus is positively charged, whereas the atom is neutral.

99. because they are charged

100. Moseley discovered that each element contains a unique positive charge (or number of protons) in its nucleus. Thus, the number of protons in an atom's nucleus uniquely identifies it as an atom of a particular element.

101. mass number = 39; charge = 0

102. B-11 must occur in greater abundance because the atomic weight of bromine is much closer to the mass of B-11 than to the mass of B-10.

103. $^{28}_{14}Si$, $^{29}_{14}Si$, $^{30}_{14}Si$

ASSESSMENT

MIXED REVIEW

90. Determine what was wrong with Dalton's theory and provide the most recent version of the atomic structure.

91. Cathode-Ray Tube Describe a cathode-ray tube and how it operates.

92. Subatomic Particles Explain how J. J. Thomson's determination of the charge-to-mass ratio of the electron led to the conclusion that atoms were composed of subatomic particles.

93. Gold Foil Experiment How did the actual results of Rutherford's gold foil experiment differ from the results he expected?

94. If a nucleus contains 12 protons, how many electrons are in the neutral atom? Explain.

95. An atom's nucleus has 92 protons and its mass number is 235. How many neutrons are in the nucleus? What is the name of the atom?

96. Complete **Table 8.**

Table 8 Composition of Various Isotopes

Isotope			Zn-64		
Atomic number				9	11
Mass number	32				23
Number of protons	16				
Number of neutrons		24		10	
Number of electrons		20			

97. Approximately how many times greater is the diameter of an atom than the diameter of its nucleus? Knowing that most of an atom's mass is contained in the nucleus, what can you conclude about the density of the nucleus?

98. Is the charge of a nucleus positive, negative, or zero? The charge of an atom?

99. Why are electrons in a cathode-ray tube deflected by electric fields?

100. What was Henry Moseley's contribution to the modern understanding of the atom?

101. What is the mass number of potassium-39? What is the isotope's charge?

102. Boron-10 and boron-11 are the naturally occurring isotopes of elemental boron. If boron has an atomic mass of 10.81 amu, which isotope occurs in greater abundance?

103. Semiconductors Silicon is important to the semiconductor manufacturing industry. The three naturally occurring isotopes of silicon are silicon-28, silicon-29, and silicon-30. Write the symbol for each.

104. Titanium Use **Table 9** to calculate the atomic mass of titanium.

Table 9 Titanium Isotopes

Isotope	Atomic Mass (amu)	Relative Abundance (%)
Ti-46	45.953	8.00
Ti-47	46.952	7.30
Ti-48	47.948	73.80
Ti-49	48.948	5.50
Ti-50	49.945	5.40

105. Describe how each type of radiation affects an atom's atomic number and mass number.

106. Relative Abundances Magnesium constitutes about 2% of Earth's crust and has three naturally occurring isotopes. Suppose you analyze a mineral and determine that it contains the three isotopes in the following proportions: Mg-24 (abundance = 79%), Mg-25 (abundance = 10%), and Mg-26 (abundance = 11%). If your friend analyzes a different mineral containing magnesium, do you expect her to obtain the same relative abundances for each magnesium isotope? Explain your reasoning.

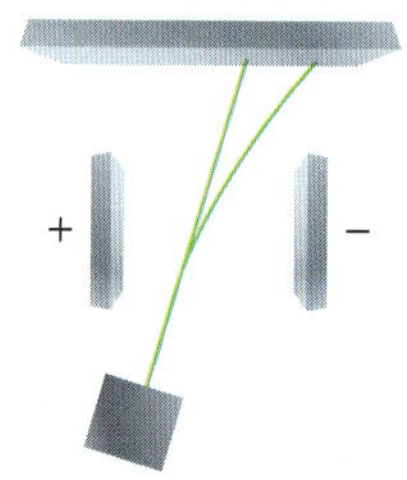

■ **Figure 24**

107. Radiation Identify the two types of radiation shown in **Figure 24.** Explain your reasoning.

THINK CRITICALLY

108. Formulate How were scientific methods used to determine the model of the atom? Why is the model considered a theory?

109. Discuss What experiment led to the dispute of J. J. Thomson's plum pudding atomic model? Explain your answer.

110. Apply Which is greater, the number of compounds or the number of elements? The number of elements or the number of isotopes? Explain.

104. 47.89 amu

105. α: atomic number decreases by 2, mass number decreases by 4; β: atomic number increases by 1, mass number unchanged; γ: atomic number and mass number are unchanged

106. Yes. The isotopic abundance of an element is the same no matter where the element comes from.

107. The deflected beam is alpha radiation because it is deflected toward the negatively charged plate. The undeflected beam must be neutral γ rays.

THINK CRITICALLY

108. Experiments were performed to explain observations and test hypotheses. It is a theory because it is still subject to modifications as more data become available.

109. Rutherford's thin gold foil experiment. His experimental results were inconsistent with the plum pudding atomic model.

110. The number of compounds is greater than the number of elements because compounds are combinations of elements and the elements can be combined in many ways. The number of isotopes is greater than the number of atoms because each element has only one type of atom but may have more than one isotope.

Chapter Self-Check

111. Analyze An element has three naturally occurring isotopes. What other information must you know in order to calculate the element's atomic mass?

112. Apply If atoms are primarily composed of empty space, explain why you cannot pass your hand through a solid object.

113. Formulate Sketch a modern atomic model of a typical atom and identify where each type of subatomic particle would be located.

114. Apply Indium has two naturally occurring isotopes and an atomic mass of 114.818 amu. In-113 has a mass of 112.904 amu and an abundance of 4.3%. What is the identity and percent abundance of indium's other isotope?

115. Infer Sulfur's average atomic mass is close to the whole number 32. Chlorine's average atomic mass is 35.453, which is not a whole number. Suggest a possible reason for this difference.

Challenge Problem

116. Magnesium Isotopes Compute the mass number, *X*, of the third isotope of magnesium given that the respective abundances of the naturally occurring isotopes are: 79.0%, 10%, and 11% for $^{24}_{12}Mg$, $^{25}_{12}Mg$, and $^{X}_{12}Mg$. The atomic mass of magnesium is 24.305 amu.

CUMULATIVE REVIEW

117. How is a qualitative observation different from a quantitative observation? Give an example of each.

118. A 1.0-cm^3 block of gold can be flattened to a thin sheet that averages 3.0×10^{-8} cm thick. What is the area (in cm^2) of the flattened gold sheet?

119. A piece of paper has an area of 603 cm^2. How many sheets of paper would the sheet of gold mentioned in problem 118 cover?

120. Classify each mixture as heterogeneous or homogeneous.
a. salt water
b. vegetable soup
c. 14-K gold
d. concrete

121. Determine whether each change is physical or chemical.
a. Water boils.
b. A match burns.
c. Sugar dissolves in water.
d. Sodium reacts with water.
e. Ice cream melts.

122. Television and Computer Screens Describe how cathode rays are used to generate television and computer monitor images.

123. The Standard Model The standard model of particle physics describes all of the known building blocks of matter. Research the particles included in the standard model. Write a short report describing the known particles and those thought to exist but not yet detected experimentally.

124. STM Individual atoms can be seen using a sophisticated device known as a scanning tunneling microscope. Write a short report on how the scanning tunneling microscope works and create a gallery of this microscope's images from sources such as books, magazines, and the Internet.

DBQ Document-Based Questions

Zirconium *is a lustrous, gray-white metal. Because of its high resistance to corrosion and its low cross section for neutron absorption, it is often used in nuclear reactors. It can also be processed to produce gems that look like diamonds that is used in jewelry.*

Table 10 *shows the relative abundances of zirconium isotopes.*

Table 10 Relative Abundances of Zirconium Isotopes

Element	Mass (amu)	Abundance (%)
Zirconium-90	89.905	51.45
Zirconium-91	90.906	11.22
Zirconium-92	91.905	17.15
Zirconium-94	93.906	17.38
Zirconium-96	95.908	2.80

Data obtained from: Lide, David R., ed. 2005. *CRC Handbook of Chemistry and Physics.* Boca Raton: CRC Press.

125. What is the mass number of each zirconium isotope?

126. Compute the number of protons and neutrons for each zirconium isotope.

127. Does the number of protons or neutrons remain the same for all isotopes? Explain.

128. Based on the relative abundances of each isotope, predict to which isotope's mass the average atomic mass of zirconium is going to be closest.

129. Calculate the weighted average atomic mass of zirconium.

111. You also need to know the mass and percent abundance of each isotope.

112. Atoms are extremely small and very close together in comparison to our hand. Atoms in a solid object are bonded together by electrical forces—bonds that are not easily broken.

113. Sketches should look similar to Figure 14.

114. The other isotope is In-115. Its percent abundance is 95.7%.

115. Sulfur has an isotope with a very high abundance. Chlorine has more than one isotope with a large percent abundance.

CHALLENGE PROBLEM

116. 26 amu

CUMULATIVE REVIEW

117. A qualitative observation does not involve measurement (water is hot). A quantitative observation involves measurement (the water is 42°C).

118. area = 3.3×10^7 cm^2

119. sheets = 55,000

120.a. homogeneous
b. heterogeneous
c. homogeneous
d. heterogeneous

121.a. physical change
b. chemical change
c. physical change
d. chemical change
e. physical change

WRITING IN ▶ Chemistry

Rubric

122. The back of a computer monitor screen or television screen is coated with a phosphorescent material that glows when a beam of electrons strikes on it. Phosphorescent means that the material emits a light of a different color.

123. known particles: protons, neutrons, electrons, quarks, pions
not detected yet: Higgs boson

124. A point moves across a sample and the electrons in the point interact with the electrons surrounding the superficial atoms in the sample. This interaction is recorded electronically.

DBQ Document-Based Questions

Data obtained from: Lide, David R., ed. 2005. *CRC Handbook of Chemistry and Physics.* Boca Raton: CRC Press.

125. refer to table
126. refer to table
127. protons; isotopes are atoms of the same element with different mass numbers.
128. zirconium-90
129. 91.23 amu

Element	Mass Number	Number of Neutrons	Number of Protons
Zirconium-90	90	50	40
Zirconium-91	91	51	40
Zirconium-92	92	52	40
Zirconium-94	94	54	40
Zirconium-96	96	56	40

MULTIPLE CHOICE

1. B
2. A
3. D
4. C
5. B
6. D
7. B
8. B
9. B
10. D

MULTIPLE CHOICE

1. Which describes an atom of plutonium?
   - **A.** It can be divided into smaller particles that retain all the properties of plutonium.
   - **B.** It cannot be divided into smaller particles that retain all the properties of plutonium.
   - **C.** It does not possess all the properties of a larger quantity of plutonium.
   - **D.** It has an atomic number of 244.

2. Neptunium's only naturally occurring isotope, $^{237}_{93}Np$, decays by emitting one alpha particle, one beta particle, and one gamma ray. What is the new atom formed from this decay?
   - **A.** $^{233}_{92}U$
   - **B.** $^{241}_{93}Np$
   - **C.** $^{233}_{90}Th$
   - **D.** $^{241}_{92}U$

3. Which type of matter has a definite composition throughout and is made of more than one type of element?
   - **A.** heterogeneous mixture
   - **B.** homogeneous mixture
   - **C.** element
   - **D.** compound

Use the diagram below to answer Question 4.

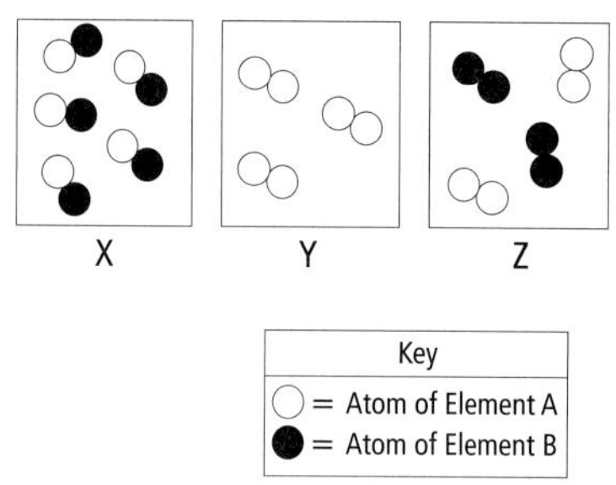

4. Which diagram shows a mixture?
   - **A.** X
   - **B.** Y
   - **C.** Z
   - **D.** both X and Z

5. The Moon is approximately 384,400 km from Earth. What is this value in scientific notation?
   - **A.** 384.4×10^3 km
   - **B.** 3.844×10^5 km
   - **C.** 3.844×10^{-5} km
   - **D.** 3844×10^{-2} km

6. Why does an atom have no net electric charge?
   - **A.** Its subatomic particles carry no electric charges.
   - **B.** The positively charged protons cancel out the negatively charged neutrons.
   - **C.** The positively charged neutrons cancel out the negatively charged electrons.
   - **D.** The positively charged protons cancel out the negatively charged electrons.

7. How many neutrons, protons, and electrons does $^{126}_{52}Te$ have?
   - **A.** 126 neutrons, 52 protons, and 52 electrons
   - **B.** 74 neutrons, 52 protons, and 52 electrons
   - **C.** 52 neutrons, 74 protons, and 74 electrons
   - **D.** 52 neutrons, 126 protons, and 126 electrons

Use the figure below to answer Question 8.

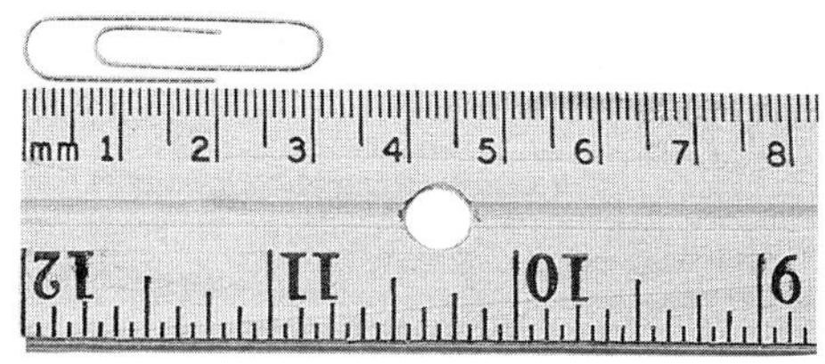

8. Record the length of this paper clip to the appropriate number of significant digits.
   - **A.** 31 mm
   - **B.** 31.1 mm
   - **C.** 30.1 mm
   - **D.** 31.15 mm

9. Element X has an unstable nucleus due to an over-abundance of neutrons. All are likely to occur EXCEPT
   - **A.** element X will undergo radioactive decay.
   - **B.** element X will remain an unstable radioactive element.
   - **C.** element X will gain more protons to balance the neutrons it possesses.
   - **D.** element X will spontaneously lose energy.

10. What makes up most of the volume of an atom?
    - **A.** protons
    - **B.** neutrons
    - **C.** electrons
    - **D.** empty space

Matt Meadows

SHORT ANSWER

11. A 36.41-g sample of calcium carbonate ($CaCO_3$) contains 14.58 g of calcium and 4.36 g of carbon. What is the mass of oxygen contained in the sample? What is the percent by mass of each element in this compound?

Use the table below to answer Questions 12 and 13.

Characteristics of Naturally Occurring Neon Isotopes			
Isotope	Atomic Number	Mass (amu)	Percent Abundance
^{20}Ne	10	19.992	90.48
^{21}Ne	10	20.994	0.27
^{22}Ne	10	21.991	9.25

12. For each isotope listed above, write the number of protons, electrons, and neutrons it contains.

13. Using the data in the table above, calculate the average atomic mass of neon.

EXTENDED RESPONSE

14. Assume that Element Q has the following three isotopes: ^{248}Q, ^{252}Q, and ^{259}Q. If the atomic mass of Q is 258.63, which of its isotopes is most abundant? Explain your answer.

15. Iodine-131 undergoes radioactive decay to form an isotope with 54 protons and 77 neutrons. What type of decay occurs in this isotope? Explain how you can tell.

16. You are given an aluminum cube. Your measurements show that its sides are 2.14 cm and its mass is 25.1 g. Explain how you would find its density. If the density of aluminum is known to be 2.70 g/cm^3, what is your percent error?

SAT SUBJECT TEST: CHEMISTRY

For each question below, indicate whether Statement I is true or false and indicate whether Statement II is true or false. If Statement II is a *correct explanation* of Statement I, write CE on your paper.

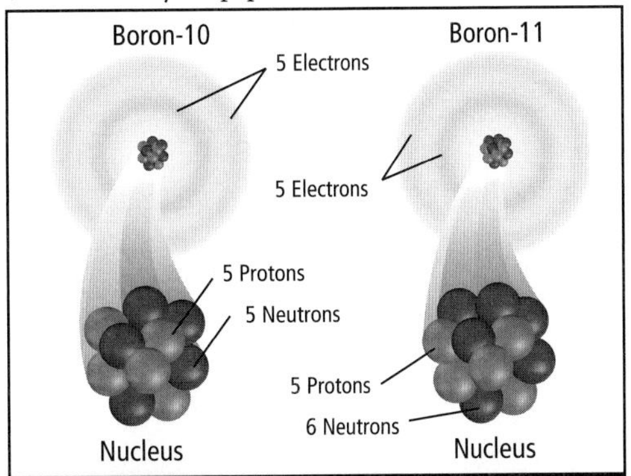

Statement I		Statement II
17. The two atoms of boron pictured above are isotopes	BECAUSE	they have the same number of protons but a different number of neutrons.
18. Most alpha particles shot at a piece of gold foil travel through it	BECAUSE	an atom has a large nucleus compared to its overall size.
19. A beam of neutrons is attracted to the charged plates surrounding it	BECAUSE	neutrons have no charge.
20. Carbon and oxygen can form either CO or CO_2	BECAUSE	carbon and oxygen obey the law of definite composition.
21. A mixture of sand and water is heterogeneous	BECAUSE	water is a compound formed from hydrogen and oxygen.

NEED EXTRA HELP?																					
If You Missed Question . . .	1	2	3	4	5	6	7	8	9	10	11	12	13	14	15	16	17	18	19	20	21
Review Section . . .	4.2	4.4	3.4	4.2	4.3	3.1	2.2	2.3	2.2	4.2	3.4	4.3	4.3	4.3	4.4	2.3	4.3	4.2	4.2	4.1	3.3

SHORT ANSWER

11. mass of oxygen = 17.47 g
40.04% calcium, 12.0% carbon, 47.98% oxygen.

12. ^{20}Ne: 10p, 10n, 10e
^{21}Ne: 10p, 11n, 10e
^{22}Ne: 10p, 12n, 10e

13. average atomic mass = 20.18 amu

EXTENDED RESPONSE

14. ^{259}Q is probably the most abundant isotope because the average atomic mass is close to 259. If a different isotope were more abundant, the average atomic mass would be much lower.

15. beta decay; The atomic number changes from 53 (iodine) to 54 (xenon), whereas the mass number does not change at all (131 for iodine, 54 + 77 for xenon).

16. Use the formula for volume of a cube (L × W × H), then use the density formula (D = m/v) to solve for density.
volume = 2.56 g/cm^3
percent error = 5.18% error

SAT SUBJECT TEST: CHEMISTRY

17. T, T, CE
18. T, F
19. F, T
20. T, F
21. T, T (Although both statements are true, Statement II is NOT a correct explanation of Statement I.)

CHAPTER 5 Electrons in Atoms

BIGIDEA **The atoms of each element have a unique arrangement of electrons.**

ESSENTIAL QUESTIONS	RESOURCES TO ASSESS MASTERY
SECTION 1 Light and Quantized Energy **1.** How do the wave and particle natures of light compare? **2.** What is a quantum of energy and how is it related to an energy change of matter? **3.** How do continuous electromagnetic spectra and atomic emission spectra compare and contrast? 2 sessions 1 block	**Progress Monitoring** Formative Assessment, pp. 141, 145 Reading Check, pp. 139, 141, 142, 144 Section Review, p. 145
SECTION 2 Quantum Theory and the Atom **1.** How do the Bohr and quantum mechanical models of the atom compare? **2.** What is the impact of de Broglie's wave-particle duality and the Heisenberg uncertainty principle on the current view of electrons in atoms? **3.** What are the relationships among a hydrogen atom's energy levels, sublevels, and atomic orbitals? 2 sessions 1 block	**Progress Monitoring** Formative Assessment, pp. 147, 150, 154 Reading Check, pp. 148, 151, 152, 153, 154 Section Review, p. 155
SECTION 3 Electron Configuration **1.** How are the Pauli exclusion principle, the aufbau principle, and Hund's rule used to write electron configurations with orbital diagrams and electron configuration notation? **2.** What are valence electrons, and how do electron-dot structures represent an atom's valence electrons? 1 session 0.5 block	**Progress Monitoring** Formative Assessment, pp. 157, 159 Reading Check, pp. 157, 159 Section Review, p. 162 **Summative Assessment** Chapter Assessment, p. 166 *eAssessment* Chapter Test (scaffolded)

Go online!

	LEVELED RESOURCES	LAB MATERIALS
	Science Notebook 5.1 OL ***FAST FILE Chapter Resources:*** MiniLab Worksheet, p. 12 OL ChemLab Worksheet, p. 4 OL Study Guide, p. 14 BL	LaunchLAB p. 134 / **10 min** wrapped box MiniLAB p. 144 / **30 min** cotton swabs, lithium chloride, Bunsen burner, sodium chloride, potassium chloride, calcium chloride, strontium chloride, unknown solution ChemLAB p. 164 / **45 min** ring stand with clamp, 40-W tubular lightbulb, light socket with grounded power cord, 275-mL polystyrene culture flask, Flinn C-Spectra® or similar diffraction grating, red food coloring, green food coloring, blue food coloring, yellow food coloring, set of colored pencils, spectrum tubes, spectrum-tube power supply
	Science Notebook 5.2 OL ***FAST FILE Chapter Resources:*** Study Guide, p. 16 BL	
	Science Notebook 5.3 OL ***FAST FILE Chapter Resources:*** Study Guide, p. 18 BL	

ADDITIONAL RESOURCES

- eAssessment
- ConnectED Student Center
- ConnectED Teacher Center
- Lesson Presentations
- eSolutions
- Videos
- Animations
- Online Quizzes
- Virtual Investigations
- Multilingual eGlossary
- Personal Tutors
- WebQuests
- Applying Practices Worksheets
- PBLs

Practice:

- Challenge Problems AL
- Supplemental Problems BL OL
- Chapter Test (Scaffolded)
- Standardized Test Practice

Additional Resources:

- Blueprints for Success
- Solving Problems: A Chemistry Handbook

Lab Resources:

- Laboratory Manual OL
- CBL Laboratory Manual OL
- Small-Scale Laboratory Manual OL
- Forensics Laboratory Manual OL

BL Below Level OL On Level AL Advanced Learners EL English Learners COOP LEARN Cooperative Learning

CHAPTER 5

LaunchLAB

Rubric

How do you know what is inside an atom?

Purpose Students will make observations using all the senses except sight.

Safety Precautions Approve lab safety forms before work begins.

Disposal Keep boxes for use next year.

Teaching Strategies

- Try to use objects in the box that are simple, but challenging.
- When students are through with the lab, you might want to identify the objects, or, to demonstrate that chemists cannot always see what they are looking for, you might want to leave the object's identity a mystery!

Expected Results Results will vary. Students should try to use senses other than sight to determine the relative size, mass, shape, and number of objects.

CHAPTER 5

Electrons in Atoms

BIGIDEA The atoms of each element have a unique arrangement of electrons.

SECTIONS

1 Light and Quantized Energy

2 Quantum Theory and the Atom

3 Electron Configuration

LaunchLAB

How do you know what is inside an atom?

Imagine that it is your birthday, and there is one wrapped present that is different from all the rest. Unlike the other gifts that you can open, you can only guess what is inside this package. In trying to determine the structure of the atom, early chemists had a similar experience. How good are your skills of observation and deduction? In this lab, you will try to determine what is inside a box through observation methods.

FOLDABLES Study Organizer

Electron Configuration

Make a concept-map book. Label it as shown. Use it to help you summarize the three rules that define how electrons are arranged in an atom.

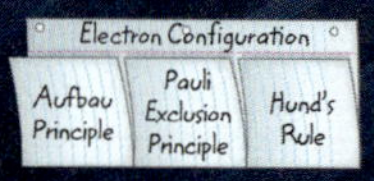

Scientists use stellar absorption spectra to gain information such as a star's temperature and elemental composition. They determined from stellar spectra that stars are made of the same elements as those found on Earth.

Jerry Lodriguss/Science Source

Procedure

1. Read and complete the lab safety form.
2. Obtain a **wrapped box** from your instructor.
3. Using as many observation methods as you can, and without unwrapping or opening the box, try to determine what is inside the box.
4. Record the observations you make throughout this discovery process.

Analysis

1. **Describe** how you were able to determine characteristics such as the size, shape, and composition of the object in the box. Answers will vary, but may include determining whether the object inside has the same size and shape as the box, lifting the box to estimate its weight, shaking the box to sense any motion of the object inside the box,trying to detect odors from the box, and listening for sounds from inside the box.
2. **Indicate** what senses you used to make your observations. Answers will likely include hearing, touch, and smell.
3. **Discuss** why it is hard to determine what type of object is in the box without opening it. Students will determine that observations typically rely heavily upon sight, although touch and hearing are somewhat useful.

Inquiry After reading the chapter, design another investigation that illustrates the difficulties associated with the study of subatomic particles. Answers will vary depending on what students choose to investigate.

Introduce the BIGIDEA

Electron Arrangements and Properties To introduce this chapter Big Idea, have students give the symbols and names of the elements with atomic numbers 17, 18, and 19. 17, Cl, chlorine; 18, Ar, argon; 19, K, potassium Compare and contrast the radically different properties of the three elements and point out that each successive element has one more proton and one more electron than the previous element. Chlorine is a yellow-green gas at room temperature and is very reactive. Argon, a gas used in incandescent bulbs, is very unreactive. Potassium, a metal at room temperature, is so reactive that it must be stored in kerosene or oil to prevent reaction with the oxygen and water in the air. Explain that the different properties of the three elements can be attributed to the arrangements of electrons in their atoms.

Tie to Previous Knowledge

Have students review the following concepts before studying this chapter.

- atomic structure

Use the Photo

Spectra Have students compare and contrast the Betelgeuse and Rigel emission spectra with the visible portion of the electromagnetic spectrum shown in Figure 5. The stars' emission spectra contain only certain colors, while the visible portion of the electromagnetic spectrum contains all the colors.

SECTION 1

1 Focus

MAINIDEA

Electromagnetic Radiation's Dual Nature Ask students to envision the way that water molecules on a lake's surface behave as a water wave travels across the lake. The water molecules move up and down as the wave travels across the lake's surface. Then, ask what happens when the water wave reaches the lake's shore. Some of the wave's energy is transferred to the particles that make up the shore, often shifting or moving them. Explain that as a ray of light (a form of electromagnetic radiation) travels from place to place, electric and magnetic fields move from side to side and up and down. Also, explain that when light transfers some of its energy to matter–for example, when sunlight touches one's dark clothing–the energy is absorbed in certain amounts, called quanta. In other words, light seems to travel as a wave but is emitted and absorbed by matter in only certain, definite amounts. OL

2 Teach

Concept Development

Matter Explain the concept that matter is made up of atoms. For example, water contains two atoms of hydrogen for every atom of oxygen and the two elements are always in the same proportion in the compound. Point out, however, that something well beyond this concept must account for the vastly different chemical behaviors of hydrogen, oxygen, and the other chemical elements.

SECTION 1

Light and Quantized Energy

MAINIDEA Light, a form of electromagnetic radiation, has characteristics of both a wave and a particle.

Essential Questions

- How do the wave and particle natures of light compare?
- What is a quantum of energy and how is it related to an energy change of matter?
- How do continuous electromagnetic spectra and atomic emission spectra compare and contrast?

Review Vocabulary

radiation: the rays and particles—alpha particles, beta particles, and gamma rays—that are emitted by radioactive material

New Vocabulary

electromagnetic radiation
wavelength
frequency
amplitude
electromagnetic spectrum
quantum
Planck's constant
photoelectric effect
photon
atomic emission spectrum

CHEM 4 YOU Have you ever come inside on a cold day, headed for the kitchen, and popped a cold snack into the microwave oven? When the microwaves reached your snack, small packets of energy warmed it in practically no time at all.

The Atom and Unanswered Questions

After discovering three subatomic particles in the early 1900s, scientists continued their quest to understand atomic structure and the arrangement of electrons within atoms.

Rutherford proposed that all of an atom's positive charge and virtually all of its mass are concentrated in a nucleus that is surrounded by fast-moving electrons. The model did not explain how the atom's electrons are arranged in the space around the nucleus. Nor did it address the question of why the negatively charged electrons are not pulled into the atom's positively charged nucleus. Rutherford's nuclear model did not begin to account for the differences and similarities in chemical behavior among the various elements.

For example, consider the elements lithium, sodium, and potassium, which are found in different periods on the periodic table but have similar chemical behaviors. All three elements appear metallic in nature, and their atoms react vigorously with water to liberate hydrogen gas. In fact, as shown in **Figure 1,** both sodium and potassium react so violently that the hydrogen gas can ignite and even explode.

In the early 1900s, scientists began to unravel the puzzle of chemical behavior. They observed that certain elements emitted visible light when heated in a flame. Analysis of the emitted light revealed that an element's chemical behavior is related to the arrangement of the electrons in its atoms. To understand this relationship and the nature of atomic structure, it will be helpful to first understand the nature of light.

Figure 1 Different elements can have similar reactions with water.

DIFFERENTIATED INSTRUCTION

Hearing Impaired Help students visualize the characteristics of various types of electromagnetic waves by writing the following three items (A–C) and questions (1–3) on the board:

A. Visible light

B. Microwaves

C. Radio waves

1. Which moves through air at such a speed that it travels around Earth approximately seven times in one second? A, B, and C. Note: Visible light, microwaves, and radio waves are three forms of electromagnetic radiation. All three travel at a speed of 3.00×10^8 m/s in a vacuum and almost the same speed in air.
2. Which might have a wavelength equal to the length of three football fields? C, radio waves
3. Which might have a wavelength equal to the width of a pencil? B, microwaves OL

The Wave Nature of Light

Visible light is a type of **electromagnetic radiation**–a form of energy that exhibits wavelike behavior as it travels through space. Other examples of electromagnetic radiation include microwaves that cook your food, X rays that doctors and dentists use to examine bones and teeth, and waves that carry radio and television programs into homes.

Characteristics of waves All waves can be described by several characteristics, a few of which might be familiar to you from everyday experience. You might have seen concentric waves when dropping an object into water, as shown in **Figure 2a.**

The **wavelength** (represented by λ, the Greek letter lambda) is the shortest distance between equivalent points on a continuous wave. For example, in **Figure 2b,** the wavelength is measured from crest to crest or from trough to trough. Wavelength is usually expressed in meters, centimeters, or nanometers ($1 \text{ nm} = 1 \times 10^{-9} \text{ m}$).

The **frequency** (represented by ν, the Greek letter nu) is the number of waves that pass a given point per second. One hertz (Hz), the SI unit of frequency, equals one wave per second. In calculations, frequency is expressed with units of waves per second, (1/s) or (s^{-1}); the term *waves* is understood. A particular frequency can be expressed in the following ways: 652 Hz = 652 waves/second = 652/s = 652 s^{-1}.

The **amplitude** of a wave is the wave's height from the origin to a crest, or from the origin to a trough, as illustrated in **Figure 2b.** Wavelength and frequency do not affect the amplitude of a wave.

All electromagnetic waves, including visible light, travel at a speed of 3.00×10^8 m/s in a vacuum. Because the speed of light is such an important and universal value, it is given its own symbol, *c*. The speed of light is the product of its wavelength (λ) and its frequency (ν).

APPLYING PRACTICES

Use Mathematical Representations Go to the resources tab in ConnectED to find the Applying Practices worksheet *Wave Characteristics.*

Electromagnetic Wave Relationship

$$c = \lambda\nu$$

c is the speed of light in a vacuum.
λ is the wavelength.
ν is the frequency.

The speed of light in a vacuum is equal to the product of the wavelength and the frequency.

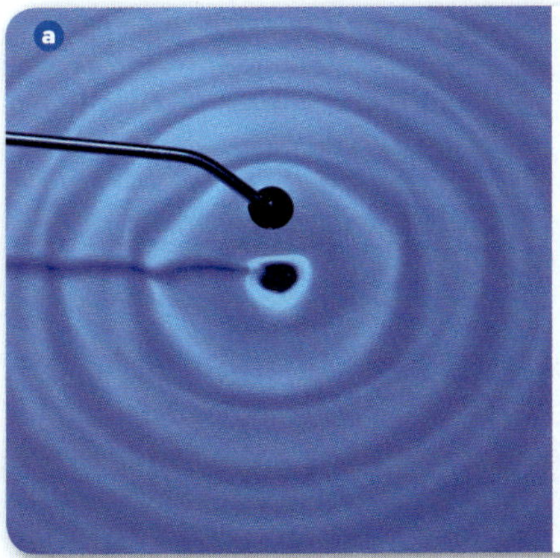

■ **Figure 2** **a.** The concentric waves in the water show the characteristic properties of all waves. **b.** Amplitude, wavelength, and frequency are the main characteristics of waves.

Identify *a crest, a trough, and one wavelength in the photo.*

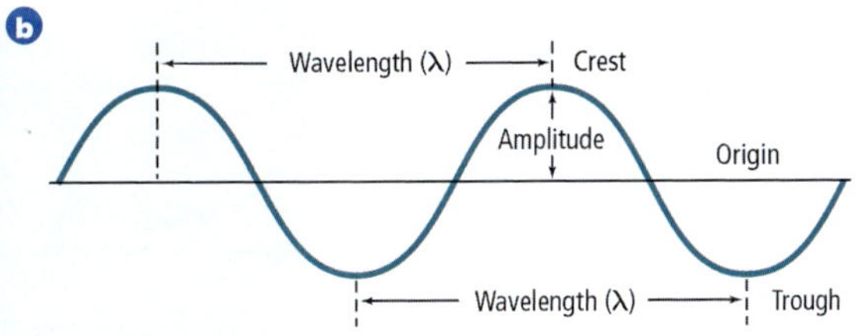

Quick Demo

Wave Characteristics Bring in a coiled spring toy and attach it securely to an object on one side of the room. Demonstrate wave characteristics, frequency, and energy–by generating standing waves. Start with a half wave, showing the longest wavelength, lowest frequency, and least energy. Work up to two or two and one-half standing waves. It will be obvious that more energy is needed as the number of standing waves increases. With each increase in the number of waves, ask students what is happening to frequency and wavelength, and how energy is changing. Frequency is increasing, wavelength is decreasing, and energy is increasing. BL OL EL

■ **Caption Question Fig. 2** Students should point to the correct parts of the photo.

Content Background

Value of C In the formula $c = \lambda\nu$, ν (the frequency of an electromagnetic wave) can be measured precisely with lasers and atomic clocks. The λ (wavelength) of an electromagnetic wave, however, cannot be measured with such extreme precision. Accordingly, the International Commission of Weights and Measures decided in 1983 to make the speed of light a defined quantity. c, the speed of light in a vacuum, is defined as exactly 299,792,458 m/s. However, $c = 3.00 \times 10^8$ is sufficiently accurate for most purposes.

CHEMISTRY PROJECT

Classical Physics and Electrons in Atoms Have students research and explain how electrons in atoms should behave according to classical physics. Have them draw diagrams illustrating their findings. Negatively charged electrons orbiting the nucleus should spiral into the positively charged nucleus, giving off energy in the process. OL EL

■ **Caption Question Fig. 3**
Wavelength and frequency do not affect the amplitude of a wave.

Math in Chemistry

Wavelength and Frequency
Explain that when two quantities are related mathematically in such a way that the increase in one quantity is proportional to the decrease in the other quantity, the two quantities are said to be inversely proportional. Point out that the relationship $c = \lambda\nu$ is valid because the quantities λ and ν are inversely related.

Visual Learning

Figure 3 Have students count the number of wavelengths shown in the two waves, both in the same total length. One wave has four wavelengths; the other has seven wavelengths. Ask how the wavelength of the higher-frequency wave compares with that of the lower-frequency wave. The higher-frequency wavelength is 4/7 that of the lower-frequency wavelength. Ask how the frequency of the higher-frequency wave compares with that of the lower-frequency wave. The higher frequency is 7/4 that of the lower-frequency wave. Use these answers to reinforce the inverse relationship between wavelength and frequency. **OL**

■ **Figure 3** These waves illustrate the relationship between wavelength and frequency. As wavelength increases, frequency decreases.
Infer ***Does frequency or wavelength affect amplitude?***

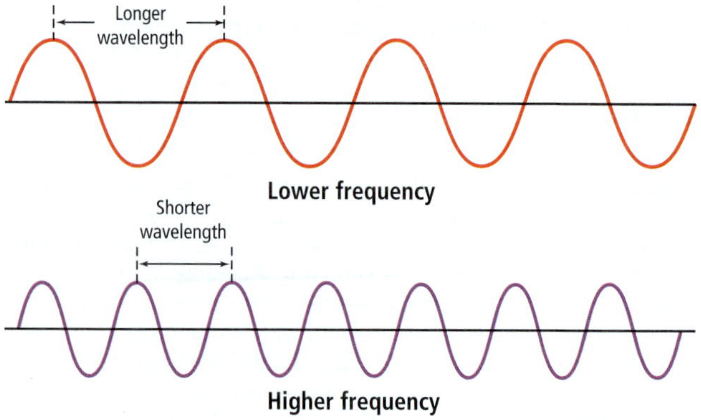

Although the speed of all electromagnetic waves in a vacuum is the same, waves can have different wavelengths and frequencies. As you can see from the equation on the previous page, wavelength and frequency are inversely related; in other words, as one quantity increases, the other decreases. To better understand this relationship, examine the two waves illustrated in **Figure 3.** Although both waves travel at the speed of light, you can see that the red wave has a longer wavelength and lower frequency than the violet wave.

Electromagnetic spectrum Sunlight, which is one example of white light, contains a nearly continuous range of wavelengths and frequencies. White light passing through a prism separates into a continuous spectrum of colors similar to the spectrum in **Figure 4.** These are the colors of the visible spectrum. The spectrum is called continuous because each point of it corresponds to a unique wavelength and frequency. You might be familiar with the colors of the visible spectrum. If you have ever seen a rainbow, you have seen all of the visible colors at once. A rainbow is formed when tiny drops of water in the air disperse the white light from the Sun into its component colors, producing a spectrum that arches across the sky.

■ **Figure 4** When white light passes through a prism, it is separated into a continuous spectrum of its different components—red, orange, yellow, green, blue, indigo, and violet light.

CHEMISTRY JOURNAL

Frequencies and Daily Living In order to reinforce the concept of frequency, have students think of and describe at least five phenomena they encounter that recur or occur at given frequencies in their daily lives. Have them describe these phenomena and, when possible, quantify the frequencies. **BL** **OL**

The visible spectrum of light shown in **Figure 4,** however, comprises only a small portion of the complete electromagnetic spectrum, which is illustrated in **Figure 5.** The **electromagnetic spectrum,** also called the EM spectrum, includes all forms of electromagnetic radiation, with the only differences in the types of radiation being their frequencies and wavelengths. Note in **Figure 4** that the bend varies with the wavelengths as they pass through the prism, resulting in the sequence of the colors red, orange, yellow, green, blue, indigo, and violet. In examining the energy of the radiation shown in **Figure 5,** note that energy increases with increasing frequency. Thus, looking back at **Figure 3,** the violet light, with its greater frequency, has more energy than the red light. This relationship between frequency and energy will be explained in the next section.

Because all electromagnetic waves travel at the same speed in a given medium, you can use the formula $c = \lambda\nu$ to calculate the wavelength or frequency of any wave.

☑ READING CHECK **State** the relationship between the energy and the frequency of electromagnetic radiation.

Connection to Physics Electromagnetic radiation from diverse origins constantly bombards us. In addition to the radiation from the Sun, human activities also produce radiation which include radio and TV signals, phone relay stations, lightbulbs, medical X-ray equipment, and particle accelerators. Natural sources on Earth, such as lightning, natural radioactivity, and even the glow of fireflies, also contribute. Our knowledge of the universe is based on electromagnetic radiation emitted by distant objects and detected with instruments on Earth.

CAREERS IN CHEMISTRY

Spectroscopist Spectroscopy is the study of the spectra absorbed or emitted by matter. Because each element's spectrum is unique, it is like a fingerprint. Astrophysicists use spectroscopy to investigate what a star, such as the Sun, is made of. A star's absorption spectrum shows many dark lines, which allow spectroscopists to identify the elements present in the star.

WebQuest

■ **Figure 5** The electromagnetic spectrum covers a wide range of frequencies. The visible-light section of the spectrum is very narrow. As frequency and energy increase, wavelength decreases.

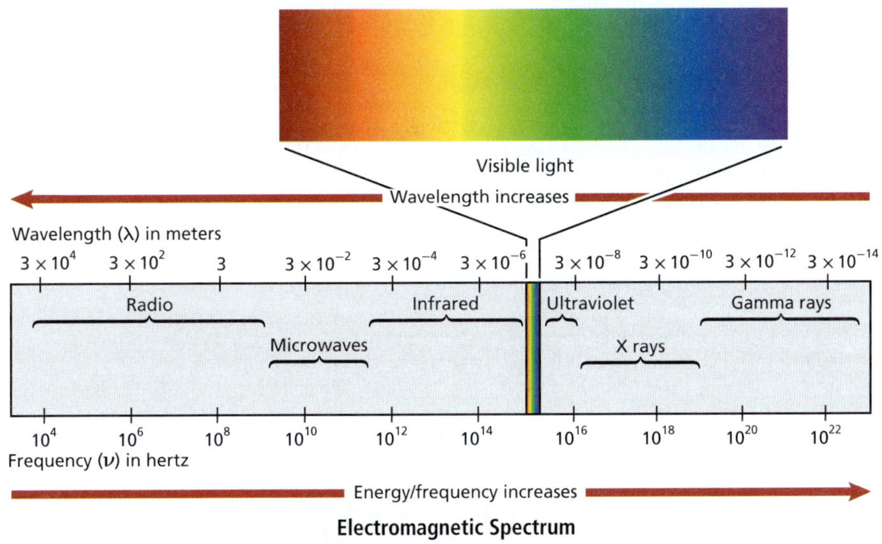

Quick Demo

Reflection and Refraction

Project the beam from a high-intensity projector onto the side of a large beaker of water. Darken the room and adjust the arrangement so students can see the visible portion of the electromagnetic spectrum on a wall or screen. Explain that reflection and refraction separate the component colors of white light from the projector as they pass through the beaker and water. Point out that rainbows are formed in much the same manner when the colors in sunlight separate as they are reflected and refracted by raindrops.

CAREERS IN CHEMISTRY

WebQuest

☑ READING CHECK The energy increases when the frequency increases.

Content Background

Electromagnetic Waves An electromagnetic wave consists of oscillating electric and magnetic fields. The two fields oscillate at right angles to one another. For example, if the electric field oscillates up and down, the magnetic field oscillates from side to side. Both the electric and magnetic fields oscillate at right angles to the direction of propagation of the electromagnetic wave.

CHEMISTRY PROJECT

Electromagnetic Waves and Uses Have students research and discuss the many ways humans use electromagnetic waves to transmit information and carry energy from place to place. OL

IN-CLASS Example

Question The red-colored light in a fireworks display might be produced when strontium salts are heated. What is the frequency of such red light with a wavelength of 6.50×10^{-7} m?

Answer $4.62 \times 10^{14}\ s^{-1}$
$\nu = (3.00 \times 10^{8}\ m/s) \div 6.50 \times 10^{-7}\ m$
$= 4.62 \times 10^{14}\ s^{-1}$

PRACTICE Problems

Have students refer to the Selected Solutions appendix for complete solutions to odd-numbered problems. The complete solutions for all problems can be found in the Solutions Manual.

1. 6.12×10^{14} Hz
2. 2.61×10^{18} Hz
3. 3.00×10^{8} m/s
4. FM: $\lambda = 3.17$ m; AM: $\lambda = 370$ m
   FM: drawing b; AM: drawing a

Reinforcement

Electromagnetic waves When the people in a stadium do the "wave," the wave travels around the stadium as individual persons move their bodies and arms up and down. Point out, however, that each person transmitting the wave remains in the same place. Thus, the energy travels around the stadium but not the people. In the same way, an electromagnetic wave carries energy but not matter.

EXAMPLE Problem 1

Find help with **Algebraic Equations.** Math Handbook

CALCULATING WAVELENGTH OF AN EM WAVE Microwaves are used to cook food and transmit information. What is the wavelength of a microwave that has a frequency of 3.44×10^{9} Hz?

1 ANALYZE THE PROBLEM

You are given the frequency of a microwave. You also know that because microwaves are part of the electromagnetic spectrum, their speeds, frequencies, and wavelengths are related by the formula $c = \lambda\nu$. The value of c is a known constant. First, solve the equation for wavelength, then substitute the known values and solve.

Known	Unknown
$\nu = 3.44 \times 10^{9}$ Hz	$\lambda = ?$ m
$c = 3.00 \times 10^{8}$ m/s	

2 SOLVE FOR THE UNKNOWN

Solve the equation relating the speed, frequency, and wavelength of an electromagnetic wave for wavelength (λ).

$c = \lambda\nu$ — State the electromagnetic wave relationship.

$\lambda = c/\nu$ — Solve for λ.

$\lambda = \dfrac{3.00 \times 10^{8}\ m/s}{3.44 \times 10^{9}\ Hz}$ — Substitute $c = 3.00 \times 10^{8}$ m/s and $\nu = 3.44 \times 10^{9}$ Hz.

Note that hertz is equivalent to 1/s or s^{-1}.

$\lambda = \dfrac{3.00 \times 10^{8}\ m/\cancel{s}}{3.44 \times 10^{9}\ \cancel{s^{-1}}}$ — Divide numbers and units.

$\lambda = 8.72 \times 10^{-2}$ m

3 EVALUATE THE ANSWER

The answer is correctly expressed in a unit of wavelength (m). Both of the known values in the problem are expressed with three significant figures, so the answer should have three significant figures, which it does. The value for the wavelength is within the wavelength range for microwaves shown in **Figure 5.**

PRACTICE Problems

Do additional problems. Online Practice

1. Objects get their colors from reflecting only certain wavelengths when hit with white light. Light reflected from a green leaf is found to have a wavelength of 4.90×10^{-7} m. What is the frequency of the light?
2. X-rays can penetrate body tissues and are widely used to diagnose and treat disorders of internal body structures. What is the frequency of an X-ray with a wavelength of 1.15×10^{-10} m?
3. After careful analysis, an electromagnetic wave is found to have a frequency of 7.8×10^{6} Hz. What is the speed of the wave?
4. **Challenge** While an FM radio station broadcasts at a frequency of 94.7 MHz, an AM station broadcasts at a frequency of 820 kHz. What are the wavelengths of the two broadcasts? Which of the drawings below corresponds to the FM station? To the AM station?

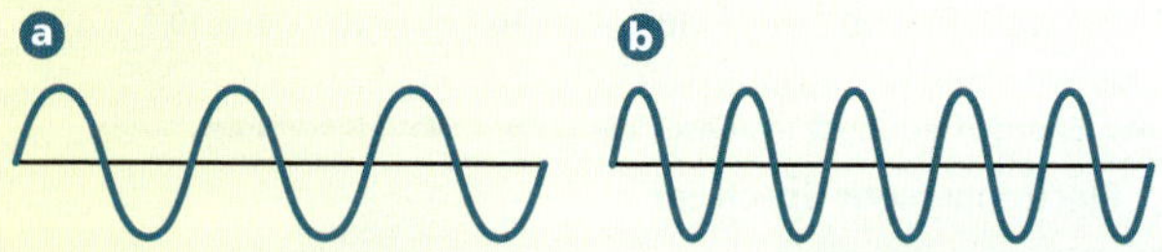

DIFFERENTIATED INSTRUCTION

English Learners Have English learners look up and then explain the meanings of several key English terms used in this section: *radiation, spectrum, constant, effect, emission, quantum*. Then, ask them to use the terms in a paragraph about waves. EL BL

The Particle Nature of Light

While considering light as a wave explains much of its everyday behavior, it fails to adequately describe important aspects of light's interactions with matter. The wave model of light cannot explain why heated objects emit only certain frequencies of light at a given temperature, or why some metals emit electrons when light of a specific frequency shines on them. Scientists realized that a new model or a revision of the wave model of light was needed to address these phenomena.

The quantum concept When objects are heated, they emit glowing light. **Figure 6** illustrates this phenomenon with iron. A piece of iron appears dark gray at room temperature, glows red when heated sufficiently, and turns orange, then bluish in color at even higher temperatures. As you will learn in later chapters, the temperature of an object is a measure of the average kinetic energy of its particles. As the iron gets hotter, it possesses a greater amount of energy and emits different colors of light. These different colors correspond to different frequencies and wavelengths.

The wave model could not explain the emission of these different wavelengths. In 1900, German physicist Max Planck (1858–1947) began searching for an explanation of this phenomenon as he studied the light emitted by heated objects. His study led him to a startling conclusion: matter can gain or lose energy only in small, specific amounts called quanta. A **quantum** is the minimum amount of energy that can be gained or lost by an atom.

VOCABULARY
ACADEMIC VOCABULARY
Phenomenon
an observable fact or event
During rainstorms, electric currents often pass from the sky to Earth—a phenomenon we call lightning.

☑ READING CHECK **Explain** why the color of heated objects changes with temperature.

Planck and other physicists of the time thought the concept of quantized energy was revolutionary, and some found it disturbing. Prior experience had led scientists to think that energy could be absorbed and emitted in continually varying quantities, with no minimum limit to the amount. For example, think about heating a cup of water in a microwave oven. It seems that you can add any amount of thermal energy to the water by regulating the power and duration of the microwaves. Instead, the water's temperature increases in infinitesimal steps as its molecules absorb quanta of energy. Because these steps are so small, the temperature seems to rise in a continuous, rather than a stepwise, manner.

©Mark Karrass/Corbis

■ **Figure 6** The wavelength of the light emitted by heated metal, such as the iron at left, depends on the temperature. At room temperature, iron is gray. When heated, it first turns red, then glowing orange.
Identify *the color of the piece of iron with the greatest kinetic energy.*

ASSESSMENT

Performance Have students develop an investigation or demonstration that illustrates the quantum concept. They might use a balance and small objects having nearly equal masses, such as paper clips. Or, they might use a graduated cylinder and small objects having nearly equal volumes, such as marbles or ball bearings. OL

Enrichment

Optical Pyrometer Have interested students research and make a class presentation or report on the operation of an optical pyrometer—a device that measures extremely high temperatures by the wavelengths of light emitted by the objects. OL

☑ READING CHECK The temperature of an object is a measure of the average kinetic energy of its particles. As the object becomes hotter, it emits light at higher frequencies, and therefore of different colors.

■ **Caption Question Fig. 6** glowing orange

CHEMISTRY JOURNAL

What's a quantum? Have students research the reactions of Planck's contemporaries to his quantum concept. Have them list and explain the reactions of Planck's contemporaries in their chemistry journals. OL

Concept Development

Chemical Behavior Explain to students that they might think of the light emitted by an atom as a "window into the atom." Explain further that the chemical behaviors of the elements are related to the arrangement of electrons within their atoms.

Build a Model

Photoelectric Effect Have student groups build a setup that models the photoelectric effect. For example, the setup might show that impacting small magnets attached to a heavy iron object with lightweight and low-energy objects, such as marshmallows, will not displace the magnets. Then, the setup could show that heavier objects with greater energy displace the magnets. Have students draw the analogy between the marshmallows and low-energy photons and between the heavier objects and high-energy photons. **OL** **EL** **COOP LEARN**

☑ READING CHECK The photoelectric effect is the phenomenon of electron emission from the surface of metals in the presence of light whose frequency is at or above a certain value.

RealWorld CHEMISTRY

The Photoelectric Effect

SOLAR ENERGY is sometimes used to power road signs. Photovoltaic cells use the photoelectric effect to convert the energy of light into electric energy.

Planck proposed that the energy emitted by hot objects was quantized. He then went further and demonstrated mathematically that a relationship exists between the energy of a quantum and the frequency of the emitted radiation.

Energy of a Quantum

$$E_{\text{quantum}} = h\nu$$

$E_{quantum}$ represents energy.
h is Planck's constant.
ν represents frequency.

The energy of a quantum is given by the product of Planck's constant and the frequency.

Planck's constant has a value of 6.626×10^{-34} J•s, where J is the symbol for joule, the SI unit of energy. The equation shows that the energy of radiation increases as the radiation's frequency, ν, increases.

According to Planck's theory, for a given frequency, ν, matter can emit or absorb energy only in whole-number multiples of $h\nu$; that is, $1h\nu$, $2h\nu$, $3h\nu$, and so on. A useful analogy for this concept is that of a child building a wall of wooden blocks. The child can add to or take away height from the wall only in increments of whole numbers of blocks. Similarly, matter can have only certain amounts of energy—quantities of energy between these values do not exist.

The photoelectric effect Scientists also knew that the wave model of light could not explain a phenomenon called the photoelectric effect. In the **photoelectric effect,** electrons, called photoelectrons, are emitted from a metal's surface when light of a certain frequency, or higher than a certain frequency, shines on the surface, as shown in **Figure 7.**

The wave model predicts that given enough time, even low-energy, low-frequency light would accumulate and supply enough energy to eject photoelectrons from a metal. In reality, a metal will not eject photoelectrons below a specific frequency of incident light. For example, no matter how intensely or how long it shines, light with a frequency less than 1.14×10^{15} Hz does not eject photoelectrons from silver. But even dim light with a frequency equal to or greater than 1.14×10^{15} Hz ejects photoelectrons from silver.

☑ READING CHECK **Describe** the photoelectric effect.

Figure 7 The photoelectric effect occurs when light of a certain frequency strikes a metal surface and ejects electrons. When the intensity of the light increases, the number of electrons ejected increases. When the frequency (energy) of the light increases, the energy of the ejected electrons increases.

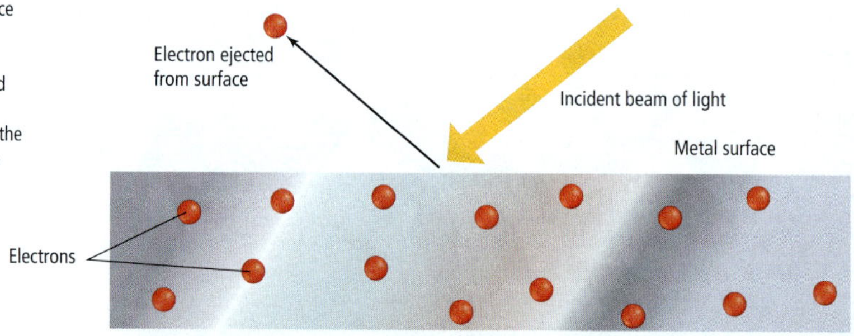

©Andrew Fox/Corbis

CHEMISTRY PROJECT

Sunscreens Because the molecules in sunscreens vibrate and absorb certain frequencies of ultraviolet (UV) light, the sunscreens can help protect people from the harmful effects of sun exposure. Have students research and write a description of the newest types of sunscreens, the molecules involved in each, and the wavelengths of UV light absorbed. **OL**

Light's dual nature To explain the photoelectric effect, Albert Einstein proposed in 1905 that light has a dual nature. A beam of light has wavelike and particlelike properties. It can be thought of as a beam of bundles of energy called photons. A **photon** is a massless particle that carries a quantum of energy. Extending Planck's idea of quantized energy, Einstein calculated that a photon's energy depends on its frequency.

Energy of a Photon

$$E_{photon} = h\nu$$

E_{photon} represents energy.
h is Planck's constant.
ν represents frequency.

The energy of a photon is given by the product of Planck's constant and the frequency.

Einstein also proposed that the energy of a photon must have a certain threshold value to cause the ejection of a photoelectron from the surface of the metal. Thus, even small numbers of photons with energy above the threshold value will cause the photoelectric effect. Einstein won the Nobel Prize in Physics in 1921 for this work.

APPLYING PRACTICES

Evaluate Claims, Evidence, and Reasoning Go to the resources tab in ConnectED to find the Applying Practices worksheet *Is light a wave or a particle?*

EXAMPLE Problem 2

Find help with **scientific notation.** Math Handbook

CALCULATE THE ENERGY OF A PHOTON Every object gets its color by reflecting a certain portion of incident light. The color is determined by the wavelength of the reflected photons, thus by their energy. What is the energy of a photon from the violet portion of the Sun's light if it has a frequency of 7.230×10^{14} s^{-1}?

1 ANALYZE THE PROBLEM

Known	Unknown
$\nu = 7.230 \times 10^{14}$ s^{-1}	E_{photon} = ? J
$h = 6.626 \times 10^{-34}$ J•s	

2 SOLVE FOR THE UNKNOWN

$E_{photon} = h\nu$ — State the equation for the energy of a photon.

$E_{photon} = (6.626 \times 10^{-34}\ \text{J•s})(7.230 \times 10^{14}\ \text{s}^{-1})$ — Substitute $h = 6.626 \times 10^{-34}$ J•s and $\nu = 7.230 \times 10^{14}$ s^{-1}.

$E_{photon} = 4.791 \times 10^{-19}$ J — Multiply and divide numbers and units.

3 EVALUATE THE ANSWER

As expected, the energy of a single photon of light is extremely small. The unit is joules, an energy unit, and there are four significant figures.

EXAMPLE PROBLEM

PRACTICE Problems

Do additional problems. Online Practice

5. Calculate the energy possessed by a single photon of each of the following types of electromagnetic radiation.
   a. 6.32×10^{20} s^{-1} b. 9.50×10^{13} Hz c. 1.05×10^{16} s^{-1}
6. The blue color in some fireworks occurs when copper(I) chloride is heated to approximately 1500 K and emits blue light of wavelength 4.50×10^{2} nm. How much energy does one photon of this light carry?
7. **Challenge** The microwaves used to heat food have a wavelength of 0.125 m. What is the energy of one photon of the microwave radiation?

PRACTICE PROBLEMS

IN-CLASS Example

Question A new sunscreen is reputed to protect against the UV-A waves that can cause skin cancer. How much energy is possessed by a single photon of UV-A electromagnetic radiation with frequency 9.231×10^{14} s^{-1}?

Answer 6.116×10^{-19} J

$E_{photon} = (6.626 \times 10^{-34}\ \text{J•s})(9.231 \times 10^{14}\ \text{s}^{-1})$

$= 6.116 \times 10^{-19}$ J

PRACTICE Problems

Have students refer to the Selected Solutions appendix for complete solutions to odd-numbered problems. The complete solutions for all problems can be found in the Solutions Manual.

5. a. 4.19×10^{-13} J
   b. 6.29×10^{-20} J
   c. 6.96×10^{-18} J
6. 4.42×10^{-19} J
7. 1.59×10^{-24} J

DIFFERENTIATED INSTRUCTION

Advanced Learners Have students research and perhaps explain to the class how astrophysicists determine which elements make up Earth's Sun and other stars. In general, because a star is made of hot, glowing gases, its emitted light can be gathered by a telescope and analyzed. From the atomic emission and absorption spectra of the light, the elements present in the star can be determined. AL

☑ READING CHECK When excited, atoms return to their ground state by emitting light, which corresponds to a certain transition of electrons between orbitals. The lines in the emission spectrum of an element correspond to the transitions.

MiniLAB

Rubric

Purpose Students will observe the colors of light emitted when certain compounds are burned in a flame.

Process Skills classify, compare and contrast, observe and infer

Safety Precautions Approve lab safety forms before work begins. Remind students to use caution with the flame. Although the wet swab will not burn easily, have a beaker of tap water set out for students to extinguish the hot swabs. Review MSDS's for all hazardous chemicals used in this lab.

Disposal Swabs should be thrown in the trash, not into the sink. Check local regulations to determine if the chemicals used in the lab are permitted in the school trash. If not, waste must be sent to a landfill site approved for the disposal of chemical and hazardous wastes.

Teaching Strategies

- Remind students to avoid touching the Bunsen burner with the swabs (to avoid possible contamination).

Expected Results
See data table below:

Compound	Flame color
lithium chloride	red
sodium chloride	yellow
potassium chloride	violet
calcium chloride	red-orange
strontium chloride	bright red
unknown	depends on compound

MiniLAB

Identify Compounds

How do flame colors vary for different elements?

Procedure

1. Read and complete the lab safety form.
2. Dip one of six **cotton swabs** into the **lithium chloride** solution. Put the swab into the flame of a **Bunsen burner.** Observe the color of the flame, and record it in your data table.
3. Repeat Step 2 for each of the metallic chloride solutions **(sodium chloride, potassium chloride, calcium chloride, and strontium chloride).** Record the color of each flame in your data table.
4. Compare your results to the flame tests shown in the Elements Handbook.
5. Repeat Step 2 using a sample of **unknown solution** obtained from your teacher. Record the color of the flame produced.
6. Dispose of the used cotton swabs as directed by your teacher.

Analysis

1. **Suggest** a reason why each compound produced a flame of a different color, even though they each contain chlorine.
2. **Explain** how an element's flame test might be related to its atomic emission spectrum.
3. **Infer** the identity of the unknown crystals. Explain your reasoning.

Atomic Emission Spectra

Have you ever wondered how light is produced in the glowing tubes of neon signs? This process is another phenomenon that cannot be explained by the wave model of light. The light of the neon sign is produced by passing electricity through a tube filled with neon gas. Neon atoms in the tube absorb energy and become excited. These excited atoms return to their stable state by emitting light to release that energy. If the light emitted by the neon is passed through a glass prism, neon's atomic emission spectrum is produced. The **atomic emission spectrum** of an element is the set of frequencies of the electromagnetic waves emitted by atoms of the element. Neon's atomic emission spectrum consists of several individual lines of color corresponding to the frequencies of the radiation emitted by the atoms of neon. It is not a continuous range of colors, as in the visible spectrum of white light.

☑ READING CHECK **Explain** how an emission spectrum is produced.

Each element's atomic emission spectrum is unique and can be used to identify an element or determine whether that element is part of an unknown compound. For example, when a platinum wire is dipped into a strontium nitrate solution and then inserted into a burner flame, the strontium atoms emit a characteristic red color. You can perform a series of flame tests by doing the MiniLab.

Figure 8 shows an illustration of the characteristic purple-pink glow produced by excited hydrogen atoms and the visible portion of hydrogen's emission spectrum responsible for producing the glow. Note how the line nature of hydrogen's atomic emission spectrum differs from that of a continuous spectrum.

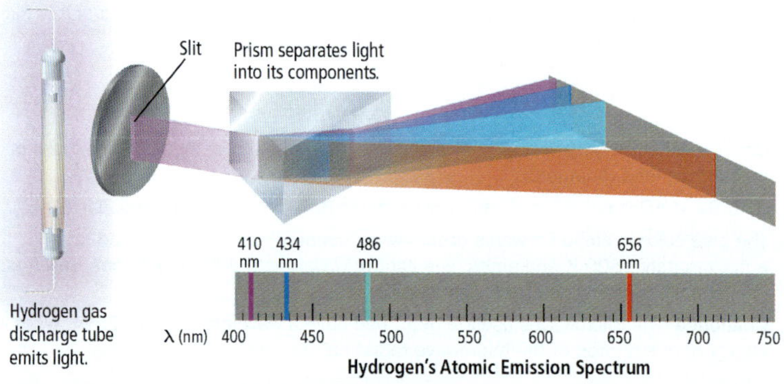

■ **Figure 8** The purple light emitted by hydrogen can be separated into its different components using a prism. Hydrogen has an atomic emission spectrum that comprises four lines of different wavelengths.
Determine *Which line has the highest energy?*

Analysis

1. The colors are due primarily to electron transitions of the metal atoms. The colors are characteristic of lithium, sodium, potassium, calcium, and strontium.
2. The colors are a composite of each element's visible spectrum.
3. Answers will vary depending on the identity of the unknown sample.

■ **Caption Question Fig. 8** The line corresponding to the 410 nm wavelength has the highest energy.

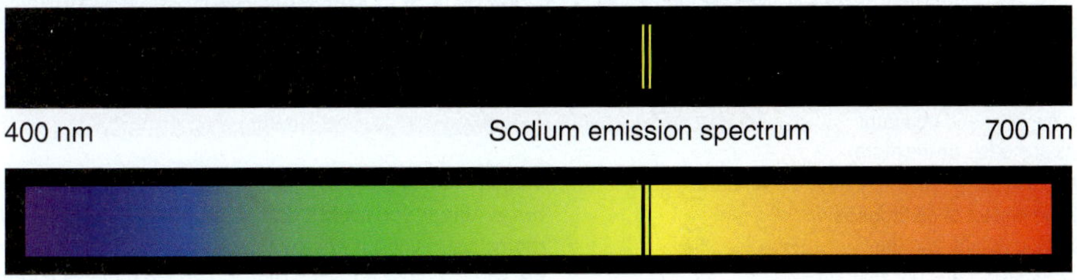

■ **Figure 9** The bottom spectrum is an absorption spectrum. It is composed of black lines on a continuous spectrum. The black lines correspond to certain frequencies absorbed by a given element, sodium in this case. They can be matched to the colored lines present in sodium's emission spectrum, shown above the absorption spectrum.

Connection to Astronomy An atomic emission spectrum is characteristic of the element being examined and can be used to identify that element. The fact that only certain colors appear in an element's atomic emission spectrum means that only specific frequencies of light are emitted. Because those emitted frequencies are related to energy by the formula $E_{photon} = h\nu$, only photons with specific energies are emitted. This was not predicted by the laws of classical physics. Scientists had expected to observe the emission of a continuous series of colors as excited electrons lost energy. Elements absorb the same specific frequencies of light as the frequencies they emit, thus creating an absorption spectrum. In an absorption spectrum, the absorbed frequencies appear as black lines, as shown in **Figure 9.** By comparing the black lines to the emission spectrum of elements, scientists are able to determine the composition of the outer layers of stars.

SECTION 1 REVIEW

Section Self-Check

Section Summary

- All waves are defined by their wavelengths, frequencies, amplitudes, and speeds.
- In a vacuum, all electromagnetic waves travel at the speed of light.
- All electromagnetic waves have both wave and particle properties.
- Matter emits and absorbs energy in quanta.
- White light produces a continuous spectrum. An element's emission spectrum consists of a series of lines of individual colors.

8. MAINIDEA **Compare** the dual nature of light.
9. **Describe** the phenomena that can be explained only by the particle model of light.
10. **Compare and contrast** continuous spectrum and emission spectrum.
11. **Assess** Employ quantum theory to assess the amount of energy that matter gains and loses.
12. **Discuss** the way in which Einstein utilized Planck's quantum concept to explain the photoelectric effect.
13. **Calculate** Heating 235 g of water from 22.6°C to 94.4°C in a microwave oven requires 7.06×10^4 J of energy. If the microwave frequency is $2.88 \times 10^{10}\ s^{-1}$, how many quanta are required to supply the 7.06×10^4 J?
14. **Interpret Scientific Illustrations** Use **Figure 5** and your knowledge of electromagnetic radiation to match the numbered items with the lettered items. The numbered items may be used more than once or not at all.

a. longest wavelength	**1.** gamma ray
b. highest frequency	**2.** infrared wave
c. greatest energy	**3.** radio waves

3 Assess

Check for Understanding

Ask students to explain why chemists found Rutherford's nuclear model of the atom lacking. It did not explain or account for the differences in the chemical behavior of the elements. OL

Reteach

Reinforce the concept that red light has less energy than blue light. Explaining that you are making a solution of a fluorescent substance, prepare a solution of 10 g fluorescein in 100 mL water in a 150 mL beaker. Darken the room and shine a flashlight's beam through a transparent red cellophane sheet into the fluorescein solution. When the flashlight is turned off, the solution will not fluoresce. Then, repeat the process, but use a blue cellophane sheet rather than a red one. The solution will fluoresce when the flashlight is turned off. Ask students to explain the results. The blue light waves have a higher frequency, shorter wavelength, and greater energy than the red light waves. The solution can be flushed down a drain with water. OL

ChemLAB

The ChemLab located at the end of the chapter can be used at this point in the lesson.

ASSESSMENT

Knowledge Ask students to compare the wavelengths, frequencies, and energies of microwaves and X-rays. Microwaves have longer wavelengths, lower frequencies, and lower energies than X-rays. OL

SECTION 1 REVIEW

8. Light exhibits wavelike behavior as it travels through space. Light exhibits particlelike behavior as it interacts with matter.
9. The particle model must be used to explain the photoelectric effect, the color of hot objects, and atomic emission spectra.
10. A continuous spectrum shows the colors of all the wavelengths. An emission spectrum shows only the wavelengths corresponding to a given element.
11. A single quantum is the minimum amount of energy that can be lost or gained by an atom. Therefore, matter loses or gains energy only in multiples of one quantum.
12. Einstein proposed that electromagnetic radiation has a wave-particle nature; that the energy of a quantum, or photon, depends on the frequency of the radiation; that the energy of the photon is given by the formula $E_{photon} = h\nu$; and that only photons whose energy exceeds a threshold value will cause emission of photoelectrons.
13. 3.70×10^{27}
14. a: 3, b: 1, c: 1

SECTION 2

1 Focus

MAINIDEA

Waves and Quantized Energies Draw a circle with a dot in the middle on the board and explain that it is one way of representing the circular orbit of an electron around an atomic nucleus. Then, explain that moving particles, such as electrons, have wavelike properties. Challenge three students to come to the board, to divide the circles into three, four, and five equal segments, respectively, and to draw the same number of standing waves on the circles. The standing waves should resemble those in Figure 13c. Ask which of the wave patterns represents the shortest wavelength, highest frequency, and greatest energy. The one with five complete wavelengths. Point out that when an electron with wavelike motion is restricted to a circular orbit of fixed radius, only certain wavelengths, frequencies, and energies are possible. OL AL EL

2 Teach

Visual Learning

Table 1 Ask students to examine the table's Relative Energy column and determine Bohr's formula relating the hydrogen atom's relative energy to the electron's Bohr atomic orbit (n). $E_n = n^2E_1$ OL

SECTION 2

Quantum Theory and the Atom

MAINIDEA Wavelike properties of electrons help relate atomic emission spectra, energy states of atoms, and atomic orbitals.

Essential Questions

- How do the Bohr and quantum mechanical models of the atom compare?
- What is the impact of de Broglie's wave-particle duality and the Heisenberg uncertainty principle on the current view of electrons in atoms?
- What are the relationships among a hydrogen atom's energy levels, sublevels, and atomic orbitals?

Review Vocabulary

atom: the smallest particle of an element that retains all the properties of that element; is composed of electrons, protons, and neutrons

New Vocabulary

ground state
quantum number
de Broglie equation
Heisenberg uncertainty principle
quantum mechanical model of the atom
atomic orbital
principal quantum number
principal energy level
energy sublevel

CHEM 4 YOU Imagine climbing a ladder and trying to stand between the rungs. Unless you could stand on air, it would not work. When atoms are in various energy states, electrons behave in much the same way as a person climbing up the rungs of a ladder.

Bohr's Model of the Atom

The dual wave-particle model of light accounted for several previously unexplainable phenomena, but scientists still did not understand the relationships among atomic structure, electrons, and atomic emission spectra. Recall that hydrogen's atomic emission spectrum is discontinuous; that is, it is made up of only certain frequencies of light. Why are the atomic emission spectra of elements discontinuous rather than continuous? Niels Bohr, a Danish physicist working in Rutherford's laboratory in 1913, proposed a quantum model for the hydrogen atom that seemed to answer this question. Bohr's model also correctly predicted the frequencies of the lines in hydrogen's atomic emission spectrum.

Energy states of hydrogen Building on Planck's and Einstein's concepts of quantized energy, Bohr proposed that the hydrogen atom has only certain allowable energy states. The lowest allowable energy state of an atom is called its **ground state.** When an atom gains energy, it is said to be in an excited state.

Bohr also related the hydrogen atom's energy states to the electron within the atom. He suggested that the electron in a hydrogen atom moves around the nucleus in only certain allowed circular orbits. The smaller the electron's orbit, the lower the atom's energy state, or energy level. Conversely, the larger the electron's orbit, the higher the atom's energy state, or energy level. Thus, a hydrogen atom can have many different excited states, although it contains only one electron. Bohr's idea is illustrated in **Figure 10.**

Figure 10 The figure shows an atom that has one electron. Note that the illustration is not to scale. In its ground state, the electron is associated with the lowest energy level. When the atom is in an excited state, the electron is associated with a higher energy level.

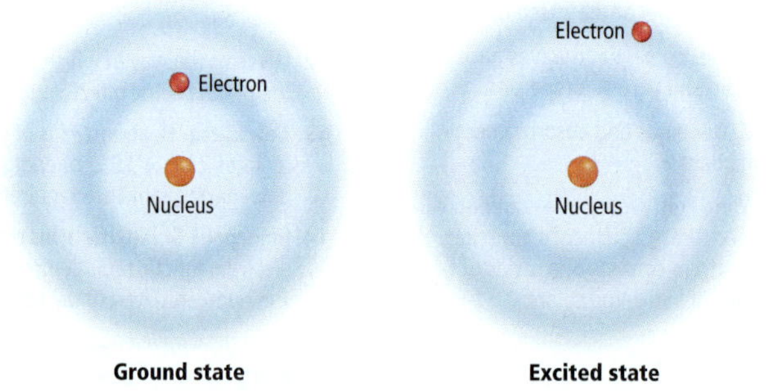

DIFFERENTIATED INSTRUCTION

Struggling Learners Demonstrate the electron transitions associated with energy-level changes. Tell students that a book on the floor represents an electron in an atom's lowest-energy orbit. Raise the book to a higher energy level (their chair). Ask if energy is required. yes Ask what happens when the book returns to the floor. Energy is released. Explain the analogy between the book's energy levels and an electron's transitions between atomic orbits. Point out that the energy needed to raise an electron to a higher-energy orbit is exactly the same as the energy released when the electron returns to its original orbit. BL EL

Table 1 Bohr's Description of the Hydrogen Atom

Bohr's Atomic Orbit	Quantum Number	Orbit Radius (nm)	Corresponding Atomic Energy Level	Relative Energy
First	$n = 1$	0.0529	1	E_1
Second	$n = 2$	0.212	2	$E_2 = 4E_1$
Third	$n = 3$	0.476	3	$E_3 = 9E_1$
Fourth	$n = 4$	0.846	4	$E_4 = 16E_1$
Fifth	$n = 5$	1.32	5	$E_5 = 25E_1$
Sixth	$n = 6$	1.90	6	$E_6 = 36E_1$
Seventh	$n = 7$	2.59	7	$E_7 = 49E_1$

In order to complete his calculations, Bohr assigned a number, n, called a **quantum number,** to each orbit. He also calculated the radius of each orbit. For the first orbit, the one closest to the nucleus, $n = 1$ and the orbit radius is 0.0529 nm; for the second orbit, $n = 2$ and the orbit radius is 0.212 nm; and so on. Additional information about Bohr's description of hydrogen's allowed orbits and energy levels is given in **Table 1.**

The hydrogen line spectrum Bohr suggested that the hydrogen atom is in the ground state, also called the first energy level, when its single electron is in the $n = 1$ orbit. In the ground state, the atom does not radiate energy. When energy is added from an outside source, the electron moves to a higher-energy orbit, such as the $n = 2$ orbit shown in **Figure 11.** Such an electron transition raises the atom to an excited state. When the atom is in an excited state, the electron can drop from the higher-energy orbit to a lower-energy orbit. As a result of this transition, the atom emits a photon corresponding to the energy difference between the two levels.

$$\Delta E = E_{\text{higher-energy orbit}} - E_{\text{lower-energy orbit}} = E_{\text{photon}} = h\nu$$

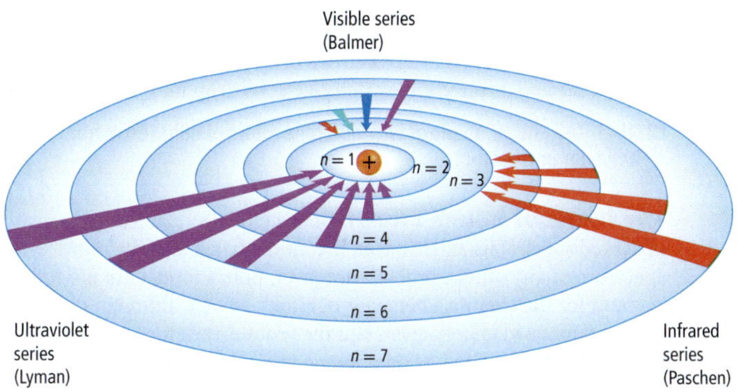

Figure 11 When an electron drops from a higher-energy orbit to a lower-energy orbit, a photon is emitted. The ultraviolet (Lyman), visible (Balmer), and infrared (Paschen) series correspond to electrons dropping to $n = 1$, $n = 2$, and $n = 3$, respectively.

View an **animation about the Balmer Series.**

Concepts In Motion

ASSESSMENT

Performance Have students make a large copy of the hydrogen atom's Bohr orbits (as shown in Figure 11) on a piece of construction paper. Have them put a coin in the lowest orbit to represent the orbital occupancy related to hydrogen's lowest energy state. Then, have them move the coin between the appropriate orbits to simulate the following orbit transitions and spectral lines in hydrogen's atomic emission spectrum: violet (6→2), blue-violet (5→2), blue-green (4→2), and red (3→2). OL EL

DIFFERENTIATED INSTRUCTION

Advanced Learners Ask advanced students to work through Bohr's use of Newton's second law ($F = ma$), Coulomb's constant (K), and Bohr's own concept of quantized angular momenta to derive the relationship $r_n = (h^2n^2)/(4\pi^2 Kmq^2)$. Then, have them use the equation to calculate the radii of the hydrogen atom's first four Bohr orbits. AL

☑ **READING CHECK** When an electron returns to its ground state from an excited state, the atom emits a photon whose frequency corresponds to the energy difference between the two energy levels. Each frequency corresponds to a certain color.

Concept Development

The Real-World-Reading-Link analogy between electron energy states and rungs on a ladder is useful in helping students understand the Bohr and Quantum Mechanical Models of the atom. Remind students, however, that the analogy has limitations. Ask students what is correct and what is incorrect about the ladder analogy. Correct: The atom has only certain allowable energy states or energy levels. Incorrect: Atomic energy states are not evenly spaced, as are the rungs on a ladder. **BL** **OL**

Quick Demo

WARNING: ***Wear safety goggles and conduct this demo behind an explosion shield. Do not allow the forks to touch. Do not allow anyone to come near the demo or touch any part of the setup.*** Insert two forks into the ends of a large dill pickle. With the cord unplugged, attach leads from a 110-V current to the two forks. Plug in the cord. The electric current will produce a yellow glow in the pickle. Ask students what may account for the yellow glow. The pickle has been brined with salt, and the electric current excites sodium ions in the pickle. The excited sodium ions produce the yellow emission spectrum of sodium when they drop back to lower energy states. Allow students to use hand-held diffraction gratings to examine the emission spectrum.

■ **Figure 12** Only certain energy levels are allowed. The energy levels are similar to the rungs of a ladder. The four visible lines correspond to electrons dropping from a higher n to the orbit $n = 2$. As n increases, the hydrogen atom's energy levels are closer to each other.

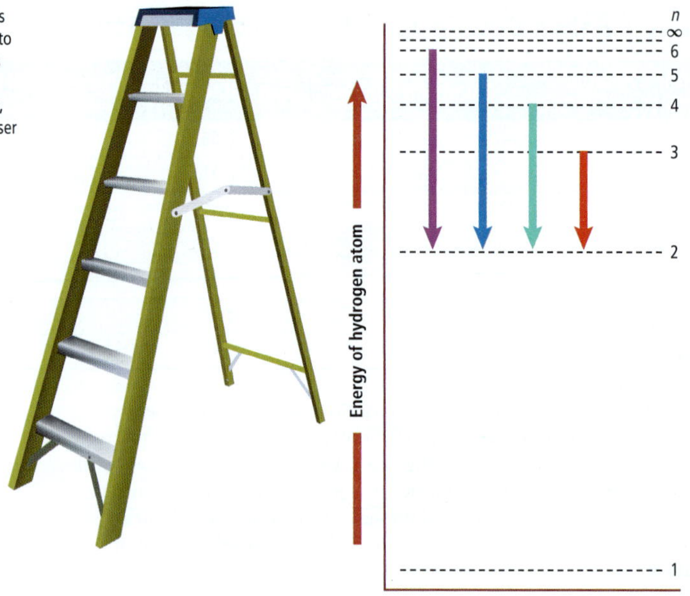

View an **animation about electron transitions.**

Concepts In Motion

Because only certain atomic energies are possible, only certain frequencies of electromagnetic radiation can be emitted. You might compare hydrogen's atomic energy states to rungs on a ladder. A person can climb up or down the ladder only from rung to rung. Similarly, the hydrogen atom's electron can move only from one allowable orbit to another, and therefore, can emit or absorb only certain amounts of energy, corresponding to the energy difference between the two orbits.

Figure 12 shows that, unlike rungs on a ladder, however, the hydrogen atom's energy levels are not evenly spaced. **Figure 12** also illustrates the four electron transitions that account for visible lines in hydrogen's atomic emission spectrum, shown in **Figure 8.** Electron transitions from higher-energy orbits to the second orbit account for all of hydrogen's visible lines, which form the Balmer series. Other electron transitions have been measured that are not visible, such as the Lyman series (ultraviolet), in which electrons drop into the $n = 1$ orbit, and the Paschen series (infrared), in which electrons drop into the $n = 3$ orbit.

☑ READING CHECK **Explain** why different colors of light result from electron behavior in the atom.

The limits of Bohr's model Bohr's model explained hydrogen's observed spectral lines. However, the model failed to explain the spectrum of any other element. Moreover, Bohr's model did not fully account for the chemical behavior of atoms. In fact, although Bohr's idea of quantized energy levels laid the groundwork for atomic models to come, later experiments demonstrated that the Bohr model was fundamentally incorrect. The movements of electrons in atoms are not completely understood even now; however, substantial evidence indicates that electrons do not move around the nucleus in circular orbits.

DIFFERENTIATED INSTRUCTION

English Learners Have English learners look up and then explain the meanings of several key English terms used in this section: *state* (as in ground state), *uncertainty, principal, level* (noun). Ask students to use each term in a sentence or a paragraph. **EL** **BL**

The Quantum Mechanical Model of the Atom

Scientists in the mid-1920s, by then convinced that the Bohr atomic model was incorrect, formulated new and innovative explanations of how electrons are arranged in atoms. In 1924, a French graduate student in physics named Louis de Broglie (1892–1987) proposed an idea that eventually accounted for the fixed energy levels of Bohr's model.

Electrons as waves De Broglie had been thinking that Bohr's quantized electron orbits had characteristics similar to those of waves. For example, as **Figures 13a** and **13b** show, only multiples of half-wavelengths are possible on a plucked harp string because the string is fixed at both ends. Similarly, de Broglie saw that only odd numbers of wavelengths are allowed in a circular orbit of fixed radius, as shown in **Figure 13c.** He also reflected on the fact that light—at one time thought to be strictly a wave phenomenon—has both wave and particle characteristics. These thoughts led de Broglie to pose a new question: If waves can have particlelike behavior, could the opposite also be true? That is, can particles of matter, including electrons, behave like waves?

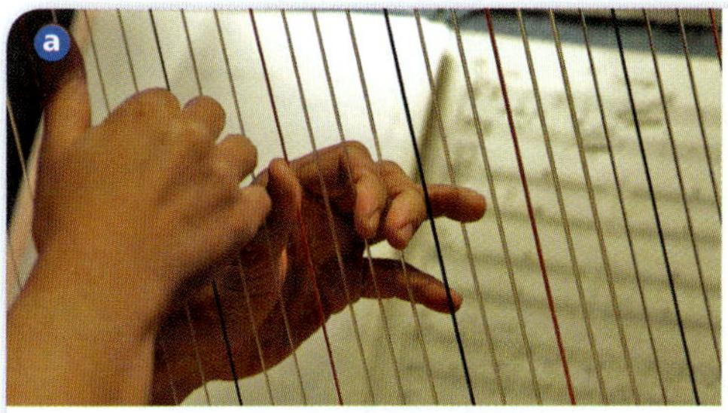

■ **Figure 13 a.** The string on the harp vibrates between two fixed endpoints. **b.** The vibrations of a string between the two fixed endpoints labeled *A* and *B* are limited to multiples of half-wavelengths. **c.** Electrons on circular orbits can only have odd numbers of wavelengths.

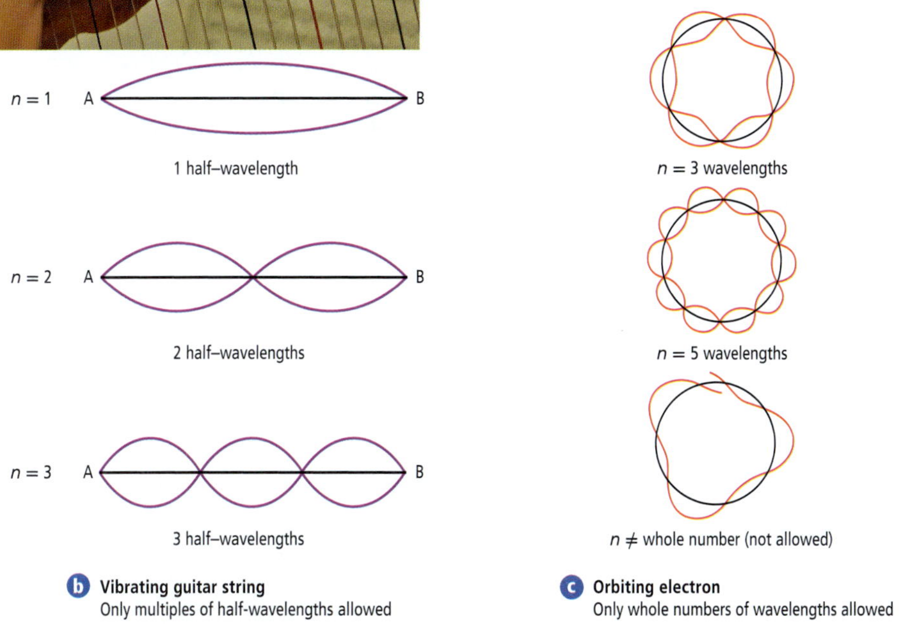

Quick Demo

Electrons Have a fan rotating at high speed when students enter the classroom so that they will not have seen the fan's blades in a stopped position. As soon as the class period begins, ask them to describe the fan's blades. They will be able to describe the blades' approximate length and little else. Explain that scientists experience somewhat the same situation in trying to describe electrons in atoms. The electrons move about the nucleus and appear to fill the entire volume, yet occupy very little volume themselves. Explain that due to the motion of the electrons and certain limitations in our ability to view them (as described by Heisenberg's uncertainty principle), we are unable to simultaneously describe exactly where the electrons are and where they are going.

Content Background

The Bohr Model and the Quantum Mechanical Model
In the Bohr model of the hydrogen atom, each possible electron orbit has a definite radius, as described in Table 1. The quantum mechanical (QM) model, however, predicts only the probability of finding the electron at a specific location in the atom. Interestingly, the QM model's most probable distance of the electron from the nucleus in the hydrogen atom matches the radius of the Bohr orbit.

CHEMISTRY JOURNAL

Gases for IR and UV Have students research the types of gases used to emit infrared and ultraviolet electromagnetic radiation. Have them summarize their findings in their chemistry journals. OL

Problem-Solving LAB

Purpose Students will learn how to relate emission-spectrum wavelengths to electron-orbit transitions in the Bohr atomic model, calculate the wavelengths using the equation $\frac{1}{\lambda} = (1.09678 \times 10^7\ \mathrm{m}^{-1})\left(\frac{1}{n_f^2} - \frac{1}{n_i^2}\right)$, and determine the energy per quantum for a given wavelength with the equation $E = \frac{hc}{\lambda}$.

Process Skills interpret scientific illustrations, use numbers, think critically

Teaching Strategies

- Balmer found his famous series for hydrogen in 1886. Point out to students that experimental constraints at that time limited his ability to determine wavelengths to those in the visible and near ultraviolet, from approximately 250 nm to 700 nm. Thus, all Balmer's lines lie in that region.
- You may wish to run through a set of example calculations with students to ensure that they can perform the calculations properly.
- Emphasize to students that the absorption and emission spectra of hydrogen are particularly important in astronomy because most of the universe is made of hydrogen.

Think Critically

1. a. $\lambda = 6.56465 \times 10^{-7}$ m
   b. $\lambda = 4.86273 \times 10^{-7}$ m
   c. $\lambda = 4.34171 \times 10^{-7}$ m
   d. $\lambda = 4.10292 \times 10^{-7}$ m
2. $n_i = 3, n_f = 2$: calculated wavelength of 6.56465×10^{-7} m closely matches experimental wavelength of 6562 Å; $n_i = 4, n_f = 2$: calculated wavelength of 4.86273×10^{-7} m closely matches experimental wavelength of 4861 Å; $n_i = 5, n_f = 2$: calculated wavelength of 4.34171×10^{-7} m closely matches experimental wavelength of 4340 Å; $n_i = 6, n_f = 2$: calculated wavelength of 4.10292×10^{-7} m closely matches experimental wavelength of 4101 Å

The **de Broglie equation** predicts that all moving particles have wave characteristics. It also explains why it is impossible to notice the wavelength of a fast-moving car. An automobile moving at 25 m/s and having a mass of 910 kg has a wavelength of 2.9×10^{-38} m, a wavelength far too small to be seen or detected. By comparison, an electron moving at the same speed has the easily measured wavelength of 2.9×10^{-5} m. Subsequent experiments have proven that electrons and other moving particles do indeed have wave characteristics. De Broglie knew that if an electron has wavelike motion and is restricted to circular orbits of fixed radius, only certain wavelengths, frequencies, and energies are possible. Developing his idea, de Broglie derived the following equation.

Particle Electromagnetic–Wave Relationship

$$\lambda = \frac{h}{m\nu}$$

λ represents wavelength.
***h* is Planck's constant.**
***m* represents mass of the particle.**
ν represents velocity.

The wavelength of a particle is the ratio of Planck's constant and the product of the particle's mass and its velocity.

Problem-Solving LAB

Interpret Scientific Illustrations

What electron transitions account for the Balmer series? Hydrogen's emission spectrum comprises three series of lines. Some wavelengths are ultraviolet (Lyman series) and infrared (Paschen series). Visible wavelengths comprise the Balmer series. The Bohr atomic model attributes these spectral lines to transitions from higher-energy states with electron orbits in which $n = n_i$ to lower-energy states with smaller electron orbits in which $n = n_f$.

Analysis

The image at right illustrates some of the transitions in hydrogen's Balmer series. These Balmer lines are designated H_α (6562 Å), H_β (4861 Å), H_γ (4340 Å), and H_δ (4101 Å). Each wavelength (λ) is related to an electron transition within a hydrogen atom by the following equation, in which $1.09678 \times 10^7\ \mathrm{m}^{-1}$ is known as the Rydberg constant.

$$\frac{1}{\lambda} = 1.09678 \times 10^7\left(\frac{1}{n_f^2} - \frac{1}{n_i^2}\right)\mathrm{m}^{-1}$$

For hydrogen's Balmer series, electron orbit transitions occur from larger orbits to the $n = 2$ orbit; that is, $n_f = 2$.

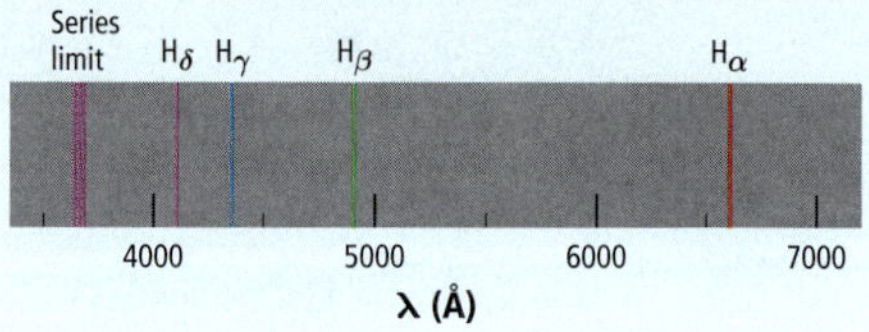

Think Critically

1. **Calculate** the wavelengths for the following electron orbit transitions.
   a. $n_i = 3$; $n_f = 2$
   b. $n_i = 4$; $n_f = 2$
   c. $n_i = 5$; $n_f = 2$
   d. $n_i = 6$; $n_f = 2$
2. **Relate** the Balmer-series wavelengths you calculated in Question 1 to those determined experimentally. Allowing for experimental error and calculation uncertainty, do the wavelengths match? Explain your answer. One angstrom (Å) equals 10^{-10} m.
3. **Apply** the formula $E = hc/\lambda$ to determine the energy per quantum for each of the orbit transitions in Question 1.
4. **Extend** the Bohr model by calculating the wavelength and energy per quantum for the electron orbit transition for which $n_f = 3$ and $n_i = 5$. This transition accounts for a spectral line in hydrogen's Paschen series.

3. a. 3.027×10^{-19} J
   b. 4.087×10^{-19} J
   c. 4.577×10^{-19} J
   d. 4.844×10^{-19} J
4. 1.549×10^{-19} J

ASSESSMENT

Skill Have the students extend the ideas presented here to make a prediction concerning the spectrum that would be emitted from hydrogen-like atoms, such as He^+ or Li^{2+}. Or, have them predict what would happen to the continuous spectrum of light if it passed through a cell containing hydrogen gas. OL

The Heisenberg uncertainty principle Step by step, scientists such as Rutherford, Bohr, and de Broglie had been unraveling the mysteries of the atom. However, a conclusion reached by the German theoretical physicist Werner Heisenberg (1901–1976) proved to have profound implications for atomic models.

Heisenberg showed that it is impossible to take any measurement of an object without disturbing the object. Imagine trying to locate a hovering, helium-filled balloon in a darkened room. If you wave your hand about, you can locate the balloon's position when you touch it. However, when you touch the balloon, you transfer energy to it and change its position. You could also detect the balloon's position by turning on a flashlight. Using this method, photons of light reflected from the balloon would reach your eyes and reveal the balloon's location. Because the balloon is a macroscopic object, the effect of the rebounding photons on its position is very small and not observable.

Imagine trying to determine an electron's location by "bumping" it with a high-energy photon. Because such a photon has about the same energy as an electron, the interaction between the two particles changes both the wavelength of the photon and the position and velocity of the electron, as shown in **Figure 14.** In other words, the act of observing the electron produces a significant, unavoidable uncertainty in the position and motion of the electron. Heisenberg's analysis of interactions, such as those between photons and electrons, led him to his historic conclusion. The **Heisenberg uncertainty principle** states that it is fundamentally impossible to know precisely both the velocity and position of a particle at the same time.

☑ READING CHECK **Explain** the Heisenberg uncertainty principle.

Although scientists of the time found Heisenberg's principle difficult to accept, it has been proven to describe the fundamental limitations of what can be observed. The interaction of a photon with a macroscopic object such as a helium-filled balloon has so little effect on the balloon that the uncertainty in its position is too small to measure. But that is not the case with an electron moving at 6×10^6 m/s near an atomic nucleus. The uncertainty of the electron's position is at least 10^{-9} m, about 10 times greater than the diameter of the entire atom.

The Heisenberg uncertainty principle also means that it is impossible to assign fixed paths for electrons like the circular orbits in Bohr's model. The only quantity that can be known is the probability for an electron to occupy a certain region around the nucleus.

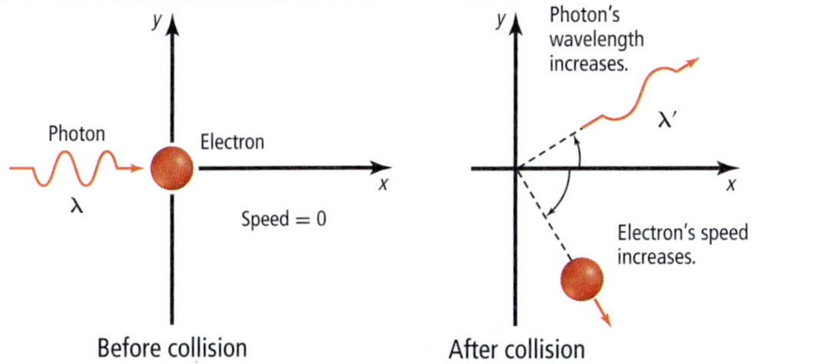

■ **Figure 14** When a photon interacts with an electron at rest, both the velocity and the position of the electron are modified. This illustrates the Heisenberg uncertainty principle. It is impossible to know at the same time the position and the velocity of a particle.
Explain ***Why has the photon's energy changed?***

Enrichment

Uncertainty Principle Make a sign labeled *Heisenberg Might Have Slept Here.* Show it to students, and ask how the uncertainty about whether or not Heisenberg slept in a given location is analogous to an electron's position in an atom. Heisenberg's principle states that it is fundamentally impossible to know both a particle's motion (momentum) and position at the same time. OL

☑ READING CHECK The Heisenberg uncertainty principle states that it is not possible to know precisely the velocity and the position of a particle at the same time.

■ **Caption Question Fig. 14** It transferred some energy to the electron.

Quick Demo

Photons and Electrons
Give a heavy ball to a blindfolded student in the middle of a ring of students (about a 1.5-m radius). Quietly, set a 50-mL, plastic graduated cylinder in the place of a student in the ring. Instruct the student to gently roll the ball in various directions until the cylinder is located. When the ball finally hits the cylinder, it knocks the cylinder from its original position. Then, ask the students if the information gained from rolling the ball gives the cylinder's position after impact. The cylinder is no longer where it was before being hit with the ball. Then describe the analogy to the photon and electron. BL EL

DIFFERENTIATED INSTRUCTION

Advanced Learners Have students investigate and report on whether quantum mechanics invalidates the laws and models of classical physics. In general, classical physics laws and models are valid approximations of the laws of quantum mechanics. As such, they accurately describe and predict behavior at the macroscopic level. However, quantum mechanics is needed to accurately describe and explain atomic and subatomic behavior. AL

The Schrödinger wave equation In 1926, Austrian physicist Erwin Schrödinger (1887–1961) furthered the wave-particle theory proposed by de Broglie. Schrödinger derived an equation that treated the hydrogen atom's electron as a wave. Schrödinger's new model for the hydrogen atom seemed to apply equally well to atoms of other elements—an area in which Bohr's model failed. The atomic model in which electrons are treated as waves is called the wave mechanical model of the atom or, the **quantum mechanical model of the atom.** Like Bohr's model, the quantum mechanical model limits an electron's energy to certain values. However, unlike Bohr's model, the quantum mechanical model makes no attempt to describe the electron's path around the nucleus.

☑ READING CHECK **Compare and contrast** Bohr's model and the quantum mechanical model.

The Schrödinger wave equation is too complex to be considered here. However, each solution to the equation is known as a wave function, which is related to the probability of finding the electron within a particular volume of space around the nucleus. Recall from your study of mathematics that an event with a high probability is more likely to occur than one with a low probability.

Electron's probable location The wave function predicts a three-dimensional region around the nucleus, called an **atomic orbital,** which describes the electron's probable location. An atomic orbital is like a fuzzy cloud in which the density at a given point is proportional to the probability of finding the electron at that point. **Figure 15a** illustrates the probability map that describes the electron in the atom's lowest energy state. The probability map can be thought of as a time-exposure photograph of the electron moving around the nucleus, in which each dot represents the electron's location at an instant in time. The high density of dots near the nucleus indicates the electron's most probable location. However, because the cloud has no definite boundary, it is also possible that the electron might be found at a considerable distance from the nucleus.

☑ READING CHECK **Describe** where electrons are located in an atom.

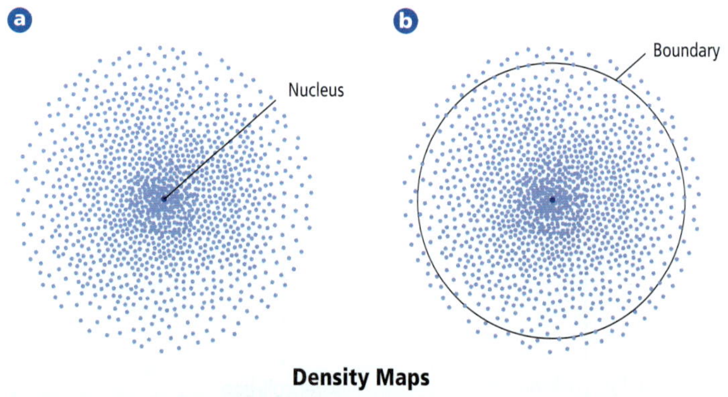

Figure 15 The density map represents the probability of finding an electron at a given position around the nucleus. **a.** The higher density of points near the nucleus shows that the electron is more likely to be found close to the nucleus. **b.** At any given time, there is a 90% probability of finding the electron within the circular region shown. This surface is sometimes chosen to represent the boundary of the atom. In this illustration, the circle corresponds to a projection of the 3-dimensional sphere that contains the electrons.

Apply Chemistry

Lasers A photon striking an atom in an excited state stimulates it to make a transition to a lower-energy state and emit a second photon coherent with the first. Coherent means that the photons have the same associated wavelengths and are in phase (crest-to-crest and trough-to-trough). In a laser, photons from many atoms are reflected back and forth until they build to an intense, small beam—typically about 0.5 mm in diameter.

Medical lasers can be engineered to produce pulses of varying wavelength, intensity, and duration. For example, ophthalmologists can reshape corneas by removing tissue with 10-ns pulses from a 193-nm wavelength argon laser.

Because laser beams can be focused to small diameters, they can be used for internal surgeries, destroying target tissue without adversely affecting surrounding tissue. In addition, by channeling laser beams through optical fibers, doctors can perform surgeries in previously unreachable parts of the body. For example, bundles of optical fibers threaded through arteries can carry laser beams that destroy blockages.

☑ READING CHECK Both models limit an electron's energy to certain values. Unlike the Bohr model, the quantum mechanical model makes no attempt to describe the electron's path around the nucleus.

☑ READING CHECK Electrons are located around the nucleus at a position that can be described only by a probability map. A boundary surface is chosen to contain the region that the electron can be expected to occupy 90% of the time.

DIFFERENTIATED INSTRUCTION

English Learners Have English Learners investigate two acronyms related to the quantum mechanical model and emission spectra, MASER and LASER. MASER is an acronym for **M**icrowave **A**mplification by **S**timulated **E**mission of **R**adiation (Masers differ from lasers in that maser emissions involve vibrational energy states of molecules and laser emissions involve energy-level transitions of atoms.). LASER is an acronym for **L**ight **A**mplification by **S**timulated **E**mission of **R**adiation. Demonstrate a low-powered laser that is safe to use in the classroom by aiming the beam in a safe direction and clapping chalkboard erasers over the beam. The beam becomes visible as the light reflects from chalk particles in the air. EL

Hydrogen's Atomic Orbitals

Because the boundary of an atomic orbital is fuzzy, the orbital does not have an exact defined size. To overcome the inherent uncertainty about the electron's location, chemists arbitrarily draw an orbital's surface to contain 90% of the electron's total probability distribution. This means that the probability of finding the electron within the boundary is 0.9 and the probability of finding it outside the boundary is 0.1. In other words, it is more likely to find the electron close to the nucleus and within the volume defined by the boundary, than to find it outside the volume. The circle shown in **Figure 15b** encloses 90% of the lowest-energy orbital of hydrogen.

Principal quantum number Recall that the Bohr atomic model assigns quantum numbers to electron orbits. Similarly, the quantum mechanical model assigns four quantum numbers to atomic orbitals. The first one is the **principal quantum number** (n) and indicates the relative size and energy of atomic orbitals. As n increases, the orbital becomes larger, the electron spends more time farther from the nucleus, and the atom's energy increases. Therefore, n specifies the atom's major energy levels. Each major energy level is called a **principal energy level.** An atom's lowest principal energy level is assigned a principal quantum number of 1. When the hydrogen atom's single electron occupies an orbital with $n = 1$, the atom is in its ground state. Up to 7 energy levels have been detected for the hydrogen atom, giving n values ranging from 1 to 7.

Watch a **video about fireworks.**

Video

Energy sublevels Principal energy levels contain **energy sublevels.** Principal energy level 1 consists of a single sublevel, principal energy level 2 consists of two sublevels, principal energy level 3 consists of three sublevels, and so on. To better understand the relationship between the atom's energy levels and sublevels, picture the seats in a wedge-shaped section of a theater, as shown in **Figure 16.** As you move away from the stage, the rows become higher and contain more seats. Similarly, the number of energy sublevels in a principal energy level increases as n increases.

☑ READING CHECK **Explain** the relationship between energy levels and sublevels.

■ **Figure 16** Energy levels can be thought of as rows of seats in a theater. The rows that are higher up and farther from the stage contain more seats. Similarly, energy levels related to orbitals farther from the nucleus contain more sublevels.

Identify Misconceptions

Students might think that the hydrogen atom's energy levels are evenly spaced.

Uncover the Misconception Have students compare hydrogen's energy levels shown in Figure 12 with the rungs on a ladder. Unlike the rungs on a ladder, hydrogen's energy levels are not evenly spaced.

Demonstrate the Concept Have students calculate and compare the ratios E_n/E_{n-1} from E_2 through E_7 using information from the *Relative Energy* of Table 1.
$E_2/E_1 = 4$, $E_3/E_2 = 2.25$, $E_4/E_3 = 1.78$, $E_5/E_4 = 1.56$, $E_6/E_5 = 1.44$, $E_7/E_6 = 1.36$

Assess New Knowledge Have students use their calculated energy ratios from Demonstrate the Concept to make their own energy diagrams for hydrogen's energy levels similar to the one in Figure 12. Their energy diagrams should show clearly that hydrogen's energy levels become more closely spaced as n increases. OL

☑ READING CHECK The number of energy sublevels in a principal energy level increases as n increases.

CHEMISTRY PROJECT

Models of the Atom Have students research and explain the experimental evidence accompanying the evolution of models of the atom. Ask them to include Thomson's plum-pudding model, Rutherford's nuclear model, the Bohr model, and the quantum mechanical model. OL

What's CHEMISTRY Got To Do With It?

Video

Lighting Up the Night Sky

READING CHECK *s orbitals are spherical. p orbitals are dumbbell-shaped.*

Enrichment

Sublevels Students might think that the letters s, p, d, and f, which represent sublevels, are arbitrary and perhaps mysterious. Explain that the letters originated from descriptions of spectral lines as sharp, principal, diffuse, and fundamental.

Concept Development

Hydrogen's Orbitals Show students the radial probability distributions of hydrogen's 1s, 2s, and 3s orbitals. Explain that such distributions show where the electron is most likely to be found.

Point out that the 1s orbital's radial probability distribution (rpd) has a single maximum near the nucleus. This maximum represents a region of high electron density. The larger 2s orbital's rpd has two regions of electron density, the more distant region having the higher density.

The two regions are separated by a spherical node where the probability of finding the electron is zero. The 3s orbital's rpd has three regions of electron density and two nodes. As with the 2s orbital, the region of highest electron density is the one farthest from the nucleus.

ASSESSMENT

Knowledge Ask students which hydrogen energy-level transition accounts for the violet line in its emission spectrum. *$n = 6 \rightarrow n = 2$* **OL**

Shapes of orbitals Sublevels are labeled *s, p, d,* or *f* according to the shapes of the atom's orbitals. All s orbitals are spherical, and all p orbitals are dumbbell-shaped; however, not all d or f orbitals have the same shape. Each orbital can contain, at most, two electrons. The single sublevel in principal energy level 1 corresponds to a spherical orbital called the 1s orbital. The two sublevels in principal energy level 2 are designated 2s and 2p. The 2s sublevel corresponds to the 2s orbital, which is spherical like the 1s orbital but larger in size, as shown in **Figure 17a.** The 2p sublevel corresponds to three dumbbell-shaped p orbitals designated $2p_x$, $2p_y$, and $2p_z$. The subscripts *x, y,* and *z* merely designate the orientations of p orbitals along the *x, y,* and *z* coordinate axes, as shown in **Figure 17b.** Each of the p orbitals related to an energy sublevel has the same energy.

READING CHECK **Describe** the shapes of s and p orbitals.

Principal energy level 3 consists of three sublevels designated 3s, 3p, and 3d. Each d sublevel relates to five orbitals of equal energy. Four of the d orbitals have identical shapes but different orientations along the *x, y,* and *z* coordinate axes. However, the fifth orbital, d_{z^2}, is shaped and oriented differently than the other four. The shapes and orientations of the five d orbitals are illustrated in **Figure 17c.** The fourth principal energy level ($n = 4$) contains a fourth sublevel, called the 4f sublevel, which relates to seven f orbitals of equal energy. The f orbitals have complex, multilobed shapes.

Figure 17 The shapes of atomic orbitals describe the probable distribution of electrons in energy sublevels.

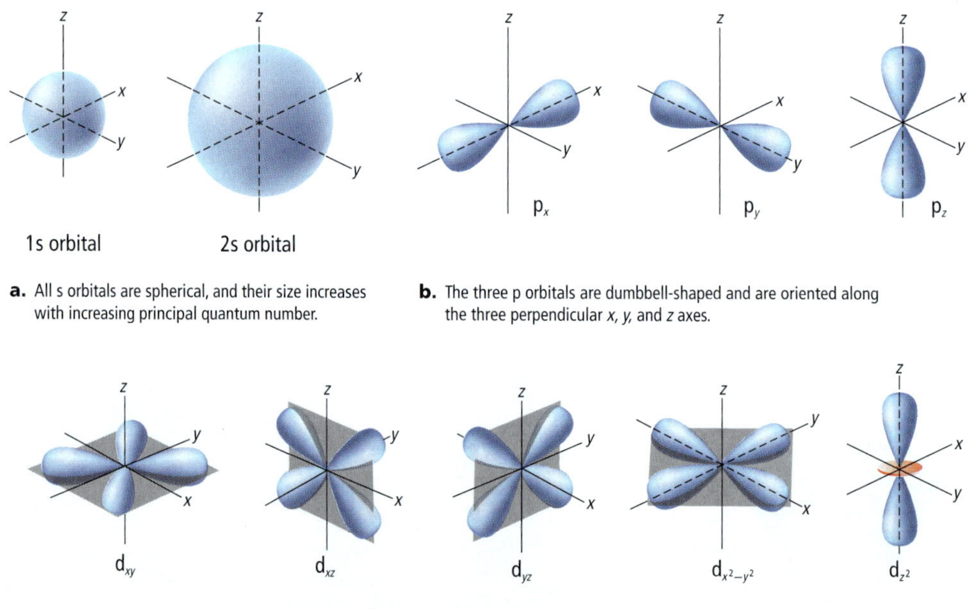

a. All s orbitals are spherical, and their size increases with increasing principal quantum number.

b. The three p orbitals are dumbbell-shaped and are oriented along the three perpendicular *x, y,* and *z* axes.

c. Four of the five d orbitals have the same shape but lie in different planes. The d_{z^2} orbital has its own unique shape.

CHEMISTRY JOURNAL

Orbital Shapes Have students sketch the shapes and orientations of hydrogen's 3s, 3p, and 3d orbitals. Have them label the orbital sketches and include them in their chemistry journals. **OL** **EL**

Table 2 Hydrogen's First Four Principal Energy Levels

Principal Quantum Number (*n*)	Sublevels (Types of Orbitals) Present	Number of Orbitals Related to Sublevel	Total Number of Orbitals Related to Principal Energy Level (n^2)
1	s	1	1
2	s p	1 3	4
3	s p d	1 3 5	9
4	s p d f	1 3 5 7	16

Hydrogen's first four principal energy levels, sublevels, and related atomic orbitals are summarized in **Table 2.** Note that the number of orbitals related to each sublevel is always an odd number, and that the maximum number of orbitals related to each principal energy level equals n^2.

At any given time, the electron in a hydrogen atom can occupy just one orbital. You can think of the other orbitals as unoccupied spaces —spaces available should the atom's energy increase or decrease. For example, when the hydrogen atom is in the ground state, the electron occupies the 1s orbital. However, when the atom gains a quantum of energy, the electron is excited to one of the unoccupied orbitals. Depending on the amount of energy available, the electron can move to the 2s orbital, to one of the three 2p orbitals, or to any other orbital that is vacant.

SECTION 2 REVIEW

Section Self-Check

Section Summary

- Bohr's atomic model attributes hydrogen's emission spectrum to electrons dropping from higher-energy to lower-energy orbits.
- The de Broglie equation relates a particle's wavelength to its mass, its velocity, and Planck's constant.
- The quantum mechanical model assumes that electrons have wave properties.
- Electrons occupy three-dimensional regions of space called atomic orbitals.

15. MAINIDEA **Explain** the reason, according to Bohr's atomic model, why atomic emission spectra contain only certain frequencies of light.

16. Differentiate between the wavelength of visible light and the wavelength of a moving soccer ball.

17. Enumerate the sublevels contained in the hydrogen atom's first four energy levels. What orbitals are related to each s sublevel and each p sublevel?

18. Explain why the location of an electron in an atom is uncertain using the Heisenberg uncertainty principle and de Broglie's wave-particle duality. How is the location of electrons in atoms defined?

19. Calculate Use the information in **Table 1** to calculate how many times larger the hydrogen atom's seventh Bohr radius is than its first Bohr radius.

20. Compare and contrast Bohr's model and the quantum mechanical model of the atom.

3 Assess

Check for Understanding

Ask students to explain why higher energy levels are made of sublevels associated with more electrons than lower energy levels. Higher energy levels are associated with larger volumes, which can contain more orbitals than smaller volumes. It is reasonable, therefore, that more electrons can be contained in the greater number of orbitals associated with higher energy levels. **OL**

Reteach

Explain that an electron's position and velocity within an atomic orbital are not known. Reiterate that at a given instant, there is a 10% probability that the electron is outside the orbital's 90% probability surface. **OL**

Extension

According to quantum mechanics, each electron in an atom can be described by four quantum numbers. Three of these (n, l, and m_l) are related to the probability of finding the electron at various points in space. The fourth (m_s) is related to the direction of electron spin—either clockwise or counterclockwise. The principal quantum number, n, specifies the atom's energy level associated with the electron. l specifies the energy sublevel and describes the shape of the region of space in which the electron moves. m_l specifies the orientation in space of the orbital containing the electron. m_s specifies the orientation of the electron's spin axis. **AL**

SECTION 2 REVIEW

15. Because only certain atomic energies are possible, only certain frequencies of radiation can be emitted from an atom.

16. The wavelength of the moving soccer ball is much smaller than the wavelengths of visible light. The moving soccer ball's wavelength is too small to see or detect.

17. First energy level, s; second energy level, s and p; third energy level, s, p, and d; fourth energy level, s, p, d, and f. Each s sublevel is related to a spherical s orbital. Each p sublevel is related to three dumbbell-shaped orbitals (p_x, p_y, and p_z).

18. An electron has wave-particle characteristics and does not have a single, definite location in space. The Heisenberg uncertainty principle states that it is fundamentally impossible to know precisely both the velocity and position of a particle at the same time.

19. $n = 7$: radius: 2.59 nm; $n = 1$: radius: 0.0529 nm
2.59 nm ÷ 0.0529 nm = 49.0 times larger

20. Bohr model: the electron is a particle; the hydrogen atom has only certain allowable energy states. Quantum mechanical model: the electron is a wave-particle phenomenon; an electron's energy is limited to certain values. Also, the quantum mechanical model makes no assertions regarding the electron's path around the nucleus.

SECTION 3

1 Focus

MAINIDEA

Learning the Aufbau Sequence Draw a sublevel diagram on the chalkboard, listing each sublevel and drawing the diagonal arrows that reflect the aufbau sequence. Ask students to write the sequence of sublevels by following the arrows in sequence from top to bottom. 1s, 2s, 2p, 3s, 3p, 4s, 3d, 4p, 5s, 4d, 5p, 6s, 4f, 5d, 6p, 7s, 5f, 6d, 7p Point out that they can construct and use such a sublevel diagram whenever they need to follow the aufbau sequence in determining an atom's electron arrangement. OL

2 Teach

Use Science Terms

Aufbau Principle Explain that the name *aufbau* is derived from the German *aufbauen,* which means *to build up.*

- **Caption Question Fig. 18** 5p

SECTION 3

Essential Questions

- How are the Pauli exclusion principle, the aufbau principle, and Hund's rule used to write electron configurations using orbital diagrams and electron configuration notation?
- What are valence electrons, and how do electron-dot structures represent an atom's valence electrons?

Review Vocabulary

electron: a negatively charged, fast-moving particle with an extremely small mass that is found in all forms of matter and moves through the empty space surrounding an atom's nucleus

New Vocabulary

electron configuration
aufbau principle
Pauli exclusion principle
Hund's rule
valence electron
electron-dot structure

Electron Configuration

MAINIDEA A set of three rules can be used to determine electron arrangement in an atom.

CHEM 4 YOU As students board a bus, they each sit in a separate bench seat until they are all full. Then, they begin sharing seats. Electrons fill atomic orbitals in a similar way.

Ground-State Electron Configuration

When you consider that atoms of the heaviest elements contain more than 100 electrons, the idea of determining electron arrangements in atoms with many electrons seems daunting. Fortunately, all atoms can be described with orbitals similar to hydrogen's. This allows us to describe arrangements of electrons in atoms using a few specific rules.

The arrangement of electrons in an atom is called the atom's **electron configuration.** Because low-energy systems are more stable than high-energy systems, electrons in an atom tend to assume the arrangement that gives the atom the lowest energy possible. The most stable, lowest-energy arrangement of the electrons is called the element's ground-state electron configuration. Three rules, or principles—the aufbau principle, the Pauli exclusion principle, and Hund's rule—define how electrons can be arranged in an atom's orbitals.

The aufbau principle The **aufbau principle** states that each electron occupies the lowest energy orbital available. Therefore, your first step in determining an element's ground-state electron configuration is learning the sequence of atomic orbitals from lowest energy to highest energy. This sequence, known as an aufbau diagram, is shown in **Figure 18.** In the diagram, each box represents an atomic orbital.

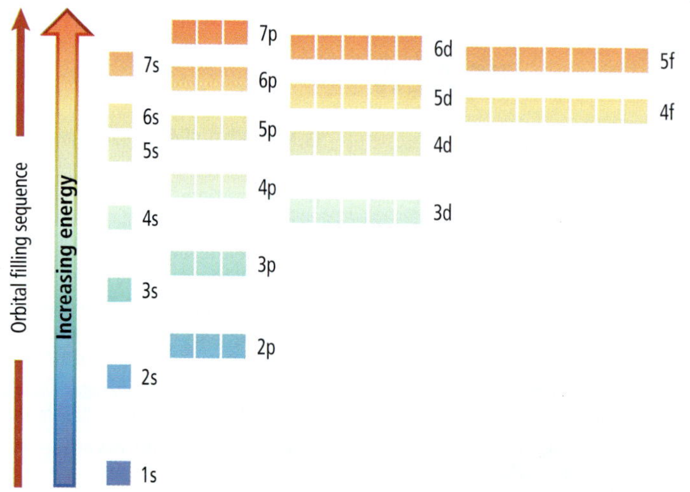

■ **Figure 18** The aufbau diagram shows the energy of each sublevel relative to the energy of other sublevels. Each box on the diagram represents an atomic orbital.

Determine *Which sublevel has the greater energy, 4d or 5p?*

Demonstration

Emission Spectra

Purpose
to illustrate the relationship between the electron configurations of nonmetals and their emission spectra

Materials
Spectrum tubes (H and Ne); spectrum tube supply; Flinn C-Spectra® diffraction grating; colored pencils or chalk

Safety Precautions
Use care around the spectrum tube high-voltage power supply. Spectrum tubes become hot when used.

Procedure
An inexpensive alternative to a spectroscope can be created by taping a small piece of the Flinn C-Spectra® diffraction grating to a 3 × 5 inch card. Have students view the spectrum emitted from the lights in the classroom. Then, darken the room and have them view the excited neon atoms in the powered neon spectrum tubes. Use colored pencils to record the emission spectrum of neon as seen through their diffraction gratings. Remind students that neon contains 10 electrons. Now repeat the process using a hydrogen spectrum tube.

Table 3 Features of the Aufbau Diagram

Feature	Example
All orbitals related to an energy sublevel are of equal energy.	All three 2p orbitals are of equal energy.
In a multi-electron atom, the energy sublevels within a principal energy level have different energies.	The three 2p orbitals are of higher energy than the 2s orbital.
In order of increasing energy, the sequence of energy sublevels within a principal energy level is s, p, d, and f.	If $n = 4$, then the sequence of energy sublevels is 4s, 4p, 4d, and 4f.
Orbitals related to energy sublevels within one principal energy level can overlap orbitals related to energy sublevels within another principal level.	The orbital related to the atom's 4s sublevel has a lower energy than the five orbitals related to the 3d sublevel.

Table 3 summarizes several features of the aufbau diagram. Although the aufbau principle describes the sequence in which orbitals are filled with electrons, it is important to know that atoms are not built up electron by electron.

The Pauli exclusion principle Electrons in orbitals can be represented by arrows in boxes. Each electron has an associated spin, similar to the way a top spins on its point. Like the top, the electron is able to spin in only one of two directions. An arrow pointing up [↑] represents the electron spinning in one direction, and an arrow pointing down [↓] represents the electron spinning in the opposite direction. An empty box [ ] represents an unoccupied orbital, a box containing a single up arrow [↑] represents an orbital with one electron, and a box containing both up and down arrows [↑↓] represents a filled orbital.

The **Pauli exclusion principle** states that a maximum of two electrons can occupy a single atomic orbital, but only if the electrons have opposite spins. Austrian physicist Wolfgang Pauli (1900–1958) proposed this principle after observing atoms in excited states. An atomic orbital containing paired electrons with opposite spins is written as [↑↓]. Because each orbital can contain, at most, two electrons, the maximum number of electrons related to each principal energy level equals $2n^2$.

Hund's rule The fact that negatively charged electrons repel each other has an important impact on the distribution of electrons in equal-energy orbitals. **Hund's rule** states that single electrons with the same spin must occupy each equal-energy orbital before additional electrons with opposite spins can occupy the same orbitals. For example, let the boxes below represent the 2p orbitals. One electron enters each of the three 2p orbitals before a second electron enters any of the orbitals. The sequence in which six electrons occupy three p orbitals is shown below.

1. [↑][ ][ ] 2. [↑][↑][ ] 3. [↑][↑][↑]

4. [↑↓][↑][↑] 5. [↑↓][↑↓][↑] 6. [↑↓][↑↓][↑↓]

READING CHECK **State** the three rules that define how electrons are arranged in atoms.

VOCABULARY

WORD ORIGIN

Aufbau
comes from the German word *aufbauen,* which means *to build up* or *arrange.*

FOLDABLES®
Incorporate information from this section into your Foldable.

READING CHECK The aufbau principle states that each electron occupies the lowest energy orbital available. The Pauli exclusion principle states that a maximum of two electrons can occupy a single atomic orbital. Hund's rule states that single electrons with the same spin must occupy each equal-energy orbital before additional electrons with opposite spin can occupy the same orbital.

FOLDABLES® Rubric

Build a Model

Use this activity to help students understand how neon's lowest two energy levels relate to one another. Have students work in groups to build a model that shows the atomic orbitals of a neon atom. Have them use plastic foam models for neon's 1s, $2p_x$, $2p_y$, and $2p_z$ orbitals and metal rods to connect the orbital models. Have them use a transparent material such as netting for the 2s orbital. Ask each group to demonstrate its model for the class. Be sure to point out that an actual atomic orbital does not have the well-defined boundaries of such a model. OL COOP LEARN

Because hydrogen has 1 electron, ask students to predict if there will be more or fewer lines in hydrogen's spectrum.

Results

The red-orange spectrum of neon also contains some green lines. Usually, only 3 of the 4 lines of hydrogen are visible.

Analysis

1. Write the electron configurations of neon and hydrogen. Ne: $1s^22s^22p^6$, H: $1s^1$
2. What is the appearance of neon in the excited state? In the ground state, neon is a clear, colorless gas. In the excited state it gives off a red-orange light.
3. Of the two spectra viewed, did hydrogen or neon have more lines? Explain why. Neon has more lines than hydrogen because its ten electrons have a greater number of possible energy transitions.

ASSESSMENT

Skill Have students view the excited spectral tube of another element such as mercury. Ask them to predict if Hg will have more lines than neon and hydrogen because it has 80 electrons. No, Hg has fewer lines in the visible spectrum. However, there are many additional lines in mercury's IR and UV spectra. OL

▪ **Caption Question Fig. 19** 10

Visual Learning

Table 4 Have students write an electron configuration notation that shows the orbital occupancy related to a phosphorus atom's 3p sublevel. $3p_x^{1}3p_y^{1}3p_z^{1}$ A chlorine atom's 3s and 3p sublevels. $3s^{2}3p_x^{2}3p_y^{2}3p_z^{1}$ OL

Concept Development

Hund's Rule Ask students to think about and explain the analogy between Hund's rule and the behavior of total strangers as they board an empty bus. Passengers tend to sit in separate rows until people occupy all rows. Only when no more empty rows are available do two passengers occupy a single row. For electrons, the situation is much the same as they occupy orbitals related to a sublevel. Chemistry's bus principle is known as Hund's rule. BL OL

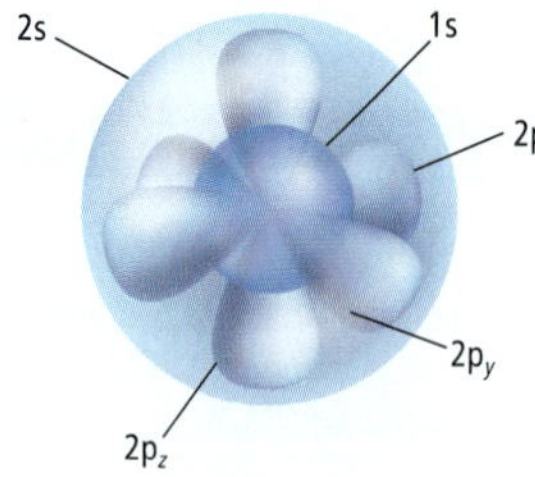

Neon atom

[↑↓] [↑↓] [↑↓][↑↓][↑↓]
1s 2s 2p

▪ **Figure 19** The 1s, 2s, and 2p orbitals of a neon atom overlap.
Determine how many electrons a neon atom has.

Electron Arrangement

You can represent an atom's electron configuration using one of two convenient methods: orbital diagrams or electron configuration notation.

Orbital diagrams As mentioned earlier, electrons in orbitals can be represented by arrows in boxes. Each box is labeled with the principal quantum number and sublevel associated with the orbital. For example, the orbital diagram for a ground-state carbon atom, shown below, contains two electrons in the 1s orbital, two electrons in the 2s orbital, and one electron in two of three separate 2p orbitals. The third 2p orbital remains unoccupied.

[↑↓] [↑↓] [↑][↑][ ]
1s 2s 2p

Electron configuration notation The electron configuration notation designates the principal energy level and energy sublevel associated with each of the atom's orbitals and includes a superscript representing the number of electrons in the orbital. For example, the electron configuration notation of a ground-state carbon atom is written $1s^{2}2s^{2}2p^{2}$. Orbital diagrams and electron configuration notations for the elements in periods one and two of the periodic table are shown in **Table 4. Figure 19** illustrates how the 1s, 2s, $2p_x$, $2p_y$, and $2p_z$ orbitals illustrated previously in **Figure 17** overlap in the neon atom.

Explore **electron configurations and orbital diagrams with an interactive table.** Concepts In Motion

Table 4 Electron Configurations and Orbital Diagrams for Elements 1–10

Element	Atomic Number	Orbital Diagram 1s	2s	$2p_x$	$2p_y$	$2p_z$	Electron Configuration Notation
Hydrogen	1	[↑]					$1s^{1}$
Helium	2	[↑↓]					$1s^{2}$
Lithium	3	[↑↓]	[↑]				$1s^{2}\ 2s^{1}$
Beryllium	4	[↑↓]	[↑↓]				$1s^{2}\ 2s^{2}$
Boron	5	[↑↓]	[↑↓]	[↑]			$1s^{2}\ 2s^{2}\ 2p^{1}$
Carbon	6	[↑↓]	[↑↓]	[↑]	[↑]		$1s^{2}\ 2s^{2}\ 2p^{2}$
Nitrogen	7	[↑↓]	[↑↓]	[↑]	[↑]	[↑]	$1s^{2}\ 2s^{2}\ 2p^{3}$
Oxygen	8	[↑↓]	[↑↓]	[↑↓]	[↑]	[↑]	$1s^{2}\ 2s^{2}\ 2p^{4}$
Fluorine	9	[↑↓]	[↑↓]	[↑↓]	[↑↓]	[↑]	$1s^{2}\ 2s^{2}\ 2p^{5}$
Neon	10	[↑↓]	[↑↓]	[↑↓]	[↑↓]	[↑↓]	$1s^{2}\ 2s^{2}\ 2p^{6}$

DIFFERENTIATED INSTRUCTION

Struggling Learners Look for ways to encourage and help these students as they learn to write electron configuration notations. One way is to have each of them make an aufbau diagram on one side of a 4 × 6 or 5 × 7 card and a sublevel diagram on the other side. Increase students' confidence by having them use the cards as they practice writing electron configurations for various elements. BL

Note that the electron configuration notation does not usually show the orbital distributions of electrons related to a sublevel. It is understood that a designation such as nitrogen's $2p^3$ represents the orbital occupancy $2p_x{}^1 2p_y{}^1 2p_z{}^1$.

For sodium, the first ten electrons occupy 1s, 2s, and 2p orbitals. Then, according to the aufbau sequence, the eleventh electron occupies the 3s orbital. The electron configuration notation and orbital diagram for sodium are written as follows.

$$1s^2 2s^2 2p^6 3s^1$$

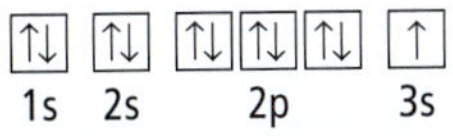

Noble-gas notation is a method of representing electron configurations of noble gases. Noble gases are the elements in the last column of the periodic table. They have eight electrons in their outermost orbital and they are unusually stable. You will learn more about noble gases in a later chapter. The noble-gas notation uses bracketed symbols. For example, [He] represents the electron configuration for helium, $1s^2$, and [Ne] represents the electron configuration for neon, $1s^2 2s^2 2p^6$. Compare the electron configuration for neon with sodium's configuration above. Note that the inner-level configuration for sodium is identical to the electron configuration for neon. Using noble-gas notation, sodium's electron configuration can be shortened to the form $[Ne]3s^1$. The electron configuration for an element can be represented using the noble-gas notation for the noble gas in the previous period and the electron configuration for the additional orbitals being filled. The complete and abbreviated (using noble-gas notation) electron configurations of the period 3 elements are shown in **Table 5.**

☑ READING CHECK **Explain** how to write the noble-gas notation for an element. What is the noble-gas notation for calcium?

Explore **electron configurations with an interactive table.** Concepts In Motion

Table 5 Electron Configurations for Elements 11–18

Element	Atomic Number	Complete Electron Configuration	Electron Configuration Using Noble Gas
Sodium	11	$1s^2 2s^2 2p^6 3s^1$	$[Ne]3s^1$
Magnesium	12	$1s^2 2s^2 2p^6 3s^2$	$[Ne]3s^2$
Aluminum	13	$1s^2 2s^2 2p^6 3s^2 3p^1$	$[Ne]3s^2 3p^1$
Silicon	14	$1s^2 2s^2 2p^6 3s^2 3p^2$	$[Ne]3s^2 3p^2$
Phosphorus	15	$1s^2 2s^2 2p^6 3s^2 3p^3$	$[Ne]3s^2 3p^3$
Sulfur	16	$1s^2 2s^2 2p^6 3s^2 3p^4$	$[Ne]3s^2 3p^4$
Chlorine	17	$1s^2 2s^2 2p^6 3s^2 3p^5$	$[Ne]3s^2 3p^5$
Argon	18	$1s^2 2s^2 2p^6 3s^2 3p^6$	$[Ne]3s^2 3p^6$ or [Ar]

Get help with **electron configuration notation.**

Personal Tutor

VOCABULARY

SCIENCE USAGE V. COMMON USAGE

Period

Science usage: a horizontal row of elements in the current periodic table

There are seven periods in the current periodic table.

Common usage: an interval of time determined by some recurring phenomenon

The period of Earth's orbit is one year.

ASSESSMENT

Knowledge Ask students to write electron configurations and construct orbital notations and electron dot structures for atoms of all the elements in the third period on the periodic table. **OL**

Content Background

More Exceptions to Predicted Configurations The exceptions to predicted ground-state electron configurations become more numerous among the transition elements in periods 5 and 6. Period 5: niobium, $[Kr]5s^1 4d^4$; molybdenum, $[Kr]5s^1 4d^5$; ruthenium, $[Kr]5s^1 4d^7$; rhodium, $[Kr]5s^1 4d^8$; palladium, $[Kr]4d^{10}$; silver, $[Kr]5s^1 4d^{10}$. Period 6: lanthanum, $[Xe]6s^2 4f^0 5d^1$; platinum, $[Xe]6s^1 4f^{14} 5d^9$; gold, $[Xe]6s^1 4f^{14} 5d^{10}$.

☑ **READING CHECK** The noble-gas notation uses bracketed symbols for the noble gas in the previous period followed by the electron configuration for the additional orbitals being filled. Ca: $[Ar]4s^2$

CULTURAL DIVERSITY

The Development of Fireworks Explain that the Chinese may have first used fireworks about the second century B.C. After inventing explosive black powder, which they called gung pow, the Chinese developed black-powder "crackers" that produced loud explosions. Most scholars believe that the Chinese used these crackers to frighten off evil spirits and to celebrate weddings, births, battle victories, and eclipses of the Moon. Fireworks became much more interesting and colorful in the 1830s, when Italian pyrotechnics experts added potassium chlorate to the mix. The potassium chlorate provided more oxygen for the chemical reaction, making it burn faster and hotter. This enabled the Italians to include various inorganic compounds that burn at high temperatures and create spectacular colors. The colors in fireworks are due to energy-level transitions of electrons in the metal atoms of these inorganic compounds.

Reinforcement

Energy-Level Sequence Point out that some textbooks, reference books, and periodic tables show electron configurations written in energy-level sequence rather than in aufbau sequence. Reinforce that using the energy-level sequence for electron configurations does not render the aufbau sequence invalid.

PRACTICE Problems

Have students refer to the Selected Solutions appendix for complete solutions to odd-numbered problems. The complete solutions for all problems can be found in the Solutions Manual.

21. a. $[Ar]4s^23d^{10}4p^5$
b. $[Kr]5s^2$
c. $[Kr]5s^24d^{10}5p^3$
d. $[Xe]6s^24f^{14}5d^5$
e. $[Xe]6s^24f^9$
f. $[Ar]4s^23d^2$

22. 5; 11

23. 6

24. indium

25. $[Xe]6s^2$; barium

PROBLEM-SOLVING STRATEGIES

Apply the Strategy

$[Kr]5s^24d^2$

Explore **electron configurations.**

Virtual Investigations

Exceptions to predicted configurations You can use the aufbau diagram to write correct ground-state electron configurations for all elements up to and including vanadium, atomic number 23. However, if you were to proceed in this manner, your configurations for chromium, $[Ar]4s^23d^4$, and copper, $[Ar]4s^23d^9$, would be incorrect. The correct configurations for these two elements are $[Ar]4s^13d^5$ for chromium and $[Ar]4s^13d^{10}$ for copper. The electron configurations for these two elements, as well as those of several other elements, illustrate the increased stability of half-filled and filled sets of s and d orbitals.

PROBLEM-SOLVING STRATEGY

Filling Atomic Orbitals

By drawing a sublevel diagram and following the arrows, you can write the ground-state electron configuration for any chemical element.

1. Sketch the sublevel diagram on a blank piece of paper.
2. Determine the number of electrons in one atom of the element for which you are writing the electron configuration. The number of electrons in a neutral atom equals the element's atomic number.
3. Starting with 1s, write the aufbau sequence of atomic orbitals by following the diagonal arrows from the top of the sublevel diagram to the bottom. When you complete one line of arrows, move to the right, to the beginning of the next line of arrows. As you proceed, add superscripts indicating the numbers of electrons in each set of atomic orbitals. Continue only until you have sufficient atomic orbitals to accommodate the total number of electrons in one atom of the element.
4. Apply noble-gas notation.

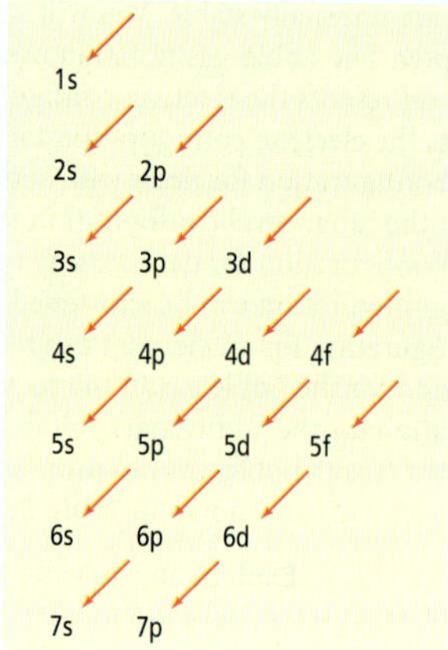

The sublevel diagram shows the order in which the orbitals are usually filled.

Apply the Strategy

Write the ground-state electron configuration for zirconium.

PRACTICE Problems

Do additional problems. Online Practice

21. Write ground-state electron configurations for the following elements.

a. bromine (Br) c. antimony (Sb) e. terbium (Tb)
b. strontium (Sr) d. rhenium (Re) f. titanium (Ti)

22. A chlorine atom in its ground state has a total of seven electrons in orbitals related to the atom's third energy level. How many of the seven electrons occupy p orbitals? How many of the 17 electrons in a chlorine atom occupy p orbitals?

23. When a sulfur atom reacts with other atoms, electrons in orbitals related to the atom's third energy level are involved. How many such electrons does a sulfur atom have?

24. An element has the ground-state electron configuration $[Kr]5s^24d^{10}5p^1$. It is part of some semiconductors and used in various alloys. What element is it?

25. **Challenge** In its ground state, an atom of an element has two electrons in all orbitals related to the atom's highest energy level for which $n = 6$. Using noble-gas notation, write the electron configuration for this element, and identify the element.

DIFFERENTIATED INSTRUCTION

Visually Impaired Make or purchase cellular polystyrene or papier-mâché models of s, p, and d orbitals. Allow visually impaired students to feel the models and trace their contours to gain a better appreciation of their shapes and orientations. BL EL

Valence Electrons

Only certain electrons, called valence electrons, determine the chemical properties of an element. **Valence electrons** are defined as electrons in the atom's outermost orbitals—generally those orbitals associated with the atom's highest principal energy level. For example, a sulfur atom contains 16 electrons, only six of which occupy the outermost 3s and 3p orbitals, as shown by sulfur's electron configuration. Sulfur has six valence electrons.

$$\text{S} \quad [\text{Ne}]3s^2 3p^4$$

Similarly, although a cesium atom contains 55 electrons, it has just one valence electron, the 6s electron shown in cesium's electron configuration.

$$\text{Cs} \quad [\text{Xe}]6s^1$$

Electron-dot structures Because valence electrons are involved in forming chemical bonds, chemists often represent them visually using a simple shorthand method, called electron-dot structure. An atom's **electron-dot structure** consists of the element's symbol, which represents the atomic nucleus and inner-level electrons, surrounded by dots representing all of the atom's valence electrons. American chemist G. N. Lewis (1875–1946) devised the method while teaching a college chemistry class in 1902.

In writing an atom's electron-dot structure, dots representing valence electrons are placed one at a time on the four sides of the symbol (they may be placed in any sequence) and then paired up until all are shown. The ground-state electron configurations and electron-dot structures for the elements in the second period are shown in **Table 6.**

Explore **electron-dot structures** with an interactive table. Concepts In Motion

Table 6 Electron Configurations and Dot Structures

Element	Atomic Number	Electron Configuration	Electron-Dot Structure
Lithium	3	$1s^2 2s^1$	Li·
Beryllium	4	$1s^2 2s^2$	·Be·
Boron	5	$1s^2 2s^2 2p^1$	·Ḃ·
Carbon	6	$1s^2 2s^2 2p^2$	·Ċ· (with dot below)
Nitrogen	7	$1s^2 2s^2 2p^3$	·N̈· (with dot below)
Oxygen	8	$1s^2 2s^2 2p^4$	:Ö· (with dot below)
Fluorine	9	$1s^2 2s^2 2p^5$	:F̈· (with pair below)
Neon	10	$1s^2 2s^2 2p^6$	:N̈e: (with pair below)

Use Science Terms

Valence Explain to students that some textbooks and reference books use the term *valence* in place of oxidation number. For example, some books might indicate that oxygen has a valence of 2–.

Content Background

Valence Electrons Explain to students that some inner-level d electrons are often considered valence electrons for transition elements. For example, although an atom of iron has just two electrons in its outermost (4s) orbitals, an additional electron associated with one of the atom's 3d orbitals is often involved in bonding. And, in an atom of manganese, as many as five 3d-orbital electrons can be involved in bonding.

CHEMISTRY JOURNAL

Another Solar System—What if? Ask students to write essays for their journals in which they speculate about flying a spacecraft to a planet in a different solar system. In the new solar system, they discover that each atomic orbital of the planet's solid, liquid, and gaseous matter might contain up to three electrons rather than just two. Their speculation should focus on the characteristics of the elements on this new planet. OL

IN-CLASS Example

Question How many electrons are shown in the electron-dot structures for atoms of lithium and potassium? Atoms of fluorine and iodine? Atoms of boron and thallium?

Answer Lithium and potassium have one dot. Fluorine and iodine have seven dots. Boron and thallium have three dots.

PRACTICE Problems

Have students refer to the Selected Solutions appendix for complete solutions to odd-numbered problems. The complete solutions for all problems can be found in the Solutions Manual.

26. Refer to the Solutions Manual.
27. aluminum; 3 electrons
28. beryllium

3 Assess

Check for Understanding

Ask students to predict the maximum number of electrons that can exist in orbitals related to an atom's fourth and fifth energy levels—assuming an element existed that contained enough electrons. Give students the formula $2n^2$, which can be used to calculate the number of electrons related to each value of n. 32 and 50 electrons, respectively OL

Reteach

Have students write the electron-dot structure of strontium. The structure includes the symbol Sr and two dots. Ask what the two dots represent. They represent the two electrons in a strontium atom's outermost, 5s, orbital. Then, ask what the electron-dot structure does not communicate about the strontium atom's electrons. It does not specify which orbital contains the two electrons, nor does it give any information about strontium's inner level electrons. OL

Extension

Ask students to identify the elements that have the following ground-state electron configurations.

$[Ar]4s^2 3d^5$ manganese

$[Xe]6s^2 4f^{14} 5d^{10} 6p^3$ bismuth OL

EXAMPLE Problem 3

EXAMPLE PROBLEM

ELECTRON-DOT STRUCTURES Some toothpastes contain stannous fluoride, a compound of tin and fluorine. What is tin's electron-dot structure?

1 ANALYZE THE PROBLEM

Consult the periodic table to determine the total number of electrons in a tin atom. Write out tin's electron configuration, and determine its number of valence electrons. Then use the rules for electron-dot structures to draw the electron-dot structure for tin.

2 SOLVE FOR THE UNKNOWN

Tin has an atomic number of 50. Thus, a tin atom has 50 electrons.

$$[Kr]5s^2 4d^{10} 5p^2$$

Write out tin's electron configuration using noble-gas notation. The closest noble gas is Kr.

The two 5s and the two 5p electrons (the electrons in the orbitals related to the atom's highest principal energy level) represent tin's four valence electrons. Draw the four valence electrons around tin's chemical symbol (Sn) to show tin's electron-dot structure. 

3 EVALUATE THE ANSWER

The correct symbol for tin (Sn) has been used, and the rules for drawing electron-dot structures have been correctly applied.

PRACTICE Problems

PRACTICE PROBLEMS

Do additional problems. Online Practice

26. Draw electron-dot structures for atoms of the following elements.
a. magnesium **b.** thallium **c.** xenon

27. An atom of an element has a total of 13 electrons. What is the element, and how many electrons are shown in its electron-dot structure?

28. Challenge This element exists in the solid state at room temperature and at normal atmospheric pressure and is found in emerald gemstones. It is known to be one of the following elements: carbon, germanium, sulfur, cesium, beryllium, or argon. Identify the element based on the electron-dot structure at right.

·X·

SECTION 3 REVIEW

Section Self-Check

Section Summary

- The arrangement of electrons in an atom is called the atom's electron configuration.
- Electron configurations are defined by the aufbau principle, the Pauli exclusion principle, and Hund's rule.
- An element's valence electrons determine the chemical properties of the element.
- Electron configurations can be represented using orbital diagrams, electron configuration notation, and electron-dot structures.

29. MAINIDEA **Apply** the Pauli exclusion principle, the aufbau principle, and Hund's rule to write out the electron configuration and draw the orbital diagram for each of the following elements.
a. silicon **b.** fluorine **c.** calcium **d.** krypton

30. Define *valence electron.*

31. Illustrate and describe the sequence in which ten electrons occupy the five orbitals related to an atom's d sublevel.

32. Extend the aufbau sequence through an element that has not yet been identified, but whose atoms would completely fill 7p orbitals. How many electrons would such an atom have? Write its electron configuration using noble-gas notation for the previous noble gas, radon.

33. Interpret Scientific Illustrations Which is the correct electron-dot structure for an atom of selenium? Explain.

a. 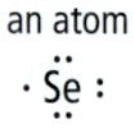 **b.** 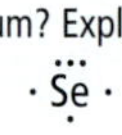 **c.** 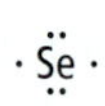 **d.** 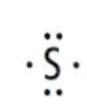

SECTION 3 REVIEW

29. Refer to the Solutions Manual for orbital diagrams.
a. Si $1s^2 2s^2 2p^6 3s^2 3p^2$
b. F $1s^2 2s^2 2p^5$
c. Ca $1s^2 2s^2 2p^6 3s^2 3p^6 4s^2$
d. Kr $1s^2 2s^2 2p^6 3s^2 3p^6 4s^2 3d^{10} 4p^6$

30. electrons in an atom's outermost orbital

31. Refer to the Solutions Manual for illustrations. Single electrons with the same spin occupy each equal-energy orbital before additional electrons with opposite spins occupy the same orbital.

32. 118 electrons. $[Rn]7s^2 5f^{14} 6d^{10} 7p^6$

33. c is correct; a shows three two-electron orbitals; b shows one three-electron orbital; d has the wrong symbol

CHEMISTRY & health

Tiny Tweezers

Peering through a microscope, a cell biologist can grasp a single cell with a pair of "tweezers." But these are not the kind of tweezers you might find in a medicine cabinet. These tweezers are made from two laser beams and can hold very tiny things such as cells and even individual atoms.

You might have heard that lasers can be used to cut things. Laser "scissors" are used in some surgeries. But surprisingly, lasers can also trap living cells and other microscopic objects in their beams without damaging them. How can beams of light hold things in place?

Gripping with light When light rays pass through a cell, they change direction slightly. This is similar to how light rays bend when passing through water in an aquarium. When light rays are bent, they exert a force. Large objects, such as aquariums, are too massive to be affected by this miniscule force, but tiny cells respond to the force. If the light rays are positioned in just the right way, they can hold a small object in place, as shown in **Figure 1.**

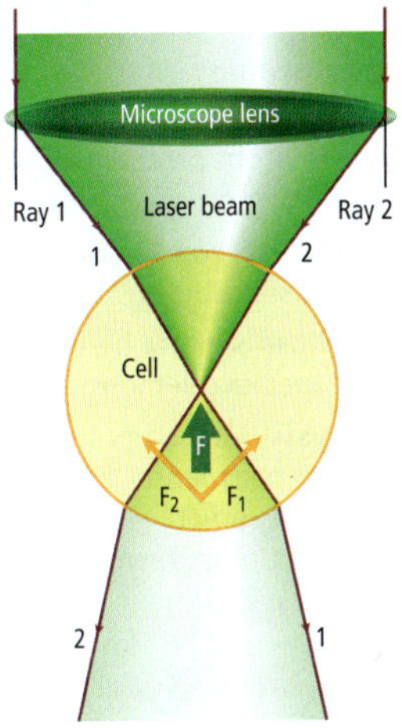

Figure 1 The light rays in the laser beam bend as they pass through the cell membranes. The bending of the light rays creates forces on the cell. The combined forces hold the cell in place.

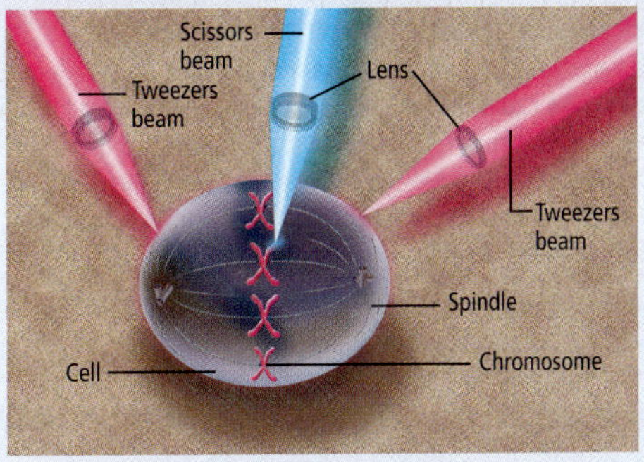

Figure 2 Organelles found within living cells are accessible to the smallest lasers.

Lasers and cancer So what use do scientists have for these tiny tweezers? One group of scientists is using them to study cell organelles. They are studying the forces exerted by mitotic spindles—the grouping of microtubules that coordinates cell division. The spindles guide replicated chromosomes to opposite sides of the cell—a key role in cell division. However, scientists do not know exactly how the spindles perform this function.

Tiny laser scissors have been used to cut off pieces of chromosomes during cell division. Laser tweezers were then used to move the pieces around the cell and the spindles, as shown in **Figure 2.** Knowing the force with which the tweezers grasp the chromosomes, scientists can measure the opposing force exerted by the spindles. Scientists hope that learning how spindles function during cell division will help them learn more about diseases related to cell division, such as cancer—a disease in which cells divide uncontrollably.

WRITING IN ▶ Chemistry

Laser Light Lasers can be found in a wide variety of everyday settings. Research the different types of lasers you might encounter daily, and find out what kind of light each laser uses. Summarize the results of your research in a presentation.

WebQuest

CHEMISTRY & health

Purpose

Students will learn how lasers can be used as microscopic tweezers and scissors in the smallest of surgeries.

Background

Laser is an acronym for Light Amplification by the Stimulated Emission of Radiation. Laser light can be contrasted with other sorts of colored light, such as that produced by neon lightbulbs. In a laser beam, all the bits of light (called photons) are the same wavelength (wavelength determines the color of the light) and are in phase with one another. This property, called coherence, is the crucial difference between laser light and other light sources.

Teaching Strategies

Model the size of a cell, a molecule, and an atom as a classroom exercise. For instance, if an atom were the size of a dot, perhaps 1 cm in diameter, then a typical sugar molecule would be about 10 cm in diameter and an average human cell would be around 200,000 cm (or 2 km) wide. Such a comparison might make students think that cells are large; on the contrary, it shows that atoms are amazingly small. On this same scale, a human being would be about 150,000 km long, stretching nearly halfway from Earth to the Moon.

WRITING IN ▶ Chemistry

Rubric WebQuest

Summarize Some of the lasers students might list from their everyday lives include: laser scanners found in stores and libraries to read barcodes; CD players and CD-ROMs; laser pointers; police officers use invisible infrared laser beams rather than radar systems to watch for speeding motorists.

ChemLAB

Rubric

Preparation

Time Allotment one class period

Process Skills compare and contrast, predict, think critically, classify, observe and infer, sequence

Safety Precautions Approve lab safety forms before work begins. Do not let students handle the spectrum power supplies or tubes. Warn students not to touch the gas spectrum tubes during use because they are very hot and can cause burns. Exercise caution around the spectrum power supplies, as they present a significant electrical shock hazard. Plug into wall receptacles only protected by GFI circuits.

Disposal The flasks of food coloring solutions can be reused.

Preparation of Materials

- Set up light sockets with lightbulbs prior to class and have them plugged in.
- Set up spectrum power supplies and tubes prior to class.

Procedure

- Have several groups of students start their observations of the gas discharge tubes first, so that the area does not become crowded by the end of the class period.
- The Flinn C-Spectra is much easier to use than a spectroscope for viewing spectra. It can be ordered from:
  Flinn Scientific, Inc.
  P.O. Box 219
  Batavia, IL 60510
  www.flinnsci.com
- A possible source for the gas spectrum tubes and power supplies would be the physics lab.

ChemLAB

Analyze Line Spectra

Background: Emission spectra are produced when excited atoms return to a more stable state by emitting radiation of specific wavelengths. When white light passes through a sample, atoms in the sample absorb specific wavelengths. This produces dark lines in the continuous spectrum of white light and is called an absorption spectrum.

Question: *What absorption and emission spectra do various substances produce?*

Materials

ring stand with clamp
40-W tubular lightbulb
light socket with grounded power cord
275-mL polystyrene culture flask
Flinn C-Spectra® or similar diffraction grating
red, green, blue, and yellow food coloring
set of colored pencils
spectrum tubes (hydrogen, neon, and sodium)
spectrum-tube power supply (3)

Safety Precautions

WARNING: ***Use care around the spectrum-tube power supplies. Spectrum tubes will become hot when used.***

Procedure

1. Read and complete the lab safety form.
2. Use a Flinn C-Spectra® or similar diffraction grating to view an incandescent lightbulb. Draw the observed spectrum using colored pencils.
3. Use the Flinn C-Spectra® to view the emission spectra from tubes of gaseous hydrogen, neon, and sodium. Use colored pencils to draw the observed spectra.
4. Fill a 275-mL culture flask with about 100 mL of water. Add two or three drops of red food coloring to the water. Shake the solution.
5. Repeat Step 4 for the green, blue, and yellow food coloring.
6. Set up the 40-W lightbulb so that it is near eye level. Place the flask with red food coloring about 8 cm from the lightbulb so that you are able to see light from the bulb above the solution and light from the bulb projecting through the solution.
7. With the room lights darkened, view the light using the Flinn C-Spectra®. The top spectrum viewed will be a continuous spectrum from the white lightbulb. The bottom spectrum will be the absorption spectrum of the red solution. Use colored pencils to make a drawing of the absorption spectra you observe.
8. Repeat Steps 6 and 7 using the green, blue, and yellow solutions.
9. **Cleanup and Disposal** Turn off the light and spectrum-tube power supplies. Wait several minutes for the lightbulb and spectrum tubes to cool. Dispose of the liquids and store the lightbulb and spectrum tubes as directed by your teacher.

Analyze and Conclude

1. **Think Critically** How can the single electron in a hydrogen atom produce all of the lines found in its emission spectrum?
2. **Predict** How can you predict the absorption spectrum of a solution by looking at its color?
3. **Apply** How can spectra be used to identify the presence of specific elements in a substance?
4. **Error Analysis** Name a potential source of error in this experiment. Choose one of the elements you observed, and research its absorption spectrum. Compare your findings with the results of your experiment.

INQUIRY EXTENSION

Hypothesize What would happen if you mixed more than one color of food coloring with water and repeated the experiment? Design an experiment to test your hypothesis.

Matt Meadows

Analyze and Conclude

1. At any given time, the electron occupies a single orbital. However, it can move into other vacant orbitals as the atom absorbs or releases energy.
2. The color of a solution is due to the color of light it transmits. The colors not transmitted are absorbed, and these colors comprise the absorption spectrum.
3. The spectrum of each element is unique. Thus, the presence of a unique atomic spectrum indicates the presence of that element.
4. Answers will vary.

INQUIRY EXTENSION

Answers will vary.

Vocabulary Practice

BIGIDEA The atoms of each element have a unique arrangement of electrons.

SECTION 1 Light and Quantized Energy

MAINIDEA Light, a form of electromagnetic radiation, has characteristics of both a wave and a particle.

- All waves are defined by their wavelengths, frequencies, amplitudes, and speeds.

$$c = \lambda\nu$$

- In a vacuum, all electromagnetic waves travel at the speed of light.
- All electromagnetic waves have both wave and particle properties.
- Matter emits and absorbs energy in quanta.

$$E_{quantum} = h\nu$$

- White light produces a continuous spectrum. An element's emission spectrum consists of a series of lines of individual colors.

VOCABULARY
- electromagnetic radiation
- wavelength
- frequency
- amplitude
- electromagnetic spectrum
- quantum
- Planck's constant
- photoelectric effect
- photon
- atomic emission spectrum

SECTION 2 Quantum Theory and the Atom

MAINIDEA Wavelike properties of electrons help relate atomic emission spectra, energy states of atoms, and atomic orbitals.

- Bohr's atomic model attributes hydrogen's emission spectrum to electrons dropping from higher-energy to lower-energy orbits.

$$\Delta E = E_{\text{higher-energy orbit}} - E_{\text{lower-energy orbit}} = E_{\text{photon}} = h\nu$$

- The de Broglie equation relates a particle's wavelength to its mass, its velocity, and Planck's constant.

$$\lambda = h / m\nu$$

- The quantum mechanical model assumes that electrons have wave properties.
- Electrons occupy three-dimensional regions of space called atomic orbitals.

VOCABULARY
- ground state
- quantum number
- de Broglie equation
- Heisenberg uncertainty principle
- quantum mechanical model of the atom
- atomic orbital
- principal quantum number
- principal energy level
- energy sublevel

SECTION 3 Electron Configuration

MAINIDEA A set of three rules can be used to determine electron arrangement in an atom.

- The arrangement of electrons in an atom is called the atom's electron configuration.
- Electron configurations are defined by the aufbau principle, the Pauli exclusion principle, and Hund's rule.
- An element's valence electrons determine the chemical properties of the element.
- Electron configurations can be represented using orbital diagrams, electron configuration notation, and electron-dot structures.

VOCABULARY
- electron configuration
- aufbau principle
- Pauli exclusion principle
- Hund's rule
- valence electron
- electron-dot structure

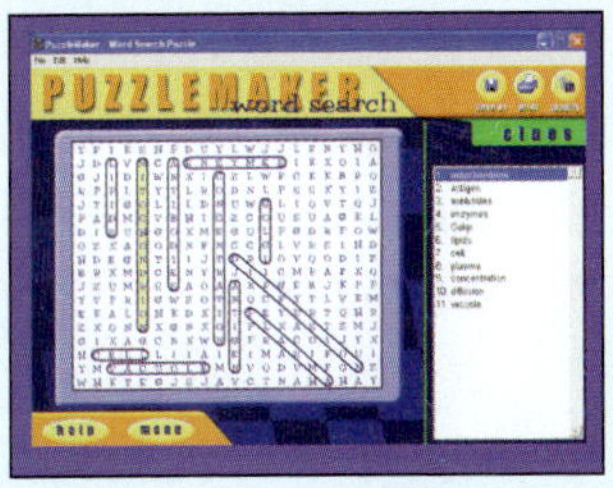

Vocabulary Practice

For additional practice with vocabulary, have students access the Vocabulary PuzzleMaker.

CHAPTER 5 STUDY GUIDE

Use the Vocabulary

To reinforce chapter vocabulary, have students write a sentence using each term. OL EL

Review Strategies

- Ask students to write the equation that relates frequency and wavelength. OL
- Have students write the equation that relates the energy of a quantum to the associated frequency. OL
- Ask students to relate Heisenberg's uncertainty principle to electrons in atoms. OL
- Have students explain the relationship between an atom's orbitals and its energy levels. OL
- Problems from the Supplemental Practice Problems appendix or the Supplemental Problems booklet can be used for review. OL

Students can review with these online resources.

Vocabulary Practice

Vocabulary eGames
Vocabulary eFlashcards
Vocabulary PuzzleMaker

Section Self-Check

Chapter Self-Check

Online Test Practice

Use *eAssessment* to:

- create multiple versions of tests
- edit existing questions and add your own questions
- build tests aligned with select state curriculum standards using built-in state curriculum tags
- track students' progress

SECTION 1

Mastering Concepts

34. a. Frequency is the number of waves that pass a given point per second.

b. Wavelength is the shortest distance between equivalent points on a continuous wave.

c. A quantum is the minimum amount of energy that can be lost or gained by an atom.

d. An atom's ground state is its lowest allowable energy state.

35. d. X-rays, a. ultraviolet light, b. microwaves, c. radio waves

36. 2.88×10^{21} gamma-ray electromagnetic waves of this frequency pass a given point per second.

37. a phenomenon in which a metal emits electrons when light of a sufficient frequency shines on it

38. The light from a neon sign contains only certain visible colors, while sunlight contains the full spectrum of colors.

39. According to Planck, for a given frequency, ν, matter can emit or absorb energy only in discrete quanta that are whole-number multiples of $h\nu$, where h is Planck's constant.

40. He proposed that photons must have a certain minimum energy level, or threshold, value to cause the ejection of a photoelectron.

41. The red waves have a longer wavelength and a lower frequency.

42. The color of the light changes as the object acquires more energy.

43. The wave model does not explain the photoelectric effect, atomic emission spectra, or why matter emits different frequencies of light at different temperatures.

44. Both types of waves travel at the same speed in a vacuum, 3.00×10^8 m/s. Radio waves have a longer wavelength and lower frequency than ultraviolet waves.

CHAPTER 5 ASSESSMENT

SECTION 1

Mastering Concepts

34. Define the following terms.
- **a.** frequency
- **b.** wavelength
- **c.** quantum
- **d.** ground state

35. Arrange the following types of electromagnetic radiation in order of increasing wavelength.
- **a.** ultraviolet light
- **b.** microwaves
- **c.** radio waves
- **d.** X-rays

36. A gamma ray has a frequency of 2.88×10^{21} Hz. What does this mean?

37. What is the photoelectric effect?

38. Neon Sign How does light emitted from a neon sign differ from sunlight?

39. Explain Planck's quantum concept as it relates to energy lost or gained by matter.

40. How did Einstein explain the photoelectric effect?

41. Rainbow What are two differences between the red and green electromagnetic waves in a rainbow?

42. Temperature What happens to the light emitted by a heated, glowing object as its temperature increases?

43. What are three deficiencies of the wave model of light related to light's interaction with matter?

44. How are radio waves and ultraviolet waves similar? How are they different?

Mastering Problems

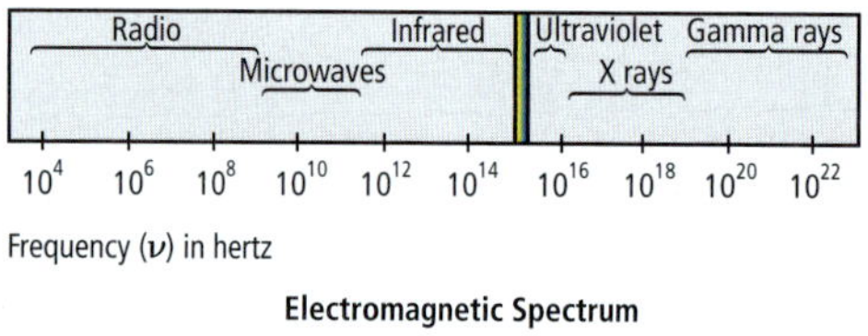

■ **Figure 20**

45. Radiation Use **Figure 20** to determine the following types of radiation.
- **a.** radiation with a frequency of $8.6 \times 10^{11}\ s^{-1}$
- **b.** radiation with a wavelength of 4.2 nm
- **c.** radiation with a frequency of 5.6 MHz
- **d.** radiation that travels at a speed of 3.00×10^8 m/s

46. What is the wavelength of electromagnetic radiation with a frequency of 5.00×10^{12} Hz? What kind of electromagnetic radiation is this?

47. What is the frequency of electromagnetic radiation with a wavelength of 3.33×10^{-8} m? What type of electromagnetic radiation is this?

48. What is the speed of an electromagnetic wave with a frequency of 1.33×10^{17} Hz and a wavelength of 2.25 nm?

49. What is the energy of a photon of red light that has a frequency of 4.48×10^{14} Hz?

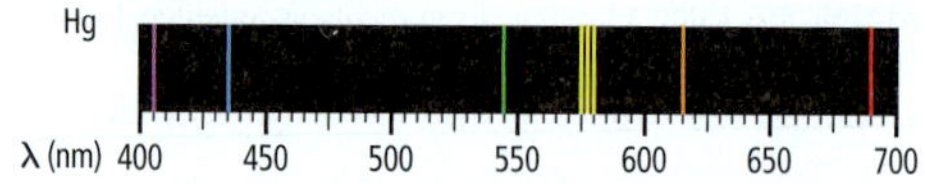

■ **Figure 21**

50. Mercury Mercury's atomic emission spectrum is shown in **Figure 21.** Estimate the wavelength of the orange line. What is its frequency? What is the energy of a photon corresponding to the orange line emitted by the mercury atom?

51. What is the energy of an ultraviolet photon that has a wavelength of 1.18×10^{-8} m?

52. A photon has an energy of 2.93×10^{-25} J. What is its frequency? What type of electromagnetic radiation is the photon?

53. A photon has an energy of 1.10×10^{-13} J. What is the photon's wavelength? What type of electromagnetic radiation is it?

54. Spacecraft How long does it take a radio signal from the *Voyager* spacecraft to reach Earth if the distance between *Voyager* and Earth is 2.72×10^9 km?

55. Radio Waves If your favorite FM radio station broadcasts at a frequency of 104.5 MHz, what is the wavelength of the station's signal in meters? What is the energy of a photon of the station's electromagnetic signal?

56. Platinum What minimum frequency of light is needed to eject a photoelectron from atoms of platinum, which require at least 9.08×10^{-19} J/photon?

57. Eye Surgery The argon fluoride (ArF) laser used in some refractive eye surgeries emits electromagnetic radiation of 193.3 nm wavelength. What is the frequency of the ArF laser's radiation? What is the energy of a single quantum of the radiation?

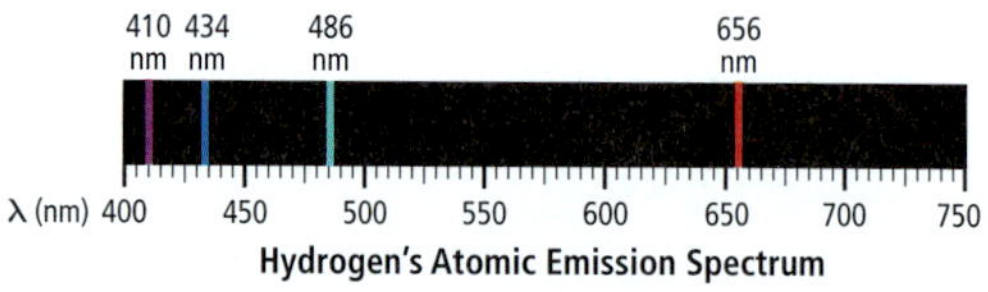

■ **Figure 22**

58. Hydrogen One line in hydrogen's emission spectrum has a wavelength of 486 nm. Examine **Figure 22** to determine the line's color. What is the line's frequency?

Mastering Problems

45. a. infrared

b. X-ray

c. AM radio

d. any EM wave

46. $\lambda = 6.00 \times 10^{-5}$ m; infrared radiation

47. $\nu = 9.01 \times 10^{15}\ s^{-1}$; ultraviolet radiation

48. $\nu = 3.00 \times 10^8$ m/s

49. $E_{photon} = 2.97 \times 10^{-19}$ J

50. $\lambda = 615$ nm, $\nu = 4.88 \times 10^{14}\ s^{-1}$, $E_{photon} = 3.23 \times 10^{-19}$ J

51. $E_{photon} = 1.68 \times 10^{-17}$ J

52. $\nu = 4.42 \times 10^8\ s^{-1}$; TV or FM wave

53. $\lambda = 1.81 \times 10^{-12}$ m; an X-ray or gamma-ray

54. t = 9070 s, or 151 min

55. $\lambda = 2.871$ m, $E_{photon} = 6.924 \times 10^{-26}$ J

56. 1.37×10^{15} Hz

57. $\nu = 1.55 \times 10^{15}\ s^{-1}$; $E = 1.03 \times 10^{-18}$ J

58. The line is blue-green. Its frequency is $6.17 \times 10^{14}\ s^{-1}$.

Chapter Self-Check

SECTION 2

Mastering Concepts

59. According to the Bohr model, how do electrons move in atoms?

60. What does *n* designate in Bohr's atomic model?

61. What is the difference between an atom's ground state and an excited state?

62. What is the name of the atomic model in which electrons are treated as waves? Who first wrote the electron wave equations that led to this model?

63. What is an atomic orbital?

64. What does *n* represent in the quantum mechanical model of the atom?

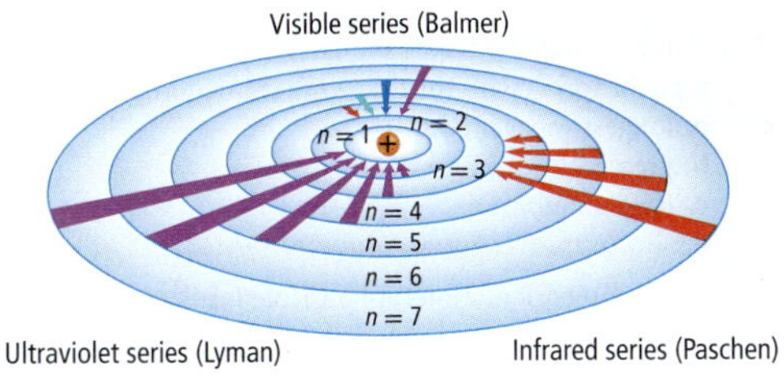

■ **Figure 23**

65. Electron Transition According to the Bohr model shown in **Figure 23,** what type of electron-orbit transitions produce the ultraviolet lines in hydrogen's Lyman series?

66. How many energy sublevels are contained in each of the hydrogen atom's first three energy levels?

67. What atomic orbitals are related to a p sublevel?

68. What do the sublevel designations s, p, d, and f specify with respect to the atom's orbitals?

69. How are the five orbitals related to an atom's d sublevel designated?

70. What is the maximum number of electrons an orbital can contain?

71. Describe the relative orientations of the orbitals related to an atom's 2p sublevel.

72. How many electrons can be contained in all the orbitals related to an argon atom's third energy level?

73. How does the quantum mechanical model of the atom describe the paths of an atom's electrons?

74. Macroscopic Objects Why do we not notice the wavelengths of moving objects such as automobiles?

75. Why is it impossible to know precisely the velocity and position of an electron at the same time?

SECTION 3

Mastering Concepts

76. In what sequence do electrons fill the atomic orbitals related to a sublevel?

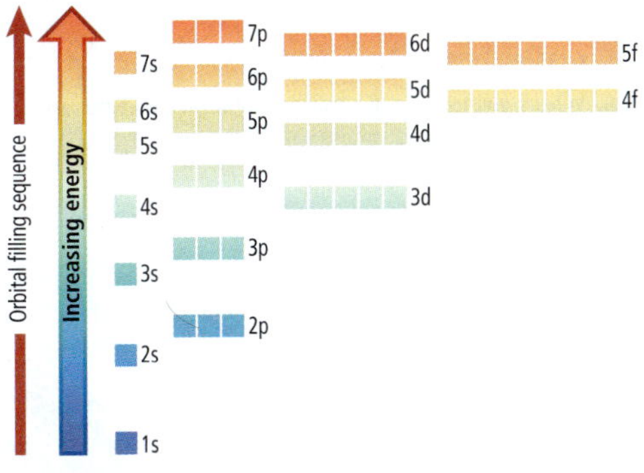

■ **Figure 24**

77. Rubidium Using **Figure 24,** explain why one electron in a rubidium atom occupies a 5s orbital rather than a 4d or 4f orbital.

78. What are valence electrons? How many of a magnesium atom's 12 electrons are valence electrons?

79. Light is said to have a dual wave-particle nature. What does this statement mean?

80. Describe the difference between a quantum and a photon.

81. How many electrons are shown in each element's electron-dot structure?

a. carbon
b. iodine
c. calcium
d. gallium

82. When writing the electron configuration notation for an atom, what three principles or rules should you follow?

83. Write the electron configuration and draw the orbital notation for atoms of oxygen and sulfur.

Mastering Problems

84. List the aufbau sequence of orbitals from 1s to 7p.

85. Write each element's orbital notation and complete electron configuration.

a. beryllium
b. aluminum
c. nitrogen
d. sodium

86. Use noble-gas notation to describe the electron configurations of the elements represented by the following symbols.

a. Kr
b. P
c. Zr
d. Pb

SECTION 2

Mastering Concepts

59. Electrons move in circular orbits around the nucleus.

60. The quantum number *n* specifies the electron's orbit.

61. An atom's ground state is its lowest energy state, while any energy state higher than the ground state is an excited state.

62. the quantum mechanical model of the atom; Erwin Schrödinger

63. a three-dimensional region around the nucleus describing an electron's probable location

64. *n* represents an orbital's principal quantum number, which indicates the relative size and energy of the orbital.

65. The Lyman series is caused by electron transitions from Bohr's higher energy orbits to the $n = 1$ orbit.

66. Energy level 1 has one sublevel, energy level 2 has two sublevels, and energy level 3 has three sublevels.

67. p_x, p_y, and p_z orbitals

68. their shapes

69. xy, xz, yz, $x^2 - y^2$, z^2

70. two electrons

71. Lying along the x, y, and z coordinate axes, the three p orbitals are mutually perpendicular.

72. eight electrons

73. The quantum mechanical model gives no description of the electrons' paths.

74. Their wavelengths are too small to be seen.

75. The photon required to measure an electron's velocity or position changes both the position and velocity of the electron.

SECTION 3

Mastering Concepts

76. Each orbital must contain a single electron before any orbital contains two electrons.

77. The orbital related to the 5s sublevel has a lower energy than orbitals related to the 4d and 4f sublevels.

78. Valence electrons are the electrons in an atom's outermost orbitals; 2.

79. Light exhibits wavelike behavior in some situations and particlelike behavior in others.

80. A quantum is the minimum amount of energy that can be lost or gained by an atom, while a photon is a particle of light that carries a quantum of energy.

81. a: 4; **b:** 7; **c:** 2; **d:** 3

82. Pauli exclusion principle, aufbau principle, and Hund's rule

83. oxygen: $1s^22s^22p^4$; The orbital diagram has five boxes with two arrows in the first three and single arrows in the last two. sulfur: $[Ne]3s^23p^4$; The orbital diagram has nine boxes with two arrows in the first seven and single arrows in the last two.

Mastering Problems

84. 1s, 2s, 2p, 3s, 3p, 4s, 3d, 4p, 5s, 4d, 5p, 6s, 4f, 5d, 6p, 7s, 5f, 6d, 7p

85. a. Be $1s^22s^2$

↑↓ ↑↓
1s 2s

b. Al $1s^22s^22p^63s^23p^1$

↑↓ ↑↓ ↑↓ ↑↓ ↑↓ ↑↓ ↑
1s 2s 2p 3s 3p

c. N $1s^22s^22p^3$

↑↓ ↑↓ ↑ ↑ ↑
1s 2s 2p

d. Na $1s^22s^22p^63s^1$

↑↓ ↑↓ ↑↓ ↑↓ ↑↓ ↑
1s 2s 2p 3s

86. a. Kr $[Ar]4s^2 3d^{10} 4p^6$
b. P $[Ne]3s^2 3p^3$
c. Zr $[Kr]5s^2 4d^2$
d. Pb $[Xe]6s^2 4f^{14} 5d^{10} 6p^2$

87. a. F
b. Ca
c. Nd
d. Te
e. Md
f. Br

88. d

89. b

90. a. ·Ċ· (4 dots)
b. ·Äs· (5 dots)
c. :P̈o· (6 dots)
d. ·K
e. Ḃa·

91. 18; 15; 4

92. b

93. $[Kr]5s^2 4d^{10} 5p^2$ ·Ṡn·

MIXED REVIEW

94. a. 18 c. 18
b. 32 d. 8

95. $\nu = 5.20 \times 10^{-7}$ m

96. a. longest wavelength: 4
b. greatest frequency: 3
c. largest amplitude: 1 and 3
d. shortest wavelength: 3

97. a. 1 c. 5
b. 3 d. 7

98. helium, calcium, cobalt, barium

99. $n = 4 \longrightarrow n = 2$

100. The two dots are the atom's two 4s valence electrons.

101. $\nu = 4.54 \times 10^{15}\ s^{-1}$; $\lambda = 6.60 \times 10^{-8}$ m

102. francium

103. Bohr proposed that atoms emit light of certain wavelengths and energies when electrons move from higher-energy orbits to lower-energy orbits.

104. 3.10×10^{19} photons

105. Its wavelength decreases.

106. The energy of the atom increases as one or more electrons move into higher energy orbitals.

87. What element is represented by each electron configuration?
a. $1s^2 2s^2 2p^5$
b. $[Ar]4s^2$
c. $[Xe]6s^2 4f^4$
d. $[Kr]5s^2 4d^{10} 5p^4$
e. $[Rn]7s^2 5f^{13}$
f. $1s^2 2s^2 2p^6 3s^2 3p^6 4s^2 3d^{10} 4p^5$

88. Which electron configuration notation describes an atom in an excited state?
a. $[Ar]4s^2 3d^{10} 4p^2$
b. $[Ne]3s^2 3p^5$
c. $[Kr]5s^2 4d^1$
d. $[Ar]4s^2 3d^8 4p^1$

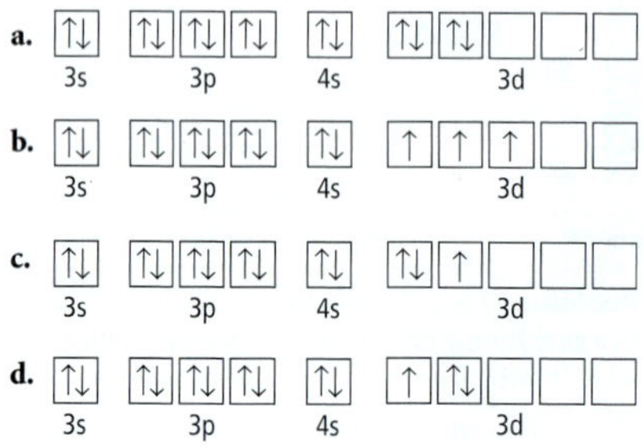

■ **Figure 25**

89. Which orbital diagram in **Figure 25** is correct for an atom in its ground state?

90. Draw an electron-dot structure for an atom of each element.
a. carbon
b. arsenic
c. polonium
d. potassium
e. barium

91. Arsenic An atom of arsenic has how many electron-containing orbitals? How many of the orbitals are completely filled? How many of the orbitals are associated with the atom's $n = 4$ principal energy level?

■ **Figure 26**

92. Which element could have the ground-state electron-dot notation shown in **Figure 26?**
a. manganese **c.** calcium
b. antimony **d.** samarium

93. For an atom of tin in the ground state, write the electron configuration using noble-gas notation, and draw its electron-dot structure.

MIXED REVIEW

94. What is the maximum number of electrons that can be contained in an atom's orbitals having the following principal quantum numbers?
a. 3 **c.** 6
b. 4 **d.** 7

95. What is the wavelength of light with a frequency of 5.77×10^{14} Hz?

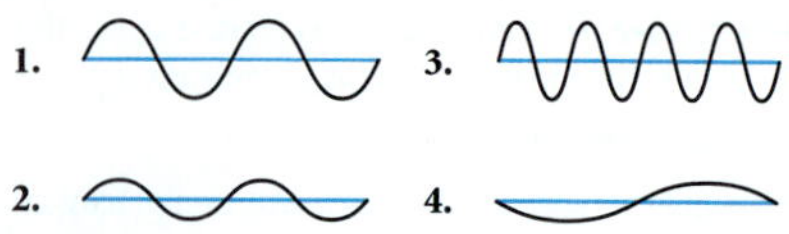

■ **Figure 27**

96. Waves Using the waves shown in **Figure 27**, identify the wave or waves with the following characteristics.
a. longest wavelength
b. greatest frequency
c. largest amplitude
d. shortest wavelength

97. How many orientations are possible for the orbitals related to each sublevels?
a. s **c.** d
b. p **d.** f

98. Which elements have only two electrons in their electron-dot structures: hydrogen, helium, lithium, aluminum, calcium, cobalt, bromine, krypton, or barium?

99. In Bohr's atomic model, what electron-orbit transition produces the blue-green line in hydrogen's atomic emission spectrum?

100. Zinc A zinc atom contains a total of 18 electrons in its 3s, 3p, and 3d orbitals. Why does its electron-dot structure show only two dots?

101. X-Ray An X-ray photon has an energy of 3.01×10^{-18} J. What is its frequency and wavelength?

102. Which element has the ground-state electron configuration represented by the noble-gas notation $[Rn]7s^1$?

103. How did Bohr explain atomic emission spectra?

104. Infrared Radiation How many photons of infrared radiation with a frequency of 4.88×10^{13} Hz are required to provide an energy of 1.00 J?

105. Light travels slower in water than it does in air; however, its frequency remains the same. How does the wavelength of light change as it travels from air to water?

106. According to the quantum mechanical model of the atom, what happens when an atom absorbs a quantum of energy?

Chapter Self-Check

THINK CRITICALLY

107. Compare and Contrast Briefly discuss the difference between an orbit in Bohr's model of the atom and an orbital in the quantum mechanical view of the atom.

108. Calculate It takes 8.17×10^{-19} J of energy to remove one electron from a gold surface. What is the maximum wavelength of light capable of causing this effect?

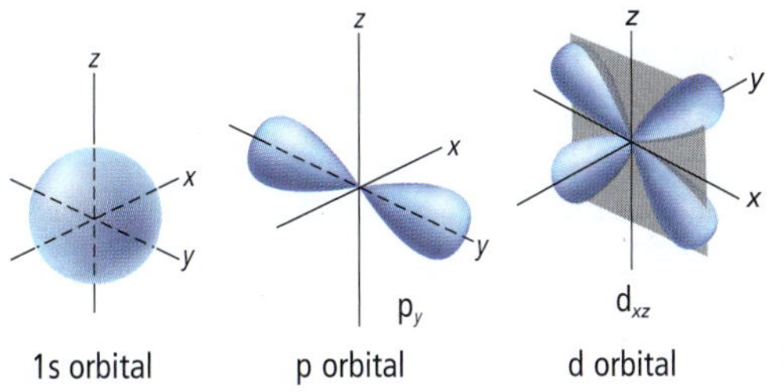

■ **Figure 28**

109. Describe the shapes of the atomic orbitals shown in **Figure 28.** Specify their orientations and relate each orbital to a particular type of energy sublevel.

110. Infer Suppose that you live in a universe in which the Pauli exclusion principle states that a maximum of three, rather than two, electrons can occupy a single atomic orbital. Evaluate and explain the new chemical properties of the elements lithium and phosphorus.

CHALLENGE PROBLEM

111. Hydrogen Atom The hydrogen atom's energy is -6.05×10^{-20} J when the electron is in the $n = 6$ orbit and -2.18×10^{-18} J when the electron is in the $n = 1$. Calculate the wavelength of the photon emitted when the electron drops from the $n = 6$ orbit to the $n = 1$ orbit. Use the following values: $h = 6.626 \times 10^{-34}$ J•s and $c = 3.00 \times 10^8$ m/s.

CUMULATIVE REVIEW

112. Round 20.56120 g to three significant figures.

113. Identify whether each statement describes a chemical property or a physical property.

a. Mercury is a liquid at room temperature.
b. Sucrose is a white, crystalline solid.
c. Iron rusts when exposed to moist air.
d. Paper burns when ignited.

114. An atom of gadolinium has an atomic number of 64 and a mass number of 153. How many electrons, protons, and neutrons does it contain?

WRITING IN ▶ Chemistry

115. Neon Signs To make neon signs emit different colors, manufacturers often fill the signs with gases other than neon. Write an essay about the use of gases in neon signs and the colors produced by the gases.

116. Rutherford's Model Imagine that you are a scientist in the early twentieth century, and you have just learned the details of a new, nuclear model of the atom proposed by the prominent English physicist Ernest Rutherford. After analyzing the model, you discern what you believe to be important limitations. Write a letter to Rutherford in which you express your concerns regarding his model. Use diagrams and examples of specific elements to help you make your point.

DBQ Document-Based Questions

Sodium Vapor *When sodium metal is vaporized in a gas-discharge lamp, two closely spaced, bright yellow-orange lines are produced. Because sodium vapor lamps are electrically efficient, they are used widely for outdoor lighting, such as streetlights and security lighting.*

Figure 29 *shows the emission spectrum of sodium metal. The entire visible spectrum is shown for comparison.*

Data obtained from: Volland, W. March 2005. *Spectroscopy: Element Identification and Emission Spectra.*

■ **Figure 29**

117. Differentiate between the two spectra shown above.

118. Sodium's two bright lines have wavelengths of 588.9590 nm and 589.9524 nm. What is the ground-state electron configuration notation for sodium, and how does sodium's electron configuration relate to the lines?

119. Calculate the energies of photons related to the two lines using the relationships expressed in the following equations.

$$E_{photon} = h\nu;\ c = \lambda\nu;\ E = hc/\lambda$$

DBQ Document-Based Questions

Data obtained from: Dr. Walt Volland. March, 2005. *Spectroscopy: Element Identification and Emission Spectra*

117. One shows all the colors of the complete visible spectrum; the other shows just certain colors emitted by sodium atoms, known as sodium's atomic emission spectrum.

118. $1s^22s^22p^63s^1$; The two lines are produced when sodium atoms drop from excited states to lower-energy states. This occurs when electrons in higher-energy orbitals drop into lower-energy orbitals.

119. 3.38×10^{-19} J and 3.37×10^{-19} J

THINK CRITICALLY

107. In the Bohr model, an orbit is a circular path taken by an electron as it moves around the atomic nucleus. In the quantum mechanical model, an orbital is a three-dimensional region around the nucleus that describes the electron's probable location.

108. $\lambda = 2.43 \times 10^{-7}$ m

109. The first orbital is spherical and related to an s sublevel. The second orbital is dumbbell-shaped, oriented along the *y*-axis, and related to a p sublevel. The third orbital consists of two, perpendicular dumbbell-shaped parts, lies in the *xz* plane, and is related to a d sublevel.

110. Both lithium and phosphorus would be noble gases. Lithium, with an electron configuration notation of $1s^3$, would be analogous to helium ($1s^2$). Phosphorus, with an electron configuration notation of $1s^32s^32p^9$, would be analogous to neon ($1s^2$, $2s^2$, $2p^6$).

CHALLENGE PROBLEM

111. $\lambda = 9.38 \times 10^{-8}$ m

CUMULATIVE REVIEW

112. 20.6 g

113. a. physical property
b. physical property
c. chemical property
d. chemical property

114. 64 electrons, 64 protons, 89 neutrons

WRITING IN ▶ Chemistry

Rubric

115. Students answers might include the following elements and colors: helium (yellow); neon (orange-red); sodium (yellow); argon (lavender); krypton (white); xenon (blue).

116. Answers will vary.

MULTIPLE CHOICE

1. D
2. C
3. B
4. C
5. D
6. C
7. D
8. A
9. D

MULTIPLE CHOICE

1. Cosmic rays are high-energy radiation from outer space. What is the frequency of a cosmic ray that has a wavelength of 2.67×10^{-13} m when it reaches Earth? (The speed of light is 3.00×10^{8} m/s.)

A. $8.90 \times 10^{-22}\ s^{-1}$
B. $3.75 \times 10^{12}\ s^{-1}$
C. $8.01 \times 10^{-5}\ s^{-1}$
D. $1.12 \times 10^{21}\ s^{-1}$

2. Which is the electron-dot structure for indium?

A. · In
B. · In ·
C. $\cdot\,\dot{\text{In}}\,\cdot$
D. $\cdot\,\underset{\cdot}{\dot{\text{In}}}\,\cdot$

Use the figure below to answer Questions 3 and 4.

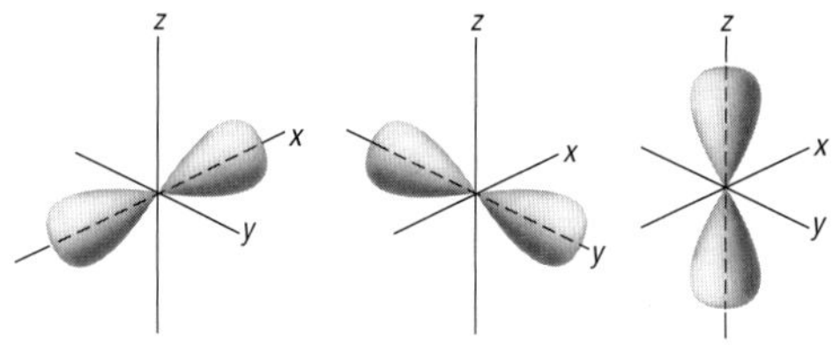

3. To which sublevel do all of these orbitals belong?

A. s
B. p
C. d
D. f

4. How many electrons total can reside in this sublevel?

A. 2
B. 3
C. 6
D. 8

5. What is the maximum theoretical number of electrons related to the fifth principal energy level of an atom?

A. 2
B. 8
C. 18
D. 32

Use the periodic table and the table below to answer Questions 6 to 8.

Electron Configurations for Selected Transition Metals

Element	Symbol	Atomic Number	Electron Configuration
Vanadium	V	23	$[Ar]4s^2 3d^3$
Yttrium	Y	39	$[Kr]5s^2 4d^1$
			$[Xe]6s^2 4f^{14} 5d^6$
Scandium	Sc	21	$[Ar]4s^2 3d^1$
Cadmium	Cd	48	

6. Using noble-gas notation, what is the ground-state electron configuration of Cd?

A. $[Kr]4d^{10}4f^{2}$
B. $[Ar]4s^{2}3d^{10}$
C. $[Kr]5s^{2}4d^{10}$
D. $[Xe]5s^{2}4d^{10}$

7. What is the element that has the ground-state electron configuration $[Xe]6s^{2}4f^{14}5d^{6}$?

A. La
B. Ti
C. W
D. Os

8. What is the complete electron configuration of a scandium atom?

A. $1s^{2}2s^{2}2p^{6}3s^{2}3p^{6}4s^{2}3d^{1}$
B. $1s^{2}2s^{2}2p^{7}3s^{2}3p^{7}4s^{2}3d^{1}$
C. $1s^{2}2s^{2}2p^{5}3s^{2}3p^{5}4s^{2}3d^{1}$
D. $1s^{2}2s^{1}2p^{7}3s^{1}3p^{7}4s^{2}3d^{1}$

9. Which is NOT evidence that a chemical change has occurred?

A. The properties of the substances involved in the reaction have changed.
B. An odor is produced.
C. The composition of the substances involved in the reaction have changed.
D. The total mass of all substances involved has changed.

Online Test Practice

SHORT ANSWER

Use the data below to answer Questions 10 to 13.

Temperature of Water with Heating	
Time (s)	Temperature (°C)
0	16.3
30	19.7
60	24.2
90	27.8
120	32.0
150	35.3
180	39.6
210	43.3
240	48.1

10. Make a graph showing temperature versus time.

11. Is the heating of this sample of water a linear process? Explain how you can tell.

12. Use your graph to find the approximate rate of heating in degrees per second. What is this value in degrees per minute?

13. Show the equation to convert the temperature at 180 s from degrees Celsius to Kelvin and to degrees Fahrenheit.

EXTENDED RESPONSE

14. Compare the information provided in an electron-dot structure with the information in an electron configuration.

15. Explain why $1s^22s^22p^63s^23p^64s^24d^{10}4p^2$ is not the correct electron configuration for germanium (Ge). Write the correct electron configuration for germanium.

SAT SUBJECT TEST: CHEMISTRY

Use the diagram below to answer Questions 16 and 17.

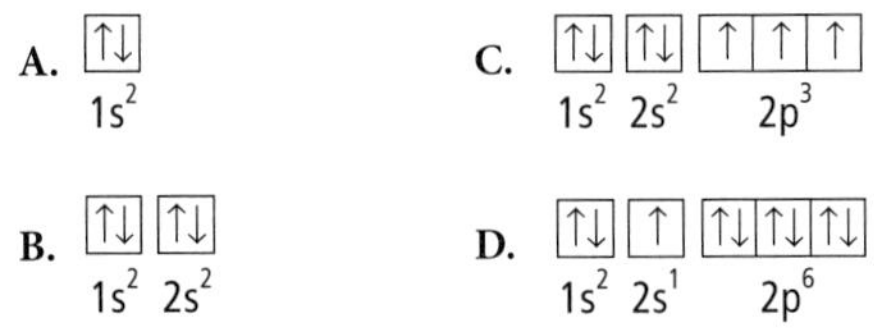

16. Which shows an orbital diagram that violates the aufbau principle?
A. A
B. B
C. C
D. D
E. none

17. Which shows the orbital diagram for the element beryllium?
A. A
B. B
C. C
D. D
E. none

18. A student performs an experiment to measure the boiling point of pentane and measures it at 37.2°C. The literature reports this value as 36.1°C. What is the student's percent error?
A. 97.0%
B. 2.95%
C. 1.1%
D. 15.5%
E. 3.05%

19. Which method of separating components of a mixture depends on the different boiling points of the components of the mixture?
A. chromatography
B. filtration
C. crystallization
D. distillation
E. sublimation

NEED EXTRA HELP?																			
If You Missed Question . . .	1	2	3	4	5	6	7	8	9	10	11	12	13	14	15	16	17	18	19
Review Section . . .	5.1	5.3	5.2	5.2	5.3	5.3	5.3	5.3	3.2	2.4	2.4	2.4	2.1	5.3	5.3	5.3	5.3	2.3	3.3

SHORT ANSWER

10. Refer to the Solutions Manual. Check that the graph is approximately linear, with time on the *x*-axis and temperature on the *y*-axis.

11. This process is occurring at a constant rate. One can tell because there is only one constant slope; the best-fit line is linear.

12. Use slope = rise/run to find the slope; students should select two points on the line of best fit to compare (not two points from the data table) such as (45 s, 22°) and (220 s, 45°). With these points, slope = (45 − 22) ÷ (220 − 45) = 0.13 degrees per second. Multiplying by 60 seconds per minute to convert this value into degrees per minute gives 7.8°/min.

13. 40°C + 273 = 313 K; (9/5)(40°C) + 32 = 104°F

EXTENDED RESPONSE

14. An electron-dot structure provides information about how many valence or outer energy level electrons are in the atom, whereas the electron configuration shows the energy level and sublevel for all electrons in the atom.

15. The electrons in the d sublevel are in the third energy level, not the fourth. The correct electron configuration would be $1s^22s^22p^63s^23p^64s^23d^{10}4p^2$.

SAT SUBJECT TEST: CHEMISTRY

16. D
17. B
18. E
19. D

CHAPTER 6 The Periodic Table and Periodic Law

BIGIDEA **Periodic trends in the properties of atoms allow us to predict physical and chemical properties.**

ESSENTIAL QUESTIONS	RESOURCES TO ASSESS MASTERY
SECTION 1 **Development of the Modern Periodic Table** **1.** How was the periodic table developed? **2.** What are the key features of the periodic table? 1 session 0.5 block	**Progress Monitoring** Formative Assessment, pp. 177, 179 Reading Check, pp. 176, 177 Section Review, p. 181
SECTION 2 **Classification of the Elements** **1.** Why do elements in the same group have similar properties? **2.** Based on their electron configurations, what are the four blocks of the periodic table? 1 session 0.5 block	**Progress Monitoring** Formative Assessment, pp. 183, 185 Reading Check, p. 185 Section Review, p. 186
SECTION 3 **Periodic Trends** **1.** What are the period and group trends of different properties? **2.** How are period and group trends in atomic radii related to electron configuration? 2 sessions 1 block	**Progress Monitoring** Formative Assessment, pp. 188, 190, 191 Reading Check, pp. 188, 191, 192 Graph Check, p. 191 Section Review, p. 194 **Summative Assessment** Chapter Assessment, p. 198 *eAssessment* Chapter Test (scaffolded)

Go online!

	LEVELED RESOURCES	LAB MATERIALS
	Science Notebook 6.1 OL ***FAST FILE Chapter Resources:*** ChemLab Worksheet, p. 30 OL Study Guide, p. 42 BL	LaunchLAB p. 172 / **15 min** bolts, screws, nails, ruler, balance ChemLAB p. 196 / **60 min** stoppered test tubes and plastic dishes containing small samples of elements, conductivity apparatus, 1.0*M* HCl, small hammer, test tubes, test-tube rack, 10-mL graduated cylinder, spatula, glass-marking pencil
	Science Notebook 6.2 OL ***FAST FILE Chapter Resources:*** Study Guide, p. 44 BL	
	Science Notebook 6.3 OL ***FAST FILE Chapter Resources:*** MiniLab Worksheet, p. 28 OL Study Guide, p. 46 BL	MiniLAB p. 193 / **40 min** index cards, pencil

ADDITIONAL RESOURCES

eAssessment
ConnectED Student Center
ConnectED Teacher Center
Lesson Presentations
eSolutions
Videos
Animations
Online Quizzes
Virtual Investigations
Multilingual eGlossary
Personal Tutors
WebQuests
Applying Practices Worksheets
PBLs

Practice:

Challenge Problems AL
Supplemental Problems BL OL
Chapter Test (Scaffolded)
Standardized Test Practice

Additional Resources:

Blueprints for Success
Solving Problems: A Chemistry Handbook

Lab Resources:

Laboratory Manual OL
CBL Laboratory Manual OL
Small-Scale Laboratory Manual OL
Forensics Laboratory Manual OL

BL Below Level OL On Level AL Advanced Learners EL English Learners COOP LEARN Cooperative Learning

CHAPTER 6

LaunchLAB

Rubric

How can you recognize trends?

Purpose Students will use the organization of everyday items to help visualize the periodic trends on the table of elements.

Teaching Strategies
- Try to select fasteners that are different enough to lead to more trends.
- Buttons, food items, and postage stamps could be used as alternatives in this lab.
- To make the lab authentic, try to pick samples so that one or two of the fasteners do not seem to fit the trends.
- Extend this lab by having the students design a periodic table of items found in their home.

Expected Results The students will have created a periodic table of fasteners in a 4 × 3 grid.

Procedure
1. Read and complete the lab safety form.
2. Obtain a sample of **fasteners,** including **bolts, screws,** and **nails.**
3. Measure the length of each fastener with a **ruler.**
4. Use a **balance** to measure the mass of each fastener.
5. Place the nails in a series from smallest to largest.
6. Continue to arrange a series of screws and a series of bolts that also correspond to the series of nails created in Step 5.

Analysis
1. **Make a table** listing the length and mass of each fastener. Answers will vary. Sample answer:

Item Length Mass	Nail 2.6 cm 0.295 g	Nail 3.1 cm 0.648 g	Nail 4.1 cm 0.860 g	Nail 7.3 cm 4.302 g
Item Length Mass	Screw 1.2 cm 0.819 g	Screw 1.9 cm 1.607 g	Screw 3.1 cm 1.765 g	Screw 3.2 cm 3.926 g
Item Length Mass	Bolt 1.8 cm 1.502 g	Bolt 2.8 cm 1.723 g	Bolt 4.0 cm 2.596 g	Bolt 8.0 cm 5.192 g

2. **Describe** the trend in mass as you go from left to right across each row of the table. Answers will vary based on samples provided. Generally, the mass will increase from left to right across the row.
3. **Describe** the trend in mass as you go down each column of the table. Answers will vary based on samples provided. For the order nail, screw, bolt, generally, the mass will increase from top to bottom in the column.

CHAPTER 6
The Periodic Table and Periodic Law

BIGIDEA Periodic trends in the properties of atoms allow us to predict physical and chemical properties.

SECTIONS
1. Development of the Modern Periodic Table
2. Classification of the Elements
3. Periodic Trends

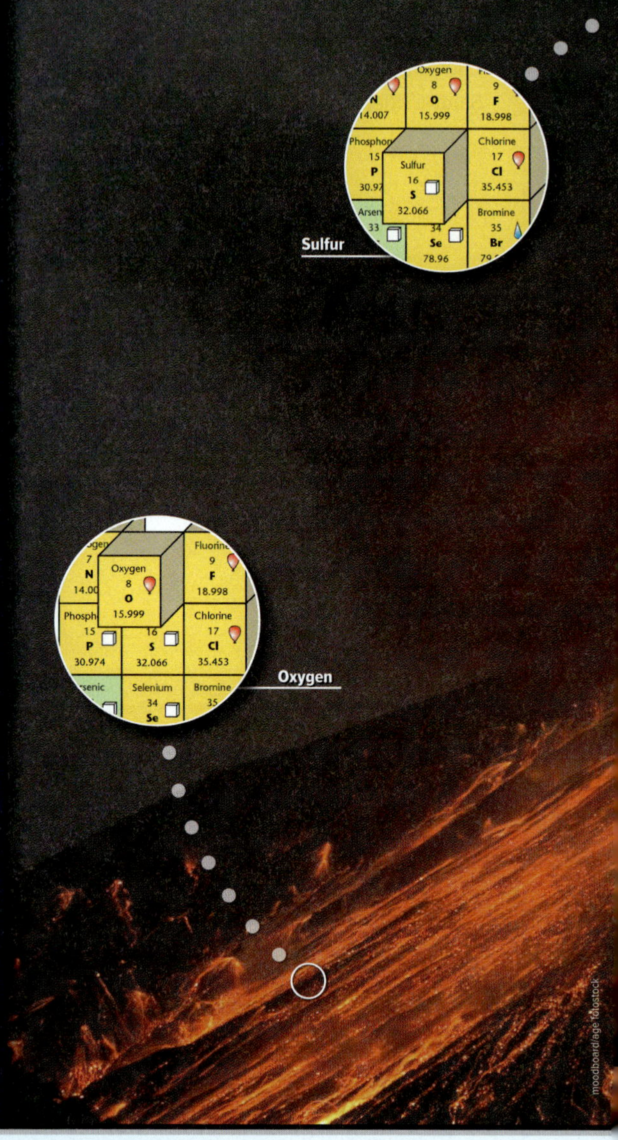

LaunchLAB

How can you recognize trends?

The periodic table of the elements is arranged so that the properties of the elements repeat in a regular way. Such an arrangement can also be used for common items. In this lab, you will study ways to organize data according to trends.

FOLDABLES Study Organizer

Periodic Trends

Make a folded chart. Label it as shown. Use it to organize information about periodic trends.

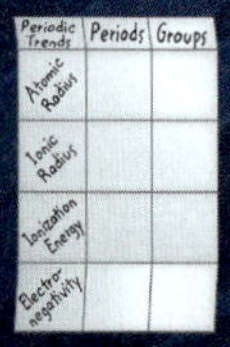

There are currently 117 elements in the periodic table. Approximately 90 of them occur naturally. By mass, oxygen is the most abundant element in Earth's crust (just under 50%).

4. **Analyze** your organization of the fasteners, and explain any other trends that you find in the table. Answers will vary based on samples provided. Explanations could include statements like "The fasteners were shinier across the row and more dull down the column."

Inquiry Create a periodic table of carbonated beverages in a manner similar to this lab. What properties did you use? Answers will vary. Generally, the students will describe the brand, the amount of calories, the amount of sodium, and the color of the beverage.

Introduce the BIGIDEA

Trends To introduce this chapter's Big Idea, draw the following pattern of shapes on the board: □□□■■●●●○▲▲▲□□□■■●●●○▲

Ask students to predict what the next shape will be? black triangle Point out that trends, such as the pattern of shapes, help you predict things and events. Explain that in this chapter they will study trends that are useful for predicting properties of the elements.

Tie to Previous Knowledge

Have students review the following concepts before studying this chapter.

- atomic structure
- electron configurations, valence electrons, electron-dot structures

Use the Photo

Properties of Elements Have students describe the photograph. The photograph shows an erupting volcano, and its lava. It also shows some elements from the periodic table associated with the lava, crust, and gases. Ask them what the purpose of the photograph is. The photograph shows some of the elements that exist in volcanoes.

SECTION 1

1 Focus

MAINIDEA

Organizing the Elements Have students observe and compare some copper foil and some aluminum foil. Ask them to describe ways that the two are alike. Possible answer: They are both metal and are both malleable. Point out that these are just some of the many properties that scientists considered when developing the periodic table. As scientists learned more about the elements, they organized the elements in different ways. OL

2 Teach

Quick Demo

Arranging Elements Randomly display about 30 stock bottles of elements and compounds. Ask students whether such an arrangement facilitates being able to locate a particular substance. Have students suggest better ways to arrange the substances. OL

SECTION 1

Essential Questions

- How was the periodic table developed?
- What are the key features of the periodic table?

Review Vocabulary

atomic number: the number of protons in an atom

New Vocabulary

periodic law
group
period
representative element
transition element
metal
alkali metal
alkaline earth metal
transition metal
inner transition metal
lanthanide series
actinide series
nonmetal
halogen
noble gas
metalloid

Development of the Modern Periodic Table

MAINIDEA The periodic table evolved over time as scientists discovered more useful ways to compare and organize the elements.

CHEM 4 YOU Imagine grocery shopping if all the apples, pears, oranges, and peaches were mixed into one bin at the grocery store. Organizing things according to their properties is often useful. Scientists organize the many different types of chemical elements in the periodic table.

Development of the Periodic Table

In the late 1700s, French scientist Antoine Lavoisier (1743–1794) compiled a list of all elements that were known at the time. The list, shown in **Table 1,** contained 33 elements organized in four categories. Many of these elements, such as silver, gold, carbon, and oxygen, have been known since prehistoric times. The 1800s brought a large increase in the number of known elements. The advent of electricity, which was used to break down compounds into their components, and the development of the spectrometer, which was used to identify the newly isolated elements, played major roles in the advancement of chemistry. The industrial revolution of the mid-1800s also played a major role, which led to the development of many new chemistry-based industries, such as the manufacture of petrochemicals, soaps, dyes, and fertilizers. By 1870, there were over 60 known elements.

Along with the discovery of new elements came volumes of new scientific data related to the elements and their compounds. Chemists of the time were overwhelmed with learning the properties of so many new elements and compounds. What chemists needed was a tool for organizing the many facts associated with the elements. A significant step toward this goal came in 1860, when chemists agreed upon a method for accurately determining the atomic masses of the elements. Until this time, different chemists used different mass values in their work, making the results of one chemist's work hard to reproduce by another. With newly agreed-upon atomic masses for the elements, the search for relationships between atomic mass and elemental properties, and a way to organize the elements began in earnest.

Table 1 Lavoisier's Table of Simple Substances (Old English Names)

Gases	light, heat, dephlogisticated air, phlogisticated gas, inflammable air
Metals	antimony, silver, arsenic, bismuth, cobalt, copper, tin, iron, manganese, mercury, molybdena, nickel, gold, platina, lead, tungsten, zinc
Nonmetals	sulphur, phosphorus, pure charcoal, radical muriatique*, radical fluorique*, radical boracique*
Earths	chalk, magnesia, barote, clay, siliceous earth

*no English name

CHEMISTRY PROJECT

Law of Octaves Have students work in small groups to create a poster that shows and describes a musical octave and describes the relationship that Newlands noticed between the musical octave and repetition in properties of elements. If possible, have at least one student in each group that can read music. OL COOP LEARN

John Newlands In 1864, English chemist John Newlands (1837–1898) proposed an organizational scheme for the elements. He noticed that when the elements were arranged by increasing atomic mass, their properties repeated every eighth element. A pattern such as this is called periodic because it repeats in a specific manner. Newlands named the periodic relationship that he observed in chemical properties the *law of octaves,* after the musical octave in which notes repeat every eighth tone. **Figure 1** shows how Newlands organized 14 of the elements known in the mid-1860s. Acceptance of the law of octaves was hampered because the law did not work for all of the known elements. Also, the use of the word *octave* was harshly criticized by fellow scientists, who thought that the musical analogy was unscientific. While his law was not generally accepted, the passage of a few years would show that Newlands was basically correct; the properties of elements do repeat in a periodic way.

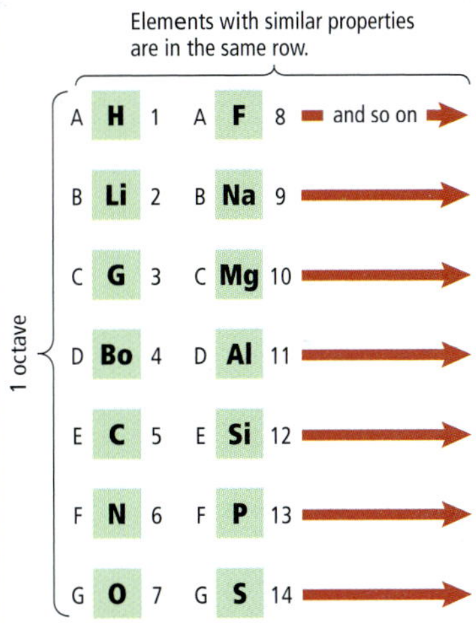

Figure 1 John Newlands noticed that the properties of elements repeated every eighth element, in the same way musical notes repeat every eighth note and form octaves.

Meyer and Mendeleev In 1869, German chemist Lothar Meyer (1830–1895) and Russian chemist Dmitri Mendeleev (1834–1907) each demonstrated a connection between atomic mass and the properties of elements. Mendeleev, however, is generally given more credit than Meyer because he published his organizational scheme first. Like Newlands several years earlier, Mendeleev noticed that when the elements were ordered by increasing atomic mass, there was a periodic pattern in their properties. By arranging the elements in order of increasing atomic mass into columns with similar properties, Mendeleev organized the elements into a periodic table. Mendeleev's table, shown in **Figure 2,** became widely accepted because he predicted the existence and properties of undiscovered elements that were later found. Mendeleev left blank spaces in the table where he thought the undiscovered elements should go. By noting trends in the properties of known elements, he was able to predict the properties of the yet-to-be-discovered elements scandium, gallium, and germanium.

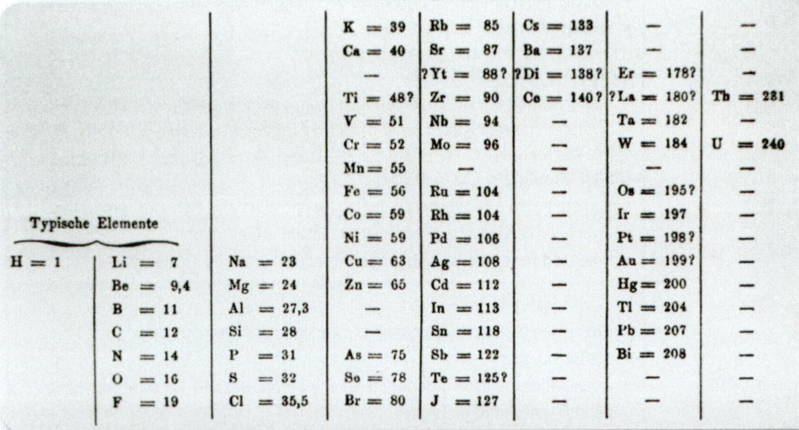

Typische Elemente							
			K = 39	Rb = 85	Cs = 133	—	—
			Ca = 40	Sr = 87	Ba = 137	—	—
			—	?Yt = 88?	?Di = 138?	Er = 178?	—
			Ti = 48?	Zr = 90	Ce = 140?	?La = 180?	Tb = 281
			V = 51	Nb = 94	—	Ta = 182	—
			Cr = 52	Mo = 96	—	W = 184	U = 240
			Mn = 55	—	—	—	—
			Fe = 56	Ru = 104	—	Os = 195?	—
			Co = 59	Rh = 104	—	Ir = 197	—
			Ni = 59	Pd = 106	—	Pt = 198?	—
H = 1	Li = 7	Na = 23	Cu = 63	Ag = 108	—	Au = 199?	—
	Be = 9,4	Mg = 24	Zn = 65	Cd = 112	—	Hg = 200	—
	B = 11	Al = 27,3	—	In = 113	—	Tl = 204	—
	C = 12	Si = 28	—	Sn = 118	—	Pb = 207	—
	N = 14	P = 31	As = 75	Sb = 122	—	Bi = 208	—
	O = 16	S = 32	Se = 78	Te = 125?	—	—	—
	F = 19	Cl = 35,5	Br = 80	J = 127	—	—	—

Figure 2 In the first version of his table, published in 1869, Mendeleev arranged elements with similar chemical properties horizontally. He left empty spaces for elements that were not yet discovered.

Build a Model

Periodic Table Have your class create a classroom periodic table that describes the characteristics of each of the students. Each student will be a cell in the classroom periodic table. Provide each student with a one-foot square poster board. In the center of the square, each student should write his or her two-letter student symbol (capped initial of last name followed by lower-case initial of first name). Pick student-based characteristics such as shoe size, hair color, sports interest, favorite school subject, and so on. Define a location on the card for each characteristic. Have each student complete his or her card; then, have the class determine an appropriate method for organizing the squares, so that, if possible, a periodic pattern is developed. OL EL

Extension

Have students research the biography of the main contributors to the development of the periodic table. They should include information on some of their other contributions to chemistry. OL AL

DIFFERENTIATED INSTRUCTION

Advanced Learners Mendeleev predicted properties of several elements that had not been discovered at the time he published his periodic table. Provide students with several of the properties he predicted, and have them research the properties of the actual element that was discovered. One example is ekasilicon, which was named gallium once it was discovered. Mendeleev predicted that ekasilicon would have an atomic mass of 68 amu, a low melting point, a density of 5.9 g/cm^3, and an oxide formula of Ea_2O_3. Have students evaluate how close Mendeleev's predictions were to the actual properties. Germanium has an atomic mass of 72.61 amu and a density of 5.60 g/cm^3. Ask students what they think the prefix *eka-* means. The prefix *eka-* means *one* in Sanskrit. Germanium was named ekasilicon originally because it was one place over from silicon in Mendeleev's periodic table. AL

Identify Misconceptions

Students often have difficulty understanding the vast amount of information that can be inferred by an element's position on the periodic table.

Uncover the Misconception First, have students create a list of phenomena or events that are periodic or cyclical. This list might include seasons of the year, phases of the moon, the school year, days of the week, octaves in music, and math functions, such as sine or cosine. Then, ask them to explain why the word periodic is appropriate to the periodic table.

Demonstrate the Concept Play a game of Twenty Questions or Who am I? Choose an element and tell students that they will be provided with clues (properties) about the element's identity. Start with several elements with familiar properties. Increase the difficulty of the game by selecting elements with less familiar properties. Suggest that there are too many elements and properties for scientists to remember. The periodic table helps scientists organize seemingly unrelated properties and chemical facts so that general trends can be recognized.

Assess New Knowledge Pick five elements that represent periodicity. Using a separate file card for each element, list several properties of the element. Leave one of the properties for one of the elements as an unknown. Have students organize the five cards and predict the missing information. Have them explain how they were able to predict the missing information. OL

READING CHECK Mendeleev organized the elements by increasing atomic mass. Moseley organized them by increasing atomic number.

Moseley Mendeleev's table, however, was not completely correct. After several new elements were discovered and the atomic masses of the known elements were more accurately determined, it became apparent that several elements in his table were not in the correct order. Arranging the elements by mass resulted in several elements being placed in groups of elements with differing properties.

The reason for this problem was determined in 1913 by English chemist Henry Moseley (1887–1915). As you might recall, Moseley discovered that atoms of each element contain a unique number of protons in their nuclei—the number of protons being equal to the atom's atomic number. By arranging the elements in order of increasing atomic number, the problems with the order of the elements in the periodic table were solved. Moseley's arrangement of elements by atomic number resulted in a clear periodic pattern of properties. The statement that there is a periodic repetition of chemical and physical properties of the elements when they are arranged by increasing atomic number is called the **periodic law.**

VOCABULARY
WORD ORIGIN
Periodic
comes from the Greek word *periodos,* meaning *way around, circuit*

READING CHECK **Compare and contrast** the ways in which Mendeleev and Moseley organized the elements.

Table 2 summarizes the contributions of Newlands, Meyer, Mendeleev, and Moseley to the development of the periodic table. The periodic table brought order to seemingly unrelated facts and became a significant tool for chemists. It is a useful reference for understanding and predicting the properties of elements and for organizing knowledge of atomic structure. Do the Problem-Solving Lab later in this chapter to see how the periodic law can be used to predict unknown elemental properties.

Table 2 Contributions to the Classification of Elements

John Newlands (1837–1898) • arranged elements by increasing atomic mass • noticed the repetition of properties every eighth element • created the law of octaves
Lothar Meyer (1830–1895) • demonstrated a connection between atomic mass and elements' properties • arranged the elements in order of increasing atomic mass
Dmitri Mendeleev (1834–1907) • demonstrated a connection between atomic mass and elements' properties • arranged the elements in order of increasing atomic mass • predicted the existence and properties of undiscovered elements
Henry Moseley (1887–1915) • discovered that atoms contain a unique number of protons called the atomic number • arranged elements in order of increasing atomic number, which resulted in a periodic pattern of properties

CHEMISTRY JOURNAL

Student News Correspondents Have students pretend they are newspaper reporters interviewing Mendeleev and Moseley. They should conduct some background research in order to be effective reporters. In addition to asking about their subjects' respective chemical discoveries, suggest that some of the interview questions relate to the subjects' experiences during the time period in which they lived, their educational backgrounds, and the countries in which they lived. OL

The Modern Periodic Table

The modern periodic table consists of boxes, each containing an element name, symbol, atomic number, and atomic mass. A typical box from the table is shown in **Figure 3.** The boxes are arranged in order of increasing atomic number into a series of columns, called **groups** or families, and rows, called **periods.** The table is shown in **Figure 5** on the next page and on the inside back cover of your textbook.

☑ READING CHECK **Define** *groups* and *periods.*

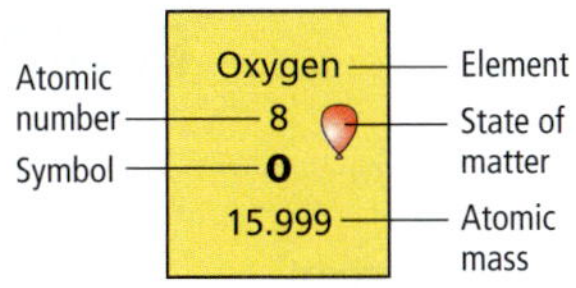

Figure 3 A typical box from the periodic table contains the element's name, its chemical symbol, its atomic number, its atomic mass, and its state.

Beginning with hydrogen in period 1, there are a total of seven periods. Each group is numbered 1 through 18. For example, period 4 contains potassium and calcium. Scandium (Sc) is in the third column from the left, which is group 3. Oxygen is in group 16. The elements in groups 1, 2, and 13 to 18 possess a wide range of chemical and physical properties. For this reason, they are often referred to as the main group, or **representative elements.** The elements in groups 3 to 12 are referred to as the **transition elements.** Elements are classified as metals, nonmetals, and metalloids.

Metals Elements that are generally shiny when smooth and clean, solid at room temperature, and good conductors of heat and electricity are called **metals.** Most metals are also malleable and ductile, meaning that they can be pounded into thin sheets and drawn into wires, respectively. Most representative elements and all transition elements are metals. If you look at boron (B) in column 13, you will see a heavy stairstep line that zigzags down to astatine (At) at the bottom of group 17. This stairstep line is a visual divider between the metals and the nonmetals on the table. In **Figure 5,** metals are represented by the blue boxes.

Alkali metals Except for hydrogen, all of the elements on the left side of the table are metals. The group 1 elements (except for hydrogen) are known as the **alkali metals.** Because they are so reactive, alkali metals usually exist as compounds with other elements. Two familiar alkali metals are sodium (Na), one of the components of salt, and lithium (Li), often used in batteries.

Alkaline earth metals The **alkaline earth metals** are in group 2. They are also highly reactive. Calcium (Ca) and magnesium (Mg), two minerals important for your health, are examples of alkaline earth metals. Because magnesium is solid and relatively light, it is used in the fabrication of electronic devices, such as the laptop shown in **Figure 4.**

Figure 4 Because magnesium is light and strong, it is often used in the production of electronic devices. For instance, this laptop case is made of magnesium.

Extension

The Discovery of Elements

Provide students with a list of element names and their dates of discovery, or have students research these dates. Give each student a set of colored pencils and a basic periodic table that contains only element names or symbols. Have students color each of the following sets of elements a different color.

- elements known by the year 100
- elements discovered from 101 to 1600
- elements discovered from 1601 to 1799
- elements discovered from 1800 through the time Mendeleev published his first periodic table (~1870)
- elements discovered from 1871 to 1980
- elements discovered since 1981

OL EL

ASSESSMENT

Knowledge Point out that the placement of several pairs of elements on Mendeleev's table were incorrect. Argon and potassium are one pair whose correct positions were reversed. Ask students to suggest reasons why Mendeleev used atomic mass instead of atomic number to organize the elements. The property of atomic number was not discovered until the early 1900s, thus making it impossible for Mendeleev to base his periodic table on it. OL

☑ **READING CHECK** groups: columns in the periodic table; periods: rows in the periodic table.

CHEMISTRY PROJECT

Group Names Have students research the history of current and past names used for sets of elements on the periodic table, such as alkali metals, alkaline earth metals, rare earth metals, halogens, and noble gases (inert gases). Students should consider whether the names accurately reflect properties that are now known of the elements. OL EL

Apply Chemistry

Gemstone Chemistry Gemstones often owe their color to atoms of transition elements that are substituted into a crystal structure. For example, consider the mineral corundum (Al_2O_3). If chromium atoms replace a few aluminum atoms in the crystal structure, the resulting crystal has a brilliant red color. This crystal is the gemstone known as a ruby. The substitution of iron atoms results in the gemstone known as topaz. The substitution of titanium atoms results in a sapphire. Another example occurs in the mineral known as beryl ($Be_3Al_2Si_6O_{18}$). If a few of the aluminum atoms in the crystal are replaced with chromium atoms, the result is a brilliant green emerald. The substitution of one atom for another occurs because the atoms have similar atomic radii and valence electrons.

Figure 5

PERIODIC TABLE OF THE ELEMENTS

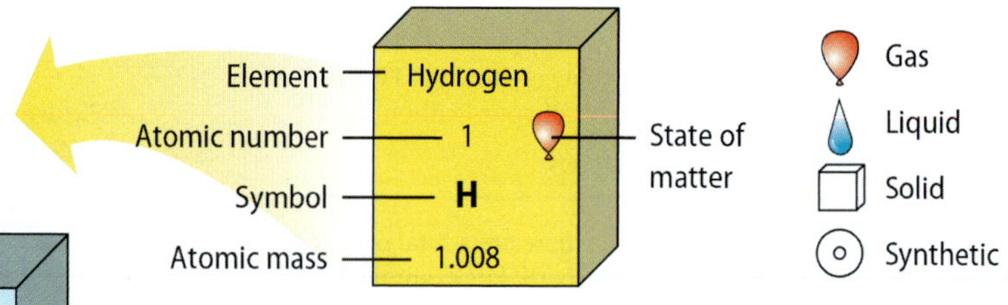

	1	2	3	4	5	6	7	8	9
1	Hydrogen 1 H 1.008								
2	Lithium 3 Li 6.941	Beryllium 4 Be 9.012							
3	Sodium 11 Na 22.990	Magnesium 12 Mg 24.305							
4	Potassium 19 K 39.098	Calcium 20 Ca 40.078	Scandium 21 Sc 44.956	Titanium 22 Ti 47.867	Vanadium 23 V 50.942	Chromium 24 Cr 51.996	Manganese 25 Mn 54.938	Iron 26 Fe 55.847	Cobalt 27 Co 58.933
5	Rubidium 37 Rb 85.468	Strontium 38 Sr 87.62	Yttrium 39 Y 88.906	Zirconium 40 Zr 91.224	Niobium 41 Nb 92.906	Molybdenum 42 Mo 95.94	Technetium 43 Tc (98)	Ruthenium 44 Ru 101.07	Rhodium 45 Rh 102.906
6	Cesium 55 Cs 132.905	Barium 56 Ba 137.327	Lanthanum 57 La 138.905	Hafnium 72 Hf 178.49	Tantalum 73 Ta 180.948	Tungsten 74 W 183.84	Rhenium 75 Re 186.207	Osmium 76 Os 190.23	Iridium 77 Ir 192.217
7	Francium 87 Fr (223)	Radium 88 Ra (226)	Actinium 89 Ac (227)	Rutherfordium 104 Rf (261)	Dubnium 105 Db (262)	Seaborgium 106 Sg (266)	Bohrium 107 Bh (264)	Hassium 108 Hs (277)	Meitnerium 109 Mt (268)

The number in parentheses is the mass number of the longest lived isotope for that element.

Lanthanide series	Cerium 58 Ce 140.115	Praseodymium 59 Pr 140.908	Neodymium 60 Nd 144.242	Promethium 61 Pm (145)	Samarium 62 Sm 150.36	Europium 63 Eu 151.965
Actinide series	Thorium 90 Th 232.038	Protactinium 91 Pa 231.036	Uranium 92 U 238.029	Neptunium 93 Np (237)	Plutonium 94 Pu (244)	Americium 95 Am (243)

CULTURAL DIVERSITY

International Tables Have students research and collect periodic tables from a variety of countries, such as Japan, Russia, Germany, Spain, and Mexico. Display these periodic tables and have students examine them for similarities and differences. For example, a Japanese periodic table uses the same symbol (Na) for sodium as an English table, but the element name is written in Kanji or Japanese calligraphy. If time permits, have students further investigate how the English translation for the element name compares to the English name for the element. While there are alternative forms of the periodic table, students will find that the shape of the periodic table is relatively universal.

View an **animation about the periodic table.**

Concepts In Motion

Explore **updates to the periodic table.**

Periodic Table

- Metal
- Metalloid
- Nonmetal
- Recently observed

10	11	12	13	14	15	16	17	18
								Helium 2 **He** 4.003
			Boron 5 **B** 10.811	Carbon 6 **C** 12.011	Nitrogen 7 **N** 14.007	Oxygen 8 **O** 15.999	Fluorine 9 **F** 18.998	Neon 10 **Ne** 20.180
			Aluminum 13 **Al** 26.982	Silicon 14 **Si** 28.086	Phosphorus 15 **P** 30.974	Sulfur 16 **S** 32.066	Chlorine 17 **Cl** 35.453	Argon 18 **Ar** 39.948
Nickel 28 **Ni** 58.693	Copper 29 **Cu** 63.546	Zinc 30 **Zn** 65.39	Gallium 31 **Ga** 69.723	Germanium 32 **Ge** 72.61	Arsenic 33 **As** 74.922	Selenium 34 **Se** 78.96	Bromine 35 **Br** 79.904	Krypton 36 **Kr** 83.80
Palladium 46 **Pd** 106.42	Silver 47 **Ag** 107.868	Cadmium 48 **Cd** 112.411	Indium 49 **In** 114.82	Tin 50 **Sn** 118.710	Antimony 51 **Sb** 121.757	Tellurium 52 **Te** 127.60	Iodine 53 **I** 126.904	Xenon 54 **Xe** 131.290
Platinum 78 **Pt** 195.08	Gold 79 **Au** 196.967	Mercury 80 **Hg** 200.59	Thallium 81 **Tl** 204.383	Lead 82 **Pb** 207.2	Bismuth 83 **Bi** 208.980	Polonium 84 **Po** 208.982	Astatine 85 **At** 209.987	Radon 86 **Rn** 222.018
Darmstadtium 110 **Ds** (281)	Roentgenium 111 **Rg** (272)	Copernicium 112 **Cn** (285)	Ununtrium * 113 **Uut** (284)	Flerovium 114 **Fl** (289)	Ununpentium * 115 **Uup** (288)	Livermorium 116 **Lv** (293)	Ununseptium * 117 **Uus** (294)	Ununoctium * 118 **Uuo** (294)

* The names and symbols for elements 113, 115, 117 and 118 are temporary. Final names will be approved by IUPAC (International Union of Pure and Applied Chemistry).

Gadolinium 64 **Gd** 157.25	Terbium 65 **Tb** 158.925	Dysprosium 66 **Dy** 162.50	Holmium 67 **Ho** 164.930	Erbium 68 **Er** 167.259	Thulium 69 **Tm** 168.934	Ytterbium 70 **Yb** 173.04	Lutetium 71 **Lu** 174.967
Curium 96 **Cm** (247)	Berkelium 97 **Bk** (247)	Californium 98 **Cf** (251)	Einsteinium 99 **Es** (252)	Fermium 100 **Fm** (257)	Mendelevium 101 **Md** (258)	Nobelium 102 **No** (259)	Lawrencium 103 **Lr** (262)

Quick Demo

Properties of Metals

Demonstrate to the class the malleability of several different metals. Using similar thicknesses of sheets of copper, tin, lead, and iron, have student volunteers test and compare the malleability with a hammer. **WARNING:** ***Students should wear safety goggles and be instructed on the safe use of the hammers.*** Have students rank the metals in order of decreasing malleability. BL EL

ASSESSMENT

Skill Display samples of elements that represent metals, nonmetals, and metalloids. Have students record their observations of each sample and classify it as a metal, nonmetal, or metalloid. If possible, students should identify each element and state its group. OL

Reinforcement

Skill Give each student a blank periodic table. Ask them to draw an arrow on their table showing the direction of increasing metallic properties across the table and another showing increasing metallic properties within a group. OL EL

DIFFERENTIATED INSTRUCTION

English Learners Have students create a dot-to-dot puzzle using chemical symbols as clues. Draw the outline of a piece of lab equipment, a lab setup, or some other chemistry-related picture using a dark felt tip marker. Place another sheet of paper over the first and mark dots along the image, especially at critical direction changes. Have students label the dots with chemical symbols instead of numbers. The atomic number of each chemical symbol represents the number of each dot (H is 1, He is 2, Li is 3, and so on.). Students should exchange and complete each other's puzzles. EL OL

Problem-Solving LAB

Purpose Students will use the periodic law to determine the melting point, boiling point, and heat of vaporization of francium.

Process Skills predict, identify variables, make and use graphs, interpret data, observe and infer, apply concepts

Teaching Strategies

- Ask students how the periodic law can be used to predict the melting point, boiling point, and heat of vaporization of francium. By plotting a graph of these properties versus atomic number for the known alkali metals, francium's values can be extrapolated from the graph.
- Explain why there is so little francium in Earth's crust. Francium-223 is the only naturally occurring isotope of the element. It is produced from the alpha decay of actinium-227, which itself is produced from the decay of uranium-238. For one ton of U-238, only 0.2 mg of Ac-227 is formed. Ac-227, with a half-life of 22 years, produces only 3.8×10^{-10} g of Fr-223. The unstable Fr-223 decays very quickly, having a half-life of only 22 minutes.

Think Critically

1. A graph of each property versus atomic number is the best approach. By extending the data curve through to francium's atomic number of 87, its radius, melting point, and boiling point can be determined. R = 270 pm, MP = 27°C, and BP ≈ 670°C.
2. Francium is probably a liquid at room temperature. Its melting point is 27°C, and should be close to this value according to the trend shown in the table.
3. The radius prediction is most inaccurate. The effect of the principal energy level on the radius is harder to extrapolate accurately because it varies from period to period.
4. Even 1 million atoms collected together as a solid are microscopic. For comparison, a grain of salt contains about 10^{15} sodium atoms.

Problem-Solving LAB

Analyze Trends

Francium—solid, liquid, or gas?
Francium was discovered in 1939, but its existence was predicted by Mendeleev in the 1870s. It is the least stable of the first 101 elements: Its most stable isotope has a half-life of just 22 minutes! Use your knowledge about the properties of other alkali metals to predict some of francium's properties.

Analysis

In the spirit of Dmitri Mendeleev's prediction of the properties of then-undiscovered elements, use the given information about the known properties of the alkali metals to devise a method for determining the corresponding property of francium.

Alkali Metals Data

Element	Melting Point (°C)	Boiling Point (°C)	Radius (pm)
Lithium	180.5	1342	152
Sodium	97.8	883	186
Potassium	63.4	759	227
Rubidium	39.30	688	248
Cesium	28.4	671	265
Francium	?	?	?

Think Critically

1. **Devise** an approach that clearly displays the trends for each of the properties given in the table and allows you to extrapolate a value for francium. Use the periodic law as a guide.
2. **Predict** whether francium is a solid, a liquid, or a gas. How can you support your prediction?
3. **Infer** which column of data presents the greatest possible error in making a prediction. Explain.
4. **Determine** why producing 1 million francium atoms per second is not enough to make measurements, such as density or melting point.

Transition and inner transition metals The transition elements are divided into **transition metals** and **inner transition metals.** The two sets of inner transition metals, known as the **lanthanide series** and **actinide series,** are located along the bottom of the periodic table. The rest of the elements in groups 3 to 12 make up the transition metals. Elements from the lanthanide series are used extensively as phosphors, substances that emit light when struck by electrons. Because it is strong and light, the transition metal titanium is used to make frames for bicycles and eyeglasses.

Connection to Biology **Nonmetals** Nonmetals occupy the upper-right side of the periodic table. They are represented by the yellow boxes in **Figure 5.** **Nonmetals** are elements that are generally gases or brittle, dull-looking solids. They are poor conductors of heat and electricity. The only nonmetal that is a liquid at room temperature is bromine (Br). The most abundant element in the human body is the nonmetal oxygen, which constitutes 65% of the body mass.

Group 17 is comprised of highly reactive elements that are known as **halogens.** Like the group 1 and group 2 elements, the halogens are often part of compounds. Compounds made with the halogen fluorine (F) are commonly added to toothpaste and drinking water to prevent tooth decay. The extremely unreactive group 18 elements are commonly called the **noble gases** and are used in lasers, a variety of light bulbs, and neon signs.

VOCABULARY
SCIENCE USAGE VS. COMMON USAGE
Conductor
Science usage: a substance or body capable of transmitting electricity, heat, or sound
Copper is a good conductor of heat.
Common usage: a person who conducts an orchestra, chorus, or other group of musical performers
The new conductor helped the orchestra perform at its best.

■ **Figure 6** Scientists developing submarine technology created this robot that looks and swims like a real fish. Its body is made of a silicon resin that softens in water.

Metalloids The elements in the green boxes bordering the stairstep line in **Figure 5** are called metalloids, or semimetals. **Metalloids** have physical and chemical properties of both metals and nonmetals. Silicon (Si) and germanium (Ge) are two important metalloids, used extensively in computer chips and solar cells. Silicon is also used to make prosthetics or in lifelike applications, as shown in **Figure 6.**

This introduction to the periodic table touches only the surface of its usefulness. You can refer to the Elements Handbook at the end of your textbook to learn more about the elements and their various groups.

SECTION 1 REVIEW

Section Self-Check

Section Summary

- The elements were first organized by increasing atomic mass, which led to inconsistencies. Later, they were organized by increasing atomic number.
- The periodic law states that when the elements are arranged by increasing atomic number, there is a periodic repetition of their chemical and physical properties.
- The periodic table organizes the elements into periods (rows) and groups or families (columns); elements with similar properties are in the same group.
- Elements are classified as either metals, nonmetals, or metalloids.

1. **MAIN**IDEA **Describe** the development of the modern periodic table. Include contributions made by Lavoisier, Newlands, Mendeleev, and Moseley.
2. **Sketch** a simplified version of the periodic table, and indicate the location of metals, nonmetals, and metalloids.
3. **Describe** the general characteristics of metals, nonmetals, and metalloids.
4. **Identify** each of the following as a representative element or a transition element.
   a. lithium (Li) **b.** platinum (Pt) **c.** promethium (Pm) **d.** carbon (C)
5. **Compare** For each of the given elements, list two other elements with similar chemical properties.
   a. iodine (I) **b.** barium (Ba) **c.** iron (Fe)
6. **Compare** According to the periodic table, which two elements have an atomic mass less than twice their atomic number?
7. **Interpret Data** A company plans to make an electronic device. They need to use an element that has chemical behavior similar to that of silicon (Si) and lead (Pb). The element must have an atomic mass greater than that of sulfur (S), but less than that of cadmium (Cd). Use the periodic table to determine which element the company could use.

ChemLAB

The ChemLab located at the end of the chapter can be used at this point in the lesson.

3 Assess

Check for Understanding

Have students write a paragraph using each of the vocabulary terms listed for the section. OL

Reteach

Organize students into pairs. Have one student select an element that the other student must classify using the terms introduced in this section. Have students reverse roles so that both students have the opportunity to respond. OL COOP LEARN

Extension

Have students predict the properties for element 117. They should offer supporting reasons for their predictions. OL

SECTION 1 REVIEW

1. Lavoisier organized a list of the known elements of his day as four categories. Newlands was the first to organize the elements and show that properties repeated in a periodic way. Mendeleev and Meyer proposed periodic tables showing a relationship between atomic mass and elemental properties. Moseley organized the elements by atomic number instead of atomic mass.
2. Simplified tables should resemble Figure 5 with the groups and periods labeled. Refer to the Solutions Manual.
3. metals: shiny, ductile, malleable, good conductors of heat and electricity; nonmetals: dull, brittle, poor conductors of heat and electricity; metalloids: properties of both metals and nonmetals
4. **a.** representative; **b.** transition; **c.** transition; **d.** representative
5. **a.** any other group 17 element; **b.** any other group 2 element; **c.** any other group 8 element
6. hydrogen and oxygen
7. germanium (Ge)

SECTION 2

1 Focus

MAINIDEA

Organizing by Blocks Have students think about the last time they went shopping in a department store. Ask them how they knew where to find the items they wanted. The store is divided into departments. What are some of the departments? shoes, clothes for teen girls, clothes for teen boys, household items, and so on Point out that chemical elements are also arranged into departments called *groups* on the periodic table. OL

2 Teach

Concept Development

Electron Configuration

Emphasize that electron configuration is a periodic trend and that it determines an element's chemical properties.

SECTION 2

Essential Questions

- Why do elements in the same group have similar properties?
- Based on their electron configurations, what are the four blocks of the periodic table?

Review Vocabulary

valence electron: electron in an atom's outermost orbital; determines the chemical properties of an atom

Classification of the Elements

MAINIDEA Elements are organized into different blocks in the periodic table according to their electron configurations.

CHEM 4 YOU A house number is not enough to deliver a letter to the correct address. More information, such as street name, city, and state, is necessary to deliver the letter. Similarly, chemical elements are identified according to details about the arrangement of their electrons.

Organizing the Elements by Electron Configuration

As you learned previously, electron configuration determines the chemical properties of an element. Writing out electron configurations using the aufbau diagram can be tedious. Fortunately, you can determine an atom's electron configuration and its number of valence electrons from its position on the periodic table. The electron configurations for some of the group 1 elements are listed in **Table 3.** All four configurations have a single electron in their outermost orbitals.

Valence electrons Recall that electrons in the highest principal energy level of an atom are called valence electrons. Each of the group 1 elements has one electron in its highest energy level; thus, each element has one valence electron. The group 1 elements have similar chemical properties because they all have the same number of valence electrons. This is one of the most important relationships in chemistry; atoms in the same group have similar chemical properties because they have the same number of valence electrons. Each group 1 element has a valence electron configuration of s^1. Each group 2 element has a valence electron configuration of s^2. Each column in groups 1, 2, and 13 to 18 on the periodic table has its own valence electron configuration.

Valence electrons and period The energy level of an element's valence electrons indicates the period on the periodic table in which it is found. For example, lithium's valence electron is in the second energy level and lithium is found in period 2. Now look at gallium, with its electron configuration of $[Ar]4s^23d^{10}4p^1$. Gallium's valence electrons are in the fourth energy level, and gallium is found in the fourth period.

Table 3 Electron Configuration for the Group 1 Elements

Period 1	hydrogen	$1s^1$	$1s^1$
Period 2	lithium	$1s^22s^1$	$[He]2s^1$
Period 3	sodium	$1s^22s^22p^63s^1$	$[Ne]3s^1$
Period 4	potassium	$1s^22s^22p^63s^23p^64s^1$	$[Ar]4s^1$

DIFFERENTIATED INSTRUCTION

Visually Impaired Have a visually impaired student work with a sighted student to study the organization of elements on the periodic table by electron configuration. Have the sighted student construct a model of an s-block element by cutting a two-cup section from an egg carton. The visually impaired student can model a group 1 element by placing a marble in one cup, or a group 2 element by placing a marble in each of the two cups. Models for p, d, and f-group elements can similarly be made and described by both students. OL BL

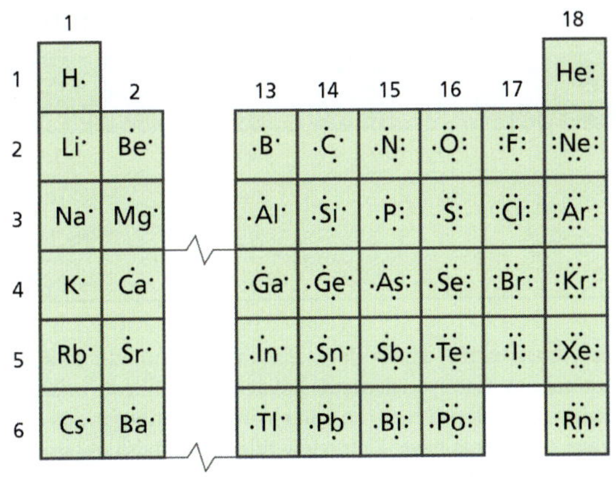

■ **Figure 7** The figure shows the electron-dot structure of most representative elements.

Observe ***How does the number of valence electrons vary within a group?***

Valence electrons of the representative elements

Elements in group 1 have one valence electron; group 2 elements have two valence electrons. Group 13 elements have three valence electrons, group 14 elements have four, and so on. The noble gases in group 18 each have eight valence electrons, with the exception of helium, which has only two valence electrons. **Figure 7** shows how the electron-dot structures you studied previously illustrate the connection between group number and number of valence electrons. Notice that the number of valence electrons for the elements in groups 13 to 18 is ten less than their group number.

The s-, p-, d-, and f-Block Elements

The periodic table has columns and rows of varying sizes. The reason behind the table's odd shape becomes clear if it is divided into sections, or blocks, representing the atom's energy sublevel being filled with valence electrons. Because there are four different energy sublevels (s, p, d, and f), the periodic table is divided into four distinct blocks, as shown in **Figure 8.**

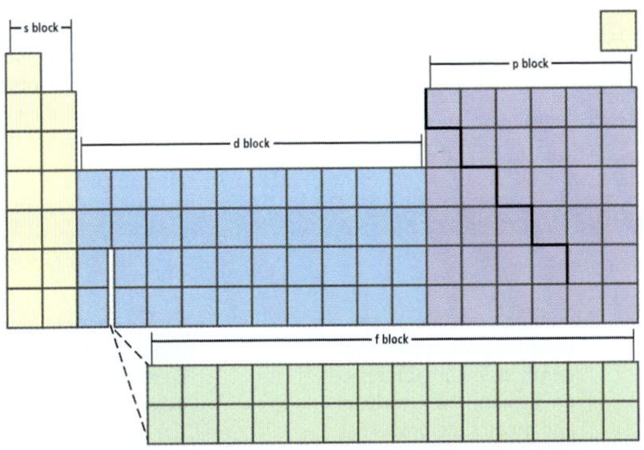

■ **Figure 8** The periodic table is divided into four blocks—s, p, d, and f.

Analyze ***What is the relationship between the maximum number of electrons an energy sublevel can hold and the number of columns in that block on the diagram?***

ASSESSMENT

Knowledge Both helium and beryllium have two valence electrons. One of the elements is relatively reactive; the other is nonreactive. Have students explain the difference in chemical reactivity in spite of the similar electron configurations. OL

■ **Caption Question Fig. 7**
It remains the same.

■ **Caption Question Fig. 8**
The number of columns in the block is equal to the maximum number of electrons the energy sublevel can hold.

DIFFERENTIATED INSTRUCTION

Hearing Impaired Assign each student a representative element. Have them construct a three-dimensional cube using poster board and artistically label one side of the cube with the name of the element. Another side of the cube should show the element's group, while another side provides the element's electron configuration. They should use the remaining sides to illustrate properties of the element, its uses, and some of its important compounds. Hang these cubes from the ceiling in the classroom. BL EL

Concept Development

Scientific Knowledge Most scientific innovations, like the development of the periodic table, build on knowledge gained by many scientists over time. Tell students to pair with another student to brainstorm an invention or technology you use regularly. What critical knowledge or concepts had to be in place to develop the technology? If one concept had not been understood, was developed later, or was not applied, how might the invention or technology be different? While answers will vary based on the technology selected, students should understand that acquiring scientific knowledge is a building process that incorporates the work of many over time. **AL** **OL**

Content Background

IUPAC The International Union of Pure and Applied Chemistry (IUPAC) was formed in 1919 by chemists from both industry and the academic world. Group goals included standardizing weights, measures, names, and symbols, and fostering international scientific communication. During the Cold War, a period of global tension that lasted from the end of World War II until the fall of the Soviet Union in 1991, IUPAC played a key role in maintaining technical communication among scientists worldwide. One of the current roles of IUPAC is approving names of newly discovered elements, which can be named after a mythological concept, a mineral, a place or country, a property, or a scientist.

Explore **noble gas configuration** with an **interactive table**. Concepts In Motion

Table 4 Noble Gas Electron Configuration

Period	Principal Energy Level	Element	Electron Configuration
1	$n = 1$	helium	$1s^2$
2	$n = 2$	neon	$[He]2s^22p^6$
3	$n = 3$	argon	$[Ne]3s^23p^6$
4	$n = 4$	krypton	$[Ar]4s^23d^{10}4p^6$

Watch a **video about elements and the periodic table.** Video

VOCABULARY

ACADEMIC VOCABULARY

Structure
something made up of more-or-less interdependent elements or parts
Many scientists were involved in the discovery of the structure of the atom.

s-Block elements The s-block consists of groups 1 and 2, and the element helium. Group 1 elements have partially filled s orbitals containing one valence electron and electron configurations ending in s^1. Group 2 elements have completely filled s orbitals containing two valence electrons and electron configurations ending in s^2. Because s orbitals hold two electrons at most, the s-block spans two groups.

p-Block elements After the s sublevel is filled, the valence electrons next occupy the p sublevel. The p-block, comprised of groups 13 through 18, contains elements with filled or partially filled p orbitals. There are no p-block elements in period 1 because the p sublevel does not exist for the first principal energy level ($n = 1$). The first p-block element is boron (B), in the second period. The p-block spans six groups because the three p orbitals can hold a maximum of six electrons. The group 18 elements (noble gases) are unique members of the p-block. Their atoms are so stable that they undergo virtually no chemical reactions. The electron configurations of the first four noble gas elements are shown in **Table 4.** Here, both the s and p orbitals corresponding to the period's principal energy level are completely filled. This arrangement of electrons results in an unusually stable atomic structure. Together, the s- and p-blocks comprise the representative elements.

■ **Figure 9**
History of the Periodic Table
The modern periodic table is the result of the work of many scientists over the centuries who studied elements and discovered periodic patterns in their properties.

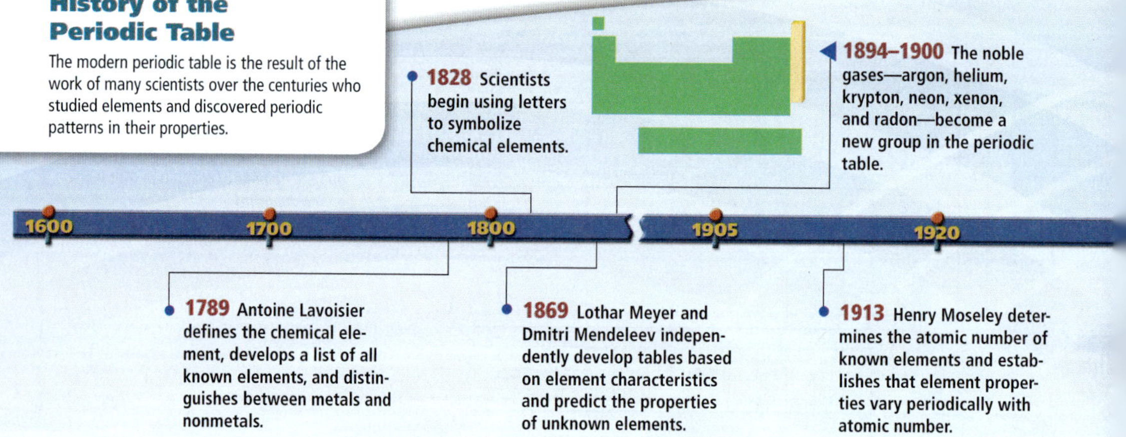

CHEMISTRY JOURNAL

Noble Gases Ask students to research how and when the noble gases were discovered. They should also research the uses of noble gases, their relative reactivities, and if they form any compounds. Students should write a paragraph describing why the noble gases belong on the far right-hand side of the periodic table. Have them add their research and paragraphs to their journals. **OL**

What's CHEMISTRY Got To Do With It?

Video

It's Elemental

d-Block elements The d-block contains the transition metals and is the largest of the blocks. Although there are a number of exceptions, d-block elements are usually characterized by a filled outermost s orbital of energy level *n*, and filled or partially filled d orbitals of energy level $n-1$. As you move across a period, electrons fill the d orbitals. For example, scandium (Sc), the first d-block element, has an electron configuration of $[Ar]4s^2 3d^1$. Titanium, the next element on the table, has an electron configuration of $[Ar]4s^2 3d^2$. Note that titanium's filled outermost s orbital has an energy level of $n = 4$, while the d orbital, which is partially filled, has an energy level of $n = 3$. As you learned previously, the aufbau principle states that the 4s orbital has a lower energy level than the 3d orbital. Therefore, the 4s orbital is filled before the 3d orbital. The five d orbitals can hold a total of ten electrons; thus, the d-block spans ten groups on the periodic table.

f-Block elements The f-block contains the inner transition metals. Its elements are characterized by a filled, or partially filled outermost s orbital, and filled or partially filled 4f and 5f orbitals. The electrons of the f sublevel do not fill their orbitals in a predictable manner. Because there are seven f orbitals holding up to a maximum of 14 electrons, the f-block spans 14 columns of the periodic table.

Therefore, the s-, p-, d-, and f-blocks determine the shape of the periodic table. As you proceed down through the periods, the principal energy level increases, as does the number of orbitals containing electrons. Note that period 1 contains only s-block elements, periods 2 and 3 contain both s- and p-block elements, periods 4 and 5 contain s-, p-, and d-block elements, and periods 6 and 7 contain s-, p-, d-, and f-block elements.

The development of the periodic table took many years and is still an ongoing project as new elements are synthetized. Refer to **Figure 9** to learn more about the history of the periodic table and the work of the many scientists who contributed to its development.

☑ READING CHECK **Summarize** how each block of the periodic table is defined.

CAREERS IN CHEMISTRY

Research Chemist Some nuclear chemists specialize in studying the newest and heaviest elements. To produce heavy elements, a nuclear chemist works with a large team, including physicists, engineers, and technicians. Heavy elements are produced by collisions in a particle accelerator. The nuclear chemist analyzes the data from these collisions to identify the elements and understand their properties.

WebQuest

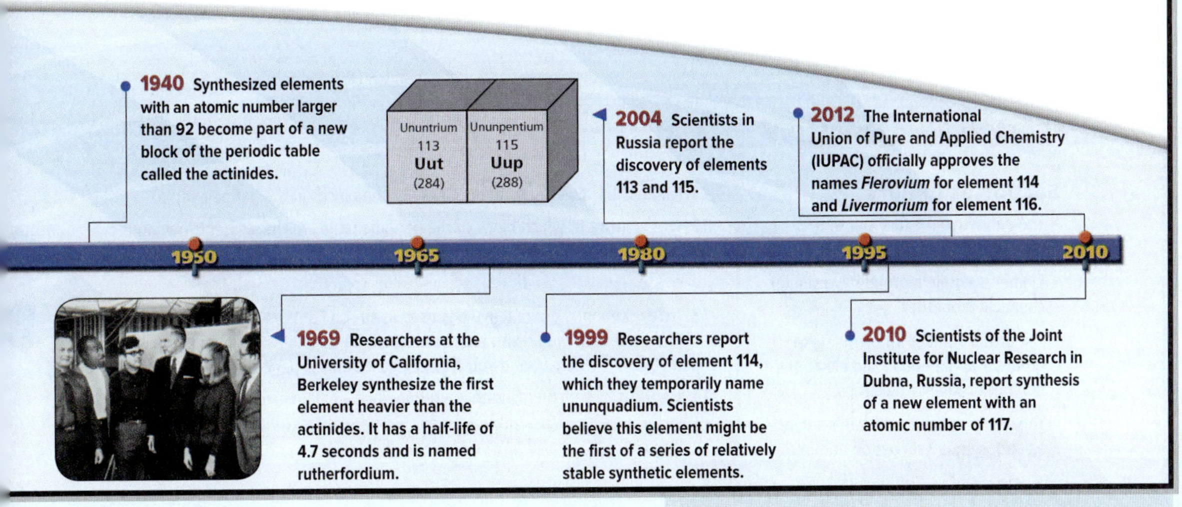

ASSESSMENT

Skill A student is given clues about three elements. One of the elements is an alkali metal, one is from the carbon family, and one is from the oxygen family.

- Element A is metallic, shiny, a good conductor, and reacts slowly with HCl to form H_2 gas.
- Element B is a yellow solid and is not a good conductor.
- Element C exhibits a metallic luster, conducts electricity, and forms a white powder when exposed to air.

Have students propose identities for each element, support their choices, and write electron configurations for each. OL

A: most likely lithium because it is an alkali metal that, in its group, reacts the slowest with HCl; $[He]2s^1$

B: must be sulfur because of its yellow color, nonconductive properties, and the fact that it is in the same group as oxygen; $[Ne]3s^2 3p^4$

C: must be lead because it is gray, malleable, and in the same group as carbon; $[Xe]6s^2 4f^{14} 5d^{10} 6p^2$

CAREERS IN CHEMISTRY

WebQuest

Apply Chemistry

Synthetic Elements Students might wonder why scientists continue trying to produce new synthetic elements. Part of the reason is for the sake of research itself, but occasionally the newly synthesized radioactive elements do find commercial applications. One example is americium, element 95, discovered in 1944. Americium is used in smoke detectors and other high-precision measuring devices. Scientists are also hoping that element 96, curium, could eventually have commercial applications as a small, portable, power generator.

☑ READING CHECK They are defined according to the energy sublevels that are being filled.

CHEMISTRY PROJECT

The Newest Elements Have students research some of the newest elements on the periodic table (ununquadium, ununhexium, ununtrium, and ununpentium) and an element discovered but not yet confirmed (ununseptium). Students should include details about when and where the elements were discovered as well as the status of naming the elements. OL

IN-CLASS Example

Question Without using the periodic table, determine the group, period, and block of atoms with the following electron configurations:

1. $[Kr]5s^1$
2. $[Ar] 4s^2 3d^{10} 4p^5$
3. $[He]2s^2 2p^5$

Answer

1. group 1, period 5, s-block
2. group 17, period 4, p-block
3. group 17, period 2, p-block

PRACTICE Problems

Have students refer to the Selected Solutions appendix for complete solutions to odd-numbered problems. The complete solutions for all problems can be found in the Solutions Manual.

8. a. group 2; period 3; block s
   b. group 2; period 2; block s
   c. group 17; period 5; block p
9. a. Sc, Y, La, Ac
   b. N, P, As, Sb, Bi
   c. Ne, Ar, Kr, Xe, Rn
10. a. $1s^2 2s^2 2p^6 3s^2 3p^6 4s^2$
    b. $1s^2 2s^2 2p^6 3s^2 3p^6 4s^2 3d^{10}$
    c. $1s^2 2s^2 2p^6 3s^2 3p^6 4s^2 3d^{10} 4p^6 5s^2 4d^{10} 5p^6$
    d. $1s^2 2s^2 2p^4$

3 Assess

Check for Understanding

Give blank periodic tables to students and have them label the periods, groups, and blocks. OL EL

Reteach

Have student groups make four flashcards. Each card should identify an element by its period number, group number, valence electrons, and family name, respectively. Student groups should challenge one another to identify the elements. OL COOP LEARN

Extension

Have students brainstorm ways to help them recall element names and numbers. Explain to them that associations with interesting ideas, items, or concepts can help students remember important information. OL

EXAMPLE Problem 1

EXAMPLE PROBLEM

ELECTRON CONFIGURATION AND THE PERIODIC TABLE Strontium, which is used to produce red fireworks, has an electron configuration of $[Kr]5s^2$. Without using the periodic table, determine the group, period, and block of strontium.

1 ANALYZE THE PROBLEM

You are given the electron configuration of strontium.

Known	**Unknown**
Electron configuration = $[Kr]5s^2$	Group = ?
	Period = ?
	Block = ?

2 SOLVE FOR THE UNKNOWN

The s^2 indicates that strontium's valence electrons fill the s sublevel. Thus, strontium is in group 2 of the **s-block.**

For representative elements, the number of valence electrons can indicate the group number.

The 5 in $5s^2$ indicates that strontium is in **period 5.**

The number of the highest energy level indicates the period number.

3 EVALUATE THE ANSWER

The relationships between electron configuration and position on the periodic table have been correctly applied.

PRACTICE Problems

Do additional problems. Online Practice

PRACTICE PROBLEMS

8. Without using the periodic table, determine the group, period, and block of an atom with the following electron configurations.
   a. $[Ne]3s^2$ b. $[He]2s^2$ c. $[Kr]5s^2 4d^{10} 5p^5$
9. What are the symbols for the elements with the following valence electron configurations?
   a. $s^2 d^1$ b. $s^2 p^3$ c. $s^2 p^6$
10. **Challenge** Write the electron configuration of the following elements.
    a. the group 2 element in the fourth period
    b. the group 12 element in the fourth period
    c. the noble gas in the fifth period
    d. the group 16 element in the second period

SECTION 2 REVIEW

Section Self-Check

Section Summary

- The periodic table has four blocks (s, p, d, f).
- Elements within a group have similar chemical properties.
- The group number for elements in groups 1 and 2 equals the element's number of valence electrons.
- The energy level of an atom's valence electrons equals its period number.

11. **MAIN IDEA Explain** what determines the blocks in the periodic table.
12. **Determine** in which block of the periodic table are the elements having the following valence electron configurations.
    a. $s^2 p^4$ b. s^1 c. $s^2 d^1$ d. $s^2 p^1$
13. **Infer** Xenon, a nonreactive gas used in strobe lights, is a poor conductor of heat and electricity. Would you expect xenon to be a metal, a nonmetal, or a metalloid? Where would you expect it to be on the periodic table? Explain.
14. **Explain** why elements within a group have similar chemical properties.
15. **Model** Make a simplified sketch of the periodic table, and label the s-, p-, d-, and f-blocks.

SECTION 2 REVIEW

11. The energy sublevel being filled defines the blocks in the periodic table.
12. a. p-block
    b. s-block
    c. d-block
    d. p-block
13. nonmetal; The nonreactive gases are noble gases in group 18 on the right side of the periodic table.
14. because they have the same valence electron configuration
15. Sketches should look similar to Figure 8. Refer to the Solutions Manual for a sample sketch.

Essential Questions

- What are the period and group trends of different properties?
- How are period and group trends in atomic radii related to electron configuration?

Review Vocabulary

principal energy level: the major energy level of an atom

New Vocabulary

ion
ionization energy
octet rule
electronegativity

Periodic Trends

MAINIDEA Trends among elements in the periodic table include their sizes and their abilities to lose or attract electrons.

CHEM 4 YOU A calendar is a useful tool for keeping track of activities. The pattern of days, from Sunday to Saturday, is repeated week after week. If you list an activity many weeks ahead, you can tell from the day of the week what else might happen on that day. In much the same way, the organization of the periodic table tells us about the behavior of many of the elements.

Atomic Radius

Many properties of the elements tend to change in a predictable way, known as a trend, as you move across a period or down a group. Atomic size is one such periodic trend. The sizes of atoms are influenced by electron configuration.

Recall that the electron cloud surrounding a nucleus does not have a clearly defined edge. The outer limit of an electron cloud is defined as the spherical surface within which there is a 90% probability of finding an electron. However, this surface does not exist in a physical way, as the outer surface of a golf ball does. Atomic size is defined by how closely an atom lies to a neighboring atom. Because the nature of the neighboring atom can vary from one substance to another, the size of the atom itself also tends to vary somewhat from substance to substance.

For metals such as sodium, the atomic radius is defined as half the distance between adjacent nuclei in a crystal of the element, as shown in **Figure 10.** For elements that commonly occur as molecules, such as many nonmetals, the atomic radius is defined as half the distance between nuclei of identical atoms that are chemically bonded together. The atomic radius of a nonmetal diatomic hydrogen molecule (H_2) is shown in **Figure 10.**

Figure 10 Atomic radii depend on the type of bonds that atoms form.

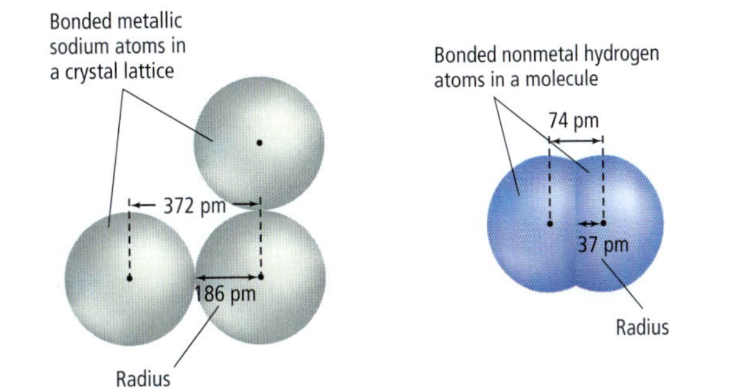

The radius of a metal atom is one-half the distance between two adjacent atoms in the crystal.

The radius of a nonmetal atom is often determined from a molecule of two identical atoms.

SECTION 3

1 Focus

MAINIDEA

Trends Show students a group of ten different-sized balls of modeling clay made of various colors. Ask them to describe two ways they could arrange the balls. by size or by color Point out that size of the atoms is one way elements are arranged on the periodic table. Other properties are also important. BL

2 Teach

Concept Development

Periodic Trends Help students develop the concept of periodic trends by providing them with opportunities to plot specific data and analyze the resulting graphs. For instance, they can plot the phases of the moon as a function of days. OL

DIFFERENTIATED INSTRUCTION

Struggling Learners Draw an atomic model of sodium on the board. The atom should have eleven protons in the nucleus, two electrons in the first shell, eight electrons in the second shell, and one electron in the third shell. Ask students what happens to the size of the atom if the outermost valence electron is removed. Have students draw atomic models of other elements within the same group or period. Make sure they understand that as they move across a period, the increases in the nuclear charges have greater impact on the atomic radii than the increasing numbers of electrons around the nuclei. This results in the trend of decreasing atomic radii. BL EL

☑ READING CHECK The nuclear charge increases, while electrons are being added to the same energy level. Because the amount of shielding remains the same, the nucleus pulls more strongly on the valence electrons, which decreases the radius.

Caption Question Fig. 11 Electrons occupy larger, higher-energy orbitals; inner core electrons shield valence electrons from the increased charge in the nucleus.

ASSESSMENT

Skill Students often have difficulty interpreting graphs and generalizing the information.

- Provide students with several graphs and ask them to write several sentences describing any trends they see in the data. The graphs they analyze do not have to be of periodic trends; select a wide variety of graphs, such as business trends, seasonal temperature, population over time, crop yields, and rainfall.
- Select a new scientific graph and ask students specific questions that require them to read and analyze data points on the graph. Ask them to summarize any trends shown in the graph. Then, ask a question that requires them to apply the trends in order to answer. OL

FOLDABLES® Rubric

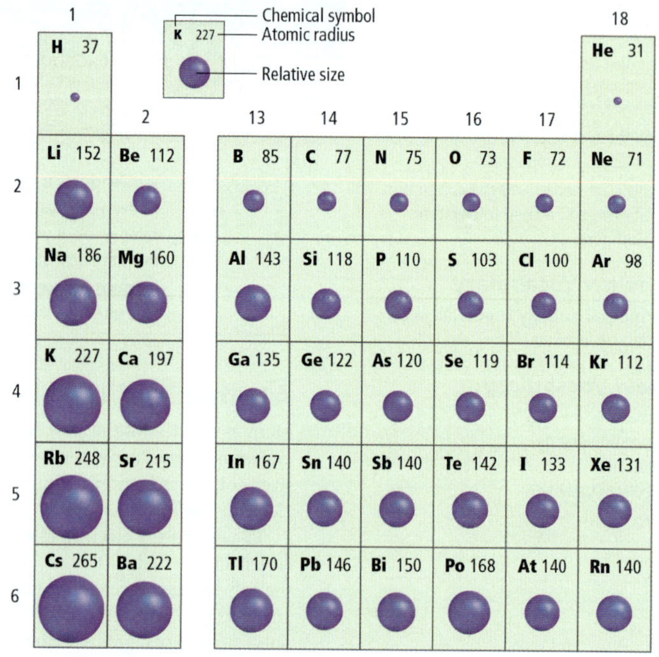

■ **Figure 11** The atomic radii of the representative elements, given in picometers (10^{-12} m), vary as you move from left to right within a period and down a group.

Infer ***why the atomic radii increase as you move down a group.***

View an **animation about trends in atomic radii.**

Concepts In Motion

FOLDABLES® Incorporate information from this section into your Foldable.

Trends within periods In general, there is a decrease in atomic radii as you move from left to right across a period. This trend, shown in **Figure 11,** is caused by the increasing positive charge in the nucleus and the fact that the principal energy level within a period remains the same. Each successive element has one additional proton and electron, and each additional electron is added to orbitals corresponding to the same principal energy level. Moving across a period, no additional electrons come between the valence electrons and the nucleus. Thus, the valence electrons are not shielded from the increased nuclear charge, which pulls the outermost electrons closer to the nucleus.

☑ READING CHECK Discuss how the fact that the principal energy level remains the same within a period explains the decrease in the atomic radii across a period.

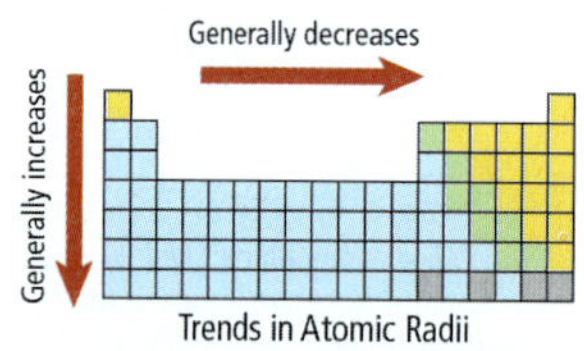

■ **Figure 12** Atomic radii generally decrease from left to right in a period and generally increase as you move down a group.

Trends within groups Atomic radii generally increase as you move down a group. The nuclear charge increases, and electrons are added to orbitals corresponding to successively higher principal energy levels. However, the increased nuclear charge does not pull the outer electrons toward the nucleus to make the atom smaller.

Moving down a group, the outermost orbital increases in size along with the increasing principal energy level; thus, the atom becomes larger. The larger orbital means that the outer electrons are farther from the nucleus. This increased distance offsets the pull of the increased nuclear charge. Also, as additional orbitals between the nucleus and the outer electrons are occupied, these electrons shield the outer electrons from the nucleus. **Figure 12** summarizes the group and period trends.

CHEMISTRY PROJECT

Comparison of Atomic Radii Have students work in small groups to make scaled models of the atoms using the atomic radii shown in Figure 11. Models might be made using modeling clay or foam sheets. Some students may prefer to use graphics software to model the atoms. OL COOP LEARN

EXAMPLE Problem 2

INTERPRET TRENDS IN ATOMIC RADII Which has the largest atomic radius: carbon (C), fluorine (F), beryllium (Be), or lithium (Li)? Answer without referring to **Figure 11.** Explain your answer in terms of trends in atomic radii.

1 ANALYZE THE PROBLEM

You are given four elements. First, determine the groups and periods the elements occupy. Then apply the general trends in atomic radii to determine which has the largest atomic radius.

2 SOLVE FOR THE UNKNOWN

From the periodic table, all the elements are found to be in period 2. — *Determine the periods.*

Ordering the elements from left-to-right across the period yields: Li, Be, C, and F.

The first element in period 2, lithium, has the largest radius. — *Apply the trend of decreasing radii across a period.*

3 EVALUATE THE ANSWER

The period trend in atomic radii has been correctly applied. Checking radii values in **Figure 11** verifies the answer.

PRACTICE Problems

Do additional problems. Online Practice

Answer the following questions using your knowledge of group and period trends in atomic radii. Do not use the atomic radii values in Figure 11 to answer the questions.

16. Which has the largest atomic radius: magnesium (Mg), silicon (Si), sulfur (S), or sodium (Na)? The smallest?

17. The figure on the right shows helium, krypton, and radon. Which one is krypton? How can you tell?

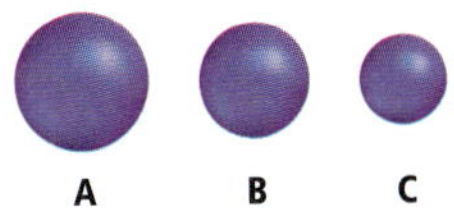

18. Can you determine which of two unknown elements has the larger radius if the only known information is that the atomic number of one of the elements is 20 greater than the other? Explain.

19. Challenge Determine which element in each pair has the largest atomic radius:

a. the element in period 2, group 1; or the element in period 3, group 18

b. the element in period 5, group 2; or the element in period 3, group 16

c. the element in period 3, group 14; or the element in period 6, group 15

d. the element in period 4, group 18; or the element in period 2, group 16

Ionic Radius

Atoms can gain or lose one or more electrons to form ions. Because electrons are negatively charged, atoms that gain or lose electrons acquire a net charge. Thus, an **ion** is an atom or a bonded group of atoms that has a positive or negative charge. You will learn about ions later, but for now, consider how the formation of an ion affects the size of an atom.

When atoms lose electrons and form positively charged ions, they always become smaller. The reason is twofold. The electron lost from the atom will almost always be a valence electron. The loss of a valence electron can leave a completely empty outer orbital, which results in a smaller radius. Furthermore, the electrostatic repulsion between the now-fewer number of remaining electrons decreases. As a result, they experience a greater nuclear charge allowing these remaining electrons to be pulled closer to the positively charged nucleus.

IN-CLASS Example

Question Which has the largest radius: sodium (Na), phosphorus (P), neon (Ne), or rubidium (Rb)?

Answer rubidium

PRACTICE Problems

Have students refer to the Selected Solutions appendix for complete solutions to odd-numbered problems. The complete solutions for all problems can be found in the Solutions Manual.

16. largest: Na; smallest: S

17. B. The atomic radius increases when going down a group, so helium is the smallest and radon is the biggest.

18. No. If all that is known is that the atomic number of one element is 20 greater than that of the other, then you will be unable to determine the specific groups and periods that the elements are in. Without this information, you cannot apply the periodic trends in atomic size to determine which element has the larger radius.

19. a. element in period 2, group 1

b. element in period 5, group 2

c. element in period 6, group 15

d. element in period 4, group 18

CHEMISTRY JOURNAL

Comparison of Ionic Radii Ask students to consider the electronic structure of the ions O^{2-} and Mg^{2+}. Point out that although the oxygen ion was formed by the addition of two electrons, and the magnesium ion was formed by the loss of two electrons, they both now have the electronic structure of neon, which has 10 electrons. Have students write a description in their journals of why the ionic radius of O^{2-} (140 pm) and the ionic radius of Mg^{2+} (72 pm) are so different. Mg^{2+} has 12 protons, and O^{2-} has 8 protons, therefore Mg^{2+} has a greater nuclear charge than O^{2-}. The greater attraction pulls the electrons in closer to the nucleus and makes the ionic radius of Mg^{2+} smaller. OL

ASSESSMENT

Performance Ask students to draw a metal atom and its ion, showing relative sizes. The ion will be smaller. Ask students to repeat the procedure with a halogen atom and its ion. The ion will be larger. **OL** **EL**

Math in Chemistry

Atomic and Ionic Radii Have students compare the radii listed in Figure 11 and Figure 14. Write the following problem comparing the radii for cesium on the board:

$$\frac{\text{ionic radius}}{\text{atomic radius}} = \frac{167}{265} \approx 0.63$$

The radius of Cs^+ is 63% the radius of Cs. Have students make similar comparisons for Li, Mg, and F. Note that for fluorine, students should divide atomic radius by ionic radius. The radius of Li^+ is 50% the radius of Li. The radius of Mg^{2+} is 45% the radius of Mg. The radius of F is 54% the radius of F^-. **OL**

■ **Caption Question Fig. 14** because the outer electrons are in orbitals corresponding to higher principal energy levels

FOLDABLES® Rubric

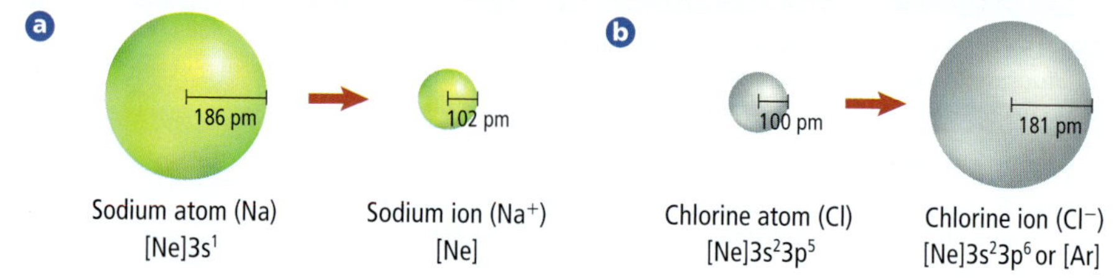

■ **Figure 13** The size of atoms varies greatly when they form ions.
a. Positive ions are smaller than the neutral atoms from which they form.
b. Negative ions are larger than the neutral atoms from which they form.

FOLDABLES® Incorporate information from this section into your Foldable.

When atoms gain electrons and form negatively charged ions, they become larger. The addition of an electron to an atom increases the electrostatic repulsion between the atom's outer electrons, forcing them to move farther apart. The increased distance between the outer electrons results in a larger radius.

Figure 13a illustrates how the radius of sodium decreases when sodium atoms form positive ions, and **Figure 13b** shows how the radius of chlorine increases when chlorine atoms form negative ions.

Trends within periods The ionic radii of most of the representative elements are shown in **Figure 14.** Note that elements on the left side of the table form smaller positive ions, and elements on the right side of the table form larger negative ions. In general, as you move from left to right across a period, the size of the positive ions gradually decreases. Then, beginning in group 15 or 16, the size of the much-larger negative ions also gradually decreases.

■ **Figure 14** The ionic radii of most of the representative elements are shown in picometers (10^{-12} m).
Explain *why the ionic radii increase for both positive and negative ions as you move down a group.*

Period	1	2	13	14	15	16	17
2	Li 76, 1+	Be 31, 2+	B 20, 3+	C 15, 4+	N 146, 3-	O 140, 2-	F 133, 1-
3	Na 102, 1+	Mg 72, 2+	Al 54, 3+	Si 41, 4+	P 212, 3-	S 184, 2-	Cl 181, 1-
4	K 138, 1+	Ca 100, 2+	Ga 62, 3+	Ge 53, 4+	As 222, 3-	Se 198, 2-	Br 196, 1-
5	Rb 152, 1+	Sr 118, 2+	In 81, 3+	Sn 71, 4+	Sb 62, 5+	Te 221, 2-	I 220, 1-
6	Cs 167, 1+	Ba 135, 2+	Tl 95, 3+	Pb 84, 4+	Bi 74, 5+		

Key: K 138, 1+ — Ionic radius (138); Chemical symbol (K); Charge (1+); Relative size (circle)

Demonstration

Activity of Alkali Metals

Purpose
to demonstrate that chemical reactivity follows a predictable pattern

Materials
Overhead projector; explosion shield; 600-mL beakers (3); phenolphthalein indicator (10 drops); clear plastic wrap; cubes of Li, Na, and K, (~ 2 mm on a side); wire screen (10 cm × 10 cm) See the Teacher Handbook for preparation of solutions.

Safety Precautions
Wear safety goggles and an apron. Use an explosion shield. Make sure there are no open flames or possible ignition sources.

Disposal Neutralize the solutions formed using acetic acid or dilute HCl. Flush the neutralized solutions down a drain with water.

Procedure
Cover the stage and lower lens of an overhead

Trends within groups As you move down a group, an ion's outer electrons are in orbitals corresponding to higher principal energy levels, resulting in a gradual increase in ionic size. Thus, the ionic radii of both positive and negative ions increase as you move down a group. The group and period trends in ionic radii are summarized in **Figure 15.**

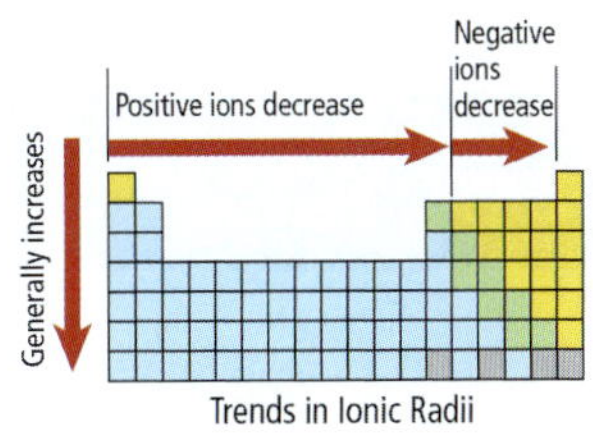

Figure 15 The diagram summarizes the general trends in ionic radii.

Ionization Energy

To form a positive ion, an electron must be removed from a neutral atom. This requires energy. The energy is needed to overcome the attraction between the positive charge of the nucleus and the negative charge of the electron. **Ionization energy** is defined as the energy required to remove an electron from a gaseous atom. For example, 8.64×10^{-19} J is required to remove an electron from a gaseous lithium atom. The energy required to remove the first outermost electron from an atom is called the first ionization energy. The first ionization energy of lithium equals 8.64×10^{-19} J. The loss of the electron results in the formation of a Li^+ ion. The first ionization energies of the elements in periods 1 through 5 are plotted on the graph in **Figure 16.**

READING CHECK **Define** *ionization energy.*

Think of ionization energy as an indication of how strongly an atom's nucleus holds onto its valence electrons. A high ionization energy value indicates the atom has a strong hold on its electrons. Atoms with large ionization energy values are less likely to form positive ions. Likewise, a low ionization energy value indicates an atom loses an outer electron easily. Such atoms are likely to form positive ions. Lithium's low ionization energy, for example, is important for its use in lithium-ion computer backup batteries, where the ability to lose electrons easily makes a battery that can quickly provide a large amount of electrical power.

Get help with **identifying trends.**

Personal Tutor

APPLYING PRACTICES

Use the Periodic Table Go to the resources tab in ConnectED to find the Applying Practices worksheet *Electron Patterns in Atoms.*

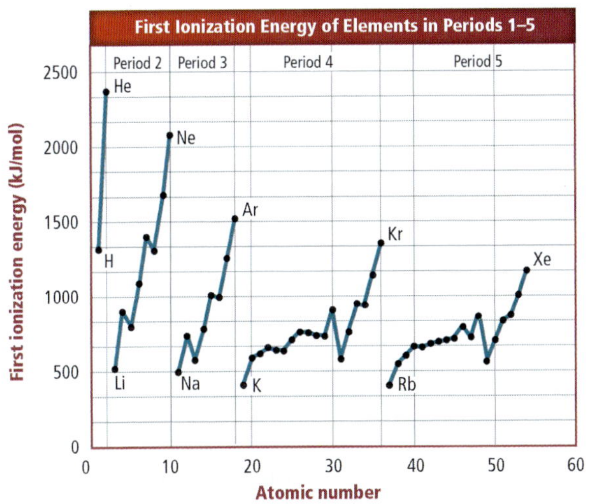

Figure 16 The first ionization energies for elements in periods 1 through 5 are shown as a function of the atomic number.

GRAPH CHECK

Describe the trend in first ionization energies within a group.

Math in Chemistry

Trends in Element Properties

Assign a trend to each group of students and have them create a table of data for that trend for the first 40 elements of the periodic table. Assigned trends should include ionization energy, electronegativity, atomic radii, ionic radii, density, and melting point. Students might need a variety of resources, such as the text, a *CRC Handbook of Chemistry and Physics,* a periodic table, internet sites, or other reference materials, to complete their data table.

Have students enter the data as a set of lists into a graphing calculator. The first list should be the atomic numbers 1 through 40. These data should be graphed on the *x*-axis. The specific data related to the trend should be graphed on the *y*-axis.

Have students display and print the graph. On the printed graph, have them draw a vertical line through each alkali metal (explain that the interval between each vertical line represents a period). They should then color-code each data point that represents an alkali metal. A different color should be used for noble gases. Repeat this process with a third color for the halogens. Finally, have students write a paragraph describing the trend that results when comparing elements within a period and elements in the same group. OL

COOP LEARN

READING CHECK It is the energy required to remove an electron from a gaseous atom.

GRAPH CHECK It decreases.

projector with clear plastic wrap. Place a 600-mL beaker with 100 mL of water on the projector. **WARNING:** ***Place explosion shield around the projector.*** Darken the room. Drop a small piece of lithium into the water. **WARNING:** ***Quickly cover the beaker with a wire screen. The gas produced is flammable.*** In separate beakers, repeat the procedure using Na and K.

Results

Metals skim across the water with speeds related to their reactivity. Li: slow, least; Na: fast, active; K: flammable, very active.

Analysis

1. Which metal reacts the fastest? K
2. Which metal has its outer-level electron farthest from the nucleus? K
3. How does the element's position in the column relate to its reactivity? Explain. Li is first and least reactive; K is third and most reactive. K is the most reactive because its valence electrons are in a higher energy level. They experience less pull towards the nucleus and can be removed more easily.

ASSESSMENT

Knowledge Have students apply the trend in reactivities they observed to predict how the activity of Rb and Cs will compare with that of Li, Na, and K. Rb and Cs will be more reactive, with Cs being the most reactive. OL

Visual Learning

Table 5 Have students study the table for the period 2 elements. Ask them to predict how they could use the chart of ionization energies to predict the number of electrons an atom would lose when forming an ion. OL EL

Quick Demo

Reactivity Ask students to predict which element, magnesium or calcium, will be more reactive. Add several drops of phenolphthalein to several milliliters of water in a test tube. Drop a small piece of magnesium into the test tube. Have students record their observations. Repeat this procedure with a small piece of calcium metal in a second test tube of phenolphthalein and water. Make sure to use a fresh piece of calcium that has not started to oxidize. After students record their observations, discuss what the color change indicates. a change in pH Students should notice that calcium is noticeably more reactive than the magnesium. A slow reaction between the magnesium and the water does occur, as evidenced by the small bubbles and slight color change. Repeat the reaction with hot water. Students should note that the reaction occurs more vigorously for both metals. Neutralize the solutions created and flush them down the drain with water. OL

☑ READING CHECK four

Table 5 Successive Ionization Energies for the Period 2 Elements

Element	Valence Electrons	Ionization Energy (kJ/mol)*								
		1st	2nd	3rd	4th	5th	6th	7th	8th	9th
Li	1	520	7300	11,810						
Be	2	900	1760	14,850	21,010					
B	3	800	2430	3660	25,020	32,820				
C	4	1090	2350	4620	6220	37,830	47,280			
N	5	1400	2860	4580	7480	9440	53,270	64,360		
O	6	1310	3390	5300	7470	10,980	13,330	71,870	84,080	
F	7	1680	3370	6050	8410	11,020	15,160	17,870	92,040	106,430
Ne	8	2080	3950	6120	9370	12,180	15,240	20,000	23,070	115,380

*mol is an abbreviation for mole, a quantity of matter.

RealWorld CHEMISTRY

Ionization Energy

SCUBA DIVING The increased pressure that scuba divers experience far below the water's surface can cause too much oxygen to enter their blood, which would result in confusion and nausea. To avoid this, divers sometimes use a gas mixture called *heliox*—oxygen diluted with helium. Helium's high ionization energy ensures that it will not react chemically in the bloodstream.

Each set of connected points on the graph in **Figure 16** represents the elements in a period. The group 1 metals have low ionization energies. Thus, group 1 metals (Li, Na, K, Rb) are likely to form positive ions. The group 18 elements (He, Ne, Ar, Kr, Xe) have high ionization energies and are unlikely to form ions. The stable electron configuration of gases of group 18 greatly limits their reactivity.

Removing more than one electron After removing the first electron from an atom, it is possible to remove additional electrons. The amount of energy required to remove a second electron from a 1+ ion is called the second ionization energy, the amount of energy required to remove a third electron from a 2+ ion is called the third ionization energy, and so on. **Table 5** lists the first through ninth ionization energies for elements in period 2.

Reading across **Table 5** from left to right, you will see that the energy required for each successive ionization always increases. However, the increase in energy does not occur smoothly. Note that for each element there is an ionization for which the required energy increases dramatically. For example, the second ionization energy of lithium (7300 kJ/mol) is much greater than its first ionization energy (520 kJ/mol). This means that a lithium atom is likely to lose its first valence electron but extremely unlikely to lose its second.

☑ READING CHECK **Infer** how many electrons carbon is likely to lose.

If you examine the table, you will notice that the ionization at which the large increase in energy occurs is related to the atom's number of valence electrons. Lithium has one valence electron and the increase occurs after the first ionization energy. Lithium easily forms the common lithium 1+ ion but is unlikely to form a lithium 2+ ion. The increase in ionization energy shows that atoms hold onto their inner core electrons much more strongly than they hold onto their valence electrons.

CHEMISTRY JOURNAL

Group Numbering Systems Have students investigate and report on different numbering systems used on periodic tables. Have students summarize their findings in their journals. OL

Trends within periods As shown in **Figure 16** and by the values in **Table 5,** first ionization energies generally increase as you move from left to right across a period. The increased nuclear charge of each successive element produces an increased hold on the valence electrons.

Trends within groups First ionization energies generally decrease as you move down a group. This decrease in energy occurs because atomic size increases as you move down the group. Less energy is required to remove the valence electrons farther from the nucleus. **Figure 17** summarizes the group and period trends in first ionization energies.

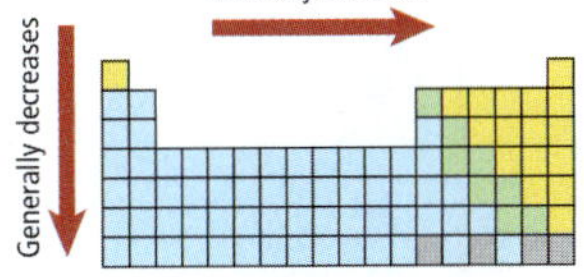

Figure 17 Ionization energies generally increase from left to right in a period and generally decrease as you move down a group.

Octet rule When a sodium atom loses its single valence electron to form a 1+ sodium ion, its electron configuration changes as shown below.

Sodium atom $1s^22s^22p^63s^1$ Sodium ion $1s^22s^22p^6$

Note that the sodium ion has the same electron configuration as neon ($1s^22s^22p^6$), a noble gas. This observation leads to one of the most important principles in chemistry, the octet rule. The **octet rule** states that atoms tend to gain, lose, or share electrons in order to acquire a full set of eight valence electrons. This reinforces what you learned earlier, that the electron configuration of filled s and p orbitals of the same energy level (consisting of eight valence electrons) is unusually stable. Note that the first-period elements are an exception to the rule, as they are complete with only two valence electrons.

The octet rule is useful for determining the type of ions likely to form. Elements on the right side of the periodic table tend to gain electrons in order to acquire the noble gas configuration; therefore, these elements tend to form negative ions. In a similar manner, elements on the left side of the table tend to lose electrons and form positive ions.

FOLDABLES® Incorporate information from this section into your Foldable.

Explore the **periodic properties of the elements.**

Virtual Investigations

MiniLAB

Organize Elements

Can you find the pattern?

Procedure

1. Read and complete the lab safety form.
2. Make a set of element cards based on the information in the chart at right.
3. Organize the cards by increasing mass, and start placing them into a 4 column × 3 row grid.
4. Place each card based on its properties, and leave gaps when necessary.

Symbol	Mass (g)	State	Color
Ad	52.9	solid/liquid	orange
Ax	108.7	ductile solid	light blue
Bp	69.3	gas	red
Cx	112.0	brittle solid	light green
Lq	98.7	ductile solid	blue
Pk	83.4	brittle solid	green
Qa	68.2	ductile solid	dark blue
Rx	106.2	liquid	yellow
Tu	64.1	brittle solid	hunter
Xn	45.0	gas	crimson

Analysis

1. **Make a table** listing the placement of each element.
2. **Describe** the period (across) and group (down) trends for the color in your new table.
3. **Describe** the period and group trends for the mass in your new table. Explain your placement of any elements that do not fit the trends.
4. **Predict** the placement of a newly found element, Ph, that is a fuchsia gas. What would be an expected range for the mass of Ph?
5. **Predict** the properties for the element that would fill the last remaining gap in the table.

MiniLAB

Rubric

Purpose Students will organize a periodic table of unknown elements and predict the properties of missing elements.

Process Skills formulate models, make and use tables, analyze patterns, use models

Teaching Strategies

- Having premade cards can shorten the time necessary to complete the lab.
- Extend this lab by having the students predict the whole next row of elements.

Expected Results Students will have organized the known cards according to the following grid.

Xn	Ad	Tu	Qa
Bp		Pk	Lq
	Rx	Cx	Ax

Analysis

1. Refer to the table in Expected Results.
2. The color wavelength decreases across the period and becomes lighter down the group.
3. The mass increases across the period and down the group. Cx does not fit the period trend for mass, but it fits in the third column with other green, brittle solids.
4. Ph would fit in the third period, first column based on color and stated trends. The mass would be between 69.3 g and 106.9 g.
5. The remaining gap would be a yellow-colored liquid with a mass most likely between 69.3 g and 83.4 g.

DIFFERENTIATED INSTRUCTION

Struggling Learners Help students visualize the octet rule by giving them a bowl with slanted sides and a small rubber ball. Point out that the ball does not move when it is placed in the center of the bowl. When students place the ball on the inner side of the bowl, forces cause it to be unstable and to fall toward the center, its most stable place. Relate this to the way forces cause an atom to be unstable unless it has a full set of valence electrons. BL

FOLDABLES® Rubric

■ **Caption Question Fig. 18** They are so stable that they are unlikely to form compounds.

3 Assess

Check for Understanding

Informally quiz students on periodic trends by giving them pairs of elements and asking them to identify which is larger, smaller, more reactive, or more electronegative. **OL**

Reteach

Have students label the trends on an outline of a periodic table by using arrows pointing in the direction that the trend increases. Repeat this labeling process with the other trends studied. **OL**

Extension

Have students research information on Linus Pauling and how the value of 3.98 was determined. **AL**

Increasing electronegativity →

Decreasing electronegativity ↓

- electronegativity < 1.0
- 1.0 ≤ electronegativity < 2.0
- 2.0 ≤ electronegativity < 3.0
- 3.0 ≤ electronegativity < 4.0

1 **H** 2.20																	2 **He**
3 **Li** 0.98	4 **Be** 1.57											5 **B** 2.04	6 **C** 2.55	7 **N** 3.04	8 **O** 3.44	9 **F** 3.98	10 **Ne**
11 **Na** 0.93	12 **Mg** 1.31											13 **Al** 1.61	14 **Si** 1.90	15 **P** 2.19	16 **S** 2.58	17 **Cl** 3.16	18 **Ar**
19 **K** 0.82	20 **Ca** 1.00	21 **Sc** 1.36	22 **Ti** 1.54	23 **V** 1.63	24 **Cr** 1.66	25 **Mn** 1.55	26 **Fe** 1.83	27 **Co** 1.88	28 **Ni** 1.91	29 **Cu** 1.90	30 **Zn** 1.65	31 **Ga** 1.81	32 **Ge** 2.01	33 **As** 2.18	34 **Se** 2.55	35 **Br** 2.96	36 **Kr**
37 **Rb** 0.82	38 **Sr** 0.95	39 **Y** 1.22	40 **Zr** 1.33	41 **Nb** 1.6	42 **Mo** 2.16	43 **Tc** 2.10	44 **Ru** 2.2	45 **Rh** 2.28	46 **Pd** 2.20	47 **Ag** 1.93	48 **Cd** 1.69	49 **In** 1.78	50 **Sn** 1.96	51 **Sb** 2.05	52 **Te** 2.1	53 **I** 2.66	54 **Xe**
55 **Cs** 0.79	56 **Ba** 0.89	57 **La** 1.1	72 **Hf** 1.3	73 **Ta** 1.5	74 **W** 1.7	75 **Re** 1.9	76 **Os** 2.2	77 **Ir** 2.2	78 **Pt** 2.2	79 **Au** 2.4	80 **Hg** 1.9	81 **Tl** 1.8	82 **Pb** 1.8	83 **Bi** 1.9	84 **Po** 2.0	85 **At** 2.2	86 **Rn**
87 **Fr** 0.70	88 **Ra** 0.90	89 **Ac** 1.1	104 **Rf**	105 **Db**	106 **Sg**	107 **Bh**	108 **Hs**	109 **Mt**	110 **Ds**	111 **Rg**	112 **Cn**	113 **Uut**	114 **Fl**	115 **Uup**	116 **Lv**	117 **Uus**	118 **Uuo**

Electronegativity Values in Paulings

■ **Figure 18** The electronegativity values for most of the elements are shown. The values are given in Paulings.

Infer *why electronegativity values are not listed for the noble gases.*

View an **animation about trends in electronegativity.**

Concepts In Motion

Electronegativity

The **electronegativity** of an element indicates the relative ability of its atoms to attract electrons in a chemical bond. As shown in **Figure 18,** electronegativity generally decreases as you move down a group. **Figure 18** also indicates that electronegativity generally increases as you move from left to right across a period.

Electronegativity values are expressed in terms of a numerical value of 3.98 or less. The units of electronegativity are arbitrary units called Paulings, named after American scientist Linus Pauling (1901–1994). Fluorine is the most electronegative element, with a value of 3.98, and cesium and francium are the least electronegative elements, with values of 0.79 and 0.70, respectively. In a chemical bond, the atom with the greater electronegativity more strongly attracts the bond's electrons. Note that because the noble gases form very few compounds, they do not have electronegativity values.

SECTION 3 REVIEW

Section Self-Check

Section Summary

- Atomic and ionic radii decrease from left to right across a period, and increase as you move down a group.
- Ionization energies generally increase from left to right across a period, and decrease as you move down a group.
- The octet rule states that atoms gain, lose, or share electrons to acquire a full set of eight valence electrons.
- Electronegativity generally increases from left to right across a period, and decreases as you move down a group.

20. MAINIDEA **Explain** how the period and group trends in atomic radii are related to electron configuration.

21. Indicate whether fluorine or bromine has a larger value for each of the following properties.

a. electronegativity **c.** atomic radius
b. ionic radius **d.** ionization energy

22. Explain why it takes more energy to remove the second electron from a lithium atom than it does to remove the fourth electron from a carbon atom.

23. Calculate Determine the differences in electronegativity, ionic radius, atomic radius, and first ionization energy for oxygen and beryllium.

24. Make and Use Graphs Graph the atomic radii of the representative elements in periods 2, 3, and 4 versus their atomic numbers. Connect the points of elements in each period, so that there are three separate curves on the graph. Summarize the trends in atomic radii shown on your graph. Explain.

SECTION 3 REVIEW

20. Atomic radii increase down a group as electrons are added to higher energy levels and inner core electrons shield the valence electrons from the increased nuclear charge. Atomic radii decrease across a period as increased nuclear charge coupled with unchanging shielding by inner core electrons pulls the valence electrons (being added to the same energy level) closer to the nucleus.

21. a. fluorine **c.** bromine
b. bromine **d.** fluorine

22. Lithium's second removed electron is an inner core electron, not a valence electron. Carbon's fourth removed electron is still a valence electron.

23. electronegativity: oxygen is 1.87 Paulings more electronegative; ionic radius: oxygen is 109 pm bigger; atomic radius: Beryllium is 39 pm bigger; first ionization energy: oxygen requires 410 kJ/mol more energy

24. In general, atomic radii decrease across a period due to increased nuclear charge; they increase down a group due to valence electrons placed in larger orbitals belonging to higher principal energy levels. As a result, there is an increase in the amount of shielding by the inner core electrons.

CHEMISTRY & health

Elements of the Body

Every time you eat a sandwich or take a breath, you are taking in elements your body needs to function normally. These elements have specific properties, depending on their location on the periodic table. **Figure 1** shows the percent by mass composition of cells in the human body.

Oxygen In an adult body, there are more than 14 billion billion billion oxygen atoms! Without a constant input of oxygen into the blood, the human body could die in just a few minutes.

Carbon Carbon can form strong bonds with itself and other elements. Carbon forms the long-chained carbon backbones that are an essential part of organic molecules such as carbohydrates, proteins, and lipids. The DNA molecule that determines your physical features relies on the versatility of carbon and its ability to bond with many different elements.

Hydrogen There are more hydrogen atoms in the body than atoms of all the other elements combined, although hydrogen represents only 10% of the composition by mass because of their significantly lower mass. The human body, requires hydrogen not in its elemental form, but in a variety of essential compounds, like water. With oxygen and carbon, hydrogen is also a crucial part of carbohydrates and other organic molecules that your body needs for energy.

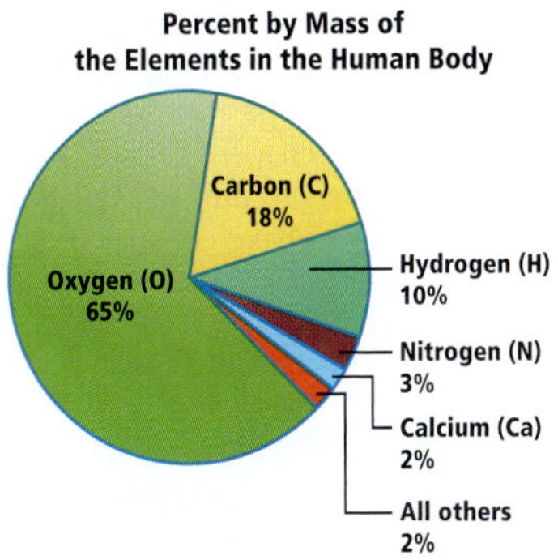

Figure 1 The human body is composed of many different elements.

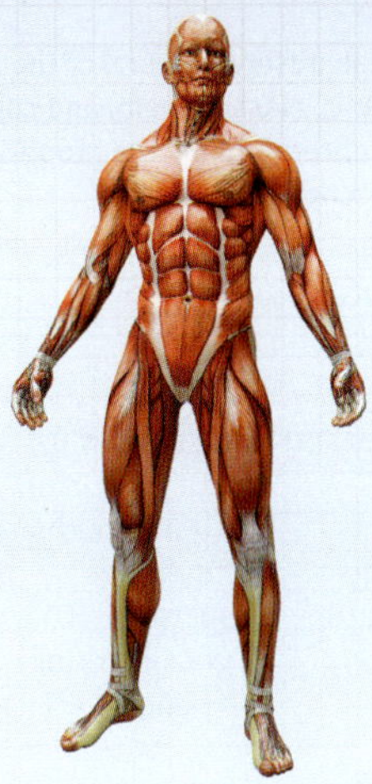

Figure 2 The entire human body is covered with muscles.

Nitrogen As shown in **Figure 2,** the human body is entirely covered with muscle. Nitrogen atoms are found in compounds that make up the proteins your body needs to build muscle.

Other elements in the body Oxygen, carbon, hydrogen, and nitrogen are the most abundant elements in your body but only a few of the elements that your body needs to live and grow. Trace elements, which together make up less than 2% of the body's mass, are a critical part of your body. Your bones and teeth could not grow without the constant intake of calcium. Although sulfur comprises less than 1 percent of the human body by mass, it is an essential component and is found in the proteins in your fingernails for instance. Sodium and potassium are crucial for the transmission of electrical signals in your brain.

WRITING IN ▶ Chemistry

Can you get all of the trace elements you need by eating only pre-packaged food? Why are trace elements necessary? Research trace elements and design a graphic novel that teaches elementary students about these nutrients.

WebQuest

WRITING IN ▶ Chemistry

Rubric WebQuest

Research Students should realize that some pre-packaged foods might not contain all the nutrients required by your body. Also, they should think about the fact that although many elements are present in very small amounts in the human body, they play crucial roles in the normal function of the body. One example of this would be iron. By mass, iron is present only in tiny amounts in your body, but without the iron in hemoglobin, blood could not transport oxygen.

CHEMISTRY & health

Purpose

Students will learn about how various elements are necessary for the normal function of the human body.

Background

Nitrogen fixation is the process by which bacteria in the soil convert atmospheric nitrogen into compounds usable by other living things. Humans can only consume nitrogen in the form of these compounds. Much of the hydrogen and oxygen in the body is present in the form of water. Because about 60% of the human body, by weight, is water and water has two hydrogen atoms to each oxygen atom, most of the atoms in the body are hydrogen.

Teaching Strategies

- Ask students if they use iodized salt, and point out that the iodine in iodized salt is required in trace amounts for the proper function of our thyroid gland which regulates, among other things, the body's metabolism.
- Some students might take a daily multivitamin. Encourage them to look at the back of the bottle for the mass and percentage of different elements contained in the pill. Students might also wish to research why certain elements are needed by the body.
- Have students make a circle graph detailing the percentages of certain elements in the human body.

ChemLAB

Rubric

Preparation

Time Allotment one and a half class periods

Process Skills observe and infer, interpret data, classify, compare and contrast

Safety Precautions Approve lab safety forms before work begins. Review the MSDS for hazardous materials in the lab with students before starting the activity. Students must wear aprons and goggles because the elements could shatter when struck with the hammer. Also caution students about the risk of hydrochloric acid to eyes and clothes. Remind students that they should never taste a chemical substance.

Disposal Neutralize the HCl solution with sodium carbonate. Then, flush it down a drain with a large amount of water.

Preparation of Materials

- The following element samples should be obtained for use in this lab. Test tubes samples: carbon, nitrogen, oxygen, magnesium, aluminum, silicon, red phosphorus, sulfur, chlorine, calcium, selenium, tin, iodine, and lead.
- Dish samples: carbon, magnesium, aluminum, silicon, sulfur, and tin
- The lab can be performed even if some of the listed elements are not available.
- If other elements are substituted, make sure they do not produce hazardous reactions.
- See the Teacher Handbook for preparation of all solutions.

Procedure

Set up stations at various locations in the lab.

ChemLAB

Investigate Descriptive Chemistry

Background: **You can observe several of the representative elements, classify them, and compare their properties. The observation of the properties of elements is called descriptive chemistry.**

Question: *What is the pattern of properties of the representative elements?*

Materials

- stoppered test tubes and plastic dishes containing small samples of elements
- conductivity apparatus
- 1.0*M* HCl
- small hammer
- test tubes (6)
- test-tube rack
- 10-mL graduated cylinder
- spatula
- glass-marking pencil

Safety Precautions

WARNING: Never test chemicals by tasting. 1.0M HCl is harmful to eyes and clothing. Brittle samples might shatter into sharp pieces.

Procedure

1. Read and complete the lab safety form.
2. Observe and record the appearance (physical state, color, luster, texture, and so on) of the element sample in each test tube without removing the stoppers.
3. Remove a small sample of each of the elements contained in a plastic dish and place it on a hard surface. Gently tap each element sample with a small hammer. If the element is malleable, it will flatten. If it is brittle, it will shatter. Record your observations.
4. Use the conductivity tester to determine which elements conduct electricity. Clean the electrodes with water, and dry them before testing each element.
5. Label each test tube with the symbol for one of the elements in the plastic dishes. Using a graduated cylinder, add 5 mL of water to each test tube.
6. Use a spatula to put a small amount of each element into the corresponding test tubes. Using a graduated cylinder, add 5 mL of 1.0*M* HCl to each test tube. Observe each tube for at least 1 minute. The formation of bubbles is evidence of a reaction between the acid and the element. Record your observations.

Observation of Elements

Classification	Properties
Metals	• malleable • good conductor of electricity • lustrous • silver or white in color • many react with acids
Nonmetals	• solids, liquids, or gases • do not conduct electricity • do not react with acids • likely brittle if solid
Metalloids	• combine properties of metals and nonmetals

7. **Cleanup and Disposal** Dispose of all materials as instructed by your teacher.

Analyze and Conclude

1. **Interpret Data** Using the table above and your observations, list the element samples that display the general characteristics of metals.
2. **Interpret Data** Using the table above and your observations, list the element samples that display the general characteristics of nonmetals.
3. **Interpret Data** Using the table above and your observations, list the element samples that display the general characteristics of metalloids.
4. **Model** Construct a periodic table, and label the representative elements by group (1 through 17). Using your results and the periodic table presented in this chapter, record the identities of elements observed during the lab in the periodic table you have constructed.
5. **Infer** Describe any trends among the elements you observed in the lab.

INQUIRY EXTENSION

Investigate Were there any element samples that did not fit into one of the three categories? What additional investigations could you conduct to learn even more about these elements' characteristics?

Analyze and Conclude

1.–4. Answers will vary depending on the samples provided to students.

5. Students might note that the metallic characteristic increases from right-to-left and from top-to-bottom.

INQUIRY EXTENSION

Answers will vary.

CHAPTER 6 STUDY GUIDE

Vocabulary Practice

BIGIDEA Periodic trends in the properties of atoms allow us to predict physical and chemical properties.

SECTION 1 Development of the Modern Periodic Table

MAINIDEA The periodic table evolved over time as scientists discovered more useful ways to compare and organize the elements.

- The elements were first organized by increasing atomic mass, which led to inconsistencies. Later, they were organized by increasing atomic number.
- The periodic law states that when the elements are arranged by increasing atomic number, there is a periodic repetition of their chemical and physical properties.
- The periodic table organizes the elements into periods (rows) and groups or families (columns); elements with similar properties are in the same group.
- Elements are classified as either metals, nonmetals, or metalloids.

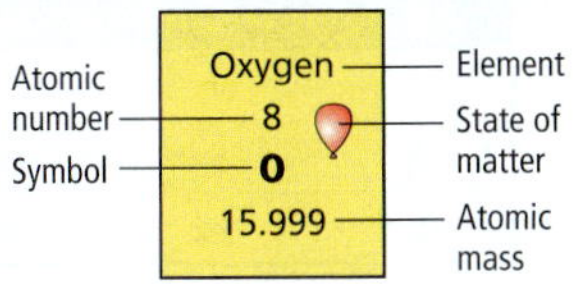

VOCABULARY
- periodic law
- group
- period
- representative element
- transition element
- metal
- alkali metal
- alkaline earth metal
- transition metal
- inner transition metal
- lanthanide series
- actinide series
- nonmetal
- halogen
- noble gas
- metalloid

SECTION 2 Classification of the Elements

MAINIDEA Elements are organized into different blocks in the periodic table according to their electron configurations.

- The periodic table has four blocks (s, p, d, f).
- Elements within a group have similar chemical properties.
- The group number for elements in groups 1 and 2 equals the element's number of valence electrons.
- The energy level of an atom's valence electrons equals its period number.

SECTION 3 Periodic Trends

MAINIDEA Trends among elements in the periodic table include their sizes and their abilities to lose or attract electrons.

- Atomic and ionic radii decrease from left to right across a period, and increase as you move down a group.
- Ionization energies generally increase from left to right across a period, and decrease as you move down a group.
- The octet rule states that atoms gain, lose, or share electrons to acquire a full set of eight valence electrons.
- Electronegativity generally increases from left to right across a period, and decreases as you move down a group.

VOCABULARY
- ion
- ionization energy
- octet rule
- electronegativity

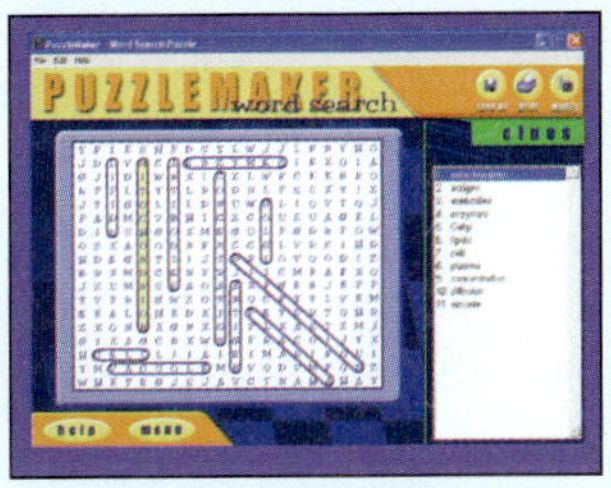

Vocabulary Practice

For additional practice with vocabulary, have students access the Vocabulary PuzzleMaker.

Use the Vocabulary

To reinforce chapter vocabulary, have students write a sentence using each term. OL EL

Review Strategies

- Have students work in groups to create a puzzle of scrambled words using the vocabulary terms. When complete, the puzzle should be handed to another group, which will unscramble the words and provide definitions. OL
- Have students label a blank periodic table with the trends as they have studied. OL
- Problems from the Supplemental Practice Problems appendix or the Supplemental Problems booklet can be used for review. OL

Students can review with these online resources.

Vocabulary Practice

Vocabulary eGames
Vocabulary eFlashcards
Vocabulary PuzzleMaker

Section Self-Check

Chapter Self-Check

Online Test Practice

Use *eAssessment* to:
- create multiple versions of tests
- edit existing questions and add your own questions
- build tests aligned with select state curriculum standards using built-in state curriculum tags
- track students' progress

SECTION 1

Mastering Concepts

25. Mendeleev used atomic mass instead of atomic number to order the elements. This resulted in some elements being out of order.
26. Newlands introduced the idea of periodically repeating properties.
27. Mendeleev's work was published first, he did more to show periodic trends, and he predicted properties of several yet-to-be-discovered elements.
28. When the elements are arranged by increasing atomic number, there is a periodic repetition of their chemical and physical properties.
29. Metals are generally dense, solid, shiny, ductile, malleable, and good conductors of heat and electricity.
30. Metalloids have properties intermediate between metals and nonmetals.
31. a. nonmetal
    b. metal
    c. metalloid
    d. metal
32. a. 2
    b. 4
    c. 3
    d. 1
33. Refer to the Solutions Manual for a sample table.
34. The line indicates that the lanthanide and actinide series of elements would be at that location if there were enough horizontal room on the page.
35. a. Hg
    b. Rn
    c. Sn
    d. Ni
36. The halogen would have atomic number 117. The noble gas would have atomic number 118.

Mastering Problems

37. Argon and potassium would be switched. Cobalt and nickel would be switched. Tellurium and iodine would be switched.
38. group 17, period 7; metalloid
39. ununseptium, ununoctium, ununennium, unbinilium
40. a. Si
    b. In
    c. W

CHAPTER 6 ASSESSMENT

SECTION 1

Mastering Concepts

25. Explain how Mendeleev's periodic table was in error.

26. Explain the contribution of Newlands's law of octaves to the development of the modern periodic table.

27. Lothar Meyer and Dmitri Mendeleev both proposed similar periodic tables in 1869. Why is Mendeleev generally given credit for the periodic table?

28. What is the periodic law?

29. Describe the general characteristics of metals.

30. What are the general properties of a metalloid?

31. Identify each of the following as a metal, a nonmetal, or a metalloid.
a. oxygen
b. barium
c. germanium
d. iron

32. Match each item on the left with its corresponding group on the right.
a. alkali metals — **1.** group 18
b. halogens — **2.** group 1
c. alkaline earth metals — **3.** group 2
d. noble gases — **4.** group 17

33. Sketch a simplified periodic table, and use labels to identify the alkali metals, alkaline earth metals, transition metals, inner transition metals, noble gases, and halogens.

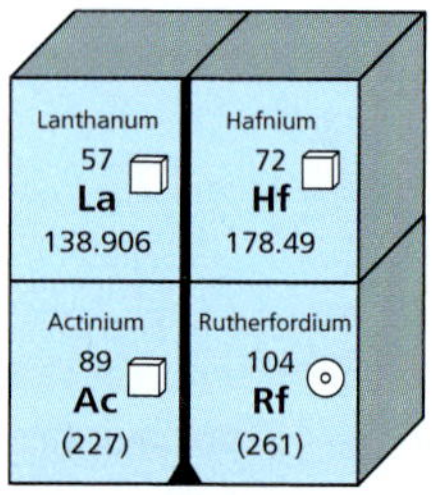

Figure 19

34. Explain what the dark line running down the middle of **Figure 19** indicates.

35. Give the chemical symbol of each of the following elements.
a. a metal used in thermometers
b. a radioactive gas used to predict earthquakes; the noble gas with the greatest atomic mass
c. a coating for food cans; it is the metal in group 14 with the lowest atomic mass
d. transition metal that is used to make burglar-proof vaults; also the name of a coin

36. If a new halogen and a new noble gas were discovered, what would be their atomic numbers?

Mastering Problems

37. If the periodic table were arranged by atomic mass, which of the first 55 elements would be ordered differently than they are in the existing table?

38. New Heavy Element Scientists recently reported an element with 117 protons. What is its group and period? Would it be a metal, a metalloid, or a nonmetal?

39. Naming New Elements Recently discovered elements that have not been fully verified are given temporary names using the prefix words in **Table 6.** Based on this system, write names for elements 117 to 120.

Table 6 Prefixes

0	1	2	3	4
nil	un	b(i)	tr(i)	quad
5	6	7	8	9
pent	hex	sept	oct	en(n)

40. Give the chemical symbol for each element.
a. the element in period 3 that can be used in making computer chips because it is a metalloid
b. the group 13, period 5 metal used in making flat screens for televisions
c. an element used as a filament in lightbulbs; has the highest atomic mass of the natural elements in group 6

SECTION 2

Mastering Concepts

41. Household Products Why do the elements chlorine, used in laundry bleach, and iodine, a nutrient added to table salt, have similar chemical properties?

42. How is the energy level of an atom's valence electrons related to its period in the periodic table?

43. How many valence electrons does each noble gas have?

44. What are the four blocks of the periodic table?

45. What electron configuration has the greatest stability?

46. Explain how an atom's valence electron configuration determines its place in the periodic table.

47. Write the electron configuration for the element fitting each of the following descriptions.
a. the metal in group 15 that is part of compounds often found in cosmetics
b. the halogen in period 3 that is part of a bleaching compound used in paper production
c. the transition metal that is a liquid at room temperature; is sometimes used in outdoor security lights

SECTION 2

Mastering Concepts

41. They have the same valence electron configuration (s^2p^5).
42. The energy level of an atom's valence electrons equals its period number.
43. All noble gases have eight valence electrons, except for helium, which has two.
44. s-, p-, d-, and f-block
45. ns^2np^6, where n is the energy level
46. Elements in a given column have the same number of valence electrons. The energy level of an atom's valence electrons determines its period.
47. a. Bi: $[Xe]6s^24f^{14}5d^{10}6p^3$
    b. Cl: $[Ne]3s^23p^5$
    c. Hg: $[Xe]6s^24f^{14}5d^{10}$
48. a. 3, period 5, d-block
    b. 15, period 4, p-block
    c. 18, period 2, p-block
    d. 13, period 3, p-block
49. larger
50. The first energy level has only the s sublevel. The second and third energy levels have only the s and p sublevels. The fourth and fifth energy levels have s, p, and d sublevels.

Chapter Self-Check

48. Determine the group, period, and block in which each of the following elements is located in the periodic table.
a. $[Kr]5s^24d^1$ **c.** $[He]2s^22p^6$
b. $[Ar]4s^23d^{10}4p^3$ **d.** $[Ne]3s^23p^1$

49. Given any two elements within a group, is the element with the larger atomic number likely to have a larger or smaller atomic radius than the other element?

50. **Table 7** shows the number of elements in the first five periods of the periodic table. Explain why some of the periods have different numbers of elements.

Table 7 Number of Elements in Periods 1–5

Period	1	2	3	4	5
Number of elements	2	8	8	18	18

51. **Coins** One of the transition groups is often called the coinage group because at one time many coins were made of these metals. Which group is this? What element in this group is still used in many U.S. coins today?

52. Do any of the halogens have their valence electrons in orbitals of the same energy level? Explain.

53. The transition elements have their valence electrons in orbitals of more than one energy level, but the representative elements have their valence electrons in orbitals of only one energy level. Show this by using the electron configurations of a transition element and a representative element as examples.

Mastering Problems

54. **Fireworks** Barium is a metal that gives a green color to fireworks. Write the electron configuration for barium. Classify it according to group, period, and block in the periodic table.

55. **Headphones** Neodymium magnets can be used in stereo headphones because they are powerful and lightweight. Write the electron configuration for neodymium. In which block of the periodic table is it?

56. **Soda Cans** The metal used to make soda cans has the electron configuration $[Ne]3s^23p^1$. Identify the metal and give its group, period, and block.

57. Identify each missing part of **Table 8.**

Table 8 Electron Configuration

Period	Group	Element	Electron Configuration
3		Mg	$[Ne]3s^2$
4	14	Ge	
	12	Cd	$[Kr]5s^24d^{10}$
2	1		$[He]2s^1$

SECTION 3

Mastering Concepts

58. What is ionization energy?

59. An element forms a negative ion when ionized. On what side of the periodic table is the element located? Explain.

60. Of the elements magnesium, calcium, and barium, which forms the ion with the largest radius? The smallest? What periodic trend explains this?

61. Explain why each successive ionization of an electron requires a greater amount of energy.

62. How does the ionic radius of a nonmetal compare with its atomic radius? Explain the change in radius.

63. Explain why atomic radii decrease as you move from left to right across a period.

64. Which element has the larger ionization energy?
a. Li, N **b.** Kr, Ne **c.** Cs, Li

65. Explain the octet rule. Why are hydrogen and helium exceptions to the octet rule?

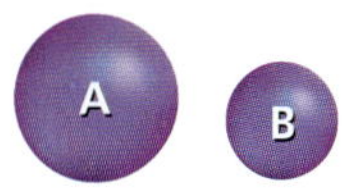

■ **Figure 20**

66. Use **Figure 20** to answer each of the following questions. Explain your reasoning for each answer.
a. If A is an ion and B is an atom of the same element, is the ion a positive or negative ion?
b. If A and B represent the atomic radii of two elements in the same period, what is their order?
c. If A and B represent the ionic radii of two elements in the same group, what is their order?

67. How many valence electrons do elements in group 1 have? In group 18?

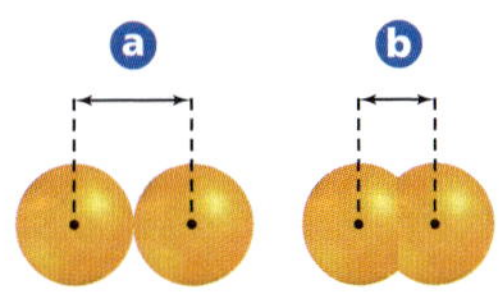

■ **Figure 21**

68. **Figure 21** shows two ways to define an atomic radius. Describe each method. When is each method used?

69. **Chlorine** The electron configuration of a chlorine atom is $[Ne]3s^23p^5$. When it gains an electron and becomes an ion, its electron configuration changes to $[Ne]3s^23p^6$, or [Ar], the electron configuration for argon. Has the chlorine atom changed to an argon atom? Explain.

51. group 11; Standard U.S. coinage is still made with some copper.

52. No, because each halogen is in a different period, each has its valence electrons in orbitals of a different energy level.

53. Possible answer: Chlorine's electron configuration is $[Ne]3s^23p^5$, which has its valence electrons in orbitals of the third energy level. Iron's electron configuration is $[Ar]4s^23d^6$, which has its valence electrons in orbitals of both the third and fourth energy levels.

Mastering Problems

54. Its electron configuration is $[Xe]6s^2$. It is in group 2, period 6, s block.

55. Its electron configuration is $[Xe]6s^24f^4$. It is in the f block of elements.

56. The metal is aluminum. It is in group 13, period 3, p block.

57. **a.** 2
b. $[Ar]4s^23d^{10}4p^2$
c. 5
d. Li

SECTION 3

Mastering Concepts

58. Ionization energy is the energy needed to remove an electron from a neutral atom in its gaseous state.

59. Elements on the right side of the periodic table gain electrons to gain a stable octet.

60. Ba^{2+} is the largest; Mg^{2+} is the smallest; ionic size increases down a group.

61. With each removed electron, there are fewer electrons to shield the remaining electrons from the electrostatic force of attraction of the nucleus. The increased nuclear attraction makes it more difficult to remove subsequent electrons.

62. The ionic radius of a nonmetal is larger than its neutral atom. Nonmetals tend to gain electrons in the atom's current energy level; these additional electrons repel each other and increase the size of the ion.

63. Atomic radii decrease left-to-right because the nuclear charge increases as the shielding of inner core electrons remains constant. The increased attraction of the nucleus for its electrons pulls the electrons inward, resulting in a decreased atomic size.

64. a. N; b. Ne; c. Li

65. The ns^2np^6 electron configuration, known as the octet configuration, contains eight electrons and generally has the lowest energy and is the most stable. Atoms gain, lose, or share electrons in order to obtain the stable octet configuration. Hydrogen and helium have a complete energy level with just two valence electrons.

66. **a.** The ion is negative. A negative ion is always larger than its atom.
b. A is to the left of B. Atomic radius in a period decreases left-to-right.
c. A is below B. Ionic radius increases down a group.

67. 1; 8

68. Method A is used for a metal. The atomic radius is half the distance between two adjacent atoms in a crystal of the metal. Method B is used for a nonmetal that occurs as a molecule. The atomic radius is half the distance between nuclei of identical atoms that are bonded together.

69. No; the electron configurations of a chlorine ion and an argon atom are the same, but the chlorine ion still has 17 protons and retains its identity as chlorine.

Mastering Problems

70. a. chlorine, carbon, oxygen
b. chlorine, oxygen, carbon

71. Oxygen is more electronegative than silicon. When silicon and oxygen bond, silicon atoms become smaller as its electrons are pulled closer to the oxygen atoms. The oxygen atoms become larger as a result.

72. Carbon atoms decrease in size. Nitrogen and oxygen atoms increase in size.

MIXED REVIEW

73. An ion is an atom that has lost or gained one or more electrons.

74. because an atom has no definite boundary

75. boron

76. Bromine; electronegativity tends to increase from left to right across the periodic table.

77. A is the s-block elements, with full or partly-filled s orbitals. B is the p-block elements, with full or partly-filled p orbitals. C is the d-block elements, with full or partly-filled d orbitals. D is the f-block elements, with full or partly filled f orbitals.

78. a. As **b.** N
c. Be

79. The s block represents the filling of the s orbital, which holds a maximum of two electrons. The p-block represents the filling of the three p orbitals, which hold a maximum of six electrons. The d-block represents the filling of the five d orbitals, which hold a maximum of ten electrons.

80. Scientists have refined their methods for measuring atomic masses.

81. The order is O, S, Se, and Te. This is an example of a group trend.

82. The element calcium is in group 2, period 4, s-block.

83. The p orbital does not exist for energy level 1. The first energy level consists only of a single s orbital that holds a maximum of two electrons.

84. copper and silver

85. platinum

ASSESSMENT

Mastering Problems

70. Sport Bottles Some sports bottles are made of Lexan, a plastic containing a compound of the elements chlorine, carbon, and oxygen. Order these elements from greatest to least according to atomic radius and ionic radius.

71. Contact Lenses Soft contact lenses are made of silicon and oxygen atoms bonded together. Create a table listing the atomic and ionic electron configurations, and the atomic and ionic radii for silicon and oxygen. When silicon bonds with oxygen, which atoms become larger and which become smaller? Why?

72. Artificial Sweetener Some diet sodas contain the artificial sweetener aspartame, a compound containing carbon, nitrogen, oxygen, and other atoms. Create a table showing the atomic and ionic radii of carbon, nitrogen, and oxygen. (Assume the ionization states shown in **Figure 14.**) Use the table to predict whether the sizes of carbon, nitrogen, and oxygen atoms increase or decrease in size when they form bonds in aspartame.

MIXED REVIEW

73. Define an ion.

74. Explain why the radius of an atom cannot be measured directly.

75. What is the metalloid in period 2 of the periodic table that is part of compounds used as water softeners?

76. Do you expect cesium, a group 1 element used in infrared lamps, or bromine, a halogen used in firefighting compounds, to have the greatest electronegativity? Why?

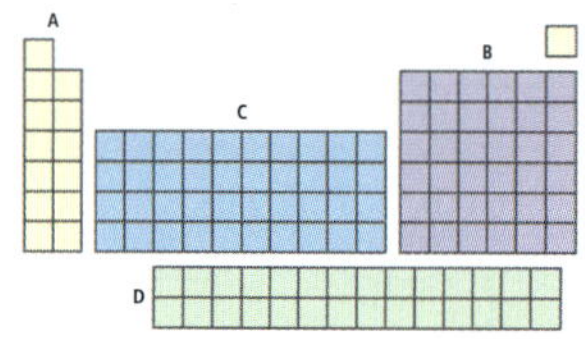

■ **Figure 22**

77. Figure 22 shows different sections of the periodic table. Give the name of each section, and explain what the elements in each section have in common.

78. Which element in each pair is more electronegative?
a. K, As **b.** N, Sb **c.** Sr, Be

79. Explain why the s-block of the periodic table is two-groups wide, the p-block is six-groups wide, and the d-block is ten-groups wide.

80. Most of the atomic masses in Mendeleev's table are different from today's values. Explain why.

81. Arrange the elements oxygen, sulfur, tellurium, and selenium in order of increasing atomic radii. Is your order an example of a group trend or a period trend?

82. Milk The element with the electron configuration [Ar]$4s^2$ is an important mineral in milk. Identify this element's group, period, and block in the periodic table.

83. Why are there no p-block elements in the first period?

84. Jewelry What are the two transition metals that are used in making jewelry and are the group 11 elements with the lowest atomic masses?

85. Which has the largest ionization energy, platinum, an element sometimes used in dental crowns, or cobalt, an element that provides a bright blue color to pottery?

THINK CRITICALLY

86. Apply Sodium forms a 1+ ion, while fluorine forms a 1− ion. Write the electron configuration for each ion. Why don't these two elements form 2+ and 2− ions, respectively?

87. Make and Use Graphs The densities of the group 15 elements are given in **Table 9.** Plot density versus atomic number, and state any trends you observe.

Table 9 Group 15 Density Data

Element	Atomic Number	Density (g/cm^3)
Nitrogen	7	1.14×10^{-3}
Phosphorus	15	1.82
Arsenic	33	5.22
Antimony	51	6.53
Bismuth	83	10.05

88. Generalize The outer-electron configurations of elements in group 1 can be written as ns^1, where n refers to the element's period and its principal energy level. Develop a similar notation for all the other groups of the representative elements.

89. Identify A period 3 representative element is part of the rough material on the side of a match box used for lighting matches. **Table 10** shows the ionization energies for this element. Use the information in the table to infer the identity of the element. Explain.

Table 10 Ionization Energies in kJ/mol

Number	1st	2nd	3rd	4th	5th	6th
Ionization energy	1010	1905	2910	4957	6265	21,238

THINK CRITICALLY

86. Both ions have the configuration $1s^2\,2s^2\,2p^6$, a stable, noble gas configuration. Sodium's second ionization energy is much greater than its first, showing that it holds onto its inner core electrons much more strongly than its valence electron. Similarly, the energy required for fluorine to accept a second electron is too great.

87. The graph should show density increasing with increasing atomic number. Note that the density of nitrogen is so low because it is the only element that exists as a gas (the others are solids). Refer to the Solutions Manual for graph.

88. group 2, ns^2; group 13, ns^2np^1; group 14, ns^2np^2; group 15, ns^2np^3; group 16, ns^2np^4; group 17, ns^2np^5; group 18, ns^2np^6

89. Phosphorus; the jump in ionization energy after the 5th level indicates that the element has five valence electrons.

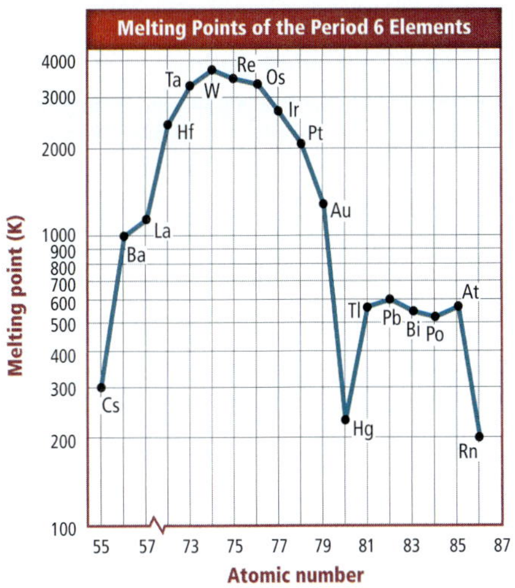

■ **Figure 23**

90. Interpret Data The melting points of the period 6 elements are plotted versus atomic number in **Figure 23.** Determine the trends in melting point and the orbital configurations of the elements. Form a hypothesis that explains the trends.

CHALLENGE PROBLEM

91. Ionization energies are expressed in kilojoules per mole (one mole contains 6.02×10^{23} atoms), but the energy to remove an electron from a gaseous atom is expressed in joules. Use the values in **Table 5** to calculate the energy, in joules, required to remove the first electron from an atom of Li, Be, B, and C. Then, use the relationship 1 eV $= 1.60 \times 10^{-19}$ J to convert the values to electron volts.

CUMULATIVE REVIEW

92. Define *matter.* Identify whether or not each of the following is a form of matter.
- **a.** microwaves
- **b.** helium inside a balloon
- **c.** heat from the Sun
- **d.** velocity
- **e.** a speck of dust
- **f.** the color blue

93. Convert the following mass measurements as indicated.
- **a.** 1.1 cm to meters
- **b.** 76.2 pm to millimeters
- **c.** 11 mg to kilograms
- **d.** 7.23 µg to kilograms

94. How is the energy of a quantum of emitted radiation related to the frequency of the radiation?

95. What element has the ground-state electron configuration of $[Ar]4s^23d^6$?

Chapter Self-Check

WRITING IN ▶ Chemistry

96. Triads In the early 1800s, German chemist J. W. Döbereiner proposed that some elements could be classified into sets of three, called triads. Research and write a report on Döbereiner's triads. What elements comprised the triads? How were the properties of elements within a triad similar?

97. Affinity Electron affinity is another periodic property of the elements. Write a summary on what electron affinity is, and describe its group and period trends.

DBQ Document-Based Questions

Mendeleev's original periodic table is remarkable given the knowledge of elements at that time, and yet it is different from the modern version. Compare Mendeleev's table, shown in **Table 11,** *with the modern periodic table shown in* **Figure 5.**

Data obtained from: Dmitrii Mendeleev, *The Principles of Chemistry,* 1891.

Table 11 Groups of Elements

Series	0	I	II	III	IV	V	VI	VII	VIII
1	—	H	—	—	—		—	—	
2	He	Li	Be	B	C	N	O	F	
3	Ne	Na	Mg	Al	Si	P	S	Cl	
4	Ar	K	Ca	So	Ti	V	Cr	Mn	Fe
5		Cu	Zn	Ga	Ge	As	Se	Br	Co Ni (Cu)
6	Kr	Rb	Sr	Y	Zr	Nb	Mo	—	Ru
7		Ag	Cad	In	Sn	Sb	Te	I	Rh Pd (Ag)
8	Xe	Cs	Ba	La	—	—	—	—	—
9		—	—	—	—	—	—	—	—
10	—	—	—	Yb	—	Ta	W	—	Os
11		Au	Hg	Tl		Bi	—	—	Ir Pt (Au)
12	—	—	Rd	—	Th	—	U		

98. Mendeleev placed the noble gases on the left of his table. Why does placement on the right of the modern table make more sense?

99. Which block on Mendeleev's table was most like today's placement? Which block was least like today's placement? Why?

100. Most of the atomic masses in Mendeleev's table differ from today's values. Why do you think this is so?

90. For the d-block elements, the highest values occur for half-filled and near half-filled d orbitals. (Re with a configuration of $5d^5$ has the highest melting point.) Relating to Hund's rule, it seems that metallic bonding strengthens as the number of unpaired electrons increases, reaching a maximum when the orbital is just over half-filled. Note that Hg and Rn have no unpaired electrons and substantially lower melting points. For the p-block elements (81–86), again the elements with unpaired p electrons tend to have higher melting points.

CHALLENGE PROBLEM

91. Li: 8.64×10^{-19} J, or 5.40 eV
Be: 1.50×10^{-18} J, or 9.38 eV
B: 1.33×10^{-18} J, or 8.31 eV
C: 1.81×10^{-18} J, or 11.3 eV

CUMULATIVE REVIEW

92. Matter is anything that has mass and takes up space.
- **a.** no
- **b.** yes
- **c.** no
- **d.** no
- **e.** yes
- **f.** no

93.
- **a.** 1.1×10^{-2} m
- **b.** 7.62×10^{-8} mm
- **c.** 1.1×10^{-5} kg
- **d.** 7.23×10^{-9} kg

94. The energy of a quantum equals the frequency times Planck's constant.

95. iron

WRITING IN ▶ Chemistry

Rubric

96. Döbereiner noticed that the atomic weight of strontium fell midway between the atomic weights of calcium and barium, elements that possessed similar chemical properties. He also studied the halogen triad composed of chlorine, bromine, and iodine and the alkali-metal triad made up of lithium, sodium, and potassium. Döbereiner proposed that nature contained triads of elements in which the middle element (when ordered by atomic weight) had properties that were an average of the other two members.

97. Students will find that electron affinity (EA) is the energy change that accompanies one mole of electrons being added to one mole of gaseous atoms or ions. With many irregularities (and excluding the noble gases), first electron affinity (EA_1) generally decreases from top to bottom within a group and increases from left to right within a period.

DBQ Document-Based Questions

Data obtained from: Dmitrii Mendeleev, 2004. *Principles of Chemistry,* 1891.

98. Placing them on the right has the representative elements ordered from left to right as orbitals of the energy levels are filled, ending on the right with noble gases, which have filled outer orbitals.

99. He had the s block most like today's table, and the f-block least like today's table. The s-block elements were most known in his time, but little was known about the f-block elements.

100. Scientists have refined their methods for measuring atomic masses.

MULTIPLE CHOICE

1. A
2. A
3. D
4. A
5. B
6. D
7. D
8. B
9. D
10. D

MULTIPLE CHOICE

1. Elements in the same group of the periodic table have the same
A. number of valence electrons.
B. physical properties.
C. number of electrons.
D. electron configuration.

2. Which statement is NOT true?
A. The atomic radius of Na is less than the atomic radius of Mg.
B. The electronegativity of C is greater than the electronegativity of B.
C. The ionic radius of Br^- is greater than the atomic radius of Br.
D. The first ionization energy of K is greater than the first ionization energy of Rb.

3. What is the group, period, and block of an atom with the electron configuration $[Ar]4s^2 3d^{10} 4p^4$?
A. group 14, period 4, d-block
B. group 16, period 3, p-block
C. group 14, period 4, p-block
D. group 16, period 4, p-block

Use the table below to answer Questions 4 and 5.

Characteristics of Elements

Element	Block	Characteristic
X	s	soft solid; reacts readily with oxygen
Y	p	gas at room temperature; forms salts
Z	—	inert gas

4. In which group does Element X most likely belong?
A. 1
B. 17
C. 18
D. 4

5. In which block is Element Z most likely found?
A. s-block
B. p-block
C. d-block
D. f-block

Use the table below to answer Questions 6 and 7.

Percent Composition By Mass of Selected Nitrogen Oxides

Compound	Percent Nitrogen	Percent Oxygen
N_2O_4	30.4%	69.6%
N_2O_3	?	?
N_2O	63.6%	36.4%
N_2O_5	25.9%	74.1%

6. What is the percent nitrogen in the compound N_2O_3?
A. 44.7%
B. 46.7%
C. 28.1%
D. 36.8%

7. A sample of a nitrogen oxide contains 1.29 g of nitrogen and 3.71 g of oxygen. Which compound is this most likely to be?
A. N_2O_4
B. N_2O_3
C. N_2O
D. N_2O_5

8. On the modern periodic table, metalloids are found only in
A. the d-block.
B. groups 13 through 17.
C. the f-block.
D. groups 1 and 2.

9. Which group is composed entirely of nonmetals?
A. 1
B. 13
C. 15
D. 18

10. It can be predicted that element 118 would have properties similar to a(n)
A. alkali earth metal.
B. halogen.
C. metalloid.
D. noble gas.

Online Test Practice

SHORT ANSWER

11. Write the electron configuration for the element arsenic (As).

12. Write the nuclear decay equation for the beta decay of iodine-131.

13. Two students are identifying a sample of tap water. Student A says that tap water is a mixture, while Student B says that it is a compound. Which student is correct? Justify your answer.

EXTENDED RESPONSE

Use the table below to answer Questions 14 and 15.

Successive Ionization Energies for Selected Period 2 Elements, in kJ/mol

Element	Li	Be	B	C
Valence e–	1	2	3	4
First ionization energy	520	900	800	1090
Second ionization energy	7300	1760	2430	2350
Third ionization energy		14,850	3660	4620
Fourth ionization energy			25,020	6220
Fifth ionization energy				37,830

14. Correlate the biggest jump in ionization energy to the number of valence electrons in each atom.

15. Predict which ionization energy will show the largest jump for magnesium. Explain your answer.

SAT SUBJECT TEST: CHEMISTRY

For Questions 16 to 19, answer true or false for the first statement, and true or false for the second statement. If the second statement is a correct explanation of the first statement, write CE.

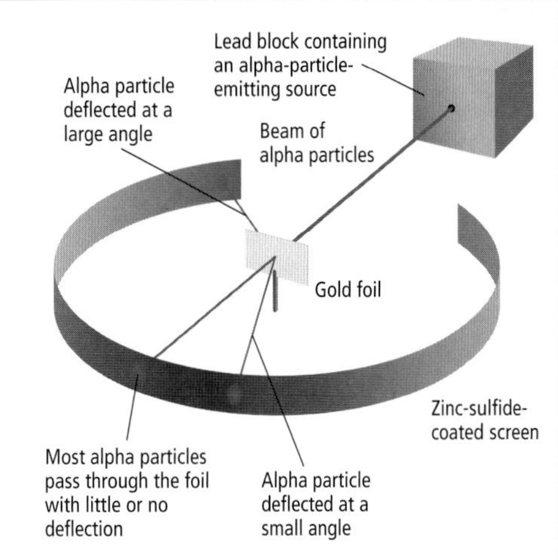

Statement I		Statement II
16. Some particles bounce off the gold foil	BECAUSE	the nucleus is negatively charged.
17. Some particles bounce off the gold foil	BECAUSE	they hit protons in the nucleus.
18. Many particles pass through the gold foil	BECAUSE	atoms are made of protons, neutrons, and electrons.
19. The symbol for an alpha particle is ${}^{4}_{2}He$	BECAUSE	protons and neutrons have about the same mass.

NEED EXTRA HELP?																			
If You Missed Question . . .	1	2	3	4	5	6	7	8	9	10	11	12	13	14	15	16	17	18	19
Review Section . . .	6.2	6.3	6.2	6.2	6.2	3.4	3.4	6.2	6.2	6.3	5.3	4.4	3.3	6.3	6.3	4.2	4.2	4.2	4.4

SHORT ANSWER

11. $1s^22s^22p^63s^23p^64s^23d^{10}4p^3$

12. ${}^{131}_{53}I \longrightarrow {}^{131}_{54}Xe + {}^{0}_{-1}\beta$

13. Because this is tap water, Student A is correct. Tap water contains the compound water along with dissolved gases and minerals. Only pure H_2O would be considered a compound.

EXTENDED RESPONSE

14. It is easier to remove valence electrons from a partially filled shell. For lithium, it requires much more energy to remove the second electron than the first because the second electron is part of a completely filled outer shell. Its removal causes the atom to become unstable, and therefore requires a great deal of energy to remove.

15. Magnesium will have its largest increase in ionization energy for the third ionization energy. The first two ionization energies are how much energy it requires to remove magnesium's two valence electrons. The third ionization energy will disrupt a complete octet, and therefore requires a lot more energy.

SAT SUBJECT TEST: CHEMISTRY

16. T, F
17. T, T, CE
18. T, T
19. T, T

CHAPTER 7 Ionic Compounds and Metals

BIGIDEA **Atoms in ionic compounds are held together by chemical bonds formed by the attraction of oppositely charged ions.**

ESSENTIAL QUESTIONS	RESOURCES TO ASSESS MASTERY
SECTION 1 Ion Formation **1.** What holds atoms together in a chemical bond? **2.** How do positive and negative ions form? **3.** How does ion formation relate to electron configuration? 1 session 0.5 block	**Progress Monitoring** Formative Assessment, pp. 208, 209 Reading Check, p. 207 Section Review, p. 209
SECTION 2 Ionic Bonds and Ionic Compounds **1.** How do ionic bonds form and how are the ions arranged in an ionic compound? **2.** What can you conclude about the strength of ionic bonds based on the physical properties of ionic compounds? **3.** Is ionic bond formation exothermic or endothermic? 1 session 0.5 block	**Progress Monitoring** Formative Assessment, pp. 215, 217 Reading Check, pp. 213, 214 Section Review, p. 217
SECTION 3 Names and Formulas for Ionic Compounds **1.** What is a formula unit and how does it relate to an ionic compound's composition? **2.** How do you write the formulas for compounds formed from different ions and oxyanions? **3.** What are the naming conventions for ionic compounds and oxyanions? 2 sessions 1 block	**Progress Monitoring** Formative Assessment, pp. 220, 222, 223, 224 Reading Check, p. 219 Section Review, p. 224
SECTION 4 Metallic Bonds and the Properties of Metals **1.** What are the characteristics of a metallic bond? **2.** How does the electron sea model account for the physical properties of metals? **3.** What are alloys, and how can they be categorized? 2 sessions 1 block	**Progress Monitoring** Formative Assessment, p. 228 Reading Check, p. 226 Section Review, p. 228 **Summative Assessment** Chapter Assessment, p. 232 *eAssessment* Chapter Test (Scaffolded)

Go online!

LEVELED RESOURCES	LAB MATERIALS
Science Notebook 7.1 OL ***FAST FILE* Chapter Resources:** Study Guide, p. 72 BL	LaunchLAB p. 204 / **15 min** table salt, well plate, disposable pipette, conductivity tester, sugar, distilled water
Science Notebook 7.2 OL ***FAST FILE* Chapter Resources:** ChemLab Worksheet, p. 58 OL Study Guide, p. 73 BL	ChemLAB p. 230 / **45 min** magnesium ribbon, crucible, ring stand and ring, clay triangle, Bunsen burner, stirring rod, crucible tongs, centigram balance, 100-mL beaker, distilled water, conductivity tester
Science Notebook 7.3 OL ***FAST FILE* Chapter Resources:** Study Guide, p. 75 BL	
Science Notebook 7.4 OL ***FAST FILE* Chapter Resources:** MiniLab Worksheet, p. 56 OL Study Guide, p. 77 BL	MiniLAB p. 227 / **25 min** hairpins, forceps, laboratory burner, 250-mL beaker, cold water

ADDITIONAL RESOURCES

eAssessment
ConnectED Student Center
ConnectED Teacher Center
Lesson Presentations
eSolutions
Videos
Animations
Online Quizzes
Virtual Investigations
Multilingual eGlossary
Personal Tutors
WebQuests
Applying Practices Worksheets
PBLs

Practice:

Challenge Problems AL
Supplemental Problems BL OL
Chapter Test (Scaffolded)
Standardized Test Practice

Additional Resources:

Blueprints for Success
Solving Problems: A Chemistry Handbook

Lab Resources:

Laboratory Manual OL
CBL Laboratory Manual OL
Small-Scale Laboratory Manual OL
Forensics Laboratory Manual OL

BL Below Level   OL On Level   AL Advanced Learners   EL English Learners   COOP LEARN Cooperative Learning

CHAPTER 7

LaunchLAB

Rubric

What compounds conduct electricity in solution?

Purpose Students will use a conductivity tester to observe that some compounds conduct electricity in aqueous solution.

Safety Precautions Approve lab safety forms before work begins.

Disposal Solutions and liquids can be safely flushed into the sanitary water system.

Teaching Strategies

- Alternate ionic and covalent compounds can be selected based on availability.
- Extend the lab using common solutions. For example, sport drinks often mention electrolytes on their labels. Most students have heard of electrolytes but do not understand their function. If sport drinks are used, remind students not to drink in the lab.

Expected Results The ionic compounds dissociate in solution and are conductors in aqueous solution. The covalent compounds do not dissociate, and therefore do not conduct electricity.

CHAPTER 7
Ionic Compounds and Metals

BIGIDEA Ionic compounds are held together by chemical bonds formed by the attraction of oppositely charged ions.

SECTIONS

LaunchLAB

What compounds conduct electricity in solution?

For a material to conduct an electric current, it must contain charged particles that can move throughout the substance. Electrical conductivity is a property of matter that tells you something about bonding.

FOLDABLES Study Organizer

Ionic Compounds

Make a trifold book. Label it as shown. Use it to help you organize information about ionic compounds.

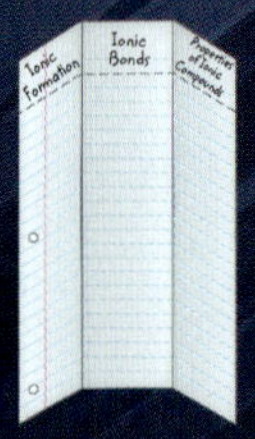

Procedure

1. Read and complete the lab safety form.
2. Make a data table to record your observations.
3. Fill an open well in a **well plate** with **table salt (NaCl).**
4. Use a **disposable pipette** to transfer approximately 1 mL of **table salt (NaCl) solution** to an open well in the well plate.
5. Place the probes of a **conductivity tester** in the well plate containing the solid table salt. If the light is illuminated, the table salt conducts electricity. Repeat with the solution.
6. Repeat Steps 3 to 5 using **sugar ($C_{12}H_{22}O_{11}$)** instead of table salt.
7. Repeat Steps 3 to 5 using **distilled water** instead of tap water.

Analysis

1. **Organize** Make a table listing the compounds and the results of the conductivity tests.

Substance	Conductivity Result
NaCl (solid)	no
NaCl (aq)	yes
$C_{12}H_{22}O_{11}$ (solid)	no
$C_{12}H_{22}O_{11}$ (aq)	no
Distilled water	no

2. **Explain** your results. Table salt formed charged particles when dissolved in water.

Inquiry Create a model to describe how compounds that conduct electricity in solution differ from compounds that do not conduct electricity in solution. The compounds that conduct electricity in solution dissociate into separate charged ions, and this enables the flow of electricity. The compounds do not conduct electricity in solution and do not separate into charged particles.

Metals and ionic compounds surround this scuba diver. For example, the metal tank holds the oxygen gas that keeps the diver alive under water. And the ionic compound calcium carbonate makes up much of the reef the diver is exploring. Even the water itself contains ionic compounds—there are about 35 g of ionic salts dissolved in every liter of sea water.

Introduce the BIGIDEA

Chemical Bonds To introduce this chapter's Big Idea, burn a small strip of magnesium ribbon. Caution students not to look directly at the burning magnesium. Once the reaction is complete, ask students what they observed. Students observe that the compound formed is different from the original magnesium. Ask students what reacted with the magnesium. air, which is composed mainly of nitrogen and oxygen Have students locate magnesium, oxygen, and nitrogen on the periodic table and predict what type of ions they form. Magnesium is located in group 2 and will form a 2+ ion. Oxygen is found in group 16 and will form a 2− ion. Nitrogen is located in group 15 and will form a 3− ion. Tell students that they will create a similar reaction later in the ChemLab.

Tie to Previous Knowledge

Have students review the following concepts before studying this chapter.

- atomic structure
- electron configuration
- periodic trends

Use the Photo

Chemical Bonds in Nature

Have students observe the opening photo and ask students what the coral reef is composed of. calcium carbonate Have students notice the drawing of calcium carbonate. Ask them what they observe. The calcium carbonate contains positive calcium ions and negative carbonate ions. For every 2+ calcium ions there is a 2− carbonate ion. Ask students to contrast the aluminum drawing with the calcium carbonate drawing. They should notice a lack of negatively charged particles in the aluminum. Tell them that the aluminum contains a metallic bond, while the calcium carbonate contains an ionic bond. Explain that the aluminum contains negatively charged particles that are not shown in the illustration. Both bonds are discussed in this chapter.

SECTION 1

1 Focus

MAINIDEA

Electron Configuration Draw the electron configurations of calcium and argon on the board. Ask students to compare the two electron configurations. The electron configuration for calcium has two more valence electrons. Remind students that argon is a noble gas and has a stable electron configuration. Now, draw the electron configuration for chlorine on the board and ask students to compare it to that of argon. Chlorine has one less electron than argon. Ask students what calcium must do to have the electron configuration of argon. Calcium must lose the two outer electrons forming a 2+ ion. What must chlorine do to have the electron configuration of argon? Chlorine must gain one outer electron forming a 1− ion. OL

2 Teach

Concept Development

Electron Configurations On the board, write the complete electron configurations of potassium and argon. Ask students to compare and contrast them. They are the same except that potassium also has a $4s^1$ electron. Ask students why argon is particularly stable. It is a noble gas and has eight valence electrons. Ask students how potassium could achieve a noble gas configuration. The potassium must lose the $4s^1$ electron. Remind students that potassium becomes 1+ due to the extra proton in the nucleus. Continue by discussing the configurations of calcium and gallium. OL

SECTION 1

Essential Questions

- What holds atoms together in a chemical bond?
- How do positive and negative ions form?
- How does ion formation relate to electron configuration?

Review Vocabulary

octet rule: atoms tend to gain, lose, or share electrons in order to acquire eight valence electrons

New Vocabulary

chemical bond
cation
anion

Ion Formation

MAINIDEA Ions form when atoms gain or lose valence electrons to achieve a stable octet electron configuration.

CHEM 4 YOU Imagine that you and a group of friends go to a park to play soccer. There, you meet a larger group that also wants to play. To form even teams, one group loses members and the other group gains members. Atoms sometimes behave in a similar manner to form compounds.

Valence Electrons and Chemical Bonds

Imagine going on a scuba dive, diving below the ocean's surface and observing the awe-inspiring world below. You might explore the colorful and exotic organisms teeming around a coral reef, such as the one shown in **Figure 1.** The coral is formed from a compound called calcium carbonate, which is just one of thousands of compounds found on Earth. How do so many compounds form from the relatively few elements known to exist? The answer to this question involves the electron structure of atoms and the nature of the forces between atoms.

In previous chapters, you learned that elements within a group on the periodic table have similar properties. Many of these properties depend on the number of valence electrons the atom has. These valence electrons are involved in the formation of chemical bonds between two atoms. A **chemical bond** is the force that holds two atoms together. Chemical bonds can form by the attraction between the positive nucleus of one atom and the negative electrons of another atom, or by the attraction between positive ions and negative ions. This chapter discusses chemical bonds formed by ions, atoms that have acquired a positive or negative charge. You will learn about bonds that form from the sharing of electrons in a later chapter.

Figure 1 As carbon dioxide dissolves in ocean water, carbonate ions are produced. Coral polyps capture these carbonate ions, producing crystals of calcium carbonate, which they secrete as an exoskeleton. Over time, the coral reef forms. A coral reef is a complex habitat that supports coral, algae, mollusks, echinoderms, and a variety of fishes.

David Nardini/Getty Images

Explore electron-dot structures with an interactive table. Concepts In Motion

Table 1 Electron-Dot Structures

Group	1	2	13	14	15	16	17	18
Diagram	Li·	·Be·	·Ḃ·	·Ċ·	·N̈·	·Ö:	:F̈:	:N̈e:

Valence electrons Recall that an electron-dot structure is a type of diagram used to keep track of valence electrons. Electron-dot structures are especially helpful when used to illustrate the formation of chemical bonds. **Table 1** shows several examples of electron-dot structures. For example, carbon, with an electron configuration of $1s^22s^22p^2$, has four valence electrons in the second energy level. These valence electrons are represented by the four dots around the symbol C in the table.

Also, recall that ionization energy refers to how easily an atom loses an electron and that electron affinity indicates how much attraction an atom has for electrons. Noble gases, which have high ionization energies and low electron affinities, show a general lack of chemical reactivity. Other elements on the periodic table react with each other, forming numerous compounds. The difference in reactivity is directly related to the valence electrons.

The difference in reactivity involves the octet—the stable arrangement of eight valence electrons in the outer energy level. Unreactive noble gases have electron configurations that have a full outermost energy level. This level is filled with two electrons for helium ($1s^2$) and eight electrons for the other noble gases (ns^2np^6). Elements tend to react to acquire the stable electron structure of a noble gas.

FOLDABLES® Incorporate information from this section into your Foldable.

Positive Ion Formation

A positive ion forms when an atom loses one or more valence electrons in order to attain a noble gas configuration. A positively charged ion is called a **cation.** To understand the formation of a positive ion, compare the electron configurations of the noble gas neon (atomic number 10) and the alkali metal sodium (atomic number 11).

Neon atom (Ne)	$1s^22s^22p^6$
Sodium atom (Na)	$1s^22s^22p^63s^1$

Note that the sodium atom has one 3s valence electron; it differs from the noble gas neon by that single valence electron. When sodium loses this outer valence electron, the resulting electron configuration is identical to that of neon. **Figure 2** shows how a sodium atom loses its valence electron to become a sodium cation.

By losing an electron, the sodium atom acquires the stable outer-electron configuration of neon. It is important to understand that although sodium now has the electron configuration of neon, it is not neon. It is a sodium ion with a single positive charge. The 11 protons that establish the character of sodium still remain within its nucleus.

READING CHECK **Identify** the number of electrons in the outermost energy level that are associated with maximum stability.

Figure 2 In the formation of a positive ion, a neutral atom loses one or more valence electrons. The atom is neutral because it contains equal numbers of protons and electrons; the ion, however, contains more protons than electrons and has a positive charge.

Analyze ***Does the removal of an electron from a neutral atom require energy or release energy?***

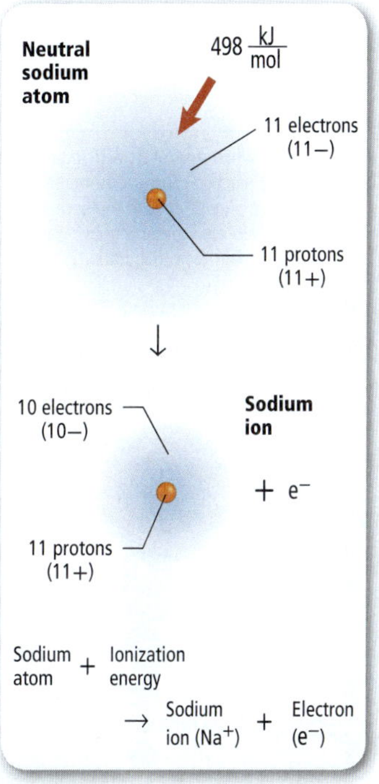

Visual Learning

Table 1 Tell students that G. N. Lewis developed electron dot diagrams as a simple way to show and keep track of valence electrons. Have students look at Table 1 and note that up to two dots can be placed on each of the four sides of a symbol. Tell students that all four sides of the symbol are equivalent. Have students write and compare the electron-dot diagrams for several elements. OL

Reinforcement

Electron-Dot Structures Give students blank periodic tables. Ask them to locate several elements and to draw electron-dot diagrams for atoms of those elements. Remind students that valence electrons can only be s and p electrons. OL EL

FOLDABLES® Rubric

READING CHECK Eight electrons, also known as an octet, are associated with maximum stability.

Caption Question Fig. 2 It requires energy.

DIFFERENTIATED INSTRUCTION

Struggling Learners Show that when an atom becomes stable, it does not become a noble gas. Place a circle of transparency film on an overhead projector. In the center, place 11 small beads to represent the protons in the sodium nucleus. Ask students to determine the charge on the nucleus. 11+ Represent the energy levels in the atom by placing 2 beads, 8 beads, and 1 bead, respectively, in inner, middle, and outer concentric circles positioned around the nucleus. Determine the charge of the electron cloud. 11− Remove the outermost bead, which represents the valence electron, and have students determine the charge of the ion. 1+ Remind students that the number of protons, not the number of electrons, determines the identity of the atom. BL EL

Concept Development

Transition Elements The electron configurations of some transition metals show a limitation of the octet rule. Transition metals generally do not form ions with noble-gas configurations. In forming ions, transition metals lose s electrons first, but they can also lose d electrons. Write the electron configuration for the element iron. $[Ar]\,3d^6 4s^2$ Iron forms two ions, Fe^{2+} and Fe^{3+}. When forming Fe^{2+}, the two 4s electrons are lost, resulting in $[Ar]3d^6$. This configuration is only somewhat stable, and the ion will readily lose a 3d electron, forming Fe^{3+}. The electron configuration for Fe^{3+} is $[Ar]3d^5$. This half-full d sublevel is not as stable as an octet, but it is more stable than the original configuration for iron or for Fe^{2+}. **OL**

ASSESSMENT

Skill Have students write the electron configuration for each of the following ions and determine which ions possess noble gas configurations and which possess pseudo-noble gas configurations: Ti^{4+}, Se^{2-}, Ge^{3+}, Sc^{3+}. All but Ge^{3+} form noble gas configurations. **OL**

Table 2 Group 1, 2, and 13 Ions

Group	Configuration	Charge of Ion Formed
1	[noble gas] ns^1	1+ when the s^1 electron is lost
2	[noble gas] ns^2	2+ when the s^2 electrons are lost
13	[noble gas] ns^2np^1	3+ when the s^2p^1 electrons are lost

Metal ions Metals atoms are reactive because they lose valence electrons easily. The group 1 and 2 metals are the most reactive metals on the periodic table. For example, potassium and magnesium, group 1 and 2 elements, respectively, form K^+ and Mg^{2+} ions. Some group 13 atoms also form ions. The ions formed by metal atoms in groups 1, 2, and 13 are summarized in **Table 2.**

Transition metal ions Recall that, in general, transition metals have an outer energy level of ns^2. Going from left to right across a period, atoms of each element fill an inner d sublevel. When forming positive ions, transition metals commonly lose their two valence electrons, forming 2+ ions. However, it is also possible for d electrons to be lost. Thus, transition metals also commonly form ions of 3+ or greater, depending on the number of d electrons in the electron structure. It is difficult to predict the number of electrons that will be lost. For example, iron (Fe) forms both Fe^{2+} and Fe^{3+} ions. A useful rule of thumb for these metals is that they form ions with a 2+ or a 3+ charge.

Pseudo-noble gas configurations Although the formation of an octet is the most stable electron configuration, other electron configurations can also provide some stability. For example, elements in groups 11–14 lose electrons to form an outer energy level containing full s, p, and d sublevels. These relatively stable electron arrangements are referred to as pseudo-noble gas configurations. In **Figure 3,** the zinc atom has the electron configuration of $1s^2 2s^2 2p^6 3s^2 3p^6 4s^2 3d^{10}$. When forming an ion, the zinc atom loses the two 4s electrons in the outer energy level, and the stable configuration of $1s^2 2s^2 2p^6 3s^2 3p^6 3d^{10}$ results in a pseudo-noble gas configuration.

Figure 3 When zinc reacts with iodine, the heat of the reaction causes solid iodine to sublimate into a purple vapor. At the bottom of the tube, ZnI_2 is formed containing Zn^{2+} ions with a pseudo-noble gas configuration.

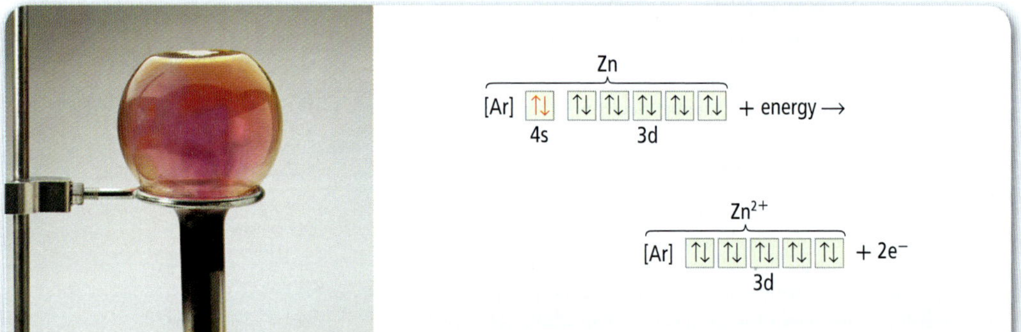

When the two 4s valence electrons are lost, a stable pseudo-noble gas configuration consisting of filled s, p, and d sublevels is achieved. Note that the filled 3s and 3p orbitals exist as part of the [Ar] configuration.

CHEMISTRY PROJECT

Group Numbering Have students draw a simplified periodic table and label the column headings (groups). Instruct them to draw, in each cell on the table, the Lewis structure for each element in period 2. Ask them to study each structure and its number of valence electrons and relate these to the Group 1 to 8 numbering system. **OL**

CHEMISTRY JOURNAL

Summary of Ion Formation Have students make charts in their chemistry journals with the following column heads: *Symbol of Element, Electron Configuration, Electron-dot Diagram, Ion with Charge, Number of Electrons Lost or Gained,* and *Name of Ion.* Using the elements cesium, barium, arsenic, iodine, zinc, and phosphorus, students should complete the charts. **OL**

Table 3 Group 15–17 Ions

Group	Configuration	Charge of Ion Formed
15	[noble gas] ns^2np^3	3– when three electrons are gained
16	[noble gas] ns^2np^4	2– when two electrons are gained
17	[noble gas] ns^2np^5	1– when one electron is gained

Negative Ion Formation

Nonmetals, which are located on the right side of the periodic table, easily gain electrons to attain a stable outer electron configuration. Examine **Figure 4.** To attain a noble-gas configuration, chlorine gains one electron, forming an ion with a 1– charge. After gaining the electron, the chloride ion has the electron configuration of an argon atom.

Chlorine atom (Cl)	$1s^22s^22p^63s^23p^5$
Argon atom (Ar)	$1s^22s^22p^63s^23p^6$
Chloride ion (Cl^-)	$1s^22s^22p^63s^23p^6$

An **anion** is a negatively charged ion. To designate an anion, the ending *-ide* is added to the root name of the element. Thus, a chlorine atom becomes a chloride anion. What is the name of the nitrogen anion?

Nonmetal ions As shown in **Table 3,** nonmetals gain the number of electrons that, when added to their valence electrons, equals 8. For example, consider phosphorus, with five valence electrons. To form a stable octet, the atom gains three electrons and forms a phosphide ion with a 3– charge. Likewise, oxygen, with six valence electrons, gains two electrons and forms an oxide ion with a 2– charge.

Some nonmetals can lose or gain other numbers of electrons to form an octet. For example, in addition to gaining three electrons, phosphorus can lose five. However, in general, group 15 elements gain three electrons, group 16 elements gain two, and group 17 elements gain one to achieve an octet.

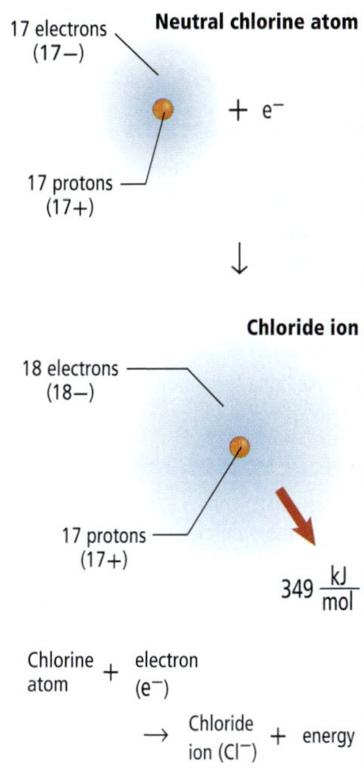

Figure 4 During the formation of the negative chloride ion, a neutral atom gains one electron. The process releases 349 kJ/mol of energy.

Compare *How do the energy changes accompanying positive ion and negative ion formation compare?*

SECTION 1 REVIEW

Section Self-Check

Section Summary

- A chemical bond is the force that holds two atoms together.
- Some atoms form ions to gain stability. A stable configuration involves a complete outer energy level, usually consisting of eight valence electrons.
- Ions are formed by the loss or gain of valence electrons.
- The number of protons remains unchanged during ion formation.

1. MAINIDEA **Compare** the stability of a lithium atom with that of its ion, Li^+.
2. **Describe** two different causes of the force of attraction in a chemical bond.
3. **Apply** Why are all of the elements in group 18 relatively unreactive, whereas those in group 17 are very reactive?
4. **Summarize** ionic bond formation by correctly pairing these terms: *cation, anion, electron gain,* and *electron loss.*
5. **Apply** Write out the electron configuration for each atom. Then, predict the change that must occur in each to achieve a noble-gas configuration.
   a. nitrogen **b.** sulfur **c.** barium **d.** lithium
6. **Model** Draw models to represent the formation of the positive calcium ion and the negative bromide ion.

- **In-Text Question** nitride
- **Caption Question Fig. 4** cation formation requires energy; anion formation releases energy

3 Assess

Check for Understanding

Ask students to determine the ions formed from strontium, aluminum, and sulfur. Sr^{2+}, Al^{3+}, S^{2-} OL EL

Reteach

Write the symbol for fluorine on the board, and have students determine its electron configuration. $1s^22s^22p^5$ Have students calculate the number of valence electrons. 7 Ask students what ion will form when fluorine reacts. F^- What is the name of this ion? fluoride OL EL

Extension

On a blank periodic table, have students write the formulas for ions formed from various elements. OL

ASSESSMENT

Knowledge Have each student write five representative elements that have an atomic number of 54 or lower on a piece of paper. Another student should write the electron configuration for each element, draw the electron dot diagram, and determine what type of ion is most likely formed. OL

SECTION 1 REVIEW

1. The Li^+ ion is more stable because it has a complete octet.
2. The two causes are the attraction between the positive nucleus of one atom and the negative electrons of another atom, and the attraction between positive ions and negative ions.
3. The group 18 elements, known as noble gases, have complete outer energy levels and do not easily form ions. The group 17 elements are highly reactive because they need only to gain a single electron to form an octet.
4. anion, electron gain, cation, electron loss
5. **a.** [He]$2s^22p^3$; gain 3 electrons (3– ion) or lose 5 electrons (5+ ion)
   b. [Ne]$3s^23p^4$; gain 2 electrons (2– ion)
   c. [Xe]$6s^2$; lose 2 electrons (2+ ion)
   d. [He]$2s^1$; lose 1 electron (1+ ion)
6. Models should show that the calcium atom loses two electrons, forming Ca^{2+}, and that bromine gains one electron, forming Br^-. The models should also show the additional energy to form the calcium ion, and the release of energy to form the bromide ion.

SECTION 2

1 Focus

MAINIDEA

Charges on Ions Bring a balloon into class, rub it on your hair, and stick it to the wall. Ask students why the balloon sticks to the wall. The wall has a charge opposite that of the balloon and oppositely charged substances attract. Draw the electron configuration for calcium on the board. Ask students to explain how it forms a 2+ ion. The calcium atom loses the two outer electrons. Ask students what happens to the positively charged calcium ion. It will attract a negatively charged particle. OL

2 Teach

Extension

Ionic Compounds in Paint Have students check the label of ingredients on paint cans, either from their home or a neighborhood paint supplier. With students working in groups, ask them to determine which of the ingredients are ionic. Answers will vary; ionic compounds present are usually titanium dioxide, calcium carbonate, calcium silicate, and aluminum silicate. OL COOP LEARN

SECTION 2

Ionic Bonds and Ionic Compounds

MAINIDEA Oppositely charged ions attract each other, forming electrically neutral ionic compounds.

CHEM 4 YOU Have you ever tried to separate sheets of plastic wrap that are stuck together? The hard-to-separate layers attract each other due to their oppositely charged surfaces.

Essential Questions

- How do ionic bonds form and how are the ions arranged in an ionic compound?
- What can you conclude about the strength of ionic bonds based on the physical properties of ionic compounds?
- Is ionic bond formation exothermic or endothermic?

Review Vocabulary

compound: a chemical combination of two or more different elements

New Vocabulary

ionic bond
ionic compound
crystal lattice
electrolyte
lattice energy

Formation of an Ionic Bond

What do the reactions shown in **Figure 5** have in common? In both cases, elements react with each other to form a compound. **Figure 5a** shows the reaction between the elements sodium and chlorine. During this reaction, a sodium atom transfers its valence electron to a chlorine atom and becomes a positive ion. The chlorine atom accepts the electron into its outer energy level and becomes a negative ion. The oppositely charged ions attract each other, forming the compound sodium chloride. The electrostatic force that holds oppositely charged particles together in an ionic compound is referred to as an **ionic bond.** Compounds that contain ionic bonds are **ionic compounds.** If ionic bonds occur between metals and the nonmetal oxygen, oxides form. Most other ionic compounds are called salts.

Binary ionic compounds Thousands of compounds contain ionic bonds. Many ionic compounds are binary, which means that they contain only two different elements. Binary ionic compounds contain a metallic cation and a nonmetallic anion. Sodium chloride (NaCl) is a binary compound because it contains two different elements, sodium and chlorine. Magnesium oxide (MgO), the reaction product shown in **Figure 5b,** is also a binary ionic compound.

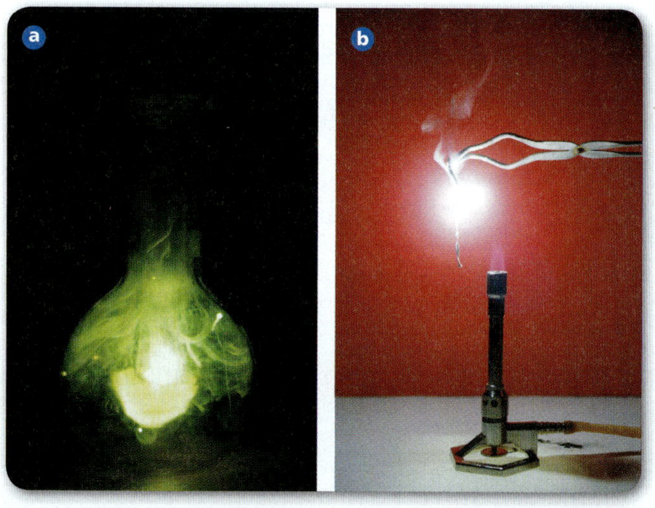

(l)Andrew Lambert Photography/Photo Researchers, (r)Charles D. Winters/Photo Researchers

Figure 5 Each of these chemical reactions produces an ionic compound while releasing a large amount of energy. **a.** The reaction that occurs between elemental sodium and chlorine gas produces a white crystalline solid. **b.** When a ribbon of magnesium metal burns in air, it forms the ionic compound magnesium oxide.

CHEMISTRY PROJECT

Form an Ionic Bond Have students write a story or create a model about the formation of an ionic bond. Then, have them read their story or present their models to the class. OL EL

Compound formation and charge What role does ionic charge play in the formation of ionic compounds? To answer this question, examine how calcium fluoride forms. Calcium has the electron configuration $[Ar]4s^2$, and needs to lose two electrons to attain the stable configuration of argon. Fluorine has the configuration $[He]2s^22p^5$, and must gain one electron to attain the stable configuration of neon. Because the number of electrons lost and gained must be equal, two fluorine atoms are needed to accept the two electrons lost from the calcium atom.

$$1 \text{ Ca }\cancel{\text{ion}}\left(\frac{2+}{\text{Ca }\cancel{\text{ion}}}\right) + 2 \text{ F }\cancel{\text{ions}}\left(\frac{1-}{\text{F }\cancel{\text{ion}}}\right) = (1)(2+) + (2)(1-) = 0$$

As you can see, the overall charge of one unit of calcium fluoride (CaF_2) is zero. **Table 4** summarizes several ways in which the formation of an ionic compound such as sodium chloride can be represented.

FOLDABLES® Incorporate information from this section into your Foldable.

View an **animation of ionic bond formation in sodium chloride.** Concepts In Motion

Table 4 Formation of Sodium Chloride

Chemical Equation

$Na + Cl \rightarrow Na^+ + Cl^- + \text{energy}$

Electron Configurations

One electron is transferred.

$\underbrace{[Ne]3s^1}_{Na} + \underbrace{[Ne]3s^23p^5}_{Cl} \rightarrow \underbrace{[Ne]}_{Na^+} + \underbrace{[Ar]}_{Cl^-} + \text{energy}$

Orbital Notation

One electron is transferred.

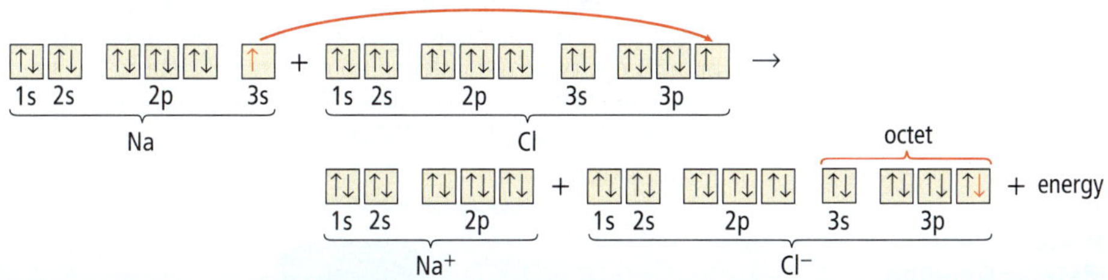

Electron-Dot Structures

One electron is transferred.

Na· + ·C̈l: → [Na]⁺ + [:C̈l:]⁻ + energy

Atomic Models

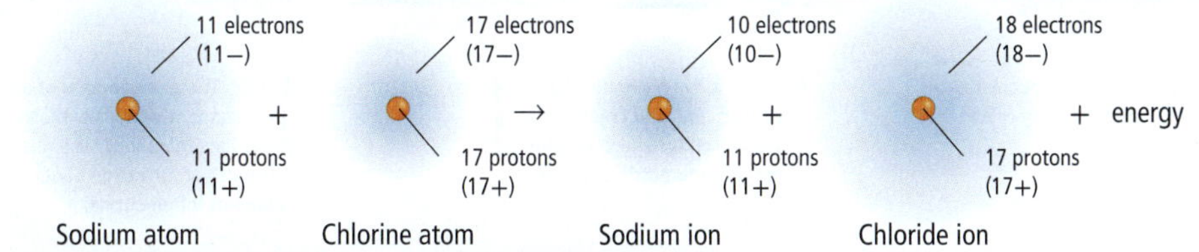

Quick Demo

Conductivity of Ionic Compounds Demonstrate that ionic compounds do not conduct an electrical current in the solid state, but that they do conduct an electrical current in both the molten state and in solution. Place some sodium chloride crystals in a watch glass. Insert a conductivity apparatus and have the students observe. The bulb will not light, indicating no flow of electricity. Remove the tester and dissolve the sodium chloride in the water. Insert the conductivity apparatus and observe. The bulb will light. Ask students why the sodium chloride did not conduct in the crystalline form but does in solution. In the solid phase, the ions are unable to move, whereas in solution, they do. Use the conductivity tester to confirm that solid lithium chloride does not conduct a current, but molten lithium chloride does. If your laboratory heat source will not melt the compound, discuss with students what would happen. OL EL

FOLDABLES® Rubric

DIFFERENTIATED INSTRUCTION

Struggling Learners Give each student between one and seven marbles. Explain that the number of marbles they have corresponds to the number of electrons in the highest energy levels of an element. Ask them to find another student that has the correct number of marbles that would enable both to complete their energy levels and form ionic bonds. BL EL COOP LEARN

PRACTICE Problems

Have students refer to the Selected Solutions appendix for complete solutions to odd-numbered problems. The complete solutions for all problems can be found in the Solutions Manual.

7. Three Na atoms each lose 1 e−, forming 1+ ions. One N atom gains 3 e−, forming a 3− ion. The ions attract, forming Na_3N. The overall charge on one formula unit of Na_3N is zero.
8. Two Li atoms each lose 1 e−, forming 1+ ions. One O atom gains 2 e−, forming a 2− ion. The ions attract, forming Li_2O. The overall charge on one formula unit of Li_2O is zero.
9. One Sr atom loses 2 e−, forming a 2+ ion. Two F atoms each gain 1 e−, forming 1− ions. The ions attract, forming SrF_2. The overall charge on one formula unit of SrF_2 is zero.
10. Two Al atoms each lose 3 e−, forming 3+ ions. Three S atoms gain 2 e− each, forming 2− ions. The ions attract, forming Al_2S_3. The overall charge on one formula unit of Al_2S_3 is zero.
11. Three group 1 atoms loses 1 e−, forming 1+ ions. One group 15 atom gains 3 e−, forming a 3− ion. The ions attract, forming X_3Y, where *X* represents a group 1 atom and *Y* represents a group 15 atom.

Next, consider aluminum oxide, the whitish coating that forms on aluminum chairs. To acquire a noble-gas configuration, each aluminum atom loses three electrons and each oxygen atom gains two electrons. Thus, three oxygen atoms are needed to accept the six electrons lost by two aluminum atoms. The neutral compound formed is aluminum oxide (Al_2O_3).

$$2\text{ Al }\cancel{\text{ions}}\left(\frac{3+}{\text{Al }\cancel{\text{ion}}}\right) + 3\text{ O }\cancel{\text{ions}}\left(\frac{2-}{\text{O }\cancel{\text{ion}}}\right) = 2(3+) + 3(2-) = 0$$

PRACTICE Problems

Do additional problems. Online Practice

Explain how an ionic compound forms from these elements.

7. sodium and nitrogen
8. lithium and oxygen
9. strontium and fluorine
10. aluminum and sulfur
11. **Challenge** Explain how elements in the two groups shown on the periodic table at the right combine to form an ionic compound.

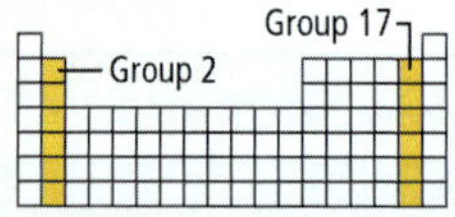

Properties of Ionic Compounds

The chemical bonds in a compound determine many of its properties. For ionic compounds, the ionic bonds produce unique physical structures, unlike those of other compounds. The physical structures of ionic compounds also contribute to their physical properties. These properties have been used in many applications, as discussed in **Figure 6.**

Physical structure In an ionic compound, large numbers of positive ions and negative ions exist together in a ratio determined by the number of electrons transferred from the metal atom to the nonmetal atom. These ions are packed into a regular repeating pattern that balances the forces of attraction and repulsion between the ions.

CBW/Alamy

■ **Figure 6**

Milestones in Ionic and Metallic Bonding

A series of discoveries helped scientists understand the properties of ionic and metallic substances—leading to the creation of new tools and materials.

1897 J. J. Thomson speculates that electrons play a key role in chemical bonding.

1900

1910

1913 X-ray crystallography reveals sodium ions and chloride ions in sodium chloride are arranged in regular geometric patterns.

1916 Gilbert Lewis proposes a bonding theory based on the interaction of electrons among atoms.

1930

1932 The development of an electronegativity scale allows scientists to quantify the relative strength of attraction of each element for electrons.

CIRCA 1940 Metallurgists develop alloys that perform under extreme temperature, pressure, and centrifugal force. Such alloys are later used in jet engines and spacecraft.

CHEMISTRY PROJECT

Shape Memory Alloys Provide supplemental readings describing the properties and current applications of shape memory alloys (SMA). After reading, have students divide into small groups to brainstorm additional SMA applications. AL OL

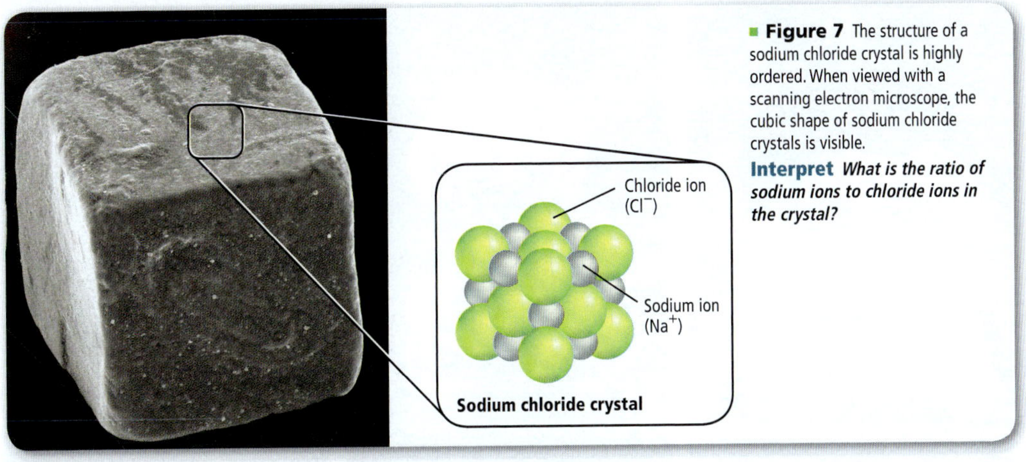

■ **Figure 7** The structure of a sodium chloride crystal is highly ordered. When viewed with a scanning electron microscope, the cubic shape of sodium chloride crystals is visible.

Interpret *What is the ratio of sodium ions to chloride ions in the crystal?*

Examine the pattern of the ions in the sodium chloride crystal shown in **Figure 7.** Note the highly organized nature of an ionic crystal—the consistent spacing of the ions and the uniform pattern formed by them. Although the ion sizes are not the same, each sodium ion in the crystal is surrounded by six chloride ions, and each chloride ion is surrounded by six sodium ions. What shape would you expect a large crystal of this compound to be? As shown in **Figure 7,** the one-to-one ratio of sodium and chloride ions produces a highly ordered cubic crystal. As in all ionic compounds, in NaCl, no single unit consisting of only one sodium ion and one chloride ion is formed. Instead, large numbers of sodium ions and chloride ions exist together. If you can, obtain a magnifying lens and use it to examine some crystals of table salt (NaCl). What is the shape of these small salt crystals?

☑ READING CHECK **Explain** what determines the ratio of positive ions to negative ions in an ionic crystal.

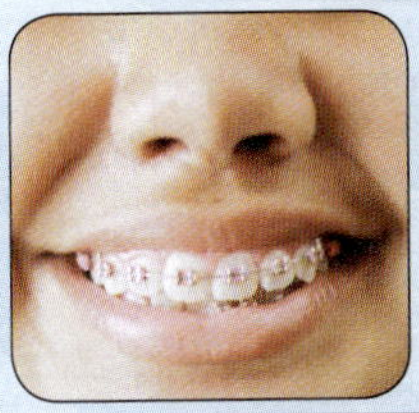

1962 A nickel-titanium alloy with "shape memory" is discovered. The alloy reverts to its original shape after being deformed. Dental braces are one of many applications.

2004 Scientists develop a nickel-gadolinium alloy that absorbs radioactive neutrons emitted by nuclear waste. Applications include transport and storage of highly radioactive fuel.

1970 — 1990 — 2000 — 2010

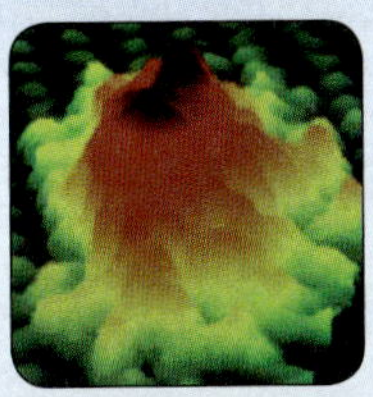

1981 Invention of the scanning tunneling microscope allows researchers to study atomic-scale images in three dimensions.

2009 U.S. aluminum can recycling rate reaches an all-time high of 57%. Recycling these cans saves roughly 95% of the energy required to extract aluminum from its ore.

DIFFERENTIATED INSTRUCTION

Advanced Learners Have students choose one possible application of SMA. Design a strategy for developing and marketing this application. The strategy should include a design sketch, a list of potential users, and marketing suggestions. Answers will vary, but might include applications in medical, defense, and aerospace industries, as well as products related to health and safety and home comfort and efficiency. AL

Content Background

Linus Pauling Linus Pauling is the only person ever to receive two unshared Nobel Prizes, for Chemistry in 1954 and for Peace in 1962. His prolific scientific research was not limited to the nature of chemical bonds, but included important discoveries in genetic diseases, immunology, brain function, and nutritional therapy. Pauling became a well-known public figure in America as a proponent of the health benefits of vitamin C and an activist against the development, testing, and abuse of nuclear weapons after World War II. Linus Pauling died in 1994.

■ **Caption Question Fig. 7**
one-to-one

☑ **READING CHECK** The ratio of ions in the crystal depends on the charges of the ions making up the compound.

Apply Chemistry

Common Ionic Compounds
Some common household items are made from ionic compounds. The following list gives examples of household items and the ionic compounds that are used to produce them.

- Carpet: calcium carbonate
- Ceramics/Glass: calcium carbonate, lithium borate, magnesium silicate, sodium carbonate
- Glossy Paper: aluminum silicate, calcium carbonate, sodium sulfate, calcium oxide, sodium carbonate, titanium(IV) oxide
- Toothpaste: calcium carbonate, sodium carbonate, sodium-hydrogen carbonate, sodium fluoride

Math in Chemistry

Lattice Energy and Interionic Distance Lattice energy is a result of the electrostatic attractions between oppositely charged ions. There is a relationship between lattice energy and the interionic distance, which is the sum of the ionic radii of a positive and a negative ion. Have students determine the interionic distance for LiF, LiCl, and LiI using the following ionic radii: $Li^+ = 68$ pm, $F^- = 133$ pm, $Cl^- = 181$ pm, and $I^- = 220$ pm. The lattice energy for each of these compounds is: LiF = 1030 kJ/mol, LiCl = 834 kJ/mol, and LiI = 730 kJ/mol. Have them graph the interionic radii on the x-axis and lattice energy as a positive value on the y-axis. What does the graph indicate? An inverse relationship; as interionic distance increases, lattice energy decreases. Use the graph to estimate the lattice energy for LiBr. The ionic radius of $Br^- = 196$ pm. 788 kJ/mol AL

☑ READING CHECK beryl is a silicate; aragonite is a carbonate

Figure 8 Aragonite ($CaCO_3$), barite ($BaSO_4$), and beryl ($Be_3Al_2Si_6O_{18}$) are examples of minerals that are ionic compounds. The ions that form them are bonded together in a crystal lattice. Differences in ion size and charge result in different ionic crystal shapes, a topic that will be discussed later.

The strong attractions among the positive ions and the negative ions in an ionic compound result in the formation of a crystal lattice. A **crystal lattice** is a three-dimensional geometric arrangement of particles. In a crystal lattice, each positive ion is surrounded by negative ions, and each negative ion is surrounded by positive ions. Ionic crystals vary in shape due to the sizes and relative numbers of the ions bonded, as shown by the minerals in **Figure 8.**

Connection to Earth Science The minerals shown in **Figure 8** are just a few of the types studied by mineralogists, scientists who study minerals. They make use of several classification schemes to organize the thousands of known minerals. Color, crystal structure, hardness, chemical, magnetic, and electric properties, and numerous other characteristics are used to classify minerals. The types of anions minerals contain can also be used to identify them. For example, more than one-third of all known minerals are silicates, which are minerals that contain an anion that is a combination of silicon and oxygen. Halides contain fluoride, chloride, bromide, or iodide ions. Other mineral classes include boron-containing anions known as borates and carbon-oxygen containing anions known as carbonates.

☑ READING CHECK **Identify** the mineral shown in **Figure 8** that is a silicate. Identify the mineral that is a carbonate.

Physical properties Melting point, boiling point, and hardness are physical properties of matter that depend on how strongly the particles that make up the matter are attracted to one another. Another property—the ability of a material to conduct electricity—depends on the availability of freely moving charged particles. Ions are charged particles, so whether they are free to move determines whether an ionic compound conducts electricity. In the solid state, the ions in an ionic compound are locked into fixed positions by strong attractive forces. As a result, ionic solids do not conduct electricity.

CHEMISTRY JOURNAL

Ionic Compound Using the methods shown in Table 4, have students show how an ionic compound between calcium and phosphorus forms by drawing diagrams in their journals. OL EL

Table 5 Melting and Boiling Points of Some Ionic Compounds

Compound	Melting Point (°C)	Boiling Point (°C)
NaI	660	1304
KBr	734	1435
NaBr	747	1390
$CaCl_2$	782	>1600
NaCl	801	1413
MgO	2852	3600

The situation changes dramatically, however, when an ionic solid melts to become a liquid or is dissolved in solution. The ions—previously locked in position—are now free to move and conduct an electric current. Both ionic compounds in solution and in the liquid state are excellent conductors of electricity. An ionic compound whose aqueous solution conducts an electric current is called an **electrolyte.** You will learn more about solutions of electrolytes later.

Because ionic bonds are relatively strong, ionic crystals require a large amount of energy to be broken apart. Thus, ionic crystals have high melting points and high boiling points, as shown in **Table 5.** Many crystals, including gemstones, have brilliant colors. These colors are due to the presence of transition metals in the crystal lattices.

Ionic crystals are also hard, rigid, brittle solids due to the strong attractive forces that hold the ions in place. When an external force is applied to the crystal—a force strong enough to overcome the attractive forces holding the ions in position within the crystal—the crystal cracks or breaks apart, as shown in **Figure 9.** The crystal breaks apart because the applied force repositions the like-charged ions next to each other; the resulting repulsive force breaks apart the crystal.

VOCABULARY

SCIENCE USAGE V. COMMON USAGE

Conduct

Science usage: the ability to transmit light, heat, sound, or electricity

The material did not conduct electricity well.

Common usage: to guide or lead

It was the manager's job to conduct the training session.

Figure 9 Strong attractive forces hold the ions in place until a force strong enough to overcome the attraction is applied.

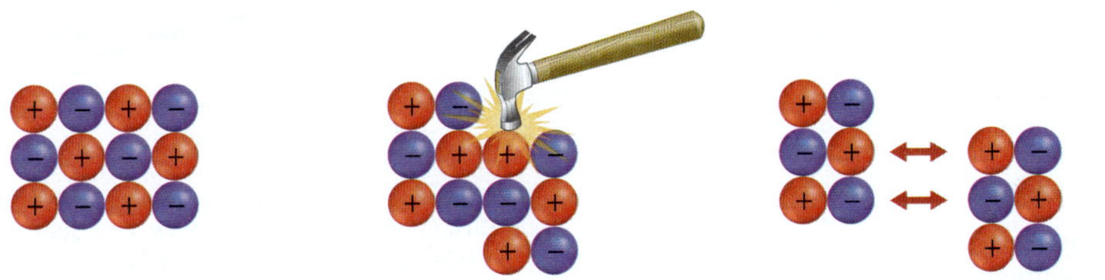

Undisturbed ionic crystal
Before the force is applied, the crystal has a uniform pattern of ions.

Applied force realigns particles.
If the applied force is strong enough, it pushes the ions out of alignment.

Forces of repulsion break crystal apart.
A repulsive force created by nearby like-charged ions breaks apart the crystal.

ASSESSMENT

Performance Using the following data, have students design a graph or table to show trends in lattice energy.

Compound	Lattice energy (kJ/mol)
LiF	1030
LiCl	834
LiBr	788
LiI	730
NaF	910
NaCl	769
NaBr	732
NaI	682
KF	808
KCl	701
KBr	671
KI	632

Have students predict the shape of the graph of the lattice energies of RbF, RbCl, RbBr, and RbI. Have them make graphs that test their predictions. OL EL

ChemLAB

The ChemLab located at the end of the chapter can be used at this point in the lesson.

CHEMISTRY PROJECT

Ions in Biology Many ions play key roles in biological processes. These ions include the ions of sodium, potassium, magnesium, calcium, chloride, iodide, and fluoride. Have students research the importance of these ions and prepare a presentation. OL

Data Analysis LAB

About the Lab

- Scientists embedded silver bromide into the polymer NPVP in hopes of developing antimicrobial materials and surfaces that can be used in personal hygiene, biomedical devices, and the food industry.
- Silver ions exhibit broad spectrum biocidal activity in fighting different bacteria, fungi and viruses. Silver containing materials are widely used in biomedical industries because of their biocidal properties.
- The ionic compound silver bromide is embedded into a polymer to determine if it has antimicrobial properties against both gram positive and gram negative bacteria.
- The polymer NPVP captures and kills bacteria which adhere to the surface. The surface of the polymer is embedded with nanoparticles of silver bromide. The Ag^+ nanoparticles continue to kill bacteria after the surface of the polymer is completely covered with dead bacteria.
- Students are to analyze graphs of different materials with antimicrobial properties over a period of time.
- Also see "New Nanotechnology Receives FDA Approval—Setting Stage For New Era In Battle Against Hospital Related Infections," December 12, 2005. *Nano Techwire;* and "Nano Silver Fights Infections," August 1, 2005, © 2005 by United Press International.

Think Critically

1. Both 21% and 43% NPVP show antimicrobial properties. The addition of silver bromide ions greatly increases the antimicrobial properties of 21% NPVP and 43% NPVP.
2. Silver nitrate ($AgNO_3$) and the two composites (AgBr/43% NPVP and AgBr/21% NPVP) were able to reduce the bacteria population to zero. The time needed to reduce the bacteria population to zero for each substance is as follows: $AgNO_3$, 50 minutes; AgBr/21% NPVP, ~85 minutes; AgBr/43% NPVP, ~200 minutes.
3. Yes, the two composites, AgBr/43% NPVP and AgBr/21% NPVP, both show significant antimicrobial properties. Both composites produced zero or nearly zero bacteria populations after 50 minutes.

Energy and the Ionic Bond

During every chemical reaction, energy is either absorbed or released. If energy is absorbed during a chemical reaction, the reaction is endothermic. If energy is released, it is exothermic.

The formation of ionic compounds from positive ions and negative ions is always exothermic. The attraction of the positive ion for the negative ions close to it forms a more stable system that is lower in energy than the individual ions. If the amount of energy released during bond formation is reabsorbed, the bonds holding the positive ions and negative ions together will break apart.

Lattice energy Because the ions in an ionic compound are arranged in a crystal lattice, the energy required to separate 1 mol of the ions of an ionic compound is referred to as the **lattice energy.** The strength of the forces holding ions in place is reflected by the lattice energy. The greater the lattice energy, the stronger the force of attraction.

Lattice energy is directly related to the size of the ions bonded. Smaller ions form compounds with more closely spaced ionic charges. Because the electrostatic force of attraction between opposite charges increases as the distance between the charges decreases, smaller ions produce stronger interionic attractions and greater lattice energies. For example, the lattice energy of a lithium compound is greater than that of a potassium compound containing the same anion because the lithium ion is smaller than the potassium ion.

Data Analysis LAB

Based on Real Data*

Interpret Data

Can embedding nanoparticles of silver into a polymer give the polymer antimicrobial properties? Researchers tested the antimicrobial properties of a new composite material—the polymer poly(4-vinyl-N-hexylpyridinium bromide), known as NPVP, which attracts cations. It is known that silver ions from silver bromide and silver nitrate exhibit antimicrobial activity. Silver bromide was embedded into the NPVP polymer. Scientists tested the antimicrobial properties of the composite material. Their results, illustrated in the graph, show the growth of *E. coli* bacteria over a period of approximately four hours. Each line represents the *E. coli* population in response to the introduction of a particular substance.

Data and Observations

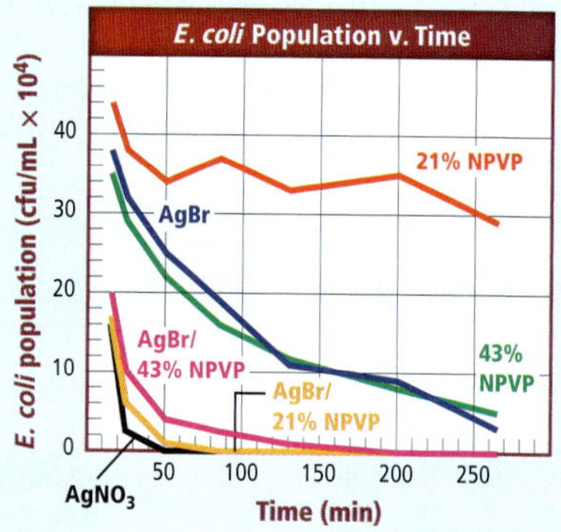

Think Critically

1. **Interpret** Does the addition of silver bromide (AgBr) ions to NPVP improve the antimicrobial properties of the composite?
2. **Interpret** Which composite reduced the *E. coli* population to zero? How long does it take for each substance to reduce the bacteria population to zero?
3. **Conclude** Does a composite polymer containing NPVP and silver bromide show antimicrobial properties? Explain your answer.

*Data obtained from: Sambhy, V., et al. Published on the Web 7/7/2006. Silver Bromide Nanoparticle/Polymer Composites. *Journal of the American Chemical Society.*

Table 6 Lattice Energies of Some Ionic Compounds

Compound	Lattice Energy (kJ/mol)	Compound	Lattice Energy (kJ/mol)
KI	632	KF	808
KBr	671	AgCl	910
RbF	774	NaF	910
NaI	682	LiF	1030
NaBr	732	$SrCl_2$	2142
NaCl	769	MgO	3795

The value of lattice energy is also affected by the charge of the ion. The ionic bond formed from the attraction of ions with larger positive or negative charges generally has a greater lattice energy. The lattice energy of MgO is almost four times greater than that of NaF because the charge of the ions in MgO is greater than the charge of the ions in NaF. The lattice energy of $SrCl_2$ is between the lattice energies of MgO and NaF because $SrCl_2$ contains ions with both higher and lower charges.

☑ READING CHECK **Summarize** the charges on each ion in the following ionic compounds: MgO, NaF, $SrCl_2$.

Table 6 shows the lattice energies of some ionic compounds. Examine the lattice energies of RbF and KF. Because K^+ has a smaller ionic radius than Rb^+, KF has a greater lattice energy than RbF. This confirms that lattice energy is related to ion size. Notice the lattice energies of $SrCl_2$ and AgCl. How do they show the relationship between lattice energy and the charge of the ions involved?

SECTION 2 REVIEW

Section Self-Check

Section Summary

- Ionic compounds contain ionic bonds formed by the attraction of oppositely charged ions.
- Ions in an ionic compound are arranged in a repeating pattern known as a crystal lattice.
- Ionic compound properties are related to ionic bond strength.
- Ionic compounds are electrolytes; they conduct an electric current in the liquid phase and in aqueous solution.
- Lattice energy is the energy needed to remove 1 mol of ions from its lattice.

12. MAINIDEA **Explain** how an ionic compound made up of charged particles can be electrically neutral.
13. **Describe** the energy change associated with ionic bond formation, and relate it to stability.
14. **Identify** three physical properties of ionic compounds that are associated with ionic bonds, and relate them to bond strength.
15. **Explain** how ions form bonds, and describe the structure of the resulting compound.
16. **Relate** lattice energy to ionic-bond strength.
17. **Apply** Use electron configurations, orbital notation, and electron-dot structures to represent the formation of an ionic compound from the metal strontium and the nonmetal chlorine.
18. **Design** a concept map that shows the relationships among ionic bond strength, physical properties of ionic compounds, lattice energy, and stability.

■ **In-Text Question** Sr^{2+} has the greater charge. The lattice energy of $SrCl_2$ is greater than AgCl.

3 Assess

Check for Understanding

Ask students to determine the ionic compound formed from Mg and Cl and explain how it was formed. $MgCl_2$; magnesium loses two valence electrons, while chlorine gains only one valence electron. The ratio of Mg^{2+} to Cl^- is 1:2. OL

Reteach

Using the elements aluminum and oxygen, have students determine the formula for the ionic compound formed. Al_2O_3 OL EL

Extension

Tell students that iron can form both 2+ and 3+ ions. Ask students to write the ionic compounds formed when these ions combine with oxygen to form oxides. Fe_2O_3 and Fe_3O_2 OL EL

ASSESSMENT

Knowledge Have each student list a metal and a nonmetal. Have another student determine the ionic compound formed between them. OL COOP LEARN

☑ READING CHECK Mg^{2+}, O^{2-}, Na^+, F^-, Sr^{2+}, Cl^-

SECTION 2 REVIEW

12. The total positive charge of the cations in the compound equals the total negative charge of the anions in the compound.
13. Ionic bond formation is exothermic; the lower-energy product is more stable than the original reactants.
14. properties: exist as crystals; high melting and boiling points; hard, rigid, brittle solids. These properties are due to the strength of the electrostatic attraction of opposite charged ions in close proximity.
15. Electrons are transferred between atoms forming ions. Electrostatic forces hold the ions together in the ionic compound. The ions are arranged in a regular repeating pattern in an ionic crystal.
16. As lattice energy becomes larger, the stronger the attraction between the ions and thus, the stronger the ionic bond.
17. Drawings should include one Sr atom losing 2 e^- and forming an Sr^{2+} ion, and two Cl atoms each gaining 1 e^- and forming two Cl^- ions. These ions attract, forming $SrCl_2$.
18. Concepts maps will vary but should correlate greater bond strength to increased stability and a higher lattice energy, and should show that physical properties such as high melting and boiling points, brittleness, and conductivity are due to the strength of ionic bonds.

SECTION 3

1 Focus

MAINIDEA

Naming Compounds Write the full name of a student on the board. Ask students how they normally write their names. first name followed by last name Write the name *sodium chloride* on the board. Ask students which substance lost electrons. sodium Ask students which substance gained electrons. chlorine Write the term *calcium carbonate* on the board, and have the student predict which substance gained electrons and which substance lost electrons. Carbonate gained electrons and calcium lost electrons. OL

2 Teach

Concept Development

Balance Charges Using the oxidation numbers for monatomic ions in Table 7 and Table 8, work through several examples following Example Problem 1. Avoid using the technique that crosses oxidation numbers to get subscripts. This technique is quick and accurate if subscripts are reduced to the simplest ratios, but it does not lead to an understanding of balanced charges.

- **In-Text Question** Be^{2+}; I^-; N^{3-}

SECTION 3

Essential Questions

- What is a formula unit and how does it relate to an ionic compound's composition?
- How do you write the formulas for compounds formed from different ions and oxyanions?
- What are the naming conventions for ionic compounds and oxyanions?

Review Vocabulary

nonmetal: an element that is generally a gas or a dull, brittle solid and is a poor conductor of heat and electricity

New Vocabulary

formula unit
monatomic ion
polyatomic ion
oxyanion

Names and Formulas for Ionic Compounds

MAINIDEA In written names and formulas for ionic compounds, the cation appears first, followed by the anion.

CHEM 4 YOU Although people have a wide range of names, most have both a first name and a last name. Ionic compound names are similar, in that they also consist of two parts.

Formulas for Ionic Compounds

Because chemists around the world need to be able to communicate with one another, they have developed a set of rules for naming compounds. Using this standardized naming system, you can write a chemical formula from a compound's name and name a compound given its chemical formula.

Recall that an ionic compound is made up of ions arranged in a repeating pattern. The chemical formula for an ionic compound, called a **formula unit,** represents the simplest ratio of the ions involved. For example, the formula unit of magnesium chloride is $MgCl_2$ because the magnesium and chloride ions exist in a 1:2 ratio.

The overall charge of a formula unit is zero because the formula unit represents the entire crystal, which is electrically neutral. The formula unit for $MgCl_2$ contains one Mg^{2+} ion and two Cl^- ions, for a total charge of zero.

Monatomic ions Binary ionic compounds are composed of positively charged monatomic ions of a metal and negatively charged monatomic ions of a nonmetal. A **monatomic ion** is a one-atom ion, such as Mg^{2+} or Br^-. **Table 7** indicates the charges of common monatomic ions according to their location on the periodic table. What is the formula for the beryllium ion? The iodide ion? The nitride ion?

Transition metals, which are in groups 3 through 12, and metals in groups 13 and 14 are not included in **Table 7** because of the variance in ionic charges of atoms in the groups. Most transition metals and metals in groups 13 and 14 can form several different positive ions.

Table 7 Common Monatomic Ions

Group	Atoms that Commonly Form Ions	Charge of Ions
1	H, Li, Na, K, Rb, Cs	1+
2	Be, Mg, Ca, Sr, Ba	2+
15	N, P, As	3−
16	O, S, Se, Te	2−
17	F, Cl, Br, I	1−

Demonstration

Demonstrate Oxidation States of Vanadium

Purpose
to demonstrate the colorful ions formed as the oxidation number of vanadium changes from 5+ to 4+, 3+, and 2+

Materials
Granular zinc-mercury(II) chloride(aq), ammonium metavanadate (NH_4VO_3) solution, 500-mL Erlenmeyer flask with stopper, powder funnel, 250-mL graduated cylinder. See the Teacher Handbook for preparation of solutions.

Safety Precautions

Disposal Hazardous waste as described by EPA regulators will be produced and should be disposed of accordingly.

Procedure
The zinc-mercury amalgam is stored under water. After pouring off the water, pour 63 g of the zinc-mercury amalgam into a 500-mL

Table 8 Monatomic Metal Ions

Group	Common Ions
3	Sc^{3+}, Y^{3+}, La^{3+}
4	Ti^{2+}, Ti^{3+}
5	V^{2+}, V^{3+}
6	Cr^{2+}, Cr^{3+}
7	Mn^{2+}, Mn^{3+}, Tc^{2+}
8	Fe^{2+}, Fe^{3+}
9	Co^{2+}, Co^{3+}
10	Ni^{2+}, Pd^{2+}, Pt^{2+}, Pt^{4+}
11	Cu^{+}, Cu^{2+}, Ag^{+}, Au^{+}, Au^{3+}
12	Zn^{2+}, Cd^{2+}, Hg_2^{2+}, Hg^{2+}
13	Al^{3+}, Ga^{2+}, Ga^{3+}, In^{+}, In^{2+}, In^{3+}, Tl^{+}, Tl^{3+}
14	Sn^{2+}, Sn^{4+}, Pb^{2+}, Pb^{4+}

Oxidation numbers The charge of a monatomic ion is equal to its oxidation number, or oxidation state. As shown in **Table 8,** most transition metals and group 13 and 14 metals have more than one possible ionic charge. Note that the ionic charges given in **Table 8** are the most common ones, not the only ones possible.

The oxidation number of an element in an ionic compound equals the number of electrons transferred from the atom to form the ion. For example, a sodium atom transfers one electron to a chlorine atom to form sodium chloride. This results in Na^+ and Cl^-. Thus, the oxidation number of sodium in the compound is 1+ because one electron was transferred from the sodium atom. Because an electron is transferred to the chlorine atom, its oxidation number is 1−.

Formulas for binary ionic compounds In the chemical formula for any ionic compound, the symbol of the cation is always written first, followed by the symbol of the anion. Subscripts, which are small numbers to the lower right of a symbol, represent the number of ions of each element in an ionic compound. If no subscript is written, it is assumed to be one. You can use oxidation numbers to write formulas for ionic compounds. Recall that ionic compounds have no charge. If you add the oxidation number of each ion multiplied by the number of these ions in a formula unit, the total must be zero.

Suppose you need to determine the formula for one formula unit of the compound that contains sodium and fluoride ions. Start by writing the symbol and charge for each ion: Na^+ and F^-. The ratio of ions in a formula unit of the compound must show that the number of electrons lost by the metal equals the number of electrons gained by the nonmetal. This occurs when one sodium atom transfers one electron to the fluorine atom; the formula unit is NaF.

☑ READING CHECK **Relate** the charge of an ion to its oxidation number.

CAREERS IN CHEMISTRY

Food Scientist Have you ever thought about the science behind the food you eat? Food scientists are concerned about the effects of processing on the appearance, aroma, taste, and the vitamin and mineral content of food. They also develop and improve foods and beverages. Food scientists often maintain "tasting notebooks" as they learn the characteristics of individual and blended flavors.

WebQuest

VOCABULARY

ACADEMIC VOCABULARY

Transfer

to cause to pass from one to another

Carlos had to transfer to a new school when his parents moved to a new neighborhood.

☑ READING CHECK The charge of an ion is equal to its oxidation number.

Identify Misconceptions

Students often think that formula units represent actual numbers of atoms.

Uncover the Misconception Display crystals of different ionic compounds. Each ion in the crystal is surrounded by ions oppositely charged. Students should define formula units as the ratios of these ions.

Demonstrate the Concept Set up a microscope and allow students to view the cubic crystals of NaCl. Using toothpicks and small and large marshmallows representing Na^+ and Cl^-, respectively, form the NaCl crystal lattice. Point out that one Na^+ is surrounded by six Cl^- and one Cl^- is surrounded by six Na^+. The ratio of Na^+ to Cl^- is one-to-one, and the formula unit is NaCl.

Assess New Knowledge Have students determine the cation to anion ratio for the formula units that represent Ag_2S, $MgCl_2$, Al_2O_3, SnF_4, and FeO. 2:1, 1:2, 2:3, 1:4, 1:1 OL

CAREERS IN CHEMISTRY

WebQuest

Erlenmeyer flask through a powder funnel. Add 140 mL of ammonium metavanadate solution to the zinc-mercury amalgam in the flask. **WARNING:** ***The solutions are toxic and corrosive.*** Note the color change as vanadium with oxidation number 5+ (VO_2^+, yellow) is converted to vanadium with oxidation number 4+ (VO^{2+}, blue). The yellow and blue of the two ions mix to give a green color. Stopper the flask, and shake gently until the color of the solution changes from green to blue to blue-green as vanadium with oxidation number 4+ (VO^{2+}, blue) is converted to V^{3+} (blue-green). Finally, shake the stoppered flask vigorously to change V^{3+} to V^{2+} (violet).

Results

Equilibrium reactions account for the mixtures of colors.

$$Zn + 2VO_2^+ + 4H^+ \leftrightharpoons 2VO^{2+} + Zn^{2+} + 2H_2O$$
$$Zn + 2VO^{2+} + 4H^+ \leftrightharpoons 2V^{3+} + Zn^{2+} + 2H_2O$$
$$2V^{3+} + Zn \leftrightharpoons 2V^{2+} + Zn^{2+}$$

Analysis

1. Where is vanadium found on the periodic table? group 5, period 4, among the transition elements
2. What is the electron configuration of vanadium with a 5+ oxidation number? [Ar]

IN-CLASS Example

Question A compound formed between magnesium and fluorine is transparent over a wide range of wavelengths. Lenses and windows can be made of this compound. Determine the correct formula for the ionic compound formed between magnesium and fluorine

Answer Magnesium from group 2 forms a Mg^{2+} ion, and fluorine, a group 17 element, forms an F^- ion. Because the charges are not the same, subscripts must be used to indicate the ratio of positive to negative ions in the compound. One magnesium atom loses two electrons, and each fluorine atom gains one electron. Thus, for the number of electrons lost to equal the number of electrons gained, one magnesium atom must react with two fluorine atoms. This yields MgF_2.

IN-CLASS Example

Question Stibnite is a mineral formed from sulfur and antimony with a +3 oxidation number. Determine the correct formula for the ionic compound formed between antimony and sulfur.

Answer The +3 oxidation number indicates that antimony has lost 3 electrons to form Sb^{3+}. Sulfur, a group 16 element, gains 2 electrons to form S^{2-}. Because the charges are not the same, subscripts must be used to indicate the ratio of positive to negative ions.

One antimony atom loses three electrons and each sulfur atom gains 2 electrons. The smallest number both 2 and 3 divide into evenly is 6. Therefore, a total of six electrons must be transferred. Sb_2S_3 is the correct formula.

Visual Learning

Table 9 Remind students that each of the listed ions contains at least two atoms but acts as a single ion. Have students write examples of ionic compounds containing polyatomic ions. **OL**

RealWorld CHEMISTRY

Ionic Compounds

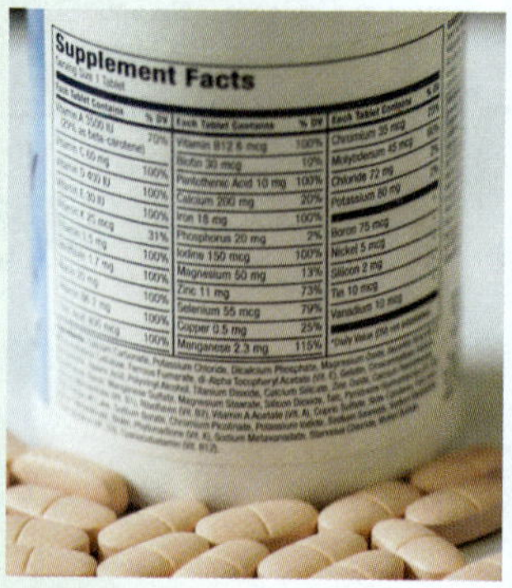

MINERAL SUPPLEMENTS To function properly, your body requires a daily intake of many different minerals. To ensure they are getting what they need, many people take a daily multivitamin and a mineral supplement. The minerals in these supplements come from a variety of ionic compounds. In fact, the majority of minerals found in mineral supplements come from ground-up rocks.

EXAMPLE Problem 1

FORMULA FOR AN IONIC COMPOUND Determine the formula for the ionic compound formed from potassium and oxygen.

1 ANALYZE THE PROBLEM

You are given that potassium and oxygen ions form an ionic compound; the formula for the compound is the unknown. First, write out the symbol and oxidation number for each ion involved in the compound. Potassium, from group 1, forms 1+ ions, and oxygen, from group 16, forms 2− ions.

K^+ O^{2-}

Because the charges are not the same, you need to determine the subscripts to use to indicate the ratio of positive ions to negative ions.

2 SOLVE FOR THE UNKNOWN

A potassium atom loses one electron, while an oxygen atom gains two electrons. If combined in a one-to-one ratio, the number of electrons lost by potassium will not balance the number of electrons gained by oxygen. Thus, two potassium ions are needed for each oxide ion. The formula is **K_2O.**

3 EVALUATE THE ANSWER

The overall charge of the compound is zero.

$$2\ \cancel{\text{K ions}}\left(\frac{1+}{\cancel{\text{K ion}}}\right) + 1\ \cancel{\text{O ion}}\left(\frac{2-}{\cancel{\text{O ion}}}\right) = 2(1+) + 1(2-) = 0$$

EXAMPLE Problem 2

FORMULA FOR AN IONIC COMPOUND Determine the formula for the compound formed from aluminum ions and sulfide ions.

1 ANALYZE THE PROBLEM

You are given that aluminum and sulfur form an ionic compound; the formula for the ionic compound is the unknown. First, determine the charges of each ion. Aluminum, from group 13, forms 3+ ions, and sulfur, from group 16, forms 2− ions.

Al^{3+} S^{2-}

Each aluminum atom loses three electrons, while each sulfur atom gains two electrons. The number of electrons lost must equal the number of electrons gained.

2 SOLVE FOR THE UNKNOWN

The smallest number that can be divided evenly by both 2 and 3 is 6. Therefore, six electrons are transferred. Three sulfur atoms accept the six electrons lost by two aluminum atoms. The correct formula, **Al_2S_3,** shows two aluminum ions bonded to three sulfide ions.

3 EVALUATE THE ANSWER

The overall charge of one formula unit of this compound is zero.

$$2\ \cancel{\text{Al ions}}\left(\frac{3+}{\cancel{\text{Al ion}}}\right) + 3\ \text{S}\ \cancel{\text{ions}}\left(\frac{2-}{\cancel{\text{S ion}}}\right) = 2(3+) + 3(2-) = 0$$

ASSESSMENT

Knowledge Have students explain which electrons are lost as vanadium's oxidation number changes from 0 to 5+. the 4s and 3d electrons **OL**

DIFFERENTIATED INSTRUCTION

Advanced Learners Have capable students research the Born-Haber cycle for the energy changes in sodium chloride formation and design a Born-Haber cycle for the formation of potassium fluoride. **AL**

PRACTICE Problems

Do additional problems. Online Practice

Write formulas for the ionic compounds formed by the following ions.

19. potassium and iodide

20. magnesium and chloride

21. aluminum and bromide

22. cesium and nitride

23. Challenge Write the general formula for the ionic compound formed by elements from the two groups shown on the periodic table at the right.

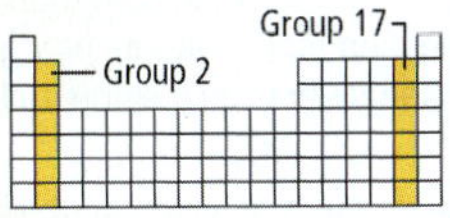

Formulas for polyatomic ionic compounds Many ionic compounds contain **polyatomic ions,** which are ions made up of more than one atom. **Table 9** and **Figure 10** list some common polyatomic ions. Also, refer to **Table R-5** in the Student Resources. A polyatomic ion acts as an individual ion in a compound and its charge applies to the entire group of atoms. Thus, the formula for a polyatomic compound follows the same rules used for a binary compound.

Because a polyatomic ion exists as a unit, never change subscripts of the atoms within the ion. If more than one polyatomic ion is needed, place parentheses around the ion and write the appropriate subscript outside the parentheses. For example, consider the compound formed from the ammonium ion (NH_4^+) and the oxide ion (O^{2-}). To balance the charges, the compound must have two ammonium ions for each oxide ion. To add a subscript to ammonium, enclose it in parentheses, then add the subscript. The correct formula is $(NH_4)_2O$.

Table 9 Common Polyatomic Ions

Ion	Name	Ion	Name
NH_4^+	ammonium	IO_4^-	periodate
NO_2^-	nitrite	$C_2H_3O_2^-$	acetate
NO_3^-	nitrate	$H_2PO_4^-$	dihydrogen phosphate
OH^-	hydroxide	CO_3^{2-}	carbonate
CN^-	cyanide	SO_3^{2-}	sulfite
MnO_4^-	permanganate	SO_4^{2-}	sulfate
HCO_3^-	hydrogen carbonate	$S_2O_3^{2-}$	thiosulfate
ClO^-	hypochlorite	O_2^{2-}	peroxide
ClO_2^-	chlorite	CrO_4^{2-}	chromate
ClO_3^-	chlorate	$Cr_2O_7^{2-}$	dichromate
ClO_4^-	perchlorate	HPO_4^{2-}	hydrogen phosphate
BrO_3^-	bromate	PO_4^{3-}	phosphate
IO_3^-	iodate	AsO_4^{3-}	arsenate

Figure 10 Ammonium and phosphate ions are polyatomic; that is, they are made up of more than one atom. Each polyatomic ion, however, acts as a single unit and has one charge.

Identify ***What are the charges of the ammonium ion and phosphate ion, respectively?***

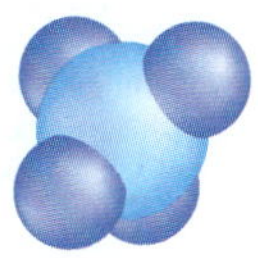

Ammonium ion (NH_4^+)

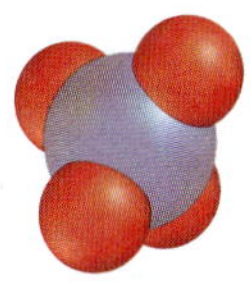

Phosphate ion (PO_4^{3-})

PRACTICE Problems

Have students refer to the Selected Solutions appendix for complete solutions to odd-numbered problems. The complete solutions for all problems can be found in the Solutions Manual.

19. KI

20. $MgCl_2$

21. $AlBr_3$

22. Cs_3N

23. The general formula is XY_2, where X represents the group 2 element and Y represents the group 17 element.

Caption Question Fig. 10 ammonium ion, 1+; phosphate ion, 3−

Quick Demo

Polyatomic Ions This demo shows that polyatomic ions act as a single unit. Place a conductivity probe in distilled water. (A CBL may be used, but is not required.) Add 1 drop of 1.0*M* NaCl solution and record the conductivity reading. Continue adding the solution dropwise and recording data until ten drops have been added. Have students graph the data and remind them that NaCl is composed of two ions. Repeat the procedure using $NaNO_3$. Overlap the graphs and ask students to compare them. Repeat the procedure using $MgCl_2$, then use $Mg(NO_3)_2$. If time permits, repeat with $AlCl_3$, followed by $Al(NO_3)_3$. Overlap all the chloride graphs, and ask students to relate slope to the number of ions present. Slope increases as the number of ions increases. Then, overlap all the nitrate graphs to confirm that the nitrate ion acted as a single ion. OL

CULTURAL DIVERSITY

Pottery One of the most ancient arts is the making of pottery. The oldest known pottery dates back to 400 B.C. in Japan. Chinese, Middle Eastern, and Native American potters also produced a variety of pottery. Basically, the chemical composition of pottery is that of clay or clay blends, $Al_2O_3{\cdot}2SiO_2{\cdot}2H_2O$. When formed to the desired shape, the clay piece is baked until it is hard and the intensity of the heat determines the degree of hardness. Pottery is often coated with glaze to render it waterproof and add decoration. Pottery glaze has a composition similar to that of pottery clay, as well as a mixture of ionically bonded metal oxides. The addition of metal oxides to the glaze results in many properties, including hardness, color, surface character, crystal development, and chemical stability. Adding Fe_2O_3 produces a rich red color, while CuO produces a clear green color. The addition of Al_2O_3 adds hardness and durability to the pottery.

IN-CLASS Example

Question A transparent, colorless mineral found at Mount Vesuvius is called mascagnite. This is the name given to the ionic compound formed between the ammonium ion and the sulfate ion. Determine the correct formula for this ionic compound.

Answer Ammonium, NH_4^+, has lost an electron. Sulfate, SO_4^{2-} has gained two electrons. Two ammonium ions must lose the two electrons gained by one sulfate ion. $(NH_4)_2SO_4$

PRACTICE Problems

Have students refer to the Selected Solutions appendix for complete solutions to odd-numbered problems. The complete solutions for all problems can be found in the Solutions Manual.

24. $NaNO_3$
25. $Ca(ClO_3)_2$
26. $Al_2(CO_3)_3$
27. Answers will vary; example: $MgCO_3$

ASSESSMENT

Skill Play a memory game using names and formulas for ionic compounds. Draw a four-by-four chart and project the image. Place the names for eight compounds in eight of the squares and the formulas for the same compounds in the other eight squares. Cover each square, and have students turn the covers over two at a time to match each formula with its name. OL

EXAMPLE Problem 3

EXAMPLE PROBLEM

FORMULA FOR A POLYATOMIC IONIC COMPOUND A compound formed by calcium ions and phosphate ions is often used in fertilizers. Write the compound's formula.

1 ANALYZE THE PROBLEM

You know that calcium and phosphate ions form an ionic compound; the formula for the compound is the unknown. First, write each ion along with its charge. Calcium, from group 2, forms 2+ ions, and the polyatomic phosphate acts as a single unit with a 3− charge.

Ca^{2+} $\quad$ PO_4^{3-}

Each calcium atom loses two electrons, while each polyatomic phosphate group gains three electrons. The number of electrons lost must equal the number of electrons gained.

2 SOLVE FOR THE UNKNOWN

The smallest number evenly divisible by both charges is 6. Thus, a total of six electrons are transferred. The negative charge from two phosphate ions equals the positive charge from three calcium ions. In the formula, place the polyatomic ion in parentheses and add a subscript to the outside. The correct formula for the compound is $Ca_3(PO_4)_2$.

3 EVALUATE THE ANSWER

The overall charge of one formula unit of calcium phosphate is zero.

$$3\ \text{Ca}\ \cancel{\text{ions}}\left(\frac{2+}{\text{Ca}\ \cancel{\text{ion}}}\right) + 2\ \text{PO}_4\ \cancel{\text{ions}}\left(\frac{3-}{\text{PO}_4\ \cancel{\text{ion}}}\right) = 3(2+) + 2(3-) = 0$$

PRACTICE Problems

Do additional problems. Online Practice

Write formulas for ionic compounds composed of the following ions.

24. sodium and nitrate
25. calcium and chlorate
26. aluminum and carbonate
27. Challenge Write the formula for an ionic compound formed by ions from a group 2 element and polyatomic ions composed of only carbon and oxygen.

Get help **naming ionic compounds.** Personal Tutor

Watch a **video about compound formation.** Video

Names for Ions and Ionic Compounds

Scientists use a systematic approach when naming ionic compounds. Because ionic compounds have both cations and anions, the naming system accounts for both of these ions.

Naming an oxyanion An **oxyanion** is a polyatomic ion composed of an element, usually a nonmetal, bonded to one or more oxygen atoms. More than one oxyanion exists for some nonmetals, such as nitrogen and sulfur. These ions are easily named using the rules in **Table 10.**

Table 10 Oxyanion Naming Conventions for Sulfur and Nitrogen

• Identify the ion with the greatest number of oxygen atoms. This ion is named using the root of the nonmetal and the suffix *-ate*.
• Identify the ion with fewer oxygen atoms. This ion is named using the root of the nonmetal and the suffix *-ite*. Examples: NO_3^- nitrate, NO_2^- nitrite, SO_4^{2-} sulfate, SO_3^{2-} sulfite

Video Lab

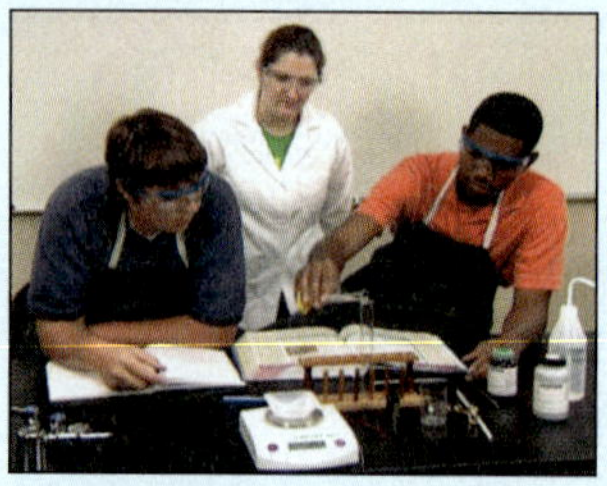

Video

Formation of Compounds

CHEMISTRY JOURNAL

Composition of Ocean Water Ocean water contains potassium, sodium, magnesium, hydrogen carbonate, sulfate, chloride, and bromide ions. Have students write and name all the formula units that can be composed from a combination of these ions. OL

As shown in **Table 11,** chlorine forms four oxyanions that are named according to the number of oxygen atoms present. Names of similar oxyanions formed by other halogens follow the rules used for chlorine. For example, bromine forms the bromate ion (BrO_3^-), and iodine forms the periodate ion (IO_4^-) and the iodate ion (IO_3^-).

Naming ionic compounds Chemical nomenclature is a systematic way of naming compounds. Now that you are familiar with chemical formulas, you can use the following five rules to name ionic compounds.

1. Name the cation followed by the anion. Remember that the cation is always written first in the formula.
2. For monatomic cations, use the element name.
3. For monatomic anions, use the root of the element name plus the suffix *-ide.*
   Example:

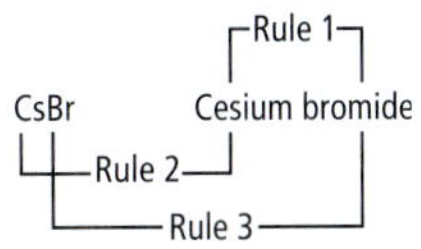

4. To distinguish between multiple oxidation numbers of the same element, the name of the chemical formula must indicate the oxidation number of the cation. The oxidation number is written as a Roman numeral in parentheses after the name of the cation.
   Note: This rule applies to the transition metals and metals on the right side of the periodic table, which often have more than one oxidation number. See **Table 8.** It does not apply to group 1 and group 2 cations, as they have only one oxidation number.
   Examples:
   Fe^{2+} and O^{2-} ions form FeO, known as iron(II) oxide.
   Fe^{3+} and O^{2-} ions form Fe_2O_3, known as iron(III) oxide.
5. When the compound contains a polyatomic ion, simply use the name of the polyatomic ion in place of the anion or cation.
   Examples:
   The name for NaOH is sodium hydroxide.
   The name for $(NH_4)_2S$ is ammonium sulfide.

Table 11 Oxyanion Naming Conventions for Chlorine

- The oxyanion with the greatest number of oxygen atoms is named using the prefix *per-*, the root of the nonmetal, and the suffix *-ate.*
- The oxyanion with one fewer oxygen atom is named using the root of the nonmetal and the suffix *-ate.*
- The oxyanion with two fewer oxygen atoms is named using the root of the nonmetal and the suffix *-ite.*
- The oxyanion with three fewer oxygen atoms is named using the prefix *hypo-*, the root of the nonmetal, and the suffix *-ite.*

Examples:

ClO_4^- perchlorate	ClO_3^- chlorate
ClO_2^- chlorite	ClO^- hypochlorite

Explore **ionic compounds.**

Virtual Investigations

PRACTICE Problems — Do additional problems. Online Practice

Name the following compounds.

28. NaBr

29. $CaCl_2$

30. KOH

31. $Cu(NO_3)_2$

32. Ag_2CrO_4

33. Challenge The ionic compound NH_4ClO_4 is a key reactant used in solid rocket boosters, such as those that power the Space Shuttle into orbit. Name this compound.

ASSESSMENT

Knowledge Have students draw a five-by-five chart on a piece of paper. Ask them to write five metal ions in the left margin, including the ammonium ion, and five nonmetal ions at the top, including at least two polyatomic anions. Have them trade charts with another student. Have each student write the correct formulas for the ionic compounds formed in the square. Hand the completed chart to another student to evaluate the responses. OL

Reinforcement

Visual Aids Have students make flash cards to help them remember the names, formulas, and charges of polyatomic ions. BL EL

PRACTICE Problems

Have students refer to the Selected Solutions appendix for complete solutions to odd-numbered problems. The complete solutions for all problems can be found in the Solutions Manual.

28. sodium bromide
29. calcium chloride
30. potassium hydroxide
31. copper(II) nitrate
32. silver chromate
33. ammonium perchlorate

DIFFERENTIATED INSTRUCTION

Struggling Learners Students might have trouble determining formula units from the ions involved. Use puzzle-like pieces to represent ions. Pieces representing positive ions have one wedge-shaped tab for each positive charge, and negative ions will have a matching notch for each negative charge. Allow students to manipulate the pieces to discover that, for example, two 1+ ions are needed for each 2− ion. Have students write the correct formula units for the ionic compounds formed. BL EL

■ **In-Text Question** Answers will vary but should show a logical, systematic approach.

3 Assess

Check for Understanding

Use the ingredients list from a bottle of vitamins to write on the board the names or formulas of ionic compounds that make up the vitamin tablets. Have students determine the correct formula unit or name for each compound. Compounds often present in vitamins are calcium phosphate, magnesium phosphate, copper(II) oxide, zinc oxide, calcium carbonate, magnesium oxide, titanium(IV) oxide, potassium chloride, potassium iodide, tin(II) chloride, and chromium(III) chloride. **OL**

Reteach

Review naming ionic compounds and writing their formula units. Tell students that artificial ocean water, often used in saltwater aquariums, is made by mixing the following: 24.72 g NaCl, 0.67 g KCl, 1.36 g $CaCl_2$, 4.66 g $MgCl_2$, 6.29 g $MgSO_4$, and 0.18 g $NaHCO_3$. Have students name each ionic compound present. **OL** **EL**

Extension

Have students design quiz show questions and answers for categories such as oxidation numbers, formula units for ionic compounds, and names for ionic compounds. **OL**

ASSESSMENT

Knowledge Write formulas for several ions, both positive and negative, on a transparency. Point out one positive and one negative ion, and have students determine the correct formula and name of the ionic compound formed. **OL**

PROBLEM-SOLVING STRATEGY

Naming Ionic Compounds

Naming ionic compounds is easy if you follow this naming-convention flowchart.

Apply the Strategy

Name the compounds KOH and Ag_2CrO_4 using this flowchart.

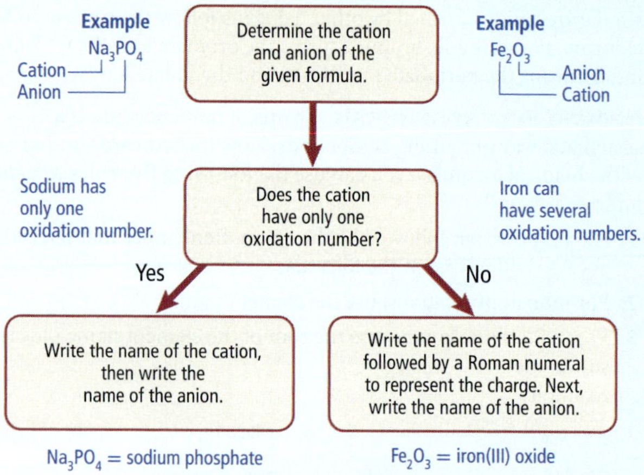

APPLYING PRACTICES

Construct and Revise an Explanation Go to the resources tab in ConnectED to find the Applying Practices worksheet *Electron States and Simple Chemical Reactions.*

The Problem-Solving Strategy above reviews the steps used in naming ionic compounds if the formula is known. Naming ionic compounds is important in communicating the cation and anion present in a crystalline solid or aqueous solution. How might you change the diagram to help you write the formulas for ionic compounds if you know their names?

The ion-containing substances you have investigated so far have been ionic compounds. In the next section, you will learn how ions relate to the structure and properties of metals.

SECTION 3 REVIEW

Section Self-Check

Section Summary

- A formula unit gives the ratio of cations to anions in the ionic compound.
- A monatomic ion is formed from one atom. The charge of a monatomic ion is equal to its oxidation number.
- Roman numerals indicate the oxidation number of cations having multiple possible oxidation states.
- Polyatomic ions consist of more than one atom and act as a single unit.
- To indicate more than one polyatomic ion in a chemical formula, place parentheses around the polyatomic ion and use a subscript.

34. MAINIDEA **State** the order in which the ions associated with a compound composed of potassium and bromine would be written in the chemical formula and the compound name.

35. Describe the difference between a monatomic ion and a polyatomic ion, and give an example of each.

36. Apply Ion X has a charge of 2+, and ion Y has a charge of 1−. Write the formula unit of the compound formed from the ions.

37. State the name and formula for the compound formed from Mg and Cl.

38. Write the name and formula for the compound formed from sodium ions and nitrite ions.

39. Analyze What subscripts would you most likely use if the following substances formed an ionic compound?

a. an alkali metal and a halogen

b. an alkali metal and a nonmetal from group 16

c. an alkaline earth metal and a halogen

d. an alkaline earth metal and a nonmetal from group 16

SECTION 3 REVIEW

34. The cation (potassium) is stated first, followed by the anion (bromide).

35. monatomic ions are one-atom ions, Cl^-; polyatomic ions are two or more atoms grouped together having a net charge, ClO_3^-

36. XY_2

37. magnesium chloride, $MgCl_2$

38. sodium nitrite, $NaNO_2$

39. a. 1, 1
b. 2, 1
c. 1, 2
d. 1, 1

SECTION 4

Essential Questions

- What are the characteristics of a metallic bond?
- How does the electron sea model account for the physical properties of metals?
- What are alloys, and how can they be categorized?

Review Vocabulary

physical property: a characteristic of matter that can be observed or measured without altering the sample's composition

New Vocabulary

electron sea model
delocalized electron
metallic bond
alloy

Metallic Bonds and the Properties of Metals

MAINIDEA Metals form crystal lattices and can be modeled as cations surrounded by a "sea" of freely moving valence electrons.

CHEM 4 YOU Imagine a buoy in the ocean, bobbing by itself surrounded by a vast expanse of open water. Though the buoy stays in the same area, the ocean water freely flows past. In some ways, this description also applies to metallic atoms and their electrons.

Metallic Bonds

Although metals are not ionic, they share several properties with ionic compounds. The bonding in both metals and ionic compounds is based on the attraction of particles with unlike charges. Metals often form lattices in the solid state. These lattices are similar to the ionic crystal lattices discussed earlier. In such a lattice, 8 to 12 other metal atoms closely surround each metal atom.

A sea of electrons Although metal atoms always have at least one valence electron, they do not share these valence electrons with neighboring atoms, nor do they lose their valence electrons. Instead, within the crowded lattice, the outer energy levels of the metal atoms overlap. This unique arrangement is described by the electron sea model. The **electron sea model** proposes that all the metal atoms in a metallic solid contribute their valence electrons to form a "sea" of electrons. This sea of electrons surrounds the metal cations in the lattice.

The electrons present in the outer energy levels of the bonding metallic atoms are not held by any specific atom and can move easily from one atom to the next. Because they are free to move, they are often referred to as **delocalized electrons.** When the atom's outer electrons move freely throughout the solid, a metallic cation is formed. Each such ion is bonded to all neighboring metal cations by the sea of valence electrons, as shown in **Figure 11.** A **metallic bond** is the attraction of a metallic cation for delocalized electrons.

■ **Figure 11** The valence electrons in metals (shown as a blue cloud of minus signs) are evenly distributed among the metallic cations (shown in red). Attractions between positive cations and the negative "sea" hold the metal atoms together in a lattice.
Explain *Why are electrons in metals known as delocalized electrons?*

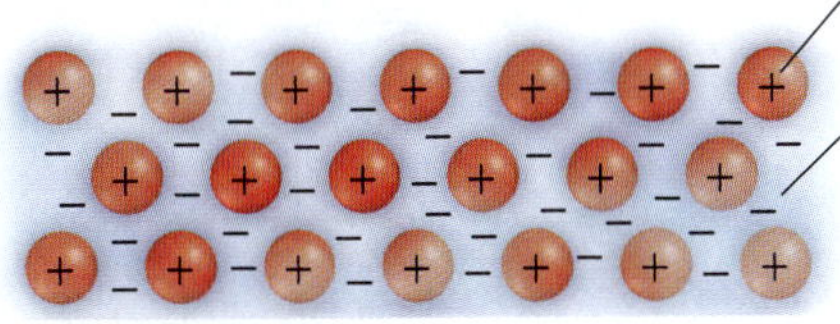

SECTION 4

1 Focus

MAINIDEA

Model Electron Motion Ask students to walk around and between the desks in the room while staying within the boundaries of the outermost desks. Tell students they represent the "sea of electrons" found in a metallic solid, such as copper, and that the desks represent positive copper ions. Tell them that every desk should have two students around it and that two desks can claim the same student. Ask students what they notice. The number of electrons and number of copper ions do not change but the electrons (students) are free to move between the copper ions (desks). Ask students what would happen if the outer row of desks were moved in. The electrons (students) would move in because they are confined by the outer row of copper ions (desks).

Ask students what would happen if the desks were put into a long row. The students (electrons) would move along with the desks (copper ions). This represents what happens in a metallic bond. OL

2 Teach

Visual Learning

Figure 11 Have students observe the figure and discuss how metallic bonds have metal cations surrounded by a "sea" of mobile electrons. Using a conductivity device, touch the surface of a piece of zinc or copper. The bulb will light up. Ask students why the bulb lights up. mobile electrons Point out that unlike ionic compounds, in which bonding electrons are attached to the ions, bonding electrons in metals are free to move. OL

■ **Caption Question Fig. 11**
because they are free to move through the metal

DIFFERENTIATED INSTRUCTION

Advanced Learners Approve lab safety forms before work begins. Have students research the thermal conductivity of metals. Obtain different metal wires, such as Al, Zn, Cu, and Fe, of equal gauge and length. On one end of each wire, place a small bead of wax. Using tongs, have students heat the opposite end of the wire and determine and compare the times needed to melt the wax. AL EL

READING CHECK metals dent when hit, whereas ionic compounds shatter

Quick Demo

Alloys In an evaporating dish, dissolve 5.0 g of sodium hydroxide in 25 mL of water. Add 2.0 g of zinc to the solution, and place on a hot plate. Heat the mixture to just below the boiling point. Place some new pennies into the evaporating dish, and turn the pennies several times with forceps. When the pennies are silver in color, take the evaporating dish off the hot plate and remove the pennies. Rinse and dry the pennies. Ask students what has happened. The pennies are coated with zinc. Place one of the pennies on the hot plate. The penny will turn golden. Ask students what happened. The zinc and the copper on the surface of the penny mixed. Discuss alloys, and tell students that the heat increased the motion of the atoms and allowed them to mix faster. Allow the students to examine a new penny, a zinc-coated penny, and a brass-coated penny. Ask what type of alloy was formed. A substitutional alloy was formed because copper and zinc atoms are almost identical in size.

Disposal Wash remaining zinc with cold water, drain, and place in the trash. Dilute the NaOH solution and flush down a drain. OL

Table 12 Melting and Boiling Points

Element	Melting Point (°C)	Boiling Point (°C)
Lithium	180	1342
Tin	232	2602
Aluminum	660	2519
Barium	727	1897
Silver	962	2162
Copper	1085	2562

Properties of metals The physical properties of metals can be explained by metallic bonding. These properties provide evidence of the strength of metallic bonds.

Melting and boiling points The melting points of metals vary greatly. Mercury is a liquid at room temperature, which makes it useful in scientific instruments such as thermometers and barometers. On the other hand, tungsten has a melting point of 3422°C. Lightbulb filaments are usually made from tungsten, as are certain spacecraft parts.

In general, metals have moderately high melting points and high boiling points, as shown in **Table 12.** The melting points are not as extreme as the boiling points because the cations and electrons are mobile in a metal. It does not take an extreme amount of energy for them to be able to move past each other. However, during boiling, atoms must be separated from the group of cations and electrons, which requires much more energy.

Malleability, ductility, and durability Metals are malleable, which means they can be hammered into sheets, and they are ductile, which means they can be drawn into wire. **Figure 12** shows how the mobile particles involved in metallic bonding can be pushed or pulled past each other. Metals are generally durable. Although metallic cations are mobile in a metal, they are strongly attracted to the electrons surrounding them and are not easily removed from the metal.

Thermal conductivity and electrical conductivity The movement of mobile electrons around positive metallic cations makes metals good conductors. The delocalized electrons move heat from one place to another much more quickly than the electrons in a material that does not contain mobile electrons. Mobile electrons easily move as part of an electric current when an electric potential is applied to a metal. These same delocalized electrons interact with light, absorbing and releasing photons, thereby creating the property of luster in metals.

Hardness and strength The mobile electrons in transition metals consist not only of the two outer s electrons but also of the inner d electrons. As the number of delocalized electrons increases, so do the properties of hardness and strength. For example, strong metallic bonds are found in transition metals such as chromium, iron, and nickel, whereas alkali metals are considered soft because they have only one delocalized electron, ns^1.

READING CHECK **Contrast** the behavior of metals and ionic compounds when each is struck by a hammer.

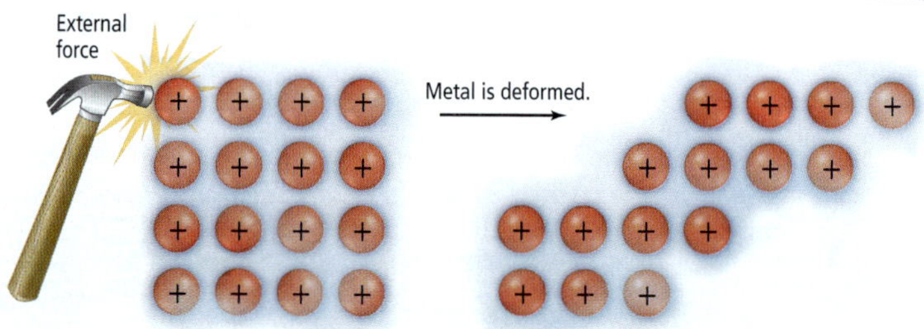

Figure 12 An applied force causes metal ions to move through delocalized electrons, making metals malleable and ductile.

CHEMISTRY JOURNAL

Types of Alloys Have students diagram microscopic views of both a substitutional alloy and an interstitial alloy in their journals. OL EL

Metal Alloys

Due to the nature of metallic bonds, it is relatively easy to introduce other elements into the metallic crystal, forming an alloy. An **alloy** is a mixture of elements that has metallic properties. Because of their unique blend of properties, alloys have a wide range of commercial applications. Stainless steel, brass, and cast iron are a few of the many useful alloys.

Properties of alloys The properties of alloys differ somewhat from the properties of the elements they contain. For example, steel is iron mixed with at least one other element. Some properties of iron are present, but steel has additional properties, such as increased strength. Some alloys vary in properties, depending on how they are manufactured. In the case of some metals, different properties can result based on heating and cooling.

VOCABULARY

WORD ORIGIN

Alloy
comes from the Latin word *alligare,* which means *to bind.*

MiniLAB

Observe Properties

How do the properties of steel change when it is subjected to different types of heat treatment?
For centuries, people have treated metals with heat to change their properties. The final properties of the metal depend on the temperature to which the metal is heated and the rate at which it cools.

Procedure

1. Read and complete the lab safety form.
2. Examine a property of spring steel by trying to bend open one of three **hairpins.** Record your observations.
3. Next, hold each end of the hairpin with a pair of **forceps.** Place the curved central loop portion of the hairpin in the top of the blue flame from a **laboratory burner.** When the metal turns red, pull the hairpin open to form a straight piece of metal. Allow it to cool as you record your observations. Repeat Step 3 for the remaining two hairpins.
WARNING: ***Do not touch the hot metal. Do not hold your hand above the flame of the laboratory burner.***
4. To make softened steel, use a pair of forceps to hold all three hairpins vertically in the flame from the laboratory burner until the hairpins are glowing red all over. Slowly raise the three hairpins straight up and out of the flame so they cool slowly. Slow cooling results in the formation of large crystals.
5. After cooling, bend each of the three hairpins into the shape of the letter J. Record how the metal feels as you bend it.
6. To harden the steel, use the forceps to hold two of the bent hairpins in the flame until they are glowing red all over. Quickly plunge the hot metals into a **250-mL beaker** containing approximately 200 mL of **cold water.** Quick cooling causes the crystal size to be small.
7. Attempt to straighten one of the bends. Record your observations.
8. To temper the steel, use the forceps to hold the remaining hardened metal bend above the flame for a brief period of time. Slowly move the metal back and forth just above the flame until the gray metal turns to an iridescent blue-gray color. Do not allow the metal to become hot enough to glow red. Slowly cool the metal, and then try to unbend it using the end of your finger. Record your observations.

Analysis

1. **Analyze** your results, and identify the two types of steel that appear to have their properties combined in tempered steel.
2. **Hypothesize** how the different observed properties relate to crystal size.
3. **State** a use for spring steel that takes advantage of its unique properties.
4. **Infer** the advantages and disadvantages of using softened steel for body panels on automobiles.
5. **Apply** What is a major disadvantage of hardened steel? Do you think hardened steel would be wear-resistant and retain a sharpened edge? Explain your reasoning.

CHEMISTRY PROJECT

Chemistry at Work Have students research the alloys used in the manufacturing of medical implants and prepare a newspaper article. OL

MiniLAB

Rubric

Purpose Students will observe how the properties of a metal change as the rate of cooling changes and will hypothesize how the size and type of crystal formed affect properties.

Process Skills observe and infer, recognize cause and effect, hypothesize

Safety Precautions Approve lab safety forms before work begins. Have students wear an apron and goggles. Remind them to be careful around the burner flame and to allow time for the hot metal to cool before handling. Use caution when selecting hairpins, as many are plastic coated and will emit toxic fumes when held in a flame.

Disposal Students should place the metal into a designated waste container.

Teaching Strategies

- Remind students that small crystals form when a rapid change in temperature occurs.
- Point out that the hairpins cool quickly, usually in two to three minutes.

Analysis

1. spring and hardened steel
2. Possible hypothesis: Soft steel has large crystals; tempered steel has intermediate-sized crystals.
3. Answers might include pop-up tent, spring toys, and clips.
4. Smooth curves are possible, but they dent easily.
5. it is brittle and breaks easily; yes

3 Assess

Check for Understanding

Ask students to compare and contrast ionic compounds and metallic solids. Accept any reasonable answer. OL EL

Reteach

Show students several metallic and ionic solids. Demonstrate the differences in their physical properties. OL EL

Extension

Have interested students research industrial alloys and report to the class. Ask students to find out if industries in their area use the alloys they researched. AL

ASSESSMENT

Skill Have students complete a chart that compares metallic bonding to ionic bonding. Charts might include types of particles, bond formation, structure, physical properties, and possible laboratory identification. OL

Figure 13 Bicycle frames are sometimes made of 3/2.5 titanium alloy, an alloy of titanium containing 3% aluminum and 2.5% vanadium.

Table 13 Commercial Alloys

Common Name	Composition	Uses
Alnico	Fe 50%, Al 20%, Ni 20%, Co 10%	magnets
Brass	Cu 67–90%, Zn 10–33%	plumbing, hardware, lighting
Bronze	Cu 70–95%, Zn 1–25%, Sn 1–18%	bearings, bells, medals
Cast iron	Fe 96–97%, C 3–4%	casting
Gold, 10-carat	Au 42%, Ag 12–20%, Cu 37.46%	jewelry
Lead shot	Pb 99.8%, As 0.2%	shotgun shells
Pewter	Sn 70–95%, Sb 5–15%, Pb 0–15%	tableware
Stainless steel	Fe 73–79%, Cr 14–18%, Ni 7–9%	instruments, sinks
Sterling silver	Ag 92.5%, Cu 7.5%	tableware, jewelry

Table 13 lists some commercially important alloys and their uses. An alloy of titanium and vanadium is used for the bicycle frame shown in **Figure 13.** Alloys such as this are classified into one of two basic types, substitutional alloys and interstitial alloys.

Substitutional alloys In a substitutional alloy, some of the atoms in the original metallic solid are replaced by other metals of similar atomic size. Sterling silver is an example of a substitutional alloy. In sterling silver, copper atoms replace some of the silver atoms in the metallic crystal. The resulting solid has properties of both silver and copper.

Interstitial alloys An interstitial alloy is formed when the small holes (interstices) in a metallic crystal are filled with smaller atoms. The best-known interstitial alloy is carbon steel. Holes in the iron crystal are filled with carbon atoms, and the physical properties of iron are changed. Iron is relatively soft and malleable. However, the presence of carbon makes the solid harder, stronger, and less ductile than pure iron.

SECTION 4 REVIEW

Section Self-Check

Section Summary

- A metallic bond forms when metal cations attract freely moving, delocalized valence electrons.
- In the electron sea model, electrons move through the metallic crystal and are not held by any particular atom.
- The electron sea model explains the physical properties of metallic solids.
- Metal alloys are formed when a metal is mixed with one or more other elements.

40. MAINIDEA **Contrast** the structures of ionic compounds and metals.
41. **Explain** how the conductivity of electricity and the high boiling points of metals are explained by metallic bonding.
42. **Contrast** the cause of the attraction in ionic bonds and metallic bonds.
43. **Summarize** alloy types by correctly pairing these terms and phrases: *substitutional, interstitial, replaced,* and *filled in.*
44. **Design an experiment** that could be used to distinguish between a metallic solid and an ionic solid. Include at least two different methods for comparing the solids. Explain your reasoning.
45. **Model** Draw a model to represent the physical property of metals known as ductility, or the ability to be drawn into a wire. Base your drawing on the electron sea model shown in **Figure 11.**

Greg Huglin/SuperStock

SECTION 4 REVIEW

40. Ions in ionic compounds are arranged in a repeating pattern of alternating charges, whereas metals consist of fixed cations surrounded by a sea of mobile, or delocalized electrons.
41. Delocalized electrons can move through the solid to conduct an electric current. The number of delocalized electrons and the strength of a metallic bond determine the melting point.
42. Ionic bonds are held together by the electrostatic force of attraction between ions, whereas a metallic bond is due to the attraction of metallic cations for delocalized electrons.
43. substitutional, replaced, interstitial, filled in
44. A typical student experiment might involve using a conductivity tester to test solids and solutions, and using a hammer to check malleability and brittleness.
45. Diagrams should show metal ions being moved into a longer, thinner form through a sea of electrons.

everydayCHEMISTRY

Killer Fashion

Shiny and colorful, costume jewelry can be inexpensive and fun. But is it safe? Usually the answer is yes. But some costume jewelry, particularly pieces made in developing countries, such as China and India, might pose a danger due to high levels of the toxic element lead (Pb).

Poisoned plumbing When lead gets wet, a certain amount of it dissolves, becoming lead (Pb^{2+}) ions. Inside the body, these ions can replace calcium (Ca^{2+}) ions. Other than their similar electric charges, lead and calcium are different (for one thing, lead ions are much heavier than calcium ions), and the presence of lead can cause learning disabilities, coma, or even death.

It might be surprising, then, to learn that lead was used by the Romans in, of all things, their water pipes! In fact, the symbol for lead—Pb—comes from the Latin word *plumbum,* which still appears in English as the root of the word *plumber,* one who works with pipes.

Toxic pottery While lead is not found in modern plumbing, it can still be found in other things. The pot shown in **Figure 1** was created with lead glaze and fired using traditional Mexican techniques to give it its distinctive black color. Glazes containing lead compounds can also create vibrant colors when fired under different conditions.

Figure 1 Lead compounds in pottery glaze give this pot its distinctive look.

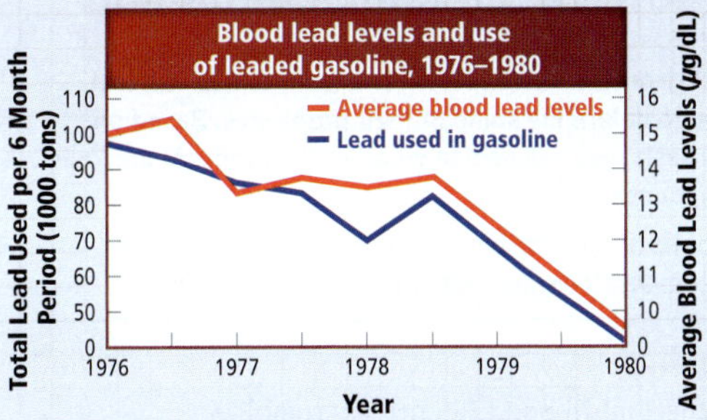

Figure 2 Lead levels in Americans' blood dropped as leaded gasoline was phased out.

A useful poison Before it was known to be highly toxic, lead had a number of applications beyond pottery and plumbing. Lead has been used in paint and even gasoline, where its presence reduced "knock"—the tendency of gasoline to explode at the wrong time within the engine block. In the 1970s, when leaded gasoline was phased out in the United States, blood lead levels dropped immediately (see **Figure 2**).

But other avenues, such as jewelry or toys manufactured in other countries, can still contain lead. A lead-rich piece of costume jewelry might rest harmlessly against the skin until the metal finds its way into the mouth of a curious child or a day-dreaming teenager.

Chelation Children are particularly susceptible to lead poisoning, due to their smaller body sizes and rapid rates of development. In serious cases, a process called chelation therapy might be the only way to save the child's life. Chelation therapy reverses one important effect of lead poisioning, replacing toxic lead with beneficial calcium in the body.

WRITING IN ▶ Chemistry

Sense of Danger Our sense of taste can detect certain toxins found naturally in plants. Research other modern toxins, such as lead and antifreeze, to find out why they don't elicit a negative response from our taste buds.

WebQuest

WRITING IN ▶ Chemistry

Rubric WebQuest

Research Student research should cite speculation by scientists that because humans and their ancestors did not often encounter elemental lead (or other modern toxins) in the natural environment, we have evolved having no natural aversion to eating the toxin. In fact, certain lead compounds found in paint (such as lead acetate) have a sweet taste. Research might also discuss how, contrary to the situation with lead, many plant-produced toxins have a bitter taste. This is likely due to the fact that humans and plants have coevolved for millions of years.

everydayCHEMISTRY

Purpose

Students will learn how lead exposure can affect their health and of several sources of lead exposure.

Background

Exposure to lead sources cannot be avoided. In addition to the sources mentioned in the article, old peeling and flaking lead-containing house paint is a major source of lead exposure. Even though lead was banned from use in house paints in 1978, the problems related to old house paint will continue for many years.

A common student misconception is that the graphite in pencils is the element lead. Tell them that this is not so. There are two reasons why the graphite in pencils is referred to as "lead." First, elemental lead was used for writing in ancient times. Second, when graphite (a type of carbon) was discovered and its usefulness for writing was recognized, it was at first believed to be a form of lead.

Teaching Strategies

- Examine the graph shown in Figure 2. Discuss with your class why, with so many lead sources in the environment, (including lead from paint, lead from pottery, and lead from the costume jewelry), blood lead levels were so greatly affected when lead was banned from use in gasoline. Bring up the idea of exposure level, methods of ingestion and absorption, and typical (average) exposure across a large population.
- Ask students to infer what this graph means for the blood lead levels of people living in parts of the world where leaded gasoline is still used. Their blood lead levels will be elevated.

ChemLAB

Rubric

Preparation

Time Allotment one class period

Process Skills classify, observe and infer, use numbers, compare and contrast, predict

Safety Precautions Approve lab safety forms before work begins. Have students wear an apron and goggles. Tell students not to look directly at the light emitted by the burning magnesium. Advise them that the crucible will remain hot enough to burn the skin for about five minutes. Review the MSDS for magnesium with students prior to the lab activity. Make sure fire suppression equipment for metals is available (type D extinguisher) or a bucket of sand. If possible, perform lab under a fume hood.

Disposal Place the magnesium product in a receptacle for solid waste.

Procedure

- The lengths of magnesium should be cut in advance.
- Adjust lab burners to produce the hottest possible flame.
- Tell students that at high temperatures oxygen and nitrogen in the air will react with magnesium.
- Troubleshooting To observe the burning magnesium indirectly, place a piece of shiny metal behind the crucible and a piece of cardboard in front of it.

Expected Results

Sample Data

Mass of empty crucible: 7.56 g
Mass of crucible + Mg ribbon before heating: 7.85 g
Mass of Mg ribbon: 0.29 g
Mass of crucible + Mg ribbon after heating: 7.93 g
Mass of Mg products: 0.37 g

ChemLAB

Synthesize an Ionic Compound

Background: You will form two compounds and test them to determine some of their properties. Based on your tests, you will decide whether the products are ionic compounds.

Question: *Can the physical properties of a compound indicate that they have ionic bonds?*

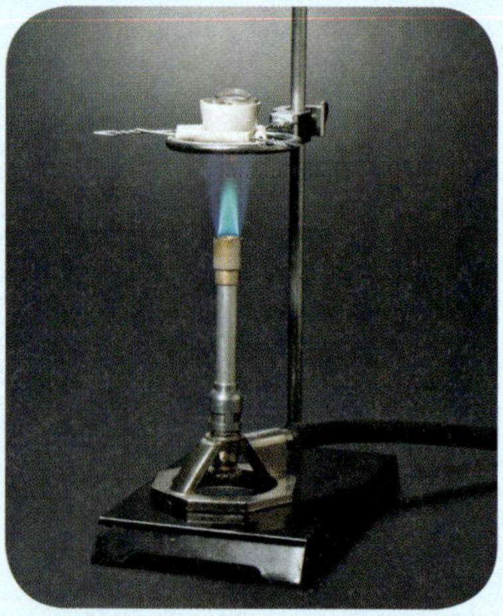

Matt Meadows

Materials

magnesium ribbon (25 cm)
ring stand and ring
Bunsen burner
crucible tongs
100-mL beaker
conductivity tester
crucible
clay triangle
stirring rod
centigram balance
distilled water

Safety Precautions

WARNING: ***Do not look directly at the burning magnesium; the intensity of the light can damage your eyes. Avoid handling heated materials until they have cooled.***

Procedure

1. Read and complete the lab safety form.
2. Record all measurements in your data table.
3. Position the ring on the ring stand about 7 cm above the top of the Bunsen burner. Place the clay triangle on the ring.
4. Measure the mass of the clean, dry crucible.
5. Roll 25 cm of magnesium ribbon into a loose ball. Place it in the crucible. Measure the mass of the magnesium and crucible together.
6. Place the crucible on the triangle, and heat it with a hot flame (flame tip should be near the crucible).
7. Turn off the burner as soon as the magnesium ignites and begins to burn with a bright white light. Allow it to cool, and measure the mass of the magnesium product and the crucible.
8. Place the dry, solid product in the beaker.
9. Add 10 mL of distilled water to the beaker, and stir. Check the mixture with a conductivity tester.
10. **Cleanup and Disposal** Dispose of the product as directed by your teacher. Wash out the crucible with water. Return all lab equipment to its proper place.

Analyze and Conclude

1. **Analyze Data** Calculate the mass of the ribbon and the product. Record these masses in your table.
2. **Classify** the forms of energy released. What can you conclude about the stability of products?
3. **Infer** Does the magnesium react with the air?
4. **Predict** the ionic formulas for the two binary products formed, and write their names.
5. **Analyze and Conclude** The product of the magnesium-oxygen reaction is white, whereas the product of the magnesium-nitrogen reaction is yellow. Which compound makes up most of the product?
6. **Analyze and Conclude** Did the magnesium compounds conduct a current when in solution? Do these results verify that the compounds are ionic?
7. **Error Analysis** If the results show that the magnesium lost mass instead of gaining mass, cite possible sources of the error.

INQUIRY EXTENSION

Design an Experiment If the magnesium compounds conduct a current in solution, can you affect how well they conduct electricity? If they did not conduct a current, could they? Design an experiment to find out.

Analyze and Conclude

1. Refer to Expected Results.
2. heat and light; It is more stable than the reacting elements.
3. There is an increase in mass from 0.29 g to 0.37 g.
4. MgO, magnesium oxide; Mg_3N_2, magnesium nitride
5. MgO; The product appears white.
6. yes; Yes, because ionic compounds conduct an electric current in solution.
7. Possible answers include that some of the product blew away or that the reaction was incomplete.

INQUIRY EXTENSION

Student experimental designs will vary. However, the basic point students should investigate is that more concentrated ionic solutions are more conductive than less concentrated ones.

CHAPTER 7 STUDY GUIDE

Vocabulary Practice

BIGIDEA Ionic compounds are held together by chemical bonds formed by the attraction of oppositely charged ions.

SECTION 1 Ion Formation

MAINIDEA Ions form when atoms gain or lose valence electrons to achieve a stable octet electron configuration.

- A chemical bond is the force that holds two atoms together.
- Some atoms form ions to gain stability. This stable configuration involves a complete outer energy level, usually consisting of eight valence electrons.
- Ions are formed by the loss or gain of valence electrons.
- The number of protons remains unchanged during ion formation.

VOCABULARY
- chemical bond
- cation
- anion

SECTION 2 Ionic Bonds and Ionic Compounds

MAINIDEA Oppositely charged ions attract each other, forming electrically neutral ionic compounds.

- Ionic compounds contain ionic bonds formed by the attraction of oppositely charged ions.
- Ions in an ionic compound are arranged in a repeating pattern known as a crystal lattice.
- Ionic compound properties are related to ionic bond strength.
- Ionic compounds are electrolytes; they conduct an electric current in the liquid phase and in aqueous solution.
- Lattice energy is the energy needed to remove 1 mol of ions from its lattice.

VOCABULARY
- ionic bond
- ionic compound
- crystal lattice
- electrolyte
- lattice energy

SECTION 3 Names and Formulas for Ionic Compounds

MAINIDEA In written names and formulas for ionic compounds, the cation appears first, followed by the anion.

- A formula unit gives the ratio of cations to anions in the ionic compound.
- A monatomic ion is formed from one atom. The charge of a monatomic ion is equal to its oxidation number.
- Roman numerals indicate the oxidation number of cations having multiple possible oxidation states.
- Polyatomic ions consist of more than one atom and act as a single unit.
- To indicate more than one polyatomic ion in a chemical formula, place parentheses around the polyatomic ion and use a subscript.

VOCABULARY
- formula unit
- monatomic ion
- polyatomic ion
- oxyanion

SECTION 4 Metallic Bonds and the Properties of Metals

MAINIDEA Metals form crystal lattices and can be modeled as cations surrounded by a "sea" of freely moving valence electrons.

- A metallic bond forms when metal cations attract freely moving, delocalized valence electrons.
- In the electron sea model, electrons move through the metallic crystal and are not held by any particular atom.
- The electron sea model explains the physical properties of metallic solids.
- Metal alloys are formed when a metal is mixed with one or more other elements.

VOCABULARY
- electron sea model
- delocalized electron
- metallic bond
- alloy

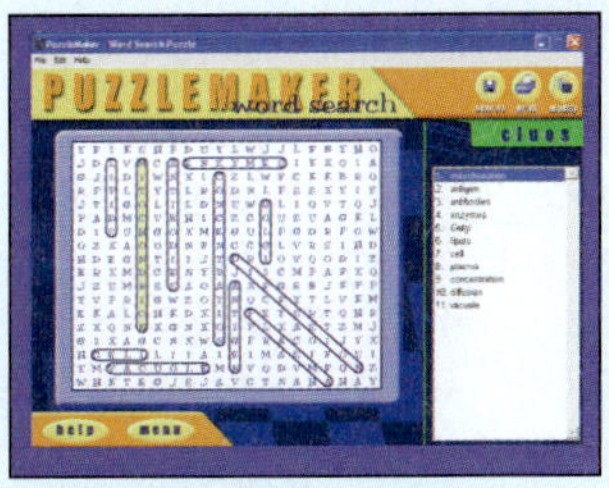

Vocabulary Practice

For additional practice with vocabulary, have students access the Vocabulary PuzzleMaker.

Use the Vocabulary

To reinforce chapter vocabulary, have students write a sentence using each term. OL EL

Review Strategies

- Have students describe the formation of cations and anions, and give an example of each. OL
- Have students discuss the nature of ionic bonds and identify physical properties associated with ionic bonds. OL
- Have students summarize a metallic bond, and identify physical properties associated with metallic bonding. OL
- Problems from the Supplemental Practice Problems appendix can be used for review. OL

Students can review with these online resources.

Vocabulary Practice

Vocabulary eGames
Vocabulary eFlashcards
Vocabulary PuzzleMaker

Section Self-Check

Chapter Self-Check

Online Test Practice

Use *eAssessment* to:

- create multiple versions of tests
- edit existing questions and add your own questions
- build tests aligned with select state curriculum standards using built-in state curriculum tags
- track students' progress

CHAPTER 7 ASSESSMENT

SECTION 1

Mastering Concepts

46. How do positive ions and negative ions form?

47. When do chemical bonds form?

48. Why are halogens and alkali metals likely to form ions? Explain your answer.

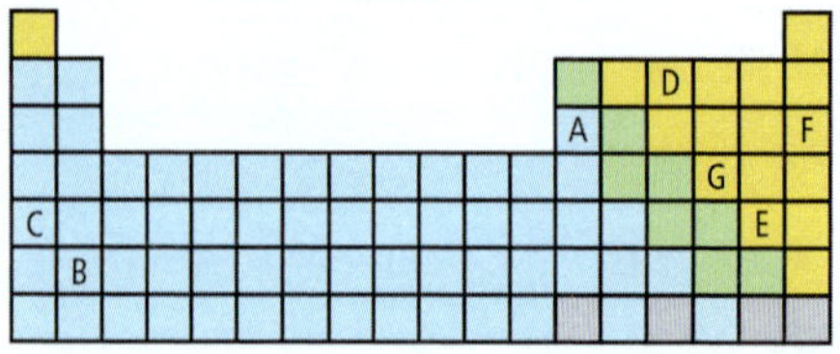

Figure 14

49. The periodic table shown in **Figure 14** contains elements labeled *A–G*. For each labeled element, state the number of valence electrons and identify the ion that will form.

50. Discuss the importance of electron affinity and ionization energy in the formation of ions.

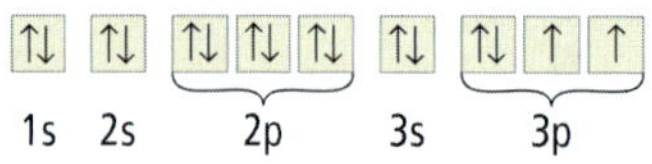

Figure 15

51. The orbital notation of sulfur is shown in **Figure 15.** Explain how sulfur forms its ion.

Mastering Problems

52. Give the number of valence electrons in an atom of each element.
- **a.** cesium
- **b.** rubidium
- **c.** gallium
- **d.** zinc
- **e.** strontium

53. Explain why noble gases are not likely to form chemical bonds.

54. Discuss the formation of the barium ion.

55. Explain how an anion of nitrogen forms.

56. The more reactive an atom, the higher its potential energy. Which atom has higher potential energy, neon or fluorine? Explain.

57. Explain how the iron atom can form both an iron 2+ ion and an iron 3+ ion.

58. Predict the reactivity of each atom based on its electron configuration.
- **a.** potassium
- **b.** fluorine
- **c.** neon

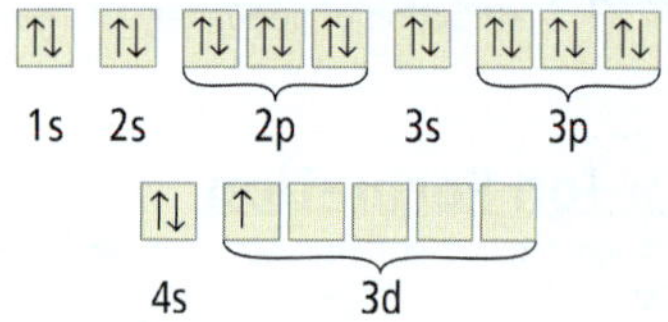

Figure 16

59. Discuss the formation of a 3+ scandium ion using its orbital notation, shown in **Figure 16.**

SECTION 2

Mastering Concepts

60. What does the term *electrically neutral* mean when discussing ionic compounds?

61. Discuss the formation of ionic bonds.

62. Explain why potassium does not bond with neon to form a compound.

63. Briefly discuss three physical properties of ionic solids that are linked to ionic bonds.

64. Describe an ionic crystal, and explain why ionic crystals for different compounds might vary in shape.

65. How does lattice energy change with a change in the size of an ion?

66. In **Figure 14,** the element labeled *B* is barium, and the element labeled *E* is iodine. Explain why the compound formed between these elements will not be BaI.

Mastering Problems

67. Determine the ratio of cations to anions in each.
- **a.** potassium chloride, a salt substitute
- **b.** calcium fluoride, used in the steel industry
- **c.** calcium oxide, used to remove sulfur dioxide from power-plant exhaust
- **d.** strontium chloride, used in fireworks

68. Look at **Figure 14;** describe the ionic compound that forms from the elements represented by C and D.

69. Discuss the formation of an ionic bond between zinc and oxygen.

70. Using orbital notation, diagram the formation of an ionic bond between aluminum and fluorine.

71. Using electron configurations, diagram the formation of an ionic bond between barium and nitrogen.

72. Conductors Under certain conditions, ionic compounds conduct an electric current. Describe these conditions, and explain why ionic compounds are not always used as conductors.

SECTION 1

Mastering Concepts

46. An atom gains or loses electrons to achieve a stable electron configuration.

47. when a positive nucleus attracts electrons of another atom, or oppositely charged ions attract

48. Halogens need to gain only one electron to have a noble gas electron configuration. Alkali metals need to lose one.

49. A: three valence electrons, Al^{3+}; B: two valence electrons, Ba^{2+}; C: one valence electron, Rb^{+}; D: five valence electrons N^{3-}; E: seven valence electrons, I^{-}; F: eight valence electrons, no ion formed, G: six valence electrons Se^{2-}

50. high electron affinity: atom easily gains an electron; low ionization energy: atom easily loses an electron

51. Sulfur gains 2 electrons in the 3p sublevel, forming a complete octet.

Mastering Problems

52. a. 1 **b.** 1 **c.** 3 **d.** 2 **e.** 2

53. They have a full outer energy level.

54. Ba loses 2 e^- and forms Ba^{2+} which has the stable electron configuration of Xe.

55. N gains 3 e^-, forming N^{3-} which has the stable electron configuration of Ne.

56. F; because it will gain one more electron to fill its outer energy level

57. Iron has the electron configuration $[Ar]4s^2 3d^6$. To form the 2+ ion, the iron atom loses the $4s^2$ electrons. When forming the 3+ ion, the iron atom loses, the $4s^2$ electrons and one 3d electron.

58. a. very reactive, loses 1 e^-, forms 1+ ion
b. reactive, gains 1 e^-, forms 1− ion
c. unreactive, has an octet

59. Scandium, $[Ar]4s^2 3d^1$, loses its $4s^2$ and $3d^1$ electrons to form the 3+ ion.

SECTION 2

Mastering Concepts

60. The number of electrons lost is equal to the number of electrons gained.

61. A positive ion is attracted to a negative ion and lattice energy is released.

62. Neon has an octet; it is already stable.

63. are crystals; high mp and bp

64. a geometric arrangement of ions; shape varies due to ion size and number of ions

65. As the size of an ion increases, the lattice energy decreases.

66. Ba forms Ba^{2+}; I forms I^-; 1 Ba ion and 2 I ions are needed to form an electrically neutral compound.

Mastering Problems

67. a. 1:1 **b.** 1:2 **c.** 1:1 **d.** 1:2

68. C represents Rb, which forms Rb^+; D represents N, which forms N^{3-}; 3 Rb and 1 N ions form Rb_3N.

69. Zn forms Zn^{2+}; O forms O^{2-}; ions attract, form ZnO

70. Refer to Solutions Manual; ions attract to form AlF_3

71. Refer to Solutions Manual. To form a compound, a total of six electrons must transfer from three barium atoms to two nitrogen atoms.

72. Ionic compounds will conduct electric current in the molten state or when dissolved in water but are nonconducting solids at room temperature.

73. CaKr, because Kr is a noble gas; $BaCl_3$ and MgF, because charges are not balanced

74. MgO; It has the highest lattice energy.

75. KCl: K smaller than Cs, both are 1+; the smaller the ion, the greater the lattice energy. CaO: Ca ion is 2+ while K is 1+; the greater an ion's charge, the greater its lattice energy.

Chapter Self-Check

73. Which of these compounds are not likely to occur: CaKr, Na_2S, $BaCl_3$, MgF? Explain your choices.

74. Use **Table 6** to determine which ionic compound has the highest melting point: MgO, KI, or AgCl. Explain your answer.

75. Which has the greater lattice energy, CsCl or KCl? K_2O or CaO? Explain your choices.

SECTION 3

Mastering Concepts

76. What information do you need to write a correct chemical formula to represent an ionic compound?

77. When are subscripts used in formulas for ionic compounds?

78. Discuss how an ionic compound is named.

79. Using oxidation numbers, explain why the formula NaF_2 is incorrect.

80. Explain what the name chromium(III) oxide means in terms of electrons lost and gained, and identify the correct formula.

Mastering Problems

81. Give the formula for each ionic compound.
 a. calcium iodide
 b. silver bromide
 c. copper(II) chloride
 d. potassium periodate
 e. silver acetate

82. Name each of the following ionic compounds.
 a. K_2O
 b. $CaCl_2$
 c. Mg_3N_2
 d. NaClO
 e. KNO_3

83. Complete **Table 14** by placing the symbols, formulas, and names in the blanks.

Table 14 Identifying Ionic Compounds

Cation	Anion	Name	Formula
		ammonium sulfate	
			PbF_2
		lithium bromide	
			Na_2CO_3
Mg^{2+}	PO_4^{3-}		

84. **Chrome** Chromium, a transition metal used in chrome plating, forms both the Cr^{2+} and Cr^{3+} ions. Write the formulas for the ionic compounds formed when these ions react with fluorine ions and with oxygen ions.

85. Which are correct formulas for ionic compounds? For those that are not correct, give the correct formula and justify your answer.
 a. AlCl
 b. Na_3SO_4
 c. $BaOH_2$
 d. Fe_2O

86. Write the formulas for all of the ionic compounds that can be formed by combining each of the cations with each of the anions listed in **Table 15.** Name each compound formed.

Table 15 List of Cations and Anions

Cations	Anions
K^+	SO_3^{2-}
NH_4^+	I^-
Fe^{3+}	NO_3^-

SECTION 4

Mastering Concepts

87. Describe a metallic bond.

88. Briefly explain why metallic alloys are made.

89. Briefly describe how malleability and ductility of metals are explained by metallic bonding.

90. Compare and contrast the two types of metal alloys.

91. Explain how a metallic bond is similar to an ionic bond.

92. **Brass** Copper and zinc are used to form brass, an alloy. Briefly explain why these two metals form a substitutional alloy and not an interstitial alloy.

Mastering Problems

93. How is a metallic bond different from an ionic bond?

94. **Silver** Briefly explain why silver is a good conductor of electricity.

95. **Steel** Briefly explain why steel, an alloy of iron, is used to build the supporting structure of many buildings.

96. The melting point of beryllium is 1287°C, while that of lithium is 180°C. Explain the large difference in values.

97. Titanium has a boiling point of 3287°C, and copper has a boiling point of 2562°C. Explain why there is a difference in the boiling points of these two metals.

98. **Alloys** Describe the difference between the metal alloy sterling silver and carbon steel in terms of the types of alloys involved.

SECTION 3

Mastering Concepts

76. the metallic ion and the nonmetallic ion, including their charges

77. when more than one unit of an ion is in the simplest ratio of the ions

78. name cation; name anion; monatomic cations use element name; monatomic anions use root plus *-ide*; transition metals and some metals have multiple oxidation states; if there's a polyatomic ion, name ion

79. 1+ and 1− ions should be in a 1:1 ratio; the correct formula is NaF

80. The (III) indicates Cr has lost 3 e^-; oxide indicates O atom gained 2 e^-; Cr_2O_3

Mastering Problems

81. a. CaI_2 b. AgBr c. $CuCl_2$ d. KIO_4 e. $AgC_2H_3O_2$

82. a. potassium oxide
 b. calcium chloride
 c. magnesium nitride
 d. sodium hypochlorite
 e. potassium nitrate

83. NH_4^+; SO_4^{2-}; ammonium sulfate; $(NH_4)_2SO_4$
 Pb^{2+}; F^-; lead(II) fluoride; PbF_2
 Li^+; Br^-; lithium bromide; LiBr
 Na^+; CO_3^{2-}; sodium carbonate; Na_2CO_3
 Mg^{2+}; PO_4^{3-}; magnesium phosphate; $Mg_3(PO_4)_2$

84. fluorine: CrF_2, CrF_3; oxygen: CrO, Cr_2O_3

85. a. $AlCl_3$; 1 Al^{3+} ion bonds to 3 Cl^- ions
 b. Na_2SO_4; 2 Na^+ ions bond to SO_4^{2-}
 c. $Ba(OH)_2$; parentheses needed
 d. Fe_2O_3 or FeO; iron forms Fe^{2+} or Fe^{3+}

86. K_2SO_3, potassium sulfite; KI, potassium iodide; KNO_3, potassium nitrate; $(NH_4)_2SO_3$, ammonium sulfite; NH_4I, ammonium iodide; NH_4NO_3, ammonium nitrate; $Fe_2(SO_3)_3$, iron(III) sulfite; FeI_3, iron(III) iodide; $Fe(NO_3)_3$, iron(III) nitrate

SECTION 4

Mastering Concepts

87. Each positive metal ion is attracted to delocalized valence electrons.

88. Alloys have slightly different properties than those of the pure metal they are mixed from. Some alloys have increased strength and are harder than the pure metal.

89. When a force is applied to a metallic solid, the metal ions move as well as the delocalized electrons.

90. substitutional alloy: metal atoms similar in size; interstitial alloy: two differently sized atoms

91. The bonds are similar because they are formed by the attraction of oppositely charged particles. Ionic bonds are between oppositely charged ions and metallic bonds are between a metallic ion and negative valence electrons.

92. Cu and Zn ions are about the same size and can replace each other, forming a substitutional alloy.

Mastering Problems

93. metallic: attraction between a positive metal ion and free valence electrons; ionic bond: attraction between a positive metallic ion and a negative nonmetallic ion

94. It has delocalized electrons that are free to move.

95. Iron forms a strong metallic bond, giving solid iron hardness and strength.

96. Be has 2 delocalized electrons per atom; lithium has 1; as number of delocalized electrons increases, lattice energy increases, thus raising the melting point

97. Ti: up to 4 delocalized e^-; Cu: up to 2 delocalized e^-; metallic bonding is greater in Ti

98. Sterling silver is substitutional (Ag and Cu). Carbon steel is interstitial (Fe and C).

Mixed Review

99. 6, 6, 5, 5, and 7, respectively

100. Ca, $[Ar]4s^2$, loses 2 e^-; if it loses an inner 3p electron, it is unstable

101. $MgCl_2$; lattice energy increases with increased charge.

102. **a.** Na_2S **b.** $FeCl_3$ **c.** Na_2SO_4 **d.** $Ca_3(PO_4)_2$ **e.** $Zn(NO_3)_2$

103. CoO, cobalt(II) oxide; Co_2O_3, cobalt(III) oxide

104. Selenium; 6; Se^{2-}
Tin; 4; Sn^{2+}
Iodine; 7; I^-
Argon; 8; none

105. Delocalized electrons allow it to conduct. It is malleable and ductile.

106. Nickel, $[Ar]3d^84s^2$, will lose the two outer 4s electrons.

107. Both are composed of S and O. Sulfate (SO_4) has more oxygen atoms than sulfite (SO_3). The suffix *–ite* is used to indicate one less oxygen atom. Both ions have the same oxidation number, 2−.

108. K loses 1 e^-; I gains 1 e^-; KI forms; see Solutions Manual

109. A Mg atom loses 2 e^- and forms Mg^{2+}. An O atom gains 2 e^- and forms O^{2-}. 1 Mg ion attracts 1 O ion, forming MgO. 3 Mg atoms each lose 2 e^-, forming Mg^{2+}. 2 N atoms each gain 3 e^-, forming N^{3-}. The ions form Mg_3N_2.

110. Sodium metal contains metallic bonds. Sodium chloride is an ionic solid.

111. **a.** calcium oxide
b. barium sulfide
c. aluminum phosphate
d. barium hydroxide
e. strontium nitrate

THINK CRITICALLY

112. Concept maps will vary.

113. **a.** NaCl; smaller ion size
b. Cu; it is smaller
c. MgO; Mg has a greater charge

114. cations: lose electrons, have + charge
anions: gain electrons, have − charge

115. **a.** metal is copper(II) or copper(I)
b. formula unit not simplest ratio
c. Pb cannot have 5+ oxidation state
d. prefixes not used in ionic compounds
e. If a polyatomic ion requires a subscript, use parentheses.
Flowcharts will vary.

MIXED REVIEW

99. Give the number of valence electrons for atoms of oxygen, sulfur, arsenic, phosphorus, and bromine.

100. Explain why calcium can form a Ca^{2+} ion but not a Ca^{3+} ion.

101. Which ionic compound would have the greatest lattice energy: NaCl, KCl, or $MgCl_2$? Explain your answer.

102. Give the formula for each ionic compound.
a. sodium sulfide
b. iron(III) chloride
c. sodium sulfate
d. calcium phosphate
e. zinc nitrate

103. Cobalt, a transition metal, forms both the Co^{2+} and Co^{3+} ions. Write the correct formulas, and give the name for the oxides formed by the two different ions.

104. Complete **Table 16.**

Table 16 Element, Electron, and Ion Data

Element	Valence Electrons	Ion Formed
Selenium		
Tin		
Iodine		
Argon		

105. Gold Briefly explain why gold can be used both in jewelry and as a conductor in electronic devices.

106. Discuss the formation of the nickel ion with a 2+ oxidation number.

107. Compare the oxyanions sulfate and sulfite.

108. Using electron-dot structures, diagram the formation of an ionic bond between potassium and iodine.

109. Magnesium forms both an oxide and a nitride when burned in air. Discuss the formation of magnesium oxide and magnesium nitride when magnesium atoms react with oxygen and nitrogen atoms.

110. An external force easily deforms sodium metal, while sodium chloride shatters when the same amount of force is applied. Why do these two solids behave so differently?

111. Name each ionic compound.
a. CaO
b. BaS
c. $AlPO_4$
d. $Ba(OH)_2$
e. $Sr(NO_3)_2$

THINK CRITICALLY

112. Design a concept map to explain the physical properties of both ionic compounds and metallic solids.

113. Predict which solid in each pair will have the higher melting point. Explain your answers.
a. NaCl or CsCl
b. Ag or Cu
c. Na_2O or MgO

114. Compare and contrast cations and anions.

115. Observe and Infer Identify the mistakes in the incorrect formulas and formula names, and design a flowchart to prevent the mistakes.
a. copper acetate
b. Mg_2O_2
c. Pb_2O_5
d. disodium oxide
e. $Mg_3(PO_4)_2$

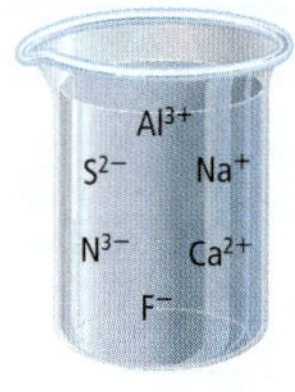

■ **Figure 17**

116. Apply Examine the ions in the beaker shown in **Figure 17.** Identify two compounds that could form using the available ions, and explain why this is possible.

117. Apply Praseodymium is a lanthanide element that reacts with hydrochloric acid, forming praseodymium(III) chloride. It also reacts with nitric acid, forming praseodymium(III) nitrate. Praseodymium has the electron configuration $[Xe]4f^36s^2$.
a. Examine the electron configuration, and explain how praseodymium forms a 3+ ion.
b. Write the correct formulas for both compounds formed by praseodymium.

118. Hypothesize Look at the locations of potassium and calcium on the periodic table. Form a hypothesis to explain why the melting point of calcium is considerably higher than the melting point of potassium.

119. Assess Explain why the term *delocalized* is an appropriate term for the electrons involved in metallic bonding.

120. Apply All uncharged atoms have valence electrons. Explain why elements such as iodine and sulfur do not have metallic bonds.

121. Analyze Explain why lattice energy is released upon formation of an ionic compound.

116. The compounds that can be formed are: Al_2S_3, AlN, AlF_3, Na_2S, Na_3N, NaF, CaS, Ca_3N_2, and CaF_2. Students should explain about electrons transferred from the atom to form a positive ion as well as electrons gained by atoms to form negative ions. They should also discuss the attraction between positive ions and negative ions and the forming of an electrically neutral compound.

117. **a.** Praseodymium must lose the outer $6s^2$ electrons and one of the 4f electrons to form the 3+ ion.
b. The compounds formed are $PrCl_3$ and $Pr(NO_3)_3$.

118. Calcium has two delocalized electrons for every one for potassium. Thus, calcium has a higher melting point.

119. They are free to move; the electrons are not held to any specific atom.

120. They gain electrons. Thus, their electrons are not delocalized.

121. Bonded ions have lower potential energy than separated ions. The difference in their energies is the lattice energy released when an ionic bond forms.

Chapter Self-Check

CHALLENGE PROBLEM

122. Ionic Compounds Chrysoberyl is a transparent or translucent mineral that is sometimes opalescent. It is composed of beryllium aluminum oxide, $BeAl_2O_4$. Identify the oxidation numbers of each of the ions found in this compound. Explain the formation of this ionic compound.

CUMULATIVE REVIEW

123. You are given a liquid of unknown density. The mass of a graduated cylinder containing 2.00 mL of the liquid is 34.68 g. The mass of the empty graduated cylinder is 30.00 g. Given this information, determine the density of the liquid.

124. In the laboratory, students used a balance and a graduated cylinder to collect the data shown in **Table 17.** Calculate the density of the sample. If the accepted value of this sample is 7.01 g/mL, calculate the percent error.

Table 17 Volume and Mass Data

Mass of sample	19.21 g
Volume of water alone	39.0 mL
Volume of water + sample	43.1 mL

125. A mercury atom drops in energy from 1.413×10^{-18} J to 1.069×10^{-18} J.

- **a.** What is the energy of the photon emitted by the mercury atom?
- **b.** What is the frequency of the photon emitted by the mercury atom?
- **c.** What is the wavelength of the photon emitted by the mercury atom?

126. Which element has the greater ionization energy, chlorine or carbon?

127. Compare and contrast the ways in which metals and nonmetals form ions, and explain why they are different.

128. What are transition elements?

129. Write the symbol and name of the element that fits each description.

- **a.** the second-lightest of the halogens
- **b.** the metalloid with the lowest period number
- **c.** the only group 16 element that is a gas at room temperature
- **d.** the heaviest of the noble gases
- **e.** the group 15 nonmetal that is a solid at room temperature

WRITING IN ▶ Chemistry

130. Free Radicals Many researchers believe that free radicals are responsible for the effects of aging and cancer. Research free radicals, and write about the cause and what can be done to prevent free radicals.

131. Growing Crystals Crystals of ionic compounds can be easily grown in the laboratory setting. Research the growth of crystals, and design an experiment to grow a crystal in the laboratory.

DBQ Document-Based Questions

Oceans *As part of an analysis of the world's oceans, scientists summarized the ion-related data shown in* **Table 18.**

Data from: Royal Society of Chemistry, *All at sea? The chemistry of the oceans.*

Table 18 The Twelve Most-Common Ions in the Sea

Ion	Concentration (mg/dm^3)	% by mass (of total dissolved solids)
Cl^-	19,000	55.04
Na^+	10,500	30.42
SO_4^{2-}	2655	7.69
Mg^{2+}	1350	3.91
Ca^{2+}	400	1.16
K^+	380	1.10
CO_3^{2-}	140	0.41
Br^-	65	0.19
BO_3^{3-}	20	0.06
SiO_3^{2-}	8	0.02
Sr^{2+}	8	0.02
F^-	1	0.003

132. Identify the anions and cations listed in **Table 18.**

133. Create a bar graph of each ion's concentration. Explain why this is a difficult graph to draw.

134. Sodium chloride is not the only ionic compound that forms from sea water. Identify four other compounds that could be formed that contain the sodium ion. Write both the formula and the name for each compound.

CHALLENGE PROBLEM

122. Be, a group 2 element, forms a 2+ ion, Al, a group 13 element, forms a 3+ ion, and O, a group 16 element, forms a 2− ion. There are two electrons lost from one atom of beryllium and six electrons lost from two atoms of aluminum. Four oxygen atoms gain a total of eight electrons with each oxygen atom gaining two electrons. The positive ions attract the negative ions forming an electrically neutral compound.

CUMULATIVE REVIEW

123. density = 2.34 g/mL

124. density = 4.7 g/mL; % error = 33%

125. a. $\Delta E = 3.440 \times 10^{-19}$ J

b. $\nu = 5.192 \times 10^{14}\ s^{-1}$

c. $\lambda = 5.77 \times 10^{-7}$ m or 577 nm

126. chlorine

127. Metals lose electrons to form cations; nonmetals gain electrons to form anions. Both form ions to gain stability.

128. the d block elements

129. a. Cl, chlorine

b. B, boron

c. O, oxygen

d. Rn, radon

e. P, phosphorus

WRITING IN ▶ Chemistry

Rubric

130. Student answers will vary. Students should discuss reduction and oxidation (gain and loss of electrons) in forming free radicals as well as antioxidants, Vitamin E and Vitamin C.

131. Student answers will vary. Students should include the use of supersaturated solutions and that the evaporation of water from the solution allows crystals to grow large over a period of time.

DBQ Document-Based Questions

Data obtained from: Royal Society of Chemistry, *All at sea? The chemistry of the oceans.*

132. anions: chloride, Cl^-, sulfate, SO_4^{2-}, carbonate, CO_3^{2-}, bromide, Br^-, borate, BO_3^{3-}, silicate, SiO_3^{2-}, fluoride, F^-
cations: sodium, Na^+, magnesium, Mg^{2+}, strontium, Sr^{2+}, calcium, Ca^{2+}, potassium, K^+

133. Bar charts should correspond to concentration data from Table 18. The graph is difficult to draw because the range of the data is so great; some data are very small and some data are very large.

134. Students should identify four of the following compounds: sodium chloride, NaCl; sodium sulfate, Na_2SO_4; sodium carbonate, Na_2CO_3; sodium bromide, NaBr; sodium borate, Na_3BO_3; sodium silicate, Na_2SO_3; sodium fluoride, NaF.

MULTIPLE CHOICE

1. B
2. C
3. A
4. D
5. B
6. C
7. C
8. D
9. B

MULTIPLE CHOICE

Use the figure below to answer Question 1.

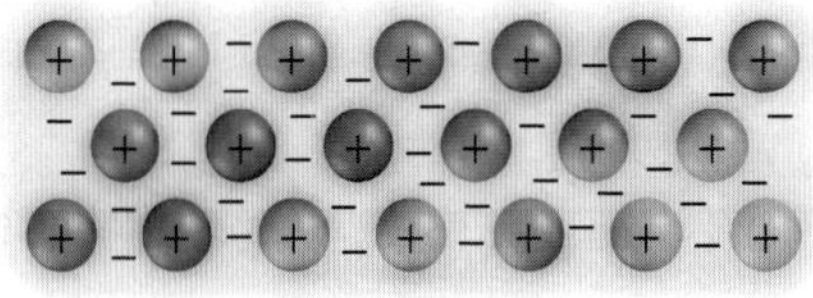

1. Which description is supported by the model shown?
A. Metals are shiny, reflective substances.
B. Metals are excellent conductors of heat and electricity.
C. Ionic compounds are malleable compounds.
D. Ionic compounds are good conductors of electricity.

2. Which is NOT true of the Sc^{3+} ion?
A. It has the same electron configuration as Ar.
B. It is a scandium ion with three positive charges.
C. It is considered to be a different element than a neutral Sc atom.
D. It was formed by the loss of 4s and 3d electrons.

3. Of the salts below, which would require the most energy to break the ionic bonds?
A. $BaCl_2$
B. LiF
C. NaBr
D. KI

4. The high strength of its ionic bonds results in all of the following properties of NaCl EXCEPT
A. hard crystals.
B. high boiling point.
C. high melting point.
D. low solubility.

5. Which is the correct formula for the compound chromium(III) sulfate?
A. Cr_3SO_4
B. $Cr_2(SO_4)_3$
C. $Cr_3(SO_4)_2$
D. $Cr(SO_4)_3$

Use the table below to answer Questions 6–8.

Physical Properties of Selected Compounds			
Compound	Bond Type	Melting Point (°C)	Boiling Point (°C)
F_2	Nonpolar covalent	−220	−188
CH_4	Nonpolar covalent	−183	−162
NH_3	Polar covalent	−78	−33
CH_3Cl	Polar covalent	−64	61
KBr	Ionic	730	1435
Cr_2O_3	Ionic	?	4000

6. A compound is discovered to have a melting point of −100°C. Which could be true of this compound?
A. It definitely has an ionic bond.
B. It definitely has a polar covalent bond.
C. It has either a polar covalent bond or a nonpolar covalent bond.
D. It has either a polar covalent bond or an ionic bond.

7. Which could NOT be the melting point of Cr_2O_3?
A. 2375°C
B. 950°C
C. 148°C
D. 3342°C

8. Which is supported by the data in the table?
A. Nonpolar covalent bonds have high boiling points.
B. Polar covalent bonds have high melting points.
C. Ionic bonds have low melting points.
D. Ionic bonds have high boiling points.

9. Which is the correct orbital diagram for the third and fourth principal energy levels of vanadium?

A. 3s [↑↓] 3p [↑↓][↑↓][↑↓] 4s [↑↓] 3d [↑↓][↑↓][ ][ ][ ]

B. 3s [↑↓] 3p [↑↓][↑↓][↑↓] 4s [↑↓] 3d [↑][↑][↑][ ][ ]

C. 3s [↑↓] 3p [↑↓][↑↓][↑↓] 4s [↑↓] 3d [↑↓][↑][ ][ ][ ]

D. 3s [↑↓] 3p [↑↓][↑↓][↑↓] 4s [↑↓] 3d [↑][↑↓][ ][ ][ ]

SHORT ANSWER

Use the table below to answer Questions 10–12.

Lutetium is a rare-earth element that can be used to speed up the chemical reactions involved in petroleum processing. It has two naturally occurring isotopes.

Isotope	Form of Decay	Percent Abundance
$^{175}_{71}Lu$	none	97.41
$^{176}_{71}Lu$	beta	2.59

10. Show the setup and calculate the average atomic mass of lutetium.

11. Identify the product when lutetium-176 goes through nuclear decay.

12. Compare the number of protons and neutrons in each of these isotopes.

EXTENDED RESPONSE

13. Relate the change in atomic radius to the changes in atomic structure that occur across the periodic table.

Use the diagram below to answer Question 14.

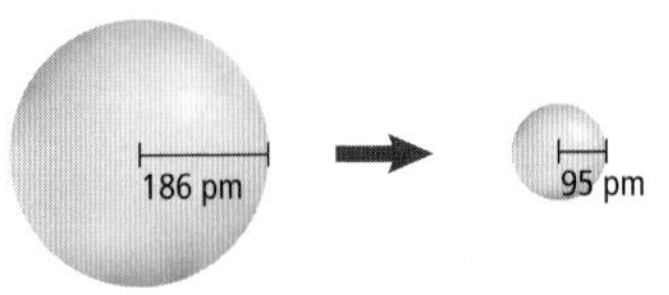

Sodium atom (Na) [Ne]3s[1] — Sodium ion (Na^+) [Ne]

14. Relate the change in ionic radius to the changes in ion formation that occur across the periodic table.

SAT SUBJECT TEST: CHEMISTRY

Use the diagram below to answer Question 15.

15. Which describes the state of matter shown?
- **A.** solid, because the particles are tightly packed against one another
- **B.** gas, because the particles are flowing past one another
- **C.** liquid, because the particles are able to move freely
- **D.** solid, because there is a regular pattern to the particles
- **E.** liquid, because the particles are in constant contact as they flow past each other

Use the list of elements below to answer Questions 16–20.

- **A.** sodium
- **B.** chromium
- **C.** boron
- **D.** argon
- **E.** chlorine

16. Which has its outermost electrons in an s-sublevel?

17. Which has seven valence electrons?

18. Which is a transition metal?

19. Which has an electron configuration of $1s^22s^22p^63s^23p^5$?

20. Which is a noble gas?

NEED EXTRA HELP?																				
If You Missed Question . . .	1	2	3	4	5	6	7	8	9	10	11	12	13	14	15	16	17	18	19	20
Review Section . . .	7.4	7.1	7.2	7.2	7.3	7.3	7.3	7.3	5.3	4.3	4.4	4.3	6.3	6.3	3.1	5.3	5.3	6.2	5.3	6.2

SHORT ANSWER

10. $(0.9741)(175) + (0.0259)(176)$
$= 175$ amu

11. It goes through beta decay to form $^{176}_{72}Hf$.

12. Both isotopes have 71 protons. Lu-175 has 104 neutrons, while Lu-176 has 105 neutrons.

EXTENDED RESPONSE

13. Atomic radii generally decrease across a given period because the increasing positive charge in the nucleus tends to contract the outer electron orbitals that are being filled at the same principal energy level. Atomic radii generally increase down a given group because new principal energy levels (at larger radii) are being added to the atom. Increasing positive charge in the nucleus is not sufficient to overcome this effect.

14. Cations are formed from their corresponding neutral atoms by releasing outer valence electrons to achieve a stable noble gas configuration. The ionic radius is smaller than the neutral atomic radius because all of the valence electrons comprising the highest principal energy level are being released.

SAT SUBJECT TEST: CHEMISTRY

15. E
16. A
17. E
18. B
19. E
20. D

CHAPTER 8 Covalent Bonding

BIGIDEA **Covalent bonds form when atoms share electrons.**

ESSENTIAL QUESTIONS	RESOURCES TO ASSESS MASTERY
SECTION 1 The Covalent Bond **1.** How does the octet rule apply to atoms that form covalent bonds? **2.** Why do atoms form single, double, and triple covalent bonds? **3.** What are sigma and pi bonds and how do they contrast? **4.** How are the strength of a covalent bond, its bond length, and its bond dissociation energy related? 2 sessions 1 block	**Progress Monitoring** Formative Assessment, p. 245 Reading Check, pp. 243, 244, 246 Section Review, p. 247
SECTION 2 Naming Molecules **1.** What rules do you follow to name a binary molecular compound from its molecular formula? **2.** How are acidic solutions named? 1 session 0.5 block	**Progress Monitoring** Formative Assessment, pp. 249, 250, 251 Reading Check, p. 249 Section Review, p. 252
SECTION 3 Molecular Structures **1.** What are the basic steps used to draw Lewis structures? **2.** Why does resonance occur and what are some resonance structures? **3.** Which molecules are exceptions to the octet rule, and why do these exceptions occur? 2 sessions 1 block	**Progress Monitoring** Formative Assessment, pp. 257, 258 Reading Check, p. 259 Section Review, p. 260
SECTION 4 Molecular Shapes **1.** What is the VSEPR bonding theory? **2.** How can you use the VSEPR model to predict the shape of, and the bond angles in, a molecule? **3.** What is hybridization? 1 session 0.5 block	**Progress Monitoring** Formative Assessment, p. 263 Reading Check, p. 262 Section Review, p. 264
SECTION 5 Electronegativity and Polarity **1.** How is electronegativity used to determine bond type? **2.** How do polar and nonpolar covalent bonds and polar and nonpolar molecules compare? How do they contrast? **3.** What are the characteristics of covalently bonded compounds? 1 session 0.5 block	**Progress Monitoring** Formative Assessment, pp. 267, 270 Reading Check, pp. 266, 267 Graph Check, p. 266 Section Review, p. 270 **Summative Assessment** Chapter Assessment, p. 274 *eAssessment* Chapter Test (Scaffolded)

Go online!

LEVELED RESOURCES	LAB MATERIALS
Science Notebook 8.1 OL ***FAST FILE Chapter Resources:*** MiniLab Worksheet, p. 86 OL Study Guide, p. 100 BL	LaunchLAB p. 238 / **15 min** paper towels, lab gloves, paper cup, graduated cylinder, food coloring, wood splint, sodium silicate solution, ethanol MiniLAB p. 242 / **20 min** permanent marker, 9-inch aluminum pie pan, hot plate, sugar crystals, salt crystals, paraffin, hot mitt
Science Notebook 8.2 OL ***FAST FILE Chapter Resources:*** Study Guide, p. 101 BL	
Science Notebook 8.3 OL ***FAST FILE Chapter Resources:*** Study Guide, p. 102 BL	
Science Notebook 8.4 OL ***FAST FILE Chapter Resources:*** Study Guide, p. 103 BL	
Science Notebook 8.5 OL ***FAST FILE Chapter Resources:*** ChemLab Worksheet, p. 88 OL Study Guide, p. 104 BL	ChemLAB p. 272 / **45 min** molecular model kit

ADDITIONAL RESOURCES

- eAssessment
- ConnectED Student Center
- ConnectED Teacher Center
- Lesson Presentations
- eSolutions
- Videos
- Animations
- Online Quizzes
- Virtual Investigations
- Multilingual eGlossary
- Personal Tutors
- WebQuests
- Applying Practices Worksheets
- PBLs

Practice:

- Challenge Problems AL
- Supplemental Problems BL OL
- Chapter Test (Scaffolded)
- Standardized Test Practice

Additional Resources:

- Blueprints for Success
- Solving Problems: A Chemistry Handbook

Lab Resources:

- Laboratory Manual OL
- CBL Laboratory Manual OL
- Small-Scale Laboratory Manual OL
- Forensics Laboratory Manual OL

BL Below Level   OL On Level   AL Advanced Learners   EL English Learners   COOP LEARN Cooperative Learning

CHAPTER 8

LaunchLAB

Rubric

What type of compound is used to make a super ball?

Purpose Students will build a polymer from silicon and ethanol that can bounce when shaped into a sphere.

Safety Precautions Approve lab safety forms before work begins. Review MSDS for ethanol and sodium silicate with students prior to doing the lab. Ethanol is very flammable and its vapors are explosive–use caution and appropriate ventilation. Sodium silicate is very alkaline and is a skin irritant. Students must wear safety goggles and gloves when handling the polymer.

Disposal Wrap the excess product and wooden splints in newspaper. Place these pieces of newspaper in a box or plastic bag and seal it. Have the sealed box or bag disposed of in a landfill that takes chemical or hazardous waste materials.

Teaching Strategies

- Have groups vary the amount of ethanol used in making the ball, using between 5 and 15 mL. Have students observe the effects this has on the outcome and hypothesize about the cause.
- The ball might become brittle and crumble. Ask students to determine the cause of this.

Expected Results Students should be able to form a sphere that bounces. The product might be brittle and might crumble but can be reformed by shaping it in gloved hands.

CHAPTER 8
Covalent Bonding

BIGIDEA Covalent bonds form when atoms share electrons.

SECTIONS

LaunchLAB

What type of compound is used to make a super ball?

Super balls are often made of a silicon compound called organosilicon oxide ($Si(OCH_2CH_3)_2O$). In this lab, you will compare the properties of organosilicon oxide to the properties of ionic compounds, which you studied previously.

FOLDABLES Study Organizer

Bond Character

Make a layered-look book. Label it as shown. Use it to help you organize your study of the three major types of bonding.

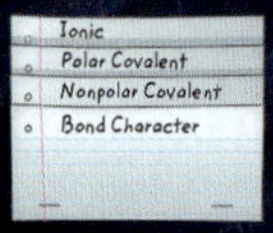

Procedure

1. Read and complete the lab safety form.
2. Spread several **paper towels** across your desk or lab work area. Put on **lab gloves.** Place a **paper cup** on the paper towels.
3. Using a **graduated cylinder,** measure 20.0 mL of **sodium silicate solution,** and pour it into the cup. Add one drop of **food coloring** and 10.0 mL of **ethanol** to the cup. Stir the mixture clockwise with a **wooden splint** for 3 s. **WARNING:** ***Keep ethanol away from flame and spark sources, as its vapors can be explosive.***
4. Working over paper towels, pour the mixture onto one of your glove-covered palms. Gently squeeze out excess liquid as the mixture solidifies.
5. Roll the solid between glove-covered hands and form a ball. Drop it on the floor and observe what happens.
6. Store the ball in an airtight container. You will need to reshape the ball before using it again.

Analysis

1. **Describe** the properties of the ball that you observed. The ball loses shape when it sits, but when shaped as a ball, it bounces. As the ball dries out, it becomes brittle and crumbles.
2. **Compare** the properties you observed with those of an ionic compound. Ionic compounds form crystals, dissolve in water, and have high melting points. The ball was made of two liquids at room temperature, it bounces, does not dissolve in water, and loses its shape over time.

Inquiry How many electrons do silicon and oxygen atoms need to form octets? If both atoms must gain electrons, how can they form a bond with each other? Silicon has 4 valence electrons and oxygen has 6 valence electrons. To form octets, silicon must gain 4 electrons and oxygen must gain 2 electrons. To form a bond, these atoms must share electrons.

Introduce the BIGIDEA

Sharing Electrons To introduce this chapter's Big Idea, have students draw the Lewis structures for hydrogen and oxygen atoms. Ask them how many additional electrons a hydrogen atom needs in order to have the same structure as helium. 1 additional electron Ask students how many electrons an atom of oxygen needs to obtain the noble gas arrangement of neon. 2 additional electrons Ask students how hydrogen and oxygen atoms are able to bond to form water, H_2O. They must share electrons. Two hydrogen atoms share an electron each with 1 oxygen atom.

Tie to Previous Knowledge

Have students review the following concepts before studying this chapter.

- atomic structure
- electron configuration
- periodic trends
- periodic properties of elements
- ionic bonding

Use the Photo

Establishing Relationships
The art in the chapter opener shows several different ways of describing or illustrating a water molecule: a Lewis structure, a ball-and-stick model, and a space-filling model, finally showing a group of such molecules in a water droplet. Ask students to describe the differences between each method of illustrating the molecule. Answers will vary, but should focus on the different purposes served by each approach.

SECTION 1

1 Focus

MAINIDEA

Covalent Bonds Draw the Lewis structures for fluorine, chlorine, oxygen, sulfur, and bromine on the board. Ask students to locate these elements on the periodic table and identify what they have in common. They are all nonmetals and gain electrons to form negative ions. If two of these atoms form a compound, what must they do to form an octet? They must share electrons. Draw carbon's Lewis structure and show students an atomic model of carbon. On both, point out the four places carbon can form a single covalent bond. Bond a hydrogen atom into each of the places. Ask students what is represented. The model represents the electrons shared between a central carbon atom and four terminal hydrogen atoms in a CH_4 molecule. OL

2 Teach

Quick Demo

Potential Energy Use this demo to relate potential energy to stability. Make a pendulum using a foam-rubber ball and a string or spring. Ask where the pendulum has the greatest potential energy. when the pendulum is pulled up to its maximum height Ask where the pendulum has the least potential energy. when it hangs straight down Explain that the lowest potential energy occurs when the pendulum is at the lowest point in its motion. Have students note that eventually a swinging pendulum comes to a stop in the lowest energy position. Relate the pendulum's higher-energy positions to instability. OL

SECTION 1

The Covalent Bond

MAINIDEA Atoms gain stability when they share electrons and form covalent bonds.

Essential Questions

- How does the octet rule apply to atoms that form covalent bonds?
- Why do atoms form single, double, and triple covalent bonds?
- What are sigma and pi bonds and how do they contrast?
- How are the strength of a covalent bond, its bond length, and its bond dissociation energy related?

Review Vocabulary

chemical bond: the force that holds two atoms together

New Vocabulary

covalent bond
molecule
Lewis structure
sigma bond
pi bond
endothermic reaction
exothermic reaction

CHEM 4 YOU Have you ever run in a three-legged race? Each person in the race shares one of their legs with a teammate to form a single three-legged team. In some ways, a three-legged race mirrors how atoms share electrons and join together as a unit.

Why Do Atoms Bond?

Understanding the bonding in compounds is essential to developing new chemicals and technologies. To understand why new compounds form, recall what you know about elements that do not tend to form new compounds—the noble gases. You learned that all noble gases have stable electron arrangements. This stable arrangement consists of a full outer energy level and has lower potential energy than other electron arrangements. Because of their stable configurations, noble gases seldom form compounds.

Gaining stability The stability of an atom, ion, or compound is related to its energy; that is, lower energy states are more stable. From your study of ionic bonds, you know that metals and nonmetals can gain stability by transferring (gaining or losing) electrons to form ions. The resulting ions have stable noble-gas electron configurations. From the octet rule you know that atoms with a complete octet, a configuration of eight valence electrons, are stable. In this chapter, you will learn that the sharing of valence electrons is another way atoms can acquire the stable electron configuration of noble gases. The water droplets shown in **Figure 1** consist of water molecules formed when hydrogen and oxygen atoms share electrons.

■ **Figure 1** Each water droplet is made up of water molecules. Each water molecule is made up of two hydrogen atoms and one oxygen atom that have bonded by sharing electrons. The shapes of the drops are due to intermolecular forces acting on the water molecules.

CHEMISTRY JOURNAL

Bonding Limerick Read students the following limerick:

Bond formations can be unwell
(Electrons join up in a shell)
For it doesn't seem fine
To say they combine
When we know very well they repel!

Have students develop limericks, poems, or rhymes that depict some aspect of covalent bonds. OL

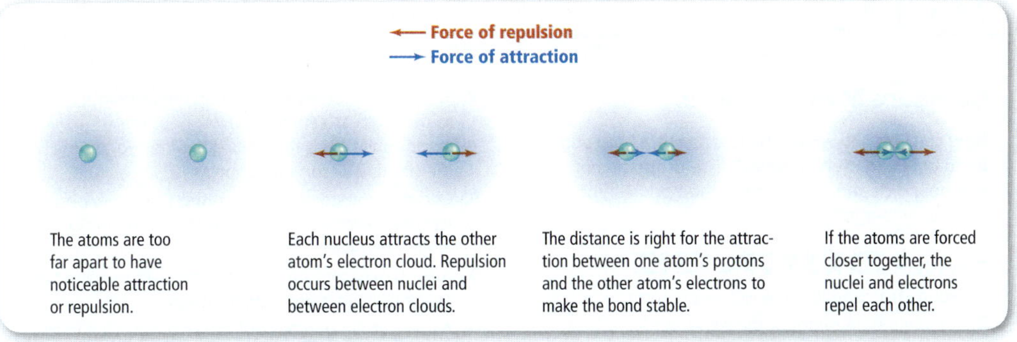

■ **Figure 2** The arrows in this diagram show the net forces of attraction and repulsion acting on two fluorine atoms as they move toward each other. The overall force between two atoms is the result of electron-electron repulsion, nucleus-nucleus repulsion, and nucleus-electron attraction. At the position of maximum net attraction, a covalent bond forms.

Relate ***How is the stability of the bond related to the forces acting on the atoms?***

What is a covalent bond?

You just read that atoms can share electrons to form stable electron configurations. How does this occur? Are there different ways in which electrons can be shared? How are the properties of these compounds different from those formed by ions? Read on to answer these questions.

Shared electrons Atoms in nonionic compounds share electrons. The chemical bond that results from sharing valence electrons is a **covalent bond.** A **molecule** is formed when two or more atoms bond covalently. In a covalent bond, the shared electrons are considered to be part of the outer energy levels of both atoms involved. Covalent bonding generally can occur between elements that are near each other on the periodic table. The majority of covalent bonds form between atoms of nonmetallic elements.

Covalent bond formation Diatomic molecules, such as hydrogen (H_2), nitrogen (N_2), oxygen (O_2), fluorine (F_2), chlorine (Cl_2), bromine (Br_2), and iodine (I_2), form when two atoms of each element share electrons. They exist this way because the two-atom molecules are more stable than the individual atoms.

Consider fluorine, which has an electron configuration of $1s^22s^22p^5$. Each fluorine atom has seven valence electrons and needs another electron to form an octet. As two fluorine atoms approach each other, several forces act, as shown in **Figure 2.** Two repulsive forces act on the atoms, one from each atom's like-charged electrons and one from each atom's like-charged protons. A force of attraction also acts, as one atom's protons attract the other atom's electrons. As the fluorine atoms move closer, the attraction of the protons in each nucleus for the other atom's electrons increases until a point of maximum net attraction is achieved. At that point, the two atoms bond covalently and a molecule forms. If the two nuclei move closer, the repulsion forces increase and exceed the attractive forces.

The most stable arrangement of atoms in a covalent bond exists at some optimal distance between nuclei. At this point, the net attraction is greater than the net repulsion. Fluorine exists as a diatomic molecule because the sharing of one pair of electrons gives each fluorine atom a stable noble-gas configuration. As shown in **Figure 3,** each fluorine atom in the fluorine molecule has one pair of electrons that are covalently bonded (shared) and three pairs of electrons that are unbonded (not shared). Unbonded pairs are also known as lone pairs.

■ **Figure 3** Two fluorine atoms share a pair of electrons to form a covalent bond. Note that the shared electron pair gives each atom a complete octet.

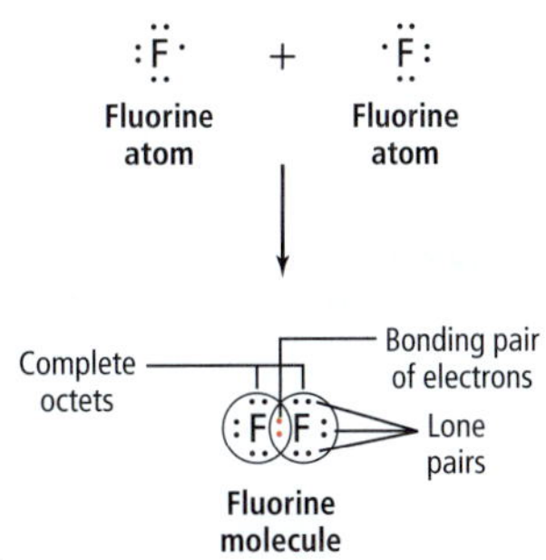

Visual Learning

Figure 2 Have students look at the figure and discuss the potential energy change as the fluorine atoms approach each other in terms of attractive forces and repulsive forces. At a certain optimal point, attractive and repulsive forces balance, and a bond forms. If the repulsive force is greater, no bond forms and the atoms stay separated. **OL**

■ **Caption Question Fig. 2**
A stable bond is formed at the location of maximum net attraction.

Concept Development

Electron Nature It is important for students to understand that electrons are identical, regardless of which atom or energy level they are in. Students must also understand that electrons in the electron cloud are not static. They move at nearly the speed of light, yet are attracted to the positively charged nucleus. When a covalent bond forms, nuclei of both atoms contribute to the force that attracts the electrons. A single atom does not have sole ownership of the shared electrons.

Reinforcement

Dot Diagrams Review electron dot diagrams. Point out that covalent bonds can be represented in a variety of ways.

CHEMISTRY JOURNAL

Why Covalent Bonds Form Have students write a paragraph explaining why a covalent bond can form between a chlorine atom and a fluorine atom. **OL**

MiniLAB

Rubric

Purpose Students will compare the melting points of compounds with covalent bonds and compounds with ionic bonds.

Process Skills observe and infer, classify, compare and contrast, interpret data

Safety Precautions Approve lab safety forms before work begins. Do not directly inhale the vapors of the substances. Make sure the area is well ventilated or students are working under a fume hood.

Disposal Carefully remove the lid and place in a waste container.

Teaching Strategies Students should work in groups to do this lab. A candle can be substituted for the Bunsen burner.

Expected Result

- paraffin melts first
- sugar melts second, then burns
- sodium chloride does not melt

Analysis

1. The paraffin melted first. The salt crystals did not melt.
2. paraffin, low; sugar, medium; salt crystals, very high
3. ionic bonds: salt covalent bonds: paraffin and sugar
4. Ionic compounds have higher melting points than covalently bonded compounds.

MiniLAB

Compare Melting Points

How can you determine the relationship between bond type and melting point? The properties of a compound depend on whether the bonds in the compound are ionic or covalent.

Procedure

1. Read and complete the lab safety form.
2. Create a data table for the experiment.
3. Using a **permanent marker,** draw three lines on the inside bottom of a **disposable, 9-inch aluminum pie pan** to create three, equal wedges. Label the wedges, ***A, B,*** and ***C.***
4. Set the pie pan on a **hot plate.**
   WARNING: ***Hot plate and metal pie pan will burn skin—handle with care.***
5. Obtain samples of the following from your teacher and deposit them onto the labeled wedges as follows: **sugar crystals** ($C_{12}H_{22}O_{11}$), *A*; **salt crystals** (NaCl) *B*; **paraffin** ($C_{23}H_{48}$), *C*.
6. Predict the order in which the compounds will melt.
7. Turn the temperature knob on the hot plate to the highest setting. You will heat the compounds for 5 min. Assign someone to time the heating of the compounds.
8. Observe the compounds during the 5-min period. Record which compounds melt and the order in which they melt.
9. After 5 min, turn off the hot plate and remove the pie pan using a **hot mitt** or **tongs.**
10. Allow the pie pan to cool, and then place it in the proper waste container.

Analysis

1. **State** Which solid melted first? Which solid did not melt?
2. **Apply** Based on your observations and data, describe the melting point of each solid as low, medium, high, or very high.
3. **Infer** Which compounds are bonded with ionic bonds? Which are bonded with covalent bonds?
4. **Summarize** how the type of bonding affects the melting points of compounds.

Single Covalent Bonds

When only one pair of electrons is shared, such as in a hydrogen molecule, it is a single covalent bond. The shared electron pair is often referred to as the bonding pair. For a hydrogen molecule, shown in **Figure 4,** each covalently bonded atom equally attracts the pair of shared electrons. Thus, the two shared electrons belong to each atom simultaneously, which gives each hydrogen atom the noble-gas configuration of helium ($1s^2$) and lower energy. The hydrogen molecule is more stable than either hydrogen atom is by itself.

Recall that electron-dot diagrams can be used to show valence electrons of atoms. In a **Lewis structure,** they can represent the arrangement of electrons in a molecule. A line or a pair of vertical dots between the symbols of elements represents a single covalent bond in a Lewis structure. For example, a hydrogen molecule is written as H–H or H:H.

■ **Figure 4** When two hydrogen atoms share a pair of electrons, each hydrogen atom is stable because it has a full outer energy level.

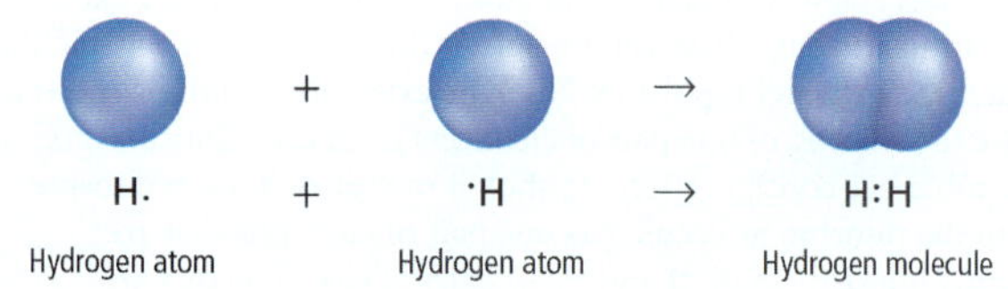

Group 17 and single bonds The halogens—the group 17 elements, such as fluorine—have seven valence electrons. To form an octet, one more electron is needed. Therefore, atoms of group 17 elements form single covalent bonds with atoms of other nonmetals, such as carbon. You have already read that the atoms of some group 17 elements form covalent bonds with identical atoms. For example, fluorine exists as F_2 and chlorine exists as Cl_2.

Group 16 and single bonds An atom of a group 16 element can share two electrons and can form two covalent bonds. Oxygen is a group 16 element with an electron configuration of $1s^22s^22p^4$. Water is composed of two hydrogen atoms and one oxygen atom. Each hydrogen atom has the noble-gas configuration of helium when it shares one electron with oxygen. Oxygen, in turn, has the noble-gas configuration of neon when it shares one electron with each hydrogen atom. **Figure 5a** shows the Lewis structure for a molecule of water. Notice that the oxygen atom has two single covalent bonds and two unshared pairs of electrons.

Group 15 and single bonds Group 15 elements form three covalent bonds with atoms of nonmetals. Nitrogen is a group 15 element with the electron configuration of $1s^22s^22p^3$. Ammonia (NH_3) has three single covalent bonds. Three nitrogen electrons bond with the three hydrogen atoms leaving one pair of unshared electrons on the nitrogen atom. **Figure 5b** shows the Lewis structure for an ammonia molecule. Nitrogen also forms similar compounds with atoms of group 17 elements, such as nitrogen trifluoride (NF_3), nitrogen trichloride (NCl_3), and nitrogen tribromide (NBr_3). Each atom of these group 17 elements and the nitrogen atom share an electron pair.

Group 14 and single bonds Atoms of group 14 elements form four covalent bonds. A methane molecule (CH_4) forms when one carbon atom bonds with four hydrogen atoms. Carbon, a group 14 element, has an electron configuration of $1s^22s^22p^2$. With four valence electrons, carbon needs four more electrons for a noble gas configuration. Therefore, when carbon bonds with other atoms, it forms four bonds. Because a hydrogen atom, a group 1 element, has one valence electron, it takes four hydrogen atoms to provide the four electrons needed by a carbon atom. The Lewis structure for methane is shown in **Figure 5c.** Carbon also forms single covalent bonds with other nonmetal atoms, including those in group 17.

☑ READING CHECK **Describe** how a Lewis structure shows a covalent bond.

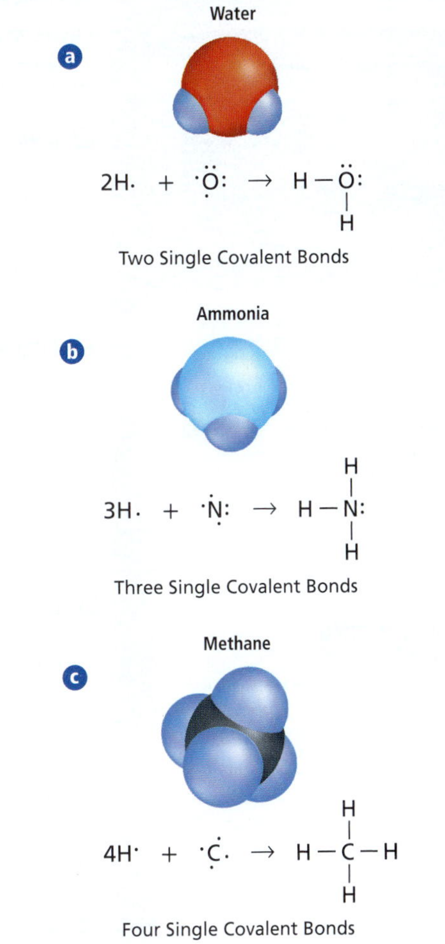

Figure 5 These chemical equations show how atoms share electrons and become stable. As shown by the Lewis structure for each molecule, all atoms in each molecule achieve a full outer energy level.

Describe *For the central atom in each molecule, describe how the octet rule is met.*

Caption Question Fig. 5 water: two electrons from each bond with hydrogen and two pairs of lone electrons; ammonia: two electrons from each bond with hydrogen and one pair of lone electrons; methane: two electrons from each bond with hydrogen

Build a Model

Lewis Structures Give students four index cards that have been cut in half to form eight pieces. Have them write the following Lewis dot diagrams on the cards: one for C, two for H, one for S, and four for F. Tell students they will use these cards to line up unpaired electrons from one atom with the unpaired electrons from other atoms to form single covalent bonds. Have students form the compounds CF_4 and H_2S. BL EL

☑ READING CHECK A covalent bond is shown as a pair of dots or as a short line.

DIFFERENTIATED INSTRUCTION

Struggling Learners Using a piece of acetate, draw circles to represent atoms. In each circle, represent a single atom by using its electron dot diagram. Show students two F atoms, and ask what each one needs to become stable. one electron Overlap the circles so that each F atom has eight electrons, and ask students how many electrons each atom has. eight Tell students that the shared electrons form a covalent bond. Overlap two H atoms and one O atom. Ask students how many electrons O has. eight How many electrons does each H atom have? two Ask students if H is stable with two electrons. Yes, because it has the noble gas configuration of He. Ask students how many bonds are formed. two single covalent bonds BL

IN-CLASS Example

Question Hydrogen selenide, H_2Se, is a toxic gas with properties similar to hydrogen sulfide. Draw the Lewis structure for this molecule.

Answer

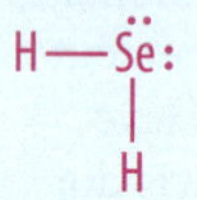

PRACTICE Problems

Have students refer to the Selected Solutions appendix for complete solutions to odd-numbered problems. The complete solutions for all problems can be found in the Solutions Manual.

1. 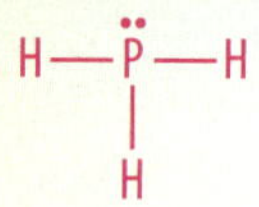

2. 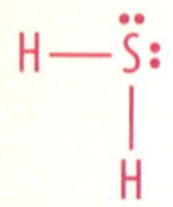

3. 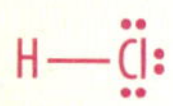

4. 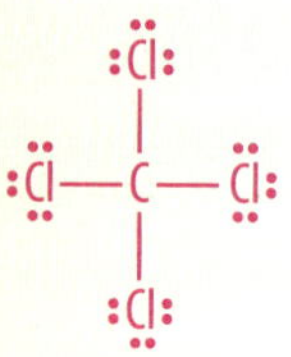

5. 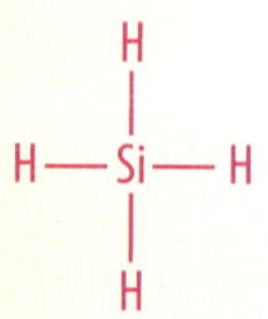

6. Using 1 and 16 to represent atoms of groups 1 and 16, respectively, the generic structure is:

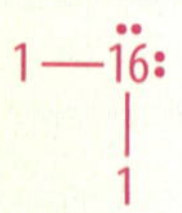

☑ **READING CHECK** Sigma bonds can form from the overlap of an s orbital with another s orbital, an s orbital with a p orbital, or a p orbital end-to-end with another p orbital.

■ **Figure 6** The frosted-looking portions of this glass were chemically etched using hydrogen fluoride (HF), a weak acid. Hydrogen fluoride reacts with silica, the major component of glass, and forms gaseous silicon tetrafluoride (SiF_4) and water.

EXAMPLE Problem 1

LEWIS STRUCTURE OF A MOLECULE The pattern on the glass shown in **Figure 6** was made by chemically etching its surface with hydrogen fluoride (HF). Draw the Lewis structure for a molecule of hydrogen fluoride.

1 ANALYZE THE PROBLEM

You are given the information that hydrogen and fluorine form the molecule hydrogen fluoride. An atom of hydrogen, a group 1 element, has only one valence electron. It can bond with any nonmetal atom when they share one pair of electrons. An atom of fluorine, a group 17 element, needs one electron to complete its octet. Therefore, a single covalent bond forms when atoms of hydrogen and fluorine bond.

2 SOLVE FOR THE UNKNOWN

To draw a Lewis structure, first draw the electron-dot diagram for each of the atoms. Then, rewrite the chemical symbols and draw a line between them to show the shared pair of electrons. Finally, add dots to show the unshared electron pairs.

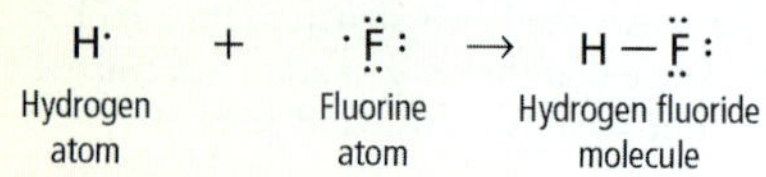

3 EVALUATE THE ANSWER

Each atom in the new molecule now has a noble-gas configuration and is stable.

PRACTICE Problems

Do additional problems. Online Practice

Draw the Lewis structure for each molecule.

1. PH_3
2. H_2S
3. HCl
4. CCl_4
5. SiH_4
6. **Challenge** Draw a generic Lewis structure for a molecule formed between atoms of group 1 and group 16 elements.

APPLYING PRACTICES

Construct and Revise an Explanation
Go to the resources tab in ConnectED to find the Applying Practices worksheet *Electron States and Simple Chemical Reactions.*

The sigma bond Single covalent bonds are also called **sigma bonds,** represented by the Greek letter sigma (σ). A sigma bond occurs when the pair of shared electrons is in an area centered between the two atoms. When two atoms share electrons, their valence atomic orbitals overlap end-to-end, concentrating the electrons in a bonding orbital between the two atoms. A bonding orbital is a localized region where bonding electrons will most likely be found. Sigma bonds can form when an s orbital overlaps with another s orbital or a p orbital, or two p orbitals overlap end-to-end. Water (H_2O), ammonia (NH_3), and methane (CH_4) have sigma bonds, as shown in **Figure 7.**

VOCABULARY
ACADEMIC VOCABULARY
Overlap
to occupy the same area in part
The two driveways overlap at the street forming a common entrance.

☑ **READING CHECK** **List** the orbitals that can form sigma bonds in a covalent compound.

CHEMISTRY PROJECT

Chemistry in Medicine Ask students to research the use of perfluorooctylbromide ($C_8F_{17}Br$) as an artificial oxygen carrier in synthetic blood. Have them report on how the compound works in the body to supply oxygen to tissues. Ask them to report on the compound's molecular structure. What types of bonds does the molecule contain? all bonds are sigma bonds (single covalent bonds) How many electron pairs do the C, F, and Br atoms in the compound share? Carbon shares four electron pairs, forming C-C bonds, C-F bonds, and C-Br bonds. Fluorine shares one electron pair and bromine shares one electron pair. OL

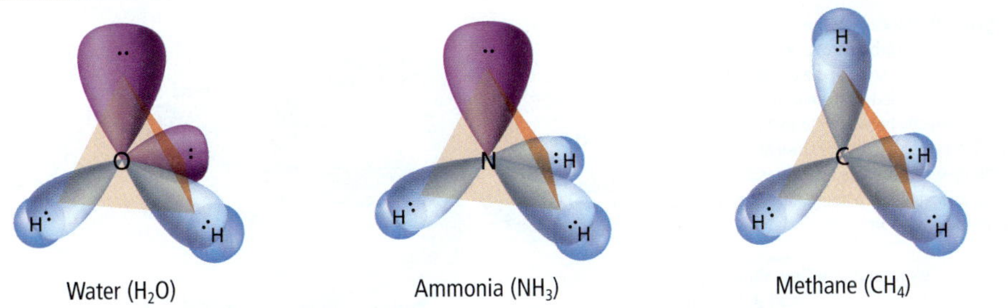

■ **Figure 7** Sigma bonds formed in each of these molecules when the atomic orbital of each hydrogen atom overlapped end-to-end with the orbital of the central atom.

Interpret ***Identify the number of sigma bonds in each molecule.***

Multiple Covalent Bonds

In some molecules, atoms have noble-gas configurations when they share more than one pair of electrons with one or more atoms. Sharing multiple pairs of electrons forms multiple covalent bonds. A double covalent bond and a triple covalent bond are examples of multiple bonds. Carbon, nitrogen, oxygen, and sulfur atoms often form multiple bonds with other nonmetals. How do you know if two atoms will form a multiple bond? In general, the number of valence electrons needed to form an octet equals the number of covalent bonds that can form.

Double bonds A double covalent bond forms when two pairs of electrons are shared between two atoms. For example, atoms of the element oxygen only exist as diatomic molecules. Each oxygen atom has six valence electrons and must obtain two additional electrons for a noble-gas configuration, as shown in **Figure 8a.** A double covalent bond forms when each oxygen atom shares two electrons; a total of two pairs of electrons are shared between the two atoms.

Triple bonds A triple covalent bond forms when three pairs of electrons are shared between two atoms. Diatomic nitrogen (N_2) molecules contain a triple covalent bond. Each nitrogen atom shares three electron pairs, forming a triple bond with the other nitrogen atom as shown in **Figure 8b.**

The pi bond A multiple covalent bond consists of one sigma bond and at least one pi bond. A **pi bond,** represented by the Greek letter pi (π), forms when parallel orbitals overlap and share electrons. The shared electron pair of a pi bond occupies the space above and below the line that represents where the two atoms are joined together.

FOLDABLES® Incorporate information from this section into your Foldable.

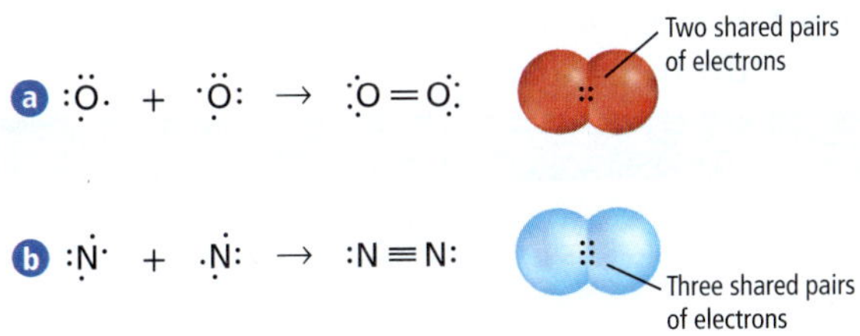

■ **Figure 8** Multiple covalent bonds form when two atoms share more than one pair of electrons. **a.** Two oxygen atoms form a double bond. **b.** A triple bond forms between two nitrogen atoms.

Get help with **multiple covalent bonds.** Personal Tutor

■ **Caption Question Fig. 7** water, 2; ammonia, 3; methane, 4

Concept Development

Electron Sharing Make sure students understand that an orbital can contain only two electrons at a time. An orbital with a single electron can only share its electron with another single-electron orbital. A double covalent bond involves two orbitals sharing four electrons between two atoms. A triple covalent bond involves three orbitals sharing six electrons between two atoms.

ASSESSMENT

Skill Divide students into groups. Hand each group a set of cards containing electron dot diagrams of atoms. The set should include 2 C atoms, 2 N atoms, 2 O atoms, 6 H atoms, and 4 Cl atoms. Place a hydrogen card next to a chlorine card. Show students that by sharing a pair of electrons each atom obtains a stable outer electron configuration. Have students determine the Lewis structure for as many molecules as possible by combining the cards. Remind students that multiple bonds are possible with carbon, nitrogen, and oxygen atoms. OL EL COOP LEARN

FOLDABLES® Rubric

Content Background

Exceptional Oxygen Spins of unpaired electrons in atoms or ions cause a substance to be paramagnetic, that is, attracted to an external magnetic field. A substance is diamagnetic, unaffected by the magnetic field, if all of the electrons are paired. Although the Lewis structure and bond strength for diatomic oxygen indicates there is a double bond between the two atoms, oxygen exhibits exceptional behavior in the laboratory. It is paramagnetic. The electrons associated with the π bond have unpaired, or parallel, spins.

Concept Development

Bond Strength An important factor in determining the chemical reactivity of a molecule is the strength of the covalent bonds holding the molecule together. When comparing two molecules with similar composition, the one with weaker bonds is usually more reactive. Have students compare the bond energies of the molecules CH_3Cl and CH_2Cl_2. 1567 kJ/mol, 1482 kJ/mol Ask students which molecule is more likely to react. CH_2Cl_2 **OL**

Enrichment

Bond Energy Use the equation shown below to calculate the change in enthalpy of a chemical reaction.

$\Delta H = \sum$(bond energies of broken bonds) $- \sum$(bond energies of formed bonds)

Have students determine the total enthalpy change when Cl_2 reacts with CH_4 to form CH_3Cl and HCl. The enthalpy change is −104 kJ. Ask students what the negative indicates about the energy of the reaction and the strength of the bonds formed. A negative enthalpy change indicates that the reaction is exothermic, and that the bonds in the product molecules are stronger than the bonds in the reactant molecules. Have them draw the Lewis structures of the molecules and predict which bonds are broken and which are formed. bonds broken are one Cl—Cl and one C—H; bonds formed are one H—Cl and one C—Cl **AL**

READING CHECK As the bond type increases from single covalent, to double covalent, to triple covalent, the bond length decreases.

■ **Figure 9** Notice how the multiple bond between the two carbon atoms in ethene (C_2H_4) consists of a sigma bond and a pi bond. The sigma bond is formed by the end-to-end overlap of orbitals directly between the two carbon atoms. The carbon atoms are close enough that the side-by-side p orbitals overlap and form the pi bond. This results in a doughnut-shaped cloud around the sigma bond.

View an **animation about sigma and pi bonding.**

Concepts In Motion

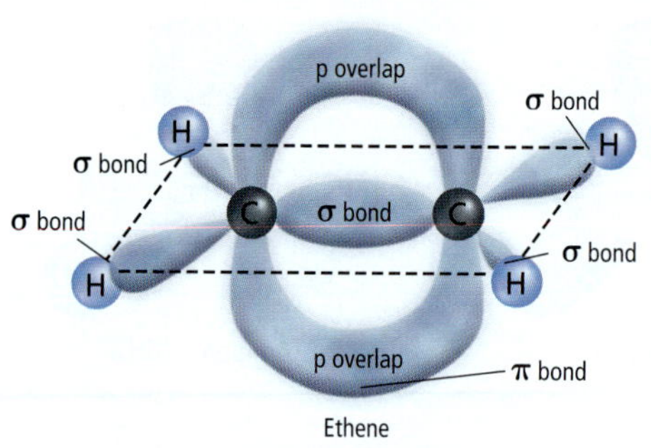

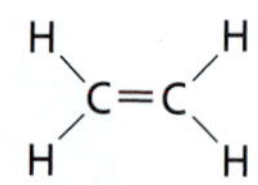

It is important to note that molecules having multiple covalent bonds contain both sigma and pi bonds. A double covalent bond, as shown in **Figure 9,** consists of one pi bond and one sigma bond. A triple covalent bond consists of two pi bonds and one sigma bond.

The Strength of Covalent Bonds

Recall that a covalent bond involves attractive and repulsive forces. In a molecule, nuclei and electrons attract each other, but nuclei repel other nuclei, and electrons repel other electrons. When this balance of forces is upset, a covalent bond can be broken. Because covalent bonds differ in strength, some bonds break more easily than others. Several factors influence the strength of covalent bonds.

Bond length The strength of a covalent bond depends on the distance between the bonded nuclei. The distance between the two bonded nuclei at the position of maximum attraction is called bond length, as shown in **Figure 10.** It is determined by the sizes of the two bonding atoms and how many electron pairs they share. Bond lengths for molecules of fluorine (F_2), oxygen (O_2), and nitrogen (N_2) are listed in **Table 1.** Notice that as the number of shared electron pairs increases, the bond length decreases.

Bond length and bond strength are also related: the shorter the bond length, the stronger the bond. Therefore, a single bond, such as that in F_2, is weaker than a double bond, such as that in O_2. Likewise, the double bond in O_2 is weaker than the triple bond in N_2.

READING CHECK **Relate** covalent bond type to bond length.

■ **Figure 10** Bond length is the distance from the center of one nucleus to the center of the other nucleus of two bonded atoms.

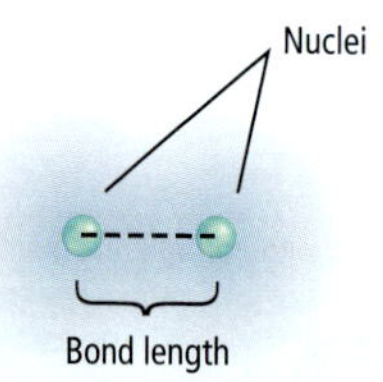

Table 1 Covalent Bond Type and Bond Length

Molecule	Bond Type	Bond Length
F_2	single covalent	1.43×10^{-10} m
O_2	double covalent	1.21×10^{-10} m
N_2	triple covalent	1.10×10^{-10} m

DIFFERENTIATED INSTRUCTION

Struggling Learners Have students place a single toothpick between two gumdrops to form a single bond. Then, have them break 1 cm off each of two toothpicks, and place them between two gumdrops to form a double bond. Have them break 2 cm off each of three toothpicks, and place them between two gumdrops to represent a triple bond. Have students identify the shortest and longest bonds. Ask students to carefully break the bonds to determine which bond is easiest and which is the most difficult to break. A triple bond is the shortest, is the most difficult to break, and has the highest bond energy. The single bond is the longest, is the easiest to break, and has the lowest bond energy. **BL** **EL**

Table 2 Bond-Dissociation Energy

Molecule	Bond-Dissociation Energy
F_2	159 kJ/mol
O_2	498 kJ/mol
N_2	945 kJ/mol

Bonds and energy An energy change occurs when a bond between atoms in a molecule forms or breaks. Energy is released when a bond forms, but energy must be added to break a bond. The amount of energy required to break a specific covalent bond is called bond-dissociation energy and is always a positive value. The bond-dissociation energies for the covalent bonds in molecules of fluorine, oxygen, and nitrogen are listed in **Table 2.**

Bond-dissociation energy also indicates the strength of a chemical bond because of the inverse relationship between bond energy and bond length. As indicated in **Table 1** and **Table 2,** the smaller the bond length is, the greater the bond-dissociation energy. The sum of the bond-dissociation energy values for all of the bonds in a molecule is the amount of chemical potential energy in a molecule of that compound.

The total energy change of a chemical reaction is determined from the energy of the bonds broken and formed. An **endothermic reaction** occurs when a greater amount of energy is required to break the existing bonds in the reactants than is released when the new bonds form in the products. An **exothermic reaction** occurs when more energy is released during product bond formation than is required to break bonds in the reactants. **Figure 11** illustrates a common exothermic reaction. You will study exothermic and endothermic reactions in much greater detail when you study the energy changes in chemical reactions.

Figure 11 Breaking the C–C bonds in charcoal and the O–O bonds in the oxygen in air requires an input of energy. Energy is released as heat and light when bonds form, producing CO_2. Thus, the burning of charcoal is an exothermic reaction.

SECTION 1 REVIEW

Section Self-Check

Section Summary

- Covalent bonds form when atoms share one or more pairs of electrons.
- Sharing one pair, two pairs, and three pairs of electrons forms single, double, and triple covalent bonds, respectively.
- Orbitals overlap directly in sigma bonds. Parallel orbitals overlap in pi bonds. A single covalent bond is a sigma bond but multiple covalent bonds are made of both sigma and pi bonds.
- Bond length is measured nucleus-to-nucleus. Bond-dissociation energy is needed to break a covalent bond.

7. MAINIDEA **Identify** the type of atom that generally forms covalent bonds.
8. **Describe** how the octet rule applies to covalent bonds.
9. **Illustrate** the formation of single, double, and triple covalent bonds using Lewis structures.
10. **Compare and contrast** ionic bonds and covalent bonds.
11. **Contrast** sigma bonds and pi bonds.
12. **Apply** Create a graph using the bond-dissociation energy data in **Table 2** and the bond-length data in **Table 1.** Describe the relationship between bond length and bond-dissociation energy.
13. **Predict** the relative bond-dissociation energies needed to break the bonds in the structures below.

a. $H-C\equiv C-H$ **b.**

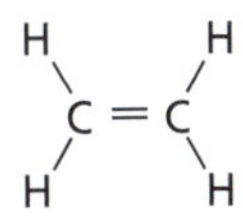

3 Assess

Check for Understanding

Ask students to draw the Lewis structure for the molecule H_2S. OL

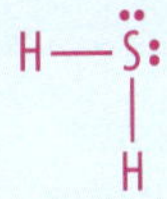

Reteach

Draw the structural formulas for C_2H_6, C_2H_4, and C_2H_2. Ask students to identify all sigma and pi bonds. Using a table of bond energies and bond length, compare the bond lengths and bond energies of C—C, C=C, and C≡C. Remind students that as the number of bonds between two atoms increases, the bond length becomes shorter and the bond itself becomes stronger. BL

Extension

Have interested students research trinitrotoluene (TNT) and determine the available bond energy if all bonds in a molecule of this compound were broken. AL

SECTION 1 REVIEW

7. Most covalent bonds form between nonmetallic elements.
8. Atoms share valence electrons; the shared electrons complete the octet of each atom.
9. Student Lewis structures should show the sharing of a single pair of electrons, two pairs of electrons, and three pairs of electrons, respectively, for single, double, and triple covalent bonds.
10. Valence electrons are involved in both types of bonds. In covalent bonds, atoms share electrons, whereas in ionic bonds, electrons are transferred between atoms.
11. A sigma bond is a single covalent bond formed from the direct overlap of orbitals. A pi bond is the parallel overlap of p orbitals.
12. Student graphs should show that as bond length decreases, the bond dissociation energy increases.
13. **a.** C—H: less energy than C≡C
    b. C—H: less energy than C=C

SECTION 2

1 Focus

MAINIDEA

Naming Molecules Ask students how they would identify or name their grandmother's sister in terms of their familial relationship. Their grandmother's sister is their great aunt. Ask them to explain how they determined that name. They should describe a systematic way of naming relatives. Mother's or father's brothers are uncles. Mother's or father's sisters are aunts. Explain that molecules are named using a systematic approach. OL BL

2 Teach

Visual Learning

Table 3 Review oxidation numbers, naming ionic compounds, and writing formula units. Unlike ionic compounds, the number of each atom that makes up a molecule is identified by a prefix. Write several molecular formulas on the board and have students use Table 3 to practice naming molecules. OL

SECTION 2

Essential Questions

- What rules do you follow to name a binary molecular compound from its molecular formula?
- How are acidic solutions named?

Review Vocabulary

oxyanion: a polyatomic ion in which an element (usually a nonmetal) is bonded to one or more oxygen atoms

New Vocabulary

oxyacid

Naming Molecules

MAINIDEA Specific rules are used when naming binary molecular compounds, binary acids, and oxyacids.

CHEM 4 YOU *You probably know that your mother's mother is your grandmother, and that your grandmother's sister is your great-aunt. But what do you call your grandmother's brother's daughter? Naming molecules requires a set of rules, just as naming family relationships requires rules.*

Naming Binary Molecular Compounds

Many molecular compounds have common names, but they also have scientific names that reveal their composition. To write the formulas and names of molecules, you will use processes similar to those described for ionic compounds.

Start with a binary molecular compound. Note that a binary molecular compound is composed only of two nonmetal atoms—not metal atoms or ions. An example is dinitrogen monoxide (N_2O), a gaseous anesthetic that is more commonly known as nitrous oxide or laughing gas. The naming of N_2O is explained in the following rules.

1. The first element in the formula is always named first, using the entire element name. **N is the symbol for *nitrogen*.**
2. The second element in the formula is named using its root and adding the suffix *-ide*. **O is the symbol for oxygen so the second word is *oxide*.**
3. Prefixes are used to indicate the number of atoms of each element that are present in the compound. **Table 3** lists the most common prefixes used. **There are two atoms of nitrogen and one atom of oxygen, so the first word is *dinitrogen* and second word is *monoxide*.**

There are exceptions to using the prefixes shown in **Table 3.** The first element in the compound name never uses the *mono-* prefix. For example, CO is carbon monoxide, not monocarbon monoxide. Also, if using a prefix results in two consecutive vowels, one of the vowels is usually dropped to avoid an awkward pronunciation. For example, notice that the oxygen atom in CO is called monoxide, not monooxide.

Explore **naming covalent compounds with an interactive table.** Concepts In Motion

Table 3 Prefixes in Covalent Compounds

Number of Atoms	Prefix	Number of Atoms	Prefix
1	mono-	6	hexa-
2	di-	7	hepta-
3	tri-	8	octa-
4	tetra-	9	nona-
5	penta-	10	deca-

Demonstration

Forming Bonds

Purpose

to demonstrate the formation of ionic and covalent bonds

Materials

Magnesium ribbon (5 cm); roll sulfur (2 g); tongs; large metal can; deflagrating spoon; laboratory burner

Safety Precautions

Disposal Dispose of the solid in a landfill approved to receive chemical waste.

Procedure

Place a large can in the sink. Hold the magnesium ribbon with tongs and light it. Hold the burning ribbon in the can. If the room lights are turned off, students can safely observe the light. **WARNING**: ***Wear goggles when burning magnesium. Do not look directly at the burning metal.*** In a darkened

EXAMPLE Problem 2

NAMING BINARY MOLECULAR COMPOUNDS Name the compound P_2O_5, which is used as a drying and dehydrating agent.

1 ANALYZE THE PROBLEM

You are given the formula for a compound. The formula contains the elements and the number of atoms of each element in one molecule of the compound. Because only two different elements are present and both are nonmetals, the compound can be named using the rules for naming binary molecular compounds.

2 SOLVE FOR THE UNKNOWN

First, name the elements involved in the compound.

phosphorus — **The first element, represented by P, is phosphorus.**

oxide — **The second element, represented by O, is oxygen. Add the suffix *-ide* to the root of oxygen, *ox-*.**

phosphorus oxide — **Combine the names.**

Now modify the names to indicate the number of atoms present in a molecule.

diphosphorus pentoxide — **From the formula P_2O_5, you know that two phosphorus atoms and five oxygen atoms make up a molecule of the compound. From Table 3, you know that *di-* is the prefix for two and *penta-* is the prefix for five. The *a* in *penta-* is not used because *oxide* begins with a vowel.**

3 EVALUATE THE ANSWER

The name diphosphorus pentoxide shows that a molecule of the compound contains two phosphorus atoms and five oxygen atoms, which agrees with the compound's chemical formula, P_2O_5.

EXAMPLE PROBLEM

PRACTICE Problems

Do additional problems. Online Practice

Name each of the binary covalent compounds listed below.

14. CO_2

15. SO_2

16. NF_3

17. CCl_4

18. Challenge What is the formula for diarsenic trioxide?

Common names for some molecular compounds Have you ever enjoyed an icy, cold glass of dihydrogen monoxide on a hot day? You probably have but you most likely called it by its common name, water. Recall that many ionic compounds have common names in addition to their scientific ones. For example, baking soda is sodium hydrogen carbonate and common table salt is sodium chloride.

Many binary molecular compounds, such as nitrous oxide and water, were discovered and given common names long before the present-day naming system was developed. Other binary covalent compounds that are generally known by their common names rather than their scientific names are ammonia (NH_3), hydrazine (N_2H_4), and nitric oxide (NO).

☑ READING CHECK **Apply** What are the scientific names for ammonia, hydrazine, and nitric oxide?

IN-CLASS Example

Question PCl_5 is a colorless to pale, yellow solid with a pungent odor mainly used as a gentle chlorinating agent in manufacturing dyes and pharmaceuticals. Name this compound.

Answer The first element, P, is phosphorus. The second element, Cl, is chlorine. Add the suffix *-ide* to the root of chlorine to make chloride. Now modify the name to indicate the number of atoms present in a molecule. There are no subscripts given for phosphorus, indicating that there is only one atom of this element present. Chloride has the subscript 5, the prefix for which is *penta-*. Hence, the compound is called *phosphorus pentachloride*.

PRACTICE Problems

Have students refer to the Selected Solutions appendix for complete solutions to odd-numbered problems. The complete solutions for all problems can be found in the Solutions Manual.

14. carbon dioxide

15. sulfur dioxide

16. nitrogen trifluoride

17. carbon tetrachloride

18. As_2O_3

Reinforcement

Molecule Mobile Have students make a mobile that can be used to name molecules. Display the mobile in the classroom. BL EL

☑ READING CHECK nitrogen trihydride, dinitrogen tetrahydride, nitrogen monoxide

room, place a small piece of roll sulfur in a deflagrating spoon and briefly heat it in a burner flame. The sulfur will ignite and burn with a low blue flame. **WARNING**: ***Use a fume hood; SO_2 vapor is toxic.***

Results

Magnesium burns with a bright, white-hot flame. Sulfur burns with a low blue flame. The stability of MgO is evidenced by the large amount of heat and light produced when it is formed, compared to the heat and light produced in the formation of SO_2. Thus, the ionic bonds in MgO are stronger (more stable) than the covalent bonds in SO_2.

Analysis

Ask these questions:

1. In the demonstration, with what element do magnesium and sulfur react? oxygen
2. Write balanced chemical equations for the chemical reactions. $2Mg(s) + O_2(g) \longrightarrow 2MgO(s)$; $S(s) + O_2(g) \longrightarrow SO_2(g)$

ASSESSMENT

Knowledge If Mg forms an ionic bond, and S forms a covalent bond, which bond type is the stronger bond? Why? The ionic bond is stronger because more energy is released during its formation. OL

ASSESSMENT

Knowledge Show students a chart that contains molecular formulas and have students name each molecule or acid. Use a second chart that contains names of both molecules and acids. Have students determine the correct molecular formula for each substance. **OL**

Quick Demo

Acid Formation Add 2 drops of bromothymol blue indicator to 50.0 mL of distilled water in a flask. Place a straw in the water and have a student blow through the straw. The indicator will turn from blue to yellow because the carbon dioxide exhaled from the student reacts with the water forming an acidic solution.

$CO_2 + H_2O \rightarrow H_2CO_3$

Write the formula for the acid formed (H_2CO_3) on the board, and ask students if it is a binary acid or an oxyacid. oxyacid Ask students to name the acid. carbonic acid **OL**

Naming Acids

Water solutions of some molecules are acidic and are named as acids. Acids are important compounds with specific properties. If a compound produces hydrogen ions (H^+) in solution, it is an acid. For example, HCl produces H^+ in solution and is an acid. Two common types of acids exist—binary acids and oxyacids.

Naming binary acids A binary acid contains hydrogen and one other element. The naming of the common binary acid known as hydrochloric acid is explained in the following rules.

1. The first word has the prefix *hydro-* to name the hydrogen part of the compound. The rest of the first word consists of a form of the root of the second element plus the suffix *-ic*. **HCl (hydrogen and chlorine) becomes *hydrochloric*.**
2. The second word is always *acid*. **Thus, HCl in a water solution is called *hydrochloric acid*.**

Although the term *binary* indicates exactly two elements, a few acids that contain more than two elements are named according to the rules for naming binary acids. If no oxygen is present in the formula for the acidic compound, the acid is named in the same way as a binary acid, except that the root of the second part of the name is the root of the polyatomic ion that the acid contains. For example, HCN, which is composed of hydrogen and the cyanide ion, is called *hydrocyanic acid* in solution.

Naming oxyacids An acid that contains both a hydrogen atom and an oxyanion is referred to as an **oxyacid.** Recall that an oxyanion is a polyatomic ion containing one or more oxygen atoms. The following rules explain the naming of nitric acid (HNO_3), an oxyacid.

1. First, identify the oxyanion present. The first word of an oxyacid's name consists of the root of the oxyanion and the prefix *per-* or *hypo-* if it is part of the oxyanion's name. The first word of the oxyacid's name also has a suffix that depends on the oxyanion's suffix. If the oxyanion's name ends with the suffix *-ate,* replace it with the suffix *-ic*. If the name of the oxyanion ends with the suffix *-ite,* replace it with the suffix *-ous*. **NO_3^-, the nitrate ion, becomes *nitric*.**
2. The second word of the name is always *acid*. **HNO_3 (hydrogen and the nitrate ion) becomes *nitric acid*.**

Table 4 shows how the names of several oxyacids follow these rules. Notice that the hydrogen in an oxyacid is not part of the name.

Table 4 Naming Oxyacids

Compound	Oxyanion	Acid Suffix	Acid Name
$HClO_3$	chlorate	-ic	chloric acid
$HClO_2$	chlorite	-ous	chlorous acid
HNO_3	nitrate	-ic	nitric acid
HNO_2	nitrite	-ous	nitrous acid

DIFFERENTIATED INSTRUCTION

Advanced Learners Have interested students research and write a report or prepare a presentation on the chemical and physical properties, sources, and uses of compounds that contain only nitrogen and oxygen. These compounds are nitrogen monoxide, nitrogen dioxide, dinitrogen monoxide, dinitrogen trioxide, dinitrogen tetroxide, and dinitrogen pentoxide. **AL**

CHEMISTRY PROJECT

Molecular Compounds Using a 2 × 2 grid, have students determine all possible molecules formed when either carbon or hydrogen reacts with either oxygen or fluorine. Have them use the oxidation states of 4+ for carbon, 1+ for hydrogen, 2− for oxygen, and 1− for fluorine. Write the correct formula and name for each compound. **OL**

Explore **naming covalent compounds** with an interactive table. Concepts In Motion

Table 5 Formulas and Names of Some Covalent Compounds

Formula	Common Name	Molecular Compound Name
H_2O	water	dihydrogen monoxide
NH_3	ammonia	nitrogen trihydride
N_2H_4	hydrazine	dinitrogen tetrahydride
HCl	muriatic acid	hydrochloric acid
$C_9H_8O_4$	aspirin	2-(acetyloxy)benzoic acid

You have learned that naming covalent compounds follows different sets of rules depending on the composition of the compound. **Table 5** summarizes the formulas and names of several covalent compounds. Note that an acid, whether a binary acid or an oxyacid, can have a common name in addition to its compound name.

PRACTICE Problems Do additional problems. Online Practice

Name the following acids. Assume each compound is dissolved in water.

19. HI **20.** $HClO_3$ **21.** $HClO_2$ **22.** H_2SO_4 **23.** H_2S

24. Challenge What is the formula for periodic acid?

Writing Formulas from Names

The name of a molecular compound reveals its composition and is important in communicating the nature of the compound. Given the name of any binary molecule, you should be able to write the correct chemical formula. The prefixes used in a name indicate the exact number of each atom present in the molecule and determine the subscripts used in the formula. If you are having trouble writing formulas from the names for binary compounds, you might want to review the naming rules listed on pages at the beginning of this section.

The formula for an acid can also be derived from the name. It is helpful to remember that all binary acids contain hydrogen and one other element. For oxyacids—acids containing oxyanions—you will need to know the names of the common oxyanions. If you need to review oxyanion names, see **Table 9** in the previous chapter.

PRACTICE Problems Do additional problems. Online Practice

Give the formula for each compound.

25. silver chloride
26. dihydrogen monoxide
27. chlorine trifluoride
28. diphosphorus trioxide
29. disulfur decafluoride
30. Challenge What is the formula for carbonic acid?

PRACTICE Problems

Have students refer to the Selected Solutions appendix for complete solutions to odd-numbered problems. The complete solutions for all problems can be found in the Solutions Manual.

19. hydroiodic acid
20. chloric acid
21. chlorous acid
22. sulfuric acid
23. hydrosulfuric acid
24. HIO_4
25. AgCl
26. H_2O
27. ClF_3
28. P_2O_3
29. S_2F_{10}
30. H_2CO_3

ASSESSMENT

Knowledge Have students write molecular formulas, names of molecules, names of acids, or formulas of acids on the front of note cards. Have another student write the correct formula or name on the back of each note card. Use these note cards for review. EL

CHEMISTRY JOURNAL

Form Compounds Give students the following chart with the answers left blank. Have them write the formula and name the molecule that forms in their journals. Remind students that usually the least electronegative atom is written first in a compound. OL

	Br	F	Cl
C	CBr_4 carbon tetrabromide	CF_4 carbon tetrafluoride	CCl_4 carbon tetrachloride
P	PBr_3 phosphorus tribromide	PF_3 phosphorus trifluoride	PCl_3 phosphorus trichloride
S	SBr_2 sulfur dibromide	SF_2 sulfur difluoride	SCl_2 sulfur dichloride

■ **Caption Question Fig. 12**
H_2SO_3 is an oxyacid; HBr is a binary acid.

3 Assess

Check for Understanding

Show students a 4 × 4 grid containing eight molecules and their corresponding names, randomly arranged. Cover each square and play a game by matching the formula of the molecule with its correct name. BL EL

Reteach

Many common air pollutants form acids in water. Have students create tables that have the following column heads: Formula of Pollutant, Name of Molecule, Formula of Acid, and Name of Acid. For each of the following pollutants, provide information for one column and have students complete the table: SO_2 (sulfur dioxide, H_2SO_3, sulfurous acid), CO_2 (carbon dioxide, H_2CO_3, carbonic acid), SO_3 (sulfur trioxide, H_2SO_4, sulfuric acid), NO_2 (nitrogen dioxide, HNO_3, nitric acid). OL

Extension

Have students write answers and questions for a game using the categories Molecular Formulas, Names of Molecules, Names of Acids, and Formulas for Acids. OL

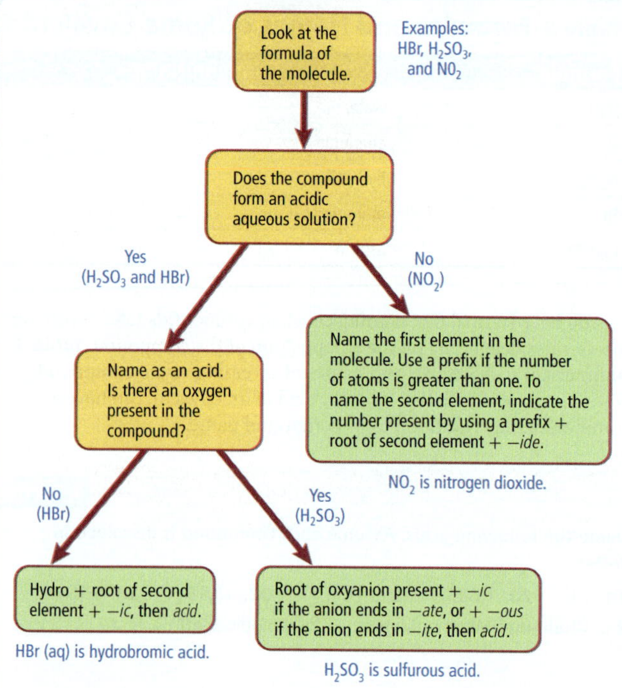

■ **Figure 12** Use this flowchart to name molecular compounds when their formulas are known.
Apply ***Which compound above is an oxyacid? Which is a binary acid?***

The flowchart in **Figure 12** can help you determine the name of a molecular covalent compound. To use the chart, start at the top and work downward by reading the text contained in the colored boxes. Apply the text at each step of the flowchart to the formula of the compound that you wish to name.

SECTION 2 REVIEW

Section Self-Check

Section Summary

- Names of covalent molecular compounds include prefixes for the number of each atom present. The final letter of the prefix is dropped if the element name begins with a vowel.
- Molecules that produce H^+ in solution are acids. Binary acids contain hydrogen and one other element. Oxyacids contain hydrogen and an oxyanion.

31. MAINIDEA **Summarize** the rules for naming binary molecular compounds.

32. Define a binary molecular compound.

33. Describe the difference between a binary acid and an oxyacid.

34. Apply Using the system of rules for naming binary molecular compounds, describe how you would name the molecule N_2O_4.

35. Apply Write the molecular formula for each of these compounds: *iodic acid, disulfur trioxide, dinitrogen monoxide,* and *hydrofluoric acid.*

36. State the molecular formula for each compound listed below.

a. dinitrogen trioxide
b. nitrogen monoxide
c. hydrochloric acid
d. chloric acid
e. sulfuric acid
f. sulfurous acid

SECTION 2 REVIEW

31. Name the first element in the formula first. Name the second element using its root plus the suffix –ide. Add prefixes to indicate the number of atoms of each element present

32. a molecule composed of only two nonmetal elements

33. A binary acid contains hydrogen and one other element. An oxyacid contains hydrogen, another element, and oxygen.

34. There are two atoms of nitrogen; use the prefix *di–* with the name nitrogen. There are four atoms of oxygen, so use the prefix *tetra–* plus the root of oxygen plus the ending *–ide*. The name is dinitrogen tetroxide.

35. HIO_3, S_2O_3, N_2O, HF

36. **a.** N_2O_3
b. NO
c. HCl
d. $HClO_3$
e. H_2SO_4
f. H_2SO_3

SECTION 3

Essential Questions

- What are the basic steps used to draw Lewis structures?
- Why does resonance occur, and what are some resonance structures?
- Which molecules are exceptions to the octet rule, and why do these exceptions occur?

Review Vocabulary

ionic bond: the electrostatic force that holds oppositely charged particles together in an ionic compound

New Vocabulary

structural formula
resonance
coordinate covalent bond

Molecular Structures

MAINIDEA Structural formulas show the relative positions of atoms within a molecule.

CHEM 4 YOU As a child, you might have played with plastic building blocks that connected only in certain ways. If so, you probably noticed that the shape of the object you built depended on the limited ways the blocks interconnected. Building molecules out of atoms works in a similar way.

Structural Formulas

You have already studied the structure of ionic compounds—substances formed from ionic bonds. The covalent molecules you have read about in this chapter have structures that are different from those of ionic compounds. In studying the molecular structures of covalent compounds, models are used as representations of the molecule.

The molecular formula, which shows the element symbols and numerical subscripts, tells you the type and number of each atom in a molecule. As shown in **Figure 13,** there are several different models that can be used to represent a molecule. Note that in the ball-and-stick and space-filling molecular models, atoms of each specific element are represented by spheres of a representative color, as shown in **Table R-1** in the Student Resources. These colors are used for identifying the atoms if the chemical symbol of the element is not present.

One of the most useful molecular models is the **structural formula,** which uses letter symbols and bonds to show relative positions of atoms. You can predict the structural formula for many molecules by drawing the Lewis structure. You have already seen some simple examples of Lewis structures, but more involved structures are needed to help you determine the shapes of molecules.

■ **Figure 13** All of these models can be used to show the relative locations of atoms and electrons in the phosphorus trihydride (phosphine) molecule.
Compare and contrast *the types of information contained in each model.*

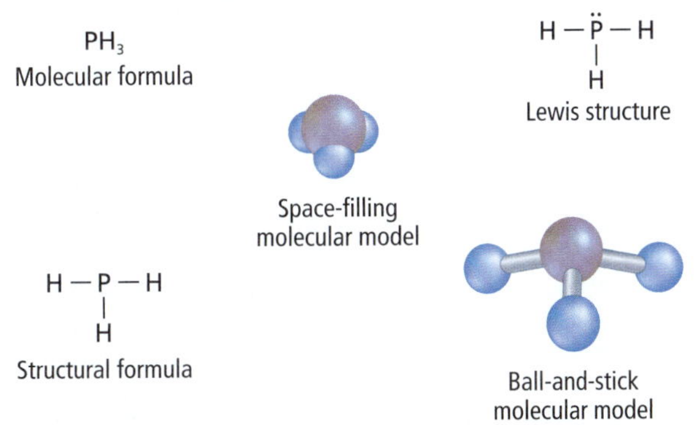

SECTION 3

1 Focus

MAINIDEA

Molecular Structures Give the students a set of snap-together toy blocks and ask them to make a pyramid. Ask them if there was a set pattern they followed. The shape of the blocks determined how they could build the pyramid. Tell the students that molecular shapes are determined by the way atoms share electrons. OL

2 Teach

Quick Demo

Octet Models Place a circle with the electron dot diagram of a sulfur atom and two circles with the electron dot diagram of oxygen on a felt board. Ask students how these atoms could be arranged to satisfy the octet rule. Students will use trial-and-error methods. Tell students that a systematic approach can be used to determine the Lewis structure of molecules, including those that show exceptions to the octet rule. OL EL

■ **Caption Question Fig. 13** All models show the number and type of each atom; Lewis structures, structural formulas, ball-and-stick models, and space-filling models show geometric shape; Lewis structures show distribution of valence electrons among bonding pairs and lone pairs. Space-filling models show the relative sizes of the atoms.

Visual Learning

Build Models Using a molecular model kit or marshmallows and toothpicks, have students build models of NH_4^+, CH_2O, BCl_3, and HF. Have students create a chart showing the ball-and-stick model, Lewis structure, and structural formula for each substance. OL EL

Apply Chemistry

Dyes Organic dyes are organic molecules that absorb select wavelengths of visible light and reflect all other wavelengths. These dyes are used to produce colorful fabrics. They are also used in recordable compact discs. Recordable CDs have a layer of colored, organic dye sandwiched between a clear rigid backing and a reflective surface. A laser is used to burn data onto the CD. The bonds in the organic dye molecule absorb the laser light and change the structure of the molecule, producing an opaque substance. This change in structure is irreversible; therefore, data can be written only once on any part of the compact disc.

Concept Development

Lewis Structure Using the systematic approach, have students draw Lewis structures for several molecules. Remind them that carbon atoms can form chains with other carbon atoms. The carbon atoms will bond together with a single, double, or triple bond, depending on the available electron pairs. Have students draw the Lewis structure for ethane (C_2H_6), ethene (C_2H_4), and ethyne (C_2H_2), using the systematic technique. OL

Reinforcement

Representing Molecules Point out that both Lewis structures and structural formulas are used to show the arrangement and ratio of atoms in a compound. Lewis structures can be used to represent both ionic and covalent compounds. Structural formulas are used only to represent molecules. Structural formulas use a line to represent a shared pair of electrons. However, unshared, lone, pairs of electrons are still represented by electron dots.

Lewis structures Although it is fairly easy to draw Lewis structures for most compounds formed by nonmetals, it is a good idea to follow a regular procedure. Whenever you need to draw a Lewis structure, follow the steps outlined in this Problem-Solving Strategy.

PROBLEM-SOLVING STRATEGY

Drawing Lewis Structures

1. Predict the location of certain atoms.

   The atom that has the least attraction for shared electrons will be the central atom in the molecule. This element is usually the one closer to the left side of the periodic table. The central atom is located in the center of the molecule; all other atoms become terminal atoms.

   Hydrogen is always a terminal, or end, atom. Because it can share only one pair of electrons, hydrogen can be connected to only one other atom.

2. Determine the number of electrons available for bonding.

   This number is equal to the total number of valence electrons in the atoms that make up the molecule.

3. Determine the number of bonding pairs.

   To do this, divide the number of electrons available for bonding by two.

4. Place the bonding pairs.

   Place one bonding pair (single bond) between the central atom and each of the terminal atoms.

5. Determine the number of electron pairs remaining.

   To do this, subtract the number of pairs used in Step 4 from the total number of bonding pairs determined in Step 3. These remaining pairs include lone pairs as well as pairs used in double and triple bonds. Place lone pairs around each terminal atom (except H atoms) bonded to the central atom to satisfy the octet rule. Any remaining pairs will be assigned to the central atom.

6. Determine whether the central atom satisfies the octet rule.

   Is the central atom surrounded by four electron pairs? If not, it does not satisfy the octet rule. To satisfy the octet rule, convert one or two of the lone pairs on the terminal atoms into a double bond or a triple bond between the terminal atom and the central atom. These pairs are still associated with the terminal atom as well as with the central atom. Remember that carbon, nitrogen, oxygen, and sulfur often form double and triple bonds.

Apply the Strategy

Study Example Problems 3 through 5 to see how the steps in the Problem-Solving Strategy are applied.

CHEMISTRY PROJECT

Gilbert Lewis Have students research Gilbert Newton Lewis and make posters depicting his original Lewis structures. Students should share the posters with the class. EL

EXAMPLE Problem 3

Find help with **dimensional analysis.** Math Handbook

LEWIS STRUCTURE FOR A COVALENT COMPOUND WITH SINGLE BONDS Ammonia is a raw material used in the manufacture of many products, including fertilizers, cleaning products, and explosives. Draw the Lewis structure for ammonia (NH_3).

1 ANALYZE THE PROBLEM

Ammonia molecules consist of one nitrogen atom and three hydrogen atoms. Because hydrogen must be a terminal atom, nitrogen is the central atom.

2 SOLVE FOR THE UNKNOWN

Find the total number of valence electrons available for bonding.

$$1\text{ N atom} \times \frac{5\text{ valence electrons}}{1\text{ N atom}} + 3\text{ H atoms} \times \frac{1\text{ valence electron}}{1\text{ H atom}}$$
$$= 8\text{ valence electrons}$$

There are 8 valence electrons available for bonding.

$$\frac{8\text{ electrons}}{2\text{ electrons/pair}} = 4\text{ pairs}$$

Determine the total number of bonding pairs. To do this, divide the number of available electrons by two.

Four pairs of electrons are available for bonding.

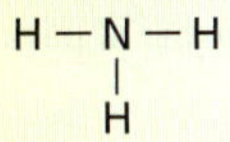

Place a bonding pair (a single bond) between the central nitrogen atom and each terminal hydrogen atom.

Determine the number of bonding pairs remaining.

4 pairs total − 3 pairs used
= 1 pair available

Subtract the number of pairs used in these bonds from the total number of pairs of electrons available.

The remaining pair—a lone pair—must be added to either the terminal atoms or the central atom. Because hydrogen atoms can have only one bond, they have no lone pairs.

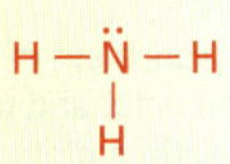

Place the remaining lone pair on the central nitrogen atom.

3 EVALUATE THE ANSWER

Each hydrogen atom shares one pair of electrons, as required, and the central nitrogen atom shares three pairs of electrons and has one lone pair, providing a stable octet.

PRACTICE Problems

Do additional problems. Online Practice

37. Draw the Lewis structure for BH_3.

38. Challenge A nitrogen trifluoride molecule contains numerous lone pairs. Draw its Lewis structure.

CHEMISTRY JOURNAL

Lewis Structures Have students predict the Lewis structures for the following common molecules: carbon monoxide (CO), carbon dioxide (CO_2), sulfur dioxide (SO_2), sulfur trioxide (SO_3), hydrogen cyanide (HCN), dinitrogen monoxide (N_2O), nitrogen dioxide (NO_2), and nitrogen monoxide (NO). **OL**

IN-CLASS Example

Question Silicon tetrahydride is a gas at room temperature that undergoes spontaneous combustion in air. What is the Lewis structure for SiH_4?

Answer

$$1\text{ Si atom} \times \frac{4\text{ valence electrons}}{\text{Si atom}}$$
$$+ 4\text{ H atoms} \times \frac{1\text{ valence electron}}{\text{H atom}}$$
$$= \frac{8\text{ total electrons}}{2} = 4\text{ bonding pairs}$$

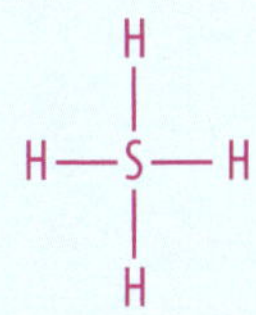

Use all bonding pairs. Silicon has an octet and hydrogen only shares one pair. This Lewis structure satisfies the octet rule.

Content Background

Space Molecules In dust clouds throughout the universe, dust particles provide a surface for chemical reactions to take place. Atoms join together to create larger molecules. As these molecules tumble through space, they change energy and give off radiation in the form of radio waves. In May 2000, scientists discovered the simple sugar molecule ribose floating in the dust cloud called Sagittarius B2. Astronomers have identified more than 120 space molecules.

PRACTICE Problems

Have students refer to the Selected Solutions appendix for complete solutions to odd-numbered problems. The complete solutions for all problems can be found in the Solutions Manual.

37.

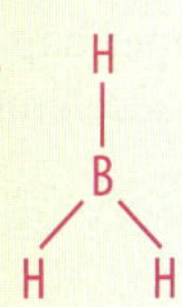

38.

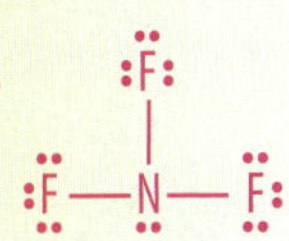

IN-CLASS Example

Question Selenium trioxide is a yellowish-white crystalline powder that decomposes on heating and is a very toxic compound. What is the Lewis structure for SeO_3?

Answer

$$1 \text{ Se atom} \times \frac{6 \text{ valence electrons}}{\text{Se atom}} + 3 \text{ O atoms} \times \frac{6 \text{ valence electrons}}{\text{O atom}} = \frac{24 \text{ valence electrons}}{2} = 12 \text{ electron pairs}$$

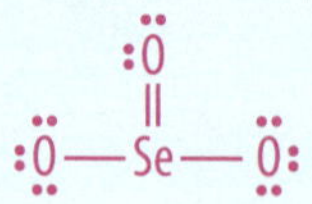

All bonded atoms have an octet.

PRACTICE Problems

Have students refer to the Selected Solutions appendix for complete solutions to odd-numbered problems. The complete solutions for all problems can be found in the Solutions Manual.

39. 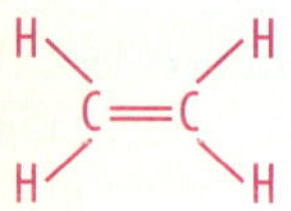

40. 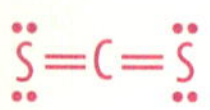

Apply Chemistry

Ripening Fruits and vegetables naturally produce ethylene gas (C_2H_4) as they ripen. For some fruits and vegetables, the ripening process is accelerated when they are exposed to ethylene. The effect of ethylene on fruits and vegetables is dependent on the sensitivity of the item. Tell students that they can accelerate the ripening of fruit at home by placing unripe fruit in a closed paper bag. The trapped ethylene gas accelerates the ripening process.

Get help with **greatest common factors.**

Personal Tutor

EXAMPLE Problem 4

LEWIS STRUCTURE FOR A COVALENT COMPOUND WITH MULTIPLE BONDS Carbon dioxide is a product of all cellular respiration. Draw the Lewis structure for carbon dioxide (CO_2).

1 ANALYZE THE PROBLEM

The carbon dioxide molecule consists of one carbon atom and two oxygen atoms. Because carbon has less attraction for shared electrons, carbon is the central atom, and the two oxygen atoms are terminal.

2 SOLVE FOR THE UNKNOWN

Find the total number of valence electrons available for bonding.

$$1 \text{ C atom} \times \frac{4 \text{ valence electrons}}{1 \text{ C atom}} + 2 \text{ O atoms} \times \frac{6 \text{ valence electrons}}{1 \text{ O atom}}$$
$$= 16 \text{ valence electrons}$$

There are 16 valence electrons available for bonding.

$$\frac{16 \text{ electrons}}{2 \text{ electrons/pair}} = 8 \text{ pairs}$$

Determine the total number of bonding pairs by dividing the number of available electrons by two.

Eight pairs of electrons are available for bonding.

O—C—O

Place a bonding pair (a single bond) between the central carbon atom and each terminal oxygen atom.

Determine the number of electron pairs remaining. Subtract the number of pairs used in these bonds from the total number of pairs of electrons available.

8 pairs total − 2 pairs used
= 6 pairs available

Subtract the number of pairs used in these bonds from the total number of pairs of electrons available.

:Ö—C—Ö:

Add three lone pairs to each terminal oxygen atom.

Determine the number of electron pairs remaining.

6 pairs available − 6 pairs used
= 0 pairs available

Subtract the lone pairs from the pairs available.

Examine the incomplete structure above (showing the placement of the lone pairs). Note that the carbon atom does not have an octet and that there are no more electron pairs available. To give the carbon atom an octet, the molecule must form double bonds.

Ö=C=Ö

Use a lone pair from each O atom to form a double bond with the C atom.

3 EVALUATE THE ANSWER

Both carbon and oxygen now have an octet, which satisfies the octet rule.

PRACTICE Problems

Do additional problems. Online Practice

39. Draw the Lewis structure for ethylene, C_2H_4.

40. Challenge A molecule of carbon disulfide contains both lone pairs and multiple-covalent bonds. Draw its Lewis structure.

DIFFERENTIATED INSTRUCTION

Advanced Learners Have students make posters that list the steps of the systematic approach as they determine the Lewis structure of one of the following molecules or ions: OH^-, PF_6^-, BF_3, ClO_3^-, SF_3N, or CH_2Cl_2. **AL**

Lewis structures for polyatomic ions Although the unit acts as anion, the atoms within a polyatomic ion are covalently bonded. The procedure for drawing Lewis structures for polyatomic ions is similar to drawing them for covalent compounds. The main difference is in finding the total number of electrons available for bonding. Compared to the number of valence electrons present in the atoms that make up the ion, more electrons are present if the ion is negatively charged and fewer are present if the ion is positive. To find the total number of electrons available for bonding, first find the number available in the atoms present in the ion. Then, subtract the ion charge if the ion is positive, and add the ion charge if the ion is negative.

EXAMPLE Problem 5

LEWIS STRUCTURE FOR A POLYATOMIC ION Draw the correct Lewis structure for the polyatomic ion phosphate (PO_4^{3-}).

1 ANALYZE THE PROBLEM

You are given that the phosphate ion consists of one phosphorus atom and four oxygen atoms and has a charge of 3−. Because phosphorus has less attraction for shared electrons than oxygen, phosphorus is the central atom and the four oxygen atoms are terminal atoms.

2 SOLVE FOR THE UNKNOWN

Find the total number of valence electrons available for bonding.

$$1 \text{ P atom} \times \frac{5 \text{ valence electrons}}{\text{P atom}} + 4 \text{ O atoms} \times \frac{6 \text{ valence electrons}}{\text{O atom}}$$

$$+ 3 \text{ electrons from the negative charge} = 32 \text{ valence electrons}$$

$$\frac{32 \text{ electrons}}{2 \text{ electrons/pair}} = 16 \text{ pair}$$

Determine the total number of bonding pairs.

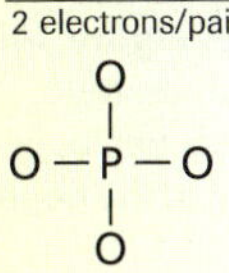

Draw single bonds from each terminal oxygen atom to the central phosphorus atom.

16 pairs total − 4 pairs used
= 12 pairs available

Subtract the number of pairs used from the total number of pairs of electrons available.

Add three lone pairs to each terminal oxygen atom.
12 pairs available − 12 lone pairs used = 0

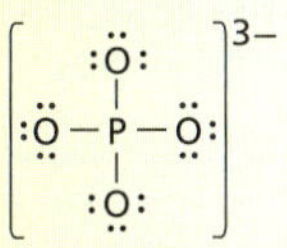

Subtracting the lone pairs used from the pairs available verifies that there are no electron pairs available for the phosphorus atom. The Lewis structure for the phosphate ion is shown.

3 EVALUATE THE ANSWER

All of the atoms have an octet, and the group has a net charge of 3−.

PRACTICE Problems

Do additional problems. Online Practice

41. Draw the Lewis structure for the NH_4^+ ion.

42. **Challenge** The ClO_4^- ion contains numerous lone pairs. Draw its Lewis structure.

RealWorld CHEMISTRY

Phosphorus and Nitrogen

Suzanne Long/Alamy

ALGAL BLOOMS Phosphorus and nitrogen are nutrients required for algae growth. Both can enter lakes and streams from discharges of sewage and industrial waste, and in fertilizer runoff. If these substances build up in a body of water, a rapid growth of algae, known as an algal bloom, can occur, forming a thick layer of green slime over the water's surface. When the algae use up the supply of nutrients, they die and decompose. This process reduces the amount of dissolved oxygen in the water that is available to other aquatic organisms.

ASSESSMENT

Skill On a note card, write the chemical formula for a molecule or polyatomic ion. Ask students to draw the Lewis structure on the other side of the card. OL

Concept Development

Model Resonance Choose three students to represent atoms in the polyatomic ion NO_2^-. Arrange the three students in a sequence with the nitrogen atom as the central atom. One bond is a double bond, and the other bond is a single bond. Ask students to determine the resonance structures. OL EL

IN-CLASS Example

Question What is the correct Lewis structure for the polyatomic ion BrO_4^-?

Answer

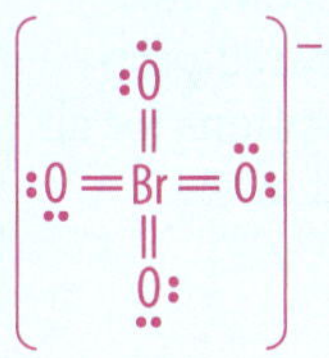

PRACTICE Problems

Have students refer to the Selected Solutions appendix for complete solutions to odd-numbered problems. The complete solutions for all problems can be found in the Solutions Manual.

41. 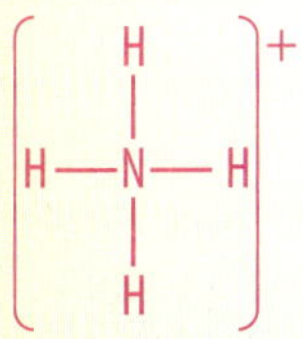

42. 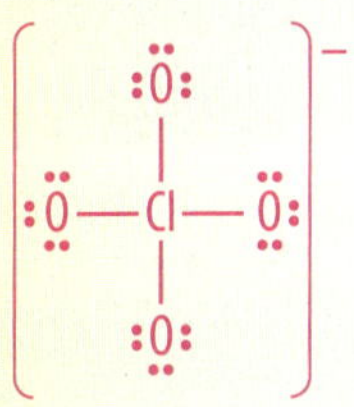

Enrichment

Chemistry of Chocolate The Mayans and Aztecs consumed chocolate, as do many people throughout the world today. Cocoa and chocolate products contain theobromine, which is chemically related to caffeine, though lacking caffeine's stimulant effect. OL

1. Write the molecular formula for theobromine. $C_7H_8N_4O_2$
2. Identify the number of sigma and pi bonds present in this molecule. 22 sigma bonds and 4 pi bonds

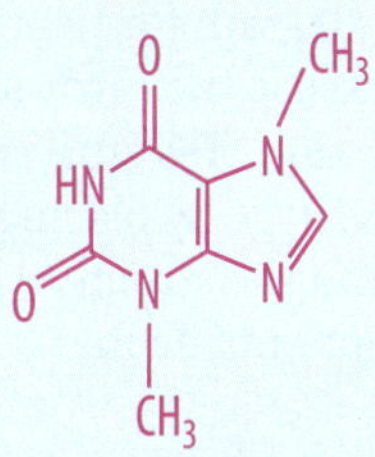

theobromine

Content Background

Exceptions to the Octet Rule

Many compounds form that do not have eight electrons around each atom. Compounds that contain beryllium and boron as the central atom are often deficient in electrons. Expanded octets, those having more than eight electrons, occur only around a central nonmetallic atom from period 3 or higher because they have d orbitals available.

ASSESSMENT

Knowledge Phosgene is a colorless, highly toxic gas used in warfare. The formula for this molecule is CCl_2O. Have students draw two resonance structures for this molecule. **OL**

PRACTICE Problems

Have students refer to the Selected Solutions appendix for complete solutions to odd-numbered problems. The complete solutions for all problems can be found in the Solutions Manual.

43. 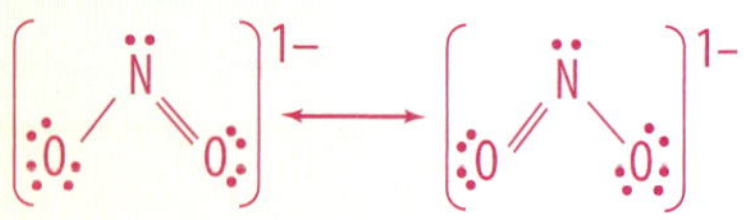

44. 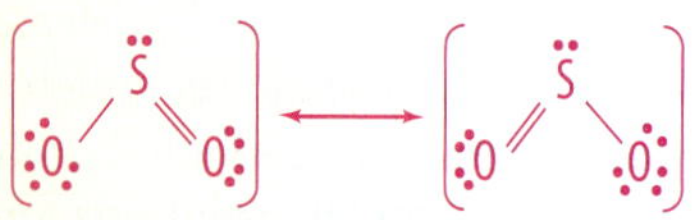

45. 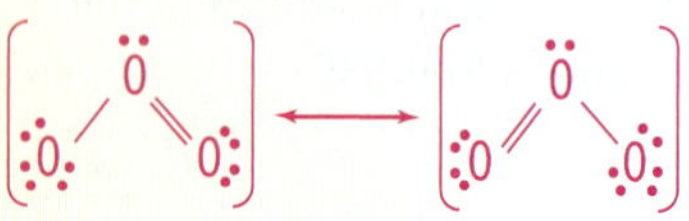

46. 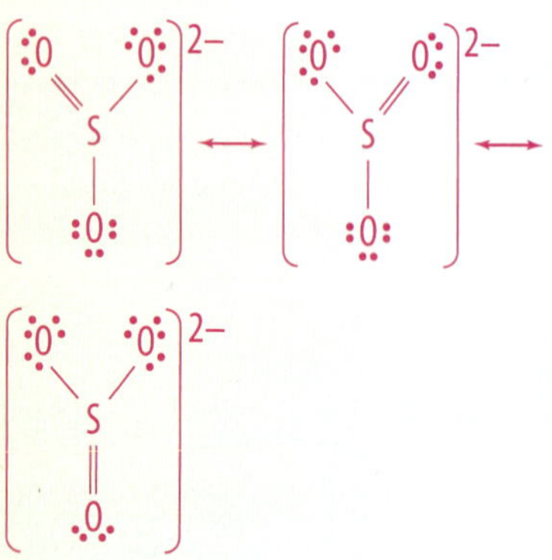

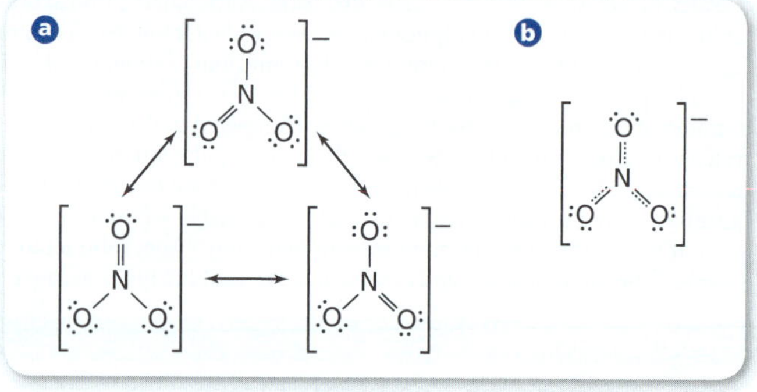

■ **Figure 14** The nitrate ion (NO_3^-) exhibits resonance. **a.** These resonance structures differ only in the location of the double bond. The locations of the nitrogen and oxygen atoms stay the same. **b.** The actual nitrate ion is like an average of the three resonance structures in **a.** The dotted lines indicate possible locations of the double bond.

VOCABULARY

SCIENCE USAGE V. COMMON USAGE

Resonance

Science usage: a phenomenon related to the stability of a molecule; a large vibration in a mechanical system caused by a small periodic stimulus

The new molecule had several resonance structures.

Common usage: a quality of richness or variety

The sound of the orchestra had resonance.

Resonance Structures

Using the same sequence of atoms, it is possible to have more than one correct Lewis structure when a molecule or polyatomic ion has both a double bond and a single bond. Consider the polyatomic ion nitrate (NO_3^-), shown in **Figure 14a.** Three equivalent structures can be used to represent the nitrate ion.

Resonance is a condition that occurs when more than one valid Lewis structure can be written for a molecule or ion. The two or more correct Lewis structures that represent a single molecule or ion are referred to as resonance structures. Resonance structures differ only in the position of the electron pairs, never the atom positions. The location of the lone pairs and bonding pairs differs in resonance structures. The molecule O_3 and the polyatomic ions NO_3^-, NO_2^-, SO_3^{2-}, and CO_3^{2-} all exhibit resonance.

It is important to note that each molecule or ion that exhibits resonance behaves as if it has only one structure. Refer to **Figure 14b.** Experimentally measured bond lengths show that the bonds are identical to each other. They are shorter than single bonds but longer than double bonds. The actual bond length is an average of the bonds in the resonance structures.

PRACTICE Problems — Do additional problems. Online Practice

Draw the Lewis resonance structures for the following molecules.

43. NO_2^- **44.** SO_2 **45.** O_3

46. Challenge Draw the Lewis resonance structure for the ion SO_3^{2-}.

■ **Figure 15** The central nitrogen atom in this NO_2 molecule does not satisfy the octet rule; the nitrogen atom has only seven electrons in its outer energy level.

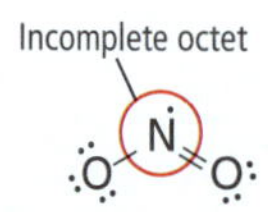

Exceptions to the Octet Rule

Generally, atoms attain an octet when they bond with other atoms. Some molecules and ions, however, do not obey the octet rule. There are several reasons for these exceptions.

Odd number of valence electrons First, a small group of molecules might have an odd number of valence electrons and be unable to form an octet around each atom. For example, NO_2 has five valence electrons from nitrogen and 12 from oxygen, totaling 17, which cannot form an exact number of electron pairs. See **Figure 15.** ClO_2 and NO are other examples of molecules with odd numbers of valence electrons.

DIFFERENTIATED INSTRUCTION

Struggling Learners Using index cards that have been cut in half, have students write the electron dot diagram for C on one card and the electron dot diagram for O on three cards. Ask students to use the cards to form the polyatomic CO_3^{2-} ion. Tell students to indicate the ion's charge by placing large square brackets around the structure and writing a 2– superscript outside the brackets. Student structures should show a central carbon atom, three terminal oxygen atoms, two single bonds, and one double bond. Ask students if the atoms could be arranged another way and have octets for all atoms. Two other resonance structures are possible. The location of the double bond determines the resonance structure. **BL** **EL**

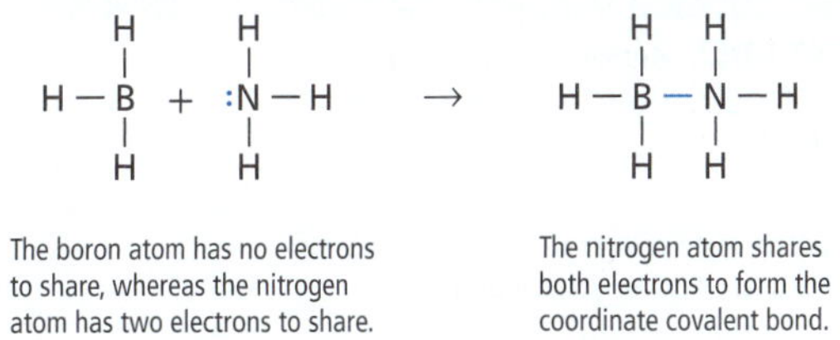

■ **Figure 16** In this reaction between boron trihydride (BH_3) and nitrogen (NH_3), the nitrogen atom donates both electrons that are shared by boron and nitrogen, forming a coordinate covalent bond.

Interpret ***Does the coordinate covalent bond in the product molecule satisfy the octet rule?***

Suboctets and coordinate covalent bonds Another exception to the octet rule is due to a few compounds that form suboctets—stable configurations with fewer than eight electrons present around an atom. This group is relatively rare, and BH_3 is an example. Boron, a group 13 metalloid, forms three covalent bonds with other nonmetallic atoms.

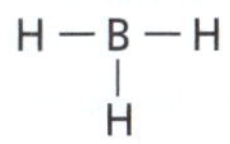

The boron atom shares only six electrons—too few to form an octet. Such compounds tend to be reactive and can share an entire pair of electrons donated by another atom.

A **coordinate covalent bond** forms when one atom donates both of the electrons to be shared with an atom or ion that needs two electrons to form a stable electron arrangement with lower potential energy. Refer to **Figure 16.** Atoms or ions with lone pairs often form coordinate covalent bonds with atoms or ions that need two more electrons.

Expanded octets The third group of compounds that does not follow the octet rule has central atoms that contain more than eight valence electrons. This electron arrangement is referred to as an expanded octet. An expanded octet can be explained by considering the d orbitals that occur in the energy levels of elements in period three or higher. An example of an expanded octet, shown in **Figure 17,** is the bond formation in the molecule PCl_5. Five bonds are formed with ten electrons shared in one s orbital, three p orbitals, and one d orbital. Another example is the molecule SF_6, which has six bonds sharing 12 electrons in an s orbital, three p orbitals, and two d orbitals. When you draw the Lewis structures for these compounds, either extra lone pairs are added to the central atom or more than four bonding atoms are present in the molecule.

☑ READING CHECK **Summarize** three reasons why some molecules do not conform to the octet rule.

■ **Figure 17** Prior to the reaction of PCl_3 and Cl_2, every reactant atom follows the octet rule. After the reaction, the product, PCl_5, has an expanded octet containing ten electrons.

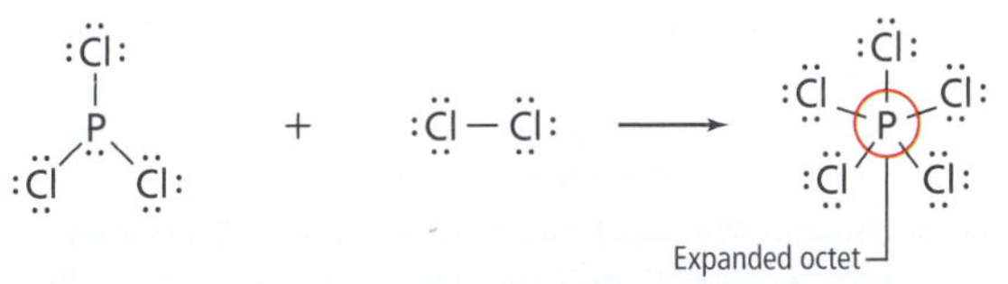

■ **Caption Question Fig. 16** Yes, each atom involved in the bond has a complete octet.

Visual Learning

Expanded Octets Have students locate the nonmetals on the periodic table located in period 3 or higher. These elements can have expanded octets. Explain that phosphorus, a group 15 nonmetal with 5 valence electrons, can share all of its electrons. This expands its octet to sp^3d, a configuration capable of holding 10 electrons. Using marshmallows to represent atoms and toothpicks to represent single covalent bonds, have students make models of PF_5, SCl_6, and ClF_5. **AL**

☑ READING CHECK the molecule has an odd number of valence electrons; the molecule is stable with less than eight valence electrons; the molecule is stable with more than eight valence electrons (expanded octet)

3 Assess

Check for Understanding

Write a molecular formula on the board. Have students draw the Lewis structure for the molecule. **OL**

Reteach

Review the steps used in drawing Lewis structures using the molecules N_2O and BeF_2. Have students determine whether these molecules are exceptions to the octet rule or if they exhibit resonance. **OL**

Extension

Have students research the compounds formed from the noble gas xenon. Have them draw the Lewis structure that represents each molecule formed. **OL**

DIFFERENTIATED INSTRUCTION

Struggling Learners Use the following analogy to clarify coordinate covalent bonds to students. You borrow a CD from a friend. While using the CD, you treat the CD as belonging to you, yet at the same time your friend counts that CD as part of his or her CD collection. **BL**

IN-CLASS Example

Question Selenium hexafluoride is a toxic gas used as an electrical insulator. What is the Lewis structure for this compound?

Answer Each fluorine has an octet and the central Se atom has an expanded octet.

PRACTICE Problems

Have students refer to the Selected Solutions appendix for complete solutions to odd-numbered problems. The complete solutions for all problems can be found in the Solutions Manual.

47.

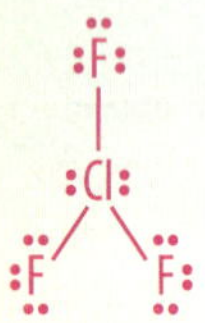

48.

49.

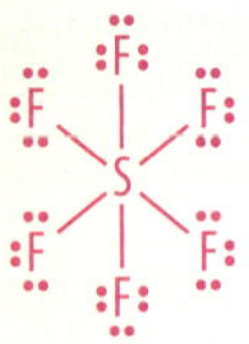

EXAMPLE Problem 6

EXAMPLE PROBLEM

LEWIS STRUCTURE: EXCEPTION TO THE OCTET RULE Xenon is a noble gas that will form a few compounds with nonmetals that strongly attract electrons. Draw the correct Lewis structure for xenon tetrafluoride (XeF_4).

1 ANALYZE THE PROBLEM

You are given that a molecule of xenon tetrafluoride consists of one xenon atom and four fluorine atoms. Xenon has less attraction for electrons, so it is the central atom.

2 SOLVE FOR THE UNKNOWN

First, find the total number of valence electrons.

$$1 \text{ Xe atom} \times \frac{8 \text{ valence electrons}}{1 \text{ Xe atom}} + 4 \text{ F atoms} \times \frac{7 \text{ valence electrons}}{1 \text{ F atom}} = 36 \text{ valence electrons}$$

$$\frac{36 \text{ electrons}}{2 \text{ electrons/pair}} = 18 \text{ pairs}$$

Determine the total number of bonding pairs.

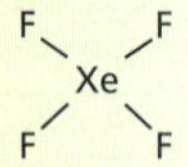

Use four bonding pairs to bond the four F atoms to the central Xe atom.

18 pairs available − 4 pairs used = 14 pairs available

Determine the number of remaining pairs.

$$14 \text{ pairs} - 4 \text{ F atoms} \times \frac{3 \text{ pairs}}{1 \text{ F atom}} = 2 \text{ pairs unused}$$

Add three pairs to each F atom to obtain an octet. Determine how many pairs remain.

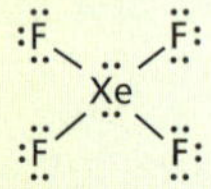

Place the two remaining pairs on the central Xe atom.

3 EVALUATE THE ANSWER

This structure gives xenon 12 total electrons, an expanded octet. Xenon compounds, such as the XeF_4 shown here, are toxic because they are highly reactive.

PRACTICE Problems

Do additional problems. Online Practice

Draw the expanded octet Lewis structure for each molecule.

47. ClF_3

48. PCl_5

49. Challenge Draw the Lewis structure for the molecule formed when six fluorine atoms and one sulfur atom bond covalently.

SECTION 3 REVIEW

Section Self-Check

Section Summary

- Different models can be used to represent molecules.
- Resonance occurs when more than one valid Lewis structure exists for the same molecule.
- Exceptions to the octet rule occur in some molecules.

50. MAINIDEA Describe the information contained in a structural formula.

51. State the steps used to draw Lewis structures.

52. Summarize exceptions to the octet rule by correctly pairing these molecules and phrases: odd number of valence electrons, PCl_5, ClO_2, BH_3, expanded octet, less than an octet.

53. Evaluate A classmate states that a binary compound having only sigma bonds displays resonance. Could the classmate's statement be true?

54. Draw the resonance structures for the dinitrogen oxide (N_2O) molecule.

55. Draw the Lewis structures for CN^-, SiF_4, HCO_3^-, and, AsF_6^-.

SECTION 3 REVIEW

50. types and number of atoms; rough molecular shape

51. determine central atom and terminal atoms; determine number of bonding electrons; determine bonding pairs; connect terminal atoms to the central atom with single bonds; determine remaining number of bonding pairs; apply octet rule and form double or triple bonds if needed

52. expanded octet, PCl_5; odd number of valence electrons, ClO_2; less than an octet, BH_3

53. No, a molecule or polyatomic ion must have both a single bond and a double bond in order to display resonance.

54.

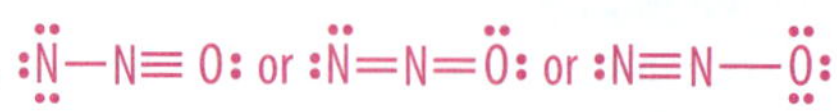

55.

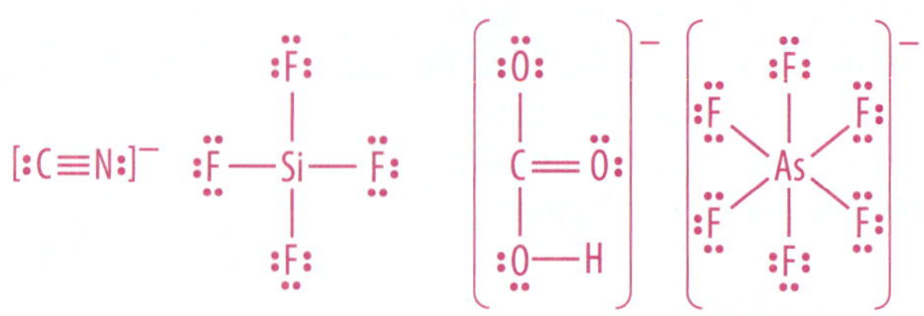

SECTION 4

Molecular Shapes

MAINIDEA The VSEPR model is used to determine molecular shape.

Essential Questions

- What is the VSEPR bonding theory?
- How can you use the VSEPR model to predict the shape of, and the bond angles in, a molecule?
- What is hybridization?

Review Vocabulary

atomic orbital: the region around an atom's nucleus that defines an electron's probable location

New Vocabulary

VSEPR model
hybridization

CHEM 4 YOU Have you ever rubbed two balloons in your hair to create a static electric charge on them? If you brought the balloons together, their like charges would cause them to repel each other. Molecular shapes are also affected by the forces of electric repulsion.

VSEPR Model

The shape of a molecule determines many of its physical and chemical properties. Often, shapes of reactant molecules determine whether or not they can get close enough to react. Electron densities created by the overlap of the orbitals of shared electrons determine molecular shape. Theories have been developed to explain the overlap of bonding orbitals and can be used to predict the shape of the molecule.

The molecular geometry, or shape, of a molecule can be determined once a Lewis structure is drawn. The model used to determine the molecular shape is referred to as the **V**alence **S**hell **E**lectron **P**air **R**epulsion model, or **VSEPR model.** This model is based on an arrangement that minimizes the repulsion of shared and unshared electron pairs around the central atom.

Bond angle To understand the VSEPR model better, imagine balloons that are inflated to similar sizes and tied together, as shown in **Figure 18.** Each balloon represents an electron-dense region. The repulsive force of this electron-dense region keeps other electrons from entering this space. When a set of balloons is connected at a central point, which represents a central atom, the balloons naturally form a shape that minimizes interactions between the balloons.

The electron pairs in a molecule repel one another in a similar way. These forces cause the atoms in a molecule to be positioned at fixed angles relative to one another. The angle formed by two terminal atoms and the central atom is a bond angle. Bond angles predicted by VSEPR are supported by experimental evidence.

Unshared pairs of electrons are also important in determining the shape of the molecule. These electrons occupy a slightly larger orbital than shared electrons. Therefore, shared bonding orbitals are pushed together by unshared pairs.

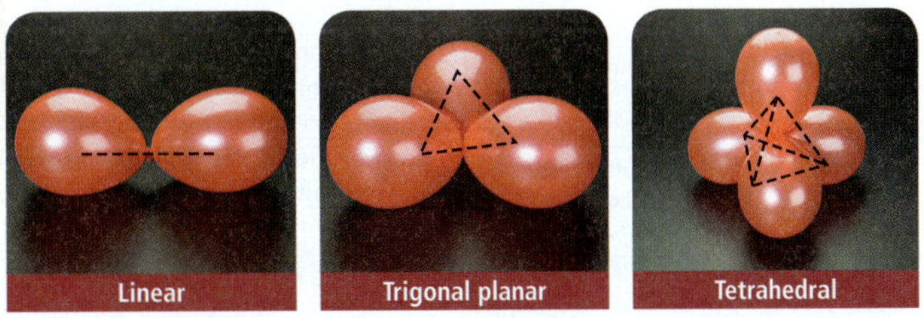

Figure 18 Electron pairs in a molecule are located as far apart as they can be, just as these balloons are arranged. Two pairs form a linear shape. Three pairs form a trigonal planar shape. Four pairs form a tetrahedral shape.

Matt Meadows

SECTION 4

1 Focus

MAINIDEA

Molecular Shape Bring four inflated balloons to class and tell students the balloons represent bonding pairs of electrons. Hold the ends of two balloons together in the center and ask students to predict the shape of the molecule. linear Hold three balloons together in the center and ask students to predict the shape of the molecule. trigonal planar (a triangle) Finally, hold all four balloons together in the center and ask for the shape. tetrahedral OL

2 Teach

Quick Demo

Model Electron Sharing Using small balloons to represent shared pairs of electrons and larger balloons to represent unshared pairs of electrons, make models to represent HCl, H_2O, NH_3, and CH_4. Use the following balloon combinations for the models: HCl, one small and three large; H_2O, two small and two large; NH_3, three small and one large; CH_4, four small. Tie the balloons together and ask students to notice the shape of the resultant molecules. BL EL

Math in Chemistry

Geometry Using a molecular model set, construct molecules, beginning with H_2. Ask students why a hydrogen molecule is considered linear. A straight line can be drawn through the two atoms. Next, construct BeH_2, and ask students to determine its shape. linear Construct BH_3, and have students notice that all four atoms lie in the same plane. Cut a triangle from transparency film, place it on top of the molecule, and have students determine the bond angle. 120° The molecular shape is referred to as trigonal planar. Construct a CH_4 molecule, and have students cut four equilateral triangles and assemble them over the molecular model, forming a geometric model. Ask students to describe the shape. tetrahedron Construct NH_3 and have students assemble a geometric shape to encompass the molecule. Tell students the shape is known as trigonal pyramidal. Students should notice that there are no interior points in this molecular shape, and each atom aligns with a vertex. Construct an H_2O molecule, and ask students to determine the shape. bent Point out that this molecular shape is planar, and the atoms form an obtuse angle. OL EL

Extension

Complex Geometry Repeat the activity above with a PCl_5 molecule and the SF_6 molecule. Have students determine the geometric shapes of the molecules. trigonal bipyramidal, octahedral AL EL

■ **Caption Question Fig. 19** four

☑ READING CHECK one

VOCABULARY
WORD ORIGIN
Trigonal planar
comes from the Latin words *trigonum*, which means *triangular*, and *plan-*, which means *flat*

Explore **molecular shapes.**
Virtual Investigations

Connection to Biology The shape of food molecules is important to our sense of taste. The surface of your tongue is covered with taste buds, each of which contains from 50 to 100 taste receptor cells. Taste receptor cells can detect five distinct tastes—sweet, bitter, salty, sour, and umami (the taste of MSG, monosodium glutamate)—but each receptor cell responds best to only one taste.

The shapes of food molecules are determined by their chemical structures. When a molecule enters a taste bud, it must have the correct shape for the nerve in each receptor cell to respond and send a message to the brain. The brain then interprets the message as a certain taste. When such molecules bind to sweet receptors, they are sensed as sweet. The greater the number of food molecules that fit a sweet receptor cell, the sweeter the food tastes. Sugars and artificial sweeteners are not the only sweet molecules. Some proteins found in fruits are also sweet molecules. Some common molecular shapes are illustrated in **Table 6.**

Hybridization

A hybrid occurs when two things are combined and the result has characteristics of both. For example, a hybrid automobile uses both gas and electricity as energy sources. During chemical bonding, different atomic orbitals undergo hybridization. To understand this, consider the bonding involved in the methane molecule (CH_4). The carbon atom has four valence electrons with the electron configuration$[He]2s^22p^2$. You might expect the two unpaired p electrons to bond with other atoms and the 2s electrons to remain an unshared pair. However, carbon atoms undergo **hybridization,** a process in which atomic orbitals mix and form new, identical hybrid orbitals.

The hybrid orbitals in a carbon atom are shown in **Figure 19.** Note that each hybrid orbital contains one electron that it can share with another atom. The hydrid orbital is called an sp^3 orbital because the four hybrid orbitals form from one s orbital and three p orbitals. Carbon is the most common element that undergoes hybridization.

The number of atomic orbitals that mix and form the hybrid orbital equals the total number of pairs of electrons, as shown in **Table 6.** In addition, the number of hybrid orbitals formed equals the number of atomic orbitals mixed. For example, $AlCl^3$ has a total of three pairs of electrons and VSEPR predicts a trigonal planar molecular shape. This shape results when one s and two p orbitals on the central atom, Al, mix and form three identical sp^2 hybrid orbitals.

Lone pairs also occupy hybrid orbitals. Compare the hybrid orbitals of $BeCl_2$ and H_2O in **Table 6.** Both compounds contain three atoms. Why does an H_2O molecule contain sp^3 orbitals? There are two lone pairs on the central oxygen atom in H_2O. Therefore, there must be four hybrid orbitals—two for bonding and two for the lone pairs.

Recall that multiple covalent bonds consist of one sigma bond and one or more pi bonds. Only the two electrons in the sigma bond occupy hybrid orbitals such as sp and sp^2. The remaining unhybridized p orbitals overlap to form pi bonds. It is important to note that single, double, and triple covalent bonds contain only one hybrid orbital. Thus, CO_2, with two double bonds, forms sp hybrid orbitals.

☑ READING CHECK **State** the number of electrons that are available for bonding in a hybrid sp^3 orbital.

■ **Figure 19** A carbon atom's 2s and 2p electrons occupy the hybrid sp^3 orbitals. Notice that the hybrid orbitals have an intermediate amount of potential energy when compared with the energy of the original s and p orbitals. According to VSEPR theory, a tetrahedral shape minimizes repulsion between the hybrid orbitals in a CH_4 molecule.

Identify *How many faces does the tetrahedral shape formed by the sp^3 orbitals have?*

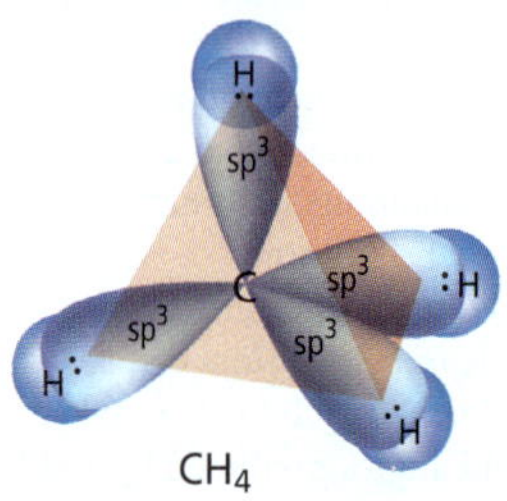

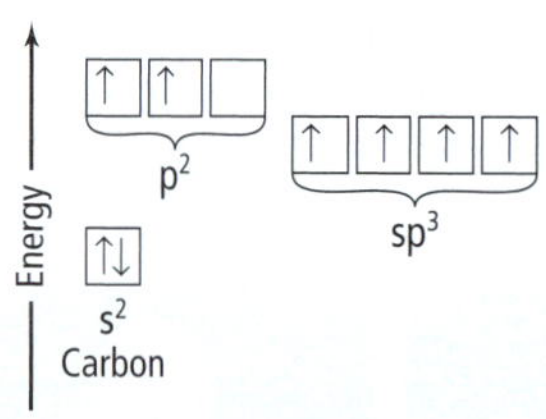

View an **animation about molecular shapes.**
Concepts In Motion

DIFFERENTIATED INSTRUCTION

Advanced Learners Have students research the molecular orbital theory. This theory describes electrons in molecules by using molecular orbitals instead of atomic orbitals. Students should use diagrams and report their findings to the class. AL

Explore **molecular shapes with an interactive table.** Concepts In Motion

Table 6 Molecular Shapes

Molecule	Total Pairs	Shared Pairs	Lone Pairs	Hybrid Orbitals	Molecular Shape*	
$BeCl_2$	2	2	0	sp	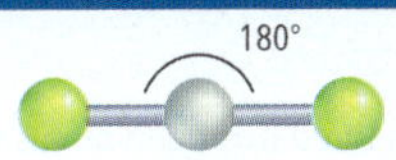 Linear	The $BeCl_2$ molecule contains only two pairs of electrons shared with the central Be atom. These bonding electrons have the maximum separation, a bond angle of 180°, and the molecular shape is linear.
$AlCl_3$	3	3	0	sp^2	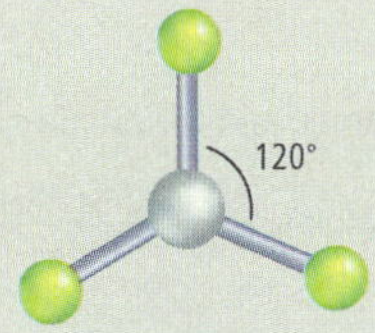 Trigonal planar	The three bonding electron pairs in $AlCl_3$ have maximum separation in a trigonal planar shape with 120° bond angles.
CH_4	4	4	0	sp^3	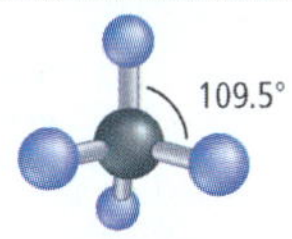 Tetrahedral	When the central atom in a molecule has four pairs of bonding electrons, as CH_4 does, the shape is tetrahedral. The bond angles are 109.5°.
NH_3	4	3	1	sp^3	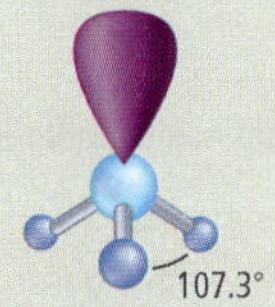 Trigonal pyramidal	NH_3 has three single covalent bonds and one lone pair. The lone pair takes up a greater amount of space than the shared pairs. There is stronger repulsion between the lone pair and the bonding pairs than between two bonding pairs. The resulting geometry is trigonal pyramidal, with 107.3° bond angles.
H_2O	4	2	2	sp^3	 Bent	Water has two covalent bonds and two lone pairs. Repulsion between the lone pairs causes the angle to be 104.5°, less than both tetrahedral and trigonal pyramid. As a result, water molecules have a bent shape.
$NbBr_5$	5	5	0	sp^3d	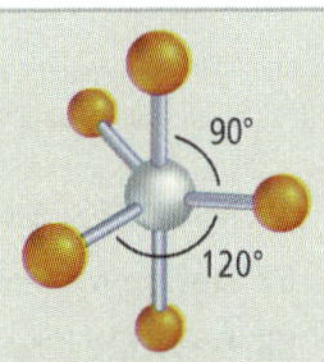 Trigonal bipyramidal	The $NbBr_5$ molecule has five pairs of bonding electrons. The trigonal bipyramidal shape minimizes the repulsion of these shared electron pairs.
SF_6	6	6	0	sp^3d^2	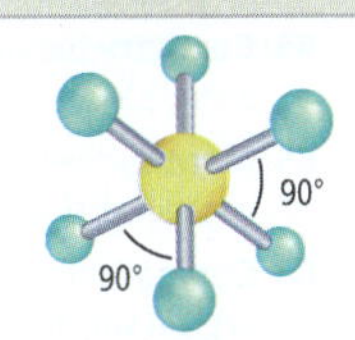 Octahedral	As with $NbBr_5$, SF_6 has no unshared electron pairs on the central atom. However, six shared pairs arranged about the central atom result in an octahedral shape.

*Balls represent atoms, sticks represent bonds, and lobes represent lone pairs of electrons.

CHEMISTRY PROJECT

Sulforaphane The compound sulforaphane is an ingredient in broccoli and has been discovered to be a potent anticancer agent. Have students research sulforaphane ($C_5H_{11}OS$) and draw models to represent this molecule. **AL**

CHEMISTRY JOURNAL

Molecular Shape Have students predict the molecular shape and bond angle of the common molecules CO_2, NO_2, CH_2O, CCl_4, H_2S, and HF. **OL**

ASSESSMENT

Performance Using the VSEPR model, have students build models that have these shapes: linear, trigonal planer, trigonal bipyramidal, and octahedral. **OL**

3 Assess

Check for Understanding

Have students use the Lewis structure for AsH_3 to predict the shape of the molecule, the bond angle, and the hybrid orbitals of arsenic. **OL**

Reteach

Have students use balls of modeling clay to represent nuclei and nonvalence electrons, colored toothpicks to represent bonded electrons, and noncolored toothpicks to represent lone pairs. Construct CH_4, and have students determine the hybrid orbitals of carbon and the shape of the molecule. sp^3, tetrahedral Repeat for NH_3. sp^3, trigonal pyramidal Continue with H_2O. sp^3, bent **OL**

Extension

Ask students to explain how a carbon atom has four bonding sites when its electron configuration of $[He]2s^22p^2$ suggests it should only have two bonding sites. The $2s^22p^2$ orbitals undergo hybridization, forming four sp^3 bonding orbitals.

ASSESSMENT

Performance Have students write the formula of a simple molecule on a sheet of paper. Have another student draw a model of the molecule, predict the geometry, the hybridization, and the bond angle. **OL**

IN-CLASS Example

Question Sulfur dichloride is a cherry-red liquid at room temperature. What is the shape of sulfur dichloride? Determine the bond angle, and identify the type of hybrid.

Answer All atoms have an octet. The shape is bent, with a bond angle of 104.5°, and it is an sp^3 hybrid.

PRACTICE Problems

Have students refer to the Selected Solutions appendix for complete solutions to odd-numbered problems. The complete solutions for all problems can be found in the Solutions Manual.

56. trigonal planar, 120°, sp^2
57. bent, 104.5°, sp^3
58. linear, 180°, sp
59. tetrahedral, 109°, sp^3
60. tetrahedral, 109°, sp^3

EXAMPLE Problem 7

EXAMPLE PROBLEM

FIND THE SHAPE OF A MOLECULE Phosphorus trihydride, a colorless gas, is produced when organic materials, such as fish flesh, rot. What is the shape of a phosphorus trihydride molecule? Predict the bond angle and identify hybrid orbitals.

1 ANALYZE THE PROBLEM

A phosphorus trihydride molecule has three hydrogen atoms bonded to a central phosphorus atom.

2 SOLVE FOR THE UNKNOWN

Find the total number of valence electrons and the number of electron pairs.

$$1 \text{ P atom} \times \frac{5 \text{ valence electrons}}{1 \text{ P atom}} + 3 \text{ H atoms} \times \frac{1 \text{ valence electron}}{1 \text{ F atom}} = 8 \text{ valence electrons}$$

$$\frac{8 \text{ electrons}}{2 \text{ electrons/pair}} = 4 \text{ pairs}$$

Determine the total number of bonding pairs.

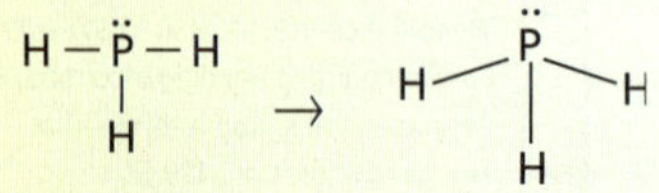

Lewis structure → Molecular shape

Draw the Lewis structure, using one pair of electrons to bond each H atom to the central P atom and assigning the lone pair to the P atom.

The molecular shape is trigonal pyramidal with a predicted 107° bond angle and sp^3 hybrid orbitals.

3 EVALUATE THE ANSWER

All electron pairs are used and each atom has a stable electron configuration.

PRACTICE Problems

Do additional problems. Online Practice

Determine the molecular shape, bond angle, and hybrid orbitals for each molecule.

56. BF_3

57. OCl_2

58. BeF_2

59. CF_4

60. Challenge For a NH_4^+ ion, identify its molecular shape, bond angle, and hybrid orbitals.

SECTION 4 REVIEW

Section Self-Check

Section Summary

- VSEPR model theory states that electron pairs repel each other and determine both the shape of and bond angles in a molecule.
- Hybridization explains the observed shapes of molecules by the presence of equivalent hybrid orbitals.

61. MAINIDEA **Summarize** the VSEPR bonding theory.

62. Define the term *bond angle*.

63. Describe how the presence of a lone electron pair affects the spacing of shared bonding orbitals.

64. Compare the size of an orbital that has a shared electron pair with one that has a lone pair.

65. Identify the type of hybrid orbitals present and bond angles for a molecule with a tetrahedral shape.

66. Compare the molecular shapes and hybrid orbitals of PF_3 and PF_5 molecules. Explain why their shapes differ.

67. List in a table, the Lewis structure, molecular shape, bond angle, and hybrid orbitals for molecules of CS_2, CH_2O, H_2Se, CCl_2F_2, and NCl_3.

SECTION 4 REVIEW

61. VSEPR theory determines molecular geometry based on the repulsive nature of electron pairs around a central atom.
62. The bond angle is the angle formed by any two terminal atoms and the central atom.
63. A lone pair occupies more space than a shared electron pair, thus, the presence of a lone pair pushes the bonding pairs closer together.
64. The orbital containing a lone electron pair occupies more space than a shared electron pair.
65. sp^3 and 109°
66. PF_3 is trigonal pyramidal with sp^3 hybrid orbitals. PF_5 is trigonal bipyramidal with sp^3d hybrid orbitals. Shape is determined by the type of hybrid orbital.
67. Refer to the Solutions Manual for Lewis structures.
    CS_2: linear, 180°, sp
    CH_2O: trigonal planar, 120°, sp^2
    H_2Se: bent, 104.5°, sp^3
    CCl_2F_2: tetrahedral, 109°, sp^3
    NCl_3: trigonal pyramidal, 107°, sp^3

SECTION 5

Essential Questions

- How is electronegativity used to determine bond type?
- How do polar and nonpolar covalent bonds and polar and nonpolar molecules compare? How do they contrast?
- What are the characteristics of covalently bonded compounds?

Review Vocabulary

electronegativity: the relative ability of an atom to attract electrons in a chemical bond

New Vocabulary

polar covalent bond

Electronegativity and Polarity

MAINIDEA A chemical bond's character is related to each atom's attraction for the electrons in the bond.

CHEM 4 YOU **The stronger you are, the more easily you can do pull-ups. Just as people have different abilities for doing pull-ups, atoms in chemical bonds have different abilities to attract (pull) electrons.**

Electronegativity and Bond Character

The type of bond formed during a reaction is related to each atom's attraction for electrons. Electron affinity is a measure of the tendency of an atom to accept an electron. Excluding noble gases, electron affinity increases with increasing atomic number within a period and decreases with increasing atomic number within a group. The scale of electronegativities—shown in **Figure 20**—allows chemists to evaluate the electron affinity of specific atoms in a compound. Recall that electronegativity indicates the relative ability of an atom to attract electrons in a chemical bond. Note that electronegativity values were assigned, whereas electron affinity values were measured.

Electronegativity The version of the periodic table of the elements shown in **Figure 20** lists electronegativity values. Note that fluorine has the greatest electronegativity value (3.98), while francium has the least (0.7). Because noble gases do not generally form compounds, individual electronegativity values for helium, neon, and argon are not listed. However, larger noble gases, such as xenon, sometimes bond with highly electronegative atoms, such as fluorine.

Figure 20 Electronegativity values are derived by comparing an atom's attraction for shared electrons to that of a fluorine atom's attraction for shared electrons. Note that the electronegativity values for the lanthanide and actinide series, which are not shown, range from 1.12 to 1.7.

Electronegativity Values for Selected Elements

Legend: Metal, Metalloid, Nonmetal

1 H 2.20																
3 Li 0.98	4 Be 1.57											5 B 2.04	6 C 2.55	7 N 3.04	8 O 3.44	9 F 3.98
11 Na 0.93	12 Mg 1.31											13 Al 1.61	14 Si 1.90	15 P 2.19	16 S 2.58	17 Cl 3.16
19 K 0.82	20 Ca 1.00	21 Sc 1.36	22 Ti 1.54	23 V 1.63	24 Cr 1.66	25 Mn 1.55	26 Fe 1.83	27 Co 1.88	28 Ni 1.91	29 Cu 1.90	30 Zn 1.65	31 Ga 1.81	32 Ge 2.01	33 As 2.18	34 Se 2.55	35 Br 2.96
37 Rb 0.82	38 Sr 0.95	39 Y 1.22	40 Zr 1.33	41 Nb 1.6	42 Mo 2.16	43 Tc 2.10	44 Ru 2.2	45 Rh 2.28	46 Pd 2.20	47 Ag 1.93	48 Cd 1.69	49 In 1.78	50 Sn 1.96	51 Sb 2.05	52 Te 2.1	53 I 2.66
55 Cs 0.79	56 Ba 0.89	57 La 1.10	72 Hf 1.3	73 Ta 1.5	74 W 1.7	75 Re 1.9	76 Os 2.2	77 Ir 2.2	78 Pt 2.2	79 Au 2.4	80 Hg 1.9	81 Tl 1.8	82 Pb 1.8	83 Bi 1.9	84 Po 2.0	85 At 2.2
87 Fr 0.7	88 Ra 0.9	89 Ac 1.1														

Video Lab

Video

Chemical Bonding

SECTION 5

1 Focus

MAINIDEA

Electronegativity and Polarity Have two students help with a quick demo. Ask both students to pull on a rope with equal strength. Tell the class the rope represents a shared pair of electrons. Ask the class what this represents when atoms share electrons. equal sharing of electrons Ask one student to pull harder than the other student. The second student should be pulled toward the first student. Ask the class what this represents when atoms share electrons. unequal sharing of electrons Have students identify which atoms have a greater tendency to gain electrons. the ones with the greatest pull on the electrons Ask students what type of bond is represented if the electron is completely pulled away from one atom. an ionic bond OL

2 Teach

Quick Demo

Polar Water Rub a rubber rod with wool or silk to induce a charge and bring it near a stream of water from an opened buret. Have students observe the water move toward the charged rod. Repeat the procedure using paint thinner. nothing happens Have students try to explain the results. When polarity is discussed in this section, refer to this Quick Demo and point out that water is a polar molecule, whereas paint thinner is nonpolar. Partial charges on water molecules are attracted to the rubber rod, which has a negative charge. OL

Identify Misconceptions

A common misconception is that the bond between two atoms is purely covalent or purely ionic.

Uncover the Misconception

Have students write analogies to represent nonpolar covalent bonds, polar covalent bonds, and ionic bonds.

Demonstrate the Concept

Have students imagine they are eating at a restaurant with another person and each orders a different type of sandwich to be shared. If they share equally, what would they do? They would cut the sandwich in half, and each would get a half. Ask students what unequal sharing would be. The shared portions are not equal. If one person eats his or her entire sandwich and then takes some of the other person's sandwich, what bond type does this represent? ionic

Assess New Knowledge

Using the graph of electronegativities and percent ionic character in Figure 21, have students determine the bond type of the several pairs of atoms. Ask them to also identify which atom is more likely to have the electron. H and Br polar covalent, Br; C and O polar covalent, O; Li and F ionic, F; O and O nonpolar covalent, the electrons are shared equally. **OL**

In-Text Question

What percent ionic character is a bond between two atoms that have an electronegativity difference of 2.00? About 60%

Where would LiBr be plotted on the graph? just to the left of NaBr on the curve, at an electronegativity difference of 1.98

READING CHECK 0%

GRAPH CHECK approximately 74%

Table 7 EN Difference and Bond Character

Electronegativity Difference	Bond Character
> 1.7	mostly ionic
0.4 – 1.7	polar covalent
< 0.4	mostly covalent
0	nonpolar covalent

Bond character A chemical bond between atoms of different elements is never completely ionic or covalent. The character of a bond depends on how strongly each of the bonded atoms attracts electrons. As shown in **Table 7,** the character and type of a chemical bond can be predicted using the electronegativity difference of the elements that bond. Electrons in bonds between identical atoms have an electronegativity difference of zero—meaning that the electrons are equally shared between the two atoms. This type of bond is considered nonpolar covalent, or a pure covalent bond. On the other hand, because different elements have different electronegativities, the electron pairs in a covalent bond between different atoms are not shared equally. Unequal sharing results in a **polar covalent bond.** When there is a large difference in the electronegativity between bonded atoms, an electron is transferred from one atom to the other, which results in bonding that is primarily ionic.

Watch a **video about chemical bonding.** Video

Bonding is not often clearly ionic or covalent. An electronegativity difference of 1.70 is considered 50 percent covalent and 50 percent ionic. As the difference in electronegativity increases, the bond becomes more ionic in character. Generally, ionic bonds form when the electronegativity difference is greater than 1.70. However, this cutoff is sometimes inconsistent with experimental observations of two nonmetals bonding together. **Figure 21** summarizes the range of chemical bonding between two atoms. What percent ionic character is a bond between two atoms that have an electronegativity difference of 2.00? Where would LiBr be plotted on the graph?

READING CHECK Analyze What is the percent ionic character of a pure covalent bond?

Figure 21 This graph shows that the difference in electronegativity between bonding atoms determines the percent ionic character of the bond. Above 50% ionic character, bonds are mostly ionic.

GRAPH CHECK
Determine the percent ionic character of calcium oxide.

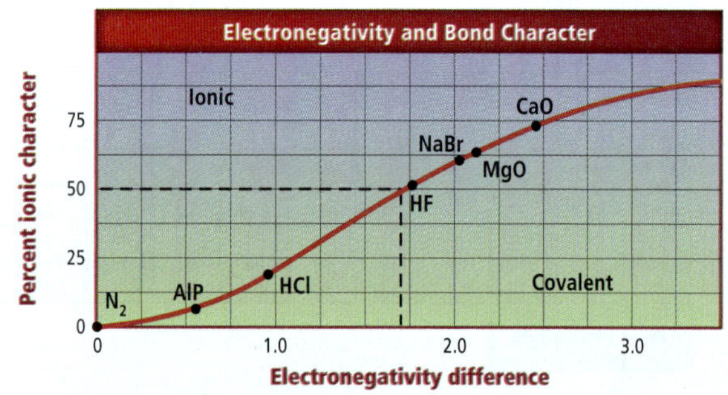

DIFFERENTIATED INSTRUCTION

Struggling Learners Have a simulated tug-of-war with seven people on one side and one on the other to represent transfer of electrons. Have two equally sized people represent a nonpolar molecule. Finally, have one large person and one small person represent a polar bond. **BL** **EL**

Electronegativity	Cl	= 3.16
Electronegativity	H	= 2.20
Difference		= 0.96

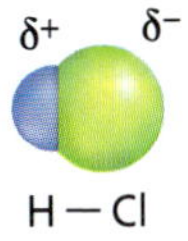

View an **animation about bond types.**

Concepts In Motion

■ **Figure 22** Chlorine's electronegativity is higher than that of hydrogen. Therefore, in a molecule containing hydrogen and chlorine, the shared pair of electrons is with the chlorine atom more often than it is with the hydrogen atom. Symbols are used to indicate the partial charge at each end of the molecule from this unequal sharing of electrons.

Polar Covalent Bonds

As you just learned, polar covalent bonds form because not all atoms that share electrons attract them equally. A polar covalent bond is similar to a tug-of-war in which the two teams are not of equal strength. Although both sides share the rope, the stronger team pulls more of the rope toward its side. When a polar bond forms, the shared electron pair or pairs are pulled toward one of the atoms. Thus, the electrons spend more time around that atom than the other atom. This results in partial charges at the ends of the bond.

The Greek letter delta (δ) is used to represent a partial charge. In a polar covalent bond, δ^- represents a partial negative charge and δ^+ represents a partial positive charge. As shown in **Figure 22,** δ^- and δ^+ can be added to a molecular model to indicate the polarity of the covalent bond. The more-electronegative atom is at the partially negative end, while the less-electronegative atom is at the partially positive end. The resulting polar bond often is referred to as a dipole (two poles).

Molecular polarity Covalently bonded molecules are either polar or nonpolar; which type depends on the location and nature of the covalent bonds in the molecule. A distinguishing feature of nonpolar molecules is that they are not attracted by an electric field. Polar molecules, however, are attracted by an electric field. Because polar molecules are dipoles with partially charged ends, they have an uneven electron density. This results in the tendency of polar molecules to align with an electric field.

Polarity and molecular shape You can learn why some molecules are polar and some are not by comparing water (H_2O) and carbon tetrachloride (CCl_4) molecules. Both molecules have polar covalent bonds. According to the data in **Figure 20,** the electronegativity difference between a hydrogen atom and an oxygen atom is 1.24. The electronegativity difference between a chlorine atom and a carbon atom is 0.61. Although these electronegativity differences vary, a H—O bond and a C—Cl bond are considered to be polar covalent.

$$\overset{\delta^+}{\text{H}}\text{—}\overset{\delta^-}{\text{O}} \qquad \overset{\delta^+}{\text{C}}\text{—}\overset{\delta^-}{\text{Cl}}$$

According to their molecular formulas, both molecules have more than one polar covalent bond. However, only the water molecule is polar. Why might one molecule with polar covalent bonds be polar, while a second molecule with polar covalent bonds is nonpolar? The answer lies in the shapes of the molecules.

CAREERS IN CHEMISTRY

Flavor Chemist A flavor chemist, or flavorist, must know how chemicals react and change in different conditions. A degree in chemistry is an asset, but is not required. Most flavorists work for companies that supply flavors to the food and beverage industries. A certified flavorist trains for five years in a flavor laboratory, passes an oral examination, and then works under supervision for another two years.

WebQuest

Quick Demo

Polar Molecules Place two or three stirring bars with non-stick coatings on a flat surface. The stirring bars should be spaced away from each other and have random orientations. Bring a magnet near the stirring bars. The stirring bars will align themselves with the magnetic field. Tell students that a polar molecule behaves much the same way in an electric field because it has a positive pole and a negative pole.

CAREERS IN CHEMISTRY

WebQuest

ASSESSMENT

Performance Show students molecular models of common molecules. Have them determine the bond type present and then determine the type of molecule based on its symmetry. The molecules SO_2, CS_2, CH_3Cl, CH_2O, HCl, and H_2S can be used. OL

CHEMISTRY JOURNAL

Bond Type Using the electronegativity difference, have students determine if the bonds in the following molecules are polar covalent, nonpolar covalent, or ionic: H_2S, CH_4, NH_3, KCl, N_2, CaO, CO_2, HCl. KCl and CaO are ionic, N_2 is nonpolar covalent, and the rest are polar covalent. OL

CHEMISTRY PROJECT

Chromatography Have student groups research the separation technique known as paper chromatography. Ask them to describe how polarity differences are used to separate a mixture of substances. Students should outline a lab activity to show how chromatography works and perform the lab for the class. OL

Quick Demo

Polarity Made Apparent
In four labeled beakers, put 100 mL of methanol, ethanol, 2-propanol (isopropyl alcohol) and acetone. To each beaker, add a few crystals of Reichardt's dye and stir. Reichardt's dye is a compound that indicates the polarity of solvents by color. It should be green-blue in methanol, blue-violet in ethanol, red-violet in 2-propanol, and orange in acetone. As the polarity of the solvent increases, the light absorbed by the dye changes. Yellow is most polar and green-blue is the least polar.

Caption Question Fig. 24
No, detergents must be used because the polar water molecules cannot dissolve the nonpolar oil molecules.

Figure 23 A molecule's shape determines its polarity.

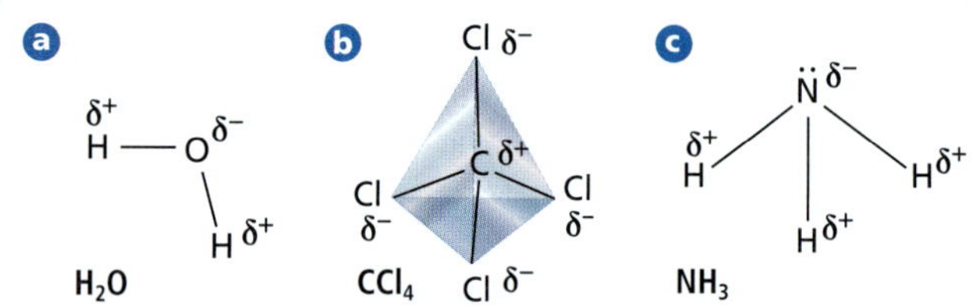

a. The bent shape of a water molecule makes it polar.

b. The symmetry of a CCl_4 molecule results in an equal distribution of charge, and the molecule is nonpolar.

c. The asymmetric shape of an ammonia molecule results in an unequal charge distribution and the molecule is polar.

The shape of an H_2O molecule, as determined by VSEPR, is bent because the central oxygen atom has lone pairs of electrons, as shown in **Figure 23a.** Because the polar H—O bonds are asymmetric in a water molecule, the molecule has a definite positive end and a definite negative end. Thus, it is polar.

A CCl_4 molecule is tetrahedral, and therefore, symmetrical, as shown in **Figure 23b.** The electric charge measured at any distance from its center is identical to the charge measured at the same distance to the opposite side. The average center of the negative charge is located on each chlorine atom. The positive center is also located on the carbon atom. Because the partial charges are balanced, CCl_4 is a nonpolar molecule. Note that symmetric molecules are usually nonpolar, and molecules that are asymmetric are polar as long as the bond type is polar.

Is the molecule of ammonia (NH_3), shown in **Figure 23c,** polar? It has a central nitrogen atom and three terminal hydrogen atoms. Its shape is trigonal pyramidal because of the lone pair of electrons present on the nitrogen atom. Using **Figure 20,** you can find that the electronegativity difference of hydrogen and nitrogen is 0.84, making each N—H bond polar covalent. The charge distribution is unequal because the molecule is asymmetric. Thus, the molecule is polar.

Solubility of polar molecules The physical property known as solubility is the ability of a substance to dissolve in another substance. The bond type and the shape of the molecules present determine solubility. Polar molecules and ionic compounds are usually soluble in polar substances, but nonpolar molecules dissolve only in nonpolar substances, as shown in **Figure 24.**

Figure 24 Symmetric covalent molecules, such as oil and most petroleum products, are nonpolar. Asymmetric molecules, such as water, are usually polar. As shown in this photo, polar and nonpolar substances usually do not mix.
Infer *Will water alone clean oil from a fabric?*

Tony Craddock/Photo Researchers

CHEMISTRY JOURNAL

Like Dissolves Like Add 20 mL of carbon tetrachloride (CCl_4) and 20 mL of water to a beaker. The liquids form layers and do not mix. CCl_4 is more dense than water and forms the lower layer. In a second beaker, add 20 ml of CCl_4 only. In a third beaker, add 20 mL of water only. Add a few crystals of iodine and and a few crystals of potassium permanganate to the two beakers. The iodine dissolves in the nonpolar solvent CCl_4. The potassium permanganate dissolved in the polar water. Water is polar and attracts the ions present in the ionic compound potassium permanganate. Have students draw a water molecule and identify the portion of the molecule that attracts the potassium ion and the portion that attracts the permanganate ion. The central oxygen atom has a partial negative charge and attracts the potassium ion. The two hydrogen atoms create a partial positive region that attracts the permanganate ion. OL

Properties of Covalent Compounds

Table salt, an ionic solid, and table sugar, a covalent solid, are similar in appearance. However, these compounds behave differently when heated. Salt does not melt, but sugar melts at a relatively low temperature. Does the type of bonding in a compound affect its properties?

Explore **covalent compounds.**

Virtual Investigations

Intermolecular forces Differences in properties are a result of differences in attractive forces. In a covalent compound, the covalent bonds between atoms in molecules are strong, but the attraction forces between molecules are relatively weak. These weak attraction forces are known as intermolecular forces, or van der Waals forces. Intermolecular forces vary in strength but are weaker than the bonds that join atoms in a molecule or ions in an ionic compound.

There are different types of intermolecular forces. Between nonpolar molecules, the force is weak and is called a dispersion force, or induced dipole. The force between oppositely charged ends of two polar molecules is called a dipole-dipole force. The more polar the molecule, the stronger the dipole-dipole force. The third force, a hydrogen bond, is especially strong. It forms between the hydrogen end of one dipole and a fluorine, oxygen, or nitrogen atom on another dipole. You will study intermolecular forces in more detail when you study states of matter.

Data Analysis LAB

Based on Real Data*

Interpret Data

How does the polarity of the mobile phase affect chromatograms? Chromatography is a technique in which a moving phase transports and separates the components of a mixture. A chromatograph is created by recording the intensity of each component carried in the moving phase versus time. The peak areas on the chromatograph indicate the amount of each component present in the mixture.

High-performance liquid chromatography, or HPLC, is used by analytical chemists to separate mixtures of solutes. During HPLC, components that are strongly attracted to the extracting solvent are retained longer by the moving phase and tend to appear early on a chromatograph. Several scientists performed HPLC using a methanol-water mixture as the extracting solvent to separate a phenol-benzoic acid mixture. Their results are shown in the graph to the right.

Think Critically

1. **Explain** the different retention times shown on the chromatograms.
2. **Infer** from the graph the component, phenol or benzoic acid, that is in excess. Explain your answer.

Data and Observations

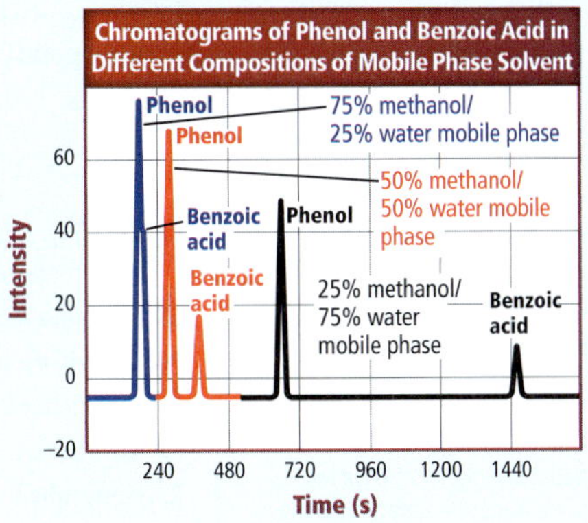

*Data obtained from: Joseph, Seema M. and Palasota, John A. 2001. The combined effects of pH and percent methanol on the HPLC separation of benzoic acid and phenol. *Journal of Chemical Education* 78:1381.

3. **Infer** which component of the mixture has more polar molecules.
4. **Determine** the most effective composition of the mobile phase (of those tested) for separating phenol from benzoic acid. Explain.

CULTURAL DIVERSITY

Medical Research The use of herbal remedies has been well documented in China over the past 2500 years. Western scientists and pharmaceutical companies are looking at covalent compounds contained in these remedies as sources of possible drugs for the treatment of cancer and AIDS. The National Cancer Institute recently compiled data that catalogs the medicinal effects of Chinese herbs. In 1992, the National Institute of Health created an Office of Alternative Medicine to evaluate a variety of medical therapies, including five related to Chinese herbal medicine. Several U.S. pharmaceutical companies have formed over the past few years to develop drugs from Chinese plants.

ChemLAB

The ChemLab located at the end of the chapter can be used at this point in the lesson.

Data Analysis LAB

About the Lab

- Phenol and benzoic acid have similar structures as shown below.

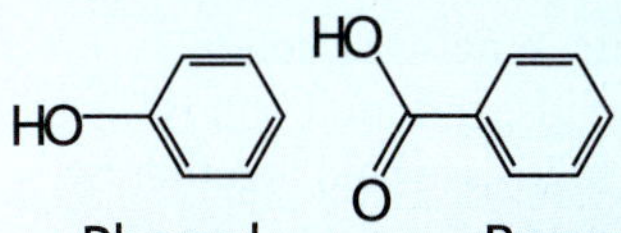

Phenol Benzoic acid

- For additional details about original experiment see: Orna, Mary V. Chemistry, Color, and Art. *Journal of Chemical Education.* 78, no. 10 (2001).
- Note that both the peak intensities and the time elapsed before each peaks occurs (for phenol and benzoic acid) become more equal as the concentration of water in the mobile phase decreases.

Think Critically

1. Phenol has a lower retention time relative to benzoic acid because it has a greater attraction to the mobile phase compared to benzoic acid. Because both phenol and benzoic acid are organic compounds, they are considerably more soluble in methanol (a polar organic molecule). Therefore, their retention times decrease as percent methanol increases.
2. Phenol is in excess because it has the largest peak. The area under the peak corresponds to the amount of the component present in the mixture.
3. Phenol is the more polar compound. It has the greater affinity for the polar solvent.
4. Answers and reasoning will vary. A 50% methanol/50% water mobile-phase solvent mixture extracts more of the mixture components (phenol and benzoic acid) than the other mixtures, has good separation between the two peak intensities, and requires less time.

3 Assess

Check for Understanding

Ask students to compare the characteristics of ionic solids to those of covalent and network solids. **OL**

Reteach

Have students make a chart and discuss the properties of ionic, metallic, covalent, and network solids. Point out that a high melting point is a result of a three-dimensional attraction of an atom or ion for electrons. This occurs in network solids, metallic solids, and ionic solids. Covalent solids are held together by intermolecular attractions and have lower melting points. **OL**

Extension

Have students investigate intermolecular bonds. Students should use drawings to represent each type of van der Waals force and give an example. Display students' work. **OL**

ASSESSMENT

Knowledge Write several chemical formulas on the board. Have students predict the general physical properties of each compound based on its bonding. **OL**

Figure 25 Network solids are often used in cutting tools because of their extreme hardness. Here, a diamond-tipped saw blade cuts through stone.

Forces and properties The properties of covalent molecular compounds are related to the relatively weak intermolecular forces holding the molecules together. These weak forces result in the relatively low melting and boiling points of molecular substances compared with those of ionic substances. That is why, when heated moderately, sugar melts but salt does not.

Weak intermolecular forces also explain why many molecular substances exist as gases or vaporize readily at room temperature. Oxygen (O_2), carbon dioxide (CO_2), and hydrogen sulfide (H_2S) are examples of covalent compounds that are gases at room temperature. Because the hardness of a substance depends on the intermolecular forces between individual molecules, many covalent molecules are relatively soft solids. Paraffin, found in candles and other products, is a common example of a covalent solid.

In the solid phase, molecules align to form a crystal lattice. This molecular lattice is similar to that of an ionic solid, but with less attraction between particles. The structure of the lattice is affected by molecular shape and the type of intermolecular force. Most molecular information has been determined by studying molecular solids.

Covalent Network Solids

There are some solids, often called covalent network solids, that are composed only of atoms interconnected by a network of covalent bonds. Quartz and diamond are two common examples of network solids. In contrast to molecular solids, network solids are typically brittle, nonconductors of heat or electricity, and extremely hard. Analyzing the structure of a diamond explains some of its properties. In a diamond, each carbon atom is bonded to four other carbon atoms. This tetrahedral arrangement, which is shown in **Figure 25,** forms a strongly bonded crystal system that is extremely hard and has a very high melting point.

SECTION 5 REVIEW

Section Self-Check

Section Summary

- The electronegativity difference determines the character of a bond between atoms.
- Polar bonds occur when electrons are not shared equally forming a dipole.
- The spatial arrangement of polar bonds in a molecule determines the overall polarity of a molecule.
- Molecules attract each other by weak intermolecular forces. In a covalent network solid, each atom is covalently bonded to many other atoms.

68. MAINIDEA **Summarize** how electronegativity difference is related to bond character.
69. **Describe** a polar covalent bond.
70. **Describe** a polar molecule.
71. **List** three properties of a covalent compound in the solid phase.
72. **Categorize** bond types using electronegativity difference.
73. **Generalize** Describe the general characteristics of covalent network solids.
74. **Predict** the type of bond that will form between the following pair of atoms:
    - **a.** H and S
    - **b.** C and H
    - **c.** Na and S
75. **Identify** each molecule as polar or nonpolar: SCl_2, CS_2, and CF_4.
76. **Determine** whether a compound made of hydrogen and sulfur atoms is polar or nonpolar.
77. **Draw** the Lewis structures for the molecules SF_4 and SF_6. Analyze each structure to determine whether the molecule is polar or nonpolar.

David Gowans/Alamy

SECTION 5 REVIEW

68. The greater the EN difference, the greater the ionic nature of the bond.
69. A polar covalent bond has unequal sharing of electrons. The electrons are pulled toward one of the atoms, generating partial charges on the ends.
70. It has a greater electron density on one side of the molecule.
71. The solid state of a molecule is crystalline. A molecular solid is a nonconductor, is soft, and has a low melting point.
72. If the difference is zero, the bond is considered nonpolar covalent; if between 0.4 and 1.7, polar covalent; if greater than 1.7, ionic.
73. brittle, nonconductors of heat and electricity, extremely hard
74. **a.** mostly covalent **b.** mostly covalent **c.** polar covalent
75. SCl_2, polar; CS_2, nonpolar; CF_4, nonpolar
76. polar; because it is asymmetric
77.

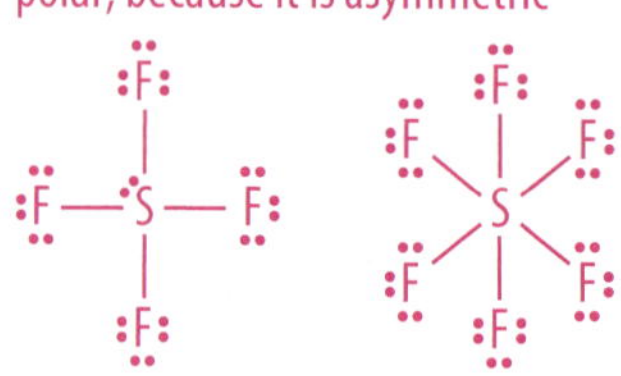

polar nonpolar

STICKY FEET: How Geckos Grip

For a gecko, hanging from a wall or a ceiling is no great feat. The key to a gecko's amazing grip is found on each of its toes. Researchers have determined that a gecko's grip depends on the sticking power of atoms themselves.

1 Gecko toe The bottom of a gecko's toe is covered with millions of tiny hairs, called setae, arranged in rows.

2 Setae Setae are complex structures. The end of each seta has microscopic branches called spatulae.

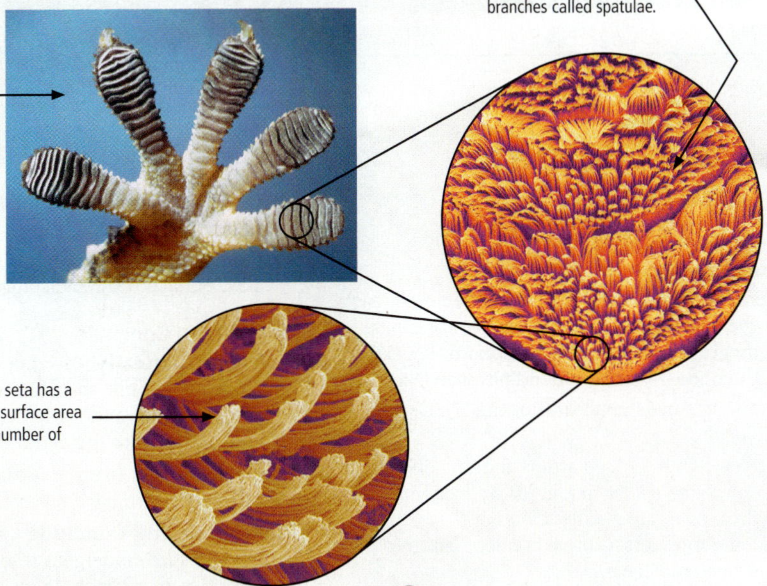

3 Spatulae Each seta has a relatively enormous surface area because of its vast number of spatulae.

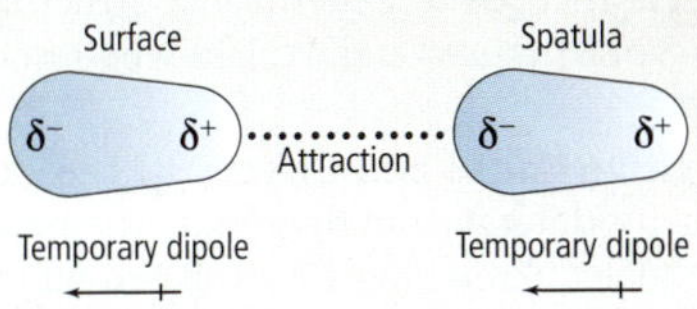

4 Sticking Van der Waals forces form between a surface and a gecko's spatulae. When multiplied by the spatulae's vast surface area, the sum of the weak van der Waals forces is more than enough to balance the pull of gravity and hold a gecko in place.

5 Letting go A gecko simply curls its toes when it wants to move. This reduces the amount of surface contact and the van der Waals forces, and the gecko loses its grip.

WRITING IN ▶ Chemistry

Invent Using their knowledge of how geckos stick to surfaces, scientists are developing applications for geckolike materials. Some possible applications include mini-robots that climb walls and tape that sticks even under water. What uses for a new sticky geckolike material can you think of?

WebQuest

WRITING IN ▶ Chemistry

Rubric WebQuest

Invent Some other applications that have been discussed include less painful adhesive bandages, super-grip athletic equipment, and sticky tools for nanosurgery.

HOW IT works

Purpose

Students will learn how van der Waals forces provide the sticking power for geckos' toes.

Background

Gecko toes look like suction cups, but they do not function that way. Suction was ruled out as an explanation for how they stick to surfaces by testing gecko feet in a vacuum. Other ideas, such as static cling or water-based attraction, have also been ruled out by experiment.

Teaching Strategies

The diagram has five numbered text blocks that explain the source of the gecko's phenomenal sticking power.

- Number 1 shows a gecko's toe magnified, showing rows containing hundreds of thousands of tiny hairs, or setae.
- Numbers 2 and 3 explain how the setae branch out into even smaller structures called spatulae. The angle at which setae meet the surface is an important factor in determining the holding power.
- Number 4 explains how large surface area and van der Waals forces act to create a significant gripping strength. Van der Waals forces are intermolecular forces; that is, they act between molecules.
- Number 5 explains how a gecko lets go by curling its toes. Curling the toes reduces the number of setae making contact with the surface at the optimal angle. This results in decreased van der Waals forces and reduced gripping strength.

ChemLAB

Rubric

Preparation

Time Allotment one class period

Process Skills applying concepts, comparing and contrasting, formulating models

Safety Precautions Approve lab safety forms before work begins.

Procedure

- Prominently display the color code key used to identify atoms of each element.
- Review the molecular geometry examples in Table 6 with students prior to beginning the lab, or ask them to review the table on their own before beginning.

Analyze and Conclude

1. increasing flexibility: triple, double, single; increasing strength: single, double, triple
2. The H_2O molecule has two bonds and two lone pairs around the central atom. The lone pairs take up space around the central atom and repel the bonding electrons, causing the bent shape. The CO_2 molecule has two double bonds with no lone pairs. The bonding electrons repel to form the linear shape, which maximizes the distance between electron densities.
3. The SO_3 molecule exhibits resonance. It has a central S and three terminal O atoms. One of the terminal O atoms forms a double bond. Three resonance structures exist, one for each possible location of the double bond.
4. The following molecules are polar: H_2O, PH_3, HCN, and CO. All others are nonpolar.

ChemLAB

Model Molecular Shapes

Background: **Covalent bonding occurs when atoms share valence electrons. In the Valence Shell Electron Pair Repulsion (VSEPR) theory, the way in which valence electrons of bonding atoms are positioned is the basis for predicting a molecule's shape. This method of visualizing shape is also based on the molecule's Lewis structure.**

Question: *How do the Lewis structure and the positions of valence electrons affect the shape of the covalent compound?*

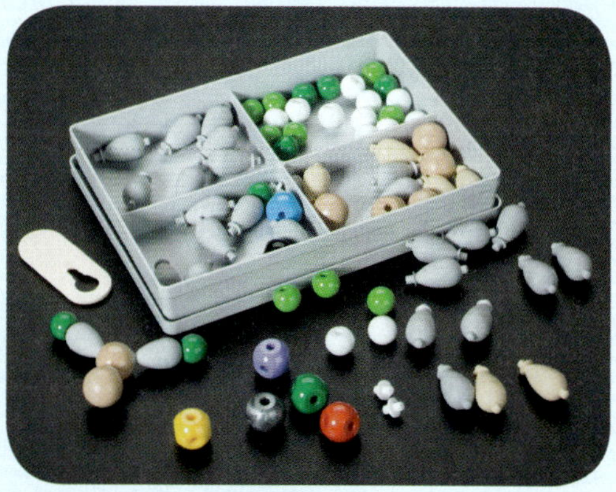

Matt Meadows

Materials

molecular model kit

Safety Precautions

Procedure

1. Read and complete the lab safety form.
2. Create a table to record your data.
3. Note and record the color used to represent each of the following atoms in the molecular model kit: hydrogen (H), oxygen (O), phosphorus (P), carbon (C), fluorine (F), sulfur (S), and nitrogen (N).
4. Draw the Lewis structures of the H_2, O_2, and N_2 molecules.
5. Obtain two hydrogen atoms and one connector from the molecular model kit, and assemble a hydrogen (H_2) molecule. Observe that your model represents a single-bonded diatomic hydrogen molecule.
6. Obtain two oxygen atoms and two connectors from the molecular model kit, and assemble an oxygen (O_2) molecule. Observe that your model represents a double-bonded diatomic oxygen molecule.
7. Obtain two nitrogen atoms and three connectors from the molecular model kit, and assemble one nitrogen (N_2) molecule. Observe that your model represents a triple-bonded diatomic nitrogen molecule.
8. Recognize that diatomic molecules such as those formed in this lab are always linear. Diatomic molecules are made up of only two atoms and two points (atoms) can only be connected by a straight line.
9. Draw the Lewis structure of water (H_2O), and construct its molecule.
10. Classify the shape of the H_2O molecule using information in **Table 6.**
11. Repeat Steps 9 and 10 for the PH_3, CF_4, CO_2, SO_3, HCN, and CO molecules.

Analyze and Conclude

1. **Think Critically** Based on the molecular models you built and observed in this lab, rank single, double, and triple bonds in order of increasing flexibility and increasing strength.
2. **Observe and Infer** Explain why H_2O and CO_2 molecules have different shapes.
3. **Analyze and Conclude** One of the molecules from this lab undergoes resonance. Identify the molecule that has three resonance structures, draw the structures, and explain why resonance occurs.
4. **Recognize Cause and Effect** Use the electronegativity difference to determine the polarity of the molecules in Steps 9–11. Based on their calculated bond polarities and the models constructed in this lab, determine the molecular polarity of each structure.

INQUIRY EXTENSION

Model Use a molecular model kit to build the two resonance structures of ozone (O_3). Then, use Lewis structures to explain how you can convert between the two resonance structures by interchanging a lone pair for a covalent bond.

INQUIRY EXTENSION

Students should assemble two models, each with a central oxygen atom joined to two terminal oxygen atoms. One terminal atom is joined by a single bond, the other by a double bond. The location of these two bonds determines the two resonance structures. The Lewis structures should show that you can convert between the two resonance structures by swapping the position of a lone pair and a covalent bond.

CHAPTER 8 STUDY GUIDE

Vocabulary Practice

BIGIDEA Covalent bonds form when atoms share electrons.

SECTION 1 The Covalent Bond

MAINIDEA Atoms gain stability when they share electrons and form covalent bonds.

- Covalent bonds form when atoms share one or more pairs of electrons.
- Sharing one pair, two pairs, and three pairs of electrons forms single, double, and triple covalent bonds, respectively.
- Orbitals overlap directly in sigma bonds. Parallel orbitals overlap in pi bonds. A single covalent bond is a sigma bond but multiple covalent bonds are made of both sigma and pi bonds.
- Bond length is measured nucleus-to-nucleus. Bond dissociation energy is needed to break a covalent bond.

VOCABULARY
- covalent bond
- molecule
- Lewis structure
- sigma bond
- pi bond
- endothermic reaction
- exothermic reaction

SECTION 2 Naming Molecules

MAINIDEA Specific rules are used when naming binary molecular compounds, binary acids, and oxyacids.

- Names of covalent molecular compounds include prefixes for the number of each atom present. The final letter of the prefix is dropped if the element name begins with a vowel.
- Molecules that produce H^+ in solution are acids. Binary acids contain hydrogen and one other element. Oxyacids contain hydrogen and an oxyanion.

VOCABULARY
- oxyacid

SECTION 3 Molecular Structures

MAINIDEA Structural formulas show the relative positions of atoms within a molecule.

- Different models can be used to represent molecules.
- Resonance occurs when more than one valid Lewis structure exists for the same molecule.
- Exceptions to the octet rule occur in some molecules.

VOCABULARY
- structural formula
- resonance
- coordinate covalent bond

SECTION 4 Molecular Shapes

MAINIDEA The VSEPR model is used to determine molecular shape.

- VSEPR model theory states that electron pairs repel each other and determine both the shape of and bond angles in a molecule.
- Hybridization explains the observed shapes of molecules by the presence of equivalent hybrid orbitals.

VOCABULARY
- VSEPR model
- hybridization

SECTION 5 Electronegativity and Polarity

MAINIDEA A chemical bond's character is related to each atom's attraction for the electrons in the bond.

- The electronegativity difference determines the character of a bond between atoms.
- Polar bonds occur when electrons are not shared equally forming a dipole.
- The spatial arrangement of polar bonds in a molecule determines the overall polarity of a molecule.
- Molecules attract each other by weak intermolecular forces. In a covalent network solid, each atom is covalently bonded to many other atoms.

VOCABULARY
- polar covalent bond

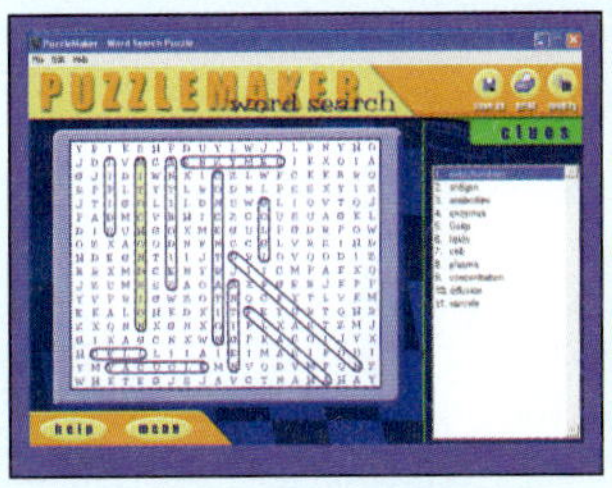

Vocabulary Practice

For additional practice with vocabulary, have students access the Vocabulary PuzzleMaker.

Using the Vocabulary

To reinforce chapter vocabulary, have students write a sentence using each term. OL EL

Review Strategies

- Have students name and write formulas and Lewis structures for molecular compounds and acids. OL
- Have students identify the hybrid orbitals present in a molecule.
- Discuss resonance hybrids and exceptions to the octet rule. OL
- Have students determine molecular shapes and bond angles. OL
- Have students use electronegativity difference to determine bond type and polarity. OL

Students can review with these online resources.

Vocabulary Practice

Vocabulary eGames
Vocabulary eFlashcards
Vocabulary PuzzleMaker

Section Self-Check

Chapter Self-Check

Online Test Practice

Use *eAssessment* to:

- create multiple versions of tests
- edit existing questions and add your own questions
- build tests aligned with select state curriculum standards using built-in state curriculum tags
- track students' progress

SECTION 1

Mastering Concepts

78. Atoms lose, gain, or share electrons to achieve an octet. Covalent bonding occurs when atoms share electrons to achieve an octet.
79. The nucleus of one atom attracts the electrons of the other atom, and they share one or more pairs of electrons.
80. Molecules bond covalently.
81. As the atoms approach, the net force of attraction increases. At a certain optimal distance between atoms, the net attractive force is maximized. If the atoms move closer than the optimal distance, repulsive force exceeds attractive force.
82. single covalent bond: sigma bond; double bond: a sigma bond and a pi bond; triple bond: one sigma and two pi bonds

Mastering Problems

83. N: 5, 3; As: 5, 3; Br: 7, 1; Se: 6, 2
84. **a.** single bonds: sigma bonds; double bond: one sigma bond and one pi bond
 b. single bonds: sigma bonds; triple bond: one sigma and two pi bonds
85. The triple bond in CO is the shortest and the strongest.
86. The triple bond in C ≡ N⁻ is shorter and stronger.
87. a, c, b

SECTION 2

Mastering Concepts

88. Answers should agree with Figure 12.
89. when it releases H^+ in water solution
90. Sulfur hexafluoride: 1 S atom bonded with 6 F atoms. Disulfur tetrafluoride: 2 S atoms bonded with 4 F atoms.
91. The name *silicon* indicates 1 Si atom. The prefix *di-* means 2 and *oxide* indicates oxygen. The correct formula is SiO_2.

Mastering Problems

92. $HClO_2$: chlorous acid; H_3PO_4: phosphoric acid; H_2Se: hydroselenic acid; $HClO_3$: chloric acid
93. **a.** nitrogen trifluoride
 b. nitrogen monoxide
 c. sulfur trioxide
 d. silicon tetrafluoride

CHAPTER 8 ASSESSMENT

SECTION 1

Mastering Concepts

78. What is the octet rule, and how is it used in covalent bonding?

79. Describe the formation of a covalent bond.

80. Describe the bonding in molecules.

81. Describe the forces, both attractive and repulsive, that occur as two atoms move closer together.

82. How could you predict the presence of a sigma or pi bond in a molecule?

Mastering Problems

83. Give the number of valence electrons in N, As, Br, and Se. Predict the number of covalent bonds needed for each of these elements to satisfy the octet rule.

84. Locate the sigma and pi bonds in each of the molecules shown below.

a. 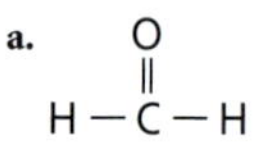

b. H — C ≡ C — H

85. In the molecules CO, CO_2, and CH_2O, which C—O bond is the shortest? Which C—O bond is the strongest?

86. Consider the carbon-nitrogen bonds shown below:

C≡N⁻ and 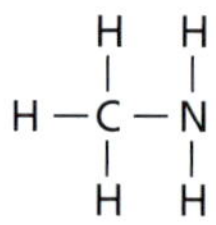

Which bond is shorter? Which is stronger?

87. Rank each of the molecules below in order of the shortest to the longest sulfur-oxygen bond length.
a. SO_2 **b.** SO_3^{2-} **c.** SO_4^{2-}

SECTION 2

Mastering Concepts

88. Explain how molecular compounds are named.

89. When is a molecular compound named as an acid?

90. Explain the difference between sulfur hexafluoride and disulfur tetrafluoride.

91. **Watches** The quartz crystals used in watches are made of silicon dioxide. Explain how you use the name to determine the formula for silicon dioxide.

Mastering Problems

92. Complete **Table 8.**

Table 8 Acid Names

Formula	Name
$HClO_2$	
H_3PO_4	
H_2Se	
$HClO_3$	

93. Name each molecule.
a. NF_3 **c.** SO_3
b. NO **d.** SiF_4

94. Name each molecule.
a. SeO_2 **c.** N_2F_4
b. SeO_3 **d.** S_4N_4

95. Write the formula for each molecule.
a. sulfur difluoride **c.** carbon tetrafluoride
b. silicon tetrachloride **d.** sulfurous acid

96. Write the formula for each molecule.
a. silicon dioxide **c.** chlorine trifluoride
b. bromous acid **d.** hydrobromic acid

SECTION 3

Mastering Concepts

97. What must you know in order to draw the Lewis structure for a molecule?

98. **Doping Agent** Material scientists are studying the properties of polymer plastics doped with AsF_5. Explain why the compound AsF_5 is an exception to the octet rule.

99. **Reducing Agent** Boron trihydride (BH_3) is used as reducing agent in organic chemistry. Explain why BH_3 often forms coordinate covalent bonds with other molecules.

100. Antimony and chlorine can form antimony trichloride or antimony pentachloride. Explain how these two elements can form two different compounds.

Mastering Problems

101. Draw three resonance structures for the polyatomic ion CO_3^{2-}.

102. Draw the Lewis structures for these molecules, each of which has a central atom that does not obey the octet rule.
a. PCl_5 **c.** ClF_5
b. BF_3 **d.** BeH_2

94. **a.** selenium dioxide **c.** dinitrogen tetrafluoride
 b. selenium trioxide **d.** tetrasulfur tetranitride
95. **a.** SF_2 **b.** $SiCl_4$ **c.** CF_4 **d.** H_2SO_3
96. **a.** SiO_2 **b.** $HBrO_2$ **c.** ClF_3 **d.** HBr

SECTION 3

Mastering Concepts

97. the number of valence electrons for each atom
98. Arsenic has five bonding positions with a total of 10 shared electrons. This is greater than the eight electrons that occupy an octet.
99. BH_3 only has six electrons and does not have an electron arrangement with a low amount of potential energy. It will share a lone pair with another molecule to form this electron arrangement.
100. Antimony has five valence electrons, one lone pair, and three positions where it can share one electron with a chlorine atom. This will form $SbCl_3$. Antimony can also expand its octet and bond with all five valence electrons, forming $SbCl_5$.

Mastering Problems

101. Refer to the Solutions Manual for structures.
102. Refer to the Solutions Manual for structures.

Chapter Self-Check

103. Draw two resonance structures for the polyatomic ion HCO_2^-.

104. Draw the Lewis structure for a molecule of each of these compounds and ions.
a. H_2S
b. BF_4^-
c. SO_2
d. $SeCl_2$

105. Which elements in the list below are capable of forming molecules in which one of its atoms has an expanded octet? Explain your answer.
a. B
b. C
c. P
d. O
e. Se

SECTION 4

Mastering Concepts

106. What is the basis of the VSEPR model?

107. What is the maximum number of hybrid orbitals a carbon atom can form?

108. What is the molecular shape of each molecule? Estimate the bond angle for each molecule, assuming that there is not a lone pair.
a. A—B
b. A—B—A
c. 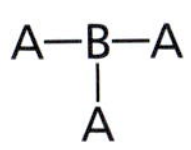
d. 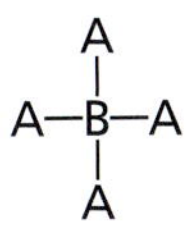

109. Parent Compound PCl_5 is used as a parent compound to form many other compounds. Explain the theory of hybridization and determine the number of hybrid orbitals present in a molecule of PCl_5.

Mastering Problems

110. Complete **Table 9** by identifying the expected hybrid on the central atom. You might find drawing the molecule's Lewis structure helpful.

Table 9 Structures

Formula	Hybrid Orbital	Lewis Structure
XeF_4		
TeF_4		
KrF_2		
OF_2		

111. Predict the molecular shape of each molecule.
a. COS
b. CF_2Cl_2

112. For each molecule listed below, predict its molecular shape and bond angle, and identify the hybrid orbitals. Drawing the Lewis structure might help you.
a. SCl_2
b. NH_2Cl
c. HOF
d. BF_3

SECTION 5

Mastering Concepts

113. Describe electronegativity trends in the periodic table.

114. Explain the difference between nonpolar molecules and polar molecules.

115. Compare the location of bonding electrons in a polar covalent bond with those in a nonpolar covalent bond. Explain your answer.

116. What is the difference between a covalent molecular solid and a covalent network solid? Do their physical properties differ? Explain your answer.

Mastering Problems

117. For each pair, indicate the more polar bond by circling the negative end of its dipole.
a. C—S, C—O
b. C—F, C—N
c. P—H, P—Cl

118. For each of the bonds listed, tell which atom is more negatively charged.
a. C—H
b. C—N
c. C—S
d. C—O

119. Predict which bond is the most polar.
a. C—O
b. Si—O
c. C—Cl
d. C—Br

120. Rank the bonds according to increasing polarity.
a. C—H
b. N—H
c. Si—H
d. O—H
e. Cl—H

121. Refrigerant The refrigerant known as freon-14 is an ozone-damaging compound with the formula CF_4. Why is the CF_4 molecule nonpolar even though it contains polar bonds?

122. Determine if these molecules and ion are polar. Explain your answers.
a. H_3O^+
b. PCl_5
c. H_2S
d. CF_4

123. Use Lewis structures to predict the molecular polarities for sulfur difluoride, sulfur tetrafluoride, and sulfur hexafluoride.

Mastering Problems

103. Refer to the Solutions Manual for structures.

104. Refer to the Solutions Manual for structures.

105. P and Se because they are period 3 and higher and have a d sublevel available

SECTION 4

Mastering Concepts

106. the repulsive nature of electron pairs around a central atom

107. four

108. **a.** linear, 180°
b. linear, 180°
c. trigonal planar, 120°
d. tetrahedral, 109°

109. The theory of hybridization explains the shapes of molecules by the formation of identical hybrid orbitals from the atomic orbitals of the atoms in the molecule; five identical sp^3d orbitals

Mastering Problems

110. XeF_4: sp^3d^2; TeF_4: sp^3d; KrF_2: sp^3d; OF_2: sp^3

111. **a.** linear **b.** tetrahedral

112. **a.** bent, 104.5°, sp^3
b. trigonal pyramidal, 107°, sp^3
c. bent, 104.5°, sp^3
d. trigonal planar, 120°, sp^2

SECTION 5

Mastering Concepts

113. It increases left to right in a period and decreases top to bottom in a group.

114. A nonpolar molecule has a symmetric distribution of charge, while a polar molecule has a concentration of electrons on one side of the molecule.

115. Electrons in a polar bond are closer to the more electronegative atom because of unequal sharing. Those in a nonpolar bond are shared equally.

116. A covalent molecular solid is soft and has a low melting point because of weak intermolecular forces. A covalent network solid has a high melting point and is very hard because of the strength of the network of covalent bonds.

Mastering Problems

117. The following are circled:
a. O **b.** F **c.** Cl

118. **a.** C **b.** N **c.** S **d.** O

119. Si—O

120. in order of increasing polarity: c, a, b, e, d

121. equal distribution of charge in a symmetrical molecule

122. **a.** polar, asymmetrical
b. nonpolar, symmetrical
c. polar, asymmetrical
d. nonpolar, symmetrical

123. SF_2 and SF_4 are asymmetrical molecules and are polar. SF_6 is a symmetrical molecule and is nonpolar.

MIXED REVIEW

124. **a.** ClO **c.** PCl_5
b. H_3AsO_4 **d.** H_2S

125. **a.** phosphorus trichloride
b. dichlorine heptoxide
c. tetraphosphorus hexoxide
d. nitrogen monoxide

126. **a.**
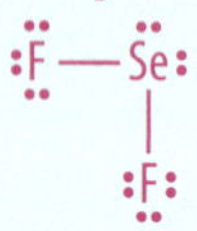

b.
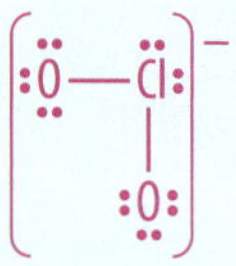

c.
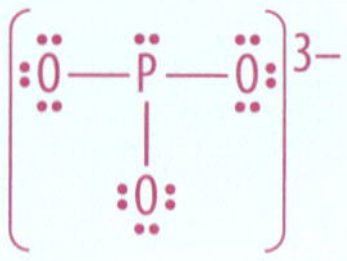

d.
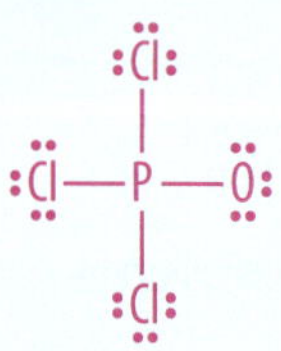

e.
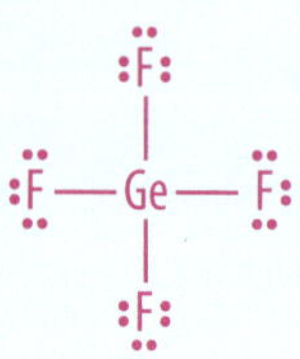

127. The polar molecules are CH_3Cl, ClF, and NCl_3 because each molecule is asymmetric and the charge is not distributed evenly.

128. in order of least to greatest bond character: e, d, a, c, b

129.
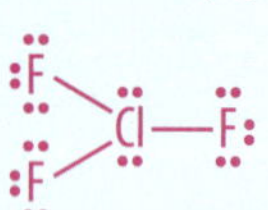

sp^3d

130. Single covalent: 2 shared electrons, any halogen or group 17 element
Double covalent: 4 shared electrons, group 16 elements
Triple covalent: 6 shared electrons, group 15 elements

THINK CRITICALLY

131. Concept maps will vary.

132. The name arsenic(III) oxide states that arsenic has an oxidation number of 3+ and oxide is 2–. The correct formula is As_2O_3. The name diarsenic trioxide states that there are two atoms of arsenic and three atoms of oxygen. The correct formula is As_2O_3. Even though they are named differently, they both represent the same formula.

133. Column 2: the electrostatic attraction of a positive ion for a negative ion; the sharing of electrons between two atoms; the attraction of a positive ion for delocalized electrons; atoms covalently bonded to many other atoms in a crystal lattice. Column 3: hard, rigid, brittle, crystalline, high melting point, nonconductor in the solid state; soft, low melting point, nonconductor in the solid state; a crystal that conducts heat and electricity, malleable, ductile, high melting point; crystal is hard, rigid, brittle, nonconductor. Sample answers for Column 4: NaCl, CO_2, Ag, diamond.

134. Sigma bonds are the N—H bonds and the C—N bonds as well as one of the C—O bonds. The other C—O bond is a pi bond. The lone pairs are located on both N atoms and on the O atom.

135. **a.** polar **b.** nonpolar **c.** polar

136. The first carbon (bonded to three H atoms and one C atom) atom is a sp^3 hybrid because it has 4 bonding positions. The second carbon atom (bonded to one C atom and one N atom) is an sp hybrid because it has two bonding positions.

ASSESSMENT

MIXED REVIEW

124. Write the formula for each molecule.
a. chlorine monoxide
b. arsenic acid
c. phosphorus pentachloride
d. hydrosulfuric acid

125. Name each molecule.
a. PCl_3
b. Cl_2O_7
c. P_4O_6
d. NO

126. Draw the Lewis structure for each molecule or ion.
a. SeF_2
b. ClO_2^-
c. PO_3^{3-}
d. $POCl_3$
e. GeF_4

127. Determine which of the molecules are polar. Explain your answers.
a. CH_3Cl
b. ClF
c. NCl_3
d. BF_3
e. CS_2

128. Arrange the bonds in order of least to greatest polar character.
a. C—O
b. Si—O
c. Ge—O
d. C—Cl
e. C—Br

129. **Rocket Fuel** In the 1950s, the reaction of hydrazine with chlorine trifluoride (ClF_3) was used as a rocket fuel. Draw the Lewis structure for ClF_3 and identify the hybrid orbitals.

130. Complete **Table 10,** which shows the number of electrons shared in a single covalent bond, a double covalent bond, and a triple covalent bond. Identify the group of atoms that will form each of these bonds.

Table 10 Shared Pairs

Bond Type	Number of Shared Electrons	Atoms that Form the Bond
Single covalent		
Double covalent		
Triple covalent		

THINK CRITICALLY

131. **Organize** Design a concept map that explains how VSEPR model theory, hybridization theory, and molecular shape are related.

132. **Compare and contrast** the two covalent compounds identified by the names arsenic(III) oxide and diarsenic trioxide.

133. **Make and Use Tables** Use your knowledge of ionic, metallic, and covalent bonds to complete **Table 11.**

Table 11 Properties and Bonding

Solid	Bond Description	Characteristic of Solid	Example
Ionic			
Covalent molecular			
Metallic			
Covalent network			

134. **Apply** Urea, whose structure is shown below, is a compound used in manufacturing plastics and fertilizers. Identify the sigma bond, pi bonds, and lone pairs present in a molecule of urea.

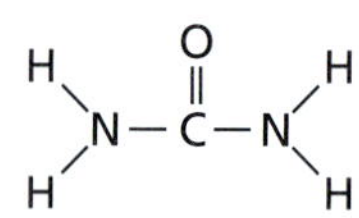

135. **Analyze** For each of the characteristics listed below, identify the polarity of a molecule with that characteristic.
a. solid at room temperature
b. gas at room temperature
c. attracted to an electric current

136. **Apply** The structural formula for acetonitrile, CH_3CN, is shown below.

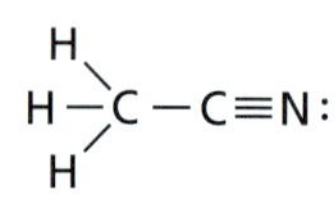

Examine the structure of the acetonitrile molecule. Determine the number of carbon atoms in the molecule, identify the hybrid present in each carbon atom, and explain your reasoning.

Chapter Self-Check

CHALLENGE PROBLEM

137. Examine the bond-dissociation energies for the various bonds listed in **Table 12.**

Table 12 Bond-Dissociation Energies

Bond	Bond-Dissociation Energy (kJ/mol)	Bond	Bond-Dissociation Energy (kJ/mol)
C—C	348	O—H	467
C=C	614	C—N	305
C≡C	839	O=O	498
N—N	163	C—H	416
N=N	418	C—O	358
N≡N	945	C=O	745

a. Draw the correct Lewis structures for C_2H_2 and HCOOH.
b. Determine the amount of energy needed to break apart each of these molecules.

CUMULATIVE REVIEW

138. Table 13 lists a liquid's mass and volume data. Create a line graph of this data with the volume on the *x*-axis and the mass on the *y*-axis. Calculate the slope of the line. What information does the slope give you?

Table 13 Mass v. Volume

Volume	Mass
4.1 mL	9.36 g
6.0 mL	14.04 g
8.0 mL	18.72 g
10.0 mL	23.40 g

139. Write the correct chemical formula for each compound.
a. calcium carbonate
b. potassium chlorate
c. silver acetate
d. copper(II) sulfate
e. ammonium phosphate

140. Write the correct chemical name for each compound.
a. NaI
b. $Fe(NO_3)_3$
c. $Sr(OH)_2$
d. $CoCl_2$
e. $Mg(BrO_3)_2$

WRITING IN ▶ Chemistry

141. Antifreeze Research ethylene glycol to learn its chemical formula. Explain how its structure makes it a useful antifreeze and coolant.

142. Detergents Choose a laundry detergent to research and write an essay about its chemical composition. Explain how it removes oil and grease from fabrics.

DBQ Document-Based Questions

Luminol *Crime-scene investigators often use the covalent compound luminol to find blood evidence. The reaction between luminol, certain chemicals, and hemoglobin, a protein in blood, produces light.* **Figure 26** *shows a ball-and-stick model of luminol.*

Data obtained from: Fleming, Declan., 2002. The Chemiluminescence of Luminol, Exemplarchem, *Royal Society of Chemistry.*

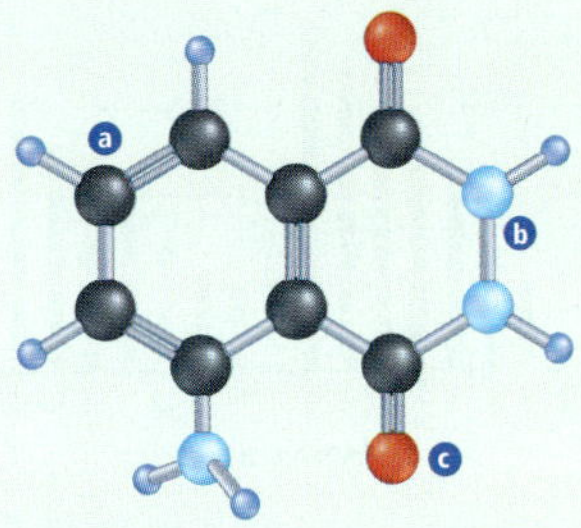

■ **Figure 26**

143. Determine the molecular formula for luminol and draw its Lewis structure.

144. Indicate the hybrid present on the atoms labeled *a, b,* and *c* in **Figure 26.**

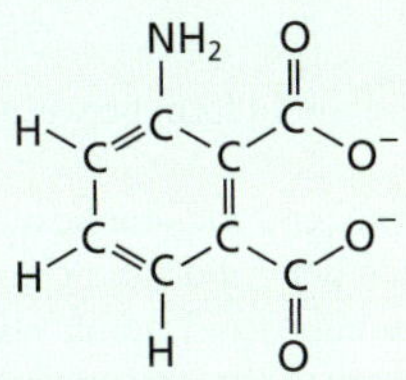

APA ion

■ **Figure 27**

145. When luminol comes in contact with the iron ion in hemoglobin, it reacts to produce Na_2APA, water, nitrogen, and light energy. Given the structural formula of the APA ion in **Figure 27,** write the chemical formula for the polyatomic APA ion.

CHALLENGE PROBLEM

137. a. H—C≡C—H

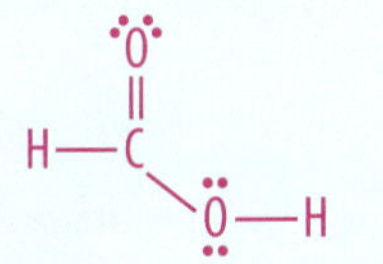

b. C_2H_2 = 1671 kJ/mol; HCOOH = 1986 kJ/mol

CUMULATIVE REVIEW

138. 2.34 g/mL; density
139. a. $CaCO_3$
b. $KClO_3$
c. $AgC_2H_3O_2$
d. $CuSO_4$
e. $(NH_4)_3PO_4$
140. a. sodium iodide
b. iron(III) nitrate
c. strontium hydroxide
d. cobalt(II) chloride
e. magnesium bromate

WRITING IN ▶ Chemistry

Rubric

141.

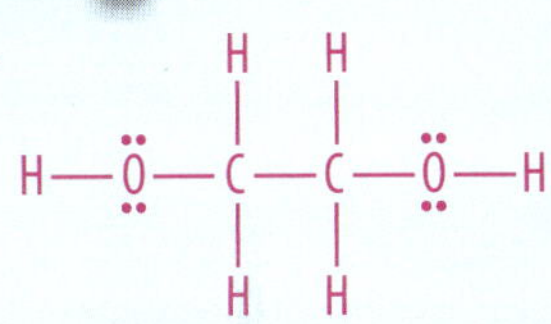

Answers will vary. Students might note that the presence of –OH groups make ethylene glycol miscible in water and contribute to its relatively high boiling point and relatively low freezing point.

142. Answers should include a discussion of the nonpolar end of a detergent molecule and the polar end of the same molecule, which allow it to attract both water and oil.

DBQ Document-Based Questions

Data obtained from: Fleming, Declan., 2002. The Chemiluminescence of Luminol, Exemplarchem, *Royal Society of Chemistry.*

143. $C_8H_7O_2N_3$

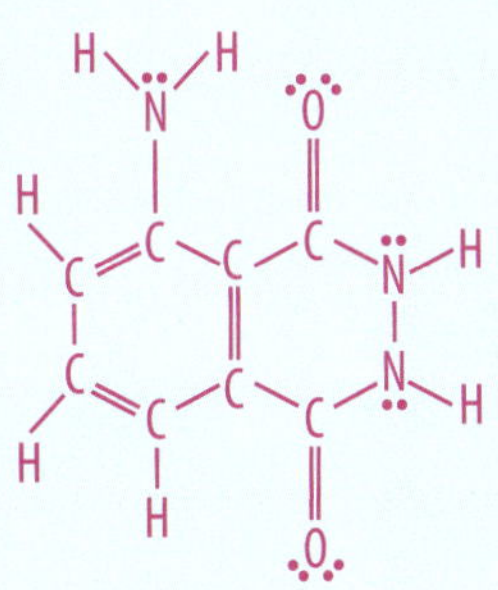

144. A, sp^2; B, sp^3; C, sp^2
145. $C_8H_5NO_4^{\ 2-}$

CUMULATIVE STANDARDIZED TEST PRACTICE

MULTIPLE CHOICE

1. D
2. A
3. B
4. D
5. B
6. C
7. D
8. D
9. A
10. B

CHAPTER 8 STANDARDIZED TEST PRACTICE

MULTIPLE CHOICE

1. The common name of SiI_4 is tetraiodosilane. What is its molecular compound name?
A. silane tetraiodide
B. silane tetraiodine
C. silicon iodide
D. silicon tetraiodide

2. Which compound contains at least one pi bond?
A. CO_2
B. $CHCl_3$
C. AsI_3
D. BeF_2

Use the graph below to answer Questions 3 and 4.

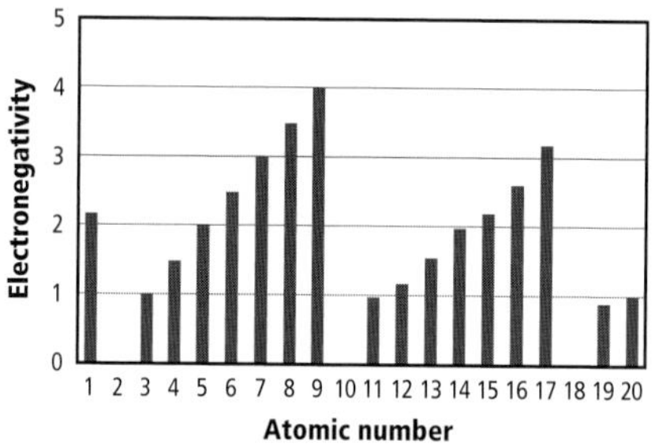

3. What is the electronegativity of the element with atomic number 14?
A. 1.5
B. 1.8
C. 2.0
D. 2.2

4. An ionic bond would form between which pairs of elements?
A. atomic number 3 and atomic number 4
B. atomic number 7 and atomic number 8
C. atomic number 4 and atomic number 18
D. atomic number 8 and atomic number 12

5. Which is the Lewis structure for silicon disulfide?
A. :S::Si::S:
B. S::Si::S (with additional dots)
C. S:Si:S (with additional dots)
D. :S:Si:S: (with additional dots)

6. The central selenium atom in selenium hexafluoride forms an expanded octet. How many electron pairs surround the central Se atom?
A. 4
B. 5
C. 6
D. 7

Use the table below to answer Questions 7 and 8.

Bond Dissociation Energies at 298 K			
Bond	kJ/mol	Bond	kJ/mol
Cl–Cl	242	N≡N	945
C–C	345	O–H	467
C–H	416	C–O	358
C–N	305	C=O	745
H–I	299	O=O	498
H–N	391		

7. Which diatomic gas has the shortest bond between its two atoms?
A. HI
B. O_2
C. Cl_2
D. N_2

8. Approximately how much energy will it take to break all the bonds present in the molecule below?

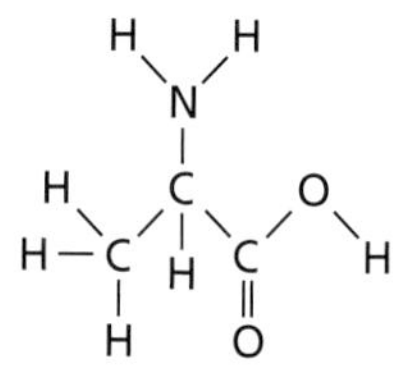

A. 3024 kJ/mol
B. 4318 kJ/mol
C. 4621 kJ/mol
D. 5011 kJ/mol

9. Which compound does NOT have a bent molecular shape?
A. BeH_2
B. H_2S
C. H_2O
D. SeH_2

10. Which compound is nonpolar?
A. H_2S
B. CCl_4
C. SiH_3Cl
D. AsH_3

Online Test Practice

SHORT ANSWER

11. Oxyacids contain hydrogen and an oxyanion. There are two different oxyacids that contain hydrogen, nitrogen, and oxygen. Identify these two oxyacids. How can they be distinguished on the basis of their names and formulas?

Use the atomic emission spectrum below to answer Questions 12 and 13.

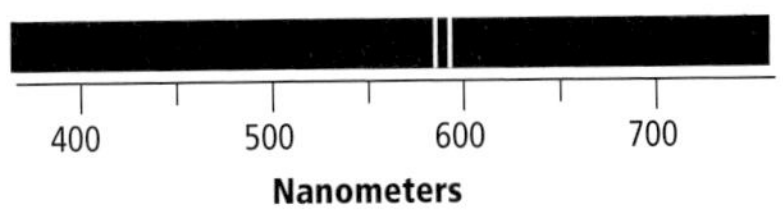

12. Estimate the wavelength of the photons being emitted by this element.

13. Find the frequency of the photons being emitted by this element.

EXTENDED RESPONSE

Use the table below to answer Question 14.

Percent Abundance of Silicon Isotopes		
Isotope	Mass	Percent Abundance
^{28}Si	27.98 amu	92.21 %
^{29}Si	28.98 amu	4.70 %
^{30}Si	29.97 amu	3.09 %

14. Your lab partner calculates the average atomic mass of these three silicon isotopes. His average atomic mass value is 28.98 amu. Explain why your lab partner is incorrect, and show how to calculate the correct average atomic mass.

SAT SUBJECT TEST: CHEMISTRY

Use the list of separation techniques below to answer Questions 15 to 17.

A. filtration
B. distillation
C. crystallization
D. chromatography
E. sublimation

15. Which technique separates components of a mixture with different boiling points?

16. Which technique separates components of a mixture based on the size of its particles?

17. Which technique is based on the stronger attraction some components have for the stationary phase compared to the mobile phase?

Use the table below to answer Questions 18 to 19.

Electron-Dot Structures								
Group	1	2	13	14	15	16	17	18
Diagram	Li·	·Be·	·B̈·	·C̈·	·N̈·	·Ö:	:F̈:	:N̈e:

18. Based on the Lewis structures shown, which elements will combine in a 2:3 ratio?
A. lithium and carbon
B. beryllium and fluorine
C. beryllium and nitrogen
D. boron and oxygen
E. boron and carbon

19. How many electrons will beryllium have in its outer energy level after it forms an ion to become chemically stable?
A. 0
B. 2
C. 4
D. 6
E. 8

NEED EXTRA HELP?																			
If You Missed Question . . .	1	2	3	4	5	6	7	8	9	10	11	12	13	14	15	16	17	18	19
Review Section . . .	8.2	8.1	8.5	8.5	8.3	8.3	8.1	8.1	8.4	8.5	8.2	5.2	5.2	4.3	3.3	3.3	3.3	7.2	7.2

SHORT ANSWER

11. Nitric acid (HNO_3) and nitrous acid (HNO_2). The *-ic* suffix indicates the larger number of oxygen atoms; the *-ous* suffix indicates the lower number of oxygen atoms.
12. 580 nm
13. 5.2×10^{14} Hz

EXTENDED RESPONSE

14. The lab partner is incorrect because he has taken a simple average of the masses of each isotope. An average atomic mass must take into account the percent abundance of each isotope and find a weighted average. The correct answer (28.09 amu) is found as follows: (27.98 amu × 92.21%) + (28.98 amu × 4.70%) + (29.97 amu × 3.09%).

SAT SUBJECT TEST: CHEMISTRY

15. B
16. A
17. D
18. D
19. B

CHAPTER 9 Chemical Reactions

BIGIDEA **Millions of chemical reactions in and around you transform reactants into products, resulting in the absorption or release of energy.**

ESSENTIAL QUESTIONS	RESOURCES TO ASSESS MASTERY
SECTION 1 **Reactions and Equations** **1.** What is evidence of chemical change? **2.** How are chemical reactions represented? **3.** Why do chemical equations need to be balanced and how is this accomplished? 1 session 0.5 block	**Progress Monitoring** Formative Assessment, pp. 285, 287, 288 Section Review, p. 288
SECTION 2 **Classifying Chemical Reactions** **1.** How are chemical reactions classified? **2.** What are the characteristics of different classes of chemical reactions? 2 sessions 1 block	**Progress Monitoring** Formative Assessment, p. 294 Reading Check, pp. 294, 297 Section Review, p. 298
SECTION 3 **Reactions in Aqueous Solutions** **1.** What are aqueous solutions? **2.** How are complete ionic and net ionic equations written for chemical reactions in aqueous solutions? **3.** How can you predict whether reactions in aqueous solutions will produce a precipitate, water, or a gas? 2 sessions 1 block	**Progress Monitoring** Formative Assessment, p. 307 Reading Check, pp. 301, 303, 307 Section Review, p. 308 **Summative Assessment** Chapter Assessment, p. 312 *eAssessment* Chapter Test (scaffolded)

Go online!

LEVELED RESOURCES	LAB MATERIALS
Science Notebook 9.1 OL ***FAST FILE* Chapter Resources:** Study Guide, p. 20 BL	LaunchLAB p. 280 / **15 min** distilled water, 100-mL beaker, 25-mL graduated cylinder, stirring rod, pipette, thermometer, 0.1*M* ammonia, universal indicator effervescent tablet
Science Notebook 9.2 OL ***FAST FILE* Chapter Resources:** ChemLab Worksheet, p. 3 OL Study Guide, p. 22 BL	ChemLAB p. 310 / **45 min** 1.0*M* $Zn(NO_3)_2$, Al wire, Mg ribbon, Zn metal strips, 1.0*M* $Al(NO_3)_2$, 1.0*M* $Cu(NO_3)_2$ 1.0*M* $Mg(NO_3)_2$, Cu wire, Emery cloth or sandpaper, pipettes, 24-well microscale reaction plate, wire cutters
Science Notebook 9.3 OL ***FAST FILE* Chapter Resources:** MiniLab Worksheet, p. 2 OL Study Guide, p. 24 BL	MiniLAB p. 301 / **30 min** distilled water, NaOH pellets, Epsom salts ($MgSO_4$), balance, 150-mL beaker, stirring rod, 100-mL graduated cylinder

ADDITIONAL RESOURCES

- eAssessment
- ConnectED Student Center
- ConnectED Teacher Center
- Lesson Presentations
- eSolutions
- Videos
- Animations
- Online Quizzes
- Virtual Investigations
- Multilingual eGlossary
- Personal Tutors
- WebQuests
- Applying Practices Worksheets
- PBLs

Practice:

- Challenge Problems AL
- Supplemental Problems BL OL
- Chapter Test (Scaffolded)
- Standardized Test Practice

Additional Resources:

- Blueprints for Success
- Solving Problems: A Chemistry Handbook

Lab Resources:

- Laboratory Manual OL
- CBL Laboratory Manual OL
- Small-Scale Laboratory Manual OL
- Forensics Laboratory Manual OL

BL Below Level　OL On Level　AL Advanced Learners　EL English Learners　COOP LEARN Cooperative Learning

CHAPTER 9

LaunchLAB

Rubric

How do you know when a chemical change has occurred?

Purpose Students will classify an observed change as either chemical or physical and identify the evidence of a chemical change.

Safety Precautions Approve lab safety forms before work begins. Review the ammonia MSDS with students. **WARNING**: ***Ammonia vapors are extremely irritating.*** Cover the desktop with an absorbent cloth that can catch any overflow from the reaction.

Disposal Contents of the test tube can be flushed down a drain.

Teaching Strategies
- Make sure the effervescent tablet contains a carbonate or hydrogen carbonate and a weak acid such as citric acid.
- See the Teacher Handbook for preparation of solutions.

Expected Results Citric acid in the tablet dissolves in water, turning the indicator from blue to red. The acid reacts with sodium hydrogen carbonate, producing carbon dioxide gas. As the gas escapes, the solution is buffered and the acidity decreases. The red color quickly changes to orange, then yellow, and after standing, to green. The contents of the test tube foam and overflow. The temperature of the solution decreases 2°C.

CHAPTER 9
Chemical Reactions

BIGIDEA **Millions of chemical reactions in and around you transform reactants into products, resulting in the absorption or release of energy.**

SECTIONS

1 Reactions and Equations

2 Classifying Chemical Reactions

3 Reactions in Aqueous Solutions

LaunchLAB

How do you know when a chemical change has occurred?

There are different ways to know when a chemical change has occurred, including temperature change, color change, and formation of a gas or solid. In this lab, you will observe evidence of a chemical change.

FOLDABLES® Study Organizer

Chemical Reactions

Make a concept-map book. Label it as shown. Use it to help you organize information about how chemical reactions are classified.

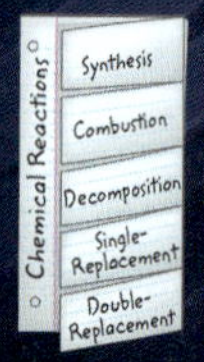

Procedure

1. Read and complete the lab safety form.
2. Measure 10.0 mL of **distilled water** in a **25-mL graduated cylinder,** and pour it into a **100-mL beaker.** Using a **pipette,** add one drop of **0.1*M* ammonia** to the water. **WARNING:** ***Ammonia vapors are extremely irritating.***
3. Stir 15 drops of **universal indicator** into the solution with a **stirring rod.** Observe the solution's color. Measure its temperature with a **thermometer.**
4. Drop an **effervescent tablet** into the solution. Observe what happens. Record your observations, including any temperature change.

Analysis

1. **Describe** any changes in the color or temperature of the solution. The color of the solution changed from blue to red, then orange, then yellow, then green. The temperature of the solution decreased 2°C.
2. **Explain** Was a gas produced? If so, what did you observe to support this conclusion? Yes, a gas was produced. The contents foamed.
3. **Analyze** Did a physical change or a chemical change occur? Explain. Each of the characteristics of the reaction is evidence of a chemical change, therefore, a chemical change is observed.

Before wood can burn, the water inside it boils off. Wood bursts into flames once it reaches 260°C. The smoke released during the reaction contains over 100 chemical substances.

Inquiry What does the universal indicator tell you about the solution? Design an experiment to support your prediction. The universal indicator tells you that the pH of the solution is changing (a chemical reaction is taking place). Individual experimental designs will vary.

Introduce the BIGIDEA

Physical and Chemical Changes To introduce this chapter's Big Idea, lead students in a discussion about wood burning. Examples might include a forest fire or fireplace. Ask students to distinguish between physical and chemical changes. Physical change does not alter the chemical composition of a substance, whereas chemical change does. Ask students to list physical changes and chemical changes of wood. Physical changes might include cutting wood into logs. Chemical changes might include photosynthesis to make food for growth and the formation of carbon when burned. Ask students if the burning of any substance leads to a chemical change, and have them give examples. yes; paper, charcoal, wood and cars Ask students to determine whether physical or chemical change occurs when wood is burned, and justify their responses. Chemical change occurs because the wood is converted to ashes.

Tie to Previous Knowledge

Have students review the following concepts before studying this chapter.

- writing formulas for ionic compounds
- writing formulas for molecular compounds

Use the Photo

Combustion of a Substance Ask students to predict what would happen to trees in a forest fire. Trees would burn and go through a chemical change, that is, wood is converted into gases and ashes. Ask students to list ways in which a forest fire might impact the environment and living organisms in the forest. Answers will vary but might include destruction of animal food sources, water pollution, soil regeneration, and the germination of some types of seeds.

SECTION 1

1 Focus

MAINIDEA

Balance Refer students to the MiniLab–Model Isotopes–and ask them to recall how they used the balance to determine the mass of 10 pennies. Suppose each penny has an identical mass, what will happen to the mass if, first, 10 pennies are placed on the balance and, second only 5 pennies are placed on the balance? The overall mass will decrease by one half for 5 pennies. Ask students to make the analogy of this activity with reactants and products in a chemical reaction. More reactant produces more product; and less reactant produces less product. OL

2 Teach

Concept Development

Amounts of Reactants Ask students to write a word equation for preparing a cake made from scratch, a process that requires energy. Student answers will vary but might include: energy + salt + sugar + baking powder + eggs + water + shortening + flour → cake. Emphasize that the precise amounts of reactants and products in a chemical reaction (and equation) are critical. For example, 2.5 kg salt, 5 g sugar, 4 L water, 4 doz eggs, and so on, will likely not produce a proper cake. OL

SECTION 1

Essential Questions

- What is evidence of chemical change?
- How are chemical reactions represented?
- Why do chemical equations need to be balanced and how is this accomplished?

Review Vocabulary

chemical change: a process involving one or more substances changing into a new substance

New Vocabulary

chemical reaction
reactant
product
chemical equation
coefficient

Reactions and Equations

MAINIDEA Chemical reactions are represented by balanced chemical equations.

CHEM 4 YOU When you purchase bananas from a grocery store, they might be green. Within a few days, the bananas turn yellow. This color change is one of the ways you can tell a chemical reaction has occurred.

Chemical Reactions

Do you know that the foods you eat, the fibers in your clothes, and the plastic in your CDs have something in common? Foods, fibers, and plastics are produced when the atoms in substances are rearranged to form different substances. Atoms are rearranged during the forest fire shown in the photo at the beginning of the chapter. They were also rearranged when you dropped the effervescent tablet into the beaker of water and indicator during the Launch Lab.

The process by which the atoms of one or more substances are rearranged to form different substances is called a **chemical reaction.** A chemical reaction is another name for a chemical change, which you read about previously. Chemical reactions affect every part of your life. They break down your food, producing the energy you need to live. Chemical reactions in the engines of cars and buses provide the energy to power the vehicles. They produce natural fibers, such as cotton and wool, in plants and animals. In factories, they produce synthetic fibers such as nylon, shown in **Figure 1.**

Evidence of a chemical reaction How can you tell when a chemical reaction has taken place? Although some chemical reactions are hard to detect, many reactions provide physical evidence that they have occurred. A temperature change can indicate a chemical reaction. Many reactions, such as those that occur during the burning of wood, release energy in the form of heat and light. Other chemical reactions absorb heat.

Figure 1 When adipoyl chloride in dichloromethane reacts with hexanediamine, nylon is formed. Nylon is used in many products, including carpeting, clothing, sports equipment, and tires.

Charles D. Winters/Photo Researchers

CHEMISTRY JOURNAL

Observing Chemical Change Have students keep a running list in their chemistry journals of chemical reactions they use or observe. Have students also describe and include the evidence for each listed reaction. OL

■ **Figure 2** Each of these photos illustrates evidence of a chemical reaction.
Describe *the evidence in each photo that tells you a chemical reaction has occurred.*

In addition to a temperature change, other types of evidence might indicate that a chemical reaction has occurred. One indication of a chemical reaction is a color change. For example, you might have noticed that the color of some nails that are left outside changes from silver to orange-brown in a short time. The color change is evidence that a chemical reaction occurred between the iron in the nail and the oxygen in air. A banana changing from green to yellow is another example of a color change indicating that a chemical reaction has occurred. Odor, gas bubbles, and the formation of a solid are other indications of chemical change. Each of the photographs in **Figure 2** shows evidence of a chemical reaction.

Representing Chemical Reactions

Chemists use statements called equations to represent chemical reactions. Equations show a reaction's **reactants,** which are the starting substances, and **products,** which are the substances formed during the reaction. Chemical equations do not express numerical equalities as mathematical equations do because during chemical reactions the reactants are used up as the products form. Instead, the equations used by chemists show the direction in which the reaction progresses. Therefore, an arrow rather than an equal sign is used to separate the reactants from the products. You read the arrow as *react to produce* or *yield*. The reactants are written to the left of the arrow, and the products are written to the right of the arrow. When there are two or more reactants, or when there are two or more products, a plus sign separates each reactant or each product. These elements of equation notation are shown below.

$$\text{Reactant 1} + \text{Reactant 2} \rightarrow \text{Product 1} + \text{Product 2}$$

In equations, symbols are used to show the physical states of the reactants and products. Reactants and products can exist as solids, liquids, and gases. When they are dissolved in water, they are said to be aqueous. It is important to show the physical states of a reaction's reactants and products in an equation because the physical states provide clues about how the reaction occurs. Some basic symbols used in equations are shown in **Table 1.**

Table 1 Symbols Used in Equations

Symbol	Purpose
$+$	separates two or more reactants or products
$\rightarrow$	separates reactants from products
$\rightleftharpoons$	separates reactants from products and indicates a reversible reaction
(s)	identifies a solid state
(l)	identifies a liquid state
(g)	identifies a gaseous state
(aq)	identifies a water solution

Math in Chemistry

Oxidation Numbers Remind students that when writing the formula for a compound, the oxidation numbers of the component elements or ions must add to zero. For example:
H_2O $(2 \times (1+)) + (2-) = 0$;
$(NH_4)_2S$ $(2 \times (1+)) + (2-) = 0$;
$Ti(NO_2)_4$ $(1 \times (4+)) + (4 \times (1-)) = 0$.

Extension

Chemical Shorthand Ask students to list possible reasons why symbols are used in chemistry. Shorthand way of describing complex reactions and ideas, chemical symbols allow scientists speaking different languages to communicate more easily. Ask students to list what other disciplines use symbols to communicate more effectively. Answers may vary but should include math. OL

■ **Caption Question Fig. 2**
The apple is turning brown (changing color). The effervescent tablet shows a gas (bubbles) being formed. The pizza baking shows a change in color, production of a gas (to raise the dough), and production of an odor.

CHEMISTRY PROJECT

Writing Chemical Equations Ask students to consider some of the chemical reactions they have described in their journals. Have them write word, skeleton, and balanced chemical equations for any reactions they can. Have students make 3 dimensional models or drawings of the reactions. OL

Reinforcement

Writing Formulas Emphasize the importance of writing correct formulas for all reactants and products in the skeleton equation and not changing the formula during the balancing process. Ask students to write the chemical formula for water and then for hydrogen peroxide. H_2O and H_2O_2. Ask them what might happen in a chemical reaction if these two chemical formulas were changed in the chemical equation. The changed reaction would be a totally different chemical reaction than the original. Point out that once correct formulas are written, only coefficients can be added and changed to balance the equation. OL EL

PRACTICE Problems

Have students refer to the Selected Solutions appendix for complete solutions to odd-numbered problems. The complete solutions for all problems can be found in the Solutions Manual.

1. $H_2(g) + Br_2(g) \longrightarrow HBr(g)$
2. $CO(g) + O_2(g) \longrightarrow CO_2(g)$
3. potassium chlorate(s) ⟶ potassium chloride(s) + oxygen(g); $KClO_3(s) \longrightarrow KCl(s) + O_2(g)$

■ **Figure 3** Science, like all other disciplines, has a specialized language that allows specific information to be communicated in a uniform manner. This reaction between aluminum and bromine can be described by a word equation, a skeleton equation, or a balanced chemical equation.

Word equations You can use statements called word equations to indicate the reactants and products of chemical reactions. The word equation below describes the reaction between aluminum (Al) and bromine (Br), which is shown in **Figure 3.** Aluminum is a solid, and bromine is a liquid. The brownish-red cloud in the photograph is excess bromine. The reaction's product, which is solid particles of aluminum bromide ($AlBr_3$), settles on the bottom of the beaker.

Reactant 1 + Reactant 2 ⟶ Product 1

aluminum(s) + bromine(l) ⟶ aluminum bromide(s)

This word equation reads, "Aluminum and bromine react to produce aluminum bromide."

Skeleton equations Although word equations help to describe chemical reactions, they lack important information. A skeleton equation uses chemical formulas rather than words to identify the reactants and the products. For example, the skeleton equation for the reaction between aluminum and bromine uses the formulas for aluminum, bromine, and aluminum bromide in place of words.

$$Al(s) + Br_2(l) \longrightarrow AlBr_3(s)$$

How would you write the skeleton equation that describes the reaction between carbon and sulfur to form carbon disulfide? Carbon and sulfur are solids. First, write the chemical formulas for the reactants to the left of the arrow. Then, separate the reactants with a plus sign and indicate their physical states.

$$C(s) + S(s) \longrightarrow$$

Finally, write the chemical formula for the product, liquid carbon disulfide, to the right of the arrow and indicate its physical state. The result is the skeleton equation for the reaction.

$$C(s) + S(s) \longrightarrow CS_2(l)$$

This skeleton equation tells us that carbon in the solid state reacts with sulfur in the solid state to produce carbon disulfide in the liquid state.

VOCABULARY
ACADEMIC VOCABULARY
Formula
an expression using chemical symbols to represent a chemical reaction *The chemical formula for water is H_2O.*

PRACTICE Problems

Do additional problems. Online Practice

Write skeleton equations for the following word equations.

1. Hydrogen and bromine gases react to yield hydrogen bromide.
   hydrogen(g) + bromine(g) ⟶ hydrogen bromide(g)
2. When carbon monoxide and oxygen react, carbon dioxide forms.
   carbon monoxide(g) + oxygen(g) ⟶ carbon dioxide(g)
3. **Challenge** Write the word equation and the skeleton equation for the following reaction: when heated, solid potassium chlorate yields solid potassium chloride and oxygen gas.

Demonstration

A Metal and a Nonmetal React

Purpose
to combine a metal and nonmetal into a stable compound by releasing energy

Materials
Zinc dust (1 g); iodine crystals (3 g); dry test tube with stopper; evaporating dish; glass bottle with stopper; dropper

Safety Precautions

Always perform this demonstration in a fume hood. Use caution when working with chemicals and extreme temperatures.

Disposal This demonstration produces hazardous waste as described by the EPA and should be disposed of accordingly.

Procedure
Put 1 g of zinc dust and 3 g of iodine crystals into a dry test tube. Stopper the test tube and mix by shaking. Transfer the mixture to an

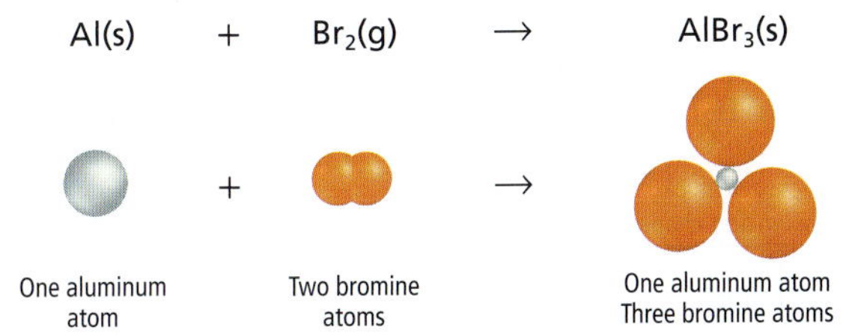

Figure 4 The information conveyed by skeleton equations is limited. In this case, the skeleton equation is correct, but it does not show the exact number of atoms that interact. Refer to **Table R-1** in the Student Resources for a key to atom color conventions.

Chemical equations Like word equations, skeleton equations lack some information about reactions. Recall that the law of conservation of mass states that in a chemical change, matter is neither created nor destroyed. Chemical equations must show that matter is conserved during a reaction. Skeleton equations lack that information.

Look at **Figure 4.** The skeleton equation for the reaction between aluminum and bromine shows that one aluminum atom and two bromine atoms react to produce a substance containing one aluminum atom and three bromine atoms. Was a bromine atom created in the reaction? Atoms are not created in chemical reactions, and to accurately show what happened, more information is needed.

To accurately represent a chemical reaction by an equation, the equation must show equal numbers of atoms of each reactant and each product on both sides of the arrow. Such an equation is called a balanced chemical equation. A **chemical equation** is a statement that uses chemical formulas to show the identities and relative amounts of the substances involved in a chemical reaction.

Balancing Chemical Equations

The balanced equation for the reaction between aluminum and bromine, shown in **Figure 5,** reflects the law of conservation of mass. To balance an equation, you must find the correct coefficients for the chemical formulas in the skeleton equation. A **coefficient** in a chemical equation is the number written in front of a reactant or product. Coefficients are usually whole numbers and are not usually written if the value is one. The coefficients in a balanced equation describe the lowest whole number ratio of the amounts of all of the reactants and products.

Get help with **balancing chemical equations.**

Personal Tutor

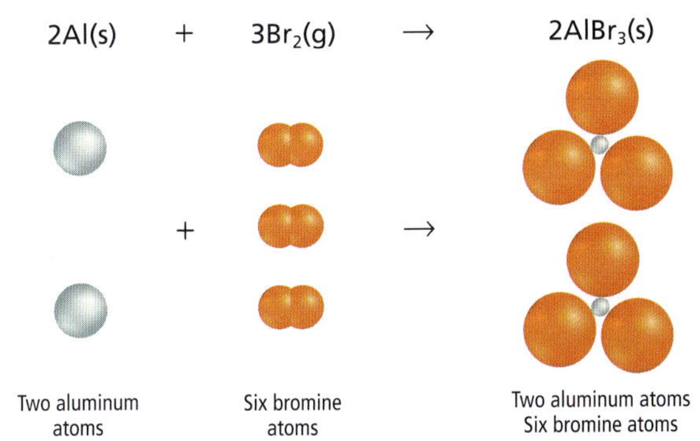

Figure 5 In a balanced chemical equation, the number of particles on the reactant side of the equation equals the number of particles on the product side of the equation. In this case, two aluminum atoms and six bromine atoms are needed on both sides of the equation.

Quick Demo

Silver Iodide Formation

WARNING: ***Silver nitrate is toxic and will stain skin and clothing.*** Pour approximately 5 mL of 0.1*M* silver nitrate solution into one small test tube and 5 mL of 0.1*M* potassium iodide solution into another. Carefully lower both test tubes into a small flask so that the tubes remain upright. Stopper the flask. Weigh the flask and contents. Tip the flask so that the test tubes spill their contents into the flask, causing the formation of yellow, insoluble, silver iodide. Ask students to predict any change in mass. Because a dense solid formed, students might predict that the mass increased. Then, weigh the flask and contents to demonstrate the law of conservation of mass. OL

evaporating dish. In the fume hood, carefully add a few drops of water to the mixture using a dropper. **WARNING:** ***This demonstration should be done only in a fume hood. I_2 vapors are toxic.*** The dish will become hot. Collect the toxic, violet-colored iodine vapors in a wide-mouth glass bottle that is inverted and clamped over the evaporating dish. Immediately stopper the bottle.

Results

Violet fumes are given off from the reacting mixture as an exothermic reaction proceeds.

Analysis

In forming zinc iodide, one zinc atom reacts with two iodine atoms. Use this information and the dot diagrams to explain how the reaction occurred. Zinc shares one electron with each iodine atom.

ASSESSMENT

Skill Ask students to mark on a blank periodic table the position of the dividing line between metals and nonmetals. OL

Steps for balancing equations Most chemical equations can be balanced by following the steps given in **Table 2.** For example, you can use these steps to write the chemical equation for the reaction between hydrogen (H_2) and chlorine (Cl_2) that produces hydrogen chloride (HCl).

Explore **balancing chemical equations with an interactive table.** Concepts In Motion

Table 2 **Steps for Balancing Equations**

Step	Process	Example
1	*Write the skeleton equation for the reaction.* Make sure that the chemical formulas correctly represent the substances. An arrow separates the reactants from the products, and a plus sign separates multiple reactants and products. Show the physical states of all reactants and products.	$H_2(g)$ + $Cl_2(g)$ → HCl(g)  Two hydrogen atoms; Two chlorine atoms; One hydrogen atom One chlorine atom
2	*Count the atoms of the elements in the reactants.* If a reaction involves identical polyatomic ions in the reactants and products, count each polyatomic ion as a single element. This reaction does not involve any polyatomic ions. Two atoms of hydrogen and two atoms of chlorine are reacting.	H_2 + Cl_2 → 2 atoms H 2 atoms Cl
3	*Count the atoms of the elements in the products.* One atom of hydrogen and one atom of chlorine are produced.	HCl 1 atom H + 1 atom Cl
4	*Change the coefficients to make the number of atoms of each element equal on both sides of the equation.* Never change a subscript in a chemical formula to balance an equation because doing so changes the identity of the substance.	H_2 + Cl_2 → 2HCl 2 atoms H; 2 atoms Cl; 2 atoms H + 2 atoms Cl  Two hydrogen atoms; Two chlorine atoms; Two hydrogen atoms Two chlorine atoms
5	*Write the coefficients in their lowest possible ratio.* The coefficients should be the smallest possible whole numbers. The ratio 1 hydrogen to 1 chlorine to 2 hydrogen chloride (1:1:2) is the lowest-possible ratio because the coefficients cannot be reduced further and still remain whole numbers.	$H_2(g) + Cl_2(g) \longrightarrow 2HCl(g)$ 1:1:2 1 H_2 to 1 Cl_2 to 2 HCl
6	*Check your work.* Make sure that the chemical formulas are written correctly. Then, check that the number of atoms of each element is equal on both sides of the equation.	H_2 + Cl_2 → 2HCl 2 atoms H 2 atoms Cl 2 atoms H + 2 atoms Cl There are two hydrogen atoms and two chlorine atoms on both sides of the equation.

Identify Misconceptions

Students might think that the coefficients in a chemical equation must balance.

Uncover the Misconception Have students write the balanced equation for the reaction between carbon (graphite) and gaseous oxygen yielding gaseous carbon dioxide. $C(s) + O_2(g) \longrightarrow CO_2(g)$ Ask them to compare the sum of the coefficients on the two sides of the balanced equation. The sum on the reactant side is two, and the sum on the product side is one, showing that the sums do not have to be equal.

Demonstrate the Concept Use ball-and-stick molecular models to model the reaction between gaseous acetylene (C_2H_2) and gaseous oxygen yielding gaseous carbon dioxide and liquid water: $2C_2H_2(g) + 5O_2(g) \longrightarrow 4CO_2(g) + 2H_2O(l)$. Show that the number of atoms of each element in the products equals the number in the reactants, but the sum of reactant coefficients (7) does not equal the sum of product coefficients (6).

Assess New Knowledge Have students write the balanced equation for the reaction between solid phosphorus (P_4) and liquid bromine yielding solid phosphorus pentabromide. $P_4(s) + 10Br_2(l) \longrightarrow 4PBr_5(s)$ Emphasize the point by having them compare the sum of reactant coefficients (11) with the sum of product coefficients (4). OL

DIFFERENTIATED INSTRUCTION

Struggling Learners Matter is conserved in a chemical reaction. If you have 9 g of reactants, the total mass of products is 9 g. Reacting 1 g of hydrogen with 8 g of oxygen produces 9 g of water. Show students the word equation with the masses of oxygen and hydrogen for this reaction. Ask students how many grams of water are produced. 9 g BL

EXAMPLE Problem 1

Get help with **ratios.** Math Handbook

WRITING A BALANCED CHEMICAL EQUATION Write the balanced chemical equation for the reaction in which aqueous sodium hydroxide and aqueous calcium bromide react to produce solid calcium hydroxide and aqueous sodium bromide.

1 ANALYZE THE PROBLEM

You are given the reactants and products in a chemical reaction. Start with a skeleton equation, and use the steps given in **Table 2** for balancing chemical equations.

2 SOLVE FOR THE UNKNOWN

Write the skeleton equation for the chemical reaction. Be sure to put the reactants on the left side of the arrow and the products on the right. Separate the substances with plus signs, and indicate their physical states.

$$NaOH(aq) + CaBr_2(aq) \rightarrow Ca(OH)_2(s) + NaBr(aq)$$

1 Na, 1 O, 1 H, 1 Ca, 2 Br	**Count the atoms of each element in the reactants.**
1 Na, 2 O, 2 H, 1 Ca, 1 Br	**Count the atoms of each element in the products.**
$2NaOH + CaBr_2 \rightarrow Ca(OH)_2 + NaBr$	**Insert the coefficient 2 in front of NaOH to balance the hydroxide ions.**
$2NaOH + CaBr_2 \rightarrow Ca(OH)_2 + 2NaBr$	**Insert the coefficient 2 in front of NaBr to balance the Na and Br atoms.**
The ratio of the coefficients is 2:1:1:2.	**Write the coefficients in their lowest-possible ratio.**
Reactants: 2 Na, 2 OH, 1 Ca, 2 Br Products: 2 Na, 2 OH, 1 Ca, 2 Br.	**Check to make sure that the number of atoms of each element is equal on both sides of the equation.**

3 EVALUATE THE ANSWER

The chemical formulas for all substances are written correctly. The number of atoms of each element is equal on both sides of the equation. The coefficients are written in the lowest possible ratio. The balanced chemical equation for the reaction is

$$2NaOH(aq) + CaBr_2(aq) \rightarrow Ca(OH)_2(s) + 2NaBr(aq)$$

PRACTICE Problems

Do additional problems. Online Practice

Write chemical equations for each of the following reactions.

4. In water, iron(III) chloride reacts with sodium hydroxide, producing solid iron(III) hydroxide and sodium chloride.

5. Liquid carbon disulfide reacts with oxygen gas, producing carbon dioxide gas and sulfur dioxide gas.

6. **Challenge** A piece of zinc metal is added to a solution of dihydrogen sulfate. This reaction produces a gas and a solution of zinc sulfate.

Marilyn Genter/The Image Works

RealWorld CHEMISTRY

Calcium Hydroxide

REEF AQUARIUMS An aqueous solution of calcium hydroxide is used in reef aquariums to provide calcium for animals such as snails and corals. Calcium hydroxide reacts with the carbon dioxide in the water to produce calcium and bicarbonate ions. Reef animals use the calcium to grow shells and strong skeletal systems.

IN-CLASS Example

Question Write a balanced chemical equation for the reaction of carbon monoxide (CO) and oxygen (O_2) that produces carbon dioxide (CO_2).

Answer $2CO(g) + O_2(g) \rightarrow 2CO_2(g)$

PRACTICE Problems

Have students refer to the Selected Solutions appendix for complete solutions to odd-numbered problems. The complete solutions for all problems can be found in the Solutions Manual.

4. $FeCl_3(aq) + 3NaOH(aq) \rightarrow Fe(OH)_3(s) + 3NaCl(aq)$

5. $CS_2(l) + 3O_2(g) \rightarrow CO_2(g) + 2SO_2(g)$

6. $Zn(s) + H_2SO_4(aq) \rightarrow H_2(g) + ZnSO_4(aq)$

ASSESSMENT

Knowledge Have students determine whether or not each of the following chemical equations is balanced. For any equation that is not balanced, have students change coefficients to correct the imbalance.

a. $2Al + 6HCl \rightarrow 3AlCl_3 + 3H_2$
The equation is not balanced. The correct equation is $2Al + 6HCl \rightarrow 2AlCl_3 + 3H_2$.

b. $Ba(ClO_3)_2 \rightarrow BaCl_2 + 2O_2$
The equation is not balanced. The correct equation is $Ba(ClO_3)_2 \rightarrow BaCl_2 + 3O_2$.

c. $3Sn(SO_4)_2 + 4K_3PO_4 \rightarrow Sn_3(PO_4)_4 + 6K_2SO_4$
The equation is balanced.

d. $C_5H_{12} + 11O_2 \rightarrow 5CO_2 + 12H_2O$
The equation is not balanced. The correct equation is $C_5H_{12} + 8O_2 \rightarrow 5CO_2 + 6H_2O$. **OL**

CULTURAL DIVERSITY

The Language of Chemistry Ask students if they can tell what subject is being discussed in the following passage from a Spanish textbook: "El gas cloro se representa por la fórmula Cl_2. Otros seis elementos comunes que existen normalmente como moléculas diatómicas son el hidrógeno (H_2), el nitrógeno (N_2), el oxígeno (O_2) y el flúor (F_2). El bromo (Br_2) existe en su estado gaseoso a una temperatura sobre 58.8°C. El yodo (I_2) se transforma en gas a la temperatura de 184°C." On the basis of the chemical formulas, students will probably notice that the subject is chemistry. Explain that although scientists in different countries communicate with each other using various written and spoken languages, scientists the world over represent chemical reactions using the same chemical symbols and equations.

3 Assess

Check for Understanding

On the board, write $N_2(g) + O_2(g) \rightarrow N_2O_5(s)$. Ask students why it is incorrect to balance the equation by changing the formula for the reacting oxygen to O_5. Although the equation would be balanced, it would be incorrect because oxygen is diatomic (O_2), not pentatomic (O_5). OL

Reteach

Have each student review the reactions that they have recorded in their chemistry journal and write a balanced equation for any of those reactions that are not already balanced. Have students share reactions with the class. OL

Extension

For reactions in which polyatomic ions obviously remain intact, students might consider them as units to simplify the balancing process. For example, in the reaction $2Li_3PO_4(aq) + 3CaCl_2(aq) \rightarrow 6LiCl(aq) + Ca_3(PO_4)_2(s)$, two phosphate ions on each side of the equation balance phosphorus and oxygen. Have students balance several equations containing intact polyatomic ions. OL

ASSESSMENT

Knowledge Ask students to write the balanced chemical equation for the reaction of lead nitrate and potassium iodide. $Pb(NO_3)_2(aq) + 2KI(aq) \rightarrow PbI_2(s) + 2KNO_3(aq)$ OL

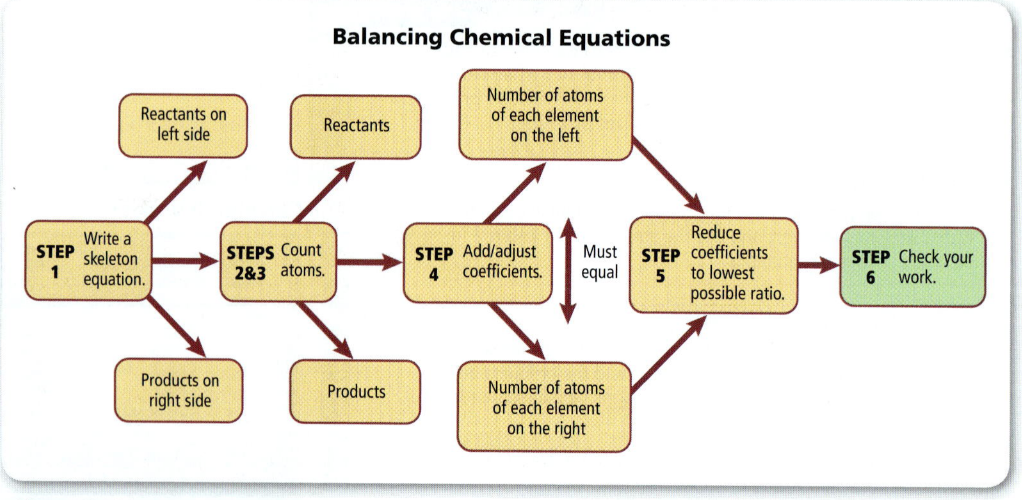

■ **Figure 6** It is imperative to your study of chemistry to be able to balance chemical equations. Use this flowchart to help you master the skill. Notice that the numbered steps correspond to the steps in **Table 2.**

Explore **balancing chemical reactions.**

Virtual Investigations

Obeying the law of conservation of mass Probably the most fundamental concept of chemistry is the law of conservation of mass that you encountered previously. All chemical reactions obey the law that matter is neither created nor destroyed. Therefore, it is also fundamental that the equations that represent chemical reactions include sufficient information to show that the reaction obeys the law of conservation of mass.

You have learned how to show the conservation of mass with balanced chemical equations. The flowchart shown in **Figure 6** summarizes the steps for balancing equations. You will probably find that some chemical equations can be balanced easily, whereas others are more difficult to balance. Most chemical equations can be balanced by the process you learned in this section.

SECTION 1 REVIEW

Section Self-Check

Section Summary

- Some physical changes are evidence that indicate a chemical reaction has occurred.
- Word equations and skeleton equations provide important information about a chemical reaction.
- A chemical equation gives the identities and relative amounts of the reactants and products that are involved in a chemical reaction.
- Balancing an equation involves adjusting the coefficients until the number of atoms of each element is equal on both sides of the equation.

7. MAINIDEA **Explain** why it is important that a chemical equation be balanced.
8. **List** three types of physical evidence that indicate a chemical reaction has occurred.
9. **Compare and contrast** a skeleton equation and a chemical equation.
10. **Explain** why it is important to reduce coefficients in a balanced equation to the lowest-possible whole-number ratio.
11. **Analyze** When balancing a chemical equation, can you adjust the subscript in a formula? Explain.
12. **Assess** Is the following equation balanced? If not, correct the coefficients to balance the equation.

$$2K_2CrO_4(aq) + Pb(NO_3)_2(aq) \rightarrow 2KNO_3(aq) + PbCrO_4(s)$$

13. **Evaluate** Aqueous phosphoric acid and aqueous calcium hydroxide react to form solid calcium phosphate and water. Write a balanced chemical equation for this reaction.

SECTION 1 REVIEW

7. Because mass is neither created nor destroyed in chemical reactions, the numbers of atoms of all elements must be equal on both sides of the reaction arrow.
8. Answers might include release or absorption of energy, change in color, change in odor, formation of a gas, or formation of a solid.
9. The skeleton equation includes the formulas of reactants and products. The chemical equation gives the relative amounts of reactants and products.
10. Coefficients in the lowest ratio most clearly indicate the relative amounts of substances in a reaction.
11. No. Doing so changes the identity of the substance.
12. No. The correct equation is $K_2CrO_4(aq) + Pb(NO_3)_2(aq) \rightarrow 2KNO_3(aq) + PbCrO_4(s)$.
13. $2H_3PO_4(aq) + 3Ca(OH)_2(aq) \rightarrow Ca_3(PO_4)_2(s) + 6H_2O(aq)$

Essential Questions

- How are chemical reactions classified?
- What are the characteristics of different classes of chemical reactions?

Review Vocabulary

metal: an element that is a solid at room temperature, a good conductor of heat and electricity, and is generally shiny

New Vocabulary

synthesis reaction
combustion reaction
decomposition reaction
single-replacement reaction
double-replacement reaction
precipitate

Classifying Chemical Reactions

MAINIDEA There are four types of chemical reactions: synthesis, combustion, decomposition, and replacement reactions.

CHEM 4 YOU It could take you a long time to find a specific novel in an unorganized bookstore. Bookstores classify and organize books into different categories to make your search easier. Chemical reactions are also classified and organized into different categories.

Types of Chemical Reactions

Chemists classify chemical reactions in order to organize the many reactions that occur daily. Knowing the categories of chemical reactions can help you remember and understand them. It can also help you recognize patterns and predict the products of many chemical reactions. One way chemists classify reactions is to distinguish among the four types: synthesis, combustion, decomposition, and replacement reactions. Some reactions fit into more than one of these types.

Synthesis Reactions

In **Figure 7,** sodium and chlorine react to produce sodium chloride. This reaction is a **synthesis reaction**—a chemical reaction in which two or more substances (A and B) react to produce a single product (AB).

$$A + B \rightarrow AB$$

When two elements react, the reaction is always a synthesis reaction.

Two compounds can also combine to form one compound. For example, the reaction between calcium oxide (CaO) and water (H_2O) to form calcium hydroxide ($Ca(OH)_2$) is a synthesis reaction.

$$CaO(s) + H_2O(l) \rightarrow Ca(OH)_2(s)$$

Another type of synthesis reaction involves a reaction between a compound and an element, as happens when sulfur dioxide gas (SO_2) reacts with oxygen gas (O_2) to form sulfur trioxide (SO_3).

$$2SO_2(g) + O_2(g) \rightarrow 2SO_3(g)$$

Figure 7 In this synthesis reaction, two elements, sodium and chlorine, react to produce one compound, sodium chloride.

$$2Na(s) + Cl_2(g) \rightarrow 2NaCl(s)$$

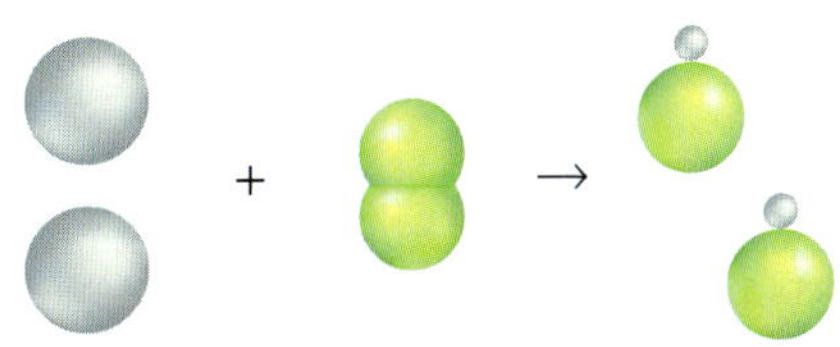

1 Focus

MAINIDEA

Classify Reactions Ask students why it might be important to organize chemical reactions according to different types or categories. to allow for easier study and understanding of reactions Ask students what type of reaction occurs when wood is burned. combustion reaction Water is produced by the reaction of hydrogen gas and oxygen gas. Ask students what type of reaction this is. synthesis reaction OL

2 Teach

Use Science Terms

Make or Break Have students write statements explaining the meaning of the terms *synthesize* and *decompose*. Ask them to write four illustrative sentences–two that include each word. OL EL

Content Background

Real-World Connection The London smog disaster of 1952 was triggered by a temperature inversion, a weather condition in which cool, static air is trapped close to the ground. The smog in December, 1952 contained unusually high amounts of soot, tar, and sulfur dioxide from coal burned in homes and industrial sources. Because most deaths were the result of pneumonia, bronchitis, tuberculosis and heart failure, common causes of death, few realized that the death rate had increased so dramatically. Dr. Robert Waller of St. Bartholomew's Hospital stated in an interview for the BBC, "The interesting thing is that no one realized at the time that the number of deaths was increasing ...until florists began to run out of flowers and undertakers began to run out of coffins."

Use Science Terms

Nanoscale Have student research the meaning of the prefix *nano*. 10^{-9} Have students research the meaning of the term *nanoscale engineering* and some of possible future applications of nanoscale engineering. Possible applications include colloidal science, microprocessors, and microelectromechanical systems. OL EL

■ **Figure 8** The light produced by a sparkler is the result of a combustion reaction between oxygen and different metals.

VOCABULARY
WORD ORIGIN
Combustion
comes from the Latin word *comburere,* meaning *to burn.*

Combustion Reactions

The synthesis reaction between sulfur dioxide and oxygen can also be classified as a combustion reaction. In a **combustion reaction,** such as the one shown in **Figure 8,** oxygen combines with a substance and releases energy in the form of heat and light. Oxygen can combine in this way with many different substances, making combustion reactions common. To learn more about the discovery of the chemical reaction for combustion and other reactions, review **Figure 9.**

A combustion reaction occurs between hydrogen and oxygen when hydrogen is heated, as illustrated in **Figure 10.** Water is formed during the reaction, and a large amount of energy is released. Another important combustion reaction occurs when coal is burned to produce energy. Coal is called a fossil fuel because it contains the remains of plants that lived long ago. It is composed primarily of the element carbon. Coal-burning power plants generate electric power in many parts of the United States. The primary reaction that occurs in these plants is between carbon and oxygen.

$$C(s) + O_2(g) \rightarrow CO_2(g)$$

■ **Figure 9**
Real-World Chemical Reactions
Throughout history, people have worked to understand and apply the power of chemical reactions to solve problems.

1600 — 1700 — 1800 — 1905 — 1920

1635 America's first chemical plant opens in Boston. Products include saltpeter, a component of gunpowder, and alum, a chemical used in tanning animal skins.

1775 Antoine Lavoisier demonstrates that combustion is an exothermic chemical reaction involving oxygen.

CIRCA 1800 Experiments with plants result in the discovery of the balanced chemical equation for photosynthesis.

1885 The internal combustion engine is invented. It later becomes the prototype for the modern gas engine.

1909–1910 German chemists Fritz Haber and Carl Bosch develop the Haber-Bosch process for synthesizing ammonia.

CHEMISTRY JOURNAL

International Chemistry In the early 1900's, demand for nitrogen-based fertilizers exceeded supply. The largest source of chemicals for fertilizer production was a huge guano deposit in Chile which was rapidly being consumed. The Haber-Bosch process, developed in Germany around World War I, ultimately freed world agriculture from dependence on Chile. Ask students to compare world dependence on Chilean ammonia in 1900 with world dependence on oil from a few sources today. How does dependence on a resource affect international relations? Can the desire to overcome resource dependence spur scientific discovery? While answers will vary, students should recognize that practical applications of chemistry can have far-reaching effects. Have students record their responses in their chemistry journals. OL

$2H_2(g)$ + $O_2(g)$ → $2H_2O(g)$

■ **Figure 10** During a combustion reaction between oxygen and hydrogen, water is formed.

Analyze *Why is this chemical reaction both a synthesis reaction and a combustion reaction?*

Note that the combustion reactions just mentioned are also synthesis reactions. However, not all combustion reactions are synthesis reactions. For example, the reaction involving methane gas (CH_4) and oxygen illustrates a combustion reaction in which one substance replaces another in the formation of products.

$$CH_4(g) + 2O_2(g) \rightarrow CO_2(g) + 2H_2O(g)$$

Methane, which belongs to a group of substances called hydrocarbons, is the major component of natural gas. All hydrocarbons contain carbon and hydrogen and burn in oxygen to yield carbon dioxide and water. You will learn more about hydrocarbons in a later chapter.

PRACTICE Problems

Do additional problems. Online Practice

Write chemical equations for the following reactions. Classify each reaction into as many categories as possible.

14. The solids aluminum and sulfur react to produce aluminum sulfide.

15. Water and dinitrogen pentoxide gas react to produce aqueous hydrogen nitrate.

16. The gases nitrogen dioxide and oxygen react to produce dinitrogen pentoxide gas.

17. Challenge Sulfuric acid (H_2SO_4) and sodium hydroxide solutions react to produce aqueous sodium sulfate and water.

1974–1978 Researchers demonstrate that chlorofluorocarbons (CFCs) can deplete the ozone layer. The use of CFCs as spray propellants is banned in the United States.

2004 Scientists discover that migrating birds are guided by chemical reactions in their bodies that are influenced by Earth's magnetic field.

1950 — 1965 — 1980 — 1995 — 2010

1952 A heavy smog—sulfur dioxide and other coal-burning products—settles over London for five days in December, causing 4000 deaths.

1995 Researchers use the atomic force microscope to create and observe chemical reactions as they occur molecule by molecule, paving the way for nanoscale engineering.

2010 Deborah Jin and colleagues discover that chemical reactions can still occur at temperatures near absolute zero when they observed the formation of KRb from a reaction of diatomic potassium and rubidium molecules.

■ **Caption Question Fig. 10** It is a synthesis reaction because two elements combine to form one product, and it is a combustion reaction because oxygen is combining with another substance and releasing energy.

PRACTICE Problems

Have students refer to the Selected Solutions appendix for complete solutions to odd-numbered problems. The complete solutions for all problems can be found in the Solutions Manual.

14. $2Al(s) + 3S(s) \rightarrow Al_2S_3(s)$; synthesis

15. $H_2O(l) + N_2O_5(g) \rightarrow 2HNO_3(aq)$; synthesis

16. $4NO_2(g) + O_2(g) \rightarrow 2N_2O_5(g)$; synthesis and combustion

17. $H_2SO_4(aq) + 2NaOH(aq) \rightarrow Na_2SO_4(aq) + 2H_2O(l)$; synthesis

Content Background

Products of Decomposition Reactions Most decomposition reactions require the input of energy, often in the form of heat or electricity. Students will find it possible to predict the products of many decomposition reactions by recognizing these five general types.

1. metal carbonate $\rightarrow$ metal oxide $+ CO_2$
2. metal chlorate $\rightarrow$ metal chloride $+ O_2$
3. metal hydroxide $\rightarrow$ metal oxide $+ H_2O$
4. metal oxide $\rightarrow$ metal $+$ oxygen
5. oxyacid $\rightarrow$ nonmetal oxide $+ H_2O$

PRACTICE Problems

Have students refer to the Selected Solutions appendix for complete solutions to odd-numbered problems. The complete solutions for all problems can be found in the Solutions Manual.

18. $2Al_2O_3(s) \rightarrow 4Al(s) + 3O_2(g)$
19. $Ni(OH)_2(s) \rightarrow NiO(s) + H_2O(l)$
20. $2NaHCO_3(s) \rightarrow Na_2CO_3(aq) + CO_2(g) + H_2O(l)$

FOLDABLES® Rubric

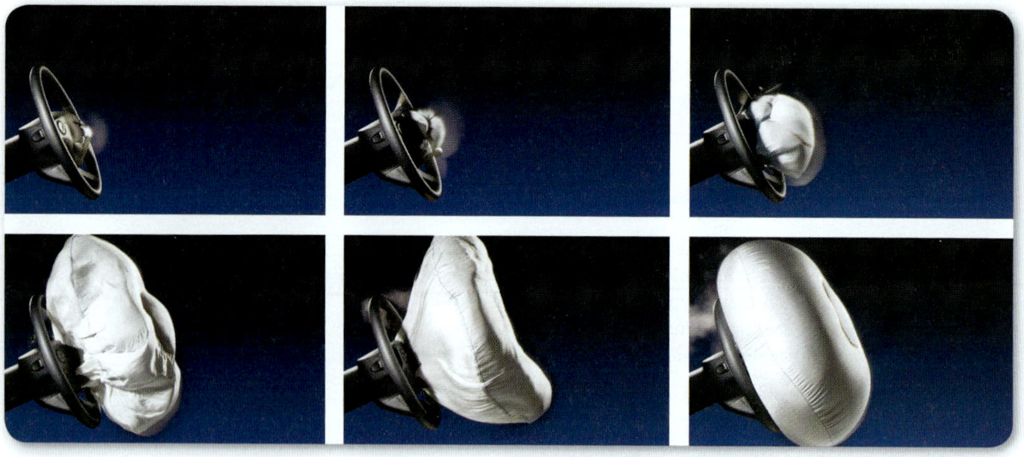

■ **Figure 11** The decomposition of sodium azide, which produces a gas, is the chemical reaction that inflates air bags.

FOLDABLES® Incorporate information from this section into your Foldable.

Decomposition Reactions

Some chemical reactions are essentially the opposite of synthesis reactions. These reactions are classified as decomposition reactions. A **decomposition reaction** is one in which a single compound breaks down into two or more elements or new compounds. In generic terms, decomposition reactions can be represented as follows.

$$AB \rightarrow A + B$$

Decomposition reactions often require an energy source, such as heat, light, or electricity, to occur. For example, ammonium nitrate breaks down into dinitrogen monoxide and water when the reactant is heated to a high temperature.

$$NH_4NO_3(s) \rightarrow N_2O(g) + 2H_2O(g)$$

Notice that this decomposition reaction involves one reactant breaking down into more than one product.

The outcome of another decomposition reaction is shown in **Figure 11.** Automobile safety air bags inflate rapidly as sodium azide pellets decompose. A device that can provide an electric signal to start the reaction is packaged inside air bags along with the sodium azide pellets. When the device is activated, sodium azide decomposes, producing nitrogen gas that quickly inflates the air bag.

$$2NaN_3(s) \rightarrow 2Na(s) + 3N_2(g)$$

PRACTICE Problems

Do additional problems. Online Practice

Write chemical equations for the following decomposition reactions.

18. Aluminum oxide(s) decomposes when electricity passes through it.

19. Nickel(II) hydroxide(s) decomposes to produce nickel(II) oxide(s) and water.

20. Challenge Heating sodium hydrogen carbonate(s) produces sodium carbonate(aq) and water. Carbon dioxide gas is also produced.

CHEMISTRY PROJECT

Combustion of Carbohydrates Ask students to research the catabolic pathway of carbohydrates, including the reaction of glucose with oxygen. Students should also place this reaction in one of the categories studied in this chapter. The reaction of glucose with oxygen: $C_6H_{12}O_6(s) + 6O_2(g) \rightarrow 6CO_2(g) + 6H_2O(l) +$ energy. This is a combustion reaction. **OL**

Figure 12 In a single-replacement reaction, the atoms of one element replace the atoms of another element in a compound.

Replacement Reactions

In contrast to synthesis, combustion, and decomposition reactions, many chemical reactions are replacement reactions and involve the replacement of an element in a compound. These reactions are also called displacement reactions. There are two types of replacement reactions: single-replacement reactions and double-replacement reactions.

Single-replacement reactions The reaction between lithium and water is shown in **Figure 12.** The following chemical equation shows that a lithium atom replaces one of the hydrogen atoms in a water molecule.

$$2Li(s) + 2H_2O(l) \longrightarrow 2LiOH(aq) + H_2(g)$$

A reaction in which the atoms of one element replace the atoms of another element in a compound is called a **single-replacement reaction.**

$$A + BX \longrightarrow AX + B$$

Metal replaces hydrogen or another metal The reaction between lithium and water is one type of single-replacement reaction, in which a metal replaces a hydrogen atom in a water molecule. Another type of single-replacement reaction occurs when one metal replaces another metal in a compound dissolved in water. **Figure 12** shows a single-replacement reaction occurring when a bar of pure copper is placed in aqueous silver nitrate. The crystals that are accumulating on the copper bar are the silver atoms that the copper atoms replaced.

$$Cu(s) + 2AgNO_3(aq) \longrightarrow 2Ag(s) + Cu(NO_3)_2(aq)$$

A metal will not always replace another metal in a compound dissolved in water because metals differ in their reactivities. Reactivity is the ability to react with another substance. An activity series of some metals is shown in **Figure 13.** This series orders metals by reactivity with other metals. Single-replacement reactions are used to determine a metal's position on the list. The most active metals are at the top of the list. The least active metals are at the bottom. Similarly, the reactivity of each halogen has been determined and listed, as shown in **Figure 13.**

(l)Charles D. Winters/Photo Researchers, (r)McGraw-Hill Education

Figure 13 An activity series, similar to the series shown here for various metals and halogens, is a useful tool for determining whether a chemical reaction will occur and for determining the result of a single-replacement reaction.

Most active ↓ **Least active**

METALS
Lithium
Rubidium
Potassium
Calcium
Sodium
Magnesium
Aluminum
Manganese
Zinc
Iron
Nickel
Tin
Lead
Copper
Silver
Platinum
Gold

Most active ↓ **Least active**

HALOGENS
Fluorine
Chlorine
Bromine
Iodine

Quick Demo

Single-Replacement Reaction Obtain a small piece of galvanized iron from a hardware store or from a heating/cooling contractor's scrap. At the beginning of a class period, immerse the piece of galvanized iron in a 100-mL beaker containing 80 mL of dilute hydrochloric acid in a fume hood. Have students describe the reaction that occurs. Bubbles form rapidly for a time. Then, the bubbling slows considerably. Explain that a single-replacement reaction has occurred between the hydrochloric acid and the galvanized iron's zinc coating. Ask students to write the chemical equation for the reaction and identify the gas produced. $Zn(s) + 2HCl(aq) \longrightarrow ZnCl_2(aq) + H_2(g)$ The gas is hydrogen. Ask students why the iron metal under the zinc coating reacts more slowly than the zinc. Zinc is more reactive than iron. Wash the zinc compounds down a drain with large quantities of water. **OL**

CHEMISTRY JOURNAL

Single-Replacement Reactions Have students write chemical equations for the single-replacement reactions that occur when aluminum metal is placed in aqueous solutions of nickel(II) nitrate, sodium nitrate, lead(IV) nitrate, and gold(III) nitrate. If no reaction occurs, have them write NR in place of the products. Have students include the equations in their chemistry journals. $2Al(s) + 3Ni(NO_3)_2(aq) \longrightarrow 2Al(NO_3)_3(aq) + 3Ni(s)$, $Al(s) + NaNO_3(aq) \longrightarrow NR$, $4Al(s) + 3Pb(NO_3)_4(aq) \longrightarrow 4Al(NO_3)_3(aq) + 3Pb(s)$, $Al(s) + Au(NO_3)_3(aq) \longrightarrow Al(NO_3)_3(aq) + Au(s)$ **OL**

READING CHECK The atoms of one element replace the atoms of another element in a compound.

Problem-Solving LAB

Purpose Students will correlate the properties of metals with their chemical activity.

Process Skills recognize cause and effect, apply concepts, classify, predict

Teaching Strategies

- Ask students what properties should be considered in determining the relative reactivity of the metals. Electronegativity, oxidation states, atomic radius, ionization energy, and effective nuclear charge are properties that exhibit periodic trends.
- Ask students to explore how each of their suggestions would help predict the reactivity of a particular element. Generally, metals with a low ionization energy and large radius react more easily. Further consideration must be given to the stability of the shell structure.

Think Critically

1. Student graphs should accurately reflect the data in the table.
2. With increasing atomic number, atomic radius increases, ionization energy decreases, and electronegativity decreases.
3. These trends correlate with the decrease in reactivity down the period.
4. Astatine is a group 17 element and follows the trends described for the halogens. Therefore, At would be at the bottom of the activity series on the basis of its periodic trends.

ASSESSMENT

Knowledge Students should write an account of their Problem-Solving Lab findings and explain how the interpretations of their findings were used to place the other given elements in the reactive series. OL

You can use the activity series to predict whether or not certain reactions will occur. A specific metal can replace any metal listed below it that is in a compound. It cannot replace any metal listed above it. For example, copper atoms replace silver atoms in a solution of silver nitrate. However, if you place a silver wire in aqueous copper(II) nitrate, the silver atoms will not replace the copper. Silver is listed below copper in the activity series, so no reaction occurs. The letters NR (no reaction) are commonly used to indicate that a reaction will not occur.

$$Ag(s) + Cu(NO_3)_2(aq) \rightarrow NR$$

Nonmetal replaces nonmetal A third type of single-replacement reaction involves the replacement of a nonmetal in a compound by another nonmetal. Halogens are frequently involved in these types of reactions. Like metals, halogens exhibit different activity levels in single-replacement reactions. The reactivities of halogens, determined by single-replacement reactions, are also shown in **Figure 13.** The most active halogen is fluorine, and the least active is iodine. A more reactive halogen replaces a less reactive halogen that is part of a compound dissolved in water. For example, fluorine replaces bromine in water containing dissolved sodium bromide. However, bromine does not replace fluorine in water containing dissolved sodium fluoride.

$$F_2(g) + 2NaBr(aq) \rightarrow 2NaF(aq) + Br_2(l)$$
$$Br_2(g) + 2NaF(aq) \rightarrow NR$$

READING CHECK **Explain** how a single-replacement reaction works.

Problem-Solving LAB

Analyze Trends

How can you explain the reactivities of halogens? The location of all the halogens in group 17 in the periodic table tells you that halogens have common characteristics. Indeed, halogens are all nonmetals and have seven electrons in their outermost orbitals. However, each halogen also has its own characteristics, such as the ability to react with other substances.

Analysis

Examine the accompanying data table. It includes data about the atomic radii, ionization energies, and electronegativities of the halogens.

Properties of Halogens

Halogen	Atomic Radius (pm)	Ionization Energy (kJ/mol)	Electronegativity
Fluorine	72	1681	3.98
Chlorine	100	1251	3.16
Bromine	114	1140	2.96
Iodine	133	1008	2.66
Astatine	140	920	2.2

Think Critically

1. **Make graphs** Use the information in the data table to make three line graphs.
2. **Describe** any periodic trends that you identify in the data.
3. **Relate** any periodic trends that you identify among the halogens to the activity series of halogens shown in **Figure 13.**
4. **Predict** the location of the element astatine in the activity series of halogens. Explain.

DIFFERENTIATED INSTRUCTION

Advanced Learners Ask students to devise and explain an experimental procedure they would use to place an unknown metal in the activity series of metals. Typical student procedures will involve putting samples of the metal in aqueous solutions of the salts of other, known metals. Such experiments provide data that allow students to place the unknown metal above metals it replaces and below metals it does not replace. AL

EXAMPLE Problem 2

SINGLE-REPLACEMENT REACTIONS Predict the products that will result when these reactants combine, and write a balanced chemical equation for each reaction.

a. $Fe(s) + CuSO_4(aq) \rightarrow$

b. $Br_2(l) + MgCl_2(aq) \rightarrow$

c $Mg(s) + AlCl_3(aq) \rightarrow$

1 ANALYZE THE PROBLEM

You are given three sets of reactants. Using **Figure 13,** you must first determine if each reaction occurs. Then, if a reaction is predicted, you can determine the product(s) of the reaction. With this information you can write a skeleton equation for the reaction. Finally, you can use the steps for balancing chemical equations to write the complete balanced chemical equation.

2 SOLVE FOR THE UNKNOWN

a. Iron is listed above copper in the activity series. Therefore, the first reaction will occur because iron is more reactive than copper. In this case, iron will replace copper. The skeleton equation for this reaction is

$$Fe(s) + CuSO_4(aq) \rightarrow FeSO_4(aq) + Cu(s)$$

This equation is balanced.

b. In the second reaction, chlorine is more reactive than bromine because bromine is listed below chlorine in the activity series. Therefore, the reaction will not occur. The skeleton equation for this situation is

$$Br(l) + MgCl_2(aq) \rightarrow NR$$

No balancing is required.

c. Magnesium is listed above aluminum in the activity series. Therefore, the third reaction will occur because magnesium is more reactive than aluminum. In this case, magnesium will replace aluminum. The skeleton equation for this reaction is

$$Mg(s) + AlCl_3(aq) \rightarrow Al(s) + MgCl_2(aq)$$

This equation is not balanced. The balanced equation is

$$3Mg(s) + 2AlCl_3(aq) \rightarrow 2Al(s) + 3MgCl_2(aq)$$

3 EVALUATE THE ANSWER

The activity series shown in **Figure 13** supports the reaction predictions. The chemical equations are balanced because the number of atoms of each substance is equal on both sides of the equation.

PRACTICE Problems

Do additional problems. Online Practice

Predict whether the following single-replacement reactions will occur. If a reaction occurs, write a balanced equation for the reaction.

21. $K(s) + ZnCl_2(aq) \rightarrow$

22. $Cl_2(g) + HF(aq) \rightarrow$

23. $Fe(s) + Na_3PO_4(aq) \rightarrow$

24. Challenge $Al(s) + Pb(NO_3)_2(aq) \rightarrow$

StudioSource/Alamy

RealWorld CHEMISTRY

Single-Replacement Reactions

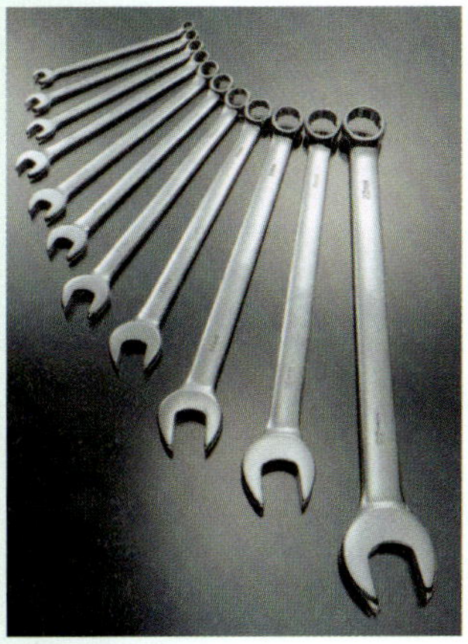

ZINC PLATING Tools made of steel are often covered with a layer of zinc to prevent corrosion. Zinc is more reactive than the lead in steel. During zinc plating, the zinc replaces some of the surface lead, coating the steel.

IN-CLASS Example

Question Predict if the following reaction will occur and indicate products formed.

1. $Li(s) + NaOH(aq) \rightarrow$

2. $F_2(g) + HCl(aq) \rightarrow$

3. $Ag(s) + AlCl_3(aq) \rightarrow$

Answer

1. $Li(s) + NaOH(aq) \rightarrow Na(s) + LiOH(aq)$

2. $F_2(g) + 2HCl(aq) \rightarrow Cl_2(g) + 2HF(aq)$

3. $Ag(s) + AlCl_3(aq) \rightarrow$ no reaction

PRACTICE Problems

Have students refer to the Selected Solutions appendix for complete solutions to odd-numbered problems. The complete solutions for all problems can be found in the Solutions Manual.

21. Yes. K is above Zn in the activity series. $2K(s) + ZnCl_2(aq) \rightarrow Zn(s) + 2KCl(aq)$

22. No. Cl is below F in the activity series.

23. No. Fe is below Na in the activity series.

24. Yes. Al is above Pb in the activity series. $2Al(s) + 3Pb(NO_3)_2(s) \rightarrow 3Pb(s) + 2Al(NO_3)_3(aq)$

ChemLAB

The ChemLab located at the end of the chapter can be used at this point in the lesson.

CHEMISTRY JOURNAL

An Analogy for Reactions Have students compare reaction types with dance activities and record their comparisons in their chemistry journals. Synthesis—two people come together to dance. Decomposition—two people part when the dance is over. Single-replacement—another person cuts in and replaces one of the two partners. Double-replacement—two couples switch dance partners. OL

Extension

Activity Series of Metals Explain to students that the reactivities of metals in the activity series can be described in greater detail than is shown in **Figure 13.** Metals from magnesium down to iron react with steam (but not cold water). Metals from nickel to lead react with neither water nor steam. All metals above silver react with oxygen, forming oxides, while silver, platinum, and gold form oxides only indirectly.

Apply Chemistry

Jewelry Ask students if they have ever had an allergic reaction to metal jewelry. Then, ask them to consider the position of nickel, gold, silver, and platinum on the activity series. Ask students which is most active and which is least active. most active: nickel, least active: gold Ask students which of these metals might be most likely to cause skin reactions when used in jewelry, nickel and which type of metal jewelry might be the best selection for a person with allergies to jewelry. gold, platinum, or silver OL

■ **Figure 14** The color-coding in the generic equation for a double-replacement reaction and in the equation for the reaction between calcium hydroxide and hydrochloric acid shows the anions changing places.

$$AX + BY \rightarrow AY + BX$$

$$Ca(OH)_2(aq) + 2HCl(aq) \rightarrow CaCl_2(aq) + 2H_2O(l)$$

Double-replacement reactions The final type of replacement reaction, which involves an exchange of ions between two compounds, is called a **double-replacement reaction.**

In the generic equation in **Figure 14,** A and B represent positively charged ions (cations), and X and Y represent negatively charged ions (anions). Notice that the anions have switched places and are now bonded to the other cations in the reaction. In other words, X replaces Y and Y replaces X—a double replacement. More simply, the positive and negative ions of two compounds switch places.

The reaction between calcium hydroxide and hydrochloric acid is a double-replacement reaction.

$$Ca(OH)_2(aq) + 2HCl(aq) \rightarrow CaCl_2(aq) + 2H_2O(l)$$

The ionic components of the reaction are Ca^{2+}, OH^-, H^+, and Cl^-. Knowing this, you can now see the two replacements of the reaction. The anions (OH^- and Cl^-) have changed places and are now bonded to the other cations (Ca^{2+} and H^+), as shown in **Figure 14.**

The reaction between sodium hydroxide and copper(II) chloride in solution is also a double-replacement reaction.

$$2NaOH(aq) + CuCl_2(aq) \rightarrow 2NaCl(aq) + Cu(OH)_2(s)$$

In this case, the anions (OH^- and Cl^-) changed places and bonded to the other cations (Na^+ and Cu^{2+}). **Figure 15** shows that the result of this reaction is a solid product, copper(II) hydroxide. A solid produced during a chemical reaction in a solution is called a **precipitate.**

■ **Figure 15** When aqueous sodium hydroxide is added to a solution of copper(II) chloride, the anions (OH^- and Cl^-) change places. The resulting products are sodium chloride, which remains in solution, and copper(II) hydroxide, the blue solid in the test tube.

View an **animation about precipitate formation.**

Concepts In Motion

Andrew Lambert Photography/Photo Researchers

CHEMISTRY PROJECT

Chemistry in Industry Ask students to research how scrubbers remove sulfur and nitrogen oxides from power plant emissions. Ask them to make a flowchart illustrating their findings, write equations for any chemical reactions involved, and classify each reaction. Have them present the flowcharts to the class. OL EL

Table 3 Guidelines for Writing Double-Replacement Reactions

Step	Example
1. Write the components of the reactants in a skeleton equation.	$Al(NO_3)_3 + H_2SO_4$
2. Identify the cations and the anions in each compound.	$Al(NO_3)_3$ has Al^{3+} and NO_3^- H_2SO_4 has H^+ and SO_4^{2-}
3. Pair up each cation with the anion from the other compound.	Al^{3+} pairs with SO_4^{2-} H^+ pairs with NO_3^-
4. Write the formulas for the products using the pairs from Step 3.	$Al_2(SO_4)_3$ HNO_3
5. Write the complete equation for the double-replacement reaction.	$Al(NO_3)_3 + H_2SO_4 \rightarrow Al_2(SO_4)_3 + HNO_3$
6. Balance the equation.	$2Al(NO_3)_3 + 3H_2SO_4 \rightarrow Al_2(SO_4)_3 + 6HNO_3$

Products of double-replacement reactions One of the key characteristics of double-replacement reactions is the type of product that is formed when the reaction takes place. All double-replacement reactions produce either water, a precipitate, or a gas. Refer back to the two double-replacement reactions previously discussed in this section. The reaction between calcium hydroxide and hydrochloric acid produces water. A precipitate is produced in the reaction between sodium hydroxide and copper(II) chloride. An example of a double-replacement reaction that forms a gas is that of potassium cyanide and hydrobromic acid.

$$KCN(aq) + HBr(aq) \rightarrow KBr(aq) + HCN(g)$$

It is important to be able to evaluate the chemistry of double-replacement reactions and predict the products of these reactions. The basic steps to write double-replacement reactions are given in **Table 3.**

☑ READING CHECK **Describe** what happens to the anions in a double-replacement reaction.

PRACTICE Problems

Do additional problems. Online Practice

Write the balanced chemical equations for the following double-replacement reactions.

25. The two substances at right react to produce solid silver iodide and aqueous lithium nitrate.

26. Aqueous barium chloride and aqueous potassium carbonate react to produce solid barium carbonate and aqueous potassium chloride.

27. Aqueous sodium oxalate and aqueous lead(II) nitrate react to produce solid lead(II) oxalate and aqueous sodium nitrate.

28. Challenge Acetic acid (CH_3COOH) and potassium hydroxide react to produce potassium acetate and water.

PRACTICE PROBLEMS

DIFFERENTIATED INSTRUCTION

English Learners Have English learners describe the Quick Demo from this section of the chapter, or another classroom demonstration relating to the section, both in their native language and in English. If time permits, allow these students to enhance their English skills by comparing and discussing their English descriptions of the demonstration. EL OL

☑ READING CHECK The anions change places. The anion for compound A bonds with the cation for compound B, and the anion for compound B bonds with the cation for compound A. Two new compounds are formed.

PRACTICE Problems

Have students refer to the Selected Solutions appendix for complete solutions to odd-numbered problems. The complete solutions for all problems can be found in the Solutions Manual.

25. $LiI(aq) + AgNO_3(aq) \rightarrow AgI(s) + LiNO_3(aq)$

26. $BaCl_2(aq) + K_2CO_3(aq) \rightarrow BaCO_3(s) + 2KCl(aq)$

27. $Na_2C_2O_4(aq) + Pb(NO_3)_2(aq) \rightarrow PbC_2O_4(s) + 2NaNO_3(aq)$

28. $CH_3COOH(aq) + KOH(aq) \rightarrow CH_3COOK(aq) + H_2O(l)$

Build a Model

Double Replacement Have students build a model, using a model kit, of each of the reactants in the double-replacement reaction $KCN(aq) + HBr(aq) \rightarrow KBr(aq) + HCN(g)$.

Ask students to manipulate the models so that the reactants produce the products. Have students describe in writing how the models were manipulated so the reactants formed products. BL EL

3 Assess

Check for Understanding

Ask students to explain why reaction 1 occurs but reaction 2 does not.

1. $2KBr(aq) + Cl_2(g) \rightarrow 2KCl(aq) + Br_2(l)$
2. $2KBr(aq) + I_2(s) \rightarrow 2KI(aq) + Br_2(l)$

Chlorine is above bromine in the activity series of halogens; however, iodine is below bromine in the series. OL

Reteach

Ask students to classify each of the following reactions:

1. $Ti(s) + 2Cl_2(g) \rightarrow TiCl_4(s)$ synthesis
2. $AgNO_3(aq) + LiI(aq) \rightarrow AgI(s) + LiNO_3(aq)$ double replacement
3. $3Mg(s) + 2Cr(NO_3)_3(aq) \rightarrow 3Mg(NO_3)_2(aq) + 2Cr(s)$ single replacement
4. $2C_2H_2(g) + 5O_2(g) \rightarrow 4CO_2(g) + 2H_2O(l)$ combustion
5. $2H_2O_2(aq) \rightarrow 2H_2O(l) + O_2(g)$ decomposition OL

Extension

Ask students to write the balanced chemical equation for the reaction between sucrose ($C_{12}H_{22}O_{11}(s)$) and oxygen, yielding gaseous carbon dioxide and liquid water. $C_{12}H_{22}O_{11}(s) + 12O_2(g) \rightarrow 12CO_2(g) + 11H_2O(l)$ Point out that a chemical equation represents a ratio of the enormous amount of atoms, molecules, and ions in a reaction. OL

Explore **types of chemical reactions with an interactive table.** Concepts In Motion

Table 4 Predicting Products of Chemical Reactions

Type of Reaction	Reactants	Probable Products	Generic Equation
Synthesis	• two or more substances	• one compound	$A + B \rightarrow AB$
Combustion	• a metal and oxygen • a nonmetal and oxygen • a compound and oxygen	• the oxide of the metal • the oxide of the nonmetal • two or more oxides	$A + O_2 \rightarrow AO$
Decomposition	• one compound	• two or more elements and/or compounds	$AB \rightarrow A + B$
Single-replacement	• a metal and a compound • a nonmetal and a compound	• a new compound and the replaced metal • a new compound and the replaced non-metal	$A + BX \rightarrow AX + B$
Double-replacement	• two compounds	• two different compounds, one of which is a solid, water, or a gas	$AX + BY \rightarrow AY + BX$

Table 4 summarizes the various types of chemical reactions. Use the table to help you organize the reactions, so that you can identify each and predict its products. For example, how would you determine what type of reaction occurs when solid calcium oxide and carbon dioxide gas react to produce solid calcium carbonate? First, write the chemical equation.

$$CaO(s) + CO_2(g) \rightarrow CaCO_3(s)$$

Second, determine what is happening in the reaction. In this case, two substances are reacting to form one compound. Third, use the table to identify the type of reaction. This particular reaction is a synthesis reaction. Fourth, check your answer by comparing the chemical equation to the generic equation provided in the table for that type of reaction.

$$CaO(s) + CO_2(g) \rightarrow CaCO_3(s)$$
$$A \quad + \quad B \quad \rightarrow \quad AB$$

Synthesis Reaction

SECTION 2 REVIEW

Section Self-Check

Section Summary

- Classifying chemical reactions makes them easier to understand, remember, and recognize.
- Activity series of metals and halogens can be used to predict if single-replacement reactions will occur.

29. MAINIDEA **Describe** the four types of chemical reactions and their characteristics.
30. **Explain** how an activity series of metals is organized.
31. **Compare and contrast** single-replacement reactions and double-replacement reactions.
32. **Describe** the result of a double-replacement reaction.
33. **Classify** What type of reaction is most likely to occur when barium reacts with fluorine? Write the chemical equation for the reaction.
34. **Interpret Data** Could the following reaction occur? Explain your answer.

$$3Ni + 2AuBr_3 \rightarrow 3NiBr_2 + 2Au$$

SECTION 2 REVIEW

29. Synthesis: two substances react to yield a single product. Combustion: a substance reacts with oxygen, producing heat and light. Decomposition: a single compound breaks down into two or more elements or new compounds. Replacement: the atoms of one element replace the atoms of another element in a compound (single-replacement), or positive ions are exchanged between two compounds (double-replacement).
30. An activity series of metals orders metals by their reactivity with other metals. The most active metals are on the top of the list, and the least active metals are at the bottom of the list.
31. In a single-replacement reaction, atoms of one element replace atoms of another element in a compound. In a double-replacement reaction, two compounds dissolved in water exchange positive ions.
32. Double-replacement reactions produce two different compounds, one being a solid precipitate, water, or gas.
33. A synthesis reaction will likely occur. $Ba + F_2 \rightarrow BaF_2$
34. The reaction does occur because nickel is more reactive than gold.

SECTION 3

Reactions in Aqueous Solutions

MAINIDEA Double-replacement reactions occur between substances in aqueous solutions and produce precipitates, water, or gases.

Essential Questions

- What are aqueous solutions?
- How are complete ionic and net ionic equations written for chemical reactions in aqueous solutions?
- How can you predict whether reactions in aqueous solutions will produce a precipitate, water, or a gas?

Review Vocabulary

solution: a uniform mixture that might contain solids, liquids, or gases

New Vocabulary

aqueous solution
solute
solvent
complete ionic equation
spectator ion
net ionic equation

CHEM 4 YOU One way to make lemonade involves using a powdered drink mix and water. When the powdered drink mix is added to the water, the flavor crystals dissolve in the water, forming a solution. The solution is lemonade-flavored drink.

Aqueous Solutions

You read previously that a solution is a homogeneous mixture. Many of the reactions discussed in the previous section involve substances dissolved in water. When a substance dissolves in water, a solution forms. An **aqueous solution** contains one or more substances called **solutes** dissolved in the water. In this case, water is the **solvent**—the most plentiful substance in the solution.

Molecular compounds in solution Although water is always the solvent in aqueous solutions, there are many possible solutes. Some solutes, such as sucrose (table sugar) and ethanol (grain alcohol), are molecular compounds that exist as molecules in aqueous solutions. Other solutes are molecular compounds that form ions when they dissolve in water. For example, the molecular compound hydrogen chloride forms hydrogen ions and chloride ions when it dissolves in water, as shown in **Figure 16.** An equation can be used to show this ionization process.

$$HCl(aq) \rightarrow H^+(aq) + Cl^-(aq)$$

Compounds such as hydrogen chloride that produce hydrogen ions in aqueous solution are acids. In fact, an aqueous solution of hydrogen chloride is often referred to as hydrochloric acid.

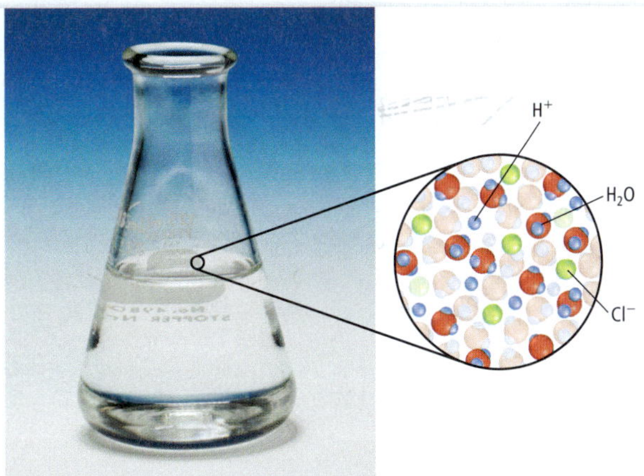

Figure 16 In water, hydrogen chloride (HCl) breaks apart into hydrogen ions (H^+) and chloride ions (Cl^-).

SECTION 3

1 Focus

MAINIDEA

Reaction Products Ask students to give examples of aqueous solution reactions that produce precipitates, water, or gases. Perform an in-class Quick Demo of one of the reactions, for example:
$NaCl(aq) + AgNO_3(aq) \rightarrow AgCl(s) + NaNO_3(aq)$
Ionic form of equation is:
$Ag^+(aq) + Cl^-(aq) \rightarrow AgCl(s)$ OL

2 Teach

Visual Learning

Figures 16 and 17 Briefly explain and use models to show students how compounds form ions when they dissolve in water. Show how molecular compounds such as HCl form ions as they dissolve, a process known as ionization. Also, demonstrate how the positive and negative ions that make up ionic compounds such as NaOH merely separate as the compound dissolves in water, a process known as dissociation.

Concept Development

Precipitates Remind students that a precipitate is a solid formed from a chemical reaction in a solution. A precipitate separates from the solution, unlike a solid dissolved in a solution. Point out that the residue left behind when water evaporates is not a precipitate because the solid residue was not formed by a chemical reaction.

Quick Demo

Formulas for Precipitate Formation Add 0.27 g of sodium carbonate to 5 mL of water in a test tube. Stopper and shake the tube until all the solid dissolves. Add 0.41 g of calcium nitrate to 5 mL of water in another test tube. As before, stopper and shake the tube until all the solid dissolves. Have students observe evidence of reaction when you mix the contents of the two tubes. (Dry the $CaCO_3$ precipitate, and put it in a waste container. Flush $NaNO_3$ down a drain.) A white precipitate of calcium carbonate forms. Have students write the balanced chemical equation, complete ionic equation, and net ionic equation for the reaction.

$Na_2CO_3(aq) + Ca(NO_3)_2 \longrightarrow 2NaNO_3(aq) + CaCO_3(s)$

$2Na^+(aq) + CO_3^{2-}(aq) + Ca^{2+}(aq) + 2NO_3^{2-}(aq) \longrightarrow 2Na^+(aq) + 2NO_3^{2-}(aq) + CaCO_3(s)$

$CO_3^{2-}(aq) + Ca^{2+}(aq) \longrightarrow CaCO_3(s)$ **OL**

VOCABULARY

SCIENCE USAGE V. COMMON USAGE

Compound

Science usage: a chemical combination of two or more different elements

Salt is a compound comprised of the elements sodium and chlorine.

Common usage: a word that consists of two or more words

Two compound words are basketball *and* textbook.

Ionic compounds in solution In addition to molecular compounds, ionic compounds might be solutes in aqueous solutions. Recall that ionic compounds consist of positive ions and negative ions held together by ionic bonds. When ionic compounds dissolve in water, their ions can separate—a process called dissociation. For example, an aqueous solution of the ionic compound sodium chloride contains Na^+ and Cl^- ions.

Types of Reactions in Aqueous Solutions

When two aqueous solutions that contain ions as solutes are combined, the ions might react with one another. These reactions are always double-replacement reactions. The solvent molecules, which are all water molecules, do not usually react. Three types of products can form from the double-replacement reaction: a precipitate, water, or a gas.

Reactions that form precipitates Some reactions that occur in aqueous solutions produce precipitates. For example, recall from Section 2 that when aqueous solutions of sodium hydroxide and copper(II) chloride are mixed, a double-replacement reaction occurs in which the precipitate copper(II) hydroxide forms.

$$2NaOH(aq) + CuCl_2(aq) \rightarrow 2NaCl(aq) + Cu(OH)_2(s)$$

Note that the chemical equation does not show some details of this reaction. Sodium hydroxide and copper(II) chloride are ionic compounds. Therefore, in aqueous solutions they exist as Na^+, OH^-, Cu^{2+}, and Cl^- ions, as shown in **Figure 17.** When their solutions are combined, Cu^{2+} ions in one solution and OH^- ions in the other solution react to form the precipitate copper(II) hydroxide, $Cu(OH)_2(s)$. The Na^+ and Cl^- ions remain dissolved in the new solution.

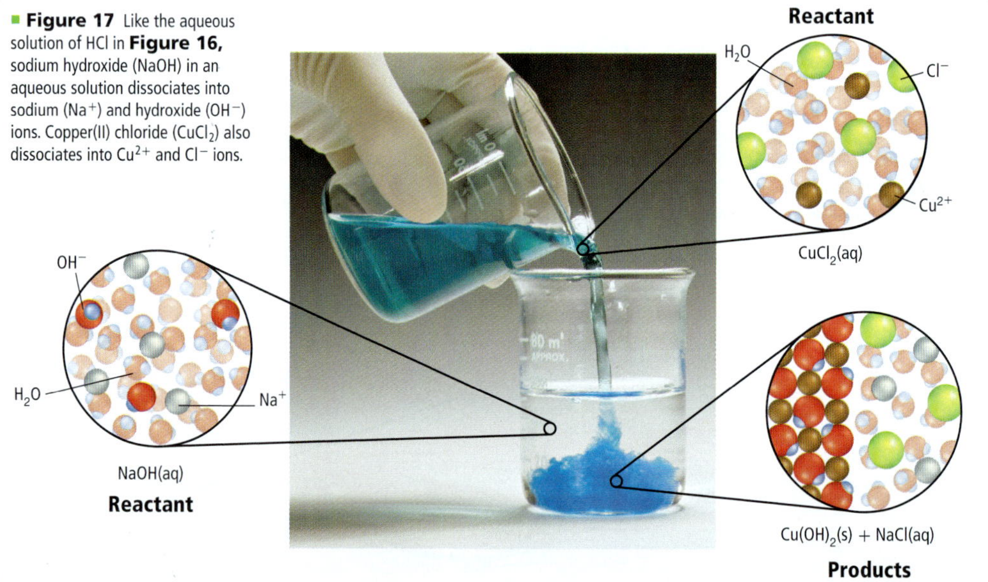

Figure 17 Like the aqueous solution of HCl in **Figure 16,** sodium hydroxide (NaOH) in an aqueous solution dissociates into sodium (Na^+) and hydroxide (OH^-) ions. Copper(II) chloride ($CuCl_2$) also dissociates into Cu^{2+} and Cl^- ions.

Matt Meadows

DIFFERENTIATED INSTRUCTION

Advanced Learners Approve lab safety forms before work begins. Assemble an apparatus from a 20 mm × 150 mm test tube, a one-hole stopper that fits the test tube and has a short glass tube in the hole, a 30 cm length of rubber tubing attached to the short glass tube, and a 10 cm length of glass tubing attached to the other end of the rubber tubing. Have students pour 5 mL of vinegar into the test tube and pour 100 mL of limewater (saturated calcium hydroxide solution) into a 150-mL beaker. Then, have students add 2.5 g of baking soda to the vinegar and quickly insert the one-hole stopper into the test tube so that the evolved gas passes through the tubing and bubbles through the limewater. Have them make observations. As the carbon dioxide gas produced by the reaction bubbles through the limewater, a cloud of insoluble calcium carbonate forms—a test for carbon dioxide. Ask students to write chemical equations for the reaction between the acetic acid in vinegar and sodium hydrogen carbonate and the reaction between carbon dioxide and aqueous calcium hydroxide to form insoluble calcium carbonate. $HC_2H_3O_2(aq) + NaHCO_3(s) \longrightarrow H_2O(l) + CO_2(g) + NaC_2H_3O_2(aq)$; $CO_2(g) + Ca(OH)_2(aq) \longrightarrow CaCO_3(s) + H_2O(l)$ **AL**

MiniLAB

Observe a Precipitate-Forming Reaction

How do two liquids form a solid?

Procedure

1. Read and complete the lab safety form.
2. Place 50 mL **distilled water** in a **150-mL beaker.**
3. Measure about 4 g **NaOH pellets** on a **balance.** Add the NaOH pellets to the beaker one at a time. Mix with a **stirring rod** until each NaOH pellet dissolves before adding the next pellet.
4. Measure about 6 g **Epsom salts ($MgSO_4$)** and place it in another 150-mL beaker. Add 50 mL distilled water to the Epsom salts. Mix with another stirring rod until the Epsom salts dissolve.
5. Slowly pour the Epsom salts solution into the NaOH solution. Record your observations.
6. Stir the new solution. Record your observations.
7. Allow the precipitate to settle, then decant the liquid from the solid into a **100-mL graduated cylinder.**
8. Dispose of the solid as instructed by your teacher.

Analysis

1. **Write** a balanced chemical equation for the reaction between the NaOH and $MgSO_4$. Note that most sulfate compounds exist as ions in aqueous solutions.
2. **Write** the complete ionic equation for this reaction.
3. **Determine** which ions are spectator ions, then write the net ionic equation for this reaction.

Ionic equations To show the details of reactions that involve ions in aqueous solutions, chemists use ionic equations. Ionic equations differ from chemical equations in that substances that are ions in solution are written as ions in the equation. Look again at the reaction between aqueous solutions of sodium hydroxide and copper(II) chloride. To write the ionic equation for this reaction, you must show the reactants, NaOH(aq) and $CuCl_2$(aq), and the product, NaCl(aq), as ions.

$$2Na^+(aq) + 2OH^-(aq) + Cu^{2+}(aq) + 2Cl^-(aq) \rightarrow 2Na^+(aq) + 2Cl^-(aq) + Cu(OH)_2(s)$$

An ionic equation that shows all of the particles in a solution as they exist is called a **complete ionic equation.** Note that the sodium ions and the chloride ions are both reactants and products. Because they are both reactants and products, they do not participate in the reaction. Ions that do not participate in a reaction are called **spectator ions** and are not usually shown in ionic equations. **Net ionic equations** are ionic equations that include only the particles that participate in the reaction. Net ionic equations are written from complete ionic equations by removing all spectator ions. For example, a net ionic equation is what remains after the sodium and chloride ions are crossed out of this complete ionic equation.

$$\cancel{2Na^+(aq)} + 2OH^-(aq) + Cu^{2+}(aq) + \cancel{2Cl^-(aq)} \rightarrow \cancel{2Na^+(aq)} + \cancel{2Cl^-(aq)} + Cu(OH)_2(s)$$

Only the hydroxide and copper ions are left in the net ionic equation shown below.

$$2OH^-(aq) + Cu^{2+}(aq) \rightarrow Cu(OH)_2(s)$$

☑ READING CHECK **Compare** How are ionic equations different from chemical equations?

MiniLAB

Rubric

Purpose Students will write a balanced chemical equation, a complete ionic equation, and a net ionic equation for a chemical reaction that produces a precipitate.

Process Skills classify, observe, infer

Safety Precautions Approve lab safety forms before work begins. Be sure students wear aprons and goggles. NaOH is corrosive to skin. Epsom salts are a skin irritant.

Disposal The magnesium hydroxide precipitate can be placed in the trash. Do not put it down a drain.

Teaching Strategies
Show students how to decant a liquid that contains a solid.

Expected Results The reaction will form a white precipitate, magnesium hydroxide. Sodium sulfate is also formed, but it is soluble in water and is not seen in the solution.

Analysis

1. $MgSO_4(aq) + 2NaOH(aq) \rightarrow Mg(OH)_2(s) + Na_2SO_4(aq)$
2. $Mg^{2+} + SO_4^{2-} + 2Na^+ + 2OH^- \rightarrow Mg(OH)_2(s) + 2Na^+ + SO_4^{2-}$
3. Spectator ions: SO_4^{2-} and $2Na^+$; $Mg^{2+} + 2OH^- \rightarrow Mg(OH)_2(s)$

☑ READING CHECK In an ionic equation, the substances that are ions in solution are written as ions in the equation.

IN-CLASS Example

Question Complete the following chemical equation
$KCl(aq) + AgNO_3(s) \rightarrow$

Answer
$KCl(aq) + AgNO_3(s) \rightarrow AgCl(s) + KNO_3(aq)$

PRACTICE Problems

Have students refer to the Selected Solutions appendix for complete solutions to odd-numbered problems. The complete solutions for all problems can be found in the Solutions Manual.

35. chemical equation: $KI(aq) + AgNO_3(aq) \rightarrow KNO_3(aq) + AgI(s)$
complete ionic equation: $\cancel{K^+}(aq) + I^-(aq) + Ag^+(aq) + \cancel{NO_3^-}(aq) \rightarrow \cancel{K^+}(aq) + \cancel{NO_3^-}(aq) + AgI(s)$
net ionic equation: $I^-(aq) + Ag^+(aq) \rightarrow AgI(s)$

36. chemical equation: $2(NH_4)_3PO_4(aq) + 3Na_2SO_4(aq) \rightarrow 3(NH_4)_2SO_4(aq) + 2Na_3PO_4(aq)$
complete ionic equation: $\cancel{6NH_4^+}(aq) + \cancel{2PO_4^{3-}}(aq) + \cancel{6Na^+}(aq) + \cancel{3SO_4^{2-}}(aq) \rightarrow \cancel{6NH_4^+}(aq) + \cancel{3SO_4^{2-}}(aq) + \cancel{6Na^+}(aq) + \cancel{2PO_4^{3-}}(aq)$
No reaction occurs; therefore, there is no net ionic equation

37. chemical equation: $AlCl_3(aq) + 3NaOH(aq) \rightarrow Al(OH)_3(s) + 3NaCl(aq)$
complete ionic equation: $Al^{3+}(aq) + \cancel{3Cl^-}(aq) + \cancel{3Na^+}(aq) + 3OH^-(aq) \rightarrow Al(OH)_3(s) + \cancel{3Na^+}(aq) + \cancel{3Cl^-}(aq)$
net ionic equation: $Al^{3+}(aq) + 3OH^-(aq) \rightarrow Al(OH)_3(s)$

38. chemical equation: $Li_2SO_4(aq) + Ca(NO_3)_2(aq) \rightarrow 2LiNO_3(aq) + CaSO_4(s)$
complete ionic equation: $\cancel{2Li^+}(aq) + SO_4^{2-}(aq) + Ca^{2+}(aq) + \cancel{2NO_3^-}(aq) \rightarrow \cancel{2Li^+}(aq) + \cancel{2NO_3^-}(aq) + CaSO_4(s)$
net ionic equation: $SO_4^{2-}(aq) + Ca^{2+}(aq) \rightarrow CaSO_4(s)$

39. chemical equation: $5Na_2CO_3(aq) + 2MnCl_5(aq) \rightarrow 10NaCl(aq) + Mn_2(CO_3)_5(s)$
complete ionic equation: $\cancel{10Na^+}(aq) + 5CO_3^{2-}(aq) + 2Mn^{5+}(aq) + \cancel{10Cl^-}(aq) \rightarrow \cancel{10Na^+}(aq) + \cancel{10Cl^-}(aq) + Mn_2(CO_3)_5(s)$
net ionic equation: $5CO_3^{2-}(aq) + 2Mn^{5+}(aq) \rightarrow Mn_2(CO_3)_5(s)$

EXAMPLE Problem 3

REACTIONS THAT FORM A PRECIPITATE Write the chemical, complete ionic, and net ionic equations for the reaction between aqueous solutions of barium nitrate and sodium carbonate that forms the precipitate barium carbonate.

1 ANALYZE THE PROBLEM

You are given the word equation for the reaction between barium nitrate and sodium carbonate. You must determine the chemical formulas and relative amounts of all reactants and products to write the balanced chemical equation. To write the complete ionic equation, you need to show the ionic states of the reactants and products. By crossing out the spectator ions from the complete ionic equation, you can write the net ionic equation. The net ionic equation will include fewer substances than the other equations.

2 SOLVE FOR THE UNKNOWN

Write the correct chemical formulas and physical states for all substances involved in the reaction.

$Ba(NO_3)_2(aq) + Na_2CO_3(aq) \rightarrow BaCO_3(s) + NaNO_3(aq)$

$Ba(NO_3)_2(aq) + Na_2CO_3(aq) \rightarrow BaCO_3(s) + 2NaNO_3(aq)$ — **Balance the skeleton equation.**

$Ba^{2+}(aq) + 2NO_3^-(aq) + 2Na^+(aq) + CO_3^{2-}(aq) \rightarrow BaCO_3(s) + 2Na^+(aq) + 2NO_3^-(aq)$ — **Show the ions of the reactants and the products.**

$Ba^{2+}(aq) + \cancel{2NO_3^-(aq)} + \cancel{2Na^+(aq)} + CO_3^{2-}(aq) \rightarrow BaCO_3(s) + \cancel{2Na^+(aq)} + \cancel{2NO_3^-(aq)}$ — **Cross out the spectator ions from the complete ionic equation.**

$Ba^{2+}(aq) + CO_3^{2-}(aq) \rightarrow BaCO_3(s)$ — **Write the net ionic equation.**

3 EVALUATE THE ANSWER

The net ionic equation includes fewer substances than the other equations because it shows only the reacting particles. The particles composing the solid precipitate that is the result of the reaction are no longer ions.

PRACTICE Problems

Do additional problems. Online Practice

Write chemical, complete ionic, and net ionic equations for each of the following reactions that might produce a precipitate. Use *NR* to indicate that no reaction occurs.

35. Aqueous solutions of potassium iodide and silver nitrate are mixed, forming the precipitate silver iodide.

36. Aqueous solutions of ammonium phosphate and sodium sulfate are mixed. No precipitate forms and no gas is produced.

37. Aqueous solutions of aluminum chloride and sodium hydroxide are mixed, forming the precipitate aluminum hydroxide.

38. Aqueous solutions of lithium sulfate and calcium nitrate are mixed, forming the precipitate calcium sulfate.

39. **Challenge** When aqueous solutions of sodium carbonate and manganese(V) chloride are mixed, a precipitate forms. The precipitate is a compound containing manganese.

CHEMISTRY JOURNAL

Ions in Groundwater Have students research the types of ions found in groundwater in various parts of your county or state. Determine which ions, if any, are beneficial and which ions are usually removed (or partially removed) in water-treatment plants. Ask students to summarize their findings in their chemistry journals. OL

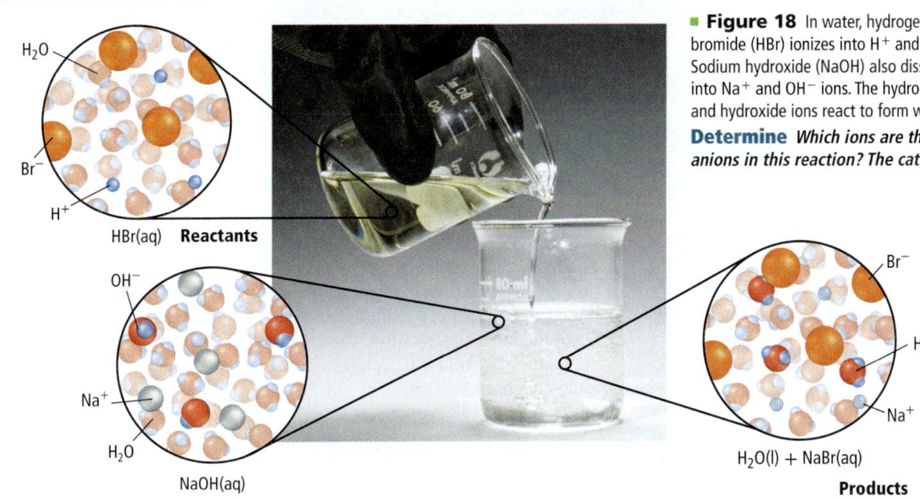

Figure 18 In water, hydrogen bromide (HBr) ionizes into H^+ and Br^- ions. Sodium hydroxide (NaOH) also dissociates into Na^+ and OH^- ions. The hydrogen ions and hydroxide ions react to form water.
Determine *Which ions are the anions in this reaction? The cations?*

Reactions that form water Another type of double-replacement reaction that occurs in an aqueous solution produces water molecules. The water molecules produced in the reaction increase the number of solvent particles. Unlike reactions in which a precipitate forms, no evidence of a chemical reaction is observable because water is colorless, odorless, and already makes up most of the solution. For example, when you mix hydrobromic acid (HBr) with a sodium hydroxide solution (NaOH), as shown in **Figure 18,** a double-replacement reaction occurs and water is formed. The chemical equation for this reaction is shown below.

$$HBr(aq) + NaOH(aq) \rightarrow H_2O(l) + NaBr(aq)$$

In this case, the reactants and the product sodium bromide exist as ions in an aqueous solution. The complete ionic equation for this reaction shows these ions.

$$H^+(aq) + Br^-(aq) + Na^+(aq) + OH^-(aq) \rightarrow H_2O(l) + Na^+(aq) + Br^-(aq)$$

Look carefully at the complete ionic equation. The reacting solute ions are the hydrogen ions and hydroxide ions because the sodium ions and bromine ions are both spectator ions. If you cross out the spectator ions, you are left with the ions that take part in the reaction.

$$H^+(aq) + \cancel{Br^-(aq)} + \cancel{Na^+(aq)} + OH^-(aq) \rightarrow H_2O(l) + \cancel{Na^+(aq)} + \cancel{Br^-(aq)}$$

This equation is the net ionic equation for the reaction.

$$H^+(aq) + OH^-(aq) \rightarrow H_2O(l)$$

☑ READING CHECK **Analyze** In the reaction between hydrobromic acid and sodium hydroxide, why are the sodium ions and bromine ions called spectator ions?

Caption Question Fig. 18 anions: Br^- and OH^-; cations: H^+ and Na^+

☑ **READING CHECK** They are the ions that do not take part in the reaction.

Use Science Terms

Spectator Have students research the meaning of the term *spectator* when speaking about a sporting event and relate the term to its use in chemistry. OL EL

Extension

Water Softeners Ask students if they have hard or soft water at home. Explain that hard water has a high calcium or magnesium content and will form soap scum on the shower or tub, will not totally rinse soap from clothes, and will leave a deposit in a pot that is boiled dry. Explain that water can be softened, by ion exchange. Calcium or magnesium ions from the water are exchanged with sodium ions in an ion exchanger or water softener. Ask students why sodium will replace calcium and magnesium ions. Na is more active than Ca or Mg Ask students if the replacement of calcium or magnesium ions with sodium ions might be a health risk for a person. If that person has high blood pressure, the increased sodium intake from softened water might be a health risk. OL

CHEMISTRY JOURNAL

What's a hydrogen to do? Have students write journal entries that describe the "activities" and "travels" of a hydrogen atom during the following reactions and processes: a gaseous hydrogen molecule reacts with a gaseous iodine molecule to form gaseous hydrogen iodide; gaseous hydrogen iodide dissolves in water to form hydroiodic acid; hydroiodic acid reacts with aqueous lithium sulfide to form hydrogen sulfide gas. Encourage students to look up the physical and chemical properties of iodine and hydrogen sulfide and possibly include these properties in their stories. OL

IN-CLASS Example

Question Complete the following chemical equation
$HNO_3(aq) + KOH(aq) \rightarrow$

Answer
$HNO_3(aq) + KOH(aq) \rightarrow H_2O(l) + KNO_3(aq)$

PRACTICE Problems

Have students refer to the Selected Solutions appendix for complete solutions to odd-numbered problems. The complete solutions for all problems can be found in the Solutions Manual.

40. chemical equation: $H_2SO_4(aq) + 2KOH(aq) \rightarrow 2H_2O(l) + K_2SO_4(aq)$
complete ionic equation: $2H^+(aq) + \cancel{SO_4^{2-}}(aq) + \cancel{2K^+}(aq) + 2OH^-(aq) \rightarrow 2H_2O(l) + \cancel{2K^+}(aq) + \cancel{SO_4^{2-}}(aq)$
net ionic equation: $2H^+(aq) + 2OH^-(aq) \rightarrow 2H_2O(l)$ or $H^+(aq) + OH^-(aq) \rightarrow H_2O(l)$

41. chemical equation: $2HCl(aq) + Ca(OH)_2(aq) \rightarrow 2H_2O(l) + CaCl_2(aq)$
complete ionic equation: $2H^+(aq) + \cancel{2Cl^-}(aq) + \cancel{Ca^{2+}}(aq) + 2OH^-(aq) \rightarrow 2H_2O(l) + \cancel{Ca^{2+}}(aq) + \cancel{2Cl^-}(aq)$
net ionic equation: $H^+(aq) + OH^-(aq) \rightarrow H_2O(l)$

42. chemical equation: $HNO_3(aq) + NH_4OH(aq) \rightarrow H_2O(l) + NH_4NO_3(aq)$
complete ionic equation: $H^+(aq) + \cancel{NO^{3-}}(aq) + \cancel{NH_4^+}(aq) + OH^-(aq) \rightarrow H_2O(l) + \cancel{NH_4^+}(aq) + \cancel{NO_3^-}(aq)$
net ionic equation: $H^+(aq) + OH^-(aq) \rightarrow H_2O(l)$

43. chemical equation: $H_2S(aq) + 1\,Ca(OH)_2(aq) \rightarrow 2H_2O(l) + CaS(aq)$
complete ionic equation: $2H^+(aq) + \cancel{S^{2-}}(aq) + \cancel{Ca^{2+}}(aq) + 2OH^-(aq) \rightarrow 2H_2O(l) + \cancel{Ca^{2+}}(aq) + \cancel{S^{2-}}(aq)$
net ionic equation: $H^+(aq) + OH^-(aq) \rightarrow H_2O(l)$

44. chemical equation: $2C_6H_5COOH(aq) + Mg(OH)_2(aq) \rightarrow Mg(C_6H_5COO)_2(aq) + 2H_2O(l)$
complete ionic equation:
$\cancel{2C_6H_5COO^-(aq)} + 2H^+(aq) + \cancel{Mg^{+}(aq)} + 2OH^-(aq) \rightarrow \cancel{2C_6H_5COO^-(aq)} + \cancel{Mg^{+}(aq)} + 2H_2O(l)$
net ionic equation:
$2H^+(aq) + 2OH^-(aq) \rightarrow 2H_2O(l)$

EXAMPLE Problem 4

REACTIONS THAT FORM WATER Write the chemical, complete ionic, and net ionic equations for the reaction between hydrochloric acid and aqueous lithium hydroxide. This reaction produces water and aqueous lithium chloride.

1 ANALYZE THE PROBLEM

You are given the word equation for the reaction that occurs between hydrochloric acid and aqueous lithium hydroxide to produce water and aqueous lithium chloride. You must determine the chemical formulas for and relative amounts of all reactants and products to write the balanced chemical equation. To write the complete ionic equation, you need to show the ionic states of the reactants and products. By crossing out the spectator ions from the complete ionic equation, you can write the net ionic equation.

2 SOLVE FOR THE UNKNOWN

Write the skeleton equation for the reaction and balance it.

$HCl(aq) + LiOH(aq) \rightarrow H_2O(l) + LiCl(aq)$

$H^+(aq) + Cl^-(aq) + Li^+(aq) + OH^-(aq) \rightarrow H_2O(l) + Li^+(aq) + Cl^-(aq)$

Show the ions of the reactants and the products.

$H^+(aq) + \cancel{Cl^-(aq)} + \cancel{Li^+(aq)} + OH^-(aq) \rightarrow H_2O(l) + \cancel{Li^+(aq)} + \cancel{Cl^-(aq)}$

Cross out the spectator ions from the complete ionic equation.

$H^+(aq) + OH^-(aq) \rightarrow H_2O(l)$

Write the net ionic equation.

3 EVALUATE THE ANSWER

The net ionic equation includes fewer substances than the other equations because it shows only those particles involved in the reaction that produces water. The particles that compose the product water are no longer ions.

PRACTICE Problems

Do additional problems. Online Practice

Write chemical, complete ionic, and net ionic equations for the reactions between the following substances, which produce water.

40. Mixing sulfuric acid (H_2SO_4) and aqueous potassium hydroxide produces water and aqueous potassium sulfate.

41. Mixing hydrochloric acid (HCl) and aqueous calcium hydroxide produces water and aqueous calcium chloride.

42. Mixing nitric acid (HNO_3) and aqueous ammonium hydroxide produces water and aqueous ammonium nitrate.

43. Mixing hydrosulfuric acid (H_2S) and aqueous calcium hydroxide produces water and aqueous calcium sulfide.

44. Challenge When benzoic acid (C_6H_5COOH) and magnesium hydroxide are mixed, water and magnesium benzoate are produced.

DIFFERENTIATED INSTRUCTION

Struggling Learners Have struggling students build chemical models—using a model kit—to demonstrate a synthesis reaction and replacement reactions. The models might be presented to their classmates. Use these examples of chemical reactions for the models.

1. $H_2 + O_2 \rightarrow 2H_2O$ (synthesis reaction)
2. $2Na + 2H_2O \rightarrow 2NaOH + H_2$ (single replacement reaction)
3. $CO_2 + 4HCl \rightarrow CCl_4 + 2H_2O$ (double replacement reaction) BL EL

Reactions that form gases A third type of double-replacement reaction that occurs in aqueous solutions results in the formation of a gas. Some gases commonly produced in these reactions are carbon dioxide, hydrogen cyanide, and hydrogen sulfide.

A gas-producing reaction occurs when you mix hydroiodic acid (HI) with an aqueous solution of lithium sulfide. Bubbles of hydrogen sulfide gas form in the container during the reaction. Lithium iodide is also produced in this reaction and remains dissolved in the solution.

$$2HI(aq) + Li_2S(aq) \rightarrow H_2S(g) + 2LiI(aq)$$

The reactants hydroiodic acid and lithium sulfide exist as ions in aqueous solution. Therefore, you can write an ionic equation for this reaction. The complete ionic equation includes all of the substances in the solution.

$$2H^+(aq) + 2I^-(aq) + 2Li^+(aq) + S^{2-}(aq) \rightarrow H_2S(g) + 2Li^+(aq) + 2I^-(aq)$$

Note that there are many spectator ions in the equation. When the spectator ions are crossed out, only the substances involved in the reaction remain in the equation.

$$2H^+(aq) + \cancel{2I^-(aq)} + \cancel{2Li^+(aq)} + S^{2-}(aq) \rightarrow H_2S(g) + \cancel{2Li^+(aq)} + \cancel{2I^-(aq)}$$

This is the net ionic equation.

$$2H^+(aq) + S^{2-}(aq) \rightarrow H_2S(g)$$

If you completed the Launch Lab at the beginning of this chapter, you observed another gas-producing reaction. In that reaction, carbon dioxide gas was produced and bubbled out of the solution. Another reaction that produces carbon dioxide gas occurs in your kitchen when you mix vinegar and baking soda. Vinegar is an aqueous solution of acetic acid and water. Baking soda essentially consists of sodium hydrogen carbonate. Rapid bubbling occurs when vinegar and baking soda are combined. The bubbles are carbon dioxide gas escaping from the solution. You can see this reaction occurring in **Figure 19.**

A reaction similar to the one between vinegar and baking soda occurs when you combine any acidic solution and sodium hydrogen carbonate. In all cases, two reactions must occur almost simultaneously in the solution to produce the carbon dioxide gas. One of these is a double-replacement reaction and the other is a decomposition reaction.

For example, when you dissolve sodium hydrogen carbonate in hydrochloric acid, a gas-producing double-replacement reaction occurs. The hydrogen in the hydrochloric acid and the sodium in the sodium hydrogen carbonate replace each other.

$$HCl(aq) + NaHCO_3(aq) \rightarrow H_2CO_3(aq) + NaCl(aq)$$

Sodium chloride is an ionic compound, and its ions remain separate in the aqueous solution. However, as the carbonic acid (H_2CO_3) forms, it decomposes immediately into water and carbon dioxide.

$$H_2CO_3(aq) \rightarrow H_2O(l) + CO_2(g)$$

Figure 19 When vinegar and baking soda (sodium hydrogen carbonate, $NaHCO_3$) combine, the result is a vigorous bubbling that releases carbon dioxide (CO_2).

Charles D. Winters/Photo Researchers

Apply Chemistry

Chemistry in Baking Explain to students that some biscuits and other baked goods rise during baking because of trapped carbon dioxide gas produced by the reaction between baking powder and water. Baking powder is a mixture of baking soda (sodium hydrogen carbonate, $NaHCO_3$), a dry acid such as tartaric acid ($H_2C_4H_4O_6$), and usually starch to keep the mixture dry. Water added to baking powder dissolves the baking soda and tartaric acid, which facilitates the following chemical reaction: $H_2C_4H_4O_6(aq) + 2NaHCO_3(aq) \rightarrow 2H_2O(l) + 2CO_2(g) + Na_2C_4H_4O_6(aq)$. Demonstrate the reaction by adding some water to a small sample of baking powder in a watch glass and project the image.

CHEMISTRY PROJECT

Hard-Water Cleanup Have students research why vinegar can be used in the home to remove hard water deposits. Have the students compare the cost of using vinegar to the cost of using other hard water, lime dissolving chemicals. Have students review safety precautions for use of commercial products. Have students poll class members for products used in their own homes. Results can be reported to the class. OL

IN-CLASS Example

Question Write a chemical, complete ionic and net-ionic equation for the chemical reaction below.

$HCl(aq) + Zn(s) \rightarrow$

Answer

$2HCl(aq) + Zn(s) \rightarrow H_2(g) + ZnCl_2(aq)$

$2H^+(aq) + 2Cl^-(aq) + Zn(s) \rightarrow H_2(g) + Zn^{2+}(aq) + 2Cl^-(aq)$

$2H^+(aq) + Zn(s) \rightarrow H_2(g) + Zn^{2+}(aq)$

PRACTICE Problems

Have students refer to the Selected Solutions appendix for complete solutions to odd-numbered problems. The complete solutions for all problems can be found in the Solutions Manual.

45. chemical equation: $2HClO_4(aq) + K_2CO_3(aq) \rightarrow H_2O(l) + CO_2(g) + 2KClO_4(aq)$
complete ionic equation: $2H^+(aq) + \cancel{2ClO_4^-}(aq) + \cancel{2K^+}(aq) + CO_3^{2-}(aq) \rightarrow H_2O(l) + CO_2(g) + \cancel{2K^+}(aq) + \cancel{2ClO_4^-}(aq)$
net ionic equation: $2H^+(aq) + CO_3^{2-}(aq) \rightarrow H_2O(l) + CO_2(g)$

46. chemical equation: $H_2SO_4(aq) + 2NaCN(aq) \rightarrow 2HCN(g) + Na_2SO_4(aq)$
complete ionic equation: $2H^+(aq) + \cancel{SO_4^{2-}}(aq) + \cancel{2Na^+}(aq) + 2CN^-(aq) \rightarrow 2HCN(g) + \cancel{2Na^+}(aq) + \cancel{SO_4^{2-}}(aq)$
net ionic equation: $2H^+(aq) + 2CN^-(aq) \rightarrow 2HCN(g)$ or $H^+(aq) + CN^-(aq) \rightarrow HCN(g)$

47. chemical equation: $2HBr(aq) + (NH_4)_2CO_3(aq) \rightarrow H_2O(l) + CO_2(g) + 2NH_4Br(aq)$
complete ionic equation: $2H^+(aq) + \cancel{2Br^-}(aq) + \cancel{2NH_4^+}(aq) + CO_3^{2-}(aq) \rightarrow H_2O(l) + CO_2(g) + \cancel{2NH_4^+}(aq) + \cancel{2Br^-}(aq)$
net ionic equation: $2H^+(aq) + CO_3^{2-}(aq) \rightarrow H_2O(l) + CO_2(g)$

48. chemical equation: $2HNO_3(aq) + KRbS(aq) \rightarrow H_2S(g) + KRb(NO_3)_2(aq)$
complete ionic equation: $2H^+(aq) + \cancel{2NO_3^-}(aq) + \cancel{K^+}(aq) + \cancel{Rb^+}(aq) + S^{2-}(aq) \rightarrow H_2S(g) + \cancel{K^+}(aq) + \cancel{Rb^+}(aq) + \cancel{2NO_3^-}(aq)$
net ionic equation: $2H^+(aq) + S^{2-}(aq) \rightarrow H_2S(g)$

EXAMPLE Problem 5

REACTIONS THAT FORM GASES Write the chemical, complete ionic, and net ionic equations for the reaction between hydrochloric acid and aqueous sodium sulfide, which produces hydrogen sulfide gas.

1 ANALYZE THE PROBLEM

You are given the word equation for the reaction between hydrochloric acid (HCl) and sodium sulfide (Na_2S). You must write the skeleton equation and balance it. To write the complete ionic equation, you need to show the ionic states of the reactants and products. By crossing out the spectator ions in the complete ionic equation, you can write the net ionic equation.

2 SOLVE FOR THE UNKNOWN

Write the correct skeleton equation for the reaction.

$HCl(aq) + Na_2S(aq) \rightarrow H_2S(g) + NaCl(aq)$

$2HCl(aq) + Na_2S(aq) \rightarrow H_2S(g) + 2NaCl(aq)$ — **Balance the skeleton equation.**

$2H^+(aq) + 2Cl^-(aq) + 2Na^+(aq) + S^{2-}(aq) \rightarrow H_2S(g) + 2Na^+(aq) + 2Cl^-(aq)$ — **Show the ions of the reactants and the products.**

$2H^+(aq) + \cancel{2Cl^-(aq)} + \cancel{2Na^+(aq)} + S^{2-}(aq) \rightarrow H_2S(g) + \cancel{2Na^+(aq)} + \cancel{2Cl^-(aq)}$ — **Cross out the spectator ions from the complete ionic equation.**

$2H^+(aq) + S^{2-}(aq) \rightarrow H_2S(g)$ — **Write the net ionic equation in its smallest whole-number ratio.**

3 EVALUATE THE ANSWER

The net ionic equation includes fewer substances than the other equations because it shows only those particles involved in the reaction that produce hydrogen sulfide. The particles that compose the product are no longer ions.

PRACTICE Problems

Do additional problems. Online Practice

Write chemical, complete ionic, and net ionic equations for these reactions.

45. Perchloric acid ($HClO_4$) reacts with aqueous potassium carbonate, forming carbon dioxide gas and water.

46. Sulfuric acid (H_2SO_4) reacts with aqueous sodium cyanide, forming hydrogen cyanide gas and aqueous sodium sulfate.

47. Hydrobromic acid (HBr) reacts with aqueous ammonium carbonate, forming carbon dioxide gas and water.

48. Nitric acid (HNO_3) reacts with aqueous potassium rubidium sulfide, forming hydrogen sulfide gas.

49. Challenge Aqueous potassium iodide reacts with lead nitrate in solution, forming solid lead iodide.

49. chemical equation: $2KI(aq) + Pb(NO_3)_2(aq) \rightarrow 2KNO_3(aq) + PbI_2(s)$
complete ionic equation: $\cancel{2K^+(aq)} + 2I^-(aq) + Pb^{2+}(aq) + \cancel{2NO_3^-(aq)} \rightarrow \cancel{2K^+(aq)} + \cancel{2NO_3^-(aq)} + PbI_2(s)$
net ionic equation: $Pb^{2+}(aq) + 2I^-(aq) \rightarrow PbI_2(s)$

CHEMISTRY PROJECT

Identifying Reaction Types Ask students to cite chemical reactions that do not fit into the categories discussed in this chapter. Have them write the equations for these reactions and, if they are able, specify the reaction type. Examples include combinations of these categories plus reactions such as redox, saponification, and polymerization. OL

Double-replacement reaction

$$AX + BY \rightarrow AY + BX$$

$$HCl(aq) + NaHCO_3(aq) \rightarrow H_2CO_3(aq) + NaCl(aq)$$

$$AB \rightarrow A + B$$

$$H_2CO_3(aq) \rightarrow H_2O(l) + CO_2(g)$$

Decomposition reaction

■ **Figure 20** When HCl is combined with $NaHCO_3$, a double-replacement reaction takes place, followed immediately by a decomposition reaction.

Watch a **video about chemical reactions and equations.**

Video

Overall equations Recall that when you combine an acidic solution, such as hydrochloric acid, and sodium hydrogen carbonate, two reactions occur—a double-replacement reaction and a decomposition reaction. These reactions are shown in **Figure 20.** The two reactions can be combined and represented by one chemical equation in a process similar to adding mathematical equations. An equation that combines two reactions is called an overall equation. To write an overall equation, the reactants in the two reactions are written on the reactant side of the combined equation, and the products of the two reactions are written on the product side. Then, any substances that are on both sides of the equation are crossed out.

Reaction 1	$HCl(aq) + NaHCO_3(aq) \rightarrow H_2CO_3(aq) + NaCl(aq)$
Reaction 2	$H_2CO_3(aq) \rightarrow H_2O(l) + CO_2(g)$
Combined equation	$HCl(aq) + NaHCO_3(aq) + H_2\cancel{CO_3}(aq) \rightarrow H_2\cancel{CO_3}(aq) + NaCl(aq) + H_2O(l) + CO_2(g)$
Overall equation	$HCl(aq) + NaHCO_3(aq) \rightarrow H_2O(l) + CO_2(g) + NaCl(aq)$

In this case, the reactants in the overall equation exist as ions in aqueous solutions. Therefore, a complete ionic equation can be written for the reaction.

$$H^+(aq) + Cl^-(aq) + Na^+(aq) + HCO_3^-(aq) \rightarrow H_2O(l) + CO_2(g) + Na^+(aq) + Cl^-(aq)$$

Note that the sodium and chloride ions are the spectator ions. When you cross them out, only the substances that take part in the reaction remain.

$$H^+(aq) + \cancel{Cl^-(aq)} + \cancel{Na^+(aq)} + HCO_3^-(aq) \rightarrow H_2O(l) + CO_2(g) + \cancel{Na^+(aq)} + \cancel{Cl^-(aq)}$$

The net ionic equation shows that both water and carbon dioxide gas are produced in this reaction.

$$H^+(aq) + HCO_3^-(aq) \rightarrow H_2O(l) + CO_2(g)$$

☑ READING CHECK **Describe** What is an overall equation?

☑ **READING CHECK** An equation that combines two reactions.

ASSESSMENT

Skill Have students create tables that summarize the types of reactions discussed in this section. Ask students to give two example equations for each type of reaction included in the table. OL

ASSESSMENT

Knowledge Ask students to predict the formula of the missing product (X) in each of the following reactions:

1. $BaCl_2(aq) + K_2SO_4(aq) \rightarrow 2KCl(aq) + X$ X is $BaSO_4$, an insoluble product.
2. $HI(aq) + LiOH(aq) \rightarrow LiI(aq) + X$ X is H_2O.
3. $2HCl(aq) + (NH_4)_2S(aq) \rightarrow 2NH_4Cl(aq) + X$ X is H_2S, a gas. OL

Video Lab

Video

Chemical Reactions and Equations

3 Assess

Check for Understanding

Ask students to predict which double-replacement reactions occur. Have them explain their reasoning.

1. $CuCl_2(aq) + 2NaNO_3(aq) \rightarrow Cu(NO_3)_2(aq) + 2NaCl(aq)$
2. $CuCl_2(aq) + Na_2CO_3(aq) \rightarrow CuCO_3(s) + 2NaCl(aq)$
3. $2HCl(aq) + Na_2CO_3(aq) \rightarrow H_2O(l) + CO_2(g) + 2NaCl(aq)$

Reactions 2 and 3 occur. Reaction 2 forms a solid, and reaction 3 forms water and a gas. **OL**

Reteach

Ask students to identify any spectator ions in the following complete ionic equation. Then, have them write the net ionic equation.

$H^+(aq) + BrO_3^-(aq) + K^+(aq) + OH^-(aq) \rightarrow H_2O(l) + BrO_3^-(aq) + K^+(aq)$ Spectator ions are BrO_3^- and K^+. The net ionic equation is $H^+(aq) + OH^-(aq) \rightarrow H_2O(l)$. **BL**

Extension

Invite a chemistry professor or graduate student to the classroom to discuss the importance of balanced equations in graduate-level research. Prepare some equations representing chemical reactions and ask the guest speaker to demonstrate how he or she balances chemical equations. **OL**

CAREERS IN CHEMISTRY

WebQuest

Figure 21 After a bicarbonate ion (HCO_3^-) enters a red blood cell, it reacts with a hydrogen ion (H^+) to form water and carbon dioxide (CO_2). The CO_2 is exhaled from the lungs during respiration.

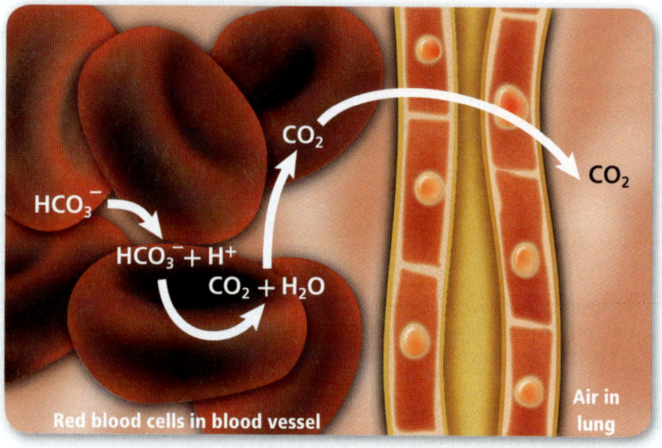

CAREERS IN CHEMISTRY

Biochemist A biochemist is a scientist who studies the chemical processes of living organisms. A biochemist might study functions of the human body or research how food, drugs, and other substances affect living organisms.

WebQuest

Connection to Biology The reaction between hydrogen ions and bicarbonate ions to produce water and carbon dioxide is an important one in your body. This reaction is occurring in the blood vessels of your lungs as you read these words. As shown in **Figure 21,** the carbon dioxide gas produced in your cells is transported in your blood in the form of bicarbonate ions (HCO_3^-). In the blood vessels of your lungs, the HCO_3^- ions combine with H^+ ions to produce CO_2, which you exhale.

This reaction also occurs in products that are made with baking soda, which contains sodium bicarbonate. Sodium bicarbonate makes baked goods rise. It is used as an antacid and in deodorants to absorb moisture and odors. Baking soda can be added to toothpaste to whiten teeth and freshen breath. As a paste, sodium bicarbonate can be used in cleaning and scrubbing. It is also used as a fire-suppression agent in some fire extinguishers.

SECTION 3 REVIEW

Section Self-Check

Section Summary

- In aqueous solutions, the solvent is always water. There are many possible solutes.
- Many molecular compounds form ions when they dissolve in water. When some ionic compounds dissolve in water, their ions separate.
- When two aqueous solutions that contain ions as solutes are combined, the ions might react with one another. The solvent molecules do not usually react.
- Reactions that occur in aqueous solutions are double-replacement reactions.

50. **MAIN IDEA List** three common types of products produced by reactions that occur in aqueous solutions.
51. **Describe** solvents and solutes in an aqueous solution.
52. **Distinguish** between a complete ionic equation and a net ionic equation.
53. **Write** complete ionic and net ionic equations for the reaction between sulfuric acid (H_2SO_4) and calcium carbonate ($CaCO_3$).
$$H_2SO_4(aq) + CaCO_3(s) \rightarrow H_2O(l) + CO_2(g) + CaSO_4(aq)$$
54. **Analyze** Complete and balance the following equation.
$$CO_2(g) + HCl(aq) \rightarrow$$
55. **Predict** What type of product would the following reaction be most likely to produce? Explain your reasoning.
$$Ba(OH)_2(aq) + 2HCl(aq) \rightarrow$$
56. **Formulate Equations** A reaction occurs when nitric acid (HNO_3) is mixed with an aqueous solution of potassium hydrogen carbonate. Aqueous potassium nitrate is produced. Write the chemical and net ionic equations for the reaction.

SECTION 3 REVIEW

50. precipitates, water, and gases
51. A solvent is the most plentiful substance in a solution, and a solute is the substance dissolved in the solvent.
52. In a complete ionic equation, all dissolved ionic compounds and highly ionized molecular compounds are shown as free ions. A net ionic equation includes only the particles that take part in the reaction.
53. Complete: $2H^+(aq) + SO_4^{2-}(aq) + CaCO_3(s) \rightarrow H_2O(l) + CO_2(g) + Ca^{2+}(aq) + SO_4^{2-}(aq)$
Net: $2H^+(aq) + CaCO_3(s) \rightarrow H_2O(l) + CO_2(g) + Ca^{2+}(aq)$
54. $CO_2(g) + 4HCl(aq) \rightarrow CCl_4(l) + 2H_2O(l)$
55. water: The reactants would break down into these ions in solution: $Ba^+ + OH^- + H^+ + Cl^-$. The barium and chloride ions are spectator ions, so the ions that take part in the reaction are the OH^- and H^+ ions, which form water.
56. Chemical: $HNO_3(aq) + KHCO_3(aq) \rightarrow H_2O(l) + CO_2(g) + KNO_3(aq)$
Net: $H^+(aq) + HCO_3^-(aq) \rightarrow H_2O(l) + CO_2(g)$

HOW IT works

LIGHTING UP THE NIGHT: Bioluminescence

In the gathering darkness, a male firefly announces his presence by sending a signal in yellow-green light. A female near the ground answers his call, and he descends. The result might be a successful mating, or, if the female of another firefly species has fooled the male, he might be greedily devoured. The production of light by the firefly is the result of a chemical process called bioluminescence. This process is a strategy used by a wide variety of living things in many different environments. How does it work?

1 Flashy Beetles Fireflies (or lightning bugs) are not flies at all, but a group of beetles that flash their mating signals. They also use their light to lure their prey. The yellow-green light comes from cells in their lower abdomen. The wavelength for this light is between 510 and 670 nm.

2 Bioluminescence The glow of the firefly is the result of a chemical reaction. The reactants are oxygen and luciferin, a light-emitting substance found in some organisms. An enzyme, luciferase, speeds up the reaction. The products of this reaction are oxyluciferin and energy, in the form of light.

3 Glowing Discoveries Research into bioluminescence led to the discovery of green fluorescent protein (GFP), which is found in some species of jellyfish. GFP emits a green light when exposed to UV light. Researchers have inserted GFP into various organisms, such as mice, for research purposes. Examples of what scientists are using GFP to study include cancer, malaria, and cellular processes.

(l)Darwin Dale/Photo Researchers, (r)Eye of Science/Photo Researchers, (bkgd)QuizChang/Getty Images

WRITING IN ▶ Chemistry

Research Identify different life forms that use bioluminescence and create a pamphlet showing how bioluminescence is effective in each of these organisms.

WebQuest

WRITING IN ▶ Chemistry

Rubric WebQuest

Research Some other organisms that use bioluminescence include deep-sea creatures. With so little light available at these depths, the bioluminescence is used mostly as a courtship signal. However, other creatures use their bioluminescent body parts as a lure to attract prey. Some female lightning bugs use bioluminescent signaling to lure a male of another species. The unsuspecting male is then devoured.

HOW IT works

Purpose

Students will learn how fireflies and other organisms use chemical reactions to generate bioluminescence.

Background

Fireflies use nitric oxide (NO) to control the production of light within their abdomens.

Nitric oxide is formed at high temperature in a nitrogen-oxygen atmosphere. The gas can contribute to ozone breakdown and can lead to acid precipitation. It is also an important compound in many living systems.

Mitochondria are known as the powerhouses of the cell. They take in oxygen and use it to provide the cell with energy. Most events that deprive the mitochondria of oxygen are harmful, even fatal to cells. In this case, though, turning off the mitochondria is the reason for the fireflies' bioluminescent reaction.

Teaching Strategies

- The diagram shows a number of fireflies, as well as one up close. It also shows some applications for green fluorescent protein (GFP).
- Explain to students that the luciferin must return to its pre-flashing state before the firefly can flash again. Therefore, this is a cyclical reaction that ends where it started. Many biological reactions are similarly cyclical. Discuss in class why cyclical chemical reactions occur in nature. List some other examples. digestion and respiration

SMALL SCALE ChemLAB

Rubric

Preparation

Time Allotment one class period

Process Skills observe and infer, sequence, compare and contrast, apply concepts, predict

Safety Precautions Approve lab safety forms before work begins. Be sure students wear an apron and goggles, and wash hands with soap after the lab. Review MSDS with students. Account for all Mg ribbon.

Disposal Unreacted metal can be reclaimed and reused. The contents of the well plate can be emptied into a waste container and evaporated to dryness in the fume hood. Dispose of the solid waste at an approved chemical disposal site.

Preparation of Solutions Use the following masses of nitrate salts to prepare 1.0 L of a 1.0*M* solution:
298 g $Zn(NO_3)_2 \cdot 6H_2O$
375 g $Al(NO_3)_3 \cdot 9H_2O$
242 g $Cu(NO_3)_2 \cdot 3H_2O$
256 g $Mg(NO_3)_2 \cdot 6H_2O$
See the Teacher Handbook for preparation of all solutions.

Alternate Materials Other metals and their nitrate salts can be substituted for the ones listed.

Procedure

- A well in the reaction plate will be half filled by 2 mL of solution.
- The metal must be cleaned with emery cloth or sandpaper and have a luster.
- **Troubleshooting** Remind students that if a shiny piece of metal darkens, that is evidence of a chemical change.

SMALL SCALE ChemLAB

Develop an Activity Series

Background: Some metals are more reactive than others. By comparing how different metals react with the known ions in aqueous solutions, an activity series for the tested materials can be developed. The activity series will reflect the relative reactivity of the tested metals.

Question: *How is an activity series developed?*

Materials

1.0*M* $Zn(NO_3)_2$	Al wire
1.0*M* $Al(NO_3)_3$	Mg ribbon
1.0*M* $Cu(NO_3)_2$	Zn metal strips (4)
1.0*M* $Mg(NO_3)_2$	Emery cloth or sandpaper
pipettes (4)	24-well microscale reaction plate
wire cutters	
Cu wire	

Safety Precautions

Procedure

1. Read and complete the lab safety form.
2. Create a table to record your data.
3. Use a pipette to fill the four wells in column 1 of the reaction plate with 2 mL of 1.0*M* $Al(NO_3)_3$ solution.
4. Repeat the procedure in Step 3 to fill the four wells in column 2 with 2 mL of 1.0*M* $Mg(NO_3)_2$.
5. Repeat the procedure in Step 3 to fill the four wells in column 3 with 2 mL of 1.0*M* $Zn(NO_3)_2$.
6. Repeat the procedure in Step 3 to fill the four wells in column 4 with 2 mL of 1.0*M* $Cu(NO_3)_2$.
7. With the emery cloth or sandpaper, polish 10 cm of aluminum wire until it is shiny. Use wire cutters to carefully cut the aluminum wire into four 2.5-cm pieces. Place a piece of the aluminum wire in each well of row A containing solution.
8. Repeat the procedure in Step 7 using 10 cm of magnesium ribbon. Place a piece of Mg ribbon in each well of row B containing solution.
9. Use the emery cloth or sandpaper to polish each small strip of zinc metal. Place a piece of Zn metal in each well of row C containing solution.
10. Observe what happens in each well. After 5 minutes, record your observations in the data table you made.

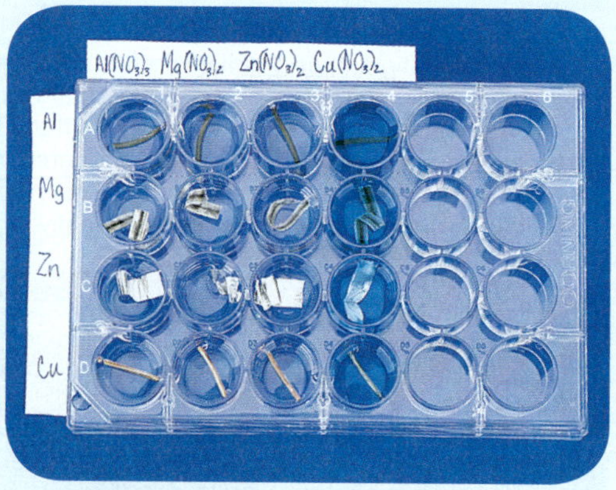

11. **Cleanup and Disposal** Dispose of the chemicals, solutions, and pipettes as directed by your teacher. Wash and return all lab equipment to the designated location. Wash your hands thoroughly.

Analyze and Conclude

1. **Observe and Infer** In which wells of the reaction plate did chemical reactions occur? Which metal reacted with the most solutions? Which metal reacted with the fewest solutions? Which metal is the most reactive?
2. **Sequence** The most-active metal reacted with the most solutions. The least-active metal reacted with the fewest solutions. Order the four metals from most active to least active.
3. **Apply** Write a chemical equation for each single-replacement reaction that occurred on your reaction plate.
4. **Real-World Chemistry** Under what circumstances might it be important to know the activity tendencies of a series of elements?
5. **Error Analysis** How does your answer from Question 2 above compare with the activity series in **Figure 13?** What could account for the differences?

INQUIRY EXTENSION

Design an Experiment Think of three "what if" questions about this investigation that might affect your results. Design an experiment to test one of them.

Matt Meadows

Analyze and Conclude

1. A3, A4, B1, B3, B4, C4
   Mg – 3 reactions
   Cu – 0 reactions
   Magnesium is the most reactive.
2. Mg, Al, Zn, Cu
3. A3: $2Al + 3Zn(NO_3)_2 \rightarrow 2Al(NO_3)_3 + 3Zn$
   A4: $2Al + 3Cu(NO_3)_2 \rightarrow 2Al(NO_3)_3 + 3Cu$
   B1: $3Mg + 2Al(NO_3)_2 \rightarrow 3Mg(NO_3)_2 + 2Al$
   B3: $Mg + Zn(NO_3)_2 \rightarrow Mg(NO_3)_2 + Zn$
   B4: $Mg + Cu(NO_3)_2 \rightarrow Mg(NO_3)_2 + Cu$
   C4: $Zn + Cu(NO_3)_2 \rightarrow Zn(NO_3)_2 + Cu$
4. Answers may vary. For example, industrial chemists could use the information to fine-tune a product or process.
5. Answers will vary. If the procedure was not followed or if the contents of two wells mixed due to over-filling, then the activity series would not agree.

INQUIRY EXTENSION

Students might include checking the concentrations of the solutions, changing the temperature at which the reaction occurs, or conducting the experiment without polishing wires. Experiments should be designed to test only one variable at a time. See the Solutions Manual.

CHAPTER 9 STUDY GUIDE

Vocabulary Practice

BIGIDEA Millions of chemical reactions in and around you transform reactants into products, resulting in the absorption or release of energy.

SECTION 1 Reactions and Equations

MAINIDEA Chemical reactions are represented by balanced chemical equations.

- Some physical changes are evidence that indicate a chemical reaction has occurred.
- Word equations and skeleton equations provide important information about a chemical reaction.
- A chemical equation gives the identities and relative amounts of the reactants and products that are involved in a chemical reaction.
- Balancing an equation involves adjusting the coefficients until the number of atoms of each element is equal on both sides of the equation.

H_2 + Cl_2 → 2HCl

2 atoms H + 2 atoms Cl → 2 atoms H + 2 atoms Cl

Two hydrogen atoms + Two chlorine atoms → Two hydrogen atoms Two chlorine atoms

VOCABULARY
- chemical reaction
- reactant
- product
- chemical equation
- coefficient

SECTION 2 Classifying Chemical Reactions

MAINIDEA There are four types of chemical reactions: synthesis, combustion, decomposition, and replacement reactions.

- Classifying chemical reactions makes them easier to understand, remember, and recognize.
- Activity series of metals and halogens can be used to predict if single-replacement reactions will occur.

VOCABULARY
- synthesis reaction
- combustion reaction
- decomposition reaction
- single-replacement reaction
- double-replacement reaction
- precipitate

SECTION 3 Reactions in Aqueous Solutions

MAINIDEA Double-replacement reactions occur between substances in aqueous solutions and produce precipitates, water, or gases.

- In aqueous solutions, the solvent is always water. There are many possible solutes.
- Many molecular compounds form ions when they dissolve in water. When some ionic compounds dissolve in water, their ions separate.
- When two aqueous solutions that contain ions as solutes are combined, the ions might react with one another. The solvent molecules do not usually react.
- Reactions that occur in aqueous solutions are double-replacement reactions.

VOCABULARY
- aqueous solution
- solute
- solvent
- complete ionic equation
- spectator ion
- net ionic equation

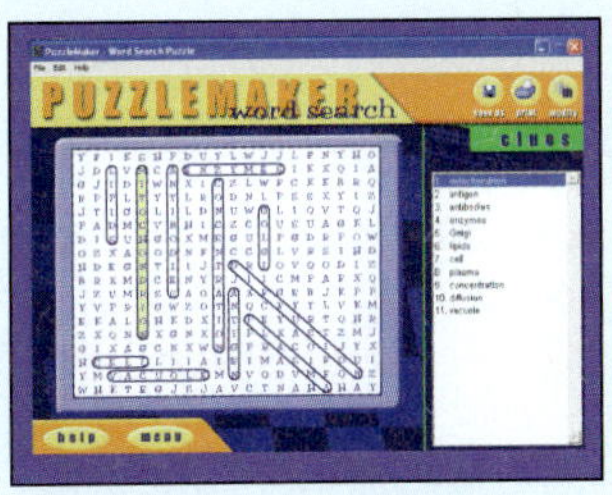

Vocabulary Practice

For additional practice with vocabulary, have students access the Vocabulary PuzzleMaker.

Use the Vocabulary

To reinforce chapter vocabulary, have students write a sentence using each term. OL EL

Review Strategies

- Have students list the various factors that might indicate a chemical reaction has occurred. OL
- Have students describe the steps involved in writing word equations, skeleton equations, and chemical equations. OL
- Ask students to describe each of the five classes of chemical reactions studied in this chapter. OL
- Problems from the Supplemental Practice Problems appendix or the Supplemental Problems booklet can be used for review. OL

Students can review with these online resources.

Vocabulary Practice

Vocabulary eGames
Vocabulary eFlashcards
Vocabulary PuzzleMaker

Section Self-Check

Chapter Self-Check

Online Test Practice

Use *eAssessment* to:
- create multiple versions of tests
- edit existing questions and add your own questions
- build tests aligned with select state curriculum standards using built-in state curriculum tags
- track students' progress

SECTION 1

Mastering Concepts

57. Representation of a chemical reaction using chemical symbols and numbers to indicate the reactants and products.
58. A chemical reaction occurs when reactants are converted into products.
59. Reactants are the initial components and products are the resultant components.
60. Arrows separate reactants from products and specify direction. Coefficients specify the relative amount of the components.
61. Yes.
62. **a.** $NO_2(g)$ **c.** $BaCl_2(aq)$
    b. $Ga(l)$ **d.** $(NH_4)_2CO_3(s)$
63. $K(s)$ and $Zn(NO_3)_2(aq)$
64. $2H_2S(g) + 3O_2(g) \longrightarrow 2SO_2(g) + 2H_2O(g)$
65. **a.** copper(s) + oxygen(g) ⟶ copper(II) oxide(s)
    b. potassium(s) + water(l) ⟶ potassium hydroxide(aq) + hydrogen(g)
    c. calcium chloride(aq) + sodium sulfate(aq) ⟶ calcium sulfate(s) + sodium chloride(aq)
66. **a.** $(NH_4)_2Cr_2O_7(s) \longrightarrow Cr_2O_3(s) + N_2(g) + 4H_2O(g)$
    b. $6CO_2(g) + 6H_2O(l) \longrightarrow C_6H_{12}O_6(s) + 6O_2(g)$

Mastering Problems

67. $HI(g) \longrightarrow H_2(g) + I_2(g)$
68. **a.** $Na_2CO_3(s) \longrightarrow Na_2O(s) + CO_2(g)$
    b. $Al(s) + I_2(s) \longrightarrow AlI_3(s)$
    c. $FeO(s) + O_2(g) \longrightarrow Fe_2O_3(s)$
69. **a.** $C_4H_{10}(l) + O_2(g) \longrightarrow CO_2(g) + H_2O(l)$
    b. $Al_2(CO_3)_3(s) \longrightarrow Al_2O_3(s) + CO_2(g)$
    c. $AgNO_3(aq) + Na_2S(aq) \longrightarrow Ag_2S(s) + NaNO_3(aq)$
70. $Li(s) + Cl_2(g) \longrightarrow LiCl(s)$
71. **a.** $Fe(s) + F_2(g) \longrightarrow FeF_3(s)$
    b. $SO_3(g) + H_2O(l) \longrightarrow H_2SO_4(aq)$
    c. $Na(s) + MgI_2(aq) \longrightarrow NaI(aq) + Mg(s)$
    d. $V(s) + O_2(g) \longrightarrow V_2O_5(s)$
72. **a.** $Li(s) + AuCl_3(aq) \longrightarrow LiCl(aq) + Au(s)$
    b. $Fe(s) + Sn(NO_3)_4(aq) \longrightarrow Fe(NO_3)_3(aq) + Sn(s)$
    c. $NiCl_2(s) + O_2(g) \longrightarrow NiO(s) + Cl_2O_5(g)$
    d. $Li_2CrO_4(aq) + BaCl_2(aq) \longrightarrow LiCl(aq) + BaCrO_4(s)$

CHAPTER 9 ASSESSMENT

SECTION 1

Mastering Concepts

57. Define *chemical equation.*

58. Distinguish between a chemical reaction and a chemical equation.

59. Explain the difference between reactants and products.

60. What do the arrows and coefficients in equations communicate?

61. Does a conversion of a substance into a new substance always indicate that a chemical reaction has occurred? Explain.

62. Write formulas for the following substances and designate their physical states.
- **a.** nitrogen dioxide gas
- **b.** liquid gallium
- **c.** barium chloride dissolved in water
- **d.** solid ammonium carbonate

63. Identify the reactants in the following reaction: When potassium is dropped into aqueous zinc nitrate, zinc and aqueous potassium nitrate form.

64. Balance the reaction of hydrogen sulfide with atmospheric oxygen gas.

$$H_2S(g) + O_2(g) \rightarrow SO_2(g) + H_2O(g)$$

65. Write word equations for the following skeleton equations.
- **a.** $Cu(s) + O_2(g) \rightarrow CuO(s)$
- **b.** $K(s) + H_2O(l) \rightarrow KOH(aq) + H_2(g)$
- **c.** $CaCl_2(aq) + Na_2SO_4(aq) \rightarrow CaSO_4(s) + NaCl(aq)$

66. Balance the following reactions.
- **a.** $(NH_4)_2Cr_2O_7(s) \rightarrow Cr_2O_3(s) + N_2(g) + H_2O(g)$
- **b.** $CO_2(g) + H_2O(l) \rightarrow C_6H_{12}O_6(s) + O_2(g)$

Mastering Problems

67. Hydrogen iodide gas breaks down into hydrogen gas and iodine gas during a decomposition reaction. Write a skeleton equation for this reaction.

68. Write skeleton equations for these reactions.
- **a.** sodium carbonate(s) → sodium oxide(s) + carbon dioxide(g)
- **b.** aluminum(s) + iodine(s) → aluminum iodide(s)
- **c.** iron(II) oxide(s) + oxygen(g) → iron(III) oxide(s)

69. Write skeleton equations for these reactions.
- **a.** butane (C_4H_{10})(l) + oxygen(g) → carbon dioxide(g) + water(l)
- **b.** aluminum carbonate(s) → aluminum oxide(s) + carbon dioxide(g)
- **c.** silver nitrate(aq) + sodium sulfide(aq) → silver sulfide(s) + sodium nitrate(aq)

70. Write a skeleton equation for the reaction between lithium(s) and chlorine gas to produce lithium chloride(s).

71. Write skeleton equations for these reactions.
- **a.** iron(s) + fluorine(g) → iron(III) fluoride(s)
- **b.** sulfur trioxide(g) + water(l) → sulfuric acid(aq)
- **c.** sodium(s) + magnesium iodide(aq) → sodium iodide(aq) + magnesium(s)
- **d.** vanadium(s) + oxygen(g) → vanadium(V) oxide(s)

72. Write skeleton equations for these reactions.
- **a.** lithium(s) + gold(III) chloride(aq) → lithium chloride(aq) + gold(s)
- **b.** iron(s) + tin(IV) nitrate(aq) → iron(III) nitrate(aq) + tin(s)
- **c.** nickel(II) chloride(s) + oxygen(g) → nickel(II) oxide(s) + dichlorine pentoxide(g)
- **d.** lithium chromate(aq) + barium chloride(aq) → lithium chloride(aq) + barium chromate(s)

73. Balance the skeleton equations for the reactions described in Question 71.

74. Balance the skeleton equations for the reactions described in Question 72.

75. Write chemical equations for these reactions.
- **a.** When solid naphthalene ($C_{10}H_8$) burns in air, the reaction yields gaseous carbon dioxide and liquid water.
- **b.** Bubbling hydrogen sulfide gas through manganese(II) chloride dissolved in water results in the formation of the precipitate manganese(II) sulfide and hydrochloric acid.
- **c.** Solid magnesium reacts with nitrogen gas to produce solid magnesium nitride.
- **d.** Heating oxygen difluoride gas yields oxygen gas and fluorine gas.

SECTION 2

Mastering Concepts

76. List each of the four types of chemical reactions and give an example for each type.

77. How would you classify a chemical reaction between two reactants that produces one product?

78. Under what conditions does a precipitate form in a chemical reaction?

79. Will a metal always replace another metal in a compound dissolved in water? Explain.

80. In each of the following pairs, which element will replace the other in a reaction?
- **a.** tin and sodium
- **b.** fluorine and iodine
- **c.** lead and silver
- **d.** copper and nickel

73. **a.** $2Fe(s) + 3F_2(g) \longrightarrow 2FeF_3(s)$
    b. $SO_3(g) + H_2O(l) \longrightarrow H_2SO_4(aq)$
    c. $2Na(s) + MgI_2(aq) \longrightarrow 2NaI(aq) + Mg(s)$
    d. $4V(s) + 5O_2(g) \longrightarrow 2V_2O_5(s)$
74. **a.** $3Li(s) + AuCl_3(aq) \longrightarrow 3LiCl(aq) + Au(s)$
    b. $4Fe(s) + 3Sn(NO_3)_4(aq) \longrightarrow 4Fe(NO_3)_3(aq) + 3Sn(s)$
    c. $NiCl_2(s) + 3O_2(g) \longrightarrow NiO(s) + Cl_2O_5(g)$
    d. $Li_2CrO_4(aq) + BaCl_2(aq) \longrightarrow 2LiCl(aq) + BaCrO_4(s)$
75. **a.** $C_{10}H_8(s) + 12O_2(g) \longrightarrow 10CO_2(g) + 4H_2O(l)$
    b. $H_2S(g) + MnCl_2(aq) \longrightarrow MnS(s) + 2HCl(aq)$
    c. $3Mg(s) + N_2(g) \longrightarrow Mg_3N_2(s)$
    d. $2OF_2(g) \longrightarrow O_2(g) + 2F_2(g)$

SECTION 2

Mastering Concepts

76. Refer to the Solutions Manual.
77. It is a synthesis reaction.
78. When the reaction occurs in aqueous solution and the product of the reaction is insoluble.
79. No. The most active metal will replace the least active metal, but the opposite will not occur.
80. **a.** Na replaces Sn
    b. F replaces I
    c. Pb replaces Ag
    d. Ni replaces Cu

Chapter Self-Check

Mastering Problems

81. Classify each of the reactions represented by the chemical equations in Question 71.

82. Classify each of the reactions represented by the chemical equations in Question 72.

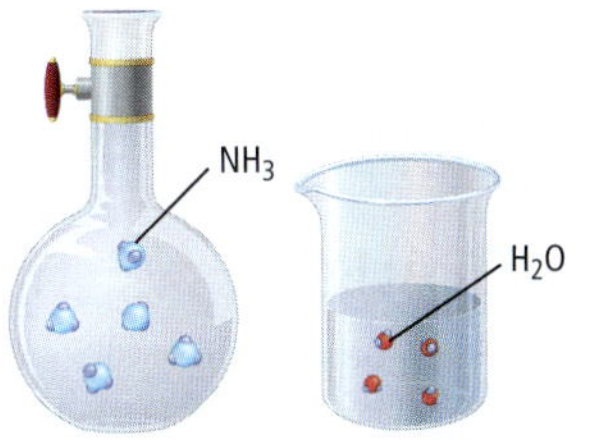

■ **Figure 22**

83. Use **Figure 22** to answer the following questions.
  a. Write a chemical equation for the reaction between the two compounds shown in the figure.
  b. Classify this reaction.

84. Write a balanced chemical equation for the combustion of liquid methanol (CH_3OH).

85. Write chemical equations for each of the following synthesis reactions.
  a. boron + fluorine →
  b. germanium + sulfur →
  c. zirconium + nitrogen →
  d. tetraphosphorus decoxide + water → phosphoric acid

86. Combustion Write a chemical equation for the combustion of each of the following substances. If a compound contains carbon and hydrogen, assume that carbon dioxide gas and liquid water are produced.
  a. solid barium
  b. solid boron
  c. liquid acetone (C_3H_6O)
  d. liquid octane (C_8H_{18})

87. Write chemical equations for each of the following decomposition reactions. One or more products might be identified.
  a. magnesium bromide →
  b. cobalt(II) oxide →
  c. titanium(IV) hydroxide → titanium(IV) oxide + water
  d. barium carbonate → barium oxide + carbon dioxide

88. Write chemical equations for the following single-replacement reactions that might occur in water. If no reaction occurs, write *NR* in place of the products.
  a. nickel + magnesium chloride →
  b. calcium + copper(II) bromide →
  c. potassium + aluminum nitrate →
  d. magnesium + silver nitrate →

SECTION 3

Mastering Concepts

89. Complete the following word equation.

Solute + Solvent →

90. Define each of the following terms: *solution, solvent,* and *solute.*

91. When reactions occur in aqueous solutions, what common types of products are produced?

92. Compare and contrast chemical equations and ionic equations.

93. What is a net ionic equation? How does it differ from a complete ionic equation?

94. Define *spectator ion.*

95. Write the net ionic equation for a chemical reaction that occurs in an aqueous solution and produces water.

Mastering Problems

96. Complete the following chemical equations.
  a. $Na(s) + H_2O(l) \rightarrow$
  b. $K(s) + H_2O(l) \rightarrow$

97. Complete the following chemical equation.

$$CuCl_2(s) + Na_2SO_4(aq) \rightarrow$$

98. Write complete ionic and net ionic equations for the chemical reaction in Question 97.

99. Write complete ionic and net ionic equations for each of the following reactions.
  a. $K_2S(aq) + CoCl_2(aq) \rightarrow 2KCl(aq) + CoS(s)$
  b. $H_2SO_4(aq) + CaCO_3(s) \rightarrow H_2O(l) + CO_2(g) + CaSO_4(s)$
  c. $2HClO(aq) + Ca(OH)_2(aq) \rightarrow 2H_2O(l) + Ca(ClO)_2(aq)$

100. A reaction occurs when hydrosulfuric acid (H_2S) is mixed with an aqueous solution of iron(III) bromide. The reaction produces solid iron(III) sulfide and aqueous hydrogen bromide. Write the chemical and net ionic equations for the reaction.

101. Write complete ionic and net ionic equations for each of the following reactions.
  a. $H_3PO_4(aq) + 3RbOH(aq) \rightarrow 3H_2O(l) + Rb_3PO_4(aq)$
  b. $HCl(aq) + NH_4OH(aq) \rightarrow H_2O(l) + NH_4Cl(aq)$
  c. $2HI + (NH_4)_2S(aq) \rightarrow H_2S(g) + 2NH_4I(aq)$
  d. $HNO_3(aq) + KCN(aq) + HCN(g) + KNO_3(aq)$

102. Paper A reaction occurs when sulfurous acid (H_2SO_3) is mixed with an aqueous solution of sodium hydroxide. The reaction produces aqueous sodium sulfite, a chemical used in manufacturing paper. Write the chemical and net ionic equations for the reaction.

Mastering Problems

81. a. synthesis
  b. synthesis
  c. single replacement
  d. combustion and synthesis

82. a. single replacement
  b. single replacement
  c. combustion
  d. double replacement

83. a. $NH_3(g) + H_2O(l) \rightarrow NH_4^+(aq) + OH^-(aq)$
  b. single-replacement reaction

84. $2CH_3OH(l) + 3O_2(g) \rightarrow 2CO_2(g) + 4H_2O(g)$

85. a. $2B(s) + 3F_2(g) \rightarrow 2BF_3(g)$
  b. $Ge(s) + 2S(s) \rightarrow GeS_2(s)$
  c. $3Zr(s) + 2N_2(g) \rightarrow Zr_3N_4(s)$
  d. $P_4O_{10}(s) + 6H_2O(l) \rightarrow 4H_3PO_4(aq)$

86. a. $2Ba(s) + O_2(g) \rightarrow 2BaO(s)$
  b. $4B(s) + 3O_2(g) \rightarrow 2B_2O_3(s)$
  c. $C_3H_6O(l) + 4O_2(g) \rightarrow 3CO_2(g) + 3H_2O(l)$
  d. $2C_8H_{18}(l) + 25O_2(g) \rightarrow 16CO_2(g) + 18H_2O(l)$

87. a. $MgBr_2(s) \rightarrow Mg(s) + Br_2(l)$
  b. $2CoO(s) \rightarrow 2Co(s) + O_2(g)$
  c. $Ti(OH)_4(s) \rightarrow TiO_2(s) + 2H_2O(l)$
  d. $BaCO_3(s) \rightarrow BaO(s) + CO_2(g)$

88. a. $Ni(s) + MgCl_2(aq) \rightarrow NR$
  b. $Ca(s) + CuBr_2(aq) \rightarrow Cu(s) + CaBr_2(aq)$
  c. $3K(s) + Al(NO_3)_3(aq) \rightarrow Al(s) + 3KNO_3(aq)$
  d. $Mg(s) + 2AgNO_3(aq) \rightarrow 2Ag(s) + Mg(NO_3)_2(aq)$

SECTION 3

Mastering Concepts

89. Solute + Solvent → Solution

90. Refer to the Solutions Manual.

91. solids, water, and gases

92. Refer to the Solutions Manual.

93. Refer to the Solutions Manual.

94. Refer to the Solutions Manual.

95. $H^+(aq) + OH^-(aq) \rightarrow H_2O(l)$

Mastering Problems

96. a. $2Na(s) + 2H_2O(l) \rightarrow 2NaOH(aq) + H_2$
  b. $2K(s) + 2H_2O(l) \rightarrow 2KOH(aq) + H_2(g)$

97. $CuCl_2(s) + Na_2SO_4(aq) \rightarrow CuSO_4(aq) + 2NaCl(aq)$

98. ionic equation: $CuCl_2(s) + 2Na^+(aq) + SO_4^{2-}(aq) \rightarrow Cu^{2+}(aq) + SO_4^{2-}(aq) + 2Na^+(aq) + 2Cl^-(aq)$
net ionic equation: $CuCl_2(s) \rightarrow Cu^{2+}(aq) + 2Cl^-(aq)$

99. a. Complete: $2K^+(aq) + S^{2-}(aq) + Co^{2+}(aq) + 2Cl^-(aq) \rightarrow 2K^+(aq) + 2Cl^-(aq) + CoS(s)$
Net: $S^{2-}(aq) + Co^{2+}(aq) \rightarrow CoS(s)$
  b. Complete: $2H^+(aq) + SO_4^{2-}(aq) + CaCO_3(s) \rightarrow H_2O(l) + CO_2(g) + CaSO_4(s)$
Net: $2H^+(aq) + SO_4^{2-}(aq) + CaCO_3(s) \rightarrow H_2O(l) + CO_2(g) + CaSO_4(s)$
  c. Complete: $2H^+(aq) + 2ClO^-(aq) + Ca^{2+}(aq) + 2OH^-(aq) \rightarrow 2H_2O(l) + Ca^{2+}(aq) + 2ClO^-(aq)$
Net: $H^+(aq) + OH^-(aq) \rightarrow H_2O(l)$

100. Chemical: $3H_2S(aq) + 2FeBr_3(aq) \rightarrow 6HBr(aq) + Fe_2S_3(s)$
Net: $3S^{2-}(aq) + 2Fe^{3+}(aq) \rightarrow Fe_2S_3(s)$

101. a. Complete: $3H^+(aq) + PO_4^{3-}(aq) + 3Rb^+(aq) + 3OH^-(aq) \rightarrow 3H_2O(l) + 3Rb^+(aq) + PO_4^{3+}(aq)$
Net: $H^+(aq) + OH^-(aq) \rightarrow H_2O(l)$
  b. Complete: $H^+(aq) + Cl^-(aq) + NH_4^+(aq) + OH^-(aq) \rightarrow H_2O(l) + NH_4^+(aq) + Cl^-(aq)$
Net: $H^+(aq) + OH^-(aq) \rightarrow H_2O(l)$
  c. Complete: $2H^+(aq) + 2I^-(aq) + 2NH_4^+(aq) + S^{2-}(aq) \rightarrow H_2S(g) + 2NH_4^+(aq) + 2I^-(aq)$
Net: $2H^+(aq) + S^{2-}(aq) \rightarrow H_2S(g)$
  d. Complete: $H^+(aq) + NO_3^-(aq) + K^+(aq) + CN^-(aq) \rightarrow HCN(g) + K^+(aq) + NO_3^-(aq)$
Net: $H^+(aq) + CN^-(aq) \rightarrow HCN(g)$

102. See the Solutions Manual

MIXED REVIEW

103. glucose and oxygen

104. An aqueous sucrose solution contains water molecules and sucrose molecules. An aqueous hydrogen chloride solution contains water molecules, hydrogen ions, and chloride ions.

105. a. benzene(l) + oxygen(g) ⟶ carbon dioxide(g) + water (l)
b. carbon monoxide(g) + oxygen(g) ⟶ carbon dioxide(g)
c. chlorine(g) + sodium bromide(s) ⟶ sodium chloride(s) + bromine(g)
d. calcium carbonate(s) ⟶ calcium oxide(s) + carbon dioxide(g)

106. a. combustion
b. synthesis or combustion
c. single-replacement
d. decomposition

107. a. $(NH_4)_3PO_4(aq) + CrBr_3(aq) \longrightarrow NH_4Br(aq) + CrPO_4(s)$
b. $Cr(OH)_6(s) \longrightarrow CrO_3(s) + H_2O(l)$
c. $Al(s) + CuCl(aq) \longrightarrow AlCl_3(aq) + Cu(s)$
d. $KI(aq) + HgNO_3(aq) \longrightarrow KNO_3(aq) + HgI(s)$

108. Refer to the Solutions Manual.

109. Refer to the Solutions Manual.

110. a. no reaction
b. reaction occurs
c. reaction occurs

111. $3NaOH(aq) + AlCl_3(s) \longrightarrow 3NaCl(aq) + Al(OH)_3(aq)$

112. a. the beaker with silver nitrate
b. silver chloride
c. $HCl(aq) + AgNO_3(aq) \longrightarrow AgCl(s) + HNO_3(aq)$
d. double-replacement reaction

113. Skeleton equation: $Fe(s) + Cl_2(g) \longrightarrow FeCl_3(s)$
Balanced chemical equation: $2Fe(s) + 3Cl_2(g) \longrightarrow 2FeCl_3(s)$

114. $2H_2O(l) \longrightarrow 2H_2(g) + O_2(g)$; Hydrogen gas and oxygen gas

115. Refer to the Solutions Manual.

116. Refer to the Solutions Manual.

THINK CRITICALLY

117. Refer to the Solutions Manual.

118. Refer to the Solutions Manual.

119. Refer to the Solutions Manual.

120. Refer to the Solutions Manual.

121. $2Li(s) + 2H_2O(l) \longrightarrow 2LiOH(aq) + H_2(g)$
$2Na(s) + 2H_2O(l) \longrightarrow 2NaOH(aq) + H_2(g)$
$2K(s) + 2H_2O(l) \longrightarrow 2KOH(aq) + H_2(g)$
Sodium would release the least amount of heat, followed by potassium, then lithium.

122. a. $KNO_3(aq) + CsCl(aq) \longrightarrow KCl(aq) + CsNO_3(aq)$
$K^+(aq) + NO_3^-(aq) + Cs^+(aq) + Cl^-(aq) \longrightarrow K^+(aq) + Cl^-(aq) + Cs^+(aq) + NO_3^-(aq)$
Ions remain in solution. No reaction occurs.
b. $Ca(OH)_2(aq) + 2KCN(aq) \longrightarrow Ca(CN)_2(aq) + 2KOH(aq)$
$Ca^{2+}(aq) + 2OH^-(aq) + 2K^+(aq) + 2CN^-(aq) \longrightarrow Ca^{2+}(aq) + 2CN^-(aq) + 2K^+(aq) + 2OH^-(aq)$
Ions remain in solution. No reaction occurs.
c. Chemical: $2Li_3PO_4(aq) + 3MgSO_4(aq) \longrightarrow 3Li_2SO_4(aq) + Mg_3(PO_4)_2(s)$
Net ionic: $2PO_4^{3-}(aq) + 3Mg^{2+}(aq) \longrightarrow Mg_3(PO_4)_2(s)$
d. Chemical: $HBrO(aq) + NaOH(aq) \longrightarrow H_2O(l) + NaBrO(aq)$
Net ionic: $H^+(aq) + OH^-(aq) \longrightarrow H_2O(l)$

123. A nail is composed of iron, Fe. Iron atoms react with oxygen gas to form rust, iron(III) oxide. Iron atoms do not react with nitrogen gas.

124. $4Al(s) + 3O_2(g) \longrightarrow 2Al_2O_3(s)$

ASSESSMENT

MIXED REVIEW

103. Photosynthesis Identify the products in the following reaction that occurs in plants: Carbon dioxide and water react to produce glucose and oxygen.

104. How will aqueous solutions of sucrose and hydrogen chloride differ?

105. Write the word equation for each of these skeleton equations. C_6H_6 is the formula for benzene.
a. $C_6H_6(l) + O_2(g) \rightarrow CO_2(g) + H_2O(l)$
b. $CO(g) + O_2(g) \rightarrow CO_2(g)$
c. $Cl_2(g) + NaBr(s) \rightarrow NaCl(s) + Br_2(g)$
d. $CaCO_3(s) \rightarrow CaO(s) + CO_2(g)$

106. Classify each of the reactions represented by the chemical equations in Question 105.

107. Write skeleton equations for the following reactions.
a. ammonium phosphate(aq) + chromium(III) bromide(aq) → ammonium bromide(aq) + chromium(III) phosphate(s)
b. chromium(VI) hydroxide(s) → chromium(VI) oxide(s) + water(l)
c. aluminum(s) + copper(I) chloride(aq) → aluminum chloride(aq) + copper(s)
d. potassium iodide(aq) + mercury(I) nitrate(aq) → potassium nitrate(aq) + mercury(I) iodide(s)

108. Balance the skeleton equations for the reactions described in Question 107.

109. Classify each of the reactions represented by the chemical equations in Question 108.

110. Predict whether each of the following reactions will occur in aqueous solutions. If you predict that a reaction will not occur, explain your reasoning. *Note: Barium sulfate and silver bromide precipitate in aqueous solutions.*
a. sodium hydroxide + ammonium sulfate →
b. niobium(V) sulfate + barium nitrate →
c. strontium bromide + silver nitrate →

111. Complete the missing information in the following skeleton equation and balance the chemical equation:

$$NaOH(aq) + \_\_\_ \rightarrow 3NaCl(aq) + Al(OH)_3(aq)$$

112. Precipitate Formation The addition of hydrochloric acid to beakers containing solutions of either sodium chloride (NaCl) or silver nitrate ($AgNO_3$) causes a white precipitate in one of the beakers.
a. Which beaker contains a precipitate?
b. What is the precipitate?
c. Write a chemical equation showing the reaction.
d. Classify the reaction.

113. Write the skeleton equation and the balanced chemical equation for the reaction between iron and chlorine.

114. Write a chemical equation representing the decomposition of water into two gaseous products. What are the products?

115. Distinguish between an ionic compound and a molecular compound dissolved in water. Do all molecular compounds ionize when dissolved in water? Explain.

116. Classify the type of reactions that occur in aqueous solutions, and give an example to support your answer.

THINK CRITICALLY

117. Explain how an equation can be balanced even if the number of reactant particles differs from the number of product particles.

118. Apply Describe the reaction of aqueous solutions of sodium sulfide and copper(II) sulfate, producing the precipitate copper(II) sulfide.

119. Predict A piece of aluminum metal is placed in aqueous KCl. Another piece of aluminum is placed in an aqueous $AgNO_3$ solution. Explain why a chemical reaction does or does not occur in each instance.

120. Design an Experiment You suspect that the water in a lake close to your school might contain lead in the form of Pb^{2+}(aq) ions. Formulate your suspicion as a hypothesis and design an experiment to test your theory. Write the net ionic equations for the reactions of your experiment. *(Hint: In aqueous solution, Pb^{2+} forms solid compounds with Cl^-, Br^-, I^-, and SO_4^{2-} ions.)*

121. Predict When sodium metal reacts with water, it produces sodium hydroxide, hydrogen gas, and heat. Write balanced chemical equations for Li, Na, and K reacting with water. Use **Figure 13** to predict the order of the amount of heat released from least to most amount of heat released.

122. Apply Write the chemical equations and net ionic equations for each of the following reactions that might occur in aqueous solutions. If a reaction does not occur, write *NR* in place of the products. Magnesium phosphate precipitates in an aqueous solution.
a. $KNO_3 + CsCl \rightarrow$
b. $Ca(OH)_2 + KCN \rightarrow$
c. $Li_3PO_4 + MgSO_4 \rightarrow$
d. $HBrO + NaOH \rightarrow$

123. Analyze Explain why a nail exposed to air forms rust, whereas the same nail exposed to a pure nitrogen environment does not form rust.

124. Evaluate Write a balanced chemical equation for the reaction of aluminum with oxygen to produce aluminum oxide.

Chapter Self-Check

CHALLENGE PROBLEM

125. A single-replacement reaction occurs between copper and silver nitrate. When 63.5 g of copper reacts with 339.8 g of silver nitrate, 215.8 g of silver is produced. Write a balanced chemical equation for this reaction. What other product formed? What is the mass of the second product?

CUMULATIVE REVIEW

126. Complete the following problems in scientific notation. Round off to the correct number of significant figures.
- **a.** $(5.31 \times 10^{-2}\ cm) \times (2.46 \times 10^{5}\ cm)$
- **b.** $(6.42 \times 10^{-2}\ g) \div (3.21 \times 10^{-3}\ g)$
- **c.** $(9.87 \times 10^{4}\ g) - (6.2 \times 10^{3}\ g)$

127. Distinguish between a mixture, a solution, and a compound.

128. Data from chromium's four naturally occurring isotopes is provided in **Table 5.** Calculate chromium's atomic mass.

Table 5 Chromium Isotope Data

Isotope	Percent Abundance	Mass (amu)
Cr-50	4.35%	49.946
Cr-52	83.79%	51.941
Cr-53	9.50%	52.941
Cr-54	2.36%	53.939

129. Differentiate between electron configuration and electron-dot structure.

130. Identify the elements by their electron configuration.
- **a.** $1s^2 2s^2 2p^6 3s^2 3p^6 4s^2 3d^{10} 4p^5$
- **b.** $[Ne]3s^2 3p^4$
- **c.** $[Xe]6s^2$

131. Write the electron configuration for the element fitting each description.
- **a.** a metalloid in group 13
- **b.** a nonmetal in group 15, period 3

132. Describe the formation of positive and negative ions.

133. Write the formula for the compounds made from each of the following pairs of ions.
- **a.** copper(I) and sulfite
- **b.** tin(IV) and fluoride
- **c.** gold(III) and cyanide
- **d.** lead(II) and sulfide

WRITING IN ▶ Chemistry

134. Kitchen Chemistry Make a poster describing chemical reactions that occur in the kitchen.

135. Mathematical Equations Write a report that compares and contrasts chemical equations and mathematical equations.

136. Balance Equations Create a flowchart describing how to balance a chemical equation.

DBQ Document-Based Questions

Solubility *Scientists, in determining whether a precipitate will occur in a chemical reaction, use a solubility rules chart.* **Table 6** *lists the solubility rules for ionic compounds in water.*

Data obtained from: Van Der Sluys, W.G. 2001, *J. Chem. Ed.* 78:111–115

Table 6 Solubility Rules for Ionic Compound in Water

Ionic Compound	Rule
Soluble salts	Group 1 cations and NH_4^+ ions form soluble salts.
	All nitrates are soluble.
	Most halides are soluble, except those of Pb^{2+}, Hg_2^{2+}, Ag^+, and Cu^+.
	Most sulfates are soluble, with the exception of those of Ba^{2+}; Sr^{2+}, Pb^{2+}, Ag^+, Ca^{2+}, and Hg_2^{2+} form slightly soluble sulfates.
Insoluble salts	Hydroxides, oxides, and sulfides are usually insoluble, except that those of group 1 ions and NH_4^+ are soluble and those of group 2 ions are slightly soluble.
	Chromates, phosphates, and carbonates are usually insoluble, except that those of group 1 ions and NH_4^+ are soluble.

Using the solubility rules provided in the table above, complete the following chemical equations. Indicate whether a precipitate forms or not. Identify the precipitate. If no reaction occurs, write *NR*.

137. $Ca(NO_3)_2(aq) + Na_2CO_3(aq) \rightarrow$

138. $Mg(s) + NaOH(aq) \rightarrow$

139. $PbS(s) + LiNO_3(aq) \rightarrow$

CHALLENGE PROBLEM

125. $2AgNO_3 + Cu \rightarrow Cu(NO_3)_2 + 2Ag$; 187.5 g of $Cu(NO_3)_2$

CUMULATIVE REVIEW

126. a. $1.31 \times 10^4\ cm^2$
b. 2.00×10^1
c. $9.2 \times 10^4\ g$

127. A mixture is a physical blend of two or more pure substances in any proportion in which each substance retains its individual properties. A solution is a uniform mixture that may contain solids, liquids, or gases. A compound is a chemical combination of two or more different elements to form a substance with new properties. The components of mixtures and solutions can be separated by physical means. A compound can be broken down only by chemical means.

128. 51.99 amu

129. Electron configuration is the arrangement of electrons in an atom. An electron-dot structure consists of the element's symbol and valence electrons.

130. a. Br
b. S
c. Ba

131. a. B: $[He]2s^2 2p^1$
b. P: $[Ne]3s^2 3p^3$

132. Positive ions form when atoms lose valence electrons. Negative ions form when valence electrons are added to an atom.

133. a. Cu_2SO_3
b. SnF_4
c. $Au(CN)_3$
d. PbS

135. Student answers might include the following: chemical equations contain reaction arrows, while mathematical equations contain equal signs; chemical equations contain formulas for substances, while mathematical equations contain variables; because chemical equations contain formulas that represent kinds and numbers of atoms, they must be balanced; chemical equations might contain energy terms as well as formulas.

136. Student flowcharts should be similar to Figure 6.

DBQ Document-Based Questions

Data obtained from: Van Der Sluys, W.G. 2001. *J. Chem. Ed.* 78:111–115.

137. $Ca(NO_3)_2(aq) + Na_2CO_3(aq) \rightarrow CaCO_3(s) + 2NaNO_3(aq)$
$CaCO_3(s)$ is the precipitate that forms.

138. $Mg(s) + NaOH(aq) \rightarrow NR$

139. $PbS(s) + 2LiNO_3(aq) \rightarrow Li_2S(aq) + Pb(NO)_2(aq)$
No precipitate forms.

WRITING IN ▶ Chemistry

Rubric

134. Answers will vary. Examples might include the reaction of yeast when bread dough rises, frying an egg, baking a cake, or making jello.

MULTIPLE CHOICE

1. D
2. A
3. A
4. C
5. B
6. D
7. A
8. B

MULTIPLE CHOICE

1. What type of reaction is described by the following equation?

$Cs(s) + H_2O(l) \rightarrow CsOH(aq) + H_2(g)$

A. synthesis
B. combustion
C. decomposition
D. single-replacement

Use the figure below to answer Question 2.

Activity Series of Halogens

Most active ↓ **Least active**

Fluorine
Chlorine
Bromine
Iodine

2. Which reaction between halogens and halide salts will occur?
A. $F_2(g) + FeI_2(aq) \rightarrow FeF_2(aq) + I_2(l)$
B. $I_2(s) + MnBr_2(aq) \rightarrow MnI_2(aq) + Br_2(g)$
C. $Cl_2(s) + SrF_2(aq) \rightarrow SrCl_2(aq) + F_2(g)$
D. $Br_2(l) + CoCl_2(aq) \rightarrow CoBr_2(aq) + Cl_2(g)$

3. Which is the electron configuration for iron?
A. $1s^22s^22p^63s^23p^64s^23d^6$
B. $[Ar]3d^6$
C. $1s^22p^63p^63d^6$
D. $[Ar]4s^24d^6$

4. Which is a description of a pattern displayed by elements in the periodic table?
A. repetition of their physical properties when arranged by increasing atomic radius
B. repetition of their chemical properties when arranged by increasing atomic mass
C. periodic repetition of their properties when arranged by increasing atomic number
D. periodic repetition of their properties when arranged by increasing atomic mass

5. When moving down a group on the periodic table, which two atomic properties follow the same trend?
A. atomic radius and ionization energy
B. ionic radius and atomic radius
C. ionization energy and ionic radius
D. ionic radius and electronegativity

Use the table below to answer Questions 6 to 8.

Physical Properties of Select Ionic Compounds

Compound	Name	State at 25°C	Soluble in Water?	Melting Point (°C)
$NaClO_3$	sodium chlorate	solid	yes	248
Na_2SO_4	sodium sulfate	solid	yes	884
$NiCl_2$	nickel(II) chloride	solid	yes	1031
$Ni(OH)_2$	nickel(II) hydroxide	solid	no	230
$AgNO_3$	silver nitrate	solid	yes	210

6. An aqueous solution of nickel(II) sulfate is mixed with aqueous sodium hydroxide. Will a visible reaction occur?
A. No, solid nickel(II) hydroxide is soluble in water.
B. No, solid sodium sulfate is soluble in water.
C. Yes, solid sodium sulfate will precipitate out of the solution.
D. Yes, solid nickel(II) hydroxide will precipitate out of the solution.

7. What happens when $AgClO_3(aq)$ and $NaNO_3(aq)$ are mixed?
A. No visible reaction occurs.
B. Solid $NaClO_3$ precipitates out of the solution.
C. NO_2 gas is released during the reaction.
D. Solid Ag metal is produced.

8. Finely ground nickel(II) hydroxide is placed in a beaker of water. It sinks to the bottom of the beaker and remains unchanged. An aqueous solution of hydrochloric acid (HCl) is then added to the beaker, and the $Ni(OH)_2$ disappears. Which equation best describes what occurred in the beaker?
A. $Ni(OH)_2(s) + HCl(aq) \rightarrow NiO(aq) + H_2(g) + HCl(aq)$
B. $Ni(OH)_2(s) + 2HCl(aq) \rightarrow NiCl_2(aq) + 2H_2O(l)$
C. $Ni(OH)_2(s) + 2H_2O(l) \rightarrow NiCl_2(aq) + 2H_2O(l)$
D. $Ni(OH)_2(s) + 2H_2O(l) \rightarrow NiCl_2(aq) + 3H_2O(l) + O_2(g)$

SHORT ANSWER

Use the diagram below to answer Questions 9 and 10.

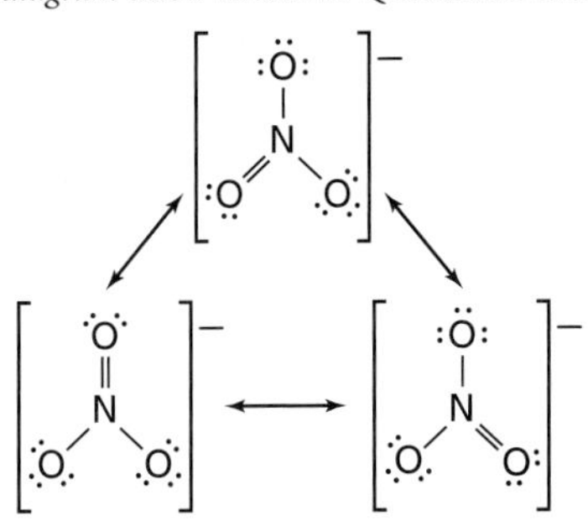

9. What is the name for the multiple Lewis structures shown in the diagram?

10. Why do these structures form?

11. Write the balanced chemical equation for the reaction of solid calcium with water to form calcium hydroxide in solution and hydrogen gas.

EXTENDED RESPONSE

Use the partial chemical equation below to answer Questions 12 and 13.

$$AlCl_3(aq) + Fe_2O_3(aq) \rightarrow$$

12. What type of reaction will this be? Explain how you can tell from the reactants.

13. Predict what the products of this reaction will be. Use evidence from the reaction to support your answer.

14. What is the electron configuration for the ion P^{3-}? Explain how this configuration is different from the configuration for the neutral atom of phosphorus.

SAT SUBJECT TEST: CHEMISTRY

15. Chloroform ($CHCl_3$) was one of the first anesthetics used in medicine. The chloroform molecule contains 26 valence electrons total. How many of these valence electrons are part of covalent bonds?
    A. 26
    B. 13
    C. 8
    D. 4
    E. 2

16. Which is NOT true of an atom obeying the octet rule?
    A. obtains a full set of eight valence electrons
    B. acquires the valence configuration of a noble gas
    C. electron configuration is unusually stable
    D. has an s^2p^6 valence configuration
    E. will lose electrons

Use the figure below to answer Question 17.

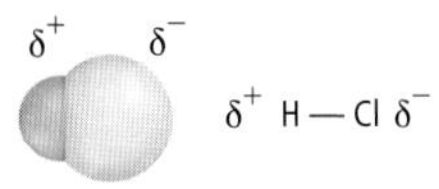

17. Which statement does NOT correctly describe the model of HCl shown above?
    A. A nonpolar bond exists between these atoms.
    B. Chlorine has a stronger attraction for electrons than does hydrogen.
    C. The electrons in the bond are shared unequally.
    D. This compound dissolves in a polar substance.
    E. Chlorine is the more electronegative atom.

18. The combustion of ethanol (C_2H_6O) produces carbon dioxide and water vapor. What equation best describes this process?
    A. $C_2H_6O(l) + O_2(g) \rightarrow CO_2(g) + H_2O(l)$
    B. $C_2H_6O(l) \rightarrow 2CO_2(g) + 3H_2O(l)$
    C. $C_2H_6O(l) + 3O_2(g) \rightarrow 2CO_2(g) + 3H_2O(g)$
    D. $C_2H_6O(l) \rightarrow 3O_2(l) + 2CO_2(g) + 3H_2O(l)$
    E. $C_2H_6O(l) \rightarrow 2CO_2(g) + 3H_2O(g)$

NEED EXTRA HELP?

If You Missed Question . . .	1	2	3	4	5	6	7	8	9	10	11	12	13	14	15	16	17	18
Review Section . . .	9.2	9.2	5.3	6.3	6.3	9.3	9.3	9.3	8.3	8.3	9.1	9.2	9.2	5.3	8.3	8.3	8.5	9.2

SHORT ANSWER

9. These are resonance structures.
10. A pair of electrons can form a double bond in several different positions. This means that there are several equivalent Lewis structures for the same molecule.
11. $Ca(s) + 2H_2O(l) \longrightarrow Ca(OH)_2(aq) + H_2(g)$

EXTENDED RESPONSE

12. This will be a double-replacement reaction. There are two compounds reacting together. In the other types of reactions, a single element participates as one of the reactants, but not here.
13. The products will be $FeCl_3$ and Al_2O_3. In a double-replacement reaction, the positive and negative ions exchange places so that each positive ion forms a compound with a new negative ion.
14. $1s^22s^22p^63s^23p^6$ is the electron configuration for P^{3-}. This differs from the configuration of the neutral phosphorus atom because it has three extra valence electrons, completing its valence shell and making it resemble a noble gas in its configuration. A neutral atom of P has the configuration of $1s^22s^22p^63s^23p^3$.

SAT SUBJECT TEST: CHEMISTRY

15. C
16. E
17. A
18. C

CHAPTER 10 The Mole

BIGIDEA **The mole represents a large number of extremely small particles.**

ESSENTIAL QUESTIONS	RESOURCES TO ASSESS MASTERY
SECTION 1 Measuring Matter 1. How is a mole used to indirectly count the number of particles of matter? 2. What is a common everyday counting unit to which the mole can be related? 3. How can moles be converted to number of representative particles and vice versa? 1 session 0.5 block	**Progress Monitoring** Formative Assessment, pp. 321, 324 Reading Check, p. 323 Section Review, p. 324
SECTION 2 Mass and the Mole 1. Why can the mass of an atom be related to the mass of a mole of atoms? 2. How can the number of moles be converted to the mass of an element and vice versa? 3. How can the number of moles be converted to the number of atoms of an element and vice versa? 2 sessions 1 block	**Progress Monitoring** Formative Assessment, pp. 327, 329, 331 Reading Check, p. 327 Section Review, p. 332
SECTION 3 Moles of Compounds 1. What are the mole relationships shown by a chemical formula? 2. How is the molar mass of a compound calculated? 3. How can the number of moles be converted to the mass of a compound and vice versa? 4. What conversion factors are applied to determine the number of atoms or ions in a known mass of a compound? 2 sessions 1 block	**Progress Monitoring** Formative Assessment, pp. 335, 336, 337, 339 Section Review, p. 340
SECTION 4 Empirical and Molecular Formulas 1. What is meant by the percent composition of a compound? 2. How can the empirical and molecular formulas for a compound be determined from mass percent and actual mass data? 2 sessions 1 block	**Progress Monitoring** Formative Assessment, pp. 341, 345, 348, 350 Reading Check, p. 344 Section Review, p. 350
SECTION 5 Formulas of Hydrates 1. What is a hydrate and how does its name relate to its composition? 2. How is the formula of a hydrate determined from laboratory data? 1 session 0.5 block	**Progress Monitoring** Formative Assessment, p. 354 Reading Check, p. 352 Section Review, p. 354 **Summative Assessment** Chapter Assessment, p. 358 *eAssessment* Chapter Test (scaffolded)

Go online!

LEVELED RESOURCES	LAB MATERIALS
Science Notebook 10.1 OL ***FAST FILE Chapter Resources:*** Study Guide, p. 44 BL	LaunchLAB p. 318 / **15 min** paper clip, gumdrop or marshmallow, ruler
Science Notebook 10.2 OL ***FAST FILE Chapter Resources:*** Study Guide, p. 45 BL	
Science Notebook 10.3 OL ***FAST FILE Chapter Resources:*** Study Guide, p. 46 BL	
Science Notebook 10.4 OL ***FAST FILE Chapter Resources:*** MiniLab Worksheet, p. 34 OL Study Guide, p. 47 BL	MiniLAB p. 342 / **25 min** chewing gum, weighing paper, balance, tap water, 250-mL beaker, stirring rod, paper towels, scissors, 10-cm × 10-cm piece of window screen
Science Notebook 10.5 OL ***FAST FILE Chapter Resources:*** ChemLab Worksheet, p. 36 OL Study Guide, p. 49 BL	ChemLAB p. 356 / **45 min** Bunsen burner, ring stand and ring, crucible and lid, clay triangle, Epsom salts (hydrated $MgSO_4$), balance, crucible tongs, spatula, spark lighter or matches

ADDITIONAL RESOURCES

eAssessment
ConnectED Student Centre
ConnectED Teacher Centre
Lesson Presentations
eSolutions
Videos
Animations
Online Quizzes
Virtual Investigations
Multilingual eGlossary
Personal Tutors
WebQuests
Applying Practices Worksheets
PBLs

Practice:

Challenge Problems AL
Supplemental Problems BL OL
Chapter Test (Scaffolded)
Standardized Test Practice

Additional Resources:

Blueprints for Success
Solving Problems: A Chemistry Handbook

Lab Resources:

Laboratory Manual OL
CBL Laboratory Manual OL
Small-Scale Laboratory Manual OL
Forensics Laboratory Manual OL

BL Below Level   OL On Level   AL Advanced Learners   EL English Learners   COOP LEARN Cooperative Learning

CHAPTER 10

LaunchLAB

Rubric

How much is a mole?

Purpose Students will develop a mental picture of the size of a mole.

Safety Precautions Approve lab safety forms before work begins.

Teaching Strategies

- Have students review conversions of metric units and scientific notation.
- Discuss the meaning of a light year–distance traveled by light in one year in a vacuum. The speed of light is 3.00×10^8 m/sec.
- Have students calculate the actual distance traveled by light in one year. $(3.00 \times 10^8 \text{ m/s}) \times (60 \text{ s/min}) \times (60 \text{ min/hr}) \times (24 \text{ hr/day}) \times (365 \text{ days/yr}) = 9.46 \times 10^{15}$ m
- Students can convert the distance of one light year in meters to the distance in miles. The conversion factor is 9.61×10^3 m/mile. $(9.46 \times 10^{15} \text{ m}) \times (1 \text{ mile}/9.61 \times 10^3 \text{ m}) = 9.84 \times 10^{11}$ miles
- Ask students to compare the miles traveled by light in one year to the distance across the United States (3000 miles) or the circumference of Earth (25,000 miles). $(9.84 \times 10^{11} \text{ miles})/(3000 \text{ miles}) = 400{,}000{,}000$ times across the United States. $(9.84 \times 10^{11} \text{ miles})/(25{,}000 \text{ miles}) = 39{,}000{,}000$ Earth circumferences
- Objects of any uniform length might be used.

Expected Results See Analysis section.

Procedure

1. Read and complete the lab safety form.
2. Select an item to measure, such as a **paper clip, gum drop,** or **marshmallow,** from the choices provided by your teacher. **WARNING:** ***Do not eat or taste any items used in the lab.***
3. Use a **ruler** to measure the length of your item to the nearest 0.1 cm.

CHAPTER 10

The Mole

BIGIDEA **The mole represents a large number of extremely small particles.**

SECTIONS

1 Measuring Matter

2 Mass and the Mole

3 Moles of Compounds

4 Empirical and Molecular Formulas

5 Formulas of Hydrates

LaunchLAB

How much is a mole?

Counting large numbers of items is easier when you use counting units, such as decades or dozens. Chemists use a counting unit called the mole. In this lab, you will use a common item to help you understand how much a mole really is.

FOLDABLES® Study Organizer

Conversion Factors

Make a two-tab Book. Label it as shown. Use it to help you organize information about conversion factors.

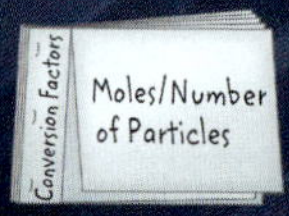

Analysis

1. **Calculate** If a mole is 6.02×10^{23} items, how far will a mole of your items, placed end-to-end lengthwise, extend into space? Express your answer in meters. Answers depend on the object chosen. For paper clips:
$$(6.02 \times 10^{23} \text{ paper clips})\left(\frac{3.2 \text{ cm}}{1 \text{ paperclip}}\right)\left(\frac{1 \text{ m}}{100 \text{ cm}}\right) = 1.9 \times 10^{22} \text{ m}$$
2. **Calculate** Convert the distance in Question 1 to light-years (ly). ($1 \text{ ly} = 9.46 \times 10^{15}$ m) Answers depend on the object chosen. For paper clips:
$$(1.9 \times 10^{22} \text{ m})\left(\frac{1 \text{ ly}}{9.46 \times 10^{15} \text{ m}}\right) = 2.0 \times 10^6 \text{ ly}$$
3. **Compare** the distance you calculated in Question 2 with these astronomical distances:
   a. distance to nearest star (other than the Sun) = 4.3 ly Answers depend on the object chosen. For paper clips, the distance calculated is much larger.

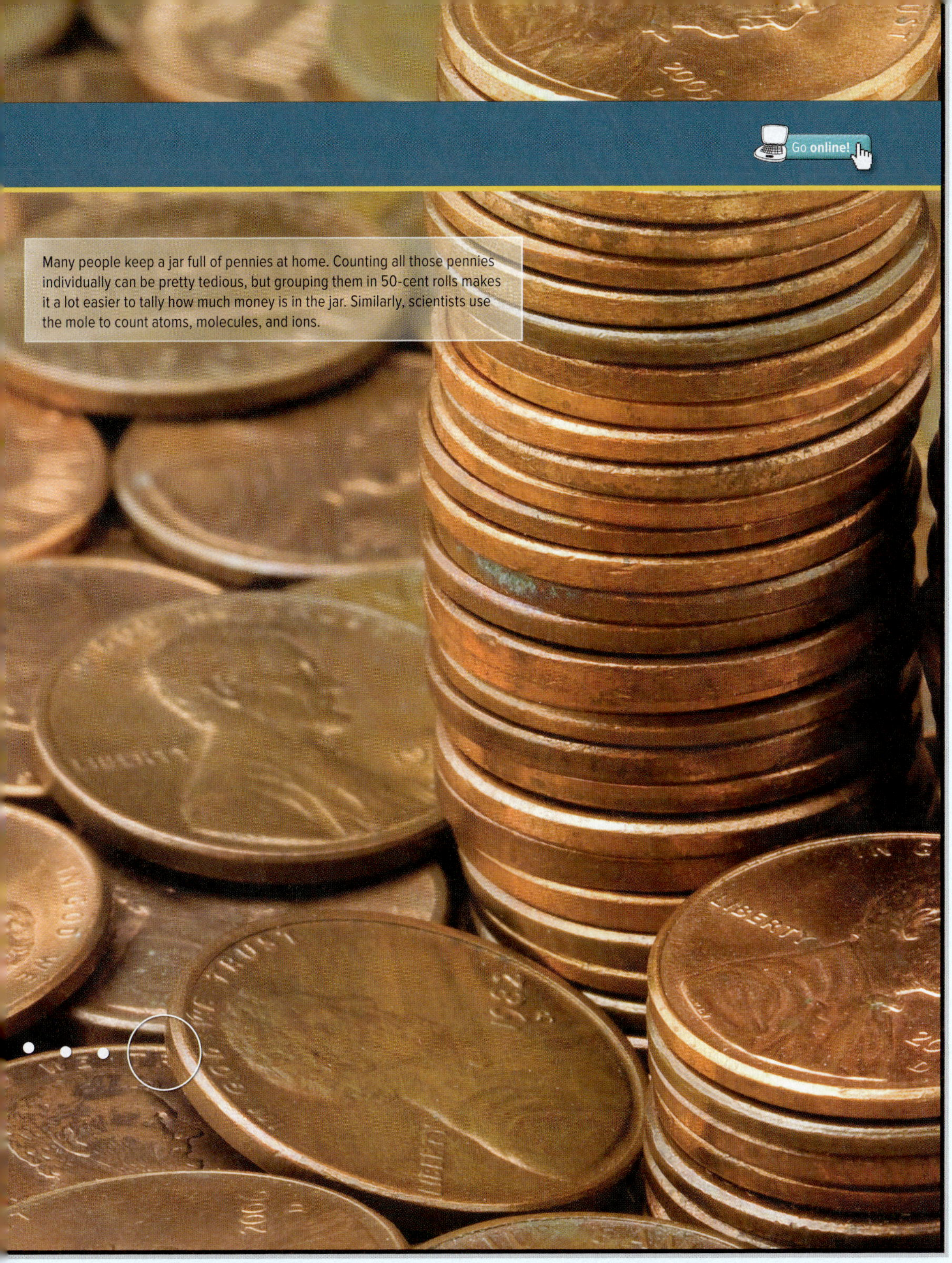

Many people keep a jar full of pennies at home. Counting all those pennies individually can be pretty tedious, but grouping them in 50-cent rolls makes it a lot easier to tally how much money is in the jar. Similarly, scientists use the mole to count atoms, molecules, and ions.

b. distance to the center of our galaxy = 30,000 ly Answers will vary depending on the object chosen. For paper clips, the distance calculated is much larger.

c. distance to nearest galaxy = 2×10^6 ly Answers depend on the object chosen. For paper clips, the distance calculated is equal to the distance to the nearest galaxy.

Inquiry Compare your item to another used by one of your classmates. Would a mole of your item have the same mass as a mole of the other item? Design an investigation to determine if there is a relationship between mass and moles. Student investigations will vary. One mole of different items will have different masses. Answers should show that there is a relationship between the number of moles of an item and the total mass (a greater number of moles of a given item results in greater mass), but no relationship between mass and moles (the mass of three moles depends on the item).

Introduce the BIGIDEA

Count by Group To introduce this chapter's Big Idea, bring rolls of nickels and quarters into class. Ask students what they have in common. They contain the same number of coins. Ask students what is different between the two rolls other than the monetary value. the mass of each roll Ask students why coins are placed in rolls. so they can quickly be counted by group instead of individually Ask students other ways individual items are counted. Eggs are counted by the dozen, pencils by the gross, and shoes by the pair. Tell students that chemists use a very large unit called a *mole* to count very small atoms, molecules, and formula units.

Tie to Previous Knowledge

Have students review the following concepts before studying this chapter.

- scientific notation, significant figures
- dimensional analysis
- average atomic mass

Use the Photo

Groups and Units Have students examine the opening photo. The number of pennies in the photo is difficult to determine. Ask students how they could determine the quantity of pennies without counting them one at a time. Answers will vary. They might begin by estimating the pennies in the stacks and then multiplying to get a total. An easy way to count large quantities of pennies is to count them in rolls, which are groups of 50.

SECTION 1

1 Focus

MAINIDEA

Count the Very Small Have students discuss results from the Launch Lab. Ask why the mole is a convenient unit for scientists dealing with extremely small objects such as atoms. It allows them to count everyday quantities of atoms without using extremely large numbers. To illustrate just how large a number a mole is, tell students that one mole of marshmallows would cover the United States to a depth of about 5.01×10^7 miles or 8.07×10^7 km. OL

2 Teach

Math in Chemistry

Scientific Notation Because Avogadro's number (6.02×10^{23}) is used extensively in this chapter, it is important that students understand basic math operations involving scientific notation. Tell students that when numbers in scientific notation are multiplied, only the decimal portion is multiplied–the exponents are added. Ask students to solve the following problems: (25.6 mol)(6.02×10^{23} atoms/mol) and (1.2×10^2 mol)(6.02×10^{23} ions/mol) 1.54×10^{25} atoms and 7.2×10^{25} ions, respectively Remind students that when numbers in scientific notation are divided, only the decimal portion is divided–the exponents are subtracted. Ask students to solve the following problems:
(8.2×10^{28})/(6.02×10^{23})
(15×10^{-1})/(6.02×10^{23})
1.4×10^5 and 2.5×10^{-24}, respectively. OL

■ **Caption Question Fig. 1**
Answers might include trio, quartet, six-pack, and so on.

SECTION 1

Essential Questions

- How is a mole used to indirectly count the number of particles of matter?
- What is a common everyday counting unit to which the mole can be related?
- How can moles be converted to number of representative particles and vice versa?

Review Vocabulary

molecule: two or more atoms that covalently bond together to form a unit

New Vocabulary

mole
Avogadro's number

Measuring Matter

MAINIDEA Chemists use the mole to count atoms, molecules, ions, and formula units.

CHEM 4 YOU Has your class ever had a contest to guess how many pennies or jelly beans were in a jar? You might have noticed that the smaller the object is, the harder it is to count.

Counting Particles

If you were buying a bouquet of roses for a special occasion, you probably would not ask for 12 or 24; you would ask for one or two dozen. Similarly, you might buy a pair of gloves, a ream of paper for your printer, or a gross of pencils. Each of the units shown in **Figure 1**—a pair, a dozen, a gross, and a ream—represents a specific number of items. These units make counting objects easier. It is easier to buy and sell paper by the ream—500 sheets—than by the individual sheet.

Each of the counting units shown in **Figure 1** is appropriate for certain kinds of objects, depending primarily on their size and function. But regardless of the object—gloves, eggs, pencils, or paper—the number that the unit represents is always constant. Chemists also need a convenient method for accurately counting the number of atoms, molecules, or formula units in a sample of a substance. However, atoms are so small and there are so many of them in even the smallest sample that it is impossible to count them directly. Because of this, chemists created a counting unit called the mole. In the Launch Lab, you probably found that a mole of any object is an enormous number of items.

■ **Figure 1** Different units are used to count different types of objects. A pair is two objects, a dozen is 12, a gross is 144, and a ream is 500.
List *What other counting units are you familiar with?*

Matt Meadows

CHEMISTRY JOURNAL

Counting Particles Ask students to develop a diagram that can be used to compare a dozen to a mole. Tell students to think of other words that represent a counted value, such as a trio, a quartet, a gross, and so on, and to come up with another analogy for a mole. BL EL

The mole The **mole,** abbreviated mol, is the SI base unit used to measure the amount of a substance. A mole is defined as the number of carbon atoms in exactly 12 g of pure carbon-12. Through years of experimentation, it has been established that a mole of anything contains 6.0221367×10^{23} representative particles. A representative particle is any kind of particle, such as an atom, a molecule, a formula unit, an electron, or an ion. If you write out Avogadro's number, it looks like this.

$$602{,}213{,}670{,}000{,}000{,}000{,}000{,}000$$

The number 6.0221367×10^{23} is called **Avogadro's number,** in honor of the Italian physicist and lawyer Amedeo Avogadro, who, in 1811, determined the volume of 1 mol of a gas. In this book, Avogadro's number is rounded to three significant figures, 6.02×10^{23}.

To count extremely small particles, such as atoms, Avogadro's number must be an enormous quantity. As you might imagine, Avogadro's number would not be convenient for measuring a quantity of marbles. Avogadro's number of marbles with a five-eighth inch diameter would cover the surface of Earth to a depth of more than fourteen kilometers! **Figure 2,** however, shows that it is convenient to use the mole to measure amounts of substances. One-mole quantities of water, copper, and salt are shown. Each one has a different representative particle. The representative particle in a mole of water is the water molecule, the representative particle in a mole of copper is the copper atom, and the representative particle in a mole of sodium chloride is the NaCl formula unit.

Vocabulary

Science usage v. Common usage

Mole

Science usage: an SI base unit used to measure the quantity of matter

The chemist measured out a mole of the compound.

Common usage: a small burrowing animal

The damage to the lawn was caused by a mole.

Figure 2 The amount of each substance shown is 6.02×10^{23}, or 1 mol, of representative particles. The representative particle for each substance is shown in a box. Refer to **Table R-1** in the Student Resources for a key to atom color conventions.

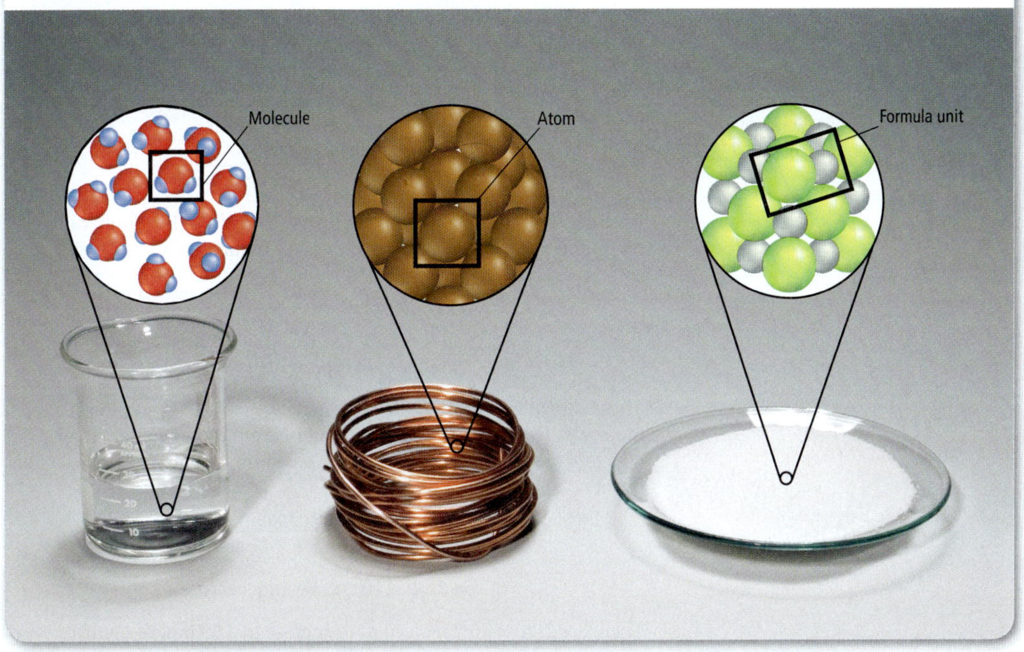

Quick Demo

Compare Quantities Place 1 mol quantities of substances such as Al, Cu, NaCl, $C_{12}H_{22}O_{11}$ (sucrose), H_2O, C (charcoal), Zn, and C_2H_5OH (ethanol) into sealed bottles. Label each with its name and the words *One Mole*. Ask students how these samples and those in Figure 2 differ from one another and what they have in common. The samples differ in appearance, size, and volume, but they each contain the same number of representative particles (one mole). Ask how many particles are represented in each sample. 6.02×10^{23} particles Have students account for the differences among the samples. Each is composed of different atoms or molecules. OL

ASSESSMENT

Skill Assign a particular substance and number of moles of the substance to each student. The substances could be elements, molecules, or ionic compounds. Have each student name the representative particle and determine the number of representative particles contained in the molar amount. OL

CHEMISTRY JOURNAL

Quantities and Units Ask students to go to their neighborhood grocery stores to find out how different common food items are sold. In their journals, have them explain why the items are sold in different quantities and units. OL EL

Concept Development

Quantities and Qualities

Measure 27.0 g of aluminum foil (one mole) and show it to the class. Tell students it is one mole and ask how many representative particles are present. 1 mol of Al contains 6.02×10^{23} atoms of Al Crumple the foil and ask what changed and what stayed the same. Students should understand that although the size and shape have changed, the crumpled foil still contains 1 mol of Al atoms. Tear the foil into two pieces and ask what changed. Each piece contains less than 1 mol of Al atoms. OL EL

☑ READING CHECK other units than those desired will remain

FOLDABLES® Rubric

12 roses = 1 dozen roses

■ **Figure 3** A key to using dimensional analysis is correctly identifying the mathematical relationship between the units you are converting. The relationship shown here, 12 roses = 1 dozen roses, can be used to write two conversion factors.

FOLDABLES® Incorporate information from this section into your Foldable.

Converting Between Moles and Particles

Suppose you buy three-and-one-half dozen roses and want to know how many roses you have. Recall what you have learned about conversion factors. You can multiply the known quantity (3.5 dozen roses) by a conversion factor to express the quantity in the units you want (number of roses). First, identify the mathematical relationship that relates the given unit with the desired unit. **Figure 3** shows the relationship.

Relationship: 1 dozen roses = 12 roses

By dividing each side of the equality by the other side, you can write two conversion factors from the relationship.

$$\text{Conversion factors: } \frac{12 \text{ roses}}{1 \text{ dozen roses}} \text{ and } \frac{1 \text{ dozen roses}}{12 \text{ roses}}$$

Then choose the conversion factor that, when multiplied by the known quantity, results in the desired unit. When set up correctly, all units cancel except those required for the answer.

$$\text{Conversion: } 3.5 \cancel{\text{dozen roses}} \times \frac{12 \text{ roses}}{1 \cancel{\text{dozen roses}}} = 42 \text{ roses}$$

Here, *dozens of roses* cancels, leaving *roses* as the desired unit.

☑ READING CHECK **Describe** how you can tell if the wrong conversion factor has been used.

Moles to particles Now suppose you want to determine how many particles of sucrose are in 3.50 mol of sucrose. The relationship between moles and representative particles is given by Avogadro's number.

1 mol of representative particles = 6.02×10^{23} representative particles

Using this relationship, you can write two different conversion factors that relate representative particles and moles.

$$\frac{6.02 \times 10^{23} \text{ representative particles}}{1 \text{ mol}}$$

$$\frac{1 \text{ mol}}{6.02 \times 10^{23} \text{ representative particles}}$$

By using the correct conversion factor, you can find the number of representative particles in a given number of moles.

$$\text{number of moles} \times \frac{6.02 \times 10^{23} \text{ representative particles}}{1 \text{ mol}}$$

$$= \text{number of representative particles}$$

As shown in **Figure 4,** the representative particle of sucrose is a molecule. To obtain the number of sucrose molecules contained in 3.50 mol of sucrose, you need to use Avogadro's number as a conversion factor.

$$3.50 \cancel{\text{mol sucrose}} \times \frac{6.02 \times 10^{23} \text{ molecules sucrose}}{1 \cancel{\text{mol sucrose}}}$$

$$= 2.11 \times 10^{24} \text{ molecules sucrose}$$

There are 2.11×10^{24} molecules of sucrose in 3.50 mol of sucrose.

CHEMISTRY PROJECT

Mole of Copper Ask students to determine the mass of one mole of copper. 63.5 g Using copper pennies, have students construct a pile of pennies that contains one mole of copper. Tell students to assume that the pennies are 100% copper (though this is not the case). mass of penny = 2.5 g
63.5 g/2.5 g = 25 pennies OL

PRACTICE Problems

Do additional problems. Online Practice

1. Zinc (Zn) is used to form a corrosion-inhibiting surface on galvanized steel. Determine the number of Zn atoms in 2.50 mol of Zn.
2. Calculate the number of molecules in 11.5 mol of water (H_2O).
3. Silver nitrate ($AgNO_3$) is used to make several different silver halides used in photographic films. How many formula units of $AgNO_3$ are there in 3.25 mol of $AgNO_3$?
4. **Challenge** Calculate the number of oxygen atoms in 5.00 mol of oxygen molecules. Oxygen is a diatomic molecule, O_2.

Particles to moles Now suppose you want to find out how many moles are represented by a certain number of representative particles. To do this, you can use the inverse of Avogadro's number as a conversion factor.

$$\text{number of representative particles} \times \frac{1 \text{ mol}}{6.02 \times 10^{23} \text{ representative particles}} = \text{number of moles}$$

For example, if instead of knowing how many moles of sucrose you have, suppose you knew that a sample contained 2.11×10^{24} molecules of sucrose. To convert this number of molecules of sucrose to moles of sucrose, you need a conversion factor that has moles in the numerator and molecules in the denominator.

$$2.11 \times 10^{24} \cancel{\text{molecules sucrose}} \times \frac{1 \text{ mol}}{6.02 \times 10^{23} \cancel{\text{molecules sucrose}}} = 3.50 \text{ mol sucrose}$$

Thus, 2.11×10^{24} molecules of sucrose is 3.50 mol of sucrose.

You can convert between moles and number of representative particles by multiplying the known quantity by the proper conversion factor. Example Problem 1 further illustrates the conversion process.

READING CHECK **List** the two conversion factors that can be written from Avogadro's number.

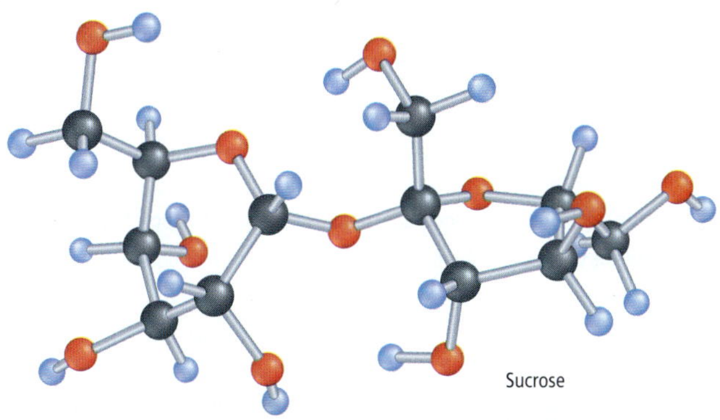

Figure 4 The representative particle of sucrose is a molecule. The ball-and-stick model shows that a molecule of sucrose is a single unit made up of carbon, hydrogen, and oxygen.

Analyze *Use the ball-and-stick model of sucrose to write the chemical formula for sucrose.*

DIFFERENTIATED INSTRUCTION

Struggling Learners The use of conversion factors might confuse some students. Have students make an index card with 6.02×10^{23} particles/1 mol on one side and 1 mol/6.02×10^{23} particles on the opposite side. When analyzing a problem, have students flip the card over as needed to determine which conversion factor is needed to cancel the known unit and give the desired unit for the answer. BL EL

PRACTICE Problems

Have students refer to the Selected Solutions appendix for complete solutions to odd-numbered problems. The complete solutions for all problems can be found in the Solutions Manual.

1. 1.51×10^{24} atoms
2. 6.92×10^{24} molecules
3. 1.96×10^{24} formula units
4. 6.02×10^{24} atom

Caption Question Fig. 4
$C_{12}H_{22}O_{11}$ OL

READING CHECK

$$\frac{1 \text{ mol}}{6.02 \times 10^{23} \text{ molecules}} \text{ and } \frac{6.02 \times 10^{23} \text{ molecules}}{1 \text{ mol}}$$

3 Assess

Check for Understanding

Weigh one mole of a substance, such as copper shot (63.5 g). Ask students how many atoms of copper are on the balance. 6.02×10^{23} atoms Cu Reduce the mass by one-half and ask students how much copper is present in terms of moles and representative particles. 0.50 mol Cu or 3.01×10^{23} atoms Cu OL

Reteach

Ask students why it is important to have a counting unit as large as the mole. A large counting unit like the mole allows you to count many small particles. OL

Extension

Ask students to invent a new counting unit that might be useful for counting marbles, CDs, baseballs, or books in the library. OL EL

IN-CLASS Example

Question Platinum (Pt) is a metal often used in jewelry, but it is also used in laboratory equipment and emission control devices in automobiles. Calculate the moles of platinum if you have a ring that contains 5.50×10^{22} atoms of platinum.

Answer

$$5.50 \times 10^{22} \text{ Pt atoms} \times \frac{1 \text{ mol}}{6.02 \times 10^{23} \text{ Pt atoms}} = 0.914 \text{ mol}$$

ASSESSMENT

Knowledge Have students write on a card a problem requiring a conversion from moles of a substance to representative particles. Each student will pass the card to a neighboring student to solve. The neighbor will solve the conversion and write a problem requiring a conversion from representative particles to moles and pass it back for solution. Both students must agree on the two answers. OL COOP LEARN

PRACTICE Problems

Have students refer to the Selected Solutions appendix for complete solutions to odd-numbered problems. The complete solutions for all problems can be found in the Solutions Manual.

5. a. 9.55 mol
   b. 4.15×10^{-4} mol
6. a. a molecule; 6.23 mol CO_2
   b. a formula unit; 0.595 mol $ZnCl_2$

EXAMPLE Problem 1

Find help with **dimensional analysis.** Math Handbook

EXAMPLE PROBLEM

PARTICLES-TO-MOLES CONVERSION Zinc (Zn) is used as a corrosion-resistant coating on iron and steel. It is also an essential trace element in your diet. Calculate the number of moles of zinc that contain 4.50×10^{24} atoms.

1 ANALYZE THE PROBLEM

You are given the number of atoms of zinc and must find the equivalent number of moles. If you compare 4.50×10^{24} atoms Zn with 6.02×10^{23}, the number of atoms in 1 mol, you can predict that the answer should be less than 10 mol.

Known	**Unknown**
number of atoms = 4.50×10^{24} atoms Zn	moles Zn = ? mol
1 mol Zn = 6.02×10^{23} atoms Zn	

2 SOLVE FOR THE UNKNOWN

Use a conversion factor—the inverse of Avogadro's number—that relates moles to atoms.

$$\text{number of atoms} \times \frac{1 \text{ mol}}{6.02 \times 10^{23} \text{ atoms}} = \text{number of moles}$$

Apply the conversion factor.

$$4.50 \times 10^{24} \text{ atoms Zn} \times \frac{1 \text{ mol Zn}}{6.02 \times 10^{23} \text{ atoms Zn}} = 7.48 \text{ mol Zn}$$

Substitute number of Zn atoms = 4.50×10^{24}. Multiply and divide numbers and units.

3 EVALUATE THE ANSWER

Both the number of Zn atoms and Avogadro's number have three significant figures. Therefore, the answer is expressed correctly with three digits. The answer is less than 10 mol, as predicted, and has the correct unit, moles.

PRACTICE Problems

Do additional problems. Online Practice

5. How many moles contain each of the following?
   a. 5.75×10^{24} atoms Al
   b. 2.50×10^{20} atoms Fe
6. **Challenge** Identify the representative particle for each formula, and convert the given number of representative particles to moles.
   a. 3.75×10^{24} CO_2
   b. 3.58×10^{23} $ZnCl_2$

SECTION 1 REVIEW

Section Self-Check

Section Summary

- The mole is a unit used to count particles of matter indirectly. One mole of a pure substance contains Avogadro's number of representative particles.
- Representative particles include atoms, ions, molecules, formula units, electrons, and other similar particles.
- One mole of carbon-12 atoms has a mass of exactly 12 g.
- Conversion factors written from Avogadro's relationship can be used to convert between moles and number of representative particles.

7. MAINIDEA **Explain** why chemists use the mole.
8. **State** the mathematical relationship between Avogadro's number and 1 mol.
9. **List** the conversion factors used to convert between particles and moles.
10. **Explain** how a mole is similar to a dozen.
11. **Apply** How does a chemist count the number of particles in a given number of moles of a substance?
12. **Calculate** the mass in atomic mass units of 0.25 mol of carbon-12 atoms.
13. **Calculate** the number of representative particles of each substance.
    a. 11.5 mol Ag
    b. 18.0 mol H_2O
    c. 0.150 mol NaCl
    d. 1.35×10^{-2} mol CH_4
14. **Arrange** these three samples from smallest to largest in terms of number of representative particles: 1.25×10^{25} atoms of zinc (Zn), 3.56 mol of iron (Fe), and 6.78×10^{22} molecules of glucose ($C_6H_{12}O_6$).

SECTION 1 REVIEW

7. Chemists use the mole because it is a convenient way of knowing how many representative particles are in a sample.
8. One mole contains Avogadro's number (6.02×10^{23}) of representative particles.
9. $\frac{6.02 \times 10^{23} \text{ representative particles}}{1 \text{ mol}}, \frac{1 \text{ mol}}{6.02 \times 10^{23} \text{ representative particles}}$
10. The mole is a unit for counting 6.02×10^{23} representative particles. The dozen is used to count 12 items.
11. They multiply the number of moles by Avogadro's number.
12. 1.8×10^{24} amu carbon-12
13. a. 6.92×10^{24} atoms Ag
    b. 1.08×10^{25} molecules H_2O
    c. 9.03×10^{22} formula units NaCl
    d. 8.13×10^{21} molecules CH_4
14. From smallest to largest: 6.78×10^{22} molecules glucose, 2.14×10^{24} atoms Fe, 1.25×10^{25} atoms Zn.

SECTION 2

Mass and the Mole

MAINIDEA A mole always contains the same number of particles; however, moles of different substances have different masses.

Essential Questions

- Why can the mass of an atom be related to the mass of a mole of atoms?
- How can the number of moles be converted to the mass of an element and vice versa?
- How can the number of moles be converted to the number of atoms of an element and vice versa?

Review Vocabulary

conversion factor: a ratio of equivalent values used to express the same quantity in different units

New Vocabulary

molar mass

CHEM 4 YOU When purchasing a dozen eggs, you can pick from several sizes—medium, large, and extra-large. The size of the egg does not affect how many come in the carton. A similar situation exists with the size of the atoms that make up a mole.

The Mass of a Mole

You would not expect a dozen limes to have the same mass as a dozen eggs. Because eggs and limes differ in size and composition, it is not surprising that they have different masses, as shown in **Figure 5.** One-mole quantities of two different substances have different masses for the same reason—the substances have different compositions. For example, if you put one mole of carbon and one mole of copper on separate balances, you would see a difference in mass, just as you do for the eggs and the limes. This occurs because carbon atoms differ from copper atoms. Thus, the mass of 6.02×10^{23} carbon atoms does not equal the mass of 6.02×10^{23} copper atoms.

Recall that each atom of carbon-12 has a mass of 12 amu. The atomic masses of all other elements are established relative to carbon-12. For example, an atom of hydrogen-1 has a mass of approximately 1 amu, one-twelfth the mass of a carbon-12 atom. The mass of an atom of helium-4 is approximately 4 amu, one-third the mass of one atom of carbon-12.

You might have noticed, however, that the atomic-mass values given on the periodic table are not exact integers. For example, you will find 12.011 amu for carbon, 1.008 amu for hydrogen, and 4.003 amu for helium. These noninteger values occur because the values are weighted averages of the masses of all the naturally occurring isotopes of each element.

■ **Figure 5** A dozen limes has approximately twice the mass of a dozen eggs. The difference in mass is reasonable because limes are different from eggs in composition and size.

Matt Meadows

SECTION 2

1 Focus

MAINIDEA

Mass and Number Show students two beakers, A and B. Tell them that beaker A contains 58.5 g of table salt (NaCl), and beaker B contains 342 g of table sugar ($C_{12}H_{22}O_{11}$). Tell students that both beakers contain the same number of particles–1 mole. Ask students to explain why the mass is different when the number of particles is the same. The mass of a formula unit of salt is different than the mass of a molecule of sugar. Thus, the same number of particles (formula units and molecules) of each have different masses. OL

2 Teach

Quick Demo

Extrapolate Data Place a sack of pre-1982 pennies on a balance and determine the mass. The average mass of a pre-1982 penny is 3.10 g. Ask students how they can determine the number of pennies in the bag. Divide the total mass by 3.10 g. After students have calculated the number of pennies, have them verify their results by counting. OL

CHEMISTRY PROJECT

How many moles? Give groups of students two bags containing different numbers of iron nails, a bag containing one mole of iron filings, and an empty bag. Ask how they could determine the molar mass of iron and the number of moles of iron in each bag of nails. Have students carry out the procedure and write a newspaper article. OL COOP LEARN

Caption Question Fig. 6 63.546 g

Problem-Solving LAB

Purpose Students will determine the relationship among atomic mass, molar mass, and Avogadro's number.

Process Skills formulate models, draw conclusions, use numbers

Teaching Strategies

- Although the masses of protons and neutrons are known to eight or ten significant digits, the figures used in this lab are rounded to three. Students should expect to calculate approximate values.
- Students might ask how 12 nuclear particles, each with a mass of 1.007 amu, can be in the nucleus of a carbon atom with a mass of exactly 12 amu. The answer involves the mass equivalent of the binding energy of the electrons, which is being neglected in these calculations.

Think Critically

1. $\frac{1.672 \times 10^{-24}\ \text{g}}{1\ \text{nuclear particle}} \times \frac{4\ \text{nuclear particles}}{1\ \text{He atom}}$
$= 6.680 \times 10^{-24}$ g/He atom
2. A model of C-12 should have six protons and six neutrons.
$\frac{1.672 \times 10^{-24}\ \text{g}}{1\ \text{nuclear particle}} \times \frac{12\ \text{nuclear particles}}{1\ \text{C-12 atom}}$
$= 2.004 \times 10^{-23}$ g/C-12 atom
The carbon atom = 12 amu.
3. $1.007\ \text{g} \times \frac{1\ \text{H-1 atom}}{1.672 \times 10^{-24}\ \text{g}}$
$= 6.023 \times 10^{23}$ H-1 atoms
4. $\frac{6.680 \times 10^{-24}\ \text{g}}{\text{He atom}} \times (6.023 \times 10^{23}\ \text{He atoms})$
$= 4.023$ g
$\frac{2.004 \times 10^{-23}\ \text{g}}{\text{C atom}} \times (6.023 \times 10^{23}\ \text{C atoms})$
$= 12.07$ g
5. The mass of one mole of any atom is the same value in grams as the mass of one atom in amu.

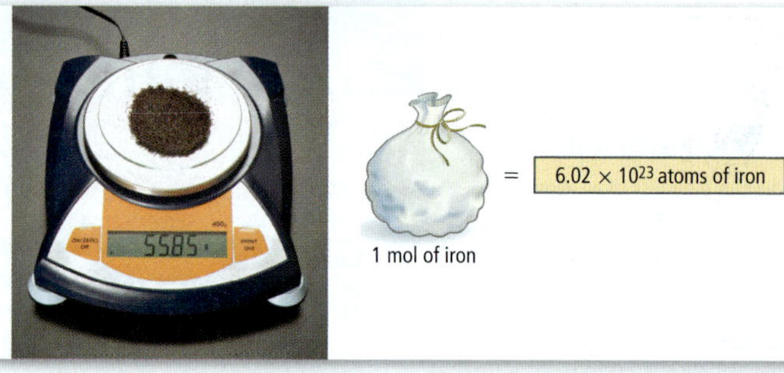

Figure 6 One mole of iron, represented by a bag of particles, contains Avogadro's number of atoms and has a mass equal to its atomic mass in grams.

Apply *What is the mass of one mole of copper?*

View an **animation about molar mass.**

Concepts In Motion

Molar Mass How does the mass of one atom relate to the mass of one mole of that atom? Recall that the mole is defined as the number of carbon-12 atoms in exactly 12 g of pure carbon-12. Thus, the mass of one mole of carbon-12 atoms is 12 g. Whether you are considering a single atom or Avogadro's number of atoms (a mole), the masses of all atoms are established relative to the mass of carbon-12. The mass in grams of one mole of any pure substance is called its **molar mass.**

The molar mass of any element is numerically equal to its atomic mass and has the units g/mol. As given on the periodic table, an atom of iron has an atomic mass of 55.845 amu. Thus, the molar mass of iron is 55.845 g/mol, and 1 mol (or 6.02×10^{23} atoms of iron) has a mass of 55.845 g. Note that by measuring 55.845 g of iron, you indirectly count out 6.02×10^{23} atoms of iron. **Figure 6** shows the relationship between molar mass and one mole of an element.

Problem-Solving LAB

Formulate a Model

How are molar mass, Avogadro's number, and the atomic nucleus related?
A nuclear model of mass can provide a simple picture of the connections among the mole, molar mass, and the number of representative particles in a mole.

Analysis

The diagram to the right shows the space-filling models of hydrogen-1 and helium-4 nuclei. The hydrogen-1 nucleus contains one proton with a mass of 1.007 amu. The mass of a proton, in grams, has been determined experimentally to be 1.672×10^{-24} g. The helium-4 nucleus contains two protons and two neutrons and has a mass of approximately 4 amu.

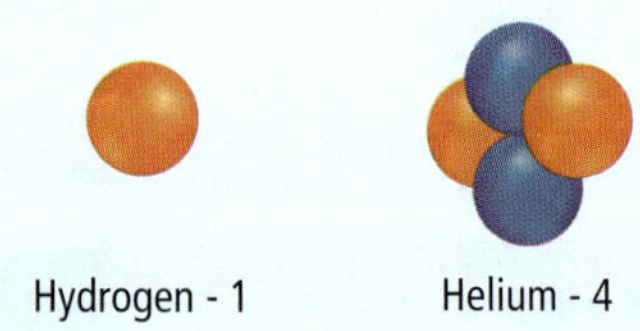

Think Critically

1. **Apply** What is the mass in grams of one helium atom? (The mass of a neutron is approximately the same as the mass of a proton.)
2. **Draw** Carbon-12 contains six protons and six neutrons. Draw the carbon-12 nucleus and calculate the mass of one atom in amu and g.
3. **Apply** How many atoms of hydrogen-1 are in a 1.007-g sample? Recall that 1.007 amu is the mass of one atom of hydrogen-1. Round your answer to two significant digits.
4. **Apply** If you had samples of helium and carbon that contained the same number of atoms as you calculated in Question 3, what would be the mass in grams of each sample?
5. **Conclude** What can you conclude about the relationship between the number of atoms and the mass of each sample?

Matt Meadows

CHEMISTRY JOURNAL

Molar Mass Students should determine the chemical formula and molar mass of each of the following compounds. OL

Substance	Chemical Formula	Molar Mass
Hydrogen peroxide	H_2O_2	34.014 g/mol
Acetic acid	CH_3COOH	60.052 g/mol
Calcium phosphate	$Ca_3(PO_4)_2$	310.174 g/mol
Silver nitrate	$AgNO_3$	169.872 g/mol
Ethanol	C_2H_6O	46.069 g/mol

Using Molar Mass

Imagine that your class bought jelly beans in bulk to sell by the dozen at a candy sale. You soon realize that it is too much work to count out each dozen, so you instead decide to measure the jelly beans by mass. You find that the mass of 1 dozen jelly beans is 35 g. This relationship and the conversion factors that stem from it are as follows:

$$1 \text{ dozen jelly beans} = 35 \text{ g jelly beans}$$

$$\frac{35 \text{ g jelly beans}}{1 \text{ dozen jelly beans}} \text{ and } \frac{1 \text{ dozen jelly beans}}{35 \text{ g jelly beans}}$$

What mass of jelly beans should you measure if a customer wants 5 dozen jelly beans? To determine this mass, you would multiply the number of dozens of jelly beans to be sold by the correct conversion factor. Select the conversion factor with the units you are converting to in the numerator (g) and the units you are converting from in the denominator (dozen).

$$5 \cancel{\text{ dozen jelly beans}} \times \frac{35 \text{ g jelly beans}}{1 \cancel{\text{ dozen jelly beans}}} = 175 \text{ g jelly beans.}$$

A quantity of 5 dozen jelly beans has a mass of 175 g.

READING CHECK **Compare** How are the jelly bean conversion factors used above similar to the molar mass of a compound?

Get help with **conversion factors.**

Personal Tutor

Moles to mass Now suppose that while working in a chemistry lab, you need 3.00 mol of copper (Cu) for a chemical reaction. How would you measure that amount? Like the 5 dozen jelly beans, the number of moles of copper can be converted to an equivalent mass and measured on a balance.

To calculate the mass of a given number of moles, simply multiply the number of moles by the molar mass.

$$\text{number of moles} \times \frac{\text{mass in grams}}{1 \text{ mole}} = \text{mass}$$

If you check the periodic table, you will find that copper, element 29, has an atomic mass of 63.546 amu. You know that the molar mass of an element (in g/mol) is equal to its atomic mass (given in amu). Thus, copper has a molar mass of 63.546 g/mol. By using the molar mass, you can convert 3.00 mol of copper to grams of copper.

$$3.00 \cancel{\text{ mol Cu}} \times \frac{63.546 \text{ g Cu}}{1 \cancel{\text{ mol Cu}}} = 191 \text{ g Cu}$$

So, as shown in **Figure 7,** you can measure the 3.00 mol of copper needed for the reaction by using a balance to measure out 191 g of copper. The reverse conversion—from mass to moles—also involves the molar mass as a conversion factor, but it is the inverse of the molar mass that is used. Can you explain why?

FOLDABLES® Incorporate information from this section into your Foldable.

Figure 7 To measure 3.00 mol of copper, place a weighing paper on a balance, tare the balance, and then add the 191 g of copper.

Matt Meadows

Connection to Biology Cellular biologists continually discover new biologic proteins. After a new biomolecule is discovered, biologists determine the molar mass of the compound using a technique known as mass spectrometry. In addition to the molar mass, mass spectrometry also provides additional information that helps the biologist reveal the compound's composition.

ASSESSMENT

Performance Have students, working in groups of three, make a presentation of their understanding of the use of the nuclear model to explain the mass of a mole of an element. To make each presentation unique and to illustrate the fact that the molar mass of an element is the weighted average of all the isotopes of the element, have each group explain the atomic mass of a different element having a number of different isotopes, such as chlorine. OL COOP LEARN

READING CHECK Both express a ratio of mass to some other counting unit (dozen or mole).

In-Text Question The reverse conversion—from mass to moles—also involves the molar mass as a conversion factor, but it is the inverse of the molar mass that is used. Can you explain why? To convert a mass in grams to moles, the inverse molar mass is used so that the unit of grams cancel.

FOLDABLES® Rubric

DIFFERENTIATED INSTRUCTION

Advanced Learners Point out that determining the slope of a line plotted on a graph is often used to determine the value of a constant. Have students determine the masses of three samples of copper, 25.0 g, 50.0 g, and 75.0 g, and calculate the corresponding numbers of moles 0.393 mol, 0.787 mol, and 1.18 mol, respectively. Have them graph mass versus number of moles using this data. If a graphing calculator is available, have students use the linear regression function and determine the slope of the line. Compare the calculated value to the molar mass. Ask students to explain their results. The slope of the graph is 63.5 g/mol, which is the molar mass of copper. As mass increases, the number of moles of copper in each sample increases, a direct relationship. As an extension, ask students to determine the number of copper atoms in each sample. AL

IN-CLASS Example

Question Zirconium (Zr) is a metal used in nuclear reactors because it is resistant to corrosion. What is the mass of a sample that contains 4.05 mol of zirconium?

Answer
$$4.05\ \cancel{\text{mol Zr}} \times \frac{91.22\text{ g Zr}}{1\ \cancel{\text{mol Zr}}} = 369\text{ g Zr}$$

PRACTICE Problems

Have students refer to the Selected Solutions appendix for complete solutions to odd-numbered problems. The complete solutions for all problems can be found in the Solutions Manual.

15. a. 96.3 g Al
b. 1.20×10^3 g Si
16. a. 2.03×10^4 g Co
b. 1.60 g Zn

RealWorld CHEMISTRY

The Importance of Chromium

CHROMIUM What gives these rims their mirrorlike finish? The metal alloy rim has been plated, or coated, with a thin layer of chromium. Chrome plating has been used in the automobile industry for decades because of its beauty and its corrosion resistance.

EXAMPLE Problem 2

MOLE-TO-MASS CONVERSION Chromium (Cr), a transition element, is a component of chrome plating. Chrome plating is used on metals and in steel alloys to control corrosion. Calculate the mass in grams of 0.0450 mol Cr.

1 ANALYZE THE PROBLEM

You are given the number of moles of chromium and must convert it to an equivalent mass using the molar mass of chromium from the periodic table. Because the sample is less than one-tenth of a mole, the answer should be less than one-tenth of the molar mass.

Known	Unknown
number of moles = 0.0450 mol Cr	mass Cr = ? g
molar mass Cr = 52.00 g/mol Cr	

2 SOLVE FOR THE UNKNOWN

Use a conversion factor—the molar mass—that relates grams of chromium to moles of chromium. Write the conversion factor with moles of chromium in the denominator and grams of chromium in the numerator. Substitute the known values into the equation and solve.

$$\text{moles Cr} \times \frac{\text{grams Cr}}{1\text{ mol Cr}} = \text{grams Cr}$$
Apply the conversion factor.

$$0.0450\ \cancel{\text{mol Cr}} \times \frac{52.00\text{ g Cr}}{1\ \cancel{\text{mol Cr}}} = 2.34\text{ g Cr}$$
Substitute 0.450 mol for moles Cr and 52.00 g/mol for molar mass of Cr. Multiply and divide numbers and units.

3 EVALUATE THE ANSWER

The known number of moles of chromium has the smallest number of significant figures, three, so the answer is correctly stated with three digits. The answer is less than one-tenth the mass of 1 mol, as predicted, and is in grams, a mass unit.

PRACTICE Problems

Do additional problems. Online Practice

15. Determine the mass in grams of each of the following.
a. 3.57 mol Al
b. 42.6 mol Si

16. Challenge Convert each given quantity in scientific notation to mass in grams expressed in scientific notation.
a. 3.45×10^2 mol Co
b. 2.45×10^{-2} mol Zn

Explore **moles, mass, and molecules.** Virtual Investigations

Many of the values for atomic mass given in the periodic table have five significant figures. However, in Example Problem 2, the periodic table value of 51.996 g/mol Cr was rounded to 52.00 g/mol. It is generally okay to round reference values as long as you keep one more significant figure than the answer will have. In the case of Example Problem 2, 52.00 has one more significant figure than 0.0450 mol Cr, which limits the answer to three significant figures.

Jeff Greenberg/PhotoEdit

CHEMISTRY JOURNAL

Conversion Factors Have students begin to compile a list of conversion factors that are being introduced for making conversions between moles, mass, and number of paticles. Encourage them to update their lists as they continue their study of moles and stoichiometry, and refer to them when they are solving problems. BL OL EL

EXAMPLE Problem 3

MASS-TO-MOLE CONVERSION Calcium (Ca), the fifth most-abundant element on Earth, is always found combined with other elements because of its high reactivity. How many moles of calcium are in 525 g Ca?

1 ANALYZE THE PROBLEM

You must convert the mass of calcium to moles of calcium. The mass of calcium is more than ten times larger than the molar mass. Therefore, the answer should be greater than 10 mol.

Known
mass = 525 g Ca
molar mass Ca = 40.08 g/mol Ca

Unknown
number of moles Ca = ? mol

2 SOLVE FOR THE UNKNOWN

Use a conversion factor—the inverse of molar mass—that relates moles of calcium to grams of calcium. Substitute the known values and solve.

$$\text{mass Ca} \times \frac{1 \text{ mol Ca}}{\text{grams Ca}} = \text{moles Ca}$$

Apply the conversion factor.

$$525 \cancel{\text{g Ca}} \times \frac{1 \text{ mol Ca}}{40.08 \cancel{\text{g Ca}}} = 13.1 \text{ mol Ca}$$

Substitute mass Ca = 525 g, and inverse molar mass of Ca = 1 mol/40.08 g. Multiply and divide numbers and units.

3 EVALUATE THE ANSWER

The mass of calcium has the fewest significant figures, three, so the answer is expressed correctly with three digits. As predicted, the answer is greater than 10 mol and has the expected unit.

PRACTICE Problems

Do additional problems. Online Practice

17. Determine the number of moles in each of the following.
a. 25.5 g Ag **b.** 300.0 g S

18. Challenge Convert each mass to moles. Express the answer in scientific notation.
a. 1.25×10^3 g Zn **b.** 1.00 kg Fe

Converting between mass and atoms So far, you have learned how to convert mass to moles and moles to mass. You can go one step further and convert mass to the number of atoms. Recall the jelly beans you were selling at the candy sale. At the end of the day, you find that 550 g of jelly beans is left unsold. Without counting, can you determine how many jelly beans that is? You know that 1 dozen jelly beans has a mass of 35 g and that 1 dozen is 12 jelly beans. Thus, you can first convert the 550 g to dozens of jelly beans by using the conversion factor that relates dozens and mass.

$$550 \cancel{\text{g jelly beans}} \times \frac{1 \text{ dozen jelly beans}}{35 \cancel{\text{g jelly beans}}} = 16 \text{ dozen jelly beans}$$

Next, you can determine how many jelly beans are in 16 dozen by multiplying by the conversion factor that relates number of particles (jelly beans) and dozens.

CHEMISTRY JOURNAL

Mass, Moles, and Particles Ask students to write statements that relate mass to moles, moles to mass, moles to particles, and particles to moles. OL

ASSESSMENT

Skills Ask students to make a table with columns labeled *Mass, Moles,* and *Number of Representative Particles*. The table should have seven rows, one for each of the elements listed below. One piece of information is given for each element. Students must calculate the remaining data to complete the table. The given data are as follows: 2.5 g Au, 4.95 mol Zn, 3.95×10^{23} atoms C, 12.5 mol Cu, 8.75×10^{21} atoms Sc, 49.6 g Se, 3.21 mol U, and 4.93×10^{25} atoms Ba. OL

Mass	Moles	Number of Particles
2.5 g Au	0.013	7.6×10^{21}
324 g Zn	4.95	2.98×10^{24}
7.88 g C	0.656	3.95×10^{23}
794 g Cu	12.5	7.53×10^{24}
0.652 g Sc	0.0145	8.75×10^{21}
49.6 g Se	0.628	3.78×10^{23}
764 g U	3.21	1.93×10^{24}
11,200 g Ba	81.9	4.93×10^{25}

IN-CLASS Example

Question Titanium (Ti) is a metal that is often used to produce strong, lightweight alloys. If an alloy contains 645 g of titanium, how many moles of titanium are present?

Answer

$$654 \cancel{\text{g Ti}} \times \frac{1 \text{ mol Ti}}{47.87 \cancel{\text{g Ti}}} = 13.7 \text{ mol Ti}$$

PRACTICE Problems

Have students refer to the Selected Solutions appendix for complete solutions to odd-numbered problems. The complete solutions for all problems can be found in the Solutions Manual.

17. a. 0.236 mol Ag
b. 9.355 mol S

18. a. 1.91×10^1 mol Zn
b. 1.79×10^1 mol Fe

IN-CLASS Example

Question Scandium (Sc) is a metal found in rare minerals that are mined in Scandinavia. If a mineral sample contains 25.6 g of scandium, how many atoms of scandium are in the sample?

Answer

$$25.6\ \cancel{g\ Sc} \times \frac{1\ mol\ Sc}{44.96\ \cancel{g\ Sc}} = 0.569\ mol\ Sc$$

$$0.569\ \cancel{mol}\ Sc \times \frac{6.02 \times 10^{23}\ atoms}{1\ \cancel{mol}}$$

$$= 3.43 \times 10^{23}\ atoms\ of\ Sc$$

The conversion factor relating number of jelly beans and dozens is 12 jelly beans/dozen. Applying it yields the answer in jelly beans.

$$16\ \cancel{dozen} \times \frac{12\ jelly\ beans}{1\ \cancel{dozen}} = 192\ jelly\ beans$$

The 550 g of leftover jelly beans is equal to 192 jelly beans.

Just as you could not make a direct conversion from the mass of jelly beans to the number of jelly beans, you cannot make a direct conversion from the mass of a substance to the number of representative particles of that substance. You must first convert mass to moles by multiplying by a conversion factor that relates moles and mass. That conversion factor is the molar mass. The number of moles must then be multiplied by a conversion factor that relates the number of representative particles to moles. For this conversion, you will use Avogadro's number. This two-step process is shown in Example Problem 4.

EXAMPLE Problem 4

Find help with **dimensional analysis.** Math Handbook

EXAMPLE PROBLEM

MASS-TO-ATOMS CONVERSION Gold (Au) is one of a group of metals called the coinage metals (copper, silver, and gold). How many atoms of gold are in a U.S. Eagle, a gold alloy bullion coin with a mass of 31.1 g Au?

1 ANALYZE THE PROBLEM

You must determine the number of atoms in a given mass of gold. Because you cannot convert directly from mass to the number of atoms, you must first convert the mass to moles using the molar mass. Then, convert moles to the number of atoms using Avogadro's number. The given mass of the gold coin is about one-sixth the molar mass of gold (196.97 g/mol), so the number of gold atoms should be approximately one-sixth Avogadro's number.

Known	**Unknown**
mass = 31.1 g Au	**number of atoms Au = ?**
molar mass Au = 196.97 g/mol Au	

2 SOLVE FOR THE UNKNOWN

Use a conversion factor—the inverse of the molar mass—that relates moles of gold to grams of gold.

$$mass\ Au \times \frac{1\ mol\ Au}{grams\ Au} = moles\ Au$$ **Apply the conversion factor.**

$$31.1\ \cancel{g\ Au} \times \frac{1\ mol\ Au}{196.97\ \cancel{g\ Au}} = 0.158\ mol\ Au$$ **Substitute mass Au = 31.1 g and the inverse molar mass of Au = 1 mol/196.97 g. Multiply and divide numbers and units.**

To convert the calculated moles of gold to atoms, multiply by Avogadro's number.

$$moles\ Au \times \frac{6.02 \times 10^{23}\ atoms\ Au}{1\ mol\ Au} = \mathbf{atoms\ Au}$$ **Apply the conversion factor.**

$$0.158\ \cancel{mol\ Au} \times \frac{6.02 \times 10^{23}\ atoms\ Au}{1\ \cancel{mol\ Au}} = \mathbf{9.51 \times 10^{22}\ atoms\ Au}$$ **Substitute moles Au = 0.158 mol, and solve.**

3 EVALUATE THE ANSWER

The mass of gold has the smallest number of significant figures, three, so the answer is expressed correctly with three digits. The answer is approximately one-sixth Avogadro's number, as predicted, and the correct unit, atoms, is obtained.

DIFFERENTIATED INSTRUCTION

Advanced Learners Using the content label from a bottle of vitamins and minerals, have capable students determine the number of moles and atoms of each element—for example, copper, zinc, magnesium, calcium, selenium—present in one tablet. If a person were to take two tablets three times a day, how many moles of each element would be taken? How many atoms of each element would be taken per day? AL

EXAMPLE Problem 5

Find help with **significant figures.** Math Handbook

ATOMS-TO-MASS CONVERSION Helium (He) is an unreactive noble gas often found in underground deposits mixed with methane. The mixture is separated by cooling the gaseous mixture until all but the helium has liquefied. A party balloon contains 5.50×10^{22} atoms of helium gas. What is the mass, in grams, of the helium?

1 ANALYZE THE PROBLEM

You are given the number of atoms of helium and must find the mass of the gas. First, convert the number of atoms to moles, then convert moles to grams.

Known	Unknown
number of atoms He = 5.50×10^{22} atoms He	mass = ? g He
molar mass He = 4.00 g/mol He	

2 SOLVE FOR THE UNKNOWN

Use a conversion factor—the inverse of Avogadro's number—that relates moles to number of atoms.

$$\text{atoms He} \times \frac{1 \text{ mol He}}{6.02 \times 10^{23} \text{ atoms He}} = \text{moles He}$$

Apply the conversion factor.

$$5.50 \times 10^{22} \cancel{\text{atoms He}} \times \frac{1 \text{ mol He}}{6.02 \times 10^{23} \cancel{\text{atoms He}}} = 0.0914 \text{ mol He}$$

Substitute atoms He = 5.50×10^{22} atoms. Multiply and divide numbers and units.

Next, apply a conversion factor—the molar mass of helium—that relates mass of helium to moles of helium.

$$\text{moles He} \times \frac{\text{grams He}}{1 \text{ mol He}} = \text{grams He}$$

Apply the conversion factor.

$$0.0914 \cancel{\text{mol He}} \times \frac{4.00 \text{ g He}}{1 \cancel{\text{mol He}}} = 0.366 \text{ g He}$$

Substitute moles He = 0.0914 mol, molar mass He = 4.00 g/mol, and solve.

3 EVALUATE THE ANSWER

The answer is expressed correctly with three significant figures and is in grams, a mass unit.

PRACTICE Problems

Do additional problems. Online Practice

19. How many atoms are in each of the following samples?

- **a.** 55.2 g Li
- **b.** 0.230 g Pb
- **c.** 11.5 g Hg

20. What is the mass in grams of each of the following?

- **a.** 6.02×10^{24} atoms Bi
- **b.** 1.00×10^{24} atoms Mn
- **c.** 3.40×10^{22} atoms He
- **d.** 1.50×10^{15} atoms N
- **e.** 1.50×10^{15} atoms U

21. Challenge Convert each given mass to number of representative particles. Identify the type of representative particle, and express the number in scientific notation.

- **a.** 4.56×10^{3} g Si
- **b.** 0.120 kg Ti

IN-CLASS Example

Question Neon is the unreactive noble gas found in neon lamps. The gas gives off a reddish glow in these lamps. If a neon bulb contains 2.69×10^{22} atoms of neon, what is the mass, in grams, of the neon?

Answer

$$2.69 \times 10^{22} \cancel{\text{atoms Ne}} \times \frac{1 \cancel{\text{mol Ne}}}{6.02 \times 10^{23} \cancel{\text{atoms Ne}}} \times \frac{20.18 \text{ g Ne}}{1 \cancel{\text{mol Ne}}} = 0.902 \text{ g Ne}$$

PRACTICE Problems

Have students refer to the Selected Solutions appendix for complete solutions to odd-numbered problems. The complete solutions for all problems can be found in the Solutions Manual.

19. a. 4.79×10^{24} atoms Li
b. 6.68×10^{20} atoms Pb
c. 3.45×10^{22} atoms Hg

20. a. 2.09×10^{3} g Bi
b. 91.3 g Mn
c. 0.226 g He
d. 3.49×10^{-8} g N
e. 5.93×10^{-7} g U

21. a. 9.77×10^{25} atoms Si
b. 1.51×10^{24} atoms Ti

ASSESSMENT

Skill Have students solve the following problem. What is the mass of 5.25×10^{23} atoms of potassium? How many moles of potassium is this? 34.1 g, 0.872 mol OL

CHEMISTRY JOURNAL

Conversion Map Have students design a concept map that can be used to convert from atoms to moles, moles to atoms, moles to mass, mass to moles, mass to atoms, and atoms to mass. OL EL

3 Assess

Check for Understanding

Give groups of students test tubes containing different masses of lead shot. Have each group determine the mass, moles, and number of atoms of lead in their test tube. Ask students what would change if the test tubes contained the same mass of iron instead of lead. Because the molar mass of iron is 55.85 g/mol and the molar mass of lead is 207.2 g/mol, test tubes containing iron will have 3.71 times as many moles as test tubes containing the same mass of lead. **OL** **COOP LEARN**

Reteach

Use small candies as models for representative particles. Have students work in pairs to determine the mass of one piece of candy, such as candy corn or jellybeans. Using this value as the mass of one representative particle, students can determine the mass of one mole of candies. Given a bag of candy, students can find the number of moles in one bag of candy (neglect the mass of the empty bag), the number of bags needed for one mole, and the cost of one mole. **OL** **COOP LEARN**

Extension

Have students build mobiles illustrating the mole concept. Include stems for particles and for mass. Molar mass, Avogadro's number, and conversion factors should be included. This visual reference will help students as they convert between moles, mass, and number of representative particles. **OL** **EL**

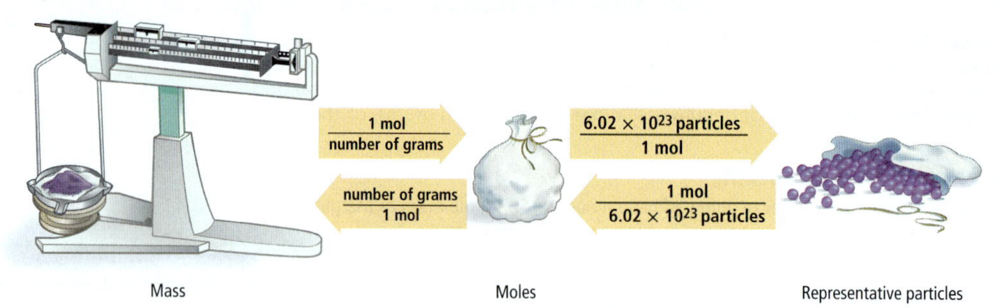

■ **Figure 8** The mole is at the center of conversions between mass and particles (atoms, ions, or molecules). In the figure, mass is represented by a balance, moles by a bag of particles, and representative particles by the contents that are spilling out of the bag. Two steps are needed to convert from mass to representative particles or the reverse.

Now that you have practiced conversions between mass, moles, and representative particles, you probably realize that the mole is at the center of these calculations. Mass must always be converted to moles before being converted to atoms, and atoms must similarly be converted to moles before calculating their mass. **Figure 8** shows the steps to follow as you complete these conversions. In the Example Problems, two steps were used to convert either mass to moles to atoms, or atoms to moles to mass. Instead of two separate steps, these conversions can be made in one step. Suppose you want to find out how many atoms of oxygen (O) are in 1.00 g of oxygen. This calculation involves two conversions—mass to moles and then moles to atoms. You could set up one equation like this.

$$1.00\ \cancel{\text{g O}} \times \frac{1\ \cancel{\text{mol O}}}{15.999\ \cancel{\text{g O}}} \times \frac{6.02 \times 10^{23}\ \text{atoms O}}{1\ \cancel{\text{mol O}}} = 3.76 \times 10^{22}\ \text{atoms O}$$

SECTION 2 REVIEW

Section Self-Check

Section Summary

- The mass in grams of one mole of any pure substance is called its molar mass.
- The molar mass of an element is numerically equal to its atomic mass.
- The molar mass of any substance is the mass in grams of Avogadro's number of representative particles of the substance.
- Molar mass is used to convert from moles to mass. The inverse of molar mass is used to convert from mass to moles.

22. MAINIDEA **Summarize** in terms of particles and mass, one-mole quantities of two different monatomic elements.
23. **State** the conversion factor needed to convert between mass and moles of the atom fluorine.
24. **Explain** how molar mass relates the mass of an atom to the mass of a mole of atoms.
25. **Describe** the steps used to convert the mass of an element to the number of atoms of the element.
26. **Arrange** these quantities from smallest to largest in terms of mass: 1.0 mol of Ar, 3.0×10^{24} atoms of Ne, and 20 g of Kr.
27. **Identify** the quantity that is calculated by dividing the molar mass of an element by Avogadro's number.
28. **Design** a concept map that shows the conversion factors needed to convert between mass, moles, and number of particles.

SECTION 2 REVIEW

22. Each one-mole quantity has 6.02×10^{23} particles (atoms), but they will have different masses.
23. Mass-to-mole conversions use the conversion factor 1 mol/18.998 g. Moles-to-mass conversions use the conversion factor 18.998 g/1 mol.
24. Molar mass is the mass in grams of one mole of any pure substance.
25. Multiply the mass by the inverse of molar mass, and then multiply by Avogadro's number.
26. 20 g Kr, 1.0 mol Ar, 3.0×10^{24} atoms Ne
27. Because the molar mass is a ratio of grams per mole and Avogadro's number is a ratio of particles per mole, dividing the molar mass of an element by Avogadro's number yields the mass of a single representative particle of that element.
28. Student concept maps will vary but should show the correct sets of conversion factors needed to convert among mass, moles, and number of particles. See Solutions Manual.

Essential Questions

- What are the mole relationships shown by a chemical formula?
- How is the molar mass of a compound calculated?
- How can the number of moles be coverted to the mass of a compound and vice versa?
- What conversion factors are applied to determine the number of atoms or ions in a known mass of a compound?

Review Vocabulary

representative particle: an atom, molecule, formula unit, or ion

Moles of Compounds

MAINIDEA The molar mass of a compound can be calculated from its chemical formula and can be used to convert from mass to moles of that compound.

CHEM 4 YOU Imagine checking two pieces of luggage at the airport, only to find out that one of them is over the weight limit. Because the weight of each suitcase depends on the combination of the items packed inside, changing the combination of the items in the two suitcases changes the weight of each.

Chemical Formulas and the Mole

You have learned that different kinds of representative particles are counted using the mole. In the last section, you read how to use molar mass to convert among moles, mass, and number of particles of an element. Can you make similar conversions for compounds and ions? Yes, you can, but to do so you will need to know the molar mass of the compounds and ions involved.

Recall that a chemical formula indicates the numbers and types of atoms contained in one unit of the compound. Consider the compound dichlorodifluoromethane with the chemical formula CCl_2F_2. The subscripts in the formula indicate that one molecule of CCl_2F_2 consists of one carbon (C) atom, two chlorine (Cl) atoms, and two fluorine (F) atoms. These atom are chemically bonded together. The C-Cl-F ratio in CCl_2F_2 is 1:2:2.

Now suppose you had a mole of CCl_2F_2. The representative particles of the compound are molecules, and a mole of CCl_2F_2 contains Avogadro's number of molecules. The C-Cl-F ratio in one mole of CCl_2F_2 would still be 1:2:2, as it is in one molecule of the compound. **Figure 9** illustrates this principle for a dozen CCl_2F_2 molecules. Check for yourself that a dozen CCl_2F_2 molecules contains one dozen carbon atoms, two dozen chlorine atoms, and two dozen fluorine atoms. The chemical formula CCl_2F_2 not only represents an individual molecule of CCl_2F_2, it also represents a mole of the compound.

Figure 9 A dozen molecules of CCl_2F_2 contains one dozen carbon atoms, two dozen chlorine atoms, and two dozen fluorine atoms.

Interpret *How many of each kind of atom—carbon, chlorine, and fluorine—are contained in 1 mol of CCl_2F_2?*

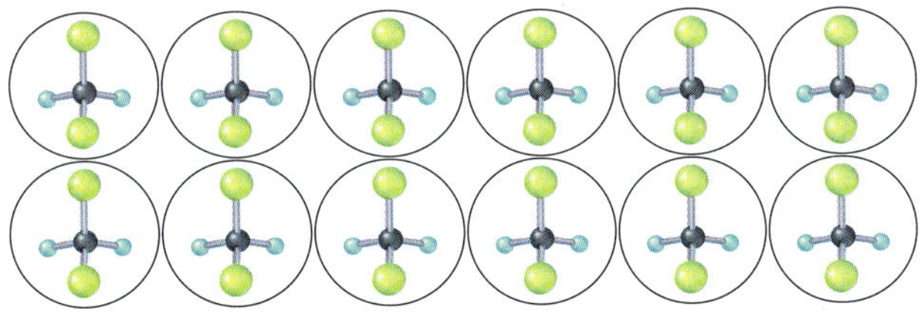

CHEMISTRY PROJECT

Chemical Formulas and the Mole Ask students to research the chemical formulas for the following compounds. For each compound, have them list the mole relationships among the component elements. OL

Substance	Chemical Formula	Mole Relationships
Butanol	$C_4H_{10}O$	4 mol C:10 mol H:1 mol O
Hydrazine	N_2H_4	2 mol N:4 mol H
Muriatic acid	HCl	1 mol H:1 mol Cl
Blue vitriol	$CuSO_4$	1 mol Cu:1 mol S:4 mol O

SECTION 3

1 Focus

MAINIDEA

Count Compounds Bring two beakers to class, one containing 342 g of table sugar and the second containing 180 g of glucose. Tell students that both beakers contain one mole and each sample contains carbon atoms, hydrogen atoms and oxygen atoms. Ask students why the beakers contain different amounts. The beakers contain different compounds, each of which is composed of carbon, hydrogen, and oxygen. Tell students that the first beaker contains $C_{12}H_{22}O_{11}$ and the second beaker contains $C_6H_{12}O_6$. Discuss how the mass of a compound is calculated using the molar mass of the component elements. OL

2 Teach

Content Background

Aluminum Example Problem 6 involves aluminum oxide, or alumina, which is the source material for the production of aluminum. Aluminum, which is a very common metal today, was once rare and extremely expensive. Though bauxite is the principle ore containing alumina, bauxite is not used directly to make aluminum. It is first refined into alumina. The Bayer refining process, developed in 1888 by the Austrian chemist Karl Bayer, is used to refine bauxite into alumina. The white powder product of the process (alumina) is transformed into aluminum metal in the smelting process.

Caption Question Fig. 9 one mole C atoms, two moles Cl atoms, two moles F atoms

Identify Misconceptions

Students might recognize that a mole of a compound represents 6.02×10^{23} particles, but they might not realize that there are other molar relationships that can be defined in that mole. For example, the number of moles of fluorine atoms in one mole of freon molecules.

Uncover the Misconception Analogies can be very helpful. Ask students how many legs a cat has. four How many legs do a dozen cats have? 48 How many cats are needed to have 200 legs? 50 How many legs would a mole of cats have? 2.41×10^{24} Then, ask how many atoms of fluorine are in one freon molecule. two Finally, how many atoms of fluorine are in one mole of freon-12? 1.20×10^{24}

Demonstrate the Concept Make a model of CH_4. Take the model apart and ask students the number of hydrogen atoms per methane molecule. four Ask how many hydrogen atoms would be in one mole of methane.

$$1\ \text{mol } CH_4 \times \frac{6.02 \times 10^{23}\ \text{molecules}}{1\ \text{mol}} \times \frac{4\text{H atoms}}{1\ \text{molecule } CH_4} = 2.41 \times 10^{24}$$

Assess New Knowledge Have students write mole ratios for the methane molecule. OL

IN-CLASS Example

Question Hematite is the mineral form of iron(III) oxide. Determine the number of moles of oxide ions, O^{2-}, contained in 6.25 mol of Fe_2O_3.

Answer

$$6.25\ \cancel{\text{mol } Fe_2O_3} \times \frac{3\ \text{mol O}}{1\ \cancel{\text{mol } Fe_2O_3}} = 18.8\ \text{mol } O^{2-}\ \text{ions}$$

VOCABULARY

ACADEMIC VOCABULARY

Ratio
the relationship in size or quantity of two or more things; proportion
The test results showed his LDL-to-HDL cholesterol ratio was too high.

In some chemical calculations, you might need to convert between moles of a compound and moles of individual atoms in the compound. The following ratios, or conversion factors, can be written for use in these calculations for the molecule CCl_2F_2.

$$\frac{1\ \text{mol C atoms}}{1\ \text{mol } CCl_2F_2} \qquad \frac{2\ \text{mol Cl atoms}}{1\ \text{mol } CCl_2F_2} \qquad \frac{2\ \text{mol F atoms}}{1\ \text{mol } CCl_2F_2}$$

To find out how many moles of fluorine atoms are in 5.50 moles of freon, you multiply the moles of freon by the conversion factor relating moles of fluorine atoms to moles of freon.

$$\text{moles } CCl_2F_2 \times \frac{\text{moles F atoms}}{1\ \text{mol } CCl_2F_2} = \text{moles F atoms}$$

$$5.50\ \cancel{\text{mol } CCl_2F_2} \times \frac{2\ \text{mol F atoms}}{1\ \cancel{\text{mol } CCl_2F_2}} = 11.0\ \text{mol F atoms}$$

Conversion factors such as the one just used for fluorine can be written for any element in a compound. The number of moles of the element that goes in the numerator of the conversion factor is the subscript for that element in the chemical formula.

EXAMPLE Problem 6

MOLE RELATIONSHIPS FROM A CHEMICAL FORMULA Aluminum oxide (Al_2O_3), often called alumina, is the principal raw material for the production of aluminum (Al). Alumina occurs in the minerals corundum and bauxite. Determine the moles of aluminum ions (Al^{3+}) in 1.25 mol of Al_2O_3.

1 ANALYZE THE PROBLEM

You are given the number of moles of Al_2O_3 and must determine the number of moles of Al^{3+} ions. Use a conversion factor based on the chemical formula that relates moles of Al^{3+} ions to moles of Al_2O_3. Every mole of Al_2O_3 contains 2 mol of Al^{3+} ions. Thus, the answer should be two times the number of moles of Al_2O_3.

Known	Unknown
number of moles = 1.25 mol Al_2O_3	number of moles = ? mol Al^{3+} ions

2 SOLVE FOR THE UNKNOWN

Use the relationship that 1 mol of Al_2O_3 contains 2 mol of Al^{3+} ions to write a conversion factor.

$$\frac{2\ \text{mol } Al^{3+}\ \text{ions}}{1\ \text{mol } Al_2O_3}$$ Create a conversion factor relating moles of Al^{3+} ions to moles of Al_2O_3.

To convert the known number of moles of Al_2O_3 to moles of Al^{3+} ions, multiply by the ions-to-moles conversion factor.

$$\text{moles } Al_2O_3 \times \frac{2\ \text{mol } Al^{3+}\ \text{ions}}{1\ \text{mol } Al_2O_3} = \text{moles } Al^{3+}\ \text{ions}$$ Apply the conversion factor.

$$1.25\ \cancel{\text{mol } Al_2O_3} \times \frac{2\ \text{mol } Al^{3+}\ \text{ions}}{1\ \cancel{\text{mol } Al_2O_3}} = 2.50\ \text{mol } Al^{3+}\ \text{ions}$$ Substitute moles Al_2O_3 = 1.25 mol Al_2O_3 and solve.

3 EVALUATE THE ANSWER

Because the conversion factor is a ratio of whole numbers, the number of significant digits is based on the moles of Al_2O_3. Therefore, the answer is expressed correctly with three significant figures. As predicted, the answer is twice the number of moles of Al_2O_3.

DIFFERENTIATED INSTRUCTION

Struggling Learners On the board, write the chemical formulas of several different compounds that each contain atoms of carbon, hydrogen, and oxygen. Ask students to create a table with column headings titled Chemical Formula, Number of Carbon Atoms, Molar Mass of Carbon, Mass of Carbon, Number of Oxygen Atoms, Molar Mass of Oxygen, Mass of Oxygen, Number of Hydrogen Atoms, Molar Mass of Hydrogen, Mass of Hydrogen, and Molar Mass of Compound. Walk students through calculating the mass of each component element and then summing the mass contributions to determine the molar mass of the compound. BL

PRACTICE Problems

Do additional problems. Online Practice

29. Zinc chloride ($ZnCl_2$) is used in soldering flux, an alloy used to join two metals together. Determine the moles of Cl^- ions in 2.50 mol $ZnCl_2$.

30. Plants and animals depend on glucose ($C_6H_{12}O_6$) as an energy source. Calculate the number of moles of each element in 1.25 mol $C_6H_{12}O_6$.

31. Iron(III) sulfate [$Fe_2(SO_4)_3$] is sometimes used in the water purification process. Determine the number of moles of sulfate ions present in 3.00 mol of $Fe_2(SO_4)_3$.

32. How many moles of oxygen atoms are present in 5.00 mol of diphosphorus pentoxide (P_2O_5)?

33. Challenge Calculate the number of moles of hydrogen atoms in 1.15×10^1 mol of water. Express the answer in scientific notation.

The Molar Mass of Compounds

The mass of your backpack is the sum of the mass of the pack and the masses of the books, notebooks, pencils, lunch, and miscellaneous items you put into it. You could find its mass by determining the mass of each item separately and adding them together. Similarly, the mass of a mole of a compound equals the sum of the masses of all the particles that make up the compound.

Suppose you want to determine the molar mass of the compound potassium chromate (K_2CrO_4). Start by looking up the molar mass of each element present in K_2CrO_4. Then, multiply each molar mass by the number of moles of that element in the chemical formula. Adding the masses of each element yields the molar mass of K_2CrO_4.

$$2\ \cancel{\text{mol K}} \times \frac{39.10\text{ g K}}{1\ \cancel{\text{mol K}}} = 78.20\text{ g}$$

$$1\ \cancel{\text{mol Cr}} \times \frac{52.00\text{ g Cr}}{1\ \cancel{\text{mol Cr}}} = 52.00\text{ g}$$

$$4\ \cancel{\text{mol O}} \times \frac{16.00\text{ g O}}{1\ \cancel{\text{mol O}}} = 64.00\text{ g}$$

$$\text{molar mass } K_2CrO_4 = 194.20\text{ g}$$

The molar mass of a compound demonstrates the law of conservation of mass; the total mass of the reactants that reacted equals the mass of the compound formed. **Figure 10** shows equivalent masses of one mole of potassium chromate, sodium chloride, and sucrose.

PRACTICE Problems

Do additional problems. Online Practice

34. Determine the molar mass of each ionic compound.

a. NaOH **b.** $CaCl_2$ **c.** $KC_2H_3O_2$

35. Calculate the molar mass of each molecular compound.

a. C_2H_5OH **b.** HCN **c.** CCl_4

36. Challenge Identify each substance as a molecular compound or an ionic compound, and then calculate its molar mass.

a. $Sr(NO_3)_2$ **b.** $(NH_4)_3PO_4$ **c.** $C_{12}H_{22}O_{11}$

■ **Figure 10** Because each substance contains different numbers and kinds of atoms, their molar masses are different. The molar mass of each compound is the sum of the masses of all the elements contained in the compound.

Potassium chromate (K_2CrO_4)

Sodium chloride (NaCl)

Sucrose ($C_{12}H_{22}O_{11}$)

PRACTICE Problems

Have students refer to the Selected Solutions appendix for complete solutions to odd-numbered problems. The complete solutions for all problems can be found in the Solutions Manual.

29. 5.00 mol Cl^-
30. 7.50 mol C; 15.0 mol H; 7.50 mol O
31. 9.00 mol SO_4^{2-}
32. 25.0 mol O
33. 2.30×10^1 mol H

ASSESSMENT

Knowledge Give each student a different chemical formula and have them identify the number of each atom per single formula unit or molecule. Then, have them identify the number of moles of each atom per mole of the same substance. OL

PRACTICE Problems

Have students refer to the Selected Solutions appendix for complete solutions to odd-numbered problems. The complete solutions for all problems can be found in the Solutions Manual.

34. a. 40.00 g/mol
b. 110.98 g/mol
c. 98.14 g/mol
35. a. 46.07 g/mol
b. 27.03 g/mol
c. 153.81 g/mol
36. a. ionic; 211.64 g/mol
b. ionic; 149.10 g/mol
c. molecular; 342.30 g/mol

CHEMISTRY JOURNAL

Molar Masses Ask students to calculate the molar masses of $CaCO_3$, $C_{12}H_{22}O_{11}$ (sucrose), NaCl, KNO_3, and C_8H_{18} (octane) and to record their calculations and answers in their journals. The molar mass values of these compounds will be used in future journal entries. OL

Reinforcement

Significant Digits To help students arrive at the same answers provided in the text, have them use the molar masses of the elements expressed to the second decimal place. Have students determine the molar mass of sodium thiosulfate ($Na_2S_2O_3$) using molar masses from the periodic table. 158.12 g/mol Remind students how to apply the rules for significant figures to the addition of molar mass values. OL

ASSESSMENT

Knowledge Give students individual vials of common substances labeled with their chemical formulas. Have students determine the masses of the substances in their vials, the molar masses of the substances, the number of moles of the substances present, and the number of moles of all ions or atoms present. OL

IN-CLASS Example

Question Baking soda, $NaHCO_3$ is a compound used in many recipes. What is the mass of 1.25 moles of $NaHCO_3$?

Answer First determine the molar mass of $NaHCO_3$ [84.01 g $NaHCO_3$/mol]

$$1.25 \text{ mol } NaHCO_3 \times \frac{84.01 \text{ g } NaHCO_3}{1 \text{ mol } NaHCO_3} = 105 \text{ g } NaHCO_3$$

PRACTICE Problems

Have students refer to the Selected Solutions appendix for complete solutions to odd-numbered problems. The complete solutions for all problems can be found in the Solutions Manual.

37. 319 g
38. 5.93 g
39. $KMnO_4$; 403 g

Converting Moles of a Compound to Mass

Suppose you need to measure a certain number of moles of a compound for an experiment. First, you must calculate the mass in grams that corresponds to the necessary number of moles. Then, you can measure that mass on a balance. In Example Problem 2, you learned how to convert the number of moles of elements to mass using molar mass as the conversion factor. The procedure is the same for compounds, except that you must first calculate the molar mass of the compound.

EXAMPLE Problem 7

Find help **with dimensional analysis.** Math Handbook

EXAMPLE PROBLEM

MOLE-TO-MASS CONVERSION FOR COMPOUNDS The characteristic odor of garlic is due to allyl sulfide [$(C_3H_5)_2S$]. What is the mass of 2.50 mol of $(C_3H_5)_2S$?

1 ANALYZE THE PROBLEM

You are given 2.50 mol of $(C_3H_5)_2S$ and must convert the moles to mass using the molar mass as a conversion factor. The molar mass is the sum of the molar masses of all the elements in $(C_3H_5)_2S$.

Known	Unknown
number of moles = 2.50 mol $(C_3H_5)_2S$	molar mass = ? g/mol $(C_3H_5)_2S$
	mass = ? g $(C_3H_5)_2S$

2 SOLVE FOR THE UNKNOWN

Calculate the molar mass of $(C_3H_5)_2S$.

$1 \text{ mol S} \times \frac{32.07 \text{ g S}}{1 \text{ mol S}} = 32.07 \text{ g S}$ — Multiply the moles of S in the compound by the molar mass of S.

$6 \text{ mol C} \times \frac{12.01 \text{ g C}}{1 \text{ mol C}} = 72.06 \text{ g C}$ — Multiply the moles of C in the compound by the molar mass of C.

$10 \text{ mol H} \times \frac{1.008 \text{ g H}}{1 \text{ mol H}} = 10.08 \text{ g H}$ — Multiply the moles of H in the compound by the molar mass of H.

molar mass = (32.07 g + 72.06 g + 10.08 g) = 114.21 g/mol $(C_3H_5)_2S$ — Total the mass values.

Use a conversion factor—the molar mass—that relates grams to moles.

$\text{moles } (C_3H_5)_2S \times \frac{\text{grams } (C_3H_5)_2S}{1 \text{ mol } (C_3H_5)_2S} = \text{mass } (C_3H_5)_2S$ — Apply the conversion factor.

$2.50 \text{ mol } (C_3H_5)_2S \times \frac{114.21 \text{ g } (C_3H_5)_2S}{1 \text{ mol } (C_3H_5)_2S} = 286 \text{ g } (C_3H_5)_2S$ — Substitute moles $(C_3H_5)_2S$ = 2.5 mol, molar mass $(C_3H_5)_2S$ = 114.21 g/mol, and solve.

PRACTICE Problems

Do additional problems. Online Practice

37. The United States chemical industry produces more sulfuric acid (H_2SO_4), in terms of mass, than any other chemical. What is the mass of 3.25 mol of H_2SO_4?

38. What is the mass of 4.35×10^{-2} mol of zinc chloride ($ZnCl_2$)?

39. **Challenge** Write the chemical formula for potassium permanganate, and then calculate the mass in grams of 2.55 mol of the compound.

DIFFERENTIATED INSTRUCTION

Struggling Learners Have groups of students make 12 molecular models of CO_2 using molecular model kits or gumdrops and toothpicks. Have them put their CO_2 models in an empty egg carton. Ask students to write down the number of CO_2 molecules. Then, have students take apart the molecules and place the carbon atoms in the egg carton. Have them record how many carbon atoms are present. Students can then put the oxygen atoms in two cartons and record the number of oxygen atoms present. Ask how many dozen carbon atoms and oxygen atoms are present in one dozen CO_2 molecules. Finally, ask how many moles of carbon atoms and oxygen atoms are present in one mole of CO_2. BL EL COOP LEARN

Converting the Mass of a Compound to Moles

Imagine that an experiment you are doing in the laboratory produces 5.55 g of a compound. How many moles is this? To find out, you calculate the molar mass of the compound and determine it to be 185.0 g/mol. The molar mass relates grams and moles, but this time you need the inverse of the molar mass as the conversion factor.

$$5.50 \text{ g compound} \times \frac{1 \text{ mol compound}}{185.0 \text{ g compound}} = 0.0297 \text{ mol compound}$$

EXAMPLE Problem 8

Find help with significant figures. Math Handbook

MASS-TO-MOLE CONVERSION FOR COMPOUNDS Calcium hydroxide [$Ca(OH)_2$] is used to remove sulfur dioxide from the exhaust gases emitted by power plants and for softening water by the elimination of Ca^{2+} and Mg^{2+} ions. Calculate the number of moles of calcium hydroxide in 325 g of the compound.

1 ANALYZE THE PROBLEM

You are given 325 g of $Ca(OH)_2$ and must solve for the number of moles of $Ca(OH)_2$. You must first calculate the molar mass of $Ca(OH)_2$.

Known	Unknown
mass = 325 g $Ca(OH)_2$	molar mass = ? g/mol $Ca(OH)_2$
	number of moles = ? mol $Ca(OH)_2$

2 SOLVE FOR THE UNKNOWN

Determine the molar mass of $Ca(OH)_2$.

$1 \text{ mol Ca} \times \frac{40.08 \text{ g Ca}}{1 \text{ mol Ca}} = 40.08 \text{ g}$ — Multiply the moles of Ca in the compound by the molar mass of Ca.

$2 \text{ mol O} \times \frac{16.00 \text{ g O}}{1 \text{ mol O}} = 32.00 \text{ g}$ — Multiply the moles of O in the compound by the molar mass of O.

$2 \text{ mol H} \times \frac{1.008 \text{ g H}}{1 \text{ mol H}} = 2.016 \text{ g}$ — Multiply the moles of H in the compound by the molar mass of H.

molar mass = (40.08 g + 32.00 g + 2.016 g) = **74.10 g/mol $Ca(OH)_2$** — Total the mass values.

Use a conversion factor—the inverse of the molar mass—that relates moles to grams.

$325 \text{ g } Ca(OH)_2 \times \frac{1 \text{ mol } Ca(OH)_2}{74.10 \text{ g } Ca(OH)_2} = 4.39 \text{ mol } Ca(OH)_2$ — Apply the conversion factor. Substitute mass Ca = 325 g, inverse molar mass $Ca(OH)_2$ = 1 mol/74.10 g, and solve.

3 EVALUATE THE ANSWER

To check the reasonableness of the answer, round the molar mass of $Ca(OH)_2$ to 75 g/mol and the given mass of $Ca(OH)_2$ to 300 g. Seventy-five is contained in 300 four times. Thus, the answer is reasonable. The unit, moles, is correct, and there are three significant figures.

EXAMPLE PROBLEM

PRACTICE Problems

Do additional problems. Online Practice

40. Determine the number of moles present in each compound.

a. 22.6 g $AgNO_3$ **b.** 6.50 g $ZnSO_4$ **c.** 35.0 g HCl

41. Challenge Identify each as an ionic or molecular compound and convert the given mass to moles. Express your answers in scientific notation.

a. 2.50 kg Fe_2O_3 **b.** 25.4 mg $PbCl_4$

ASSESSMENT

Knowledge Have students determine the mass of 0.200 mol of sodium chloride, 0.100 mol of potassium nitrate, and 0.250 mol of hydrogen chloride. 11.7 g NaCl, 10.1 g KNO_3, 9.11 g HCl OL

IN-CLASS Example

Question Titanium(IV) oxide is a common compound used as a white pigment. White road marking paint contains TiO_2. Calculate the moles of titanium(IV) oxide in 43.5 g of the compound.

Answer Calculate the molar mass of TiO_2: 79.88 g/mol

$$43.5 \text{ g } TiO_2 \times \frac{1 \text{ mol } TiO_2}{79.87 \text{ g } TiO_2} = 0.545 \text{ mol } TiO_2$$

PRACTICE Problems

Have students refer to the Selected Solutions appendix for complete solutions to odd-numbered problems. The complete solutions for all problems can be found in the Solutions Manual.

40. a. 0.133 mol
b. 0.0403 mol
c. 0.960 mol

41. a. ionic; 1.57×10^1 mol
b. ionic; 7.28×10^{-5} mol

CHEMISTRY JOURNAL

Know Your Antacid Have students examine the label on a bottle of antacid tablets (choose a brand that contains calcium carbonate). Ask them to record in their journals all information needed to determine the mass of calcium carbonate per tablet. Then, have them determine the moles of calcium carbonate per tablet, the moles of calcium ions per tablet, and the moles of carbonate ions per tablet. OL

Quick Demo

Calculate Sucrose Using a regular (non-diet) can of a soft drink as an example, point out to students that the drink contains 13% sugar by mass and has a mass of 355 g. Write the formula for sugar (sucrose) on the board: $C_{12}H_{22}O_{11}$. Have students calculate the number of moles of sucrose in the soft drink. **OL**

IN-CLASS Example

Question Silicon dioxide is usually found in the form of quartz. A sample of silicon dioxide has a mass of 42.7 g.

a. Determine the number of silicon atoms in the sample.

b. Determine the number of oxygen atoms in the sample.

c. Determine the mass of one molecule of silicon dioxide.

Answer

a. Use the molar mass of SiO_2 [60.09 g/mol] to calculate mol SiO_2. [0.711 mol SiO_2] Determine the number of SiO_2 molecules present. [4.28×10^{23} molecules of SiO_2] Calculate the number of Si atoms.

$$4.28 \times 10^{23} \cancel{\text{molecules } SiO_2} \times \frac{1 \text{ atom Si}}{1 \cancel{SiO_2}} = 4.28 \times 10^{23} \text{ atoms Si}$$

b. $$4.28 \times 10^{23} \cancel{\text{molecules } SiO_2} \times \frac{2 \text{ atoms O}}{1 \cancel{SiO_2}} = 8.56 \times 10^{23} \text{ atoms O}$$

c. $$\left(\frac{60.09 \text{ g } SiO_2}{1 \cancel{\text{mol}}\ SiO_2}\right) \times \left(\frac{1 \cancel{\text{mol}}}{6.02 \times 10^{23} \text{ molecules}}\right) = 9.98 \times 10^{-23} \text{ g/molecule of } SiO_2$$

FOLDABLES® Rubric

FOLDABLES® Incorporate information from this section into your Foldable.

Converting the Mass of a Compound to Number of Particles

Example Problem 8 illustrated how to find the number of moles of a compound contained in a given mass. Now, you will learn how to calculate the number of representative particles—molecules or formula units—contained in a given mass and, in addition, the number of atoms or ions.

Recall that no direct conversion is possible between mass and number of particles. You must first convert the given mass to moles by multiplying by the inverse of the molar mass. Then, you can convert moles to the number of representative particles by multiplying by Avogadro's number. To determine numbers of atoms or ions in a compound, you will need conversion factors that are ratios of the number of atoms or ions in the compound to 1 mol of compound. These are based on the chemical formula. Example Problem 9 provides practice in solving this type of problem.

EXAMPLE Problem 9

Find help with **significant fitures.** Math Handbook

CONVERSION FROM MASS TO MOLES TO PARTICLES Aluminum chloride ($AlCl_3$) is used in refining petroleum and manufacturing rubber and lubricants. A sample of aluminum chloride has a mass of 35.6 g.

a. How many aluminum ions are present?

b. How many chloride ions are present?

c. What is the mass, in grams, of one formula unit of aluminum chloride?

1 ANALYZE THE PROBLEM

You are given 35.6 g of $AlCl_3$ and must calculate the number of Al^{3+} ions, the number of Cl^- ions, and the mass in grams of one formula unit of $AlCl_3$. Molar mass, Avogadro's number, and ratios from the chemical formula are the necessary conversion factors. The ratio of Al^{3+} ions to Cl^- ions in the chemical formula is 1:3. Therefore, the calculated numbers of ions should be in that same ratio. The mass of one formula unit in grams will be an extremely small number.

Known

mass = 35.6 g $AlCl_3$

Unknown

number of ions = ? Al^{3+} ions

number of ions = ? Cl^- ions

mass = ? g/formula unit $AlCl_3$

2 SOLVE FOR THE UNKNOWN

Determine the molar mass of $AlCl_3$.

$$1 \cancel{\text{mol Al}} \times \frac{26.98 \text{ g Al}}{1 \cancel{\text{mol Al}}} = 26.98 \text{ g Al}$$

Multiply the moles of Al in the compound by the molar mass of Al.

$$3 \cancel{\text{mol Cl}} \times \frac{35.45 \text{ g Cl}}{1 \cancel{\text{mol Cl}}} = 106.35 \text{ g Cl}$$

Multiply the moles of Cl in the compound by the molar mass of Cl.

$$\text{molar mass} = (26.98 \text{ g} + 106.35 \text{ g}) = 133.33 \text{ g/mol } AlCl_3$$

Total the molar mass values.

Use a conversion factor—the inverse of the molar mass—that relates moles to grams.

$$\text{mass } AlCl_3 \times \frac{1 \text{ mol } AlCl_3}{\text{grams } AlCl_3} = \text{moles } AlCl_3$$

Apply the conversion factor.

$$35.6 \cancel{\text{ g } AlCl_3} \times \frac{1 \text{ mol } AlCl_3}{133.33 \cancel{\text{ g } AlCl_3}} = 0.267 \text{ mol } AlCl_3$$

Substitute mass $AlCl_3$ = 35.6 g and inverse molar mass $AlCl_3$ = 1 mol/133.33 g, and solve.

CHEMISTRY PROJECT

Avogadro's Number Have students research the history of Avogadro's number and make class reports. When Avogadro proposed his hypothesis, there were no data to confirm it. Several other scientists were involved in determining the number of particles in one mole: Robert Brown, Jean-Baptiste Perrin, Albert Einstein, R. A. Millikan, and Stanislao Cannizzaro. **OL**

Use Avogadro's number.

$0.267\ \text{mol AlCl}_3 \times \frac{6.02 \times 10^{23}\ \text{formula units}}{1\ \text{mol AlCl}_3}$ — Multiply and divide numbers and units.

$= 1.61 \times 10^{23}$ formula units $AlCl_3$

To calculate the number of Al^{3+} and Cl^- ions, use the ratios from the chemical formula as conversion factors.

$1.61 \times 10^{23}\ \text{AlCl}_3\ \text{formula units} \times \frac{1\ Al^{3+}\ \text{ion}}{1\ \text{AlCl}_3\ \text{formula unit}}$ — Multiply and divide numbers and units.

$= 1.61 \times 10^{23}$ Al^{3+} ions

$1.61 \times 10^{23}\ \text{AlCl}_3\ \text{formula units} \times \frac{3\ Cl^-\ \text{ions}}{1\ \text{AlCl}_3\ \text{formula unit}}$ — Multiply and divide numbers and units.

$= 4.83 \times 10^{23}$ Cl^- ions

Calculate the mass in grams of one formula unit of $AlCl_3$. Use the inverse of Avogadro's number as a conversion factor.

$\frac{133.33\ \text{g AlCl}_3}{1\ \text{mol}} \times \frac{1\ \text{mol}}{6.02 \times 10^{23}\ \text{formula units}}$ — Substitute mass $AlCl_3 = 133.33$ g, and solve.

$= 2.21 \times 10^{-22}$ g $AlCl_3$/formula unit

3 EVALUATE THE ANSWER

A minimum of three significant figures is used in each value in the calculations. Therefore, the answers have the correct number of digits. The number of Cl^- ions is three times the number of Al^{3+} ions, as predicted. The mass of a formula unit of $AlCl_3$ can be checked by calculating it in a different way. Divide the sample mass of $AlCl_3$ (35.6 g) by the number of formula units contained in the mass (1.61×10^{23} formula units) to obtain the mass of one formula unit. The two answers are the same.

PRACTICE Problems

Do additional problems. Online Practice

42. Ethanol (C_2H_5OH), a domestically produced fuel source, is often blended with gasoline. A sample of ethanol has a mass of 45.6 g.
- **a.** How many carbon atoms does the sample contain?
- **b.** How many hydrogen atoms are present?
- **c.** How many oxygen atoms are present?

43. A sample of sodium sulfite (Na_2SO_3) has a mass of 2.25 g.
- **a.** How many Na^+ ions are present?
- **b.** How many SO_3^{2-} ions are present?
- **c.** What is the mass in grams of one formula unit of Na_2SO_3?

44. A sample of carbon dioxide (CO_2) has a mass of 52.0 g.
- **a.** How many carbon atoms are present?
- **b.** How many oxygen atoms are present?
- **c.** What is the mass in grams of one molecule of CO_2?

45. What mass of sodium chloride (NaCl) contains 4.59×10^{24} formula units?

46. Challenge A sample of silver chromate has a mass of 25.8 g.
- **a.** Write the formula for silver chromate.
- **b.** How many cations are present in the sample?
- **c.** How many anions are present in the sample?
- **d.** What is the mass in grams of one formula unit of silver chromate?

PRACTICE PROBLEMS

PRACTICE Problems

Have students refer to the Selected Solutions appendix for complete solutions to odd-numbered problems. The complete solutions for all problems can be found in the Solutions Manual.

42. a. 1.19×10^{24} C atoms
b. 3.58×10^{24} H atoms
c. 5.96×10^{23} O atoms
43. a. 2.16×10^{22} Na^+ ions
b. 1.08×10^{22} SO_3^{2-} ions
c. 2.09×10^{-22} g Na_2SO_3/formula unit
44. a. 7.10×10^{23} C atoms
b. 1.42×10^{24} O atoms
c. 7.31×10^{-23} g CO_2/molecule
45. 445 g NaCl
46. a. Ag_2CrO_4
b. 9.36×10^{22} Ag^+ ions
c. 4.68×10^{22} CrO_4^{2-} ions
d. 5.51×10^{-22} g Ag_2CrO_4/formula unit

ASSESSMENT

Knowledge Tell students a box of macaroni and cheese contains 0.5600 g of sodium chloride. Ask them to determine the moles of sodium chloride, the number of sodium chloride formula units, and the number of atoms of sodium. OL
0.009582 mol NaCl
5.77×10^{21} formula units NaCl
5.77×10^{21} atoms Na

CHEMISTRY JOURNAL

Moles of Gasoline Have students find out and record the price of one gallon of gasoline. As they study Section 3, have them answer the following questions.

1. A gallon of gasoline is 3.78 L and has a density of 0.700 g/mL. What is the mass of one gallon of gas? 2650 g
2. If the formula for gasoline is C_8H_{18}, determine the moles of gasoline in one gallon. 23.2 mol
3. How many atoms of carbon are in one gallon of gasoline? 1.12×10^{26} C atoms
4. What is the price per mole of gasoline? answer depends on price per gallon
5. If you purchased $10.00 worth of gas, how many moles did you put into the tank? answer depends on price per gallon OL

3 Assess

Check for Understanding

Ask students to determine the molar mass of sugar (sucrose, $C_{12}H_{22}O_{11}$). 342.3 g/mol Have them determine the mass of one cup of sugar and the number of moles of sugar in one cup. about 192 g; about 0.561 mol Ask how many atoms of carbon, hydrogen, and oxygen are in one cup of sugar. about 4.05×10^{24} C atoms; about 7.43×10^{24} H atoms; about 3.71×10^{24} O atoms **OL**

Reteach

Write the formulas NaCl, $Al(NO_3)_3$, and CH_3CH_2OH on the board. Ask students how many moles of each atom or ion are present in one mole of each compound. NaCl: 1 mol Na^+ ion and 1 mol Cl^- ion; $Al(NO_3)_3$: 1 mol Al^{3+} ion and 3 mol NO_3^- ions; CH_3CH_2OH: 2 mol C atoms, 6 mol H atoms, 1 mol O atoms **OL**

Extension

Ask students to determine the number of moles of table sugar (sucrose) in a five-pound bag. (1 lb = 454 g) 6.63 mol sucrose Ask students to use the label of their favorite candy bar to determine the number of moles of sugar it contains, assuming the sugar in the candy bar is all sucrose. Answers will depend on the candy bar but will usually be about 0.050 mol sugar. **OL**

■ **Figure 11** Note the central position of the mole. To go from the left, right, or top of the diagram to any other place, you must go through the mole. The conversion factors on the arrows provide the means for making the conversions.

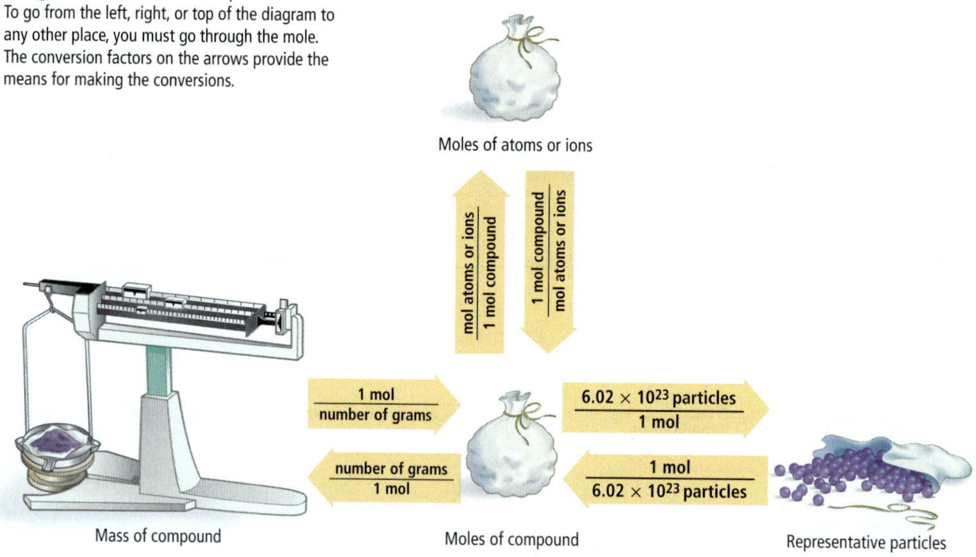

Conversions between mass, moles, and the number of particles are summarized in **Figure 11.** Note that molar mass and the inverse of molar mass are conversion factors between mass and number of moles. Avogadro's number and its inverse are the conversion factors between moles and the number of representative particles. To convert between moles and the number of moles of atoms or ions contained in the compound, use the ratio of moles of atoms or ions to 1 mole of compound or its inverse, which are shown on the upward and downward arrows in **Figure 11.** These ratios are derived from the subscripts in the chemical formula.

SECTION 3 REVIEW

Section Self-Check

Section Summary

- Subscripts in a chemical formula indicate how many moles of each element are present in 1 mol of the compound.
- The molar mass of a compound is calculated from the molar masses of all the elements in the compound.
- Conversion factors based on a compound's molar mass are used to convert between moles and mass of a compound.

47. **MAIN**IDEA **Describe** how to determine the molar mass of a compound.
48. **Identify** the conversion factors needed to convert between the number of moles and the mass of a compound.
49. **Explain** how you can determine the number of atoms or ions in a given mass of a compound.
50. **Apply** How many moles of K, C, and O atoms are there in 1 mol of $K_2C_2O_4$?
51. **Calculate** the molar mass of $MgBr_2$.
52. **Calculate** Calcium carbonate is the calcium source for many vitamin tablets. The recommended daily allowance of calcium is 1000 mg of Ca^{2+} ions. How many moles of Ca^{2+} does 1000 mg represent?
53. **Design** a bar graph that will show the number of moles of each element present in 500 g of a particular form of dioxin ($C_{12}H_4Cl_4O_2$), a powerful poison.

SECTION 3 REVIEW

47. Multiply the mass of one mole of each element by the number of moles of that element in the chemical formula. Add the resulting masses.
48. $\frac{\text{number of grams}}{1 \text{ mol}}, \frac{1 \text{ mol}}{\text{number of grams}}$
49. Convert the mass to moles, multiply the number of moles by the ratio of the number of atoms or ions to one mole, and multiply by Avogadro's number.
50. 2 mol K, 2 mol C, 4 mol O
51. 184.113 g/mol
52. 0.02 mol Ca^{2+}
53. Student bar graphs should show the following molar quantities: 24 mol C, 8 mol H, 8 mol Cl, and 4 mol O.

SECTION 4

Empirical and Molecular Formulas

MAINIDEA A molecular formula of a compound is a whole-number multiple of its empirical formula.

Essential Questions

- What is meant by the percent composition of a compound?
- How can the empirical and molecular formulas for a compound be determined from mass percent and actual mass data?

Review Vocabulary

percent by mass: the ratio of the mass of each element to the total mass of the compound expressed as a percent

New Vocabulary

percent composition
empirical formula
molecular formula

CHEM 4 YOU You might have noticed that some beverage bottles and food packages contain two or more servings instead of the single serving you expect. How would you determine the total number of calories contained in the package?

Percent Composition

Chemists, such as those shown in **Figure 12,** are often involved in developing new compounds for industrial, pharmaceutical, and home uses. After a synthetic chemist (one who makes new compounds) has produced a new compound, an analytical chemist analyzes the compound to provide experimental proof of its composition and its chemical formula.

It is the analytical chemist's job to identify the elements a compound contains and determine their percents by mass. Gravimetric and volumetric analyses are experimental procedures based on the measurement of mass for solids and liquids, respectively.

Percent composition from experimental data For example, consider a 100-g sample of a compound that contains 55 g of Element X and 45 g of Element Y. The percent by mass of any element in a compound can be found by dividing the mass of the element by the mass of the compound and multiplying by 100.

$$\text{percent by mass (element)} = \frac{\text{mass of element}}{\text{mass of compound}} \times 100$$

Figure 12 New compounds are first made on a small scale by a synthetic chemist like the one shown on the left. Then, an analytical chemist, like the one shown on the right, analyzes the compound to verify its structure and percent composition.

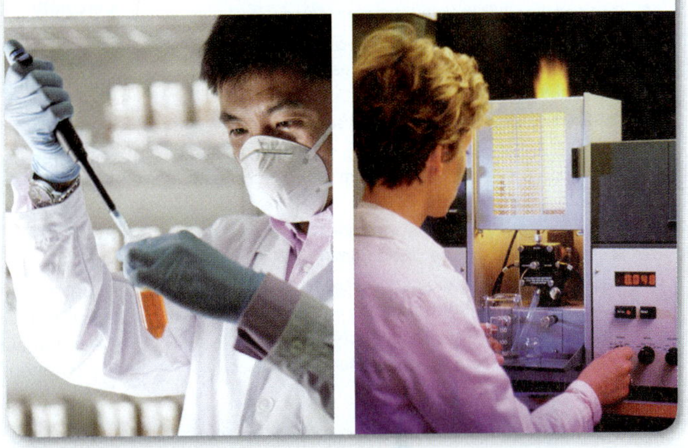

SECTION 4

1 Focus

MAINIDEA

Calculate Amounts Bring in a large bag of potato chips and have students guess the number of servings in the bag. Answers will vary. Ask students if they could determine the number of servings if they were given the total mass of chips and the mass of one serving. Yes, the number of servings would be equal to the total mass divided by the mass of an individual serving. Tell students that the empirical formula can be represented by a single serving, while the molecular formula can be represented by the total number of servings. BL OL

2 Teach

ASSESSMENT

Performance Have students measure the mass of a bag of microwave popcorn before it is heated and again after popping. Caution students not to open the microwaved popcorn until the steam has had time to dissipate. Have them calculate the percent mass lost to steam during popping. OL

DIFFERENTIATED INSTRUCTION

Struggling Learners In the *Journal of Chemical Education,* 1988, Joel S. Thompson wrote a rhyme that can be used to determine a simple chemical formula.

Percent to mass; Mass to mole; Divide by small; Multiply 'til whole

Give students the percent composition of a compound, 43.6% phosphorus and 56.4% oxygen. Ask students to write each line of the rhyme and show how each line is used to determine the empirical formula of this compound.

Percent to mass: 43.6% P = 43.6 g P; 56.4% g O = 56.4 g O
Mass to mole: 43.6 g P = 1.41 mol P; 56.4 g O = 3.52 mol O
Divide by small: 1.41 mol P/1.41 mol = 1 P; 3.52 mol O/1.41 mol = 2.5 O
Multiply 'til whole: 2(1 P) = 2 P; 2(2.5 O) = 5 O; P_2O_5 BL

MiniLAB

Rubric

Purpose Students will calculate the percent composition of solid sweeteners in gum to infer whether the sweetener is a coating or is mixed throughout the gum.

Process Skills observe and infer, measure

Safety Precautions Approve lab safety forms before work begins. Students should not place the gum in their mouths. Do not use sharp-pointed scissors.

Disposal Discard the gum in a waste container.

Teaching Strategies

- Avoid using hard, candy-coated gums.
- Ask students if the percentage of mass dissolved would be different for a chunky piece of gum compared to a long flat piece of gum. Yes, because the surface areas are different.
- Have students compare the masses of all the pieces of gum. Ask them if they think the gum is machine-made, and why. The gum is likely machine made as the mass of each piece of gum is almost identical.

Expected Results The gum will lose mass after being placed in water because the sweeteners are water-soluble. The cut piece with the larger surface area will expose more of the soluble chemicals to the water and will have a greater loss of mass.

Analysis

1. mass dissolved = initial mass − final mass = 3.11 g − 2.84 g = 0.27 g
2. mass dissolved = initial mass − final mass = 3.11 g − 2.75 g = 0.36 g
3. whole: (0.27 g/3.11 g) × 100 = 8.7% pieces: (0.36 g/3.11 g) × 100 = 12%
4. Because more mass dissolves when the surface area is larger, you can infer that the sweeteners and flavors are mixed throughout the gum.

CAREERS IN CHEMISTRY

Medicinal Chemist Much like a chef trying to perfect a recipe, the medicinal chemist works to perfect the best combination of reactants. They use their knowledge of the effects of toxins and medicines on the human body to synthesize new molecules that target disease.

WebQuest

Because percent means parts per 100, the percents by mass of all the elements of a compound must always add up to 100.

$$\frac{55\ \cancel{g}\ \text{element X}}{100\ \cancel{g}\ \text{compound}} \times 100 = 55\%\ \text{element X}$$

$$\frac{45\ \cancel{g}\ \text{element Y}}{100\ \cancel{g}\ \text{compound}} \times 100 = 45\%\ \text{element Y}$$

Thus, the compound is 55% X and 45% Y. The percent by mass of each element in a compound is the **percent composition** of a compound.

Percent composition from the chemical formula The percent composition of a compound can also be determined from its chemical formula. To do this, assume you have exactly 1 mol of the compound and use the chemical formula to calculate the compound's molar mass. Then, determine the mass of each element in a mole of the compound by multiplying the element's molar mass by its subscript in the chemical formula. Finally, use the equation below to find the percent by mass of each element.

Percent by Mass from the Chemical Formula

$$\text{percent by mass} = \frac{\text{mass of element in 1 mol of compound}}{\text{molar mass of compound}} \times 100$$

The percent by mass of an element in a compound is the mass of the element in 1 mol of the compound divided by the molar mass of the compound, multiplied by 100.

Example Problem 10 covers calculating percent composition.

MiniLAB

Analyze Chewing Gum

Are sweetening and flavoring added as a coating or mixed throughout chewing gum?

Procedure

1. Read and complete the lab safety form.
2. Unwrap two pieces of **chewing gum.** Place each piece on a **weighing paper.** Measure and record each mass using a **balance.**
   WARNING: *Do not eat any items used in the lab.*
3. Add 150 mL of cold **tap water** to a **250-mL beaker.** Place one piece of chewing gum in the water, and stir with a **stirring rod** for 2 min.
4. Pat the gum dry using **paper towels.** Measure and record the mass of the dried gum.
5. Use **scissors** to cut the second piece of gum into small pieces. Repeat Step 3 using fresh water. Keep the pieces from clumping together.
   WARNING: *Use caution with scissors.*
6. Use a 10-cm × 10-cm piece of **window screen** to strain the water from the gum. Pat the gum dry using paper towels. Measure and record the mass of the dried gum.

Analysis

1. **Calculate** For the uncut piece of gum, calculate the mass of sweeteners and flavorings that dissolved in the water. The mass of sweeteners and flavorings is the difference between the original mass of the gum and the mass of the dried gum.
2. **Calculate** For the gum cut into small pieces, calculate the mass of dissolved sweeteners and flavorings.
3. **Apply** For each piece of gum, determine the percent of the original mass from the soluble sweeteners and flavorings.
4. **Infer** What can you infer from the two percentages? Is the gum sugar-coated or are the sweeteners and flavorings mixed throughout?

CAREERS IN CHEMISTRY

WebQuest

CHEMISTRY JOURNAL

Percent by Mass Have each student write a chemical formula in their journal. Ask them determine the percent by mass of each element in the compound. Have students exchange journals with a partner and repeat the process. Partners should compare answers and resolve discrepancies. Answers will vary. Possible formulas and answers are NaCl: 39.3% Na, 60.7% Cl; H_2O: 11.2% H, 88.8% O; CO_2: 27.3% C, 72.7% O; NaOH: 57.5% Na, 40.0% O, 2.5% H OL COOP LEARN

EXAMPLE Problem 10

Find help with **percents.** Math Handbook

CALCULATING PERCENT COMPOSITION Sodium hydrogen carbonate ($NaHCO_3$), also called baking soda, is an active ingredient in some antacids used for the relief of indigestion. Determine the percent composition of $NaHCO_3$.

1 ANALYZE THE PROBLEM

You are given only the chemical formula. Assume you have 1 mol of $NaHCO_3$. Calculate the molar mass and the mass of each element in 1 mol to determine the percent by mass of each element in the compound. The sum of all percents should be 100, although your answer might vary slightly due to rounding.

Known	Unknown
formula = $NaHCO_3$	percent Na = ?
	percent H = ?
	percent C = ?
	percent O = ?

2 SOLVE FOR THE UNKNOWN

Determine the molar mass of $NaHCO_3$ and each element's contribution.

$$1\ \cancel{\text{mol Na}} \times \frac{22.99\ \text{g Na}}{1\ \cancel{\text{mol Na}}} = 22.99\ \text{g Na}$$

Multiply the molar mass of Na by the number of Na atoms in the compound.

$$1\ \cancel{\text{mol H}} \times \frac{1.008\ \text{g H}}{1\ \cancel{\text{mol H}}} = 1.008\ \text{g H}$$

Multiply the molar mass of H by the number of H atoms in the compound.

$$1\ \cancel{\text{mol C}} \times \frac{12.01\ \text{g C}}{1\ \cancel{\text{mol C}}} = 12.01\ \text{g C}$$

Multiply the molar mass of C by the number of C atoms in the compound.

$$3\ \cancel{\text{mol O}} \times \frac{16.00\ \text{g O}}{1\ \cancel{\text{mol O}}} = 48.00\ \text{g O}$$

Multiply the molar mass of O by the number of O atoms in the compound.

$$\text{molar mass} = (22.99\ \text{g} + 1.008\ \text{g} + 12.01\ \text{g} + 48.00\ \text{g})$$
$$= 84.01\ \text{g/mol NaHCO}_3$$

Total the mass values.

Use the percent by mass equation.

$$\%\ \text{mass element} = \frac{\text{mass of element in 1 mol of compound}}{\text{molar mass of compound}} \times 100$$

State the equation.

$$\text{percent Na} = \frac{22.99\ \text{g/mol}}{84.01\ \text{g/mol}} \times 100 = 27.37\%\ \text{Na}$$

Substitute mass of Na in 1 mol compound = 22.99 g/mol and molar mass $NaHCO_3$ = 84.01 g/mol. Calculate % Na.

$$\text{percent H} = \frac{1.008\ \text{g/mol}}{84.01\ \text{g/mol}} \times 100 = 1.200\%\ \text{H}$$

Substitute mass of H in 1 mol compound = 1.008 g/mol and molar mass $NaHCO_3$ = 84.01 g/mol. Calculate % H.

$$\text{percent C} = \frac{12.01\ \text{g/mol}}{84.01\ \text{g/mol}} \times 100 = 14.30\%\ \text{C}$$

Substitute mass of C in 1 mol compound = 12.01 g/mol and molar mass $NaHCO_3$ = 84.01 g/mol. Calculate % C.

$$\text{percent O} = \frac{48.00\ \text{g/mol}}{84.01\ \text{g/mol}} \times 100 = 57.14\%\ \text{O}$$

Substitute mass of O in 1 mol compound = 48.00 g/mol and molar mass $NaHCO_3$ = 84.01 g/mol. Calculate % O.

$NaHCO_3$ is 27.37% Na, 1.200% H, 14.30% C, and 57.14% O.

3 EVALUATE THE ANSWER

All masses and molar masses contain four significant figures. Therefore, the percents are correctly stated with four significant figures. When rounding error is accounted for, the sum of the mass percents is 100%, as required.

EXAMPLE PROBLEM

IN-CLASS Example

Question The chemical formula of caffeine is $C_8H_{10}N_4O_2$. Determine the percent composition of the molecule.

Answer Formula $C_8H_{10}N_4O_2$: 96.08 g C, 10.08 g H, 56.04 g N, 32.00 g O
Molar mass = 194.20 g/mol $C_8H_{10}N_4O_2$: 49.47% C, 5.19% H, 28.86% N, 16.48% O

Quick Demo

Reactions Affect Mass

Scratch the edge of a post-1982 penny to expose the zinc core. Measure and record the mass of the penny. Place the penny into 6.0*M* HCl and have students observe the reaction of zinc with HCl.
WARNING: *6.0M HCl will burn skin and irritate nasal linings. Do not inhale vapors. If spilled on skin, wash the area with soap and water, and neutralize with a baking soda paste.* When the reaction stops, remove the penny with plastic forceps, rinse several times with water, and then with alcohol. Allow the penny to dry. Remass the penny. Have students calculate what mass of the penny was zinc (the part that dissolved) and what mass was copper (the remaining penny). Determine the mass percent of zinc and the mass percent of copper in the penny. Disposal: Dilute the acid solution with more water and flush it down a drain. Post-1982 pennies are 2.5% Cu and 97.5% Zn. Pre-1982 pennies are 95% Cu and 5% zinc. OL COOP LEARN

CHEMISTRY JOURNAL

Percent Composition of Coins Ask students to take the coins they have in their pockets and use the data below to determine the percent by mass of Ag, Cu, Zn, and Ni in each coin. Quarters and dimes minted before 1965 are 90% silver and 10% copper. Quarters and dimes minted after 1965 contain 75% Cu and 25% Ni by mass. Nickels are 75% Cu and 25% Ni by mass. Pre-1982 pennies are 95% Cu and 5% Zn, and post-1982 pennies are 2.5% Cu and 97.5% Zn. Have students mass their coins and write the calculations in their journals. OL

PRACTICE Problems

Have students refer to the Selected Solutions appendix for complete solutions to odd-numbered problems. The complete solutions for all problems can be found in the Solutions Manual.

54. 3.08% H; 31.61% P; 65.31% O

55. H_2SO_3

56. 36.11% Ca; 63.89% Cl

57. a. sodium, sulfur, and oxygen; Na_2SO_4

b. ionic

c. 32.37% Na; 22.58% S; 45.05% O

☑ **READING CHECK** 1) Assume that the total mass of the compound is 100.00 g. The percent by mass of each element is equal to the mass of that element in grams. 2) Convert the mass of each element to moles using the molar mass as a conversion factor. 3) Divide each molar amount by the smallest mole value. 4) If needed, multiply each by an integer to determine the smallest whole-number ratio. 5) Write the empirical formula using the smallest whole-number ratio.

Concept Development

Use Ratios Give each student, or group of students, a box containing different numbers of small and large wooden beads, but maintain the ratio of two small beads to one large bead in each box. Tell students the beads represent moles of two different atoms. Have the students determine the mass of all the small beads, the mass of all the large beads, and the total mass of the beads. Then, have students determine the mass of one small bead, one large bead, and the sum of the two masses. Have them record the data in a data table. Students can then calculate the percent by mass of the small and large beads and record the results. When the class results are examined, point out that the mass percent is approximately the same for all boxes. Using the mass of a small bead and the mass of a large bead, determine the ratio of small to large beads. two small beads for every one large bead Ask students if their box had that ratio. They should notice that all beads were in a 2:1 ratio. **OL** **COOP LEARN**

PRACTICE Problems

Do additional problems. Online Practice

54. What is the percent composition of phosphoric acid (H_3PO_4)?

55. Which has the larger percent by mass of sulfur, H_2SO_3 or $H_2S_2O_8$?

56. Calcium chloride ($CaCl_2$) is sometimes used as a de-icer. Calculate the percent by mass of each element in $CaCl_2$.

57. Challenge Sodium sulfate is used in the manufacture of detergents.

a. Identify each of the component elements of sodium sulfate, and write the compound's chemical formula.

b. Identify the compound as ionic or covalent.

c. Calculate the percent by mass of each element in sodium sulfate.

Empirical Formula

When a compound's percent composition is known, its formula can be calculated. First, determine the smallest whole-number ratio of the moles of the elements in the compound. This ratio gives the subscripts in the empirical formula. The **empirical formula** for a compound is the formula with the smallest whole-number mole ratio of the elements. The empirical formula might or might not be the same as the actual molecular formula. If the two formulas are different, the molecular formula will always be a simple multiple of the empirical formula. The empirical formula for hydrogen peroxide is HO; the molecular formula is H_2O_2. In both formulas, the ratio of oxygen to hydrogen is 1:1.

Percent composition or masses of the elements in a given mass of compound can be used to determine the formula for the compound. If percent composition is given, assume the total mass of the compound is 100.00 g and that the percent by mass of each element is equal to the mass of that element in grams. This can be seen in **Figure 13,** where 100.00 g of the 40.05% S and 59.95% O compound contains 40.05 g of S and 59.95 g of O. The mass of each element is then converted to moles.

$$40.05 \cancel{g\,S} \times \frac{1 \text{ mol S}}{32.07 \cancel{g\,S}} = 1.249 \text{ mol S}$$

$$59.95 \cancel{g\,O} \times \frac{1 \text{ mol O}}{16.00 \cancel{g\,O}} = 3.747 \text{ mol O}$$

Thus, the mole ratio of S atoms to O atoms in the oxide is 1.249:3.747.

When the values in a mole ratio are not whole numbers, they cannot be used as subscripts in a chemical formula. You can convert the ratio to whole numbers by recognizing that the element with the smallest number of moles might have the smallest subscript possible, 1. To make the mole value of sulfur equal to 1, divide both mole values by the moles of sulfur (1.249). This does not change the ratio between the two elements because both are divided by the same number.

$$\frac{1.249 \text{ mol S}}{1.249} = 1 \text{ mol S} \qquad \frac{3.747 \text{ mol O}}{1.249} = 3 \text{ mol O}$$

The simplest whole-number mole ratio of S to O is 1:3. Thus, the empirical formula is SO_3. Sometimes, dividing by the smallest mole value does not yield whole numbers. In such cases, each mole value must then be multiplied by the smallest factor that will make it a whole number. This is shown in Example Problem 11.

☑ **READING CHECK** **List** the steps needed to calculate the empirical formula from percent composition data.

■ **Figure 13** Keep this figure in mind when doing problems using percent composition. You can always assume that you have a 100-g sample of the compound and use the percents of the elements as masses of the elements.

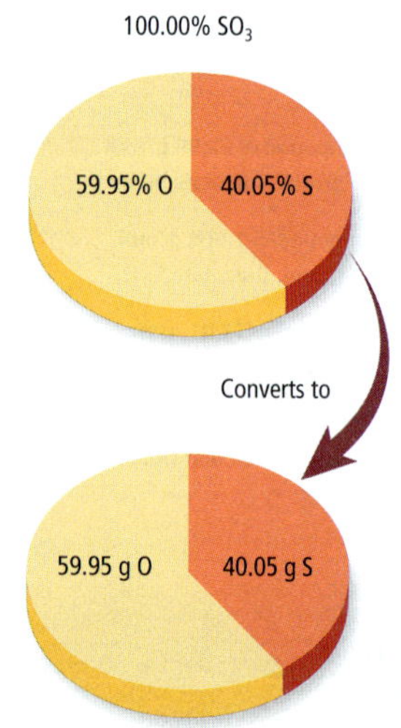

Demonstration

Empirical Formulas

Purpose

to emphasize the information needed to determine an empirical formula

Materials

test tube; 2-hole stopper; glass and rubber tubing; Bunsen burner; ring stand; test tube holder; copper(II) oxide; scupula

Safety Precautions

Disposal Copper product can be reused.

Procedure

Set up equipment as shown. Close off the burner's air supply. Turn on the gas and light the burner. When the flame is yellow, adjust the air control to get a hot, blue flame. Heat the sample gently, then strongly. Heat for 10 minutes and note changes. When the reaction is complete, remove the burner from the test tube. Do not turn off the burner; gas must be in contact with the product as it cools. Turn off the burner when the test tube is cool.

EXAMPLE Problem 11

Find help with **ratios.** Math Handbook

EMPIRICAL FORMULA FROM PERCENT COMPOSITION Methyl acetate is a solvent commonly used in some paints, inks, and adhesives. Determine the empirical formula for methyl acetate, which has the following chemical analysis: 48.64% carbon, 8.16% hydrogen, and 43.20% oxygen.

1 ANALYZE THE PROBLEM

You are given the percent composition of methyl acetate and must find the empirical formula. Because you can assume that each percent by mass represents the mass of the element in a 100.00-g sample, the percent sign can be replaced with the unit grams. Then, convert from grams to moles and find the smallest whole-number ratio of moles of the elements.

Known	Unknown
percent by mass C = 48.64% C	empirical formula = ?
percent by mass H = 8.16% H	
percent by mass O = 43.20% O	

2 SOLVE FOR THE UNKNOWN

Convert each mass to moles using a conversion factor—the inverse of the molar mass—that relates moles to grams.

$$48.64 \cancel{g\,C} \times \frac{1 \text{ mol C}}{12.01 \cancel{g\,C}} = 4.050 \text{ mol C}$$ **Substitute mass C = 48.64 g, inverse molar mass C = 1 mol/12.01 g, and calculate moles of C.**

$$8.16 \cancel{g\,H} \times \frac{1 \text{ mol H}}{1.008 \cancel{g\,H}} = 8.10 \text{ mol H}$$ **Substitute mass H = 8.16 g, inverse molar mass H = 1 mol/1.008 g, and calculate moles of H.**

$$43.20 \cancel{g\,O} \times \frac{1 \text{ mol O}}{16.00 \cancel{g\,O}} = 2.700 \text{ mol O}$$ **Substitute mass O = 43.20 g, inverse molar mass O = 1 mol/16.00 g, and calculate moles of O.**

Methyl acetate has a mole ratio of (4.050 mol C):(8.10 mol H):(2.700 mol O).

Next, calculate the simplest ratio of moles of elements by dividing the moles of each element by the smallest value in the calculated mole ratio.

$$\frac{4.050 \text{ mol C}}{2.700} = 1.500 \text{ mol C} = 1.5 \text{ mol C}$$ **Divide moles of C by 2.700.**

$$\frac{8.10 \text{ mol H}}{2.700} = 3.00 \text{ mol H} = 3 \text{ mol H}$$ **Divide moles of H by 2.700.**

$$\frac{2.700 \text{ mol O}}{2.700} = 1.000 \text{ mol O} = 1 \text{ mol O}$$ **Divide moles of O by 2.700.**

The simplest mole ratio is (1.5 mol C):(3 mol H):(1 mol O). Multiply each number in the ratio by the smallest number—in this case 2—that yields a ratio of whole numbers.

$2 \times 1.5 \text{ mol C} = 3 \text{ mol C}$ **Multiply moles of C by 2 to obtain a whole number.**

$2 \times 3 \text{ mol H} = 6 \text{ mol H}$ **Multiply moles of H by 2 to obtain a whole number.**

$2 \times 1 \text{ mol O} = 2 \text{ mol O}$ **Multiply moles of O by 2 to obtain a whole number.**

The simplest whole-number ratio of atoms is (3 atoms C):(6 atoms H):(2 atoms O). Thus, the empirical formula of methyl acetate is **$C_3H_6O_2$**.

3 EVALUATE THE ANSWER

The calculations are correct, and significant figures have been observed. To check that the formula is correct, calculate the percent composition represented by the formula. The percent composition checks exactly with the data given in the problem.

IN-CLASS Example

Question Cinnamon contains cinnamaldehyde. A molecule of cinnamaldehyde contains 81.79% C, 6.10% H, and 12.11% O. Determine the molecule's empirical formula.

Answer

6.81 mol C, 6.05 mol H, 0.757 mol O

C: 6.81 mol C/0.757 mol = 9

H: 6.05 mol H/0.757 mol = 8

O: 0.757 mol O/0.757 mol = 1

Thus, the formula is C_9H_8O.

Concept Development

Ratios and Percentages

Continue the activity described in the Concept Development on the previous page by giving students another box to represent another compound. This box should have the same ratio of items, but different substances, for example, two wooden beads and one glass marble. Have students calculate the percent by mass of each item and the simplest whole-number ratio of items. Ask students if the boxes represent the same compound. Students should notice that although the empirical formula is the same, the mass percent is different, so the boxes must contain models of different compounds. **OL**

Results

The black copper(II) oxide will turn to reddish-brown copper metal.

Analysis

1. What is the product of this reaction? copper metal
2. From the results of the experiment, can you identify the formula for the copper oxide? Why or why not? No, you do not know the amount of oxygen that was produced.

ASSESSMENT

Performance Have students design an experiment to determine the empirical formula for the copper oxide. Determine the mass of the copper oxide before heating and determine the mass of the product. Calculate the empirical formula from this data. **OL**

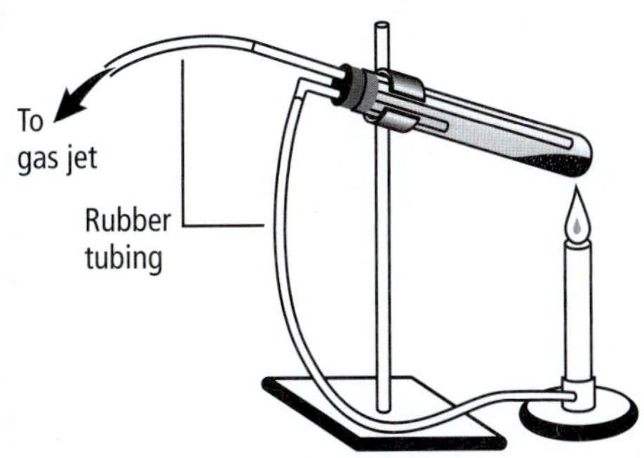

PRACTICE Problems

Have students refer to the Selected Solutions appendix for complete solutions to odd-numbered problems. The complete solutions for all problems can be found in the Solutions Manual.

58. N_2O_3
59. Al_2S_3
60. CH_3
61. $C_9H_8O_4$

Visual Learning

Figure 14 Point out that acetylene is a gas and, therefore, is different from benzene, which is a liquid solvent. Both are 92.25% carbon and 7.75% hydrogen. Therefore, both have the empirical formula CH. Ask students what must be different other than their chemical and physical properties. the chemical formulas If their chemical formulas are different, can they have the same molar mass? No, the molar masses must be different. Ask how many CH units are in one benzene molecule. six Have students determine the molar mass of benzene. 78.11 g/mol Have them determine the mass of the empirical formula CH. 13.02 g/mol Have them divide the mass of benzene by the mass of the CH formula. six OL

PRACTICE Problems

PRACTICE PROBLEMS

Do additional problems. Online Practice

58. The circle graph at the right gives the percent composition for a blue solid. What is the empirical formula for this solid?

59. Determine the empirical formula for a compound that contains 35.98% aluminum and 64.02% sulfur.

60. Propane is a hydrocarbon, a compound composed only of carbon and hydrogen. It is 81.82% carbon and 18.18% hydrogen. What is the empirical formula?

61. Challenge Aspirin is the world's most-often used medication. The chemical analysis of aspirin indicates that the molecule is 60.00% carbon, 4.44% hydrogen, and 35.56% oxygen. Determine the empirical formula for aspirin.

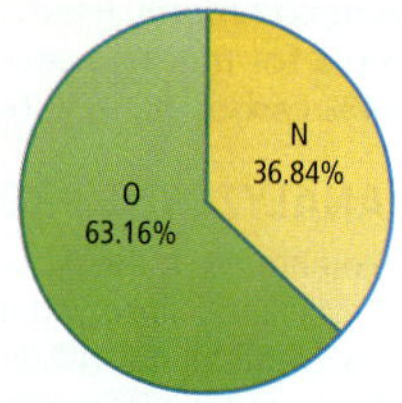

Molecular Formula

Would it surprise you to learn that substances with distinctly different properties can have the same percent composition and the same empirical formula? How is this possible? Remember that the subscripts in an empirical formula indicate the simplest whole-number ratio of moles of the elements in the compound. However, the simplest ratio does not always indicate the actual ratio in the compound. To identify a new compound, a chemist determines the **molecular formula,** which specifies the actual number of atoms of each element in one molecule or formula unit of the substance. **Figure 14** shows an important use of the gas acetylene. It has the same percent composition and the same empirical formula (CH) as benzene, which is a liquid. Yet chemically and structurally, acetylene and benzene are very different.

To determine the molecular formula for a compound, the molar mass of the compound must be determined through experimentation and compared with the mass represented by the empirical formula. For example, the molar mass of acetylene is 26.04 g/mol, and the mass of the empirical formula (CH) is 13.02 g/mol. Dividing the actual molar mass by the mass of the empirical formula indicates that the molar mass of acetylene is two times the mass of the empirical formula.

$$\frac{\text{experimentally determined molar mass of acetylene}}{\text{mass of empirical formula}} = \frac{26.04\ \cancel{\text{g/mol}}}{13.02\ \cancel{\text{g/mol}}} = 2.000$$

Because the molar mass of acetylene is two times the mass represented by the empirical formula, the molecular formula of acetylene must contain twice the number of carbon and hydrogen atoms as represented by the empirical formula.

Figure 14 Acetylene is a gas used for welding because of the high-temperature flame produced when it is burned with oxygen.

Tony Freeman/PhotoEdit

CULTURAL DIVERSITY

Is it spicy? Hot peppers are commonly used by cooks in Mexico, Spain, Thailand, South American, India, and China. All of these countries have hot climates. The spicy foods from these countries make you sweat. The evaporation of sweat absorbs heat and cools you off. Capsaicinoids, which are found in the white tissues inside peppers, are the class of compounds responsible for the "heat" in spicy foods. To determine the concentration of capsaicinoids in peppers, analytical chemists test an extract from the pepper in a device called a high performance liquid chromatograph. The chromatographic results are measured in ASTM pungency units and reported in Scoville units, the unit used to measure the hotness of a pepper. Scoville units range from zero for bell peppers to 300,000 for habanero peppers to over 15 million for pure capsaicin.

Similarly, when the experimentally determined molar mass of benzene, 78.12 g/mol, is compared with the mass of the empirical formula, the molar mass of benzene is found to be six times the mass of the empirical formula.

$$\frac{\text{experimentally determined molar mass of benzene}}{\text{mass of the empirical formula CH}} = \frac{78.12\ \text{g mol}}{13.02\ \text{g mol}} = 6.000$$

The molar mass of benzene is six times the mass represented by the empirical formula, so the molecular formula for benzene must represent six times the number of carbon atoms and hydrogen atoms shown in the empirical formula. You can conclude that the molecular formula for acetylene is 2 × CH, or C_2H_2, and the molecular formula for benzene is 6 × CH, or C_6H_6.

A molecular formula can be represented as the empirical formula multiplied by an integer *n*.

$$\text{molecular formula} = (\text{empirical formula})n$$

The integer *n* is the factor (6 in the example of benzene above) by which the subscripts in the empirical formula must be multiplied to obtain the molecular formula.

The steps in determining empirical and molecular formulas from percent composition or mass data are outlined in **Figure 15.** As in other calculations, the route leads from mass through moles because formulas are based on the relative numbers of moles of elements in each mole of compound.

Figure 15 Use this flowchart to guide you through the steps in determining the empirical and molecular formulas for compounds.
Describe *How is the integer* n *related to the empirical and molecular formulas?*

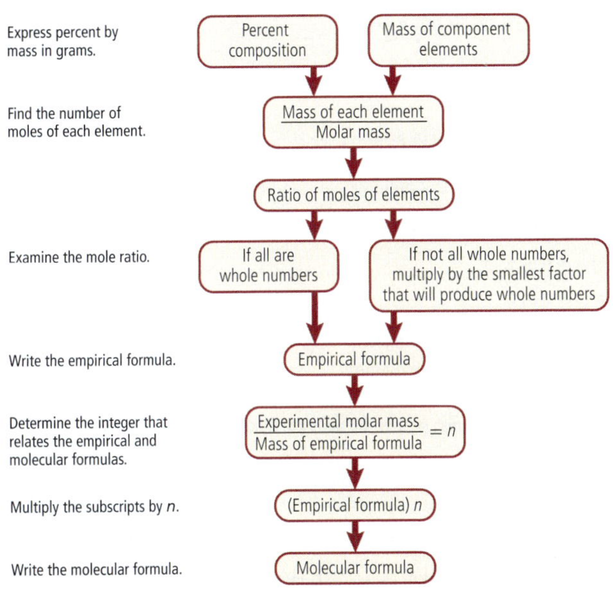

Concept Development

Calculation Have students refer to the data they obtained from the boxes of beads in the Concept Developments on the previous pages. Tell them that each box represents a different molecule. Ask them what is the same for each molecule? The ratio of small to large beads (atoms) is two to one. What is different for each box? the total mass Using the mass of two small beads and one large bead, ask students to calculate the number of 2:1 units in their boxes. Answers will vary depending upon the number of beads in each box. Have students multiply the numbers in the ratio (2:1) by the number of units in their boxes and ask if the sum of the numbers in the new ratio is the same as the number of beads in their boxes. They should be the same. **OL**

Extension

Gas Chromatography Have students investigate the operation of a gas chromatograph. Have them write a report or make a class presentation that explains how the instrument can be used to determine the concentrations of the capsa-icinoids discussed in Cultural Diversity on the previous page. **AL**

Caption Question Fig. 15 The integer *n* is the factor by which the empirical formula must be multiplied in order to obtain the molecular formula. **OL**

IN-CLASS Example

Question A molecule of nicotine contains 74.0% C, 8.65% H, and 17.35% N. The molar mass of nicotine is 162.26 g/mol. Determine the empirical and molecular formula of nicotine.

Answer
6.16 mol C
8.58 mol H
1.238 mol N
C: 6.16 mol C/1.238 mol = 5
H: 8.58 mol H/1.238 mol = 7
1.238 mol N/1.238 mol = 1
Thus, the empirical formula = C_5H_7N.
Empirical mass = 81.12 g/mol
n = (molar mass of nicotine)/(molar mass of empirical formula)
n = (162.26 g/mol)/(81.12 g/mol) = 2
$(C_5H_7N) \times 2 = C_{10}H_{14}N_2$

ASSESSMENT

Knowledge Have students determine the empirical formula and molecular formula for a compound with the following percent composition, 26.7% P, 12.1% N, and 61.2% Cl. The molar mass is 695 g/mol. $PNCl_2$; $P_6N_6Cl_{12}$ OL

Concept Development

Determine Composition
Have students refer to the data they obtained from the boxes of beads in Concept Development activities on the previous pages. Using the masses of the two different sets of beads and the total masses, determine the percent composition of the two modeled compounds. OL

EXAMPLE Problem 12

Find help with **ratios.** Math Handbook

EXAMPLE PROBLEM

DETERMINING A MOLECULAR FORMULA Succinic acid is a substance produced by lichens. Chemical analysis indicates it is composed of 40.68% carbon, 5.08% hydrogen, and 54.24% oxygen and has a molar mass of 118.1 g/mol. Determine the empirical and molecular formulas for succinic acid.

1 ANALYZE THE PROBLEM

You are given the percent composition. Assume that each percent by mass represents the mass of the element in a 100.00-g sample. You can compare the given molar mass with the mass represented by the empirical formula to find *n*.

Known	**Unknown**
percent by mass C = 40.68% C	empirical formula = ?
percent by mass H = 5.08% H	molecular formula = ?
percent by mass O = 54.24% O	
molar mass = 118.1 g/mol succinic acid	

2 SOLVE FOR THE UNKNOWN

Use the percents by mass as masses in grams, and convert grams to moles by using a conversion factor—the inverse of molar mass—that relates moles to mass.

$$40.68\ \cancel{g\,C} \times \frac{1\ \text{mol C}}{12.01\ \cancel{g\,C}} = 3.387\ \text{mol C}$$

Substitute mass C = 40.68 g, inverse molar mass C = 1 mol/12.01 g, and solve for moles of C.

$$5.08\ \cancel{g\,H} \times \frac{1\ \text{mol H}}{1.008\ \cancel{g\,H}} = 5.04\ \text{mol H}$$

Substitute mass H = 5.08 g, inverse molar mass H = 1 mol/1.008 g, and solve for moles of H.

$$54.24\ \cancel{g\,O} \times \frac{1\ \text{mol O}}{16.00\ \cancel{g\,O}} = 3.390\ \text{mol O}$$

Substitute mass O = 54.24 g, inverse molar mass O = 1 mol/16.00 g, and solve for moles of O.

The mole ratio in succinic acid is (3.387 mol C):(5.04 mol H):(3.390 mol O).
Next, calculate the simplest ratio of moles of elements by dividing the moles of each element by the smallest value in the calculated mole ratio.

$$\frac{3.387\ \text{mol C}}{3.387} = 1\ \text{mol C}$$

Divide moles of C by 3.387.

$$\frac{5.04\ \text{mol H}}{3.387} = 1.49\ \text{mol H} \approx 1.5\ \text{mol H}$$

Divide moles of H by 3.387.

$$\frac{3.390\ \text{mol O}}{3.387} = 1.001\ \text{mol O} \approx 1\ \text{mol O}$$

Divide moles of O by 3.387.

The simplest mole ratio is 1:1.5:1. Multiply all mole values by 2 to obtain whole numbers.

$2 \times 1\ \text{mol C} = 2\ \text{mol C}$ Multiply moles of C by 2.

$2 \times 1.5\ \text{mol H} = 3\ \text{mol H}$ Multiply moles of H by 2.

$2 \times 1\ \text{mol O} = 2\ \text{mol O}$ Multiply moles of O by 2.

The simplest whole-number mole ratio is 2:3:2. The empirical formula is $C_2H_3O_2$.
Calculate the empirical formula mass using the molar mass of each element.

$$2\ \cancel{\text{mol C}} \times \frac{12.01\ \text{g C}}{1\ \cancel{\text{mol C}}} = 24.02\ \text{g C}$$

Multiply the molar mass of C by the moles of C atoms in the compound.

$$3\ \cancel{\text{mol H}} \times \frac{1.008\ \text{g H}}{1\ \cancel{\text{mol H}}} = 3.024\ \text{g H}$$

Multiply the molar mass of H by the moles of H atoms in the compound.

$$2\ \cancel{\text{mol O}} \times \frac{16.00\ \text{g O}}{1\ \cancel{\text{mol O}}} = 32.00\ \text{g O}$$

Multiply the molar mass of O by the moles of O atoms in the compound.

molar mass $C_2H_3O_2$ = (24.02 g + 3.024 g + 32.00 g) = 59.04 g/mol Total the mass values.

CHEMISTRY PROJECT

Medicines from Plants Many medications, such as digoxin, quinine, digitalis, and ephedrine, have been obtained from plant-derived extracts. Ask students to research one of these substances and write an article to share with the class. OL

Divide the experimentally determined molar mass of succinic acid by the mass of the empirical formula to determine *n*.

$$n = \frac{\text{molar mass of succinic acid}}{\text{molar mass of } C_2H_3O_2} = \frac{118.1 \text{ g/mol}}{59.04 \text{ g/mol}} = 2.000$$

Multiply the subscripts in the empirical formula by 2 to determine the actual subscripts in the molecular formula.

$2 \times (C_2H_3O_2) = C_4H_6O_4$

The molecular formula for succinic acid is $C_4H_6O_4$.

3 EVALUATE THE ANSWER

The calculation of the molar mass from the molecular formula gives the same result as the given, experimentally-determined molar mass.

EXAMPLE Problem 13

Find help with **ratios.** Math Handbook

EXAMPLE PROBLEM

CALCULATING AN EMPIRICAL FORMULA FROM MASS DATA The mineral ilmenite is usually mined and processed for titanium, a strong, light, and flexible metal. A sample of ilmenite contains 5.41 g of iron, 4.64 g of titanium, and 4.65 g of oxygen. Determine the empirical formula for ilmenite.

1 ANALYZE THE PROBLEM

You are given the masses of the elements found in a known mass of ilmenite and must determine the empirical formula of the mineral. Convert the known masses of each element to moles, then find the smallest whole-number ratio of the moles of the elements.

Known	Unknown
mass of iron = 5.41 g Fe	empirical formula = ?
mass of titanium = 4.64 g Ti	
mass of oxygen = 4.65 g O	

2 SOLVE FOR THE UNKNOWN

Convert each known mass to moles by using a conversion factor—the inverse of molar mass—that relates moles to grams.

$$5.41 \text{ g Fe} \times \frac{1 \text{ mol Fe}}{55.85 \text{ g Fe}} = 0.0969 \text{ mol Fe}$$

Multiply mass Fe = 5.41 g by the inverse molar mass Fe = 1 mol/55.85 g, and calculate moles of Fe.

$$4.64 \text{ g Ti} \times \frac{1 \text{ mol Ti}}{47.87 \text{ g Ti}} = 0.0969 \text{ mol Ti}$$

Multiply mass Ti = 4.64 g by the inverse molar mass Ti = 1 mol/47.87 g, and calculate moles of Ti.

$$4.65 \text{ g O} \times \frac{1 \text{ mol O}}{16.00 \text{ g O}} = 0.291 \text{ mol O}$$

Multiply mass O = 4.65 g by the inverse molar mass O = 1 mol/16.00 g, and calculate moles of O.

The mineral ilmenite has a mole ratio of (0.0969 mol Fe):(0.0969 mol Ti):(0.291 mol O).
Calculate the simplest ratio by dividing each mole value by the smallest value in the ratio.

$$\frac{0.0969 \text{ mol Fe}}{0.0969} = 1 \text{ mol Fe}$$

Divide the moles of Fe by 0.0969.

$$\frac{0.0969 \text{ mol Ti}}{0.0969} = 1 \text{ mol Ti}$$

Divide the moles of Ti by 0.0969.

$$\frac{0.291 \text{ mol O}}{0.0969} = 3 \text{ mol O}$$

Divide the moles of O by 0.0969.

Because all the mole values are whole numbers, the simplest whole-number mole ratio is (1 mol Fe):(1 mol Ti):(3 mol O). The empirical formula for ilmenite is $FeTiO_3$.

CHEMISTRY PROJECT

What Is a mass spectrometer?

Have interested students research mass spectrometers and write a report on this analytical instrument and what information it can provide. Students can share their findings with the class. OL

3 Assess

Check for Understanding

Have each student determine the percent composition and molar mass of an assigned compound. Pairs of students should then exchange data and determine the empirical and molecular formulas of the compounds. The pairs of students should work out any discrepancies in the calculations. OL COOP LEARN

Reteach

Ask students to outline the procedure for determining an empirical formula. OL

Extension

Challenge students to find the molecular formulas for sets of compounds that have the same empirical formula. A good place to look is in the Molecular Formulas Index of Organic Compounds in the *Handbook of Chemistry and Physics.* AL

IN-CLASS Example

Question Chemical analysis shows that a compound commonly used as a bleaching agent is composed of 13.79 g Na, 21.27 g Cl and 9.60 g O. Determine the empirical formula of this compound.

Answer

Na: 0.600 mol Na/0.600 mol = 1

Cl: 0.600 mol Cl/0.600 mol = 1

O: 0.600 mol O/0.600 mol = 1

Thus, the empirical formula is NaClO.

PRACTICE Problems

Have students refer to the Selected Solutions appendix for complete solutions to odd-numbered problems. The complete solutions for all problems can be found in the Solutions Manual.

62. C_4H_{10}
63. N_2O_2
64. K_2O
65. $C_6H_6O_2$
66. $C_{17}H_{19}O_3N$

ASSESSMENT

Knowledge When heated in an atmosphere of pure nitrogen, magnesium reacts to form a nitride. Give students the following data and have them determine the formula of the compound.

Mass of empty crucible = 15.05 g
Mass of crucible + Mg ribbon = 17.45 g
Mass of crucible + product = 18.37 g Mg_3N_2 OL

3 EVALUATE THE ANSWER

The mass of iron is slightly greater than the mass of titanium, but the molar mass of iron is also slightly greater than that of titanium. Thus, it is reasonable that the numbers of moles of iron and titanium are equal. The mass of titanium is approximately the same as the mass of oxygen, but the molar mass of oxygen is about one-third that of titanium. Thus, a 3:1 ratio of oxygen to titanium is reasonable.

PRACTICE Problems

Do additional problems. Online Practice

62. A compound was found to contain 49.98 g of carbon and 10.47 g of hydrogen. The molar mass of the compound is 58.12 g/mol. Determine the molecular formula.

63. A colorless liquid composed of 46.68% nitrogen and 53.32% oxygen has a molar mass of 60.01 g/mol. What is the molecular formula?

64. When an oxide of potassium is decomposed, 19.55 g of K and 4.00 g of O are obtained. What is the empirical formula for the compound?

65. Challenge Analysis of a chemical used in photographic developing fluid yielded the percent composition data shown in the circle graph to the right. If the chemical's molar mass is 110.0 g/mol, what is its molecular formula?

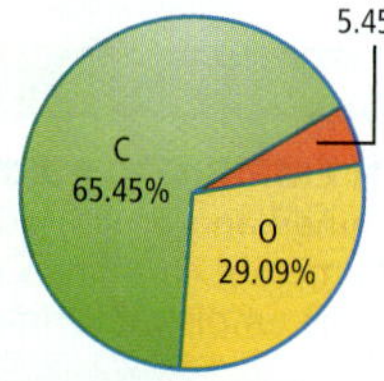

66. Challenge Analysis of the pain reliever morphine yielded the data shown in the table. Determine the empirical formula of morphine.

Element	Mass (g)
carbon	17.900
hydrogen	1.680
oxygen	4.225
nitrogen	1.228

SECTION 4 REVIEW

Section Self-Check

Section Summary

- The percent by mass of an element in a compound gives the percentage of the compound's total mass due to that element.
- The subscripts in an empirical formula give the smallest whole-number ratio of moles of elements in the compound.
- The molecular formula gives the actual number of atoms of each element in a molecule or formula unit of a substance.
- The molecular formula is a whole-number multiple of the empirical formula.

67. MAINIDEA **Assess** A classmate tells you that experimental data shows a compound's molecular formula to be 2.5 times its empirical formula. Is he correct? Explain.

68. Calculate Analysis of a compound composed of iron and oxygen yields 174.86 g of Fe and 75.14 g of O. What is the empirical formula for this compound?

69. Calculate An oxide of aluminum contains 0.545 g of Al and 0.485 g of O. Find the empirical formula for the oxide.

70. Explain how percent composition data for a compound are related to the masses of the elements in the compound.

71. Explain how you can find the mole ratio in a chemical compound.

72. Apply The molar mass of a compound is twice that of its empirical formula. How are the compound's molecular and empirical formulas related?

73. Analyze Hematite (Fe_2O_3) and magnetite (Fe_3O_4) are two ores used as sources of iron. Which ore provides the greater percent of iron per kilogram?

SECTION 4 REVIEW

67. No, he is in error because the molecular formula must be a whole-number multiple of the empirical formula.

68. Fe_2O_3

69. Al_2O_3

70. Percent composition is numerically equal to the mass in grams of each element in a 100.0 g sample.

71. The mole ratio is determined by calculating the moles of each element in the compound and dividing each number of moles by the smallest number of moles. It is sometimes necessary to multiply the ratio by the smallest integer that will give whole numbers.

72. The molecular formula is equal to twice the empirical formula.

73. Hematite is 69.94% Fe, magnetite is 72.36% Fe. Magnetite contains a greater percentage of iron per kilogram than hematite.

SECTION 5

Formulas of Hydrates

MAINIDEA Hydrates are solid ionic compounds in which water molecules are trapped.

Essential Questions

- What is a hydrate and how does its name relate to its composition?
- How is the formula of a hydrate determined from laboratory data?

Review Vocabulary

crystal lattice: a three-dimensional geometric arrangement of particles

New Vocabulary

hydrate

CHEM 4 YOU Some products, such as electronic equipment, are boxed with small packets labeled *dessicant.* These packets control moisture by absorbing water. Some contain ionic compounds called hydrates.

Naming Hydrates

Have you ever watched crystals slowly form from a water solution? Sometimes, water molecules adhere to the ions as the solid forms. The water molecules that become part of the crystal are called waters of hydration. Solid ionic compounds in which water molecules are trapped are called hydrates. A **hydrate** is a compound that has a specific number of water molecules bound to its atoms. **Figure 16** shows the beautiful gemstone known as opal, which is hydrated silicon dioxide (SiO_2). The unusual coloring is the result of water in the mineral.

In the formula of a hydrate, the number of water molecules associated with each formula unit of the compound is written following a dot—for example, $Na_2CO_3 \cdot 10H_2O$. This compound is called sodium carbonate decahydrate. In the word *decahydrate,* the prefix *deca-* means *ten* and the root word *hydrate* refers to *water.* A decahydrate has ten water molecules associated with one formula unit of compound. The mass of water associated with a formula unit is included in molar mass calculations. The number of water molecules associated with hydrates varies widely. Some common hydrates are listed in **Table 1.**

Explore **naming hydrates** with an interactive table. Concepts In Motion

Table 1 Formulas of Hydrates

Prefix	Molecules H_2O	Formula	Name
Mono-	1	$(NH_4)_2C_2O_4 \cdot H_2O$	ammonium oxalate monohydrate
Di-	2	$CaCl_2 \cdot 2H_2O$	calcium chloride dihydrate
Tri-	3	$NaC_2H_3O_2 \cdot 3H_2O$	sodium acetate trihydrate
Tetra-	4	$FePO_4 \cdot 4H_2O$	iron(III) phosphate tetrahydrate
Penta-	5	$CuSO_4 \cdot 5H_2O$	copper(II) sulfate pentahydrate
Hexa-	6	$CoCl_2 \cdot 6H_2O$	cobalt(II) chloride hexahydrate
Hepta-	7	$MgSO_4 \cdot 7H_2O$	magnesium sulfate heptahydrate
Octa-	8	$Ba(OH)_2 \cdot 8H_2O$	barium hydroxide octahydrate
Deca-	10	$Na_2CO_3 \cdot 10H_2O$	sodium carbonate decahydrate

Figure 16 The presence of water and various mineral impurities accounts for the variety of different-colored opals. Further changes in color occur when opals are allowed to dry out.

SECTION 5

1 Focus

MAINIDEA

Hydrates and Hydration Show the class slices of fresh and dehydrated fruit. Ask students what the difference is between the two. The fresh fruit slices contain water, whereas the dehydrated slices do not. Ask students how they could find out how much water was removed from the fruit when it was dehydrated. Mass the fresh fruit and place it in an oven to dehydrate. Remass the dehydrated fruit after removing it from the oven. The difference in mass equals the mass of water removed through dehydration. Tell students that many compounds contain water in the crystalline structure known as water of hydration, and this water can be removed by heating the substance. OL

2 Teach

Visual Learning

Figure 16 Some opals with orange-red coloring are referred to as fire opals. The greater the water concentration in the opal, the greater the fire. Some fire opals have up to 10% water by mass. Ask students to infer why opals are often stored on damp cotton. to maintain their water concentration Opals often crack if heated. Ask students to infer an explanation. Heating an opal creates water vapor which expands and cracks the opal. BL

DIFFERENTIATED INSTRUCTION

Advanced Learners Ask students to measure and record the mass of a quantity of plaster of paris. Add water and have students mold the paste into a shape. Allow the product to dry and then measure its mass again. Ask students to determine how much water was absorbed by the calcium sulfate. Have them determine the formula for the compound formed and compare their results with the formula for gypsum. Plaster of paris has one water molecule per two formula units of $CaSO_4$. Gypsum is calcium sulfate dihydrate. AL

Quick Demo

Measure Water Content

Add 5 mL of cooking oil to a clean flask. Make an aluminum foil lid with holes for the flask. Mass 20 kernels of popcorn and add them to the flask. Place the aluminum lid on the flask and measure and record the mass of the entire assembly. Heat the flask over a burner or on a hot plate until the popcorn is popped. **WARNING:** ***The flask will be hot.*** Measure and record the final mass when the flask is cool. Ask students to determine the percent water in the popcorn. Results will vary. Point out that this type of experiment can be used to determine the amount of water in hydrates. **OL**

Concept Development

Calculate Hydrates Discuss how to calculate the molar mass of a hydrate. Remind students that the dot in the formula of a hydrate represents the addition of water molecules and should not be confused with the dot used in mathematics to mean multiplication. Thus, the formula $CaSO_4 \cdot 2H_2O$ indicates 1 mole of $CaSO_4$ and 2 moles of water. The molar mass is 172.18 g/mol.

■ **In-Text Question** barium chloride dihydrate

ChemLAB

The ChemLab located at the end of the chapter can be used at this point in the lesson.

☑ **READING CHECK** The dot signifies that the water molecules are trapped in the compound and are not chemically bonded.

The hydrate cobalt(II) chloride hexahydrate is pink.

The hydrate can be heated to drive off the water of hydration.

Anhydrous cobalt(II) chloride is blue.

■ **Figure 17** Water of hydration can be removed by heating a hydrate, producing an anhydrous compound that can look very different from its hydrated form.

Matt Meadows

Analyzing a Hydrate

When a hydrate is heated, water molecules are driven off leaving an anhydrous compound, or one "without water." See **Figure 17.** The series of photos show that when pink cobalt(II) chloride hexahydrate is heated, blue anhydrous cobalt(II) chloride is produced.

VOCABULARY
WORD ORIGIN
Anhydrous
comes from the Greek prefix *an–*, meaning *not* or *without*, and *–hydrous*, from the Greek root *hydro* meaning *water*

How can you determine the formula of a hydrate? You must find the number of moles of water associated with 1 mol of the hydrate. Suppose you have a 5.00-g sample of a hydrate of barium chloride. You know that the formula is $BaCl_2 \cdot xH_2O$. You must determine *x*, the coefficient of H_2O in the hydrate formula that indicates the number of moles of water associated with 1 mol of $BaCl_2$. To find *x*, you would heat the sample of the hydrate to drive off the water of hydration. After heating, the dried substance, which is anhydrous $BaCl_2$, has a mass of 4.26 g. The mass of the water of hydration is the difference between the mass of the hydrate (5.00 g) and the mass of the anhydrous compound (4.26 g).

$$5.00 \text{ g } BaCl_2 \text{ hydrate} - 4.26 \text{ g anhydrous } BaCl_2 = 0.74 \text{ g } H_2O$$

You now know the masses of $BaCl_2$ and H_2O in the sample. You can convert these masses to moles using the molar masses. The molar mass of $BaCl_2$ is 208.23 g/mol, and the molar mass of H_2O is 18.02 g/mol.

$$4.26 \cancel{\text{g BaCl}_2} \times \frac{1 \text{ mol BaCl}_2}{208.23 \cancel{\text{g BaCl}_2}} = 0.0205 \text{ mol BaCl}_2$$

$$0.74 \cancel{\text{g H}_2\text{O}} \times \frac{1 \text{ mol H}_2\text{O}}{18.02 \cancel{\text{g H}_2\text{O}}} = 0.041 \text{ mol H}_2\text{O}$$

Now that the moles of $BaCl_2$ and H_2O have been determined, you can calculate the ratio of moles of H_2O to moles of $BaCl_2$ which is *x*, the coefficient that precedes H_2O in the formula for the hydrate.

$$x = \frac{\text{moles H}_2\text{O}}{\text{moles BaCl}_2} = \frac{0.041 \text{ mol H}_2\text{O}}{0.0205 \text{ mol BaCl}_2} = \frac{2.0 \text{ mol H}_2\text{O}}{1.00 \text{ mol BaCl}_2} = \frac{2}{1}$$

The ratio of moles of H_2O to moles of $BaCl_2$ is 2:1, so 2 mol of water is associated with 1 mol of barium chloride. The value of the coefficient *x* is 2 and the formula of the hydrate is $BaCl_2 \cdot 2H_2O$. What is the name of the hydrate? The ChemLab at the end of this chapter will give you practice in experimentally determining the formula of a hydrate.

☑ **READING CHECK** **Explain** why a dot is used in writing the formula of a hydrate.

CHEMISTRY JOURNAL

Hydrates Have students pick one of the hydrates in Table 1 for investigation. Have them determine the molar mass and the percent composition in terms of the percent anhydrous compound and the percent water. Students can record their calculations in their journals. **OL**

EXAMPLE Problem 14

Find help with **ratios.** Math Handbook

Determining the Formula of a Hydrate A mass of 2.50 g of blue, hydrated copper sulfate ($CuSO_4 \cdot xH_2O$) is placed in a crucible and heated. After heating, 1.59 g of white anhydrous copper sulfate ($CuSO_4$) remains. What is the formula for the hydrate? Name the hydrate.

1 ANALYZE THE PROBLEM

You are given a mass of hydrated copper sulfate. The mass after heating is the mass of the anhydrous compound. You know the formula for the compound, except for *x*, the number of moles of water of hydration.

Known

mass of hydrated compound = 2.50 g $CuSO_4 \cdot xH_2O$
mass of anhydrous compound = 1.59 g $CuSO_4$
molar mass H_2O = 18.02 g/mol H_2O
molar mass $CuSO_4$ = 159.6 g/mol $CuSO_4$

Unknown

formula of hydrate = ?
name of hydrate = ?

2 SOLVE FOR THE UNKNOWN

Determine the mass of water lost.

mass of hydrated copper sulfate	2.50 g
mass of anhydrous copper sulfate	−1.59 g
mass of water lost	0.91 g

Subtract the mass of anhydrous $CuSO_4$ from the mass of $CuSO_4 \cdot xH_2O$.

Convert the known masses of H_2O and anhydrous $CuSO_4$ to moles using a conversion factor—the inverse of molar mass—that relates moles and mass.

$$1.59\ \cancel{\text{g CuSO}_4} \times \frac{1\ \text{mol CuSO}_4}{159.6\ \cancel{\text{g CuSO}_4}} = 0.00996\ \text{mol CuSO}_4$$

Substitute mass $CuSO_4$ = 1.59 g, inverse molar mass $CuSO_4$ = 1 mol/159.6 g, and solve.

$$0.91\ \cancel{\text{g H}_2\text{O}} \times \frac{1\ \text{mol H}_2\text{O}}{18.02\ \cancel{\text{g H}_2\text{O}}} = 0.050\ \text{mol H}_2\text{O}$$

Substitute mass H_2O = 0.91 g, inverse molar mass H_2O = 1 mol/18.02 g, and solve.

$$x = \frac{\text{moles H}_2\text{O}}{\text{moles CuSO}_4}$$

State the ratio of moles of H_2O to moles of $CuSO_4$.

$$x = \frac{0.050\ \text{mol H}_2\text{O}}{0.00996\ \text{mol CuSO}_4} \approx \frac{5.0\ \text{mol H}_2\text{O}}{1\ \text{mol CuSO}_4} = 5$$

Substitute moles of H_2O = 0.050 mol, moles of $CuSO_4$ = 0.00996 mol. Divide numbers, and cancel units to determine the simplest whole-number ratio.

The ratio of H_2O to $CuSO_4$ is 5:1, so the formula for the hydrate is **$CuSO_4 \cdot 5H_2O$.**
The name of the hydrate is **copper(II) sulfate pentahydrate.**

3 EVALUATE THE ANSWER

Copper(II) sulfate pentahydrate is a common hydrate listed in **Table 1.**

PRACTICE Problems

Do additional problems. Online Practice

74. The composition of a hydrate is given in the circle graph shown at the right. What is the formula and name of this hydrate?

75. Challenge An 11.75-g sample of a common hydrate of cobalt(II) chloride is heated. After heating, 0.0712 mol of anhydrous cobalt chloride remains. What is the formula and the name of this hydrate?

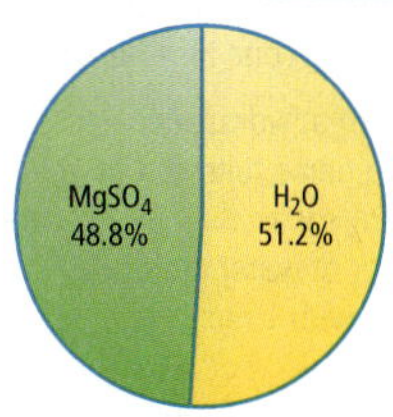

CHEMISTRY PROJECT

Analyzing a Hydrate Soil samples found on Mars by NASA have indicated the presence of magnesium sulfate heptahydrate, calcium sulfate dihydrate, and lithium sulfate monohydrate. These hydrates might prove to benefit future exploration on Mars because they represent stored water that can be extracted from the rocks on the surface of Mars. Have students determine the mass of water present in one mole of each of the above compounds and describe how the water could be extracted and stored for future use. Ask students to use diagrams to support their answers. OL

IN-CLASS Example

Question The hydrated form of nickel(II) chloride is a green compound commonly found in the chemistry laboratory. If a 5.00 g sample of hydrated nickel(II) chloride loses 2.27 g of water when heated, what is the formula of the hydrate? Name the hydrate.

Answer
Mass of hydrated compound = 5.00 g
Mass of water lost = 2.27 g
Mass of anhydrous compound = 2.73 g
0.0211 mol $NiCl_2$
0.126 mol H_2O
0.126 mol H_2O/0.0211 mol $NiCl_2$ =
6 $NiCl_2 \cdot 6H_2O$
Nickel(II) chloride hexahydrate

Apply Chemistry

Gypsum, Heat, and Water

Calcium sulfate dihydrate ($CaSO_4 \cdot 2H_2O$) is also known as gypsum. When gypsum is heated and ground into a powder, plaster of paris [$CaSO_4 \cdot (1/2)H_2O$] results. When water is added to plaster of paris, the resulting gypsum paste can be molded into a desired shape and allowed to harden. As the excess water evaporates, hard gypsum forms. Ask students what happens when water is added to the plaster of paris. It hydrates and turns back into gypsum. Anhydrous calcium sulfate is commonly called anhydrite, meaning *without water.* Although large deposits of anhydrite exist, they are of little economic value. However, experiments are ongoing to convert these deposits to valuable gypsum. Ask students what task the scientists are trying to do. They are trying to hydrate the anhydrous compound. OL

PRACTICE Problems

Have students refer to the Selected Solutions appendix for complete solutions to odd-numbered problems. The complete solutions for all problems can be found in the Solutions Manual.

74. $MgSO_4 \cdot 7H_2O$; magnesium sulfate heptahydrate.
75. $CoCl_2 \cdot 2H_2O$; cobalt(II) chloride dihydrate

3 Assess

Check for Understanding

Ask students why a desiccant such as $CaCl_2$ is heated from time to time. to remove absorbed water OL

Reteach

Soak two pieces of filter paper in a strong solution of cobalt chloride and allow the paper to dry. **WARNING:** ***Wear gloves. $CoCl_2$ is a skin irritant and moderately toxic if ingested.*** Place one piece of the dried paper in a humid environment. It will turn pink. Place another piece in a beaker and heat it for a few minutes. It will turn blue. Ask students to explain their observations. The cobalt chloride forms a pink hydrate when it absorbs water. When heated, the hydrate loses water and turns blue. OL

Extension

Tell students that equal molar amounts of barium hydroxide and a hydrate of barium hydroxide have masses of 21.4 g and 39.4 g, respectively. Ask them to find the formula for the hydrate. $Ba(OH)_2{\cdot}8H_2O$ OL

ASSESSMENT

Performance Give each student a formula for a hydrate from Table 1 and have them determine the percent ionic compound and the percent water in the compound. Have each student write a problem providing the calculated data and asking for the formula for the hydrate. Students can exchange problems. OL COOP LEARN

Figure 18 Calcium chloride, in the bottom of the desiccator, keeps the air inside the desiccator dry. In the chemistry lab, calcium chloride can also be packed into glass tubes called drying tubes. Drying tubes protect reactions from atmospheric moisture, but allow gases produced by reactions to escape.

Uses of Hydrates

Anhydrous compounds have important applications in the chemistry laboratory. Calcium chloride forms three hydrates—a monohydrate, a dihydrate, and a hexahydrate. As shown in **Figure 18,** anhydrous calcium chloride is placed in the bottom of tightly sealed containers called desiccators. The calcium chloride absorbs moisture from the air inside the desiccator, creating a dry atmosphere in which other substances can be kept dry. Calcium sulfate is often added to solvents such as ethanol and ethyl ether to keep them free of water.

The ability of the anhydrous form of a hydrate to absorb water also has some important commercial applications. Electronic and optical equipment, particularly equipment that is transported overseas by ship, is often packaged with packets of desiccant. Desiccants prevent moisture from interfering with the sensitive electronic circuitry. While some types of desiccant simply absorb moisture, other types bond with moisture from the air and form hydrates.

Some hydrates, sodium sulfate decahydrate ($Na_2SO_4{\cdot}10H_2O$) for example, are used to store solar energy. When the Sun's energy heats the hydrate to a temperature greater than 32°C, the single formula unit of Na_2SO_4 in the hydrate dissolves in the 10 mol of water of hydration. In the process, energy is absorbed by the hydrate. This energy is released when the temperature decreases and the hydrate crystallizes again.

SECTION 5 REVIEW

Section Self-Check

Section Summary

- The formula of a hydrate consists of the formula of the ionic compound and the number of water molecules associated with one formula unit.
- The name of a hydrate consists of the compound name followed by the word *hydrate* with a prefix indicating the number of water molecules associated with 1 mol of the compound.
- Anhydrous compounds are formed when hydrates are heated.

76. MAINIDEA **Summarize** the composition of a hydrate.
77. **Name** the compound that has the formula $SrCl_2{\cdot}6H_2O$.
78. **Describe** the experimental procedure for determining the formula of a hydrate. Explain the reason for each step.
79. **Apply** A hydrate contains 0.050 mol of H_2O to every 0.00998 mol of ionic compound. Write a generalized formula of the hydrate.
80. **Calculate** the mass of the water of hydration if a hydrate loses 0.025 mol of H_2O when heated.
81. **Arrange** these hydrates in order of increasing percent water content: $MgSO_4{\cdot}7H_2O$, $Ba(OH)_2{\cdot}8H_2O$, and $CoCl_2{\cdot}6H_2O$.
82. **Apply** Explain how the hydrate in **Figure 17** might be used as a means of roughly determining the probability of rain.

Matt Meadows

SECTION 5 REVIEW

76. A hydrate is a solid ionic compound in which water molecules are trapped.
77. strontium chloride hexahydrate
78. Mass an empty crucible. Add some hydrate and remass. Heat the crucible to drive out the water. Cool and remass. Determine the moles of the anhydrous compound. Subtract the mass of the crucible after heating from the mass of the crucible with the hydrate. The difference is the mass of the water lost. Determine the moles of water. Determine the simplest whole-number ratio of moles of water to moles of anhydrous compound, which will yield the formula of the hydrate.
79. $XY{\cdot}5H_2O$, where XY represents the ionic compound
80. water of hydration = 0.45 g H_2O
81. $CoCl_2{\cdot}6H_2O$; $Ba(OH)_2{\cdot}8H_2O$; $MgSO_4{\cdot}7H_2O$
82. The hydrate is pink in moist air.

everyday CHEMISTRY

History In a Glass of Water

Recall the last glass of water you drank. Although it seems unbelievable, that glass of water almost certainly contained water molecules that were also consumed by Albert Einstein, Joan of Arc, or Confucius! Just how can two glasses of water poured at different times in history contain some of the same molecules? Avogadro's number and molar calculations tell the story.

Oceans and moles The total mass of the water in Earth's oceans and from a variety of other sources is approximately 1.4×10^{24} g. In contrast, an 8-fluid ounce glass of water contains about 2.3×10^2 g, or 230 g, of water. Using this data, you can calculate the total number of glasses of water available on Earth to drink, and the total number of water molecules contained in those glasses.

You know that one mol of water has a mass of about 18 g. Using dimensional analysis you can convert the grams of water in a glass to moles.

$$\frac{230\ \cancel{\text{g water}}}{\text{glass}} \times \frac{1\ \text{mol water}}{18\ \cancel{\text{g water}}} \approx 13\ \text{mol water/glass}$$

Thus, one glass of water contains around 13 moles of water. Now convert moles of water to molecules of water by using Avogadro's number.

$$\frac{13\ \cancel{\text{mol water}}}{\text{glass}} \times \frac{6 \times 10^{23}\ \text{molecules water}}{1\ \cancel{\text{mol water}}} \approx 8 \times 10^{24}\ \text{molecules water/glass}$$

Because you know the total mass of water and the mass of water per glass, you can calculate the total number of glasses of water available for drinking.

$$1.4 \times 10^{24}\ \cancel{\text{g water}} \times \frac{1\ \text{glass}}{230\ \cancel{\text{g water}}} \approx 6 \times 10^{21}\ \text{glasses}$$

So, there are 8×10^{24} molecules in a single glass of water and there are 6×10^{21} glasses of water on Earth. Comparing these numbers, you can see that there are about 1000 times more molecules in a single glass of water than there are glasses of water on Earth!

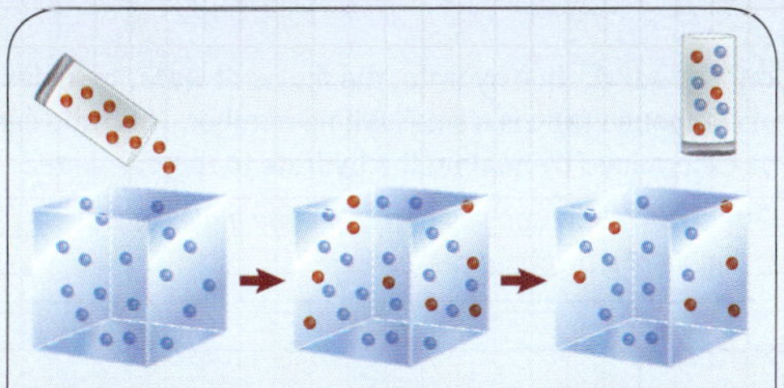

Figure 1 Molecules from the first glass of water (red) are poured back into a container that holds all of Earth's water molecules (blue). A second glass of water taken from the container contains a small number of water molecules that were also in the first glass.

Giant container Suppose all the water on Earth was stored in a single, cube-shaped container. It would be enormous, with sides about 1100 km long! Imagine filling your glass with water from the container. Pour the water back into the container and wait for the water to mix completely. Then refill your glass. Would any of the molecules from the first glass be found in the second glass?

As shown in **Figure 1,** it is likely that the two glasses will share some number of water molecules. Why? Because there are 1000 times more molecules in a glass than there are glasses in the container, on average, the second glass will contain about 1000 molecules that were also in the first glass. This is true for any two glasses.

The power of big numbers Now, consider the amount of water—much more than a single glass—that passed through Einstein, Joan of Arc, or Confucius in their lifetimes. Assuming the molecules of water mixed evenly throughout the entire volume of Earth's water, you can understand how every glass of water must contain some of those same molecules.

WRITING IN ▸ Chemistry

Estimate The estimating process used in this article is sometimes called a "back-of-the-envelope" calculation. Use this method to estimate the total mass of all of the students in your school.

WebQuest

everyday CHEMISTRY

Purpose

Students will see how Avogadro's number and the power of estimation can lead to some startling conclusions.

Background

Avogadro's number can be seen as a portal linking the microscopic world of atoms and molecules to the macroscopic world of paperclips, spoons, and glasses of water. It is interesting to note that one mole of any "large" object (even something as small as a bacterium) would be too large to deal with on a practical basis. One mole of bacteria, for instance, would weigh several million metric tons.

Teaching Strategies

- Scientists often use scientific notation to express very large numbers. To help students understand the need for scientific notation, try writing out some of the numbers discussed in the article (for instance, the molecules in a glass of water) first without, and then with, scientific notation.
- Try several "back of the envelope" calculations in class. Some examples might be: the amount of oil changed at an oil and lube shop in a year, the number of oxygen molecules you breathe in each day, or the amount of graphite used on all the standardized tests taken in a school year.

WRITING IN ▸ Chemistry

Rubric WebQuest

Estimate Suppose a school has a student body of 500, and that the average mass of each student is about 60 kg. The total mass of the students is calculated as follows. 500 students $\times$ 60 kg/student $=$ 30,000 kg

ChemLAB

Rubric

Preparation

Time Allotment one class period

Process Skills acquire and analyze information, interpret data, hypothesize, use numbers

Safety Precautions

- Approve lab safety forms before work begins.
- Advise students to use Bunsen burners with care and that a crucible can be hot even though it does not look hot.
- Epsom salts is a mild skin irritant and an eye and respiratory irritant. Among the alternative materials, $BaCl_2$ is highly toxic. $CuSO_4$ is toxic and is a skin and respiratory irritant. Review the MSDS with students prior to the activity.

Disposal Anhydrous Epsom salts can be put into the trash. The Epsom salts can also be reclaimed by allowing the crystals to hydrate with water vapor from the air.

Alternative Materials A hot plate and an evaporating dish can be used in place of the Bunsen burner and crucible; however, this method will not remove all the water from the Epsom salts. The likely result of this method is $MgSO_4{\cdot}3H_2O$. Other hydrates, such as $BaCl_2{\cdot}2H_2O$, $Na_2CO_3{\cdot}H_2O$, and $CuSO_4{\cdot}5H_2O$, can be used. Epsom salts does not produce fumes with heating.

Procedure

- Demonstrate heating the crucible with a low flame and gradually increase the intensity.
- Epsom salts do not require the crucible to get red hot, but heating should be enough to completely remove the water of hydration.
- Students will be able to see water boiling out of the Epsom salts. They will know the water is removed when the boiling stops.

ChemLAB

Determine the Formula of a Hydrate

Background: **In a hydrate, the moles of water to moles of compound ratio is a small whole number. This ratio can be determined by heating the hydrate to remove water.**

Question: *How can you determine the moles of water in a mole of a hydrated compound?*

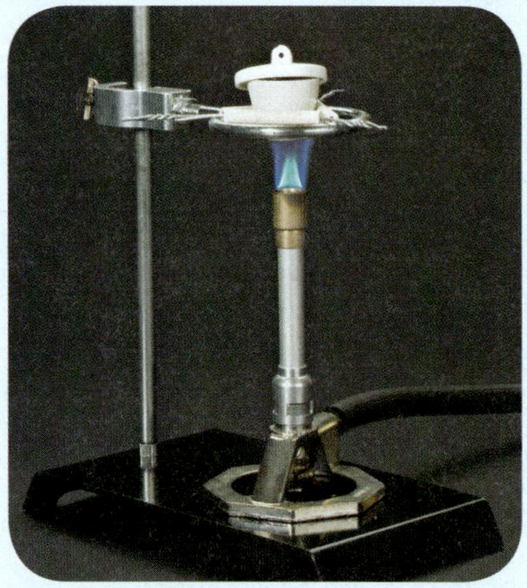

Materials

Bunsen burner
ring stand and ring
crucible and lid
clay triangle
crucible tongs
balance
Epsom salts (hydrated $MgSO_4$)
spatula
spark lighter or matches

Safety Precautions

WARNING: ***Turn off the Bunsen burner when not in use. Crucible, lid, and triangle will be hot and can burn skin. Do not inhale fumes–they are respiratory irritants.***

Procedure

1. Read and complete the lab safety form.
2. Prepare a data table.
3. Measure the mass of the crucible and its lid to the nearest 0.01 g.
4. Add about 3 g hydrated $MgSO_4$ to the crucible. Measure the mass of the crucible, lid, and hydrate to the nearest 0.01 g.
5. Record your observations of the hydrate.
6. Place the triangle on the ring of the ring stand. Adjust the ring stand so the triangle will be positioned near the tip of the Bunsen burner's flame. Do not light the Bunsen burner yet.
7. Carefully place the crucible in the triangle with its lid slightly ajar.
8. Begin heating with a low flame, then gradually progress to a stronger flame. Heat for about 10 min, then turn off the burner.
9. Use tongs to carefully remove the crucible from the triangle. Use tongs to place the lid on the crucible. Allow everything to cool.
10. Measure the mass of the crucible, lid, and $MgSO_4$.
11. Record your observations of the anhydrous $MgSO_4$.
12. **Cleanup and Disposal** Discard the anhydrous $MgSO_4$ as directed by your teacher. Return all lab equipment to its proper place and clean your station.

Analyze and Conclude

1. **Calculate** Use your experimental data to calculate the formula for hydrated $MgSO_4$.
2. **Observe and Infer** How do the appearances of the hydrated and anhydrous $MgSO_4$ crystals compare? How are they different?
3. **Conclude** Why might the method used not be suitable for determining the water of hydration for all hydrates?
4. **Error Analysis** If the hydrate's formula is $MgSO_4{\cdot}7H_2O$, what is the percent error in your formula for hydrated $MgSO_4$? What are the possible sources for the error? What procedural changes could you make to reduce the error?
5. **Predict** the result of leaving the anhydrous crystals uncovered overnight.

INQUIRY EXTENSION

Design an experiment to test whether a compound is hydrated or anhydrous.

Analyze and Conclude

1. $MgSO_4{\cdot}7H_2O$
2. The hydrated $MgSO_4$ is shiny and translucent. The anhydrous $MgSO_4$ is opaque and bright white.
3. Some hydrates may decompose when heated.
4. Answers will vary.
$$\left(\frac{(7.00 - 6.96)}{7.00}\right)100 = 0.571\%\ \text{error}$$
A possible source of error could be that not all of the water was completely removed from the hydrated $MgSO_4$.
5. The anhydrous $MgSO_4$ may absorb water.

INQUIRY EXTENSION

Answers will vary, but student experiments will likely involve measuring initial compound mass, heating, and measuring the mass again after heating. Accept all reasonable answers.

CHAPTER 10 STUDY GUIDE

Vocabulary Practice

BIGIDEA The mole represents a large number of extremely small particles.

SECTION 1 Measuring Matter

MAINIDEA Chemists use the mole to count atoms, molecules, ions, and formula units.

- The mole is a unit used to count particles of matter indirectly. One mole of a pure substance contains Avogadro's number of representative particles.
- Representative particles include atoms, ions, molecules, formula units, electrons, and other similar particles.
- One mole of carbon-12 atoms has a mass of exactly 12 g.
- Conversion factors written from Avogadro's relationship can be used to convert between moles and number of representative particles.

VOCABULARY
- mole
- Avogadro's number

SECTION 2 Mass and the Mole

MAINIDEA A mole always contains the same number of particles; however, moles of different substances have different masses.

- The mass in grams of 1 mol of any pure substance is called its molar mass.
- The molar mass of an element is numerically equal to its atomic mass.
- The molar mass of any substance is the mass in grams of Avogadro's number of representative particles of the substance.
- Molar mass is used to convert from moles to mass. The inverse of molar mass is used to convert from mass to moles.

VOCABULARY
- molar mass

SECTION 3 Moles of Compounds

MAINIDEA The molar mass of a compound can be calculated from its chemical formula and can be used to convert from mass to moles of that compound.

- Subscripts in a chemical formula indicate how many moles of each element are present in 1 mol of the compound.
- The molar mass of a compound is calculated from the molar masses of all of the elements in the compound.
- Conversion factors based on a compound's molar mass are used to convert between moles and mass of a compound.

SECTION 4 Empirical and Molecular Formulas

MAINIDEA A molecular formula of a compound is a whole-number multiple of its empirical formula.

- The percent by mass of an element in a compound gives the percentage of the compound's total mass due to that element.
- The subscripts in an empirical formula give the smallest whole-number ratio of moles of elements in the compound.
- The molecular formula gives the actual number of atoms of each element in a molecule or formula unit of a substance.
- The molecular formula is a whole-number multiple of the empirical formula.

VOCABULARY
- percent composition
- empirical formula
- molecular formula

SECTION 5 Formulas of Hydrates

MAINIDEA Hydrates are solid ionic compounds in which water molecules are trapped.

- The formula of a hydrate consists of the formula of the ionic compound and the number of water molecules associated with one formula unit.
- The name of a hydrate consists of the compound name and the word *hydrate* with a prefix indicating the number of water molecules in 1 mol of the compound.
- Anhydrous compounds are formed when hydrates are heated.

VOCABULARY
- hydrate

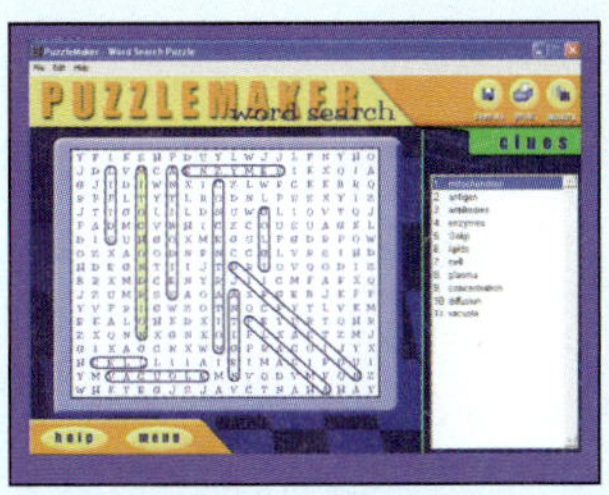

Vocabulary Practice

For additional practice with vocabulary, have students access the Vocabulary PuzzleMaker.

Use the Vocabulary

To reinforce chapter vocabulary, have students write a sentence using each term. OL EL

Review Strategies

- Have students give an example of each type of mole conversion. OL
- Have students summarize how to calculate the percent composition of a compound, how to determine the empirical and molecular formulas of a compound, and how to determine the formula of a hydrate. OL
- Problems from the Supplemental Practice Problems appendix or the Supplemental Problems booklet can be used for review. OL

Students can review with these online resources.

Vocabulary Practice

Vocabulary eGames
Vocabulary eFlashcards
Vocabulary PuzzleMaker

Section Self-Check

Chapter Self-Check

Online Test Practice

Use *eAssessment* to:

- create multiple versions of tests
- edit existing questions and add your own questions
- build tests aligned with select state curriculum standards using built-in state curriculum tags
- track students' progress

SECTION 1

Mastering Concepts

83. 6.02×10^{23}

84. 6.02×10^{23} atoms

85. There are 6.02×10^{23} particles of both elements. Silver contains 47 protons, 47 electrons and 61 neutrons per atom, while platinum contains 78 protons, 78 electrons and 117 neutrons per atom. A mole of platinum would contain more protons, electrons and neutrons but the same number of atoms.

86. A mole allows a chemist to accurately measure the number of atoms, molecules, or formula units in a substance.

87. One roll of pennies always contains 50 pennies, and one roll of dimes always contains 50 dimes. Each roll contains a specific number of coins. A mole also contains a specific number of particles, 6.02×10^{23} particles.

88. Avogadro's number is the number of particles in one mole of a substance. It can be used to convert particles to moles or moles to particles.

89. Flowcharts will vary but should clearly show the use of proper conversion factors.

Mastering Problems

90. a. 1.51×10^{23} atoms
b. 5.15×10^{21} formula units
c. 2.13×10^{25} molecules
d. 2.56×10^{23} molecules

91. a. 2.68×10^{24} molecules
b. 1.51×10^{23} formula units
c. 1.35×10^{24} molecules
d. 5.76×10^{24} atoms

92. a. 8.13×10^{23} molecules
b. 1.53×10^{23} molecules
c. 7.53×10^{23} molecules
d. 9.030×10^{25} molecules

93. a. 5.40×10^{-4} mol
b. 8.24 mol
c. 2.59×10^{-1} mol
d. 2.08×10^{1} mol

94. a. 2.51×10^{-9} mol
b. 2.56×10^{22} molecules
c. 1.49×10^{2} mol
d. 3.55×10^{24} atoms

CHAPTER 10 ASSESSMENT

SECTION 1

Mastering Concepts

83. What is the numerical value of Avogadro's number?

84. How many atoms of potassium does 1 mol of potassium contain?

85. Compare a mole of Ag-108 and a mole of Pt-195 using atoms, protons, electrons, and neutrons.

86. Why is the mole an important unit to chemists?

87. Currency Examine the information in **Table 2** and explain how rolls used to count pennies and dimes are similar to moles.

Table 2 Rolled-Coin Values

Coin	Value of a Roll of Coins
Penny	\$0.50
Dime	\$5.00

88. Explain how Avogadro's number is used as a conversion factor.

89. Conversion Design a flowchart that could be used to help convert particles to moles or moles to particles.

Mastering Problems

90. Determine the number of representative particles in each substance.
a. 0.250 mol of silver
b. 8.56×10^{-3} mol of sodium chloride
c. 35.3 mol of carbon dioxide
d. 0.425 mol of nitrogen (N_2)

91. Determine the number of representative particles in each substance.
a. 4.45 mol of $C_6H_{12}O_6$
b. 0.250 mol of KNO_3
c. 2.24 mol of H_2
d. 9.56 mol of Zn

92. How many molecules are contained in each compound?
a. 1.35 mol of carbon disulfide (CS_2)
b. 0.254 mol of diarsenic trioxide (As_2O_3)
c. 1.25 mol of water
d. 150.0 mol of HCl

93. Determine the number of moles in each substance.
a. 3.25×10^{20} atoms of lead
b. 4.96×10^{24} molecules of glucose
c. 1.56×10^{23} formula units of sodium hydroxide
d. 1.25×10^{25} copper(II) ions

94. Perform the following conversions.
a. 1.51×10^{15} atoms of Si to mol of Si
b. 4.25×10^{-2} mol of H_2SO_4 to molecules of H_2SO_4
c. 8.95×10^{25} molecules of CCl_4 to mol of CCl_4
d. 5.90 mol of Ca to atoms of Ca

95. How many moles contain the given quantity?
a. 1.25×10^{15} molecules of carbon dioxide
b. 3.59×10^{21} formula units of sodium nitrate
c. 2.89×10^{27} formula units of calcium carbonate

96. RDA of Selenium The recommended daily allowance (RDA) of selenium in your diet is 6.97×10^{-7} mol. How many atoms of selenium is this?

■ **Figure 19**

97. The two solutions shown in **Figure 19** are mixed. What is the total number of metal ions in the mixture?

98. Jewelry A bracelet containing 0.200 mol metal atoms is 75% gold. How many particles of gold atoms are in the bracelet?

99. Snowflakes A snowflake contains 1.9×10^{18} molecules of water. How many moles of water does it contain?

100. If you could count two atoms every second, how long would it take you to count a mole of atoms? Assume that you counted continually for 24 hours every day. How does the time you calculated compare with the age of Earth, which is estimated to be 4.5×10^{9} years old?

101. Chlorophyll The green color of leaves is due to the presence of chlorophyll, $C_{55}H_{72}O_5N_4Mg$. A fresh leaf was found to have 1.5×10^{-5} mol of chlorophyll per cm^2. How many chlorophyll molecules are in 1 cm^2?

SECTION 2

Mastering Concepts

102. Explain the difference between atomic mass (amu) and molar mass (g).

103. Which contains more atoms, a mole of silver atoms or a mole of gold atoms? Explain your answer.

104. Which has more mass, a mole of potassium or a mole of sodium? Explain your answer.

105. Explain how you would convert from number of atoms of a specific element to its mass.

106. Discuss the relationships that exist between the mole, molar mass, and Avogadro's number.

107. Barbed Wire Barbed wire is often made of steel, which is primarily iron, and coated with zinc. Compare the number of particles and the mass of 1 mol of each.

95. a. 2.08×10^{-9} mol
b. 5.96×10^{-3} mol
c. 4.80×10^{3} mol

96. 4.20×10^{17} atoms

97. 2.29×10^{23} Cu^{2+} and Ca^{2+} ions

98. 9.03×10^{22} atoms Au

99. 3.2×10^{-6} mol

100. 9.5×10^{15} yr; about 2 million times longer

101. 9.03×10^{18} chlorophyll molecules/cm^2

SECTION 2

Mastering Concepts

102. Atomic mass (amu) is the mass of an individual particle (atom, molecule). Molar mass (grams) is the mass of a mole of particles.

103. They both contain the same number of atoms because a mole of anything contains 6.02×10^{23} representative particles.

104. The molar mass of K is 39.098 g/mol; the molar mass of Na is 22.990 g/mol. Thus, a mole of K has a greater mass.

Chapter Self-Check

Mastering Problems

108. Calculate the mass of each element.
a. 5.22 mol of He **c.** 2.22 mol of Ti
b. 0.0455 mol of Ni **d.** 0.00566 mol of Ge

109. Perform the following conversions.
a. 3.50 mol of Li to g of Li
b. 7.65 g of Co to mol of Co
c. 5.62 g of Kr to mol of Kr
d. 0.0550 mol of As to g of As

110. Determine the mass in grams of each element.
a. 1.33×10^{22} mol of Sb **c.** 1.22×10^{23} mol of Ag
b. 4.75×10^{14} mol of Pt **d.** 9.85×10^{24} mol of Cr

111. Complete **Table 3.**

Table 3 Mass, Mole, and Particle Data

Mass	Moles	Particles
	3.65 mol Mg	
29.54 g Cr		
		3.54×10^{25} atoms P
	0.568 mol As	

112. Convert each to mass in grams.
a. 4.22×10^{15} atoms U
b. 8.65×10^{25} atoms H
c. 1.25×10^{22} atoms O
d. 4.44×10^{23} atoms Pb

113. Calculate the number of atoms in each element.
a. 25.8 g of Hg **c.** 150 g of Ar
b. 0.0340 g of Zn **d.** 0.124 g of Mg

114. Arrange from least to most in moles: 3.00×10^{24} atoms Ne, 4.25 mole Ar, 2.69×10^{24} atoms Xe, 65.96 g Kr.

115. Balance Precision A sensitive electronic balance can detect masses of 1×10^{-8} g. How many atoms of silver would be in a sample having this mass?

116. A sample of a compound contains 3.86 g of sulfur and 4.08 g of vanadium. How many atoms of sulfur and vanadium does the compound contain?

117. Which has more atoms, 10.0 g of C or 10.0 g of Ca? How many atoms does each have?

118. Which has more atoms, 10.0 mol of C or 10.0 mol of Ca? How many atoms does each have?

119. A mixture contains 0.250 mol of Fe and 1.20 g of C. What is the total number of atoms in the mixture?

120. Respiration Air contains several gases. Argon makes up 0.934% of the air. If a person takes a breath that contains 0.600 g of air, calculate the number of argon atoms inhaled.

SECTION 3

Mastering Concepts

121. What information is provided by the formula for potassium chromate (K_2CrO_4)?

122. In the formula for sodium phosphate (Na_3PO_4), how many moles of sodium are represented? How many moles of phosphorus? How many moles of oxygen?

123. Explain how you determine the molar mass of a compound.

124. Insect Repellent Many insect repellents use DEET as the active ingredient. DEET was patented in 1946 and is effective against many biting insects. What must you know to determine the molar mass of DEET?

125. Why can molar mass be used as a conversion factor?

126. List three conversion factors used in molar conversions.

127. Which of these contains the most moles of carbon atoms per mole of the compound: ascorbic acid ($C_6H_8O_6$), glycerin ($C_3H_8O_3$), or vanillin ($C_8H_8O_3$)? Explain.

Mastering Problems

128. How many moles of oxygen atoms are contained in each compound?
a. 2.50 mol of $KMnO_4$
b. 45.9 mol of CO_2
c. 1.25×10^{-2} mol of $CuSO_4{\cdot}5H_2O$

129. How many carbon tetrachloride (CCl_4) molecules are in 3.00 mol of CCl_4? How many carbon atoms? How many chlorine atoms? How many total atoms?

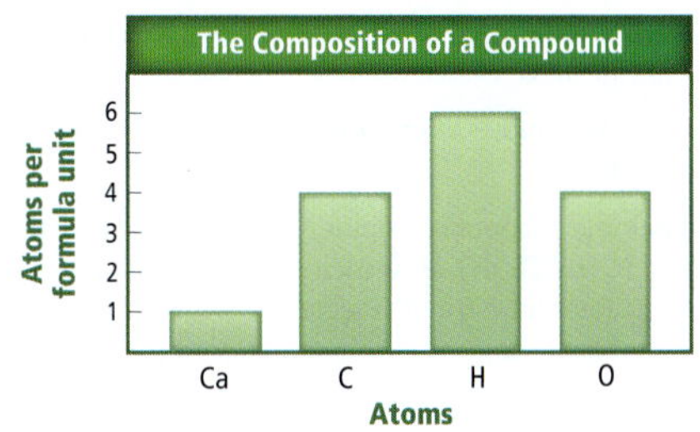

■ **Figure 20**

130. The graph in **Figure 20** shows the numbers of atoms of each element in a compound. What is the compound's formula? What is its molar mass?

131. Determine the molar mass of each compound.
a. nitric acid (HNO_3)
b. ammonium nitrate (NH_4NO_3)
c. zinc oxide (ZnO)
d. cobalt chloride ($CoCl_2$)

105. Convert the number of atoms to moles, then multiply moles by the molar mass of the element.

106. Molar mass is the mass in grams of one mole of any pure substance. Avogadro's number is the number of representative particles in one mole. The mass of 6.02×10^{23} representative particles of a substance is the molar mass of the substance.

107. The mole of iron and the mole of zinc both contain 6.02×10^{23} particles, but a mole of iron has a mass of 55.85 g/mol, and a mole of zinc has a mass of 65.39 g/mol.

Mastering Problems

108. a. 20.9 g **c.** 106 g
b. 2.67 g **d.** 0.411 g

109. a. 24.3 g **c.** 0.0671 mol
b. 0.130 mol **d.** 4.12 g

110. a. 1.62×10^{24} g
b. 9.27×10^{16} g
c. 1.32×10^{25} g
d. 5.12×10^{26} g

111.

Table 3	Mass, Mole, and Particle Data	
Mass	Moles	Particles
88.7 g Mg	3.65 mol Mg	2.20×10^{24} atoms Mg
29.54 g Cr	0.5681 mol Cr	3.420×10^{23} atoms Cr
1820 g P	58.8 mol P	3.54×10^{25} atoms P
42.6 g As	0.568 mol As	3.42×10^{23} atoms As

112. a. 1.98×10^{-6} g **c.** 0.332 g
b. 145 g **d.** 153 g

113. a. 7.74×10^{22} atoms
b. 3.23×10^{20} atoms
c. 2.3×10^{24} atoms
d. 3.07×10^{21} atoms

114. Kr, Ar, Xe, Ne

115. 6×10^{13} atoms

116. 7.25×10^{22} atoms S
4.82×10^{22} atoms V

117. 5.01×10^{23} atoms C; 1.50×10^{23} atoms Ca; 10.0 g C contains more atoms

118. Both samples have 6.02×10^{24} atoms.

119. 2.11×10^{23} total atoms

120. 8.44×10^{19} atoms Ar per breath

SECTION 3

Mastering Concepts

121. One mole of K_2CrO_4 contains two moles of K^+ ions and one mole of CrO_4^{2-} ions.

122. 3 mol Na, 1 mol P, 4 mol O

123. Molar mass is determined by multiplying the molar mass of each element present in the compound by its subscript and then adding these values.

124. You must know the chemical formula of the substance to determine the molar mass.

125. Molar mass is the mass of one mole of a compound. It can be used to convert moles of the compound to mass or mass of the compound to moles.

126. 1 mol/number of grams, 6.02×10^{23} representative particles/1 mol, 1 mol/6.02×10^{23} representative particles, number of grams/1 mol

127. vanillin ($C_8H_8O_3$); eight C atoms/molecule

Mastering Problems

128. a. 10.0 mol
b. 91.8 mol
c. 0.113 mol

129. 1.81×10^{24} molecules CCl_4; 1.81×10^{24} atoms C; 7.24×10^{24} atoms Cl; 9.05×10^{24} total atoms

130. $CaC_4H_6O_4$ [$Ca(C_2H_3O_2)_2$]; 158.17 g/mol

131. a. 63.02 g/mol **c.** 81.39 g/mol
b. 80.05 g/mol **d.** 129.83 g/mol

132. 114.21 g/mol
133. a. 2.27 mol
b. 3.12 mol
134. a. 6.05 g
b. 9.26×10^3 g
135. 8.48 g
136. 1650 g
137. 5.48×10^{-4} g
138. a. 2.35×10^{-4} mol ions
b. 0.300 mol ions
c. 8.75×10^{-3} mol ions
d. 3.00×10^{-9} mol ions
139. 1.082×10^{24} formula units
140. 1.07×10^{22} atoms
141. 1080 g
142. 6.14×10^{23} molecules
143. 21.1 g
144. 1295 mol
145. 1.25×10^{22} molecules
146. 4.904 mol
147. 1.359×10^{23} ions
148. 2×10^{21} molecules
149. 2.58×10^{23} ions
150. 6.84×10^{23} atoms
151. 4.12×10^{-3} mol C
5.15×10^{-3} mol H
2.06×10^{-3} mol N
1.03×10^{-3} mol O
152. 18.3 cm^3

SECTION 4

Mastering Concepts

153. Percent composition is the percent by mass of each element in a compound.
154. the percent composition of the compound
155. the percent composition of the compound and the molar mass
156. An empirical formula is the smallest whole-number ratio of elements that make up a compound (CH). A molecular formula specifies the actual number of atoms of each element in one molecule or formula unit of the substance (C_6H_6).
157. They are the same when the subscripts for each element present are the same. Na_2O is both the empirical and molecular formula for sodium oxide.
158. The chemist must have either an analysis of the elements that compose the molecule or the percent composition of the compound to determine the chemical formula.
159. NO_2 is the empirical formula and N_2O_4 is the molecular formula of the same compound. $(NO_2)_2 = N_2O_4$
160. Yes, for every pure substance, the percent by mass of each element is the same regardless of the size of the sample.

Mastering Problems

161. $BaCO_3$
162. hematite

132. Garlic Determine the molar mass of allyl sulfide, the compound responsible for the smell of garlic. The chemical formula of allyl sulfide is $(C_3H_5)_2S$.

133. How many moles are in 100.0 g of each compound?
a. dinitrogen oxide (N_2O)
b. methanol (CH_3OH)

134. What is the mass of each compound?
a. 4.50×10^{-2} mol of $CuCl_2$
b. 1.25×10^2 mol of $Ca(OH)_2$

135. Acne Benzoyl peroxide ($C_{14}H_{10}O_4$) is a substance used as an acne medicine. What is the mass in grams of 3.50×10^{-2} mol $C_{14}H_{10}O_4$?

136. Glass Etching Hydrofluoric acid is a substance used to etch glass. Determine the mass of 4.95×10^{25} HF molecules.

137. What is the mass of a mole of electrons if one electron has a mass of 9.11×10^{-28} g?

138. How many moles of ions are in each compound?
a. 0.0200 g of $AgNO_3$
b. 0.100 mol of K_2CrO_4
c. 0.500 g of $Ba(OH)_2$
d. 1.00×10^{-9} mol of Na_2CO_3

139. How many formula units are present in 500.0 g of lead(II) chloride?

140. Determine the number of atoms in 3.50 g of gold.

141. Calculate the mass of 3.62×10^{24} molecules of glucose ($C_6H_{12}O_6$).

142. Determine the number of molecules of ethanol (C_2H_5OH) in 47.0 g.

143. What mass of iron(III) chloride contains 2.35×10^{23} chloride ions?

144. How many moles of iron can be recovered from 100.0 kg of Fe_3O_4?

145. Cooking A common cooking vinegar is 5.0% acetic acid (CH_3COOH). How many molecules of acetic acid are present in 25.0 g of vinegar?

146. Calculate the moles of aluminum ions present in 250.0 g of aluminum oxide (Al_2O_3).

147. Determine the number of chloride ions in 10.75 g of magnesium chloride.

148. Pain Relief Acetaminophen, a common aspirin substitute, has the formula $C_8H_9NO_2$. Determine the number of molecules of acetaminophen in a 500-mg tablet.

149. Calculate the number of sodium ions present in 25.0 g of sodium chloride.

150. Determine the number of oxygen atoms present in 25.0 g of carbon dioxide.

151. Espresso There is 1.00×10^2 mg of caffeine in a shot of espresso. The chemical formula of caffeine is $C_8H_{10}N_4O_2$. Determine the moles of each element present in the caffeine in one shot of espresso.

152. The density of lead (Pb) is 11.3 g/cm^3. Calculate the volume of 1 mol of Pb.

SECTION 4

Mastering Concepts

153. Explain what is meant by percent composition.

154. What information must a chemist obtain in order to determine the empirical formula of an unknown compound?

155. What information must a chemist have to determine the molecular formula for a compound?

156. What is the difference between an empirical formula and a molecular formula? Provide an example.

157. When can the empirical formula be the same as the molecular formula?

158. Antibacterial Soap Triclosan is an antibacterial agent included in detergents, dish soaps, laundry soaps, deodorants, cosmetics, lotions, creams, toothpastes, and mouthwashes. The chemical formula for triclosan is $C_{12}H_7Cl_3O_2$. What information did the chemist need to determine this formula?

159. Which of the following formulas—NO, N_2O, NO_2, N_2O_4, and N_2O_5—represent the empirical and molecular formulas of the same compound? Explain your answer.

160. Do all pure samples of a given compound have the same percent composition? Explain.

Mastering Problems

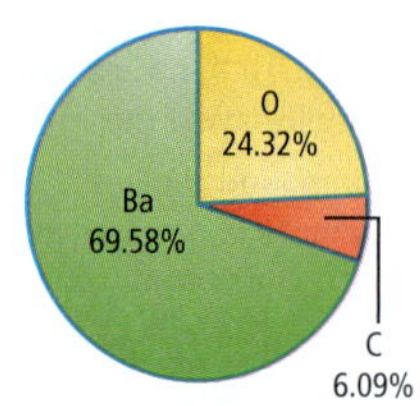

■ **Figure 21**

161. The circle graph in **Figure 21** shows the percent composition of a compound containing barium, carbon, and oxygen. What is the empirical formula of this compound?

162. Iron Three naturally occurring iron compounds are pyrite (FeS_2), hematite (Fe_2O_3), and siderite ($FeCO_3$). Which contains the greatest percentage of iron?

Chapter Self-Check

163. Express the composition of each compound as the mass percent of its elements (percent composition).
a. sucrose ($C_{12}H_{22}O_{11}$) **c.** magnetite (Fe_3O_4)
b. aluminum sulfate ($Al_2(SO_4)_3$)

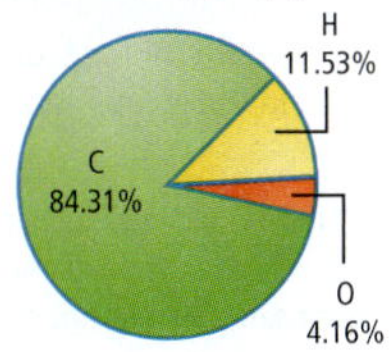

Molar mass = 384 g/mol

■ **Figure 22**

164. Vitamin D_3 Your body's ability to absorb calcium is aided by vitamin D_3. Chemical analysis of vitamin D_3 yields the data shown in **Figure 22.** What are the empirical and molecular formulas for vitamin D_3?

165. When a 30.98-g sample of phosphorus reacts with oxygen, a 71.00-g sample of phosphorus oxide is formed. What is the percent composition of the compound? What is the empirical formula for this compound?

166. Cholesterol Heart disease is linked to high blood cholesterol levels. What is the percent composition of the elements in a molecule of cholesterol ($C_{27}H_{45}OH$)?

167. Determine the empirical formula for each compound.
a. ethylene (C_2H_4)
b. ascorbic acid ($C_6H_8O_6$)
c. naphthalene ($C_{10}H_8$)

168. Caffeine The stimulant effect of coffee is due to caffeine, $C_8H_{10}N_4O_2$. Calculate the molar mass of caffeine. Determine its percent composition.

169. Which titanium-containing mineral, rutile (TiO_2) or ilmenite ($FeTiO_3$), has the larger percentage of titanium?

170. Vitamin E Many plants contain vitamin E ($C_{29}H_{50}O_2$), a substance that some think slows the aging process in humans. What is the percent composition of vitamin E?

171. Artificial Sweetener Determine the percent composition of aspartame ($C_{14}H_{18}N_2O_5$), an artificial sweetener.

172. MSG Monosodium glutamate, known as MSG, is sometimes added to food to enhance flavor. Analysis determined this compound to be 35.5% C, 4.77% H, 8.29% N, 13.6% Na, and 37.9% O. What is its empirical formula?

173. What is the empirical formula of a compound that contains 10.52 g Ni, 4.38 g C, and 5.10 g N?

174. Patina The Statue of Liberty has turned green because of the formation of a patina. Two copper compounds, $Cu_3(OH)_4SO_4$ and $Cu_4(OH)_6SO_4$, form this patina. Find the mass percentage of copper in each compound.

SECTION 5

Mastering Concepts

175. What is a hydrated compound? Use an example to illustrate your answer.

176. Explain how hydrates are named.

177. Desiccants Why are certain electronic devices transported with desiccants?

178. In a laboratory setting, how would you determine if a compound was a hydrate?

179. Write the formula for the following hydrates.
a. nickel(II) chloride hexahydrate
b. cobalt(II) chloride hexahydrate
c. magnesium carbonate pentahydrate
d. sodium sulfate decahydrate

Mastering Problems

180. Determine the mass percent of anhydrous sodium carbonate (Na_2CO_3) and water in sodium carbonate decahydrate ($Na_2CO_3{\cdot}10H_2O$).

181. Table 4 shows data from an experiment to determine the formulas of hydrated barium chloride. Determine the formula for the hydrate and its name.

Table 4 Data for $BaCl_2{\cdot}xH_2O$

Mass of empty crucible	21.30 g
Mass of hydrate + crucible	31.35 g
Initial mass of hydrate	
Mass after heating 5 min	29.87 g
Mass of anhydrous solid	

182. Chromium(III) nitrate forms a hydrate that is 40.50% water by mass. What is its chemical formula?

183. Determine the percent composition of $MgCO_3{\cdot}5H_2O$ and draw a pie graph to represent the hydrate.

184. What is the formula and name of a hydrate that is 85.3% barium chloride and 14.7% water?

185. Gypsum is hydrated calcium sulfate. A 4.89-g sample of this hydrate was heated. After the water was removed, 3.87 g anhydrous calcium sulfate remained. Determine the formula for this hydrate and name the compound.

186. A 1.628-g sample of a hydrate of magnesium iodide is heated until its mass is reduced to 1.072 g and all water has been removed. What is the formula of the hydrate?

187. Borax Hydrated sodium tetraborate ($Na_2B_4O_7{\cdot}xH_2O$) is commonly called borax. Chemical analysis indicates that this hydrate is 52.8% sodium tetraborate and 47.2% water. Determine the formula and name the hydrate.

163. a. 42.10% C; 6.480% H; 51.42% O
b. 15.77% Al; 28.12% S; 56.11% O
c. 72.37% Fe; 27.63% O

164. The empirical formula is equal to the molecular formula, $C_{27}H_{44}O$.

165. 43.63% P, 56.37% O; P_2O_5

166. 83.86% C; 11.99% H; 4.138% O

167. a. CH_2
b. $C_3H_4O_3$
c. C_5H_4

168. 194.20 g/mol; 49.47% C, 5.191% H, 28.86% N, 16.48% O

169. TiO_2

170. 80.87% C; 11.70% H; 7.430% O

171. 57.12% C; 6.164% H; 9.521% N; 27.18% O

172. $C_5H_8NO_4Na$

173. $Ni(CN)_2$

174. $Cu_3(OH)_4SO_4$ = 53.72% Cu
$Cu_4(OH)_6SO_4$ = 56.20% Cu

SECTION 5

Mastering Concepts

175. A hydrated compound is a compound that has a specific number of water molecules associated with its atoms, for example, $Na_2CO_3{\cdot}10H_2O$ and $CuSO_4{\cdot}5H_2O$.

176. First, name the compound. Then, add a prefix (*mono-, di-, tri-*) before the word *hydrate,* that indicates how many water molecules are associated with one formula unit of the compound.

177. Dessicants are anhydrous forms of a compound that absorb water from the air and keep it off of electronic devices.

178. Mass a sample of the compound before heating. Heat the compound, allow it to cool, and remass the sample. A change in mass might indicate that the compound is a hydrate.

179. a. $NiCl_2{\cdot}6\,H_2O$
b. $CoCl_2{\cdot}6\,H_2O$
c. $MgCO_3{\cdot}5\,H_2O$
d. $Na_2SO_4{\cdot}10\,H_2O$

Mastering Problems

180. 37.03% Na_2CO_3; 62.96% H_2O

181. $BaCl_2{\cdot}2H_2O$, barium chloride dihydrate

182. $Cr(NO_3)_3{\cdot}9H_2O$

183. Circle graph should show the following percentages: 13.93% Mg, 6.89% C, 27.52% O, 51.66% H_2O

184. $BaCl_2{\cdot}2H_2O$, barium chloride dihydrate

185. $CaSO_4{\cdot}2H_2O$, calcium sulfate dihydrate

186. $MgI_2{\cdot}8H_2O$

187. $Na_2B_4O_7{\cdot}10H_2O$, sodium tetraborate decahydrate

MIXED REVIEW

188. D < A < B < C

189. 52.42 g O, 22.95 g N, 19.68 g C, 4.96 g H

190. Avogadro's number is the number of particles in one mole. You must mass out 237.9 g of $CoCl_2{\cdot}6H_2O$ to have one mole of particles of this compound.

191. calcium

192. $C_5H_{10}S_2$

193. **a.** 0.00490 mol
b. 0.00406 mol
c. 0.0227 mol

194. 85.7 % C, 14.3 % H; CH_2; C_3H_6

195. Al_2O_3

196. C_5H_4; $C_{10}H_8$

197. $C_4H_{10}O$ and $C_9H_8O_4$

THINK CRITICALLY

198. Chalcopyrite ($CuFeS_2$) is 34.62% copper by mass (determined from percent composition) and chalcocite (Cu_2S) is 79.84% copper by mass. Chalcocite would yield the greater quantity of copper because the ore has the greater percentage copper by mass.

199. The rocks contain two different compounds of lead and sulfur. Sample 1 has the formula PbS and Sample 2 has the formula PbS_2.

200. The bar graph should show values of 0.00570 mol Y, 0.00950 mol Al, and 0.0228 mol O.

201. A molecule of TNT contains 21 atoms but a mole of TNT contains 6.02×10^{23} molecules. The statement is false as stated. "Trinitrotoluene, TNT, has the formula $C_7H_5N_3O_6$ which contains 21 atoms but a mole of TNT contains 6.02×10^{23} molecules."

202. Determine and record the mass of an empty evaporating dish. Add about 2 g of the hydrate. Measure and record the mass. Heat the evaporating dish gently for 5 minutes, and strongly for another 5 minutes to evaporate all the water. Allow the dish to cool, and measure and record the mass. Determine the masses of the anhydrous solid and the water lost. Calculate the number of moles of anhydrous compound and water. Determine the ratio of moles of water to moles of anhydrous compound. Use the whole-number ratio of the moles as the coefficient of H_2O in the formula.

203. Concept maps will vary but should show logical connections between Avogadro's number of particles, a mole, and molar mass, and also show how the molecular formula is a whole-number multiple of the empirical formula.

MIXED REVIEW

188. Rank samples A–D from least number of atoms to greatest number of atoms. A: 1.0 mol of H_2; B: 0.75 mol of H_2O; C: 1.5 mol of NaCl; D: 0.50 mol of Ag_2S

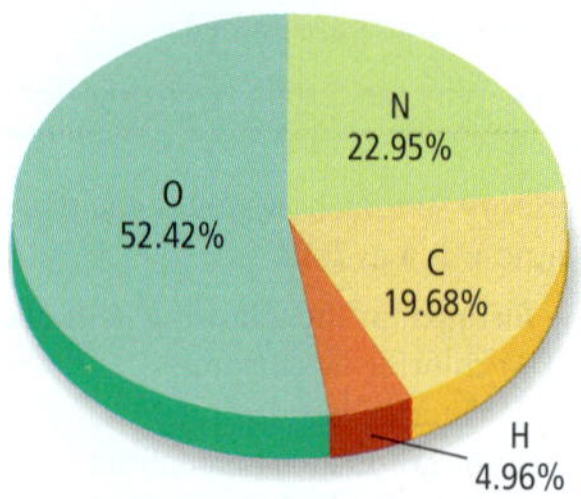

■ **Figure 23**

189. The graph in **Figure 23** shows the percent composition of a compound containing carbon, hydrogen, oxygen, and nitrogen. How many grams of each element are present in 100.0 g of the compound?

190. How many grams of $CoCl_2{\cdot}6H_2O$ must you measure out in a container to have exactly Avogadro's number of particles?

191. One atom of an unknown element has a mass of 6.66×10^{-23} g. What is the identity of this element?

192. **Skunks** Analysis of skunk spray yields a molecule with 44.77% C, 7.46% H and 47.76% S. What is the chemical formula for this molecule found in the spray from skunks that scientists think is partly responsible for the strong odor?

193. How many moles are present in 1.00 g of each compound?
a. l-tryptophan ($C_{11}H_{12}N_2O_2$), an essential amino acid
b. magnesium sulfate heptahydrate, also known as Epsom salts
c. propane (C_3H_8), a fuel

194. A compound contains 6.0 g of carbon and 1.0 g of hydrogen, and has a molar mass of 42.0 g/mol. What are the compound's percent composition, empirical formula, and molecular formula?

195. Which of these compounds has the greatest percent of oxygen by mass: TiO_2, Fe_2O_3, or Al_2O_3?

196. **Mothballs** Naphthalene, commonly found in mothballs, is composed of 93.7% carbon and 6.3% hydrogen. The molar mass of naphthalene is 128 g/mol. Determine the empirical and molecular formulas for naphthalene.

197. Which of these molecular formulas are also empirical formulas: ethyl ether ($C_4H_{10}O$), aspirin ($C_9H_8O_4$), butyl dichloride ($C_4H_8O_2$), glucose ($C_6H_{12}O_6$)?

THINK CRITICALLY

198. **Apply Concepts** A mining company has two possible sources of copper: chalcopyrite ($CuFeS_2$) and chalcocite (Cu_2S). If the mining conditions and the extraction of copper from the ore were identical for each of the ores, which ore would yield the greater quantity of copper? Explain your answer.

199. **Analyze and Conclude** On a field trip, students collected rock samples. Analysis of the rocks revealed that two of the rock samples contained lead and sulfur. **Table 5** shows the percent lead and sulfur in each of the rocks. Determine the empirical formula of each rock. What can the students conclude about the rock samples?

Table 5 Lead and Sulfur Content

Rock Sample	% Lead	% Sulfur
1	86.6%	13.4%
2	76.4%	23.6%

200. **Graph** A YAG, or yttrium aluminum garnet ($Y_3Al_5O_{12}$), is a synthetic gemstone which has no counterpart in nature. Design a bar graph to indicate the moles of each element present in a 5.67 carat yttrium aluminum garnet. (1 carat = 0.20 g)

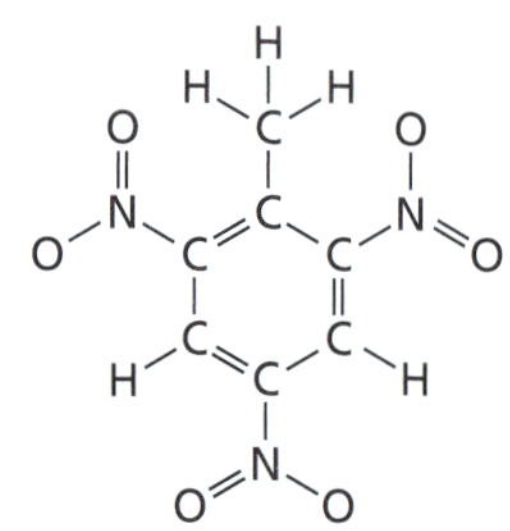

■ **Figure 24**

201. **Assess** The structure of the TNT molecule is shown in **Figure 24.** Critique the statement "Trinitrotoluene, TNT, contains 21 atoms per mole." What is correct about the statement and what is incorrect? Rewrite the statement.

202. **Design an Experiment** Design an experiment that can be used to determine the amount of water in alum ($KAl(SO_4)_2{\cdot}xH_2O$).

203. Design a concept map that illustrates the mole concept. Include the terms *moles, Avogadro's number, molar mass, number of particles, percent composition, empirical formula, and molecular formula.*

Chapter Self-Check

CHALLENGE PROBLEM

204. Two different compounds are composed of Elements X and Y. The formulas of the compounds are X_2Y_3 and XY. A 0.25 mol sample of XY has a mass of 17.96 g, and a 0.25 mol sample of X_2Y_3 has a mass of 39.92 g.
a. What are the atomic masses of elements X and Y?
b. What are the formulas for the compounds?

CUMULATIVE REVIEW

205. Express each answer with the correct number of significant figures.
a. $18.23 - 456.7$
b. $4.233 \div 0.0131$
c. $(82.44 \times 4.92) + 0.125$

206. Making Candy A recipe for pralines calls for the candy mixture to be heated until it reaches the "soft ball" stage, at about 236°F. Can a Celsius thermometer with a range of −10 to 110°C be used to determine when the "soft ball" stage is reached?

207. Contrast atomic number and mass number. Compare these numbers for isotopes of an element.

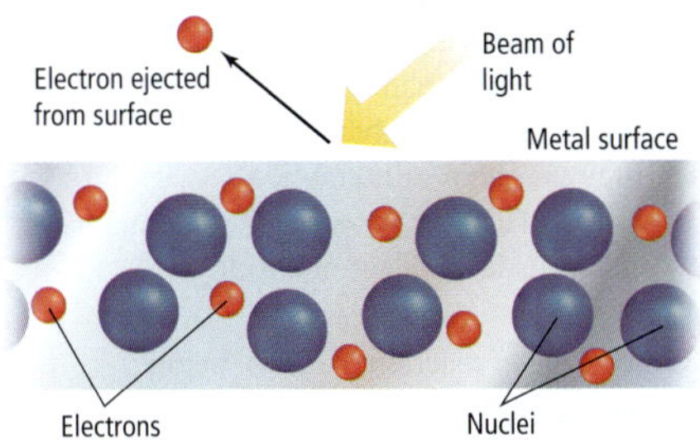

■ **Figure 25**

208. Describe the phenomenon in **Figure 25.** Why are the electrons not bound to the nuclei?

209. Given the elements Ar, Cs, Br, and Ra, identify those that form positive ions. Explain your answer.

210. Write the formula and name the compound formed when each pair of elements combine.
a. barium and chlorine
b. aluminum and selenium
c. calcium and phosphorus

211. Write balanced equations for each reaction.
a. Magnesium metal and water combine to form solid magnesium hydroxide and hydrogen gas.
b. Dinitrogen tetroxide gas decomposes into nitrogen dioxide gas.
c. Aqueous solutions of sulfuric acid and potassium hydroxide undergo a double-replacement reaction.

WRITING IN ▶ Chemistry

212. Natural Gas Natural gas hydrates are chemical compounds known as clathrate hydrates. Research natural gas hydrates and prepare an educational pamphlet for consumers. The pamphlet should discuss the composition and structure of the compounds, the location of the hydrates, their importance to consumers, and the environmental impact of using the hydrates.

213. Avogadro Research and report on the life of Italian chemist Amedeo Avogadro (1776–1856) and how his work led scientists to determine the number of particles in a mole.

214. Luminol Crime-scene investigators use luminol to visualize blood residue. Research luminol and determine its chemical formula and percent composition.

DBQ Document-Based Questions

Space Shuttle Propellants *At liftoff, the orbiter and an external fuel tank carry 3,164,445 L of the liquid propellants hydrogen, oxygen, hydrazine, monomethylhydrazine, and dinitrogen tetroxide. Their total mass is 727,233 kg. Data for the propellants carried at liftoff are given in* **Table 6.**

Data obtained from: "Space Shuttle Use of Propellants and Fluids." September 2001. *NASA Fact Sheet.*

Table 6 Space Shuttle Liquid Propellants

Propellants	Molecular Formula	Mass (kg)	Moles	Molecules
Hydrogen	H_2		5.14×10^7	
Oxygen	O_2			1.16×10^{31}
Hydrazine		493		
Monomethyl-hydrazine	CH_3NHNH_2	4909		
Dinitrogen tetroxide	N_2O_4		8.64×10^4	

215. Hydrazine contains 87.45% nitrogen and 12.55% hydrogen, and has a molar mass of 32.04 g/mol. Determine hydrazine's molecular formula. Record the molecular formula in **Table 6.**

216. Complete **Table 6** by calculating the number of moles, mass in kilograms, or molecules for each propellant. Give all answers to three significant figures.

CHALLENGE PROBLEM

204. a. X = 55.84 g/mol; Y = 16 g/mol
b. FeO and Fe_2O_3

CUMULATIVE REVIEW

205. a. −438.5
b. 323
c. 406.125

206. No, the Celsius thermometer could not be used because 113°C is beyond the range of the thermometer.

207. Atomic number equals the number of protons. Mass number equals the number of protons plus the number of neutrons. Two isotopes of an element will have the same atomic number but different mass numbers.

208. The illustration shows the photoelectric effect: the emission of electrons from a metal surface when light of sufficient energy strikes the metal. The electrons in metals are not bound to nuclei; they are free to move.

209. Cs and Ra can form positive ions. The configuration of Cs is [Xe]$6s^1$; Cs loses 1 valence electron to form a 1^+ ion. The configuration of Ra is [Rn]$7s^2$; Ra loses 2 electrons to form a 2^+ ion. Argon is a noble gas and does not lose or gain electrons. Br gains one electron to form a 1^- ion.

210. a. $BaCl_2$; barium chloride
b. Al_2Se_3; aluminum selenide
c. Ca_3P_2; calcium phosphide

211. a. $Mg(s) + 2H_2O(l) \longrightarrow Mg(OH)_2(s) + H_2(g)$
b. $N_2O_4(g) \longrightarrow 2NO_2(g)$
c. $H_2SO_4(aq) + 2KOH(aq) \longrightarrow K_2SO_4(aq) + 2H_2O(l)$

213. Answers will vary. Students should mention Avogadro's hypothesis. Avogadro formulated his hypothesis as an explanation for earlier works by Gay-Lussac and Ritter. His ideas were rejected by chemists of his day but were revived later by the Italian chemist Stanislao Cannizzaro. Avogadro died before seeing his ideas accepted.

214. Luminol ($C_8H_7O_2N_3$) reacts with compounds in blood and releases energy in the form of light. When crime scene investigators see the light, they know blood is at the scene. The percent composition of luminol is 54.23% C, 3.98% H, 23.72% N, and 18.06% O.

DBQ Document-Based Questions

Data obtained from: "Space Shuttle Use of Propellants and Fluids." September 2001. *NASA Fact Sheet.*

215. N_2H_4

216. Refer to the Solutions Manual.

WRITING IN ▶ Chemistry

Rubric

212. Answers will vary. Natural gas hydrates are crystalline solids with a water-ice lattice within which light hydrocarbon molecules are trapped. They naturally form in Arctic permafrost regions and on the sea floor. Gas hydrates may be a new clean energy source. Removal of gas hydrates might trigger the release of huge amounts of methane. Large methane releases might have caused climatic warming in the geologic past.

MULTIPLE CHOICE

1. B
2. D
3. C
4. B
5. A
6. C
7. C
8. D
9. C
10. C

MULTIPLE CHOICE

Use the graph below to answer Questions 1 to 4.

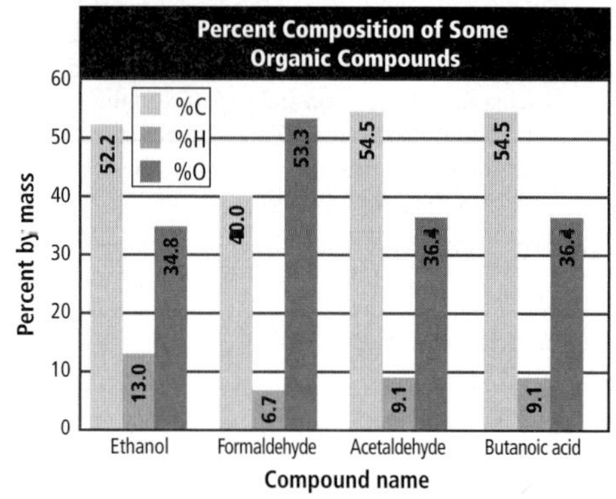

1. Acetaldehyde and butanoic acid must have the same
A. molecular formula.
B. empirical formula.
C. molar mass.
D. chemical properties.

2. If the molar mass of butanoic acid is 88.1 g/mol, what is its molecular formula?
A. $C_3H_4O_3$
B. C_2H_4O
C. $C_5H_{12}O_1$
D. $C_4H_8O_2$

3. What is the empirical formula of ethanol?
A. C_4HO_3
B. $C_2H_6O_2$
C. C_2H_6O
D. $C_4H_{13}O_2$

4. The empirical formula of formaldehyde is the same as its molecular formula. How many grams are in 2.000 mol of formaldehyde?
A. 30.00 g
B. 60.06 g
C. 182.0 g
D. 200.0 g

5. Which does NOT describe a mole?
A. a unit used to count particles directly
B. Avogadro's number of molecules of a compound
C. the number of atoms in exactly 12 g of pure C-12
D. the SI unit for the amount of a substance

Use the graph below to answer Question 6.

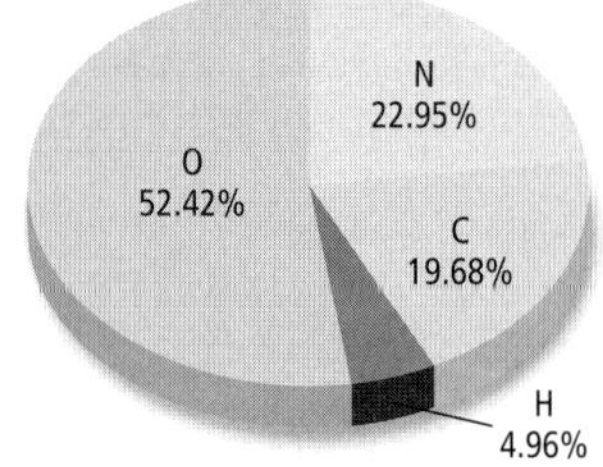

6. What is the empirical formula for this compound?
A. $C_6H_2N_6O_3$
B. $C_4HN_5O_{10}$
C. CH_3NO_2
D. CH_5NO_3

7. Which is NOT true of molecular compounds?
A. Triple bonds are stronger than single bonds.
B. Electrons are shared in covalent bonds.
C. All atoms have eight valence electrons when they are chemically stable.
D. Lewis structures show the arrangements of electrons in covalent molecules.

8. Which type of reaction is shown below?

$$2HI + (NH_4)_2S \rightarrow H_2S + 2NH_4I$$

A. synthesis
B. decomposition
C. single replacement
D. double replacement

9. How many atoms are in 0.625 moles of Ge (atomic mass = 72.59 amu)?
A. 2.73×10^{25}
B. 6.99×10^{25}
C. 3.76×10^{23}
D. 9.63×10^{23}

10. What is the mass of one molecule of barium hexafluorosilicate ($BaSiF_6$)?
A. 1.68×10^{26} g
B. 2.16×10^{21} g
C. 4.64×10^{-22} g
D. 6.02×10^{-23} g

SHORT ANSWER

Use the table below to answer Question 11.

Charges of Some Ions	
Ion	Formula
Sulfide	S^{2-}
Sulfite	$SO_3{}^{2-}$
Sulfate	$SO_4{}^{2-}$
Thiosulfate	$S_2O_3{}^{2-}$
Copper(I)	Cu^+
Copper(II)	Cu^{2+}

11. How many possible compounds can be made from the ions of copper, sulfur, and oxygen in the table above? Write their names and formulas.

EXTENDED RESPONSE

Use the figure below to answer Question 12.

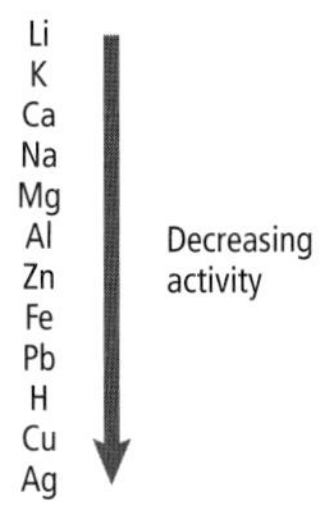

You have been asked to identify a sample of a metal. You have aqueous solutions of KCl, $AlCl_3$, $FeCl_3$, and $CuCl_2$ available. Could the metal be either lithium, zinc, or lead?

12. Explain how you would use these solutions to identify what metal your sample is made of.

SAT SUBJECT TEST: CHEMISTRY

13. It takes 2 iron atoms and 6 chlorine atoms to make 2 iron(III) chloride particles. How many chlorine atoms are required to make 18 iron(III) chloride particles?

A. 9
B. 18
C. 27
D. 54
E. 72

14. What is the molar mass of fluorapatite ($Ca_5(PO_4)_3F$)?

A. 314 g/mol
B. 344 g/mol
C. 442 g/mol
D. 504 g/mol
E. 524 g/mol

15. Which is not a correct formula for an ionic compound?

A. $CaCl_2$
B. Na_2SO_4
C. Al_3S_2
D. $Mg(NO_3)_2$
E. NaCl

Use the table below to answer question 16.

Percent composition of selected hydrocarbons			
Compound	%C	%H	%O
$C_4H_{10}O$	64.81	13.60	21.59
$C_6H_{12}O_4$	48.64	8.108	43.24
$C_7H_{16}O_3$	56.76	10.81	32.43
$C_5H_8O_5$	40.54	5.405	54.05

16. A 25.0-g sample of an unknown hydrocarbon is composed of 12.16 g carbon, 2.027 g hydrogen, and 10.81 g oxygen. If its molar mass is 148 g/mol, what is the molecular formula for this compound?

A. $C_4H_{10}O$
B. $C_6H_{12}O_4$
C. $C_7H_{16}O_3$
D. $C_5H_8O_5$
E. $C_8H_5O_5$

NEED EXTRA HELP?																
If You Missed Question . . .	1	2	3	4	5	6	7	8	9	10	11	12	13	14	15	16
Review Section . . .	10.4	10.4	10.4	10.4	10.1	10.4	8.3	9.2	10.1	10.2	7.3	9.2	10.3	10.3	7.3	10.4

SHORT ANSWER

11. Eight compounds are possible:
Copper(I) thiosulfate: $Cu_2S_2O_3$
Copper(II) thiosulfate: CuS_2O_3
Copper(I) sulfate: Cu_2SO_4
Copper(II) sulfate: $CuSO_4$
Copper(I) sulfite: Cu_2SO_3
Copper(II) sulfite: $CuSO_3$
Copper(I) sulfide: Cu_2S
Copper(II) sulfide: CuS

EXTENDED RESPONSE

12. You would need to test each sample of metal for a reaction with each of the solutions in turn. If the sample is lithium, it is more reactive than any of the four aqueous chloride solutions and will display a reaction when placed into any of them. Lead is less reactive than any of the aqueous solutions except for the copper; therefore, if the metal reacts with the copper solution but none of the others, it is likely to be lead. Zinc is more reactive than iron and copper but less reactive than potassium and aluminum, so if the unidentified metal is zinc it will react with the iron and copper solutions but not with the potassium or aluminum solutions.

SAT SUBJECT TEST: CHEMISTRY

13. D
14. D
15. C
16. B

CHAPTER 11 Stoichiometry

BIGIDEA **Mass relationships in chemical reactions confirm the law of conservation of mass.**

ESSENTIAL QUESTIONS	RESOURCES TO ASSESS MASTERY
SECTION 1 **Defining Stoichiometry** **1.** Which relationships can be derived from a balanced chemical equation? **2.** How are mole ratios written from a balanced chemical equation? 1 session 0.5 block	**Progress Monitoring** Formative Assessment, p. 371 Reading Check, pp. 369, 371 Section Review, p. 372
SECTION 2 **Stoichiometric Calculations** **1.** What is the sequence of steps used in solving stoichiometric problems? **2.** How are these steps applied to solve stoichiometric problems? 2 sessions 1 block	**Progress Monitoring** Formative Assessment, p. 374 Section Review, p. 378
SECTION 3 **Limiting Reactants** **1.** In a chemical reaction, which reactant is the limiting reactant? **2.** How much of the excess reactant remains after the reaction is complete? **3.** How do you calculate the mass of a product when the amounts of more than one reactant are given? 1 session 0.5 block	**Progress Monitoring** Formative Assessment, pp. 381, 383, 384 Reading Check, p. 380 Section Review, p. 384
SECTION 4 **Percent Yield** **1.** What is the theoretical yield of a chemical reaction? **2.** How do you calculate the percent yield for a chemical reaction? 1 session 0.5 block	**Progress Monitoring** Formative Assessment, p. 388 Section Review, p. 388 **Summative Assessment** Chapter Assessment, p. 392 *eAssessment* Chapter Test (Scaffolded)

Go online!

LEVELED RESOURCES	LAB MATERIALS
Science Notebook 11.1 OL ***FAST FILE Chapter Resources:*** Study Guide, p. 72 BL	LaunchLAB p. 366 / **15 min** 10-mL graduated cylinder, 0.01*M* potassium permanganate, 100-mL beaker, 0.01*M* sodium hydrogen sulfite solution, stirring rod
Science Notebook 11.2 OL ***FAST FILE Chapter Resources:*** MiniLab Worksheet, p. 58 OL ChemLab Worksheet, p. 60 OL Study Guide, p. 74 BL	MiniLAB p. 378 / **25 min** balance, crucible, sodium hydrogen carbonate ($NaHCO_3$), ring stand, ring, clay triangle, Bunsen burner, crucible tongs ChemLAB p. 390 / **45 min** copper(II) sulfate pentahydrate, iron metal filings (20 mesh), distilled water, hot plate, beaker tongs, balance, stirring rod, 150-mL beaker, 400-mL beaker, 100-mL graduated cylinder, weighing paper
Science Notebook 11.3 OL ***FAST FILE Chapter Resources:*** Study Guide, p. 76 BL	
Science Notebook 11.4 OL ***FAST FILE Chapter Resources:*** Study Guide, p. 77 BL	

ADDITIONAL RESOURCES

eAssessment
ConnectED Student Center
ConnectED Teacher Center
Lesson Presentations
eSolutions
Videos
Animations
Online Quizzes
Virtual Investigations
Multilingual eGlossary
Personal Tutors
WebQuests
Applying Practices Worksheets
PBLs

Practice:

Challenge Problems AL
Supplemental Problems BL OL
Chapter Test (Scaffolded)
Standardized Test Practice

Additional Resources:

Blueprints for Success
Solving Problems: A Chemistry Handbook

Lab Resources:

Laboratory Manual OL
CBL Laboratory Manual OL
Small-Scale Laboratory Manual OL
Forensics Laboratory Manual OL

BL Below Level   OL On Level   AL Advanced Learners   EL English Learners   COOP LEARN Cooperative Learning

CHAPTER 11

Stoichiometry

BIGIDEA **Mass relationships in chemical reactions confirm the law of conservation of mass.**

SECTIONS

1 Defining Stoichiometry

2 Stoichiometric Calculations

3 Limiting Reactants

4 Percent Yield

LaunchLAB

What evidence can you observe that a reaction has stopped?

During a chemical reaction, reactants are consumed as new products form. In this lab, you will look for signs a chemical reaction has stopped.

FOLDABLES Study Organizer

Steps in Stoichiometric Calculations

Make a four-tab book. Label the tabs with the steps in stoichiometric calculations. Use it to help you summarize the steps in solving a stoichiometric problem.

CHAPTER 11

LaunchLAB

Rubric

What evidence can you observe that a reaction has stopped?

Purpose Students will observe a chemical reaction in which one of the reactants is completely used up and the other is in excess.

Safety Precautions Approve lab safety forms before work begins. Review the MSDS for potassium permanganate and sodium hydrogen sulfite with students prior to doing the lab. Do not allow students to prepare the solutions required for the lab, as $KMnO_4$ is a dangerous oxidizer in powder and crystal form. Remind students to wear aprons and goggles.

Disposal The $KMnO_4$ solution must be processed or saved as chemical waste–do not pour the solution down the drain.

Teaching Strategies

- Place five test tubes and ten rubber stoppers on the desk. Place one stopper into each test tube, and ask students which item is in excess.
- See the Teacher Handbook for preparation of solutions.

Expected Results Initially, the $KMnO_4$ solution is dark purple. The solution gradually becomes colorless as the $NaHSO_3$ solution is added. Expect that 5 mL $KMnO_4$ solution will require 27 mL $NaHSO_3$ solution to change it to a clear, colorless solution.

Procedure

1. Read and complete the lab safety form.
2. Use a **10-mL graduated cylinder** to measure out 5.0 mL **0.01*M* potassium permanganate** ($KMnO_4$). Add the solution to a **100-mL beaker.**
3. Clean and dry the graduated cylinder, and then use it to measure 5.0 mL **0.01*M* sodium hydrogen sulfite solution** ($NaHSO_3$). Slowly add this solution to the beaker while stirring with a **stirring rod.** Record your observations.
4. Repeat Step 3 until the $KMnO_4$ solution in the beaker turns colorless. Stop adding the $NaHSO_3$ solution as soon as you obtain a colorless solution. Record your observations.

Analysis

1. **Identify** the evidence you observed that a chemical reaction was occurring. As colorless sodium hydrogen sulfite was added to the purple potassium permanganate solution, a color change from purple to colorless was observed.

Green plants make their own food through photosynthesis, a biological process that involves a series of chemical reactions. These reactions begin with carbon dioxide and water. On a summer day, one acre of corn produces enough oxygen through photosynthesis to meet the needs of 130 people.

2. **Explain** why slowly adding the $NaHSO_3$ solution while stirring is a better experimental technique than adding 5.0 mL of the solution all at once. Adding the solution all at once could lead to an error in the volume of $NaHSO_3$ solution needed to change the purple $KMnO_4$ solution to a colorless solution. The error could be as large as 5.0 mL.

Inquiry Would anything more have happened if you continued to add $NaHSO_3$ solution to the beaker? Explain. Nothing more will happen because the solution is colorless, meaning that there is no more potassium permanganate to react.

Introduce the BIGIDEA

Reactive Substances To introduce this chapter's Big Idea, ask four students to each add 5.0 mL of water to an empty film canister. Tell them to divide a sodium bicarbonate tablet into four unequal pieces–small, medium, medium-large, and large. Have them place each piece of the tablet on the lid of a film canister. When you say so, students should snap the lid onto the canister, invert the canister on the table, and back away from the table. Film canisters will pop and be launched into the air. Ask students which canister had the biggest pop. The canister with the largest piece of the tablet will likely have the biggest pop. Ask students if the size of the sodium bicarbonate in the canister determined how high the canister traveled. Results will likely correlate to the size of the tablet.

Tie to Previous Knowledge

Have students review the following concepts before studying this chapter.

- scientific notation, significant figures, dimensional analysis
- average atomic mass
- mole conversions

Use the Photo

Conservation of Mass Write the balanced basic equation for photosynthesis on the board. Remind students of the law of the conservation of mass. Ask them to verify that mass is conserved during photosynthesis. Students should verify that the total number of each type of atom, and therefore the mass, is identical on both sides of the equation.

SECTION 1

1 Focus

MAINIDEA

Reactants and Products Place a lump of clay on a watch glass, and insert a birthday candle into the clay. Place the watch glass and candle into a bowl with water that covers the bottom of the clay. Light the candle. Ask students to predict what will happen if you lower a beaker over the candle. Post all predictions on the board. Place the beaker over the candle. Ask students to form a hypothesis that explains why the candle went out. The hypothesis might include the following: The candle was undergoing a combustion reaction. It needs oxygen to burn. Placing the beaker over the burning candle limited the amount of oxygen available for the reaction. Tell students in this chapter they will study the relationships between quantities of reactants and products. OL

SECTION 1

Essential Questions

- Which relationships can be derived from a balanced chemical equation?
- How are mole ratios written from a balanced chemical equation?

Review Vocabulary

reactant: the starting substance in a chemical reaction

New Vocabulary

stoichiometry
mole ratio

Defining Stoichiometry

MAINIDEA The amount of each reactant present at the start of a chemical reaction determines how much product can form.

CHEM 4 YOU **Have you ever watched a candle burning? You might have watched the candle burn out as the last of the wick was used up. Or, maybe you used a candle snuffer to put out the flame. Either way, the combustion reaction ended when one of the reactants was used up.**

Particle and Mole Relationships

In doing the Launch Lab, were you surprised when the purple color of potassium permanganate disappeared as you added sodium hydrogen sulfite? If you concluded that the potassium permanganate had been used up and the reaction had stopped, you are right. Chemical reactions stop when one of the reactants is used up. When planning the reaction of potassium permanganate and sodium hydrogen sulfite, a chemist might ask, "How many grams of potassium permanganate are needed to react completely with a known mass of sodium hydrogen sulfite?" Or, when analyzing a photosynthesis reaction, you might ask, "How much oxygen and carbon dioxide are needed to form a known mass of sugar?" Stoichiometry is the tool for answering these questions.

Stoichiometry The study of quantitative relationships between the amounts of reactants used and amounts of products formed by a chemical reaction is called **stoichiometry.** Stoichiometry is based on the law of conservation of mass. Recall that the law states that matter is neither created nor destroyed in a chemical reaction. In any chemical reaction, the amount of matter present at the end of the reaction is the same as the amount of matter present at the beginning. Therefore, the mass of the reactants equals the mass of the products. Note the reaction of powdered iron (Fe) with oxygen (O_2) shown in **Figure 1.** Although iron reacts with oxygen to form a new compound, iron(III) oxide (Fe_2O_3), the total mass is unchanged.

Figure 1 The balanced chemical equation for this reaction between iron and oxygen provides the relationships between amounts of reactants and products.

Charles D. Winters/Photo Researchers

DIFFERENTIATED INSTRUCTION

Struggling Learners Using a molecular model kit, make a model of CH_4 and two O_2 molecules. Ask students to identify the models. Make models of CO_2 and two H_2O molecules and ask students to identify these models. Tell students that the two sets of models represent the reactants and the products of the combustion of methane. Arrange the reactants and products on opposite sides of a cardboard arrow. Ask students to count the number of each atom before and after the reaction. one carbon, four hydrogen, and four oxygen atoms Remind them that coefficients indicate the number of moles as well as the number of molecules. Have them interpret the equation in moles. $1 \text{ mol } CH_4 + 2 \text{ mol } O_2 \rightarrow 1 \text{ mol } CO_2 + 2 \text{ mol } H_2O$ BL EL

Explore balanced chemical equations with an interactive table. **Concepts In Motion**

Table 1 Relationships Derived from a Balanced Chemical Equation

$4Fe(s)$	+	$3O_2(g)$	→	$2Fe_2O_3(s)$
iron	+	oxygen	→	iron(III) oxide
4 atoms Fe	+	3 molecules O_2	→	2 formula units Fe_2O_3
4 mol Fe	+	3 mol O_2	→	2 mol Fe_2O_3
223.4 g Fe	+	96.00 g O_2	→	319.4 g Fe_2O_3
319.4 g reactants			→	319.4 g products

The balanced chemical equation for the chemical reaction shown in **Figure 1** is as follows.

$$4Fe(s) + 3O_2(g) \rightarrow 2Fe_2O_3(s)$$

You can interpret this equation in terms of representative particles by saying that four atoms of iron react with three molecules of oxygen to produce two formula units of iron(III) oxide. Remember that coefficients in an equation represent not only numbers of individual particles but also numbers of moles of particles. Therefore, you can also say that four moles of iron react with three moles of oxygen to produce two moles of iron(III) oxide.

The chemical equation does not directly tell you anything about the masses of the reactants and products. However, by converting the known mole quantities to mass, the mass relationships become obvious. Recall that moles are converted to mass by multiplying by the molar mass. The masses of the reactants are as follows.

$$4\ \cancel{\text{mol Fe}} \times \frac{55.85\text{ g Fe}}{1\ \cancel{\text{mol Fe}}} = 223.4\text{ g Fe}$$

$$3\ \cancel{\text{mol O}_2} \times \frac{32.00\text{ g O}_2}{1\ \cancel{\text{mol O}_2}} = 96.00\text{ g O}_2$$

The total mass of the reactants is: (223.4 g + 96.00 g) = 319.4 g

Similarly, the mass of the product is calculated as follows:

$$2\ \cancel{\text{mol Fe}_2\text{O}_3} \times \frac{159.7\text{ g Fe}_2\text{O}_3}{1\ \cancel{\text{mol Fe}_2\text{O}_3}} = 319.4\text{ g}$$

Note that the mass of the reactants equals the mass of the product.

$$\text{mass of reactants} = \text{mass of products}$$
$$319.4\text{ g} = 319.4\text{ g}$$

As predicted by the law of conservation of mass, the total mass of the reactants equals the mass of the product. The relationships that can be determined from a balanced chemical equation are summarized in **Table 1.**

VOCABULARY

WORD ORIGIN

Stoichiometry
comes from the Greek words *stoikheion,* which means element, and *metron,* which means to measure

☑ READING CHECK **List** the types of relationships that can be derived from the coefficients in a balanced chemical equation.

CHEMISTRY JOURNAL

Interpret Chemical Equations Have students interpret the following equations in terms of particles, moles, and mass.

$2KClO_3 \rightarrow 2KCl + 3O_2$
$Fe_2O_3 + 3H_2 \rightarrow 2Fe + 3H_2O$

Have them demonstrate in their journals that the law of conservation of mass is observed. **OL**

Visual Learning

Table 1 Ask students to use the table as a guide for interpreting the equation for the combustion of methane. Make a drawing of the reaction to help students. **OL** **EL**

☑ READING CHECK The coefficients in a balanced equation represent the number of representative particles as well as the number of moles of particles. Thus, mole and number of particle relationships can be written from a balanced chemical equation. Because masses can be derived from the coefficients by converting the known mole quantities to mass, mass relationships can also be written.

Quick Demo

Quantities of Reactants

Take four 125 mL flasks and four balloons. To each flask, add 10 mL of 2.0*M* acetic acid. Use a balance to obtain 0.5, 1.0, 2.5, and 5.0 g samples of baking soda using weighing paper to protect the balance. Ask students what will happen when the baking soda is added to the vinegar. A reaction will take place that releases a gas. Ask students if all flasks will produce the same amount of carbon dioxide. Answers will vary. Place the baking soda into the acetic acid in the flasks and quickly cover the top of each flask with the balloon. Ask students why all of the balloons were not the same size. Each reaction produced a different amount of carbon dioxide, because of the differing amounts of acetic acid and baking soda present. **BL** **EL**

IN-CLASS Example

Question Geologists test for calcium carbonate in rocks using a hydrochloric acid solution. Bubbling indicates the release of carbon dioxide from the limestone according to the following reaction: $CaCO_3$ + 2HCl → $CaCl_2$ + H_2O + CO_2 Interpret the equation for this reaction in terms of representative particles, moles, and mass. Show that the law of conservation of mass is observed.

Answer 1 formula unit of $CaCO_3$ + 2 molecules of HCl → 1 formula unit of $CaCl_2$ + 1 molecule of H_2O + 1 molecule of CO_2; 1 mol $CaCO_3$ + 2 mol HCl → 1 mol $CaCl_2$ + 1 mol H_2O + 1 mol CO_2; 100 g $CaCO_3$ + 72.9 g HCl → 111 g $CaCl_2$ + 18.0 g H_2O + 44.0 g
175 g reactants = 175 g products.
Mass is conserved.

PRACTICE Problems

Have students refer to the Selected Solutions appendix for complete solutions to odd-numbered problems. The complete solutions for all problems can be found in the Solutions Manual.

1. a. 34.062 g reactants = 34.062 g products
   b. 92.566 g reactants = 92.566 g products
   c. 80.608 g reactants = 80.608 g products
2. a. 2Na(s) + $2H_2O$(l) → 2NaOH(aq) + $1H_2$(g)
   82.01 g reactants = 82.01 g products
   b. 4Zn(s) + $10HNO_3$(aq) → $4Zn(NO_3)_2$(aq) + $1N_2O$(g) + $5H_2O$(l)
   891.68 g reactants = 891.68 g products

EXAMPLE Problem 1

Find help with rounding. Math Handbook

EXAMPLE PROBLEM

INTERPRETING CHEMICAL EQUATIONS The combustion of propane (C_3H_8) provides energy for heating homes, cooking food, and soldering metal parts. Interpret the equation for the combustion of propane in terms of representative particles, moles, and mass. Show that the law of conservation of mass is observed.

1 ANALYZE THE PROBLEM

The coefficients in the balanced chemical equation shown below represent both moles and representative particles, in this case molecules. Therefore, the equation can be interpreted in terms of molecules and moles. The law of conservation of mass will be verified if the masses of the reactants and products are equal.

Known

$C_3H_8(g) + 5O_2(g) \rightarrow 3CO_2(g) + 4H_2O(g)$

Unknown

Equation interpreted in terms of molecules = ?
Equation interpreted in terms of moles = ?
Equation interpreted in terms of mass = ?

2 SOLVE FOR THE UNKNOWN

The coefficients in the chemical equation indicate the number of molecules.

1 molecule C_3H_8 + 5 molecules O_2 → 3 molecules CO_2 + 4 molecules H_2O

The coefficients in the chemical equation also indicate the number of moles.

1 mol C_3H_8 + 5 mol O_2 → 3 mol CO_2 + 4 mol H_2O

To verify that mass is conserved, first convert moles of reactant and product to mass by multiplying by a conversion factor—the molar mass—that relates grams to moles.

$$\text{moles of reactant or product} \times \frac{\text{grams reactant or product}}{\text{1 mol reactant or product}} = \text{grams of reactant or product}$$

$$1 \cancel{\text{mol } C_3H_8} \times \frac{44.09 \text{ g } C_3H_8}{1 \cancel{\text{mol } C_3H_8}} = 44.09 \text{ g } C_3H_8$$ Calculate the mass of the reactant C_3H_8.

$$5 \cancel{\text{mol } O_2} \times \frac{32.00 \text{ g } O_2}{1 \cancel{\text{mol } O_2}} = 160.0 \text{ g } O_2$$ Calculate the mass of the reactant O_2.

$$3 \cancel{\text{mol } CO_2} \times \frac{44.01 \text{ g } CO_2}{1 \cancel{\text{mol } CO_2}} = 132.0 \text{ g } CO_2$$ Calculate the mass of the product CO_2.

$$4 \cancel{\text{mol } H_2O} \times \frac{18.02 \text{ g } H_2O}{1 \cancel{\text{mol } H_2O}} = 72.08 \text{ g } H_2O$$ Calculate the mass of the product H_2O

44.09 g C_3H_8 + 160.0 g O_2 = **204.1 g reactants** — Add the masses of the reactants.

132.0 g CO_2 + 72.08 g H_2O = **204.1 g products** — Add the masses of the products.

204.1 g reactants = 204.1 g products — The law of conservation of mass is observed.

3 EVALUATE THE ANSWER

The sums of the reactants and the products are correctly stated to the first decimal place because each mass is accurate to the first decimal place. The mass of reactants equals the mass of products, as predicted by the law of conservation of mass.

CHEMISTRY PROJECT

Stoichiometry in Space Have students research the role of lithium hydroxide in the removal of the carbon dioxide exhaled by astronauts. Ask them to discuss why the carbon dioxide must be removed, and how stoichiometric calculations are used to minimize the weight of lithium hydroxide carried into space. Ask advanced students to perform a rough calculation of the amount of carbon dioxide exhaled by a person in a day. OL AL

DIFFERENTIATED INSTRUCTION

Hearing Impaired Have students view a video clip of a shuttle launch. Show on the board that the fuel for the launch is a mixture of hydrogen and oxygen. Have students write a paragraph describing what factors engineers have to consider when using these fuels. Paragraphs should include such factors as how much hydrogen is present and how much oxygen is needed. OL

PRACTICE Problems

Do additional problems. Online Practice

1. Interpret the following balanced chemical equations in terms of particles, moles, and mass. Show that the law of conservation of mass is observed.
   a. $N_2(g) + 3H_2(g) \rightarrow 2NH_3(g)$
   b. $HCl(aq) + KOH(aq) \rightarrow KCl(aq) + H_2O(l)$
   c. $2Mg(s) + O_2(g) \rightarrow 2MgO(s)$
2. **Challenge** For each of the following, balance the chemical equation; interpret the equation in terms of particles, moles, and mass; and show that the law of conservation of mass is observed.
   a. ___Na(s) + ___$H_2O(l)$ → ___NaOH(aq) + ___$H_2(g)$
   b. ___Zn(s) + ___$HNO_3(aq)$ → ___$Zn(NO_3)_2(aq)$ + ___$N_2O(g)$ + ___$H_2O(l)$

PRACTICE PROBLEMS

Get help with **ratios.** Personal Tutor

Mole ratios You have read that the coefficients in a chemical equation indicate the relationships between moles of reactants and products. You can use the relationships between coefficients to derive conversion factors called mole ratios. A **mole ratio** is a ratio between the numbers of moles of any two of the substances in a balanced chemical equation. Consider the reaction between potassium (K) and bromine (Br_2) to form potassium bromide (KBr). The product of the reaction, the ionic salt potassium bromide, is prescribed by veterinarians, like the one in **Figure 2,** as an antiepileptic medication for dogs.

$$2K(s) + Br_2(l) \rightarrow 2KBr(s)$$

What mole ratios can be written for this reaction? Starting with the reactant potassium, you can write a mole ratio that relates the moles of potassium to each of the other two substances in the equation. Thus, one mole ratio relates the moles of potassium used to the moles of bromine used. The other mole ratio relates the moles of potassium used to the moles of potassium bromide formed.

$$\frac{2 \text{ mol K}}{1 \text{ mol Br}_2} \text{ and } \frac{2 \text{ mol K}}{2 \text{ mol KBr}}$$

Two other mole ratios show how the moles of bromine relate to the moles of the other two substances in the equation—potassium and potassium bromide.

$$\frac{1 \text{ mol Br}_2}{2 \text{ mol K}} \text{ and } \frac{1 \text{ mol Br}_2}{2 \text{ mol KBr}}$$

Similarly, two ratios relate the moles of potassium bromide to the moles of potassium and bromine.

$$\frac{2 \text{ mol KBr}}{2 \text{ mol K}} \text{ and } \frac{2 \text{ mol KBr}}{1 \text{ mol Br}_2}$$

These six ratios define all the mole relationships in this equation. Each of the three substances in the equation forms a ratio with the two other substances.

☑ READING CHECK **Identify** the source from which a chemical reaction's mole ratios are derived.

■ **Figure 2** Potassium metal and liquid bromine react vigorously to form the ionic compound potassium bromide. Bromine is one of the two elements that are liquids at room temperature (mercury is the other). Potassium is a highly reactive metal. Potassium bromide is an ionic salt that is used to treat epilepsy in dogs.

ASSESSMENT

Skill Have each student write a balanced equation on a sheet of paper and give it to another student, who will write the mole ratios found in the equation. OL

3 Assess

Check for Understanding

Place a small piece of calcium into a test tube containing 3.0*M* HCl. In a second test tube, collect the hydrogen gas. Demonstrate the presence of hydrogen in the test tube using a burning splint. Note the characteristic pop. On the board, write the word equation for the reaction: calcium plus hydrochloric acid produces calcium chloride and hydrogen. Ask students to write the balanced equation. $Ca + 2HCl \rightarrow CaCl_2 + H_2$ Have students write at least two mole ratios for the equation. OL

Reteach

Pour 10 mL of 0.10*M* $Ba(OH)_2$ into a test tube. Add 10 mL of 0.10*M* sulfuric acid. Ask students to write the equation for the reaction that occurs and all the possible mole ratios. $Ba(OH)_2 + H_2SO_4 \rightarrow BaSO_4 + 2H_2O$ OL

Extension

Ask students to explain how mole ratios derived from a balanced chemical equation can be used to relate the masses of reactants and products in a reaction. Because molar quantities are related to mass by Avogadro's number, mole ratios can be converted to mass ratios. OL

☑ READING CHECK A chemical reaction's mole ratios are derived from the relationships between coefficients in a balanced chemical equation. A mole ratio is a ratio between the numbers of moles of any two substances in the equation.

PRACTICE Problems

Do additional problems. Online Practice

3. Determine all possible mole ratios for the following balanced chemical equations.
   - a. $4Al(s) + 3O_2(g) \rightarrow 2Al_2O_3(s)$
   - b. $3Fe(s) + 4H_2O(l) \rightarrow Fe_3O_4(s) + 4H_2(g)$
   - c. $2HgO(s) \rightarrow 2Hg(l) + O_2(g)$
4. **Challenge** Balance the following equations, and determine the possible mole ratios.
   - a. $ZnO(s) + HCl(aq) \rightarrow ZnCl_2(aq) + H_2O(l)$
   - b. butane (C_4H_{10}) + oxygen → carbon dioxide + water

VOCABULARY

ACADEMIC VOCABULARY

Derive
to obtain from a specified source
The researcher was able to derive the meaning of the illustration from ancient texts.

The decomposition of potassium chlorate ($KClO_3$) is sometimes used to obtain small amounts of oxygen in the laboratory.

$$2KClO_3(s) \rightarrow 2KCl(s) + 3O_2(g)$$

The mole ratios that can be written for this reaction are as follows.

$$\frac{2 \text{ mol } KClO_3}{2 \text{ mol } KCl} \text{ and } \frac{2 \text{ mol } KClO_3}{3 \text{ mol } O_2}$$

$$\frac{2 \text{ mol } KCl}{2 \text{ mol } KClO_3} \text{ and } \frac{2 \text{ mol } KCl}{3 \text{ mol } O_2}$$

$$\frac{3 \text{ mol } O_2}{2 \text{ mol } KClO_3} \text{ and } \frac{3 \text{ mol } O_2}{2 \text{ mol } KCl}$$

Note that the number of mole ratios you can write for a chemical reaction involving a total of n substances is $(n)(n-1)$. Thus, for reactions involving four and five substances, you can write 12 and 20 moles ratios, respectively.

Four substances: $(4)(3) = 12$ mole ratios

Five substances: $(5)(4) = 20$ mole ratios

SECTION 1 REVIEW

Section Self-Check

Section Summary

- Balanced chemical equations can be interpreted in terms of moles, mass, and representative particles (atoms, molecules, formula units).
- The law of conservation of mass applies to all chemical reactions.
- Mole ratios are derived from the coefficients of a balanced chemical equation. Each mole ratio relates the number of moles of one reactant or product to the number of moles of another reactant or product in the chemical reaction.

5. **MAINIDEA Compare** the mass of the reactants and the mass of the products in a chemical reaction, and explain how these masses are related.
6. **State** how many mole ratios can be written for a chemical reaction involving three substances.
7. **Categorize** the ways in which a balanced chemical equation can be interpreted.
8. **Apply** The general form of a chemical reaction is $xA + yB \rightarrow zAB$. In the equation, A and B are elements, and x, y, and z are coefficients. State the mole ratios for this reaction.
9. **Apply** Hydrogen peroxide (H_2O_2) decomposes to produce water and oxygen. Write a balanced chemical equation for this reaction, and determine the possible mole ratios.
10. **Model** Write the mole ratios for the reaction of hydrogen gas and oxygen gas, $2H_2(g) + O_2(g) \rightarrow 2H_2O$. Make a sketch of six hydrogen molecules reacting with the correct number of oxygen molecules. Show the water molecules produced.

PRACTICE Problems

Have students refer to the Selected Solutions appendix for complete solutions to odd-numbered problems. The complete solutions for all problems can be found in the Solutions Manual.

3. a. $\frac{4 \text{ mol Al}}{3 \text{ mol } O_2}$ $\frac{3 \text{ mol } O_2}{2 \text{ mol } Al_2O_3}$ $\frac{2 \text{ mol } Al_2O_3}{4 \text{ mol Al}}$

$\frac{3 \text{ mol } O_2}{4 \text{ mol Al}}$ $\frac{2 \text{ mol } Al_2O_3}{3 \text{ mol } O_2}$ $\frac{4 \text{ mol Al}}{2 \text{ mol } Al_2O_3}$

b. $\frac{3 \text{ mol Fe}}{4 \text{ mol } H_2O}$ $\frac{3 \text{ mol Fe}}{4 \text{ mol } H_2}$ $\frac{3 \text{ mol Fe}}{1 \text{ mol } Fe_3O_4}$

$\frac{4 \text{ mol } H_2O}{3 \text{ mol Fe}}$ $\frac{4 \text{ mol } H_2}{3 \text{ mol Fe}}$ $\frac{1 \text{ mol } Fe_3O_4}{3 \text{ mol Fe}}$

$\frac{1 \text{ mol } Fe_3O_4}{4 \text{ mol } H_2}$ $\frac{1 \text{ mol } Fe_3O_4}{4 \text{ mol } H_2O}$ $\frac{4 \text{ mol } H_2O}{4 \text{ mol } H_2}$

$\frac{4 \text{ mol } H_2}{1 \text{ mol } Fe_3O_4}$ $\frac{4 \text{ mol } H_2O}{1 \text{ mol } Fe_3O_4}$ $\frac{4 \text{ mol } H_2}{4 \text{ mol } H_2O}$

c. $\frac{2 \text{ mol HgO}}{2 \text{ mol Hg}}$ $\frac{1 \text{ mol } O_2}{2 \text{ mol Hg}}$ $\frac{1 \text{ mol } O_2}{2 \text{ mol HgO}}$

$\frac{2 \text{ mol Hg}}{2 \text{ mol HgO}}$ $\frac{2 \text{ mol Hg}}{1 \text{ mol } O_2}$ $\frac{2 \text{ mol HgO}}{1 \text{ mol } O_2}$

4. a. $ZnO + 2HCl \rightarrow ZnCl_2 + H_2O$

$\frac{1 \text{ mol ZnO}}{2 \text{ mol HCl}}$ $\frac{1 \text{ mol ZnO}}{1 \text{ mol } ZnCl_2}$ $\frac{1 \text{ mol ZnO}}{1 \text{ mol } H_2O}$

$\frac{2 \text{ mol HCl}}{1 \text{ mol ZnO}}$ $\frac{2 \text{ mol HCl}}{1 \text{ mol } ZnCl_2}$ $\frac{2 \text{ mol HCl}}{1 \text{ mol } H_2O}$

$\frac{1 \text{ mol } ZnCl_2}{1 \text{ mol ZnO}}$ $\frac{1 \text{ mol } ZnCl_2}{2 \text{ mol HCl}}$ $\frac{1 \text{ mol } ZnCl_2}{1 \text{ mol } H_2O}$

$\frac{1 \text{ mol } H_2O}{1 \text{ mol ZnO}}$ $\frac{1 \text{ mol } H_2O}{2 \text{ mol HCl}}$ $\frac{1 \text{ mol } H_2O}{1 \text{ mol } ZnCl_2}$

b. $2C_4H_{10} + 13O_2 \rightarrow 8CO_2 + 10H_2O$

$\frac{2 \text{ mol } C_4H_{10}}{13 \text{ mol } O_2}$ $\frac{2 \text{ mol } C_4H_{10}}{8 \text{ mol } CO_2}$ $\frac{2 \text{ mol } C_4H_{10}}{10 \text{ mol } H_2O}$

$\frac{13 \text{ mol } O_2}{2 \text{ mol } C_4H_{10}}$ $\frac{8 \text{ mol } CO_2}{2 \text{ mol } C_4H_{10}}$ $\frac{10 \text{ mol } H_{20}}{2 \text{ mol } C_4H_{10}}$

$\frac{10 \text{ mol } H_2}{13 \text{ mol } O_2}$ $\frac{10 \text{ mol } H_2}{8 \text{ mol } CO_2}$ $\frac{8 \text{ mol } CO_2}{13 \text{ mol } O_2}$

$\frac{13 \text{ mol } O_2}{10 \text{ mol } H_2O}$ $\frac{8 \text{ mol } CO_2}{10 \text{ mol } H_2O}$ $\frac{13 \text{ mol } O_2}{8 \text{ mol } CO_2}$

SECTION 1 REVIEW

5. The coefficients indicate the molar relationship between each pair of reactants and products. The masses of reactants and products are equal.
6. $(3)(2) = 6$ ratios
7. particles (atoms, molecules, formula units), moles, and mass
8. xA/yB and xA/zAB, yB/xA and yB/zAB, zAB/xA and zAB/yB
9. $2H_2O_2 \rightarrow 2H_2O + O_2$
   2 mol H_2O_2/2 mol H_2O, 2 mol H_2O_2/1 mol O_2, 2 mol H_2O/2 mol H_2O_2, 2 mol H_2O/1 mol O_2, 1 mol O_2/2 mol H_2O_2, 1 mol O_2/2 mol H_2O
10. $2H_2/O_2$ and $2H_2/2H_2O$, $O_2/2H_2$ and $O_2/2H_2O$, $2H_2O/2H_2$ and $2H_2O/O_2$
    Student sketches should show six hydrogen molecules reacting with three oxygen molecules to form six water molecules.

Essential Questions

- What is the sequence of steps used in solving stoichiometric problems?
- How are these steps applied to solve stoichiometric problems?

Review Vocabulary

chemical reaction: a process in which the atoms of one or more substances are rearranged to form different substances

Stoichiometric Calculations

MAINIDEA The solution to every stoichiometric problem requires a balanced chemical equation.

CHEM 4 YOU Baking requires accurate measurements. That is why it is necessary to follow a recipe when baking cookies from scratch. If you need to make more cookies than a recipe yields, what must you do?

Using Stoichiometry

What tools are needed to perform stoichiometric calculations? All stoichiometric calculations begin with a balanced chemical equation. Mole ratios based on the balanced chemical equation are needed, as well as mass-to-mole conversions.

Stoichiometric mole-to-mole conversion The vigorous reaction between potassium and water is shown in **Figure 3.** The balanced chemical equation is as follows.

$$2K(s) + 2H_2O(l) \longrightarrow 2KOH(aq) + H_2(g)$$

FOLDABLES® Incorporate information from this section into your Foldable.

From the balanced equation, you know that two moles of potassium yield one mole of hydrogen. But how much hydrogen is produced if only 0.0400 mol of potassium is used? To answer this question, identify the given, or known, substance and the substance that you need to determine. The given substance is 0.0400 mol of potassium. The unknown is the number of moles of hydrogen. Because the given substance is in moles and the unknown substance to be determined is also in moles, this problem involves a mole-to-mole conversion.

To solve the problem, you need to know how the unknown moles of hydrogen are related to the known moles of potassium. Previously, you learned to derive mole ratios from the balanced chemical equation. Mole ratios are used as conversion factors to convert the known number of moles of one substance to the unknown number of moles of another substance in the same reaction. Several mole ratios can be written from the equation, but how do you choose the correct one?

■ **Figure 3** Potassium metal reacts vigorously with water, releasing so much heat that the hydrogen gas formed in the reaction catches fire.

Stephen Frisch/McGraw-Hill Education

1 Focus

MAINIDEA

Adjust for Quantitative Differences A recipe for a banana split calls for four ingredients, each in a particular quantity: one banana, two scoops of ice cream, 50 mL of chocolate syrup and 50 mL of strawberry syrup. Write the ingredients on the board in the form of a chemical equation: __banana + _ice cream + __chocolate syrup + __strawberry syrup → __banana split. Then, "balance" the equation by giving the quantities: banana + 2 ice cream + 50 chocolate syrup + 50 strawberry syrup → 1 banana split. Ask students what is needed to make five banana splits following the same recipe. 5 bananas, 10 scoops of ice cream, 250 mL of chocolate syrup, and 250 mL of strawberry syrup. BL OL

2 Teach

Concept Development

Using Moles Point out to students that moles are always involved when solving stoichiometry problems. Remind them that the mole ratio is needed to convert from one substance in the balanced equation to another substance. If they are in doubt about how to proceed in solving a problem, remind them to "go to moles first." OL

FOLDABLES® Rubric

CHEMISTRY PROJECT

Soap Making Soap making is a multistep process involving several chemical reactions. Have students research the process, identify the compounds and substances involved, and summarize the basic reactions. In particular, have students write a balanced chemical equation for a reaction used to produce lye. From the equation, ask them to write all of the mole ratios. AL

Quick Demo

Balancing Equations Ask students to write the balanced equations for the formation of carbon disulfide from carbon and sulfur and to write the meaning of each formula. 4C (4 atoms of carbon) + $1S_8$ (1 molecule of sulfur) $\rightarrow$ $4CS_2$ (4 molecules of carbon disulfide) Ask students to write another equation including the masses of the reactants and the product. 48 u C + 256 u S $\rightarrow$ 304 u CS_2 Explain that individual atoms and molecules are too small to be massed. It is easier to make carbon disulfide by reacting four molar masses of carbon, 48 g, rather than four atomic masses, and one molar mass of S_8, 256 g, rather than one molecular mass. Ask students how much carbon disulfide would be produced. one molar mass of carbon disulfide, 304 g **OL** **EL**

ASSESSMENT

Knowledge Have students consider the reaction of 2.50 mol calcium hydride with excess water according to this equation.
$CaH_2 + 2H_2O \rightarrow Ca(OH)_2 + 2H_2$
Ask the following questions: What mass of calcium hydroxide will be produced? 185 g What mass of water is needed? 90.1 g What mass of hydrogen is produced? 10.1 g What is the mass of 2.50 mol calcium hydride? 105 g Have students verify the law of conservation of mass. 105 g CaH_2 + 90.1 g H_2O = 185 g $Ca(OH)_2$ + 10.1 g H_2 **OL**

As shown below, the correct mole ratio, 1 mol H_2 to 2 mol K, has moles of unknown in the numerator and moles of known in the denominator. Using this mole ratio converts the moles of potassium to the unknown number of moles of hydrogen.

$$\text{moles of known} \times \frac{\text{moles of unknown}}{\text{moles of known}} = \text{moles of unknown}$$

$$0.0400 \text{ mol K} \times \frac{1 \text{ mol } H_2}{2 \text{ mol K}} = 0.0200 \text{ mol } H_2$$

The following Example Problems show mole-to-mole, mole-to-mass, and mass-to-mass stoichiometry problems. The process used to solve these problems is outlined in the Problem-Solving Strategy below.

PROBLEM-SOLVING STRATEGY

Mastering Stoichiometry

The flowchart below outlines the steps used to solve mole-to-mole, mole-to-mass, and mass-to-mass stoichiometric problems.

1. Complete Step 1 by writing the balanced chemical equation for the reaction.
2. To determine where to start your calculations, note the unit of the given substance.
   - If mass (in grams) of the given substance is the starting unit, begin your calculations with Step 2.
   - If amount (in moles) of the given substance is the starting unit, skip Step 2 and begin your calculations with Step 3.
3. The end point of the calculation depends on the desired unit of the unknown substance.
   - If the answer must be in moles, stop after completing Step 3.
   - If the answer must be in grams, stop after completing Step 4.

Apply the Strategy

Apply the Problem-Solving Strategy to Example Problems 2, 3, and 4.

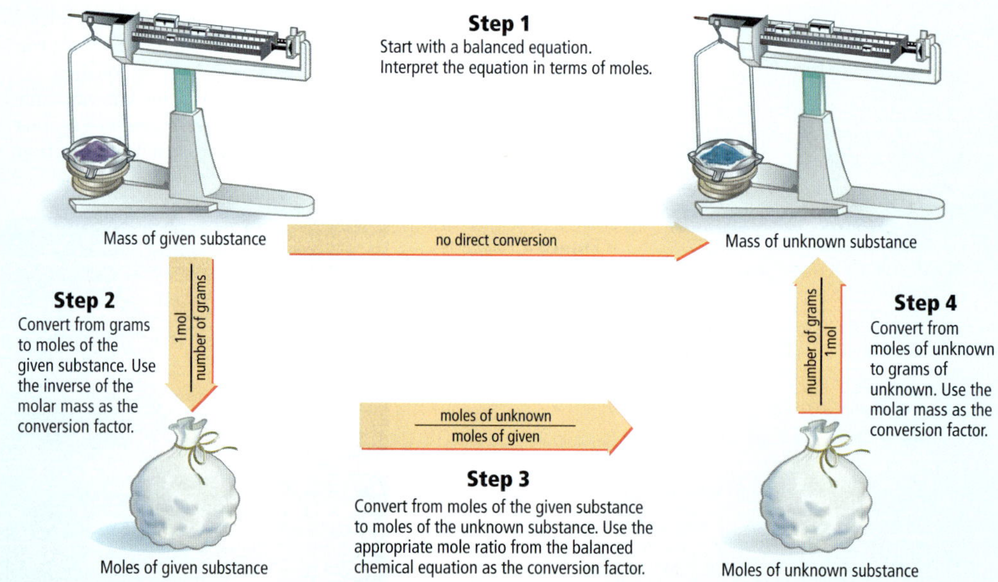

DIFFERENTIATED INSTRUCTION

Struggling Learners Stress the importance of solving stoichiometric problems step by step and understanding the reason for each step. Ask students why the balanced chemical equation is important. It determines the mole ratios. Give them an equation, and ask them to explain how they would convert from moles of one substance to moles of another substance in the same equation. Use the mole ratio: moles of unknown substance divided by moles of known substance. This might be a point of confusion. Have students write a mole ratio on one side of an index card and the inverse of the same ratio on the other side. As they solve a problem, they can look at the index card to determine which ratio gives the correct unit for the answer. **BL**

EXAMPLE Problem 2

Find help with **ratios.** Math Handbook

MOLE-TO-MOLE STOICHIOMETRY One disadvantage of burning propane (C_3H_8) is that carbon dioxide (CO_2) is one of the products. The released carbon dioxide increases the concentration of CO_2 in the atmosphere. How many moles of CO_2 are produced when 10.0 mol of C_3H_8 are burned in excess oxygen in a gas grill?

1 ANALYZE THE PROBLEM

You are given moles of the reactant, C_3H_8 and must find the moles of the product, CO_2. First write the balanced chemical equation, then convert from moles of C_3H_8 to moles of CO_2. The correct mole ratio has moles of unknown substance in the numerator and moles of known substance in the denominator.

Known	**Unknown**
moles C_3H_8 = 10.0 mol C_3H_8	**moles CO_2 = ? mol CO_2**

2 SOLVE FOR THE UNKNOWN

Write the balanced chemical equation for the combustion of C_3H_8. Use the correct mole ratio to convert moles of known (C_3H_8) to moles of unknown (CO_2).

10.0 mol **? mol**

$$C_3H_8(g) + 5O_2(g) \rightarrow 3CO_2(g) + 4H_2O(g)$$

$$\text{Mole ratio: } \frac{3 \text{ mol } CO_2}{1 \text{ mol } C_3H_8}$$

$$10.0 \cancel{\text{ mol } C_3H_8} \times \frac{3 \text{ mol } CO_2}{1 \cancel{\text{ mol } C_3H_8}} = \mathbf{30.0 \text{ mol } CO_2}$$

Burning 10.0 moles of C_3H_8 produces 30.0 moles CO_2.

3 EVALUATE THE ANSWER

Because the given number of moles has three significant figures, the answer also has three figures. The balanced chemical equation indicates that 1 mol of C_3H_8 produces 3 mol of CO_2. Thus, 10.0 mol of C_3H_8 produces three times as many moles of CO_2, or 30.0 mol.

PRACTICE Problems

Do additional problems. Online Practice

11. Methane and sulfur react to produce carbon disulfide (CS_2), a liquid often used in the production of cellophane.

$$\_\_CH_4(g) + \_\_S_8(s) \rightarrow \_\_CS_2(l) + \_\_H_2S(g)$$

a. Balance the equation.

b. Calculate the moles of CS_2 produced when 1.50 mol S_8 is used.

c. How many moles of H_2S are produced?

12. Challenge Sulfuric acid (H_2SO_4) is formed when sulfur dioxide (SO_2) reacts with oxygen and water.

a. Write the balanced chemical equation for the reaction.

b. How many moles of H_2SO_4 are produced from 12.5 moles of SO_2?

c. How many moles of O_2 are needed?

RealWorld CHEMISTRY

Outdoor Cooking

GAS GRILLS Using outdoor grills is a popular way to cook. Gas grills burn either natural gas or propane that is mixed with air. The initial spark is provided by a grill starter. Propane is more commonly used for fuel because it can be supplied in liquid form in a portable tank. Combustion of liquid propane also releases more energy than natural gas.

Burke/Triolo Productions/Photolibrary/Getty Images

IN-CLASS Example

Question Butane, C_4H_{10}, is the gas burned in disposable lighters. How many moles of oxygen are needed to burn 5.0 mol of butane in a lighter to produce carbon dioxide and water?

Answer

$$2\,C_4H_{10} + 13\,O_2 \rightarrow 8\,CO_2 + 10\,H_2O$$

$$\text{mole ratio: } \frac{13 \text{ mol } O_2}{2 \text{ mol } C_4H_{10}}$$

$$5.0 \cancel{\text{ mol } C_4H_{10}} \times \frac{13 \text{ mol } O_2}{2 \cancel{\text{ mol } C_4H_{10}}}$$

$$= 32 \text{ mol } O_2$$

PRACTICE Problems

Have students refer to the Selected Solutions appendix for complete solutions to odd-numbered problems. The complete solutions for all problems can be found in the Solutions Manual.

11. a. $2CH_4(g) + S_8(s) \rightarrow 2CS_2(l) + 4H_2S(g)$

b. $1.50 \cancel{\text{ mol } S_8} \times \frac{2 \text{ mol } CS_2}{1 \cancel{\text{ mol } S_8}} = 3.00 \text{ mol } CS_2$

c. $1.50 \cancel{\text{ mol } S_8} \times \frac{4 \text{ mol } H_2S}{1 \cancel{\text{ mol } S_8}} = 6.00 \text{ mol } H_2S$

12. a. $2SO_2(g) + O_2(g) + 2H_2O(l) \rightarrow 2H_2SO_4(aq)$

b. $12.5 \cancel{\text{ mol } SO_2} \times \frac{2 \text{ mol } H_2SO_4}{2 \cancel{\text{ mol } SO_2}} = 12.5 \text{ mol } H_2SO_4 \text{ produced}$

c. $12.5 \cancel{\text{ mol } SO_2} \times \frac{1 \text{ mol } O_2}{2 \cancel{\text{ mol } SO_2}} = 6.25 \text{ mol } O_2 \text{ needed}$

CHEMISTRY JOURNAL

How much acid is neutralized? Have students find the active ingredients on the label of a container of antacid tablets and identify the compound that neutralizes stomach acid. Have them note the recommended dosage and the amount of the neutralizing ingredient per tablet. Ask students to determine the mass of the active ingredient per dose. If the active ingredient is a metal hydroxide, the hydroxide reacts with HCl to produce a metal chloride and water. If the active ingredient is a metal carbonate, the products are a metal chloride, carbon dioxide, and water. Have students write the equation of the reaction of the antacid. Using the mass of the active ingredient per dose, have them determine the mass of hydrochloric acid neutralized per dose. **OL**

IN-CLASS Example

Question What is the mass of hydrogen produced when a 0.200 mol sample of sodium reacts with an excess of water to produce hydrogen and sodium hydroxide?

Answer

$2\,Na + 2\,H_2O \longrightarrow 2\,NaOH + H_2$

$\frac{1\ mol\ H_2}{2\ mol\ Na}$

$0.200\ \cancel{mol\ Na} \times \frac{1\ mol\ H_2}{2\ \cancel{mol\ Na}}$

$= 0.100\ mol\ H_2$

$0.100\ \cancel{mol\ H_2} \times \frac{2.016\ g\ H_2}{1\ \cancel{mol\ H_2}}$

$= 0.202\ g\ H_2$ will be produced

PRACTICE Problems

Have students refer to the Selected Solutions appendix for complete solutions to odd-numbered problems. The complete solutions for all problems can be found in the Solutions Manual.

13. $88.6\ g\ Cl_2$

14. **a.** $177\ g\ Cl_2$

b. $15.0\ g\ C$

c. $292\ g$

15. $64.64\ g\ N_2$

16. $2SO_2(g) + O_2(g) + 2H_2O(l) \longrightarrow 2H_2SO_4(aq)$; $3.83\ g\ H_2SO_4$

Stoichiometric mole-to-mass conversion Now, suppose you know the number of moles of a reactant or product in a reaction and you want to calculate the mass of another product or reactant. This is an example of a mole-to-mass conversion.

EXAMPLE Problem 3

Find help with **significant figures.** Math Handbook

EXAMPLE PROBLEM

MOLE-TO-MASS STOICHIOMETRY Determine the mass of sodium chloride (NaCl), commonly called table salt, produced when 1.25 mol of chlorine gas (Cl_2) reacts vigorously with excess sodium.

1 ANALYZE THE PROBLEM

You are given the moles of the reactant, Cl_2, and must determine the mass of the product, NaCl. You must convert from moles of Cl_2 to moles of NaCl using the mole ratio from the equation. Then, you need to convert moles of NaCl to grams of NaCl using the molar mass as the conversion factor.

Known	**Unknown**
moles of chlorine = 1.25 mol Cl_2	**mass of sodium chloride = ? g NaCl**

2 SOLVE FOR THE UNKNOWN

$$\begin{matrix} & 1.25\ mol & ?\ g \\ 2Na(s) + & Cl_2(g) \longrightarrow & 2NaCl(s)\end{matrix}$$

Write the balanced chemical equation, and identify the known and the unknown values.

Mole ratio: $\frac{2\ mol\ NaCl}{1\ mol\ Cl_2}$

$$1.25\ \cancel{mol\ Cl_2} \times \frac{2\ mol\ NaCl}{1\ \cancel{mol\ Cl_2}} = 2.50\ mol\ NaCl$$

Multiply moles of Cl_2 by the mole ratio to get moles of NaCl.

$$2.50\ \cancel{mol\ NaCl} \times \frac{58.44\ g\ NaCl}{1\ \cancel{mol\ NaCl}} = \mathbf{146\ g\ NaCl}$$

Multiply moles of NaCl by the molar mass to get grams of NaCl.

3 EVALUATE THE ANSWER

Because the given number of moles has three significant figures, the mass of NaCl also has three. To quickly assess whether the calculated mass value for NaCl is correct, perform the calculations in reverse: divide the mass of NaCl by the molar mass of NaCl, and then divide the result by 2. You will obtain the given number of moles of Cl_2.

PRACTICE Problems

Do additional problems. Online Practice

PRACTICE PROBLEMS

13. Sodium chloride is decomposed into the elements sodium and chlorine by means of electrical energy. How much chlorine gas, in grams, is obtained from the process diagrammed at right?

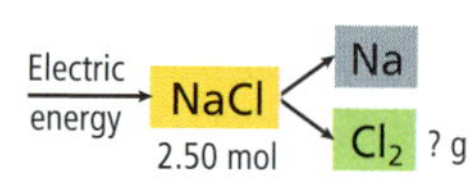

14. **Challenge** Titanium is a transition metal used in many alloys because it is extremely strong and lightweight. Titanium tetrachloride ($TiCl_4$) is extracted from titanium oxide (TiO_2) using chlorine and coke (carbon).

$$TiO_2(s) + C(s) + 2Cl_2(g) \longrightarrow TiCl_4(s) + CO_2(g)$$

a. What mass of Cl_2 gas is needed to react with 1.25 mol of TiO_2?

b. What mass of C is needed to react with 1.25 mol of TiO_2?

c. What is the mass of all of the products formed by reaction with 1.25 mol of TiO_2?

CULTURAL DIVERSITY

Stoichiometry in Soap Making Soap making dates to 2800 B.C. in Babylonia. During the 17th century, soap was a luxury item for the wealthy. However, as materials became more readily available, soap was made in almost every household. Crude stoichiometric techniques were used by all early soap-making cultures. Soap making requires three steps, making lye, rendering (or cleaning) fat, and then heating the lye-fat mixture. Sodium chloride was added to make the soap hard. Scents were also added, and the soap was rolled into balls or cut into bars to be stored. Deer fat, whale blubber, tallow, and olive oil have been used as sources of fat in soap depending upon which was available in the area. Countries bordering the Mediterranean Sea used the ashes of local shrubs instead of potash (K_2CO_3). Ashes from these plants are rich in sodium carbonate, and were a valuable trade item called barilla.

Stoichiometric mass-to-mass conversion If you were preparing to carry out a chemical reaction in the laboratory, you would need to know how much of each reactant to use in order to produce the mass of product you required. Example Problem 4 demonstrates how you can use a measured mass of the known substance, the balanced chemical equation, and mole ratios from the equation to find the mass of the unknown substance. The ChemLab at the end of this chapter will provide you with laboratory experience in determining a mole ratio.

Watch a **video about stoichiometry.** Video

EXAMPLE Problem 4

Find help with **dimensional analysis.** Math Handbook

MASS-TO-MASS STOICHIOMETRY Ammonium nitrate (NH_4NO_3), an important fertilizer, produces dinitrogen monoxide (N_2O) gas and H_2O when it decomposes. Determine the mass of H_2O produced from the decomposition of 25.0 g of solid NH_4NO_3.

1 ANALYZE THE PROBLEM

You are given a description of the chemical reaction and the mass of the reactant. You need to write the balanced chemical equation and convert the known mass of the reactant to moles of the reactant. Then, use a mole ratio to relate moles of the reactant to moles of the product. Finally, use the molar mass to convert from moles of the product to the mass of the product.

Known
mass of ammonium nitrate = 25.0 g NH_4NO_3

Unknown
mass of water = ? g H_2O

2 SOLVE FOR THE UNKNOWN

$$\overset{25.0\text{ g}}{NH_4NO_3(s)} \rightarrow N_2O(g) + \overset{?\text{ g}}{2H_2O(g)}$$

Write the balanced chemical equation, and identify the known and unknown values.

$$25.0\ \cancel{\text{g NH}_4\text{NO}_3} \times \frac{1\text{ mol NH}_4\text{NO}_3}{80.04\ \cancel{\text{g NH}_4\text{NO}_3}} = 0.312\text{ mol NH}_4\text{NO}_3$$

Multiply grams of NH_4NO_3 by the inverse of molar mass to get moles of NH_4NO_3.

$$\text{Mole ratio: } \frac{2\text{ mol H}_2\text{O}}{1\text{ mol NH}_4\text{NO}_3}$$

$$0.312\ \cancel{\text{mol NH}_4\text{NO}_3} \times \frac{2\text{ mol H}_2\text{O}}{1\ \cancel{\text{mol NH}_4\text{NO}_3}} = 0.624\text{ mol H}_2\text{O}$$

Multiply moles of NH_4NO_3 by the mole ratio to get moles of H_2O.

$$0.624\ \cancel{\text{mol H}_2\text{O}} \times \frac{18.02\text{ g H}_2\text{O}}{1\ \cancel{\text{mol H}_2\text{O}}} = \mathbf{11.2\text{ g H}_2\text{O}}$$

Multiply moles of H_2O by the molar mass to get grams of H_2O.

3 EVALUATE THE ANSWER

The number of significant figures in the answer, three, is determined by the given grams of NH_4NO_3. To verify that the mass of H_2O is correct, perform the calculations in reverse.

EXAMPLE PROBLEM

PRACTICE Problems

Do additional problems. Online Practice

15. One of the reactions used to inflate automobile air bags involves sodium azide (NaN_3): $2NaN_3(s) \rightarrow 2Na(s) + 3N_2(g)$. Determine the mass of N_2 produced from the decomposition of NaN_3 shown at right.

16. Challenge In the formation of acid rain, sulfur dioxide (SO_2) reacts with oxygen and water in the air to form sulfuric acid (H_2SO_4). Write the balanced chemical equation for the reaction. If 2.50 g of SO_2 reacts with excess oxygen and water, how much H_2SO_4, in grams, is produced?

Video Lab

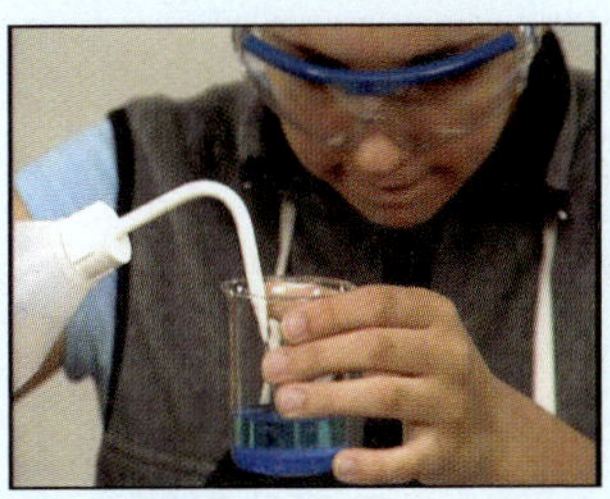

Video

Stoichiometry

IN-CLASS Example

Question The Titan rocket uses a combination of hydrazine, N_2H_4, and dinitrogen tetroxide, N_2O_4 as rocket fuel according to the **reaction:**
$2N_2H_4 + N_2O_4 \rightarrow 3N_2 + 4H_2O$
If 200 g of hydrazine is used, how many grams of nitrogen is produced?

Answer $2N_2H_4 + N_2O_4 \rightarrow 3N_2 + 4H_2O$

$$200\ \cancel{\text{g N}_2\text{H}_4} \times \frac{1\text{ mol N}_2\text{H}_4}{32.05\ \cancel{\text{g N}_2\text{H}_4}} = 6.24\text{ mol N}_2\text{H}_4$$

$$6.24\ \cancel{\text{mol N}_2\text{H}_4} \times \frac{4\text{ mol H}_2\text{O}}{2\ \cancel{\text{mol N}_2\text{H}_4}} = 12.5\text{ mol H}_2\text{O}$$

$$12.5\ \cancel{\text{mol H}_2\text{O}} \times \frac{18.02\text{ g H}_2\text{O}}{1\ \cancel{\text{mol H}_2\text{O}}} = 225\text{ g H}_2\text{O}$$

3 Assess

Check for Understanding

Write the balanced equation for the reaction between iron and sulfur on the board: $16Fe + 3S_8 \rightarrow 8Fe_2S_3$. Have students determine the mass of iron(III) sulfide produced if 7.00 g of iron is used. 13.0 g Fe_2S_3 **OL**

Reteach

Write this equation on the board: $6CO_2 + 6H_2O \rightarrow C_6H_{12}O_6 + 6O_2$. Ask students how many grams of glucose form when 88.0 g of CO_2 reacts? Before they pick up their pencils and calculators, have them verbalize each step in the solution and explain why each step is needed. Once they can explain each step, they will be able to solve the problem. 60.0 g $C_6H_{12}O_6$ **OL**

Extension

Give student groups a stoichiometric problem to solve. Ask them to illustrate the solution using a poster similar to the Problem-Solving Strategy *Mastering Stoichiometry* graphic. **OL**

ChemLAB

The ChemLab located at the end of the chapter can be used at this point in the lesson.

MiniLAB

Rubric

Purpose Predict the mass of a product and compare it with the experimental mass.

Process Skills collect and interpret data, predict, use numbers

Safety Precautions Approve lab safety forms before beginning. Warn students about the hot crucible.

Disposal The sodium carbonate product can be placed in the trash.

Teaching Strategies
Have students calculate the final mass of the product before starting.

Expected Results 3.12 g of baking soda yields about 1.90 g Na_2CO_3

Analysis

1. During heating, the product initially looked "wet" and bubbles appeared. In time, the product "dried out."
2. The two masses should be similar.
3. Assuming accepted and actual values are 1.97 g and 1.90 g, respectively; error = −0.07 g; % error = 3.55%.
4. errors from each mass measurement, mass of moisture absorbed by crucible OL

MiniLAB

Apply Stoichiometry

How much sodium carbonate (Na_2CO_3) is produced when baking soda decomposes? Baking soda is used in many baking recipes because it makes batter rise, which results in a light and fluffy texture. This occurs because baking soda, sodium hydrogen carbonate ($NaHCO_3$), decomposes upon heating to form carbon dioxide gas according to the following equation.

$$2NaHCO_3 \rightarrow Na_2CO_3 + CO_2 + H_2O$$

Procedure

1. Read and complete the lab safety form.
2. Create a data table to record your experimental data and observations.
3. Use a **balance** to measure the mass of a clean, dry **crucible.** Add about 3.0 g of **sodium hydrogen carbonate ($NaHCO_3$),** and measure the combined mass of the crucible and $NaHCO_3$. Record both masses in your data table, and calculate the mass of the $NaHCO_3$.
4. Use this starting mass of $NaHCO_3$ and the balanced chemical equation to calculate the mass of Na_2CO_3 that will be produced.
5. Set up a **ring stand** with a **ring** and **clay triangle** for heating the crucible.
6. Heat the crucible with a **Bunsen burner,** slowly at first and then with a stronger flame, for 7–8 min. Record your observations during the heating.
7. Turn off the burner, and use **crucible tongs** to remove the hot crucible.
   WARNING: ***Do not touch the hot crucible with your hands.***
8. Allow the crucible to cool, and then measure the mass of the crucible and Na_2CO_3.

Analysis

1. **Describe** what you observed during the heating of the baking soda.
2. **Compare** your calculated mass of Na_2CO_3 with the actual mass you obtained from the experiment.
3. **Calculate** Assume that the mass of Na_2CO_3 that you calculated in Step 4 is the accepted value for the mass of product that will form. Calculate the error and percent error associated with the experimentally measured mass.
4. **Identify** sources of error in the procedure that led to errors calculated in Question 3.

SECTION 2 REVIEW

Section Self-Check

Section Summary

- Chemists use stoichiometric calculations to predict the amounts of reactants used and products formed in specific reactions.
- The first step in solving stoichiometric problems is writing the balanced chemical equation.
- Mole ratios derived from the balanced chemical equation are used in stoichiometric calculations.
- Stoichiometric problems make use of mole ratios to convert between mass and moles.

17. MAINIDEA **Explain** why a balanced chemical equation is needed to solve a stoichiometric problem.
18. **List** the four steps used in solving stoichiometric problems.
19. **Describe** how a mole ratio is correctly expressed when it is used to solve a stoichiometric problem.
20. **Apply** How can you determine the mass of liquid bromine (Br_2) needed to react completely with a given mass of magnesium?
21. **Calculate** Hydrogen reacts with excess nitrogen as follows:
    $$N_2(g) + 3H_2(g) \rightarrow 2NH_3(g)$$
    If 2.70 g of H_2 reacts, how many grams of NH_3 is formed?
22. **Design** a concept map for the following reaction.
    $$CaCO_3(s) + 2HCl(aq) \rightarrow CaCl_2(aq) + H_2O(l) + CO_2(g)$$
    The concept map should explain how to determine the mass of $CaCl_2$ produced from a given mass of HCl.

SECTION 2 REVIEW

17. The coefficients in the balanced equation indicate the molar relationship between each pair of reactants and products.
18. balance the equation; convert the mass of the known substance to moles of known substance; use the mole ratio to convert from moles of the known to moles of the unknown; convert moles of unknown to mass of the unknown
19. moles of unknown/moles of known
20. Write a balanced equation. Convert the given mass of Mg to moles. Use the mole ratio from the balanced equation to convert moles of Mg to moles of Br_2. Convert from moles of Br_2 to mass of Br_2.
21. 15.2 g NH_3
22. Concept maps will vary, but all should show the use of these conversion factors: the inverse of molar mass, the mole ratio, the molar mass.

SECTION 3

Essential Questions

- In a chemical reaction, which reactant is the limiting reactant?
- How much of the excess reactant remains after the reaction is complete?
- How do you calculate the mass of a product when the amounts of more than one reactant are given?

Review Vocabulary

molar mass: the mass in grams of one mole of any pure substance

New Vocabulary

limiting reactant
excess reactant

Limiting Reactants

MAINIDEA A chemical reaction stops when one of the reactants is used up.

CHEM 4 YOU **If there are more boys than girls at a school dance, some boys will be left without dance partners. The situation is much the same for the reactants in a chemical reaction—excess reactants cannot participate.**

Why do Reactions Stop?

Rarely in nature are the reactants present in the exact ratios specified by the balanced chemical equation. Generally, one or more reactants are in excess and the reaction proceeds until all of one reactant is used up. When a reaction is carried out in the laboratory, the same principle applies. Usually, one or more of the reactants are in excess, while one is limited. The amount of product depends on the reactant that is limited.

Limiting and excess reactants Recall the reaction from the Launch Lab. After the colorless solution formed, adding more sodium hydrogen sulfite had no effect because there was no more potassium permanganate available to react with it. Potassium permanganate was a limiting reactant. As the name implies, the **limiting reactant** limits the extent of the reaction and, thereby, determines the amount of product formed. A portion of all the other reactants remains after the reaction stops. Reactants leftover when a reaction stops are **excess reactants.**

To help you understand limiting and excess reactants, consider the analogy in **Figure 4.** From the available tools, four complete sets consisting of a pair of pliers, a hammer, and two screwdrivers can be assembled. The number of sets is limited by the number of available hammers. Pliers and screwdrivers remain in excess.

Figure 4 Each tool set must have one hammer, so only four sets can be assembled.
Interpret *How many more hammers are required to complete a fifth set?*

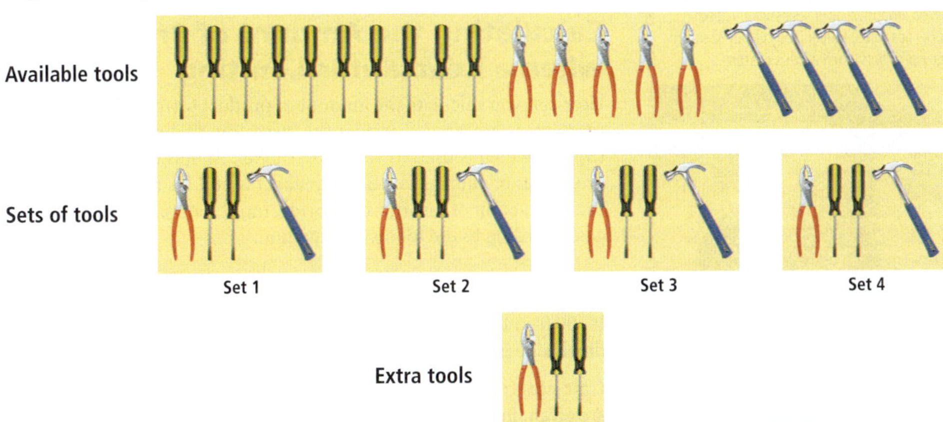

Aaron Haupt

SECTION 3

1 Focus

MAINIDEA

Limited Reactants Refer students to the Launch Lab and ask what happened to the potassium permanganate. The solution turned from purple to colorless. How much sodium hydrogen sulfite was added before the color change occurred? Answers will vary. It should be around 8 mL of $NaHSO_3$ added to change the color. Ask students what happened if they added another drop of sodium hydrogen sulfite. Nothing, the solution remained colorless. Ask if they can explain why that was the case. All the potassium permanganate had reacted, and there was no more to react with the additional drop. OL

2 Teach

Quick Demo

Adding Reactants Using a straw, exhale into a test tube of limewater. Explain that the limewater becomes cloudy because insoluble calcium carbonate is formed. Write the equation for the reaction on the board or project the image: $CO_2(g) + CaO(aq) \rightarrow CaCO_3(s)$. Ask students to determine the mass of calcium carbonate that would be produced if you exhaled 0.0900 mol CO_2 into the test tube. 9.00 g $CaCO_3$ OL

Caption Question Fig. 4 one additional hammer is needed BL EL

CHEMISTRY JOURNAL

Stoichiometric Steps In their chemistry journals, have students write the heading "Steps in Solving Limiting Reactant Stoichiometric Problems." As they study Section 3, have students list the steps needed for calculating the amounts of products formed when the masses of all reactants are given. Students should explain the purpose for each step. OL

READING CHECK six molecules

Identify Misconceptions

Students often have difficulty determining the limiting reactant when given the mass of more than one reactant.

Uncover the Misconception

Ask how much magnesium oxide is produced when 48.6 g of magnesium reacts with 64.0 g of oxygen according to this equation: $2Mg + O_2 \rightarrow 2MgO$. Have students determine the moles of each of the reactants given in the problem. 2.00 mol Mg, 2.00 mol O_2 Ask which is the limiting reactant. Mg Ask what is the mass of magnesium oxide produced. 80.6 g

Demonstrate the Concept

Use nuts and bolts to represent moles of atoms. Represent 2 mol of Mg using two bolts and 2 mol of O_2 using four nuts. Have students identify the mole ratio between O_2 molecules and Mg atoms. 1 mol O_2 molecules/2 mol Mg atoms, or 2 nuts/2 bolts Ask what is in excess. 1 mol O_2, or 2 nuts Represent 1 mol of MgO using one bolt and one nut. Ask how many MgO formula units can be formed. 2 units of 1 nut and 1 bolt with 2 nuts left over

Assess New Knowledge

Give students this problem. A 54.0-g sample of aluminum reacts with 508 g of iodine according to this equation: $2Al + 3I_2 \rightarrow 2AlI_3$. Have them determine the number of moles of each reactant. 2.00 mol Al, 2.00 mol I_2 Ask which is the limiting reactant. I_2 Ask how many moles of AlI_3 are formed. 1.33 mol OL BL EL

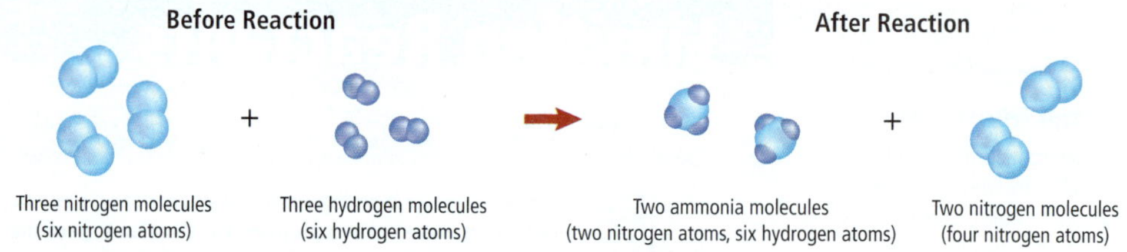

Figure 5 If you check all the atoms present before and after the reaction, you will find that some of the nitrogen molecules are unchanged. These nitrogen molecules are the excess reactant.

View an **animation about limiting reactants.**

Concepts In Motion

Determining the limiting reactant The calculations you did in the previous section were based on having the reactants present in the ratio described by the balanced chemical equation. When this is not the case, the first thing you must do is determine which reactant is limiting.

Consider the reaction shown in **Figure 5,** in which three molecules of nitrogen (N_2) and three molecules of hydrogen (H_2) react to form ammonia (NH_3). In the first step of the reaction, all the nitrogen molecules and hydrogen molecules are separated into individual atoms. These atoms are available for reassembling into ammonia molecules, just as the tools in **Figure 4** are available to be assembled into tool kits. How many molecules of ammonia can be produced from the available atoms? Two ammonia molecules can be assembled from the hydrogen atoms and nitrogen atoms because only six hydrogen atoms are available—three for each ammonia molecule. When the hydrogen is gone, two unreacted molecules of nitrogen remain. Thus, hydrogen is the limiting reactant and nitrogen is the excess reactant. It is important to know which reactant is the limiting reactant because, as you have just read, the amount of product formed depends on this reactant.

READING CHECK Extend How many more hydrogen molecules would be needed to completely react with the excess nitrogen molecules shown in **Figure 5?**

Figure 6 Natural rubber, which is soft and very sticky, is hardened in a chemical process called vulcanization. During vulcanization, molecules become linked together, forming a durable material that is harder, smoother, and less sticky. These properties make vulcanized rubber ideal for many products, such as this caster.

Calculating the Amount of Product when a Reactant Is Limiting

How can you calculate the amount of product formed when one of the reactants is limiting? Consider the formation of disulfur dichloride (S_2Cl_2), which is used to vulcanize rubber. As shown in **Figure 6,** the properties of vulcanized rubber make it useful for many products. In the production of disulfur dichloride, molten sulfur reacts with chlorine gas according to the following equation.

$$S_8(l) + 4Cl_2(g) \rightarrow 4S_2Cl_2(l)$$

If 200.0 g of sulfur reacts with 100.0 g of chlorine, what mass of disulfur dichloride is produced?

Calculating the limiting reactant The masses of both reactants are given. First, determine which one is the limiting reactant, because the reaction stops producing product when the limiting reactant is used up.

joey333/iStock/Getty Images

CHEMISTRY PROJECT

Maximum Yield of Carbon Dioxide Have student groups design an experiment to determine the maximum yield of carbon dioxide produced by the reaction of baking soda with vinegar. Suggested equipment includes a plastic soft drink bottle, vinegar, baking soda, a balloon to cover the mouth of the bottle, and a teaspoon. One teaspoonful of baking soda is approximately 10 g and one teaspoonful of vinegar is approximately 4.9 mL. Vinegar is 5% acetic acid. Suggest that students use a constant amount of baking soda (one teaspoonful) and vary the amount of vinegar added. Ask students to find out if there is a point at which no more product is produced. OL COOP LEARN

Moles of reactants Identifying the limiting reactant involves finding the number of moles of each reactant. You can do this by converting the masses of chlorine and sulfur to moles. Multiply each mass by a conversion factor that relates moles and mass–the inverse of molar mass.

$$100.0\ \cancel{g\ Cl_2} \times \frac{1\ mol\ Cl_2}{70.91\ \cancel{g\ Cl_2}} = 1.410\ mol\ Cl_2$$

$$200.0\ \cancel{g\ S_8} \times \frac{1\ mol\ S_8}{256.5\ \cancel{g\ S_8}} = 0.7797\ mol\ S_8$$

Using mole ratios The next step involves determining whether the two reactants are in the correct mole ratio, as given in the balanced chemical equation. The coefficients in the balanced chemical equation indicate that 4 mol of chlorine is needed to react with 1 mol of sulfur. This 4:1 ratio from the equation must be compared with the actual ratio of the moles of available reactants just calculated above. To determine the actual ratio of moles, divide the number of available moles of chlorine by the number of available moles of sulfur.

$$\frac{1.410\ mol\ Cl_2\ available}{0.7797\ mol\ S_8\ available} = \frac{1.808\ mol\ Cl_2\ available}{1\ mol\ S_8\ available}$$

Only 1.808 mol of chlorine is available for every 1 mol of sulfur, instead of the 4 mol of chlorine required by the balanced chemical equation. Therefore, chlorine is the limiting reactant.

Calculating the amount of product formed After determining the limiting reactant, the amount of product in moles can be calculated by multiplying the given number of moles of the limiting reactant (1.410 mol Cl_2) by the mole ratio relating disulfur dichloride and chlorine. Then, moles of S_2Cl_2 are converted to grams of S_2Cl_2 by multiplying by the molar mass. These calculations can be combined as shown.

$$1.410\ \cancel{mol\ Cl_2} \times \frac{4\ \cancel{mol\ S_2Cl_2}}{4\ \cancel{mol\ Cl_2}} \times \frac{135.0\ g\ S_2Cl_2}{1\ \cancel{mol\ S_2Cl_2}} = 190.4\ g\ S_2Cl_2$$

Thus, 190.4 g S_2Cl_2 forms when 1.410 mol Cl_2 reacts with excess S_8.

Analyzing the excess reactant Now that you have determined the limiting reactant and the amount of product formed, what about the excess reactant, sulfur? How much of it reacted?

Moles reacted You need to make a mole-to-mass calculation to determine the mass of sulfur needed to react completely with 1.410 mol of chlorine. First, obtain the number of moles of sulfur by multiplying the moles of chlorine by the S_8-to-Cl_2 mole ratio.

$$1.410\ \cancel{mol\ Cl_2} \times \frac{1\ mol\ S_8}{4\ \cancel{mol\ Cl_2}} = 0.3525\ mol\ S_8$$

Mass reacted Next, to obtain the mass of sulfur needed, multiply 0.3525 mol of S_8 by its molar mass.

$$0.3525\ \cancel{mol\ S_8} \times \frac{265.5\ g\ S_8}{1\ \cancel{mol\ S_8}} = 90.42\ g\ S_8\ needed$$

Excess remaining Knowing that 200.0 g of sulfur is available and that only 90.42 g of sulfur is needed, you can calculate the amount of sulfur left unreacted when the reaction ends.

$$200.0\ g\ S_8\ available - 90.42\ g\ S_8\ needed = 109.6\ g\ S_8\ in\ excess$$

CAREERS IN CHEMISTRY

Pharmacist Knowledge of drug composition, modes of action, and possible harmful interactions with other substances allows a pharmacist to counsel patients on their care. Pharmacists also mix chemicals to form powders, tablets, ointments, and solutions.

WebQuest

APPLYING PRACTICES

Use Mathematical Representations Go to the resources tab in ConnectED to find the Applying Practices worksheet *Conservation of Mass.*

VOCABULARY

SCIENCE USAGE V. COMMON USAGE

Product

Science usage: a new substance formed during a chemical reaction

The sole reaction product was a colorless gas.

Common usage: something produced

The cosmetics counter in the department store had hundreds of products from which to choose.

Explore **limiting reactants.**

Virtual Investigations

DIFFERENTIATED INSTRUCTION

Advanced Learners Divide students into groups and tell them they are to form their own cookie company. They have the following ingredients available for making the cookies described in the Assessment on this page. Note, 1 teaspoon = 4.9 mL.

1000 c flour	10 doz eggs	40 sticks butter
1.00 L baking soda	100 oz chocolate chips	500 c brown sugar
500 c granulated sugar	1.00 L salt	1.00 L vanilla

Which of the ingredients is the limiting ingredient? butter How many dozen cookies can be made? 80 dozen If a dozen cookies sells for $4.00, how much money would the company gross? $320.00 AL COOP LEARN

CAREERS IN CHEMISTRY

WebQuest

ASSESSMENT

Knowledge Give students this recipe for chocolate chip cookies.

$2\frac{1}{4}$ c flour
1 tsp salt
1 tsp baking soda
1 c softened butter (2 sticks)
$\frac{3}{4}$ c granulated sugar
$\frac{3}{4}$ c brown sugar
2 eggs
2 tsp vanilla
3 oz chocolate chips
Yields 4 dozen cookies

Ask students how they could make twice as many cookies. double the ingredients Ask how they could make these cookies with only one egg. cut the recipe in half Ask students to relate this to limiting reactants. The number of eggs limits the number of cookies that can be made. Likewise, the amounts of reactants determine the quantity of products. OL

Content Background

Testing for Chemicals The primary pigment in lead-based paints was $PbCO_3{\cdot}Pb(OH)_2$ prior to 1940. Today, we know that lead in household products poses a health hazard, but old paint is still found in many homes. The level of lead in homes can be determined using a lead detection kit. One lead test kit uses a swab containing potassium rhodizonate ($C_6K_2O_6$). The swab is rubbed over the surface to be tested. If lead is present, a pink color appears on the swab due to the presence of lead(II) rhodizonate. The reaction is a single replacement described by the following equation: $Pb^{2+} + C_6K_2O_6 \rightarrow C_6PbO_6 + 2K^+$ The larger the amount of lead, the more red the swab becomes. This kit can detect amounts of lead as low as 0.006 g.

IN-CLASS Example

Question When silver (Ag) reacts with sulfur (S_8), the compound silver sulfide (Ag_2S) is formed.
$16Ag + S_8 \rightarrow 8Ag_2S$

a. When 4.00 g of silver reacts with 4.00 g of sulfur, what mass of silver sulfide is produced?

b. How much of the excess reactant remains after the reaction stops?

Answer

a. $4.00\ \cancel{g\ Ag} \times \frac{1\ mol}{107.9\ \cancel{g\ Ag}}$

$= 0.0371\ mol\ Ag$

$4.00\ \cancel{g\ S_8} \times \frac{1\ mol}{256.5\ \cancel{g\ S_8}}$

$= 0.0156\ mol\ S_8$

$0.0371\ \cancel{mol\ Ag} \times \frac{8\ mol\ Ag_2S}{16\ \cancel{mol\ Ag}}$

$= 0.0186\ mol\ Ag_2S$

$0.0186\ \cancel{mol\ Ag_2S} \times \frac{247.9\ g\ Ag_2S}{1\ \cancel{mol\ Ag_2S}}$

$= 4.60\ g\ Ag_2S$ produced.

b. $0.0371\ \cancel{mol\ Ag} \times \frac{1\ mol\ S_8}{16\ \cancel{mol\ Ag}}$

$= 0.00232\ mol\ S_8$ needed

$0.00232\ \cancel{mol\ S_8} \times \frac{256.5\ g\ S_8}{1\ \cancel{mol\ S_8}}$

$= 0.595\ g\ S_8$

$4.00\ g - 0.595\ g = 3.40\ g\ S_8$ in excess

EXAMPLE Problem 5

Find help with **dimensional analysis.** Math Handbook

EXAMPLE PROBLEM

DETERMINING THE LIMITING REACTANT The reaction between solid white phosphorus (P_4) and oxygen produces solid tetraphosphorus decoxide (P_4O_{10}). This compound is often called diphosphorus pentoxide because its empirical formula is P_2O_5.

a. Determine the mass of P_4O_{10} formed if 25.0 g of P_4 and 50.0 g of oxygen are combined.

b. How much of the excess reactant remains after the reaction stops?

1 ANALYZE THE PROBLEM

You are given the masses of both reactants, so you must identify the limiting reactant and use it to find the mass of the product. From moles of the limiting reactant, the moles of the excess reactant used in the reaction can be determined. The number of moles of the excess reactant that reacted can be converted to mass and subtracted from the given mass to find the amount in excess.

Known	**Unknown**
mass of phosphorus = 25.0 g P_4	**mass of tetraphosphorus decoxide = ? g P_4O_{10}**
mass of oxygen = 50.0 g O_2	**mass of excess reactant = ? g excess reactant**

2 SOLVE FOR THE UNKNOWN

Determine the limiting reactant.

$$\underset{25.0\ g}{P_4(s)} + \underset{50.0\ g}{5O_2(g)} \rightarrow \underset{?\ g}{P_4O_{10}(s)}$$

Write the balanced chemical equation, and identify the known and the unknown.

Determine the number of moles of the reactants by multiplying each mass by the conversion factor that relates moles and mass—the inverse of molar mass.

$$25.0\ \cancel{g\ P_4} \times \frac{1\ mol\ P_4}{123.9\ \cancel{g\ P_4}} = 0.202\ mol\ P_4$$

Calculate the moles of P_4.

$$50.0\ \cancel{g\ O_2} \times \frac{1\ mol\ O_2}{32.00\ \cancel{g\ O_2}} = 1.56\ mol\ O_2$$

Calculate the moles of O_2.

Calculate the actual ratio of available moles of O_2 and available moles of P_4.

$$\frac{1.56\ mol\ O_2}{0.202\ mol\ P_4} = \frac{7.72\ mol\ O_2}{1\ mol\ P_4}$$

Calculate the ratio of moles of O_2 to moles of P_4.

Determine the mole ratio of the two reactants from the balanced chemical equation.

Mole ratio: $\frac{5\ mol\ O_2}{mol\ P_4}$

Because 7.72 mol of O_2 is available but only 5 mol is needed to react with 1 mol of P_4, O_2 is in excess and P_4 is the limiting reactant. Use the moles of P_4 to determine the moles of P_4O_{10} that will be produced. Multiply the number of moles of P_4 by the mole ratio of P_4O_{10} (the unknown) to P_4 (the known).

$$0.202\ \cancel{mol\ P_4} \times \frac{1\ mol\ P_4O_{10}}{1\ \cancel{mol\ P_4}} = 0.202\ mol\ P_4O_{10}$$

Calculate the moles of product (P_4O_{10}) formed.

To calculate the mass of P_4O_{10}, multiply moles of P_4O_{10} by the conversion factor that relates mass and moles—molar mass.

$$0.202\ \cancel{mol\ P_4O_{10}} \times \frac{283.9\ g\ P_4O_{10}}{1\ \cancel{mol\ P_4O_{10}}} = \mathbf{57.3\ g\ P_4O_{10}}$$

Calculate the mass of the product P_4O_{10}.

Demonstration

Limiting Reactants

Purpose

to observe the effect of a limiting reactant on a chemical reaction

Materials

Burets (2); metric ruler; 25 mL graduated cylinder; stirring rods; 18 × 150 mm test tubes (8); test-tube rack; 0.10*M* $CuSO_4$; 0.20*M* KOH; centrifuge; distilled water; See the Teacher Handbook for preparation of solutions.

Safety Precautions

Disposal

$CuSO_4$ and KOH solutions can be evaporated to dryness in a fume hood. The solid can be disposed of at an approved waste facility.

Procedure

Fill one buret with 0.20*M* KOH and the other with 0.10*M* $CuSO_4$. Label eight test tubes according to the amount of KOH to be added: 2.0 mL, 4.0 mL, 6.0 mL, 8.0 mL, 10.0 mL, 12.0 mL, 14.0 mL, and 16.0 mL. Dispense 10.0 mL $CuSO_4$ into all eight labeled test tubes. Add KOH according to the volume on the test tube label. Add enough water to each test tube so that each will have 26.0 mL. Stir and place each test tube in the centrifuge for about one minute. Remove each test tube from the centrifuge. Measure the height of the

Because O_2 is in excess, only part of the available O_2 is consumed. Use the limiting reactant, P_4, to determine the moles and mass of O_2 used.

$$0.202 \cancel{\text{mol } P_4} \times \frac{5 \text{ mol } O_2}{1 \cancel{\text{mol } P_4}} = 1.01 \text{ mol } O_2$$

Multiply the moles of limiting reactant by the mole ratio to determine moles of excess reactant needed.

Convert moles of O_2 consumed to mass of O_2 consumed.

$$1.01 \cancel{\text{mol } O_2} \times \frac{32.00 \text{ g } O_2}{1 \cancel{\text{mol } O_2}} = 32.3 \text{ g } O_2$$

Multiply the moles of O_2 by the molar mass.

Calculate the amount of excess O_2.

50.0 g O_2 available − 32.3 g O_2 consumed = **17.7 g O_2 in excess**

Subtract the mass of O_2 used from the mass available.

3 EVALUATE THE ANSWER

All values have a minimum of three significant figures, so the mass of P_4O_{10} is correctly stated with three digits. The mass of excess O_2 (17.7 g) is found by subtracting two numbers that are accurate to the first decimal place. Therefore, the mass of excess O_2 correctly shows one decimal place. The sum of the O_2 that was consumed (32.3 g) and the given mass of P_4 (25.0 g) is 57.3 g, the calculated mass of the product P_4O_{10}.

PRACTICE Problems

Do additional problems. Online Practice

23. The reaction between solid sodium and iron(III) oxide is one in a series of reactions that inflates an automobile airbag: $6Na(s) + Fe_2O_3(s) \rightarrow 3Na_2O(s) + 2Fe(s)$. If 100.0 g of Na and 100.0 g of Fe_2O_3 are used in this reaction, determine the following.

a. limiting reactant

b. reactant in excess

c. mass of solid iron produced

d. mass of excess reactant that remains after the reaction is complete

24. Challenge Photosynthesis reactions in green plants use carbon dioxide and water to produce glucose ($C_6H_{12}O_6$) and oxygen. A plant has 88.0 g of carbon dioxide and 64.0 g of water available for photosynthesis.

a. Write the balanced chemical equation for the reaction.

b. Determine the limiting reactant.

c. Determine the excess reactant.

d. Determine the mass in excess.

e. Determine the mass of glucose produced.

PRACTICE PROBLEMS

Connection to Biology Your body needs vitamins, minerals, and elements in small amounts to facilitate normal metabolic reactions. A lack of these substances can lead to abnormalities in growth, development, and the functioning of your body's cells. Phosphorus, for example, is an essential element in living systems; phosphate groups occur regularly in strands of DNA. Potassium is needed for proper nerve function, muscle control, and blood pressure. A diet low in potassium and high in sodium might be a factor in high blood pressure. Another example is vitamin B-12. Without adequate vitamin B-12, the body is unable to synthesize DNA properly, affecting the production of red blood cells.

ASSESSMENT

Knowledge Have students write the equation for the reaction of lithium with bromine to produce lithium bromide. Have them determine the following, given that 25.0 g of lithium and 25.0 g of bromine are present at the beginning of the reaction: the limiting reactant, the mass of lithium bromide produced, the excess reactant, the mass of the excess reactant. Bromine is the limiting reactant; 27.1 g LiBr; lithium is the excess reactant; 22.8 g excess. **OL**

PRACTICE Problems

Have students refer to the Selected Solutions appendix for complete solutions to odd-numbered problems. The complete solutions for all problems can be found in the Solutions Manual.

23. a. Fe_2O_3
b. Na
c. 69.92 g Fe
d. 13.6 g Na

24. a. $6CO_2(g) + 6H_2O(l) \rightarrow C_6H_{12}O_6(aq) + 6O_2(g)$
b. CO_2
c. H_2O
d. 28.0 g
e. 60.0 g

precipitate and note the color of the supernatant in each test tube.

Results

Test tubes labeled *2, 4, 6, 8* have blue supernatant and increasing amounts of precipitate. Test tubes labeled *10, 12, 14, 16* have the same amount of precipitate and clear colorless supernatant.

Analysis

1. What is the balanced equation for the reaction between KOH and $CuSO_4$? $2KOH(aq) + CuSO_4(aq) \rightarrow Cu(OH)_2(s) + K_2SO_4(aq)$.
2. Explain the blue color of the supernatant in some test tubes. $CuSO_4$ is in excess in these tubes.
3. In which test tube(s) did the maximum amount of precipitate occur? in those with 10, 12, 14, and 16 mL of potassium hydroxide

ASSESSMENT

Knowledge Explain why the amount of precipitate levels off in the test tubes that have 10.0 mL or more of KOH. KOH is in excess and $CuSO_4$ is the limiting reactant. **OL**

3 Assess

Check for Understanding

Write the equation for coal gasification, a process that converts coal into methane gas, on the board or project the image: $2C + 2H_2O \rightarrow CH_4 + CO_2$. Have students determine the yield of methane if 2000 g of carbon and 2000 g of water are used. 900 g OL

Reteach

Ask students to determine the limiting reactant and the excess reactant in these situations.

1. charcoal briquettes burning in an outdoor grill Charcoal is limiting; oxygen is in excess.
2. a candle burning in a closed container Oxygen is limiting; candle wax is in excess.

Ask students what data should be collected and what calculations must be done to confirm their choices. The number of moles of each reactant, the actual mole ratio, and the ratio according to the balanced equation. OL

Extension

Students can design an experiment that could be used to determine the maximum rise of a cupcake using different amounts of baking powder. AL

ASSESSMENT

Skill Ask each student to write a balanced equation with at least two reactants and provide masses for each. Working in pairs, students should solve their two problems for the limiting reactant, the excess reactant, and the amount of each product. OL COOP LEARN

Figure 7 With insufficient oxygen, the burner on the left burns with a yellow, sooty flame. The burner on the right burns hot and clean because an excess of oxygen is available to react completely with the methane gas.

Why use an excess of a reactant? Many reactions stop while portions of the reactants are still present in the reaction mixture. Because this is inefficient and wasteful, chemists have found that by using an excess of one reactant—often the least expensive one—reactions can be driven to continue until all of the limiting reactant is used up. Using an excess of one reactant can also speed up a reaction.

Figure 7 shows an example of how controlling the amount of a reactant can increase efficiency. Your lab likely uses the type of Bunsen burner shown in the figure. If so, you know that this type of burner has a control that lets you adjust the amount of air that mixes with the methane gas. How efficiently the burner operates depends on the ratio of oxygen to methane gas in the fuel mixture. When the air is limited, the resulting flame is yellow because of glowing bits of unburned fuel. This unburned fuel leaves soot (carbon) deposits on glassware. Fuel is wasted because the amount of energy released is less than the amount that could have been produced if enough oxygen were available. When sufficient oxygen is present in the combustion mixture, the burner produces a hot, intense blue flame. No soot is deposited because the fuel is completely converted to carbon dioxide and water vapor.

SECTION 3 REVIEW

Section Self-Check

Section Summary

- The limiting reactant is the reactant that is completely consumed during a chemical reaction. Reactants that remain after the reaction stops are called excess reactants.
- To determine the limiting reactant, the actual mole ratio of the available reactants must be compared with the ratio of the reactants obtained from the coefficients in the balanced chemical equation.
- Stoichiometric calculations must be based on the limiting reactant.

25. **MAIN IDEA Describe** the reason why a reaction between two substances comes to an end.
26. **Identify** the limiting and the excess reactant in each reaction.
    - **a.** Wood burns in a campfire.
    - **b.** Airborne sulfur reacts with the silver plating on a teapot to produce tarnish (silver sulfide).
    - **c.** Baking powder in batter decomposes to produce carbon dioxide.
27. **Analyze** Tetraphosphorus trisulphide (P_4S_3) is used in the match heads of some matches. It is produced in the reaction $8P_4 + 3S_8 \rightarrow 8P_4S_3$. Determine which of the following statements are incorrect, and rewrite the incorrect statements to make them correct.
    - **a.** 4 mol P_4 reacts with 1.5 mol S_8 to form 4 mol P_4S_3.
    - **b.** Sulfur is the limiting reactant when 4 mol P_4 and 4 mol S_8 react.
    - **c.** 6 mol P_4 reacts with 6 mol S_8, forming 1320 g P_4S_3.

Matt Meadows

SECTION 3 REVIEW

25. one of the reactants is used up
26. **a.** The wood limits the reaction. Oxygen is in excess. The fire will burn only while wood is present.
    b. Silver is the limiting reactant. Sulfur is in excess. When a layer of tarnish covers the silver surface, it prevents the sulfur in the air from reacting.
    c. A decomposition reaction usually has only one reactant. The reaction is limited by the amount of baking powder present.
27. **a.** correct
    b. Phosphorus is the limiting reactant.
    c. correct

SECTION 4

Essential Questions

- What is the theoretical yield of a chemical reaction?
- How do you calculate the percent yield for a chemical reaction?

Review Vocabulary

process: a series of actions or operations

New Vocabulary

theoretical yield
actual yield
percent yield

Percent Yield

MAINIDEA Percent yield is a measure of the efficiency of a chemical reaction.

CHEM 4 YOU Imagine that you are practicing free throws and you take 100 practice shots. Theoretically, you could make all 100 shots. In actuality, however, you know you will not make all of the shots. Chemical reactions also have theoretical and actual outcomes.

How Much Product?

While solving stoichiometric problems in this chapter, you might have concluded that chemical reactions always proceed in the laboratory according to the balanced equation and produce the calculated amount of product. This, however, is not the case. Just as you are unlikely to make 100 out of 100 free throws during basketball practice, most reactions never succeed in producing the predicted amount of product. Reactions do not go to completion or yield as expected for a variety of reasons. Liquid reactants and products might adhere to the surfaces of their containers or evaporate. In some instances, products other than the intended ones might be formed by competing reactions, thus reducing the yield of the desired product. Or, as shown in **Figure 8,** some amount of any solid product is usually left behind on filter paper or lost in the purification process. Because of these problems, chemists need to know how to gauge the yield of a chemical reaction.

Theoretical and Actual Yields In many of the stoichiometric calculations you have performed, you have calculated the amount of product produced from a given amount of reactant. The answer you obtained is the theoretical yield of the reaction. The **theoretical yield** is the maximum amount of product that can be produced from a given amount of reactant.

A chemical reaction rarely produces the theoretical yield of product. A chemist determines the actual yield of a reaction through a careful experiment in which the mass of the product is measured. The **actual yield** is the amount of product produced when the chemical reaction is carried out in an experiment.

■ **Figure 8** Silver chromate is formed when potassium chromate is added to silver nitrate. Note that not all of the precipitate can be removed from the filter paper. Still more of the precipitate is lost because it adheres to the sides of the beaker.

SECTION 4

1 Focus

MAINIDEA

Actual and Theoretical Yield Pop a bag of popcorn and ask students if all of the popcorn popped. Some of the corn kernels will remain unpopped. Tell students that not all reactants produce products in a chemical reaction. The number of popped kernels in the bag is the actual yield. They can use stoichiometry to determine the theoretical yield. BL OL

2 Teach

Math in Chemistry

Calculate Percentages Students might have learned that to obtain a percent, they should divide the part by the whole and multiply by 100. Help them to see that when they divide the part by the whole, they are finding out the fraction of the whole that the part is, or part per one. When they multiply by 100, they are finding the parts per 100, or percent. Ask what they would do to find parts per million. Multiply part per one by 10^6. BL

CHEMISTRY PROJECT

Evaluate Choices Have students develop six multiple-choice questions with four answer choices, one being the correct answer. These questions, with the correct answers, can be used in a class game. OL

CHEMISTRY JOURNAL

Batting Averages If students do not know what is meant by a baseball batting average, have them research this subject and write in their journals how a batting average is similar to percent yield. BL EL

IN-CLASS Example

Question Zinc can be removed from bronze by placing bronze in hydrochloric acid. The zinc reacts with the hydrochloric acid producing zinc chloride and hydrogen gas, and leaving the copper behind.

a. If 25.0 g of zinc are in a sample of bronze, determine the theoretical yield of hydrogen gas. $Zn + 2\,HCl \rightarrow ZnCl_2 + H_2$

b. If the reaction yields 0.680 g H_2, determine the percent yield.

Answer

a. $25.0\ \text{g Zn} \times \frac{1\ \text{mol Zn}}{65.39\ \text{g Zn}} = 0.382\ \text{mol Zn}$

$\frac{1\ \text{mol } H_2}{1\ \text{mol Zn}}$

$0.382\ \text{mol Zn} \times \frac{1\ \text{mol } H_2}{1\ \text{mol Zn}} = 0.382\ \text{mol } H_2$

$0.382\ \text{mol } H_2 \times \frac{2.016\ \text{g } H_2}{1\ \text{mol } H_2} = 0.771\ \text{g of } H_2$ is the theoretical yield

b. $\frac{0.680\ \text{g } H_2\ \text{(actual yield)}}{0.771\ \text{g } H_2\ \text{(theoretical yield)}} \times 100 = 88.2\%$ yield of H_2

Quick Demo

Calculating Yields Add about 5.0 mL of cooking oil to a flask. Place 20 kernels of popcorn in the flask and cover with a piece of aluminum foil with holes punched in the top. Place the flask on a hot plate and pop the popcorn. Pour out the popcorn and count the number of kernels that popped as the actual yield. The theoretical yield is 20 kernels. Ask students to determine the percent yield using the results. Answers will vary depending on the data collected. BL EL

Percent yield Chemists need to know how efficient a reaction is in producing the desired product. One way of measuring efficiency is by means of percent yield. **Percent yield** of product is the ratio of the actual yield to the theoretical yield expressed as a percent.

Percent Yield

$$\text{percent yield} = \frac{\text{actual yield}}{\text{theoretical yield}} \times 100$$

The actual yield divided by the theoretical yield multiplied by 100 is the percent yield.

EXAMPLE Problem 6

Find help with **percents.** Math Handbook

PERCENT YIELD Solid silver chromate (Ag_2CrO_4) forms when excess potassium chromate (K_2CrO_4) is added to a solution containing 0.500 g of silver nitrate ($AgNO_3$). Determine the theoretical yield of Ag_2CrO_4. Calculate the percent yield if the reaction yields 0.455 g of Ag_2CrO_4.

1 ANALYZE THE PROBLEM

You know the mass of a reactant and the actual yield of the product. Write the balanced chemical equation, and calculate theoretical yield by converting grams of $AgNO_3$ to moles of $AgNO_3$, moles of $AgNO_3$ to moles of Ag_2CrO_4, and moles of Ag_2CrO_4 to grams of Ag_2CrO_4. Calculate the percent yield from the actual yield and the theoretical yield.

Known	Unknown
mass of silver nitrate = 0.500 g $AgNO_3$	theoretical yield = ? g Ag_2CrO_4
actual yield = 0.455 g Ag_2CrO_4	percent yield = ? % Ag_2CrO_4

2 SOLVE FOR THE UNKNOWN

$$\underset{0.500\ \text{g}}{2AgNO_3(aq)} + K_2CrO_4(aq) \rightarrow \underset{?\ \text{g}}{Ag_2CrO_4(s)} + 2KNO_3(aq)$$

Write the balanced chemical equation, and identify the known and the unknown.

$$0.500\ \text{g } AgNO_3 \times \frac{1\ \text{mol } AgNO_3}{169.9\ \text{g } AgNO_3} = 2.94 \times 10^{-3}\ \text{mol } AgNO_3$$

Use molar mass to convert grams of $AgNO_3$ to moles of $AgNO_3$.

$$2.94 \times 10^{-3}\ \text{mol } AgNO_3 \times \frac{1\ \text{mol } Ag_2CrO_4}{2\ \text{mol } AgNO_3} = 1.47 \times 10^{-3}\ \text{mol } Ag_2CrO_4$$

Use the mole ratio to convert moles of $AgNO_3$ to moles of Ag_2CrO_4.

$$1.47 \times 10^{-3}\ \text{mol } Ag_2CrO_4 \times \frac{331.7\ \text{g } Ag_2CrO_4}{1\ \text{mol } Ag_2CrO_4} = 0.488\ \text{g } Ag_2CrO_4$$

Calculate the theoretical yield.

$$\frac{0.455\ \text{g } Ag_2CrO_4}{0.488\ \text{g } Ag_2CrO_4} \times 100 = 93.2\%\ Ag_2CrO_4$$

Calculate the percent yield.

3 EVALUATE THE ANSWER

The quantity with the fewest significant figures has three, so the percent is correctly stated with three digits. The molar mass of Ag_2CrO_4 is about twice the molar mass of $AgNO_3$, and the ratio of moles of $AgNO_3$ to moles of Ag_2CrO_4 in the equation is 2:1. Therefore, 0.500 g of $AgNO_3$ should produce about the same mass of Ag_2CrO_4. The actual yield of Ag_2CrO_4 is close to 0.500 g, so a percent yield of 93.2% is reasonable.

DIFFERENTIATED INSTRUCTION

Advanced Learners Have capable students design a spreadsheet that can be used to determine the theoretical yield of a product from a given mass of reactant and the percent yield of the reaction when given the actual yield. First, they must determine the theoretical yield of the reaction when given a mass of one of the reactants. Second, they must be able to enter an actual yield and calculate the percent yield of the reactions. The spreadsheet can be used as a tutorial for other students. AL

PRACTICE Problems

Do additional problems. Online Practice

28. Aluminum hydroxide ($Al(OH)_3$) is often present in antacids to neutralize stomach acid (HCl). The reaction occurs as follows: $Al(OH)_3(s) + 3HCl(aq) \rightarrow AlCl_3(aq) + 3H_2O(l)$. If 14.0 g of $Al(OH)_3$ is present in an antacid tablet, determine the theoretical yield of $AlCl_3$ produced when the tablet reacts with HCl.
29. Zinc reacts with iodine in a synthesis reaction: $Zn + I_2 \rightarrow ZnI_2$.
    a. Determine the theoretical yield if 1.912 mol of zinc is used.
    b. Determine the percent yield if 515.6 g of product is recovered.
30. **Challenge** When copper wire is placed into a silver nitrate solution ($AgNO_3$), silver crystals and copper(II) nitrate ($Cu(NO_3)_2$) solution form.
    a. Write the balanced chemical equation for the reaction.
    b. If a 20.0-g sample of copper is used, determine the theoretical yield of silver.
    c. If 60.0 g of silver is recovered from the reaction, determine the percent yield of the reaction.

Data Analysis LAB

Based on Real Data[1, 2]

Analyze and Conclude

Can rocks on the Moon provide an effective oxygen source for future lunar missions?
Although the Moon has no atmosphere and thus no oxygen, its surface is covered with rocks and soil made from oxides. Scientists, looking for an oxygen source for future long-duration lunar missions, are researching ways to extract oxygen from lunar soil and rock. Analysis of samples collected during previous lunar missions provided scientists with the data shown in the table. The table identifies the oxides in lunar soil as well as each oxide's percent-by-weight of the soil.

Think Critically

1. **Calculate** For each of the oxides listed in the table, determine the mass (in grams) that would exist in 1.00 kg of lunar soil.
2. **Apply** Scientists want to release the oxygen from its metal oxide using a decomposition reaction: metal oxide → metal + oxygen. To assess the viability of this idea, determine the amount of oxygen per kilogram contained in each of the oxides found in lunar soil.
3. **Identify** What oxide would yield the most oxygen per kilogram? The least?
4. **Determine** the theoretical yield of oxygen from the oxides present in a 1.00-kg sample of lunar soil.

Data and Observation

Moon-Rock Data[1]

Oxide	% Mass of Soil
SiO_2	47.3%
Al_2O_3	17.8%
CaO	11.4%
FeO	10.5%
MgO	9.6%
TiO_2	1.6%
Na_2O	0.7%
K_2O	0.6%
Cr_2O_3	0.2%
MnO	0.1%

[1]Data obtained from: McKay, et al. 1994. JSC-1: A new lunar soil stimulant. *Engineering, Construction, and Operations in Space* IV: 857–866, American Society of Civil Engineers.

[2]Data obtained from: Berggren, et al. 2005. Carbon monoxide silicate reduction system. *Space Resources Roundtable* VII.

5. **Calculate** Using methods currently available, scientists can produce 15 kg of oxygen from 100 kg of lunar soil. What is the percent yield of the process?

CULTURAL DIVERSITY

Norbert Rillieux was born in New Orleans in 1806. His father was a French plantation owner and his mother was an African-American slave. When Norbert was born, his father had the choice of declaring him free or, as was usually the custom, a slave. Fortunately for Norbert, he was declared free. Even though he was a free man, Norbert experienced racial discrimination throughout his life.

Taking advantage of the excellent educational opportunities available to him, Norbert studied science and math and became a chemical engineer. Having grown up observing the dangerous and labor-intensive process of refining sugar, he developed a vastly improved process. His invention, the multiple-effect pan evaporator, revolutionized sugar refining and was eventually patented. His new method could produce a higher-quality sugar at nearly half the cost.

PRACTICE Problems

Have students refer to the Selected Solutions appendix for complete solutions to odd-numbered problems. The complete solutions for all problems can be found in the Solutions Manual.

28. 23.9 g of AlCl3 is the theoretical yield.
29. a. 610.3 g ZnI_2
    b. 84.48% yield of ZnI_2
30. a. $Cu(s) + 2AgNO_3(aq) \rightarrow 2Ag(s) + Cu(NO_3)_2(aq)$
    b. 68.0 g of Ag
    c. 88.2% yield

Data Analysis LAB

About the Lab

- This lab asks students to determine the amount of available oxygen in the lunar soil, the theoretical yield of oxygen, and the percent yield using an achievable actual yield.
- Soil on the Moon is composed of many different oxides that are rich in oxygen. Scientists are studying if the oxygen in Moon rocks can provide astronauts with oxygen.
- NASA has simulated the composition of lunar soil. Scientists are trying to achieve a cost-effective way of extracting the oxygen from this soil.
- Refer to Gregory Mone, "Mining the Moon," *Popular Science*, July 2006, 38–39, for details.

Think Critically

1. TiO_2: 16 g; Al_2O_3: 178 g; SiO_2: 473 g; FeO: 105 g; MgO: 96 g; CaO: 114 g; Na_2O: 7 g; K_2O: 6 g; MnO: 1 g; Cr_2O_3: 2 g
2. TiO_2: 0.00641 kg O_2; Al_2O_3: 0.0838 kg O_2; SiO_2: 0.252 kg O_2; FeO: 0.0234 kg O_2; MgO: 0.0381 kg O_2; CaO: 0.0325 kg O_2; Na_2O: 0.00181 kg O_2; K_2O: 0.000988 kg O_2; MnO: 0.000225 kg O_2; Cr_2O_3: 0.000632 kg O_2;
3. SiO_2 yields the most; MnO the least
4. 0.439 kg of O_2/1.00 kg lunar soil
5. $0.15 \text{ kg}/0.439 \times 100 = 34\%$

ASSESSMENT

Performance Have each student write four multiple choice questions involving percent yield calculations. Have them exchange their questions with a classmate and each answer the other's questions. Have them resolve any disagreements about the correct answer. OL

3 Assess

Check for Understanding

Write the following equation on the board: $CH_4 + 2O_2 \rightarrow CO_2 + 2H_2O$. Ask students to determine the percent yield if 10.0 g of methane is burned producing 19.5 g of water.
86.7% yield OL

Reteach

Discuss the burning of gasoline in an internal combustion engine: $2C_8H_{18} + 25O_2 \rightarrow 16CO_2 + 18H_2O$. Ask students to determine the percent yield of carbon dioxide if 700.0 g of octane produces 1800.0 g of carbon dioxide. Have students explain each step of the calculation.
83.4% yield CO_2 OL

Extension

Have students research internal combustion engines and what is being done to increase the efficiency of automobiles and reduce air pollution. OL

■ **Figure 9** Sulfur, such as these piles at Vancouver Harbor, can be extracted from petroleum products by a chemical process. Sulfur is also mined by forcing hot water into underground deposits and pumping the liquid sulfur to the surface.

Percent Yield in the Marketplace

Percent yield is important in the cost effectiveness of many industrial manufacturing processes. For example, the sulfur shown in **Figure 9** is used to make sulfuric acid (H_2SO_4). Sulfuric acid is an important chemical because it is a raw material used to make products such as fertilizers, detergents, pigments, and textiles.

The cost of sulfuric acid affects the cost of many of the consumer items you use every day. The first two steps in the manufacturing process are shown below.

Step 1 $\quad S_8(s) + 8O_2(g) \rightarrow 8SO_2(g)$

Step 2 $\quad 2SO_2(g) + O_2(g) \rightarrow 2SO_3(g)$

In the final step, SO_3 combines with water to produce H_2SO_4.

Step 3 $\quad SO_3(g) + H_2O(l) \rightarrow H_2SO_4(aq)$

The first step, the combustion of sulfur, produces an almost 100% yield. The second step also produces a high yield if a catalyst is used at the relatively low temperature of 400°C. A catalyst is a substance that speeds a reaction but does not appear in the chemical equation. Under these conditions, the reaction is slow. Raising the temperature increases the reaction rate but decreases the yield.

To maximize yield and minimize time in the second step, engineers have devised a system in which the reactants, O_2 and SO_2, are passed over a catalyst at 400°C. Because the reaction releases a great deal of heat, the temperature gradually increases with an accompanying decrease in yield. Thus, when the temperature reaches approximately 600°C, the mixture is cooled and then passed over the catalyst again. A total of four passes over the catalyst with cooling between passes results in a yield greater than 98%.

SECTION 4 REVIEW

Section Self-Check

Section Summary

- The theoretical yield of a chemical reaction is the maximum amount of product that can be produced from a given amount of reactant. Theoretical yield is calculated from the balanced chemical equation.
- The actual yield is the amount of product produced. Actual yield must be obtained through experimentation.
- Percent yield is the ratio of actual yield to theoretical yield expressed as a percent. High percent yield is important in reducing the cost of every product produced through chemical processes.

31. MAIN IDEA **Identify** which type of yield—theoretical yield, actual yield, or percent yield—is a measure of the efficiency of a chemical reaction.
32. **List** several reasons why the actual yield from a chemical reaction is not usually equal to the theoretical yield.
33. **Explain** how percent yield is calculated.
34. **Apply** In an experiment, you combine 83.77 g of iron with an excess of sulfur and then heat the mixture to obtain iron(III) sulfide.
$$2Fe(s) + 3S(s) \rightarrow Fe_2S_3(s)$$
What is the theoretical yield, in grams, of iron(III) sulfide?
35. **Calculate** the percent yield of the reaction of magnesium with excess oxygen: $2Mg(s) + O_2(g) \rightarrow 2MgO(s)$

Reaction Data	
Mass of empty crucible	35.67 g
Mass of crucible and Mg	38.06 g
Mass of crucible and MgO (after heating)	39.15 g

SECTION 4 REVIEW

31. percent yield
32. Not all reactions go to completion. Some of the reactants or products stick to the surface of the container and are not massed or transferred. Other unexpected products form from competing reactions.
33. Divide the actual yield by the theoretical yield and multiply the quotient by 100.
34. 155.9 g Fe_2S_3
35. 87.9% yield of MgO

CHEMISTRY & health

Battling Resistant Strains

The human immunodeficiency virus (HIV), the virus that causes AIDS, has proven to be among the most incurable foes ever faced by modern medical science. One reason for this is the virus's remarkable ability to adapt. Resistant strains of the virus appear quickly, rendering obsolete the newest and most powerful AIDS drugs. Now some researchers are using the virus's adaptability as a way to fight it.

Selecting resistance PA-457 is a promising new anti-HIV drug synthesized from betulinic acid, an organic compound derived from some plants, including the bark of birch trees. To find out just what PA-457 does to HIV, known as the drug's mechanism of action, researchers took what might seem a strange step: they encouraged samples of HIV to develop resistance to PA-457.

Researchers subjected HIV samples to small doses of PA-457. Using a low dose made it more likely that some of the virus would survive the treatment and possibly develop resistance. Those viruses that survived exposure were collected, and their genetic sequences were examined. The surviving viruses were found to have a mutation in the genes that control how the virus builds a structure called a capsid, shown in **Figure 1.**

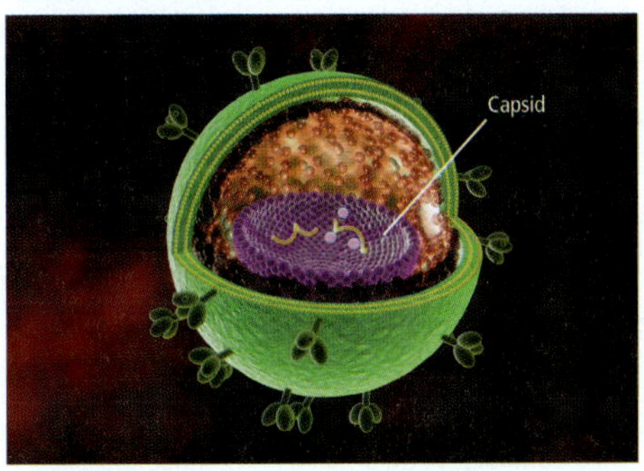

Figure 1 In a normal HIV virus, the capsid forms a protective coating around the genetic material.

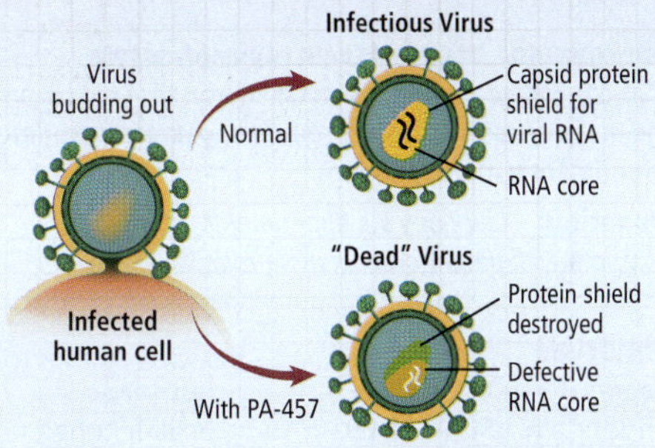

Figure 2 When treated with PA-457, the HIV capsid becomes misshapen and collapses, resulting in the death of the virus.

Surprise attack This finding was surprising, because it showed that, unlike most drugs, PA-457 attacks the HIV structure, rather than the enzymes that help HIV reproduce, as illustrated in **Figure 2.** This makes PA-457 among the first of a new class of HIV drugs known as maturation inhibitors—drugs that can prevent the virus from maturing during the late stages in its development.

Slowing evolution The hope is that because PA-457 and other maturation inhibitors attack the HIV structure, resistance will be slower to develop. Even so, maturation inhibitors will likely be prescribed in combination with other AIDS drugs that attack HIV at different stages of its life cycle.

This practice, called multidrug therapy, makes it harder for HIV to develop resistance because any surviving virus would need to have multiple mutations—at least one for each anti-HIV drug. These mutations are less likely to occur at the same time.

WRITING IN ▶ Chemistry

Research how scientists determine the safe dosing level for an experimental drug. Debate how a drug's effectiveness must be balanced with its potential toxicity and side effects.

WebQuest

WRITING IN ▶ Chemistry

Rubric WebQuest

Research Answers will vary. Students should find that the dose-response relationship for therapeutic drugs varies. Some drugs are very predictable, requiring little monitoring, but others require much closer monitoring. Disease, the patient's age, and drug interactions can all affect the level of a drug in the patient's system. Additionally, some drugs become toxic even at concentrations close to the therapeutic dose and need to be monitored very closely.

CHEMISTRY & health

Purpose

Students will learn how the adaptability of HIV can be used to learn how new HIV drugs work.

Background

Drug resistance is conferred by mutations of HIV's genetic sequence. HIV does not possess DNA (deoxyribonucleic acid), but rather stores its genetic information on a related molecule called RNA (ribonucleic acid). Researchers map the RNA sequence of HIV to discover which part of the RNA controls each part of the virus. When mutations are discovered, researchers can go back to that map to determine which part of the virus is altered by the mutation.

Teaching Strategies

- Discuss examples of mutations that provide resistance. These might include pesticide resistance in insects, herbicide resistance in weeds, and antibiotic resistance in disease-causing bacteria.
- Researchers encouraged resistance to PA-457 by administering low doses of the drug. Why not high doses? High doses might kill those viruses with only partial resistance. Compare this with the need to complete full courses of antibiotics to prevent the development of resistant bacteria.
- Note the crucial difference between viruses and bacteria: viruses are not affected by antibiotics. Have students find out why.

ChemLAB

Rubric

Preparation

Time Allotment one class period

Process Skills observe and infer, measure, collect data, interpret data, use numbers, compare and contrast, apply concepts

Safety Precautions Approve lab safety forms before work begins. Review the MSDS for copper(II) sulfate pentahydrate and iron with students before beginning the lab. Copper(II) sulfate is toxic. Use only a GFCI outlet for the hot plate. Be sure students wear aprons, goggles, and plastic gloves. Students must wash their hands with soap after the lab.

Disposal The copper(II) sulfate and iron(II) sulfate solutions can be evaporated to dryness in the fume hood. The solid can then be disposed of at a designated waste facility.

Procedure

- Decanting might be a new lab technique for students. Show them how to slowly pour the liquid from the beaker down the stirring rod and into another beaker.
- Tell students that the copper metal will not stick to the stirring rod when it is completely dry.
- **Troubleshooting** For best results, use iron filings that are free of oxidation.

ChemLAB

Determine the Mole Ratio

Background: Iron reacts with copper(II) sulfate ($CuSO_4$). By measuring the mass of iron that reacts and the mass of copper metal produced, you can calculate the experimental mole ratio.

Question: *How does the experimental mole ratio compare with the theoretical mole ratio?*

Materials

copper(II) sulfate pentahydrate ($CuSO_4 \cdot 5H_2O$)	hot plate
iron metal filings (20 mesh)	beaker tongs
distilled water	balance
150-mL beaker	stirring rod
100-mL graduated cylinder	400-mL beaker
	weighing paper

Safety Precautions

WARNING: ***Hot plates can cause burns. Turn off hot plates when not in use. Use only GFCI-protected circuits.***

Procedure

1. Read and complete the lab safety form.
2. Measure the mass of a clean, dry 150-mL beaker. Record all measurements in a data table.
3. Place approximately 12 g $CuSO_4 \cdot 5H_2O$ into the 150-mL beaker, and measure the combined mass.
4. Add 50 mL of distilled water to the $CuSO_4 \cdot 5H_2O$. Place the mixture on a hot plate set at medium, and stir until all of the solid dissolves (do not boil). Using tongs, remove the beaker from the hot plate.
5. Measure about 2 g of iron filings onto a piece of weighing paper. Measure the mass of the filings.
6. While stirring, slowly add the iron filings to the hot copper(II) sulfate solution. Be careful not to splash the hot solution.
7. Allow the reaction mixture to sit for 5 min.
8. Use the stirring rod to decant (pour off) the liquid into a 400-mL beaker. Be careful to decant only the liquid–leave the solid copper metal behind.
9. Add 15 mL of distilled water to the copper solid, and carefully swirl the beaker to wash the copper. Decant the liquid into the 400-mL beaker.
10. Repeat Step 9 two more times.
11. Place the beaker containing the wet copper on the hot plate. Use low heat to dry the copper.
12. After the copper is dry, use tongs to remove the beaker from the hot plate and allow it to cool.
13. Measure the mass of the beaker and the copper.
14. **Cleanup and Disposal** The dry copper can be placed in a waste container. Moisten any residue that sticks to the beaker, and wipe it out using a paper towel. Pour the unreacted copper(II) sulfate and iron(II) sulfate solutions into a large beaker. Return all lab equipment to its proper place.

Analyze and Conclude

1. **Apply** Write a balanced chemical equation for the reaction and calculate the mass of copper (Cu) that should have formed from the sample of iron (Fe) used. This mass is the theoretical yield.
2. **Interpret Data** Using your data, determine the mass and the moles of copper produced. Calculate the moles of iron used, and determine the whole-number iron-to-copper mole ratio and percent yield.
3. **Compare and Contrast** Compare the theoretical iron-to-copper mole ratio to the mole ratio you calculated using the experimental data.
4. **Error Analysis** Identify sources of the error that resulted in deviation from the mole ratio given in the balanced chemical equation.

INQUIRY EXTENSION

Compare your results with those of other lab teams. Form a hypothesis to explain any differences. How would you test your hypothesis?

Analyze and Conclude

Numerical answers based on sample data.

1. $Fe(s) + CuSO_4(aq) \rightarrow Cu(s) + FeSO_4(aq)$; 2.30 g Cu
2. 2.26 g Cu; 0.0356 mol Cu; 0.0362 mol Fe; Fe:Cu mole ratio = 1.02:1; percent yield = 98.3%
3. The ratio of Fe to Cu in the equation is 1:1, almost the same as the experimental ratio.
4. Copper was not completely dry; some copper could have oxidized if heated too much; copper could have been lost.

INQUIRY EXTENSION

Student answers will vary but should discuss possible error sources and their impact on the results.

CHAPTER 11 STUDY GUIDE

Vocabulary Practice

BIGIDEA Mass relationships in chemical reactions confirm the law of conservation of mass.

SECTION 1 Defining Stoichiometry

MAINIDEA The amount of each reactant present at the start of a chemical reaction determines how much product can form.

- Balanced chemical equations can be interpreted in terms of moles, mass, and representative particles (atoms, molecules, formula units).
- The law of conservation of mass applies to all chemical reactions.
- Mole ratios are derived from the coefficients of a balanced chemical equation. Each mole ratio relates the number of moles of one reactant or product to the number of moles of another reactant or product in the chemical reaction.

VOCABULARY
- stoichiometry
- mole ratio

SECTION 2 Stoichiometric Calculations

MAINIDEA The solution to every stoichiometric problem requires a balanced chemical equation.

- Chemists use stoichiometric calculations to predict the amounts of reactants used and products formed in specific reactions.
- The first step in solving stoichiometric problems is writing the balanced chemical equation.
- Mole ratios derived from the balanced chemical equation are used in stoichiometric calculations.
- Stoichiometric problems make use of mole ratios to convert between mass and moles.

SECTION 3 Limiting Reactants

MAINIDEA A chemical reaction stops when one of the reactants is used up.

- The limiting reactant is the reactant that is completely consumed during a chemical reaction. Reactants that remain after the reaction stops are called excess reactants.
- To determine the limiting reactant, the actual mole ratio of the available reactants must be compared with the ratio of the reactants obtained from the coefficients in the balanced chemical equation.
- Stoichiometric calculations must be based on the limiting reactant.

VOCABULARY
- limiting reactant
- excess reactant

SECTION 4 Percent Yield

MAINIDEA Percent yield is a measure of the efficiency of a chemical reaction.

- The theoretical yield of a chemical reaction is the maximum amount of product that can be produced from a given amount of reactant. Theoretical yield is calculated from the balanced chemical equation.
- The actual yield is the amount of product produced. Actual yield must be obtained through experimentation.
- Percent yield is the ratio of actual yield to theoretical yield expressed as a percent. High percent yield is important in reducing the cost of every product produced through chemical processes.

$$\text{Percent yield} = \frac{\text{actual yield}}{\text{theoretical yield}} \times 100$$

VOCABULARY
- theoretical yield
- actual yield
- percent yield

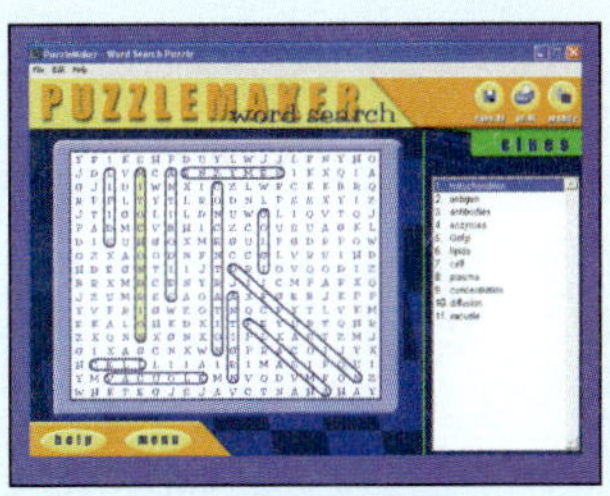

Vocabulary Practice

For additional practice with vocabulary, have students access the Vocabulary PuzzleMaker.

Use the Vocabulary

To reinforce chapter vocabulary, have students write a sentence using each term. OL EL

Review Strategies

- Have students explain why a balanced chemical equation is necessary in stoichiometry. OL
- Have students list the steps in stoichiometric problem solving. OL
- Have students explain how a limiting reactant affects a stoichiometric problem. OL
- Have students explain the percent yield of a reaction. OL
- Problems from the Supplemental Practice Problems appendix or the Supplemental Problems booklet can be used for review. OL

Students can review with these online resources.

Vocabulary Practice

Vocabulary eGames
Vocabulary eFlashcards
Vocabulary PuzzleMaker

Section Self-Check

Chapter Self-Check

Online Test Practice

Use *eAssessment* to:

- create multiple versions of tests
- edit existing questions and add your own questions
- build tests aligned with select state curriculum standards using built-in state curriculum tags
- track students' progress

SECTION 1

Mastering Concepts

36. Mole ratios are determined by the coefficients in a balanced equation. If the equation is not balanced, the relationship between reactants and products cannot be determined.

37. relationships among particles, moles, and mass for all reactants and products

38. Mole ratios allow for the conversion from moles of one substance in a balanced chemical equation to moles of another substance in the same equation.

39. moles B/moles A

40. The coefficients in the balanced chemical equation show the numbers of representative particles involved in a reaction. Subscripts give the numbers of different kinds of atoms within a molecule or formula unit.

41. The mass of the reactants will always equal the mass of the products.

42. 1 mol $(NH_4)_2Cr_2O_7$/1 mol N_2 and inverse; 1 mol $(NH_4)_2Cr_2O_7$/1 mol Cr_2O_3 and inverse; 1 mol $(NH_4)_2Cr_2O_7$/4 mol H_2O and inverse

43. $2M_2N \longrightarrow M_4 + N_2$; 1 mol N_2/2 mol M_2N, 1 mol N_2/1 mol M_4 2 mol M_2N/1 mol M_4, 2 mol M_2N/1 mol N_2, 1 mol M_4/1 mol N_2, 1 mol M_4/2 mol M_2N

Mastering Problems

44. 4 atoms Al + 3 molecules $O_2 \longrightarrow$ 2 formula units Al_2O_3; 4 mol Al + 3 mol $O_2 \longrightarrow$ 2 mol Al_2O_3; 107.93 g Al + 95.99 g $O_2 \longrightarrow$ 203.92 g Al_2O_3

45. 1 formula unit SnO_2 + 2 atoms C $\longrightarrow$ 1 atom Sn + 2 molecules CO; 1 mol SnO_2 + 2 mol C $\longrightarrow$ 1 mol Sn + 2 mol CO; 150.71 g SnO_2 + 24.02 g C $\longrightarrow$ 118.71 g Sn + 56.02 g CO

46. $Cu(s) + 4HNO_3(aq) \longrightarrow Cu(NO_3)_2(aq) + 2NO_2(g) + 2H_2O(l)$; answers may include 1 mol Cu/4 mol HNO_3 and inverse, 1 mol Cu/1 mol $Cu(NO_3)_2$ and inverse, 1 mol Cu/2 mol NO_2 and inverse, 1 mol Cu/2 mol H_2O and inverse, 4 mol HNO_3/1 mol $Cu(NO_3)_2$ and inverse, 4 mol HNO_3/2 mol NO_2 and inverse, 4 mol HNO_3/2 mol H_2O and inverse, 1 mol $Cu(NO_3)_2$/2 mol NO_2 and inverse, 1 mol $Cu(NO_3)_2$/2 mol H_2O and inverse, 2 mol NO_2/2 mol H_2O and inverse

CHAPTER 11 ASSESSMENT

SECTION 1

Mastering Concepts

36. Why must a chemical equation be balanced before you can determine mole ratios?

37. What relationships can be determined from a balanced chemical equation?

38. Explain why mole ratios are central to stoichiometric calculations.

39. What is the mole ratio that can convert from moles of A to moles of B?

40. Why are coefficients used in mole ratios instead of subscripts?

41. Explain how the conservation of mass allows you to interpret a balanced chemical equation in terms of mass.

42. When heated by a flame, ammonium dichromate decomposes, producing nitrogen gas, solid chromium(III) oxide, and water vapor.

$$(NH_4)_2Cr_2O_7 \rightarrow N_2 + Cr_2O_3 + 4H_2O$$

Write the mole ratios for this reaction that relate ammonium dichromate to the products.

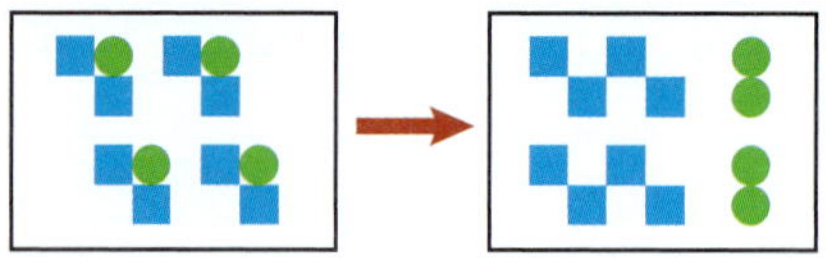

■ **Figure 10**

43. **Figure 10** depicts an equation with squares representing Element M and circles representing Element N. Write a balanced equation to represent the picture shown, using smallest whole-number ratios. Write mole ratios for this equation.

Mastering Problems

44. Interpret the following equation in terms of particles, moles, and mass.

$$4Al(s) + 3O_2(g) \rightarrow 2Al_2O_3(s)$$

45. **Smelting** When tin(IV) oxide is heated with carbon in a process called smelting, the element tin can be extracted.

$$SnO_2(s) + 2C(s) \rightarrow Sn(l) + 2CO(g)$$

Interpret the chemical equation in terms of particles, moles, and mass.

46. When solid copper is added to nitric acid, copper(II) nitrate, nitrogen dioxide, and water are produced. Write the balanced chemical equation for the reaction. List six mole ratios for the reaction.

47. When hydrochloric acid solution reacts with lead(II) nitrate solution, lead(II) chloride precipitates and a solution of nitric acid is produced.

a. Write the balanced chemical equation for the reaction.

b. Interpret the equation in terms of molecules and formula units, moles, and mass.

48. When aluminum is mixed with iron(III) oxide, iron metal and aluminum oxide are produced, along with a large quantity of heat. What mole ratio would you use to determine moles of Fe if moles of Fe_2O_3 is known?

$$Fe_2O_3(s) + 2Al(s) \rightarrow 2Fe(s) + Al_2O_3(s) + \text{heat}$$

49. Solid silicon dioxide, often called silica, reacts with hydrofluoric acid (HF) solution to produce the gas silicon tetrafluoride and water.

a. Write the balanced chemical equation for the reaction.

b. List three mole ratios, and explain how you would use them in stoichiometric calculations.

50. **Chrome** The most important commercial ore of chromium is chromite ($FeCr_2O_4$). One of the steps in the process used to extract chromium from the ore is the reaction of chromite with coke (carbon) to produce ferrochrome ($FeCr_2$).

$$2C(s) + FeCr_2O_4(s) \rightarrow FeCr_2(s) + 2CO_2(g)$$

What mole ratio would you use to convert from moles of chromite to moles of ferrochrome?

51. **Air Pollution** The pollutant SO_2 is removed from the air in a reaction that also involves calcium carbonate and oxygen. The products of this reaction are calcium sulfate and carbon dioxide. Determine the mole ratio you would use to convert moles of SO_2 to moles of $CaSO_4$.

52. Two substances, W and X, react to form the products Y and Z. **Table 2** shows the moles of the reactants and products involved when the reaction was carried out. Use the data to determine the coefficients that will balance the equation W + X → Y + Z.

Table 2 Reaction Data

Moles of Reactants		Moles of Products	
W	X	Y	Z
0.90	0.30	0.60	1.20

53. **Antacids** Magnesium hydroxide is an ingredient in some antacids. Antacids react with excess hydrochloric acid in the stomach to relieve indigestion.

$$\_\_Mg(OH)_2 + \_\_HCl \rightarrow \_\_MgCl_2 + \_\_H_2O$$

a. Balance the reaction of $Mg(OH)_2$ with HCl.

b. Write the mole ratio that would be used to determine the number of moles of $MgCl_2$ produced when HCl reacts with $Mg(OH)_2$.

47. a. $2HCl(aq) + Pb(NO_3)_2(aq) \longrightarrow PbCl_2(s) + 2HNO_3(aq)$

b. 2 molecules HCl + 1 formula unit $Pb(NO_3)_2 \longrightarrow$ 1 formula unit $PbCl_2$ + 2 molecules HNO_3; 2 mol HCl + 1 mol $Pb(NO_3)_2 \longrightarrow$ 1 mol $PbCl_2$ + 2 mol HNO_3; 72.9 g HCl + 331.2 g $Pb(NO_3)_2 \longrightarrow$ 278.1 g $PbCl_2$ + 126.0 g HNO_3

48. 2 mol Fe/1 mol Fe_2O_3

49. a. $SiO_2(s) + 4HF(aq) \longrightarrow SiF_4(g) + 2H_2O(l)$

b. Students may write any 3 of the 12 ratios. Explanations should correctly describe how the ratio can be used as a conversion factor.

50. 1 mol $FeCr_2$/1 mol $FeCr_2O_4$

51. 2 mol $CaSO_4$/2 mol SO_2

52. $3W + X \longrightarrow 2Y + 4Z$

53. a. $1Mg(OH)_2 + 2HCl \longrightarrow 1MgCl_2 + 2H_2O$

b. 1 mol $MgCl_2$/1 mol $Mg(OH)_2$ or 1 mol $MgCl_2$/2 mol HCl

Chapter Self-Check

SECTION 2

Mastering Concepts

54. What is the first step in all stoichiometric calculations?

55. What information does a balanced equation provide?

56. On what law is stoichometry based, and how do the calculations support this law?

57. How is molar mass used in some stoichiometric calculations?

58. What information must you have in order to calculate the mass of product formed in a chemical reaction?

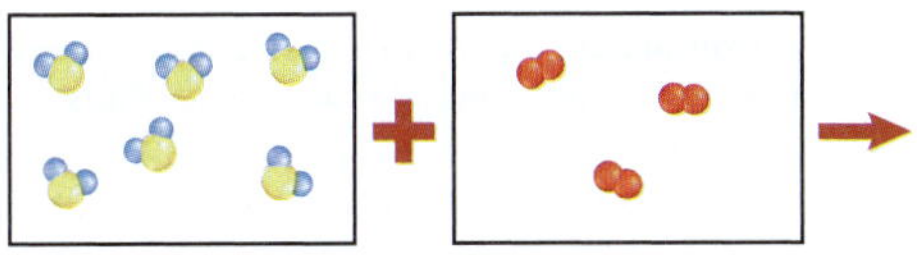

■ **Figure 11**

59. Each box in **Figure 11** represents the contents of a flask. One flask contains hydrogen sulfide, and the other contains oxygen. When the contents of the flasks are mixed, a reaction occurs and water vapor and sulfur are produced. In the figure, the red circles represent oxygen, the yellow circles represent sulfur, and blue circles represent hydrogen.

a. Write the balanced chemical equation for the reaction.

b. Using the same color code, sketch a representation of the flask after the reaction occurs.

Mastering Problems

60. Ethanol (C_2H_5OH), also known as grain alcohol, can be made from the fermentation of sugar ($C_6H_{12}O_6$). The unbalanced chemical equation for the reaction is shown below.

$$\_\_C_6H_{12}O_6 \rightarrow \_\_C_2H_5OH + \_\_CO_2$$

Balance the chemical equation and determine the mass of C_2H_5OH produced from 750 g of $C_6H_{12}O_6$.

61. Welding If 5.50 mol of calcium carbide (CaC_2) reacts with an excess of water, how many moles of acetylene (C_2H_2), a gas used in welding, will be produced?

$$CaC_2(s) + 2H_2O(l) \rightarrow Ca(OH)_2(aq) + C_2H_2(g)$$

62. Antacid Fizz When an antacid tablet dissolves in water, the fizz is due to a reaction between sodium hydrogen carbonate ($NaHCO_3$), also called sodium bicarbonate, and citric acid ($H_3C_6H_5O_7$).

$$3NaHCO_3(aq) + H_3C_6H_5O_7(aq) \rightarrow 3CO_2(g) + 3H_2O(l) + Na_3C_6H_5O_7(aq)$$

How many moles of $Na_3C_6H_5O_7$ can be produced if one tablet containing 0.0119 mol of $NaHCO_3$ is dissolved?

63. Esterification The process in which an organic acid and an alcohol react to form an ester and water is known as esterification. Ethyl butanoate ($C_3H_7COOC_2H_5$), an ester, is formed when the alcohol ethanol (C_2H_5OH) and butanoic acid (C_3H_7COOH) are heated in the presence of sulfuric acid.

$$C_2H_5OH(l) + C_3H_7COOH(l) \rightarrow C_3H_7COOC_2H_5(l) + H_2O(l)$$

Determine the mass of ethyl butanoate produced if 4.50 mol of ethanol is used.

64. Greenhouse Gas Carbon dioxide is a greenhouse gas that is linked to global warming. It is released into the atmosphere through the combustion of octane (C_8H_{18}) in gasoline. Write the balanced chemical equation for the combustion of octane and calculate the mass of octane needed to release 5.00 mol of CO_2.

65. A solution of potassium chromate reacts with a solution of lead(II) nitrate to produce a yellow precipitate of lead(II) chromate and a solution of potassium nitrate.

a. Write the balanced chemical equation.

b. Starting with 0.250 mol of potassium chromate, determine the mass of lead chromate formed.

66. Rocket Fuel The exothermic reaction between liquid hydrazine (N_2H_4) and liquid hydrogen peroxide (H_2O_2) is used to fuel rockets. The products of this reaction are nitrogen gas and water.

a. Write the balanced chemical equation.

b. How much hydrazine, in grams, is needed to produce 10.0 mol of nitrogen gas?

67. Chloroform ($CHCl_3$), an important solvent, is produced by a reaction between methane and chlorine.

$$CH_4(g) + 3Cl_2(g) \rightarrow CHCl_3(g) + 3HCl(g)$$

How much CH_4, in grams, is needed to produce 50.0 grams of $CHCl_3$?

68. Oxygen Production The Russian Space Agency uses potassium superoxide (KO_2) for the chemical oxygen generators in their space suits.

$$4KO_2 + 2H_2O + 4CO_2 \rightarrow 4KHCO_3 + 3O_2$$

Complete **Table 3**.

Table 3 Oxygen Generation Reaction Data

Mass KO_2	Mass H_2O	Mass CO_2	Mass $KHCO_3$	Mass O_2
				380 g

69. Gasohol is a mixture of ethanol and gasoline. Balance the equation, and determine the mass of CO_2 produced from the combustion of 100.0 g of ethanol.

$$C_2H_5OH(l) + O_2(g) \rightarrow CO_2(g) + H_2O(g)$$

SECTION 2

Mastering Concepts

54. Write a balanced chemical equation for the reaction.

55. The balanced equation provides the relationship between reactants and products, and the coefficients in the equation are used to write mole ratios relating reactants and products.

56. Stoichiometry is based on the law of conservation of mass. The calculations are used to determine the mass of reactants and products. Once found, the sum of reactants will equal the sum of products, verifying the law of conservation of mass.

57. Molar mass is a conversion factor for converting moles of a given substance to mass or mass of a given substance to moles.

58. You must have the balanced chemical equation and know the quantity of one substance in the reaction other than the product you are to determine.

59. a. $8H_2S(g) + 4O_2(g) \rightarrow 8H_2O(g) + S_8(s)$

b. Student sketches should show the formation of six water molecules (H_2O) and six sulfur atoms (S).

Mastering Problems

60. $C_6H_{12}O_6 \rightarrow 2C_2H_5OH + 2CO_2$; 390 g C_2H_5OH

61. 5.50 mol of C_2H_2

62. 0.00397 mol $Na_3C_6H_5O_7$

63. 523 g $C_3H_7COOC_2H_5$

64. $2C_8H_{18}(l) + 25O_2(g) \rightarrow 16CO_2(g) + 18H_2O(l)$; 71.4 g C_8H_{18}

65. a. $K_2CrO_4(aq) + Pb(NO_3)_2(aq) \rightarrow PbCrO_4(s) + 2KNO_3(aq)$

b. 80.8 g $PbCrO_4$

66. a. $N_2H_4(l) + 2H_2O_2(l) \rightarrow N_2(g) + 4H_2O(g)$

b. 3.20×10^2 g N_2H_4

67. 6.72 g CH_4

68. mass KO_2 = 1100 g, mass H_2O = 140 g, mass $CO_2 = 7.0 \times 10^2$ g, mass $KHCO_3$ = 1600 g

69. $C_2H_5OH(l) + 3O_2(g) \rightarrow 2CO_2(g) + 3H_2O(l)$; 191.0 g CO_2

70. a. $Pb(s) + PbO_2(s) + 2H_2SO_4(aq) \longrightarrow 2PbSO_4(aq) + 2H_2O(l)$
b. 73.2 g $PbSO_4$

71. a. 50.2 g Au
b. 33.5% gold in ore

72. 0.587 g $Na_3Ag(S_2O_3)_2$

SECTION 3

Mastering Concepts

73. The actual mole ratio of reactants from the chemical equation is compared to the mole ratio determined from the given quantities.

74. The limiting reactant is the reactant that produces the lowest number of moles of product. Mass does not determine the limiting reactant but the number of moles.

75. a. $3M_2 + N_2 \longrightarrow 2M_3N$
b. 6 moles of element M (in the form of 3 moles of M_2) and 6 moles of element N (likewise, 3 moles of N_2)
c. 2 moles of M_3N form with 2 moles of N_2 unreacted (4 total moles of element N)
d. M_2 is the limiting reactant and N_2 is the excess reactant.

Mastering Problems

76. Hydrogen is limiting; ethyne is the excess reactant. One mol of ethyne is left over.

77. 4.0 mol $Fe(OH)_2$

78. 10.0 mol $CsXeF_7$

79. 1120 g Fe

80. Cl_2 is the limiting reactant; phosphorus is in excess.

81. a. MnO_2 is the limiting reactant.
b. 17.1 g $Zn(OH)_2$

82. $2Li(s) + Br_2(l) \longrightarrow 2LiBr(s)$
a. Br_2
b. 27.1 g LiBr
c. Li, 22.8 g

70. Car Battery Car batteries use lead, lead(IV) oxide, and a sulfuric acid solution to produce an electric current. The products of the reaction are lead(II) sulfate in solution and water.
a. Write the balanced equation for the reaction.
b. Determine the mass of lead(II) sulfate produced when 25.0 g of lead reacts with an excess of lead(IV) oxide and sulfuric acid.

71. To extract gold from its ore, the ore is treated with sodium cyanide solution in the presence of oxygen and water.

$$4Au(s) + 8NaCN(aq) + O_2(g) + 2H_2O(l) \rightarrow 4NaAu(CN)_2(aq) + 4NaOH(aq)$$

a. Determine the mass of gold that can be extracted if 25.0 g of sodium cyanide is used.
b. If the mass of the ore from which the gold was extracted is 150.0 g, what percentage of the ore is gold?

72. Film Photographic film contains silver bromide in gelatin. Once exposed, some of the silver bromide decomposes, producing fine grains of silver. The unexposed silver bromide is removed by treating the film with sodium thiosulfate. Soluble sodium silver thiosulfate ($Na_3Ag(S_2O_3)_2$) is produced.

$$AgBr(s) + 2Na_2S_2O_3(aq) \rightarrow Na_3Ag(S_2O_3)_2(aq) + NaBr(aq)$$

Determine the mass of $Na_3Ag(S_2O_3)_2$ produced if 0.275 g of AgBr is removed.

SECTION 3

Mastering Concepts

73. How is a mole ratio used to find the limiting reactant?

74. Explain why the statement, "The limiting reactant is the reactant with the lowest mass" is incorrect.

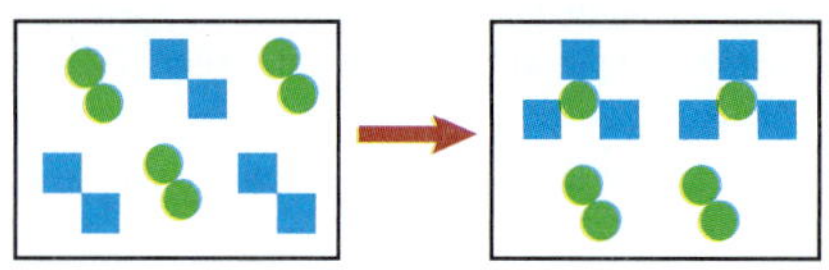

■ **Figure 12**

75. Figure 12 uses squares to represent Element M and circles to represent Element N.
a. Write the balanced equation for the reaction.
b. If each square represents 1 mol of M and each circle represents 1 mol of N, how many moles of M and N were present at the start of the reaction?
c. How many moles of product form? How many moles of Element M and Element N are unreacted?
d. Identify the limiting reactant and the excess reactant.

Mastering Problems

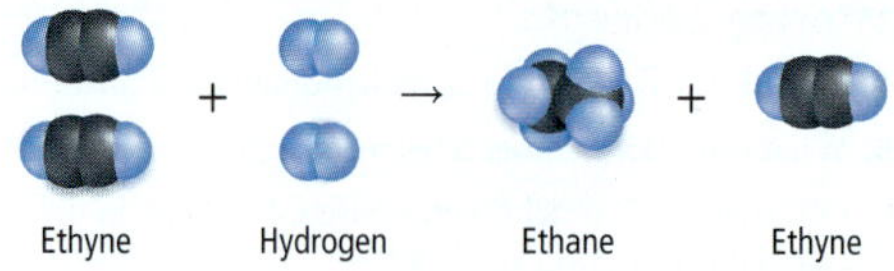

■ **Figure 13**

76. The reaction between ethyne (C_2H_2) and hydrogen (H_2) is illustrated in **Figure 13.** The product is ethane (C_2H_6). Which is the limiting reactant? Which is the excess reactant? Explain.

77. Nickel-Iron Battery In 1901, Thomas Edison invented the nickel-iron battery. The following reaction takes place in the battery.

$$Fe(s) + 2NiO(OH)(s) + 2H_2O(l) \rightarrow Fe(OH)_2(s) + 2Ni(OH)_2(aq)$$

How many mol of $Fe(OH)_2$ is produced when 5.00 mol of Fe and 8.00 mol of NiO(OH) react?

78. One of the few xenon compounds that form is cesium xenon heptafluoride ($CsXeF_7$). How many moles of $CsXeF_7$ can be produced from the reaction of 12.5 mol of cesium fluoride with 10.0 mol of xenon hexafluoride?

$$CsF(s) + XeF_6(s) \longrightarrow CsXeF_7(s)$$

79. Iron Production Iron is obtained commercially by the reaction of hematite (Fe_2O_3) with carbon monoxide. How many grams of iron is produced when 25.0 mol of hematite reacts with 30.0 mol of carbon monoxide?

$$Fe_2O_3(s) + 3CO(g) \longrightarrow 2Fe(s) + 3CO_2(g)$$

80. The reaction of chlorine gas with solid phosphorus (P_4) produces solid phosphorus pentachloride. When 16.0 g of chlorine reacts with 23.0 g of P_4, which reactant is limiting? Which reactant is in excess?

81. Alkaline Battery An alkaline battery produces electrical energy according to this equation.

$$Zn(s) + 2MnO_2(s) + H_2O(l) \rightarrow Zn(OH)_2(s) + Mn_2O_3(s)$$

a. Determine the limiting reactant if 25.0 g of Zn and 30.0 g of MnO_2 are used.
b. Determine the mass of $Zn(OH)_2$ produced.

82. Lithium reacts spontaneously with bromine to produce lithium bromide. Write the balanced chemical equation for the reaction. If 25.0 g of lithium and 25.0 g of bromine are present at the beginning of the reaction, determine
a. the limiting reactant.
b. the mass of lithium bromide produced.
c. the excess reactant and the excess mass.

SECTION 4

Mastering Concepts

83. What is the difference between actual yield and theoretical yield?

84. How are actual yield and theoretical yield determined?

85. Can the percent yield of a chemical reaction be more than 100%? Explain your answer.

86. What relationship is used to determine the percent yield of a chemical reaction?

87. What experimental information do you need in order to calculate both the theoretical and the percent yield of any chemical reaction?

88. A metal oxide reacts with water to produce a metal hydroxide. What additional information would you need to determine the percent yield of metal hydroxide from this reaction?

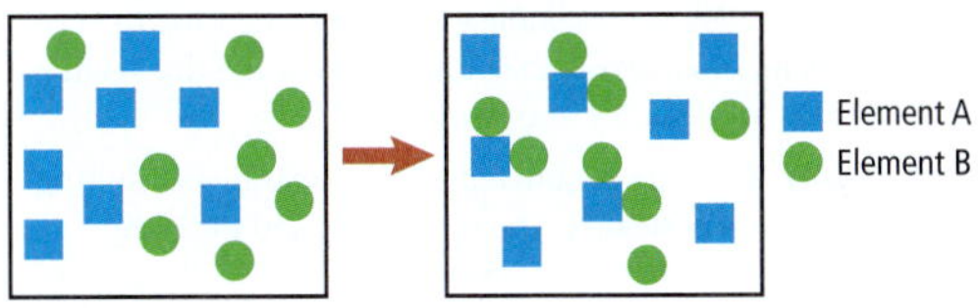

■ **Figure 14**

89. Examine the reaction represented in **Figure 14.** Determine if the reaction went to completion. Explain your answer, and calculate the percent yield of the reaction.

Mastering Problems

90. Ethanol (C_2H_5OH) is produced from the fermentation of sucrose ($C_{12}H_{22}O_{11}$) in the presence of enzymes.

$$C_{12}H_{22}O_{11}(aq) + H_2O(g) \rightarrow 4C_2H_5OH(l) + 4CO_2(g)$$

Determine the theoretical yield and the percent yield of ethanol if 684 g of sucrose undergoes fermentation and 349 g of ethanol is obtained.

91. Lead(II) oxide is obtained by roasting galena, lead(II) sulfide, in air. The unbalanced equation is:

$$PbS(s) + O_2(g) \rightarrow PbO(s) + SO_2(g)$$

a. Balance the equation, and determine the theoretical yield of PbO if 200.0 g of PbS is heated.

b. What is the percent yield if 170.0 g of PbO is obtained?

92. Upon heating, calcium carbonate ($CaCO_3$) decomposes to calcium oxide (CaO) and carbon dioxide (CO_2).

a. Determine the theoretical yield of CO_2 if 235.0 g of $CaCO_3$ is heated.

b. What is the percent yield of CO_2 if 97.5 g of CO_2 is collected?

93. Hydrofluoric acid solutions cannot be stored in glass containers because HF reacts readily with silicon dioxide in glass to produce hexafluorosilicic acid (H_2SiF_6).

$$SiO_2(s) + 6HF(aq) \rightarrow H_2SiF_6(aq) + 2H_2O(l)$$

40.0 g SiO_2 and 40.0 g HF react to yield 45.8 g H_2SiF_6.

a. What is the limiting reactant?

b. What is the mass of the excess reactant?

c. What is the theoretical yield of H_2SiF_6?

d. What is the percent yield?

94. Van Arkel Process Pure zirconium is obtained using the two-step Van Arkel process. In the first step, impure zirconium and iodine are heated to produce zirconium iodide (ZrI_4). In the second step, ZrI_4 is decomposed to produce pure zirconium.

$$ZrI_4(s) \rightarrow Zr(s) + 2I_2(g)$$

Determine the percent yield of zirconium if 45.0 g of ZrI_4 is decomposed and 5.00 g of pure Zr is obtained.

95. Methanol, wood alcohol, is produced when carbon monoxide reacts with hydrogen gas.

$$CO + 2H_2 \rightarrow CH_3OH$$

When 8.50 g of carbon monoxide reacts with an excess of hydrogen, 8.52 g of methanol is collected. Complete **Table 4,** and calculate the percent yield.

Table 4 Methanol Reaction Data

	CO(g)	$CH_3OH(l)$
Mass	8.50 g	
Molar mass	28.01 g/mol	32.05 g/mol
Moles		

96. Phosphorus (P_4) is commercially prepared by heating a mixture of calcium phosphate ($Ca_3(PO_4)_2$), sand (SiO_2), and coke (C) in an electric furnace. The process involves two reactions.

$$2Ca_3(PO_4)_2(s) + 6SiO_2(s) \rightarrow 6CaSiO_3(l) + P_4O_{10}(g)$$

$$P_4O_{10}(g) + 10C(s) \rightarrow P_4(g) + 10CO(g)$$

The P_4O_{10} produced in the first reaction reacts with an excess of coke (C) in the second reaction. Determine the theoretical yield of P_4 if 250.0 g of $Ca_3(PO_4)_2$ and 400.0 g of SiO_2 are heated. If the actual yield of P_4 is 45.0 g, determine the percent yield of P_4.

97. Chlorine forms from the reaction of hydrochloric acid with manganese(IV) oxide. The balanced equation is:

$$MnO_2 + 4HCl \rightarrow MnCl_2 + Cl_2 + 2H_2O$$

Calculate the theoretical yield and the percent yield of chlorine if 86.0 g of MnO_2 and 50.0 g of HCl react. The actual yield of Cl_2 is 20.0 g.

SECTION 4

Mastering Concepts

83. Actual yield is the amount of product obtained experimentally. Theoretical yield is the amount of product predicted by a stoichiometric calculation.

84. Actual yield is determined through experimentation. Theoretical yield is calculated from a given reactant or the limiting reactant.

85. No, you cannot produce more product than the theoretical yield, which is determined from the starting reactants.

86. (actual yield/theoretical yield) × 100 = percent yield

87. The quantity of one reactant and the actual yield of the product.

88. The mass of one substance in the reaction and the actual mass of metal hydroxide produced.

89. The reaction did not go to completion. Using squares to represent Element A and circles to represent Element B, the initial products would have yielded four AB_2 particles, but only three were produced. There are enough remaining unreacted A and B particles to produce one more AB_2 particle. The percent yield is 75%.

Mastering Problems

90. theoretical yield = 369 g C_2H_5OH
percent yield C_2H_5OH = 94.6%

91. a. $2PbS + 3O_2 \rightarrow 2PbO + 2SO_2$
theoretical yield = 186.6 g PbO

b. percent yield PbO = 91.10%

92. a. theoretical yield CO_2 = 103 g

b. percent yield CO_2 = 94.7%

93. a. HF is the limiting reactant.

b. SiO_2 is the excess reactant, 20.0 g.

c. theoretical yield = 48.0 g H_2SiF_6

d. percent yield H_2SiF_6 = 95.4%

94. 73.0% yield Zr

95. mass $CH_3OH(l)$ = 9.71 g
moles CO(g) = 0.303 mol
moles $CH_3OH(l)$ = 0.303 mol
percent yield = 87.7%

96. theoretical yield = 49.92 g P_4
percent yield P_4 = 90.1%

97. theoretical yield = 24.3 g Cl_2
percent yield Cl_2 = 82.3%

MIXED REVIEW

98. 2 mol NH_4NO_3/1 mol CuS

99. 401 g CaNCN

100. 25.6 g Cu

101. **a.** $2NO(g) + O_2(g) \longrightarrow 2NO_2(g)$

b. 2 mol NO_2/2 mol NO

102. theoretical yield = 4.04 g H_2
percent yield H_2 = 94.1%

103. The graph levels off because the oxygen limits the reaction at that point. fixed amount = 0.329 mol O_2

THINK CRITICALLY

104. No, percent yields cannot be greater than 100%. High results could mean the product was not completely dry, or it was contaminated.

105. **a.** No, because there is only one reactant.

b. Yes, because there are two reactants. Not enough information is given to identify which is the limiting reactant.

106. Example Experiment: Obtain and record the mass of an empty evaporating dish. Add 2.00 g of copper(II) sulfate pentahydrate and obtain and record the mass of the hydrate and evaporating dish. Heat the dish gently for 5 minutes, then strongly for 5 minutes to drive off the water. Cool the dish and remeasure the mass. Record. Determine the mass of the anhydrous copper sulfate. Using the equation $CuSO_4{\cdot}5H_2O \longrightarrow CuSO_4 + 5H_2O$ and the initial mass of the copper(II) sulfate pentahydrate, determine the theoretical yield of copper(II) sulfate. Determine the actual yield of copper(II) sulfate. Divide the actual yield by the theoretical yield and multiply by 100 to determine percent yield of copper(II) sulfate.

107. When you fan the flame, additional oxygen is added and the remaining coals can burn.

108. **a.** $2Na_3PO_4(aq) + 3Co(NO_3)_2(aq) \longrightarrow Co_3(PO_4)_2(s) + 6NaNO_3(aq)$

b. Trial 1: limiting reactant is Na_3PO_4, excess reactant is $Co(NO_3)_2$.
Trials 2 through 4: limiting reactant is $Co(NO_3)_2$, excess reactant is Na_3PO_4.

MIXED REVIEW

98. Ammonium sulfide reacts with copper(II) nitrate in a double replacement reaction. What mole ratio would you use to determine the moles of NH_4NO_3 produced if the moles of CuS are known?

99. Fertilizer The compound calcium cyanamide (CaNCN) is used as a nitrogen source for crops. To obtain this compound, calcium carbide is reacted with nitrogen at high temperatures.

$$CaC_2(s) + N_2(g) \rightarrow CaNCN(s) + C(s)$$

What mass of CaNCN can be produced if 7.50 mol of CaC_2 reacts with 5.00 mol of N_2?

100. When copper(II) oxide is heated in the presence of hydrogen gas, elemental copper and water are produced. What mass of copper can be obtained if 32.0 g of copper(II) oxide is used?

101. Air Pollution Nitrogen monoxide, which is present in urban air pollution, immediately converts to nitrogen dioxide as it reacts with oxygen.

a. Write the balanced chemical equation for the formation of nitrogen dioxide from nitrogen monoxide.

b. What mole ratio would you use to convert from moles of nitrogen monoxide to moles of nitrogen dioxide?

102. Electrolysis Determine the theoretical and percent yield of hydrogen gas if 36.0 g of water undergoes electrolysis to produce hydrogen and oxygen and 3.80 g of hydrogen is collected.

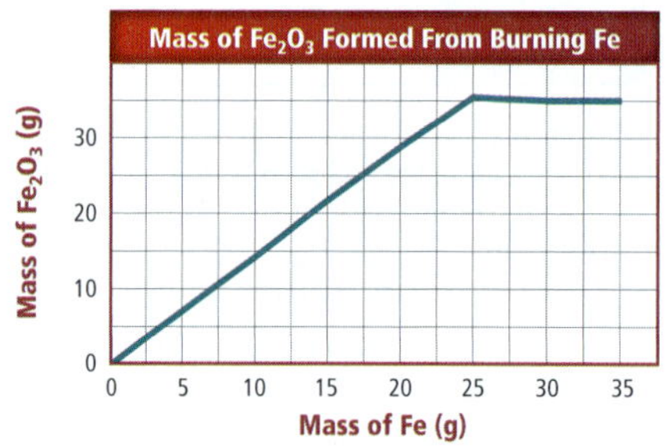

■ **Figure 15**

103. Iron reacts with oxygen as shown.

$$4Fe(s) + 3O_2(g) \rightarrow 2Fe_2O_3(s)$$

Different amounts of iron were burned in a fixed amount of oxygen. For each mass of iron burned, the mass of iron(III) oxide formed was plotted on the graph shown in **Figure 15.** Why does the graph level off after 25.0 g of iron is burned? How many moles of oxygen are present in the fixed amount?

THINK CRITICALLY

104. Analyze and Conclude In an experiment, you obtain a percent yield of product of 108%. Is such a percent yield possible? Explain. Assuming that your calculation is correct, what reasons might explain such a result?

105. Observe and Infer Determine whether each reaction depends on a limiting reactant. Explain why or why not, and identify the limiting reactant.

a. Potassium chlorate decomposes to form potassium chloride and oxygen.

b. Silver nitrate and hydrochloric acid react to produce silver chloride and nitric acid.

106. Design an Experiment Design an experiment that can be used to determine the percent yield of anhydrous copper(II) sulfate when copper(II) sulfate pentahydrate is heated to remove water.

107. Apply When a campfire begins to die down and smolder, you can rekindle the flame by fanning the fire. Explain, in terms of stoichiometry, why the fire again begins to flare up when fanned.

108. Apply Students conducted a lab to investigate limiting and excess reactants. The students added different volumes of sodium phosphate solution (Na_3PO_4) to a beaker. They then added a constant volume of cobalt(II) nitrate solution ($Co(NO_3)_2$), stirred the contents, and allowed the beakers to sit overnight. The next day, each beaker had a purple precipitate at the bottom. The students decanted the supernatant from each beaker, divided it into two samples, and added one drop of sodium phosphate solution to one sample and one drop of cobalt(II) nitrate solution to the second sample. Their results are shown in **Table 5.**

a. Write a balanced chemical equation for the reaction.

b. Based on the results, identify the limiting reactant and the excess reactant for each trial.

Table 5 Reaction Data for $Co(NO_3)_2$ and Na_3PO_4

Trial	Volume Na_3PO_4	Volume $Co(NO_3)_2$	Reaction with Drop of Na_3PO_4	Reaction with Drop of $Co(NO_3)_2$
1	5.0 mL	10.0 mL	purple precipitate	no reaction
2	10.0 mL	10.0 mL	no reaction	purple precipitate
3	15.0 mL	10.0 mL	no reaction	purple precipitate
4	20.0 mL	10.0 mL	no reaction	purple precipitate

Chapter Self-Check

CHALLENGE PROBLEM

109. When 9.59 g of a certain vanadium oxide is heated in the presence of hydrogen, water and a new oxide of vanadium are formed. This new vanadium oxide has a mass of 8.76 g. When the second vanadium oxide undergoes additional heating in the presence of hydrogen, 5.38 g of vanadium metal forms.
- **a.** Determine the empirical formulas for the two vanadium oxides.
- **b.** Write balanced equations for the steps of the reaction.
- **c.** Determine the mass of hydrogen needed to complete the steps of this reaction.

CUMULATIVE REVIEW

110. You observe that sugar dissolves more quickly in hot tea than in iced tea. You state that higher temperatures increase the rate at which sugar dissolves in water. Is this statement a hypothesis or a theory? Why?

111. Write the electron configuration for each of the following atoms.
- **a.** fluorine
- **b.** aluminum
- **c.** titanium
- **d.** radon

112. Explain why the gaseous nonmetals exist as diatomic molecules, but other gaseous elements exist as single atoms.

113. Write a balanced equation for the reaction of potassium with oxygen.

114. What is the molecular mass of UF_6? What is the molar mass of UF_6?

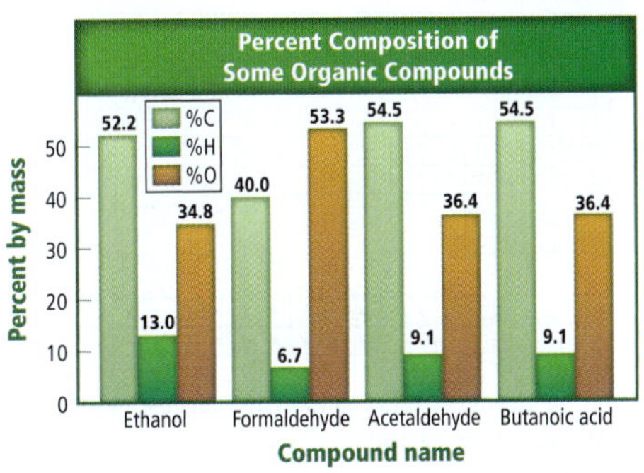

■ **Figure 16**

115. **Figure 16** gives percent composition data for several organic compounds.
- **a.** How are the molecular and empirical formulas of acetaldehyde and butanoic acid related?
- **b.** What is the empirical formula of butanoic acid?

WRITING IN ▶ Chemistry

116. Air Pollution Research the air pollutants produced by combustion of gasoline in internal combustion engines. Discuss the common pollutants and the reaction that produces them. Show, through the use of stoichiometry, how each pollutant could be reduced if more people used mass transit.

117. Haber Process The percent yield of ammonia produced when hydrogen and nitrogen are combined under ordinary conditions is extremely small. However, the Haber Process combines the two gases under a set of conditions designed to maximize yield. Research the conditions used in the Haber Process, and find out why the development of the process was of great importance.

DBQ Document-Based Question

Chemical Defense *Many insects secrete hydrogen peroxide (H_2O_2) and hydroquinone $C_6H_4(OH)_2$. Bombardier beetles take this a step further by mixing these chemicals with a catalyst. The result is an exothermic chemical reaction and a spray of hot, irritating chemicals for any would-be predator. Researchers hope to use a similar method to reignite aircraft turbine engines.*

Figure 17 *below shows the unbalanced chemical reaction that results in the bombardier beetle's defensive spray.*

Data obtained from: Becker, Bob. April 2006. *ChemMatters.* 24: no. 2.

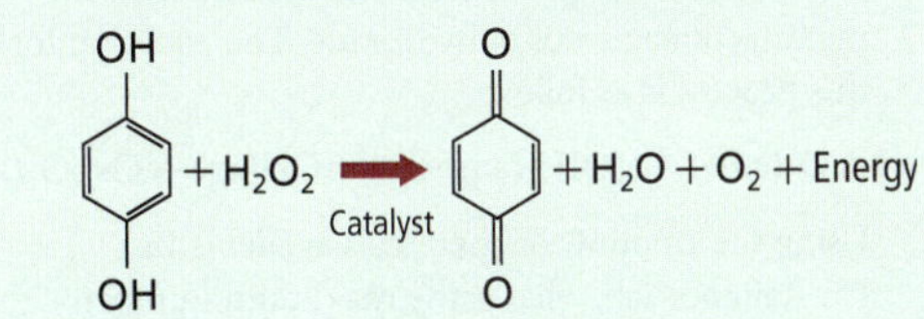

$C_6H_4(OH)_2$ Hydroquinone — $C_6H_4O_2$ Benzoquinone

■ **Figure 17**

118. Balance the equation in **Figure 17.** If the bombardier beetle stores 100.0 mg of hydroquinone ($C_6H_4(OH)_2$) along with 50.0 mg of hydrogen peroxide (H_2O_2), what is the limiting reactant?

119. What is the excess reactant and how many milligrams are in excess?

120. How many milligrams of benzoquinone will be produced?

CHALLENGE PROBLEM

109. a. V_2O_5 and VO_2
b. $V_2O_5 + H_2 \longrightarrow 2VO_2 + H_2O$
$VO_2 + 2H_2 \longrightarrow V + 2H_2O$
c. 0.532 g H_2

CUMULATIVE REVIEW

110. a. hypothesis, because it is based only on observation, not on data

111. a. $[He]2s^22p^5$
b. $[Ne]3s^23p^1$
c. $[Ar]4s^23d^2$
d. $[Xe]6s^24f^{14}5d^{10}6p^6$

112. Diatomic molecules achieve a noble gas electron configuration by forming covalent bonds. The monoatomic gases already have noble gas electron configurations.

113. $4K(s) + O_2(g) \longrightarrow 2K_2O(s)$

114. 352.02 amu; 352.02 g/mol

115. a. They are related by a whole-number multiplier.
b. C_2H_4O

WRITING IN ▶ Chemistry

Rubric

116. Answers will vary. Common pollutants are SO_2, NO, NO_2 and O_3. Check the stoichiometry and be sure it accounts for a decrease in the pollutant.

117. Answers will vary. Make sure the following equation is included.
$N_2(g) + 3H_2(g) \leftrightharpoons 2NH_3(g) + 92$ kJ
The aim of the Haber process was to control a reaction so that a large amount of a useful product was yielded quickly. The process was of great importance because the Germans had to come up with a nitrogen compound that could be produced in large amounts.

DBQ Document-Based Questions

Data obtained from: Becker, Bob. April 2006. *ChemMatters.* 24: no. 2.

118. H_2O_2 is the limiting reactant.
119. 19.1 mg of $C_6H_4(OH)_2$ excess
120. 79.4 mg $C_6H_4O_2$

MULTIPLE CHOICE

1. D
2. A
3. C
4. B
5. C
6. D
7. C
8. B

MULTIPLE CHOICE

1. Stoichiometry is based on the law of
A. constant mole ratios.
B. Avogadro's constant.
C. conservation of energy.
D. conservation of mass.

Use the graph below to answer Questions 2 to 5.

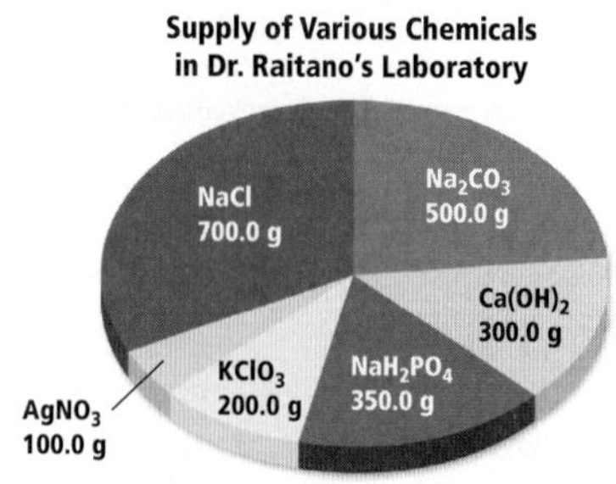

2. Pure silver metal can be made using the reaction shown below.

$$Cu(s) + 2AgNO_3(aq) \rightarrow 2Ag(s) + Cu(NO_3)_2(aq)$$

How many grams of copper metal will be needed to use up all of the $AgNO_3$ in Dr. Raitano's laboratory?
A. 18.70 g
B. 37.30 g
C. 74.70 g
D. 100.0 g

3. The LeBlanc process is the traditional method of manufacturing sodium hydroxide. The equation for this process is as follows.

$$Na_2CO_3(aq) + Ca(OH)_2(aq) \rightarrow 2NaOH(aq) + CaCO_3(s)$$

Using the amounts of chemicals available in Dr. Raitano's lab, what is the maximum number of moles of NaOH that can be produced?
A. 4.050 mol
B. 4.720 mol
C. 8.097 mol
D. 9.430 mol

4. Pure O_2 gas can be generated from the decomposition of potassium chlorate ($KClO_3$):

$$2KClO_3(s) \rightarrow 2KCl(s) + 3O_2(g)$$

If half of the $KClO_3$ in the lab is used and 12.8 g of oxygen gas is produced, what is the percent yield of this reaction?
A. 12.8%
B. 32.7%
C. 65.6%
D. 98.0%

5. Sodium dihydrogen pyrophosphate ($Na_2H_2P_2O_7$), more commonly known as baking powder, is manufactured by heating NaH_2PO_4 to a high temperature.

$$2NaH_2PO_4(s) \rightarrow Na_2H_2P_2O_7(s) + H_2O(g)$$

If 444.0 g of $Na_2H_2P_2O_7$ is needed, how much more NaH_2PO_4 will Dr. Raitano have to buy to make enough $Na_2H_2P_2O_7$?
A. 0.000 g
B. 94.00 g
C. 130.0 g
D. 480.0 g

6. Red mercury(II) oxide decomposes at high temperatures to form mercury metal and oxygen gas.

$$2HgO(s) \rightarrow 2Hg(l) + O_2(g)$$

If 3.55 mol of HgO decomposes to form 1.54 mol of O_2 and 618 g of Hg, what is the percent yield of this reaction?
A. 13.2%
B. 42.5%
C. 56.6%
D. 86.8%

Use the diagram below to answer Questions 7 and 8.

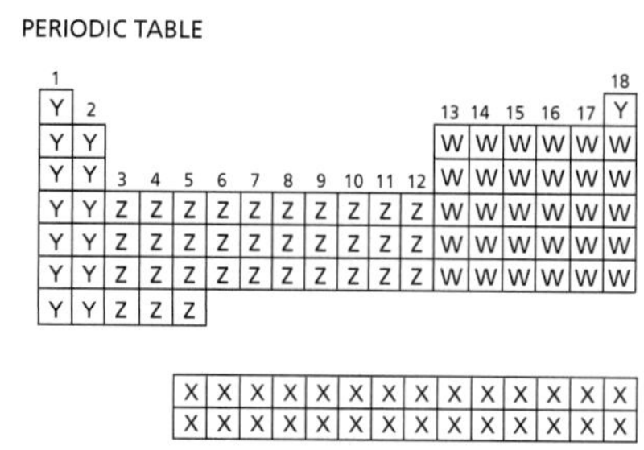

7. Which elements tend to have the largest atomic radius in their periods?
A. W
B. X
C. Y
D. Z

8. Elements labeled *W* have their valence electrons in which sublevel?
A. s
B. p
C. d
D. f

SHORT ANSWER

9. Dimethyl hydrazine $(CH_3)_2N_2H_2$ ignites on contact with dinitrogen tetroxide (N_2O_4).

$$(CH_3)_2N_2H_2(l) + 2N_2O_4(l) \rightarrow 3N_2(g) + 4H_2O(g) + 2CO_2(g)$$

Because this reaction produces an enormous amount of energy from a small amount of reactants, it was used to drive the rockets on the Lunar Excursion Modules (LEMs) of the Apollo space program. If 18.0 mol of dinitrogen tetroxide is consumed in this reaction, how many moles of nitrogen gas will be released?

EXTENDED RESPONSE

Use the table below to answer Questions 10 and 11.

First Ionization Energy of Period 3 Elements		
Element	**Atomic Number**	**1st Ionization Energy, kJ/mol**
Sodium	11	496
Magnesium	12	736
Aluminum	13	578
Silicon	14	787
Phosphorus	15	1012
Selenium	16	1000
Chlorine	17	1251
Argon	18	1521

10. Plot the data from this data table. Place atomic numbers on the *x*-axis.

11. Summarize the general trend in ionization energy. How does ionization energy relate to the number of valence electrons in an element?

SAT SUBJECT TEST: CHEMISTRY

12. How much cobalt(III) titanate (Co_2TiO_4), in moles, is in 7.13 g of the compound?
   A. 2.39×10^1 mol
   B. 3.10×10^{-2} mol
   C. 3.22×10^1 mol
   D. 4.17×10^{-2} mol
   E. 2.28×10^{-2} mol

Use the pictures below to answer Questions 13 to 17.

A.
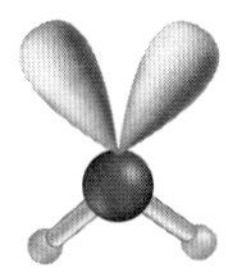

B.
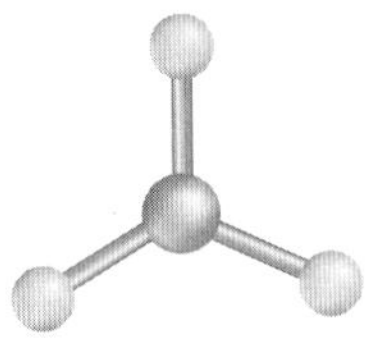

C.
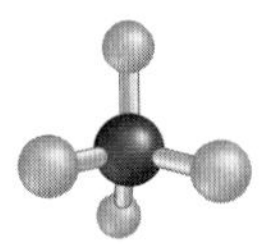

D.
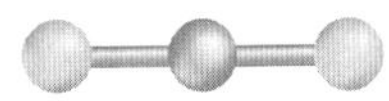

E.
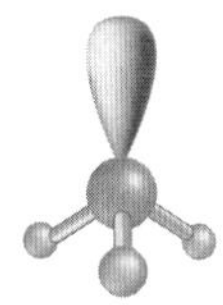

13. Hydrogen sulfide displays this molecular shape.

14. Molecules with this shape have four shared pairs of electrons and no lone pairs of electrons.

15. This molecular shape is known as *trigonal planar.*

16. Carbon dioxide displays this molecular shape.

17. This molecular shape displays sp^2 hybridization.

NEED EXTRA HELP?																	
If You Missed Question . . .	1	2	3	4	5	6	7	8	9	10	11	12	13	14	15	16	17
Review Section . . .	11.1	11.2	11.2	11.4	11.3	11.4	6.3	5.3	11.2	6.3	6.3	10.3	8.4	8.4	8.4	8.4	8.4

SHORT ANSWER

9. 27 moles N_2

EXTENDED RESPONSE

10. Data should form an approximately linear relationship with a few jagged edges, similar to Figure 16 on page 191.

11. Ionization energy generally increases as you move across a period or row in the periodic table. Elements in the first few families have only 1 or 2 valence electrons, which are relatively easy to remove because this will result in a complete outer shell. Elements on the right side of the periodic table have very high ionization energies because of their smaller sizes and steadily increasing nuclear charges.

SAT SUBJECT TEST: CHEMISTRY

12. B
13. A
14. C
15. B
16. D
17. B

CHAPTER 12 States of Matter

BIGIDEA **Kinetic-molecular theory explains the different properties of solids, liquids, and gases.**

ESSENTIAL QUESTIONS	RESOURCES TO ASSESS MASTERY
SECTION 1 Gases 1. How is the kinetic-molecular theory used to explain the behavior of gases? 2. Why does mass affect the rates of diffusion and effusion? 3. How is gas pressure measured and how is the partial pressure of a gas calculated? 2 sessions 1 block	**Progress Monitoring** Formative Assessment, pp. 405, 407, 409 Reading Check, p. 405 Section Review, p. 410
SECTION 2 Forces of Attraction 1. What are intramolecular forces? 2. How do intermolecular forces compare? How do they contrast? 1 session 0.5 block	**Progress Monitoring** Formative Assessment, pp. 412, 413 Reading Check, pp. 412, 413 Section Review, p. 414
SECTION 3 Liquids and Solids 1. How do the arrangements of particles in liquids and solids differ? 2. What are the factors that affect viscosity? 3. How are the unit cell and crystal lattice related? 2 sessions 1 block	**Progress Monitoring** Formative Assessment, pp. 419, 421, 422, 424 Reading Check, pp. 418, 420, 422 Section Review, p. 424
SECTION 4 Phase Changes 1. How can the addition and removal of energy cause a phase change? 2. What is a phase diagram? 1 session 0.5 block	**Progress Monitoring** Formative Assessment, pp. 427, 429 Reading Check, p. 428 Graph Check, pp. 426, 429, 430 Section Review, p. 430 **Summative Assessment** Chapter Assessment, p. 434 *eAssessment* Chapter Test (Scaffolded)

Go online!

LEVELED RESOURCES	LAB MATERIALS
Science Notebook 12.1 OL ***FAST FILE Chapter Resources:*** Study Guide, p. 96 BL	LaunchLAB p. 400 / **15 min** 100-mL graduated cylinder, water, ruler, ball bearing, stopwatch, vegetable oil
Science Notebook 12.2 OL ***FAST FILE Chapter Resources:*** Study Guide, p. 98 BL	
Science Notebook 12.3 OL ***FAST FILE Chapter Resources:*** MiniLab Worksheet, p. 86 OL Study Guide, p. 99 BL	MiniLAB p. 423 / **20 min** soda straws, 22- or 26-gauge wire, scissors
Science Notebook 12.4 OL ***FAST FILE Chapter Resources:*** ChemLab Worksheet p. 88 OL Study Guide, p. 100 BL	ChemLAB p. 432 / **45 min** distilled water, ethanol, isopropyl alcohol, acetone, household ammonia, droppers, small plastic cups, grease pencil or masking tape and marking pen, paper towel, wax paper, stopwatch

ADDITIONAL RESOURCES

eAssessment
ConnectED Student Center
ConnectED Teacher Center
Lesson Presentations
eSolutions
Videos
Animations
Online Quizzes
Virtual Investigations
Multilingual eGlossary
Personal Tutors
WebQuests
Applying Practices Worksheets
PBLs

Practice:

Challenge Problems AL
Supplemental Problems BL OL
Chapter Test (Scaffolded)
Standardized Test Practice

Additional Resources:

Blueprints for Success
Solving Problems: A Chemistry Handbook

Lab Resources:

Laboratory Manual OL
CBL Laboratory Manual OL
Small-Scale Laboratory Manual OL
Forensics Laboratory Manual OL

BL Below Level OL On Level AL Advanced Learners EL English Learners COOP LEARN Cooperative Learning

CHAPTER 12

LaunchLAB

Rubric

How do different liquids affect the speed of a sinking ball bearing?

Purpose Students will observe viscosity in liquids.

Safety Precautions Approve lab safety forms before work begins.

Disposal Do not pour vegetable oil down the drain. Collect oil in a container and place it in the trash.

Teaching Strategies Students need to be quick to record ball-drop times. Tell students to be prepared to record the time before the ball is dropped.

Expected Results The ball drops at a slower rate in the vegetable oil.

Procedure

1. Read and complete the lab safety form.
2. Fill a **100-mL graduated cylinder** with **water.** Be sure to fill it exactly to the 100-mL mark.
3. Place the end of a **ruler** on the tabletop. Drop a **ball bearing** (or other small, round object) from a mark on the ruler just above the surface of the water. Use a **stopwatch** to time the ball bearing as it sinks to the bottom. Record this time in a data table.
4. Repeat Steps 2 and 3 two more times, dropping the object from the same height each time. Calculate the average drop time of your three trials.
5. Repeat Steps 2–4 using **vegetable oil** instead of water.

CHAPTER 12

States of Matter

BIGIDEA Kinetic-molecular theory explains the different properties of solids, liquids, and gases.

SECTIONS

1 Gases

2 Forces of Attraction

3 Liquids and Solids

4 Phase Changes

LaunchLAB

How do different liquids affect the speed of a sinking ball bearing?

You've probably noticed that different liquids might have vastly different properties. For example, liquids such as maple syrup, corn oil, and vegetable oil are much thicker than liquids such as water. In this lab, you will study how the speed of an object is affected by the viscosity of two different liquids.

FOLDABLES Study Organizer

States of Matter

Make a three-tab book. Label it as shown. Use it to help you summarize information about three common states of matter.

The iodine thermometer contains a few grams of iodine inside a sealed sphere. As the outdoor temperature increases, the iodine changes from a solid directly to a gas. The deeper the violet color, the higher the temperature.

Analysis

1. **Compare** the average drop time for the two liquids. The average drop time for the water should be less than the average drop time for the vegetable oil.
2. **Infer** the relationship between the times that you recorded and how easily the liquid flows as you pour it. Vegetable oil has more resistance to flow (greater viscosity) than water, and the drop times are longer for vegetable oil than for water.

Inquiry How does temperature affect the speed with which a ball bearing sinks in a liquid? Develop a hypothesis, and design an experiment to test your hypothesis. As temperature increases, resistance to flow (viscosity) decreases, and the drop times also decrease. Experimental designs will vary.

Introduce the BIGIDEA

Kinetic-Molecular Theory To introduce this chapter's Big Idea, ask students what they remember about the kinetic-molecular theory from previous science courses. Responses will vary. Write the correct student responses on the board and correct any misconceptions that students might have. Ask a student volunteer to name the four states of matter. solid, liquid, gas, and plasma Tell students that only three states of matter will be covered in this chapter: solid, liquid, and gas.

Tie to Previous Knowledge

Have students review the following concepts before studying this chapter.

- density
- physical properties

Use the Photo

Iodine Thermometer Ask students what the circles in the photo are. iodine thermometers Ask students why the thermometers are different shades of purple. The photos of the thermometers were taken at different times of the day and the temperatures varied.

Ask students what made the early morning thermometer light purple and the late afternoon thermometer dark purple. The outdoor temperature was warmer in the late afternoon and more of the solid iodine had converted to a gas, which made the late-afternoon thermometer darker purple than the early-morning thermometer. Ask students if they can tell from the photo what the container of the iodine thermometers is. a sealed, round-bottom flask Ask students to look carefully and see if they can find the mouth of the flask.

1 Focus

MAINIDEA

Particles and Gases Bring an inflatable plastic toy, such as a beach ball, to class. Begin to inflate the toy. Ask a student volunteer to explain what occurs inside the toy when the air particles enter. Accept all reasonable responses. The air particles move around in random motion and exert a force on the inside of the toy. Ask students what happens as more and more air enters the toy. The air particles exert more and more pressure on the sides of the toy as the amount of air inside the toy increases. Blow up the toy partially and ask students if the air particles are compressible. yes **OL**

2 Teach

Content Background

Nationalities The two scientists that developed the kinetic molecular theory were from two different countries. Ludwig Boltzmann was from Austria and James Maxwell was from Scotland.

▪ **Question p. 402** Accept all reasonable responses. Correct responses should include details using the kinetic-molecular theory.

SECTION 1

Gases

MAINIDEA Gases expand, diffuse, exert pressure, and can be compressed because they are in a low-density state consisting of tiny, constantly moving particles.

CHEM 4 YOU If you have gone camping, you might have slept on an air-filled mattress. How did the mattress compare to lying on the ground? It was probably warmer and more comfortable. The properties of the air mattress are due to the particles that make up the air inside it.

Essential Questions

- How is the kinetic-molecular theory used to explain the behavior of gases?
- Why does mass affect the rates of diffusion and effusion?
- How is gas pressure measured and how is the partial pressure of a gas calculated?

Review Vocabulary

kinetic energy: energy due to motion

New Vocabulary

kinetic-molecular theory
elastic collision
temperature
diffusion
Graham's law of effusion
pressure
barometer
pascal
atmosphere
Dalton's law of partial pressures

The Kinetic-Molecular Theory

You have learned that composition (the types of atoms present) and structure (their arrangement) determine the chemical properties of matter. Composition and structure also affect the physical properties of matter. Based solely on physical appearance, you can distinguish between the solids and liquids, as shown in **Figure 1.** By contrast, substances that are gases at room temperature usually display similar physical properties despite their different compositions. Why is there so little variation in behavior among gases? Why are the physical properties of gases different from those of liquids and solids?

By the eighteenth century, scientists knew how to collect gaseous products by displacing water. Now, they could observe and measure properties of individual gases. About 1860, chemists Ludwig Boltzmann and James Maxwell, who were working in different countries, each proposed a model to explain the properties of gases. That model is the kinetic-molecular theory. Because all of the gases known to Boltzmann and Maxwell contained molecules, the name of the model refers to molecules. The word *kinetic* comes from a Greek word meaning *to move*. Objects in motion have energy called kinetic energy. The **kinetic-molecular theory** describes the behavior of matter in terms of particles in motion. The model makes several assumptions about the size, motion, and energy of gas particles.

▪ **Figure 1** You can distinguish some materials by looking at them, but this is not true for many gases.

Gold

Graphite

Mercury

(l)©Don Mason/Corbis; (c)©Lester V. Bergman/Corbis; (r)Dirk Wiersma/Photo Researchers

DIFFERENTIATED INSTRUCTION

Struggling Learners Remind students that kinetic energy is directly related to the mass and velocity of an object. The object can be as large as a passenger aircraft or as small as a subatomic particle. Remind students that the equation for determining the kinetic energy of an object is $KE = \frac{1}{2} mv^2$. **BL**

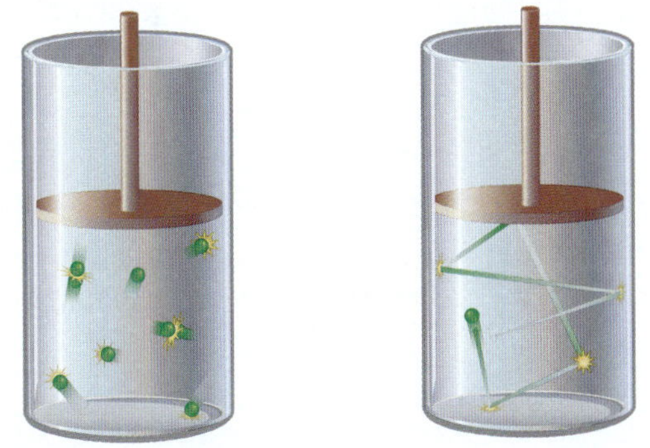

■ **Figure 2** Kinetic energy can be transferred between gas particles during an elastic collision. Between collisions, the particles move in straight lines.

Explain ***the influence that gas particles have on each other, both in terms of collisions and what happens to particles between collisions.***

Particle size Gases consist of small particles that are separated from one another by empty space. The volume of the particles is small compared with the volume of the empty space. Because gas particles are far apart, they experience no significant attractive or repulsive forces.

Particle motion Gas particles are in constant, random motion. Particles move in a straight line until they collide with other particles or with the walls of their container, as shown in **Figure 2.** Collisions between gas particles are elastic. An **elastic collision** is one in which no kinetic energy is lost. Kinetic energy can be transferred between colliding particles, but the total kinetic energy of the two particles does not change.

Particle energy Two factors determine the kinetic energy of a particle: mass and velocity. The kinetic energy of a particle can be represented by the following equation.

$$KE = \frac{1}{2}mv^2$$

KE is kinetic energy, m is the mass of the particle, and v is its velocity. Velocity reflects both the speed and the direction of motion. In a sample of a single gas, all particles have the same mass, but all particles do not have the same velocity. Therefore, all particles do not have the same kinetic energy. **Temperature** is a measure of the average kinetic energy of the particles in a sample of matter.

Explaining the Behavior of Gases

The kinetic-molecular theory helps explain the behavior of gases. For example, the constant motion of gas particles allows a gas to expand until it fills its container, such as when you inflate a beach ball. As you blow air into the ball, the air particles spread out and fill the inside of the container—the beach ball.

Vocabulary
Word origin
Gas
comes from the Latin word *chaos,* which means *space*

Low density Remember that density is mass per unit volume. The density of chlorine gas is 2.898×10^{-3} g/mL at 20°C; the density of solid gold is 19.3 g/mL. Gold is more than 6700 times as dense as chlorine. This large difference cannot be due only to the difference in mass between gold atoms and chlorine molecules (about 3:1). As the kinetic-molecular theory states, a great deal of space exists between gas particles. Thus, there are fewer chlorine molecules than gold atoms in the same volume.

Quick Demo

Displacement Place an inverted funnel in a beaker of water and push its cone into the water, keeping the funnel end open. Ask students to explain what is happening. Water moves into the funnel, displacing air. Next, plug the opening of the funnel stem with your finger as you again push the funnel into the water. Ask students to explain the results. Only a small amount of water moves into the funnel because the funnel-stem opening is blocked, preventing the displacement of gas particles in the air by the water. This experiment, first performed hundreds of years ago, demonstrated that air is not empty space. OL EL

■ **Caption Question Fig. 2**
Between collisions, gas particles are so far apart that they experience no significant attractive or repulsive forces. When gas particles collide, the collision is elastic and the total kinetic energy remains the same.

CHEMISTRY PROJECT

Boltzmann and Maxwell Ask students to research Ludwig Boltzmann and James Maxwell, whose work on gases led to the kinetic-molecular theory. Ask them to write a report including a synopsis of the contributions of both men and an evaluation of which scientist's work made a stronger contribution to the theory. OL

Enrichment

Effusion Rate Assign students to groups of three or four, and ask each group to calculate the ratio of effusion rates for different pairs of noble gases. OL COOP LEARN

- **Caption Question Fig. 3** The density of the particles decreases as you move from left to right because the amount of empty space increases.

- **In-Text Question** The gas particles effuse, and the object deflates as the gas particles escape.

Content Background

Effusion The complete definition of *effusion* includes the concept *from an area of high pressure to an area of lower pressure,* which cannot be introduced because pressure has not yet been defined. Revisit this definition once *pressure* has been introduced and defined.

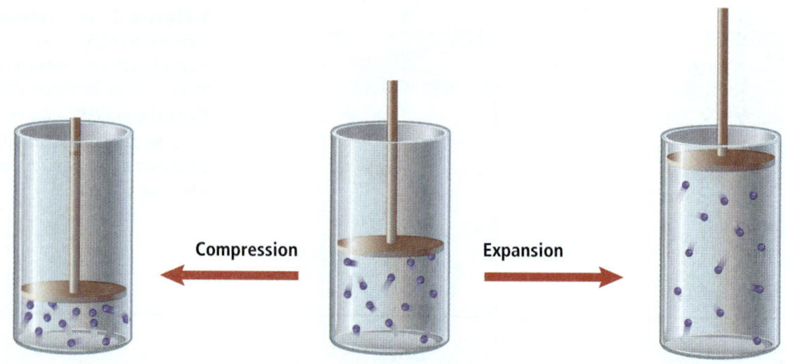

Figure 3 In a closed container, compression and expansion change the volume occupied by a constant mass of particles.
Relate *the change in volume to the density of the gas particles in each cylinder.*

Compression and expansion If you squeeze a pillow made of foam, you can compress it; that is, you can reduce its volume. Air, which is a mixture of gases, is also compressible. The large amount of empty space between the particles in the air allows air to be squeezed into a smaller volume. When the volume of a container is made larger, the random motion of the particles fills the available space. **Figure 3** illustrates what happens to the density of a gas in a container as it is compressed and as it is allowed to expand.

Diffusion and effusion According to the kinetic-molecular theory, there are no significant forces of attraction between gas particles. Thus, gas particles can flow easily past each other. Often, the space into which a gas flows is already occupied by another gas. The random motion of the gas particles causes the gases to mix until they are evenly distributed. **Diffusion** is the term used to describe the movement of one material through another. The term might be new, but you are probably familiar with the process. If food is cooking in the kitchen, you can smell it throughout the house because the gas particles diffuse. Particles diffuse from an area of high concentration (the kitchen) to one of low concentration (the other rooms in the house).

Effusion is a process related to diffusion. During effusion, a gas escapes through a tiny opening. What happens when you puncture a container, such as a balloon or a tire? In 1846, Thomas Graham conducted experiments to measure the rates of effusion for different gases at the same temperature. Graham designed his experiments so that the gases effused into a vacuum—space containing no matter. He discovered an inverse relationship between effusion rates and molar mass. **Graham's law of effusion** states that the rate of effusion for a gas is inversely proportional to the square root of its molar mass.

Graham's Law

$$\text{Rate of effusion} \propto \frac{1}{\sqrt{\text{molar mass}}}$$

The rate of diffusion or effusion of a gas is inversely proportional to the square root of its molar mass.

DIFFERENTIATED INSTRUCTION

Struggling Learners Some students will find it easier to solve problems that use Graham's law of effusion if they first estimate what the answer will be and then compare their calculated answer with their prediction. For example, if they are given the rate of effusion for hydrogen and asked to use it to calculate the rate for chlorine, they should first compare the molar masses of the two gases. If they realize that chlorine molecules are much more massive than hydrogen molecules, they will predict that the effusion rate of chlorine will be much slower than that of hydrogen. BL EL

The rate of diffusion depends mainly on the mass of the particles involved. Lighter particles diffuse more rapidly than heavier particles. Recall that different gases at the same temperature have the same average kinetic energy as described by the equation KE = ½ mv^2. However, the mass of gas particles varies from gas to gas. For lighter particles to have the same average kinetic energy as heavier particles, they must have, on average, a greater velocity.

Graham's law also applies to rates of diffusion, which is logical because heavier particles diffuse more slowly than lighter particles at the same temperature. Using Graham's law, you can set up a proportion to compare the diffusion rates for two gases.

$$\frac{\text{Rate}_A}{\text{Rate}_B} = \sqrt{\frac{\text{molar mass}_B}{\text{molar mass}_A}}$$

READING CHECK **Explain** why the rate of diffusion depends on the mass of the particles.

Explore **kinetic theory.**

Virtual Investigations

EXAMPLE Problem 1

Find help with **square and cube roots.** Math Handbook

GRAHAM'S LAW Ammonia has a molar mass of 17.0 g/mol; hydrogen chloride has a molar mass of 36.5 g/mol. What is the ratio of their diffusion rates?

1 ANALYZE THE PROBLEM

You are given the molar masses for ammonia and hydrogen chloride. To find the ratio of the diffusion rates for ammonia and hydrogen chloride, use the equation for Graham's law of effusion.

Known	Unknown
molar mass_{HCl} = 36.5 g/mol	ratio of diffusion rates = ?
molar mass_{NH_3} = 17.0 g/mol	

2 SOLVE FOR THE UNKNOWN

$$\frac{\text{Rate}_{NH_3}}{\text{Rate}_{HCl}} = \sqrt{\frac{\text{molar mass}_{HCl}}{\text{molar mass}_{NH_3}}}$$

State the ratio derived from Graham's law.

$$= \sqrt{\frac{36.5\ \text{g/mol}}{17.0\ \text{g/mol}}} = 1.47$$

Substitute molar mass_{HCl} = 36.5 g/mol and molar mass_{NH_3} = 17.0 g/mol.

The ratio of diffusion rates is 1.47.

3 EVALUATE THE ANSWER

A ratio of roughly 1.5 is logical because molecules of ammonia are about half as massive as molecules of hydrogen chloride. Because the molar masses have three significant figures, the answer also does. Note that the units cancel, and the answer is stated correctly without any units.

EXAMPLE PROBLEM

PRACTICE Problems

Do additional problems. Online Practice

1. Calculate the ratio of effusion rates for nitrogen (N_2) and neon (Ne).
2. Calculate the ratio of diffusion rates for carbon monoxide and carbon dioxide.
3. **Challenge** What is the rate of effusion for a gas that has a molar mass twice that of a gas that effuses at a rate of 3.6 mol/min?

ASSESSMENT

Skill Ask students to design an experiment to compare the rates of diffusion of substances found in different perfumes. Suggest a format for students to follow with the problem, objectives, materials, safety precautions, procedure, and data collection. OL

READING CHECK Lighter particles diffuse more rapidly because they have greater velocities than heavier particles with equal energy.

IN-CLASS Example

Question Calculate the ratio of effusion rates for helium and argon.

Answer

$$\frac{\text{Rate}_{He}}{\text{Rate}_{Ar}} = \sqrt{\frac{\text{atomic mass}_{Ar}}{\text{atomic mass}_{He}}}$$

$$= \sqrt{\frac{39.948\ \text{g/mol}}{4.003\ \text{g/mol}}}$$

$$= 3.159$$

PRACTICE Problems

Have students refer to the Selected Solutions appendix for complete solutions to odd-numbered problems. The complete solutions for all problems can be found in the Solutions Manual.

1. $R_{N_2}/R_{Ne} = 0.8488$
2. 1.253
3. 2.5 mol/min

CHEMISTRY JOURNAL

Applying Pressure Have students write these questions down and answer them from their study of Section 1.

1. Why is stepping on a nail more painful than lying on a bed of nails? When lying on a bed of nails, the force exerted by your mass is spread out over a larger area than when you step on a nail, so you feel less pressure, and thus less pain at any given point.
2. Why can you ski across deep snow, whereas you would sink into it wearing boots? With skis, the force exerted is spread out over a larger area, so the pressure on any given area of the snow is reduced. OL

Math in Chemistry

Square Roots When students use Graham's law to solve problems involving the relationship between the molar masses of gases and their rates of effusion, remind them that square roots are involved. Many students make the common mistake of ignoring the square root symbol and treating the equation like a direct proportion.

Use Science Terms

Temperature and Pressure Have students write definitions of the terms *temperature* and *pressure;* then, have them write sentences that include the terms. OL EL

■ **Caption Question Fig. 4**

at the heel because it is the smallest area

Content Background

Single-Breath Dives Records dating to 4500 B.C. in Mesopotamia reveal that people were making single-breath, deep dives into the sea in search of the precious mother of pearl. These "free-dives" have recently become a competition of who can dive the deepest. On June 14, 2007, Austrian diver Herbert Nitsch dove to a new record depth of 214 m. At these depths, the pressure of the water pushing down on the body is comparable to the weight of a full-grown elephant standing on one foot.

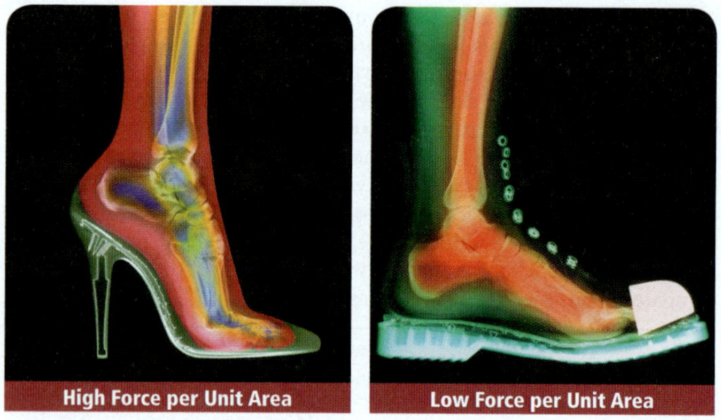

■ **Figure 4** High-heeled shoes increase the pressure on a surface because the area touching the floor is reduced. In flatter-heeled shoes, such as boots, the force is applied over a larger area.

Infer *where the highest pressure is located between the floor and high-heel shoe.*

H. Turvey/Photo Researchers

Gas Pressure

Have you watched someone try to walk across snow, mud, or hot asphalt in high heels? If so, you might have noticed that the heels sank into the soft surface. **Figure 4** shows why a person sinks when wearing high heels but does not sink when wearing boots. In each case, the force pressing down on the soft surface is related to the person's mass. With boots, the force is spread out over a larger area. **Pressure** is defined as force per unit area. The area of the bottom of a boot is much larger than the area of the bottom of a high-heeled shoe. So, the pressure on the soft surface is less with a boot than it is with high heels.

Gas particles also exert pressure when they collide with the walls of their container. Because an individual gas particle has little mass, it can exert little pressure. However, a liter-sized container could hold 10^{22} gas particles. With this many particles colliding, the pressure can be high.

Air pressure Earth is surrounded by an atmosphere that extends into space for hundreds of kilometers. Because the particles in air move in every direction, they exert pressure in all directions. This pressure is called atmospheric pressure, or air pressure. Air pressure varies at different points on Earth. Because gravity is greater at the surface of Earth, there are more particles than at higher altitudes where the force of gravity is less. Fewer particles at higher elevations exert less force than the greater concentration of particles at lower altitudes. Therefore, air pressure is less at higher altitudes than it is at sea level. At sea level, atmospheric pressure is about one-kilogram per square centimeter.

■ **Figure 5** Torricelli was the first to design equipment to show that the atmosphere exerted pressure.

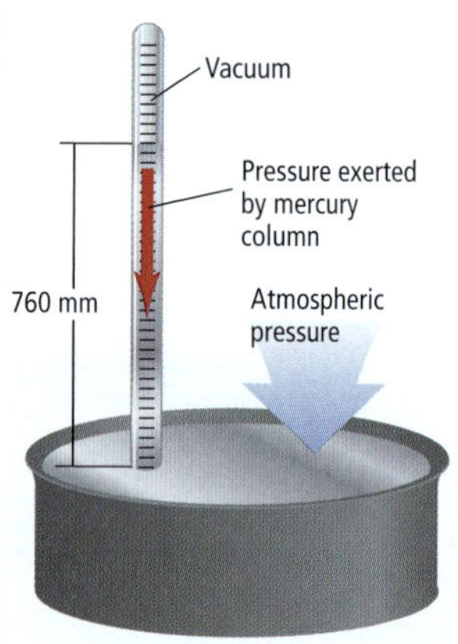

Measuring air pressure Italian physicist Evangelista Torricelli (1608–1647) was the first to demonstrate that air exerted pressure. He noticed that water pumps were unable to pump water higher than about 10 m. He hypothesized that the height of a column of liquid would vary with the density of the liquid. To test this idea, Torricelli designed the equipment shown in **Figure 5.** He filled a thin glass tube that was closed at one end with mercury. While covering the open end so that air could not enter, he inverted the tube and placed it (open end down) in a dish of mercury. The height of the mercury column fell to about one-fourteenth of a similar water column. This validated Terricelli's hypothesis because mercury is approximately fourteen times more dense than water.

CHEMISTRY JOURNAL

Torricelli Ask students to write an epitaph for Evangelista Torricelli that briefly describes the major events in his life and work. OL

Barometers The device that Torricelli invented is called a barometer. A **barometer** is an instrument used to measure atmospheric pressure. As Torricelli demonstrated, at sea level the height of the mercury in a barometer is usually about 760 mm. The exact height of the mercury is determined by two forces. Gravity exerts a constant downward force on the mercury. This force is opposed by an upward force exerted by air pressing down on the surface of the mercury. Changes in air temperature or humidity cause air pressure to vary.

Manometers A manometer is an instrument used to measure gas pressure in a closed container. In a manometer, a flask is connected to a U-tube that contains mercury, as shown in **Figure 6.** When the valve between the flask and the U-tube is opened, gas particles diffuse out of the flask into the U-tube. The released gas particles push down on the mercury in the tube. The difference in the height of the mercury in the two arms is used to calculate the pressure of the gas in the flask.

Units of pressure The SI unit of pressure is the pascal (Pa). It is named for Blaise Pascal (1623–1662), a French mathematician and philosopher. The pascal is derived from the SI unit of force, the newton (N). One **pascal** is equal to a force of one newton per square meter: 1 Pa equals 1 N/m^2. Many fields of science still use more traditional units of pressure. For example, engineers often report pressure as pounds per square inch (psi). The pressures measured by barometers and manometers can be reported in millimeters of mercury (mmHg).
There is also a unit called the torr and another unit called a bar.

At sea level, the average air pressure is 101.3 kPa when the temperature is 0°C. Air pressure is often reported in a unit called an atmosphere (atm). One **atmosphere** is equal to 760 mmHg or 760 torr or 101.3 kilopascals (kPa). **Table 1** compares different units of pressure. Because the units 1 atm, 760 mmHg, and 760 torr are defined units, they should have as many significant figures as needed when used in calculations.

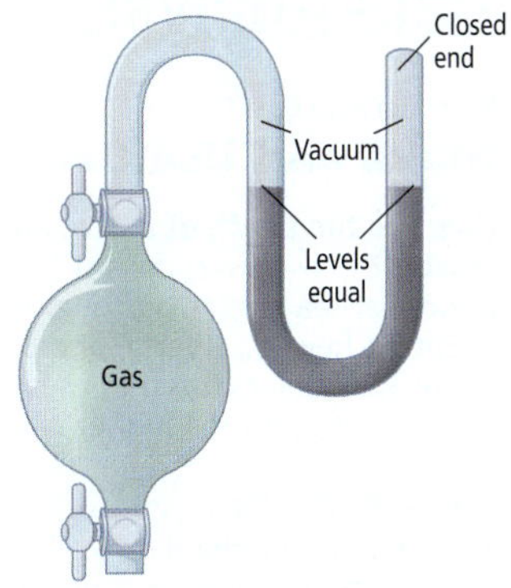

Before gas is released into the U-tube, the mercury is at the same height in each arm.

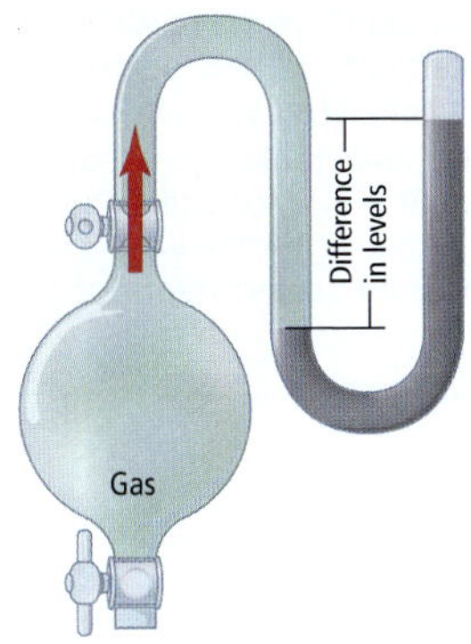

After gas is released into the U-tube, the heights in the two arms are no longer equal.

Figure 6 A manometer measures the pressure of an enclosed gas.

Table 1 Comparison of Pressure Units

Unit	Number Equivalent to 1 atm	Number Equivalent to 1 kPa
Kilopascal (kPa)	101.3 kPa	—
Atmosphere (atm)	—	0.009869 atm
Millimeters of mercury (mmHg)	760 mmHg	7.501 mmHg
Torr	760 torr	7.501 torr
Pounds per square inch (psi or lb/in^2)	14.7 psi	0.145 psi
Bar	1.01 bar	0.01 bar

ASSESSMENT

Knowledge Ask students to write a test question on the material in this section. Then, collect the questions and have the class discuss the answers, or have students exchange their questions and quiz each other. OL COOP LEARN

Content Background

Measuring Air Pressure Ask students if they think that, by changing the diameter or length of the tube in Torricelli's apparatus, the level of mercury in the column would vary. Explain that Torricelli found that it did not matter what diameter or length of tube he used, the mercury always fell to a height of about 760 mm. Review with students the two forces that determine the exact height of the mercury. gravity and air pressure

Extension

Barometers Ask interested students to research the operation and uses of aneroid barometers. Student groups can prepare classroom presentations about their findings. OL COOP LEARN

DIFFERENTIATED INSTRUCTION

Advanced Learners Ask interested students to design, build, and test their own barometers. AL

Data Analysis LAB

About the Lab

- Not all divers dive at altitudes at or near sea level. Some divers dive at higher altitudes, such as in Canada, northwestern United States, Colorado, and Switzerland.
- For pressure gauges that do not correct for altitude variations, correction factors must be manually added to dive depths.
- Divers need to know the depth in which they are diving to determine safe diving times.
- Also see Sawatzky, D. "Diving at Altitude Part II." *Diver Magazine,* August 2000.

Think Critically

1. Refer to the Solutions Manual for a sample graph.
2. 20 m
3. The amount of time that it is safe to stay under water is directly related to the diving depth. If you do not know your actual diving depth, you cannot determine how long it is safe to stay at a particular diving depth.

Caption Question Fig. 7
They will have equal pressure; they are equal moles of gas at the same temperature and volume.

Data Analysis LAB

Based on Real Data*

Make and Use Graphs

How are the depth of a dive and altitude related? Most divers dive at locations that are at or near sea level in altitude. However, divers in Saskatchewan, Alberta, and British Columbia, Canada, as well as much of the northwestern United States, dive at higher altitudes.

Think Critically

1. **Compare** Use the data in the table to make a graph of atmospheric pressure versus altitude.
2. **Calculate** What is your actual diving depth if your depth gauge reads 18 m, but you are at an altitude of 1800 m and your gauge does not compensate for altitude?
3. **Infer** Dive tables are used to determine how long it is safe for a diver to stay under water at a specific depth. Why is it important to know the correct depth of the dive?

Data and Observations

The table shows the pressure gauge correction factor for high altitude underwater diving.

Altitude Diving Correction Factors

Altitude (m)	Atmospheric Pressure (atm)	Pressure Gauge Correction Factor (m)
0	1.000	0.0
600	0.930	0.7
1200	0.864	1.4
1800	0.801	2.0
2400	0.743	2.7
3000	0.688	3.2

*Data obtained from: Sawatzky, D. 2000. Diving at Altitude Part I. *Diver Magazine.* June 2000.

Dalton's law of partial pressures When Dalton studied the properties of gases, he found that each gas in a mixture exerts pressure independently of the other gases present. Illustrated in **Figure 7, Dalton's law of partial pressures** states that the total pressure of a mixture of gases is equal to the sum of the pressures of all the gases in the mixture. The portion of the total pressure contributed by a single gas is called its partial pressure. The partial pressure of a gas depends on the number of moles of gas, the size of the container, and the temperature of the mixture. It does not depend on the identity of the gas. At a given temperature and pressure, the partial pressure of 1 mol of any gas is the same. Dalton's law of partial pressures can be summarized by the equation at the top of the next page.

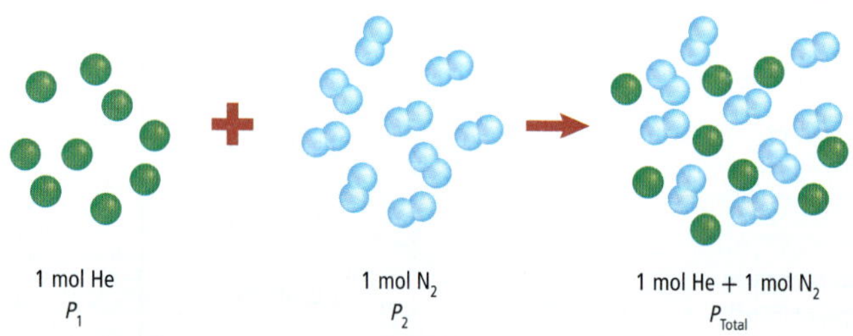

Figure 7 When gases mix, the total pressure of the mixture is equal to the sum of the partial pressures of the individual gases.
Determine *How do the partial pressures of nitrogen gas and helium gas compare when a mole of nitrogen gas and a mole of helium gas are in the same closed container?*

CHEMISTRY JOURNAL

Marine Animals Have students research how marine animals can withstand the huge pressures that are present under the sea and explain their findings in their chemistry journals. OL

Dalton's Law of Partial Pressures

$$P_{total} = P_1 + P_2 + P_3 + \ldots P_n$$

P_{total} represents total pressure. P_1, P_2, and P_3 represent the partial pressures of each gas up to the final gas, P_n.

To calculate the total pressure of a mixture of gases, add the partial pressures of each of the gases in the mixture.

Get help with **Dalton's law of partial pressures.**

Personal Tutor

Look again at **Figure 7.** What happens when 1 mol of helium and 1 mol of nitrogen are combined in a single closed container? Because neither the volume nor the number of particles changes, the total pressure equals the sum of the two partial pressures.

EXAMPLE Problem 2

Find help with **significant figures.** Math Handbook

THE PARTIAL PRESSURE OF A GAS A mixture of oxygen (O_2), carbon dioxide (CO_2), and nitrogen (N_2) has a total pressure of 0.97 atm. What is the partial pressure of O_2 if the partial pressure of CO_2 is 0.70 atm and the partial pressure of N_2 is 0.12 atm?

1 ANALYZE THE PROBLEM

You are given the total pressure of a mixture and the partial pressure of two gases in the mixture. To find the partial pressure of the third gas, use the equation that relates partial pressures to total pressure.

Known	Unknown
$P_{N_2} = 0.12$ atm	$P_{O_2} = ?$ atm
$P_{CO_2} = 0.70$ atm	
$P_{total} = 0.97$ atm	

2 SOLVE FOR THE UNKNOWN

$P_{total} = P_{N_2} + P_{CO_2} + P_{O_2}$ — State Dalton's law of partial pressures.

$P_{O_2} = P_{total} - P_{CO_2} - P_{N_2}$ — Solve for P_{O_2}.

$P_{O_2} = 0.97\text{ atm} - 0.70\text{ atm} - 0.12\text{ atm}$ — Substitute $P_{N_2} = 0.12$ atm, $P_{CO_2} = 0.70$ atm, and $P_{total} = 0.97$ atm.

$P_{O_2} = 0.15$ **atm**

3 EVALUATE THE ANSWER

Adding the calculated value for the partial pressure of oxygen to the known partial pressures gives the total pressure, 0.97 atm. The answer has two significant figures to match the data.

EXAMPLE PROBLEM

PRACTICE Problems

Do additional problems. Online Practice

4. What is the partial pressure of hydrogen gas in a mixture of hydrogen and helium if the total pressure is 600 mmHg and the partial pressure of helium is 439 mmHg?
5. Find the total pressure for a mixture that contains four gases with partial pressures of 5.00 kPa, 4.56 kPa, 3.02 kPa, and 1.20 kPa.
6. Find the partial pressure of carbon dioxide in a gas mixture with a total pressure of 30.4 kPa if the partial pressures of the other two gases in the mixture are 16.5 kPa and 3.7 kPa.
7. **Challenge** Air is a mixture of gases. By percentage, it is roughly 78 percent nitrogen, 21 percent oxygen, and 1 percent argon. (There are trace amounts of many other gases in air.) If the atmospheric pressure is 760 mmHg, what are the partial pressures of nitrogen, oxygen, and argon in the atmosphere?

PRACTICE PROBLEMS

ASSESSMENT

Knowledge Ask students to solve the following problem and to explain their answers. Compare the speed of particles in the air at the top of the troposphere, where the temperature is about −50°C, to their speed at sea level, where the temperature is about −20°C. (Have students predict which particles will move faster based on what they know about the kinetic-molecular theory.) The particles in the air at sea level will move faster, as they are at a higher temperature and have greater kinetic energy. (Average kinetic energy is directly proportional to temperature.) OL

IN-CLASS Example

Question A mixture of oxygen (O_2), dinitrogen monoxide (N_2O), and argon (Ar) has a total pressure of 0.98 atm. What is the partial pressure of N_2O, if the partial pressure of O_2 is 0.48 atm and the partial pressure of Ar is 0.15 atm?

Answer

$P_{total} = P_{O_2} + P_{N_2O} + P_{Ar}$

$P_{N_2O} = P_{total} - P_{O_2} - P_{Ar}$

$P_{N_2O} = 0.98\text{ atm} - 0.48\text{ atm} - 0.15\text{ atm}$

$P_{N_2O} = 0.35\text{ atm}$

PRACTICE Problems

Have students refer to the Selected Solutions appendix for complete solutions to odd-numbered problems. The complete solutions for all problems can be found in the Solutions Manual.

4. 161 mmHg
5. 13.78 kPa
6. 10.2 kPa
7. $N_2 = 590$ mmHg; $O_2 = 160$ mmHg; Ar = 8 mmHg

■ Caption Question **Fig. 8** 97.7 kPa

FOLDABLES® Rubric

3 Assess

Check for Understanding

Ask students to compare the processes of diffusion and effusion and to describe situations in which each takes place. Diffusion is the movement of one substance through another and takes place whenever gas particles, such as perfume or air freshener, move through a room. Effusion is the escape of a gas through a small hole and occurs when air slowly seeps out of microscopic openings in balloons or tires. **OL**

Reteach

To emphasize that pressure depends on the force per unit area (not just the total force exerted by an object) ask students which will cause more wear and tear on a carpeted floor over a set length of time: a 125-pound person who walks back and forth on the carpet wearing shoes with spike heels, or a 125-pound person who makes the same number of trips wearing flat-soled shoes? The shoes with spiked heels will place more pressure on the carpet because all of the person's mass will be focused on just two small areas. The flat shoes allow the mass to be distributed over a larger area, so they exert less pressure per unit area. **BL**

Extension

Ask students to use the definition of pressure to explain how people doing karate can break thick boards with the sides of their hands. They are exerting a large force over a small area, placing enough pressure on the board to break it. **OL**

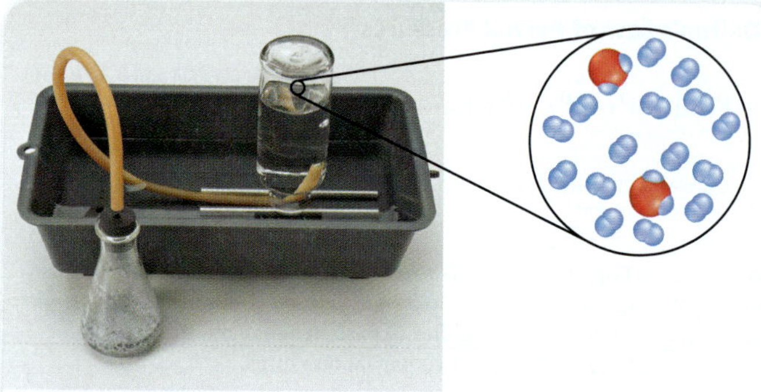

■ **Figure 8** In the flask, sulfuric acid (H_2SO_4) reacts with zinc to produce hydrogen gas. The hydrogen is collected at 20°C.

Calculate *the partial pressure of hydrogen at 20°C if the total pressure of the hydrogen and water vapor mixture is 100.0 kPa.*

FOLDABLES® Incorporate information from this section into your Foldable.

Using Dalton's law Partial pressures can be used to determine the amount of gas produced by a reaction. The gas produced is bubbled into an inverted container of water, as shown in **Figure 8.** As the gas collects, it displaces the water. The gas collected in the container will be a mixture of hydrogen and water vapor. Therefore, the total pressure inside the container will be the sum of the partial pressures of hydrogen and water vapor.

The partial pressures of gases at the same temperature are related to their concentration. The partial pressure of water vapor has a fixed value at a given temperature. You can look up the value in a reference table. At 20°C, the partial pressure of water vapor is 2.3 kPa. You can calculate the partial pressure of hydrogen by subtracting the partial pressure of water vapor from the total pressure.

As you will read later, knowing the pressure, volume, and temperature of a gas allows you to calculate the number of moles of the gas. Temperature and volume can be measured during an experiment. Once the temperature is known, the partial pressure of water vapor is used to calculate the pressure of the gas. The known values for volume, temperature, and pressure are then used to find the number of moles.

SECTION 1 REVIEW

Section Self-Check

Section Summary

- The kinetic-molecular theory explains the properties of gases in terms of the size, motion, and energy of their particles.
- Dalton's law of partial pressures is used to determine the pressures of individual gases in gas mixtures.
- Graham's law is used to compare the diffusion rates of two gases.

8. MAINIDEA **Explain** Use the kinetic theory to explain the behavior of gases.
9. **Describe** how the mass of a gas particle affects its rate of effusion and diffusion.
10. **Explain** how gas pressure is measured.
11. **Explain** why the container of water must be inverted when a gas is collected by displacement of water.
12. **Calculate** Suppose two gases in a container have a total pressure of 1.20 atm. What is the pressure of Gas B if the partial pressure of Gas A is 0.75 atm?
13. **Infer** whether or not temperature has any effect on the diffusion rate of a gas. Explain your answer.

SECTION 1 REVIEW

8. Gases consist of small particles in random motion, which experience elastic collisions.
9. The rate of effusion and diffusion decreases as mass increases.
10. Atmospheric pressure is measured using a barometer. The gas pressure in a closed container is measured using a manometer.
11. If the container is not inverted, the gas, which is less dense than water, will rise through the water and escape from the opening of the container.
12. 0.45 atm
13. As temperature increases, the velocity of the particles increase and the particles will diffuse faster.

Forces of Attraction

Essential Questions

- What are intramolecular forces?
- How do intermolecular forces compare? How do they contrast?

Review Vocabulary

polar covalent: a type of bond that forms when electrons are not shared equally

New Vocabulary

dispersion force
dipole-dipole force
hydrogen bond

MAINIDEA Intermolecular forces—including dispersion forces, dipole-dipole forces, and hydrogen bonds—determine a substance's state at a given temperature.

CHEM 4 YOU You might be aware that water is a substance that is found as a solid, a liquid, and a gas at temperatures and pressures common on Earth. This unique property, along with others that enable life as we understand it to exist, stems from the forces that exist between water molecules.

Intermolecular Forces

If all particles of matter at room temperature have the same average kinetic energy, why are some materials gases while others are liquids or solids? The answer lies with the attractive forces within and between particles. The attractive forces that hold particles together in ionic, covalent, and metallic bonds are called intramolecular forces. The prefix *intra-* means *within*. For example, intramural sports are competitions among teams from within a single school or district. The term *molecular* can refer to atoms, ions, or molecules. **Table 2** summarizes what you read previously about intramolecular forces.

Intramolecular forces do not account for all attractions between particles. There are forces of attraction called intermolecular forces. The prefix *inter-* means *between* or *among*. For example, an interview is a conversation between two people. These forces can hold together identical particles, such as water molecules in a drop of water, or two different types of particles, such as carbon atoms in graphite and the cellulose particles in paper. The three intermolecular forces that will be discussed in this section are dispersion forces, dipole-dipole forces, and hydrogen bonds. Although some intermolecular forces are stronger than others, all intermolecular forces are weaker than the intramolecular forces involved in bonding.

Table 2 Comparison of Intramolecular Forces

Force	Model	Basis of Attraction	Example
Ionic	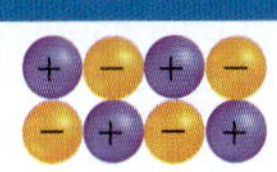	cations and anions	NaCl
Covalent	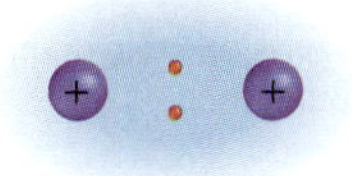	positive nuclei and shared electrons	H_2
Metallic	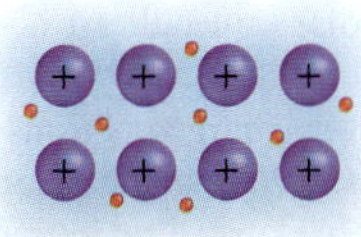	metal cations and mobile electrons	Fe

1 Focus

MAINIDEA

Intermolecular Forces Tell students that the prefix *inter-* means between or among. Ask students what they think *intermolecular forces* means. forces among or between molecules Ask students to infer how the forces between molecules would determine the state of matter. Accept all reasonable responses, but correct any misconceptions. OL

2 Teach

Quick Demo

Molecular Attraction Use hook-and-loop tape to model how the strength of intermolecular attraction is affected by the size of molecules. Have students observe the separation of double strips of hook-and-loop tape in different lengths. Ask them whether they think it is easier to separate longer or shorter molecules. BL

DIFFERENTIATED INSTRUCTION

Struggling Learners Students who have trouble understanding the additive nature of interparticle forces might benefit from making their own models of small and large molecules using cutouts of adhesive-backed shelf paper. Have students cut out different-sized circles of the shelf paper, remove the backing, and stick them together. Then, students should try to separate the circles, comparing how long it takes to separate circles of different sizes. BL EL

ASSESSMENT

Skill Have students make a table of dispersion forces, dipole-dipole forces, and hydrogen bonds, listing their relative strength, how each force forms, and examples of the types of molecules in which they form. **OL**

Caption Question Fig. 9
They represent a partial positive and a partial negative charge, respectively.

READING CHECK They result from temporary shifts in the density of electrons in electron clouds.

READING CHECK Astatine should be a solid like iodine and for the same reasons. The larger halogen molecules have more electrons and greater dispersion forces. The dispersion forces hold the molecules close together creating solids at room temperature.

Content Background

Fritz London London was a German-American physicist. He was born in Breslau, Germany (now Poland) and educated in Germany and France. In 1939, he immigrated to the United States. He was a professor of chemistry at Duke University in Durham, North Carolina. He not only studied hydrogen bonding, but he studied superfluids and superconductivity as well.

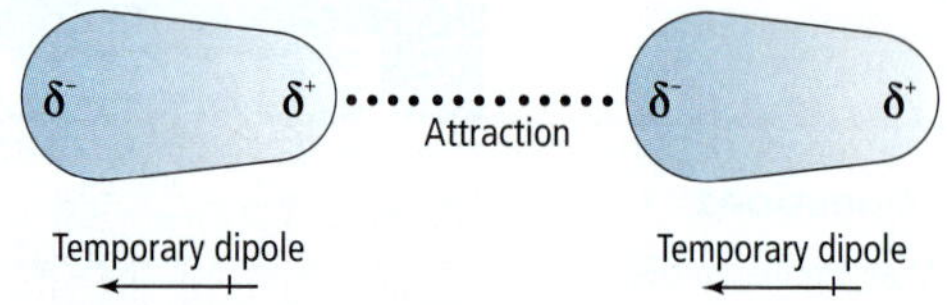

Figure 9 When two molecules are close together, the electron clouds repel each other, creating temporary dipoles. The δ sign represents an area of partial charge on the molecule.
Explain *what the δ+ and δ− signs on a temporary dipole represent.*

Dispersion forces Recall that oxygen molecules are nonpolar because electrons are evenly distributed between the equally electronegative oxygen atoms. Under the right conditions, however, oxygen molecules can be compressed into a liquid. For oxygen to condense, there must be some force of attraction between its molecules.

The force of attraction between oxygen molecules is called a dispersion force. **Dispersion forces** are weak forces that result from temporary shifts in the density of electrons in electron clouds. Dispersion forces are sometimes called London forces after the German-American physicist who first described them, Fritz London.

Remember that the electrons in an electron cloud are in constant motion. When two molecules are in close contact, especially when they collide, the electron cloud of one molecule repels the electron cloud of the other molecule. The electron density around each nucleus is, for a moment, greater in one region of each cloud. Each molecule forms a temporary dipole. When temporary dipoles are close together, a weak dispersion force exists between oppositely charged regions of the dipoles, as shown in **Figure 9.**

READING CHECK **Explain** why dispersion forces form.

Dispersion forces exist between all particles. Dispersion forces are weak for small particles, and these forces have an increasing effect as the number of electrons involved increases. Thus, dispersion forces tend to become stronger as the size of the particles increase. For example, fluorine, chlorine, bromine, and iodine exist as diatomic molecules. Recall that the number of nonvalence electrons increases from fluorine to chlorine to bromine to iodine. Because the larger halogen molecules have more electrons, there can be a greater difference between the positive and negative regions of their temporary dipoles and, thus, stronger dispersion forces. This difference in dispersion forces explains why fluorine and chlorine are gases, bromine is a liquid, and iodine is a solid at room temperature.

READING CHECK **Infer** the physical state of the element astatine at room temperature and explain your reasoning.

VOCABULARY
ACADEMIC VOCABULARY
Orient
to arrange in a specific position; to align in the same direction
The blooms of the flowers were all oriented toward the setting Sun.

Dipole-dipole forces Polar molecules contain permanent dipoles; that is, some regions of a polar molecule are always partially negative and some regions of the molecule are always partially positive. These attractions between oppositely charged regions of polar molecules are called **dipole-dipole forces.** Neighboring polar molecules orient themselves so that oppositely charged regions align.

CHEMISTRY PROJECT

Intermolecular Forces Have students create illustrated posters that explain the various types of intermolecular forces. If possible, display the posters in the classroom. **BL** **EL**

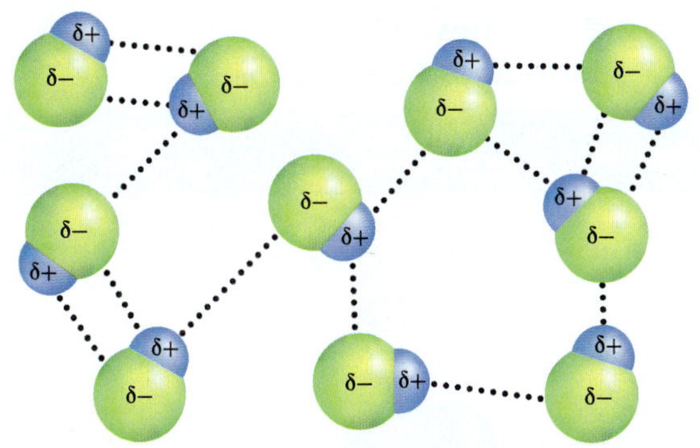

Figure 10 Neighboring polar molecules orient themselves so that oppositely charged regions align.
Identify *the types of forces that are represented in this figure.*

When hydrogen-chloride gas molecules approach, the partially positive hydrogen atom in one molecule is attracted to the partially negative chlorine atom in another molecule. **Figure 10** shows multiple attractions among hydrogen-chloride molecules. Because the dipoles are permanent, you might expect dipole-dipole forces to be stronger than dispersion forces. This prediction holds true for small polar molecules with large dipoles. However, for many polar molecules, including the HCl molecules in **Figure 10,** dispersion forces dominate dipole-dipole forces.

☑ READING CHECK **Compare** dipole-dipole forces and dispersion forces.

Hydrogen bonds One special type of dipole-dipole attraction is called a hydrogen bond. A **hydrogen bond** is a dipole-dipole attraction that occurs between molecules containing a hydrogen atom bonded to a small, highly electronegative atom with at least one lone electron pair. Hydrogen bonds typically dominate both dispersion forces and dipole-dipole forces. For a hydrogen bond to form, hydrogen must be bonded to either a fluorine, oxygen, or nitrogen atom. These atoms are electronegative enough to cause a large partial positive charge on the hydrogen atom, yet small enough that their lone pairs of electrons can come close to hydrogen atoms. For example, in a water molecule, the hydrogen atoms have a large partial positive charge and the oxygen atom has a large partial negative charge. When water molecules approach, a hydrogen atom on one molecule is attracted to the lone pair of electrons on the oxygen atom on the other molecule, as shown in **Figure 11.**

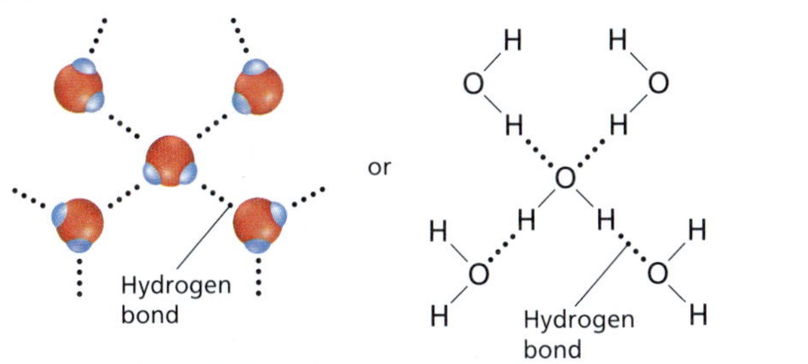

Figure 11 The hydrogen bonds between water molecules are stronger than typical dipole-dipole attractions because the bond between hydrogen and oxygen is highly polar.

Apply Chemistry

Attraction and Repulsion How well different fabrics absorb or repel water depends on their chemical structure. Cotton is formed from cellulose, a polymer of glucose. Each glucose molecule contains multiple hydroxyl (–OH) groups, each of which can hydrogen bond to more than one water molecule. Therefore, cotton absorbs water well and is often used to make towels. In hot weather, cotton clothing can cool because the body's heat energy is used to evaporate water (from sweat) that is bonded to the cotton. Nylon's structure includes only a few groups that can hydrogen bond to water, and most of the nylon polymer is made of $-CH_2$ groups that can form only weaker dispersion forces. Therefore, nylon repels water and is used in rain jackets and umbrellas.

ASSESSMENT

Skill Ask students to research the chemical structures of several different types of fabrics. Students should use the structures to rank the fabrics in terms of the number of hydrogen bonds that the molecules of each fabric should form with water, and then test their rankings by performing an absorbency test with water. OL

Caption Question Fig. 10
dipole-dipole forces

☑ READING CHECK Dipole-dipole forces exist between permanent dipoles; dispersion forces exist between temporary dipoles.

CHEMISTRY JOURNAL

Linus Pauling Ask students to write a short summary of the work of Linus Pauling as it relates to hydrogen bonds. OL

3 Assess

Check for Understanding

Ask students what kind(s) of interparticle forces must be overcome to:

a. melt ice? hydrogen bonds
b. boil water? hydrogen bonds
c. melt NaCl? ionic bonds
d. sublime I_2? dispersion forces OL

Reteach

Ask students to rank the intermolecular forces in order of increasing strength. dispersion forces ⟶ dipole-dipole forces ⟶ hydrogen bonds OL

Extension

Ask students which of the following compounds can form dipole-dipole forces. Cl_2; CO; NO; CH_4 CO, NO OL

Table 3 Properties of Three Molecular Compounds

Compound	Molecular Structure	Molar Mass (g)	Boiling Point (°C)
Water (H_2O)	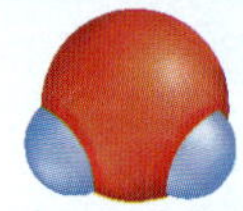	18.0	100
Methane (CH_4)	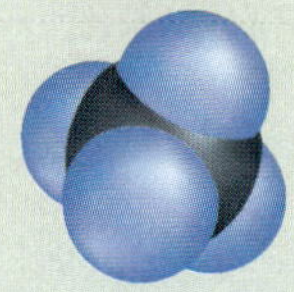	16.0	−161.5
Ammonia (NH_3)	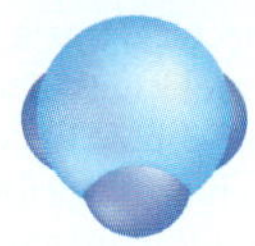	17.0	−33.3

Hydrogen bonds explain why water is a liquid at room temperature, while compounds of comparable mass are gases. Look at the data in **Table 3.** The difference between methane and water is easy to explain. Because methane molecules are nonpolar, the only forces holding the molecules together are relatively weak dispersion forces. The difference between ammonia and water is not as obvious. Molecules of both compounds can form hydrogen bonds. Yet, ammonia is a gas at room temperature, which indicates that the attractive forces between ammonia molecules are not as strong. Because oxygen atoms are more electronegative than nitrogen atoms, the O–H bonds in water are more polar than the N–H bonds in ammonia. As a result, the hydrogen bonds between water molecules are stronger than the hydrogen bonds between ammonia molecules.

SECTION 2 REVIEW

Section Self-Check

Section Summary

- Intramolecular forces are stronger than intermolecular forces.
- Dispersion forces are intermolecular forces between temporary dipoles.
- Dipole-dipole forces occur between polar molecules.

14. MAINIDEA **Explain** what determines a substance's state at a given temperature.

15. **Compare and contrast** intermolecular forces and describe intramolecular forces.

16. **Evaluate** Which of the molecules listed below can form hydrogen bonds? For which of the molecules would dispersion forces be the only intermolecular force? Give reasons for your answers.

a. H_2 **b.** H_2S **c.** HCl **d.** HF

17. **Intepret Data** In a methane molecule (CH_4), there are four single covalent bonds. In an octane molecule (C_8H_{18}), there are 25 single covalent bonds. How does the number of bonds affect the dispersion forces in samples of methane and octane? Which compound is a gas at room temperature? Which is a liquid?

SECTION 2 REVIEW

14. The intermolecular forces between the particles determine the state of a substance. In a solid, the intermolecular forces are very strong and hold the particles together. In a liquid, the intermolecular forces are weaker and in a gas, the particles no longer experience intermolecular forces.

15. Intermolecular forces occur between particles. Intramolecular forces hold particles together.

16. Hydrogen bonds: **d**; only dispersion forces: **a; d** is a polar molecule with an oxygen, nitrogen, or fluorine atom bonded to hydrogen. Molecule **a** is nonpolar.

17. More bonds mean more electrons to form temporary dipoles, which means greater dispersion forces. Methane is a gas; octane is a liquid.

SECTION 3

Liquids and Solids

MAIN IDEA The particles in solids and liquids have a limited range of motion and are not easily compressed.

Essential Questions

- How do the arrangements of particles in liquids and solids differ?
- What are the factors that affect viscosity?
- How are the unit cell and crystal lattice related?

Review Vocabulary

meniscus: the curved surface of a column of liquid

New Vocabulary

viscosity
surface tension
surfactant
crystalline solid
unit cell
allotrope
amorphous solid

CHEM 4 YOU Did you ever wonder why syrup that is stored in the refrigerator is harder to pour than syrup stored in the pantry? You probably know that warming syrup makes it pour more easily. But why does an increase in temperature help?

Liquids

Although the kinetic-molecular theory was developed to explain the behavior of gases, the model also applies to liquids and solids. When applying the kinetic-molecular theory to the solid and liquid states of matter, you must consider the forces of attraction between particles as well as their energy of motion.

Previously, you read that a liquid can take the shape of its container but its volume is fixed. In other words, the particles can flow to adjust to the shape of a container, but the liquid cannot expand to fill its container, as shown in **Figure 12.** According to the kinetic-molecular theory, individual particles do not have fixed positions in the liquid. Forces of attraction between particles in the liquid limit their range of motion so that the particles remain closely packed in a fixed volume.

Density and compression At 25°C and 1 atm of air pressure, liquids are much denser than gases. The density of a liquid is much greater than that of its vapor at the same conditions. For example, liquid water is about 1250 times denser than water vapor at 25°C and 1 atm of pressure. Because they are at the same temperature, both gas and liquid particles have the same average kinetic energy. Thus, the higher density of liquids is due to the intermolecular forces that hold particles together.

Unlike gases, liquids are considered incompressible in many applications. The change in volume for liquids is much smaller because liquid particles are already tightly packed. An enormous amount of pressure must be applied to reduce the volume of a liquid by a very small amount.

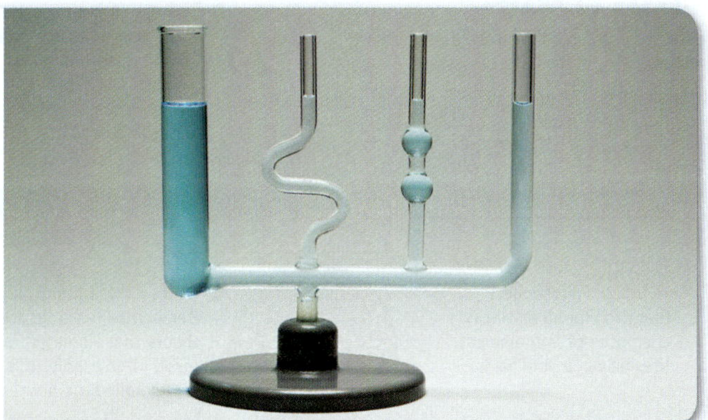

Figure 12 Liquids flow and take the shape of their container, but they do not expand to fill their container like gases.
Infer *the reason that the liquid is at the same level in each of the interconnected tubes.*

DIFFERENTIATED INSTRUCTION

Advanced Learners Ask interested students to design an apparatus or an experiment to test their predictions about the relative viscosities of several different liquids. Challenge them further by asking them to include in their designs a study of the effect of temperature on viscosity. AL

SECTION 3

1 Focus

MAIN IDEA

Range of Motion Fill a small beaker to the top with BBs. Place a stiff card over the top of the beaker, and gently shake the beaker. (The BBs should have very limited movement because the beaker is full.) Ask students to describe the movement of the BBs and why the movement is restricted. The BBs are tightly packed, and there is no place for them to go. Tell students that this is similar to the movement of particles in a solid. The particles are tightly packed, and there is little room for movement.

Remove some of the BBs. The BBs should have some movement. Place the stiff card over the top of the beaker and gently shake the beaker. Ask students to describe the current movement of the BBs. The movement of the BBs is less restricted, and they have more room in which to move. Tell students that this is similar to the movement of particles in a liquid. The particles are not as tightly packed as in a solid, and they have some room for movement. BL OL

2 Teach

Build a Model

Phases of Matter Ask students to create a model, collage, poster, or diagrammed chart of the three phases of a substance, such as water. Ask students to display their work in the classroom. OL EL

Caption Question Fig. 12
Pressure depends on the depth of the liquid. The pressure at the bottom of the connected tubes is constant, resulting in the fluid columns of equal height.

Content Background

Atoms Democritus was a Greek philosopher who lived from about 460 B.C. to 370 B.C. He reasoned that all matter is made of *atomos,* tiny pieces that are indivisible, and that the atomos of different things differed in size and shape. He also thought that atomos were eternal and could not be created or destroyed. Later, Democritus concluded that hard objects are made of atomos that are close together and soft objects are made of atomos that are farther apart. Because of Aristotle's rejection of these ideas, Democritus' theories were disregarded for almost two thousand years.

Figure 13 Gases and liquids have the ability to flow and diffuse. These photos show one liquid diffusing through another liquid.

Fluidity Gases and liquids are classified as fluids because they can flow and diffuse. **Figure 13** shows one liquid diffusing through another liquid. Liquids usually diffuse more slowly than gases at the same temperature, because intermolecular attractions interfere with the flow. Thus, liquids are less fluid than gases. A comparison between water and natural gas can illustrate this difference. When there is a leak in a basement water pipe, the water remains in the basement unless the amount of water released exceeds the volume of the basement.

A gas will not stay in the basement. For example, natural gas, or methane, is a fuel burned in gas furnaces, hot-water heaters, and stoves. Gas that leaks from a gas pipe diffuses throughout the house. Because natural gas is odorless, companies that supply the fuel include a compound with a distinct odor. Adding odor to natural gas warns the homeowner of the leak. The customer has time to shut off the gas supply, open windows to allow the gas to diffuse, and call the gas company to report the leak.

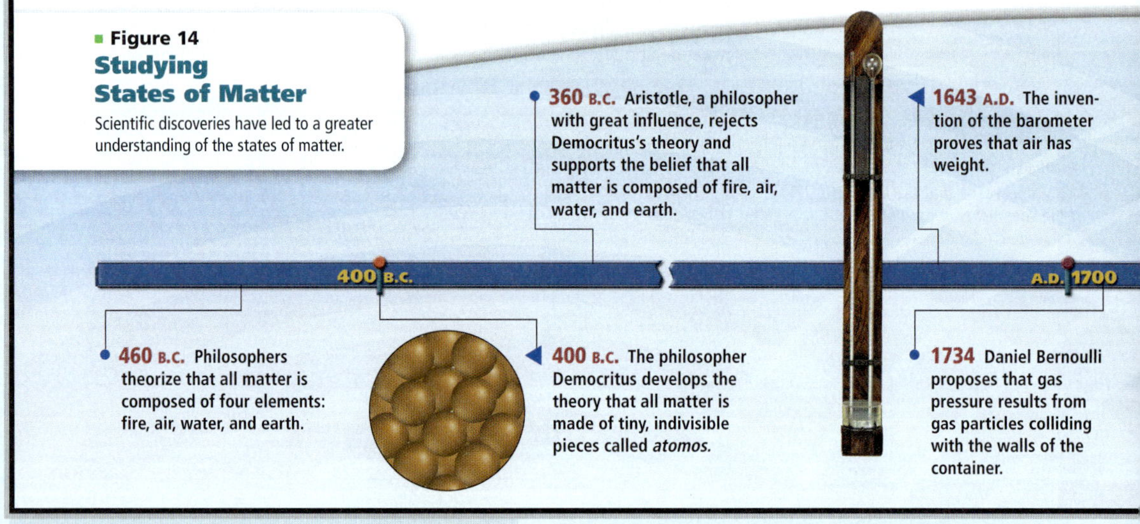

Figure 14
Studying States of Matter
Scientific discoveries have led to a greater understanding of the states of matter.

CHEMISTRY PROJECT

Superfluids Have students research the properties of superfluids. Have students compare and contrast the properties of superfluids with the properties of ordinary fluids (liquids and gases). Both superfluids and ordinary fluids lack a definite shape and have the ability to flow. Properties observed only in superfluids include frictionless flow, superleaks (flowing through materials), the fountain effect, and quantized vortices (many small whirlpools instead of one). **OL**

Viscosity You are already familiar with viscosity if you have ever tried to get honey out of a bottle. **Viscosity** is a measure of the resistance of a liquid to flow. The particles in a liquid are close enough for attractive forces to slow their movement as they flow past one another. The viscosity of a liquid is determined by the type of intermolecular forces in the liquid, the size and shape of the particles, and the temperature.

You should note that not all liquids have viscosity. Scientists discovered superfluids in 1937. Scientists cooled liquid helium below −270.998°C and discovered that the properties of the liquid changed. The superfluid helium lost viscosity—the resistance to flow. The discovery of superfluidity and other milestones in our understanding of states of matter are shown in **Figure 14.**

Attractive forces In typical liquids, the stronger the intermolecular attractive forces, the higher the viscosity. If you have used glycerol in the laboratory to help insert a glass tube into a rubber stopper, you know that glycerol is a viscous liquid. **Figure 15** uses structural formulas to show the hydrogen bonding that makes glycerol so viscous. The hydrogen atoms attached to the oxygen atoms in each glycerol molecule are able to form hydrogen bonds with other glycerol molecules. The red dots in **Figure 15** show where the hydrogen bonds form between molecules.

Particle size and shape The size and shape of particles also affect viscosity. Recall that the overall kinetic energy of a particle is determined by its mass and velocity. Suppose the attractive forces between molecules in Liquid A and Liquid B are similar. If the molecules in Liquid A are more massive than the molecules in Liquid B, Liquid A will have a greater viscosity. Liquid A's molecules will, on average, move more slowly than the molecules in Liquid B. Molecules with long chains, such as cooking oils and motor oil, have a higher viscosity than shorter, more-compact molecules, assuming the molecules exert the same type of attractive forces. Within the long chains, there is less distance between atoms on neighboring molecules and, thus, a greater chance for attractions between atoms.

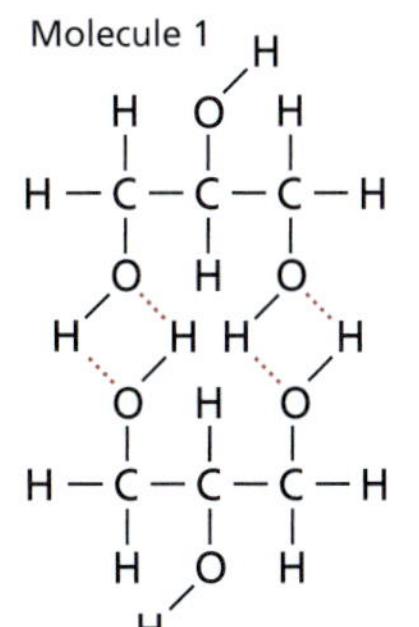

Figure 15 This diagram shows two glycerol molecules and the hydrogen bonds between them.

Determine *the possible number of hydrogen bonds a glycerol molecule can form with a second molecule.*

1808 John Dalton proposes that all matter is composed of tiny particles.

1937 Scientists discover superfluids—unusual fluids with properties not observed in ordinary matter.

2003 Deborah S. Jin creates the first fermionic condensate—a superfluid considered to be a sixth state of matter.

1800 1900 2000

1927 The term *plasma* is first used to describe a fourth state of matter, which is found in lightning.

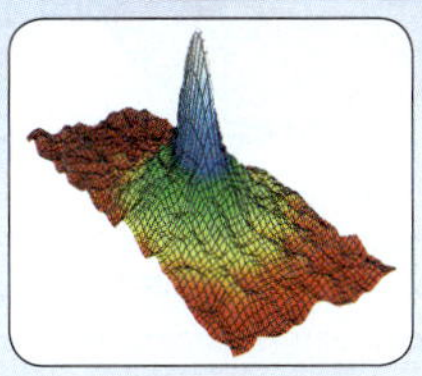

1995 A fifth state of matter, a gaseous superfluid called a Bose-Einstein condensate, is created and named after Satyendra Nath Bose and Albert Einstein.

Caption Question Fig. 15 four

Apply Chemistry

Supercritical Fluid Supercritical fluid extraction is a relatively new technique for separating compounds and mixtures. A supercritical fluid is a substance that, at its critical temperature and pressure, has some properties of liquids and some properties of gases. Supercritical fluids can dissolve as much solute as a good liquid solvent, yet they can be more easily removed than most liquid solvents once the process is finished. Because it is inexpensive and nontoxic and usually has good selectivity and capacity, carbon dioxide is one of the most widely used supercritical fluids. Supercritical carbon dioxide is now used to extract caffeine from coffee beans when making decaffeinated coffee, replacing previously used toxic liquid solvents, such as dichloromethane.

CHEMISTRY JOURNAL

Viscosity Have students observe everyday examples of viscosity. They should record a description of each situation and provide an explanation of what is occurring. Then, they should summarize what is common to the events and connect this information to what they have learned about intermolecular forces. OL

Temperature Viscosity decreases with temperature. When you pour a small amount of cooking oil into a frying pan, the oil tends not to spread across the bottom of the pan until you heat it. With the increase in temperature, there is an increase in the average kinetic energy of the oil molecules. The added energy makes it easier for the molecules to overcome the intermolecular forces that keep the molecules from flowing.

Another example of the effects of temperature on viscosity is motor oil. Motor oil keeps the moving parts of an internal combustion engine lubricated. Because temperature changes affect the viscosity of motor oil, people once used different motor-oil blends in winter and summer. The motor oil used in winter was designed to flow at low temperatures. The motor oil used in summer was more viscous so that it could maintain sufficient viscosity on extremely hot days or during long trips. Today, additives in motor oil help adjust the viscosity so that the same oil blend can be used all year. Molecules in the additives are compact spheres with relatively low viscosity at cool temperatures. At high temperatures, the shape of the additive molecules changes to long strands. These strands get tangled with the oil molecules, which increases the viscosity of the oil.

☑ READING CHECK **Infer** why it is important for motor oil to remain viscous.

Surface tension Intermolecular forces do not have an equal effect on all particles in a liquid, as shown in **Figure 16.** Particles in the middle of the liquid can be attracted to particles above them, below them, and to either side. For particles at the surface of the liquid, there are no attractions from above to balance the attractions from below. Thus, there is a net attractive force pulling down on particles at the surface. The surface tends to have the smallest possible area and to act as though it is stretched tight like the head of a drum. For the surface area to increase, particles from the interior must move to the surface. It takes energy to overcome the attractions holding these particles in the interior. The energy required to increase the surface area of a liquid by a given amount is called **surface tension.** Surface tension is a measure of the inward pull by particles in the interior.

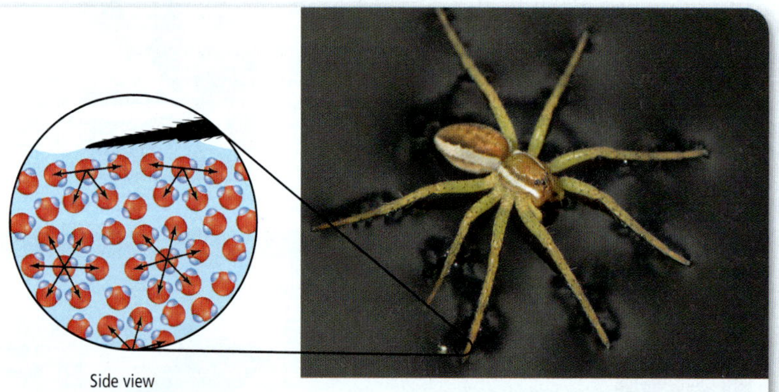

Figure 16 At the surface of water, the particles are drawn toward the interior until attractive and repulsive forces are balanced.

Side view
Intermolecular forces just below the surface of the water create surface tension.

The surface tension of the water allows this spider to walk on the surface of the water.

Quick Demo

Surface Tension Fill a beaker with water until it is full but not overflowing. Ask students to predict how many pennies can be added to the beaker without causing the water to overflow. Students might predict a low number, but 25–35 pennies should be able to be added. Carefully add pennies, one at a time, into the water, dropping them in edge down. Ask students to explain why all the pennies do not cause the water to flow out. Surface tension causes the surface of the water to bulge, as if it were a stretched membrane, rather than to flow out. Repeat this demonstration using a different liquid, such as rubbing alcohol, and ask students to predict how the results will differ from those observed using water. OL EL

☑ READING CHECK To keep the engine parts well lubricated so that they do not stick together or over heat.

Use Science Terms

Fluidity Have students write statements explaining the meaning of the terms *viscosity* and *surfactant.* OL EL

Demonstration

Surface Tension

Purpose

to observe surface tension

Materials

1-pint canning jar with ring lid; wire screen; distilled water with food coloring added (1 L); card stock (1 piece)

Safety Precautions

Disposal Items can be cleaned and reused.

Procedure

Cut a piece of wire window screen to fit inside the ring lid of a 1-pint canning jar. Screw on the lid with the wire screen in place. Pour the colored water through the screen to fill the canning jar to overflowing. Have students predict what will happen when the jar is inverted. Cover the top of the jar with a piece of card stock. Turn the jar upside down over a sink. Remove the card stock by sliding it off to

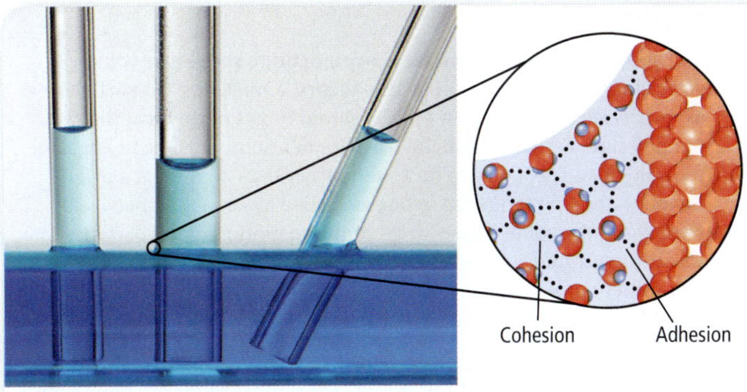

The force of attraction between the water molecules and the silicon dioxide in the glass causes the water molecules to creep up the glass.

Water molecules are attracted to each other—cohesion—and to the silicon dioxide molecules in the glass—adhesion.

■ **Figure 17** Water molecules have cohesive and adhesive properties.
Infer ***why the water level is higher in the smaller diameter tube.***

In general, the stronger the attractions between particles, the greater the surface tension. Water has a high surface tension because its molecules can form multiple hydrogen bonds. Drops of water are shaped like spheres because the surface area of a sphere is smaller than the surface area of any other shape of similar volume. Water's high surface tension is what allows the spider in **Figure 16** to walk on the surface of the pond.

The same forces that allow the spider to stay dry on the surface of a pond also makes it difficult to use water alone to remove dirt from skin and clothing. Because dirt particles cannot penetrate the surface of the waterdrops, water alone cannot remove the dirt. Soaps and detergents decrease the surface tension of water by disrupting the hydrogen bonds between water molecules. When the hydrogen bonds are broken, the water spreads out allowing the dirt to be carried away by the water. Compounds that lower the surface tension of water are called surface-active agents or **surfactants.**

Watch a **video about enzymes and surfactants.**

Video

Cohesion and adhesion When water is placed into a narrow container, such as the glass tubes in **Figure 17,** you can see that the surface of the water is not straight. The surface forms a concave meniscus; that is, the surface dips in the center. **Figure 17** models what is happening to the water at the molecular level. There are two types of forces at work: cohesion and adhesion. Cohesion describes the force of attraction between identical molecules. Adhesion describes the force of attraction between molecules that are different. Because the adhesive forces between water molecules and the silicon dioxide in glass are greater than the cohesive forces between water molecules, the water rises along the inner walls of the cylinder.

Capillary action If the cylinder is extremely narrow, a thin film of water will be drawn upward. Narrow tubes are called capillary tubes. This movement of a liquid such as water is called capillary action, or capillarity. Capillary action helps explain how paper towels can absorb large amounts of water. The water is drawn into the narrow spaces between the cellulose fibers in paper towels by capillary action. In addition, the water molecules form hydrogen bonds with cellulose molecules.

VOCABULARY
SCIENCE USAGE V. COMMON USAGE

Force

Science usage: a push or a pull, having both magnitude and direction, that is exerted on an object
The gravitational force exists between any two objects with mass and is directly proportional to their masses.

Common usage: a group of people who have the power to work toward a desired outcome
The U.S. labor force increased its productivity last year.

■ **Caption Question Fig. 17**
The ratio of the adhesive force to the fluid weight is greater in the smaller diameter tube than in the larger diameter tube.

Apply Chemistry

Thermal Expansion Most solids exhibit positive thermal expansion–they expand when heated and contract when cooled. However, the crystalline solid zirconium tungstate (ZrW_2O_8), discovered in 1996, has negative thermal expansion and contracts when heated. In fact, it shrinks uniformly in all three dimensions over a temperature range from 0.3 K to 1050 K (its decomposition point). Because of these properties, it has the potential to be used in materials where low thermal expansion is needed, such as in spacecraft, circuit boards, telescopes, and even dentistry. Have students research the structure of this compound to find out why it has negative thermal expansion and research applications where it is used. **AL**

What's CHEMISTRY Got To Do With It?

Video

Squeaky Clean.

the side. Have a student look up into the jar and report what he or she observes. While holding the inverted jar over the sink, slowly tip the jar sideways until the water flows through the wire screen into the sink.

Results
The surface tension is sufficient to form a film that holds the water in the jar when it is inverted.

Analysis

1. Why does the water remain in the inverted jar? The surface tension of the water forms a film that covers the holes in the window screen.
2. Why does water flow out of the jar when it is tilted? As air enters the jar, it displaces the water and pushes it out of the jar.

ASSESSMENT

Performance Have students design and conduct experiments to determine whether the results of this demonstration depend on the size of the jar and the size of the holes in the wire screen. **OL**

Content Background

Crystalline Solids Students might be intrigued by the very narrow range (just a few degrees Celsius) that separates the melting points and boiling points of the noble gases. Explain to students that Table 5 describes the solid phases of substances that can exist as gases or liquids at room temperature.

READING CHECK Hydrogen bonding in ice creates a less dense structure than the structure in water. Therefore, ice floats on water because of its lower density.

Extension

Icebergs The largest iceberg known to exist broke off the Ross Ice Shelf in Antarctica on March 20, 2000. It was about the size of the state of Connecticut. It measured 295 km long and 37 km wide. It extended about 200 m below the surface of the ocean and rose about 30 m above the surface of the ocean.

CAREERS IN CHEMISTRY

WebQuest

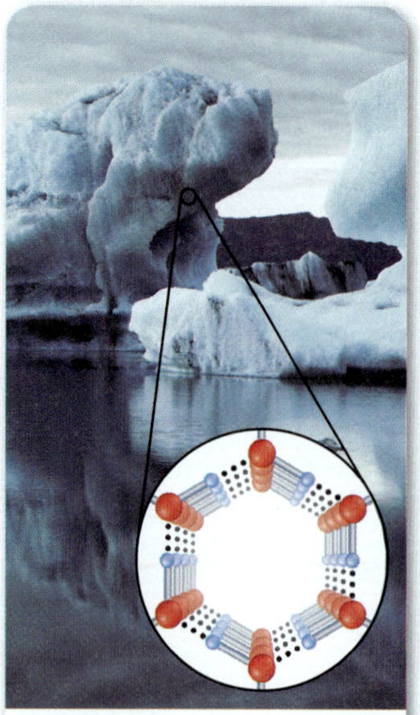

Figure 18 An iceberg can float because the rigid, three-dimensional structure of ice keeps water molecules farther apart than they are in liquid water. This open, symmetrical structure of ice results from hydrogen bonding.

Solids

Did you ever wonder why solids have a definite shape and volume? According to the kinetic-molecular theory, a mole of solid particles has as much kinetic energy as a mole of liquid or gas particles at the same temperature. By definition, the particles in a solid must be in constant motion. For a substance to be a solid rather than a liquid at a given temperature, there must be strong attractive forces acting between particles in the solid. These forces limit the motion of the particles to vibrations around fixed locations in the solid. Thus, there is more order in a solid than in a liquid. Because of this order, solids are not fluid. Only gases and liquids are classified as fluids.

Density of solids In general, the particles in a solid are more closely packed than those in a liquid. Thus, most solids are more dense than most liquids. When the liquid and solid states of a substance coexist, the solid almost always sinks in the liquid. Solid cubes of benzene sink in liquid benzene because solid benzene is more dense than liquid benzene. There is about a 10% difference in density between the solid and liquid states of most substances. Because the particles in a solid are closely packed, ordinary amounts of pressure will not change the volume of a solid.

You cannot predict the relative densities of ice and liquid water based on benzene. Ice cubes and icebergs float because water is less dense as a solid than it is as a liquid. **Figure 18** shows the reason for the exception. As water freezes, each H_2O molecule can form hydrogen bonds with up to four neighboring molecules. As a result, the water molecules in ice are less-closely packed together than in liquid water.

READING CHECK **Describe** in your own words why ice floats in water.

Crystalline solids Although ice is unusual in its density, ice is typical of most solids in that its molecules are packed together in a predictable way. A **crystalline solid** is a solid whose atoms, ions, or molecules are arranged in an orderly, geometric structure. The locations of particles in a crystalline solid can be represented as points on a framework called a crystal lattice. **Figure 19** shows three ways that particles in a crystal lattice can be arranged to form a cube.

CAREERS IN CHEMISTRY

Metallurgist Metallurgists are engineers who are involved in all stages of processing metals, from extracting and refining to casting the final product. At each stage, metallurgists must understand the physical and chemical properties of metals. A college degree is necessary to become a metallurgist, and many go on to earn advanced degrees.

WebQuest

Figure 19 These drawings show three of the ways particles are arranged in crystal lattices. Each sphere represents a particle. **a.** Particles are arranged only at the corners of the cube. **b.** There is a particle in the center of the cube. **c.** There are particles in the center of each of the six cubic faces but no particle in the center of the cube itself.

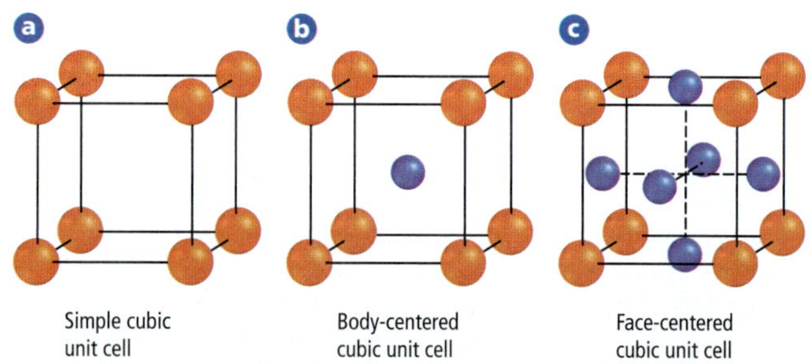

CULTURAL DIVERSITY

The Agony of the Feet Every year, firewalkers tread barefoot across glowing coals to open and close the three-day celebrations for the May 21 Feast Day of Saints Constantine and Helen in Aghia Eleni, Greece. This Greek Orthodox rite has never been fully sanctioned by the church, but references to firewalking can be traced back to Dionysus, the Greek god of revelry. How can firewalkers walk on hot coals for several minutes without burning their feet? The low thermal conductivity of the mostly molecular solids found in wood and charcoal makes them transfer heat so slowly that if you walk across the burning embers quickly you will not be injured. This is the same reason a wooden boardwalk is not as hot as sand, which contains compounds with a higher thermal conductivity.

A **unit cell** is the smallest arrangement of atoms in a crystal lattice that has the same symmetry as the whole crystal. Like the formula unit that you read about previously, a unit cell is a small, representative part of a larger whole. The unit cell can be thought of as a building block whose shape determines the shape of the crystal.

Table 4 shows seven categories of crystals based on shape. Crystal shapes differ because the surfaces, or faces, of unit cells do not always meet at right angles, and the edges of the faces vary in length. In **Table 4,** the edges are labeled *a, b,* and *c;* the angles at which the faces meet are labeled α, β, and γ.

Explore **unit cells** with an interactive table. **Concepts In Motion**

Table 4 Unit Cells

 Halite (rock salt)	 Vesuvianite	 Aragonite
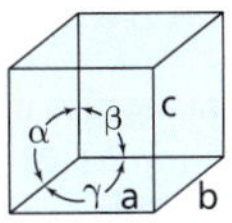 a = b = c α = β = γ = 90° Cubic	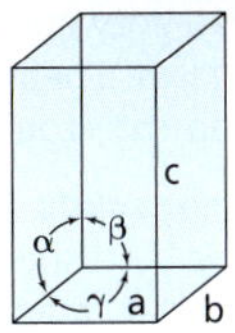 a = b ≠ c α = β = γ = 90° Tetragonal	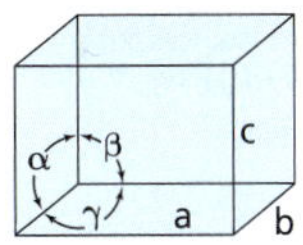 a ≠ b ≠ c α = β = γ = 90° Orthorhombic

 Microcline	 Beryl (emerald)	 Tourmaline	 Crocoite
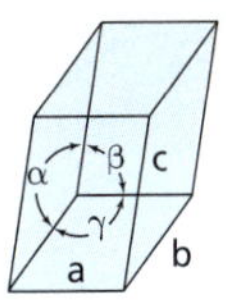 a ≠ b ≠ c α ≠ β ≠ γ ≠ 90° Triclinic	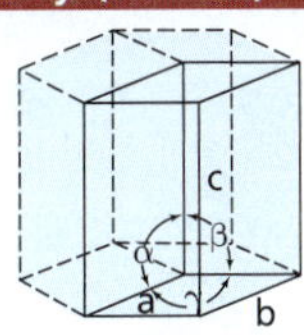 a = b ≠ c α = β = 90°, γ = 120° Hexagonal	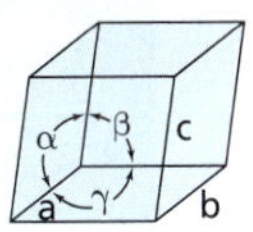 a = b = c α = β = γ ≠ 90° Rhombohedral	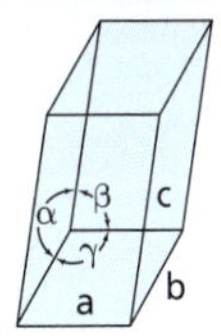 a ≠ b ≠ c α = γ = 90° ≠ β Monoclinic

ASSESSMENT

Knowledge Have students research the career field of gemology by interviewing a local jeweler and preparing a brief report of their findings. A certified gemologist can be invited to speak to the class about gems–their structure, characteristic features, and qualities. **OL**

Enrichment

Liquid Crystals Students might be interested to know that substances exist that have some properties of both solids and liquids. Liquid crystals can flow, but they also have some crystalline structure. Have students research liquid crystals and their applications. Students should present a short oral report to the class that includes some type of visual aid. **AL**

DIFFERENTIATED INSTRUCTION

Advanced Learners Ask students to construct and display models of the seven crystal systems: cubic, tetragonal, hexagonal, rhombohedral, orthorhombic, monoclinic, and triclinic. **AL**

ASSESSMENT

Performance Have groups of two or three students build models of one of the three types of cubic unit cells. Have each group share its model with the class. Quiz students to see if they can label the models correctly. OL COOP LEARN

Content Background

Quartz Quartz is the second most abundant mineral on Earth. It exists in two forms: alpha-quartz, which has a trigonal structure, and beta-quartz, which has a hexagonal structure. Some types of quartz are used as gemstones, such as amethyst, citrine, smoky quartz, and rose quartz. Sandstone, which contains quartz, is used as a building material. Quartz is also used in other applications, such as in the manufacturing of sandpaper, glass, and ceramics, and in grindstones.

Explore **types of crystalline solids with an interactive table.** Concepts In Motion

Table 5 Types of Crystalline Solids

Type	Unit Particles	Characteristics of Solid Phase	Examples
Atomic	atoms	soft to very soft; very low melting points; poor conductivity	group 18 elements
Molecular	molecules	fairly soft; low to moderately high melting points; poor conductivity	I_2, H_2O, NH_3, CO_2, $C_{12}H_{22}O_{11}$ (table sugar)
Covalent network	atoms connected by covalent bonds	very hard; very high melting points; often poor conductivity	diamond (C) and quartz (SiO_2)
Ionic	ions	hard; brittle; high melting points; poor conductivity	NaCl, KBr, $CaCO_3$
Metallic	atoms surrounded by mobile valence electrons	soft to hard; low to very high melting points; malleable and ductile; excellent conductivity	all metallic elements

APPLYING PRACTICES

PBL Go to the resources tab in ConnectED to find the PBL *Touching the Future*.

Categories of crystalline solids Crystalline solids can be classified into five categories based on the types of particles that they contain and how thoses particles are bonded together: atomic solids, molecular solids, covalent network solids, ionic solids, and metallic solids. **Table 5** summarizes the general characteristics of each category and provides examples. The only atomic solids are noble gases. Their properties reflect the weak dispersion forces between the atoms.

Molecular solids In molecular solids, the molecules are held together by dispersion forces, dipole-dipole forces, or hydrogen bonds. Most molecular compounds are not solids at room temperature. Even water, which can form strong hydrogen bonds, is a liquid at room temperature. Molecular compounds such as sugar are solids at room temperature because of their large molar masses. With larger molecules, many weak attractions can combine to hold the molecules together. Because they contain no ions, molecular solids are poor conductors of heat and electricity.

Figure 20 The most common kind of quartz has a hexagonal crystal structure.

MiniLAB

Model Crystal Unit Cells

How can you make physical models that illustrate the structures of crystals?

Procedure

1. Read and complete the lab safety form.
2. Cut **four soda straws** into thirds. Wire the straw pieces together to make a cube using **22- or 26-gauge wire.** Use **scissors** to cut the wire. Refer to **Table 4** for a guide to crystal shapes.
3. To model a rhombohedral crystal, deform the cube from Step 2 until no angles are 90°.
4. To model a hexagonal crystal, make the base angle $\gamma = 120°$ and the other two angles equal 90°.
5. To model a tetragonal crystal, cut **4 straws** in half. Cut 4 of the pieces in half again. Wire the 8 shorter pieces to make 4 square ends. Use the longer pieces to connect the square ends.
6. To model the orthorhombic crystal, cut **4 straws** in half. Cut one-third off 4 of the halves, creating 4 each of three different lengths. Connect the 4 long, 4 medium, and 4 short pieces so that each side is a rectangle.
7. To model the monoclinic crystal, deform the model from Step 6 along one axis. To model the triclinic crystal, deform the model from Step 6 until it has no 90° angles.

Analysis

1. **Evaluate** Which two models have three axes of equal length? How do these models differ?
2. **Determine** which model includes a square and a rectangle.
3. **Determine** which models have three unequal axes.
4. **Infer** Do you think crystals are perfect, or do they have defects? Explain your answer.

Covalent network solids Atoms such as carbon and silicon, which can form multiple covalent bonds, are able to form covalent network solids. The covalent network structure of quartz, which contains silicon, is shown in **Figure 20.** Carbon forms three types of covalent network solids–diamond, graphite, and buckminsterfullerene. An element, such as carbon, that exists in different forms at the same state–solid, liquid, or gas–is called an **allotrope.** For more information about carbon allotropes see the Elements Handbook.

Ionic solids Remember that each ion in an ionic solid is surrounded by ions of opposite charge. The type of ions and the ratio of ions determine the structure of the lattice and the shape of the crystal. The network of attractions that extends throughout an ionic crystal gives these compounds their high melting points and hardness. Ionic crystals are strong, but brittle. When ionic crystals are struck, the cations and anions are shifted from their fixed positions. Repulsions between ions of like charge cause the crystal to shatter.

Metallic solids Recall that metallic solids consist of positive metal ions surrounded by a sea of mobile electrons. The strength of the metallic bonds between cations and electrons varies among metals and accounts for their wide range of physical properties. For example, tin melts at 232°C, but nickel melts at 1455°C. The mobile electrons make metals malleable–easily hammered into shapes–and ductile–easily drawn into wires. When force is applied to a metal, the electrons shift and thereby keep the metal ions bonded in their new positions. Mobile electrons make metals good conductors of heat and electricity. As shown in **Figure 21,** metal wiring is used to carry electricity to businesses and homes.

Figure 21 Homes, business, and equipment of all types use metal wiring to carry electricity. The metal is usually copper, but other metals are used in special applications.

☑ READING CHECK **Describe** the properties of metals that make them useful for making jewelry.

☑ READING CHECK Metals are malleable, and thin sheets can be formed into various decorative designs. Metals are ductile, and they can be used to make earring posts, earring wires, and necklaces.

MiniLAB

Rubric

Purpose Students will construct and compare crystal unit cell models.

Process Skills classify, compare and contrast, formulate models

Safety Precautions Approve lab safety forms before work begins. Wear apron and goggles. Be careful using scissors. The end of the wire is sharp.

Disposal Have students place the crystal models into a designated container.

Teaching Strategies

- Students should refer to Table 4 while constructing their crystals.
- Students can make the crystal models at home or work in teams.
- Collect all of the cubic unit cells of one type and demonstrate how they fit together to make a large crystal lattice.
- Remind students that they have learned about only seven basic unit cells. There are more types to learn about in advanced courses.

Expected Results Students will construct the seven unit cells shown in Table 4.

Analysis

1. The two crystal models that have all three axes of equal length are cubic and rhombohedral. They differ in that the cubic has only 90-degree angles; the rhombohedral has no 90-degree angles.
2. the tetragonal crystal model
3. The three crystal models that have three axes of unequal length are orthorhombic, monoclinic, and triclinic.
4. They are not perfectly uniform, as they can be affected by external forces and conditions.

ASSESSMENT

Knowledge Ask students why there is confusion over the classification of the structure of glass. Using X-ray diffraction glass does not appear to have a crystalline structure, but when neutrons are used there does seem to be some orderly pattern in some areas. Ask them to define the term *amorphous solid.* OL

FOLDABLES® Rubric

3 Assess

Check for Understanding

Ask students to describe how a liquid with high viscosity differs in appearance from one with low viscosity. A viscous liquid often appears thick and moves very slowly when poured. A nonviscous liquid appears thinner and moves rapidly when poured. OL

Reteach

Demonstrate surface tension using a beaker of water and a thumbtack. Ask students to draw a diagram showing how the molecules of water are arranged around the thumbtack so that the tack stays on the surface. Refer to Figure 16. OL EL

Extension

Ask students to design an experiment in which they test the thermal and electrical conductivity of different types of crystalline solids. OL

Figure 22 Native Americans used the glass-like amorphous rock obsidian to make arrowheads and knives, because it can form sharp edges when broken. Obsidian rock forms when lava cools too quickly to form crystals.

FOLDABLES® Incorporate information from this section into your Foldable.

Amorphous solids An **amorphous solid** is one in which the particles are not arranged in a regular, repeating pattern. It does not contain crystals. The term *amorphous* is derived from a Greek word that means *without shape.* An amorphous solid often forms when a molten material cools too quickly to allow enough time for crystals to form. **Figure 22** shows an example of an amorphous solid.

Glass, rubber, and many plastics are amorphous solids. Recent studies have shown that glass might have some structure. When X-ray diffraction is used to study glass, there appears to be no pattern to the distribution of atoms. When neutrons are used instead, an orderly pattern of silicate units can be detected in some regions. Researchers hope to use this new information to control the structure of glass for optical applications and to produce glass that can conduct electricity.

SECTION 3 REVIEW

Section Self-Check

Section Summary

- The kinetic-molecular theory explains the behavior of solids and liquids.
- Intermolecular forces in liquids affect viscosity, surface tension, cohesion, and adhesion.
- Crystalline solids can be classified by their shape and composition.

18. MAINIDEA **Contrast** the arrangement of particles in solids and liquids.
19. **Describe** the factors that affect viscosity.
20. **Explain** why soap and water are used to clean clothing instead of water alone.
21. **Compare** a unit cell and a crystal lattice.
22. **Describe** the difference between a molecular solid and a covalent network solid.
23. **Explain** why water forms a meniscus when it is in a graduated cylinder.
24. **Infer** why the surface of mercury in a thermometer is convex; that is, the surface is higher at the center.
25. **Predict** which solid is more likely to be amorphous—one formed by allowing a molten material to cool slowly to room temperature or one formed by quickly cooling the same material in an ice bath.
26. **Design** an experiment to compare the relative abilities of water and isopropyl alcohol to support skipping stones. Include a prediction about which liquid will be better, along with a brief explanation of your prediction.

SECTION 3 REVIEW

18. The particles are closer together in solids than in liquids because of intermolecular attractions.
19. Viscosity of a liquid is determined by the type of intermolecular forces in the liquid, the size and shape of the particles, and the temperature.
20. Soaps and detergents decrease the surface tension of water by breaking the hydrogen bonds, which allows the dirt to be carried away by the water.
21. Unit cells are the building blocks of the crystal lattice.
22. A molecular solid is formed from molecules held together by intermolecular forces that are weaker than covalent bonds; covalent network solids are formed from molecules held together by covalent bonds.
23. Because the adhesive forces between water molecules and the silicon dioxide in glass are greater than the cohesive forces between water molecules, the water rises along the inner walls of the cylinder.
24. The cohesion between mercury atoms is stronger than the adhesion between mercury and glass.
25. The one that cooled quickly in an ice bath is more likely to be amorphous because amorphous solids often form when molten material cools too quickly for crystals to form.
26. Make sure students use correct scientific procedures in their experimental designs.

SECTION 4

Essential Questions

- How can the addition and removal of energy cause a phase change?
- What is a phase diagram?

Review Vocabulary

phase change: a change from one state of matter to another

New Vocabulary

melting point
vaporization
evaporation
vapor pressure
boiling point
freezing point
condensation
deposition
phase diagram
triple point

Phase Changes

MAINIDEA Matter changes phase when energy is added or removed.

CHEM 4 YOU Have you ever wondered where the matter in a solid air freshener goes? The day it is opened and put in a room, it is a solid, fragrant mass. Day-by-day, the solid gets smaller and smaller. Finally, almost nothing is left and it is time to put a new one out. You never observe a puddle of liquid like you would see if it had melted.

Phase Changes That Require Energy

Most substances can exist in three states depending on the temperature and pressure. A few substances, such as water, exist in all three states under ordinary conditions. States of a substance are referred to as phases when they coexist as physically distinct parts of a mixture. Ice water is a heterogeneous mixture with two phases, solid ice and liquid water. When energy is added or removed from a system, one phase can change into another, as shown in **Figure 23.** Because you are familiar with the phases of water—ice, liquid water, and water vapor—and have observed changes between those phases, we can use water as the primary example in the discussion of phase changes.

Melting What does happen to ice cubes in a glass of ice water? When ice cubes are placed in water, the water is at a higher temperature than the ice. Heat flows from the water to the ice. Heat is the transfer of energy from an object at a higher temperature to an object at a lower temperature. At ice's melting point, the energy absorbed by the ice is not used to raise the temperature of the ice. Instead, it disrupts the hydrogen bonds holding the water molecules together in the ice crystal. When molecules on the surface of the ice absorb enough energy to break the hydrogen bonds, they move apart and enter the liquid phase. As molecules are removed, the ice cube shrinks. The process continues until all of the ice melts. If a tray of ice cubes is left on a counter, where does the energy to melt the cubes come from?

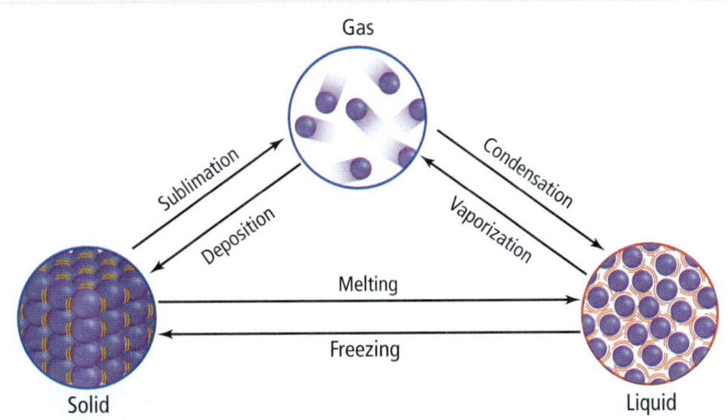

■ **Figure 23** The diagram shows the six possible transitions between phases.
Determine *what phase changes occur between solids and liquids.*

SECTION 4

1 Focus

MAINIDEA

Phase Change Place a beaker of water on a hot plate. Bring the water to a boil. Ask students to describe the change in energy between the hot plate and the water. Thermal energy is being added to the water by the hot plate. Ask students to describe the changes in the water. Water is going from the liquid phase to the gaseous state. Point out to students that particles can move from one phase of matter to another when energy is added or removed. OL

2 Teach

Extension

Electric Power Generation
Students might be interested to know that steam is used to generate electricity. Water is converted to steam by using energy from fossil fuels and nuclear power. Pressurized steam is used to turn giant turbines that generate electricity.

■ **Caption Question Fig. 23** melting, freezing

■ **In-Text Question** thermal energy from the surroundings

CHEMISTRY JOURNAL

Observing Phase Changes Ask students to keep a daily listing in their journals for one week in which they describe their observations of phase changes taking place. They should also name each phase change process that they observe. OL

Identify Misconceptions

Students might have misconceptions about the activity of molecules during phase changes.

Uncover the Misconception Ask students what happens to molecules in a solid as it melts. They might respond that the molecules themselves break apart. They might also describe boiling and sublimation in the same way.

Demonstrate the Concept Use drawings and models to show how intermolecular forces, rather than intramolecular bonds break as a substance is heated.

Assess New Knowledge Ask students to draw several water molecules, using solid lines for intramolecular bonds and dashed lines for intermolecular forces. Then, ask them to redraw the molecules as the water is heated. Intermolecular forces are overcome; the intramolecular bonds remain intact. **OL**

Caption Question Fig. 24 The peak of the graph would shift to the right because of the increased kinetic energy.

GRAPH CHECK The particles enter the vapor phase.

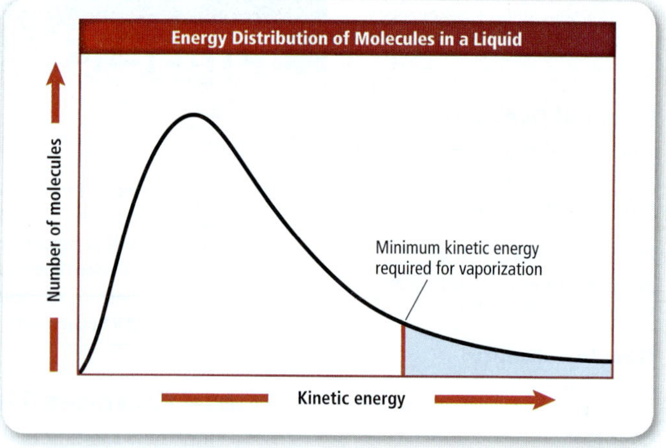

Figure 24 This graph shows a typical distribution of kinetic energy of molecules in a liquid at 25°C. The most probable amount of kinetic energy for a molecule lies at the peak of the curve.

Describe *how the curve would look for the same liquid at 30°C.*

The amount of energy required to melt 1 mol of a solid depends on the strength of the forces keeping the particles together in the solid. Because hydrogen bonds between water molecules are strong, a relatively large amount of energy is required. However, the energy required to melt ice is much less than the energy required to melt table salt because the ionic bonds in sodium chloride are much stronger than the hydrogen bonds in ice.

The temperature at which the liquid phase and the solid phase of a given substance can coexist is a characteristic physical property of many solids. The **melting point** of a crystalline solid is the temperature at which the forces holding its crystal lattice together are broken and it becomes a liquid. It is difficult to specify an exact melting point for an amorphous solid because they tend to melt over a temperature range.

Vaporization When an ice cube melts, the temperature of the ice and the water produced remains constant. Once all of the ice has melted, additional energy added to the system increases the kinetic energy of the liquid molecules. The temperature of the system begins to rise. In liquid water, some molecules will have more kinetic energy than other molecules. **Figure 24** shows how energy is distributed among the molecules in a liquid at 25°C. The shaded portion indicates those molecules that have the energy required to overcome the forces of attraction holding the molecules together in the liquid.

GRAPH CHECK **Describe** what happens to the particles in the shaded portion on the graph.

RealWorld CHEMISTRY
Evaporation

PERSPIRATION Evaporation is one way your body controls its temperature. When you become hot, your body releases sweat from glands in your skin. Water molecules in sweat can absorb heat energy from your skin and evaporate. Excess heat is carried from all parts of your body to your skin by your blood.

Particles that escape from the liquid enter the gas phase. For a substance that is ordinarily a liquid at room temperature, the gas phase is called a vapor. **Vaporization** is the process by which a liquid changes to a gas or vapor. If the input of energy is gradual, the molecules tend to escape from the surface of the liquid. Remember that molecules at the surface are attracted to fewer other molecules than are molecules in the interior. When vaporization occurs only at the surface of a liquid, the process is called **evaporation.** Even at cold temperatures, some water molecules have enough energy to evaporate. As the temperature rises, more and more molecules enter the gas phase.

CHEMISTRY PROJECT

Body Temperature Ask students to research the mechanisms by which the human body regulates its temperature and the health consequences when these mechanisms fail. Have students write a report or prepare an oral presentation on the subject. **AL**

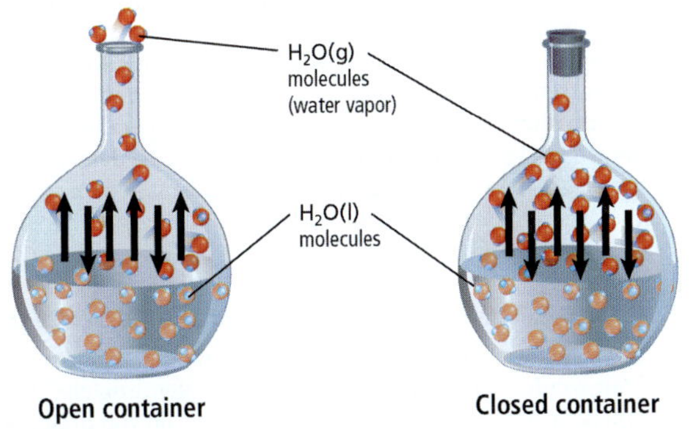

■ **Figure 25** Evaporation occurs in both open and closed containers. In an open container, water molecules that evaporate can escape from the container. Water vapor collects above the liquid in a closed container.

Figure 25 compares evaporation in an open container with evaporation in a closed container. If water is in an open container, all the molecules will eventually evaporate. The time it takes for them to evaporate depends on the amount of water and the available energy. In a partially filled, closed container, the situation is different. Water vapor collects above the liquid and exerts pressure on the surface of the liquid. The pressure exerted by a vapor over a liquid is called **vapor pressure.**

Boiling The temperature at which the vapor pressure of a liquid equals the external or atmospheric pressure is called the **boiling point.** Use **Figure 26** to compare what happens to a liquid at temperatures below its boiling point with what happens to a liquid at its boiling point. At the boiling point, molecules throughout the liquid have enough energy to vaporize. Bubbles of vapor collect below the surface of the liquid and rise to the surface.

APPLYING PRACTICES

Plan and Conduct an Investigation Go to the resources tab in ConnectED to find the Applying Practices worksheet *Investigate Interparticle Forces.*

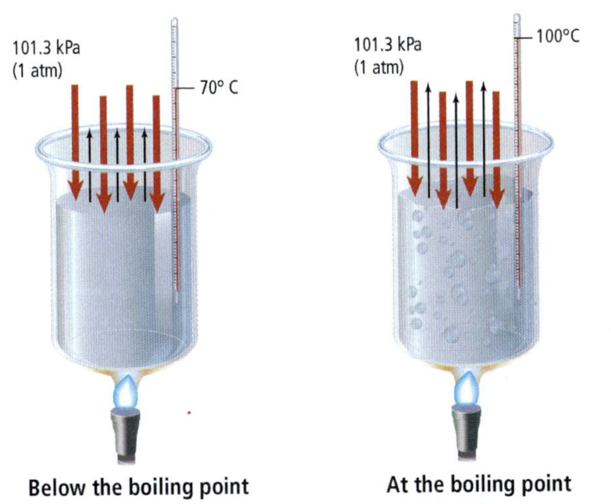

■ **Figure 26** As temperature increases, water molecules gain kinetic energy. Vapor pressure increases (black arrows) but is less than atmospheric pressure (red arrows). A liquid has reached its boiling point when its vapor pressure is equal to atmospheric pressure. At sea level, the boiling point of water is 100°C.

Extension

Melting Points Ask students to research the melting points of ten compounds for which the data are not included in their textbook. Tell students to summarize their findings in a table for their journals. OL

ASSESSMENT

Performance Ask students to design an experiment (including the setup) to measure the amount of gas produced by the sublimation of a solid air freshener. These designs should incorporate a sealed apparatus with tubing running from the air freshener container into an inverted graduated cylinder placed in a dish of water. OL

ChemLAB

The ChemLab located at the end of the chapter can be used at this point in the lesson.

CHEMISTRY PROJECT

Museum Specimens Ask students to research and present a report comparing the newer freeze-drying process to older taxidermy methods used to preserve animal specimens for museums. If possible, ask students to bring in specimens that have been preserved by both methods. OL

Use Science Terms

Phase Changes Have students write statements explaining the meaning of the terms *melting, vaporization, evaporation, sublimation, condensation, deposition* and *freezing.* BL EL

■ **Caption Question Fig. 27**
Dry ice sublimes instead of forming a liquid, which could damage the food products.

☑ **READING CHECK** Water in the air condenses on condensation nuclei when it rises or passes over cooler land or water.

■ **Figure 27** Steaks, seafood, and other highly perishable food products often are shipped in a container with dry ice to keep the food cold during shipping.
Explain *why dry ice is preferred over regular ice for shipping steaks and other food products.*

Sublimation Many substances have the ability to change directly from the solid phase to the gas phase. Recall that sublimation is the process by which a solid changes directly to a gas without first becoming a liquid. Solid iodine and solid carbon dioxide (dry ice) sublime at room temperature. Dry ice, shown in **Figure 27,** keeps objects that could be damaged by melting water cold during shipping. Mothballs, which contain the compounds naphthalene or *p*-dichlorobenzene, also sublime, as do solid air fresheners.

Phase Changes That Release Energy

Have you ever awakened on a chilly morning to see frost on your windows or the grass covered with water droplets? When you set a glass of ice water on a picnic table, do you notice beads of water on the outside of the glass? These events are examples of phase changes that release energy into the surroundings.

Freezing Suppose you place liquid water in an ice tray into a freezer. As heat is removed from the water, the molecules lose kinetic energy and their velocity decreases. The molecules are less likely to flow past one another. When enough energy has been removed, the hydrogen bonds between water molecules keep the molecules fixed, or frozen, into set positions. Freezing is the reverse of melting. The **freezing point** is the temperature at which a liquid is converted into a crystalline solid.

Condensation When a water vapor molecule loses energy, its velocity decreases. The water vapor molecule is more likely to form a hydrogen bond with another water molecule. The formation of a hydrogen bond releases thermal energy and indicates a change from the vapor phase to the liquid phase. The process by which a gas or a vapor becomes a liquid is called **condensation.** Condensation is the reverse of vaporization.

Different factors contribute to condensation. However, condensation always involves the transfer of thermal energy. For example, water vapor molecules can come in contact with a cold surface, such as the side of a glass of ice water. Thermal energy transfers from the water vapor molecules to the cool glass, causing condensation on the outside of the glass. A similar process can occur during the night when water vapor in the air condenses and dew forms on blades of grass.

Connection to Earth Science Precipitation, clouds, and fog all result from condensation. They form as air cools when it rises or passes over cooler land or water. Their formations require a second factor, microscopic particles suspended in the air called condensation nuclei. These can be particles, such as soot and dust, or aerosols, such as sulfur dioxide and nitrogen oxide, on which water vapor condenses. In some circumstances, warm air can settle on top of cooler air, which is called a temperature inversion. **Figure 28** shows fog trapped in a mountain valley by such an inversion.

■ **Figure 28** Normally, air becomes cooler as elevation increases. A temperature inversion occurs when the situation is reversed and the air becomes warmer at higher elevations. Inversions can trap smog over cities and fog in mountain valleys.

☑ **READING CHECK** **Describe** the condensation of water vapor in the atmosphere.

CHEMISTRY JOURNAL

Morning Dew When poets want to evoke the essence of the morning, they often make a reference to dew–the moisture on grass that disappears as the Sun and the temperature rise. But what causes dew? On a warm day with high humidity, the air becomes saturated with moisture. There is only so much water the air can hold at any given temperature, and that amount decreases when the temperature drops. The coolest nighttime temperatures tend to be in the hours just before sunrise. When the temperature drops, the air cannot hold as much moisture as it did earlier in the night. Excess moisture then condenses on anything the air touches, typically on the grass. Eventually, warmth from the rising sun's rays evaporates this moisture, and the dew is gone. Have students write a poem or song about dew, including scientific information in their creations. OL

Deposition When water vapor comes in contact with a cold window in winter, it forms a solid deposit on the window called frost. **Deposition** is the process by which a substance changes from a gas or vapor to a solid without first becoming a liquid. Deposition is the reverse of sublimation. Snowflakes form when water vapor high up in the atmosphere changes directly into solid ice crystals. Energy is released as the crystals form.

Phase Diagrams

There are two variables that combine to control the phase of a substance: temperature and pressure. These variables can have opposite effects on a substance. For example, a temperature increase causes more liquid to vaporize, but an increase in pressure causes more vapor to condense. A **phase diagram** is a graph of pressure versus temperature that shows in which phase a substance exists under different conditions of temperature and pressure.

Figure 29 shows the phase diagram for water. You can use this graph to predict what phase water will be in for any combination of temperature and pressure. Note that there are three regions representing the solid, liquid, and vapor phases of water and three curves that separate the regions from one another. At points that fall along the curves, two phases of water can coexist. The short, yellow curve shows the temperature and pressure conditions under which solid water and water vapor can coexist. The long, blue curve shows the temperature and pressure conditions under which liquid water and water vapor can coexist. The red curve shows the temperature and pressure conditions under which solid water and liquid water can coexist.

Point A on the phase diagram of water—the point where the yellow, blue, and red curves meet—is the triple point for water. The **triple point** is the point on a phase diagram that represents the temperature and pressure at which three phases of a substance can coexist. All six phase changes can occur at the triple point: freezing and melting; evaporation and condensation; sublimation and deposition. Point B is called the critical point. This point indicates the critical pressure and critical temperature above which water cannot exist as a liquid. If water vapor is at the critical temperature, an increase in pressure will not change the vapor into a liquid.

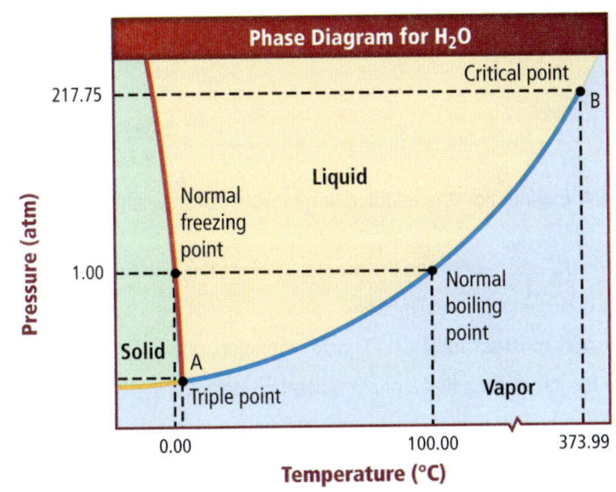

Figure 29 This phase diagram shows the phase of water at different temperatures and pressures.

GRAPH CHECK

Determine the phase of water at 2.00 atm and 100.00°C.

Quick Demo

Sublimation and Condensation Place a small lump of solid air freshener in the bottom of a beaker. Put a watch glass full of ice on top of the beaker, and place this setup in a shallow pan of warm water. Ask students to observe what happens and name the process that takes place. The heat from the warm water causes the solid air freshener to undergo sublimation, and the cold surface of the watch glass causes the vaporized air freshener to condense into a solid again. If the air freshener is colored, ask students to explain why the material that sublimes is white. The dye is a separate compound from the air freshener and does not undergo sublimation itself. OL EL

ASSESSMENT

Skill Ask students to create a poster showing the movement of the particles of a substance among its three phases. In their drawings, students should show the relative spacing of particles in all phases and indicate whether energy is absorbed or released during each transition. OL EL

Reinforcement

Kinetic Energy Explain that the average kinetic energy of particles determines their temperature. So even at temperatures below 100°C, water in an open container will evaporate. Liquids do not have to reach a full boil to be converted to gases.

GRAPH CHECK liquid

DIFFERENTIATED INSTRUCTION

Struggling Learners Some students might have difficulty interpreting phase diagrams. Suggest that they interpret the diagrams one axis at a time. To identify the solid, liquid, and gas phases, students should first consider temperature. For example, students have learned that water is a solid at temperatures below 0°C, and as the temperature increases, it first becomes a liquid, then a gas. So, except for certain substances with unusual properties, the section on the left of the diagram represents the solid phase, the middle section the liquid phase, and the right section the gas phase. When they observe the pressure axis, they should make the same conclusion, knowing that at very high pressures, materials compress into solids. The section at the highest pressure represents the solid phase, the next section the liquid phase, and at the lowest pressure, the gas phase. BL

GRAPH CHECK At high pressures, water will form a solid at lower temperatures. This is not true for carbon dioxide.

3 Assess

Check for Understanding

Ask students to identify the following processes as exothermic or endothermic: melting, condensation, freezing, sublimation, deposition, and vaporization. Endothermic: melting, sublimation, vaporization; exothermic: condensation, freezing, deposition. **OL**

Reteach

Ask students what is happening at the molecular level when, during the winter, snow disappears gradually without first melting. Under the right conditions, solid water can undergo sublimation. **OL**

Extension

Ask students to use Figure 30 to determine one temperature and pressure combination at which carbon dioxide is in the solid phase, one combination for the liquid phase, and one for the gas phase. Answers will vary but should agree with the figure. **OL**

View an **animation about a phase diagram.**

Concepts In Motion

GRAPH CHECK

Contrast the slope of the red line in water's phase diagram with that of the red line in carbon dioxide's phase diagram. How do water and carbon dioxide differ in their reaction to increased pressure at the solid/liquid boundary?

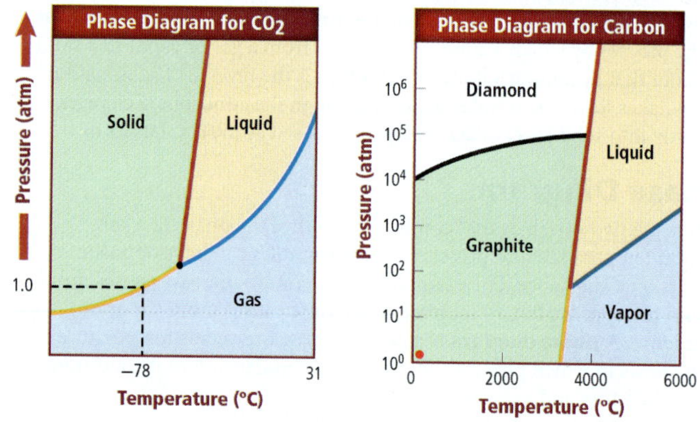

■ **Figure 30** Phase diagrams show useful information, such as why carbon dioxide sublimes at normal conditions and the existence of two forms of solid carbon.

The phase diagram for each substance is different because the normal boiling and freezing points of substances are different. However, each diagram will supply the same type of data for the phases, including a triple point. Of course, the range of temperatures chosen will vary to reflect the physical properties of the substance.

Phase diagrams can provide important information for substances. For example, the phase diagram for carbon dioxide in **Figure 30** shows why carbon dioxide sublimes at normal conditions. Find 1.0 atm on the carbon dioxide graph and follow the dashed line to the yellow line. The graph shows that carbon dioxide changes from a solid to a gas at 1 atm. If you extend the dashed line past the yellow line, the graph shows that carbon dioxide does not liquefy as temperature increases. It remains a gas.

The diagram on the right is a phase diagram for carbon. Notice that the graph contains two allotropes of carbon in the solid region. Graphite is the standard state of carbon at normal temperatures and pressures, designated by a red dot. Diamond is more stable at higher temperatures and pressures. Diamonds that exist at normal room conditions originally formed at high temperature and pressure.

SECTION 4 REVIEW

Section Self-Check

Section Summary

- States of a substance are referred to as phases when they coexist as physically distinct parts of a mixture.
- Energy changes occur during phase changes.
- Phase diagrams show how different temperatures and pressures affect the phase of a substance.

27. MAINIDEA **Explain** how the addition or removal of energy can cause a phase change.
28. **Explain** the difference between the processes of melting and freezing.
29. **Compare** deposition and sublimation.
30. **Compare and contrast** sublimation and evaporation.
31. **Describe** the information that a phase diagram supplies.
32. **Explain** what the triple point and the critical point on a phase diagram represent.
33. **Determine** the phase of water at 75.00°C and 3.00 atm using **Figure 29.**

SECTION 4 REVIEW

27. The addition of energy increases the kinetic energy of the particles, which reduces the intermolecular forces between the particles. As energy is removed, the kinetic energy of the particles decreases and intermolecular forces increase.
28. Freezing occurs when a liquid becomes a solid and energy is released. Melting requires an input of energy to convert a solid to a liquid.
29. A substance undergoes deposition when it goes from the vapor phase to the solid phase without going through the liquid phase. A substance undergoes sublimation when it goes from the solid phase directly to the vapor phase without going through the liquid phase.
30. In both processes, the substances become a vapor. During sublimation, the substance goes from the solid phase directly to the vapor phase. During evaporation, particles in a liquid gain enough energy to enter the vapor phase.
31. the combinations of temperature and pressure under which a given substance exists as a solid, liquid, and/or gas
32. Triple point: the temperature at which the three phases of a substance can coexist; critical point: the pressure and temperature above which a substance cannot exist as a liquid.
33. liquid

everyday CHEMISTRY

Cocoa Chemistry

Chocolate is a food product that is native to Central America and Mexico. The Aztec ruler Montezuma served the bitter cocoa-bean drink to Hernan Cortéz in 1519. Cortéz took the cocoa beans and the recipe for the chocolate beverage to Spain where it became a very popular, but expensive beverage. Chocolate remained a food product for the wealthy until the mid-nineteenth century, when the price of chocolate became affordable and processing techniques improved. The chocolate served today bears little resemblance to the chocolate served in Montezuma's court. Processing techniques as well as additives create the smooth, sweet, delightful treat that you enjoy today.

Melts in your mouth Chocolate is a mixture of cocoa, cocoa butter, and other ingredients. This mixture is a solid at room temperature but melts in your mouth. Why? Because one of the main ingredients in chocolate—cocoa butter—is a fat that melts at or near body temperature.

Particle size Chocolate is a liquid during the mixing process. The cocoa butter in the melted chocolate coats the solid particles of cocoa, sugar, and milk solids. The solid particles in the mixture must not be too large, or the chocolate will have a gritty texture. Generally, the particles are ground to a maximum diameter of 2.0×10^{-5} to 3.0×10^{-5} m.

Controlling flow As you can see in **Figure 1,** a large number of small particles has a larger surface area than a single particle of the same mass.

Figure 1 Although the mass of each particle or group of particles is the same, increasing the surface area allows more cocoa butter to coat the particles, which improves the flow of the chocolate.

Peter Scholey/Photographer's Choice RF/Getty Images

Figure 2 Chocolate is carefully processed so that the proper crystal structure forms in the chocolate. These crystals give chocolate the characteristics found in popular chocolate bars.

Smaller particles in the chocolate requires more cocoa butter to coat the solid surfaces. It is the excess cocoa butter *between* the solid particles that allows chocolate to flow.

Smooth texture If the chocolate contains too little cocoa butter between the particles, the chocolate will be too thick to flow into a mold. To improve the flow of the chocolate without increasing particle size, manufacturers can either add more fat to the mixture or add an emulsifier, such as lecithin. Lecithin is a fat often obtained from soybeans that helps keep the fat molecules evenly suspended, or emulsified, in the chocolate.

Crystallization Another important process in chocolate manufacturing is tempering. During the tempering of the chocolate, the temperature of the chocolate is carefully controlled to ensure that the desired crystals form. When chocolate is not properly tempered, crystals form that create poor-quality chocolate. The desired crystals make the chocolate in **Figure 2** glossy and firm, and allow it to snap well and melt near body temperature.

WRITING IN ▶ Chemistry

Research to find out more about chocolate and design a timeline that summarizes its changes over time. WebQuest

everyday CHEMISTRY

Purpose

Students will learn how particle size and fat content work together to affect the viscosity of liquid chocolate.

Background

Chocolate can be purchased at a wide array of prices and with varying quality. Considering the information in the article, and noting that cocoa butter is one of the more expensive ingredients in chocolate, discuss how changing particle size and fat quantity can change the quality and cost of chocolate.

Teaching Strategies

Chocolate is a good subject for laboratory analysis. Many possible projects are described in *The Science of Chocolate* by Stephen T. Beckett, published in 2000 by the Royal Society of Chemistry.

WRITING IN ▶ Chemistry

Rubric WebQuest

Research The reports might contain a variety of information, such as where the beans are grown, how they are processed, or how the final product is distributed and sold.

SMALL SCALE ChemLAB

Rubric

Preparation

Time Allotment one class period

Process Skills acquire and analyze information, classify, draw conclusions, observe and infer, recognize cause and effect

Safety Precautions Approve lab safety forms before work begins. Review MSDS for all hazardous chemicals with students before beginning the lab. **WARNING:** ***Acetone, ethanol, and isopropyl alcohol are flammable–vapors may be explosive.*** Keep open flames and sparks away from the vapors of these liquids–this lab should not be performed without adequate ventilation. These liquids can also dry the skin or irritate the eyes.

Disposal Instruct students to return each unused liquid to the proper container for reuse.

Procedure

- Tell students that if the drop of liquid is still visible after 5 minutes, they should record the data as 300 seconds.
- Ethanol can be warmed in a warm water bath on a hot plate.

Expected Results

Liquid	Evap. time (s)	Obs: Shape of liquid drop (top, side views)
Dist. water	>300	Top: spherical; Side: domed
Ethanol	140	Top: jagged circular; Side: flat
Ethanol (warm)	<140	Top: jagged circular; Side: flat
Isopropyl alcohol	146	Same as ethanol
Acetone	8.4	Top: small circular; Side: flat
Ammonia	>300	Same as water

SMALL SCALE ChemLAB

Compare Rates of Evaporation

Background: **Several factors determine how fast a sample of liquid will evaporate. The volume of the sample is a key factor. A drop of water takes less time to evaporate than a liter of water. The amount of energy supplied to the sample is another factor.**

Question: *How do intermolecular forces affect the evaporation rates of liquids?*

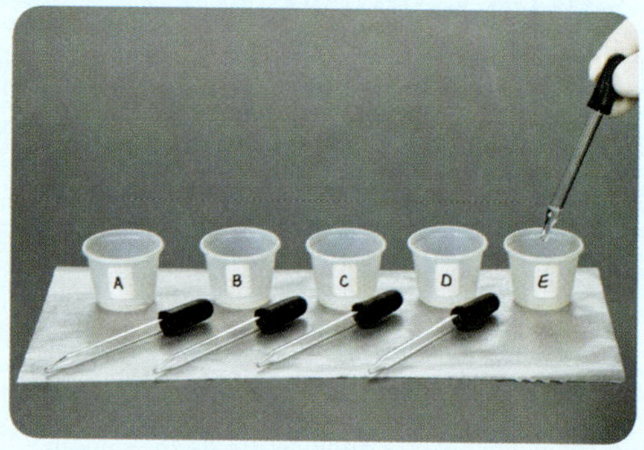

Matt Meadows

Materials

distilled water
ethanol
isopropyl alcohol
acetone
household ammonia
droppers (5)
small plastic cups (5)
grease pencil or masking tape and a marking pen
paper towel
square of waxed paper
stopwatch

Safety Precautions

Procedure

1. Read and complete the lab safety form.
2. Make a data table to record data.
3. Use a grease pencil or masking tape to label each of 5 small plastic cups. Use *A* for distilled water, *B* for ethanol, *C* for isopropyl alcohol, *D* for acetone, and *E* for household ammonia. Place the plastic cups on a paper towel.
4. Use a dropper to collect about 1 mL of distilled water and place the water in the cup labeled *A*. Place the dropper on the paper towel directly in front of the cup. Repeat with the other liquids.
5. Place a square of waxed paper on your lab surface. Plan where on the waxed paper you will place each of the five drops that you will test to avoid mixing.
6. Have your stopwatch ready. Collect some water in your water dropper and place a single drop on the waxed paper. Begin timing. Time how long it takes for the drop to completely evaporate. While you wait, make a top-view and side-view drawing of the drop. If the drop takes longer than 5 min to evaporate, record > 300 s in your data table.
7. Repeat Step 6 with the four other liquids.
8. Use the above procedure to design an experiment in which you can observe the effect of temperature on the rate of evaporation of ethanol. Your teacher will provide a sample of warm ethanol.
9. **Cleanup and Disposal** Clean up lab materials as instructed by your teacher.

Analyze and Conclude

1. **Classify** Which liquids evaporated quickly? Which liquids were slow to evaporate?
2. **Evaluate** Based on your data, in which liquid(s) are the attractive forces between molecules most likely dispersion forces?
3. **Consider** What is the relationship between surface tension and the shape of a liquid drop? What are the attractive forces that increase surface tension?
4. **Assess** The isopropyl alcohol you used was a mixture of isopropyl alcohol and water. Would pure isopropyl alcohol evaporate more quickly or more slowly compared to the alcohol and water mixture? Give a reason for your answer.
5. **Evaluate** Household ammonia is a mixture of ammonia and water. Based on the data you collected, is there more ammonia or more water in the mixture? Explain.
6. **Evaluate** How does the rate of evaporation of warm ethanol compare to ethanol at room temperature?
7. **Error Analysis** How could you change the procedure to make it more precise?

INQUIRY EXTENSION

Design an Experiment How would different surfaces affect your results? Design an experiment to test your hypothesis.

Analyze and Conclude

1. acetone, fastest; water and ammonia, slowest
2. in the liquid that is nonpolar and that evaporates the fastest (acetone)
3. The greater the surface tension, the more spherical or dome-shaped the drop. Intermolecular forces increase surface tension.
4. Pure alcohol would evaporate more quickly because of reduced hydrogen bonding.
5. More water; the evaporation rate is similar to that of water.
6. Warm ethanol evaporates faster than ethanol at room temperature because of increased kinetic energy.
7. by determining the exact volume of each liquid tested

INQUIRY EXTENSION

Make sure students use correct scientific procedures in their experimental designs.

CHAPTER 12 STUDY GUIDE

Vocabulary Practice

BIGIDEA Kinetic-molecular theory explains the different properties of solids, liquids, and gases.

SECTION 1 Gases

MAINIDEA Gases expand, diffuse, exert pressure, and can be compressed because they are in a low-density state consisting of tiny, constantly-moving particles.

- The kinetic-molecular theory explains the properties of gases in terms of the size, motion, and energy of their particles.
- Dalton's law of partial pressures is used to determine the pressures of individual gases in gas mixtures.
- Graham's law is used to compare the diffusion rates of two gases.

$$\frac{\text{Rate}_A}{\text{Rate}_B} = \sqrt{\frac{\text{molar mass}_B}{\text{molar mass}_A}}$$

VOCABULARY
- kinetic-molecular theory
- elastic collision
- temperature
- diffusion
- Graham's law of effusion
- pressure
- barometer
- pascal
- atmosphere
- Dalton's law of partial pressures

SECTION 2 Forces of Attraction

MAINIDEA Intermolecular forces—including dispersion forces, dipole-dipole forces, and hydrogen bonds—determine a substance's state at a given temperature.

- Intramolecular forces are stronger than intermolecular forces.
- Dispersion forces are intermolecular forces between temporary dipoles.
- Dipole-dipole forces occur between polar molecules.

VOCABULARY
- dispersion force
- dipole-dipole force
- hydrogen bond

SECTION 3 Liquids and Solids

MAINIDEA The particles in solids and liquids have a limited range of motion and are not easily compressed.

- The kinetic-molecular theory explains the behavior of solids and liquids.
- Intermolecular forces in liquids affect viscosity, surface tension, cohesion, and adhesion.
- Crystalline solids can be classified by their shape and composition.

VOCABULARY
- viscosity
- surface tension
- surfactant
- crystalline solid
- unit cell
- allotrope
- amorphous solid

SECTION 4 Phase Changes

MAINIDEA Matter changes phase when energy is added or removed.

- States of a substance are referred to as phases when they coexist as physically distinct parts of a mixture.
- Energy changes occur during phase changes.
- Phase diagrams show how different temperatures and pressures affect the phase of a substance.

VOCABULARY
- melting point
- vaporization
- evaporation
- vapor pressure
- boiling point
- freezing point
- condensation
- deposition
- phase diagram
- triple point

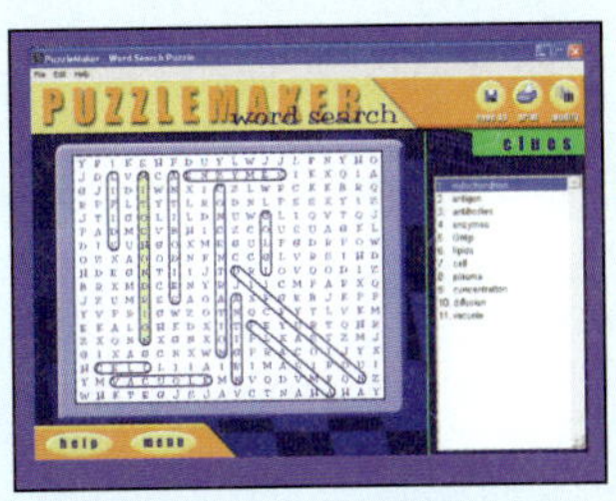

Vocabulary Practice

For additional practice with vocabulary, have students access the Vocabulary PuzzleMaker.

Use the Vocabulary

To reinforce chapter vocabulary, have students write a sentence using each term. OL EL

Review Strategies

- Have students make a chart comparing the characteristics of solids, liquids, and gases. OL
- Have students learn these pairs of terms that have opposite meanings: *vaporization/condensation, sublimation/deposition,* and *melting/freezing.* OL
- Problems from the Supplemental Practice Problems appendix or the Supplemental Problems booklet can be used for review. OL

Students can review with these online resources.

Vocabulary Practice

Vocabulary eGames
Vocabulary eFlashcards
Vocabulary PuzzleMaker

Section Self-Check

Chapter Self-Check

Online Test Practice

Use *eAssessment* to:

- create multiple versions of tests
- edit existing questions and add your own questions
- build tests aligned with select state curriculum standards using built-in state curriculum tags
- track students' progress

SECTION 1

Mastering Concepts

34. one in which no kinetic energy is lost

35. It is directly proportional to their temperature.

36. Because of the space between gas particles, gases are easily compressed when pushed into a smaller volume. When the pressure is removed, their random motion enables gases to expand.

37. (1). Matter is composed of small particles.
(2). The particles are in constant motion and undergo elastic collisions.
(3). The particles have kinetic energy and the average kinetic energy of the particles is temperature.

38. Gases have low density, can be compressed, will expand to fill all available space, and can undergo diffusion and effusion.

39. Both involve the movement of gas particles. Diffusion is the movement of one substance through another; effusion is when a substance under pressure escapes through a tiny opening. Effusion and diffusion rates are inversely related to molecular mass of a gas.

40. Density decreases because the gas particles occupy more volume per unit mass.

41. Because of the variation in air pressure with elevation; At high elevations, reduced air pressure results in a lower boiling point for water and cooking time is longer.

Mastering Problems

42. 36.0 g/mol

43. $Rate_{Kr}/Rate_{Ne} = 0.4907$

44. 3.56 g/mol

45. 0.01 atm

46. 0.99 atm

47. 0.332 atm; It is about one-third of the 1-atm pressure at sea level.

48. 84.0 kPa = 0.829 atm and 6.30×10^2 torr

49. 8.4 atm = 8.5×10^2 kPa and 6.4×10^3 mmHg

50. The gases will diffuse until both bulbs are filled with the same gas mixtures.

CHAPTER 12 ASSESSMENT

SECTION 1

Mastering Concepts

34. What is an elastic collision?

35. How does the kinetic energy of particles vary as a function of temperature?

36. Use the kinetic-molecular theory to explain the compression and expansion of gases.

37. List the three basic assumptions of the kinetic-molecular theory.

38. Describe the common properties of gases.

39. Compare diffusion and effusion. Explain the relationship between the rates of these processes and the molar mass of a gas.

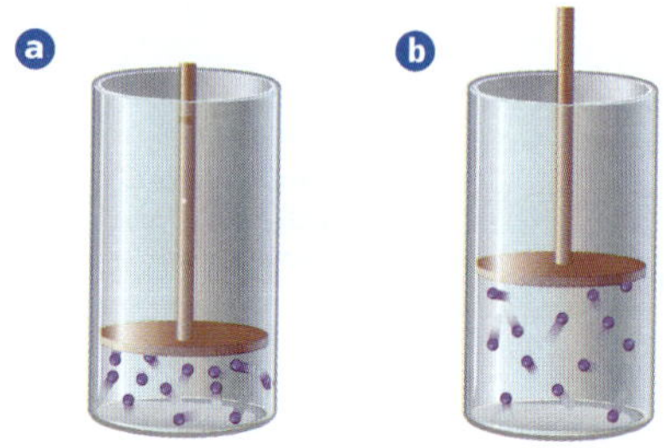

■ **Figure 31**

40. In **Figure 31,** what happens to the density of gas particles in the cylinder as the piston moves from Position A to Position B?

41. Baking Explain why the baking instructions on a box of cake mix are different for high and low elevations. Would you expect to have a longer or a shorter cooking time at a high elevation?

Mastering Problems

42. What is the molar mass of a gas that takes three times longer to effuse than helium?

43. What is the ratio of effusion rates of krypton and neon at the same temperature and pressure?

44. Calculate the molar mass of a gas that diffuses three times faster than oxygen under similar conditions.

45. What is the partial pressure of water vapor in an air sample when the total pressure is 1.00 atm, the partial pressure of nitrogen is 0.79 atm, the partial pressure of oxygen is 0.20 atm, and the partial pressure of all other gases in air is 0.0044 atm?

46. What is the total gas pressure in a sealed flask that contains oxygen at a partial pressure of 0.41 atm and water vapor at a partial pressure of 0.58 atm?

47. Mountain Climbing The pressure atop the world's highest mountain, Mount Everest, is usually about 33.6 kPa. Convert the pressure to atmospheres. How does the pressure compare with the pressure at sea level?

48. High Altitude The atmospheric pressure in Denver, Colorado, is usually about 84.0 kPa. What is this pressure in atm and torr units?

49. At an ocean depth of 76.2 m, the pressure is about 8.4 atm. Convert the pressure to mmHg and kPa units.

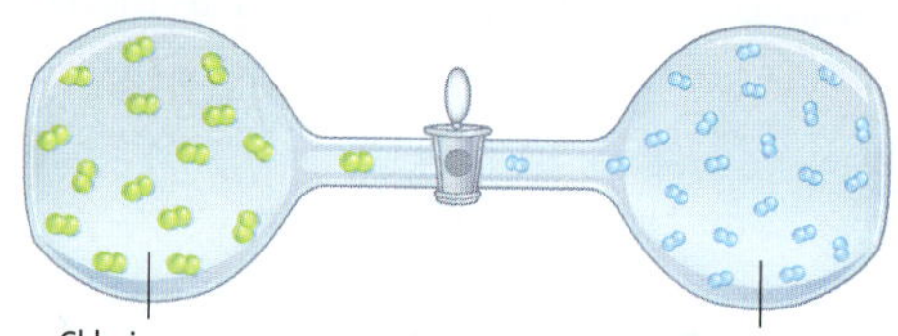

■ **Figure 32**

50. Figure 32 represents an experimental set-up in which the left bulb is filled with chlorine gas and the right bulb is filled with nitrogen gas. Describe what happens when the stopcock is opened. Assume that the temperature of the system is held constant during the experiment.

SECTION 2

Mastering Concepts

51. Explain the difference between a temporary dipole and a permanent dipole.

52. Why are dispersion forces weaker than dipole-dipole forces?

53. Explain why hydrogen bonds are stronger than most dipole-dipole forces.

54. Compare intramolecular and intermolecular forces.

55. Hypothesize why long, nonpolar molecules would interact more strongly with one another than spherical nonpolar molecules of similar composition.

Mastering Problems

56. Polar Molecules Use relative differences in electronegativity to label the ends of the polar molecules listed as partially positive or partially negative.
a. HF **b.** HBr **c.** NO **d.** CO

57. Draw the structure of the dipole-dipole interaction between two molecules of carbon monoxide.

58. Decide which of the substances listed can form hydrogen bonds.
a. H_2O **b.** H_2O_2 **c.** HF **d.** NH_3

SECTION 2

Mastering Concepts

51. A temporary dipole forms when one molecule is close to another molecule, and the electrons repel each other creating a greater electron density in one part of the molecule. Permanent dipoles are found in polar molecules in which some regions of the molecule are always partially positive and partially negative.

52. Dispersion forces are between temporary dipoles. Dipole-dipole forces are between permanent dipoles.

53. A hydrogen bond involves a large difference in electronegativity between the hydrogen atom and the atom it is attached to (O, N, or F), making the bond extremely polar.

54. Intramolecular forces hold atoms together in a molecule while intermolecular forces hold different molecules together.

55. Because long molecules have greater surface areas, more intermolecular forces can exist.

59. Decide which one of the molecules listed below can form intermolecular hydrogen bonds, and then draw it, showing several molecules attached together by hydrogen bonds.
a. NaCl **b.** $MgCl_2$ **c.** H_2O_2 **d.** CO_2

SECTION 3

Mastering Concepts

60. What is surface tension, and what conditions must exist for it to occur?
61. Explain why the surface of water in a graduated cylinder is curved.
62. Which liquid is more viscous at room temperature, water or molasses? Explain.
63. Explain how two different forces play a role in capillary action.

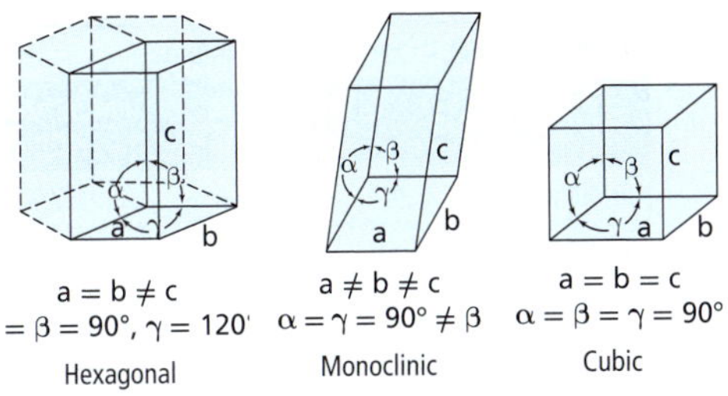

■ **Figure 33**

64. Use the drawings in **Figure 33** to compare the cubic, monoclinic, and hexagonal crystal systems.
65. What is the difference between a network solid and an ionic solid?
66. Explain why most metals bend when struck but most ionic solids shatter.
67. List the types of crystalline solids that are usually good conductors of heat and electricity.
68. How does the strength of a liquid's intermolecular forces affect its viscosity?
69. Explain why water has a higher surface tension than benzene, whose molecules are nonpolar.
70. Compare the numbers of particles in **Figure 19** that form the basic shapes of one unit cell for each of the following.
a. simple cubic
b. body-centered cubic
71. Predict which solid is more likely to be amorphous—one formed by cooling a molten material over 4 h at room temperature or one formed by cooling a molten material quickly in an ice bath.
72. **Conductivity** Predict which solid will conduct electricity better—sugar or salt.
73. Explain why ice floats in water but solid benzene sinks in liquid benzene. Which behavior is more "normal"?

Mastering Problems

74. Given edge lengths and face angles, predict the shape of each of the following crystals.
a. $a = 3$ nm, $b = 3$ nm, $c = 3$ nm; $\alpha = 90°$, $\beta° = 90$, $\gamma = 90°$
b. $a = 4$ nm, $b = 3$ nm, $c = 5$ nm; $\alpha = 90°$, $\beta° = 100$, $\gamma = 90°$
c. $a = 3$ nm, $b = 3$ nm, $c =$ 5nm; $\alpha = 90°$, $\beta° = 90$, $\gamma = 90°$
d. $a = 3$ nm, $b = 3$ nm, $c = 5$ nm; $\alpha = 90°$, $\beta° = 90$, $\gamma = 120°$

SECTION 4

Mastering Concepts

75. How does sublimation differ from deposition?
76. Compare boiling and evaporation.
77. Define the term *melting point.*
78. Explain the relationships among vapor pressure, atmospheric pressure, and boiling point.
79. Explain why dew forms on cool mornings.
80. **Snow** Why does a pile of snow slowly shrink even on days when the temperature never rises above the freezing point of water?

Mastering Problems

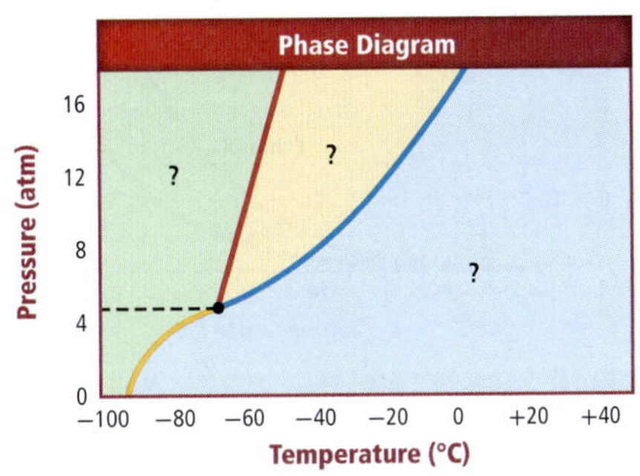

■ **Figure 34**

81. Copy and label the solid, liquid, and gas phases, triple point, and critical point on **Figure 34.**
82. Why does it take more energy to boil 10 g of liquid water than to melt an equivalent mass of ice?

Mastering Problems

56. a. $H^+—F^-$; b. $H^+—Br^-$; c. $N^+—O^-$; d. $C^+—O^-$
57. Refer to Figure 9. The drawing should show two CO molecules, with the C partially positive and the O partially negative. The C of each molecule should be bonded to the O of the other.
58. All of the substances can form hydrogen bonds.
59. H_2O_2 can form hydrogen bonds. Refer to the Solutions Manual for drawing.

SECTION 3

Mastering Concepts

60. The energy needed to increase the surface area of a liquid by a given amount; strong interparticle forces must exist between the liquid particles.
61. Adhesion between water and glass is greater than cohesion between water molecules.
62. Molasses; stronger intermolecular forces keep molasses from flowing.
63. Capillary action results from the opposing forces of adhesion and cohesion. Adhesive forces between water molecules and molecules in the glass of a capillary tube are stronger than the cohesive forces holding water molecules together, so water moves up the walls of the capillary tube.
64. In the cubic crystal system, all of the sides are an equal length and the angles are all 90°. In the monoclinic crystal system, a, b, and c are unequal in length and α and γ are 90° angles, but β is not. In the hexagonal crystal system, a and b are equal in length, but c is not. The angles α and β are equal to 90°, but γ is equal to 120°.
65. Network solid is held together by covalent bonds; ionic solid by electrostatic attraction
66. The sea of electrons that holds metal ions together can easily move to accommodate outside forces; in ionic solids, a strong force can separate the solid along the plane, bonding groups of atoms together.
67. Metallic solids; ionic solids when molten or dissolved in an aqueous solution.
68. Stronger intermolecular forces result in higher viscosity because the forces hold the particles together too tightly for them to flow easily.
69. Surface tension increases with strength of interparticle forces. Water molecules are held together by strong hydrogen bonds, resulting in higher surface tension; the weaker dispersion forces between benzene molecules result in lower surface tension.
70. a. 8; b. 9
71. The one cooled quickly in an ice bath because amorphous solids often form when molten material cools too quickly for crystals to form.
72. Salt, because it contains ions while sugar is a molecular solid without any ions.
73. Ice floats in water because strong hydrogen bonds enforce a unique hexagonal structure among water molecules in the solid phase. As a result of this structure, ice is less dense than liquid water. Benzene has no such forces and becomes more dense as it freezes. Benzene is more "normal".

Mastering Problems

74. **a.** cubic
b. monoclinic
c. tetragonal
d. hexagonal

SECTION 4

Mastering Concepts

75. Sublimation occurs when a solid is converted to a gas; deposition occurs when a gas is converted to a solid.

76. Evaporation: conversion of a liquid to a gas at the liquid's surface; boiling: when vapor pressure is equal to external atmospheric pressure; it occurs at and below the surface where bubbles form.

77. The temperature at which the crystal lattice of a solid disintegrates and it becomes a liquid.

78. Boiling point is the temperature at which vapor pressure, exerted by liquid molecules escaping from the surface of a sample, equals the atmospheric pressure on the surface of the liquid.

79. When water vapor in the air comes in contact with a cool object, it condenses on the object.

80. Some of the snow sublimes.

Mastering Problems

81. Refer to the Solutions Manual.

82. Melting does not require as much energy because the particles in a solid do not have to move far apart or gain much movement to form a liquid.

MIXED REVIEW

83. Because the particles in gases and liquids are held together by fewer attractive forces than in solids, allowing them to flow.

84. Oxygen molecules are nonpolar and held together by dispersion forces, making them easy to separate. Water molecules are held together by stronger hydrogen bonds, making them harder to separate. As a result, water has a higher boiling point.

85. Because particles in gases are farther apart than particles in liquids or solids, there is more space for the particles to be compressed.

86. Metallic bonds holding mercury atoms together are stronger than hydrogen bonds holding water molecules together, so the mercury atoms are more closely packed, resulting in greater mass per unit volume.

87. The container with twice the pressure has twice the number of particles.

MIXED REVIEW

83. Use the kinetic-molecular theory to explain why both gases and liquids are fluids.

84. Use intermolecular forces to explain why oxygen is a gas at room temperature and water is a liquid.

85. Use the kinetic-molecular theory to explain why gases are easier to compress than liquids or solids.

86. At 25°C and a pressure of 760 mmHg, the density of mercury is 13.5 g/mL; water at the same temperature and pressure has a density of 1.00 g/mL. Explain this difference in terms of intermolecular forces and the kinetic-molecular theory.

87. If two identical containers each hold the same gas at the same temperature but the pressure inside one container is exactly twice that of the other container, what must be true about the amount of gas inside each container?

88. List three types of intermolecular forces.

89. When solid sugar crystals are dissolved in a glass of water, they form a clear homogeneous solution in which the crystals are not visible. If the beaker is left out at room temperature for a few days, the crystals reappear in the bottom and on the sides of the glass. Is this an example of freezing?

THINK CRITICALLY

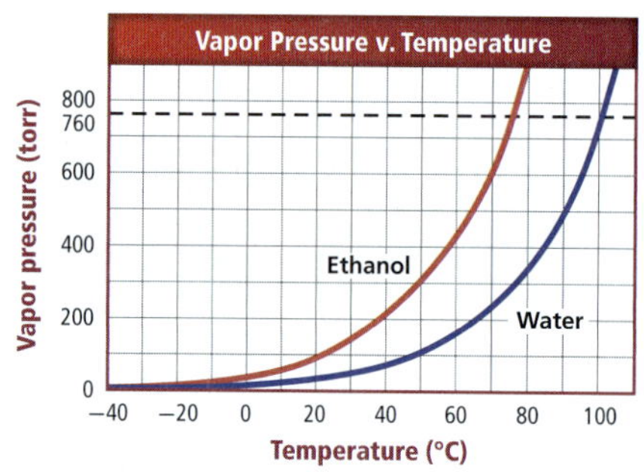

■ **Figure 35**

90. Interpret Graphs Examine **Figure 35**, which plots vapor pressure versus temperature for water and ethanol.
- **a.** What is the boiling point of water at 1 atm?
- **b.** What is the boiling point of ethanol at 1 atm?
- **c.** Estimate the temperature at which water will boil when the atmospheric pressure is 0.80 atm.

91. Hypothesize What type of crystalline solid do you predict would best suit the following needs?
- **a.** a material that can be melted and reformed at a low temperature
- **b.** a material that can be drawn into long, thin wires
- **c.** a material that conducts electricity when molten
- **d.** an extremely hard material that is nonconductive

92. Compare and Contrast An air compressor uses energy to squeeze air particles together. When the air is released, it expands, allowing the energy to be used for purposes such as gently cleaning surfaces without using a more abrasive liquid or solid. Hydraulic systems use pressurized fluids to transmit and multiply force. What do you think are some advantages and disadvantages of these two types of technology?

93. Graph Use **Table 6** to construct a phase diagram for ammonia.

Table 6 Phase Diagram for Ammonia

Selected Points	Pressure (atm)	Temperature (°C)
Triple point	0.060	−77.7
Critical point	112	132.2
Normal boiling point	1.0	−33.5
Normal freezing point	1.0	−77.7

94. Apply A solid being heated stays at a constant temperature until it is completely melted. What happens to the heat energy put into the system during that time?

95. Communicate Which process—effusion or diffusion—is responsible for your being able to smell perfume from an open bottle that is located across the room from you? Explain.

96. Infer A laboratory demonstration involves pouring bromine vapors, which are a deep red color, into a flask of air and then tightly sealing the top of the flask. The bromine is observed to first sink to the bottom of the beaker. After several hours have passed, the red color is distributed equally throughout the flask.
- **a.** Is bromine gas more or less dense than air?
- **b.** Would liquid bromine diffuse more or less quickly than gaseous bromine after you pour it into another liquid?

97. Analyze Use your knowledge of intermolecular forces to predict whether ammonia (NH_3) or methane (CH_4) will be more soluble in water.

98. Evaluate List three changes that require energy and three that release energy.

88. hydrogen bonding, dispersion forces, dipole-dipole forces

89. No. The only phase change that occurred is liquid water evaporating to form water vapor. The sugar was always in the solid phase, even when not visible. The crystals became large enough to see with the unaided eye over time.

THINK CRITICALLY

90. **a.** 100°C
b. 78.5°C
c. 94°C

91. **a.** molecular solid
b. metallic solid
c. ionic solid
d. covalent network solid

92. Air compressors produce rapid outputs of a large amount of energy because gases can be greatly compressed. Hydraulic systems involve liquids, which cannot be compressed as much, and are more useful for a slow, steady output of energy.

93. Refer to the Solutions Manual.

94. The energy is used to break the intermolecular forces holding the particles of the solid together.

95. Diffusion; gas particles in the perfume are mixing with air particles.

Chapter Self-Check

99. Evaluate Supercritical carbon dioxide is a liquid form of CO_2 used in the food industry to decaffeinate tea, coffee, and colas, as well as in the pharmaceutical industry to form polymer microparticles used in drug delivery systems. Use **Figure 36** to determine what conditions must be used to form supercritical carbon dioxide.

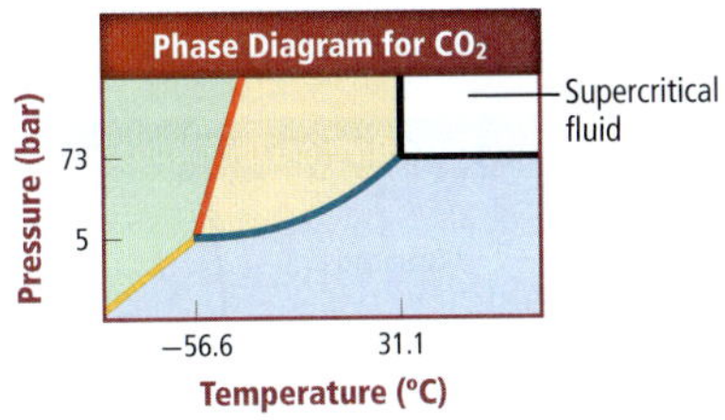

■ **Figure 36**

Challenge Problem

100. You have a solution containing 135.2 g of dissolved KBr in 2.3 L of water. What volume of this solution, in mL, would you use to make 1.5 L of a 0.1 mol/L KBr solution? What is the boiling point of this new solution?

CUMULATIVE REVIEW

101. Identify each of the following as an element, a compound, a homogeneous mixture, or a heterogeneous mixture.

a. air
b. blood
c. antimony
d. ammonia
e. mustard
f. water

102. You are given two clear, colorless aqueous solutions. You are told that one solution contains an ionic compound, and one contains a covalent compound. How could you determine which is an ionic solution and which is a covalent solution?

103. Which branch of chemistry would most likely study matter and phase changes?

a. biochemistry
b. organic chemistry
c. physical chemistry
d. polymer chemistry

104. What type of reaction is the following?

$$K_2CO_3(aq) + BaCl_2(aq) \rightarrow 2KCl(aq) + BaCO_3(s)$$

a. combustion
b. double-replacement
c. single-replacement
d. synthesis

105. Which chemist produced the first widely used and accepted periodic table?

a. Dmitri Mendeleev
b. Henry Moseley
c. John Newlands
d. Lothar Meyer

WRITING IN ▶ Chemistry

106. Musk is the basic ingredient of many perfumes, soaps, shampoos, and even foods such as chocolates, licorice, and hard candies. Both synthetic and natural musk molecules have high molecular weights compared to other perfume ingredients, and as a result, have a slower rate of diffusion, assuring a slow, sustained release of fragrance. Write a report on the chemistry of perfume ingredients, emphasizing the importance of diffusion rate as a property of perfume.

107. Birthstones Find out what your birthstone is and write a brief report about the chemistry of that gem. Find out its chemical composition, which category its unit cell is in, how hard and durable it is, and what its approximate cost is at present.

108. Propane gas is a commonly used heating fuel for gas grills and homes. However, it is not packaged as a gas. It is liquefied and referred to as liquid propane or "LP gas." Make a poster explaining the advantages and disadvantages of storing and transporting propane as a liquid rather than a gas.

109. Other States of Matter Research and prepare an oral report about one of the following topics: plasma, superfluids, fermionic condensate, or Bose-Einstein condensate. Share your report with your classmates and prepare a visual aid that can be used to explain your topic.

DBQ Document-Based Questions

Iodine *Solid iodine that is left at room temperature sublimates from a solid to a gas. But when heated quickly, a different process takes place, as described here.*

"About 1 g of iodine crystals is placed in a sealed glass ampoule and gently heated on a hot plate. A layer of purple gas is formed at the bottom, and the iodine liquefies. If one tilts the tube, this liquid flows along the wall as a narrow stream and solidifies very quickly."

Data obtained from: Leenson, 2005. Sublimination of Iodine at Various Pressures: Multipurpose Experiments in Inorganic and Physical Chemistry. *Journal of Chemical Education* 82(2):241–245.

110. Why does solid iodine sublime readily? Use your knowledge of intermolecular forces to explain.

111. Why is liquid iodine not usually visible if crystals are heated in the open air?

112. Why is it necessary to use a sealed ampoule in this investigation?

113. Infer why the iodine solidifies when the tube is tilted.

96. a. more dense
b. less quickly

97. Ammonia will be more soluble in water than methane because ammonia and water can form relatively strong intermolecular attractions (hydrogen bonds) between them. Methane and water will form only intermolecular dispersive forces which are weaker than hydrogen bonds.

98. requires energy—melting, sublimation, vaporization; releases energy—freezing, deposition, condensation

99. A pressure above 73 bars and a temperature above 31.1°C.

CHALLENGE PROBLEM

100. 300 mL

CUMULATIVE REVIEW

101. a. homogeneous mixture
b. heterogeneous mixture
c. element
d. compound
e. heterogeneous mixture
f. compound

102. They could be tested for conductivity. The ionic compound in solution should conduct electricity; the covalent compound should not.

103. c

104. b

105. a

WRITING IN ▶ Chemistry

Rubric

106. Student reports should point out that a slow effusion rate results in a longer lasting fragrance.

107. Check student reports for accuracy.

108. A liquid requires a smaller container than a gas, and the container can hold more fuel as a liquid rather than a solid.

109. Check student reports.

DBQ Document-Based Questions

Data obtained from: Leenson, 2005. Sublimination of Iodine at Various Pressures: Multipurpose Experiments in Inorganic and Physical Chemistry. *Journal of Chemical Education* 82(2):241–245.

110. The dispersive forces that hold iodine molecules together in its solid crystal are relatively weak so when those bonds break, the atoms can move apart directly from the solid to the gas state.

111. Iodine melts at 112.9°C, boils at 183.0°C, and its vapor pressure is 100 mmHg at 116.5°C. When being heated rapidly or in a closed vessel solid iodine will melt. A vapor pressure of about 100 mmHg must be attained to allow liquid iodine to form and prevent sublimation. In containers open to the air, the crystals will usually sublime completely before melting.

112. If the ampoule were not sealed, the iodine vapor would escape into the room.

113. The iodine cools quickly as it is tilted.

MULTIPLE CHOICE

1. C
2. B
3. B
4. D
5. D
6. C
7. C
8. B

MULTIPLE CHOICE

1. What is the ratio of diffusion rates for nitric oxide (NO) and nitrogen tetroxide (N_2O_4)?
   A. 0.326
   B. 0.571
   C. 1.751
   D. 3.066

2. Which is NOT an assumption of the kinetic-molecular theory?
   A. Collisions between gas particles are elastic.
   B. All the gas particles in a sample have the same velocity.
   C. A gas particle is not significantly attracted or repelled by other gas particles.
   D. All gases at a given temperature have the same average kinetic energy.

3. A sealed flask contains neon, argon, and krypton gas. If the total pressure in the flask is 3.782 atm, the partial pressure of Ne is 0.435 atm, and the partial pressure of Kr is 1.613 atm, what is the partial pressure of Ar?
   A. 2.048 atm
   B. 1.734 atm
   C. 1556 atm
   D. 1318 atm

Use the figure below to answer Question 4.

4. Hydrogen and nitrogen react as shown to form ammonia (NH_3). What is true of this reaction?
   A. Three ammonia molecules are formed, with zero molecules remaining.
   B. Two ammonia molecules are formed, with two hydrogen molecules remaining.
   C. Six ammonia molecules are formed, with zero molecules remaining.
   D. Two ammonia molecules are formed, with two nitrogen molecules remaining.

5. Which does not affect the viscosity of a liquid?
   A. intermolecular attractive forces
   B. size and shape of molecules
   C. temperature of the liquid
   D. capillary action

Use the graph below to answer Questions 6 to 8.

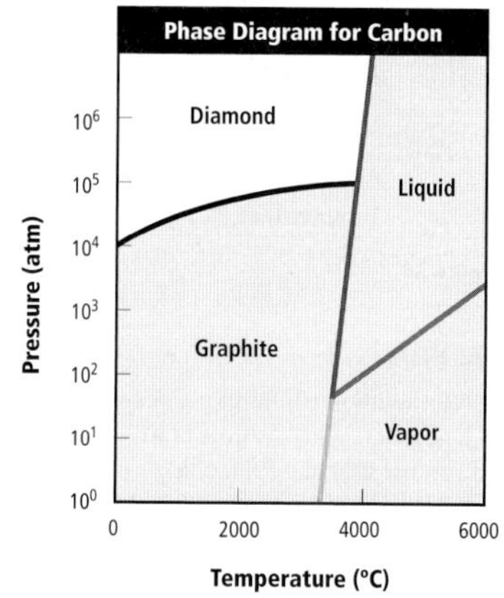

6. Under what conditions is diamond most likely to form?
   A. temperatures > 5000°C and pressures < 100 atm
   B. temperatures > 6000°C and pressures < 25 atm
   C. temperatures < 3500°C and pressures > 10^5 atm
   D. temperatures < 4500°C and pressures < 10 atm

7. Find the point on the graph at which carbon exists in three phases: solid graphite, solid diamond, and liquid carbon. What are the approximate temperature and pressure at that point?
   A. 4700°C and 10^6 atm
   B. 3000°C and 10^3 atm
   C. 4000°C and 10^5 atm
   D. 3500°C and 80 atm

8. In what form or forms does carbon exist at 6000°C and 10^5 atm?
   A. diamond only
   B. liquid carbon only
   C. diamond and liquid carbon
   D. liquid carbon and graphite

Online Test Practice

SHORT ANSWER

Use the table below to answer Questions 9 and 10.

Properties of Single Bonds		
Bond	**Strength (kJ/mol)**	**Length (pm)**
H – H	435	74
Br – Br	192	228
C – C	347	154
C – H	393	104
C – N	305	147
C – O	356	143
Cl – Cl	243	199
I – I	151	267
S – S	259	208

9. Create a graph to show how bond length varies with bond strength. Place bond strength on the x-axis.

10. Summarize the relationship between bond strength and bond length.

EXTENDED RESPONSE

Use the table below to answer Question 11.

Geometry of $AlCl_3$ and PCl_3		
Compound	$AlCl_3$	PCl_3
Molecular Shape	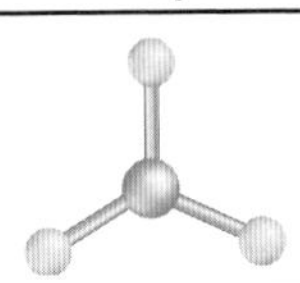	

11. What are the names of the shapes of the molecules for each compound? Explain how the atomic arrangements in each compound result in their different shapes despite their similar formulas.

SAT SUBJECT TEST: CHEMISTRY

12. Potassium chromate and lead(II) acetate are both dissolved in a beaker of water, where they react to form solid lead(II) chromate. What is the balanced net ionic equation describing this reaction?
    A. $Pb^{2+}(aq) + C_2H_3O_2^-(aq) \rightarrow Pb(C_2H_3O_2)_2(s)$
    B. $Pb^{2+}(aq) + 2CrO_4^-(aq) \rightarrow Pb(CrO_4)_2(s)$
    C. $Pb^{2+}(aq) + CrO_4^{2-}(aq) \rightarrow PbCrO_4(s)$
    D. $Pb^+(aq) + C_2H_3O_2^-(aq) \rightarrow PbC_2H_3O_2(s)$
    E. $Pb^{2+}(aq) + CrO_4^-(aq) \rightarrow PbCrO_5(s)$

13. The solid phase of a compound has a definite shape and volume because its particles
    A. are not in constant motion.
    B. are always more tightly packed in the liquid phase.
    C. can vibrate only around fixed points.
    D. are held together by strong intramolecular forces.
    E. have no intermolecular forces.

Use the table below to answer Questions 14 and 15.

Properties of Sulfuric Acid	
Formula	H_2SO_4
Molar mass	98.08 g/mol
Density	1.834 g/mL

14. What is the mass of 75.0 mL of sulfuric acid?
    A. 40.9 g
    B. 138 g
    C. 98.08 g
    D. 180 g
    E. 198.4 g

15. How many atoms of oxygen are present in 2.4 mol of sulfuric acid?
    A. 940 atoms
    B. 230 atoms
    C. 1.5×10^{24} atoms
    D. 5.8×10^{24} atoms
    E. 6.02×10^{23} atoms

NEED EXTRA HELP?															
If You Missed Question . . .	1	2	3	4	5	6	7	8	9	10	11	12	13	14	15
Review Section . . .	12.1	12.1	12.1	11.1	12.3	12.4	12.4	12.4	8.1	8.1	8.4	9.3	12.3	2.1	10.3

SHORT ANSWER

9. Refer to the Solutions Manual for graph.
10. As the bond strength increases, the bond length decreases.

EXTENDED RESPONSE

11. $AlCl_3$ has a trigonal planar shape, while PCl_3 has trigonal pyramidal shape. The difference in their shapes is due to the number of unpaired electron pairs. Because aluminum has three valence electrons, when it bonds to chlorine, all of those electrons are used in bonding and no unpaired electrons are left, therefore giving it a planar shape. PCl_3 has a pyramidal shape because its five valence electrons are not all used in bonding to three chlorine atoms, leaving one pair of electrons, which repel the bonding electrons to make a three-dimensional shape.

SAT SUBJECT TEST: CHEMISTRY

12. C
13. C
14. B
15. D

CHAPTER 13 Gases

BIGIDEA **Gases respond in predictable ways to changes in pressure, temperature, volume, and number of particles.**

ESSENTIAL QUESTIONS	RESOURCES TO ASSESS MASTERY
SECTION 1 **The Gas Laws** **1.** What are the relationships between pressure, temperature, and volume of a constant amount of gas? **2.** How can you use the gas laws to solve problems involving the pressure, temperature, and volume of a constant amount of gas? 2 sessions 1 block	**Progress Monitoring** Formative Assessment, pp. 442, 443, 445, 448, 451 Reading Check, p. 449 Graph Check, pp. 442, 445, 447 Section Review, p. 451
SECTION 2 **The Ideal Gas Law** **1.** How does Avogadro's principle relate the number of particles of gas to the gas's volume? **2.** How is the amount of gas present related to its pressure, temperature, and volume by the ideal gas law? **3.** What are the properties of real gases and of ideal gases? 2 sessions 1 block	**Progress Monitoring** Formative Assessment, pp. 453, 454, 456, 458 Reading Check, pp. 454, 457 Section Review, p. 459
SECTION 3 **Gas Stoichiometry** **1.** What stoichiometric relationships can be determined for gaseous reactants and products from balanced chemical equations? **2.** How are the amounts of gaseous reactants and products in a chemical reaction calculated? 1 session 0.5 block	**Progress Monitoring** Formative Assessment, p. 464 Section Review, p. 464 **Summative Assessment** Chapter Assessment, p. 468 *eAssessment* Chapter Test (scaffolded)

LEVELED RESOURCES	LAB MATERIALS
Science Notebook 13.1 OL ***FAST FILE* Chapter Resources:** Study Guide, p. 18 BL	**LaunchLAB** p. 440 / **15 min** bucket, ice, round balloons, stirring rod, water, string, paper towels
Science Notebook 13.2 OL ***FAST FILE* Chapter Resources:** MiniLab Worksheet, p. 2 OL ChemLab Worksheet, p. 3 OL Study Guide, p. 19 BL	**MiniLAB** p. 457 / **20 min** thermometer, barometer or weather radio, candle, masking tape, matches, baking soda ($NaHCO_3$), vinegar (5% CH_3COOH), beaker, aluminum foil **ChemLAB** p. 466 / **45 min** popcorn kernels, 10-mL graduated cylinder, vegetable oil, 250-mL beaker, wire gauze squares, beaker tongs, Bunsen burner, balance, ring stand, distilled water, small iron ring, paper towels.
Science Notebook 13.3 OL ***FAST FILE* Chapter Resources:** Study Guide, p. 23 BL	

eAssessment
ConnectED Student Center
ConnectED Teacher Center
Lesson Presentations
eSolutions
Videos
Animations
Online Quizzes
Virtual Investigations
Multilingual eGlossary
Personal Tutors
WebQuests
Applying Practices Worksheets
PBLs

Practice:

Challenge Problems AL
Supplemental Problems BL OL
Chapter Test (Scaffolded)
Standardized Test Practice

Additional Resources:

Blueprints for Success
Solving Problems: A Chemistry Handbook

Lab Resources:

Laboratory Manual OL
CBL Laboratory Manual OL
Small-Scale Laboratory Manual OL
Forensics Laboratory Manual OL

BL Below Level   OL On Level   AL Advanced Learners   EL English Learners   COOP LEARN Cooperative Learning

CHAPTER 13

LaunchLAB

Rubric

How does temperature affect the volume of a gas?

Purpose Students will demonstrate the relationship between temperature and volume of a gas in a closed system.

Safety Precautions Approve lab safety forms before work begins. **WARNING:** ***Spilled water is a slipping hazard. Some students might be allergic to latex.*** Survey students before the lab and find an alternative to latex for students with the allergy. Remind students to wear goggles and keep the balloons in their possession.

Teaching Strategies

- Supply enough ice for each student to make an ice bath.
- Remind students to measure around the widest part of the balloon. Students can use a marker to note where the initial circumference was measured.
- The circumference of the cold balloon must be measured as soon as it is taken from the bucket.

CHAPTER 13

Gases

BIGIDEA Gases respond in predictable ways to changes in pressure, temperature, volume, and number of particles.

SECTIONS

1 The Gas Laws

2 The Ideal Gas Law

3 Gas Stoichiometry

LaunchLAB

How does temperature affect the volume of a gas?

In a hot-air balloon, the burners raise the temperature of the air inside the balloon to keep it aloft. In this lab, you will examine how temperature changes affect the volume of a balloon.

FOLDABLES® Study Organizer

The Gas Laws

Make a layered-look book. Label it with the gas laws. Summarize your notes on each tab.

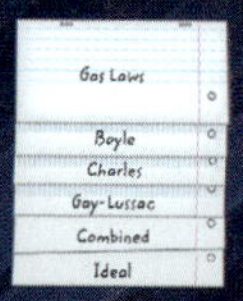

Procedure

1. Read and complete the lab safety form.
2. Inflate a **round balloon,** and tie it closed.
3. Pour cold **water** into a **bucket** until it is half full, then add **ice.** Use **paper towels** to wipe up any spilled water.
4. Use **string** to measure the circumference of the balloon.
5. Use a **stirring rod** to stir the water in the bucket to equalize the temperature. Submerge the balloon in the ice water for 15 min.
6. Remove the balloon from the water. Measure the circumference again.

Analysis

1. **Describe** what happened to the size of the balloon when its temperature decreased. The size of the balloon decreased when the temperature decreased.
2. **Predict** what might happen to the balloon's size if the bucket contained warm water. The size of the balloon will increase if the temperature increases.

Inquiry What do you think would happen if you filled the balloon with helium instead of air and repeated the experiment? For an ideal gas, like helium, the size of the balloon will decrease when the temperature is decreased and the size of the balloon will increase if the temperature is increased. As a follow-up question, you could ask how the temperature might affect how fast a helium-filled balloon would rise. It would rise faster at higher temperatures and slower at lower temperatures.

In the nineteenth century, Joseph Gay-Lussac used hot air balloon flights for research and experimentation. His work, along with that of other scientists, contributed to the development of the gas laws.

Introduce the BIGIDEA

Predictable Responses To introduce this chapter's Big Idea, lead students in a discussion about things that happen in a predictable way. Ask students what happens when you step on a car's gas pedal. The car accelerates. Ask students what happens when you change the pressure applied to the gas pedal. If you press lightly, the car accelerates slowly. If you press down harder, the car accelerates faster. Another way to demonstrate this concept would be to place a thermometer in hot and cold water and have students predict what will happen to the temperature reading on the thermometer. The temperature reading will rise in the hot water and decrease in the cold water.

Tie to Previous Knowledge

Have students review the following concepts before studying this chapter.

- density
- balancing equations
- stoichiometry of reactions
- kinetic-molecular theory
- gas pressure

Use the Photo

Hot Air Balloons Ask students to predict why a hot air balloon needs a burner to stay aloft. The burner is used to heat the gas in the balloon. As temperature increases, both the volume and the pressure of the gas increase, enabling the balloon to stay aloft. In Section 1, students will read about Jacques Charles and Joseph Gay-Lussac, whose work will help students answer this question.

SECTION 1

1 Focus

MAINIDEA

Change in Variables Refer students to this section's Launch Lab and ask them to recall what happened to the balloon when it was placed in cold water. It decreased in size. Ask if the change in temperature affected the volume of the gas in the balloon and how they know. Yes, because the size of the balloon decreased, the volume of the gas decreased. Ask if the change in temperature affected the pressure of the gas in the balloon. Have them explain their answers. Yes. Because the sides of the balloon were less firm, the pressure on the gas decreased. BL OL AL

2 Teach

Concept Development

Gas Remind students that most of a gas is empty space. The volume of a sample of gas is much greater than the total volume of its particles. BL

ASSESSMENT

Performance Ask students to use a beaker, a graduated cylinder, a piece of rubber tubing, and water to design a procedure that can be used to measure the volume of gas in a balloon. One way students could measure the volume is by filling the cylinder with water, inverting it into a beaker full of water, releasing the air from the balloon into the cylinder using the tubing, and then recording the volume of water displaced in the cylinder. OL

☑ GRAPH CHECK FIG. 1 4 L

SECTION 1

The Gas Laws

MAINIDEA For a fixed amount of gas, a change in one variable—pressure, temperature, or volume—affects the other two.

Essential Questions

- What are the relationships between pressure, temperature, and volume of a constant amount of gas?
- How can you use the gas laws to solve problems involving the pressure, temperature, and volume of a constant amount of gas?

Review Vocabulary

scientific law: describes a relationship in nature that is supported by many experiments

New Vocabulary

Boyle's law
absolute zero
Charles's law
Gay-Lussac's law
combined gas law

CHEM 4 YOU What might happen to the gas in a balloon if you decreased its volume by squeezing it? You would feel increasing resistance as you squeeze and might see part of the balloon bulge.

Boyle's Law

As the balloon example illustrates, the pressure of a gas and its volume are related. Robert Boyle (1627–1691), an Irish chemist, described this relationship between the pressure and the volume of a gas.

How are pressure and volume related? Boyle designed experiments like the one shown in **Figure 1.** He showed that if the temperature and the amount of gas are constant, doubling the pressure decreases the volume by one-half. On the other hand, reducing the pressure by one-half doubles the volume. A relationship in which one variable increases proportionally as the other variable decreases is known as an inversely proportional relationship.

Boyle's law states that the volume of a fixed amount of gas held at a constant temperature varies inversely with the pressure. Look at the graph in **Figure 1,** in which volume versus pressure is plotted for a gas. The plot of an inversely proportional relationship results in a downward curve.

■ **Figure 1** As the external pressure on the cylinder's piston increases, the volume inside the cylinder decreases. The graph shows the inverse relationship between pressure and volume.

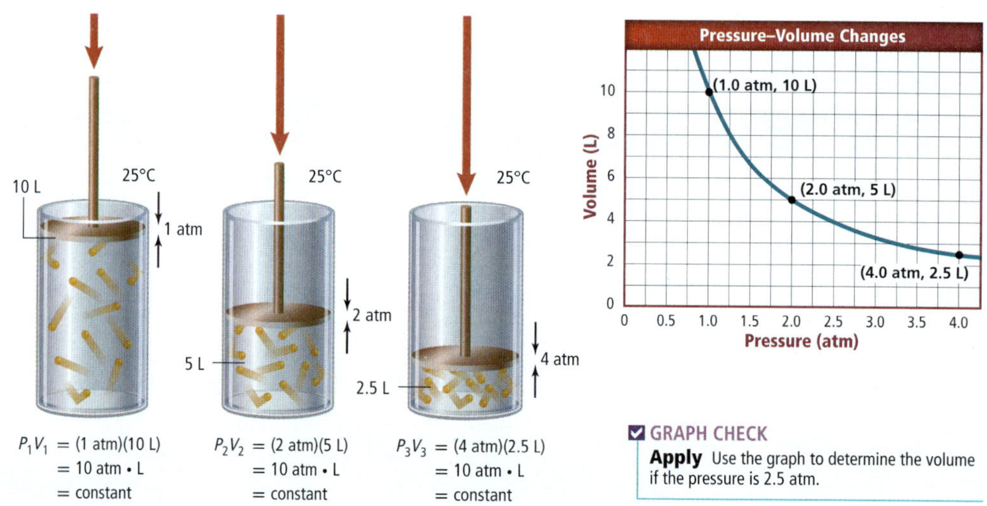

☑ GRAPH CHECK

Apply Use the graph to determine the volume if the pressure is 2.5 atm.

Demonstration

Demonstrate Boyle's Law

Purpose

to demonstrate Boyle's law

Materials

Empty, colorless, 2-L soft drink bottle with cap; dropper; food coloring

Safety Precautions

Disposal Bottles and droppers can be reused or recycled. Rinse water and dye down a drain.

Procedure

Remove the label from an empty, colorless, 2-L soft drink bottle. Add water up to 4 cm from the top. Fill a dropper halfway with water colored with vegetable dye. Put the dropper into the 2-L bottle, and screw the cap on tightly. Tell students to observe closely.

Squeeze the sides of the bottle firmly until the dropper descends. Then, stop squeezing the bottle.

Note that the product of the pressure and the volume for each point in **Figure 1** is 10 atm·L. Boyle's law can be expressed mathematically as follows.

Boyle's Law

$$P_1V_1 = P_2V_2$$ ***P* represents pressure. *V* represents volume.**

For a given amount of gas held at constant temperature, the product of pressure and volume is a constant.

P_1 and V_1 represent the initial conditions, and P_2 and V_2 represent new conditions. If you know any three of these values, you can solve for the fourth by rearranging the equation.

EXAMPLE Problem 1

Find help with **inverse relationships.** Math Handbook

BOYLE'S LAW A diver blows a 0.75-L air bubble 10 m under water. As it rises to the surface, the pressure goes from 2.25 atm to 1.03 atm. What will be the volume of air in the bubble at the surface?

1 ANALYZE THE PROBLEM

According to Boyle's law, the decrease in pressure on the bubble will result in an increase in volume, so the initial volume should be multiplied by a pressure ratio greater than 1.

Known	Unknown
$V_1 = 0.75$ L	$V_2 = ?$ L
$P_1 = 2.25$ atm	
$P_2 = 1.03$ atm	

2 SOLVE FOR THE UNKNOWN

Use Boyle's law. Solve for V_2, and calculate the new volume.

$P_1V_1 = P_2V_2$ — **State Boyle's law.**

$V_2 = V_1\left(\frac{P_1}{P_2}\right)$ — **Solve for V_2.**

$V_2 = 0.75\text{ L}\left(\frac{2.25\text{ atm}}{1.03\text{ atm}}\right)$ — **Substitute $V_1 = 0.75$ L, $P_1 = 2.25$ atm, and $P_2 = 1.03$ atm.**

$V_2 = 0.75\text{ L}\left(\frac{2.25\text{ }\cancel{\text{atm}}}{1.03\text{ }\cancel{\text{atm}}}\right) = 1.6\text{ L}$ — **Multiply and divide numbers and units.**

3 EVALUATE THE ANSWER

The pressure decreases by roughly half, so the volume should roughly double. The answer is expressed in liters, a unit of volume, and correctly contains two significant figures.

EXAMPLE PROBLEM

PRACTICE Problems

Do additional problems. Online Practice

Assume that the temperature and the amount of gas are constant in the following problems.

1. The volume of a gas at 99.0 kPa is 300.0 mL. If the pressure is increased to 188 kPa, what will be the new volume?
2. The pressure of a sample of helium in a 1.00-L container is 0.988 atm. What is the new pressure if the sample is placed in a 2.00-L container?
3. **Challenge** Air trapped in a cylinder fitted with a piston occupies 145.7 mL at 1.08 atm pressure. What is the new volume when the piston is depressed, increasing the pressure by 25%?

PRACTICE PROBLEMS

Enrichment

Properties of Gases Ask groups of three or four students to research the properties of an element that is a gas at normal atmospheric pressure and room temperature. Have student groups prepare classroom presentations about their findings. OL COOP LEARN

IN-CLASS Example

Question A sample of helium gas in a balloon is compressed from 4.0 L to 2.5 L at a constant temperature. If the pressure of the gas in the 4.0-L volume is 210 kPa, what will the pressure be at 2.5 L?

Answer 340 kPa

$P_2 = 210\text{ kPa}\left(\frac{4.0\cancel{L}}{2.5\cancel{L}}\right)$

$= 340\text{ kPa}$

PRACTICE Problems

Have students refer to the Solutions to Selected Practice Problems appendix for solutions to odd-numbered problems. The solutions for all problems can be found in the Solutions Manual.

1. 158 mL
2. 0.494 atm
3. 117 mL

Results

The level of the colored water in the dropper rises, the dropper becomes more dense, and it descends.

Analysis

Ask these questions.

1. As the pressure is increased on the air in the dropper, does the volume increase or decrease? The volume decreases.
2. As the pressure is decreased on the air in the dropper, does the volume increase or decrease? The volume increases.
3. Is the relationship between gas volume and pressure an inverse or direct relationship? inverse relationship

ASSESSMENT

Knowledge Have students write a statement of Boyle's law. The pressure and volume of a confined gas at constant temperature are inversely proportional. OL

Problem-Solving Lab

Purpose Students will apply Boyle's law to lung function in the human body.

Process Skills recognize cause and effect, analyze and interpret illustrations, apply concepts

Teaching Strategies

- Use a plastic syringe without a needle to demonstrate lung function. With the plunger depressed completely and a hand under, but not in contact with, the end of the barrel, pull the stopper up and have students describe what they feel. Next, depress the plunger and have students describe what they feel.
- Have students compare parts of the syringe to components of the lungs, and describe how volume changes inside the syringe resulted in pressure changes and air movement.

Think Critically

1. Boyle's Law states that at constant temperature the volume of a constant amount of gas varies inversely with pressure. When you inhale, lung volume increases. Pressure in the lungs decreases and air moves in. When you exhale, lung volume decreases. Pressure in the lungs increases and air moves out.
2. When someone is hit in the abdomen, the diaphragm is temporarily paralyzed. When the diaphragm does not move up and down, lung volume does not change. If the volume does not change, pressure inside the lungs does not change, and air is not pushed into and out of the lungs.
3. If parts of the lungs lose elasticity, it is more difficult to change lung volume. With less change in volume, there is less difference in pressure. With less difference in pressure, it is more difficult for air to enter or exit the lungs.
4. As a scuba diver ascends toward the water's surface, pressure decreases. A decrease in pressure results in an increase in volume. If a diver holds his or her breath while ascending, air volume in the lungs will increase.

Problem-Solving LAB

Apply Scientific Explanations

What does Boyle's law have to do with breathing? You take a breath about 20 times per minute, exchanging carbon dioxide gas for life-sustaining oxygen. How do pressure and volume change in your lungs as you breathe?

Analysis

The spongy, elastic tissue that makes up your lungs allows them to expand and contract in response to movement of the diaphragm, a strong muscle beneath the lungs. As your diaphragm moves downward, increasing lung volume, you inhale. As your diaphragm moves upward, decreasing lung volume, you exhale.

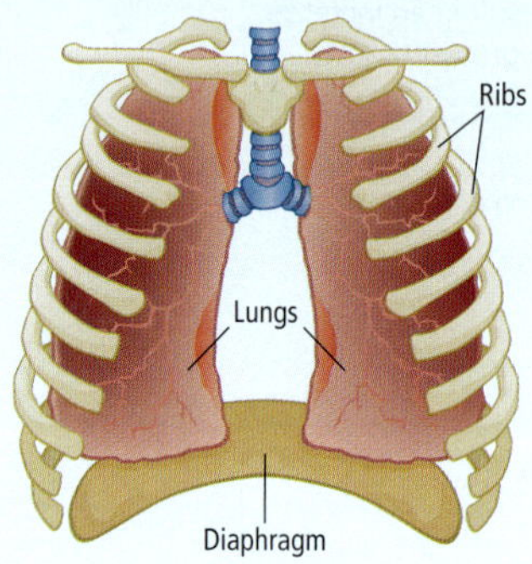

Think Critically

1. **Apply** Boyle's law to explain why air enters your lungs when you inhale and leaves when you exhale.
2. **Explain** what happens inside the lungs when a blow to the abdomen knocks the wind out of a person. Use Boyle's law to determine your answer.
3. **Infer** Parts of the lungs lose elasticity and become enlarged when a person has emphysema. From what you know about Boyle's law, why does this condition affect breathing?
4. **Explain** why beginning scuba divers are taught never to hold their breath while ascending from deep water.

Charles's Law

In the Launch Lab, you observed that a balloon's circumference decreased after the balloon was submerged in ice water. Why did this happen? After a cool evening, a rubber pool raft can appear partially inflated. During a sunny afternoon, the same raft can appear fully inflated. Why did the appearance of the raft change? These questions can be answered by applying a second gas law—Charles's law.

How are temperature and volume related? Jacques Charles (1746–1823), a French physicist, studied the relationship between volume and temperature. He observed that as temperature increases, so does the volume of a gas sample when the amount of gas and the pressure remain constant. This property is explained by the kinetic-molecular theory: as temperature increases, gas particles move faster, striking the walls of their container more frequently and with greater force. Because pressure depends on the frequency and force with which gas particles strike the walls of their container, this would increase the pressure. For the pressure to stay constant, volume must increase so that the particles have farther to travel before striking the walls. Having to travel farther decreases the frequency with which the particles strike the walls of the container.

The cylinders in **Figure 2** show how the volume of a fixed amount of gas changes as the gas is heated. Unlike **Figure 1**, where pressure in addition to that of the atmosphere was applied to the piston, the piston in **Figure 2** is free to float. This means that the piston will be supported by the gas inside the cylinder at a level where the pressure of the gas exactly matches that of the atmosphere. As you can see, the volume occupied by a gas at 1 atm increases as the temperature in the cylinder increases. The distance the piston moves is a measure of the increase in volume of the gas as it is heated.

Graphing the relationship of temperature and volume **Figure 2** also shows graphs of the relationship between the temperature and the volume of a fixed amount of gas at constant pressure. The plot of volume versus temperature is a straight line. Note that you can predict the temperature at which the volume will reach 0 L by extrapolating the line to temperatures below the values that were measured.

In the first graph, the temperature that corresponds to 0 L is −273.15°C. This relationship is linear, but it is not a direct proportion. For example, you can see that the graph of the line does not pass through the origin and that doubling the temperature from 25°C to 50°C does not double the volume.

CHEMISTRY JOURNAL

Pressure Relationships Have students write the following questions in their chemistry journals and answer them as soon as they have learned enough from their study of Section 1.

1. Why do breads and muffins rise when baked? The yeast or baking powder in the dough produces carbon dioxide gas, which expands as it is heated.
2. Why do helium balloons rise? Helium is less dense than air.
3. Why do tennis balls come in pressurized cans? Tennis balls contain trapped gases that are inserted under pressure, which gives them greater bounce. Once the balls are out of the pressurized cans, the gas gradually escapes. OL

■ **Figure 2** When the cylinder is heated, the kinetic energy of the gas particles increases, causing them to push the piston outward. The graphs show the relationship of volume to Celsius and Kelvin temperature.

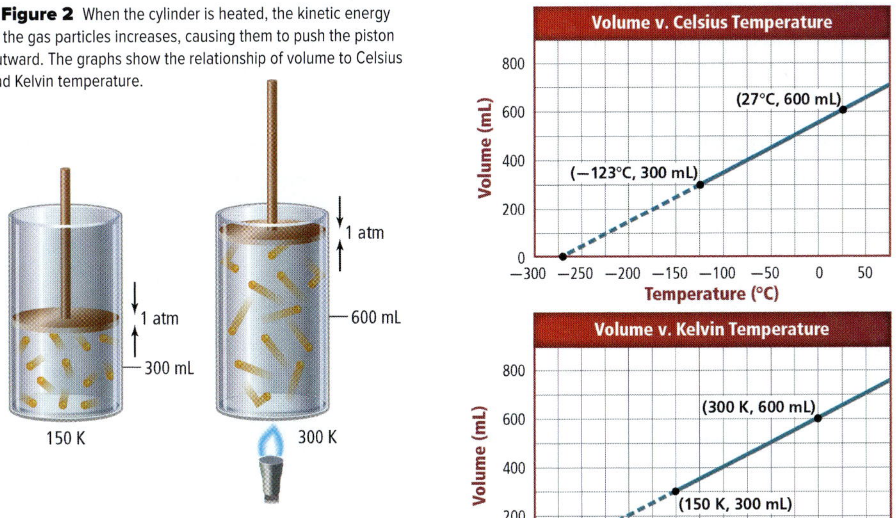

$\frac{V_1}{T_1} = \frac{300 \text{ mL}}{150 \text{ K}}$
$= 2$ mL/K
= constant

$\frac{V_2}{T_2} = \frac{600 \text{ mL}}{300 \text{ K}}$
$= 2$ mL/K
= constant

The second graph in **Figure 2,** which plots the Kelvin (K) temperature against volume, does show a direct proportion. A temperature of 0 K corresponds to 0 mL, and doubling the temperature doubles the volume. Zero on the Kelvin scale is also known as **absolute zero.** Absolute zero represents the lowest possible theoretical temperature. At absolute zero, the atoms are all in the lowest possible energy state.

☑ GRAPH CHECK **Explain** why the second graph in **Figure 2** shows a direct proportion, but the first graph does not.

Using Charles's law **Charles's law** states that the volume of a given amount of gas is directly proportional to its Kelvin temperature at constant pressure. Charles's law can be expressed as follows.

FOLDABLES® Incorporate information from this section into your Foldable.

Charles's Law

$$\frac{V_1}{T_1} = \frac{V_2}{T_2}$$

V represents volume.
T represents temperature.

For a given amount of gas at constant pressure, the quotient of the volume and Kelvin temperature is a constant.

In the equation above, V_1 and T_1 represent initial conditions, while V_2 and T_2 are new conditions. As with Boyle's law, if you know three of the values, you can calculate the fourth.

The temperature must be expressed in kelvins when using the equation for Charles's law. To convert a temperature from Celsius degrees to kelvins, add 273 to the Celsius temperature:

$$T_K = 273 + T_C$$

DIFFERENTIATED INSTRUCTION

Struggling Learners Some students will find solving gas law problems easier if they first predict what their answers will be, then solve the problem, and finally compare the calculated answers with the predictions. For example, if students are using Boyle's law to solve for the final pressure of a gas after being given the initial pressure and the initial and final volumes, ask them whether they think the pressure will increase or decrease *before* they try to do the calculation. If their predictions do not match their answers, they know to double check their calculations. BL

■ **In-Text Question p. 444** In the Launch Lab, the ice water caused the temperature of the gas inside the balloon to decrease, which decreased the volume. In the raft, the sun warmed the gas inside the raft, increasing the volume.

Content Background

Industrial Production of Gases

Five of the top ten industrial chemicals in the United States are gases–oxygen, nitrogen, ethylene, propylene, and chlorine. The greatest industrial use for oxygen is in the steel industry, where it reacts with impurities in crude iron. Nitrogen is extracted from air and then reacted with hydrogen to form ammonia, which is used in fertilizers and explosives. In a process called cracking, ethylene and propylene are produced by heating petroleum to approximately 500°C in the presence of a catalyst. Both gases are used in making plastics, particularly polyethylene and polypropylene. Chlorine is formed by the electrolysis of table salt (NaCl).

ASSESSMENT

Knowledge Ask each student to write a question about either Boyle's law or Charles's law. Have students exchange questions and quiz each other. OL

☑ GRAPH CHECK In the second graph, 0 K corresponds to 0 mL, and doubling the temperature doubles the volume. In the first graph, 0°C does not correspond to 0 mL, and doubling the temperature does not double the volume.

FOLDABLES® Rubric

IN-CLASS Example

Question What is the volume of the air in a balloon that occupies 0.620 L at 25°C if the temperature is lowered to 0.00°C?

Answer 0.57 L

$T_1 = 25°C + 273 = 298\ K$

$T_2 = 0.00°C + 273 = 273\ K$

$$V_2 = \frac{V_1T_2}{T_1} = \frac{(0.620\ L)(273\ \cancel{K})}{298\ \cancel{K}} = 0.57\ L$$

PRACTICE Problems

Have students refer to the Solutions to Selected Practice Problems appendix for complete solutions to odd-numbered problems. The complete solutions for all problems can be found in the Solutions Manual.

4. 3.1 L
5. 330°C
6. 2.58 L
7. 190 K

Reinforcement

Baking Ask students why baking directions are different for high altitudes. The atmospheric pressure at high altitudes is less than it is at low altitudes, which makes the water in foods boil at a lower temperature. As a result, cooking time must be increased or oven temperatures must be increased. OL BL

EXAMPLE Problem 2

Find help with **significant digits.** Math Handbook

CHARLES'S LAW A helium balloon in a closed car occupies a volume of 2.32 L at 40.0°C. If the car is parked on a hot day and the temperature inside rises to 75.0°C, what is the new volume of the balloon, assuming the pressure remains constant?

1 ANALYZE THE PROBLEM

Charles's law states that as the temperature of a fixed amount of gas increases, so does its volume, assuming constant pressure. Therefore, the volume of the balloon will increase. The initial volume should be multiplied by a temperature ratio greater than 1.

Known
$T_2 = 40.0°C$
$V_1 = 2.32\ L$
$T_2 = 75.0°C$

Unknown
$V_2 = ?\ L$

2 SOLVE FOR THE UNKNOWN

Convert degrees Celsius to kelvins.

$T_K = 273 + T_C$ — **Apply the conversion factor.**

$T_1 = 273 + 40.0°C = 313.0\ K$ — **Substitute $T_1 = 40.0°C$.**

$T_2 = 273 + 75.0°C = 348.0\ K$ — **Substitute $T_2 = 75.0°C$.**

Use Charles's law. Solve for V_2, and substitute the known values into the rearranged equation.

$\frac{V_1}{T_1} = \frac{V_2}{T_2}$ — **State Charles's law.**

$V_2 = V_1\left(\frac{T_2}{T_1}\right)$ — **Solve for V_2.**

$V_2 = 2.32\ L\left(\frac{348.0\ K}{313.0\ K}\right)$ — **Substitute $V_1 = 2.32\ L$, $T_1 = 313.0\ K$, and $T_2 = 348.0\ K$.**

$V_2 = 2.32\ L\left(\frac{348.0\ \cancel{K}}{313.0\ \cancel{K}}\right) = \mathbf{2.58\ L}$ — **Multiply and divide numbers and units.**

3 EVALUATE THE ANSWER

The increase in kelvins is relatively small, so the volume should show a small increase. The unit of the answer is liters, a volume unit, and there are three significant figures.

PRACTICE Problems

Do additional problems. Online Practice

Assume that the pressure and the amount of gas remain constant in the following problems.

4. What volume will the gas in the balloon at right occupy at 250 K?

5. A gas at 89°C occupies a volume of 0.67 L. At what Celsius temperature will the volume increase to 1.12 L?

6. The Celsius temperature of a 3.00-L sample of gas is lowered from 80.0°C to 30.0°C. What will be the resulting volume of this gas?

7. **Challenge** A gas occupies 0.67 L at 350 K. What temperature is required to reduce the volume by 45%?

CHEMISTRY JOURNAL

Practical Gases Ask students to keep a record in their journals of how they use gases as part of their everyday activities. Have them do this every day for two weeks. Ask them to note for each instance whether the gas was compressed. OL

Gay-Lussac's Law

In the Launch Lab, you saw Charles's law in action as the balloon's volume changed in response to temperature. What would have happened if the balloon's shape were rigid? If volume is constant, is there a relationship between temperature and pressure? The answer to that question is found in Gay-Lussac's law.

How are temperature and pressure of a gas related?

Pressure is a direct result of collisions between gas particles and the walls of their container. An increase in temperature increases collision frequency and energy, so raising the temperature should also raise the pressure if the volume is not changed. Joseph Gay-Lussac (1778–1850) found that a direct proportion exists between Kelvin temperature and pressure, as illustrated in **Figure 3.** **Gay-Lussac's law** states that the pressure of a fixed amount of gas varies directly with the Kelvin temperature when the volume remains constant. It can be expressed mathematically as follows.

Gay-Lussac's Law

$$\frac{P_1}{T_1} = \frac{P_2}{T_2}$$

P represents pressure.
T represents temperature.

For a given amount of gas held at constant volume, the quotient of the pressure and the Kelvin temperature is a constant.

As with Boyle's and Charles's laws, if you know any three of the four variables, you can calculate the fourth using this equation. Remember that temperature must be in kelvins whenever it is used in a gas law equation.

CAREERS IN CHEMISTRY

Meteorologist Relationships among pressure, temperature, and volume of air help meteorologists understand and predict the weather. For example, winds and fronts result from pressure changes caused by the uneven heating of Earth's atmosphere by the Sun.

WebQuest

Figure 3 When the cylinder is heated, the kinetic energy of the particles increases, increasing both the frequency and energy of the collisions with the container wall. The volume of the cylinder is fixed, so the pressure exerted by the gas increases.

View an **animation of the gas laws.**

Concepts In Motion

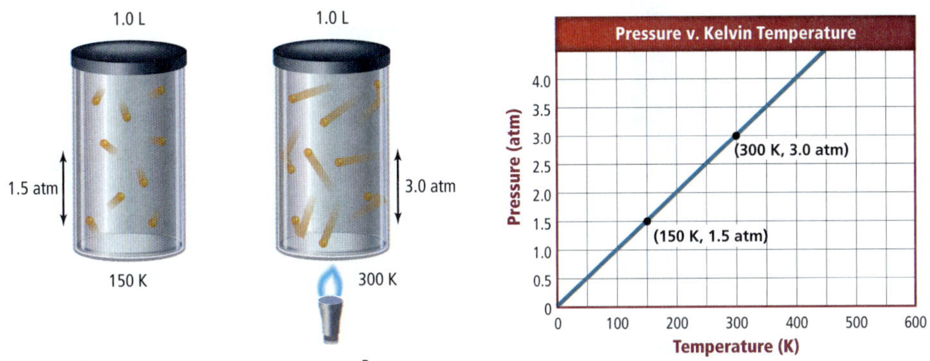

$\frac{P_1}{T_1} = \frac{1.5 \text{ atm}}{150 \text{ K}}$
$= 0.01 \text{ atm/K}$
$=$ constant

$\frac{P_2}{T_2} = \frac{3.0 \text{ atm}}{300 \text{ K}}$
$= 0.01 \text{ atm/K}$
$=$ constant

☑ GRAPH CHECK

Compare and contrast the graphs in **Figures 2** and **3.**

CAREERS IN CHEMISTRY

WebQuest

Identify Misconceptions

Students might think air fills in the space between gas particles.

Uncover the Misconception Have students draw a picture of the inside of a balloon, representing all gas particles as circles. Students should notice both the particles present and the distances between them.

Demonstrate the Concept Have each student use a finger to plug the end of a syringe without a needle. Have them use the plungers to try to compress the air in the syringes. Have them repeat this procedure with water in the syringes. Students will observe the compressibility of air versus that of water. Discuss with students that empty space must exist between gas particles for the air to be compressed.

Assess New Knowledge Ask students to draw an illustration of what happens to the particles in air in the balloon as you squeeze it. The drawings should show particles in the balloon starting out far apart but moving closer together as the balloon is compressed. OL

☑ GRAPH CHECK All three graphs are straight lines. The two graphs that use the temperature in kelvins show direct proportions. The graphs in Figure 2 relate volume to temperature. The graph in Figure 3 relates pressure to temperature.

IN-CLASS Example

Question A gas in a sealed container has a pressure of 125 kPa at a temperature of 30.0°C. If the pressure in the container is increased to 201 kPa, what is the new temperature?

Answer 214°C

$T_1 = 30.0°C + 273 = 303\ K$

$T_2 = \frac{T_1P_2}{P_1} = \frac{(303\ K)(201\ kPa)}{125\ kPa} = 487\ K$

$487\ K - 273 = 214°C$

PRACTICE Problems

Have students refer to the Solutions to Selected Practice Problems appendix for complete solutions to odd-numbered problems. The complete solutions for all problems can be found in the Solutions Manual.

8. 1.96 atm
9. −138°C
10. 273°C

Build a Model

Gas Laws Have students work in groups to build a model of Boyle's law, Charles's law, or Gay-Lussac's law. Ask each group to demonstrate its model to the class. OL COOP LEARN

ASSESSMENT

Knowledge Ask students which gas law they need to use to solve the following problem, then have them solve it. A gas at 20.0°C occupies 1.00 L. Assuming pressure remains constant, what volume will it occupy at 30.0°C? Students need to use Charles's law, which relates the temperature and volume of gases. Answer = 1.03 L OL

RealWorld CHEMISTRY

Gay-Lussac's Law

PRESSURE COOKERS A pressure cooker is a pot with a lid that locks into place. This seals the container, which keeps its volume constant. Heating the pot increases the pressure in the cooker. As pressure increases, the temperature continues to increase and foods cook faster.

EXAMPLE Problem 3

GAY-LUSSAC'S LAW The pressure of the oxygen gas inside a canister is 5.00 atm at 25.0°C. The canister is located at a camp high on Mount Everest. If the temperature there falls to −10.0°C, what is the new pressure inside the canister?

1 ANALYZE THE PROBLEM

Gay-Lussac's law states that if the temperature of a gas decreases, so does its pressure when volume is constant. Therefore, the pressure in the oxygen canister will decrease. The initial pressure should be multiplied by a temperature ratio less than 1.

Known	Unknown
$P_1 = 5.00$ atm	$P_2 = ?$ atm
$T_1 = 25.0°C$	
$T_2 = -10.0°C$	

2 SOLVE FOR THE UNKNOWN

Convert degrees Celsius to kelvins.

$T_K = 273 + T_C$ — Apply the conversion factor.

$T_1 = 273 + 25.0°C = 298.0\ K$ — Substitute $T_1 = 25.0°C$.

$T_2 = 273 + (-10.0°C) = 263.0\ K$ — Substitute $T_2 = -10.0°C$.

Use Gay-Lussac's law. Solve for P_2, and substitute the known values into the rearranged equation.

$\frac{P_1}{T_1} = \frac{P_2}{T_2}$ — State Gay-Lussac's law.

$P_2 = P_1\left(\frac{T_2}{T_1}\right)$ — Solve for P_2.

$P_2 = 5.00\ atm\left(\frac{263.0\ K}{298.0\ K}\right)$ — Substitute $P_1 = 5.00$ atm, $T_1 = 298.0$ K, and $T_2 = 263.0$ K.

$P_2 = 5.00\ atm\left(\frac{263.0\ \cancel{K}}{298.0\ \cancel{K}}\right) = 4.41\ atm$ — Multiply and divide numbers and units.

3 EVALUATE THE ANSWER

Kelvin temperature decreases, so the pressure should decrease. The unit is atm, a pressure unit, and there are three significant figures.

PRACTICE Problems

Do additional problems. Online Practice

Assume that the volume and the amount of gas are constant in the following problems.

8. The pressure in an automobile tire is 1.88 atm at 25.0°C. What will be the pressure if the temperature increases to 37.0°C?

9. Helium gas in a 2.00-L cylinder is under 1.12 atm pressure. At 36.5°C, that same gas sample has a pressure of 2.56 atm. What was the initial temperature in degrees Celsius of the gas in the cylinder?

10. Challenge If a gas sample has a pressure of 30.7 kPa at 0.00°C, by how many degrees Celsius does the temperature have to increase to cause the pressure to double?

CHEMISTRY PROJECT

Weather and Atmospheric Pressure Ask students to log for two weeks both the daily barometric pressure and the weather conditions. Ask them to correlate weather conditions with barometric pressure levels and to summarize their correlations. Have students share results and discuss how barometric pressure is involved in weather prediction. OL

Figure 4 Tethers attached at the sides of a weather balloon hold it in place while it is being filled with helium or hydrogen gas. Weather balloons carry instruments that send data, such as air temperature, pressure, and humidity, to receivers on the ground. As the balloon rises, its volume responds to changes in temperature and pressure, expanding until the sides burst. A small parachute returns the instruments to Earth.

The Combined Gas Law

In a number of applications involving gases, such as the weather balloon in **Figure 4,** pressure, temperature, and volume might all change. Boyle's, Charles's, and Gay-Lussac's laws can be combined into a single law. This **combined gas law** states the relationships between pressure, temperature, and volume of a fixed amount of gas. All three variables have the same relationship to each other as they have in the other gas laws: pressure is inversely proportional to volume and directly proportional to temperature, and volume is directly proportional to temperature. The combined gas law can be expressed mathematically as follows.

Explore the **gas laws.**

Virtual Investigations

The Combined Gas Law

$$\frac{P_1V_1}{T_1} = \frac{P_2V_2}{T_2}$$

P represents pressure. *V* represents volume. *T* represents temperature.

For a given amount of gas, the product of pressure and volume, divided by the Kelvin temperature, is a constant.

Using the combined gas law The combined gas law enables you to solve problems involving changes in more than one variable. It also provides a way for you to remember the other three laws without memorizing each equation. If you can write out the combined gas law equation, equations for the other laws can be derived from it by remembering which variable is held constant in each case.

For example, if temperature remains constant as pressure and volume vary, then $T_1 = T_2$. After simplifying the combined gas law under these conditions, you are left with $P_1V_1 = P_2V_2$, which you should recognize as the equation for Boyle's law.

Get help with the **combined gas law.**

Personal Tutor

☑ READING CHECK **Derive** Charles's and Gay-Lussac's laws from the combined gas law.

Apply Chemistry

Tire Pressure Inflate a bicycle tire to the recommended pressure, and ask students to predict what will happen to the pressure as the temperature changes. Let them propose a method for testing their predictions and then run the tests. Students can test their predictions by placing the tire at locations with different temperatures and then measuring the pressure using a gauge. The pressure will increase as the temperature increases and decrease as the temperature decreases. OL EL

Quick Demo

Mass Record the mass of approximately 100 popcorn kernels. Pop the kernels in an air popper, and then record the mass of the resulting popcorn and unpopped kernels. Ask the students to explain how the difference in mass relates to the gas laws. Students will have an opportunity to do a similar activity with the ChemLab at a later time. The popcorn loses mass because water vapor escapes as the kernels break open during popping. Water trapped in the kernels vaporizes as it is heated; the pressure and volume of the trapped gases increase as more heat is added, resulting in the popping observed. OL

☑ READING CHECK The combined gas law can be expressed as $P_1V_1/T_1 = P_2V_2/T_2$. For Charles's Law, the pressure is constant, so $P_1 = P_2$. Divide both sides of the combined gas law by *P* to get Charles's Law ($V_1/T_1 = V_2/T_2$). For Gay-Lussac's Law, the volume is constant, so $V_1 = V_2$. Divide both sides of the combined gas law by *V* to get Gay-Lussac's Law ($P_1/T_1 = P_2/T_2$).

DIFFERENTIATED INSTRUCTION

Hearing Impaired Students who learn visually will be able to understand the combined gas law better if they "discover" it themselves, step-by-step. Write the symbols P_1, V_1, T_1, P_2, V_2, T_2, and "=" on index cards, and let students arrange the cards to form Boyle's, Charles's, and Gay-Lussac's laws. Then, ask them to arrange all of the cards into an equation that still keeps the relationships in all three gas laws correctly set up as direct or inverse proportions. OL COOP LEARN

IN-CLASS Example

Question An unopened, cold 2.00-L bottle of soda contains 46.0 mL of gas confined at a pressure of 1.30 atm at a temperature of 5.0°C. If the bottle is dropped into a lake and sinks to a depth at which the pressure is 1.52 atm and the temperature is 2.09°C, what will be the volume of the gas in the bottle?

Answer 39 mL

$T_1 = 5.0°C + 273 = 278.0\ K$

$T_2 = 2.09°C + 273 = 275.09\ K$

$$V_2 = \frac{P_1T_2V_1}{P_2T_1} = \frac{(1.30\ \cancel{atm})(275.09\ \cancel{K})(46.0\ mL)}{(1.52\ \cancel{atm})(278.0\ \cancel{K})} = 39\ mL$$

PRACTICE Problems

Have students refer to the Solutions to Selected Practice Problems appendix for complete solutions to odd-numbered problems. The complete solutions for all problems can be found in the Solutions Manual.

11. 0.214 mL
12. 72 mL
13. down

Use Science Terms

Gas Law Terms Have students write statements explaining the meanings of the terms *molar volume* and *combined gas law.* **BL** **EL**

Reinforcement

Sharing Solutions Have students work in groups to solve different combined gas law problems and ask each group to present its solution to the rest of the class. **OL** **COOP LEARN**

EXAMPLE Problem 4

THE COMBINED GAS LAW A gas at 110 kPa and 30.0°C fills a flexible container with an initial volume of 2.00 L. If the temperature is raised to 80.0°C and the pressure increases to 440 kPa, what is the new volume?

1 ANALYZE THE PROBLEM

Both pressure and temperature change, so you will need to use the combined gas law. The pressure quadruples, but the temperature does not increase by such a large factor. Therefore, the new volume will be smaller than the starting volume.

Known
$P_1 = 110$ kPa $P_2 = 440$ kPa
$T_1 = 30.0°C$ $T_2 = 80.0°C$
$V_1 = 2.00$ L

Unknown
$V_2 = ?$ L

2 SOLVE FOR THE UNKNOWN

Convert degrees Celsius to kelvins.

$T_K = 273 + T_C$ — **Apply the conversion factor.**

$T_1 = 273 + 30.0°C = 303.0\ K$ — **Substitute $T_1 = 30.0°C$.**

$T_2 = 273 + 80.0°C = 353.0\ K$ — **Substitute $T_2 = 80.0°C$.**

Use the combined gas law. Solve for V_2, and substitute the known values into the rearranged equation.

$\frac{P_1V_1}{T_1} = \frac{P_2V_2}{T_2}$ — **State the combined gas law.**

$V_2 = V_1\left(\frac{P_1}{P_2}\right)\left(\frac{T_2}{T_1}\right)$ — **Solve for V_2.**

$V_2 = 2.00\ L\left(\frac{110\ kPa}{440\ kPa}\right)\left(\frac{353.0\ K}{303.0\ K}\right)$ — **Substitute $V_1 = 2.00$ L, $P_1 = 110$ kPa, $P_2 = 440$ kPa, $T_2 = 353.0$ K, and $T_1 = 303.0$ K.**

$V_2 = 2.00\ L\left(\frac{110\ \cancel{kPa}}{440\ \cancel{kPa}}\right)\left(\frac{353.0\ \cancel{K}}{303.0\ \cancel{K}}\right) = 0.58\ L$ — **Multiply and divide numbers and units.**

3 EVALUATE THE ANSWER

Because the pressure change is much greater than the temperature change, the volume undergoes a net decrease. The unit is liters, a volume unit, and there are two significant figures.

PRACTICE Problems

Do additional problems. Online Practice

Assume that the amount of gas is constant in the following problems.

11. A sample of air in a syringe exerts a pressure of 1.02 atm at 22.0°C. The syringe is placed in a boiling-water bath at 100.0°C. The pressure is increased to 1.23 atm by pushing the plunger in, which reduces the volume to 0.224 mL. What was the initial volume?

12. A balloon contains 146.0 mL of gas confined at a pressure of 1.30 atm and a temperature of 5.0°C. If the pressure doubles and the temperature decreases to 2.0°C, what will be the volume of gas in the balloon?

13. Challenge If the temperature in the gas cylinder at right increases to 30.0°C and the pressure increases to 1.20 atm, will the cylinder's piston move up or down?

DIFFERENTIATED INSTRUCTION

Advanced Learners Ask interested students to use the following equipment to design an apparatus to test predictions made using the combined gas law: inflatable spherical ball, pump, tire gauge, thermometer, and tape measure. Ask them to run a test in which they make predictions and then use their apparatus to test them. **AL** **COOP LEARN**

Explore the **gas laws** with an interactive table. **Concepts In Motion**

Table 1 The Gas Laws

Law	Boyle's	Charles's	Gay-Lussac's	Combined
Formula	$P_1V_1 = P_2V_2$	$\frac{V_1}{T_1} = \frac{V_2}{T_2}$	$\frac{P_1}{T_1} = \frac{P_2}{T_2}$	$\frac{P_1V_1}{T_1} = \frac{P_2V_2}{T_2}$
What is constant?	amount of gas, temperature	amount of gas, pressure	amount of gas, volume	amount of gas
Graphic organizer	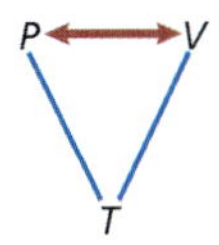	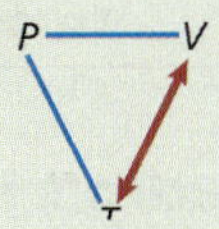	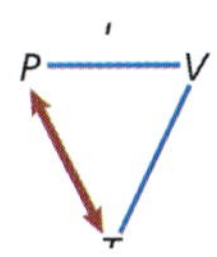	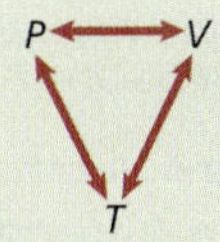

Temperature scales and the gas laws You might have noticed that the work done by Charles and Gay-Lussac preceded the development of the Kelvin scale, yet their laws require the use of temperature in kelvins. In the 1700s and early 1800s, scientists worked with several different scales. For example, a scale called the Réaumur scale was often used in France around Charles's time. On this scale—or any scale not based on absolute zero—the expression for Charles's law is more complex, requiring two constants in addition to *V* and *T*. The Kelvin scale simplified matters, resulting in the familiar gas laws presented here.

You have now seen how pressure, temperature, and volume affect a gas sample. You can use the gas laws, summarized in **Table 1,** as long as the amount of gas remains constant. But what happens if the amount of gas changes? In the next section, you will add the fourth variable, amount of gas present, to the gas laws.

SECTION 1 REVIEW

Section Self-Check

Section Summary

- Boyle's law states that the volume of a fixed amount of gas is inversely proportional to its pressure at constant temperature.
- Charles's law states that the volume of a fixed amount of gas is directly proportional to its Kelvin temperature at constant pressure.
- Gay-Lussac's law states that the pressure of a fixed amount of gas is directly proportional to its Kelvin temperature at constant volume.
- The combined gas law relates pressure, temperature, and volume in a single statement.

14. **MAIN**IDEA **State** the relationships between pressure, temperature, and volume of a fixed amount of gas.
15. **Explain** Which of the three variables that apply to equal amounts of gases are directly proportional? Which are inversely proportional?
16. **Analyze** A weather balloon is released into the atmosphere. You know the initial volume, temperature, and air pressure. What information will you need to predict its volume when it reaches its final altitude? Which law would you use to calculate this volume?
17. **Infer** why gases such as the oxygen used at hospitals are compressed. Why must compressed gases be shielded from high temperatures? What must happen to compressed oxygen before it can be inhaled?
18. **Calculate** A rigid plastic container holds 1.00 L of methane gas at 660 torr pressure when the temperature is 22.0°C. How much pressure will the gas exert if the temperature is raised to 44.6°C?
19. **Design a concept map** that shows the relationships between pressure, volume, and temperature in Boyle's, Charles's, and Gay-Lussac's laws.

ASSESSMENT

Skill Have students determine what additional information is needed to solve this problem. If a sample of gas at 12.0°C and 1.06 atm pressure is moved to a 2.30-L container at 24.9°C, what is the final pressure of the gas? To use the combined gas law to solve this problem, students need to know the initial volume of the gas. **OL**

3 Assess

Check for Understanding

Have students write explanations that compare and contrast direct and inverse relationships. **OL**

Reteach

Ask students to explain why hot-air balloons rise. The heated air is less dense than the surrounding air and thus has less mass per unit volume. **OL**

Extension

Ask students to explain why the standard pressure of 1 atm is not always equal to the atmospheric pressure. Atmospheric pressure varies due to varying altitude and weather conditions. **OL**

SECTION 1 REVIEW

14. This relationship is given by the combined gas law: $P_1V_1/T_1 = P_2V_2/T_2$. For example: when the temperature increases, either the volume or pressure increases (or both).
15. *P* and *V* are directly proportional to *T*, and *P* and *V* are inversely proportional to each other.
16. You would need to know the final temperature and final pressure to calculate the final volume. Use the combined gas law.
17. A greater mass confined to a smaller volume makes transporting and storing of gases easier. Increasing temperature increases pressure, and the cylinders might explode. Before compressed oxygen can be breathed, it must be decompressed.
18. 711 torr
19. The concept map should show how *P*, *V*, and *T* are proportional to one another. Students should also label each pair of variables used in the gas laws. Refer to the Solutions Manual.

SECTION 2

1 Focus

MAINIDEA

The Ideal Gas Law states the interrelationships among the volume, temperature, pressure, and amount of a gas. Ask students to list all of the ways they can think of to increase the volume of a balloon. Possible answers include add more air, increasing the temperature or decreasing the pressure. Ask students how they might be able to demonstrate the effect of decreasing pressure. Place the balloon in a vacuum chamber. **BL** **OL** **AL**

2 Teach

Reinforcement

Numbers of Gas Particles The converse of Avogadro's principle is that equal numbers of gas particles at the same temperature and pressure have the same volume. This principle can be used to reinforce the mole concept because it helps students understand that equal numbers of molecules of different gases can have the same volume, yet different masses. For example, 1 mol, or 6.02×10^{23} molecules, of hydrogen gas at STP has the same volume as 1 mol, or 6.02×10^{23} molecules, of nitrogen gas, but the mass of the hydrogen is 2 g, while the mass of the nitrogen is 28 g.

▪ **Caption Question Fig. 13.5** Even though the particles making up different gases can vary greatly in size, the kinetic-molecular theory states that the particles in a gas are usually far enough apart that the actual size of the particle is not significant in determining the volume of a gas. However, in liquids and solids, molecules are close enough together to make a difference in their volume.

SECTION 2

The Ideal Gas Law

MAINIDEA The ideal gas law relates the number of particles to pressure, temperature, and volume.

Essential Questions

- How does Avogadro's principle relate the number of particles of gas to the gas's volume?
- How is the amount of gas present related to its pressure, temperature, and volume by the ideal gas law?
- What are the properties of real gases and of ideal gases?

Review Vocabulary

mole: an SI base unit used to measure the amount of a substance; the amount of a pure substance that contains 6.02×10^{23} representative particles

New Vocabulary

Avogadro's principle
molar volume
standard temperature and pressure (STP)
ideal gas constant (R)
ideal gas law

CHEM 4 YOU You know that adding air to a tire causes the pressure in the tire to increase. But did you know that the recommended pressure for car tires is specified for cold tires? As tires roll over the road, friction causes their temperatures to increase. This also causes the pressure to increase.

Avogadro's Principle

The particles that make up different gases can vary greatly in size. However, kinetic-molecular theory assumes that the particles in a gas sample are far enough apart that size has very little influence on the volume occupied by a gas. For example, 1000 relatively large krypton gas particles occupy the same volume as 1000 smaller helium gas particles at the same temperature and pressure. It was Avogadro who first proposed this idea in 1811. **Avogadro's principle** states that equal volumes of gases at the same temperature and pressure contain equal numbers of particles. **Figure 5** shows equal volumes of carbon dioxide, helium, and oxygen.

Volume and moles Recall that one mole of a substance contains 6.02×10^{23} particles. The **molar volume** of a gas is the volume that 1 mol occupies at 0.00°C and 1.00 atm pressure. The conditions of 0.00°C and 1.00 atm are known as **standard temperature and pressure (STP).** Avogadro showed experimentally that 1 mol of any gas occupies a volume of 22.4 L at STP. Because the volume of 1 mol of a gas at STP is 22.4 L, you can use 22.4 L/mol as a conversion factor whenever a gas is at STP.

For example, suppose you want to find the number of moles in a sample of gas that has a volume of 3.72 L at STP. Use the molar volume to convert from volume to moles.

$$3.72\ \text{L} \times \frac{1\ \text{mol}}{22.4\ \text{L}} = 0.166\ \text{mol}$$

▪ **Figure 5** Gas tanks of equal volume that are at the same pressure and temperature contain equal numbers of gas particles, regardless of which gas they contain.
Infer *Why doesn't Avogadro's principle apply to liquids and solids?*

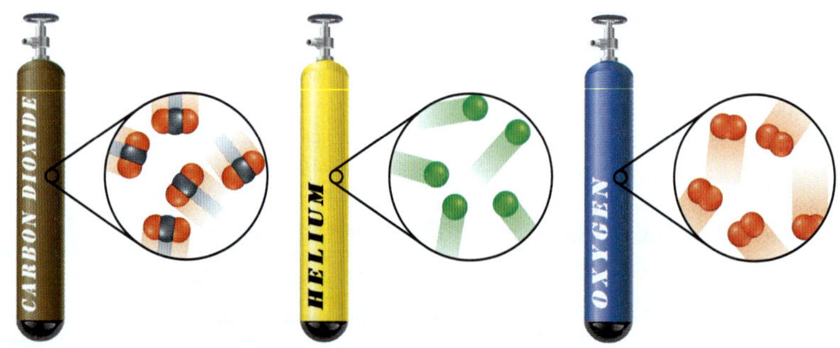

CULTURAL DIVERSITY

High-Altitude Adaptations Many populations of people throughout the world are native to high altitudes, including Andean peoples in South America and Himalayan peoples in Asia. The air at 5400 m at the Mount Everest Base Camp in the Himalayas has a partial pressure of oxygen about one-half that found at sea level. Many mountain climbers must use oxygen tanks on Everest, while their local guides often are less dependent on additional oxygen. For the past 100 years, scientists have been studying the genetic and phenotypical adaptations of these populations to determine how they differ from people adapted to living at lower altitudes. Many native highlanders have high hemoglobin concentrations in their blood, which enables them to carry what little oxygen there is more efficiently. Most people who adapt to a higher altitude before puberty have relatively large chest dimensions, which allow for large lung capacities. Alveoli in their lungs might increase in both number and size

EXAMPLE Problem 5

Find help with unit conversion. Math Handbook

MOLAR VOLUME The main component of natural gas used for home heating and cooking is methane (CH_4). Calculate the volume that 2.00 kg of methane gas will occupy at STP.

1 ANALYZE THE PROBLEM

The number of moles can be calculated by dividing the mass of the sample, *m*, by its molar mass, M. The gas is at STP (0.00°C and 1.00 atm pressure), so you can use the molar volume to convert from the number of moles to the volume.

Known	Unknown
m = 2.00 kg	V = ? L
T = 0.00°C	
P = 1.00 atm	

2 SOLVE FOR THE UNKNOWN

Determine the molar mass for methane.

$$M = 1\text{ C atom}\left(\frac{12.01\text{ amu}}{1\text{ C atom}}\right) + 4\text{ H atoms}\left(\frac{1.01\text{ amu}}{1\text{ H atom}}\right)$$

Determine the molecular mass.

$$= 12.01\text{ amu} + 4.04\text{ amu} = 16.05\text{ amu}$$
$$= 16.05\text{ g/mol}$$

Express the molecular mass as g/mol to arrive at the molar mass.

Determine the number of moles of methane.

$$2.00\text{ kg}\left(\frac{1000\text{ g}}{1\text{ kg}}\right) = 2.00 \times 10^3\text{ g}$$

Convert the mass from kg to g.

$$\frac{m}{M} = \frac{2.00 \times 10^3\text{ g}}{16.05\text{ g/mol}} = 125\text{ mol}$$

Divide mass by molar mass to determine the number of moles.

Use the molar volume to determine the volume of methane at STP.

$$V = 125\text{ mol} \times \frac{22.4\text{ L}}{1\text{ mol}} = 2.80 \times 10^3\text{ L}$$

Use the molar volume, 22.4 L/mol, to convert from moles to the volume.

3 EVALUATE THE ANSWER

The amount of methane present is much more than 1 mol, so you should expect a large volume, which is in agreement with the answer. The unit is liters, a volume unit, and there are three significant figures.

PRACTICE Problems

Do additional problems. Online Practice

20. What size container do you need to hold 0.0459 mol of N_2 gas at STP?

21. How much carbon dioxide gas, in grams, is in a 1.0-L balloon at STP?

22. What volume in milliliters will 0.00922 g of H_2 gas occupy at STP?

23. What volume will 0.416 g of krypton gas occupy at STP?

24. Calculate the volume that 4.5 kg of ethylene gas (C_2H_4) will occupy at STP.

25. **Challenge** A flexible plastic container contains 0.860 g of helium gas in a volume of 19.2 L. If 0.205 g of helium is removed at constant pressure and temperature, what will be the new volume?

when compared with their sea-level counterparts. This knowledge could help in treating the chronic medical conditions often seen in high altitude populations, such as pulmonary hypertension, swelling, and chronic mountain sickness. It might also help determine what medical intervention should be offered to mountain climbers, rescue workers, and other newcomers to high altitudes who experience difficulties.

IN-CLASS Example

Question Give the following scenario to students: 11.2 L of an unknown gas at STP has a mass of 22 g. What is the molecular mass of the unknown gas? First, use the molar volume to calculate the number of moles. Then, calculate the molar mass.

Answer 1 mol/22.4 L × 11.2 L = 0.5 mol
22 g/0.5 mol = 44 g/mol

PRACTICE Problems

Have students refer to the Solutions to Selected Practice Problems appendix for complete solutions to odd-numbered problems. The complete solutions for all problems can be found in the Solutions Manual.

20. 1.03 L
21. 2.0 g
22. 102 mL
23. 0.111 L
24. 3.6×10^3 L
25. 14.6 L

ASSESSMENT

Knowledge Put 1 L of several different gases, such as hydrogen, methane, propane, and nitrogen, in separate balloons and write the chemical symbol for the contained gas on each. If the gases are unavailable, draw the balloons on the board. Ask students whether the volumes of the gases are equal. yes The number of molecules? yes The masses? no The number of moles? yes OL

Quick Demo

Gas and Volume Place 5 mL of water in a balloon. Add a small piece of calcium carbide (CaC_2) to the balloon. Tie off the balloon and ask students to explain what they observe in terms of the variables involved. The volume of the gas in the balloon increases because the amount of gas increases. Temperature and pressure stay constant. OL

ASSESSMENT

Knowledge Have students solve the following problem. A sample of dry gas weighing 2.1025 g is found to occupy 2.850 L at 22.0°C and 740.0 mm Hg. How many moles of gas are present? 0.115 mol OL

READING CHECK Volume and pressure are directly proportional to the number of moles.

FOLDABLES® Rubric

ChemLAB

The ChemLab located at the end of the chapter can be used at this point in the lesson.

■ **Figure 6** The volume and temperature of this tire stay the same as air is added. However, the pressure in the tire increases as the amount of air present increases.

The Ideal Gas Law

Avogadro's principle and the laws of Boyle, Charles, and Gay-Lussac can be combined into a single mathematical statement that describes the relationships between pressure, volume, temperature, and number of moles of a gas. This formula works best for gases that obey the assumptions of the kinetic-molecular theory. Known as ideal gases, their particles occupy a negligible volume and are far enough apart that they exert minimal attractive or repulsive forces on one another.

From the combined gas law to the ideal gas law The combined gas law relates the variables of pressure, volume, and temperature for a given amount of gas.

$$\frac{P_1V_1}{T_1} = \frac{P_2V_2}{T_2}$$

For a specific sample of gas, this relationship of pressure, volume, and temperature is always the same. You could rewrite the relationship represented in the combined gas law as follows.

$$\frac{PV}{T} = \text{constant}$$

As **Figure 6** illustrates, increasing the amount of gas present in a sample will raise the pressure if temperature and volume are constant. Likewise, if pressure and temperature remain constant, the volume will increase as more particles of a gas are added. In fact, we know that both volume and pressure are directly proportional to the number of moles, n, so n can be incorporated into the combined gas law as follows.

$$\frac{PV}{nT} = \text{constant}$$

FOLDABLES® Incorporate information from this section into your Foldable.

Experiments using known values of P, T, V, and n have determined the value of this constant. It is called the **ideal gas constant,** and it is represented by the symbol R. If pressure is in atmospheres, the value of R is 0.0821 L·atm/mol·K. Note that the units for R are simply the combined units for each of the four variables. **Table 2** shows the numerical values for R in different units of pressure.

READING CHECK **Explain** why the number of moles, n, was added to the denominator of the equation above.

Substituting R for the constant in the equation above and rearranging the variables gives the most familiar form of the ideal gas law. The **ideal gas law** describes the physical behavior of an ideal gas in terms of the pressure, volume, temperature, and number of moles of gas present.

Table 2 Values of R

Value of R	Units of R
0.0821	$\frac{L \cdot atm}{mol \cdot K}$
8.314	$\frac{L \cdot kPa}{mol \cdot K}$
62.4	$\frac{L \cdot mm\ Hg}{mol \cdot K}$

The Ideal Gas Law

$$PV = nRT$$

P represents pressure. V represents volume. n represents number of moles. R is the ideal gas constant. T represents temperature.

For a given amount of gas held at constant temperature, the product of pressure and volume is a constant.

If you know any three of the four variables, you can rearrange the equation to solve for the unknown.

CHEMISTRY JOURNAL

History of Chemistry Have students write a summary in their journals about their research on the work of the alchemists of the Middle Ages. Ask them to investigate what the alchemists discovered about the nature of gases. Also, ask them to include a drawing of an early experimental apparatus used by those early chemists. OL

EXAMPLE Problem 6

Find help with **significant figures.** Math Handbook

THE IDEAL GAS LAW Calculate the number of moles of ammonia gas (NH_3) contained in a 3.0-L vessel at 3.00×10^2 K with a pressure of 1.50 atm.

1 ANALYZE THE PROBLEM

You are given the volume, temperature, and pressure of a gas sample. Use the ideal gas law, and select the value of R that contains the pressure units given in the problem. Because the pressure and temperature are close to STP, but the volume is much smaller than 22.4 L, it would make sense if the calculated answer were much smaller than 1 mol.

Known

$V = 3.0$ L

$T = 3.00 \times 10^2$ K

$P = 1.50$ atm

$R = 0.0821 \frac{\text{L·atm}}{\text{mol·K}}$

Unknown

$n = ?$ mol

2 SOLVE FOR THE UNKNOWN

Use the ideal gas law. Solve for n, and substitute the known values.

$PV = nRT$ — **State the ideal gas law.**

$n = \frac{PV}{RT}$ — **Solve for n.**

$$n = \frac{(1.50\ \text{atm})(3.0\ \text{L})}{\left(0.0821\ \frac{\text{L·atm}}{\text{mol·K}}\right)(3.00 \times 10^2\ \text{K})}$$

Substitute $V = 3.0$ L, $T = 3.00 \times 10^2$ K, $P = 1.50$ atm, and $R = 0.0821$ L·atm/mol·K.

$$n = \frac{(1.50\ \cancel{\text{atm}})(3.0\ \cancel{\text{L}})}{\left(0.0821\ \frac{\cancel{\text{L·atm}}}{\text{mol·}\cancel{\text{K}}}\right)(3.00 \times 10^2\ \cancel{\text{K}})} = 0.18\ \text{mol}$$

Multiply and divide numbers and units.

3 EVALUATE THE ANSWER

The answer agrees with the prediction that the number of moles present will be significantly less than 1 mol. The unit of the answer is the mole, and there are two significant figures.

PRACTICE Problems

Do additional problems. Online Practice

26. Determine the Celsius temperature of 2.49 mol of a gas contained in a 1.00-L vessel at a pressure of 143 kPa.

27. Calculate the volume of a 0.323-mol sample of a gas at 265 K and 0.900 atm.

28. What is the pressure, in atmospheres, of a 0.108-mol sample of helium gas at a temperature of 20.0°C if its volume is 0.505 L?

29. If the pressure exerted by a gas at 25°C in a volume of 0.044 L is 3.81 atm, how many moles of gas are present?

30. Challenge An ideal gas has a volume of 3.0 L. If the number of moles of gas and the temperature are doubled, while the pressure remains constant, what is the new volume?

Quick Demo

Moles of a Gas Ask a student volunteer to pump up a deflated ball or tire. As the other students watch, ask how they can calculate the number of moles of gas in the ball or tire. The number of moles, n, equals PV/RT. R is a constant, and they can use a thermometer to measure the air temperature and a tire gauge to measure air pressure in the tire. To find the volume of air, students can either measure the tire or ball and calculate the volume, or, can empty the air into a large, water-filled graduated cylinder that is inverted in a bucket of water and then measure the volume of displaced water. **OL**

IN-CLASS Example

Question Determine the Kelvin temperature required for 0.0140 mol of gas to fill a balloon to 1.20 L under 0.988 atm pressure.

Answer 1030 K

$$T = \frac{PV}{nR} = \frac{(0.988\ \cancel{\text{atm}})(1.20\ \cancel{\text{L}})}{(0.0140\ \cancel{\text{mol}})\left(0.0821\ \frac{\cancel{\text{L·atm}}}{\cancel{\text{mol}}\text{·K}}\right)}$$

$= 1030$ K

PRACTICE Problems

Have students refer to the Solutions to Selected Practice Problems appendix for complete solutions to odd-numbered problems. The complete solutions for all problems can be found in the Solutions Manual.

26. −266°C

27. 7.81 L

28. 5.14 atm

29. 6.9×10^{-3} mol

30. 12 L

CHEMISTRY JOURNAL

Atmospheric Ozone Ask students to write about the journey of a molecule of chlorofluorocarbon (CFC). CFCs include the freons now banned in the United States but still used in many other countries. Students should follow the atoms in the CFC as they travel from an aerosol can into the atmosphere where the chlorine is thought to react with ozone. **OL**

DIFFERENTIATED INSTRUCTION

Advanced Learners Ask students to research how the ideal gas law can work for real gases. The van der Waals equation can be used with constants a and b put in for each gas: $(P + n^2a/V^2)(V - nb) = nRT$. The first part of this equation corrects for attractions between real gas particles, and the second corrects for volume of real gas particles. Note that a and b are constants that are experimentally determined for each particular gas. **AL**

ASSESSMENT

Skill Write the combined gas law in words on the board, and ask students to use it to write the equation for the law. The pressure of a fixed amount of a gas is inversely proportional to its volume and directly proportional to its temperature, and its volume is directly proportional to its temperature. $P_1 \times V_1/T_1 = P_2 \times V_2/T_2$ **OL**

Enrichment

Carbonation Take students on a short field trip to the school cafeteria, concession stand, or a nearby restaurant that has a soda machine connected to tanks of carbon dioxide gas for carbonation. Have an employee show the students how the tank is set up to carbonate the soda. **OL**

■ **Caption Question Fig 7** Carbon dioxide gas is more dense than the oxygen in air. As a result, the heavier carbon dioxide displaces the oxygen from the surrounding area.

Video Lab

Video

Gases

VOCABULARY
WORD ORIGIN
Mole
comes from the German word *Mol*, which is short for *Molekulargewicht*, meaning *molecular weight*

Watch a **video about gases.**

Video

The Ideal Gas Law—Molar Mass and Density

The ideal gas law can be used to solve for the value of any one of the four variables *P*, *V*, *T*, or *n* if the values of the other three are known. However, you can also rearrange the $PV = nRT$ equation to calculate the molar mass and density of a gas sample.

Molar mass and the ideal gas law To find the molar mass of a gas sample, the mass, temperature, pressure, and volume of the gas must be known. Recall that the number of moles of a gas (n) is equal to the mass (m) divided by the molar mass (M). Therefore, the n in the equation can be replaced by m/M.

$$PV = nRT \quad \text{substitute } n = \frac{m}{M} \quad PV = \frac{mRT}{M}$$

You can rearrange the new equation to solve for the molar mass.

$$M = \frac{mRT}{PV}$$

Density and the ideal gas law Recall that the density (D) of a substance is defined as mass (m) per unit volume (V). After rearranging the ideal gas equation to solve for molar mass, you can substitute D for m/V.

$$M = \frac{mRT}{PV} \quad \text{substitute } \frac{m}{V} = D \quad M = \frac{DRT}{P}$$

You can rearrange the new equation to solve for density.

$$D = \frac{MP}{RT}$$

Why might you need to know the density of a gas? Consider the requirements to fight a fire. One way to put out a fire is to remove its oxygen source by covering it with another gas that will neither burn nor support combustion, as shown in **Figure 7.** This gas must have a greater density than oxygen so that it will displace the oxygen at the source of the fire. You can observe a similar application of density by doing the MiniLab on the next page.

■ **Figure 7** To extinguish a fire, you need to take away fuel, oxygen, or heat. The fire extinguisher at right contains carbon dioxide, which displaces oxygen but does not burn. It also has a cooling effect due to the rapid expansion of the carbon dioxide as it is released from the nozzle.
Explain *Why does carbon dioxide displace oxygen?*

CULTURAL DIVERSITY

A Bit of Bubbly Champagne is produced in the la Champagne region of northern France. Seventeenth-century monks, including cellarmaster Dom Perignon, tried for years to get rid of the bubbles, which were considered a sign of poor wine-making. These tiny bubbles form because the conditions of the region result in a short growing season and a late grape harvest. The cool weather stops the fermentation process before the sugar in the wine has all been changed to alcohol. In spring, the fermentation starts again in the bottled wine, creating trapped carbon dioxide. Eventually, the monks gave up trying to get rid of the bubbles and perfected the making of sparkling wines. By law, since 1927, only sparkling beverages produced in that region of France, which includes 312 villages, can be labeled *champagne.*

MiniLAB

Model a Fire Extinguisher

Why is carbon dioxide used in fire extinguishers?

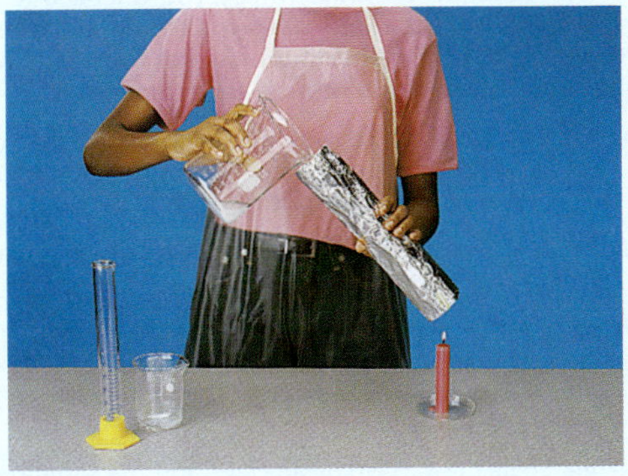

Procedure

1. Read and complete the lab safety form.
2. Measure the temperature with a **thermometer.** Obtain the air pressure with a **barometer** or **weather radio.** Record your data.
3. Roll a 23-cm × 30-cm piece of **aluminum foil** into a cylinder that is 30 cm long and roughly 6 cm in diameter. Tape the edges with **masking tape.**
4. Use **matches** to light a **candle.** **WARNING:** ***Run water over the extinguished match before throwing it away. Keep hair and clothing away from the flame.***
5. Place 30 g of **baking soda ($NaHCO_3$)** in a large **beaker.** Add 40 mL of **vinegar (5% CH_3COOH).**
6. Quickly position the foil cylinder at about 45° up and away from the top of the candle flame. **WARNING:** ***Do not touch the end of the aluminum tube that is near the burning candle.***
7. While the reaction in the beaker is actively producing carbon dioxide gas, carefully pour the gas, but not the liquid, out of the beaker and into the top of the foil tube. Record your observations.

Analysis

1. **Apply** Calculate the molar volume of carbon dioxide gas (CO_2) at room temperature and atmospheric pressure.
2. **Calculate** the room-temperature densities in grams per liter of carbon dioxide, oxygen, and nitrogen gases. Recall that you will need to calculate the molar mass of each gas in order to calculate densities.
3. **Interpret** Do your observations and calculations support the use of carbon dioxide gas to extinguish fires? Explain.

Real Versus Ideal Gases

What does the term *ideal gas* mean? Ideal gases follow the assumptions of the kinetic-molecular theory. According to this theory, an ideal gas is one whose particles take up no space. Ideal gases experience no intermolecular attractive forces, nor are they attracted or repelled by the walls of their containers. The particles of an ideal gas are in constant, random motion, moving in straight lines until they collide with each other or with the walls of the container. Additionally, these collisions are perfectly elastic, which means that the kinetic energy of the system does not change. An ideal gas follows the gas laws under all conditions of temperature and pressure.

In reality, no gas is truly ideal. All gas particles have some volume, however small, and are subject to intermolecular interactions. Also, the collisions that particles make with each other and with the container are not perfectly elastic. Despite that, most gases will behave like ideal gases at a wide range of temperatures and pressures. Under the right conditions, calculations made using the ideal gas law closely approximate experimental measurements.

☑ READING CHECK **Explain** the relationship between the kinetic-molecular theory and an ideal gas.

Matt Meadows

☑ READING CHECK An ideal gas follows all the assumptions of the kinetic-molecular theory: the particles take up no space, the particles experience no intermolecular attractive forces, and the collisions of the particles are perfectly elastic.

MiniLAB

Rubric

Purpose Students will use Avogadro's principle and the combined gas law to calculate and compare the density of three gases at room temperature and determine barometric pressure.

Process Skills use numbers, hypothesize

Safety Precautions Approve lab safety forms before work begins. Review MSDS for hazardous chemicals. Have students wear aprons and goggles. Remind them to be careful around the candle flame and the hot aluminum.

Disposal Students should flush the contents of the beaker down a drain and properly dispose of burned matches.

Teaching Strategies

- See the Teacher Handbook for preparation of solutions.
- Use a weather radio to determine the local barometric pressure if you do not have a barometer.
- Have students calculate the volume of CO_2 gas needed to fill the cylinder. $V = \pi r^2 h$; approximately 850 mL
- Remind students to be certain the end of the tube is positioned so that the gas will pour out directly onto the flame.

Expected Results The CO_2 gas, which is heavier than the oxygen in air, will flow down the foil cylinder and extinguish the candle flame.

Analysis

1. Using $V_2 = P_1V_1T_2/(T_1P_2)$ at 298 K and 98.6 kPa, the molar volume is 25.1 L.
2. CO_2 44.0 g/25.1 L = 1.75 g/L; O_2 32.0 g/25.1 L = 1.27 g/L; N_2 28.0 g/25.1 L = 1.12 g/L
3. Yes, the heavier CO_2 gas moves down the cylinder, displacing the air and extinguishing the candle flame.

PROBLEM-SOLVING STRATEGIES

Apply the Strategy
Students should use the strategy to show the derivation from the ideal gas law to Boyle's law ($P_1V_1 = P_2V_2$), Gay-Lussac's law ($P_1/T_1 = P_2/T_2$), and the combined gas law ($P_1V_1/T_1 = P_2V_2/T_2$).

ASSESSMENT

Skill Ask students to demonstrate their skill at reading a barometer. Check the pressure first, explaining to the students how to read the instrument. Then ask each student to read the atmospheric pressure, and record each reading on a piece of paper. OL BL EL

Math in Chemistry

Calculating Percent Change
Some gas law problems involve calculating a percent increase or percent decrease in temperature, pressure, or volume. In these cases, students will need to calculate the ratio of the final/initial values. To calculate the percent increase or decrease, they can use the following relationship: % change = $[(X_2/X_1) - 1] \times 100$
X can represent temperature in kelvins, pressure, volume or amount. In-class example problem for Gay-Lussac's Law: You accidentally left a 1.00-L soda bottle in a sunny location. The temperature increased from 20°C to 31°C. The bottle is closed and made of rigid plastic. So, the volume cannot increase. Ask students if the pressure increased or decreased. increased Calculate the percentage increase.
$P_2/P_1 = T_2/T_1 = 1.038$. The pressure increased by 3.8 percent. OL

PROBLEM-SOLVING STRATEGIES

Deriving Gas Laws

If you master the following strategy, you will need to remember only one gas law—the ideal gas law. Consider the example of a fixed amount of gas held at constant pressure. You need Charles's law to solve problems involving volume and temperature.

1. Use the ideal gas law to write two equations that describe the gas sample at two different volumes and temperatures. (Quantities that do not change are shown in **red**.)
2. Isolate volume and temperature—the two conditions that vary—on the same side of each equation.
3. Because n, R, and P are constant under these conditions, you can set the volume and temperature conditions equal, deriving Charles's law.

$$PV_1 = nRT_1 \qquad PV_2 = nRT_2$$

$$\frac{V_1}{T_1} = \frac{nR}{P} \qquad \frac{V_2}{T_2} = \frac{nR}{P}$$

$$\frac{V_1}{T_1} = \frac{V_2}{T_2}$$

Apply the Strategy

Derive Boyle's law, Gay-Lussac's law, and the combined gas law based on the above example.

Watch a **video about gases, pressure, and scuba diving.**

Video

Extreme pressure and temperature When is the ideal gas law not likely to work for a real gas? Real gases deviate most from ideal gas behavior at high pressures and low temperatures. The nitrogen gas in the tanks shown in **Figure 8** behaves as a real gas. Lowering the temperature of nitrogen gas results in less kinetic energy of the gas particles, which means their intermolecular attractive forces are strong enough to affect their behavior. When the temperature is low enough, this real gas condenses to form a liquid. The propane gas in the tanks shown in **Figure 8** also behaves as a real gas. Increasing the pressure on a gas forces the gas particles closer together until the volume occupied by the gas particles themselves is no longer negligible. Real gases such as propane will liquefy if enough pressure is applied.

Figure 8 Real gases do not follow the ideal gas law at all pressures and temperatures.

Nitrogen gas turns to liquid at −196°C. At this temperature, scientists can preserve biological specimens, such as body tissues, for future research or medical procedures.

About 270 times more propane can be stored as a liquid than as a gas in the same amount of space. Your family might use small tanks of liquid propane as fuel for your barbecue grill or larger tanks for heating and cooking.

(l)Pasquale Sorrentino/Photo Researchers; (r)Paul Broadbent/Alamy

CHEMISTRY PROJECT

Joseph Priestly Have students research the work of British scientist Joseph Priestley, who was given credit for the discovery of oxygen. Point out that Priestley, like many scientists, was active in politics. Because he was persecuted in Britain for his support of the American Revolution, he came to America in 1794. OL

What's CHEMISTRY Got To Do With It?

Video

Leagues Under the Sea

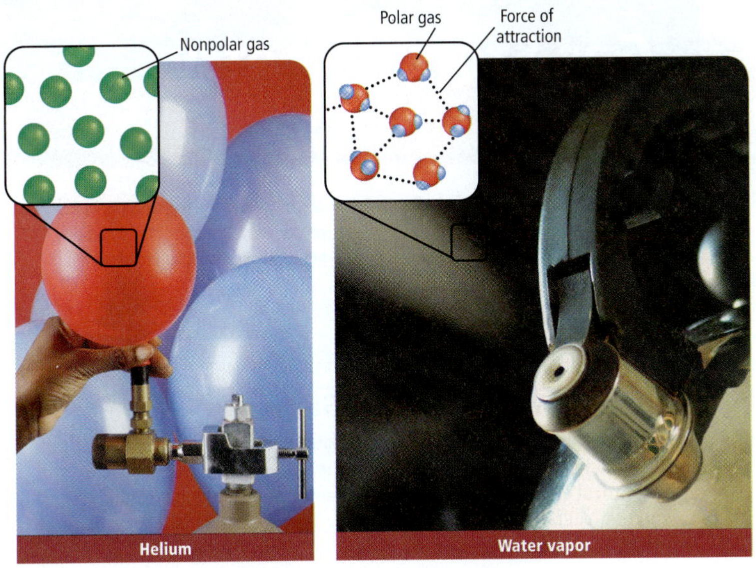

■ **Figure 9** In a nonpolar gas, there is minimal attraction between particles. However, polar gases, such as water vapor, experience forces of attraction between particles.

Infer *Assuming the volume of the particles is negligible, how will the measured pressure for a sample of gas that experiences significant intermolecular attractive forces compare to the pressure predicted by the ideal gas law?*

Polarity and size of particles The nature of the particles making up a gas also affects how ideally the gas behaves. For example, polar gas molecules, such as water vapor, generally have larger attractive forces between their particles than nonpolar gases, such as helium. The oppositely charged ends of polar molecules are pulled together through electrostatic forces, as shown in **Figure 9.** Therefore, polar gases do not behave as ideal gases. Also, the particles of gases composed of larger nonpolar molecules, such as butane (C_4H_{10}), occupy more actual volume than an equal number of smaller gas particles in gases such as helium (He). Therefore, larger gas particles tend to exhibit a greater departure from ideal behavior than do smaller gas particles.

SECTION 2 REVIEW

Section Self-Check

Section Summary

- Avogadro's principle states that equal volumes of gases at the same pressure and temperature contain equal numbers of particles.
- The ideal gas law relates the amount of a gas present to its pressure, temperature, and volume.
- The ideal gas law can be used to find molar mass if the mass of the gas is known or the density of the gas if its molar mass is known.
- At very high pressures and very low temperatures, real gases behave differently than ideal gases.

31. MAINIDEA **Explain** why Avogadro's principle holds true for ideal gases that have small particles and for ideal gases that have large particles.
32. **State** the equation for the ideal gas law.
33. **Analyze** how the ideal gas law applies to real gases using the kinetic-molecular theory.
34. **Predict** the conditions under which a real gas might deviate from ideal behavior.
35. **List** common units for each variable in the ideal gas law.
36. **Calculate** A 2.00-L flask is filled with propane gas (C_3H_8) at a pressure of 1.00 atm and a temperature of $-15.0°C$. What is the mass of the propane in the flask?
37. **Make and Use Graphs** For every 6°C drop in temperature, the air pressure in a car's tires goes down by about 1 psi (14.7 psi = 1.00 atm). Make a graph illustrating the change in tire pressure from 20°C to $-20°C$ (assume 30.0 psi at 20°C).

■ **Caption Question Fig. 9**
It will be less than the pressure predicted by the ideal gas law.

Visual Learning

Figure 9 Have students explain why nonpolar gases behave more like ideal gases than polar gases. Polar gases have more interparticle attractions due to their oppositely charged poles that can attract those of other identical molecules. Ask students to draw diagrams of the following gas molecules: CO, N_2, and NH_3. Have them label each gas as polar or nonpolar. For each polar gas, ask them to label the positive and negative poles. N_2 is nonpolar; CO and NH_3 are polar. Positive poles are the C in CO and the H in NH_3. Negative poles are the O in CO and N in NH_3. OL EL

3 Assess

Check for Understanding

Ask groups of students to write and solve two ideal gas law problems. OL

Reteach

Ask students to determine how to calculate how many moles of gas are contained in an inflated tire. Students can use the ideal gas law to do this calculation if they have a tire gauge, a thermometer, and a tape measure. OL

Extension

Have students conduct research to identify noble gases that will react under extreme conditions and write the equations for those reactions. AL

SECTION 2 REVIEW

31. The size of any gas particle is so small compared to the volume of the gas, it is assumed that no particle has any volume of its own.
32. $PV = nRT$
33. A real gas behaves most like an ideal gas under conditions that increase the distance and reduce the attractions among gas particles. The best conditions for that are high temperature and low pressure.
34. A real gas might deviate from ideal behavior under conditions that decrease the distance and increase the attractions among gas particles, such as low temperature and high pressure.
35. *P*: atm, mm Hg, torr, kPa; *V*: L, mL; *T*: K; *n*: mol
36. 4.16 g
37. Graph should show air pressure graphed with relation to temperature; the resulting plot will be a straight line showing a direct relationship between the variables.

SECTION 3

1 Focus

MAINIDEA

Balance Gas Reactions Use paper models to illustrate the balanced chemical reaction between hydrogen gas and oxygen gas to form water vapor: $2H_2(g) + O_2(g) \rightarrow 2H_2O(g)$. White paper circles represent hydrogen atoms and red paper circles represent oxygen atoms. Ask students to recite the balanced chemical equation aloud in two different ways: Two moles of hydrogen gas plus one mole of oxygen gas form two moles of water vapor. Two volumes of hydrogen gas plus one volume of oxygen gas form two volumes of water vapor. BL EL OL

2 Teach

Build a Model

Gas Molecules Ask students to use ball-and-stick models of gas molecules to illustrate what happens when a real gas encounters an extremely high pressure and a low temperature. Students should show the molecules moving slowly, gathering close together, and experiencing interparticle attractions. BL EL

Reinforcement

Algebra Review Review with students the algebraic steps required to solve the ideal gas equation for each of the variables *P*, *V*, *n*, and *T*. Use dimensional analysis to demonstrate that each variable has the appropriate units.

SECTION 3

Essential Questions

- What stoichiometric ratios can be determined for gaseous reactants and products from balanced chemical equations?
- How are the amounts of gaseous reactants and products in a chemical reaction calculated?

Review Vocabulary

coefficient: the number written in front of a reactant or product in a chemical equation, which tells the smallest number of particles of the substance involved in the reaction

Gas Stoichiometry

MAINIDEA When gases react, the coefficients in the balanced chemical equation represent both molar amounts and relative volumes.

CHEM 4 YOU To make a cake, it is important to add the ingredients in the correct proportions. In a similar way, the correct proportions of reactants are needed in a chemical reaction to yield the desired products.

Stoichiometry of Reactions Involving Gases

The gas laws can be applied to calculate the stoichiometry of reactions in which gases are reactants or products. Recall that the coefficients in chemical equations represent molar amounts of substances taking part in the reaction. For example, hydrogen gas can react with oxygen gas to produce water vapor.

$$2H_2(g) + O_2(g) \rightarrow 2H_2O(g)$$

From the balanced chemical equation, you know that 2 mol of hydrogen gas reacts with 1 mol of oxygen gas, producing 2 mol of water vapor. This tells you the molar ratios of substances in this reaction. Avogadro's principle states that equal volumes of gases at the same temperature and pressure contain equal numbers of particles. Thus, for gases, the coefficients in a balanced chemical equation represent not only molar amounts but also relative volumes. Therefore, 2 L of hydrogen gas would react with 1 L of oxygen gas to produce 2 L of water vapor.

Stoichiometry and Volume–Volume Problems

To find the volume of a gaseous reactant or product in a reaction, you must know the balanced chemical equation for the reaction and the volume of at least one other gas involved in the reaction. Examine the reaction in **Figure 10,** which shows the combustion of methane. This reaction takes place every time you light a Bunsen burner.

Because the coefficients represent volume ratios for gases taking part in the reaction, you can determine that it takes 2 L of oxygen to react completely with 1 L of methane. The complete combustion of 1 L of methane will produce 1 L of carbon dioxide and 2 L of water vapor.

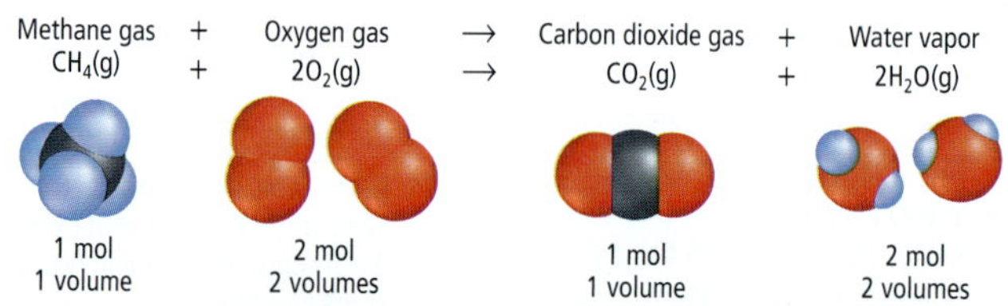

Figure 10 The coefficients in a balanced equation show the relationships between numbers of moles of all reactants and products, and the relationships between volumes of any gaseous reactants or products. From these coefficients, volume ratios can be set up for any pair of gases in the reaction.

DIFFERENTIATED INSTRUCTION

Struggling Learners Some students understand stoichiometry problems better if they can visualize reactions using ball-and-stick models of reactants and products. Ask students to build models of the molecules involved in the combustion of methane, and use those models to find the mole ratio of: **a.** methane to oxygen, 1:2; **b.** oxygen to carbon dioxide, 2:1; **c.** methane to carbon dioxide, 1:1; **d.** methane to water 1:2. BL EL

Note that no conditions of temperature and pressure are listed. They are not needed as part of the calculation because after mixing, both gases are at the same temperature and pressure. The temperature of the entire reaction might change during the reaction, but a change in temperature would affect all gases in the reaction the same way. Therefore, you do not need to consider pressure and temperature conditions.

EXAMPLE Problem 7

Find help with ratios. Math Handbook

VOLUME–VOLUME PROBLEMS What volume of oxygen gas is needed for the complete combustion of 4.00 L of propane gas (C_3H_8)? Assume that pressure and temperature remain constant.

1 ANALYZE THE PROBLEM

You are given the volume of a gaseous reactant in a chemical reaction. Remember that the coefficients in a balanced chemical equation provide the volume relationships of gaseous reactants and products.

Known	Unknown
$V_{C_3H_8} = 4.00\ L$	$V_{O_2} = ?\ L$

2 SOLVE FOR THE UNKNOWN

Use the balanced equation for the combustion of C_3H_8. Find the volume ratio for O_2 and C_3H_8, then solve for V_{O_2}.

$C_3H_8(g) + 5O_2(g) \rightarrow 3CO_2(g) + 4H_2O(g)$ — Write the balanced equation.

$\frac{5\ \text{volumes}\ O_2}{1\ \text{volume}\ C_3H_8}$ — Find the volume ratio for O_2 and C_3H_8.

$V_{O_2} = (4.00\ L\ C_3H_8) \times \frac{5\ \text{volumes}\ O_2}{1\ \text{volume}\ C_3H_8}$ — Multiply the known volume of C_3H_8 by the volume ratio to find the volume of O_2.

$= 20.0\ L\ O_2$

3 EVALUATE THE ANSWER

The coefficients in the combustion equation show that a much larger volume of O_2 than C_3H_8 is used up in the reaction, which is in agreement with the calculated answer. The unit of the answer is liters, a unit of volume, and there are three significant figures.

PRACTICE Problems

Do additional problems. Online Practice

38. How many liters of propane gas (C_3H_8) will undergo complete combustion with 34.0 L of oxygen gas?
39. Determine the volume of hydrogen gas needed to react completely with 5.00 L of oxygen gas to form water.
40. What volume of oxygen is needed to completely combust 2.36 L of methane gas (CH_4)?
41. **Challenge** Nitrogen and oxygen gases react to form dinitrogen monoxide gas (N_2O). What volume of O_2 is needed to produce 34 L of N_2O?

RealWorld CHEMISTRY

Using Stoichiometry

KILNS Correct proportions of gases are needed for many chemical reactions. Although many pottery kilns are fueled by methane, a precise mixture of propane and air can be used to fuel a kiln if methane is unavailable.

IN-CLASS Example

Question In-class example for volume-volume problems: Ethanol burns to form carbon dioxide gas and water vapor according to the following balanced chemical equation: $CH_3CH_2OH(g) + 3O_2(g) \rightarrow 2CO_2(g) + 3H_2O(g)$. What volume of carbon dioxide gas will be formed if the reaction uses 6 L of oxygen?

Answer From the coefficients of the balanced chemical equation, the volume ratio of O_2 to CO_2 is 3/2. Use this ratio to determine the volume of CO_2 = 4 L.

PRACTICE Problems

Have students refer to the Solutions to Selected Practice Problems appendix for complete solutions to odd-numbered problems. The complete solutions for all problems can be found in the Solutions Manual.

38. 6.80 L C_3H_8
39. 10.0 L H_2
40. 4.72 L O_2
41. 17 L O_2

Concept Development

Concept Map Describe the four steps required to solve a stoichiometry problem and work with students to make a concept map to represent the process. (1) Write the balanced equation for the reaction. (2) Convert the given mass or volume of a reactant or product to moles. (3) Use the coefficients in the balanced equation to set up the appropriate mole ratio and multiply to calculate moles of the desired reactant or product. (4) Convert moles of the desired reactant or product to mass or volume as required.

CHEMISTRY PROJECT

Gas Law Application Ask interested students to watch the movie *Around the World in 80 Days,* and write a brief review of the movie in which they focus on the way the gas laws are used in flying the balloon. Have students include the reviews in their portfolios. OL

Enrichment

Compressed Gas Ask a representative of a company that sells compressed gases to give a classroom presentation on uses for the gases and how to handle compressed gas tanks.

Content Background

Discovery of Helium The element helium was observed in the Sun using spectroscopy in 1868, 23 years before it was found in uranium ore on Earth. Its discovery in the Sun is why its name comes from the Greek word for the Sun, *Helios.* In 1900, helium was still rare, but the 1905 analysis of gas from a natural gas well in Kansas showed that it contained about 2% helium and the rest was mostly nitrogen. Today, the private industries with the largest helium production in the U.S. are in Kansas and Texas.

Quick Demo

Decomposition of H_2O_2

Place a dry 500 mL Erlenmeyer flask on a balance and set the tare to "0". Add 100 g of 3% hydrogen peroxide (available from a grocery store or pharmacy). Add about 0.5 g of dry yeast as a catalyst to the hydrogen peroxide. Record the mass. The mixture should bubble as the hydrogen peroxide decomposes. Record the change in mass after the bubbles subside. Ask students what accounts for the loss in mass. O_2 was lost to the surroundings. Ask students what role the yeast played in the reaction. The yeast was a catalyst. OL

■ **Figure 11** Ammonia is essential in the production of fertilizers containing nitrogen. Proper levels of soil nitrogen lead to increased crop yields.

PhotoLink/Getty Images

Stoichiometry and Volume–Mass Problems

Connection to Biology What you have learned about stoichiometry can be applied to the production of ammonia (NH_3) from nitrogen gas (N_2). Fertilizer manufacturers use ammonia to make nitrogen-based fertilizers. Nitrogen is an essential element for plant growth. Natural sources of nitrogen in soil, such as nitrogen fixation by plants, the decomposition of organic matter, and animal wastes, do not always supply enough nitrogen for optimum crop yields. **Figure 11** shows a farmer applying fertilizer rich in nitrogen to the soil. This enables the farmer to produce a crop with a higher yield.

Example Problem 8 shows how to use a volume of nitrogen gas to produce a certain amount of ammonia. In doing this type of problem, remember that the balanced chemical equation allows you to find ratios for only moles and gas volumes, not for masses. All masses given must be converted to moles or volumes before being used as part of a ratio. Also, remember that the temperature units used must be kelvins.

VOCABULARY

ACADEMIC VOCABULARY

Ratio
the relationship in quantity between two things
In a water molecule, the ratio of hydrogen to oxygen is 2:1.

EXAMPLE Problem 8

EXAMPLE PROBLEM

VOLUME–MASS PROBLEMS Ammonia is synthesized from hydrogen and nitrogen.

$$N_2(g) + 3H_2(g) \rightarrow 2NH_3(g)$$

If 5.00 L of nitrogen reacts completely with hydrogen at a pressure of 3.00 atm and a temperature of 298 K, how much ammonia, in grams, is produced?

1 ANALYZE THE PROBLEM

You are given the volume, pressure, and temperature of a gas sample. The mole and volume ratios of gaseous reactants and products are given by the coefficients in the balanced chemical equation. Volume can be converted to moles and thus related to mass by using molar mass and the ideal gas law.

Known	Unknown
$V_{N_2} = 5.00$ L	$m_{NH_3} = ?$ g
$P = 3.00$ atm	
$T = 298$ K	

2 SOLVE FOR THE UNKNOWN

Determine how many liters of gaseous ammonia will be made from 5.00 L of nitrogen gas.

$$\frac{1 \text{ volume } N_2}{2 \text{ volumes } NH_3}$$

Find the volume ratio for N_2 and NH_3 using the balanced equation.

$$5.00 \text{ L } \cancel{N_2}\left(\frac{2 \cancel{\text{volumes}} NH_3}{1 \cancel{\text{volume}} \cancel{N_2}}\right) = 10.0 \text{ L } NH_3$$

Multiply the known volume of N_2 by the volume ratio to find the volume of NH_3.

Use the ideal gas law. Solve for n, and calculate the number of moles of NH_3.

$$PV = nRT$$

State the ideal gas law.

$$n = \frac{PV}{RT}$$

Solve for n.

$$n = \frac{(3.00 \text{ atm})(10.0 \text{ L})}{\left(0.0821 \frac{\text{L·atm}}{\text{mol·K}}\right)(298 \text{ K})}$$

Substitute $P = 3.00$ atm, $V_{NH_3} = 10.0$ L, and $T = 298$ K.

$$n = \frac{(3.00 \cancel{\text{atm}})(10.0 \cancel{\text{L}})}{\left(0.0821 \frac{\cancel{\text{L·atm}}}{\text{mol·}\cancel{\text{K}}}\right)(298 \cancel{\text{K}})} = 1.23 \text{ mol } NH_3$$

Multiply and divide numbers and units.

$$M = \left(\frac{1 \cancel{\text{N atom}} \times 14.01 \text{ amu}}{1 \cancel{\text{N atom}}}\right) + \left(\frac{3 \cancel{\text{H atoms}} \times 1.01 \text{ amu}}{1 \cancel{\text{H atom}}}\right)$$

$$= 17.04 \text{ amu}$$

Find the molecular mass of NH_3.

$$M = 17.04 \text{ g/mol}$$

Express molar mass in units of g/mol.

Convert moles of ammonia to grams of ammonia.

$$1.23 \cancel{\text{mol } NH_3} \times \frac{17.04 \text{ g } NH_3}{1 \cancel{\text{mol } NH_3}} = \mathbf{21.0 \text{ g } NH_3}$$

Use the molar mass of ammonia as a conversion factor.

3 EVALUATE THE ANSWER

To check your answer, calculate the volume of reactant nitrogen at STP. Then, use molar volume and the mole ratio between N_2 and NH_3 to determine how many moles of NH_3 were produced. The unit of the answer is grams, a unit of mass. There are three significant figures.

PRACTICE Problems

Do additional problems. Online Practice

42. Ammonium nitrate is a common ingredient in chemical fertilizers. Use the reaction shown to calculate the mass of solid ammonium nitrate that must be used to obtain 0.100 L of dinitrogen monoxide gas at STP.

$$NH_4NO_3(s) \rightarrow N_2O(g) + 2H_2O(g)$$

43. When solid calcium carbonate ($CaCO_3$) is heated, it decomposes to form solid calcium oxide (CaO) and carbon dioxide gas (CO_2). How many liters of carbon dioxide will be produced at STP if 2.38 kg of calcium carbonate reacts completely?

44. When iron rusts, it undergoes a reaction with oxygen to form iron(III) oxide.

$$4Fe(s) + 3O_2(g) \rightarrow 2Fe_2O_3(s)$$

Calculate the volume of oxygen gas at STP that is required to completely react with 52.0 g of iron.

45. Challenge An excess of acetic acid is added to 28 g of sodium bicarbonate at 25°C and 1 atm pressure. During the reaction, the gas cools to 20°C. What volume of carbon dioxide will be produced? The balanced equation for the reaction is shown below.

$$NaHCO_3(aq) + CH_3COOH(aq) \rightarrow NaCH_3COO(aq) + CO_2(g) + H_2O(l)$$

PRACTICE PROBLEMS

IN-CLASS Example

Question Ethene gas, also called ethylene, can polymerize to form polyethylene plastic according to the following chemical equation: $n(C_2H_4)(g) \rightarrow (CH_2 - CH_2)n(s)$ Start with 2240 L of ethylene gas at STP. (The actual reaction takes place at 220 °C and 1200 atm.) If the reaction converts all of the ethene gas to polyethylene, how many grams of polyethylene will be produced?

Answer 1 mole of gas at STP conditions has a volume of 22.4 L. The molar mass of a monomer unit of polyethylene is 26 g/mol.

$$2240 \cancel{\text{L}} \times \frac{1 \cancel{\text{mol}}}{22.4 \cancel{\text{L}}} \times \frac{26 \text{ g}}{\cancel{\text{mol}}} = 2600 \text{ g}$$

PRACTICE Problems

Have students refer to the Solutions to Selected Practice Problems appendix for complete solutions to odd-numbered problems. The complete solutions for all problems can be found in the Solutions Manual.

42. 0.357 g NH_4NO_3
43. 533 L CO_2
44. 15.6 L O_2
45. 7.9 L CO_2

CHEMISTRY JOURNAL

Importance of Ammonia Have students research non-agricultural uses for ammonia. Ask students to include the chemical reactions that are involved in each example they find. Have students share their findings with the class in the form of a graphic or presentation. Uses include the paper industry, pharmaceutical industry, petroleum industry, and household cleaners and detergents. **OL**

3 Assess

Check for Understanding

Place a 10.0-g ball of aluminum foil in 38.0 mL of concentrated hydrochloric acid in a well-ventilated area. Ask students to explain what they see. Bubbles of hydrogen gas form as the acid reacts with the metal. Ask students to write a balanced equation for this reaction and use it to predict how many liters of hydrogen gas will be produced at STP if the 10.0 g of aluminum reacts completely. $2Al(s) + 6HCl(aq) \rightarrow 3H_2(g) + 2AlCl_3(aq)$; 12.5 L For disposal, filter the product. Pour the liquid down a drain with plenty of water. Discard the solid waste in the trash can. **OL**

Reteach

Explain to students that the combustion reactions in their bodies that break down the foods they eat are similar to those that take place when methane is burned. Ask students to write a balanced equation for the combustion of sucrose ($C_{12}H_{22}O_{11}$) and to calculate the number of liters of oxygen gas needed to combust a one-ounce packet (30.0 g) of sugar at STP. $C_{12}H_{22}O_{11}(g) + 12O_2(g) \rightarrow 12CO_2(g) + 11H_2O(g)$; 23.6 L **OL**

Extension

Ask students to research how a small amount of nitroglycerin can produce a big explosion. Liquid nitroglycerin explodes to form extremely large volumes of several different gases. $4C_3H_5N_3O_9(l) \rightarrow 6N_2(g) + O_2(g) + 12CO_2(g) + 10H_2O(g)$ **OL**

ASSESSMENT

Knowledge Ask students to explain why the conditions of temperature and pressure must be known to solve a problem in which the volume of a gaseous reactant is used to determine the number of moles of a solid product. The volume of the gas must be converted to moles because reaction stoichiometry is based on mole ratios. Moles of a gas will occupy different volumes depending on the temperature and pressure. **OL**

Figure 12 To effectively manufacture a product, such as many of these plastics, it is essential to answer the following questions. How much of a reactant should be purchased? How much of a product will be produced?

Comstock/Getty Images

Stoichiometric problems, such as the ones in this section, are considered in industrial processes. For example, ethene gas (C_2H_4), also called ethylene, is the raw material for making polyethylene polymers. Polyethylene is produced when numerous ethene molecules join together in chains of repeating $—CH_2–CH_2—$ units. These polymers are used to make many everyday items, such as the ones shown in **Figure 12.** The general formula for this polymerization reaction is shown below. In this formula, n is the number of units used.

$$n(C_2H_4)(g) \rightarrow —(CH_2-CH_2)_n—(s)$$

If you were a process engineer for a polyethylene manufacturing plant, you would need to know about the properties of ethene gas and the polymerization reaction. Knowledge of the gas laws would help you calculate both the mass and volume of raw material needed under different temperature and pressure conditions to make different types of polyethylene.

SECTION 3 REVIEW

Section Self-Check

Section Summary

- The coefficients in a balanced chemical equation specify volume ratios for gaseous reactants and products.
- The gas laws can be used along with balanced chemical equations to calculate the amount of a gaseous reactant or product in a reaction.

46. MAINIDEA Explain When fluorine gas combines with water vapor, the following reaction occurs.

$$2F_2(g) + 2H_2O(g) \rightarrow O_2(g) + 4HF(g)$$

If the reaction starts with 2 L of fluorine gas, how many liters of water vapor react with the fluorine, and how many liters of oxygen and hydrogen fluoride are produced?

47. Analyze Is the volume of a gas directly or inversely proportional to the number of moles of a gas at constant temperature and pressure? Explain.

48. Calculate One mole of a gas occupies a volume of 22.4 L at STP. Calculate the temperature and pressure conditions needed to fit 2 mol of a gas into a volume of 22.4 L.

49. Interpret Data Ethene gas (C_2H_4) reacts with oxygen to form carbon dioxide and water. Write a balanced equation for this reaction, then find the mole ratios of substances on each side of the equation.

SECTION 3 REVIEW

46. 2 L H_2O, 1 L O_2, and 4 L HF

47. Directly proportional; as the amount of gas increases, so does volume.

48. Student answers will vary. Temperature can be halved or pressure doubled or a combination of lowering temperature and increasing pressure.

49. $C_2H_4(g) + 3O_2(g) \rightarrow 2CO_2 + 2H_2O$
The mole ratio of ethene to oxygen is 1:3; the mole ratio of carbon dioxide to water is 2:2.

CHEMISTRY&health

Health Under Pressure

You live, work, and play in air that is generally about 1 atm in pressure and 21% oxygen. Have you ever wondered what might happen if the pressure and the oxygen content of the air were greater? Would you recover from illness or injury more quickly? These questions are at the heart of hyperbaric medicine.

Hyperbaric medicine The prefix *hyper-* means *above* or *excessive*, and a bar is a unit of pressure equal to 100 kPa, roughly normal atmospheric pressure. Thus, the term *hyperbaric* refers to pressure that is greater than normal. Patients receiving hyperbaric therapy are exposed to pressures greater than the pressure of the atmosphere at sea level.

The oxygen connection Greater pressure is most often combined with an increase in the concentration of oxygen a patient receives. The phrase *hyperbaric oxygen therapy* (HBOT) refers to treatment with 100% oxygen. **Figure 1** shows a chamber that might be used for HBOT. Inside the hyperbaric chamber, pressures can reach five to six times normal atmospheric pressure. At hyperbaric therapy centers across the country, HBOT is used to treat a wide range of conditions, including burns, decompression sickness, slow-healing wounds, anemia, and some infections.

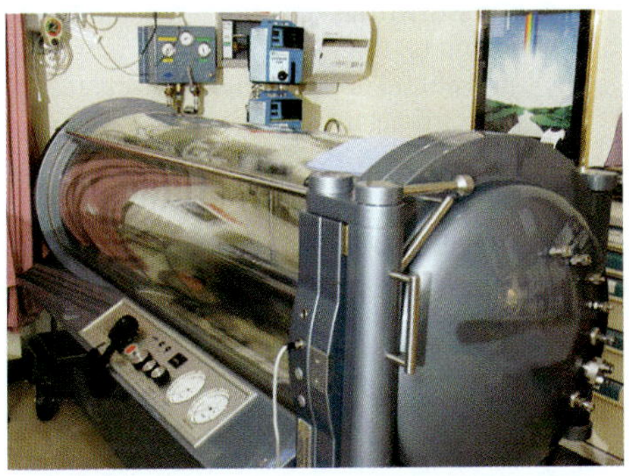

Figure 1 During HBOT, the patient lies in a hyperbaric chamber. A technician controls the pressure and oxygen levels.

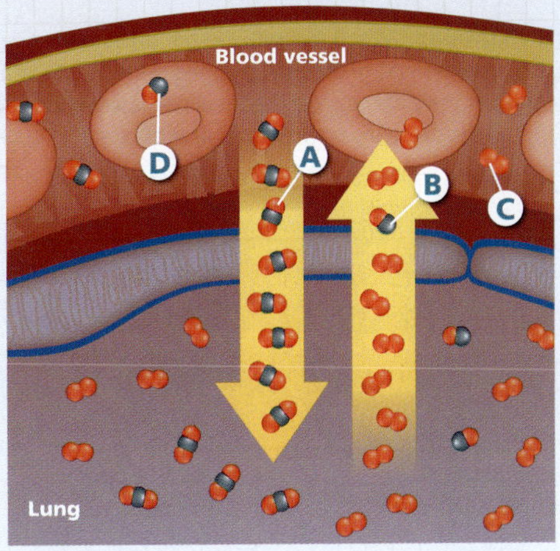

Figure 2 Gases are exchanged between the lungs and the circulatory system.

Carbon-monoxide poisoning Use **Figure 2** to help you understand how HBOT aids in the treatment of carbon-monoxide poisoning.

Normal gas exchange Oxygen (O_2) moves from the lungs to the blood and binds to the hemoglobin in red blood cells. Carbon dioxide (CO_2) is released, as shown by **A**.

Abnormal gas exchange If carbon monoxide (CO) enters the blood, as shown by **B**, it, instead of oxygen, binds to the hemoglobin. Cells in the body begin to die from oxygen deprivation.

Oxygen in blood plasma In addition to the oxygen carried by hemoglobin, oxygen is dissolved in the blood plasma, as shown by **C**. HBOT increases the concentration of dissolved oxygen to an amount that can sustain the body.

Eliminating carbon monoxide Pressurized oxygen also helps remove any carbon monoxide bound to hemoglobin, as shown by **D**.

WRITING IN ▸ Chemistry

Research and prepare an informational pamphlet about the use of HBOT to treat slow-healing wounds.

WebQuest

CHEMISTRY&health

Purpose

Students will investigate the principles behind hyperbaric oxygen therapy (HBOT) and explore how HBOT is used to treat carbon-monoxide poisoning.

Background

Patients undergoing HBOT often report sensations similar to those felt when ascending and descending during airplane flight. As pressure slowly builds, patients might feel increased ear pressure, which can be alleviated by yawning or swallowing. As pressure is released at the end of the treatment, ears typically "pop" automatically. Conditions of temperature and humidity are carefully controlled in a hyperbaric chamber.

Teaching Strategies

Use Boyle's law to compare the volume of gases in a scuba diver's lungs at the ocean surface and at a depth of 30 meters. Pressure and volume are inversely related. Because the pressure is greater at 30 meters, the volume of gases in the scuba diver's lungs will be less than at the surface. Have students apply their knowledge of hyperbaric therapy and Boyle's law to consider how HBOT helps treat divers who have ascended too quickly, resulting in nitrogen gas coming rapidly out of solution in body fluids and tissues and forming gaseous bubbles. Boyle's law is applied when using HBOT to treat divers for decompression sickness. As the pressure applied to the chamber increases, the volume of the nitrogen bubbles decreases.

WRITING IN ▸ Chemistry

Rubric WebQuest

Research After completing informational pamphlets, students should be able to describe the use of HBOT in wound treatment. Provide an assessment rubric that allows students to self- or peer-evaluate pamphlets on a scale of 1 to 5. Rubric items might include: The information in the pamphlet is presented clearly. The pamphlet describes conditions during a typical therapy session. The pamphlet describes advantages and potential dangers associated with HBOT. The pamphlet would be effective in helping patients choose whether or not to undergo HBOT.

ChemLAB

Rubric

Preparation

Time Allotment one class period

Process Skills collect and organize data, hypothesize, observe

Safety Precautions Approve lab safety forms before work begins. Do not allow the students to eat the popcorn that is made with the laboratory glassware. Warn students that hot oil can splatter and burn skin.

Disposal Have students dispose of the popcorn in a waste container. The oil can be washed from the beaker with soap and water.

Procedure

- The additional piece of wire gauze should be sufficient to hold the popcorn in the beaker.
- Some popcorn kernels will not pop.
- It is important that the students do not burn their popcorn. Wire gauze with a ceramic circle is recommended to distribute the burner heat.

Sample Data

Water volume	5.0 mL
Total volume of water and popcorn	7.8 mL
Mass of beaker, oil, and kernels before popping	110.943 g
Mass of beaker, oil, and popcorn after popping	109.952 g

ChemLAB

Determine Pressure in Popcorn Kernels

Background: When the water vapor pressure inside a popcorn kernel is great enough, the kernel bursts and releases the water vapor. The ideal gas law can be used to find the pressure in the kernel as it bursts.

Question: *How much pressure is required to burst a kernel of popcorn?*

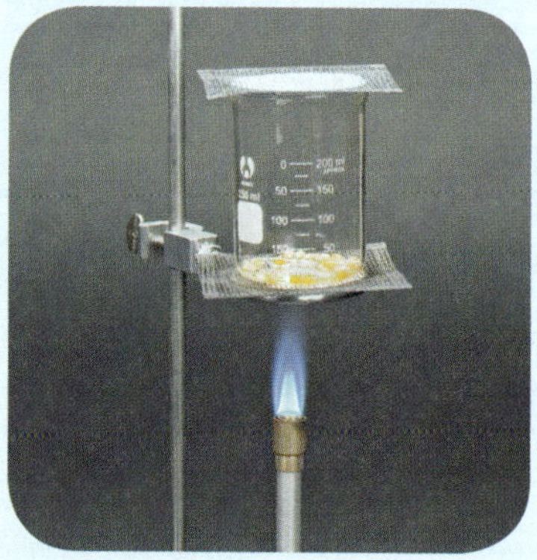

Matt Meadows

Materials

popcorn kernels (18–20)
10-mL graduated cylinder
vegetable oil (1.5 mL)
250 mL beaker
wire gauze squares (2)
beaker tongs
Bunsen burner
balance
ring stand
distilled water
small iron ring
paper towels

Safety Precautions

Procedure

1. Read and complete the lab safety form.
2. Create a table to record your data.
3. Place approximately 5 mL of distilled water in the graduated cylinder, and record the volume.
4. Place 18–20 popcorn kernels in the graduated cylinder with the water. Tap the cylinder to force any air bubbles off the kernels. Record the new volume.
5. Remove the kernels from the graduated cylinder, and dry them.
6. Place the dry kernels and 1.0–1.5 mL of vegetable oil into the beaker.
7. Measure the total mass of the beaker, oil, and kernels.
8. Set up a Bunsen burner with a ring stand, ring, and wire gauze.
9. Place the beaker on the wire gauze and ring. Place another piece of wire gauze on top of the beaker.
10. Gently heat the beaker with the burner. Move the burner back and forth to heat the oil evenly.
11. Observe the changes in the kernels and oil while heating, then turn off the burner when the popcorn has popped and before any burning occurs.
12. Using the beaker tongs, remove the beaker from the ring and allow it to cool completely.
13. Measure the final mass of the beaker, oil, and popcorn once cooling is complete.
14. **Cleanup and Disposal** Dispose of the popcorn and oil as directed by your teacher. Wash and return all lab equipment to its designated location.

Analyze and Conclude

1. **Calculate** the volume of the popcorn kernels, in liters, by the difference in the volumes of distilled water before and after adding popcorn.
2. **Calculate** the total mass of water vapor released using the mass measurements of the beaker, oil, and popcorn before and after popping.
3. **Convert** Use the molar mass of water to find the number of moles of water released.
4. **Use Formulas** Use the temperature of the boiling oil (225°C) as your gas temperature and the volume of popcorn, and calculate the pressure of the gas using the ideal gas law.
5. **Compare and contrast** atmospheric pressure to the pressure of the water vapor in the kernel.
6. **Infer** why all the popcorn kernels did not pop.
7. **Error Analysis** Identify a potential source of error for this lab, and suggest a method to correct it.

INQUIRY EXTENSION

Design an experiment that tests the amount of pressure necessary to burst different types of popcorn kernels.

Analyze and Conclude

1. Sample calculation: 7.8 mL − 5.0 mL = 2.8 mL; 2.8 mL = 0.0028 L
2. Sample calculation: 110.943 g − 109.952 g = 0.991 g water vapor
3. Sample calculation: 0.991 g H_2O × (1 mol H_2O)/(18.02 g H_2O) = 0.0550 mol H_2O
4. Sample calculation: $P = nRT/V$ = ((0.0550 mol) (0.0821 (L•atm)/(mol•K))(498 K)/0.0028 L) = 803 atm
5. The calculated value is much higher, and therefore the kernel exploded outward.
6. The unpopped kernels could be due to lower water content. This would result in less pressure as the water becomes a vapor.
7. Answers will vary. Sources of error include the presence of water vapor on the beaker, loss of some of the oil, and the fact that some kernels do not pop.

INQUIRY EXTENSION

Students might discover that different types of popcorn require different pressures to pop. The amount of water content will have the biggest effect on the rate of popping.

CHAPTER 13 STUDY GUIDE

Vocabulary Practice

BIGIDEA Gases respond in predictable ways to changes in pressure, temperature, volume, and number of particles.

SECTION 1 The Gas Laws

MAINIDEA For a fixed amount of gas, a change in one variable—pressure, temperature, or volume—affects the other two.

- Boyle's law states that the volume of a fixed amount of gas is inversely proportional to its pressure at constant temperature.

$$P_1V_1 = P_2V_2$$

- Charles's law states that the volume of a fixed amount of gas is directly proportional to its Kelvin temperature at constant pressure.

$$\frac{V_1}{T_1} = \frac{V_2}{T_2}$$

- Gay-Lussac's law states that the pressure of a fixed amount of gas is directly proportional to its Kelvin temperature at constant volume.

$$\frac{P_1}{T_1} = \frac{P_2}{T_2}$$

- The combined gas law relates pressure, temperature, and volume in a single statement.

$$\frac{P_1V_1}{T_1} = \frac{P_2V_2}{T_2}$$

VOCABULARY
- Boyle's law
- absolute zero
- Charles's law
- Gay-Lussac's law
- combined gas law

SECTION 2 The Ideal Gas Law

MAINIDEA The ideal gas law relates the number of particles to pressure, temperature, and volume.

- Avogadro's principle states that equal volumes of gases at the same pressure and temperature contain equal numbers of particles.
- The ideal gas law relates the amount of a gas present to its pressure, temperature, and volume.

$$PV = nRT$$

- The ideal gas law can be used to find molar mass if the mass of the gas is known or the density of the gas if its molar mass is known.

$$\text{M} = \frac{mRT}{PV} \qquad D = \frac{MP}{RT}$$

- At very high pressures and very low temperatures, real gases behave differently than ideal gases.

VOCABULARY
- Avogadro's principle
- molar volume
- standard temperature and pressure (STP)
- ideal gas constant (R)
- ideal gas law

SECTION 3 Gas Stoichiometry

MAINIDEA When gases react, the coefficients in the balanced chemical equation represent both molar amounts and relative volumes.

- The coefficients in a balanced chemical equation specify volume ratios for gaseous reactants and products.
- The gas laws can be used along with balanced chemical equations to calculate the amount of a gaseous reactant or product in a reaction.

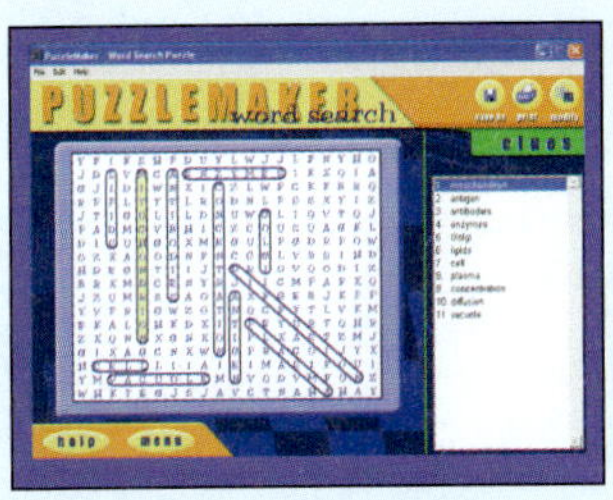

Vocabulary Practice

For additional practice with vocabulary, have students access the Vocabulary PuzzleMaker.

CHAPTER 13 STUDY GUIDE

Use the Vocabulary

To reinforce chapter vocabulary, have students write a sentence using each term. OL EL

Review Strategies

- Have students list the SI and one other common unit for volume, pressure, and temperature. OL
- Have students summarize the gas laws by listing the values that are held constant and those that vary for each law. OL
- Problems from the Supplemental Practice Problems appendix or the Supplemental Problems booklet can be used for review. OL

Students can review with these online resources.

Vocabulary Practice

Vocabulary eGames
Vocabulary eFlashcards
Vocabulary PuzzleMaker

Section Self-Check

Chapter Self-Check

Online Test Practice

Use *eAssessment* to:

- create multiple versions of tests
- edit existing questions and add your own questions
- build tests aligned with select state curriculum standards using built-in state curriculum tags
- track students' progress

SECTION 1

Mastering Concepts

50. Boyle's law: the volume of a given amount of gas held at a constant temperature varies inversely with pressure, $P_1V_1 = P_2V_2$; Charles's law: the volume of a given mass of gas is directly proportional to its kelvin temperature at constant pressure, $V_1/T_1 = V_2/T_2$; Gay-Lussac's law: the pressure of a given mass of gas varies directly with the kelvin temperature when the volume remains constant, $P_1/T_1 = P_2/T_2$; combined gas law: states the relationships between pressure, volume, and temperature of a fixed amount of gas, $P_1V_1/T_1 = P_2V_2/T_2$

51. One variable always decreases as the other increases.

52. One variable always increases as the other increases.

53. $T = 0.00°C$ (273 K) and $P = 1.00$ atm

54. atm for pressure, kelvins for temperature, and L for volume

Mastering Problems

55. Charles's law states that the volume of a given mass of gas is directly proportional to temperature. The graphed data follow this law because doubling the temperature doubles the volume. Therefore, the data are accurate.

56. 5.91×10^4 L

57. **a.** 1.6 atm
b. 200 K
c. 510 mm Hg

58. 2.68 L

59. 1.17 atm

60. 74.8 kPa

CHAPTER 13 ASSESSMENT

SECTION 1

Mastering Concepts

50. State Boyle's law, Charles's law, Gay-Lussac's law, and the combined gas law in words and equations.

51. If two variables are inversely proportional, what happens to the value of one as the value of the other increases?

52. If two variables are directly proportional, what happens to the value of one as the value of the other increases?

53. List the standard conditions for gas measurements.

54. Identify the units most commonly used for P, V, and T.

Mastering Problems

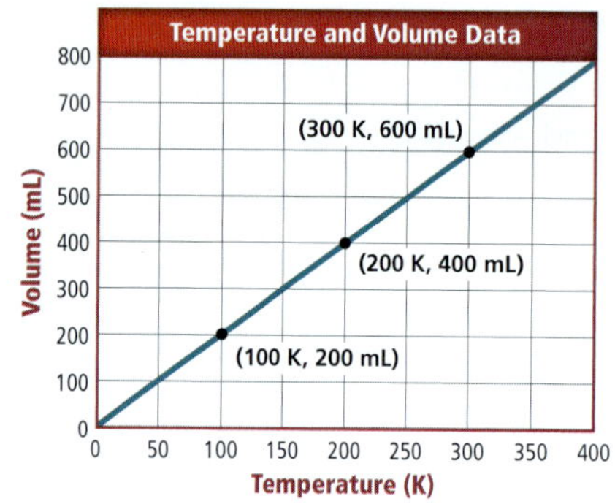

■ **Figure 13**

55. Use Charles's law to determine the accuracy of the data plotted in **Figure 13.**

56. Weather Balloons A weather balloon is filled with helium that occupies a volume of 5.00×10^4 L at 0.995 atm and 32.0°C. After it is released, it rises to a location where the pressure is 0.720 atm and the temperature is −12.0°C. What is the volume of the balloon at the new location?

57. Use Boyle's, Charles's, or Gay-Lussac's law to calculate the missing value in each of the following.
a. $V_1 = 2.0$ L, $P_1 = 0.82$ atm, $V_2 = 1.0$ L, $P_2 = ?$
b. $V_1 = 250$ mL, $T_1 = ?$, $V_2 = 400$ mL, $T_2 = 298$ K
c. $V_1 = 0.55$ L, $P_1 = 740$ mm Hg, $V_2 = 0.80$ L, $P_2 = ?$

58. Hot-Air Balloons A sample of air occupies 2.50 L at a temperature of 22.0°C. What volume will this sample occupy inside a hot-air balloon at a temperature of 43.0°C? Assume that the pressure inside the balloon remains constant.

59. What is the pressure of a fixed volume of hydrogen gas at 30.0°C if it has a pressure of 1.11 atm at 15.0°C?

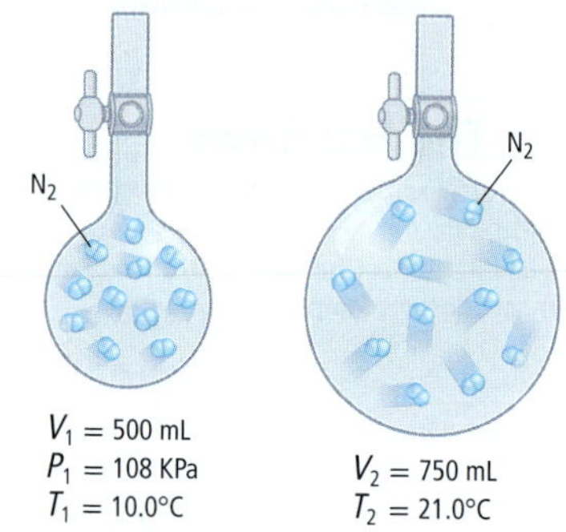

■ **Figure 14**

60. A sample of nitrogen gas is transferred to a larger flask, as shown in **Figure 14.** What is the pressure of nitrogen in the second flask?

SECTION 2

61. State Avogadro's principle.

62. State the ideal gas law.

63. What volume is occupied by 1 mol of a gas at STP? What volume does 2 mol occupy at STP?

64. Define the term *ideal gas,* and explain why there are no true ideal gases in nature.

65. List two conditions under which a gas is least likely to behave ideally.

66. What units must be used to express the temperature in the equation for the ideal gas law? Explain.

Mastering Problems

67. Home Fuel Propane (C_3H_8) is a gas commonly used as a home fuel for cooking and heating.
a. Calculate the volume that 0.540 mol of propane occupies at STP.
b. Think about the size of this volume and the amount of propane that it contains. Why do you think propane is usually liquefied before it is transported?

68. Careers in Chemistry A physical chemist measured the lowest pressure achieved in a laboratory—about 1.0×10^{-15} mm Hg. How many molecules of gas are present in a 1.00-L sample at that pressure if the sample's temperature is 22.0°C?

69. Calculate the number of moles of O_2 gas held in a sealed, 2.00-L tank at 3.50 atm and 25.0°C. How many moles would be in the tank if the temperature was raised to 49.0°C and the pressure remained constant?

SECTION 2

Mastering Concepts

61. At a fixed temperature and pressure, equal volumes of any ideal gas contain equal numbers of particles.

62. The ideal gas law describes the physical behavior of an ideal gas in terms of the pressure, volume, temperature, and number of moles of gas present: $PV = nRT$.

63. 22.4 L; 44.8 L

64. An ideal gas is one whose particles take up no space and have no intermolecular attractive forces and it follows the gas laws under all conditions of temperature and pressure. No gas is truly ideal because all gas particles have some volume and are subject to intermolecular interactions.

65. high pressure and low temperature

66. kelvins; V is not directly proportional to Celsius temperature.

Mastering Problems

67. **a.** 12.1 L
b. Propane occupies a much smaller volume when liquefied.

68. 3.3×10^4 molecules

69. 0.286 mol; 0.265 mol

Chapter Self-Check

70. **Perfumes** Geraniol is a compound found in rose oil that is used in perfumes. What is the molar mass of geraniol if its vapor has a density of 0.480 g/L at a temperature of 260.0°C and a pressure of 0.140 atm?

71. Find the volume that 42 g of carbon monoxide gas occupies at STP.

72. Determine the density of chlorine gas at 22.0°C and 1.00 atm.

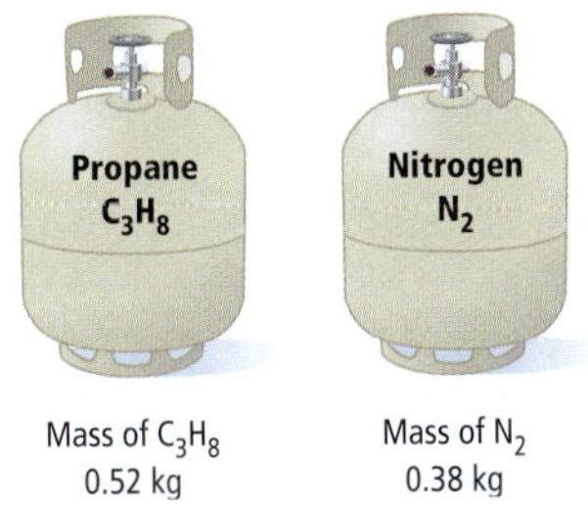

■ **Figure 15**

73. Which of the gases in **Figure 15** occupies the greatest volume at STP? Explain your answer.

74. If the containers in **Figure 15** each hold 4.00 L, what is the pressure inside each? Assume ideal behavior.

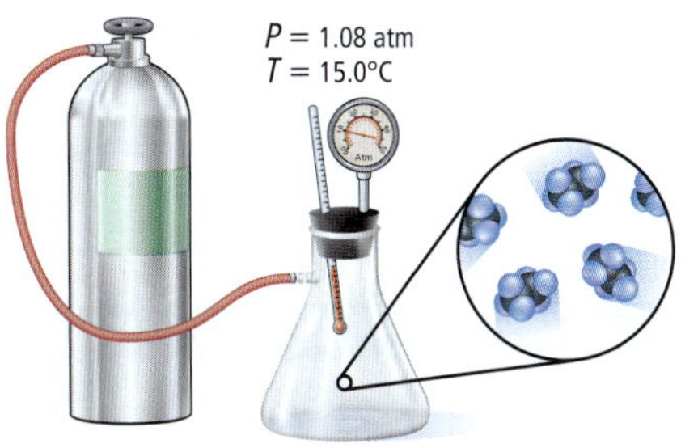

■ **Figure 16**

75. A 2.00-L flask is filled with ethane gas (C_2H_6) from a small cylinder, as shown in **Figure 16.** What is the mass of the ethane in the flask?

76. What is the density of a sample of nitrogen gas (N_2) that exerts a pressure of 5.30 atm in a 3.50-L container at 125°C?

77. How many moles of helium gas (He) would be required to fill a 22-L container at a temperature of 35°C and a pressure of 3.1 atm?

78. Before a reaction, two gases share a container at a temperature of 200 K. After the reaction, the product is in the same container at a temperature of 400 K. If both V and P are constant, what must be true of n?

SECTION 3

Mastering Concepts

79. Why must an equation be balanced before using it to determine the volumes of gases involved in a reaction?

80. It is not necessary to consider temperature and pressure when using a balanced equation to determine relative gas volume. Why?

81. What information do you need to solve a volume-mass problem that involves gases?

82. Explain why the coefficients in a balanced chemical equation represent not only molar amounts but also relative volumes for gases.

83. Do the coefficients in a balanced chemical equation represent volume ratios for solids and liquids? Explain.

Mastering Problems

84. **Ammonia Production** Ammonia is often formed by reacting nitrogen and hydrogen gases. How many liters of ammonia gas can be formed from 13.7 L of hydrogen gas at 93.0°C and a pressure of 40.0 kPa?

85. A 6.5-L sample of hydrogen sulfide is treated with a catalyst to promote the reaction shown below.

$$2H_2S(g) + O_2(g) \rightarrow 2H_2O(g) + 2S(s)$$

If the H_2S reacts completely at 2.0 atm and 290 K, how much water vapor, in grams, is produced?

86. To produce 15.4 L of nitrogen dioxide at 310 K and 2.0 atm, how many liters of nitrogen gas and oxygen gas are required?

87. Use the reaction shown below to answer these questions.

$$2CO(g) + 2NO(g) \rightarrow N_2(g) + 2CO_2(g)$$

a. What is the volume ratio of carbon monoxide to carbon dioxide in the balanced equation?

b. If 42.7 g of CO is reacted completely at STP, what volume of N_2 gas will be produced?

88. When 3.00 L of propane gas is completely combusted to form water vapor and carbon dioxide at 350°C and 0.990 atm, what mass of water vapor results?

89. When heated, solid potassium chlorate ($KClO_3$) decomposes to form solid potassium chloride and oxygen gas. If 20.8 g of potassium chlorate decomposes, how many liters of oxygen gas will form at STP?

90. **Welding** The gas acetylene, often used for welding, burns according to the following equation.

$$2C_2H_2(g) + 5O_2(g) \rightarrow 2H_2O(g) + 4CO_2(g)$$

If you have a 10.0-L tank of acetylene at 25.0°C and 1.00 atm pressure, how many moles of CO_2 will be produced if you burn all the acetylene in the tank?

70. 1.50×10^2 g/mol
71. 34 L
72. 2.93 g/L
73. The N_2 occupies the greatest volume at STP. The N_2 occupies 310 L, while the C_3H_8 occupies only 260 L.
74. propane: 66.1 atm; nitrogen: 76.2 atm
75. 2.75 g
76. 4.55 g/L
77. 2.7 mol
78. With a constant volume and pressure and doubling of the temperature, the number of moles would be half the initial number of moles

SECTION 3

Mastering Concepts

79. Equation coefficients represent ratios between gas volumes in the reaction.
80. Temperature and pressure are the same for each gas involved in the reaction. These conditions affect each gas in the same way.
81. balanced equation, at least one mass or volume value for a reactant or product, and P and T conditions under which gas volumes have been measured
82. Avogadro's principle states that equal volumes of gases at the same temperature and pressure contain equal numbers of particles (or an equal number of moles). Therefore, the coefficients also represent the relative volumes of the gases.
83. No, this relationship only applies to gases that behave like ideal gases.

Mastering Problems

84. 9.13 L
85. 9.9 g
86. N_2 gas: 7.7 L; O_2 gas: 15.4 L
87. **a.** 1:1
    b. 17.1 L
88. 4.2 g
89. 5.71 L
90. 0.817 mol

MIXED REVIEW

91. a. $CH_4(g) + 2O_2(g) \longrightarrow CO_2(g) + 2H_2O(g)$
b. 1:2

92. 3.9 g

93. 4.94×10^{-6} atm

94. a. 4.48 L
b. 2.37 L
c. 4.06 L

95. 2.63 g

96. 1.17 atm

97. 0.19 mol

98. a. 1.2 L
b. 0.75 L

THINK CRITICALLY

99. 0.56 g

100. a. 4.55 g/L
b. 1.51×10^{22} molecules per liter

101. 380 L

102. 20.4 atm

103. 1850 L

104.

Table 3 Volume of H_2 Collected

Trial	T(°C)	V(mL)
1	300	48
2	175	37
3	110	32
4	0	22
5	−100	15
6	−150	11

Extrapolating the graph to a volume of 0 mL intersects the temperature axis at approximately −273°C, the temperature called absolute zero.

105. 8.314×10^6 cm^3• Pa/(mol•K)

106. At high pressures and low temperatures, the ideal gas law will calculate a pressure that is higher than the gas actually exerts. Under these conditions, the effects of intermolecular forces become more important. Attractions between particles will lower the force of the collisions with the container wall, resulting in an actual pressure that is lower than the pressure calculated by the ideal gas law.

MIXED REVIEW

91. Gaseous methane (CH_4) undergoes complete combustion by reacting with oxygen gas to form carbon dioxide and water vapor.
a. Write a balanced equation for this reaction.
b. What is the volume ratio of methane to water in this reaction?

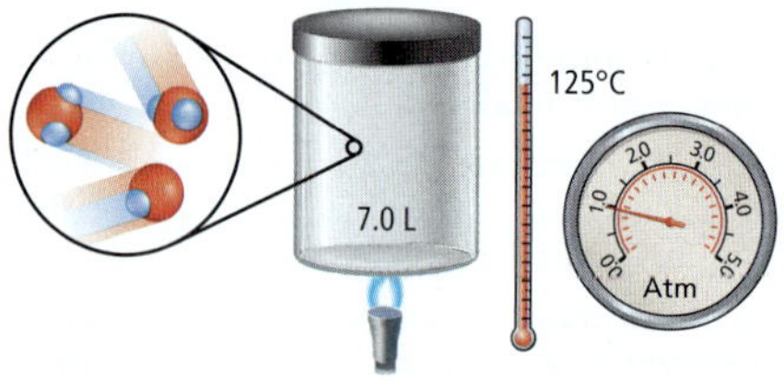

■ **Figure 17**

92. Calculate the amount of water vapor, in grams, contained in the vessel shown in **Figure 17.**

93. Television Determine the pressure inside a television picture tube with a volume of 3.50 L that contains 2.00×10^{-5} g of nitrogen gas at 22.0°C.

94. Determine how many liters 8.80 g of carbon dioxide gas would occupy at:
a. STP
b. 160°C and 3.00 atm
c. 288 K and 118 kPa

95. Oxygen Consumption If 5.00 L of hydrogen gas, measured at a temperature of 20.0°C and a pressure of 80.1 kPa, is burned in excess oxygen to form water, what mass of oxygen will be consumed? Assume temperature and pressure remain constant.

96. A fixed amount of oxygen gas is held in a 1.00-L tank at a pressure of 3.50 atm. The tank is connected to an empty 2.00-L tank by a tube with a valve. After this valve has been opened and the oxygen is allowed to flow freely between the two tanks at a constant temperature, what is the final pressure in the system?

97. If 2.33 L of propane at 24°C and 67.2 kPa is completely burned in excess oxygen, how many moles of carbon dioxide will be produced?

98. Respiration A human breathes about 0.50 L of air during a normal breath. Assume the conditions are at STP.
a. What is the volume of one breath on a cold day atop Mt. Everest? Assume −60°C and 253 mm Hg pressure.
b. Air normally contains about 21% oxygen. If the O_2 content is about 14% atop Mt. Everest, what volume of air does a person need to breathe to supply the body with the same amount of oxygen?

THINK CRITICALLY

99. Apply An oversized helium balloon in a floral shop must have a volume of at least 3.8 L to rise. When 0.1 mol is added to the empty balloon, its volume is 2.8 L. How many grams of He must be added to make it rise? Assume constant T and P.

100. Calculate A toy manufacturer uses tetrafluoroethane ($C_2H_2F_4$) at high temperatures to fill plastic molds for toys.
a. What is the density (in g/L) of $C_2H_2F_4$ at STP?
b. Find the molecules per liter of $C_2H_2F_4$ at 220°C and 1.0 atm.

101. Analyze A solid brick of dry ice (CO_2) weighs 0.75 kg. Once the brick has fully sublimated into CO_2 gas, what would its volume be at STP?

102. Apply Calculate the pressure of 4.67×10^{22} molecules of CO gas mixed with 2.87×10^{24} molecules of N_2 gas in a 6.00-L container at 34.8°C.

103. Analyze When nitroglycerin ($C_3H_5N_3O_9$) explodes, it decomposes into the following gases: CO_2, N_2, O_2, and H_2O. If 239 g of nitroglycerin explodes, what volume will the mixture of gaseous products occupy at 1.00 atm pressure and 2678°C?

104. Make and Use Graphs The data in **Table 3** show the volume of hydrogen gas collected at several different temperatures. Illustrate these data with a graph. Use the graph to complete the table. Determine the temperature at which the volume will reach a value of 0 mL. What is this temperature called?

Table 3 Volume of H_2 Collected

Trial	T (°C)	V (mL)
1	300	48
2	175	37
3	110	
4	0	22
5		15
6	−150	11

105. Apply What is the numerical value of the ideal gas constant (R) in $\frac{cm^3{\cdot}Pa}{mol{\cdot}K}$?

106. Infer At very high pressures, will the ideal gas law calculate a pressure that is higher or lower than the actual pressure exerted by a sample of gas? How will the calculated pressure compare to the actual pressure at low temperatures? Explain your answers.

Chapter Self-Check

CHALLENGE PROBLEM

107. **Baking** A baker uses baking soda as the leavening agent for his pumpkin-bread recipe. The baking soda decomposes according to two possible reactions.

$$2NaHCO_3(s) \rightarrow Na_2CO_3(s) + H_2O(l) + CO_2(g)$$

$$NaHCO_3(s) + H^+(aq) \rightarrow H_2O(l) + CO_2(g) + Na^+(aq)$$

Calculate the volume of CO_2 that forms per gram of $NaHCO_3$ by each reaction process. Assume the reactions take place at 210°C and 0.985 atm.

CUMULATIVE REVIEW

108. Convert each mass measurement to its equivalent in kilograms.
- **a.** 247 g
- **b.** 53 mg
- **c.** 7.23 mg
- **d.** 975 mg

109. Write the electron configuration for each atom.
- **a.** iodine
- **b.** boron
- **c.** chromium
- **d.** krypton
- **e.** calcium
- **f.** cadmium

110. For each element, tell how many electrons are in each energy level and write the electron dot structure.
- **a.** Kr
- **b.** Sr
- **c.** P
- **d.** B
- **e.** Br
- **f.** Se

111. How many atoms of each element are present in five formula units of calcium permanganate?

112. You are given two clear, colorless aqueous solutions. One solution contains an ionic compound, and one contains a covalent compound. How could you determine which solution is the ionic solution and which solution is the covalent solution?

113. Write a balanced equation for the following reactions.
- **a.** Zinc displaces silver in silver chloride.
- **b.** Sodium hydroxide and sulfuric acid react to form sodium sulfate and water.

114. Terephthalic acid is an organic compound used in the formation of polyesters. It contains 57.8% C, 3.64% H, and 38.5% O. The molar mass is approximately 166 g/mol. What is the molecular formula of terephthalic acid?

115. The particles of which gas have the highest average speed? The lowest average speed?
- **a.** carbon monoxide at 90°C
- **b.** nitrogen trifluoride at 30°C
- **c.** methane at 90°C
- **d.** carbon monoxide at 30°C

WRITING IN ▶ Chemistry

116. **Hot-Air Balloons** Many early balloonists dreamed of completing a trip around the world in a hot-air balloon, a goal not achieved until 1999. Write about what you imagine a trip in a balloon would be like, including a description of how manipulating air temperature would allow you to control altitude.

117. **Scuba** Investigate and explain the function of the regulators on the air tanks used by scuba divers.

DBQ Document-Based Questions

The Haber Process *Ammonia (NH_3) is used in the production of fertilizer, refrigerants, dyes, and plastics. The Haber process is a method of producing ammonia through a reaction of molecular nitrogen and hydrogen. The equation for the reversible reaction is:*

$$N_2(g) + 3H_2(g) \rightleftharpoons 2NH_3(g) + 92\ kJ$$

Figure 18 *shows the effect of temperature and pressure on the amount of ammonia produced by the Haber process.*

Data obtained from: Smith, M. 2004. *Science* 39:1021–1034.

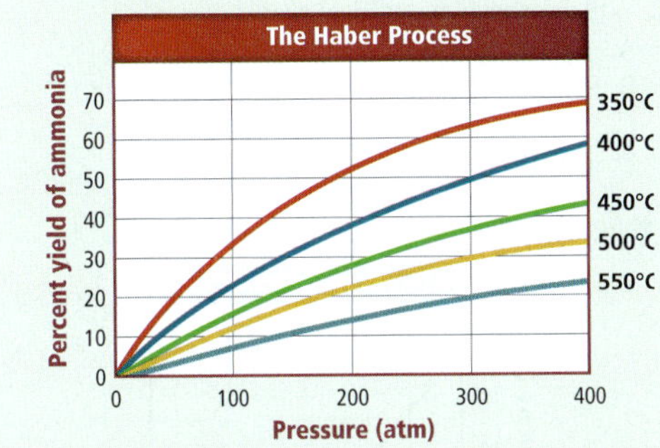

■ **Figure 18**

118. Explain how the percent yield of ammonia is affected by pressure and temperature.

119. The Haber process is typically run at 200 atm and 450°C, a combination proven to yield a substantial amount of ammonia in a short time.
- **a.** If the temperature goes from 450°C to 500°C, what pressure is necessary to maintain the typical percent yield?
- **b.** How do you think lowering the temperature of this reaction below 450°C would affect the amount of time required to produce ammonia?

CHALLENGE PROBLEM

107. $2NaHCO_3(s) \rightarrow Na_2CO_3(s) + H_2O(l) + CO_2(g)$ = 0.24 L per gram. $NaHCO_3(s) + H^+(aq) \rightarrow H_2O(l) + CO_2(g) + Na^+(aq)$ = 0.48 L per gram

CUMULATIVE REVIEW

108.
- **a.** 0.247 kg
- **b.** 5.3×10^{-5} kg
- **c.** 7.23×10^{-6} kg
- **d.** 9.75×10^{-4} kg

109.
- **a.** $[Kr]4d^{10}5s^25p^5$
- **b.** $[He]2s^22p^1$
- **c.** $[Ar]3d^54s^1$
- **d.** $[Ar]3d^{10}4s^24p^6$
- **e.** $[Ar]4s^2$
- **f.** $[Kr]4d^{10}5s^2$

110.
- **a.** 2,8,18,8 :Kr:
- **b.** 2,8,18,8,2 •Sr•
- **c.** 2,8,5 •P•
- **d.** 2,3 •B•
- **e.** 2,8,18,7 :Br•
- **f.** 2,8,18,6 :Se•

111. 5 formula units $Ca(MnO_4)_2$: 5 Ca, 10 Mn, 40 O

112. The ionic solution will conduct electricity. The covalent solution will not conduct electricity.

113.
- **a.** $Zn(s) + 2AgCl(aq) \rightarrow ZnCl_2(aq) + 2Ag(s)$
- **b.** $2NaOH(aq) + H_2SO_4(aq) \rightarrow Na_2SO_4(aq) + 2H_2O(l)$

114. $C_8H_6O_4$

115. c, b; The average speed is greater at higher temperature and lower at greater molar mass.

WRITING IN ▶ Chemistry

Rubric

116. Student answers should include a description of how the difference in density of hot air and cold air is what allows hot-air balloons to stay aloft, and how manipulating the rate of heating of air allows a balloonist to ascend and descend.

117. Student answers should include the overall function of the air tank regulator, a device that changes air pressure levels and delivers air. The first stage regulator is the regulator attached to the scuba tank, and it lowers the tank pressure to ambient pressure plus a predetermined pressure (e.g., ambient + 140 psi). The second stage regulator follows the first stage regulator in line, and it delivers compressed air to the diver.

DBQ Document-Based Questions

Data obtained from: Smith, M. 2004. Science 39:1021–1034.

118. Ammonia yield is increased by very high pressures. Yield decreases at very high temperatures.

119.
- **a.** At 200 atm and 450°C, the percent yield is around 28%. To maintain this yield a pressure of ~280 atm is needed.
- **b.** Lowering the temperature of this reaction slows the rate of reaction, increasing the amount of time required to produce ammonia.

MULTIPLE CHOICE

1. D
2. D
3. C
4. A
5. C
6. A
7. B

MULTIPLE CHOICE

Use the graph below to answer Questions 1 and 2.

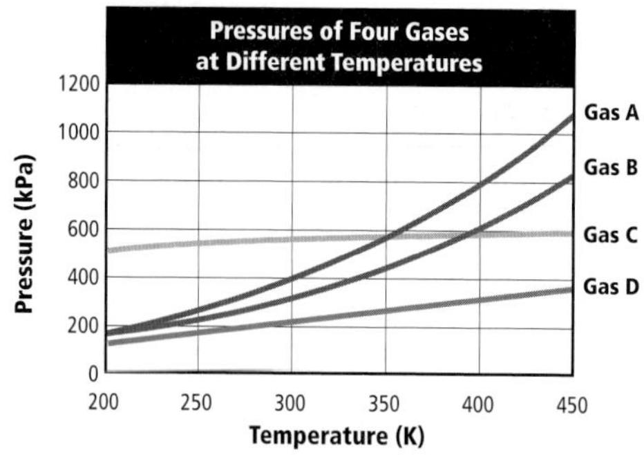

1. Which is evident in the graph above?
A. As temperature increases, pressure decreases.
B. As pressure increases, volume decreases.
C. As temperature increases, the number of moles decreases.
D. As pressure decreases, temperature decreases.

2. Which behaves as an ideal gas?
A. Gas A
B. Gas B
C. Gas C
D. Gas D

Use the graph below to answer Question 3.

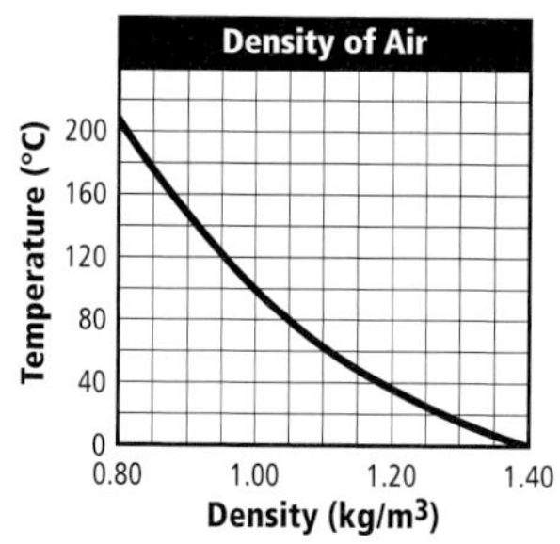

3. The graph shows data from an experiment that analyzed the relationship between temperature and air density. What is the independent variable in the experiment?
A. density
B. mass
C. temperature
D. time

4. Hydrofluoric acid (HF) is used in the manufacture of electronics equipment. It reacts with calcium silicate ($CaSiO_3$), a component of glass. What type of property prevents hydrofluoric acid from being transported or stored in glass containers?
A. chemical property
B. extensive physical property
C. intensive physical property
D. quantitative property

5. Sodium hydroxide (NaOH) is a strong base found in products used to clear clogged plumbing. What is the percent composition of sodium hydroxide?
A. 57.48% Na, 60.00% O, 2.52% H
B. 2.52% Na, 40.00% O, 57.48% H
C. 57.48% Na, 40.00% O, 2.52% H
D. 40.00% Na, 2.52% O, 57.48% H

Use the circle graph below to answer Question 6.

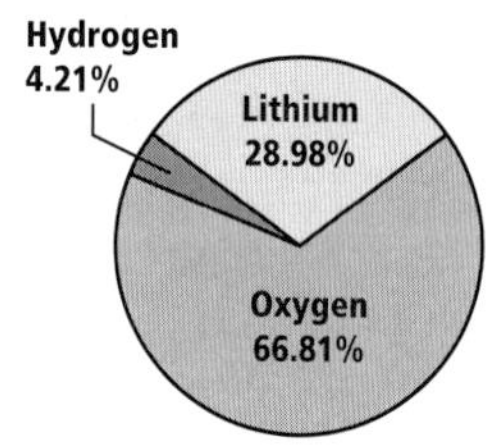

6. What is the empirical formula for this compound?
A. LiOH
B. Li_2OH
C. Li_3OH
D. $LiOH_2$

7. While it is on the ground, a blimp is filled with 5.66×10^6 L of He gas. The pressure inside the grounded blimp, where the temperature is 25°C, is 1.10 atm. Modern blimps are nonrigid, which means that their volumes can change. If the pressure inside the blimp remains the same, what will be the volume of the blimp at a height of 2300 m, where the temperature is 12°C?
A. 2.72×10^6 L
B. 5.40×10^6 L
C. 5.66×10^6 L
D. 5.92×10^6 L

SHORT ANSWER

8. Describe several observations that provide evidence that a chemical change has occurred.

9. Identify five diatomic molecules that occur naturally, and explain why the atoms in these molecules share a pair, or pairs, of electrons.

10. The diagram below shows the Lewis structure for the polyatomic ion nitrate (NO_3^-). Define the term *polyatomic ion,* and give examples of other ions of this type.

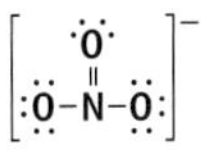

EXTENDED RESPONSE

Use the table below to answer Question 11.

Radon Levels August 2004 through July 2005

Date	Radon Level (pCi/L)	Date	Radon Level (pCi/L)
8/04	0.26	2/05	0.087
9/04	0.052	3/05	0.087
10/04	0.087	4/05	0.10
11/04	0.052	5/05	0.22
12/04	0.069	6/05	0.087
1/05	0.035	7/05	0.16

11. Radon is a radioactive gas produced when radium in soil and rock decays. It is a known carcinogen. The data above show radon levels measured in a fictitious community. Select a method for graphing these data. Explain the reasons for your choice, and graph the data.

SAT SUBJECT TEST: CHEMISTRY

12. Which diagram shows the relationship between volume and pressure for a gas at constant temperature?

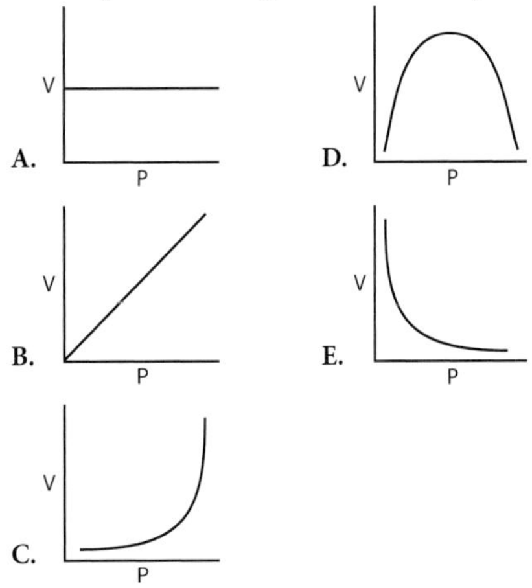

13. The reaction that provides blowtorches with their intense flame is the combustion of acetylene (C_2H_2) with oxygen to form carbon dioxide and water vapor. Assuming that the pressure and temperature of the reactants are the same, what volume of oxygen gas is required to completely burn 5.60 L of acetylene?
    - A. 2.24 L
    - B. 5.60 L
    - C. 8.20 L
    - D. 11.2 L
    - E. 14.0 L

14. Assuming ideal behavior, how much pressure will 0.0468 g of ammonia (NH_3) gas exert on the walls of a 4.00-L container at 35.0°C?
    - A. 0.0174 atm
    - B. 0.296 atm
    - C. 0.0126 atm
    - D. 0.00198 atm
    - E. 0.278 atm

NEED EXTRA HELP?

If You Missed Question . . .	1	2	3	4	5	6	7	8	9	10	11	12	13	14
Review Section . . .	13.1	13.2	1.3	3.1	10.4	10.4	13.1	3.2	8.1	7.3	2.4	13.1	13.1	13.1

SHORT ANSWER

8. Evidence of chemical change includes temperature change, color change, the production of an odor or a gas, and the precipitation of a solid.
9. Hydrogen (H_2), oxygen (O_2), nitrogen (N_2), fluorine (F_2), chlorine (Cl_2), bromine (Br_2), and iodine (I_2) occur naturally. By sharing a pair, or pairs, of electrons, both atoms achieve noble gas configurations, resulting in atomic stability.
10. A polyatomic ion is an ion made up of more than one atom that acts as a single unit with a net charge. Other examples include hydroxide (OH^-), chlorite (ClO_2^-), and cyanide (CN^-).

EXTENDED RESPONSE

11. Students should choose a bar or line graph for the data, with justification that each data point can be represented on the graph.

SAT SUBJECT TEST: CHEMISTRY

12. E
13. E
14. A

CHAPTER 14 Mixtures and Solutions

BIGIDEA **Nearly all of the gases, liquids, and solids that make up our world are mixtures.**

ESSENTIAL QUESTIONS	RESOURCES TO ASSESS MASTERY
SECTION 1 **Types of Mixtures** **1.** How do the properties of suspensions, colloids, and solutions compare? **2.** What are the types of colloids and types of solutions? **3.** How are the electrostatic forces in colloids described? 2 sessions 1 block	**Progress Monitoring** Formative Assessment, p. 479 Reading Check, p. 477 Section Review, p. 479
SECTION 2 **Solution Concentration** **1.** How can concentration be described using different units? **2.** How are the concentrations of solutions determined? **3.** What is the molarity of a solution and how can it be calculated? 1 session 0.5 block	**Progress Monitoring** Formative Assessment, pp. 485, 487 Reading Check, p. 482 Section Review, p. 488
SECTION 3 **Factors Affecting Solvation** **1.** How do intermolecular forces affect solvation? **2.** What is solubility? **3.** Which factors affect solubility? 1 session 0.5 block	**Progress Monitoring** Formative Assessment, pp. 489, 493, 494, 496 Reading Check, p. 492 Graph Check, p. 493 Section Review, p. 497
SECTION 4 **Colligative Properties of Solutions** **1.** What are colligative properties? **2.** What are four colligative properties of solutions? **3.** How are the boiling point elevation and freezing point depression of a solution determined? 2 sessions 1 block	**Progress Monitoring** Reading Check, p. 499 Formative Assessment, pp. 500, 502 Graph Check, p. 501 Section Review, p. 504 **Summative Assessment** Chapter Assessment, p. 508 *eAssessment* Chapter Test (Scaffolded)

Go online!

LEVELED RESOURCES	LAB MATERIALS
Science Notebook 14.1 OL ***FAST FILE Chapter Resources:*** Study Guide, p. 44 BL	LaunchLAB p. 474 / **20 min** ammonium chloride (NH_4Cl), balance, 100-mL beaker, water, 50-mL graduated cylinder, stirring rod, calcium chloride ($CaCl_2$)
Science Notebook 14.2 OL ***FAST FILE Chapter Resources:*** Study Guide, p. 45 BL	
Science Notebook 14.3 OL ***FAST FILE Chapter Resources:*** ChemLab Worksheet, p. 33 OL Study Guide, p. 47 BL	ChemLAB p. 506 / **45 min** copper(II) sulfate pentahydrate, distilled water, test tubes, 25-mL graduated cylinder, glass stirring rod, tweezers, test tube rack, mortar and pestle, spatula, clock
Science Notebook 14.4 OL ***FAST FILE Chapter Resources:*** MiniLab Worksheet, p. 32 OL Study Guide, p. 49 BL	MiniLAB p. 502 / **20 min** 400-mL beakers, crushed ice, cold tap water, nonmercury thermometer, stirring rod, rock salt (NaCl)

ADDITIONAL RESOURCES

eAssessment
ConnectED Student Center
ConnectED Teacher Center
Lesson Presentations
eSolutions
Videos
Animations
Online Quizzes
Virtual Investigations
Multilingual eGlossary
Personal Tutors
WebQuests
Applying Practices Worksheets
PBLs

Practice:

Challenge Problems AL
Supplemental Problems BL OL
Chapter Test (Scaffolded)
Standardized Test Practice

Additional Resources:

Blueprints for Success
Solving Problems: A Chemistry Handbook

Lab Resources:

Laboratory Manual OL
CBL Laboratory Manual OL
Small-Scale Laboratory Manual OL
Forensics Laboratory Manual OL

BL Below Level OL On Level AL Advanced Learners EL English Learners COOP LEARN Cooperative Learning

CHAPTER 14

LaunchLAB

Rubric

How does energy change when solutions form?

Purpose Students will investigate whether the dissolution of two different solids in water is an exothermic or endothermic process.

Safety Precautions Approve lab safety forms before work begins. Review the MSDS for hazardous chemicals used in this lab with students.

Disposal Dispose of solutions by rinsing them down a drain with excess water.

Teaching Strategies

- After completing the lab, students might want to investigate using other solids. Two other suggested solids to investigate are magnesium sulfate heptahydrate (Epsom salts) and anhydrous sodium borate (fused borax). Do not use plain borax, which is a decahydrate.
- For more precise results, students can use stirring rods and thermometers.

Expected Results Calcium chloride: temperature increases
Ammonium chloride: temperature decreases

(Optional solids)
Epsom salts: temperature decreases
Fused borax: temperature increases

CHAPTER 14

Mixtures and Solutions

BIGIDEA Nearly all of the gases, liquids, and solids that make up our world are mixtures.

SECTIONS

1 Types of Mixtures

2 Solution Concentration

3 Factors Affecting Solvation

4 Colligative Properties of Solutions

LaunchLAB

How does energy change when solutions form?

When a solution forms, the forces of attraction between the solvent particles and the solute particles disrupt the intermolecular forces in the solute. In this lab, you will discover whether this interaction affects the energy of the solution.

FOLDABLES Study Organizer

Concentration

Make a bound book. Use it to help you organize information about the concentration of solutions.

Procedure

1. Read and complete the lab safety form.
2. Measure 10 g of **ammonium chloride (NH_4Cl)** using a **balance,** and place it in a **100-mL beaker.**
3. Measure 30 mL of **water** with a **50-mL graduated cylinder,** and add to the N H_4Cl, stirring with your **stirring rod.**
4. Feel the bottom of the beaker, and record your observations.
5. Repeat Steps 2–4 using **calcium chloride ($CaCl_2$).**
6. Dispose of the solutions by flushing them down a drain with water.

Analysis

1. **Compare** Which dissolving process is exothermic, and which is endothermic? Calcium chloride dissolves exothermically. Ammonium chloride dissolves endothermically.
2. **Infer** What are everyday applications for dissolving processes that are exothermic? Endothermic? Answers will vary, but might include hot packs and cold packs used for medical purposes.

Inquiry If you wanted a greater temperature change, would you add more solute or more solvent? Explain. To a point, adding more solute would result in a greater temperature change of the solution. Answers will vary.

Two mixtures used in the construction of this building are steel and concrete. Iron is the major component of steel, but other elements, such as nickel, manganese, chromium, vanadium, and tungsten, can be added depending on the desired properties. Cement is used to make concrete and mortar to form building materials that are strong and can withstand everyday environmental stresses.

Introduce the BIGIDEA

Everyday Mixtures To introduce this chapter's Big Idea, have students draw the foods found in their favorite meal. Examples will vary and might include pizza and soda. Have students draw arrows to and label the different parts of each food. Students could indicate that pizza is made of crust, sauce, and toppings while soda is made of primarily sugar, water, and artificial coloring. Explain to students that these items are mixtures because they contain two or more items that retain their properties. Challenge students to identify foods that are not mixtures. Students should recognize that the task is difficult because almost all foods are mixtures. One example of a pure substance (non-mixture) could be water. Extend the discussion to items beyond food to help students understand that mixtures are everywhere.

Tie to Previous Knowledge

Have students review the following concepts before studying this chapter.

- solutions, alloys, heterogeneous and homogeneous mixtures
- ionic compounds
- covalent compounds
- determining moles and molar mass

Use the Photo

Mixtures Explain to students that mixtures are found almost everywhere. Have students look at the photo and identify what material provides the foundational structure for the building shown. steel Ask students whether or not steel is a mixture. Steel is a mixture of iron, carbon, and other elements. Ask students to identify other building materials in the photo that they think are mixtures. Brick and mortar are mixtures. Have students explain in their science journals why they categorized their chosen materials as mixtures. In Section 1 students will learn what a mixture is and how to identify different types of mixtures.

SECTION 1

1 Focus

MAINIDEA

Types of Mixtures Refer students to Table 2. Ask students what type of mixtures the table represents. The table provides examples of solutions, which are homogeneous mixtures. Ask students to take one of the solutions listed and propose a way to make it into a heterogeneous mixture (suspension or colloid). Student answers will vary. One example might be to add vinegar to oil to make a suspension. Another example might be fog (a colloid of water droplets in air). Ask students to explain how they know that their examples are heterogeneous. Student responses will vary. The vinegar-oil mixture will settle out over time. Fog exhibits the Tyndall effect. OL

2 Teach

Concept Development

Mixtures v. Solutions After developing the concept that many mixtures are solutions, it is important to remind students that not all mixtures are solutions. Many important mixtures are suspensions or colloids. OL

SECTION 1

Types of Mixtures

MAINIDEA Mixtures can be either heterogeneous or homogeneous.

Essential Questions

- How do the properties of suspensions, colloids, and solutions compare?
- What are the types of colloids and types of solutions?
- How are the electrostatic forces in colloids described?

Review Vocabulary

solute: a substance dissolved in a solution

New Vocabulary

suspension
colloid
Brownian motion
Tyndall effect
soluble
miscible
insoluble
immiscible

CHEM 4 YOU **If you have ever filled a pail with ocean water, you might have noticed that some of the sediment settles to the bottom of the pail. However, the water will be salty no matter how long you let the pail sit. Why do some substances settle out but others do not?**

Heterogeneous Mixtures

Recall that a mixture is a combination of two or more pure substances in which each pure substance retains its individual chemical properties. Heterogeneous mixtures do not blend smoothly throughout, and the individual substances remain distinct. Two types of heterogeneous mixtures are suspensions and colloids.

Suspensions A **suspension** is a mixture containing particles that settle out if left undisturbed. The muddy water shown in **Figure 1** is a suspension. Pouring a liquid suspension through a filter will also separate out the suspended particles.

Thixotropic mixtures Some suspensions separate into a solidlike mixture on the bottom and water on the top. When the solidlike mixture is stirred or agitated, it flows like a liquid. Substances that behave in this way are said to be thixotropic. For example, toothpaste is thixotropic—it acts as a liquid when it is squeezed from the tube but as a solid when it sits on your brush. Some paints are thixotropic—you can stir them in the paint can yet they don't flow down the stirring stick or brush when you hold them up. Builders in earthquake zones must be aware that some clays are thixotropic. These clays form liquids in response to the agitation of an earthquake, which causes structures built on them to collapse.

Figure 1 A suspension can be separated by allowing it to sit for a period of time. A liquid suspension can also be separated by pouring it through a filter.

DIFFERENTIATED INSTRUCTION

Visually Impaired Prepare several aqueous cornstarch solutions, and let the cornstarch settle out so it forms a thick layer on the bottom of each container. Place a metal spoon or other unbreakable stirring device into each solution. Have students use the spoon to probe the bottom of the container, and note the solid feel of the cornstarch phase. Then, have them stir the cornstarch, and note how the cornstarch quickly breaks apart and becomes easy to stir. Discuss the thixotropic behavior of the cornstarch. BL EL

Table 1 Types of Colloids

Category	Example	Dispersed Particles	Dispersing Medium
Solid sol	colored gems	solid	solid
Sol	blood, gelatin	solid	liquid
Solid emulsion	butter, cheese	liquid	solid
Emulsion	milk, mayonnaise	liquid	liquid
Solid foam	marshmallow, soaps that float	gas	solid
Foam	whipped cream, beaten egg white	gas	liquid
Solid aerosol	smoke, dust in air	solid	gas
Liquid aerosol	spray deodorant, fog, clouds	liquid	gas

Colloids Particles in a suspension are much larger than atoms and can settle out of solution. A heterogeneous mixture of intermediate-sized particles (between atomic-scale size of solution particles and the size of suspension particles) is a **colloid.** Colloid particles are between 1 nm and 1000 nm in diameter and do not settle out. Milk is a colloid. The components of homogenized milk cannot be separated by settling or by filtration.

The most abundant substance in the mixture is the dispersion medium. Colloids are categorized according to the phases of their dispersed particles and dispersing mediums. Milk is a colloidal emulsion because liquid particles are dispersed in a liquid medium. Other types of colloids are described in **Table 1.**

The dispersed particles in a colloid are prevented from settling out because they often have polar or charged atomic groups on their surfaces. These areas on their surfaces attract the positively or negatively charged areas of the dispersing-medium particles. This results in the formation of electrostatic layers around the particles, as shown in **Figure 2.** The layers repel each other when the dispersed particles collide; thus, the particles remain in the colloid.

If you interfere with the electrostatic layering, colloid particles will settle out of the mixture. For example, if you stir an electrolyte into a colloid, the dispersed particles clump together, destroying the colloid. Heating also destroys a colloid because it gives colliding particles enough kinetic energy to overcome the electrostatic forces and settle out.

Brownian motion The dispersed particles of liquid colloids make jerky, random movements. This erratic movement of colloid particles is called **Brownian motion.** It was first observed by, and later named for, the Scottish botanist Robert Brown (1773–1858), who noticed the random movements of pollen grains dispersed in water. Brownian motion results from collisions of particles of the dispersion medium with the dispersed particles. These collisions help to prevent the colloid particles from settling out of the mixture.

☑ READING CHECK **Describe** two reasons why particles in a colloid do not settle out.

Figure 2 The dispersing medium particles form charged layers around the colloid particles. These charged layers repel each other and keep the particles from settling out.

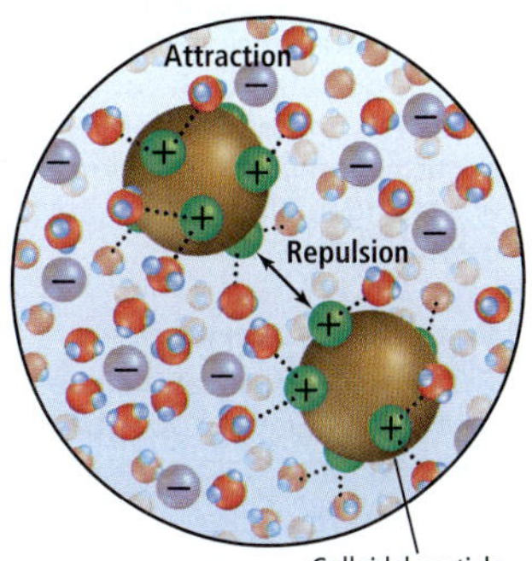

Identify Misconceptions

Many students will think that all mixtures are solutions.

Uncover the Misconception Ask students to classify the following mixtures: blood, fog, paint, and mayonnaise.

Demonstrate the Concept Using three mixtures (carbonated orange drink, muddy water, and motor oil), have students produce a list of properties exhibited by these different mixtures. For example, the drink has color but is transparent, the muddy water is opaque, and the motor oil is somewhat cloudy. The muddy water begins to settle, but the solute particles in the motor oil and orange drink do not settle. Show that only the mud can be separated from the water by filtration. Discuss how particle size affects the classification of mixtures.

Assess New Knowledge Have students draw three beakers containing three solutions. Have them illustrate how particle size would affect one or more of the properties of a solution, colloid, or suspension. Students should then explain their illustrations to other students in the classroom. OL

Quick Demo

Pour some water into a beaker. Add several milliliters of milk and stir thoroughly. Darken the room. Use a laser pointer or a flashlight as a light source and point it so that the beam of light travels through the mixture. Have students look at the beaker through the side and note the cone of light. This is the Tyndall effect. **WARNING:** ***Remind students that laser pointers should not be aimed at a person's eyes.***

CHEMISTRY JOURNAL

Kitchen Chemistry Have students research cooking techniques and discuss how solution chemistry is applied in the kitchen. What kinds of ingredients are added to foods as emulsifiers to help keep them mixed? What ingredients are used as thickening agents? How do the thickening agents work? Have students record their findings in their journals. OL

☑ READING CHECK Because dispersed particles in a colloid often have polar groups, they attract the positively or negatively charged areas of the dispersing-medium particles. Brownian motion also keeps particles in a colloid from settling out.

■ **Caption Question Fig. 3** The beaker on the left in which the beam of light can be seen is the colloid.

Data Analysis LAB

Purpose Students will design an experiment to measure the relationship between turbidity and visibility.

Process Skills design an experiment, identify variables, think critically, analyze and conclude

Teaching Strategies

- Discuss the details of conducting an experiment that is designed to find the relationship between an independent variable (the one varied by the experimenter) and the dependent variable (the result that occurs when the independent variable is changed). Ask students what other factors might affect the results and how these factors can be controlled. Light intensity must be kept constant, the colloid must be uniform, and the same technique should be used for each measurement.
- When the Tyndall effect is used to monitor water clarity, the measurement is called turbidity. The transmittance of light through a given depth of water is measured and compared to the same reading for pure water.
- Suggest students use milk mixed in varying ratios with water as a basis for determining their turbidity scales. Discuss how the number of particles in the milk can be related to the turbidity of the solution.

Think Critically

1. Answers will vary. The amount of turbidity can be equated with the number of milk drops used to create the colloid. The concentration of the milk-water colloid is the independent variable. The distance at which the mark on the bottom of the graduated cylinder can be seen might be the dependent variable. The control is pure water.
2. Through research, students could estimate the volume of molecules in a standard milk sample based on its composition. The colloid particles at each concentration could then be extrapolated from the standard sample.
3. Answers will vary. Students might mention water safety. Answers also could include precautions, such as gloves and goggles, to avoid exposure to potential contaminants.
4. Possible materials include milk, water, a ruler, a dropper, and a graduated cylinder with a mark on its bottom (for determining the depth through which the mark can no longer be seen).

■ **Figure 3** Particles in a colloid scatter light, unlike particles in a solution. Called the Tyndall effect, the beam of light is visible in the colloid because of light scattering.
Determine *which mixture is the colloid.*

Tyndall effect Concentrated colloids are often cloudy or opaque. Dilute colloids sometimes appear as clear as solutions. Dilute colloids appear to be homogeneous solutions because their dispersed particles are so small. However, dispersed colloid particles scatter light, a phenomenon known as the **Tyndall effect.** In **Figure 3,** a beam of light is shone through two unknown mixtures. You can observe that dispersed colloid particles scatter the light, unlike particles in the solution. Suspensions also exhibit the Tyndall effect, but solutions never exhibit the Tyndall effect. You have observed the Tyndall effect if you have observed rays of sunlight passing through smoke-filled air, or viewed lights through fog. The Tyndall effect can be used to determine the amount of colloid particles in suspension.

Homogeneous Mixtures

Cell solutions, ocean water, and steel might appear dissimilar, but they share certain characteristics. You learned earlier that solutions are homogeneous mixtures that contain two or more substances called the solute and the solvent. The solute is the substance that dissolves. The solvent is the dissolving medium. When you look at a solution, it is not possible to distinguish the solute from the solvent.

Types of solutions A solution might exist as a gas, a liquid, or a solid, depending on the state of its solvent, as shown in **Table 2.** Air is a gaseous solution, and its solvent is nitrogen gas. Braces that you wear on your teeth might be made of nitinol, a solid solution of titanium in nickel. Most solutions, however, are liquids. You read previously that reactions can take place in aqueous solutions, or solutions in which the solvent is water. Water is the most common solvent among liquid solutions.

Data Analysis LAB

Based on Real Data*

Design an Experiment

How can you measure turbidity? The National Primary Drinking Water Regulations set the standards for public water systems. Turbidity—a measure of the cloudiness of water that results from the suspension of solids in the water—is often associated with contamination from viruses, parasites, and bacteria. Most of these colloid particles come from erosion, industrial and human waste, algae blooms from fertilizers, and decaying organic matter.

Data and Observation

The Tyndall effect can be used to measure the turbidity of water. Your goal is to plan a procedure and develop a scale to interpret data.

Think Critically

1. **Identify** the variables that can be used to relate the ability of light to pass through the liquid and the number of the colloid particles present. What will you use as a control?
2. **Relate** the variables used in the experiment to the actual number of colloid particles that are present.
3. **Analyze** What safety precautions must be considered?
4. **Determine** the materials you need to measure the Tyndall effect. Select technology to collect or interpret data.

*Data obtained from U.S. Environmental Protection Agency. 2006. *The Office of Groundwater and Drinking Water.*

Explore types of solutions with an interactive table. Concepts In Motion

Table 2 Types and Examples of Solutions

Type of Solution	Example	Solvent	Solute
Gas	air	nitrogen (gas)	oxygen (gas)
Liquid	carbonated water	water (liquid)	carbon dioxide (gas)
	ocean water	water (liquid)	oxygen gas (gas)
	antifreeze	water (liquid)	ethylene glycol (liquid)
	vinegar	water (liquid)	acetic acid (liquid)
	ocean water	water (liquid)	sodium chloride (solid)
Solid	dental amalgam	silver (solid)	mercury (liquid)
	steel	iron (solid)	carbon (solid)

Just as solutions can exist in different forms, the solutes in the solutions can be gases, liquids, or solids, also shown in **Table 2.** Solutions, such as ocean water, can contain more than one solute.

Forming solutions Some combinations of substances readily form solutions, and others do not. A substance that dissolves in a solvent is said to be **soluble** in that solvent. For example, sugar is soluble in water—a fact you might have learned by dissolving sugar in flavored water to make a sweetened beverage, such as tea or lemonade. Two liquids that are soluble in each other in any proportion, such as those that form the antifreeze listed in **Table 2,** are said to be **miscible.** A substance that does not dissolve in a solvent is said to be **insoluble** in that solvent. Sand is insoluble in water. The liquids in a bottle of oil and vinegar separate shortly after they are mixed. Oil is insoluble in vinegar. Two liquids that can be mixed together but separate shortly after are said to be **immiscible.**

SECTION 1 REVIEW

Section Self-Check

Section Summary

- The individual substances in a heterogeneous mixture remain distinct.
- Two types of heterogeneous mixtures are suspensions and colloids.
- Brownian motion is the erratic movement of colloid particles.
- Colloids exhibit the Tyndall effect.
- A solution can exist as a gas, a liquid, or a solid, depending on the solvent.
- Solutes in a solution can be gases, liquids, or solids.

1. MAINIDEA **Explain** Use the properties of seawater to describe the characteristics of mixtures.
2. **Distinguish** between suspensions and colloids.
3. **Identify** the various types of solutions. Describe the characteristics of each type of solution.
4. **Explain** Use the Tyndall effect to explain why it is more difficult to drive through fog using high beams than using low beams.
5. **Describe** different types of colloids.
6. **Explain** Why do dispersed colloid particles stay dispersed?
7. **Summarize** What causes Brownian motion?
8. **Compare and Contrast** Make a table that compares the properties of suspensions, colloids, and solutions.

3 Assess

Check for Understanding

Ask students to compare the properties of a heterogeneous mixture and a colloid. How would they distinguish between them? A heterogeneous mixture and a colloid can be distinguished by shining a beam of light through the solutions. OL

Reteach

Have students collect pictures of the common household mixtures described in Table 1 and paste them on a poster board. Under each picture, have them describe the dispersed particles and dispersing medium for the mixture. Label the picture according to the proper classification *sol, emulsion,* or *aerosol.* BL EL

Extension

Have students classify various health and beauty products, such as liquid makeup, shampoos, and shaving creams, based on what type of colloid each is. OL EL

ASSESSMENT

Performance Assign each student or group of students a mixture and have them design an experiment that would allow them to classify the mixture as a solution, suspension, or a colloid. OL

SECTION 1 REVIEW

1. Answers will vary but might include that seawater is a heterogeneous mixture with dirt and mud particles, and it is a homogeneous mixture with dissolved substances.
2. Suspension particles are larger than colloidal particles. Suspension particles settle out of the mixture, whereas colloidal particles do not.
3. All solutions are homogeneous mixtures containing two or more substances. Solutions might be liquid, solid, or gas. Solution types are identified in Table 2.
4. High beams are aimed farther down the road than low beams. Because the fog scatters light, there is less light from the high beams to illuminate the road than from the low beams. Also, because the high beams are aimed more directly into the fog, more of their light is reflected back toward the driver, making it more difficult to see.
5. Refer to Table 1 for descriptions of colloid types.
6. The particles do not settle out because they have polar or charged layers surrounding them. These layers repel each other and prevent the particles from settling or separating.
7. Collisions of particles of the dispersion medium with the dispersed particles result in Brownian motion.
8. Student tables will vary, but should include particle size, if the particles settle out, and if the particles display the Tyndall effect.

SECTION 2

1 Focus

MAINIDEA

Percents and Moles Ask students what a percentage score on a test means. Students should indicate that a percentage represents the proportion of answers a person got right on a test. Explain to students that, in the same way, a percentage value of a solution indicates the proportion of solute present in the whole solution. Refer students to Figure 4. Ask students to predict which picture represents a solution with a higher percentage of solute. The darker tea contains a higher percentage of solute and, therefore, has a higher concentration. Show students the formulas for percent by mass and percent by volume. Ask students to guess why molarity might be a better way to express the concentration for a solution such as tea or sugar water. Molarity contains measurement units that are readily used when working with a solid solute and a liquid solvent. Discuss with students the need for various representations of concentration in order to accurately accommodate laboratory calculations and material comparisons. OL

2 Teach

Concept Development

Ratios Introduce students to the different units of solution concentration as ratios. Ask students to define population density. Discuss the difference between the number of people in a large city and a rural community. Remind them that it is not enough to describe just the total number of people, it is also important to know something about the amount of space the population occupies. Suggest that concentration ratios are similar. It is not enough to describe just the amount of solute in solution, it is also important to know how much solvent is present. OL

SECTION 2

Solution Concentration

MAINIDEA Concentration can be expressed in terms of percent or in terms of moles.

Essential Questions

- How can the concentration be described using different units?
- How are the concentrations of solutions determined?
- What is the molarity of a solution and how can it be calculated?

Review Vocabulary

solvent: the substance that dissolves a solute to form a solution

New Vocabulary

concentration
molarity
molality
mole fraction

CHEM 4 YOU **Have you ever tasted a glass of iced tea and found it too strong or too bitter? To adjust the taste, you could add water to dilute the tea, or you could add sugar to sweeten it. Either way, you are changing the concentration of the particles dissolved in the water.**

Expressing Concentration

The **concentration** of a solution is a measure of how much solute is dissolved in a specific amount of solvent or solution. Concentration can be described qualitatively using the words *concentrated* or *dilute.* Notice the containers of tea in **Figure 4.** One of the tea solutions is more concentrated than the other. In general, a concentrated solution contains a large amount of solute. The darker tea has more tea particles than the lighter tea. Conversely, a dilute solution contains a small amount of solute. The lighter tea in **Figure 4** is dilute and contains less tea particles than the darker tea.

Although qualitative descriptions of concentration can be useful, solutions are more often described quantitatively. Some commonly used quantitative descriptions are percent by mass, percent by volume, molarity, and molality. These descriptions express concentration as a ratio of measured amounts of solute and solvent or solution. **Table 3** lists each ratio's description.

Which quantitative description should be used? The description used depends on the type of solution analyzed and the reason for describing it. For example, a chemist working with a reaction in an aqueous solution most likely refers to the molarity of the solution because he or she needs to know the number of particles involved in the reaction.

■ **Figure 4** The strength of the tea corresponds to its concentration. The darker container of tea is more concentrated than the lighter container.

Table 3 Concentration Ratios

Concentration Description	Ratio
Percent by mass	$\frac{\text{mass of solute}}{\text{mass of solution}} \times 100$
Percent by volume	$\frac{\text{volume of solute}}{\text{volume of solution}} \times 100$
Molarity	$\frac{\text{moles of solute}}{\text{liter of solution}}$
Molality	$\frac{\text{moles of solute}}{\text{kilogram of solvent}}$
Mole fraction	$\frac{\text{moles of solute}}{\text{moles of solute} + \text{moles of solvent}}$

DIFFERENTIATED INSTRUCTION

Struggling Learners Provide students with an equation sheet. Make sure that one of the first steps in solving problems in this section is to identify on paper what is known and unknown. Knowing this, students should select the appropriate equation. BL

Percent by mass The percent by mass is the ratio of the solute's mass to the solution's mass expressed as a percent. The mass of the solution equals the sum of the masses of the solute and the solvent.

Percent by Mass

$$\text{percent by mass} = \frac{\text{mass of solute}}{\text{mass of solution}} \times 100$$

Percent by mass equals the mass of the solute divided by the mass of the whole solution, multiplied by 100.

FOLDABLES® Incorporate information from this section into your Foldable.

EXAMPLE Problem 1

Find help with **percents.** Math Handbook

CALCULATE PERCENT BY MASS In order to maintain a sodium chloride (NaCl) concentration similar to ocean water, an aquarium must contain 3.6 g NaCl per 100.0 g of water. What is the percent by mass of NaCl in the solution?

1 ANALYZE THE PROBLEM

You are given the amount of sodium chloride dissolved in 100.0 g of water. The percent by mass of a solute is the ratio of the solute's mass to the solution's mass, which is the sum of the masses of the solute and the solvent.

Known	Unknown
mass of solute = 3.6 g NaCl	percent by mass = ?
mass of solvent = 100.0 g H_2O	

2 SOLVE FOR THE UNKNOWN

Find the mass of the solution.

mass of solution = grams of solute + grams of solvent

mass of solution = 3.6 g + 100.0 g = 103.6 g — Substitute mass of solute = 3.6 g, and mass of solvent = 100.0 g.

Calculate the percent by mass.

$$\text{percent by mass} = \frac{\text{mass of solute}}{\text{mass of solution}} \times 100$$ — State the equation for percent by mass.

$$\text{percent by mass} = \frac{3.6\text{ g}}{103.6\text{ g}} \times 100 = 3.5\%$$ — Substitute mass of solute = 3.6 g, and mass of solution = 103.6 g.

3 EVALUATE THE ANSWER

Because only a small mass of sodium chloride is dissolved per 100.0 g of water, the percent by mass should be a small value, which it is. The mass of sodium chloride was given with two significant figures; therefore, the answer is also expressed with two significant figures.

EXAMPLE PROBLEM

PRACTICE Problems

Do additional practice. Online Practice

9. What is the percent by mass of $NaHCO_3$ in a solution containing 20.0 g of $NaHCO_3$ dissolved in 600.0 mL of H_2O?

10. You have 1500.0 g of a bleach solution. The percent by mass of the solute sodium hypochlorite (NaOCl) is 3.62%. How many grams of NaOCl are in the solution?

11. In Question 10, how many grams of solvent are in the solution?

12. **Challenge** The percent by mass of calcium chloride in a solution is found to be 2.65%. If 50.0 g of calcium chloride is used, what is the mass of the solution?

DIFFERENTIATED INSTRUCTION

Struggling Learners Have students role play solution concentration. Tape the perimeter of a large square on the floor. The square represents the amount of solvent. Have three students stand in the square. Add three more students and ask them to describe what happens to the numerical value of the concentration. Repeat the scenario with three students in the square, then decrease the size of the square. Ask what happens to the concentration value. BL EL

IN-CLASS Example

Question A 650-mg multivitamin contains 1.0 mg folic acid. What is the percent by mass of folic acid in the vitamin?

Answer 0.15%

$$\text{Percent by mass} = \frac{\text{mass of solute}}{\text{mass of solution}} \times 100$$

$$= \frac{1.0\text{ mg}}{650\text{ mg}} \times 100$$

$$= 0.15\%$$

PRACTICE Problems

Have students refer to the Selected Solutions appendix for complete solutions to odd-numbered problems. The complete solutions for all problems can be found in the Solutions Manual.

9. 3.23%
10. 54.3 g
11. 1445.7 g
12. 1.89×10^3 g

FOLDABLES® Rubric

Concept Development

Percentages Review the percent-by-volume and percent-by-mass equations. Remind students that an important difference when solving percent-by-mass and percent-by-volume problems is that the unit used for the amount of solution is different. BL

☑ READING CHECK Percent by mass is a comparison between the mass of solute and the total mass of solution. Percent by volume is a comparison between the volume of the solute and the total volume of the solution.

PRACTICE Problems

Have students refer to the Selected Solutions appendix for complete solutions to odd-numbered problems. The complete solutions for all problems can be found in the Solutions Manual.

13. 18%
14. 2.1%
15. 120 mL

Quick Demo

Volume Demonstrate that two separate liquid volumes do not always add together mathematically to give the new solution volume. Mix 50.0 mL water with 50.0 mL isopropyl alcohol in a 100-mL graduated cylinder. Have students observe the volume of the resulting solution. Note that it is less than 100.0 mL. Have students calculate the percent by volume based on the volume of the solution measured after the two liquids were mixed together. When finished with the demo, the solution can be flushed down a drain. OL

☑ READING CHECK The molar concentration is 0.5*M*.

■ **Figure 5** B20 is 20% by volume biodiesel and 80% by volume petroleum diesel. Biodiesel is an alternative fuel that can be produced from renewable resources, such as vegetable oil.

Percent by volume Percent by volume usually describes solutions in which both solute and solvent are liquids. The percent by volume is the ratio of the volume of the solute to the volume of the solution, expressed as a percent. The volume of the solution is the sum of the volumes of the solute and the solvent. Calculations of percent by volume are similar to those involving percent by mass.

Percent by Volume

$$\text{percent by volume} = \frac{\text{volume of solute}}{\text{volume of solution}} \times 100$$

Percent by volume equals the volume of solute divided by the volume of the solution, multiplied by 100.

Biodiesel, shown in **Figure 5,** is a clean-burning alternative fuel that is produced from renewable resources. Biodiesel can be used in diesel engines with little or no modifications. Biodiesel is simple to use, biodegradable, nontoxic, and it does not contain some of the pollutants found in regular gasoline. It does not contain petroleum, but it can be blended with petroleum diesel to create a biodiesel blend. B20 is 20% by volume biodiesel, 80% by volume petroleum diesel.

☑ READING CHECK **Compare** percent mass and percent volume.

PRACTICE Problems

Do additional problems. Online Practice

13. What is the percent by volume of ethanol in a solution that contains 35 mL of ethanol dissolved in 155 mL of water?

14. What is the percent by volume of isopropyl alcohol in a solution that contains 24 mL of isopropyl alcohol in 1.1 L of water?

15. Challenge If 18 mL of methanol is used to make an aqueous solution that is 15% methanol by volume, how many milliliters of solution is produced?

Molarity Percent by volume and percent by mass are only two of the commonly used ways to quantitatively describe the concentrations of liquid solutions. One of the most common units of solution concentration is molarity. **Molarity** *(M)* is the number of moles of solute dissolved per liter of solution. Molarity is also known as molar concentration, and the unit *M* is read as molar. A liter of solution containing 1 mol of solute is a 1*M* solution, which is read as a one-molar solution. A liter of solution containing 0.1 mol of solute is a 0.1*M* solution. To calculate a solution's molarity, you must know the volume of the solution in liters and the amount of dissolved solute in moles.

Molarity

$$\text{molarity } (M) = \frac{\text{moles of solute}}{\text{liters of solution}}$$

The molarity of a solution equals the moles of solute divided by the liters of solution.

☑ READING CHECK **Determine** What is the molar concentration of a liter solution with 0.5 mol of solute?

CHEMISTRY JOURNAL

Compare Concentration Have students develop a list of concentration units related to household and daily situations. Examples might include medication dosages in mg/kg (compare adult and child dosages of cough syrup), percent hydrogen peroxide (compare antiseptic to hydrogen peroxide used to bleach hair), and concentrated orange juice or insecticides (compare concentrate and dilute forms). OL

EXAMPLE Problem 2

CALCULATING MOLARITY A 100.5-mL intravenous (IV) solution contains 5.10 g of glucose ($C_6H_{12}O_6$). What is the molarity of this solution? The molar mass of glucose is 180.16 g/mol.

1 ANALYZE THE PROBLEM

You are given the mass of glucose dissolved in a volume of water. The molarity of the solution is the ratio of moles of solute per liter of solution.

Known

mass of solute = 5.10 g $C_6H_{12}O_6$

molar mass of $C_6H_{12}O_6$ = 180.16 g/mol

volume of solution = 100.5 mL

Unknown

solution concentration = ? *M*

2 SOLVE FOR THE UNKNOWN

Calculate the number of moles of $C_6H_{12}O_6$.

$$(5.10\ \cancel{g\ C_6H_{12}O_6})\left(\frac{1\ \text{mol}\ C_6H_{12}O_6}{180.16\ \cancel{g\ C_6H_{12}O_6}}\right)$$

$$= 0.0283\ \text{mol}\ C_6H_{12}O_6$$

Multiply grams of $C_6H_{12}O_6$ by the molar mass of $C_6H_{12}O_6$.

Convert the volume of H_2O to liters.

$$(100.5\ \cancel{mL}\ \text{solution})\left(\frac{1\ \text{L}}{1000\ \cancel{mL}}\right) = 0.1005\ \text{L solution}$$

Use the conversion factor 1 L/1000 mL.

Solve for the molarity.

$$M = \frac{\text{moles of solute}}{\text{liters of solutions}}$$

State the molarity equation.

$$M = \left(\frac{0.0283\ \text{mol}\ C_6H_{12}O_6}{0.1005\ \text{L solution}}\right)$$

Substitute moles of $C_6H_{12}O_6$ = 0.0283 and volume of solution = liters of solution = 0.1005 L.

$$M = \left(\frac{0.0282\ \text{mol}\ C_6H_{12}O_6}{1\ \text{L solution}}\right) = 0.282M$$

Divide numbers and units.

3 EVALUATE THE ANSWER

The molarity value will be small because only a small mass of glucose was dissolved in the solution. The mass of glucose used in the problem has three significant figures; therefore, the value of the molarity also has three significant figures.

PRACTICE Problems

Do additional problems. Online Practice

16. What is the molarity of an aqueous solution containing 40.0 g of glucose ($C_6H_{12}O_6$) in 1.5 L of solution?

17. Calculate the molarity of 1.60 L of a solution containing 1.55 g of dissolved KBr.

18. What is the molarity of a bleach solution containing 9.5 g of NaOCl per liter of bleach?

19. Challenge How much calcium hydroxide ($Ca(OH)_2$), in grams, is needed to produce 1.5 L of a 0.25*M* solution?

CAREERS IN CHEMISTRY

Pharmacy Technician Most pharmacists rely on pharmacy technicians to prepare the proper medications to fill prescriptions. These technicians read patient charts and prescriptions in order to prepare the proper concentration, or dose, of medication that is to be administered to patients.

WebQuest

IN-CLASS Example

Question A student dissolves 2.75 g of sodium hydroxide (NaOH) in water to make 250 mL of solution. What is the molarity of the solution? The molar mass of NaOH is 39.997 g/mol.

Answer

$$\text{mol NaOH} = 2.75\ \text{g NaOH} \times \frac{1\ \text{mol NaOH}}{39.997\ \text{g NaOH}}$$

$$= 0.0688\ \text{mol NaOH}$$

$$\text{L of solution} = 250\ \text{mL solution} \times \frac{1\ \text{L}}{1000\ \text{mL}}$$

$$= 0.25\ \text{L}$$

$$\text{Molarity}\ (M) = \frac{\text{moles of solute}}{\text{liters of solution}}$$

$$= \frac{0.0688\ \text{mol NaOH}}{0.25\ \text{L solution}}$$

$$= 0.28M$$

CAREERS IN CHEMISTRY

WebQuest

Math in Chemistry

Moles v. Volume Have students solve for the missing quantities in the following chart for a 0.149*M* solution of Na_2SO_4.

Mass (g)	Moles	Volume (mL)
5.29	0.0373	250
11.1	0.0782	525
14.3	0.101	675
17.7	0.125	839
60.3	0.425	2850

Using a graphing calculator, enter the values for moles in one list and the values from volume in a second list. Graph moles versus volume in liters. Find the best-fit line and use the linear regression function of the calculator to determine the equation of the resulting straight line. Ask what the graph's slope means. The slope is equal to the molarity. OL

CHEMISTRY JOURNAL

IV Solutions Intravenous solutions are important in medicine to help stabilize the fluid and electrolyte levels of patients. Solutions of sodium chloride and glucose are commonly used. Have students investigate why medical establishments typically use percent concentration instead of molarity. Work should be recorded in student journals. OL

PRACTICE Problems

Have students refer to the Selected Solutions appendix for complete solutions to odd-numbered problems. The complete solutions for all problems can be found in the Solutions Manual.

16. 0.15*M*

17. $8.13 \times 10^{-3}M$

18. 0.13*M*

19. 28 g

■ **Caption Question Fig. 6** Copper(II) sulfate takes up space and will add volume to the solution.

PRACTICE Problems

Have students refer to the Selected Solutions appendix for complete solutions to odd-numbered problems. The complete solutions for all problems can be found in the Solutions Manual.

20. 11 g
21. 11 g
22. 3.0×10^1 g
23. 0.87 mL

Quick Demo

Dilution Before class begins, using colored stock such as copper(II) sulfate, nickel(II) nitrate, or food coloring in water, prepare four dilutions of known concentration. Fill test tubes with 15 to 20 mL of each solution and put them in a test-tube rack in order of decreasing concentration. **WARNING:** ***Copper(II) sulfate is toxic.*** Display the test tubes in class, and ask students what happens to the color of the solution as concentration decreases. In another test tube, have a solution labeled *unknown*. Ask students to use the samples displayed to determine the approximate concentration of the unknown by matching its color to a known color. Explain to students that chemists have developed instrumentation such as colorimeters and spectrophotometers that depend on color to help determine concentration. When finished, allow water to evaporate and save crystals so that they can be used again. **OL**

■ **Figure 6** Accurately preparing a solution of copper(II) sulfate involves several steps.
Explain ***why you cannot add 375 g of copper(II) sulfate directly to 1 L of water to make a 1.5M solution.***

Step 1: The mass of the solute is measured.

Step 2: The solute is placed in a volumetric flask of the correct volume.

Step 3: Distilled water is added to the flask to bring the solution level up to the calibration mark.

Preparing molar solutions Now that you know how to calculate the molarity of a solution, how do you think you would prepare 1 L of a 1.50*M* aqueous solution of copper(II) sulfate pentahydrate ($CuSO_4 \cdot 5H_2O$)? A 1.50*M* aqueous solution of $CuSO_4 \cdot 5H_2O$ contains 1.50 mol of $CuSO_4 \cdot 5H_2O$ dissolved in 1 L of solution. The molar mass of $CuSO_4 \cdot 5H_2O$ is about 249.70 g. Thus, 1.50 mol of $CuSO_4 \cdot 5H_2O$ has a mass of 375 g, an amount that you can measure on a balance.

$$\frac{1.50 \text{ mol } CuSO_4 \cdot 5H_2O}{1 \text{ L solution}} \times \frac{249.7 \text{ g } CuSO_4 \cdot 5H_2O}{1 \text{ mol } CuSO_4 \cdot 5H_2O} = \frac{375 \text{ g } CuSO_4 \cdot 5H_2O}{1 \text{ L solution}}$$

You cannot simply add 375 g of $CuSO_4 \cdot 5H_2O$ to 1 L of water to make the 1.50*M* solution. Like all substances, $CuSO_4 \cdot 5H_2O$ takes up space and will add volume to the solution. Therefore, you must use slightly less than 1 L of water to make 1 L of solution, as shown in **Figure 6.**

You will often do experiments that call for small quantities of solution. For example, you might need only 100 mL of a 1.50*M* $CuSO_4 \cdot 5H_2O$ solution for an experiment. Look again at the definition of molarity. As calculated above, a 1.50*M* solution of $CuSO_4 \cdot 5H_2O$ contains 1.50 mol of $CuSO_4 \cdot 5H_2O$ per 1 L of solution. Therefore, 1 L of solution contains 375 g of $CuSO_4 \cdot 5H_2O$.

This relationship can be used as a conversion factor to calculate how much solute you need for your experiment.

$$100 \text{ mL} \times \frac{1 \text{ L}}{1000 \text{ mL}} \times \frac{375 \text{ g } CuSO_4 \cdot 5H\ 2O}{1 \text{ L solution}} = 37.5 \text{ g } CuSO_4 \cdot 5H_2O$$

Thus, you would need to measure out 37.5 g of $CuSO_4 \cdot 5H_2O$ to make 100 mL of a 1.50*M* solution.

PRACTICE Problems

Do additional problems. Online Practice

20. How many grams of $CaCl_2$ would be dissolved in 1.0 L of a 0.10*M* solution of $CaCl_2$?

21. How many grams of $CaCl_2$ should be dissolved in 500.0 mL of water to make a 0.20*M* solution of $CaCl_2$?

22. What mass of NaOH is in 250 mL of a 3.0*M* NaOH solution?

23. **Challenge** What volume of ethanol (C_2H_5OH) is in 100.0 mL of 0.15*M* solution? The density of ethanol is 0.7893 g/mL.

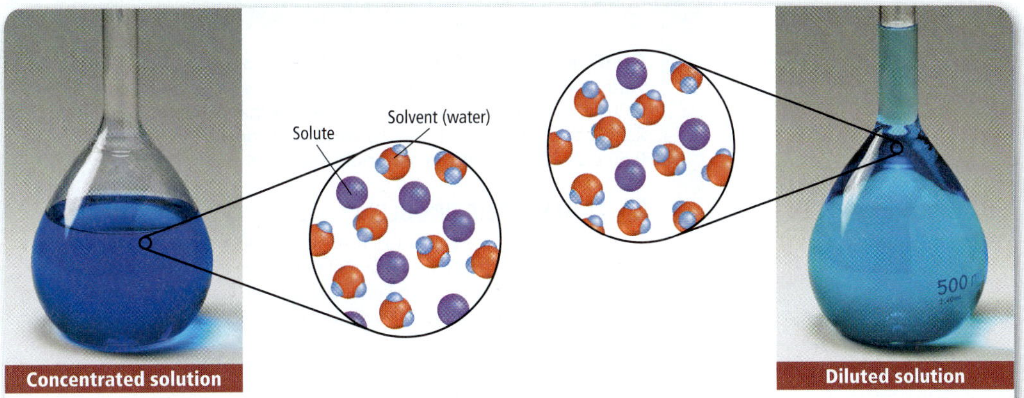

■ **Figure 7** A concentrated solution can be diluted by adding solvent. The number of moles of solute does not change when a concentrated solution is diluted.

Diluting molar solutions In the laboratory, you might use concentrated solutions of standard molarities, called stock solutions. For example, concentrated hydrochloric acid (HCl) is $12M$. Recall that a concentrated solution has a large amount of solute. You can prepare a less-concentrated solution by diluting the stock solution with additional solvent. When you add solvent, you increase the number of solvent particles among which the solute particles move, as shown in **Figure 7,** thereby decreasing the solution's concentration.

How do you determine the volume of stock solution you must dilute? You can rearrange the expression of molarity to solve for moles of solute.

$$\text{molarity } (M) = \frac{\text{moles of solute}}{\text{liters of solution}}$$

$$\text{moles of solute} = \text{molarity} \times \text{liters of solution}$$

Because the total number of moles of solute does not change during dilution,

$$\text{moles of solute in the stock solution} = \text{moles of solute after dilution.}$$

Substituting moles of solute with molarity times liters of solution, the relationship can be expressed in the dilution equation.

Dilution Equation

$$M_1V_1 = M_2V_2$$

M represents molarity.
V represents volume.

For a given amount of solute, the product of the molarity and volume of the stock solution equals the product of the molarity and the volume of the dilute solution.

M_1 and V_1 represent the molarity and volume of the stock solution, and M_2 and V_2 represent the molarity and volume of the dilute solution. Before dilution, a concentrated solution contains a fairly high ratio of solute particles to solvent particles. After adding more solvent, the ratio of solute particles to solvent particles has decreased.

VOCABULARY

ACADEMIC VOCABULARY

Concentrated
less dilute or diffuse
We added more water to the lemonade because it was too concentrated.

CHEMISTRY PROJECT

Allowable Contaminants Have student groups investigate the EPA's allowable limits for various contaminants found in drinking water. Local, state, and national allowable limits might vary. Have groups compare different regions of the country and speculate why regulations might vary. OL

ASSESSMENT

Performance Assign each student or pair of students a dilution to prepare from a stock solution. Give them a file card with the following information: solution name, stock solution concentration, final concentration, and final volume.

Have students use simulated stock solutions prepared from food coloring and water. (Use a solution of yellow food coloring for $0.250M$ potassium chromate; use a solution of green food coloring for $0.355M$ nickel(II) nitrate; use a solution of blue food coloring for $0.175M$ copper(II) acetate; and use a solution of red and blue food coloring for $0.865M$ chromium(III) nitrate).

Students can use volumetric flasks or graduated cylinders depending on their lab skill level. Provide each student an appropriate container in which to prepare the solution and have students provide a complete description of the calculations and steps used to prepare the solution. OL

Enrichment

Normality Ask students to research the definition of normality. Have them determine how molarity can be converted to units of normality.

IN-CLASS Example

Question If a student used 75 mL of a concentrated hydrochloric acid (HCl) stock solution to make 1.5 L of a 0.50M HCl solution, what was the original concentration of the stock solution?

Answer 11M

$M_1V_1 = M_2V_2$

$M_1 = M_2\left(\frac{V_2}{V_1}\right)$

$= 0.50M\left(\frac{1.5\text{ L}}{0.075\text{ L}}\right)$

$= 1.0 \times 10^1 M$

PRACTICE Problems

Have students refer to the Selected Solutions appendix for complete solutions to odd-numbered problems. The complete solutions for all problems can be found in the Solutions Manual.

24. 125 mL
25. 5.0 mL
26. 91 g

EXAMPLE Problem 3

DILUTING STOCK SOLUTIONS If you know the concentration and volume of the solution you want to prepare, you can calculate the volume of stock solution you will need. What volume, in milliliters, of 2.00M calcium chloride ($CaCl_2$) stock solution would you use to make 0.50 L of 0.300M calcium chloride solution?

1 ANALYZE THE PROBLEM

You are given the molarity of a stock solution of $CaCl_2$ and the volume and molarity of a dilute solution of $CaCl_2$. Use the relationship between molarities and volumes to find the volume, in liters, of the stock solution required. Then, convert the volume to milliliters.

Known	Unknown
$M_1 = 2.00M\ CaCl_2$	$V_1 = ?$ mL 2.00M $CaCl_2$
$M_2 = 0.300M$	
$V_2 = 0.50$ L	

2 SOLVE FOR THE UNKNOWN

Solve the molarity-volume relationship for the volume of the stock solution V_1.

$M_1V_1 = M_2V_2$ — **State the dilution equation.**

$V_1 = V_2\left(\frac{M_2}{M_1}\right)$ — **Solve for V_1.**

$V_1 = (0.50\text{ L})\left(\frac{0.300M}{2.00M}\right)$ — **Substitute $M_1 = 2.00M$, $M_2 = 0.300M$, and $V_2 = 0.50$ L.**

$V_1 = (0.50\text{ L})\left(\frac{0.300\cancel{M}}{2.00\cancel{M}}\right) = 0.075\text{ L}$ — **Multiply and divide numbers and units.**

$V_1 = (0.075\text{ L})\left(\frac{1000\text{ mL}}{1\text{ L}}\right) = 75\text{ mL}$ — **Convert to milliliters using the conversion factor 1000 mL/1 L.**

To make the dilution, measure out 75 mL of the stock solution and dilute it with enough water to make the final volume 0.50 L.

3 EVALUATE THE ANSWER

The volume V_1 was calculated, and then its value was converted to milliliters. This volume should be less than the final volume of the dilute solution, and it is. Of the given information, V_2 had the fewest number of significant figures, with two. Thus, the volume V_1 should also have two significant figures, and it does.

PRACTICE Problems

Do additional problems. Online Practice

24. What volume of a 3.00M KI stock solution would you use to make 0.300 L of a 1.25M KI solution?

25. How many milliliters of a 5.0M H_2SO_4 stock solution would you need to prepare 100.0 mL of 0.25M H_2SO_4?

26. **Challenge** If 0.50 L of 5.00M stock solution of HCl is diluted to make 2.0 L of solution, how much HCl, in grams, is in the solution?

CULTURAL DIVERSITY

Fluoridated Water Research has shown a link between the presence of fluoride ions in drinking water and a decrease in dental cavities. Cavities form in teeth when common sugars ferment in the mouth, forming lactic acid. This acid reacts with the enamel on the tooth, causing a loss of enamel. Introducing fluoride levels as low as 5 ppm (5 mg fluoride/kg of water) in drinking water inhibits this enamel-decaying reaction in the mouth. Some regions of the world have this level of fluoride naturally contained in their drinking water. However, too much fluoride is not good; excess levels of fluoride in water can result in a condition known as fluorosis. Ask students why this condition might occur more frequently in regions of the world where the climate is very hot or where people consume more water.

Molality The volume of a solution changes with temperature as it expands or contracts. This change in volume alters the molarity of the solution. Masses, however, do not change with temperature. It is sometimes more useful to describe solutions in terms of how many moles of solute are dissolved in a specific mass of solvent. Such a description is called **molality**—the ratio of the number of moles of solute dissolved in 1 kg of solvent. The unit *m* is read as molal. A solution containing 1 mol of solute per kilogram of solvent is a one-molal solution.

Get help with **calculating molarity and molality.**

Personal Tutor

Molality

$$\text{molality } (m) = \frac{\text{moles of solute}}{\text{kg of solvent}}$$

The molality of a solution equals the moles of solute divided by kilograms of solvent.

EXAMPLE Problem 4

CALCULATING MOLALITY In the lab, a student adds 4.5 g of sodium chloride (NaCl) to 100.0 g of water. Calculate the molality of the solution.

1 ANALYZE THE PROBLEM

You are given the mass of solute and solvent. Determine the number of moles of solute. Then, you can calculate the molality.

Known
mass of water (H_2O) = 100.0 g
mass of sodium chloride (NaCl) = 4.5 g

Unknown
m = ? mol/kg

2 SOLVE FOR THE UNKNOWN

$$4.5 \text{ g NaCl} \times \frac{1 \text{ mol NaCl}}{58.44 \text{ g NaCl}} = 0.077 \text{ mol NaCl}$$

Calculate the number of moles of solute.

$$100.0 \text{ g } H_2O \times \frac{1 \text{ kg } H_2}{1000 \text{ g } H_2O} = 0.1000 \text{ kg } H_2O$$

Convert the mass of H_2O from grams to kilograms using the factor 1 kg/1000 g.

Substitute the known values into the expression for molality, and solve.

$$m = \frac{\text{moles of solute}}{\text{kilograms of solvent}}$$

Write the equation for molality.

$$m = \frac{0.077 \text{ mol NaCl}}{0.1000 \text{ kg } H_2O} = 0.77 \text{ mol/kg}$$

Substitute moles of solute = 0.077 mol NaCl, kilograms of solvent = 0.1000 kg H_2O.

3 EVALUATE THE ANSWER

Because there was less than one-tenth of a mole of solute present in one-tenth of a kilogram of water, the molality should be less than one, and it is. The mass of sodium chloride was given with two significant figures; therefore, the molality is also expressed with two significant figures.

PRACTICE Problems

Do additional problems. Online Practice

27. What is the molality of a solution containing 10.0 g of Na_2SO_4 dissolved in 1000.0 g of water?

28. Challenge How much ($Ba(OH)_2$), in grams, is needed to make a 1.00*m* aqueous solution?

The addition of fluoride to drinking water varies from one locale to another and from one country to another. Adding low levels of fluoride to drinking water is a controversial issue in many countries. Some people think adding fluoride ions to public drinking water is a form of compulsory medication. Debate reasons why people might object to adding fluoride in spite of the positive benefit of decreased tooth decay. Some reasons might include accidental overdose, individual toxicity and sensitivity, and exposure to fluoride through toothpastes and rinses.

Reinforcement

Concentration Units Students often confuse *molarity* and *molality*. Have them write the terms as headings to two columns. Beneath the appropriate head, have them write each term's definition, the equation used to calculate it, and reasons why that concentration unit might be used. OL

ASSESSMENT

Skill Assign each student a problem to solve from the problems at the end of the chapter or from the Supplemental Practice Problems appendix. OL

IN-CLASS Example

Question An experiment requires that 3.2 grams of calcium hydroxide ($Ca(OH)_2$) be added to 1.0 kg of water. What is the molality of the resultant solution? The molar mass of $Ca(OH)_2$ is 74.093 g/mol.

Answer

$$\text{moles} = \frac{3.2 \text{ g}}{74.093 \text{ g/mol}} = 0.043 \text{ mol}$$

$$\text{Molality } (m) = \frac{\text{moles of solute}}{\text{kg of solvent}} = \frac{0.043 \text{ mol } Ca(OH)_2}{1.0 \text{ kg } H_2O} = 0.043 \text{ mol/kg}$$

PRACTICE Problems

Have students refer to the Selected Solutions appendix for complete solutions to odd-numbered problems. The complete solutions for all problems can be found in the Solutions Manual.

27. 0.0704*m*
28. 171 g per kg of water

PRACTICE Problems

Have students refer to the Selected Solutions appendix for complete solutions to odd-numbered problems. The complete solutions for all problems can be found in the Solutions Manual.

29. 0.118
30. 43.8%

3 Assess

Check for Understanding

Divide the class into groups of four. In each group, assign each student one problem: molarity, molality, mole fraction, or percent solution. Each student should solve the assigned problem and then explain the solution to the other members of the group. OL COOP LEARN

Reteach

Have students create a chart that lists the steps needed to determine concentration in each of the units discussed in this section. Then, solve the following problem for at least two different measures: 115.0 grams acetic acid ($HC_2H_3O_2$) in 1500.0 mL water. Remember that for water, 1 mL equals 1 g. Remind students to pay careful attention to units. OL

Extension

Parts per Million A commonly used environmental unit of concentration is ppm, or parts per million. One ppm is the equivalent of 1 mg/liter. Compute the molarity and molality of an aqueous solution that is 5 ppm benzene (C_6H_6). $6.4 \times 10^{-5}M$; $6.4 \times 10^{-5}m$ OL AL

Hydrochloric Acid in Aqueous Solution

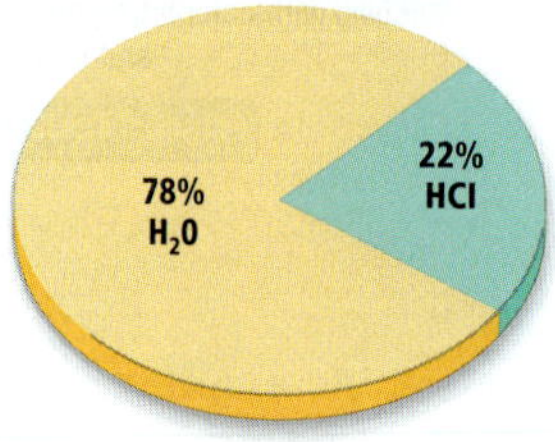

$$X_{HCl} + X_{H_2O} = 1.00$$
$$0.22 + 0.78 = 1.00$$

Figure 8 The mole fraction expresses the number of moles of solute and solvent relative to the total number of moles of solution. Each mole fraction can be thought of as a percent. For example, the mole fraction of water (X_{H_2O}) is 0.78, which is equivalent to saying the solution contains 78% water (on a mole basis).

Mole fraction If you know the number of moles of solute and solvent, you can also express the concentration of a solution as a **mole fraction**—the ratio of the number of moles of solute or solvent in solution to the total number of moles of solute and solvent, as shown in **Figure 8.**

The symbol X is commonly used for mole fraction, with a subscript to indicate the solvent or solute. The mole fraction for the solvent (X_A) and the mole fraction for the solute (X_B) can be expressed as follows.

Mole Fraction

$$X_A = \frac{n_A}{n_A + n_B} \qquad X_B = \frac{n_B}{n_A + n_B}$$

X_A and X_B represent the mole fractions of each substance.
n_A and n_B represent the number of moles of each substance.

A mole fraction equals the number of moles of solute or solvent in a solution divided by the total number of moles of solute and solvent.

For example, suppose a hydrochloric acid solution contains 36 g of HCl and 64 g of H_2O. To convert these masses to moles, you would use the molar masses as conversion factors.

$$n_{HCl} = 36 \cancel{\text{g HCl}} \times \frac{1 \text{ mol HCl}}{36.5 \cancel{\text{g HCl}}} = 0.99 \text{ mol HCl}$$

$$n_{H_2O} = 64 \cancel{\text{g H}_2\text{O}} \times \frac{1 \text{ mol H}_2\text{O}}{18.0 \cancel{\text{g H}_2\text{O}}} = 3.6 \text{ mol H}_2\text{O}$$

The mole fractions of HCl and water can be expressed as follows.

$$X_{HCl} = \frac{n_{HCl}}{n_{HCl} + n_{H_2O}} = \frac{0.99 \text{ mol HCl}}{0.99 \text{ mol HCl} + 3.6 \text{ mol H}_2\text{O}} = 0.22$$

$$X_{H_2O} = \frac{n_{H_2O}}{n_{HCl} + n_{H_2O}} = \frac{3.6 \text{ mol H}_2\text{O}}{0.99 \text{ mol HCl} + 3.6 \text{ mol H}_2\text{O}} = 0.78$$

PRACTICE Problems

Do additional problems. Online Practice

29. What is the mole fraction of NaOH in an aqueous solution that contains 22.8% NaOH by mass?

30. Challenge If the mole fraction of sulfuric acid (H_2SO_4) in an aqueous solution is 0.325, what is the percent by mass of H_2SO_4?

SECTION 2 REVIEW

Section Self-Check

Section Summary

- Concentrations can be measured qualitatively and quantitatively.
- Molarity is the number of moles of solute dissolved per liter of solution.
- Molality is the ratio of the number of moles of solute dissolved in 1 kg of solvent.
- The number of moles of solute does not change during a dilution.

31. MAINIDEA **Compare and contrast** five quantitative ways to describe the composition of solutions.

32. Explain the similarities and differences between a 1*M* solution of NaOH and a 1*m* solution of NaOH.

33. Calculate A can of chicken broth contains 450 mg of sodium chloride in 240.0 g of broth. What is the percent by mass of sodium chloride in the broth?

34. Solve How much ammonium chloride (NH_4Cl), in grams, is needed to produce 2.5 L of a 0.5*M* aqueous solution?

35. Outline the laboratory procedure for preparing a specific volume of a dilute solution from a concentrated stock solution.

SECTION 2 REVIEW

31. Molarity, molality, and mole fraction are based on moles of solute per some other quantity; percent by volume and molarity are defined on a per volume of solution basis; molality and mole fraction are based on a per quantity of solvent basis; percent by mass and percent by volume are the only ratios involving percentages.

32. Both solutions contain NaOH (solute) dissolved in water (solvent). The 1*m* solution contains 1 mole of NaOH per kilogram of water; the 1*M* solution contains 1 mole of NaOH per liter of solution.

33. 0.19%

34. 70 g NH_4Cl

35. Calculate the volume of stock solution needed and add it to a volumetric flask. Add water up to the flask's calibration line.

SECTION 3

Factors Affecting Solvation

MAINIDEA Factors such as temperature, pressure, and polarity affect the formation of solutions.

Essential Questions

- How do intermolecular forces affect solvation?
- What is solubility?
- Which factors affect solubility?

Review Vocabulary

exothermic: a chemical reaction in which more energy is released than is required to break bonds in the initial reactants

New Vocabulary

solvation
heat of solution
unsaturated solution
saturated solution
supersaturated solution
Henry's law

CHEM 4 YOU If you have ever made microwavable soup from a dry mix, you added cold water to the dry mix and stirred. At first, only a small amount of the powdered mix dissolves in the cold water. After heating it in the microwave and stirring again, all of the powdered mix dissolves and you have soup.

The Solvation Process

Why are some substances soluble in each other, while others are not? To form a solution, solute particles must separate from one another and the solute and solvent particles must mix. Recall that attractive forces exist among the particles of all substances. Attractive forces exist between the pure solute particles, between the pure solvent particles, and between the solute and solvent particles. When a solid solute is placed in a solvent, the solvent particles completely surround the surface of the solid solute. If the attractive forces between the solvent and solute particles are greater than the attractive forces holding the solute particles together, the solvent particles pull the solute particles apart and surround them. These surrounded solute particles then move away from the solid solute and out into the solution.

The process of surrounding solute particles with solvent particles to form a solution is called **solvation,** as shown in **Figure 9.** Solvation in water is called hydration. "Like dissolves like" is the general rule used to determine whether solvation will occur in a specific solvent. To determine whether a solvent and solute are alike, you must examine the bonding and polarity of the particles and the intermolecular forces among particles.

Figure 9 Salt begins to separate when it is dropped into water. The solute particles are pulled from the solid and surrounded by solvent particles.

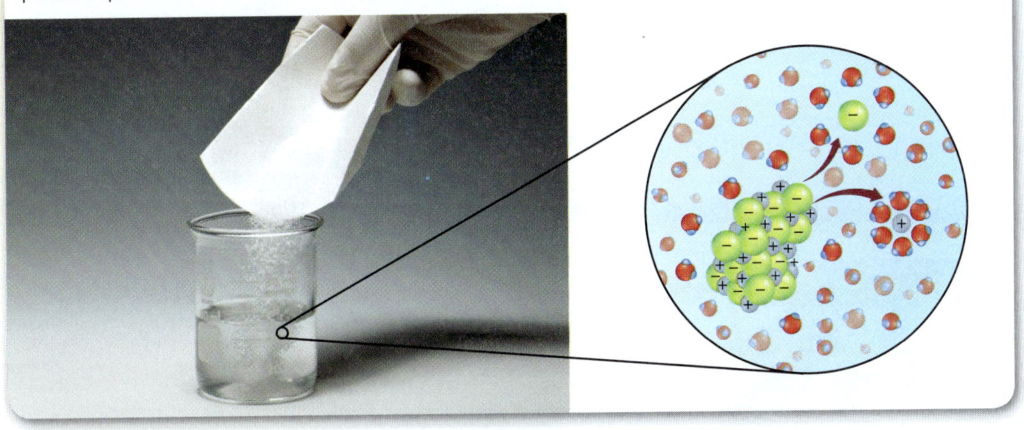

Matt Meadows

DIFFERENTIATED INSTRUCTION

English Learners Have students create a poster or diagram illustrating solvation. They should label the positive and negative ions and the polar ends of the water molecule. Have them share their posters or diagrams with the rest of the class and explain the process. EL OL

SECTION 3

1 Focus

MAINIDEA

Formation Factors Refer students to Figure 9. Ask students to summarize what is happening. Salt is dissolving in water. Ask students to identify the solute and solvent. Students should be able to identify the salt and the water molecules in the particles figure. Students should also recognize that salt is the solute and water is the solvent. Ask students to look again at the figure and explain what is happening to the solute. The solute is being pulled apart and surrounded by the solvent. Explain to students that this is the solvation process and that there are three main factors that affect solvation: temperature, pressure (gases), and polarity. Remind students again that in order for solvation to occur, the solute must be pulled apart by the solvent. Have students predict how each factor affects dissolving. Students should then use these predictions as a point of reference for their readings. OL

2 Teach

Concept Development

Common Solutions Ask students to generate a list of solutions that they use regularly. Then, review the list in class to verify that students have correctly listed only solutions. OL

ASSESSMENT

Knowledge Have students write an explanation of the general rule "like dissolves like" in their own words, citing examples from their own experiences. OL

Identify Misconceptions

Students often think that water is the solvent in all solutions.

Uncover the Misconception Provide the class with as many common solutions as possible. Solutions can include window cleaning products; brass, steel, and pewter; rubbing alcohol; tincture of iodine; carbonated soda in a closed container; and sport drinks. Ask groups of students to describe the composition of each solution, and if possible, state the solvent, the solute, and the states of matter.

Demonstrate the Concept Write the column headings *mixture, solute,* and *solvent* on the board and have student groups list the components of the mixture. Help students label the components as *solid, liquid,* or *gas*. Students should start to realize that solutions are not just solids dissolved in liquids, and that water is not always the solvent.

Assess New Knowledge Call on students to cite examples of different types of solution systems. OL

Figure 10 Sodium chloride dissolves in water as the water molecules surround the sodium and chloride ions. Note how the polar water molecules orient themselves differently around the positive and negative ions.

View an **animation about the dissolution of compounds.**

Concepts In Motion

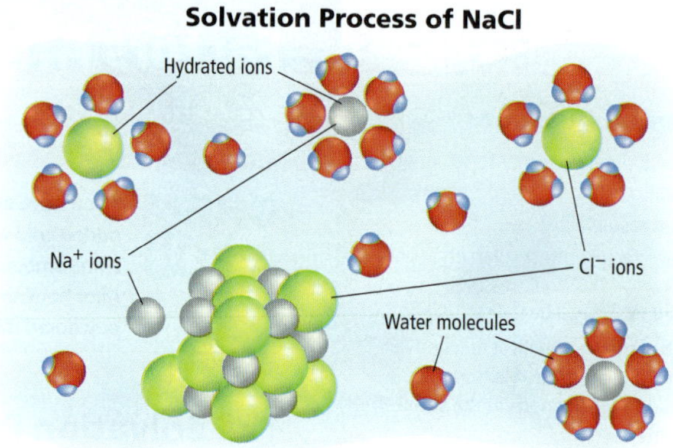

Aqueous solutions of ionic compounds Recall that water molecules are polar molecules and are in constant motion, as described by the kinetic-molecular theory. When a crystal of an ionic compound, such as sodium chloride (NaCl), is placed in water, the water molecules collide with the surface of the crystal. The charged ends of the water molecules attract the positive sodium ions and negative chloride ions. This attraction between the dipoles and the ions is greater than the attraction among the ions in the crystal, so the ions break away from the surface. The water molecules surround the ions, and the solvated ions move into the solution, shown in **Figure 10,** exposing more ions on the surface of the crystal. Solvation continues until the entire crystal has dissolved.

Not all ionic substances are solvated by water molecules. Gypsum is insoluble in water because the attractive forces between the ions in gypsum are so strong that they cannot be overcome by the attractive forces of the water molecules. As shown in **Figure 11,** the discoveries of specific solutions and mixtures, such as plaster made out of gypsum, have contributed to the development of many products and processes.

Figure 11
Milestones in Solution Chemistry

Scientists working with solutions have contributed to the development of products and processes in fields including medical technology, food preparation and preservation, and public health and safety.

1866 The invention of celluloid, a solution of camphor and cellulose, marks the beginning of the plastics industry.

1883 The first successful centrifuge uses the force created by a high rate of spin to separate components of a mixture.

1899 Newly patented technology reduces the size of fat globules dispersed in raw milk, preventing formation of a cream layer—a process called homogenization.

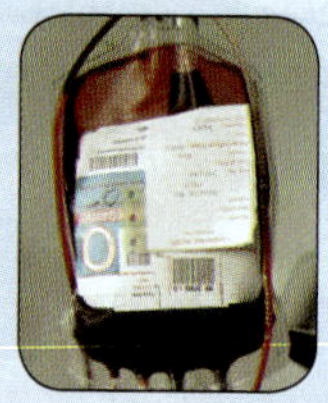

1916 Doctors develop a glycerol solution that allows blood to be stored for up to several weeks after collection for use in transfusions.

1870 1890 1910

CHEMISTRY PROJECT

Sport Drinks Have students research sport drinks and compare the nutritional information and the ingredients of different brands. Is there a benefit to drinking a sports drink if you are an athlete or have just completed a strenuous workout? Is a sports drink significantly different from other noncarbonated, flavored beverages? Have students compare the cost of various sports drinks to other flavored noncarbonated beverages. OL

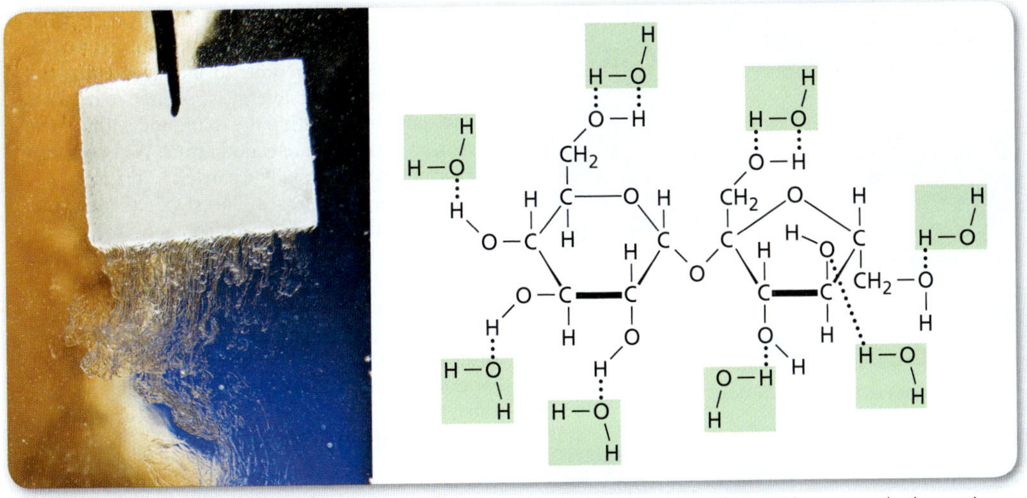

■ **Figure 12** Sucrose molecules contain eight O–H bonds and are polar. Polar water molecules form hydrogen bonds with the O–H bonds, which pulls the sucrose into solution.

Aqueous solutions of molecular compounds Water is also a good solvent for many molecular compounds. Table sugar is the molecular compound sucrose. As shown in **Figure 12,** sucrose molecules are polar and have several O–H bonds. As soon as the sugar crystals contact the water, water molecules collide with the outer surface of the crystal. Each O–H bond becomes a site for hydrogen bonding with water. The attractive forces among sucrose molecules are overcome by the attractive forces between polar water molecules and polar sucrose molecules. Sucrose molecules leave the crystal and become solvated by water molecules.

Oil is a substance made up primarily of carbon and hydrogen. It does not form a solution with water. There is little attraction between the polar water molecules and the nonpolar oil molecules. However, oil spills can be cleaned up with a nonpolar solvent because nonpolar solutes are more readily dissolved in nonpolar solvents.

Watch a **video about water and its solutions.**

Video

1943 The first artificial kidney removes toxins dissolved in a patient's blood.

1950

1964 Stephanie Kwolek discovers a synthetic fiber, formed from liquid crystals in solution, that is stronger than steel and lighter than fiberglass.

1970

1980 A type of gypsum board is developed as a firewall system to separate townhome and condominium units.

1990

2003 Scientists develop chemical packets that remove toxic metals and pesticides and kill pathogens in drinking water. They can be distributed to survivors of natural disasters.

2010

2012 Initial work indicates the bacterium GFAJ-1 can use arsenic in place of phosphorus in biomolecules. However, further research shows that the bacterium, while arsenic tolerant, takes in phosphorus from solution to build these molecules.

Content Background

Safe Drinking Water Natural disasters, including floods, hurricanes, and earthquakes can affect drinking water systems. Water can become contaminated by sewage, agricultural waste, or industrial waste. Bacterial infection in humans can result in diseases such as cholera, typhoid, and dysentery. Water purification packets, developed in 2003, contain calcium hypochlorite that kills many microorganisms, and chemicals that cause precipitation of heavy metals and pesticides. These chemicals are dissolved into water from a local source and then filtered through a cloth to remove impurities, producing safe and clear drinking water.

Apply Chemistry

Detergents Soaps and laundry detergents are an application of the rule "like dissolves like." Long hydrocarbon chains make up most soaps and detergents. These chains are nonpolar except for the carboxyl group at one end. The long nonpolar hydrocarbon chains are attracted to the grease, grime, and oil in clothing and tend to surround the stain particles. The polar ends of the soap molecule orient themselves toward the water molecules. As a result, the nonpolar stain that is surrounded by soap molecules can be washed out of the fabric.

CHEMISTRY JOURNAL

Discoveries Have students discuss the people who might have been most affected by each discovery or development in the time line. How might our society be different today if this discovery had not occurred? What individuals might be most affected and how? Have students record their responses in their journals. OL

Video Lab

Video

Water and Its Solutions

☑ **READING CHECK** Energy is required to overcome the attractive forces within a solute and within the solvent. When solute and solvent particles attract each other, energy is released. The net energy change for these two processes determines whether the solution absorbs or releases energy as it forms.

Concept Development

Solvation Process Explain that the solvation mechanism can be thought of as the sum of three processes: (1) breaking up of the solute into separate component particles, (2) overcoming of interparticle forces in the solvent to make room for the solute (expanding the solvent), and (3) interaction of the solute and solvent particles to form the solution. **OL**

A sugar cube in iced tea will dissolve slowly, but stirring will make the sugar cube dissolve more quickly.

Granulated sugar dissolves more quickly in iced tea than a sugar cube, and stirring will make the granulated sugar dissolve even more quickly.

Granulated sugar dissolves very quickly in hot tea.

Figure 13 Agitation, surface area, and temperature affect the rate of solvation.

Heat of solution During the process of solvation, the solute must separate into particles. Solvent particles must also move apart in order to allow solute particles to come between them. Energy is required to overcome the attractive forces within the solute and within the solvent, so both steps are endothermic. When solute and solvent particles mix, the particles attract each other and energy is released. This step in the solvation process is exothermic. The overall energy change that occurs during the solution formation process is called the **heat of solution.**

As you observed in the Launch Lab at the beginning of this chapter, some solutions release energy as they form, whereas others absorb energy during formation. For example, after ammonium nitrate dissolves in water, its container feels cool. In contrast, after calcium chloride dissolves in water, its container feels warm.

☑ READING CHECK **Explain** why some solutions absorb energy during formation, while others release energy during formation.

Factors That Affect Solvation

Solvation occurs only when the solute and solvent particles come in contact with each other. There are three common ways, shown in **Figure 13,** to increase the collisions between solute and solvent particles and thus increase the rate at which the solute dissolves: agitation, increasing the surface area of the solute, and increasing the temperature of the solvent.

Agitation Stirring or shaking—agitation of the mixture—moves dissolved solute particles away from the contact surfaces more quickly and thereby allows new collisions between solute and solvent particles to occur. Without agitation, solvated particles move away from the contact areas slowly.

Surface area Breaking the solute into small pieces increases its surface area. A greater surface area allows more collisions to occur. This is why a teaspoon of granulated sugar dissolves more quickly than an equal amount of sugar in cube form.

Temperature The rate of solvation is affected by temperature. For example, sugar dissolves more quickly in hot tea, shown in **Figure 13,** than it does in iced tea. Additionally, hotter solvents generally can dissolve more solid solute. Hot tea can hold more dissolved sugar than the iced tea. Most solids act in the same way as sugar—as temperature increases, the rate of solvation also increases. Solvation of other substances, such as gases, decreases at higher temperatures. For example, a carbonated soft drink will lose its fizz (carbon dioxide) faster at room temperature than when cold.

Demonstration

Temperature and Solubility

Purpose
to demonstrate the effect of temperature on solubility

Materials
125-mL Erlenmeyer flasks (2); hot plate; rods (2); $Ca(CH_3COO)_2$ (20 g); KNO_3 (20 g); water

Safety Precautions

Disposal Rinse all chemicals down a drain with at least 100 times their volume in water.

Procedure
Measure out 20 g each of calcium acetate ($Ca(CH_3COO)_2$) and potassium nitrate (KNO_3). Place each solid in a separate 125-mL Erlenmeyer flask containing 50 mL of water. Have students note that the calcium acetate dissolves, whereas the potassium nitrate is only partially soluble. Have students predict what will happen if the solutions are heated. Place both flasks on a hot plate and heat the solutions, stirring occasionally.

Results
The dissolved calcium acetate will crystallize, and the undissolved potassium nitrate will dissolve.

Solubility

Just as solvation can be understood at the particle level, so can solubility. The solubility of a solute also depends on the nature of the solute and solvent. When a solute is added to a solvent, solvent particles collide with the solute's surface particles; solute particles begin to mix randomly among the solvent particles. At first, the solute particles are carried away from the crystal. However, as the number of solvated particles increases, the same random mixing results in increasingly frequent collisions between solvated solute particles and the remaining crystal. Some colliding solute particles rejoin the crystal, or crystallize, as illustrated in **Figure 14.** As solvation continues, the crystallization rate increases, while the solvation rate remains constant. As long as the solvation rate is greater than the crystallization rate, the net effect is continuing solvation.

Depending on the amount of solute present, the rates of solvation and crystallization might eventually equalize. No more solute appears to dissolve and a state of dynamic equilibrium exists between crystallization and solvation (as long as the temperature remains constant).

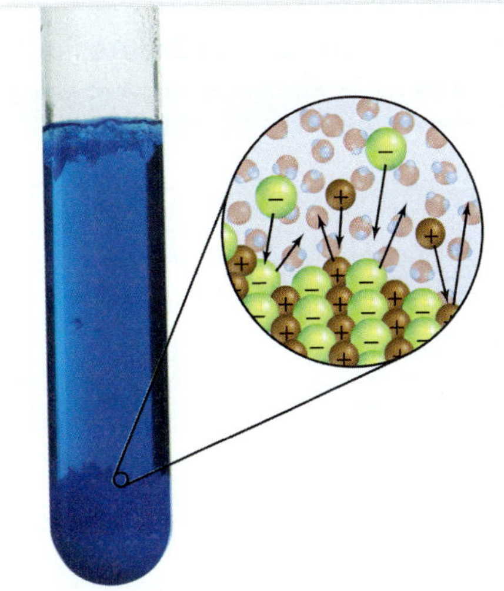

■ **Figure 14** In a saturated solution, the rate of solvation equals the rate of crystallization. The amount of dissolved solute does not change.

Unsaturated solutions An **unsaturated solution** is one that contains less dissolved solute for a given temperature and pressure than a saturated solution. In other words, more solute can be dissolved in an unsaturated solution.

Saturated solutions Although solute particles continue to dissolve and crystallize in solutions that reach equilibrium, the overall amount of dissolved solute in the solution remains constant. Such a solution, illustrated in **Figure 14,** is said to be a **saturated solution;** it contains the maximum amount of dissolved solute for a given amount of solvent at a specific temperature and pressure.

Temperature and supersaturated solutions Solubility is affected by raising the temperature of the solvent because the kinetic energy of its particles is increased, resulting in more-frequent collisions and collisions with greater energy than those that occur at lower temperatures. The fact that many substances are more soluble at high temperatures is demonstrated in **Figure 15.** For example, calcium chloride ($CaCl_2$) has a solubility of about 64 g $CaCl_2$ per 100 g H_2O at 10°C. Increasing the temperature to approximately 27°C increases the solubility by almost 50%, to 100 g $CaCl_2$ per 100 g H_2O. For other substances, such as cerium sulfate, $Ce_2(SO_4)_3$, solubility initially decreases rapidly as temperature increases, but then levels off and remains constant.

■ **Figure 15** The solubilities of several substances as a function of temperature are shown in this graph.

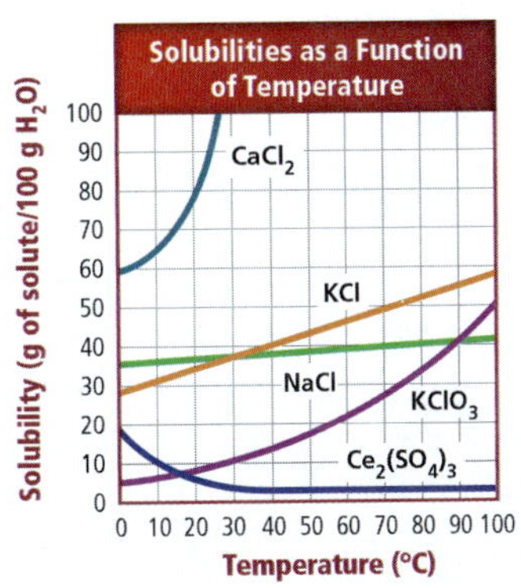

GRAPH CHECK

Interpret What is the solubility of NaCl at 80°C?

GRAPH CHECK 40 g per 100 g H_2O

Concept Development

Supersaturated Solution Use the following analogy to help students understand the terms *saturated, unsaturated,* and *supersaturated.* Suggest that the seats in the classroom represent the amount of solvent. Students represent the solute. If there are 20 students and 25 seats, there is still room to add more students; similarly, additional solute can be added to an unsaturated solution. With 25 students and 25 chairs, the classroom holds the maximum number of students it can accommodate; similarly, a saturated solution can accommodate no more solute. With 30 students squeezing into 25 chairs, the classroom holds more students than it normally can; similarly, a supersaturated solution contains more solute than the solution can normally hold. Have students note that students in the supersaturated class are not comfortable, just as a supersaturated solution is comparatively unstable. When the uncomfortable extra students get tired of being squeezed, they stand up and vacate the chairs. This is analogous to excess solute crystallizing out of a supersaturated solution. BL EL

ChemLAB

The ChemLab located at the end of the chapter can be used at this point in the lesson.

Solubility in g/100 mL H_2O		
Solid	**0°C**	**100°C**
$Ca(CH_3COO)_2$	37.4	29.7
KNO_3	13.3	24.7

Because the solubility of calcium acetate decreases as the temperature increases, it begins to crystallize out of the solution at higher temperatures. The solubility of potassium nitrate increases dramatically at higher temperatures.

Analysis

1. Do the observations agree with your predictions? Most students will predict that both substances will dissolve when heated.
2. Is the slope of potassium nitrate's solubility curve positive or negative? positive
3. Is the slope of the calcium acetate's solubility curve positive or negative? negative

ASSESSMENT

Knowledge Ask students to use this demonstration to preview Le Châtelier's principle as it relates to solution equilibrium. Potassium nitrate has a positive enthalpy of solution (endothermic) and the graph of solubility versus temperature has a positive slope. The opposite is true for calcium acetate. OL AL

Visual Learning

Table 4 Ask students questions based on the data in the table. What happens to the solubility of aluminum sulfate as temperature increases? Sodium chloride? Potassium chloride? Their solubilities increase. Is there a trend or pattern that can be suggested concerning solubility of a substance as temperature increases? The solubility of most non-gas substances increases with increasing temperature. Are there any exceptions? calcium hydroxide, ammonia, carbon dioxide, oxygen Have students note that with the exception of calcium hydroxide, the other three substances are gases that obey Henry's law. OL

Quick Demo

Supersaturated Before class, prepare a supersaturated solution. To do this, slowly heat approximately 10 g of sodium acetate with about 8 mL of distilled water in a small Erlenmeyer flask until the solid is completely dissolved. Let the solution cool to room temperature, being careful not to disturb it while cooling. Add one or two crystals of sodium acetate to the solution and have students observe the sodium acetate crystals form. The sodium acetate can be reused after the water is allowed to evaporate.

ASSESSMENT

Knowledge Ask students to write a paragraph describing what they would expect to observe when one or two crystals of a solute are added to unsaturated, saturated, and supersaturated solutions. OL AL

Table 4 Solubilities of Solutes in Water at Various Temperatures

Substance	Formula	Solubility (g/100 g H_2O)*			
		0°C	20°C	60°C	100°C
Aluminum sulfate	$Al_2(SO_4)_3$	31.2	36.4	59.2	89.0
Barium hydroxide	$Ba(OH)_2$	1.67	3.89	20.94	–
Calcium hydroxide	$Ca(OH)_2$	0.189	0.173	0.121	0.076
Lithium sulfate	Li_2SO_4	36.1	34.8	32.6	–
Potassium chloride	KCl	28.0	34.2	45.8	56.3
Sodium chloride	NaCl	35.7	35.9	37.1	39.2
Silver nitrate	$AgNO_3$	122	216	440	733
Sucrose	$C_{12}H_{22}O_{11}$	179.2	203.9	287.3	487.2
Ammonia*	NH_3	1130	680	200	–
Carbon dioxide*	CO_2	1.713	0.878	0.359	–
Oxygen*	O_2	0.048	0.031	0.019	–

* L/1 L H_2O of gas at standard pressure (101 kPa)

VOCABULARY
WORD ORIGIN
Saturated
comes from the Latin *saturatus* meaning *to fill*

The effect of temperature on solubility is also illustrated by the data in **Table 4.** Notice in **Table 4** that at 20°C, 203.9 g of sucrose ($C_{12}H_{22}O_{11}$) dissolves in 100 g of water. At 100°C, 487.2 g of sucrose dissolves in 100 g of water, a nearly 140% increase in solubility.

The fact that solubility changes with temperature and that some substances become more soluble with increasing temperature is the key to forming supersaturated solutions. A **supersaturated solution** contains more dissolved solute than a saturated solution at the same temperature. To make a supersaturated solution, a saturated solution is formed at a high temperature and then cooled slowly. The slow cooling allows the excess solute to remain dissolved in solution at the lower temperature.

■ **Figure 16** When a seed crystal is added to a supersaturated solution, the excess solute crystallizes out of the solution.

Stephen Frisch/McGraw-Hill Education

CHEMISTRY JOURNAL

Make Solutions Have students write a paragraph explaining how a solubility graph can be used in making saturated, unsaturated, and supersaturated solutions. Then, have students exchange paragraphs and use them to make each type of solution. OL

■ **Figure 17** Hot spring mineral deposits are an example of crystals that formed from supersaturated solutions.

Supersaturated solutions are unstable. If a tiny amount of solute, called a seed crystal, is added to a supersaturated solution, the excess solute precipitates quickly, as illustrated in **Figure 16.** Crystallization can also occur if the inside of the container is scratched or the supersaturated solution undergoes a physical shock, such as stirring or tapping the container. Using crystals of silver iodide (AgI) to seed air that is supersaturated with water vapor causes the water particles to come together and form droplets that might fall to Earth as rain. This technique is called cloud seeding. Rock candy and mineral deposits at the edges of mineral springs, such as those shown in **Figure 17,** are both formed from supersaturated solutions.

Solubility of gases In **Table 4** you can see that the gases oxygen and carbon dioxide are less soluble at higher temperatures than at lower temperatures. This is a predictable trend for all gaseous solutes in liquid solvents. Can you explain why? Recall that the kinetic energy of gas particles allows them to escape from a solution more readily at higher temperatures. Thus, as a solution's temperature increases, the solubility of a gaseous solute decreases.

Pressure and Henry's law Pressure affects the solubility of gaseous solutes in solutions. The solubility of a gas in any solvent increases as its external pressure (the pressure above the solution) increases. Carbonated beverages depend on this fact. Carbonated beverages contain carbon dioxide gas dissolved in an aqueous solution. In bottling or canning the beverage, carbon dioxide is dissolved in the solution at a pressure higher than atmospheric pressure. When the beverage container is opened, the pressure of the carbon dioxide gas in the space above the liquid decreases. As a result, bubbles of carbon dioxide gas form in the solution, rise to the top, and escape. Unless the container is sealed, the process will continue until the solution loses almost all of its carbon dioxide gas and goes flat. The decreased solubility of the carbon dioxide contained in the beverage after it is opened can be described by Henry's law.

Explore **salts and solubility.**

Virtual Investigations

Essential Standards Self-Check

Distinguish between unsaturated, saturated, and supersaturated solutions. What factors influence saturation?

VOCABULARY

SCIENCE USAGE V. COMMON USAGE

Pressure

Science usage: the force exerted over an area

As carbon dioxide escapes the solution, the pressure in the closed bottle increases.

Common usage: The burden of physical or mental stress

There is a lot of pressure to do well on exams.

Quick Demo

Sugar Crystals Bring 250 mL of water to boil in a beaker. Add sugar to the water until no more sugar dissolves, then pour the solution into a jar. Tie a heavy piece of string or twine to one end of a wood splint and use the splint to suspend the string into the solution. Place the jar in a corner of the room where it will remain undisturbed for several days. Have students observe the formation of sugar crystals on the string. OL

Enrichment

Gaseous Element Ask groups of three or four students to research the properties of an element that is a gas at normal atmospheric pressure and room temperature. Have student groups prepare classroom presentations about their findings. BL EL

Visual Learning

Figure 17 Ask students to brainstorm other examples of supersaturated solutions. Point out the figure and the description in the text of cloud seeding to promote rainfall. An example of a supersaturated solution in cooking is the making of fudge. BL

ASSESSMENT

Knowledge Ask students to explain why a carbonated soda, once opened, will go flat more quickly at room temperature than when placed in a refrigerator. Because carbon dioxide is more soluble at lower temperatures, less CO_2 will escape the solution when placed in a refrigerator. OL

■ **Caption Question Fig. 18** When the cap is removed, the decreased pressure above the solution results in the decreased solubility of the carbon dioxide.

Extension

Dissolved Oxygen Ask students to hypothesize about the oxygen needs of fish found in different regions of the world with different water temperatures. Discuss the terms *dissolved oxygen (DO)* and *biological oxygen demand (BOD)*. Make a connection about what happens to the amount of dissolved oxygen as the temperature increases. Have students draw a solubility graph for the amount of oxygen dissolved based on ocean water temperature. OL

Henry's law states that at a given temperature, the solubility (S) of a gas in a liquid is directly proportional to the pressure (P) of the gas above the liquid. When the bottle of soda is closed, as illustrated in **Figure 18,** the pressure above the solution keeps carbon dioxide from escaping the solution. You can express this relationship in the following way.

Henry's Law

$$\frac{S_1}{P_1} = \frac{S_2}{P_2}$$

S represents solubility.
P represents pressure.

At a given temperature, the quotient of solubility of a gas and its pressure is constant.

You will often use Henry's law to determine the solubility S_2 at a new pressure P_2, where P_2 is known. The basic rules of algebra can be used to solve Henry's law for any one specific variable. To solve for S_2, begin with the standard form of Henry's law.

$$\frac{S_1}{P_1} = \frac{S_2}{P_2}$$

Cross multiplying yields the following expression.

$$S_1P_2 = P_1S_2$$

Dividing both sides of the equation by P_1 yields the desired result–the equation solved for S_2.

$$\frac{S_1P_2}{P_1} = \frac{\cancel{P_1}S_2}{\cancel{P_1}} \qquad S_2 = \frac{S_1P_2}{P_1}$$

■ **Figure 18** Carbon dioxide (CO_2) is dissolved in soda. Some CO_2 also is found in the gas above the liquid.

Explain ***Why does the carbon dioxide escape from the solution when the cap is removed?***

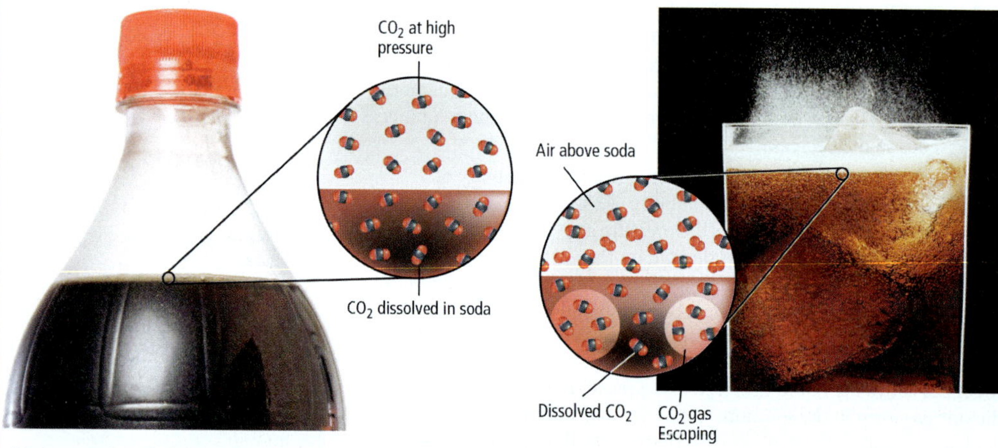

The pressure above the solution of a closed soda bottle keeps excess carbon dioxide from escaping the solution.

The pressure above the solution decreases when the cap is removed, which decreases the solubility of the carbon dioxide.

CHEMISTRY JOURNAL

Air and Health Have students research the historical development of the composition of air used in scuba tanks and diving bells. They should include information about decompression sickness, or the bends, and discuss its cause and remedy. Have student use Henry's law to discuss the bends. Another topic for research is the use of hyperbaric chambers in hospitals. The chambers use increased oxygen concentration in the air in order to increase the amount of oxygen that reaches a patient's blood and brain. The chambers are commonly used to promote healing of burn victims as the increased oxygen supply helps to regenerate cells. OL

EXAMPLE Problem 5

HENRY'S LAW If 0.85 g of a gas at 4.0 atm of pressure dissolves in 1.0 L of water at 25°C, how much will dissolve in 1.0 L of water at 1.0 atm of pressure and the same temperature?

1 ANALYZE THE PROBLEM

You are given the solubility of a gas at an initial pressure. The temperature of the gas remains constant as the pressure changes. Because decreasing pressure reduces a gas's solubility, less gas should dissolve at the lower pressure.

Known
$S_1 = 0.85$ g/L
$P_1 = 4.0$ atm
$P_2 = 1.0$ atm

Unknown
$S_2 = ?$ g/L

2 SOLVE FOR THE UNKNOWN

$$\frac{S_1}{P_1} = \frac{S_2}{P_2}$$ State Henry's law.

$$S_2 = S_1\left(\frac{P_2}{P_1}\right)$$ Solve Henry's law for S_2.

$$S_2 = \left(\frac{0.85\text{ g}}{1.0\text{ L}}\right)\left(\frac{1.0\text{ atm}}{4.0\text{ atm}}\right) = 0.21\text{ g/L}$$ Substitute $S_1 = 0.85$ g/L, $P_1 = 4.0$ atm, and $P_2 = 1.0$ atm. Multiply and divide numbers and units.

3 EVALUATE THE ANSWER

The solubility decreased as expected. The pressure on the solution was reduced from 4.0 atm to 1.0 atm, so the solubility should be reduced to one-fourth its original value, which it is. The unit g/L is a solubility unit, and there are two significant figures.

EXAMPLE PROBLEM

PRACTICE Problems

Do additional problems. Online Practice

36. If 0.55 g of a gas dissolves in 1.0 L of water at 20.0 kPa of pressure, how much will dissolve at 110.0 kPa of pressure?

37. A gas has a solubility of 0.66 g/L at 10.0 atm of pressure. What is the pressure on a 1.0-L sample that contains 1.5 g of gas?

38. Challenge The solubility of a gas at 7.0 atm of pressure is 0.52 g/L. How many grams of the gas would be dissolved per 1.0 L if the pressure increased 40.0 percent?

SECTION 3 REVIEW

Section Self-Check

Section Summary

- The process of solvation involves solute particles surrounded by solvent particles.
- Solutions can be unsaturated, saturated, or supersaturated.
- Henry's law states that at a given temperature, the solubility (S) of a gas in a liquid is directly proportional to the pressure (P) of the gas above the liquid.

39. MAINIDEA **Describe** factors that affect the formation of solutions.

40. Define *solubility.*

41. Describe how intermolecular forces affect solvation.

42. Explain on a particle basis why the vapor pressure of a solution is lower than that of the pure solvent.

43. Summarize If a seed crystal is added to a supersaturated solution, how would you characterize the resulting solution?

44. Make and Use Graphs Use the information in **Table 4** to graph the solubilities of aluminum sulfate, lithium sulfate, and potassium chloride at 0°C, 20°C, 60°C, and 100°C. Which substance's solubility is most affected by increasing temperature?

3 Assess

Check for Understanding

Ask students to identify at least one solid solute and one gaseous solute in ocean water. Ask students to compare the degree of solubility of oxygen or carbon dioxide (two common gases dissolved in ocean water) in warmer regions (near the equator) with ocean water in colder regions of the world. Have students describe the difference in solubility using Henry's law. OL

Reteach

Have students write the equation for Henry's law and explain the equation in their own words. OL

IN-CLASS Example

Question If 1.2 g of a gas at 3.5 atm of pressure dissolves in 1.0 L of water at 25°C, how much pressure is needed to dissolve 2.4 g of the gas in 1.0 L of water at the same temperature?

Answer 7.0 atm

$$\frac{S_1}{P_1} = \frac{S_2}{P_2}$$

$$P_2 = \left(\frac{S_2}{S_1}\right)P_1 = \left(\frac{2.4\text{ g/L}}{1.2\text{ g/L}}\right)3.5\text{ atm} = 7.0\text{ atm}$$

PRACTICE Problems

Have students refer to the Selected Solutions appendix for complete solutions to odd-numbered problems. The complete solutions for all problems can be found in the Solutions Manual.

36. 3.0 g/L
37. 23 atm
38. 0.73 g/L

SECTION 3 REVIEW

39. Surface area, temperature, and pressure affect the formation of solutions.

40. Solubility refers to the maximum amount of solute that can dissolve in a given amount of solvent at a particular temperature and pressure.

41. The attractive forces between solute and solvent particles overcome the forces holding the solute particles together, thus pulling the solute particles apart.

42. When a solvent contains a solute, fewer solvent particles occupy the surface. Fewer particles escape into the gaseous state.

43. After the excess solute particles crystallize out of solution, the solution is saturated.

44. Refer to the Solution Manual for the graph. Aluminum sulfate shows the greatest change in solubility over the temperature range.

SECTION 4

1 Focus

MAINIDEA

Solute Dependency Ask students to recall the formula for molality. Students should remember that molality is defined as the moles of solute per kg of solvent. Refer students to the formulas for boiling point elevation and freezing point depression. Ask students what they notice is the exact same in each formula. The presence of a molality value. Write the two formulas on the board next to one another. Covering up the K_f variable, ask students what the relationship is between boiling point elevation and molality. Students should recognize that they are directly related. Next, cover the K_b variable and repeat questioning. Students should recognize that they are directly related. Remind students that molality is determined by the moles of solute per kg of solvent. Therefore, it can also be said that colligative properties, such as boiling point elevation and freezing point depression, are directly related to the amount of solute particles found in solution. OL

2 Teach

Enrichment

Electrolytes Have students research the role of electrolytes in the body. Research should include the role of electrolytes in nerve and muscle function. Students should also include information on the issues that arise when electrolytes are imbalanced in the body. OL

SECTION 4

Colligative Properties of Solutions

MAINIDEA Colligative properties depend on the number of solute particles in a solution.

Essential Questions

- What are colligative properties?
- What are four colligative properties of solutions?
- How are the boiling point elevation and freezing point depression of a solution determined?

Review Vocabulary

ion: an atom that is electrically charged

New Vocabulary

colligative property
vapor pressure lowering
boiling point elevation
freezing point depression
osmosis
osmotic pressure

CHEM 4 YOU **If you live in an area that experiences cold winters, you have probably noticed people spreading salt to melt icy sidewalks and roads. How does salt help make a winter's drive safer?**

Electrolytes and Colligative Properties

Solutes affect some of the physical properties of their solvents. Early researchers were puzzled to discover that the effects of a solute on a solvent depended only on how many solute particles were in the solution, not on the specific solute dissolved. Physical properties of solutions that are affected by the number of particles but not by the identity of dissolved solute particles are called **colligative properties.** The word *colligative* means *depending on the collection.* Colligative properties include vapor pressure lowering, boiling point elevation, freezing point depression, and osmotic pressure.

Electrolytes in aqueous solution You read previously that ionic compounds are called electrolytes because they dissociate in water to form a solution that conducts electric current, as shown in **Figure 19.** Some molecular compounds ionize in water and are also electrolytes. Electrolytes that produce many ions in a solution are called strong electrolytes; those that produce only a few ions in a solution are called weak electrolytes.

View an **animation about strong, weak, and nonelectrolytes.**

Concepts In Motion

■ **Figure 19** Sodium chloride conducts electricity well because it is an electrolyte. Sucrose does not conduct electricity because it is not an electrolyte.

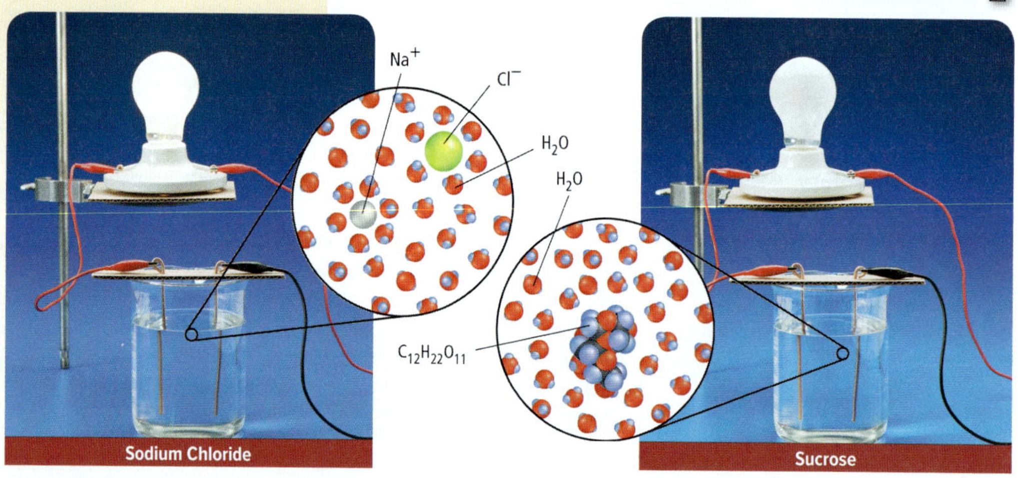

Sodium chloride is a strong electrolyte. It dissociates in solution, producing Na^+ and Cl^- ions.

$$NaCl(s) \rightarrow Na^+(aq) + Cl^-(aq)$$

Dissolving 1 mol of NaCl in 1 kg of water would not yield a 1*m* solution of ions. Rather, there would be 2 mol of solute particles in solution—1 mol each of Na^+ and Cl^- ions.

Nonelectrolytes in aqueous solution Many molecular compounds dissolve in solvents but do not ionize. Such solutions do not conduct an electric current, as shown in **Figure 19,** and the solutes are called nonelectrolytes. Sucrose is an example of a nonelectrolyte. A 1*m* sucrose solution contains only 1 mol of sucrose particles.

☑ READING CHECK **Infer** Which compound would have the greater effect on colligative properties, sodium chloride or sucrose?

Vapor Pressure Lowering

You learned previously that vapor pressure is the pressure exerted in a closed container by particles that have escaped the liquid's surface and entered the gaseous state. In a closed container at constant temperature and pressure, the solvent particles of a pure solvent reach a state of dynamic equilibrium, escaping and reentering the liquid state at the same rate.

Experiments show that adding a nonvolatile solute (one that has little tendency to become a gas) to a solvent lowers the solvent's vapor pressure. The particles that produce vapor pressure escape the liquid phase at its surface. When a solvent is pure, as shown in **Figure 20,** its particles occupy the entire surface area. When the solvent contains solute, as also shown in **Figure 20,** a mix of solute and solvent particles occupies the surface area. With fewer solvent particles at the surface, fewer particles enter the gaseous state, and the vapor pressure is lowered. The greater the number of solute particles in a solvent, the lower the resulting vapor pressure. Thus, **vapor pressure lowering** is due to the number of solute particles in solution and is a colligative property of solutions.

You can predict the relative effect of a solute on vapor pressure based on whether the solute is an electrolyte or a nonelectrolyte. For example, 1 mol each of the solvated nonelectrolytes glucose, sucrose, and ethanol molecules has the same relative effect on the vapor pressure. However, 1 mol each of the solvated electrolytes sodium chloride (NaCl), sodium sulfate (Na_2SO_4), and aluminum chloride ($AlCl_3$) has an increasingly greater effect on vapor pressure because of the increasing number of ions each produces in solution.

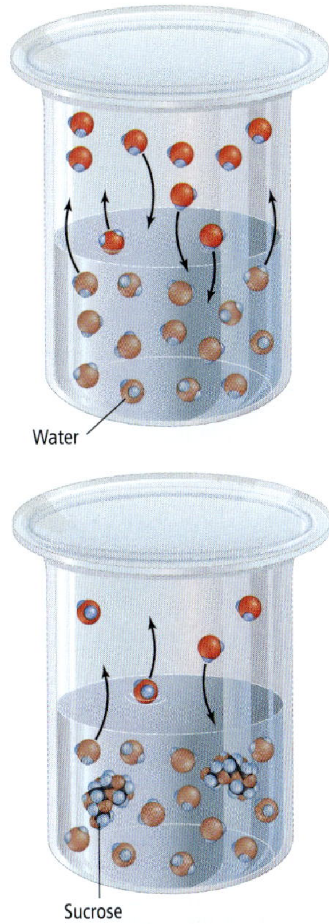

Figure 20 The vapor pressure of a pure solvent is greater than the vapor pressure of a nonvolatile solution.

Identify Misconceptions

Most students will think adding a solute will not change the normal boiling point or freezing point of a solution. Remind students that for a pure solvent, a substance's boiling and condensing points occur at the same temperature, and its freezing and melting points also occur at the same temperature (but at a temperature different from that of the boiling/condensation point).

Uncover the Misconception Ask students to brainstorm practical examples of situations in which a freezing point is lowered or a boiling point is increased. Examples might include adding salt to water when cooking, adding salt to ice on roadways in the winter, adding salt to water when making ice cream, and spraying a liquid such as ethylene glycol on a plane to melt the ice before the plane takes off.

Demonstrate the Concept Put 100 mL of water in a clear, plastic bottle, such as an empty soda container, and replace the cap. Repeat with a solution of rubbing alcohol that is 70% alcohol and 30% water. Label the plastic containers clearly and place them on their sides in a freezer at the beginning of the class period. At the end of the class period, retrieve the bottles and compare the two solutions. The water should show more evidence of freezing. Have students explain why.

Assess New Knowledge Have students describe the effect of vapor pressure on the normal boiling point and freezing point of a solvent when a solute is added. OL

DIFFERENTIATED INSTRUCTION

Advanced Learners Have students explain why the temperatures of both solutions must be the same in order to compare the vapor pressure of two solutions. At higher temperatures, molecules can escape from the liquid at a faster rate.

☑ READING CHECK Sodium chloride will have the greater effect because it will dissociate, forming twice as many particles.

ASSESSMENT

Knowledge Have students apply their knowledge of boiling-point elevation and freezing-point depression to describe why antifreeze is both a coolant in the summer and a preventative to engine freeze-up in the winter. OL

Apply Chemistry

Antifreeze Some animals have adaptations that allow them to survive in subfreezing temperatures. Fishes that live in the Antarctic produce proteins and glycoproteins that serve as antifreeze, and some reptiles use glucose and glycerol. OL

ASSESSMENT

Knowledge Have students solve the following problem and include a written explanation of the steps they used to arrive at the solution. What mass of glucose ($C_6H_{12}O_6$) will produce the same boiling-point elevation as 171 g of sucrose ($C_{12}H_{22}O_{11}$)? 90.0 g glucose OL

Boiling Point Elevation

Because a nonvolatile solute lowers a solvent's vapor pressure, it also affects the boiling point of the solvent. Recall that liquid in a pot on a stove boils when its vapor pressure equals the atmospheric pressure. When the temperature of a solution containing a nonvolatile solute is raised to the boiling point of the pure solvent, the resulting vapor pressure is still less than the atmospheric pressure and the solution will not boil. Thus, the solution must be heated to a higher temperature to supply the additional kinetic energy needed to raise the vapor pressure to atmospheric pressure. The temperature difference between a solution's boiling point and a pure solvent's boiling point is called the **boiling point elevation.**

For nonelectrolytes, the value of the boiling point elevation, which is symbolized ΔT_b, is directly proportional to the solution's molality.

Boiling Point Elevation

$$\Delta T_b = K_b m$$

ΔT_b represents the boiling point elevation.
K_b represents the molal boiling elevation constant.
m represents molality.

The temperature difference is equal to the molal boiling point elevation constant multiplied by the solution's molality.

The molal boiling point elevation constant, K_b, is the difference in boiling points between a $1m$ nonvolatile, nonelectrolyte solution and a pure solvent. Boiling point elevation is expressed in units of °C/m and varies for different solvents. Values of K_b for several common solvents are found in **Table 5.** Note that water's K_b value is 0.512°C/m. This means that a $1m$ aqueous solution containing a nonvolatile, nonelectrolyte solute boils at 100.512°C—a temperature just 0.512°C higher than pure water's boiling point of 100.0°C.

Explore **colligative properties.**

Virtual Investigations

Like vapor pressure lowering, boiling point elevation is a colligative property. The value of the boiling point elevation is directly proportional to the solution's solute molality; that is, the greater the number of solute particles in the solution, the greater the boiling point elevation. Because it is related to mole fraction, which involves the number of solute particles, molality is used as the concentration. Molality also uses mass of solvent rather than volume, and therefore is not affected by temperature changes. Examine **Figure 21** and notice that the curve for a solution lies below the curve for the pure solvent at any temperature.

Table 5 Molal Boiling Point Elevation Constants (K_b)

Solvent	Boiling Point (°C)	K_b (°C/m)
Water	100.0	0.512
Benzene	80.1	2.53
Carbon tetrachloride	76.7	5.03
Ethanol	78.5	1.22
Chloroform	61.7	3.63

CHEMISTRY PROJECT

Melting Ice Have students design a plan to investigate melting ice in pure water and in saltwater. Would the rate of melting be the same in both solutions? Students might also investigate the effect of different concentrations of salt in the saltwater solutions on melting ice. Have students explain any differences they find. OL

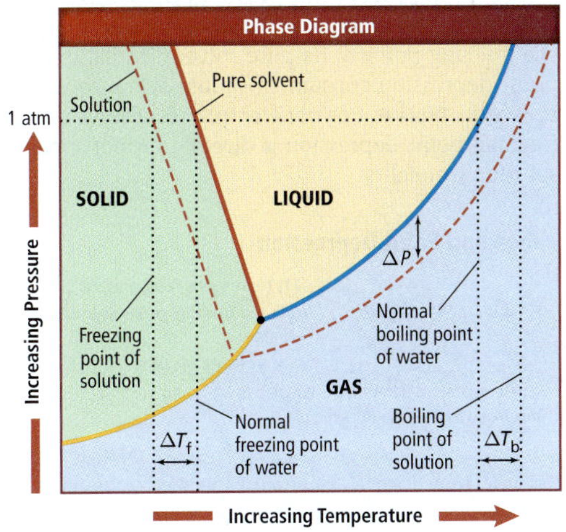

■ **Figure 21** Temperature and pressure affect solid, liquid, and gas phases of a pure solvent (solid lines) and a solution (dashed line).

☑ GRAPH CHECK
Describe how the difference between the solid lines and dashed line corresponds to vapor pressure lowering, boiling point elevation, and freezing point depression. Use specific data from the graph to support your answer.

Freezing Point Depression

At a solvent's freezing point temperature, the particles no longer have sufficient kinetic energy to overcome the interparticle attractive forces; the particles form into a more organized structure in the solid state. In a solution, the solute particles interfere with the attractive forces among the solvent particles. This prevents the solvent from entering the solid state at its normal freezing point.

The freezing point of a solution is always lower than that of a pure solvent. **Figure 21** shows the differences in boiling and melting points of pure water and an aqueous solution. By comparing the solid and dashed lines, you can see that the temperature range over which the aqueous solution exists as a liquid is greater than that of pure water. Two common applications of freezing point depression, shown in **Figure 22,** use salt to lower the freezing point of a water solution.

■ **Figure 22** By adding salts to the ice on a road, the freezing point of the ice is lowered, which results in the ice melting. Adding salt to ice when making ice cream lowers the freezing point of the ice, allowing the resulting water to freeze the ice cream.

☑ GRAPH CHECK The difference between the solid and dashed line at the liquid-gas boundary corresponds to vapor pressure lowering (ΔP). Boiling point elevation (ΔT_b) is the difference between the normal boiling point on the solid line and the corresponding spot on the dashed line; freezing point depression (ΔT_f) is similarly the difference at the normal freezing point of water

Visual Learning

Phase Diagram Have students interpret the phase diagram using each axis separately. When considering the temperature axis, have students explain the diagram in terms of the properties of water. The left side of the diagram represents the solid phase, the middle represents the liquid phase, and the right side represents the gas phase. When considering the pressure axis, have students explain the diagram in terms of what happens to substances when exposed to very high pressures. The section of the diagram at high pressure represents the solid phase, the next section represents the liquid phase, and the section with the lowest pressure represents the gas phase. OL

Quick Demo

Ice Cream Bring in an ice-cream maker and make ice cream during the class period. Emphasize the importance of adding rock salt to the ice.

CHEMISTRY JOURNAL

Salting Roads Rock salt is used extensively to melt the ice on roadways in the winter. Have students compare the positive factors (reducing accidents) and negative factors (corrosion of cars and bridges, the accelerated deterioration of roadways, and the killing of roadside plant life) of using salt. OL

MiniLAB

Rubric

Purpose Students will investigate the effect of a solute on the freezing point of a solution.

Process Skills measure, compare and contrast, predict

Safety Precautions Approve lab safety forms before work begins. Wear a lab apron and safety goggles during the lab.

Teaching Strategies

- Students should read Section 4 before performing this laboratory investigation.
- Use the results of this investigation to initiate a detailed discussion of the chemistry involved in the making of homemade ice cream.
- Remind students not to use the thermometers as stirring rods.

Expected Results The salt will lower the freezing point of the water 4° to 6°C.

Analysis

1. The freezing point of the water should be lowered 4°–6°C by the addition of the salt. The ions interfere with the attractive forces between water molecules, thus preventing the water from freezing at its normal freezing temperature of 0°C.
2. The beaker containing only ice serves as the control.
3. The number of particles in the solution affects the colligative properties of a solution. Because 1 mol of sodium chloride produces 2 mol of ions in solution, it has a greater effect on the freezing point and boiling point than a solute that produces only 1 mol of particles in solution.
4. Fine table salt is a better choice because it dissolves more quickly in the cold water than coarse rock salt thus producing the greatest freezing point depression as quickly as possible.

Table 6 Molal Freezing Point Depression Constants (K_f)

Solvent	Freezing Point (°C)	K_f (°C/m)
Water	0.0	1.86
Benzene	5.5	5.12
Carbon tetrachloride	−23.0	29.8
Ethanol	−114.1	1.99
Chloroform	−63.5	4.68

A solution's **freezing point depression,** ΔT_f, is the difference in temperature between its freezing point and the freezing point of its pure solvent. Molal freezing point depression constants (K_f) for several solvents are shown in **Table 6.** For nonelectrolytes, the value of the freezing point depression is directly proportional to the solution's molality.

Freezing Point Depression

$$\Delta T_f = K_f m$$

ΔT_f represents temperature.
K_f is the freezing point depression constant.
m represents molality.

The temperature difference is equal to the freezing point depression constant multiplied by the solution's molality.

As with K_b values, K_f values are specific to their solvents. With water's K_f value of 1.86°C/m, a 1m aqueous solution containing a nonvolatile, nonelectrolye solute freezes at −1.86°C rather than at pure water's freezing point of 0.0°C. Glycerol is a nonelectrolyte solute produced by many fish and insects to keep their blood from freezing during cold winters. Antifreeze and many de-icer solutions contain the nonelectrolyte solute ethylene glycol.

Notice that the equations for boiling point elevation and freezing point depression specify the molality of a nonelectrolyte. For electrolytes, you must make sure to use the effective molality of the solution. Example Problem 6 illustrates this point.

MiniLAB

Examine Freezing Point Depression

How do you measure freezing point depression?

Procedure

1. Read and complete the lab safety form.
2. Fill two **400-mL beakers** with **crushed ice.** Add 50 mL of **cold tap water** to each beaker.
3. Measure the temperature of each beaker using a **nonmercury thermometer.**
4. Stir the contents of each beaker with a **stirring rod** until both beakers are at a constant temperature—approximately 1 min. Record the temperature.
5. Add 75 g of **rock salt (NaCl)** to one of the beakers. Continue stirring both beakers. Some of the salt will dissolve.
6. When the temperature in each beaker is constant, record the final readings.
7. To clean up, flush the contents of each beaker down the drain with excess water.

Analysis

1. **Compare** your readings taken for the ice water and the salt water. How do you explain the observed temperature change?
2. **Explain** why salt was added to only one of the beakers.
3. **Explain** Salt is a strong electrolyte that produces two ions, Na^+ and Cl^-, when it dissociates in water. Explain why this is important to consider when calculating the colligative property of freezing point depression.
4. **Predict** whether it would be better to use coarse rock salt or fine table salt when making homemade ice cream. Explain.

ASSESSMENT

Knowledge Have students calculate the freezing point of a 2.5M aqueous solution of glucose (a nonelectrolyte). −4.68°C OL

EXAMPLE Problem 6

Find help with **algebraic equations.** Math Handbook

CHANGES IN BOILING AND FREEZING POINTS Sodium chloride (NaCl) is often used to prevent icy roads and to freeze ice cream. What are the boiling and freezing points of a 0.029*m* aqueous solution of sodium chloride?

1 ANALYZE THE PROBLEM

You are given the molality of an aqueous sodium chloride solution. First, calculate ΔT_b and ΔT_f based on the number of particles in the solution. Then, to determine the elevated boiling point and the depressed freezing point, add ΔT_b to the normal boiling point and subtract ΔT_f from the normal freezing point of water.

Known	Unknown
	boiling point = ?°C
molality of solution = 0.029*m*	freezing point = ?°C
K_b = 0.512°C/m	
K_f = 1.86°C/m	

2 SOLVE FOR THE UNKNOWN

Determine the molality of the particles.

particle molality = 2 × 0.029*m* = 0.058*m*

Determine ΔT_b and ΔT_f

$\Delta T_b = K_b m$

$\Delta T_f = K_f m$

State the boiling point elevation and freezing point depression formulas.

$\Delta T_b = (0.512°C/\cancel{m})(0.058\cancel{m}) = 0.030°C$

$\Delta T_f = (1.86°C/\cancel{m})(0.058\cancel{m}) = 0.11°C$

Substitute K_b = 0.512°C/*m*, K_f = 1.86°C/*m*, and *m* = 0.058*m*.

Determine the elevated boiling point and depressed freezing point of the solution.

boiling point = 100.000°C + 0.030°C
= **100.030°C**

freezing point = 0.00°C − 0.11°C
= **−0.11°C**

Add ΔT_b to the normal boiling point and subtract ΔT_f from the normal freezing point.

3 EVALUATE THE ANSWER

The boiling point is higher and the freezing point is lower, as expected. Because the molality of the solution has two significant figures, both ΔT_b and ΔT_f have two significant figures. Because the normal boiling point and freezing point are exact values, they do not affect the number of significant figures in the final answer.

PRACTICE Problems

Do additional problems. Online Practice

45. What are the boiling point and freezing point of a 0.625*m* aqueous solution of any nonvolatile, nonelectrolyte solute?

46. What are the boiling point and freezing point of a 0.40*m* solution of sucrose in ethanol?

47. Challenge A 0.045*m* solution (consisting of a nonvolatile, nonelectrolyte solute) is experimentally found to have a freezing point depression of 0.080°C. What is the freezing point depression constant (K_f)? Which is most likely to be the solvent: water, ethanol, or chloroform?

RealWorld CHEMISTRY

Freezing Point Depression

SALTWATER FISH Maintaining the proper saline (salt) concentration is important to the health of saltwater fish. In the ocean, the presence of salt in arctic areas keeps the water from freezing, allowing aquatic life to be sustained.

IN-CLASS Example

Question What are the boiling point and freezing point of a 0.16*m* aqueous solution of calcium chloride ($CaCl_2$)?

Answer

boiling point = 100.25°C;
freezing point = −0.89°C

Particle molality = 3 × 0.16*m* = 0.48*m*

$\Delta T_b = K_b m$
= (0.512°C/*m*)(0.48*m*)
= 0.25°C

Boiling point = 100.0°C + 0.25°C
= 100.25°C

$\Delta T_f = K_f m$
= (1.86°C/*m*)(0.48*m*)
= 0.89°C

Freezing point = 0.0°C − 0.89°C
= −0.89°C

PRACTICE Problems

Have students refer to the Selected Solutions appendix for complete solutions to odd-numbered problems. The complete solutions for all problems can be found in the Solutions Manual.

45. 100.320°C; −1.16°C
46. 79.0°C; −114.9°C
47. 1.8°C/*m*; water

DIFFERENTIATED INSTRUCTION

Advanced Learners Have students research the role of the kidneys in the body as well as the medical process known as dialysis. Have them write a paragraph explaining how osmosis, a colligative property, is used in this medical procedure. **AL**

3 Assess

Check for Understanding

Have students list the four colligative properties and explain them in terms of vapor pressure. OL

Reteach

Have students work in groups of two to quiz each other on the colligative properties, their definitions, and practical applications. OL

Extension

Have students write a letter to their local, county, or state highway department asking what substances are used on roadways during the winter. Find out if the substances and methods used have changed over time, and if so, for what reasons. OL

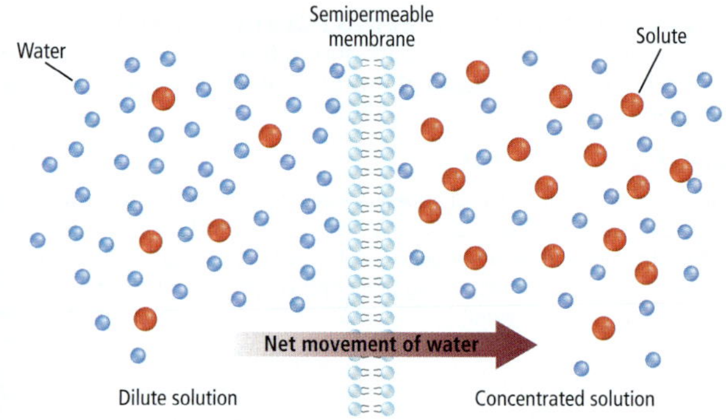

Figure 23 Due to osmosis, solvents diffuse from a lower solute concentration to a higher solute concentration through semipermeable membranes.

View an **animation about osmosis.**

Concepts In Motion

Osmotic Pressure

Connection to Biology Recall that diffusion is the mixing of gases or liquids resulting from their random motions. **Osmosis** is the diffusion of a solvent through a semipermeable membrane. Semipermeable membranes are barriers that allow some, but not all, particles to cross. The membranes surrounding all living cells are semipermeable membranes. Osmosis plays an important role in many biological systems, such as the uptake of nutrients by plants.

Examine a system in which a dilute solution is separated from a concentrated solution by a semipermeable membrane, illustrated in **Figure 23.** During osmosis, water molecules move in both directions across the membrane, but the solute molecules cannot cross it. Water molecules diffuse across the membrane from the dilute solution to the concentrated solution. The amount of additional pressure caused by the water molecules that moved into the concentrated solution is called the **osmotic pressure.** Osmotic pressure depends on the number of solute particles in a given volume of solution and is a colligative property of solutions.

SECTION 4 REVIEW

Section Self-Check

Section Summary

- Nonvolatile solutes lower the vapor pressure of a solution.
- Boiling point elevation is directly related to the solution's molality.
- A solution's freezing point depression is always lower than that of the pure solvent.
- Osmotic pressure depends on the number of solute particles in a given volume.

48. MAINIDEA **Explain** the nature of colligative properties.
49. **Describe** four colligative properties of solutions.
50. **Explain** why a solution has a higher boiling point than that of the pure solvent.
51. **Solve** An aqueous solution of calcium chloride ($CaCl_2$) boils at 101.3°C. How many kilograms of calcium chloride were dissolved in 1000.0 g of the solvent?
52. **Calculate** the boiling point elevation of a solution containing 50.0 g of glucose ($C_6H_{12}O_6$) dissolved in 500.0 g of water. Calculate the freezing point depression for the same solution.
53. **Investigate** A lab technician determines the boiling point elevation of an aqueous solution of a nonvolatile, nonelectrolyte to be 1.12°C. What is the solution's molality?

SECTION 4 REVIEW

48. Colligative properties depend on the number of solute particles in a solution.
49. Vapor pressure lowering: the decrease in vapor pressure with increasing solute particles in solution; boiling point elevation: the increase in boiling point with increasing solute particles in solution; freezing point depression: the decrease in freezing point with increasing solute particles in solution; osmotic pressure: the change in osmotic pressure with increasing solute particles in solution.
50. Solute particles in solution decrease the vapor pressure above the solution. Because a solution boils when its vapor pressure equals the external pressure, this decrease in vapor results in the need for a higher temperature in order for the solution to boil.
51. 0.0936 kg
52. $\Delta T_b = 0.285°C$; $\Delta T_f = 1.03°C$
53. 2.19*m*

CHEMISTRY & careers

Career: Environmental Chemist

A CO_2 Solution

Geologic records indicate that the levels of atmospheric carbon dioxide (CO_2) are likely higher today than in the past 20 million years. Anthropogenic (an thruh pah JEN ihk) CO_2, which means CO_2 from human-made sources, has contributed to this high level. CO_2 does not remain in the atmosphere indefinitely. Oceans naturally contain CO_2 that comes from the atmosphere and from living organisms. Oceans have absorbed nearly 50% of anthropogenic CO_2. Some scientists think that over the next thousand years, as much as 90% of anthropogenic CO_2 will dissolve in the oceans.

Collecting CO_2 data The rate at which CO_2 dissolves into the oceans is influenced by many factors including temperature, concentration of CO_2 in the air and in the water, and the mixing of air and water due to wind and waves. A team of researchers spent years collecting and analyzing CO_2 data from thousands of collection points throughout the world's oceans. The data, shown in **Figure 1,** indicate that the North Atlantic has the most anthropogenic CO_2 per square meter of ocean surface. The combination of temperature, depth, and current make the North Atlantic an efficient absorber of anthropogenic CO_2.

CO_2 capture and storage One way to reduce the amount of CO_2 released into the atmosphere would be to capture and store the CO_2 produced when fossil fuels are burned. Researchers are investigating the possibility of directly injecting captured CO_2 into the ocean to speed up the dissolution process. This could reduce the greenhouse effect of CO_2 gas. However, upsetting the natural balance of dissolved CO_2 can have profound effects on water chemistry, which can harm or even kill marine life. For example, coral reefs throughout the world already show signs of stress due to increasing levels of dissolved CO_2.

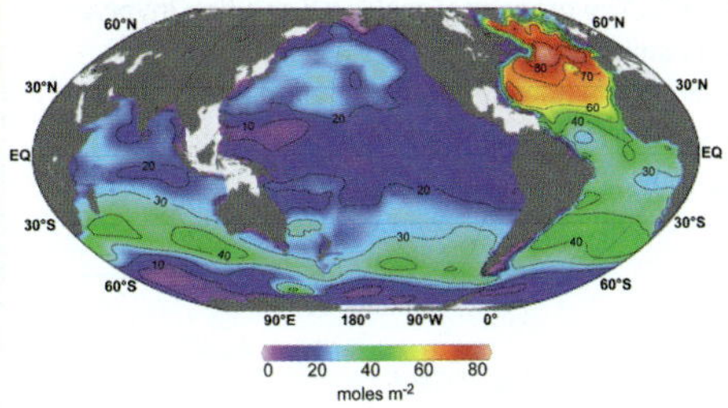

Figure 1 The red, yellow, and green regions represent areas where high levels of anthropogenic CO_2 are dissolved in the water.

Data obtained from: Sabine et al. 2004. The oceanic sink for anthropogenic CO_2. *Science* 305: 367–371.

Deep ocean sequestration A proposal that might reduce atmospheric CO_2 and protect life in the upper ocean is to liquefy the CO_2 and pump it deep under water, a process known as deep ocean sequestration. It is thought that the extreme pressure at depths greater than 3000 m will cause the CO_2 to form a hydrate. The hydrate will dissolve into the deep ocean water, but the CO_2 will remain trapped for hundreds of years far from the upper ocean and atmosphere.

Ongoing research Scientists are working on many of the unanswered questions about deep ocean sequestration, such as the effect of CO_2 on deep-sea organisms. There are still many technological problems involving capturing, storing, and transporting large quantities of liquid CO_2. If the technological problems can be solved, the public as well as government officials will have to consider the relative dangers of releasing CO_2 into the air and into the ocean.

WRITING IN ▶ Chemistry

Brainstorm a list of questions that must be addressed through research before deep ocean sequestration is attempted.

WebQuest

WRITING IN ▶ Chemistry

Rubric WebQuest

Brainstorm Answers will vary, but might include: What are specific short- and long-term effects of elevated CO_2 exposure on deep sea organisms and the entire marine ecosystem? How long will liquefied CO_2 remain trapped, and what are potential consequences if large levels of CO_2 do not remain at depths below 3000 m indefinitely? How will decreasing levels of atmospheric CO_2 affect world climate? How might geologic events (plate movement, earthquakes) affect areas of CO_2 sequestration?

CHEMISTRY & careers

Purpose

Students will identify world ocean locations where CO_2 is dissolved in the greatest quantities. Students will describe deep-ocean sequestration and evaluate questions related to the idea that require additional research.

Background

Ocean sequestration research includes two major approaches. Questions remain on both fronts. While technology exists for the first approach—directly injecting CO_2 into deep ocean locations—insufficient data exists to evaluate the effectiveness of the approach and potential environmental effects, including increased acidity in marine ecosystems. A second approach involves enhancing natural carbon sequestration by adding nutrients to ocean surface water, an action that would stimulate phytoplankton growth. Increased numbers of phytoplankton would consume greater amounts of CO_2, allowing more CO_2 to be drawn from the atmosphere. Potential environmental effects have not been fully determined.

Teaching Strategies

- Have students analyze Figure 1 to rank ocean locations based on CO_2 concentration. Ask students to look for patterns in the data. Students should consider how observed patterns might relate to specific factors affecting the rate of CO_2 dissolution in ocean water.
- After completing the feature activity, ask students to form an opinion about the merits of deep ocean sequestration of CO_2 on a large scale versus the potential environmental issues the process could raise.

ChemLAB

Rubric

Preparation

Time Allotment one class period

Process Skills observe and infer, compare and contrast, recognize cause and effect, think critically, collect and organize data, interpret data, measure, experiment, analyze and conclude

Safety Precautions Review the MSDS for copper(II) sulfate pentahydrate with students prior to doing the lab.

Disposal Small amounts of copper(II) sulfate pentahydrate solution are normally considered safe to wash down the drain if it is connected to a sanitary water system. Check your local and state requirements.

Alternate Disposal Evaporate the water and dispose of the dry waste in a landfill.

Preparation of Materials
A copper(II) sulfate pentahydrate crystal size of ~ 0.2 g is best.

Procedure

- Another ionic solid (like rock salt) could be used.
- Be sure to provide approximately equal sized crystals.
- Have the students note changes in the solid and the solution.
- A white piece of paper held behind the test tubes will help in spotting small differences in the color.
- **Troubleshooting** It is important that students do not disturb the other test tubes while agitating the second and fourth tubes.

ChemLAB

Investigate Factors Affecting Solubility

Background: The process of making a solution involves the solvent coming in contact with the solute particles. When you add a soluble compound to water, several factors affect the rate of solution formation.

Question *How do factors affect the rate of solution formation?*

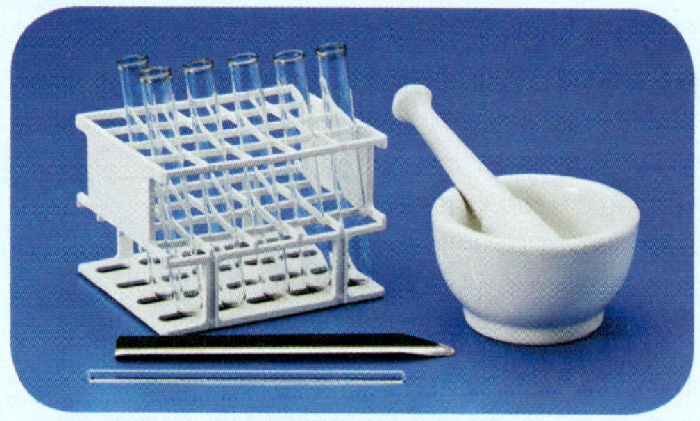

Materials

copper(II) sulfate pentahydrate
distilled water
test tubes (6)
25-mL graduated cylinder
glass stirring rod
tweezers
test tube rack
mortar and pestle
spatula
clock

Safety Precautions

Procedure

1. Read and complete the lab safety form.
2. Create a table to record your data.
3. Write a hypothesis that uses what you know about reaction rates to explain what you might observe during the procedure.
4. Place the 6 test tubes in the test tube rack.
5. Place one crystal of copper(II) sulfate pentahydrate in each of the first two test tubes.
6. For the remaining test tubes, use the mortar and pestle to crush the crystals. Use the spatula to scrape the crystals into the remaining test tubes.
7. Measure 15-mL of room-temperature distilled water. Pour the water into the first test tube and record the time.
8. Observe the solution in the test tube just after adding the water and after 15 min.
9. Leave the first test tube undisturbed in the rack.
10. Repeat Steps 7 and 8 for the third and fourth test tubes.
11. Use the glass stirring rod to agitate the second test tube for 1 to 2 min.
12. Leave the third test tube undisturbed.
13. Agitate the fourth test tube with the glass stirring rod for 1 to 2 min.
14. Repeat Steps 7 and 8 for the fifth test tube using cold water. Leave the fifth test tube undisturbed.
15. Repeat Steps 7 and 8 for the sixth test tube using hot water. Leave the sixth test tube undisturbed.
16. **Cleanup and Disposal** Dispose of the remaining solids and solutions as directed by your teacher. Wash and return all lab equipment to the designated locations.

Analyze and Conclude

1. **Compare and Contrast** What effect did you observe due to the agitation of the second and fourth test tubes versus the solutions in the first and third test tubes?
2. **Observe and Infer** What factor caused the more rapid solution formation in the fourth test tube in comparison to the second test tube?
3. **Recognize Cause and Effect** Why do you think the results for the third, fifth, and sixth test tubes were different?
4. **Discuss** whether or not your data supported your hypothesis.
5. **Error Analysis** Identify a major potential source of error for this lab, and suggest an easy method to correct it.

INQUIRY EXTENSION

Think Critically The observations in this lab were macroscopic in nature. Propose a submicroscopic explanation to account for these factors that affected the rate of solution formation. At the molecular level, what is occurring to speed solution formation in each case?

Analyze and Conclude

1. The test tubes that were agitated showed darker solutions than the first and third tubes that were left undisturbed. The fourth test tube contained the darkest solution.
2. The fourth test tube had a much faster rate of dissolving due to the crushing of the crystal. By crushing the solid, the surface area of the solid is increased and this causes the solid to dissolve more readily.
3. The solid in the sixth test tube dissolved the fastest due to the increased temperature. The cold water in the fifth test tube was very slow to dissolve the solid.
4. Answers will vary.
5. The major potential source of error is the size of the crystal. If the crystals are very different in size, the results can be altered.

INQUIRY EXTENSION

The process of solution formation is based on the interactions of the solvent and solute. Any process that increases these interactions will increase the rate of solution formation.

CHAPTER 14 STUDY GUIDE

Vocabulary Practice

BIGIDEA Nearly all of the gases, liquids, and solids that make up our world are mixtures.

SECTION 1 Types of Mixtures

MAINIDEA Mixtures can be either heterogeneous or homogeneous.

- The individual substances in a heterogeneous mixture remain distinct.
- Two types of heterogeneous mixtures are suspensions and colloids.
- Brownian motion is the erratic movement of colloid particles.
- Colloids exhibit the Tyndall effect.
- A solution can exist as a gas, a liquid, or a solid, depending on the solvent.
- Solutes in a solution can be gases, liquids, or solids.

VOCABULARY
- suspension
- colloid
- Brownian motion
- Tyndall effect
- soluble
- miscible
- insoluble
- immiscible

SECTION 2 Solution Concentration

MAINIDEA Concentration can be expressed in terms of percent or in terms of moles.

- Concentrations can be measured qualitatively and quantitatively.
- Molarity is the number of moles of solute dissolved per liter of solution.

$$\text{molarity } (M) = \frac{\text{moles of solute}}{\text{liters of solution}}$$

- Molality is the ratio of the number of moles of solute dissolved in 1 kg of solvent.

$$\text{molality } (m) = \frac{\text{moles of solute}}{\text{kilograms of solvent}}$$

- The number of moles of solute does not change during a dilution.

$$M_1V_1 = M_2V_2$$

VOCABULARY
- concentration
- molarity
- molality
- mole fraction

SECTION 3 Factors Affecting Solvation

MAINIDEA Factors such as temperature, pressure, and polarity affect the formation of solutions.

- The process of solvation involves solute particles surrounded by solvent particles.
- Solutions can be unsaturated, saturated, or supersaturated.
- Henry's law states that at a given temperature, the solubility (S) of a gas in a liquid is directly proportional to the pressure (P) of the gas above the liquid.

$$\frac{S_1}{P_1} = \frac{S_2}{P_2}$$

VOCABULARY
- solvation
- heat of solution
- unsaturated solution
- saturated solution
- supersaturated solution
- Henry's law

SECTION 4 Colligative Properties of Solutions

MAINIDEA Colligative properties depend on the number of solute particles in a solution.

- Nonvolatile solutes lower the vapor pressure of a solution.
- Boiling point elevation is directly related to the solution's molality.

$$\Delta T_b = K_b m$$

- A solution's freezing point depression is always lower than that of the pure solvent.

$$\Delta T_f = K_f m$$

- Osmotic pressure depends on the number of solute particles in a given volume.

VOCABULARY
- colligative property
- vapor pressure lowering
- boiling point elevation
- freezing point depression
- osmosis
- osmotic pressure

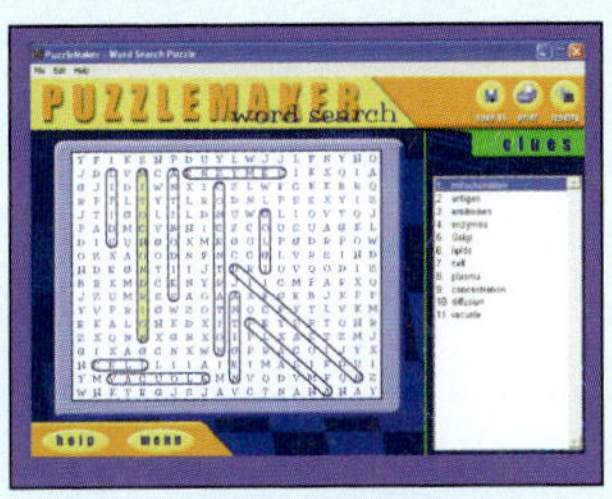

Vocabulary Practice

For additional practice with vocabulary, have students access the Vocabulary PuzzleMaker.

Use the Vocabulary

To reinforce chapter vocabulary, have students write a sentence using each term. OL EL

Review Strategies

- Have students organize the terms in this chapter in groups–*supersaturated, saturated,* and *unsaturated solutions; concentrated* and *dilute; molarity* and *molality;* and *percent by mass* and *percent by volume.* Students should understand the similarities and differences among the terms. OL
- Problems from the Supplemental Practice Problems appendix or the Supplemental Problems booklet can be used for review. OL

Students can review with these online resources.

Vocabulary Practice

Vocabulary eGames
Vocabulary eFlashcards
Vocabulary PuzzleMaker

Section Self-Check

Chapter Self-Check

Online Test Practice

Use *eAssessment* to:

- create multiple versions of tests
- edit existing questions and add your own questions
- build tests aligned with select state curriculum standards using built-in state curriculum tags
- track students' progress

CHAPTER 14 ASSESSMENT

SECTION 1

Mastering Concepts

54. Explain what is meant by the statement "not all mixtures are solutions."

55. What is the difference between a solute and a solvent?

56. What is a suspension, and how does it differ from a colloid?

57. How can the Tyndall effect be used to distinguish between a colloid and a solution? Why?

58. Name a colloid formed from a gas dispersed in a liquid.

■ **Figure 24**

59. Salad dressing What type of heterogenous mixture is shown in **Figure 24?** What characteristic is most useful in classifying the mixture?

60. What causes the Brownian motion observed in liquid colloids?

61. Aerosol sprays are categorized as colloids. Identify the phases of an aerosol spray.

SECTION 2

Mastering Concepts

62. What is the difference between percent by mass and percent by volume?

63. What is the difference between molarity and molality?

64. What factors must be considered when creating a dilute solution from a stock solution?

65. How do 0.5*M* and 2.0*M* aqueous solutions of NaCl differ?

66. Under what conditions might a chemist describe a solution in terms of molality? Why?

Mastering Problems

67. According to lab procedure, you stir 25.0 g of $MgCl_2$ into 550 mL of water. What is the percent by mass of $MgCl_2$ in the solution?

68. How many grams of LiCl are in 275 g of a 15% aqueous solution of LiCl?

69. You need to make a large quantity of a 5% solution of HCl but have only 25 mL HCl. What volume of 5% solution can be made from this volume of HCl?

70. Calculate the percent by volume of a solution created by adding 75 mL of acetic acid to 725 mL of water.

71. Calculate the molarity of a solution that contains 15.7 g of $CaCO_3$ dissolved in enough water to make 275 mL of solution.

72. What is the volume of a 3.00*M* solution made with 122 g of LiF?

73. How many moles of BaS would be used to make 1.5×10^3 mL of a 10.0*M* solution?

74. How much $CaCl_2$, in grams, is needed to make 2.0 L of a 3.5*M* solution?

75. Stock solutions of HCl with various molarities are frequently prepared. Complete **Table 7** by calculating the volume of concentrated, or 12*M*, hydrochloric acid that should be used to make 1.0 L of HCl solution with each molarity listed.

Table 7 HCl Solutions

Molarity of HCl Desired	Volume of 12*M* HCl Stock Solution Needed (mL)
0.50	
1.0	
1.5	
2.0	
5.0	

76. How much 5.0*M* nitric acid (HNO_3), in milliliters, is needed to make 225 mL of 1.0*M* HNO_3?

77. Experiment In the lab, you dilute 55 mL of a 4.0*M* solution to make 250 mL of solution. Calculate the molarity of the new solution.

78. How many milliliters of 3.0*M* phosphoric acid (H_3PO_4) can be made from 95 mL of a 5.0*M* H_3PO_4 solution?

79. If you dilute 20.0 mL of a 3.5*M* solution to make 100.0 mL of solution, what is the molarity of the dilute solution?

80. What is the molality of a solution that contains 75.3 g of KCl dissolved in 95.0 g of water?

81. How many grams of Na_2CO_3 must be dissolved into 155 g of water to create a solution with a molality of 8.20 mol/kg?

82. What is the molality of a solution containing 30.0 g of naphthalene ($C_{10}H_8$) dissolved in 500.0 g of toluene?

83. What are the molality and mole fraction of solute in a 35.5 percent by mass aqueous solution of formic acid (HCOOH)?

SECTION 1

Mastering Concepts

54. Solutions are homogeneous mixtures that are uniform in composition with a single phase. Mixtures can also be heterogeneous, where the substances that make them up remain distinct.

55. A solute is the substance being dissolved. The solvent is the substance in which the solute dissolves.

56. A suspension is a heterogeneous mixture that settles out if left undisturbed. The particles dispersed in a colloid are much smaller than those in a suspension and do not settle out.

57. A beam of light is visible in a colloid but not in a solution. Dispersed colloid particles are large enough to scatter light (Tyndall effect).

58. Students' answers might include whipped cream or beaten egg whites.

59. The mixture is a suspension. Left undisturbed, the mixture components settle out.

60. The random particle movements in liquid colloids result from collisions between particles in the mixture.

61. The most abundant mixture component is in the gas phase. The dispersed particles are in the liquid phase.

SECTION 2

Mastering Concepts

62. Percent by mass is a comparison between the mass of solute and the total mass of the solution. Percent by volume is a comparison between the volume of the solute and the total volume of the solution.

63. Molarity is solution concentration expressed as the moles of solute per volume of solution. Molality expresses concentration as moles of solute per kilogram of solvent. Molality does not depend upon the temperature of the solution.

64. The molarity and volume of both stock solution and dilute solution are required in the formula $M_1V_1 = M_2V_2$.

65. The 2*M* solution contains more moles of NaCl per volume of water than the 0.5*M* solution.

66. Under conditions of changing temperature. Because molality is based on mass, it does not change with temperature.

Mastering Problems

67. 4.3%
68. 41 g
69. 500 mL
70. 9.4%
71. 0.571*M*
72. 1.57 L
73. 15 mol
74. 770 g
75. 42 mL; 83 mL; 130 mL; 170 mL; 420 mL
76. 45 mL
77. 0.88*M*
78. 160 mL
79. 0.70*M*
80. 10.6 mol/kg
81. 135 g
82. 0.468*m*
83. 12.0*m*; 0.177

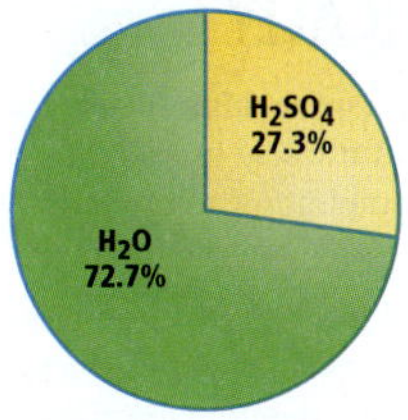

■ **Figure 25**

84. What is the mole fraction of H_2SO_4 in a solution containing the percentage of sulfuric acid and water shown in **Figure 25?**

85. Calculate the mole fraction of $MgCl_2$ in a solution created by dissolving 132.1 g of $MgCl_2$ in 175 mL of water.

SECTION 3

Mastering Concepts

86. Describe the process of solvation.

87. What are three ways to increase the rate of solvation?

88. Explain the difference between saturated and unsaturated solutions.

Mastering Problems

89. At a pressure of 1.5 atm, the solubility of a gas is 0.54 g/L. Calculate the solubility when the pressure is doubled.

90. At 4.5 atm of pressure, the solubility of a gas is 9.5 g/L. How much gas, in grams, will dissolve in 1 L if the pressure is reduced by 3.5 atm?

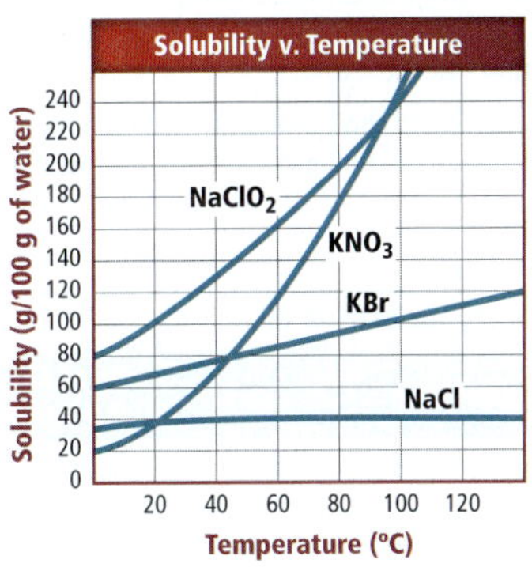

■ **Figure 26**

91. Using **Figure 26,** compare the solubility of potassium bromide (KBr) and potassium nitrate (KNO_3) at 80°C.

92. The solubility of a gas at 37.0 kPa is 1.80 g/L. At what pressure will the solubility reach 9.00 g/L?

93. Use Henry's law to complete **Table 8.**

Table 8 Solubility and Pressure

Solubility (g/L)	Pressure (kPa)
2.9	?
3.7	32
?	39

94. Soft Drinks The partial pressure of CO_2 inside a bottle of soft drink is 4.0 atm at 25°C. The solubility of CO_2 is 0.12 mol/L. When the bottle is opened, the partial pressure drops to 3.0×10^{-4} atm. What is the solubility of CO_2 in the open drink? Express your answer in grams per liter.

SECTION 4

Mastering Concepts

95. Define the term *colligative property.*

96. Use the terms *dilute* and *concentrated* to compare the solution on both sides of a semipermeable membrane.

97. Identify each variable in the following formula:

$$\Delta T_b = K_b m$$

98. Define the term *osmotic pressure,* and explain why it is considered a colligative property.

Mastering Problems

99. Calculate the freezing point of a solution of 12.1 g of naphthalene ($C_{10}H_8$) dissolved in 0.175 kg of benzene (C_6H_6). Refer to **Table 6** for needed data.

100. In the lab, you dissolve 179 g of $MgCl_2$ into 1.00 L of water. Use **Table 6** to find the freezing point of the solution.

101. Cooking A cook prepares a solution for boiling by adding 12.5 g of NaCl to a pot holding 0.750 L of water. At what temperature should the solution in the pot boil? Refer to **Table 5** for needed data.

102. The boiling point of ethanol (C_2H_5OH) changes from 78.5°C to 85.2°C when an amount of naphthalene ($C_{10}H_8$) is added to 1.00 kg of ethanol. How much naphthalene, in grams, is required to cause this change? Refer to **Table 5** for needed data.

103. Ice Cream A rock salt (NaCl), ice, and water mixture is used to cool milk and cream to make homemade ice cream. How many grams of rock salt must be added to the water to lower the freezing point by 10.0°C?

84. 0.0650

85. 0.125

SECTION 3

Mastering Concepts

86. A solute introduced into a solvent is surrounded by solvent particles. Due to the attraction between solute and solvent particles, solute particles are pulled apart and surrounded by solvent particles. Once separated, solute particles disperse into solution.

87. increase the temperature of the solvent, increase the surface area of the solute, agitation

88. A saturated solution contains the maximum amount of solute under a given set of conditions. An unsaturated solution contains less than the maximum amount.

Mastering Problems

89. 1.1 g/L

90. 2.1 g

91. The solubility of KBr is 95 g/100 g H_2O. The solubility of KNO_3 is nearly twice as high at the same temperature, at nearly 170 g/100 g H_2O.

92. 185 kPa

93. 25 kPa; 4.5 g/L

94. 4.0×10^{-4} g/L

SECTION 4

Mastering Concepts

95. A physical property of a solution that is affected by the number of solute particles but not their nature. Examples: vapor pressure lowering, boiling point elevation, freezing point depression, osmotic pressure.

96. If there is a concentration gradient, the solution is more dilute on one side of the membrane and more concentrated on the other side of the membrane.

97. ΔT_b represents the difference between the boiling points of a solution and the pure solvent; K_b is the molal boiling point elevation constant; *m* represents the solution molality.

98. Osmotic pressure is the pressure exerted by water molecules that move into solution through osmosis. Osmotic pressure is a colligative property because it depends on the number of solute particles dissolved in solution.

Mastering Problems

99. 2.7°C

100. −10.5°C

101. 100.29°C

102. 7.0×10^2 g

103. 157 g NaCl per 1 kg H_2O

MIXED REVIEW

104. $MgCl_2$(s) in H_2O(l): Yes. NH_3(l) in C_6H_6(l): No. H_2(g) in H_2O(l): No. I_2(l) in Br_2(l): Yes. Predictions are based on the general rule "like dissolves like." A polar solvent like water will dissolve a polar solute like magnesium chloride, and a nonpolar solvent like liquid bromine will dissolve a nonpolar solute like liquid iodine. Ammonia is a polar molecule, while benzene is nonpolar. Water is a polar molecule, while diatomic hydrogen is nonpolar.

105. When a colloid is exposed to heat, suspended particles can settle out. Paint should be stored in a cool location where it cannot freeze, and away from direct sunlight and objects like water heaters or furnaces that generate heat.

106. 50.0 grams $SrCl_2$ has the greatest effect; T_b $SrCl_2$ solution $= 100.000°C + 0.484°C = 100.484°C$; T_b $C_6H_{12}O_6$ solution $= 100.000°C + 0.426°C = 100.426°C$

107. For the gases, solubility decreases as temperature increases. For most solids, solubility increases as temperature increases. $Ca(OH)_2$ and Li_2SO_4 do not follow the general trend for solids.

108. $X_{N_2} = 0.804$; $X_{O_2} = 0.189$; $X_{Ar} = 0.00723$

109. Unsaturated; the solubility of KCl in water increases with temperature. A solution at 50°C holds more solute than one at 25°C.

110. 246 g

111. 0.501*m*

THINK CRITICALLY

112. 50 mL sulfuric acid needed. Subtract the volume of sulfuric acid from the total solution volume to determine a volume of 950 mL H_2O needed. Dissolve 50 mL sulfuric acid in somewhat less than 950 mL H_2O. Add water until the volume of the solution is 1000 mL.

ASSESSMENT

MIXED REVIEW

104. Apply your knowledge of polarity and solubility to predict whether solvation is possible in each situation shown in **Table 9.** Explain your answers.

Table 9 Is solvation possible?

Solute	Solvent
solid $MgCl_2$	liquid H_2O
liquid NH_3	liquid C_6H_6
gaseous H_2	liquid H_2O
liquid I_2	liquid Br_2

105. Household Paint Some types of paint are colloids composed of pigment particles dispersed in oil. Based on what you know about colloids, recommend an appropriate location for storing cans of leftover household paint. Justify your recommendation.

106. Which solute has the greatest effect on the boiling point of 1.00 kg of water: 50.0 g of strontium chloride ($SrCl_2$) or 150.0 g of glucose ($C_6H_{12}O_6$)? Justify your answer.

107. Analyze solubility and temperature data in **Table 4** to determine the general trend followed by the gases in the chart. Compare this trend to the trend followed by most of the solids in the chart. Identify the solids listed that do not follow the general trend.

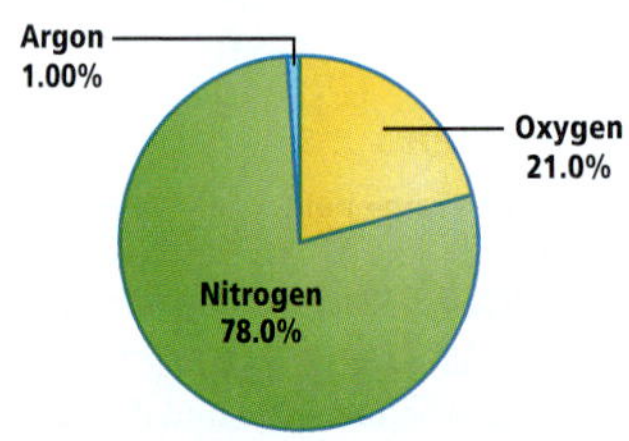

■ **Figure 27**

108. An air sample yields the mass percent composition shown in **Figure 27.** Calculate the mole fraction for each gas present in the sample.

109. If you prepared a saturated aqueous solution of potassium chloride at 25°C and then heated it to 50°C, would you describe the solution as unsaturated, saturated, or supersaturated? Explain.

110. How many grams of calcium nitrate ($Ca(NO_3)_2$) would you need to prepare 3.00 L of a 0.500*M* solution?

111. What would be the molality of the solution described in the previous problem? The density of the $Ca(NO_3)_2$ solution is 1.08 kg/L.

THINK CRITICALLY

112. Develop a plan for making 1000 mL of a 5% by volume solution of sulfuric acid in water. Your plan should describe the amounts of solute and solvent necessary, as well as the steps involved in making the solution.

113. Compare and Infer Study the phase diagram in **Figure 21.** Compare the dotted lines surrounding ΔT_f and ΔT_b, and describe the differences you observe. How might these lines be positioned differently for solutions of electrolytes and nonelectrolytes? Why?

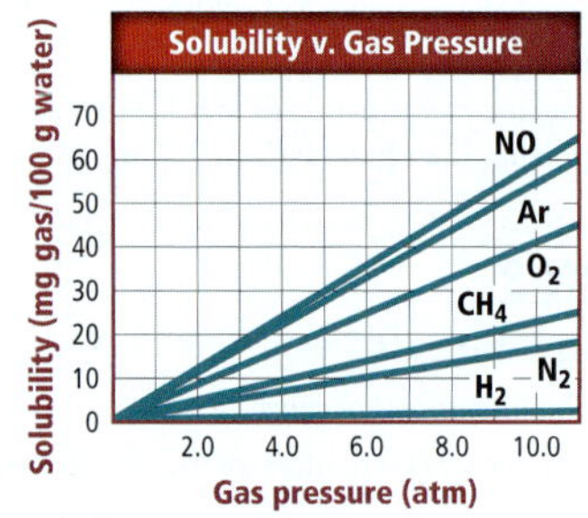

■ **Figure 28**

114. Extrapolate The solubility of argon in water at various pressures is shown in **Figure 28.** Extrapolate the data to 15 atm. Use Henry's law to verify the solubility determined by your extrapolation.

115. Infer Dehydration occurs when more fluid is lost from the body than is taken in. Scuba divers are advised to hydrate their bodies before diving. Use your knowledge of the relationship between pressure and gas solubility to explain the importance of hydration prior to a dive.

116. Graph Table 10 shows solubility data that was collected in an experiment. Plot a graph of the molarity of KI versus temperature. What is the solubility of KI at 55°C?

Table 10 Solubility of KI

Temperature (°C)	Grams of KI per 100.0 g Solution
20	144
40	162
60	176
80	192
100	206

113. The freezing point of the solution is below the normal freezing point of water, while the boiling point of the solution is above the normal boiling point of water. ΔT_f and ΔT_b would be larger for electrolytes than nonelectrolytes. Electrolytes dissociate in water, resulting in a larger number of particles in solution.

114. 82 mg/100 g H_2O

115. As pressure increases with water depth during a dive, gas concentration in the blood increases. If blood (solvent) volume is low, the gas (solute) concentration will be higher than normal levels at specific depths. A well-hydrated diver has a greater amount of solvent in which gases can be dissolved.

116. Molarity equals 8.67*M*, 9.76*M*, 10.6*M*, 11.6*M*, and 12.4*M* at 20°C, 40°C, 60°C, 80°C, and 100°C, respectively. The graph increases from left to right. The solubility of KI at 55°C is about 10.4*M*.

Chapter Self-Check

117. Design an Experiment You are given a sample of a solid solute and three aqueous solutions containing that solute. How would you determine which solution is saturated, unsaturated, and supersaturated?

118. Compare Which of the following solutions has the highest concentration? Rank the solutions from the greatest to the smallest boiling point elevation. Explain your answer.
a. 0.10 mol NaBr in 100.0 mL solution
b. 2.1 mol KOH in 1.00 L solution
c. 1.2 mol $KMnO_4$ in 3.00 L solution

CHALLENGE PROBLEMS

119. Interpret the solubility data in **Table 11** using the concept of Henry's law.

Table 11 Measurements of Solubility of a Gas

Measurement	Solubility
1	0.225
2	0.45
3	0.9
4	1.8
5	3.6

120. You have a solution containing 135.2 g of dissolved KBr in 2.3 L of water. What volume of this solution, in mL, would you use to make 1.5 L of a 0.10*M* KBr solution? What is the boiling point of this new solution?

CUMULATIVE REVIEW

121. The radius of an argon atom is 94 pm. Assuming the atom is spherical, what is the volume of an argon atom in cubic nanometers? $V = 4/3\pi r^3$

122. Identify which molecule is polar.
a. SiH_4
b. NO_2
c. SiO_2
d. HBr

123. Name the following compounds.
a. NaBr
b. $Pb(CH_3COO)_2$
c. $(NH_4)_2CO_3$

124. A 12.0-g sample of an element contains 5.94×10^{22} atoms. What is the unknown element?

125. Pure bismuth can be produced by the reaction of bismuth oxide with carbon at high temperatures.

$$2Bi_2O_3 + 3C \rightarrow 4Bi + 3CO_2$$

How many moles of Bi_2O_3 reacted to produce 12.6 mol of CO_2?

WRITING IN ▶ Chemistry

126. Homogenized Milk The first homogenized milk was sold in the United States around 1919. Today, almost all milk sold in this country is homogenized in the form of a colloidal emulsion. Research the homogenization process. Write a brief article describing the process. The article should include a flowchart or diagram of the process, as well as a discussion of the reputed benefits and drawbacks associated with drinking homogenized milk.

DBQ Document-Based Questions

Annual Mean Dissolved Oxygen *The data in* **Figure 29** *shows the average dissolved oxygen values, in milliliters per liter, in ocean-surface waters during a one-month period in 2001. Longitude is indicated horizontally, and latitude is indicated vertically.*

Data obtained from: National Oceanographic Data Center. 2002. *World Ocean Atlas 2001 Figures.*

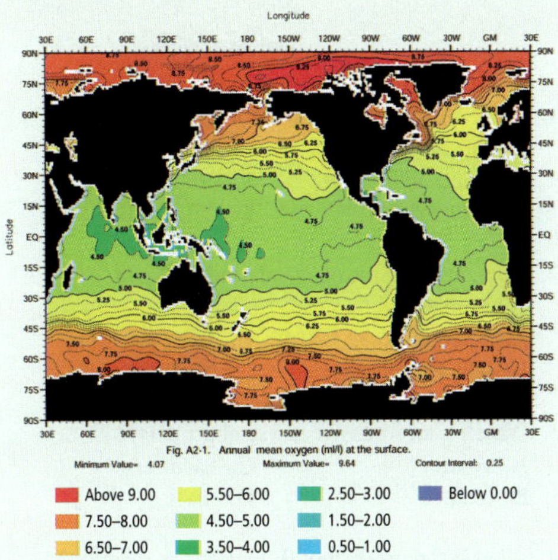

■ **Figure 29**

127. Are dissolved oxygen values most closely related to latitude or longitude? Why do you think this is true?

128. At what latitude are average dissolved oxygen values the lowest?

129. Describe the general trend defined by the data. Relate the trend to the relationship between gas solubility and temperature.

117. Add a pinch of solute to each container. If the solution is supersaturated, crystallization will occur; saturated, no solute will dissolve; unsaturated, solute will dissolve.

118. The molarities are 1.0*M* NaBr, 2.1*M* KOH, and 0.40*M* $KMnO_4$. Because the KOH solution contributes the greatest concentration of particles to solution, it has the greatest boiling point elevation; $KMnO_4$ has the lowest concentration of particles and the smallest boiling point elevation. Boiling point elevation depends only upon concentration.

CHALLENGE PROBLEM

119. In Henry's law, solubility is directly proportional to pressure. In this example, each measurement indicates a doubling of the solubility value. This indicates that the pressure is also doubling between measurements. An additional observation might include that from Measurement 1 to Measurement 5, the solubility has increased by a factor of 16. Therefore, the pressure would do the same.

120. 3.0×10^2 mL; 100.10°C

CUMULATIVE REVIEW

121. 3.5×10^{-3} nm^3

122. SiH_4 nonpolar; NO_2 polar; H_2S polar; NCl_3 polar

123. **a.** sodium bromide, **b.** lead (II) acetate, **c.** ammonium carbonate

124. The atomic mass is 122 amu. The element is antimony.

125. 8.40 mol Bi_2O_3

DBQ Document-Based Questions

Data obtained from: World Ocean Atlas 2001 Figures. June, 2002. *National Oceanographic Data Center.*

127. Dissolved oxygen values are most closely related to latitude. Surface land and water temperatures are more closely correlated to latitude than longitude.

128. Values are lowest near the equator.

129. In general, dissolved oxygen in surface ocean waters increases as latitude increases towards both north and south. Surface water temperatures are greatest near the equator. Surface water temperature decreases toward the poles. As temperature decreases, gas solubility generally increases.

WRITING IN ▶ Chemistry

Rubric

126. Students' answers will vary. Students should note that raw milk contains fat dispersed throughout. If left to stand, the fat separates out, leaving a cream layer and a skim milk layer. The process of homogenization breaks the fat globules into smaller sizes and reduces their tendency to form a cream layer.

MULTIPLE CHOICE

1. C
2. D
3. C
4. B
5. C
6. C
7. D
8. C
9. B
10. A

MULTIPLE CHOICE

Use the graph below to answer Questions 1 and 2.

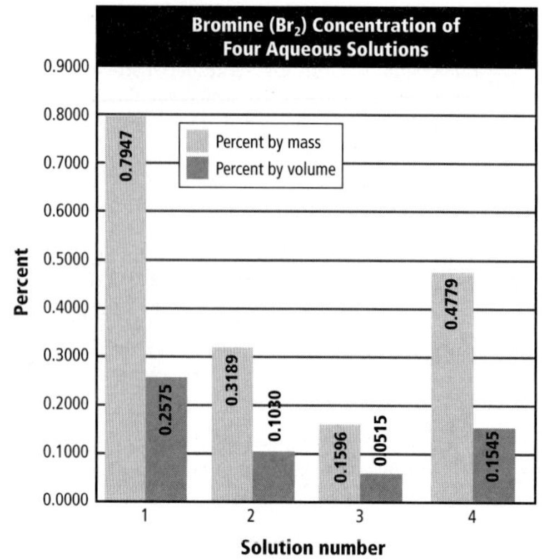

1. What is the volume of bromine (Br_2) in 7.000 L of Solution 1?
   A. 55.63 mL
   B. 8.808 mL
   C. 18.03 mL
   D. 27.18 mL

2. How many grams of Br_2 are in 55.00 g of Solution 4?
   A. 3.560 g
   B. 0.08498 g
   C. 1.151 g
   D. 0.2628 g

3. Which is an intensive physical property?
   A. volume
   B. length
   C. hardness
   D. mass

4. What is the product of this synthesis reaction?
   $Cl_2(g) + 2NO(g) \rightarrow$?
   A. NCl_2
   B. $2NOCl$
   C. N_2O_2
   D. $2ClO$

5. If 1 mol of each of the solutes listed below is dissolved in 1 L of water, which solute will have the greatest effect on the vapor pressure of its respective solution?
   A. KBr
   B. $C_6H_{12}O_6$
   C. $MgCl_2$
   D. $CaSO_4$

Use the diagram below to answer Question 6.

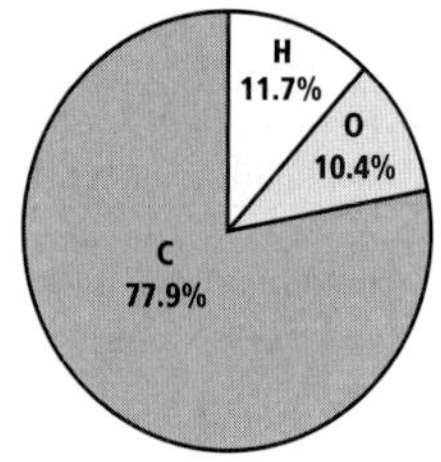

6. What is the empirical formula for this substance?
   A. CH_2O
   B. C_8HO
   C. $C_{10}H_{18}O$
   D. $C_7H_{12}O$

7. What is the correct chemical formula for the ionic compound formed by the calcium ion (Ca^{2+}) and the acetate ion ($C_2H_3O_2^-$)?
   A. $CaC_2H_3O_2$
   B. $CaC_4H_6O_3$
   C. $(Ca)_2C_2H_3O_2$
   D. $Ca(C_2H_3O_2)_2$

Use the reaction below to answer Questions 8 and 9.

$$Fe_3O_4(s) + 4H_2(g) \rightarrow 3\ Fe(s) + 4\ H_2O(l)$$

8. If 16 mol of H_2 are used, how many moles of Fe will be produced?
   A. 6
   B. 3
   C. 12
   D. 9

9. If 7 mol of Fe_3O_4 are mixed with 30 mol of H_2, what will be true?
   A. There will be no reactants left.
   B. 2 mol of hydrogen gas will be left over.
   C. 30 mol of water will be produced.
   D. 7 mol of Fe will be produced.

10. What is the molar mass of Fe_3O_4?
    A. 231.54 g/mol
    B. 71.85 g/mol
    C. 287.40 g/mol
    D. 215.56 g/mol

Online Test Practice

SHORT ANSWER

Use the graph below to answer Questions 11 to 13.

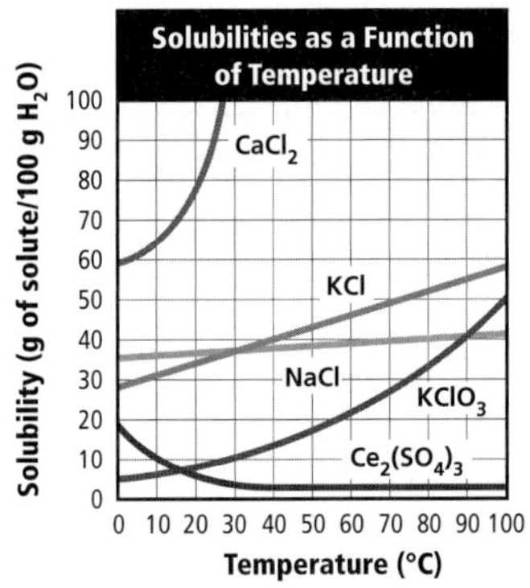

11. How many moles of $KClO_3$ can be dissolved in 100 g of water at 60°C?

12. Which can hold more solute at 20°C: a solution of NaCl or KCl? How does this compare to their solubilities at 80°C?

13. How many moles of $KClO_3$ would be required to make 1 L of a saturated solution of $KClO_3$ at 75°C?

EXTENDED RESPONSE

Use information below to answer Questions 14 and 15.

The electron configuration for silicon is $1s^22s^22p^63s^23p^2$.

14. Explain how this configuration demonstrates the Aufbau principle.

15. Draw the orbital diagram for silicon. Explain how Hund's rule and the Pauli exclusion principle are used in constructing the orbital diagram.

SAT SUBJECT TEST: CHEMISTRY

16. What volume of a 0.125*M* $NiCl_2$ solution contains 3.25 g of $NiCl_2$?
   - **A.** 406 mL
   - **B.** 32.5 mL
   - **C.** 38.5 mL
   - **D.** 26.0 mL
   - **E.** 201 mL

17. Which is NOT a colligative property?
   - **A.** boiling point elevation
   - **B.** freezing point depression
   - **C.** vapor pressure lowering
   - **D.** osmotic pressure
   - **E.** solubility

Use the data table below to answer Questions 18 and 19.

Electronegativities of Selected Elements						
H						
2.20						
Li	Be	B	C	N	O	F
0.98	1.57	2.04	2.55	3.04	3.44	3.98
Na	Mg	Al	Si	P	S	Cl
0.93	1.31	1.61	1.90	2.19	2.58	3.16

18. What is the electronegativity difference in Li_2O?
   - **A.** 1.48
   - **B.** 2.46
   - **C.** 3.40
   - **D.** 4.42
   - **E.** 5.19

19. Which bond has the greatest polarity?
   - **A.** C–H
   - **B.** Si–O
   - **C.** Mg–Cl
   - **D.** Al–N
   - **E.** H–Cl

NEED EXTRA HELP?																			
If You Missed Question . . .	1	2	3	4	5	6	7	8	9	10	11	12	13	14	15	16	17	18	19
Review Section . . .	14.2	14.2	3.1	9.2	14.4	10.4	7.3	11.2	11.3	10.3	14.3	14.3	14.3	5.3	5.3	14.2	14.4	8.5	8.5

SHORT ANSWER

11. 21 grams

12. At 20°C, the NaCl solution can hold more solute. At 80°C, the solubilities are reversed and KCl is more soluble than NaCl.

13. $(3.0 \times 10^1 \text{ g/L})(1 \text{ mol}/122.55 \text{ g } KClO_3) = 0.24 \text{ mol } KClO_3$ in 1 liter.

EXTENDED RESPONSE

14. The aufbau principle dictates that electrons must fill the lowest available energy levels before filling any higher energy levels.

15. Hund's rule mandates that the last two electrons will be placed in separate p-orbitals. The Pauli exclusion principal determines that shared electrons in any given orbital must have opposite spins, as shown by up and down arrows.

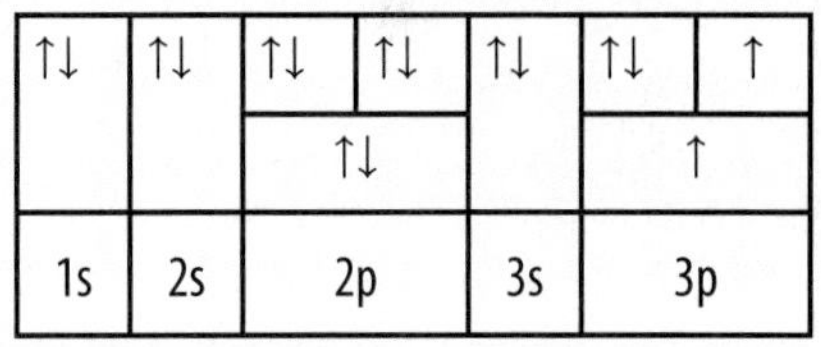

SAT SUBJECT TEST: CHEMISTRY

16. E
17. E
18. B
19. C

CHAPTER 15 Energy and Chemical Change

BIGIDEA **Chemical reactions usually absorb or release energy.**

ESSENTIAL QUESTIONS	RESOURCES TO ASSESS MASTERY
SECTION 1 Energy **1.** What is energy? **2.** How do potential and kinetic energy differ? **3.** How can chemical potential energy be related to the heat lost or gained in chemical reactions? **4.** How is the amount of heat absorbed or released by a substance calculated as its temperature changes? 1 session 0.5 block	**Progress Monitoring** Formative Assessment, pp. 517, 519 Reading Check, p. 517 Section Review, p. 522
SECTION 2 Heat **1.** How is a calorimeter used to measure energy that is absorbed or released? **2.** What do enthalpy and enthalpy change mean in terms of chemical reactions and processes? 2 sessions 1 block	**Progress Monitoring** Formative Assessment, pp. 524, 525, 527 Reading Check, p. 524 Section Review, p. 528
SECTION 3 Thermochemical Equations **1.** How are thermochemical equations for chemical reactions and other processes written? **2.** How is energy lost or gained during changes of state? **3.** How is the heat that is absorbed or released in a chemical reaction calculated? 1 session 0.5 block	**Progress Monitoring** Formative Assessment, p. 532 Reading Check, p. 531 Section Review, p. 533
SECTION 4 Calculating Enthalpy Change **1.** How is Hess's law applied to calculate the enthalpy change for a reaction? **2.** What is the basis for the table of standard enthalpies of formation? **3.** How is ΔH_{rxn} calculated using thermochemical equations? **4.** What is the enthalpy change for a reaction using standard enthalpies of formation data? 2 sessions 1 block	**Progress Monitoring** Formative Assessment, pp. 539, 541 Reading Check, p. 536 Section Review, p. 541
SECTION 5 Reaction Spontaneity **1.** What is the difference between spontaneous and nonspontaneous processes? **2.** How do changes in entropy and free energy determine the spontaneity of chemical reactions and other processes? 1 session 0.5 block	**Progress Monitoring** Formative Assessment, pp. 546, 547 Section Review, p. 548 **Summative Assessment** Chapter Assessment, p. 552 *eAssessment* Chapter Test (scaffolded)

LEVELED RESOURCES	LAB MATERIALS
Science Notebook 15.1 OL ***FAST FILE* Chapter Resources:** Study Guide, p. 74 BL	LaunchLAB p. 514 / **45 min** calcium chloride ($CaCl_2$), ammonium nitrate (NH_4NO_3), graduated cylinder, nonmercury thermometer, distilled water, balance, test tubes, potassium nitrate (KNO_3)
Science Notebook 15.2 OL ***FAST FILE* Chapter Resources:** MiniLab Worksheet, p. 58 OL ChemLab Worksheet, p. 59 OL Study Guide, p. 75 BL	MiniLAB p. 526 / **30 min** hot plate, metal cylinder, polystyrene coffee cup, nonmercury thermometer, distilled water, 250-mL beaker, crucible tongs, graduated cylinder, balance ChemLAB p. 550 / **45 min** large potato chip or other snack food, 250-mL beakers, 100-mL graduated cylinder, evaporating dish, nonmercury thermometer, ring stand with ring, wire gauze, matches, stirring rod, balance
Science Notebook 15.3 OL ***FAST FILE* Chapter Resources:** Study Guide, p. 76 BL	
Science Notebook 15.4 OL ***FAST FILE* Chapter Resources:** Study Guide, p. 77 BL	
Science Notebook 15.5 OL ***FAST FILE* Chapter Resources:** Study Guide, p. 79 BL	

ADDITIONAL RESOURCES

eAssessment
ConnectED Student Center
ConnectED Teacher Center
Lesson Presentations
eSolutions
Videos
Animations
Online Quizzes
Virtual Investigations
Multilingual eGlossary
Personal Tutors
WebQuests
Applying Practices Worksheets
PBLs

Practice:

Challenge Problems AL
Supplemental Problems BL OL
Chapter Test (Scaffolded)
Standardized Test Practice

Additional Resources:

Blueprints for Success
Solving Problems: A Chemistry Handbook

Lab Resources:

Laboratory Manual OL
CBL Laboratory Manual OL
Small-Scale Laboratory Manual OL
Forensics Laboratory Manual OL

BL Below Level   OL On Level   AL Advanced Learners   EL English Learners   COOP LEARN Cooperative Learning

CHAPTER 15

LaunchLAB

Rubric

How can you make a cold pack?

Purpose Students will observe exothermic and endothermic solutions and predict real-world uses for the chemicals.

Safety Precautions

Approve lab safety forms before work begins. Review the MSDS of hazardous chemicals with students. **WARNING:** ***Ammonium nitrate is very reactive and should be handled with caution. Use chemical splash goggles. Keep away from heat sources.***

Disposal Solutions and solids can be safely flushed down a drain with water.

Teaching Strategies

- Actual volumes and masses are not critical as long as they are similar. Quantities can be adjusted as needed.
- MSDS formats and information will vary. Student exposure to these different formats can lead to interesting discussion of the information required.
- Premeasure ammonium nitrate for students.

Expected Results

1. The potassium nitrate should produce the best endothermic solution. The calcium chloride makes an exothermic solution that is used for the ice-melt pellets in "driveway heat."
2. The ammonium nitrate would make the best chemical cold pack due to the largest temperature reduction.
3. The calcium chloride could be used in a chemical hot pack or to melt ice.

CHAPTER 15

Energy and Chemical Change

BIGIDEA **Chemical reactions usually absorb or release energy.**

SECTIONS

1. Energy
2. Heat
3. Thermochemical Equations
4. Calculating Enthalpy Change
5. Reaction Spontaneity

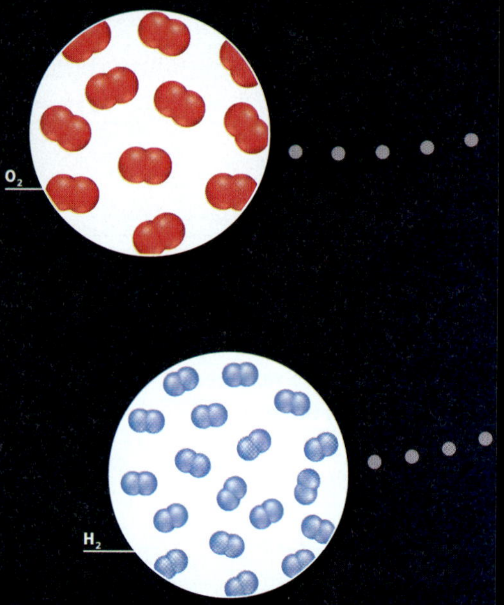

The three main engines of the space shuttle use more than 547,000 kg of liquid oxygen and approximately 92,000 kg of liquid hydrogen to lift a total mass of 2.04×10^6 kg.

LaunchLAB

How can you make a cold pack?

Chemical cold packs are used for fast relief of pain due to injury. Some chemical cold packs contain two separate compounds that are combined in a process that absorbs heat. In this lab, you will test three chemicals to determine which would make the best chemical cold pack.

FOLDABLES® Study Organizer

Gibbs Free Energy Equation

Make a concept-map book. Label it as shown. Use it to organize your study of the energy equation.

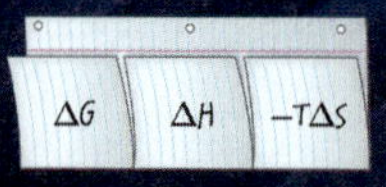

Procedure

1. Read and complete the lab safety form.
2. Use a **graduated cylinder** to place 15 mL of **distilled water** into each of **three test tubes.**
3. Use a **nonmercury thermometer** to find the temperature of the distilled water. Record the initial temperature of the water in a data table.
4. Use a **balance** to measure the mass of 1.0 g of **potassium nitrate (KNO_3).** Add the KNO_3 to Test Tube 1. **WARNING:** ***Keep all chemicals used in this lab away from heat sources.***
5. Mix, and record the maximum or minimum temperature reached by the solution.
6. Repeat Steps 4 and 5 with samples of **calcium chloride ($CaCl_2$)** and **ammonium nitrate (NH_4NO_3).**

Analysis

1. **Analyze and Conclude** Which is the best chemical for a chemical cold pack? potassium nitrate

Go online!

2. **Describe** an alternate use better suited for one of the other chemicals used in the lab. The calcium chloride could be used as a hot pack.

Inquiry Investigate a change that you could make in the procedure that would increase the temperature change. Add a larger mass of the compound to the same amount of water.

Introduce the BIGIDEA

Endothermic and Exothermic Changes To introduce this chapter's Big Idea, have students brainstorm to think of various chemical reactions and physical changes that absorb or release energy in the form of heat. List the reactions and changes under two categories on the board labeled *Endothermic and Exothermic*. Choose an endothermic physical change such as the vaporization of water and have students write the equation and include energy as a reactant. energy + $H_2O(l) \rightarrow H_2O(g)$ Ask students to write the equation in reverse. $H_2O(g) \rightarrow H_2O(l)$ + energy Point out that the second equation represents the exothermic condensation of water vapor to liquid water.

Tie to Previous Knowledge

Have students review the following concepts before studying this chapter.

- writing chemical equations
- mole concept
- kinetic theory

Use the Photo

Rocket Reaction Have students examine the chapter opener photo and describe the chemical reaction that powers the space shuttle's booster rockets. Liquid hydrogen reacts with liquid oxygen. Have students examine the blowouts showing the reactants and product and challenge them to write the balanced equation for the reaction, $2H_2 + O_2 \rightarrow 2H_2O$. Then, ask what the reaction produces, or liberates, in addition to water vapor. The reaction liberates a large amount of energy. Explain to students that they will learn in this chapter how to write a type of chemical equation that includes the change in energy. OL EL

1 Focus

MAINIDEA

Energy Forms and Conversions Hold a book at arm's length and let it drop to the desktop or floor. Ask students to describe the energy conversion that took place. When the book drops, potential energy of position is converted to energy of motion, called kinetic energy. When the book stops, kinetic energy is mainly converted to heat and sound. Light a wooden splint, and ask students to describe the energy conversion that resulted. Chemical potential energy in the wood splint is converted to heat and light. Point out to students that although energy is converted, the total amount of energy remains unchanged. **OL**

2 Teach

Concept Development

Energy Exchanges Industrial chemical reactions are very important to the production of many products. Determining which reactions will be optimal for production can include calculating amount of reactants needed, whether the reaction occurs without adding energy, production of sufficient amounts of product, and where energy transfers occur in the reactions. Thermodynamics can help follow the flow of energy in reactions, calculate the energy requirements of the reactions, production of excess heat that must be eliminated or conserved, and give manufacturers the information necessary to have adequate supplies of energy for the manufacturing process. The three laws of thermodynamics allow for the study of each phase of reactions and tracing energy transfers and changes in entropy.

SECTION 1

Energy

MAINIDEA Energy can change form and flow, but it is always conserved.

Essential Questions

- What is energy?
- How do potential and kinetic energy differ?
- How can chemical potential energy be related to the heat lost or gained in chemical reactions?
- How is the amount of heat absorbed or released by a substance calculated as its temperature changes?

Review Vocabulary

temperature: a measure of the average kinetic energy of the particles in a sample of matter

New Vocabulary

energy
law of conservation of energy
chemical potential energy
heat
calorie
joule
specific heat

CHEM 4 YOU **Have you ever watched a roller coaster zoom up and down a track or experienced the thrill of a coaster ride? Each time a coaster climbs a steep grade or plunges down the other side, its energy changes from one form to another.**

The Nature of Energy

You are probably familiar with the term *energy.* Perhaps you have heard someone say, "I just ran out of energy," after a strenuous game or a difficult day. Solar energy, nuclear energy, energy-efficient automobiles, and other energy-related topics are often discussed in the media.

Energy cooks the food you eat and propels the vehicles that transport you. If the day is especially hot or cold, energy from burning fuels helps maintain a comfortable temperature in your home and school. Electric energy provides light and powers devices from computers and TV sets to cellular phones, MP3 players, and calculators. Energy was involved in the manufacture and delivery of every material and device in your home. Your every movement and thought requires energy. In fact, you can think of each cell in your body as a miniature factory that runs on energy derived from the food you eat.

What is energy? **Energy** is the ability to do work or produce heat. It exists in two basic forms: potential energy and kinetic energy. Potential energy is energy due to the composition or position of an object. A macroscopic example of potential energy of position is a downhill skier poised at the starting gate for a race, as shown in **Figure 1a.** After the starting signal is given, the skier's potential energy changes to kinetic energy during the speedy trip to the finish line, as shown in **Figure 1b.** Kinetic energy is energy of motion. You can observe kinetic energy in the motion of objects and people all around you.

Figure 1 At the top of the course, the skier in **a** has high potential energy because of her position. In **b,** the skier's potential energy changes to kinetic energy.
Compare ***How is the potential energy of the skier different at the starting gate and at the finish line?***

Caption Question Fig. 1 The higher elevation at the starting gate compared to the finish line means that her potential energy was greater at the starting gate.

■ **Figure 2** Energy can change from one form to another but is always conserved.
In **a,** the potential energy of water is converted to kinetic energy of motion as it falls through the intake from its high position in the reservoir. The rushing water spins the turbine to generate electric energy. In **b,** the potential energy stored in the bonds of propane molecules is converted to heat.

Chemical systems contain both kinetic energy and potential energy. Recall that the kinetic energy of a substance is directly related to the constant random motion of its representative particles and is proportional to temperature. As temperature increases, the motion of submicroscopic particles increases. The potential energy of a substance depends on its composition: the type of atoms in the substance, the number and type of chemical bonds joining the atoms, and the particular way the atoms are arranged.

Law of conservation of energy When water rushes through turbines in the hydroelectric plant shown in **Figure 2a,** some of the water's kinetic energy is converted to electric energy. Propane (C_3H_8) is an important fuel for cooking and heating. In **Figure 2b,** propane gas combines with oxygen to form carbon dioxide and water. Potential energy stored in the propane bonds is given off as heat. In both of these examples, energy changes from one form to another, but energy is conserved–the total amount of energy remains constant. To better understand the conservation of energy, suppose you have money in two accounts at a bank and you transfer funds from one account to the other. Although the amount of money in each account has changed, the total amount of your money in the bank remains the same. When applied to energy, this analogy embodies the law of conservation of energy. The **law of conservation of energy** states that in any chemical reaction or physical process, energy can be converted from one form to another, but it is neither created nor destroyed. This is also known as the first law of thermodynamics.

Chemical potential energy The energy that is stored in a substance because of its composition is called **chemical potential energy.** Chemical potential energy plays an important role in chemical reactions. For example, the chemical potential energy of propane results from the arrangement of the carbon and hydrogen atoms and the strength of the bonds that join them.

☑ READING CHECK **State** the law of conservation of energy in your own words.

APPLYING PRACTICES

Develop and Use Models Go to the resources tab in ConnectED to find the Applying Practices worksheet *Modeling Energy in Chemical Reactions.*

Alan Sirulnikoff/Photo Researchers

ASSESSMENT

Performance Have students make a poster or chart that depicts United States energy use from approximately 1800 to the present in each of these categories: wood, coal, petroleum and natural gas, hydroelectric, and nuclear. Discuss the possible reasons for and likely consequences of trends shown by the chart. OL EL

☑ READING CHECK Energy may change forms but is never gained or lost.

CHEMISTRY PROJECT

Energy Value of Foods Have students look up the caloric content of at least ten food items including all the foods consumed in a single meal, such as a lunch or dinner. Then, have students make posters or construct a table of class data. OL EL

CHEMISTRY JOURNAL

Life Without Petroleum Ask students to write a short story in their journals describing what their lives might be like living without petroleum. OL

IN-CLASS Example

Question Glucose is a simple sugar found in fruit. Burning 1.00 g of glucose releases 15.6 kJ of energy. How many Calories are released?

Answer 3.73 Calories
Calories = 15.6 kJ × (1000 J/ kJ) × (1 cal/4.184 J) × (1 Calorie/1000 cal) = 3.73 Calories

Quick Demo

Endothermic and Exothermic Reactions

Measure 100 mL of water into each of two 250-mL beakers. Put a Celsius thermometer in each beaker and allow students to read the water's temperature. Record the temperatures and remove the thermometers. While you add approximately 9 g of anhydrous borax to one beaker, have your student add approximately 13 g of Epsom salts to the second beaker. When dissolved, read and record the temperatures of the solutions. Temperature increased with the dissolving of anhydrous borax and decreased with the dissolving of Epsom salts. Ask students which process is exothermic and which is endothermic. The dissolving of anhydrous borax is exothermic. The dissolving of Epsom salts is endothermic. Flush solutions down the drain with plenty of water. OL

Table 1 Relationships Among Energy Units

Relationship	Conversion Factors
1 J = 0.2390 cal	$\frac{1\ J}{0.2390\ cal}$ $\frac{0.2390\ cal}{1\ J}$
1 cal = 4.184 J	$\frac{1\ cal}{4.184\ J}$ $\frac{4.184\ J}{1\ cal}$
1 Calorie = 1 kcal	$\frac{1\ Calorie}{1000\ cal}$ $\frac{1000\ cal}{1\ Calorie}$

Heat The principal component of gasoline is octane (C_8H_{18}). When gasoline burns in an automobile's engine, some of octane's chemical potential energy is converted to the work of moving the pistons, which ultimately moves the wheels and propels the automobile. However, much of the chemical potential energy of octane is released as heat. The symbol q is used to represent **heat,** which is energy that is in the process of flowing from a warmer object to a cooler object. When the warmer object loses energy, its temperature decreases. When the cooler object absorbs energy, its temperature rises.

Measuring Heat

The flow of energy and the resulting change in temperature are clues to how heat is measured. In the metric system of units, the amount of energy required to raise the temperature of one gram of pure water by one degree Celsius (1°C) is defined as a **calorie** (cal). When your body breaks down sugars and fats to form carbon dioxide and water, these exothermic reactions generate heat that can be measured in Calories. Note that the nutritional Calorie is capitalized. That is because one nutritional Calorie equals 1000 calories, or one kilocalorie (kcal). Recall that the prefix *kilo-* means 1000. For example, one tablespoon of butter contains approximately 100 Calories. This means that if the butter was burned completely to produce carbon dioxide and water, 100 kcal (100,000 cal) of heat would be released.

The SI unit of of energy and of heat is the **joule** (J). One joule is the equivalent of 0.2390 calories, and one calorie equals 4.184 joules. **Table 1** summarizes the relationships between calories, nutritional Calories, joules, and kilojoules (kJ) and the conversion factors you can use to convert from one unit to another.

EXAMPLE Problem 1

Find help with **unit conversion.** Math Handbook

EXAMPLE PROBLEM

CONVERT ENERGY UNITS A breakfast of cereal, orange juice, and milk might contain 230 nutritional Calories. Express this energy in joules.

1 ANALYZE THE PROBLEM

You are given an amount of energy in nutritional Calories. You must convert nutritional Calories to calories and then convert calories to joules.

Known
amount of energy = 230 Calories

Unknown
amount of energy = ? J

2 SOLVE FOR THE UNKNOWN

Convert nutritional Calories to calories.

$230\ \cancel{Calories} \times \frac{1000\ cal}{1\ \cancel{Calorie}} = 2.3 \times 10^5\ cal$ Apply the relationship 1 Calorie = 1000 cal.

Convert calories to joules.

$2.3 \times 10^5\ \cancel{cal} \times \frac{4.184\ J}{1\ \cancel{cal}} = 9.6 \times 10^5\ J$ Apply the relationship 1 cal = 4.184 J.

3 EVALUATE THE ANSWER

The minimum number of significant figures used in the conversion is two, and the answer correctly has two digits. A value of the order of 10^5 or 10^6 is expected because the given number of kilocalories is of the order of 10^2 and it must be multiplied by 10^3 to convert it to calories. Then, the calories must be multiplied by a factor of approximately 4. Therefore, the answer is reasonable.

Demonstration

A Thermite Reaction

Purpose
to demonstrate an exothermic reaction

Materials
Thermite mixture; thermite starter; 11-cm round piece of filter paper; bucket of dry sand; crucible tongs; 8 cm length of Mg ribbon; burner; hammer; magnet

Safety Precautions

A safety shield must be used. Do not allow students to come closer than 24–30 ft. or to look directly at the burning Mg ribbon.

Disposal Reuse the sand. Discard the glass in a waste container.

Procedure
Fill a large bucket with dry sand and place it in a sink. Form a large, funnel-shaped depression in the sand. Use 11-cm filter paper to form a paper cone. Place the paper cone in the sand near the center of the bucket at the bottom of the depression. There must be at least 10 cm of sand around and under the cone. Place enough thermite mixture in the paper cone to almost fill it and mix in 5 mL of the starter. Make a slight depression in the center of the thermite mixture and fill with thermite starter. Insert 4 cm of an 8-cm length

PRACTICE Problems

Do additional problems. Online Practice

1. A fruit-and-oatmeal bar contains 142 nutritional Calories. Convert this energy to calories.
2. An exothermic reaction releases 86.5 kJ. How many kilocalories of energy are released?
3. **Challenge** Define a new energy unit, named after yourself, with a magnitude of one-tenth of a calorie. What conversion factors relate this new unit to joules? To Calories?

Specific Heat

You have read that one calorie, or 4.184 J, is required to raise the temperature of one gram of pure water by one degree Celsius (1°C). That quantity, 4.184 J/(g·°C), is defined as the specific heat (c) of water. The **specific heat** of any substance is the amount of heat required to raise the temperature of one gram of that substance by one degree Celsius. Because different substances have different compositions, each substance has its own specific heat.

To raise the temperature of water by one degree Celsius, 4.184 J must be absorbed by every gram of water. Much less energy is required to raise the temperature of an equal mass of concrete by one degree Celsius. You might have noticed that concrete sidewalks get hot during a sunny summer day. How hot depends on the specific heat of concrete, but other factors are also important. The specific heat of concrete is 0.84 J/(g·°C), which means that the temperature of concrete increases roughly five times more than water's temperature when equal masses of concrete and water absorb the same amount of energy. As shown in **Figure 3,** people who have been walking on hot concrete surfaces might want to cool their feet in the water of a fountain.

Figure 3 The cooler waters of the fountain are welcome after walking on the hot concrete sidewalk. Water must absorb five times the energy of an equal mass of concrete to reach an equivalent temperature.

Infer ***How would the temperature change of the concrete compare to that of the water over the course of a cool night.***

(l)Stephen Chernin/Getty Images; (r)©Bob Krist/Corbis

PRACTICE Problems

Have students refer to the Selected Solutions appendix for complete solutions to odd-numbered problems. The complete solutions for all problems can be found in the Solutions Manual.

1. 142 000 cal
2. 20.7 kcal
3. Unit X = 0.1 cal;
   1 cal = 4.184 J;
   X = 0.4184 J
   1 cal = 0.001 Calorie
   X = 0.0001 Calorie

Content Background

Another Unit of Heat Ask students how the heating and cooling capacities of most furnaces and air conditioners in the United States are rated. in BTU, British thermal units Explain that 1 BTU = 1055 J and is the energy required to raise the temperature of one pound of water one degree Fahrenheit, from 63°F to 64°F. OL

Enrichment

Energy to Heat Explain that even a well-tuned automobile is able to convert only about 25% of the chemical energy in gasoline to useful mechanical energy that moves the car and its occupants. Point out that the remaining 75% of energy is lost to the surroundings as heat.

Caption Question Fig. 3 The concrete cools more than the water during the night.

of Mg ribbon into the starter and thermite. Allow the remaining 4 cm of ribbon to stick out to be used as a fuse. Place a safety shield between the class and the sink. **WARNING:** ***Wear safety goggles.*** Move students to the back of the room. Darken the room. Bend the end of the Mg ribbon over to form a fuse and light it with a burner. Move away quickly. The heat generated produces white-hot molten iron and sparks. After the reaction stops, have students look at the molten sand and metal while the room is dark. After about 15 min., use crucible tongs to hold the metal in a stream of water. When completely cooled, use a hammer to break off the melted sand around the iron core. Test the pieces with a magnet.

Results

Al reacted with Fe_2O_3 to produce Fe.

Analysis

1. Write the balanced chemical equation. $Fe_2O_3 + 2Al \longrightarrow Al_2O_3 + 2Fe + \text{energy}$
2. What was used to supply the high activation energy required for the reaction? the heat of the burning Mg ribbon

ASSESSMENT

Knowledge Have students explain how the thermite reaction can be used for practical applications. underwater welding is one example OL

Identify Misconceptions

Students might think that heat and temperature are the same thing.

Uncover the Misconception Ask students which has a higher temperature, 100.0 g of ethanol at 25.0°C or 10.0 g of ethanol at 30.0°C. 10.0 g of ethanol at 30.0°C Then, ask which can transfer more heat as it cools by 10.0°C. 100.0 g of ethanol at 25.0°C

Demonstrate the Concept Demonstrate for students how to calculate the amounts of heat involved in the previous example.

$q_1 = 2.44 \text{ J/(g·°C)} \times 100.0 \text{ g} \times 10.0\text{°C} = 2440 \text{ J}$

$q_2 = 2.44 \text{ J/(g·°C)} \times 10.0 \text{ g} \times 10.0\text{°C} = 244 \text{ J}$

Assess New Knowledge Give students the following information: a 5.00-g piece of lead at 85.0°C and a 20.00-g piece of aluminum at 65.0°C are placed in a container of cold water. After the two metals have attained thermal equilibrium, a thermometer in the water reads 25.0°C. Ask students which metal had the higher initial temperature. lead Ask which had the greater ΔT as it cooled? lead Have students compare the heat lost by the two metals and explain the difference.

$q_{Pb} = 0.129 \text{ J/(g·°C)} \times 5.00 \text{ g} \times 60.0\text{°C} = 38.7 \text{ J}$

$q_{Al} = 0.897 \text{ J/(g·°C)} \times 5.00 \text{ g} \times 40.0\text{°C} = 179 \text{ J}$

Aluminum started at a lower temperature and had a smaller ΔT, but it lost more heat because it has a higher specific heat. **OL**

Table 2 Specific Heats at 298 K (25°C)

Substance	Specific heat J/(g·°C)
Water(l)	4.184
Ethanol(l)	2.44
Water(s)	2.03
Water(g)	2.01
Beryllium(s)	1.825
Magnesium(s)	1.023
Aluminum(s)	0.897
Concrete(s)	0.84
Granite(s)	0.803
Calcium(s)	0.647
Iron(s)	0.449
Strontium(s)	0.301
Silver(s)	0.235
Barium(s)	0.204
Lead(s)	0.129
Gold(s)	0.129

Calculating heat absorbed Suppose that the temperature of a 5.00×10^3-g block of concrete sidewalk increased by 6.0°C. Would it be possible to calculate the amount of heat it had absorbed? Recall that the specific heat of a substance tells you the amount of heat that must be absorbed by 1 g of a substance to raise its temperature 1°C. **Table 2** shows the specific heats for some common substances. The specific heat of concrete is 0.84 J/(g·°C), so 1 g of concrete absorbs 0.84 J when its temperature increases by 1°C. To determine the heat absorbed by 5.00×10^3 g of concrete you must multiply the 0.84 J by 5.00×10^3. Then, because the concrete's temperature changed by 6.0°C, you must multiply the product of the mass and the specific heat by 6.0°C.

Equation for Calculating Heat

$$q = c \times m \times \Delta T$$

q represents the heat absorbed or released. *c* represents the specific heat of the substance. *m* represents the mass of the sample in grams. ΔT is the change in temperature in °C, or $T_{final} - T_{initial}$.

The quantity of heat absorbed or released by a substance is equal to the product of its specific heat, the mass of the substance, and the change in its temperature.

You can use this equation to calculate the heat absorbed by the concrete block.

$$q = c \times m \times \Delta T$$

$$q_{concrete} = \frac{0.84 \text{ J}}{(\text{g·°C})} \times (5.00 \times 10^3 \text{ g}) \times 6.0\text{°C} = 25{,}000 \text{ J or } 25 \text{ kJ}$$

The total amount of heat absorbed by the concrete block is 25,000 J or 25 kJ.

For comparison, how much heat would be absorbed by 5.00×10^3 g of the water in the fountain when its temperature is increased by 6.0°C? The calculation for q_{water} is the same as it is for concrete except that you must use the specific heat of water, 4.184 J/(g·°C).

$$q_{water} = \frac{4.184 \text{ J}}{(\text{g·°C})} \times (5.00 \times 10^3 \text{ g}) \times 6.0\text{°C} = 1.3 \times 10^5 \text{ J or } 130 \text{ kJ}$$

If you divide the heat absorbed by the water (130 kJ) by the heat absorbed by the concrete (25 J), you will find that for the same change in temperature, the water absorbed more than five times the amount of heat absorbed by the concrete block.

Calculating heat released Substances can both absorb and release heat. The same equation for *q*, the quantity of heat, can be used to calculate the energy released by substances when they cool off. Suppose the 5.00×10^3-g piece of concrete reached a temperature of 74.0°C during a sunny day and cooled down to 40.0°C at night. How much heat was released? First calculate ΔT.

$$\Delta T = 74.0\text{°C} - 40.0\text{°C} = 34.0\text{°C}$$

Then, use the equation for quantity of heat.

$$q = c \times m \times \Delta T$$

$$q_{concrete} = \frac{0.84 \text{ J}}{(\text{g·°C})} \times (5.00 \times 10^3 \text{ g}) \times 34.0\text{°C} = 140{,}000 \text{ J or } 140 \text{ kJ}$$

EXAMPLE Problem 2

CALCULATE SPECIFIC HEAT In the construction of bridges and skyscrapers, gaps must be left between adjoining steel beams to allow for the expansion and contraction of the metal due to heating and cooling. The temperature of a sample of iron with a mass of 10.0 g changed from 50.4°C to 25.0°C with the release of 114 J. What is the specific heat of iron?

1 ANALYZE THE PROBLEM

You are given the mass of the sample, the initial and final temperatures, and the quantity of heat released. You can calculate the specific heat of iron by rearranging the equation that relates these variables to solve for *c*.

Known

energy released = 114 J — $T_i = 50.4°C$

mass of iron = 10.0 g Fe — $T_f = 25.0°C$

Unknown

specific heat of iron, *c* = ? J/(g·°C)

2 SOLVE FOR THE UNKNOWN

Calculate ΔT.

$$\Delta T = 50.4°C - 25.0°C = 25.4°C$$

Write the equation for calculating the quantity of heat.

$$q = c \times m \times \Delta T$$ State the equation for calculating heat.

$$\frac{c \times \cancel{m} \times \cancel{\Delta T}}{\cancel{m} \times \cancel{\Delta T}} = \frac{q}{m \times \Delta T}$$ Solve for *c*.

$$c = \frac{q}{m \times \Delta T}$$

$$c = \frac{114\ J}{(10.0\ g)(25.4°C)}$$ Substitute $q = 114$ J, $m = 10.0$ g, and $\Delta T = 25.4°C$.

$$c = 0.449\ J/(g\cdot°C)$$ Multiply and divide numbers and units.

3 EVALUATE THE ANSWER

The values used in the calculation have three significant figures, so the answer is correctly stated with three digits. The value of the denominator of the equation is approximately two times the value of the numerator, so the final result, which is approximately 0.5, is reasonable. The calculated value is the same as that recorded for iron in **Table 2.**

PRACTICE Problems

Do additional problems. Online Practice

4. If the temperature of 34.4 g of ethanol increases from 25.0°C to 78.8°C, how much heat has been absorbed by the ethanol? Refer to **Table 2.**
5. A 155-g sample of an unknown substance was heated from 25.0°C to 40.0°C. In the process, the substance absorbed 5696 J of energy. What is the specific heat of the substance? Identify the substance among those listed in **Table 2.**
6. **Challenge** A 4.50-g nugget of pure gold absorbed 276 J of heat. The initial temperature was 25.0°C. What was the final temperature?

Matt Meadows

RealWorld CHEMISTRY

Specific Heat

ABSORBING HEAT You might have wrapped your hands around a cup of hot chocolate to stay warm at a fall football game. In much the same way, long ago, children sometimes walked to school on wintry days carrying hot, baked potatoes in their pockets. The potatoes provided warmth for cold hands, but by the time the school bell rang, the potatoes had cooled off. At lunchtime, the cold potatoes might have been placed in or on the schoolhouse stove to warm them again for eating.

Explore **phase changes.**

Virtual Investigations

IN-CLASS Example

Question A 38.8-g piece of metal alloy absorbs 181 J as its temperature increases from 25.0°C to 36.0°C. What is the alloy's specific heat?

Answer 0.424 J/(g·°C)
$181\ J = c \times 38.8\ g \times 11.0°C$;
$c = 0.424\ J/(g\cdot°C)$

PRACTICE Problems

Have students refer to the Selected Solutions appendix for complete solutions to odd-numbered problems. The complete solutions for all problems can be found in the Solutions Manual.

4. 4.52×10^3 J
5. 2.45 J/(g·°C) The specific heat is very close to the value for ethanol.
6. 5.00×10^2 °C

Concept Development

Specific Heat Ask students which process loses more energy: the cooling of 1 mol of water from its boiling point to its freezing point or the condensation of 1 mol of steam at 100°C to 1 mol of water at the same temperature. Students will probably think more energy is lost in the cooling process because the temperature change is 100°C. Demonstrate the calculation for the cooling: 18.0 g × 4.18 J/g°C × 100°C = 7520 J. Explain that the value is much less than that for the condensation of steam at 100°C. 40,700 J OL

DIFFERENTIATED INSTRUCTION

English Learners Ask students to prepare flash cards for the vocabulary terms and phrases in this section. Each card should have a term or phrase on one side and the definition on the reverse side. Then, have students work in pairs using the cards until they have mastered the terms. EL OL

3 Assess

Check for Understanding

Ask students to decide if each of the following changes is exothermic or endothermic. 1. Tea is warmed in a microwave oven. endothemic 2. Natural gas is burned in the furnace of a home. exothermic 3. Ice on the pond melts in the sunlight. endothermic OL EL

Reteach

Ask students to explain the difference between temperature and heat. Temperature is the measure of the average kinetic energy of the particles that make up a substance. Heat is a form of energy that flows from a warmer substance to a cooler substance. OL

Extension

Explain to students that the study of energy in chemical reactions is known as thermodynamics and that thermodynamics has its own specialized vocabulary. In their chemistry journals, ask them to keep a list of thermodynamic terms and definitions. OL EL

Figure 4 Each photoelectric cell on this panel absorbs the Sun's radiation and converts it to electricity quietly and without causing pollution.

Using the Sun's energy Because of its high specific heat, water is sometimes used to harness the energy of the Sun. After water has been heated by solar radiation, the hot water can be circulated in homes and businesses to provide heat. Radiation from the Sun could supply all the energy needs of the world and reduce or eliminate the use of carbon dioxide-producing fuels, but several factors have delayed the development of solar technologies. For example, the Sun shines for only a part of each day. In some areas, clouds often reduce the amount of available radiation. Because of this variability, effective methods for storing energy are critical.

A more promising approach to the use of solar energy is the development of photovoltaic cells, such as those shown in **Figure 4.** These devices convert solar radiation directly to electricity. Photovoltaic cells supply power for astronauts in space, but they are not used extensively for ordinary energy needs. That is because the cost of supplying electricity by means of photovoltaic cells is high compared to the cost of burning coal or oil.

SECTION 1 REVIEW

Section Self-Check

Section Summary

- Energy is the capacity to do work or produce heat.
- Chemical potential energy is energy stored in the chemical bonds of a substance by virtue of the arrangement of the atoms and molecules.
- Chemical potential energy is released or absorbed as heat during chemical processes or reactions.

7. MAINIDEA **Explain** how energy changes from one form to another in an exothermic reaction. In an endothermic reaction.
8. **Distinguish** between kinetic and potential energy in the following examples: two separated magnets; an avalanche of snow; books on library shelves; a mountain stream; a stock-car race; separation of charge in a battery.
9. **Explain** how the light and heat of a burning candle are related to chemical potential energy.
10. **Calculate** the amount of heat absorbed when 5.50 g of aluminum is heated from 25.0°C to 95.0°C. The specific heat of aluminum is 0.897 J/(g·°C).
11. **Interpret Data** Equal masses of aluminum, gold, iron, and silver were left to sit in the Sun at the same time and for the same length of time. Use **Table 2** to arrange the four metals according to the increase in their temperatures from largest increase to smallest.

SECTION 1 REVIEW

7. Chemical potential energy changes to heat in exothermic reactions, and the heat is released. In endothermic reactions, heat is absorbed and changed to chemical potential energy.
8. Two separated magnets illustrate potential energy. In a snow avalanche, positional potential energy changes to kinetic energy. Books on a shelf illustrate positional potential energy. As water races down a mountain stream, positional potential energy changes to kinetic energy. In a stock car race, chemical potential energy changes to kinetic energy. The separation of charge in a battery illustrates electrical potential energy.
9. Chemical potential energy, contained in the candle, is changed to energy in the form of light and heat and released as the chemical combustion reaction takes place.
10. 345 J
11. The temperature change is inversely proportional to the specific heat: aluminum, iron, silver, gold.

SECTION 2

Heat

MAINIDEA The enthalpy change for a reaction is the enthalpy of the products minus the enthalpy of the reactants.

CHEM 4 YOU Think about standing under a hot shower, relaxing as your body absorbs heat from the water. When you jump into a cold pool, you might shiver as your body loses heat. In a similar way, some chemical reactions absorb heat whereas others release heat.

Essential Questions

- How is a calorimeter used to measure energy that is absorbed or released?
- What do enthalpy and enthalpy change mean in terms of chemical reactions and processes?

Review Vocabulary

pressure: force applied per unit area

New Vocabulary

calorimeter
thermochemistry
system
surroundings
universe
enthalpy
enthalpy (heat) of reaction

Calorimetry

Have you ever wondered how food chemists obtain the Calorie information that appears on packaged food? The packages record the results of combustion reactions carried out in calorimeters. A **calorimeter** is an insulated device used for measuring the amount of heat absorbed or released during a chemical or physical process. A known mass of water is placed in an insulated chamber to absorb the energy released from the reacting system or to provide the energy absorbed by the system. The data to be collected is the change in temperature of this mass of water. **Figure 5** shows the kind of calorimeter, called a bomb calorimeter, that is used by food chemists.

Determining specific heat Satisfactory results can be obtained in your calorimetry experiments using the much simpler foam-cup calorimeter. These calorimeters are open to the atmosphere, so reactions carried out in them occur at constant pressure. You can use them to determine the specific heat of an unknown metal.

Suppose you put 125 g of water into a foam-cup calorimeter and find that its initial temperature is 25.60°C. Then you heat a 50.0-g sample of the unknown metal to 115.0°C and put the metal sample into the water. Heat flows from the hot metal to the cooler water, and the temperature of the water rises. The flow of heat stops only when the temperature of the metal and the water are equal.

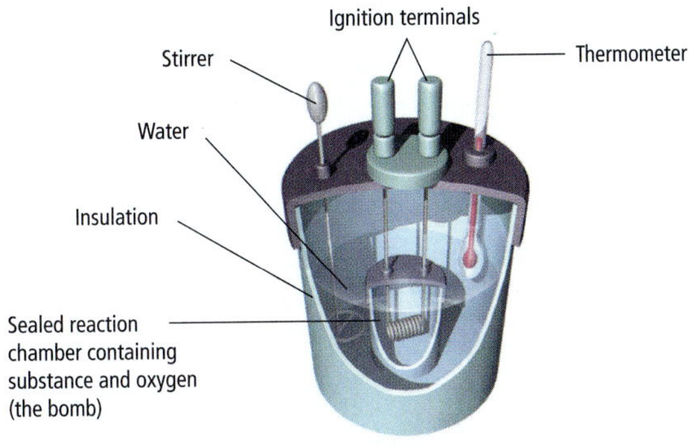

Figure 5 A sample is positioned in a steel inner chamber called the bomb, which is filled with oxygen at high pressure. Surrounding the bomb is a measured mass of water stirred by a low-friction stirrer to ensure uniform temperature. The reaction is initiated by a spark, and the temperature is recorded until it reaches its maximum.

Infer *Why is it important that the stirrer does not create friction?*

View an **animation about calorimetry.**

Concepts In Motion

CHEMISTRY JOURNAL

The Bomb Calorimeter Have students research how a bomb calorimeter is used to determine the caloric content of foods. Have them draw a diagram of the calorimeter and explain how it works. Students should include their diagrams and written explanations in their chemistry journals. OL EL

SECTION 2

1 Focus

MAINIDEA

An Analogy for Enthalpy Change Use the following analogy to help students understand how enthalpy change is calculated: Suppose that your bank account, for a given month, has a beginning balance of $1000 and an ending balance of $850. If you calculate the change, ΔB, as $B_{ending} - B_{beginning}$, the calculation is $\Delta B = \$850 - \$1000 = -\$150$. Explain that enthalpy change, ΔH, is calculated as follows: $\Delta H = H_{products} - H_{reactants}$. Point out that for a reaction that loses enthalpy (exothermic), like a bank account that loses money, $H_{products}$ is less than $H_{reactants}$ and ΔH has a negative sign. Then, explain the similarity between a bank account that gains money and a reaction that gains enthalpy.

2 Teach

Concept Development

Heat of Combustion Explain that when a substance is burned in a calorimeter, the energy released per gram of each substance is called the heat of combustion. If 1 mol of a substance is completely burned, the energy evolved is known as the molar heat of combustion.

Caption Question Fig. 5 Friction creates heat, which would add heat to the water and introduce error into the measured change in temperature.

ASSESSMENT

Skill Have students devise a procedure to determine the specific heat of a metal object of unknown composition using a thermometer, beaker, hot plate, balance, water, graduated cylinder, and foam cup calorimeter. If time permits, give interested students a metal object and have them perform the experiment. AL

Reinforcement

Energy Changes Have students compare energy changes in chemical reactions to profits and losses in a business. Each month the business has receipts (positive dollar amounts) and expenses (negative dollar amounts). If receipts exceed expenditures, a positive dollar amount, or profit, occurs. If expenditures are greater than receipts, money is lost and a negative dollar amount results. OL

ChemLAB

The ChemLab located at the end of the chapter can be used at this point in the lesson.

☑ **READING CHECK** q is quantity of heat, m is mass in grams, ΔT is the change in temperature and c is the specific heat.

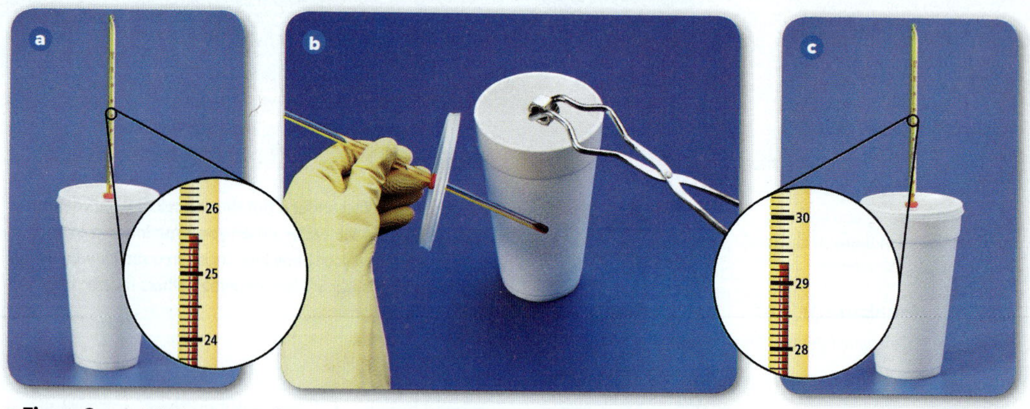

Figure 6 **a.** An initial temperature of 25.60°C is recorded for the 125 g of water in the calorimeter. **b.** A 50.0-g sample of an unknown metal is heated to 115.0°C and placed in the calorimeter. **c.** The metal transfers heat to the water until metal and water are at the same temperature. The final temperature is 29.30°C.

Explore **calorimetry.**

Virtual Investigations

APPLYING PRACTICES

Plan and Conduct an Investigation Go to the resources tab in ConnectED to find the Applying Practices worksheet *Coffee Cup Calorimetry.*

Figure 6 shows an experimental procedure. Note that the temperature in the calorimeter becomes constant at 29.30°C, which is the final temperature attained by both the water and the metal. Assuming no heat is lost to the surroundings, the heat gained by the water is equal to the heat lost by the metal. This quantity of heat can be calculated using the equation you learned in Section 1.

$$q = c \times m \times \Delta T$$

☑ **READING CHECK** **Define** the four variables in the equation above.

First, calculate the heat gained by the water. To do this, you need the specific heat of water, 4.184 J/(g•°C).

$$q_{water} = 4.184\ \text{J/(g•°C)} \times 125\ \text{g} \times (29.30°\text{C} - 25.60°\text{C})$$

$$q_{water} = 4.184\ \text{J/}(\cancel{\text{g}}\cdot\cancel{°\text{C}}) \times 125\ \cancel{\text{g}} \times 3.70\cancel{°\text{C}}$$

$$q_{water} = 1940\ \text{J}$$

The heat gained by the water, 1940 J, equals the heat lost by the metal, q_{metal}, so you can write this equation.

$$q_{metal} = q_{water}$$

$$q_{metal} = -1940\ \text{J}$$

$$c_{metal} \times m \times \Delta T = -1940\ \text{J}$$

Now, solve the equation for the specific heat of the metal, c_{metal}, by dividing both sides of the equation by $m \times \Delta T$.

$$c_{metal} = \frac{-1940\ \text{J}}{m \times \Delta T}$$

The change in temperature for the metal, ΔT, is the difference between the final temperature of the water and the initial temperature of the metal (29.30°C − 115.0°C = −85.7°C). Substitute the known values of m and ΔT (50.0 g and −85.7°C) into the equation and solve.

$$c_{metal} = \frac{-1940\ \text{J}}{(50.0\ \text{g})(-85.7°\text{C})} = 0.453\ \text{J/(g•°C)}$$

The unknown metal has a specific heat of 0.453 J/(g•°C). **Table 2** shows that the metal could be iron.

CHEMISTRY PROJECT

Building Materials Have students research how the specific heat of masonry building materials is important in the energy requirements of buildings. Students might present their findings to the class with posters or presentations. OL

EXAMPLE Problem 3

Find help with algebraic equations. Math Handbook

EXAMPLE PROBLEM

USING SPECIFIC HEAT A piece of metal with a mass of 4.68 g absorbs 256 J of heat when its temperature increases by 182°C. What is the specific heat of the metal? Could the metal be one of the alkaline earth metals listed in **Table 2?**

1 ANALYZE THE PROBLEM

You are given the mass of the metal, the amount of heat it absorbs, and the temperature change. You must calculate the specific heat. Use the equation for *q*, the quantity of heat, but solve for specific heat, *c*.

Known	Unknown
mass of metal, $m = 4.68$ g	**specific heat, c = ? J/(g·°C)**
quantity of heat absorbed, $q = 256$ J	
$\Delta T = 182°C$	

2 SOLVE FOR THE UNKNOWN

$q = c \times m \times \Delta T$ — State the equation for the quantity of heat, *q*.

$c = \frac{q}{m \times \Delta T}$ — Solve for *c*.

$c = \frac{256\ \text{J}}{(4.68\ \text{g})(182°\text{C})} = \mathbf{0.301\ J/(g{\cdot}°C)}$ — Substitute $q = 256$ J, $m = 4.68$ g, and $\Delta T = 182°C$.

Table 2 indicates that the metal could be strontium.

3 EVALUATE THE ANSWER

The three quantities used in the calculation have three significant figures, and the answer is correctly stated with three digits. The calculations are correct and yield the expected unit.

PRACTICE Problems

Do additional problems. Online Practice

PRACTICE PROBLEMS

12. A 90.0-g sample of an unknown metal absorbed 25.6 J of heat as its temperature increased 1.18°C. What is the specific heat of the metal?

13. The temperature of a sample of water increases from 20.0°C to 46.6°C as it absorbs 5650 J of heat. What is the mass of the sample?

14. How much heat is absorbed by a 2.00×10^3 g granite boulder ($c_{granite} = 0.803$ J/(g·°C)) as its temperature changes from 10.0°C to 29.0°C?

15. Challenge If 335 g of water at 65.5°C loses 9750 J of heat, what is the final temperature of the water?

Chemical Energy and the Universe

Virtually every chemical reaction and change of physical state either releases or absorbs heat. **Thermochemistry** is the study of heat changes that accompany chemical reactions and phase changes. The burning of fuels always produces heat. Some products have been engineered to produce heat on demand. For example, soldiers in the field use a highly exothermic reaction to heat their meals. You might have used a heat pack to warm your hands on a cold day. The energy released by a heat pack is produced by the following reaction and is shown in the equation as one of the products.

$$4Fe(s) + 3O_2(g) \rightarrow 2Fe_2O_3(s) + 1625\ \text{kJ}$$

IN-CLASS Example

Question A 124-g sample of ethanol at an initial temperature of 30.0°C absorbs 1560 J of heat. Given ethanol's specific heat, 2.44 J/(g·°C), what is the final temperature of the ethanol?

Answer $T_f = 35.2°C$
$1{,}560\ \text{J} = 2.44\ \text{J/(g·°C)} \times 124\ \text{g} \times \Delta T$;
$\Delta T = 5.16°C$;
$5.16°C = T_f - 30.0°C$;
$T_f = 35.2°C$

PRACTICE Problems

Have students refer to the Selected Solutions appendix for complete solutions to odd-numbered problems. The complete solutions for all problems can be found in the Solutions Manual.

12. 0.241 J/(g·°C)
13. 50.8 g
14. 30,500 J
15. 58.5°C

ASSESSMENT

Knowledge Ask students why more heat is required to raise the temperature of a given mass of water a given number of degrees than is needed to raise the same mass of ethanol by the same number of degrees. Water has a higher specific heat than ethanol. OL

DIFFERENTIATED INSTRUCTION

Hearing Impaired Use ball-and-stick molecular models to depict the exothermic reaction $2H_2(g) + O_2(g) \rightarrow 2H_2O(l)$. As you construct the reactant molecules, draw them on the board with the energy diagram at a high level (toward the top of the board). Then, construct the product molecules and draw them on the energy diagram at a lower energy level (toward the bottom of the board). Indicate how the energy changed by drawing an arrow pointing downward from reactants to products. OL EL

MiniLAB

Rubric

Purpose Students will experimentally determine the specific heat of a metal and compare it to the accepted value.

Process Skills describe, make and use tables, compare, calculate, use formulas

Safety Precautions

Approve lab safety forms before work begins. Warn students to be careful with the hot plate and boiling water. Use only GFI protected electrical receptacles.

Teaching Strategies

- Provide different metallic samples to the groups so that they can share the data and complete additional calculations.
- Some repetition of samples is useful so that students can combine and compare data.
- A burner, ring, and stand can be used instead of a hot plate.

Expected Results The values of the specific heat might be slightly inaccurate because some of the boiling water will stay on the metal during the transfer.

Sample Data

Cylinder height	2.5 cm
Cylinder radius	0.9 cm
Cylinder mass	65.637 g
Volume of water	90.0 mL
Cool water	28.0 °C
Hot water	98.5 °C
Max temp of mixture	32.0 °C

MiniLAB

Determine Specific Heat

How can you determine the specific heat of a metal? You can use a coffee-cup calorimeter to determine the specific heat of a metal.

Procedure

1. Read and complete the lab safety form.
2. Make a table to record your data.
3. Pour approximately 150 mL of **distilled water** into a **250-mL beaker.** Place the beaker on a **hot plate** set on high.
4. Use a **balance** to find the mass of a **metal cylinder.**
5. Using **crucible tongs,** carefully place the metal cylinder in the beaker on the hot plate.
6. Measure 90.0 mL of distilled water using a **graduated cylinder.**
7. Pour the water into a **polystyrene coffee cup** nested in a second **250-mL beaker.**
8. Measure and record the temperature of the water using a **nonmercury thermometer.**
9. When the water on the hot plate begins to boil, measure and record the temperature as the initial temperature of the metal.
10. Carefully add the hot metal to the cool water in the coffee cup with the crucible tongs. Do not touch the hot metal with your hands.
11. Stir, and measure the maximum temperature of the water after the metal was added.

Analysis

1. **Calculate** the heat gained by the water. The specific heat of H_2O is 4.184 J/g•°C. Because the density of water is 1.0 g/mL, use the volume of water as the mass.
2. **Calculate** the specific heat of your metal. Assume that the heat absorbed by the water equals the heat lost by the metal.
3. **Compare** this experimental value to the accepted value for your metal.
4. **Describe** major sources of error in this lab. What modifications could you make in this experiment to reduce the error?

Because you are interested in the heat given off by the chemical reaction going on inside the pack, it is convenient to think of the pack and its contents as the system. In thermochemistry, the **system** is the specific part of the universe that contains the reaction or process you wish to study. Everything in the universe other than the system is considered the **surroundings.** Therefore, the **universe** is defined as the system plus the surroundings.

$$\text{universe} = \text{system} + \text{surroundings}$$

What kind of energy transfer occurs during the exothermic heat-pack reaction? Heat produced by the reaction flows from the heat pack (the system) to your cold hands (part of the surroundings).

What happens in an endothermic reaction or process? The flow of heat is reversed. Heat flows from the surroundings to the system. When barium hydroxide and ammonium thiocyanate crystals, shown in **Figure 7,** are placed in a beaker and mixed, a highly endothermic reaction occurs. Placing the beaker on a wet board allows heat to flow from the water and board (the surroundings) into the beaker (the system). The temperature change is great enough that the beaker freezes to the board.

■ **Figure 7** In this endothermic reaction, the reacting mixture draws enough energy from the water and the wet board to cause the wet board to freeze to the beaker.

Matt Meadows

Enthalpy and enthalpy changes The total amount of energy a substance contains depends on many factors, some of which are still not completely understood. Therefore, it is impossible to know the total energy content of a substance. Fortunately, chemists are usually more interested in changes in energy during reactions than in the absolute amounts of energy contained in the reactants and products.

Analysis

1. –3. Student results will vary depending on the metal used.

4. Students might cite loss of heat to the surroundings, hot water clinging to the metal or loss of heat from the metal in the process of transfer, measurement error, and spillage. Students should suggest that the greatest source of error, the loss of heat from the calorimeter, can be reduced by improving the insulation.

CHEMISTRY PROJECT

Cataloging Everyday Reactions

Ask students to think about two exothermic and two endothermic reactions or processes they encounter in their daily lives. Ask them to record the following information: a description of each reaction or process, the direction of heat flow between the system and the surroundings, the change in enthalpy of the system, the change in enthalpy of the surroundings, and the sign (positive or negative) of ΔH_{rxn}. OL

For many reactions, the amount of energy lost or gained can be measured conveniently in a calorimeter at constant pressure, as shown in the experiment in **Figure 6.** The foam cup is not sealed, so the pressure is constant. Many reactions take place at constant atmospheric pressure; for example, those that occur in living organisms on Earth's surface, in lakes and oceans, and those that take place in open beakers and flasks in the laboratory. The energy released or evolved from reactions carried out at constant pressure is sometimes given the symbol q_p. To more easily measure and study the energy changes that accompany such reactions, chemists have defined a property called enthalpy. **Enthalpy** (H) is the heat content of a system at constant pressure.

Although you cannot measure the actual energy or enthalpy of a substance, you can measure the change in enthalpy, which is the heat absorbed or released in a chemical reaction. The change in enthalpy for a reaction is called the **enthalpy (heat) of reaction** (ΔH_{rxn}). You have already learned that a symbol preceded by the Greek letter delta (Δ) means a change in the property. Thus, ΔH_{rxn} is the difference between the enthalpy of the substances that exist at the end of the reaction and the enthalpy of the substances present at the start.

$$\Delta H_{rxn} = H_{final} - H_{initial}$$

Because the reactants are present at the beginning of the reaction and the products are present at the end, ΔH_{rxn} is defined by this equation.

$$\Delta H_{rxn} = H_{products} - H_{reactants}$$

The sign of the enthalpy of reaction Recall the heat-pack reaction.

$$4Fe(s) + 3O_2(g) \rightarrow 2Fe_2O_3(s) + 1625 \text{ kJ}$$

According to the equation, the reactants in this exothermic reaction lose heat. Therefore, $H_{products} < H_{reactants}$. When $H_{reactants}$ is subtracted from the smaller $H_{products}$, a negative value for ΔH_{rxn} results. Enthalpy changes for exothermic reactions are always negative. The equation for the heat-pack reaction and its enthalpy change are usually written as shown.

$$4Fe(s) + 3O_2(g) \rightarrow 2Fe_2O_3(s) \quad \Delta H_{rxn} = -1625 \text{ kJ}$$

A diagram of the enthalpy change is shown in **Figure 8.**

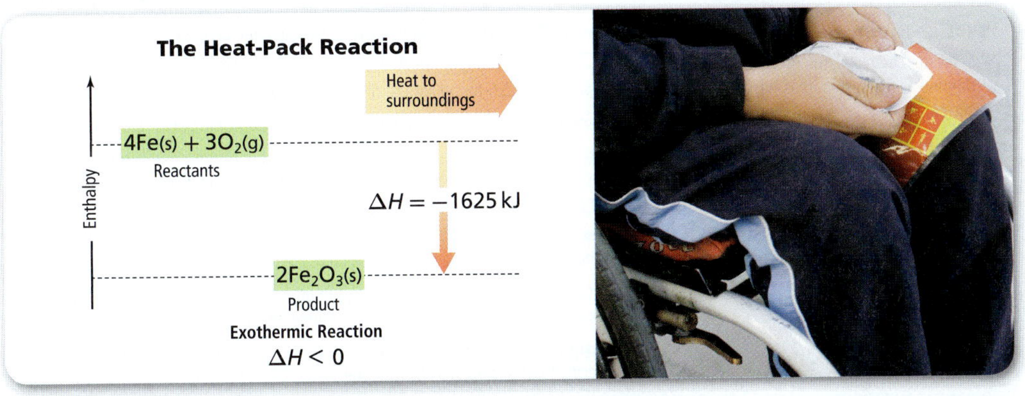

Tim Fuller

CAREERS IN CHEMISTRY

Heating and Cooling Specialist
Heating and cooling system mechanics install, maintain, and repair refrigeration and heating equipment in homes and in industry. They must understand how heat is exchanged by means of exothermic and endothermic processes. They must be able to read blueprints and use a wide range of tools, from pipe cutters to computerized diagnostic devices. Such mechanics might specialize in one aspect of this field, or become proficient in all areas.

WebQuest

■ **Figure 8** The downward arrow shows that 1625 kJ of heat is released to the surroundings in the reaction between iron and oxygen to form Fe_2O_3. A heat pack utilizing this reaction of iron and oxygen provides energy for warming cold hands.

Explain *how the diagram shows that the reaction is exothermic.*

CAREERS IN CHEMISTRY

WebQuest

Math in Chemistry

Signs of Energy Changes Help students understand the + and − signs assigned to changes in energy during a chemical reaction by pointing out that the sign always reflects the system's point of view. A + sign for ΔH indicates that the energy of the system has increased while a − sign indicates that the energy of the system has decreased. OL

Concept Development

Enthalpy Explain that enthalpy is a state function—a property of a system that depends only on its present state. Explain that an important characteristic of a state function, such as H, is that a change in H (ΔH) is independent of the pathway between the initial and final states. $\Delta H_{rxn} = H_{products} - H_{reactants}$

■ **Caption Question Fig. 8** The reaction is exothermic because the energy of the reactants is higher on the energy scale than the products.

ASSESSMENT

Skill Have students write the thermochemical equation for a reaction between two solid substances, A and B, which yields two gaseous substances, C and D, and releases 128 kJ of heat. $A(s) + B(s) \rightarrow C(g) + D(g) \; \Delta H = -128 \text{ kJ}$ OL

Visual Learning

Figure 7 Explain to students that not all of the energy released in a reaction is absorbed by the water in the calorimeter. Some energy is absorbed by the calorimeter itself, the amount depending on the materials used in the calorimeter and how it is constructed. Therefore, new calorimeters are calibrated individually to determine a calorimeter constant, which is then applied to all calculations made using that calorimeter.

3 Assess

Check for Understanding

Give students the following oral assignment: Define the universe and give three examples. After some discussion, point out that, in thermodynamics, the universe is simply defined as the sum of the system and surroundings (universe = system + surroundings). Therefore, an example of the universe could be any combination of system and surroundings. **OL** **EL**

Reteach

Ask students to define and compare *energy, heat,* and *enthalpy.* Energy is the ability to do work or produce heat. Heat is a form of energy that flows from a warmer object or material to a colder object or material. Enthalpy is the heat content of a system at constant pressure. **OL**

Extension

Arrange a field trip to a nearby electrical power plant or ask a power company representative to make a class presentation on the production of electricity. Alternatively, an interested student might interview the representative and make a class presentation of his or her findings. **OL** **EL**

■ **Caption Question Fig. 9** 27 kJ

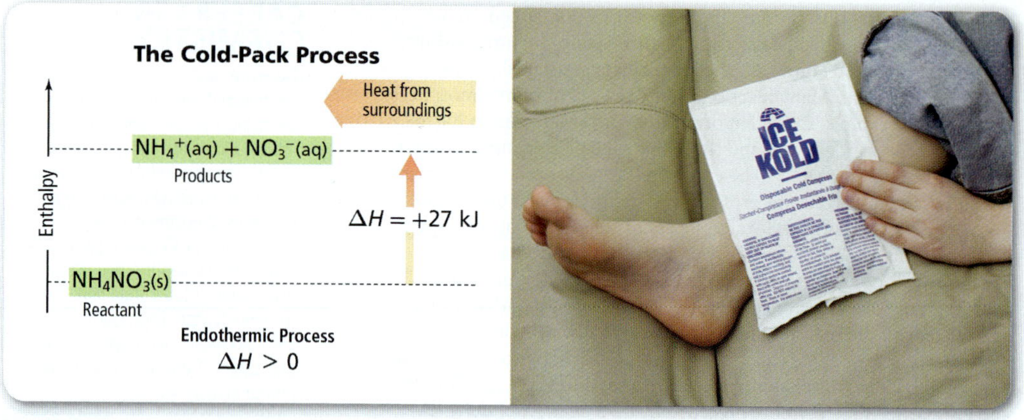

■ **Figure 9** The upward arrow shows that 27 kJ of heat is absorbed from the surroundings in the process of dissolving NH_4NO_3. This reaction is the basis for the cold pack. When the cold pack is placed on a person's elbow, his elbow supplies the required heat and is itself cooled.

Determine ***How much energy of ammonium nitrate is absorbed when a cold pack is activated?***

Matt Meadows/Getty Images

Now, recall the cold-pack process.

$$27 \text{ kJ} + NH_4NO_3(s) \rightarrow NH_4^+(aq) + NO_3^-(aq)$$

For this endothermic process, $H_{products} > H_{reactants}$. Therefore, when $H_{reactants}$ is subtracted from the larger $H_{products}$, a positive value for ΔH_{rxn} is obtained. Chemists write the equation for the cold-pack process and its enthalpy change in the following way.

$$NH_4NO_3(s) \rightarrow NH_4^+(aq) + NO_3^-(aq) \quad \Delta H_{rxn} = 27 \text{ kJ}$$

Figure 9 shows the energy change for the cold-pack process. In this process, the enthalpy of the products is 27 kJ greater than the enthalpy of the reactant because energy is absorbed. Thus, the sign of ΔH_{rxn} for this and all endothermic reactions and processes is positive. Recall that the sign of ΔH_{rxn} for all exothermic reactions is negative.

The enthalpy change, ΔH, is equal to q_p, the heat gained or lost in a reaction or process carried out at constant pressure. Because all reactions presented in this textbook occur at constant pressure, you can assume that $q = \Delta H_{rxn}$.

SECTION 2 REVIEW

Section Self-Check

Section Summary

- In thermochemistry, the universe is defined as the system plus the surroundings.
- The heat lost or gained by a system during a reaction or process carried out at constant pressure is called the change in enthalpy (ΔH).
- When ΔH is positive, the reaction is endothermic. When ΔH is negative, the reaction is exothermic.

16. MAINIDEA **Describe** how you would calculate the amount of heat absorbed or released by a substance when its temperature changes.

17. Explain why ΔH for an exothermic reaction always has a negative value.

18. Explain why a measured volume of water is an essential part of a calorimeter.

19. Explain why you need to know the specific heat of a substance in order to calculate how much heat is gained or lost by the substance as a result of a temperature change.

20. Describe what the system means in thermodynamics, and explain how the system is related to the surroundings and the universe.

21. Calculate the specific heat in J/(g•°C) of an unknown substance if a 2.50-g sample releases 12.0 cal as its temperature changes from 25.0°C to 20.0°C.

22. Design an Experiment Describe a procedure you could follow to determine the specific heat of a 45-g piece of metal.

SECTION 2 REVIEW

16. The heat absorbed or released equals the specific heat of the substance times its mass times its change in temperature.

17. $\Delta H_{rxn} = H_{products} - H_{reactants}$ and $H_{products} < H_{reactants}$.

18. You need the mass of the water in order to calculate the heat absorbed or released ($q = cm\Delta T$).

19. The specific heat of a substance tells you the number of joules that are lost or gained for every degree change in temperature and for every gram of the substance.

20. The system contains the process being studied. The surroundings are everything except the system, and the universe is the system and its surroundings.

21. 4.02 J/(g•°C)

22. Put a known mass of water into a calorimeter; measure its temperature. Heat a 45-g metal sample to 100°C and add the sample to water. Wait until the temperature of water is constant. Measure the final temperature of the water and calculate the specific heat of the metal.

SECTION 3

Essential Questions

- How are thermochemical equations for chemical reactions and other processes written?
- How is energy lost or gained during changes of state?
- How is the heat that is absorbed or released in a chemical reaction calculated?

Review Vocabulary

combustion reaction: a chemical reaction that occurs when a substance reacts with oxygen, releasing energy in the form of heat and light

New Vocabulary

thermochemical equation
enthalpy (heat) of combustion
molar enthalpy (heat) of vaporization
molar enthalpy (heat) of fusion

Thermochemical Equations

MAINIDEA Thermochemical equations express the amount of heat released or absorbed by chemical reactions.

CHEM 4 YOU Have you ever been exhausted after a hard race or other strenuous activity? If you felt as if your body had less energy than before the event, you were right. That tired feeling relates to combustion reactions that occur in the cells of your body, the same combustion you might observe in a burning campfire.

Writing Thermochemical Equations

The change in energy is an important part of chemical reactions, so chemists include ΔH as part of many chemical equations. The heat-pack and cold-pack equations are called thermochemical equations when they are written as follows.

$$4Fe(s) + 3O_2(g) \rightarrow 2Fe_2O_3(s) \quad \Delta H = -1625 \text{ kJ}$$

$$NH_4NO_3(s) \rightarrow NH_4^+(aq) + NO_3^-(aq) \quad \Delta H = 27 \text{ kJ}$$

A **thermochemical equation** is a balanced chemical equation that includes the physical states of all reactants and products and the energy change, usually expressed as the change in enthalpy, ΔH.

The highly exothermic combustion of glucose ($C_6H_{12}O_6$) occurs in the body as food is metabolized to produce energy. The thermochemical equation for the combustion of glucose is shown below.

$$C_6H_{12}O_6(s) + 6O_2(g) \rightarrow 6CO_2(g) + 6H_2O(l) \quad \Delta H_{comb} = -2808 \text{ kJ}$$

The **enthalpy (heat) of combustion** (ΔH_{comb}) of a substance is the enthalpy change for the complete burning of one mole of the substance. Standard enthalpies of combustion for several substances are given in **Table 3.** Standard enthalpy changes have the symbol ΔH°. The zero superscript tells you that the enthalpy changes were determined with all reactants and products at standard conditions. Standard conditions are 1 atm pressure and 298 K (25°C) and should not be confused with standard temperature and pressure (STP).

Table 3 Standard Enthalpies of Combustion

Substance	Formula	ΔH°_{comb} (kJ/mol)
Sucrose (table sugar)	$C_{12}H_{22}O_{11}(s)$	−5644
Octane (a component of gasoline)	$C_8H_{18}(l)$	−5471
Glucose (a simple sugar found in fruit)	$C_6H_{12}O_6(s)$	−2808
Propane (a gaseous fuel)	$C_3H_8(g)$	−2219
Methane (a gaseous fuel)	$CH_4(g)$	−891

CHEMISTRY PROJECT

Write Thermochemical Equations Have students write the thermochemical equations for the combustion of one mole of each of the substances listed in Table 3. AL

SECTION 3

1 Focus

MAINIDEA

Write a Thermochemical Equation Refer students back to the chapter opener photo and have them recall the chemical reaction that powers the space shuttle's booster rockets. $2H_2 + O_2 \rightarrow 2H_2O$ Point out that a similar, but less energetic, reaction occurs when hydrogen gas burns in air (reacts with oxygen) to produce liquid water. Explain that this exothermic reaction releases 286 kJ per mole of water formed. Then, lead students through the process of writing the thermochemical equation for the reaction, being sure to include physical states of both reactants and products.
$H_2(g) + 1/2O_2(g) \rightarrow H_2O(l)$
$\Delta H = -286$ kJ
or $2H_2(g) + O_2(g) \rightarrow 2H_2O(l)$
$\Delta H = -572$ kJ OL

2 Teach

Use Science Terms

Terms to Know Have students use a dictionary to find and explain the meanings of several key English words used in Sections 1, 2, and 3: *potential, conservation, capacity, calorimetry, system, surroundings, universe, thermochemical, combustion, vaporization, and fusion.* OL EL

Quick Demo

An Exothermic Change

Place approximately 20 g sodium thiosulfate pentahydrate ($Na_2S_2O_3{\cdot}5H_2O$) in a large, glass heat-resistant test tube and add 3 or 4 drops of water. Heat the tube and contents in a hot-water bath until the entire solid has melted or dissolved. Remove the tube, place a thermometer in the liquid, and allow it to cool to room temperature. Have a student volunteer touch the tube and contents while you use forceps to add a $Na_2S_2O_3{\cdot}5H_2O$ seed crystal. If the supersaturated solution does not crystalize with one seed crystal, gently agitate the test tube or add a second crystal. The liquid solidifies rapidly, releasing considerable heat. Ask what energy transfer has occurred and the sign of ΔH for the change. The system has lost energy to the surroundings. ΔH is negative. OL

Table 4 Standard Enthalpies of Vaporization and Fusion

Substance	Formula	ΔH°_{vap} (kJ/mol)	ΔH°_{fus} (kJ/mol)
Water	H_2O	40.7	6.01
Ethanol	C_2H_5OH	38.6	4.94
Methanol	CH_3OH	35.2	3.22
Acetic acid	CH_3COOH	23.4	11.7
Ammonia	NH_3	23.3	5.66

Changes of State

Many processes other than chemical reactions absorb or release heat. For example, think about what happens when you step out of a hot shower. You shiver as water evaporates from your skin. That is because your skin provides the heat needed to vaporize the water.

As heat is taken from your skin to vaporize the water, you cool down. The heat required to vaporize one mole of a liquid is called its **molar enthalpy (heat) of vaporization** (ΔH_{vap}). Similarly, if you want a glass of cold water, you might drop an ice cube into it. The water cools as it provides the heat to melt the ice. The heat required to melt one mole of a solid substance is called its **molar enthalpy (heat) of fusion** (ΔH_{fus}). Because vaporizing a liquid and melting a solid are endothermic processes, their ΔH values are positive. Standard molar enthalpies of vaporization and fusion for five common compounds are shown in **Table 4.**

Thermochemical equations for changes of state The vaporization of water and the melting of ice can be described by the following equations.

$$H_2O(l) \rightarrow H_2O(g) \quad \Delta H_{vap} = 40.7 \text{ kJ}$$
$$H_2O(s) \rightarrow H_2O(l) \quad \Delta H_{fus} = 6.01 \text{ kJ}$$

The first equation indicates that 40.7 kJ of energy is absorbed when one mole of water is converted to one mole of water vapor. The second equation indicates that 6.01 kJ of energy is absorbed when one mole of ice melts to form one mole of liquid water.

What happens in the reverse processes, when water vapor condenses to liquid water or liquid water freezes to ice? The same amounts of energy are released in these exothermic processes as are absorbed in the endothermic processes of vaporization and melting. Thus, the molar enthalpy (heat) of condensation (ΔH_{cond}) and the molar enthalpy of vaporization have the same numerical value but opposite signs. Similarly, the molar enthalpy (heat) of solidification (ΔH_{solid}) and the molar enthalpy of fusion have the same numerical value but opposite signs.

$$\Delta H_{vap} = -\Delta H_{cond}$$
$$\Delta H_{fus} = -\Delta H_{solid}$$

These relationships are illustrated in **Figure 10.**

Figure 10 The upward arrows show that the energy of the system increases as ice melts and then vaporizes. The downward arrows show that the energy of the system decreases as water vapor condenses and then solidifies.

Phase Changes for Water

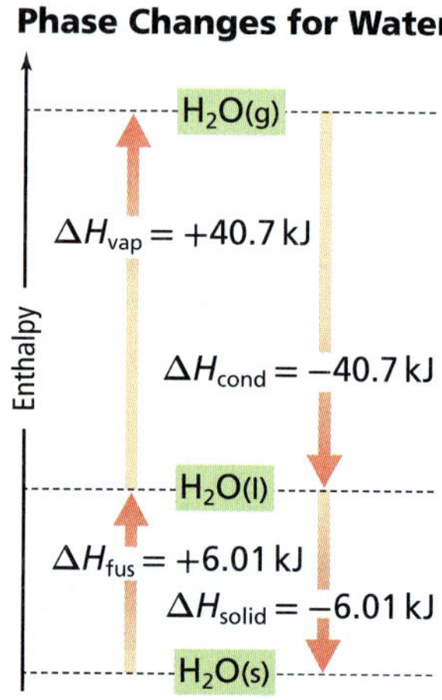

View an **animation about heat flow in endothermic and exothermic reactions.**

Concepts In Motion

CHEMISTRY JOURNAL

Gasoline Additives The requirement that gasoline in some areas of the country contain oxygenating compounds is under close scrutiny because of its added cost and some negative environmental effects. Have students research the effects—on mileage and on the environment—of adding oxygenating compounds such as ethanol and methyl tertbutyl ether (MTBE) to gasoline. Have them record their findings in their chemistry journals. OL

Compare the following equations for the condensation and freezing of water with the equations on the previous page for the vaporization and melting of water.

$$H_2O(g) \rightarrow H_2O(l) \qquad \Delta H_{cond} = -40.7 \text{ kJ}$$

$$H_2O(l) \rightarrow H_2O(s) \qquad \Delta H_{solid} = -6.01 \text{ kJ}$$

Some farmers make use of the heat of fusion of water to protect fruit and vegetables from freezing. If the temperature is predicted to drop to freezing, they flood their orchards or fields with water. When the water freezes, energy (ΔH_{fus}) is released and often warms the surrounding air enough to prevent frost damage. In the Problem-Solving Lab that follows, you will draw the heating curve of water and interpret it using the heats of fusion and vaporization.

☑ READING CHECK **Categorize** condensation, solidification, vaporization, and fusion as exothermic or endothermic processes.

Problem-Solving LAB

Make and Use Graphs

How can you derive a heating curve? Water molecules have a strong attraction to one another because they are polar and they form hydrogen bonds. The polarity of water accounts for its high specific heat and relatively high enthalpies of fusion and vaporization.

Analysis

Use the data in the table to plot a heating curve of temperature versus time for a 180-g sample of water as it is heated at a constant rate from −20°C to 120°C. Draw a best-fit line through the points. Note the time required for water to pass through each segment of the graph.

Think Critically

1. **Analyze** each of the five regions of the graph, which are distinguished by an abrupt change in slope. Indicate how the absorption of heat changes the energy (kinetic and potential) of the water molecules.
2. **Calculate** the amount of heat required to pass through each region of the graph (180 g H_2O = 10 mol H_2O, ΔH_{fus} = 6.01 kJ/mol, ΔH_{vap} = 40.7 kJ/mol, $c_{H_2O(s)}$ = 2.03 J/(g·°C), $c_{H_2O(\ell)}$ = 4.184 J/(g·°C), $c_{H_2O(g)}$ = 2.01 J/(g·°C). How does the length of time needed to pass through each region relate to the amount of heat absorbed?

Time and Temperature Data for Water

Time (min)	Temperature (°C)	Time (min)	Temperature (°C)
0.0	−20	13.0	100
1.0	0	14.0	100
2.0	0	15.0	100
3.0	9	16.0	100
4.0	26	17.0	100
5.0	42	18.0	100
6.0	58	19.0	100
7.0	71	20.0	100
8.0	83	21.0	100
9.0	92	22.0	100
10.0	98	23.0	100
11.0	100	24.0	100
12.0	100	25.0	100

3. **Infer** What would the heating curve of ethanol look like? Ethanol melts at −114°C and boils at 78°C. Sketch ethanol's curve from −120°C to 90°C. What factors determine the lengths of the flat regions of the graph and the slope of the curve between the flat regions?

DIFFERENTIATED INSTRUCTION

Advanced Learners Explain that the ΔH value of −286 kJ for the formation of 1 mol of liquid water from its constituent elements in their standard states is known as the heat of formation of liquid water. Heats of formation can be used to calculate ΔH values for many reactions by subtracting the sum of the heats of formation of the reactants from the sum of the heats of formation of the products: $\Delta H_{(reaction)} = \Sigma\, \Delta H_{f(products)} - \Sigma\, \Delta H_{f(reactants)}$. Have students use a table of standard heats of formation to determine $\Delta H_{(reaction)}$ for the following reactions. Point out that the table values are molar heats of formation, so they must be multiplied by the number of moles in the reaction. AL

a. $2NO(g) + O_2(g) \rightarrow 2NO_2$ [−116.2 kJ]
b. $Fe_2O_3 + 3CO(g) \rightarrow 3CO_2(g) + 2Fe(s)$ [−24.8 kJ]

Problem-Solving LAB

Purpose Students will plot the heating curve of water and infer the internal energy changes that take place.

Process Skills make and use graphs, apply concepts, interpret data, recognize cause and effect

Teaching Strategies

- Review the relationship between temperature and the average kinetic energy of the molecules.
- Use magnets to model the potential energy associated with strong attractive dipole forces between water molecules.
- Student's graphs will have three steps. The first slope will increase from −20°C to 0°C and plateau at 0°C. The second slope increases from 0°C to 100°C and plateaus at 100°C. The third slope increases from 100°C to 120°C.

Think Critically

1. From −20°C to 0°C, kinetic energy increases. At the 0°C plateau, potential energy increases. From 0°C to 100°C, kinetic energy increases. At the 100°C plateau, potential energy increases. From 100°C to 120°C, kinetic energy increases.
2. For −20°C to 0°C, q = 7 kJ
   Melting the ice, q = 60 kJ
   For 0°C to 100°C, q = 80 kJ
   Vaporizing the water, q = 400 kJ. For 100°C to 120°C, q = 7 kJ.
3. From −120 °C to −114°C the curve rises regularly. At −114°C it becomes horizontal for a time and then rises again until it reaches 78°C, where it becomes horizontal again. After a time, the curve rises again to 90°C. The lengths of the flat regions depend on the amount of ethanol being heated and the amount of heat being added with time. Those factors and the specific heat of the substance determine the slope of the upward curve between the flat regions.

☑ **READING CHECK** condensation, exothermic; solidification, exothermic; vaporization, endothermic; fusion, endothermic

IN-CLASS Example

Question How much heat is required to vaporize 166 g of water at its boiling point?

Answer 375 kJ
mol of water = 166 g ÷ 18.0 g/mol = 9.22 mol
9.22 mol × 40.7 kJ/mol = 375 kJ

PRACTICE Problems

Have students refer to the Selected Solutions appendix for complete solutions to odd-numbered problems. The complete solutions for all problems can be found in the Solutions Manual.

23. 2.58 kJ
24. 376 kJ
25. 232 g CH_4

ASSESSMENT

Skill Ask students to write the thermochemical equation for the complete oxidation of sucrose ($C_{12}H_{22}O_{11}$) to gaseous carbon dioxide and liquid water.
$C_{12}H_{22}O_{11}(s) + 12O_2(g) \rightarrow 12CO_2(g) + 11H_2O(l)$; $\Delta H = -5644$ kJ **OL**

EXAMPLE Problem 4

Find help with **dimensional analysis.** Math Handbook

EXAMPLE PROBLEM

THE ENERGY RELEASED IN A REACTION A bomb calorimeter is useful for measuring the energy released in combustion reactions. The reaction is carried out in a constant-volume bomb with a high pressure of oxygen. How much heat is evolved when 54.0 g glucose ($C_6H_{12}O_6$) is burned according to this equation?

$$C_6H_{12}O_6(s) + 6O_2(g) \rightarrow 6CO_2(g) + 6H_2O(l) \quad \Delta H_{comb} = -2808 \text{ kJ}$$

1 ANALYZE THE PROBLEM

You are given a mass of glucose, the equation for the combustion of glucose, and ΔH_{comb}. You must convert grams of glucose to moles of glucose. Because the molar mass of glucose is more than three times the mass of glucose burned, you can predict that the energy evolved will be less than one-third ΔH_{comb}.

Known	**Unknown**
mass of glucose = 54.0 g $C_6H_{12}O_6$	q = ? kJ
ΔH_{comb} = −2808 kJ	

2 SOLVE FOR THE UNKNOWN

Convert grams of $C_6H_{12}O_6$ to moles of $C_6H_{12}O_6$.

$$54.0 \text{ g } C_6H_{12}O_6 \times \frac{1 \text{ mol } C_6H_{12}O_6}{180.18 \text{ g } C_6H_{12}O_6} = 0.300 \text{ mol } C_6H_{12}O_6$$

Multiply by the inverse of molar mass, $\frac{1 \text{ mol}}{180.18 \text{ g}}$.

Multiply moles of $C_6H_{12}O_6$ by the enthalpy of combustion, ΔH_{comb}.

$$0.300 \text{ mol } C_6H_{12}O_6 \times \frac{2808 \text{ kJ}}{1 \text{ mol } C_6H_{12}O_6} = 842 \text{ kJ}$$

Multiply moles of glucose by $\frac{2808 \text{ kJ}}{1 \text{ mol } C_6H_{12}O_6}$.

3 EVALUATE THE ANSWER

All values in the calculation have at least three significant figures, so the answer is correctly stated with three digits. As predicted, the released energy is less than one-third ΔH_{comb}.

PRACTICE Problems

Do additional problems. Online Practice

23. Calculate the heat required to melt 25.7 g of solid methanol at its melting point. Refer to **Table 4.**

24. How much heat evolves when 275 g of ammonia gas condenses to a liquid at its boiling point? Use **Table 4** to determine ΔH_{cond}.

25. Challenge What mass of methane (CH_4) must be burned in order to liberate 12,880 kJ of heat? Refer to **Table 3**.

Connection to Biology When a mole of glucose is burned in a bomb calorimeter, 2808 kJ of energy is released. The same amount of energy is produced in your body when an equal mass of glucose is metabolized in the process of cellular respiration. The process takes place in every cell of your body in a series of complex steps in which glucose is broken down and carbon dioxide and water are released. These are the same products produced by the combustion of glucose in a calorimeter. The energy released is stored as chemical potential energy in the bonds of molecules of adenosine triphosphate (ATP). When energy is needed by any part of the body, molecules of ATP release their energy.

CHEMISTRY JOURNAL

Thermochemical Equations After studying Section 3, have students write thermochemical equations in their chemistry journals for three or more chemical reactions they studied in previous chapters. Have them obtain the needed heat of formation values from Table R-11. **OL**

DIFFERENTIATED INSTRUCTION

Advanced Learners Have students research the role of thermochemisty in the field of materials science engineering. Tell students that one place they might start their research is in the area of adhesives such as those used on tapes and in glues. **AL**

Combustion Reactions

Combustion is the reaction of a fuel with oxygen. In biological systems, food is the fuel. **Figure 11** illustrates some of the many foods that contain glucose as well as other foods that contain carbohydrates that are readily converted to glucose in your body. You also depend on other combustion reactions to keep you warm or cool, and to transport you in vehicles. One way you might heat your home or cook your food is by burning methane gas. The combustion of one mole of methane produces 891 kJ according to this equation.

$$CH_4(g) + 2O_2(g) \rightarrow CO_2(g) + 2H_2O(l) + 891 \text{ kJ}$$

Most vehicles—cars, airplanes, boats, and trucks—run on the combustion of gasoline, which is mostly octane (C_8H_{18}). **Table 3** shows that the burning of one mole of octane produces 5471 kJ. The equation for the combustion of gasoline is as follows.

$$C_8H_{18}(l) + \frac{25}{2}O_2(g) \rightarrow 8CO_2(g) + 9H_2O(l) + 5471 \text{ kJ}$$

Another combustion reaction is the reaction between hydrogen and oxygen.

$$H_2(g) + \frac{1}{2}O_2(g) \rightarrow H_2O(l) + 286 \text{ kJ}$$

The combustion of hydrogen provides the energy to lift the shuttle into space, as illustrated on the opening page of this chapter.

Figure 11 These foods are fuels for the body. They provide the glucose that is burned to produce 2808 kJ/mol to carry on the activities of life.

SECTION 3 REVIEW

Section Self-Check

Section Summary

- A thermochemical equation includes the physical states of the reactants and products and specifies the change in enthalpy.
- The molar enthalpy (heat) of vaporization, ΔH_{vap}, is the amount of energy required to evaporate one mole of a liquid.
- The molar enthalpy (heat) of fusion, ΔH_{fus}, is the amount of energy needed to melt one mole of a solid.

26. MAIN IDEA Write a complete thermochemical equation for the combustion of ethanol (C_2H_5OH). $\Delta H_{comb} = -1367$ kJ/mol

27. Determine Which of the following processes are exothermic? Endothermic?

a. $C_2H_5OH(l) \rightarrow C_2H_5OH(g)$

b. $Br_2(l) \rightarrow Br_2(s)$

c. $C_5H_{12}(g) + 8O_2(g) \rightarrow 5CO_2(g) + 6H_2O(l)$

d. $NH_3(g) \rightarrow NH_3(l)$

e. $NaCl(s) \rightarrow NaCl(l)$

28. Explain how you could calculate the heat released in freezing 0.250 mol water.

29. Calculate How much heat is released by the combustion of 206 g of hydrogen gas? $\Delta H_{comb} = -286$ kJ/mol

30. Apply The molar heat of vaporization of ammonia is 23.3 kJ/mol. What is the molar heat of condensation of ammonia?

31. Interpret Scientific Illustrations The reaction A → C is shown in the enthalpy diagram at right. Is the reaction exothermic or endothermic? Explain your answer.

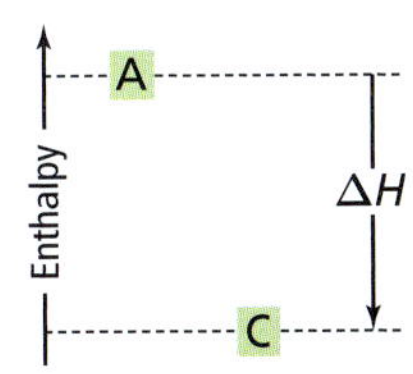

3 Assess

Check for Understanding

Have students classify the following changes of state as exothermic or endothermic.

$C_3H_8(g) \rightarrow C_3H_8(l)$ exothermic

$CO_2(s) \rightarrow CO_2(g)$ endothermic

$C_{10}H_8(s) \rightarrow C_{10}H_8(l)$ endothermic

$H_2O(l) \rightarrow H_2O(s)$ exothermic

OL EL

Reteach

Have students write and compare the chemical and thermochemical equations for the combustion of one mole of methane.

$CH_4 + 2O_2 \rightarrow CO_2 + 2H_2O$

$CH_4(g) + 2O_2(g) \rightarrow CO_2(g) + 2H_2O(l); \Delta H = -891$ kJ

The thermochemical equation includes the physical state of the reactants and products and the change in enthalpy. OL

Extension

Explain to students that some solid substances go directly from the solid state to the gas phase without becoming liquids. This process is called sublimation. Ask students how the energy involved in the sublimation of a mole of water compares to the energy required to melt and then vaporize a mole of water. The same amount of energy would be needed. OL

SECTION 3 REVIEW

26. $C_2H_5OH(l) + 3O_2(g) \rightarrow 2CO_2(g) + 3H_2O(l)\ \Delta H_{comb} = -1367$ kJ/mol
27. Reactions **b, c,** and **d** are exothermic. Reactions **a** and **e** are endothermic.
28. Multiply 0.250 mol times the molar heat of fusion of water, 6.01 kJ/mol.
29. 29,300 kJ
30. −23.3 kJ/mol
31. The reaction is exothermic because the product (C) has a lower energy than the reactant (A).

SECTION 4

1 Focus

MAINIDEA

Use Hess's Law To ensure that students will not be intimidated by the seeming complexity of using Hess's Law when they first read the section, lead them through its use with an example such as the one presented in the first few pages.

a. $S(s) + O_2(g) \rightarrow SO_2(g)$ $\Delta H = -297$ kJ

b. $2SO_3(g) \rightarrow 2SO_2(g) + O_2(g)$ $\Delta H = 198$ kJ

c. $2S(s) + 2O_2(g) \rightarrow 2SO_2(g)$ $\Delta H = -594$ kJ

d. $2SO_2(g) + O_2(g) \rightarrow 2SO_3(g)$ $\Delta H = -198$ kJ

Ask what formulas cancel when equations **c** and **d** are added to produce the desired equation. $2SO_2(g)$ cancels on both sides of the final equation. OL

2 Teach

Concept Development

Hess's Law Help students understand Hess's law with this analogy. Suppose you are going on a long trip by automobile. If you drive 800 km on Day 1, 720 km on Day 2, 600 km on Day 3, and 680 km on Day 4, you can calculate your total trip mileage by adding the day totals. (800 km + 720 km + 600 km + 680 km = 2800 km) Explain that, if a reaction is the sum of two or more reactions, the sum of the enthalpy changes for the two or more reactions is the enthalpy change for the overall reaction. OL

SECTION 4

Calculating Enthalpy Change

MAINIDEA The enthalpy change for a reaction can be calculated using Hess's law.

Essential Questions

- How is Hess's law applied to calculate the enthalpy change for a reaction?
- What is the basis for the table of standard enthalpies of formation?
- How is ΔH_{rxn} calculated using thermochemical equations?
- What is the enthalpy change for a reaction using standard enthalpies of formation data?

Review Vocabulary

allotrope: one of two or more forms of an element with different structures and properties when they are in the same state

New Vocabulary

Hess's law
standard enthalpy (heat) of formation

CHEM 4 YOU **Maybe you have watched a two-act play or a two-part TV show. Each part tells some of the story, but you have to see both parts to understand the entire story. Like such a play or show, some reactions are best understood when you view them as the sum of two or more simpler reactions.**

Hess's Law

Sometimes it is impossible or impractical to measure the ΔH of a reaction by using a calorimeter. Consider the reaction in **Figure 12,** the conversion of carbon in its allotropic form, diamond, to carbon in its allotropic form, graphite.

$$C(s, \text{diamond}) \rightarrow C(s, \text{graphite})$$

This reaction occurs so slowly that measuring the enthalpy change is impossible. Other reactions occur under conditions difficult to duplicate in a laboratory. Still others produce products other than the desired ones. For these reactions, chemists use a theoretical way to determine ΔH.

Suppose you are studying the formation of sulfur trioxide in the atmosphere. You would need to determine ΔH for this reaction.

$$2S(s) + 3O_2(g) \rightarrow 2SO_3(g) \quad \Delta H = ?$$

Unfortunately, laboratory experiments to produce sulfur trioxide and determine its ΔH result in a mixture of products that is mostly sulfur dioxide (SO_2). In situations such as this, you can calculate ΔH by using Hess's law of heat summation. **Hess's law** states that if you can add two or more thermochemical equations to produce a final equation for a reaction, then the sum of the enthalpy changes for the individual reactions is the enthalpy change for the final reaction.

■ **Figure 12** The expression "diamonds are forever" suggests the durability of diamonds and tells you that the conversion of diamond to graphite is so slow that it would be impossible to measure its enthalpy change.

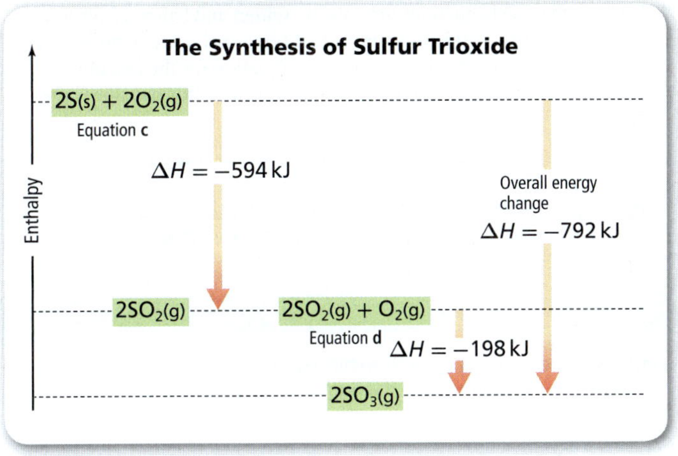

Figure 13 The arrow on the left indicates the release of 594 kJ as S and O_2 react to form SO_2 (Equation **c**). Then, SO_2 and O_2 react to form SO_3 (Equation **d**) with the release of 198 kJ (middle arrow). The overall energy change (the sum of the two processes) is shown by the arrow on the right.
Determine *the enthalpy change for the decomposition of SO_3 to S and O_2.*

Applying Hess's law How can Hess's law be used to calculate the energy change for the reaction that produces SO_3?

$$2S(s) + 3O_2(g) \rightarrow 2SO_3(g) \quad \Delta H = ?$$

Step 1 Chemical equations are needed that contain the substances found in the desired equation and have known enthalpy changes. The following equations contain S, O_2, and SO_3.

a. $S(s) + O_2(g) \rightarrow SO_2(g) \quad \Delta H = -297 \text{ kJ}$

b. $2SO_3(g) \rightarrow 2SO_2(g) + O_2(g) \quad \Delta H = 198 \text{ kJ}$

Step 2 The desired equation shows two moles of sulfur reacting, so rewrite Equation **a** for two moles of sulfur by multiplying the coefficients by two. Double the enthalpy change ΔH because twice the energy will be released if two moles of sulfur react. With these changes, Equation **a** becomes the following (Equation **c**).

c. $2S(s) + 2O_2(g) \rightarrow 2SO_2(g) \quad \Delta H = 2(-297 \text{ kJ}) = -594 \text{ kJ}$

Step 3 In the desired equation, sulfur trioxide is a product rather than a reactant, so reverse Equation **b.** When you reverse an equation, you must also change the sign of its ΔH. Equation **b** then becomes Equation **d.**

d. $2SO_2(g) + O_2(g) \rightarrow 2SO_3(g) \quad \Delta H = -198 \text{ kJ}$

Step 4 Add Equations **c** and **d** to obtain the desired reaction. Add the corresponding ΔH values. Cancel any terms that are common to both sides of the combined equation.

$$2S(s) + 2O_2(g) \rightarrow 2SO_2(g) \qquad \Delta H = -594 \text{ kJ}$$
$$2SO_2(g) + O_2(g) \rightarrow 2SO_3(g) \qquad \Delta H = -198 \text{ kJ}$$
$$\cancel{2SO_2(g)} + 2S(s) + 3O_2(g) \rightarrow \cancel{2SO_2(g)} + 2SO_3(g) \quad \Delta H = -792 \text{ kJ}$$

The thermochemical equation for the burning of sulfur to form sulfur trioxide is as follows. **Figure 13** diagrams the energy changes.

$$2S(s) + 3O_2(g) \rightarrow 2SO_3(g) \quad \Delta H = -792 \text{ kJ}$$

Concept Development

More Hess's Law Illustrate the application of Hess's law by using these general equations A + B → C; C + D → E + B

Show how C and B can be eliminated from the overall equation because they are present in equal amounts on both sides. Show that the sum of the two reactions is A + D → E.

Caption Question Fig. 13 +792 kJ

CHEMISTRY JOURNAL

Balancing a Checkbook Have students describe the similarities between balancing a checkbook and calculating ΔH for a chemical reaction using Hess's law. If they do not have a checking account or do not know how to balance one, provide an example and show how to balance the account. Have them include their descriptions in their chemistry journals. OL

IN-CLASS Example

Question Determine ΔH for the reaction $C(s, graphite) + 1/2\ O_2(g) \rightarrow CO(g)$ using the following thermochemical equations.

A. $C(s, graphite) + O_2(g) \rightarrow CO_2(g)$ $\Delta H = -394\ kJ$

B. $CO(g) + 1/2\ O_2(g) \rightarrow CO_2(g)$ $\Delta H = -283\ kJ$

Answer $\Delta H = -111\ kJ$

Produce Equation C by reversing Equation B and changing the sign of its ΔH.

C. $CO_2(g) \rightarrow CO(g) + 1/2\ O_2(g)$ $\Delta H = 283\ kJ$

Add Equations A and C to produce the desired Equation, $C(s, graphite) + 1/2\ O_2(g) \rightarrow CO(g)$. Adding the ΔH values for Equations A and C yields −111 kJ.

☑ **READING CHECK** The equation above is written for the formation of 1 mol of SO_3 using a fractional coefficient. The equation on the previous page was written for the formation of two moles of SO_3. Thus, ΔH for the equation on the previous page is twice as great as ΔH for the equation above.

Quick Demo

Exothermic Decomposition

WARNING: ***30% hydrogen peroxide is a strong oxidizing agent and can cause severe burns. Wear eye protection, a laboratory apron, and rubber gloves. Perform the demonstration in a fume hood or behind a safety shield.*** Set a glass 100-mL graduated cylinder in a trough or sink. Add 5 mL 30% hydrogen peroxide (H_2O_2) and 5 mL liquid dishwashing detergent. Use a scoop to add approximately 1 g of potassium iodide (KI) to the cylinder. (KI catalyzes the decomposition of H_2O_2.) The 30% H_2O_2 decomposes rapidly to H_2O and O_2, releasing much heat and forming O_2-containing suds.

Thermochemical equations are usually written and balanced for one mole of product. Often, that means that fractional coefficients must be used. For example, the thermochemical equation for the reaction between sulfur and oxygen to form one mole of sulfur trioxide is the following.

$$S(s) + \frac{3}{2}O_2(g) \rightarrow SO_3(g)\ \ \Delta H = -396\ kJ$$

☑ **READING CHECK** **Compare** the equation above with the thermochemical equation developed on the previous page. How are they different?

EXAMPLE Problem 5

EXAMPLE PROBLEM

HESS'S LAW Use thermochemical Equations **a** and **b** below to determine ΔH for the decomposition of hydrogen peroxide (H_2O_2), a compound that has many uses ranging from bleaching hair to powering rocket engines.

$2H_2O_2(l) \rightarrow 2H_2O(l) + O_2(g)$

a. $2H_2(g) + O_2(g) \rightarrow 2H_2O(l)$ $\Delta H = -572\ kJ$

b. $H_2(g) + O_2(g) \rightarrow H_2O_2(l)$ $\Delta H = -188\ kJ$

1 ANALYZE THE PROBLEM

You have been given two chemical equations and their enthalpy changes. These two equations contain all the substances found in the desired equation.

Known

a. $2H_2(g) + O_2(g) \rightarrow 2H_2O(l)$ $\Delta H = -572\ kJ$

b. $H_2(g) + O_2(g) \rightarrow H_2O_2(l)$ $\Delta H = -188\ kJ$

Unknown

$\Delta H = ?\ kJ$

2 SOLVE FOR THE UNKNOWN

H_2O_2 is a reactant.

$H_2O_2(aq) \rightarrow H_2(g) + O_2(g)$ $\Delta H = 188\ kJ$ — Reverse Equation **b** and change the sign of ΔH.

Two moles of H_2O_2 are needed.

c. $2H_2O_2(aq) \rightarrow 2H_2(g) + 2O_2(g)$ — Multiply the reversed Equation **b** by two to obtain Equation **c.**

ΔH for Equation **c** = (188 kJ)(2) — Multipy 188 kJ by two to obtain ΔH for Equation **c.**

= 376 kJ

c. $2H_2O_2(aq) \rightarrow 2H_2(g) + 2O_2(g)$ $\Delta H = 376\ kJ$ — Write Equation **c** and ΔH.

Add Equations **a** and **c,** canceling any terms common to both sides of the combined equation. Add the enthalpies of Equations **a** and **c**.

a. $2H_2(g) + O_2(g) \rightarrow 2H_2O(l)$ $\Delta H = -572\ kJ$ — Write Equation **a.**

c. $2H_2O_2(l) \rightarrow 2H_2(g) + 2O_2(g)$ $\Delta H = 376\ kJ$ — Write Equation **c.**

$2H_2O_2(l) \rightarrow 2H_2O(l) + O_2(g)$ **$\Delta H = -196\ kJ$** — Add Equations **a** and **c.** Add the enthalpies.

3 EVALUATE THE ANSWER

The two equations produce the desired equation. All values are accurate to the ones place, so ΔH is correctly stated.

CHEMISTRY JOURNAL

Thermochemical Formation Equations Have students write the thermochemical formation equations for the following compounds and place them in their journals: silver cyanide(s), $\Delta H_f^\circ = 146\ kJ/mol$; magnesium hydroxide(s), $\Delta H_f^\circ = -925\ kJ/mol$; phosphorus pentafluoride(g), $\Delta H_f^\circ = -1578\ kJ/mol$; $CH_3COOH(l)$, $\Delta H_f^\circ = -484\ kJ/mol$. OL

PRACTICE Problems

Do additional problems. Online Practice

32. Use Equations **a** and **b** to determine ΔH for the following reaction.

$2CO(g) + 2NO(g) \rightarrow 2CO_2(g) + N_2(g)$ $\Delta H = ?$

a. $2CO(g) + O_2(g) \rightarrow 2CO_2(g)$ $\Delta H = -566.0$ kJ

b. $N_2(g) + O_2(g) \rightarrow 2NO(g)$ $\Delta H = -180.6$ kJ

33. Challenge ΔH for the following reaction is -1789 kJ. Use this and Equation **a** to determine ΔH for Equation **b.**

$4Al(s) + 3MnO_2(s) \rightarrow 2Al_2O_3(s) + 3Mn(s)$ $\Delta H = -1789$ kJ

a. $4Al(s) + 3O_2(g) \rightarrow 2Al_2O_3(s)$ $\Delta H = -3352$ kJ

b. $Mn(s) + O_2(g) \rightarrow MnO_2(s)$ $\Delta H = ?$

Standard Enthalpy (Heat) of Formation

Hess's law allows you to calculate unknown ΔH values using known reactions and their experimentally determined ΔH values. However, recording ΔH values for all known chemical reactions would be a huge and unending task. Instead, scientists record and use enthalpy changes for only one type of reaction—a reaction in which a compound is formed from its elements in their standard states. The standard state of a substance means the normal physical state of the substance at 1 atm and 298 K (25°C). For example, in their standard states, iron is a solid, mercury is a liquid, and oxygen is a diatomic gas.

The ΔH value for such a reaction is called the standard enthalpy (heat) of formation of the compound. The **standard enthalpy (heat) of formation** (ΔH_f°) is defined as the change in enthalpy that accompanies the formation of one mole of the compound in its standard state from its elements in their standard states. A typical standard heat of formation reaction is the formation of one mole of SO_3 from its elements.

$$S(s) + \frac{3}{2}O_2(g) \rightarrow SO_3(g) \quad \Delta H_f^\circ = -396 \text{ kJ}$$

The product of this equation is SO_3, a suffocating gas that produces acid rain when mixed with moisture in the atmosphere. The destructive results of acid precipitation are shown in **Figure 14.**

Figure 14 Sulfur trioxide combines with water in the atmosphere to form sulfuric acid (H_2SO_4), a strong acid, which reaches Earth as acid precipitation. Acid precipitation slowly destroys trees and property.

PRACTICE Problems

Have students refer to the Selected Solutions appendix for complete solutions to odd-numbered problems. The complete solutions for all problems can be found in the Solutions Manual.

32. -385.4 kJ

33. Because the direction of Equation *b* was changed, ΔH for Equation *b* $= -521$ kJ.

Quick Demo

Burning Sugar Demonstrate the exothermic combustion of table sugar by first dabbing some fine ashes from a wood splint on a sugar cube. The wood ashes act as a catalyst for the combustion. Use tongs to hold the cube in a burner flame. The resulting carbon can be put in the trash when cool. Have students write the thermochemical equation for the combustion of sucrose to gaseous carbon dioxide and liquid water.

$C_{12}H_{22}O_{11}(s) + 12O_2(g) \rightarrow 12CO_2(g) + 11H_2O(l)$; $\Delta H = -5644$ kJ

DIFFERENTIATED INSTRUCTION

Advanced Learners Have students use the enthalpy of combustion of sucrose (−5644 kJ/mol) and the enthalpies of formation of gaseous carbon dioxide and liquid water to calculate ΔH_f° for sucrose.

$C_{12}H_{22}O_{11}(s) + 12O_2(g) \rightarrow 12CO_2(g) + 11H_2O(l)$

$\Delta H^\circ_{rxn} = S\Delta H_f^\circ(\text{products}) - S\Delta H_f^\circ(\text{reactants})$

$-5644 \text{ kJ} = (12(-394 \text{ kJ}) + 11(-286 \text{ kJ})) - \Delta H_f^\circ \text{ sucrose}$

ΔH_f° sucrose $= -2230$ kJ **AL**

Apply Chemistry

Heat Pumps Explain that many homes in the United States are heated and cooled by heat pumps, which utilize the endothermic vaporization and exothermic condensation of coolant compounds called hydrochlorofluorocarbons (HCFCs), which are made up of the elements hydrogen, chlorine, fluorine, and carbon. HCFC's have high enthalpies of vaporization and condensation. Explain that a heat pump is like a refrigerator that can be run in two directions. In summer, the heat pump makes use of the endothermic vaporization of the coolant to absorb heat from the home and the exothermic condensation of the coolant to expel the heat into the outside air. In winter, the heat pump uses the endothermic vaporization of the coolant to absorb heat from the cold outside air and its exothermic condensation to transfer the heat into the home. HCFC's will be phased out of use due to the ozone depleting properties, but other compounds are being developed to replace HCFC's.

Caption Question Fig. 15
H_2O should be placed below 0.0 kJ about three-quarters of the distance to −396 kJ.

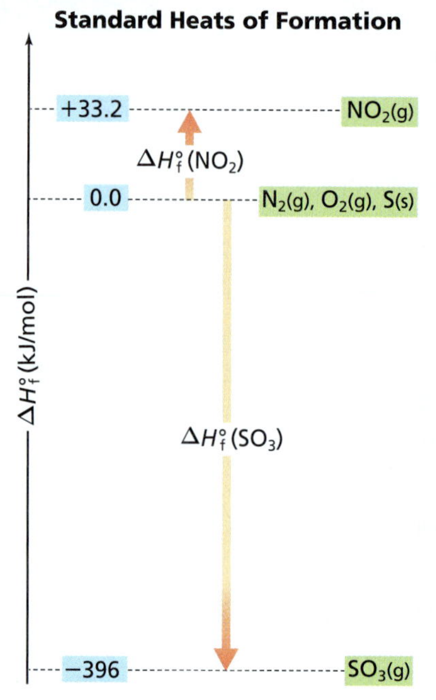

Figure 15 ΔH_f° for the elements N_2, O_2, and S is 0.0 kJ. When N_2 and O_2 react to form 1 mole of NO_2, 33.2 kJ is absorbed. Thus, ΔH_f° for NO_2 is +33.2 kJ/mol. When S and O_2 react to form one mole of SO_3, 396 kJ is released. Therefore, ΔH_f° for SO_3 is −396 kJ/mol.

Predict *Describe the approximate location of water on the scale. The heat of formation for the reaction* $H_2(g) + \frac{1}{2}O_2(g) \rightarrow H_2O(l)$ *is* $\Delta H_f^\circ = -286$ *kJ/mol.*

Where do standard heats of formation come from? When you state the height of a mountain, you do so relative to some point of reference—usually sea level. In a similar way, standard enthalpies of formation are stated based on the following assumption: Elements in their standard state have a ΔH_f° of 0.0 kJ. With zero as the starting point, the experimentally determined enthalpies of formation of compounds can be placed on a scale above and below the elements in their standard states. Think of the zero of the enthalpy scale as being similar to the arbitrary assignment of 0.0°C to the freezing point of water. All substances warmer than freezing water have a temperature above zero. All substances colder than freezing water have a temperature below zero.

Enthalpies of formation from experiments
Standard enthalpies of formation of many compounds have been measured experimentally. For example, consider the equation for the formation of nitrogen dioxide.

$$\frac{1}{2}N_2(g) + O_2(g) \rightarrow NO_2(g) \quad \Delta H_f^\circ = +33.2 \text{ kJ}$$

The elements nitrogen and oxygen are diatomic gases in their standard states, so their standard enthalpies of formation are zero. When nitrogen and oxygen gases react to form one mole of nitrogen dioxide, the experimentally determined ΔH for the reaction is +33.2 kJ. That means that 33.2 kJ of energy is absorbed in this endothermic reaction. The energy content of the product NO_2 is 33.2 kJ greater than the energy content of the reactants. On a scale on which ΔH_f° of reactants is 0.0 kJ, ΔH_f° of NO_2 is +33.2 kJ. **Figure 15** shows that on the scale of standard enthalpies of formation, NO_2 is placed 33.2 kJ above the elements from which it was formed. Sulfur trioxide (SO_3) is placed 396 kJ below zero on the scale because the formation of $SO_3(g)$ is an exothermic reaction. The energy content of the sulfur trioxide, ΔH_f°, is −396 kJ. **Table 5** lists standard enthalpies of formation for some common compounds. A more complete list is in **Table R-11**.

Table 5 Standard Enthalpies of Formation

Compound	Formation Equation	ΔH_f°(kJ/mol)
$H_2S(g)$	$H_2(g) + S(s) \rightarrow H_2S(g)$	−21
$HF(g)$	$\frac{1}{2}H_2(g) + \frac{1}{2}F_2(g) \rightarrow HF(g)$	−273
$SO_3(g)$	$S(s) + \frac{3}{2}O_2(g) \rightarrow SO_3(g)$	−396
$SF_6(g)$	$S(s) + 3F_2(g) \rightarrow SF_6(g)$	−1220

CHEMISTRY PROJECT

Germain Hess Have students research the life of Germain Hess and present their findings in a report, time line or graphic. Ask students to include the several major political and scientific events of the time in which Hess lived. Ask students how those events may have had an impact on the scientific research in Hess's time. OL AL

■ **Figure 16** Sulfur hexafluoride is used to etch minute and sometimes intricate patterns on silicon wafers in the production semiconductor devices. Semiconductors are important components of modern electronic equipment, including computers, cell phones, and MP3 players.

Using standard enthalpies of formation Standard enthalpies of formation can be used to calculate the enthalpies of many reactions under standard conditions ΔH°_{rxn} using Hess's law. Suppose you want to calculate ΔH°_{rxn} for a reaction that produces sulfur hexafluoride. Sulfur hexafluoride is a stable, unreactive gas with some interesting applications, one of which is shown in **Figure 16.**

$$H_2S(g) + 4F_2(g) \rightarrow 2HF(g) + SF_6(g) \quad \Delta H^\circ_{rxn} = ?$$

Step 1 Refer to **Table 5** to find an equation for the formation of each of the three compounds in the desired equation—HF, SF_6, and H_2S.

a. $\frac{1}{2}H_2(g) + \frac{1}{2}F_2(g) \rightarrow HF(g) \quad \Delta H^\circ_f = -273 \text{ kJ}$

b. $S(s) + 3F_2(g) \rightarrow SF_6(g) \quad \Delta H^\circ_f = -1220 \text{ kJ}$

c. $H_2(g) + S(s) \rightarrow H_2S(g) \quad \Delta H^\circ_f = -21 \text{ kJ}$

Step 2 Equations **a** and **b** describe the formation of the products HF and SF_6 in the desired equation, so use Equations **a** and **b** in the direction in which they are written.

Equation **c** describes the formation of a product, H_2S, but in the desired equation, H_2S is a reactant. Reverse Equation **c** and change the sign of its ΔH°_f.

$$H_2S(g) \rightarrow H_2(g) + S(s) \quad \Delta H^\circ_f = 21 \text{ kJ}$$

Step 3 Two moles of HF are required. Multiply Equation **a** and its enthalpy change by two.

$$H_2(g) + F_2(g) \rightarrow 2HF(g) \quad \Delta H^\circ_f = 2(-273) = -546 \text{ kJ}$$

Step 4 Add the three equations and their enthalpy changes. The elements H_2 and S cancel.

$$\cancel{H_2(g)} + F_2(g) \rightarrow 2HF(g) \quad \Delta H^\circ_f = -546 \text{ kJ}$$
$$\cancel{S(s)} + 3F_2(g) \rightarrow SF_6(g) \quad \Delta H^\circ_f = -1220 \text{ kJ}$$
$$H_2S(g) \rightarrow \cancel{H_2(g)} + \cancel{S(s)} \quad \Delta H^\circ_f = 21 \text{ kJ}$$
$$H_2S(g) + 4F_2(g) \rightarrow 2HF(g) + SF_6(g) \quad \Delta H^\circ_{rxn} = -1745 \text{ kJ}$$

CHEMISTRY JOURNAL

Chemical to Thermochemical Have students choose five chemical equations from the Practice Problems in the Chemical Reactions chapter. Have them look up the enthalpies of formation of the reactants and products and calculate ΔH for each of the reactions. Finally, have students convert each chemical equation to a thermochemical equation. Students should summarize their results in tables and include the tables in their chemistry journals. OL

ASSESSMENT

Performance Have students form teams (pro and con) to research and debate the use of fossil fuels as a source of energy for generating electric power in the United States. Be sure the teams include both economic and environmental considerations in their debate. OL COOP LEARN

Visual Learning

Table 5 Have students pick three reactions, other than the formation of $NO_2(g)$, which is shown in Figure 15, from the list in Table 5 and construct diagrams that show the energy changes involved. Have them use Figure 15 as a model for their diagrams. OL

IN-CLASS Example

Question Use standard enthalpies of formation from Table R-11 to calculate ΔH°_{rxn} for the following reaction: $CH_4(g) + 2Cl_2(g) \rightarrow CCl_4(l) + 2H_2(g)$

Answer $\Delta H^\circ_{rxn} = -53.6$ kJ
$\Delta H^\circ_{rxn} = -128.2 \text{ kJ} - (-74.6 \text{ kJ})$
$= -53.6$ kJ

PRACTICE Problems

Have students refer to the Selected Solutions appendix for complete solutions to odd-numbered problems. The complete solutions for all problems can be found in the Solutions Manual.

34. a. $2NO(g) + O_2(g) \rightarrow 2NO_2(g)$
Formation of NO: $N_2 + O_2 \rightarrow 2NO$
Formation of NO_2: $N_2 + 2O_2 \rightarrow 2NO_2$
NO is a reactant in the problem, so add the reversed NO formation equation to the NO_2 formation equation:
$2NO + N_2 + 2O_2 \rightarrow$
$N_2 + O_2 + 2NO_2$
$2NO + O_2 \rightarrow 2NO_2$

b. $H_2(g) + S(s) + 2O_2(g) \rightarrow H_2SO_4(l)$
$SO_3(g) \rightarrow S(s) + \frac{3}{2}O_2(g)$
$H_2O(l) \rightarrow H_2(g) + \frac{1}{2}O_2(g)$
$SO_3(g) + H_2O(l) \rightarrow H_2SO_4(l)$

35. −1398 kJ

36. −2186 kJ

37. 33.2 kJ

The summation equation The stepwise procedure you have just read about shows how standard heats of formation equations combine to produce the desired equation and its ΔH°_{rxn}. The procedure can be summed up in the following formula.

Summation Equation

$$\Delta H^\circ_{rxn} = \Sigma\Delta H^\circ_f(\text{products}) - \Sigma\Delta H^\circ_f(\text{reactants})$$

ΔH°_{rxn} represents the standard enthalpy of the reaction.
Σ represents the sum of the terms.
ΔH°_f(products) and ΔH°_f(reactants) represent the standard enthalpies of formation of all the products and all the reactants.

ΔH°_{rxn} is obtained by subtracting the sum of heats of formation of the reactants from the sum of the heats of formation of the products.

You can see how this formula applies to the reaction between hydrogen sulfide and fluorine.

$$H_2S(g) + 4F_2(g) \rightarrow 2HF(g) + SF_6(g)$$

$$\Delta H^\circ_{rxn} = [(2)\Delta H^\circ_f(HF) + \Delta H^\circ_f(SF_6)] - [\Delta H^\circ_f(H_2S) + (4)\Delta H^\circ_f(F_2)]$$

$$\Delta H^\circ_{rxn} = [(2)(-273 \text{ kJ}) + (-1220 \text{ kJ})] - [-21 \text{ kJ} + (4)(0.0 \text{ kJ})]$$

$$\Delta H^\circ_{rxn} = -1745 \text{ kJ}$$

EXAMPLE Problem 6

EXAMPLE PROBLEM

ENTHALPY CHANGE FROM STANDARD ENTHALPIES OF FORMATION Use standard enthalpies of formation to calculate ΔH°_{rxn} for the combustion of methane.

$$CH_4(g) + 2O_2(g) \rightarrow CO_2(g) + 2H_2O(l)$$

1 ANALYZE THE PROBLEM

You are given an equation and asked to calculate the change in enthalpy. The formula $\Delta H^\circ_{rxn} = \Sigma\Delta H^\circ_{rxn}(\text{products}) - \Sigma\Delta H^\circ_f(\text{reactants})$ can be used with data from **Table R-11**.

Known	**Unknown**
$\Delta H^\circ_f(CO_2) = -394$ kJ	$\Delta H^\circ_{rxn} = ?$ kJ
$\Delta H^\circ_f(H_2O) = -286$ kJ	
$\Delta H^\circ_f(CH_4) = -75$ kJ	
$\Delta H^\circ_f(O_2) = 0.0$ kJ	

2 SOLVE FOR THE UNKNOWN

Use the formula $\Delta H^\circ_{rxn} = \Sigma\Delta H^\circ_f(\text{products}) - \Sigma\Delta H^\circ_f(\text{reactants})$.

Expand the formula to include a term for each reactant and product. Multiply each term by the coefficient of the substance in the balanced chemical equation.

$\Delta H^\circ_{rxn} = [\Delta H^\circ_f(CO_2) + (2)\Delta H^\circ_f(H_2O)] - [\Delta H^\circ_f(CH_4) + (2)\Delta H^\circ_f(O_2)]$

Substitute CO_2 and H_2O for the products, CH_4 and O_2 for the reactants. Multiply H_2O and O_2 by two.

$\Delta H^\circ_{rxn} = [(-394 \text{ kJ}) + (2)(-286 \text{ kJ})] - [(-75 \text{ kJ}) + (2)(0.0 \text{ kJ})]$

Substitute $\Delta H^\circ_f(CO_2) = -394$ kJ, $\Delta H^\circ_f(H_2O) = -286$ kJ, $\Delta H^\circ_f(CH_4) = -75$ kJ, and $\Delta H^\circ_f(O_2) = 0.0$ kJ into the equation.

$\Delta H^\circ_{rxn} = [-966 \text{ kJ}] - [-75 \text{ kJ}] = -966 \text{ kJ} + 75 \text{ kJ} = \mathbf{-891 \text{ kJ}}$

The combustion of 1 mol CH_4 releases 891 kJ.

3 EVALUATE THE ANSWER

All values are accurate to the ones place. Therefore, the answer is correct as stated. The calculated value is the same as that given in **Table 3.** You can check your answer by using the stepwise procedure shown previously.

PRACTICE Problems

Do additional problems. Online Practice

PRACTICE PROBLEMS

34. Show how the sum of enthalpy of formation equations produces each of the following reactions. You do not need to look up and include ΔH values.

a. $2NO(g) + O_2(g) \rightarrow 2NO_2(g)$

b. $SO_3(g) + H_2O(l) \rightarrow H_2SO_4(l)$

35. Use standard enthalpies of formation from **Table R-11** to calculate ΔH°_{rxn} for the following reaction.

$4NH_3(g) + 7O_2(g) \rightarrow 4NO_2(g) + 6H_2O(l)$

36. Determine ΔH°_{comb} for butanoic acid, $C_3H_7COOH(l) + 5O_2(g) \rightarrow 4CO_2(g) + 4H_2O(l)$. Use data in **Table R-11** and the following equation.

$4C(s) + 4H_2(g) + O_2(g) \rightarrow C_3H_7COOH(l)\ \ \Delta H = -534\ kJ$

37. Challenge Two enthalpy of formation equations, **a** and **b,** combine to form the equation for the reaction of nitrogen oxide and oxygen. The product of the reaction is nitrogen dioxide: $NO(g) + \frac{1}{2}O_2(g) \rightarrow NO_2(g)\ \ \Delta H^\circ_{rxn} = -58.1\ kJ$

a. $\frac{1}{2}N_2(g) + \frac{1}{2}O_2(g) \rightarrow NO(g)\ \ \Delta H^\circ_f = 91.3\ kJ$

b. $\frac{1}{2}N_2(g) + O_2(g) \rightarrow NO_2(g)\ \ \Delta H^\circ_f = ?$

What is ΔH°_f for Equation **b**?

SECTION 4 REVIEW

Section Self-Check

Section Summary

- The enthalpy change for a reaction can be calculated by adding two or more thermochemical equations and their enthalpy changes.
- Standard enthalpies of formation of compounds are determined relative to the assigned enthalpy of formation of the elements in their standard states.

38. MAINIDEA Explain what is meant by Hess's law and how it is used to determine ΔH°_{rxn}.

39. Explain in words the formula that can be used to determine ΔH°_{rxn} when using Hess's law.

40. Describe how the elements in their standard states are defined on the scale of standard enthalpies of formation.

41. Examine the data in **Table 5.** What conclusion can you draw about the stabilities of the compounds listed relative to the elements in their standard states? Recall that low energy is associated with stability.

42. Calculate Use Hess's law to determine ΔH for the reaction $NO(g) + O(g) \rightarrow NO_2(g)\ \ \Delta H = ?$ given the following reactions. Show your work.

$O_2(g) \rightarrow 2O(g)\ \ \Delta H = +495\ kJ$

$2O_3(g) \rightarrow 3O_2(g)\ \ \Delta H = -427\ kJ$

$NO(g) + O_3(g) \rightarrow NO_2(g) + O_2(g)\ \ \Delta H = -199\ kJ$

43. Interpret Scientific Illustrations Use the data below to draw a diagram of standard heats of formation similar to **Figure 15** and use your diagram to determine the heat of vaporization of water at 298 K.

Liquid water: $\Delta H^\circ_f = -285.8$ kJ/mol

Gaseous water: $\Delta H^\circ_f = -241.8$ kJ/mol

3 Assess

Check for Understanding

Ask students if elements in their standard states possess zero energy. no Ask, "If that's so, why are elements in their standard states assigned enthalpies of zero?" They are assigned enthalpies of zero in order to establish a standard value that can be compared with the enthalpies of compounds or the enthalpies of elements in nonstandard states. OL EL

Reteach

Ask students how they might use enthalpy of formation data to calculate ΔH for a reaction they cannot perform. Write the equation for the reaction. Then, use the formula $\Delta H^\circ_{rxn} = \Sigma\Delta H^\circ_f(\text{products}) - \Sigma\Delta H^\circ_f(\text{reactants})$. OL

Extension

Give students this data: ΔH°_f(carbon, graphite) = 0.0 kJ/mol, ΔH°_f (carbon, diamond) = 1.9 kJ/mol. Ask them to write the equation for the conversion of graphite to diamond and determine ΔH_{rxn}. $C(graphite) \rightarrow C(diamond)$; $\Delta H_{rxn} = 1.9\ kJ - 0.0\ kJ = 1.9\ kJ$. Ask what is the standard state for carbon. graphite OL

ASSESSMENT

Knowledge Have students write the formation equation for solid copper(II) nitrate. $Cu(s) + N_2(g) + 3O_2(g) \rightarrow Cu(NO_3)_2(s)$ Ask students what the enthalpy change for the equation is called. the enthalpy of formation of solid copper(II) nitrate OL

SECTION 4 REVIEW

38. Refer to the Solutions Manual. Hess's law says that if two or more equations add up to an overall equation, the ΔH°_{rxn} of the overall equation is the sum of the ΔH°_{rxn} values of the equations that were combined.

39. $\Delta H^\circ_{rxn} = \Sigma\Delta H^\circ_f(\text{products}) - \Sigma\Delta H^\circ_f(\text{reactants})$. The enthalpy of reaction under standard conditions (1 atm and 298 K) equals the sum of the standard enthalpies of formation of the products minus the sum of the standard enthalpies of formation of the reactants.

40. They are assigned enthalpies of formation of zero.

41. All compounds listed in Table 5 are more stable than their constituent elements.

42. Refer to the Solutions Manual. $NO(g) + O(g) \rightarrow NO_2(g)$ $\Delta H = -233\ kJ$

43. Diagrams will show liquid water at 285.8 kJ/mol below 0.0 kJ and gaseous water at 241.8 kJ/mol below 0.0 kJ. The heat of vaporization is the energy difference between the two lines or $-241.8\ kJ - (-285.8\ kJ) = 44.0\ kJ$.

SECTION 5

1 Focus

MAINIDEA

How ΔH_{system} Affects the Surroundings From Section 2, have students recall the definition of the universe and how endothermic and exothermic processes affect the surroundings. universe = system + surroundings; In an endothermic process, heat flows from the surroundings into the system. In an exothermic process, heat flows from the system into the surroundings. Explain that a system's ΔH, which is positive for an endothermic process and negative for an exothermic process, plays a significant role in determining whether or not a process is spontaneous because of how it affects the surroundings. OL

2 Teach

Visual Learning

Figure 17 Ask students if painting an iron object makes the rusting reaction shown in the photo non-spontaneous. no Ask how painting an iron object enables it to exist in wet conditions without rusting. A coat of paint prevents the iron from coming into contact with oxygen in the air. OL

SECTION 5

Reaction Spontaneity

MAINIDEA Changes in enthalpy and entropy determine whether a process is spontaneous.

Essential Questions

- What is the difference between spontaneous and nonspontaneous processes?
- How do changes in entropy and free energy determine the spontaneity of chemical reactions and other processes?

Review Vocabulary

vaporization: the energy-requiring process by which a liquid changes to a gas or vapor

New Vocabulary

spontaneous process
entropy
second law of thermodynamics
free energy

CHEM 4 YOU How is it that some newer buildings appear to be falling apart when others that are much older seem to stand forever? It might be the level of maintenance and work put into them. Similarly, in chemistry, without a constant influx of energy, there is a natural tendency toward disorder.

Spontaneous Processes

In **Figure 17** you can see a familiar picture of what happens to an iron object when it is left outdoors in moist air. Iron rusts slowly according to the same chemical equation that describes what happens in the heat pack you read about earlier in the chapter.

$$4Fe(s) + 3O_2(g) \rightarrow 2Fe_2O_3(s) \quad \Delta H = -1625\ kJ$$

The heat pack goes into action the moment you activate it. Similarly, unprotected iron objects rust whether you want them to or not.

Rusting is spontaneous. Any physical or chemical change that once begun, occurs with no outside intervention is a **spontaneous process.** However, for many spontaneous processes, some energy from the surroundings must be supplied to get the process started. For example, you might use a match to light a Bunsen burner in your school lab.

Suppose you reverse the direction of the equation for the rusting of iron. Recall that when you change the direction of a reaction, the sign of ΔH changes. The reaction becomes endothermic.

$$2Fe_2O_3(s) \rightarrow 4Fe(s) + 3O_2(g) \quad \Delta H = 1625\ kJ$$

Reversing the equation will not make rust decompose spontaneously into iron and oxygen under ordinary conditions. The equation represents a reaction that is not spontaneous.

Figure 17 Left unattended, with abundant water and oxygen in the air, the iron in this boat spontaneously converts to rust (Fe_2O_3).

CULTURAL DIVERSITY

Wheat-Straw Ethanol A company in Babilafuente (Salamanca), Spain was the first in the world to implement continuous production of cellulosic ethanol from wheat straw on a commercial scale, roughly five million liters per year. Cellulosic ethanol is ethanol produced from cellulose. Enzymatic catalysts play an important role in the process, prior to fermentation and ethanol recovery. Production of ethanol from cellulose holds great promise as a means of reducing the use of petroleum-based fuels and reducing emissions of greenhouse gases.

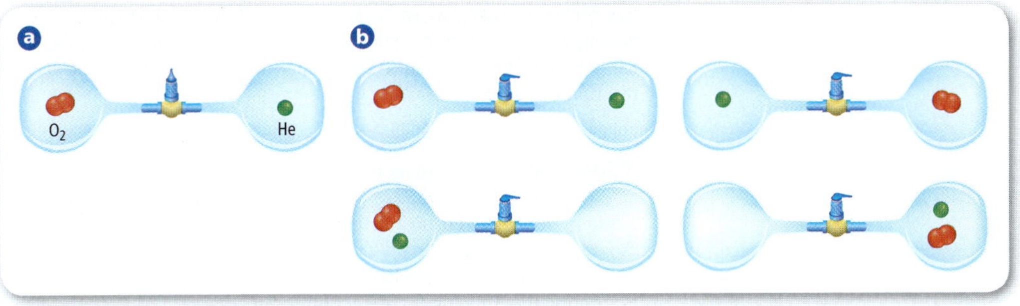

■ **Figure 18** In **a,** an oxygen molecule and a helium atom are each confined to a single bulb. When the stopcock is opened in **b,** the gas particles move freely into the double volume available. Four arrangements of the particles, which represent an increase in entropy, are possible at any given time.

The formation of rust on iron is an exothermic and spontaneous reaction. The reverse reaction is endothermic and nonspontaneous. You might conclude that all exothermic processes are spontaneous and all endothermic processes are nonspontaneous. But remember that ice melting at room temperature is a spontaneous, endothermic process. Something other than ΔH plays a role in determining whether a chemical process occurs spontaneously under a given set of conditions. That something is called entropy.

What is entropy? You're probably not surprised when the smell of brownies baking in the kitchen wafts to wherever you are in your home. And you know that gases tend to spread throughout Earth's atmosphere. Why do gases behave this way? When gases spread out, a system reaches a state of maximum entropy. **Entropy** (S) is a measure of the number of possible ways that the energy of a system can be distributed, and this is related to the freedom of the system's particles to move and number of ways they can be arranged.

Consider the two bulbs in **Figure 18.** When the stopcock is closed, one bulb contains a single molecule of oxygen. The other contains one atom of helium. When the stopcock is opened, the gas particles pass freely between the bulbs. Each gas particle can spread out into twice its original volume. The particles might be found in any of the four arrangements shown. The entropy of the system is greater with the stopcock open because the number of possible arrangements of the particles and the distribution of their energies is increased.

As the number of particles increases, the number of possible arrangements for a group of particles increases dramatically. If the two bulbs contained a total of ten particles, the number of possible arrangements would be 1024 times more than if the particles were confined to a single bulb. In general, the number of possible arrangements available to a system increases under the following conditions: when volume increases, when energy increases, when the number of particles increases, or when the particles' freedom of movement increases.

The second law of thermodynamics The tendency toward increased entropy is summarized in the **second law of thermodynamics,** which states that spontaneous processes always proceed in such a way that the entropy of the universe increases. Entropy is sometimes considered to be a measure of the disorder or randomness of the particles that make up a system. Particles that are more spread out are said to be more disordered, causing the system to have greater entropy than when the particles are closer together.

VOCABULARY

SCIENCE USAGE V. COMMON USAGE

System

Science usage: the particular reaction or process being studied

The universe consists of the system and the surroundings.

Common usage: an organized or established procedure

She worked out a system in which everyone would have an equal opportunity.

Get help with **probability.**

Personal Tutor

Extension

Molecular Motion Ask an interested student to make a poster that depicts the molecular arrangements and motions that result during the following changes, all of which result in higher entropy states: a solid melting, a solid dissolving in a liquid, a liquid evaporating, and a gas at a lower temperature heated to a higher temperature at constant volume. Discuss each of the changes with the class. OL EL

Use Science Terms

Entropy Have students write statements that describe several reactions and processes. Each statement should include the term *entropy.* OL EL

DIFFERENTIATED INSTRUCTION

Visually Impaired Put several drops of peppermint oil in one evaporating dish and several drops of wintergreen in another. Put the dish with the peppermint oil on a hot plate set on low, and put the dish with the oil of wintergreen in a beaker of crushed ice. Put the two dishes equidistant from the class. The odor of peppermint oil reaches students first and is much stronger. OL

Quick Demo

Probability and Disorder Demonstrate the relationship between the number of arrangements and probability by rolling a pair of dice 72 times. After each roll, have a student record the total number of dots on the tops of the two dice in a table with columns for 2 through 12 dots. Calculate the percent occurrence of each of the totals. Because 72 rolls are relatively few trials, your data might only approximate the following theoretical distribution: 2 (2 rolls, 2.8%); 3 (4 rolls, 5.6%); 4 (6 rolls, 8.3%); 5 (8 rolls, 11.1%); 6 (10 rolls, 13.9%); 7 (12 rolls, 16.7%); 8 (10 rolls, 13.9%); 9 (8 rolls, 11.1%); 10 (6 rolls, 8.3%); 11 (4 rolls, 5.6%); 12 (2 rolls, 2.8%). Point out that rolling two dice simultaneously yields 36 discrete outcomes, and the probable percent occurrence of each total depends upon the number of arrangements of the two dice that can yield that total. For example, 7 can occur by six combinations:
die #1 = 6 and die #2 = 1;
die #1 = 5 and die #2 = 2;
die #1 = 4 and die #2 = 3;
die #1 = 3 and die #2 = 4;
die #1 = 2 and die #2 = 5;
die #1 = 1 and die #2 = 6.
Explain that disordered systems of atoms, molecules, and ions have many more possible arrangements than ordered ones. Accordingly, higher-entropy arrangements are more probable than lower-entropy ones. OL

Predicting changes in entropy Recall that the change in enthalpy for a reaction is equal to the enthalpy of the products minus the enthalpy of the reactants. The change in entropy (ΔS) during a reaction or process is similar.

$$\Delta S_{system} = S_{products} - S_{reactants}$$

If the entropy of a system increases during a reaction or process, $S_{products} > S_{reactants}$ and ΔS_{system} is positive. Conversely, if the entropy of a system decreases during a reaction or process, $S_{products} < S_{reactants}$ and ΔS_{system} is negative.

You can sometimes predict if ΔS_{system} is positive or negative by examining the equation for a reaction or process.

1. *Entropy changes associated with changes in state can be predicted.* In solids, molecules have limited movement. In liquids, they have some freedom to move, and in gases, molecules can move freely within their container. Thus, entropy increases as a substance changes from a solid to a liquid and from a liquid to a gas. ΔS_{system} is positive as water vaporizes and methanol melts.

$$H_2O(l) \rightarrow H_2O(g) \quad \Delta S_{system} > 0$$

$$CH_3OH(s) \rightarrow CH_3OH(l) \quad \Delta S_{system} > 0$$

2. *The dissolving of a gas in a solvent always results in a decrease in entropy.* Gas particles have more entropy when they can move freely than when they are dissolved in a liquid or solid that limits their movements and randomness. ΔS_{system} is negative for the dissolving of oxygen in water as shown in **Figure 19.**

$$O_2(g) \rightarrow O_2(aq) \quad \Delta S_{system} < 0$$

3. *Assuming no change in physical state occurs, the entropy of a system usually increases when the number of gaseous product particles is greater than the number of gaseous reactant particles.* For the following reaction, ΔS_{system} is positive because two molecules of gas react and three molecules of gas are produced.

$$2SO_3(g) \rightarrow 2SO_2(g) + O_2(g) \quad \Delta S_{system} > 0$$

VOCABULARY
WORD ORIGIN
Random
comes from the Germanic word *rinnan,* meaning *to run;* a haphazard course

Figure 19 In the bubbles, the nitrogen and oxygen gas molecules that make up most of the air can move more freely than when dissolved in the aquarium water.

DIFFERENTIATED INSTRUCTION

Struggling Learners Illustrate the increase in entropy that occurs when sodium chloride dissolves in water by first showing students a model of the low-entropy, face-centered cubic NaCl crystal lattice. Then, draw a diagram on the board of a beaker of water showing the Na^+ and Cl^- ions distributed randomly throughout the beaker. Explain that the highly disordered arrangement of the ions in solution represents a higher state of entropy than the rigidly organized ions in the solid state. BL EL

■ **Figure 20** Sodium chloride and liquid water are pure substances, each with a degree of orderliness. When sodium chloride dissolves in water, the entropy of the system increases because sodium ions, chloride ions, and water molecules mix together to create a large number of random arrangements.

4. *With some exceptions, entropy increases when a solid or a liquid dissolves in a solvent.* The solute particles, which are together before dissolving, become dispersed throughout the solvent. The solute particles have more freedom of movement, as shown in **Figure 20** for the dissolving of sodium chloride in water. ΔS_{system} is positive.

$$NaCl(s) \rightarrow Na^{+}(aq) + Cl^{-}(aq) \quad \Delta S_{system} > 0$$

5. *The random motion of the particles of a substance increases as its temperature increases.* Recall that the kinetic energy of molecules increases with temperature. Increased kinetic energy means faster movement and more possible arrangements of particles. Therefore, the entropy of any substance increases as its temperature increases. ΔS_{system} is positive.

PRACTICE Problems Do additional problems. Online Practice

44. Predict the sign of ΔS_{system} for each of the following changes.

a. $ClF(g) + F_2(g) \rightarrow ClF_3(g)$

b. $NH_3(g) \rightarrow NH_3(aq)$

c. $CH_3OH(l) \rightarrow CH_3OH(aq)$

d. $C_{10}H_8(l) \rightarrow C_{10}H_8(s)$

45. Challenge Comment on the sign of ΔS_{system} for the following reaction.
$Fe(s) + Zn^{2+}(aq) \rightarrow Fe^{2+}(aq) + Zn(s)$

Connection to Earth Science Earth's spontaneous processes Volcanoes, fumaroles, hot springs, and geysers are evidence of geothermal energy in Earth's interior. Volcanoes are vents in Earth's crust from which molten rock (magma), steam, and other materials flow. When surface water moves downward through Earth's crust, it can interact with magma and/or hot rocks. Water that comes back to the surface in hot springs is heated to temperatures much higher than the surrounding air temperatures. Geysers are hot springs that spout hot water and steam into the air. Fumaroles emit steam and other gases, such as hydrogen sulfide. These geothermal processes are obviously spontaneous. Can you identify increases in entropy in these Earth processes?

DIFFERENTIATED INSTRUCTION

Advanced Learners The entropy of a system usually increases when a solid or liquid dissolves to form a solution. However, explain to capable students that some solutes form highly charged ions in solution, which can result in extensive clustering of water molecules in ordered patterns around the ions and a resulting decrease in entropy. AL

PRACTICE Problems

Have students refer to the Selected Solutions appendix for complete solutions to odd-numbered problems. The complete solutions for all problems can be found in the Solutions Manual.

44. a. negative
b. negative
c. negative
d. negative

45. The states of the two reactants are the same on both sides of the equation, so it is impossible from the equation alone to predict the sign of ΔS_{system}.

Enrichment

State of Disorder Explain to students that the tendencies of reactions to occur from states of high to low energy and from states of low entropy to high entropy are manifestations of the second law of thermodynamics. The second law of thermodynamics states that in any spontaneous process, the entropy of the universe always increases. Point out that the tendency of the reaction states to change from high to low energy exists because the energy lost by the system is gained by the surroundings, thereby increasing the thermal entropy of the surroundings.

■ **In-Text Question** Volcanoes: When the magma flows and steam escapes from volcano vents, the particles have more freedom of motion, increasing the system entropy. Hot springs and geysers: When surface water makes its way down to magma and hot rocks, the water is heated and possibly changes state to a vapor, increasing its thermal energy and the system entropy.

ASSESSMENT

Skill Give students the data in the following table.

Ice Melting at Two Temperatures		
T (K)	ΔH_{fus} (J/mol)	ΔS_{fus} (J/mol)
272	6011	22.06
274	6011	22.06

The table lists quantities associated with ice melting at −1°C (272 K) and +1°C (274 K). Ask why the sign of ΔH_{fus} is positive. because the process is endothermic Ask why the sign of ΔS_{fus} is positive. because the system's entropy increases during melting Have students use the formula $\Delta G = \Delta H - T\Delta S$ to calculate ΔG for ice melting at the two temperatures. At 272 K, $\Delta G = 11$ J/mol. At 274 K, $\Delta G = -33$ J/mol. Have students use their results to explain why ice melts at 274 K but not at 272 K. At 274 K, the negative ΔG means that the entropy of the universe increases; therefore, the process is spontaneous. At 272 K, the positive ΔG means that the entropy of the universe decreases; therefore, the process is not spontaneous. **OL**

FOLDABLES® Rubric

Figure 21 It is difficult to recognize this ancient Greek sculpture as the head of a lion. The particles of limestone that are loosened by wind and weather or dissolved by rain disperse randomly, destroying the precise representation of the image and increasing the entropy of the universe.

FOLDABLES® Incorporate information from this section into your Foldable.

Entropy, the Universe, and Free Energy

If you happen to break an egg, you know you cannot reverse the process and again make the egg whole. Similarly, an abandoned barn gradually disintegrates into a pile of decaying wood and a statue dissolves slowly in rainwater and disperses into the ground, as shown in **Figure 21.** Order turns to disorder in these processes, and the entropy of the universe increases.

What effect does entropy have on reaction spontaneity? Recall that the second law of thermodynamics states that the entropy of the universe must increase as a result of a spontaneous reaction or process. Therefore, the following is true for any spontaneous process.

$$\Delta S_{\text{universe}} > 0$$

Because the universe equals the system plus the surroundings, any change in the entropy of the universe is the sum of changes occurring in the system and surroundings.

$$\Delta S_{\text{universe}} = \Delta S_{\text{system}} + \Delta S_{\text{surroundings}}$$

In nature, $\Delta S_{\text{universe}}$ tends to be positive for reactions and processes under the following conditions.

1. *The reaction or process is exothermic, which means ΔH_{system} is negative.* The heat released by an exothermic reaction raises the temperature of the surroundings and thereby increases the entropy of the surroundings. $\Delta S_{\text{surroundings}}$ is positive.
2. *The entropy of the system increases, so ΔS_{system} is positive.*

Thus, exothermic chemical reactions accompanied by an increase in entropy are all spontaneous.

Free energy Can you definitely determine if a reaction is spontaneous? In 1878, J. Willard Gibbs, a physicist at Yale University, defined a combined enthalpy-entropy function called Gibbs free energy that answers that question. For reactions or processes that take place at constant pressure and temperature, Gibbs free energy (G_{system}), commonly called **free energy,** is energy that is available to do work. Thus, free energy is useful energy. In contrast, some entropy is associated with energy that is spread out into the surroundings as, for example, random molecular motion, and cannot be recovered to do useful work. The free energy change (ΔG_{system}) is the difference between the system's change in enthalpy (ΔH_{system}) and the product of the kelvin temperature and the change in entropy ($T\Delta S_{\text{system}}$).

Gibbs Free Energy Equation

$$\Delta G_{\text{system}} = \Delta H_{\text{system}} - T\Delta S_{\text{system}}$$

ΔG_{system} represents the free energy change. ΔH_{system} represents the change in enthalpy. T is temperature. ΔS_{system} represents the change in entropy.

The free energy released or absorbed in a chemical reaction is equal to the difference between the enthalpy change and the product of the change in entropy (in joules per kelvin) and the temperature (in kelvins).

To calculate Gibbs free energy, it is usually necessary to convert units because ΔS is usually expressed in J/K, whereas ΔH is expressed in kJ.

CHEMISTRY JOURNAL

The Work of J. Willard Gibbs Have students research the work of J. Willard Gibbs (1839–1903), an American scientist born in New Haven, Connecticut. In their chemistry journals, have them summarize Gibbs's contributions to our understanding of energy and entropy associated with processes and reactions. **OL**

The sign of free energy When a reaction or process occurs under standard conditions (298 K and 1 atm), the standard free energy change can be expressed as follows.

$$\Delta G^\circ_{system} = \Delta H^\circ_{system} - T\Delta S^\circ_{system}$$

If the sign of the free energy change (ΔG°_{system}) is negative, the reaction is spontaneous. If the sign of the free energy change is positive, the reaction is nonspontaneous.

Recall that free energy is energy that is available to do work. In contrast, energy related to entropy is useless because it is dispersed and cannot be harnessed to do work.

Calculating free energy change How do changes in enthalpy and entropy affect free energy change and spontaneity for the reaction between nitrogen and hydrogen to form ammonia?

$$N_2(g) + 3H_2(g) \rightarrow 2NH_3(g)$$

$$\Delta H^\circ_{system} = -91.8 \text{ kJ} \quad \Delta S^\circ_{system} = -197 \text{ J/K}$$

The entropy of the system decreases because 4 mol of gaseous molecules react and only 2 mol of gaseous molecules are produced. Therefore, ΔS°_{system} is negative. A decrease in the entropy of the system tends to make the reaction nonspontaneous, but the reaction is exothermic (ΔH°_{system} is negative), which tends to make the reaction spontaneous. To determine which of the two tendencies predominates, you must calculate ΔG°_{system} for the reaction. First, convert ΔS°_{system} to kilojoules.

$$\Delta S^\circ_{system} = -197 \cancel{J}/K \times \frac{1 \text{ kJ}}{1000 \cancel{J}} = -0.197 \text{ kJ/K}$$

Now, substitute ΔH°_{system}, T, and ΔS°_{system} into the equation for ΔG°_{system}.

$$\Delta G^\circ_{system} = \Delta H^\circ_{system} - T\Delta S^\circ_{system}$$

$$\Delta G^\circ_{system} = -91.8 \text{ kJ} - (298 \cancel{K})(-0.197 \text{ kJ}/\cancel{K})$$

$$\Delta G^\circ_{system} = -91.8 \text{ kJ} + 58.7 \text{ kJ} = -33.1 \text{ kJ}$$

ΔG°_{system} for this reaction is negative, so the reaction is spontaneous.

The reaction between nitrogen and hydrogen demonstrates that the entropy of a system can decrease during a spontaneous process. However, it can do so only if the entropy of the surroundings increases more than the entropy of the system decreases. Thus, the entropy of the universe (system + surroundings) always increases in any spontaneous process. **Table 6** shows how reaction spontaneity depends on the signs of ΔH_{system} and ΔS_{system}.

VOCABULARY

ACADEMIC VOCABULARY

Demonstrate
to show clearly
People are standing by to demonstrate how the device works.

Explore **reaction spontaneity with an interactive table.** Concepts In Motion

Table 6 Reaction Spontaneity $\Delta G_{system} = \Delta H_{system} - T\Delta S_{system}$

ΔH_{system}	ΔS_{system}	ΔG_{system}	Reaction Spontaneity
negative	positive	always negative	always spontaneous
negative	negative	negative or positive	spontaneous at lower temperatures
positive	positive	negative or positive	spontaneous at higher temperatures
positive	negative	always positive	never spontaneous

IN-CLASS Example

Question A reaction has a ΔH value of -28.8 kJ and a ΔS value of -54.0 J/K. Is the reaction spontaneous at 298 K?

Answer The reaction is spontaneous at 298 K.
$(-28.8 \text{ kJ})(1000 \text{ J/kJ}) = -28{,}800 \text{ J}$
$\Delta G_{system} = -28{,}800 \text{ J} - (298 \text{ K})(-54.0 \text{ J/K}) = -12{,}700 \text{ J}$. Because ΔG_{system} is negative, the reaction is spontaneous at 298 K.

ASSESSMENT

Knowledge Ask students to define the circumstances under which an exothermic reaction with a negative ΔS_{system} is spontaneous. The reaction is spontaneous only if the sum of ΔH_{system} and $-T\Delta S_{system}$ yields a negative value. OL

CHEMISTRY PROJECT

Predicting Entropy Changes Have students write a balanced chemical equation for each of the following types of changes and chemical reactions. Have them create a table in which they include the type of change or reaction, the balanced equation, the predicted change in entropy, and the predicted sign of ΔS_{system}.

1. A solid dissolves in a solvent. entropy increases; ΔS_{system} is +
2. A gas dissolves in a solvent. entropy decreases; ΔS_{system} is −
3. The number of gaseous reactant particles exceeds the number of gaseous product particles. entropy decreases; ΔS_{system} is −
4. A substances changes from the liquid state to the gaseous state. entropy increases; ΔS_{system} is + OL

PRACTICE Problems

Have students refer to the Selected Solutions appendix for complete solutions to odd-numbered problems. The complete solutions for all problems can be found in the Solutions Manual.

46. a. spontaneous reaction
b. nonspontaneous reaction
c. nonspontaneous reaction
d. nonspontaneous reaction

47. spontaneous above 3910 K

3 Assess

Check for Understanding

Give students this equation for the formation of silicon carbide from sand and coke: 624.7 kJ + $SiO_2(s)$ + $3C(s) \rightarrow SiC(s) + 2CO(g)$. Ask them to write the system's enthalpy change and predict the sign of the entropy change. $\Delta H^\circ_{rxn} = 624.7$ kJ, ΔS_{system} is positive. **OL** **EL**

Reteach

Burn a wood splint and ask students how the entropies of the system and the surroundings change. Both increase. The reaction is exothermic, increasing the thermal entropy of the surroundings. The ashes have more entropy than the wood. **OL**

Extension

Have students observe chemical and physical processes in their surroundings, list and classify each process as spontaneous or nonspontaneous, the flow of energy in the process and whether or not the process increases the entropy of the universe. **OL** **EL**

EXAMPLE Problem 7

EXAMPLE PROBLEM

DETERMINE REACTION SPONTANEITY For a process, ΔH_{system} = 145 kJ and ΔS_{system} = 322 J/K. Is the process spontaneous at 382 K?

1 ANALYZE THE PROBLEM

You must calculate ΔG_{system} to determine spontaneity.

Known	Unknown
T = 382 K	sign of ΔG_{system} = ?
ΔH_{system} = 145 kJ	
ΔS_{system} = 322 J/K	

2 SOLVE FOR THE UNKNOWN

Convert ΔS_{system} to kJ/K

$322\ \text{J/K} \times \frac{1\ \text{kJ}}{1000\ \text{J}} = 0.322\ \text{kJ/K}$ — Convert ΔS_{system} to kJ/K.

Solve the free energy equation.

$\Delta G_{system} = \Delta H_{system} - T\Delta S_{system}$ — State the Gibbs free energy equation.

$\Delta G_{system} = 145\ \text{kJ} - (382\ \text{K})(0.322\ \text{kJ/K})$ — Substitute T = 382 K, ΔH_{system} = 145 kJ, and ΔS_{system} = 0.322 kJ/K

$\Delta G_{system} = 145\ \text{kJ} - 123\ \text{kJ} = 22\ \text{kJ}$ — Multiply and subtract numbers.

Because ΔG_{system} is positive, the reaction is nonspontaneous.

3 EVALUATE THE ANSWER

Because ΔH is positive and the temperature is not high enough to make the second term of the equation greater than the first, ΔG_{system} is positive. The significant figures are correct.

PRACTICE Problems

Do additional problems. Online Practice

46. Determine whether each of the following reactions is spontaneous.

a. $\Delta H_{system} = -75.9$ kJ, T = 273 K, ΔS_{system} = 138 J/K
b. $\Delta H_{system} = -27.6$ kJ, T = 535 K, $\Delta S_{system} = -55.2$ J/K
c. ΔH_{system} = 365 kJ, T = 388 K, $\Delta S_{system} = -55.2$ J/K
d. ΔH_{system} = 452 kJ, T = 165 K, ΔS_{system} = 55.7 J/K

47. Challenge Given $\Delta H_{system} = -144$ kJ and $\Delta S_{system} = -36.8$ J/K for a reaction, determine the lowest temperature in kelvins at which the reaction would be spontaneous.

SECTION 5 REVIEW

Section Self-Check

Section Summary

- Entropy is a measure of the disorder or randomness of a system.
- Spontaneous processes always result in an increase in the entropy of the universe.
- Free energy is the energy available to do work. The sign of the free energy change indicates whether the reaction is spontaneous.

48. MAIN IDEA Compare and contrast spontaneous and nonspontaneous reactions.

49. Describe how a system's entropy changes if the system becomes more disordered during a process.

50. Decide Does the entropy of a system increase or decrease when you dissolve a cube of sugar in a cup of tea? Define the system, and explain your answer.

51. Determine whether the system $\Delta H_{system} = -20.5$ kJ, T = 298 K, and $\Delta S_{system} = -35.0$ J/K is spontaneous or nonspontaneous.

52. Outline Use the blue and red headings to outline the section. Under each heading, summarize the important ideas discussed.

SECTION 5 REVIEW

48. A reaction occurs spontaneously only when the temperature, entropy change within the system, and energy exchanged between the system and surroundings cause the entropy of the universe to increase.

49. The system's entropy increases.

50. The system's entropy increases. The system consists of the sugar and tea. Randomness or disorder increases as sugar molecules, which were originally locked into position in the solid structure of the sugar cube, disperse throughout the tea.

51. The system is spontaneous.

52. Students outlines should include all important ideas expressed in the Section Summary.

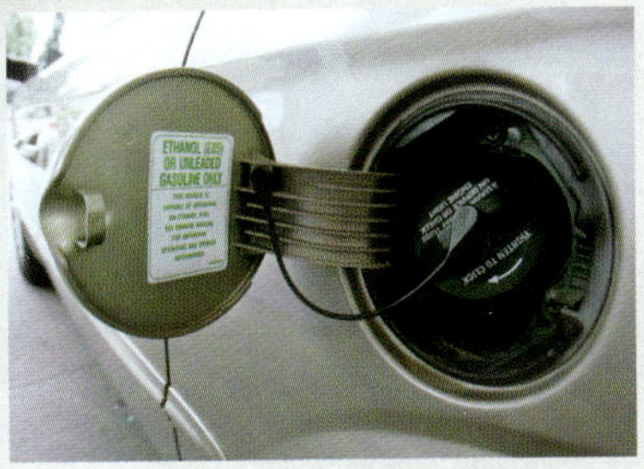

DRIVING THE FUTURE: Flexible Fuel Vehicles

The service stations of the not-too-distant future will not only deliver various grades of gasoline, but they will also pump a fuel called E85. This fuel can be used in a flexible-fuel vehicle, or FFV. Conventional vehicles operate on 100% gasoline or on a blend of 10% ethanol and 90% gasoline. FFVs, however, operate on all these blends and E85, which is 85% ethanol. E85 has the advantage of not being highly dependent on fossil fuels.

1 Renewable Resource E85 is 15% gasoline and 85% ethanol by volume. Ethanol (C_2H_5OH) is a renewable fuel that can be produced domestically.

2 Environmental Benefit Compared with gasoline, burning E85 can reduce emissions of greenhouse gases such as carbon dioxide and nitrogen oxides.

3 Combustion Requirements The FFV engine burning E85 requires a richer mixture (more fuel, less air) than for an equal volume of gasoline. The FFV fuel injectors, therefore, must be able to inject up to 30% more fuel.

4 Damage Prevention The ethanol content of E85 is high enough to damage some of the material used in the construction of conventional vehicles. Therefore, the FFV fuel tank is made of stainless steel. The fuel lines are also made of stainless steel or lined with nonreactive materials.

WRITING IN ▶ Chemistry

Write thermochemical equations for the complete combustion of 1 mol octane (C_8H_{18}), a component of gasoline, and 1 mol ethanol (ΔH_{comb} of C_8H_{18} = −5471 kJ/mol; ΔH_{comb} of C_2H_5OH = −1367 kJ/mol). Which releases the greater amount of energy per mole of fuel? Which releases more energy per kilogram of fuel? Discuss the significance of your findings.

WebQuest

WRITING IN ▶ Chemistry

Rubric WebQuest

Combustion of ethanol: $C_2H_5OH(l) + 3O_2(g) \rightarrow 2CO_2(g) + 3H_2O(g)$ $\Delta H°_{comb} = -1367$ kJ;
Combustion of octane: $C_8H_{18}(l) + 12.5O_2(g) \rightarrow 8CO_2(g) + 9H_2O(g)$ $\Delta H°_{comb} = -5471$ kJ;
Octane releases more energy per mole. Ethanol's molar mass is 46.07g. (−1367 kJ ÷ 0.04607 kg = −29,670 kJ/kg); Octane's molar mass is 114.23 g. (−5471 kJ ÷ 0.11423 kg = −47,890 kJ/kg); Octane releases more energy per kilogram. Because the combustion of ethanol releases less energy than the combustion of an equal mass (and volume) of gasoline, the FFV's miles per gallon (mpg) will be less on E85 than on 100% gasoline.

HOW IT works

Purpose

Students will learn how flexible-fuel vehicles (FFVs) can utilize fuels ranging from 100% gasoline to E85, which is 15% gasoline and 85% ethanol by volume.

Background

Gasoline is a mixture of hydrocarbons derived from petroleum. Ethanol, also called grain alcohol, is water-soluble and biodegradable. E85 spills, therefore, pose less threat to the environment than gasoline spills.

Ethanol can damage the fuel systems of conventional vehicles. To avoid this damage, FFV fuel lines are made of stainless steel or lined with a plastic–polytetrofluoroethylene (PTFE).

In most FFVs, an oxygen sensor called a lambda probe analyzes the exhaust gases and sends the information to an on-board computer that adjusts the fuel-to-air ratio and spark timing for optimum operation.

Estimated reductions of greenhouse gases when using E85 vary widely and depend on vehicle technologies, ethanol feedstocks, and ethanol production methods.

Teaching Strategies

- Have students interpret the illustration to answer the following question: How are FFV fuel injectors different from those in conventional vehicles? FFV fuel injectors must be able to inject more fuel.
- Lead students in a discussion of the economic and environmental factors related to burning E85.
- Use this feature to review stoichiometry. Have students calculate the fuel-oxygen mass ratios for the two combustion reactions. The mass ratio of octane to oxygen for complete combustion is 1:3.502; the mass ratio of ethanol to oxygen is 1:2.0838.

ChemLAB

Rubric

Preparation

Time Allotment one class period

Process Skills observe, measure, infer

Safety Precautions Approve lab safety forms before work begins. Students should wear aprons and goggles. The exothermic reaction produces flames. Students with long hair should tie their hair back.

Disposal The ash can be placed in a receptacle for solid waste. The soot can be washed from the beaker with soap and water.

Alternative Materials Similar snack products, such as corn chips, can be used instead of potato chips.

Procedure

- Point out that the soot on the beaker is unburned carbon. It accounts for a large percent error.
- **Troubleshooting** Flames will rise 10 cm above the top of the potato chip.

Analyze and Conclude

1. The reaction is exothermic because heat and light were visible and the temperature of the water increased.
2. The potato chip reacts with oxygen from the air to produce carbon dioxide gas, water vapor, and unburned carbon. The chip was not completely consumed as indicated by the presence of soot and ash.
3. 50.26 g water, 15.3°C, 3220 J/chip
4. 0.770 Calories
5. 28 g per serving (0.770 Calories/1 chip) × (1 chip/1.63 g) × 28 g/1 serving = 13.2 Calories
6. 75 Calories per serving, but answers will vary according to the chips used (75 Calories − 13.2 Calories)/75 Calories × 100 = 82% error.

ChemLAB

Measure Calories

Background: The burning of a potato chip releases heat stored in the substances contained in the chip. Using calorimetry, you will approximate the amount of energy contained in a potato chip.

Question: *How many Calories are in a potato chip?*

Materials

large potato chip or other snack food
250-mL beaker
100-mL graduated cylinder
evaporating dish
nonmercury thermometer
ring stand with ring
wire gauze
matches
stirring rod
balance

Safety Precautions

WARNING: ***Hot objects might not appear to be hot. Do not heat broken, chipped, or cracked glassware. Tie back long hair. Do not eat any items used in the lab.***

Procedure

1. Read and complete the lab safety form.
2. Measure the mass of a potato chip and record it in a data table.
3. Place the potato chip in an evaporating dish on the metal base of the ring stand. Position the ring and wire gauze so that they will be 10 cm above the top of the potato chip.
4. Measure the mass of an empty 250-mL beaker and record it in your data table.
5. Using a graduated cylinder, measure 50 mL of water and pour it into the beaker. Measure the mass of the beaker and water and record it in your data table.
6. Measure and record the initial temperature of the water.
7. Place the beaker on the wire gauze on the ring stand. Use a match to ignite the bottom of the potato chip.
8. Gently stir the water in the beaker while the chip burns. Measure and record the highest temperature attained by the water.
9. **Cleanup and Disposal** Wash all lab equipment and return it to its designated place.

Analyze and Conclude

1. **Classify** Is the reaction exothermic or endothermic? Explain how you know.
2. **Observe and Infer** Describe the reactant and products of the chemical reaction. Was the reactant (potato chip) completely consumed? What evidence supports your answer?
3. **Calculate** Determine the mass of the water and its temperature change. Use the equation $q = c \times m \times \Delta T$ to calculate how much heat, in joules, was transferred to the water by the burning of the chip.
4. **Calculate** Convert the quantity of heat from joules/chip to Calories/chip.
5. **Calculate** From the information on the chip container, determine the mass in grams of one serving. Determine how many Calories are contained in one serving. Use your data to calculate the number of Calories released by the combustion of one serving.
6. **Error Analysis** Compare your calculated Calories per serving with the value on the chip's container. Calculate the percent error.
7. **Calculate** a class average to compare to the chip container. Why would more data lead to more precise results?

INQUIRY EXTENSION

Predict Do all potato chips have the same number of calories? Make a plan to test several different brands of chips.

INQUIRY EXTENSION

Answers will vary. All experiments should include the elements of the ChemLab including safety and testing precautions.

CHAPTER 15 STUDY GUIDE

Vocabulary Practice

BIGIDEA Chemical reactions usually absorb or release energy.

SECTION 1 Energy

MAINIDEA Energy can change form and flow, but it is always conserved.

- Energy is the capacity to do work or produce heat.
- Chemical potential energy is energy stored in the chemical bonds of a substance by virtue of the arrangement of the atoms and molecules.
- Chemical potential energy is released or absorbed as heat during chemical processes or reactions.

$$q = c \times m \times \Delta T$$

VOCABULARY
- energy
- law of conservation of energy
- chemical potential energy
- heat
- calorie
- joule
- specific heat

SECTION 2 Heat

MAINIDEA The enthalpy change for a reaction is the enthalpy of the products minus the enthalpy of the reactants.

- In thermochemistry, the universe is defined as the system plus the surroundings.
- The heat lost or gained by a system during a reaction or process carried out at constant pressure is called the change in enthalpy (ΔH).
- When ΔH is positive, the reaction is endothermic. When ΔH is negative, the reaction is exothermic.

VOCABULARY
- calorimeter
- thermochemistry
- system
- surroundings
- universe
- enthalpy
- enthalpy (heat) of reaction

SECTION 3 Thermochemical Equations

MAINIDEA Thermochemical equations express the amount of heat released or absorbed by chemical reactions.

- A thermochemical equation includes the physical states of the reactants and products and specifies the change in enthalpy.
- The molar enthalpy (heat) of vaporization, ΔH_{vap}, is the amount of energy required to evaporate one mole of a liquid.
- The molar enthalpy (heat) of fusion, ΔH_{fus}, is the amount of energy needed to melt one mole of a solid.

VOCABULARY
- thermochemical equation
- enthalpy (heat) of combustion
- molar enthalpy (heat) of vaporization
- molar enthalpy (heat) of fusion

SECTION 4 Calculating Enthalpy Change

MAINIDEA The enthalpy change for a reaction can be calculated using Hess's law.

- The enthalpy change for a reaction can be calculated by adding two or more thermochemical equations and their enthalpy changes.
- Standard enthalpies of formation of compounds are determined relative to the assigned enthalpy of formation of the elements in their standard states.

$$\Delta H^\circ_{rxn} = \Sigma\ \Delta H^\circ_f(\text{products}) - \Sigma \Delta H^\circ_f(\text{reactants})$$

VOCABULARY
- Hess's law
- standard enthalpy (heat) of formation

SECTION 5 Reaction Spontaneity

MAINIDEA Changes in enthalpy and entropy determine whether a process is spontaneous.

- Entropy is a measure of the disorder or randomness of a system.
- Spontaneous processes always result in an increase in the entropy of the universe.
- Free energy is the energy available to do work. The sign of the free energy change indicates whether the reaction is spontaneous.

$$\Delta G_{system} = \Delta H_{system} - T\Delta S_{system}$$

VOCABULARY
- spontaneous process
- entropy
- second law of thermodynamics
- free energy

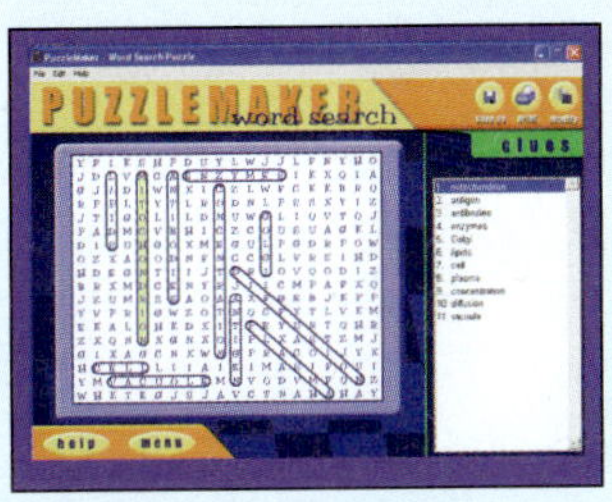

Vocabulary Practice

For additional practice with vocabulary, have students access the Vocabulary PuzzleMaker.

CHAPTER 15 STUDY GUIDE

Use the Vocabulary

To reinforce chapter vocabulary, have students write a sentence using each term. OL EL

Review Strategies

- Ask students to relate chemical potential energy to exothermic and endothermic reactions. OL
- Have students describe the steps involved in writing thermochemical equations. OL
- Have students explain the significance and use of Hess's law in determining the enthalpies of chemical reactions. OL

Students can review with these online resources.

Vocabulary Practice

Vocabulary eGames
Vocabulary eFlashcards
Vocabulary PuzzleMaker

Section Self-Check

Chapter Self-Check

Online Test Practice

Use *eAssessment* to:
- create multiple versions of tests
- edit existing questions and add your own questions
- build tests aligned with select state curriculum standards using built-in state curriculum tags
- track students' progress

CHAPTER 15 ASSESSMENT

SECTION 1

Mastering Concepts

53. Compare and contrast temperature and heat.

54. How does the chemical potential energy of a system change during an endothermic reaction?

55. Describe a situation that illustrates potential energy changing to kinetic energy.

56. Cars How is the energy in gasoline converted and released when it burns in an automobile engine?

57. Nutrition How does the nutritional Calorie compare with the calorie? What is the relationship between the Calorie and a kilocalorie?

58. What quantity has the units J/(g•°C)?

■ **Figure 22**

59. Describe what might happen in **Figure 22** when the air above the surface of the lake is colder than the water.

60. Ethanol has a specific heat of 2.44 J/(g•°C). What does this mean?

61. Explain how the amount of energy required to raise the temperature of an object is calculated.

Mastering Problems

62. Nutrition A food item contains 124 nutritional Calories. How many calories does the food item contain?

63. How many joules are absorbed in a process that absorbs 0.5720 kcal?

64. Transportation Ethanol is being used as an additive to gasoline. The combustion of 1 mol of ethanol releases 1367 kJ of energy. How many Calories are released?

65. To vaporize 2.00 g of ammonia, 656 calories are required. How many kilojoules are required to vaporize the same mass of ammonia?

66. One mole of ethanol releases 326.7 Calories of energy during combustion. How many kilojoules are released?

67. Metallurgy A 25.0-g bolt made of an alloy absorbed 250 J of heat as its temperature changed from 25.0°C to 78.0°C. What is the specific heat of the alloy?

SECTION 2

Mastering Concepts

68. Why is a foam cup used in a student calorimeter rather than a typical glass beaker?

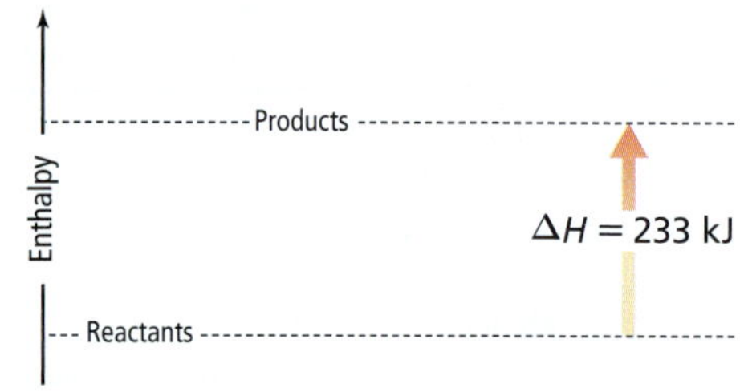

■ **Figure 23**

69. Is the reaction shown in **Figure 23** endothermic or exothermic? How do you know?

70. Give two examples of chemical systems and define the universe in terms of those examples.

71. Under what condition is the heat (q) evolved or absorbed in a chemical reaction equal to a change in enthalpy (ΔH)?

72. The enthalpy change for a reaction, ΔH, is negative. What does this indicate about the chemical potential energy of the system before and after the reaction?

73. What is the sign of ΔH for an exothermic reaction? An endothermic reaction?

Mastering Problems

74. How many joules of heat are lost by 3580 kg of granite as it cools from 41.2°C to −12.9°C? The specific heat of granite is 0.803 J/(g•°C).

75. Swimming Pool A swimming pool measuring 20.0 m × 12.5 m is filled with water to a depth of 3.75 m. If the initial temperature is 18.4°C, how much heat must be added to the water to raise its temperature to 29.0°C? Assume that the density of water is 1.000 g/mL.

76. How much heat is absorbed by a 44.7-g piece of lead when its temperature increases by 65.4°C?

77. Food Preparation When 10.2 g of canola oil at 25.0°C is placed in a wok, 3.34 kJ of heat is required to heat it to a temperature of 196.4°C. What is the specific heat of the canola oil?

78. Alloys When a 58.8-g piece of hot alloy is placed in 125 g of cold water in a calorimeter, the temperature of the alloy decreases by 106.1°C, while the temperature of the water increases by 10.5°C. What is the specific heat of the alloy?

Wesley Hitt/Alamy

SECTION 1

Mastering Concepts

53. Heat is a form of energy that flows from a warmer object to a cooler object. Temperature is a measure of the average kinetic energy of the particles in a sample of matter.

54. It increases.

55. Student answers will vary. A typical answer is: During an avalanche, the potential energy of snow at a higher altitude is converted to kinetic energy as the snow cascades down a mountain.

56. Some is converted to work to move pistons in the engine; much is released as heat.

57. One nutritional Calorie equals 1000 calories. One nutritional Calorie equals 1 kilocalorie.

58. specific heat

59. If the air is cool enough, water vapor from the lake might condense and form fog. Heat will be transferred from the warmer water to the cooler air. The air immediately above the water will be slightly warmer than the surrounding air, and the fog might appear to rise off the lake somewhat like steam.

60. It means that 2.44 J is required to raise the temperature of 1 g of ethanol by one degree Celsius.

61. The amount of energy required equals the product of the object's specific heat, its mass, and its change in temperature.

Mastering Problems

62. 124,000 calories

63. 2,393 J

64. 327 Calories

65. 2.74 kJ

66. 1367 kJ

67. 0.189 J/g•°C

SECTION 2

Mastering Concepts

68. The foam cup is better insulated than a glass beaker, so that a minimal amount of heat is transferred into or out of the calorimeter.

69. The reaction is endothermic because the enthalpy of the products is 233 kJ higher than the enthalpy of the reactants.

70. universe = system + surroundings; Student answers will vary.

71. when the reaction is carried out at constant pressure

72. The system's chemical potential energy is less after the reaction than before the reaction.

73. ΔH is negative for an exothermic reaction and positive for an endothermic reaction.

Mastering Problems

74. 1.56×10^{8} J

75. 4.16×10^{10} J

76. 377 J

77. 1.91 J/(g•°C)

78. 0.880 J/(g•°C)

Chapter Self-Check

SECTION 3

Mastering Concepts

79. Write the sign of ΔH_{system} for each of the following changes in physical state.
- **a.** $C_2H_5OH(s) \rightarrow C_2H_5OH(l)$
- **b.** $H_2O(g) \rightarrow H_2O(l)$
- **c.** $CH_3OH(l) \rightarrow CH_3OH(g)$
- **d.** $NH_3(l) \rightarrow NH_3(s)$

80. The molar enthalpy of fusion of methanol is 3.22 kJ/mol. What does this mean?

81. Explain how perspiration can help cool your body.

82. Write the thermochemical equation for the combustion of methane. Refer to **Table 3.**

Mastering Problems

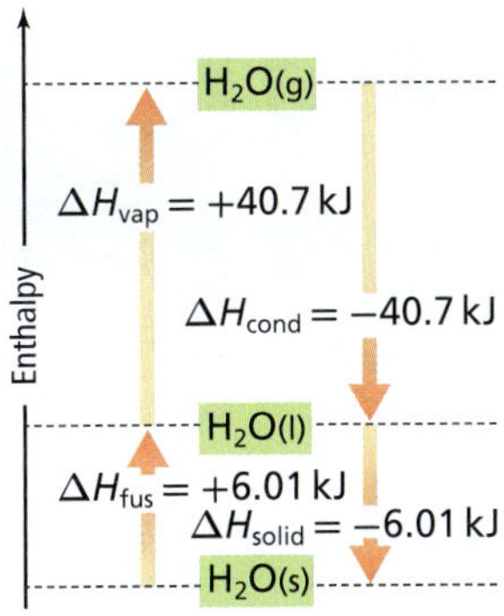

■ **Figure 24**

83. Use information from **Figure 24** to calculate how much heat is required to vaporize 4.33 mol of water at 100°C.

84. Agriculture Water is sprayed on oranges during a frosty night. If an average of 11.8 g of water freezes on each orange, how much heat is released?

85. Grilling What mass of propane (C_3H_8) must be burned in a barbecue grill to release 4560 kJ of heat? The ΔH_{comb} of propane is −2219 kJ/mol.

86. Heating with Coal How much heat is liberated when 5.00 kg of coal is burned if the coal is 96.2% carbon by mass and the other materials in the coal do not react? ΔH_{comb} of carbon is −394 kJ/mol.

87. How much heat is evolved when 1255 g of water condenses to a liquid at 100°C?

88. A sample of ammonia ($\Delta H_{solid} = -5.66$ kJ/mol) liberates 5.66 kJ of heat as it solidifies at its melting point. What is the mass of the sample?

SECTION 4

Mastering Concepts

89. For a given compound, what does the standard enthalpy of formation describe?

90. How does ΔH for a thermochemical equation change when the amounts of all substances are tripled and the equation is reversed?

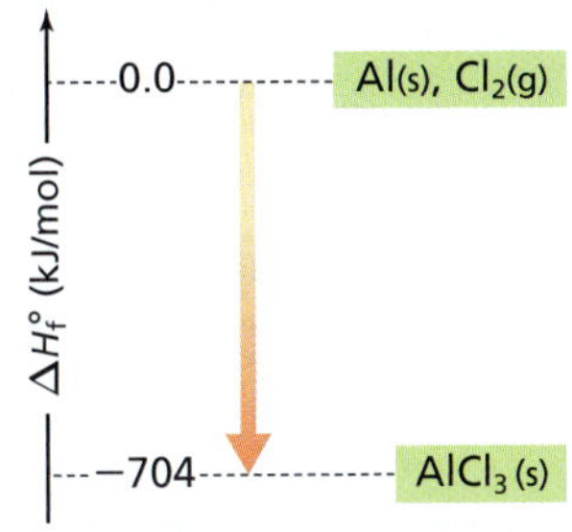

■ **Figure 25**

91. Use **Figure 25** to write the thermochemical equation for the formation of 1 mol of aluminum chloride (a solid in its standard state) from its constituent elements in their standard states.

Mastering Problems

92. Use standard enthalpies of formation from **Table R-11** to calculate ΔH°_{rxn} for the following reaction.

$$P_4O_6(s) + 2O_2(g) \rightarrow P_4O_{10}(s)$$

93. Use Hess's law and the following thermochemical equations to produce the thermochemical equation for the reaction C(s, diamond) → C(s, graphite). What is ΔH for the reaction?
- **a.** $C(s, graphite) + O_2(g) \rightarrow CO_2(g) \quad \Delta H = -394$ kJ
- **b.** $C(s, diamond) + O_2(g) \rightarrow CO_2(g) \quad \Delta H = -396$ kJ

94. Use Hess's law and the changes in enthalpy for the following two generic reactions to calculate ΔH for the reaction $2A + B_2C_3 \rightarrow 2B + A_2C_3$.

$$2A + \frac{3}{2}C_2 \rightarrow A_2C_3 \quad \Delta H = -1874 \text{ kJ}$$

$$2B + \frac{3}{2}C_2 \rightarrow B_2C_3 \quad \Delta H = -285 \text{ kJ}$$

SECTION 5

Mastering Concepts

95. Under what conditions is an endothermic chemical reaction in which the entropy of the system increases likely to be spontaneous?

SECTION 3

Mastering Concepts

79. a. positive.
b. negative.
c. positive.
d. negative.

80. It means that 3.22 kJ of energy is required to melt one mole of methanol.

81. Your body is cooled as it supplies the heat required to vaporize water from your skin.

82. $CH_4(g) + 2O_2(g) \rightarrow CO_2(g) + 2H_2O(l)$
$\Delta H = -891$ kJ

Mastering Problems

83. 176 kJ
84. −3.94 kJ
85. 90.60 g
86. −158,000 kJ
87. 2830 kJ
88. 17.03 g

SECTION 4

Mastering Concepts

89. Standard enthalpy of formation describes the change in enthalpy when one mole of the compound in its standard state is formed from its constituent elements in their standard states.

90. ΔH is tripled, and its sign is changed.

91. $Al(s) + 3/2Cl_2(g) \rightarrow AlCl_3(s)$
$\Delta H^\circ_f = -704$ kJ

Mastering Problems

92. −1343.9 kJ
93. −2 kJ
94. −1589 kJ

SECTION 5

Mastering Concepts

95. Such a reaction is likely to be spontaneous only at higher temperatures.

96. Because a gaseous product is formed, it is likely that the system's entropy increases.

97. For a spontaneous reaction, ΔG_{system} must be negative as calculated in the expression $\Delta G^\circ_{system} = \Delta H^\circ_{system} - T\Delta S^\circ_{system}$. Reactions **a** and **c** both have a positive ΔH_{system}. However, both reactions also have more moles of gaseous products than gaseous reactants, which suggests that entropy increases as products form. So, higher temperatures will tend to make these reactions spontaneous (a negative value for ΔG_{system}). On the other hand, reaction **b** has fewer moles of gas on the products side, which means entropy decreases as products form. But because ΔH_{system} is negative for this reaction, it will tend to be spontaneous at lower temperatures.

98. The heat released by an exothermic reaction increases the entropy of the surroundings. Such a reaction decreases ΔG_{system} because ΔH_{system} is negative in the equation $\Delta G^\circ_{system} = \Delta H^\circ_{system} - T\Delta S^\circ_{system}$

Mastering Problems

99. **a.** 87.9 kJ nonspontaneous
b. −270 kJ spontaneous
c. 84.1 kJ nonspontaneous

100. 88.4 K

101. +0.119 kJ/K

102. Yes. The reaction is spontaneous under standard conditions.

103. 266 K

96. Predict how the entropy of the system changes for the reaction $CaCO_3(s) \rightarrow CaO(s) + CO_2(g)$. Explain.

97. Which of these reactions would you expect to be spontaneous at relatively high temperatures? At relatively low temperatures? Explain.

a. $2NH_3(g) \rightarrow N_2(g) + 3H_2(g)$ $\quad \Delta H_{system} = 92$ kJ
b. $2NO_2(g) \rightarrow N_2O_4(g)$ $\quad \Delta H_{system} = -58$ kJ
c. $CaCO_3(g) \rightarrow CaO(s) + CO_2(g)$ $\quad \Delta H_{system} = 178$ kJ

98. Explain how an exothermic reaction changes the entropy of the surroundings. Does the enthalpy change for such a reaction increase or decrease ΔG_{system}? Explain.

Mastering Problems

99. Calculate ΔG_{system} for each process, and state whether the process is spontaneous or nonspontaneous.

a. $\Delta H_{system} = 145$ kJ, $T = 293$ K, $\Delta S_{system} = 195$ J/K
b. $\Delta H_{system} = -232$ kJ, $T = 273$ K, $\Delta S_{system} = 138$ J/K
c. $\Delta H_{system} = -15.9$ kJ, $T = 373$ K, $\Delta S_{system} = -268$ J/K

100. Calculate the temperature at which $\Delta G_{system} = 0$ if $\Delta H_{system} = 4.88$ kJ and $\Delta S_{system} = 55.2$ J/K.

101. For the change $H_2O(l) \rightarrow H_2O(g)$, ΔG°_{system} is 8.557 kJ and ΔH°_{system} is 44.01 kJ. What is ΔS°_{system} for the change?

102. Is the reaction to convert copper(II) sulfide to copper(II) sulfate spontaneous under standard conditions? $CuS(s) + 2O_2(g) \rightarrow CuSO_4(s)$. $\Delta H^\circ_{rxn} = -718.3$ kJ, and $\Delta S^\circ_{rxn} = -368$ J/K. Explain.

103. Calculate the temperature at which $\Delta G_{system} = -34.7$ kJ if $\Delta H_{system} = -28.8$ kJ and $\Delta S_{system} = 22.2$ J/K.

MIXED REVIEW

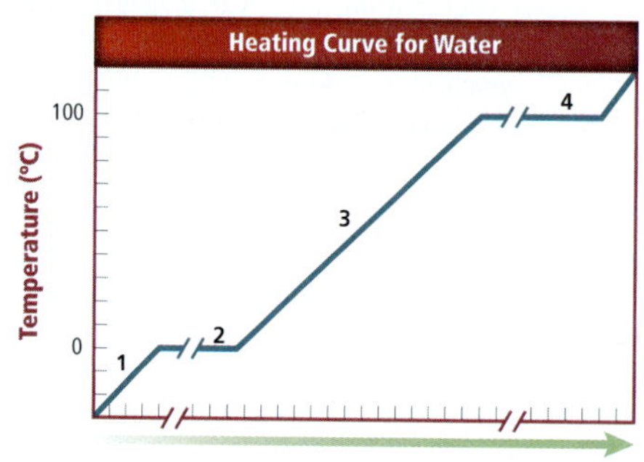

■ **Figure 26**

104. Heat was added consistently to a sample of water to produce the heating curve in **Figure 26.** Identify what is happening in Sections 1, 2, 3, and 4 on the curve.

105. Bicycling Describe the energy conversions that occur when a bicyclist coasts down a long grade, then struggles to ascend a steep grade.

106. Hiking Imagine that on a cold day you are planning to take a thermos of hot soup with you on a hike. Explain why you might fill the thermos with hot water first before filling it with the hot soup.

107. Differentiate between the enthalpy of formation of $H_2O(l)$ and $H_2O(g)$. Why is it necessary to specify the physical state of water in the following thermochemical equation $CH_4(g) + 2O_2(g) \rightarrow CO_2(g) + 2H_2O(l \text{ or } g)$ $\Delta H = ?$

THINK CRITICALLY

■ **Figure 27**

108. Analyze both of the images in **Figure 27** in terms of potential energy of position, chemical potential energy, kinetic energy, and heat.

109. Apply Phosphorus trichloride is a starting material for the preparation of organic phosphorous compounds. Demonstrate how thermochemical equations **a** and **b** can be used to determine the enthalpy change for the reaction $PCl_3(l) + Cl_2(g) \rightarrow PCl_5(s)$.

a. $P_4(s) + 6Cl_2(g) \rightarrow 4PCl_3(l)$ $\quad \Delta H = -1280$ kJ
b. $P_4(s) + 10Cl_2(g) \rightarrow 4PCl_5(s)$ $\quad \Delta H = -1774$ kJ

110. Calculate Suppose that two pieces of iron, one with a mass exactly twice the mass of the other, are placed in an insulated calorimeter. If the original temperatures of the larger piece and the smaller piece are 90.0°C and 50.0°C, respectively, what is the temperature of the two pieces when thermal equilibrium has been established? Refer to **Table 2** for the specific heat of iron.

111. Predict which of the two compounds, methane gas (CH_4) or methanal vapor (CH_2O), has the greater molar enthalpy of combustion. Explain your answer. *(Hint: Write and compare the balanced chemical equations for the two combustion reactions.)*

MIXED REVIEW

104. Section 1: The kinetic energy of the water (ice) is increasing as the temperature rises.
Section 2: Potential energy is increasing as the system absorbs energy in the process of melting.
Section 3: The kinetic energy of the water is increasing as the temperature rises.
Section 4: Potential energy is increasing as the system absorbs energy in the process of evaporating.

105. Descending a long grade, potential energy of position is converted to kinetic energy of motion. Ascending a steep grade, chemical potential energy and kinetic energy are converted to potential energy.

106. The hot water will transfer energy to the thermos in the form of heat, raising the temperature of the thermos to nearly that of the hot soup. Because the temperatures of the thermos and soup are similar, the soup will lose little heat to the thermos when placed inside.

107. ΔH°_f for $H_2O(l)$ and $H_2O(g)$ differ by approximately the enthalpy of vaporization of water. Because water in the liquid state has an enthalpy of formation that differs from that of water in the gaseous state, the enthalpy change for the reaction depends upon the physical states of all reactants and products.

THINK CRITICALLY

108. The snow has positional potential energy, which changes to kinetic energy of motion. Wood has chemical potential, which can be converted to heat, light, and kinetic energy of motion.

109. refer to Solutions Manual $\Delta H = -124$ kJ

110. $T_f = 76.7$°C

111. Methane likely has the greater molar enthalpy of combustion.

Chapter Self-Check

CHALLENGE PROBLEM

112. A sample of natural gas is analyzed and found to be 88.4% methane (CH_4) and 11.6% ethane (C_2H_6) by mass. The standard enthalpy of combustion of methane to gaseous carbon dioxide (CO_2) and liquid water (H_2O) is −891 kJ/mol. Write the equation for the combustion of gaseous ethane to carbon dioxide and water. Calculate the standard enthalpy of combustion of ethane using standard enthalpies of formation from **Table R-11.** Using that result and the standard enthalpy of combustion of methane in **Table 3,** calculate the energy released by the combustion of 1 kg of natural gas.

CUMULATIVE REVIEW

113. Why is it necessary to perform repeated experiments in order to support a hypothesis?

114. Phosphorus has the atomic number 15 and an atomic mass of 31 amu. How many protons, neutrons, and electrons are in a neutral phosphorus atom?

115. What element has the electron configuration $[Ar]4s^1 3d^5$?

116. Name the following molecular compounds.
a. S_2Cl_2
b. CS_2
c. SO_3
d. P_4O_{10}

117. Determine the molar mass for the following compounds.
a. $Co(NO_3)_2 \cdot 6H_2O$
b. $Fe(OH)_3$

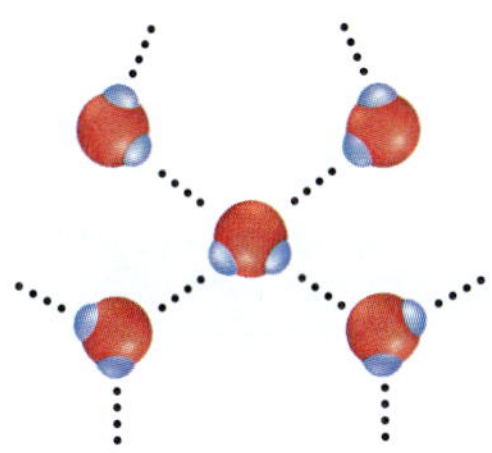

■ **Figure 28**

118. What kind of chemical bond is represented by the dotted lines in **Figure 28?**

119. A sample of oxygen gas has a volume of 20.0 cm^3 at −10.0°C. What volume will this sample occupy if the temperature rises to 110 °C?

120. What is the molarity of a solution made by dissolving 25.0 g of sodium thiocyanate (NaSCN) in enough water to make 500 mL of solution?

121. List three colligative properties of solutions.

WRITING IN ▶ Chemistry

122. Alternate Fuels Use various sources to explain how hydrogen might be produced, transported, and used as a fuel for automobiles. Summarize the benefits and drawbacks of using hydrogen as an alternative fuel for internal combustion engines.

123. Wind Power Research the use of wind as a source of electrical power. Explain the possible benefits, disadvantages, and limitations of its use.

DBQ Document-Based Questions

Cooking Oil *A university research group burned four cooking oils in a bomb calorimeter to determine if a relationship exists between the enthalpy of combustion and the number of double bonds in an oil molecule. Cooking oils typically contain long chains of carbon atoms linked by either single or double bonds. A chain with no double bonds is said to be saturated. Oils with one or more double bonds are unsaturated. The enthalpies of combustion of the four oils are shown in* **Table 7.** *The researchers calculated that the results deviated by only 0.6% and concluded that a link between saturation and enthalpy of combustion could not be detected by the experimental procedure used.*

Data obtained from: http: Heat of Combustion Oils. April 1998. *University of Pennsylvania.*

Table 7 Combustion Results for Oils

Type of Oil	ΔH_{comb} (kJ/g)
Soy oil	40.81
Canola oil	41.45
Olive oil	39.31
Extra-virgin olive oil	40.98

124. Which of the oils tested provided the greatest amount of energy per unit mass when burned?

125. According to the data, how much energy would be liberated by burning 0.554 kg of olive oil?

126. Assume that 12.2 g of soy oil is burned and that all the energy released is used to heat 1.600 kg of water, initially at 20.0°C. What is the final temperature of the water?

127. Oils can be used as fuels. How many grams of canola oil would have to be burned to provide the energy to vaporize 25.0 g of water? (ΔH_{vap} = 40.7 kJ/mol).

CHALLENGE PROBLEM

112. −55,200 kJ

CUMULATIVE REVIEW

113. Experiments must be repeated to be sure that they yield similar results each time.

114. number of protons = 15; number of electrons = 15; number of neutrons = mass number - number of protons = 16

115. chromium

116. a. disulfur dichloride
b. carbon disulfide
c. sulfur trioxide
d. tetraphosphorus decoxide

117. a. 291.07 g/mol
b. 106.88 g/mol

118. Hydrogen bonds

119. 29.1 cm^3

120. 0.616*M*

121. vapor pressure lowering, boiling point elevations, freezing point depression

WRITING IN ▶ Chemistry

Rubric

122. Students may write that hydrogen could best be used as an automobile fuel in fuel cells. Much of the technology now used for handling methane and propane gases could be adapted for use with hydrogen. Much of the hydrogen now available is a byproduct of the petrochemical industry. For full-scale use of hydrogen as a fuel for automobiles and for other energy needs, hydrogen would probably be produced by the electrolysis of water using renewable sources of energy such as wind power or solar energy. The only product of the combustion of hydrogen is water, so it is a nonpolluting source of power.

123. Student will note that the wind is not a steady source of energy and there will always be a need for a backup. Wind power is nonpolluting. People might object because windmills disturb the natural beauty of the landscape. Windmills can also kill birds. When windmills are located offshore, fish could be adversely affected by the structures.

DBQ Document-Based Questions

Data obtain from: Argeros, A., Pincas, D., Shinar, Z., and Sultenfuss, A. April, 1998. Heat of Combustion of Oils. *University of Pennsylvania*.

124. canola oil: 41.45 kJ/g
125. 21,800 kJ
126. 94.4° C
127. 1.37 g canola oil

MULTIPLE CHOICE

1. C
2. A
3. D
4. A
5. D
6. D
7. C
8. A
9. B

MULTIPLE CHOICE

Use the graph below to answer Questions 1 to 3.

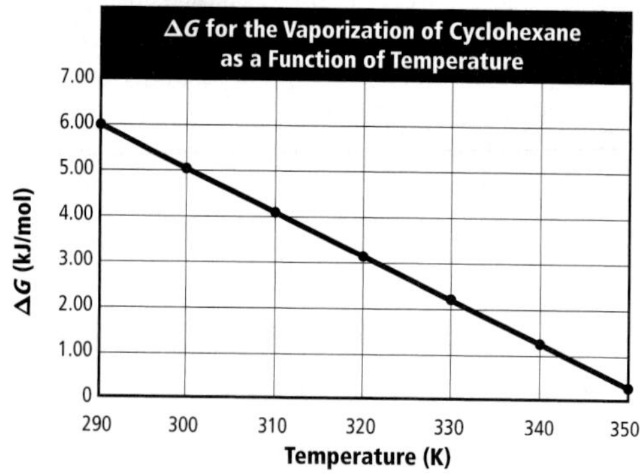

1. In the range of temperatures shown, the vaporization of cyclohexane
A. does not occur at all.
B. will occur spontaneously.
C. is not spontaneous.
D. occurs only at high temperatures.

2. What is the standard free energy of vaporization, ΔG°_{vap}, of cyclohexane at 300 K?
A. 5.00 kJ/mol
B. 4.00 kJ/mol
C. 3.00 kJ/mol
D. 2.00 kJ/mol

3. When ΔG°_{vap} is plotted as a function of temperature, the slope of the line equals ΔS°_{vap} and the y-intercept of the line equals ΔH°_{vap}. What is the approximate standard entropy of the vaporization of cyclohexane?
A. −50.0 J/mol•K
B. −10.0 J/mol•K
C. −5.0 J/mol•K
D. −100 J/mol•K

4. The metal yttrium, atomic number 39, forms
A. positive ions.
B. negative ions.
C. both positive and negative ions.
D. no ions at all.

5. Given the reaction $2Al + 3FeO \rightarrow Al_2O_3 + 3Fe$, what is the mole-to-mole ratio between iron(II) oxide and aluminum oxide?
A. 2:3
B. 1:1
C. 3:2
D. 3:1

Use the table below to answer Question 6.

Electronegativity of Selected Elements

H						
2.20						
Li	Be	B	C	N	O	F
0.98	1.57	2.04	2.55	3.04	3.44	3.98
Na	Mg	Al	Si	P	S	Cl
0.93	1.31	1.61	1.90	2.19	2.58	3.16

6. Which bond is the most electronegative?
A. H–H
B. H–C
C. H–N
D. H–O

7. Element Q has an oxidation number of +2, while Element M has an oxidation number of −3. Which is the correct formula for a compound made of elements Q and M?
A. Q_2M_3
B. M_2Q_3
C. Q_3M_2
D. M_3Q_2

8. Wavelengths of light shorter than about 4.00×10^{-7} m are not visible to the human eye. What is the energy of a photon of ultraviolet light having a frequency of 5.45×10^{16} s^{-1}? (Planck's constant is 6.626×10^{-34} J•s.)
A. 3.61×10^{-17} J
B. 1.22×10^{-50} J
C. 8.23×10^{49} J
D. 3.81×10^{-24} J

Use the graph below to answer Question 9.

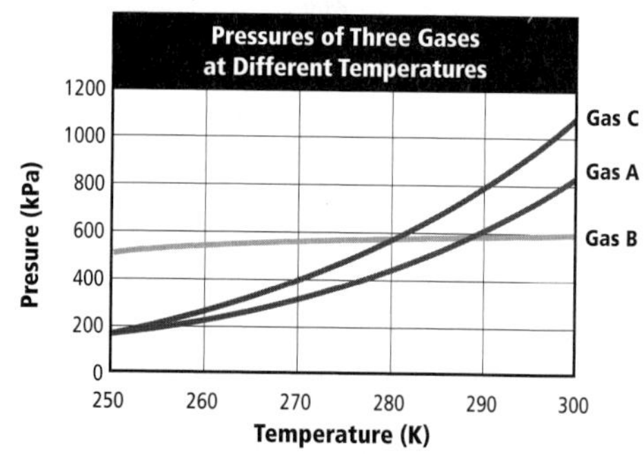

9. What is the pressure of Gas B at 295 K?
A. 500 kPa
B. 600 kPa
C. 700 kPa
D. 900 kPa

SHORT ANSWER

Use the figure below to answer Questions 10 to 12.

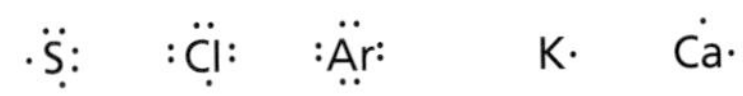

10. Explain why argon is not likely to form a compound.
11. What is the chemical formula for calcium chloride? Explain the formation of this ionic compound using the electron-dot structures above.
12. Use electron-dot models to explain what charge sulfur will most likely have when it forms an ion.

EXTENDED RESPONSE

Use the information below to answer Questions 13 and 14.

A sample of gas occupies a certain volume at a pressure of 1 atm. If the pressure remains constant, heating causes the gas to expand, as shown below.

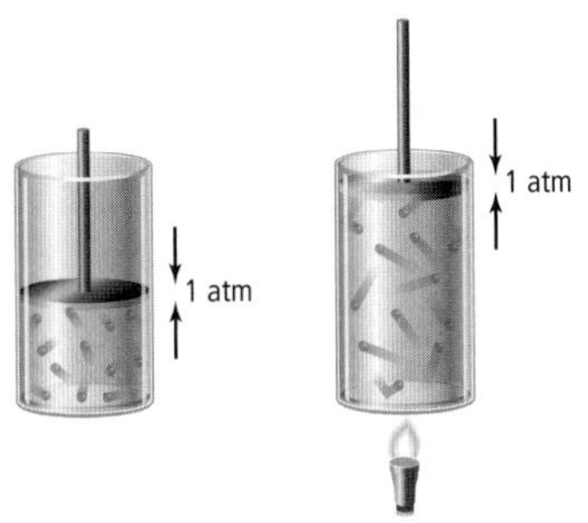

13. State the gas law that describes why the gas in the second canister occupies a greater volume than the gas in the first canister.
14. If the volume in the first container is 2.1 L at a temperature of 300.0 K, to what temperature must the second canister be heated to reach a volume of 5.4 L? Show your setup and the final answer.

SAT SUBJECT TEST: CHEMISTRY

15. The specific heat of ethanol is 2.44 J/g•°C. How many kilojoules of energy are required to heat 50.0 g of ethanol from −20.0°C to 68.0°C?
   A. 10.7 kJ
   B. 8.30 kJ
   C. 2.44 kJ
   D. 1.22 kJ
   E. 5.86 kJ

16. If 3.00 g of aluminum foil, placed in an oven and heated from 20.0°C to 662.0°C, absorbs 1728 J of heat, what is the specific heat of aluminum?
   A. 0.131 J/g•°C
   B. 0.870 J/g•°C
   C. 0.897 J/g•°C
   D. 2.61 J/g•°C
   E. 0.261 J/g•°C

Use the table below to answer Questions 17 and 18.

Density and Electronegativity Data for Elements

Element	Density (g/ml)	Electronegativity
Aluminum	2.698	1.6
Fluorine	1.696×10^{-3}	4.0
Sulfur	2.070	2.6
Copper	8.960	1.9
Magnesium	1.738	1.3
Carbon	3.513	2.6

17. A sample of metal has a mass of 9.250 g and occupies a volume of 5.250 mL. Which metal is it?
   A. aluminum
   B. magnesium
   C. carbon
   D. copper
   E. sulfur

18. Which pair is most likely to form an ionic bond?
   A. carbon and sulfur
   B. aluminum and magnesium
   C. copper and sulfur
   D. magnesium and fluorine
   E. aluminum and carbon

NEED EXTRA HELP?																		
If You Missed Question . . .	1	2	3	4	5	6	7	8	9	10	11	12	13	14	15	16	17	18
Review Section . . .	15.5	15.5	15.5	8.5	11.1	8.5	7.3	5.1	13.1	5.3	5.3	5.3	13.1	13.1	15.2	15.2	2.1	8.5

SHORT ANSWER

10. Argon already has a full outer energy level (eight valence electrons) and is not likely to form an ion. It does not need to gain or lose any electrons in order to become chemically stable.
11. $CaCl_2$; a calcium atom becomes Ca^{2+}, losing its two valence electrons to two chlorine atoms, which each become Cl^-.
12. Sulfur has six valence electrons. Because atoms are more stable when they have 8 valence electrons completing their outer energy levels, sulfur tends to gain two electrons to become the ion S^{2-}.

EXTENDED RESPONSE

13. This is Charles's law: at a constant pressure, the volume of a given mass of gas is directly proportional to its kelvin temperature.
14. $T_1/V_1 = T_2/V_2$

$$T_2 = \frac{T_1V_2}{V_1} = \frac{(300.0\ K)(5.4\ L)}{2.1\ L} = 770\ K$$

SAT SUBJECT TEST: CHEMISTRY

15. A
16. C
17. B
18. D

CHAPTER 16 Reaction Rates

BIGIDEA **Every chemical reaction proceeds at a definite rate, but can be speeded up or slowed down by changing the conditions of the reaction.**

ESSENTIAL QUESTIONS	RESOURCES TO ASSESS MASTERY
SECTION 1 **A Model for Reaction Rates** **1.** How can average rates of chemical reactions be calculated from experimental data? **2.** How are the rates of chemical reactions related to collisions between reacting particles? 2 sessions 1 block	**Progress Monitoring** Formative Assessment, p. 560, 561, 565, 566 Reading Check, p. 563 Graph Check, p. 565 Section Review, p. 567
SECTION 2 **Factors Affecting Reaction Rates** **1.** What are the factors that affect the rates of chemical reactions? **2.** What is the role of a catalyst? 2 sessions 1 block	**Progress Monitoring** Formative Assessment, pp. 569, 572, 573 Reading Check, p. 569 Graph Check, pp. 570, 572 Section Review, p. 573
SECTION 3 **Reaction Rate Laws** **1.** What is the relationship between reaction rate and concentration? **2.** How are reaction orders determined using the method of initial rates? 1 session 0.5 block	**Progress Monitoring** Formative Assessment, pp. 574, 577 Reading Check, pp. 575, 576 Graph Check, p. 575 Section Review, p. 577
SECTION 4 **Instantaneous Reaction Rates and Reaction Mechanisms** **1.** How are instantaneous rates of chemical reactions calculated? **2.** What substances and steps are involved in a reaction mechanism? **3.** How is the instantaneous rate of a complex reaction related to its reaction mechanism? 1 session 0.5 block	**Progress Monitoring** Formative Assessment, pp. 581, 582 Reading Check, p. 581 Graph Check, pp. 578, 582 Section Review, p. 582 **Summative Assessment** Chapter Assessment, p. 586 *eAssessment* Chapter Test (Scaffolded)

Go online!

LEVELED RESOURCES	LAB MATERIALS
Science Notebook 16.1 OL ***FAST FILE Chapter Resources:*** Study Guide, p. 100 BL	LaunchLAB p. 558 / **15 min** baker's yeast, hydrogen peroxide, beaker or cup, toothpick
Science Notebook 16.2 OL ***FAST FILE Chapter Resources:*** MiniLab Worksheet, p. 88 OL ChemLab Worksheet, p. 89 OL Study Guide, p. 102 BL	MiniLAB p. 571 / **15 min** effervescent tablets, 250-mL beaker, hot plate, non-mercury thermometer, water, stopwatch or clock with second hand, balance ChemLAB p. 584 / **45 min** 10-mL graduated pipette, safety pipette filler, 6*M* hydrochloric acid, distilled water, 25-mm x 150-mm test tubes labeled 1–4, test-tube rack, magnesium ribbon, emery cloth or fine sandpaper, scissors, plastic ruler, tongs, watch with second hand or stopwatch, stirring rod.
Science Notebook 16.3 OL ***FAST FILE Chapter Resources:*** Study Guide, p. 103 BL	
Science Notebook 16.4 OL ***FAST FILE Chapter Resources:*** Study Guide, p. 104 BL	

ADDITIONAL RESOURCES

eAssessment
ConnectED Student Center
ConnectED Teacher Center
Lesson Presentations
eSolutions
Videos
Animations
Online Quizzes
Virtual Investigations
Multilingual eGlossary
Personal Tutors
WebQuests
Applying Practices Worksheets
PBLs

Practice:

Challenge Problems AL
Supplemental Problems BL OL
Chapter Test (Scaffolded)
Standardized Test Practice

Additional Resources:

Blueprints for Success
Solving Problems: A Chemistry Handbook

Lab Resources:

Laboratory Manual OL
CBL Laboratory Manual OL
Small-Scale Laboratory Manual OL
Forensics Laboratory Manual OL

BL Below Level   OL On Level   AL Advanced Learners   EL English Learners   COOP LEARN Cooperative Learning

CHAPTER 16

LaunchLAB

Rubric

How can you accelerate a reaction?

Purpose Students will observe how an enzyme increases the rate of a chemical reaction.

Safety Precautions Approve lab safety forms before work begins. **WARNING: *hydrogen peroxide is corrosive.***

Disposal The spent hydrogen peroxide/yeast mixture can be flushed down a drain with water.

Teaching Strategies

- Describe yeast metabolism, and explain how yeast causes bread to rise.
- Continue to develop the concept of enzymes acting as catalysts in biological functions. Also discuss chemical catalysts in other every-day situations.
- See the Teacher Handbook for preparation of solutions.

Expected Results The hydrogen peroxide is clear and colorless. When yeast is added, bubbles form. After a short while, the mixture also starts to foam.

Procedure

1. Read and complete the lab safety form.
2. Create a *Before and After* table to record your observations.
3. Pour about 10 mL of **hydrogen peroxide** into a small **beaker or cup.** Observe the hydrogen peroxide. Complete the *Before* column with your initial observations.
   WARNING: *Hydrogen peroxide is corrosive. Avoid contact with skin and eyes.*
4. Add 0.1 g of **baker's yeast** to the hydrogen peroxide. Stir gently with a **toothpick,** and observe the mixture again. Complete the *After* column with your observations.

CHAPTER 16

Reaction Rates

BIGIDEA Every chemical reaction proceeds at a definite rate, but can be speeded up or slowed down by changing the conditions of the reaction.

SECTIONS

1. A Model for Reaction Rates
2. Factors Affecting Reaction Rates
3. Reaction Rate Laws
4. Instantaneous Reaction Rates and Reaction Mechanisms

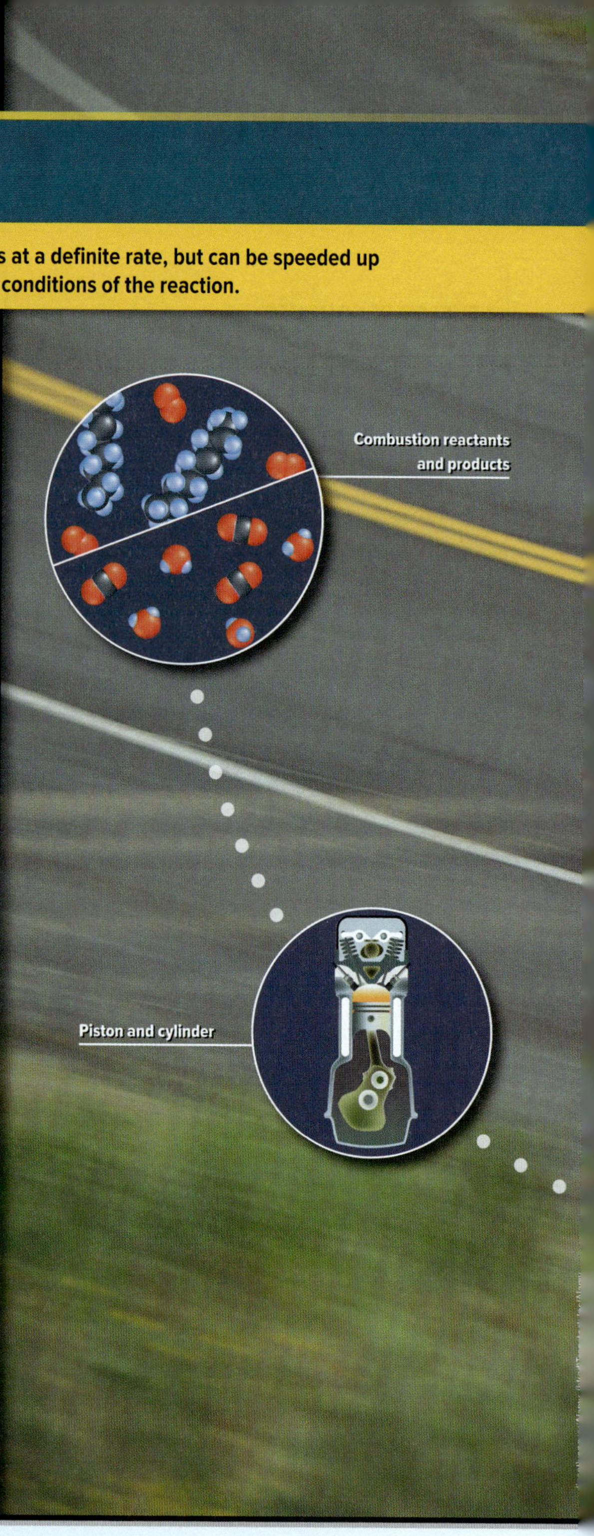

LaunchLAB

How can you accelerate a reaction?

Some chemical reactions go so slowly that nothing seems to be happening. In this lab, you will study one way of speeding up a slow reaction.

FOLDABLES® Study Organizer

Reaction Rates

Make a layered-look book. Label it as shown. Use it to help you organize information about factors affecting reaction rates.

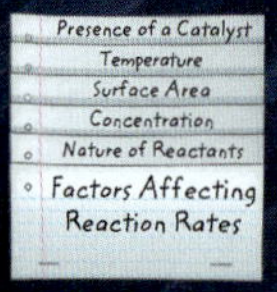

Analysis

1. **Identify** the two products formed when hydrogen peroxide decomposes. oxygen and water
2. **Explain** why bubbles are produced in Step 4 but not in Step 3. Yeast contains an enzyme that acts as a catalyst for the decomposition of hydrogen peroxide.

Inquiry What would happen if you added more or less yeast? What if you did not stir the mixture? Design an experiment to test one of these variables. Answers will vary. Students might hypothesize that more yeast will increase the rate of reaction, not stirring the mixture might reduce the rate of reaction, or changing the temperature of the reactant might change the rate of reaction.

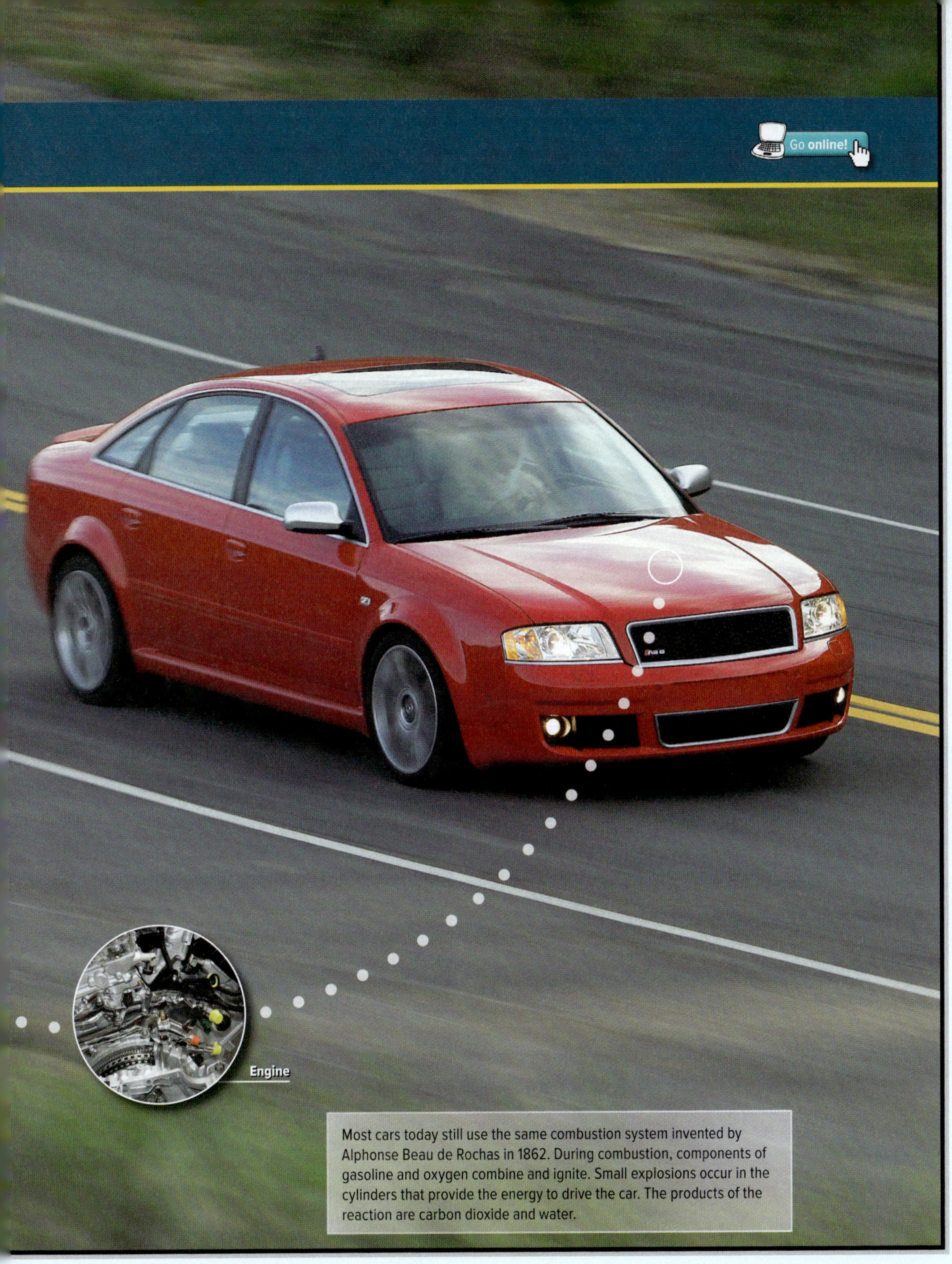

Most cars today still use the same combustion system invented by Alphonse Beau de Rochas in 1862. During combustion, components of gasoline and oxygen combine and ignite. Small explosions occur in the cylinders that provide the energy to drive the car. The products of the reaction are carbon dioxide and water.

Introduce the BIGIDEA

Fast and Slow Reactions To introduce this chapter's Big Idea have students brainstorm to think of chemical reactions that occur at different rates. List the reactions under three categories on the board: extremely fast, medium, extremely slow. Choose a medium or extremely slow reaction and ask students how they think the rate might be increased. Example: The rate at which a wooden log burns might be increased by reducing it to fine shavings or using a bellows to increase its contact with oxygen.

Tie to Previous Knowledge

Have students review the following concepts before studying this chapter.

- chemical reactions, equations for chemical reactions
- gas laws
- molarity

Use the Photo

Compare Reaction Rates
Have students describe the reaction that takes place in the cylinders of an automobile engine. A spark causes gasoline vapors to react with oxygen in a combustion reaction. Ask students to compare the rate of this reaction with the rate at which an iron nail rusts when left outside in inclement weather. The rate of the gasoline combustion reaction is much faster than the rate at which the nail rusts.

SECTION 1

1 Focus

MAINIDEA

Collisions and Reaction Rate
Show students a photo of a space shuttle launching and ask them why NASA engineers use liquid rather than gaseous reactants (hydrogen and oxygen) to power the shuttle's booster rockets. The two elements are much denser in the liquid state than in the gaseous state. Therefore, many more H_2 and O_2 molecules collide per unit time, greatly increasing the rate at which energy is released to propel the shuttle into orbit. BL OL AL

2 Teach

ASSESSMENT

Knowledge Write the equation $CH_3CH_2I + OH^- \rightarrow CH_3CH_2IOH^- \rightarrow CH_3CH_2OH + I^-$ on the board and have students identify the activated complex, the products, and the reactants. Activated complex is $CH_3CH_2IOH^-$. Products are CH_3CH_2OH and I^-. Reactants are CH_3CH_2I and OH^-. OL

SECTION 1

A Model for Reaction Rates

MAINIDEA Collision theory is the key to understanding why some reactions are faster than others.

Essential Questions

- How can average rates of chemical reactions be calculated from experimental data?
- How are the rates of chemical reactions related to collisions between reacting particles?

Review Vocabulary

energy: the ability to do work or produce heat; it exists in two basic forms: potential energy and kinetic energy

New Vocabulary

reaction rate
collision theory
activated complex
activation energy

CHEM 4 YOU Which is faster: walking to school, or riding in a bus or car? Determining how fast a person can get to school is not all that different from calculating the rate of a chemical reaction. Either way, you are measuring change over time.

Expressing Reaction Rates

In the Launch Lab, you discovered that the decomposition of hydrogen peroxide can be a fast reaction, or it can be a slow one. However, *fast* and *slow* are inexact terms. Chemists, engineers, chefs, welders, concrete mixers, and others often need to be more specific. For example, a chef must know the rate at which a roast cooks to determine when it will be ready to serve. The person mixing the concrete must know the rate of mixing water, sand, gravel, and cement so that the resulting concrete can be poured at the correct consistency. Delaying pouring can result in concrete that is not strong enough for its purpose.

Think about how you express the speed or rate of a moving object. The speedometer of the speeding racer in **Figure 1** shows that the car is moving at 320 km/h. The speed of a sprinter on a track team might be expressed in meters per second (m/s). Generally, the average rate of an action or process is defined as the change in a given quantity during a specific period of time. Recall from your study of math that the Greek letter *delta* (Δ) before a quantity indicates a change in the quantity. In equation form, average rate or speed is written as follows.

$$\text{average rate} = \frac{\Delta\text{quantity}}{\Delta t}$$

Figure 1 The speedometer of the racer shows its speed in km/h or mph, both of which are the change in distance divided by the change in time. The sprinter's speed might be measured in m/s.

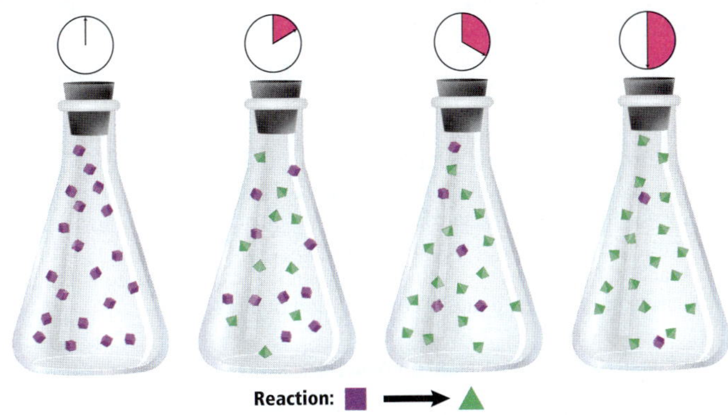

■ **Figure 2** Over time, the reactant changes to a product. The rate of a chemical reaction can be expressed as a change in the number of moles of reactant or product during an interval of time.

Calculate *the rate of change for each interval.*

Figure 2 shows how a reaction proceeds from reactant to product over time. Notice that the amount of the reactant decreases as the amount of product increases. If you know the change in a product or a reactant during a segment of time, you can calculate the average rate of the reaction. Most often, chemists are concerned with changes in the molar concentration (mol/L, M) of a reactant or product during a reaction. Therefore, the **reaction rate** of a chemical reaction is the change in concentration of a reactant or product per unit of time, generally expressed as mol/(L • s). Brackets around the formula for a substance denote its molar concentration. For example, $[NO_2]$ represents the molar concentration of NO_2.

Reaction rates are determined experimentally by measuring the concentrations of reactants and/or products as an actual chemical reaction proceeds. Reaction rates cannot be calculated from balanced equations.

Suppose you wish to express the average rate of the following reaction during the time period beginning at time t_1 and ending at time t_2.

$$CO(g) + NO_2(g) \rightarrow CO_2(g) + NO(g)$$

Calculating the rate at which the products of the reaction are produced results in a reaction rate with a positive value. The rate calculation based on the production of NO has the following form.

$$\text{Average reaction rate} = \frac{[NO] \text{ at time } t_2 - [NO] \text{ at time } t_1}{t_2 - t_1} = \frac{\Delta[NO]}{\Delta t}$$

For example, if the concentration of NO is $0.000M$ at time $t_1 = 0.00$ s and $0.010M$ two seconds after the reaction begins, the following calculation gives the average rate of the reaction expressed as moles of NO produced per liter per second.

$$\text{Average reaction rate} = \frac{0.010M - 0.000M}{2.00 \text{ s} - 0.00 \text{ s}}$$
$$= \frac{0.010M}{2.00 \text{ s}} = 0.0050 \text{ mol/(L·s)}$$

Notice how the units work out:

$$\frac{M}{s} = \frac{\text{mol}}{\text{L}} \cdot \frac{1}{\text{s}} = \frac{\text{mol}}{(\text{L}\cdot\text{s})}$$

VOCABULARY

SCIENCE USAGE V. COMMON USAGE

Concentration

Science usage: quantitative measure of the amount of solute in a given amount of solvent or solution

The solution has a concentration of six moles per liter.

Common usage: the focus of attention on a single object or purpose

The concentration of the audience was completely on the performer.

CHEMISTRY JOURNAL

Fast and Slow Ask students to list at least five fast reactions or processes and five slow reactions or processes that they encounter in their daily lives. Have students list reactants and products for at least one of the reactions or processes. If possible, have them obtain information about one reaction or process sufficient to calculate and express an average reaction rate. OL

■ **Caption Question Fig. 2** For production of green:

$$0\text{–}10 \text{ s: rate} = \frac{10 \text{ particles} - 0 \text{ particles}}{10\text{s}} = 1 \text{ particle/s}$$

$$10\text{–}20 \text{ s: rate} = \frac{15 \text{ particles} - 10 \text{ particles}}{10\text{s}} = 0.5 \text{ particles/s}$$

$$20\text{–}30 \text{ s: rate} = \frac{18 \text{ particles} - 15 \text{ particles}}{10\text{s}} = 0.3 \text{ particles/s}$$

Quick Demo

Popping Rate Bring a portable microwave oven and a bag of uncooked, unopened microwave popcorn into the classroom. (If you cannot bring a microwave oven into the classroom to perform the demonstration, bring a bag of popped corn and give students the popping time.) Following the instructions on the oven and on the bag of popcorn, pop the corn and have students time the process from the first "pop" to the last. After the popped corn has cooled, help students understand the nature of rates by having them open the bag, count the number of popped kernels, and calculate the average popping rate expressed in kernels per second. OL

ASSESSMENT

Knowledge For the reaction $A + 2B \rightarrow 3C$ at time = 0.000s, [A] = 2.00 mol/L, [B] = 4.00 mol/L, and [C] = 0.00 mol/L. At time = 3.00 minutes, A has dropped to 0.50 mol/L. Have students calculate and express the rate of the reaction over the 3.00-minute time interval in mol A consumed/(L•min), mol B consumed/(L•min), and mol C produced/(L•min).
Rate = 0.500 mol A/(L•min);
Rate = 1.00 mol B/(L•min);
Rate = 1.50 mol C/(L•min) OL

IN-CLASS Example

Question At a given temperature, N_2O_5 decomposes to form NO_2 and O_2. Calculate the average rate of the reaction if $[N_2O_5]$ decreases from 0.1000 mol/L at $t = 0.000$ s to 0.0500 mol/L at $t = 80.0$ s.

Answer 6.25×10^{-4} mol/(L•s)

Average rate

$$= -\frac{(0.0500 \text{ mol/L} - 0.1000 \text{ mol/L})}{(80.0 \text{ s} - 0.0 \text{ s})}$$

$$= 6.25 \times 10^{-4} \text{ mol/(L•s)}$$

You can also choose to state the rate of the reaction as the rate at which CO is consumed, as shown below.

$$\text{average reaction rate} = \frac{[\text{CO}] \text{ at time } t_2 - [\text{CO}] \text{ at time } t_1}{t_2 - t_1} = \frac{\Delta[\text{CO}]}{\Delta t}$$

Do you predict a positive or a negative value for this reaction rate? In this case, a negative value indicates that the concentration of CO decreases as the reaction proceeds. However, reaction rates must always be positive. When the rate is measured by the consumption of a reactant, scientists apply a negative sign to the calculation to get a positive reaction rate. Thus, the following form of the average rate equation is used to calculate the rate of consumption of a reactant.

Average Reaction Rate Equation

$$\text{average reaction rate} = -\frac{\Delta[\text{reactant}]}{\Delta t}$$

Δ[reactant] represents the change in concentration of a reactant.
Δ*t* represents the change in time.

The average reaction rate for the consumption of a reactant is the negative change in the concentration of the reactant divided by the elapsed time.

EXAMPLE Problem 1

Find help with **algebraic equations.** Math Handbook

EXAMPLE PROBLEM

CALCULATE AVERAGE REACTION RATES In a reaction between butyl chloride (C_4H_9Cl) and water, the concentration of C_4H_9Cl is 0.220*M* at the beginning of the reaction. At 4.00 s, the concentration of C_4H_9Cl is 0.100*M*. Calculate the average reaction rate over the given time period expressed as moles of C_4H_9Cl consumed per liter per second.

1 ANALYZE THE PROBLEM

You are given the initial and final concentrations of the reactant C_4H_9Cl and the initial and final times. You can calculate the average reaction rate of the chemical reaction using the change in concentration of butyl chloride in four seconds.

Known	Unknown
$t_1 = 0.00$ s	Average reaction rate = ? mol/(L·s)
$t_2 = 4.00$ s	
$[C_4H_9Cl]$ at $t_1 = 0.220M$	
$[C_4H_9Cl]$ at $t_2 = 0.100M$	

2 SOLVE FOR THE UNKNOWN

$$\text{Average reaction rate} = -\frac{[C_4H_9Cl] \text{ at } t_2 - [C_4H_9Cl] \text{ at } t_1}{t_2 - t_1}$$

State the average reaction rate equation.

$$= -\frac{0.100M - 0.220M}{4.00 \text{ s} - 0.00 \text{ s}}$$

Substitute $t_2 = 4.00$ s, $t_1 = 0.00$ s, $[C_4H_9Cl]$ at $t_2 = 0.100$ *M*, and $[C_4H_9Cl]$ at $t_1 = 0.220M$.

$$= -\frac{0.100 \text{ mol/L} - 0.220 \text{ mol/L}}{4.00 \text{ s} - 0.00 \text{ s}}$$

Substitute mol/L for *M* and perform the calculations. Note that the minus sign cancels out.

$$\text{Average reaction rate} = \frac{0.120 \text{ mol/L}}{4.00 \text{ s}} = \mathbf{0.0300\ mol/(L•s)}$$

3 EVALUATE THE ANSWER

The average reaction rate of 0.0300 moles C_4H_9Cl consumed per liter per second is reasonable based on the starting and ending amounts. The answer is correctly expressed in three significant figures.

Get help with **reaction rate problems.** Personal Tutor

CHEMISTRY JOURNAL

Kilometers per Liter and Kilometers per Dollar To give students experience in stating the change of one quantity in proportion to the change of another, ask them to keep track of the miles driven, gallons of gasoline consumed, and cost of gasoline for an automobile for one week. Then, have them convert English units to SI units and calculate the automobile's mileage in kilometers driven per liter consumed and the cost to drive in kilometers driven per dollar spent (gasoline only). If a student's family has more than one automobile, ask him or her to keep records for all the vehicles and compare the mileage and cost-to-drive figures. OL

PRACTICE Problems

Do additional problems. Online Practice

PRACTICE PROBLEMS

Use the data in the following table to calculate the average reaction rates.

Experimental Data for $H_2 + Cl_2 \rightarrow 2HCl$			
Time (s)	$[H_2]$ (*M*)	$[Cl_2]$ (*M*)	[HCl] (*M*)
0.00	0.030	0.050	0.000
4.00	0.020	0.040	

1. Calculate the average reaction rate expressed in moles H_2 consumed per liter per second.
2. Calculate the average reaction rate expressed in moles Cl_2 consumed per liter per second.
3. **Challenge** If the average reaction rate for the reaction, expressed in moles of HCl formed, is 0.0050 mol/L•s, what concentration of HCl would be present after 4.00 s?

Collision Theory

Have you ever watched children trying to break a piñata? Each hit with a stick can result in emptying the piñata of its contents, as shown in **Figure 3.** The reactants in a chemical reaction must also collide in order to form products. **Figure 3** also represents a reaction between the molecules A_2 and B_2 to form AB. The reactant molecules must come together in a collision in order to react and produce molecules of AB. The figure is an illustration of **collision theory,** which states that atoms, ions, and molecules must collide in order to react.

☑ READING CHECK **Predict** why a collision between two particles is necessary for a reaction to occur.

Look at the reaction between carbon monoxide (CO) gas and nitrogen dioxide (NO_2) gas at a temperature above 500 K.

$$CO(g) + NO_2(g) \rightarrow CO_2(g) + NO(g)$$

The reactant molecules collide to produce carbon dioxide (CO_2) gas and nitrogen monoxide (NO) gas. However, calculations of the number of molecular collisions per second yield a puzzling result: only a small fraction of collisions produce reactions.

Figure 3 Just as a stick must hit the piñata hard enough to break it open, particles in chemical reactions must collide with a sufficient amount of energy for a reaction to occur.

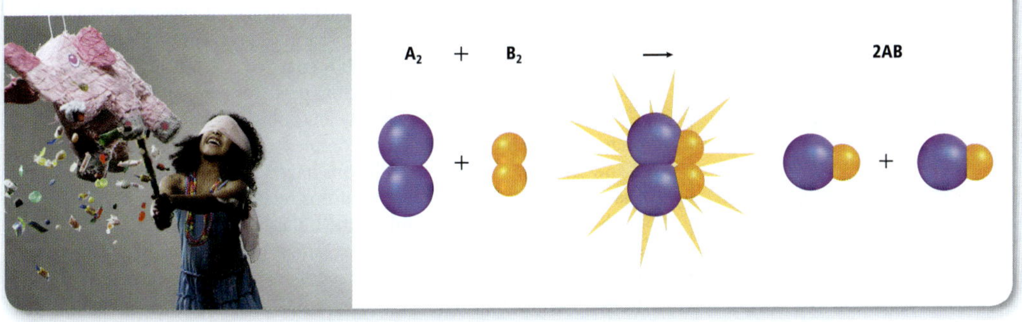

CULTURAL DIVERSITY

Ultrafast Reaction Rates Ahmed Zewail, an Egyptian-born professor of chemistry and physics at California Institute of Technology, won the 1999 Nobel Prize in Chemistry for his development of laser techniques that enable scientists to measure the rates of ultrafast reactions and see the movements of atoms. Zewail used ultrafast laser pulses that, acting like a strobe light, captured changes in molecular structure at intervals just under 10 femtoseconds. With this technique, scientists can directly observe phenomena such as bond breaking, bond making, short-lived intermediates, and transition states. Zewail's techniques can be applied to gases, liquids, solids, clusters of atoms and molecules, surfaces, catalysts, polymers, and biological systems such as proteins and DNA.

PRACTICE Problems

Have students refer to the Selected Solutions appendix for complete solutions to odd-numbered problems. The complete solutions for all problems can be found in the Solutions Manual.

1. 0.0025 mol/(L•s)
2. 0.0025 mol/(L•s)
3. 0.020*M*

☑ **READING CHECK** Particles that are to react must come close to each other so that new bonds in the product can form, while old bonds in the reactants break.

Content Background

Types of Catalysts Catalysts are classified as either homogeneous or heterogeneous. A homogeneous catalyst is one that is present in the same physical state, or phase, as the reactants. Gaseous nitrogen monoxide (NO), for example, is a homogeneous catalyst for the formation of ozone in the troposphere, the part of the atmosphere closest to Earth's surface. A heterogeneous catalyst is one that exists in a phase different from the reactants, often a solid. For example, solid beads in an automobile's catalytic converter help convert unburned hydrocarbons to carbon dioxide and water.

Quick Demo

Activation Energy Hold up a strike-anywhere-match and ask students to explain why the match is not burning even though all the necessary reactants are present and the air contains an adequate supply of oxygen. To initiate the reaction, activation energy must be supplied in the form of friction between the match and the matchbook's friction strip. OL

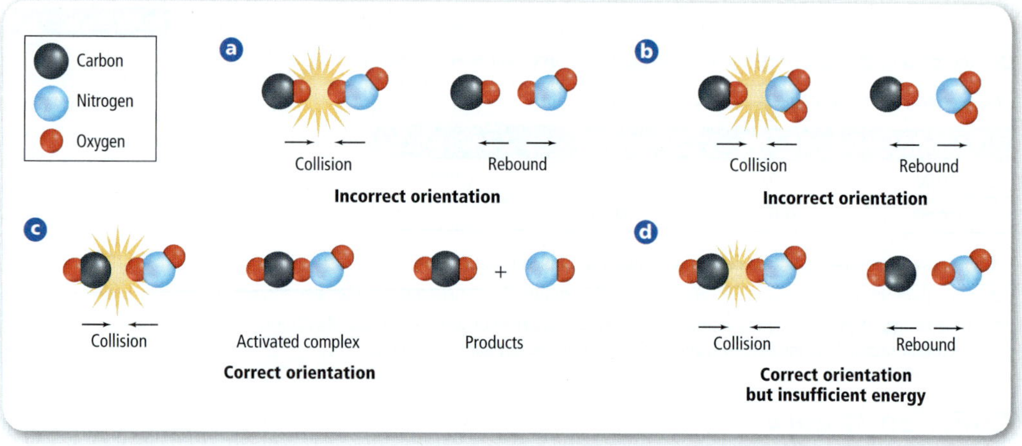

Figure 4 This figure shows four different collision orientations between CO molecules and NO_2 molecules. The collisions in **a** and **b** do not result in a reaction because the molecules are not in position to form bonds. The molecules in **c** collide in the correct orientation, and a reaction occurs. Although the molecules in **d** are also in the correct orientation, they have insufficient energy to react.

View an **animation about molecular orientation and collision effectiveness.**

Concepts In Motion

Collision orientation and the activated complex Why do most collisions fail to produce products? What other factors must be considered? **Figure 4a** and **b** show one possible answer to this question. These illustrations indicate that in order for a collision to lead to a reaction, the carbon atom in a CO molecule must contact an oxygen atom in an NO_2 molecule at the instant of impact. This is the only way in which a temporary bond can form between the carbon atom and an oxygen atom. The collisions shown in **Figure 4a** and **b** do not lead to reactions because the molecules collide in unfavorable orientations. A carbon atom does not contact an oxygen atom at the instant of impact, so the molecules simply rebound.

When the orientation of colliding molecules is correct, as shown in **Figure 4c,** a reaction can occur. An oxygen atom is transferred from an NO_2 molecule to a CO molecule. When this occurs, a short-lived entity called an activated complex is formed, in this case OCONO. An **activated complex,** sometimes called a transition state, is a temporary, unstable arrangement of atoms in which old bonds are breaking and new bonds are forming. As a result, the activated complex might form products or might break apart to re-form the reactants.

Activation energy and reaction rate The collision depicted in **Figure 4d** does not lead to a reaction for a different reason—insufficient energy. Just as the piñata does not break open unless it is hit hard enough, no reaction occurs between the CO and NO_2 molecules unless they collide with sufficient energy. The minimum amount of energy that reacting particles must have to form the activated complex and lead to a reaction is called the **activation energy** (E_a). **Table 1** summarizes the conditions under which colliding particles can react.

A high E_a means that relatively few collisions have the required energy to produce the activated complex, and the reaction rate is slow. A low E_a means that more collisions have sufficient energy to react, and the reaction rate is faster. Think of this relationship in terms of a person pushing a heavy cart up a hill. If the hill is high, a substantial amount of energy is required to move the cart, and it might take a long time to get it to the top. If the hill is low, less energy is required and the task might be accomplished faster.

Table 1 Collision Theory Summary
1. Reacting substances (atoms, ions, or molecules) must collide.
2. Reacting substances must collide in the correct orientation.
3. Reacting substances must collide with sufficient energy to form an activated complex.

DIFFERENTIATED INSTRUCTION

Struggling Learners Have students represent the reaction between CO and NO_2 with ball-and-stick molecular models. Have them use the models to show the favorable collision orientation, and also the unfavorable collision orientation shown in Figure 4. Ask them to explain why each depicted collision either leads to reaction or does not initiate reaction. BL EL

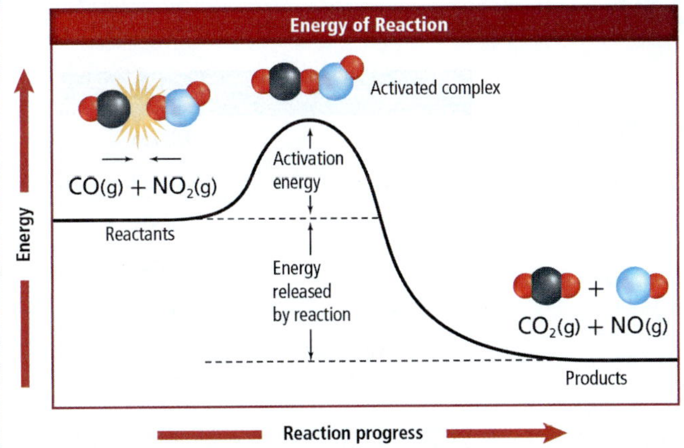

■ **Figure 5** When an exothermic reaction occurs, molecules collide with enough energy to overcome the activation energy barrier. They form an activated complex, then release energy and form products at a lower energy level.

☑ GRAPH CHECK
Explain how you can tell from the graph that the reaction described is an exothermic reaction.

Figure 5 shows the energy diagram for the progress of the reaction between carbon monoxide and nitrogen dioxide. Does this energy diagram look somewhat different from those you previously studied for thermochemistry? Why? This diagram shows the activation energy of the reaction. Activation energy can be thought of as a barrier the reactants must overcome in order to form the products. In this case, the CO and NO_2 molecules collide with enough energy to overcome the barrier, and the products formed lie at a lower energy level. Recall that reactions that lose energy are called exothermic reactions.

For many reactions, the process from reactants to products is reversible. **Figure 6** illustrates the reverse endothermic reaction between CO_2 and NO to re-form CO and NO_2. In this reaction, the reactants lie at a low energy level. They must overcome a significant activation energy to re-form CO and NO_2. This requires a greater input of energy than the forward reaction. If this reverse reaction is achieved, CO and NO_2 again lie at a higher energy level.

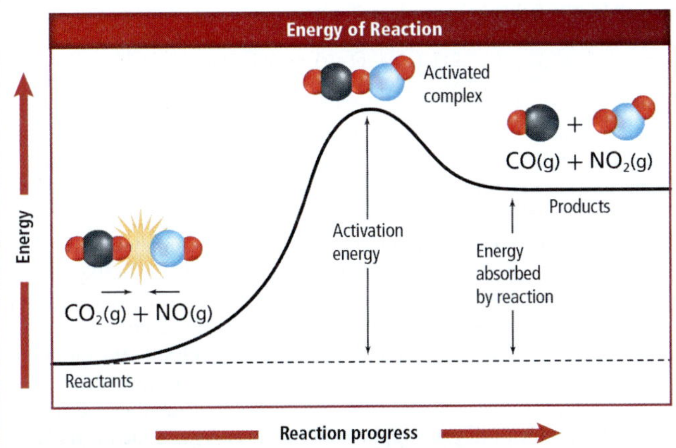

■ **Figure 6** In the reverse reaction, which is endothermic, the reactant molecules are at a lower energy than the products. To react, the reactants must absorb enough energy to overcome the activation energy barrier and form higher-energy products.

☑ GRAPH CHECK
Compare Figures 5 and **6** to determine whether the activation energy for the forward reaction is larger or smaller than the activation energy for the reverse reaction.

ASSESSMENT

Skill Have students write a hypothesis stating what they expect the results of adding more energy to a system would do to the movement of particles in the system. Instruct students to make drawings and illustrations describing the activity of the particles. OL

☑ **GRAPH CHECK FIG. 5** The reactants are at a higher energy level than the products. The energy is released as heat.

☑ **GRAPH CHECK FIG. 6** The activation energy for the reverse reaction is larger than the activation energy for the forward reaction.

Visual Learning

Reaction Direction Point out to students that the energy diagram shown in Figure 6 is simply the reverse of the reaction described in Figure 5 written in the forward direction. The higher energy barrier in Figure 6 indicates that the reverse reaction's activation energy is greater than that of the forward reaction. Explain that this factor favors the forward reaction over the reverse reaction. Also, remind them that the forward reaction is exothermic, which, as they learned previously, also favors the forward reaction. OL

Problem-Solving LAB

Purpose Students will devise a method to illustrate how the kinetic-molecular theory is related to the collision theory and chemical reactions.

Process Skills Analyze and conclude, designing an experiment, interpreting scientific illustrations

Teaching Strategies

- Review with students kinetic-molecular theory. Review also the definition of the collision theory.
- Use this lab and discussion to introduce the concept of factors that affect reaction rates, which are presented in the next section.

Think Critically

1. average rate $= \Delta[N_2O_5]/\Delta t$
   0–20 min:
   rate $= 4.12 \times 10^{-4}$ mol/(L•min)
   40–60 min:
   rate $= 1.18 \times 10^{-4}$ mol/(L•min)
   80–100 min:
   rate $= 3.65 \times 10^{-5}$ mol/(L•min)
2. average rate $= \Delta[NO_2]/\Delta t$
   0–20 min:
   rate $= 8.24 \times 10^{-4}$ mol/(L•min)
   40–60 min:
   rate $= 2.36 \times 10^{-4}$ mol/(L•min)
   80–100 min:
   rate $= 7.30 \times 10^{-5}$ mol/(L•min)
   Because 2 mol of NO_2 are produced for each mol of N_2O_5 consumed, the rates of production of NO_2 are twice the rates of consumption of N_2O_5.
3. The rate of decomposition of N_2O_5 decreases over time.
4. Collision theory states that the reacting atoms, molecules, or ions must collide in order to react. According to that model, as the concentration of N_2O_5 decreases over the time of the experiment, fewer available molecules result in fewer intermolecular collisions per unit time and, therefore, a decreasing reaction rate.

Problem-Solving LAB

Interpret Data

How does the rate of decomposition vary over time? The compound dinitrogen pentoxide (N_2O_5) decomposes in air according to the equation

$$2N_2O_5(g) \rightarrow 4NO_2(g) + O_2(g)$$

Knowing the rate of decomposition allows the concentration of dinitrogen pentoxide to be determined at any time.

Analysis

The table shows the results of an experiment in which the concentration of N_2O_5 was measured over time at normal atmospheric pressure and a temperature of 45°C.

Time (min)	$[N_2O_5]$ (mol/L)
0	0.01756
20.0	0.00933
40.0	0.00531
60.0	0.00295
80.0	0.00167
100.0	0.00094

Think Critically

1. **Calculate** the average reaction rate for each time interval: 0–20 min, 40–60 min, and 80–100 min. Express each rate as a positive number and in moles of N_2O_5 consumed per liter per minute.
2. **Express** the average reaction rate for each time interval in moles of NO_2 produced per liter per minute. Use the reaction equation to explain the relationship between these rates and those calculated in Question 1.
3. **Interpret** the data and your calculations in describing how the average rate of decomposition of N_2O_5 varies over time.
4. **Apply** collision theory to infer why the reaction rate varies as it does.

Spontaneity and Reaction Rate

Recall that reaction spontaneity is related to change in free energy (ΔG). If ΔG is negative, the reaction is spontaneous under the conditions specified. If ΔG is positive, the reaction is not spontaneous. Now consider whether spontaneity has any effect on reaction rates. Are more spontaneous reactions faster than less spontaneous ones?

To investigate the relationship between spontaneity and reaction rate, consider the following gas-phase reaction between hydrogen and oxygen.

$$2H_2(g) + O_2(g) \rightarrow 2H_2O(g)$$

Here, $\Delta G = -458$ kJ at 298 K (25°C) and 1 atm pressure. Because ΔG is negative, the reaction is spontaneous. For the same reaction, $\Delta H = -484$ kJ, which means that the reaction is highly exothermic. You can examine the speed of this reaction by filling a tape-wrapped soda bottle with stoichiometric quantities of the two gases—two volumes hydrogen and one volume oxygen. A thermometer in the stopper allows you to monitor the temperature inside the bottle. As you watch for evidence of a reaction, the temperature remains constant for hours. Have the gases escaped, or have they failed to react?

If you remove the stopper and hold a burning splint to the mouth of the bottle, a reaction occurs explosively. Clearly, the hydrogen and oxygen gases have not escaped from the bottle. Yet, they did not react noticeably until you supplied additional energy in the form of a lighted splint.

VOCABULARY
ACADEMIC VOCABULARY
Investigate
to observe by study or close examination
They decided to investigate how the mice were getting into the house.

ASSESSMENT

Performance Have students draw general energy diagrams for an endothermic reaction, being sure to include an activation energy barrier. Ask them which reaction (forward or reverse) has the greater activation energy. The forward reaction has the greater activation energy. BL EL

DIFFERENTIATED INSTRUCTION

Hearing Impaired Have hearing-impaired students use molecular models to depict the reactants, activated complex, and products for the favorable collision orientation between a CO molecule and a NO_2 molecule. Also, have them correlate the models of reactants, activated complex, and products with the appropriate locations on Figure 4. OL EL

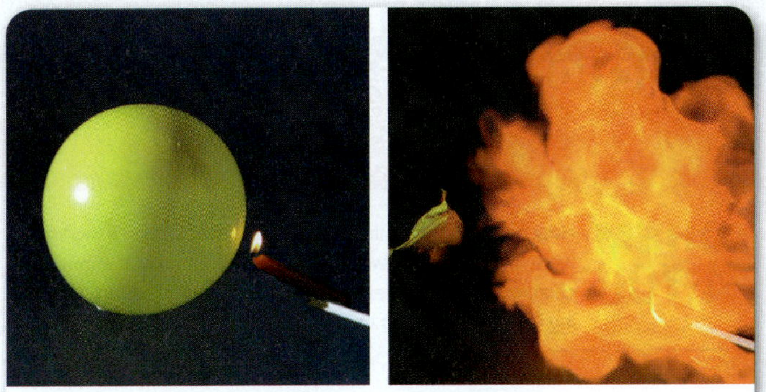

■ **Figure 7** The hydrogen and oxygen in the balloon do not react until the balloon is touched by a flame. Then, an explosive reaction occurs.
Explain *the role of the flame.*

Figure 7 illustrates the reaction between hydrogen and oxygen in a similar way. The balloon is filled with a mixture of hydrogen gas and oxygen gas that appears not to react. When the lighted candle introduces additional energy, an explosive reaction occurs between the gases. Similarly, the air-fuel mixture in the cylinders of a car show little sign of reaction until a spark from a spark plug initiates a small explosion which produces energy to move the car. Logs on the forest floor combine slowly with oxygen in the air as they decompose, but they also combine with oxygen and burn rapidly in a forest fire once they are ignited.

As these examples show, reaction spontaneity in the form of ΔG implies nothing about the speed of the reaction; ΔG indicates only the natural tendency for a reaction or process to proceed. Factors other than spontaneity, however, do affect the rate of a chemical reaction. You will learn about these factors in the next section.

SECTION 1 REVIEW

Section Self-Check

Section Summary

- The rate of a chemical reaction is expressed as the rate at which a reactant is consumed or the rate at which a product is formed.
- Reaction rates are generally calculated and expressed in moles per liter per second (mol/(L • s)).
- In order to react, the particles in a chemical reaction must collide.
- The rate of a chemical reaction is unrelated to the spontaneity of the reaction.

4. **MAIN**IDEA **Relate** collision theory to reaction rate.
5. **Explain** what the reaction rate indicates about a particular chemical reaction.
6. **Compare** the concentrations of the reactants and products during the course of a chemical reaction (assuming no additional reactants are added).
7. **Compare** the average reaction rate measured over an initial, short time interval to one measured over a long time interval.
8. **Describe** the relationship between activation energy and the rate of a reaction.
9. **Summarize** what happens during the brief existence of an activated complex.
10. **Apply** collision theory to explain why collisions between two reacting particles do not always result in the formation of a product.
11. **Interpret** how the speed of a chemical reaction is related to the spontaneity of the reaction.
12. **Calculate** the average rate of a reaction between hypothetical molecules A and B if the concentration of A changes from 1.00*M* to 0.50*M* in 2.00 s.

■ **Caption Question Fig. 7** The flame provides the activation energy needed to initiate the reaction.

3 Assess

Check for Understanding

Ask students to identify which condition of the collision theory is related to the reaction's activation energy and relate how that condition explains why H_2 and O_2 molecules might not react. Activation energy is related to the fact that the molecules must collide with sufficient energy to form the activated complex. **OL**

Reteach

Have students relate the following three reasons why a dart thrown at a target board might not score a 10 (by analogy) to the three reasons that H_2 and O_2 molecules might not react.

1. The dart misses the 10 area on the target. The molecules might not collide.
2. The fins of the dart (rather than the point) hit the target. The molecules might collide with an incorrect orientation.
3. The dart's speed is so low that it bounces off the target. The molecules might collide with insufficient energy to form the activated complex. **OL** **BL**

Extension

Ask a group of students to devise a setup that demonstrates activation energy using small model cars and a flexible track. **OL**

SECTION 1 REVIEW

4. In order for a reaction to occur, molecules, atoms, or ions must collide. The frequency, orientation, and energy of those collisions determine the rate of the overall reaction.
5. The reaction rate indicates the rate of change of the concentration of a reactant or product in mol/(L•s).
6. The concentrations of the reactants decrease, and the concentrations of the products increase at the same rate.
7. As the concentration of the reactants decreases, the reaction rate decreases. Thus, the longer the time period, the smaller the average reaction rate.
8. The higher the activation energy, the slower the rate of the reaction.
9. Bonds in the reactants are in the process of breaking, while new bonds are beginning to form to produce the products.
10. The collision must be in a correct orientation and have sufficient energy to form the activated complex.
11. The speed and spontaneity of a chemical reaction are not related.
12. Rate = 0.25 mol/(L•s)

SECTION 2

1 Focus

MAINIDEA

Surface Area and Rate Measure the mass of a sizeable chunk of $CuSO_4 \cdot 5H_2O$. Measure out an equal mass of small crystals of the same compound. Place both samples in beakers of equal size with magnetic stirring bars, and add equal volumes of water. Place both beakers on magnetic stirrers set on the same speed, and have students observe and compare the rates at which the $CuSO_4$ samples dissolve, as evidenced by the developing intensities of blue color. Ask students to explain the difference. The small crystals dissolve faster because they present a greater surface area in contact with molecules of water. Point out that surface area is another factor that affects chemical reaction rates. BL OL AL

2 Teach

Reinforcement

Electron Configuration Remind students that, in general, an element's reactivity is related to its electron configuration. For example, elements whose atoms have one electron in their outermost energy levels and elements whose atoms lack a single electron tend to be extremely reactive. OL

Caption Question Fig. 8

a. $Cu + 2AgNO_3 \rightarrow 2Ag + Cu(NO_3)_2$

b. $Zn + 2AgNO_3 \rightarrow 2Ag + Zn(NO_3)_2$

SECTION 2

Factors Affecting Reaction Rates

MAINIDEA Factors such as reactivity, concentration, temperature, surface area, and catalysts affect the rate of a chemical reaction.

Essential Questions

- What are the factors that affect the rates of chemical reactions?
- What is the role of a catalyst?

Review Vocabulary

concentration: a quantitative measure of the amount of solute in a given amount of solvent or solution

New Vocabulary

catalyst
inhibitor
heterogeneous catalyst
homogeneous catalyst

CHEM 4 YOU How quickly do you think a forest fire would spread if the trees were far apart or the wood were damp? Similarly, the rate of a chemical reaction is dependent on a number of factors, including the concentrations and physical properties of the reactants.

The Nature of Reactants

Some substances react more readily than others. For example, copper and zinc are both metals and they have similar physical properties because of their relative positions on the periodic table. But they react at different rates when placed in aqueous silver nitrate solutions of equal concentration.

When a copper strip is placed in 0.05*M* silver nitrate, as shown in **Figure 8a,** the copper and silver nitrate react to form silver metal and aqueous copper(II) nitrate. When a zinc strip is placed in 0.05*M* silver nitrate, as shown in **Figure 8b,** the zinc and silver nitrate react to form silver metal and aqueous zinc nitrate. You can see that the reactions are similar. However, compare the amounts of silver formed in the two photographs, which were taken after the same number of minutes had elapsed. **Figure 8** shows that more silver formed in the reaction of zinc and silver nitrate than in the reaction of copper and silver nitrate. The reaction of zinc with silver nitrate occurs faster because zinc is higher on the activity series than copper.

Copper strip in silver nitrate

Zinc strip in silver nitrate

Figure 8 Zinc is more reactive than copper, so it reacts with silver nitrate faster than copper does.
Write *the balanced equations for the reactions at right.*

Demonstration

Surface Area and Reaction Rate

Purpose
to demonstrate that increasing the surface area of a reactant increases the rate of a chemical reaction

Materials
Lycopodium powder (2 g); ceramic pad; laboratory burner or butane lighter

Safety Precautions
WARNING: *The falling dust will produce a roaring flame. Perform this experiment under a fume hood.*

Disposal Dispose of the product in the trash.

Procedure
Dim the room lights. Using a burner or lighter, attempt to ignite about 2 g of lycopodium powder placed in a small pile

The concentration of oxygen in the air surrounding the candle is about 20%.

The candle burns more rapidly because the jar contains almost 100% oxygen.

Figure 9 The brighter flame in the jar containing a greater amount of oxygen indicates an increase in reaction rate. The higher oxygen concentration accounts for the faster reaction.

Concentration

One way chemists can change the rate of a reaction is by changing the concentrations of the reactants. Remember that collision theory states that particles must collide in order to react. The more particles that are present, the more often collisions occur. Think about bumper cars at an amusement park. When more cars are in operation, the number of collisions increases. The same is true for a reaction in which Reactant A combines with Reactant B. At given concentrations of A and B, molecules of A and B collide to produce AB at a particular rate. What happens if the concentration of B is increased? Molecules of A collide with molecules of B more frequently because more molecules of B are available. More collisions ultimately increase the rate of reaction.

☑ READING CHECK **Predict** what would happen to the rate of the reaction if the concentration of A was increased.

Look at the reactions shown in **Figure 9.** The wax in the candle undergoes combustion. In the first photo, the candle burns in air. How does this compare with the second photo, in which the burning candle is placed inside a jar containing nearly 100% oxygen–approximately five times the concentration of oxygen in air? According to collision theory, the higher concentration of oxygen increases the collision frequency between the wax molecules in the candle and oxygen molecules. As a result, the rate of the reaction increases, resulting in a larger, brighter flame.

Surface Area

Now suppose you lowered a red-hot chunk of steel into a flask of oxygen gas and a red-hot bundle of steel wool into another flask of oxygen gas. What might be different? The oxygen would react with the chunk of steel much more slowly than it would with the steel wool. Using what you know about collision theory, can you explain why? You are correct if you said that, for the same mass of iron, steel wool has more surface area than the chunk of steel. The greater surface area of the steel wool allows oxygen molecules to collide with many more iron atoms per unit of time.

FOLDABLES®
Incorporate information from this section into your Foldable.

APPLYING PRACTICES

Apply Scientific Principles and Evidence Go to the resources tab in ConnectED to find the Applying Practices worksheet *Concentration, Temperature, and Reaction Rates.*

Quick Demo

Surface Area Pour 25 mL of household vinegar into each of two petri dishes, and project an image of the dishes. With the projector on, add 0.1 g of magnesium turnings simultaneously to one dish and 0.1 g of 20-mesh granular magnesium to the other dish. Because the granular magnesium presents a greater surface area than the magnesium turnings, the granular magnesium's bubbling reaction occurs at a higher rate. The residue from this reaction can be flushed down a drain. OL

ASSESSMENT

Knowledge Ask students to think of at least one (each) highly reactive and at least one (each) relatively unreactive solid, liquid, and gas. Highly reactive materials: solids —TNT, dynamite, gunpowder; liquids—nitroglycerin, gasoline; gases—hydrogen, methane, propane. Relatively unreactive materials: solids—teflon, treated wood; liquids—water, liquid nitrogen; gases— helium, neon, nitrogen. OL

☑ READING CHECK The reaction rate would increase because more molecules of A would be present to collide with B.

FOLDABLES® Rubric

on a ceramic pad. Have students observe that there is no reaction. Then, lift the pad containing the powder about 40 cm above the burner flame, and turn the pad over. Have students discuss the effect of increased surface area on reaction rate. Relate this to dust explosions in grain elevators.

Results
The lycopodium powder does not ignite until dispersed above the flame. The falling dust ignites explosively.

Analysis

1. Summarize this demonstration. Reaction rate increases when surface area increases.
2. Predict which would dissolve first, a sugar cube or the same mass of granulated sugar. the granulated sugar
3. What do you think happens to the collision rate between oxygen molecules and dust particles as the surface area increases? The collision rate increases as surface area increases.

ASSESSMENT

Performance Have interested students photograph or shoot video of the demonstration; then write a narrative describing the demonstration at the molecular level. OL EL

ChemLAB

The ChemLab located at the end of the chapter can be used at this point in the lesson.

Identify Misconceptions

Students often confuse catalysts and intermediates in reaction mechanisms.

Uncover the Misconception Have students examine the three elementary reactions that comprise the mechanism for the complex reaction shown in Figure 20. Ask if N_2O_2 and N_2O are catalysts or intermediates. Both are intermediates because they are produced in one elementary step and consumed in a subsequent step.

Demonstrate the Concept Have students examine the following general reaction mechanism and identify catalysts and intermediates.

$A + B + C \rightarrow X$

$X + D \rightarrow Y + A$

$Y \rightarrow E$

X and Y are intermediates. A is a catalyst.

Assess New Knowledge Have students write the overall reaction for the previous reaction. $B + C + D \rightarrow E$. Ask students to explain why A is a catalyst whereas B and C are reactants. A is a catalyst because it is not consumed. However, B and C are reactants because they are rearranged into different molecules in the first elementary step and are not reformed. OL

GRAPH CHECK 23

Figure 10 The greater surface area of the steel wool means that more collisions can occur between the metal and oxygen.

For the same mass, many small particles have more total surface area than one large particle. For example, observe the reactions shown in **Figure 10.** The hot nail glows in oxygen in **Figure 10a,** but the same mass of steel wool in **Figure 10b** bursts into flames. Increasing the surface area of a reactant speeds up the rate of reaction by increasing the collision rate between reacting particles.

Temperature

Increasing the temperature of a reaction generally increases the rate of a reaction. For example, you know that the reactions that cause foods to spoil occur faster at room temperature than when the foods are refrigerated. The graph in **Figure 11** illustrates that increasing the temperature by 10 K can approximately double the rate of a reaction. How can such a small increase in temperature have such a significant effect?

Recall that increasing the temperature of a substance increases the average kinetic energy of the particles that make up the substance. For that reason, reacting particles collide more frequently at higher temperatures than at lower temperatures. However, that fact alone does not account for the increase in reaction rate with increasing temperature.

To better understand how reaction rate varies with temperature, examine the second graph in **Figure 11.** This graph compares the numbers of particles that have sufficient energy to react at temperatures T_1 and T_2, where T_2 is greater than T_1. The dotted line indicates the activation energy (E_a) for the reaction. The shaded area under each curve represents the number of collisions that have energy equal to or greater than the activation energy. How do the shaded areas compare? The number of high-energy collisions at the higher temperature, T_2, is greater than the number at the lower temperature, T_1. Therefore, as the temperature increases, more collisions result in a reaction.

GRAPH CHECK **Determine** the relative reaction rate at 325 K.

Figure 11 Increasing the temperature of a reaction increases the frequency of collisions and therefore the rate of the reaction. Increasing the temperature also raises the kinetic energy of the particles. More of the collisions at high temperatures have enough energy to overcome the activation energy barrier and react.

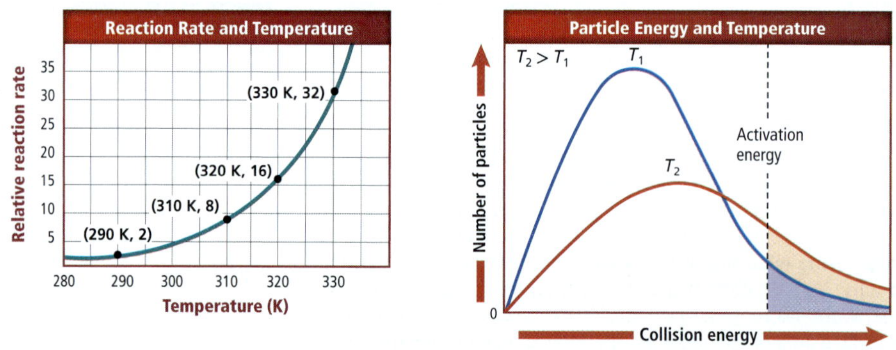

CHEMISTRY PROJECT

Reaction Pathways Have students make a general energy diagram for each of the following types of chemical reactions: exothermic, endothermic. The diagrams should show both uncatalyzed and catalyzed reaction pathways. OL

CHEMISTRY JOURNAL

Kitchen Reaction Rates Ask students to keep notes while they watch or assist in the preparation of at least five meals. Their notes should describe ways in which conditions in their kitchens are varied or controlled to influence the rates of reactions during food storage and preparation. OL EL

MiniLAB

Examine Reaction Rate and Temperature

What is the effect of temperature on a common chemical reaction?

Procedure

1. Read and complete the lab safety form.
2. Break a single **effervescent tablet** into four equal pieces.
3. Use a **balance** to measure the mass of one piece of the tablet. Measure 50 mL of room-temperature **water** (approximately 20°C) into a **250-mL beaker.** Use a **nonmercury thermometer** to measure the temperature of the water.
4. With a **stopwatch** or a **clock with a second hand** ready, add the piece of tablet to the water. Record the amount of time elapsed between when the tablet hits the water and when all of the solid has dissolved.
5. Repeat Steps 3 and 4, this time gradually warming the 50 mL of water to about 50°C on a **hot plate.** Maintain the temperature (equilibrate) throughout the run.

Analysis

1. **Identify** the initial mass, the final mass, and t_1 and t_2 for each trial run.
2. **Calculate** the reaction rate by finding the mass of reactant consumed per second for each run.
3. **Describe** the relationship between reaction rate and temperature for this reaction.
4. **Predict** what the reaction rate would be if the reaction were carried out at 40°C and explain the basis for your prediction. To test your prediction, repeat the reaction at 40°C using another piece of tablet.
5. **Evaluate** how well your prediction for the reaction rate at 40°C compares to the measured reaction rate.

Catalysts and Inhibitors

The temperature and the concentration of reactants affect the rate of a reaction, but an increase in temperature is not always the best, or most practical, thing to do. For example, suppose that you want to increase the rate of the decomposition of glucose in a living cell. Increasing the temperature and/or the concentration of reactants is not an option because doing so might harm or kill the cell.

Catalysts Many chemical reactions in living organisms would not occur quickly enough to sustain life at normal living temperatures if it were not for the presence of enzymes. An enzyme is a type of **catalyst,** a substance that increases the rate of a chemical reaction without being consumed in the reaction. Catalysts are used extensively in manufacturing because quickly producing more of a product reduces its cost. A catalyst does not yield more product and is not included in either the reactants or the products of the reaction. Thus, catalysts are not included in chemical equations.

Inhibitors Another type of substance that affects reaction rates is called an inhibitor. Unlike a catalyst, which speeds up reaction rates, an **inhibitor** is a substance that slows down reaction rates, or inhibits reactions. Some inhibitors prevent a reaction from happening at all.

How catalysts and inhibitors work A catalyst lowers the activation energy required for a reaction to take place at a given temperature. Recall that a low activation energy means that more of the collisions between particles will have sufficient energy to overcome the activation energy barrier and bring about a reaction. By lowering the activation energy, a catalyst increases the average reaction rate.

RealWorld CHEMISTRY

Excluding Oxygen

FOOD PRESERVATION Foods often spoil because they react with oxygen. Many methods of food preservation maintain product freshness by excluding oxygen. For example, apples stored in an atmosphere of carbon dioxide can be kept fresh long after harvest. Foods such as popcorn and crackers are often packaged in an atmosphere of an unreactive gas such as nitrogen or argon.

DIFFERENTIATED INSTRUCTION

Advanced Learners Have students research and write descriptions of how changes in temperature affect the rate law for a chemical reaction. The form of the rate law does not change, but changing the temperature changes the value of the rate constant *k*. The rate constant, *k*, is proportional to $e^{-E_a/RT}$, where $e^{-E_a/RT}$ reflects the fraction of collisions with sufficient energy to initiate a reaction. AL

MiniLAB

Rubric

Purpose Students will determine the effect of temperature on the rate of a chemical reaction.

Process Skills collect and interpret data, make and use graphs, observe, predict recognize cause and effect

Safety Precautions Approve Lab Safety Forms before work begins. Hot plate surface will be hot. Use only a GFI protected receptacle for the hot plate.

Teaching Strategies

- Use an effervescent tablet that contains sodium hydrogen carbonate and citric acid or another weak acid.
- Describe the chemical reaction for the tablet: acetyl salicylic acid (aspirin) + citric acid + sodium hydrogen carbonate → sodium citrate + sodium acetyl salicylate + water + carbon dioxide (produces the bubbles).

Expected Results As the temperature increases the rate of reaction will increase.

Analysis

1. Data will vary.
2. The reaction rate will vary, but the reaction rate should be faster at 50°C than it is at 20°C.
3. The reaction proceeds faster at the higher temperature.
4. A ten degree increase in temperature should result in doubling the rate of the reaction. Thus, a reasonable prediction is that the reaction rate would quadruple.
5. Possible answer: The prediction was not validated by the experimental data possibly because of differences in the masses of the tablets used. But the rate at 40°C was higher than the rate at 20°C.

ASSESSMENT

Skill Instead of using paper for graphing, have students analyze data from the MiniLab using a graphing calculator or a computer graphing program. AL

Apply Chemistry

Ozone Consuming Vehicles Explain to students that in the future, automobiles and air conditioners might purify our air, consuming ozone and carbon dioxide as they operate. A company that manufactures catalytic converters for automobiles, has developed a catalyst that catalyzes the conversion of ozone (O_3) to diatomic oxygen (O_2) and the conversion of carbon monoxide (CO) to carbon dioxide (CO_2). Because the catalyst works well at warm temperatures, it could be painted on automobile radiators, air conditioner compressors, and other devices in which fans move large volumes of air for cooling purposes.

Use Science Terms

Names for Systems Have students explain the meanings of the terms *homogeneous* and *heterogeneous,* and give example of homogeneous and heterogeneous systems. OL EL

GRAPH CHECK FIG. 12 The catalyst changes the activation energy only; it does not affect the energy level of either the reactants or products. Therefore, the energy released is not affected.

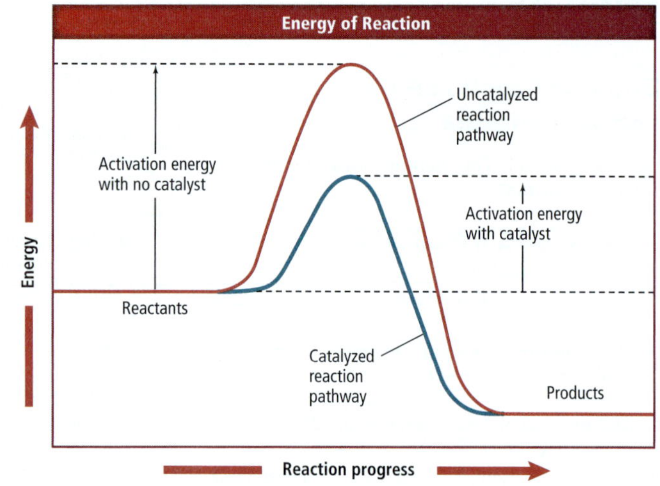

Figure 12 The activation energy of the catalyzed reaction is lower than that of the uncatalyzed reaction. Thus the catalyzed reaction produces products at a faster rate than the uncatalyzed reaction does.

GRAPH CHECK
Determine from the graph how the use of a catalyst affects the energy released in the reaction.

Figure 12 shows the energy diagram for an exothermic chemical reaction. The red line represents the reaction pathway with no catalyst present. The blue line represents the catalyzed reaction pathway. Note that the activation energy for the catalyzed reaction is much lower than for the uncatalyzed reaction. You can think of the reaction's activation energy as an obstacle to be cleared, as shown in **Figure 13.** In this analogy, much less energy is required for the horse and rider to clear the lower barrier than to jump the higher hurdle.

Inhibitors can act in a variety of ways. Some block lower energy pathways and thus raise the activation energy of a reaction. Others react with the catalyst and destroy it or prevent it from performing its function. In biological reactions, an inhibitor might bind the enzyme that catalyzes a reaction and prevent the reaction from occurring. In the food industry, inhibitors are called preservatives or antioxidants. Preservatives are safe to eat and give food longer shelf lives.

Figure 13 A higher activation energy means that reacting particles must have more energy in order to react. The horse and rider exert little energy jumping the low barrier. Greater speed and energy are needed to clear the higher hurdle.

DIFFERENTIATED INSTRUCTION

Visually Impaired Help visually impaired students understand the effect of a catalyst on activation energy. Have them simulate an exothermic reaction with a large activation energy by first rolling a ball up an incline until it reaches the top and falls or rolls down to a lower level (such as a ramp or hill on school property). Explain that the incline represents the activation energy (an energy barrier that must be overcome to initiate the reaction). The high point on the incline represents the reactants' energy, and the low point represents the products' energy. Then, to simulate the lower activation energy resulting from a catalyst, have students perform the same activity with a lower incline. OL EL

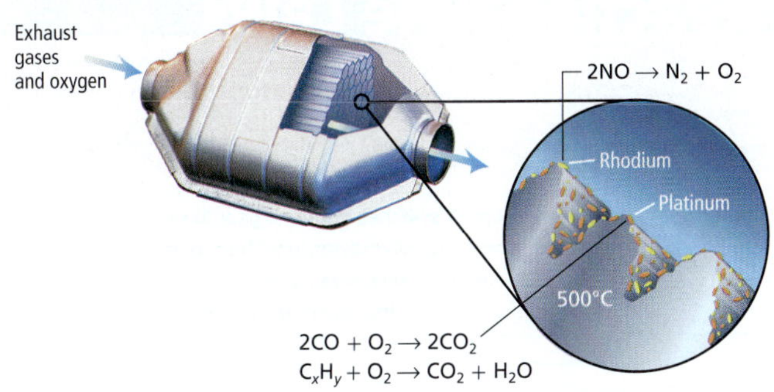

Figure 14 The inside of a catalytic converter is coated with particles of rhodium and platinum. At 500°C, rhodium catalyzes the conversion of nitrogen oxide (NO) to nitrogen (N_2) and oxygen (O_2). Platinum catalyzes the conversion of carbon monoxide (CO) to carbon dioxide (CO_2) and converts any unburned gasoline, represented by C_xH_y, to carbon dioxide and water vapor (H_2O).

Heterogeneous and homogeneous catalysts Today's automobiles are required by law to be equipped with catalytic converters. **Figure 14** shows the reactions within a catalytic converter that convert harmful exhaust gases to acceptable substances. Nitrogen monoxide is converted to nitrogen and oxygen, carbon monoxide to carbon dioxide, and unburned gasoline to carbon dioxide and water. The most effective catalysts for this application are transition metal oxides and metals such as rhodium and platinum. Because the catalysts in a catalytic converter are solids and the reactions they catalyze are gaseous, the catalysts are called heterogeneous catalysts. A **heterogeneous catalyst** exists in a physical state different than that of the reaction it catalyzes. A catalyst that exists in the same physical state as the reaction it catalyzes is called a **homogeneous catalyst.** In the Launch Lab, you used a heterogenous catalyst (yeast) to speed up the decomposition of hydrogen peroxide. The same result can be obtained by using a potassium iodide (KI) solution. Iodide ions (I^- (aq)), present in the same physical state as the hydrogen peroxide molecules, act as a homogeneous catalyst in the decomposition.

Explore **reaction rates.**

Virtual Investigations

SECTION 2 REVIEW

Section Self-Check

Section Summary

- Key factors that influence the rate of chemical reactions include reactivity, concentration, surface area, temperature, and catalysts.
- Raising the temperature of a reaction generally increases the rate of the reaction by increasing the collision frequency and the number of collisions that form an activated complex.
- Catalysts increase the rates of chemical reactions by lowering activation energies.

13. **MAIN**IDEA **Explain** why magnesium metal reacts with hydrochloric acid (HCl) at a faster rate than iron does.
14. **Explain** how collision theory accounts for the effect of concentration on reaction rate.
15. **Explain** the difference between a catalyst and an inhibitor.
16. **Describe** the effect on the rate of a reaction if one of the reactants is ground to a powder rather than used as a single chunk.
17. **Infer** If increasing the temperature of a reaction by 10 K approximately doubles the reaction rate, what would be the effect of increasing the temperature by 20 K?
18. **Research** how catalysts are used in industry, in agriculture, or in the treatment of contaminated soil, waste, or water. Write a short report summarizing your findings about the role of a catalyst in one of these applications.

3 Assess

Check for Understanding

Provide students with a general energy curve for an endothermic reaction—both catalyzed and uncatalyzed. Check for understanding by asking them to label the following: products, reactants, activated complex, catalyzed reaction pathway, and uncatalyzed reaction pathway. OL

Reteach

Ask students to explain the two factors that cause the rate of a chemical reaction to increase with increasing temperature. Increasing the temperature increases both the reacting particles' collision frequency and collision energy. OL

Extension

Ask students to investigate the reactions used to develop and print photographic film. Have them pay particular attention to how photographers control the rates of these reactions. You might want to have a student or a group of students make a class presentation of their findings. OL COOP LEARN

ASSESSMENT

Skill Ask students to create a table that summarizes the effects of the following factors on the rates of chemical reactions: reactivity, concentration, surface area, temperature, and catalysts. OL

SECTION 2 REVIEW

13. Magnesium is a more active metal than iron, which means that magnesium is more reactive. Thus, the reaction of magnesium with HCl is faster than the reaction of iron.
14. Increasing reactant concentration increases collision frequency between reactant particles.
15. A catalyst speeds up the rate of a reaction by lowering the activation energy. An inhibitor slows or even stops a reaction by interfering with the reactants or with the catalyst.
16. The rate of the reaction increases because more surface area is available for reaction.
17. The rate would quadruple.
18. Answers will vary depending upon the application chosen, but all reports should have in common that the catalyst increases the rate of the reaction and that it is not consumed in the reaction.

SECTION 3

1 Focus

MAINIDEA

Direct Proportions and Rate Laws Draw a rectangle on the board, and ask students what formula they would use to calculate the area. $A = l \times w$. Then, double the length of the rectangle to $2l$, leaving the width unchanged. Ask students why and how the area has changed. Because the length has changed to $2l$, the new area is twice the original area. Point out that a chemical reaction's rate law is similar in that the reaction's rate is related mathematically to the concentrations of the reactants. If the reaction is first order in a reactant, for example, doubling its concentration doubles the rate. BL OL AL

2 Teach

ASSESSMENT

Performance Have students design and carry out an investigation in which they evaluate the effect of vinegar concentration on the rate of the reaction between vinegar, $CH_3COOH(aq)$, and baking soda, $NaHCO_3(s)$. Student experiments could involve diluting vinegar and comparing the rates at which $CO_2(g)$ is produced from given concentrations of vinegar for a given mass of baking soda. OL

SECTION 3

Reaction Rate Laws

MAINIDEA The reaction rate law is an experimentally determined mathematical relationship that relates the speed of a reaction to the concentrations of the reactants.

Essential Questions

- What is the relationship between reaction rate and concentration?
- How are reaction orders determined using the method of initial rates?

Review Vocabulary

reactant: the starting substance in a chemical reaction

New Vocabulary

rate law
specific rate constant
reaction order
method of initial rates

CHEM 4 YOU When a bicyclist switches from first gear to second gear, the bicycle travels a greater distance with each revolution of the pedals. In the same way, when a chemist increases the concentration of a reactant, the rate of the reaction increases.

Writing Reaction Rate Laws

In Section 1, you learned how to calculate the average rate of a chemical reaction. The word *average* is important because most chemical reactions slow down as the reactants are consumed and fewer particles are available to collide. Chemists quantify the results of collision theory in an equation called a rate law. A **rate law** expresses the relationship between the rate of a chemical reaction and the concentration of reactants at a given temperature. For example, the reaction A → B is a one-step reaction. The rate law for this reaction is expressed as follows.

One-Step Reaction Rate Law

$$\text{rate} = k[\text{A}]$$

[A] represents the concentration of a reactant; *k* is a constant.

The rate of a one-step reaction is the product of the concentration of the reactant and a constant.

The symbol k is the **specific rate constant,** a numerical value that relates the reaction rate and the concentrations of reactants at a given temperature. The specific rate constant is unique for every reaction and can have a variety of units including L/(mol•s), L^2/(mol^2•s), and s^{-1}. A rate law must be determined experimentally as illustrated in **Figure 15.**

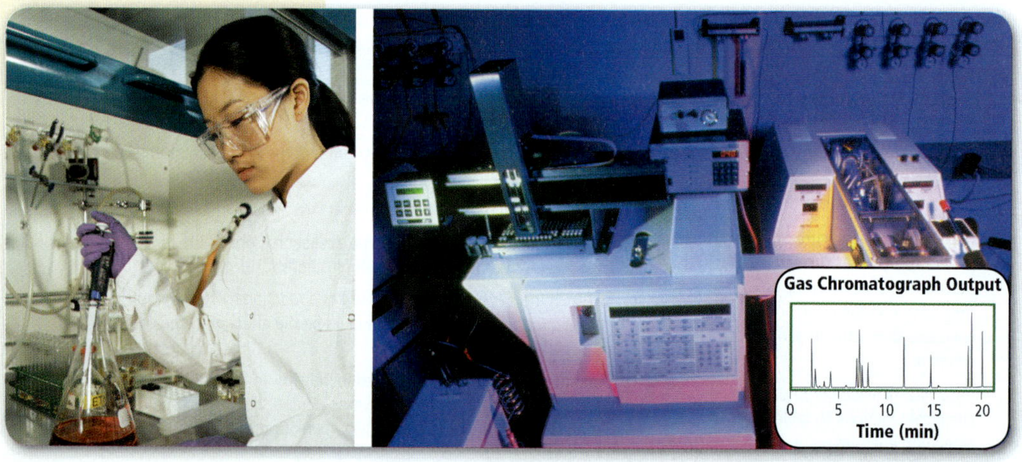

Figure 15 Gas chromatography is one technique that can be used for reaction rate determination. To determine the rate of a reaction, samples of the reaction mixture are withdrawn at regular intervals while the reaction is proceeding. The samples are immediately injected into a gas chromatograph, which separates the components and helps identify them.

DIFFERENTIATED INSTRUCTION

English Learners Have English learners look up in a dictionary the following key terms and write a sentence using each: *instantaneous, elementary, complex, mechanism,* and *intermediate.* Also, ask these students to teach the class how to speak and write these words in their native languages. EL BL

The rate law shows that the reaction rate is directly proportional to the molar concentration of A. The specific rate constant, k, does not change with concentration; however, k does change with temperature. A large value of k means that A reacts rapidly to form B.

First-order reaction rate laws In the expression $Rate = k[A]$, it is understood that the notation [A] means the same as $[A]^1$. For reactant A, the understood exponent 1 is called the reaction order. The **reaction order** for a reactant defines how the rate is affected by the concentration of that reactant. For example, the rate law for the decomposition of H_2O_2 is expressed by the following equation.

$$\text{Rate} = k[H_2O_2]$$

Because the reaction rate is directly proportional to the concentration of H_2O_2 raised to the first power ($[H_2O_2]^1$), the decomposition of H_2O_2 is said to be first order in H_2O_2. Because the reaction is first order in H_2O_2, the reaction rate changes in the same proportion that the concentration of H_2O_2 changes. So, if the H_2O_2 concentration decreases to one-half its original value, the reaction rate is also reduced by one-half.

Recall that reaction rates are determined from experimental data. Because reaction order is based on reaction rates, it follows that reaction order is also determined experimentally. Finally, because the rate constant, k, describes the reaction rate, it must also be determined experimentally. The graph in **Figure 16** shows how the initial reaction rate for the decomposition of H_2O_2 changes with the concentration of H_2O_2.

☑ READING CHECK **Infer** If the reaction order for a reactant is first order, how will the rate of the reaction change if the concentration of the reactant is tripled?

Figure 16 The graph shows a direct relationship between the concentration of H_2O_2 and its rate of decomposition.

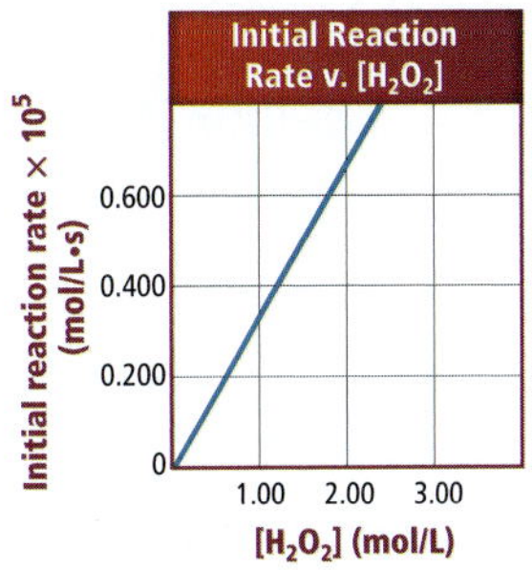

☑ GRAPH CHECK

Apply Using the graph, determine the initial reaction rate when $[H_2O_2]$ is 1.50 mol/L.

Other-order reaction rate laws The overall reaction order of a chemical reaction is the sum of the orders for the individual reactants in the rate law. Many chemical reactions, particularly those that have more than one reactant, are not first-order. Consider the general form for a chemical reaction with two reactants. In this chemical equation, a and b are coefficients.

$$aA + bB \rightarrow \text{products}$$

The general rate law for such a reaction is described below.

The General Rate Law

$$\text{rate} = k[A]^m[B]^n$$

[A] and [B] represent the concentrations of reactants A and B. The exponents m and n are the reaction orders.

The rate of a reaction is equal to the product of k and the concentrations of the reactants each raised to a power (order) that is determined experimentally.

Only if the reaction between A and B occurs in a single step (and with a single activated complex) does $m = a$ and $n = b$. That is unlikely, however, because single-step reactions are not common. For example, consider the multi-step reaction between nitrogen monoxide (NO) and hydrogen (H_2), which is described by the following equation.

$$2NO(g) + 2H_2(g) \rightarrow N_2(g) + 2H_2O(g)$$

Quick Demo

Oxygen and Combustion Rate Measure approximately 0.4 g of MnO_2 into a 400-mL beaker. Pour 40 mL of 6% H_2O_2 into the beaker, and allow oxygen gas to accumulate by covering the beaker with a watch glass or glass plate. Wearing goggles and an apron, hold a wooden laboratory splint with forceps or tongs and ignite the tip of the splint. After observing the reaction rate in air, blow out the flame, remove the cover, and immediately lower the glowing splint into the beaker. The splint will burst into flame and burn brightly in the oxygen gas, demonstrating the presence of oxygen concentration in the rate law for the combustion of wood. Filter the liquid from the MnO_2, and dispose of the MnO_2 as solid waste. Flush the filtrate down a drain.

☑ GRAPH CHECK 0.460×10^5 or 4.60×10^4 mol/(L•s)

☑ READING CHECK The rate of reaction will be tripled.

DIFFERENTIATED INSTRUCTION

Advanced Learners Have students create a data table showing initial molar concentrations of Substances A and B for three trials in the reaction $A + B \rightarrow C$. Give them the rate law $\text{Rate} = k[A]^2[B]^3$, and have them express the reaction rate as moles of C produced per liter per second. Data should show that when [A] is doubled, the rate increases by a factor of 4; when [B] is doubled, the rate increases by a factor of 8. AL

Math in Chemistry

Exponents Lead students through several examples of mathematical operations to make sure they understand how to use exponents in rate laws. Use 0.10 for the rate constant k and initial $[A] = 0.50$ in the following examples. In each example, [A] is doubled in Trial 2.

- $Rate = k[A]$ (understood exponent is 1) Trial 1: $Rate = 0.10(0.50) = 0.050$; Trial 2: $Rate = 0.10(1.00) = 0.10$; Point out that doubling [A] doubled the rate because $2^1 = 2$.
- $Rate = k[A]^2$ Trial 1: $Rate = 0.10(0.50)^2 = 0.025$; Trial 2: $Rate = 0.10(1.00)^2 = 0.10$; point out that doubling [A] increased the rate by a factor of 4 because $2^2 = 4$.
- $Rate = k[A]^3$ Trial 1: $Rate = 0.10(0.50)^3 = 0.0125$; Trial 2: $Rate = 0.10(1.00)^3 = 0.10$; point out that doubling [A] increased the rate by a factor of 8 because $2^3 = 8$. OL

☑ READING CHECK The overall order is the sum of the exponents on the terms in the rate equation.

PRACTICE Problems

Have students refer to the Selected Solutions appendix for complete solutions to odd-numbered problems. The complete solutions for all problems can be found in the Solutions Manual.

19. $Rate = k[A]^3$
20. $Rate = k[O_2][NO]^2$
21. Examining trials 1 and 2, doubling [A] has no effect on the rate; therefore, the reaction is zero order in A. Examining trials 2 and 3, doubling [B] doubles the rate; therefore, the reaction is first order in B. $Rate = k[A]^0[B] = k[B]$
22. Examining trials 1 and 2, doubling $[CH_3CHO]$ increases the rate by a factor of four. Examining trials 2 and 3, $[CH_3CHO]$ is again doubled so the rate again must increase by a factor of four. Therefore, the rate for trial 3 is 43.2×10^{-11} mol/(L•s).

This reaction has the following rate law.

$$rate = k[NO]^2[H_2]$$

The rate law was determined from experimental data that indicate that the rate depends on the concentration of the reactants as follows: If [NO] doubles, the rate quadruples; if $[H_2]$ doubles, the rate doubles. The reaction is described as second order in NO, first order in H_2, and third order overall. The overall order is the sum of the orders for the individual reactants (the sum of the exponents), which is (2 + 1), or 3.

☑ READING CHECK **Explain** how you can determine the overall order of the reaction from the rate equation.

VOCABULARY

WORD ORIGIN

Initial
adjective from Latin *initium,* meaning *of or relating to the beginning*

Determining Reaction Order

One common experimental method of evaluating reaction order is called the method of initial rates. The **method of initial rates** determines reaction order by comparing the initial rates of a reaction carried out with varying reactant concentrations. The initial rate measures how fast the reaction proceeds at the moment at which the reactants are mixed and the concentrations of the reactants are known. To understand how this method works, consider the general reaction $aA + bB \rightarrow$ products. Suppose that the reaction is carried out three times with varying concentrations of A and B and yields the initial reaction rates shown in **Table 2.** Recall that the general rate law for this type of reaction is as follows.

$$rate = k[A]^m[B]^n$$

To determine m, the exponent of [A], compare the concentrations and reaction rates in Trials 1 and 2. As you can see from the data, while the concentration of B remains constant, the concentration of A in Trial 2 is twice that of Trial 1. Note that the initial rate in Trial 2 is twice that of Trial 1. Because doubling [A] doubles the rate, the reaction must be first order in A. Since you are doubling the concentration, consider how 2^m affects the rate. The rate also doubles, so $2^m = 2$. Because $2^m = 2$, m must equal 1. The same method is used to determine n, the exponent of [B], except this time Trials 2 and 3 are compared. Doubling the concentration of B causes the rate to increase by four times. Because $2^n = 4$, n must equal 2. Thus the reaction is second order in B, giving the following overall rate law.

$$rate = k[A]^1[B]^2$$

The overall reaction order is third order (sum of exponents 2 + 1 = 3).

Table 2 Experimental Initial Rates for $aA + bB \rightarrow$ products

Trial	Initial [A](*M*)	Initial [B](*M*)	Initial Rate (mol/(L · s))
1	0.100	0.100	2.00×10^{-3}
2	0.200	0.100	4.00×10^{-3}
3	0.200	0.200	16.00×10^{-3}

CHEMISTRY JOURNAL

Reaction Rate Laws Have students create tables in their journals in which they describe the effect of doubling, tripling, and quadrupling [A] on the overall rate of chemical reactions that have the following rate laws: $Rate = k[A]^0$; $Rate = k[A]$; $Rate = k[A]^2$; $Rate = k[A]^3$. OL

PRACTICE Problems

Do additional problems. Online Practice

19. Write the rate law for the reaction $aA \rightarrow bB$ if the reaction is third order in A. [B] is not part of the rate law.

20. The rate law for the reaction $2NO(g) + O_2(g) \rightarrow 2NO_2(g)$ is first order in O_2 and third order overall. What is the rate law for the reaction?

21. Given the experimental data below, use the method of initial rates to determine the rate law for the reaction $aA + bB \rightarrow$ products. *(Hint: Any number to the zero power equals one. For example, $(0.22)^0 = 1$ and $(55.6)^0 = 1$.)*

Practice Problem 21 Experimental Data

Trial	Initial [A](*M*)	Initial [B](*M*)	Initial Rate (mol/(L·s))
1	0.100	0.100	2.00×10^{-3}
2	0.200	0.100	2.00×10^{-3}
3	0.200	0.200	4.00×10^{-3}

22. Challenge The rate law for the reaction $CH_3CHO(g) \rightarrow CH_4(g) + CO(g)$ is Rate $= k[CH_3CHO]^2$. Use this information to fill in the missing experimental data below.

Practice Problem 22 Experimental Data

Trial	Initial $[CH_3CHO]$(*M*)	Initial Rate (mol/(L·s))
1	2.00×10^{-3}	2.70×10^{-11}
2	4.00×10^{-3}	10.8×10^{-11}
3	8.00×10^{-3}	

SECTION 3 REVIEW

Section Self-Check

Section Summary

- The mathematical relationship between the rate of a chemical reaction at a given temperature and the concentrations of reactants is called the rate law.
- The rate law for a chemical reaction is determined experimentally using the method of initial rates.

23. MAINIDEA Explain what the rate law for a chemical reaction tells you about the reaction.

24. Apply the rate-law equations to show the difference between a first-order reaction with a single reactant and a second-order reaction with a single reactant.

25. Explain the function of the specific rate constant in a rate-law equation.

26. Explain Under what circumstance is the specific rate constant (k) not a constant? What does the size of k indicate about the rate of a reaction?

27. Suggest a reason why, when given the rate of a chemical reaction, it is important to know that the reaction rate is an average reaction rate.

28. Explain how the exponents in the rate equation for a chemical reaction relate to the coefficients in the chemical equation.

29. Determine the overall reaction order for a reaction between A and B for which the rate law is rate $= k[A]^2[B]^2$.

30. Design an Experiment Explain how you would design an experiment to determine the rate law for the general reaction $aA + bB \rightarrow$ products using the method of initial rates.

3 Assess

Check for Understanding

Ask students to give the individual reaction orders for each substance and the overall reaction order for the following rate law: Rate $= k[NO_2]^2[Cl_2]$. The reaction is second order in NO_2, first order in Cl_2, and third order overall. OL

Reteach

Have students write the rate law for the general reaction involving the reactants A, B, and C if the reaction is second order in A, zero order in B, and first order in C. Rate $= k[A]^2[C]$. OL

Extension

Ask interested students to prepare and perform a class demonstration of an iodine-clock reaction. Reactions involving a potassium iodate solution and a sodium bisulfite/starch solution are described in many laboratory manuals and books of chemistry demonstrations. OL

ASSESSMENT

Knowledge Ask students to determine how the initial rate changes (assuming constant temperature) in each of the following experiments.

a. The reaction is first order in Reactant A, and [A] is halved. The rate is halved.

b. The reaction is second order in Reactant B, and [B] is tripled. The rate increases by a factor of nine. OL

SECTION 3 REVIEW

23. The rate law expresses the mathematical relationship between the reaction's rate and the concentrations of reactants.

24. First order reaction: Rate $= k[A]$; Second-order reaction: Rate $= k[A]^2$

25. The specific rate constant (k) relates reaction rate and concentration at a specific temperature.

26. k changes with temperature. The larger the value of k the faster the reaction.

27. The rate of a reaction decreases over time as reactant concentrations decrease. Therefore, the rate is an average over time rather than the rate at a given instant.

28. There is no relationship in general. In the relatively rare case of a single-step reaction with a single activated complex, the exponents are equal to the coefficients.

29. The overall reaction is fourth.

30. Determine the order of reactant A by measuring the reaction rate for several trials in which [A] is varied while [B] remains constant. Determine the order of reactant B by measuring the reaction rate for several trials in which [B] is varied while [A] remains constant.

SECTION 4

1 Focus

MAINIDEA

Rate-Determining-Step Analogy Make a paper-clip chain with one link made of paper and tape. Pull the chain from both ends until the paper-and-tape link breaks. Ask students if they are reminded of a common expression. A chain is no stronger than its weakest link. Point out that most chemical reactions, also have a weakest link in the sense that a multiple-step reaction can proceed no faster than the slowest of its elementary steps. BL OL AL

2 Teach

Visual Learning

Slope of a Line Figure 17 Review with students how to determine the slope of a straight line, which is constant. That is, $\Delta y/\Delta x$ (in the case of reaction rate, $\Delta[A]/\Delta t$) is the same for any line segment. However, the plot of concentration of a reactant or product over an appreciable time interval will be a curve. Point out that, because the slope of the curve varies, the reaction rate varies. Students can determine the instantaneous reaction rate at any point on the curve by drawing a straight line tangent to the curve at that point and determining the slope of the line. OL EL

☑ GRAPH CHECK $[H_2O_2]$ in mol/L is plotted on the y-axis; relative time in seconds is plotted on the x-axis.

SECTION 4

Essential Questions

- How are instantaneous rates of chemical reactions calculated?
- What substances and steps are involved in a reaction mechanism?
- How is the instantaneous rate of a complex reaction related to its reaction mechanism?

Review Vocabulary

decomposition reaction: a chemical reaction that occurs when a single compound breaks down into two or more elements or new compounds

New Vocabulary

instantaneous rate
complex reaction
reaction mechanism
intermediate
rate-determining step

Instantaneous Reaction Rates and Reaction Mechanisms

MAINIDEA The slowest step in a sequence of steps determines the rate of the overall chemical reaction.

CHEM 4 YOU Buying lunch in the cafeteria is a series of steps: picking up a tray and tableware, choosing food items, and paying the cashier. The first two steps might go rapidly, but a long line at the cashier will slow down the whole experience. Similarly, a reaction can go no faster than its slowest step.

Instantaneous Reaction Rates

Chemists often need to know more than the average reaction rate. A pharmacist developing a new drug treatment might need to know the progress of a reaction at an exact instant. Consider the decomposition of hydrogen peroxide (H_2O_2), which is represented as follows.

$$2H_2O_2(aq) \rightarrow 2H_2O(l) + O_2(g)$$

For this reaction, the decrease in H_2O_2 concentration over time is shown in **Figure 17.** The curved line shows how the reaction rate decreases as the reaction proceeds. The **instantaneous rate** is the reaction rate at a specific time. It is the slope of the straight line tangent to the curve at a specific time. The expression $\Delta[H_2O_2]/\Delta t$ is one way to express the reaction rate. In other words, the rate of change in H_2O_2 concentration relates to one specific point (or instant) on the graph.

You can determine the instantaneous rate for a reaction in another way if you are given the reactant concentrations at a given temperature and know the experimentally determined rate law and the specific rate constant at that temperature.

■ **Figure 17** The instantaneous rate for a specific point in the reaction progress can be determined from the tangent to the curve that passes through that point.

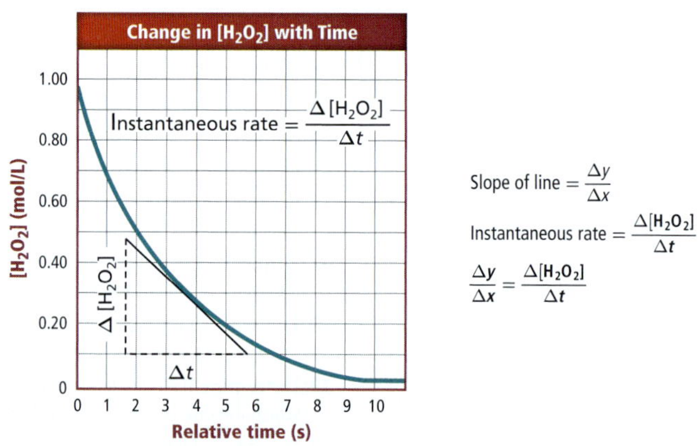

☑ GRAPH CHECK
Identify the variables that are plotted on the y-axis and on the x-axis.

Consider the decomposition of dinitrogen pentoxide (N_2O_5) into nitrogen dioxide (NO_2) and oxygen (O_2), which proceeds as follows.

$$2N_2O_5(g) \rightarrow 4NO_2(g) + O_2(g)$$

The experimentally determined rate law for this reaction is

$$\text{rate} = k[N_2O_5]$$

where $k = 1.0 \times 10^{-5}\ s^{-1}$. If $[N_2O_5] = 0.350M$, the instantaneous reaction rate would be calculated as

$$\text{rate} = (1.0 \times 10^{-5}\ s^{-1})(0.350\ mol/L) = 3.5 \times 10^{-6}\ mol/(L \cdot s)$$

EXAMPLE Problem 2

Find help with **dimensional analysis.** Math Handbook

CALCULATE INSTANTANEOUS REACTION RATES The following reaction is first order in H_2 and second order in NO with a rate constant of $2.90 \times 10^2\ (L^2/(mol^2 \cdot s))$.

$$2NO(g) + H_2(g) \rightarrow N_2O(g) + H_2O(g)$$

Calculate the instantaneous rate when the reactant concentrations are $[NO] = 0.00200M$ and $[H_2] = 0.00400M$.

1 ANALYZE THE PROBLEM

The rate law can be expressed by rate $= k[NO]^2[H_2]$. Therefore, the instantaneous reaction rate can be determined by inserting reactant concentrations and the specific rate constant into the rate law equation.

Known	Unknown
$[NO] = 0.00200M$	rate = ? mol/(L • s)
$[H_2] = 0.00400M$	
$k = 2.90 \times 10^2\ (L^2/(mol^2 \cdot s))$	

2 SOLVE FOR THE UNKNOWN

rate $= k[NO]^2[H_2]$ — State the rate law.

rate $= (2.90 \times 10^2\ L^2/(mol^2 \cdot s))(0.00200\ mol/L)^2(0.00400\ mol/L)$ — Substitute $k = 2.90 \times 10^2\ (L^2/(mol^2 \cdot s))$, $[NO] = 0.00200M$, and $[H_2] = 0.00400M$.

rate $= 4.64 \times 10^{-6}\ mol/(L \cdot s)$ — Multiply the numbers and units.

3 EVALUATE THE ANSWER

Units in the calculation cancel to give mol/(L • s), which is a common unit for reaction rates. A magnitude of approximately 10^{-6} mol/(L • s) fits with the quantities given and the rate law equation. The answer is correctly expressed with three significant figures.

EXAMPLE PROBLEM

PRACTICE Problems

Do additional problems. Online Practice

Use the rate law in Example Problem 2 and the concentrations given in Practice Problems 31 and 32 to calculate the instantaneous rate for the reaction between NO and H_2.

31. $[NO] = 0.00500M$ and $[H_2] = 0.00200M$

32. $[NO] = 0.0100M$ and $[H_2] = 0.00125M$

33. Challenge Calculate [NO] for the reaction in Example Problem 2 if the rate is 9.00×10^{-5} mol/(L • s) and $[H_2]$ is $0.00300M$.

Extension

Determining Rates Many reaction rates are determined experimentally using spectrometric, conductometric, or manometric methods. Ask interested students to research these methods and make a class presentation explaining their uses. AL

IN-CLASS Example

Question The following reaction is second order in $Fe(OH)^{2+}$, and the rate constant at 298 K is 4.5×10^2 (L/mol•s).

$2[FeOH]^{2+}(aq) \rightarrow [Fe_2(OH)_2]^{4+}(aq)$

Calculate the instantaneous rate when the concentration of $Fe(OH)^{2+}$ is 0.040 mol/L.

Answer 7.2×10^{-1} mol/(L•s)

Rate $= 4.5 \times 10^2\ (L/mol \cdot s)(0.040\ mol/L)^2$
$= 7.2 \times 10^{-1}$ mol/(L•s)

PRACTICE Problems

Have students refer to the Selected Solutions appendix for complete solutions to odd-numbered problems. The complete solutions for all problems can be found in the Solutions Manual.

31. Rate $= 1.45 \times 10^{-5}$ mol/(L•s)

32. Rate $= 3.63 \times 10^{-5}$ mol/(L•s)

33. $[NO] = 1.02 \times 10^{-2}M$

Quick Demo

Limiting a Reaction Clamp three funnels that have different flow rates to a ring stand, one above the other, placing the funnel with the smallest flow rate in the middle. Place a large beaker under the bottom funnel, then pour dry sand into the topmost funnel. Just as the rate-limiting step in a complex reaction limits the overall reaction rate, sand builds up in the middle, or "rate-determining," funnel.

CAREERS IN CHEMISTRY

WebQuest

■ **Caption Question Fig. 18** Cl is a catalyst in the decomposition of ozone.

Apply Chemistry

Air Bags Air bags are designed to slow the speed of the passenger's body to zero without damage to the body. The air bag has only a limited time to initiate and complete the chemical reaction needed to inflate the bag. The inflation system is the reaction of sodium azide (NaN_3) with potassium nitrate (KNO_3) to produce nitrogen gas (N_2). Hot blasts of nitrogen inflate the air bag. There is not much time between a car contacting an object, the sensor activation of the air bag, and the inflation of the bag. The whole process occurs in 0.04 s. Have students discuss how reaction rate, concentrations of reactants, and limiting reactants might all play a part in the action of the airbag. Students might suggest that reactant amounts compromise the timing of the reaction or whether the reaction will occur. OL

CAREERS IN CHEMISTRY

Chemical Engineer An understanding of reaction mechanisms is vital to chemical engineers. Their jobs often include scaling up a laboratory synthesis of a substance to large-scale production in a manufacturing plant. They must design the production facility and monitor its safe and efficient operation.

WebQuest

Reaction Mechanisms

Most chemical reactions consist of sequences of two or more simpler reactions. For example, recent evidence indicates that the reaction $2O_3 \rightarrow 3O_2$ occurs in three steps after intense ultraviolet radiation from the Sun liberates chlorine atoms from certain compounds in Earth's stratosphere. Steps 1 and 2 in this reaction might occur simultaneously or in reverse order.

1. Chlorine atoms decompose ozone according to the equation $Cl + O_3 \rightarrow O_2 + ClO$.
2. Ultraviolet radiation causes the decomposition reaction $O_3 \rightarrow O + O_2$.
3. ClO produced in the reaction in Step 1 reacts with O produced in Step 2 according to the equation $ClO + O \rightarrow Cl + O_2$.

Each of the reactions described in Steps 1 through 3 is called an elementary step. These elementary steps, illustrated in **Figure 18,** comprise the complex reaction $2O_3 \rightarrow 3O_2$. A **complex reaction** is one that consists of two or more elementary steps. A **reaction mechanism** is the complete sequence of elementary steps that makes up a complex reaction. Adding elementary Steps 1 through 3 and canceling formulas that occur in equal amounts on both sides of the reaction arrow produce the net equation for the complex reaction as shown.

Elementary step:	$\cancel{Cl}(g) + O_3(g) \rightarrow \cancel{ClO}(g) + O_2(g)$
Elementary step:	$O_3(g) \rightarrow \cancel{O}(g) + O_2(g)$
Elementary step:	$\cancel{ClO}(g) + \cancel{O}(g) \rightarrow \cancel{Cl}(g) + O_2(g)$
Complex reaction:	$2O_3 \rightarrow 3O_2$

Because chlorine atoms react in Step 1 and are re-formed in Step 3, chlorine is said to catalyze the decomposition of ozone. Because ClO and O are formed in Steps 1 and 2, respectively, and are consumed in the reaction in Step 3, they are called intermediates. An **intermediate** is a substance produced in one elementary step of a complex reaction and consumed in a subsequent elementary step. Like catalysts, intermediates do not appear in the net chemical equation.

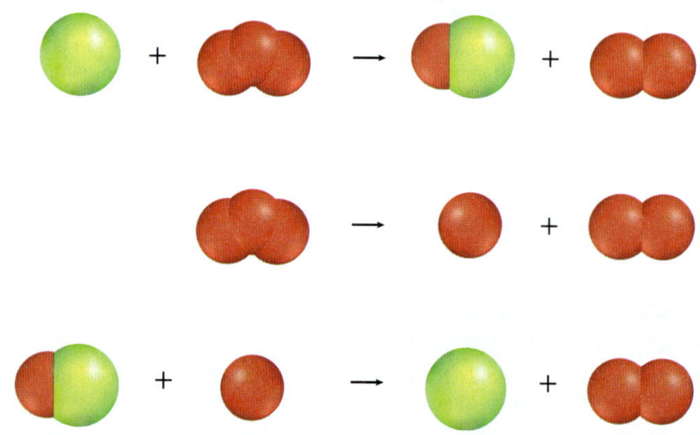

■ **Figure 18** ClO and O are intermediates in the three elementary steps of the complex reaction producing oxygen gas (O_2) from ozone (O_3).
Infer *What is the function of chlorine (Cl) in the complex reaction?*

CHEMISTRY JOURNAL

Reaction Rates—How Poetic Have a contest in which students write creative, poems that incorporate several facets of reaction rates. For example, you could require that the first stanza begin with the line: *Reaction rates, reaction rates....* Have students include the poems in their chemistry journals. Ask a colleague to help you judge the poems and choose winners. After awarding prizes, read the winning poems to the class. OL

Connection to Physics **Investigating reaction mechanisms** How is it possible to discover the presence of intermediates and determine their role in a chemical reaction? Learning how particles change their identities in the course of a chemical reaction means detecting evidence of bonds breaking and bonds forming. These processes take an extremely short period of time—time measured in femtoseconds. A femtosecond (fs) is one-millionth of a billionth of a second (0.000000000000001 second). Until recently, scientists could only calculate and imagine the actual atomic activity that occurs when bonds are broken and new bonds are made.

In 1999, Dr. Ahmed Zewail of the California Institute of Technology won a Nobel Prize for his achievements in the field of femtochemistry. Zewail developed an ultrafast laser device that can record pictures of chemical reactions as they happen. The laser "flashes" every 10 femtoseconds to record the movements of particles just as if they were being recorded on frames taken by a movie camera. Thus, a femtosecond recording of molecular motion could have as many as 10^{14} frames per second. The molecular motion corresponds to bond formation and breakage and can be related to the various possible intermediates and the products that are formed during a reaction. Zewail was able to witness an interaction between benzene (C_6H_6) and iodine (I_2) over a period of 1500 fs. A collision of iodine with benzene resulted in the breaking of the bond between the iodine atoms, after which the two atoms moved apart from one another. Technology such as this allows chemists to test their hypotheses about possible intermediates and reaction mechanisms.

READING CHECK **Explain** the importance of the methods of femtochemistry to the study of reaction mechanisms.

Rate-determining step Every complex reaction has an elementary step that is slower than all the other steps. The slowest elementary step in a complex reaction is called the **rate-determining step.** A reaction cannot go faster than its slowest elementary step. An analogy for the rate-determining step is shown in **Figure 19.**

Stephen Wilkes/Getty Images

Figure 19 At highway toll booths, drivers must slow down and stop as tolls are paid. Although they can resume their speeds after paying the toll, the pause affects their overall rate of travel. In a similar way, the overall rate of a chemical reaction is dependent on how fast the slowest elementary step proceeds.

ASSESSMENT

Performance Have students use ball-and-stick molecular models to simulate the three-step reaction between monatomic chlorine and ozone. Ask students to demonstrate and explain why Cl is said to catalyze the reaction and why ClO and O are said to be reaction intermediates. OL EL

READING CHECK Chemists can observe chemical interactions and identify intermediates.

Extension

Air Bag Chemistry There are several steps in the chemical reactions that occur during airbag deployment. Have students research these reactions and write a balanced equation for the initial reaction. Have students also research the amounts of reactants used in most air bags. Some students may also choose to research the environmental hazards of undeployed air bags in junk vehicles. Students should find the reaction $2NaN_3 \longrightarrow 2Na + 3N_2$; 130 g of sodium azide will produce 67 L of N_2, the amount filling an air bag. Sodium azide is explosive at high temperature, which might be reached in a scrap metal shredder. OL AL

DIFFERENTIATED INSTRUCTION

Struggling Learners Use one or more concrete examples to help students better understand how the rate-determining step of a complex reaction limits the overall reaction rate. For example, you might describe simplified steps for the factory production of bicycles. In this example, the steps that occur in Departments 1 and 2 occur simultaneously.

Dept. 1: production of frames; rate = 10 per hour

Dept. 2: production of wheels; rate = 20 per hour

Dept. 3: assembly of bicycle (1 frame + 2 wheels + misc.); rate = 8 per hour

Point out that although the steps that occur in Departments 1 and 2 would enable a factory output of 10 bicycles per hour, the step that occurs in Department 3 limits the output to 8 per hour. Therefore, Step 3 is the rate-determining step. BL

☑ GRAPH CHECK The reaction is exothermic because the energy level of the reactants is higher than that of the products.

3 Assess

Check for Understanding

Ask students to explain why the rate of a chemical reaction generally decreases over time. The rate decreases because the reactant's concentration(s) decrease(s). OL

Reteach

Reinforce the concept that most reactions occur in a series of steps called a reaction mechanism. Point out that a reaction might have several different mechanisms. In particular, explain that the mechanism for a complex, catalyzed reaction always differs from that of the uncatalyzed reaction. OL

Extension

Physicians and other health care personnel must have a good understanding of metabolic reactions and the conditions that affect them. Ask a health care professional to make a class presentation on this topic. OL

ASSESSMENT

Knowledge Ask students to differentiate between an elementary reaction and a complex reaction and to give one example of each. An elementary reaction occurs in a single step; a complex reaction occurs in two or more elementary steps. Examples will vary. OL

Figure 20 The three peaks in this energy diagram correspond to activation energies to form the activated complexes. The middle peak represents the highest energy barrier to overcome; therefore, the reaction involving $N_2O_2 + H_2$ is the rate-determining step.

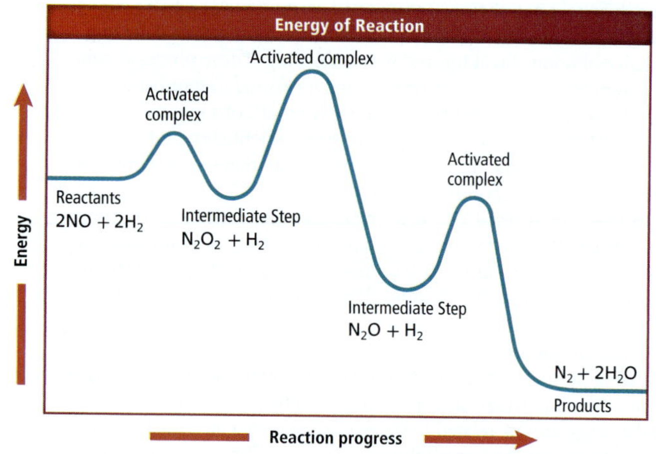

☑ GRAPH CHECK

Determine from the graph whether the overall reaction is exothermic or endothermic.

To see how the rate-determining step affects reaction rate, consider again the gas-phase reaction from Example Problem 2. A mechanism for this reaction consists of the following elementary steps.

$$2NO \rightarrow \cancel{N_2O_2} \quad \text{(fast)}$$
$$\cancel{N_2O_2} + H_2 \rightarrow \cancel{N_2O} + H_2O \quad \text{(slow)}$$
$$\cancel{N_2O} + H_2 \rightarrow N_2 + H_2O \quad \text{(fast)}$$
$$\overline{2NO(g) + 2H_2(g) \rightarrow N_2(g) + 2H_2O(g)}$$

The first and third elementary steps occur relatively fast, so the slow middle step is the rate-determining step.

Figure 20 shows how energy changes as this complex reaction proceeds. Each step of the reaction has its own activation energy, represented by a peak in the graph. Activation energy for Step 2 is higher than for Steps 1 and 3, which is why Step 2 is the rate-determining step.

SECTION 4 REVIEW

Section Self-Check

Section Summary

- The reaction mechanism of a chemical reaction must be determined experimentally.
- For a complex reaction, the rate-determining step limits the instantaneous rate of the overall reaction.

34. MAINIDEA **Compare and contrast** an elementary chemical reaction with a complex chemical reaction.
35. **Explain** how the rate law for a chemical reaction is used to determine the instantaneous rate of the reaction.
36. **Define** a reaction mechanism and an intermediate.
37. **Distinguish** between an intermediate and an activated complex.
38. **Relate** the size of the activation energy of an elementary step in a complex reaction to the rate of that step.
39. **Calculate** A reaction between A and B to form AB is first order in A and first order in B. The rate constant, k, equals 0.500 mol/(L • s). What is the rate of the reaction when $[A] = 2.00 \times 10^{-2} M$ and $[B] = 1.50 \times 10^{-2} M$?

SECTION 4 REVIEW

34. An elementary reaction occurs in a single step. A complex reaction consists of two or more elementary steps.
35. Insert actual experimental rate constant and reactant concentrations into the rate law, then calculate.
36. A reaction mechanism is the complete sequence of elementary steps that make up a complex reaction. An intermediate is a substance produced in one elementary step and consumed in a subsequent step.
37. An intermediate is a species that exists for a measurable length of time during the course of the steps in a multistep reaction mechanism. An activated complex only exists for an instant, while the reactants are colliding.
38. The higher the activation energy of an elementary step, the slower the rate of that step.
39. Rate $= 1.50 \times 10^{-4}$ mol/(L•s)

CHEMISTRY & health

Reaction Rate and Body Temperature

Imagine that you're late for school and rush outside without putting on your jacket. It's a chilly day, and soon you begin to shiver. Shivering is an automatic response by your body that helps maintain your normal body temperature, which is important.

What is normal body temperature?

Normal human body temperature is approximately 37°C, but it can vary with age, gender, time of day, and level of activity. Your temperature goes up when you engage in strenuous activities or when the temperature of the air around you is high. It can also go down when you take a cold shower or forget to wear your jacket in cold weather.

Chemical reactions heat the body Inside each cell of the body, food is metabolized to produce energy that is either used or stored in large molecules called adenosine triphosphate (ATP). When energy is needed, ATP splits into adenosine diphosphate (ADP) and a phosphate group (P_i) and energy is released.

$$ATP \rightarrow ADP + P_i + Energy$$

Reactions such as this require enzymes that regulate their rates. These enzymes are protein catalysts that are most efficient within the range of normal human body temperatures. Without the help of enzymes and a temperature near 37°C, reactions such as this one could not occur at a rate that would meet the needs of the body. Outside this temperature range, reaction rates are slower, as shown in **Figure 1.**

Regulating body temperature The area of the brain called the hypothalamus regulates body temperature by a complex feedback system. The system maintains a balance between the thermal energy released by chemical reactions within the body and the thermal energy exchanged between the body and the environment.

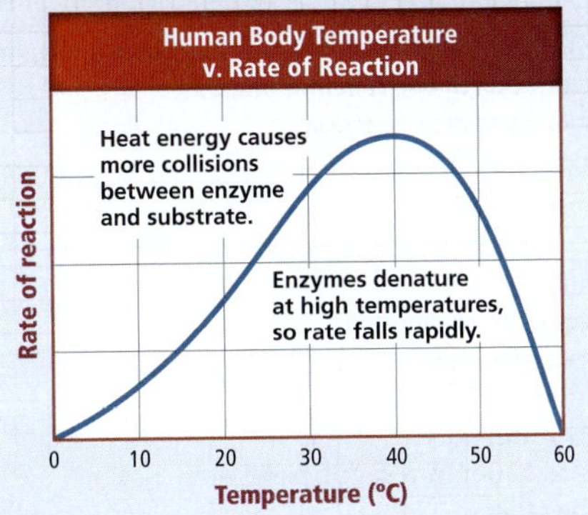

Figure 1 Optimum temperature for humans is close to 37°C. Excessive thermal energy results in the breakdown of a protein's structure, preventing it from functioning as it should.

Hypothermia—low body temperature

When hypothermia is detected, the hypothalamus begins actions that increase the release of thermal energy. Shivering is the rapid contractions of muscles that result from chemical reactions that release thermal energy. The body also begins actions to conserve thermal energy, including reducing blood flow to the skin.

Hyperthermia—high body temperature

Excessive thermal energy, either from the environment or because of increased chemical reactions within the body, causes the body to respond by sweating. Blood vessels near the skin's surface dilate, and heart and lung functions increase. These actions result in an increase in the release of thermal energy to the environment. The entire system of temperature control is designed to keep reactions within the body occurring at the optimal rate.

WRITING IN ▶ Chemistry

Research Write a patient-information brochure about the medical treatment of hypothermia and hyperthermia. Describe who might be at risk, any long term effects these conditions might have, and how they might be prevented.

WebQuest

CHEMISTRY & health

Purpose

Students will explore the relationship between body temperature and chemical reactions that control body functions.

Background

The decomposition of ATP ($C_{10}H_{16}N_5O_{13}P_3$) into ADP ($C_{10}H_{15}N_5O_{10}P_2$) and an inorganic phosphate is an exothermic reaction. This reaction is analogous to the function of rechargeable batteries: as batteries are used, they give up potential energy until it is completely converted into electric energy and heat. The batteries can only be used again after additional energy is added. In this analogy, ATP is the recharged battery. ADP is the used battery. There is roughly 0.1 mol of ATP in the human body. Under normal energy consumption, each ATP molecule is recycled between 2000 and 3000 times daily to meet the body's energy needs.

Teaching Strategies

- Have students work in pairs to analyze Figure 1. Ask students to describe the relationship between graphed variables.
- Ask students to consider the body's energy needs under conditions of hypothermia and hyperthermia. Have students predict which condition would require the greatest use of ATP.

WRITING IN ▶ Chemistry

Rubric WebQuest

Research Hypothermia would require the greatest use of ATP. Hypothermia symptoms include drowsiness, weakness and loss of coordination, pale, cold skin, confusion, uncontrollable shivering, and slowed breathing and heart rate. Victims should be warmed using blankets, body heat, and/or warm compresses. CPR and/or rescue breathing may be required. Hyperthermia symptoms include cool, clammy skin, rapid heart rate and weakened pulse, muscle spasms, and faintness. Victims should lie in a cool place, sponge off with cool water, and drink nonalcoholic, non-caffeinated beverages.

ChemLAB

Rubric

Preparation

Time Allotment one class period

Process Skills observe and infer, sequence, collect data, interpret data, hypothesize

Safety Precautions Approve lab safety forms before work begins. Remind students: Wear an apron, gloves and goggles. Always add acid to water. Never pipette any chemical by mouth. Use a safety pipette filler. Hand out Mg allotment to students.

Disposal Neutralize the acid from the test tubes with baking soda ($NaHCO_3$) before rinsing it down a drain with water. $MgCl_2$ solution can be flushed down a drain.

Preparation of Solutions Concentrated hydrochloric acid (HCl(aq)) is 12*M*. Dilute it with an equal volume of distilled water to produce approximately 6*M* HCl. Always add acid to water, not the reverse.

Procedure

- Clean the Mg metal with emery cloth or sandpaper until it shines.
- Remind students to carefully measure the volume of the acid and water that is transferred from each test tube.

ChemLAB

Observe How Concentration Affects Reaction Rate

Background: Collision theory describes how a change in concentration of one reactant affects the rate of a chemical reaction.

Question: *How does the concentration of a reactant affect the reaction rate?*

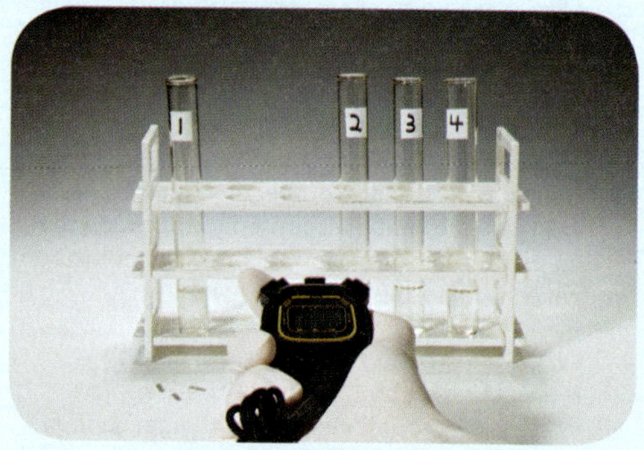

Materials

10-mL graduated pipette
safety pipette filler
6*M* hydrochloric acid
distilled water
25-mm × 150-mm test tubes, labeled *1–4*
test-tube rack
magnesium ribbon
emery cloth or fine sandpaper
scissors
plastic ruler
tongs
watch with second hand or stopwatch
stirring rod

Safety Precautions

WARNING: ***Never pipette any chemical by mouth. Hydrochloric acid is corrosive. Avoid contact with skin and eyes.***

Procedure

1. Read and complete the lab safety form.
2. Use a safety pipette to draw 10 mL of 6.0*M* hydrochloric acid (HCl) into a 10-mL graduated pipette.
3. Dispense the 10 mL of 6.0*M* HCl into Test Tube 1.
4. Draw 5.0 mL of the 6.0*M* HCl from Test Tube 1 with the pipette. Dispense this acid into Test Tube 2. Use the pipette to add an additional 5.0 mL of distilled water. Mix with the stirring rod. This solution is 3.0*M* HCl.
5. Draw 5.0 mL of the 3.0*M* HCl from Test Tube 2 and dispense it into Test Tube 3. Add 5.0 mL of distilled water and stir. This solution is 1.5*M* HCl.
6. Draw 5.0 mL of the 1.5*M* HCl from Test Tube 3 and dispense it into Test Tube 4. Add 5.0 mL of distilled water and stir. This solution is 0.75*M* HCl.
7. Draw 5.0 mL of the 0.75*M* HCl from Test Tube 4. Neutralize and discard it in the sink.
8. Using tongs, place a 1-cm length of magnesium ribbon into Test Tube 1. Record in your data table the time in seconds it takes for the bubbling to stop.
9. Repeat Step 8 using the remaining three test tubes. Record the time in seconds it takes for the bubbling to stop in each test tube.
10. **Cleanup and Disposal** Place acid solutions in an acid discard container. Thoroughly wash all test tubes and lab equipment. Discard other materials as directed by your teacher. Return all lab equipment to its proper place.

Analyze and Conclude

1. **Make a Graph** Plot the concentration of the acid on the *x*-axis and the reaction time on the *y*-axis. Draw a smooth curve through the data points.
2. **Conclude** Based on your graph, what is the relationship between the acid concentration and the reaction rate?
3. **Hypothesize** Write a hypothesis using collision theory, reaction rate, and reactant concentration to explain your results.
4. **Error Analysis** Compare your experimental results with those of other students in the laboratory. Explain the differences.

INQUIRY EXTENSION

Design an Experiment Based on your observations and results, would temperature variations affect reaction rates? Plan an experiment to test your hypothesis.

Matt Meadows

Analyze and Conclude

1. The graph should show that the reaction time decreases with an increase in concentration.
2. Reaction time decreases with an increase in acid concentration.
3. As the concentration of the HCl reactant increases in the test tube, the probability of an effective collision with the Mg metal increases. These collisions displace H_2 gas from the acid, increasing the rate of the reaction as is evidenced by increased bubbling.
4. Students' results should be close. Discrepancies could arise from inaccurate dilutions of the acid samples and variability in the length of the magnesium ribbon.

INQUIRY EXTENSION

Students' experiments might involve choosing one dilution and repeating the experiment at two or three different temperatures.

CHAPTER 16 STUDY GUIDE

Vocabulary Practice

BIGIDEA Every chemical reaction proceeds at a definite rate, but can be speeded up or slowed down by changing the conditions of the reaction.

SECTION 1 A Model for Reaction Rates

MAINIDEA Collision theory is the key to understanding why some reactions are faster than others.

- The rate of a chemical reaction is expressed as the rate at which a reactant is consumed or the rate at which a product is formed.

$$\text{average reaction rate} = -\frac{\Delta[\text{reactant}]}{\Delta t} = \frac{\Delta[\text{product}]}{\Delta t}$$

- Reaction rates are generally calculated and expressed in moles per liter per second (mol/(L•s)).
- In order to react, the particles in a chemical reaction must collide.
- The rate of a chemical reaction is unrelated to the spontaneity of the reaction.

VOCABULARY
- reaction rate
- collision theory
- activated complex
- activation energy

SECTION 2 Factors Affecting Reaction Rates

MAINIDEA Factors such as reactivity, concentration, temperature, surface area, and catalysts affect the rate of a chemical reaction.

- Key factors that influence the rate of chemical reactions include reactivity, concentration, surface area, temperature, and catalysts.
- Raising the temperature of a reaction generally increases the rate of the reaction by increasing the collision frequency and the number of collisions that form an activated complex.
- Catalysts increase the rates of chemical reactions by lowering activation energies.

VOCABULARY
- catalyst
- inhibitor
- heterogeneous catalyst
- homogeneous catalyst

SECTION 3 Reaction Rate Laws

MAINIDEA The reaction rate law is an experimentally determined mathematical relationship that relates the speed of a reaction to the concentrations of the reactants.

- The mathematical relationship between the rate of a chemical reaction at a given temperature and the concentrations of reactants is called the rate law.

$$\text{rate} = k[\text{A}]$$

$$\text{rate} = k[\text{A}]^m[\text{B}]^n$$

- The rate law for a chemical reaction is determined experimentally using the method of initial rates.

VOCABULARY
- rate law
- specific rate constant
- reaction order
- method of initial rates

SECTION 4 Instantaneous Reaction Rates and Reaction Mechanisms

MAINIDEA The slowest step in a sequence of steps determines the rate of the overall chemical reaction.

- The reaction mechanism of a chemical reaction must be determined experimentally.
- For a complex reaction, the rate-determining step limits the instantaneous rate of the overall reaction.

VOCABULARY
- instantaneous rate
- complex reaction
- reaction mechanism
- intermediate
- rate-determining step

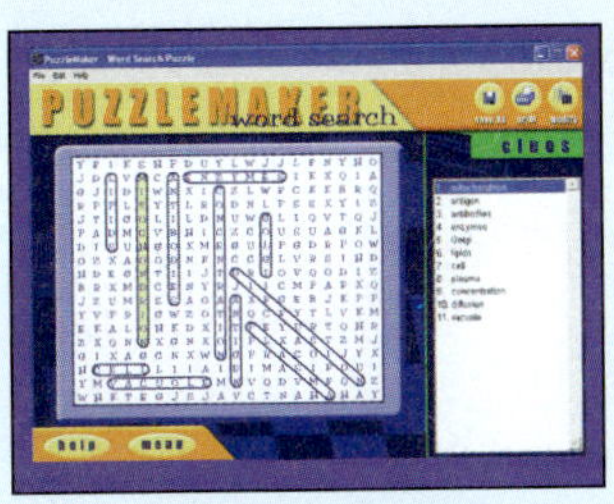

Vocabulary Practice

For additional practice with vocabulary, have students access the Vocabulary PuzzleMaker.

Use the Vocabulary

To reinforce chapter vocabulary, have students write a sentence using each term. OL EL

Review Strategies

- Have students summarize how reaction rates are expressed and explain how the collision model relates to reaction rates. OL
- Have students list and explain the conditions and factors that affect reaction rates. OL
- Have students give an example of a reaction rate law. OL
- Have students explain the difference between an average reaction rate and an instantaneous reaction rate. OL

Students can review with these online resources.

Vocabulary Practice

Vocabulary eGames
Vocabulary eFlashcards
Vocabulary PuzzleMaker

Section Self-Check

Chapter Self-Check

Online Test Practice

Use *eAssessment* to:

- create multiple versions of tests
- edit existing questions and add your own questions
- build tests aligned with select state curriculum standards using built-in state curriculum tags
- track students' progress

SECTION 1

Mastering Concepts

40. The concentration of reactants decreases and the concentration of products increases.

41. The average rate of a reaction is the change in the concentration of a reactant or product that occurs during a specified time interval.

42. The rate may be expressed as the decrease in [A] per unit time: Rate $= \Delta[A]/\Delta t$. Numerically, the two rates would be equal; however, $\Delta[A]$ would be negative, showing a decrease in concentration of the reactant as the reaction proceeds, and $\Delta[B]$ would be positive, showing an increase in concentration of the product as the reaction proceeds.

43. The activated complex is a transition state between reactants and products.

44. The molecules do not react if they collide with insufficient energy or lack favorable orientations at the instant of impact.

45. **a.** reactants 2
b. activated complex 3
c. products 4
d. activation energy 1

46. The activation energy for the forward reaction is less than the activation energy for the reverse reaction.

Mastering Problems

47. The average reaction rate is 0.0250 mol I_2 consumed per liter per minute.

48. Average rate $= 0.0617$ mol/3.00 min $= 2.06 \times 10^{-2}$ mol/min

49. Rate $= 3.75 \times 10^{-4}$ mol/(L•min)

SECTION 2

Mastering Concepts

50. The rate of a reaction depends on the reactivity of the reactants. Under a given set of conditions, the greater the reactivity of the reactants, the faster the reaction will proceed.

51. Increasing the concentration increases the rate, and decreasing the concentration decreases the rate.

52. Increasing the concentration of a reactant means more particles are packed into the same space. As a result, the frequency of collisions between reacting particles increases.

CHAPTER 16 ASSESSMENT

SECTION 1

Mastering Concepts

40. What happens to the concentrations of the reactants and products during the course of a chemical reaction?

41. Explain what is meant by the average rate of a reaction.

42. How would you express the rate of the chemical reaction $A \rightarrow B$ based on the concentration of Reactant A? How would that rate compare with the reaction rate based on the Product B?

43. What is the role of the activated complex in a chemical reaction?

44. Suppose two molecules that can react collide. Under what circumstances do the colliding molecules not react?

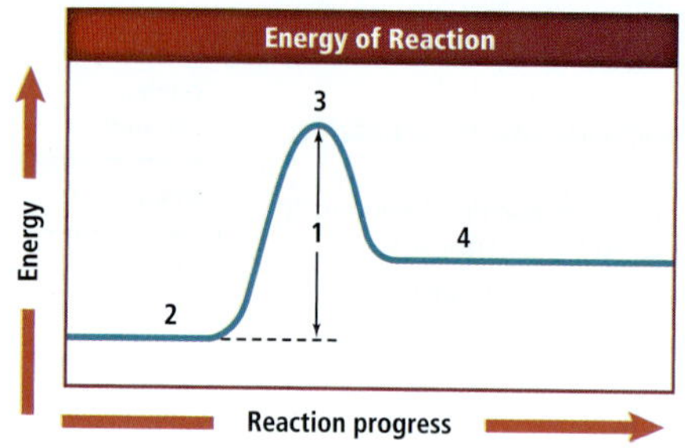

■ **Figure 21**

45. **Figure 21** is an energy diagram for a reaction. Match the appropriate number with the quantity it represents.
a. reactants
b. activated complex
c. products
d. activation energy

46. If $A \rightarrow B$ is exothermic, how does the activation energy for the forward reaction compare with the activation energy for the reverse reaction $(A \leftarrow B)$?

Mastering Problems

47. In the gas-phase reaction, $I_2 + Cl_2 \rightarrow 2ICl$, $[I_2]$ changes from 0.400*M* at 0.00 min to 0.300*M* at 4.00 min. Calculate the average reaction rate in moles of I_2 consumed per liter per minute.

48. In a reaction $Mg(s) + 2HCl(aq) \rightarrow H_2(g) + MgCl_2(aq)$, 6.00 g of Mg was present at 0.00 min. After 3.00 min, 4.50 g of Mg remained. Express the average rate as mol Mg consumed/min.

49. If a chemical reaction occurs at the rate of 2.25×10^{-2} moles per liter per second at 322 K, what is the rate expressed in moles per liter per minute?

SECTION 2

Mastering Concepts

50. What role does the reactivity of the reactants play in determining the rate of a chemical reaction?

51. In general, what is the relationship between reaction rate and reactant concentration?

52. Apply collision theory to explain why increasing the concentration of a reactant usually increases the reaction rate.

53. Explain why a crushed solid reacts with a gas more quickly than a large chunk of the same solid.

54. **Food Preservation** Apply collision theory to explain why foods usually spoil more slowly when refrigerated than at room temperature.

55. Apply collision theory to explain why powdered zinc reacts to form hydrogen gas faster than large pieces of zinc when both are placed in hydrochloric acid solution.

56. Hydrogen peroxide decomposes to water and oxygen gas more rapidly when manganese dioxide is added. The manganese dioxide is not consumed in the reaction. Explain the role of the manganese dioxide.

Mastering Problems

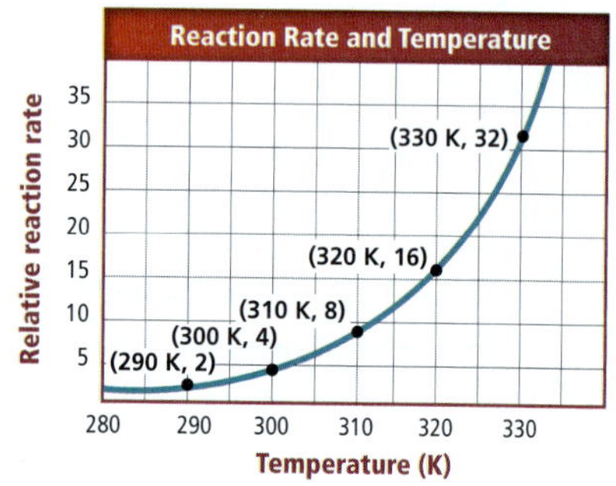

■ **Figure 22**

57. Examine **Figure 22,** which relates relative reaction rate and temperature. Approximately how does the reaction rate change for each increase of 10 K?

58. Suppose that a large volume of 3% hydrogen peroxide decomposes to produce 12 mL of oxygen gas in 100 s at 298 K. Estimate how much oxygen gas would be produced by an identical solution in 100 s at 308 K.

59. Using the information in Question 58, estimate how much oxygen gas would be produced in an identical solution in 100 seconds at 318 K. Estimate the time needed to produce 12 mL of oxygen gas at 288 K.

53. The finely divided solid presents a larger surface area than the large chunk.

54. Lowering the temperature decreases the frequency and the energy of collisions between reactant particles in the food, therefore lowering the rate of chemical reactions leading to spoilage.

55. The greater surface area of powdered zinc allows more zinc atoms to collide with reacting particles in the acid per unit time.

56. Manganese dioxide catalyzes the decomposition reaction by lowering the reaction's activation energy.

Mastering Problems

57. For every 10 K change in temperature, the relative reaction rate doubles.

58. For each increase of 10 K, the rate approximately doubles. Therefore, approximately 24 mL of oxygen would be produced.

59. At 318 K (a 20 degree increase in temperature): The rate would be quadrupled, so 48 mL of oxygen would be produced. At 288 K: For each decrease of 10 K, the rate is approximately halved. Therefore, approximately twice as much time, 200 seconds, would be required to produce the same volume of oxygen.

Chapter Self-Check

SECTION 3

Mastering Concepts

60. In the method of initial rates used to determine the rate law for a chemical reaction, what is the significance of the word *initial?*

61. Why must the rate law for a chemical reaction be based on experimental evidence rather than the balanced equation for the reaction?

62. Assume that the rate law for a generic chemical reaction is rate = $[A][B]^3$. What is the reaction order in A, the reaction order in B, and the overall reaction order?

63. Consider the generic chemical reaction: $A + B \rightarrow AB$. Based on experimental data, the reaction is second order in Reactant A. If the concentration of A is halved, and all other conditions remain unchanged, how does the reaction rate change?

Mastering Problems

64. The instantaneous rate data in **Table 3** were obtained for the reaction $H_2(g) + 2NO(g) \rightarrow H_2O(g) + N_2O(g)$ at a given temperature and concentration of NO. How does the instantaneous rate of this reaction change as the initial concentration of H_2 is changed? Based on the data, is $[H_2]$ part of the rate law? Explain.

Table 3 Reaction Between $H_2(g)$ and NO(g)

$[H_2]$ (mol/L)	Instantaneous Rate (mol/L·s)
0.18	6.00×10^{-3}
0.32	1.07×10^{-2}
0.58	1.93×10^{-2}

65. Suppose that a generic chemical reaction has the rate law of rate = $[A]^2[B]^3$ and that the reaction rate under a given set of conditions is 4.5×10^{-4} mol/(L•min). If the concentrations of both A and B are doubled and all other reaction conditions remain constant, how will the reaction rate change?

66. The experimental data in **Table 4** were obtained for the decomposition of azomethane ($CH_3N_2CH_3$) at a particular temperature according to the equation $CH_3N_2CH_3(g) \rightarrow C_2H_6(g) + N_2(g)$. Use the data to determine the reaction's experimental rate law.

Table 4 Decomposition of Azomethane

Experiment Number	Initial $[CH_3N_2CH_3]$	Initial Reaction Rate
1	0.012*M*	2.5×10^{-6} mol/(L•s)
2	0.024*M*	5.0×10^{-6} mol/(L•s)

67. Use the data in **Table 4** to calculate the value of the specific rate constant, *k*.

68. At the same temperature, predict the reaction rate when the initial concentration of $CH_3N_2CH_3$ is 0.048*M*. Use the data in **Table 4.**

SECTION 4

Mastering Concepts

69. Distinguish between a simple and a complex reaction, an elementary step, and a reaction mechanism.

70. Suppose that a chemical reaction takes place in a two-step mechanism.

Step 1 (fast) $A + B \rightarrow C$

Step 2 (slow) $C + D \rightarrow E$

Which step in the reaction mechanism is the rate-determining step? Explain.

71. In the reaction described in Question 70, what are Steps 1 and 2 called? What is substance C called?

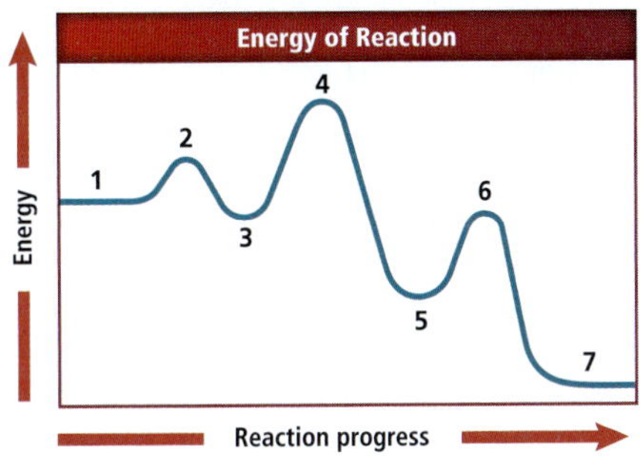

■ **Figure 23**

72. In **Figure 23,** identify each of the labels 1, 2, 3, 4, 5, 6, and 7 as one of the following: activated complex, intermediate, reactants, or products.

Mastering Problems

73. Dinitrogen pentoxide decomposes in chloroform at a rate of 2.48×10^{-4} mol/(L•min) at a particular temperature according to the equation $2N_2O_5 \rightarrow 4NO_2 + O_2$. The reaction is first order in N_2O_5. Given an initial concentration 0.400 mol/L, what is the rate constant for the reaction? What is the approximate $[N_2O_5]$ after the reaction proceeds for 1.30 h?

74. Radioactive decay is first order in the decaying isotope. For example, strontium-90 contained in fallout from nuclear explosions decays to yttrium-90 and a beta particle. Write the rate law for the decay of strontium-90.

SECTION 3

Mastering Concepts

60. The initial rate means the moment at which the reactants are mixed at the stated concentrations; however, the rate begins to decrease the instant the reaction starts.

61. Most chemical reactions occur in more than one step.

62. The reaction is first order in A, third order in B, and fourth order overall.

63. The rate decreases to one-fourth its initial value.

Mastering Problems

64. As the concentration of H_2 increases, the reaction rate increases. Yes; the fact that the rate increases as $[H_2]$ increases indicates that $[H_2]$ appears in the rate law and is raised to a positive exponent. The data show that the ratios between two successive $[H_2]$ values are equal to the ratios between the corresponding instantaneous rate values. Therefore, the positive exponent of $[H_2]$ in the rate law is one.

65. Because $(2^2)(2^3) = 32$, the reaction rate will increase by a factor of 32 to 1.4×10^{-2} mol/(L•min).

66. In experiment 2, doubling the concentration of azomethane, $CH_3N_2CH_3$, doubles the reaction rate. Therefore, Rate = $k[CH_3N_2CH_3]$.

67. $k = 2.1 \times 10^{-4}\ s^{-1}$

68. Rate = 1.0×10^{-5} mol/(L•s)

SECTION 4

Mastering Concepts

69. A simple reaction occurs in a single step; a complex reaction is made up of two or more elementary steps; a reaction mechanism consists of the complete sequence of elementary steps that make up a complex reaction.

70. step 2; The reaction can proceed no faster than the slowest step.

71. elementary steps; an intermediate

72. 1, reactants; 2, activated complex; 3, intermediate; 4, activated complex; 5, intermediate; 6, activated complex; 7, products

Mastering Problems

73. Rate = $k[N_2O_5]$
$k = 6.2 \times 10^{-4}\ min^{-1}$
Final $[N_2O_5] = 0.381$ mol/L

74. Rate = $k[^{90}_{38}Sr]$

MIXED REVIEW

75. The statement is not valid. Because most chemical reactions occur is a sequence of elementary steps, the rate law must be determined by experimental methods.

76. Average reaction rate $= 4.00 \times 10^{-3}$ mol/L•min

77. Volume of $H_2 = 1.38$ L H_2

78. Average rate $= 4.70 \times 10^{-3}$ mol/(L•min)

79. The overall reaction is $2NO_2Cl(g) \longrightarrow 2NO_2(g) + Cl_2(g)$ The species, Cl, is an intermediate in the overall reaction because it is formed in the first step and consumed in the second step.

80. The energy diagram for assumption A has a higher hump for the first activated complex. The energy diagram for assumption B has a higher hump for the second activated complex.

81. Initial rate $= 1.2 \times 10^{-5}$ mol/(L•s)

82. Rate, mol/(L•s); Rate (mol/L•s) $= k(\text{mol/L})^3$; k, L^2/mol^2•s

THINK CRITICALLY

83. The activation energy of the forward reaction is greater than that of the reverse reaction.

84. The greater shaded area under the curve representing the higher temperature, T_2, demonstrates that the number of collisions that result in reaction is greater at the higher temperature.

85. Student data should show that the change in the initial reaction rate is proportional to the square of the change in concentration of Reactant X.

86. Such a complex reaction might have more than one rate-determining step if two or more of the elementary steps occur at identical rates.

87. The number of A–B collision combinations increased from four to eight, a factor of two. Because the reaction rate depends upon the number of molecular collisions, the rate probably doubles.

88. Increasing the temperature by 10 K increases the average speed of the reacting particles. Therefore, the collision frequency increases. In addition, the number of collisions possessing sufficient energy to form the activated complex increases, often by a factor of approximately two.

MIXED REVIEW

75. Evaluate the validity of this statement: You can determine the rate law for a chemical reaction by examining the mole ratio of reactants in the balanced equation. Explain your answer.

76. The concentration of Reactant A decreases from 0.400 mol/L at 0.00 min to 0.384 mol/L at 4.00 min. Calculate the average reaction rate during this time period. Express the rate in mol/(L • min).

77. The mass of a sample of magnesium is measured and the sample is placed in a container of hydrochloric acid. A chemical reaction occurs according to the equation $Mg(s) + 2HCl(aq) \rightarrow H_2(g) + MgCl_2(aq)$. Use the data in **Table 5** to calculate the volume of hydrogen gas produced at STP during the 3.00-min reaction. *(Hint: 1 mol of an ideal gas occupies 22.4 L at STP)*

Table 5 Reaction of Magnesium and Hydrochloric Acid

Time (min)	Mass of Magnesium (g)	Volume of Hydrogen at STP (L)
0.00	6.00	0.00
3.00	4.50	?

78. If the concentration of a reaction product increases from 0.0882 mol/L to 0.1446 mol/L in 12.0 minutes, what is the average reaction rate during the time interval?

79. A two-step mechanism has been proposed for the decomposition of nitryl chloride (NO_2Cl).

Step 1: $NO_2Cl(g) \rightarrow NO_2(g) + Cl(g)$

Step 2: $NO_2Cl(g) + Cl(g) \rightarrow NO_2(g) + Cl_2(g)$

What is the overall reaction? Identify any intermediates in the reaction sequence, and explain why they are called intermediates.

80. Compare and contrast the reaction energy diagrams for the overall decomposition of nitryl chloride by the mechanism in Problem 79 under two assumptions: A—that the first step is slower; B—that the second step is slower.

81. Automobile Engine The following reaction takes place in an automobile's engine and exhaust system.

$$NO_2(g) + CO(g) \rightarrow NO(g) + CO_2(g)$$

The reaction's rate law at a particular temperature is Rate $= 0.50$ L/(mol • s)$[NO_2]^2$. What is the reaction's initial, instantaneous rate when $[NO_2] = 0.0048$ mol/L?

82. The concentrations in a chemical reaction are expressed in moles per liter and time is expressed in seconds. If the overall rate law is third-order, what are the units for the rate and the rate constant?

THINK CRITICALLY

83. Visualize the reaction energy diagram for a one-step, endothermic chemical reaction. Compare the heights of the activation energies for the forward and reverse reactions.

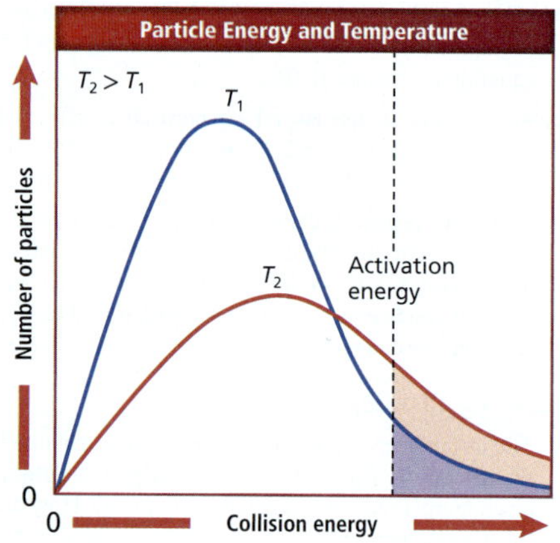

■ **Figure 24**

84. Differentiate between the shaded areas in **Figure 24** at temperatures T_1 and T_2 on the basis of the number of collisions per unit time that might occur with energy equal to or greater than the activation energy.

85. Apply the method of initial rates to determine the order of a chemical reaction with respect to Reactant X. Create a set of hypothetical experimental data that would lead you to conclude that the reaction is second order in X.

86. Formulate a rationale to explain how a complex chemical reaction might have more than one rate-determining elementary step.

87. Construct a diagram that shows all of the possible collision combinations between two molecules of Reactant A and two molecules of Reactant B. Now, increase the number of molecules of A from two to four and sketch each possible A-B collision combination. By what factor did the number of collision combinations increase? What does this imply about the reaction rate?

88. Apply collision theory to explain two reasons why increasing the temperature of a reaction by 10 K often doubles the reaction rate.

89. Create a table of concentrations, starting with 0.100M concentrations of all reactants, that you would propose in order to establish the rate law for the reaction $aA + bB + cD \rightarrow$ products using the method of initial rates.

89. The table should include sets of trials in which the concentrations of two reactants are held constant, while that of the third reactant is varied. For three reactants, at least four trials would be needed.

Chapter Self-Check

CHALLENGE PROBLEM

90. Hydrocarbons Heating cyclopropane (C_3H_6) converts it to propene ($CH_2{=}CHCH_3$). The rate law is first order in cyclopropane. If the rate constant at a particular temperature is $6.22 \times 10^{-4}\ s^{-1}$ and the concentration of cyclopropane is held at 0.0300 mol/L, what mass of propene is produced in 10.0 min in a volume of 2.50 L?

CUMULATIVE REVIEW

91. For the following categories of elements, state the possible number(s) of electrons in their outermost orbitals in the ground state.
- **a.** p-block elements
- **b.** nitrogen-group elements
- **c.** d-block elements
- **d.** noble-gas elements
- **e.** s-block elements

92. Classify each of the following elements as a metal, nonmetal, or metalloid.
- **a.** molybdenum
- **b.** bromine
- **c.** arsenic
- **d.** neon
- **e.** cerium

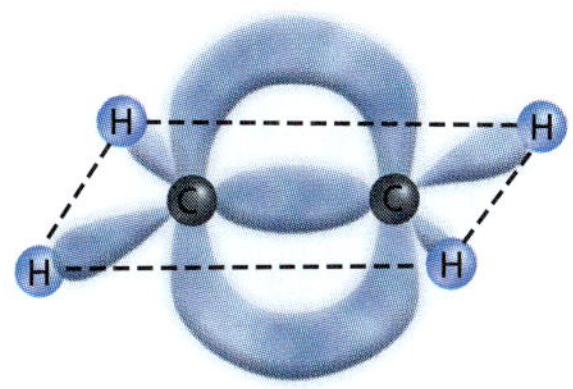

Ethene

Figure 25

93. Using **Figure 25,** determine how many sigma and pi bonds are contained in a single ethene molecule.

94. Balance the following equations.
- **a.** $Sn(s) + NaOH(aq) \rightarrow Na_2SnO_2(aq) + H_2(g)$
- **b.** $C_8H_{18}(l) + O_2(g) \rightarrow CO_2(g) + H_2O(l)$
- **c.** $Al(s) + H_2SO_4(aq) \rightarrow Al_2(SO_4)_3(aq) + H_2(g)$

95. What mass of iron(III) chloride is needed to prepare 1.00 L of a 0.255*M* solution?

96. What must you know to calculate the boiling point elevation of a solution of hexane in benzene?

97. ΔH for a reaction is negative. Compare the energy of the products and the reactants. Is the reaction endothermic or exothermic?

WRITING IN ▶ Chemistry

98. Pharmaceuticals Imagine that your nation is experiencing an influenza epidemic. Fortunately, scientists have recently discovered a new catalyst that increases the rate of production of an effective flu medicine. Write a newspaper article describing how the catalyst works. Include a reaction energy diagram and an explanation detailing the importance of the discovery.

99. Lawn Care Write an advertisement that explains that Company A's fertilizer works better than Company B's fertilizer because it has smaller sized granules. Include applicable diagrams.

DBQ Document-Based Questions

Chemical Indicators *Phenolphthalein is a chemical indicator used to show the presence of a base. The data in* **Table 6** *presents the decrease in phenolphthalein concentration with time when a 0.0050M phenolphthalein solution is added to a solution that has a concentration of hydroxide ion equal to 0.61M.*

Table 6 Reaction Between Phenolphthalein and Excess Base

Concentration of Phenolphthalein (*M*)	Time (s)
0.0050	0.0
0.0040	22.3
0.0020	91.6
0.0010	160.9
0.00050	230.3
0.00015	350.7

Data obtained from: Bodner Research Web. 2006. "Chemical Kinetics," *General Chemistry Help.*

100. What is the average rate of the reaction in the first 22.3 s expressed in moles of phenolphthalein consumed per liter per second?

101. What is the average rate of the reaction as the phenolphthalein concentration decreases from 0.00050*M* to 0.00015*M*?

102. The rate law is rate = *k*[phenolphthalein]. If the rate constant for the reaction is $1.0 \times 10^{-2}\ s^{-1}$, what is the instantaneous rate of reaction when the concentration of phenolphthalein is 0.0025*M*?

CHALLENGE PROBLEM

90. 1.18 g

CUMULATIVE REVIEW

91. **a.** 3–8
b. 5
c. 1–10
d. 8
e. 1 or 2

92. **a.** metal
b. nonmetal
c. metalloid
d. nonmetal
e. metal

93. five sigma bonds and one pi bond

94. **a.** $Sn(s) + 2NaOH(aq) \rightarrow Na_2SnO_2(aq) + H_2(g)$
b. $2C_8H_{18}(l) + 25O_2(g) \rightarrow 16CO_2(g) + 18H_2O(l)$
c. $2Al(s) + 3H_2SO_4(aq) \rightarrow Al_2(SO_4)_3(aq) + 3H_2(g)$

95. 41.4 g $FeCl_3$

96. the molarity of the solution and the boiling point elevation constant for benzene

97. The reaction is exothermic because ΔH is negative. The reactants have more energy than the products.

WRITING IN ▶ Chemistry

Rubric

98. Answers will vary but should include a description of how a catalyst works, a reaction energy diagram and an explanation detailing the importance of the discovery.

99. Students should note that the larger surface area afforded by the smaller granules enhances the rate of nutrient delivery to the soil.

DBQ Document-Based Questions

Data obtained from: Chemical Kinetics. *Division of Chemistry Education.* Purdue University.

100. Average rate $= 4.5 \times 10^{-5}$ mol/(L•s)
101. Average rate $= 2.9 \times 10^{-6}$ mol/(L•s)
102. Rate $= 2.5 \times 10^{-5}$ mol/(L•s)

MULTIPLE CHOICE

1. D
2. D
3. B
4. C
5. B
6. B
7. D
8. C
9. B
10. B

MULTIPLE CHOICE

1. The rate of a chemical reaction is all of the following EXCEPT

A. the speed at which a reaction takes place.
B. the change in concentration of a reactant per unit time.
C. the change in concentration of a product per unit time.
D. the amount of product formed in a certain period of time.

2. How can colloids be distinguished from solutions?

A. Dilute colloids have particles that can be seen with the naked eye.
B. Colloid particles are much smaller than solvated particles.
C. Colloid particles that are dispersed will settle out of the mixture in time.
D. Colloids will scatter light beams that are shone through them.

Use the graph below to answer Questions 3 and 4.

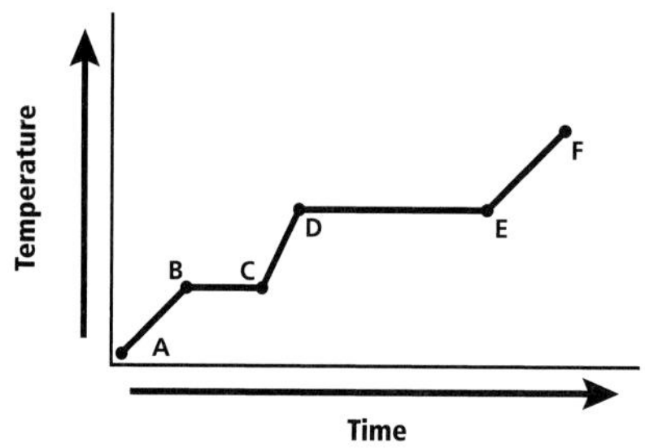

3. During which segment is this substance undergoing melting?

A. segment AB
B. segment BC
C. segment CD
D. segment DE

4. As the substance heats from point C to point D, which is true of the substance?

A. potential energy increases, kinetic energy decreases
B. potential energy increases, kinetic energy increases
C. potential energy remains constant, kinetic energy increases
D. potential energy decreases, kinetic energy remains constant

5. How much water must be added to 6.0 mL of a 0.050*M* stock solution to dilute it to 0.020*M*?

A. 15 mL
B. 9.0 mL
C. 6.0 mL
D. 2.4 mL

6. Which is NOT an acceptable unit for expressing a reaction rate?

A. *M*/min
B. L/s
C. mol/(mL•h)
D. mol/(L•min)

7. Which is the strongest type of intermolecular bond?

A. ionic bond
B. dipole-dipole force
C. dispersion force
D. hydrogen bond

Use the diagram below to answer Questions 8 and 9.

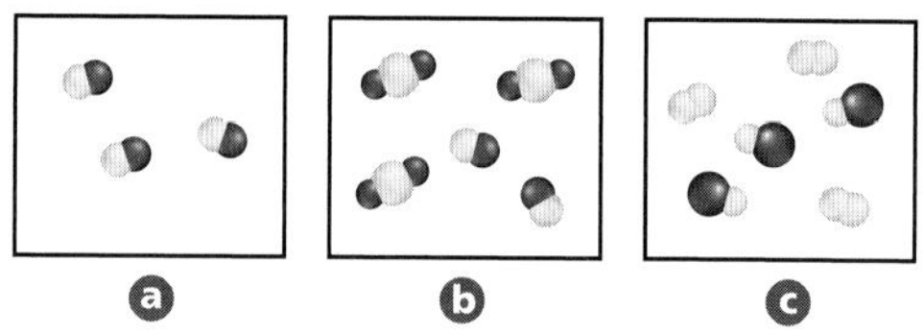

8. Which sample could contain particles of oxygen gas?

A. a
B. b
C. c
D. Both a and b

9. Which sample could contain particles of carbon dioxide?

A. a
B. b
C. c
D. Both a and b

10. How many moles are in 4.30×10^2 g of calcium phosphate ($Ca_3(PO_4)_2$)?

A. 0.721 moles
B. 1.39 moles
C. 1.54 moles
D. 3.18 moles

Online Test Practice

SHORT ANSWER

Use the following information to answer Question 11.

The complete dissociation of acid H_3A takes place in three steps:	
$H_3A(aq) \rightarrow H_2A^-(aq) + H^+(aq)$	rate $= k_1[H_3A]$ $k_1 = 3.2 \times 10^2\ s^{-1}$
$H_2A^-(aq) \rightarrow HA^{2-}(aq) + H^+(aq)$	rate $= k_2[H_2A^-]$ $k_2 = 1.5 \times 10^2\ s^{-1}$
$HA^{2-}(aq) \rightarrow A^{3-}(aq) + H^+(aq)$	rate $= k_3[HA^{2-}]$ $k_3 = 0.8 \times 10^2\ s^{-1}$
overall reaction: $H_3A(aq) \rightarrow A^{3-}(aq) + 3H^+(aq)$	

11. When the reactant concentrations are $[H_3A] = 0.100M$, $[H_2A^-] = 0.500M$, and $[HA^{2-}] = 0.200M$, which reaction is the rate-determining step? Explain how you can tell.

12. The rate law for $A + B + C \rightarrow$ products is: rate $= k[A]^2[C]$. If $k = 6.92 \times 10^{-5}\ L^2/(mol^2 \cdot s)$, $[A] = 0.175M$, $[B] = 0.230M$, and $[C] = 0.315M$, what is the instantaneous reaction rate?

EXTENDED RESPONSE

Use the following reaction to answer Questions 13 to 15.

Sodium nitride (Na_3N) breaks down to form sodium metal and nitrogen gas.

13. Write the balanced chemical equation for the reaction.

14. Classify the type of reaction. Explain your answer.

15. Show the steps to determine the amount of nitrogen gas that can be produced from 32.5 grams of sodium nitride.

SAT SUBJECT TEST: CHEMISTRY

Use the table below to answer Questions 16 to 18.

Reaction: $SO_2Cl_2(g) \rightarrow SO_2(g) + Cl_2(g)$			
Experimental Data Collected for Reaction			
Time (min)	$[SO_2Cl_2]$ (M)	$[SO_2]$ (M)	$[Cl_2]$ (M)
0.0	1.00	0.00	0.00
100.0	0.87	0.13	0.13
200.0	0.74	?	?

16. What is the average reaction rate for this reaction, expressed in moles SO_2Cl_2 consumed per liter per minute?
    - **A.** 1.3×10^{-3} mol/(L•min)
    - **B.** 2.6×10^{-1} mol/(L•min)
    - **C.** 7.4×10^{-3} mol/(L•min)
    - **D.** 8.7×10^{-3} mol/(L•min)
    - **E.** 2.6×10^{-3} mol/(L•min)

17. On the basis of the average reaction rate, what will the concentrations of SO_2 and Cl_2 be at 200.0 min?
    - **A.** 0.13*M*
    - **B.** 0.26*M*
    - **C.** 0.39*M*
    - **D.** 0.52*M*
    - **E.** 0.87*M*

18. How long will it take for half of the original amount of SO_2Cl_2 to decompose at the average reaction rate?
    - **A.** 285 min
    - **B.** 335 min
    - **C.** 385 min
    - **D.** 401 min
    - **E.** 516 min

19. A sample of argon gas is compressed into a volume of 0.712 L by a piston exerting 3.92 atm of pressure. The piston is released until the pressure of the gas is 1.50 atm. What is the new volume of the gas?
    - **A.** 0.272 L
    - **B.** 3.67 L
    - **C.** 5.86 L
    - **D.** 4.19 L
    - **E.** 1.86 L

NEED EXTRA HELP?

If You Missed Question . . .	1	2	3	4	5	6	7	8	9	10	11	12	13	14	15	16	17	18	19
Review Section . . .	16.1	14.1	12.4	12.4	14.2	16.1	12.2	3.4	3.4	10.3	16.4	16.4	9.1	9.2	11.2	16.3	16.3	16.3	13.1

SHORT ANSWER

11. Use the rate law expressions and *k* values given to find the rate for each step. Because the third step has the lowest rate, it is the rate-determining step.

12. Rate $= 6.68 \times 10^{-7}$ mol/(L•s)

EXTENDED RESPONSE

13. $2Na_3N(s) \rightarrow 6Na(s) + N_2(g)$

14. Decomposition; The reactants are breaking into smaller particles.

15. $32.5\ g\ Na_3N \left(\frac{1\ mol\ Na_3N}{82.977\ g\ Na_3N}\right)\left(\frac{1\ mol\ N_2}{2\ mol\ Na_3N}\right)\left(\frac{28.014\ g\ N_2}{1\ mol\ N_2}\right) = 5.49\ g\ N_2$

SAT SUBJECT TEST: CHEMISTRY

16. A
17. B
18. C
19. E

CHAPTER 17 Chemical Equilibrium

BIGIDEA **Many reactions and processes reach a state of chemical equilibrium in which both reactants and products are formed at equal rates.**

ESSENTIAL QUESTIONS	RESOURCES TO ASSESS MASTERY
SECTION 1 A State of Dynamic Balance **1.** What are the characteristics of chemical equilibrium? **2.** How are equilibrium expressions written for systems that are at equilibrium? **3.** How are equilibrium constants calculated from concentration data? 1 session 0.5 block	**Progress Monitoring** Formative Assessment, pp. 595, 599, 605 Reading Check, pp. 597, 604 Graph Check, p. 595 Section Review, p. 605
SECTION 2 Factors Affecting Chemical Equilibrium **1.** Which factors affect chemical equilibrium? **2.** How does Le Châtelier's principle apply to equilibrium systems? 2 sessions 1 block	**Progress Monitoring** Formative Assessment, pp. 607, 609, 610 Reading Check, p. 608 Section Review, p. 611
SECTION 3 Using Equilibrium Constants **1.** How are the equilibrium concentrations of reactants and products determined? **2.** How is the solubility of a compound calculated from its solubility product constant? **3.** Why is the common ion effect important? 2 sessions 1 block	**Progress Monitoring** Formative Assessment, pp. 612, 614, 618, 620, 621 Reading Check, p. 618 Graph Check, p. 620 Section Review, p. 622 **Summative Assessment** Chapter Assessment, p. 626 *eAssessment* Chapter Test (Scaffolded)

Go online!

LEVELED RESOURCES	LAB MATERIALS
Science Notebook 17.1 OL ***FAST FILE Chapter Resources:*** Study Guide, p. 14 BL	LaunchLAB p. 592 / **15 min** graduated cylinder, 100-mL beaker, food coloring, water, glass tubes of equal diameter
Science Notebook 17.2 OL ***FAST FILE Chapter Resources:*** MiniLab Worksheet, p. 2 OL Study Guide, p. 16 BL	MiniLAB p. 611 / **15 min** 0.1*M* $CoCl_2$ solution, concentrated HCl, water, table salt, ice bath, nonmercury thermometer, test tubes, hot water bath
Science Notebook 17.3 OL ***FAST FILE Chapter Resources:*** ChemLab Worksheet, p. 3 OL Study Guide, p. 18 BL	ChemLAB p. 624 / **45 min** $AgNO_3$ solution, NaCl solution, Na_2S solution, 24-well microplate; thin-stem pipettes

ADDITIONAL RESOURCES

eAssessment
ConnectED Student Center
ConnectED Teacher Center
Lesson Presentations
eSolutions
Videos
Animations
Online Quizzes
Virtual Investigations
Multilingual eGlossary
Personal Tutors
WebQuests
Applying Practices Worksheets
PBLs

Practice:

Challenge Problems AL
Supplemental Problems BL OL
Chapter Test (Scaffolded)
Standardized Test Practice

Additional Resources:

Blueprints for Success
Solving Problems: A Chemistry Handbook

Lab Resources:

Laboratory Manual OL
CBL Laboratory Manual OL
Small-Scale Laboratory Manual OL
Forensics Laboratory Manual OL

BL Below Level　OL On Level　AL Advanced Learners　EL English Learners　COOP LEARN Cooperative Learning

CHAPTER 17

LaunchLAB

Rubric

What is equal about equilibrium?

Purpose Students will discover that equilibrium does not mean equal amounts of reactant and product.

Safety Precautions

Approve lab safety forms before work begins.

Teaching Strategies

- The two glass tubes should be longer than the graduated cylinder is tall and have the same internal diameters.
- Demonstrate how to transfer the water from the beaker to the graduated cylinder and back to the beaker. Pairs of students can make simultaneous transfers.
- Have students write their observations on a lab form or in their journals.

Expected Results Initially, the height of the 20 mL of water in the graduated cylinder is greater than the height of the 20 mL of water in the beaker. Therefore, water will reach a higher level in the glass tube placed in the graduated cylinder than it will in the tube placed in the beaker. As time passes, the heights of the water in the tubes will even out. At the end, 30 mL is in the beaker and 10 mL is in the graduated cylinder.

CHAPTER 17

Chemical Equilibrium

BIGIDEA Many reactions and processes reach a state of chemical equilibrium in which both reactants and products are formed at equal rates.

SECTIONS

1 A State of Dynamic Balance

2 Factors Affecting Chemical Equilibrium

3 Using Equilibrium Constants

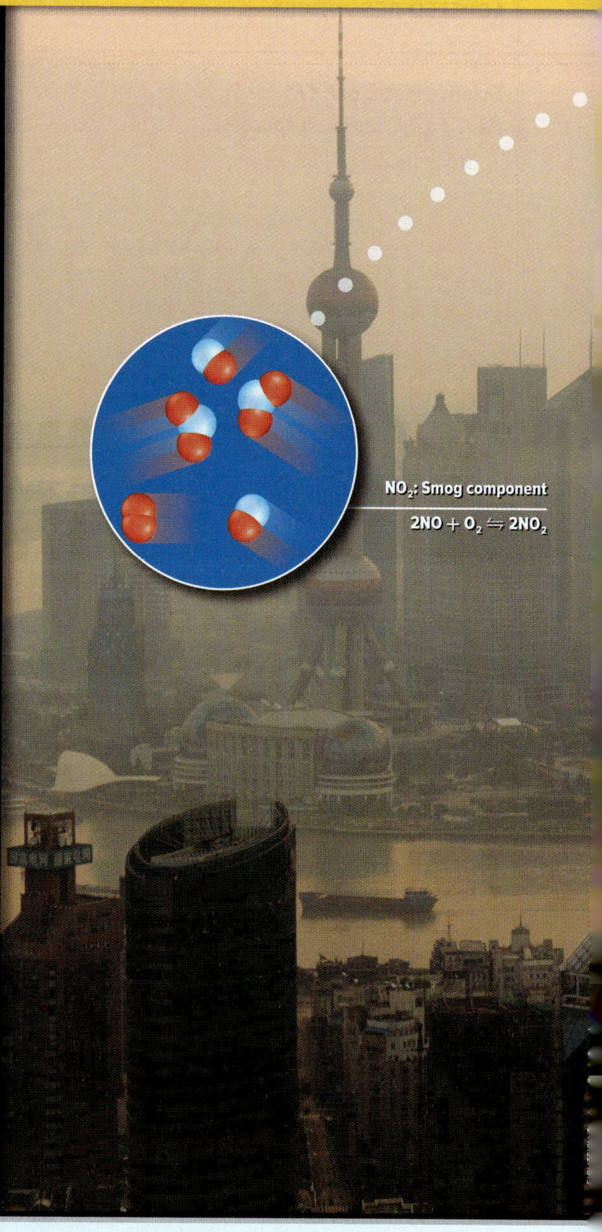

LaunchLAB

What is equal about equilibrium?

Equilibrium is a point of balance in which opposing changes cancel each other. In this lab, you will transfer different volumes of water from one container to another until equilibrium is reached.

FOLDABLES Study Organizer

Changes Affecting Equilibrium

Make a folded chart. Label it as shown. Use it to help you organize information about the factors that affect equilibrium.

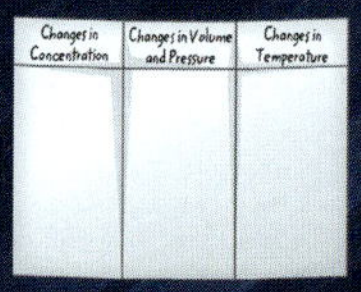

Procedure

1. Read and complete the lab safety form.
2. Measure 20 mL of water in a **graduated cylinder** and pour it into a **100-mL beaker.** Fill the graduated cylinder to the 20-mL mark with water. Add two drops of **food coloring** to the water in each container.
3. Obtain **two glass tubes of equal diameter.** Place one tube in the graduated cylinder and the other in the beaker.
4. Work with a partner. With the ends of the tubes at the bottoms of their containers, cover the open ends of the glass tubes with your index fingers so that water becomes trapped in the tubes. Simultaneously, move each tube to the other container and release your fingers to release the water.
5. Repeat the transfer process about 25 times. Record your observations.

Analysis

1. **Describe** your observations during the transfer process. As transfers continued, the level of water in the beaker grew higher, and the level in the graduated cylinder dropped lower until the levels were equal. The tube from the graduated cylinder could transfer more water than the tube from the beaker because the water in the graduated cylinder was deeper than the water in the beaker.
2. **Explain** Would the final result be different if you had continued the transfer process for a longer

No other human activity causes as much air pollution as the use of motor vehicles. Over the last decade catalytic converters and changes in gasoline additives have made cars 40% cleaner.

time? Once the levels were equal, the amount of water transferred from the beaker was equal to the amount of water transferred from the graduated cylinder, so no further change would occur in the levels of water.

Inquiry Could you illustrate equilibrium using glass tubes of different diameters? Explain. No, the amount of water transferred should depend only on the height of the water in the container from which it was taken. If the diameters of the tubes were different, the amount of water taken with each transfer would also depend on the diameter of the tube being used.

Introduce the BIGIDEA

Reversible Reactions To introduce this chapter's Big Idea, have students write the balanced equation for the chemical reaction in which sulfur dioxide and oxygen produce sulfur trioxide. $2SO_2 + O_2 \rightarrow 2SO_3$ Then, ask students to write the equation for the decomposition of sulfur trioxide to sulfur dioxide and oxygen. $2SO_3 \rightarrow 2SO_2 + O_2$ Ask how the two reactions are related. The second reaction is the reverse of the first reaction. Explain that reactions that can occur in both the forward and the reverse reactions are called reversible reactions. The forward and reverse reactions can occur at equal rates, called chemical equilibrium.

Tie to Previous Knowledge

Have students review the following concepts before studying this chapter.

- chemical reactions, equations for chemical reactions
- reaction rates, reaction rate laws

Use the Photo

Reaction Equilibrium Focus students' attention on the brown haze in the chapter opener photo, as well as the two chemical equations: $N_2 + O_2 \rightarrow 2NO$ and $2NO + O_2 \rightarrow 2NO_2$. Explain that the two reactions play roles in the formation of photochemical smog. Ask where the reactants in the first equation might be found. Nitrogen and oxygen gases make much of the atmosphere near Earth's surface. Ask why photochemical smog forms most often in and around cities with significant automobile traffic. The first reaction occurs more readily when the nitrogen and oxygen in air are heated to high temperatures, as in car engines. Explain that many chemical reactions occur in two directions, forward and reverse, and that the two reactions can exist in a state of balance called an equilibrium state.

SECTION 1

1 Focus

MAINIDEA

***K* Values Matter** Write the equation for the following reversible reaction on the board, explaining that CH_4 (methane) and H_2S (hydrogen sulfide) are two components of natural gas: $CH_4(g) + 2H_2S(g) \rightleftharpoons CS_2(g) + 4H_2(g)$. Then, work with students to help them set up the equilibrium constant expression for the equilibrium.

$$K = \frac{[CS_2][H_2]^4}{[CH_4][H_2S]^2}$$

Explain that $K = 0.036$ at 960°C, and ask students to explain why keeping natural gas at that temperature would not be a good way to be sure that the equilibrium mixture contains large amounts of hydrogen gas. The small *K* value means that the equilibrium mixture kept at that temperature would contain relatively small amounts of the products, which include hydrogen gas. **OL**

2 Teach

Concept Development

Law of Mass Action The law of chemical equilibrium is often referred to as the law of mass action. The law was proposed in 1864 by Cato Guldberg and Peter Waage, who worked at the University of Oslo. The law of mass action applies to a closed system. In a closed system, there is no energy or mass transferred into or out of the system. In a closed system, changes continue, but there is no net change, representing an equilibrium state.

SECTION 1

A State of Dynamic Balance

MAINIDEA Chemical equilibrium is described by an equilibrium constant expression that relates the concentrations of reactants and products.

Essential Questions

- What are the characteristics of chemical equilibrium?
- How are equilibrium expressions written for systems that are at equilibrium?
- How are equilibrium constants calculated from concentration data?

Review Vocabulary

free energy: the energy that is available to do work—the difference between the change in enthalpy and the product of the entropy change and the absolute temperature

New Vocabulary

reversible reaction
chemical equilibrium
law of chemical equilibrium
equilibrium constant
homogeneous equilibrium
heterogeneous equilibrium

CHEM 4 YOU Imagine a tug-of-war between two teams. Because the rope between them is not moving, it might seem that neither team is pulling. In fact, both teams are pulling, but the forces exerted by the two teams are equal and opposite, so they are in complete balance.

What is equilibrium?

Often, chemical reactions reach a point of balance or equilibrium. If you performed the Launch Lab, you found that a point of balance was reached in the transfer of water from the beaker to the graduated cylinder and from the graduated cylinder to the beaker.

Consider the reaction for the formation of ammonia from nitrogen and hydrogen that you read about previously.

$$N_2(g) + 3H_2(g) \rightarrow 2NH_3(g) \quad \Delta G° = -33.1 \text{ kJ}$$

Ammonia is important in agriculture as a fertilizer and an additive to animal feed grains. In industry, it is a raw material for the manufacture of many products such as nylon, as shown in **Figure 1.**

The equation for the production of ammonia has a negative standard free energy, $\Delta G°$. Recall that a negative sign for $\Delta G°$ indicates that the reaction is spontaneous under standard conditions, defined as 298 K and 1 atm, but spontaneous reactions are not always fast. When carried out under standard conditions, this ammonia-forming reaction is much too slow. To produce ammonia at a rate that is practical, the reaction must be carried out at a much higher temperature and pressure.

Figure 1 Ammonia reacts with both ends of a six-carbon molecule to form a diamine (1,6-diaminohexane). This is one step in the formation of the polymer nylon. Here nylon fibers, to be used in tire manufacturing, are being wound onto a spool.

Ted Horowitz/Alamy

Demonstration

A Model for Equilibrium

Purpose
to model an equilibrium system

Materials
Two water troughs or glass aquariums; 250-mL beaker; 600-mL beaker

Safety Precautions
Approve lab safety forms before work begins.

Disposal Pour water down a drain.

Procedure
Ask for two student volunteers, and have the remaining students watch the demonstration. Fill the two troughs or aquariums, with unequal depths of water. Have one student scoop water from trough 1 to trough 2 with the 250-mL beaker. Simultaneously, have the second student scoop water from trough 2 to trough 1 with the 600-mL beaker. Allow the process to continue until the depths of water in the two troughs no longer change.

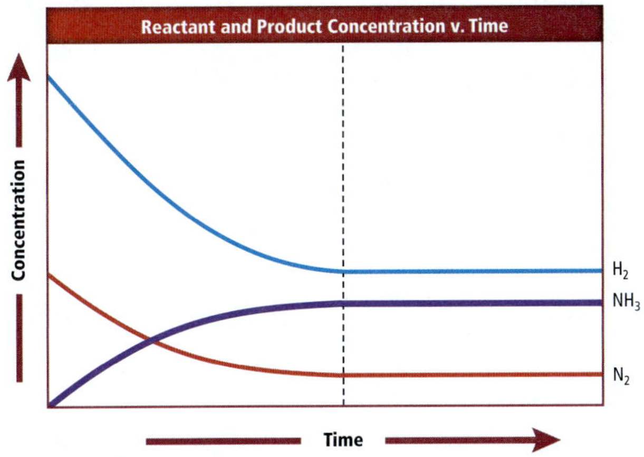

Figure 2 The concentrations of the reactants (H_2 and N_2) decrease at first, while the concentration of the product (NH_3) increases. Then, before the reactants are used up, all concentrations become constant.

☑ GRAPH CHECK
Explain how the graph shows that the concentrations of the reactants and products become constant.

What happens when 1 mol of nitrogen and 3 mol of hydrogen, the number of moles shown as coefficients in the chemical equation, are placed in a closed reaction vessel at 723 K? Because the reaction is spontaneous, nitrogen and hydrogen react. **Figure 2** illustrates the progress of the reaction. Note that the concentration of the product, NH_3, is zero at the start and gradually increases with time. The reactants, H_2 and N_2, are consumed in the reaction, so their concentrations gradually decrease. After a period of time, however, the concentrations of H_2, N_2, and NH_3 no longer change. All concentrations become constant, as shown by the horizontal lines on the right side of the diagram. The concentrations of H_2 and N_2 are not zero, so not all of the reactants were converted to product, even though $\Delta G°$ for this reaction is negative.

☑ GRAPH CHECK **Describe** the slopes of the curves for the reactants and for the product on the left of the vertical dotted line. How do the slopes differ on the right of the dotted line?

Reversible reactions and chemical equilibrium When a reaction results in an almost complete conversion of reactants to products, chemists say that the reaction goes to completion—but most reactions do not go to completion. The reactions appear to stop because they are reversible. A **reversible reaction** is a chemical reaction that can occur in both the forward and the reverse directions.

$$\text{Forward: } N_2(g) + 3H_2(g) \rightarrow 2NH_3(g)$$
$$\text{Reverse: } N_2(g) + 3H_2(g) \leftarrow 2NH_3(g)$$

Chemists combine these two equations into a single equation that uses a double arrow to show that both reactions occur.

$$N_2(g) + 3H_2(g) \rightleftharpoons 2NH_3(g)$$

The reactants in the forward reaction are on the left of the arrows. The reactants in the reverse reaction are on the right of the arrows. In the forward reaction, hydrogen and nitrogen combine to form the product ammonia. In the reverse reaction, ammonia decomposes into the products hydrogen and nitrogen.

VOCABULARY
ACADEMIC VOCABULARY
Convert
to change from one form or function to another
She converted a spare bedroom into an office.

Visual Learning

Figure 2 Help students appreciate how the graph relates to the equation for the production of ammonia, and use it to introduce the idea of reversible reactions. Ask students to look at the labeling of the *x*- and *y*-axes. Ask what is changing with time. the concentrations of the reactants and products Ask which quantities are decreasing and which are increasing. H_2 and N_2 are decreasing and NH_3 is increasing. Ask if, on the left side of the graph, H_2 and N_2 are decreasing at the same rate. H_2 is decreasing faster than N_2. Write the equation for the reaction on the board. Ask if the equation confirms their answer. Yes, in the equation, three moles of H_2 react for every mole of N_2, so three moles of H_2 are used up for every one mole of N_2. Ask students what is happening to the concentration of NH_3. The concentration of NH_3 is increasing. Ask what happens on the right side of the vertical dashed line. The concentrations of H_2, N_2, and NH_3 do not continue to change. Ask them to draw an inference from that observation. The reaction has reached a state of equilibrium. Ask if all the H_2 and N_2 have been depleted. No, the concentration of the two reactants are not zero. **OL**

☑ GRAPH CHECK On the left, the reactant curves slope downward as the reactants are depleted. The product curve slopes upward. On the right, the slopes of all curves, reactants, and products, have a slope of zero, which means that they are neither increasing or decreasing.

Results
A state of equilibrium is attained. One trough will have more water than the other when the less-than-full 600-mL beaker can transfer only the same volume of water as the 250-mL beaker.

Analysis
Ask these questions.

1. Before the transfer process began, what did you predict would happen as the process continued? The student with the larger, 600-mL beaker would empty one trough.
2. Why did your predicted outcome not take place? Water levels became constant when the less-than-full 600-mL beaker could transfer only the same volume of water as the 250-mL beaker.
3. Were the water levels in the two troughs equal when the levels became constant? no

ASSESSMENT

Knowledge Have students describe the analogy between the water at equilibrium and the reactants and products in a system at equilibrium. Just as the amounts of water in the two troughs were unequal, the amounts of reactants and products in a system at equilibrium do not have to be equal. Equilibrium requires only that the rates of the forward and reverse reactions occur at equal rates. **OL**

Visual Learning

Figure 3 Ask students why the number of molecules of ammonia increases from Figure 3a through Figure 3c, while the numbers of molecules of hydrogen and nitrogen decrease. Ammonia is a product of the reaction in which hydrogen and nitrogen are consumed. Ask what happens to the amounts of the three substances as the reaction progresses from Figure 3c to Figure 3d. They remain constant. Point out that this constancy does not mean that the reactions have stopped, but that a state of equilibrium has been established. **OL**

Reinforcement

Concentrations Be sure to point out that although a state of equilibrium is illustrated for the hydrogen-nitrogen-ammonia system, the amounts and concentrations of the three substances at equilibrium are not equal.

■ Caption Question Fig. 3

a. Only reactants are present.

b. More than two ammonia molecules are present, so they can react in the reverse direction.

c. The number of reactant and product molecules are the same in **c** and **d,** so equilibrium is established.

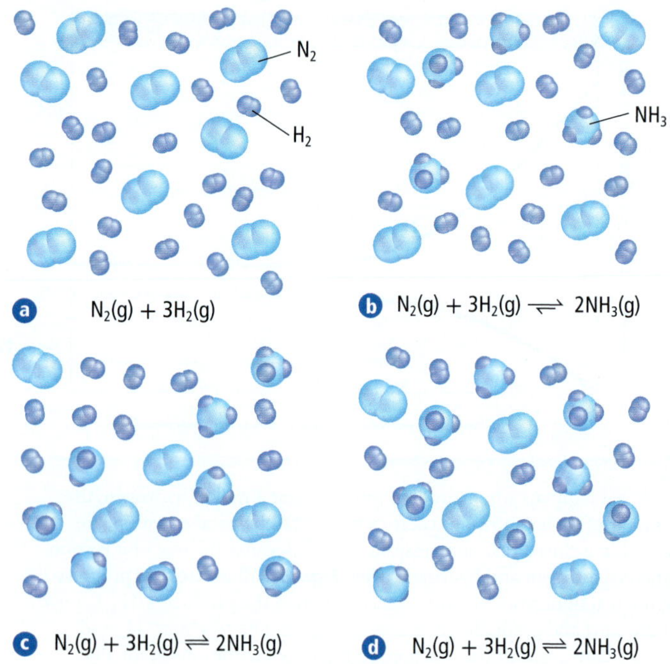

■ **Figure 3** The progress of a reaction to produce ammonia from hydrogen and nitrogen is shown in **a.** through **d.**

Interpret *Study the diagrams to answer the following questions. In* **a,** *how do you know that the reaction has not yet begun? In* **b,** *what evidence indicates that the reverse reaction has begun? Compare* **c** *with* **d.** *How do you know that equilibrium has been reached?*

How does the reversibility of this reaction affect the production of ammonia? **Figure 3a** shows a mixture of nitrogen and hydrogen just as the reaction begins at a definite, initial rate. No ammonia is present, therefore only the forward reaction can occur.

$$N_2(g) + 3H_2(g) \rightarrow 2NH_3(g)$$

As hydrogen and nitrogen combine to form ammonia, their concentrations decrease, as shown in **Figure 3b.** You learned previously that the rate of a reaction depends on the concentration of the reactants. The decrease in the concentration of the reactants causes the rate of the forward reaction to slow. As soon as ammonia is present, the reverse reaction can occur, slowly at first, but at an increasing rate as the concentration of ammonia increases.

$$N_2(g) + 3H_2(g) \leftarrow 2NH_3(g)$$

As the reaction proceeds, the rate of the forward reaction continues to decrease and the rate of the reverse reaction continues to increase until the two rates are equal. At that point, ammonia is produced at the same rate it is decomposed, so the concentrations of N_2, H_2, and NH_3 remain constant, as shown in **Figures 3c** and **3d.** The system has reached a state of balance or equilibrium. The word *equilibrium* means that opposing processes are in balance. **Chemical equilibrium** is a state in which the forward and reverse reactions balance each other because they take place at equal rates.

$$\text{Rate}_{\text{forward reaction}} = \text{Rate}_{\text{reverse reaction}}$$

CHEMISTRY JOURNAL

Law of Mass Action Have students research the work of the Norwegian chemists Cato Maximilian Guldberg and Peter Waage that led them to propose the law of mass action. Have them describe how the law of mass action results in the formatting of equilibrium constant expressions. **AL**

You can recognize that the ammonia-forming reaction reaches a state of chemical equilibrium because its chemical equation is written with a double arrow like this.

$$N_2(g) + 3H_2(g) \rightleftharpoons 2NH_3(g)$$

At equilibrium, the concentrations of reactants and products are constant, as shown in **Figures 3c** and **3d.** However, that doesn't mean that the amounts or concentrations of reactants and products are equal. That is seldom the case. In fact, it is not unusual for the equilibrium concentrations of a reactant and product to differ by a factor of one million or more.

READING CHECK **Explain** the meaning of a double arrow in chemical equations.

Connection to Physics **The dynamic nature of equilibrium** A push or pull on an object is a force. When you push on a door or pull on a dog's leash, you exert a force. When two or more forces are exerted on the same object in the same direction, they add together. One force subtracts from the other if the forces are in opposite directions. Thus, in a tug-of-war, when two teams pull on a rope with equal force, the resulting force has a magnitude of zero and the rope does not move. The system is said to be in equilibrium. Similarly, the people on the seesaw in **Figure 4a** represent a system in equilibrium. The equal-and-opposite forces on both ends of the seesaw are called balanced forces. If, instead, one force is greater in magnitude than the other, the combined force is greater than zero and is called an unbalanced force. An unbalanced force causes an object to accelerate, which is what has happened in **Figure 4b.**

Figure 4 In **a,** all the forces are in perfect balance, so the position of the seesaw remains steady. In **b,** the unbalanced force on the left causes the seesaw to change its position.
Explain *this analogy in terms of chemical equilibrium.*

READING CHECK The double arrow means that a reaction is at equilibrium and that the forward reaction and the reverse reaction are occurring at the same rate.

Caption Question Fig. 4
In a reaction at equilibrium, the rates of the forward and reverse reactions are equal as are the forces operating on the seesaw. They are in balance so nothing appears to be happening. However, if something upsets the balance, action begins again.

Concept Development

Dynamic Equilibrium Help students understand the dynamic nature of equilibrium with this example. Draw on the board a bridge that connects a city of 10,000 vehicles with a city of 50,000 vehicles. Draw arrows of equal length and opposite directions above and below the bridge and label both arrows with *500 vehicles per hour.* Point out that a kind of equilibrium state has been achieved in the sense that the number of vehicles crossing the bridge hourly in one direction is equal to the number crossing in the opposite direction. Emphasize the fact that the numbers of vehicles in the two cities do not have to be equal.

DIFFERENTIATED INSTRUCTION

Struggling Learners Illustrate the dynamic nature of equilibrium for students by describing the following situation. A person is rowing a boat upstream at exactly the same speed as the current is pushing it downstream. Point out that the boat has no net movement in relation to the stream's banks, but a state of dynamic, physical equilibrium exists. BL

Identify Misconceptions

Students might think that equilibrium systems must contain equal amounts of reactants and products.

Uncover the Misconception Ask students to describe the relative amounts of reactants and products in a system at equilibrium. If they answer 50% reactants and 50% products, they display the misconception.

Demonstrate the Concept Have students estimate the relative equilibrium concentrations of N_2, H_2, and NH_3 shown in Figure 2. Ask if they are equal. Approximately $1N_2$: $4H_2$: $3NH_3$; They are not equal.

Assess New Knowledge Remind students that the value of the equilibrium constant for the equilibrium system $H_2 + I_2 \leftrightharpoons 2HI$ at 731 K is given as 49.7. Ask them what the value of K_{eq} would be if all concentrations were 1.0*M*. $K_{eq} = 1.0$ Point out that the K_{eq} value of 49.7 clearly shows that the equilibrium system contains more products than reactants. OL

Caption Question Fig. 5 If people are walking back and forth, the inhabitants of each building are not the same from moment to moment. The same is true of a chemical reaction at equilibrium. The atoms that are part of a product at one time might be part of a reactant at another time.

Figure 5 Suppose a certain number of people are confined to the two buildings connected by this walkway and that people can walk back and forth between the buildings. The number of people in each building will remain constant only if the same number of people cross the bridge in one direction as cross in the opposite direction.

Decide *whether the same people will always be in the same building. How does your answer apply to chemical equilibrium?*

Like equal forces opposing each other, equilibrium is a state of action, not inaction. For example, consider this analogy: The glassed-in walkway, shown in **Figure 5,** connects two buildings. Suppose that all entrances and exits for the buildings, except the walkway, are closed for a day. And suppose that the same number of persons cross the walkway in each direction every hour. Given these circumstances, the number of persons in each building remains constant even though people continue to cross between the two buildings. Note that the numbers of persons in the two buildings do not have to be equal. Equilibrium requires only that the number of persons crossing the walkway in one direction is equal to the number crossing in the opposite direction.

The dynamic nature of chemical equilibrium can be illustrated by placing equal masses of iodine crystals in two interconnected flasks, as shown in **Figure 6a.** The flask on the left contains iodine molecules made up entirely of the nonradioactive isotope I-127. The flask on the right contains iodine molecules made up of the radioactive isotope I-131. The radiation counters indicate the difference in the levels of radioactivity within each flask.

Each flask is a closed system. No reactant or product can enter or leave. At 298 K and 1 atm, this equilibrium is established in both flasks.

$$I_2(s) \rightleftharpoons I_2(g)$$

In the forward process, called sublimation, iodine molecules change directly from the solid phase to the gas phase. In the reverse process, gaseous iodine molecules return to the solid phase. A solid-vapor equilibrium is established in each flask.

When the stopcock in the tube connecting the two flasks is opened, as in **Figure 6b,** iodine vapor can travel back and forth between the two flasks. After a period of time, the readings on the radiation counters indicate that the flask on the left contains as many radioactive I-131 molecules as the flask on the right in both the vapor and the solid phases.

The evidence suggests that iodine molecules constantly change from the solid phase to the gas phase according to the forward process, and that gaseous iodine molecules convert back to the solid phase according to the reverse process. The constant readings on both radiation detectors indicate that equilibrium has been established in the combined volume of the two flasks.

DIFFERENTIATED INSTRUCTION

Visually Impaired Have two students sit on opposite sides of a table, rest their right (or left) elbows on the table, and lock hands palm-to-palm. Have them push gently with equal force so that their hands do not move. Ask them if the lack of movement means that no forces are being exerted. no Point out that opposing forces cancel each other, and make the analogy with the equilibrium state. Equilibrium is a state in which the concentrations of reactants and products do not change because two opposing processes are occurring at equal rates. BL EL

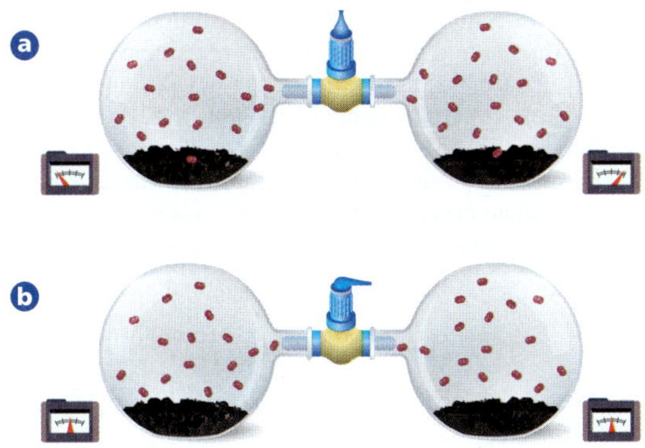

Figure 6a. Radioactive iodine molecules in the flask on the right are separated from nonradioactive iodine in the flask on the left. Note the readings on the radiation monitors. **b.** After the stopcock has been open for a time, the radiation monitors show that radioactive molecules are in both flasks. The particles must have moved back and forth between the flasks and between the solid and the gaseous phases.

Equilibrium Expressions

Some chemical systems have little tendency to react. Others go to completion. The majority of reactions reach a state of equilibrium with some of the reactants unconsumed. If the reactants are not all consumed, then the amount of products produced is less than the amount predicted by the balanced chemical equation. According to the equation for the ammonia-producing reaction, 2 mol of ammonia should be produced when 1 mol of nitrogen and 3 mol of hydrogen react. However, because the reaction reaches a state of equilibrium, less than 2 mol of ammonia are obtained.

VOCABULARY
WORD ORIGIN
Completion
comes from the Latin verb *completus,* which means *having all necessary parts, elements, or steps.*

The law of chemical equilibrium In 1864, Norwegian chemists Cato Maximilian Guldberg and Peter Waage jointly proposed and developed the **law of chemical equilibrium,** which states that at a given temperature, a chemical system might reach a state in which a particular ratio of reactant and product concentrations has a constant value. The general equation for a reaction at equilibrium is as follows.

$$aA + bB \rightleftharpoons cC + dD$$

If the law of chemical equilibrium is applied to this reaction, the following ratio is obtained.

The Equilibrium Constant Expression

$$K_{eq} = \frac{[C]^c[D]^d}{[A]^a[B]^b}$$

[A] and [B] are the molar concentrations of the reactants. [C] and [D] are the molar concentrations of the products.

The exponents *a, b, c,* and *d,* are the coefficients in the balanced equation.

The equilibrium constant expression is the ratio of the molar concentrations of the products to the molar concentrations of the reactants with each concentration raised to a power equal to its coefficient in the balanced chemical equation.

The **equilibrium constant,** K_{eq}, is the numerical value of the ratio of product concentrations to reactant concentrations, with each concentration raised to the power equal to its coefficient in the balanced equation. The value of K_{eq} is constant only at a specified temperature.

ASSESSMENT

Knowledge Ask students to write the equation for the equilibrium system that yields the equilibrium constant expression
$K_{eq} = [C]^2[D]/[A][B]^3$
$A + 3B \rightleftharpoons 2C + D$ OL

Math in Chemistry

Units for K_{eq} Point out that although it is common to see K_{eq} values given without units, equilibrium constants based on observed concentrations might (or might not) have units. Because concentration units are mol/L, only when the sums of exponents in the numerator and denominator are equal do the units cancel, making K_{eq} unitless. If the sums are not equal, however, units such as mol/L or L^2/mol^2 result. Explain, for example, that units for K_{eq} in the expression $K_{eq} = [C]^2[D]/[A][B]^3$ are L/mol.

DIFFERENTIATED INSTRUCTION

English Learners Have English learners look up and then explain the meanings of several key English terms used in this section: *reversible, equilibrium, homogeneous, heterogeneous.* Have them use each term in a sentence or write a paragraph using all of the terms. EL BL

Concept Development

Molar Concentration Lead students through the process of calculating the molar concentration of liquid ethanol. 1000 mL CH_3CH_2OH/1 L $CH_3CH_2OH \times 0.789$ g CH_3CH_2OH/1 mL $CH_3CH_2OH \times 1$ mol CH_3CH_2OH/46.07 g $CH_3CH_2OH = 17.1$ mol/L Point out that unless the density of liquid ethanol changes due to a change in temperature, 17.1 mol/L might be considered a constant for the substance. OL

Reinforcement

The value of K_{eq} Ask students how the value of K_{eq}, is related to the relative amounts of reactants and products present at equilibrium. If K_{eq} is greater than 1, there are more products than reactants. If K_{eq} is less than one, there are more reactants than products. BL EL

How can you interpret the size of the equilibrium constant? Recall that a fraction with a numerator greater than its denominator has a value greater than 1. And a fraction with a numerator less than its denominator has a value less than 1. For example, compare the ratios 5/1 and 1/5. Five is a larger number than one-fifth. Because the product concentrations are in the numerator of the equilibrium expression, a numerically large K_{eq} means that the equilibrium mixture contains more products than reactants. Similarly, a K_{eq} less than 1 means that the equilibrium mixture contains more reactants than products.

$K_{eq} > 1$: Products are favored at equilibrium.

$K_{eq} < 1$: Reactants are favored at equilibrium.

Expressions for homogeneous equilibria Gaseous hydrogen iodide is produced by the equilibrium reaction of hydrogen gas with iodine. Iodine and some of its compounds have important uses in medicine, as illustrated in **Figure 7.** How would you write the equilibrium constant expression for this reaction in which hydrogen and iodine react to form hydrogen iodide?

$$H_2(g) + I_2(g) \rightleftharpoons 2HI(g)$$

This reaction is a **homogeneous equilibrium,** which means that all the reactants and products are in the same physical state. All participants are gases. First, place the product concentration in the numerator and the reactant concentrations in the denominator.

$$\frac{[HI]}{[H_2][I_2]}$$

The expression becomes equal to K_{eq} when you add the coefficients from the balanced chemical equation as exponents.

$$K_{eq} = \frac{[HI]^2}{[H_2][I_2]}$$

K_{eq} for this equilibrium at 731 K is 49.7. Note that 49.7 has no units. When writing equilibrium constant expressions, it is customary to omit units.

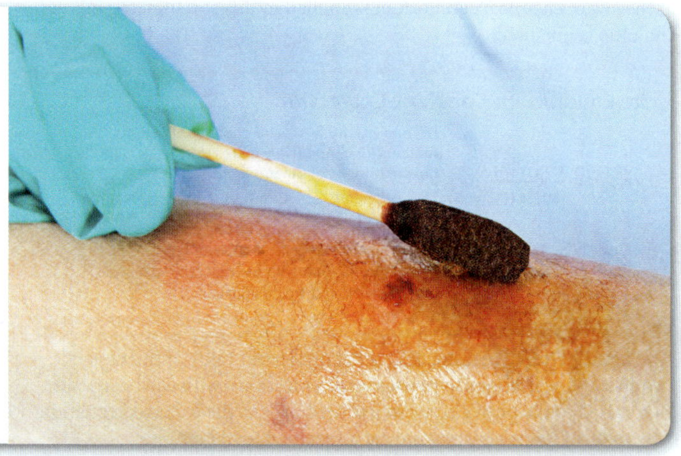

Figure 7 Because of iodine's antibacterial properties, solutions of iodine and iodine compounds are used externally as antiseptics. Some iodine compounds are used internally. For example, doctors use potassium iodide (KI) in the treatment of goiter, a condition characterized by the enlargement of the thyroid gland.

CHEMISTRY JOURNAL

Concentrations of Solids and Liquids Ask students to explain why concentrations of solids and liquids are omitted from the equilibrium constant expressions for heterogeneous equilibria. Have them write the chemical equations and equilibrium constant expressions for two heterogeneous equilibria that involve solids and two that involve liquids. OL

EXAMPLE Problem 1

EQUILIBRIUM CONSTANT EXPRESSIONS FOR HOMOGENEOUS EQUILIBRIA Millions of tons of ammonia (NH_3) are produced each year for use in the manufacture of products such as explosives, fertilizers, and synthetic fibers. You might have used ammonia in your home as a household cleaner, which is particularly useful for cleaning glass. Ammonia is manufactured from its elements, hydrogen and nitrogen, using the Haber process. Write the equilibrium constant expression for the following reaction.

$$N_2(g) + 3H_2(g) \rightleftharpoons 2NH_3(g)$$

1 ANALYZE THE PROBLEM

The equation for the reaction provides the information needed to write the equilibrium constant expression. The equilibrium is homogeneous because the reactants and product are in the same physical state.

The general form of the equilibrium constant expression is

$$K_{eq} = \frac{[C]^c}{[A]^a[B]^b}$$

Known
$[A] = [N_2]$, coefficient $N_2 = 1$
$[B] = [H_2]$, coefficient $H_2 = 3$
$[C] = [NH_3]$, coefficient $NH_3 = 2$

Unknown
$K_{eq} = ?$

2 SOLVE FOR THE UNKNOWN

Form a ratio of product concentration to reactant concentrations.

$$K_{eq} = \frac{[C]^c}{[A]^a[B]^b}$$ State the general form of the equilibrium constant expression.

$$K_{eq} = \frac{[NH_3]^c}{[N_2]^a[H_2]^b}$$ Substitute $A = N_2$, $B = H_2$, and $C = NH_3$.

$$K_{eq} = \frac{[NH_3]^2}{[N_2][H_2]^3}$$ Substitute $a = 1$, $b = 3$, and $c = 2$.

3 EVALUATE THE ANSWER

The product concentration is in the numerator and the reactant concentrations are in the denominator. Product and reactant concentrations are raised to powers equal to their coefficients.

PRACTICE Problems

Do additional problems. Online Practice

1. Write equilibrium constant expressions for these equilibria.
   a. $N_2O_4(g) \rightleftharpoons 2NO_2(g)$
   b. $2H_2S(g) \rightleftharpoons 2H_2(g) + S_2(g)$
   c. $CO(g) + 3H_2(g) \rightleftharpoons CH_4(g) + H_2O(g)$
   d. $4NH_3(g) + 5O_2(g) \rightleftharpoons 4NO(g) + 6H_2O(g)$
   e. $CH_4(g) + 2H_2S(g) \rightleftharpoons CS_2(g) + 4H_2(g)$
2. **Challenge** Write the chemical equation that has the equilibrium constant expression $K_{eq} = \frac{[CO]^2[O_2]}{[CO_2]^2}$.

RealWorld CHEMISTRY

Thyroid Health

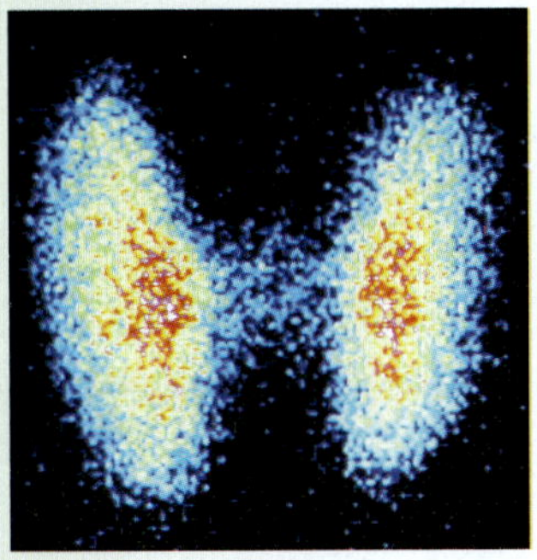

NUCLEAR MEDICINE Iodine-131 is a radioactive isotope that is absorbed by the thyroid gland. It is used in medicine to diagnose and treat diseases of the thyroid. When iodine-131 is administered to a patient, radiation from the isotope creates an image of the gland on film that reveals abnormalities. The image above can be used to diagnose and treat people with thyroid diseases or abnormalities, such as Graves' disease. Graves' disease is a treatable disease that is a common cause of an overactive thyroid gland.

IN-CLASS Example

Question The decomposition of phosgene, a chemical warfare agent outlawed by the Geneva Convention, is described by the equilibrium $COCl_2(g) \rightleftharpoons CO(g) + Cl_2(g)$. What is the equilibrium constant expression for the equilibrium?

Answer $K = \frac{[CO][Cl_2]}{[COCl_2]}$

PRACTICE Problems

Have students refer to the Selected Solutions appendix for complete solutions to odd-numbered problems. The complete solutions for all problems can be found in the Solutions Manual.

1. a. $K_{eq} = [NO_2]^2/[N_2O_4]$
   b. $K_{eq} = [H_2]^2[S_2]/[H_2S]^2$
   c. $K_{eq} = [CH_4][H_2O]/[CO][H_2]^3$
   d. $K_{eq} = [NO]^4[H_2O]^6/[NH_3]^4[O_2]^5$
   e. $K_{eq} = [CS_2][H_2]^4/[CH_4][H_2S]^2$
2. $2CO_2(g) \rightleftharpoons 2CO(g) + O_2(g)$

CHEMISTRY PROJECT

Reversible Occurrences and Processes Ask students to list at least five everyday occurrences or processes that are reversible. Have students share their findings to the class in graphic form or oral report. Have students discuss the importance of the reversible reactions they listed. OL

Quick Demo

Equilibrium Constant Fill a small flask approximately half-full of water. Stopper the flask securely, and show it to students. The next day, ask if the amount of water in the flask is changing. no Then, ask if that means that no reactions or processes are occurring in the flask. no Point out that the amount of liquid water remains constant even though two opposing processes are in equilibrium. Ask students to write the chemical equation and equilibrium constant expression for the heterogeneous equilibrium inside the flask.
$H_2O(l) \leftrightharpoons H_2O(g)$, $K_{eq} = [H_2O(g)]$ **OL**

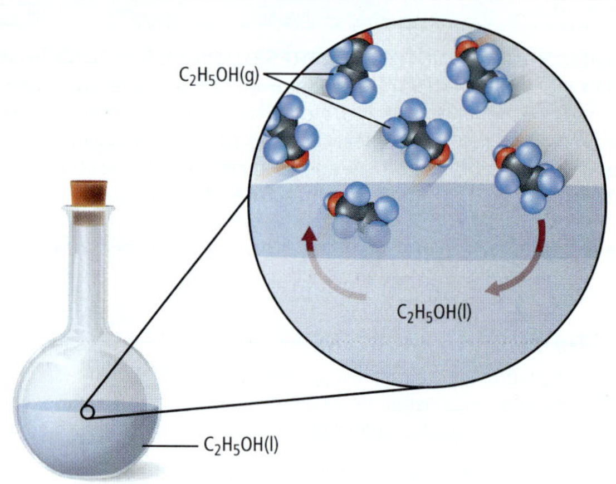

■ **Figure 8** At equilibrium the rate of evaporation of ethanol (C_2H_5OH) equals the rate of condensation. This two-phase equilibrium is called a heterogeneous equilibrium. K_{eq} depends only on $[C_2H_5OH(g)]$.

Expressions for heterogeneous equilibria You have learned to write K_{eq} expressions for homogeneous equilibria, those in which all reactants and products are in the same physical state. When the reactants and products are present in more than one physical state, the equilibrium is called a **heterogeneous equilibrium.** When ethanol is placed in a closed flask, a liquid-vapor equilibrium is established, as illustrated in **Figure 8.**

$$C_2H_5OH(l) \rightleftharpoons C_2H_5OH(g)$$

To write the equilibrium constant expression for this process, you would form a ratio of the product to the reactant. At a given temperature, the ratio would have a constant value K.

$$K = \frac{[C_2H_5OH(g)]}{[C_2H_5OH(l)]}$$

Note that the concentration of liquid ethanol is in the denominator. Liquid ethanol is a pure substance, so its concentration is its density expressed in moles per liter. Recall that at any given temperature, density is constant. No matter how much or how little C_2H_5OH is present, its concentration remains constant. Therefore, the term in the denominator is a constant and can be combined with K in the expression for K_{eq}.

$$K[C_2H_5OH(l)] = [C_2H_5OH(g)] = K_{eq}$$

The equilibrium constant expression for this phase change is

$$K_{eq} = [C_2H_5OH(g)]$$

Solids are also pure substances with unchanging concentrations, so equilibria involving solids are simplified in the same way. Recall the experiment involving the sublimation of iodine crystals in **Figure 6.**

$$I_2(s) \rightleftharpoons I_2(g)$$
$$K_{eq} = [I_2(g)]$$

The equilibrium constant, K_{eq}, depends only on the concentration of gaseous iodine in the system.

DIFFERENTIATED INSTRUCTION

Advanced Learners Have capable students use algebra to determine the number of molecules of each substance present at equilibrium for the reaction $A_3(g) + AB(g) \leftrightharpoons A_2(g) + A_2B(g)$, $K_{eq} = 16$. Initially, the system contains nine A_3 molecules and 12 AB molecules. $16 = ((x)A_2 \times (x)A_2B)/((9 - x)A_3 \times (12 - x)AB)$ Solving the equation for x gives a value of 8. Therefore, the equilibrium mixture contains 1 molecule of A_3, 4 molecules of AB, 8 molecules of A_2, and 8 molecules of A_2B. **AL**

EXAMPLE Problem 2

EQUILIBRIUM CONSTANT EXPRESSIONS FOR HETEROGENEOUS EQUILIBRIA In addition to its uses in baking and as an antacid and cleaning agent, baking soda is often placed in open boxes in refrigerators to freshen the air as shown in **Figure 9.** Write the equilibrium constant expression for the decomposition of baking soda (sodium hydrogen carbonate).

$$2NaHCO_3(s) \rightleftharpoons Na_2CO_3(s) + CO_2(g) + H_2O(g)$$

1 ANALYZE THE PROBLEM

You are given a heterogeneous equilibrium involving gases and solids. Solids are omitted from the equilibrium constant expression.

Known

$[C] = [Na_2CO_3]$, coefficient $Na_2CO_3 = 1$
$[D] = [CO_2]$, coefficient $CO_2 = 1$
$[E] = [H_2O]$, coefficient $H_2O = 1$
$[A] = [NaHCO_3]$, coefficient $NaHCO_3 = 2$

Unknown

equilibrium constant expression = ?

2 SOLVE FOR THE UNKNOWN

Form a ratio of product concentrations to reactant concentrations.

$$K_{eq} = \frac{[C]^c[D]^d[E]^e}{[A]^a[B]^b}$$ **State the general form of the equilibrium constant expression.**

$$K_{eq} = \frac{[Na_2CO_3]^c[CO_2]^d[H_2O]^e}{[NaHCO_3]^a}$$ **Substitute A = $NaHCO_3$, C = Na_2CO_3, D = CO_2, and E = H_2O.**

$$K_{eq} = \frac{[Na_2CO_3]^1[CO_2]^1[H_2O]^1}{[NaHCO_3]^2}$$ **Substitute a = 2, c = 1, d = 1, and e = 1.**

$$K_{eq} = [CO_2][H_2O]$$ **Omit terms involving solid substances.**

3 EVALUATE THE ANSWER

The expression correctly applies the law of chemical equilibrium to the equation.

PRACTICE Problems

Do additional problems. Online Practice

3. Write equilibrium constant expressions for these heterogeneous equilibria.
   a. $C_{10}H_8(s) \rightleftharpoons C_{10}H_8(g)$
   b. $H_2O(l) \rightleftharpoons H_2O(g)$
   c. $CaCO_3(s) \rightleftharpoons CaO(s) + CO_2(g)$
   d. $C(s) + H_2O(g) \rightleftharpoons CO(g) + H_2(g)$
   e. $FeO(s) + CO(g) \rightleftharpoons Fe(s) + CO_2(g)$
4. **Challenge** Solid iron reacts with chlorine gas to form solid iron(III) chloride ($FeCl_3$). Write the balanced equation and the equilibrium constant expression for the reaction.

Figure 9 Sodium hydrogen carbonate (baking soda) absorbs odors and freshens the air in a refrigerator. It is also a key ingredient in some toothpastes.

Get help with **equilibrium constant expressions.**
Personal Tutor

Quick Demo

Equilibrium in Gases
Obtain a bottle of smelling salts and pass it around the classroom for students to smell. Explain that smelling salts are primarily ammonium carbonate, which decomposes to ammonium hydrogen carbonate and ammonia according to the equation $(NH_4)_2CO_3(s) \rightarrow NH_4HCO_3(s) + NH_3(g)$. The ammonia produced in the reaction acts as a heart stimulant when used as smelling salts. Ask students to write the equilibrium constant expression for the heterogeneous chemical equilibrium that exists inside the closed bottle of smelling salts. $K_{eq} = [NH_3(g)]$ OL

IN-CLASS Example

Question Write the equilibrium constant expression for the following heterogeneous equilibrium: $MgCO_3(s) \leftrightharpoons MgO(s) + CO_2(g)$

Answer $K = [CO_2(g)]$

PRACTICE Problems

Have students refer to the Selected Solutions appendix for complete solutions to odd-numbered problems. The complete solutions for all problems can be found in the Solutions Manual.

3. a. $K_{eq} = [C_{10}H_8(g)]$
   b. $K_{eq} = [H_2O(g)]$
   c. $K_{eq} = [CO_2(g)]$
   d. $K_{eq} = [CO(g)][H_2(g)]/[H_2O(g)]$
   e. $K_{eq} = [CO_2(g)]/[CO(g)]$
4. $2Fe(s) + 3Cl_2(g) \leftrightharpoons 2FeCl_3(s)$
   $K_{eq} = 1/[Cl_2]^3$

DIFFERENTIATED INSTRUCTION

Struggling Learners Have students work together in small groups and write down several examples of systems in nature or their surroundings that involve equilibrium. Have students represent the system and the dynamics of the equilibrium in a graphic format. BL EL COOP LEARN

☑ READING CHECK At equilibrium, the rates of the forward and reverse reactions are equal. If one or more of the reactants or products were not present, the two opposing reactions could not take place.

CAREERS IN CHEMISTRY

WebQuest

IN-CLASS Example

Question What is the value of K_{eq} for the following equilibrium if $[N_2] = 0.20$ mol/L, $[O_2] = 0.15$ mol/L, and $[NO] = 0.0035$ mol/L?
$N_2(g) + O_2(g) \rightleftharpoons 2NO(g)$

Answer $K_{eq} = 4.1 \times 10^{-4}$

$$K_{eq} = \frac{(0.0035)^2}{(0.20)(0.15)} = 4.1 \times 10^{-4}$$

PRACTICE Problems

Have students refer to the Selected Solutions appendix for complete solutions to odd-numbered problems. The complete solutions for all problems can be found in the Solutions Manual.

5. $K_{eq} = 0.213$
6. $K_{eq} = 3.93$
7. $[COCl_2] = 0.27M$

CAREERS IN CHEMISTRY

Science Writer To convey scientific information to the nonscientific reader, a writer must have a broad knowledge of science and the ability to write clear, concise, and understandable prose. Science writers make complex subjects, such as chemical equilibrium, accessible to readers with no prior knowledge of the subject.

WebQuest

Equilibrium Constants

For a given reaction at a given temperature, K_{eq} will always be the same regardless of the initial concentrations of reactants and products. To test this statement, three experiments were carried out using the following reaction.

$$H_2(g) + I_2(g) \rightleftharpoons 2HI(g)$$

The results are summarized in **Table 1.** In Trial 1, 1.0000 mol H_2 and 2.0000 mol I_2 were placed in a 1.0000-L vessel. No HI was present at the beginning of Trial 1. In Trial 2, only HI was present at the start of the experiment. In Trial 3, each of the three substances had the same initial concentration. The reactions were carried out at 731 K.

Equilibrium concentrations When equilibrium was established, the concentration of each substance was determined experimentally. Note that the equilibrium concentrations are not the same in the three trials, yet when each set of equilibrium concentrations is put into the equilibrium constant expression, the value of K_{eq} is the same. Each set of equilibrium concentrations represents an equilibrium position.

The value of K_{eq} Although an equilibrium system has only one value for K_{eq} at a particular temperature, it has an unlimited number of equilibrium positions. Equilibrium positions depend on the initial concentrations of the reactants and products. The large value of K_{eq} for the reaction $H_2(g) + I_2(g) \rightleftharpoons 2HI(g)$ means that at equilibrium the product is present in larger amount than the reactants. However, many equilibria have small K_{eq} values. For the equilibrium $N_2(g) + O_2(g) \rightleftharpoons 2NO(g)$, K_{eq} equals 4.6×10^{-31} at 298 K. A K_{eq} this small means that the product, NO, is practically nonexistent at equilibrium.

Equilibrium characteristics You might have noticed certain characteristics of all chemical reactions that reach equilibrium. First, the reaction must take place in a closed system—no reactant or product can enter or leave the system. Second, the temperature must remain constant. Third, all reactants and products are present, and they are in constant dynamic motion. This means that equilibrium is dynamic, not static.

☑ READING CHECK **Explain** why it is important that all reactants and products be present at equilibrium.

Table 1 Experimental Data for HI Reaction Equilibrium

	Initial Concentrations			Equilibrium Concentrations			K_{eq}
Trial	$[H_2]_0$ (M)	$[I_2]_0$ (M)	$[HI]_0$ (M)	$[H_2]_{eq}$ (M)	$[I_2]_{eq}$ (M)	$[HI]_{eq}$ (M)	$\frac{[HI]^2}{[H_2][I_2]} = K_{eq}$
1	1.0000	2.0000	0	0.06587	1.0659	1.8682	$\frac{[1.8682]^2}{[0.06587][1.0659]} = 49.70$
2	0	0	5.0000	0.5525	0.5525	3.8950	$\frac{[3.8950]^2}{[0.5525][0.5525]} = 49.70$
3	1.0000	1.0000	1.0000	0.2485	0.2485	1.7515	$\frac{[1.7515]^2}{[0.2485][0.2485]} = 49.70$

DIFFERENTIATED INSTRUCTION

Advanced Learners Challenge advanced learners to add two more equilibrium positions (lines) to Table 1, columns 5 to 8, with equilibrium concentrations different from the ones already in the table. One position should have equal equilibrium concentrations of H_2 and I_2; the other position should have unequal concentrations of H_2 and I_2. The concentrations in both lines should yield K_{eq} values of 49.70. AL

EXAMPLE Problem 3

Find help solving algebraic equations. Math Handbook

EXAMPLE PROBLEM

THE VALUE OF EQUILIBRIUM CONSTANTS Calculate the value of K_{eq} for the equilibrium constant expression $K_{eq} = \frac{[NH_3]^2}{[N_2][H_2]^3}$ given concentration data at one equilibrium position: $[NH_3] = 0.933$ mol/L, $[N_2] = 0.533$ mol/L, $[H_2] = 1.600$ mol/L.

1 ANALYZE THE PROBLEM

You have been given the equilibrium constant expression and the concentration of each reactant and product. You must calculate the equilibrium constant.

Known

$K_{eq} = \frac{[NH_3]^2}{[N_2][H_2]^3}$

$[N_2] = 0.533$ mol/L

$[NH_3] = 0.933$ mol/L

$[H_2] = 1.600$ mol/L

Unknown

$K_{eq} = ?$

2 SOLVE FOR THE UNKNOWN

$$K_{eq} = \frac{[0.933]^2}{[0.533][1.600]^3} = 0.399$$

Substitute $[NH_3] = 0.933$ mol/L, $[N_2] = 0.533$ mol/L, and $[H_2] = 1.600$ mol/L.

3 EVALUATE THE ANSWER

The answer is correctly stated with three digits. The largest concentration value is in the denominator and raised to the third power, so a value less than 1 is reasonable.

PRACTICE Problems

Do additional problems. Online Practice

5. Calculate K_{eq} for the equilibrium in Practice Problem 1a using the data $[N_2O_4] = 0.0185$ mol/L and $[NO_2] = 0.0627$ mol/L.
6. Calculate K_{eq} for the equilibrium in Practice Problem 1c using the data $[CO] = 0.0613$ mol/L, $[H_2] = 0.1839$ mol/L, $[CH_4] = 0.0387$ mol/L, and $[H_2O] = 0.0387$ mol/L.
7. **Challenge** The reaction $COCl_2(g) \rightleftharpoons CO(g) + Cl_2(g)$ reaches equilibrium at 900 K. K_{eq} is 8.2×10^{-2}. If the equilibrium concentrations of CO and Cl_2 are 0.150*M*, what is the equilibrium concentration of $COCl_2$?

SECTION 1 REVIEW

Section Self-Check

Section Summary

- A reaction is at equilibrium when the rate of the forward reaction equals the rate of the reverse reaction.
- The equilibrium constant expression is a ratio of the molar concentrations of the products to the molar concentrations of the reactants with each concentration raised to a power equal to its coefficient in the balanced chemical equation.
- The value of the equilibrium constant expression, K_{eq}, is a constant for a given temperature.

8. **MAIN**IDEA **Explain** how the size of the equilibrium constant relates to the amount of product formed at equilibrium.
9. **Compare** homogeneous and heterogeneous equilibria.
10. **List** three characteristics a reaction mixture must have if it is to attain a state of chemical equilibrium.
11. **Calculate** Determine the value of K_{eq} at 400 K for this equation: $PCl_5(g) \rightleftharpoons PCl_3(g) + Cl_2(g)$ if $[PCl_5] = 0.135$ mol/L, $[PCl_3] = 0.550$ mol/L, and $[Cl_2] = 0.550$ mol/L.
12. **Interpret Data** The table below shows the value of the equilibrium constant for a reaction at three different temperatures. At which temperature is the concentration of the products the greatest? Explain your answer.

K_{eq} and Temperature

263 K	273 K	373 K
0.0250	0.500	4.500

3 Assess

Check for Understanding

Ask students if a liquid-vapor equilibrium can be established over a liquid in a container open to the atmosphere. Generally, no, however, an equilibrium can be established if the atmosphere is saturated or becomes saturated with the vapor. OL

Reteach

Ask students to use the density of water at 277 K (1.000 g/mL) to calculate the molar concentration of liquid water at that temperature. 55.5 mol/L OL

Extension

Have interested students use the ideal gas law ($PV = nRT$) to work out the relationship between K_{eq} and K_p, the equilibrium constant expressed in partial pressures of gases rather than molar concentrations. $K_{eq} = K_p(1/RT)^{\Delta n}$ where Δn = moles of gaseous products − moles of gaseous reactants. AL

ASSESSMENT

Knowledge Ask students to write equilibrium constant expressions for the following equilibria.

$2SO_2(g) + O_2(g) \leftrightharpoons 2SO_3(g)$

$K_{eq} = [SO_3]^2/[SO_2]^2[O_2]$

$Ni(OH)_2(s) \leftrightharpoons NiO(s) + H_2O(g)$

$K_{eq} = [H_2O(g)]$ OL

SECTION 1 REVIEW

8. The larger the equilibrium constant the greater the amount of products formed at equilibrium.
9. In a homogenous equilibrium, all the reactants and products are in the same physical state. In a heterogeneous equilibrium, reactants and products are in different physical states.
10. To attain equilibrium, a reaction mixture must be in a closed container, it must be at constant temperature, and all reactants and products must be present.
11. $K_{eq} = 2.24$
12. 373 K; The products are in the numerator of the ratio, so the larger the K_{eq} is, the greater the concentration of the products.

SECTION 2

1 Focus

MAINIDEA

CO_2 Equilibrium An important environmental equilibrium involves the carbon dioxide in Earth's atmosphere and dissolved in its oceans. Ask students to write the equilibrium equation. $CO_2(g) \leftrightharpoons CO_2(aq)$ Ask what happens to $CO_2(g)$ in Earth's atmosphere as a result of the combustion of fossil fuels. $CO_2(g)$ increases. Explain that such a change increases the rate of the reaction to the right. Ask what happens as a result? $CO_2(aq)$ increases until the rate at which the oceans release CO_2 equals the rate at which it dissolves—in other words, until a new equilibrium position is established. Point out, however, that the increased concentration of CO_2 in Earth's oceans changes the pH (acidity) of the water, which might affect marine life in negative ways. OL

2 Teach

Reinforcement

Le Châtelier's Principle Emphasize that although Le Châtelier's principle is useful in determining how changes in conditions affect many chemical reactions, it can be applied only to systems at equilibrium.

Concept Development

Heat Explain that heat can be thought of as a reactant in an endothermic reaction and a product in an exothermic reaction.

SECTION 2

Factors Affecting Chemical Equilibrium

MAINIDEA When changes are made to a system at equilibrium, the system shifts to a new equilibrium position.

Essential Questions

- Which various factors affect chemical equilibrium?
- How does Le Châtelier's principle apply to equilibrium systems?

Review Vocabulary

reaction rate: the change in concentration of a reactant or product per unit time, generally calculated and expressed in moles per liter per second.

New Vocabulary

Le Châtelier's principle

CHEM 4 YOU When demand for a product equals the available supply, the price remains constant. If demand exceeds supply, the price of the product increases. The price becomes constant again when supply and demand regain a state of balance. Systems at equilibrium behave in a similar way.

Le Châtelier's Principle

Suppose the by-products of an industrial process are the gases carbon monoxide and hydrogen, and a company chemist believes these gases can be combined to produce the fuel methane (CH_4). When CO and H_2 are placed in a closed vessel at 1200 K, this exothermic reaction ($\Delta H = -206.5$ kJ) establishes equilibrium (Equilibrium Position 1).

$$\underset{0.30000M}{CO(g)} + \underset{0.10000M}{3H_2(g)} \rightleftharpoons \underset{0.05900M}{CH_4(g)} + \underset{0.02000M}{H_2O(g)} \quad \Delta H^\circ = -206.5 \text{ kJ}$$

Inserting these concentrations into the equilibrium expression gives an equilibrium constant equal to 3.933.

$$K_{eq} = \frac{[CH_4][H_2O]}{[CO][H_2]^3} = \frac{(0.05900)(0.02000)}{(0.30000)(0.10000)^3} = 3.933$$

Unfortunately, a methane concentration of 0.05900 mol/L in the equilibrium mixture is too low to be of any practical use. Could the chemist change the equilibrium position and thereby increase the amount of methane? An analogy might be the runner on a treadmill shown in **Figure 10.** If the runner increases the speed of the treadmill, she must also increase her speed to restore equilibrium.

Figure 10 A runner gradually increases the speed of the treadmill. With each change, she must increase her running speed in order to restore her equilibrium at the new treadmill setting. Similarly, a chemist can change the conditions of a reaction at equilibrium in order to increase the amount of product.

Demonstration

Hydrogen Bonding

Purpose

to demonstrate hydrogen bonding in the context of equilibrium

Materials

$CoCl_2 \cdot 2H_2O$ (1.4 g); acetone (60 mL); 100-mL graduated cylinder; H_2O (30 mL)

Safety Precautions

Disposal Waste should be disposed of as described by EPA regulations.

Procedure

Dissolve 1.4 g of $CoCl_2 \cdot 2H_2O$ in 30 mL of H_2O. Place this solution in a 100-mL graduated cylinder. Slowly pour 60 mL of acetone down the inside of the graduated cylinder. **WARNING:** ***The cobalt chloride solution is toxic and acetone is flammable.*** Have students observe the blue layer on top.

In 1888, French chemist Henri-Louis Le Châtelier discovered that there are ways to control equilibria to make reactions more productive. He proposed what is now called **Le Châtelier's principle:** If a stress is applied to a system at equilibrium, the system shifts in the direction that relieves the stress. A stress is any kind of change in a system at equilibrium that upsets the equilibrium.

FOLDABLES® Incorporate information from this section into your Foldable.

Applying Le Châtelier's Principle

How could the industrial chemist apply Le Châtelier's principle to increase her yield of methane? She will need to adjust any factors that will shift the equilibrium to the product side of the reaction.

Changes in concentration Adjusting the concentrations of either the reactants or the products puts a stress on the equilibrium. Earlier, you read about collision theory, which states that particles must collide in order to react. The number of collisions between reacting particles depends on the concentration of the particles, so perhaps the chemist can change the equilibrium by changing concentrations.

Adding reactants Suppose additional carbon monoxide is injected into the reaction vessel, raising the concentration of carbon monoxide from 0.30000*M* to 1.00000*M*. The higher carbon monoxide concentration immediately increases the number of effective collisions between CO and H_2 molecules and upsets the equilibrium. The rate of the forward reaction increases, as indicated by the longer arrow to the right.

$$CO(g) + 3H_2(g) \rightleftharpoons CH_4(g) + H_2O(g)$$

In time, the rate of the forward reaction slows down as the concentrations of CO and H_2 decrease. Simultaneously, the rate of the reverse reaction increases as more CH_4 and H_2O molecules are produced. Eventually, a new equilibrium position (Position 2) is established.

$$\begin{array}{ccccccc} CO(g) & + & 3H_2(g) & \rightleftharpoons & CH_4(g) & + & H_2O(g) \\ 0.99254M & & 0.07762M & & 0.06648M & & 0.02746M \end{array}$$

$$K_{eq} = \frac{[CH_4][H_2O]}{[CO][H_2]^3} = \frac{(0.06648)(0.02746)}{(0.99254)(0.07762)^3} = 3.933$$

Note that although K_{eq} has not changed, the new equilibrium position results in the desired effect–an increased concentration of methane. The results of this experiment are summarized in **Table 2.**

Could you have predicted this result using Le Châtelier's principle? Yes. Think of the increased concentration of CO as a stress on the equilibrium. The equilibrium system reacts to the stress by consuming CO at an increased rate. This response, called a shift to the right, forms more CH_4 and H_2O. Any increase in the concentration of a reactant results in a shift to the right and additional product.

VOCABULARY
SCIENCE USAGE V. COMMON USAGE

Stress

Science usage: any kind of change in a system at equilibrium that upsets the equilibrium

The stress of the addition of more reactant to the reaction mixture caused the rate of the forward reaction to increase.

Common usage: physical or mental strain or pressure

He felt that the stress of taking on another task would be too great.

APPLYING PRACTICES

PBL Go to the resources tab in ConnectED to find the PBL *Food for Thought.*

Table 2 At Equilibrium: $CO(g) + 3H_2(g) \rightleftharpoons CH_4(g) + H_2O(g)$

Equilibrium position	$[CO]_{eq}$ (*M*)	$[H_2]_{eq}$ (*M*)	$[CH_4]_{eq}$ (*M*)	$[H_2O]_{eq}$ (*M*)	K_{eq}
1	0.30000	0.10000	0.05900	0.02000	3.933
2	0.99254	0.07762	0.06648	0.02746	3.933

Quick Demo

Stress on Equilibrium Tell students that chemical equilibrium will help simulate the changing of seasons in this Quick Demo. Place a few crystals, about 0.3 g, of crushed $CoCl_2 \cdot 6H_2O$ in a test tube. **WARNING: *There should be no open flames in the room.*** Add 10 mL ethanol and shake or stir vigorously until most of the solid has dissolved. If the solution is not pink, add water until the solution just turns light pink. Make a simple painting of a winter scene on a piece of white construction paper, painting the snow with the cobalt chloride solution. To simulate a change in seasons, warm the painting over a hot plate or hot-air register. Explain that the snow turns bluish-green because of the stress created by the heat.

FOLDABLES® Rubric

Results

The acetone forms hydrogen bonds with water and decreases the effective concentration of water in the top layer of the solution. The equilibrium shifts to the right to form the blue chloro complex in the following reaction.

$$\underset{\text{(pink)}}{Co(H_2O)_6^{2+}} + 4Cl^- \leftrightharpoons \underset{\text{(blue)}}{CoCl_4^{2-}} + 6H_2O$$

Analysis

1. Why does the acetone layer form on top of the water? Acetone is less dense than water.
2. What do chemists call the formation of an attractive force between the hydrogen atoms of one water molecule and the oxygen atom of another water molecule? hydrogen bonding
3. Using the chemical equation for the equilibrium, which direction is favored when the water is attracted to the acetone and its effective concentration is thereby lowered? the right, favoring the formation of the blue, $CoCl_4^{2-}$

ASSESSMENT

Knowledge Ask students to explain the difference between a hydrogen bond and a covalent bond. A covalent bond is the sharing of one or more pairs of electrons. A hydrogen bond is an attractive force between a hydrogen atom and an electronegative element having unshared electrons. **OL**

■ **Caption Question Fig. 11**
In the same way that a storekeeper can keep selling products by keeping stock on the shelves, in a chemical reaction, products will continue to be produced if reactants are added. The added reactants put a stress on the equilibrium, which moves in a way to relieve the stress. This movement is a shift to the right, and causes the production of more products.

■ **Caption Question Fig. 12**
If you added more H_2, the equilibrium would shift to the right. If you removed CH_4, the equilibrium would shift to the right.

☑ READING CHECK It would shift to the left toward making more reactant.

■ **In-Text Question** The equilibrium shifts in the direction of the products.

Visual Learning

Figure 12 Explain that a longer equilibrium arrow to the left or right means that one reaction occurs faster than the other *temporarily.* Emphasize that once a new equilibrium position is established, the two reactions again occur at equal rates.

■ **Figure 11** Storekeepers know that all products should be available at all times, so when stocks get low, they must be replaced.
Explain *this analogy in terms of Le Châtelier's principle.*

Removing products Suppose that rather than injecting more reactant, the chemist decides to remove a product (H_2O) by adding a desiccant to the reaction vessel. Recall that a desiccant is a substance that absorbs water. What does Le Châtelier's principle predict the equilibrium will do in response to a decrease in the concentration of water? The equilibrium shifts in the direction that will tend to bring the concentration of water back up. That is, the equilibrium shifts to the right and results in additional product.

Think about how supermarket shelves are kept stocked, as shown in **Figure 11.** As customers buy items from the shelves, it is someone's job to replace whatever is removed. Similarly, the equilibrium reaction restores some of the lost water by producing more water. In any equilibrium, the removal of a product results in a shift to the right and the production of more product.

Adding products The equilibrium position can also be shifted to the left, toward the reactants. Le Châtelier's principle predicts that if additional product is added to a reaction at equilibrium, the reaction will shift to the left. The stress is relieved by converting products to reactants. If one of the reactants is removed, a similar shift to the left will occur.

When predicting the results of a stress on an equilibrium using Le Châtelier's principle, have the equation for the reaction in view. The effects of changing concentrations are summarized in **Figure 12.**

☑ READING CHECK **Describe** how an equilibrium shifts if a reactant is removed.

Changes in volume and pressure Consider again the reaction for making methane from by-product gases.

$$CO(g) + 3H_2(g) \rightleftharpoons CH_4(g) + H_2O(g)$$

Can this reaction be forced to produce more methane by changing the volume of the reaction vessel? Suppose the volume can be changed using a pistonlike device similar to the one shown in **Figure 13.** If the piston is forced downward, the volume of the system decreases. Recall that Boyle's law states that decreasing the volume at constant temperature increases the pressure. The increased pressure is a stress on the reaction at equilibrium. How does the equilibrium respond to the disturbance and relieve the stress?

■ **Figure 12** The addition or removal of a reactant or product shifts the equilibrium in the direction that relieves the stress. Note the unequal arrows, which indicate the direction of the shift.
Describe *how the reaction would shift if you added H_2. If you removed CH_4.*

$$CO(g) + 3H_2(g) \rightleftharpoons CH_4(g) + H_2O(g)$$

Equilibrium shifts to the right.

1. $CO(g) + 3H_2(g) \rightleftharpoons CH_4(g) + H_2O(g)$ — $CO(g)$ Add a reactant.
2. $CO(g) + 3H_2(g) \rightleftharpoons CH_4(g) + H_2O(g)$ — Remove a product. ($H_2O(g)$)

Equilibrium shifts to the left.

3. $CO(g) + 3H_2(g) \rightleftharpoons CH_4(g) + H_2O(g)$ — Remove a reactant. ($CO(g)$)
4. $CO(g) + 3H_2(g) \rightleftharpoons CH_4(g) + H_2O(g)$ — Add a product. $H_2O(g)$

Jochen Sand/Digital Vision/Getty Images

DIFFERENTIATED INSTRUCTION

Struggling Learners Help students visualize Le Châtelier's principle with this analogy. Hens lay eggs in order to incubate and hatch them. If their eggs are removed after laying, hens lay more eggs in order to fulfill their purpose. Farmers encourage hens to lay more eggs (the reaction) by removing the eggs (product) as soon as they are laid. BL

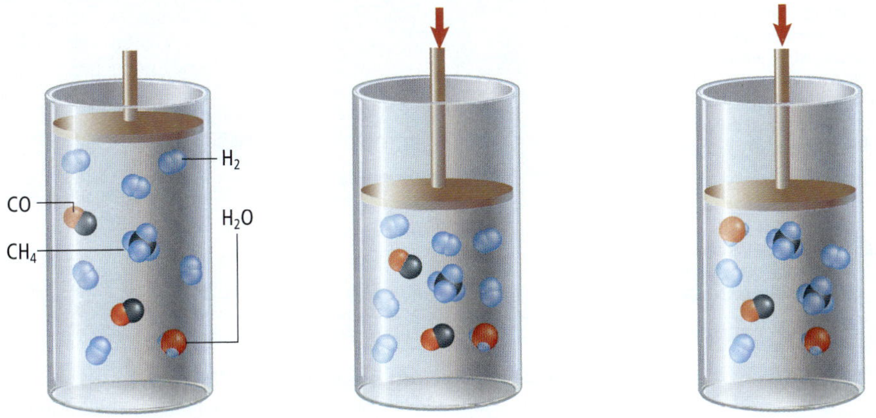

The reaction between CO and H_2 is at equilibrium.

Lowering the piston decreases the volume and increases the pressure.

As a result, more molecules of the products form. Their formation relieves the stress on the system.

Figure 13 For the reaction between CO and H_2 at constant temperature, changing the volume of the reaction vessel changes the concentrations of gaseous reactants and products. Increasing the pressure shifts the equilibrium to the right and increases the amount of product.

Compare *the numbers of product molecules on the left with the numbers on the right.*

The pressure exerted by an ideal gas depends on the number of gas particles that collide with the walls of the vessel. The more gas particles contained in the vessel, the greater the pressure will be. If the number of gas particles is increased at constant temperature, the pressure of the gas increases. If the number of gas particles is decreased, the pressure decreases. How does this relationship between numbers of gas particles and pressure apply to the reaction for making methane?

Moles of reactant versus moles of product Compare the number of moles of gaseous reactants in the equation to the number of moles of gaseous products. For every two moles of gaseous products, four moles of gaseous reactants are consumed, a net decrease of two moles. If you apply Le Châtelier's principle, you can see that the equilibrium can relieve the stress of increased pressure by shifting to the right. **Figure 13** shows that this shift decreases the total number of moles of gas, and thus the pressure inside the reaction vessel decreases. Although the shift to the right does not reduce the pressure to its original value, it has the desired effect—more methane is produced.

Changing the volume (and pressure) of an equilibrium system shifts the equilibrium only if the number of moles of gaseous reactants is different from the number of moles of gaseous products. If the number of moles of gas is the same on both sides of the equation, changes in volume and pressure have no effect on the equilibrium.

Changes in temperature A change in temperature alters both the equilibrium position and the equilibrium constant. Recall that virtually every chemical reaction is either endothermic or exothermic. The reaction for making methane has a negative $\Delta H°$, which means that the forward reaction is exothermic and the reverse reaction is endothermic.

$$CO(g) + 3H_2(g) \rightleftharpoons CH_4(g) + H_2O(g) \quad \Delta H° = -206.5 \text{ kJ}$$

In this case, you can think of heat as a product in the forward reaction and a reactant in the reverse reaction.

$$CO(g) + 3H_2(g) \rightleftharpoons CH_4(g) + H_2O(g) + \text{heat}$$

CHEMISTRY JOURNAL

Le Châtelier in Verse Have a contest in which students write creative stories or poems that illustrate Le Châtelier's principle. Have students include the stories or poems in their chemistry journals. Ask a colleague to help you choose winners, and award prizes. Read the best stories or poems to the class. OL

ASSESSMENT

Performance The equilibrium between NO_2 and N_2O_4 is described by the equation $2NO_2(g) \rightleftharpoons N_2O_4(g) + 57.2$ kJ
(brown) (colorless)
Have students make a poster that illustrates how the relative numbers of molecules of NO_2 and N_2O_4 in a closed container change and how the color of the equilibrium mixture changes when the system is subjected to two stresses: (1) the volume of the container is increased; (2) the temperature is lowered. When the container's volume is increased, the number of NO_2 molecules increases and the color becomes darker. When the temperature is decreased, the number of NO_2 molecules increases, and the color darkens. OL

Apply Chemsitry

Synthesis of Ethyl Acetate The ester ethyl acetate may be synthesized by heating ethanol with ethanoic acid in the presence of a catalytic amount of a strong acid such as sulfuric acid. In this process, an equilibrium forms and lies slightly on the ester side. $CH_3COOH + CH_3CH_2OH \rightleftharpoons CH_3COOCH_2CH_3 + H_2O$ Applying Le Chatelier's principle, yields can be improved by adding excess alcohol or by removing water—either as an azeotropic mixture with benzene or with a drying agent.

Caption Question Fig. 13 On the left, at lower pressure, there are two product molecules (and eight reactant molecules). On the right, at higher pressure, there are four product molecules (and four reactant molecules). The overall effect of the increase in pressure is a reduction of the number of molecules in the reaction chamber from 10 to 8.

In-Text Question As pressure is added to the system, the number of molecules of methane produced increases.

3 Assess

Check for Understanding

Ask students to explain why the addition of a catalyst does not change an equilibrium position. A catalyst speeds up the forward and reverse reactions equally. OL

Reteach

Summarize the factors that can cause a change in equilibrium position: concentration; volume (pressure); temperature. Ask student volunteers to explain how equilibria might shift in response to each stress. If the concentration of a substance is changed, the equilibrium shifts in the direction that tends to restore the original concentration. If the volume (pressure) of the system is changed (and the reactant and product sides of the equilibrium equation have different numbers of gaseous molecules), the equilibrium shifts in the direction that tends to restore the original pressure. If the temperature is changed, the equilibrium shifts in the direction that tends to restore the original temperature. OL

Extension

Explain that adding a noble gas to an equilibrium system does not cause the equilibrium to shift to the right or left because, although the total pressure in the container increases, the partial pressures or concentrations of the reactants and products do not change. OL

ASSESSMENT

Knowledge Ask students to explain how the following stresses affect the equilibrium system.

$PF_3(g) + F_2(g) \rightleftharpoons PF_5(g) + \text{heat}$ (1) Volume is decreased. The equilibrium shifts to the right. (2) Temperature is increased. The equilibrium shifts to the left. OL

■ **Figure 14** When placed in a boiling-water bath (right), the equilibrium shifts in the endothermic direction, which produces more reddish-brown NO_2. The mixture becomes lighter in color when placed in ice (left) because the equilibrium shifts in the exothermic direction, in which more NO_2 is converted to colorless N_2O_4.

View an **animation about equilibrium shifts.**

Concepts In Motion

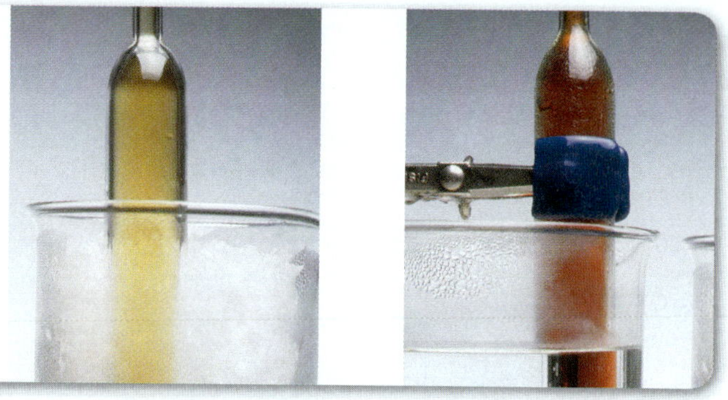

Charles D. Winters/Science Source

Heat and equilibrium position According to Le Châtelier's principle, if heat is added to an equilibrium system, the equilibrium shifts in the direction in which heat is used up; that is, the equilibrium shifts to the left and decreases the concentration of methane (CH_4). Lowering the temperature shifts the equilibrium to the right because the forward reaction liberates heat and relieves the stress. In shifting to the right, the equilibrium produces more methane.

Temperature and K_{eq} Any change in temperature results in a change in K_{eq}. Recall that the larger the value of K_{eq}, the more product is found in the equilibrium mixture. Thus, for the methane-producing reaction, K_{eq} increases in value when the temperature is lowered and decreases in value when the temperature is raised.

The conversion between dinitrogen tetroxide (N_2O_4) and nitrogen dioxide (NO_2) responds to changes in temperature in an observable way. This endothermic equilibrium is described by the following equation.

$$N_2O_4(g) \rightleftharpoons 2NO_2(g) \quad \Delta H° = 57.2 \text{ kJ}$$

N_2O_4 is a colorless gas; NO_2 is a reddish-brown gas. **Figure 14** shows that the color of the equilibrium mixture in ice is much lighter than when the mixture is heated in boiling water. The removal of heat by cooling shifts the equilibrium to the left and creates more colorless N_2O_4. Adding heat shifts the equilibrium to the right and creates more reddish-brown NO_2. **Figure 15** shows the effects of heating and cooling on the reactions you have been reading about.

■ **Figure 15** For the exothermic reaction between CO and H_2, raising the temperature shifts the equilibrium to the left (Equation 1). Lowering the temperature results in a shift to the right (Equation 2). The opposite is true for the endothermic reaction involving NO and N_2O_4 (Equations 3 and 4).

Exothermic Reaction

Equilibrium shifts to the left. (heat — Raise the temperature.)

1. $CO(g) + 3H_2(g) \rightleftharpoons CH_4(g) + H_2O(g) + \text{heat}$

Equilibrium shifts to the right. (Lower the temperature.)

2. $CO(g) + 3H_2(g) \rightleftharpoons CH_4(g) + H_2O(g) + \text{heat}$

Endothermic Reaction

Equilibrium shifts to the right.

3. $\text{heat} + N_2O_4(g) \rightleftharpoons 2\,NO_2(g)$ (heat — Raise the temperature.)

Equilibrium shifts to the left.

4. $\text{heat} + N_2O_4(g) \rightleftharpoons 2\,NO_2(g)$ (Lower the temperature.)

CHEMISTRY PROJECT

The Haber Process Have students research the Haber Process and the effect that the development of this process has had on agriculture. The Haber Process was invented by Fritz Haber, a German research chemist, by applying Le Châtelier's principle. OL

MiniLAB

Observe Shifts in Equilibrium

If a stress is placed on a reaction at equilibrium, how will the system shift to relieve the stress?

Procedure

1. Read and complete the lab safety form.
2. Place about 2 mL of **0.1*M* $CoCl_2$ solution** in a **test tube.** Record the color of the solution.
3. Add about 3 mL of **concentrated HCl** to the test tube. Record the color of the solution. **WARNING:** HCl can burn skin and clothing.
4. Add enough **water** to the test tube to make a color change occur. Record the color.
5. Add about 2 mL of 0.1*M* $CoCl_2$ to another test tube. Add concentrated HCl a drop at a time until the solution turns purple. If the solution becomes blue, add water until it turns purple.
6. Place the test tube in an **ice bath** that has had some **table salt** sprinkled into the ice water. Record the color of the solution in the test tube.
7. Place the test tube in a **hot water bath.** Use a **nonmercury thermometer** to determine that the temperature is at least 70°C. Record the solution's color.

Analysis

1. **Interpret** Use the equation for the reaction you just observed to explain your observations of color in Steps 2–4. The equation is as follows.
$$\underset{\text{pink}}{Co(H_2O)_6^{2+}} + 4Cl^- \rightleftharpoons \underset{\text{blue}}{CoCl_4^{2-}} + 6H_2O$$
2. **Describe** how the equilibrium shifts when energy is added or removed.
3. **Interpret** From your observations of color in Steps 6 and 7, determine whether the reaction is exothermic or endothermic.

Catalysts and equilibrium Changes in concentration, volume, and temperature make a difference in the amount of product formed in a reaction. Can a catalyst also affect product concentration? A catalyst speeds up a reaction, but it does so equally in both directions. Therefore, a catalyzed reaction reaches equilibrium more quickly but with no change in the amount of product formed.

SECTION 2 REVIEW

Section Self-Check

Section Summary

- Le Châtelier's principle describes how an equilibrium system shifts in response to a stress or a disturbance.
- When an equilibrium shifts in response to a change in concentration or volume, the equilibrium position changes but K_{eq} remains constant. A change in temperature, however, alters both the equilibrium position and the value of K_{eq}.

13. **MAIN**IDEA **Explain** how a system at equilibrium responds to a stress and list factors that can be stresses on an equilibrium system.
14. **Explain** how decreasing the volume of the reaction vessel affects each equilibrium.
 a. $2SO_2(g) + O_2(g) \rightleftharpoons 2SO_3(g)$ **b.** $H_2(g) + Cl_2(g) \rightleftharpoons 2HCl(g)$
15. **Decide** whether higher or lower temperatures will produce more CH_3CHO in the following equilibrium. $C_2H_2(g) + H_2O(g) \rightleftharpoons CH_3CHO(g)$ $\Delta H° = -151$ kJ
16. **Demonstrate** The table below shows the concentrations of Substances A and B in two reaction mixtures. A and B react according to the equation $2A \rightleftharpoons B$; $K_{eq} = 200$. Are the two mixtures at different equilibrium positions?

Concentration Data in mol/L

Reaction	[A]	[B]
1	0.0100	0.0200
2	0.0500	0.500

17. **Design** a concept map that shows ways in which Le Châtelier's principle can be applied to increase the products in a system at equilibrium and to increase the reactants in such a system.

MiniLAB

Rubric

Purpose Students will observe visible evidence of shifts in equilibrium when stresses are placed on a reaction at equilibrium.

Process Skills analyze, compare and contrast, draw a conclusion, recognize cause and effect

Safety Precautions

Approve lab safety forms before work begins. Concentrated HCl is very corrosive to skin and eyes. **WARNING: *Cobalt chloride is a possible carcinogen. Do not work with cobalt chloride dust.*** Review MSDS for HCl with the students.

Disposal Neutralize solutions with sodium hydrogen carbonate before flushing down a drain.

Teaching Strategies

- See the Teacher Handbook for preparation of solutions.
- Reinforce the idea that complex ions are responsible for the colored solutions.
- Provide HCl in dropper bottles. HCl should be dispensed in a fume hood.

Expected Results Addition of HCl to the pink solution produces a blue solution. When water is added to the blue solution, it turns pink. The purple solution will turn pink in the ice bath but turns blue in the hot water bath.

Analysis

1. Excess chloride ions pushes the equilibrium toward the blue ion; water pushes toward the pink ion.
2. Heating shifts the equilibrium to the blue, cooling shifts it to the pink.
3. The reaction is endothermic.

SECTION 2 REVIEW

13. If possible, the equilibrium shifts in the direction that relieves the stress. Changes in concentration, pressure (volume), and temperature can be stresses.
14. a. The equilibrium shifts to the right.
 b. The stress has no effect on the equilibrium.
15. lower temperatures
16. The two mixtures are at the same equilibrium position.
17. Concept maps should show that products can be increased by increasing the concentration of the reactants, by removing products, or by raising or lowering the temperature depending upon whether the reaction is exothermic or endothermic.

SECTION 3

1 Focus

MAINIDEA

The Barium Sulfate Equilibrium Ask students if they or anyone they know has had an X-ray of the digestive tract. Explain that persons who have such X-rays might be asked to ingest barium sulfate ($BaSO_4$) to increase the X-ray's definition. Point out that barium ions are toxic, and ask how it is possible that ingesting $BaSO_4$ is not hazardous. The solubility of $BaSO_4$ is so low that relatively few Ba^{2+} form. Lead students in writing the solubility equation and solubility constant expression for $BaSO_4$. $BaSO_4(s) \rightleftharpoons Ba^{2+}(aq) + SO_4^{2-}(aq)$; $K_{sp} = [Ba^{2+}][SO_4^{2-}]$ Have students look up the $BaSO_4$ K_{sp} value in Table 3 and lead them through the calculation to find the actual barium-ion concentration in a saturated solution at 298 K. $K_{sp} = 1.1 \times 10^{-10}$; $[Ba^{2+}] = 1.0 \times 10^{-5}M$ OL

2 Teach

Content Background

Using Stresses An industrial chemist might apply Le Châtelier's principle to the reaction between H_2 and CO by removing one or both of the product gases as they are formed and simultaneously replenishing the reactants. Both are stresses on the equilibrium that produce more product.

SECTION 3

Essential Questions

- How are the equilibrium concentrations of reactants and products determined?
- How is the solubility of a compound calculated from its solubility product constant?
- Why is the common ion effect important?

Review Vocabulary

solubility: the maximum amount of solute that will dissolve in a given amount of solvent at a specific temperature and pressure

New Vocabulary

solubility product constant
common ion
common ion effect

Using Equilibrium Constants

MAINIDEA Equilibrium constant expressions can be used to calculate concentrations and solubilities.

CHEM 4 YOU **If you have ever tried to squeeze yourself into the backseat of a car already occupied by several of your friends, you know there is a limit to how many people the seat can hold. An ionic compound encounters a similar situation when being dissolved in a solution.**

Calculating Equilibrium Concentrations

How can the equilibrium constant expression be used to calculate the concentration of a product? The K_{eq} for the reaction that forms CH_4 from H_2 and CO is 3.933 at 1200 K. If the concentrations of H_2, CO, and H_2O are known, the concentration of CH_4 can be calculated.

$$\underset{0.850M}{CO(g)} + \underset{1.333M}{3H_2(g)} \rightleftharpoons \underset{?M}{CH_4(g)} + \underset{0.286M}{H_2O(g)}$$

$$K_{eq} = \frac{[CH_4][H_2O]}{[CO][H_2]^3}$$

Solve the expression for the unknown $[CH_4]$ by multiplying both sides of the equation by $[CO][H_2]^3$ and dividing both sides by $[H_2O]$.

$$[CH_4] = K_{eq} \times \frac{[CO][H_2]^3}{[H_2O]}$$

Substitute the known concentrations and the value of K_{eq} (3.933).

$$[CH_4] = 3.933 \times \frac{(0.850)(1.333)^3}{(0.286)} = 27.7 \text{ mol/L}$$

The equilibrium concentration of CH_4 is 27.7 mol/L.

Is a yield of 27.7 mol/L sufficient to make the conversion of waste CO and H_2 to methane practical? That depends on the cost of methane. **Figure 16** shows a tanker transporting natural gas, which is primarily methane, to ports around the world.

Figure 16 New port terminals are being planned to accommodate tankers, which carry increasing amounts of natural gas around the world to meet both industrial and home needs. Natural gas, which is primarily methane, is used for heating and cooking.

MsLightBox/E+/Getty Images

ASSESSMENT

Knowledge Have students research the fluoridation of drinking water. Have them look more closely into the effect of salivary acid on hydroxyapatite and apply what they learn to the question of fluoridation. OL

EXAMPLE Problem 4

Find help with **square and cube roots.** Math Handbook

CALCULATING EQUILIBRIUM CONCENTRATIONS At a temperature of 1405 K, hydrogen sulfide, which has a foul odor resembling rotten eggs, decomposes to form hydrogen and diatomic sulfur. The equilibrium constant for the reaction is 2.27×10^{-3}.

$$2H_2S(g) \rightleftharpoons 2H_2(g) + S_2(g)$$

What is the concentration of hydrogen gas if $[S_2] = 0.0540$ mol/L and $[H_2S] = 0.184$ mol/L?

1 ANALYZE THE PROBLEM

You have been given K_{eq} and two of the three variables in the equilibrium constant expression. The equilibrium expression can be solved for $[H_2]$. K_{eq} is less than one, so more reactants than products are in the equilibrium mixture. Thus, you can predict that $[H_2]$ will be less than 0.184 mol/L, the concentration of the reactant H_2S.

Known	**Unknown**
$K_{eq} = 2.27 \times 10^{-3}$	$[H_2] = ?$ mol/L
$[S_2] = 0.0540$ mol/L	
$[H_2S] = 0.184$ mol/L	

2 SOLVE FOR THE UNKNOWN

$$\frac{[H_2]^2[S_2]}{[H_2S]^2} = K_{eq}$$ State the equilibrium constant expression.

Solve the equation for $[H_2]$.

$$[H_2]^2 = K_{eq} \times \frac{[H_2S]^2}{[S_2]}$$ Multiply both sides by $[H_2S]^2$. Divide both sides by $[S_2]$.

$$[H_2] = \sqrt{K_{eq} \times \frac{[H_2S]^2}{[S_2]}}$$ Take the square root of both sides.

$$[H_2] = \sqrt{(2.27 \times 10^{-3}) \times \frac{(0.184)^2}{(0.0540)}}$$ Substitute $K_{eq} = 2.27 \times 10^{-3}$, $[H_2S] = 0.184$ mol/L, and $[S_2] = 0.0540$ mol/L.

$$[H_2] = 0.0377 \text{ mol/L}$$ Multiply, divide, and take the square root.

The equilibrium concentration of H_2 is 0.0377 mol/L.

3 EVALUATE THE ANSWER

The answer is correctly stated with three significant figures. As predicted, the equilibrium concentration of H_2 is less than 0.184 mol/L.

EXAMPLE PROBLEM

PRACTICE Problems

Do additional problems. Online Practice

18. At a certain temperature, $K_{eq} = 10.5$ for the equilibrium $CO(g) + 2H_2(g) \rightleftharpoons CH_3OH(g)$. Calculate the following concentrations:

a. [CO] in an equilibrium mixture containing 0.933 mol/L H_2 and 1.32 mol/L CH_3OH

b. $[H_2]$ in an equilibrium mixture containing 1.09 mol/L CO and 0.325 mol/L CH_3OH

c. $[CH_3OH]$ in an equilibrium mixture containing 0.0661 mol/L H_2 and 3.85 mol/L CO

19. Challenge In a generic reaction $A + B \rightleftharpoons C + D$, 1.00 mol of A and 1.00 mol of B are allowed to react in a 1-L flask until equilibrium is established. If the equilibrium concentration of A is 0.450 mol/L, what is the equilibrium concentration of each of the other substances? What is K_{eq}?

PRACTICE PROBLEMS

CHEMISTRY JOURNAL

Electrolytes Have students research the ionic solution present in human blood plasma. Explain that the solution is called an electrolyte. Also, have students research the effects of unusually high or low electrolyte concentrations in the blood and how the body normally maintains appropriate levels of these ions. Ask them to summarize their findings in their chemistry journals. **OL**

IN-CLASS Example

Question If K_{eq} for the following equilibrium is 0.110 at a particular temperature, $[I_2] = 0.0330$, and $[Cl_2] = 0.220$, what is the equilibrium concentration of ICl? $2ICl(g) \leftrightarrows I_2(g) + Cl_2(g)$

Answer $[ICl] = 0.257$ mol/L

$$0.110 = \frac{(0.0330)(0.220)}{[ICl]^2}$$

$[ICl]^2 = 0.066$

$[ICl] = 0.257$ mol/L

PRACTICE Problems

Have students refer to the Selected Solutions appendix for complete solutions to odd-numbered problems. The complete solutions for all problems can be found in the Solutions Manual.

18. a. $[CO] = 0.144M$
b. $[H_2] = 0.169M$
c. $[CH_3OH] = 0.177M$

19. $K_{eq} = 1.49$

Math in Chemistry

Equilibrium Constant Point out that solving equilibrium constant expressions for some concentrations might involve calculating roots. For example, explain that solving for $[H_2]$ in the expression

$$K_{eq} = \frac{[CH_4][H_2O]}{[CO][H_2]^3}$$ will involve taking the cube root of $$\frac{[CH_4][H_2O]}{[CO]\cdot K_{eq}}$$

$$[H_2]^3 = \frac{[CH_4][H_2O]}{[CO]K_{eq}}$$

$$[H_2] = \sqrt[3]{\frac{[CH_4][H_2O]}{[CO]K_{eq}}}$$

Refer to the Math Handbook for further review.

Reinforcement

Le Châtelier's Principle Ask students to explain how adding additional sulfate ions to a saturated solution of barium sulfate would affect the concentration of barium ions. According to Le Châtelier's principle, the equilibrium would shift to lower the sulfate ion concentration, and favor the formation of $BaSO_4$. Such a shift would also lower the concentration of barium ions. OL

Concept Development

Scientific Notation Some students might have difficulty comparing numerical values expressed in scientific notation. Help them clarify their understanding by asking them to rank the following quantities from largest to smallest: (a) 5.55×10^{-10}; (b) 8.40×10^{-5}; (c) 1.43×10^{-25}; (d) 2.76×10^{4}; (e) 3.44×10^{-3}; (f) 9.81×10^{-33}; (g) 9.81×10^{-10}; (h) 2.50×10^{-10}. d, e, b, g, a, h, c, f Refer to the Math Handbook for further review. BL EL

ASSESSMENT

Skill Refer students to Table 3 and have them predict what might happen when solutions of $Pb(NO_3)_2$ and Na_2SO_4 are mixed. Ask what might happen if a solution of NaI is added. Have them write equations and solubility product constant expressions for the equilibria involved. OL

Figure 17 The water of the Great Salt Lake is much saltier than sea water. The high concentration of salt makes the water dense enough that most people can float in it. The Salar de Uyuni, or Uyuni Salt Flats, at right, were left behind when a similar prehistoric lake dried.

The Solubility Product Constant

Some ionic compounds, such as sodium chloride, dissolve readily in water, and some, such as barium sulfate ($BaSO_4$) barely dissolve at all. On dissolving, all ionic compounds dissociate into ions.

$$NaCl(s) \rightarrow Na^{+}(aq) + Cl^{-}(aq)$$

Connection to Earth Science Because of the high solubility of NaCl, the oceans and some lakes contain large amounts of salt. **Figure 17** shows the Great Salt Lake next to one of the Uyuni flats in Bolivia, which were left behind when a prehistoric lake dried.

Sometimes low solubility is also important. Although barium ions are toxic to humans, patients must ingest barium sulfate prior to having an X-ray of the digestive tract taken. Can patients safely ingest $BaSO_4$?

Barium sulfate dissociates in water according to this equation.

$$BaSO_4(s) \rightarrow Ba^{2+}(aq) + SO_4^{2-}(aq)$$

As soon as the first product ions form, the reverse reaction begins.

$$BaSO_4(s) \leftarrow Ba^{2+}(aq) + SO_4^{2-}(aq)$$

In time, equilibrium is established.

$$BaSO_4(s) \rightleftharpoons Ba^{2+}(aq) + SO_4^{2-}(aq)$$

For sparingly soluble compounds such as $BaSO_4$, the rates become equal when the concentrations of the aqueous ions are exceedingly small. Nevertheless, the solution at equilibrium is a saturated solution.

Writing solubility product constant expressions The equilibrium constant expression for the dissolving of a sparingly soluble compound is called the **solubility product constant,** K_{sp}. The solubility product constant expression is the product of the concentrations of the dissolved ions, each raised to the power equal to the coefficient of the ion in the chemical equation. Recall that the concentration of a pure substance is its density in moles per liter, which is constant at a given temperature. Therefore, in heterogeneous equilibria, pure solids and liquids are omitted from equilibrium expressions.

CHEMISTRY PROJECT

Lead-Based Paints Have students research the insoluble lead compounds that were used as paint pigments for many years until it was discovered that people—mostly children—were being poisoned by exposure to the lead in these paints. Have interested students illustrate the brilliant colors of these compounds on poster paper. OL COOP LEARN

Now you can write the solubility product constant expression for the dissolving of barium sulfate ($BaSO_4$) in water. The K_{sp} for the process is 1.1×10^{-10} at 298 K.

$$K_{sp} = [Ba^{2+}][SO_4^{2-}] = 1.1 \times 10^{-10}$$

The small value of K_{sp} for $BaSO_4$ indicates that products are not favored at equilibrium. The concentration of barium ions at equilibrium is only $1.0 \times 10^{-5}M$, and a patient, such as the one shown in **Figure 18,** can safely ingest a barium sulfate solution.

The solubility product constant for the antacid magnesium hydroxide ($Mg(OH)_2$) provides another example.

$$Mg(OH)_2(s) \rightleftharpoons Mg^{2+}(aq) + 2OH^-(aq)$$
$$K_{sp} = [Mg^{2+}][OH^-]^2$$

K_{sp} depends only on the concentrations of the ions in the saturated solution. However, some of the undissolved solid, no matter how small the amount, must be present in the equilibrium mixture.

The solubility product constants for some ionic compounds are listed in **Table 3.** Note that they are all small numbers. Solubility product constants are measured and recorded only for sparingly soluble compounds.

Using solubility product constants The solubility product constants in **Table 3** have been determined through careful experiments. K_{sp} values are important because they can be used to determine the solubility of a sparingly soluble compound. Recall that the solubility of a compound in water is the amount of the substance that will dissolve in a given volume of water at a given temperature.

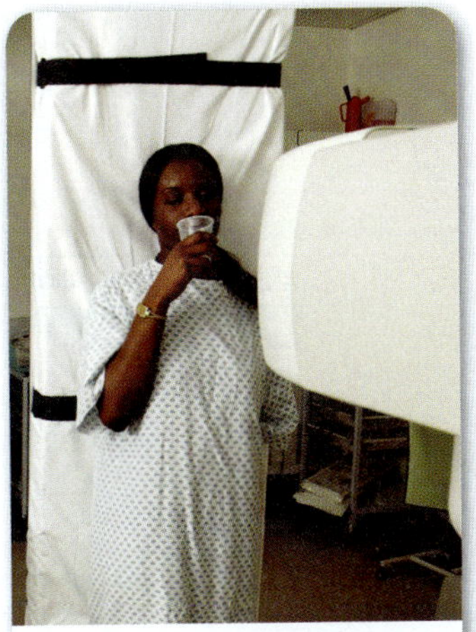

■ **Figure 18** Greater definition is possible in a gastrointestinal X-ray when patients drink a thick mixture containing barium sulfate. Barium sulfate is a poisonous substance, but it has such low solubility that only a minimal amount can dissolve in the patient's body.

Universal Images Group/Getty Images

Table 3 Solubility Product Constants at 298 K

Compound	K_{sp}	Compound	K_{sp}	Compound	K_{sp}
Carbonates		**Halides**		**Hydroxides**	
$BaCO_3$	2.6×10^{-9}	CaF_2	3.5×10^{-11}	$Al(OH)_3$	4.6×10^{-33}
$CaCO_3$	3.4×10^{-9}	$PbBr_2$	6.6×10^{-6}	$Ca(OH)_2$	5.0×10^{-6}
$CuCO_3$	2.5×10^{-10}	$PbCl_2$	1.7×10^{-5}	$Cu(OH)_2$	2.2×10^{-20}
$PbCO_3$	7.4×10^{-14}	PbF_2	3.3×10^{-8}	$Fe(OH)_2$	4.9×10^{-17}
$MgCO_3$	6.8×10^{-6}	PbI_2	9.8×10^{-9}	$Fe(OH)_3$	2.8×10^{-39}
Ag_2CO_3	8.5×10^{-12}	AgCl	1.8×10^{-10}	$Mg(OH)_2$	5.6×10^{-12}
$ZnCO_3$	1.5×10^{-10}	AgBr	5.4×10^{-13}	$Zn(OH)_2$	3×10^{-17}
Hg_2CO_3	3.6×10^{-17}	AgI	8.5×10^{-17}	**Sulfates**	
Chromates		**Phosphates**		$BaSO_4$	1.1×10^{-10}
$BaCrO_4$	1.2×10^{-10}	$AlPO_4$	9.8×10^{-21}	$CaSO_4$	4.9×10^{-5}
$PbCrO_4$	2.3×10^{-13}	$Ca_3(PO_4)_2$	2.1×10^{-33}	$PbSO_4$	2.5×10^{-8}
Ag_2CrO_4	1.1×10^{-12}	$Mg_3(PO_4)_2$	1.0×10^{-24}	Ag_2SO_4	1.2×10^{-5}

Reinforcement

Constant Concentrations Point out that, as with solubility product constants, incorporating constant concentrations into equilibrium constants is not unusual. Remind students that in Section 1, concentrations of solids and liquids were similarly incorporated when writing equilibrium constant expressions for heterogeneous equilibria.

Concept Development

Solubility Equilibrium Emphasize to students that when writing chemical equations for solubility equilibria, the solid is always written as the reactant—to the left of the equilibrium arrows—and the ions are always written as the products—to the right of the arrows.

Math in Chemistry

Calculator Practice Students might need extra practice in using their calculators to solve problems involving scientific notation. In particular, students commonly make the mistake of using the times (×) sign when entering scientific-notation numbers. Point out that the exponent key ([EXP] on most calculators, [EEX] or [EE] on others) represents "× 10." To help students with this process, lead them through entering several numbers in scientific-notation and carrying out calculations with the numbers. BL EL

DIFFERENTIATED INSTRUCTION

Struggling Learners Guide students with special needs through the steps necessary to calculate roots using their calculators—particularly roots involving scientific notation. Conduct sufficient practice so that students become comfortable with this type of calculation. BL EL

IN-CLASS Example

Question At 298 K, K_{sp} for lead(II) iodide is 9.8×10^{-9}. What is the molar solubility of PbI_2 at 298 K?

Answer $s = 1.3 \times 10^{-3}$ mol/L

$$9.8 \times 10^{-9} = s(2s)^2 = 4s^3$$
$$s^3 = 2.5 \times 10^{-9}$$
$$s = \sqrt[3]{2.5 \times 10^{-9}}$$
$$s = 1.3 \times 10^{-3} \text{ mol/L}$$

PRACTICE Problems

Have students refer to the Selected Solutions appendix for complete solutions to odd-numbered problems. The complete solutions for all problems can be found in the Solutions Manual.

20. **a.** $s = 4.8 \times 10^{-7} M$
b. $s = 1.3 \times 10^{-5} M$
c. $s = 5.8 \times 10^{-5} M$

21. $s = 7.27 \times 10^{-5}$ g/L

ChemLAB

The ChemLab located at the end of the chapter can be used at this point in the lesson.

Suppose you wish to determine the solubility of silver iodide (AgI) in mol/L at 298 K. The equilibrium equation and solubility product constant expression are as follows.

$$\text{AgI(s)} \rightleftharpoons \text{Ag}^+\text{(aq)} + \text{I}^-\text{(aq)}$$
$$K_{sp} = [\text{Ag}^+][\text{I}^-] = 8.5 \times 10^{-17} \text{ at 298 K}$$

Explore **salts and solubility.**
Virtual Investigations

It is convenient to let *s* represent the solubility of AgI, that is, the number of moles of AgI that dissolves in one liter of solution. The equation indicates that for every mole of AgI that dissolves, an equal number of moles of Ag^+ ions forms in solution. Therefore, $[Ag^+]$ equals *s*. Every Ag^+ has an accompanying I^- ion, so $[I^-]$ also equals *s*. Substituting s for $[Ag^+]$ and $[I^-]$, the K_{sp} expression becomes the following.

$$[\text{Ag}^+][\text{I}^-] = (s)(s) = s^2 = 8.5 \times 10^{-17}$$
$$s = \sqrt{8.5 \times 10^{-17}} = 9.2 \times 10^{-9} \text{ mol/L}$$

The solubility of AgI is 9.2×10^{-9} mol/L at 298 K.

EXAMPLE Problem 5

EXAMPLE PROBLEM

CALCULATING MOLAR SOLUBILITY Use the K_{sp} value from **Table 3** to calculate the solubility in mol/L of copper(II) carbonate ($CuCO_3$) at 298 K.

1 ANALYZE THE PROBLEM

You have been given the solubility product constant for $CuCO_3$. The copper and carbonate ion concentrations are in a one-to-one relationship with the molar solubility of $CuCO_3$. Use *s* to represent the molar solubility of $CuCO_3$. Then use the solubility product constant expression to solve for the solubility. Because K_{sp} is of the order of 10^{-10}, you can predict that the solubility will be the square root of K_{sp}, or about 10^{-5}.

Known	Unknown
K_{sp} ($CuCO_3$) = 2.5×10^{-10}	s = ? mol/L

2 SOLVE FOR THE UNKNOWN

$CuCO_3(s) \rightleftharpoons Cu^{2+}(aq) + CO_3^{2-}(aq)$ — State the balanced chemical equation for the solubility equilibrium.

$K_{sp} = [Cu^{2+}][CO_3^{2-}] = 2.5 \times 10^{-10}$ — State the solubility product constant expression.

$s = [Cu^{2+}] = [CO_3^{2-}]$ — Relate $[Cu^{2+}]$ and $[CO_3^{2-}]$ to the solubility of $CuCO_3$, *s*.

$(s)(s) = s^2 = 2.5 \times 10^{-10}$ — Substitute *s* for $[Cu^{2+}]$ and $[CO_3^{2-}]$ in the expression for K_{sp}.

$s = \sqrt{2.5 \times 10^{-10}} = \mathbf{1.6 \times 10^{-5}}$ **mol/L** — Solve for *s*, and calculate the answer.

The molar solubility of $CuCO_3$ in water at 298 K is 1.6×10^{-5} mol/L.

3 EVALUATE THE ANSWER

The K_{sp} value has two significant figures, so the answer is correctly expressed with two digits. As predicted, the molar solubility of $CuCO_3$ is approximately 10^{-5} mol/L.

PRACTICE Problems

Do additional problems. Online Practice

20. Use the data in **Table 3** to calculate the solubility in mol/L of the following ionic compounds at 298 K.

a. $PbCrO_4$ **b.** AgCl **c.** $CaCO_3$

21. Challenge The K_{sp} of lead(II) carbonate ($PbCO_3$) is 7.40×10^{-14} at 298 K. What is the solubility of lead(II) carbonate in g/L?

DIFFERENTIATED INSTRUCTION

Struggling Learners Students with below-level abilities will be better able to solve solubility problems if they are shown the steps they need to follow to solve the problems. In particular, be sure to show such students how to use their calculators to solve solubility problems involving scientific notation, squares, square roots, cubes, and cube roots. BL

You have read that the solubility product constant can be used to determine the molar solubility of an ionic compound. You can apply this information as you perform the ChemLab at the end of this chapter. K_{sp} can also be used to find the concentrations of the ions in a saturated solution.

EXAMPLE Problem 6

CALCULATING ION CONCENTRATION Magnesium hydroxide is a white solid obtained from seawater and used in the formulation of many medications, in particular those whose function is to neutralize excess stomach acid. Determine the hydroxide ion concentration in a saturated solution of $Mg(OH)_2$ at 298 K. The K_{sp} equals 5.6×10^{-12}.

1 ANALYZE THE PROBLEM

You have been given the K_{sp} for $Mg(OH)_2$. The moles of Mg^{2+} ions in solution equal the moles of $Mg(OH)_2$ that dissolved, but the moles of OH^- ions in solution are two times the moles of $Mg(OH)_2$ that dissolved. You can use these relationships to write the solubility product constant expression in terms of one unknown. Because the equilibrium expression is a third-power equation, you can predict that $[OH^-]$ will be approximately the cube root of 10^{-12}, or approximately 10^{-4}.

Known

$K_{sp} = 5.6 \times 10^{-12}$

Unknown

$[OH^-] = ?$ mol/L

2 SOLVE FOR THE UNKNOWN

$Mg(OH)_2(s) \rightleftharpoons Mg^{2+}(aq) + 2OH^-(aq)$ — State the equation for the solubility equilibrium.

$K_{sp} = [Mg^{2+}][OH^-]^2 = 5.6 \times 10^{-12}$ — State the K_{sp} expression.

Let $x = [Mg^{2+}]$. Because there are two OH^- ions for every Mg^{2+} ion, $2x = [OH^-]$.

$(x)(2x)^2 = 5.6 \times 10^{-12}$ — Substitute $x = [Mg^{2+}]$ and $2x = [OH^-]$

$(x)(4)(x)^2 = 5.6 \times 10^{-12}$ — Square the terms.

$4x^3 = 5.6 \times 10^{-12}$ — Combine the terms.

$x^3 = \frac{5.6 \times 10^{-12}}{4} = 1.4 \times 10^{-12}$ — Divide.

$x = [Mg^{2+}] = \sqrt[3]{1.4 \times 10^{-12}} = 1.1 \times 10^{-4}$ mol/L — Use your calculator to determine the cube root.

Multiply $[Mg^{2+}]$ by 2 to obtain $[OH^-]$.

$[OH^-] = 2[Mg^{2+}] = 2(1.1 \times 10^{-4}$ mol/L$) =$ **2.2×10^{-4} mol/L**

3 EVALUATE THE ANSWER

The given K_{sp} has two significant figures, so the answer is correctly stated with two digits. As predicted, $[OH^-]$ is about 10^{-4} mol/L.

EXAMPLE PROBLEM

PRACTICE Problems

Do additional problems. Online Practice

22. Use K_{sp} values from **Table 3** to calculate the following.
    - **a.** $[Ag^+]$ in a solution of AgBr at equilibrium
    - **b.** $[F^-]$ in a saturated solution of CaF_2
    - **c.** $[Ag^+]$ in a solution of Ag_2CrO_4 at equilibrium
23. Calculate the solubility of Ag_3PO_4 ($K_{sp} = 2.6 \times 10^{-18}$).
24. **Challenge** The solubility of silver chloride (AgCl) is 1.86×10^{-4} g/100 g of H_2O at 298 K. Calculate the K_{sp} for AgCl.

IN-CLASS Example

Question Given that K_{sp} for $ZnCO_3$ at 298 K is 1.5×10^{-10} mol/L, what is the carbonate ion concentration in a saturated $ZnCO_3$ solution at 298 K?

Answer $[CO_3^{2-}] = 1.2 \times 10^{-5}$ mol/L

$K_{sp} = [Zn^{2+}][CO_3^{2-}] = s^2$

$1.5 \times 10^{-10} = s^2$

$s = \sqrt{1.5 \times 10^{-10}}$

$s = 1.2 \times 10^{-5}$ mol/L $= [CO_3^{2-}]$

PRACTICE Problems

Have students refer to the Selected Solutions appendix for complete solutions to odd-numbered problems. The complete solutions for all problems can be found in the Solutions Manual.

22. **a.** $[Ag^+] = 7.3 \times 10^{-7} M$
    b. $[F^-] = 4.2 \times 10^{-4} M$
    c. $[Ag^+] = 1.3 \times 10^{-4} M$
23. $s = 1.8 \times 10^{-5}$ mol/L
24. $K_{sp} = 1.7 \times 10^{-10}$

CHEMISTRY PROJECT

Molar Solubility from K_{sp} Have students write the equation for the solubility equilibrium for silver carbonate. Then, have them calculate the compound's molar solubility in water at 298 K. **OL**

$Ag_2CO_3(s) \rightleftharpoons 2Ag^+(aq) + CO_3^{2-}(aq)$

$s =$ mol/L Ag_2CO_3 that dissolves

$[CO_3^{2-}] = s$ mol/L, $[Ag^+] = 2s$ mol/L

$K_{sp} = [Ag^+]^2[CO_3^{2-}]$

$8.5 \times 10^{-12} = (2s)^2(s) = 4s^3$

$s = 1.3 \times 10^{-4}$ mol/L

ASSESSMENT

Skill Have students look up five compounds that have widely varying K_{sp} values. Have them use the K_{sp} values to calculate the molar solubilities of the compounds and the solubilities in grams per liter of solution. **OL**

ASSESSMENT

Skill To assess how well students understand how to calculate ion concentrations when solutions are mixed, ask them to determine all final ion concentrations in the solution that results from mixing equal volumes of 0.100*M* NaCl solution and 0.100*M* $BaCl_2$ solution. $[Na^+] = 0.050M$, $[Ba^{2+}] = 0.050M$, $[Cl^-] = 0.150M$ **OL**

READING CHECK A precipitate will form if the concentration of Fe^{3+} and $Fe(CN)_6^{4-}$ are greater than those that can exist in a saturated solution of $Fe_4(Fe(CN)_6)_3$.

Table 4 Ion Concentrations

Original Solutions (mol/L)	Mixture (mol/L)
$[Fe^{3+}] = 0.10$	$[Fe^{3+}] = 0.050$
$[Cl^-] = 0.30$	$[Cl^-] = 0.15$
$[K^+] = 0.40$	$[K^+] = 0.20$
$[Fe(CN)_6^{4-}] = 0.10$	$[Fe(CN)_6^{4-}] = 0.050$

Predicting precipitates Suppose equal volumes of 0.10*M* aqueous solutions of iron(III) chloride ($FeCl_3$) and potassium hexacyanoiron(II) ($K_4Fe(CN)_6$) are combined. Will a precipitate form as shown in **Figure 19?** The following double-replacement reaction might occur.

$$4FeCl_3 + 3K_4Fe(CN)_6 \rightarrow 12KCl + Fe_4(Fe(CN)_6)_3$$

You can use K_{sp} to predict whether a precipitate will form when any two ionic solutions are mixed.

For the reaction above, a precipitate is likely to form only if either product, KCl or $Fe_4(Fe(CN)_6)_3$, has low solubility. You might know that KCl is a soluble compound and would be unlikely to precipitate. But K_{sp} for $Fe_4(Fe(CN)_6)_3$ is a very small number, 3.3×10^{-41}, which suggests that $Fe_4(Fe(CN)_6)_3$ might precipitate if the concentrations of its ions are large enough. How large is large enough?

The following equilibrium is possible between solid $Fe_4(Fe(CN)_6)_3$—a precipitate—and its ions in solution, Fe^{3+} and $Fe(CN)_6^{4-}$.

$$Fe_4(Fe(CN)_6)_3(s) \rightleftharpoons 4Fe^{3+}(aq) + 3Fe(CN)_6^{4-}(aq)$$

When the $FeCl_3$ and $Fe_4(Fe(CN)_6)_3(s)$ solutions are mixed, if the concentrations of the ions Fe^{3+} and $Fe(CN)_6^{4-}$ are greater than those that can exist in a saturated solution of $Fe_4(Fe(CN)_6)_3$, the equilibrium will shift to the left and $Fe_4(Fe(CN)_6)_3(s)$ will precipitate. To predict whether a precipitate will form when the two solutions are mixed, you must first calculate the concentrations of the ions.

READING CHECK Explain the conditions under which you would predict that a precipitate would form.

Calculating ion concentrations **Table 4** shows the concentrations of the ions of reactants and products in the original solutions (0.10*M* $FeCl_3$ and 0.10*M* $K_4Fe(CN)_6$) and in the mixture immediately after equal volumes of the two solutions were mixed. Note that $[Cl^-]$ is three times as large as $[Fe^{3+}]$ because the ratio of Cl^- to Fe^{3+} in $FeCl_3$ is 3:1. Also note that $[K^+]$ is four times as large as $[Fe(CN)_6^{4-}]$ because the ratio of K^+ to $Fe(CN)_6^{4-}$ in $K_4Fe(CN)_6$ is 4:1. In addition, note that the concentration of each ion in the mixture is one-half its original concentration. This is because when equal volumes of two solutions are mixed, the same number of ions are dissolved in twice as much solution. Therefore, the concentration is reduced by one-half.

You can now use the data in the table to make a trial to see if the concentrations of Fe^{3+} and $Fe(CN)_6^{4-}$ in the mixed solution exceed the value of K_{sp} when substituted into the solubility product constant expression.

$$K_{sp} = [Fe^{3+}]^4[Fe(CN)_6^{4-}]^3$$

Remember that you have not determined whether the solution is saturated. When you make this substitution, it will not necessarily give the solubility product constant. Instead, it provides a number called the ion product (Q_{sp}). Q_{sp} is a trial value that can be compared with K_{sp}.

$$Q_{sp} = [Fe^{3+}]^4[Fe(CN)_6^{4-}]^3 = (0.050)^4(0.050)^3 = 7.8 \times 10^{-10}$$

You can now compare Q_{sp} and K_{sp}. This comparison can have one of three outcomes: Q_{sp} can be less than K_{sp}, equal to K_{sp}, or greater than K_{sp}.

Figure 19 Because its ion-product constant (Q_{sp}) is greater than K_{sp}, you could predict that this precipitate of $Fe_4(Fe(CN)_6)_3$ would form.

View an **animation of a precipitation reaction.**

Concepts In Motion

McGraw-Hill Education

CULTURAL DIVERSITY

Prussian Blue The common name for highly insoluble, deep blue iron(III) ferrocyanide $Fe[Fe^{3+}Fe^{2+}(CN)_6]_3$ is Prussian blue. The compound's original name came from its use as a dye in German army uniforms. A German color maker and dyer in Berlin named Diesbach accidentally discovered Prussian blue's deep blue color in 1704, mistakenly thinking—because it was made from cattle blood—that the pigment would be red. Prussian blue was the earliest of the modern synthetic colors. After its discovery in 1704, the manufacturing process of this pigment was passed to Diesbach's pupil, de Pierre, who in turn began manufacturing Prussian blue in Paris. Simon Eikenlenberg, a Dutch painter, wrote on the knowledge of Prussian blue in his manuscript *Notes on Paint and Painting* in 1722. By 1724, the manufacturing process of the pigment had spread to England.

1. If $Q_{sp} < K_{sp}$, the solution is unsaturated. No precipitate will form.
2. If $Q_{sp} = K_{sp}$, the solution is saturated, and no change will occur.
3. If $Q_{sp} > K_{sp}$, a precipitate will form, reducing the concentrations of the ions in the solution until the product of their concentrations in the K_{sp} expression equals the numerical value of K_{sp}. Then the system is in equilibrium, and the solution is saturated.

In the case of the $Fe_4(Fe(CN)_6)_3$ equilibrium, Q_{sp} (7.8×10^{-10}) is larger than K_{sp}(3.3×10^{-41}) and a deeply colored blue precipitate of $Fe_4(Fe(CN)_6)_3$ forms, as shown in **Figure 19.**

EXAMPLE Problem 7

Find help with **algebraic equations.** Math Handbook

EXAMPLE PROBLEM

PREDICTING A PRECIPITATE Predict whether a precipitate of $PbCl_2$ will form if 100 mL of $0.0100M$ NaCl is added to 100 mL of $0.0200M$ $Pb(NO_3)_2$.

1 ANALYZE THE PROBLEM

You have been given equal volumes of two solutions with known concentrations. The concentrations of the initial solutions allow you to calculate the concentrations of Pb^{2+} and Cl^- ions in the mixed solution.

Known	Unknown
100 mL $0.0100M$ NaCl	$Q_{sp} > K_{sp}$?
100 mL $0.0200M$ $Pb(NO_3)_2$	
$K_{sp} = 1.7 \times 10^{-5}$	

2 SOLVE FOR THE UNKNOWN

$PbCl_2(s) \rightleftharpoons Pb^{2+}(aq) + 2Cl^-(aq)$ — State the equation for the dissolving of $PbCl_2$.

$Q_{sp} = [Pb^{2+}][Cl^-]^2$ — State the ion product expression, Q_{sp}.

Mixing the solutions dilutes their concentrations by one-half.

$[Pb^{2+}] = \frac{0.0200M}{2} = 0.0100M$ — Divide $[Pb^{2+}]$ by 2.

$[Cl^-] = \frac{0.0100M}{2} = 0.00500M$ — Divide $[Cl^-]$ by 2.

$Q_{sp} = (0.0100)(0.00500)^2 = 2.5 \times 10^{-7}$ — Substitute $[Pb^{2+}] = 0.0100M$ and $[Cl^-] = 0.00500M$ into Q_{sp}.

$Q_{sp}\ (2.5 \times 10^{-7}) < K_{sp}\ (1.7 \times 10^{-5})$ — Compare Q_{sp} with K_{sp}.

A precipitate will not form.

3 EVALUATE THE ANSWER

Q_{sp} is less than K_{sp}. The Pb^{2+} and Cl^- ions are not present in high enough concentrations in the mixed solution to cause precipitation to occur.

PRACTICE Problems

Do additional problems. Online Practice

25. Use K_{sp} values from **Table 3** to predict whether a precipitate will form when equal volumes of the following solutions are mixed.
   a. $0.10M$ $Pb(NO_3)_2$ and $0.030M$ NaF
   b. $0.25M$ K_2SO_4 and $0.010M$ $AgNO_3$

26. Challenge Will a precipitate form when 250 mL of $0.20M$ $MgCl_2$ is added to 750 mL of $0.0025M$ NaOH?

IN-CLASS Example

Question Will a precipitate form when equal volumes of $0.020M$ $Ca(NO_3)$ and $0.0064M$ NaF solutions are mixed at 298 K? At 298 K, K_{sp} for $CaF_2 = 3.5 \times 10^{-11}$.

Answer A precipitate of CaF_2 will form.
$Q_{sp} = [Ca^{2+}][F^-]^2 = (0.010)(0.0032)^2$
$= 1.0 \times 10^{-7}$
$Q_{sp} > K_{sp}$

PRACTICE Problems

Have students refer to the Selected Solutions appendix for complete solutions to odd-numbered problems. The complete solutions for all problems can be found in the Solutions Manual.

25. a. $Q_{sp} > K_{sp}$ so a precipitate of PbF_2 will form.
b. $Q_{sp} < K_{sp}$ so no precipitate will form.
26. A precipitate will form.

Content Background

Precipitate Colors and Qualitative Analysis Explain to students that precipitate colors are often characteristic of particular compounds and can be used to confirm the presence of a particular ion in solution. For example, as a test for the presence of iron(III) ions, the addition of several drops of aqueous potassium hexacyanoferrate(II) solution to a solution containing iron(III) produces a deep blue precipitate of iron(III) hexacyanoferrate(II).

CHEMISTRY JOURNAL

Spectator Ions and Activity Have interested and capable students research the effects of spectator ions on solubility. Spectator ions increase the solubilities of ionic compounds in aqueous solutions. Have them explain how activities take these electrostatic interactions into account and replace concentrations in the K_{sp}, thus providing a more accurate equilibrium constant. Have students put their findings in their chemistry journals. AL

Quick Demo

Common Ion Effect Wear goggles and a lab apron. Demonstrate the common ion effect by adding 3 or 4 mL of concentrated HCl to 5 mL of a saturated NaCl solution in one test tube and 3 or 4 mL of 50% NaOH solution to 5 mL of saturated NaCl solution in a second test tube. **WARNING: *HCl and NaOH solutions are corrosive.*** Because both the HCl and NaOH solutions add common ions to the $NaCl(s) \leftrightharpoons Na^+(aq) + Cl^-(aq)$ equilibrium, they lower the solubility of NaCl and a precipitate forms.

ASSESSMENT

Knowledge Ask students to explain two ways in which they could use the common ion effect to lower the solubility of silver iodide in an aqueous solution. Increase the concentration of silver ions by adding a soluble silver compound, or increase the concentration of iodide ions by adding a soluble iodide compound. **OL**

GRAPH CHECK FIG. 20 The two situations represent two equilibrium positions. The concentrations are different, but in every saturated solution, the product of the ions must equal the K_{sp}.

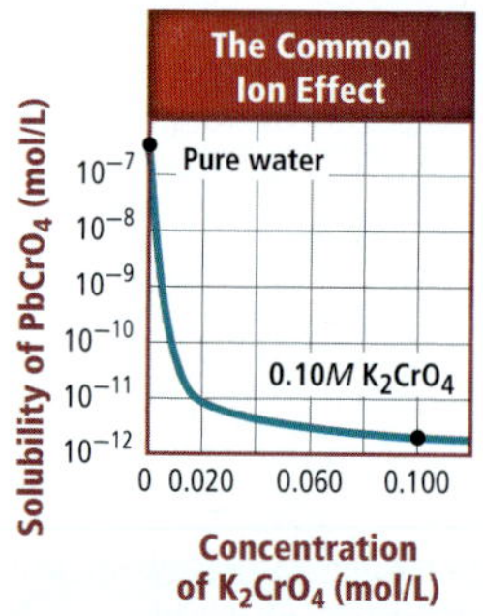

Pure water: $[Pb^{2+}] = 4.8 \times 10^{-7}$ mol/L
$[CrO_4^{2-}] = 4.8 \times 10^{-7}$ mol/L
0.10M K_2CrO_4: $[Pb^{2+}] = 2.3 \times 10^{-12}$ mol/L
$[CrO_4^{2-}] = 1.00 \times 10^{-1}$ mol/L

Figure 20 The solubility of lead chromate becomes lower as the concentration of the potassium chromate solution in which it is dissolved increases. The change is due to the presence of CrO_4^{2-} in both lead chromate and potassium chromate.

GRAPH CHECK

Verify that K_{sp} does not change as the concentration of potassium chromate increases.

The Common Ion Effect

The solubility of lead(II) chromate ($PbCrO_4$) in water is 4.8×10^{-7} mol/L at 298 K. That means you can dissolve 4.8×10^{-7} mol $PbCrO_4$ in 1.00 L of pure water. However, you cannot dissolve 4.8×10^{-7} mol $PbCrO_4$ in 1.00 L of 0.10M aqueous potassium chromate (K_2CrO_4) solution at that temperature. Why is $PbCrO_4$ less soluble in an aqueous K_2CrO_4 solution than in pure water?

The equation for the $PbCrO_4$ solubility equilibrium and the solubility product constant expression are as follows.

$$PbCrO_4(s) \rightleftharpoons Pb^{2+}(aq) + CrO_4^{2-}(aq)$$
$$K_{sp} = [Pb^{2+}][CrO_4^{2-}] = 2.3 \times 10^{-13}$$

Recall that K_{sp} is a constant at any given temperature, so if the concentration of either Pb^{2+} or CrO_4^{2-} increases when the system is at equilibrium, the concentration of the other ion must decrease. The product of the concentrations of the two ions must always equal K_{sp}. The K_2CrO_4 solution contains CrO_4^{2-} ions before any $PbCrO_4$ dissolves. In this example, the CrO_4^{2-} ion is called a common ion because it is part of both $PbCrO_4$ and K_2CrO_4. **Figure 20** shows the effect of the common ion, the CrO_4^{2-} ion, on the solubility of $PbCrO_4$. A **common ion** is an ion that is common to two or more ionic compounds. The lowering of the solubility of a substance because of the presence of a common ion is called the **common ion effect.**

Applying Le Châtelier's principle A saturated solution of lead(II) chromate ($PbCrO_4$) is shown in **Figure 21a.** Note the solid-yellow $PbCrO_4$ in the bottom of the beaker. The solution and solid are in equilibrium according to the following equation.

$$PbCrO_4(s) \rightleftharpoons Pb^{2+}(aq) + CrO_4^{2-}(aq)$$

When a solution of $Pb(NO_3)_2$ is added to the saturated $PbCrO_4$ solution, more solid $PbCrO_4$ precipitates, as shown in **Figure 21b.** The Pb^{2+} ion, common to both $Pb(NO_3)_2$ and $PbCrO_4$, reduces the solubility of $PbCrO_4$. Can this precipitation of $PbCrO_4$ be explained by Le Châtelier's principle? Adding Pb^{2+} ion to the solubility equilibrium stresses the equilibrium. To relieve the stress, the equilibrium shifts to the left to form more solid $PbCrO_4$.

Figure 21 Refer to **Figure 20** to see the effect of additional chromate ions on the solubility of lead(II) chromate. Adding Pb^{2+} ions in the form of lead nitrate ($Pb(NO_3)_2$) also affects the solubility of lead(II) chromate. **a.** $PbCrO_4(s)$ is in equilibrium with its ions in solution. **b.** The equilibrium is stressed by the addition of $Pb(NO_3)_2$ and more $PbCrO_4$ precipitate forms.

DIFFERENTIATED INSTRUCTION

Advanced Learners Have capable students calculate the molar solubility of lead(II) iodide in 0.422M potassium iodide at 298 K. Students can calculate the solubility based on the concentration of Pb^{2+}. Have them tabulate initial ion concentrations before the addition of PbI_2, the change in concentration as the result of adding PbI_2, and the equilibrium ion concentrations. The equilibrium concentrations then can be substituted into the ion product to calculate Q_{sp}, but to avoid solving a quadratic equation, a simplifying assumption can be made. Have students identify the assumption and justify its use. The assumption is that 2s is negligible compared to 0.422M. $s = 2.3 \times 10^{-4}$ mol/L. **AL**

The common ion effect also plays a role in the use of $BaSO_4$ when X-rays of the digestive system are taken. The low solubility of $BaSO_4$ helps ensure that the amount of the toxic barium ion absorbed into patient's system is small enough to be harmless. The procedure is further safeguarded by the addition of sodium sulfate (Na_2SO_4), a soluble ionic compound that provides a common ion, SO_4^{2-}.

$$BaSO_4(s) \rightleftharpoons Ba^{2+}(aq) + SO_4^{2-}(aq)$$

Le Châtelier's principle tells you that additional SO_4^{2-} from the Na_2SO_4 shifts the equilibrium to the left to produce more solid $BaSO_4$ and reduces the number of harmful Ba^{2+} ions in solution.

PROBLEM-SOLVING STRATEGY

Using Assumptions

In Example Problem 5, you calculated the molar solubility of $CuCO_3$ in pure water as 1.6×10^{-5} mol/L. But suppose that $CuCO_3$ is dissolved in a solution of 0.10*M* K_2CO_3? A common ion is in solution. If you set up the problem the same way you did in Example Problem 5, you will need to solve a quadratic equation. Solving the quadratic equation results in the correct answer, but you can make a simple assumption that streamlines the problem-solving process.

Concentration (*M*)	$CuCO_3$ (s) →	Cu^{2+} (aq) +	CO_3^{2-} (aq)
Initial	—	0	0.10
Change	—	$+s$	$+s$
Equilibrium	—	s	$0.10 + s$

Using the Quadratic Equation

1. Set up the problem
$[Cu^{2+}][CO_3^{2-}] = 2.5 \times 10^{-10}$
$(s)(0.10 + s) = 2.5 \times 10^{-10}$

2. Solve the quadratic
$0.10s + s^2 = 2.5 \times 10^{-10}$
$s^2 + 0.10s - 2.5 \times 10^{-10} = 0$

$$s = \frac{-b \pm \sqrt{b^2 - 4ac}}{2a}$$

$$= \frac{-0.10 \pm \sqrt{0.10^2 - (4)(1)(-2.5 \times 10^{-10})}}{2(1)}$$

$s = 2.5 \times 10^{-9}$ mol/L and $s = -0.10$ mol/L

Using the Simplifying Assumption

1. Set up the problem
$[Cu^{2+}][CO_3^{2-}] = 2.5 \times 10^{-10}$
$(s)(0.10 + s) = 2.5 \times 10^{-10}$

Because K_{sp} is small (2.5×10^{-10}), assume that s is negligible compared to 0.10*M*. Thus,
$0.10 + s \approx 0.10$.
$(s)(0.10) = 2.5 \times 10^{-10}$

2. Solve the problem
$(s)(0.10) = 2.5 \times 10^{-10}$

$$s = \frac{2.5 \times 10^{-10}}{(0.10)} = 2.5 \times 10^{-9} \text{ mol/L}$$

The root of the quadratic that makes sense is $s = 2.5 \times 10^{-9}$ mol/L. As you can see by comparing the two answers, the assumption gave good results more quickly and easily. However, this assumption works only for sparingly soluble compounds.

Apply the Strategy

Calculate the molar solubility of lead(II) fluoride in a 0.20*M* $Pb(NO_3)_2$ solution.

CHEMISTRY JOURNAL

Solubility and Qualitative Analysis

Have students research the way that solubility is used in qualitative analysis to separate cations into different analytical groups. Have students put their explanations, including relevant equations, in their chemistry journals. OL

3 Assess

Check for Understanding

Ask students how a solid compound's molar concentration is used in calculating the compound's molar solubility in water. It is not used. OL

Reteach

Lead students through writing the solubility product constant expressions for iron(III) hydroxide and calcium phosphate. $K_{sp} = [Fe^{3+}][OH^-]^3$; $K_{sp} = [Ca^{2+}]^3[PO_4^{3-}]^2$. OL BL

Extension

Students might be surprised to learn that spectator ions increase the solubility of an ionic compound in an aqueous solution. Explain that ions in solution can exert significant forces on each other, in effect increasing K_{sp} as ion concentration increases. OL

PROBLEM-SOLVING STRATEGIES

Apply the Strategy Calculate the molar solubility of lead(II) fluoride in a 0.20*M* $Pb(NO_3)_2$ solution.

$PbF_2(s) \longrightarrow Pb^{2+}(aq) + 2F^-(aq)$
Initial $[Pb^{2+}] = 0.20$
Change: $Pb^{2+} = +s$; $F^- = +2s$
Equilibrium: $Pb^{2+} = 0.20 + s$; $F^- = 2s$

$[Pb^{2+}][F^-]^2 = 3.3 \times 10^{-8}$
$(0.20 + s)(2s)^2 = 3.3 \times 10^{-8}$
$0.20 + s \approx 0.20$
$(0.20)(2s)^2 = 3.3 \times 10^{-8}$

$$4s^2 = \frac{3.3 \times 10^{-8}}{0.20} = 1.65 \times 10^{-7}$$

$$s^2 = \frac{1.65 \times 10^{-7}}{4}$$

$$s = \sqrt{4.12 \times 10^{-8}} = 2.0 \times 10^{-4} \text{ mol/L}$$

ASSESSMENT

Skill Ask students to use Table 3 to predict the formulas of precipitates that might form when aqueous solutions of the following compounds are mixed: (a) copper(II) chloride and sodium carbonate $CuCO_3$; (b) magnesium sulfate and phosphoric acid $Mg_3(PO_4)_2$; (c) lead(II) nitrate and lithium fluoride PbF_2. OL

Problem-Solving LAB

Purpose Students will use solubility product calculations to explore the effectiveness of fluoride in preventing tooth decay.

Process Skills use numbers, draw conclusions, think critically, apply concepts

Teaching Strategies

- Tell students how the effects of fluoride were discovered when people in areas having high concentrations of fluoride in their drinking water were found to have a low incidence of cavities. Many communities now add fluoride to public water supplies.
- Explain that if saliva were static and microbe free, solid hydroxyapatite would establish an equilibrium with its ions in solution and there would be no net loss of enamel. The flow of saliva prevents a true equilibrium from being achieved; however, remineralizing does take place.

Think Critically

1. $Ca_5(PO_4)_3OH(s) \rightleftharpoons 5Ca^{2+}(aq) + 3PO_4^{3-}(aq) + OH^-(aq)$; $K_{sp} = [Ca^{2+}]^5[PO_4^{3-}]^3[OH^-]$ The mouth is not a closed system. Saliva is being produced and swallowed regularly.
2. $Ca_5(PO_4)_3OH + NaF \rightleftharpoons NaOH + Ca_5(PO_4)_3F$
3. Solubility $Ca_5(PO_4)_3OH = 2.7 \times 10^{-5}M$ Solubility $Ca_5(PO_4)_3F = s = 6 \times 10^{-8}M$ The solubility of hydroxyapatite is 450 times greater than the solubility of fluoroapatite.
4. $Q_{sp} = 2.4 \times 10^{-40}$ The product will precipitate, remineralizing the tooth enamel.

Problem-Solving LAB

Apply Scientific Explanations

How does the fluoride ion prevent tooth decay? During the last half century, tooth decay has decreased significantly because minute quantities of fluoride ion ($6 \times 10^{-5}M$) are being added to most public drinking-water systems, and most people are using toothpastes containing sodium fluoride or tin(II) fluoride. Use what you know about the solubility of ionic compounds and reversible reactions to explore the role of the fluoride ion in maintaining cavity-free teeth.

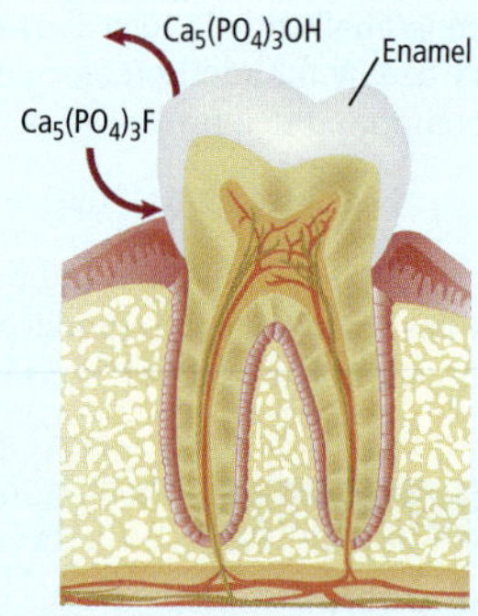

Analysis

Enamel, the hard, protective outer layer of the tooth, is 98% hydroxyapatite ($Ca_5(PO_4)_3OH$). Although insoluble in water ($K_{sp} = 6.8 \times 10^{-37}$), demineralization, which is the dissolving of hydroxyapatite, does occur, especially when the saliva contains acids. The reverse reaction, remineralization, also occurs. Remineralization is the redepositing of tooth enamel. When hydroxyapatite is in solution with fluoride ions, a double-replacement reaction can occur. A fluoride ion replaces the hydroxide ion to form fluoroapatite ($Ca_5(PO_4)_3F$), ($K_{sp} = 1 \times 10^{-60}$). Fluoroapatite remineralizes the tooth enamel, thus partially displacing hydroxyapatite. Because fluoroapatite is less soluble than hydroxyapatite, destructive demineralization is reduced.

Think Critically

1. **State** the equation for the dissolving of hydroxyapatite and its equilibrium constant expression. How do the conditions in the mouth differ from those of a true equilibrium?
2. **State** the equation that describes the double-replacement reaction that occurs between hydroxyapatite and sodium fluoride.
3. **Calculate** the solubility of hydroxyapatite and fluoroapatite in water. Compare the solubilities.
4. **Calculate** the ion product constant (Q_{sp}) for the reaction if $0.00050M$ NaF is mixed with an equal volume of $0.000015M$ $Ca_5(PO_4)_3OH$. Will a precipitate form (re-mineralization)?

SECTION 3 REVIEW

Section Self-Check

Section Summary

- Equilibrium concentrations and solubilities can be calculated using equilibrium constant expressions.
- K_{sp} describes the equilibrium between a sparingly soluble ionic compound and its ions in solution.
- If the ion product, Q_{sp}, exceeds the K_{sp} when two solutions are mixed, a precipitate will form.
- The presence of a common ion in a solution lowers the solubility of a dissolved substance.

27. MAINIDEA **List** the information you would need in order to calculate the concentration of a product in a reaction mixture at equilibrium.
28. **Explain** how to use the solubility product constant to calculate the solubility of a sparingly soluble ionic compound.
29. **Describe** how the presence of a common ion reduces the solubility of an ionic compound.
30. **Explain** the difference between K_{sp} and Q_{sp}. Is Q_{sp} an equilibrium constant?
31. **Calculate** The K_{sp} of magnesium carbonate ($MgCO_3$) is 2.6×10^{-9}. What is the solubility of $MgCO_3$ in pure water?
32. **Design an experiment** based on solubilities to demonstrate which of two ions, Mg^{2+} or Pb^{2+}, is contained in an aqueous solution. Solubility information about ionic compounds is given in **Tables R-3** and **R-8** in the Student Resources appendix.

SECTION 3 REVIEW

27. the concentrations of the reactants and all other products and K_{eq}
28. Write the equation for the solubility equilibrium and the solubility product constant expression. Let *s* equal the compound's molar solubility. Substitute the appropriate multiples of *s* into the solubility constant expression and solve the expression for *s*.
29. The presence of a common ion reduces solubility by shifting the solubility equilibrium in the direction of the solid.
30. Q_{sp} is the product of the concentrations of the ions that might be present in a solution of an ionic compound. Its value is calculated in order to compare it with K_{sp}, which is the product of the concentrations of the ions that are present in a saturated solution. K_{sp} is an equilibrium constant; Q_{sp} is not.
31. $s^2 = 5.1 \times 10^{-5}M$
32. Because magnesium chromate is soluble and lead(II) chromate is insoluble, add 10.0 mL of $0.10M$ aqueous potassium chromate solution to 100.0 mL of the unknown aqueous solution. If the unknown solution contains magnesium ion, no precipitate of $MgCrO_4$ will be produced. However, if the unknown solution contains lead(II) ion, solid, yellow $PbCrO_4$ will precipitate.

CHEMISTRY&health

Hemoglobin Rises to the Challenge

When people travel to the mountains, they often feel tired and light-headed for a time. That's because the mountain air contains fewer oxygen molecules, as shown in **Figure 1.** Over time, the fatigue lessens. The body adapts by producing more of a protein called hemoglobin.

Hemoglobin-oxygen equilibrium

Hemoglobin (Hgb) binds with oxygen molecules that enter your bloodstream, producing oxygenated hemoglobin ($Hgb(O_2)_4$). The equilibrium of Hgb and O_2 is represented as follows.

$$Hgb(aq) + 4O_2(g) \rightleftharpoons Hgb(O_2)_4(aq)$$

In the lungs When you breathe, oxygen molecules move into your blood. The equilibrium reacts to the stress by consuming oxygen molecules at an increased rate. The equilibrium shifts to the right, increasing the blood concentration of $Hgb(O_2)_4$.

$$Hgb(aq) + 4O_2(g) \rightleftharpoons Hgb(O_2)_4(aq)$$

In the tissues When the $Hgb(O_2)_4$ reaches body tissues where oxygen concentrations are low, the equilibrium shifts to the left, releasing oxygen to enable the metabolic processes that produce energy.

$$Hgb(aq) + 4O_2(g) \rightleftharpoons Hgb(O_2)_4(aq)$$

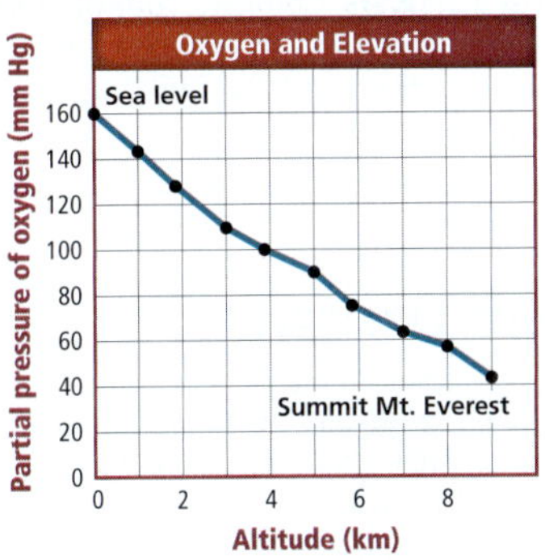

Figure 1 On the summit, the partial pressure of O_2 is much lower. Each breath a person draws contains fewer O_2 molecules.

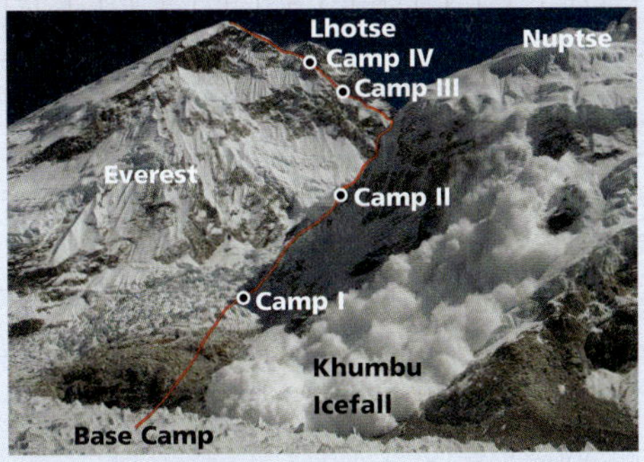

Figure 2 On Mount Everest, a climber might ascend to Camp II, descend to Base Camp, and then ascend to Camp III over the course of several days to prepare for a summit bid.

In the mountains The equilibrium reacts to the stress of thin mountain air by producing oxygen at an increased rate. The shift to the left releases oxygen molecules in your lungs, leaving less oxygenated hemoglobin in your blood.

$$Hgb(aq) + 4O_2(g) \rightleftharpoons Hgb(O_2)_4(aq)$$

The lower blood concentration of oxygenated hemoglobin means that fewer oxygen molecules are released in other parts of your body. Because less energy is produced, you feel tired.

The body adjusts Your body responds to the lower oxygen concentration by producing more hemoglobin, part of a process known as acclimatization. More hemoglobin shifts the equilibrium position back to the right.

$$Hgb(aq) + 4O_2(g) \rightleftharpoons Hgb(O_2)_4(aq)$$

The increased concentration of $Hgb(O_2)_4(aq)$ means that more oxygen molecules can be released in your body tissues. **Figure 2** shows where climbers might adjust their bodies to high elevations before beginning their summit bid.

WRITING IN ▶ Chemistry

Research the sleep disorder apnea. How would an incident of apnea affect the body's hemoglobin equilibrium?

WebQuest

CHEMISTRY&health

Purpose

Students will learn how the hemoglobin-oxygen equilibrium functions to carry oxygen from the lungs to the muscles and other tissues and how moving to a high elevation affects the equilibrium.

Background

Each hemoglobin molecule contains two α-type protein chains and two β-type protein chains, each with a heme complex that can bind with an oxygen molecule. Thus, each hemoglobin molecule can bind with four oxygen molecules to form a bright red complex, giving arterial blood its red color. When the oxygen molecules are released, their former positions are occupied by water molecules to form a bluish complex, giving venous blood its bluish tint.

In a heme complex, the central Fe^{2+} ion has coordinate covalent bonds with four nitrogen atoms of a planar porphyrin ring and with a nitrogen atom of a protein chain. Because the Fe^{2+} ion is usually six-coordinate, an oxygen molecule can form a coordinate covalent bond to occupy the sixth position.

Teaching Strategies

- Have students interpret the illustration to answer the following question: What is one way climbers can safely engage in strenuous activities in the "thinner" air at extremely high altitudes? The climbers can carry and use supplemental oxygen.
- Review coordinate covalent bonding with students.
- Show students the structure of the heme complex, and explain how an oxygen molecule can bind with the Fe^{2+} ion.

WRITING IN ▶ Chemistry

Rubric WebQuest

Research An incident of sleep apnea would reduce the amount of oxygen in the lungs and have the same effect as breathing thin mountain air. The equilibrium would produce oxygen at an increased rate. The shift to the left would release oxygen molecules in the lungs, leaving less oxygenated hemoglobin in the blood. The lower blood concentration of oxygenated hemoglobin would cause fewer oxygen molecules to be released in other parts of the body. Because less energy would be produced, the person would feel tired.

SMALL SCALE ChemLAB

Compare Two Solubility Product Constants

Background: By observing the formation of two precipitates in the same system, you can infer the relationship between the solubilities of the two ionic compounds and the numerical values of their solubility product constants (K_{sp}).

Question: *How can you use Le Châtelier's principle to evaluate the relative solubilities of two precipitates?*

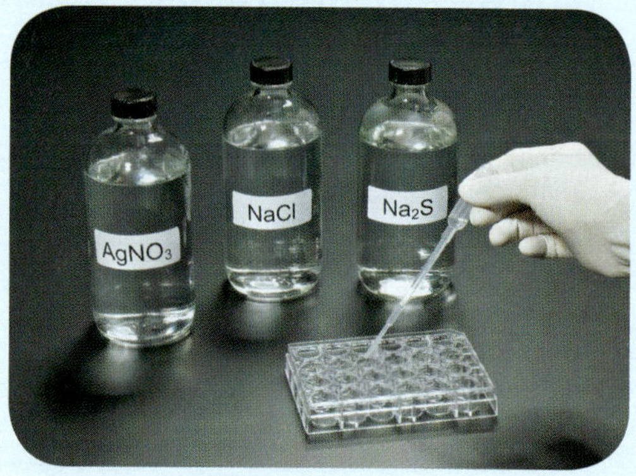

Materials

$AgNO_3$ solution
NaCl solution
Na_2S solution
24-well microplate
thin-stem pipettes (3)

Safety Precautions

WARNING: ***Silver nitrate is highly toxic and will stain skin and clothing. Sodium sulfide is a skin irritant and should be kept away from acids.***

Procedure

1. Read and complete the lab safety form.
2. Place 10 drops of $AgNO_3$ solution in Well A1 of a 24-well microplate. Place 10 drops of the same solution in Well A2.
3. Add 10 drops of NaCl solution to Well A1 and 10 drops to Well A2.
4. Allow the precipitates to form. Observe the wells from the top and the side and record your observations.
5. To Well A2, add 10 drops of Na_2S solution.
6. Allow the precipitate to form. Record your observations of the precipitate.
7. Compare the contents of Wells A1 and A2, and record your observations.
8. **Cleanup and Disposal** Use a wash bottle to transfer the contents of the well plate into a waste beaker.

Analyze and Conclude

1. **Analyze** Write the complete equation for the reaction that occurred when you mixed NaCl and $AgNO_3$ in Step 3. Write the net ionic equation.
2. **Analyze** Write the solubility product constant expression for the equilibrium established in Wells A1 and A2 in Step 3. K_{sp} (AgCl) = 1.8×10^{-10}.
3. **Analyze** Write the equation for the equilibrium that was established in Well A2 when you added Na_2S.K_{sp} (Ag_2S) = 8×10^{-48}.
4. **Identify** the two precipitates by color.
5. **Compare** the K_{sp} values for the two precipitates. Which of the two ionic compounds is more soluble?
6. **Recognize Cause and Effect** Use Le Châtelier's principle to explain how the addition of Na_2S in Step 5 affected the equilibrium established in Well A2.
7. **Calculate** the molar solubilities of the two precipitates using the K_{sp} values. Which of the precipitates is more soluble?
8. **Identify** What evidence from this experiment supports your answer to Question 7? Explain.
9. **Error Analysis** Compare your observations of the well plate from the side with your observations from the top. What did you notice?
10. **Research** how industries use precipitation to remove hazardous chemicals from wastewater.

INQUIRY EXTENSION

Soluble v. Insoluble The reactants that you used in this ChemLab are all soluble ionic compounds, and the precipitates are insoluble. How does soluble N a$_2$S differ from insoluble Ag_2S? How does soluble NaCl differ from insoluble AgCl? Use this information, K_{sp} data from **Table 3,** and other reference sources to develop general rules for solubility.

SMALL SCALE ChemLAB

Rubric

Preparation

Time Allotment one class period

Process Skills observe and infer, compare and contrast, draw a conclusion, use numbers

Safety Precautions Approve lab safety forms before work begins. Students should wear aprons and goggles and wash their hands after completing the laboratory investigation. Review MSDS for Na_2S and $AgNO_3$ with students. **WARNING: *Na_2S is a skin irritant and liberates toxic H_2S when interacting with acids.***

Disposal Use a wash bottle to transfer the contents of the microplate into a large waste beaker. Evaporate the contents of the beaker to dryness. Dispose of the solid at a facility that is licensed to receive hazardous waste and recover silver.

Preparation of Solutions
See the Teacher Handbook for preparation of all solutions.

Procedure

- Place small amounts of each solution in labeled containers throughout the laboratory.
- **Troubleshooting** Pipettes should be used for one solution only. Contamination of the solutions results in precipitation.

Analyze and Conclude

1. $AgNO_3(aq) + NaCl(aq) \longrightarrow AgCl(s) + NaNO_3(aq)$
   $Ag^+(aq) + Cl^-(aq) \longrightarrow AgCl(s)$
2. $K_{sp} = [Ag^+][Cl^-] = 1.8 \times 10^{-10}$
3. $2AgCl(s) + Na_2S(aq) \longrightarrow Ag_2S(s) + 2NaCl(aq)$
   $Ag_2S(s) \longrightarrow 2Ag^+(aq) + S^{2-}(aq)$
4. AgCl: white Ag_2S: black
5. AgCl is more soluble.
6. Silver ions are in equilibrium with solid, white AgCl. On the addition of Na_2S, S^{2-} ions combine with Ag^+ ions to form solid, black Ag_2S. According to Le Châtelier's principle, the removal of the Ag^+ ion shifts the AgCl equilibrium to the right, thus dissolving AgCl as Ag_2S forms.
7. Solubility AgCl = $1.3 \times 10^{-5}M$; Solubility $Ag_2S = 1 \times 10^{-16}M$; AgCl is more soluble.
8. The concentration of Ag^+ ions that could exist in equilibrium with solid AgCl was larger than the A g$^+$ concentration that could exist in equilibrium with solid Ag_2S. If that were not the case, there would have been no precipitate of Ag_2S.
9. After standing undisturbed for a time, the solid layers are on the bottom with the black on top of the white layer and a clear layer of solution above.
10. Answers will vary. Students might mention hydroxide and sulfide precipitation to remove heavy metals.

INQUIRY EXTENSION

Students should find that all group 1 metals are soluble. Compounds formed with the acetate, ammonium, and nitrate ions are soluble. The smaller the value of K_{sp}, the lower the molar solubility of the ionic compound.

Vocabulary Practice

BIGIDEA Many reactions and processes reach a state of chemical equilibrium in which both reactants and products are formed at equal rates.

SECTION 1 A State of Dynamic Balance

MAINIDEA Chemical equilibrium is described by an equilibrium constant expression that relates the concentrations of reactants and products.

- A reaction is at equilibrium when the rate of the forward reaction equals the rate of the reverse reaction.
- The equilibrium constant expression is a ratio of the molar concentrations of the products to the molar concentrations of the reactants with each concentration raised to a power equal to its coefficient in the balanced chemical equation.

$$K_{eq} = \frac{[C]^c[D]^d}{[A]^a[B]^b}$$

- The value of the equilibrium constant expression, K_{eq}, is a constant for a given temperature.

VOCABULARY
- reversible reaction
- chemical equilibrium
- law of chemical equilibrium
- equilibrium constant
- homogeneous equilibrium
- heterogeneous equilibrium

SECTION 2 Factors Affecting Chemical Equilibrium

MAINIDEA When changes are made to a system at equilibrium, the system shifts to a new equilibrium position.

- Le Châtelier's principle describes how an equilibrium system shifts in response to a stress or a disturbance.
- When an equilibrium shifts in response to a change in concentration or volume, the equilibrium position changes but K_{eq} remains constant. A change in temperature, however, alters both the equilibrium position and the value of K_{eq}.

VOCABULARY
- Le Châtelier's principle

SECTION 3 Using Equilibrium Constants

MAINIDEA Equilibrium constant expressions can be used to calculate concentrations and solubilities.

- Equilibrium concentrations and solubilities can be calculated using equilibrium constant expressions.
- K_{sp} describes the equilibrium between a sparingly soluble ionic compound and its ions in solution.
- If the ion product, Q_{sp}, exceeds the K_{sp} when two solutions are mixed, a precipitate will form.
- The presence of a common ion in a solution lowers the solubility of a dissolved substance.

VOCABULARY
- solubility product constant
- common ion
- common ion effect

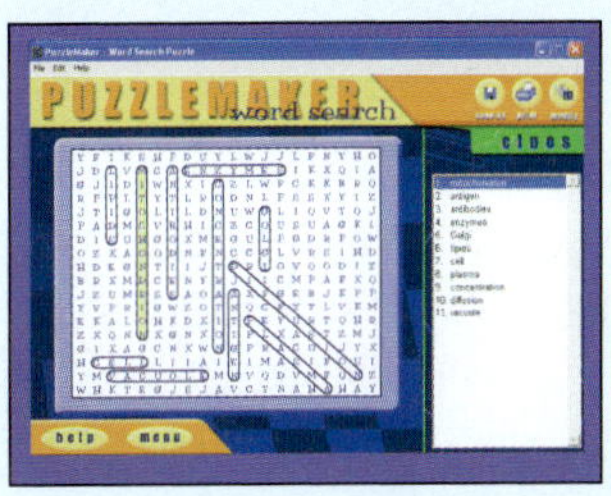

Vocabulary Practice

For additional practice with vocabulary, have students access the Vocabulary PuzzleMaker.

CHAPTER 17 STUDY GUIDE

Use the Vocabulary

To reinforce chapter vocabulary, have students write a sentence for each term. OL EL

Review Strategies

- Ask students to write the chemical equations and equilibrium constant expressions for three homogeneous equilibria. OL
- Have students write the equations to calculate the molar solubilities of AgCl and CaF_2. OL
- Ask students to explain the difference between Q_{sp} and K_{sp}. OL
- Problems from the Supplemental Practice Problems appendix or the Supplemental Problems booklet can be used for review. OL

Students can review with these online resources.

Vocabulary Practice

Vocabulary eGames
Vocabulary eFlashcards
Vocabulary PuzzleMaker

Section Self-Check

Chapter Self-Check

Online Test Practice

Use *eAssessment* to:

- create multiple versions of tests
- edit existing questions and add your own questions
- build tests aligned with select state curriculum standards using built-in state curriculum tags
- track students' progress

SECTION 1

Mastering Concepts

33. Answers might be situations such as vehicle traffic across a bridge, riding a bicycle, balancing on a seesaw, doing a handstand, and so on.

34. Reactants are continuing to form products and products are continuing to form reactants.

35. A state of physical equilibrium exists if water is being bailed from the boat at the same rate that it is entering the boat.

36. The equation represents a heterogeneous equilibrium because the reactants and products are present in more than one physical state.

37. An equilibrium position is a particular set of equilibrium concentrations.

38. the ratio of product concentrations to reactant concentrations, with each concentration raised to the power corresponding to its coefficient in the balanced equation

39. The concentrations of pure liquids and pure solids are omitted from the equilibrium constant expression.

40. A numerically large K_{eq} means that product concentrations, which are in the numerator, are greater than reactant concentrations, which are in the denominator.

41. The new value of K_{eq} is the reciprocal of its original value.

42. An equilibrium such as this can exist if the initial products formed react so quickly that the rate of the reverse reaction is quickly equal to the rate of the forward reaction. K_{eq} must have a small numerical value.

43. The concentrations of the reactants and products do not change. Individual reactant and product molecules undergo chemical change as the forward and reverse reactions proceed at equal rates.

Mastering Problems

44. a. $K_{eq} = [N_2]^3[H_2O]^4/[N_2H_4]^2[NO_2]^2$

b. $K_{eq} = [NbCl_3][NbCl_5]/[NbCl_4]^2$

45. a. $K_{eq} = [H_2O(g)][CO_2(g)]$

b. $K_{eq} = [C_6H_6(g)]$

46. $K_{eq} = [CO_2(g)]$

47. molar concentration = 135 mol/L Mn

CHAPTER 17 ASSESSMENT

SECTION 1

Mastering Concepts

33. Describe an equilibrium in everyday life that illustrates a state of balance between two opposing processes.

34. Given the fact that the concentrations of reactants and products are not changing, why is the word *dynamic* used to describe chemical equilibrium?

35. Explain how a person bailing out a row boat with a leak could represent a state of physical equilibrium.

36. Does the following equation represent a homogeneous equilibrium or a heterogeneous equilibrium? Explain. your answer.

$$H_2O(s) \rightleftharpoons H_2O(l)$$

37. What is an equilibrium position?

38. Explain how to write an equilibrium constant expression.

39. Why should you pay attention to the physical states of reactants and products when writing equilibrium constant expressions?

40. Why does a numerically large K_{eq} mean that the products are favored in an equilibrium system?

41. What happens to K_{eq} for an equilibrium system if the equation for the reaction is rewritten in the reverse?

42. How can an equilibrium system contain small and unchanging amounts of products yet have large amounts of reactants? What can you say about the relative size of K_{eq} for such an equilibrium?

43. A system, which contains only molecules as reactants and products, is at equilibrium. Describe what happens to the concentrations of the reactants and products and what happens to individual reactant and product molecules.

Mastering Problems

44. Write equilibrium constant expressions for these homogeneous equilibria.

a. $2N_2H_4(g) + 2NO_2(g) \rightleftharpoons 3N_2(g) + 4H_2O(g)$

b. $2NbCl_4(g) \rightleftharpoons NbCl_3(g) + NbCl_5(g)$

45. Write equilibrium constant expressions for these heterogeneous equilibria.

a. $2NaHCO_3(s) \rightleftharpoons Na_2CO_3(s) + H_2O(g) + CO_2(g)$

b. $C_6H_6(l) \rightleftharpoons C_6H_6(g)$

46. Heating limestone ($CaCO_3(s)$) forms quicklime ($CaO(s)$) and carbon dioxide gas. Write the equilibrium constant expression for this reversible reaction.

47. Suppose you have a cube of pure manganese metal measuring 5.25 cm on each side. You find that the mass of the cube is 1076.6 g. What is the molar concentration of manganese in the cube?

48. K_{eq} is 3.63 for the reaction $A + 2B \rightleftharpoons C$. **Table 5** shows the concentrations of the reactants and product in two different reaction mixtures at the same temperature. Determine whether both reactions are at equilibrium.

Table 5 Concentrations of A, B, and C

A (mol/L)	B (mol/L)	C (mol/L)
0.500	0.621	0.700
0.250	0.525	0.250

49. When steam is passed over iron filings, solid iron(III) oxide and gaseous hydrogen are produced in a reversible reaction. Write the balanced chemical equation and the equilibrium constant expression for the reaction, which yields iron(III) oxide and hydrogen gas.

SECTION 2

Mastering Concepts

50. What is meant by a stress on a reaction at equilibrium?

51. How does Le Châtelier's principle describe an equilibrium's response to a stress?

52. Why does removing a reactant cause an equilibrium shift to the left?

53. When an equilibrium shifts to the right, what happens to each of the following?

a. the concentration of the reactants

b. the concentration of the products

54. Carbonated Beverages Use Le Châtelier's principle to explain how a shift in the equilibrium $H_2CO_3(aq) \rightleftharpoons H_2O(l) + CO_2(g)$ causes a soft drink to go flat when its container is left open.

55. How would each of the following changes affect the equilibrium position of the system used to produce methanol from carbon monoxide and hydrogen?

$$CO(g) + 2H_2(g) \rightleftharpoons CH_3OH(g) + \text{heat}$$

a. adding CO to the system

b. cooling the system

c. adding a catalyst to the system

d. removing CH_3OH from the system

e. decreasing the volume of the system

56. Explain how a temperature increase would affect the equilibrium represented by the following equation.

$$PCl_5(g) \rightleftharpoons PCl_3(g) + Cl_2(g) + \text{heat}$$

57. A liquid solvent for chlorine is poured into a flask in which the following reaction is at equilibrium: $PCl_5(g) \rightleftharpoons PCl_3(g) + Cl_2(g) + \text{heat}$. How is the equilibrium affected when some of the chlorine gas dissolves?

48. $K_{eq} = [C]/[A][B]^2 = 3.63$
$(0.700)/(0.500)(0.621)^2 = 3.63$
$(0.250)/(0.250)(0.525)^2 = 3.63$
Both reactions are at equilibrium.

49. $2Fe(s) + 3H_2O(g) \leftrightharpoons Fe_2O_3(s) + 3H_2(g)$
$K_{eq} = [H_2(g)]^3/[H_2O(g)]^3$

SECTION 2

Mastering Concepts

50. A stress on a reaction at equilibrium is any change in concentration, volume (pressure), or temperature which causes the equilibrium to shift.

51. Le Châtelier's principle states that an equilibrium will shift in the direction that relieves a stress.

52. In order to restore the equilibrium ratio of reactants and products, the equilibrium shifts in the direction of the reactants.

53. The concentrations of the reactants decrease; the concentrations of the products increase.

Chapter Self-Check

■ **Figure 22**

58. **Figure 22** shows the following endothermic reaction at equilibrium at room temperature.
$$Co(H_2O)_6^{2+}(aq) + 4Cl^-(aq) \rightleftharpoons CoCl_4^{2-}(aq) + 6H_2O(l)$$
Given that $Co(H_2O)_6^{2+}(aq)$ is pink and $CoCl_4^{2-}(aq)$ is blue, what visual change would you expect to see if the flask were placed in an ice bath? Explain.

59. For the equilibrium described in Question 58, what visual change would you expect to see if 10 g of solid potassium chloride were added and dissolved? Explain.

60. Given two reactions at equilibrium:
a. $N_2(g) + 3H_2(g) \rightleftharpoons 2NH_3(g)$
b. $H_2(g) + Cl_2(g) \rightleftharpoons 2HCl(g)$,
explain why changing the volume of the systems alters the equilibrium position of **a** but has no effect on **b**.

61. Would you expect the numerical value of K_{eq} for the following equilibrium to increase or decrease with increasing temperature? Explain your answer.
$$PCl_5(g) \rightleftharpoons PCl_3(g) + Cl_2(g) + \text{heat}$$

62. Explain how you would regulate the pressure to favor the products in the following equilibrium system.
$$MgCO_3(s) \rightleftharpoons MgO(s) + CO_2(g)$$

63. Ethylene (C_2H_4) reacts with hydrogen to form ethane (C_2H_6).
$$C_2H_4(g) + H_2(g) \rightleftharpoons C_2H_6(g) + \text{heat}$$
How would you regulate the temperature of this equilibrium in order to accomplish each of the following?
a. increase the yield of ethane
b. decrease the concentration of ethylene
c. increase the amount of hydrogen in the system

SECTION 3

Mastering Concepts

64. What does it mean to say that two solutions have a common ion? Give an example.

65. Why are compounds such as sodium chloride usually not given K_{sp} values?

66. **X-rays** Why is barium sulfate a better choice than barium chloride for adding definition to X-rays? At 26°C, 37.5 g of $BaCl_2$ can be dissolved in 100 mL of water.

■ **Figure 23**

67. Explain what is happening in **Figure 23** in terms of Q_{sp} and K_{sp}.

68. Explain why a common ion lowers the solubility of an ionic compound.

69. Describe the solution that results when two solutions are mixed and Q_{sp} is found to equal K_{sp}. Does a precipitate form?

Mastering Problems

70. Write the K_{sp} expression for lead chromate ($PbCrO_4$), and calculate its solubility in mol/L. $K_{sp} = 2.3 \times 10^{-13}$.

71. At 350°C, $K_{eq} = 1.67 \times 10^{-2}$ for the reversible reaction $2HI(g) \rightleftharpoons H_2(g) + I_2(g)$. What is the concentration of HI at equilibrium if $[H_2]$ is $2.44 \times 10^{-3}M$ and $[I_2]$ is $7.18 \times 10^{-5}M$?

72. K_{sp} for scandium fluoride (ScF_3) at 298 K is 4.2×10^{-18}. Write the chemical equation for the solubility equilibrium of scandium fluoride in water. What concentration of Sc^{3+} ions is required to cause a precipitate to form if the fluoride-ion concentration is 0.076*M*?

73. Will a precipitate form when 62.6 mL of 0.0322*M* $CaCl_2$ and 31.3 mL of 0.0145*M* NaOH are mixed? Use data from **Table 3**. Explain your logic.

74. **Manufacturing** Ethyl acetate ($CH_3COOCH_2CH_3$), a solvent used in making varnishes and lacquers, can be produced by the reaction between ethanol and acetic acid. The equilibrium system is described by the equation
$$CH_3COOH + CH_3CH_2OH \rightleftharpoons CH_3COOCH_2CH_3 + H_2O.$$
Calculate K_{eq} using these equilibrium concentrations: $[CH_3COOCH_2CH_3] = 2.90M$, $[CH_3COOH] = 0.316M$, $[CH_3CH_2OH] = 0.313M$, and $[H_2O] = 0.114M$.

54. Because $CO_2(g)$ continually escapes from the open container, the equilibrium shifts to the right until $H_2CO_3(aq)$ is depleted.

55. **a.** The equilibrium shifts to the right.
b. The equilibrium shifts to the right.
c. The equilibrium does not shift.
d. The equilibrium shifts to the right.
e. The equilibrium shifts to the right.

56. Heating would favor the reverse (endothermic) reaction by shifting the equilibrium to the left.

57. The equilibrium shifts to the right to produce more chlorine.

58. The solution would become more pink as the equilibrium shifted to the left, releasing heat.

59. The solution would become bluer. Dissolving potassium chloride would increase the concentration of chloride ions, causing the equilibrium to shift to the right.

60. Different numbers of moles of gas are on the reactant and product sides of equation **a**, but the same number of moles of gas are on both sides of equation **b.** When the number of moles of gas is the same, changing volume has no effect on the equilibrium.

61. The numerical value of K_{eq} would decrease as the equilibrium shifted to the left, absorbing heat.

62. Decreasing the pressure would favor the formation of MgO and CO_2 because a gaseous product tends to restore the pressure.

63. **a.** lower the temperature
b. lower the temperature
c. raise the temperature

SECTION 3

Mastering Concepts

64. If two solutions have a common ion, both contain the same ion. NaCl(aq) and KCl(aq) both contain $Cl^-(aq)$.

65. Soluble compounds are usually not given K_{sp} values because the values would be large numbers. Also, such compounds are unlikely to precipitate from solutions unless ion concentrations are exceedingly high.

66. Barium ions are toxic to humans. Barium sulfate can be ingested safely only because of its exceedingly low solubility. Barium chloride's high solubility would make ingesting it extremely dangerous.

67. A precipitate forms because Q_{sp} exceeds K_{sp}.

68. If the concentration of either ion produced by the dissolving compound is increased by a common ion present in the solution, the concentration of the compound's other ion must decrease.

69. The new solution is saturated. No precipitate forms.

Mastering Problems

70. $K_{sp} = [Pb^{2+}][CrO_4^{2-}] = 2.3 \times 10^{-13}$
$s = 4.8 \times 10^{-7}$

71. $[HI] = 3.24 \times 10^{-3}M$

72. $ScF_3(s) \rightleftharpoons Sc^{3+}(aq) + 3F^-(aq)$
$[Sc^{3+}] = 9.6 \times 10^{-15}M$

73. No, a precipitate will not form. $Q_{sp} = 5.02 \times 10^{-7}$, which is less than calcium hydroxide's K_{sp} value of 5.0×10^{-6}.

74. $K_{eq} = 3.34$

MIXED REVIEW

75. Removing H_2O shifts the equilibrium to the right, producing more ethyl acetate.
76. a. The reaction would shift to the right
 b. The reaction would shift to the left
77. a. Increasing the temperature would tend to cause a shift to the left. Increasing the volume would tend to cause a shift to the right.
 b. Increasing temperature would tend to cause a shift to the right. Increasing volume would cause no shift because there are equal numbers of reactant and product particles. The equilibrium would shift to the right.
78. $s = 3.3 \times 10^{-8}$
79. The statement is not true. The value of K_{eq} reveals nothing about how fast or slowly a reaction occurs. A low value for K_{eq} merely means that the equilibrium system contains higher concentrations of reactants than products when the forward and reverse reactions are occurring at equal rates.
80. $[C_7H_6O] = 9.821$ mol/L
81. At the higher pressure (smaller volume), the equilibrium shifts in the direction that will decrease the pressure. It does so by shifting to the left, consuming more of the reddish-brown NO_2 and producing more of the colorless N_2O_4.
82. Adding potassium hydroxide increases the concentration of hydroxide ions in the solution. A constant K_{sp} requires that the concentration of aluminum ions must decrease. A shift to the left occurs demonstrating the common ion effect. $Al(OH)_3(s) \rightleftharpoons Al^{3+}(aq) + 3OH^-(aq)$; $K_{sp} = [Al^{3+}][OH^-]^3$
83. $[Cd^{2+}] = 1.8 \times 10^{-3}$ mol/L
 $[IO_3^-] = 3.6 \times 10^{-3}$ mol/L

THINK CRITICALLY

84. It is possible that the system is 50% reactants and 50% products. However, that is not necessarily the case. An equilibrium constant of 1.000 requires only that the numerical value of the ratio of product concentrations to reactant concentrations equals 1.000, with each concentration raised to the power of its coefficient in the balanced equation.
85. A continuous-flow process is better for two reasons. Both increasing reactant concentration and decreasing product concentration favor the forward reaction.
86. $BaF_2(s) \rightleftharpoons Ba^{2+}(aq) + 2F^-(aq)$
 $K_{sp} = [Ba^{2+}][F^-]^2 = 1.0 \times 10^{-6}$
 BaF_2 molar solubility $= 6.3 \times 10^{-3}$ mol/L
 $MgF_2(s) \rightleftharpoons Mg^{2+}(aq) + 2F^-(aq)$
 $K_{sp} = [Mg^{2+}][F^-]^2 = 3.7 \times 10^{-8}$
 Magnesium fluoride would precipitate first because its calculated K_{sp} of 3.7×10^{-8} is less than that of barium fluoride, 1.0×10^{-6}.
87. No, the decomposition of ammonium carbonate is endothermic, so the compound decomposes most readily at higher temperatures.
88. Sample answer: 1. Dissolve the mixture in distilled water. 2. Add an excess of a solution containing an anion such as carbonate, chromate, or sulfate that will precipitate essentially all the barium ions. 3. Filter, dry, and measure the mass of the precipitate. 4. Calculate the moles of barium compound formed, which equal the moles of barium chloride in the original mixture. 5. Calculate the mass of barium chloride in the original mixture. The remainder of the original mixture is sodium chloride.
89. Calcium phosphate has the greater solubility expressed in g/L.
 Solubility $Ca_3(PO_4)_2$ (g/L) $= 2.0 \times 10^{-4}$ g/L
 Solubility $FePO_4$ (g/L) $= 1.5 \times 10^{-9}$ g/L

ASSESSMENT

MIXED REVIEW

75. Ethyl acetate ($CH_3COOCH_2CH_3$) is produced in the equilibrium system described by the following equation.

$$CH_3COOH + CH_3CH_2OH \rightleftharpoons CH_3COOCH_2CH_3 + H_2O$$

Why does the removal of water result in the production of more ethyl acetate?

76. How would these equilibria be affected by decreasing the temperature?
a. $2O_3(g) \rightleftharpoons 3O_2(g) + \text{heat}$
b. $\text{heat} + H_2(g) + F_2(g) \rightleftharpoons 2HF(g)$

77. How would simultaneously increasing the temperature and volume of the system affect these equilibria?
a. $2O_3(g) \rightleftharpoons 3O_2(g) + \text{heat}$
b. $\text{heat} + N_2(g) + O_2(g) \rightleftharpoons 2NO(g)$

78. The solubility product constant for lead(II) arsenate ($Pb_3(AsO_4)_2$) is 4.0×10^{-36} at 298 K. Calculate the molar solubility of the compound at this temperature.

79. Evaluate this statement: A low value for K_{eq} means that both the forward and reverse reactions are occurring slowly.

80. Food Flavoring Benzaldehyde, known as artificial almond oil, is used in food flavorings. What is the molar concentration of benzaldehyde (C_7H_6O) at 298 K, when its density is 1.043 g/mL?

■ **Figure 24**

81. In the equilibrium system $N_2O_4(g) \rightleftharpoons 2NO_2(g)$, N_2O_4 is colorless and NO_2 is reddish-brown. Explain the different colors of the equilibrium system as shown in **Figure 24.**

82. Describe the process by which adding potassium hydroxide to a saturated aluminum hydroxide solution reduces the concentration of aluminum ions. Write the solubility equilibrium equation and solubility product constant expression for a saturated aqueous solution of aluminum hydroxide.

83. At 298 K, K_{sp} for cadmium iodate ($Cd(IO_3)_2$) equals 2.3×10^{-8}. What are the molar concentrations of cadmium ions and iodate ions in a saturated solution at 298 K?

THINK CRITICALLY

84. Analyze Suppose that an equilibrium system at a given temperature has a K_{eq} equal to 1.000. Evaluate the possibility that such a system is made up of 50% reactants and 50% products. Explain your answer.

85. Evaluate Imagine that you are a chemical engineer designing a production facility for a particular process. The process will utilize a reversible reaction that reaches a state of equilibrium. Analyze the merits of a continuous-flow process or a batch process for such a reaction and determine which is preferable. As a reaction proceeds in a continuous-flow process, reactants are continuously introduced into the reaction chamber and products are continuously removed from the chamber. In a batch process, the reaction chamber is charged with reactants, the reaction is allowed to occur, and the chamber is later emptied of all materials.

86. Interpret Data What compound would precipitate first if a 0.500*M* sodium fluoride solution were added gradually to a solution already containing 0.500*M* concentrations of both barium ions and magnesium ions? Use the data in **Table 6.** Write the solubility equilibrium equations and solubility product constant expressions for both compounds. Explain your answer.

Table 6 Data for Two Compounds

Compound	Molar Mass (g/mol)	Solubility at 25°C (g/L)
BaF_2	175.33	1.1
MgF_2	62.30	0.13

87. Apply Smelling salts, sometimes used to revive a person who is unconscious, are made of ammonium carbonate. The equation for the endothermic decomposition of ammonium carbonate is as follows.

$$(NH_4)_2CO_3(s) \rightleftharpoons 2NH_3(g) + CO_2(g) + H_2O(g)$$

Would you expect smelling salts to work as well on a cold winter day as on a warm summer day? Explain your answer.

88. Recognize Cause and Effect Suppose you have 12.56 g of a mixture made up of sodium chloride and barium chloride. Explain how you could use a precipitation reaction to determine how much of each compound the mixture contains.

89. Compare and Contrast Which of the two solids, calcium phosphate or iron(III) phosphate, has the greater molar solubility? $K_{sp}\ (Ca_3(PO_4)_2) = 1.2 \times 10^{-29}$; $K_{sp}\ (FePO_4) = 1.0 \times 10^{-22}$. Which compound has the greater solubility, expressed in grams per liter?

Chapter Self-Check

CHALLENGE PROBLEM

90. Synthesis of Phosgene Phosgene ($COCl_2$) is a toxic gas that is used in the manufacture of certain dyes, pharmaceuticals, and pesticides. Phosgene can be produced by the reaction between carbon monoxide and chlorine described by the equation $CO(g) + Cl_2(g) \rightleftharpoons COCl_2(g)$. Initially 1.0000 mol CO and 1.0000 mol Cl_2 are introduced into a 10.00-L reaction vessel. When equilibrium is established, both of their molar concentrations are found to be 0.0086 mol/L. What is the molar concentration of phosgene at equilibrium? What is K_{eq} for the system?

CUMULATIVE REVIEW

91. Explain the general trend in ionization energy as you go from left to right along Periods 1–5 of the periodic table.

92. How are the lengths of covalent bonds related to their strengths?

93. How are the chemical bonds in H_2, O_2, and N_2 different?

94. How can you tell if a chemical equation is balanced?

95. What mass of carbon must burn to produce 4.56 L CO_2 gas at STP?

$$C(s) + O_2(g) \rightarrow CO_2(g)$$

96. Describe a hydrogen bond. What conditions must exist for a hydrogen bond to form?

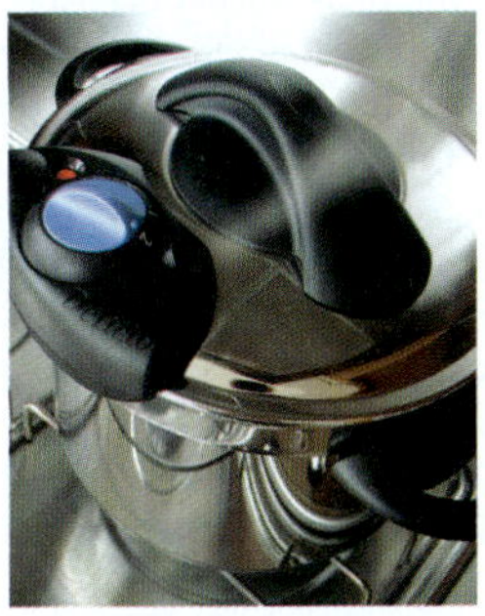

■ **Figure 25**

97. What gas law is exemplified in **Figure 25?** State the law.

98. When you reverse a thermochemical equation, why must you change the sign of ΔH?

99. What is the sign of the free energy change, $\Delta G°_{system}$, for a spontaneous reaction?

WRITING IN ▶ Chemistry

100. A New Compound Imagine that you are a scientist who has created a unique new liquid. You have named the liquid *yollane,* abbreviated *yo.* Yollane is nontoxic, inexpensive to make, and can dissolve huge volumes of gaseous carbon dioxide in the equilibrium $CO_2(g) \rightleftharpoons CO_2(yo)$, $K_{eq} = 3.4 \times 10^6$. Write a newspaper or magazine article that explains the merits of yollane in combating global warming.

101. Kidney Stones Research the role that solubility plays in the formation of kidney stones. Find out what compounds are found in kidney stones and their K_{sp} values. Summarize your findings in a health information flyer.

102. Hard Water The presence of magnesium and calcium ions in water makes the water "hard." Explain in terms of solubility why the presence of these ions is often undesirable. Find out what measures can be taken to eliminate them.

DBQ Document-Based Question

Reducing Pollution *Automobile exhausts contain the dangerous pollutants nitrogen monoxide (NO) and carbon monoxide (CO). An alloy catalyst offers a promising way to reduce the amounts of these gases in the atmosphere. When NO and CO are passed over this catalyst, the following equilibrium is established.*

$$2NO(g) + 2CO(g) \rightleftharpoons N_2(g) + 2CO_2(g)$$

The equilibrium constant is found to vary with temperature as shown in **Table 7.**

Data obtained from: Worz, et al. 2003. Cluster size-dependent mechanisms of the CO + NO reaction on small Pdn (n < or = 30) clusters on oxide surfaces. *J Am Chem Soc. 125(26): 7964–70.*

Table 7 K_{eq} v. Temperature

700 K	800 K	900 K	1000 K
9.10×10^{97}	1.04×10^{66}	4.66×10^{54}	3.27×10^{45}

103. Write the equilibrium constant expression for this equilibrium.

104. Examine the relationship between K_{eq} and temperature. Use Le Châtelier's principle to deduce whether the forward reaction is exothermic or endothermic.

105. Explain how automobile radiators plated with the alloy might help reduce the atmospheric concentrations of NO and CO.

CHALLENGE PROBLEMS

90. Initial molar concentrations of CO and Cl_2 are 1.0000 mol/10.00 L = 0.1000 mol/L. If equilibrium concentrations of CO and Cl_2 are 0.0086 mol/L, 0.1000 mol/L − 0.0086 mol/L, or 0.0914 mol/L of each, has reacted, yielding 0.0914 mol/L of $COCl_2$. $K_{eq} = 1.2 \times 10^3$

CUMULATIVE REVIEW

91. Ionizations energies tend to increase across periods 1–5.

92. A single bond is the longest and a triple bond is the shortest. Double bonds are intermediate between single and triple. A triple bond is stronger than a double bond and a double bond is stronger than a single bond.

93. In H_2, one pair of electrons is shared; in O_2, two pairs; in N_2, three pairs. All are covalent bonds.

94. The numbers and kinds of atoms are the same on both sides of the equations, and the coefficients are in the smallest possible integer ratio.

95. 2.45 g

96. A hydrogen bond is a strong dipole-dipole attraction between a hydrogen atom and a small, highly electronegative atom such as oxygen that has at least one lone pair of electrons.

97. Gay-Lussac's Law: $P_1/T_1 = P_2/T_2$

98. Reversing an exothermic thermochemical equation changes it to an endothermic equation, so instead of energy being released, energy is absorbed. Thus the sign of ΔH must change.

99. The sign of $\Delta G°_{system}$, for a spontaneous reaction is negative.

WRITING IN ▶ Chemistry

Rubric

100. Students have much positive information to use. They should recognize that the equilibrium constant for the reaction in which carbon dioxide is absorbed is a large number so the reaction should be effective in removing excess carbon dioxide from the atmosphere. Also, yollane itself is not likely to harm the environment because it is nontoxic.

101. Student reports will vary. Refer to the Solution Manual for additional information.

102. Student reports will vary. Refer to the Solution Manual for additional information.

DBQ Document-Based Questions

Data obtained from: Worz, et al, 2003. Cluster size-dependent mechanisms of the CO + NO reaction on small Pdn (N < or = 30) clusters on oxide surfaces. J. Am. 2003 *Chem Soc*; 125(26): 7964–70

103. $K_{eq} = [N_2][CO_2]^2/[NO]^2[CO]^2$

104. Because K_{eq} decreases with increasing temperature, the forward reaction is exothermic.

105. Automobile radiators operate at elevated temperatures, and automobiles effectively "sweep" large amounts of air as they travel from place to place. Therefore, automobile radiators plated with the alloy could convert significant volumes of the pollutants, NO and CO, to the less harmful substances N_2 and CO_2.

MULTIPLE CHOICE

1. D
2. B
3. B
4. B
5. A
6. A
7. C
8. C
9. A

MULTIPLE CHOICE

1. Which describes a system that has reached chemical equilibrium?
A. No new product is formed by the forward reaction.
B. The reverse reaction no longer occurs in the system.
C. The concentration of reactants in the system is equal to the concentration of products.
D. The rate at which the forward reaction occurs equals the rate of the reverse reaction.

2. The reaction between persulfate ($S_2O_8^{2-}$) and iodide (I^-) ions is often studied in student laboratories because it occurs slowly enough for its rate to be measured:

$$S_2O_8^{2-}(aq) + 2I^-(aq) \rightarrow 2SO_4^{2-}(aq) + I_2(aq)$$

This reaction has been experimentally determined to be first order in $S_2O_8^{2-}$ and first order in I^-. Therefore, what is the overall rate law for this reaction?
A. rate = $k[S_2O_8^{2-}]^2[I^-]$
B. rate = $k[S_2O_8^{2-}][I^-]$
C. rate = $k[S_2O_8^{2-}][I^-]^2$
D. rate = $k[S_2O_8^{2-}]^2[I^-]^2$

Use the diagrams below to answer Question 3.

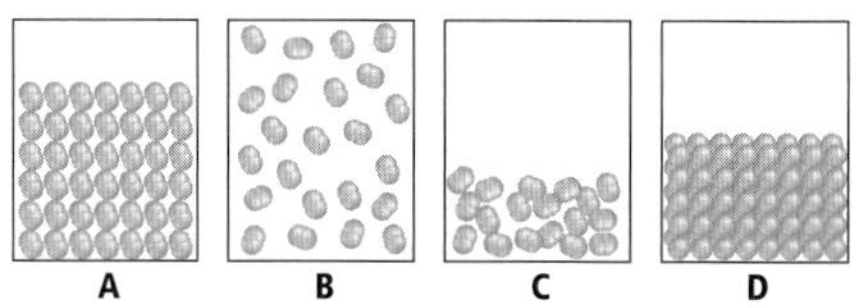

3. Which diagram shows the substance that has the weakest intermolecular forces?
A. A **C.** C
B. B **D.** D

4. Which type of intermolecular force results from a temporary imbalance in the electron density around the nucleus of an atom?
A. ionic bonds
B. London dispersion forces
C. dipole-dipole forces
D. hydrogen bonds

Use the table below to answer Questions 5 to 7.

Concentration Data for the Equilibrium System $MnCO_3(s) \rightarrow Mn^{2+}(aq) + CO_3^{2-}(aq)$ at 298 K

Trial	$[Mn^{2+}]_0$ (*M*)	$[CO_3^{2-}]_0$ (*M*)	$[Mn^{2+}]_{eq}$ (*M*)	$[CO_3^{2-}]_{eq}$ (*M*)
1	0.0000	0.00400	5.60×10^{-9}	4.00×10^{-3}
2	0.0100	0.0000	1.00×10^{-2}	2.24×10^{-9}
3	0.0000	0.0200	1.12×10^{-9}	2.00×10^{-2}

5. What is the K_{sp} for $MnCO_3$ at 298 K?
A. 2.24×10^{-11} **C.** 1.12×10^{-9}
B. 4.00×10^{-11} **D.** 5.60×10^{-9}

6. What is the molar solubility of $MnCO_3$ at 298 K?
A. $4.73 \times 10^{-6}M$ **C.** $7.48 \times 10^{-5}M$
B. $6.32 \times 10^{-2}M$ **D.** $3.35 \times 10^{-5}M$

7. A 50.0-mL volume of $3.00 \times 10^{-6}M$ K_2CO_3 is mixed with 50.0 mL of $MnCl_2$. A precipitate of $MnCO_3$ will form only when the concentration of the $MnCl_2$ solution is greater than which of the following?
A. $7.47 \times 10^{-6}M$ **C.** $2.99 \times 10^{-5}M$
B. $1.49 \times 10^{-5}M$ **D.** $1.02 \times 10^{-5}M$

8. The kinetic-molecular theory describes the microscopic behavior of gases. One main point of the theory is that within a sample of gas, the frequency of collisions between individual gas particles and between the particles and the walls of their container increases if the sample is compressed. Which gas law states this relationship in mathematical terms?
A. Gay-Lussac's law
B. Charles's law
C. Boyle's law
D. Avogadro's law

9. $AB(s) + C_2(l) \rightarrow AC(g) + BC(g)$
Which cannot be predicted about this reaction?
A. The entropy of the system decreases.
B. The entropy of the products is higher than that of the reactants.
C. The change in entropy for this reaction, ΔS_{rxn}, is positive.
D. The disorder of the system increases.

SHORT ANSWER

Use the equation below to answer Questions 10 to 12.

$$PCl_5 + H_2O \rightarrow HCl + H_3PO_4$$

10. Balance this equation, using the smallest whole-number coefficients.

11. Identify the mole ratio of water to phosphoric acid.

12. Use your balanced chemical equation to show the setup for determining the amount of hydrogen chloride produced when 25.0 g of phosphorus pentachloride is completely consumed.

EXTENDED RESPONSE

Use the graph below to answer Questions 13 to 15.

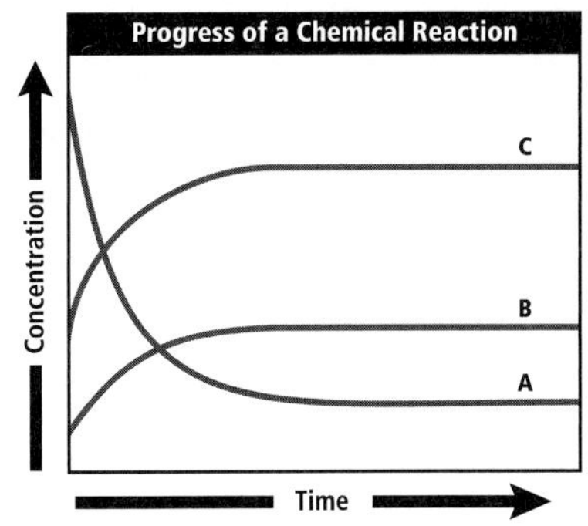

13. Describe the shape of the graph when equilibrium has been established.

14. Explain why the concentration of reactants is not zero at the end of this reaction.

15. Classify the type of chemical reaction that is represented in this graph. How do the data support your conclusion?

SAT SUBJECT TEST: CHEMISTRY

16. The formation of perchloryl fluoride (ClO_3F) has an equilibrium constant of 3.42×10^{-9} at 298 K.

$$Cl_2(g) + 3O_2(g) + F_2(g) \rightarrow 2ClO_3F(g)$$

At equilibrium, $[Cl_2] = 0.563M$, $[O_2] = 1.01M$, and $[ClO_3F] = 1.47 \times 10^{-5}M$. What is $[F_2]$?

A. $9.18 \times 10^{-2}M$
B. $3.73 \times 10^{-10}M$
C. $1.09 \times 10^{-1}M$
D. $6.32 \times 10^{-2}M$
E. $6.32 \times 10^{-7}M$

Use the graph below to answer Questions 17 and 18.

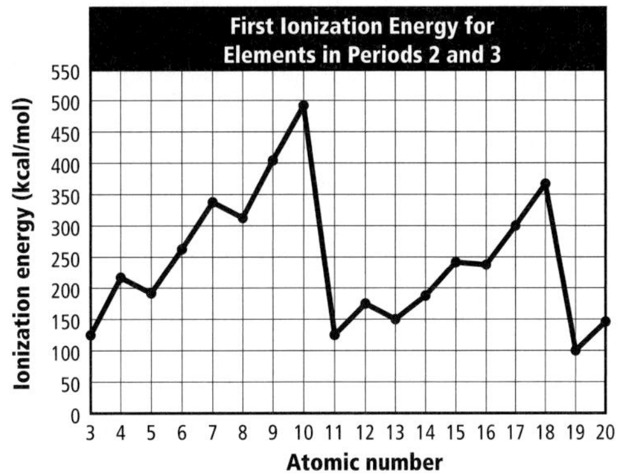

17. Which family of elements tends to have the lowest ionization energy in its period?

A. representative elements
B. transition elements
C. alkali elements
D. alkaline earth elements
E. halogens

18. Using the graph, what is the approximate ionization energy of the element with atomic number 7?

A. 300 kcal/mol
B. 310 kcal/mol
C. 325 kcal/mol
D. 340 kcal/mol
E. 390 kcal/mol

NEED EXTRA HELP?																		
If You Missed Question . . .	1	2	3	4	5	6	7	8	9	10	11	12	13	14	15	16	17	18
Review Section . . .	17.1	16.3	12.2	12.2	17.3	17.3	17.3	13.1	15.5	9.1	11.1	11.2	17.4	17.1	9.2	17.3	6.3	6.3

SHORT ANSWER

10. $PCl_5 + 4H_2O \longrightarrow 5HCl + H_3PO_4$
11. 4:1 H_2O to H_3PO_4
12. 21.9 g HCl

EXTENDED RESPONSE

13. Equilibrium has been established when the lines level off. The concentration of reactants and products remains constant from this point onward.
14. Reactants are not at a concentration of zero at the end of this reaction because the reaction is at equilibrium. The reverse reaction to make more reactant is proceeding at a rate equal to the rate of the forward reaction, which is using up reactant.
15. This reaction is most likely a decomposition reaction. There is one reactant, shown by Line A, which decreases in concentration as it gets used up. There are two products, shown by Lines B and C, which increase in concentration as they are formed from the breakdown of A.

SAT SUBJECT TEST: CHEMISTRY

16. C
17. C
18. D

CHAPTER 18 Acids and Bases

BIGIDEA **Acids and bases can be defined in terms of hydrogen ions and hydroxide ions or in terms of electron pairs.**

ESSENTIAL QUESTIONS	RESOURCES TO ASSESS MASTERY
SECTION 1 **Introduction to Acids and Bases** **1.** What are the physical and chemical properties of acids and bases? **2.** How are solutions classified as acidic, basic, or neutral? **3.** How do the Arrhenius, Brønsted-Lowry, and Lewis models of acids and bases compare? 2 sessions 1 block	**Progress Monitoring** Formative Assessment, pp. 639, 641, 643 Reading Check, p. 638 Section Review, p. 643
SECTION 2 **Strengths of Acids and Bases** **1.** How is the strength of an acid or base related to its degree of ionization? **2.** How does the strength of a weak acid compare with the strength of its conjugate base? **3.** What is the relationship between the strengths of acids and bases and the values of their ionization constants? 1 session 0.5 block	**Progress Monitoring** Formative Assessment, pp. 647, 648, 649 Reading Check, p. 646 Section Review, p. 649
SECTION 3 **Hydrogen Ions and pH** **1.** What are pH and pOH? **2.** How are pH and pOH related to the ion product constant for water? **3.** How are the pH and pOH of aqueous solutions calculated? 1 session 0.5 block	**Progress Monitoring** Formative Assessment, pp. 654, 658 Reading Check, pp. 651, 656 Section Review, p. 658
SECTION 4 **Neutralization** **1.** What do chemical equations of neutralization reactions look like? **2.** How are neutralization reactions used in acid-base titrations? **3.** How do the properties of buffered and unbuffered solutions compare? 2 sessions 1 block	**Progress Monitoring** Formative Assessment, pp. 660, 662, 667 Reading Check, p. 660 Graph Check, p. 661 Section Review, p. 668 **Summative Assessment** Chapter Assessment, p. 672 *eAssessment* Chapter Test (Scaffolded)

Go online!

	LEVELED RESOURCES	LAB MATERIALS
	Science Notebook 18.1 OL ***FAST FILE Chapter Resources:*** Study Guide, p. 44 BL	LaunchLAB p. 632 / **15 min** liquid household products, red and blue litmus paper, microplate, phenolphthalein
	Science Notebook 18.2 OL ***FAST FILE Chapter Resources:*** MiniLab Worksheet, p. 28 OL Study Guide, p. 45 BL	MiniLAB p. 648 / **30 min** 10-mL graduated cylinder, dropping pipette, 24-well microplate, conductivity tester, 6.0*M*, 1.0*M*, and 0.1*M* acetic acid solutions, glacial acetic acid
	Science Notebook 18.3 OL ***FAST FILE Chapter Resources:*** Study Guide, p. 46 BL	
	Science Notebook 18.4 OL ***FAST FILE Chapter Resources:*** ChemLab Worksheet, p. 29 OL Study Guide, p. 48 BL	ChemLAB p. 670 / **120 min** 50-mL buret, spatula, buret clamp, 250-mL Erlenmeyer flask, ring stand, 500-mL Florence flask with rubber stopper, centigram balance, wash bottle, weighing bottle, distilled water, sodium hydroxide pellets (NaOH), 250-mL beaker, potassium hydrogen phthalate ($KHC_8H_4O_4$), phenolphthalein solution

ADDITIONAL RESOURCES

eAssessment
ConnectED Student Center
ConnectED Teacher Center
Lesson Presentations
eSolutions
Videos
Animations
Online Quizzes
Virtual Investigations
Multilingual eGlossary
Personal Tutors
WebQuests
Applying Practices Worksheets
PBLs

Practice:

Challenge Problems AL
Supplemental Problems BL OL
Chapter Test (Scaffolded)
Standardized Test Practice

Additional Resources:

Blueprints for Success
Solving Problems: A Chemistry Handbook

Lab Resources:

Laboratory Manual OL
CBL Laboratory Manual OL
Small-Scale Laboratory Manual OL
Forensics Laboratory Manual OL

BL Below Level   OL On Level   AL Advanced Learners   EL English Learners   COOP LEARN Cooperative Learning

CHAPTER 18

LaunchLAB

Rubric

What is in your cupboards?

Purpose Students will use two indicators to distinguish between household products that are acidic and basic.

Safety Precautions Approve lab safety forms before work begins. **WARNING:** ***Keep phenolphthalein indicator solution away from flames and sparks.***

Disposal The products can be washed down a drain with plenty of water.

Teaching Strategies

- Distribute six to eight products to students in small labeled beakers with droppers.
- Suggested acids: white vinegar, white apple juice, colorless soft drink; suggested bases: household ammonia, colorless liquid soap, a solution of baking soda

Expected Results Acids turn blue litmus red; bases turn red litmus blue. Phenolphthalein turns pink in bases.

CHAPTER 18

Acids and Bases

BIGIDEA **Acids and bases can be defined in terms of hydrogen ions and hydroxide ions or in terms of electron pairs.**

SECTIONS

1 Introduction to Acids and Bases

2 Strengths of Acids and Bases

3 Hydrogen Ions and pH

4 Neutralization

LaunchLAB

What is in your cupboards?

You can learn something about the properties of products in your household by testing them with strips of paper called litmus paper. Can you separate household products into two groups? In this lab, you will determine whether a substance is an acid or a base.

Procedure

1. Read and complete the lab safety form.
2. Place three or four drops of several **household products** into separate wells of a **microplate.** Draw a chart to show the position of each liquid.
3. Test each product with **red and blue litmus paper.** Place two drops of **phenolphthalein** in each sample. Record your observations. **WARNING:** ***Phenolphthalein is flammable. Keep away from flames.***

Analysis

1. **Classify** the products into two groups based on your observations. Answers will vary based on what was tested. The two groups are acids and bases.
2. **Describe** how the groups differ. What can you conclude? They react with litmus and phenolphthalein based on the grouping.

Inquiry Choose one sample that reacted with the phenolphthalein. Can you reverse the reaction? Design an experiment to test your hypothesis. Answers will vary but should include the addition of an acid to neutralize the indicator.

FOLDABLES® Study Organizer

Acids and Bases

Make a two-tab book. Label it as shown. Use it to compare the main models of acids and bases.

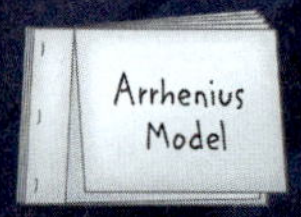

The optimal pH for an aquarium depends on the types of organisms in it. A pH of 8.2 is generally accepted as the average pH of seawater, the habitat for the fish in the photo. On the other hand, South American cichlids, which are freshwater fish, require a pH range of 6.4 to 7.0. African cichlids, also freshwater fish, thrive in water with a pH between 8.0 and 9.2.

Introduce the BIGIDEA

Acids and Bases Produce Ions To introduce this chapter's Big Idea, show students samples of solid oxalic acid ($H_2C_2O_4$) and solid sodium hydroxide (NaOH). Ask if either substance appears to be acidic or basic. No. Both are pure solids, and one cannot determine acidic or basic natures by appearance. Dissolve approximately 1 g of each solid in 100 mL of distilled water. Ask students how to test the solutions to see if they are acidic or basic. Students will probably suggest using a pH meter or litmus paper. Test the two solutions, showing that the aqueous solution of $H_2C_2O_4$ is acidic and that the aqueous solution of NaOH is basic. Then, write the chemical equations that show how $H_2C_2O_4$ produces hydrogen ions in water solution and NaOH produces hydroxide ions in water solution. $H_2C_2O_4(s) \longrightarrow H^+(aq) + HC_2O_4^-(aq)$; $NaOH(s) \longrightarrow Na^+(aq) + OH^-(aq)$ Summarize by explaining that acids produce hydrogen ions in water and bases produce hydroxide ions in water.

Tie to Previous Knowledge

Have students review the following concepts before studying this chapter.

- chemical reactions, equations for chemical reactions
- reaction rates, reaction rate laws
- chemical equilibrium

Use the Photo

Aquarium pH Ask what pH tells about the water solution in the aquarium. Refer students to the reading on the pH meter in the photo, and ask what a pH of 8.2 means. somewhat basic Ask how a solution with a pH of 8.0 compares with a solution having a pH of 8.2. The solution with a pH of 8.0 is less basic (more acidic) than the solution with a pH of 8.2. Point out that hydroxide (OH^-) ions cause both solutions to be basic and that hydroxide-ion concentration varies widely with changing pH.

SECTION 1

1 Focus

MAINIDEA

Two Acid Models Write the chemical equation for the first ionization of oxalic acid on the board. Include water as a reactant. $H_2C_2O_4(s) + H_2O(l) \rightarrow H_3O^+(aq) + HC_2O_4^-(aq)$ Point out that H_3O^+ represents a hydrated hydrogen ion, or hydronium ion. Now, rewrite the equation showing the Lewis electron-dot structures for all reactants and products. (Note: Oxalic acid is ethanedioic acid.) Ask what happens between the $H_2C_2O_4$ and H_2O molecules during the course of the reaction. A hydrogen ion, or proton, is transferred from the $H_2C_2O_4$ molecule to the H_2O molecule, where the hydrogen ion bonds with an available electron pair. Explain that one commonly-used acid-base model focuses on the $H_2C_2O_4$ molecule losing, or donating, a hydrogen ion, but that a second model focuses on the hydrogen ion from the $H_2C_2O_4$ molecule accepting an electron pair from the water molecule. Point out that both models are valid and offer different insights into how the reaction occurs. OL

2 Teach

Quick Demo

Bases in the Home Ask students how soap feels when they wash their hands. slippery Then, show them that when red litmus paper touches a wet bar of soap, the litmus paper turns blue. BL

SECTION 1

Essential Questions

- What are the physical and chemical properties of acids and bases?
- How are solutions classified as acidic, basic, or neutral?
- How do the Arrhenius, Brønsted-Lowry, and Lewis models of acids and bases compare?

Review Vocabulary

Lewis structure: a model that uses electron-dot structures to show how electrons are arranged in molecules

New Vocabulary

acidic solution
basic solution
Arrhenius model
Brønsted-Lowry model
conjugate acid
conjugate base
conjugate acid-base pair
amphoteric
Lewis model

Introduction to Acids and Bases

MAINIDEA Different models help describe the behavior of acids and bases.

CHEM 4 YOU You might not realize it, but acids and bases are two of the most common classifications of substances. You can recognize them by the tart taste of some of your favorite beverages and by the pungent odor of ammonia in some household cleaners.

Properties of Acids and Bases

When ants sense danger to their ant colony, they emit a substance called formic acid that alerts the entire colony. Acids dissolved in rainwater hollow out enormous limestone caverns and destroy valuable buildings and statues over time. Acids flavor many of the beverages and foods you like, and an acid in your stomach helps digest what you eat. Bases also play a role in your life. The soap you use and the antacid tablet you might take for an upset stomach are bases. Many household products, such as those you used in the Launch Lab, are acids and bases.

Physical properties You are probably already familiar with some of the physical properties of acids and bases. For example, you might know that acidic solutions taste sour. Carbonic and phosphoric acids give many carbonated beverages their sharp taste; citric and ascorbic acids give lemons and grapefruit their tartness; and acetic acid makes vinegar taste sour.

You might also know that basic solutions taste bitter and feel slippery. Think about how a bar of soap becomes slippery when it gets wet. You should never attempt to identify an acid or a base, or any other substance in the laboratory, by its taste or feel. **Figure 1** shows two plants growing in different soils. One grows best in acidic soil, sometimes called "sour" soil. The other thrives in basic, or alkaline, soil.

Figure 1 Rhododendrons flourish in rich, moist soil that is moderately acidic, whereas sempervivum, commonly called hen and chicks, grows best in drier, slightly basic soil.

DIFFERENTIATED INSTRUCTION

Struggling Learners Have students use drawings or models to demonstrate the structural difference between a hydroxide ion and a hydronium ion. Have students explain their models or drawings. BL EL

Acids turn blue litmus red.

Bases turn red litmus blue.

Figure 2 The strong acid hydrochloric acid (HCl), also called muriatic acid, is used to clean bricks and concrete. The strong base sodium hydroxide (NaOH) can clear clogged drains.

Electrical conductivity Another physical property of acid and base solutions is the ability to conduct electricity. Pure water is a nonconductor of electricity, but the addition of an acid or base produces ions that cause the resulting solution to become a conductor.

Chemical properties You might have already identified acids and bases by their reaction with litmus paper. Acids can also be identified by their reactions with some metals and metal carbonates.

Reactions with litmus Litmus is one of the dyes commonly used to distinguish solutions of acids and bases, as shown in **Figure 2.** Aqueous solutions of acids cause blue litmus paper to turn red. Aqueous solutions of bases cause red litmus paper to turn blue.

Reactions with metals and metal carbonates Magnesium and zinc react with aqueous solutions of acids to produce hydrogen gas. The reaction between zinc and hydrochloric acid is described by the following equation.

$$Zn(s) + 2HCl(aq) \rightarrow ZnCl_2(aq) + H_2(g)$$

Metal carbonates and hydrogen carbonates also react with aqueous solutions of acids to produce carbon dioxide (CO_2) gas. When vinegar is added to baking soda, a foaming reaction occurs between acetic acid ($HC_2H_3O_2$) dissolved in the vinegar, and sodium hydrogen carbonate ($NaHCO_3$). The production of CO_2 gas accounts for the bubbling.

$$NaHCO_3(s) + HC_2H_3O_2(aq) \rightarrow NaC_2H_3O_2(aq) + H_2O(l) + CO_2(g)$$

Geologists identify rocks as limestone (primarily $CaCO_3$) by using a hydrochloric acid solution. If a few drops of the acid produce bubbles of carbon dioxide, the rock contains limestone.

PRACTICE Problems

Do additional problems. Online Practice

1. Write balanced equations for the reactions between the following.
   a. aluminum and sulfuric acid
   b. calcium carbonate and hydrobromic acid
2. **Challenge** Write the net ionic equation for the reaction in Question 1b.

Quick Demo

Basic Indicator

WARNING: *Wear safety goggles and an apron. Perform this demo in the fume hood or behind a safety shield. No open flames in the room.*

Disposal Neutralize the NaOH solution with 1*M* HCl and flush down a drain. Drop a piece of sodium, about half the size of a pea, into an ovenproof glass beaker containing distilled water and a few drops of phenolphthalein indicator. Cover the beaker with a piece of wire gauze to prevent splattering. Explain that phenolphthalein turns pink in a basic solution and that hydrogen is produced by the reaction. Ask students why group 1 elements, such as sodium, are called alkali metals. They produce an alkaline solution in water. Have them write the formula equation. $2Na(s) + 2H_2O(l) \rightarrow 2NaOH(aq) + H_2(g)$ **OL**

PRACTICE Problems

Have students refer to the Selected Solutions appendix for complete solutions to odd-numbered problems. The complete solutions for all problems can be found in the Solutions Manual.

1. a. $2Al(s) + 3H_2SO_4(aq) \rightarrow Al_2(SO_4)_3(aq) + 3H_2(g)$
   b. $CaCO_3(s) + 2HBr(aq) \rightarrow CaBr_2(aq) + H_2O(l) + CO_2(g)$
2. $CaCO_3(s) + 2H^+(aq) + 2Br^-(aq) \rightarrow Ca^{2+}(aq) + 2Br^-(aq) + H_2O(l) + CO_2(g)$
   $CaCO_3(s) + 2H^+(aq) \rightarrow Ca^{2+}(aq) + H_2O(l) + CO_2(g)$

CHEMISTRY JOURNAL

Acids in the Home Have students check labels on foods and household products, such as cleaners and shampoos, and list in their journals any acids present in the products. **OL**

■ **Caption Question Fig. 3** at the vertical line drawn through the center of the diagram

Content Background

Acids and Bases in Industry

Acids and bases are used in many industrial processes including refining oil and sugar, electroplating, circuit board etching, wastewater neutralization, sewage treatment, and fermentation of alcohol. Products produced using acids and bases include fertilizers, explosives, plastics, inks, soap, paper, film, drugs, synthetic fabrics, dyes, latex paints, solvents, and pesticides.

■ **Figure 3** Note how $[H^+]$ and $[OH^-]$ change simultaneously. As $[H^+]$ decreases to the right, $[OH^-]$ increases to the right.
Identify *the point in the diagram at which the two ion concentrations are equal.*

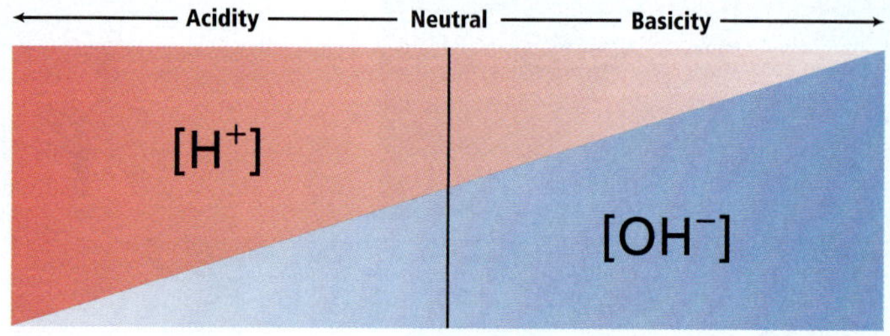

Hydronium and hydroxide ions All water solutions contain hydrogen ions (H^+) and hydroxide ions (OH^-). The relative amounts of the two ions determine whether an aqueous solution is acidic, basic, or neutral. Neutral solutions are neither acidic nor basic.

An **acidic solution** contains more hydrogen ions than hydroxide ions. A **basic solution** contains more hydroxide ions than hydrogen ions. A neutral solution contains equal concentrations of hydrogen ions and hydroxide ions. **Figure 3** illustrates these relationships. **Figure 4** describes how scientists developed an understanding of acids and bases.

Pure water produces equal numbers of H^+ ions and OH^- ions in a process called self-ionization, in which water molecules react to form a hydronium ion (H_3O^+) and a hydroxide ion.

$$\underbrace{H_2O(l) + H_2O(l)}_{\text{Water molecules}} \rightleftharpoons \underbrace{H_3O^+(aq)}_{\text{Hydronium ion}} + \underbrace{OH^-(aq)}_{\text{Hydroxide ion}}$$

The hydronium ion is a hydrogen ion that has a water molecule attached to it by a covalent bond. The symbols H^+ and H_3O^+ can be used interchangeably, as this simplified self-ionization equation shows.

$$H_2O(l) \rightleftharpoons H^+(aq) + OH^-(aq)$$

■ **Figure 4**

History of Acids and Bases

Current understanding of the structure and behavior of acids and bases is based on the contributions of chemists, biologists, environmental scientists, and inventors over the past 150 years.

1865 The introduction of an antiseptic spray containing carbolic acid marks the beginning of modern antiseptic surgery.

1869 Nucleic acids are discovered in cell nuclei. DNA and RNA are examples of nucleic acids.

1870

1883 Svante Arrhenius proposes that acids produce hydrogen ions (H^+) and bases produce hydroxide ions (OH^-) when dissolved in water.

1890

1909 The development of the pH scale allows scientists to define the acidity of a substance.

1910

1923 Scientists expand and refine the definition of acids and bases, producing the definitions currently in use.

CHEMISTRY PROJECT

Acids and Bases Have students use a state map to list several cities and towns in different parts of the state. Have students work in groups to formulate a plan for comparing the acidity of precipitation in these locations. The plan should address how and when measurements will be made, how data will be analyzed, and how results will be shared. What effects could acid precipitation have on a community? Do you think most people are concerned about acid precipitation in your community? Why or why not? Acid precipitation can make water bodies more acidic, damage trees and soil, and make buildings and outdoor sculptures decay more rapidly. **OL** **AL** **COOP LEARN**

The Arrhenius Model

If pure water itself is neutral, how does an aqueous solution become acidic or basic? The first person to answer this question was Swedish chemist Svante Arrhenius, who in 1883 proposed what is now called the Arrhenius model of acids and bases. The **Arrhenius model** states that an acid is a substance that contains hydrogen and ionizes to produce hydrogen ions in aqueous solution. A base is a substance that contains a hydroxide group and dissociates to produce a hydroxide ion in aqueous solution.

Arrhenius acids and bases As an example of the Arrhenius model of acids and bases, consider what happens when hydrogen chloride gas dissolves in water. HCl molecules ionize to form H^+ ions, which make the solution acidic.

$$HCl(g) \rightarrow H^+(aq) + Cl^-(aq)$$

When the ionic compound sodium hydroxide (NaOH) dissolves in water, it dissociates to produce OH^- ions, which make the solution basic.

$$NaOH(s) \rightarrow Na^+(aq) + OH^-(aq)$$

Although the Arrhenius model is useful in explaining many acidic and basic solutions, it has some shortcomings. For example, ammonia (NH_3) and sodium carbonate (Na_2CO_3) do not contain a hydroxide group, yet both substances produce hydroxide ions in solution and are well-known bases. Sodium carbonate is the compound that causes the alkalinity of Lake Natron, Tanzania, which is shown in **Figure 5.** Clearly, a model that includes all bases is needed.

Figure 5 Lake Natron in Africa's Great Rift Valley is a naturally basic body of water. Water, laden with dissolved sodium carbonate from surrounding volcanic rocks, drains into the lake but finds no outlet. Evaporation concentrates the mineral leaving a white crust on the surface and strongly alkaline water.

FOLDABLES® Incorporate information from this section into your Foldable.

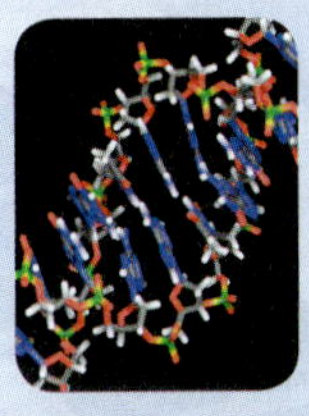

1933–34 Scientists develop portable pH meters.

1953 James Watson, Francis Crick, and Rosalind Franklin study the nucleic acid DNA, laying the framework for today's biotechnology industry.

1963 Scientists discover acid rain in North America. pH measurements show polluted rain to be 100 times more acidic than unpolluted rain.

1980s Silicon-chip pH meters have no glass component. They are now widely used in the food, cosmetic, and pharmaceutical industries.

2005 Scientists develop superacids, which are more acidic than 100% sulfuric acid. Applications include producing strong plastics and high-octane gasoline.

2010 A team of scientists at the National Institute of Standards and Technology (NIST) develops a technique to determine how a sudden change in acidity of a nanoparticle solution affects its stability. Nanoparticle production could be tailored to target tumor cells with a different acidity than that of normal cells.

1950 1970 1990 2010

FOLDABLES® Rubric

Reinforcement

Naming Acids Ask students to distinguish between HCl(g) and HCl(aq). HCl(g) is gaseous hydrogen chloride; HCl(aq) is hydrogen chloride gas dissolved in water and is called hydrochloric acid. Because the names of acids will be used often in this chapter, review with students what they learned earlier about naming binary acids and oxyacids. OL

Extension

Acid Solution Students might assume that any metallic hydroxide will produce a basic aqueous solution. Explain that dissolving highly charged metal-ion hydroxides, such as $Al(OH)_3$, in water results in acidic solutions because the positively charged metal ions attract electrons from the O–H bond, which weakens it and enables the hydrogen atom to ionize.

Content Background

Industrial Uses Acids and bases are used in many industrial processes including refining oil and sugar, electroplating, circuit board etching, wastewater neutralization, sewage treatment, and fermentation of alcohol. Products produced using acids and bases include fertilizers, explosives, plastics, inks, soap, paper, film, drugs, synthetic fabrics, dyes, latex paints, solvents, and pesticides.

The Brønsted-Lowry Model

Danish chemist Johannes Brønsted (1879–1947) and English chemist Thomas Lowry (1843–1909) proposed a more inclusive model of acids and bases—a model that focuses on the hydrogen ion (H^+). In the **Brønsted-Lowry model** of acids and bases, an acid is a hydrogen-ion donor. A base is a hydrogen-ion acceptor.

Hydrogen ion donors and acceptors The symbols X and Y represent nonmetallic elements or negative polyatomic ions. Thus, the general formula for an acid can be represented as HX or HY. When a molecule of acid HX dissolves in water, it donates a H^+ ion to a water molecule. The water molecule acts as a base and accepts the H^+ ion.

$$HX(aq) + H_2O(l) \rightleftharpoons H_3O^+(aq) + X^-(aq)$$

Upon accepting the H^+ ion, the water molecule becomes an acid, H_3O^+. The hydronium ion (H_3O^+) is an acid because it has an extra H^+ ion that it can donate. Upon donating its H^+ ion, the acid HX becomes a base, X^-. X^- is a base because it has a negative charge and can readily accept a positive hydrogen ion. Thus, an acid-base reaction in the reverse direction can occur. The acid H_3O^+ can react with the base X^- to form water and HX, establishing the following equilibrium.

$$\underset{\text{Acid}}{HX(aq)} + \underset{\text{Base}}{H_2O(l)} \rightleftharpoons \underset{\text{Conjugate acid}}{H_3O^+(aq)} + \underset{\text{Conjugate base}}{X^-(aq)}$$

Conjugate acids and bases The forward reaction is the reaction of an acid and a base. The reverse reaction is also the reaction of an acid and a base. The acid and base that react in the reverse reaction are identified under the equation as a conjugate acid and a conjugate base. A **conjugate acid** is the species produced when a base accepts a hydrogen ion. The base H_2O accepts a hydrogen ion from the acid HX and becomes the conjugate acid H_3O^+. A **conjugate base** is the species that results when an acid donates a hydrogen ion. The acid HX donates its hydrogen ion and becomes the conjugate base X^-. In the reaction shown above, the hydronium ion (H_3O^+) is the conjugate acid of the base H_2O. The X^- ion is the conjugate base of the acid HX. Brønsted-Lowry interactions involve conjugate acid-base pairs. A **conjugate acid-base pair** consists of two substances related to each other by the donating and accepting of a single hydrogen ion.

An analogy for conjugate acid-base pairs is shown in **Figure 6.** When the father has the ball in his hand, he is an acid. He throws the ball (a hydrogen ion) to his son. Now his son is the acid because he has the ball (a hydrogen ion) to give away. The father is now a base because he is available to accept the ball (a hydrogen ion). The father is the acid and the son is the base in the forward reaction. In the reverse reaction, the son has the ball and is the conjugate acid while the father is the conjugate base.

■ **Figure 6** When a father throws the ball to his son, the father is like a Brønsted-Lowry acid and the son is like a base. After the son catches the ball (hydrogen ion), he becomes like a conjugate acid.

Spencer Grant/PhotoEdit

☑ READING CHECK **Explain** how the ion HCO_3^- can be both an acid and a base

Quick Demo

Litmus Reaction

WARNING: ***Perform this demo in a well-ventilated area or under a fume hood.*** Wet a piece of red litmus paper with distilled water and hold the litmus paper with forceps over the mouth of a bottle of concentrated ammonia solution. The litmus color changes from red to blue. Ask students what this indicates regarding the interaction of ammonia with the water in the paper. A basic solution is produced. OL

☑ READING CHECK HCO_3^- has a hydrogen ion that it can donate as an acid. It also has a negative charge that will allow it to act as a base and accept a hydrogen ion.

Concept Development

Hydrogen Ion Ask students why a hydrogen ion may be called a proton. A hydrogen atom consists of a nucleus, which usually contains just one proton, and a single electron. If the electron is removed, one proton remains. OL

DIFFERENTIATED INSTRUCTION

English Learners Have English learners use a dictionary to find and then explain the meanings of several key English prefixes used in this section: *mono-*, *di-*, *tri-*, *poly-*, *conjugate*, and *polyprotic*. EL OL

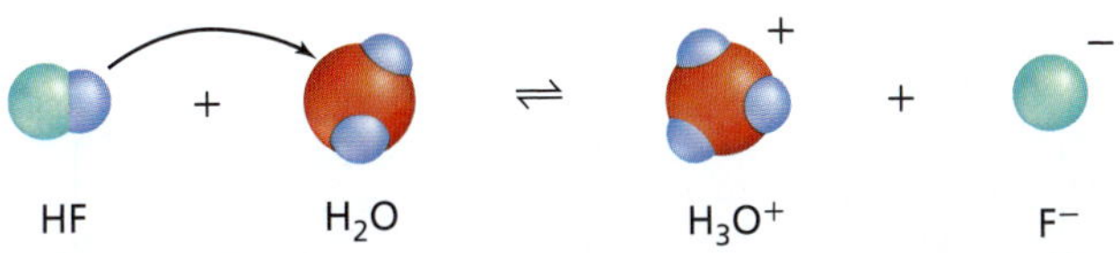

Figure 7 Hydrogen fluoride donates a hydrogen ion to a water molecule, so hydrogen fluoride is an acid.
Decide *which species is the conjugate base of hydrogen fluoride.*

Hydrogen fluoride—a Brønsted-Lowry acid Consider the equation for the ionization of hydrogen fluoride (HF) in water, shown in **Figure 7.** What are the conjugate acid-base pairs? Hydrogen fluoride, the acid in the forward reaction, produces its conjugate base F^-, the base in the reverse reaction. Water, the base in the forward reaction, produces its conjugate acid H_3O^+, the acid in the reverse reaction.

$$\underset{\text{Acid}}{HF(aq)} + \underset{\text{Base}}{H_2O(l)} \leftrightharpoons \underset{\text{Conjugate acid}}{H_3O^+(aq)} + \underset{\text{Conjugate base}}{F^-(aq)}$$

VOCABULARY
WORD ORIGIN
Conjugate
con– prefix; from Latin, meaning *with* or *together*
jugare, verb; from Latin, meaning *to join*

Hydrogen fluoride is used to manufacture a variety of fluorine-containing compounds, such as the nonstick coating on the kitchenware shown in **Figure 8.** It is an acid according to both the Arrhenius and Brønsted-Lowry definitions.

Ammonia—a Brønsted-Lowry base All of the acids and bases that fit the Arrhenius definition of acids and bases also fit the Brønsted-Lowry definition. But some other substances that lack a hydroxide group and, therefore, cannot be considered bases according to the Arrhenius definition can be classified as bases according to the Brønsted-Lowry model. One example is ammonia (NH_3).When ammonia dissolves in water, water is a Brønsted-Lowry acid in the forward reaction. Because the NH_3 molecule accepts a H^+ ion to form the ammonium ion (NH_4^+), ammonia is a Brønsted-Lowry base in the forward reaction.

$$\underset{\text{Base}}{NH_3(aq)} + \underset{\text{Acid}}{H_2O(l)} \leftrightharpoons \underset{\text{Conjugate acid}}{NH_4^+(aq)} + \underset{\text{Conjugate base}}{OH^-(aq)}$$

In the reverse reaction, the ammonium ion (NH_4^+) gives up a H^+ ion to form the molecule ammonia and thus acts as a Brønsted-Lowry acid. The ammonium ion is the conjugate acid of the base ammonia. The hydroxide ion accepts a H^+ ion to form a water molecule and is thus a Brønsted-Lowry base. The hydroxide ion is the conjugate base of the acid water.

Water—a Brønsted-Lowry acid and base Recall that when HF dissolves in water, water acts a base; when NH_3 dissolves in water, water acts as an acid. Depending on what other substances are in the solution, water can act as either an acid or a base. Water and other substances that can act as both acids and bases are said to be **amphoteric.**

Figure 8 To make the smooth, nonstick surface of this kitchenware, hydrogen fluoride is reacted with organic compounds called hydrocarbons to substitute fluorine atoms for hydrogen atoms.

Caption Question Fig. 7
The conjugate base is F^-.

ASSESSMENT

Performance Have two student volunteers plan and perform a debate involving the Arrhenius and Brønsted-Lowry models of acids and bases. Have one student assume the role of Svante Arrhenius and advocate his model while the other student plays Johannes Brønsted and supports his model. **OL**

Reinforcement

Donors and Acceptors Explain that, just as money is given by a *donor* to a charity, which is the *acceptor*, hydrogen ions or protons are given by the acid, which is the *donor*, and received by the base, which is the *acceptor*.

DIFFERENTIATED INSTRUCTION

Struggling Learners Have students make paper cutouts to represent the atoms of hydrogen, oxygen, and chlorine in the reaction between hydrogen chloride and water. They can use thumbtacks to attach the cutouts to a poster board or bulletin board, then physically transfer the H^+ from HCl to H_2O to create H_3O^+ and Cl^-. **BL**

Content Background

Common Names of Acids Some acids have common names that are different from their chemical names. For example, you might purchase muriatic acid at a hardware or home-supply store and use it to clean mortar from brick or lime deposits from tile. Muriatic acid is the commercial name for hydrochloric acid.

PRACTICE Problems

Have students refer to the Selected Solutions appendix for complete solutions to odd-numbered problems. The complete solutions for all problems can be found in the Solutions Manual.

3.

	Acid	Conjugate base	Base	Conjugate acid
a.	NH_4^+	NH_3	OH^-	H_2O
b.	HBr	Br^-	H_2O	H_3O^+
c.	H_2O	OH^-	CO_3^{2-}	HCO_3^-

4. $HSO_4^-(aq) + H_2O(l) \rightleftharpoons H_3O^+(aq) + SO_4^{2-}(aq)$.
Reactant base: H_2O;
conjugate acid: H_3O^+
Reactant acid: HSO_4^-;
conjugate base: SO_4^{2-}

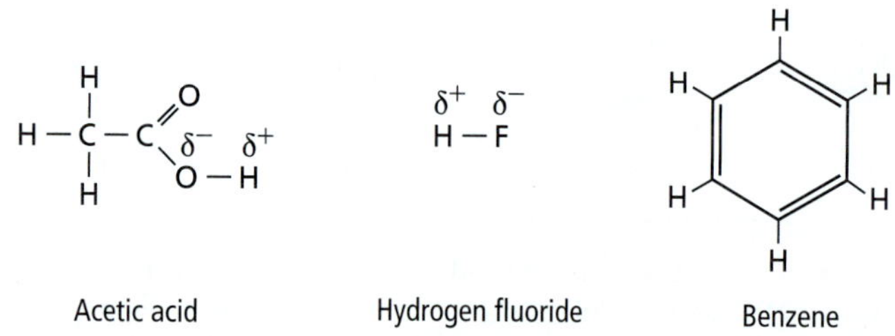

■ **Figure 9** Whether a hydrogen is ionizable depends on the polarity of its bond. In acetic acid, oxygen is more electronegative than hydrogen. The bond between oxygen and hydrogen is polar, so the hydrogen atom can ionize in solution. In hydrogen fluoride, fluorine is highly electronegative, so HF is an acid in solution. In benzene, there is little electronegativity difference between the carbon and hydrogen atoms, so benzene is not an acid.

PRACTICE Problems

Do additional problems. Online Practice

3. Identify the conjugate acid-base pairs in each reaction.

a. $NH_4^+(aq) + OH^-(aq) \rightleftharpoons NH_3(aq) + H_2O(l)$

b. $HBr(aq) + H_2O(l) \rightleftharpoons H_3O^+(aq) + Br^-(aq)$

c. $CO_3^{2-}(aq) + H_2O(l) \rightleftharpoons HCO_3^-(aq) + OH^-(aq)$

4. Challenge The products of an acid-base reaction are H_3O^+ and SO_4^{2-}. Write a balanced equation for the reaction and identify the conjugate acid-base pairs.

Monoprotic and Polyprotic Acids

From the chemical formulas of HCl and HF, you know that each acid has one hydrogen ion per molecule. An acid that can donate only one hydrogen ion is called a monoprotic acid. Other monoprotic acids are perchloric acid ($HClO_4$), nitric acid (HNO_3), hydrobromic acid (HBr), and acetic acid (CH_3COOH). Because acetic acid is a monoprotic acid, its formula is often written $HC_2H_3O_2$ to emphasize the fact that only one of the four hydrogen atoms in the molecule is ionizable.

Ionizable hydrogen atoms The difference between acetic acid's ionizable hydrogen atom and the other three hydrogen atoms is that the ionizable atom is bonded to the element oxygen, which is more electronegative than hydrogen. The difference in electronegativity makes the bond between oxygen and hydrogen polar. The structure of acetic acid is shown in **Figure 9,** along with structures of the acid HF and the nonacid benzene (C_6H_6). The hydrogen atom in hydrogen fluorine is bonded to the highly electronegative fluorine atom, so the hydrogen-fluorine bond is polar and the fluorine atom is ionizable to a certain extent. However, the hydrogen atoms in benzene are each bonded to a carbon atom. Carbon atoms have about the same electronegativity as hydrogen. These bonds are nonpolar, so benzene is not an acid.

Some acids donate more than one hydrogen ion. For example, sulfuric acid (H_2SO_4) and carbonic acid (H_2CO_3) can donate two hydrogen ions. In each compound, both hydrogen atoms are attached to oxygen atoms by polar bonds. Acids that contain two ionizable hydrogen atoms per molecule are called diprotic acids. Phosphoric acid (H_3PO_4) and boric acid (H_3BO_3) contain three ionizable hydrogen atoms per molecule. Acids with three hydrogen ions to donate are called triprotic acids. The term *polyprotic acid* can be used for any acid that has more than one ionizable hydrogen atom.

Demonstration

A Lewis Acid-Base Reaction

Purpose:
to demonstrate a Lewis acid-base reaction

Materials:
bottles of concentrated hydrochloric acid and concentrated ammonia solution

Safety Precautions
Perform this demonstration in a well-ventilated area or a fume hood.

Procedure
Place bottles of concentrated hydrochloric acid and concentrated ammonia solution side-by-side in a well-ventilated area or a fume hood. Remove the caps from both bottles and gently waft the fumes together.

Results
A white smoke of ammonium chloride (NH_4Cl) forms over the bottles.

Table 1 Some Common Acids and Their Conjugate Bases

Acid		Congugate Base	
Name	Formula	Name	Formula
Hydrochloric acid	HCl	Chloride ion	Cl^-
Nitric acid	HNO_3	Nitrate ion	NO_3^-
Sulfuric acid	H_2SO_4	Hydrogen sulfate ion	HSO_4^-
Hydrogen sulfate ion	HSO_4^-	Sulfate ion	SO_4^{2-}
Hydrofluoric acid	HF	Fluoride ion	F^-
Hydrocyanic acid	HCN	Cyanide	CN^-
Acetic acid	$HC_2H_3O_2$	Acetate ion	$C_2H_3O_2^-$
Phosphoric acid	H_3PO_4	Dihydrogen phosphate ion	$H_2PO_4^-$
Dihydrogen phosphate ion	$H_2PO_4^-$	Hydrogen phosphate ion	HPO_4^{2-}
Hydrogen phosphate ion	HPO_4^{2-}	Phosphate ion	PO_4^{3-}
Carbonic acid	H_2CO_3	Hydrogen carbonate ion	HCO_3^-
Hydrogen carbonate ion	HCO_3^-	Carbonate ion	CO_3^{2-}

All polyprotic acids ionize in steps. The three ionizations of phosphoric acid are described by the following equations.

$$H_3PO_4(aq) + H_2O(l) \rightleftharpoons H_3O^+(aq) + H_2PO_4^-(aq)$$
$$H_2PO_4^-(aq) + H_2O(l) \rightleftharpoons H_3O^+(aq) + HPO_4^{2-}(aq)$$
$$HPO_4^{2-}(aq) + H_2O(l) \rightleftharpoons H_3O^+(aq) + PO_4^{3-}(aq)$$

Table 1 shows some common monoprotic and polyprotic acids.

The Lewis Model

Notice that all substances classified as acids and bases by the Arrhenius model are classified as acids and bases by the Brønsted-Lowry model. In addition, some substances *not* classified as bases by the Arrhenius model *are* classified as bases by the Brønsted-Lowry model.

Perhaps you will not be surprised to learn that an even more general model of acids and bases was proposed by American chemist G. N. Lewis (1875–1946). Recall that Lewis developed the electron-pair theory of chemical bonding and introduced Lewis structures to keep track of the electrons in atoms and molecules. He applied his electron-pair theory of chemical bonding to acid-base reactions. Lewis proposed that an acid is an ion or molecule with a vacant atomic orbital that can accept (share) an electron pair. A base is an ion or molecule with a lone electron pair that it can donate (share). According to the **Lewis model,** a Lewis acid is an electron-pair acceptor and a Lewis base is an electron-pair donor. Note that the Lewis model includes all the substances classified as Brønsted-Lowry acids and bases and many more.

FOLDABLES® Incorporate information from this section into your Foldable.

Apply Chemistry

Acid-Paper Reactions If possible, bring in a nineteenth century book with badly decomposed pages. Explain that in the nineteenth century, paper was often sized, a process in which microscopic holes were filled with alum (aluminum sulfate) to lower absorption of moisture and reduce spreading of inks. The alum formed $Al(H_2O)_6^{3+}$ ions, which are acidic. Over time, the $Al(H_2O)_6^{3+}$ ions reacted with the wood fibers in paper, causing the paper to disintegrate and the books to fall apart. Explain that books older than the nineteenth century are generally in better condition because alum was not used for sizing at that time.

FOLDABLES® Rubric

Analysis

Ask these questions:

1. What is the chemical equation for the reaction between HCl and NH_3, which forms the white smoke (ammonium chloride, NH_4Cl)? $HCl(g) + NH_3(g) \rightarrow NH_4Cl(s)$
2. What are the Lewis structures for HCl, NH_3, and NH_4Cl? The Lewis structures show that the nitrogen atom in the NH_3 molecule has a valence-electron pair available for bonding with the hydrogen atom of HCl to form NH_4Cl.
3. Identify the Lewis acid and Lewis base in the reaction. HCl is the Lewis acid, and NH_3 is the Lewis base.

ASSESSMENT

Knowledge Have students write the balanced equation for the reaction between aqueous ammonia and water, forming aqueous ammonium hydroxide. Have them identify the Lewis acid and Lewis base in the reaction. $NH_3(g) + H_2O(l) \rightarrow NH_4^+(aq) + OH^-(aq)$; H_2O is the Lewis acid, and NH_3 is the Lewis base. OL

Concept Development

Lewis Dot Structure Have students recall that Lewis is given credit as the first scientist to discuss the significance of valence-electron pairs in chemical bonding. Review the method for writing Lewis structures for the chemical species presented in this section, including F^-, BF_3, NH_3, SO_3, and MgO. Point out that the emphasis Lewis placed on electron pairs in acid-base reactions was a natural extension of his study of the role of electron pairs in chemical bonding. **OL**

VOCABULARY

ACADEMIC VOCABULARY

Conform
to be similar or identical
Their behavior conforms to the expectations of the community.

Electron pair donors and acceptors Consider the reaction between a hydrogen ion (H^+) and a fluoride ion (F^-) to form a hydrogen fluoride (HF) molecule. The role of the electron pair is illustrated through the following Lewis structures.

In this reaction, the H^+ ion is the Lewis acid. Its vacant 1s orbital accepts an electron pair from the F^- ion. The fluoride ion is the Lewis base. It donates a lone electron pair to form the hydrogen-fluorine bond in HF. Note that this reaction also conforms to the Brønsted-Lowry model of acids and bases because H^+ can be considered a hydrogen-ion donor and F^- a hydrogen-ion acceptor.

It might surprise you to learn that the reaction of gaseous boron trifluoride (BF_3) with gaseous ammonia (NH_3) to form BF_3NH_3 is a Lewis acid-base reaction.

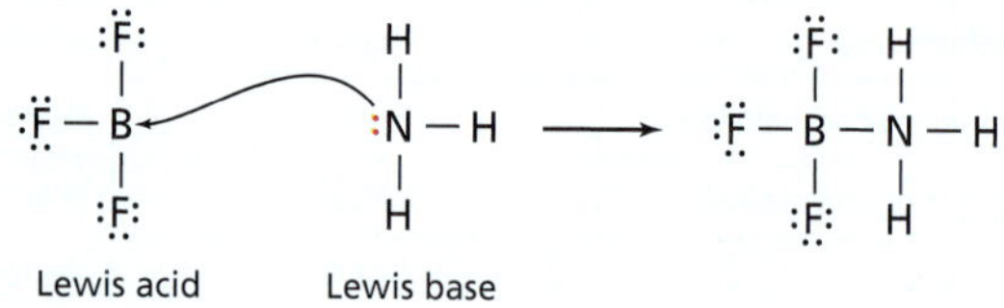

Recall that the boron atom in BF_3 has six valence electrons, so a vacant orbital can accept an electron pair from a Lewis base.

Another Lewis acid-base reaction occurs when gaseous sulfur trioxide (SO_3) is brought into contact with solid magnesium oxide (MgO).

$$SO_3(g) + MgO(s) \rightarrow MgSO_4(s)$$

The acid-base part of the reaction involves sulfur trioxide (SO_3) and the oxide ion (O^{2-}) of magnesium oxide. The product is the sulfate ion.

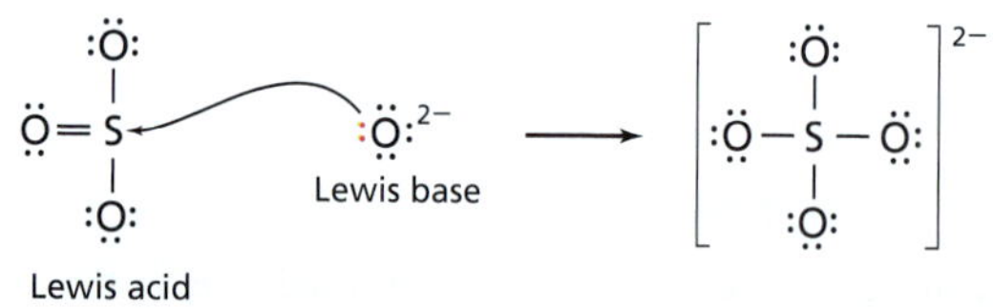

Note that the SO_3 molecule, a Lewis acid, accepts an electron pair from the O^{2-} ion, a Lewis base. The Arrhenius, Brønsted-Lowry, and Lewis acid-base models are summarized in **Table 2.**

Explore **acids and bases with an interactive table.** Concepts In Motion

Table 2 Three Models for Acids and Bases

Model	Acid Definition	Base Definition
Arrhenius	H^+ producer	OH^- producer
Brønsted-Lowry	H^+ donor	H^+ acceptor
Lewis	electron-pair acceptor	electron-pair donor

DIFFERENTIATED INSTRUCTION

Advanced Learners Challenge advanced learners to write the Lewis structures for calcium oxide (CaO), carbon dioxide (CO_2), and the calcium carbonate ($CaCO_3$) formed when CaO and CO_2 react. Have them identify the electron pair involved in the reaction, the Lewis acid, and the Lewis base. An electron pair of the oxygen in CaO is donated to the carbon atom in CO_2 to form the carbonate ion ($CO_3{}^{2-}$). Therefore, CaO is the Lewis base and CO_2 is the Lewis acid. **AL**

The reaction of SO_3 and MgO is important because it produces magnesium sulfate, a salt that forms the heptahydrate known as Epsom salt ($MgSO_4 \cdot 7H_2O$). Epsom salt has many uses, including soothing sore muscles and acting as a plant nutrient. The reaction to form magnesium sulfate also has environmental applications. When MgO is injected into the flue gases of coal-fired power plants, such as the one shown in **Figure 10,** it reacts with and removes SO_3. If SO_3 is allowed to enter the atmosphere, it can combine with water in the air to form sulfuric acid, which falls to Earth as acid precipitation.

Connection to Earth Science **Anhydrides** Like the SO_3 molecules you have been reading about, carbon dioxide gas molecules in the air also combine with water molecules in precipitation to form an acid called carbonic acid (H_2CO_3). When the acidic rainwater reaches the ground, some sinks into the soil and reaches limestone bedrock, where it slowly dissolves the limestone.

Over the course of thousands of years, the dissolution of limestone creates huge underground caverns. Within a cavern, groundwater might drip from the ceiling and deposit some of the dissolved limestone. Deposits shaped like icicles that form on the ceiling are called stalactites. Rounded masses rising from the floor are called stalagmites.

The formation of caverns occurs because carbon dioxide is an acid anhydride. An acid anhydride is an oxide that can combine with water to form an acid. Other oxides combine with water to form bases. For example, calcium oxide (CaO, lime) forms the base calcium hydroxide $Ca(OH)_2$. In general, oxides of metallic elements form bases; oxides of nonmetals form acids.

Figure 10 Sulfur trioxide, a waste gas from the burning of coal, can be removed from smokestack gases by combining it with magnesium oxide in a Lewis acid-base reaction. Note that while there is a good deal of steam coming from the cooling towers, there is little visible smoke from the smokestack.

Malcolm Fife/age fotostock

SECTION 1 REVIEW

Section Self-Check

Section Summary

- The concentrations of hydrogen ions and hydroxide ions determine whether an aqueous solution is acidic, basic, or neutral.
- An Arrhenius acid must contain an ionizable hydrogen atom. An Arrhenius base must contain an ionizable hydroxide group.
- A Brønsted-Lowry acid is a hydrogen ion donor. A Brønsted-Lowry base is a hydrogen ion acceptor.
- A Lewis acid accepts an electron pair. A Lewis base donates an electron pair.

5. **MAIN IDEA Explain** why many Lewis acids and bases are not classified as Arrhenius or Brønsted-Lowry acids and bases.
6. **Compare** the physical and chemical properties of acids and bases.
7. **Explain** how the concentrations of hydrogen ions and hydroxide ions determine whether a solution is acidic, basic, or neutral.
8. **Explain** why many compounds that contain one or more hydrogen atoms are not classified as Arrhenius acids.
9. **Identify** the conjugate acid-base pairs in the following equation.

$$HNO_2 + H_2O \rightleftharpoons NO_2^- + H_3O^+$$

10. **Write** the Lewis structure for phosphorus trichloride (PCl_3). Is PCl_3 a Lewis acid, a Lewis base, or neither?
11. **Interpret Scientific Illustrations** In the accompanying structural formula, identify any hydrogen atoms that are likely to be ionizable.

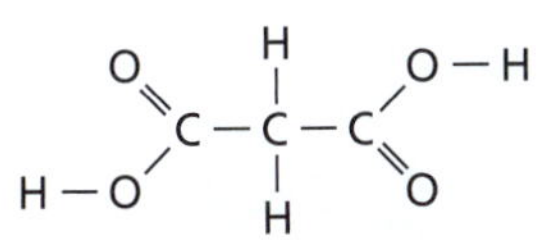

3 Assess

Check for Understanding

Ask students why hydrogen sulfide (H_2S) is generally considered an acid, while methane (CH_4) is not. Although hydrogen atoms are bonded to nonmetallic elements in both compounds, H_2S ionizes in aqueous solution to produce H^+ ions. CH_4 does not. OL

Reteach

Ask students whether the reaction $H_2O(l) + CH_3NH_2(aq) \rightarrow OH^-(aq) + CH_3NH_3^+(aq)$ is an acid-base reaction according to the Arrhenius model. no Ask if it is an acid-base reaction according to the Brønsted-Lowry model. yes Have them identify the Brønsted-Lowry acid in the forward reaction. H_2O OL

Extension

Have students write the chemical equation for the acidic reaction of $Al(H_2O)_6^{3+}$ with water. $Al(H_2O)_6^{3+}(aq) + H_2O(l) \rightleftharpoons H_3O^+(aq) + Al(H_2O)_5OH^{2+}(aq)$ OL

ASSESSMENT

Skill Ask students to write the equations for the three ionizations of boric acid (H_3BO_3). $H_3BO_3(aq) + H_2O(l) \rightleftharpoons H_3O^+(aq) + H_2BO_3^-(aq)$, $H_2BO_3^-(aq) + H_2O(l) \rightleftharpoons H_3O^+(aq) + HBO_3^{2-}(aq)$, $HBO_3^{2-}(aq) + H_2O(l) \rightleftharpoons H_3O^+(aq) + BO_3^{3-}(aq)$ OL

SECTION 1 REVIEW

5. A Lewis acid is an electron pair acceptor. A Lewis base is an electron pair donor. A Lewis acid cannot have an ionizable hydrogen ion or hydroxide ion to qualify as an Arrhenius acid or base. A Lewis acid cannot have a hydrogen ion to donate, therefore it could not qualify as a Brønsted-Lowry acid. However, all Lewis bases are Brønsted-Lowry bases because they can accept a hydrogen ion.
6. Physical properties: Acids taste sour and conduct electricity. Bases taste bitter, feel slippery, and conduct electricity. Chemical properties: Acids react with some metals to produce hydrogen gas. Acids turn blue litmus red. Bases react with acids and turn red litmus blue.
7. In an acidic solution, $[H^+] > [OH^-]$; in a neutral solution, $[H^+] = [OH^-]$; in a basic solution, $[H^+] < [OH^-]$.
8. Only compounds that have one or more ionizable hydrogen atom are Arrhenius acids.
9. HNO_2 (acid) and NO_2^- (conjugate base), H_2O (base) and H_3O^+ (conjugate acid)
10. Phosphorus in PCl_3 has three electrons, which it shares with three chlorines, and an unshared pair of electrons. The unshared pair of electrons can act as a Lewis base.
11. the two hydrogen atoms attached to oxygen atoms

SECTION 2

1 Focus

MAINIDEA

Strong and Weak Acids Write the formulas for hydrobromic acid (HBr) and carbonic acid (H_2CO_3) on the board and ask students if they can tell the relative strengths of the two acids from their formulas. Students will probably say they cannot. Some students, however, might assume that H_2CO_3 is stronger because it contains two ionizable hydrogens per molecule. That assumption is incorrect. Then, write the ionization equations for the two acids in water, with equilibrium arrows (longer arrow to the left) for carbonic acid. $HBr(g) \rightarrow H^+(aq) + Br^-(aq)$; $H_2CO_3(aq) \rightleftharpoons H^+(aq) + HCO_3^-(aq)$ Point out that the ionization of HBr is virtually 100% in dilute aqueous solution and that–as a result–the equation has a single reaction arrow. Explain, however, that the ionization of carbonic acid is less than complete and the equilibrium constant for the ionization reaction at 298 K is only 4.5×10^{-7}. OL

2 Teach

Content Background

Acid Strength and Bonding

Explain that an acid's strength relates to the strength and polarity of its H-X bond, where X represents a nonmetallic element or, in some instances, a negatively charged polyatomic ion. Point out that bond strength depends primarily upon the relative electronegativities of H and X or, for a negatively charged polyatomic ion, the electronegativities of H, X, and other elements bonded to X. In general, the weaker the H-X bond, the stronger the acid.

SECTION 2

Essential Questions

- How is the strength of an acid or base related to its degree of ionization?
- How does the strength of a weak acid compare with the strength of its conjugate base?
- What is the relationship between the strengths of acids and bases and the values of their ionization constants?

Review Vocabulary

electrolyte: an ionic compound whose aqueous solution conducts an electric current

New Vocabulary

strong acid
weak acid
acid ionization constant
strong base
weak base
base ionization constant

Strengths of Acids and Bases

MAINIDEA In solution, strong acids and bases ionize completely, but weak acids and bases ionize only partially.

CHEM 4 YOU The success of a pass in a football game depends on the passer and the receiver. How ready is the passer to pass the ball? How ready is the receiver to receive the ball? Similarly, in acid and base reactions, the progress of a reaction depends on how readily the acid donates a hydrogen ion and how readily the base accepts a hydrogen ion.

Strengths of Acids

One of the properties of acidic and basic solutions is that they conduct electricity. What can electrical conductivity tell you about the hydrogen ions and hydroxide ions in these aqueous solutions?

Suppose you test the electrical conductivities of 0.10*M* aqueous solutions of hydrochloric acid and acetic acid. The glow of the bulb in **Figure 11** indicates that the solution conducts electricity. However, if you compare the brightness of the bulb connected to the HCl solution in **Figure 11** with that of the bulb connected to the $HC_2H_3O_2$ solution in **Figure 12,** you should notice a difference. The 0.10*M* HCl solution conducts electricity better than the 0.10*M* $HC_2H_3O_2$ solution. Why is this true if the concentrations of the two acids are both 0.10*M*?

Strong acids The answer is that ions carry electric current through the solution and all the HCl molecules contained in the solution are ionized completely into hydronium ions and chloride ions. Acids that ionize completely are called **strong acids.** Because strong acids produce the maximum number of ions, they are good conductors of electricity.

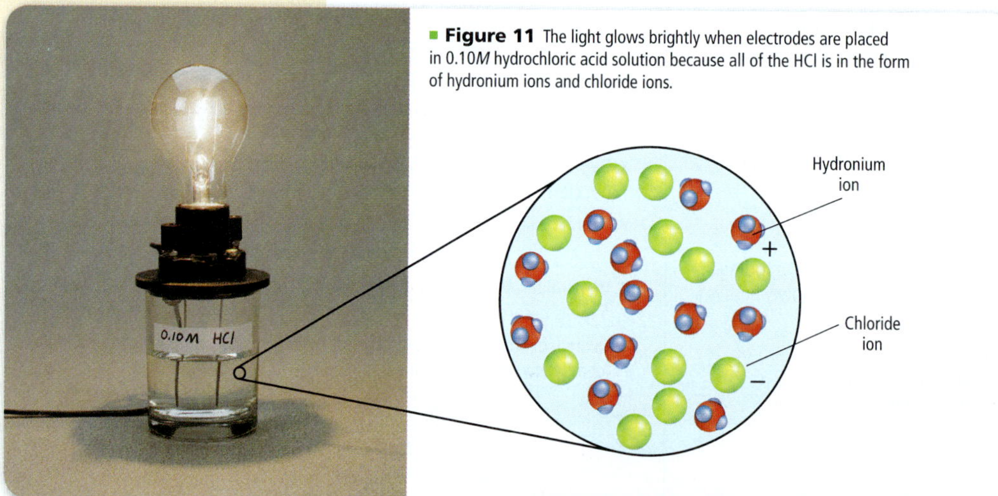

■ **Figure 11** The light glows brightly when electrodes are placed in 0.10*M* hydrochloric acid solution because all of the HCl is in the form of hydronium ions and chloride ions.

Matt Meadows

DIFFERENTIATED INSTRUCTION

Visually Impaired Have students, parents, or school personnel prepare an "acid-base buffet" made of food items that contain acids and bases. Arrange the "buffet" in the school cafeteria or the school's home economics classroom. Organize an in-school field trip in which students sample these items and report on their distinctive tastes. OL EL

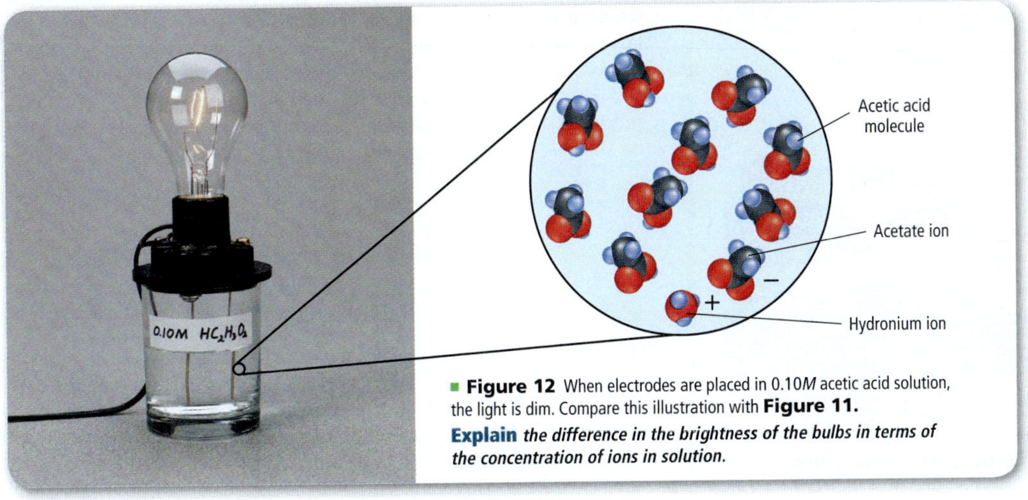

■ **Figure 12** When electrodes are placed in 0.10*M* acetic acid solution, the light is dim. Compare this illustration with **Figure 11.**
Explain *the difference in the brightness of the bulbs in terms of the concentration of ions in solution.*

Matt Meadows

The ionization of hydrochloric acid in water can be represented by the following equation, which has a single arrow pointing to the right. Recall that a single arrow means that a reaction goes to completion.

$$HCl(aq) + H_2O(l) \rightarrow H_3O^+(aq) + Cl^-(aq)$$

Because strong acids produce the maximum number of ions, their solutions are good conductors of electricity. The names and ionization equations for some strong acids are shown in **Table 3.**

Weak acids If the brightly lit bulb of the apparatus containing the HCl solution is due to the large number of ions in solution, shown in **Figure 11,** then the weakly lit bulb of the apparatus containing the $HC_2H_3O_2$ solution, shown in **Figure 12,** must mean that the acetic acid solution has fewer ions. Because the two solutions have the same molar concentrations, you can conclude that acetic acid does not ionize completely. An acid that ionizes only partially in dilute aqueous solution is a **weak acid.** Weak acids produce fewer ions and thus cannot conduct electricity as well as strong acids. **Table 3** shows ionization equations for some common weak acids.

Explore **ionization equations** with an interactive table. **Concepts In Motion**

Table 3 Ionization Equations

Strong Acids		Weak Acids	
Name	**Ionization Equation**	**Name**	**Ionization Equations**
Hydrochloric	$HCl \rightarrow H^+ + Cl^-$	Hydrofluoric	$HF \rightleftharpoons H^+ + F^-$
Hydroiodic	$HI \rightarrow H^+ + I^-$	Acetic	$HC_2H_3O_2 \rightleftharpoons H^+ + C_2H_3O_2^-$
Perchloric	$HClO_4 \rightarrow H^+ + ClO_4^-$	Hydrosulfuric	$H_2S \rightleftharpoons H^+ + HS^-$
Nitric	$HNO_3 \rightarrow H^+ + NO_3^-$	Carbonic	$H_2CO_3 \rightleftharpoons H^+ + HCO_3^-$
Sulfuric	$H_2SO_4 \rightarrow H^+ + HSO_4^-$	Hypochlorous	$HClO \rightleftharpoons H^+ + ClO^-$

■ **Caption Question Fig. 12** The light glows brighter when the concentration of ions in solution is greater. In Figure 11, all of the dissolved HCl is ionized, so the light is bright. In Figure 12, only a small amount of the $HC_2H_3O_2$ is ionized, so the light is dim.

Identify Misconceptions

Students might think that an acid's strength is directly related to the number of hydrogen atoms each molecule contains.

Uncover the Misconception
Ask students to predict which acid is stronger: HNO_3 or H_3PO_4. Students might predict that H_3PO_4 is stronger because it is triprotic.

Demonstrate the Concept
Use an electrical conductivity apparatus to show that a 0.1*M* HNO_3 solution has a higher electrical conductivity than a 0.1*M* H_3PO_4 solution, even though H_3PO_4 contains three times as many hydrogen atoms as HNO_3. Thus, HNO_3 is stronger than H_3PO_4.

Assess New Knowledge
Ask students if they can rank the acids HF, H_3BO_3, and H_2SO_3 by strength based solely on their formulas. No, the three acids have different numbers of hydrogen atoms per molecule, but their relative strengths cannot be determined from their formulas alone. **OL**

DIFFERENTIATED INSTRUCTION

Advanced Learners Ask students to research and explain the relative strengths of the oxyacids of chlorine (HOCl, $HClO_2$, and $HClO_4$). Hypochlorous acid (HOCl) is the weakest because the chlorine draws very little of the electron density and strength from the H–O bond. Perchloric acid, which can be written $HOClO_3$, is the strongest because the three oxygen atoms that are bonded to the chlorine atom draw a great deal of the electron density and strength from the Cl–O–H bond. **AL**

READING CHECK The conjugate base of a strong acid is very weak, so it has little power to capture the hydrogen ion. The conjugate base of a weak acid is stronger and competes with some level of success for the hydrogen ion.

Quick Demo

Indicators Add a few drops of methyl orange indicator solution to 50 mL of 0.1*M* acetic acid in a 100-mL beaker. The solution turns red. Then, add approximately 0.2 g of solid sodium acetate to the solution and stir. The solution turns yellow. Explain that methyl orange changes from red to yellow when the solution becomes less acidic. Ask students to write the ionization equilibrium equation for acetic acid and to explain why the solution turns yellow.
$CH_3COOH(aq) \rightleftharpoons H^+(aq) + CH_3COO^-(aq)$
The solution turns yellow because adding sodium acetate increases $[CH_3COO^-]$. According to Le Châtelier's principle, the increased concentration of acetate ion forces the equilibrium to the left. The shift to the left consumes hydrogen ions, which makes the solution less acidic. **OL**

RealWorld CHEMISTRY

Hydrogen Cyanide

A DEADLY COMPOUND
Hydrogen cyanide (HCN) is a poisonous gas found in the exhaust of vehicles, in tobacco and wood smoke, and in smoke from burning nitrogen-containing plastics. Some insects such as millipedes and burnet moths release hydrogen cyanide as a defense mechanism. A solution of hydrogen cyanide in water is called hydrocyanic acid. Fruits that have a pit, such as cherries or peaches, contain cyanohydrins, which convert to hydrocyanic acid in the digestive system if the pits are eaten. However, no hydrocyanic acid is produced in the flesh of these fruits, so the fruit can safely be eaten.

Acid strength and the Brønsted-Lowry model Can the Brønsted-Lowry model explain why HCl ionizes completely but $HC_2H_3O_2$ forms only a few ions? Consider the ionization of any strong acid, HX. Remember that the acid on the reactant side of the equation produces a conjugate base on the product side. Similarly, the base on the reactant side produces a conjugate acid.

$$\underset{\text{Acid}}{HX(aq)} + \underset{\text{Base}}{H_2O(l)} \rightarrow \underset{\text{Conjugate acid}}{H_3O^+(aq)} + \underset{\text{Conjugate base}}{X^-(aq)}$$

HX represents a strong acid and its conjugate base is weak. That is, HX is nearly 100% ionized because H_2O is a stronger base (in the forward reaction) than is the conjugate base X^- (in the reverse reaction). In other words, the ionization equilibrium lies almost completely to the right because the base H_2O has a much greater attraction for the H^+ ion than does the base X^-. Think of this as the battle of the bases: Which of the two (H_2O or X^-) has a greater attraction for the hydrogen ion? In the case of all strong acids, water is the stronger base. Notice that the equation is shown with a single arrow to the right.

How does the situation differ for any weak acid, HY?

$$\underset{\text{Acid}}{HY(aq)} + \underset{\text{Base}}{H_2O(l)} \rightleftharpoons \underset{\text{Conjugate acid}}{H_3O^+(aq)} + \underset{\text{Conjugate base}}{Y^-(aq)}$$

The ionization equilibrium for a weak acid lies far to the left because the conjugate base Y^- has a greater attraction for the H^+ ion than does the base H_2O. In the battle of the bases, the conjugate base Y^- (in the reverse reaction) is stronger than the base H_2O (in the forward reaction) and manages to capture the H^+ ion. In the case of acetic acid, the conjugate base $C_3H_2O_2^-$ (in the reverse reaction) has a stronger attraction for hydrogen ions than does the base H_2O (in the forward reaction).

$$HC_2H_3O_2(aq) + H_2O(l) \rightleftharpoons H_3O^+(aq) + C_2H_3O_2^-(aq)$$

Notice that the equation is shown with equilibrium arrows.

READING CHECK **Summarize** the important difference between strong acids and weak acids in terms of the battle of the bases.

Acid ionization constants Although the Brønsted-Lowry model helps explain acid strength, the model does not provide a quantitative way to express the strength of an acid or to compare the strengths of various acids. The equilibrium constant expression provides the quantitative measure of acid strength.

As you have read, a weak acid produces an equilibrium mixture of molecules and ions in aqueous solution. Thus, the equilibrium constant, K_{eq}, provides a quantitative measure of the degree of ionization of the acid. Consider hydrocyanic acid (HCN), also known as prussic acid which is used in dying, engraving, and tempering steel.

CHEMISTRY JOURNAL

Acids in Foods Have students evaluate the strengths of acids in foods and explain why acids such as hydrochloric acid and sulfuric acid are not used to preserve or increase the tartness of foods. Have them record their findings in their journals. **OL**

The ionization equation and equilibrium constant expression for hydrocyanic acid are as follows.

$$HCN(aq) + H_2O(l) \rightleftharpoons H_3O^+(aq) + CN^-(aq)$$

$$K_{eq} = \frac{[H_3O^+][CN^-]}{[HCN][H_2O]}$$

The concentration of liquid H_2O in the denominator of the expression is considered to be constant in dilute aqueous solutions, so it can be combined with K_{eq} to give a new equilibrium constant, K_a.

$$K_{eq}[H_2O] = K_a = \frac{[H_3O^+][CN^-]}{[HCN]} = 6.2 \times 10^{-10}$$

K_a is called the acid ionization constant. The **acid ionization constant** is the value of the equilibrium constant expression for the ionization of a weak acid. Like all equilibrium constants, the value of K_a indicates whether reactants or products are favored at equilibrium. For weak acids, the concentrations of the ions (products) in the numerator tend to be small compared to the concentration of un-ionized molecules (reactant) in the denominator. The weakest acids have the smallest K_a values because their solutions have the lowest concentrations of ions and the highest concentrations of un-ionized acid molecules. K_a values and ionization equations for several weak acids are listed in **Table 4.** Note that polyprotic acids are not necessarily strong acids for any of their ionizations. Each ionization of a polyprotic acid has a K_a value, and the values decrease for each successive ionization.

CAREERS IN CHEMISTRY

Nursery Worker The propagation and growth of plants is the primary job of a nursery worker. This involves planting, pruning, transplanting, and selling all kinds of plant material. A nursery worker must know what nutrients are needed for optimum plant growth and what soil conditions, including acidity, foster the strongest growth for each kind of plant.

WebQuest

PRACTICE Problems

Do additional problems. Online Practice

12. Write ionization equations and acid ionization constant expressions for each acid.

a. $HClO_2$ **b.** HNO_2 **c.** HIO

13. Write the first and second ionization equations for H_2SeO_3.

14. Challenge Given the expression $K_a = \frac{[AsO_4^{3-}][H_3O^+]}{[HAsO_4^{2-}]}$, write the balanced equation for the corresponding reaction.

Table 4 Ionization Constants for Weak Acids

Acid	Ionization Equation	K_a (298 K)
Hydrosulfuric, first ionization	$H_2S \rightleftharpoons H^+ + HS^-$	8.9×10^{-8}
Hydrosulfuric, second ionization	$HS^- \rightleftharpoons H^+ + S^{2-}$	1×10^{-19}
Hydrofluoric	$HF \rightleftharpoons H^+ + F^-$	6.3×10^{-4}
Hydrocyanic	$HCN \rightleftharpoons H^+ + CN^-$	6.2×10^{-10}
Acetic	$CH_3COOH \rightleftharpoons H^+ + CH_3COO^-$	1.8×10^{-5}
Carbonic, first ionization	$H_2CO_3 \rightleftharpoons H^+ + HCO_3^-$	4.5×10^{-7}
Carbonic, second ionization	$HCO_3^- \rightleftharpoons H^+ + CO_3^{2-}$	4.7×10^{-11}

CHEMISTRY PROJECT

Acid Ionization Constants Have students look up acid ionization constants in a chemistry-physics handbook for at least five weak acids not listed in Table 4. Have them write the ionization equation and K_a expression for each of the acids and rank the acids from weakest to strongest. OL

CAREERS IN CHEMISTRY

WebQuest

PRACTICE Problems

Have students refer to the Selected Solutions appendix for complete solutions to odd-numbered problems. The complete solutions for all problems can be found in the Solutions Manual.

12. a. $HClO_2(aq) + H_2O(l) \rightleftharpoons H_3O^+(aq) + ClO_2^-(aq)$
$K_a = [H_3O^+][ClO_2^-]/[HClO_2]$

b. $HNO_2(aq) + H_2O(l) \rightleftharpoons H_3O^+(aq) + NO_2^-(aq)$
$K_a = [H_3O^+][NO_2^-]/[HNO_2]$

c. $HIO(aq) + H_2O(l) \rightleftharpoons H_3O^+(aq) + IO^-(aq)$
$K_a = [H_3O^+][IO^-]/[HIO]$

13. $H_2SeO_3(aq) + H_2O(l) \rightleftharpoons HSeO_3^-(aq) + H_3O^+(aq)$
$HSeO_3^-(aq) + H_2O(l) \rightleftharpoons SeO_3^{2-}(aq) + H_3O^+(aq)$

14. $HAsO_4^{2-}(aq) + H_2O(l) \rightleftharpoons H_3O^+(aq) + AsO_4^{3-}(aq)$

ASSESSMENT

Knowledge Ask students which solution, 0.10M H_2SO_4 or 1.00M HF, is more concentrated. the HF solution Ask which is more acidic. the H_2SO_4 solution Ask them to explain. HF is a weak acid, which means that it ionizes only slightly and produces few hydrogen ions in aqueous solution. OL

MiniLAB

Rubric

Purpose Students will observe the electrical conductivity of acetic acid solutions of varying concentrations and infer the relative numbers of ions in the solutions.

Process Skills observe and infer, recognize cause and effect, hypothesize

Safety Precautions Approve lab safety forms before work begins. Acetic acid is corrosive. Use a battery-operated conductivity tester to avoid electrical shock. The Flinn conductivity meter AP1493 is recommended.

Disposal Dilute all acid solutions twentyfold and wash down a drain.

Teaching Strategies

- Have students read Section 2 before performing the investigation.
- Use the results of the investigation to initiate a discussion of proper first aid treatment if acid splashes in eyes or on skin.

Expected Results The glacial acetic acid does not conduct electricity, so neither the red nor the green LED glows. For all of the dilutions, the red LED on the tester glows brightly. For the 6.0*M* acetic acid solution, the green LED is dim; for 1.0*M* acid, the green LED is bright; for 0.1*M* acid, the green LED is dimmer than for the 6.0*M*.

Analysis

1. $CH_3COOH(aq) + H_2O(l) \rightleftharpoons H_3O^+(aq) + CH_3COO^-(aq)$
   $K_a = [CH_3COO^-][H^+]/[CH_3COOH] = 1.8 \times 10^{-5}$; Very little acetic acid is ionized.
2. Yes. Comparing 6.0*M* and 1.0*M* solutions, the 1.0*M* solution has a higher percent of ionization. The larger fraction of ions results in increased conductivity. However, continued dilution reduces the overall concentration of ions enough so that the conductivity of the 0.1M solution decreases.
3. The more dilute the weak acid becomes, the better it conducts electricity because there is a higher degree of ionization. The water acts to increase the percent of ionization. Eventually, the acid is so dilute that the conductivity decreases because there is relatively little acetic acid present.
4. A small amount of water can increase the acid's apparent strength and resulting tissue damage. For acid spills, large amounts of water should be used to quickly dilute the acid and flush it away.

MiniLAB

Compare Acid Strengths

How can you determine the relative strengths of acid solutions?

Procedure

1. Read and complete the lab safety form.
2. Use a **10-mL graduated cylinder** to measure 3 mL of **glacial acetic acid.** Use a **dropping pipette** to transfer the acid into Well A1 of a **24-well microplate.**
   WARNING: ***Glacial acetic acid is corrosive and toxic by inhalation. Handle with caution.***
3. Lower the electrodes of a **conductivity tester** into Well A1. Record your results.
4. Rinse the graduated cylinder and pipette with water. Measure 3 mL of **6.0*M* acetic acid,** and transfer it to Well A2 of the microplate. Test and record the conductivity of the solution.
5. Repeat Step 4 with **1.0*M* acetic acid** and **0.10*M* acetic acid** using Wells A3 and A4, respectively.

Analysis

1. **Write** the equation for the ionization of acetic acid in water and the equilibrium constant expression ($K_{eq} = 1.8 \times 10^{-5}$). What does the size of K_{eq} indicate about the degree of ionization?
2. **Explain** whether the following approximate percent ionizations fit your laboratory results: glacial acetic acid, 0.1%; 6.0*M* acetic acid, 0.2%; 1.0*M* acetic acid, 0.4%; 0.1*M* acetic acid, 1.3%.
3. **State** a hypothesis that explains your observations using your answer to Question 2.
4. **Utilize** your hypothesis to draw a conclusion about the need to use large amounts of water for rinsing when acid spills on living tissue.

Strengths of Bases

What you have read about acids can be applied to bases, except that OH^- ions, rather than H^+ ions, are involved. For example, the conductivity of a base depends on the extent to which the base produces OH^- ions in aqueous solution.

Strong bases A base that dissociates entirely into metal ions and hydroxide ions is known as a **strong base.** Therefore, metallic hydroxides, such as sodium hydroxide (NaOH), are strong bases.

$$NaOH(s) \rightarrow Na^+(aq) + OH^-(aq)$$

Some metallic hydroxides, such as calcium hydroxide ($Ca(OH)_2$) have low solubility and thus are poor sources of OH^- ions. Note that the solubility product constant, K_{sp}, for calcium hydroxide ($Ca(OH)_2$) is small, indicating that few OH^- ions are present in a saturated solution.

$$Ca(OH)_2(s) \rightleftharpoons Ca^{2+}(aq) + 2OH^-(aq) \quad K_{sp} = 6.5 \times 10^{-6}$$

Nevertheless, calcium hydroxide and other slightly soluble metallic hydroxides are considered strong bases because all of the compound that dissolves is completely dissociated. The dissociation equations for several strong bases are listed in **Table 5.**

Table 5 Dissociation Equations for Strong Bases

Dissociation Equations for Strong Bases
$NaOH(s) \rightarrow Na^+(aq) + OH^-(aq)$
$KOH(s) \rightarrow K^+(aq) + OH^-(aq)$
$RbOH(s) \rightarrow Rb^+(aq) + OH^-(aq)$
$CsOH(s) \rightarrow Cs^+(aq) + OH^-(aq)$
$Ca(OH)_2(s) \rightarrow Ca^{2+}(aq) + 2OH^-(aq)$
$Ba(OH)_2(s) \rightarrow Ba^{2+}(aq) + 2OH^-(aq)$

Weak bases In contrast to strong bases, a **weak base** ionizes only partially in dilute aqueous solution. For example, methylamine (CH_3NH_2) reacts with water to produce an equilibrium mixture of CH_3NH_2 molecules, $CH_3NH_3^+$ ions, and OH^- ions.

$$\underset{\text{Acid}}{CH_3NH_2(aq)} + \underset{\text{Base}}{H_2O(l)} \rightarrow \underset{\text{Conjugate acid}}{CH_3NH_3^+(aq)} + \underset{\text{Conjugate base}}{OH^-(aq)}$$

ASSESSMENT

Knowledge Ask students to predict the conductivity of 6.0*M*, 1.0*M*, and 0.1*M* solutions of the strong acid HCl. OL

Table 6 Ionization Constants of Weak Bases

Base	Ionization Equation	K_b (298 K)
Ethylamine	$C_2H_5NH_2(aq) + H_2O(l) \rightleftharpoons C_2H_5NH_3^+(aq) + OH^-(aq)$	5.0×10^{-4}
Methylamine	$CH_3NH_2(aq) + H_2O(l) \rightleftharpoons CH_3NH_3^+(aq) + OH^-(aq)$	4.3×10^{-4}
Ammonia	$NH_3(aq) + H_2O(l) \rightleftharpoons NH_4^+(aq) + OH^-(aq)$	2.5×10^{-5}
Aniline	$C_6H_5NH_2(aq) + H_2O(l) \rightleftharpoons C_6H_5NH_3^+(aq) + OH^-(aq)$	4.3×10^{-10}

This equilibrium lies far to the left because the base, CH_3NH_2, is weak and the conjugate base, OH^-, is strong. The hydroxide ion has a greater attraction for a hydrogen ion than has a molecule of methylamine.

Base ionization constants Like weak acids, weak bases also form equilibrium mixtures of molecules and ions in aqueous solution. The equilibrium constant provides a measure of the extent of the base's ionization. The equilibrium constant for the ionization of methylamine in water is defined by the following equilibrium constant expression.

$$K_b = \frac{[CH_3NH_3^+][OH^-]}{[CH_3NH_2]}$$

The **base ionization constant,** K_b, is the value of the equilibrium constant expression for the ionization of a base. The smaller the value of K_b, the weaker the base. K_b values and ionization equations for several weak bases are listed in **Table 6.**

PRACTICE Problems

Do additional problems. Online Practice

15. Write ionization equations and base ionization constant expressions for the following bases.

a. hexylamine ($C_6H_{13}NH_2$) **c.** carbonate ion (CO_3^{2-})
b. propylamine ($C_3H_7NH_2$) **d.** hydrogen sulfite ion (HSO_3^-)

16. Challenge Write an equation for a base equilibrium in which the base in the forward reaction is PO_4^{3-} and the base in the reverse reaction is OH^-.

SECTION 2 REVIEW

Section Self-Check

Section Summary

- Strong acids and strong bases are completely ionized in a dilute aqueous solution. Weak acids and weak bases are partially ionized in a dilute aqueous solution.
- For weak acids and weak bases, the value of the acid or base ionization constant is a measure of the strength of the acid or base.

17. MAINIDEA Describe the contents of dilute aqueous solutions of the strong acid HI and the weak acid HCOOH.

18. Relate the strength of a weak acid to the strength of its conjugate base.

19. Identify the conjugate acid-base pairs in each equation.

a. $HCOOH(aq) + H_2O(l) \rightleftharpoons HCOO^-(aq) + H_3O^+(aq)$

b. $NH_3(aq) + H_2O(l) \rightleftharpoons NH_4^+(aq) + OH^-(aq)$

20. Explain what the K_b for aniline ($C_6H_5NH_2$) tells you ($K_b = 4.3 \times 10^{-10}$).

21. Interpret Data Use the data in **Table 4** to put the seven acids in order according to increasing electrical conductivity.

SECTION 2 REVIEW

17. The solution of HI contains only H_3O^+ and I^- ions and water molecules. The solution of HCOOH contains H_3O^+ and $HCOO^-$ ions, and HCOOH and H_2O molecules.

18. The stronger the acid is, the weaker its conjugate base. The weaker the acid is, the stronger its conjugate base.

19. a. acid: HCOOH; conjugate base: $HCOO^-$; base: H_2O; conjugate acid: H_3O^+

b. acid: H_2O; conjugate base: OH^-; base: NH_3; conjugate acid: NH_4^+

20. The size of the K_b indicates that aniline is a weak base.

21. HS^-, HCO_3^-, HCN, H_2S, H_2CO_3, CH_3COOH, HF

PRACTICE Problems

Have students refer to the Selected Solutions appendix for complete solutions to odd-numbered problems. The complete solutions for all problems can be found in the Solutions Manual.

15. a. $C_6H_{13}NH_2(aq) + H_2O(l) \rightleftharpoons C_6H_{13}NH_3^+(aq) + OH^-(aq)$
$K_b = [C_6H_{13}NH_3^+][OH^-]/[C_6H_{13}NH_2]$

b. $C_3H_7NH_2(aq) + H_2O(l) \rightleftharpoons C_3H_7NH_3^+(aq) + OH^-(aq)$
$K_b = [C_3H_7NH_3^+][OH^-]/[C_3H_7NH_2]$

c. $CO_3^{2-}(aq) + H_2O(l) \rightleftharpoons HCO_3^-(aq) + OH^-(aq)$
$K_b = [HCO_3^-][OH^-]/[CO_3^{2-}]$

d. $HSO_3^-(aq) + H_2O(l) \rightleftharpoons H_2SO_3(aq) + OH^-(aq)$
$K_b = [H_2SO_3][OH^-]/[HSO_3^-]$

16. $PO_4^{3-}(aq) + H_2O(l) \rightleftharpoons HPO_4^{2-}(aq) + OH^-(aq)$

3 Assess

Check for Understanding

Ask students to use K_a values to rank the electrical conductivities of 0.10*M* solutions of H_2S, HCN, and H_2CO_3. $H_2CO_3 > H_2S > HCN$ **OL**

Reteach

Write on the board: 0.1*M* HCl and 1.0*M* HClO. Ask which solution is more concentrated. 1.0*M* HClO Ask which is stronger. 0.1*M* HCl **OL**

Extension

Have interested students use a college text to investigate the leveling effect of water on the strengths of strong acids and bases. Have them explain this effect in terms of the following analogy: Suppose that the five strongest people in the world were asked to lift a 50 kg weight to determine who was strongest. **AL**

ASSESSMENT

Skill Ask students to write the ionization equation and ionization constant expression for bromous acid ($HBrO_2$). $HBrO_2(aq) + H_2O(l) \rightleftharpoons H_3O^+(aq) + BrO_2^-(aq)$ or $HBrO_2(aq) \rightleftharpoons H^+(aq) + BrO_2^-(aq)$
$K_a = [H^+][BrO_2^-]/[HBrO_2]$ **OL**

SECTION 3

1 Focus

MAINIDEA

The Logarithmic Nature of pH Write the pH scale from 0 to 14 on the board and ask what pH values represent acidic, neutral, and basic solutions. Acidic solutions have pH values less than 7. Neutral solutions have a pH of 7. Basic solutions have pH values greater than 7. To illustrate the logarithmic nature of pH, ask the following questions: Based on hydrogen-ion concentration, how much more acidic is a pH 5 solution than a pH 6 solution? The pH 5 solution is ten times more acidic than the pH 6 solution. Based on hydroxide-ion concentration, how much more basic is a pH 11 solution than a pH 9 solution? The pH 11 solution is 100 times more basic than the pH 9 solution. **OL**

2 Teach

Math in Chemistry

Exponents Remind students that a change of one unit in an exponent of 10 reflects a tenfold change in the number. For example, 10^4 is ten times larger than 10^3, and 10^{-7} is ten times smaller than 10^{-6}.

Content Background

Beyond 0–14 pH Generally, the pH range is approximately 0–14, although pH values less than 0 and greater than 14 are possible. Such solutions are rare and dangerous. Also, such solutions are so concentrated that the acidic and basic substances might not be completely ionized; therefore, the ion concentrations might not be as high as expected.

SECTION 3

Essential Questions

- What are pH and pOH?
- How are pH and pOH related to the ion product constant for water?
- How are the pH and pOH of aqueous solutions calculated?

Review Vocabulary

Le Châtelier's principle: states that if a stress is applied to a system at equilibrium, the system shifts in the direction that relieves the stress

New Vocabulary

ion product constant for water
pH
pOH

Hydrogen Ions and pH

MAINIDEA pH and pOH are logarithmic scales that express the concentrations of hydrogen ions and hydroxide ions in aqueous solutions.

CHEM 4 YOU Think of two children on a seesaw. When one side of a seesaw goes up, the other side goes down. Sometimes, the seesaw is balanced in the middle. The concentrations of hydrogen ions and hydroxide ions in water solutions behave in a similar way.

Ion Product Constant for Water

Recall that pure water contains equal concentrations of H^+ and OH^- ions produced by self-ionization. **Figure 13** shows that in self-ionization, equal numbers of hydronium and hydroxide ions are formed. The equation for the equilibrium can be simplified in the following way.

$$H_2O(l) \rightleftharpoons H^+(aq) + OH^-(aq)$$

Writing K_w The double arrow indicates that this is an equilibrium. Recall that the equilibrium constant expression is written by placing the concentrations of the products in the numerator and the concentrations of the reactants in the denominator. In this case, all terms are to the first power because all the coefficients in the balanced chemical equation are 1. The concentration of pure water is constant, so $[H_2O]$ does not appear in the denominator.

The Ion Product of Water

$$K_w = [H^+][OH^-]$$

K_w is the ion product constant for water. $[H^+]$ represents the concentration of the hydrogen ion. $[OH^-]$ represents the concentration of the hydroxide ion.

In dilute aqueous solutions, the product of the concentrations of the hydrogen ion and the hydroxide ion equals K_w.

The expression for K_w is a special equilibrium constant expression that applies only to water. The constant K_w is called the ion product constant for water. The **ion product constant for water** is the value of the equilibrium constant expression for the self-ionization of water. Experiments show that in pure water at 298 K, $[H^+]$ and $[OH^-]$ are both equal to $1.0 \times 10^{-7}M$. Therefore, at 298 K, the value of K_w is 1.0×10^{-14}.

$$K_w = [H^+][OH^-] = (1.0 \times 10^{-7})(1.0 \times 10^{-7})$$
$$K_w = 1.0 \times 10^{-14}$$

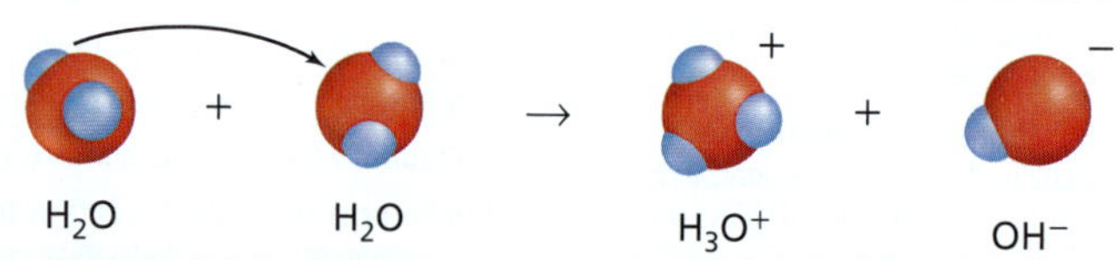

Figure 13 In the self-ionization of water, one water molecule acts as an acid, and the other acts as a base.

DIFFERENTIATED INSTRUCTION

English Learners Explain that testing acidic and basic solutions with litmus paper is so well known that related terminology is often used in the English language to describe any definitive test. Point out that such a test might be applied to determine the validity of a proposition or the ability of a person to accomplish a certain goal or successfully complete an activity. For example, one might say, "The big soccer game will be a litmus test of her goal-tending ability." **EL** **OL**

K_w and Le Châtelier's Principle The product of $[H^+]$ and $[OH^-]$ always equals 1.0×10^{-14} at 298 K. This means that if the concentration of H^+ ions increases, the concentration of OH^- ions must decrease. Similarly, an increase in the concentration of OH^- ions causes a decrease in the concentration of H^+ ions. Think about these changes in terms of Le Châtelier's principle, which you read about previously. Adding extra hydrogen ions to water at equilibrium is a stress on the system. The system reacts in a way to relieve the stress. The added H^+ ions react with OH^- ions to form more water molecules. Thus, the concentration of OH^- ions decreases. Example Problem 1 shows how you can use K_w to calculate the concentration of either H^+ or OH^- if you know the concentration of the other ion.

☑ READING CHECK **Explain** why K_w does not change when the concentration of hydrogen ions increases.

EXAMPLE Problem 1

Find help with **algebraic equations.** Math Handbook

CALCULATE $[H^+]$ AND $[OH^-]$ USING K_W At 298 K, the H^+ ion concentration in a cup of coffee is $1.0 \times 10^{-5}M$. What is the OH^- ion concentration in the coffee? Is the coffee acidic, basic, or neutral?

1 ANALYZE THE PROBLEM

You are given the concentration of the H^+ ion, and you know that K_w equals 1.0×10^{-14}. You can use the ion product constant expression to solve for $[OH^-]$. Because $[H^+]$ is greater than 1.0×10^{-7}, you can predict that $[OH^-]$ will be less than 1.0×10^{-7}.

Known
$[H^+] = 1.0 \times 10^{-5}M$
$K_w = 1.0 \times 10^{-14}$

Unknown
$[OH^-]$ = ? mol/L

2 SOLVE FOR THE UNKNOWN

Use the ion product constant expression.

$K_w = [H^+][OH^-]$ — State the ion product expression.

$[OH^-] = \frac{K_w}{[H^+]}$ — Solve for $[OH^-]$.

$[OH^-] = \frac{1.0 \times 10^{-14}}{1.0 \times 10^{-5}} = 1.0 \times 10^{-9}$ mol/L — Substitute $K_w = 1.0 \times 10^{-14}$. Substitute $[H^+] = 1.0 \times 10^{-5}M$ and solve.

Because $[H^+] > [OH^-]$, **the coffee is acidic.**

3 EVALUATE THE ANSWER

The answer is correctly stated with two significant figures because $[H^+]$ and K_w each have two significant figures. As predicted, $[OH^-]$ is less than 1.0×10^{-7} mol/L.

EXAMPLE PROBLEM

PRACTICE Problems

Do additional problems. Online Practice

22. The concentration of either the H^+ ion or the OH^- ion is given for four aqueous solutions at 298 K. For each solution, calculate $[H^+]$ or $[OH^-]$. State whether the solution is acidic, basic, or neutral.
a. $[H^+] = 1.0 \times 10^{-13}M$
b. $[OH^-] = 1.0 \times 10^{-7}M$
c. $[OH^-] = 1.0 \times 10^{-3}M$
d. $[H^+] = 4.0 \times 10^{-5}M$

23. Challenge Calculate the number of H^+ ions and the number of OH^- ions in 300 mL of pure water at 298 K.

CHEMISTRY JOURNAL

Soil pH and Plants Have interested students research plants that grow best in acidic soil and those that grow best in basic soil. Have them find out how soils can be made more acidic (sour) or more basic (sweet). Good sources of information are books on gardening, seed catalogs, florists, and people who work in plant nurseries. Students can summarize their findings in their chemistry journals. OL

☑ READING CHECK When the concentration of H^+ increases, the concentration of OH^- decreases so that the product of the two ions is always a constant.

IN-CLASS Example

Question The hydroxide-ion concentration in a detergent solution is $1.0 \times 10^{-3}M$. What is $[H^+]$ in the detergent solution?

Answer $[H^+] = 1.0 \times 10^{-11}M$
$[H^+] = (1.0 \times 10^{-14})/(1.0 \times 10^{-3})$
$= 1.0 \times 10^{-11}M$

PRACTICE Problems

Have students refer to the Selected Solutions appendix for complete solutions to odd-numbered problems. The complete solutions for all problems can be found in the Solutions Manual.

22. a. the solution is basic
b. the solution is neutral
c. the solution is basic
d. the solution is acidic

23. Number of H^+ = number of OH^- = 1.8×10^{16} ions

■ **Caption Question Fig. 14** Seawater has the higher concentration of H^+ ions; 100 times more.

Quick Demo

Carbonic Acid Put about 250 mL of 95% ethanol into a 500-mL beaker. **WARNING:** ***Make sure there are no open flames in the room.*** Add 5 or 6 drops of thymolphthalein indicator solution and sufficient 0.1*M* sodium hydroxide solution to produce a blue color. Stopper the flask until ready for the demo. Pass the flask around the class. Have each student, in turn, remove the stopper, speak a sentence into the flask, re-stopper the flask, and give the liquid a gentle swirl. At some point, enough CO_2 will have dissolved in the solution to lower the pH sufficiently that the thymolphthalein will change from blue to yellow. Ask students if they know what was special about the person's voice that caused the color to change. Some of the exhaled CO_2 dissolved, producing a weak carbonic acid solution, lowering the pH and changing the color of the indicator. **OL**

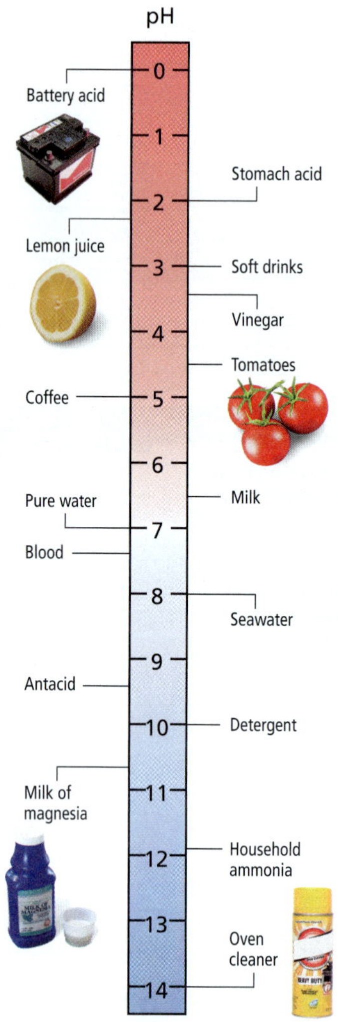

■ **Figure 14** Compare the pH values for these familiar substances.

Determine *whether seawater or detergent has a higher concentration of H^+ ions. How many times higher?*

pH and pOH

Concentrations of H^+ ions are often small numbers expressed in scientific notation. Because these numbers are cumbersome, chemists adopted an easier way to express H^+ ion concentrations.

What is pH? Chemists express the concentration of hydrogen ions using a pH scale based on common logarithms. The **pH** of a solution is the negative logarithm of the hydrogen ion concentration.

pH

$$pH = -\log [H^+]$$

$[H^+]$ represents the hydrogen ion concentration.

The pH of a solution equals the negative logarithm of the hydrogen ion concentration.

At 298 K, acidic solutions have pH values below 7. Basic solutions have pH values above 7. Thus, a solution with a pH of 0.0 is strongly acidic; a solution with a pH of 14.0 is strongly basic. The logarithmic nature of the pH scale means that a change of one pH unit represents a tenfold change in ion concentration. A solution having a pH of 3.0 has ten times the hydrogen ion concentration of a solution with a pH of 4.0. The pH scale and pH values of some common substances are shown in **Figure 14.**

What is pOH? Sometimes it is convenient to express the basicity or alkalinity of a solution on a pOH scale that mirrors the relationship between pH and $[H^+]$. The **pOH** of a solution is the negative logarithm of the hydroxide ion concentration.

pOH

$$pOH = -\log [OH^-]$$

$[OH^-]$ represents the hydroxide ion concentration.

The pOH of a solution equals the negative logarithm of the hydroxide ion concentration.

At 298 K, a solution with a pOH less than 7.0 is basic; a solution with a pOH of 7.0 is neutral; and a solution with a pOH greater than 7.0 is acidic. As with the pH scale, a change of one pOH unit expresses a tenfold change in ion concentration.

A simple relationship between pH and pOH makes it easy to calculate either quantity if the other is known.

How pH and pOH Are Related

$$pH + pOH = 14.00$$

pH represents $-\log [H^+]$. pOH represents $-\log [OH^-]$.

The sum of pH and pOH is 14.00.

Figure 15 illustrates the relationship between pH and the H^+ concentration and the relationship between pOH and OH^- concentration at 298 K.

DIFFERENTIATED INSTRUCTION

Advanced Learners Explain to students that the ion product constant of water increases with increasing temperature. For example, $K_w = 9.25 \times 10^{-14}$ at 60°C. Have students calculate the hydrogen-ion concentration and pH of pure water at this temperature. $[H^+] = 3.04 \times 10^{-7}M$, pH = 6.52 **AL**

(t)Ingram Publishing/Alamy, (cl)Sue Wilson/Alamy, (cr)foodfolio/Alamy, (bl)Eric Fowke/PhotoEdit, (br)©Ed Endicott/Alamy

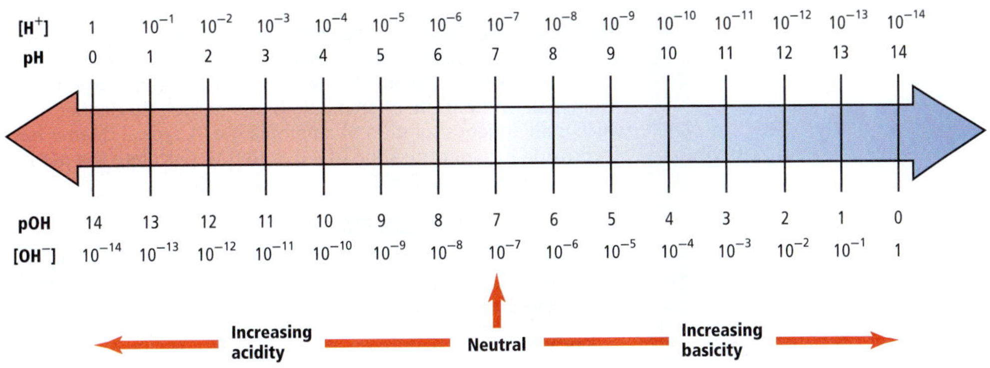

Figure 15 Study this diagram to sharpen your understanding of pH and pOH. Note that at each vertical position, the sum of pH (above the arrow) and pOH (below the arrow) equals 14. Also note that at every position, the product of $[H^+]$ and $[OH^-]$ equals 10^{-14}.

EXAMPLE Problem 2

Find help with **logarithms.** Math Handbook

CALCULATE pH FROM $[H^+]$ What is the pH of a neutral solution at 298 K?

1 ANALYZE THE PROBLEM

In a neutral solution at 298 K, $[H^+] = 1.0 \times 10^{-7}M$. You must find the negative log of $[H^+]$.

Known
$[H^+] = 1.0 \times 10^{-7}\ M$

Unknown
pH = ?

2 SOLVE FOR THE UNKNOWN

$pH = -\log [H^+]$ — State the equation for pH.

$pH = -\log (1.0 \times 10^{-7})$ — Substitute $[H^+] = 1.0 \times 10^{-7}M$.

The pH of the neutral solution at 298 K is **7.00**.

3 EVALUATE THE ANSWER

Values for pH are expressed with as many decimal places as the number of significant figures in the H^+ ion concentration. The pH is correctly stated with two decimal places.

PRACTICE Problems

Do additional problems. Online Practice

24. Calculate the pH of solutions having the following ion concentrations at 298 K.

a. $[H^+] = 1.0 \times 10^{-2}M$ **b.** $[H^+] = 3.0 \times 10^{-6}M$

25. Calculate the pH of aqueous solutions with the following $[H^+]$ at 298 K.

a. $[H^+] = 0.0055M$ **b.** $[H^+] = 0.000084M$

26. Challenge Calculate the pH of a solution having $[OH^-] = 8.2 \times 10^{-6}M$.

Get help with **logarithms.** Personal Tutor

IN-CLASS Example

Question A vinegar solution has a hydrogen-ion concentration of $4.0 \times 10^{-4}M$. What is the pH?

Answer pH = 3.40
$pH = -\log (3.98 \times 10^{-4}) = 3.40$

PRACTICE Problems

Have students refer to the Selected Solutions appendix for complete solutions to odd-numbered problems. The complete solutions for all problems can be found in the Solutions Manual.

24. a. pH = 2.00
b. pH = 5.52
25. a. pH = 2.26
b. pH = 4.08
26. pH = 8.92

Content Background

pH Although most students have heard of the term *pH*, they may wonder how the term originated. Explain that *pH* is derived from the French *pouvoir hydrogene*, which means hydrogen power.

CHEMISTRY JOURNAL

pH and Skin Have students research the pH of skin and how various products—particularly basic soaps—can interact with substances that protect the skin. Have them include their findings in their journals. OL

IN-CLASS Example

Question A particular antacid solution has a hydroxide-ion concentration of $3.2 \times 10^{-5}M$. Calculate the pOH and pH of the antacid solution.

Answer pOH = 4.49; pH = 9.51
$pOH = -\log(3.2 \times 10^{-5}) = 4.49$
$pH = 14.00 - 4.49 = 9.51$

PRACTICE Problems

Have students refer to the Selected Solutions appendix for complete solutions to odd-numbered problems. The complete solutions for all problems can be found in the Solutions Manual.

27. a. pOH = 6.00
pH = 8.00
b. pOH = 3.19
pH = 10.81
c. pH = 8.44
pOH = 5.56
d. pH = 1.60
pOH = 12.40
28. a. pOH = 4.48
pH = 9.52
b. pH = 2.02
pOH = 11.98
29. pH = 3.70
pOH = 10.30

EXAMPLE Problem 3

CALCULATE pOH AND pH FROM [OH⁻] In **Figure 16,** a cow is being fed straw and hay that has been treated with ammonia. The addition of ammonia to animal feed promotes protein growth in the animal. Another use of ammonia is as a household cleaner, which is an aqueous solution of ammonia gas. A typical cleaner has a hydroxide-ion concentration of $4.0 \times 10^{-3}M$. Calculate the pOH and pH of a cleaner at 298 K.

1 ANALYZE THE PROBLEM

You have been given the concentration of hydroxide ion and must calculate pOH and pH. First, calculate pOH using its definition. Then, calculate pH using the relationship pH + pOH = 14.00.

Known	**Unknown**
$[OH^-] = 4.0 \times 10^{-3}M$	pOH = ?
	pH = ?

2 SOLVE FOR THE UNKNOWN

$pOH = -\log[OH^-]$ — State the equation for pOH.

$pOH = -\log(4.0 \times 10^{-3})$ — Substitute $[OH^-] = 4.0 \times 10^{-3}M$.

The pOH of the solution is **2.40.**

Use the relationship between pH and pOH to find the pH.

$pH + pOH = 14.00$ — State the equation that relates pH and pOH.

$pH = 14.00 - pOH$ — Solve for pH.

$pH = 14.00 - 2.40 = 11.60$ — Substitute pOH = 2.40.

The pH of the solution is **11.60.**

3 EVALUATE THE ANSWER

The given concentration has two significant figures, so pH and pOH are correctly expressed with two decimal places. Because ammonia is a base, a small pOH value and a large pH value are reasonable.

■ **Figure 16** Farmers are able to increase the nutritional value of low-quality vegetable materials such as straw, hay, and other crop residue by immersing the materials in an atmosphere of ammonia gas for three weeks.

PRACTICE Problems

Do additional problems. Online Practice

27. Calculate the pH and pOH of aqueous solutions with the following concentrations at 298 K.
a. $[OH^-] = 1.0 \times 10^{-6}M$
b. $[OH^-] = 6.5 \times 10^{-4}M$
c. $[H^+] = 3.6 \times 10^{-9}M$
d. $[H^+] = 2.5 \times 10^{-2}M$

28. Calculate the pH and pOH of aqueous solutions with the following concentration at 298 K.
a. $[OH^-] = 0.000033M$
b. $[H^+] = 0.0095M$

29. Challenge Calculate pH and pOH for an aqueous solution containing 1.0×10^{-3} mol of HCl dissolved in 5.0 L of solution.

IMAGEMORE/age fotostock

ASSESSMENT

Knowledge Help students better understand the exponential nature of the pH scale by asking them to compare the acidity of a solution with pH = 1 with the acidity of a solution with pH = 3. pH = 1 is 10^2 or 100 times more acidic. Also, have them compare the basicity of a solution with pH = 12 to one with pH = 9. pH = 12 is 10^3 or 1000 times more basic. OL

Calculating ion concentrations from pH Sometimes, you have to calculate the concentration of H^+ ions and OH^- ions from the pH of a solution. Example Problem 4 shows how to do this.

EXAMPLE Problem 4

Find help with **antilogarithms.** Math Handbook

EXAMPLE PROBLEM

CALCULATE $[H^+]$ AND $[OH^-]$ FROM pH What are $[H^+]$ and $[OH^-]$ in a healthy person's blood that has a pH of 7.40? Assume that the temperature of the blood is 298 K.

1 ANALYZE THE PROBLEM

You have been given the pH of a solution and must calculate $[H^+]$ and $[OH^-]$. You can obtain $[H^+]$ using the equation that defines pH. Then, subtract the pH from 14.00 to obtain the pOH and use the equation that defines pOH to get $[OH^-]$.

Known	Unknown
pH = 7.40	$[H^+]$ = ? mol/L
	$[OH^-]$ = ? mol/L

2 SOLVE FOR THE UNKNOWN

Determine $[H^+]$.

$pH = -\log [H^+]$	State the equation for pH.
$-pH = \log [H^+]$	Multiply both sides of the equation by −1.
$[H^+] = \text{antilog}(-pH)$	Take the antilog of each side to solve for $[H^+]$.
$[H^+] = \text{antilog}(-7.40)$	Substitute pH = 7.40.
$[H^+] = 4.0 \times 10^{-8}M$	A calculator shows that the antilog of −7.40 is 4.0×10^{-8}.

The concentration of H^+ ions in the blood is $4.0 \times 10^{-8}M$.

Determine $[OH^-]$.

$pH + pOH = 14.00$	State the equation that relates pH and pOH.
$pOH = 14.00 - pH$	Solve for pOH.
$pOH = 14.00 - 7.40 = 6.60$	Substitute pH = 7.40.
$pOH = -\log [OH^-]$	State the equation for pOH.
$-pOH = \log [OH^-]$	Multiply both sides of the equation by −1.
$[OH^-] = \text{antilog}(-6.60)$	Take the antilog of each side and substitute pOH = 6.60.
$[OH^-] = 2.5 \times 10^{-7}M.$	A calculator shows that the antilog of −6.60 is 2.5×10^{-7}.

The concentration of OH^- ions in the blood is $2.5 \times 10^{-7}M$.

3 EVALUATE THE ANSWER

The given pH has two decimal places, so the answers must have two significant figures. A $[H^+]$ less than 10^{-7} and a $[OH^-]$ greater than 10^{-7} are reasonable, given the initial pH.

PRACTICE Problems

Do additional problems. Online Practice

30. Calculate $[H^+]$ and $[OH^-]$ in each of the following solutions.

- **a.** Milk, pH = 6.50
- **b.** Lemon juice, pH = 2.37
- **c.** Milk of magnesia, pH = 10.50
- **d.** Household ammonia, pH = 11.90

31. Challenge Calculate the $[H^+]$ and $[OH^-]$ in a sample of seawater with a pOH = 5.60.

IN-CLASS Example

Question The pH of a soft drink at 298 K is found to be 3.08. What are $[H^+]$ and $[OH^-]$ in the soft drink?

Answer $[H^+] = 8.3 \times 10^{-4}M$; $[OH^-] = 1.2 \times 10^{-11}M$

$[H^+] = \text{antilog}(-3.08) = 8.3 \times 10^{-4}M$

$pOH = 14.00 - 3.08 = 10.92$

$[OH^-] = \text{antilog}(-10.92) = 1.2 \times 10^{-11}M$

PRACTICE Problems

Have students refer to the Selected Solutions appendix for complete solutions to odd-numbered problems. The complete solutions for all problems can be found in the Solutions Manual.

30. a. $[H^+] = 3.2 \times 10^{-7}M$
$[OH^-] = 3.1 \times 10^{-8}M$
b. $[H^+] = 4.3 \times 10^{-3}M$
$[OH^-] = 2.3 \times 10^{-12}M$
c. $[H^+] = 3.2 \times 10^{-11}M$
$[OH^-] = 3.2 \times 10^{-4}M$
d. $[H^+] = 1.3 \times 10^{-12}M$
$[OH^-] = 7.9 \times 10^{-3}M$

31. $[OH^-] = 2.5 \times 10^{-6}M$
$[H^+] = 3.0 \times 10^{-9}M$

DIFFERENTIATED INSTRUCTION

Struggling Learners To reinforce that the concentration of a liquid or solid is essentially constant, lead them through the process of calculating the molar concentration of liquid water at 4°C. $1000\text{ mL } H_2O/1\text{ L } H_2O \times 1.00\text{ g } H_2O/1\text{ mL } H_2O \times 1\text{ mol } H_2O/18.0\text{ g } H_2O = 55.6\text{ mol/L}$. Point out that unless the density of water changes because of a change in temperature, 55.6 mol/L can be considered a constant for the concentration of water. BL EL

■ **Caption Question Fig. 17**
$[H^+] = 0.1M$
$[OH^-] = 0.1M$

Extension

Natural Indicators Have interested students investigate the natural indicators present in the leaves of colored plants and vegetables, such as purple cabbage. Have students boil the leaves of various plants to extract the pigments, then add small amounts of the resulting solutions to solutions of known pH. Ask them to make charts of pH colors for an extracted pigment and share their results with the class. **AL**

☑ READING CHECK A weak acid is not completely ionized. Therefore, the moles per liter of acid dissolved in the solution is not equal to the $[H^+]$.

Math in Chemistry

Help students use the antilog function on their calculators to calculate $[H^+]$ from pH and $[OH^-]$ from pOH. To calculate $[H^+]$ from pH:

1. Enter the pH in the calculator.
2. Change the sign of pH by pressing the [+/−] key.
3. Take the antilog (inverse log) of the negative pH by pressing the key [10^x] or the key sequence [inv][log]. The same three steps can be used to calculate $[OH^-]$ from pOH. **OL**

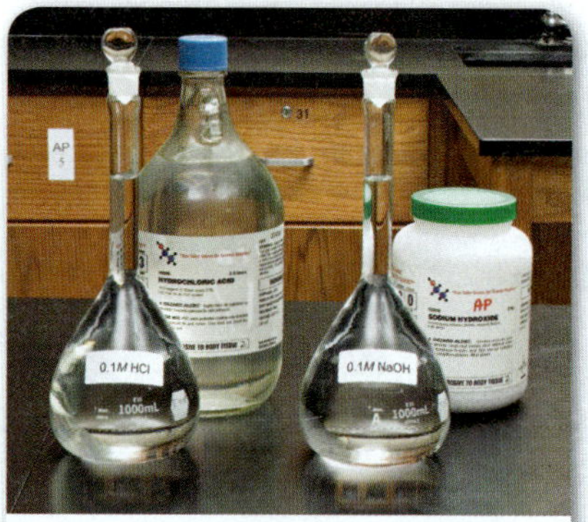

■ **Figure 17** The label on a bottle of a strong acid or a strong base tells you the concentration of hydrogen ions or hydroxide ions in the solution. That is because, in solution, strong acids and bases exist entirely as ions.
State *the $[H^+]$ in the HCl flask and the $[OH^-]$ in the NaOH flask.*

Molarity and the pH of strong acids Look at the flasks of acid and base solutions in **Figure 17.** The solutions have just been made up and are labeled with their molarity, which is the number of moles of molecules or formula units that were dissolved in 1 L of solution. One flask contains a strong acid (HCl), the other a strong base (NaOH). Recall that strong acids and bases are essentially 100% in the form of ions in solution. That means that the following reaction for the ionization of HCl goes to completion.

$$HCl(aq) \rightarrow H^+(aq) + Cl^-(aq)$$

Every HCl molecule produces one H^+ ion. The bottle labeled 0.1M HCl contains 0.1 mol of H^+ ions per liter and 0.1 mol of Cl^- ions per liter. For all strong monoprotic acids, the concentration of the acid is the concentration of H^+ ions. Thus, you can use the molarity of the acid to calculate pH.

Molarity and the pH of strong bases Similarly, the 0.1M solution of the strong base NaOH in **Figure 17** is fully ionized.

$$NaOH(aq) \rightarrow Na^+(aq) + OH^-(aq)$$

One formula unit of NaOH produces one OH^- ion. Thus, the concentration of the OH^- ions is the same as the molarity of the solution, 0.1M.

Some strong bases, such as calcium hydroxide, $Ca(OH)_2$, contain two or more OH^- ions in each formula unit. The concentration of OH^- ion in a solution of $Ca(OH)_2$ is twice the molarity of the ionic compound. For example, the concentration of hydroxide ions in a $7.5 \times 10^{-4}M$ solution of $Ca(OH)_2$ is $7.5 \times 10^{-4}M \times 2 = 1.5 \times 10^{-3}M$.

Although strong acids and strong bases are completely ionized in dilute aqueous solutions, remember that weak acids and weak bases are only partially ionized. Therefore, you must use K_a and K_b values to determine the concentrations of H^+ and OH^- ions in solutions of weak acids and bases.

☑ READING CHECK **Explain** why you cannot obtain the $[H^+]$ directly from the molarity of a weak acid solution.

Calculating K_a from pH Suppose you measured the pH of a 0.100M solution of the weak acid HF and found it to be 3.20. Would you have enough information to calculate K_a for HF?

$$HF(aq) \rightleftharpoons H^+(aq) + F^-(aq)$$

$$K_a = \frac{[H^+][F^-]}{[HF]}$$

From the pH, you could calculate $[H^+]$. Then, remember that for every mole per liter of H^+ ion there must be an equal concentration of F^- ion. That means you know two of the variables in the K_a expression. What about the third, [HF]? The concentration of HF at equilibrium is equal to the initial concentration of the acid (0.100M) minus the moles per liter of HF that dissociated, which is equal to $[H^+]$.

Matt Meadows

CHEMISTRY PROJECT

Balanced Shampoo Students might have heard or read the term "pH balanced" in ads for various shampoos. Ask them to bring in samples of the shampoos they use and work out a procedure to test the pH of these products. For example, they can dip a separate piece of pH paper into a sample of each shampoo and record the pH. Ask the class to compile a list of the ingredients in all the shampoos. Challenge them to find the ingredients that help achieve the proper pH. **OL** **COOP LEARN**

EXAMPLE Problem 5

Find help with **scientific notation.** Math Handbook

CALCULATE K_a FROM pH Formic acid is used to process latex tapped from rubber trees into natural rubber. The pH of a 0.100*M* solution of formic acid (HCOOH) is 2.38. What is K_a for HCOOH?

1 ANALYZE THE PROBLEM

You are given the pH of the formic acid solution, which allows you to calculate the concentration of the hydrogen ion.

$$HCOOH(aq) \rightleftharpoons H^+(aq) + HCOO^-(aq)$$

The balanced chemical equation shows that the concentration of $HCOO^-$ equals the concentration of H^+. The concentration of un-ionized HCOOH is the difference between the initial concentration of the acid and $[H^+]$.

Known	**Unknown**
pH = 2.38	K_a = ?
concentration of the solution = 0.100*M*	

2 SOLVE FOR THE UNKNOWN

Use the pH to calculate $[H^+]$.

$pH = -\log[H^+]$ — **Write the equation for pH.**

$[H^+] = \text{antilog}(-pH)$ — **Multiply both sides by −1 and take the antilog of each side.**

$[H^+] = \text{antilog}(-2.38)$ — **Substitute pH = 2.38.**

$[H^+] = 4.2 \times 10^{-3}M$ — **A calculator shows that the antilog of −2.38 is 4.2×10^{-3}.**

$[HCOO^-] = [H^+] = 4.2 \times 10^{-3}M$

[HCOOH] equals the initial concentration minus $[H^+]$.

$[HCOOH] = 0.100M - 4.2 \times 10^{-3}M = 0.096M$ — **Subtract $[H^+]$ from the initial [HCOOH].**

$K_a = \dfrac{[H^+][HCOO^-]}{[HCOOH]}$ — **State the acid ionization constant expression.**

$K_a = \dfrac{(4.2 \times 10^{-3})(4.2 \times 10^{-3})}{(0.096)} = 1.8 \times 10^{-4}$ — **Substitute $[H^+] = 4.2 \times 10^{-3}M$, $[HCOO^-] = 4.2 \times 10^{-3}M$, and $[HCOOH] = 0.096M$.**

The acid ionization constant for HCOOH is 1.8×10^{-4}.

3 EVALUATE THE ANSWER

The K_a is reasonable for a weak acid. The answer is correctly reported with two significant figures.

EXAMPLE PROBLEM

PRACTICE Problems

Do additional problems. Online Practice

32. Calculate the K_a for the following acids using the given information.

a. 0.220*M* solution of H_3AsO_4, pH = 1.50 **b.** 0.0400*M* solution of $HClO_2$, pH = 1.80

33. Calculate the K_a of the following acids using the given information.

a. 0.00330*M* solution of benzoic acid (C_6H_5COOH), pOH = 10.70

b. 0.100*M* solution of cyanic acid (HCNO), pOH = 11.00

c. 0.150*M* solution of butanoic acid (C_3H_7COOH), pOH = 11.18

34. Challenge Calculate the K_a of a 0.0091*M* solution of an unknown acid (HX) having a pOH of 11.32. Use **Table 4** to identify the acid.

IN-CLASS Example

Question At 298 K, the pH of a 0.150*M* solution of oxalic acid ($H_2C_2O_4$) is 1.16. What is K_a for $H_2C_2O_4$ at 298 K?

Answer $K_a = 5.9 \times 10^{-2}$

$H_2C_2O_4 \rightleftharpoons H^+ + HC_2O_4^-$

$[H^+] = \text{antilog}(-1.16) = 6.9 \times 10^{-2}M$

$[H_2C_2O_4] = 0.150M - 6.9 \times 10^{-2}M = 0.081M$

$K_a = \dfrac{(6.9 \times 10^{-2})^2}{0.081} = 5.9 \times 10^{-2}$

PRACTICE Problems

Have students refer to the Selected Solutions appendix for complete solutions to odd-numbered problems. The complete solutions for all problems can be found in the Solutions Manual.

32. a. $K_a = 5.4 \times 10^{-3}$
b. $K_a = 1.1 \times 10^{-2}$

33. a. $K_a = 8.9 \times 10^{-5}$
b. $K_a = 1.0 \times 10^{-5}$
c. $K_a = 1.5 \times 10^{-5}$

34. $K_a = 6.3 \times 10^{-4}$
The acid could be hydrofluoric acid (HF).

DIFFERENTIATED INSTRUCTION

Advanced Learners Challenge advanced learners by having them use the quadratic formula to calculate the pH of a 0.100*M* solution of hypobromous acid at 298 K. K_a for HBrO at 298 K is 2.06×10^{-9}.

	HBrO(aq)	H^+(aq)	+ BrO^-(aq)
Initial *M*	0.100	0.0	0.0
Equilibrium *M*	0.100 − *x*	*x*	*x*

Substitute the equilibrium concentrations of HBrO, H^+, and BrO^- into the acid ionization constant expression.

$$2.06 \times 10^{-9} = \frac{x^2}{0.100 - x}$$

$$x^2 + (2.06 \times 10^{-9})x - 2.06 \times 10^{-10} = 0$$

$$x = \frac{-2.06 \times 10^{-9} \pm \sqrt{(2.06 \times 10^{-9})^2 - 4(-2.06 \times 10^{-10})}}{2} = 1.43 \times 10^{-5}$$

$$pH = -\log[H^+] = -\log(1.43 \times 10^{-5}) = -\log(1.43 \times 10^{-5}) = 4.845$$ AL

3 Assess

Check for Understanding

Have students compare the [H^+] and the acidity of three solutions: solution 1 with pH = 6.00, solution 2 with pH = 3.00, solution 3 with pH = 1.00. Solution 1 is least acidic, [H^+] = 1×10^{-6}. In solution 2, [H^+] = 1×10^{-3}; therefore, solution 2 is 1000 times more acidic than solution 1. In solution 3, [H^+] = 1×10^{-1}; therefore, solution 3 is 100 times more acidic than solution 2. **OL**

Reteach

Have students use the relationship between pH and pOH and the ion product constant for water to calculate [OH^-] in a solution having pH = 2.5. [OH^-] = 3.16×10^{-12} **OL**

Extension

Advise students that it is possible for solutions to have pH values less than 0 and greater than 14. Have students calculate the pH of 1.5*M* HCl solution. pH = −0.18 **OL**

ASSESSMENT

Knowledge Ask students why S. P. L. Sørensen, the Dutch biochemist who designed the pH scale, created a scale from 1 to 14, with 7 being neutral. Because of the autoionization of water ($2H_2O(l) \rightleftharpoons H_3O^+(aq) + OH^-(aq)$), the concentrations of both hydronium and hydroxide ions are 10^{-7}*M* at 25°C. Logically, Sørensen related the pH of a solution to the exponent of the hydronium-ion concentration. **OL**

Figure 18 The approximate pH of a solution can be obtained by wetting a piece of pH paper with the solution and comparing the color of the wet paper with a set of standard colors as shown in **a.** The portable pH meter in **b,** which is being used to measure the pH of rain water, provides a more accurate measurement in the form of a digital display of the pH.

Measuring pH Perhaps you have used indicator paper to measure the pH of a solution. The litmus paper you used in the Launch Lab is an example of a kind of pH paper. All pH paper is treated with one or more substances called indicators that change color depending on the concentration of hydrogen ions in a solution. While litmus paper only indicates if a substance is an acid or a base, pH papers treated with several indicators can identify a range of pH values. Phenolphthalein, which you also used in the Launch Lab, is another example of an indicator.

When a strip of pH paper is dipped into an acidic solution or a basic solution, the color of the paper changes. To determine the pH, the new color of the paper is compared with standard pH colors on a chart, as shown in **Figure 18.** The pH meter in **Figure 18** provides a more accurate measure of pH. When electrodes are placed in a solution, the meter gives a direct analog or digital readout of pH.

SECTION 3 REVIEW

Section Self-Check

Section Summary

- The ion product constant for water, K_w, equals the product of the H^+ ion concentration and the OH^- ion concentration.
- The pH of a solution is the negative log of the hydrogen ion concentration. The pOH is the negative log of the hydroxide ion concentration. pH plus pOH equals 14.
- A neutral solution has a pH of 7.0 and a pOH of 7.0 because the concentrations of hydrogen ions and hydroxide ions are equal.

35. **MAIN**IDEA **Explain** why the pH of an acidic solution is always a smaller number than the pOH of the same solution.
36. **Describe** how you can determine the pH of a solution if you know its pOH.
37. **Explain** the significance of K_w in aqueous solutions.
38. **Explain,** using Le Châtelier's principle, what happens to the [H^+] of a 0.10*M* solution of acetic acid when a drop of NaOH solution is added.
39. **List** the information needed to calculate the K_a of a weak acid.
40. **Calculate** The pH of a tomato is approximately 4.50. What are [H^+] and [OH^-] in a tomato?
41. **Determine** the pH of a solution that contains 1.0×10^{-9} mol of OH^- ions per liter.
42. **Calculate** the pH of the following solutions.
    a. 1.0*M* HI
    b. 0.050*M* HNO_3
    c. 1.0*M* KOH
    d. 2.4×10^{-5}*M* $Mg(OH)_2$
43. **Interpret Diagrams** Refer to **Figure 15** to answer these questions: What happens to the [H^+], [OH^-], pH, and pOH as a neutral solution becomes more acidic? As a neutral solution become more basic?

SECTION 3 REVIEW

35. The sum of pH and pOH is 14.00. If a solution is acidic, its pH is less than 7.00. Therefore, pOH must be greater than 7.00.
36. Subtract the pOH from 14.00.
37. If one ion concentration is known, the other can be calculated using the K_w expression.
38. The increase in OH^- ion from the drop of NaOH shifts the self-ionization of water toward the left and increases the amount of undissociated water molecules. [OH^-] increases and [H^+] decreases.
39. the pH, pOH, or [H^+] and the initial concentration of the acid; or K_b
40. [H^+] = 3.2×10^{-5}*M*, [OH^-] = 3.2×10^{-10}*M*
41. pH = 5.00
42. **a.** pH = 0.00 **c.** pH = 14.00
    b. pH = 1.30 **d.** pH = 9.68
43. As the solution becomes more acidic, [H^+] increases from 10^{-7} to 1, [OH^-] decreases from 10^{-7} to 10^{-14}, pH changes from 7 to 0 and pOH changes from 7 to 14. As a neutral solution becomes more basic, [H^+] decreases from 10^{-7} to 10^{-14}, [OH^-] increases from 10^{-7} to 1, pH changes from 7 to 14 and pOH changes from 7 to 0.

SECTION 4

Essential Questions

- What do chemical equations of neutralization reactions look like?
- How are neutralization reactions used in acid-base titrations?
- How do the properties of buffered and unbuffered solutions compare?

Review Vocabulary

stoichiometry: the study of quantitative relationships between the amounts of reactants used and products formed by a chemical reaction; is based on the law of conservation of mass

New Vocabulary

neutralization reaction
salt
titration
titrant
equivalence point
acid-base indicator
end point
salt hydrolysis
buffer
buffer capacity

Neutralization

MAINIDEA In a neutralization reaction, an acid reacts with a base to produce a salt and water.

CHEM 4 YOU When two teams in a debate present equally convincing arguments, you might find that you are neutral—favoring neither one point of view nor the other. In a similar way, a solution is neutral when the numbers of hydrogen ions and hydroxide ions are equal.

Reactions Between Acids and Bases

If you were to experience heartburn or indigestion, you might take one of the antacids illustrated in **Figure 19** to relieve your discomfort. What kind of reaction occurs when magnesium hydroxide ($Mg(OH)_2$), the active ingredient in milk of magnesia, contacts hydrochloric acid solution (H^+ and Cl^-) produced by the stomach?

When $Mg(OH)_2$ and HCl react, a neutralization reaction occurs. A **neutralization reaction** is a reaction in which an acid and a base in an aqueous solution react to produce a salt and water. A **salt** is an ionic compound made up of a cation from a base and an anion from an acid. Neutralization is a double-replacement reaction.

Writing neutralization equations In the reaction between magnesium hydroxide and hydrochloric acid, magnesium replaces hydrogen in HCl and hydrogen replaces magnesium in $Mg(OH)_2$.

$$Mg(OH)_2(aq) + 2HCl(aq) \rightarrow MgCl_2(aq) + 2H_2O(l)$$
$$\text{Base} + \text{Acid} \rightarrow \text{Salt} + \text{Water}$$

Note that the cation from the base (Mg^{2+}) is combined with the anion from the acid (Cl^-) in the salt $MgCl_2$.

When writing neutralization equations, you must know whether all of the reactants and products in the solution exist as molecules or as formula units. For example, examine the formula equation and complete ionic equation for the reaction between hydrochloric acid and sodium hydroxide.

$$HCl(aq) + NaOH(aq) \rightarrow NaCl(aq) + H_2O(l)$$

Figure 19 A dose of any of these antacids can relieve the symptoms of acid indigestion by reacting with the acidic solution in the stomach and neutralizing it.

CHEMISTRY PROJECT

Antacids Have students examine the labels of several brands of antacids, identify the substances that neutralize acids, and investigate the possible hazards and side effects of taking these products. Have students research and compare the action of histamine receptor antagonists to antacids in reducing stomach acids. OL

SECTION 4

1 Focus

MAINIDEA

Acid-Base Neutralization Ask for a student volunteer to come to the board and write the molecular equation for the double-replacement reaction between hydrochloric acid and sodium hydroxide. $HCl(aq) + NaOH(aq) \rightarrow NaCl(aq) + H_2O(l)$ Ask what type of compound each formula represents and write the answers below the formulas in the molecular equation. an acid + a base → a salt + water Ask what kind of solution is produced when the hydrochloric acid and sodium hydroxide react in stoichiometric quantities. A neutral, salt solution is produced. Point out that such reactions are called neutralization reactions. OL

2 Teach

Concept Development

Strong and Weak Acids Explain to students that because a weak acid is only partially ionized in aqueous solution, the net ionic equation for a weak acid-strong base neutralization reaction is different from the net ionic equation for a strong acid-strong base reaction. For example, the net ionic equation for the reaction between aqueous HClO and NaOH solutions is $HClO(aq) + OH^-(aq) \rightarrow H_2O(l) + ClO^-(aq)$, whereas, the net ionic equation for the reaction between aqueous HCl and NaOH solutions is $H^+(aq) + OH^-(aq) \rightarrow H_2O(l)$.

ASSESSMENT

Performance Have interested students design an experiment to test the effectiveness of various over-the-counter antacids. After you approve their procedure, have them carry out the experiment in the laboratory. **OL**

READING CHECK $HNO_3(aq) + KOH(aq) \longrightarrow KNO_3(aq) + H_2O(l)$; $H^+(aq) + NO_3^-(aq) + K^+(aq) + OH^-(aq) \longrightarrow K^+(aq) + NO_3^-(aq) + H_2O(l)$
Eliminate spectator ions.
$H^+(aq) + OH^-(aq) \longrightarrow H_2O(l)$

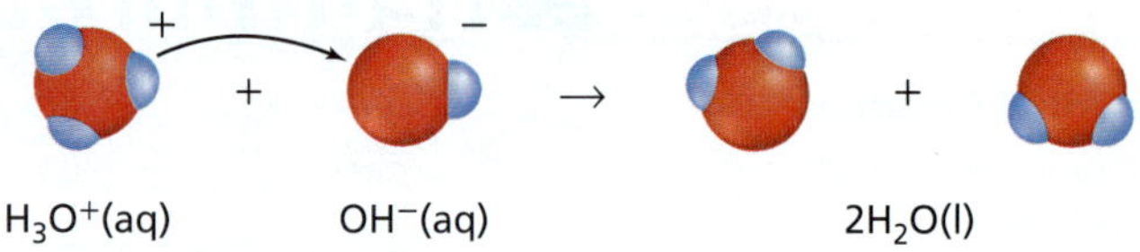

Figure 20 A hydronium ion transfers a hydrogen ion to a hydroxide ion. The loss of the hydrogen ion by H_3O^+ results in a water molecule. The gain of a hydrogen ion by OH^- also results in a water molecule.

View an **animation about a neutralization reaction.**

Concepts In Motion

Because HCl is a strong acid, NaOH a strong base, and NaCl a soluble salt, all three compounds exist as ions in an aqueous solution.

$$H^+(aq) + Cl^-(aq) + Na^+(aq) + OH^-(aq) \rightarrow Na^+(aq) + Cl^-(aq) + H_2O(l)$$

The chloride ion and the sodium ion appear on both sides of the equation, so they are spectator ions. They can be eliminated to obtain the net ionic equation for the neutralization of a strong acid by a strong base.

$$H^+(aq) + OH^-(aq) \rightarrow H_2O(l)$$

Recall that in an aqueous solution, a H^+ ion exists as a H_3O^+ ion, so the net ionic equation for an acid-base neutralization reaction is

$$H_3O^+(aq) + OH^-(aq) \rightarrow 2H_2O(l).$$

This neutralization reaction is illustrated in **Figure 20.**

READING CHECK **Demonstrate** that the equation illustrated in **Figure 20** represents the neutralization of any strong acid by a strong base by writing the complete ionic equation and the net ionic equation for the neutralization of HNO_3 by KOH.

View an **animation about titration.**

Concepts In Motion

Acid-base titration The stoichiometry of an acid-base neutralization reaction is the same as that of any other reaction that occurs in solution. In the antacid reaction described above, 1 mol of $Mg(OH)_2$ neutralizes 2 mol of HCl.

$$Mg(OH)_2(aq) + 2HCl(aq) \rightarrow MgCl_2(aq) + 2H_2O(l)$$

In the reaction of sodium hydroxide and hydrogen chloride, 1 mol of NaOH neutralizes 1 mol of HCl.

$$NaOH(aq) + HCl(aq) \rightarrow NaCl(aq) + H_2O(l)$$

Stoichiometry provides the basis for a procedure called titration, which is used to determine the concentrations of acidic and basic solutions. **Titration** is a method for determining the concentration of a solution by reacting a known volume of that solution with a solution of known concentration. If you wish to find the concentration of an acid solution, you would titrate the acid solution with a solution of a base of known concentration. You could also titrate a base of unknown concentration with an acid of known concentration. How is an acid-base titration performed? **Figure 21** illustrates one type of setup for the titration procedure outlined on the following page. In this procedure a pH meter is used to monitor the change in the pH as the titration progresses.

Figure 21 In the titration of an acid by a base, the pH meter measures the pH of the acid solution in the beaker as a solution of a base with a known concentration is added from the buret.

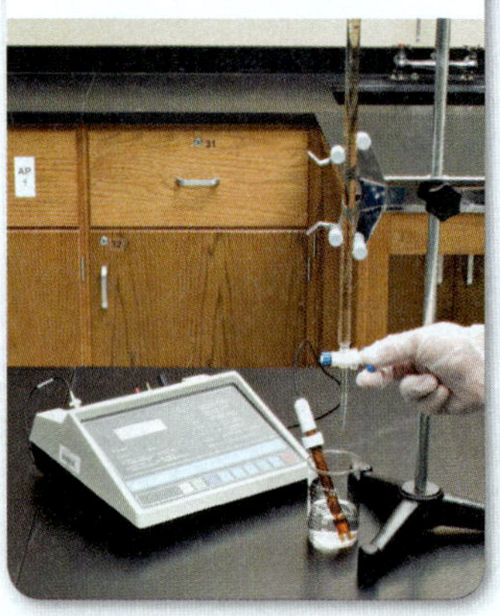

CHEMISTRY JOURNAL

What happens to hydrogen? Have students write creative essays that trace a hydrogen atom in a molecule of hydrogen chloride as the molecule dissolves in water and produces hydrochloric acid, which is then neutralized by an aqueous sodium hydroxide solution. **OL**

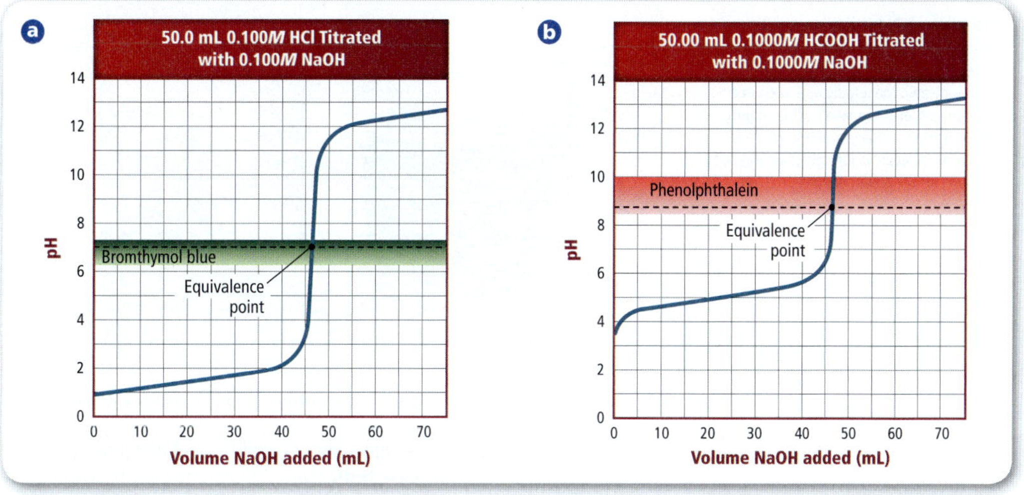

■ **Figure 22** In the titration of a strong acid by a strong base shown in **a**, a steep rise in the pH of the acid solution indicates that all of the H^+ ions from the acid have been neutralized by the OH^- ions of the base. The point at which the curve flexes (at its intersection with the dashed line) is the equivalence point of the titration. Bromthymol blue is an indicator that changes color at this equivalence point. In **b**, a weak acid (HCOOH) is titrated with a strong base (NaOH). The equivalence point is not at a pH of 7. Phenolphthalein is an indicator that changes color at this equivalence point.

Compare *the equivalence points in the two illustrations.*

Titration procedure How is an acid-base titration performed?

1. A measured volume of an acidic or basic solution of unknown concentration is placed in a beaker. The electrodes of a pH meter are immersed in this solution, and the initial pH of the solution is read and recorded.
2. A buret is filled with the titrating solution of known concentration. This is called the standard solution, or **titrant.**
3. Measured volumes of the standard solution are added slowly and mixed into the solution in the beaker. The pH is read and recorded after each addition. This process continues until the reaction reaches the **equivalence point,** which is the point at which moles of H^+ ion from the acid equal moles of OH^- ion from the base.

Figure 22a shows how the pH of the solution changes during the titration of 50.0 mL of 0.100*M* HCl, a strong acid, with 0.100*M* NaOH, a strong base. The initial pH of the 0.100*M* HCl is 1.00. As NaOH is added, the acid is neutralized and the solution's pH increases gradually. However, when nearly all of the H^+ ions from the acid have been used up, the pH increases dramatically with the addition of an exceedingly small volume of NaOH. This abrupt increase in pH occurs at the equivalence point of the titration. Beyond the equivalence point, the addition of more NaOH again results in a gradual increase in pH.

You might think that all titrations must have an equivalence point at pH 7 because that is the point at which concentrations of hydrogen ions and hydroxide ions are equal and the solution is neutral. This is not the case, however. Some titrations have equivalence points at pH values less than 7, and some have equivalence points at pH values greater than 7. These differences occur because of reactions between the newly formed salts and water, as you will read later. **Figure 22b** shows that the equivalence point for the titration of methanoic acid (a weak acid) with sodium hydroxide (a strong base) lies between pH 8 and pH 9.

☑ GRAPH CHECK **Identify** two ways in which the graphs in **Figure 22** are different.

View an **animation about neutralization reactions.**

Concepts In Motion

Apply Chemistry

Antacids Point out that the low solubilities of aluminum hydroxide and magnesium hydroxide make these compounds useful as antacids. If they were highly soluble, the high concentrations of hydroxide ions produced as the compounds dissolved would harm the tissues of the mouth, esophagus, and stomach.

■ **Caption Question Fig. 22**
In **a** the equivalence point is 7; in **b** the equivalence point is 8.2.

☑ GRAPH CHECK The equivalence points are different. In the graph of the titration of a strong acid by a strong base, the pH of the strong acid at the start of the titration is 1.00, whereas the pH of the weak acid is about 3.6. The vertical portion of the graph of a strong acid by a strong base is longer than the vertical portion of the titration of a weak acid.

CULTURAL DIVERSITY

For a Headache, Find a Willow Tree Native Americans discovered that they could relieve pain by chewing on willow bark. However, it was not until the 1800s that scientists isolated the bark's active ingredient and identified it as salicylic acid. Although salicylic acid is effective, many people do not tolerate it well—and some have allergic reactions to the compound. Fortunately, a German chemist named Felix Hoffman, who worked for the Bayer Company in the late 1800s, discovered how to modify salicylic acid so that it was less acidic, but still effective. The compound Hoffman synthesized, acetylsalicylic acid, is analgesic (anti-pain), antipyretic (anti-fever), and anti-inflammatory. Acetylsalicylic acid was first marketed as a powder called aspirin in packets and capsules in 1899.

Concept Development

Titration Explain to students that titration is a form of volumetric analysis, a laboratory procedure that can be applied to reactions other than acid-base neutralization. Point out, for example, that environmental chemists often determine the concentration of oxygen in surface waters using a titration in which a sodium thiosulfate solution is the titrant and a starch solution is the indicator.

ASSESSMENT

Performance Have students make drawings of a titration setup. Have them label and describe the function of each part of the setup, including the titrant and the solution to be titrated. OL

Content Background

Dyes Indicators must be molecules that are also able to act as both acids and bases. They must absorb light in the visible portion of the electromagnetic spectrum. The arrangement of double bonds in a molecule determines the color of light absorbed. The loss or gain of an acidic hydrogen alters the arrangement slightly and causes the molecule to change color. Molecules that impart color are often called dyes.

■ **Figure 23** The familiar dark color of tea becomes lighter when lemon juice is added. A substance contained in tea is an indicator. Most indicators are large molecules that act as weak acids. Slight differences in bonding patterns when an indicator molecule is ionized or un-ionized account for the color changes.

Acid-base indicators Chemists often use a chemical dye rather than a pH meter to detect the equivalence point of an acid-base titration. Chemical dyes whose colors are affected by acidic and basic solutions are called **acid-base indicators.** Many natural substances act as indicators. If you use lemon juice in your tea, you might have noticed that the brown color of tea gets lighter when lemon juice is added, as shown in **Figure 23.** Tea contains compounds called polyphenols that have slightly ionizable hydrogen atoms and therefore are weak acids. Adding acid in the form of lemon juice to a cup of tea depresses the ionization according to Le Châtelier's principle, and the color of the un-ionized polyphenols becomes more apparent. Many of the indicators used by chemists are shown in **Figure 24.** As shown in **Figure 22,** bromthymol blue is a good choice for a titration of a strong acid with a strong base, and phenophthalein changes color at the equivalence point of a titration of a weak acid with a strong base.

■ **Figure 24** Choosing the right indicator is important. The indicator must change color at the equivalence point of the titration which is not always at pH 7.

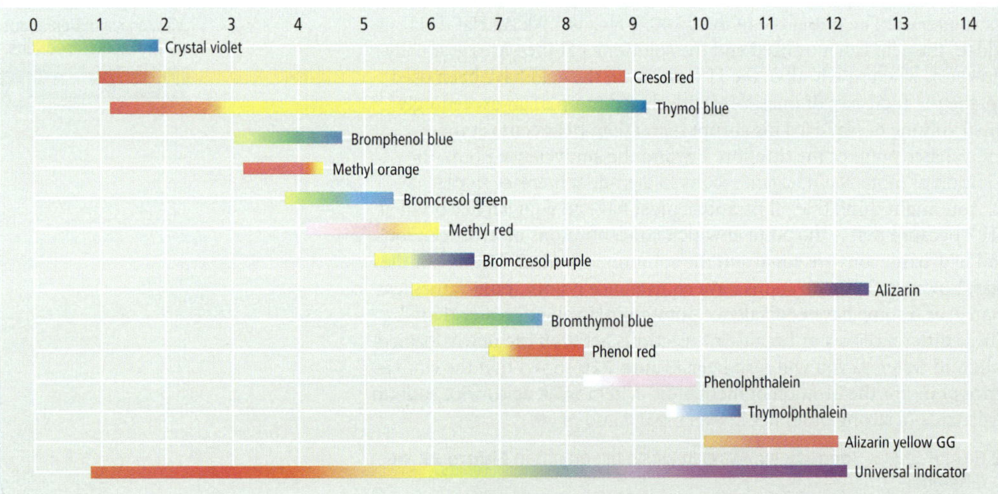

DIFFERENTIATED INSTRUCTION

English Learners Have English learners describe in their native languages and in English a Quick Demo or another classroom activity relating to this section. EL

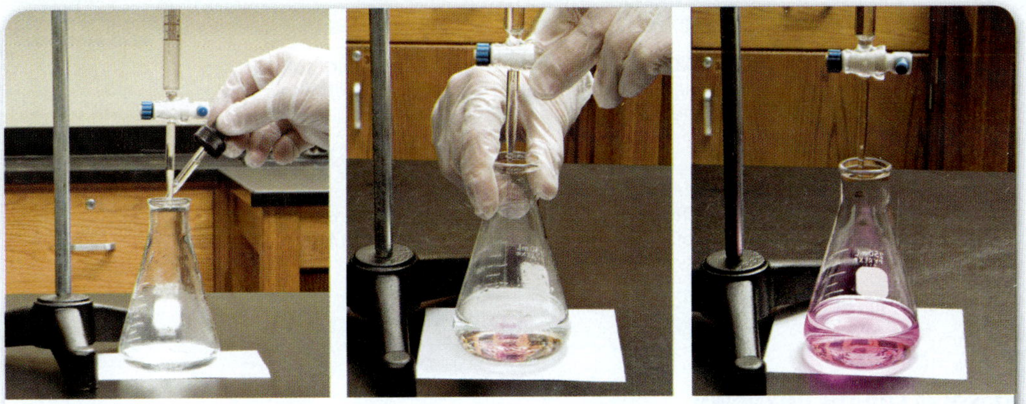

The buret contains the standard solution (0.1000M NaOH), and the flask contains 25.00 mL HCOOH solution along with a small amount of phenolphthalein indicator.

The standard solution is added slowly to the acid solution. The phenolphthalein indicator turns pink, but the color disappears upon mixing, until the end point is reached.

The end point of the titration is marked by a permanent, but very light, pink color. A careful reading of the buret reveals that 18.28 mL 0.1000M NaOH has been added.

■ **Figure 25** Titration is a precise procedure requiring practice. The white paper under the flask provides a background for viewing the indicator color change.

Indicators and titration end point Many indicators used for titration are weak acids. Each has it own particular pH or pH ranges over which it changes color. The point at which the indicator used in a titration changes color is called the **end point** of the titration. It is important to choose an indicator for a titration that will change color at the equivalence point of the titration. Remember that the role of the indicator is to indicate to you, by means of a color change, that just enough of the titrating solution has been added to neutralize the unknown solution. **Figure 25** describes the titration of an unknown solution of methanoic acid (HCOOH) with 0.1000M NaOH.

Watch a **video about acids and bases.**

Video

PROBLEM-SOLVING STRATEGY

Calculating Molarity

The balanced equation for a titration reaction is the key to calculating the unknown molarity. For example, sulfuric acid is titrated with sodium hydroxide according to this equation.

$$H_2SO_4(aq) + 2NaOH(aq) \rightarrow Na_2SO_4(aq) + 2H_2O(l)$$

1. Calculate the moles of NaOH in the standard from the titration data: molarity of the base (M_B) and the volume of the base (V_B).

   $(M_B)(V_B) = (\text{mol/L})(\text{L}) = \text{mol NaOH in standard}$

2. From the equation, you know that the mole ratio of NaOH to H_2SO_4 is 2:1. Two moles of NaOH are required to neutralize 1 mol of H_2SO_4.

   $\text{mol } H_2SO_4 \text{ titrated} = \text{mol NaOH in standard} \times \frac{1 \text{ mol } H_2SO_4}{2 \text{ mol NaOH}}$

3. M_A represents the molarity of the acid and V_A represents the volume of the acid in liters. $M_A = \frac{\text{mol } H_2SO_4 \text{ titrated}}{V_A}$

Apply this strategy as you study Example Problem 6.

Extension

Graphing Titration Use the board or projected image to illustrate the pH v. volume-of-acid curve for the titration of a polyprotic acid with sodium hydroxide. A college chemistry textbook is a good source for such a titration curve. Guide students as they write the equation for the neutralization reaction occurring during each plateau. For example, during the first plateau of the titration of H_3PO_4 with NaOH, the reaction $H_3PO_4(aq) + OH^-(aq) \rightarrow H_2PO_4^-(aq) + H_2O(l)$ occurs. During the second plateau, the reaction $H_2PO_4^-(aq) + OH^-(aq) \rightarrow HPO_4^{2-}(aq) + H_2O(l)$ occurs, and during the third plateau, the reaction $HPO_4^{2-}(aq) + OH^-(aq) \rightarrow PO_4^{3-}(aq) + H_2O(l)$ occurs.

ChemLAB

The ChemLab located at the end of the chapter can be used at this point in the lesson.

DIFFERENTIATED INSTRUCTION

Struggling Learners Have students examine the thymol blue line of Figure 24. Ask them to determine whether thymol blue would best be used as an indicator for a strong acid-strong base, a strong acid-weak base, or a weak acid-strong base reaction. The color change at pH = 2 is too low to indicate the product of neutralization. The change at pH = 8.9 might be used for the titration of a weak acid and a strong base. BL

Video Lab

Video

Acids and Bases

IN-CLASS Example

Question 28.55 mL of 0.1200*M* LiOH solution is required to neutralize 25.00 mL of a perchloric acid ($HClO_4$) solution. What is the molarity of the $HClO_4$ solution?

Answer $M_A = 1.370 \times 10^{-1}M$

$HClO_4(aq) + LiOH(aq) \longrightarrow LiClO_4(aq) + H_2O(l)$

$(0.02855\text{ L LiOH}) \times (0.1200\text{ mol LiOH/L LiOH}) = 3.426 \times 10^{-3}\text{ mol LiOH}$

$M_A = (3.426 \times 10^{-3}\text{ mol }HClO_4) \div (0.02500\text{ L }HClO_4) = 1.370 \times 10^{-1}M$

PRACTICE Problems

Have students refer to the Selected Solutions appendix for complete solutions to odd-numbered problems. The complete solutions for all problems can be found in the Solutions Manual.

44. $M_{HNO_3} = 0.2167M$

45. $M_{NH_3} = 1.178M$

46. 15.0 mL NaOH

EXAMPLE Problem 6

Find help with **dimensional analysis.** Math Handbook

EXAMPLE PROBLEM

MOLARITY FROM TITRATION DATA A volume of 18.28 mL of a standard solution of 0.1000*M* NaOH was required to neutralize 25.00 mL of a solution of methanoic acid (HCOOH). What is the molarity of the methanoic acid solution?

1 ANALYZE THE PROBLEM

You are given the molarity and volume of the NaOH solution and the volume of the methanoic acid (HCOOH) solution. The volume of base used is about four-fifths of the volume of the acid, so the molarity of the acid solution should be less than 0.1*M*.

Known	**Unknown**
$V_A = 25.00$ mL HCOOH	$M_A = ?$ mol/L
$V_B = 18.28$ mL NaOH	
$M_B = 0.1000M$	

2 SOLVE FOR THE UNKNOWN

Write the balanced formula equation for the neutralization reaction.

$$HCOOH(aq) + NaOH(aq) \rightarrow HCOONa(aq) + H_2O(l)$$

1 mol NaOH neutralizes 1 mol HCOOH. — **Write the acid to base mole relationship.**

$V_B = 18.28\ \cancel{\text{mL}} \times \frac{1\text{ L}}{1000\ \cancel{\text{mL}}} = 0.01828\text{ L}$ — **Convert volume of base from mL to L.**

Calculate moles of NaOH.

$\text{Mol NaOH} = (M_B)(V_B)$ — **Apply the relationship between moles of base, molarity of base, and volume of base.**

$\text{Mol NaOH} = (0.1000\text{ mol}/\cancel{\text{L}})(0.01828\ \cancel{\text{L}}) = 1.828 \times 10^{-3}\text{ mol NaOH}$ — **Substitute $M_B = 0.1000M$ and $V_B = 0.01828$ L.**

Calculate moles of HCOOH.

$1.828 \times 10^{-3}\ \cancel{\text{mol NaOH}} \times \frac{1\text{ mol HCOOH}}{1\ \cancel{\text{mol NaOH}}} = 1.828 \times 10^{-3}\text{ mol HCOOH}$ — **Apply the stoichiometric relationship.**

Calculate the molarity of HCOOH.

$1.828 \times 10^{-3}\text{ mol HCOOH} = (M_A)(V_A)$ — **Apply the relationship between moles of acids, molarity of acid, and volume of acid.**

$M_A = \frac{1.828 \times 10^{-3}\text{ mol HCOOH}}{V_A}$ — **Solve for M_A.**

$V_A = 25.00\ \cancel{\text{mL}} \times \frac{1\text{ L}}{1000\ \cancel{\text{mL}}} = 0.02500\text{ L HCOOH}$ — **Convert volume of acid from mL to L.**

$M_A = \frac{1.828 \times 10^{-3}\text{ mol HCOOH}}{0.02500\text{ L HCOOH}} = 7.312 \times 10^{-2}\text{ mol/L}$ — **Substitute $V_A = 0.02500$ L.**

3 EVALUATE THE ANSWER

The answer agrees with the prediction that the molarity of HCOOH is less than 0.1*M*, and is correctly recorded with four significant figures and the appropriate unit.

PRACTICE Problems

Do additional problems. Online Practice

44. What is the molarity of a nitric acid solution if 43.33 mL of 0.1000*M* KOH solution is needed to neutralize 20.00 mL of the acid solution?

45. What is the concentration of a household ammonia cleaning solution if 49.90 mL of 0.5900*M* HCl is required to neutralize 25.00 mL of the solution?

46. Challenge How many milliliters of 0.500*M* NaOH would neutralize 25.00 mL of 0.100*M* H_3PO_4?

CHEMISTRY JOURNAL

Measuring pH Have students use pH meters or acid-base indicators to investigate the pH of soil, rain, lake water, and stream water in their local area. Ask them if they can explain the pH levels that they find. Have them include their results and explanations in their journals. OL

Salt Hydrolysis

In **Figure 26,** several drops of bromthymol blue indicator solution have been added to 0.10*M* aqueous solutions of the salts ammonium chloride (NH_4Cl), sodium nitrate ($NaNO_3$), and potassium fluoride (KF). Sodium nitrate turns the indicator green, which means that the solution is neutral. The blue color of the KF solution means that the solution is basic, and the yellow color of the ammonium chloride solution indicates that the solution is acidic. Why are some aqueous salt solutions neutral, some basic, and some acidic? Many salts react with water in a process known as salt hydrolysis. In **salt hydrolysis,** the anions of the dissociated salt accept hydrogen ions from water or the cations of the dissociated salt donate hydrogen ions to water.

Figure 26 The indicator bromthymol blue provides surprising results when added to three solutions of ionic salts. An NH_4Cl solution is acidic, a $NaNO_3$ solution is neutral, and a KF solution is basic. The explanation has to do with the strengths of the acid and base from which each salt was formed.

Salts that produce basic solutions Potassium fluoride is the salt of a strong base (KOH) and a weak acid (HF). It dissociates into potassium ions and fluoride ions.

$$KF(s) \rightarrow K^+(aq) + F^-(aq)$$

The K^+ ions do not react with water, but the F^- ion is a weak Brønsted-Lowry base. Some fluoride ions establish this equilibrium with water.

$$F^-(aq) + H_2O(l) \rightleftharpoons HF(aq) + OH^-(aq)$$

Hydrogen fluoride molecules and OH^- ions are produced. The production of the OH^- ions makes the solution basic.

Salts that produce acidic solutions NH_4Cl is the salt of a weak base (NH_3) and a strong acid (HCl). When dissolved in water, the salt dissociates into ammonium ions and chloride ions.

$$NH_4Cl(s) \rightarrow NH_4^+(aq) + Cl^-(aq)$$

The Cl^- ions do not react with water, but the NH_4^+ ion is a weak Brønsted-Lowry acid. Ammonium ions react with water molecules to establish this equilibrium.

$$NH_4^+(aq) + H_2O(l) \rightleftharpoons NH_3(aq) + H_3O^+(aq)$$

Ammonia molecules and hydronium ions are produced. The presence of hydronium ions makes the solution acidic.

Salts that produce neutral solutions Sodium nitrate ($NaNO_3$) is the salt of a strong acid (HNO_3) and a strong base (NaOH). Little or no salt hydrolysis occurs because neither Na^+ nor NO_3^- react with water. Therefore, a solution of sodium nitrate is neutral.

Explore **titrations.**

Virtual Investigations

PRACTICE Problems

Do additional problems. Online Practice

47. Write equations for the salt hydrolysis reactions occurring when the following salts dissolve in water. Classify each as acidic, basic, or neutral.

a. ammonium nitrate **c.** rubidium acetate

b. potassium sulfate **d.** calcium carbonate

48. Challenge Write the equation for the reaction that occurs in a titration of ammonium hydroxide (NH_4OH) with hydrogen bromide (HBr). Will the pH at the equivalence point be greater or less than 7?

Concept Development

Acid Parent and Base Parent Help students with the process of determining a salt's "acid parent" and "base parent" by adding sufficient numbers of H^+ ions to the salt's negative ion and OH^- ions to the salt's positive ion. For example, help students work out that the parent acid and parent base of $MgSO_4$ are H_2SO_4 and $Mg(OH)_2$, respectively. OL

Reinforcement

OH^- Donor Reinforce the concept that the H_3O^+ or OH^- ions formed during salt hydrolysis do not come from the salt itself. Rather, water molecules react with either the negative or positive ions of the salt and produce additional H_3O^+ or OH^- ions.

PRACTICE Problems

Have students refer to the Selected Solutions appendix for complete solutions to odd-numbered problems. The complete solutions for all problems can be found in the Solutions Manual.

47. a. $NH_4^+(aq) + H_2O(l) \rightleftharpoons NH_3(aq) + H_3O^+(aq)$
The solution is acidic.

b. $SO_4^{2-}(aq) + H_2O(l) \rightleftharpoons HSO_4^-(aq) + OH^-(aq)$
The solution is neutral.

c. $CH_3COO^-(aq) + H_2O(l) \rightleftharpoons CH_3COOH(aq) + OH^-(aq)$
The solution is basic.

d. $CO_3^{2-}(aq) + H_2O(l) \rightleftharpoons HCO_3^-(aq) + OH^-(aq)$
The solution is basic.

48. $NH_4OH(aq) + HBr(aq) \rightarrow NH_4Br(aq) + H_2O(l)$
$NH_4^+(aq) + H_2O(aq) \rightleftharpoons H_3O^+(aq) + NH_3$
Hydronium ions are formed so the pH will be less than 7.

Caption Question Fig. 27
The jellies might sicken and die.

Reinforcement

Conjugate Acid-Base Pairs
Remind students that members of a conjugate acid-base pair are related by the presence or absence of a hydrogen ion. The acid form contains the H^{+1} ion; the base form does not. Explain that a buffer has its maximum buffering capacity in both directions when it contains equal quantities of an acid and its conjugate base or a base and its conjugate acid.

Quick Demo

Buffers Place a crushed, plain aspirin tablet in 100 mL of distilled water in a 250-mL beaker. Place a crushed, buffered aspirin tablet in 100 mL of distilled water in another 250-mL beaker. Put magnetic stirring bars in both beakers and set them on magnetic stirrers. With the stirrers set on low, place digital pH meter probes in both solutions. Record the initial pH of each solution, the pH after 5 min, and the pH after 10 min. The buffered tablet dissolves faster and produces a solution of about pH 6. The unbuffered tablet produces a solution of about pH 3. Ask students which type of aspirin is best for a person who has ulcers or is prone to acid indigestion. The buffered aspirin is better because it produces a less acidic solution. **OL**

Figure 27 To provide a healthy environment for these jellies, the pH of the aquarium water at the Monterey Bay Aquarium must be adjusted to stay within the range of 8.1 to 8.4.
Predict *what would happen if the pH were allowed to fall to 7.0.*

Sisse Brimberg/Getty Images

Buffered Solutions

It is important for the jellies shown in **Figure 27** that the aquarium water be kept within a narrow pH range. A constant pH is also important in your body. The pH of your blood must be maintained within the range of 7.1 to 7.7. The gastric juices in your stomach must have a pH between 1.6 and 1.8 to promote digestion of certain foods. Your body maintains pH values within such narrow limits by producing buffers.

What is a buffer? **Buffers** are solutions that resist changes in pH when limited amounts of acid or base are added. For example, adding 0.01 mol of HCl to 1 L of pure water lowers the pH by 5.0 units, from 7.0 to 2.0. Similarly, adding 0.01 mol of NaOH to 1 L of pure water increases the pH from 7.0 to 12.0. However, if you add the same amount of either HCl or NaOH to 1 L of a buffered solution, the pH might change by no more than 0.1 unit.

How do buffers work? A buffer is a mixture of a weak acid and its conjugate base or a weak base and its conjugate acid. The mixture of ions and molecules in a buffer solution resists changes in pH by reacting with any hydrogen ions or hydroxide ions added to the buffered solution.

Suppose that a buffer solution contains 0.1M concentrations of hydrofluoric acid (HF) and sodium fluoride (NaF). The NaF provides a 0.1M concentration of F^- ions. HF is the acid, and F^- is its conjugate base. The following equilibrium would be established.

$$HF(aq) \rightleftharpoons H^+(aq) + F^-(aq)$$

Adding an acid When an acid is added to a buffered solution, the equilibrium shifts to the left. According to Le Châtelier's principle, the added H^+ ions from the acid are a stress on the equilibrium, which is relieved by their reaction with F^- ions to form additional undissociated HF molecules.

$$HF(aq) \leftharpoons H^+(aq) + F^-(aq)$$

Equilibrium is established again with a larger amount of undissociated HF present. However, the pH of the solution has changed little because the shift to the left consumed most of the added H^+ ion.

Demonstration

Using a Buffer

Purpose
to learn how to prepare a buffer

Materials
glacial acetic acid (1 mL); 100-mL beakers (3); distilled water (60 mL); sodium acetate (4 g); large test tubes (3); indicator (10 drops); stirring rods (3); 4.0M NaOH (50 mL) (16 g in 100 mL water); dropper.

Safety Precautions

Disposal Check pH of all solutions. Adjust pH to between 3 and 10, dilute, and pour down a drain.

Procedure
A buffer system can absorb moderate amounts of acid or base without a significant change in pH. To prepare a system, add 1 mL of glacial acetic acid to 49 mL distilled H_2O. **WARNING:** ***Acid is corrosive.*** Divide the

Adding a base When a base is added to the hydrofluoric acid/fluoride ion buffer system, the added OH^- ions react with H^+ ions to form H_2O. This decreases the concentration of H^+ ions, and the equilibrium shifts to the right to replace the H^+ ions.

$$HF(aq) \rightleftharpoons H^+(aq) + F^-(aq)$$

Although the shift to the right consumes HF molecules and produces additional F^- ions, the pH remains fairly constant because the H^+ ion concentration has not changed appreciably.

A buffer solution's capacity to resist pH change can be exceeded by the addition of too much acid or base. The amount of acid or base a buffer solution can absorb without a significant change in pH is called the **buffer capacity** of the solution. The greater the concentrations of the buffering molecules and ions in the solution, the greater the solution's buffer capacity.

Choosing a buffer A buffer system is most effective when the concentrations of the conjugate acid-base pair are equal or nearly equal. Consider the $H_2PO_4^-/HPO_4^{2-}$ buffer system made by mixing equal molar amounts of NaH_2PO_4 and Na_2HPO_4.

$$H_2PO_4^- \rightleftharpoons H^+ + HPO_4^{2-}$$

What is the pH of such a buffer solution? The acid ionization constant expression for the equilibrium can provide the answer.

$$K_a = 6.2 \times 10^{-8} = \frac{[H^+][HPO_4^{2-}]}{[H_2PO^{4-}]}$$

Because the solution has been made with equal molar amounts of Na_2HPO_4 and NaH_2PO_4, $[HPO_4^{2-}]$ is equal to $[H_2PO_4^-]$. Thus, the two terms in the acid ionization expression cancel.

$$6.2 \times 10^{-8} = \frac{[H^+]\cancel{[HPO_4^{2-}]}}{\cancel{[H_2PO_4^-]}} = [H^+]$$

$$pH = -\log[H^+] = -\log(6.2 \times 10^{-8}) = 7.21$$

Thus, when equimolar amounts of each of the components are present in the $HPO_4^-/H_2PO_4^{2-}$ buffer system, the system can maintain a pH close to 7.21. Note that the pH is the negative log of K_a. **Table 7** lists several buffer systems, with the pH at which each is effective.

Vocabulary

Science usage v. Common usage

Buffer

Science usage: a solution that resists changes in pH when limited amounts of acid or base are added

The chemist decided to use a buffer consisting of equal molar amounts of formic acid and sodium formate.

Common usage: something that serves as a protective barrier

For the homes along the shore, the high seawall served as a buffer against stormy seas.

Table 7 Buffer Systems with Equimolar Components

Buffer Equilibrium	Conjugate Acid-Base Pair in Buffered Solution	Buffer pH
$HF(aq) \rightleftharpoons H^+(aq) + F^-(aq)$	HF/F^-	3.20
$CH_3COOH(aq) \rightleftharpoons H^+(aq) + CH_3COO^-(aq)$	CH_3COOH/CH_3COO^-	4.76
$H_2CO_3(aq) \rightleftharpoons H^+(aq) + HCO_3^-(aq)$	H_2CO_3/HCO_3^-	6.35
$H_2PO_4^-(aq) \rightleftharpoons H^+(aq) + HPO_4^{2-}(aq)$	$H_2PO_4^-/HPO_4^{2-}$	7.21
$NH_3(aq) + H_2O(l) \rightleftharpoons NH_4^+(aq) + OH^-(aq)$	NH_4^+/NH_3	9.4
$C_2H_5NH_2(aq) + H_2O(l) \rightleftharpoons C_2H_5NH_3^+(aq) + OH^-(aq)$	$C_2H_5NH_3^+/C_2H_5NH_2$	10.70

3 Assess

Check for Understanding

Ask students to write the formula equation, complete ionic equation, and net ionic equation for the neutralization reaction between perchloric acid and sodium hydroxide. $HClO_4(aq) + NaOH(aq) \rightarrow H_2O(l) + NaClO_4(aq)$; $H^+(aq) + ClO_4^-(aq) + Na^+(aq) + OH^- \rightarrow H_2O(l) + Na^+(aq) + ClO_4^-(aq)$; $H^+(aq) + OH^-(aq) \rightarrow H_2O(l)$. OL

Reteach

Have students choose from Table 7 a conjugate acid-base pair that could be used to produce a buffer solution of pH 9 with capacity sufficient to neutralize both added acids and bases. NH_3 (base) and NH_4^+ (conjugate acid) OL

Extension

Use the board to write the chemical equations for the reaction of both added hydrogen ion and hydroxide ion with a carbonic acid/hydrogen carbonate ion buffer solution. $H^+(aq) + HCO_3^-(aq) \rightleftharpoons H_2CO_3(aq)$ and $OH^-(aq) + H_2CO_3(aq) \rightleftharpoons H_2O(l) + HCO_3^-(aq)$ Remind students of the importance of buffers. OL

Assessment

Performance Ask students to describe the buffer's response if someone who has difficulty breathing overexerts; or ask students why a person could lose consciousness if he or she breathed in and out rapidly about 20 times, then held his or her breath. OL

acid between two 100-mL beakers. Dissolve 4 g of sodium acetate in the diluted acid in one of the beakers. Stir and label it *buffer*. Label three large test tubes *acid*, *buffer*, and *water*. Add equal amounts of the diluted acetic acid, the buffer, and distilled water separately to the appropriately labeled test tubes. Add a few drops of universal indicator to each test tube. Place a stirring rod in each test tube. Add a filled dropper of 4*M* NaOH to each solution. WARNING: ***NaOH is caustic.*** Stir and observe the color. Fill the dropper, repeat many times, and record your observations.

Results

The buffered test tube does not change color as rapidly as do the other two test tubes.

Analysis

1. Which of the three test tubes served as a control in this experiment? The control was the distilled water.
2. How was the definition of a buffer verified? Indicator color did not change as several measures of base were added.

Assessment

Knowledge Ask students if they think buffered aspirin would be better for their stomachs than unbuffered aspirin. Yes, buffered aspirin would neutralize the acid in aspirin. OL

Problem-Solving LAB

Purpose Students will use Le Châtelier's principle to explain what occurs when the components of a buffer system are changed in human blood.

Process Skills draw conclusions, recognize cause and effect, apply concepts, predict

Teaching Strategies

- Relate the carbonic acid-hydrogen carbonate buffer condition to the Brønsted-Lowry model. Show the action of carbon dioxide as an acid anhydride and the hydrogen carbonate ion as the conjugate base.
- Show how a buffer minimizes the effect of the addition of an acid or base as required by Le Châtelier's principle.

Think Critically

1. 2.0 times larger
2. A healthy body will dump acid into the blood through increased activity. The excess hydrogen carbonate ion is available to neutralize the acid, thus driving the reaction toward the production of carbon dioxide.
3. a. Vomit is acidic and will raise the pH. The buffer reaction shifts to the right. The kidneys can respond by removing hydrogen carbonate ion and the person should be kept quiet to retain CO_2.
   b. The pH rises as the hydrogen carbonate ion levels increase. This shifts the buffer reaction to the left forming more CO_2. The kidneys can respond by removing hydrogen carbonate ion and the person can breathe more rapidly to expel CO_2.

Problem-Solving LAB

Apply Scientific Explanations

How does your blood maintain its pH? Human blood contains three types of cells. Red blood cells deliver oxygen to every part of the body. White blood cells fight infections, and platelets aid in clotting when bleeding occurs. The critical functions of these cells are impaired if the pH of blood is not maintained within the narrow range of 7.1 to 7.7. Beyond this range, proteins in the body lose their structures and abilities to function. Fortunately, several buffers maintain the necessary acid/base balance. The carbonic acid/hydrogen carbonate (H_2CO_3/HCO_3^-) buffer is the most important.

$$CO_2(g) + H_2O(l) \rightleftharpoons H_2CO_3(aq) \rightleftharpoons H^+(aq) + HCO_3^-(aq)$$

As acids and bases enter the bloodstream as a result of normal activity, the blood's buffer systems shift to effectively maintain a healthful pH.

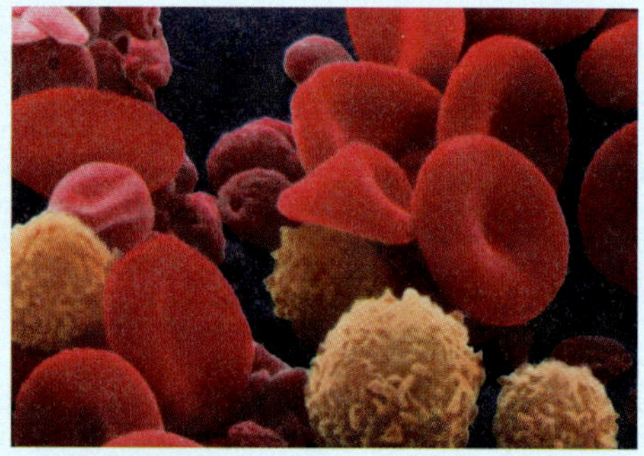

Analysis

Depending on the body's metabolic rate and other factors, the H_2CO_3/HCO_3^- equilibrium will shift according to Le Châtelier's principle. In addition, the lungs can alter the rate at which CO_2 is expelled from the body by breathing, and the kidneys can alter the rate of removal of HCO_3^- ions.

Think Critically

1. **Determine** how many times greater the $[H^+]$ is if the blood's pH changes from pH 7.4 to 7.1.
2. **Suggest** a reason why a 20:1 ratio of HCO_3^- to CO_2 in the blood is favorable for maintaining a healthy pH.
3. **Predict** whether, for each situation, the pH of the blood will rise or fall, and which way the H_2CO_3/HCO_3^- equilibrium will shift.
   a. A person with a severe stomach virus vomits many times during a 24-h period.
   b. To combat heartburn, a person takes too much ($NaHCO_3$).

SECTION 4 REVIEW

Section Self-Check

Section Summary

- In a neutralization reaction, an acid and a base react to form a salt and water.
- The net ionic equation for the neutralization of a strong acid by a strong base is $H^+(aq) + OH^-(aq) \rightarrow H_2O(l)$.
- Titration is the process in which an acid-base neutralization reaction is used to determine the concentration of a solution.
- Buffered solutions contain mixtures of molecules and ions that resist changes in pH.

49. MAINIDEA **Explain** why the net ionic equation for the neutralization reaction of any strong acid with any strong base is always the same.
50. **Explain** the difference between the equivalence point and the end point of a titration.
51. **Compare** the results of two experiments: First, a small amount of base is added to an unbuffered solution with a pH of 7. Second, the same amount of base is added to a buffered solution with a pH of 7.
52. **Calculate** the molarity of a solution of hydrobromic acid (HBr) if 30.35 mL of 0.1000*M* NaOH is required to titrate 25.00 mL of the acid to the equivalence point.
53. **Interpret** What substances could be used to make a buffer solution with a pH of 9.4. How should the amounts of the substances be related. Use **Table 7.**
54. **Design an Experiment** Describe how you would design and perform a titration in which you use 0.250*M* HNO_3 to determine the molarity of a cesium hydroxide solution. Include the formula and net ionic equations.

SECTION 4 REVIEW

49. Each neutralization reaction is the reaction of one mole of hydrogen ion with one mole of hydroxide to form one mole of water.
50. Equivalence point is the pH at which the moles of H^+ ions from the acid equal the moles of OH^- ions from the base. The end point is the point at which the indicator used in a titration changes color.
51. The pH of the unbuffered solution increases more than the pH of the buffered solution.
52. $M_A = 0.1214M$
53. Use ammonia and a salt of ammonia such as ammonium nitrate or ammonium chloride. Use equal molar amounts of the acid and its salt.
54. Place a measured volume of CsOH solution into a flask. Add an indicator such as bromothymol blue. Fill a buret with the 0.250*M* HNO_3 solution. Record the initial buret reading. Add HNO_3 solution slowly to the CsOH solution until the end point. Record the final buret reading. Calculate the volume of HNO_3 added. Use the volume and molarity of HNO_3 and the volume of CsOH to calculate the molarity of the CsOH solution. Refer to the Solutions Manual for the ionic equations.

everyday CHEMISTRY

Acid-Base Reactions on the Rise

Do you remember how much fun it was to watch a vinegar baking soda volcano erupt? The bubbles of carbon dioxide (CO_2) resulted from a decomposition reaction that quickly followed the acid-base reaction between the vinegar ($HC_2H_3O_2$), an acid, and baking soda ($NaHCO_3$), a base, as shown below.

Acid-Base Reaction

$$HC_2H_3O_2(aq) + NaHCO_3(aq) \rightarrow NaC_2H_3O_2(aq) + H_2CO_3(aq)$$

Decomposition

$$H_2CO_3(aq) \rightarrow CO_2(g) + H_2O(l)$$

The release of carbon dioxide as a result of the chemical reaction between an acid and a base, as shown in **Figure 1,** is part of the reason why baked goods rise. An ingredient that causes batter to rise when baked is called a leavening agent. The two main chemical leavening agents are baking soda and baking powder.

Baking soda Sodium hydrogen carbonate, also called sodium bicarbonate, is the chemical name for baking soda. When used in cooking, baking soda reacts with mildly acidic liquids, and carbon dioxide bubbles form. Mildly acidic liquids include vinegar, molasses, honey, citrus juice, buttermilk, and many others.

Figure 1 Carbon dioxide forms bubbles when baking soda, a base, is added to vinegar, an acid.

Figure 2 Baking traps the bubbles formed during the reaction between an acid and a base, resulting in a light, airy cake.

Baking soda must be mixed with other dry ingredients and added last to a batter so that the release of carbon dioxide is uniform throughout the batter. This acid-base reaction happens quickly. If baking soda is the only leavening agent in a recipe, the batter must be baked immediately before the bubbles have a chance to escape. Baking causes the bubbles to expand, and the cake rises. As the batter firms, the bubbles are trapped, as shown in **Figure 2.**

Baking powder If a recipe does not include an acidic liquid, baking powder is used. Most baking powder is a mixture of baking soda and two dry acids. One of the acids reacts with the baking soda when it dissolves in the batter, and the other reacts with the baking soda when heated.

Like baking soda, baking powder is mixed with other dry ingredients and added last to a batter. However, batters made with baking powder do not have to be baked immediately.

Sometimes, batters made with mildly acidic liquids include both baking powder and baking soda. Excess acid can disrupt the action of the baking powder. The baking powder provides a reliable source of carbon dioxide, and the baking soda helps to neutralize the acid.

WRITING IN ▶ Chemistry

Analyze If a recipe calls for flour, salt, sugar, bran cereal, milk, an egg, and vegetable oil, would you use baking soda or baking powder? Explain.

WebQuest

everyday CHEMISTRY

Purpose

Students will explain the purpose of a leavening agent, and will describe and compare the function of baking soda and baking powder in the baking process.

Background

Baker's yeast is another commonly used leavening agent that functions somewhat differently in the baking process than baking powder. Yeast is a unicellular fungi that consumes sugar, producing carbon dioxide and alcohol as waste products. While baking powder produces bubbles in dough within 15 minutes, yeast-leavened dough may take from two to three hours to rise.

Teaching Strategies

- Referring to an actual cake recipe, have students predict how the texture and density of the cake might differ if more or less than the specified amount of baking soda were used to prepare the batter.
- Have students use stoichiometry and the balanced equations given in the feature to calculate the amount of CO_2 gas that will be produced if given amounts of baking soda and vinegar are used in a recipe.

WRITING IN ▶ Chemistry

Rubric WebQuest

Analyze Baking soda should be used in the recipe. Milk is slightly acidic, with a pH of 6.5.

ChemLAB

Rubric

Preparation

Time Allotment three class periods

Process Skills observe and infer, collect data, interpret data, use numbers

Safety Precautions Approve lab safety forms before work begins. Students should wear safety goggles, gloves, and lab aprons. Have students wash their hands after each laboratory period. Remind students that NaOH when combined with water gets hot.

Disposal The neutralized solutions can be washed down the drain with plenty of water.

Preparation of Solutions
See the Teacher Handbook for preparation of all solutions.

Alternative Material

- Potassium hydroxide could be used instead of sodium hydroxide.

Procedure

- The 0.250*M* NaOH solution can be prepared in advance.
- The amount of potassium hydrogen phthalate for each trial can vary from 0.3 g to 0.5 g.
- Demonstrate a complete titration.
- **Troubleshooting** Students often forget to add indicator to their acid solutions.
- Have students stopper their flasks between lab periods. NaOH reacts with carbon dioxide in the air to form sodium carbonate.

ChemLAB

Standardize a Base

Background: Titration is a procedure by which the molarity of a base can be determined.

Question: *How can you determine the molarity of a solution of a base?*

Materials

50-mL buret
buret clamp
ring stand
sodium hydroxide pellets (NaOH)
potassium hydrogen phthalate ($KHC_8H_4O_4$)
distilled water
weighing bottle
spatula
250-mL Erlenmeyer flask
500-mL Florence flask with rubber stopper
250-mL beaker
centigram balance
wash bottle
phenolphthalein solution

Safety Precautions

WARNING: ***Dissolving NaOH in water generates heat. Phenolphthalein is flammable. Keep away from flames.***

Procedure

1. Read and complete the lab safety form.
2. Place about 4 g of NaOH in a 500-mL Florence flask. Add enough water to dissolve the pellets and bring the volume of the NaOH solution to about 400 mL. Stopper the flask.
3. Use the weighing bottle to mass by difference about 0.40 g of potassium hydrogen phthalate ($KHC_8H_4O_4$, molar mass = 204.32 g/mol) into a 250-mL Erlenmeyer flask. Record this mass.
4. Use a wash bottle to rinse the insides of the flask, and add about 50 mL of water. Add two drops of phenolphthalein indicator solution.
5. Rinse the buret with 10 mL of your base solution. Discard the rinse solution in a discard beaker. Attach the buret to the ring stand using the buret clamp.
6. Fill the buret with NaOH solution. The level of the liquid should be at or below the zero mark. To remove any air trapped in the tip of the buret, allow a small amount of the base to flow from the tip into the discard beaker. Read the buret to the nearest 0.02 mL, and record this initial reading.
7. Place a piece of white paper on the base of the ringstand. Swirl the flask while allowing the NaOH solution to flow slowly from the buret into the flask.

Titration Data	
	Trial 1
Mass of weighing bottle and acid	
Mass of weighing bottle	
Mass of solid acid	
Moles of acid	
Moles of base required	
Final reading of base buret	
Initial reading of base buret	
Volume of base used in mL	
Molarity of base	

8. When the pink color begins to persist longer as the flask is swirled, add the base drop-by-drop.
9. The end point is reached when one additional drop of base turns the acid pink. The pink color should persist as the flask is swirled. Record the final volume in the buret.
10. Calculate the molarity of your base using Steps 1–4 in the Analyze and Conclude section.
11. Refill the buret. Rinse the flask with water. Repeat the titration until the calculated values of the molarity for three trials show close agreement.
12. **Cleanup and Disposal** Wash the neutralized solutions down the sink with plenty of water.

Analyze and Conclude

1. **Interpret Data** For each titration, calculate the number of moles of acid used by dividing the mass of the sample by the molar mass of the acid.
2. **Infer** How many moles of base are required to react with the moles of acid you used?
3. **Calculate** Convert the volume of base to liters.
4. **Calculate** the molarity of the base by dividing the moles of base by the volume of base in liters.
5. **Error Analysis** Did your calculated molarities agree? Explain any irregularities.

INQUIRY EXTENSION

Design an Experiment Determine the concentration of a vinegar solution without using an indicator.

Analyze and Conclude

1. 0.00191 mol of $KHC_8H_4O_4$
2. The mole ratio is 1:1, so the number of moles of NaOH and $KHC_8H_4O_4$ are the same.
3. 0.00764 L
4. 0.250*M*.
5. Irregularities might include inaccurate buret readings or mass determinations, overshooting the end point, spillage, and bubbles in the buret.

INQUIRY EXTENSION

Place a measured volume of vinegar in a beaker. Titrate the vinegar with the NaOH solution (of known concentration). Use a pH meter to monitor the pH (of the vinegar). The end point is reached when a single drop of NaOH causes a large change in pH.

Vocabulary Practice

BIGIDEA Acids and bases can be defined in terms of hydrogen ions and hydroxide ions or in terms of electron pairs.

SECTION 1 Introduction to Acids and Bases

MAINIDEA Different models help describe the behavior of acids and bases.

- The concentrations of hydrogen ions and hydroxide ions determine whether an aqueous solution is acidic, basic, or neutral.
- An Arrhenius acid must contain an ionizable hydrogen atom. An Arrhenius base must contain an ionizable hydroxide group.
- A Brønsted-Lowry acid is a hydrogen ion donor. A Brønsted-Lowry base is a hydrogen ion acceptor.
- A Lewis acid accepts an electron pair. A Lewis base donates an electron pair.

VOCABULARY

- acidic solution
- basic solution
- Arrhenius model
- Brønsted-Lowry model
- conjugate acid
- conjugate base
- conjugate acid-base pair
- amphoteric
- Lewis model

SECTION 2 Strengths of Acids and Bases

MAINIDEA In solution, strong acids and bases ionize completely, but weak acids and bases ionize only partially.

- Strong acids and strong bases are completely ionized in a dilute aqueous solution. Weak acids and weak bases are partially ionized in a dilute aqueous solution.
- For weak acids and weak bases, the value of the acid or base ionization constant is a measure of the strength of the acid or base.

VOCABULARY

- strong acid
- weak acid
- acid ionization constant
- strong base
- weak base
- base ionization constant

SECTION 3 Hydrogen Ions and pH

MAINIDEA pH and pOH are logarithmic scales that express the concentrations of hydrogen ions and hydroxide ions in aqueous solutions.

- The ion product constant for water, K_w, equals the product of the H^+ ion concentration and the OH^- ion concentration.

  $K_w = [H^+][OH^-]$
- The pH of a solution is the negative log of the hydrogen ion concentration. The pOH is the negative log of the hydroxide ion concentration. pH plus pOH equals 14.

  $pH = -\log [H^+]$ $\quad$ $pOH = -\log [OH^-]$

  $pH + pOH = 14.00$
- A neutral solution has a pH of 7.0 and a pOH of 7.0 because the concentrations of hydrogen ions and hydroxide ions are equal.

VOCABULARY

- ion product constant for water
- pH
- pOH

SECTION 4 Neutralization

MAINIDEA In a neutralization reaction, an acid reacts with a base to produce a salt and water.

- In a neutralization reaction, an acid and a base react to form a salt and water.
- The net ionic equation for the neutralization of a strong acid by a strong base is $H^+(aq) + OH^-(aq) \rightarrow H_2O(l)$.
- Titration is the process in which an acid-base neutralization reaction is used to determine the concentration of a solution.
- Buffered solutions contain mixtures of molecules and ions that resist changes in pH.

VOCABULARY

- neutralization reaction
- salt
- titration
- titrant
- equivalence point
- acid-base indicator
- end point
- salt hydrolysis
- buffer
- buffer capacity

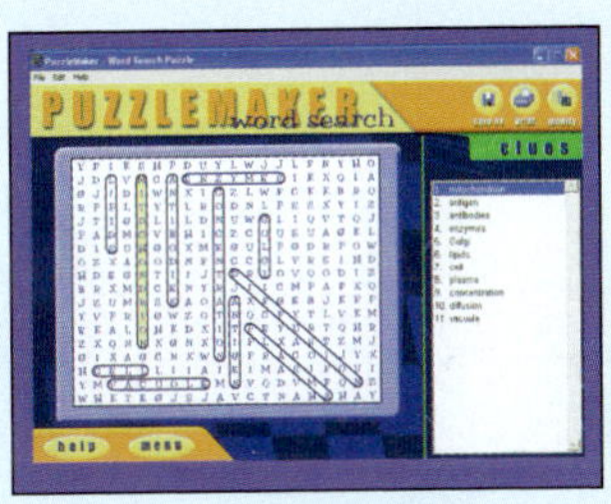

Vocabulary Practice

For additional practice with vocabulary, have students access the Vocabulary PuzzleMaker.

CHAPTER 18 STUDY GUIDE

Use the Vocabulary

To reinforce chapter vocabulary, have students write a sentence using each term. OL EL

Review Strategies

- Ask students to compare acidic, neutral, and basic solutions. OL
- Ask students to distinguish between Arrhenius and Brønsted-Lowry acids and bases and identify conjugate acid-base pairs. OL
- Ask students to distinguish between strong and weak acids. OL
- Ask students to compare the pH and pOH values in acidic, neutral, and basic solutions. OL
- Problems from the Supplemental Practice Problems appendix or the Supplemental Problems booklet can be used for review. OL

Students can review with these online resources.

Vocabulary Practice

Vocabulary eGames
Vocabulary eFlashcards
Vocabulary PuzzleMaker

Section Self-Check

Chapter Self-Check

Online Test Practice

Use *eAssessment* to:

- create multiple versions of tests
- edit existing questions and add your own questions
- build tests aligned with select state curriculum standards using built-in state curriculum tags
- track students' progress

SECTION 1

Mastering Concepts

55. An acidic solution has a larger concentration of H^+ ions than OH^- ions. A basic solution has a larger concentration of OH^- ions than H^+ ions. A neutral solution has equal concentrations of OH^- and H^+ ions.
56. $H_2O(l) + H_2O(l) \rightleftharpoons H_3O^+(aq) + OH^-(aq)$
57. **a.** acid **b.** base **c.** base **d.** acid
58. Gas is CO_2; the rock is composed of calcium carbonate.
59. The larger blue area means that basic solutions contain higher concentrations of hydroxide ions. The smaller red area means that basic solutions still contain hydrogen ions but in lower concentrations than hydroxide ions.
60. A monoprotic acid can donate one H^+ (HCl); a diprotic acid can donate two H^+ (H_2SO_4); a triprotic acid can donate three H^+ (H_3PO_4).
61. H_3O^+ is a hydrated hydrogen ion.
62. acidic: $[H^+] > [OH^-]$; neutral: $[H^+] = [OH^-]$; basic: $[H^+] < [OH^-]$
63. The Lewis model defines an acid as an electron pair acceptor, whereas in the Brønsted-Lowry model an acid is a hydrogen ion donor.

Mastering Problems

64. **a.** $Mg(OH)_2(s) \longrightarrow Mg^{2+}(aq) + 2OH^-(aq)$
    b. $Mg(s) + 2HBr(aq) \longrightarrow H_2(g) + MgBr_2(aq)$
    c. $CH_3CH_2COOH(aq) + H_2O(l) \longrightarrow H_3O^+(aq) + CH_3CH_2COO^-(aq)$
    d. $HSO_4^-(aq) + H_2O(l) \longrightarrow H_3O^+(aq) + SO_4^{2-}(aq)$

CHAPTER 18 ASSESSMENT

SECTION 1

Mastering Concepts

55. In terms of ion concentrations, distinguish between acidic, neutral, and basic solutions.
56. Write a balanced chemical equation that represents the self-ionization of water.
57. Classify each compound as an Arrhenius acid or an Arrhenius base.
    a. H_2S **c.** $Mg(OH)_2$
    b. RbOH **d.** H_3PO_4
58. **Geology** When a geologist adds a few drops of HCl to a rock, gas bubbles form. What might the geologist conclude about the nature of the gas and the rock?

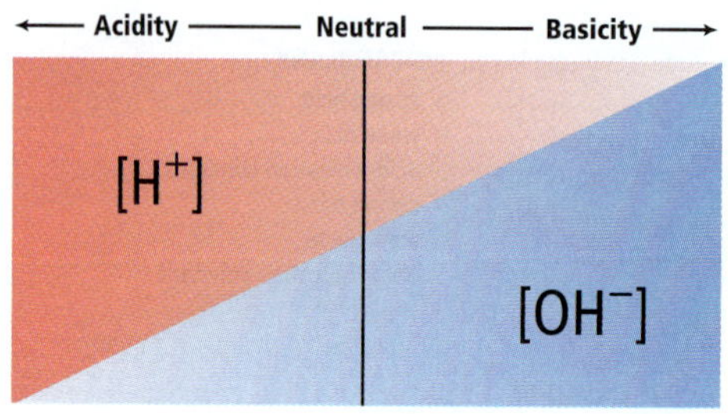

■ **Figure 28**

59. Explain the meaning of the relative sizes of the two shaded areas to the right of the dark vertical line in **Figure 28.**
60. Explain the difference between a monoprotic acid, a diprotic acid, and a triprotic acid. Give an example of each.
61. Why can H^+ and H_3O^+ be used interchangeably in chemical equations?
62. Use the symbols $<$, $>$, and $=$ to express the relationship between the concentrations of H^+ ions and OH^- ions in acidic, neutral, and basic solutions.
63. Explain how the definition of a Lewis acid differs from the definition of a Brønsted-Lowry acid.

Mastering Problems

64. Write a balanced chemical equation for each of the following.
    a. the dissociation of solid magnesium hydroxide in water
    b. the reaction of magnesium metal and hydrobromic acid
    c. the ionization of propanoic acid (CH_3CH_2COOH) in water
    d. the second ionization of sulfuric acid in water

SECTION 2

Mastering Concepts

65. Explain the difference between a strong acid and a weak acid.
66. Explain why equilibrium arrows are used in the ionization equations for some acids.

■ **Figure 29**

67. Which of the beakers shown in **Figure 29** might contain a solution of 0.1*M* hypochlorous acid? Explain your answer.
68. Explain how you would compare the strengths of two weak acids experimentally and using a table or a handbook.
69. Identify the conjugate acid-base pairs in the reaction of H_3PO_4 with water.

Mastering Problems

70. **Ammonia Cleaner** Write the chemical equation and K_b expression for the ionization of ammonia in water. How is it safe for a window cleaner to use a solution of ammonia, which is basic?
71. **Disinfectant** Hypochlorous acid is an industrial disinfectant. Write the chemical equation and the K_a expression for the ionization of hypochlorous acid in water.
72. Write the chemical equation and the K_b expression for the ionization of aniline in water. Aniline is a weak base with the formula $C_2H_5NH_2$.
73. A fictional weak base, ZaH_2, reacts with water to yield a solution with a OH^- ion concentration of 2.68×10^{-4} mol/L. The chemical equation for the reaction is $ZaH_2(aq) + H_2O(l) \leftrightharpoons ZaH_3^+(aq) + OH^-(aq)$. If $[ZaH_2]$ at equilibrium is 0.0997 mol/L, what is the value of K_b for ZaH_2?
74. Select a strong acid, and explain how you would prepare a dilute solution of the acid. Select a weak acid, and explain how you would prepare a concentrated solution of the acid.

Matt Meadows

SECTION 2

Mastering Concepts

65. In dilute aqueous solution, a strong acid ionizes completely; a weak acid ionizes slightly.
66. Equilibrium arrows are used for weak acids, which are partially ionized in water and create equilibria.
67. The beaker on the right. Because hypochlorous acid is weak, it is only slightly ionized in aqueous solution and has a low electrical conductivity.
68. Compare the conductivities of equimolar solutions of the acids. Compare acid ionization constants of the two acids.
69. acid (H_3PO_4)/conjugate base ($H_2PO_4^-$); base ($H_2O(l)$)/conjugate acid H_3O^+

Mastering Problems

70. $NH_3(aq) + H_2O(l) \rightleftharpoons NH_4^+(aq) + OH^-(aq)$
    $K_b = [NH_4^+][OH^-]/[NH_3]$;
    Ammonia is a weak base, therefore, its aqueous solution is not highly basic.
71. $HClO \rightleftharpoons H^+ + ClO^-$; $K_a = [H^+][ClO^-]/[HClO]$
72. $K_b = [C_2H_5NH_3^+][OH^-]/[C_2H_5NH_2]$
73. $K_b = 7.22 \times 10^{-7}$
74. Students might say that the dilute solution of a strong acid is prepared by dissolving a small quantity of the strong acid in a large quantity of water, and the concentrated solution of a weak acid is prepared by dissolving a large quantity of the weak acid in a small quantity of water.

SECTION 3

Mastering Concepts

75. What is the relationship between the pOH and the OH^- ion concentration of a solution?

76. Solution A has a pH of 2.0. Solution B has a pH of 5.0. Which solution is more acidic? Based on the H^+ ion concentrations in the two solutions, how many times more acidic?

77. If the concentration of H^+ ions in an aqueous solution decreases, what must happen to the concentration of OH^- ions? Why?

78. Use Le Châtelier's principle to explain what happens to the equilibrium $H_2O(l) \leftrightharpoons H^+(aq) + OH^-(aq)$ when a few drops of HCl are added to pure water.

79. Common Acids and Bases Use the data in **Table 8** to answer the following questions.

Table 8 pH values

Substance	pH
Household ammonia	11.3
Lemon juice	2.3
Antacid	9.4
Blood	7.4
Soft drinks	3.0

a. Which substance is the most basic?
b. Which substance is closest to neutral?
c. Which has a concentration of $H^+ = 4.0 \times 10^{-10}M$?
d. Which has a pOH of 11.0?
e. How many times more basic is antacid than blood?

Mastering Problems

80. What is $[OH^-]$ in an aqueous solution at 298 K in which $[H^+] = 5.40 \times 10^{-3}M$?

81. What are the pH and pOH for the solution described in Question 80?

82. If 5.00 mL of 6.00*M* HCl is added to 95.00 mL of pure water, the final volume of the solution is 100.00 mL. What is the pH of the solution?

83. Given two solutions, 0.10*M* HCl and 0.10*M* HF, which solution has the greater concentration of H^+ ions? Calculate pH values for the two solutions, given that $[H^+] = 7.9 \times 10^{-3}M$ in the 0.10*M* HF.

84. Metal Cleaner Chromic acid is used as an industrial cleaner for metals. What is K_a for the second ionization of chromic acid (H_2CrO_4) if a 0.040*M* solution of sodium hydrogen chromate has a pH of 3.946?

SECTION 4

Mastering Concepts

85. What acid and base must react to produce an aqueous sodium iodide solution?

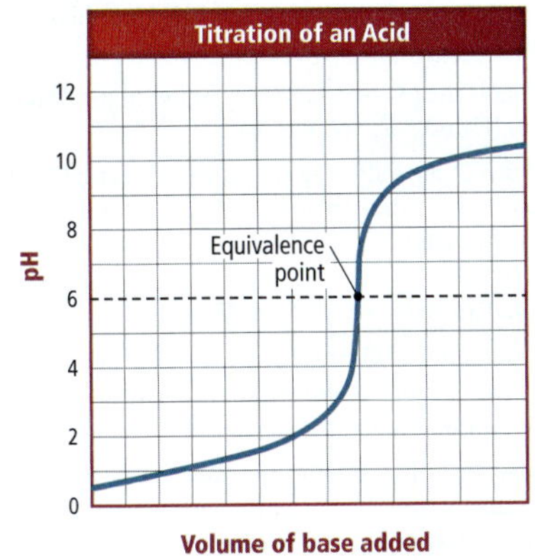

■ **Figure 30**

86. What acid-base indicators, shown in **Figure 24,** would be suitable for the neutralization reaction whose titration curve is shown in **Figure 30?** Why?

87. When might a pH meter be better than an indicator to determine the end point of an acid-base titration?

88. What happens when an acid is added to a solution containing the HF/F^- buffer system?

89. When methyl red is added to an aqueous solution, a pink color results. When methyl orange is added to the same solution, a yellow color is produced. What is the approximate pH range of the solution? Use **Figure 24.**

90. Give the name and formula of the acid and the base from which each salt was formed.
a. NaCl **b.** $KHCO_3$ **c.** NH_4NO_2 **d.** CaS

Mastering Problems

91. Write formula equations and net ionic equations for the hydrolysis of each salt in water.
a. sodium carbonate **b.** ammonium bromide

92. Air Purifier Lithium hydroxide is used to purify air by removing carbon dioxide. A 25.00-mL sample of lithium hydroxide solution is titrated to an end point by 15.22 mL of 0.3340*M* hydrochloric acid solution. What is the molarity of the LiOH solution?

93. In an acid-base titration, 45.78 mL of a sulfuric acid solution is titrated to the end point by 74.30 mL of 0.4388*M* sodium hydroxide solution. What is the molarity of the H_2SO_4 solution?

SECTION 3

Mastering Concepts

75. $pOH = -\log [OH^-]$

76. Solution A is 10^3, or 1000 times more acidic than B.

77. $[OH^-]$ increases. $[H^+][OH^-] = K_w$.

78. The HCl adds H^+ ions to the water, which causes the equilibrium to shift the left.

79. a. household ammonia
b. blood
c. antacid
d. soft drinks
e. 100 times

Mastering Problems

80. $[OH^-] = 1.85 \times 10^{-12} M$

81. pH = 2.268
pOH = 11.733

82. pH = 0.523

83. For 0.10*M* HCl, pH = 1.00.
For 0.10*M* HF, pH = 2.10.
HCl has the greater concentration of hydrogen ions because it has the greater pH.

84. $K_a = 3.20 \times 10^{-7}$

SECTION 4

Mastering Concepts

85. Hydroiodic acid and sodium hydroxide must react.

86. Bromcresol purple or alizarin would be suitable because they would change color very near a pH 6.0 equivalence point.

87. A pH meter could be used when there is no acid-base indicator that changes color at or near the equivalence point, or when such an indicator is not available.

88. The acid produces hydrogen ions, which react with F^- ions in the solution to form HF molecules. The pH will drop slightly if the buffer is within its capacity.

89. The pH is is between approximately 4.2 and 5.6.

90. a. base: sodium hydroxide (NaOH); acid: hydrochloric acid (HCl)
b. base: potassium hydroxide (KOH); acid: carbonic acid (H_2CO_3)
c. base: ammonia (NH_3); acid: nitrous acid (HNO_2)
d. base: calcium hydroxide ($Ca(OH)_2$); acid: hydrosulfuric acid (H_2S)

Mastering Problems

91. a. $Na_2CO_3(s) + H_2O(l) \longrightarrow NaHCO_3(aq) + NaOH(aq)$;
$CO_3{}^{-2}(aq) + H_2O(l) \longrightarrow HCO_3{}^-(aq) + OH^-(aq)$
b. $NH_4Br(s) + H_2O(l) \longrightarrow HBr(aq) + NH_4OH(aq)$;
$NH_4{}^+(aq) + H_2O(l) \longrightarrow NH_3(aq) + H_3O^+(aq)$

92. $M_{LiOH} = 0.2033M$

93. $M_{H_2SO_4} = 0.3561M$

MIXED REVIEW

94. $K_b = [C_2H_5NH_3^+][OH^-]/[C_2H_5NH_2]$

95. 475 mL HCl

96. pH = 4.63

97. **a** and **d** are polyprotic acids.
$H_3BO_3(aq) \rightleftharpoons H^+(aq) + H_2BO_3^-(aq)$
$H_2BO_3^-(aq) \rightleftharpoons H^+(aq) + HBO_3^{2-}(aq)$
$HBO_3^{2-}(aq) \rightleftharpoons H^+(aq) + BO_3^{3-}(aq)$
$H_2SeO_3(aq) \rightleftharpoons H^+(aq) + HSeO_3^-(aq)$
$HSeO_3^-(aq) \rightleftharpoons H^+(aq) + SeO_3^{2-}(aq)$

98. $H_2CO_3(aq) + H_2O(l) \rightleftharpoons H_3O^+(aq) + HCO_3^-(aq)$
acid (H_2CO_3); conjugate base (HCO_3^-); base ($H_2O(l)$); conjugate acid H_3O^+
$HCO_3^-(aq) + H_2O(l) \rightleftharpoons H_3O^+(aq) + CO_3^{2-}(aq)$
acid (HCO_3^-); conjugate base (CO_3^{2-}); base ($H_2O(l)$); conjugate acid H_3O^+

99. All of the $Sr(OH)_2$ that dissolves dissociates to form Sr^{2+} and OH^- ions.

100. $pOH = 1.0 \times 10^{-11}$
$pOH = 1.0 \times 10^{-8}$
$pOH = 1.0 \times 10^{-5}$
$pOH = 1.0 \times 10^{-2}$

101. $K_a = 3.1 \times 10^{-6}$

102. $OH^-(aq) + H_2PO_4^-(aq) \rightleftharpoons H_2O(l) + HPO_4^{2-}(aq)$

103. pH = 4.19

THINK CRITICALLY

104. Such a statement is misleading. If the substance dissociates or reacts with water to produce hydroxide ions in solution, it is considered a base. However, substances such as organic acids contain hydroxyl groups bonded in such as way that they donate hydrogen ions in water and produce acidic solutions.

105. All Arrhenius acids are Brønsted-Lowry acids. Most Brønsted-Lowry acids are Arrhenius acids when in aqueous solution. (Examples: HCl, H_2SO_4, H_3PO_4) Lewis acids are electron pair acceptors. Because hydrogen ion is an electron pair acceptor, all Arrhenius and Brønsted-Lowry acids are Lewis acids. Some Lewis acids are not Arrhenius or Brønsted-Lowry acids, for example, BF_3.

106. A solution having a pH of 3.0 has a hydrogen-ion concentration of $1.00 \times 10^{-3}M$. Substitute $1.00 \times 10^{-3}M$ for $[H^+]$ in the expression $K_w = [H^+][OH^-] = 1.00 \times 10^{-11}M$; pOH = 11.0

107. **a.** Lewis acids: H^+ and H_2O; Lewis base: OH^-
b. Lewis acid: BCl_3; Lewis bases: BCl_4^-, Cl^-
c. Lewis acid: SO_3; Lewis base: H_2O

108. Refer to the Solutions Manual.

109. $C_2H_5NH_3^+(aq) \rightleftharpoons H^+(aq) + C_2H_5NH_2(aq)$ When an acid is added, the buffer equilibrium shifts to the left. When a base is added, the added OH^- ions react with H^+ ions and the reaction shifts to the right.

110. Only the hydrogen atom in the COOH group is likely to be ionizable.

111. At 10°C, the pH of pure water is 7.268. At 25°C, pH = 7.000. At 40°C, pH = 6.767. It is incorrect to say that the pH of pure water is always 7.0; the pH of pure water is 7.0 only at 25°C, or 298 K.

ASSESSMENT

MIXED REVIEW

94. Write the equation for the ionization reaction and the base ionization constant expression for ethylamine ($C_2H_5NH_2$) in water.

95. How many milliliters of 0.225*M* HCl would be required to titrate 6.00 g of KOH?

96. What is the pH of a 0.200*M* solution of hypobromous acid (HBrO)? $K_a = 2.8 \times 10^{-9}$

97. Which of the following are polyprotic acids? Write successive ionization equations for the polyprotic acids in water.
a. H_3BO_3
b. CH_3COOH
c. HNO_3
d. H_2SeO_3

98. Write balanced chemical equations for the two successive ionizations of carbonic acid in water. Identify the conjugate-base pair in each of the equations.

99. **Sugar Refining** Strontium hydroxide is used in the refining of beet sugar. Only 4.1 g of strontium hydroxide can be dissolved in 1 L of water at 273 K. Given that its solubility is so low, explain how it is possible that strontium hydroxide is considered a strong base.

100. What are the concentrations of OH^- ions in solutions having pH values of 3.00, 6.00, 9.00, and 12.00 at 298 K? What are the pOH values for the solutions?

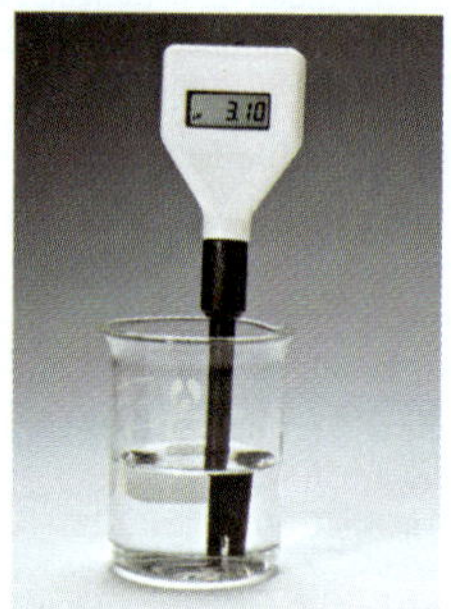

■ **Figure 31**

101. The pH probe in **Figure 31** is immersed in a 0.200*M* solution of a monoprotic acid, HA, at 303 K. What is the value of K_a for the acid at 303 K?

102. Write the chemical equation for the reaction that would occur when a base is added to a solution containing the $H_2PO_4^-/HPO_4^{2-}$ buffer system.

103. An aqueous solution buffered by benzoic acid (C_6H_5COOH) and sodium benzoate ($C_6H_5COOHNa$) is 0.0500*M* in both compounds. Given that benzoic acid's K_a equals 6.4×10^{-5}, what is the pH of the solution?

THINK CRITICALLY

104. **Critique** the following statement: "A substance whose chemical formula contains a hydroxyl group must be considered to be a base."

105. **Analyze and Conclude** Is it possible that an a Arrhenius acid is not a Brønsted-Lowry acid? Is it possible that an acid according to the Brønsted-Lowry model is not an Arrhenius acid? Is it possible that a Lewis acid could not be classified as either an Arrhenius or a Brønsted-Lowry acid? Explain and give examples.

106. **Apply Concepts** Use the ion product constant of water at 298 K to explain why a solution with a pH of 3.0 must have a pOH of 11.0.

107. **Identify** the Lewis acids and bases in the following reactions.
a. $H^+ + OH^- \leftrightharpoons H_2O$
b. $Cl^- + BCl_3 \leftrightharpoons BCl_4^-$
c. $SO_3 + H_2O \leftrightharpoons H_2SO_4$

108. **Interpret Scientific Illustrations** Sketch the shape of the approximate pH v. volume curve that would result from titrating a diprotic acid with a 0.10*M* NaOH solution.

109. **Recognize Cause and Effect** Illustrate how a buffer works using the $C_2H_5NH_3^+/C_2H_5NH_2$ buffer system. Show with equations how the weak base/conjugate acid system is affected when small amounts of acid and base are added to a solution containing this buffer system.

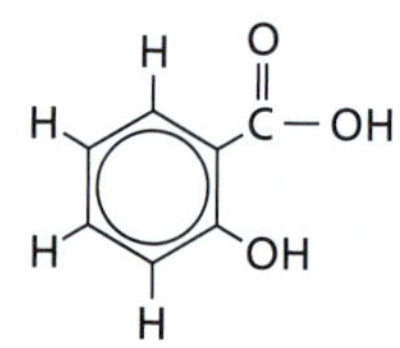

■ **Figure 32**

110. **Predict** Salicylic acid, shown in **Figure 32**, is used to manufacture acetylsalicylic acid, commonly known as aspirin. Evaluate the hydrogen atoms in the salicylic acid molecule based on your knowledge about the ionizable hydrogen in the acetic acid molecule, CH_3COOH. Predict which of salicylic acid's hydrogen atoms is likely to be ionizable.

111. **Apply Concepts** Like all equilibrium constants, the value of K_w varies with temperature. K_w equals 2.92×10^{-15} at 10°C, 1.00×10^{-14} at 25°C, and 2.92×10^{-14} at 40°C. In light of this information, calculate and compare the pH values for pure water at these three temperatures. Based on your calculations, is it correct to say that the pH of pure water is always 7.0? Explain.

Matt Meadows

Chapter Self-Check

CHALLENGE PROBLEM

112. You have 20.0 mL of a solution of a weak acid, HX, whose K_a equals 2.14×10^{-6}. The pH of the solution is found to be 3.800. How much distilled water would you have to add to the solution to increase the pH to 4.000?

CUMULATIVE REVIEW

113. What factors determine whether a molecule is polar or nonpolar?

114. What property of some liquids accounts for the meniscus that forms at the surface of a solution in a buret?

115. Which of the following physical processes are exothermic for water—freezing, boiling, condensing, subliming, evaporating?

116. Explain why an air pump gets hot when you pump air into a bicycle tire.

117. When 5.00 g of a compound was burned in a calorimeter, the temperature of 2.00 kg of water increased from 24.5°C to 40.5°C. How much heat would be released by the combustion of 1.00 mol of the compound (molar mass = 46.1 g/mol)?

118. What is the difference between an exothermic and an endothermic reaction?

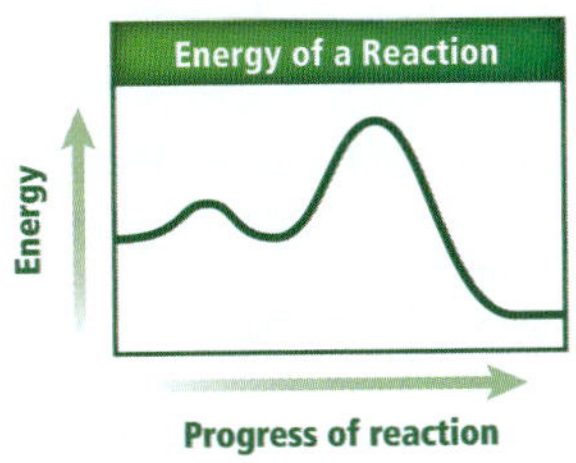

■ **Figure 33**

119. **Figure 33** shows how energy changes during the progress of a reaction.

a. Is the reaction exothermic or endothermic?

b. How many steps are in the reaction mechanism for the reaction?

c. Explain how you could use the graph to identify the rate-determining step.

120. Hydrogen and fluorine react to form HF according to the following equilibrium equation.

$$H_2(g) + F_2(g) \leftrightharpoons 2HF \quad \Delta H = -538 \text{ kJ(g)}$$

Will raising the temperature cause the amount of product to increase? Explain.

WRITING IN ▶ Chemistry

121. **Acid/Base Theories** Imagine that you are the Danish chemist Johannes Brønsted. The year is 1923, and you have formulated a new theory of acids and bases. Write a letter to Swedish chemist Svante Arrhenius in which you discuss the differences between your theory and his and point out the advantages of yours.

122. **Amino Acids** Twenty amino acids combine to form proteins in living systems. Research the structures and K_a values for five amino acids. Compare the strengths of these acids with the acids in **Table 4.**

DBQ Document-Based Questions

Rainwater **Figure 34** *shows pH measurements made from a number of the monitoring sites in New York state. The pink dot represents the average of the measurements taken at all of the sites at a particular time.*

Data obtained from: 2007 Acid Deposition Executive Summary. October, 2010. New York State Department of Environmental Conservation.

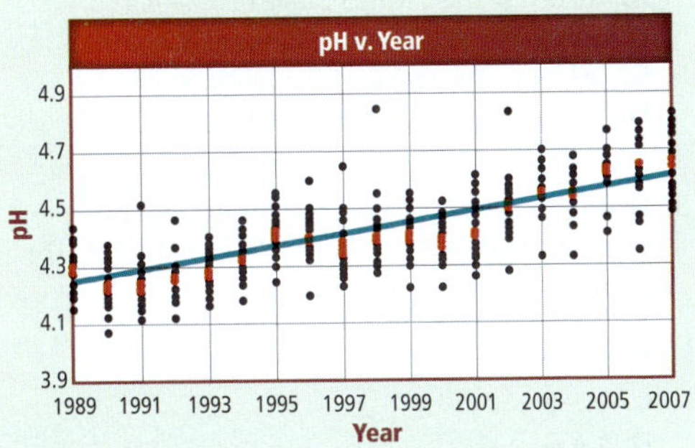

■ **Figure 34**

123. In general, what is the trend in the average pH for the years 1989 to 2007?

124. Calculate the $[H^+]$ for the lowest and the highest pH measurements recorded on the graph. How many times more acidic is the rainwater having the highest reading than the rainwater with the lowest?

125. What is the pH of the trend line in 2007? How much has the average pH changed between the years 1989 and 2007?

CHALLENGE PROBLEM

112. Add 30.1 mL of distilled water to the original 20.0 mL.

CUMULATIVE REVIEW

113. A polar molecule contains bonds in which the electrons are not shared equally because of differences in the electronegativity of the bonding atom. Also the effects of polar bonds can be cancelled or enhanced depending on the arrangement of the bonds in the molecule.

114. Adhesion and cohesions forces are at work.

115. freezing, condensing

116. Gay-Lussac's law states that when volume is a constant, temperature increases with pressure.

117. 1240 kJ/mol released

118. Energy is released in an exothermic reaction. Energy is absorbed in an endothermic reaction.

119. **a.** The reaction is exothermic because the reactants are at a higher energy than the products. **b.** Two steps because the graph shows two activation energy barriers. **c.** The rate-determining step is the one with the higher activation energy and therefore, the second step.

120. The reaction is exothermic because the sign of ΔH is negative. Therefore, heat is a product of the reaction. According to Le Châtelier's principle, addition of a product by, in this case, raising the temperature, will cause the equilibrium to shift to the left toward the reactants. Less product will be formed.

122. Students' answers will vary. For example, K_a for valine is 2.51×10^{-4} at 298 K.

DBQ Document-Based Questions

Data obtained from: 2007 Acid Deposition Executive Summary. October, 2010. New York State Department of Environmental Conservation.

123. The pH has increased gradually, from approximately 4.29 in 1989 to approximately 4.63 in 2007.

124. lowest pH(4.08), $[H^+] = 8.3 \times 10^{-5} M$; highest pH(4.85), $[H^+] = 1.4 \times 10^{-5} M$; 5.9 times more acidic

125. The trend line passes through 4.61 in 2007. The average pH changed from 4.29 in 1989 to 4.63 in 2007, a change of 0.34 .

WRITING IN ▶ Chemistry

Rubric

121. Students' letters should explain that Brønsted's theory included all the acids and bases that were defined by the Arrhenius theory. The Brønsted theory went further by explaining why some substances such as ammonia produce basic solutions but do not contain a hydroxide ion in their structures. Brønsted's theory also explains the role of water and the hydronium ion in acidic and basic solutions.

CUMULATIVE STANDARDIZED TEST PRACTICE

MULTIPLE CHOICE

1. C
2. C
3. A
4. B
5. C
6. C
7. D
8. B
9. A

CHAPTER 18 STANDARDIZED TEST PRACTICE

MULTIPLE CHOICE

Use the graph below to answer Questions 1 and 2.

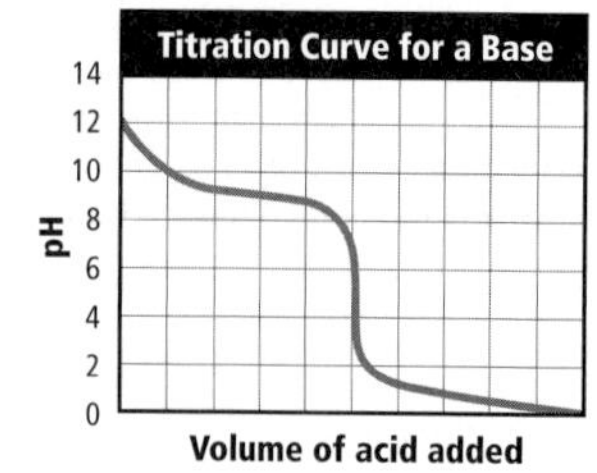

1. What is the pH at the equivalence point of this titration?
A. 10
B. 9
C. 5
D. 1

2. Which indicator would be effective for detecting the end point of this titration?
A. methyl orange, with a range of 3.2–4.4
B. phenolphthalein, with a range of 8.2–10
C. bromocresol green, with a range of 3.8–5.4
D. thymol blue, with a range of 8.0–9.6

3. Hydrogen bromide (HBr) is a strong, highly corrosive acid. What is the pOH of a 0.0375M HBr solution?
A. 12.574
B. 12.270
C. 1.733
D. 1.433

4. Cellular respiration produces about 38 mol of ATP for every mole of glucose consumed:

$$C_6H_{12}O_6 + 6O_2 \rightarrow 6CO_2 + 6H_2O + 38ATP$$

If each mole of ATP can release 30.5 kJ of energy, how much energy can be obtained from a candy bar containing 130.0 g of glucose?
A. 27.4 kJ
B. 836 kJ
C. 1159 kJ
D. 3970 kJ

Use the table below to answer Questions 5–7.

Ionization Constants and pH Data for Several Weak Organic Acids

Acid	pH of 1.000*M* Solution	K_a
Formic	1.87	1.78×10^{-4}
Cyanoacetic	?	3.55×10^{-3}
Propanoic	2.43	?
Lutidinic	1.09	7.08×10^{-3}
Barbituric	2.01	9.77×10^{-5}

5. Which acid is the strongest?
A. formic acid
B. cyanoacetic acid
C. lutidinic acid
D. barbituric acid

6. What is the hydronium ion concentration of the propanoic acid?
A. 1.4×10^{-5}
B. 2.43×10^{0}
C. 3.72×10^{-3}
D. 7.3×10^{4}

7. What is the pH of a 0.40*M* solution of cyanoacetic acid?
A. 2.06
B. 1.22
C. 2.45
D. 1.42

8. What does a value of K_{eq} greater than 1 mean?
A. More reactants than products exist at equilibrium.
B. More products than reactants exist at equilibrium.
C. The rate of the forward reaction is high at equilibrium.
D. The rate of the reverse reaction is high at equilibrium.

9. Magnesium sulfate ($MgSO_4$) is often added to water-insoluble liquid products of chemical reactions to remove unwanted water. $MgSO_4$ readily absorbs water to form two different hydrates. One of them is found to contain 13.0% H_2O and 87.0% $MgSO_4$. What is the name of this hydrate?
A. magnesium sulfate monohydrate
B. magnesium sulfate dihydrate
C. magnesium sulfate hexahydrate
D. magnesium sulfate heptahydrate

SHORT ANSWER

Use the description of an experiment below to answer Questions 10–12.

Two 0.050-mol samples of gas at 20°C are released from the end of a long tube at the same time. One gas is xenon (Xe), and the other is sulfur dioxide (SO_2).

10. Explain which gas will have traveled farther after 5 seconds. How can you tell?

11. How will increasing the temperature of this experiment affect the rate of effusion of each gas?

12. If the pressure on the xenon at the end of the experiment is 0.092 atm, what volume will it occupy?

EXTENDED RESPONSE

Use the figure below to answer Question 13.

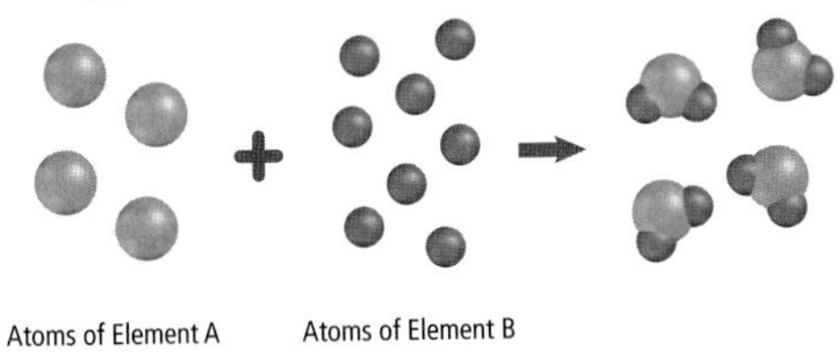

13. Explain how the chemical reaction shown in this figure demonstrates the law of conservation of mass.

14. Describe lab procedures for preparing a 0.50*M* aqueous solution of NaOH and a 0.50*m* aqueous solution of NaOH.

15. Explain how you could express the concentration of the 0.50*m* solution in Question 14 as a mole fraction.

SAT SUBJECT TEST: CHEMISTRY

16. Water has an unusually high boiling point compared to other compounds of similar molar mass because of ________.
    A. hydrogen bonding
    B. adhesive forces
    C. covalent bonding
    D. dispersion forces
    E. pi bonds

Use the graph below to answer Questions 17 and 18.

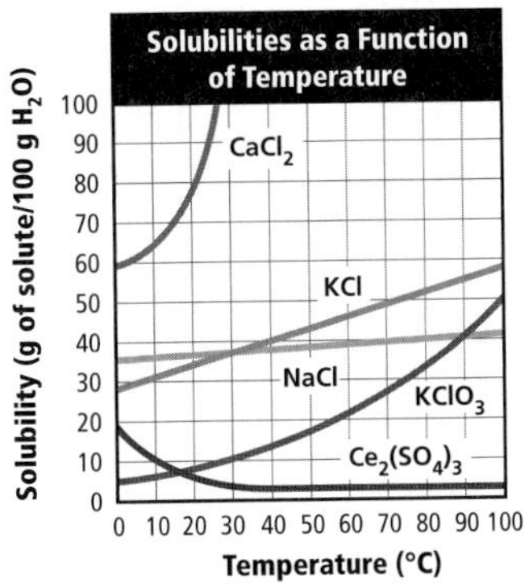

17. Which compound has a solubility of 38 g/100 g H_2O at 50°C?
    A. $CaCl_2$
    B. KCl
    C. NaCl
    D. $KClO_3$
    E. $Ce_2(SO_4)_3$

18. Which has the greatest increase in solubility as temperature increases?
    A. $Ce_2(SO_4)_3$
    B. $CaCl_2$
    C. $KClO_3$
    D. NaCl
    E. KCl

NEED EXTRA HELP?

If You Missed Question . . .	1	2	3	4	5	6	7	8	9	10	11	12	13	14	15	16	17	18
Review Section . . .	18.4	18.4	18.3	11.2	18.2	18.2	18.2	17.3	10.5	12.1	12.1	13.2	11.1	14.2	14.2	12.2	14.3	14.3

SHORT ANSWER

10. Sulfur dioxide will have traveled farther because SO_2 has the smaller mass. At 20°C, both gases have the same kinetic energy which equals $\frac{1}{2}mv^2$. Thus, the gas with the smaller mass must have the greater velocity and travel farther in 5 seconds.
11. Increasing the temperature increases the kinetic energy of the gases. The masses will not change but the velocities will and the rate of effusion will increase.
12. using $PV = nRT$, $V = 13$ L

EXTENDED RESPONSE

13. All the atoms of element A are found in the molecules of product. Similarly, all the atoms of element B are found in the molecules of the product. Thus, no atoms have been lost and no atoms have been gained.
14. To prepare a 0.50*M* solution of NaOH, calculate the mass of NaOH corresponding to one mole. Add the mass to a 1-L volumetric flask partially filled with water. Swirl to dissolve the solid. Then, fill the flask to the zero mark with water. To prepare a 0.50*m* solution of NaOH, measure 1000 mL of water, which corresponds to 1 kg. Add the mass of 0.50 mol of NaOH and swirl to dissolve.
15. To convert the unit 0.050*m* to a mole fraction, calculate the number of moles of water contained in 1.00 kg of water. The mole faction of NaOH is the moles of NaOH (0.050 mol) divided by the sum of the moles of water and the moles of NaOH. The mole fraction of water is the moles of water divided by the sum of the moles of water and the moles of NaOH.

SAT SUBJECT TEST: CHEMISTRY

16. A
17. C
18. B

CHAPTER 19 Redox Reactions

BIGIDEA **Oxidation-reduction reactions—among the most-common chemical processes in both nature and industry—involve the transfer of electrons.**

ESSENTIAL QUESTIONS	RESOURCES TO ASSESS MASTERY
SECTION 1 **Oxidation and Reduction** **1.** What are oxidation and reduction? **2.** How can oxidizing and reducing agents be identified? **3.** How is the oxidation number of an element in a compound determined? **4.** Does oxidation or reduction occur when the oxidation number increases? When it decreases? 2 sessions 1 block	**Progress Monitoring** Formative Assessment, pp. 681, 682, 686 Reading Check, p. 682 Section Review, p. 688
SECTION 2 **Balancing Redox Equations** **1.** How are changes in oxidation number related to the transfer of electrons? **2.** How can the changes in oxidation numbers be used to balance redox equations? **3.** What are half-reactions and how can they be used to balance redox equations? 2 sessions 1 block	**Progress Monitoring** Formative Assessment, pp. 691, 693, 694 Section Review, p. 696 **Summative Assessment** Chapter Assessment, p. 700 *eAssessment* Chapter Test (scaffolded)

Go online!

LEVELED RESOURCES	LAB MATERIALS
Science Notebook 19.1 OL ***FAST FILE Chapter Resources:*** MiniLab Worksheet, p. 58 OL ChemLab Worksheet, p. 60 OL Study Guide, p. 70 BL	**LaunchLAB** p. 678 / **15 min** steel wool, iron nail, 1.0*M* copper(II) sulfate ($CuSO_4$), test tube, test-tube rack **MiniLAB** p. 683 / **20 min** aluminum foil, steel wool, small tarnished object, 400-mL beaker, tap water, baking soda, table salt, beaker tongs, hot plate **ChemLAB** p. 698 / **45 min** 0.1*M* $AgNO_3$, 0.1*M* HCl, 0.1*M* $ZnSO_4$, unknown solution, Cu wire, Pb shot, Fe filings, Mg turnings, tongs or forceps, droppers, 24-well microscale reaction plate
Science Notebook 19.2 BL ***FAST FILE Chapter Resources:*** Study Guide, p. 72 BL	

ADDITIONAL RESOURCES

- eAssessment
- ConnectED Student Center
- ConnectED Teacher Center
- Lesson Presentations
- eSolutions
- Videos
- Animations
- Online Quizzes
- Virtual Investigations
- Multilingual eGlossary
- Personal Tutors
- WebQuests
- Applying Practices Worksheets
- PBLs

Practice:

- Challenge Problems AL
- Supplemental Problems BL OL
- Chapter Test (Scaffolded)
- Standardized Test Practice

Additional Resources:

- Blueprints for Success
- Solving Problems: A Chemistry Handbook

Lab Resources:

- Laboratory Manual OL
- CBL Laboratory Manual OL
- Small-Scale Laboratory Manual OL
- Forensics Laboratory Manual OL

BL Below Level   OL On Level   AL Advanced Learners   EL English Learners   COOP LEARN Cooperative Learning

LaunchLAB

Rubric

What happens when iron and copper(II) sulfate react?

Purpose Students will observe a redox reaction.

Safety Precautions Approve lab safety forms before work begins.

Disposal Solutions can be rinsed down a drain with large amounts of water. Dispose of solutions separately. Do not mix solutions in a drain. If your school is on a septic system, do not dispose of chemicals down a drain. Check with local EPA for proper disposal methods.

The copper-coated nails can be placed in the trash.

Teaching Strategies

- Prepare $CuSO_4$ in advance. See the Teacher Handbook for preparation of solutions.
- Remind students that the reaction is also a single replacement reaction. Ask them to explain why.

Expected Results The iron replaces the copper ions in solution, allowing copper metal to form on the nail. The blue copper(II) sulfate solution will turn colorless.

Procedure

1. Read and complete the lab safety form.
2. Use a piece of **steel wool** to polish the end of an **iron nail.**
3. Add about 3 mL of **1.0*M* copper(II) sulfate ($CuSO_4$)** solution to a **test tube.** Place the polished end of the nail into the $CuSO_4$ solution. Let the test tube stand in a **test-tube rack,** and observe it for about 10 min. Record your observations.

CHAPTER 19

Redox Reactions

BIGIDEA Oxidation-reduction reactions—among the most-common chemical processes in both nature and industry—involve the transfer of electrons.

SECTIONS

LaunchLAB

What happens when iron and copper(II) sulfate react?

Rust is the product of a reaction between iron and oxygen. Iron can also react with substances other than oxygen. In this lab, you will study a reaction between iron and copper(II) sulfate.

FOLDABLES® Study Organizer

Balancing Redox Equations

Make a layered-look book. Label it as shown. Use it to help you summarize information about the different methods for balancing redox equations.

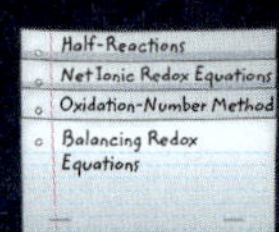

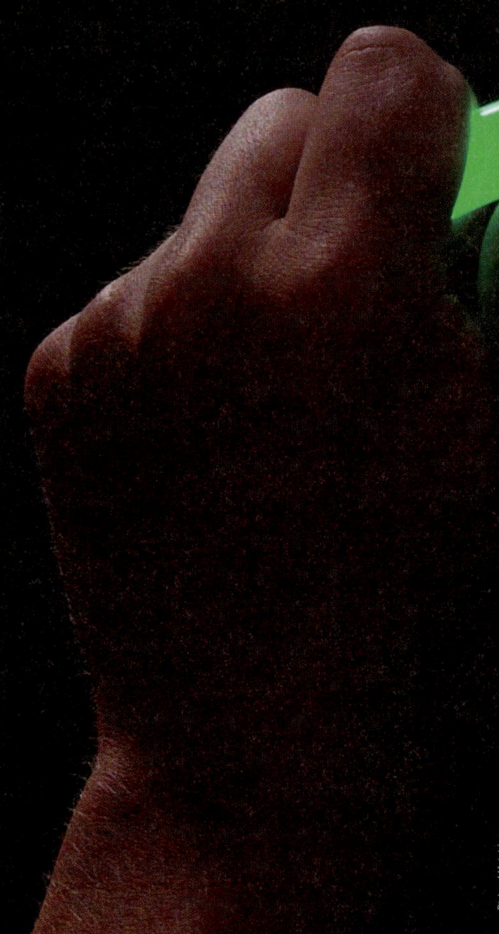

Richard Hutchings

Analysis

1. **Explain** what happened to the color of the copper(II) sulfate solution. The solution turns from blue to colorless.
2. **Identify** the substance clinging to the nail. Copper metal is found on the nail.
3. **Write** the balanced chemical equation for the reaction you observed. $Fe + CuSO_4 \rightarrow Cu + FeSO_4$

Inquiry What do you think would happen if copper was placed in an iron sulfate solution? Design an investigation to test your hypothesis. Answers will vary. When copper is placed in an iron sulfate solution, no reaction will occur. Investigations should be similar to the Launch Lab.

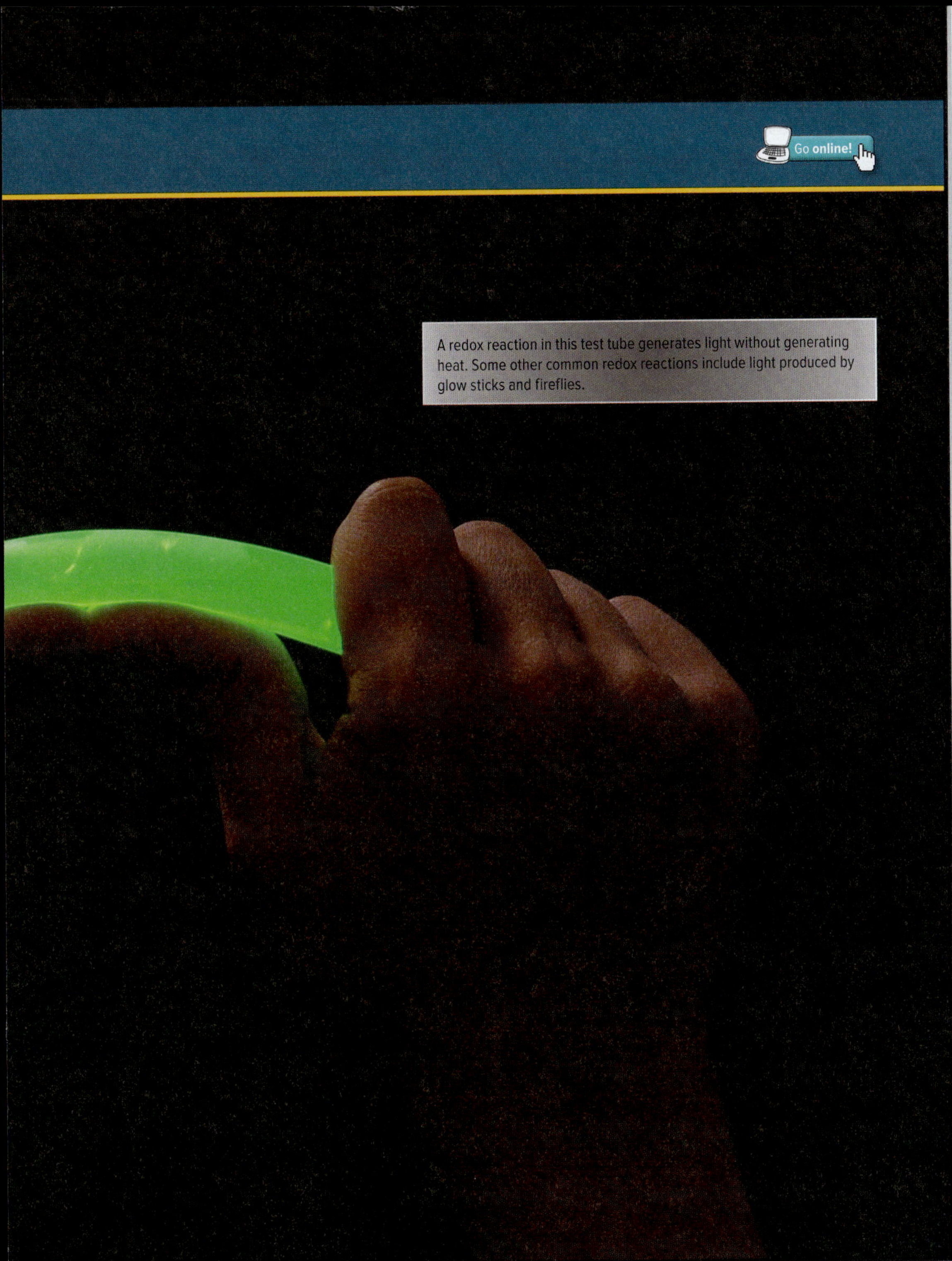

A redox reaction in this test tube generates light without generating heat. Some other common redox reactions include light produced by glow sticks and fireflies.

Introduce the BIGIDEA

Chemical Equations Review balanced chemical equations. Ask students what information is contained in a balanced chemical equation. identities of reactants and products, mole ratios, and physical states Write on the board $2Zn(s) + O_2(g) \rightarrow 2ZnO_2(s)$. Ask students what representative particles are involved in this reaction. atoms of zinc, molecules of oxygen, and formula units of zinc and oxide ions Ask students how the atoms and molecules formed ions. They gained or lost electrons. Tell students that early chemists called this an *oxidation reaction* because oxygen was a reactant. They later realized many reactions–even ones in which oxygen is not a reactant–involve the gain and loss of electrons, although the term *oxidation* is still used today.

Tie to Previous Knowledge

Have students review the following concepts before studying this chapter.

- electronegativity
- chemical reactions and types of reactions

Use the Photo

Light Stick Have students discuss situations in which they have used light sticks. Ask how light sticks work. When the light stick is bent, the glass ampoule inside breaks and two solutions inside react and give off energy. The energy excites electrons in atoms in the surrounding dye. As the atoms return to the ground state and lose energy, the dye glows. Ask students the advantage of light sticks in emergency situations. Answer might include: emits cool light, needs no external source of energy, stores easily for future use.

1 Focus

MAINIDEA

Complementary Actions Ask students to find the definition and give examples using the term *complementary*. Answers will vary but should include angles, colors, certain actions and reactions such as forces, DNA base pairs. Explain to students that oxidation and reduction reactions are complementary because one reaction cannot occur unless the other reaction occurs at the same time.

2 Teach

Quick Demo

Steel Wool Take a small piece of steel wool and fluff it until it is about twice its original size. Ask students if they think the steel wool will burn. Most students will suggest that it will get hot and possibly glow but not burn. Light a Bunsen burner and use tongs to hold the steel wool in the flame. The steel wool will burn brightly. It is a good idea to do this over a porcelain plate or dish. Model good lab practices by wearing safety goggles when performing this demonstration. Tell students that this reaction is an example of rapid oxidation in which elemental iron is reacting with the oxygen in the air. After the steel wool has cooled, it can be thrown in the trash.

■ **Caption Question Fig. 1**
The reaction is a combustion reaction.

SECTION 1

Oxidation and Reduction

MAINIDEA Oxidation and reduction are complementary—as a substance is oxidized, another substance is reduced.

Essential Questions

- What are oxidation and reduction?
- How can oxidizing and reducing agents be identified?
- How is the oxidation number of an element in a compound determined?

Review Vocabulary

spectator ion: an ion that does not participate in a reaction and is not usually shown in an ionic equation

New Vocabulary

oxidation-reduction reaction
redox reaction
oxidation
reduction
oxidation number
oxidizing agent
reducing agent

CHEM 4 YOU The light produced by a glow stick is the result of a chemical reaction. When you snap the glass capsule inside the plastic case, two chemicals are mixed and electron transfer occurs. As a result of this reaction, chemical energy is converted into light energy.

Electron Transfer and Redox Reactions

Previously, you learned that a chemical reaction can usually be classified as one of five types—synthesis, decomposition, combustion, single-replacement, or double-replacement. A defining characteristic of combustion and single-replacement reactions is that they always involve the transfer of electrons from one substance to another, as do many synthesis and decomposition reactions. For example, in the synthesis reaction in which sodium (Na) and chlorine (Cl_2) react to form the ionic compound sodium chloride (NaCl), an electron from each of two sodium atoms is transferred to the Cl_2 molecule to form two Cl^- ions.

Complete chemical equation: $2Na(s) + Cl_2(g) \rightarrow 2NaCl(s)$

Net ionic equation: $2Na(s) + Cl_2(g) \rightarrow 2Na^+ + 2Cl^-$ (ions in crystal)

An example of a combustion reaction is the burning of magnesium in air, which involves the transfer of electrons.

Complete chemical equation: $2Mg(s) + O_2(g) \rightarrow 2MgO(s)$

Net ionic equation: $2Mg(s) + O_2(g) \rightarrow 2Mg^{2+} + 2O^{2-}$ (ions in crystal)

When magnesium reacts with oxygen, as illustrated in **Figure 1**, each magnesium atom transfers two electrons to each oxygen atom. The two magnesium atoms become magnesium ions (Mg^{2+}), and the two oxygen atoms become oxide ions (O^{2-}). A reaction in which electrons are transferred from one substance to another is called an **oxidation-reduction reaction,** which is also called a **redox reaction.**

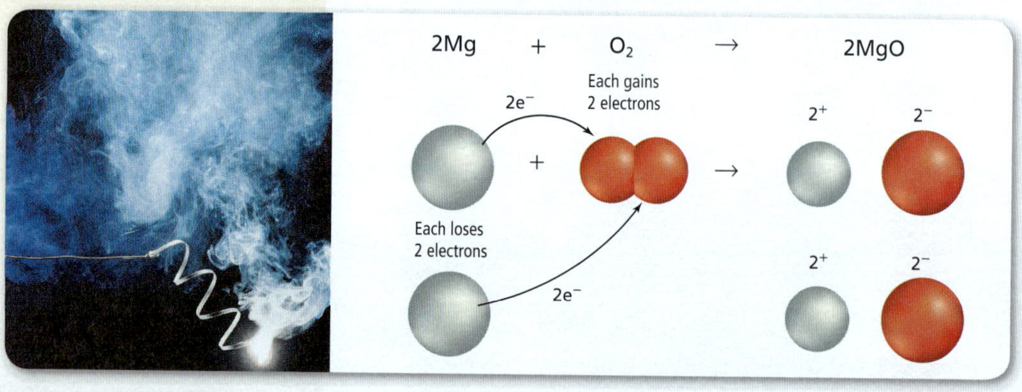

■ **Figure 1** The reaction of magnesium and oxygen involves a transfer of electrons from magnesium to oxygen. Therefore, this reaction is an oxidation-reduction reaction.
Classify *the reaction between magnesium and oxygen.*

Stephen Frisch/McGraw-Hill Education

Demonstration

Tollen's Silver Mirror Test

Purpose
to see how a reducing sugar changes the Ag ion into an Ag metal mirror

Materials
(A) $AgNO_3$ (12.8 g); **(B)** KOH(3.4 g); **(C)** dextrose (D-glucose $C_6H_{12}O_6$, 2.3 g); 15*M* NH_3 (aq); concentrated nitric acid (50 mL); beaker (600-mL); dropper; stoppered container to be silvered (1- to 2-L); graduated cylinder; stir-ring rod; distilled water (1.5 L) See the Teacher Handbook for preparation of all solutions.

Safety Precautions

WARNING: ***Nitric acid is corrosive.***

Disposal Remove Ag from inside the container with 50 mL of 1*M* HNO_3; recover $AgNO_3$ by evaporation.

Procedure
Place 150 mL of **(A)** into a 600-mL beaker.

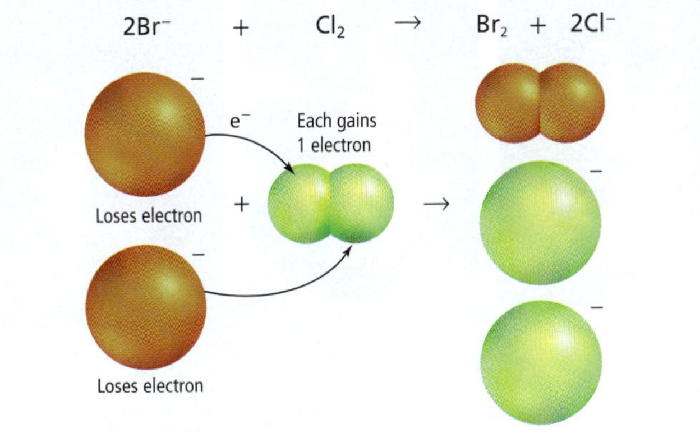

■ **Figure 2** The reaction between aqueous bromide ions and chlorine gas is a redox reaction. Here, electrons are transferred from bromide ions to chlorine.

Consider the single-replacement reaction in which chlorine in an aqueous solution reacts with bromide ions from an aqueous solution of potassium bromide, which is shown in **Figure 2.**

Complete chemical equation: $2KBr(aq) + Cl_2(aq) \rightarrow 2KCl(aq) + Br_2(aq)$

Net ionic equation: $2Br^-(aq) + Cl_2(aq) \rightarrow Br_2(aq) + 2Cl^-(aq)$

Note that the chlorine receives electrons from the bromide ions to become chloride ions. When the two bromide ions lose electrons, the two bromine atoms form a covalent bond with each other to produce Br_2 molecules.

View an **animation about redox reactions.**

Concepts In Motion

Oxidation and reduction Originally, the word *oxidation* referred only to reactions in which a substance combined with oxygen. Today, **oxidation** is defined as the complete or partial loss of electrons from a reacting substance. Look again at the net ionic equation for the reaction of sodium and chlorine. Sodium is oxidized because it loses an electron.

$$\text{Oxidation: } Na \rightarrow Na^+ + e^-$$

For oxidation to occur, the electrons lost by the substance that is oxidized must be accepted by atoms or ions of another substance. In other words, there must be an accompanying process that involves the gain of electrons. **Reduction** is the complete or partial gain of electrons by a reacting substance. Following the sodium chloride example further, the reduction reaction that accompanies the oxidation of sodium is the reduction of chlorine.

VOCABULARY
WORD ORIGIN
Reduction
comes from the Latin *re,* meaning *back,* and *ducere,* meaning *to lead*

$$\text{Reduction: } Cl_2 + 2e^- \rightarrow 2Cl^-$$

Oxidation and reduction are complementary processes: oxidation cannot occur unless reduction also occurs. It is important to recognize and distinguish between oxidation and reduction. A memory aid might help you remember the distinction. The phrase **L**oss of **E**lectrons is **O**xidation, and **G**ain of **E**lectrons is **R**eduction is shortened to **LEO GER.**

LEO the lion says **GER** or, for short, **LEO GER.**

Identify Misconceptions

One of the most common mistakes students make is to confuse the definitions of oxidation and reduction.

Uncover the Misconception Students tend to think that reduction involves a loss of electrons because students associate the word lose with reduce. Remind them that the oxidation number is reduced when an element gains electrons.

Demonstrate the Concept Be sure to reinforce this concept early by providing students with examples of half-reactions. Have them determine which element's oxidation number has increased or decreased, whether electrons are lost or gained during the process, and correctly label the half-reaction as an oxidation or reduction process.

Assess New Knowledge Have small groups of students develop an analogy that illustrates each of the following terms: *oxidation, reduction, oxidizing agent,* and *reducing agent.* Have them present their analogies. **OL** **COOP LEARN**

Add 75 mL of **(B)** to the beaker. Slowly add enough 15*M* $NH_3(aq)$ with stirring to just dissolve the precipitate. Rinse the container to be silvered with 50 mL concentrated nitric acid and then rinse thoroughly with distilled water. Pour 50 mL of **(C)** and the entire contents of the beaker ($AgNO_3 + KOH + NH_3$) into the vessel to be silvered. Rotate and tilt the stoppered container to keep all of the surfaces wet. After a minute or two, a silver mirror will begin to form. Continue to rotate and tilt the container until the mirror completely covers the inside surface. Pour out the reaction mixture and rinse the inside of the container two times with tap water.

Results

Ag^+ is reduced to Ag^0, which is deposited on the inside of the reaction vessel.

Analysis

What is the complementary oxidation reaction? The aldehyde group from the dextrose is oxidized to a carboxylic acid.

ASSESSMENT

Skill Have students research and answer these questions. What forms when an aldehyde is oxidized? a carboxylic acid What is the test called that identifies a reducing sugar? Tollen's test **OL**

ASSESSMENT

Performance Have students choose a metal to investigate. Have them research its primary ore and how the metal is extracted and purified from the ore. Oxidizing and reducing agents in each reaction should be identified. Students can give oral reports to the class. **OL**

READING CHECK chlorine

CAREERS IN CHEMISTRY

WebQuest

■ **Figure 3** The metal potassium and the nonmetal chlorine undergo a redox reaction to form potassium chloride.

Oxidation numbers A number assigned to an atom or ion to indicate its degree of oxidation or reduction is called its **oxidation number.** For example, the oxidation number of an element in an ionic compound is related to the number of electrons lost or gained by an atom of the element when it becomes an ion. The reaction of potassium metal with chlorine vapor, shown in **Figure 3,** is a redox reaction. The equation for the reaction is as follows.

Complete chemical equation: $2K(s) + Cl_2(g) \rightarrow 2KCl(s)$

Net ionic equation: $2K(s) + Cl_2(g) \rightarrow 2K^+(s) + 2Cl^-(s)$

Potassium, a group 1 element whose atoms tend to lose one electron in reactions because of its low electronegativity, is assigned an oxidation number of +1 in KCl. On the other hand, chlorine, a group 17 element whose atoms tend to gain one electron in reactions because of its high electronegativity, is assigned an oxidation number of −1 in KCl. In redox terms, you would say that potassium atoms are oxidized from 0 to the +1 state because each atom loses an electron, and chlorine atoms are reduced from 0 to the −1 state because each atom gains an electron. When an atom or ion is reduced, the numerical value of its oxidation number decreases. Conversely, when an atom or ion is oxidized, its oxidation number increases.

Oxidation numbers are tools that scientists use in written chemical equations to help them keep track of the movement of electrons in a redox reaction. Like some of the other tools you have learned about, oxidation numbers have a specific notation. Oxidation numbers are written with the positive or negative sign before the number (+3, +2), whereas ionic charge is written with the sign after the number (3+, 2+).

Oxidation number: +3 Ionic charge: 3+

READING CHECK **Determine** Which element is more likely to gain electrons, potassium or chlorine?

CAREERS IN CHEMISTRY

Potter A potter is an artist who makes pottery. He or she uses glazes containing metallic ions that have multiple oxidation states to achieve a variety of colors on ceramics. Glazes that contain copper ions produce a green-to-blue color when oxidized, and they produce a reddish color when reduced in a kiln.

WebQuest

DIFFERENTIATED INSTRUCTION

Struggling Learners Explain how mnemonics are a useful tool in remembering difficult or new concepts. Point out the mnemonic in the text and share the following mnemonic with students to help them remember what happens during the oxidation and reduction processes. The phrase *OIL RIG* reminds the students that Oxidation Is Losing electrons and Reduction Is Gaining electrons. Have the students try to think of other memory aids. **BL**

Oxidizing and Reducing Agents

The potassium-chlorine reaction in **Figure 3** can also be described by saying that "potassium is oxidized by chlorine." This description is useful because it clearly identifies both the substance that is oxidized and the substance that does the oxidizing. The substance that oxidizes another substance by accepting its electrons is called an **oxidizing agent.** This term describes the substance that is reduced. The substance that reduces another substance by losing electrons is called a **reducing agent.** A reducing agent supplies electrons to the substance being reduced (gaining electrons). The reducing agent is oxidized because it loses electrons. The reducing agent in the potassium-chlorine reaction is potassium–the substance that is oxidized.

$$\overbrace{2K(s) + Cl_2}^{\text{oxidized}}(g) \rightarrow 2KCl(s)$$
$$\text{reduced: } Cl_2(g) \rightarrow 2KCl(s)$$

Oxidizing agent: Cl_2

Reducing agent: K

A common application of redox chemistry is to remove tarnish from metal objects. Other oxidizing agents and reducing agents are useful in everyday life. For example, when you add chlorine bleach to your laundry to whiten clothes, you are using an aqueous solution of sodium hypochlorite (NaClO), an oxidizing agent. It oxidizes dyes, stains, and other materials that discolor clothes. **Table 1** summarizes the different ways to describe oxidation-reduction reactions.

Explore **redox reactions with an interactive table.** Concepts In Motion

Table 1 Summary of Redox Reactions

Process	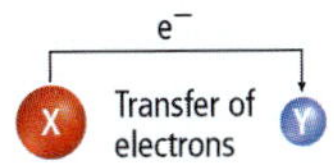
Oxidation • A reactant loses an electron. • Reducing agent is oxidized. • Oxidation number increases.	• X loses an electron. • X is the reducing agent and becomes oxidized. • The oxidation number of X increases.
Reduction • Other reactant gains an electron. • Oxidizing agent is reduced. • Oxidation number decreases.	• Y gains an electron. • Y is the oxidizing agent and becomes reduced. • The oxidation number of Y decreases.

MiniLAB

Observe a Redox Reaction

How can tarnish be removed from silver?

Procedure

1. Read and complete the lab safety form.
2. Lightly buff a piece of **aluminum foil** with **steel wool** to remove any oxide coating.
3. Wrap a **small tarnished object** in the aluminum foil, making sure that the tarnished area makes firm contact with the foil.
4. Place the wrapped object in a **400-mL** beaker and add a sufficient volume of **tap water** to cover it completely.
5. Add about 1 spoonful of **baking soda** and about 1 spoonful of **table salt** to the beaker.
6. Using beaker tongs, set the beaker and its contents on a **hot plate**, and heat until the water is almost boiling. Maintain the heat for approximately 15 min, until the tarnish disappears.

Analysis

1. **Write** the equation for the reaction of silver with hydrogen sulfide that yields silver sulfide and hydrogen.
2. **Write** the equation for the reaction of the tarnish (silver sulfide) with the aluminum foil that yields aluminum sulfide and silver.
3. **Determine** which metal, aluminum or silver, is more reactive. How do you know this from your results?
4. **Explain** why you should not use an aluminum pan to clean silver objects.

MiniLAB

Rubric

Purpose Students will use a redox reaction to clean tarnish from a silver spoon or other silver object.

Process Skills apply concepts, compare and contrast, observe and infer

Safety Precautions Approve lab safety forms before work begins. Use care when boiling the water on the hotplate. Do not let the water boil away. Add water if needed.

Disposal If your school participates in a local recycling program, the aluminum foil can be deposited in the aluminum collection bin. Otherwise, it can be disposed of in the regular trash.

Teaching Strategies

- If you do not have a tarnished silver object, spread mayonnaise or mustard, both of which contain sulfides, on a silver or a silver-plated object and leave it overnight. Rinse the object with water and wipe off all residue of the mayonnaise or mustard.
- Explain that the tarnishing compound in the environment is primarily hydrogen sulfide. It is formed from the decay of plant and animal material, and from industrial processes.

Expected Result The tarnish is removed from the entire object, not just the part in contact with the aluminum.

Analysis

1. $2Ag + H_2S \rightarrow Ag_2S + H_2$
2. $3Ag_2S + 2Al \rightarrow 6Ag + Al_2S_3$
3. Aluminum has the greater oxidation potential because it is oxidized in the reaction.
4. The aluminum pan would deteriorate.

CHEMISTRY PROJECT

Drain Cleaner Crystal drain cleaner is solid sodium hydroxide that contains aluminum. When added to water, the aluminum reacts with water in the basic solution and is oxidized. Ask students to discover what ingredients are found in liquid drain cleaners. Also, have them research the ingredients in the new foaming drain cleaners and how they work. OL

ASSESSMENT

Knowledge Consider the knowledge gained by this lab and investigate how cleaning by oxidation-reduction might apply to objects other than tarnished silver. OL

Concept Development

Prior Knowledge Review with students the concept of electronegativity. Pick several polyatomic ions with which your students are familiar, have them look up the electronegativity values, and remind them that the atom with the higher electronegativity value will be the more negative element when assigning oxidation numbers. OL

- **Caption Question Fig. 4** F; Cs

Redox and Electronegativity

The chemistry of oxidation-reduction reactions is not limited to atoms of an element changing to ions or the reverse. Some redox reactions involve changes in molecular substances or polyatomic ions in which atoms are covalently bonded to other atoms. For example, the following equation represents the redox reaction used to manufacture ammonia (NH_3).

$$N_2(g) + 3H_2(g) \rightarrow 2NH_3(g)$$

This process involves neither ions nor any obvious transfer of electrons. The reactants and products are all molecular compounds. Yet, it is still a redox reaction in which nitrogen is the oxidizing agent and hydrogen is the reducing agent.

In situations such as the formation of ammonia, where two atoms share electrons, how is it possible to say that one atom lost electrons and was oxidized, while the other atom gained electrons and was reduced? To answer this, you need to know which atom attracts electrons more strongly, or, in other words, which atom is more electronegative. You might find it helpful to review the discussion of electronegativity trends on the periodic table. **Figure 4** shows that electronegativity increases left to right across a period and generally decreases down a group. Elements with low electronegativity (Groups 1 and 2) are strong reducing agents, and those with high electronegativity (Group 17 and oxygen in Group 16) are strong oxidizing agents.

Hydrogen has an electronegativity of 2.20, and nitrogen has an electronegativity of 3.04. For the purpose of studying oxidation-reduction reactions, the more-electronegative element (in this case nitrogen) is treated as if it had been reduced by gaining electrons from the other element (hydrogen). Conversely, the less-electronegative element (hydrogen) is treated as if it had been oxidized by losing electrons to the other element (nitrogen).

$$\overbrace{N_2(g) + 3H_2(g) \rightarrow 2NH_3(g)}^{\text{reduced (partial gain of } e^-)}$$

$$\underbrace{3H_2(g) \rightarrow 2NH_3(g)}_{\text{oxidized (partial loss of } e^-)}$$

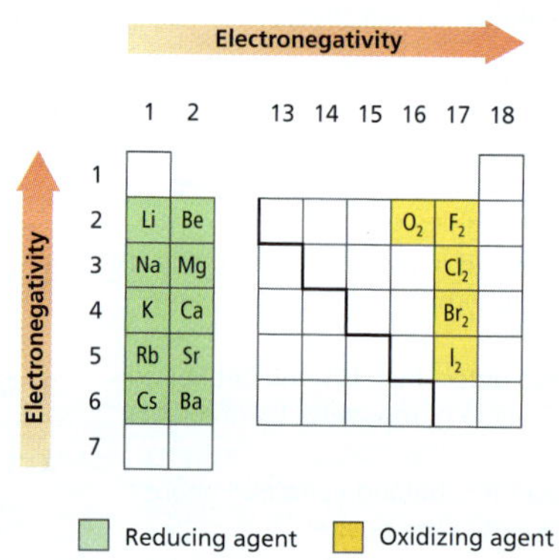

Figure 4 The electronegativity of elements increases from left to right across the periodic table, and it decreases going down a group. Elements with low electronegativity are strong reducing agents, and elements with high electronegativity are strong oxidizing agents.
Predict *which element would be the strongest oxidizing agent. Which is the strongest reducing agent?*

CHEMISTRY JOURNAL

Chemistry in Cartoons Have students draw a cartoon or brief comic strip story of an atom losing an electron to another atom. Have them include in their story captions describing which atom is oxidized and which atom is reduced. OL EL

EXAMPLE Problem 1

Identify Oxidation-Reduction Reactions The following equation represents the redox reaction of aluminum and iron.

$$2Al + 2Fe^{3+} + 3O^{2-} \rightarrow 2Fe + 2Al^{3+} + 3O^{2-}$$

Identify what is oxidized and what is reduced in this reaction. Identify the oxidizing agent and the reducing agent.

1 ANALYZE THE PROBLEM

You are given the reactants and products in the reaction. You must determine the electron transfers that occur. Then, you can apply the definitions of oxidizing agent and reducing agent to answer the question.

2 SOLVE FOR THE UNKNOWN

Identify the oxidation process and the reduction process.

$Al \rightarrow Al^{3+} + 3e^-$ (loss of e^- is oxidation) — **The aluminum atom loses three electrons and becomes an aluminum ion.**

$Fe^{3+} + 3e^- \rightarrow Fe$ (gain of e^- is reduction) — **The iron ion accepts the three electrons lost from aluminum and becomes an iron atom.**

Al is oxidized and is therefore the reducing agent. Fe^{3+} is reduced and is therefore the oxidizing agent.

3 EVALUATE THE ANSWER

In this process, aluminum atoms lost electrons and were oxidized, whereas iron ions gained electrons and were reduced. The definitions of oxidation, reduction, oxidizing agent, and reducing agent apply. Note that the ionic charge of oxygen is unchanged in this reaction; therefore, oxygen is not a key factor in this problem.

PRACTICE Problems

Do additional problems. Online Practice

1. Identify each of the following changes as either oxidation or reduction. Recall that e^- is the symbol for an electron.
   a. $I_2 + 2e^- \rightarrow 2I^-$
   b. $K \rightarrow K^+ + e^-$
   c. $Fe^{2+} \rightarrow Fe^{3+} + e^-$
   d. $Ag^+ + e^- \rightarrow Ag$
2. Identify what is oxidized and what is reduced in the following processes.
   a. $2Br^- + Cl_2 \rightarrow Br_2 + 2Cl^-$
   b. $2Ce + 3Cu^{2+} \rightarrow 3Cu + 2Ce^{3+}$
   c. $2Zn + O_2 \rightarrow 2ZnO$
   d. $2Na + 2H^+ \rightarrow 2Na^+ + H_2$
3. Identify the oxidizing agent and the reducing agent in the following equation. Explain your answer.
   $$Fe(s) + 2Ag^+(aq) \rightarrow Fe^{2+}(aq) + 2Ag(s)$$
4. **Challenge** Identify the oxidizing agent and the reducing agent in each reaction.
   a. $Mg + I_2 \rightarrow MgI_2$
   b. $H_2S + Cl_2 \rightarrow S + 2HCl$

RealWorld CHEMISTRY

Oxidation

RUST When moist air comes in contact with iron, the iron oxidizes. Iron oxide (Fe_2O_3), called rust, is common because iron combines readily with oxygen. Pure iron is uncommon in nature. Steel, a mixture that contains iron, is a commonly used form of iron. Several protective methods, such as plating, painting, and applying an enamel or plastic coating, can inhibit the production of iron oxide.

IN-CLASS Example

Question What is oxidized and what is reduced in the following equation?

$$2H^+ + 2Cl^- + Ni \rightarrow Ni^{2+} + 2Cl^- + H_2$$

Identify the oxidizing agent and the reducing agent in this equation.

Answer

reduction: $2H^+ + 2e^- \rightarrow H_2$

oxidation: $Ni \rightarrow Ni^{2+} + 2e^-$

Each H^+ gains one electron and is the species reduced. H^+ is the oxidizing agent.

Each Ni loses two electrons and is the species oxidized. Ni is the reducing agent.

PRACTICE Problems

Have students refer to the Selected Solutions appendix for complete solutions to odd-numbered problems. The complete solutions for all problems can be found in the Solutions Manual.

1. a. reduction
   b. oxidation
   c. oxidation
   d. reduction
2. a. Br^- is oxidized, Cl is reduced
   b. Ce is oxidized, Cu^{2+} is reduced
   c. Zn is oxidized, O_2 is reduced
   d. Na is oxidized, H^+ is reduced
3. Ag^+ is the oxidizing agent, Fe is the reducing agent; Ag^+ is reduced, Fe is oxidized
4. a. I_2 is the oxidizing agent, Mg is the reducing agent
   b. Cl_2 is the oxidizing agent, H_2S is the reducing agent

CHEMISTRY JOURNAL

Making Metal Have students make a list of three metal things that they use in their everyday activities. Have them research how the particular metals are made and write in their journal an explanation of the redox processes that are involved. OL

Reinforcement

Oxidation Number List the following sulfur-containing substances on the board: Na_2SO_4, H_2S, S, S_2Cl_2, SO_2, $K_2S_2O_3$. Tell the students that the sulfur in each substance has a different oxidation number. Have them organize these substances in order of increasing oxidation number of sulfur. H_2S, S, S_2Cl_2, $K_2S_2O_3$, SO_2, Na_2SO_4 **OL**

ASSESSMENT

Skill Provide students with ten compounds in sealed, clear containers. The containers should be labeled with the formula of the compound. Have the students determine the oxidation number of each element in each compound. Try to include compounds that are colored solids. Compounds could include $CuSO_4$, $K_2Cr_2O_7$, $NaNO_3$, or $Co(NO_3)_3$. Warn students that they are not to open the sealed bottles. **OL**

ChemLAB

The ChemLab located at the end of the chapter can be used at this point in the lesson.

Figure 5 Banded iron—shown in this cross-section of rock—is a result of different oxidation states of iron, which depend on which minerals are present.

Determining Oxidation Numbers

In order to understand all types of redox reactions, you must have a way to determine the oxidation number ($n_{element}$) of each element involved in a reaction. **Table 2** outlines the rules chemists use to make this determination easier.

Many elements other than those specified in the rules below, including most of the transition metals, metalloids, and nonmetals, can be found with different oxidation numbers in different compounds. For example, iron has different oxidation numbers, indicated by the different colors as shown in **Figure 5,** depending on which minerals are also present.

Table 2 Rules for Determining Oxidation Numbers

Rule	Example	$n_{element}$
1. The oxidation number of an atom of an uncombined element is zero.	Na, O_2, Cl_2, H_2	0
2. The oxidation number of a monatomic ion is equal to the charge of the ion.	Ca^{2+}	+2
	Br^-	−1
3. The oxidation number of the more-electronegative atom in a molecule or a complex ion is the same as the charge it would have if it were an ion.	N in NH_3	−3
	O in NO	−2
4. The oxidation number of the most-electronegative element, fluorine, is always −1 when it is bonded to another element.	F in LiF	−1
5. The oxidation number of oxygen in compounds is always −2 except in peroxides, such as hydrogen peroxide (H_2O_2), where it is −1. When it is bonded to fluorine, the only element more electronegative than oxygen, the oxidation number of oxygen is positive.	O in NO_2	−2
	O in H_2O_2	−1
6. The oxidation number of hydrogen in most of its compounds is +1, except in metal hydrides; then, the oxidation number is −1.	H in NaH	−1
7. The oxidation numbers of group 1 and 2 metals and aluminum are positive and equal to their number of valence electrons.	K	+1
	Ca	+2
	Al	+3
8. The sum of the oxidation numbers in a neutral compound is zero.	$CaBr_2$	(+2) + 2(−1) = 0
9. The sum of the oxidation numbers of the atoms in a polyatomic ion is equal to the charge of the ion.	SO_3^{2-}	(+4) + 3(−2) = −2

EXAMPLE Problem 2

Determine Oxidation Numbers Use the rules for determining oxidation numbers to find the oxidation number of each element in potassium chlorate ($KClO_3$) and in a sulfite ion (SO_3^{2-}).

1 ANALYZE THE PROBLEM

In the rules for determining oxidation numbers, you are given the oxidation numbers of oxygen and potassium. You are also given the overall charge of the compound or ion. Using this information and applying the rules, determine the oxidation numbers of chlorine and sulfur. (Let $n_{element}$ equal the oxidation number of the element in question.)

Known	**Unknown**
$KClO_3$	$n_{Cl} = ?$
SO_3^{2-}	$n_S = ?$
$n_O = -2$	
$n_K = +1$	

2 SOLVE FOR THE UNKNOWN

Assign the known oxidation numbers to their elements, set the sum of all oxidation numbers to zero or to the ion charge, and solve for the unknown oxidation number.

$(n_K) + (n_{Cl}) + 3(n_O) = 0$ — The sum of the oxidation numbers in a neutral compound is zero.

$(+1) + (n_{Cl}) + 3(-2) = 0$ — For group 1 metals, $n_{element} = +1$. Substitute $n_K = +1$, $n_O = -2$.

$1 + n_{Cl} + (-6) = 0$

$n_{Cl} = +5$ — Solve for n_{Cl}.

$(n_S) + 3(n_O) = -2$ — The sum of the oxidation numbers in a polyatomic ion equals the charge on the ion. Substitute $n_O = -2$.

$(n_S) + 3(-2) = -2$

$n_S + (-6) = -2$

$n_S = +4$ — Solve for n_S.

3 EVALUATE THE ANSWER

The rules for determining oxidation numbers have been correctly applied. All of the oxidation numbers in each substance add up to the proper value.

PRACTICE Problems

Do additional problems. Online Practice

5. Determine the oxidation number of the boldface element in the following formulas for compounds.
   a. Na**Cl**O_4 b. Al**P**O_4 c. H**N**O_2
6. Determine the oxidation number of the boldface element in the following formulas for ions.
   a. **N**H_4^+ b. **As**O_4^{3-} c. **Cr**O_4^{2-}
7. Determine the oxidation number of nitrogen in each of these molecules.
   a. NH_3 b. KCN c. N_2H_4
8. **Challenge** Determine the net change of oxidation number of each of the elements in these redox equations.
   a. $C + O_2 \rightarrow CO_2$
   b. $Cl_2 + ZnI_2 \rightarrow ZnCl_2 + I_2$
   c. $CdO + CO \rightarrow Cd + CO_2$

DIFFERENTIATED INSTRUCTION

Struggling Learners For students having difficulty with Practice Problems 5, 6, and 7, have them write the number of the rule for determining oxidation numbers (as given in the text), and state the reason why they chose that rule. This helps students focus on the task at hand and provides them with a more concrete method to help answer the question. BL

IN-CLASS Example

Question Find the oxidation number for each element atom in NH_4NO_3 and $Cr_2O_7^{2-}$.

Answer

NH_4NO_3

The sum of the oxidation numbers in a compound is zero.

N(ammonium) + 4H + N(nitrate) + 3O = 0

From the rules: H = +1, N (in ammonium) = −3, and O = −2. The only unknown is N (in nitrate).

−3 + 4(+1) + N(nitrate) + 3(−2) = 0

N(nitrate) + 4 − 9 = 0

N(nitrate) − 5 = 0

N(nitrate) = +5

$Cr_2O_7^{2-}$

This is an anion, so the sum of the oxidation numbers must equal the charge on the anion.

2Cr + 7O = −2

From the rules O = −2, so the unknown value is Cr.

2Cr + 7(−2) = −2

2Cr − 14 = −2

2Cr = −2 + 14 = 12

$Cr = \frac{12}{2} = +6$

PRACTICE Problems

Have students refer to the Selected Solutions appendix for complete solutions to odd-numbered problems. The complete solutions for all problems can be found in the Solutions Manual.

5. a. +7
   b. +5
   c. +3
6. a. −3
   b. +5
   c. +6
7. a. −3
   b. −3
   c. −2
8. a. C, +4; O, −2
   b. I, +1; Cl, −1; Zn, no change
   c. C, +2; Cd, −2; O, no change

3 Assess

Check for Understanding

Give students the following equation: $4Fe + 3O_2 \rightarrow 2Fe_2O_3$. Have them assign oxidation numbers and determine the oxidizing and reducing agents. Fe goes from 0 to +3. Oxygen goes from 0 to −2. Oxygen is the oxidizing agent and is reduced. Iron is the reducing agent and is oxidized. **OL**

Reteach

Have students make a chart with two columns. One column should be labeled *oxidation* and the other column should be labeled *reduction*. Have students complete the chart for the following questions.

a. What happens to the electrons when oxidation occurs? electrons are lost from particles of a substance Reduction? electrons are gained by particles of a substance

b. What happens to the charge of an atom that undergoes oxidation? It is increased. Reduction? It is decreased.

c. Which process provides the oxidizing agent? reduction Reducing agent? oxidation **OL**

Extension

Have students research the oxidation–reduction reactions that take place in a blast furnace and make a poster outlining the process. Students should point out the oxidizing and reducing agents in each reaction. **OL** **EL**

Table 3 Various Oxidation Numbers

Oxidation Number	+1	+2	+3	−1	−2
Aluminum			X		
Barium		X			
Bromine				X	
Cadmium		X			
Calcium		X			
Cesium	X				
Chlorine				X	
Fluoride				X	
Hydrogen	X			X	
Iodine				X	
Lithium	X				
Magnesium		X			
Oxygen					X
Potassium	X				
Sodium	X				
Silver	X				
Strontium		X			

Oxidation Numbers in Redox Reactions

Having studied oxidation numbers, you should be able to relate oxidation-reduction reactions to changes in oxidation number. Refer to the equation for a reaction that you saw at the beginning of this section—the replacement of bromine in aqueous potassium bromide (KBr) by chlorine (Cl_2).

$$2KBr(aq) + Cl_2(aq) \rightarrow 2KCl(aq) + Br_2(aq)$$

To learn how oxidation numbers change, start by assigning numbers, using **Table 3,** to all elements in the balanced equation. Then, review the changes, as shown in the equation below.

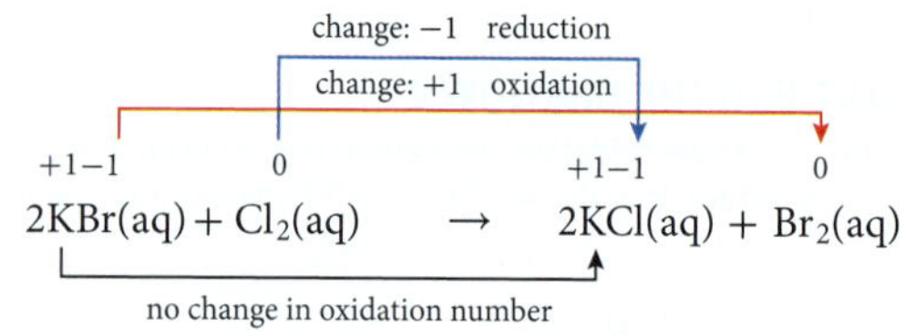

You should notice that the oxidation number of bromine changed from −1 to 0, an increase of 1. At the same time, the oxidation number of chlorine changed from 0 to −1, a decrease of 1. Therefore, chlorine is reduced and bromine is oxidized. All redox reactions follow the same pattern. When an atom is oxidized, its oxidation number increases. When an atom is reduced, its oxidation number decreases. Note that there is no change in the oxidation number of potassium. The potassium ion takes no part in the reaction and is therefore a spectator ion.

SECTION 1 REVIEW

Section Self-Check

Section Summary

- Oxidation-reduction reactions involve the transfer of electrons from one atom to another.
- When an atom or ion is reduced, its oxidation number decreases. When an atom or ion is oxidized, its oxidation number increases.
- In oxidation-reduction reactions involving molecular compounds (and polyatomic ions with covalent bonds), the more-electronegative atoms are treated as if they are reduced. The less-electronegative atoms are treated as if they are oxidized.

9. MAIN IDEA **Explain** why oxidation and reduction must always occur together.

10. Describe the roles of oxidizing agents and reducing agents in a redox reaction. How is each changed in the reaction?

11. Write the equation for the reaction of iron metal with hydrobromic acid to form aqueous iron(III) bromide and hydrogen gas. Determine the change in oxidation number for the element that is reduced and the element that is oxidized.

12. Determine the oxidation number of the boldface element in these compounds.

a. HNO_3 **c.** Sb_2O_5
b. Ca_3N_2 **d.** $CuWO_4$

13. Determine the oxidation number of the boldface element in these ions.

a. IO_4^- **c.** $B_4O_7^{2-}$
b. MnO_4^- **d.** NH_2^-

14. Make and Use Graphs Alkali metals are strong reducing agents. Make a graph showing how the reducing abilities of the alkali metals increase or decrease as you move down the family from sodium to francium.

SECTION 1 REVIEW

9. If an atom or ion loses an electron, some other species must gain the electron.

10. An oxidizing agent causes another species to be oxidized by gaining the electrons from it. A reducing agent causes another species to be reduced by losing electrons to that element.

11. $2Fe(s) + 6HBr(aq) \rightarrow 2FeBr_3(aq) + 3H_2(g)$; Fe is oxidized, H is reduced

12. a. +5
b. −3
c. +5
d. +6

13. a. +7
b. +7
c. +3
d. −3

14. In general, as you move down the periodic table within a family, the tendency to lose electrons increases, so the reducing ability increases.

SECTION 2

Balancing Redox Equations

MAINIDEA Redox equations are balanced when the total increase in oxidation numbers equals the total decrease in oxidation numbers of the atoms or ions involved in the reaction.

Essential Questions

- How are changes in oxidation number related to the transfer of electrons?
- How can changes in oxidation numbers be used to balance redox equations?
- What are half-reactions and how can they be used to balance redox equations?

Review Vocabulary

net ionic equation: an ionic equation that includes only the particles that participate in the reaction

New Vocabulary

oxidation-number method
species
half-reaction

CHEM 4 YOU When fatty substances in foods spoil, they are referred to as rancid. Large molecules are broken down through redox reactions that result in foul-smelling products. The equation for this process is complicated but can be balanced using the same rules for simpler equations.

The Oxidation-Number Method

Chemical equations must be balanced to show the correct quantities of reactants and products. Study the following unbalanced equation for the reaction that occurs when copper metal is placed in concentrated nitric acid. This reaction is shown in **Figure 6.** The brown gas that is produced is nitrogen dioxide (NO_2), from the reduction of nitrate ions (NO_3^-), and the blue solution is the result of the oxidation of copper (Cu) to copper(II) ions (Cu^{2+}).

$$Cu(s) + HNO_3(aq) \rightarrow Cu(NO_3)_2(aq) + NO_2(g) + H_2O(l)$$

Note that oxygen appears in only one reactant, HNO_3, but in all three products. Nitrogen appears in HNO_3 and in two of the products. Redox equations such as this one, in which the same element appears in several reactants and products, can be difficult to balance. As you have read, when an atom or ion loses electrons, its oxidation number increases; when an atom or ion gains electrons, its oxidation number decreases. The number of electrons transferred from atoms or ions must equal the number of electrons accepted by other atoms or ions. Therefore, the total increase in oxidation numbers (oxidation) must equal the total decrease in oxidation numbers (reduction) of the atoms or ions involved in the reaction. The balancing technique called the **oxidation-number method** is based on these principles, and is described in **Table 4.**

■ **Figure 6** Some chemical equations for redox reactions, such as the reaction between copper and nitric acid, can be difficult to balance because elements might appear more than once on each side of the equation.

Table 4 The Oxidation-Number Method

Table 4 The Oxidation-Number Method
1. Assign oxidation numbers to all chemical elements in the equation.
2. Identify the atoms or ions that are oxidized and reduced.
3. Determine the change in oxidation number for the atoms or ions that are oxidized and for the atoms or ions that are reduced.
4. Make the change in oxidation numbers equal in magnitude by adjusting coefficients in the equation.
5. If necessary, use the conventional method to balance the remainder of the equation.

DIFFERENTIATED INSTRUCTION

English Learners Have students collect pictures from advertisements for various processes that involve redox reactions or that are products of a redox reaction. Examples might include ads showing household bleach and detergents containing bleach, paints, and waxes that prevent rusting, antioxidants found in foods, or cameras and film. Have them name and describe these items in their native language and in English. Also, have them look up the derivation of the terms *oxidation* and *reduction*. To emphasize the universal language of chemistry, provide the chemical names and symbols for some of the items and point out how they do not change with local language. EL OL

SECTION 2

1 Focus

MAINIDEA

Conservation of Charge Review with students the law of conservation of matter and ask them to relate this law to chemical equations. The mass of the reactants equals the mass of the products. Tell students that charge also is conserved in a chemical reaction. OL

2 Teach

Identify Misconceptions

Students might want to think about redox only in terms of what is oxidized and what is reduced.

Uncover the Misconception Students might think that when reactions are balanced using H^+ ions, OH^- ions, and water, these substances are invented for convenience. Remind them that many redox reactions are aqueous and involve acids (H^+) or bases (OH^-).

Demonstrate the Concept Begin writing the complete, original reactions (with the acid/base conditions and water) before simplifying the reaction in its net ionic form. Then, show how to use the net ionic reaction to complete the balancing step. Finally, help students see the connection between the final balanced reaction using the methods learned in this section and the original reaction.

Assess New Knowledge Give students a reaction and have them explain why water and H^+ ions or OH^- ions are necessary to balance a redox reaction. OL

IN-CLASS Example

Question Balance the following redox equation.

$FeO + NH_3 \rightarrow N_2 + H_2O + Fe$

Answer $3FeO + 2NH_3 \rightarrow N_2 + 3H_2O + 3Fe$

PRACTICE Problems

Have students refer to the Selected Solutions appendix for complete solutions to odd-numbered problems. The complete solutions for all problems can be found in the Solutions Manual.

15. $3HCl + 2HNO_3 \rightarrow 2HOCl + 2NO + H_2O$
16. $3SnCl_4 + 2Fe \rightarrow 3SnCl_2 + 2FeCl_3$
17. $8NH_3(g) + 6NO_2(g) \rightarrow 7N_2(g) + 12H_2O(l)$
18. $SO_2 + Br_2 + 2H_2O \rightarrow 2HBr + H_2SO_4$

EXAMPLE Problem 3

Get help with **balancing redox equations.** Personal Tutor

EXAMPLE PROBLEM

THE OXIDATION-NUMBER METHOD Balance the following redox equation.

$$Cu + HNO_3 \rightarrow Cu(NO_3)_2 + NO_2 + H_2O$$

1 ANALYZE THE PROBLEM

Use the rules for determining oxidation number. The increase in oxidation number of the oxidized atoms or ions must equal the decrease in oxidation number of the reduced atoms or ions. Adjust the coefficients to balance the equation.

2 SOLVE FOR THE UNKNOWN

Assign oxidation numbers to all elements in the equation.

0 +1+5−2 +2 +5 −2 +4 −2 +1 −2

$Cu + HNO_3 \rightarrow Cu(NO_3)_2 + NO_2 + H_2O$

The oxidation number of copper increases from 0 to +2. The oxidation number of nitrogen decreases from +5 to +4.

Identify which atoms or ions are oxidized, which are reduced, and which do not change.

Cu is oxidized.
N is reduced.
H does not change.
O does not change.
N does not change in the nitrate ion (NO_3^-).

Determine the changes in oxidation number for the atoms or ions that are oxidized and for the atoms or ions that are reduced.

Changes in oxidation number:
Oxidized: Cu +2
Reduced: N −1

Copper loses electrons. It is oxidized. Nitrogen gains an electron. It is reduced.

Make the changes in oxidation numbers equal in magnitude by adjusting coefficients in the equation.

$Cu + 2HNO_3 \rightarrow Cu(NO_3)_2 + 2NO_2 + H_2O$
$2(-1) = -2$

Because the change in oxidation number for N is −1, you must add a coefficient of 2 to balance. This coefficient applies to both HNO_3 and NO_2.

Use the conventional method to balance the remainder of the equation.

$Cu + 2HNO_3 \rightarrow Cu(NO_3)_2 + 2NO_2 + H_2O$

$Cu + 4HNO_3 \rightarrow Cu(NO_3)_2 + 2NO_2 + H_2O$

The coefficient of HNO_3 must be increased from 2 to 4 to balance the four nitrogen atoms in the products.

$Cu(s) + 4HNO_3(aq) \rightarrow Cu(NO_3)_2(aq) + 2NO_2(g) + 2H_2O(l)$

Add a coefficient of 2 to H_2O to balance the four hydrogen atoms on the left.

3 EVALUATE THE ANSWER

The number of atoms of each element is equal on both sides of the equation. No subscripts have been changed.

PRACTICE Problems

Do additional problems. Online Practice

Use the oxidation-number method to balance these redox equations.

15. $HCl + HNO_3 \rightarrow HOCl + NO + H_2O$
16. $SnCl_4 + Fe \rightarrow SnCl_2 + FeCl_3$
17. $NH_3(g) + NO_2(g) \rightarrow N_2(g) + H_2O(l)$
18. **Challenge** $SO_2 + Br_2 + H_2O \rightarrow HBr + H_2SO_4$

CHEMISTRY JOURNAL

The Balance of Life Give students a free writing assignment in which they describe the tendency of our environment to be balanced. Students might make a list of things around them that appear balanced or write an essay on the importance of balance in our world. Have students describe the consequences of imbalance. OL

Balancing Net Ionic Redox Equations

Sometimes, chemists prefer to express redox reactions in the simplest possible terms—as an equation showing only the oxidation and reduction processes. Refer again to the balanced equation for the oxidation of copper by nitric acid.

$$Cu(s) + 4HNO_3(aq) \rightarrow Cu(NO_3)_2(aq) + 2NO_2(g) + 2H_2O(l)$$

Note that the reaction takes place in aqueous solution, so HNO_3, which is a strong acid, will be ionized. Likewise, copper(II) nitrate ($Cu(NO_3)_2$) will be dissociated into ions. Therefore, the equation can also be written in ionic form.

$$Cu(s) + 4H^+(aq) + 4NO_3^-(aq) \rightarrow Cu^{2+}(aq) + 2NO_3^-(aq) + 2NO_2(g) + 2H_2O(l)$$

There are four nitrate ions among the reactants, but only two of them undergo change to form two nitrogen dioxide molecules. The other two nitrate ions are only spectator ions and can be eliminated from the equation. To simplify things when writing redox equations in ionic form, chemists usually indicate hydrogen ions by H^+ with the understanding that they exist in hydrated form as hydronium ions (H_3O^+). The equation can then be rewritten showing only the substances that undergo change.

$$Cu(s) + 4H^+(aq) + 2NO_3^-(aq) \rightarrow Cu^{2+}(aq) + 2NO_2(g) + 2H_2O(l)$$

Now look at the equation in unbalanced form.

$$Cu(s) + H^+(aq) + NO_3^-(aq) \rightarrow Cu^{2+}(aq) + NO_2(g) + H_2O(l)$$

You might also see this same reaction expressed in a way that shows only the substances that are oxidized and reduced.

$$Cu(s) + NO_3^-(aq) \rightarrow Cu^{2+}(aq) + NO_2(g) \text{ (in acidic solution)}$$

In this case, the hydrogen ion and the water molecule are eliminated because neither is oxidized nor reduced. In acidic solution, hydrogen ions (H^+) and water molecules are abundant and free to participate in redox reactions as either reactants or products. Some redox reactions can occur only in basic solution. When you balance equations for these reactions, you can add hydroxide ions (OH^-) and water molecules to either side of the equation.

Data Analysis LAB

Based on Real Data*

Analyze and Conclude

How does redox lift a space shuttle? The space shuttle gains nearly 72% of its lift from its solid rocket boosters (SRBs) during the first two minutes of launch. The two pencil-shaped SRB tanks are attached to both sides of the liquid hydrogen and oxygen fuel tank. Each SRB contains approximately 499,000 kg of propellent mixture.

Data and Observations

SRB Propellent Mixture

Component	Percent Composition
Ammonium perchlorate	69.6
Aluminum	16
Catalyst	0.4
Binder	12.04
Curing agent	1.96

*Data obtained from: Dumoulin, Jim. "Solid Rocket Boosters." *NSTS Shuttle Reference Manual.* 1988

Think Critically

1. **Balance an equation** Use the oxidation-number method to balance the chemical equation for the SRB reaction.
   $NH_4ClO_4(s) + Al(s) \rightarrow Al_2O_3(g) + HCl(g) + N_2(g) + H_2O(g)$
2. **State** Which elements are reduced and which are oxidized?
3. **Infer** What are the benefits of using SRBs for the first two minutes of launch?
4. **Calculate** How many moles of water vapor are produced by one SRB?

Data Analysis LAB

About the Lab

- Review the gas laws as they apply to volume and temperature and to explosions. Also, explain the action–reaction in the rocket engine. Gases expand with temperature. When the gas hits an upper surface, it forces the shuttle upward (action), while the surface deflects the gas particles downward (reaction). The hotter the gas, the greater the force.
- Solid fuel has the densest arrangement of particles; therefore, the tanks can hold more solid fuel than fuel in another state (i.e., liquid). After the enormous amount of solid fuel is burned, the heavy containment vessels are jettisoned, making the shuttle much lighter.

Think Critically

1. $6NH_4ClO_4(s) + 10Al(s) \rightarrow 5Al_2O_3(g) + 6HCl(g) + 3N_2(g) + 9H_2O(g)$
2. Nitrogen and aluminum are oxidized; chlorine is reduced.
3. Solid fuel has the densest arrangement of particles; therefore the tanks can hold more solid fuel than fuel in another state (ie. liquid). After the solid fuel is burned, the heavy containment vessels are jettisoned, making the shuttle much lighter.
4. 4.61×10^6 mol H_2O

ASSESSMENT

Knowledge Have students balance the following spontaneous propulsion reaction that the shuttle uses in orbit. The propellant is monomethylhydrazine and the oxidizer is dinitrogen tetroxide.
$N_2O_4 + CH_3N_2H_3 \rightarrow H_2O + N_2 + CO_2$
In $CH_3N_2H_3$, the oxidation numbers for carbon and nitrogen are −4 and −1, respectively. AL

Content Background

Liftoff The space shuttle consists of the orbiter, where the astronauts reside; the external tank, which holds approximately two million liters of liquid hydrogen and oxygen; and two solid rocket boosters (SRBs) that each contain 495,000 kg of fuel. The entire system weighs approximately two million kg on the launch pad. The ammonium perchlorate–aluminum reaction is very exothermic, and the products are much more massive than the water reaction taking place in the external tank. When these heavier molecules are expelled, they provide the majority of the lift that propels the shuttle into space.

IN-CLASS Example

Question Balance the following redox equation.

$Fe^{2+}(aq) + MnO_4^-(aq) \rightarrow Fe^{3+}(aq) + Mn^{2+}(aq)$ (in acidic solution)

Answer $8H^+(aq) + 5Fe^{2+}(aq) + MnO_4^-(aq) \rightarrow 5Fe^{3+}(aq) + Mn^{2+}(aq) + 4H_2O(l)$

PRACTICE Problems

Have students refer to the Selected Solutions appendix for complete solutions to odd-numbered problems. The complete solutions for all problems can be found in the Solutions Manual.

19. $2H^+(aq) + 3H_2S(g) + 2NO_3^-(aq) \rightarrow 3S(s) + 2NO(g) + 4H_2O(l)$

20. $14H^+(aq) + Cr_2O_7^{2-}(aq) + 6I^- \rightarrow 2Cr^{3+}(aq) + 3I_2(s) + 7H_2O(l)$

21. $Zn + 2NO_3^- + 4H^+ \rightarrow Zn^{2+} + 2NO_2 + 2H_2O$

22. $6I^-(aq) + 2MnO_4^-(aq) + 4H_2O(l) \rightarrow 3I_2(s) + 2MnO_2(s) + 8OH^-(aq)$

EXAMPLE Problem 4

EXAMPLE PROBLEM

BALANCE A NET IONIC REDOX EQUATION Balance the following redox equation.

$ClO_4^-(aq) + Br^-(aq) \rightarrow Cl^-(aq) + Br_2(g)$ (in acidic solution)

1 ANALYZE THE PROBLEM

Use the rules for determining oxidation number. The increase in oxidation number of the oxidized atoms or ions must equal the decrease in oxidation number of the reduced atoms or ions. The reaction takes place under acidic conditions. Adjust the coefficients to balance the equation.

2 SOLVE FOR THE UNKNOWN

Assign oxidation numbers to all elements in the equation.

$\overset{+7\,-2}{ClO_4^-}(aq) + \overset{-1}{Br^-}(aq) \rightarrow \overset{-1}{Cl^-}(aq) + \overset{0}{Br_2}(g)$ (in acidic solution)

Use the rules in Table 2.

Identify which atoms or ions are oxidized and which are reduced.

Br is oxidized.
Cl is reduced.

The oxidation number of bromine increases from −1 to 0. The oxidation number of chlorine decreases from +7 to −1.

Determine the changes in oxidation number for the atoms or ions that are oxidized and reduced.

Changes in oxidation number:

Br +1
Cl −8

Bromine loses electrons. It is oxidized.
Chlorine gains electrons. It is reduced.

Make the changes in oxidation number equal in magnitude by adjusting coefficients in the equation.

$ClO_4^-(aq) + 8Br^-(aq) \rightarrow Cl^-(aq) + 4Br_2(g)$ (in acidic solution)

Because the change in oxidation number of Br is +1, you must add the coefficient 8 to balance the equation. $4Br_2$ represents 8 Br atoms to balance the $8Br^-$ on the left side.

Add enough hydrogen ions and water molecules to the equation to balance the oxygen atoms on both sides.

$ClO_4^-(aq) + 8Br^-(aq) + 8H^+(aq) \rightarrow Cl^-(aq) + 4Br_2(g) + 4H_2O(l)$

Because you know the reaction takes place in acid solution, you can add H^+ ions on the left side of the equation.

3 EVALUATE THE ANSWER

The number of atoms of each element is equal on both sides of the equation. As with any ionic equation, the net charge on the right equals the net charge on the left. No subscripts have been changed.

PRACTICE Problems

Do additional problems. Online Practice

Use the oxidation-number method to balance these redox equations.

19. $H_2S(g) + NO_3^-(aq) \rightarrow S(s) + NO(g)$ (in acidic solution)

20. $Cr_2O_7^{2-}(aq) + I^-(aq) \rightarrow Cr^{3+}(aq) + I_2(s)$ (in acidic solution)

21. $Zn + NO_3^- \rightarrow Zn^{2+} + NO_2$ (in acidic solution)

22. Challenge $I^-(aq) + MnO_4^-(aq) \rightarrow I_2(s) + MnO_2(s)$ (in basic solution)

CHEMISTRY PROJECT

Redox Around the House Cleaning cloths that are treated to prevent tarnishing of metal are available in the household goods sections of many stores. Have students research these types of cloths and learn how they prevent tarnish. Then, ask them to write an advertisement for the product. AL

Connection to Biology What do many deep-sea fishes and fireflies have in common with the bacterium, *Xenorhabdus luminescens?* These and other organisms emit light. Bioluminescence is the conversion of potential energy in chemical bonds into light during a redox reaction. Depending on the species, bioluminescence is produced by different chemicals and by different means. In fireflies, shown in **Figure 7,** light results from the oxidation of the molecule luciferin.

Scientists are still unraveling the mystery of bioluminescence. Some luminescent organisms emit light constantly, whereas others emit light when they are disturbed. Deep-sea fishes and some jellyfish appear to be able to control the light they emit, and one species of mushroom is known to emit light of two different colors. Zoologists have also determined that some light-emitting organisms do not produce light themselves; they produce light by harboring bioluminescent bacteria.

Figure 7 Organisms appear to use bioluminescence for different purposes. Some purposes might include attracting a mate and defense against prey. In the ocean depths, bioluminescence probably aids vision and recognition.

Balancing Redox Equations Using Half-Reactions

In chemistry, a **species** is any kind of chemical unit involved in a process. In the equilibrium equation $NH_3 + H_2O \rightarrow NH_4^+ + OH^-$, there are four species: the two molecules NH_3 and H_2O and the two ions NH_4^+ and OH^-. Oxidation-reduction reactions occur whenever a species that can give up electrons (reducing agent) comes in contact with another species that can accept them (oxidizing agent). For example, iron can reduce many species that are oxidizing agents, including chlorine.

$$2Fe + 3Cl_2 \rightarrow 2FeCl_3$$

In this reaction, each iron atom is oxidized by losing three electrons to become an Fe^{3+} ion. At the same time, each chlorine atom in Cl_2 is reduced by accepting one electron to become a Cl^- ion.

$$\text{Oxidation: } Fe \rightarrow Fe^{3+} + 3e^-$$
$$\text{Reduction: } Cl_2 + 2e^- \rightarrow 2Cl^-$$

Equations such as these represent half-reactions. A **half-reaction** is one of the two parts of a redox reaction—the oxidation half or the reduction half. **Table 5** shows a variety of reduction half-reactions that involve the oxidation of Fe to Fe^{3+}.

Vocabulary
Science usage v. Common usage
Species
Science usage: in chemistry, any kind of representative particle involved in a process
In a synthesis reaction, two distinct species combine to form a single product.
Common usage: a class of individuals having some common characteristics or qualities; a distinct sort or kind
Humans and chimpanzees are two different species.

Table 5 Redox Reactions that Oxidize Iron

Overall Reaction (unbalanced)	Oxidation Half-Reaction	Reduction Half-Reaction
$Fe + O_2 \rightarrow Fe_2O_3$		$O_2 + 4e^- \rightarrow 2O^{2-}$
$Fe + F_2 \rightarrow FeF_3$		$F_2 + 2e^- \rightarrow 2F^-$
$Fe + HBr \rightarrow H_2 + FeBr_3$	$Fe \rightarrow Fe^{3+} + 3e^-$	$2H^+ + 2e^- \rightarrow H_2$
$Fe + AgNO_3 \rightarrow Ag + Fe(NO_3)_3$		$Ag^+ + e^- \rightarrow Ag$
$Fe + CuSO_4 \rightarrow Cu + Fe_2(SO_4)_3$		$Cu^{2+} + 2e^- \rightarrow Cu$

Quick Demo

Heater Use the flameless ration heater (FRH) to show students that redox reactions are useful. The redox reaction of the heater provides heat to cook food in a situation where there is no appliance or fire for that purpose.

Demonstrate the FRH and have the students describe their observations in their chemistry journals. One of the reactions occurring in the FRH is $Mg + H_2O \rightarrow Mg(OH)_2 + H_2 +$ heat. Have students balance this reaction by the half-reaction method. OL

Assessment

Performance Have students investigate and report on one of the following oxidation reduction topics: breath test, black and white photography, use of bleach in the laundry room, hydrogen peroxide as an antiseptic, or bioluminescence. Have them explain the principles of oxidation and reduction in the context of the selected topic, and have them write the redox reaction. OL

Concept Development

Balance Organize the class into groups of two. Prepare enough file cards with a reaction written on them to give one to each group. Have each student balance the reaction individually, and then have team members discuss and compare their answers. OL COOP LEARN

Math in Chemistry

Half-Reaction Method When balancing redox reactions by the half-reaction method, note the step that is essentially solving a simultaneous equation. Step 4 Explain to students that the total number of electrons lost during the oxidation half-reaction is the same as the total number of electrons gained in the reduction half-reaction. Show this by finding the least common multiple for the two half-reactions and multiplying all substances in the half-reaction by that number. When the two half-reactions are added back together, the number of electrons lost and gained will be equal.

ASSESSMENT

Knowledge Organize the students into groups of two. Assign each group two problems from the text that they have completed for homework. Have the first student explain to the second student how he or she solved the first problem. For the second problem, have students reverse roles. **COOP LEARN**

FOLDABLES® Rubric

Figure 8 As a result of this redox reaction between iron and copper sulfate solution, solid copper metal is deposited on the iron. To balance the chemical equation for this reaction, you could use half-reactions.

FOLDABLES® Incorporate information from this section into your Foldable.

You will learn more about the importance of half-reactions when you study electrochemistry. For now, however, you can learn to use half-reactions to balance a redox equation. For example, the following unbalanced equation represents the reaction that occurs when you put an iron nail into a solution of copper(II) sulfate, as shown in **Figure 8.**

$$Fe(s) + CuSO_4(aq) \rightarrow Cu(s) + Fe_2(SO_4)_3(aq)$$

Iron atoms are oxidized as they lose electrons to the copper(II) ions. The steps for balancing redox equations by using half-reactions are shown in **Table 6.**

VOCABULARY
ACADEMIC VOCABULARY
Method:
a way of doing something
Students study for an exam using different methods.

Explore **redox titrations.**
Virtual Investigations

Table 6 The Half-Reaction Method
1. Write the unbalanced, net ionic equation for the reaction, omitting spectator ions. $Fe + Cu^{2+} + SO_4^{2-} \rightarrow Cu + 2Fe^{3+} + 3SO_4^{2-}$; $Fe + Cu^{2+} \rightarrow Cu + 2Fe^{3+}$
2. Write separate, incomplete equations for the oxidation and reduction half-reactions, including oxidation numbers. $\overset{0}{Fe} \rightarrow \overset{+3}{2Fe^{3+}}$ (oxidation) $\overset{+2}{Cu^{2+}} \rightarrow \overset{0}{Cu}$ (reduction)
3. Balance the atoms in the half-reactions. Balance the charges in each half-reaction by adding electrons as reactants or products. $2Fe \rightarrow 2Fe^{3+} + 6e^-$ $Cu^{2+} + 2e^- \rightarrow Cu$
4. Adjust the coefficients so that the number of electrons lost in oxidation equals the number of electrons gained in reduction. $2Fe \rightarrow 2Fe^{3+} + 6e^-$ $3Cu^{2+} + 6e^- \rightarrow 3Cu$
5. Add the half-reactions and cancel or reduce like terms on both sides of the equation. $2Fe + 3Cu^{2+} \rightarrow 3Cu + 2Fe^{3+}$
6. Return spectator ions, if desired. Restore state descriptions. $2Fe(s) + 3CuSO_4(aq) \rightarrow 3Cu(s) + Fe_2(SO_4)_3(aq)$

CULTURAL DIVERSITY

Redox Shapes History Have students recall various historical time periods, such as the Copper Age, Bronze Age, and Iron Age. Several of these early time periods were centered in the Middle East. In each case, the discovery of a new material by a particular cultural group or nation allowed that group to control the region for a period of time. Have students locate these key regions on a world map. What are several chemical and physical properties of copper, bronze, and iron that would provide a nation a significant advantage over neighboring countries? What is the role of redox chemistry in history?

EXAMPLE Problem 5

BALANCE A REDOX EQUATION BY USING HALF-REACTIONS
Balance the redox equation for the reaction below using half-reactions.

$$KMnO_4(aq) + SnCl_2(aq) + HCl(aq) \rightarrow MnCl_2(aq) + SnCl_4(aq) + H_2O(l) + KCl(aq)$$

1 ANALYZE THE PROBLEM

The reaction takes place in acidic solution. Use the rules for determining oxidation numbers and the steps for balancing by half-reactions to balance the equation for the reaction.

2 SOLVE FOR THE UNKNOWN

Write the unbalanced, net ionic equation for the reaction.

$MnO_4^- + Sn^{2+} \rightarrow Mn^{2+} + Sn^{4+}$

Eliminate coefficients, spectator ions, and state symbols.

Write incomplete equations for the oxidation and reduction half-reactions, including oxidation numbers.

$\overset{+2}{Sn^{2+}} \rightarrow \overset{+4}{Sn^{4+}}$ (oxidation)

$\overset{+7}{MnO_4^-} \rightarrow \overset{+2}{Mn^{2+}}$ (reduction)

Use the rules in Table 2 and Table 6.

Balance the atoms and charges in the half-reactions.

$Sn^{2+} \rightarrow Sn^{4+} + 2e^-$ (oxidation)

$5e^- + 8H^+ + MnO_4^- \rightarrow Mn^{2+} + 4H_2O$ (reduction)

In an acid solution, H_2O molecules are available in abundance and can be used to balance oxygen atoms in the half-reactions; H^+ ions are readily available and can be used to balance the charge.

Adjust the coefficients so that the number of electrons lost in oxidation (2) equals the number of electrons gained in reduction (5).

$5Sn^{2+} \rightarrow 5Sn^{4+} + 10e^-$ (oxidation)

$10e^- + 16H^+ + 2MnO_4^- \rightarrow 2Mn^{2+} + 8H_2O$ (reduction)

The least common multiple of 2 and 5 is 10. Cross-multiplying gives the balanced oxidation and reduction half-reactions.

Add the balanced half-reactions and simplify by canceling or reducing like terms on both sides of the equation.

$5Sn^{2+} + 10e^- + 16H^+ + 2MnO_4^- \rightarrow 5Sn^{4+} + 10e^- + 2Mn^{2+} + 8H_2O$

$5Sn^{2+} + 16H^+ + 2MnO_4^- \rightarrow 5Sn^{4+} + 2Mn^{2+} + 8H_2O$

Restore state descriptions and return spectator ions (K^+ and Cl^-).

$5SnCl_2(aq) + 16HCl(aq) + 2KMnO_4(aq) \rightarrow$
$5SnCl_4(aq) + 2MnCl_2(aq) + 8H_2O(l) + 2KCl(aq)$

Add K^+ ions to the two MnO_4^- ions on the left and add two K^+ ions on the right. Add Cl^- ions to the Sn^{2+} and H^+ ions on the left and to the Sn^{4+}, Mn^{2+}, and K^+ ions on the right.

3 EVALUATE THE ANSWER

A review of the balanced equation indicates that the number of atoms of each element is equal on both sides of the equation. No subscripts have been changed.

EXAMPLE PROBLEM

PRACTICE Problems

Do additional problems. Online Practice

Use the half-reaction method to balance the redox equations. Begin by writing the oxidation and reduction half-reactions. Leave the balanced equation in ionic form.

23. $Cr_2O_7^{2-}(aq) + I^-(aq) \rightarrow Cr^{3+}(aq) + I_2(s)$ (in acidic solution)

24. $Mn^{2+}(aq) + BiO_3^-(aq) \rightarrow MnO_4^-(aq) + Bi^{2+}(aq)$ (in acidic solution)

25. Challenge $N_2O(g) + ClO^-(aq) \rightarrow NO_2^-(aq) + Cl^-(aq)$ (in basic solution)

Apply Chemistry

Pressure Have students research different extraction techniques used in industry to produce pure metals, such as iron, silver, aluminum, gold, copper, and titanium. Have them investigate the cost of the raw materials of ore compared to the cost of the pure metal. They should also investigate toxicity to people and to the environment. Ask them to determine whether the process has undergone any change to provide better protection for workers and reduce problems in the environment. AL

IN-CLASS Example

Question Balance the following redox equations by using half-reactions.

$KMnO_4(aq) + KCl(aq) + H_2SO_4(aq) \rightarrow MnSO_4(aq) + K_2SO_4(aq) + H_2O(l) + Cl_2(g)$

Answer $2KMnO_4(aq) + 10KCl(aq) + 8H_2SO_4(aq) \rightarrow 2MnSO_4(aq) + 6K_2SO_4(aq) + 8H_2O(l) + 5Cl_2(g)$

PRACTICE Problems

Have students refer to the Selected Solutions appendix for complete solutions to odd-numbered problems. The complete solutions for all problems can be found in the Solutions Manual.

23. $14H^+(aq) + Cr_2O_7^{2-}(aq) + 6I^- \rightarrow 3I_2(s) + 2Cr^{3+}(aq) + 7H_2O(l)$

24. $3Mn^{2+}(aq) + 5BiO_3^-(aq) + 6H^+(aq) \rightarrow 3MnO_4^-(aq) + 5Bi^{2+}(aq) + 3H_2O(l)$

25. $N_2O(g) + 2ClO^-(aq) + 2OH^-(aq) \rightarrow 2NO_2^-(aq) + 2Cl^-(aq) + H_2O(l)$

CHEMISTRY JOURNAL

Redox in Biology Have students consult biology books and the Internet to find important biological reactions that are oxidation–reduction processes. These can be summarized in their chemistry journals. OL

PROBLEM-SOLVING STRATEGY

Apply the Strategy

$P_4 + H_2O \longrightarrow PH_3 + H_2PO_2$

$5P_4 + 24H_2O \longrightarrow 8PH_3 + 12H_2PO_2$

3 Assess

Check for Understanding

Have students explain in their own words the process of balancing a redox reaction by both the oxidationnumbermethodandthehalf-reaction method. **OL**

Reteach

One of the problems students have is organizing and keeping their work neat and orderly when balancing redox reactions by the half-reaction method. Encourage students to use plenty of space on their papers to write, check, and recheck each redox equation that they are balancing. Guide them into the habit of writing the problem correctly and including all appropriate charges. **OL**

Extension

Invite a person in your community who sells or tests water in swimming pools or hot tubs to describe the oxidation-reduction chemistry involved in maintaining the water in pools and hot tubs. **OL**

PROBLEM-SOLVING STRATEGY

Balancing Redox Equations

Determine which species is oxidized, which species is reduced, which species is the oxidizing agent, and which species is the reducing agent.

Do the oxidized and reduced species appear more than once on either side of the equation, or does the reaction occur in an acidic or basic solution?

No — **Oxidation-Number Method of Balancing Redox Equations**

- Assign oxidation numbers to all of the elements.
- Adjust the coefficients in the equation so that the oxidation numbers are equal in magnitude.
- Balance the rest of the equation by the conventional method.

Yes — **Half-Reaction Method of Balancing Redox Equations**

- Write the unbalanced, net ionic equation for the reaction, omitting the spectator ions.
- Determine the oxidation and the reduction half-reactions.
- Balance the atoms and the charges in each half-reaction.
- Adjust the coefficients so that the number of electrons lost and the number of electrons gained is equal.
- Combine the balanced half-reactions and return spectator ions.

Apply the Strategy

Balance the following equation using this flowchart.

$P_4(s) + H_2O(l) \longrightarrow PH_3(g) + H_2PO_2(aq)$

SECTION 2 REVIEW

Section Self-Check

Section Summary

- Redox equations in which the same element appears in multiple reactants and products can be difficult to balance using the conventional method.
- The oxidation-number method is based on the number of electrons transferred from atoms or ions equaling the number of electrons accepted by other atoms or ions.
- To balance equations for reactions in an acidic solution, add hydrogen ions and water molecules.
- To balance equations for reactions in a basic solution, add hydroxide ions and water molecules.
- A half-reaction is one of the two parts of a redox reaction.

26. MAINIDEA **Explain** how changes in oxidation number are related to the electrons transferred in a redox reaction. How are the changes related to the processes of oxidation and reduction?
27. **Describe** why it is important to know the conditions under which an aqueous oxidation-reduction reaction takes place in order to balance the ionic equation for the reaction.
28. **Explain** the steps of the oxidation-number method of balancing equations.
29. **State** what an oxidation half-reaction shows. What does a reduction half-reaction show?
30. **Write** the oxidation and reduction half-reactions for the redox equation.
    $Pb(s) + Pd(NO_3)_2(aq) \longrightarrow Pb(NO_3)_2(aq) + Pd(s)$
31. **Determine** The oxidation half-reaction of a redox reaction is $Sn\ 2+ \longrightarrow Sn\ 4+ + 2e\ -$, and the reduction half-reaction is $Au\ 3+ + 3e\ - \longrightarrow Au$. What minimum numbers of tin(II) ions and gold(III) ions would have to react in order to have zero electrons left over?
32. **Apply** Balance the following equations.
    a. $HClO_3(aq) \longrightarrow ClO_2(g) + HClO_4(aq) + H_2O(l)$
    b. $H_2SeO_3(aq) + HClO_3(aq) \longrightarrow H_2SeO_4(aq) + Cl_2(g) + H_2O(l)$
    c. $Cr_2O_7{}^{2-}(aq) + Fe^{2+}(aq) \longrightarrow Cr^{3+}(aq) + Fe^{3+}(aq)$ (in acidic solution)

SECTION 2 REVIEW

26. Because the nucleus (specifically, number of protons) never changes during this type of reaction, whenever there is a transfer of electrons to or from a chemical species, there is a change in the net charge of that species. Oxidation increases the oxidation number, reduction reduces it.
27. It is important to know that H_2O and either H^+ or OH^- are available to balance the equation.
28. Answers should be similar to the information in Table 4.
29. An oxidation half reaction shows how many electrons a species loses. A reduction half reaction shows how many electrons a species gains.
30. oxidation: $Pb \longrightarrow Pb^{2-} + 2e^-$
    reduction: $Pd^{2+} + 2e^- \longrightarrow Pd$
31. three Sn^{2+} ions; two Au^{3+} ions
32. **a.** $3HClO_3 \longrightarrow 2ClO_2 + HClO_4 + H_2O$
    b. $5H_2SeO_3 + 2HClO_3 \longrightarrow 5H_2SeO_4 + Cl_2 + H_2O$
    c. $Cr_2O_7{}^{2-} + 6Fe^{2+} + 14H^+ \longrightarrow 2Cr^{3+} + 6Fe^{3+} + 7H_2O$

CHEMISTRY & careers

Career: Crime-Scene Investigator

Blood That Glows

In Shakespeare's play *MacBeth*, Lady MacBeth washes the blood of King Duncan from her hands but can still see the bloodstains. In modern forensics, a chemical called luminol gives investigators similar visual ability.

Blue-green whisper Luminol oxidizes when it comes in contact with iron, as shown in **Figure 1.** In the process, the molecules release energy in the form of distinctive blue-green light. In a dark room, the faint blue glow of luminol might reveal to investigators what their eyes alone could not see—hidden traces of blood. Red blood cells consist mainly of hemoglobin—a protein that contains iron.

To use luminol, investigators mix a white powder ($C_8H_7O_3N_3$) with hydrogen peroxide (H_2O_2) and other chemicals. This creates a liquid that can be sprayed onto areas suspected of holding hidden blood evidence. If blood is present—even in quantities too small to detect with the eye—the luminol will glow. Forensic photographers then snap pictures with special cameras that can both capture the faint glow of the luminol and illuminate the surrounding area.

Figure 1 The luminol oxidizes within a beaker when an iron nail is added.

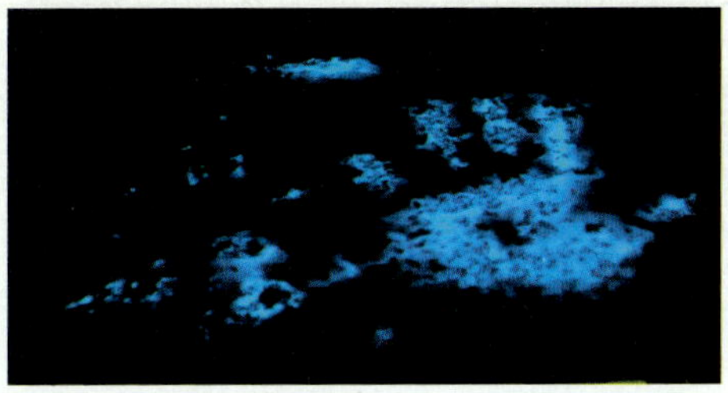

Figure 2 A luminol impression from a murder scene can be compared to a suspect's handprint.

Glowing evidence Bloodstains might reveal spatter patterns, giving clues about the type of weapon used to commit a crime. Faint luminol signals on carpet might lead investigators to much larger bloodstains. Bloody handprints, such as that in **Figure 2,** might even lead investigators to the assailant.

There are other uses for luminol besides murder investigations. In a car accident, luminol might reveal whether a victim was wearing a safety belt, even after the car has been subjected to rain, cold, or direct sunlight that can greatly alter bloodstains.

Spray of last resort Other iron-containing substances besides blood can cause luminol to glow, although experts can usually tell the difference. More importantly, luminol might interfere with other tests. For this reason, investigators normally do not use luminol until all their other investigations are complete.

WRITING IN ▸ Chemistry

News Article Research more about the use of luminol in crime scene investigations. Write a newspaper article that describes how luminol led investigators to a suspect. Describe the type of evidence that was used in the investigation.

WebQuest

CHEMISTRY & careers

Purpose

Students will learn about the light-releasing chemical luminol and its use as a forensic tool.

Background

When luminol oxidizes, it releases its energy as visible light. This process is called chemiluminescence. In this process, electrons move from a higher energy orbital to a lower one, and the color of light released is determined by the difference in energy between the two orbitals. Notice that, unlike incandescence, the color of the light released does not depend on temperature, but instead is determined by the properties of the chemicals involved.

Teaching Strategies

- Develop a list of circumstances in which bloodstains might be hidden, either deliberately or due to the circumstances of a particular crime.
- Considering that luminol requires darkness to become visible, what are some of its limitations, and how might these be overcome?
- Once blood is located via luminol, DNA tests might be performed to determine the source of the blood. Think of ways this might be valuable for law enforcement. Include in your analysis the possible presence of nonhuman blood, or blood from someone other than the crime victim.

WRITING IN ▸ Chemistry

Rubric WebQuest

News Article Answers will vary but should include that investigators could match the luminol handprint with that of the assailant. The evidence that could be used in the investigation includes the handprint, blood-spatter patterns, and other bloodstains.

SMALL SCALE ChemLAB

Forensics: Identify the Damaging Dumper

Background: Something is reacting with metals found on the hulls of many boats used on a nearby creek. The investigator has determined that there are three possible culprits, each with a different source. Your job is to test the three potential pollutants and compare them with a sample from the creek. The animals that rely on the creek as their primary water source are depending on you to solve this mystery of the damaging dumper.

Question: *How can a series of chemical reactions be used to determine what was dumped in a water supply?*

Materials

0.1*M* $AgNO_3$	Fe filings
0.1*M* HCl	Mg turnings
0.1*M* $ZnSO_4$	tongs or forceps
unknown solution	droppers (4)
Cu wire	24-well microscale reaction plate
Pb shot	

Safety Precautions

WARNING: ***Silver nitrate ($AgNO_3$) is highly toxic and will stain skin and clothing.***

Procedure

1. Read and complete the lab safety form.
2. Create a table to record your data.
3. Place the well plate on a sheet of white paper.
4. Place a piece of copper wire in four wells in the first row.
5. Repeat Step 4, by adding a small sample of iron filings to wells in the second row.
6. Repeat Step 4, by adding a piece of lead shot to wells in the third row
7. Repeat Step 4, by adding a piece of magnesium ribbon to wells in the fourth row.
8. Count 20 drops of the silver nitrate solution ($AgNO_3$) into each well in the first column.
9. Repeat Step 8, adding hydrochloric acid (HCl) in the second column.
10. Repeat Step 8, adding zinc sulfate (ZnSO 4) in the third column.

Observations

	$AgNO_3$	HCl	$ZnSO_4$	Unknown
Cu				
Fe				
Pb				
Mg				

11. Repeat Step 8, adding the unknown solution in the fourth column.
12. Allow the reactions to proceed for 5 min, and then describe the reactions. Write *NR* for any wells that do not have evidence of a reaction.
13. **Cleanup and Disposal** Dispose of the solids and solutions as directed by your teacher. Wash and return all lab equipment to its designated location.

Analyze and Conclude

1. **Summarize** the results you observed in each well. How did you know a chemical reaction occurred?
2. **Model** Write a balanced equation for each of the reactions you observed. In each one, identify the species being oxidized or reduced.
3. **Conclude** Based on your data, which solution was causing damage in the creek? Justify your answer.
4. **Use Variables, Constants, and Controls** Why was it important to compare the reactions of the unknown to more than one known solution?
5. **Research** Look up the MSDS for your chemical and report on what impact this chemical would have on the ecosystem.
6. **Extend** What would you expect if a solution of lead(II) nitrate ($Pb(NO_3)_2$) was one of the solutions?
7. **Error Analysis** Compare your results with those of other students in the laboratory. Explain any differences.

INQUIRY EXTENSION

Design an Experiment Hypothesize how you could remove this chemical from the creek without further damaging the ecology of the area. Design an experiment to test your hypothesis.

SMALL SCALE ChemLAB

Rubric

Preparation

Time Allotment one class period

Process Skills collect and organize data, design an experiment, research, formulate models, use variables, constants and controls, think critically

Disposal All solutions can be flushed down a drain with water. Solids should be discarded in the trash.

Preparation of Materials Cut samples of the metals small enough to fit into the bottom of the well-plate.

Procedure

- Use a stereo microscope to enhance the observations of the reactions.
- Use one of the standard solutions as the unknown solution.
- **Troubleshooting** It is important that the students do not mix the solutions as they will react and alter the data.

Analyze and Conclude

1. $AgNO_3$ reacts with all four metals, forming silver crystals. HCl reacts with Mg, forming gas bubbles rapidly (Mg is consumed); Fe, forming some gas bubbles; and Pb, forming few gas bubbles. HCl and Cu do not react. $ZnSO_4$ reacts with Mg, forming a layer of Zn. $ZnSO_4$ does not react with Cu, Fe, or Pb.
2. $Cu + 2AgNO_3 \rightarrow Cu(NO_3)_2 + 2Ag$; Cu oxidized; Ag^+ reduced
   $Fe + 3AgNO_3 \rightarrow Fe(NO_3)_3 + 3Ag$; Fe oxidized; Ag^+ reduced
   $Pb + 2AgNO_3 \rightarrow Pb(NO_3)_2 + 2Ag$; Pb oxidized; Ag^+ reduced
   $Mg + 2AgNO_3 \rightarrow Mg(NO_3)_2 + 2Ag$; Mg oxidized; Ag^+ reduced
   $Fe + 2HCl \rightarrow H_2 + FeCl_2$; Fe oxidized; H^+ reduced
   $Pb + 2HCl \rightarrow H_2 + PbCl_2$; Pb oxidized; H^+ reduced
   $Mg + 2HCl \rightarrow H_2 + MgCl_2$; Mg oxidized; H^+ reduced
   $Mg + ZnSO_4 \rightarrow MgSO_4 + Zn$; Mg oxidized; Zn^{2+} reduced
3. Answers will depend upon the solution used for the unknown.
4. Answers will vary. The comparison to only one solution would not provide enough data for a determination.
5. Answers will vary.
6. These reactions would occur:
   $2Fe + 3Pb(NO_3)_2 \rightarrow 2Fe(NO_3)_3 + 3Pb$
   $Mg + Pb(NO_3)_2 \rightarrow Mg(NO_3)_2 + Pb$
7. Answers will vary.

INQUIRY EXTENSION

Answers will vary. Students could propose a lab with the addition of various substances that will react with the unknown solution. Answers will depend upon the solution used for the unknown.

Vocabulary Practice

BIGIDEA Oxidation-reduction reactions—among the most-common chemical processes in both nature and industry—involve the transfer of electrons.

SECTION 1 Oxidation and Reduction

MAINIDEA Oxidation and reduction are complementary—as a substance is oxidized, another substance is reduced.

- Oxidation-reduction reactions involve the complete or partial transfer of electrons from one substance to another.
- When an atom or ion is reduced, its oxidation number is lowered. When an atom or ion is oxidized, its oxidation number is raised.
- In oxidation-reduction reactions involving molecular compounds (and polyatomic ions with covalent bonds), the more-electronegative atoms are treated as if they are reduced. The less-electronegative atoms are treated as if they are oxidized.

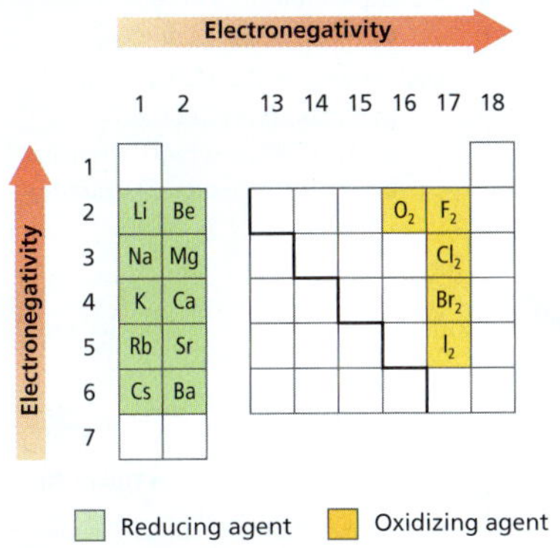

VOCABULARY
- oxidation-reduction reaction
- redox reaction
- oxidation
- reduction
- oxidation number
- oxidizing agent
- reducing agent

SECTION 2 Balancing Redox Equations

MAINIDEA Redox equations are balanced when the total increase in oxidation numbers equals the total decrease in oxidation numbers of the atoms or ions involved in the reaction.

- Redox equations in which the same element appears in several reactants and products can be difficult to balance using the conventional method.
- The oxidation-number method is based on the number of electrons transferred from atoms or ions equaling the number of electrons accepted by other atoms or ions.
- To balance equations for reactions in an acidic solution, add hydrogen ions and water molecules.
- To balance equations for reactions in a basic solution, add hydroxide ions and water molecules.
- A half-reaction is one of the two parts of a redox reaction.

VOCABULARY
- oxidation-number method
- species
- half-reaction

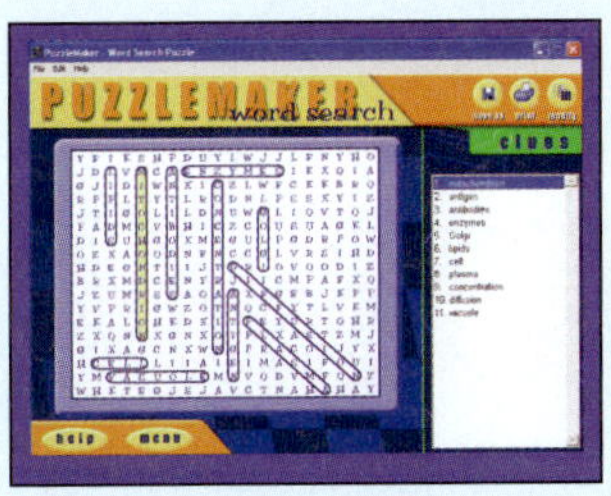

Vocabulary Practice

For additional practice with vocabulary, have students access the Vocabulary PuzzleMaker.

CHAPTER 19 STUDY GUIDE

Use the Vocabulary

To reinforce chapter vocabulary, have students write a sentence using each term. OL EL

Review Strategies

- Have students give an example for each of the rules for determining oxidation numbers. OL
- Have students create a chart or table to compare the two different methods used for balancing oxidation-reduction reactions. OL
- Problems from the Supplemental Practice Problems appendix or the Supplemental Problems booklet can be used for review. OL
- Have students summarize in their own words the steps for balancing oxidation-reduction reactions by the method of half-reactions. OL

Students can review with these online resources.

Vocabulary Practice

Vocabulary eGames
Vocabulary eFlashcards
Vocabulary PuzzleMaker

Section Self-Check

Chapter Self-Check

Online Test Practice

Use *eAssessment* to:
- create multiple versions of tests
- edit existing questions and add your own questions
- build tests aligned with select state curriculum standards using built-in state curriculum tags
- track students' progress

SECTION 1

Mastering Concepts

33. All oxidation–reduction reactions involve the transfer of electrons.

34. The word oxidation originally referred only to reactions that involved oxygen, but today it is defined as the complete or partial loss of electrons from a reacting substance.

35. Electrons are lost; electrons are gained.

36. a number assigned to an atom or ion to indicate its degree of oxidation or reduction

37. alkaline earth = +2; alkali metals = +1

38. The change in oxidation number equals the number of electrons lost in oxidation, and electrons gained in reduction.

39. Copper has different oxidation states in the two forms.

40. $2Cu(s) + O_2(g) \longrightarrow 2CuO(s)$; Cu is oxidized, O is reduced

Mastering Problems

41. a. Ga is oxidized, Br_2 is reduced.
b. Zn is oxidized, H is reduced.
c. Mg is oxidized, N_2 is reduced.

42. a. N_2 is the oxidizing agent, H_2 is the reducing agent.
b. I_2 is the oxidizing agent, Na is the reducing agent.

43. Sn

44. +7

45. a. +6
b. +6
c. +3
d. +5

46. a. oxidation
b. reduction

47. Choice **a** is not redox because none of the atoms in the reaction undergoes a change in oxidation number.

48. a. +5
b. +1
c. +3

49. a. Au, +3; Se, +6; O, −2
b. Ni, +2; C, +2; N, −3

50. SO_3^{2-} is a polyatomic ion and the oxidation number of sulfur is +4. SO_3 is a compound and the oxidation number of S in this compound is +6.

CHAPTER 19 ASSESSMENT

SECTION 1

Mastering Concepts

33. What is the main characteristic of oxidation-reduction reactions?

34. Explain why not all oxidation reactions involve oxygen.

35. In terms of electrons, what happens when an atom is oxidized? When an atom is reduced?

36. Define *oxidation number.*

37. Metals What is the oxidation number of alkaline earth metals in their compounds? Of alkali metals?

38. How does the oxidation number in an oxidation process relate to the number of electrons lost? How does the change in oxidation number in a reduction process relate to the number of electrons gained?

■ **Figure 9**

39. What probably accounts for the different forms of copper shown in **Figure 9**?

40. Copper and air Copper statues, such as the Statue of Liberty, begin to appear green after they have been exposed to air. In this redox process, copper metal reacts with oxygen to form solid copper oxide, which forms the green coating. Write the reaction for this redox process, and identify what is oxidized and what is reduced in the process.

Mastering Problems

41. Identify the species oxidized and the species reduced in each of these redox equations.
a. $3Br_2 + 2Ga \longrightarrow 2GaBr_3$
b. $2HCl + Zn \longrightarrow ZnCl_2 + H_2$
c. $3Mg + N_2 \longrightarrow Mg_3N_2$

42. Identify the oxidizing agent and the reducing agent in each of these redox equations.
a. $N_2 + 3H_2 \longrightarrow 2NH_3$
b. $2Na + I_2 \longrightarrow 2NaI$

43. What is the reducing agent in this balanced equation?

$$8H^+ + Sn + 6Cl^- + 4NO_3^- \longrightarrow SnCl_6^{2-} + 4NO_2 + 4H_2O$$

44. What is the oxidation number of manganese in $KMnO_4$?

45. Determine the oxidation number of the boldface element in these substances and ions.
a. $Ca\mathbf{Cr}O_4$ **c.** $\mathbf{N}O_2^-$
b. $NaH\mathbf{S}O_4$ **d.** $\mathbf{Br}O_3^-$

46. Identify each of these half-reactions as either oxidation or reduction.
a. $Al \longrightarrow Al^{3+} + 3e^-$
b. $Cu^{2+} + e^- \longrightarrow Cu^+$

47. Which of these equations does not represent a redox reaction? Explain your answer.
a. $LiOH + HNO_3 \longrightarrow LiNO_3 + H_2O$
b. $MgI_2 + Br_2 \longrightarrow MgBr_2 + I_2$

48. Determine the oxidation number of nitrogen in each of these molecules or ions.
a. NO_3^- **b.** N_2O **c.** NF_3

49. Determine the oxidation number of each element in these compounds or ions.
a. $Au_2(SeO_4)_3$ (gold(III) selenate)
b. $Ni(CN)_2$ (nickel(II) cyanide)

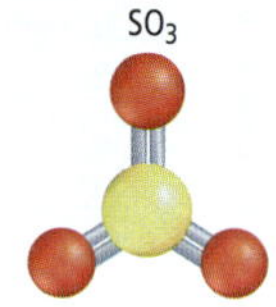

■ **Figure 10**

50. Explain how the sulfite ion (SO_3^{2-}) differs from sulfur trioxide (SO_3), shown in **Figure10.**

SECTION 2

Mastering Concepts

51. Compare and contrast balancing redox equations in acidic and basic solutions.

52. Explain why writing hydrogen ions as H^+ in redox reactions represents a simplification and not how they exist.

53. Before you attempt to balance the equation for a redox reaction, why do you need to know whether the reaction takes place in acidic or basic solution?

54. Explain what a spectator ion is.

55. Define the term *species* in terms of redox reactions.

56. Is the following equation balanced? Explain.

$$Fe(s) + Ag^+(aq) \longrightarrow Fe^{2+}(aq) + Ag(s)$$

57. Does the following equation represent a reduction or an oxidation process? Explain your answer.

$$Zn^{2+} + 2e^- \longrightarrow Zn$$

SECTION 2

Mastering Concepts

51. In a redox reaction that takes place in an acidic solution, H^- and H_2O can participate in the reaction as either reactants or products. In a basic solution, a redox reaction can involve OH^- and H_2O as either reactants or products.

52. In aqueous solution, hydrogen ions combine with water in their hydrated form, the hydronium ions (H_3O^+) and are never present as H^+. However, they are sometimes shown as H^+ to simplify the chemical equation that is written.

53. The type of solution determines whether H^+ or OH^- ions are available to balance the redox equation.

54. A spectator ion is one that is present in the same stoichiometry on both sides of a redox reaction equation. Spectator ions are not changed during a reaction, so they can be eliminated from the equation.

55. A species is any kind of chemical unit involved in the redox process. A species can be an ion, a molecule, or a free atom.

Chapter Self-Check

58. Describe what is happening to electrons in each half reaction of a redox process.

Mastering Problems

59. Use the oxidation-number method to balance these redox equations.
a. $Cl_2 + NaOH \rightarrow NaCl + HOCl$
b. $HBrO_3 \rightarrow Br_2 + H_2O + O_2$

60. Balance these net ionic equations for redox reactions.
a. $Au^{3+}(aq) + I^-(aq) \rightarrow Au(s) + I_2(s)$
b. $Ce^{4+}(aq) + Sn^{2+}(aq) \rightarrow Ce^{3+}(aq) + Sn^{4+}(aq)$

61. Use the oxidation-number method to balance the following ionic redox equations.
a. $Al + I_2 \rightarrow Al^{3+} + I^-$
b. $MnO_2 + Br^- \rightarrow Mn^{2+} + Br_2$ (in acidic solution)

62. Use the oxidation-number method to balance these redox equations.
a. $PbS + O_2 \rightarrow PbO + SO_2$
b. $NaWO_3 + NaOH + O_2 \rightarrow Na_2WO_4 + H_2O$
c. $NH_3 + CuO \rightarrow Cu + N_2 + H_2O$
d. $Al_2O_3 + C + Cl_2 \rightarrow AlCl_3 + CO$

■ Figure 11

63. **Sapphire** The mineral corundum is comprised of aluminum oxide (Al_2O_3) and is dull gray. Sapphire is mostly aluminum oxide, but it contains small amounts of Fe^{2+} and Ti^{4+}. The color of sapphire results from an electron transfer from Fe^{2+} to Ti^{4+}. Based on **Figure 11,** write an equation that describes the reaction that occurs resulting in the mineral on the right. What are the oxidizing and reducing agents?

64. Write the oxidation and reduction half-reactions represented in each of these redox equations. Write the half-reactions in net ionic form if they occur in aqueous solution.
a. $PbO(s) + NH_3(g) \rightarrow N_2(g) + H_2O(l) + Pb(s)$
b. $I_2(s) + Na_2S_2O_3(aq) \rightarrow Na_2S_2O_4(aq) + NaI(aq)$
c. $Sn(s) + 2HCl(aq) \rightarrow SnCl_2(aq) + H_2(g)$

65. Write the two half-reactions that make up the following balanced redox reaction.

$$3H_2C_2O_4 + 2HAsO_2 \rightarrow 6CO_2 + 2As + 4H_2O$$

66. Label each half-reaction as reduction or oxidation.
a. $Fe^{2+}(aq) \rightarrow Fe^{3+}(aq) + e^-$
b. $MnO_4^- + 5e^- + 8H^+ \rightarrow Mn^{2+} + 4H_2O$
c. $2H^+ + 2e^- \rightarrow H_2$
d. $F_2 \rightarrow 2F^- + 2e^-$

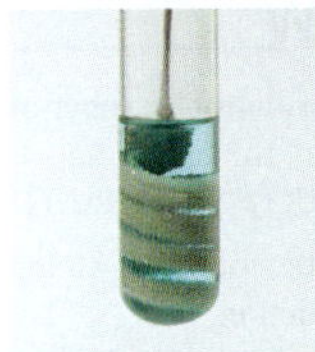

■ Figure 12

67. **Copper** When solid copper is put into a solution of silver nitrate, as shown in **Figure 12,** silver metal appears and blue copper(II) nitrate forms. Write the corresponding, unbalanced chemical equation. Next, determine the oxidation state of each element in the equation. Write the two half-reactions, labeling each as oxidation or reduction. Finally, write a balanced equation for the reaction.

68. Use the oxidation-number method to balance these ionic redox equations.
a. $MoCl_5 + S^{2-} \rightarrow MoS_2 + Cl^- + S$
b. $TiCl_6^{2-} + Zn \rightarrow Ti^{3+} + Cl^- + Zn^{2+}$

69. Use the half-reaction method to balance these equations for redox reactions. Add water molecules and hydrogen ions or hydroxide ions as needed.
a. $NH_3(g) + NO_2(g) \rightarrow N_2(g) + H_2O(l)$ (in acidic solution)
b. $Br_2 \rightarrow Br^- + BrO_3^-$ (in basic solution)

70. Balance the following redox chemical equation. Rewrite the equation in full ionic form, then derive the net ionic equation and balance by the half-reaction method. Give the final answer as it is shown below but with the balancing coefficients.

$$KMnO_4(aq) + FeSO_4(aq) + H_2SO_4(aq) \rightarrow Fe_2(SO_4)_3(aq) + MnSO_4(aq) + K_2SO_4(aq) + H_2O(l)$$

71. Use the oxidation-number method to balance these redox equations.
a. $CO + I_2O_5 \rightarrow I_2 + CO_2$
b. $SO_2 + Br_2 + H_2O \rightarrow HBr + H_2SO_4$
c. $Cu + NO_3^- \rightarrow Cu^{2+} + NO$ (in acidic solution)
d. $Zn + NO_3^- \rightarrow Zn^{2+} + NO_2$ (in acidic solution)
e. $Al + OH^- + H_2O \rightarrow H_2 + AlO_2^-$

72. Use the half-reaction method to balance these equations. Add water molecules and hydrogen ions or hydroxide ions as needed. Write the balanced equations in net ionic form.
a. $Cl^-(aq) + NO_3^-(aq) \rightarrow ClO^-(aq) + NO(g)$ (in acidic solution)
b. $IO_3^-(aq) + Br^-(aq) \rightarrow Br_2(l) + IBr(s)$ (in acidic solution)
c. $I_2(s) + Na_2S_2O_3(aq) \rightarrow Na_2S_2O_4(aq) + NaI(aq)$ (in acidic solution)

56. No. The total charge on the left-hand side does not equal the total charge on the right-hand side.
57. reduction; Electrons are gained and the oxidation number for Zn decreases.
58. Electrons are accepted by a species during the reduction half reaction, and electrons are lost from a species during an oxidation half reaction.

Mastering Problems

59. **a.** $Cl_2 + NaOH \rightarrow NaCl + HOCl$
b. $4HBrO_3 \rightarrow 2Br_2 + 2H_2O + 5O_2$
60. **a.** $2Au^{3+}(aq) + 6I^-(aq) \rightarrow 2Au(s) + 3I_2(s)$
b. $2Ce^{4+} + Sn^{2+} \rightarrow Sn^{4+} + 2Ce^{3+}$
61. **a.** $2Al + 3I_2 \rightarrow 2Al^{3+} + 6I^-$
b. $MnO_2 + 2Br^- + 4H^+ \rightarrow Mn^{2+} + Br_2 + 2H_2O$
62. **a.** $2PbS + 3O_2 \rightarrow 2PbO + 2SO_2$
b. $4NaWO_3 + O_2 + 4NaOH \rightarrow 4Na_2WO_4 + 2H_2O$
c. $2NH_3 + 3CuO \rightarrow 3Cu + N_2 + 3H_2O$
d. $Al_2O_3 + 3C + 3Cl_2 \rightarrow 2AlCl_3 + 3CO$
63. $Fe^{2+} + Ti^{4+} \rightarrow Fe^{3+} + Ti^{3+}$; Fe^{2+} is the reducing agent, Ti^{4+} is the oxidizing agent
64. **a.** $NH_3(g) \rightarrow N_2(g) + 3e^-$ oxidation
$PbO(s) + 2e^- \rightarrow Pb(s)$ reduction
b. $I_2(s) + 2e^- \rightarrow 2I^-(aq)$ reduction
$S_2O_3^{2-}(aq) \rightarrow S_2O_4^{2-}(aq) + e^-$ oxidation
c. $Sn(s) \rightarrow Sn^{2+}(aq) + 2e^-$ oxidation
$H^+(aq) + e^- \rightarrow H_2(g)$ reduction
65. $H_2C_2O_4 \rightarrow 2CO_2 + 2H^+ + 2e^-$
$HAsO_2 + 3H^+ + 3e^- \rightarrow As + 2H_2O$
66. **a.** oxidation
b. reduction
c. reduction
d. oxidation

67. $AgNO_3(aq) + Cu(s) \rightarrow Cu(NO_3)_2(aq) + Ag(s)$
oxidation: $Cu \rightarrow Cu^{2+} + 2e^-$
reduction: $e^- + Ag^+ \rightarrow Ag$
$2AgNO_3(aq) + Cu(s) \rightarrow Cu(NO_3)_2(aq) + 2Ag(s)$
68. **a.** $2MoCl_5 + 5S^{2-} \rightarrow 2MoS_2 + 10Cl^- + S$
b. $2TiCl_6^{2-} + Zn \rightarrow 2Ti^{3+} + 12Cl^- + Zn^{2+}$
69. **a.** $8NH_3 + 6NO_2 \rightarrow 7N_2 + 12H_2O$
b. $3Br_2 + 6OH^- \rightarrow 5Br^- + BrO_3^- + 3H_2O$
70. $2KMnO_4(aq) + 10FeSO_4(aq) + 8H_2SO_4(aq) \rightarrow 5Fe_2(SO_4)_3(aq) + 2MnSO_4(aq) + K_2SO_4(aq) + 8H_2O(l)$
71. **a.** $5CO + I_2O_5 \rightarrow I_2 + 5CO_2$
b. $SO_2 + Br_2 + 2H_2O \rightarrow 2HBr + H_2SO_4$
c. $3Cu + 2NO_3^- + 8H^+ \rightarrow 3Cu^{2+} + 2NO + 4H_2$
d. $Zn + 2NO_3^- + 4H^+ \rightarrow Zn^{2+} + 2NO_2 + 2H_2O$
e. $2Al + 2OH^- + 2H_2O \rightarrow 3H_2 + 2AlO_2^-$
72. **a.** $3Cl^- + 2NO_3^- + 2H^+ \rightarrow 3ClO^- + 2NO + H_2O$
b. $6H^+ + 5Br^- + IO_3^- \rightarrow 2Br_2 + IBr + 3H_2O$
c. $S_2O_3^{2-} + H_2O + I_2 \rightarrow S_2O_4^{2-} + 2H^+ + 2I^-$

MIXED REVIEW

73. a. 0, +2
b. U, +6
c. Ru, +8
d. Fe, +3

74. a. oxidation
b. oxidation
c. reduction
d. reduction

75. rule 7, −1, 0, +1, rule 8, rule 1

76. a. NH_3
b. C
c. Ir

77. a. $Fe + Te^{2+} \rightarrow Fe^{2+} + Te$
b. $3IO_4^- + 2Al + 6H^+ \rightarrow 3IO_3^- + 2Al^{3+} + 3H_2O$
c. $4I_2 + N_2O + 5H_2O \rightarrow 8I^- + 2NO_3^- + 10H^+$

78. Chromium has different oxidation states in the two forms.

79. a. $5Sb^{3+} + 2MnO_4^- + 12H_2O \rightarrow 5SbO_4^{3-} + 2Mn^{2+} + 24H^+$
b. $N_2O + 2ClO^- + 2OH^- \rightarrow 2NO_2^- + 2Cl^- + H_2O$

80. Al_2O_3; $Al_2O_3 + 2Cr^{3+} \rightarrow Cr_2O_3 + 2Al^{3+}$; It is not a redox reaction. The oxidation numbers stay the same.

81. a. $3Mg + 2Fe^{3+} \rightarrow 3Mg^{2+} + 2Fe$
b. $3SO_2 + ClO_3^- + 3H_2O \rightarrow 3SO_4^{2-} + Cl^- + 6H^+$

82. a. $3P + 2H_2O + 5HNO_3 \rightarrow 3H_3PO_4 + 5NO$
b. $2KClO_3 + 4HCl \rightarrow Cl_2 + 2ClO_2 + 2H_2O + 2KCl$

MIXED REVIEW

73. Determine the oxidation number of the boldface element in each of the following.
a. $\mathbf{O}F_2$ **b.** $\mathbf{U}O_2^{2+}$ **c.** $\mathbf{Ru}O_4$ **d.** $\mathbf{Fe}_2O_3$

74. Identify each of the following changes as either oxidation or reduction.
a. $2Cl^- \rightarrow Cl_2 + 2e^-$
b. $Na \rightarrow Na^+ + e^-$
c. $Ca^{2+} + 2e^- \rightarrow Ca$
d. $O_2 + 4e^- \rightarrow 2O^{2-}$

75. Use the rules for assigning oxidation numbers to complete **Table 7**.

Table 7 Oxidation Number Assignment

Element	Oxidation Number	Rule
K in KBr	+1	
Br in KBr		8
Cl in Cl_2		1
K in KCl		7
Cl in KCl	−1	
Br in Br_2	0	

76. Identify the reducing agents in these equations.
a. $4NH_3 + 5O_2 \rightarrow 4NO + 6H_2O$
b. $Na_2SO_4 + 4C \rightarrow Na_2S + 4CO$
c. $4IrF_5 + Ir \rightarrow 5IrF_4$

77. Write a balanced ionic redox equation using the following pairs of redox half-reactions.
a. $Fe \rightarrow Fe^{2+} + 2e^-$
$Te^{2+} + 2e^- \rightarrow Te$
b. $IO_4^- + 2e^- \rightarrow IO_3^-$
$Al \rightarrow Al^{3+} + 3e^-$ (in acidic solution)
c. $I_2 + 2e^- \rightarrow 2I^-$
$N_2O \rightarrow 2NO_3^- + 8e^-$ (in acidic solution)

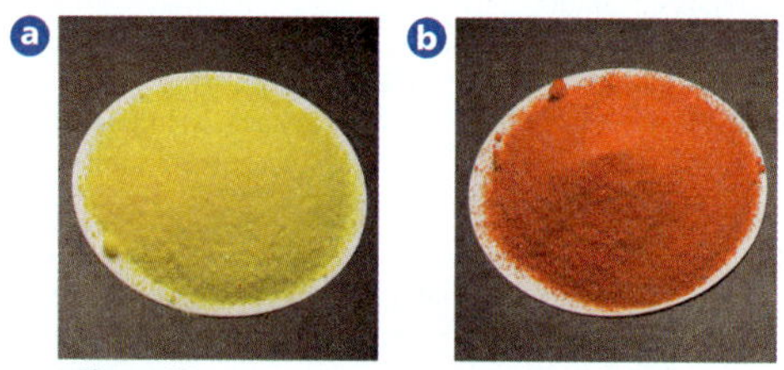

■ **Figure 13**

78. What probably accounts for the different forms of chromium shown in **Figure 13**?

79. Balance these ionic redox equations by any method.
a. $Sb^{3+} + MnO_4^- \rightarrow SbO_4^{3-} + Mn^{2+}$ (in acid solution)
b. $N_2O + ClO^- \rightarrow Cl^- + NO_2^-$ (in basic solution)

80. Gemstones Rubies are gemstones made up mainly of aluminum oxide. Their red color comes from a small amount of chromium(III) ions replacing some of the aluminum ions. Write the formula for aluminum oxide, and show the reaction in which an aluminum ion is replaced with a chromium ion. Is this a redox reaction?

81. Balance these ionic redox equations by any method.
a. $Mg + Fe^{3+} \rightarrow Mg^{2+} + Fe$
b. $ClO_3^- + SO_2 \rightarrow Cl^- + SO_4^{2-}$ (in acidic solution)

82. Balance these redox equations by any method.
a. $P + H_2O + HNO_3 \rightarrow H_3PO_4 + NO$
b. $KClO_3 + HCl \rightarrow Cl_2 + ClO_2 + H_2O + KCl$

THINK CRITICALLY

83. Apply The following equations show redox reactions that are sometimes used in the laboratory to generate pure nitrogen gas and pure dinitrogen monoxide gas (nitrous oxide, N_2O).

$$NH_4NO_2(s) \rightarrow N_2(g) + 2H_2O(l)$$
$$NH_4NO_3(s) \rightarrow N_2O(g) + 2H_2O(l)$$

a. Determine the oxidation number of each element in the two equations, and then make diagrams showing the changes in oxidation numbers that occur in each reaction.
b. Identify the atom that is oxidized and the atom that is reduced in each of the two reactions.
c. Identify the oxidizing and reducing agents in each of the two reactions.
d. Write a sentence telling how the electron transfer taking place in these two reactions differs from that taking place in the reaction below.

$$2AgNO_3 + Zn \rightarrow Zn(NO_3)_2 + 2Ag$$

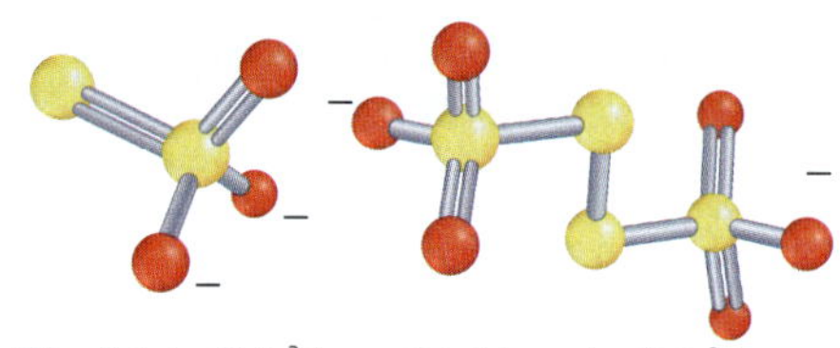

Thiosulfate ion ($S_2O_3^{2-}$) Tetrathionate ion ($S_4O_6^{2-}$)

■ **Figure 14**

84. Analyze Examine the net ionic equation below for the reaction that occurs when the thiosulfate ion ($S_2O_3^{2-}$) is oxidized to the tetrathionate ion ($S_4O_6^{2-}$). Balance the equation using the half-reaction method. **Figure 14** will help you to determine the oxidation numbers to use.

$$S_2O_3^{2-} + I_2 \rightarrow I^- + S_4O_6^{2-} \text{ (in acidic solution)}$$

THINK CRITICALLY

83. a. Refer to the Solution Manual.
b. N^{3-} to N_2 loses 3 e^- (oxidized);
N^{3+} to N_2 gains 3 e^- (reduced)
N^{3-} to N^{1+} loses 4 e^- (oxidized);
N^{5+} to N^{1+} gains 4 e^- (reduced)
c. NO_2^- and NO_3^- (oxidizing agents); NH_4^+ (reducing agent)
d. In the first two reactions, nitrogen is both oxidized and reduced. The third reaction involves redox between two different elements.

84. $2S_2O_3^{2-} + I_2 \rightarrow 2I^- + S_4O_6^{2-}$ (in acidic solution)

Chapter Self-Check

85. Predict Consider the fact that all of the following are stable compounds. What can you infer about the oxidation state of phosphorus in its compounds?

$$PH_3, PCl_3, P_2H_4, PCl_5, H_3PO_4, Na_3PO_3$$

86. Solve Potassium permanganate oxidizes chloride ions to chlorine gas. Balance the equation for this redox reaction taking place in acidic solution.

87. In the half-reaction $NO_3^- \rightarrow NH_4^+$, on which side of the equation should electrons be added? Add the correct number of electrons to the side on which they are needed, and rewrite the equation.

■ **Figure 15**

88. The redox reaction between dichromate ions and iodide ions in acidic solution is shown in **Figure 15.** Use the half-reaction method to balance the equation for this redox reaction.

CHALLENGE PROBLEM

89. For each reaction described, write the corresponding chemical equation without putting coefficients to balance it. Next, determine the oxidation state of each element in the equation. Then, write the two half-reactions, labeling which is oxidation and which is reduction. Finally, write a balanced equation for the reaction.

a. Solid mercury(II) oxide is put into a test tube and gently heated. Liquid mercury forms on the sides and in the bottom of the tube, and oxygen gas bubbles out from the test tube.

b. Solid copper pieces are put into a solution of silver nitrate. Silver metal appears and blue copper(II) nitrate forms in the solution.

CUMULATIVE REVIEW

90. A gaseous sample occupies 32.4 mL at −23°C and 0.75 atm. What volume will it occupy at STP?

91. When iron(III) chloride ($FeCl_3$) reacts in an atmosphere of pure oxygen, the following occurs:

$$4FeCl_3(s) + 3O_2(g) \rightarrow 2Fe_2O_3(s) + 6Cl_2(g)$$

If 45.0 g of $FeCl_3$ reacts and 20.5 g of iron(III) oxide is recovered, determine the percent yield.

WRITING IN ▸ Chemistry

92. Steel Research the role of oxidation-reduction reactions in the manufacture of steel. Write a summary of your findings, including appropriate diagrams and equations representing the reactions.

93. Silverware Practice your technical writing skills by writing a procedure for cleaning tarnished silverware by a redox chemical process. Be sure to include background information describing the process as well as logical steps that would enable anyone to accomplish the task.

94. Copper was a useful metal even before iron, silver, and gold metals were extracted from their ores and used as tools, utensils, jewelry, and artwork. Copper was smelted to high temperatures as early as 8000 years ago. Thousands of pieces of scrap copper have been unearthed in Virginia, where in the 1600s the colonists might have traded this material for food. Compare and contrast the processing and use of copper in those older civilizations with today.

DBQ Document-Based Questions

Glazes *The formation of color in ceramic glazes, such as in* **Figure 16,** *can be influenced by firing conditions. Metal ions such as copper that have more than one oxidation state can impart different colors to a glaze. In an oxidative firing, plenty of oxygen is allowed in the kiln, and copper ions present will make the glaze a green-to-blue color. Under reducing conditions, oxygen is limited and carbon dioxide is abundant. Copper ions in the glaze provide a reddish color.*

Data obtained from: Denio, Allen A. 2001. The joy of color in ceramic glazes with the help of redox chemistry. *Journal of Chemical Education.* 78 No 10.

■ **Figure16**

95. Write the equation for what has occurred in the pottery shown in **Figure 16.**

96. Based on the color of the pottery, what is the oxidation state of the copper that is reduced? Oxidized?

85. Phosphorus has several oxidation states (−3, +3, +5) that make phosphorus very flexible when combining with nonmetals.

86. $2MnO_4^-(aq) + 10Cl^-(aq) + 16H^+(aq) \rightarrow 2Mn^{2+}(aq) + 5Cl_2(aq) + 8H_2O(l)$

87. The oxidation state of N is reduced from +5 to −3; N must gain 8 electrons. $8e^-$ to the left side; $NO_3^- + 8e^- \rightarrow NH_4^+$

88. $6I^-(aq) + 14H^+(aq) + Cr_2O_7^{2-}(aq) \rightarrow 3I_2(s) + 7H_2O(l) + 2Cr^{3+}(aq)$

CHALLENGE PROBLEM

89. a. oxidation: $O^{2-} \rightarrow O_2 + 2e^-$
reduction $Hg^{2+} + 2e^- \rightarrow Hg$
$2HgO(s) \rightarrow O_2(g) + 2Hg(l)$

b. oxidation: $Cu \rightarrow Cu^{2+} + 2e^-$
reduction: $e^- + Ag^+ \rightarrow Ag$
$2AgNO_2(aq) + Cu(s) \rightarrow Cu(NO_3)_2(aq) + 2Ag(s)$

CUMULATIVE REVIEW

90. 27mL

91. 92.5% yield

WRITING IN ▸ Chemistry

Rubric

92. Student answers will likely include descriptions and diagrams of some or all of the following:
The chief ores of iron are its oxides, hematite (Fe_2O_3) and magnetite (Fe_3O_4), and its carbonate ($FeCO_3$). Most commonly, iron ores are reduced in a blast furnace. In the blast furnace, the important reactions are the oxidation of coke to carbon monoxide ($2C(s) + O_2(g) \rightarrow 2CO(g)$) and the reduction of iron ore by carbon monoxide, which usually occurs in steps.
$CO(g) + 3Fe_2O_3(s) \rightarrow 2Fe_3O_4(s) + CO_2(g)$
$Fe_3O_4(s) + CO(g) \rightarrow 3FeO(s) + CO_2(g)$
$FeO(s) + CO(g) \rightarrow Fe(l) + CO_2(g)$

93. Answers will vary, but students should create a logical procedure based on the concepts learned in the MiniLab for this chapter.

94. Answers will vary.

DBQ Document-Based Questions

Data obtained from: Denio, Allen A. 2001. The joy of color in ceramic glazes with the help of redox chemistry. *Journal of Chemical Education.* 78 No 10.

95. $Cu^{2+} + 2e^- \rightarrow Cu$; $Cu^{2+} + e^- \rightarrow Cu^+$

96. most reduced: 0 or +1; most oxidized: +2

MULTIPLE CHOICE

1. B
2. C
3. D
4. B
5. C
6. D
7. B
8. A

MULTIPLE CHOICE

1. Which is NOT a reducing agent in a redox reaction?
A. the substance oxidized
B. the electron acceptor
C. the less-electronegative substance
D. the electron donor

2. The reaction between nickel and copper(II) chloride is shown below.

$$Ni(s) + CuCl_2(aq) \rightarrow Cu(s) + NiCl_2(aq)$$

What are the half-reactions for this redox reaction?
A. $Ni \rightarrow Ni^{2+} + 2e^-$, $Cl_2 \rightarrow 2Cl^- + 2e^-$
B. $Ni \rightarrow Ni^{2+} + e^-$, $Cu^+ + e^- \rightarrow Cu$
C. $Ni \rightarrow Ni^{2+} + 2e^-$, $Cu^{2+} + 2e^- \rightarrow Cu$
D. $Ni \rightarrow Ni^{2+} + 2e^-$, $2Cu^+ + 2e^- \rightarrow Cu$

Use the diagram below to answer Questions 3 and 4. All four containers have a volume of 5.0 L and are at the same temperature.

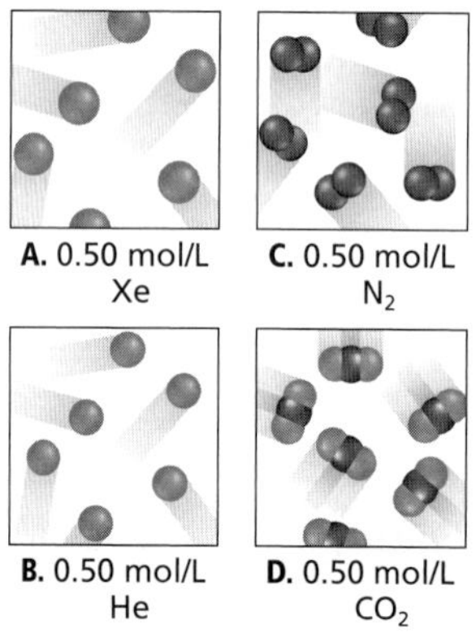

3. Which container contains 110 g of its gas?
A. A
B. B
C. C
D. D

4. If a small hole is made in each container so that the gas can escape, which container will have the fastest rate of effusion?
A. A
B. B
C. C
D. D

5. The following system is in equilibrium:

$$2S(s) + 5F_2(g) \leftrightarrows SF_4(g) + SF_6(g)$$

Which will cause the equilibrium to shift to the right?
A. increased concentration of SF_4
B. increased concentration of SF_6
C. increased pressure on the system
D. decreased pressure on the system

Use the table below to answer Question 6.

Data for the Formation of Cobalt(II) Sulfate at 25°C	
$Co(s) + S(s) + 2O_2(g) \rightarrow CoSO_4(s)$	
ΔH_f°	−888.3 kJ/mol
ΔS_f°	118.0 J/mol • K
ΔG_f°	?

6. What is the ΔG_f° for the formation of cobalt(II) sulfate from its elements?
A. −853.1 kJ/mol
B. −885.4 kJ/mol
C. −891.3 kJ/mol
D. −923.5 kJ/mol

7. Which will be the result of increasing the temperature of a reaction in a system in equilibrium where the forward reaction is endothermic?
A. The equilibrium will shift to the left.
B. The equilibrium will shift to the right.
C. The rate of the forward reaction will be decreased.
D. The rate of the reverse reaction will be decreased.

8. The reaction between sodium iodide and chlorine is shown below.

$$2NaI(aq) + Cl_2(aq) \rightarrow 2NaCl(aq) + I_2(aq)$$

The oxidation state of sodium remains unchanged for which reason?
A. Na^+ is a spectator ion.
B. Na^+ cannot be reduced.
C. Na is an uncombined element.
D. Na^+ is a monatomic ion.

Online Test Practice

SHORT ANSWER

Use the equation below to answer Questions 9 and 10.

The net ionic reaction between iodine and lead(IV) oxide is shown below.

$$I_2(s) + PbO_2(s) \rightarrow IO_3^-(aq) + Pb^{2+}(aq)$$

9. Identify the oxidation number in each participant in the reaction.

10. Explain how to identify which element is oxidized and which one is reduced.

EXTENDED RESPONSE

Use the phase diagram for CO_2 to answer Questions 11 to 13.

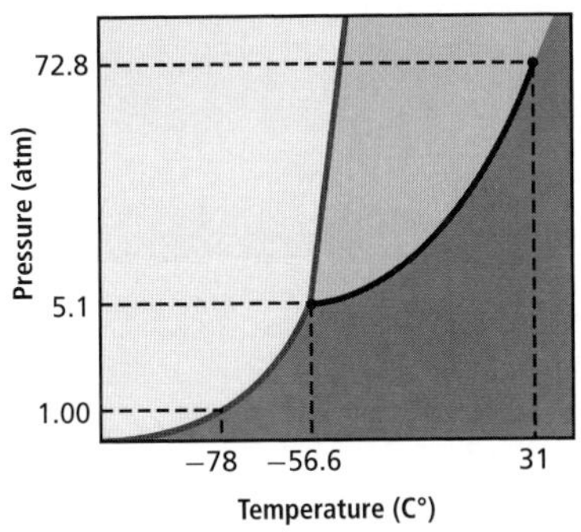

11. Explain what state or states of matter can exist at a temperature of −56.6°C and a pressure of 31.1 atm.

12. Suppose that you have a sample of CO_2 at 35°C and 83 atm. In what state of matter is the sample? Explain how you can predict this from the graph.

13. Is carbon dioxide denser in its liquid state or its solid state? Use the graph to explain.

SAT SUBJECT TEST: CHEMISTRY

14. Which statement about the common ion effect is NOT true?

A. The effects of common ions on an equilibrium system can be explained by Le Châtelier's principle.

B. The decreased solubility of an ionic compound due to the presence of a common ion is called the common ion effect.

C. The addition of NaCl to a saturated solution of AgCl will produce the common ion effect.

D. The common ion effect is due to a shift in equilibrium toward the aqueous products of a system.

E. The addition of lead(II) nitrate ($Pb(NO_3)_2$) to a saturated solution of lead(II) chromate ($PbCrO_4$) will produce the common ion effect.

Use the list below to answer Questions 15 to 18.

Five flasks contain 500 mL of a 0.250*M* aqueous solution of the indicated chemical.

A. KCl
B. CH_3OH
C. $Ba(OH)_2$
D. CH_3COOH
E. NaOH

15. Which chemical will dissociate into the greatest number of particles when in solution?

16. Which chemical has the greatest molar mass?

17. Which flask would contain 9.32 g of the labeled chemical?

18. Which flask's contents are composed of 18.7% oxygen?

NEED EXTRA HELP?

If You Missed Question . . .	1	2	3	4	5	6	7	8	9	10	11	12	13	14	15	16	17	18
Review Section . . .	19.1	19.2	13.3	12.1	17.2	15.5	17.2	19.1	19.1	19.1	12.4	12.4	12.4	17.2	14.2	10.2	10.3	10.4

SHORT ANSWER

9. $I_2(s) + PbO_2(s) \rightarrow IO_3^-(aq) + Pb^{2+}(aq)$
Oxidation numbers are:
0 +4 −2 +5 −2 +2

10. The oxidized element increases its oxidation number (I). The reduced element decreases its oxidation number (Pb).

EXTENDED RESPONSE

11. Solid
12. Liquid
13. Solid; The boundary between solid and liquid phase has a positive slope. The opposite condition holds for water which is denser in its liquid state.

SAT SUBJECT TEST: CHEMISTRY

14. D
15. C
16. C
17. A
18. C

CHAPTER 20 Electrochemistry

BIGIDEA **Chemical energy can be converted to electric energy and electric energy to chemical energy.**

ESSENTIAL QUESTIONS	RESOURCES TO ASSESS MASTERY
SECTION 1 Voltaic Cells **1.** How is electrical energy obtained from a redox reaction? **2.** What are the parts of a voltaic cell and how does each part operate? **3.** How are cell potentials calculated and the spontaneity of redox reactions determined? 2 sessions 1 block	**Progress Monitoring** Formative Assessment, pp. 713, 714, 716 Reading Check, pp. 710, 716 Graph Check, p. 714 Section Review, p. 717
SECTION 2 Batteries **1.** What are the structure, composition, and operation of the typical carbon-zinc dry-cell battery? **2.** What is the difference between primary and secondary batteries and what are two examples of each type? **3.** What is the structure of the hydrogen-oxygen fuel cell and how does it operate? **4.** What is the process of corrosion of iron and what are methods to prevent corrosion? 2 sessions 1 block	**Progress Monitoring** Formative Assessment, pp. 719, 725, 726, 727 Reading Check, pp. 719, 721, 722, 723 Section Review, p. 727
SECTION 3 Electrolysis **1.** How is it possible to reverse a spontaneous redox reaction in an electrochemical cell? **2.** How does the molten sodium chloride electrolysis reaction compare with those in the electrolysis of brine? **3.** What is the importance of electrolysis in the smelting and purification of metals? 1 session 0.5 block	**Progress Monitoring** Formative Assessment, p. 730 Reading Check, pp. 729, 730 Section Review, p. 732 **Summative Assessment** Chapter Assessment, p. 736 *eAssessment* Chapter Test (Scaffolded)

Go online!

LEVELED RESOURCES	LAB MATERIALS
Science Notebook 20.1 OL ***FAST FILE Chapter Resources:*** ChemLab Worksheet, p. 86 OL Study Guide, p. 99 BL	LaunchLAB p. 706 / **15 min** zinc strips, copper strips, lemon, voltmeter ChemLAB p. 734 / **45 min** metal strips of copper, aluminum, zinc, and magnesium, 1*M* copper(II) nitrate, 1*M* aluminum nitrate, 1*M* zinc nitrate, 1*M* magnesium nitrate, 24-well microplate, Beral-type pipette, voltmeter, filter paper, 1*M* potassium nitrate, forceps, steel wool or sandpaper, table of standard reduction potentials
Science Notebook 20.2 OL ***FAST FILE Chapter Resources:*** MiniLab Worksheet, p. 84 OL Study Guide, p. 101 BL	MiniLAB p. 726 / **30 min, 15 min each day for 2 days** sandpaper, iron nails, magnesium ribbon, copper, beakers, distilled water, salt water
Science Notebook 20.3 OL ***FAST FILE Chapter Resources:*** Study Guide, p. 103 BL	

ADDITIONAL RESOURCES

- eAssessment
- ConnectED Student Center
- ConnectED Teacher Center
- Lesson Presentations
- eSolutions
- Videos
- Animations
- Online Quizzes
- Virtual Investigations
- Multilingual eGlossary
- Personal Tutors
- WebQuests
- Applying Practices Worksheets
- PBLs

Practice:

- Challenge Problems AL
- Supplemental Problems BL OL
- Chapter Test (Scaffolded)
- Standardized Test Practice

Additional Resources:

- Blueprints for Success
- Solving Problems: A Chemistry Handbook

Lab Resources:

- Laboratory Manual OL
- CBL Laboratory Manual OL
- Small-Scale Laboratory Manual OL
- Forensics Laboratory Manual OL

BL Below Level OL On Level AL Advanced Learners EL English Learners COOP LEARN Cooperative Learning

CHAPTER 20

LaunchLAB

Rubric

How can you make a battery from a lemon?

Purpose Students will examine how a battery works.

Safety and Disposal Approve lab safety forms before work begins. Use the lemon pieces for several classes. Throw them in the trash at the end of the day. Clean and reuse the zinc and copper metal pieces.

Teaching Strategies

- Introduce the concept by connecting a battery to the voltmeter and explaining the reading.
- Metals other than zinc and copper can be used. Magnesium and iron are also good choices.

Expected Results Using zinc and copper, the voltmeter will read about 1.1 V. When one of the metals is removed from the lemon, the voltmeter will read 0 V, or whatever the default zero reading is. (The CBL multimeter setting will read about 2 V for its default reading.)

CHAPTER 20

Electrochemistry

BIGIDEA Chemical energy can be converted to electric energy and electric energy to chemical energy.

SECTIONS

LaunchLAB

How can you make a battery from a lemon?

You can purchase a handy package of portable power at any convenience store—a battery. You can also light a bulb with a lemon. In this lab, you will study how these power sources are alike.

FOLDABLES Study Organizer

Electrochemical Cells

Make a pocket book. Label it as shown. Use it to help you compare voltaic cells to electrolytic cells.

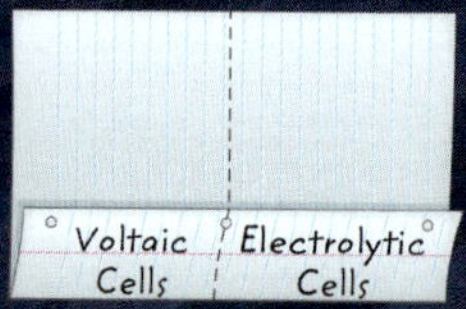

Procedure

1. Read and complete the lab safety form.
2. Insert **zinc** and **copper strips** into a **lemon,** about 2 cm apart from each other.
3. Attach the black lead from a **voltmeter** to the zinc and the red lead to the copper. Read the potential difference (voltage) shown on the voltmeter, and record your reading.
4. Remove one of the metals from the lemon, and observe what happens to the potential difference on the voltmeter.

Analysis

1. **Explain** the purpose of the zinc and copper metals. The two metals provide a potential that is measured in volts.
2. **Infer** What is the function of the lemon? The lemon acts as an electrolyte solution for ions and electrons to travel from metal to metal.

Inquiry Do you think that you can make a battery from foods other than lemons? Form a hypothesis about other foods—or types of foods—that can be made into batteries. Then, design a battery using another food. If your teacher approves your design, test your hypothesis. Foods that contain a liquid and are acidic, such as other citrus fruits, could also be used to make a battery.

A camera trap captured this image of a red fox. Camera traps are a noninvasive way to study animals by using a sensor that triggers a camera's shutter when the animal approaches. Batteries power both the camera and the sensor.

Introduce the BIGIDEA

Batteries Included To introduce this chapter's Big Idea, bring several different types of batteries to class and a piece of corroded metal. Tell students that each of the batteries involve redox reactions. If the reaction is spontaneous, the system is considered a voltaic cell. If the reaction requires energy, the system is considered an electrolytic cell.

Tie to Previous Knowledge

Have students review the following concepts before studying this chapter.

- types of reactions, activity series of elements
- energy, chemical potential energy, spontaneous process
- oxidation, reduction half-reactions

Use the Photo

Batteries Have students observe the opening photo and ask them what a battery does. converts chemical energy to electrical energy The camera trap used to take the photo used both lithium and alkaline batteries. Batteries contain electrochemical cells, which generate an electrical current. The electrodes of the lithium battery, which is found in the camera, are composed of lithium cobalt oxide, $LiCoO_2$ and carbon. The camera trap used alkaline batteries, which contain zinc and manganese oxide electrodes.

SECTION 1

1 Focus

MAINIDEA

Electron Flow Remind students of the Launch Lab's lemon battery. Ask students which direction the current seemed to flow (give a positive value). The current flowed from the zinc to the copper. Ask students what would happen if the black lead was attached to the copper and the red lead was attached to the zinc. The voltmeter would read negative or appear not to function properly. Ask students if any other food could be used in place of the lemon. They might have heard of potato clocks. OL

2 Teach

Concept Development

Got Batteries? Ask students to brainstorm a list of items that require some type of battery, such as TV remote controls. Students should be able to list items such as radios and CD players, calculators, toys, smoke detectors, clocks, cell phones, video cameras, computers, flashlights, car and lawn mower batteries. Ask them to propose reasons why certain battery types are used in some household items but not others. Some expected responses could include size of the battery, how long the battery is used at any one time, cost, whether the battery produces enough voltage to power the object. OL

SECTION 1

Voltaic Cells

MAINIDEA In voltaic cells, oxidation takes place at the anode, yielding electrons that flow to the cathode, where reduction occurs.

Essential Questions

- How is electrical energy obtained from a redox reaction?
- What are the parts of a voltaic cell, and how does each part operate?
- How are cell potentials calculated and the spontaneity of redox reactions determined?

Review Vocabulary

oxidation: the complete or partial loss of electrons from a reacting substance; increases an atom's or ion's oxidation number

reduction: the complete or partial gain of electrons by the atoms of a substance; decreases the atom's or ion's oxidation number

New Vocabulary

salt bridge
electrochemical cell
voltaic cell
half-cell
anode
cathode
reduction potential
standard hydrogen electrode

CHEM 4 YOU What could you do with half of a dollar bill? Without the other half, you cannot spend it. Voltaic cells have two half-cells, and both are required to produce energy.

Redox in Electrochemistry

Electrochemistry is the study of the redox processes by which chemical energy is converted to electrical energy and vice versa. Electrochemical processes are useful in industry and critically important for biological functioning.

You have learned that all redox reactions involve a transfer of electrons from the species that is oxidized to the species that is reduced. **Figure 1** and **Figure 2** illustrate the simple redox reaction in which zinc atoms are oxidized to form zinc (Zn^{2+}) ions. The two electrons donated from each zinc atom are accepted by a copper (Cu^{2+}) ion, which becomes an atom of copper metal. The following net ionic equation illustrates the electron transfer that occurs.

$$Zn(s) + Cu^{2+}(aq) \xrightarrow{} Zn^{2+}(aq) + Cu(s) \quad (2e^- \text{ transferred from Zn to } Cu^{2+})$$

Half-reactions Two half-reactions make up this redox process:

$$Zn \rightarrow Zn^{2+} + 2e^- \text{ (oxidation half-reaction: electrons lost)}$$
$$Cu^{2+} + 2e^- \rightarrow Cu \text{ (reduction half-reaction: electrons gained)}$$

What do you think would happen if you separated the oxidation half-reaction from the reduction half-reaction? Can a redox reaction occur? Consider **Figure 1a,** in which a zinc strip is immersed in a solution of zinc sulfate and a copper strip is immersed in a solution of copper(II) sulfate.

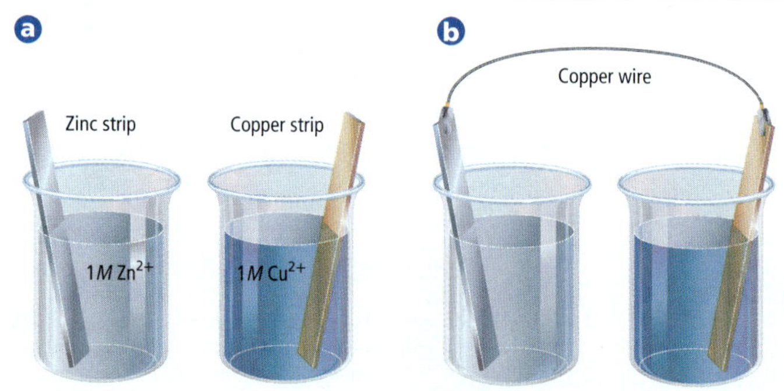

Figure 1 These containers are constructed and arranged so that zinc will be oxidized on one side, while copper ions will be reduced on the other. In **a,** zinc metal is immersed in 1 *M* zinc sulfate solution, and copper metal in 1 *M* copper sulfate. In **b,** a wire joining the zinc and copper strips provides a pathway for the flow of electrons, but the pathway is not complete. Electron transfer is still not possible.

CHEMISTRY PROJECT

Potato Clock Obtain a two-potato clock. (These are available from a variety of scientific supply stores or retail science stores.) Identify the two different electrodes as zinc and copper. Have the students diagram the two-potato clock and label the anode and cathode of the electrochemical cell. Also, ask them to label the part of the clock that is functioning as the electrolyte. OL EL

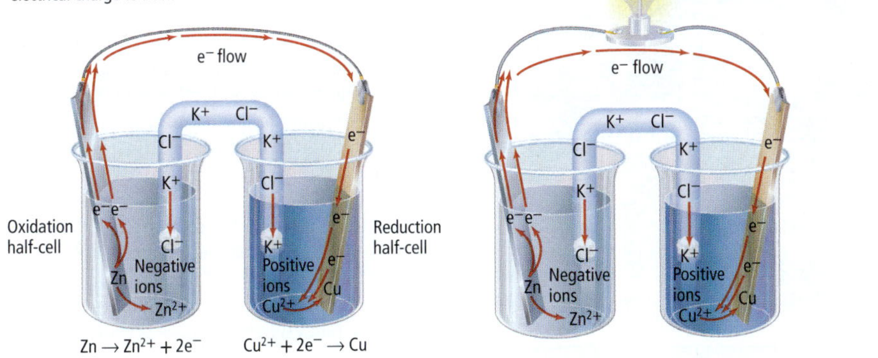

■ **Figure 2** The addition of a salt bridge completes the pathway. Negative ions move through the salt bridge to the zinc side. Positive ions move through the bridge to the copper side.

Two problems prevent a redox reaction from occurring. First, there is no way for zinc atoms to transfer electrons to copper(II) ions. This problem can be solved by connecting the zinc and copper strips with a metal wire, as shown in **Figure 1b.** The wire can serve as a pathway for electrons to flow from the zinc strip to the copper strip.

The second problem is that when the metal strips are placed in their solutions, oxidation begins at the zinc strip and reduction begins at the copper strip—but these reactions cannot continue. The reason is that as zinc is oxidized, positive zinc ions build up around the zinc electrode. As copper in the copper sulfate solution is reduced, negative sulfate ions build up around the copper electrode. The buildup of charges stops any further reaction. To solve this problem, a salt bridge must be added to the system. A **salt bridge** is a pathway to maintain solution neutrality by allowing the passage of ions from one side to another, as shown in **Figure 2.** A salt bridge consists of a tube containing a conducting solution of a soluble salt, such as KCl, held in place by an agar gel or other permeable plug. Ions can move through it, but the solutions cannot.

When the connecting metal wire and the salt bridge are in place, the spontaneous redox reaction begins. Electrons flow through the wire from the oxidation half-reaction to the reduction half-reaction, while positive and negative ions move through the salt bridge. A flow of charged particles is called an electric current. In **Figure 2,** the flow of electrons through the wire and the flow of ions through the salt bridge make up the electric current. The energy of the flowing electrons can be used to light a bulb, as shown in **Figure 2.**

Electrochemical cells The device shown in **Figure 2** is a type of electrochemical cell called a voltaic cell. An **electrochemical cell** is an apparatus that uses a redox reaction to produce electrical energy or uses electrical energy to cause a chemical reaction. A **voltaic cell** is a type of electrochemical cell that converts chemical energy to electrical energy by a spontaneous redox reaction. The voltaic cell, also shown in **Figure 3,** is named for Alessandro Volta (1745–1827), the Italian physicist who is credited with its invention in 1800.

View an **animation about a voltaic cell.**

Concepts In Motion

■ **Figure 3** This replica of one of Alessandro Volta's first cells consists of discs of zinc and copper arranged in alternating layers and separated by cloth or cardboard soaked in an acidic solution. Electric current increased with the number of metal discs used.

Identify Misconceptions

Students do not always understand that an electrical circuit requires a complete, uninterrupted flow of electrons (charged particles).

Uncover the Misconception Divide the class into small groups and provide each group with D-cell batteries, light bulbs, and single wires (with the insulation stripped from the ends). Challenge the students to light their bulbs using as many wiring pathways as possible. Ask each group to draw a diagram of the conducting pathway that lit the bulb. With each group, reinforce the concept of a complete circuit that enabled the bulb to light. Students might light bulbs individually or wire a series of batteries and bulbs to construct a more elaborate complete circuit.

Demonstrate the Concept Having shown how a complete electric circuit lights a bulb demonstrate what happens when some object or substance crosses the circuit. Have each group use the circuit they constructed to light the bulb and place a variety of substances between one end of the wire and the bottom of the battery. Explain that if the substance crossing the circuit is conductive, the circuit will conduct the current and light the bulb. If the substance is not conductive, the circuit is broken and the bulb will not light. Sample substances might include construction paper, a strip of copper, a cloth swatch soaked in salt solution or any other conducting and nonconducting materials.

Assess New Knowledge Ask students to identify how the charged particles/electrons are moving and how the circuit is completed in the electrochemical cell. OL

☑ READING CHECK The left beaker contains the anode. The anode is the zinc metal because zinc is losing electrons.

Content Background

Salt Bridges Explain that salt bridges can take various forms including the following examples: a U tube filled with salt connecting two separate beakers containing solutions, a porous cup containing one solution placed inside another container containing the second solution, or a simple paper towel soaked in an electrolyte, such as salt water, and draped over two separate beakers, thereby connecting the two solutions.

Reinforcement

Vocabulary Have students assemble in small groups and quiz each other using the vocabulary in this section. OL COOP LEARN

■ **Figure 4** The roller coaster at the top of the track has high potential energy relative to the track below because of the difference in height. Similarly, an electrochemical cell has potential energy to produce a current because there is a difference in the ability of the electrodes to move electrons from the anode to the cathode.

Chemistry of Voltaic Cells

An electrochemical cell consists of two parts called **half-cells,** in which the separate oxidation and reduction reactions take place. Each half-cell contains an electrode and a solution containing ions. An electrode is an electrically conductive material, usually a metallic strip or graphite, that conducts electrons into and out of the solution in the half cell. In **Figure 2,** the beaker with the zinc electrode is where the oxidation half of the redox reaction takes place. The beaker with the copper electrode is where the reduction half of the reaction takes place. The reaction that takes place in each half-cell is called a half-cell reaction. The electrode where oxidation takes place is called the **anode.** The electrode where reduction takes place is called the **cathode.**

☑ READING CHECK **Identify** which of the beakers in **Figure 2** contains the anode.

Voltaic cells and energy Recall that an object's potential energy is due to its position or composition. In electrochemistry, electric potential energy is a measure of the amount of current that can be generated from a voltaic cell to do work. Electric charge can flow between two points only when a difference in electric potential energy exists between the two points. In an electrochemical cell, these two points are the two electrodes. Electrons generated at the anode, the site of oxidation, are thought to be pushed or driven toward the cathode by the electromotive force (EMF). This force is due to the difference in electric potential energy between the two electrodes and is referred to as the cell potential. A volt is a unit used to measure cell potential. The electric potential difference of a voltaic cell is an indication of the energy that is available to move electrons from the anode to the cathode.

Consider the analogy illustrated in **Figure 4.** The roller coaster is almost stationary at the top of the track. Then, it plummets from its high position because of the difference in gravitational potential energy (PE) between the top and bottom of the track. The kinetic energy (KE) attained by the roller coaster is determined by the difference in height (potential energy) between the top and bottom parts of the track. Similarly, the energy of the electrons flowing from the anode to the cathode in a voltaic cell is determined by the difference in electric potential energy between the two electrodes. In redox terms, the voltage of a cell is determined by comparing the difference in the tendency of the two electrode materials to accept electrons. The greater the difference, the greater the potential energy difference between the two electrodes and the larger the voltage of the cell will be.

DIFFERENTIATED INSTRUCTION

Struggling Learners Make a visual aid to help students define and label the anode, cathode, and direction of electric current in an electrochemical cell. Draw pictures of a cell, such as one of the systems described in the text, on poster board, cut out, and attach flexible magnets to the back of the appropriate solution (ion). Now that you have defined the electrodes and solutions in each half-cell, have student volunteers label the anode and cathode, place electrons into the diagram to complete the circuit, and label the direction of the electron flow using an arrow. BL EL

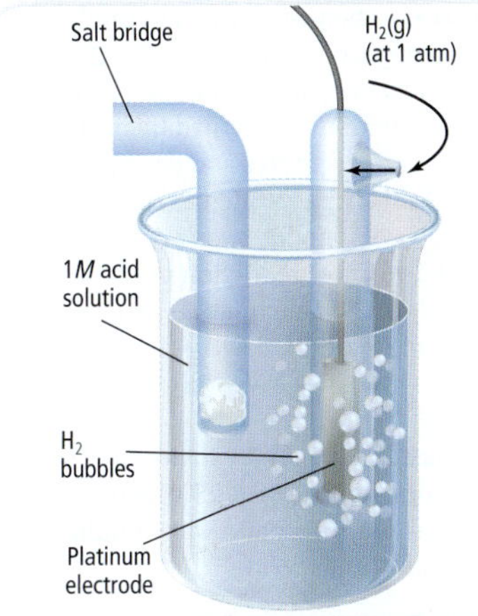

■ **Figure 5** A standard hydrogen electrode consists of a platinum electrode with hydrogen gas at 1 atm pressure bubbling into an acidic solution that is 1M in hydrogen ions. The reduction potential for this configuration is defined as 0.000 V.

Reduction:
$2H^+(aq) + 2e^- \rightarrow H_2(g)$

Oxidation:
$H_2(g) \rightarrow 2H^+(aq) + 2e^-$

The force of gravity always causes a diver to fall downward to a lower energy state, never upward to a higher energy state. When a diver steps off a diving board, his or her spontaneous motion is always downward. Similarly, in the zinc-copper cell, under standard conditions, copper(II) ions at the cathode accept electrons more readily than the zinc ions at the anode. Thus, the redox reaction occurs spontaneously only when the electrons flow from the zinc to the copper.

VOCABULARY
ACADEMIC VOCABULARY
Correspond
to be in agreement or to match
Her directions correspond with the map.

Calculating Electrochemical Cell Potentials

Recall that gaining electrons is called reduction. Building on this fact, the tendency of a substance to gain electrons is its **reduction potential.** The reduction potential of an electrode cannot be determined directly because the reduction half-reaction must be coupled with an oxidation half-reaction. When two half-reactions are coupled, the voltage generated corresponds to the difference in potential between the half-reactions. The electrical potential difference between two points is expressed in volts (V).

FOLDABLES®
Incorporate information from this section into your Foldable.

The standard hydrogen electrode Long ago, chemists decided to measure the reduction potential of all electrodes against one electrode, the standard hydrogen electrode. The **standard hydrogen electrode** consists of a small sheet of platinum immersed in a hydrochloric acid (HCl) solution that has a hydrogen-ion concentration of 1M. Hydrogen gas (H_2), at a pressure of 1 atm is bubbled in and the temperature is maintained at 25°C, as shown in **Figure 5.** The potential, also called the standard reduction potential (E^0), of this standard hydrogen electrode is defined as 0.000 V. This electrode can act as an oxidation half-reaction or a reduction half-reaction, depending on the half-cell to which it is connected. The two possible reactions at the hydrogen electrode are the following.

$$2H^+(aq) + 2e^- \underset{\text{Oxidation}}{\overset{\text{Reduction}}{\rightleftharpoons}} H_2(g) \qquad E^0 = 0.000\text{ V}$$

Extension

Chemists Have students investigate several prominent chemists whose work encompassed electrochemistry. Have them research the time period and the contributions of one scientist and write a report to share with the class. Suggest students research Galvani, Volta, Dow, Edison, or Faraday. OL

Reinforcement

Spontaneous and Nonspontaneous Reactions Have students determine which pairs represent spontaneous and which represent nonspontaneous reactions.

a. $Ni + Zn^{2+}$ nonspontaneous
b. $Fe + Cu^{2+}$ spontaneous
c. $Fe^{2+} + Cu^{2+}$ nonspontaneous OL

FOLDABLES® Rubric

Visual Learning

Memory Aids Students are likely to become confused if they always attempt to relate oxidation and reduction with positive and negative electrodes in both electrolytic and galvanic cells. Remind students that reduction always occurs at the cathode and oxidation always occurs at the anode by remembering this diagram. BL OL

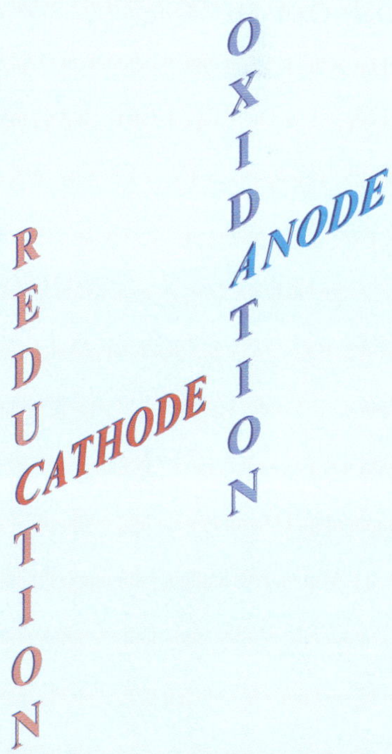

Visual Learning

Table 1 Explain to students how to use the table of standard reduction potentials, Table 1, to determine the following information.

a. the standard reduction potential for a given half-cell The standard reduction potential is found by reading the chart values for that half-cell.

b. the voltage for the voltaic cell Determine which substance is oxidized and which substance is reduced in the cell. The half-reaction that is more positive will proceed as a reduction. The voltaic cell's standard potential equals the difference. $E^0_{reduction} - E^0_{oxidation}$ OL

Apply Chemistry

Cleaning Silver from Shipwrecks Silver bars and coins that have been in wooden boxes in the ocean for long periods of time become coated with silver sulfide due to bacteria that convert sulfate ions to hydrogen sulfide gas.

When the coins are connected to the cathode of an electrolytic cell that has an unreactive anode and a sodium hydroxide electrolyte, the silver in silver sulfide is reduced to its original metallic state. Two cathodic reactions occur: $Ag_2S + 2e^- \rightarrow 2Ag + S^{2-}$ and $2H_2O + 2e^- \rightarrow H_2(g) + 2OH^-$. The anodic reaction is $2H_2O \rightarrow O_2 + 4H^+ + 4e^-$.

Half-cell potentials Over the years, chemists have measured and recorded the standard reduction potentials of many different half-cells. **Table 1** lists some common half-cell reactions in order of increasing reduction potential. The values in the table were obtained by measuring the potential when each half-cell was connected to a standard hydrogen half-cell. All of the half-reactions in **Table 1** are written as reductions. However, in any voltaic cell, which always contains two spontaneous half-reactions, the half-reaction with the lower reduction potential will proceed in the opposite direction and will be an oxidation reaction. In other words, the half-reaction that is more positive will proceed as a reduction and the half-reaction that is more negative will proceed as an oxidation.

The electrode being measured must also be under standard conditions, that is, immersed in a 1*M* solution of its ions at 25°C and 1 atm. The superscript zero in the notation E^0 is a shorthand way of indicating "measured under standard conditions."

Table 1 Standard Reduction Potentials

Half-Reaction	E^0 (V)	Half-Reaction	E^0 (V)
$Li^+ + e^- \rightleftarrows Li$	−3.0401	$Cu^{2+} + e^- \rightleftarrows Cu^+$	+0.153
$Ca^{2+} + 2e^- \rightleftarrows Ca$	−2.868	$Cu^{2+} + 2e^- \rightleftarrows Cu$	+0.3419
$Na^+ + e^- \rightleftarrows Na$	−2.71	$O_2 + 2H_2O + 4e^- \rightleftarrows 4OH^-$	+0.401
$Mg^{2+} + 2e^- \rightleftarrows Mg$	−2.372	$I_2 + 2e^- \rightleftarrows 2I^-$	+0.5355
$Be^{2+} + 2e^- \rightleftarrows Be$	−1.847	$Fe^{3+} + e^- \rightleftarrows Fe^{2+}$	+0.771
$Al^{3+} + 3e^- \rightleftarrows Al$	−1.662	$NO_3^- + 2H^+ + e^- \rightleftarrows NO_2 + H_2O$	+0.775
$Mn^{2+} + 2e^- \rightleftarrows Mn$	−1.185	$Hg_2^{2+} + 2e^- \rightleftarrows 2Hg$	+0.7973
$Cr^{2+} + 2e^- \rightleftarrows Cr$	−0.913	$Ag^+ + e^- \rightleftarrows Ag$	+0.7996
$2H_2O + 2e^- \rightleftarrows H_2 + 2OH^-$	−0.8277	$Hg^{2+} + 2e^- \rightleftarrows Hg$	+0.851
$Zn^{2+} + 2e^- \rightleftarrows Zn$	−0.7618	$2Hg^{2+} + 2e^- \rightleftarrows Hg_2^{2+}$	+0.920
$Cr^{3+} + 3e^- \rightleftarrows Cr$	−0.744	$NO_3^- + 4H^+ + 3e^- \rightleftarrows NO + 2H_2O$	+0.957
$S + 2e^- \rightleftarrows S^{2-}$	−0.47627	$Br_2(l) + 2e^- \rightleftarrows 2Br^-$	+1.066
$Fe^{2+} + 2e^- \rightleftarrows Fe$	−0.447	$Pt^{2+} + 2e^- \rightleftarrows Pt$	+1.18
$Cd^{2+} + 2e^- \rightleftarrows Cd$	−0.4030	$O_2 + 4H^+ + 4e^- \rightleftarrows 2H_2O$	+1.229
$PbI_2 + 2e^- \rightleftarrows Pb + 2I^-$	−0.365	$Cl_2 + 2e^- \rightleftarrows 2Cl^-$	+1.35827
$PbSO_4 + 2e^- \rightleftarrows Pb + SO_4^{2-}$	−0.3588	$Au^{3+} + 3e^- \rightleftarrows Au$	+1.498
$Co^{2+} + 2e^- \rightleftarrows Co$	−0.28	$MnO_4^- + 8H^+ + 5e^- \rightleftarrows Mn^{2+} + 4H_2O$	+1.507
$Ni^{2+} + 2e^- \rightleftarrows Ni$	−0.257	$Au^+ + e^- \rightleftarrows Au$	+1.692
$Sn^{2+} + 2e^- \rightleftarrows Sn$	−0.1375	$H_2O_2 + 2H^+ + 2e^- \rightleftarrows 2H_2O$	+1.776
$Pb^{2+} + 2e^- \rightleftarrows Pb$	−0.1262	$Co^{3+} + e^- \rightleftarrows Co^{2+}$	+1.92
$Fe^{3+} + 3e^- \rightleftarrows Fe$	−0.037	$S_2O_8^{2-} + 2e^- \rightleftarrows 2SO_4^{2-}$	+2.010
$2H^+ + 2e^- \rightleftarrows H_2$	**0.0000**	$F_2 + 2e^- \rightleftarrows 2F^-$	+2.866

CHEMISTRY JOURNAL

Plated and Preserved Ask students to describe in their journals how they would copperplate a pair of baby shoes or a special pair of athletic shoes. Have students include drawings in their descriptions. The shoes would be the cathode in an electrolytic cell. The anode would be copper. Once the circuit is complete, the copper ions in the solution would plate onto the surface of the shoes. OL

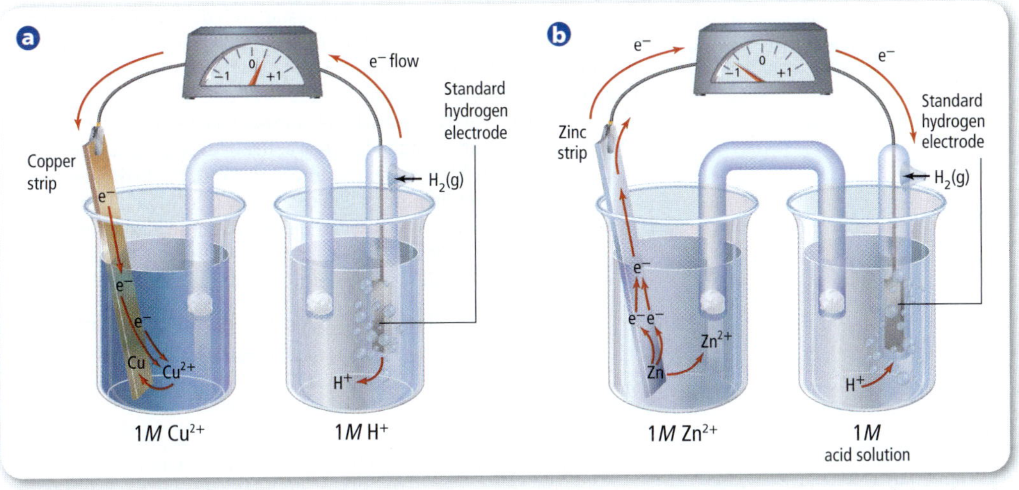

Figure 6 a. When a Cu^{2+} | Cu electrode is connected to the hydrogen electrode, electrons flow toward the copper strip and reduce Cu^{2+} ions to Cu atoms. The voltage of this reaction is +0.342 V. **b.** When a Zn | Zn^{2+} electrode is connected to the hydrogen electrode, electrons flow away from the zinc strip and zinc atoms are oxidized to Zn^{2+} ions. The voltage of this reaction is −0.762 V.

Determining electrochemical cell potentials You can use **Table 1** to calculate the electric potential of a voltaic cell consisting of a copper electrode and a zinc electrode under standard conditions. The first step is to determine the standard reduction potential for the copper half-cell (E^0_{Cu}). When the copper electrode is attached to a standard hydrogen electrode, as in **Figure 6a,** electrons flow from the hydrogen electrode to the copper electrode, and copper ions are reduced to copper metal. The E^0, measured by a voltmeter, is +0.342 V. The positive voltage indicates that Cu^{2+} ions at the copper electrode accept electrons more readily than do H^+ ions at the standard hydrogen electrode. Therefore, oxidation takes place at the hydrogen electrode and reduction takes place at the copper electrode. The oxidation and reduction half-cell reactions and the overall reaction are

$$H_2(g) \rightarrow 2H^+(aq) + 2e^- \text{ (oxidation half-cell reaction)}$$
$$Cu^{2+}(aq) + 2e^- \rightarrow Cu(s) \text{ (reduction half-cell reaction)}$$

$$H_2(g) + Cu^{2+}(aq) \rightarrow 2H^+(aq) + Cu(s) \text{ (overall redox reaction)}$$

This reaction can be written in a form called cell notation.

Reactant Product Reactant Product

$$H_2 \mid H^+ \parallel Cu^{2+} \mid Cu \qquad E^0_{Cu} = +0.342 \text{ V}$$

Oxidation half-cell Reduction half-cell

The two participants in the oxidation reaction are written first and in the order they appear in the oxidation half-reaction—reactant | product. They are followed by a double vertical line (||) representing the wire and salt bridge connecting the half-cells. Then, the two participants in the reduction reaction are written in the same reactant | product order. Note that for positive values of E^0, it is customary to place a plus sign before the voltage.

DIFFERENTIATED INSTRUCTION

Struggling Learners Have students draw a two-column table, one column labeled *Anode* and the other labeled *Cathode.* Have them place the following terms or descriptions in the appropriate column: oxidation, oxidizing agent, loses electrons, gains electrons, reduction, reducing agent. Anode: oxidation, reducing agent, loses e^-; Cathode: reduction, oxidizing agent, gains e^-. BL

ASSESSMENT

Performance Group students in teams of two or three people and ask them to copperplate a nail. Approve lab safety forms before work begins. Provide the following materials: copper(II) sulfate, an iron nail, copper wire, battery, alligator clips, and beakers. When the groups have assembled the plating apparatus, have students verbally explain their setup using proper terminology, and then have them complete the task. Have them record observations, write appropriate half-reactions, and turn in the copper-plated nail along with the written report. OL EL COOP LEARN

Quick Demo

Detect a Reaction In a beaker containing 0.1M $CuSO_4$, place a strip of zinc metal (approximately 10 cm by 5 cm). Have the students observe the reaction. In a second beaker containing 0.1M $ZnSO_4$, place a strip of copper metal (use the same size strip). Point out that there is no apparent reaction. Ask students why zinc reacts with the copper ion but copper does not react with the zinc ion. The zinc is a more reactive metal and loses electrons more easily than copper. Next, put the zinc electrode in the zinc sulfate solution and the copper electrode in the copper sulfate solution and ask students to determine whether or not a reaction is occurring. No Ask students what is needed to complete the circuit so that a reaction can occur. Students might identify a copper wire connecting the two electrodes and a salt bridge immersed in the two different solutions. OL

ASSESSMENT

Knowledge Have students determine which species in each of the following pairs is the reactant and which is the product. OL

a. $Zn|Zn^{2+} || Cu^{2+}|Cu$
b. $Zn|Zn^{2+} || Pb^{2+}|Pb$
c. $Pb|Pb^{2+} || Cu^{2+}|Cu$
d. $Cu|Cu^{2+} || Ag^{+}|Ag$

Reactants:
a. Zn, Cu^{2+}
b. Zn, Pb^{2+}
c. Pb, Cu^{2+}
d. Cu, Ag^{+}

☑ **GRAPH CHECK** Zinc

ASSESSMENT

Knowledge Have students predict the results of the Gerber Cell demonstration (see below) using different metals to see how those metals compare with copper and magnesium. OL

Content Background

Point out to students that the subscript shorthand used in the Formula for Cell Potential is written in the form of the reduction equations in Table 1. In Example Problem 1, $E^0_{cell} = E^0_{reduction} - E^0_{oxidation}$, $E^0_{cell} = E^0_{I_2|I} - E^0_{Fe^{2+}|Fe}$ where the equations for standard reduction potentials of these reactions from Table 1 are:

$I_2 + 2e^- \rightleftharpoons 2I^-$ and

$Fe^{2+} + 2e^- \rightleftharpoons Fe$.

Tell students this should not be confused with the short hand cell notation, for example $Zn|Zn^{2+}||H^+|H_2$ where the oxidation reaction is written first followed by the reduction reaction with the reactants written first for each reaction.

Explore **electrochemical cells.**

Virtual Investigations

The next step is to determine the standard reduction potential for the zinc half-cell (E^0_{Zn}). When the zinc electrode is measured against the standard hydrogen electrode under standard conditions, as in **Figure 6b,** electrons flow from the zinc electrode to the hydrogen electrode. The E^0 of the zinc half-cell, measured by a voltmeter, is –0.762 V. This means that the H^+ ions at the hydrogen electrode accept electrons more readily than do the zinc ions. Thus, the hydrogen ions have a higher reduction potential than the zinc ions. Recall that the hydrogen electrode is assigned a zero potential, so the reduction potential of the zinc electrode must have a negative value. The two half-cell reactions and the overall reaction are written as follows.

$$Zn(s) \rightarrow Zn^{2+}(aq) + 2e^- \text{ (oxidation half-cell reaction)}$$
$$2H^+(aq) + 2e^- \rightarrow H_2(g) \text{ (reduction half-cell reaction)}$$
$$Zn(s) + 2H^+(aq) \rightarrow Zn^{2+}(aq) + H_2(g) \text{ (overall redox cell reaction)}$$

This reaction can be written in the following cell notation.

Reactant Product Reactant Product

$$Zn \mid Zn^{2+} \mid\mid H^+ \mid H_2 \qquad E^0_{Zn} = -0.762 \text{ V}$$

Oxidation half-cell Reduction half-cell

The final step in calculating electrochemical cell potential is to combine the copper and zinc half-cells as a voltaic cell. This means calculating the voltaic cell's standard potential using the following formula.

Formula for Cell Potential

$$E^0_{cell} = E^0_{reduction} - E^0_{oxidation}$$

E^0_{cell} represents the overall standard cell potential.
$E^0_{reduction}$ represents the standard half-cell potential for the reduction.
$E^0_{oxidation}$ represents the standard half-cell potential for the oxidation.

The standard potential of a cell is the standard potential of the reduction half-cell minus the standard potential of the oxidation half-cell.

Because reduction occurs at the copper electrode and oxidation occurs at the zinc electrode, the E^0 values are substituted as follows.

$$\begin{aligned} E^0_{cell} &= E^0_{Cu^{2+}|Cu} - E^0_{Zn^{2+}|Zn} \\ &= +0.342 \text{ V} - (-0.762 \text{ V}) \\ &= +1.104 \text{ V} \end{aligned}$$

Notice that the negative sign in the formula automatically changes the sign of the oxidation half-reaction, so you do not reverse the sign of the standard reduction potentials listed in **Table 1** when they are used for the oxidation half-reaction.

The graph in **Figure 7** shows how the zinc half-cell with the lower reduction potential and the copper half-cell with the higher reduction potential are related.

■ **Figure 7** This simple graph illustrates how the overall cell potential is derived from the difference in reduction potential of two electrodes. Compare it to a number line. The potential difference between the zinc and copper electrodes is +1.104 V.

☑ **GRAPH CHECK**
Identify the metal, copper or zinc, that is easier to oxidize than hydrogen.

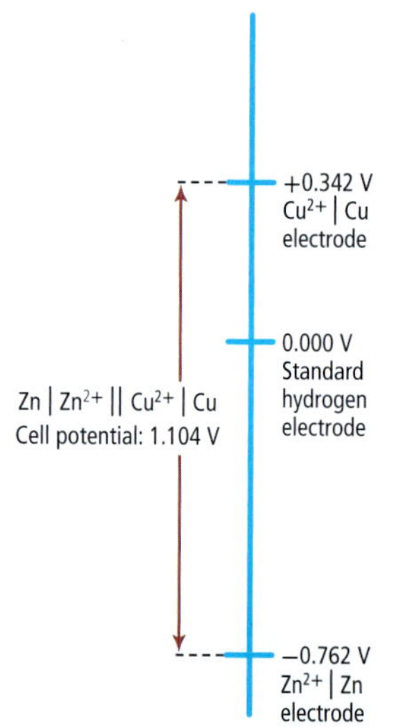

Demonstration

Gerber Cells

Purpose
to show that a voltaic cell produces a current

Materials
6 oz. baby food jar, a 50 mL beaker, dialysis tubing (2-cm by 15-cm), large stopper, 10-cm piece of magnesium ribbon, 10-cm piece of copper wire, 0.5*M* copper(II) sulfate (100 mL), 0.5*M* sodium sulfate (100 mL), 6*M* hydrochloric acid (10 mL), 6*M* nitric acid (10 mL), voltmeter, 2 connecting wires with alligator clips or voltage probe.

Safety Precautions

WARNING: *Acids are corrosive.*

Disposal Neutralize the acid with sodium bicarbonate and rinse down a drain with large quantities of water. Dispose of other chemicals and metal strips according to accepted disposal procedures.

EXAMPLE Problem 1

CALCULATE A CELL POTENTIAL The following reduction half-reactions represent the half-cells of a voltaic cell.

$$I_2(s) + 2e^- \rightarrow 2I^-(aq)$$
$$Fe^{2+}(aq) + 2e^- \rightarrow Fe(s)$$

Determine the overall cell reaction and the standard cell potential. Describe the cell using cell notation.

1 ANALYZE THE PROBLEM

You are given the half-cell equations and can find standard reduction potentials in **Table 1.** The half-reaction with the lower reduction potential will be an oxidation. With this information, you can write the overall cell reaction, calculate the standard cell potential, and describe the cell in cell notation.

Known

Standard reduction potentials for the half-cells

$E^0_{cell} = E^0_{reduction} - E^0_{oxidation}$

Unknown

overall cell reaction = ?

E^0_{cell} = ?

cell notation = ?

2 SOLVE FOR THE UNKNOWN

Find the standard reduction potentials of each half-reaction in **Table 1.**

$I_2(s) + 2e^- \rightarrow 2I^-(aq)$ $\quad E^0_{I_2|I^-} = +0.536$ V

$Fe^{2+}(aq) + 2e^- \rightarrow Fe(s)$ $\quad E^0_{Fe^{2+}|Fe} = -0.447$ V

The reduction of iodine has the higher reduction potential, so this half-reaction proceeds in the forward direction as a reduction. The iron half-reaction proceeds in the reverse direction as an oxidation.

$I_2(s) + 2e^- \rightarrow 2I^-(aq)$ (reduction half-cell reaction)

$Fe(s) \rightarrow Fe^{2+}(aq) + 2e^-$ (oxidation half-cell reaction) — **Rewrite the iron half-reaction in the correct direction.**

$I_2(s) + Fe(s) \rightarrow Fe^{2+}(aq) + 2I^-(aq)$ — **Add the two equations.**

The overall cell reaction is $I_2(s) + Fe(s) \rightarrow Fe^{2+}(aq) + 2I^-(aq)$.

Calculate the standard cell potential.

$E^0_{cell} = E^0_{reduction} - E^0_{oxidation}$ — **State the formula for cell potential.**

$E^0_{cell} = E^0_{I_2|I^-} - E^0_{Fe^{2+}|Fe}$ — **Substitute $E^0_{I_2|I^-}$ and $E^0_{Fe^{2+}|Fe}$ in the generic equation.**

$E^0_{cell} = +0.536\text{ V} - (-0.447\text{ V})$ — **Substitute $E^0_{I_2|I^-} = +0.536$ V and $E^0_{Fe^{2+}|Fe} = -0.447$ V.**

$E^0_{cell} = +0.983$ **V**

Describe the cell using cell notation.

$Fe \mid Fe^{2+}$ — **First, write the oxidation half-reaction using cell notation: reactant then product.**

$Fe \mid Fe^{2+} \parallel I_2 \mid I^-$ — **Next, write the reduction half-reaction to the right. Separate the half-cells by a double vertical line.**

Cell notation: $Fe \mid Fe^{2+} \parallel I_2 \mid I^-$

3 EVALUATE THE ANSWER

The calculated potential is reasonable given the potentials of the half-cells. E^0 is reported to the correct number of significant figures.

Find help with **calculating potentials.**

Personal Tutor

IN-CLASS Example

Question A voltaic cell is composed of a salt bridge, a silver electrode and a magnesium electrode. The reaction taking place is $2Ag^+ + Mg^0 \rightarrow 2Ag^0 + Mg^{2+}$: Draw the voltaic cell, identifying the anode, cathode and salt bridge. Identify the reaction taking place at the anode and the reaction taking place at the cathode. Determine the standard cell potential.

Answer The oxidation reaction takes place at the anode and the reduction reaction takes place at the cathode. Silver ions gain electrons and are reduced. Magnesium loses electrons and is oxidized.

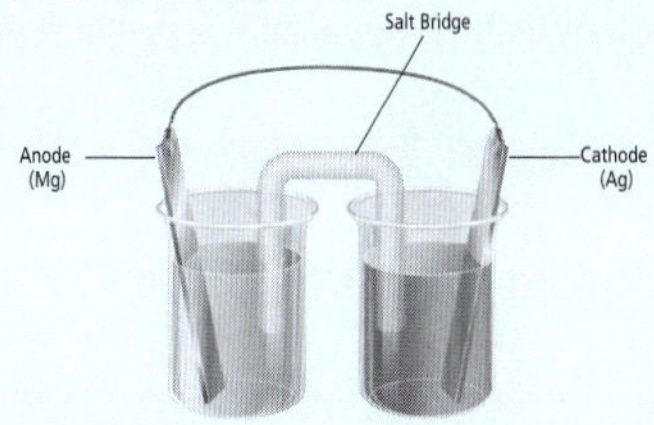

Cell potential = E^0 reduction − E^0 oxidation = 0.7996 − (−2.372) = 3.172 V

Procedure

Refer to Figure 26 for a visual aid. Clean the Cu wire with 6*M* HNO_3 and the Mg ribbon with 6*M* HCl and rinse with water. Fill the jar 2/3 full of sodium sulfate solution. Wet the dialysis tubing and knot tie one end. Fill the tube with copper(II) sulfate solution, the height of the solution slightly greater than the height of the solution in the jar. Place the copper wire in the tubing. Suspend the magnesium ribbon and dialysis tubing in the beaker, holding them in place with a large stopper. Place an alligator clip on each of the metal strips and connect to the voltmeter.

Results

A voltage will register when wires are connected. Oxidation will take place at the magnesium strip, which is the anode. $Mg \rightarrow Mg^{2+} + 2e^-$ Reduction will take place at the copper strip, which is the cathode. $Cu^{2+} + 2e^- \rightarrow Cu$

Analysis

1. Predict which metal is the anode and which metal is the cathode. Mg is the anode; Cu is the cathode.
2. Why are both copper and copper sulfate needed? Cu^{2+} ions are reduced onto copper wire, the reduction site.
3. Why is a source of magnesium ions not needed? Mg is oxidized, producing Mg ions. It is not necessary for it to start with Mg ions.

ASSESSMENT

Performance Using Table 1, have students design an electrochemical cell that would have a voltage output that you specify. Have students explain the reasons for their design and why they selected their solutions and electrodes. **OL** **EL**

PRACTICE Problems

Have students refer to the Selected Solutions appendix for complete solutions to odd-numbered problems. The complete solutions for all problems can be found in the Solutions Manual.

1. $Pt^{2+}(aq) + Sn(s) \longrightarrow Pt(s) + Sn^{2+}(aq)$
   $E^0_{cell} = +1.32\ V$
   $Sn|Sn^{2+}||Pt^{2+}|Pt$
2. $3Co^{2+}(aq) + 2Cr(s) \longrightarrow 3Co(s) + 2Cr^{3+}(aq)$
   $E^0_{cell} = +0.46\ V$
   $Cr|Cr^{3+}||Co^{2+}|Co$
3. $Hg^{2+}(aq) + Cr(s) \longrightarrow Hg(l) + Cr^{2+}(aq)$
   $E^0_{cell} = +1.764\ V$
   $Cr|Cr^{2+}||Hg^{2+}|Hg$
4. $4NO_3^-(aq) + 12OH^-(aq) + 16H^+(aq) \longrightarrow 3O_2(g) + 4NO(g) + 14H_2O(l)$
   $E^0_{cell} = +0.556\ V$
   $OH^-|O_2||NO_3^-|NO$
5. $E^0_{cell} = +0.4794\ V$
   $E^0_{cell} > 0$ spontaneous
6. $E^0_{cell} = +2.246\ V$
   $E^0_{cell} > 0$ spontaneous
7. $E^0_{cell} = -0.587\ V$
   $E^0_{cell} < 0$ not spontaneous
8. $E^0_{cell} = -2.29\ V$
   $E^0_{cell} < 0$ not spontaneous
9. $E^0_{cell} = +2.582\ V$
   Reaction is spontaneous.

READING CHECK positive

ChemLAB

The ChemLab located at the end of the chapter can be used at this point in the lesson.

PRACTICE Problems

Do additional problems. **Online Practice**

For each of these pairs of half-reactions, write the balanced equation for the overall cell reaction, and calculate the standard cell potential. Describe the reaction using cell notation. Refer to the chapter on redox reactions to review writing and balancing redox equations.

1. $Pt^{2+}(aq) + 2e^- \rightarrow Pt(s)$ and $Sn^{2+}(aq) + 2e^- \rightarrow Sn(s)$
2. $Co^{2+}(aq) + 2e^- \rightarrow Co(s)$ and $Cr^{3+}(aq) + 3e^- \rightarrow Cr(s)$
3. $Hg^{2+}(aq) + 2e^- \rightarrow Hg(l)$ and $Cr^{2+}(aq) + 2e^- \rightarrow Cr(s)$
4. **Challenge** Write the balanced equation for the cell reaction and calculate the standard cell potential for the reaction that occurs when these half-cells are connected. Describe the reaction using cell notation.
   $NO_3^- + 4H^+ + 3e^- \rightarrow NO + 2H_2O$
   $O_2 + 2H_2O + 4e^- \rightarrow 4OH^-$

Using Standard Reduction Potentials

The Example Problems showed you how to use the data from **Table 1** to calculate the standard potential (voltage) of voltaic cells. Another important use of standard reduction potentials is to determine if a proposed reaction under standard conditions will be spontaneous. How can standard reduction potentials indicate spontaneity? Electrons in a voltaic cell always flow from the half-cell with the lower standard reduction potential to the half-cell with the higher standard reduction potential, giving a positive cell voltage. To predict whether any proposed redox reaction will occur spontaneously, simply write the process in the form of half-reactions and look up the reduction potential of each. Use the values to calculate the potential of a voltaic cell operating with these two half-cell reactions. If the calculated potential is positive, the reaction is spontaneous. If the value is negative, the reaction is not spontaneous. However, the reverse of a nonspontaneous reaction will occur because it will have a positive cell voltage, which means that the reverse reaction is spontaneous.

READING CHECK **Identify** the sign of the potential of a redox reaction that occurs spontaneously.

PRACTICE Problems

Do additional problems. **Online Practice**

Calculate the cell potential to determine if each of the following balanced redox reactions is spontaneous as written. Use Table 1 to help you determine the correct half-reactions.

5. $Sn(s) + Cu^{2+}(aq) \rightarrow Sn^{2+}(aq) + Cu(s)$
6. $Mg(s) + Pb^{2+}(aq) \rightarrow Pb(s) + Mg^{2+}(aq)$
7. $2Mn^{2+}(aq) + 8H_2O(l) + 10Hg^{2+}(aq) \rightarrow 2MnO_4^-(aq) + 16H^+(aq) + 5Hg_2^{2+}(aq)$
8. $2SO_4^{2-}(aq) + Co^{2+}(aq) \rightarrow Co(s) + S_2O_8^{2-}(aq)$
9. **Challenge** Using **Table 1,** write the equation and determine the cell voltage (E^0) for the following cell. Is the reaction spontaneous?
   $Al \mid Al^{3+} \parallel Hg^{2+} \mid Hg_2^{2+}$

PROBLEM-SOLVING STRATEGY

Determining Cell Potentials

The five steps that follow summarize the procedure for calculating the potential of a voltaic cell in which a spontaneous redox reaction occurs. Suppose you must write the equation for and calculate the potential of a cell made up of these half-reactions:

$$Mn^{2+} + 2e^- \rightarrow Mn \text{ and } Fe^{3+} + 3e^- \rightarrow Fe$$

A table of reduction potentials, such as **Table 1,** is all that is required.

1. Find the two half-reactions on **Table 1.**
2. Compare the two half-cell potentials. The half-cell with the higher reduction potential is the cell in which reduction will occur. Oxidation will occur in the half-cell with the lower reduction potential.

   $Fe^{3+} + 3e^- \rightarrow Fe \quad E^0 = -0.037$ V (reduction)
   $Mn^{2+} + 2e^- \rightarrow Mn \quad E^0 = -1.185$ V (oxidation)
3. Write the equation for the reduction as it is in **Table 1.** Write the equation for the oxidation in the opposite direction.

   $Fe^{3+} + 3e^- \rightarrow Fe \qquad Mn \rightarrow Mn^{2+} + 2e^-$
4. Balance the electrons in the two half-cell equations by multiplying each by a factor. Add the equations.

   Multiply by 2. $\quad 2Fe^{3+} + 6e^- \rightarrow 2Fe$
   Multiply by 3. $\quad 3Mn \rightarrow 3Mn^{2+} + 6e^-$

   Add the equations. $\quad 2Fe^{3+} + 3Mn \rightarrow 2Fe + 3Mn^{2+}$
5. Equalizing the electrons lost and gained does not affect the E^0 for the overall reaction because E^0 is an intensive property. Use the formula: $E^0_{cell} = E^0_{reduction} - E^0_{oxidation}$ to obtain the cell potential.

   $E^0_{cell} = E^0_{Fe^{3+}|Fe} - E^0_{Mn^{2+}|Mn} = -0.037\text{ V} - (-1.185\text{ V})$
   $= +1.148$ V

Apply the Strategy

Determine E^0_{cell} for the spontaneous redox reaction that occurs between magnesium and nickel.

SECTION 1 REVIEW

Section Self-Check

Section Summary

- In a voltaic cell, oxidation and reduction take place at electrodes separated from each other.
- The standard potential of a half-cell reaction is its voltage when paired with a standard hydrogen electrode under standard conditions.
- The reduction potential of a half-cell is negative if it undergoes oxidation when connected to a standard hydrogen electrode. The reduction potential of a half-cell is positive if it undergoes reduction when connected to a standard hydrogen electrode.
- The standard potential of a voltaic cell is the difference between the standard reduction potentials of the half-cell reactions.

10. **MAIN**IDEA **Describe** the conditions under which a redox reaction causes an electric current to flow through a wire.
11. **Identify** the components of a voltaic cell. Explain the role of each component in the operation of the cell.
12. **Write** the balanced equation for the spontaneous cell reaction that occurs in a cell with these reduction half-reactions.
    a. $Ag^+(aq) + e^- \rightarrow Ag(s)$ and $Ni^{2+}(aq) + 2e^- \rightarrow Ni(s)$
    b. $Mg^{2+}(aq) + 2e^- \rightarrow Mg(s)$ and $2H^+(aq) + 2e^- \rightarrow H_2(g)$
    c. $Sn^{2+}(aq) + 2e^- \rightarrow Sn(s)$ and $Fe^{3+}(aq) + 3e^- \rightarrow Fe(s)$
    d. $PbI_2(s) + 2e^- \rightarrow Pb(s) + 2I^-(aq)$ and $Pt^{2+}(aq) + 2e^- \rightarrow Pt(s)$
13. **Determine** the standard potential for electrochemical cells in which each equation represents the overall cell reaction. Identify the reactions as spontaneous or nonspontaneous as written.
    a. $2Al^{3+}(aq) + 3Cu(s) \rightarrow 3Cu^{2+}(aq) + 2Al(s)$
    b. $Hg^{2+}(aq) + 2Cu^+(aq) \rightarrow 2Cu^{2+}(aq) + Hg(l)$
    c. $Cd(s) + 2NO_3^-(aq) + 4H^+(aq) \rightarrow Cd^{2+}(aq) + 2NO_2(g) + 2H_2O(l)$
14. **Design** a concept map for Section 1, starting with the term *electrochemical cell.* Incorporate all the new vocabulary terms in your map.

PROBLEM-SOLVING STRATEGIES

Apply the Strategy

$Mg^{2+} + 2e^- \rightarrow Mg$
$E^0 = -2.372$ V (Oxidation)
$Ni^{2+} + 2e^- \rightarrow Ni$
$E^0 = -0.257$ V (Reduction)
$Mg \rightarrow Mg^{2+} + 2e^-$
$Ni^{2+} + 2e^- \rightarrow Ni$
Overall equation:
$Mg + Ni^{2+} \rightarrow Mg^{2+} + Ni$
$E^0_{cell} = -0.257\text{ V} - (-2.372\text{ V}) = +2.115$ V

3 Assess

Check for Understanding

Provide students with an electrochemical diagram, and have them label the half-cell, anode, cathode, and use an arrow to indicate the direction of the reaction. OL EL

Reteach

Allow students to work in small groups. Have them write on index cards any questions that they still have related to Section 1. Label the file cards with a group number, collect the questions, shuffle the cards, and redistribute them to different groups. Ask students to attempt to answer the questions they were dealt. Near the end of the class period, return the questions and answers to the originating group so that they can review the answers. OL COOP LEARN

Extension

Have students explain in their own words the difference between standard reduction potentials and standard hydrogen electrodes. OL

SECTION 1 REVIEW

10. An electrochemical cell in which an oxidation half-reaction and a reduction half-reaction are connected by a salt bridge results in a flow of electrons (electric current) through a conducting wire.
11. A voltaic cell consists of an anode, a cathode, a salt bridge, and a connecting wire between the two electrodes. Oxidation takes place at the anode, reduction takes place at the cathode, the salt bridge allows movement of ions from one solution to the other, and the wire allows the passage of electrons from the anode to the cathode.
12. **a.** $2Ag^+ + Ni \rightarrow 2Ag + Ni^{2+}$
    b. $Mg + 2H^+ \rightarrow Mg^{2+} + H_2$
    c. $2Fe^{3+}(aq) + 3Sn(s) \rightarrow 2Fe(s) + 3Sn^{2+}(aq)$
    d. $Pb(s) + 2I^-(aq) + Pt^{2+}(aq) \rightarrow PbI_2(s) + Pt(s)$
13. **a.** $E^0_{cell} = -2.004$ V, nonspontaneous
    b. $E^0_{cell} = +0.698$ V, spontaneous
    c. $E^0_{cell} = +1.178$ V, spontaneous
14. Concept Maps will vary. Refer to the Solutions Manual.

SECTION 2

1 Focus

MAINIDEA

Battery Uses Bring a variety of batteries to class and ask students in which function each might be used. Answers will vary depending on the batteries. Many batteries might have more than one use, while others are used in only one way, such as a watch. Ask students what the batteries all have in common. They convert chemical energy into electrical energy. Ask students to look closely at a battery and notice the negative pole and the positive pole. Tell them this indicates the anode and cathode of the voltaic cell that makes up the battery. OL

2 Teach

Concept Development

Battery Applications Have students list all of the different kinds of batteries that they find in their households and their applications. Such a list will include: common dry cell, alkaline batteries (another type of dry cell), lithium batteries, lithium-ion batteries, NiCd batteries, NiMH batteries, and lead-acid batteries. Students can also include on their lists disposable and rechargeable batteries, AA, AAA, C, D, and 9-volt and 6-volt dry cell batteries. Have students compare the cost of the different types of batteries. OL

SECTION 2

Batteries

MAINIDEA Batteries are voltaic cells that use spontaneous reactions to provide energy for a variety of purposes.

Essential Questions

- What are the structure, composition, and operation of the typical carbon-zinc dry-cell battery?
- What is the difference between primary and secondary batteries and what are two examples of each type?
- What is the structure of the hydrogen-oxygen fuel cell and how does it operate?
- What is the process of corrosion of iron and what are methods to prevent corrosion?

Review Vocabulary

reversible reaction: a reaction that can take place in both the forward and reverse directions

New Vocabulary

battery
dry cell
primary battery
secondary battery
fuel cell
corrosion
galvanization

CHEM 4 YOU Take a moment to list some of the places where you know batteries are used. Your list might include flashlights, cars, cell phones, radios, calculators, watches, and toys, among many others. Are the batteries in these devices all the same?

Dry Cells

Some of the spontaneous cell reactions you have been reading about provide the energy of the batteries that you use every day. A **battery** is one or more voltaic cells in a single package that generates electric current. From the time of its invention in the 1860s until recently, the most commonly used voltaic cell was the zinc-carbon dry cell, shown in **Figure 8.**

Zinc-carbon dry cells A **dry cell** is an electrochemical cell in which the electrolyte is a moist paste. The paste in a zinc-carbon dry cell consists of zinc chloride, manganese(IV) oxide, ammonium chloride, and a small amount of water inside a zinc case. The zinc shell is the cell's anode, where the oxidation of zinc metal occurs according to the following equation.

$$Zn(s) \rightarrow Zn^{2+}(aq) + 2e^-$$

A carbon (graphite) rod in the center of the dry cell serves as the cathode, but the reduction half-cell reaction takes place in the paste. The carbon rod in this type of dry cell is called an inactive cathode because it is made of a material that does not participate in the redox reaction. However, the inactive electrode has the important purpose of conducting electrons. The reduction half-cell reaction for this dry cell is as follows.

$$2NH_4^+(aq) + 2MnO_2(s) + 2e^- \rightarrow Mn_2O_3(s) + 2NH_3(aq) + H_2O(l)$$

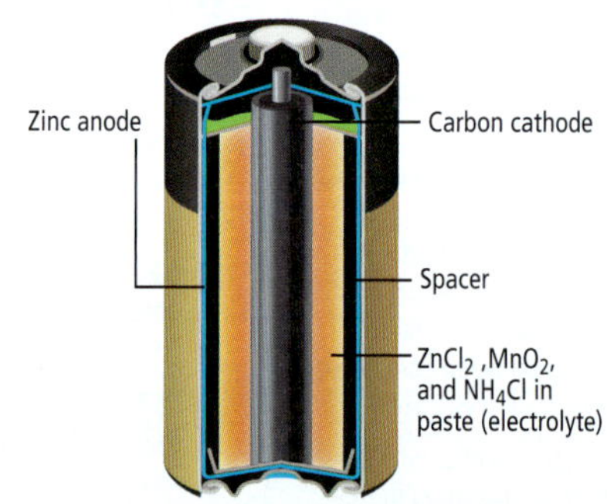

Figure 8 The so-called dry cell contains a moist paste in which the cathode half-reaction takes place. In the zinc-carbon dry cell, the zinc case acts as the anode.

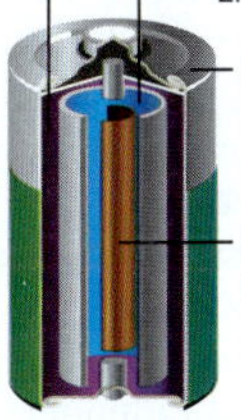

Alkaline battery

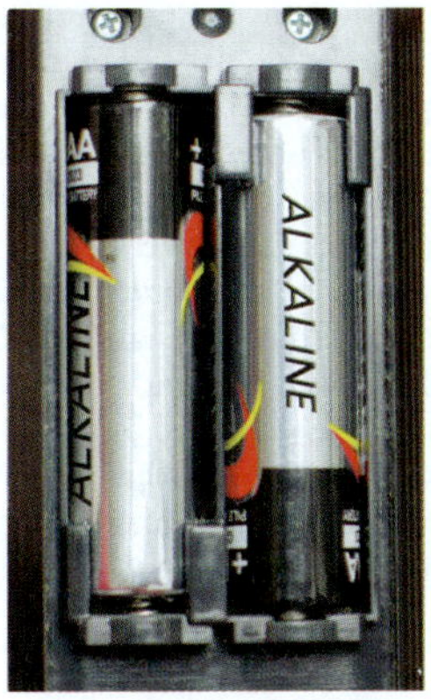

The alkaline battery uses powdered zinc in an alkaline paste as the anode.

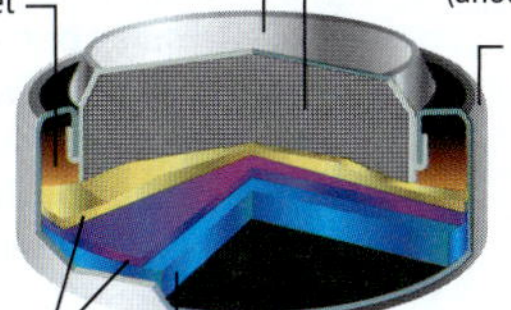

Silver button battery

Although it looks different, the silver battery is much like the alkaline battery. The silver battery uses silver oxide (Ag_2O) in the cathode instead of manganese(IV) oxide (MnO_2).

■ **Figure 9** Alkaline batteries are more efficient than zinc-carbon dry cells and are useful when smaller batteries are needed. Silver button batteries are even smaller, making them well suited to devices such as watches.

In the zinc-carbon dry cell, a spacer made of a porous material and damp from the liquid in the paste separates the paste from the zinc anode. The spacer acts as a salt bridge to allow the transfer of ions, much like the model voltaic cell you studied in Section 1. The zinc-carbon dry cell produces a voltage of 1.5 V until the reduction product, ammonia, comes out of its aqueous solution as a gas. At that point, the voltage drops to a level that makes the battery useless.

Alkaline batteries A more efficient alkaline dry cell, shown in **Figure 9,** is replacing the standard zinc-carbon dry cell in many applications. In the alkaline cell, the zinc is in a powdered form, which provides more surface area for reaction. The zinc is mixed in a paste with potassium hydroxide, a strong base, and the paste is contained in a steel case. The cathode mixture is manganese(IV) oxide, also mixed with potassium hydroxide. The anode half-cell reaction is as follows.

$$Zn(s) + 2OH^-(aq) \rightarrow ZnO(s) + H_2O(l) + 2e^-$$

The cathode half-cell reaction is as follows.

$$MnO_2(s) + 2H_2O(s) + 2e^- \rightarrow Mn(OH)_2(s) + 2OH^-(aq)$$

Alkaline batteries do not need the carbon rod cathode, so they can be made smaller and are more useful in small devices.

Silver batteries The silver battery shown in **Figure 9** is even smaller and is used to power devices such as hearing aids, watches, and cameras. The silver battery uses the same anode half-reaction as the alkaline battery, with the following cathode half-reaction.

$$Ag_2O(s) + H_2O(l) + 2e^- \rightarrow 2Ag(s) + 2OH^-(aq)$$

☑ READING CHECK **Identity** the half-reaction that occurs in both alkaline and silver batteries.

Quick Demo

Inside a Battery Using a screwdriver, dismantle a 9-volt battery to use as a display so that students can see the inside makeup of a battery. Point out the carbon and zinc electrodes. Have students count the cells in a 9-volt battery and ask them why there are more than one. **WARNING:** ***Do not get the paste wet. Intense heat will result.*** Students should realize that to produce a particular voltage, the battery must contain the number of electrochemical cells that, when multiplied by the voltage per cell, equals the total voltage. A 9-volt battery has six cells, each producing 1.5 volts, totaling 9 volts. **OL**

ASSESSMENT

Knowledge Ask students to determine the voltage output of a flashlight containing four D-size batteries. Each D battery produces 1.5 V; therefore, four D batteries produce 6 volts. **OL**

☑ READING CHECK
$Zn(s) + 2OH^-(aq) \rightarrow ZnO(s) + H_2O(l) + 2e^-$

DIFFERENTIATED INSTRUCTION

Hearing Impaired Many brand-name battery packages give an internet site address for consumer and product information. Have students access one or more of these sites and present the information these sites report about the chemistry of the batteries and performance of their products. **OL** **EL**

Extension

Battery Disposal Remind students that many batteries contain toxic metals such as cadmium, mercury, nickel, and lead. Ask them to research some of the environmental problems that can occur if these metals are disposed of improperly. Have students investigate how car batteries are disposed of in their own community. Ask them to find out if other types of batteries require special disposal in their community according to local regulations and policies. OL

Reinforcement

Test Battery Charge Ask students how they can tell when a battery is discharged. Investigate packaging that allows the consumer to test the amount of charge left in a battery. Help students discover how this test method works. Have a few batteries on hand for students to test. OL EL

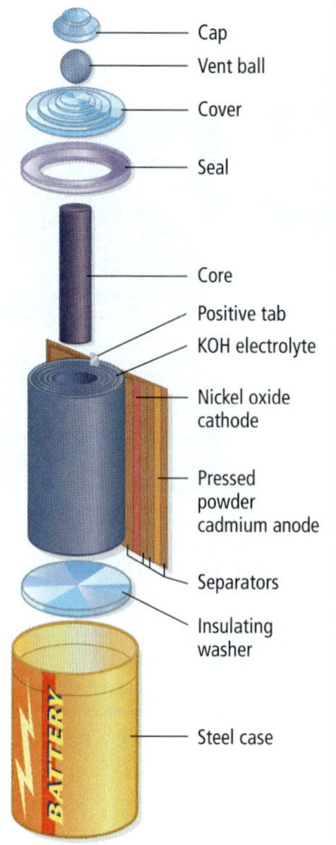

■ **Figure 10** Cordless tools and phones are often powered by rechargeable batteries, such as the NiCad battery. The battery pack is recharged by plugging it into an electric outlet, which supplies the power to drive the nonspontaneous recharge reaction.

Primary and secondary batteries Batteries are divided into two types, depending on their chemical processes. The zinc-carbon, alkaline-zinc, and silver cells are classified as primary batteries. **Primary batteries** produce electric energy by means of redox reactions that are not easily reversed. These cells deliver current until the reactants are gone, and then the battery must be discarded. Other batteries, called **secondary batteries,** depend on reversible redox reactions, so they are rechargeable. A car battery and the battery in a laptop computer are examples of secondary batteries, which are sometimes called storage batteries.

The storage batteries that power devices such as cordless drills and screwdrivers, shavers, and camcorders are usually nickel-cadmium rechargeable batteries, sometimes called NiCad batteries, as shown in **Figure 10.** For maximum efficiency, the anode and cathode are long, thin ribbons of material separated by a layer through which ions can pass. The ribbons are wound into a tight coil and packaged in a steel case. The anode reaction that occurs when the battery is used to generate electric current is the oxidation of cadmium in the presence of a base.

$$Cd(s) + 2OH^-(aq) \rightarrow Cd(OH)_2(s) + 2e^-$$

The cathode reaction is the reduction of nickel from the +3 to the +2 oxidation state.

$$NiO(OH)(s) + H_2O(l) + e^- \rightarrow Ni(OH)_2(s) + OH^-(aq)$$

When the battery is recharged, these reactions are reversed.

Lead-Acid Storage Battery

Another common storage battery is the lead-acid battery used in automobiles. Most auto batteries contain six cells that generate about 2 V each for a total output of 12 V. The anode of each cell consists of two or more grids of porous lead, and the cathode consists of lead grids filled with lead(IV) oxide. This type of battery should probably be called a lead-lead(IV) oxide battery, but the term *lead-acid* is commonly used because the battery's electrolyte is a solution of sulfuric acid. The lead-acid battery is not a dry cell.

The following equation represents the oxidation half-cell reaction at the anode where lead is oxidized from the zero oxidation state to the +2 oxidation state in $PbSO_4$.

$$Pb(s) + SO_4^{2-}(aq) \rightarrow PbSO_4(s) + 2e^-$$

The reduction of lead from the +4 to the +2 oxidation state takes place at the cathode. The half-cell reaction for the cathode is

$$PbO_2(s) + 4H^+(aq) + SO_4^{2-}(aq) + 2e^- \rightarrow PbSO_4(s) + 2H_2O(l).$$

The overall reaction is

$$Pb(s) + PbO_2(s) + 4H^+(aq) + 2SO_4^{2-}(aq) \rightarrow 2PbSO_4(s) + 2H_2O(l).$$

By looking at the half-cell reactions, you can see that lead(II) sulfate ($PbSO_4$) is the reaction product in both oxidation and reduction. Also, Pb, PbO_2, and $PbSO_4$ are solid substances, so they stay in place where they are formed. Thus, whether the battery is discharging or charging, the reactants are available where they are needed.

CHEMISTRY PROJECT

Fuel Cells Ask students to research and list some of the current applications of fuel cells. Have them research future applications of fuel cells, their limitations, and their benefits. OL

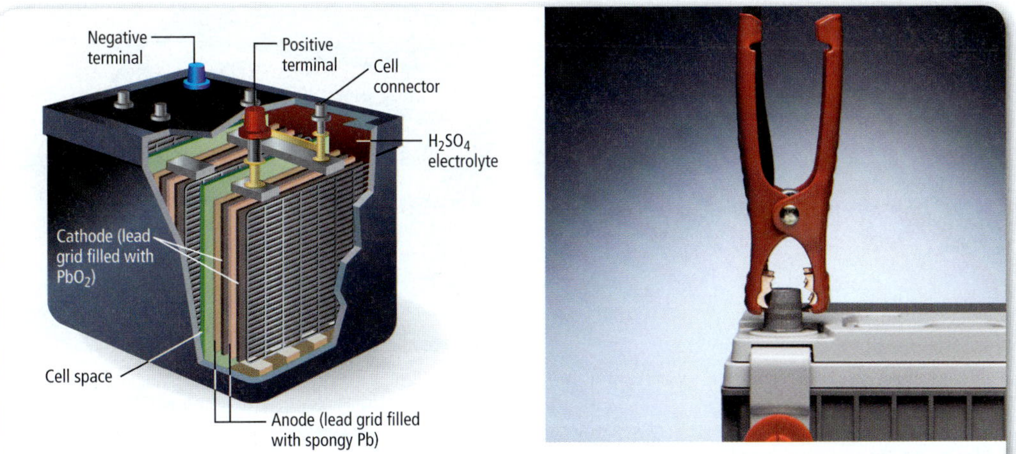

Lead-acid batteries contain lead plates and lead(IV) oxide plates. The electrolyte is a solution of sulfuric acid. When the battery is in use, the sulfuric acid is depleted and the electrolyte becomes less dense.

Low electrolyte levels can result in a dead battery. Jumper cables conduct current from a car with a good battery to start a car with a dead one.

■ **Figure 11** The lead-acid battery used in automobiles discharges when it starts the car and charges when the engine is running.

Sulfuric acid serves as the electrolyte in the battery, but, as the overall cell equation shows, it is depleted as the battery generates electric current. What happens when the battery is recharging? In this case, the reactions reverse, forming lead and lead(IV) oxide and releasing sulfuric acid, shown as $4H^+(aq) + 2SO_4^{2-}(aq)$ in the equation.

The lead-storage battery shown in **Figure 11** is a good choice for motor vehicles because it provides a large initial supply of energy to start the engine, has a long shelf life, and is reliable at low temperatures.

☑ READING CHECK **Identify** the species that is oxidized and the species that is reduced when the lead-acid battery is charging.

Lithium Batteries

Although lead-acid batteries are reliable and suitable for many applications, engineers have been working to develop batteries with less mass and higher capacity to power devices from wristwatches to electric cars. For applications in which a battery is the key component and must provide a significant amount of power, such as for the operation of an electric car, lead-acid batteries are too heavy to be feasible.

The solution is to develop lightweight batteries that store a large amount of energy for their size. Engineers have focused their attention on the element lithium for two reasons: lithium is the lightest known metal and has the lowest standard reduction potential of the metallic elements, −3.04 V, as shown in **Table 1.** A battery that oxidizes lithium at the anode can generate almost 2.3 V more than a similar battery in which zinc is oxidized.

Compare the zinc and lithium oxidation half-reactions and their standard reduction potentials.

$$Zn \rightarrow Zn^{2+} + 2e^- \qquad (E^0_{Zn^{2+}|Zn} = -0.762\ V)$$

$$Li \rightarrow Li^+ + e^- \qquad (E^0_{Li^+|Li} = -3.04\ V)$$

$$E^0_{Zn^{2+}|Zn} - E^0_{Li^+|Li} = +2.28\ V$$

VOCABULARY
WORD ORIGIN
Capacity
capac-, capax, from Latin, meaning *containing or capable of holding a great deal*

☑ READING CHECK During charging, lead in $PbSO_4$ (Pb^{2+}) is both oxidized to PbO_2 (Pb^{4+}) and reduced to Pb.

Content Backgound

Thermal Runaway Thermal runaway is a cascade effect of an exothermic reaction increasing the temperature of the media in which the reaction is taking place, which in turn, further increases the reaction rate, which produces more heat. Thermal runaway has contributed to industrial chemical accidents. When manufactured improperly, sealed lithium ion batteries can experience thermal runaway, resulting in overheating, and can explode. Cellphone and laptop batteries have been cited for causing fires and explosions. To prevent accidents from lithium ion batteries, mandatory safety devices are built in to the batteries. These include shut-down separators, tear-away tabs, and vents. Internal contamination with metal particles caused several explosions of laptop computer batteries and triggered a massive recall of laptop batteries in 2006.

CHEMISTRY JOURNAL

Rechargeable Batteries Have students research and compare the cost of rechargeable and nonrechargeable batteries for home use, comparing price to estimated hours of energy in battery operated equipment. Remind them to include the price of the charger. Have students report the data to the class and discuss the most cost effective type of battery to purchase. OL

Identify Misconceptions

Many students might think that a dry cell is dry.

Uncover the Misconception Ask students to describe how they envision the inside of a dry cell battery. What do they think is the consistency of the center substance?

Demonstrate the Concept Show students a cross section of a battery that has been dismantled. A 9-volt battery can be dismantled using a screwdriver and pliers. Enlist the help of an industrial technology teacher to cut the cross section. Cover the exposed interior of the dry cell with plastic wrap so that the paste does not dry and students will not come into contact with the substances. **WARNING:** ***Do not cut a cross section of an alkaline dry cell battery due to the corrosiveness of the substances that make up this type of battery.*** An alternative would be to refer to Figure 11 if you do not have access to cross-sections of common dry cell or 9-volt batteries.

Assess New Knowledge Ask students to explain why a dry cell is not dry. The dry cell contains an electrolytic paste made up of zinc chloride, manganese(IV) oxide, ammonium chloride, and a little water. OL EL

☑ READING CHECK They are light, long-lived, store a large amount of energy for their size, and some are rechargeable.

Lithium batteries often deliver either 3 V or 9 V and come in many sizes to fit different devices.

Lithium battery packs power this experimental car to a maximum speed of 113 km/h. The car has a range of over 320 km.

■ **Figure 12** The light weight, long life, and high potential of a lithium battery make it an excellent choice for a variety of purposes.

Lithium batteries can be either primary or secondary batteries, depending on which reduction reactions are coupled to the oxidation of lithium. For example, some lithium batteries use the same cathode reaction as zinc-carbon dry cells, the reduction of manganese(IV) oxide (MnO_2) to manganese(III) oxide (Mn_2O_3). These batteries produce an electric current of about 3 V compared to 1.5 V for zinc-carbon cells. Lithium batteries last much longer than other kinds of batteries. As a result, they are often used in watches, computers, and cameras to maintain time, date, memory, and personal settings—even when the device is turned off. **Figure 12** shows a range of available lithium batteries and a developing application.

☑ READING CHECK **List** three advantages of lithium batteries.

RealWorld CHEMISTRY

Fuel Cells

REDUCING POLLUTION One of the largest sources of air pollution in many cities is vehicles. In some European cities, experimental buses powered by hydrogen fuel cells are making a difference. Exhaust from these buses contain no carbon dioxide and no oxides of nitrogen or sulfur. Pure water is their only product.

Fuel Cells

When hydrogen burns in air, it does so explosively, with the evolution of light and heat.

$$2H_2(g) + O_2(g) \rightarrow 2H_2O(l) + \text{energy}$$

Can this reaction occur under controlled conditions inside a cell?

Connection to Physics A **fuel cell** is a voltaic cell in which the oxidation of a fuel is used to produce electric energy. Fuel cells differ from other batteries because they are provided with a continual supply of fuel from an external source. Many people think the fuel cell is a modern invention, but the first one was demonstrated in 1839 by William Grove (1811–1896), a British electrochemist. He called his cell a "gas battery." It was not until the 1950s, when scientists began working in earnest on the space program, that efficient, practical fuel cells were developed. If astronauts were to fly a space shuttle, supplies of water were needed to support their lives on board and a reliable source of electricity was needed to power the shuttle's many systems. Both of these primary needs were met with the development of the hydrogen fuel cell that controls the oxidation of hydrogen and provides both electricity and water. The cell produces no by-products to require disposal or storage on a space journey.

DIFFERENTIATED INSTRUCTION

English Learners Have students build models of cross sections of the batteries shown in the diagrams in the text or from the research that they conducted. Refer to the various figures of batteries in the text. Have students label all of the parts of the batteries in English and in the students' native languages. EL BL

How a fuel cell works As in other voltaic cells, a fuel cell has an anode and a cathode and requires an electrolyte so that ions can migrate between electrodes. A common electrolyte in a fuel cell is an alkaline solution of potassium hydroxide. Each electrode is a hollow chamber of porous carbon walls that allows contact between the inner chamber and the electrolyte surrounding it. The following oxidation half-reaction takes place at the anode.

$$2H_2(g) + 4OH^-(aq) \rightarrow 4H_2O(l) + 4e^-$$

The reaction uses the hydroxide ions that are abundant in the alkaline electrolyte and releases electrons to the anode. Electrons from the oxidation of hydrogen flow through the external circuit to the cathode where the following reduction half-reaction takes place.

$$O_2(g) + 2H_2O(l) + 4e^- \rightarrow 4OH^-(aq)$$

The electrons reduce oxygen in the presence of water to form four hydroxide ions, which replenish the hydroxide ions used up at the anode. When the two half-reactions are combined, the equation is the same as the equation for the burning of hydrogen in oxygen.

$$2H_2(g) + O_2(g) \rightarrow 2H_2O(l)$$

Because the fuel for the cell is provided from an outside source, fuel cells never run down as batteries do. They keep producing electricity as long as fuel is available.

Some fuel cells use fuels other than hydrogen. For example, methane replaces hydrogen in some cells, but has the disadvantage of producing carbon dioxide as an exhaust gas. Fuel cells such as the one shown in **Figure 13** use a plastic sheet called a proton-exchange membrane (PEM), which eliminates the need for a liquid electrolyte.

READING CHECK **Compare** fuel cells with other voltaic cells to find an important way in which they are different.

Watch a **video about hydrogen fuel cells.**

Video

CAREERS IN CHEMISTRY

Alternative Energy Specialist
If you like to invent new things and make them work, you might be interested in helping to develop new sources of energy for this increasingly energy-dependent world. Technologies include wind and solar power, geothermal energy, harnessing the energy of tides, utilizing the temperature gradients in bodies of water, and many more.

WebQuest

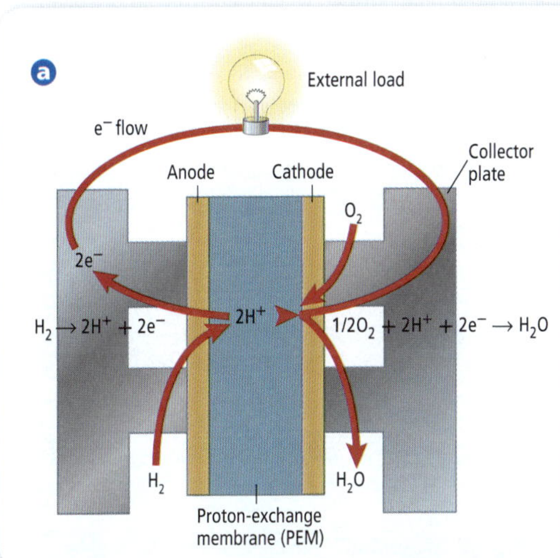

Figure 13 a. In this fuel cell, hydrogen is the fuel. The half-reactions are separated by a proton-exchange membrane so that the electrons lost in oxidation flow through an external circuit to reach the site of reduction. As electrons travel through the external circuit, they can do useful work, such as running electric motors. The by-product of this redox reaction is water. **b.** A "stack" of PEM-type cells can generate enough energy to power an electric car.

Concept Development

Corrosion Prevention Ask students if copper would make a good cathodic protector to prevent the corrosion of iron. no, because iron oxidizes easier than copper Ask a representative of a local gas or oil company to discuss with the class the use of sacrificial metal as a method of preventing the corrosion of pipeline metal. One of the questions to ask might be what elements are commonly used as cathodic protectors or sacrificial metals. OL

CAREERS IN CHEMISTRY

WebQuest

Extension

Problem Solving Have students read news accounts or watch a video of the Apollo 13 mission, during which an oxygen tank exploded and damaged the fuel cells on the space capsule. Point out how creative problem solving was key to the safe return of the astronauts. OL

READING CHECK Fuel cells need never become depleted as long as they are provided with a steady source of fuel.

What's CHEMISTRY Got To Do With It?

Video

Driving Into The Future

CULTURAL DIVERSITY

Preserving National Treasures Remind students that bridges and cars are not the only objects subject to corrosion. Many art objects are also vulnerable to corrosion. One of the best publicized restorations was that of the Statue of Liberty. Another art object that was restored was *The Thinker*, a bronze cast statue by Rodin that has been in Philadelphia since 1925. Have students research other works of art that are important to cultures around the world and that are exposed to the elements and in danger of corrosion. Have students describe the chemistry of the corrosion problem and steps that have been taken (or could be taken) to restore the treasures.

Data Analysis LAB

About the Lab

- Biofuel cells are devices that convert microbial metabolic energy into electrical energy.
- The electron mediator is required to transfer the microbial electron to the electrode in the fuel cell.
- Other articles concerning biofuel cells:
  - Bullen, R.A., Arnot, T.C., Lakeman, J.B., Walsh, F.C. "Biofuel Cells and their Development." *Biosensors and Bioelectronics* (May 2006), 2015–2045.
  - Davis, Frank, Higson, Séamus P.J. "Biofuel Cells – Recent Advances and Applications." *Biosensors and Bioelectronics* (February 2007), 1224–1235.

Think Critically

1. 15 minutes
2. Yes, the current rose significantly at 15 minutes into the experiment.
3. About 3.7 mA

Data Analysis LAB

Based on Real Data*

Interpret Graphs

How can you get electric current from microbes? Scientists have studied the use of microbes as biofuel cells. A biofuel cell directly converts microbial metabolic energy into electric current. An electron mediator facilitates transfer of electrons to an electrode. An electron mediator is a compound that taps into the electron transport chain of cells and steals the electrons that are produced.

Data and Observations

The graph shows the current produced in a biofuel cell with (blue line) and without (green line) the use of an electron mediator.

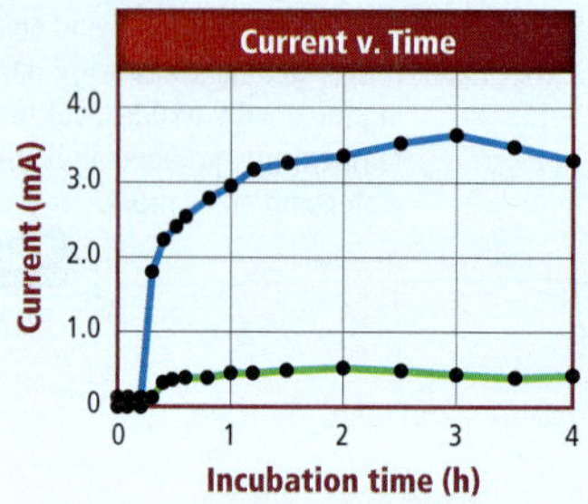

*Data obtained from: Hyun Park, Doo and J. Gregory Zeikus. April, 2000. Electricity Generation in Microbial Fuel Cells Using Neutral Red as an Electronophore. *Applied and Environmental Microbiology* 66, No. 4:1292–1297.

Think Critically

1. **Infer** the approximate time when the electron mediator was introduced.
2. **Determine** Did the introduction of the electron mediator make a difference in the current production? Explain your answer.
3. **Analyze** What is the highest current obtained by the cell?

Corrosion

In this chapter, you have examined the spontaneous redox reactions in voltaic cells. Spontaneous redox reactions also occur in nature. An example is the corrosion of iron, usually called rusting. **Corrosion** is the loss of metal resulting from an oxidation-reduction reaction of the metal with substances in the environment. Although rusting is usually thought of as a reaction between iron and oxygen, it is more complex. Both water and oxygen must be present for rusting to occur. For this reason, an iron object, such as the one shown in **Figure 14,** that has been left exposed to air and moisture is especially susceptible to rust. The portion that is in contact with the moist ground rusted first.

Rusting usually begins where there is a pit or a small break in the surface of the iron. This region becomes the anode of the cell as iron atoms begin to lose electrons as illustrated in **Figure 15.**

$$Fe(s) \rightarrow Fe^{2+}(aq) + 2e^-$$

The iron(II) ions become part of the water solution, while the electrons move through the iron to the cathode region. In effect, the piece of iron becomes the external circuit as well as the anode. The cathode is usually located at the edge of the water drop where water, iron, and air come in contact. Here, the electrons reduce oxygen from the air in the following half-reaction.

$$O_2(g) + 4H^+(aq) + 4e^- \rightarrow 2H_2O(l)$$

The supply of H^+ ions is probably furnished by carbonic acid formed when CO_2 from air dissolves in water.

Figure 14 Left unattended in the presence of air and moisture, this iron barrel is slowly being oxidized to rust (Fe_2O_3).

CHEMISTRY JOURNAL

Hardware Protection Have small groups of students visit a hardware store and list different types of nails and types of corrosion protection methods. Depending on the store, students will find nails made of iron, aluminum, and brass. Some are painted, galvanized, hot dipped, cold dipped, or electroplated. OL EL COOP LEARN

DIFFERENTIATED INSTRUCTION

Advanced Learners Have students research reasons why biofuel technology would be beneficial to the quality of life on the planet. Have students assess the limitations that might have to be overcome to make biofuel a reasonable alternative to other energy sources. AL

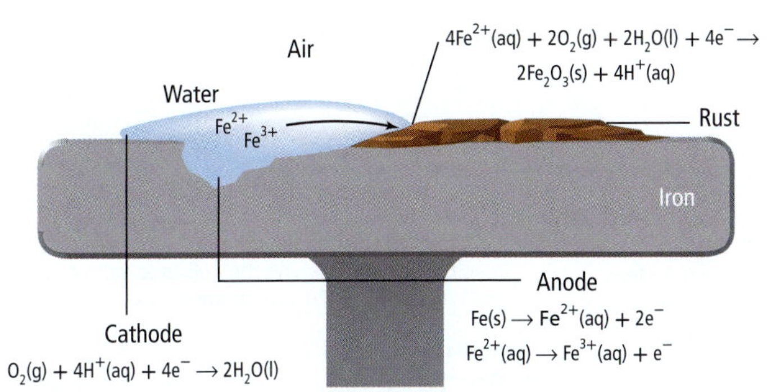

■ **Figure 15** Corrosion occurs when air, water, and iron set up a voltaic cell similar to the conditions shown at the surface of this iron I-beam. An I-beam is a large piece of iron shaped like the capital letter I and used in the construction of large buildings.
Name ***the two species that are oxidized at the anode.***

Next, the Fe^{2+} ions in solution are oxidized to Fe^{3+} ions by reacting with oxygen dissolved in the water. The Fe^{3+} ions combine with oxygen to form insoluble Fe_2O_3, rust.

$$4Fe^{2+}(aq) + 2O_2(g) + 2H_2O(l) + 4e^- \rightarrow 2Fe_2O_3(s) + 4H^+(aq)$$

Combining the three equations yields the overall cell reaction for the corrosion of iron.

$$4Fe(s) + 3O_2(g) \rightarrow 2Fe_2O_3(s)$$

Rusting is a slow process because water droplets have few ions and therefore, are not good electrolytes. However, if the water contains abundant ions, as in seawater or in regions where roads are salted in the winter, corrosion occurs much faster because the solutions are excellent electrolytes.

Preventing corrosion Corrosion of cars, bridges, ships, the structures of buildings, and other metallic objects causes more than \$100 billion in damage a year in the United States. For this reason, several means to minimize corrosion have been devised. One example is to apply a coat of paint to seal out both air and moisture, but because paint deteriorates over time, objects such as the bridge shown in **Figure 16** must be repainted often.

■ **Figure 16** Because corrosion can cause considerable damage, it is important to find ways to prevent rust and deterioration. Paint or another protective coating is one way to protect steel structures from corrosion.

DIFFERENTIATED INSTRUCTION

Hearing Impaired Assign to each student a pair of reactions that make an electrochemical cell. Have students create a poster on which they will draw and label the electrochemical reaction, incorporating as many of the vocabulary terms and concepts from this chapter as possible. OL EL

ASSESSMENT

Knowledge In a nickel–cadmium battery, the two half-reactions are $Cd + 2OH^- \rightarrow CdO + H_2O + 2e^-$ and $NiO + H_2O + 2e^- \rightarrow Ni + 2OH^-$ Have students balance the reactions using the half-reaction method. Identify what is oxidized, what is reduced, which element functions as the anode, and which element functions as the cathode. If an external source is applied during the recharging process, which element becomes the anode? The cathode?
$Cd + 2OH^- \rightarrow CdO + H_2O + 2e^-$
$NiO + H_2O + 2e^- \rightarrow Ni + 2OH^-$
Cd atoms are oxidized; Ni^{2+} ions are reduced. Cadmium is the anode; nickel is the cathode. During recharging, nickel becomes the anode and cadmium becomes the cathode. OL

■ **Caption Question Fig. 15**
Fe^{2+} and Fe

MiniLAB

Rubric

Purpose Students will observe the corrosive effects that two different metals have on an iron nail.

Process Skills classify, observe and infer, recognize cause and effect

Safety Precautions Approve lab safety forms before work begins. Remind students to be careful of the nail point. Be sure to account for all magnesium ribbon before and after the lab.

Disposal Solutions can be washed down a drain. Do not wash copper or magnesium metal down a drain. Rinse and reuse Cu and Mg.

Teaching Strategies

- See the Teacher Handbook for preparation of solutions.
- Many types of products can result from this experiment. Have students discuss the variations in their results.

Expected Results The nails wrapped with copper show corrosion. The copper still looks shiny in each. There is an orange precipitate in each of these beakers. The magnesium is corroded in each of the magnesium-wrapped nails. The nail looks unchanged where the magnesium covered it.

Analysis

1. more corrosion of the nail soaking in salt water solution than in distilled water
2. more corrosion of the Mg metal in the saltwater solution than in distilled water
3. The nail wrapped in Cu corrodes because iron is more easily oxidized. The nail wrapped in Mg does not corrode, but the Mg metal does. Mg acts as a sacrificial anode.

MiniLAB

Observe Corrosion

Which metal will corrode?

Procedure

1. Read and complete the lab safety form.
2. Use **sandpaper** to buff the surfaces of four **iron nails.** Wrap two nails with **magnesium ribbon** and two nails with **copper.** Wrap the metals tightly so that the nails do not slip out.
3. Place each of the nails in a separate **beaker.** Add **distilled water** to one of the beakers containing a copper-wrapped nail and one of the beakers containing a magnesium-wrapped nail. Add enough distilled water to just cover the wrapped nails. Add **salt water** to **two additional beakers.** Record your observations of the nails in each beaker.
4. Let the beakers stand overnight in the warmest place available. Examine the nails and solutions the next day, and record your observations.

Analysis

1. **Describe** the difference between copper-wrapped nails in the distilled water and the salt water after they have been standing overnight.
2. **Describe** the difference between the magnesium-wrapped nails in the distilled water and in the salt water.
3. **Identify** the difference in corrosion between the copper- and magnesium-wrapped nails.

The steel hulls of ships are constantly in contact with salt water, so the prevention of corrosion is vital. Although the hull can be painted, another method is used to minimize corrosion. Blocks of metals, such as magnesium, aluminum, or titanium, are placed in contact with the steel hull. These blocks oxidize more easily than iron and become the anode of the corrosion cell. They are called sacrificial anodes because they are corroded, while the iron in the hull is spared. A similar technique is used to protect underground iron pipes. Magnesium bars are attached to the pipe by wires, and these bars corrode instead of the pipe, as shown in **Figure 17.**

Figure 17 Sacrificial anodes of magnesium or other active metals are used to prevent corrosion. A magnesium rod attached to an underground iron pipe helps prevent corrosion by being oxidized itself.

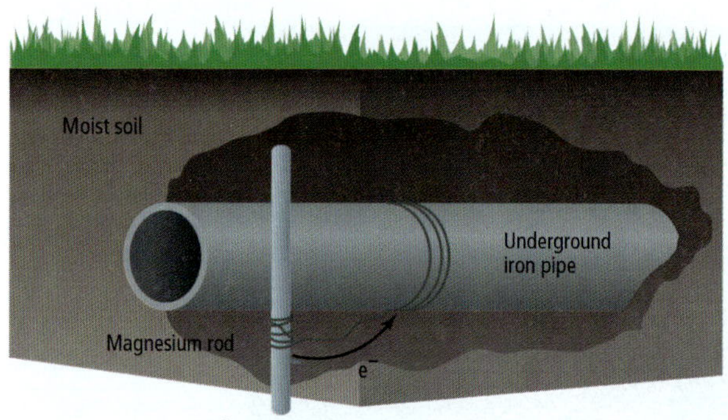

Oxidation: $Mg(s) \rightarrow Mg^{2+}(aq) + 2e^-$

Reduction: $O_2(g) + 4H^+(aq) + 4e^- \rightarrow 2H_2O(l)$

Geoff Butler

☑ **READING CHECK (p. 727)**
Methods include: application of paint to seal out air and moisture, use of a sacrificial anode, coat iron with a metal that is more resistant to corrosion such as the galvanization process

ASSESSMENT

Performance Have students test various types of nails that have protective coatings for resistance to corrosion. They should wrap these nails and place them in distilled water and saltwater solutions and make observations over a period of time. OL

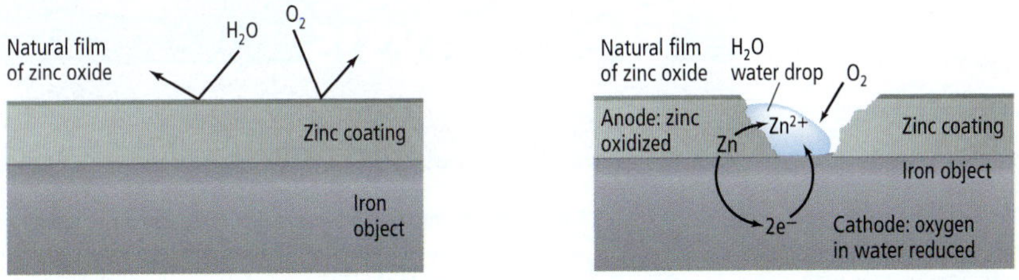

Galvanized object with zinc coating intact

The zinc coating seals the iron from air and water by forming a barrier of zinc oxide that repels water and oxygen.

Galvanized object with zinc coating broken

If the zinc coating breaks, the zinc acts as a sacrificial anode. The zinc coating, rather than the iron object, is oxidized.

Figure 18 Galvanization helps prevent corrosion in two ways.

Another approach to preventing corrosion is to coat iron with another metal that is more resistant to corrosion. In the **galvanization** process, iron is coated with a layer of zinc by either dipping the object into molten zinc or by electroplating the zinc onto it. Although zinc is more readily oxidized than iron, it is one of the self-protecting metals, a group that also includes aluminum and chromium. When exposed to air, these metals oxidize at the surface, creating a thin metal-oxide coating that seals the metal from further oxidation.

Galvanization protects iron in two ways. As long as the zinc layer is intact, water and oxygen cannot reach the iron's surface. Inevitably, the zinc coating cracks. When this happens, zinc protects iron from rapid corrosion by becoming the anode of the voltaic cell set up when water and oxygen contact iron and zinc at the same time. **Figure 18** illustrates how these two forms of corrosion protection work.

☑ READING CHECK **Summarize** methods of preventing corrosion.

SECTION 2 REVIEW

Section Self-Check

Section Summary

- Primary batteries can be used only once; secondary batteries can be recharged.
- When a battery is recharged, electric energy supplied to the battery reverses the direction of the battery's spontaneous reaction.
- Fuel cells are batteries in which the substance oxidized is a fuel from an external source.
- Methods of preventing corrosion are painting, coating with another metal, or using a sacrificial anode.

15. MAINIDEA **Identify** what is reduced and what is oxidized in the zinc-carbon dry-cell battery. What features make the alkaline dry cell an improvement over the earlier type of dry-cell battery?

16. Explain what happens when a battery is recharged.

17. Describe the half-reactions that occur in a hydrogen fuel cell, and write the equation for the overall reaction.

18. Describe the function of a sacrificial anode. How is the function of a sacrificial anode similar to galvanization?

19. Explain why lithium is a good choice for the anode of a battery.

20. Calculate Use data from **Table 1** to calculate the cell potential of the hydrogen-oxygen fuel cell.

21. Design an Experiment Use your knowledge of acids to devise a method for determining whether a lead-acid battery can deliver full charge or is beginning to run down.

3 Assess

Check for Understanding

Have the students construct a chart comparing the major types of batteries (dry cell or primary battery, lead–acid storage battery, and fuel cells) described in this chapter. Compare size, voltage produced, and lifetime of the battery. OL

Reteach

Make sure students understand the difference between a primary battery and a secondary battery. Have them classify the different batteries illustrated as primary or secondary and explain their reasons for the classification. OL

Extension

Have students research the method of retarding the rusting of metal oceangoing ships by hanging large pieces of magnesium along the hull of the ship. The more active magnesium is oxidized instead of the iron. When depleted, the magnesium strips are replaced. OL

ASSESSMENT

Knowledge Ask students some reasons why there would be a difference in the type of battery selected to supply energy to a toy compared to a battery that supplies energy to a pacemaker that is surgically implanted. The most important reason is the length of time before the battery would have to be replaced, which requires another surgical procedure. Size is a second consideration. Another concern is durability in the body and resistance to corrosion and deterioration. Cost differences are probably of least importance. OL

SECTION 2 REVIEW

15. Zn is oxidized; MnO_2 is reduced in the electrolytic paste. Zinc in a powdered form provides more surface area for reaction. Alkaline cells eliminate the inactive carbon rod cathode.

16. A power source added to the system forces the cell to operate in the non-spontaneous direction. The original materials of the cell are restored.

17. At the anode, hydrogen gas is oxidized to water molecules in the presence of hydroxide ions. At the cathode, oxygen is reduced to hydroxide ions in the presence of water molecules. Overall reaction is $2H_2(g) + O_2(g) \longrightarrow 2H_2O(l)$.

18. A sacrificial anode has a lower reduction potential than the metal it is used to preserve and preferentially corrodes. Galvanization functions in the same way. When the galvanized layer of zinc cracks or breaks, the zinc preferentially corrodes. This leaves the metal underneath free of corrosion.

19. Li is light and has the lowest reduction potential of all the metals and it also produces more voltage than a zinc half-cell would.

20. $E^0_{cell} = +1.229\ V$

21. A sample of the sulfuric acid electrolyte from the battery could be titrated with base and its molarity compared with the molarity of a sample of electrolyte taken from a new battery.

SECTION 3

1 Focus

MAINIDEA

Electrolytic Cells Bring a battery charger to class, and ask students why it is used. It recharges the battery when it loses its charge. Ask students how it works. The reactions that deplete the battery are reversed. Tell students that battery chargers work because the battery becomes an electrolytic cell instead of a voltaic cell. OL

2 Teach

Concept Development

Reversible Reactions Students do not always realize that reactions are reversible. Refer to the analogy developed earlier in the chapter with the golf ball rolling down the hill. Ask students what must happen to get the ball back up the hill. They will need to use energy to push it back up the hill. By adding energy (external electricity) to the reaction, the reaction can be forced in the opposite direction. OL

■ **Caption Question Fig. 19** In the voltaic cell, zinc is oxidized and copper ions are reduced. In the electrolytic cell, copper is oxidized and zinc ions are reduced.

SECTION 3

Electrolysis

MAINIDEA In electrolysis, a power source causes nonspontaneous reactions to occur in electrochemical cells.

Essential Questions

- How is it possible to reverse a spontaneous redox reaction in an electrochemical cell?
- How does the molten sodium chloride electrolysis reaction compare with those in the electrolysis of brine?
- What is the importance of electrolysis in the smelting and purification of metals?

Review Vocabulary

redox reaction: an oxidation-reduction reaction

New Vocabulary

electrolysis
electrolytic cell

CHEM 4 YOU When you ride a bicycle downhill, you don't have to do any work—you just coast. What is different when you ride uphill? You have to provide a lot of energy by pedaling.

Reversing Redox Reactions

When a battery generates electric current, electrons given up at the anode flow through an external circuit to the cathode, where they are used in a reduction reaction. A secondary battery is one that can be recharged by passing a current through it in the opposite direction. To help you understand the process, study the electrochemical cells in **Figure 19.** The beakers on the left contain zinc strips in solutions of zinc ions. The beakers on the right contain copper strips in solutions of copper ions. One electrochemical cell is supplying power to a light bulb by means of a spontaneous redox reaction. Electrons flow spontaneously from the zinc side to the copper side, creating an electric current. The reaction continues until the zinc strip is used up, and then the reaction stops. However, the cell can be regenerated if current is applied in the reverse direction using an external voltage source. The voltage source is required because the reverse reaction is nonspontaneous. If the voltage source is applied long enough, the cell will return to nearly its original strength.

The use of electrical energy to bring about a chemical reaction is called **electrolysis.** An electrochemical cell in which electrolysis occurs is called an **electrolytic cell.** For example, when a secondary battery is recharged, it is acting as an electrolytic cell.

■ **Figure 19** The zinc-copper electrochemical cell can be a voltaic cell or an electrolytic cell.
Infer *In each electrochemical cell, which metal is oxidized? What is reduced?*

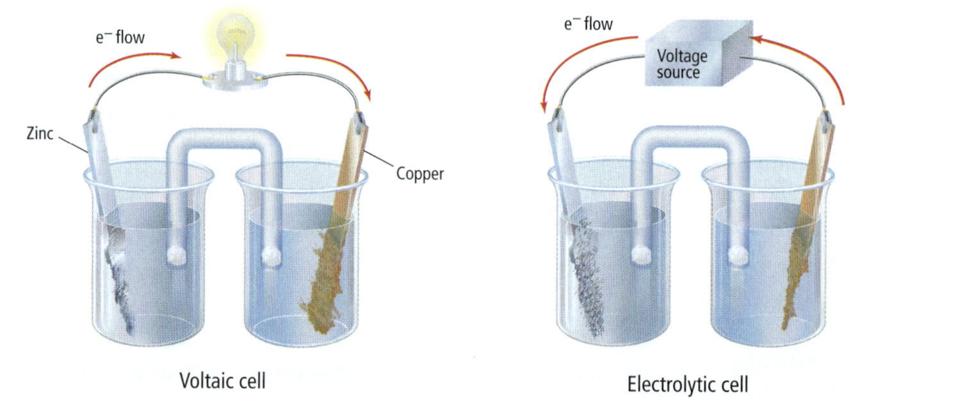

In this voltaic cell, the oxidation of zinc supplies the electrons to light the bulb and reduce copper ions. The spontaneous reaction continues until the zinc is used up.

When an outside voltage is applied, the flow of electrons is reversed and the nonspontaneous reaction occurs, which restores the conditions of the cell.

DIFFERENTIATED INSTRUCTION

Visually Impaired Ask visually impaired students to describe orally why the addition of an electrolyte to water would help with the process of electrolysis. Students should explain that water is a polar molecule and, therefore, a poor conductor of electricity. OL

Applications of Electrolysis

Recall that voltaic cells convert chemical energy to electrical energy as a result of a spontaneous redox reaction. Electrolytic cells do the opposite; they use electrical energy to drive a nonspontaneous reaction. A common example is the electrolysis of water. This reaction is the opposite of consuming hydrogen and oxygen in a fuel cell.

$$2H_2O(l) \rightarrow 2H_2(g) + O_2(g)$$

The electrolysis of water is one method by which hydrogen gas can be generated for commercial use.

Electrolysis of molten NaCl Just as electrolysis can decompose water into its elements, it can also separate molten sodium chloride into sodium metal and chlorine gas. This process is carried out in a chamber called a Down's cell, as illustrated in **Figure 20.** The electrolyte in the cell is the molten sodium chloride itself. Remember that ionic compounds can conduct electricity only when their ions are free to move, such as when they are dissolved in water or are in the molten state.

At the anode, chloride ions are oxidized to chlorine (Cl_2) gas.

$$2Cl^-(l) \rightarrow Cl_2(g) + 2e^-$$

At the cathode, sodium ions are reduced to sodium metal.

$$Na^+(l) + e^- \rightarrow Na(l)$$

The net cell reaction is the following.

$$2Na^+(l) + 2Cl^-(l) \rightarrow 2Na(l) + Cl_2(g)$$

The importance of the Down's cell can best be appreciated in terms of the important roles that both sodium and chlorine play in your life. Chlorine is used throughout the world to purify water for drinking and swimming. Many cleaning products you might use, including household bleach, contain chlorine compounds. You depend on a host of other products, such as paper, plastics, insecticides, textiles, dyes, and paints, that either contain chlorine, or chlorine was used in their production.

In its pure form, sodium is used as a coolant in nuclear reactors and in sodium vapor lamps used for outdoor lighting. In its combined form in ionic compounds, you need only look on the contents list of consumer products to find a variety of sodium salts in the products you use and the foods you eat.

☑ READING CHECK **Explain** why the sodium chloride must be molten in the Down's cell.

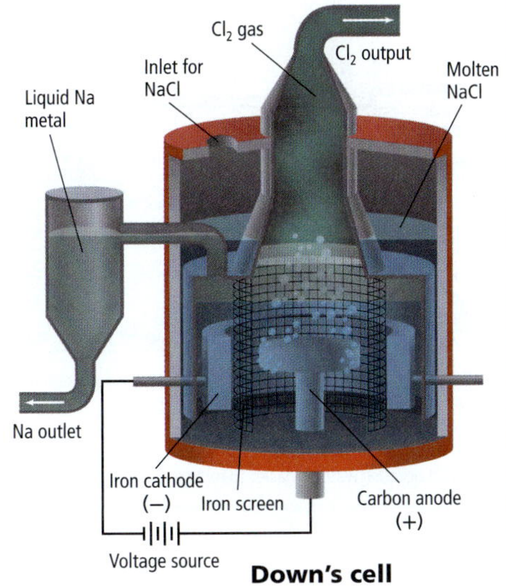

■ **Figure 20** In a Down's cell, electrons supplied by a generator are used to reduce sodium ions at the cathode. As electrons are removed from the anode, chloride ions are oxidized at the anode to chlorine gas.

FOLDABLES® Incorporate information from this section into your Foldable.

Identify Misconceptions

Students might not understand the difference between molten sodium chloride and brine.

Uncover the Misconception Have students draw a picture illustrating the difference between molten NaCl and a concentrated solution of NaCl.

Demonstrate the Concept Remind students that both sodium and chlorine are very reactive elements and are not found in nature in elemental form. Develop a detailed flowchart of the electrolysis technique that produces these elements.

Assess New Knowledge Have students list products that use chlorine or sodium as raw materials in the manufacturing process. OL

Enrichment

Art Restoration Have students investigate the process of electrophoresis as another method for restoring historical artifacts by breaking up the salt, dirt, and other particles encrusted on the pieces. Have them describe the chemistry of electrophoresis for this application. OL

☑ READING CHECK Ions can move only when they are in a molten state or in solution.

FOLDABLES® Rubric

CHEMISTRY PROJECT

Alternative Transportation Alternatives to the gasoline combustion engine are being investigated for automobiles. One possible alternative is the electric car. Have students, working in groups, investigate the cost, power output, and approximate distance traveled for both the gasoline combustion engine and the electric car; devise a spreadsheet for the research results; and write a position statement citing reasons why they would or would not purchase a vehicle powered by gasoline or electricity. Ask each group to present their position paper to the class. OL COOP LEARN

ASSESSMENT

Knowledge Have students write an explanation describing the difference between electrolysis and an electrochemical cell. OL EL

Quick Demo

Composition of a Penny

Demonstrate that a penny is a zinc core plated with a copper coating. With a file, carefully etch approximately 1/3 of the edge or rim of the penny on both sides so that some of the rim is still attached to the top and the bottom of the penny. Immerse the penny in a beaker of HCl (approximately 3*M*) and let it stand overnight. Pour the remaining hydrochloric acid down a drain upon neutralization, rinse the penny with water, and carefully pat it dry. What remains is the thin copper plating bearing the impressions of the original penny. The zinc core reacted with the hydrochloric acid. Ask students why the zinc reacted and not the copper. Did the zinc react with the hydrogen ion spontaneously? Remind students that to electroplate the copper onto the zinc core, the copper ions must be reduced at the cathode.

READING CHECK Chloride ion (Cl^-) is oxidized. Water is reduced.

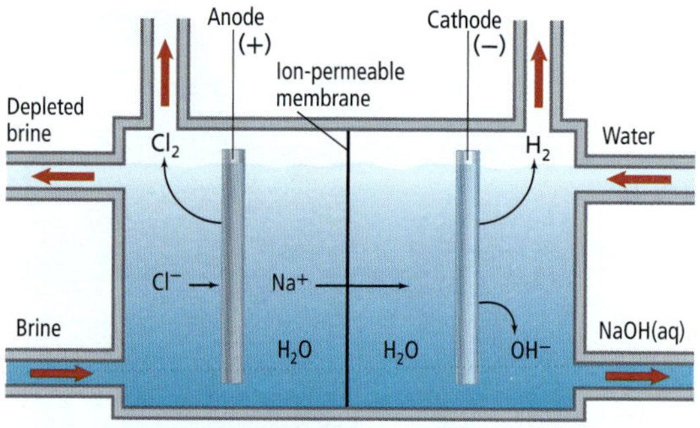

Commercial facilities use an electrolytic process to obtain hydrogen gas, chlorine gas, and sodium hydroxide from brine.

Chlorine gas is used to manufacture polyvinyl chloride products, such as these pipes for water distribution.

■ **Figure 21** In the electrolysis of brine (aqueous NaCl), sodium is not a product because water is easier to reduce.

Electrolysis of brine The decomposition of brine, an aqueous solution of sodium chloride, is also carried out by means of electrolysis. **Figure 21** shows a typical electrolytic cell and the products of the electrolysis. Two reactions are possible at the cathode: the reduction of sodium ions and the reduction of hydrogen in water molecules.

$$Na^+(aq) + e^- \rightarrow Na(s)$$

$$2H_2O(l) + 2e^- \rightarrow H_2(g) + 2OH^-(aq)$$

However, the reduction of sodium (Na^+) does not occur because water is easier to reduce, and thus is reduced preferentially.

Two reactions are also possible at the anode: the oxidation of chloride ions and the oxidation of oxygen in water molecules.

$$2Cl^-(aq) \rightarrow Cl_2(g) + 2e^-$$

$$2H_2O(l) \rightarrow O_2(g) + 4H^+(aq) + 4e^-$$

Because the desired product is chlorine (Cl_2), the concentration of chloride ions is kept high in order to favor this half-reaction. The overall cell reaction is as follows.

$$2H_2O(l) + 2NaCl(aq) \rightarrow H_2(g) + Cl_2(g) + 2NaOH(aq)$$

All three products are commercially important substances.

VOCABULARY
SCIENCE USAGE V. COMMON USAGE
Reduce
Science usage: to decrease an atom's oxidation number by the addition of electrons
Zinc reduces copper(II) ions to copper atoms by releasing two electrons.
Common usage: to diminish in size, amount, extent, or number
The number of dancers had to be reduced because the stage was too small.

READING CHECK **Name** the species that is oxidized and the species that is reduced in the electrolysis of brine.

Aluminum production Until the late nineteenth century, aluminum metal was more precious than gold because no one knew how to purify it in large quantities. In 1886, 22-year-old Charles Martin Hall (1863–1914) developed a process to produce aluminum by electrolysis. He used heat from a blacksmith forge, electricity from homemade batteries, and his mother's iron skillets as electrodes. At almost the same time, one of Le Châtelier's students, Paul L. T. Héroult (1863–1914), also 22 years old, discovered the same process. Today, it is called the Hall-Héroult process and is illustrated in **Figure 22.**

Tetra Images/Alamy

CHEMISTRY JOURNAL

Salt Brine Ask students which elements lose and which gain electrons, as well as how many are lost and gained, in the reaction that occurs in the electrolysis of salt brine. Two chloride ions lost a total of two electrons. Two hydrogen ions gain a total of two electrons. OL

In the modern version of the Hall–Héroult process, aluminum metal is obtained by electrolysis of aluminum oxide, which is refined from bauxite ore ($Al_2O_3 \cdot 2H_2O$). The aluminum oxide is dissolved at 1000°C in molten synthetic cryolite (Na_3AlF_6), another aluminum compound. The cell is lined with graphite, which forms the cathode for the reaction, as shown in **Figure 22.** Another set of graphite rods is immersed in the molten solution as an anode. The following half-reaction occurs at the cathode.

$$Al^{3+}(l) + 3e^- \rightarrow Al(l)$$

The molten aluminum settles to the bottom of the cell and is drawn off periodically. Oxide ions are oxidized at the cathode in this half-reaction.

$$2O^{2-}(aq) \rightarrow O_2(g) + 4e^-$$

Because temperatures are high, the liberated oxygen reacts with the carbon of the anode to form carbon dioxide.

$$C(s) + O_2(g) \rightarrow CO_2(g)$$

The Hall-Héroult process uses huge amounts of electrical energy. For this reason, aluminum is often produced in plants built close to large hydroelectric power stations, where electrical energy is less expensive. The vast amount of electricity needed to produce aluminum from ore is the primary reason for recycling aluminum. Recycled aluminum has already undergone electrolysis, so the only energy required to make it usable again is the heat needed to melt it in a furnace.

Purification of ores Electrolysis is also used in the purification of metals such as copper. Most copper is mined in the form of the ores chalcopyrite ($CuFeS_2$), chalcocite (Cu_2S), and malachite ($Cu_2CO_3(OH)_2$). The sulfides are most abundant and yield copper metal when heated strongly in the presence of oxygen.

$$Cu_2S(s) + O_2(g) \rightarrow 2Cu(l) + SO_2(g)$$

■ **Figure 22** The Hall-Héroult process operates at 900°C in smelters similar to this one. Note that carbon (graphite) serves as both the anode and the cathode. Recycled aluminum is often fed into the cell with the new aluminum.

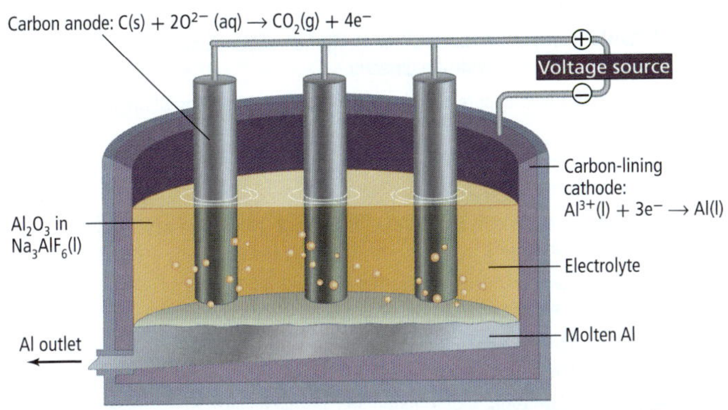

Every ton of aluminum that is recycled saves huge quantities of electrical energy that would be used to produce new aluminum from ore.

Jeff Greenberg/PhotoEdit

Quick Demo

Electrolysis Set up the apparatus so that the image can be projected. Materials needed include a clear petri dish or beaker, two pieces of insulated copper wire with about 1 cm of the copper exposed on each end, a 9-volt battery, and a solution of 0.1*M* potassium iodide (approximately 5 mL). Add an equal amount of starch solution and several drops of phenolphthalein. Stir the solution, and observe the color. Attach the wire leads to the battery and immerse the exposed ends in the solution. Allow the reaction to proceed for several minutes. Observe changes in the color of the solution and hypothesize what the color changes indicate. At one electrode, iodine is formed as indicated by the reaction with the starch to form the blue-black color. At the other electrode, a hydroxide (base) is formed as indicated by the pink-to-red color change. Have students write the two half-reactions for this electrolysis process. OL

Apply Chemistry

Plating One application of electrochemistry is in manufacturing some jewelry by electroplating a precious metal onto another metal object. Gold- or silver-plated jewelry looks similar to jewelry made of solid gold or silver. However, if the plating becomes scratched or chipped, the base metal can corrode or discolor. In addition, the thin layer of gold or silver is sometimes plated unevenly, which affects the appearance. Some artists fashion jewelry by electroplating a metal onto an object of nature, such as a pinecone or leaf.

3 Assess

Check for Understanding

Ask students which of the following metals is the most active and thus the most difficult to refine from its ore: silver, gold, iron, or copper? iron OL EL

Reteach

Take time to make sure that students recognize that many of the examples in this section relate to previously learned information such as reaction types and redox reactions. OL EL

Extension

Have students explain why the reactions discussed in this section require energy to proceed. Have students discuss the significance of the chemical reactions that are presented in this section and how these reactions affect their lives. Students should respond that the reactions require energy to proceed or that the reactions are not spontaneous without an outside energy source. Many commercial products could not be manufactured without knowledge of these specific reactions. OL

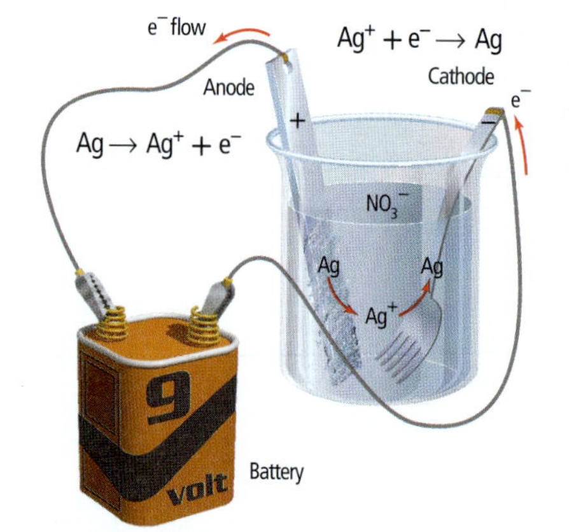

■ **Figure 23** Power is needed to oxidize silver at the anode and reduce silver at the cathode. In an electrolytic cell used for silver plating, the object to be plated is the cathode where silver ions in the electrolyte solution are reduced to silver metal and deposited on the object.

The copper obtained from this process contains many impurities and must be refined, so the molten copper is cast into large, thick plates. These plates are then used as an anode in an electrolytic cell containing a solution of copper(II) sulfate. The cathode of the cell is a thin sheet of pure copper. As current is passed through the cell, copper atoms in the impure anode are oxidized to copper(II) ions. The copper ions migrate through the solution to the cathode, where they are reduced to copper atoms. These atoms become part of the cathode, while impurities fall to the bottom of the cell.

Electroplating Objects are electroplated when a uniform coating is deposited usually as a protective or decorative layer. The substance used is typically a metal. Electroplating with a metal such as silver is accomplished with a method similar to that used to refine copper. The object to be silver-plated is the cathode of an electrolytic cell. The anode is a silver bar or sheet, as shown in **Figure 23.** At the anode, silver is oxidized to silver ions as electrons are removed by the power source. At the cathode, the silver ions are reduced to silver metal by electrons from the external power source. The silver forms a thin coating over the object being plated. Current passing through the cell must be carefully controlled in order to get a smooth, even metal coating.

Other metals are also used for electroplating. You might have costume jewelry that is electroplated with gold. Or you might admire an automobile whose steel parts such as the bumper have been made more corrosion-resistant by being electroplated first with nickel and then with chromium.

SECTION 3 REVIEW

Section Self-Check

Section Summary

- In an electrolytic cell, an outside source of power causes a nonspontaneous redox reaction to occur.
- The electrolysis of molten sodium chloride yields sodium metal and chlorine gas. The electrolysis of brine yields hydrogen gas, sodium hydroxide, and chlorine gas.
- Metals such as copper are purified in an electrolytic cell.
- Electrolysis is used to electroplate objects and to produce pure aluminum from its ore.

22. MAINIDEA **Define** *electrolysis* and relate the definition to the spontaneity of redox reactions.
23. **Explain** why the products of the electrolysis of brine and the electrolysis of molten sodium chloride are different.
24. **Describe** how impure copper obtained from the smelting of ore is purified by electrolysis.
25. **Explain,** by referring to the Hall-Héroult process, why recycling aluminum is very important.
26. **Describe** the anode and cathode of an electrolytic cell in which gold is to be plated on an object.
27. **Explain** why producing a kilogram of silver from its ions by electrolysis requires much less electric energy than producing a kilogram of aluminum from its ions.
28. **Calculate** Use **Table 1** to calculate the voltage of the Down's cell. Should the potential be positive or negative?
29. **Summarize** Write a short paragraph answering each of the three Essential Questions for Section 3 in your own words.

SECTION 3 REVIEW

22. Electrolysis is the process of using electrical energy to produce a chemical reaction. The electrolytic process is not spontaneous.
23. Electrolysis of brine involves an aqueous solution, which affects the products.
24. Cu atoms are oxidized to Cu^{2+} then subsequently reduced to pure Cu atoms, with the impurities falling away.
25. The Hall-Héroult process requires high temperatures and a large amount of electricity to separate aluminum from its ore. Recycling requires only the heat needed to melt the metal.
26. The anode is a bar of gold; the cathode is the object to be plated.
27. First, a kilogram of silver contains many fewer atoms than a kilogram of aluminum because silver has a larger molar mass. Second, silver is easier to reduce. The reduction potential for silver is +0.7996 V. The reduction potential for aluminum is −1.662 V.
28. The Down's cell is a nonspontaneous reaction, so the potential should be negative. $E^0_{cell} = -4.07$ V
29. Student paragraphs should summarize the important ideas in the section. Refer to the Solutions Manual.

HOW IT works

THE PACEMAKER: Helping a Broken Heart

Your heart is made of cardiac muscle tissue that contracts and relaxes continuously. This beating results from electric impulses moving along pathways throughout your heart. A group of specialized cells in the upper wall of the heart's right atrium—upper chamber—generates electric impulses. If these cells fail to function or the electric impulse pathways are interrupted, the heart does not beat normally. A pacemaker is an electrical device that can monitor and correct an irregular heartbeat. How does it work?

1 Leads Insulated wires called leads carry electric signals between the heart and the pacemaker. A lead is implanted into a blood vessel and then into a chamber of the heart. Pacemakers might use one, two, or three leads, each in a different chamber.

2 Pacemaker Each lead is connected to the pacemaker, which contains a battery, an antenna, and computer circuits. The pacemaker is then implanted under the skin below the collarbone. Surgery is also required to replace a pacemaker when its battery is low or the circuits fail.

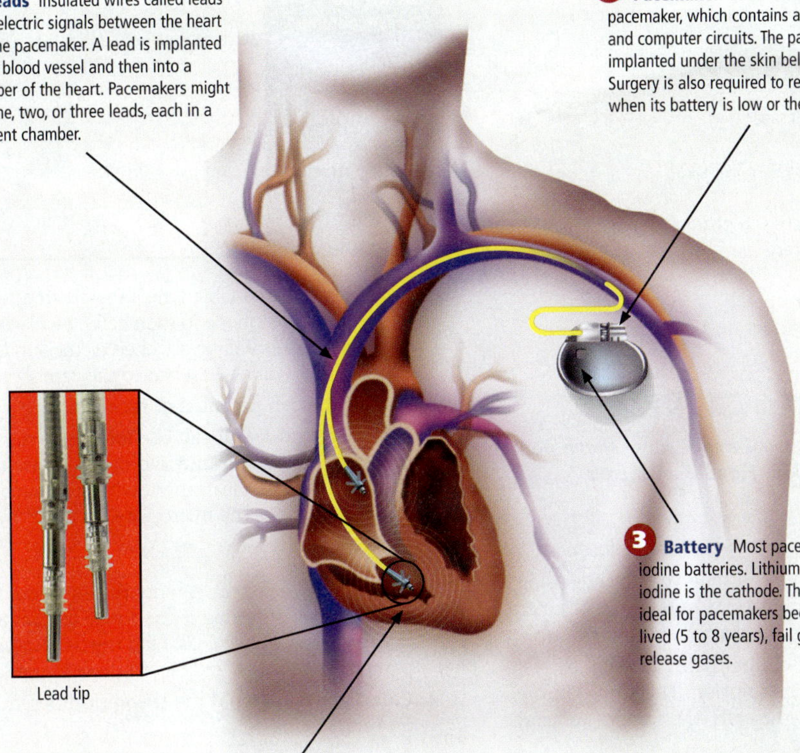

3 Battery Most pacemakers use lithium-iodine batteries. Lithium is the anode and iodine is the cathode. These batteries are ideal for pacemakers because they are long-lived (5 to 8 years), fail gradually, and do not release gases.

4 Function The tip of the lead attaches to heart tissue, transmits electric impulses, and senses the heartbeat. Pacemakers can stimulate a sluggish heartbeat, smooth an irregular beat, or activate the beating of an arrested heart.

WRITING IN ▸ Chemistry

Summarize The pacemaker is an example of an advancement in medical technology. Write a paragraph about the ways your life or that of someone you know has changed because of medical technology and how it might change in the future.

WebQuest

HOW IT works

Purpose

Students will learn how pacemakers and ICD devices can regulate heart function.

Background

Pacemakers were first used in the 1950s and have grown in reliability and functionality. They also have become smaller and lighter. Students might confuse pacemakers, devices that work with the patient's heart, with the much more complicated idea of an artificial heart. The difference might be compared with a TV remote control and the television itself.

Teaching Strategies

- The chart on this page indicates the placement of the pacemaker in relation to the heart. The leads to the heart are threaded through veins and are embedded in the walls of the heart.
- Using a schematic, detail how a lithium iodide battery for a pacemaker works. Lithium (Li) is the electron donor, producing positively-charged lithium ions in the process. Iodine (I) is the electron acceptor, forming the ionic compound lithium iodide (LiI).

WRITING IN ▸ Chemistry

Rubric WebQuest

Summarize Almost everyone has received immunizations, a technology that affects body chemistry throughout our lives. Other common technologies we might carry with us include ear tubes and tooth fillings. Contact lenses might be considered a removable body technology.

SMALL SCALE **ChemLAB**

Rubric

Preparation

Time Allotment one class period

Process Skills analyze and conclude, classify, compare and contrast

Alternative Materials Voltmeters can be used in place of the CBL. Tell the students which metals are anodes and cathodes because a negative voltage on some types of voltmeters can damage them. The CBL voltage range is ±10 volts.

Safety Precautions Approve lab safety forms before work begins. **WARNING: *Keep potassium nitrate away from heat: fire and explosion danger.*** Remind students not to use the CBL and voltage probe on AC outlets. Lab solutions are irritants. Remind students to wash hands and work area thoroughly. Review the MSDS for hazardous chemicals with students prior to doing the activity.

Disposal Rinse solutions down a drain with lots of water. Reuse metals.

Preparation of Solutions
See the Teacher Handbook for preparation of all solutions.

Procedure

- To create voltaic cells, place the four different solutions in wells A1, A2, B1, and B2. The metals should be put in the wells with the matching ion. Connect cells A1 and A2 with a salt bridge (filter paper saturated with potassium nitrate). After the potential for this cell is measured, remove the salt bridge. Use a new salt bridge to connect A1 and B1, then A1-B2, B1-B2, A2-B1, and A2-B2.
- **Troubleshooting** The CBL will read about 2 V as a default, even when nothing is being tested. This is normal.

SMALL SCALE **ChemLAB**

Measure Voltaic Cell Potentials

Background: When two different half-cells are joined, a potential difference is produced. A voltmeter measures the potential difference of combinations of half-cells.

Question: *How do measured potentials of voltaic cells compare to calculated potentials?*

Materials

metal strips of copper, aluminum, zinc, and magnesium (about 0.6 cm × 1.3 cm)
1*M* copper(II) nitrate
1*M* aluminum nitrate
1*M* zinc nitrate
1*M* magnesium nitrate
24-well microplate
Beral-type pipette (5)
voltmeter
filter paper (6 pieces 0.6 cm × 2.5 cm)
1*M* potassium nitrate
forceps
steel wool or sandpaper
table of standard reduction potentials

Safety Precautions

WARNING: ***Steel wool might have sharp points that can pierce skin. Handle with care.***

Procedure

1. Read and complete the lab safety form.
2. Plan how you will arrange voltaic cells using the four metal combinations in the 24-well microplate. Have your instructor approve your plan.
3. Soak the strips of filter paper in 2 mL of potassium nitrate solution. The strips are the salt bridges for the cells. Use forceps to handle the salt bridges.
4. Construct voltaic cells using the four metals and 1 mL of each of the solutions. Put the metals into the wells that contain the appropriate solution; for example, put the zinc metal in the solution with zinc nitrate. Use a different salt bridge for each voltaic cell. Connect the leads from the voltmeter to the metals. If you get a negative value for potential difference, switch the leads on the metals.

Voltaic Cell Potential Data			
Anode metal			
Cathode metal			
Measured cell potential (V)			
Anode half-reaction and standard potential			
Cathode half-reaction and standard potential			
Theoretical cell potential			
% Error			

5. Record in the data table which metals are the anode and the cathode in each cell. The black lead of the voltmeter will be attached to the anode. The red lead will be attached to the cathode.
6. Record the cell potential of each cell.
7. **Cleanup and Disposal** Use forceps to remove the metal strips from the microplate. Rinse them with water, then clean them with steel wool or sandpaper. Rinse the microplate.

Analyze and Conclude

1. **Apply** In the data table, write the equations for half-reactions occurring at the anode and cathode in each of the voltaic cells. Find the half-reaction potentials in **Table 1,** and record them in the table.
2. **Calculate** and record the theoretical potential for each voltaic cell.
3. **Predict** Using your data, rank the metals in order of most active to least active.
4. **Error Analysis** Calculate the percent error of the voltaic cell potential. Why is the percent error large for some voltaic cells and small for others?

INQUIRY EXTENSION

Design an Experiment that would reduce the percent error discussed in Question 4.

Analyze and Conclude

1. Refer to the Solutions Manual.
2. Refer to the Solutions Manual.
3. Refer to the Solutions Manual.
4. It is hard to get ideal conditions for each cell; some cells will get better results than others.

INQUIRY EXTENSION

Students might suggest cleaning the metal electrodes thoroughly with sandpaper or steel wool. Drying them thoroughly and checking the contacts on the voltmeter. Students might also suggest checking the molarities of the electrolyte solutions.

CHAPTER 20 STUDY GUIDE

Vocabulary Practice

BIGIDEA Chemical energy can be converted to electric energy and electric energy to chemical energy.

SECTION 1 Voltaic Cells

MAINIDEA In voltaic cells, oxidation takes place at the anode, yielding electrons that flow to the cathode, where reduction occurs.

- In a voltaic cell, oxidation and reduction take place at electrodes separated from each other.
- The standard potential of a half-cell reaction is its voltage when paired with a standard hydrogen electrode under standard conditions.
- The reduction potential of a half-cell is negative if it undergoes oxidation when connected to a standard hydrogen electrode. The reduction potential of a half-cell is positive if it undergoes reduction when connected to a standard hydrogen electrode.
- The standard potential of a voltaic cell is the difference between the standard reduction potentials of the half-cell reactions.

$$E^0_{\text{cell}} = E^0_{\text{reduction}} - E^0_{\text{oxidation}}$$

VOCABULARY
- salt bridge
- electrochemical cell
- voltaic cell
- half-cell
- anode
- cathode
- reduction potential
- standard hydrogen electrode

SECTION 2 Batteries

MAINIDEA Batteries are voltaic cells that use spontaneous reactions to provide energy for a variety of purposes.

- Primary batteries can be used only once; secondary batteries can be recharged.
- When a battery is recharged, electric energy supplied to the battery reverses the direction of the battery's spontaneous reaction.
- Fuel cells are batteries in which the substance oxidized is a fuel from an external source.
- Methods of preventing corrosion are painting, coating with another metal, or using a sacrificial anode.

VOCABULARY
- battery
- dry cell
- primary battery
- secondary battery
- fuel cell
- corrosion
- galvanization

SECTION 3 Electrolysis

MAINIDEA In electrolysis, a power source causes nonspontaneous reactions to occur in electrochemical cells.

- In an electrolytic cell, an outside source of power causes a nonspontaneous redox reaction to occur.
- The electrolysis of molten sodium chloride yields sodium metal and chlorine gas. The electrolysis of brine yields hydrogen gas, sodium hydroxide, and chlorine gas.
- Metals such as copper are purified in an electrolytic cell.
- Electrolysis is used to electroplate objects and to produce pure aluminum from its ore.

VOCABULARY
- electrolysis
- electrolytic cell

Use the Vocabulary

To reinforce chapter vocabulary, have students write a sentence using each term. OL EL

Review Strategies

- Have students work in groups to create a puzzle of scrambled words using the vocabulary from Chapter 20. OL COOP LEARN
- Have students draw and label an electrochemical cell with the appropriate terms. OL EL
- Problems from the Supplemental Practice Problems appendix or the Supplemental Problems booklet can be used for review. OL

Students can review with these online resources.

Vocabulary Practice

Vocabulary eGames
Vocabulary eFlashcards
Vocabulary PuzzleMaker

Section Self-Check

Chapter Self-Check

Online Test Practice

Use *eAssessment* to:

- create multiple versions of tests
- edit existing questions and add your own questions
- build tests aligned with select state curriculum standards using built-in state curriculum tags
- track students' progress

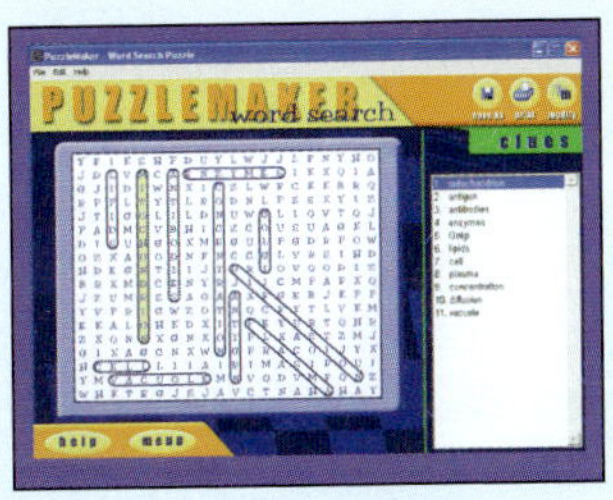

Vocabulary Practice

For additional practice with vocabulary, have students access the Vocabulary PuzzleMaker.

SECTION 1

Mastering Concepts

30. the transfer of electrons between reactants

31. oxidation of Zn(s) to $Zn^{2+} + 2e^-$

32. The salt bridge completes the electrical circuit and prevents the accumulation of positive or negative charge in the half-cells.

33. standard reduction potentials for each half-reaction

34. Al is oxidized and Cu^{2+} is reduced.

35. 25°C, 1 atm, 1*M* ion solution

36. Zinc is being oxidized; copper is the cathode.

37. The movement of the ions in the salt bridge allows the current to flow even though reactants are not in direct contact. Ions carry the electric current and prevent the buildup of positive change at the anode and negative charge at the cathode.

38. MnO_4^-, Au^+, H_2O_2

Mastering Problems

39. **a.** $Zn|Zn^{2+} \parallel H^+|H_2$
b. $H_2|H^+ \parallel Hg^{2+}|Hg$
c. $H_2|H^+ \parallel Cu^{2+}|Cu$
d. $Al\|Al^{3+} \parallel H^+|H_2$

40. **a.** $2I^- + 2Fe^{3+} \longrightarrow I_2 + 2Fe^{2+}$
b. $Sn + 2Ag^+ \longrightarrow Sn^{2+} + 2Ag$
c. $Zn + Cd^{2+} \longrightarrow Zn^{2+} + Cd$

41. **a.** +0.9258 V
b. +0.928 V
c. +0.673 V

42. **a.** The anode is zinc.
b. The cathode is silver.
c. Oxidation occurs at the zinc electrode.
d. Reduction occurs at the silver electrode.
e. The current flows from the zinc electrode to the silver electrode.
f. Positive ions flow from the anode half-cell to the cathode half-cell.
g. $E^0 = +1.5614$ V

CHAPTER 20 ASSESSMENT

SECTION 1

Mastering Concepts

30. What feature of an oxidation-reduction reaction allows it to be used to generate an electric current?

31. Describe the process that releases electrons in a zinc-copper voltaic cell.

32. What is the function of a salt bridge in a voltaic cell?

33. What information do you need in order to determine the standard voltage of a voltaic cell?

34. In a voltaic cell represented by $Al|Al^{3+} \parallel Cu^{2+}|Cu$, what is oxidized and what is reduced as the cell delivers current?

35. Under what conditions are standard reduction potentials measured?

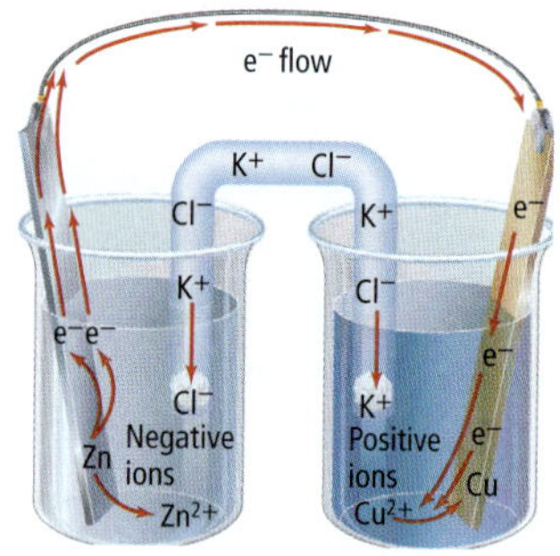

■ **Figure 24**

36. In **Figure 24,** identify the metal that is being oxidized. Identify the cathode.

37. A salt bridge is filled with KNO_3. Explain why it is necessary that the potassium ions move through the salt bridge to the cathode.

38. Recall that a reducing agent is the substance being oxidized and an oxidizing agent is the substance being reduced. Use **Table 1** to select an oxidizing agent that will convert Au to Au^{3+} but will not convert Co^{2+} to Co^{3+}.

Mastering Problems

39. Using **Table 1,** write the standard cell notation for each cell in which each of the following half-cells is connected to the standard hydrogen electrode.
a. $Zn^{2+} | Zn$ **c.** $Cu^{2+} | Cu$
b. $Hg^{2+} | Hg$ **d.** $Al^{3+} | Al$

40. Write the balanced chemical equation for the standard cell notations listed below.
a. $I^- | I_2 \parallel Fe^{3+} | Fe^{2+}$
b. $Sn | Sn^{2+} \parallel Ag^+ | Ag$
c. $Zn | Zn^{2+} \parallel Cd^{2+} | Cd$

41. Calculate the cell potentials for the following reactions.
a. $2Ag^+(aq) + Pb(s) \rightarrow Pb^{2+}(aq) + 2Ag(s)$
b. $Mn(s) + Ni^{2+}(aq) \rightarrow Mn^{2+}(aq) + Ni$
c. $I_2(aq) + Sn(s) \rightarrow 2I^-(aq) + Sn^{2+}(aq)$

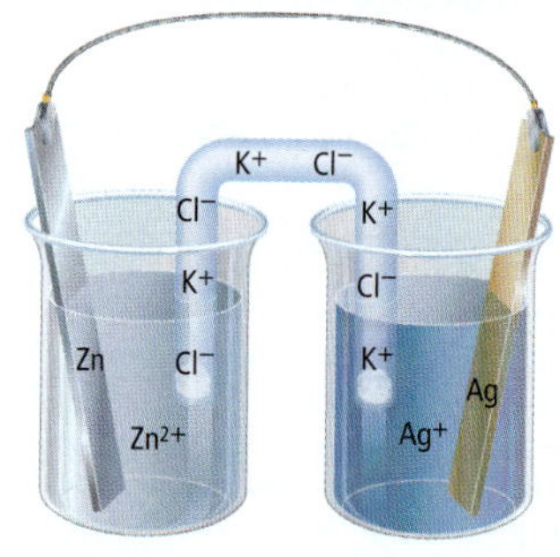

■ **Figure 25**

42. **Figure 25** illustrates a voltaic cell consisting of a strip of zinc in a 1.0*M* solution of zinc nitrate and a strip of silver in a 1.0*M* solution of silver nitrate. Use the diagram and **Table 1** to answer these questions.
a. Identify the anode.
b. Identify the cathode.
c. Where does oxidation occur?
d. Where does reduction occur?
e. In which direction is the current flowing through the connecting wire?
f. In which direction are positive ions flowing through the salt bridge?
g. What is the cell potential at 25°C and 1 atm?

SECTION 2

Mastering Concepts

43. What part of a zinc-carbon dry cell is the anode? Describe the reaction that takes place there.

44. How do primary and secondary batteries differ?

45. **Lead-Acid Battery** What substance is reduced in a lead-acid storage battery? What substance is oxidized? What substances are produced in each reaction?

46. **Biofuel Cell** At the cathode of a biofuel cell, Fe^{3+} in potassium hexacyanoferrate(III) ($K_3[Fe(CN)_6]$) is reduced to Fe^{2+} in potassium hexacyanoferrate(II) ($K_4[Fe(CN)_6]$). At the anode, reduced nicotinamide-adenine-dinucleotide (NADH) is oxidized to NAD^+. Use the following standard reduction potentials to determine the potential of the cell.

$NAD^+ + H^+ + 2e^- \rightarrow NADH \quad E^0 = -0.320$ V

$[Fe(CN)_6]^{3-} + 1e^- \rightarrow [Fe(CN)_6]^{4-} \quad E^0 = +0.36$ V

SECTION 2

Mastering Concepts

43. The anode is the zinc shell. Oxidation of Zn atoms to Zn^{2+} ions takes place.

44. Primary batteries are "throw-away"; the reaction is not easily reversed. Secondary batteries are rechargeable; the redox reaction is reversible.

45. PbO_2(s) reduced; Pb(s) oxidized; $PbSO_4$(s) and water are produced.

46. $E^0_{cell} = +0.68$ V

47. Fuel Cells List two ways in which a fuel cell differs from an ordinary battery.

48. Galvanization What is galvanization? How does galvanizing iron protect it from corrosion?

49. Batteries Explain why a lead storage battery does not produce a current when the level of H_2SO_4 is low.

50. Steel Wool is a bundle of filaments made of steel, an alloy of iron and carbon. Which would be the best way to store steel wool?
- **a.** Store it in water.
- **b.** Store it in open air.
- **c.** Store it with a desiccant.

51. Corrosion Protection List three ways metals can be protected from corrosion.

Mastering Problems

52. Half-reactions for a lead-acid storage battery are below.

$$PbO_2(s) + SO_4^{2-}(aq) + 4H^+(aq) + 2e^- \rightarrow PbSO_4(s) + 2H_2O(l) \quad E^0 = +1.685V$$

$$PbSO_4(s) + 2e^- \rightarrow Pb(s) + SO_4^{2-}(aq) \quad E^0 = -0.3588V$$

What is the standard cell potential for one cell in a car battery?

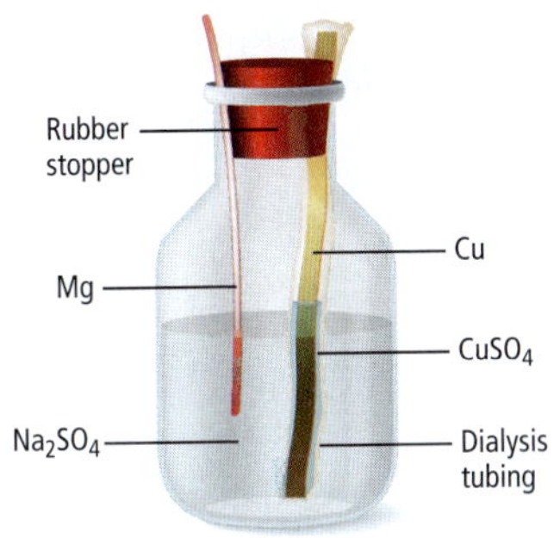

■ **Figure 26**

53. The setup in **Figure 26** acts as a battery.
- **a.** Determine the reaction that takes place at the copper strip.
- **b.** Determine the reaction that takes place at the magnesium wire.
- **c.** Identify the anode.
- **d.** Identify the cathode.
- **e.** Calculate the standard cell potential for this battery.

54. You design a battery that uses a half-cell containing Sn and Sn^{2+} and another half-cell containing Cu and Cu^{2+}. The copper electrode is the cathode, and the tin electrode is the anode. Draw the battery and write the half-reactions that occur in each half-cell. What is the maximum voltage this battery can produce?

SECTION 3

Mastering Concepts

55. How can the spontaneous redox reaction of a voltaic cell be reversed?

56. Where does oxidation take place in an electrolytic cell?

57. Down's Cell What reaction takes place at the cathode when molten sodium chloride is electrolyzed?

58. Industry Explain why the electrolysis of brine is done on a large scale at many sites around the world.

59. Recycling Explain how recycling aluminum conserves energy.

60. Describe what happens at the anode and the cathode in the electrolysis of KI(aq).

Mastering Problems

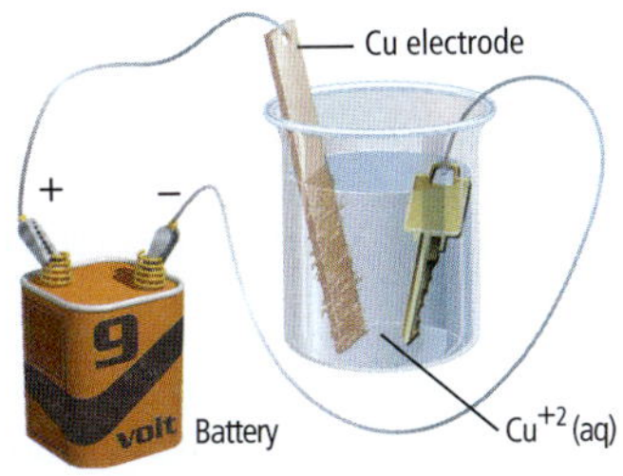

■ **Figure 27**

61. Electroplating **Figure 27** shows a key being electroplated with copper in an electrolytic cell. Where does oxidation occur? Explain your answer.

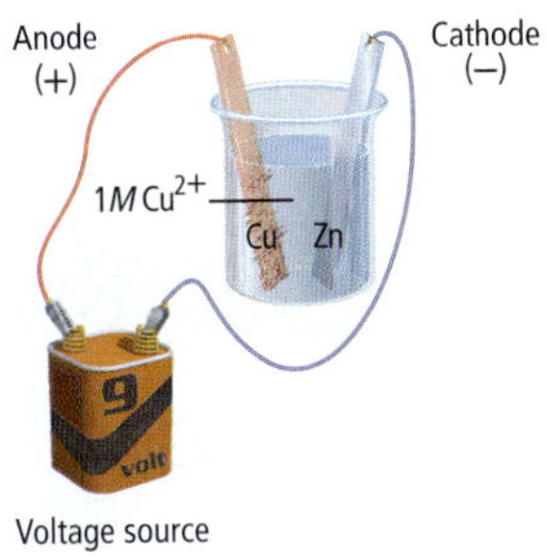

■ **Figure 28**

62. Answer the following questions based on **Figure 28.**
- **a.** Which electrode grows? Write the reaction that occurs at this electrode.
- **b.** Which electrode disappears? Write the reaction that occurs at this electrode.

63. Using **Figure 28,** explain what happens to the copper ions in solution.

47. In a fuel cell, the oxidation of a fuel is used to produce electricity. Batteries must be recharged or replaced. Current can be produced and sustained in a fuel cell as long as a fuel source is present.

48. Galvanization is the coating of corrosion-susceptible metals with "self-protecting" metals to prevent corrosion. Galvanization preserves the metal underneath by preventing air and moisture from coming in contact with it. When the galvanic coating breaks, it still protects the metal by acting as a sacrificial anode and itself being oxidized.

49. Sulfuric acid participates in the reaction. When the concentration is low, the reaction cannot take place.

50. c. Water is a reactant in the rusting process. Desiccants absorb water from the air.

51. galvanizing, painting, sacrificial anodes

Mastering Problems

52. Cell potential $= +2.044$ V

53. a. Cu^{2+} ions are reduced
$Cu^{2+} + 2e^- \longrightarrow Cu^0$

b. The magnesium is oxidized.
$Mg^0 \longrightarrow Mg^{2+} + 2e^-$

c. the magnesium wire

d. the copper strip

e. $E^0_{cell} = +2.714$ V

54. Refer to the Solutions Manual for art.
Anode: $Sn^0 \longrightarrow Sn^{2+} + 2e^-$
Cathode: $Cu^{2+} + 2e^- \longrightarrow Cu^0$
Cell potential $= +0.4794$ V

SECTION 3

Mastering Concepts

55. by passing a current through it in the opposite direction

56. at the anode

57. Na^+ ions are reduced to Na atoms.

58. The electrolysis products of brine, hydrogen gas, chlorine gas, and sodium hydroxide, are important commercial products.

59. The aluminum in cans has already been separated from its ore, which is an energy-consuming process.

60. At the cathode, potassium ions are reduced to atoms of potassium; at the anode, iodide ions are oxidized to molecules of I_2.

Mastering Problems

61. Oxidation occurs at the anode, which is the Cu electrode. Electrons move from it to the positive pole of the battery.

62. a. The zinc electrode grows; $Cu^{2+} + 2e^- \longrightarrow Cu$.

b. The copper electrode disappears; $Cu \longrightarrow Cu^{2+} + 2e^-$.

63. The copper ions migrate to the cathode and plate out on it.

MIXED REVIEW

64. In a voltaic cell, ions in solution at the cathode accept electrons more readily than the ions at the anode. When a salt bridge and wire are in place, redox occurs spontaneously and electrons flow from the anode to the cathode.

65. oxide, O^{2-}

66. oxidation: $Cr(s) \longrightarrow Cr^{3+}(aq) + 3e^-$
reduction: $Ag^+(aq) + e^- \longrightarrow Ag(s)$
Electrons flow from anode (Cr) to cathode (Ag).

67. **a.** nonspontaneous
b. nonspontaneous
c. spontaneous
d. nonspontaneous

68. **a.** $E^0_{cell} = +2.647$ V
b. $E^0_{cell} = +1.2759$ V
c. $E^0_{cell} = +0.892$ V
d. $E^0_{cell} = +0.2641$ V

69. Fe atoms oxidize to aqueous Fe^{2+} ions. Then, part of the rust-forming process involves Fe^{2+} ions dispersing through water and reacting with O_2.

70. **a.** Anode: $2H_2 + 4OH^- \longrightarrow 4H_2O + 4e^-$
Cathode: $O_2 + 2H_2O + 4e^- \longrightarrow 4OH^-$
b. Cell potential = +1.229 V

71. In a fuel cell, the oxidation of hydrogen is controlled so that most of the chemical energy is converted to electrical energy instead of heat.

72. The direction that the current is directed through the cell determines that the impure copper will be the anode.

73. chemical potential energy

74. **a.** cathode is steel; anode is Mg
b. Mg is more active and therefore more susceptible to redox which causes the magnesium to corrode before the steel.

MIXED REVIEW

64. Why do electrons flow from one electrode to the other in a voltaic cell?

65. Aluminum Production What substance is electrolyzed in the industrial process to produce aluminum metal?

66. Write the oxidation and reduction half-reactions for a silver-chromium voltaic cell. Identify the anode, cathode, and the direction of electron flow.

67. Determine whether each redox reaction is spontaneous or nonspontaneous.
a. $Mn^{2+}(aq) + 2Br^-(aq) \rightarrow Br_2(l) + Mn(s)$
b. $2Fe^{2+}(aq) + Sn^{2+}(aq) \rightarrow 2Fe^{3+}(aq) + Sn(s)$
c. $Ni^{2+}(aq) + Mg(s) \rightarrow Mg^{2+}(aq) + Ni(s)$
d. $Pb^{2+}(aq) + 2Cu^+(aq) \rightarrow Pb(s) + 2Cu^{2+}(aq)$

68. Determine the voltage of the cell in which each half-cell is connected to a $Ag^+ \mid Ag$ half-cell.
a. $Be^{2+} \mid Be$
b. $S \mid S^{2-}$
c. $Au^+ \mid Au$
d. $I_2 \mid I^-$

69. Corrosion Explain why water is necessary for the corrosion of iron.

70. Space Travel The space shuttle uses a H_2/O_2 fuel cell to produce electricity.
a. What is the reaction at the anode? At the cathode?
b. What is the standard cell potential for the fuel cell?

71. Fuel Cells Explain how the oxidation of hydrogen in a fuel cell differs from the oxidation of hydrogen when it burns in air.

72. Copper Refining In the electrolytic refining of copper, what factor determines which piece of copper is the anode and which is the cathode?

73. Storage Batteries Lead-acid batteries and other rechargeable batteries are sometimes called storage batteries. What is being stored in these batteries?

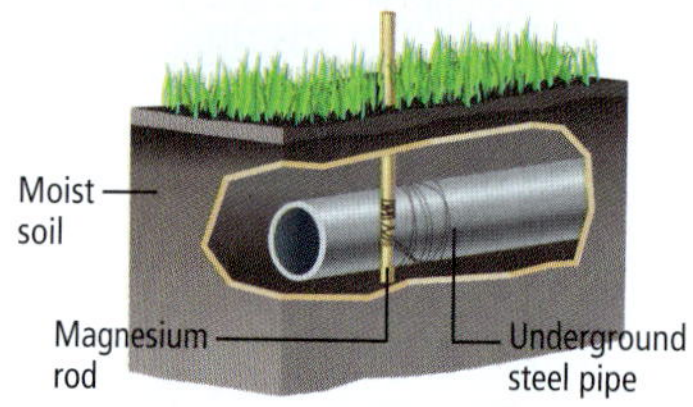

Figure 29

74. Corrosion Prevention **Figure 29** shows how buried steel pipes can be protected against corrosion. The steel pipe is connected to a more active metal that corrodes instead of the steel.
a. What is the cathode? What is the anode?
b. Describe how the magnesium metal protects the steel.

THINK CRITICALLY

75. Predict Suppose that scientists had chosen the $Cu^{2+} \mid Cu$ half-cell as a standard instead of the $H^+ \mid H_2$ half-cell? What would the potential of the hydrogen electrode be if the copper electrode were the standard? How would the relationships among the standard reduction potentials change?

76. Apply Suppose that you have a voltaic cell in which one half-cell is made up of a strip of tin immersed in a solution of tin(II) ions.
a. How could you tell by measuring voltage whether the tin strip was acting as a cathode or an anode in the cell?
b. How could you tell by simple observation whether the tin strip was acting as a cathode or an anode?

77. Hypothesize The potential of a half-cell varies with concentration of reactants and products. For this reason, standard potentials are measured at 1*M* concentration. Maintaining a pressure of 1 atm is especially important in half-cells that involve gases as reactants or products. Suggest a reason why gas pressure is critical in these cells.

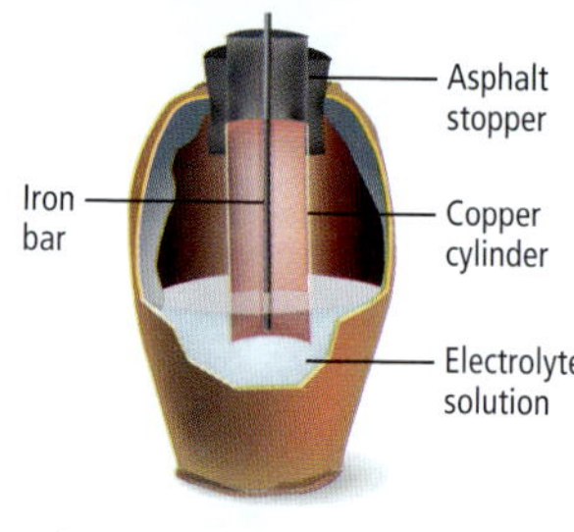

Figure 30

78. Analyze An earthen vessel was discovered in 1938 near Baghdad. This ancient vessel contained an iron bar surrounded by a copper cylinder, as shown in **Figure 30.** When filled with an electrolyte such as vinegar, this vessel might have acted as a battery.
a. Identify the cathode.
b. Identify the anode.
c. Calculate the standard cell potential of this battery.

79. Apply During electrolysis, an electrolytic cell releases bromine vapor and hydrogen gas. After electrolysis, the cell is found to contain a concentrated solution of potassium hydroxide. What was the composition of the cell before electrolysis began?

80. Hypothesize Suppose in galvanization, copper was plated on iron instead of zinc. Would copper continue to protect the iron from corrosion, as zinc does, if the copper coating became broken or cracked? Explain.

THINK CRITICALLY

75. Values in the standard reduction potential table would change by 0.3419 V. Hydrogen electrode would be −0.3419 V. Relationships would not change but voltage values would change.

76. **a.** A voltmeter indicates the flow of electrons to or from the tin strip, which shows whether the strip acts as a cathode or an anode. If the voltage is positive, tin is being oxidized.
b. A visible deposit at the cathode would indicate reduction of Sn^{2+}. A decrease in strip size would indicate oxidation of Sn at the anode.

77. Pressure is an indication of concentration; therefore, pressure is a concentration factor in half-cells involving gases.

78. **a.** the copper cylinder, $E^0 = +0.3419$ V
b. the iron bar, $E^0 = -0.447$ V
c. $Cu^{2+} + 2e^- \longrightarrow Cu$
$Fe^{2+} + 2e^- \longrightarrow Fe$
Cell potential = +0.789 V

79. potassium bromide and water

80. If the copper cracked, sites for corrosion would be exposed. No, iron is more readily oxidized than copper, so protection is diminished.

Chapter Self-Check

Challenge Problem

81. A battery is assembled using tin and mercury, which have the following reduction half-reactions:

$Sn^{2+}(aq) + 2e^- \rightarrow Sn(aq)$

$Hg^{2+}(aq) + 2e^- \rightarrow Hg(l)$

- **a.** Write a balanced equation for the cell's reaction.
- **b.** What is oxidized and what is reduced? Identify the oxidizing agent and the reducing agent.
- **c.** Which reaction occurs at the anode? At the cathode?
- **d.** What is the cell potential? Use **Table 1.**
- **e.** If sodium sulfate solution is in the salt bridge, in which direction do the sulfate ions move?

CUMULATIVE REVIEW

82. If the volume of a sample of chlorine gas is 8.2 L at 1.2 atm and 306 K, what volume will the gas occupy at STP?

83. What is meant by solvation? Explain how this process is important for the dissolving of ionic salts in water.

84. Explain how the molarity of a solution is different from its molality.

85. Define the calorie. State how the calorie is related to the Calorie and the joule.

86. Explain why you would find an aluminum chair to be hotter to sit on than a wooden bench after each had been in the sunlight for the same amount of time.

87. What does a negative sign for the free energy of a reaction tell you about the reaction? $(\Delta G_{system} = \Delta H_{system} - T\Delta S_{system})$

88. According to the collision model of chemical reactions, how is it possible that two molecules can collide but not react?

89. List five factors that can affect the rate of a reaction.

90. The decomposition reaction $A_2B \rightarrow 2A + B$ proceeds to equilibrium at 499°C. Analysis of the equilibrium mixture shows $[A_2B] = 0.855$ mol/L, $[A] = 2.045$ mol/L, and $[B] = 1.026$ mol/L. What is K_{eq}?

91. What is the solubility in mol/L of silver iodide, AgI. K_{sp} for AgI is 3.5×10^{-17}.

92. If you have a solution of a strong acid, is that the same as having a concentrated solution of the acid? Explain your answer.

93. What are the oxidation numbers for the elements in the ion PO_4^{3-}?

WRITING IN ▶ Chemistry

94. Sunken Ships Study of the sunken ocean liner *Titanic* has opened the possibility that deterioration of the steel hull might be partly due to the presence of rusticle communities. Research how the biological activity of rusticle communities results in the oxidation of iron. Write an essay that describes the role of rusticle communities in the destruction of the *Titanic.*

95. Statue of Liberty Several years ago, the supporting structure of the Statue of Liberty became so corroded that it had to be replaced entirely. Find out what the structure was made of and why it corroded so badly. Write a report that explains the chemical processes involved and include a time line of the statue, starting in France before 1886.

DBQ Document-Based Questions

Electrochemical Biological Reactions *Standard reduction potentials for some important biological reactions are given in* **Table 2.** *The strongest oxidizing agent generally available in biological systems is molecular oxygen. Consider the oxidation of reduced nicotinamide-adenine-dinucleotide (NADH) by molecular oxygen. The reaction is the following.*

$$2NADH + 2H^+ + O_2 \rightarrow 2NAD^+ + 2H_2O.$$

Data obtained from: Fromm, James Richard. 1997. "Biochemical Electrochemistry," last modified 1997, accessed September 1, 2010, http://www.3rd1000.com/chem301/chem302z.htm.

Table 2 Aqueous Standard Reduction Potentials at 25°C and pH 7.00

Electrode Couple	E^0 (V)
$2H^+ + 2e^- \rightarrow H_2(g)$	−0.4141
$NAD^+ + H^+ + 2e^- \rightarrow NADH$	−0.320
$HOOCCOCH_3$* $+ 2H^+ + 2e^- \rightarrow HOOCCHOHCH_3$**	+0.19
$Fe^{3+} + e^- \rightarrow Fe^{2+}$	+0.769
$O_2(g) + 4H^+ + 4e^- \rightarrow 2H_2O$	+0.8147

* $HOOCCOCH_3$ is pyruvic acid
** $HOOCCHOHCH_3$ is l-lactic acid.

96. Write the two half-reactions that take place in this reaction.

97. Calculate the cell potential of this reaction using **Table 2.**

98. Will NAD^+ oxidize Fe^{2+} to Fe^{3+}? Explain your answer.

CHALLENGE PROBLEM

81. a. $Hg^{2+}(aq) + Sn(s) \rightarrow Hg(l) + Sn^{2+}(aq)$

b. Mercury is reduced and tin is oxidized. The oxidizing agent is the mercury ions. The reducing agent is the tin electrode.

c. $Hg^{2+} + 2e^- \rightarrow Hg$ occurs at the cathode, and $Sn \rightarrow Sn^{2+} + 2e^-$ occurs at the anode.

d. Cell potential = +0.989 V

e. The sulfate ions move toward the tin half-cell.

CUMULATIVE REVIEW

82. 8.8 L

83. Solvation is the process in which solute particles are surrounded by solvent particles to form a solution. Polar water molecules are attracted to either positive or negative ions in an ionic solid. This attraction is greater than the attraction of the ions for each other in the solid. Water molecules surround the ions and prevent them from recombining.

84. Molarity is the number of moles of solute in a liter of solution; molality is the moles of solute in a kilogram of solvent.

85. The calorie is the amount of energy needed to raise the temperature of one gram of pure water by one degree Celsius. 1000 calories equals 1 Calorie. 1 calorie equals 4.184 J.

86. The aluminum chair has a lower specific heat than wood.

87. spontaneous as written

88. The molecules might not collide with the correct orientation, or the collision might lack the energy required to form the activated complex.

89. the nature of the reactants, temperature, concentration, surface area, and whether a catalyst is used

90. $K_{eq} = 5.02\ mol^2/L^2$

91. 5.9×10^{-9} mol/L

92. Not necessarily. A strong acid is one that ionizes completely in aqueous solution. A solution of a strong acid can be either concentrated or dilute. Whether it is concentrated or dilute depends upon the number of moles of acid in the solution.

93. P is +5; O is −2.

WRITING IN ▶ Chemistry

Rubric

94. Student papers might state that rusticle biological communities might consist of up to 35% iron compounds.

95. The most critical corrosion resulted from a naturally occurring electrochemical cell involving the iron inner structure that supports the outer copper "skin."

DBQ Document-Based Questions

Data obtained from: Fromm, James Richard. 1997. "Biochemical Electrochemistry," last modified 1997, accessed September 1, 2010, http://www.3rd1000.com/chem301/chem302z.htm.

96. $NADH \rightarrow NAD^+ + 2e^- + H^+$

$O_2 + 4H^+ + 4e^- \rightarrow 2H_2O$

97. cell potential = +1.135 V

98. No, the NAD^+ has a reduction potential of −0.320 V. The cell potential = −1.091 V, a nonspontaneous reaction.

MULTIPLE CHOICE

1. B
2. B
3. B
4. A
5. D
6. C
7. B
8. A

MULTIPLE CHOICE

Use the table below to answer Questions 1 to 4.

Selected Standard Reduction Potentials at 25°C, 1 atm, and 1*M* Ion Concentration

Half-Reaction	E^0 (V)
$Mg^{2+} + 2e^- \rightarrow Mg$	−2.372
$Al^{3+} + 3e^- \rightarrow Al$	−1.662
$Pb^{2+} + 2e^- \rightarrow Pb$	−0.1262
$Ag^+ + e^- \rightarrow Ag$	0.7996
$Hg^{2+} + 2e^- \rightarrow Hg$	0.851

1. Which metal ion is most easily reduced?
A. Mg^{2+}
B. Hg^{2+}
C. Ag^+
D. Al^{3+}

2. On the basis of the standard reduction potentials shown above, which standard cell notation correctly represents its voltaic cell?
A. $Ag \mid Ag^+ \parallel Al^{3+} \mid Al$
B. $Mg \mid Mg^{2+} \parallel H^+ \mid H_2$
C. $H_2 \mid H^+ \parallel Pb^{2+} \mid Pb$
D. $Pb \mid Pb^{2+} \parallel Al^{3+} \mid Al$

3. A voltaic cell consists of a magnesium bar dipping into a 1*M* Mg^{2+} solution and a silver bar dipping into a 1*M* Ag^+ solution. What is the standard potential of this cell?
A. 1.572 V
B. 3.172 V
C. 0.773 V
D. 3.971 V

4. Assuming standard conditions, which cell will produce a potential of 2.513 V?
A. $Al \mid Al^{3+} \parallel Hg^{2+} \mid Hg$
B. $H_2 \mid H^+ \parallel Hg^{2+} \mid Hg$
C. $Mg \mid Mg^{2+} \parallel Al^{3+} \mid Al$
D. $Pb \mid Pb^{2+} \parallel Ag^+ \mid Ag$

5. Which statement is NOT true of batteries?
A. Batteries are compact forms of voltaic cells.
B. Secondary batteries are storage batteries.
C. A battery can consist of a single cell.
D. The redox reaction in a rechargeable battery is not reversible.

6. Which is NOT a characteristic of a basic solution?
A. tastes bitter
B. conducts electricity
C. contains more H^+ ions than OH^- ions
D. feels slippery

7. A carbonated soft drink has a pH of 2.5. What is the concentration of H^+ ions in the soft drink?
A. $3 \times 10^{-12}M$
B. $3 \times 10^{-3}M$
C. $4 \times 10^{-1}M$
D. $1 \times 10^{1}M$

8. Which graph correctly shows the relationship between average kinetic energy of particles and the temperature of a sample?

A.

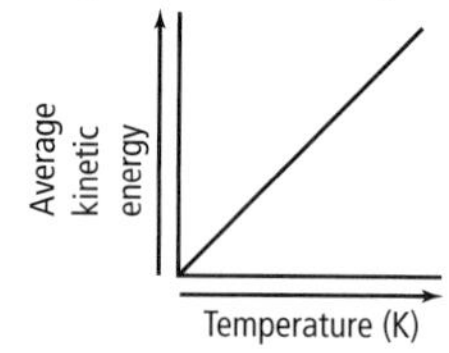

B.

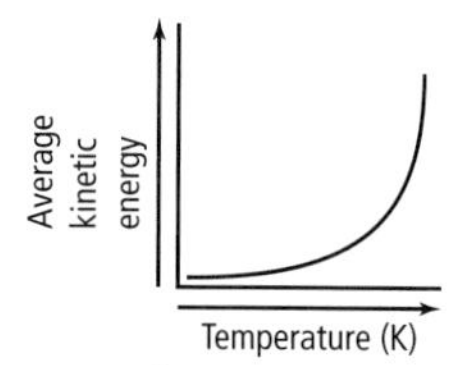

C.

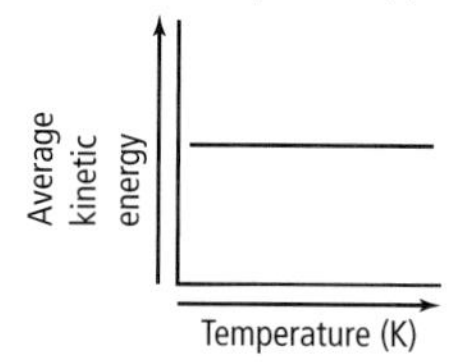

D.

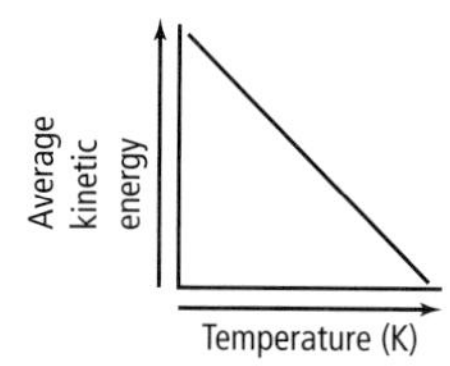

SHORT ANSWER

Use the description below to answer Questions 9 to 11.

In an experimental setup, chlorine gas and nitrogen gas are in separate containers separated by a closed stopcock. One hour after the stopcock is opened, the gases have completely mixed.

9. Five minutes after the stopcock is opened, which gas will have traveled farther, the nitrogen or the chlorine?

10. Give the ratio of the speed of nitrogen gas to the speed of chlorine gas.

11. Evaluate this statement: After one hour, the gas particles stop moving because they have completely mixed.

EXTENDED RESPONSE

Use the table below to answer Question 12.

Standard Reduction Potentials at 25°C, 1 atm, and 1*M* Solution	
$Ag^+ + e^- \rightarrow Ag$	0.7996
$Cr^{3+} + 3e^- \rightarrow Cr$	−0.744

12. Based on the standard reduction potentials given above, if a silver electrode and a chromium electrode are connected in a voltaic cell, which electrode will undergo oxidation and which will undergo reduction? Explain how you can tell.

13. Use Le Châtelier's principle to explain why the instructions for a chemical experiment sometimes instruct the chemist to cool the reaction in an ice bath.

SAT SUBJECT TEST: CHEMISTRY

14. The hydrogen sulfide produced as a by-product of petroleum refinement can be used to produce elemental sulfur: $2H_2S(g) + SO_2(g) \rightarrow 3S(l) + 2H_2O(g)$.

What is the equilibrium constant expression for this reaction?

A. $K_{eq} = \frac{[H_2O]}{[H_2S][SO_2]}$

B. $K_{eq} = \frac{[H_2S]^2[SO_2]}{[H_2O]^2}$

C. $K_{eq} = \frac{[H_2O]^2}{[H_2S]^2[SO_2]}$

D. $K_{eq} = \frac{[S]^3[H_2O]^2}{[H_2S]^2[SO_2]}$

E. $K_{eq} = \frac{[2H_2O]^2}{[2H_2S]^2[SO_2]}$

15. Which shows the correct graph of the activation energy needed for an endothermic reaction?

A.
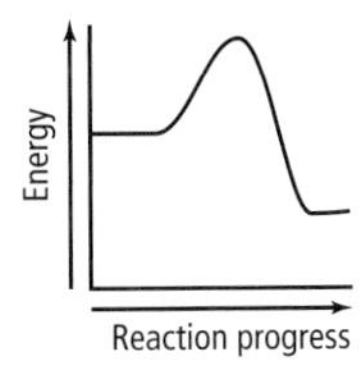

B.
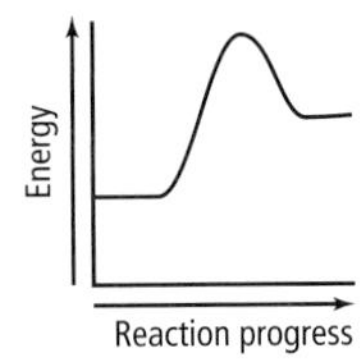

C.
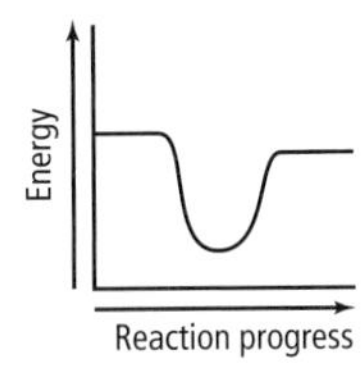

D.
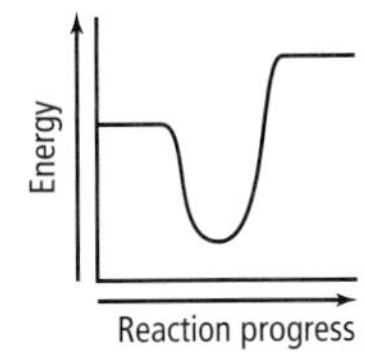

E.
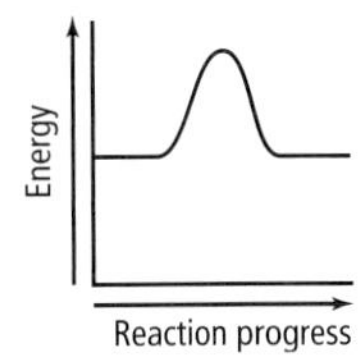

NEED EXTRA HELP?															
If You Missed Question . . .	1	2	3	4	5	6	7	8	9	10	11	12	13	14	15
Review Section . . .	20.1	20.1	20.1	20.1	20.2	18.1	18.3	15.1	12.1	12.1	12.1	8.5	17.2	17.3	15.2

SHORT ANSWER

9. The nitrogen will have traveled farther because it is made of smaller, lighter particles.
10. Rate of N_2 compared to Cl_2

$$= \sqrt{\frac{\text{molar mass } Cl_2}{\text{molar mass } N_2}} = \sqrt{\frac{70.906}{28.014}}$$

= 1.59; nitrogen diffuses 1.59 times as fast as chlorine.

11. This is not a correct statement. At equilibrium, the rates of the forward and backward processes remain constant, but they still occur.

EXTENDED RESPONSE

12. The reaction for silver has a more positive standard reduction potential, while chromium has a more negative standard reduction potential. Of any two electrodes, the one with the lower standard reduction potential will operate in the reverse direction; in this case, it is the chromium. Because it would now be losing electrons, it will be oxidized. The electrode with the more positive standard reduction potential will be reduced; in this case, it will be the silver electrode.
13. If the forward reaction is exothermic, lowering the temperature will shift the equilibrium to the right. This will, in turn, favor the forward reaction. Cooling the reaction will in this case cause more product to be formed.

SAT SUBJECT TEST: CHEMISTRY

14. C
15. B

CHAPTER 21 Hydrocarbons

BIGIDEA **Organic compounds called hydrocarbons differ by their types of bonds.**

ESSENTIAL QUESTIONS	RESOURCES TO ASSESS MASTERY
SECTION 1 **Introduction to Hydrocarbons** **1.** What do the terms *organic compound* and *organic chemistry* mean? **2.** How are hydrocarbons and the models used to represent them identified? **3.** How are saturated and unsaturated hydrocarbons distinguished? **4.** Where are hydrocarbons obtained and how are they separated? 1 session 0.5 block	**Progress Monitoring** Formative Assessment, pp. 748, 749 Reading Check, pp. 745, 746, 748 Section Review, p. 749
SECTION 2 **Alkanes** **1.** How are alkanes named when given their structures? **2.** How are alkane structures drawn when given their names? **3.** What are the properties of alkanes? 1 session 0.5 block	**Progress Monitoring** Formative Assessment, pp. 751, 753, 755, 756, 758 Reading Check, pp. 751, 752, 755 Section Review, p. 758
SECTION 3 **Alkenes and Alkynes** **1.** How do the properties of alkenes and alkynes compare with those of alkanes? **2.** How are the molecular structures of alkenes and alkynes described? **3.** How are alkenes and alkynes named when given their structures? **4.** How are alkenes and alkynes drawn when given their names? 2 sessions 1 block	**Progress Monitoring** Formative Assessment, pp. 760, 763 Reading Check, pp. 760, 762 Section Review, p. 764
SECTION 4 **Hydrocarbon Isomers** **1.** How can the two main categories of isomers—structural isomers and stereoisomers—be distinguished? **2.** How are *cis*- and *trans*- geometric isomers differentiated? **3.** What is the structural variation in molecules that results in optical isomers? 1 session 0.5 block	**Progress Monitoring** Formative Assessment, pp. 766, 767, 768, 769 Reading Check, p. 766 Section Review, p. 769
SECTION 5 **Aromatic Hydrocarbons** **1.** How do the properties of aromatic and aliphatic hydrocarbons compare and contrast? **2.** What is a carcinogen and what are some examples? 1 session 0.5 block	**Progress Monitoring** Formative Assessment, pp. 771, 772 Reading Check, pp. 771, 773 Section Review, p. 774 **Summative Assessment** Chapter Assessment, p. 778 *eAssessment* Chapter Test (Scaffolded)

Go online!

LEVELED RESOURCES	LAB MATERIALS
Science Notebook 21.1 OL ***FAST FILE Chapter Resources:*** Study Guide, p. 16 BL	LaunchLAB p. 742 / **15 min** molecular model kit
Science Notebook 21.2 OL ***FAST FILE Chapter Resources:*** ChemLab Worksheet, p. 3 OL Study Guide, p. 17 BL	ChemLAB p. 776 / **45 min** barometer, thermometer, 1-L or 2-L plastic soda bottle with cap, burner tubing, pneumatic trough, 100-mL graduated cylinder, balance (0.01g), paper towels
Science Notebook 21.3 OL ***FAST FILE Chapter Resources:*** MiniLab Worksheet, p. 2 OL Study Guide, p. 18 BL	MiniLAB p. 763 / **20 min** rubber band, wood splint, ruler, water, 150-mL beaker, dishwashing detergent, forceps, calcium carbide (CaC_2), match, stirring rod, distilled water, phenolphthalein solution
Science Notebook 21.4 OL ***FAST FILE Chapter Resources:*** Study Guide, p. 19 BL	
Science Notebook 21.5 OL ***FAST FILE Chapter Resources:*** Study Guide, p. 20 BL	

ADDITIONAL RESOURCES

eAssessment
ConnectED Student Center
ConnectED Teacher Center
Lesson Presentations
eSolutions
Videos
Animations
Online Quizzes
Virtual Investigations
Multilingual eGlossary
Personal Tutors
WebQuests
Applying Practices Worksheets
PBLs

Practice:

Challenge Problems AL
Supplemental Problems BL OL
Chapter Test (Scaffolded)
Standardized Test Practice

FAST FILE Resources:

Section Focus Transparency Masters
Math Skills Transparency Masters and Worksheets
Teaching Transparency Masters and Worksheets

Additional Resources:

Blueprints for Success
Solving Problems: A Chemistry Handbook

Lab Resources:

Laboratory Manual OL
CBL Laboratory Manual OL
Small-Scale Laboratory Manual OL
Forensics Laboratory Manual OL

BL Below Level OL On Level AL Advanced Learners EL English Learners COOP LEARN Cooperative Learning

CHAPTER 21

LaunchLAB

Rubric

How can you model simple hydrocarbons?

Purpose Students will use model-building skills to help visualize simple hydrocarbon molecules.

Safety Precautions Approve lab safety forms before work begins.

Teaching Strategies
- Provide the students with tetrahedral centers (with four holes) for the carbon atoms.
- Remind students that the carbons need to be connected in a straight-chain arrangement with no branched or cyclic structures.
- Extend this lab by having the students look for the IUPAC names for these structures in the chapter.

Expected Results The students should create structures that look like the structures in Table 1.

CHAPTER 21
Hydrocarbons

BIG IDEA Organic compounds called hydrocarbons differ by their types of bonds.

SECTIONS

1 Introduction to Hydrocarbons

2 Alkanes

3 Alkenes and Alkynes

4 Hydrocarbon Isomers

5 Aromatic Hydrocarbons

LaunchLAB

How can you model simple hydrocarbons?

Hydrocarbons are made of hydrogen and carbon atoms. Recall that carbon has four valence electrons and it can form four covalent bonds. In this lab, you will build models of hydrocarbons with two, three, four, and five carbon atoms.

Procedure
1. Read and complete the lab safety form.
2. Use a **molecular model kit** to build a structure with two carbon atoms connected by a single bond.
3. Place hydrogen atoms in all of the unoccupied positions on your model so that each carbon atom has a total of four bonds.
4. Repeat Steps 2–3 for models based on three, four, and five carbon atoms each. Be sure that each carbon atom is attached to a maximum of two other carbon atoms.

Analysis
1. **Make** a table listing the number of carbon and hydrogen atoms in each structure.

C Atoms	H Atoms
2	6
3	8
4	10
5	12

2. **Describe** the composition of each structure with a molecular formula. C_2H_6, C_3H_8, C_4H_{10}, C_5H_{12}
3. **Analyze** the pattern of the carbon-to-hydrogen ratio to develop a generic formula for hydrocarbons with single bonds. C_nH_{2n+2}

Inquiry How do you think the molecular formula would be affected if the carbon atoms were attached by double and triple bonds? There would be fewer hydrogen atoms in the molecule, and the formula would reflect fewer hydrogen atoms.

FOLDABLES Study Organizer

Hydrocarbon Compounds

Make a bound book. Label it as shown. Use it to help you organize information about hydrocarbon compounds.

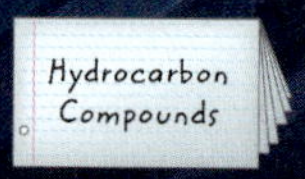

Petroleum is the primary source of hydrocarbons. Hydrocarbons are used as fuels and are the raw materials for products such as plastics, synthetic fibers, solvents, and industrial chemicals.

Introduce the BIGIDEA

Alike But Different To introduce this chapter's Big Idea, lead students in a discussion about things that are alike but different. Ask students if they have ever eaten a bag of dried mixed fruit. Ask students how the fruits were alike. Ask students how the fruits were different. Tell students that hydrocarbons are also alike but different. All hydrocarbons contain carbon and hydrogen atoms. However, they are different because they have different types of bonds, which give them different characteristics or properties.

Tie to Previous Knowledge

Have students review the following concepts before studying this chapter.

- structure of a carbon atom
- covalent bonding and molecular structure
- changes of state—boiling and melting

Use the Photo

Petroleum Ask students to identify the object in the photo. It is a pumpjack, a common device for pumping oil from the ground. Ask students what happens at the facility. Oil-field workers drill for petroleum, pump it to the surface, and transport it to a refinery. Ask students to identify uses for petroleum. Possible answers: fuel for cars, trucks, lanterns, homes, outdoor grills, and raw materials for many chemical processes, including plastics, films, and synthetic fibers.

SECTION 1

1 Focus

MAINIDEA

Carbon Building Blocks Use a molecular-model kit, and build a methane molecule (CH_4). Hold up the model and ask students if they know what molecule the model represents. Tell students the molecule is methane, which is a major component in natural gas. Explain to students that hydrocarbons are used as fuels because they combine with oxygen easily and produce a lot of heat in the process.

Build another methane molecule, and hold the model before the class. Remove a hydrogen atom from each model, and connect the two models together to create ethane. Ask students to explain how this new molecule was formed. A hydrogen was removed, and the carbon atoms bonded together. Explain to students that hydrocarbons are also used as raw materials because the molecules can be easily linked together to form long chains. The hydrogen atoms can also be removed, and a new group can be added to the molecule. The properties of the new molecule differ from the properties of the original molecule. BL OL

■ **Caption Question Fig. 1** possible answers: glucose, sucrose, or methane

SECTION 1

Introduction to Hydrocarbons

MAINIDEA Hydrocarbons are carbon-containing organic compounds that provide a source of energy and raw materials.

Essential Questions

- What do the terms *organic compound* and *organic chemistry* mean?
- How are hydrocarbons and the models used to represent them identified?
- How are saturated and unsaturated hydrocarbons distinguished?
- Where are hydrocarbons obtained and how are they separated?

Review Vocabulary

microorganism: a tiny organism, such as a bacterium or a protozoan, that cannot be seen without a microscope

New Vocabulary

organic compound
hydrocarbon
saturated hydrocarbon
unsaturated hydrocarbon
fractional distillation
cracking

CHEM 4 YOU If you have ridden in a car or a bus, you have used hydrocarbons. The gasoline and diesel fuel that are used in cars, trucks, and buses are hydrocarbons.

Organic Compounds

Chemists in the early nineteenth century knew that living things, such as the plants and panda shown in **Figure 1,** produce an immense variety of carbon compounds. Chemists referred to these compounds as *organic* compounds because they were produced by living organisms.

Once Dalton's atomic theory was accepted in the early nineteenth century, chemists began to understand that compounds, including those made by living organisms, consisted of arrangements of atoms bonded together in certain combinations. They were able to synthesize many new and useful substances. However, scientists were not able to synthesize organic compounds. Many scientists incorrectly concluded that they were unable to synthesize organic compounds because of vitalism. According to vitalism, organisms possessed a mysterious "vital force," enabling them to assemble carbon compounds.

Disproving vitalism Friedrich Wöhler (1800–1882), a German chemist, was the first scientist to realize that he had produced an organic compound, called urea, by synthesis in a laboratory. Wöhler's experiment did not immediately disprove vitalism, but it prompted a chain of similar experiments by other European chemists. Eventually, the idea that the synthesis of organic compounds required a vital force was discredited and scientists realized they could synthesize organic compounds in the laboratory.

■ **Figure 1** Living things contain, are made up of, and produce a variety of organic compounds.
Identify *two organic compounds that you have studied in a previous science course.*

Panorama Media (Beijing) Ltd./Alamy

DIFFERENTIATED INSTRUCTION

Advanced Learners Friedrich Wöhler was a German chemist whose synthesis of the organic compound urea from the inorganic substance ammonium cyanate in 1828 brought about the end of the vitalism theory. In 1835, he stated, "Organic chemistry nowadays almost drives me mad. To me it appears like a primeval tropical forest full of the most remarkable things; a dreadful endless jungle into which one does not dare to enter for there seems to be no way out." Ask students to discuss the relevancy of Wöhler's statement to today's chemistry students. AL COOP LEARN

Organic chemistry Today, the term **organic compound** is applied to all carbon-containing compounds with the primary exceptions of carbon oxides, carbides, and carbonates, which are considered inorganic. Because there are so many organic compounds, an entire branch of chemistry, called organic chemistry, is devoted to their study. Recall that carbon is an element in group 14 of the periodic table, as shown in **Figure 2.** With the electron configuration of $1s^22s^22p^2$, carbon nearly always shares its electrons and forms four covalent bonds. In organic compounds, carbon atoms are bonded to hydrogen atoms or atoms of other elements that are near carbon in the periodic table—especially nitrogen, oxygen, sulfur, phosphorus, and the halogens.

Most importantly, carbon atoms also bond to other carbon atoms and form chains from two to thousands of carbon atoms in length. Also, because carbon forms four bonds, it forms complex, branched-chain structures, ring structures, and even cagelike structures. With all of these bonding possibilities, chemists have identified millions of different organic compounds and are synthesizing more every day.

☑ READING CHECK **Explain** why carbon forms many compounds.

■ **Figure 2** Carbon is found in group 14 of the periodic table. It can bond to four other elements and form thousands of different compounds.

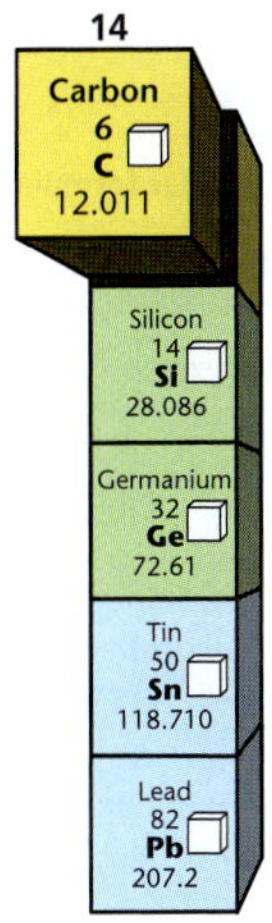

Hydrocarbons

The simplest organic compounds are **hydrocarbons,** which contain only the elements carbon and hydrogen. How many different compounds do you think two elements can form? You might guess that only a few compounds are possible. However, thousands of hydrocarbons are known, each containing only the elements carbon and hydrogen. The simplest hydrocarbon molecule, CH_4, consists of a carbon atom bonded to four hydrogen atoms. This substance, called methane, is an excellent fuel and is the main component of natural gas, as shown in **Figure 3.**

☑ READING CHECK **Name** two uses of methane or natural gas in your home or community.

A. I. Willett/Alamy

■ **Figure 3** Methane—a hydrocarbon found in natural gas—is the simplest hydrocarbon.

Identify *In addition to hydrogen, what other elements readily bond with carbon?*

2 Teach

Apply Chemistry

Coal Mines In underground coal mines, methane gas that formed along with the coal is released. If methane accumulates, a slight spark can cause an explosion. Because methane is colorless, odorless, and tasteless, it is not easy to detect.

In earlier times, miners often took a canary or other small animal underground to test the air. Because of the high metabolic rate of birds and other small animals, they are more sensitive than humans to diminished oxygen levels. When the miner's canary became unconscious or died, it signaled the miners to evacuate the mine until the methane dissipated.

Today, modern methane-detection instruments are available; however, even modern methane-detection methods must be used carefully, and they are not always 100 percent effective.

☑ **READING CHECK** Carbon forms many compounds because it can form 4 covalent bonds with other atoms, including other carbon atoms.

☑ **READING CHECK** possible answers: home heating and outdoor grilling

■ **Caption Question Fig. 3** nitrogen, oxygen, sulfur, phosphorus, and the halogens

DIFFERENTIATED INSTRUCTION

Advanced Learners Have a group of advanced learners research and create an oral presentation about vitalism. In the oral presentation, students should explain the theory, explain how the theory was disproved, and explain how the discrediting of this theory changed scientific thought. Encourage the group to make visual aids to use during their presentation. AL COOP LEARN

Quick Demo

Volume Differences Use a handheld butane lighter to demonstrate the enormous difference between the volume of a liquid hydrocarbon and the same compound in the gaseous state. Fill a large graduated cylinder with water and invert it into a pan of water. Tilt the cylinder slightly, and position the butane lighter under the water so that the butane released will enter the cylinder and displace water. Empty the contents of the lighter into the cylinder, and ask students to record the volume of gas produced. Break open the empty lighter, and measure the volume of water that it takes to fill the lighter. Ask students to compare the liquid and gas volumes. **WARNING:** ***Be sure there are no open flames in the room. Perform the Quick Demo in a well-ventilated room or under a fume hood.*** OL

☑ READING CHECK Early nineteenth-century chemists tested animal fats and plant oils to see if they reacted with bromine. Hydrocarbons that reacted with bromine were called unsaturated hydrocarbons. Hydrocarbons that did not react with bromine were called saturated hydrocarbons.

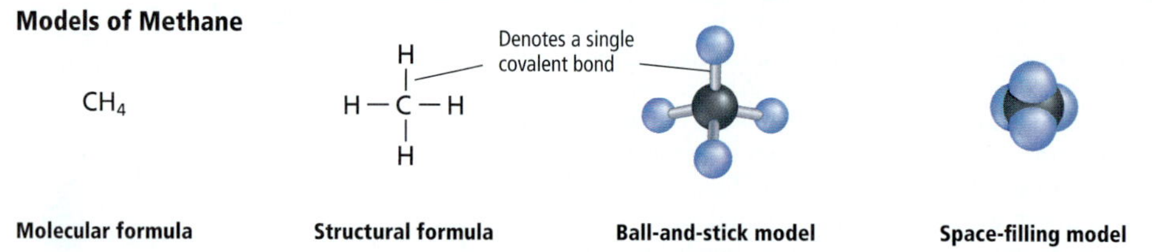

■ **Figure 4** Chemists use four different models to represent a methane (CH_4) molecule. See the Reference Tables in the Student Resources for a key to atom color conventions.

Models and hydrocarbons Chemists represent organic molecules in a variety of ways. **Figure 4** shows four different ways to represent a methane molecule. Covalent bonds are represented by a single straight line, which denotes two shared electrons. Most often, chemists use the type of model that best shows the information they want to highlight. As shown in **Figure 4,** molecular formulas give no information about the geometry of the molecule. A structural formula shows the general arrangement of atoms in the molecule but not the exact, three-dimensional geometry. The ball-and-stick model demonstrates the geometry of the molecule clearly, but the space-filling model gives a more realistic picture of what a molecule would look like if you could see it. Keep in mind as you look at the models that the atoms are held closely together by electron-sharing bonds.

■ **Figure 5** Carbon can bond to other carbon atoms in double and triple bonds. These Lewis structures and structural formulas show two ways to denote double and triple bonds.

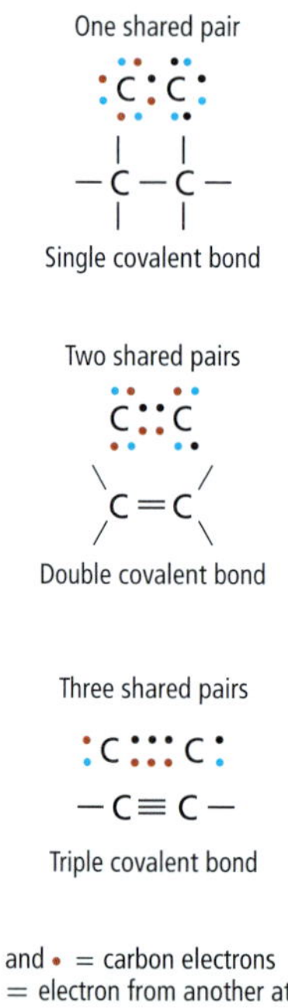

Multiple carbon-carbon bonds Carbon atoms can bond to each other not only by single covalent bonds but also by double and triple covalent bonds, as shown in **Figure 5.** Recall that in a double bond, atoms share two pairs of electrons; in a triple bond, they share three pairs of electrons.

In the nineteenth century, before chemists understood bonding and the structure of organic substances, they experimented with hydrocarbons obtained from heating animal fats and plant oils. They classified these hydrocarbons according to a chemical test in which they mixed each hydrocarbon with bromine and then measured how much reacted with the hydrocarbon. Some hydrocarbons would react with a small amount of bromine, some would react with more, and some would not react with any amount of bromine. Chemists called the hydrocarbons that reacted with bromine unsaturated hydrocarbons in the same sense that an unsaturated aqueous solution can dissolve more solute. Hydrocarbons that did not react with bromine were said to be saturated.

Present-day chemists can now explain the experimental results obtained 170 years ago. Hydrocarbons that reacted with bromine had double or triple covalent bonds. Those compounds that did not react with bromine had only single covalent bonds. Today, a hydrocarbon having only single bonds is defined as a **saturated hydrocarbon.** A hydrocarbon that has at least one double or triple bond between carbon atoms is an **unsaturated hydrocarbon.** You will learn more about these different types of hydrocarbons later in this chapter.

☑ READING CHECK **Explain** the origin of the terms *saturated* and *unsaturated hydrocarbons.*

DIFFERENTIATED INSTRUCTION

Struggling Learners Some students will gain a better understanding of the terms *saturated* and *unsaturated* as they apply to hydrocarbons if you use a simple analogy. Compare the unsaturation of a dry paper towel that can still absorb water to an alkene or alkyne that can potentially absorb more hydrogen by forming additional C–H bonds at the expense of double or triple bonds. In other words, the compound has fewer hydrogen atoms than its carbons can hold. Then, compare the saturation of a paper towel with water after a spill has been cleaned up to an alkane that already has all of the hydrogen atoms its carbons can hold. BL EL

Refining Hydrocarbons

Today, many hydrocarbons are obtained from a fossil fuel called petroleum. Petroleum formed from the remains of microorganisms that lived in Earth's oceans millions of years ago. Over time, the remains formed thick layers of mudlike deposits on the ocean floor. Heat from Earth's interior and the tremendous pressure of overlying sediments transformed this mud into oil-rich shale and natural gas. In certain kinds of geological formations, the petroleum ran out of the shale and collected in pools deep in Earth's crust. Natural gas, which formed at the same time and in the same way as petroleum, is usually found with petroleum deposits. Natural gas is composed primarily of methane, but it also contains small amounts of other hydrocarbons that have from two to five carbon atoms.

Fractional distillation Unlike natural gas, petroleum is a complex mixture containing more than a thousand different compounds. For this reason, raw petroleum, sometimes called crude oil, has little practical use. Petroleum is much more useful to humans when it is separated into simpler components or fractions. Separation is carried out in a process called **fractional distillation,** also called fractionation, which involves boiling the petroleum and collecting components or fractions as they condense at different temperatures. Fractional distillation is done in a fractionating tower similar to the one shown in **Figure 6.**

The temperature inside the fractionating tower is controlled so that it remains near 400°C at the bottom, where the petroleum is boiling, and gradually decreases toward the top. The condensation temperatures (boiling points) generally decrease as molecular mass decreases. Therefore, as the vapors travel up through the column, the hydrocarbons condense and are drawn off, as shown in **Figure 6.**

VOCABULARY

SCIENCE USAGE V. COMMON USAGE

Deposit

Science usage: a natural collection of oil or ore

There was a rich deposit of copper in the mountain.

Common usage: money placed in a bank account or the act of placing money in a bank account

The store owner placed his deposit in the after-hours slot at the bank.

Figure 6 This diagram of a fractionating tower shows that fractions with lower boiling points, such as gasoline and gaseous products, are drawn off in the cooler regions near the top of the tower. Oils and greases, having much higher boiling points, stay near the bottom of the tower and are drawn off there.

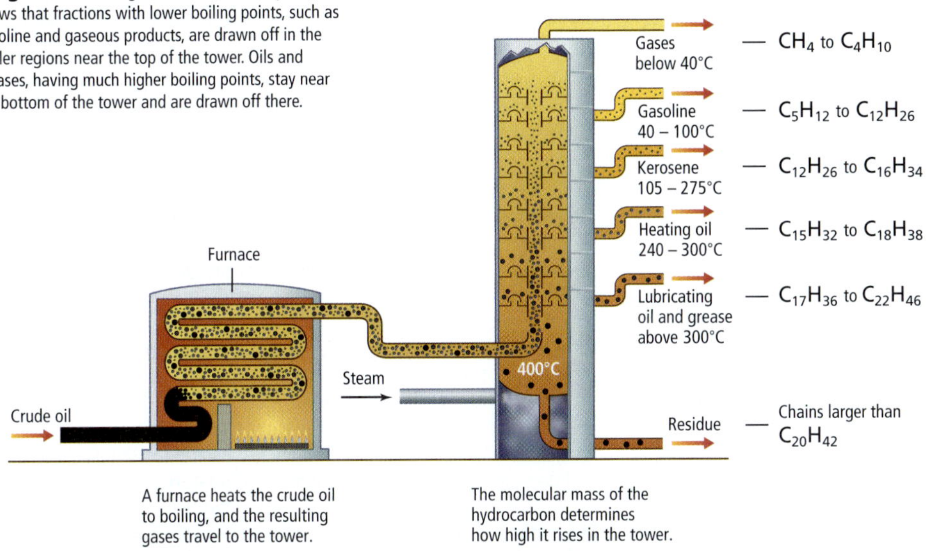

Visual Learning

Figure 6 Ask students to examine the figure that diagrams the major fractions drawn off a fractionating tower at an oil refinery. Explain that the process of simple distillation represents just the first step in the conversion of crude oil or petroleum to useful substances. Many of the fractions are further processed to obtain better separation or even to convert substances chemically into other compounds.

Some of the heavy residues from the fractionating column are distilled again under a vacuum. Lowering the pressure over a liquid will lower its boiling point. This means that the heavier fractions can be further separated without using high temperatures that might cause them to decompose. These are then used to obtain types of lubricating oils, are blended into industrial fuels, or are passed on to the cracking unit for chemical conversion to alkenes using heat and catalysts. **OL**

Quick Demo

Distillation Use a lab distillation set-up to demonstrate the separation of components from a mixture. Use salt water as the mixture and separate it into salt and distilled water. When the distillation is near completion, ask students what is left in the beaker. concentrated salt water Ask students what the composition of the distilled liquid is. distilled water **OL**

CULTURAL DIVERSITY

Distillation of Perfumes in Ancient India The knowledge of chemistry has been applied since ancient times in the distillation of perfumes and ointments in the Indus Valley region of Pakistan, India, and Afghanistan. Perfumes produced by distillation included sandalwood oil, musk tamarind, and camphor. These aromatic perfumes were sprinkled on kings during coronation ceremonies, and fragrant ointments based on sandalwood were applied during ceremonial bathing. Even today during some festivals such as Diwali, the Hindu festival of lights, aromatic slurries and pastes play an important part in rituals.

Content Background

Octane Ratings Students might be surprised to learn that octane ratings are not directly related to the 8-carbon, straight-chain alkane called octane. Octane ratings were first established by assigning a rating of zero to heptane, which was known to cause premature ignition in the cylinder, and a rating of 100 to 2,2,4-trimethylpentane, which had the best ignition qualities when tests were first performed. The compound 2,2,4-trimethylpentane was commonly called isooctane and erroneously called octane by technicians who tested gasoline. A gasoline with a rating of 90 performs about the same as a mixture of 90% isooctane and 10% heptane. Today, compounds can be added to gasoline to produce octane ratings greater than 100.

ASSESSMENT

Performance Have students research and prepare a visual aid, such as a poster or computer slide presentation, that explains octane ratings. The visual aid should contain information about tetraethyl lead in gasoline and how it affects octane ratings. BL OL AL

Caption Question Fig. 7 air, water, and soil

READING CHECK Cracking is the process in which large-chain hydrocarbons are broken into smaller-chain hydrocarbons. This process occurs in the presence of a catalyst and in the absence of oxygen.

CAREERS IN CHEMISTRY

WebQuest

Figure 7 Fractional distillation towers separate large quantities of petroleum into usable components. Thousands of products we use in our homes, for transportation, and in industry result from petroleum refining.

Infer ***What types of emissions must be controlled by refineries to protect the environment?***

Figure 6 also gives the names of the typical fractions separated from petroleum, along with their boiling points, hydrocarbon size ranges, and common uses. You might recognize some of the fractions because you use them every day. Unfortunately, fractional distillation towers, shown in **Figure 7,** do not yield fractions in the same proportions that they are needed. For example, distillation seldom yields the amount of gasoline desired. However, it yields more of the heavier oils than the market demands.

Many years ago, petroleum chemists and engineers developed a process to help match the supply with the demand. This process in which heavier fractions are converted to gasoline by breaking their large molecules into smaller molecules is called **cracking.** Cracking is done in the absence of oxygen and in the presence of a catalyst. In addition to breaking heavier hydrocarbons into molecules of the size range needed for gasoline, cracking also produces starting materials for the synthesis of many different products, including plastic products, films, and synthetic fabrics.

READING CHECK **Describe** the process in which large-chain hydrocarbons are broken into more-desirable smaller-chain hydrocarbons.

CAREERS IN CHEMISTRY

Petroleum Technician This science technician uses instruments to measure and record physical and geological information about oil or gas wells. For example, a petroleum technician might test a geological sample to determine its petroleum content and its mineral or element composition.

WebQuest

Rating gasoline None of the petroleum fractions is a pure substance. As shown in **Figure 6,** gasoline is not a pure substance, but rather a mixture of hydrocarbons. Most molecules with single covalent bonds in gasoline have 5 to 12 carbon atoms. However, the gasoline pumped into cars today is different from the gasoline used in automobiles in the early 1900s. The gasoline fraction that is distilled from petroleum is modified by adjusting its composition and adding substances to improve its performance in today's automobile engines and to reduce pollution from car exhaust.

It is critical that the gasoline-air mixture in the cylinder of an automobile engine ignite at exactly the right instant and burn evenly. If it ignites too early or too late, much energy will be wasted, fuel efficiency will drop, and the engine will wear out prematurely. Most straight-chain hydrocarbons burn unevenly and tend to ignite from heat and pressure before the piston is in the proper position and the spark plug fires. This early ignition causes a rattling or pinging noise called knocking.

CHEMISTRY JOURNAL

Oil Refining Ask students to locate the oil refinery nearest their home. Have students contact the refinery to get information about what products are produced at the company and how many barrels of crude oil the company refines per month. Students should record a synopsis of their findings in their journals. OL

CHEMISTRY PROJECT

Fuel Types Have students research the differences between fuel types. Have students include information about the various types of gasoline fuels used in cars and trucks, diesel fuel and fuels that are used in airplanes, and race cars. Encourage students to include other fuel types that they find in their research. Have students prepare a poster that explains the composition of the different fuel types. OL

■ **Figure 8** Octane ratings are used to give the antiknock rating of fuel. Mid-grade gasoline for cars has an octane rating of about 89. Aviation fuel has an octane rating of about 100. Racing fuel has an octane rating of about 110.

In the late 1920s, an antiknock, or octane rating, system for gasoline was established, resulting in the octane ratings posted on gasoline pumps like those shown in **Figure 8.** Mid-grade gasoline today has a rating of about 89, whereas premium gasoline has ratings of 91 or higher. Several factors determine which octane rating a car needs, including how much the piston compresses the air-fuel mixture and the altitude at which the car is driven.

Connection to Earth Science Since ancient times, people have found petroleum seeping from cracks in rocks. Historical records show that petroleum has been used for more than 5000 years. In the nineteenth century, as the United States entered the machine age and its population increased, the demand for petroleum products, namely kerosene for lighting and lubricants for machines, increased. In an attempt to find a reliable petroleum supply, Edwin Drake drilled the first oil well in the United States in Pennsylvania, in 1859. The oil industry flourished for a time, but when Thomas Edison introduced the electric light in 1882, investors feared that the industry was doomed. However, the invention of the automobile in the 1890s revived the industry on a massive scale.

SECTION 1 REVIEW

Section Self-Check

Section Summary

- Organic compounds contain carbon, which is able to form straight chains and branched chains.
- Hydrocarbons are organic substances composed of carbon and hydrogen.
- The major sources of hydrocarbons are petroleum and natural gas.
- Petroleum can be separated into components by the process of fractional distillation.

1. **MAIN**IDEA **Identify** three applications of hydrocarbons as a source of energy and raw materials.
2. **Name** an organic compound and explain what an organic chemist studies.
3. **Identify** what each of the four molecular models highlights about a molecule.
4. **Compare and contrast** saturated and unsaturated hydrocarbons.
5. **Describe** the process of fractional distillation.
6. **Infer** Some shortening products are described as "hydrogenated vegetable oil," which are oils that reacted with hydrogen in the presence of a catalyst. Form a hypothesis to explain why hydrogen reacted with the oils.
7. **Interpret Data** Refer to **Figure 6.** What property of hydrocarbon molecules correlates to the viscosity of a particular fraction when it is cooled to room temperature?

Rachel Epstein/PhotoEdit

ASSESSMENT

Knowledge Ask students to list the fractions that are separated from petroleum in a fractionating column and to label them in increasing order of boiling point. The fractions are petroleum gases (1 to 4 carbons); gasoline (5 to 12 carbons); kerosene (12 to 16 carbons); heating oil (15 to 18 carbons); lubricating oil (17 carbons or more); and residue (20 carbons or more). OL

3 Assess

Check for Understanding

Have students compare and contrast saturated and unsaturated hydrocarbons. OL

Reteach

Ask students why chemists use four different ways to model a hydrocarbon. The four different models give different types of information about the molecule. OL

Extension

Have students draw three different models for methane and give one advantage of using each model. OL

SECTION 1 REVIEW

1. possible applications: fuel for home heating and starting materials for the synthesis of plastic products, films, and synthetic fabrics
2. possible answer: methane; An organic chemist studies all carbon-containing compounds with the primary exceptions of carbon oxides, carbides, and carbonates.
3. The molecular formula shows the atoms in the molecule. A structural formula shows the general arrangement of the atoms. The ball-and-stick model shows the geometry. The space-filling model shows a realistic picture of what the molecule looks like.
4. Saturated hydrocarbons are hydrocarbons that contain only single bonds between carbon atoms. Unsaturated hydrocarbons are hydrocarbons that contain at least one double or triple bond between carbon atoms.
5. Fractional distillation is a process by which petroleum is separated into groups of components using boiling points as the separation mechanism.
6. Possible hypothesis: The oils react with hydrogen when double or triple bonds are broken and hydrogen atoms attach to the molecule.
7. The greater the number of carbon atoms in the chain, the greater the viscosity of the fraction.

SECTION 2

1 Focus

MAINIDEA

Single Bonds Ask students to define a single bond. the sharing of a pair of electrons between two atoms Ask students how single bonds are represented in molecular models. by a short line or bar connecting two atoms, or by a pair of dots between two atoms Tell students that a hydrocarbon that contains only single bonds in its molecular structure is called an alkane. OL

2 Teach

Quick Demo

Examples of Alkanes Find out the type of gas supplied to the lab. Most labs use either natural gas or propane. Show students two examples of alkanes. Obtain a disposable lighter and set up a Bunsen burner. Tell students that both the lighter and Bunsen burner use an alkane as a fuel. The lighter uses butane and the Bunsen burner uses either natural gas or propane. Ask students to compare the two flames. Both flames should look about the same. Differences in color are due to the oxygen/gas mix. OL

SECTION 2

Essential Questions

- How are alkanes named when given their structures?
- How are alkane structures drawn when given their names?
- What are the properties of alkanes?

Review Vocabulary

IUPAC (International Union of Pure and Applied Chemistry): an international group that aids communication between chemists by setting rules and standards in areas such as chemical nomenclature, terminology, and standardized methods

New Vocabulary

alkane
homologous series
parent chain
substituent group
cyclic hydrocarbon
cycloalkane

Alkanes

MAINIDEA Alkanes are hydrocarbons that contain only single bonds.

CHEM 4 YOU Have you ever used a Bunsen burner or an outdoor gas grill? If so, you have used an alkane. Natural gas and propane are the two most common gases used in these applications, and both are alkanes.

Straight-Chain Alkanes

Methane is the smallest member of a series of hydrocarbons known as alkanes. It is used as a fuel in homes and science labs and is a product of many biological processes. **Alkanes** are hydrocarbons that have only single bonds between atoms. Look in Section 1 to review the various models of methane. The models for ethane (C_2H_6), the second member of the alkane series, are shown in **Table 1.** Ethane consists of two carbon atoms bonded together with a single bond and six hydrogen atoms sharing the remaining valence electrons of the carbon atoms.

The third member of the alkane series, propane, has three carbon atoms and eight hydrogen atoms, giving it the molecular formula C_3H_8. The next member, butane, has four carbon atoms and the formula C_4H_{10}. Compare the structures of ethane, propane, and butane in **Table 1.**

Propane, also known as LP (liquefied propane) gas, is sold as a fuel for cooking and heating. Butane is used as fuel in small lighters and in some torches. It is also used in the manufacture of synthetic rubber.

Table 1 Simple Alkanes

Molecular Formula	Structural Formula	Ball-and-Stick Model	Space-Filling Model
Ethane (C_2H_6)	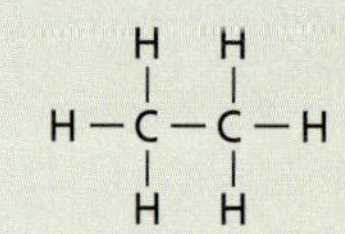	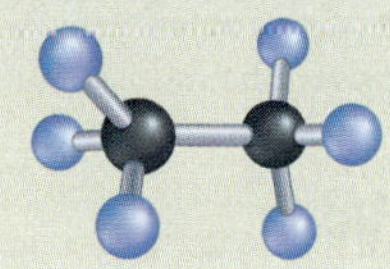	
Propane (C_3H_8)	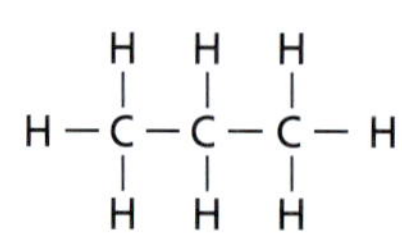	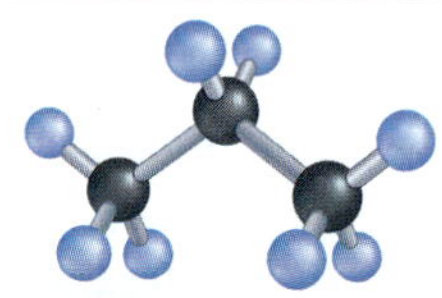	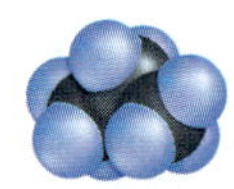
Butane (C_4H_{10})	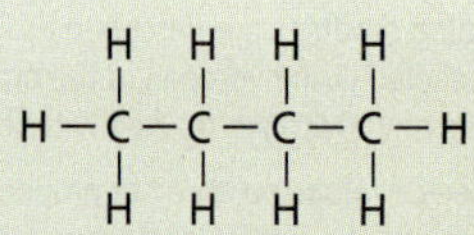	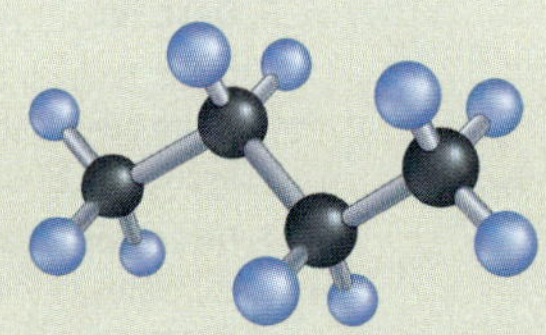	

CHEMISTRY JOURNAL

Alkane Names Point out to students that although the names for alkanes with five or more carbons have prefixes that were derived from Greek or Latin for the number of carbons, smaller alkanes were named using different sources. Ask students to research the origins of the prefixes used for alkanes that have from one to four carbons in a chain. They should include the information they find in their chemistry journals. OL

Table 2 First Ten of the Alkane Series

Name	Molecular Formula	Condensed Structural Formula
Methane	CH_4	CH_4
Ethane	C_2H_6	CH_3CH_3
Propane	C_3H_8	$CH_3CH_2CH_3$
Butane	C_4H_{10}	$CH_3CH_2CH_2CH_3$
<u>Pent</u>ane	C_5H_{12}	$CH_3CH_2CH_2CH_2CH_3$
<u>Hex</u>ane	C_6H_{14}	$CH_3CH_2CH_2CH_2CH_2CH_3$
<u>Hept</u>ane	C_7H_{16}	$CH_3CH_2CH_2CH_2CH_2CH_2CH_3$
<u>Oct</u>ane	C_8H_{18}	$CH_3(CH_2)_6CH_3$
<u>Non</u>ane	C_9H_{20}	$CH_3(CH_2)_7CH_3$
<u>Dec</u>ane	$C_{10}H_{22}$	$CH_3(CH_2)_8CH_3$

Naming straight-chain alkanes By now, you have likely noticed that names of alkanes end in *-ane.* Also, alkanes with five or more carbons in a chain have names that use a prefix derived from the Greek or Latin word for the number of carbons in each chain. For example, *pent*ane has five carbons just as a *pent*agon has five sides, and *oct*ane has eight carbons just as an *oct*opus has eight tentacles. Because methane, ethane, propane, and butane were named before alkane structures were known, their names do not have numerical prefixes. **Table 2** shows the names and structures of the first ten alkanes. Notice the underlined prefix representing the number of carbon atoms in the molecule.

In **Table 2,** you can see that the structural formulas are written in a different way from those in **Table 1.** These formulas, called condensed structural formulas, save space by not showing how the hydrogen atoms branch off from the carbon atoms. Condensed formulas can be written in several ways. In **Table 2,** the lines between carbon atoms have been eliminated to save space.

In **Table 2,** you can see that $-CH_2-$ is a repeating unit in the chain of carbon atoms. Note, for example, that pentane has one more $-CH_2-$ unit than butane. You can further condense structural formulas by writing the $-CH_2-$ unit in parentheses followed by a subscript to show the number of units, as is done with octane, nonane, and decane.

A series of compounds that differ from one another by a repeating unit is called a **homologous series.** A homologous series has a fixed numerical relationship among the numbers of atoms. For alkanes, the relationship between the numbers of carbon and hydrogen atoms can be expressed as C_nH_{2n+2}, where n is equal to the number of carbon atoms in the alkane. Given the number of carbon atoms in an alkane, you can write the molecular formula for any alkane. For example, heptane has seven carbon atoms, so its formula is $C_7H_{2(7)+2}$, or C_7H_{16}.

VOCABULARY
WORD ORIGIN
Homologous
comes from the Greek word *homologos* meaning *agreeing*

☑ READING CHECK **Write** the molecular formula for an alkane that has 13 carbon atoms in its molecular structure.

Use Science Terms

Understanding Terms Have students write statements explaining the meaning of the terms *homologous series* and *substituent group.* BL EL

ASSESSMENT

Knowledge Ask each student to write a question on the material covered at this point. Have students exchange questions and quiz each other. OL COOP LEARN

ChemLAB

The ChemLab located at the end of the chapter can be used at this point in the lesson.

ASSESSMENT

Performance After students have learned the names of the first ten straight-chain alkanes, assign each student the names of two alkanes. Ask them to draw the complete and condensed structures of the alkanes and make models of the molecules. OL EL

☑ READING CHECK $C_{13}H_{28}$

DIFFERENTIATED INSTRUCTION

Struggling Learners Some students will find it easier to learn the meaning of the general formula of alkanes (C_nH_{2n+2}) if they use it to make models of alkane structures. Assign each student an alkane. Ask them to use the general formula to predict how many carbon atoms and hydrogen atoms they will need to make a model of that alkane. Then, ask them to make the model and test their predictions. BL EL

Extension

Energy Ask students to find out the source of the energy used to heat their homes and record that information in their chemistry journals. If they burn a fuel, they should determine and record the composition of that fuel. If they use electricity, ask them to do further research to find out what type of fuel, if any, is burned to produce that electricity. Ask students to draw the structures of the major components of the fuels used. **OL**

Visual Learning

Figure 9 and Table 3 Consider using drawings of straight-chain and branched-chain hydrocarbons to introduce the concept of structural isomers. To do so, challenge students to draw structures or make models of all possible molecules having the formula C_5H_{12}. Caution them that each carbon atom must have four single covalent bonds linking it to other atoms. They should discover three possible structures, which are pictured in **Figure 17.** **OL**

☑ **READING CHECK** Butane is a straight-chain hydrocarbon and isobutane is a branched-chain hydrocarbon.

Branched-Chain Alkanes

The alkanes discussed so far in this chapter are called straight-chain alkanes because the carbon atoms are bonded to each other in a single line. Now look at the two structures in **Figure 9.** If you count the carbon and hydrogen atoms, you will discover that both structures have the same molecular formula, C_4H_{10}. Do the structures in **Figure 9** represent the same substance?

If you think that the structures represent two different substances, you are correct. The structure on the left represents butane, and the structure on the right represents a branched-chain alkane known as isobutane—a substance whose chemical and physical properties are different from those of butane. Carbon atoms can bond to one, two, three, or even four other carbon atoms. This property makes possible a variety of branched-chain alkanes.

Recall that butane is used in lighters and in torches. Isobutane is used as both an environmentally-safe refrigerant and a propellant in products such as shaving gel, as shown in **Figure 9.** In addition to these applications, both butane and isobutane are used as raw materials for many chemical processes.

☑ **READING CHECK** **Describe** the difference in the molecular structures of butane and isobutane.

Alkyl groups You have seen that both a straight-chain and a branched-chain alkane can have the same molecular formula. This fact illustrates a basic principle of organic chemistry: the order and arrangement of atoms in an organic molecule determine its identity. Therefore, the name of an organic compound must also accurately describe the molecular structure of the compound.

■ **Figure 9** Butane is a fuel used in lighters. Isobutane is used as a propellant in products such as shaving gel.

Demonstration

Combustion of Methane

Purpose

to observe the combustion of methane

Materials

400-mL beaker; 150-mL beaker; liquid detergent (25 mL); glycerin (5 mL); sucrose (5 g); rubber tubing (1-m length); small funnel; meterstick; candle; matches; newspaper; masking tape

Safety Precautions

Disposal Solution can be poured down a drain.

Procedure

Prepare a soap-bubble solution by adding 160 mL of H_2O, 25 mL of liquid detergent, and 5 mL of glycerin to the 400-mL beaker. In a separate beaker, dissolve 5 g of sucrose in 60 mL of H_2O. Gently mix the two solutions in the 400-mL beaker. Connect one end of rubber tubing to a gas outlet and the other end to a small funnel. Invert the funnel into the soap mixture, then lift the funnel from the solution. Turn on the gas briefly to make a bubble. Darken the room. Dislodge bubbles by turning the funnel sideways and gently shaking.

As each bubble rises (natural gas) or sinks (propane gas), ignite it with a burning candle taped to the end of a meterstick. It is best to have an assistant hold the meterstick

Table 3 **Common Alkyl Groups**

Name	Methyl	Ethyl	Propyl	Isopropyl	Butyl
Condensed structural formula	CH_3-	CH_3CH_2-	$CH_3CH_2CH_2-$	CH_3CHCH_3 (bond from middle C)	$CH_3CH_2CH_2CH_2-$
Structural formula	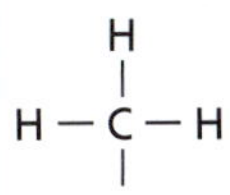	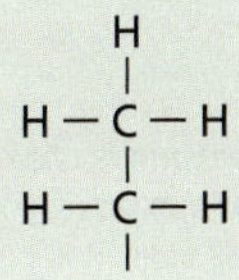	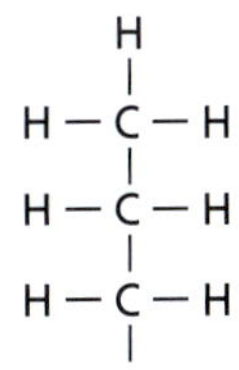	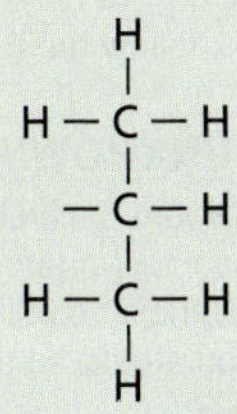	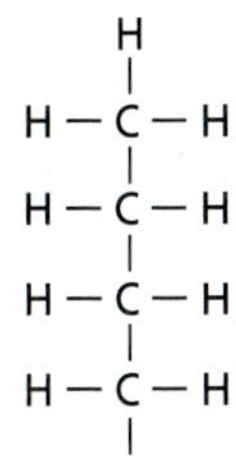

When naming branched-chain alkanes, the longest continuous chain of carbon atoms is called the **parent chain.** All side branches are called **substituent groups** because they appear to substitute for a hydrogen atom in the straight chain. Each alkane-based substituent group branching from the parent chain is named for the straight-chain alkane that has the same number of carbon atoms as the substituent. The ending *-ane* is replaced with the letters *-yl.* An alkane-based substituent group is called an alkyl group. Several alkyl groups are shown in **Table 3.**

VOCABULARY

ACADEMIC VOCABULARY

Substitute
a person or thing that takes the place of another
A substitute teacher taught chemistry class yesterday.

Naming branched-chain alkanes To name organic structures, chemists use the following systematic rules approved by the International Union of Pure and Applied Chemistry (IUPAC).

Step 1. *Count the number of carbon atoms in the longest continuous chain.* Use the name of the straight-chain alkane with that number of carbons as the name of the parent chain of the structure.

Step 2. *Number each carbon in the parent chain.* Locate the end carbon closest to a substituent group. Label that carbon *Position 1.* This step gives all the substituent groups the lowest position numbers possible.

Step 3. *Name each alkyl group substituent.* Place the name of the group before the name of the parent chain.

Step 4. *If the same alkyl group occurs more than once as a branch on the parent structure, use a prefix* (di-, tri-, tetra-, *and so on*) *before its name to indicate how many times it appears.* Then, use the number of the carbon to which each is attached to indicate its position.

Step 5. *When different alkyl groups are attached to the same parent structure, place their names in alphabetical order.* Do not consider the prefixes (*di-*, *tri-*, and so on) when determining alphabetical order.

Step 6. *Write the entire name, using hyphens to separate numbers from words and commas to separate numbers.* Do not add a space between the substituent name and the name of the parent chain.

Reinforcement

Name Alkanes Ask students if there is an alkane having the name *2-ethylbutane.* Before revealing the answer, have students draw a molecule of that name and identify the longest carbon chain. No. Following IUPAC rules, this is the name for a structure that has a 2-carbon chain attached at the second carbon of a 4-carbon chain. However, the longest continuous chain contains five carbon atoms. Ask students to give the correct name for the compound. 3-methylpentane **OL**

with the attached candle above or below the funnel before you shake bubbles free. **WARNING:** ***Do not do this demonstration near flammable materials. Place newspapers on the floor to catch wax dripping from the candle.***

Results

A flare-up occurs as each bubble bursts. The entrapped gas burns with a luminous yellow flame.

Analysis

Ask these questions.

1. The combustion you observed is a chemical reaction between what two reactants? oxygen and methane or propane
2. What did the combustion reaction produce? carbon dioxide, perhaps carbon monoxide, light, heat, water vapor, soot
3. Was the combustion reaction exothermic or endothermic? exothermic

ASSESSMENT

Knowledge Have students write the balanced equation for the combustion reaction. $CH_4 + 2O_2 \rightarrow CO_2 + 2H_2O$ or $C_3H_8 + 5O_2 \rightarrow 3CO_2 + 4H_2O$

Identify Misconceptions

Students might think that the line of carbon atoms written horizontally in a structural formula is always the chain to use in naming the substance.

Uncover the Misconception Draw the following alkane on the board: a 6-carbon chain written horizontally with a 3-carbon propyl group branching vertically off the second carbon. Ask students to write the name of this alkane.

Demonstrate the Concept Without revealing the correct IUPAC name of the compound, ask how many students named the compound 2-propylhexane and how many named it something else. Have a student who named the compound incorrectly explain his or her method for arriving at that name. Ask a student who has written the correct name, 4-*methyloctane*, to number the chain correctly on the board.

Assess New Knowledge Write carbon-chain structures for other branched alkanes, and have students find the correct numbering and write the names of the alkanes. Then, ask them to draw both skeletal and full structural formulas of compounds for which you provide names. **OL** **EL**

EXAMPLE Problem 1

EXAMPLE PROBLEM

NAMING BRANCHED-CHAIN ALKANES
Name the alkane shown.

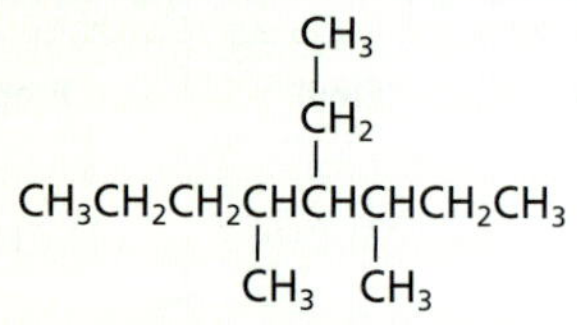

1 ANALYZE THE PROBLEM

You are given a structure. To determine the name of the parent chain and the names and locations of branches, follow the IUPAC rules.

2 SOLVE FOR THE UNKNOWN

Step 1. *Count the number of carbon atoms in the longest continuous chain.* Because structural formulas can be written with chains oriented in various ways, you need to be careful in finding the longest continuous carbon chain. In this case, it is easy. The longest chain has eight carbon atoms, so the parent name is *octane.*

Step 2. *Number each carbon in the parent chain.* Number the chain in both directions, as shown below. Numbering from the left puts the alkyl groups at Positions 4, 5, and 6. Numbering from the right puts alkyl groups at Positions 3, 4, and 5. Because 3, 4, and 5 are the lowest position numbers, they will be used in the name.

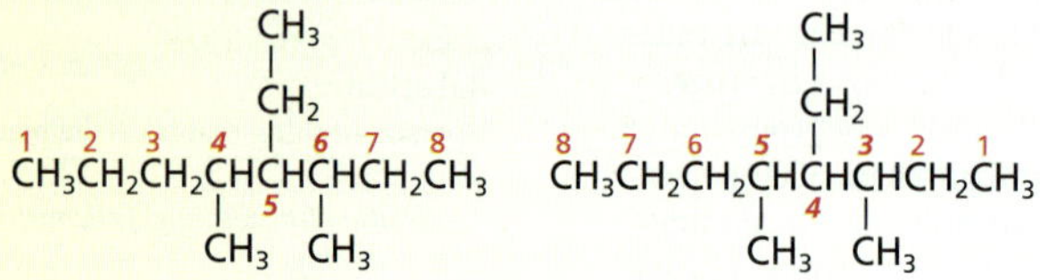

Step 3. *Name each alkyl group substituent.* Identify and name the alkyl groups branching from the parent chain. There are one-carbon methyl groups at Positions 3 and 5, and a two-carbon ethyl group at Position 4.

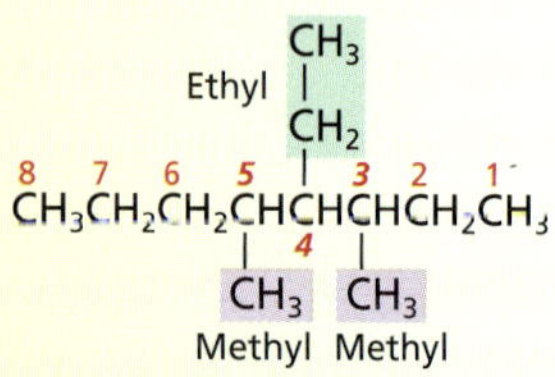

Step 4. *If the same alkyl group occurs more than once as a branch on the parent structure, use a prefix* (di-, tri-, tetra-, *and so on) before its name to indicate how many times it appears.* Look for and count the alkyl groups that occur more than once. Determine the prefix to use to show the number of times each group appears. In this example, the prefix *di-* will be added to the name *methyl* because two methyl groups are present. No prefix is needed for the one ethyl group. Then show the position of each group with the appropriate number.

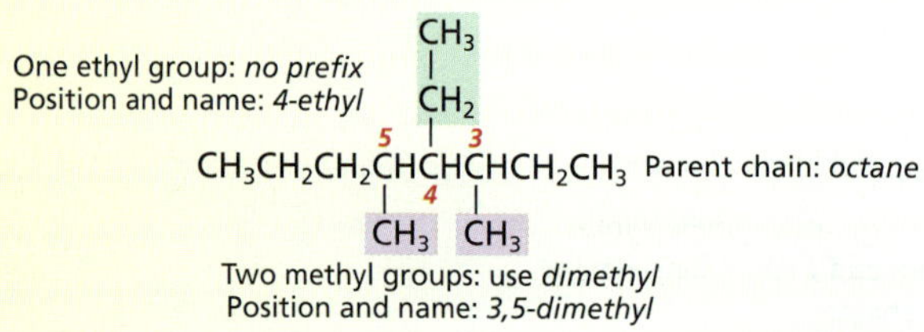

Get help **naming hydrocarbons.**

Personal Tutor

DIFFERENTIATED INSTRUCTION

English Learners Pair an English learner with another student that can communicate well with the English learner. Have the student explain how to do the Example Problems in this chapter step by step. Encourage both students to seek assistance if the need arises. **EL** **BL** **OL** **AL**

Step 5. *Whenever different alkyl groups are attached to the same parent structure, place their names in alphabetical order.* Place the names of the alkyl branches in alphabetical order, ignoring the prefixes. Alphabetical order puts the name ethyl before dimethyl.

Step 6. *Write the entire name, using hyphens to separate numbers from words and commas to separate numbers.* Write the name of the structure, using hyphens and commas as needed. The name should be written as *4-ethyl-3,5-dimethyloctane.*

3 EVALUATE THE ANSWER

The longest continuous carbon chain has been found and numbered correctly. All branches have been designated with correct prefixes and alkyl group names. Alphabetical order and punctuation are correct.

PRACTICE Problems

Do additional problems. Online Practice

8. Use the IUPAC rules to name the following structures.

a.

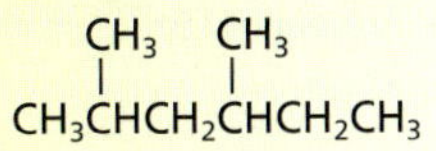

b.

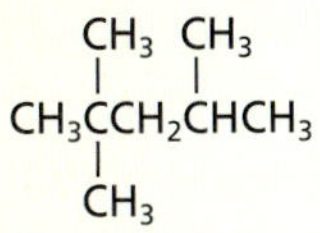

c.

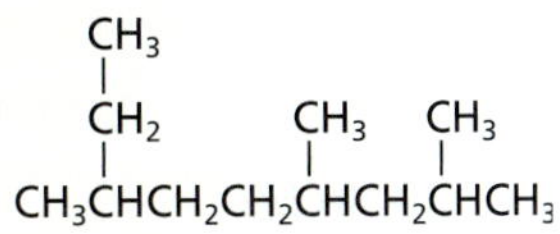

9. Challenge Draw the structures of the following branched-chain alkanes.

a. 2,3-dimethyl-5-propyldecane

b. 3,4,5-triethyloctane

PRACTICE PROBLEMS

Cycloalkanes

One of the reasons that such a variety of organic compounds exists is that carbon atoms can form ring structures. An organic compound that contains a hydrocarbon ring is called a **cyclic hydrocarbon.** To indicate that a hydrocarbon has a ring structure, the prefix *cyclo-* is used with the hydrocarbon name. Thus, cyclic hydrocarbons that contain only single bonds are called **cycloalkanes.**

Cycloalkanes can have rings with three, four, five, six, or even more carbon atoms. The name for the six-carbon cycloalkane is *cyclohexane.* Cyclohexane, which is obtained from petroleum, is used in paint and varnish removers and for extracting essential oils to make perfume. Note that cyclohexane (C_6H_{12}) has two fewer hydrogen atoms than straight-chain hexane (C_6H_{14}) because a valence electron from each of two carbon atoms is now forming a carbon-carbon bond rather than a carbon-hydrogen bond.

READING CHECK Evaluate If the prefix *cyclo-* is present in the name of an alkane, what do you know about the alkane?

As shown in **Figure 10,** cyclic hydrocarbons such as cyclohexane are represented by condensed, skeletal, and line structures. Line structures show only the carbon-carbon bonds with carbon atoms understood to be at each vertex of the structure. Hydrogen atoms are assumed to occupy the remaining bonding positions unless substituents are present. Hydrogens are also not shown in skeletal structures.

Figure 10 Cyclohexane can be represented in several ways.

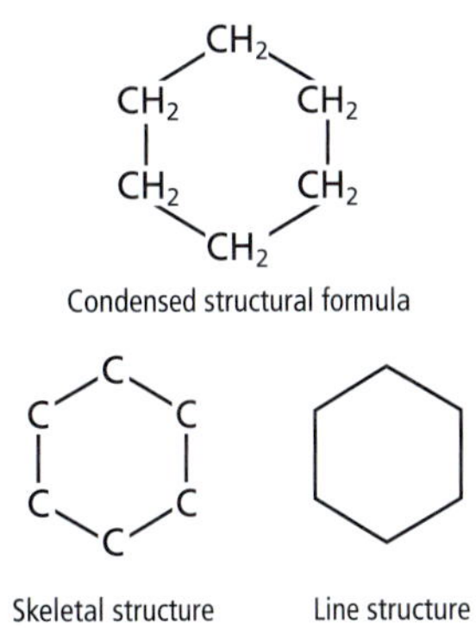

IN-CLASS Example

Question Name the alkane shown.

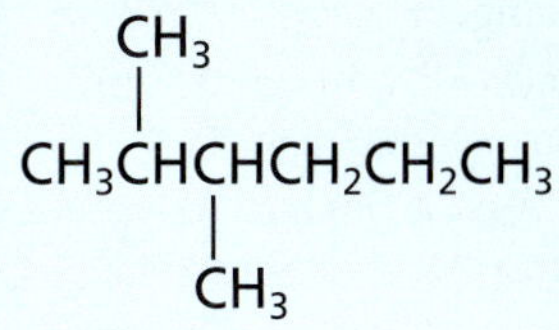

Answer 2,3-dimethylhexane

PRACTICE Problems

Have students refer to the Selected Solutions appendix for complete solutions to odd-numbered problems. The complete solutions for all problems can be found in the Solutions Manual.

8. a. 2,4-dimethylhexane
b. 2,4,7-trimethylnonane
c. 2,2,4-trimethylpentane

9. See the Solutions Manual for structures.
a. Parent chain has 10 carbon atoms, with methyl groups on carbons 2 and 3 and a propyl group on carbon 5.
b. Parent chain has 8 carbon atoms, with ethyl groups on carbons 3, 4, and 5.

ASSESSMENT

Knowledge Hectane is the IUPAC name for the straight-chain alkane that contains 100 carbon atoms. Give students that information, and then ask them the following question. What is the molecular formula of hectane? $C_{100}H_{202}$ OL

READING CHECK The alkane contains a hydrocarbon ring.

CHEMISTRY PROJECT

Cycloalkanes in Industry Have students research industrial uses of cycloalkanes and prepare a poster showing products that contain cycloalkanes or products in which cycloalkanes were used during the manufacturing process. Display the posters in the classroom as a learning device for all students. BL OL AL

Reinforcement

Naming Cycloalkanes Point out that there are six different ways to number the ring. Starting at the top, the position numbers of the substituents can be 1,2,4 or 1,4,6. Starting at the right, the numbers can be 1,2,5 or 1,3,6. Starting at the bottom, the numbers can be 1,3,4 or 1,4,5. **OL**

ASSESSMENT

Performance Have each student draw a substituted cycloalkane at the top of a sheet of paper. At the bottom of the same sheet, have students write the name of the cycloalkane. Have students fold the sheet of paper in half so that the name of the cycloalkane is not visible. Have students exchange papers. Have each student try to name the cycloalkane. After the students have named the new cycloalkane, have students check to see if the names match. If they do not match, have students work together to correctly name the cycloalkane. Encourage students to ask for help, if needed. **OL** **EL**

Enrichment

Cyclic Alkanes Have students research the sizes of cyclic alkane structures most commonly found in organic compounds, and have students determine why the size of the cyclic alkane is important. Cyclopentane and cyclohexane are the most common ring sizes found because in those structures the carbon atoms form bonds that are less strained than those in smaller or larger rings. Carbon atoms with single bonds form the strongest bonds possible when the bonds are located at angles of about 109.5° apart, as they are in methane. **AL**

Naming substituted cycloalkanes Like other alkanes, cycloalkanes can have substituent groups. Substituted cycloalkanes are named by following the same IUPAC rules used for straight-chain alkanes, but with a few modifications. With cycloalkanes, there is no need to find the longest chain because the ring is always considered to be the parent chain. Because a cyclic structure has no ends, numbering is started on the carbon that is bonded to the substituent group. When there are two or more substituents, the carbons are numbered around the ring in a way that gives the lowest-possible set of numbers for the substituents. If only one group is attached to the ring, no number is necessary. The following Example Problem illustrates the naming process for cycloalkanes.

EXAMPLE Problem 2

EXAMPLE PROBLEM

NAMING CYCLOALKANES
Name the cycloalkane shown.

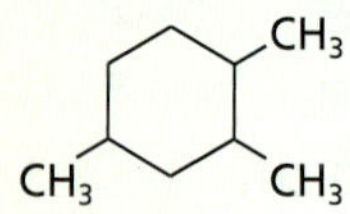

1 ANALYZE THE PROBLEM

You are given a structure. To determine the parent cyclic structure and the location of branches, follow the IUPAC rules.

2 SOLVE FOR THE UNKNOWN

Step 1. Count the carbons in the ring, and use the name of the parent cyclic hydrocarbon. In this case, the ring has six carbons, so the parent name is *cyclohexane.*

Step 2. Number the ring, starting from one of the CH_3- branches. Find the numbering that gives the lowest possible set of numbers for the branches. Here are two ways of numbering the ring.

A

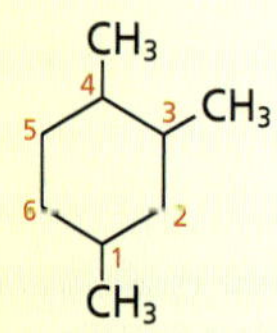

1,3,4

B

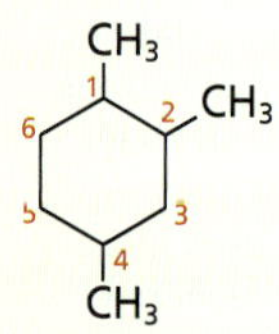

1,2,4

Numbering from the carbon atom at the bottom of the ring puts the CH_3- groups at Positions 1, 3, and 4 in Structure A. Numbering from the carbon at the top of the ring gives Positions 1, 2, and 4. All other numbering schemes place the CH_3- groups at higher position numbers. Thus, 1, 2, and 4 are the lowest possible position numbers and will be used in the name.

Step 3. Name the substituents. All three are the same—carbon methyl groups.

Step 4. Add the prefix to show the number of groups present. Three methyl groups are present, so you add the prefix *tri-* to the name *methyl* to make *trimethyl.*

Step 5. Alphabetical order can be ignored because only one type of group is present.

Step 6. Put the name together using the name of the parent cycloalkane. Use commas between separate numbers, and hyphens between numbers and words. Write the name as *1,2,4-trimethylcyclohexane.*

DIFFERENTIATED INSTRUCTION

Advanced Learners Ask students to build a model of a highly branched alkane with the formula $C_{18}H_{38}$, and then name the model using IUPAC rules. **AL**

CHEMISTRY JOURNAL

Oil and Water Ask students to write a paragraph in which they predict what will happen when they mix a small amount of motor oil or other alkane mixture with water. Students should then describe what happens when they try it. They should include these paragraphs in their chemistry journals. **OL**

3 EVALUATE THE ANSWER

The parent-ring structure is numbered to give the branches the lowest possible set of numbers. The prefix *tri-* indicates that three methyl groups are present. No alphabetization is necessary because all branches are methyl groups.

PRACTICE Problems

Do additional problems. Online Practice

PRACTICE PROBLEMS

10. Use IUPAC rules to name the following structures.

a.

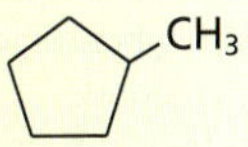

b.

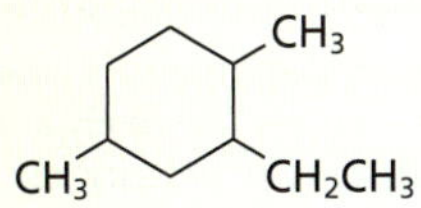

c.

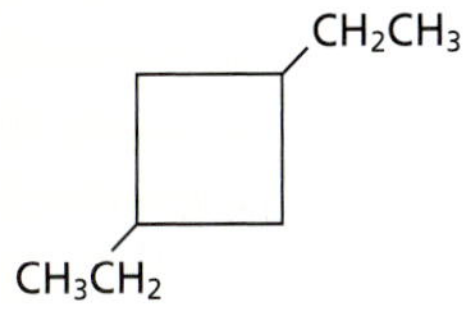

11. Challenge Draw the structures of the following cycloalkanes.

a. 1-ethyl-3-propylcyclopentane

b. 1,2,2,4-tetramethylcyclohexane

Properties of Alkanes

You have learned that the structure of a molecule affects its properties. For example, the O–H bonds in a water molecule are polar, and because the H–O–H molecule has a bent geometry, the molecule itself is polar. Thus, water molecules can form hydrogen bonds with each other. As a result, the boiling and melting points of water are much higher than those of other substances having similar molecular mass and size.

What properties would you predict for alkanes? All of the bonds in alkanes are between either a carbon atom and a hydrogen atom or between two carbon atoms. A bond between two identical atoms, such as carbon, can never be polar. Also, the C–H bonds have only a small electronegativity difference and are nonpolar. Because all of the bonds in alkanes are nonpolar, alkane molecules are nonpolar, which makes them good solvents for other nonpolar substances, as shown in **Figure 11.**

Figure 11 Many solvents—used as thinners for paints, coatings, waxes, photocopier toners, adhesives, and printer press inks—contain alkanes and cycloalkanes.

Robin Nelson/PhotoEdit

IN-CLASS Example

Question Use IUPAC rules to name the following structure.

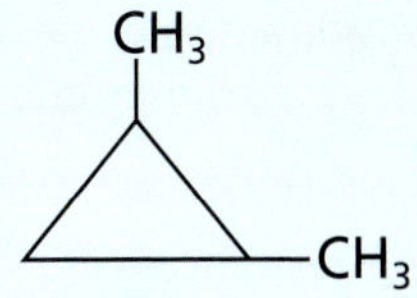

Answer 1,2-dimethylcyclopropane

PRACTICE Problems

Have students refer to the Selected Solutions appendix for complete solutions to odd-numbered problems. The complete solutions for all problems can be found in the Solutions Manual.

10. a. methylcyclopentane

b. 2-ethyl-1,4-dimethylcyclohexane

c. 1,3-diethylcyclobutane

11. See the Solutions Manual for structures.

a. Parent chain is a 5-carbon ring, with an ethyl group on carbon 1 and a propyl group on carbon 3.

b. Parent chain is a 6-carbon ring, with methyl groups on carbons 1,2,2, and 4 (four methyl groups total).

CHEMISTRY JOURNAL

Canine Arson Detectors Specially trained K-9 sniffer dogs are usually better able to find accelerants on a fire scene than humans with electronic detection devices. Hydrocarbon detectors are sensitive to gasoline components in the parts-per-million (ppm) range. Dogs, though, can pinpoint traces that escape electronic detection, often as little as 0.01 microliter of 50% evaporated gasoline. Ask students to research an actual court case in which the "testimony" of dogs was used to attempt to prove that arson was committed. OL

ASSESSMENT

Skill Ask students to write the molecular formulas for each of the following cyclic alkanes: cyclopropane, C_3H_6; cyclobutane, C_4H_8; cyclopentane, C_5H_{10}; cyclohexane, C_6H_{12}. OL

FOLDABLES® Rubric

3 Assess

Check for Understanding

Ask students to explain why cyclopropane is the smallest possible alkane. A cyclic structure cannot be formed from one or two carbon atoms. Three or more carbons must be present. OL

Reteach

Ask student volunteers to go to the board and write the structural formulas for each of the first ten alkanes in the series. OL

Extension

Ask students to learn the names of the straight-chain alkanes that contain 11 to 20 carbon atoms per molecule. AL

Table 4 Comparing Physical Properties

Substance and formula	Water (H_2O)	Methane (CH_4)
Molecular mass	18 amu	16 amu
State at room temperature	liquid	gas
Boiling point	100°C	–162°C
Melting point	0°C	–182°C

Physical properties of alkanes How do the properties of a polar and nonpolar compound compare? Refer to **Table 4,** and note that the molecular mass of methane (16 amu) is close to the molecular mass of water (18 amu). Also, water and methane molecules are similar in size. However, when you compare the melting and boiling points of methane to those of water, you can see evidence that the molecules differ in some significant way. These temperatures differ greatly because methane molecules have little intermolecular attraction compared to water molecules. This difference in attraction can be explained by the fact that methane molecules are nonpolar and do not form hydrogen bonds with each other, whereas water molecules are polar and freely form hydrogen bonds.

The difference in polarity and hydrogen bonding also explains the immiscibility of alkanes and other hydrocarbons with water. If you try to dissolve alkanes, such as lubricating oils, in water, the two liquids separate almost immediately into two phases. This separation happens because the attractive forces between alkane molecules are stronger than the attractive forces between the alkane and water molecules. Therefore, alkanes are more soluble in solvents composed of nonpolar molecules like themselves than in water, a polar solvent.

FOLDABLES® Incorporate information from this section into your Foldable.

Chemical properties of alkanes The main chemical property of alkanes is their low reactivity. Recall that many chemical reactions occur when a reactant with a full electric charge, such as an ion, or with a partial charge, such as a polar molecule, is attracted to another reactant with the opposite charge. Molecules such as alkanes, in which atoms are connected by nonpolar bonds, have no charge. As a result, they have little attraction for ions or polar molecules. The low reactivity of alkanes can also be attributed to the relatively strong C–C and C–H bonds.

SECTION 2 REVIEW

Section Self-Check

Section Summary

- Alkanes contain only single bonds between carbon atoms.
- Alkanes and other organic compounds are best represented by structural formulas and can be named using systematic rules determined by the International Union of Pure and Applied Chemistry (IUPAC).
- Alkanes that contain hydrocarbon rings are called cyclic alkanes.

12. MAINIDEA **Describe** the main structural characteristics of alkane molecules.
13. **Name** the following structures using IUPAC rules.

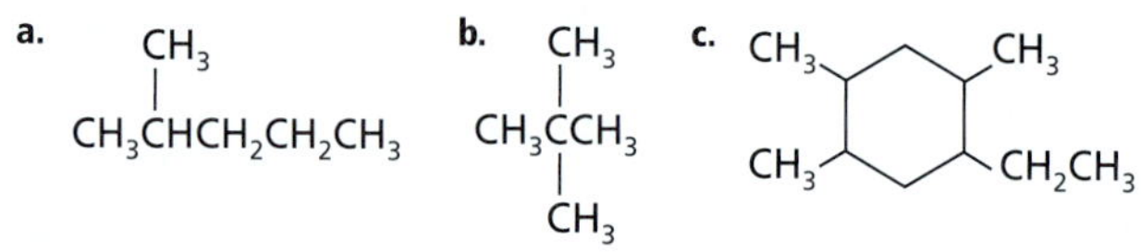

14. **Describe** the general properties of alkanes.
15. **Draw** the molecular structure for each of the following.
    a. 3,4-diethylheptane
    b. 4-isopropyl-3-methyldecane
    c. 1-ethyl-4-methylcyclohexane
    d. 1,2-dimethylcyclopropane
16. **Interpret Chemical Structures** Why is the name 3-butylpentane incorrect? Based on this name, write the structural formula for the compound. What is the correct IUPAC name for 3-butylpentane?

SECTION 2 REVIEW

12. Alkanes are chain or ring hydrocarbons containing only single carbon-carbon covalent bonds.
13. a. 2-methylpentane
    b. 2,2-dimethylpropane
    c. 1-ethyl-2,4,5-trimethylcyclohexane
14. The C-C and C-H bonds are nonpolar, making alkanes insoluble in water, a polar solvent. Alkanes are good solvents for other nonpolar substances. The bonds are also strong and stable, making alkanes relatively unreactive.
15. Refer to the Solutions Manual for drawings of the molecular structures.
16. The longest continuous carbon chain contains seven carbon atoms, not five carbon atoms. The parent chain has seven carbon atoms with an ethyl group on carbon 3: 3-ethylheptane.

SECTION 3

Alkenes and Alkynes

MAINIDEA Alkenes are hydrocarbons that contain at least one double bond, and alkynes are hydrocarbons that contain at least one triple bond.

Essential Questions

- How do the properties of alkenes and alkynes compare with those of alkanes?
- How are the molecular structures of alkenes and alkynes described?
- How are alkenes and alkynes named when given their structures?
- How are alkenes and alkynes drawn when given their names?

Review Vocabulary

hormone: chemical produced in one part of an organism and transported to another part, where it causes a physiological change

New Vocabulary

alkene
alkyne

CHEM 4 YOU Plants produce ethene as a natural ripening hormone. For efficiency in harvesting and transporting produce to market, fruits and vegetables are often picked while unripe and are exposed to ethene so they will ripen at the same time.

Alkenes

Recall that alkanes are saturated hydrocarbons because they contain only single covalent bonds between carbon atoms, and that unsaturated hydrocarbons have at least one double or triple bond between carbon atoms. Unsaturated hydrocarbons that contain one or more double covalent bonds between carbon atoms in a chain are called **alkenes.** Because an alkene must have a double bond between carbon atoms, there is no 1-carbon alkene. The simplest alkene has two carbon atoms double bonded to each other. The remaining four electrons—two from each carbon atom—are shared with four hydrogen atoms to give the molecule ethene (C_2H_4).

Alkenes with only one double bond constitute a homologous series. Recall from the previous section that a homologous series has a fixed numerical relationship among the numbers of atoms. If you study the molecular formulas for the substances shown in **Table 5,** you will see that each has twice as many hydrogen atoms as carbon atoms. The general formula for the series is C_nH_{2n}. Each alkene has two fewer hydrogen atoms than the corresponding alkane because two electrons now form the second covalent bond and are no longer available for bonding to hydrogen atoms. What are the molecular formulas for 6-carbon and 9-carbon alkenes?

Explore **alkenes** with an interactive table. Concepts In Motion

Table 5 Examples of Alkenes

Name	Ethene	Propene	1-Butene	2-Butene
Molecular formula	C_2H_4	C_3H_6	C_4H_8	C_4H_8
Structural formula	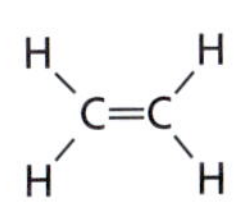	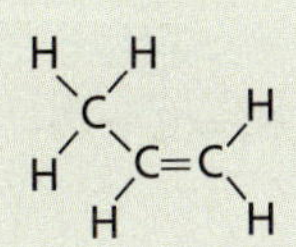	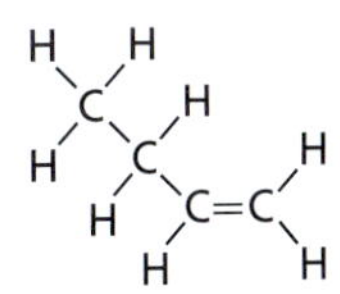	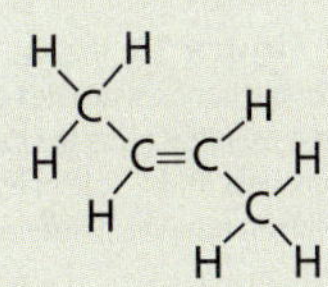
Condensed structural formula	$CH_2{=}CH_2$	$CH_3CH{=}CH_2$	$CH_3CH_2CH{=}CH_2$	$CH_3CH{=}CHCH_3$

SECTION 3

1 Focus

MAINIDEA

Alkenes and Alkynes Draw the structural formulas for ethane, ethene, and ethyne side-by-side on the board. Write the name of the structures beneath each of the drawings. Ask students to identify the differences between the structures. Students should observe that ethane has only single bonds. Ethene has one double bond, and ethyne has one triple bond. Students should also note that ethane has the highest number of hydrogen atoms, ethene has fewer hydrogen atoms, and ethyne has the least number of hydrogen atoms. Point out to students that ethene is an alkene because it has a double bond in its structure and ethyne is an alkyne because it has a triple bond in its structure. OL

2 Teach

- **In-Text Question** C_6H_{12} and C_9H_{18}

Enrichment

Old Terms Ask students to research the origin of the terms originally used to describe hydrocarbon types. Alkanes are still sometimes called paraffins, while alkenes are still occasionally referred to as olefins. Ask them to record their findings in their chemistry journals. OL

DIFFERENTIATED INSTRUCTION

Advanced Learners Ask groups of students to design chemical tests that will distinguish alkanes from alkenes and alkynes. Have them research and compare the reactivities of saturated and unsaturated hydrocarbons. AL

CHEMISTRY PROJECT

Alkenes and Alkynes Have students research and find examples of alkenes and alkynes found in nature. Have students prepare a poster showing where the alkenes or alkynes are found and the molecular structure of the compounds. OL

Quick Demo

Fruit and Ethene Place two or three pieces of overripe fruit in a bag next to a small beaker containing 20 mL of dilute aqueous iodine solution (dilute tincture of iodine or an even stronger stock solution with a light brown-yellow color that is transparent), and cover both with a larger beaker. Seal the bottom of the larger beaker to the countertop with tape. As a control, place another setup without fruit next to the first. Ask students to observe the color of the iodine solutions each day for a few days and to explain what has happened. The ethene produced by the ripening fruit is an unsaturated hydrocarbon. It reacts with iodine to form colorless alkyl halide products, so the color of the iodine solution next to the fruit lightens, while the control solution remains the same color. OL

☑ READING CHECK If the location of the double bond is not specified, it is not possible to tell, without a doubt, which compound is meant by the chemical name.

ASSESSMENT

Skill Have students determine what additional information is needed to solve the following problem.

Alkenes can be converted into alkanes using a chemical reaction called hydrogenation in which hydrogen gas is bubbled through a solution of an alkene in the presence of a catalyst. How many moles of hydrogen gas must be added to 1 mol of an alkene with more than one double bond to convert it completely to an alkane? The additional information needed is the number of double bonds present in the alkene. Each double bond requires 1 mol of hydrogen gas per mole of alkene for the conversion. OL

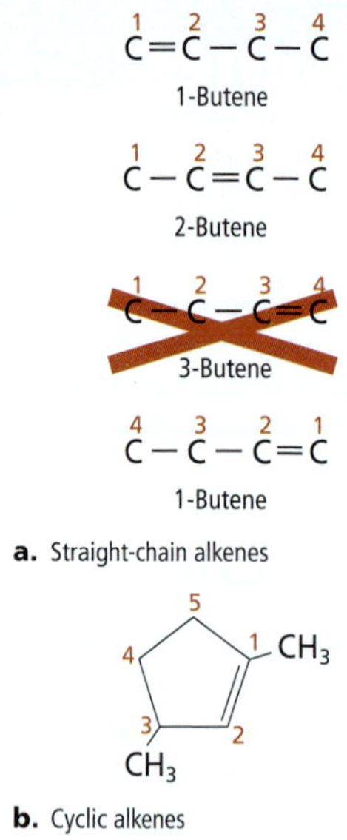

■ **Figure 12** When naming either branched or straight-chain alkenes, they must be numbered using IUPAC rules.

Naming alkenes Alkenes are named in much the same way as alkanes. Their names are formed by changing the *-ane* ending of the corresponding alkane to *-ene*. An alkane with two carbons is named *ethane,* and an alkene with two carbons is named *ethene*. Likewise, a three-carbon alkene is named propene. Ethene and propene have older, more common names: *ethylene* and *propylene,* respectively.

To name alkenes with four or more carbons in the chain, it is necessary to specify the location of the double bond, as shown in **Figure 12a.** This is done by numbering the carbons in the parent chain, starting at the end of the chain that will give the first carbon in the double bond the lowest number. Then, use only that number in the name.

Note that the third structure is not "3-butene" because it is identical to the first structure, 1-butene. It is important to recognize that 1-butene and 2-butene are two different substances, each with its own properties.

Cyclic alkenes are named in much the same way as cyclic alkanes; however, carbon number 1 must be one of the carbons connected by the double bond. In **Figure 12b,** note the numbering in the compound. The name of this compound is 1,3-dimethylcyclopentene.

☑ READING CHECK **Infer** why it is necessary to identify where the double bond is located in the name of an alkene.

Naming branched-chain alkenes When naming branched-chain alkenes, follow the IUPAC rules for naming branched-chain alkanes, but with two exceptions. First, in alkenes, the parent chain is always the longest chain that contains the double bond, whether or not it is the longest chain of carbon atoms. Second, the position of the double bond, not the branches, determines how the chain is numbered. A number specifies the location of the double bond, just as it does in straight-chain alkenes. Note that there are two 4-carbon chains in the molecule shown in **Figure 13a,** but only the one with the double bond is used as a basis for naming. This branched-chain alkene is 2-methyl-1-butene.

Some unsaturated hydrocarbons contain more than one double (or triple) bond. The number of double bonds in such molecules is shown by using a prefix (*di-*, *tri-*, *tetra-*, and so on) before the suffix *-ene*. The positions of the bonds are numbered in a way that gives the lowest set of numbers. Which numbering system would you use in the example in **Figure 13b?** Because the molecule has a seven-carbon chain, you would use the prefix *hepta-*. Because it has two double bonds, you would use the prefix *di-* before *-ene,* giving the name *heptadiene.* Adding the numbers 2 and 4 to designate the positions of the double bonds gives the name *2,4-heptadiene.*

■ **Figure 13** The positions of the double bonds in alkenes are numbered in a way that gives the lowest set of numbers. This is true of both branched and straight-chain alkenes.

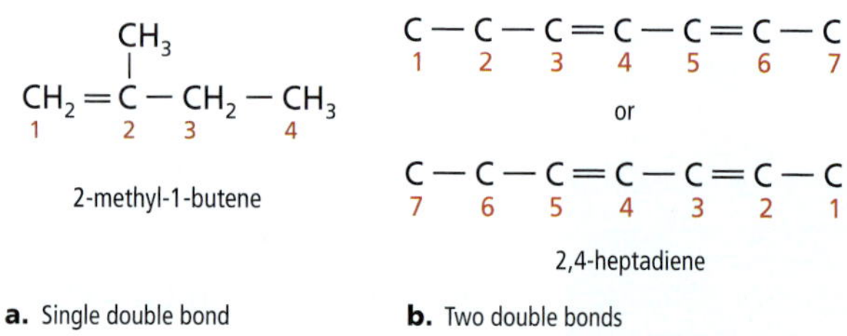

CHEMISTRY JOURNAL

Molecular Formulas of Hydrocarbons Have students write the following questions in their chemistry journals and answer them after they have read Section 3.

1. How does the molecular formula of an alkane differ from that of an alkene with the same number of carbons? The alkane contains two more hydrogens (C_nH_{2n+2}) than the alkene (C_nH_{2n}).
2. How does the molecular formula of an alkane differ from that of an alkyne with the same number of carbons? The alkane contains four more hydrogens (C_nH_{2n+2}) than the alkyne (C_nH_{2n-2}). OL

EXAMPLE Problem 3

NAMING BRANCHED-CHAIN ALKENES 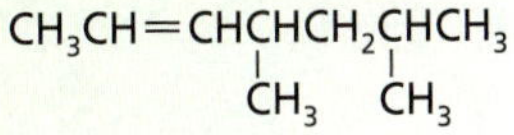

Name the alkene shown.

1 ANALYZE THE PROBLEM

You are given a branched-chain alkene that contains one double bond and two alkyl groups. Follow the IUPAC rules to name the organic compound.

2 SOLVE FOR THE UNKNOWN

Step 1. The longest continuous-carbon chain that includes the double bond contains seven carbons. The 7-carbon alkane is heptane, but the name is changed to heptene because a double bond is present.

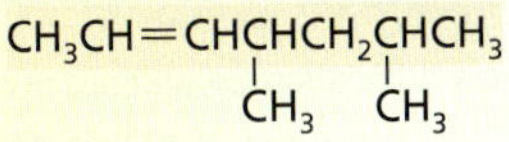

***Heptene* parent chain**

Step 2. Number the chain to give the lowest number to the double bond.

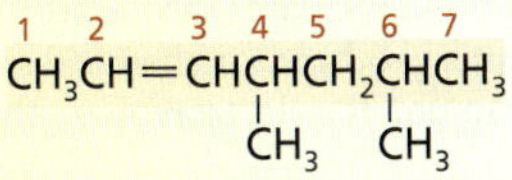

***2-Heptene* parent chain**

Step 3. Name each substituent.

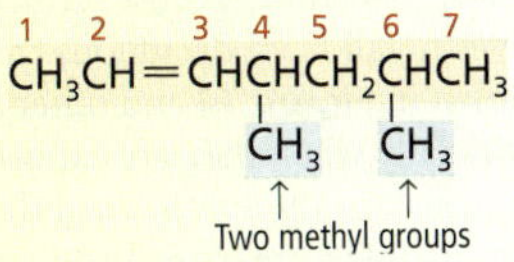

Each substituent is a methyl group.

Step 4. Determine how many of each substituent is present, and assign the correct prefix to represent that number. Then, include the position numbers to get the complete prefix.

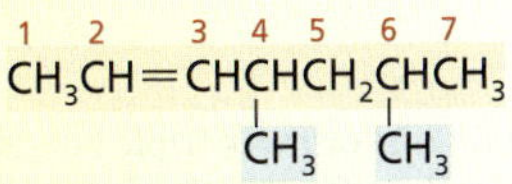

***2-Heptene* parent chain**

Two methyl groups at Positions 4 and 6

Prefix is 4,6-*dimethyl*

Step 5. The names of substituents do not have to be alphabetized because they are the same. Apply the complete prefix to the name of the parent alkene chain. Use commas between numbers, and hyphens between numbers and words. Write the name 4,6-*dimethyl-2-heptene.*

3 EVALUATE THE ANSWER

The longest carbon chain includes the double bond, and the position of the double bond has the lowest possible number. Correct prefixes and alkyl group names designate the branches.

EXAMPLE PROBLEM

PRACTICE Problems

Do additional problems. Online Practice

17. Use the IUPAC rules to name the following structures.

a. 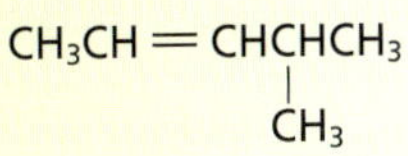

b. 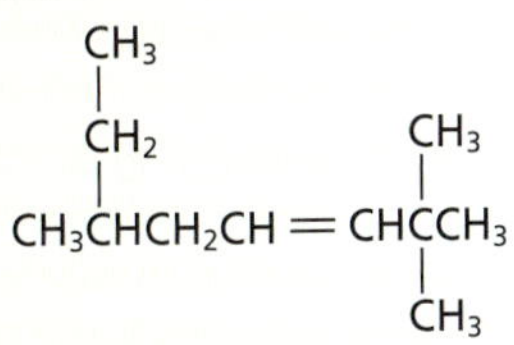

18. Challenge Draw the structure of 1,3-pentadiene.

Math in Chemistry

Hydrogenation How many moles of hydrogen gas must be added to 1 mol of an alkene with two double bonds to convert it completely to an alkane? 2 mol **OL**

IN-CLASS Example

Question Name the alkene shown.

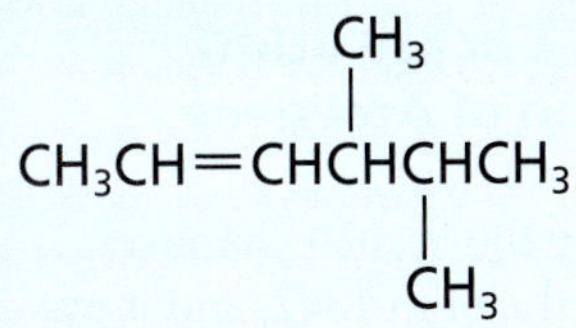

Answer 4,5-dimethyl-2-hexene

PRACTICE Problems

Have students refer to the Selected Solutions appendix for complete solutions to odd-numbered problems. The complete solutions for all problems can be found in the Solutions Manual.

17. a. 4-methyl-2-pentene

b. 2,2,6-trimethyl-3-octene

18. $CH_2{=}CHCH{=}CHCH_3$

Use Science Terms

Vocabulary Have students write statements explaining the meaning of the terms *alkene* and *alkyne*. **OL**

Concept Development

General Formulas Be sure students realize that the general formulas for alkenes and alkynes apply to those with only one unsaturated bond. They also do not apply to cyclic alkenes and alkynes. **BL** **OL**

DIFFERENTIATED INSTRUCTION

Struggling Learners Some students will be better able to understand the differences between alkanes, alkenes, and alkynes if they can compare them using models. Ask students to make models of ethane, ethene, and ethyne. Then, ask them to list similarities and differences that are apparent from the models alone. They should also discover that the two carbon atoms in ethene cannot easily rotate with respect to each other, while those in ethane can. The carbons cannot rotate in ethyne either, but this characteristic is relatively unimportant in alkynes. **BL** **EL**

■ **Caption Question Fig. 14**
All of the produce can be picked, transported to market, and sold at the same time, which maximizes profits.

☑ **READING CHECK** The triple bond has a high electron density. The cluster of electrons induces dipoles in nearby molecules, causing the nearby molecules to become unevenly charged and reactive.

Content Background

Preparation of Acetylene

Ethyne, or acetylene, was first described by the French chemist P. E. M. Berthelot in 1862, and it was first prepared on a large scale in the 1890s. First, limestone ($CaCO_3$) is heated to make quicklime, which is mostly calcium oxide (CaO).

$CaCO_3 \rightarrow CaO + CO_2$

The calcium oxide is then heated in a furnace with carbon in the form of coke, a material obtained by heating coal in the absence of oxygen. A by-product is carbon monoxide.

$CaO + 3C \rightarrow CaC_2 + CO$

The calcium carbide then reacts with water to form acetylene.

$CaC_2 + 2H_2O \rightarrow C_2H_2 + Ca(OH)_2$

Today, much acetylene is produced by the partial oxidation of methane.

$6CH_4 + O_2 \rightarrow 2C_2H_2 + 2CO + 10H_2$

■ **Figure 14** The use of ethene to ripen produce allows growers to harvest fruits and vegetables before they ripen.
Explain *why this is a benefit to growers.*

Properties and uses of alkenes Like alkanes, alkenes are nonpolar and therefore have low solubility in water as well as relatively low melting and boiling points. However, alkenes are more reactive than alkanes because the second covalent bond increases the electron density between two carbon atoms, providing a good site for chemical reactivity. Reactants that attract electrons can pull the electrons away from the double bond.

Several alkenes occur naturally in living organisms. For example, ethene is a hormone produced naturally by plants. It causes fruit to ripen and plays a part in causing leaves to fall from deciduous trees in preparation for winter. The fruits shown in **Figure 14** and other produce sold in grocery stores ripen artificially when they are exposed to ethene. Ethene is also the starting material for the synthesis of the plastic polyethylene, which is used to manufacture many products, including plastic bags, rope, and milk jugs. Other alkenes are responsible for the scents of lemons, limes, and pine trees.

Alkynes

Unsaturated hydrocarbons that contain one or more triple bonds between carbon atoms in a chain are called **alkynes.** Triple bonds involve the sharing of three pairs of electrons. The simplest and most commonly used alkyne is ethyne (C_2H_2), which is widely known by its common name *acetylene.* Study the models of ethyne in **Figure 15.**

Naming alkynes Straight-chain alkynes and branched-chain alkynes are named in the same way as alkenes. The only difference is that the name of the parent chain ends in *-yne* rather than *-ene.* Study the examples in **Table 6.** Alkynes with one triple covalent bond form a homologous series with the general formula C_nH_{2n-2}.

☑ **READING CHECK** **Infer,** by looking at the bonds in ethyne, why it is highly reactive with oxygen.

■ **Figure 15** These three molecular models represent ethyne.

Models of ethyne (acetylene)

Design Pics/David Chapman

CHEMISTRY JOURNAL

A Natural Alkyne While alkenes often occur in nature, alkynes are much harder to find. One naturally occurring alkyne is the compound dynemycin A, a complex polycyclic structure with two carbon-carbon triple bonds. It has been tested as a potential anticancer agent. Ask students to research the structure and natural sources of dynemycin A and record their findings in their chemistry journals. AL

Explore **alkynes** with an interactive table. Concepts In Motion

Table 6 Examples of Alkynes

Name	Molecular Formula	Structural Formula	Condensed Structural Formula
Ethyne	C_2H_2	$H-C\equiv C-H$	$CH\equiv CH$
Propyne	C_3H_4	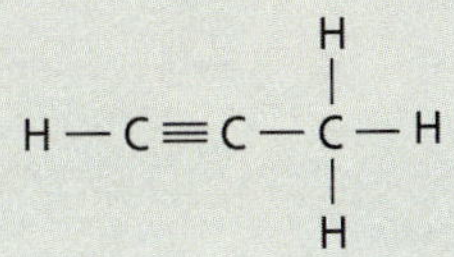	$CH\equiv CCH_3$
1-Butyne	C_4H_6	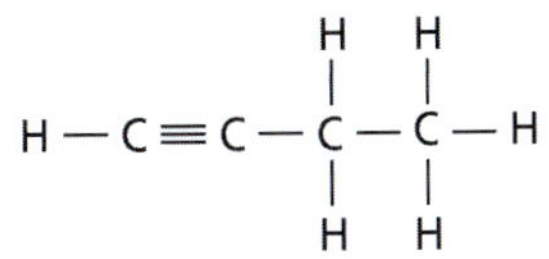	$CH\equiv CCH_2CH_3$
2-Butyne	C_4H_6	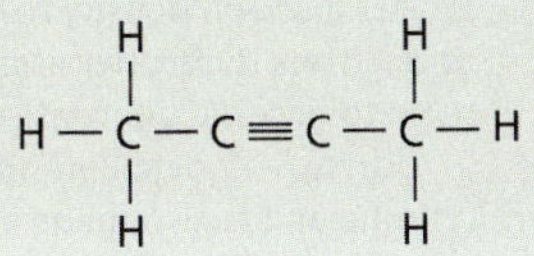	$CH_3C\equiv CCH_3$

MiniLAB

Synthesize and Observe Ethyne

Why is ethyne used in welding torches?

Procedure

1. Read and complete the lab safety form.
2. Use a **rubber band** to attach a **wood splint** to one end of a **ruler** that is about 40 cm long, so that about 10 cm of the splint extends beyond the ruler.
3. Place 120 mL **water** in a **150-mL beaker,** and add 5 mL **dishwashing detergent.** Mix thoroughly.
4. Use **forceps** to pick up a pea-sized lump of **calcium carbide** (CaC_2). Do not touch the CaC_2 with your fingers. **WARNING:** ***CaC_2 is corrosive; if CaC_2 dust touches your skin, wash it away immediately with a lot of water.*** Place the lump of CaC_2 in the beaker of detergent solution.
5. Use a **match** to light the splint while holding the ruler at the opposite end. Immediately bring the burning splint to the bubbles that have formed from the reaction in the beaker. Extinguish the splint after observing the reaction.
6. Use a **stirring rod** to dislodge a few large bubbles of ethyne. Do they float or sink in air?
7. Rinse the beaker thoroughly, then add 25 mL **distilled water** and a drop of **phenolphthalein solution.** Use forceps to place a small piece of CaC_2 in the solution. Observe the results.

Analysis

1. **Infer** What can you infer about the density of ethyne compared to the density of air?
2. **Predict** The reaction of calcium carbide with water yields two products. One is ethyne gas (C_2H_2). What is the other product? Write a balanced chemical equation for the reaction.

ASSESSMENT

Performance Have students research why calcium carbide is sometimes called miner's lamp. OL

Analysis

1. Density is slightly less than air.
2. From the color change, students should realize that a base is produced. Given that the cation present is Ca^{2+}, they might infer that the insoluble material is $Ca(OH)_2$. $CaC_2 + 2H_2O \rightarrow C_2H_2 + Ca(OH)_2$

MiniLAB

Rubric

Purpose Students will synthesize ethyne and observe a few of its properties.

Process Skills observe and infer, communicate, apply concepts

Safety Precautions Approve lab safety forms before work begins. Warn students not to lean over the beaker when lighting the bubbles. Long hair should be tied back. The somewhat caustic $Ca(OH)_2$ produced when CaC_2 reacts with moisture is the main reason for avoiding skin and eye contact.

Disposal Flush the soapy reaction solution down a drain with plenty of water. Allow any CaC_2 that has become moist to react completely with water before flushing the solution.

Teaching Strategies

- Dowels or long fireplace matches can be used rather than rulers. Be sure students extinguish the splints before they burn the ruler.
- The bubbles will be relatively pure ethyne, so they should not burn explosively when ignited as they would if they contained an air-ethyne mixture.

Expected Results When ignited, the ethyne should pop and burn in a yellow-orange ball that rises from the beaker. Soot might be deposited on the sides of the beaker as a result of incomplete combustion. Ethyne, having a molar mass of 26 g/mol, is slightly less dense than air, which has an average molar mass of about 29. Bubbles might float upward slowly but should demonstrate nearly neutral buoyancy. The phenolphthalein should turn pink as $Ca(OH)_2$ is formed.

FOLDABLES® Rubric

3 Assess

Check for Understanding

Quiz students to see if they can name straight-chain and branched alkenes using models. BL OL

Reteach

Let students work in groups to build models of branched alkenes. Ask groups to name their model and then exchange models with another group to try to name that model as well. OL EL COOP LEARN

Extension

Have students construct models of ethene and describe the geometry and shape of the molecule. They should establish that it is a flat molecule. Next, have them add methyl groups to both ends of the molecule to produce a molecule of 2-butene. Students should again describe the geometry. Some will discover that two different conformations exist. OL EL

Figure 16 Ethyne, or acetylene, reacts with oxygen in the chemical reaction $2C_2H_2 + 5O_2 \rightarrow 4CO_2 + 2H_2O$, which produces enough heat to weld metals.

Properties and uses of alkynes Alkynes have physical and chemical properties similar to those of alkenes. Alkynes undergo many of the reactions alkenes undergo. However, alkynes are generally more reactive than alkenes because the triple bonds of alkynes have even greater electron density than the double bonds of alkenes. This cluster of electrons is effective at inducing dipoles in nearby molecules, causing them to become unevenly charged and thus reactive.

FOLDABLES® Incorporate information from this section into your Foldable.

Ethyne—known commonly as acetylene—is a by-product of oil refining and is also made in large quantities by the reaction of calcium carbide (CaC_2) with water. When supplied with enough oxygen, ethyne burns with an intensely hot flame that can reach temperatures as high as 3000°C. Acetylene torches are commonly used in welding, as shown in **Figure 16.** Because the triple bond makes alkynes reactive, simple alkynes like ethyne are used as starting materials in the manufacture of plastics and other organic chemicals used in industry.

SECTION 3 REVIEW

Section Self-Check

Section Summary

- Alkenes and alkynes are hydrocarbons that contain at least one double or triple bond, respectively.
- Alkenes and alkynes are nonpolar compounds with greater reactivity than alkanes but with other properties similar to those of alkanes.

19. **MAIN IDEA Describe** how the molecular structures of alkenes and alkynes differ from the structure of alkanes.
20. **Identify** how the chemical properties of alkenes and alkynes differ from those of alkanes.
21. **Name** the structures shown using IUPAC rules.
    a. $CH \equiv CCH_2CH_3$ (with CH_3 drawn above the CH_2)
    b. $CH_3CH_2CH(CH_3)CH{=}CHCH_2CH_3$
22. **Draw** the molecular structure of 4-methyl-1,3-pentadiene and 2,3-dimethyl-2-butene.
23. **Infer** how the boiling and freezing points of alkynes compare with those of alkanes with the same number of carbon atoms. Explain your reasoning, then research data to see if it supports your idea.
24. **Predict** What geometric arrangement would you expect from the bonds surrounding the carbon atom in alkanes, alkenes, and alkynes? *(Hint: VSEPR theory can be used to predict the shape.)*

SECTION 3 REVIEW

19. Alkanes have single bonds in their structures. Alkenes have at least one double bond in their structures. Alkynes have at least one triple bond in their structures.
20. Alkenes and alkynes are much more reactive than alkanes. This is because they have regions of concentrated electron density that attract reactants of opposite charge.
21. **a.** 1-butyne
    b. 5-methyl-3-heptene
22. Refer to Solutions Manual for molecular structures.
23. Because alkynes are slightly more polar, they generally have higher melting and boiling points than alkanes. Data support this hypothesis.
24. VSEPR theory predicts the following geometric bond arrangements. alkane: tetrahedral shape; alkene: trigonal planar shape; alkyne: linear shape

SECTION 4

Essential Questions

- How can the two main categories of isomers—structural isomers and stereoisomers—be distinguished?
- How are *cis-* and *trans-* geometric isomers different?
- What is the structural variation in molecules that results in optical isomers?

Review Vocabulary

electromagnetic radiation: transverse waves that carry energy through empty space

New Vocabulary

isomer
structural isomer
stereoisomer
geometric isomer
chirality
asymmetric carbon
optical isomer
optical rotation

Hydrocarbon Isomers

MAINIDEA Some hydrocarbons have the same molecular formula but have different molecular structures.

CHEM 4 YOU Have you ever met a pair of identical twins? Identical twins have the same genetic makeup, yet they are two separate individuals with different personalities. Isomers are similar to twins—they have the same molecular formula, but different molecular structures and properties.

Structural Isomers

Examine the models of three alkanes in **Figure 17** to determine how they are similar and how they are different. All three have 5 carbon atoms and 12 hydrogen atoms, so they have the molecular formula C_5H_{12}. However, as you can see, these models represent three different arrangements of atoms and three different compounds—pentane, 2-methylbutane, and 2,2-dimethylpropane. These three compounds are isomers. **Isomers** are two or more compounds that have the same molecular formula but different molecular structures. Note that cyclopentane and pentane are not isomers because cyclopentane's molecular formula is C_5H_{10}.

There are two main classes of isomers. **Figure 17** shows compounds that are examples of structural isomers. **Structural isomers** have the same chemical formula, but their atoms are bonded in different arrangements. Structural isomers have different chemical and physical properties despite having the same formula. This observation supports one of the main principles of chemistry: The structure of a substance determines its properties. How does the trend in boiling points of C_5H_{12} isomers relate to their molecular structures?

As the number of carbons in a hydrocarbon increases, the number of possible structural isomers increases. For example, there are nine alkanes with the molecular formula C_7H_{16}. There are more than 300,000 structural isomers with the formula $C_{20}H_{42}$.

■ **Figure 17** These compounds with the same molecular formula, C_5H_{12}, are structural isomers. Note how their boiling points differ.

View an **animation about the isomers of pentane.** Concepts In Motion

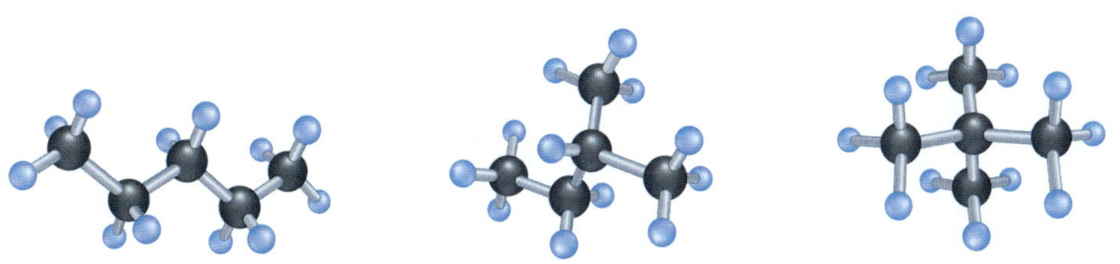

Pentane
bp = 36°C

2-Methylbutane
bp = 28°C

2,2-Dimethylpropane
bp = 9°C

SECTION 4

1 Focus

MAINIDEA

Isomers Use a molecular model kit to make the structural isomers shown in Figure 17. Display the models so that students can see them. Ask a student volunteer to give the molecular formula for each structure. C_5H_{12} Ask students how the molecular formulas are similar. The molecular formulas are the same for all three molecules. Ask a student volunteer to describe how the molecules differ. The carbon atoms are connected in different locations. Point out to students that these models represent structural isomers. OL EL

■ **In-Text Question** Students might conclude that boiling point increases as the molecule becomes less branched and more linear.

DIFFERENTIATED INSTRUCTION

Advanced Learners Ask students to draw all the structural isomers of the hydrocarbon with the formula C_7H_{16}. There are nine structural isomers: heptane, 2-methylhexane, 3-methylhexane, 2,3-dimethylpentane, 2,4-dimethylpentane, 2,2-dimethylpentane, 3,3-dimethylpentane, 3-ethylpentane, and 2,2,3-trimethylbutane. AL

2 Teach

Build a Model

***cis-* and *trans-* Isomers** Ask each group of two to three students to build models of *cis-* and *trans-*2-butene. Collect all the models and use them to demonstrate how the difference in ability for molecules to pack closely together affects their properties. OL EL

Caption Question Fig. 18 Groups of atoms attached to single-bonded carbon atoms are not fixed in space—they rotate along with the carbon atoms. Groups of atoms attached to double-bonded carbon atoms are fixed in space with respect to each other because the double bond prevents the carbon atoms from rotating.

ASSESSMENT

Knowledge Ask students to distinguish between the terms *isotope* and *isomer*. Isotopes are elements with the same atomic number but different mass numbers. Isomers are compounds with the same molecular formula but different structures. OL EL

READING CHECK Structural isomers have the same chemical formula, but their atoms are bonded in different arrangements. Geometric isomers are structural isomers that have different arrangements of groups around a double bond.

Figure 18 The single-bonded carbons in ethane are free to rotate around the bond. The double-bonded carbons in ethene resist being rotated.

Explain *How do you think this difference in ability to rotate would affect atoms or groups of atoms bonded to single-bonded and double-bonded carbon atoms?*

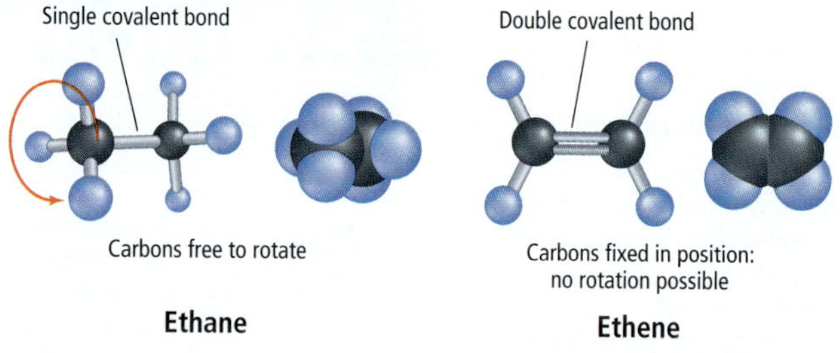

Stereoisomers

The second class of isomers involves a more subtle difference in bonding. **Stereoisomers** are isomers in which all atoms are bonded in the same order but are arranged differently in space. There are two types of stereoisomers. One type occurs in alkenes, which contain double bonds. Two carbon atoms with a single bond between them can rotate freely in relationship to each other. However, when a second covalent bond is present, the carbons can no longer rotate; they are locked in place, as shown in **Figure 18.**

Compare the two possible structures of 2-butene shown in **Figure 19.** The arrangement in which the two methyl groups are on the same side of the molecule is indicated by the prefix *cis-*. The arrangement in which the two methyl groups are on opposite sides of the molecule is indicated by the prefix *trans-*. These terms derive from Latin: *cis* means *on the same side*, and *trans* means *across from*. Because the double-bonded carbon atoms cannot rotate, the *cis-* form cannot easily change into the *trans-* form.

Isomers resulting from different arrangements of groups around a double bond are called **geometric isomers.** Note how the difference in geometry affects the isomers' physical properties, such as melting point and boiling point. Geometric isomers differ in some chemical properties as well. If the compound is biologically active, such as a drug, the *cis-* and *trans-* isomers usually have very different effects.

READING CHECK **Explain** how structural and geometric isomers differ.

Figure 19 These isomers of 2-butene differ in the arrangement in space of the two methyl groups at the ends. The double-bonded carbon atoms cannot rotate with respect to each other, so the methyl groups are fixed in one of these two arrangements.

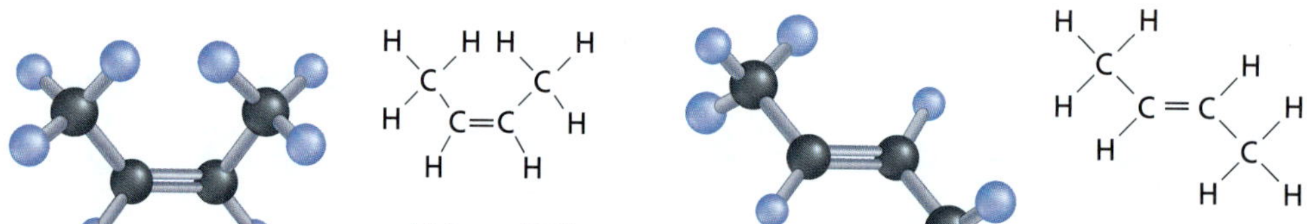

cis-2-Butene (C_4H_8)
mp = −139°C
bp = 3.7°C

trans-2-Butene (C_4H_8)
mp = −106°C
bp = 0.8°C

CHEMISTRY JOURNAL

Louis Pasteur Have students investigate Louis Pasteur's many fields of investigation and write a brief synopsis of his work in their chemistry journals. OL

■ **Figure 20** The reflection of your right hand looks the same as your left hand. However, you cannot place your hands palms down with one on top of the other and have matching parts lie on top of each other.

Chirality

Connection to Biology In 1848, the young French chemist Louis Pasteur (1822–1895) reported his discovery that crystals of the organic compound tartaric acid, existed in two shapes that were mirror images of each other. Because a person's hands are like mirror images, as shown in **Figure 20,** the crystals were called the right-handed and left-handed forms. The two forms of tartaric acid had the same chemical properties, melting point, density, and solubility in water, but only the left-handed form was produced by fermentation to make wine. In addition, bacteria were able to multiply only when they were fed the left-handed form as a nutrient.

The two crystalline forms of tartaric acid exist in the two arrangements as shown in **Figure 21.** Today, these two forms are called d-tartaric acid and l-tartaric acid. The letters d and l stand for the Latin prefixes *dextro-*, which means *to the right,* and *levo-*, which means *to the left.* The property in which a molecule exists in a right- and left-handed form is called **chirality.** Many of the substances found in living organisms, such as the amino acids that make up proteins, have this chirality. In general, living organisms make use of only one chiral form of a substance because only this form fits the active site of an enzyme.

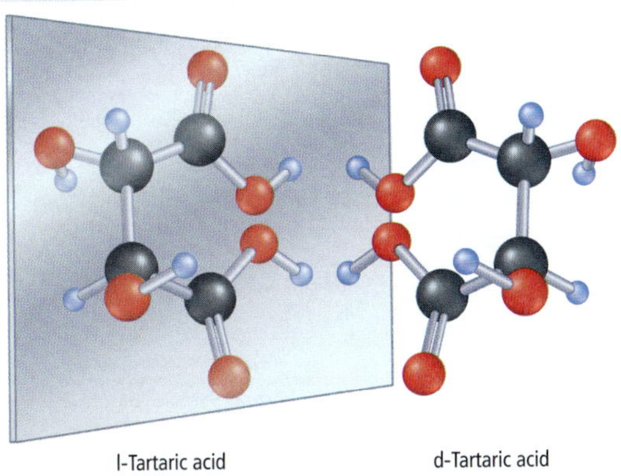

■ **Figure 21** These models represent the two forms of tartaric acid that Pasteur studied. If the model of right-handed tartaric acid (d-tartaric acid) is reflected in a mirror, its image is a model of left-handed tartaric acid (l-tartaric acid).

RealWorld CHEMISTRY

Trans fats

ISOMERS IN THE DIET Fats with *trans* isomers are called *trans* fats. Many pre-packaged foods are made with *trans* fats because they have a longer shelf life. Evidence suggests that *trans* fat increases the unhealthy form of cholesterol and decreases the healthy form, which increases the chance of heart disease.

ASSESSMENT

Knowledge Ask students to name as many common objects as they can that represent pairs of optical isomers. Possibilities include gloves, mittens, shoes, boots, right- and left-handed golf clubs, and other sports equipment. A left-handed student may be able to give examples—such as scissors and other tools—that are not as familiar to right-handed students. **BL**

Quick Demo

Optical Isomers Make models of a pair of optical isomers of a simple chiral molecule such as bromochlorofluoromethane, and hold one model up to a mirror so the students can see both it and its mirror image. Show that the mirror image has the same appearance as the model of the other isomer. Then, show that the two models cannot be rotated in any way so that they are identical, or superimposable. Let students make their own pairs of isomers and compare them. Because students have not studied organic halides, present the molecules simply as having four different groups represented by balls of four different colors.

DIFFERENTIATED INSTRUCTION

Advanced Learners Ask students to investigate the importance of chiral molecules among modern pharmaceutical agents. Correct chirality is very important. Among some pairs of optical isomers, one might have medicinal effects and the other very harmful side effects. This was first revealed when serious birth defects were caused by one optical isomer of the drug thalidomide. **AL**

ASSESSMENT

Performance Have students make and compare ball-and-stick models of *cis*-2-pentene and *trans*-2-pentene. OL

FOLDABLES® Rubric

Data Analysis LAB

About the Lab

- Chlorinated ethenes are common groundwater contaminates.
- Microbes that use ethene as a carbon and energy source under anaerobic conditions are used to clean up the ground water, but dichloroethenes (DCEs) often persist.
- This study explored the use of different reducing agents to speed the process of DCE oxidation.
- Also see Sluis, M.K. et al. 2002. *Microbiology* 148:3617–3629

Think Critically

1. lactate
2. 1,2-trans-dichloroethene

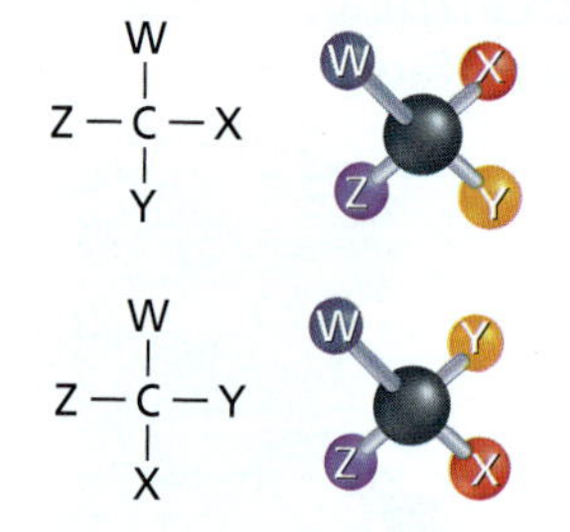

■ **Figure 22** These models represent two different molecules. Groups X and Y have switched places.

Optical Isomers

In the 1860s, chemists realized that chirality occurs whenever a compound contains an asymmetric carbon. An **asymmetric carbon** is a carbon atom that has four different atoms or groups of atoms attached to it. The four groups can always be arranged in two different ways. Suppose that groups W, X, Y, and Z are attached to the same carbon atom in the two arrangements shown in **Figure 22.** Note that the structures differ in that groups X and Y have been exchanged. You cannot rotate the two arrangements in any way that will make them identical to each other.

Now suppose that you build models of these two structures. Is there any way you could turn one structure so that it looks the same as the other? (Whether letters appear forward or backward does not matter.) You would discover that there is no way to accomplish the task without removing X and Y from the carbon atom and switching their positions. Therefore, the molecules are different even though they look very much alike.

FOLDABLES® Incorporate information from this section into your Foldable.

Isomers that result from different arrangements of four different groups around the same carbon atom represent another class of stereoisomers called optical isomers. **Optical isomers** have the same physical and chemical properties, except in chemical reactions where chirality is important, such as enzyme-catalyzed reactions in biological systems. Human cells, for example, incorporate only l-amino acids into proteins. Only the l-form of ascorbic acid is active as vitamin C. The chirality of a drug molecule can also be important. For example, only one isomer of some drugs is effective and the other isomer can be harmful.

Data Analysis LAB

Based on Real Data*

Interpret Data

What are the rates of oxidation of dichloroethene isomers? *Pseudomonas butanovora* is a bacterium that uses some alkanes, alcohols, and organic acids as sources of carbon and energy. This bacteria was tested as an agent to rid groundwater of dichloroethene (DCE) contaminants. Mixtures containing various reducing agents and butane monooxygenase in *Pseudomonas butanovora* oxidized isomers of DCE.

Data and Observations

The table shows the rate of oxidation of each compound in butane-grown *P. butanovora.*

Rates of Oxidation

	Initial Rate of Oxidation (nmol min^{-1} mg $protein^{-1}$)	
Reducing Agent	**1,2-cis DCE**	**1,2-trans DCE**
Buffer	0.9 (1.0)	1.6 (1.0)
Butyrate	6.8 (7.6)	2.0 (1.3)
Propionate	5.9 (6.6)	0.4 (0.3)
Acetate	8.5 (9.4)	3.8 (2.8)
Formate	1.4 (1.6)	1.2 (0.7)
Lactate	11 (12.2)	4.5 (2.8)

Values in parentheses represent the increase (*n*-fold) above the buffer rate.

Think Critically

1. **Compare** Which reducing agent was most useful in oxidizing each isomer?
2. **Conclude** Which isomer oxidized the slowest?

Data obtained from: Doughty, D.M. et al. 2005. Effects of dichloroethene isomers on the induction and activity of butane monooxygenase in the alkane-oxidizing Bacterium *"Pseudomonas butanovora." Applied Environmental Microbiology.* October: 6054–6059.

CHEMISTRY PROJECT

Stereochemistry and Odor Recognition Ask students to research the importance of molecular shape, particularly caused by differences in stereochemistry, on human olfactory sensation. They should write a report on this topic and present it to the class. OL

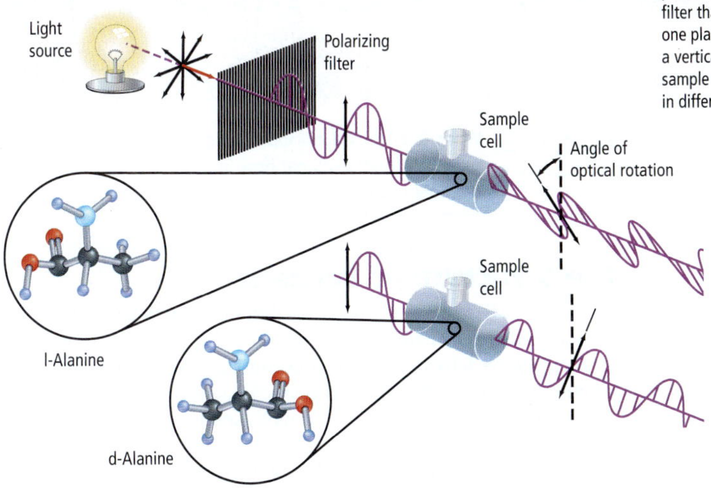

Figure 23 Polarized light can be produced by passing ordinary light through a filter that transmits light waves that lie in only one plane. Here, the filtered light waves are in a vertical plane before they pass through the sample cells. The two isomers rotate the light in different directions.

Optical rotation Mirror-image isomers are called optical isomers because they affect light passing through them. Normally, the light waves in a beam from the Sun or a lightbulb move in all possible planes. However, light can be filtered or reflected in such a way that the resulting waves all lie in the same plane. This type of light is called polarized light.

When polarized light passes through a solution containing an optical isomer, the plane of polarization is rotated to the right (clockwise, when looking toward the light source) by a d-isomer or to the left (counterclockwise) by an l-isomer, producing an effect called **optical rotation.** This effect is shown in **Figure 23.**

One optical isomer that you might have used is l-menthol. This natural isomer has a strong, minty flavor, and a cooling odor and taste. The mirror-image isomer, d-menthol, does not have the same cooling effect as l-menthol.

SECTION 4 REVIEW

Section Self-Check

Section Summary

- Isomers are two or more compounds with the same molecular formula but different molecular structures.
- Structural isomers differ in the order in which atoms are bonded to each other.
- Stereoisomers have all atoms bonded in the same order but arranged differently in space.

25. MAINIDEA **Draw** all of the structural isomers possible for the alkane with the molecular formula C_6H_{14}. Show only the carbon chains.
26. **Explain** the difference between structural isomers and stereoisomers.
27. **Draw** the structures of *cis*-3-hexene and *trans*-3-hexene.
28. **Infer** why living organisms can make use of one only chiral form of a substance.
29. **Evaluate** A certain reaction yields 80% *trans*-2-pentene and 20% *cis*-2-pentene. Draw the structures of these two geometric isomers, and develop a hypothesis to explain why the isomers form in the proportions cited.
30. **Formulate Models** Starting with a single carbon atom, draw two different optical isomers by attaching the following atoms or groups to the carbon: $-H$, $-CH_3$, $-CH_2CH_3$, and $-CH_2CH_2CH_3$.

3 Assess

Check for Understanding

Draw structures of the following pairs of molecules on the board. Ask students to determine whether or not they each represent a pair of isomers and, if they do, what kind of isomers they are.

a. pentane and 2-methylbutane
yes, structural isomers
b. *cis*- and *trans*-2-butene
yes, geometric isomers
c. butane and cyclobutane
no
d. cyclobutane and 1-butene
yes, structural isomers OL

Reteach

Ask students to draw a concept map that shows the relationships among the following terms: isomer, structural isomer, stereoisomer, geometric isomer, and optical isomer. OL

Extension

Have students research and find the molecular formulas of the l-form of ascorbic acid (Vitamin C) and the d-form of ascorbic acid. OL

ASSESSMENT

Skill Give each group of two or three students a model of a simple chiral compound. Ask each group to build a model that represents the optical isomer of the compound it was given. AL

SECTION 4 REVIEW

25. Refer to the Solutions Manual for the drawings. Answers will include 5 structural isomers: hexane, 2-methylpentane, 3-methylpentane, 2,3-dimethylbutane, and 2,2-dimethylbutane.
26. Structural isomers differ from each other in the order in which their atoms are bonded to each other, whereas stereoisomers have their atoms bonded in the same order but they are arranged differently in space.
27. Refer to the Solutions Manual for drawings. In *cis*-3-hexene, the hydrogen atoms attached to the double-bonded carbons are on the same side of the carbon chain. In the *trans* form, the hydrogen atoms are on opposite sides of the carbon chain.
28. In general, living organisms make use of only one chiral form of a substance because only this form fits the active site of an enzyme.
29. Refer to the Solutions Manual for drawings. More of the *trans* isomer forms because its structure allows the bulky methyl and ethyl groups to be farther apart than in the *cis* structure.
30. Refer to the Solutions Manual for drawings. Structures should show the listed groups attached to a single carbon atom. They should differ in that two of the attached groups should be reversed in space.

SECTION 5

1 Focus

MAINIDEA

Benzene Draw Kekulé's proposed structure of benzene on the board. Ask students what type of carbon bonds are present in this structure. alternating single and double bonds Tell students that this is what chemists once thought the structure of benzene looked like. Draw the present-day model of benzene on the board. Ask students what types of bonds are present in this structure. Accept all reasonable responses. Tell students that chemists now know that all of the electrons in benzene are shared by the six carbon nuclei in the molecule, which is represented by the circle in the middle of the model. Tell students that the sharing of electrons makes benzene very stable and unreactive. OL EL

Caption Question Fig. 24 They contain six-carbon ring structures.

SECTION 5

Aromatic Hydrocarbons

MAINIDEA Aromatic hydrocarbons are unusually stable compounds with ring structures in which electrons are shared by many atoms.

Essential Questions

- How do the properties of aromatic and aliphatic hydrocarbons compare and contrast?
- What is a carcinogen and what are some examples?

Review Vocabulary

hybrid orbitals: equivalent atomic orbitals that form during bonding by the rearrangement of valence electrons

New Vocabulary

aromatic compound
aliphatic compound

CHEM 4 YOU What do bright, colorful fabrics, asphalt roofing shingles, and essential oils for perfumes have in common? They all contain aromatic hydrocarbons.

The Structure of Benzene

Natural dyes, like those found in the fabrics in **Figure 24,** and essential oils for perfumes contain six-carbon ring structures. Compounds with these structures have been used for centuries. By the middle of the nineteenth century, chemists had a basic understanding of the structures of hydrocarbons with single, double, and triple covalent bonds. However, certain hydrocarbon ring structures remained a mystery.

The simplest example of this class of hydrocarbon is benzene, which the English physicist Michael Faraday (1791–1867) first isolated in 1825 from the gases given off when either whale oil or coal was heated. Although chemists had determined that benzene's molecular formula was C_6H_6, it was hard for them to determine what sort of hydrocarbon structure would give such a formula. After all, the formula of the saturated hydrocarbon with six carbon atoms, hexane, was C_6H_{14}. Because the benzene molecule had so few hydrogen atoms, chemists reasoned that it must be unsaturated; that is, it must have several double or triple bonds, or a combination of both. They proposed many different structures, including this one suggested in 1860.

$$CH_2{=}C{=}CH-CH{=}C{=}CH_2$$

Although this structure has a molecular formula of C_6H_6, such a hydrocarbon would be unstable and extremely reactive because of its many double bonds. However, benzene was fairly unreactive, and it did not react in the ways that alkenes and alkynes usually react. For that reason, chemists reasoned that structures such as the one shown above must be incorrect.

Figure 24 Dyes used to produce brightly-colored fabrics have been used for centuries.
Explain *What do many natural dyes and essential oils for perfumes have in common?*

R H Productions/Getty Images

DIFFERENTIATED INSTRUCTION

Struggling Learners Students might understand the delocalization of electrons that takes place in benzene better if they model it themselves. Ask six students to form a human ring in which they first model 1,3,5-cyclohexatriene (Kekulé's original model of benzene) and then benzene. Remind students that each carbon (each person) has four valence electrons, and that two of the four are involved in bonds (sigma bonds) with adjacent carbon atoms. Another electron is in a bond (also a sigma bond) with a hydrogen atom. This leaves one electron per carbon atom. In the first model, three pairs of adjacent students each raise both arms and hold two tennis balls together, representing the electrons in three stationary bonds (pi bonds). In the second model, all six students pass the six tennis balls continuously around the ring. BL EL

Kekulé's dream In 1865, the German chemist Friedrich August Kekulé (1829–1896) proposed a different kind of structure for benzene—a hexagon of carbon atoms with alternating single and double bonds. How does the molecular formula of this structure compare with that of benzene?

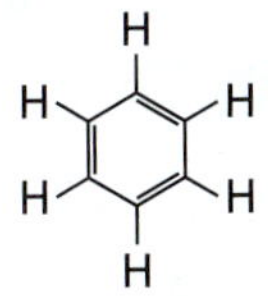

Kekulé claimed that benzene's structure came to him in a dream while he dozed in front of a fireplace in Ghent, Belgium. He said that he had dreamed of the Ouroboros, an ancient Egyptian emblem of a snake devouring its own tail, and that had made him think of a ring-shaped structure. The flat, hexagonal structure Kekulé proposed explained some of the properties of benzene, but it did not explain benzene's lack of reactivity.

A modern model of benzene Since the time of Kekulé's proposal, research has confirmed that benzene's molecular structure is indeed hexagonal. However, benzene's unreactivity could not be explained until the 1930s, when Linus Pauling proposed the theory of hybrid orbitals. When applied to benzene, this theory predicts that the pairs of electrons that form the second bond of each of benzene's double bonds are not localized between only two specific carbon atoms as they are in alkenes. Instead, the electron pairs are delocalized, which means they are shared among all six carbons in the ring. **Figure 25** shows that this delocalization makes the benzene molecule chemically stable because electrons shared by six carbon nuclei are harder to pull away than electrons held by only two nuclei. The six hydrogen atoms are usually not shown, but it is important to remember that they are there. In this representation, the circle in the middle of the hexagon symbolizes the cloud formed by the three pairs of electrons.

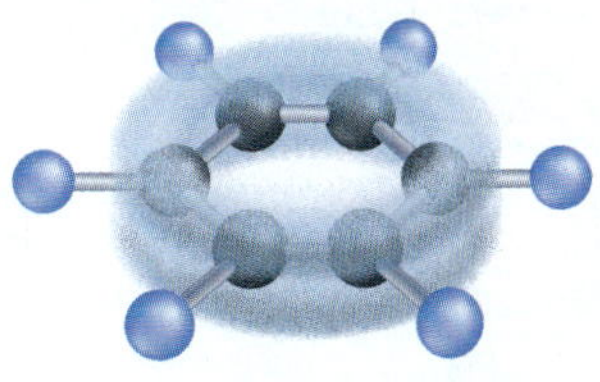

Figure 25 Benzene's bonding electrons spread evenly in a double-donut shape around the ring instead of remaining near individual atoms.

View an **animation about bonding in benzene.**

Concepts In Motion

Aromatic Compounds

Organic compounds that contain benzene rings as part of their structures are called **aromatic compounds.** The term *aromatic* was originally used because many of the benzene-related compounds known in the nineteenth century were found in pleasant-smelling oils that came from spices, fruits, and other plant parts. Hydrocarbons such as the alkanes, alkenes, and alkynes are called **aliphatic compounds** to distinguish them from aromatic compounds. The term *aliphatic* comes from the Greek word for *fat*, which is *aleiphatos*. Early chemists obtained aliphatic compounds by heating animal fats. What are some examples of animal fats that might contain aliphatic compounds?

☑ READING CHECK **Infer** why the terms *aromatic compound* and *aliphatic compound* continue to be used by chemists today.

VOCABULARY

SCIENCE USAGE V. COMMON USAGE

Aromatic

Science usage: an organic compound with increased chemical stability due to the delocalization of electrons

Benzene is an aromatic compound.

Common usage: having a strong odor or smell

The perfume was very aromatic.

CHEMISTRY PROJECT

Essential Oils Have students research essential oils and how they were acquired in ancient times. Have students write a short report explaining how they were processed, how they were used, and how the industry has changed in modern times. BL OL AL

2 Teach

ASSESSMENT

Knowledge Ask students to draw the Kekulé and present-day structures of benzene and explain the difference between the two representations. OL

In-Text Question Their molecular formulas are identical, C_6H_6

☑ **READING CHECK** These terms were consistently used through the years and became part of the language.

In-Text Question beef fat, pork fat, sheep fat, and chicken fat

Quick Demo

Burning Hydrocarbons

Aliphatic hydrocarbons generally burn cleaner in the air than do aromatic hydrocarbons, which usually give off a visible plume of black soot when burning. Demonstrate this difference with a flame test that can be used to distinguish between aliphatic and aromatic compounds. For convenience and safety, use aliphatic and aromatic polymers in this demonstration instead of liquid compounds. Hold a piece of aliphatic plastic (LDPE, HDPE, or PP) in a flame using tongs until it catches fire. Remove the burning piece from the flame, and ask students to observe. Smoke, but little or no soot is given off. Repeat, using aromatic plastic (PETE or PS). Dark, sooty smoke forms. OL EL

Reinforcement

Aromatic Compounds Show students structures or models of the following aromatic compounds that contain a benzene ring with one or more alkyl group substituents. Ask them to name the structures. If needed, review how to name substituted cycloalkanes.

1. methylbenzene (toluene)
2. 1,2-dimethylbenzene (*o*-xylene)
3. 1,3-dimethylbenzne (*m*-xylene)
4. 1,4-dimethylbenzene (*p*-xylene)

Ask students to draw structures or make models of the following compounds.

1. ethylbenzene
2. isopropylbenzene
3. 1,4-diethylbenzene AL

Apply Chemistry

Dimethyl Benzene There are three different isomers of dimethyl benzene, and they are referred to as *o*-, *m*-, and *p*-xylene. The ortho, or *o*-xylene, has a formal name of 1,2-dimethylbenzene; the meta, or *m*-xylene, has a formal name of 1,3-dimethylbenzene; and the para, or *p*-xylene, has a formal name of 1,4-dimethylbenzene. Xylene is used as a solvent in the paint, printing, rubber, glue, and leather industries; it is used as a degreaser and cleaning agent; it is a component of motor and aviation fuels; and it is used in the manufacture of some perfumes and insect repellents. Its health effects include eye damage, depression of the central nervous system, and impaired kidney and liver function.

ASSESSMENT

Knowledge Ask students to list uses for each of the following aromatic compounds.

a. *p*-xylene polyester synthesis
b. naphthalene moth repellents
c. anthracene dye and pigment synthesis OL

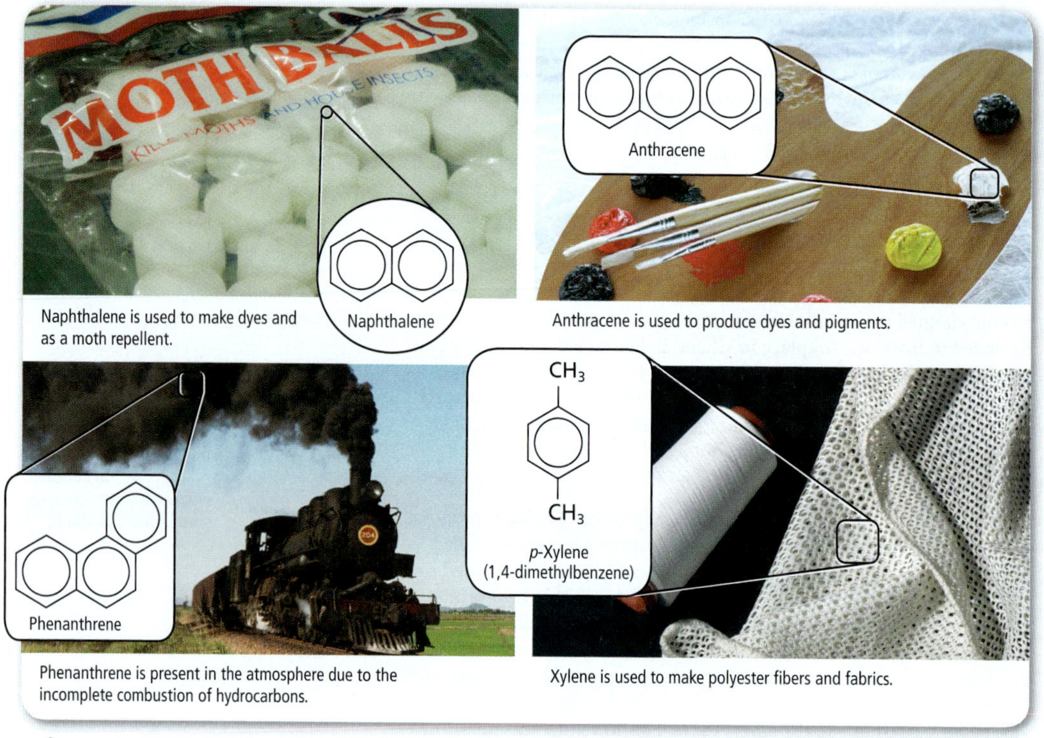

Naphthalene is used to make dyes and as a moth repellent.

Anthracene is used to produce dyes and pigments.

Phenanthrene is present in the atmosphere due to the incomplete combustion of hydrocarbons.

Xylene is used to make polyester fibers and fabrics.

Figure 26 Aromatic hydrocarbons are found in the environment due to the incomplete combustion of hydrocarbons and are used to make a variety of products.

Structures of some aromatic compounds are shown in **Figure 26.** Note that naphthalene has a structure that looks like two benzene rings arranged side by side. Naphthalene is an example of a fused-ring system, in which an organic compound has two or more cyclic structures with a common side. As in benzene, electrons are shared by the carbon atoms that make up the ring systems.

Naming substituted aromatic compounds Like other hydrocarbons, aromatic compounds can have different groups attached to their carbon atoms. For example, methylbenzene, also known as toluene, consists of a methyl group attached to a benzene ring in place of one hydrogen atom. Whenever you see something attached to an aromatic ring system, remember that the hydrogen atom is no longer there.

Substituted benzene compounds are named in the same way as cyclic alkanes. For example, ethylbenzene has a 2-carbon ethyl group attached, and 1,4-dimethylbenzene, also known as *para*-xylene, has two methyl groups attached at positions 1 and 4.

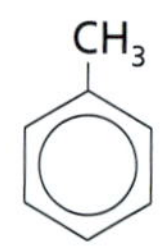

Methylbenzene (toluene)

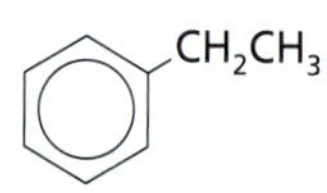

Ethylbenzene

1,4-Dimethylbenzene (*para*-xylene)

CHEMISTRY JOURNAL

Aromatic Hydrocarbons Ask students to think about why these ringed hydrocarbons are called aromatic. Have students write their response in their chemistry journals. When these hydrocarbons were first isolated in the 1860s, scientists were able to distinguish them by their odors. OL

Just as with substituted cycloalkanes, substituted benzene rings are numbered in a way that gives the lowest-possible numbers for the substituents, as shown in **Figure 27.** Numbering the ring as shown gives the numbers 1, 2, and 4 for the substituent positions. Because *ethyl* comes before *methyl* in the alphabet, it is written first in the name: 2-ethyl-1,4-dimethylbenzene.

☑ READING CHECK **Explain** what the circle means inside the six-membered ring structure in **Figure 27.**

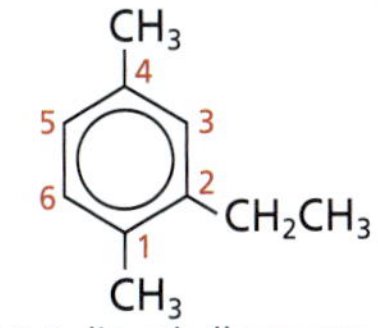

2-Ethyl-1,4-dimethylbenzene

■ **Figure 27** Substituted benzene rings are named in the same way as cyclic alkanes.

EXAMPLE Problem 4

NAMING AROMATIC COMPOUNDS Name the aromatic compound shown.

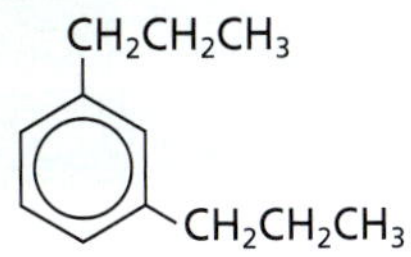

1 ANALYZE THE PROBLEM

You are given an aromatic compound. Follow the rules to name the aromatic compound.

2 SOLVE FOR THE UNKNOWN

Step 1. Number the carbon atoms to give the lowest numbers possible.

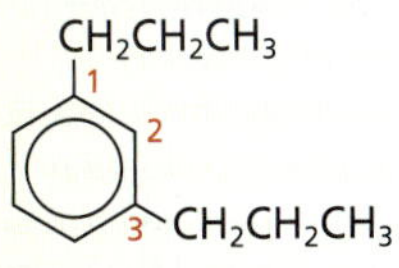

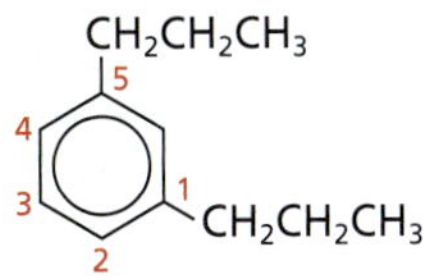

As you can see, the numbers 1 and 3 are lower than the numbers 1 and 5. So the numbers used to name the hydrocarbon should be 1 and 3.

Step 2. Determine the name of the substituents. If the same substituent appears more than once, add the prefix to show the number of groups present.

Step 3. Put the name together. Alphabetize the substituent names, and use commas between numbers and hyphens between numbers and words. Write the name as 1,3-dipropylbenzene.

3 EVALUATE THE ANSWER

The benzene ring is numbered to give the branches the lowest possible set of numbers. The names of the substituent groups are correctly identified.

PRACTICE Problems

Do additional problems. Online Practice

31. Name the following structures.

a. 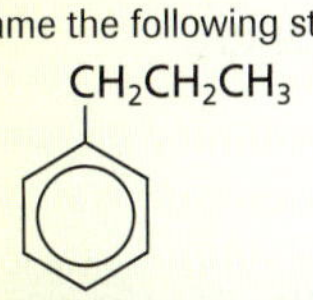

b. 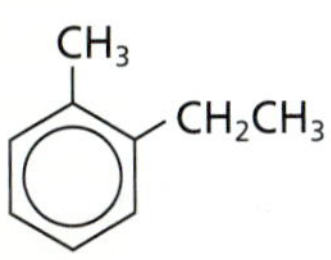

c. 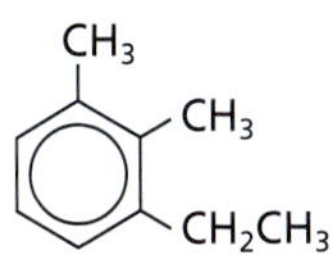

32. Challenge Draw the structure of 1,4-dimethylbenzene.

IN-CLASS Example

Question Name the following aromatic compound.

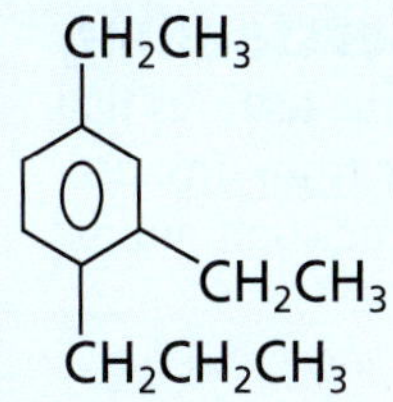

Answer 1,3-diethyl-4-propylbenzene

PRACTICE Problems

Have students refer to the Selected Solutions appendix for complete solutions to odd-numbered problems. The complete solutions for all problems can be found in the Solutions Manual.

31. a. propylbenzene
b. 1-ethyl-2-methylbenzene
c. 1-ethyl-2,3-dimethylbenzene

32. Refer to the Solutions Manual

☑ READING CHECK The electrons are shared by all six carbons in the ring.

DIFFERENTIATED INSTRUCTION

Struggling Learners Pair a below-level student with a student who understands how to name aromatic compounds. Have the student explain to the below-level student how to name an aromatic hydrocarbon step-by-step using the Example Problem. Then, have the below-level student explain the process to the student that understands, while the student is going through the process to name the compound in the In-Class Example. BL

FOLDABLES® Rubric

3 Assess

Check for Understanding

Ask students to list the features that distinguish aromatic from unsaturated aliphatic organic compounds. Aromatic compounds are cyclic hydrocarbons that are more stable and less reactive than unsaturated aliphatic compounds because the unsaturation of aromatic compounds involves a set of delocalized electrons instead of discrete double bonds. **OL**

Reteach

Draw Kekulé's structure of benzene and the present-day model next to each other on the board. Ask students which structure is a better representation of the actual structure of benzene and why it is better. The present-day model is a better model because it represents delocalized electrons that are shared among all six carbons in the ring. This model can be used to explain benzene's unusual stability better than the Kekulé structure. **OL** **EL**

Extension

Ask students to research and explain why the Kekulé structure for benzene is not an ideal representation of the actual bonding pattern in benzene. The carbon-carbon bonds in Kekulé's model have two different lengths and strengths. The double bonds would be shorter and stronger than the single bonds. This does not support the laboratory evidence as well as a model that shows all bonds with equal distribution of electrons. **AL**

Figure 28 Benzopyrene is a cancer-causing chemical that is found in soot, cigarette smoke, and car exhaust.

mcphoto/agefotostock

Carcinogens Many aromatic compounds, particularly benzene, toluene, and xylene, were once commonly used as industrial and laboratory solvents. However, tests have shown that the use of such compounds should be limited because they can affect the health of people who are exposed to them regularly. Health risks linked to aromatic compounds include respiratory ailments, liver problems, and damage to the nervous system. Beyond these hazards, some aromatic compounds are carcinogens, which are substances that can cause cancer.

FOLDABLES® Incorporate information from this section into your Foldable.

The first known carcinogen was an aromatic substance discovered around the turn of the twentieth century in chimney soot. Chimney sweeps in Great Britain were known to have abnormally high rates of cancer. Scientists discovered that the cause of the cancer was the aromatic compound benzopyrene, shown in **Figure 28.** This compound is a by-product of the burning of complex mixtures of organic substances, such as wood and coal. Some aromatic compounds found in gasoline are also known to be carcinogenic.

SECTION 5 REVIEW

Section Self-Check

Section Summary

- Aromatic hydrocarbons contain benzene rings as part of their molecular structures.
- The electrons in aromatic hydrocarbons are shared evenly over the entire benzene ring.

33. MAINIDEA **Explain** benzene's structure and how it makes the molecule unusually stable.

34. **Explain** how aromatic hydrocarbons differ from aliphatic hydrocarbons.

35. **Describe** the properties of benzene that made chemists think it was not an alkene with several double bonds.

36. **Name** the following structures.

a.

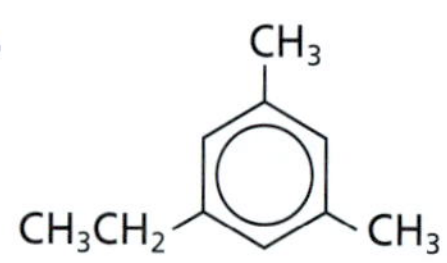

b.

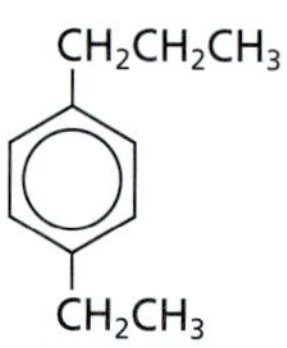

37. **Explain** why the connection between benzopyrene and cancer was significant.

SECTION 5 REVIEW

33. The electron pairs in benzene are delocalized and shared by all six carbon atoms in the ring. Benzene is relatively unreactive because the electrons are harder to pull away from six carbon atoms.

34. Aromatic compounds contain benzene rings in their structures and aliphatic hydrocarbons are straight-chain or branched-chain structures.

35. Benzene is much less reactive than alkenes with multiple double bonds, which usually are unstable. When benzene did react, the reactions were not similar to those of alkenes.

36. **a.** 1-ethyl-3,5-dimethylbenzene
b. 1-ethyl-4-propylbenzene

37. Benzopyrene was the first known carcinogen and exposure to it was occupation related. After it was discovered to be a carcinogen, measures could be taken to protect the workers. The discovery also prompted scientists and medical professionals to look for other substances that could be potentially hazardous for workers.

HOW IT works

POOCH TO POWER: How a Methane Digester Works

Officials in San Francisco are hoping the city's pet owners will contribute their animals' wastes to a pilot project that will convert organic matter into usable energy. A methane digester converts the wastes into biogas—a mixture of methane and carbon dioxide. Burning the methane provides energy for the city.

4 Gas Methane gas is collected, compressed, and either used immediately or stored. The methane can be used to heat homes or to generate electricity.

1 Bacteria Animal wastes are mixed with methane-producing bacteria in the digester. These bacteria can live only under anaerobic conditions—in an oxygen-free environment. Three different anaerobic bacteria break down the wastes, first into organic acids and then into methane gas.

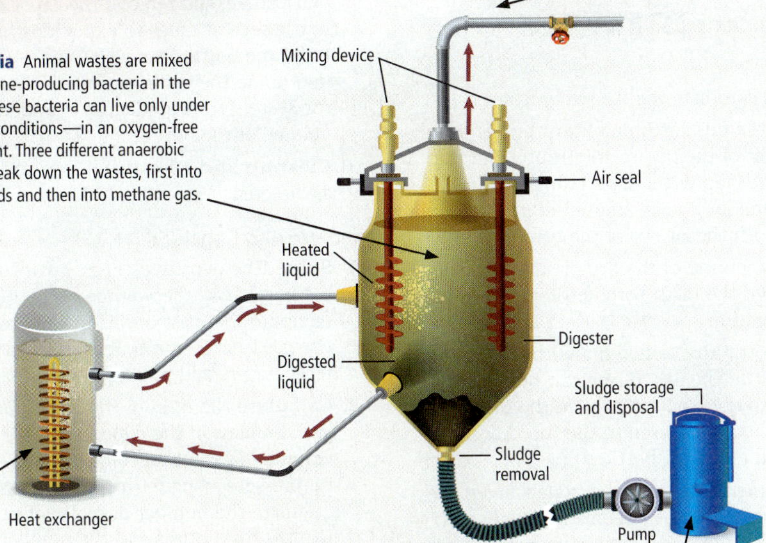

3 Sludge The bacteria cannot convert 100% of the animal wastes into methane. The remaining indigestible material, called sludge or effluent, is rich in plant nutrients and can be used as a soil conditioner.

2 Temperature As with any chemical reaction, temperature affects methane production. Like the bacteria in our own bodies, the bacteria in the digester are most efficient between 35°C and 37°C. An external heat exchanger, combined with insulation around the digester chamber, help to keep the temperature constant and within the optimal range.

WRITING IN ▶ Chemistry

Compare Research and create a pamphlet comparing the advantages of biogas production to other forms of waste disposal for agribusinesses, such as dairies and beef, pork, and poultry producers.

WebQuest

Peter Titmuss/Alamy

WRITING IN ▶ Chemistry

Rubric WebQuest

Compare Student pamphlets should address the positive attributes of producing and using biogas as an energy source. The positive attributes might include reduced air pollution, improved water quality, improved odor control, and beneficial soil additives.

HOW IT works

Purpose

Students will describe the process and assess the environmental impact of biogas production from the break-down of organic material, including pet waste.

Background

High energy costs and periodic energy shortages are prompting governments to examine alternative energy sources, including methane. Dairy farms in the United States are proving an important testing ground due to the availability of raw material. One well-fed dairy cow produces about 54 kg of manure daily, adding up to over 18,000 kg/y. Bacteria break down manure, releasing methane—a powerful greenhouse gas—to the atmosphere. A methane digester allows methane to be produced, captured, and used.

Teaching Strategies

- Use the feature diagram to identify the importance of each phase of the biogas production process.
- Have students brainstorm the benefits gained by using a methane digester as an urban energy source.

ChemLAB

Rubric

Preparation

Time Allotment one class period

Process Skills acquire and analyze information, apply concepts, measure, use numbers, analyze and conclude

Safety Precautions Approve lab safety forms before work begins. Warn students to turn off the valves after collecting the gas and to clean up water spills.

Procedure

It is inevitable that some water will remain on the sides of the bottle. Have students "flick" the bottle with a finger to get out as much water as possible.

Expected Results Natural gas in the United States averages about 85% methane, 9% ethane, 3% propane, and 3% composed of nitrogen, butane, and helium.

Analyze and Conclude

1. mass of air = density × volume Refer to sample data table.
2. Refer to sample data table. At 25°C, approximately 3% of the volume of the bottle will be occupied by water vapor because the gas was collected over water. However, the presence of water vapor can be ignored in an experiment of this precision. Students substitute values into the ideal gas equation and solve for *n*. Students should carry out the following calculation. *molar mass = measured mass of gas/calculated number of moles*
3. Results will depend on the composition of the gas.
4. Possibilities include excess water trapped in the bottle, incorrect or poor measurement techniques, or math error. A gas mixture would yield a molar mass that does not exactly equal the molar mass of any of the components.

ChemLAB

Forensics: Analyze Hydrocarbon Burner Gases

Background: A valve needs to be replaced in the science lab. The custodian says the gas used in the lab is propane, and the chemistry teacher says it is natural gas (methane). Use scientific methods to settle this dispute.

Question: *What type of alkane gas is used in the science laboratory?*

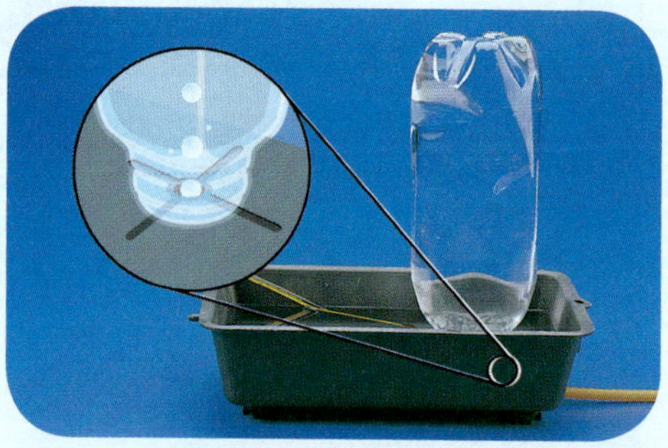

Matt Meadows

Materials

barometer	pneumatic trough
thermometer	100-mL graduated cylinder
1-L or 2-L plastic soda bottle with cap	balance (0.01g)
burner tubing	paper towels

Safety Precautions

Procedure

1. Read and complete the lab safety form.
2. Connect the burner tubing from the gas supply to the inlet of the pneumatic trough. Fill the trough with tap water. Open the gas valve slightly, and let a small amount of gas into the tank to flush the air out of the tubing.
3. Measure the mass of the dry plastic bottle and cap. Record the mass, barometric pressure, and air temperature.
4. Fill the bottle to overflowing with tap water, and screw on the cap. If some air bubbles remain, tap the bottle gently on the desktop until all the air has risen to the top. Add more water, and recap the bottle.
5. Place the thermometer in the trough. Invert the capped bottle into the pneumatic trough, and remove the cap while keeping the mouth of the bottle under water. Hold the mouth of the bottle directly over the inlet opening of the trough.
6. Slowly open the gas valve, and allow gas to enter the inverted bottle until all of the water has been displaced. Close the gas valve immediately. Record the temperature of the water.
7. While the bottle is still inverted, screw on the cap. Remove the bottle from the water, and dry the outside of the bottle.
8. Measure and record the mass of the bottle containing the burner gas.
9. Place the bottle in a fume hood, turn on the exhaust fan, and remove the cap. Compress the bottle several times to expel most of the gas. Refill the bottle to overflowing with water, and determine the volume of the bottle by pouring the water into a graduated cylinder. Record the volume of the bottle.
10. **Cleanup and Disposal** Clean your workspace as directed.

Analyze and Conclude

1. **Solve** The density of air at 1 atm and 20°C is 1.205 g/L. Use the volume of the bottle to compute the mass of the air the bottle contains. Use gas laws to compute the density of air at the temperature and pressure of your laboratory.
2. **Calculate** the mass of the empty bottle. Calculate the mass of the collected gas. Use the volume of gas, water temperature, and barometric pressure along with the ideal gas law to calculate the number of moles of gas collected. Use the mass of gas and the number of moles to calculate the molar mass of the gas.
3. **Conclude** How does your experimental molar mass compare with the molar masses of methane, ethane, and propane? Infer which gases are in the burner gas in your lab.
4. **Error Analysis** Suggest possible sources of error in the experiment.

INQUIRY EXTENSION

Design an Experiment to test how one variable, such as temperature or atmospheric pressure, affects your results.

Sample Mass and Volume Data	
Mass of bottle + air	30.49 g
Mass of air	0.82 g
Mass of "empty" bottle	29.67 g
Mass of bottle + collected burner gas	30.30 g
Mass of collected burner gas	0.63 g
Barometric pressure	1.01 atm
Temperature	24°C
Temperature	297 K
Volume of gas collected	0.630 L

INQUIRY EXTENSION

Students should find that atmospheric pressure and temperature varies so little in the lab from day to day, that it will not affect the results of this type of lab. The measurements are not precise enough to show a difference. However, students that simulate extreme temperature or atmospheric pressure changes might show a difference in the their results.

CHAPTER 21 STUDY GUIDE

Vocabulary Practice

BIGIDEA Organic compounds called hydrocarbons differ by their types of bonds.

SECTION 1 Introduction to Hydrocarbons

MAINIDEA Hydrocarbons are carbon-containing organic compounds that provide a source of energy and raw materials.

- Organic compounds contain the element carbon, which is able to form straight chains and branched chains.
- Hydrocarbons are organic substances composed of carbon and hydrogen.
- The major sources of hydrocarbons are petroleum and natural gas.
- Petroleum can be separated into components by the process of fractional distillation.

VOCABULARY
- organic compound
- hydrocarbon
- saturated hydrocarbon
- unsaturated hydrocarbon
- fractional distillation
- cracking

SECTION 2 Alkanes

MAINIDEA Alkanes are hydrocarbons that contain only single bonds.

- Alkanes contain only single bonds between carbon atoms.
- Alkanes and other organic compounds are best represented by structural formulas and can be named using systematic rules determined by the International Union of Pure and Applied Chemistry (IUPAC).
- Alkanes that contain hydrocarbon rings are called cyclic alkanes.

VOCABULARY
- alkane
- homologous series
- parent chain
- substituent group
- cyclic hydrocarbon
- cycloalkane

SECTION 3 Alkenes and Alkynes

MAINIDEA Alkenes are hydrocarbons that contain at least one double bond, and alkynes are hydrocarbons that contain at least one triple bond.

- Alkenes and alkynes are hydrocarbons that contain at least one double or triple bond, respectively.
- Alkenes and alkynes are nonpolar compounds with greater reactivity than alkanes but with other properties similar to those of alkanes.

VOCABULARY
- alkene
- alkyne

SECTION 4 Hydrocarbon Isomers

MAINIDEA Some hydrocarbons have the same molecular formula but have different molecular structures.

- Isomers are two or more compounds with the same molecular formula but different molecular structures.
- Structural isomers differ in the order in which atoms are bonded to each other.
- Stereoisomers have all atoms bonded in the same order but arranged differently in space.

VOCABULARY
- isomer
- structural isomer
- stereoisomer
- geometric isomer
- chirality
- asymmetric carbon
- optical isomer
- optical rotation

SECTION 5 Aromatic Hydrocarbons

MAINIDEA Aromatic hydrocarbons are unusually stable compounds with ring structures in which electrons are shared by many atoms.

- Aromatic hydrocarbons contain benzene rings as part of their molecular structures.
- The electrons in aromatic hydrocarbons are shared evenly over the entire benzene ring.

VOCABULARY
- aromatic compound
- aliphatic compound

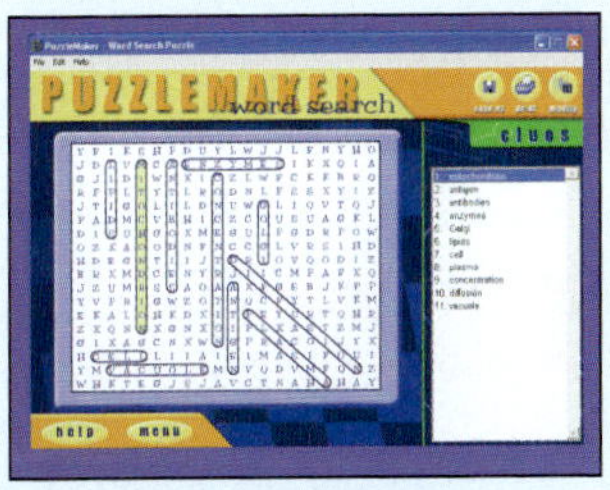

Vocabulary Practice

For additional practice with vocabulary, have students access the Vocabulary PuzzleMaker.

Use the Vocabulary

To reinforce chapter vocabulary, have students write a sentence using each term. OL EL

Review Strategies

- Have students summarize the similarities and differences between alkanes, alkenes, and alkynes. OL
- Have students make a list of the names and structures of common hydrocarbons. OL
- Problems from the Supplemental Practice Problems appendix or the Supplemental Problems booklet can be used for review. OL

Students can review with these online resources.

Vocabulary Practice

Vocabulary eGames
Vocabulary eFlashcards
Vocabulary PuzzleMaker

Section Self-Check

Chapter Self-Check

Online Test Practice

Use *eAssessment* to:

- create multiple versions of tests
- edit existing questions and add your own questions
- build tests aligned with select state curriculum standards using built-in state curriculum tags
- track students' progress

SECTION 1

Mastering Concepts

38. Chemists realized that it was possible to synthesize organic compounds without a vital force.
39. Organic compounds contain the element carbon.
40. Carbon is able to form four strong covalent bonds, including bonds with other carbon atoms.
41. petroleum and natural gas
42. boiling point
43. Saturated hydrocarbons contain all single carbon-carbon bonds. Unsaturated hydrocarbons contain one or more double or triple carbon-carbon bonds.

Mastering Problems

44. methane, propane, butane, hexane, octane
45. a. 2 electrons
    b. 4 electrons
    c. 6 electrons
46. a. structural formula and space-filling model
    b. It is an organic compound because it contains carbon and it is not one of the groups that is an exception—carbon dioxide, carbide, or carbonate.
47. Molecular models show the type of atoms in the molecule, but they do not show the geometry of the molecule. Structural formulas show the type of atoms in the molecule and the general arrangement of the atoms, but not the exact geometry. Ball-and-stick models show the types of atoms in the molecule and the general arrangement, but not the exact geometry. Space-filling models show a realistic picture of the molecule, but it is difficult to determine the types of bonds in the molecule and if the molecule is large, it is difficult to see all of the atoms in the molecule.

CHAPTER 21 ASSESSMENT

SECTION 1

Mastering Concepts

38. **Organic Chemistry** Why did Wöhler's discovery lead to the development of the field of organic chemistry?

39. What is the main characteristic of an organic compound?

40. What characteristic of carbon accounts for the large variety of organic compounds?

41. Name two natural sources of hydrocarbons.

42. Explain what physical property of petroleum compounds is used to separate them during fractional distillation.

43. Explain the difference between saturated hydrocarbons and unsaturated hydrocarbons.

Mastering Problems

44. **Distillation** Rank the compounds listed in **Table 7** in the order in which they will be distilled out of a mixture. Rank the compounds in order of first to distill to last to distill.

Table 7 Alkane Boiling Points

Compound	Boiling Point (°C)
hexane	68.7
methane	−161.7
octane	125.7
butane	−0.5
propane	−42.1

45. How many electrons are shared between two carbon atoms in each of the following carbon-carbon bonds?
    a. single bond
    b. double bond
    c. triple bond

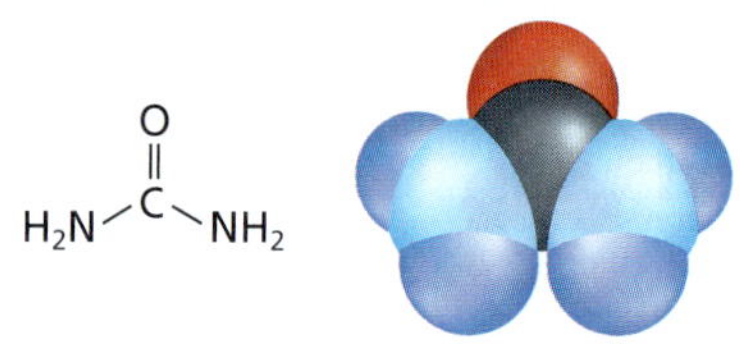

■ **Figure 29**

46. **Figure 29** shows two models of urea, a molecule that Friedrich Wöhler first synthesized in 1828.
    a. Identify the types of models shown.
    b. Is urea an organic or an inorganic compound? Explain your answer.

47. Molecules are modeled using molecular formulas, structural formulas, ball-and-stick models, and space-filling models. What are the advantages and disadvantages of each model?

SECTION 2

Mastering Concepts

48. Describe the characteristics of a homologous series of hydrocarbons.

49. **Fuels** Name three alkanes used as fuels and describe an additional application for each.

50. Draw the structural formula of each of the following.
    a. ethane
    b. hexane
    c. propane
    d. heptane

51. Write the condensed structural formulas for the alkanes in the previous question.

52. Write the name and draw the structure of the alkyl group that corresponds to each of the following alkanes.
    a. methane
    b. butane
    c. octane

53. How does the structure of a cycloalkane differ from that of a straight-chain or branched-chain alkane?

54. **Freezing and Boiling Points** Use water and methane to explain how intermolecular attractions generally affect the boiling and freezing points of a substance.

Mastering Problems

55. Name the compound represented by each of the following structural formulas.
    a. $CH_3CH_2CH_2CH_2CH_3$
    b. 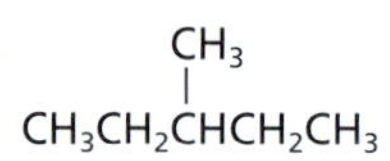
    c. 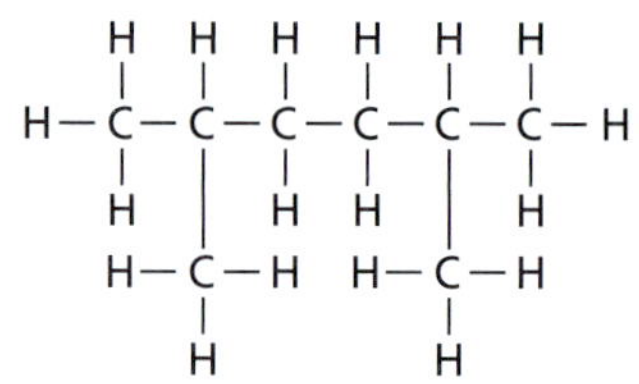
    d. 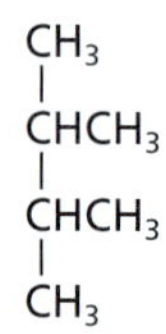

56. Draw full structural formulas for the following compounds.
    a. heptane
    b. 2-methylhexane
    c. 2,3-dimethylpentane
    d. 2,2-dimethylpropane

57. Draw condensed structural formulas for the following compounds. Use line structures for rings.
    a. 1,2-dimethylcyclopropane
    b. 1,1-diethyl-2-methylcyclopentane

SECTION 2

Mastering Concepts

48. It is a series of compounds that differ from each other by a repeating unit and have a fixed numerical relationship among the numbers of atoms.
49. Methane—fuel for cooking and heating; propane—fuel for cooking and heating; butane—small lighters and some torches
50. Refer to the Solutions Manual for structures.
51. a. CH_3CH_3
    b. $CH_3(CH_2)_4CH_3$
    c. $CH_3CH_2CH_3$
    d. $CH_3(CH_2)_5CH_3$
52. Refer to Solutions Manual for structures.
    a. methane, methyl
    b. butane, butyl
    c. octane, octyl
53. A cycloalkane contains a ring of carbon atoms.
54. Methane molecules are nonpolar and they do not form hydrogen bonds with other methane molecules. Water molecules are polar and freely form hydrogen bonds with other water molecules. Because of the attraction between molecules in water, water has a higher boiling point and a higher melting point than methane.

Chapter Self-Check

58. Name the compound represented by each of the following structural formulas.

a. 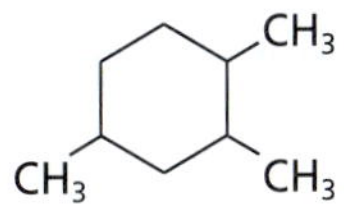

c. 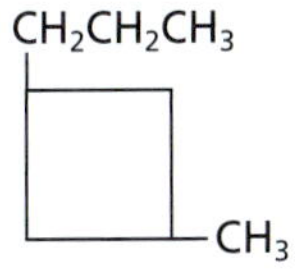

b. 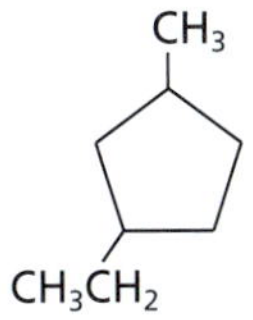

d. 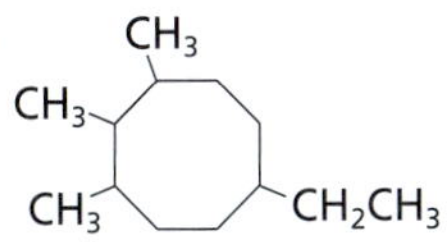

SECTION 3

Mastering Concepts

59. Explain how alkenes differ from alkanes. How do alkynes differ from both alkenes and alkanes?

60. The name of a hydrocarbon is based on the name of the parent chain. Explain how the determination of the parent chain when naming alkenes differs from the same determination when naming alkanes.

Mastering Problems

61. Name the compound represented by each of the following condensed structural formulas.

a. 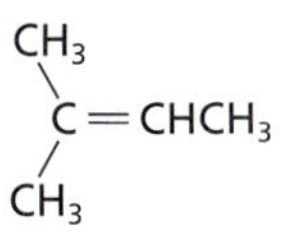

c. 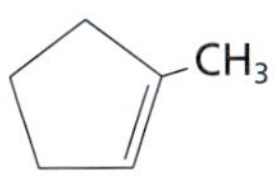

b. 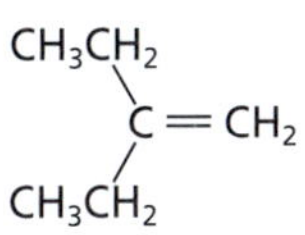

d. 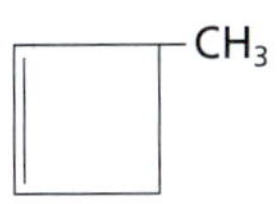

62. Draw condensed structural formulas for the following compounds. Use line structures for rings.

a. 1,4-diethylcyclohexene
b. 2,4-dimethyl-1-octene
c. 2,2-dimethyl-3-hexyne

63. Name the compound represented by the following condensed structural formula.

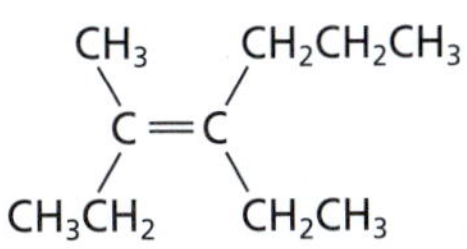

SECTION 4

Mastering Concepts

64. How are two isomers alike, and how are they different?

65. Describe the difference between *cis-* and *trans-* isomers in terms of geometrical arrangement.

66. What are the characteristics of a chiral substance?

67. Light How does polarized light differ from ordinary light, such as light from the Sun?

68. How do optical isomers affect polarized light?

Mastering Problems

69. Identify the pair of structural isomers in the following group of condensed structural formulas.

a. 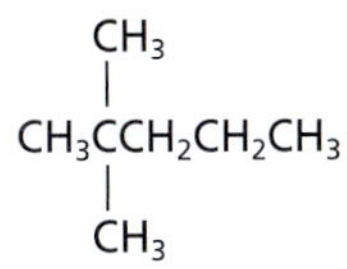

c. 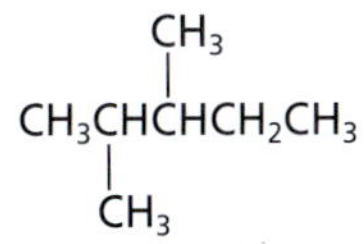

b. 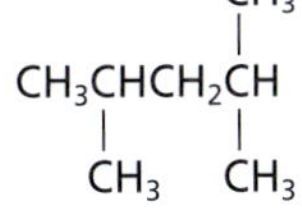

d. 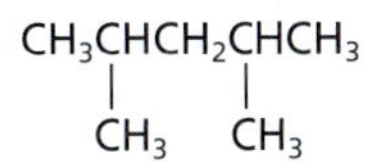

70. Identify the pair of geometric isomers among the following structures. Explain your selections. Explain how the third structure is related to the other two.

a. 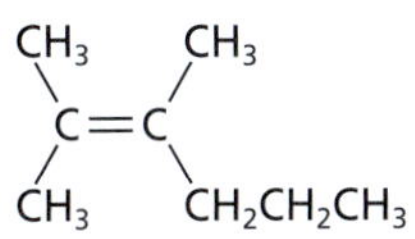

b. 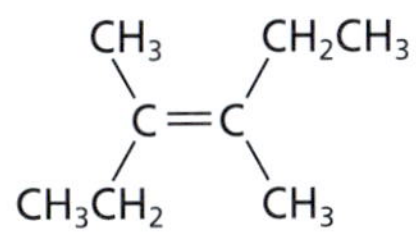

c. 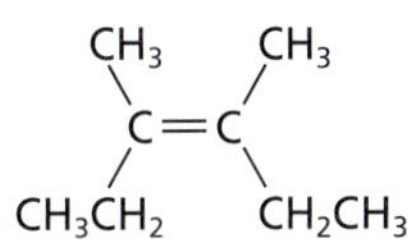

71. Draw condensed structural formulas for four different structural isomers with the molecular formula C_4H_8.

72. Draw and label the *cis-* and *trans-* isomers of the molecule represented by the following condensed formula.

$$CH_3CH=CHCH_2CH_3$$

Mastering Problems

55. a. pentane
b. 3-methylpentane
c. 2,5-dimethylhexane
d. 2,3-dimethylbutane

56. See the Solutions Manual for structures.

57. See the Solutions Manual for structures.

58. a. 1,2,4-trimethylcyclohexane
b. 1-ethyl-3-methylcyclopentane
c. 1-methyl-3-propylcyclobutane
d. 6-ethyl-1,2,3-trimethylcyclooctane

SECTION 3

Mastering Concepts

59. Alkanes contain only single bonds between carbon atoms in the molecule. Alkenes contain at least one double bond between carbon atoms in the molecule. Alkynes contain at least one triple bond between carbon atoms in a molecule.

60. When naming alkanes, the parent chain is the longest continuous carbon chain. When naming alkenes, the parent chain is the longest continuous carbon chain that includes the carbon atoms linked by a double bond.

Mastering Problems

61. a. 2-methyl-2-butene
b. 2-ethyl-1-butene
c. 1-methylcyclopentene
d. 3-methylcyclobutene

62. See the Solutions Manual for structures.

63. 4-ethyl-3-methyl-3-heptene

SECTION 4

Mastering Problems

64. Isomers have the same molecular formula but different structures. They can have different chemical and physical properties.

65. *Cis* isomers have the largest groups on the carbons in the double bond on the same side of the bond; in *trans* isomers they are on opposite sides.

66. A chiral substance has both right- and left-handed forms. Chiral substances have at least one carbon atom bonded to four different groups and are therefore asymmetrical.

67. The waves of polarized light all vibrate in one plane; in ordinary light, they vibrate in all possible planes.

68. They rotate polarized light.

Mastering Problems

69. Student answers can include any two structures except **b** and **d,** which are identical.

70. b and **c** are geometric isomers. They represent a *cis/trans* pair of isomers. **a** is a structural isomer of both **b** and **c.**

71. Refer to the Solutions Manual for structures. Student answers should show condensed structural formulas for cyclobutane, 1-butene, 2-butene, and 2-methylpropene.

72. Refer to the Solutions Manual for structures. The two hydrogen atoms bonded to the doubly-bonded carbons are on the same side of the carbon chain for the *cis* isomer and on opposite sides of the carbon chain for the *trans* isomer.

73. **a** is an optical isomer of **b, c,** and **d; a** is not superimposable with **b, c,** or **d**

SECTION 5

Mastering Concepts

74. They all have a ring structure in the molecule.
75. Carcinogens are substances that can cause cancer.

Mastering Problems

76. See Solutions Manual for structure.
77. **a.** methylbenzene (toluene)
    b. anthracene

MIXED REVIEW

78. No. They are structural isomers.
79. 20 hydrogen atoms; 18 hydrogen atoms
80. C_nH_{2n}
81. Unsaturated hydrocarbons are much more reactive.
82. No. Cyclopentane (C_5H_{10}) and pentane (C_5H_{12}) have different molecular formulas.
83. **a.** Yes
    b. No, it should be numbered from the opposite end.
    c. Yes. The double bond and methyl groups are assigned the lowest possible numbers.
    d. Yes
84. Molecular formulas cannot differentiate between isomers.
85. Structural isomers can have vastly different physical properties because they have completely different arrangements of the carbon skeleton. Stereoisomers (geometric and optical isomers) have the same carbon skeleton but different orientations in space. Geometric isomers have different properties; optical isomers differ only in the direction of rotation of polarized light and in chemical reactions that distinguish between isomers. Therefore, optical isomers would have more similar properties.
86. Numbers are needed to identify the locations of the double and triple bonds.
87. Refer to Solutions Manual for structure.

73. Three of the following structures are exactly alike, but the fourth represents an optical isomer of the other three. Identify the optical isomer, and explain how you made your choice.

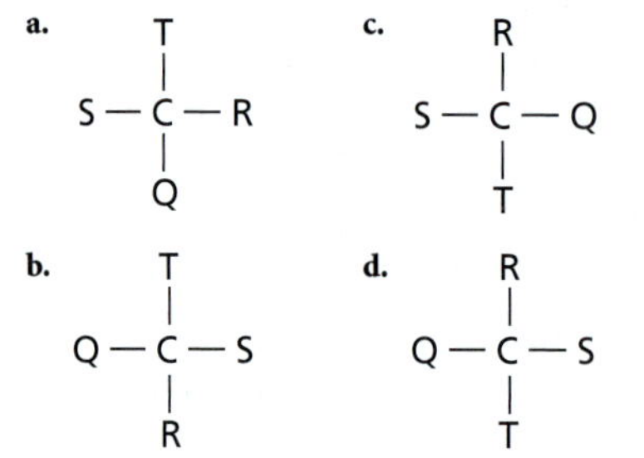

SECTION 5

Mastering Concepts

74. What structural characteristic do all aromatic hydrocarbons share?

75. What are carcinogens?

Mastering Problems

76. Draw the structural formula of 1,2-dimethylbenzene.

77. Name the compound represented by each of the following structural formulas.

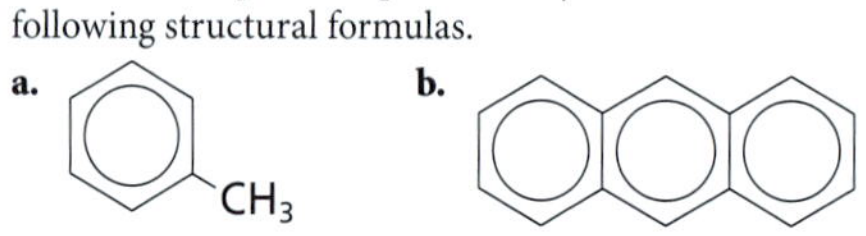

MIXED REVIEW

78. Do the following structural formulas represent the same molecule? Explain your answer.

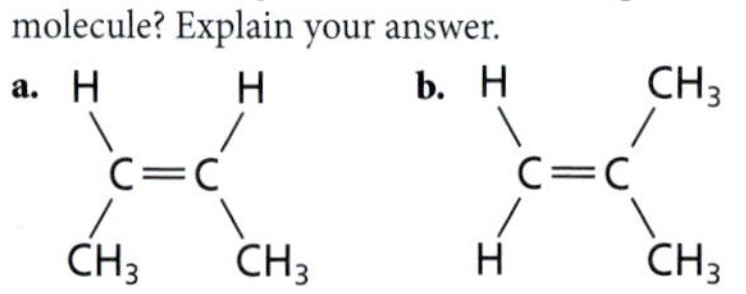

79. How many hydrogen atoms are in an alkane molecule with nine carbon atoms? How many are in an alkene with nine carbon atoms and one double bond?

80. The general formula for alkanes is C_nH_{2n+2}. Determine the general formula for cycloalkanes.

81. Manufacturing Why are unsaturated hydrocarbons more useful than saturated hydrocarbons as starting materials in chemical manufacturing?

82. Is cyclopentane an isomer of pentane? Explain your answer.

83. Determine whether each of the following structures shows the correct numbering. If the numbering is incorrect, redraw the structure with the correct numbering.

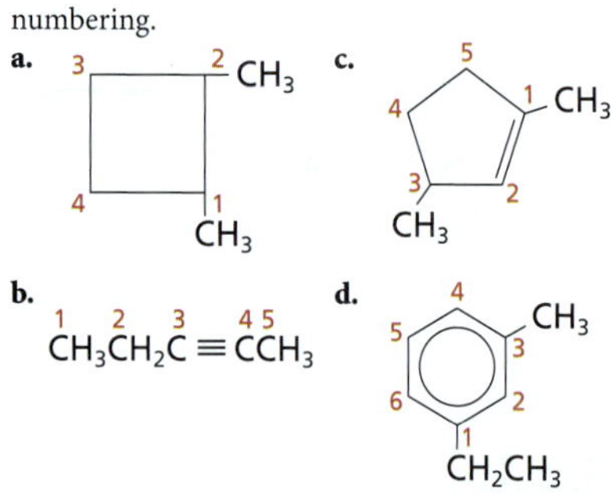

84. Why do chemists use structural formulas for organic compounds rather than molecular formulas, such as C_5H_{12}?

85. Which would you expect to have more similar physical properties, a pair of structural isomers or a pair of stereoisomers? Explain your reasoning.

86. Explain why numbers are needed in the IUPAC names of many unbranched alkenes and alkynes but not in the names of unbranched alkanes.

87. A compound with two double bonds is called a diene. The name of the structure shown is 1,4-pentadiene. Apply your knowledge of IUPAC nomenclature to draw the structure of 1,3-pentadiene.

$$H_2C{=}CH-CH_2-CH{=}CH_2$$

THINK CRITICALLY

88. Determine which two of the following names cannot be correct, and draw the structures of the molecules.
a. 2-ethyl-2-butene
b. 1,4-dimethylcyclohexene
c. 1,5-dimethylbenzene

89. Infer The sugar glucose is sometimes called dextrose because a solution of glucose is known to be dextrorotatory. Analyze the word *dextrorotatory,* and suggest what the word means.

90. Interpret Scientific Illustrations Draw Kekulé's structure of benzene, and explain why it does not truly represent the actual structure.

91. Recognize Cause and Effect Explain why alkanes, such as hexane and cyclohexane, are effective at dissolving grease, whereas water is not.

THINK CRITICALLY

88. Refer to Solutions Manual for structures.
    a. Correct name is 3-methyl-2-pentene.
    b. Name is correct.
    c. Correct name is 1,3-dimethylbenzene.
89. The prefix *dextro-* means *to the right,* and the suffix *rotatory* means *rotates.* Therefore, the natural form of glucose is chiral and rotates the plane of polarized light to the right.
90. Refer to Solutions Manual for structure. The structure shows localized electrons in the double bonds rather than delocalized electrons.
91. Grease and alkanes are both nonpolar. Water is polar. Like dissolves like.

Chapter Self-Check

92. **Explain** Use **Table 8** to construct a statement explaining the relationship between numbers of carbon atoms and boiling points of the members of the alkane series shown.

93. **Graph** the information given in **Table 8.** Predict what the boiling and melting points of the 11- and 12-carbon alkanes will be. Look up the actual values and compare your predictions to the those numbers.

Table 8 Data for Selected Alkanes

Name	Melting Point (°C)	Boiling Point (°C)
CH_4	−182	−162
C_2H_6	−183	−89
C_3H_8	−188	−42
C_4H_{10}	−138	−0.5
C_5H_{12}	−130	36
C_6H_{14}	−95	69
C_7H_{16}	−91	98
C_8H_{18}	−57	126
C_9H_{20}	−54	151
$C_{10}H_{22}$	−29	174

CHALLENGE PROBLEM

94. **Chiral Carbons** Many organic compounds have more than one chiral carbon. For each chiral carbon in a compound, a pair of stereoisomers can exist. The total number of possible isomers for the compound is equal to 2^n, where n is the number of chiral carbons. Draw each structure, and determine how many stereoisomers are possible for each compound named below.
    a. 3,5-dimethylnonane
    b. 3,7-dimethyl-5-ethyldecane

CUMULATIVE REVIEW

95. What element has the following ground-state electron configuration: $[Ar]4s^23d^6$?

96. What is the charge of an ion formed from the following families?
    a. alkali metals
    b. alkaline earth metals
    c. halogens

97. Write the chemical equations for the complete combustion of ethane, ethene, and ethyne into carbon dioxide and water.

WRITING IN ▶ Chemistry

98. **Gasoline** For many years, a principal antiknock ingredient in gasoline was the compound tetraethyllead. Research to learn about the structure of this compound, the history of its development and use, and why its use was discontinued in the United States. Find out if it is still used as a gasoline additive elsewhere in the world.

99. **Perfume** The musk used in perfumes and colognes contains many chemical compounds, including large cycloalkanes. Research and write a short report about the sources used for natural and synthetic musk compounds in these consumer products.

DBQ Document-Based Questions

Polycyclic Aromatic Hydrocarbons *PAH compounds are naturally occurring, but human activities can increase the concentrations in the environment. Soil samples were collected to study PAH compounds. The core sections were dated using radionuclides to determine when each section was deposited.*

Figure 30 *shows the concentration of polycyclic aromatic hydrocarbons (PAH) detected in Central Park in New York City.*

Data obtained from: Yan, B. et al, 2005. *Environmental Science Technology* 39 (18): 7012–7019.

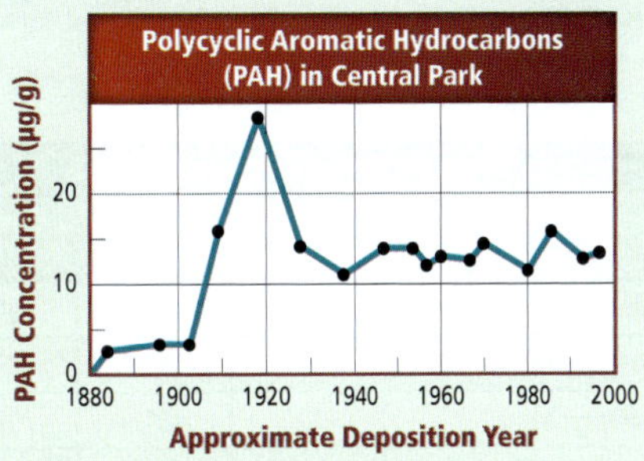

■ **Figure 30**

100. Compare the average PAH concentrations before 1905 and after 1925.

101. PAH compounds are produced in small amounts by some plants and animals, but most come from human activities, such as burning fossil fuels. Infer why the PAH levels were relatively low in the late 1800s and early 1900s.

DBQ Document-Based Questions

Data obtained from: Yan, B. et al, 2005. *Environmental Science Technology* 39 (18): 7012–7019.

100. average before 1905 about 3; average after 1925 about 13

101. The primary fuel burned by humans at this time was wood. The PAH levels began to increase when fossil fuels replaced wood as a fuel source.

92. As the number of carbon atoms increases in the chain, the boiling point increases.

93. Refer to the Solutions Manual for graph. Student predictions will vary. The actual boiling point of $C_{11}H_{24}$ is 196°C. The actual boiling point of $C_{12}H_{26}$ is 216°C.

CHALLENGE PROBLEM

94. Refer to Solutions Manual for structures.
  a. The number of chiral carbons is $2^n = 2^2 = 4$.
  b. The number of chiral carbons is $2^n = 2^3 = 8$.

CUMULATIVE REVIEW

95. iron

96. **a.** 1+
  b. 2+
  c. 1−

97. ethane:
$2C_2H_6 + 7O_2 \longrightarrow 4CO_2 + 6H_2O$;
ethene:
$C_2H_4 + 3O_2 \longrightarrow 2CO_2 + 2H_2O$;
ethyne:
$2C_2H_2 + 5O_2 \longrightarrow 4CO_2 + 2H_2O$

WRITING IN ▶ Chemistry

Rubric

98. Student answers should include a drawing of the structure of tetraethyllead ($Pb(CH_2CH_3)_4$), a discussion of when it was used in the United States, what its health hazards are, and a listing of some areas in the world where it is still added to gasoline.

99. The natural source of the musk that is used in perfumes is the male musk deer. The primary odorous compound is 3-methylcyclopentadecanone, which is synthesized in the perfume and cologne industries.

MULTIPLE CHOICE

1. C
2. D
3. A
4. B
5. D
6. A
7. A
8. A
9. A

MULTIPLE CHOICE

1. Alanine, like most amino acids, exists in two forms:

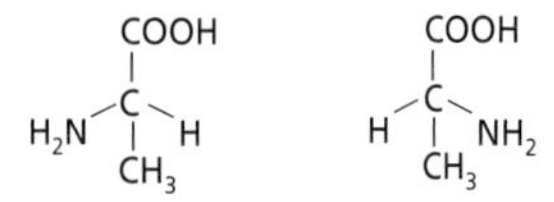

l-Alanine   d-Alanine

Almost all of the amino acids found in living organisms are in the l-form. Which term best describes l-alanine and d-alanine with respect to one another?

A. structural isomers
B. geometric isomers
C. optical isomers
D. stereoisotopes

2. Which does NOT affect reaction rate?

A. catalysts
B. surface area of reactants
C. concentration of reactants
D. reactivity of products

3. What is the molality of a solution containing 0.25 g of dichlorobenzene ($C_6H_4Cl_2$) dissolved in 10.0 g of cyclohexane (C_6H_{12})?

A. 0.17 mol/kg
B. 0.014 mol/kg
C. 0.025 mol/kg
D. 0.00017 mol/kg

Use the table below to answer Questions 4 to 6.

Data for Various Hydrocarbons

Name	Number of C Atoms	Number of H Atoms	Melting Point (°C)	Boiling Point (°C)
Heptane	7	16	−90.6	98.5
1-Heptene	7	14	−119.7	93.6
1-Heptyne	7	12	−81	99.7
Octane	8	18	−56.8	125.6
1-Octene	8	16	−101.7	121.2
1-Octyne	8	14	−79.3	126.3

4. Based on the information in the table, what type of hydrocarbon becomes a gas at the lowest temperature?

A. alkane
B. alkene
C. alkyne
D. aromatic

5. If n is the number of carbon atoms in the hydrocarbon, what is the general formula for an alkyne with one triple bond?

A. C_nH_{n+2}
B. C_nH_{2n+2}
C. C_nH_{2n}
D. C_nH_{2n-2}

6. It can be predicted from the table that nonane will have a melting point that is

A. greater than that of octane.
B. less than that of heptane.
C. greater than that of decane.
D. less than that of hexane.

7. At a pressure of 1.00 atm and a temperature of 20°C, 1.72 g CO_2 will dissolve in 1 L of water. How much CO_2 will dissolve if the pressure is raised to 1.35 atm and the temperature stays the same?

A. 2.32 g/L
B. 1.27 g/L
C. 0.785 g/L
D. 0.431 g/L

Use the diagram below to answer Question 8.

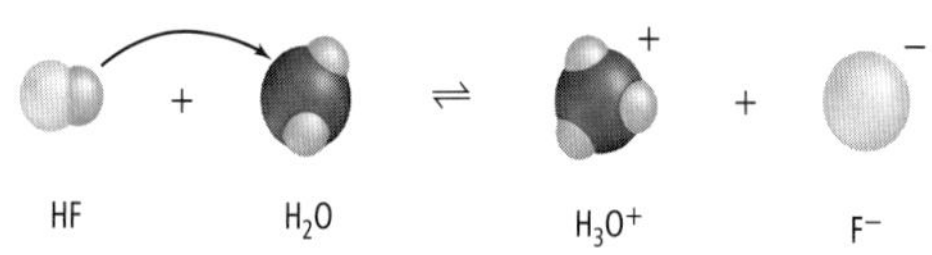

8. In the forward reaction, which substance is the Brønsted-Lowry acid?

A. HF
B. H_2O
C. H_3O^+
D. F^-

9. Which does NOT describe what happens as a liquid boils?

A. The temperature of the system rises.
B. Energy is absorbed by the system.
C. The vapor pressure of the liquid is equal to atmospheric pressure.
D. The liquid is entering the gas phase.

Online Test Practice

SHORT ANSWER

Use the diagram below to answer Questions 10 to 12.

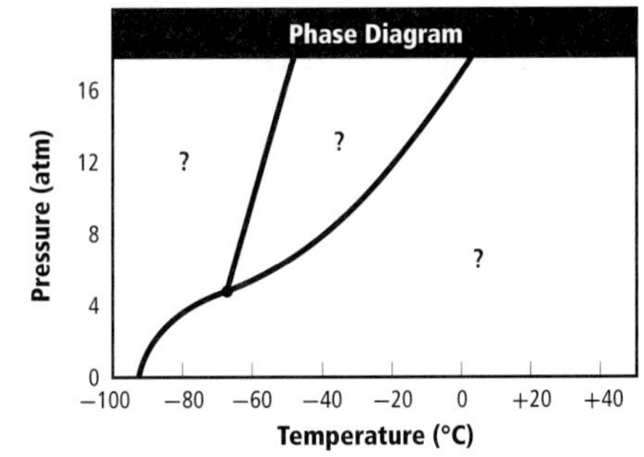

10. What state of matter is located at a temperature of −80°C and a pressure of 10 atm?

11. What are the temperature and pressure when the substance is at its triple point?

12. Describe the changes in molecular arrangement that occur when the pressure is increased from 8 atm to 16 atm, while the temperature is held constant at 0°C.

EXTENDED RESPONSE

Use the data table below to answer Questions 13 and 14.

Experimental Data for the Reaction between A and B

[A] Initial	[B] Initial	Initial rate (mol/L·s)
0.10*M*	0.10*M*	7.93
0.30*M*	0.10*M*	23.79
0.30*M*	0.20*M*	95.16

13. Find the values of *m* and *n* for the rate law expression $rate = k[A]^m[B]^n$.

14. Determine the value of *k* for this reaction.

SAT SUBJECT TEST: CHEMISTRY

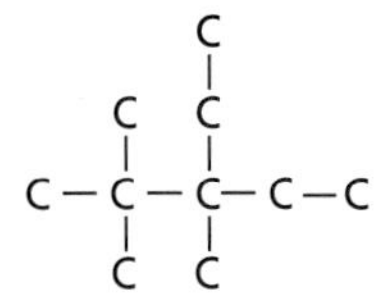

15. What is the name of the compound whose skeletal formula is shown above?

A. 2,2,3-trimethyl-3-ethylpentane
B. 3-ethyl-3,4,4-trimethylpentane
C. 2-butyl-2-ethylbutane
D. 3-ethyl-2,2,3-trimethylpentane
E. 2,2-dimethyl, 3-diethyl, 3-methylpropane

Use the graph below to answer Questions 16 and 17.

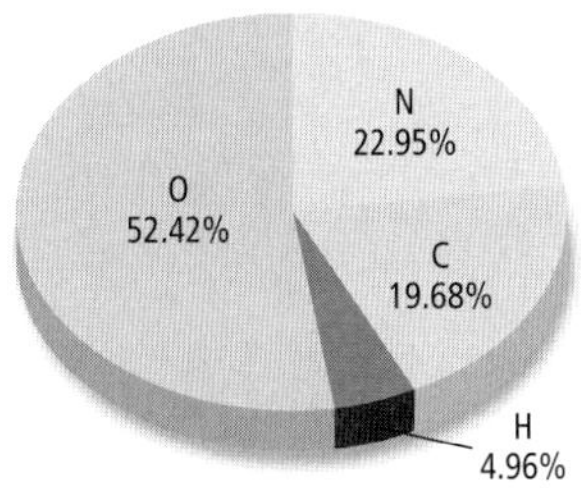

16. What is the formula for this compound?

A. $C_5H_{20}N_4O_2$
B. $C_8H_2N_9O_{11}$
C. $C_{1.6}H_5N_{1.6}O_{3.3}$
D. CH_3NO_2
E. $C_2H_5N_2O_5$

17. How many grams of nitrogen would be present in 475 g of this compound?

A. 33.9 g
B. 52.8 g
C. 67.9 g
D. 109 g
E. 120.0 g

NEED EXTRA HELP?																	
If You Missed Question . . .	1	2	3	4	5	6	7	8	9	10	11	12	13	14	15	16	17
Review Section . . .	21.4	16.2	14.2	21.3	21.3	21.3	14.1	18.1	12.4	12.4	12.4	12.4	16.3	16.3	21.2	10.4	10.4

SHORT ANSWER

10. solid

11. approximately −65°C and 4.8 atm

12. The substance is changing from a gas to a liquid as the pressure is increased; as the particles become more compact, they lose kinetic energy, become more ordered, and are spaced closer together.

EXTENDED RESPONSE

13. To solve for the exponents in the rate law, use a set of conditions in the table. Compare the ratio of concentrations of a reactant to the ratio of different rates for the same trials.
For the exponent of [A]: use the first and second trials
$Rate_2/rate_1 = ([A]_2/[A]_1)^m =$ $23.79/7.93 = 3.00; (0.30/0.10)^m =$ $3.00; 3^m = 3; m = 1$
Similarly, compare the rates of reactions when the concentration of B changes:
$Rate_3/rate_2 = ([B]_3/[B]_2)^n$ $= 95.16/23.79 = 4.00; (0.20/0.10)^n =$ $4.00; 2^n = 4; n = 2$
$Rate = k[A][B]^2$

14. $k = (7.93\ mol/L\text{-}s)/[(0.10\ mol/L)(0.10\ mol/L)^2]$
$k = 7930\ L^2/mol^2\text{-}s$
$rate = (7930\ L^2/mol^2\text{-}s)\ [A][B]^2$

SAT SUBJECT TEST: CHEMISTRY

15. D
16. D
17. D

CHAPTER 22 Substituted Hydrocarbons and Their Reactions

BIGIDEA **The substitution of different functional groups for hydrogen atoms in hydrocarbons results in a diverse group of organic compounds.**

ESSENTIAL QUESTIONS	RESOURCES TO ASSESS MASTERY
SECTION 1 **Alkyl Halides and Aryl Halides** **1.** What are functional groups, and what are some examples? **2.** How do alkyl and aryl halide structures compare and contrast? **3.** What factors affect the boiling points of organic halides? 1 session 0.5 block	**Progress Monitoring** Formative Assessment, pp. 788, 789, 790 Reading Check, pp. 788, 789, 790 Section Review, p. 791
SECTION 2 **Alcohols, Ethers, and Amines** **1.** Which functional groups define alcohols, ethers, and amines? **2.** How are the structures of alcohols, ethers, and amines drawn? **3.** What are some properties and uses of alcohols, ethers, and amines? 1 session 0.5 block	**Progress Monitoring** Formative Assessment, pp. 793, 795 Reading Check, pp. 793, 794 Section Review, p. 795
SECTION 3 **Carbonyl Compounds** **1.** Which structures identify various carbonyl compounds? **2.** What are the properties of compounds containing the carbonyl group? 2 sessions 1 block	**Progress Monitoring** Formative Assessment, pp. 798, 801 Reading Check, pp. 797, 798, 800 Section Review, p. 801
SECTION 4 **Other Reactions of Organic Compounds** **1.** How are organic reactions classified? **2.** Why is it useful to draw structural formulas when writing equations for reactions of organic compounds? **3.** How can classifying a reaction help you predict the reaction's products? 2 sessions 1 block	**Progress Monitoring** Formative Assessment, pp. 804, 806, 807 Reading Check, pp. 802, 804, 806, 807 Section Review, p. 808
SECTION 5 **Polymers** **1.** How does drawing a diagram help you understand the relationship between a polymer and the monomers from which it forms? **2.** What distinguishes addition and condensation polymerization reactions? **3.** How can you use molecular structures and the presence of functional groups to predict the properties of polymers? 1 session 0.5 block	**Progress Monitoring** Formative Assessment, pp. 811, 813 Reading Check, pp. 810, 813 Section Review, p. 814 **Summative Assessment** Chapter Assessment, p. 818 *eAssessment* Chapter Test (Scaffolded)

LEVELED RESOURCES	LAB MATERIALS
Science Notebook 22.1 OL ***FAST FILE Chapter Resources:*** Study Guide, p. 49 BL	LaunchLAB p. 784 / **20 min** graduated cylinder, 4% polyvinyl alcohol solution, disposable plastic cup, stirring rod, 4% sodium tetraborate solution, latex gloves
Science Notebook 22.2 OL ***FAST FILE Chapter Resources:*** ChemLab Worksheet, p. 32 OL Study Guide, p. 50 BL	ChemLAB p. 816 / **45 min** nonmercury thermometer, stopwatch, facial tissue, cloth towel, Beral pipettes, methanol, ethanol (95%), 2-propanol (99%), wire twist tie or small rubber band, piece of cardboard for use as a fan
Science Notebook 22.3 OL ***FAST FILE Chapter Resources:*** MiniLab Worksheet, p. 30 OL Study Guide, p. 51 BL	MiniLAB p. 800 / **20 min** tap water, 250-mL beaker, hot plate, balance, weighing paper, salicylic acid, small test tube, distilled water, 10-mL graduated cylinder, methanol, Beral pipette, concentrated sulfuric acid, test tube clamp, test tube holder, cotton ball, petri dish half
Science Notebook 22.4 OL ***FAST FILE Chapter Resources:*** Study Guide, p. 52 BL	
Science Notebook 22.5 OL ***FAST FILE Chapter Resources:*** Study Guide, p. 53 BL	

ADDITIONAL RESOURCES

- eAssessment
- ConnectED Student Centre
- ConnectED Teacher Centre
- Lesson Presentations
- eSolutions
- Videos
- Animations
- Online Quizzes
- Virtual Investigations
- Multilingual eGlossary
- Personal Tutors
- WebQuests
- Applying Practices Worksheets
- PBLs

Practice:

- Challenge Problems AL
- Supplemental Problems BL OL
- Chapter Test (Scaffolded)
- Standardized Test Practice

Additional Resources:

- Blueprints for Success
- Solving Problems: A Chemistry Handbook

Lab Resources:

- Laboratory Manual OL
- CBL Laboratory Manual OL
- Small-Scale Laboratory Manual OL
- Forensics Laboratory Manual OL

BL Below Level   OL On Level   AL Advanced Learners   EL English Learners   COOP LEARN Cooperative Learning

CHAPTER 22

LaunchLAB

Rubric

How do you make slime?

Purpose Students will observe the effect of cross-linking on polymer chains.

Safety Precautions Approve lab safety forms before work begins. Students should wear aprons and safety goggles. They should wear plastic gloves when pouring the borax solution and wash their hands after the experiment.

Disposal Contents of the paper or plastic cup can be placed in a waste container.

Teaching Strategies

- Prepare 20 mL of 4% polyvinyl alcohol for each student. To prepare solution sufficient for 50 students, slowly add, while stirring, 40 g of 98 to 100% hydrolyzed polyvinyl alcohol to 900 mL tap water. Heat this suspension to 80°C while continuing to stir. Add a few drops of food coloring. Do not overheat the solution. Cool and store the solution in a stoppered bottle.
- To prepare cross-linker (borax) solution sufficient for 50 students, dissolve 12 g of sodium tetraborate in 300 mL of warm tap water.
- The slime produced from this experiment could be a distraction in other classes during the school day. Tell students not to take their product out of the science classroom.

Expected Results The liquid turns to a solid. It will flow when draped over a student's hand, forming long strands. The strand will break if you pull it quickly.

CHAPTER 22

Substituted Hydrocarbons and Their Reactions

BIGIDEA The substitution of different functional groups for hydrogen atoms in hydrocarbons results in a diverse group of organic compounds.

SECTIONS

1 Alkyl Halides and Aryl Halides

2 Alcohols, Ethers, and Amines

3 Carbonyl Compounds

4 Other Reactions of Organic Compounds

5 Polymers

The larva of the *Cerura vinula* moth squirts formic acid when threatened. Formic acid is a substituted hydrocarbon—a type of compound that forms when one or more hydrogens in a hydrocarbon are replaced by other atoms.

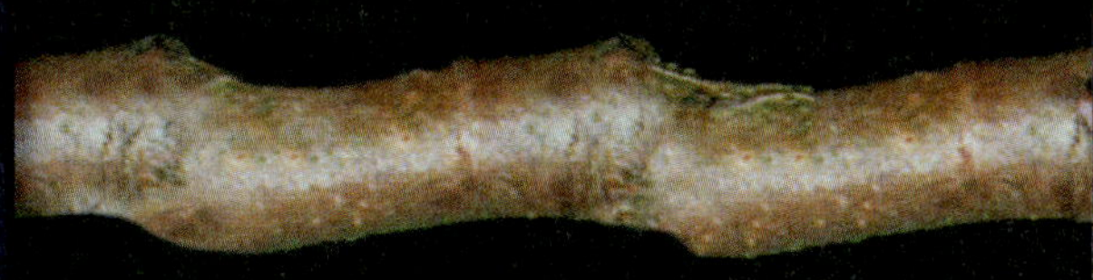

imagebroker/Alamy

LaunchLAB

How do you make slime?

In addition to carbon and hydrogen, most organic substances contain other elements that give the substances unique properties. In this lab, you will determine how the properties of substances change when groups form bonds called crosslinks between the chains.

FOLDABLES Study Organizer

Functional Groups

Make a top-tab book. Label it as shown. Use it to organize information about the functional groups of organic compounds.

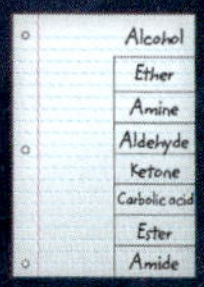

Procedure

1. Read and complete the lab safety form.
2. Use a **graduated cylinder** to measure 20 mL of **4% polyvinyl alcohol solution**. Pour the solution into a small **disposable plastic cup.** Note the viscosity of the solution as you stir it with a **stirring rod.**
3. While stirring, add 6 mL of **4% sodium tetraborate solution** to the polyvinyl alcohol solution. Continue to stir until there is no further change in the consistency of the product.
4. Use a gloved hand to scoop the material out of the cup. Knead and stretch the polymer.

Analysis

1. **Compare and contrast** the physical properties of the product and the reactants. The reactants are liquids. The product does not have all of the characteristics of a solid or a liquid.
2. **Explain** how the crosslinking of the molecular chains affected the viscosity of the solution. The cross-linking of the chains increased the viscosity of the solution.

Inquiry What is the ratio of sodium tetraborate solution to polyvinyl alcohol solution? What would you create if the ratio was changed? 6:20; a different compound.

Introduce the BIGIDEA

Substituted Hydrocarbons
To introduce this chapter's Big Idea, use methane as an example of how hydrogen atoms are substituted with other atoms or groups to create compounds with different properties. Write the molecular formula for methane (CH_4) on the board. Remind students that methane is a primary component of natural gas. Next to the molecular formula for methane, write the molecular formula for formic acid (HCOOH). Point out to students that formic acid is a substituted hydrocarbon. Three hydrogen atoms in methane are removed and a double-bonded oxygen atom and a hydroxyl group are added. Point out to students that methane and formic acid have very different properties.

Tie to Previous Knowledge
Have students review the following concepts before studying this chapter.

- combustion reactions
- interparticle forces
- hydrocarbons, isomers

Use the Photo
Formic Acid Ask students to look at the photo of the puss moth larva (*Cerura vinula*). Point out to students that this caterpillar represents an example of how a substituted hydrocarbon is used in nature. The puss moth caterpillar squirts formic acid from a gland on its body as a defense against predators. Tell students that formic acid is also a chemical compound found in the venom of stinging ants.

SECTION 1

1 Focus

MAINIDEA

Halogens Have students turn to the back inside cover of their chemistry text. Use a wall-mounted periodic table or project a periodic table on a screen or wall for students to see. Point to the halogen group in the periodic table. Ask students what this group is called. the halogens Have students find the group on their periodic tables. Tell students that when one of the halogens replaces a hydrogen atom in a hydrocarbon, an alkyl halide or aryl halide is produced. OL

Visual Learning

Figure 1 Use the photo to point out to students that functional groups are found in many common things. The rubbing alcohol contains the hydroxyl group. Vinegar contains the carboxyl group. The fruit and flowers contain esters, which give them a sweet aroma. The cheese contains several functional groups, but a ketone gives the cheese its strong aroma. The plastic toy is composed of a polymer. Polymers are also covered in this chapter. OL

SECTION 1

Essential Questions

- What are functional groups, and what are some examples?
- How do alkyl and aryl halide structures compare and contrast?
- What factors affect the boiling points of organic halides?

Review Vocabulary

aliphatic compound: a nonaromatic hydrocarbon, such as an alkane, an alkene, or an alkyne

New Vocabulary

functional group
halocarbon
alkyl halide
aryl halide
plastic
substitution reaction
halogenation

Alkyl Halides and Aryl Halides

MAINIDEA A halogen atom can replace a hydrogen atom in some hydrocarbons.

CHEM 4 YOU If you have ever played on a sports team, were individual players substituted during the game? For example, a player who is rested might substitute for a player who is tired. After the substitution, the characteristics of the team change.

Functional Groups

You read previously that in hydrocarbons, carbon atoms are linked only to other carbon atoms or hydrogen atoms. But carbon atoms can also form strong covalent bonds with other elements, the most common of which are oxygen, nitrogen, fluorine, chlorine, bromine, iodine, sulfur, and phosphorus.

Atoms of these elements occur in organic substances as parts of functional groups. In an organic molecule, a **functional group** is an atom or group of atoms that always reacts in a certain way. The addition of a functional group to a hydrocarbon structure always produces a substance with physical and chemical properties that differ from those of the parent hydrocarbon. All the items—natural and synthetic—in **Figure 1** contain functional groups that give them their individual characteristics, such as smell. Organic compounds containing several important functional groups are shown in **Table 1.** The symbols R and R′ represent carbon chains or rings bonded to the functional group. An * represents a hydrogen atom, carbon chain, or carbon ring.

Keep in mind that double and triple bonds between two carbon atoms are considered functional groups even though only carbon and hydrogen atoms are involved. By learning the properties associated with a given functional group, you can predict the properties of organic compounds for which you know the structure, even if you have never studied them.

■ **Figure 1** All of these items contain at least one of the functional groups that you will study in this chapter. For example, the fruit and flowers have sweet-smelling aromas that are due to ester molecules.

Matt Meadows

CHEMISTRY JOURNAL

Functional Groups Give students a worksheet showing the structures of several large organic molecules that contain a variety of functional groups. Explain that in this chapter they will learn to recognize and name most of these groups. Ask them to circle any groups that they know are not present in alkanes. Have students enter this information in their journals and supply the names of the compounds as they encounter them in their studies. OL

Explore **functional groups** with an interactive table. Concepts In Motion

Table 1 Organic Compounds and Their Functional Groups

Compound Type	General Formula	Functional Group
Halocarbon	R—X (X = F, Cl, Br, I)	Halogen
Alcohol	R—OH	Hydroxyl
Ether	R—O—R′	Ether
Amine	$R—NH_2$	Amino
Aldehyde	*—C(=O)—H	Carbonyl
Ketone	R—C(=O)—R′	Carbonyl
Carboxylic acid	*—C(=O)—OH	Carboxyl
Ester	*—C(=O)—O—R	Ester
Amide	*—C(=O)—N(H)—R	Amide

Organic Compounds Containing Halogens

Explore **functional groups.** Virtual Investigations

The most simple functional groups can be thought of as substituent groups attached to a hydrocarbon. Recall that a substituent group is a side branch attached to a parent chain. The elements in group 17 of the periodic table—fluorine, chlorine, bromine, and iodine—are the halogens. Any organic compound that contains a halogen substituent is called a **halocarbon.** If you replace any of the hydrogen atoms in an alkane with a halogen atom, you form an alkyl halide. An **alkyl halide** is an organic compound containing a halogen atom covalently bonded to an aliphatic carbon atom. The first four halogens—fluorine, chlorine, bromine, and iodine—are found in many organic compounds. For example, chloromethane is the alkyl halide formed when a chlorine atom replaces one of methane's four hydrogen atoms, as shown in **Figure 2.**

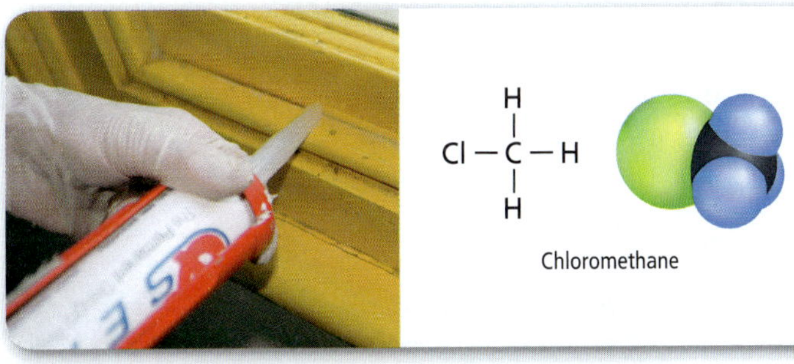

David Hoffman Photo Library/Alamy

Figure 2 Chloromethane is an alkyl halide that is used in the manufacturing process for silicone products, such as window and door sealants.

2 Teach

Quick Demo

Flame Test Perform the Beilstein flame test for students. Explain that the procedure can be a useful qualitative test for alkyl halides. **WARNING:** ***Perform the test in a fume hood or a well-ventilated room. Wear safety goggles, an apron, and a flame-proof glove or mitt.*** Dip the tip of a cool, clean, copper wire into a small sample of ethylene dichloride (1,2-dichloroethane). Place the tip of the wire in a Bunsen-burner flame. A blue or green flame demonstrates the presence of chlorine, bromine, or iodine in the compound. Advise students that chlorine-containing compounds react with copper to produce a green flame, bromine-containing compounds yield a blue-green flame, and iodine-containing compounds yield a blue flame.

DIFFERENTIATED INSTRUCTION

Advanced Learners Ask students to draw all the structural isomers of the halocarbon with the formula $C_4H_8Br_2$. There are nine different structures with that formula: 1,1-dibromobutane, 2,2-dibromobutane, 1,2-dibromobutane, 1,3-dibromobutane, 1,4-dibromobutane, 2,3-dibromobutane, 1,1-dibromo-2-methlypropane, 1,2-dibromo-2-methylpropane, and 1,3-dibromo-2-methylpropane. AL

Knowledge Show students ball-and-stick models of a number of alkyl halides and ask them to name the compounds. Be sure to include at least one structure with each of the halogens found commonly in organic compounds: fluorine, chlorine, bromine, and iodine. Also, ask students to recall all the information they can about the properties and characteristics of halogens. Halogens form anions with a 1– charge, have seven valence electrons, react with alkali metals and alkaline earth metals to form salts, and occur naturally in the diatomic form. **OL**

☑ READING CHECK Naming organic compounds must be standardized so that chemists all over the world will know exactly which compound it is.

Visual Learning

Figure 3 Use these figures and the text that describes how to name halocarbons to model how to name organic molecules. **OL**

IN-CLASS Example

Question Name the alkyl or aryl halide whose structure is shown.

a. $CH_3—CH_2—CH_2—CH_2—Br$

b.

Answer

a. 1-bromobutane

b. bromobenzene

PRACTICE Problems

Have students refer to the Selected Solutions appendix for complete solutions to odd-numbered problems. The complete solutions for all problems can be found in the Solutions Manual.

1. 2, 3-difluorobutane
2. 1-bromo-5-chloropentane
3. 1, 3-dibromo-2-chlorobenzene

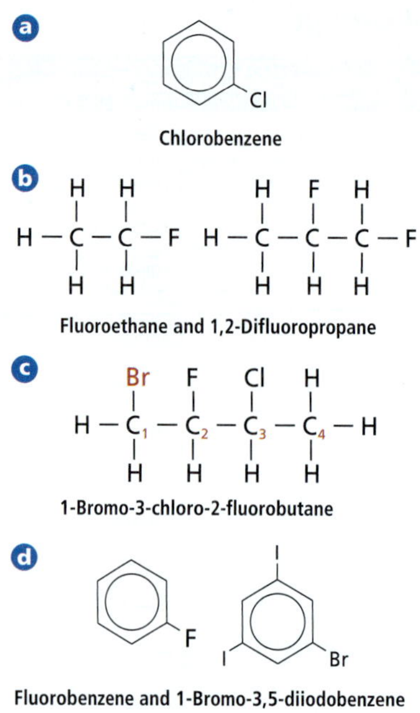

■ **Figure 3** Organic molecules containing functional groups are named based on their main-chain alkane structure using IUPAC conventions.

An **aryl halide** is an organic compound containing a halogen atom bonded to a benzene ring or other aromatic group. The structural formula for an aryl halide is created by first drawing the aromatic structure and then replacing its hydrogen atoms with the halogen atoms specified, as shown in **Figure 3a.**

Connection to Earth Science Alkyl halides are widely used as refrigerants. Until the late 1980s, alkyl halides called chlorofluorocarbons (CFCs) were widely used in refrigerators and air-conditioning systems. Recall how CFCs affect the ozone layer. CFCs have been replaced by HFCs (hydrofluorocarbons), which contain only hydrogen and fluorine atoms bonded to carbon. One of the more common HFCs is 1,1,2-trifluoroethane, also called R134a.

Naming halocarbons Organic molecules containing functional groups are given IUPAC names based on their main-chain alkane structures. For the alkyl halides, a prefix indicates which halogen is present. The prefixes are formed by changing the *-ine* at the end of each halogen name to *-o*. Thus, the prefix for fluorine is *fluoro-*, chlorine is *chloro-*, bromine is *bromo-*, and iodine is *iodo-*, as shown in **Figure 3b.**

If more than one kind of halogen atom is present in the same molecule, the atoms are listed alphabetically in the name. The chain also must be numbered in a way that gives the lowest position number to the substituent that comes first in the alphabet. Note how the alkyl halide in **Figure 3c** is named.

Similarly, the benzene ring in an aryl halide is numbered to give each substituent the lowest position number possible, as shown in **Figure 3d.**

☑ READING CHECK **Infer** why the lowest possible position number is used to name an aryl halide instead of using a randomly chosen position number.

PRACTICE Problems

Do additional problems. Online Practice

Name the alkyl or aryl halide whose structure is shown.

1.

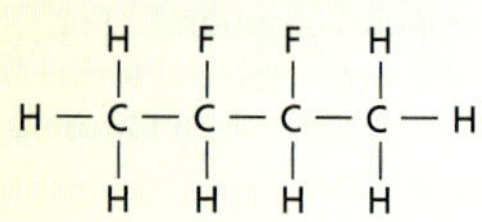

2.

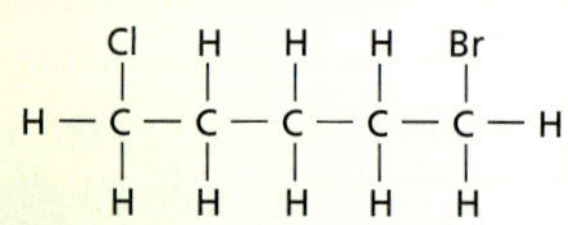

3.

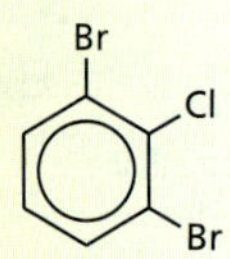

CHEMISTRY PROJECT

Chlorofluorocarbons Have students research and write a report about how some countries have banned or phased out the use of chlorofluorocarbons (CFCs) due to their suspected involvement in breaking down Earth's protective ozone layer. **OL**

Table 2 A Comparison of Alkyl Halides and Their Parent Alkanes

Structure	Name	Boiling Point (°C)	Density (g/mL) in Liquid State
CH_4	methane	−162	0.423 at −162°C (boiling point)
CH_3Cl	chloromethane	−24	0.911 at 25°C (under pressure)
$CH_3CH_2CH_2CH_2CH_3$	pentane	36	0.626
$CH_3CH_2CH_2CH_2CH_2F$	1-fluoropentane	62.8	0.791
$CH_3CH_2CH_2CH_2CH_2Cl$	1-chloropentane	108 *Increases*	0.882 *Increases*
$CH_3CH_2CH_2CH_2CH_2Br$	1-bromopentane	130	1.218
$CH_3CH_2CH_2CH_2CH_2I$	1-iodopentane	155	1.516

Properties and uses of halocarbons It is easiest to talk about properties of organic compounds containing functional groups by comparing those compounds with alkanes, whose properties you have already studied. **Table 2** lists some of the physical properties of certain alkanes and alkyl halides.

Note that each alkyl chloride has a higher boiling point and a higher density than the alkane with the same number of carbon atoms. Note also that the boiling points and densities increase as the halogen changes from fluorine to chlorine, bromine, and iodine. This trend occurs primarily because the halogens from fluorine to iodine have increasing numbers of electrons that lie farther from the halogen nucleus. These electrons shift position easily and, as a result, the halogen-substituted hydrocarbons have an increasing tendency to form temporary dipoles. Because the dipoles attract each other, the energy needed to separate the molecules also increases. Thus, the boiling points of halogen-substituted alkanes increase as the size of the halogen atom increases.

☑ READING CHECK **Explain** the relationship between the number of electrons in the halogen and the boiling point.

Organic halides are seldom found in nature, although human thyroid hormones are organic iodides. Halogen atoms bonded to carbon atoms are more reactive than the hydrogen atoms they replace. For this reason, alkyl halides are often used as starting materials in the chemical industry. Alkyl halides are also used as solvents and cleaning agents because they readily dissolve nonpolar molecules, such as greases. **Figure 4** shows an application of polytetrafluoroethene (PTFE), a plastic made from gaseous tetrafluoroethylene. A **plastic** is a polymer that can be heated and molded while relatively soft. Another plastic commonly called *vinyl* is polyvinyl chloride (PVC). It can be manufactured soft or hard, as thin sheets, or molded into objects.

☑ READING CHECK **Explain** why alkyl halides are often used in the chemical industry as starting materials instead of alkanes.

■ **Figure 4** Polytetrafluoroethene (PTFE) is made up of hundreds of units. PTFE provides a nonstick surface for many kitchen items, including bakeware.

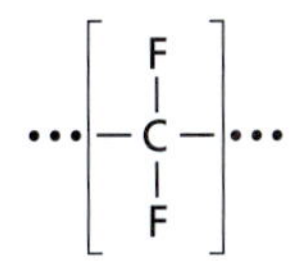

PTFE

PTFE Application

☑ READING CHECK As the number of electrons increase in a halogen-substituted alkane, the boiling point increases due to the formation of temporary dipoles between the particles.

Visual Learning

Table 2 Have students examine Table 2 and compare the boiling point and density of pentane with those of the five-carbon compounds that have functional groups in place of one of pentane's hydrogen atoms. Pentane has a boiling point of 36°C and a density of 0.626 g/mL. Based on the information in this table, substituting a functional group for a hydrogen atom in pentane results in a higher boiling point and a higher density. Boiling point values for the substituted molecules range from 62.8°C to 155°C; densities range from 0.791 g/mL to 1.516 g/mL. OL

ASSESSMENT

Knowledge Have each student write a question on the material in this section that involves naming alkyl halides. Have students exchange questions and quiz each other. OL

☑ READING CHECK The halogen atoms in the alkyl halides are more reactive than the hydrogen atoms they replaced in the alkanes.

DIFFERENTIATED INSTRUCTION

Struggling Learners Some students will find naming organic compounds that contain functional groups easier using a step-by-step approach. They should first name the compound as if no functional group were present, being sure to name the parent chain and all alkane branches correctly. Then, they should name each functional group and determine its position number. Finally, they should put all the information together to complete the compound's name. BL EL

Apply Chemistry

Synthetic and Natural Dyes

Synthetic dyes were not made until the 1800s. People that lived before that time had to be content using a limited number of natural hues to put some color in their lives. Natural purple dyes were especially hard to find, so for almost 3000 years, the dye called royal purple or Tyrian purple was one of the most precious substances on Earth. By law, Tyrian purple could be used only for coloring robes and other clothing for the Roman emperors, tabernacle curtains, and priests' vestments.

The only known source of dye was a Mediterranean sea snail, and 1200 snails had to be harvested to obtain one gram of dye. Over-harvesting almost led to the extinction of the snails, and the dye's production ceased by the fifteenth century. Today, we have numerous synthetic purple dyes that have replaced Tyrian purple.

ASSESSMENT

Skill Ask students to draw structural formulas for the following alkyl halides: 2-bromobutane; 1, 3-difluoropentane; iodocyclohexane; chlorobenzene; 1, 3-dibromobenzene; 1-bromo-3-chloro-2-fluorooctane. **OL**

READING CHECK Refer to the Solutions Manual for a drawing of each structure.

Caption Question Fig. 5
Petroleum contains alkanes that can be converted into other hydrocarbons, such as alkyl halides, alcohols, and amines, and used to make synthetic organic compounds.

Table 3 Substitution Reactions

Generic Substitution Reaction	Example of General Substitution Reaction (Halogenation)
$R{-}CH_3 + X_2 \rightarrow R{-}CH_2X + HX$ where X is fluorine, chlorine, or bromine	$C_2H_6 + Cl_2 \rightarrow C_2H_5Cl + HCl$ Ethane Chloroethane
General Alkyl Halide-Alcohol Reaction	**Example of an Alkyl Halide-Alcohol Reaction**
$R{-}X + OH^- \rightarrow R{-}OH + X^-$ Alkyl halide Alcohol	$CH_3CH_2Cl + OH^- \rightarrow CH_3CH_2OH + Cl^-$ Chloroethane Ethanol
General Alkyl Halide-Ammonia Reaction	**Example of an Alkyl Halide-Ammonia Reaction**
$R{-}X + NH_3 \rightarrow R{-}NH_2 + HX$ Alkyl halide Amine	$CH_3(CH_2)_6CH_2Br + NH_3 \rightarrow CH_3(CH_2)_6CH_2NH_2 + HBr$ 1-Bromooctane 1-Octanamine

Substitution Reactions

From where does the immense variety of organic compounds come? Amazingly enough, the ultimate source of nearly all synthetic organic compounds is petroleum. The oil-field workers shown in **Figure 5** are drilling for petroleum, which is a fossil fuel that consists almost entirely of hydrocarbons, especially alkanes. How can alkanes be converted into compounds as different as alkyl halides, alcohols, and amines?

One way is to introduce a functional group through substitution, as shown in **Table 3.** A **substitution reaction** is one in which one atom or a group of atoms in a molecule is replaced by another atom or group of atoms. With alkanes, hydrogen atoms can be replaced by atoms of halogens, typically chlorine or bromine, in a process called **halogenation.** One example of a halogenation reaction, shown in **Table 3,** is the substitution of a chlorine atom for one of ethane's hydrogen atoms. **Figure 6** shows another halogenated hydrocarbon commonly called halothane (2-bromo-2-chloro-1,1,1-trifluoroethane), which was first used as a general anesthetic in the 1950s.

Equations for organic reactions are sometimes shown in generic form. **Table 3** shows the generic form of a substitution reaction. In this reaction, *X* can be fluorine, chlorine, or bromine, but not iodine. Iodine does not react well with alkanes.

READING CHECK **Draw** the molecular structure of halothane.

Figure 5 These oil-field workers are drilling for petroleum. A single oil well can extract more than 100 barrels per day.
Explain *the relationship between petroleum and synthetic organic compounds.*

Keith Wood/Getty Images

DIFFERENTIATED INSTRUCTION

Advanced Learners Have students research and find industrial applications of the substituted reactions in Table 3. Students should share their findings with their class. **AL**

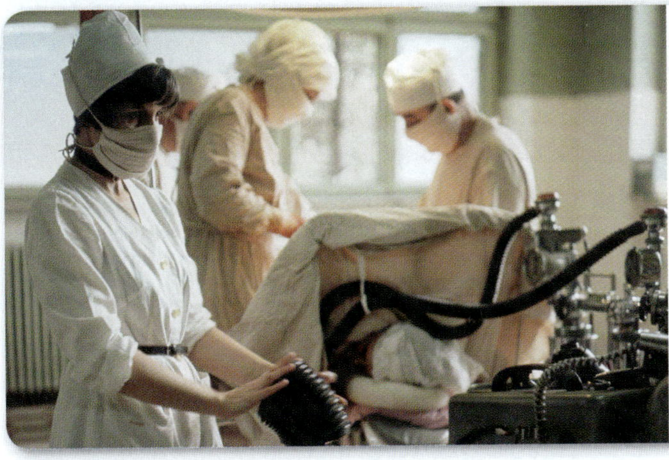

Figure 6 Halothane was introduced into medicine in the 1950s as a general anesthetic for patients undergoing surgery.

Further substitution Once an alkane has been halogenated, the resulting alkyl halide can undergo other types of substitution reactions in which the halogen atom is replaced by another atom or group of atoms. For example, reacting an alkyl halide with a basic solution results in the replacement of the halogen atom by an –OH group, forming an alcohol. An example of an alkyl halide-alcohol reaction is shown in **Table 3.** The generic form of the alkyl halide-alcohol reaction is also shown in **Table 3.**

Reacting an alkyl halide with ammonia (NH_3) replaces the halogen atom with an amino group ($-NH_2$), forming an alkyl amine, also shown in **Table 3.** The alkyl amine is one of the products produced in this reaction. Some of the newly formed amines continue to react, resulting in a mixture of amines.

FOLDABLES® Incorporate information from this section into your Foldable.

SECTION 1 REVIEW

Section Self-Check

Section Summary

- The substitution of functional groups for hydrogen in hydrocarbons creates a wide variety of organic compounds.
- An alkyl halide is an organic compound that has one or more halogen atoms bonded to a carbon atom in an aliphatic compound.

4. MAINIDEA **Compare and contrast** alkyl halides and aryl halides.
5. **Draw** structures for the following molecules.
   - **a.** 2-chlorobutane
   - **b.** 1,3-difluorohexane
   - **c.** 1,1,1-trichloroethane
   - **d.** 1-bromo-4-chlorobenzene
6. **Define** *functional group* and name the group present in each of the following structures. Name the type of organic compound each substance represents.
   - **a.** $CH_3CH_2CH_2OH$
   - **b.** CH_3CH_2F
   - **c.** $CH_3CH_2NH_2$
   - **d.** $CH_3C(=O)-OH$
7. **Evaluate** How would you expect the boiling points of propane and 1-chloropropane to compare? Explain your answer.
8. **Interpret Scientific Illustrations** Examine the pair of substituted hydrocarbons illustrated at right, and decide whether it represents a pair of optical isomers. Explain your answer.

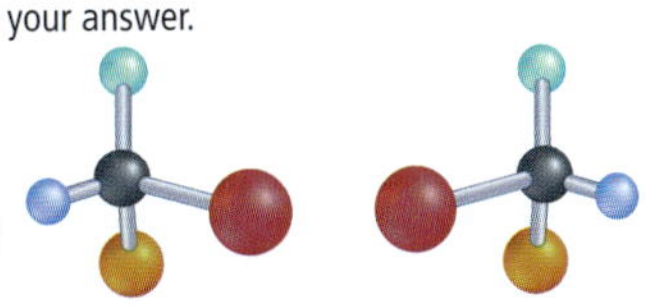

FOLDABLES® Rubric

3 Assess

Check for Understanding

Ask students if they must always indicate the position number for a functional group in an organic compound to correctly and unambiguously specify the compound's structure. No. For many compounds, only one position for a functional group is possible, or all possible positions are equivalent. OL

Reteach

Illustrate the previous point by drawing the structural formula for chloroethane on the board and asking students to name it. Point out that chlorine's position on the carbon chain need not be designated. OL

Extension

Have students compare the physical states of chloroethane and 1-chloropropane at room temperature and one atmosphere pressure. Chloroethane is a gas, and 1-chloropropane is a liquid. OL

SECTION 1 REVIEW

4. An alkyl halide is a substituted hydrocarbon that is covalently bonded to an aliphatic carbon atom. An aryl halide is a substituted hydrocarbon that contains a halogen bonded to a benzene ring or other aromatic compound.
5. Refer to the Solutions Manual for structures.
6. A functional group is an atom or group of atoms that reacts in a certain way.
   - **a.** hydroxyl group; alcohol
   - **b.** fluoro group; alkyl halide
   - **c.** amino group; amine
   - **d.** carboxyl group; carboxylic acid
7. The boiling point of 1-chloropropane should be higher than that of propane. The molecules in 1-chloropropane should form more temporary dipoles than the molecules in propane molecules.
8. They are optical isomers. They are not superimposable; however, their structures have chiral carbons and are mirror images of each other.

SECTION 2

1 Focus

MAINIDEA

Common Atoms Have students look at Table 1 in Section 1. Ask students to look at the general formulas for alcohol, ether, and amine. Ask students what are the most common atoms found in these compounds. oxygen and nitrogen BL OL

2 Teach

Quick Demo

Surprising Material Properties Thoroughly soak a one-dollar bill in methanol. Ask students if they think you are willing to throw away perfectly good money to demonstrate the volatility and flammability of an organic compound. Hold the bill with tongs. Then, use a hand-held lighter to ignite the methanol. The methanol will burn off before the paper bill catches fire. Ask students to explain what happened. Methanol burns at a temperature much lower than that required to ignite the paper. BL

SECTION 2

Alcohols, Ethers, and Amines

Essential Questions

- Which functional groups define alcohols, ethers, and amines?
- How are the structures of alcohols, ethers, and amines drawn?
- What are some properties and uses of alcohols, ethers, and amines?

Review Vocabulary

miscible: describes two liquids that are soluble in each other

New Vocabulary

hydroxyl group
alcohol
denatured alcohol
ether
amine

MAINIDEA Oxygen and nitrogen are two of the most-common atoms found in organic functional groups.

CHEM 4 YOU The last time you had a vaccination, the nurse probably disinfected your skin with an alcohol wipe before giving you the injection. Did you know that the nurse was using a substituted hydrocarbon?

Alcohols

Many organic compounds contain oxygen atoms bonded to carbon atoms. Because an oxygen atom has six valence electrons, it commonly forms two covalent bonds to gain a stable octet. An oxygen atom can form a double bond with a carbon atom, replacing two hydrogen atoms, or it can form one single bond with a carbon atom and another single bond with another atom, such as hydrogen. An oxygen-hydrogen group covalently bonded to a carbon atom is called a **hydroxyl group** (–OH). An organic compound in which a hydroxyl group replaces a hydrogen atom of a hydrocarbon is called an **alcohol.** As shown in **Table 4,** the general formula for an alcohol is ROH. **Table 4** also illustrates the relationship of the simplest alkane, methane, to the simplest alcohol, methanol.

Ethanol and carbon dioxide are produced by yeasts when they ferment sugars, such as those in grapes and bread dough. Ethanol is found in alcoholic beverages and medicinal products. Because it is an effective antiseptic, ethanol can be used to swab skin before an injection is given. It is also a gasoline additive and an important starting material for the synthesis of more complex organic compounds.

Figure 7 shows a model of an ethanol molecule and a model of a water molecule. As you compare the models, notice that the covalent bonds from the oxygen in ethanol are at roughly the same angle as the bonds around the oxygen in the water molecule. Therefore, the hydroxyl groups of alcohol molecules are moderately polar, as with water, and are able to form hydrogen bonds with the hydroxyl groups of other alcohol molecules. Due to this hydrogen bonding, alcohols have much higher boiling points than hydrocarbons of similar shape and size.

Table 4 Alcohols

General Formula	Simple Alcohol and Simple Hydrocarbon
ROH R represents carbon chains or rings bonded to the functional group	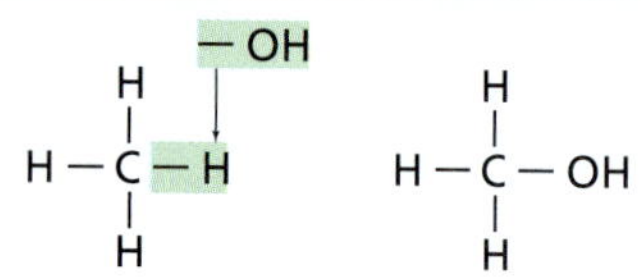 Methane (CH_4) Alkane — Methanol (CH_3OH) Alcohol

DIFFERENTIATED INSTRUCTION

Advanced Learners Ask students to research the structure of isopropanol, or rubbing alcohol, find out which industrial synthesis process is used to make it, and students report their findings to the class. AL

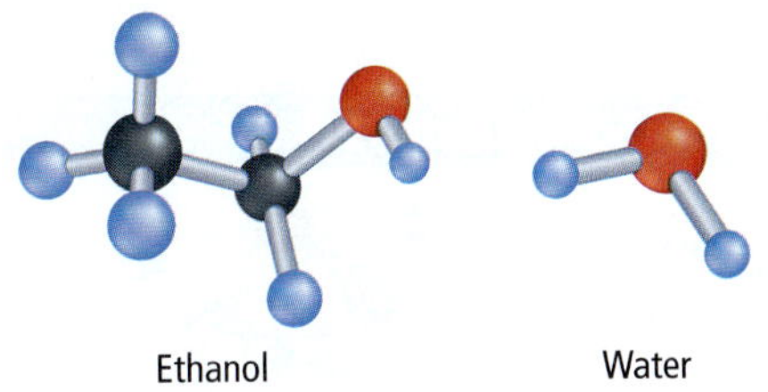

■ **Figure 7** The covalent bonds from oxygen have approximately the same bonding angle in ethanol and water.

Also, because of polarity and hydrogen bonding, ethanol is completely miscible with water. In fact, once they are mixed, it is difficult to separate water and ethanol completely. Distillation is used to remove ethanol from water, but even after that process is complete, about 5% water remains in the ethanol-water mixture.

On the shelves of drugstores, you can find bottles of ethanol labeled *denatured alcohol.* **Denatured alcohol** is ethanol to which small amounts of noxious materials, such as aviation gasoline or other organic solvents, have been added. Ethanol is denatured in order to make it unfit to drink. Because of their polar hydroxyl groups, alcohols make good solvents for other polar organic substances. For example, methanol, the smallest alcohol, is a common industrial solvent found in some paint strippers, and 2-butanol is found in some stains and varnishes.

Note that the names of alcohols are based on alkane names, like the names of alkyl halides. For example, CH_4 is methane and CH_3OH is methanol; CH_3CH_3 is ethane and CH_3CH_2OH is ethanol. When naming a simple alcohol based on an alkane carbon chain, the IUPAC rules call for naming the parent carbon chain or ring first and then changing the *-e* at the end of the name to *-ol* to indicate the presence of a hydroxyl group. In alcohols of three or more carbon atoms, the hydroxyl group can be at two or more positions. To indicate the position, a number is added, as shown in **Figure 8a** and **8b.**

☑ READING CHECK **Explain** why *4-butanol* and *3-butanol* are not the correct names for the compounds in **Figure 8a** and **8b.**

Now look at **Figure 8c.** The compound's ring structure contains six carbons with only single bonds, so you know that the parent hydrocarbon is cyclohexane. Because an –OH group is bonded to a carbon, it is an alcohol and the name will end in *-ol.* No number is necessary because all carbons in the ring are equivalent. This compound is called cyclohexanol. It is a poisonous compound used as a solvent for certain plastics and in the manufacture of insecticides.

A carbon chain can also have more than one hydroxyl group. To name these compounds, prefixes such as *di-*, *tri-*, and *tetra-* are used before the *-ol* to indicate the number of hydroxyl groups present. The full alkane name, including *-ane,* is used before the prefix.

Figure 8d shows the molecule 1,2,3-propanetriol, commonly called glycerol. It is an alcohol containing more than one hydroxyl group. Glycerol is often used as an antifreeze and as an airplane deicing fluid.

☑ READING CHECK **Explain** why numbers are not used to name the compound shown in **Figure 8c.**

■ **Figure 8** The names of alcohols are based on alkane names.

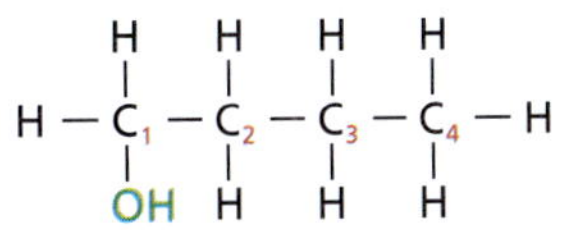

a. 1-Butanol

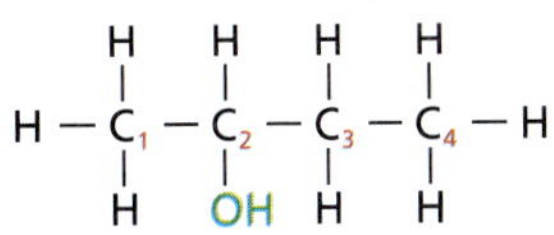

b. 2-Butanol

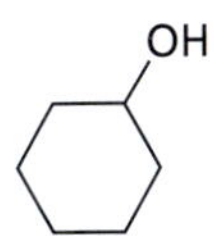

c. Cyclohexanol

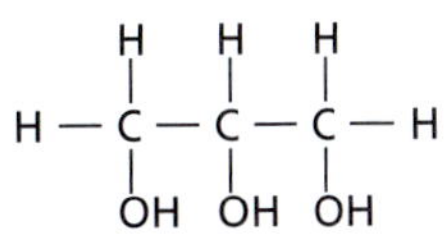

d. 1,2,3-Propanetriol (glycerol)

☑ **READING CHECK** because 3 and 4 are not the lowest possible numbers that represent the location of the functional group

☑ **READING CHECK** Numbers are not necessary because all carbons in the ring are equivalent.

Enrichment

Viscosity Have students recall the relationship between the strengths of interparticle forces and viscosities of liquids. Ask them to apply their knowledge to predict the relative viscosities of the following alcohols: ethanediol (also known as ethylene glycol); ethanol; 1, 2, 3-propanetriol (also know as glycerol). The order of viscosity from lowest to highest is: ethanol, ethanediol, and 1, 2, 3-propanetriol. The presence of more hydroxyl groups leads to more interparticle forces and higher viscosity. Chain length is also a factor. In general, a three-carbon chain has a higher viscosity than a two-carbon chain. **OL**

ASSESSMENT

Knowledge Have students write structural formulas for five alkanes containing the amino functional group, name each compound using both the suffix *-amine* and the prefixes *di-*, *tri-*, *tetra-* and so on if more than one amino group is present. The students should record the structures and names in their journals. **OL**

ChemLAB

The ChemLab located at the end of the chapter can be used at this point in the lesson.

CHEMISTRY JOURNAL

Thalidomide Ask students to research the history and write a report about the medicine thalidomide, including information about the difference in biological properties of its two optical isomers. Thalidomide is a sedative that was prescribed to many pregnant women in Europe in the 1950s and early 1960s, until it was realized that one of its optical isomers caused birth defects. Fortunately, it was never approved as a sedative in the United States. In 1998, thalidomide was approved for use in the United States as a treatment for leprosy. **OL**

FOLDABLES® Rubric

☑ READING CHECK because it is highly flammable

Identify Misconceptions

Many people believe that vitamins and medicines from natural sources are superior to those made synthetically.

Uncover the Misconception Have students use reference materials to research the structure of a vitamin, such as vitamin C. Have them draw the structure on the board and write the compound's name. Ask students if they can tell whether the displayed structure represents the vitamin's natural or its synthetic form. Students might have the misconception that a vitamin's natural and synthetic forms are different. Stress that a given vitamin structure functions properly in the human body—regardless of whether it is natural or synthetic.

Demonstrate the Concept Have students bring in labels from vitamin bottles or boxes. Have them compare claims such as "all natural" made on the labels. Point out that although the products likely contain varying amounts of contaminants from synthesis or extraction, a given vitamin is chemically identical in all preparations.

Assess New Knowledge Assign vitamins mentioned on the box labels to various students. Then, have students research the assigned vitamin structures and present them to the entire class. OL

Table 5 Ethers

General Formula	Methanol and Methyl ether	
ROR′ where R and R′ represent carbon chains or rings bonded to functional groups	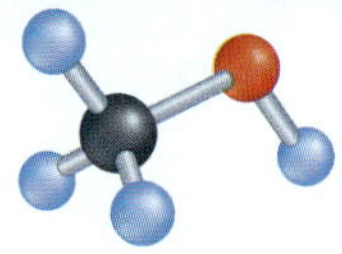 Methanol bp = 65°C	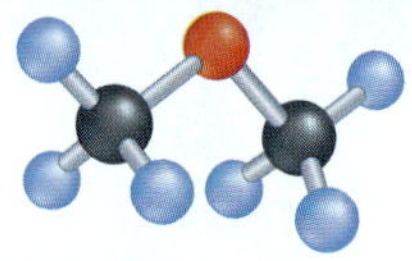 Methyl ether bp = −25°C
Examples of Ethers		
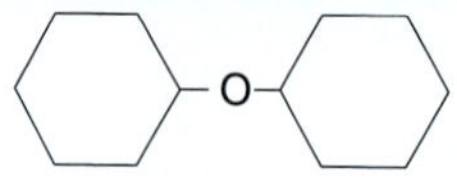 Cyclohexyl ether	$CH_3CH_2CH_2-O-CH_2CH_2CH_3$ Propyl ether	
$CH_3CH_2-O-CH_2CH_2CH_2CH_3$ Butyl ethyl ether	$CH_3CH_2-O-CH_3$ Ethyl methyl ether	

VOCABULARY
ACADEMIC VOCABULARY
Bond
to connect, bind, or join
An oxygen atom bonds to two carbon atoms in an ether.

Ethers

Ethers are another group of organic compounds in which oxygen is bonded to carbon. An **ether** is an organic compound containing an oxygen atom bonded to two carbon atoms. Ethers have the general formula ROR′, as shown in **Table 5.** The simplest ether is one in which oxygen is bonded to two methyl groups. Note the similarity between methanol and methyl ether shown in **Table 5.**

The term *ether* was first used in chemistry as a name for ethyl ether, the volatile, highly flammable substance that was commonly used as an anesthetic in surgery from 1842 until the twentieth century. As time passed, the term *ether* was applied to other organic substances having two hydrocarbon chains attached to the same oxygen atom.

FOLDABLES® Incorporate information from this section into your Foldable.

Because ethers have no hydrogen atoms bonded to the oxygen atom, their molecules cannot form hydrogen bonds with each other. Therefore, ethers are generally more volatile and have much lower boiling points than alcohols of similar size and mass. Ethers are much less soluble in water than alcohols because they have no hydrogen to donate to a hydrogen bond. However, the oxygen atom can act as a hydrogen bond receptor for the hydrogen atoms of water molecules.

☑ READING CHECK **Infer** why ethyl ether is undesirable as an anesthetic.

When naming ethers that have two identical alkyl chains bonded to oxygen, first name the alkyl group and then add the word *ether.* **Table 5** shows the structures and names of two of these symmetrical ethers, propyl ether and cyclohexyl ether. If the two alkyl groups are different, the groups are listed in alphabetical order and then followed by the word *ether.* **Table 5** contains two examples of these asymmetrical ethers, butyl ethyl ether and ethyl methyl ether.

CHEMISTRY PROJECT

Formaldehyde The single carbon aldehyde with the IUPAC name *methanal* is more commonly known as formaldehyde. Have students research the uses of formaldehyde and the health problems associated with exposure to this chemical. Ask students to write a report and present it to the class. OL

Amines

Amines contain nitrogen atoms bonded to carbon atoms in aliphatic chains or aromatic rings and have the general formula RNH_2, as shown in **Table 6.**

Chemists consider amines to be derivatives of ammonia (NH_3). Amines are considered primary, secondary, or tertiary amines depending on whether one, two, or three of the hydrogens in ammonia have been replaced by organic groups.

When naming amines, the $-NH_2$ (amino) group is indicated by the suffix *-amine.* When necessary, the position of the amino group is designated by a number, as shown in the examples in **Table 6.** When only one amino group is present, the final -e of the root hydrocarbon is dropped, as in 1-butanamine. If more than one amino group is present, the prefixes *di-*, *tri-*, *tetra-*, and so on are used to indicate the number of groups.

The amine aniline is used in the production of dyes with deep shades of color. The common name *aniline* is derived from the plant from which it was historically obtained. Cyclohexylamine and ethylamine are important in the manufacture of pesticides, plastics, pharmaceuticals, and rubber that is used to make tires.

All volatile amines have odors that humans find offensive, and amines are responsible for many of the odors characteristic of dead, decaying organisms. Two amines found in decaying human remains are putrescine and cadaverine. Specially trained dogs are used to locate human remains using these distinctive odors. Sniffer dogs are often used after catastrophic events and in forensic investigations.

Table 6 Amines

General Formula

RNH_2

where R represents a carbon chain or ring bonded to the functional group

Examples of Amines

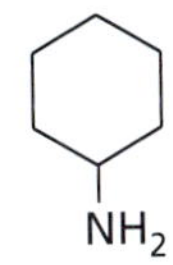

Cyclohexylamine

$CH_3CH_2-NH_2$

Ethylamine

Aniline

$NH_2CH(NH_2)CH_2CH_2CH(NH_2)NH_2$

1,1,4,4-Butanetetraamine

$NH_2CH_2CH_2CH_2NH_2$

1,3-Propanediamine

ASSESSMENT

Performance Ask students to design and set up an apparatus that can be used to compare the viscosities of ethanol, ethylene glycol (1,2-ethanediol), and glycerol (1,2, 3-propanetriol). OL

3 Assess

Check for Understanding

Ask students why propanol is not a correct name for a three-carbon alcohol. The name does not specify the location of the hydroxyl group on the three-carbon parent chain. The compound can be either 1-propanol or 2-propanol. OL

Reteach

Show students the structures of 2-propanol, propylamine (1-aminopropane), and ethyl methyl ether without the names. Have them identify the functional group present in each and give the IUPAC name for each compound. Then, ask them to determine the relationship between 2-propanol and ethyl methyl ether. They are structural isomers. OL

Extension

Have students research and make a poster showing the names, structures, and characteristics of various alkaloids, which are naturally occurring amine compounds made by plants. Common alkaloids include caffeine and nicotine. OL EL

SECTION 2 REVIEW

Section Self-Check

Section Summary

- Alcohols, ethers, and amines are formed when specific functional groups substitute for hydrogen in hydrocarbons.
- Because they readily form hydrogen bonds, alcohols have higher boiling points and higher water solubilities than other organic compounds.

9. MAINIDEA **Identify** two elements that are commonly found in functional groups.
10. **Identify** the functional group present in each of the following structures. Name the substance represented by each structure.
    a. $CH_3CH(NH_2)CH_3$
    b. 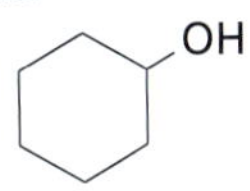
    c. $CH_3-O-CH_2CH_2CH_3$
11. **Draw** the structure for each molecule.
    a. 1-propanol
    b. 1,3-cyclopentanediol
    c. propyl ether
    d. 1,2-propanediamine
12. **Discuss** the properties of alcohols, ethers, and amines, and give one use of each.
13. **Analyze** Based on the molecular structures below, which compound would you expect to be more soluble in water? Explain your reasoning.

CH_3-O-CH_3 $\qquad$ CH_3CH_2-OH

SECTION 2 REVIEW

9. Possible answers: oxygen, nitrogen, fluorine, chlorine, bromine, iodine, sulfur, and phosphorus
10. a. $-NH_2$ represents the amino functional group; isopropylamine, 2-propanamine, or 2-aminopropane.
    b. $-OH$ represents the hydroxyl functional group; cyclohexanol.
    c. $-O-$ represents an oxygen atom in a carbon chain; methyl propyl ether.
11. a. $CH_3-CH_2-CH_2-OH$
    b. Refer to the Solutions Manual.
    c. $CH_3-CH_2-CH_2-O-CH_2-CH_2-CH_3$
    d. Refer to the Solutions Manual.
12. alcohols—moderately polar: can form hydrogen bonds with other molecules; higher boiling points than other alkanes of similar shape and size; example—ethanol; ethers—cannot form hydrogen bonds; lower boiling points and volatile; less soluble in water than alcohols; example—methyl ether; amines—some amines have volatile odors that humans find offensive; example—cyclohexylamine
13. Ethanol (on the right) is more soluble than methyl ether. Because of their molecules' greater polarities, alcohols are usually more water-soluble than ethers.

SECTION 3

1 Focus

MAINIDEA

Common Atom Have students look up the general formulas for aldehydes, ketones, carboxylic acids, esters, and amides in Table 1 in Section 1. Ask students to compare the general formulas and determine what they share. a double-bonded oxygen in the functional group BL OL

2 Teach

Content Background

Carbonyl Compounds This group contains aldehydes, ketones, carboxylic acids, esters, and amides. Carbonyl compounds contain an oxygen atom double-bonded to a carbon atom. These compounds are polar because the electron-rich oxygen atom has a partial negative charge and the electron-poor carbon atom has a partial positive charge.

SECTION 3

Essential Questions

- Which structures identify various carbonyl compounds?
- What are the properties of compounds containing the carbonyl group?

Review Vocabulary

electronegative: indicates the relative ability of an element's atoms to attract electrons in a chemical bond

New Vocabulary

carbonyl group
aldehyde
ketone
carboxylic acid
carboxyl group
ester
amide
condensation reaction

Carbonyl Compounds

MAINIDEA Carbonyl compounds contain a double-bonded oxygen in the functional group.

CHEM 4 YOU Have you ever eaten a piece of fruit-flavored candy that tasted like real fruit? Many natural fruits, such as strawberries, contain dozens of organic molecules that combine to give the distinctive aroma and flavor of fruits. The carbonyl group is found in many common types of artificial flavorings.

Organic Compounds Containing the Carbonyl Group

The arrangement in which an oxygen atom is double-bonded to a carbon atom is called a **carbonyl group.** This group is the functional group in organic compounds known as aldehydes and ketones.

Aldehydes An **aldehyde** is an organic compound in which a carbonyl group located at the end of a carbon chain is bonded to a carbon atom on one side and a hydrogen atom on the other. Aldehydes have the general formula *CHO, where * represents an alkyl group or a hydrogen atom, as shown in **Table 7.**

Aldehydes are formally named by changing the final *-e* of the name of the alkane with the same number of carbon atoms to the suffix *-al.* Thus, the formal name of the compound methanal, shown in **Table 7,** is based on the one-carbon alkane methane. Because the carbonyl group in an aldehyde always occurs at the end of a carbon chain, no numbers are used in the name unless branches or additional functional groups are present. Methanal is also commonly called formaldehyde. Ethanal has the common name *acetaldehyde.* Scientists often use the common names of organic compounds because they are familiar to chemists.

Table 7 Aldehydes

General Formula	Examples of Aldehydes		
*CHO *represents an alkyl group or a hydrogen atom	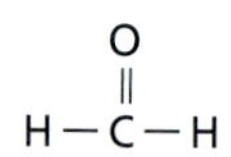 Methanal (formaldehyde)		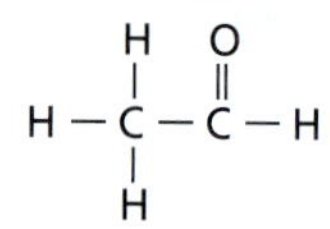 Ethanal (acetaldehyde)
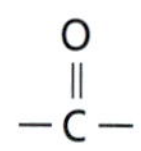 Carbonyl group	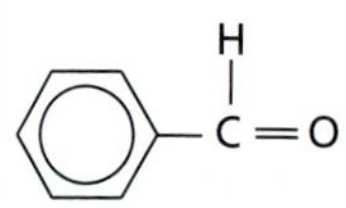 Benzaldehyde	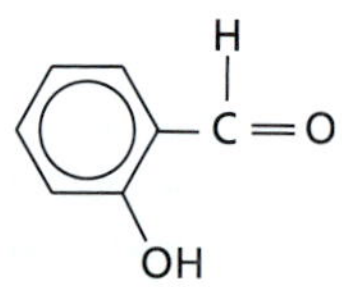 Salicylaldehyde	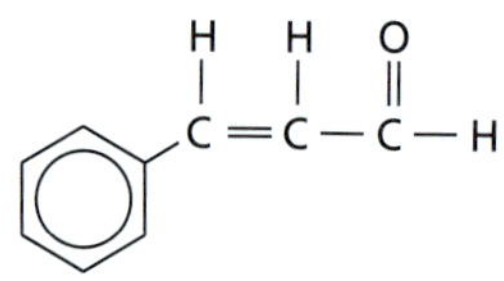 Cinnamaldehyde

CHEMISTRY PROJECT

Chemistry and Nutrition Have students browse several aisles in a supermarket and note the foods that contain palm oil, stearic acid, and benzoic acid. Have them also research the possible positive and negative effects of consuming these compounds. Have these students share their findings with the class. OL

An aldehyde molecule contains a polar, reactive structure. However, like ethers, aldehyde molecules cannot form hydrogen bonds among themselves because the molecules have no hydrogen atoms bonded to an oxygen atom. Therefore, aldehydes have lower boiling points than alcohols with the same number of carbon atoms. Water molecules can form hydrogen bonds with the oxygen atom of aldehydes, so aldehydes are more soluble in water than alkanes but not as soluble as alcohols or amines.

Formaldehyde has been used for preservation for many years, as shown in **Figure 9.** Industrially, large quantities of formaldehyde are reacted with urea to manufacture a type of grease-resistant, hard plastic used to make buttons, appliance and automotive parts, and electrical outlets, as well as the glue that holds the layers of plywood together. Benzaldehyde and salicylaldehyde, shown in **Table 7,** are two components that give almonds their natural flavor. The aroma and flavor of cinnamon, a spice that comes from the bark of a tropical tree, are produced largely by cinnamaldehyde, also shown in **Table 7.**

■ **Figure 9** A water solution of formaldehyde was used in the past to preserve biological specimens. However, formaldehyde's use has been restricted in recent years because studies indicate it might cause cancer.

Bill Aron/PhotoEdit

☑ READING CHECK **Identify** two uses for aldehydes.

Ketones A carbonyl group can also be located within a carbon chain rather than at the end. A **ketone** is an organic compound in which the carbon of the carbonyl group is bonded to two other carbon atoms. Ketones have the general formula shown in **Table 8.** The carbon atoms on either side of the carbonyl group are bonded to other atoms. The simplest ketone, commonly known as acetone, has only hydrogen atoms bonded to the side carbons, as shown in **Table 8.**

Ketones are formally named by changing the *-e* at the end of the alkane name to *-one*, and including a number before the name to indicate the position of the ketone group. In the previous example, the alkane name propane is changed to propan*one*. The carbonyl group can be located only in the center, but the prefix *2-* is usually added to the name for clarity, as shown in **Table 8.**

Ketones and aldehydes share many chemical and physical properties because their structures are similar. Ketones are polar molecules and are less reactive than aldehydes. For this reason, ketones are popular solvents for other moderately polar substances, including waxes, plastics, paints, lacquers, varnishes, and glues. Like aldehydes, ketone molecules cannot form hydrogen bonds with each other but can form hydrogen bonds with water molecules. Therefore, ketones are somewhat soluble in water. Acetone is completely miscible with water.

Table 8 Ketones

General Formula	Examples of Ketones	
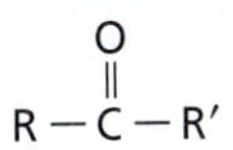 where R and R′ represent carbon chains or rings bonded to functional groups	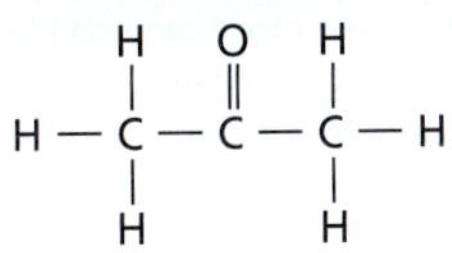 2-Propanone (acetone)	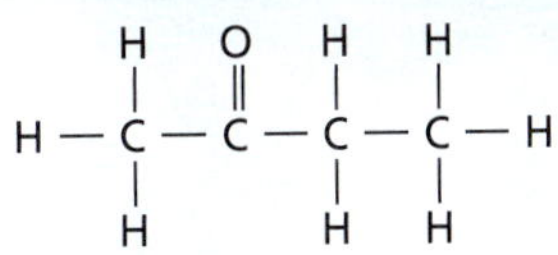 2-Butanone (methyl ethyl ketone)

Reinforcement

Substituted Hydrocarbons Have students draw from memory the functional groups found in alcohols, amines, and ethers. Then, have students differentiate these types of compounds from aldehydes, ketones, carboxylic acids, esters, and amides as they encounter them in the section. OL

Use Science Terms

Carbonyl Compounds Have students write statements explaining the meaning of the terms *ketone* and *aldehyde.* OL EL

☑ READING CHECK Possible answer: used to make buttons, appliance and automotive parts.

DIFFERENTIATED INSTRUCTION

Struggling Learners Some students will better understand differences between functional groups if they build models of compounds containing them. Ask these students to build models of butanal, 2-butanone, and butanoic acid. They should carefully examine each structure, noting similarities and differences. BL EL

Reinforcement

Nomenclature Reinforce the importance of writing the names for organic compounds carefully and correctly. Use the following examples: first write the structures on the board, then have students write correct names for the compounds: ethane; ethene; ethyne; ethanol; ethanal. **OL** **EL**

ASSESSMENT

Knowledge Ask students to match the formulas $C_7H_{14}O_2$ and $C_7H_6O_2$ with the names benzoic acid and heptanoic acid. $C_7H_6O_2$ is benzoic acid; $C_7H_{14}O_2$ is heptanoic acid. **OL**

■ **Caption Question Fig. 10** methanoic acid **OL**

☑ **READING CHECK** The formal name is formed by changing the *–ane* of the parent alkane, ethane, to *–anoic* acid.

☑ **READING CHECK** They are proton donors in solution.

Table 9 Carboxylic Acids

General Formula	Examples of Carboxylic Acids	
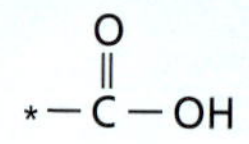 where * represents a hydrogen atom, carbon chain or ring bonded to the functional group	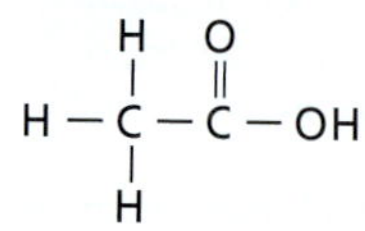 Ethanoic acid (acetic acid)	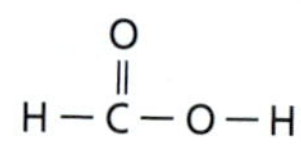 Methanoic acid (formic acid)

Carboxylic Acids

A **carboxylic acid** is an organic compound that has a carboxyl group. A **carboxyl group** consists of a carbonyl group bonded to a hydroxyl group. Thus, carboxylic acids have the general formula shown in **Table 9.** One diagram shown in **Table 9** is the structure of a familiar carboxylic acid—acetic acid, the acid found in vinegar. Although many carboxylic acids have common names, the formal name is formed by changing the *-ane* of the parent alkane to *-anoic acid*. Thus, the formal name of acetic acid is ethanoic acid.

A carboxyl group is usually represented in condensed form by writing –COOH. For example, ethanoic acid can be written as CH_3COOH. The simplest carboxylic acid consists of a carboxyl group bonded to a single hydrogen atom, HCOOH, shown in **Table 9.** Its formal name is methanoic acid, but it is more commonly known as formic acid. Some insects produce formic acid as a defense mechanism, as shown in **Figure 10.**

☑ READING CHECK **Explain** how the name *ethanoic acid* is derived.

Carboxylic acids are polar and reactive. Those that dissolve in water ionize weakly to produce hydronium ions, the anion of the acid in equilibrium with water, and the unionized acid. The ionization of ethanoic acid is an example.

$$CH_3COOH(aq) + H_2O(l) \rightleftharpoons CH_3COO^-(aq) + H_3O^+(aq)$$

Ethanoic acid (acetic acid) — Ethanoate ion (acetate ion)

Carboxylic acids can ionize in water solution because the two oxygen atoms are highly electronegative and attract electrons away from the hydrogen atom in the –OH group. As a result, the hydrogen proton can transfer to another atom that has a pair of electrons not involved in bonding, such as the oxygen atom of a water molecule. Because they ionize in water, soluble carboxylic acids turn blue litmus paper red and have a sour taste.

Some important carboxylic acids, such as oxalic acid and adipic acid, have two or more carboxyl groups. An acid with two carboxyl groups is called a dicarboxylic acid. Others have additional functional groups such as hydroxyl groups, as in the lactic acid found in yogurt. Typically, these acids are more soluble in water and often more acidic than acids with only a carboxyl group.

☑ READING CHECK **Evaluate** Using the information above, explain why carboxylic acids are classified as acids.

■ **Figure 10** Stinging ants defend themselves with a venom that contains formic acid.
Identify *another name for formic acid.*

Norm Thomas/Photo Researchers

DIFFERENTIATED INSTRUCTION

Struggling Learners Some students will need to practice naming functional groups and compounds that contain them many times before committing them to memory. Encourage students to practice by making and using flash cards, naming sample structures containing functional groups, and drawing structures from names. **BL** **EL** **COOP LEARN**

DIFFERENTIATED INSTRUCTION

Advanced Learners Ask groups of students to design a test using pH paper that can be used to distinguish carboxylic acids from other classes of organic compounds. **AL**

Table 10 **Esters**

General Formula	Example of an Ester
$-\overset{\overset{\displaystyle O}{\Vert}}{C}-O-R$ Ester group	Ethanoate group: $CH_3-\overset{\overset{\displaystyle O}{\Vert}}{C}-$; Propyl group: $-CH_2CH_2CH_3$ $CH_3-\overset{\overset{\displaystyle O}{\Vert}}{C}-O-CH_2CH_2CH_3$ Ester group: $\overset{\overset{\displaystyle O}{\Vert}}{C}-O$ Propyl ethanoate (propyl acetate)

Organic Compounds Derived from Carboxylic Acids

Several classes of organic compounds have structures in which the hydrogen or the hydroxyl group of a carboxylic acid is replaced by a different atom or group of atoms. The two most common classes are esters and amides.

Esters An **ester** is any organic compound with a carboxyl group in which the hydrogen of the hydroxyl group has been replaced by an alkyl group, producing the arrangement shown in **Table 10.** The name of an ester is formed by writing the name of the alkyl group followed by the name of the acid with the *-ic acid* ending replaced by *-ate*, as illustrated by the example shown in **Table 10.** Note how the name *propyl* results from the structural formula. The name shown in parentheses is based on the name *acetic acid*, the common name for ethanoic acid.

Esters are polar molecules and many are volatile and sweet-smelling. Many kinds of esters are found in the natural fragrances and flavors of flowers and fruits, as shown in **Figure 11.** Natural flavors, such as apple or banana, result from mixtures of many different organic molecules, including esters, but some of these flavors can be imitated by a single ester structure. Consequently, esters are manufactured for use as flavors in many foods and beverages and as fragrances in candles, perfumes, and other scented items.

VOCABULARY
SCIENCE USAGE V. COMMON USAGE
Class
Science usage: a group, set, or kind that share common traits
Esters are a class of organic molecules.
Common usage: a group of students that meet at regular intervals to study the same subject
Students meet for chemistry class during fourth period.

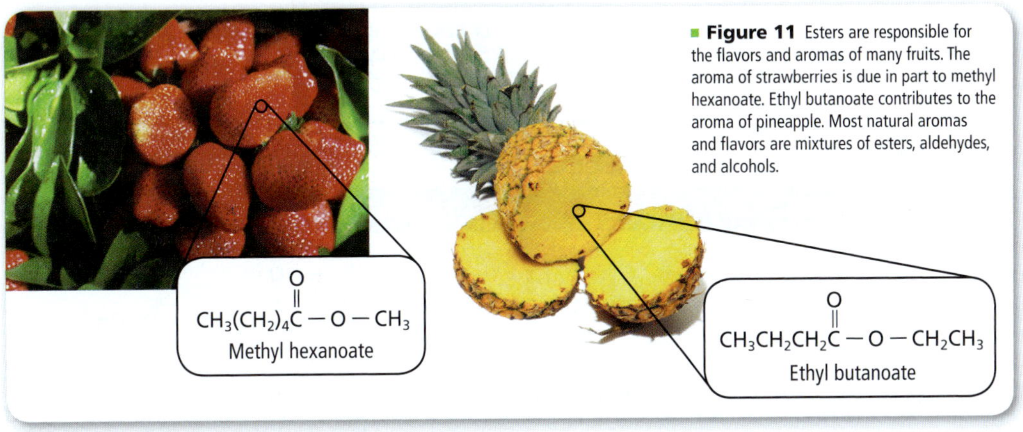

Figure 11 Esters are responsible for the flavors and aromas of many fruits. The aroma of strawberries is due in part to methyl hexanoate. Ethyl butanoate contributes to the aroma of pineapple. Most natural aromas and flavors are mixtures of esters, aldehydes, and alcohols.

Quick Demo

Sensory Experience Place samples of candies with different synthetic fruity flavors in opaque cups. Cover each cup with foil, then use a pencil to break a small hole in the foil. Pass the cups around and ask students to identify the flavors of the candies using only the scents. Some students might find this task difficult because humans detect flavors best when they can use several senses at once, including sight, smell, and taste. Explain that synthetic flavoring compounds consist of organic molecules, mostly esters and other carbonyl compounds. **OL**

Concept Development

Carboxylic Acids and Amides
To help students better understand the relationship between the formulas of carboxylic acids and amides, have them write and compare structural formulas for the following pairs of compounds: propanoic acid and propanamide; butanoic acid and butanamide. **OL**

CHEMISTRY JOURNAL

Chemistry of Fragrances Have students research and list the chemical compositions and structures of several of their favorite scents, such as rose, strawberry, lilac, or musk. They should include this list of chemical names and structures in their chemistry journals. **OL**

MiniLAB

Rubric

Purpose Students will prepare an ester by reacting an alcohol with an organic acid.

Process Skills classify, observe and infer, draw conclusions

Safety Precautions Approve lab safety forms before work begins. Students should wear aprons, gloves, and goggles while handling sulfuric acid, which is corrosive to skin and eyes. Methanol is flammable and poisonous. Students should keep it away from open flames and avoid using a laboratory burner for the hot water bath.

Disposal The cotton ball can be discarded as standard solid waste.

Teaching Strategies
So that students will be surprised at the result, you should avoid telling them about the expected odor before the experiment.

Expected Results The product will have a wintergreen odor.

Analysis

1. Student answers will vary but might include chewing gum and mint candy.
2. Student answers will vary. Advantage: synthetic esters are more efficiently and economically produced than natural esters. Disadvantage: odors of synthetic esters can differ slightly from those of natural esters, which might contain other compounds.

READING CHECK In an amide, the —OH group of a carboxylic acid is replaced by a nitrogen that is bonded to other atoms.

READING CHECK Possible answer: urea

MiniLAB

Make an Ester

How can you recognize an ester?

Procedure

1. Read and complete the lab safety form.
2. Prepare a hot-water bath by pouring 150 mL of **tap water** into a **250-mL beaker.** Place the beaker on a **hot plate** set to medium.
3. Use a **balance** and **weighing paper** to measure 1.5 g of **salicylic acid.** Place the salicylic acid in a **small test tube** and add 3 mL of **distilled water.** Use a **10-mL graduated cylinder** to measure the water. Then add 3 mL of **methanol.** Use a **Beral pipette** to add 3 drops of **concentrated sulfuric acid** to the test tube.
   WARNING: ***Concentrated sulfuric acid can cause burns. Methanol fumes are explosive—keep away from open flame. Handle chemicals with care.***
4. When the water is hot but not boiling, place the test tube in the bath for 5 min. Use a **test-tube clamp** to remove the test tube from the bath and place in a **test-tube holder** until needed.
5. Place a **cotton ball** in a **petri dish half.** Pour the contents of the test tube onto the cotton ball. Record your observation of the odor of the product.

Analysis

1. **Name** The common name of the ester that you produced is *oil of wintergreen*. Name some products that you think could contain the ester.
2. **Evaluate** the advantages and disadvantages of using synthetic esters in consumer products as compared to using natural esters.

Amides An **amide** is an organic compound in which the –OH group of a carboxylic acid is replaced by a nitrogen atom bonded to other atoms. The general structure of an amide is shown in **Table 11.** Amides are named by writing the name of the alkane with the same number of carbon atoms, and then replacing the final *-e* with *-amide.* Thus, the amide shown in **Table 11** is called ethanamide, but it can also be named acetamide from its common name, acetic acid.

READING CHECK How does an amide differ from a carboxylic acid?

The amide functional group is found repeated many times in natural proteins and some synthetic materials. For example, you might have used a nonaspirin pain reliever containing acetaminophen. In the acetaminophen structure shown in **Table 11,** notice that the amide (–NH–) group connects a carbonyl group and an aromatic group.

One important amide is caramide (NH_2CONH_2), or urea, as it is commonly known. Urea is an end product in the metabolic breakdown of proteins in mammals. It is found in the blood, bile, milk, and perspiration of mammals. When proteins are broken down, amino groups (NH_2) are removed from the amino acids. The amino groups are then converted to ammonia (NH_3) molecules that are toxic to the body. The toxic ammonia is converted to nontoxic urea in the liver. The urea is filtered out of the blood in the kidneys and passed from the body in urine.

Because of the high nitrogen content of urea and because it is easily converted to ammonia in the soil, urea is a common commercial fertilizer. Urea is also used as a protein supplement for ruminant animals, such as cattle and sheep. These animals use urea to produce proteins in their bodies.

READING CHECK **Identify** an amide that is found in the human body.

Table 11 Amides

General Formula	Examples of Amides	
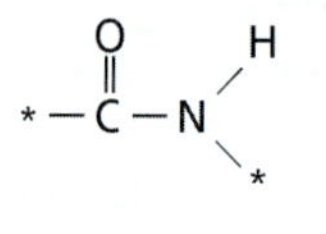 Amide group	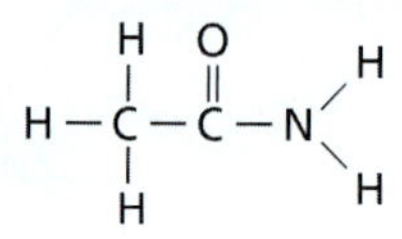 Ethanamide (acetamide)	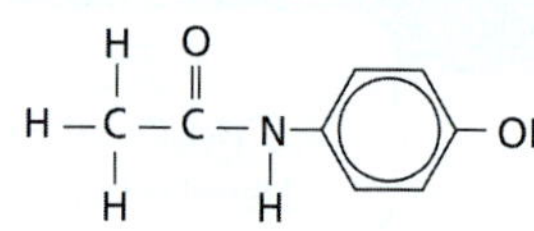 Acetaminophen

CULTURAL DIVERSITY

Early Aspirin While the German chemist Felix Hoffmann is given credit for producing the modern synthetic form of aspirin in the late 1800s, numerous ancient cultures, including Native Americans, discovered aspirin's pharmacological ancestor long before Hoffmann's time. Its past reaches to at least the fifth century B.C., when the Greek physician Hippocrates used a bitter powder he extracted from willow bark to ease aches and pains and reduce fever. The substance in the bark with medicinal properties is salicin, an organic compound similar to the modern form of aspirin.

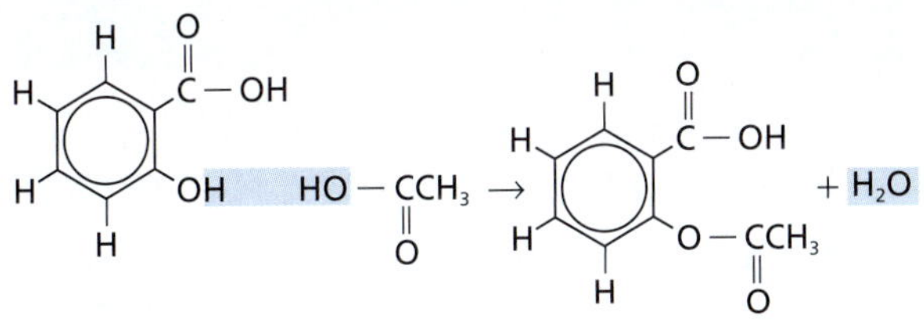

Salicylic acid | Acetic acid | Acetylsalicylic acid (aspirin) | Water

Figure 12 To synthesize aspirin, two organic molecules are combined in a condensation reaction to form a larger molecule.

Condensation Reactions

Many laboratory syntheses and industrial processes involve the reaction of two organic reactants to form a larger organic product, such as the aspirin shown in **Figure 12.** This type of reaction is known as a condensation reaction.

In a **condensation reaction,** two smaller organic molecules combine to form a more complex molecule, accompanied by the loss of a small molecule such as water. Typically, the molecule lost is formed from one particle from each of the reactant molecules. In essence, a condensation reaction is an elimination reaction in which a bond is formed between two atoms not previously bonded to each other.

The most common condensation reactions involve the combining of carboxylic acids with other organic molecules. A common way to synthesize an ester is by a condensation reaction between a carboxylic acid and an alcohol. Such a reaction can be represented by the following general equation.

$$RCOOH + R'OH \rightarrow RCOOR' + H_2O$$

FOLDABLES® Incorporate information from this section into your Foldable.

SECTION 3 REVIEW

Section Self-Check

Section Summary

- Carbonyl compounds are organic compounds that contain the C=O group.
- Five important classes of organic compounds containing carbonyl compounds are aldehydes, ketones, carboxylic acids, esters, and amides.

14. MAINIDEA Classify each of the carbonyl compounds as one of the types of organic substances you have studied in this section.

a. 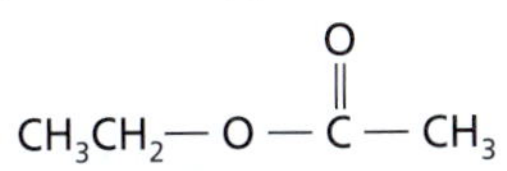

c. 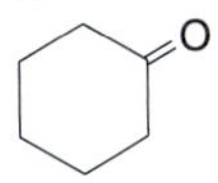

b. 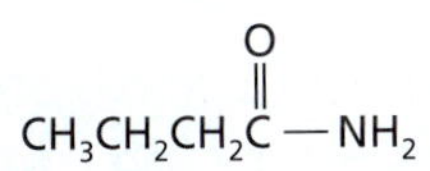

d. 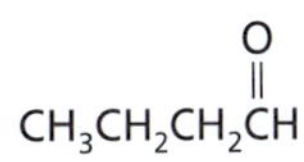

15. Describe the products of a condensation reaction between a carboxylic acid and an alcohol.

16. Determine The general formula for alkanes is C_nH_{2n+2}. Derive a general formula to represent an aldehyde, a ketone, and a carboxylic acid.

17. Infer why water-soluble organic compounds with carboxyl groups exhibit acidic properties in solutions, whereas similar compounds with aldehyde structures do not exhibit these properties.

SECTION 3 REVIEW

14. a. ester
b. amide
c. ketone
d. aldehyde

15. The products are an ester and water.

16. Aldehyde: $C_nH_{2n}O$; ketone: $C_nH_{2n}O$; carboxylic acid: $C_nH_{2n}O_2$

17. The carboxyl group can readily ionize donating an H^+ ion. However, the hydrogen atom bonded to an aldehyde's carbonyl group does not readily ionize.

FOLDABLES® Rubric

3 Assess

Check for Understanding

Ask students to give the names and structures of five compounds, one compound representing each of the following categories: an aldehyde; a ketone; a carboxylic acid; an ester; an amide. OL

Reteach

Have students list all categories of carbonyl compounds discussed in the section. Ask them to describe some of the properties of and uses for each type of compound. OL

Extension

Show students the structure of a large, complex, organic biological molecule, such as cholesterol or ATP. Ask them to circle and name all the functional groups present in the structure. OL

ASSESSMENT

Performance Have students work in small groups to build ball-and-stick models of an alkyl halide, an alcohol, an aldehyde, a ketone, a carboxylic acid, an ester, and an amide. Use the models to give a practical quiz in which all the students must name the compound represented by each model. OL EL COOP LEARN

SECTION 4

1 Focus

MAINIDEA

Predicting Products Write the following partial chemical reaction on the board: $CH_4(g) + 2O_2(g) \rightarrow$ Ask students to predict what the products will be and write the correct response on the board. $CO_2(g) + 2H_2O(g)$ Ask students how they know what the products will be. because all hydrocarbons combust to form carbon dioxide and water Tell students that chemical reactions of organic molecules can be classified into groups of similar reactions, which makes predicting the products of the reaction easier. OL

☑ **READING CHECK** Possible answer: A reaction in which atoms are removed from two adjacent carbon atoms. The carbon atoms form a double bond between them and the removed atoms form another stable molecule.

SECTION 4

Other Reactions of Organic Compounds

MAINIDEA Classifying the chemical reactions of organic compounds makes predicting products of reactions much easier.

Essential Questions

- How are organic reactions classified?
- Why is it useful to draw structural formulas when writing equations for reactions of organic compounds?
- How can classifying a reaction help you predict the reaction's products?

Review Vocabulary

catalyst: a substance that increases the rate of a chemical reaction by lowering activation energies but is not consumed in the reaction

New Vocabulary

elimination reaction
dehydrogenation reaction
dehydration reaction
addition reaction
hydration reaction
hydrogenation reaction

CHEM 4 YOU As you eat lunch, the oxidation of organic compounds is probably not on your mind. However, that is exactly what is about to occur as your cells break down the food that you eat to obtain energy for your body.

Classifying Reactions of Organic Substances

Organic chemists have discovered thousands of reactions by which organic compounds can be changed into different organic compounds. By using combinations of these reactions, chemical industries convert simple molecules from petroleum and natural gas into the large, complex organic molecules found in many useful products—including lifesaving drugs and many other consumer products as shown in **Figure 13.**

You have already read about substitution and condensation reactions. Two other important types of reactions by which organic compounds can be changed into different compounds are elimination reactions and addition reactions.

Elimination reactions One way to change an alkane into a chemically reactive substance is to form a second covalent bond between two carbon atoms, producing an alkene. Forming double bonds from single bonds between carbon atoms is an **elimination reaction,** a reaction in which a combination of atoms is removed from two adjacent carbon atoms, forming an additional bond between them. The atoms that are eliminated usually form stable molecules, such as H_2O, HCl, or H_2.

☑ **READING CHECK** **Define** *elimination reaction* in your own words.

Figure 13 Many consumer products, such as plastic containers, fibers in ropes and clothing, and oils and waxes in cosmetics, are made from petroleum and natural gas.

Cordelia Molloy/Photo Researchers

DIFFERENTIATED INSTRUCTION

Advanced Learners Ask students to research the actual steps used in the synthesis of a large organic compound, such as the anti-cancer drug taxol. As they progress through the section, ask them to label some of the steps as substitution, elimination, addition, condensation, or redox reactions. AL

■ **Figure 14** Low density polyethylene (LDPE) is made from gaseous ethene under high pressure in the presence of a catalyst. LDPE is used for playground equipment because it is easy to mold into various shapes, it is easy to dye into many colors, and it is durable. The name *polyethylene* comes from *ethylene,* which is the common name for ethene.

Ethene, the starting material for the playground equipment shown in **Figure 14,** is produced by the elimination of two hydrogen atoms from ethane. A reaction that eliminates two hydrogen atoms is called a **dehydrogenation reaction.** Note that the two hydrogen atoms form a molecule of hydrogen gas.

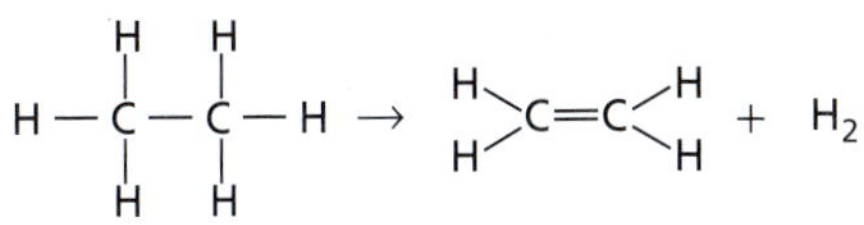

Ethane Ethene

Alkyl halides can undergo elimination reactions to produce an alkene and a hydrogen halide, as shown here.

$$\underset{\text{Alkyl halide}}{R{-}CH_2{-}CH_2{-}X} \rightarrow \underset{\text{Alkene}}{R{-}CH{=}CH_2} + \underset{\text{Hydrogen halide}}{HX}$$

Likewise, alcohols can also undergo elimination reactions by losing a hydrogen atom and a hydroxyl group to form water, as shown below. An elimination reaction in which the atoms removed form water is called a **dehydration reaction.** In the dehydration reaction, the alcohol is broken down into an alkene and water.

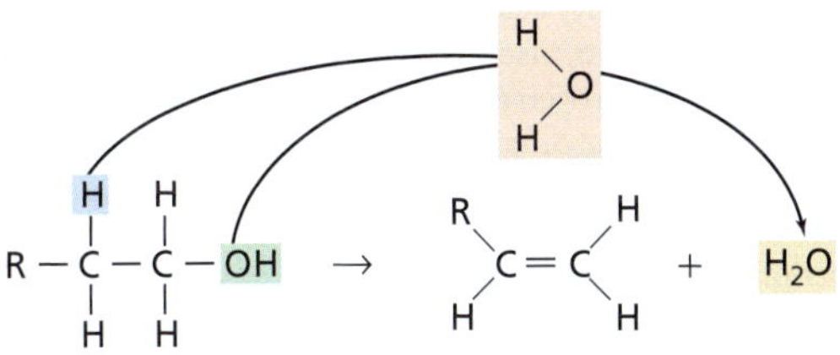

Alcohol Alkene Water

The generic form of this dehydration reaction can be written as follows.

$$R{-}CH_2{-}CH_2{-}OH \rightarrow R{-}CH{=}CH_2 + H_2O$$

Watch a **video about substituted hydrocarbons and their reactions.**

Video

Get help **identifying organic reactions.**

Personal Tutor

2 Teach

Quick Demo

Smells and Flavors Choose one of the carboxylic acid-alcohol pairs listed below under the heading *Alcohol, Acid, Flavor*. Mix together in a test tube 1 mL of each component and two drops of concentrated sulfuric acid. Place the test tube in a hot water bath for ten minutes. While waiting for the reaction to take place, ask students to use what they have learned in the previous section to predict what type of reaction will occur. a condensation reaction in which two molecules combine to form an ester Ask them how they might be able to tell when the reaction is complete. The odor will change as the reaction progresses. Due to its lower density, the ester formed might produce a visible layer on top of the original mixture. Let students examine the product and carefully waft its vapors toward their noses to try to identify its scent.

Alcohol, Acid, Flavor n-amyl alcohol, butyric acid, apricot; n-amyl alcohol, salicylic acid, pineapple; n-amyl alcohol, acetic acid, banana; octanol, acetic acid, fruity; methanol, salicylic acid, wintergreen; ethanol, butyric acid, apple; ethanol, acetic acid, fruity

Video Lab

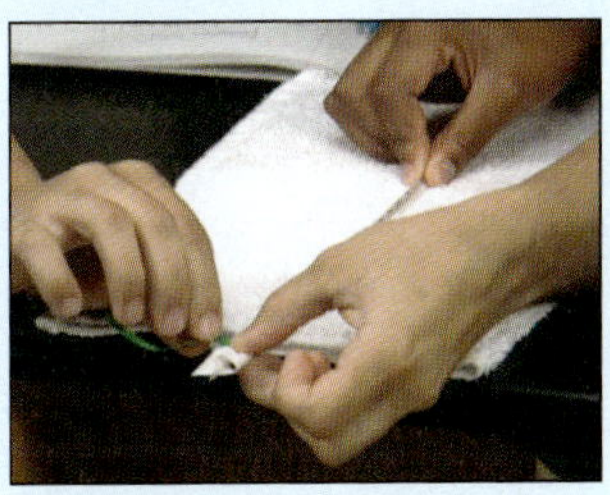

Video

Substituted Hydrocarbons and Their Reactions

Knowledge Ask students to distinguish between the terms *addition* and *elimination.* OL EL

READING CHECK dehydration reaction

Apply Chemistry

Plastic Storage Containers

In the 1930s, Earl Silas Tupper worked at the DuPont Chemical Company, which was developing plastics. Excited by the new materials, Tupper asked his supervisor if he could buy some extra plastic to use for experimentation. Tupper was given a piece of black, inflexible waste from the oil-refining process used to get starting materials for plastics. He purified it and molded it to create lightweight, unbreakable containers, bowls, and plates. Later he added airtight lids that he modeled after paint-can lids, but in reverse. Tupper founded his own company in 1938. Though his plastic storage containers are well-known today, customers needed some time to get used to the lids. In fact, his plastic storage containers were not a big success until the late 1940s.

Table 12 Summary of Addition Reactions

Reactant Alkene	Addition Reactant	Product
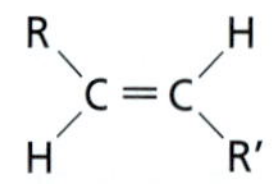	Water (hydration) H—O—H	Alcohol 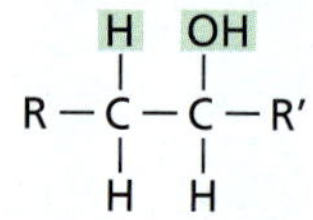
	Hydrogen (hydrogenation) H—H	Alkane 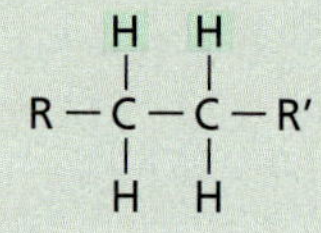
	Hydrogen halide H—X	Alkyl halide 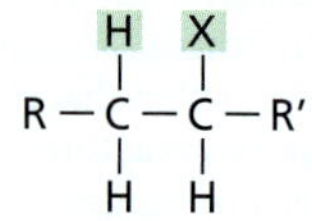
	Halogen X—X	Alkyl dihalide 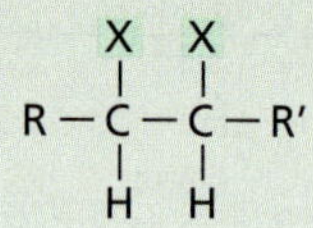

Addition reactions Another type of organic reaction appears to be an elimination reaction in reverse. An **addition reaction** results when other atoms bond to each of two atoms bonded by double or triple covalent bonds. Addition reactions typically involve double-bonded carbon atoms in alkenes or triple-bonded carbon atoms in alkynes. Addition reactions occur because double and triple bonds have a rich concentration of electrons. Therefore, molecules and ions that attract electrons tend to form bonds that use some of the electrons from the multiple bonds. The most common addition reactions are those in which H_2O, H_2, HX, or X_2 add to an alkene, as shown in **Table 12.**

A **hydration reaction,** also shown in **Table 12,** is an addition reaction in which a hydrogen atom and a hydroxyl group from a water molecule add to a double or triple bond. The generic equation shown in **Table 12** shows that a hydration reaction is the opposite of a dehydration reaction.

A reaction that involves the addition of hydrogen to atoms in a double or triple bond is called a **hydrogenation reaction.** One molecule of H_2 reacts to fully hydrogenate each double bond in a molecule. When H_2 adds to the double bond of an alkene, the alkene is converted to an alkane.

READING CHECK **Identify** the reaction that is the reverse of a hydrogenation reaction.

CHEMISTRY JOURNAL

Diels-Alder Reactions Diels-Alder reactions are fun for students to learn because they form interesting-looking cyclic products. Encourage students to research this group of reactions and include a summary of their findings in their journals. OL

Catalysts are usually needed in the hydrogenation of alkenes because the reaction's activation energy is too large without them. Catalysts such as powdered platinum or palladium provide a surface that adsorbs the reactants and makes their electrons more available to bond to other atoms.

Hydrogenation reactions are commonly used to convert the liquid unsaturated fats found in oils from plants such as soybean, corn, and peanuts into saturated fats that are solid at room temperature. These hydrogenated fats are then used to make margarine and solid shortening.

Alkynes can also be hydrogenated to produce alkenes or alkanes. One molecule of H_2 must be added to each triple bond in order to convert an alkyne to an alkene, as shown here.

$$R{-}C{\equiv}C{-}H + H_2 \rightarrow R{-}CH{=}CH_2$$

After the first molecule of H_2 is added, the alkyne is converted to an alkene. A second molecule of H_2 follows the hydrogenation reaction.

$$R{-}CH{=}CH_2 + H_2 \rightarrow R{-}CH_2{-}CH_3$$

In a similar mechanism, the addition of hydrogen halides to alkenes is an addition reaction useful to industry for the production of alkyl halides. The generic equation for this reaction is shown below.

$$R{-}CH{=}CH{-}R' + HX \rightarrow R{-}CHX{-}CH_2{-}R'$$

Data Analysis LAB

Based on Real Data*

Interpret Data

What are the optimal conditions to hydrogenate canola oil? Edible vegetable oil is hydrogenated to preserve its flavor and to alter its melting properties. Because evidence suggests that *trans*-fatty acids are associated with increased risk of heart disease and cancer, the minimum amount of *trans*-fatty acids and the maximum amount of *cis*-oleic acid are desired.

Computer models were used to simulate processing conditions and to alter eight variables to optimize the output of the desirable oil. Multiple optimal operating conditions were determined. A small-scale industrial plant was used to confirm the results of the computer simulation.

Data and Observations

The table at right shows some of the data from this investigation.

Data for Canadian Canola Oil

Trial Run	Computer Simulation		Experimental	
	trans-Fatty Acids (wt. %)	*cis*-Oleic Acid (wt. %)	*trans*-Fatty Acids (wt. %)	*cis*-Oleic Acid (wt. %)
1	4.90	69.10	5.80	70.00
2	4.79	63.75	4.61	64.00
3	4.04	68.96	4.61	67.00
4	5.99	62.80	7.10	65.00
5	4.60	68.10	5.38	66.50

Data obtained from Izadifar, M. 2005. Application of genetic algorithm for optimization of vegetable oil hydrogenation process. *Journal of Food Engineering.* 78 (2007) 1–8.

Think Critically

1. **Calculate** the percent yield for each of the trials shown in the table.
2. **Evaluate** Which trial(s) produced the highest yield of *cis*-oleic acid and the lowest yield of *trans*-fatty acids?
3. **Explain** why the techniques used in this investigation are useful in manufacturing processes.

Data Analysis LAB

About the Lab

- The primary source of *trans*-fatty acids in the diet is partially hydrogenated vegetable oils used in cooking fats and margarines.
- There are two dependent variables in this investigation: amount of *trans* isomer and amount of oleic acid produced.
- The eight variables include: reaction temperature, H_2 pressure, catalyst concentration, agitation frequency, expected iodine value, and content of initial unsaturated fatty acids including oleic, linoleic, and linolenic acid.
- Also see Izadifer, M. 2005. *Journal of Food Engineering* 66: 227–232.

Think Critically

1.

Trial Run	Percent Yield	
	trans-Fatty Acids	*cis*-Oleic Acid
1	118%	101.3%
2	96.2%	100.1%
3	114%	97.16%
4	119%	103.5%
5	117%	97.65%

2. Highest yield of *cis*-oleic acid is trial run 4. Lowest yield of *trans*-fatty acids is trial run 2.
3. Computer simulations and small-scale industrial plants are useful because they are less expensive to run than actual production lines. Chemical processes can be fine tuned with minimal expense.

CHEMISTRY PROJECT

Hydrogenated Fats Have students visit grocery-store aisles and record at least ten products that contain hydrogenated or partially hydrogenated fats or oils. Have them look up the molecular and structural formulas for one or more of the fats or oils and present their findings to the class. OL

ASSESSMENT

Performance Have students draw the general structural formulas for carboxylic acids, amides, esters, and amines. OL EL

Reinforcement

Oxidation and Reduction Have students write the chemical equation for the combustion of methane to gaseous carbon dioxide and liquid water. $CH_4(g) + 2O_2(g) \longrightarrow CO_2(g) + 2H_2O(l)$ Reinforce students' understanding of oxidation and reduction by having them determine the oxidation states of all elements, reactants, and products. CH_4: C is -4, H is $+1$, O_2: O is 0; CO_2: C is $+4$, O is -2, H_2O: H is $+1$, O is -2 Then, ask students to identify the element oxidized and the element reduced. Carbon is oxidized. Oxygen is reduced. Emphasize that although both oxidation and reduction are involved in the reaction, it is called an *oxidation reaction* because an element in methane, the organic compound, is oxidized. OL

READING CHECK methanoic acid and 2-propanone

Table 13 Oxidation-Reduction Reactions

Oxidation of an alkane to an alcohol

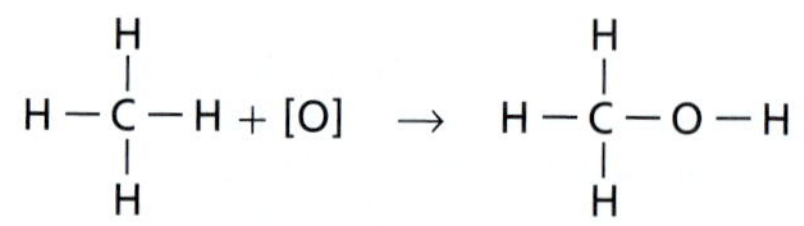

Methane → Methanol

A sequence of oxidation reactions

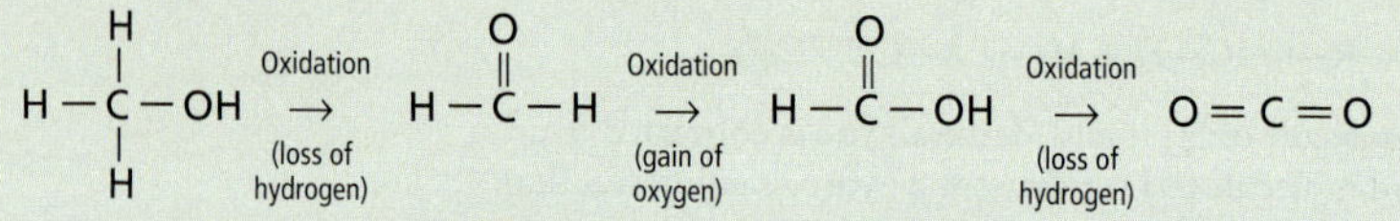

Methanol (methyl alcohol) → Methanal (formaldehyde) → Methanoic acid (formic acid) → Carbon dioxide

Oxidation of two isomers

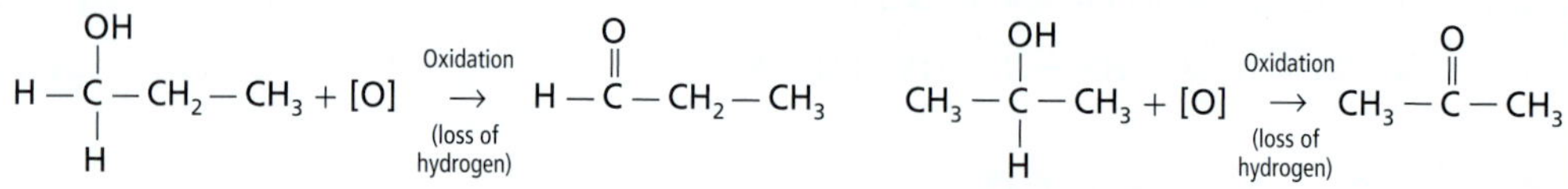

1-Propanol → Propanal; 2-Propanol → 2-Propanone

Oxidation-reduction reactions Many organic compounds can be converted to other compounds by oxidation and reduction reactions. For example, suppose you want to convert methane, the main constituent of natural gas, to methanol, a common industrial solvent and raw material for making formaldehyde and methyl esters. The conversion of methane to methanol can be represented by the equation shown in **Table 13,** in which [O] represents oxygen from an agent such as copper(II) oxide, potassium dichromate, or sulfuric acid.

What happens to methane in this reaction? Before answering, it might be helpful to review the definitions of oxidation and reduction. Oxidation is the loss of electrons, and a substance is oxidized when it gains oxygen or loses hydrogen. Reduction is the gain of electrons, and a substance is reduced when it loses oxygen or gains hydrogen. Thus, methane is oxidized as it gains oxygen and is converted to methanol. Of course, every redox reaction involves both an oxidation and a reduction; however, organic redox reactions are described based on the change in the organic compound.

Oxidizing the methanol shown in **Table 13** is the first step in the sequence of reactions that can be used to produce an aldehyde, which are also shown in **Table 13.** For clarity, oxidizing agents are omitted. Preparing an aldehyde by this method is not always a simple task because the oxidation might continue, forming the carboxylic acid.

READING CHECK **Identify** Use **Table 13** to identify two possible products that are produced when the aldehyde is further oxidized.

CULTURAL DIVERSITY

George Washington Carver Born into slavery and then freed as a child, George Washington Carver became well-known for his innovation in improving farming techniques. He also made numerous contributions to technology, many of them in the area of the chemistry of natural organic compounds. From his laboratory at Alabama's Tuskegee Institute, Carver developed more than 300 uses for peanuts and more than 100 products from the sweet potato. Other Carver innovations include synthetic marble from sawdust, plastics from wood-shavings, and writing paper from wisteria vines. Carver patented only a few of his many discoveries. Carver's epitaph summarizes his philosophy of life: "He could have added fortune to fame, but, caring for neither, found happiness and honor in being helpful to the world."

However, not all alcohols can be oxidized to aldehydes and, subsequently, carboxylic acids. To understand why, compare the oxidations of 1-propanol and 2-propanol, shown in **Table 13.** Note that oxidizing 2-propanol yields a ketone, not an aldehyde. Unlike aldehydes, ketones resist further oxidation to carboxylic acids. Thus, while the propanal formed by oxidizing 1-propanol easily oxidizes to form propanoic acid, the 2-propanone formed by oxidizing 2-propanol does not react to form a carboxylic acid.

☑ READING CHECK **Write** the equation using molecular structures like those in **Table 13** for the formation of propanoic acid.

How important are organic oxidations and reductions? You have seen that oxidation and reduction reactions can change one functional group into another. That ability enables chemists to use organic redox reactions, in conjunction with the substitution and addition reactions you read about earlier in the chapter, to synthesize a tremendous variety of useful products. On a personal note, all living systems—including you—depend on the energy released by oxidation reactions. Of course, some of the most dramatic oxidation-reduction reactions are combustion reactions. All organic compounds that contain carbon and hydrogen burn in excess oxygen to produce carbon dioxide and water. For example, the highly exothermic combustion of ethane is described by the following thermochemical equation.

$$2C_2H_6(g) + 7O_2(g) \rightarrow 4CO_2(g) + 6H_2O(l) \qquad \Delta H = -3120 \text{ kJ}$$

As you read previously, much of the world relies on the combustion of hydrocarbons as a primary source of energy. Our reliance on the energy from organic oxidation reactions is illustrated in **Figures 15.**

Predicting Products of Organic Reactions

The generic equations representing the different types of organic reactions you have learned—substitution, elimination, addition, oxidation-reduction, and condensation—can be used to predict the products of other organic reactions of the same types. For example, suppose you were asked to predict the product of an elimination reaction in which 1-butanol is a reactant. You know that a common elimination reaction involving an alcohol is a dehydration reaction.

Figure 15 People around the world depend on the oxidation of hydrocarbons to get to work and to transport products.

RealWorld CHEMISTRY

Polycyclic Aromatic Hydrocarbons (PAHs)

BIOLOGICAL MOLECULES
Hydrocarbons composed of multiple aromatic rings are called PAHs. They have been found in meteorites and identified in the material surrounding dying stars. Scientists simulated conditions in space and found that about 10% of the PAHs were converted to alcohols, ketones, and esters. These molecules can be used to form compounds that are important in biological systems.

☑ READING CHECK Refer to the Solutions Manual.

ASSESSMENT

Knowledge Ask students to predict the name of the product formed by the reaction between hydrogen and 2,3-dimethyl-1-butene. 2,3-dimethylbutane OL

Content Background

Sugars One important aldehyde that students might recognize is the open-chain form of glucose. A familiar ketone is the open-chain form of fructose. Draw the structure on the board and circle the carbonyl group.

Glucose undergoes a redox reaction during respiration when it combines with oxygen to form carbon dioxide, water, and energy. Write the balanced chemical reaction on the board for students.

Math in Chemistry

Energy Sugars, or carbohydrates, provide about 17 kJ/g of energy when consumed. If a can of soda contains 40.0 g of sugars, about how much energy is produced when the sugars are oxidized? 680 kJ

DIFFERENTIATED INSTRUCTION

Advanced Learners Have students write the chemical equations for the oxidations of 1-butanol and 2-butanol. Have them use the style, in which [O] represents oxygen from an oxidizing agent. AL

1-butanol
$CH_3CH_2CH_2CH_2OH + [O] \rightarrow CH_3CH_2CH_2CHO +$
$[O] \rightarrow CH_3CH_2CH_2COOH$
2-butanol
$CH_3CH_2CHOHCH_3 + [O] \rightarrow CH_3CH_2COCH_3$

FOLDABLES® Rubric

3 Assess

Check for Understanding

Give students the general equations for substitution, elimination, addition, condensation, and redox reactions. Ask them to use a single sentence to state what happens during each reaction. OL

Reteach

Show students the multistep reaction series below and ask them to label each individual reaction as a substitution, elimination, addition, or condensation reaction. Point out that only the major organic reactants and products are shown in the reaction series. $CH_3CH_3 \rightarrow CH_2=CH_2 \rightarrow CH_3CH_2Cl \rightarrow CH_3CH_2OH + CH_3CO_2H \rightarrow CH_3C(=O)OCH_2CH_3$. Step 1 is an elimination, Step 2 is an addition, Step 3 is a substitution, and Step 4 is a condensation. OL

Extension

Have students add the formulas of inorganic reactants or products to the equations for the reactions in the multistep series shown below.

Step 1: $CH_3CH_3 \rightarrow CH_2=CH_2 + ?$
Step 2: $CH_2=CH_2 + ? \rightarrow CH_3CH_2Cl$
Step 3: $CH_3CH_2Cl + ? \rightarrow ? + CH_3CH_2OH$
Step 4: $CH_3CH_2OH + CH_3CO_2H \rightarrow CH_3C(=O)OCH_2CH_3 + ?$
[Step 1: H_2; Step 2: HCl; Step 3: reactant is H_2O, product is HCl; Step 4: H_2O] OL

The generic equation for the dehydration of an alcohol is as follows.

$$R{-}CH_2{-}CH_2{-}OH \rightarrow R{-}CH{=}CH_2 + H_2O$$

FOLDABLES® Incorporate information from this section into your Foldable.

To determine the actual product, first draw the structure of 1-butanol. Then use the generic equation as a model to see how 1-butanol would react. The generic reaction shows that the —OH and a H— are removed from the carbon chain. Finally, draw the structure of the likely products, as shown in the following equation.

$$\underset{\text{1-Butanol}}{CH_3{-}CH_2{-}CH_2{-}CH_2{-}OH} \rightarrow \underset{\text{1-Butene}}{CH_3{-}CH_2{-}CH{=}CH_2} + H_2O$$

As another example, suppose that you wish to predict the product of the reaction between cyclopentene and hydrogen bromide. Recall that the generic equation for an addition reaction between an alkene and an alkyl halide is as follows.

$$R{-}CH{=}CH{-}R' + HX \rightarrow R{-}CHX{-}CH_2{-}R'$$

First, draw the structure for cyclopentene, the organic reactant, and add the formula for hydrogen bromide, the other reactant. From the generic equation, you can see that a hydrogen atom and a halide atom add across the double bond to form an alkyl halide. Finally, draw the formula for the likely product. If you are correct, you have written the following equation.

Cyclopentene   Hydrogen bromide   Bromocyclopentane

SECTION 4 REVIEW

Section Self-Check

Section Summary

- Most reactions of organic compounds can be classified into one of five categories: substitution, elimination, addition, oxidation-reduction, and condensation.
- Knowing the types of organic compounds reacting can enable you to predict the reaction products.

18. MAINIDEA Classify each reaction as substitution, elimination, addition, or condensation.

a. $CH_3CH=CHCH_2CH_3 + H_2 \rightarrow CH_3CH_2{-}CH_2CH_2CH_3$

b. $CH_3CH_2CH_2CH(OH)CH_3 \rightarrow CH_3CH_2CH=CHCH_3 + H_2O$

19. Identify the type of organic reaction that would best accomplish each conversion.

a. alkyl halide → alkene
b. alkene → alcohol
c. alcohol + carboxylic acid → ester
d. alkene → alkyl dihalide

20. Complete each equation by writing the condensed structural formula for the product that is most likely to form.

a. $CH_3CH=CHCH_2CH_3 + H_2 \rightarrow$

b. $CH_3CH_2CH(Cl)CH_2CH_3 + OH^- \rightarrow$

21. Predicting Products Explain why the hydration reaction involving 1-butene might yield two distinct products, whereas the hydration of 2-butene yields only one product.

SECTION 4 REVIEW

18. a. addition
 b. elimination
19. a. elimination
 b. addition
 c. condensation
 d. addition
20. a. $CH_3CH_2CH_2CH_2CH_3$
 b. $CH_3CH_2CH(OH)CH_2CH_3$
21. Hydrating 1-butene might yield 1-butanol and/or 2-butanol because hydroxyl groups might bond to carbons 1 and/or 2 of the 4-carbon chain. Hydrating 2-butene, however, yields only 2-butanol because the hydroxyl group must be on carbon 2.

SECTION 5

Essential Questions

- How does drawing a diagram help you understand the relationship between a polymer and the monomers from which it forms?
- What distinguishes addition and condensation polymerization reactions?
- How can you use molecular structures and the presence of functional groups to predict the properties of polymers?

Review Vocabulary

molecular mass: the mass of one molecule of a substance

New Vocabulary

polymer
monomer
polymerization reaction
addition polymerization
condensation polymerization
thermoplastic
thermosetting

Polymers

MAINIDEA Synthetic polymers are large organic molecules made up of repeating units that are linked together by addition or condensation reactions.

CHEM 4 YOU Think how different your life would be without plastic sandwich bags, plastic foam cups, nylon and polyester fabrics, vinyl siding on buildings, foam cushions, and a variety of other synthetic materials. These products all have at least one thing in common—they are made of polymers.

The Age of Polymers

The compact discs shown in **Figure 16** contain polycarbonate, which is made of extremely long molecules with groups of atoms that repeat in a regular pattern. This molecule is an example of a synthetic polymer. **Polymers** are large molecules consisting of many repeating structural units. In **Figure 16,** the letter *n* beside the structural unit of polycarbonate represents the number of structural units in the polymer chain. Because polymer *n* values vary widely, molecular masses of polymers range from less than 10,000 amu to more than 1,000,000 amu. A typical chain in nonstick coating on skillets has about 400 units, giving it a molecular mass of around 40,000 amu.

Before the development of synthetic polymers, people were limited to using natural substances such as stone, wood, metals, wool, and cotton. By the turn of the twentieth century, a few chemically treated natural polymers such as rubber and the first plastic, celluloid, had become available. Celluloid is made by treating cellulose from cotton or wood fiber with nitric acid.

The first synthetic polymer, synthesized in 1909, was a hard, brittle plastic called Bakelite. Because of its resistance to heat, it is still used today in stove-top appliances. Since 1909, hundreds of other synthetic polymers have been developed. Because of the widespread use of polymers, people might refer to this time as the Age of Polymers.

Figure 16 Compact discs are made of polycarbonate and contain long chains of the structural unit shown.

CHEMISTRY JOURNAL

Serendipity Many important discoveries in polymer science have been made serendipitously, which means *finding something accidentally*. Have students research how either non-stick coating or nylon was discovered and decide what essential steps led to these unexpected findings. Students should write essays in their chemistry journals summarizing their findings. OL

SECTION 5

1 Focus

MAINIDEA

Polymer Chains Ask students if they have ever created a chain to count down the days to a special event. Answers will vary. Begin creating a chain using strips of paper about 24 cm in length and a stapler. (The size of the paper strips does not matter.) As you create the chain, tell students that small molecules called monomers link together to form long chains called polymers. Tell students that these monomers link together through addition or condensation reactions. BL OL

2 Teach

Quick Demo

Polymer Formation In a well-ventilated room, demonstrate polymer formation using one or more of the following readily available products: a polyurethane foam, such as a can of insulating foam; a two-component foam used in constructing landscape items for a train set; Aerosol string-polymer; a two-component system for making a clear polymer used in flower arrangements. Ask students to carefully observe the different appearances of the monomers and the final polymer. Monomers are usually liquids or gases, while polymers are more solid than the monomers from which they are formed. OL

Content Background

Polymer Money Australia's money is made of a nonporous polymer. A protective coating is applied to keep the notes from absorbing moisture and to keep them cleaner. On average, the new bills last four to five times longer than paper money in circulation. After a note's lifespan of about 40 months, the bill is recycled to produce other plastic products. Conventional printing techniques are used to apply ink. A see-through window makes it difficult to use a scanner or copier to produce counterfeit bills.

☑ READING CHECK Both are the smallest repeating unit in a polymer. A monomer is a single molecule that forms the repeating unit in a polymer. A structural unit is repeating groups (2 or more molecules) of atoms that form a polymer.

■ **Figure 17** Polyethylene is a nontoxic, unbreakable polymer that is used to make toys for children.

Reactions Used to Make Polymers

Polymers are relatively easy to manufacture. Polymers can usually be synthesized in one step in which the major reactant is a substance consisting of small, simple organic molecules called monomers. A **monomer** is a molecule from which a polymer is made.

When a polymer is made, monomers bond together one after another in a rapid series of steps. A catalyst is usually required for the reaction to take place at a reasonable pace. With some polymers, such as polyester fabric and nylon, two or more kinds of monomers bond to each other in an alternating sequence. A reaction in which monomer units are bonded together to form a polymer is called a **polymerization reaction.** The repeating group of atoms formed by the bonding of the monomers is called the structural unit of the polymer. The structural unit of a polymer made from two different monomers has the components of both monomers.

Figure 17 shows unbreakable children's toys that are made of low-density polyethylene (LDPE), which is synthesized by polymerizing ethene under pressure. Ethene can also be made into ethylene glycol, which is the starting point for polyethylene terephthalate (PETE). PETE can be made into bottles or spun into fibers. As a fiber, it is called polyester.

Figure 18 shows highlights of polymer development that led to the Age of Polymers. Although the first synthetic polymer was developed in 1909, the industry did not flourish until after World War II.

☑ READING CHECK **Compare and contrast** a monomer and a structural unit of a polymer.

■ **Figure 18**

The Age of Polymers

Scientists working to understand the structure and properties of organic compounds have developed products that affect people everywhere. Their contributions helped usher in the Age of Polymers.

1830 — 1860 — 1890

1840s Physicians begin using ether as an anesthetic during surgery.

1865 The structure of benzene is determined. It becomes the basis for the production of aromatic compounds.

1879 Saccharin is accidentally discovered by a chemist working with coal-tar derivatives.

1899 Aspirin is widely distributed by physicians as a pain treatment. It quickly becomes the number-one selling drug worldwide.

1909 The first plastic made from synthetic polymers, Bakelite, is developed.

CHEMISTRY PROJECT

Polymer Money Circulating paper money has a limited lifespan and can be easily destroyed. It can also be counterfeited. Because of this, Australia and 23 other nations have switched to polymer banknotes. Ask students to think of another item used in our culture that could be "reinvented" in polymer form. What innovative characteristics might the polymer have that would enhance the value of the item? Are there any disadvantages to producing this item in polymer form? Have students create a brochure on the subject and present it to the class. OL

$$n\mathrm{HOOC}-(\mathrm{CH_2})_4-\mathrm{COOH} + n\mathrm{H_2N}-(\mathrm{CH_2})_6-\mathrm{NH_2} \rightarrow \left[\overset{\overset{\displaystyle \mathrm{O}}{\|}}{\mathrm{C}}-(\mathrm{CH_2})_4-\overset{\overset{\displaystyle \mathrm{O}}{\|}}{\mathrm{C}}-\mathrm{NH}-(\mathrm{CH_2})_6-\mathrm{NH}\right]_n + n\mathrm{H_2O}$$

Adipic acid — 1,6–Diamino hexane — Nylon 6,6

Figure 19 Nylon is a polymer consisting of thin strands that resemble silk.

Addition polymerization In **addition polymerization,** all of the atoms present in the monomers are retained in the polymer product. When the monomer is ethene, an addition polymerization results in the polymer polyethylene. Unsaturated bonds are broken in addition polymerization, just as they are in addition reactions. The difference is that the molecule added is a second molecule of the same substance, ethene. Note that the addition polymers in **Table 14** on the next page are similar in structure to polyethylene. That is, the molecular structure of each is equivalent to polyethylene in which other atoms or groups of atoms are attached to the chain in place of hydrogen atoms. All of these polymers are made by addition polymerization.

Condensation polymerization **Condensation polymerization** takes place when monomers containing at least two functional groups combine with the loss of a small by-product, usually water. Nylon and a type of bulletproof fabric are made this way. Nylon was first synthesized in 1931 and soon became popular because it is strong and can be drawn into thin strands resembling silk. Nylon 6,6 is the name of one type of nylon that is synthesized. One monomer is a chain, with the end carbon atoms being part of carboxyl groups, as shown in **Figure 19.** The other monomer is a chain having amino groups at both ends. These monomers undergo a condensation polymerization that forms amide groups linking the subunits of the polymer, as shown by the tinted box in **Figure 19.** Note that one water molecule is released for every new amide bond formed.

1939–1945 During World War II, nylon is allocated solely for military items such as parachutes, as shown in the photo, tents, and ponchos.

1946 Products with nonstick coating (PTFE), including bearings, bushings, gears, and cookware, become commercially available.

1959 Spandex, an elastic fiber, is commercially produced.

1988 The world's first polymer banknote is issued by the Reserve Bank of Australia. By 1996, all Australians use plastic money.

2006 Researchers develop a paper-thin, radiation-resistant, liquid-crystal polymer in which electronic circuits can be imbedded, making it useful in space applications.

2010 Scientists create a semiconducting, transparent polymer doped with fullerenes. Once the film absorbs light, it produces an electric charge. This ability could lead to applied technology, such as transparent solar cells that can be coated on windows.

1950 — 1980 — 2010

DIFFERENTIATED INSTRUCTION

Struggling Learners Some students will better understand the relationship between polymers and monomers if they can make models of monomers, join them together as a polymer, and then take them apart again. Ask students to model the process of polymerization using paper clips, safety pins, or beads that snap together. BL EL

CHEMISTRY PROJECT

Cross-Linking Have students research how cross-linking influences polymer properties such as flexibility and hardness. Have them summarize their findings, obtain samples of as many cross-linked polymers as possible, and make a class presentation summarizing their findings. AL

ASSESSMENT

Performance Have students devise and set up an apparatus for testing the relative strengths of different plastics. Have them perform similar tests on several samples. OL EL

Extension

Wrap It Up Ask students to find background information about the industrial process used to make plastic wrap. They should include a synopsis of their findings and make a poster showing a flowchart of the process. OL EL

ASSESSMENT

Performance Ask students to act out an addition polymerization reaction in which pairs of students represent monomers. In the performances, joined pairs of hands can represent two carbon atoms linked by a double bond, and releasing hands can simulate bond breaking. BL EL

Visual Learning

Table 14 Have students look around their homes and around a store that stocks home improvement materials and compile a list of polymers that they encounter in their daily lives. **OL**

What's CHEMISTRY Got To Do With It?

Video

The Plastics

Explore **polymers with an interactive table.** **Concepts In Motion**

Table 14 Common Polymers

Polymer	Applications	Structural Unit
Polyvinyl chloride (PVC)	Plastic pipes, meat wrap, upholstery, rainwear, house siding, garden hose 	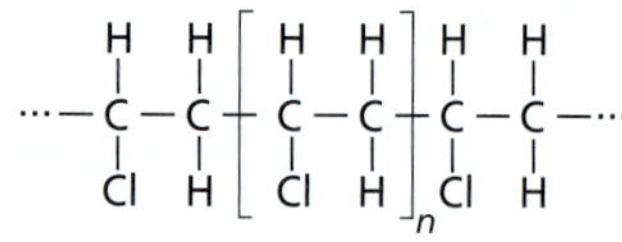 Polyvinyl chloride
Polyacrylonitrile	Fabrics for clothing and upholstery, carpet	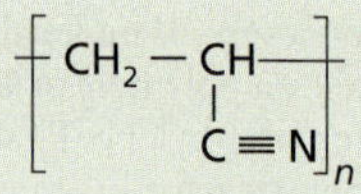
Polyvinylidene chloride	Food wrap, fabrics 	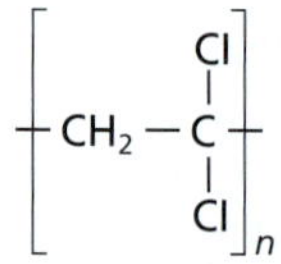
Polymethyl methacrylate	"Nonbreakable" (acrylic glass) windows, inexpensive lenses, art objects 	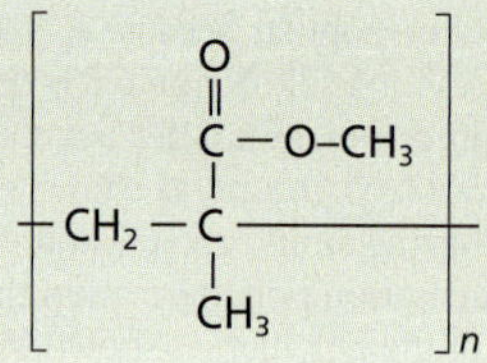
Polypropylene (PP)	Beverage containers, rope, netting, kitchen appliances	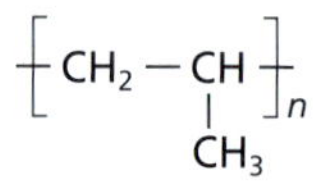
Polystyrene (PS) and styrene plastic	Foam packing and insulation, plant pots, disposable food containers, model kits 	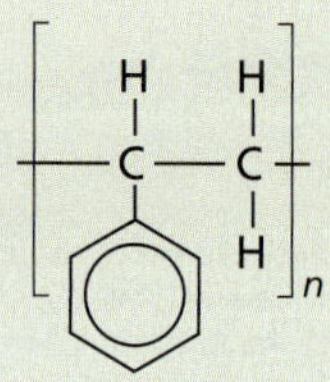
Polyethylene terephthalate (PETE)	Soft-drink bottles, tire cord, clothing, recording tape, replacements for blood vessels	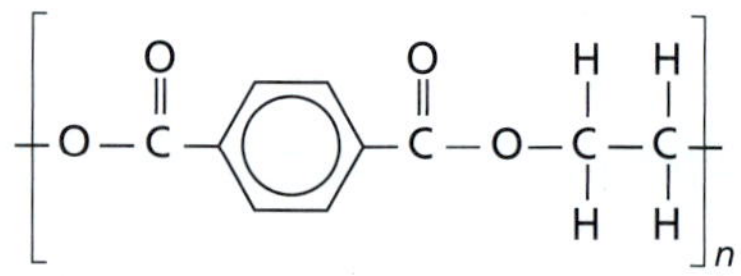
Polyurethane	Foam furniture cushions, waterproof coatings, parts of shoes	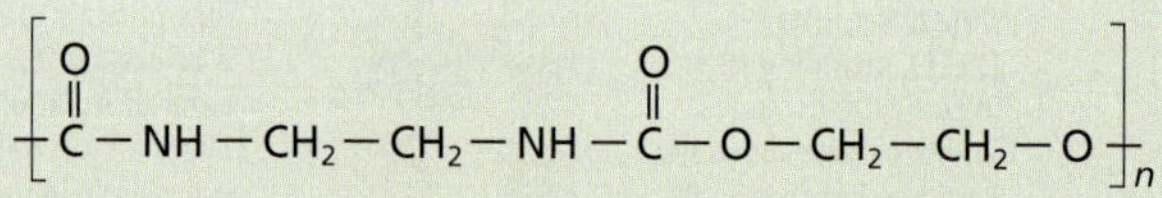

(t to b)Siede Preis/Getty Images, (2)Mark Steinmetz, (3)Dorling Kindersley/Getty Images, (4)©Lawrence Manning/Corbis

Demonstration

Polymers and Monomers

Purpose
to help students see the relationship between a polymer and a monomer

Materials
Spring-type clothespins (10), paper clips (1 box)

Safety Precautions

Disposal Reuse the materials for another demonstration.

Procedure
Clip one clothespin onto the leg of another one and continue this process to form a polymer chain. Each clothespin represents a monomer. Paper clips also can be used to represent the monomers and form a polymer chain by linking them together. If desired, make a paper-clip chain before the class period and place it in a small box; as students watch, add loose, individual paper clips to

Figure 20 Plastic lumber is made from recycled plastic, such as used soft-drink bottles, milk jugs, and other polyethylene waste.

Properties and Recycling of Polymers

Why do we use so many different polymers today? One reason is that they are easy to synthesize. Another reason is that the starting materials used to make them are inexpensive. Still another, more important, reason is that polymers have a wide range of properties. Some polymers can be drawn into fine fibers that are softer than silk, while others are as strong as steel. Polymers do not rust like steel does, and many polymers are more durable than natural materials such as wood. Fencing and decking materials made of plastic, like those shown in **Figure 20,** do not decay and do not need to be repainted.

Properties of polymers Another reason why polymers are in such great demand is that it is easy to mold them into different shapes or to draw them into thin fibers. It is not easy to do this with metals and other natural materials because they must be heated either to high temperatures, do not melt at all, or are too weak to be used to form small, thin items.

As with all substances, polymers have properties that result directly from their molecular structure. For example, polyethylene is a long-chain alkane. Thus, it has a waxy feel, does not dissolve in water, is nonreactive, and is a poor electrical conductor. These properties make it ideal for use in food and beverage containers and as an insulator in electrical wire and TV cable.

Polymers fall into two different categories, based on their melting characteristics. A **thermoplastic** polymer is one that can be melted and molded repeatedly into shapes that are retained when cooled. Polyethylene and nylon are examples of thermoplastic polymers. A **thermosetting** polymer is one that can be molded when it is first prepared, but after it cools, it cannot be remelted. This property is explained by the fact that thermosetting polymers begin to form networks of bonds in many directions when they are synthesized. By the time they have cooled, thermosetting polymers have become, in essence, a single large molecule. Bakelite is an example of a thermosetting polymer. Instead of melting, Bakelite decomposes when overheated.

READING CHECK **Compare and contrast** thermoplastic and thermosetting polymers.

CAREERS IN CHEMISTRY

Polymer Chemist Does the thought of developing new and better polymers sound inspiring and challenging to you? Polymer chemists develop new polymers and create uses or manufacturing processes for older ones.

WebQuest

VOCABULARY

WORD ORIGIN

Thermoplastic

thermo- comes from the Greek word *thermē* which means heat; *plastic* comes from the Greek word *plastikos* which means to mold or form

David R. Frazier Photolibrary, Inc.

Apply Chemistry

Polymer Foams Polyurethane foam is used as housing insulation and packaging material. One set of monomers used to make such a polymer is a polyalcohol (HO–R–OH) and a polyisocyanate (O=CN–R–NC=O). Amines or metal salts are used as catalysts in the industrial processes that produce this polymer. The polymer contains carbamate or urethane groups, which have the general structure –O–R–O–C(=O)–NH–R –NH–C(=O)–. Producing the polymer in the presence of a fluorocarbon foaming agent causes the foamy quality. The presence of multiple functional groups in the monomers results in a high level of cross-linking in the product, forming rigid foam.

CAREERS IN CHEMISTRY

WebQuest

READING CHECK Both are descriptions based on the melting characteristics of polymers. A thermoplastic can be melted and molded repeatedly. A thermosetting polymer can only be molded when it is first prepared. It cannot be remelted and remolded.

the box. Cover the box and shake it to simulate the reaction. Uncover the box and pull out the previously connected clips.

Results

The monomers link together to form a long-chain polymer.

Analysis

Ask these questions.

1. What is the generic name of the small molecules that link together? Monomers
2. If a polymer is made from ethylene monomers, what is the polymer called? Polyethylene

ASSESSMENT

Knowledge Ask students to write as many names as they can that begin with the prefix *poly*. Possible responses: polypropylene, polyethylene, polyvinyl chloride, polystyrene, polyester, polyvinylacetate OL EL

3 Assess

Check for Understanding

Ask students to draw the structures of the monomers needed to make all the addition polymers listed in Table 14. Polyvinyl chloride: $CH_2{=}CHCl$; polyacrylonitrile: $CH_2{=}CHCN$; polyvinylidene chloride: $CH_2{=}CCl_2$; polymethyl methacrylate: $CH_2{=}C(CH_3)(CO_2CH_3)$; polypropylene: $CH_3CH{=}CH_2$; polystyrene: $CH_2{=}CHC_6H_5$. **OL**

Reteach

Have students draw the structures of polymers they have not studied in this chapter. Monomers you might use include: methyl cyanoacrylate ($CH_2{=}C(CN)(CO_2CH_3)$), used to make instant glues; vinyl alcohol ($CH_2{=}CHOH$), used to make polyvinyl alcohol, a starting material for making water-soluble plastic packaging; chlorotrifluoroethylene ($CFCl{=}CF_2$), used to make a plastic explosive. **OL**

Extension

Point out that many useful polymers are made from more than one monomer structure. These polymers are referred to as copolymers. Ask students to predict the structure of polyester fiber or polymer balloons. They are made from the monomers terephthalic acid and ethylene glycol, respectively. The polyester fiber and polymer balloon contain a chain of the two monomers alternating with each other and connected by ester bonds. The formation of each ester bond was accompanied by the loss of a water molecule. **OL**

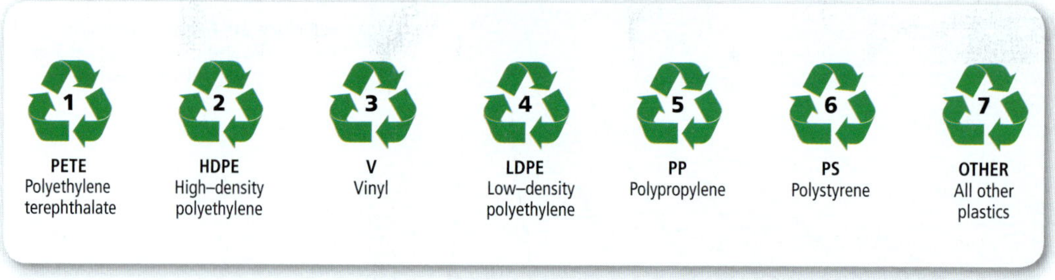

Figure 21 Codes on plastic products aid in recycling because they identify the composition of the plastic.

Watch a **video about recycling.**

Video

Recycling polymers The starting materials for the synthesis of most polymers are derived from fossil fuels. As the supply of fossil fuels becomes depleted, recycling plastics becomes more important. Recycling and buying goods made from recycled plastics decreases the amount of fossil fuels used, which conserves fossil fuels.

Currently, about 5% of the plastics used in the United States are recycled. Plastics recycling is somewhat difficult due to the large variety of different polymers found in products. Usually, the plastics must be sorted according to polymer composition before they can be reused. Thermosetting polymers are more difficult to recycle than thermoplastic polymers because only thermoplastic materials can be melted and remolded repeatedly. The task of separating plastics can be time-consuming and expensive. That is why the plastics industry and the government have tried to improve the process by providing standardized codes that indicate the composition of each plastic product. The standardized codes for plastics are shown in **Figure 21.** These codes provide a quick way for recyclers to sort plastics.

SECTION 5 REVIEW

Section Self-Check

Section Summary

- Polymers are large molecules formed by combining smaller molecules called monomers.
- Polymers are synthesized through addition or condensation reactions.
- The functional groups present in polymers can be used to predict polymer properties.

22. MAINIDEA Draw the structure for the polymer that could be produced from each of the following monomers by the method stated.

a. Addition

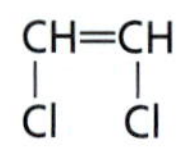

b. Condensation

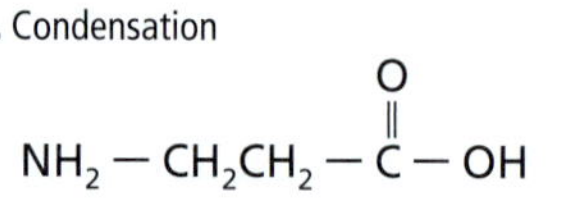

23. Label the following polymerization reaction as *addition* or *condensation*. Explain your answer.

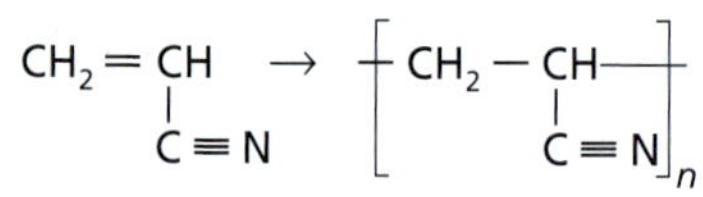

24. Identify Synthetic polymers often replace stone, wood, metals, wool, and cotton in many applications. Identify some advantages and disadvantages of using synthetic materials instead of natural materials.

25. Predict the physical properties of the polymer that is made from the following monomer. Mention solubility in water, electrical conductivity, texture, and chemical reactivity. Do you think it will be thermoplastic or thermosetting? Give reasons for your predictions.

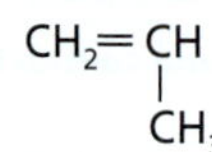

SECTION 5 REVIEW

22. **a.** Refer to Solutions Manual
    b. Refer to Solutions Manual
23. addition, because all of the atoms present in the monomer are retained in the monomer
24. synthetic materials often do not rot and decay like natural products, such as wood and cotton. Synthetic materials are easy to produce in desired shapes and sizes, such as synthetic stone. Synthetic materials usually do not rust or corrode like metals. Disadvantage: Synthetic structural products like plastic lumber are not as rigid and need more supports.
25. The polymer will have a waxy feel, low water solubility, low electrical conductivity, and low reactivity. It will be thermoplastic. It is a long-chain alkane similar to polyethylene.

everydayCHEMISTRY

Garlic: Pleasure and Pain

Did you know that the flavors of fresh and roasted garlic are very different? Fresh garlic, shown in **Figure 1,** contains substances that cause a burning sensation in your mouth. However, roasted garlic does not produce this sensation. These sensations, pleasure or pain, are results of chemical reactions.

When raw garlic is bruised, cut, or crushed, it produces a chemical called allicin, as shown in **Figure 2.** The production of allicin is a chemical defense mechanism for the garlic plant against other organisms. Allicin is an unstable compound and is converted to other compounds over time or when garlic is heated or roasted, which explains why roasted garlic does not cause the burning sensation in your mouth.

Figure 1 Fresh garlic contains a pain-producing chemical as a defense against predators.

Sensing temperature and pain Temperature and pain are sensed by neurons embedded in the skin, including the skin inside your mouth. These neurons have temperature-detecting molecules on their surfaces that are called transient receptor potential (TRP) ion channels. Different TRP channels are activated by different temperature ranges. For example, when a person touches something hot, some of the TRP ion channels open and allow charged calcium ions to enter the nerve cell. This increases the charge within the nerve cell. When the charge increases enough, an electrical signal is sent to the brain, where it is interpreted as a hot sensation.

Allicin also activates neurons. Allicin apparently acts on a pair of ion channel proteins called TRPA1 and TRPV1. When the chemical allicin is present, these channels allow ions to enter the nerve cell. The additional electric charge in the nerve cell signals the brain, where the signal is interpreted by the brain as a burning sensation.

Probing pain receptors While it is interesting to know why tasting raw garlic is painful, the understanding of how allicin causes that pain sensation is even more interesting and useful. Researchers hope that a further understanding of how these receptors work will lead to new methods for controlling chronic pain in patients.

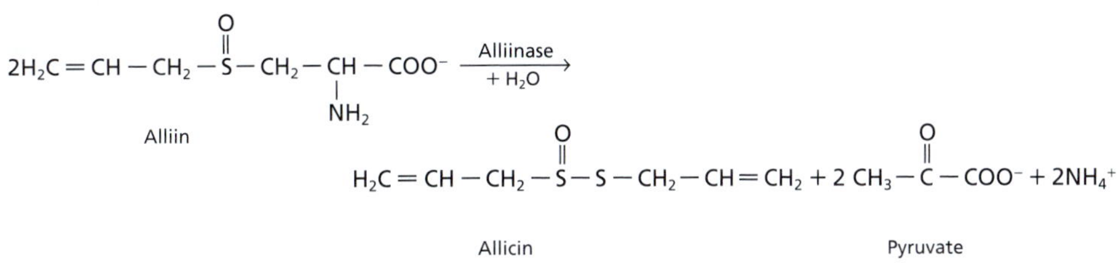

Figure 2 When garlic is bruised or damaged, alliin and the enzyme alliinase produce allicin. When you taste fresh garlic, neurons embedded in your mouth cause an electrical signal to be sent to your brain. The brain interprets the electrical signal as a burning sensation.

WRITING IN Chemistry

Research and prepare a poster that shows other chemical reactions in plants.

WebQuest

everydayCHEMISTRY

Purpose

Students will learn about recent research into the action of garlic on pain receptors in the mouth, and how that research might lead to new pain treatments.

Background

Much folklore surrounds the medicinal value of garlic, and most of that folklore is not backed by sound science. Medicinal claims range from curing the common cold to fighting cancer. While the research described in this article does suggest at least one possible benefit to garlic (the lowering of blood pressure), the researchers involved are careful to point out that the findings are preliminary and need to be investigated further.

Teaching Strategies

- Pain is an easily misunderstood topic. What is pain for? Is pain ever good? Consider different events, such as a bee sting (very painful, due to venom injected by the bee) and a tick bite (not painful, due to pain suppressors injected by the tick). Why the different response?
- Like animals, plants use many survival strategies that might involve pain or pleasure. Contrast an apple, whose taste is generally pleasant, with garlic, whose taste is undoubtedly unpleasant to most animals. Discuss why these two plants would have such different tastes.

WRITING IN Chemistry

Rubric WebQuest

Research Student posters will vary on topic and creativity. Make sure that students have adequately researched their topic and the poster clearly explains their findings.

SMALL SCALE ChemLAB

Rubric

Preparation

Time Allotment one class period

Process Skills acquire and analyze information, collect and interpret data, draw a conclusion, measure, observe and infer, predict, sequence

Safety Precautions Approve lab safety forms before work begins. Alcohols are flammable. Advise students to keep them away from open flames.

Alternate Materials

- Bathroom tissue can be used in place of facial tissue. However, do not use recycled paper products.
- Another longer, carbon-chain alcohol can be used, depending on what is available.

Procedure

- Place small marked beakers of the substances out for students.
- If time permits, students can repeat the trials to check their accuracy.

Expected Results Refer to the data table.

Evaporation Data			
Substance	Starting Temp (°C)	Temp after one minute (°C)	ΔT (°C)
Water	21	19	2
Methanol	22	8	14
Ethanol	22	13	9
2-Propanol	21	16	5

SMALL SCALE ChemLAB

Observe Properties of Alcohols

Background: Alcohols are organic compounds that contain the –OH functional group. How fast various alcohols evaporate indicates the strength of intermolecular forces in alcohols. The evaporation of a liquid is an endothermic process, absorbing energy from the surroundings. This means that the temperature will decrease as evaporation occurs.

Question: *How do intermolecular forces differ in three alcohols?*

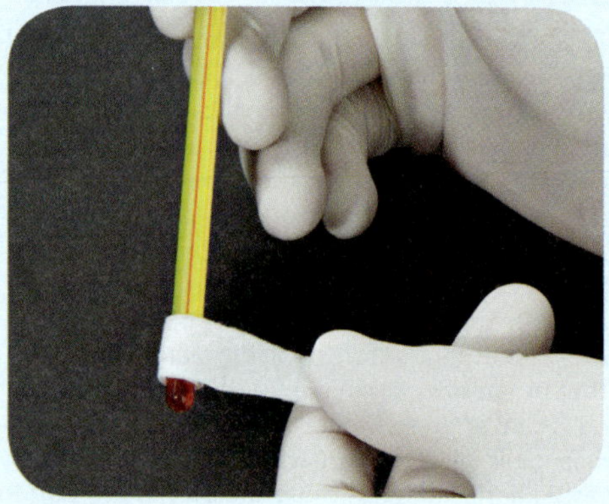

Matt Meadows

Materials

nonmercury thermometer
stopwatch
facial tissue
cloth towel
Beral pipettes (5)
methanol
ethanol (95%)
2-propanol (99%)
wire twist tie or small rubber band
piece of cardboard for use as a fan

Safety Precautions

WARNING: ***Alcohols are flammable. Keep liquids and vapors away from open flames and sparks.***

Procedure

1. Read and complete the lab safety form.
2. Prepare data tables for recording data.
3. Cut five 2-cm by 6-cm strips of tissue.
4. Place a thermometer on a folded towel lying on a flat table so that the bulb of the thermometer extends over the edge of the table. Make sure the thermometer cannot roll off the table.
5. Wrap a strip of tissue around the bulb of the thermometer. Secure the tissue with a wire twist tie placed above the bulb of the thermometer.
6. Choose one person to control the stopwatch and read the temperature on the thermometer. A second person will put a small amount of the liquid to be tested into a Beral pipette.
7. When both people are ready, squeeze enough liquid onto the tissue to completely saturate it. At the same time, the other person starts the stopwatch, reads the temperature, and records it in the data table.
8. Fan the tissue-covered thermometer bulb with a piece of cardboard or other stiff paper. After 1 min, read and record the final temperature in the data table. Remove the tissue and wipe the bulb dry.
9. Repeat Steps 5 through 8 for each of the three alcohols: methanol, ethanol, and 2-propanol.
10. Obtain the classroom temperature and humidity data from your teacher.
11. **Cleanup and Disposal** Place the used tissues in the trash. Pipettes can be reused.

Analyze and Conclude

1. **Observe and Infer** What can you conclude about the relationship between heat transfer and the differences in the temperature changes you observed?
2. **Evaluate** Molar enthalpies of vaporization (kJ/mol) for the three alcohols at 25°C are: methanol, 37.4; ethanol, 42.3; and 2-propanol, 45.4. What can you conclude about the relative strength of intermolecular forces existing in the three alcohols?
3. **Compare** Make a general statement comparing the molecular size of an alcohol in terms of the number of carbons in the carbon chain to the rate of evaporation of that alcohol.
4. **Observe and Infer** Compare your data with those of your classmates. Infer why there are differences.
5. **Error Analysis** Determine where errors might have been introduced in your procedure.

INQUIRY EXTENSION

Design an Experiment Suggest a way to make this experiment more quantitative and controlled. Design an experiment using your new method.

Analyze and Conclude

1. The greater the heat transfer during evaporation, the greater the temperature change.
2. Intermolecular forces increase as the length of the carbon chain increases. The heat of vaporization is a measure of the strength of these forces.
3. The rate of evaporation appears to decrease as the number of carbon atoms increases.
4. Differences in data might be due to differences in amount of air fanned over the thermometer, temperature and humidity in the various labs.
5. The tissue pieces might have varied in size. Air movement around the thermometer might have varied. The amount of alcohol used might have varied in each trial.

INQUIRY EXTENSION

Student might suggest that the same amount of alcohol should be added to the tissue paper for each trial. Care should be taken to ensure that the tissue papers are the same size for each trial. A small electric fan might be used to standardize the air movement around the thermometer.

CHAPTER 22 STUDY GUIDE

Vocabulary Practice

BIGIDEA The substitution of different functional groups for hydrogen atoms in hydrocarbons results in a diverse group of organic compounds.

SECTION 1 Alkyl Halides and Aryl Halides

MAINIDEA A halogen atom can replace a hydrogen atom in some hydrocarbons.

- The substitution of functional groups for hydrogen in hydrocarbons creates a wide variety of organic compounds.
- An alkyl halide is an organic compound that has one or more halogen atoms bonded to a carbon atom in an aliphatic compound.

VOCABULARY
- functional group
- halocarbon
- alkyl halide
- aryl halide
- plastic
- substitution reaction
- halogenation

SECTION 2 Alcohols, Ethers, and Amines

MAINIDEA Oxygen and nitrogen are two of the most-common atoms found in organic functional groups.

- Alcohols, ethers, and amines are formed when specific functional groups substitute for hydrogen in hydrocarbons.
- Because they readily form hydrogen bonds, alcohols have higher boiling points and higher water solubilities than other organic compounds.

VOCABULARY
- hydroxyl group
- alcohol
- denatured alcohol
- ether
- amine

SECTION 3 Carbonyl Compounds

MAINIDEA Carbonyl compounds contain a double-bonded oxygen in the functional group.

- Carbonyl compounds are organic compounds that contain the C=O group.
- Five important classes of organic compounds containing carbonyl compounds are aldehydes, ketones, carboxylic acids, esters, and amides.

VOCABULARY
- carbonyl group
- aldehyde
- ketone
- carboxylic acid
- carboxyl group
- ester
- amide
- condensation reaction

SECTION 4 Other Reactions of Organic Compounds

MAINIDEA Classifying the chemical reactions of organic compounds makes predicting products of reactions much easier.

- Most reactions of organic compounds can be classified into one of five categories: substitution, elimination, addition, oxidation-reduction, and condensation.
- Knowing the types of organic compounds reacting can enable you to predict the reaction products.

VOCABULARY
- elimination reaction
- dehydrogenation reaction
- dehydration reaction
- addition reaction
- hydration reaction
- hydrogenation reaction

SECTION 5 Polymers

MAINIDEA Synthetic polymers are large organic molecules made up of repeating units linked together by addition or condensation reactions.

- Polymers are large molecules formed by combining smaller molecules called monomers.
- Polymers are synthesized through addition or condensation reactions.
- The functional groups present in polymers can be used to predict polymer properties.

VOCABULARY
- polymer
- monomer
- polymerization reaction
- addition polymerization
- condensation polymerization
- thermoplastic
- thermosetting

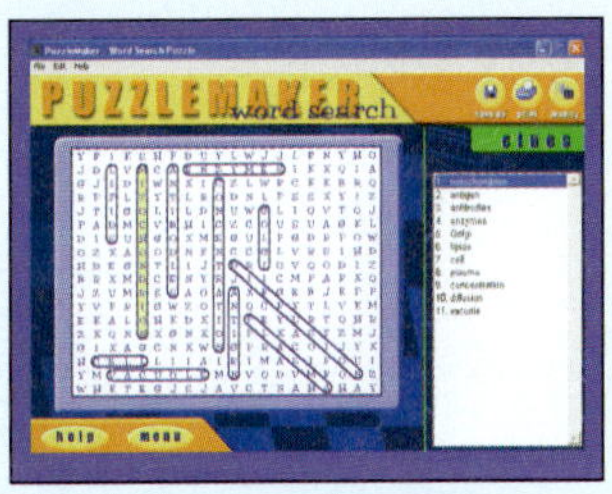

Vocabulary Practice

For additional practice with vocabulary, have students access the Vocabulary PuzzleMaker.

Use the Vocabulary

To reinforce chapter vocabulary, have students write a sentence using each term. OL EL

Review Strategies

- Have students make a list of the structures, properties, and uses of the substituted hydrocarbons studied in this chapter. OL
- Have students practice drawing the structure of a polymer given its monomer and vice versa. OL
- Problems from Supplemental Practice Problems appendix or the Supplemental Problem booklet can be used for review. OL

Students can review with these online resources.

Vocabulary Practice

Vocabulary eGames
Vocabulary eFlashcards
Vocabulary PuzzleMaker

Section Self-Check

Chapter Self-Check

Online Test Practice

Use *eAssessment* to:

- create multiple versions of tests
- edit existing questions and add your own questions
- build tests aligned with select state curriculum standards using built-in state curriculum tags
- track students' progress

SECTION 1

Mastering Concepts

26. A functional group is an atom or group of atoms in an organic molecule that always reacts in a certain way.

27. Alkyl halides have a halogen atom present on an aliphatic chain or ring of carbons, while aryl halides have a halogen atom directly bonded to a carbon on a benzene molecule or other aromatic ring.

28. bromine

29. **a.** 1-aminopentane
b. 1-aminoheptane
c. 2-aminopentane
d. 1-aminodecane

30. This trend is primarily because the halogens from fluorine to iodine have increasing numbers of electrons that lie farther from the halogen nucleus. These electrons shift easily and form temporary dipoles. The dipoles attract each other and the energy needed to separate them increases. As a result, the boiling points of halogen-substituted alkanes increase as the size of the halogen atom increases.

Mastering Problems

31. Refer to the Solutions Manual

32. Refer to the Solutions Manual

33. **a.** Refer to the Solutions Manual
b. yes; one carbon is attached to four different atoms or groups.
c. The middle carbon is chiral.

34. Refer to the Solutions Manual

35. Possible answers:
a. 1-chloropentane; 3-chloropentane
b. 1,2-difluoropropane; 1,3-difluoropropane; 2,2-difluoropropane
c. 1,2- or 1,1-dibromocyclopentane
d. 1-bromo-1-chloroethane

CHAPTER 22 ASSESSMENT

SECTION 1

Mastering Concepts

26. What is a functional group?

27. Describe and compare the structures of alkyl halides and aryl halides.

28. What reactant would you use to convert methane to bromomethane?

29. Name the amines represented by each of the condensed formulas.
a. $CH_3(CH_2)_3CH_2NH_2$
b. $CH_3(CH_2)_5CH_2NH_2$
c. $CH_3(CH_2)_2CH(NH_2)CH_3$
d. $CH_3(CH_2)_8CH_2NH_2$

30. Explain why the boiling points of alkyl halides increase in order going down the column of halides in the periodic table, from fluorine through iodine.

Mastering Problems

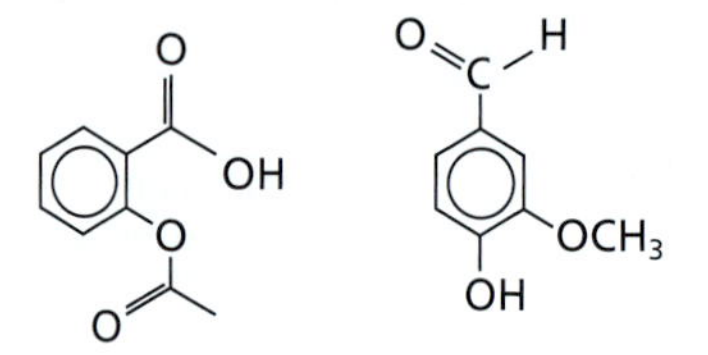

a Acetylsalicylic acid **b** Vanillin

■ **Figure 22**

31. Name and write the general formula of each of the functional groups attached to the benzene rings shown in **Figure 22.**

32. Draw structures for these alkyl and aryl halides.
a. chlorobenzene
b. 1-bromo-4-chlorohexane
c. 1,2-difluoro-3-iodocyclohexane
d. 1,3-dibromobenzene
e. 1,1,2,2-tetrafluoroethane

33. For 1-bromo-2-chloropropane:
a. Draw the structure.
b. Does the compound have optical isomers?
c. If the compound has optical isomers, identify the chiral carbon atom.

34. Draw and name all of the structural isomers possible for an alkyl halide with no branches and the molecular formula $C_5H_{10}Br_2$.

35. Name one structural isomer created by changing the position of one or more halogen atoms in each alkyl halide.
a. 2-chloropentane
b. 1,1-difluropropane
c. 1,3-dibromocyclopentane
d. 1-bromo-2-chloroethane

SECTION 2

Mastering Concepts

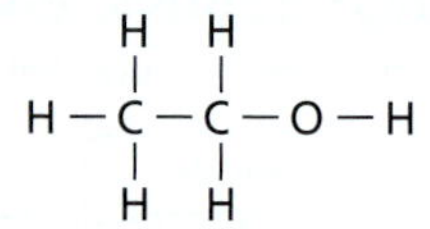

■ **Figure 23**

36. How is the compound shown in **Figure 23** denatured? What is the name of the compound?

37. Practical Applications Name one alcohol, amine, or ether that is used for each of the following purposes.
a. antiseptic
b. solvent in paint strippers
c. antifreeze
d. anesthetic
e. dye production

38. Explain why an alcohol molecule will always have a higher solubility in water than an ether molecule having an identical molecular mass.

39. Explain why ethanol has a much higher boiling point than aminoethane, even though their molecular masses are nearly equal.

Mastering Problems

40. Name one ether that is a structural isomer of each alcohol.
a. 1-butanol
b. 2-hexanol

41. Draw structures for the following alcohol, amine, and ether molecules.
a. 1,2-butanediol
b. 2-aminohexane
c. isopropyl ether
d. 2-methyl-1-butanol
e. butyl pentyl ether
f. cyclobutyl methyl ether
g. 1,3-diaminobutane
h. cyclopentanol

SECTION 3

Mastering Concepts

42. Draw the general structure for each of the following classes of organic compounds.
a. aldehyde
b. ketone
c. carboxylic acid
d. ester
e. amide

43. Common Uses Name an aldehyde, ketone, carboxylic acid, ester, or amide used for each of the following purposes.
a. preserving biological specimens
b. solvent in fingernail polish
c. acid in vinegar
d. flavoring in foods and beverages

44. What type of reaction is used to produce aspirin from salicylic acid and acetic acid?

SECTION 2

Mastering Concepts

36. Ethanol is denatured by the addition of small amounts of toxic substances, which make it unsafe to drink.

37. **a.** ethanol
b. methanol
c. ethylene glycol or propylene glycol
d. ethyl ether
e. aniline

38. Alcohols are always polar due to the asymmetrical distribution of charge around the oxygen in the –OH group. Polarity of an ether depends on its overall geometry. Alcohols are generally more soluble than corresponding ethers in water, which is a polar solvent.

39. Because O-H bonds are more polar than N-H bonds, the hydrogen bonds that form between two ethanol molecules are stronger than those that form between two aminoethane molecules. Stronger intermolecular forces result in higher boiling points.

Mastering Problems

40. **a.** ethyl ether; methyl propyl ether
b. propyl ether; isopropyl ether; butyl ethyl ether; methyl pentyl ether

41. Refer to the Solutions Manual.

Mastering Problems

45. Draw structures for each of the following carbonyl compounds.
- **a.** 2,2-dichloro-3-pentanone
- **b.** 4-methylpentanal
- **c.** isopropyl hexanoate
- **d.** octanamide
- **e.** 3-fluoro-2-methylbutanoic acid
- **f.** cyclopentanal
- **g.** hexyl methanoate

46. Name each of the following carbonyl compounds.

a.

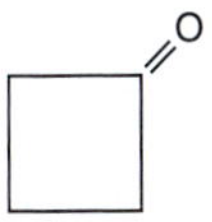

b. $CH_3-CH_2-CH_2-\overset{O}{\overset{\|}{C}}-H$

c. $CH_3\text{+}CH_2\text{+}_4\overset{O}{\overset{\|}{C}}-NH_2$

d. $CH_3\text{+}CH_2\text{+}_4\overset{O}{\overset{\|}{C}}-OH$

SECTION 4

Mastering Concepts

47. Synthetic Organic Compounds What is the starting material for making most synthetic organic compounds?

48. Explain the importance of classifying reactions.

49. List the type of organic reaction needed to perform each of the following transformations.
- **a.** alkene → alkane
- **b.** alkyl halide → alcohol
- **c.** alkyl halide → alkene
- **d.** amine + carboxylic acid → amide
- **e.** alcohol → alkyl halide
- **f.** alkene → alcohol

Mastering Problems

50. Classify each of the following organic reactions as substitution, addition, oxidation-reduction elimination, or condensation.
- **a.** 2-butene + hydrogen → butane
- **b.** propane + fluorine → 2-fluoropropane + hydrogen fluoride
- **c.** 2-propanol → propene + water
- **d.** cyclobutene + water → cyclobutanol

51. Use structural formulas to write equations for the following reactions.
- **a.** the substitution reaction between 2-chloropropane and water yielding 2-propanol and hydrogen chloride
- **b.** the addition reaction between 3-hexene and chlorine yielding 3,4-dichlorohexane

52. What type of reaction converts an alcohol into each of the following types of compounds?
- **a.** ester
- **b.** alkyl halide
- **c.** alkene
- **d.** aldehyde

53. Use structural formulas to write the equation for the condensation reaction between ethanol and propanoic acid.

SECTION 5

Mastering Concepts

54. Explain the difference between addition polymerization and condensation polymerization.

55. Which type of polymer is easier to recycle, thermosetting or thermoplastic? Explain your answer.

Mastering Problems

56. Manufacturing Polymers Draw the monomers that react to form each polymer.
- **a.** polyethylene
- **b.** polyvinyl chloride (PVC)
- **c.** polytetrafluoroethylene (PTFE)

57. Name the polymers made from the following monomers.
- **a.** $H_2C = CHCl$
- **b.** $CH_2 = CCl_2$

58. Choose the polymer of each pair that you expect to have the higher water solubility.

a.

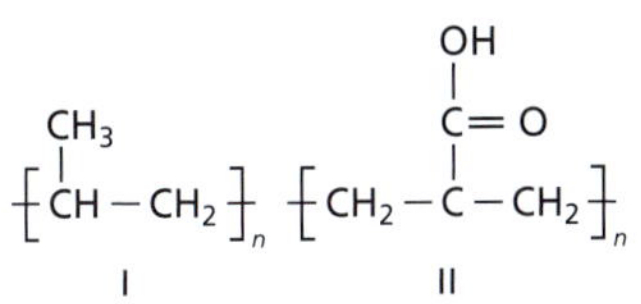

b.

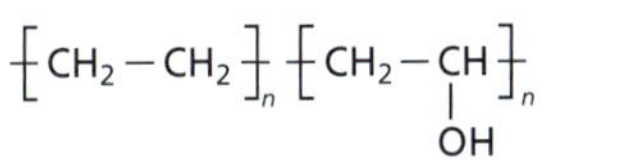

59. Examine the structures of the following polymers. Decide whether each is made by addition or condensation polymerization.
- **a.** nylon
- **b.** polyacrylonitrile
- **c.** polyurethane
- **d.** polypropylene

60. Human Hormones Which halogen is found in hormones made by a normal human thyroid gland?

SECTION 3

Mastering Concepts

42. Refer to the Solutions Manual.

43. **a.** formaldehyde
b. acetone
c. acetic acid
d. ethyl butanoate, 2-methylbutyl acetate, pentyl pentanoate, or other ester

44. condensation

Mastering Problems

45. Refer to the Solutions Manual.

46. **a.** cyclobutanone
b. butanal
c. hexanamide
d. hexanoic acid

SECTION 4

Mastering Concepts

47. fossil fuels such as petroleum

48. Because chemical reactions are so numerous, classifying them helps students and chemists better understand them, remember them, and predict the products of new reactions.

49. **a.** addition
b. substitution
c. elimination
d. condensation
e. substitution
f. addition and hydration

Mastering Problems

50. **a.** addition
b. substitution
c. elimination
d. addition

51. **a.** $CH_3CHClCH_3 + H_2O \rightarrow CH_3CH(OH)CH_3 + HCl$
b. $CH_3CH_2CH{=}CHCH_2CH_3 + Cl_2 \rightarrow CH_3CH_2CH(Cl)CH(Cl)CH_2CH_3$

52. **a.** condensation
b. substitution
c. elimination
d. oxidation

53. Refer to the Solutions Manual.

SECTION 5

Mastering Concepts

54. In addition polymerization, all the atoms in the monomers are retained in the polymer product. In condensation polymerization, monomers with at least two functional groups form the polymer, losing a small by-product such as water.

55. Thermoplastic polymers are easier to recycle because products made from them can be remelted and molded repeatedly.

Mastering Problems

56. Refer to the Solutions Manual.

57. **a.** polyvinyl chloride
b. polyvinylidene chloride

58. **a.** the second polymer
b. the second polymer

59. **a.** condensation polymerization
b. addition polymerization
c. condensation polymerization
d. addition polymerization

60. iodine

MIXED REVIEW

61. Carboxylic acids are weakly acidic, taste sour, and consist of polar molecules.

62. Refer to the Solutions Manual.

63. a. alkene
b. alkyl halide
c. alcohol
d. alcohol

64. Refer to Table 14 for answers.

65. Refer to the Solutions Manual for structures.
a. CH_3CH_2OH, ethanol
b. CH_3CH_3, ethane
c. CH_3CH_2Cl, chloroethane
d. $CH_2(F)CH_2F$, 1,2-difluoroethane

66. Refer to the Solutions Manual.

THINK CRITICALLY

67. a. carboxyl, amino, hydroxyl (2)
b. carbonyl (2), C=C

68. Ethanoic acid is soluble in water because its molecules are relatively small, form hydrogen bonds with water when un-ionized, and form ion-dipole attractions when ionized. Carboxylic acid molecules with much longer carbon chains are mostly nonpolar. These nonpolar molecules do not form strong bonds with water molecules, even though their carboxylic acid groups have a slight tendency to interact with water molecules.

69. Students should provide molecular structures for the following compounds. Refer to the Solutions Manual for structures.
a. 1-propanol, 2-propanol, ethyl methyl ether
b. 1,1-dichloroethane and 1,2-dichloroethane

70. hydroxyl (4), cyclic alkene C=C, cyclic ester

71. Check student drawings

72. Refer to the Solutions Manual.

73. a. hydrogen bonds
b. As the size of an alcohol increases, its water solubility decreases.
c. As the number of carbon atoms in an alcohol increases, the size of its nonpolar portion increases while its polar portion remains the same. As a result, its solubility in the polar water molecule decreases.

74. fractional distillation; substitution using UV light/Cl_2; elimination of HCl; addition of H_2O (hydration); oxidation

MIXED REVIEW

61. Describe the properties of carboxylic acids.

62. Draw structures of the following compounds.
a. butanone
b. propanal
c. hexanoic acid
d. heptanamide

63. Name the type of organic compound formed by each of the following reactions.
a. elimination from an alcohol
b. addition of hydrogen chloride to an alkene
c. addition of water to an alkene
d. substitution of a hydroxyl group for a halogen atom

64. List two uses for each of the following polymers.
a. polypropylene
b. polyurethane
c. polytetrafluoroethylene
d. polyvinyl chloride

65. Draw structures of and supply names for the organic compounds produced by reacting ethene with each of the following substances.
a. water
b. hydrogen
c. hydrogen chloride
d. fluorine

66. Environmentally-Safe Propellants Hydrofluoroalkanes (HFAs) are replacing chlorofluorocarbons in hand-held asthma inhalers, because of CFC damage to the ozone layer. Draw the structures of the HFAs listed below.
a. 1,1,1,2,3,3,3-heptafluoropropane
b. 1,1,1,2-tetrafluoroethane

THINK CRITICALLY

67. Interpret Scientific Illustrations List all the functional groups present in each of the following complex organic molecules.

a.

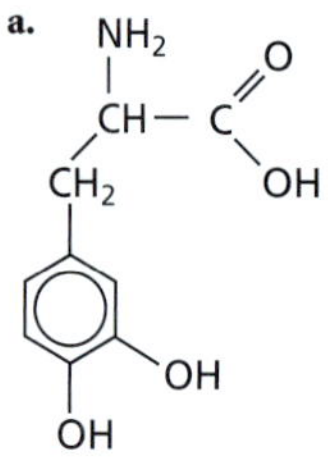

Levadopa

b.

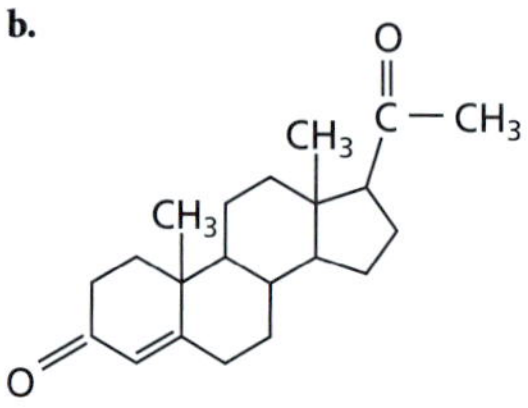

Progesterone

68. Evaluate Ethanoic acid (acetic acid) is very soluble in water. However, naturally occurring long-chain carboxylic acids, such as palmitic acid ($CH_3(CH_2)_{14}COOH$), are insoluble in water. Explain.

69. Communicate Write structural formulas for all structural isomers of molecules having the following formulas. Name each isomer.
a. C_3H_8O
b. $C_2H_4Cl_2$

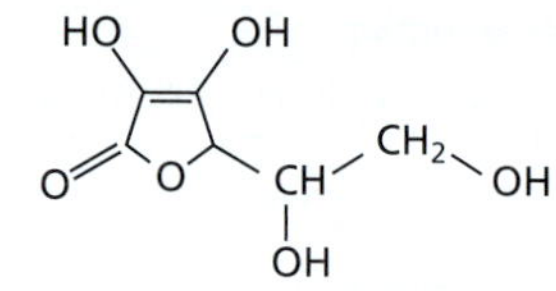

Vitamin C

Figure 24

70. Interpret Scientific Illustrations Human cells require vitamin C to properly synthesize materials that make up connective tissue such as that found in ligaments. List the functional groups present in the vitamin C molecule shown in **Figure 24.**

71. Identify Draw the structure of an example of an organic molecule that has four carbons and falls into each of the compound types listed.
a. ester
b. aldehyde
c. ether
d. alcohol

72. Predict A monohalogenation reaction describes a substitution reaction in which a single hydrogen atom is replaced by a halogen. A dihalogenation reaction is a reaction in which two hydrogen atoms are replaced by two halogen atoms.
a. Draw the structures of all the possible monohalogenation products that can form when pentane reacts with Cl_2.
b. Draw the structures of all the possible dihalogenation products that can form when pentane reacts with Cl_2.

Table 15 Alcohol Solubility in Water (mol/100 g H_2O)

Name	Alcohol	Solubility
Methanol	CH_3OH	infinite
Ethanol	C_2H_5OH	infinite
Propanol	C_3H_7OH	infinite
Butanol	C_4H_9OH	0.11
Pentanol	$C_5H_{11}OH$	0.030
Hexanol	$C_6H_{13}OH$	0.0058
Heptanol	$C_7H_{15}OH$	0.0008

73. Evaluate Examine **Table 15** comparing some alcohols and their solubility in water. Use the table to answer the following questions.
a. What type of bond forms between the –OH group of alcohols and water?
b. State a relationship between water solubility and alcohol size from the data in the table.
c. Provide an explanation for the relationship you stated in Part b.

Chapter Self-Check

74. Recognize Most useful organic molecules are made from raw materials using several steps. This is called a multistep synthesis pathway. Label the types of reaction or process taking place in each step of the multistep synthesis pathway below.

petroleum → ethane → chloroethane → ethene → ethanol → ethanoic (acetic) acid

CHALLENGE PROBLEM

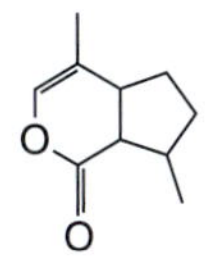

■ **Figure 25**

75. Animal Pheromones Catnip contains an organic chemical known as *nepetalactone,* shown in **Figure 25,** that is thought to mimic feline sex pheromones. Cats will rub in it, roll over it, paw at it, chew it, lick it, leap about, then purr loudly, growl, and meow for several minutes before losing interest. It takes up to two hours for the cat to "reset" and then have the same response to the catnip.

a. What type of organic compound is nepetalactone?
b. Draw the structural formula for nepetalactone on a sheet of paper and then draw in all the missing hydrogen atoms. Remember that carbon atoms must have four bonds to be stable.
c. Write the molecular formula for nepetalactone.

CUMULATIVE REVIEW

76. Explain why the concentration of ozone over Antarctica decreases at about the same time every year.

77. Why do the following characteristics apply to transition metals?
a. Ions vary in charge.
b. Many of their solids are colored.
c. Many are hard solids.

78. Determine the number of atoms in each of the following.
a. 56.1 g Al **b.** 2 moles C

79. What is a rate-determining step?

80. According to Le Châtelier's principle, how would increasing the volume of the reaction vessel affect the equilibrium $2SO_2(g) + O_2(g) \leftrightarrows 2SO_3(g)$?

81. Compare and contrast saturated and unsaturated hydrocarbons.

WRITING IN ▶ Chemistry

82. Historical Perspective Write a short story describing how your life would differ if you lived in the 1800s, before the development of synthetic polymers.

DBQ Document-Based Questions

Pharmaceutical Propellants *Many inhaled medications used to treat asthma contained chlorofluorocarbons (CFCs). However, the Montreal Protocol called for a ban of CFCs as a propellant in pharmaceutical products by 2008. Two hydrofluoroalkanes (HFAs) appear to be effective in delivering asthma medications to the lungs. However, the medication dosage had to be cut in half with the new HFA propellents.*

Figure 26 *shows the concentration after one dose of the drug beclomethasone in the blood of volunteers using a CFC or an HFA propellant in the inhaler.*

Data obtained from: Anderson, P.J. 2006. *Chest: The Cardiopulmonary and Critical Care Journal.* 120:89–93

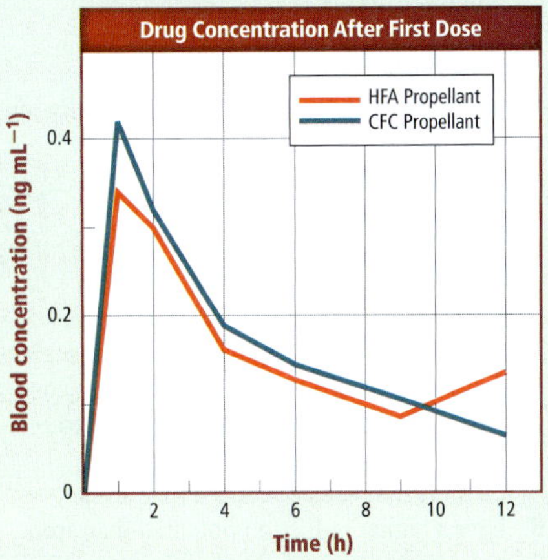

■ **Figure 26**

83. After one dose of the drug beclomethasone was given, which propellant resulted in the highest concentration of medication in the blood, HFA or CFC?

84. When does the drug reach its peak concentration?

85. Only one-half the amount of medication is needed with the HFA propellant when compared to the CFC propellant to achieve a similar blood-concentration level. Infer the advantages of using a lower dose of medication to get similar results.

CHALLENGE PROBLEM

75. a. ester
b. Refer to the Solutions Manual.
c. $C_{10}H_{14}O_2$

CUMULATIVE REVIEW

76. The same weather (temperature) patterns occur at about the same time every year.

77. a. Transition metals can lose ns^2 outer electrons forming 2+ ions. They also can lose inner d electrons and form ions of 3+ or higher.
b. Electrons in d orbitals can absorb visible light of specific wavelengths as the atom moves to an excited state.
c. Hardness is a property determined by the number of unpaired d electrons: the more unpaired d electrons, the harder the solid.

78. a. 1.25×10^{24} atoms
b. 1.20×10^{24} atoms

79. The rate-determining step is the slowest of the elementary reactions that make up a complex reaction.

80. The equilibrium will shift to the left because it has more moles of gas than the right.

81. A saturated hydrocarbon contains only single bonds. An unsaturated hydrocarbon contains at least one double or triple bond between carbon atoms.

WRITING IN ▶ Chemistry

Rubric

82. Student answers should include discussing what alternatives they would have to use in place of the many synthetic polymers in everyday use today, such as plastic bags, rubber bands, nylon and polyester fabrics, and plastic bottles.

DBQ Document-Based Questions

Data obtained from: Anderson, P.J. 2006. *Chest: The Cardiopulmonary and Critical Care Journal.* 120: 89–93.

83. HFA
84. at about 1 hour
85. If the patient uses one-half the dosage, the patient is at a lower risk of experiencing side effects from the drug.

MULTIPLE CHOICE

1. C
2. B
3. A
4. D
5. B
6. B
7. A
8. A
9. A
10. D
11. A

MULTIPLE CHOICE

1. What are the products of this reaction?

$CH_3CH_2CH_2Br + NH_3 \rightarrow ?$

A. $CH_3CH_2CH_2NH_2Br$ and H_2
B. $CH_3CH_2CH_2NH_3$ and Br_2
C. $CH_3CH_2CH_2NH_2$ and HBr
D. $CH_3CH_2CH_3$ and NH_2Br

2. What kind of reaction is this?

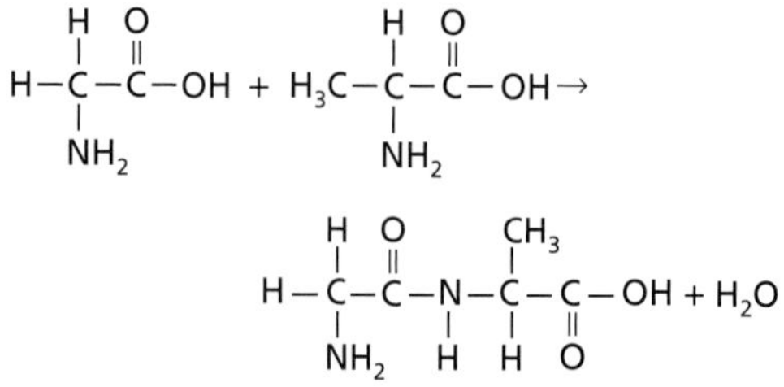

A. substitution
B. condensation
C. addition
D. elimination

3. What are the oxidation numbers of the elements in $CuSO_4$?

A. Cu = +2, S = +6, O = −2
B. Cu = +3, S = +5, O = −2
C. Cu = +2, S = +2, O = −1
D. Cu = +2, S = 0, O = −2

4. The corrosion, or rusting, of iron is an example of a naturally occurring voltaic cell. To prevent corrosion, sacrificial anodes are sometimes attached to rust-susceptible iron. Sacrificial anodes must

A. be more likely to be reduced than iron.
B. have a higher reduction potential than iron.
C. be more porous and abraded than iron.
D. lose electrons more easily than iron.

5. What type of compound does this molecule represent?

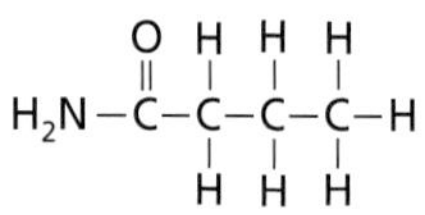

A. amine
B. amide
C. ester
D. ether

6. Diprotic succinic acid ($H_2C_4H_4O_4$) is an important part of the process that converts glucose to energy in the human body. What is the K_a expression for the second ionization of succinic acid?

A. $K_a = [H_3O^+][HC_4H_4O_4^-] / [H_2C_4H_4O_4]$
B. $K_a = [H_3O^+][C_4H_4O_4^{2-}] / [HC_4H_4O_4^-]$
C. $K_a = [H_2C_4H_4O_4] / [H_3O^+][HC_4H_4O_4^-]$
D. $K_a = [H_2C_4H_4O_4] / [H_3O^+][C_4H_4O_4^{2-}]$

Use the figure below to answer Question 7.

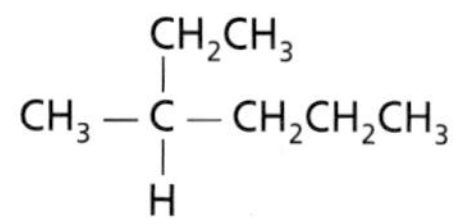

7. Which is the correct name for this compound?

A. 3-methylhexane
B. 2-ethylpentane
C. 2-propylbutane
D. 1-ethyl-1-methylbutane

8. A strip of metal X is immersed in a 1*M* solution of X^+ ions. When this half-cell is connected to a standard hydrogen electrode, a voltmeter reads a positive reduction potential. Which is true of the X electrode?

A. It accepts electrons more readily than H^+ ions.
B. It is undergoing oxidation.
C. It is adding positive X^+ ions to its solution.
D. It acts as the anode in the cell.

9. What is the mass of one formula unit of barium hexafluorosilicate ($BaSiF_6$)?

A. 4.64×10^{-22} g
B. 1.68×10^{26} g
C. 2.16×10^{21} g
D. 6.02×10^{-23} g

10. Which type of compound accepts H^+ ions?

A. an Arrhenius acid
B. an Arrhenius base
C. a Brønsted-Lowry acid
D. a Brønsted-Lowry base

11. Which substituted hydrocarbon has the general formula R–OH?

A. alcohol
B. amine
C. ketone
D. carboxylic acid

SHORT ANSWER

Use the figure below to answer Questions 12 and 13.

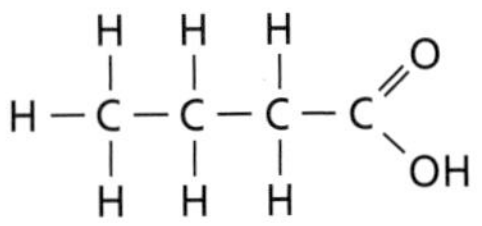

12. What is the functional group present in this compound?

13. Give the name for this compound.

EXTENDED RESPONSE

Use the graph below to answer Question 14.

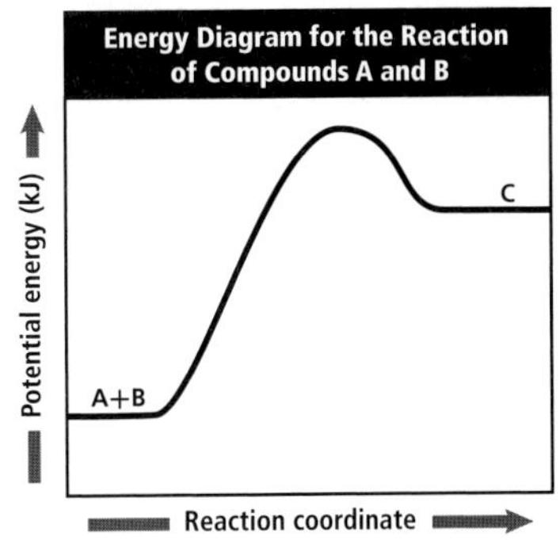

14. Discuss the reaction that results in the shape of the energy graph shown.

Use the figure below to answer Question 15.

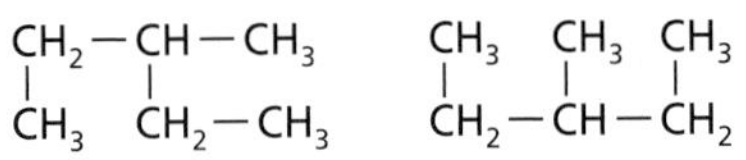

15. The two structures above both have the molecular formula C_6H_{14}. Are they isomers of one another? Explain how you can tell.

SAT SUBJECT TEST: CHEMISTRY

16. To electroplate an iron fork with silver,
- **A.** the silver electrode must have more mass than the fork.
- **B.** the iron fork must act as the anode in the cell.
- **C.** electric current must be applied to the iron fork.
- **D.** iron ions must be present in the cell solution.
- **E.** the electric current must be pulsed.

17. Which type of reaction is shown below?

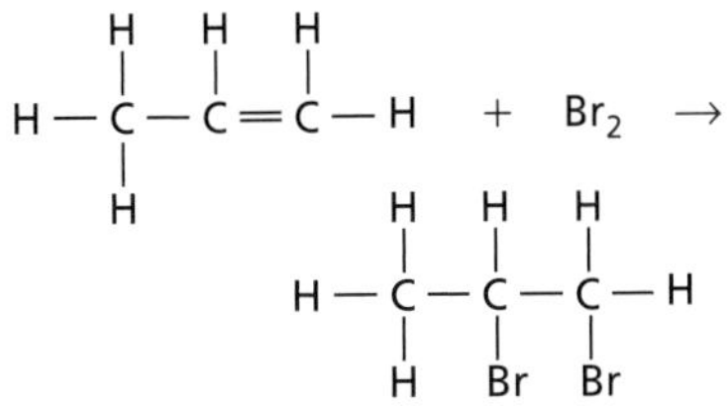

- **A.** condensation
- **B.** dehydration
- **C.** polymerization
- **D.** halogenation
- **E.** hydration

Use the table below to answer Question 18.

Experimental Data for A + B → C

Time	[A]*M*	[B]*M*	[C]*M*
0.00 sec	0.35	0.50	0.00
5.00 sec	0.15	0.30	0.40

18. Which is the rate of this reaction in terms of product produced in mol/(L•s)?
- **A.** 0.40 mol/(L•s)
- **B.** 0.85 mol/(L•s)
- **C.** 0.08 mol/(L•s)
- **D.** 0.17 mol/(L•s)
- **E.** 0.93 mol/(L•s)

NEED EXTRA HELP?

If You Missed Question . . .	1	2	3	4	5	6	7	8	9	10	11	12	13	14	15	16	17	18
Review Section . . .	22.4	22.4	19.1	20.1	22.2	18.2	21.2	20.1	10.3	18.1	22.2	22.1	22.3	16.1	21.4	20.1	22.4	16.3

SHORT ANSWER

12. carboxyl goup

13. butanoic acid

EXTENDED RESPONSE

14. This is an endothermic reaction, where energy must be added. This is shown by the energy level of the products being higher than the energy level of the reactants. The rise in the middle of the graph is due to the activation energy required to get the particles in the reaction into the proper orientation to react.

15. These are not isomers. Isomers have the same molecular formula but different geometrical forms. Although these two structures appear different, they both have the IUPAC name 3-methyl pentane. They are the same compound but viewed in different orientations.

SAT SUBJECT TEST: CHEMISTRY

16. C

17. D

18. C

CHAPTER 23 The Chemistry of Life

BIGIDEA **Biological molecules—proteins, carbohydrates, lipids, and nucleic acids—interact to carry out activities necessary to living cells.**

ESSENTIAL QUESTIONS	RESOURCES TO ASSESS MASTERY
SECTION 1 **Proteins** **1.** How can the structures of amino acids and proteins be described? **2.** What are the roles of proteins in cells? 1 session 0.5 block	**Progress Monitoring** Formative Assessment, pp. 827, 831 Reading Check, pp. 827, 828, 830 Section Review, p. 831
SECTION 2 **Carbohydrates** **1.** How can the structures of monosaccharides, disaccharides, and polysaccharides be described? **2.** What are the functions of carbohydrates in living things? 1 session 0.5 block	**Progress Monitoring** Formative Assessment, pp. 832, 833 Reading Check, p. 833 Section Review, p. 834
SECTION 3 **Lipids** **1.** How can the structures of fatty acids, triglycerides, phospholipids, and steroids be described? **2.** What are the functions of lipids in living organisms? **3.** What are some reactions that fatty acids undergo? **4.** How are the structure and function of cell membranes related? 2 sessions 1 block	**Progress Monitoring** Formative Assessment, p. 837 Reading Check, pp. 835, 836 Section Review, p. 839
SECTION 4 **Nucleic Acids** **1.** What are the structural components of nucleic acids? **2.** How is the function of DNA related to its structure? **3.** What are the structure and function of RNA? 1 session 0.5 block	**Progress Monitoring** Formative Assessment, pp. 840, 841, 842 Reading Check, p. 841 Section Review, p. 843
SECTION 5 **Metabolism** **1.** How do anabolism and catabolism compare? **2.** What is the role of ATP in metabolism? **3.** How can the processes of photosynthesis, cellular respiration, and fermentation be compared and contrasted? 1 session 0.5 block	**Progress Monitoring** Formative Assessment, pp. 846, 847, 848 Reading Check, p. 845 Section Review, p. 848 **Summative Assessment** Chapter Assessment, p. 852 *eAssessment* Chapter Test (Scaffolded)

Go online!

LEVELED RESOURCES	LAB MATERIALS
Science Notebook 23.1 OL ***FAST FILE* Chapter Resources:** ChemLab Worksheet, p. 63 OL Study Guide, p. 76 BL	LaunchLAB p. 824 / **20 min** 400-mL beaker, water, hot plate, graduated cylinder, 10% glucose solution, test tube, Benedict's solution, stirring rod, boiling chip, tongs, 10% starch solution, 10% gelatin suspension, honey suspended in water ChemLAB p. 850 / **45 min** red-skin potato pulp, 3% hydrogen peroxide solution, water, 250-mL beakers, test tubes, test-tube rack, test-tube clamp, 25-mL graduated cylinder, thermometer, ice, ruler, clock, hot plate, raw fresh liver pulp
Science Notebook 23.2 OL ***FAST FILE* Chapter Resources:** Study Guide, p. 78 BL	
Science Notebook 23.3 OL ***FAST FILE* Chapter Resources:** MiniLab Worksheet, p. 62 OL Study Guide, p. 79 BL	MiniLAB p. 837 / **30 min** 250-mL beaker, hot plate, solid vegetable shortening, 25-mL graduated cylinder, ethanol, 6.0*M* NaOH, gloves, stirring rod, tongs, cold water bath, 600-mL beaker, saturated NaCl solution, cheesecloth-lined funnel, evaporating dish
Science Notebook 23.4 OL ***FAST FILE* Chapter Resources:** Study Guide, p. 80 BL	
Science Notebook 23.5 OL ***FAST FILE* Chapter Resources:** Study Guide, p. 81 BL	

ADDITIONAL RESOURCES

- eAssessment
- ConnectEd Student Center
- ConnectEd Student Center
- Lesson Presentations
- eSolutions
- Videos
- Animations
- Online Quizzes
- Virtual Investigations
- Multilingual eGlossary
- Personal Tutors
- WebQuests
- Applying Practices Worksheets
- PBLs

Practice:

- Challenge Problems AL
- Supplemental Problems BL OL
- Chapter Test (Scaffolded)
- Standardized Test Practice

Additional Resources:

- Blueprints for Success
- Solving Problems: A Chemistry Handbook

Lab Resources:

- Laboratory Manual OL
- CBL Laboratory Manual OL
- Small-Scale Laboratory Manual OL
- Forensics Laboratory Manual OL

BL Below Level   OL On Level   AL Advanced Learners   EL English Learners   COOP LEARN Cooperative Learning

CHAPTER 23

LaunchLAB

Rubric

How do you test for simple sugars?

Purpose Students will test different foods for the presence of a simple sugar.

Safety Precautions Approve lab safety forms before work begins.

Disposal Contents of the test tube can be flushed down the sink drain with water.

Teaching Strategies

- Use this investigation to make students aware of a possible career choice as a food scientist. Show them several detailed labels that are present on today's food containers. Food scientists provide this information to inform consumers. Discuss with students why knowledge of biochemistry is useful in making informed choices about the foods they consume.
- See the Teacher Handbook for preparation of solutions.

Expected Results The glucose solution and the honey cause the blue Benedict's solution to turn yellow. This is a positive test for a monosaccharide or simple sugar.

CHAPTER 23
The Chemistry of Life

BIGIDEA Biological molecules—proteins, carbohydrates, lipids, and nucleic acids—interact to carry out activities necessary to living cells.

SECTIONS

LaunchLAB

How do you test for simple sugars?

Many different food sources supply the energy that your body uses constantly. This energy is stored in the bonds of molecules called simple sugars. In this lab, you will test several mixtures to determine if a simple sugar is present.

FOLDABLES Study Organizer

Biological Molecules

Make a concept-map book. Label it as shown. Use it to help you organize information about biological molecules.

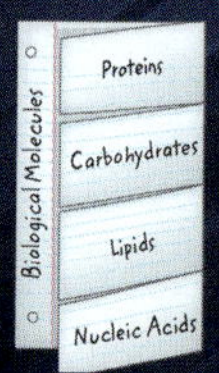

Procedure

1. Read and complete the lab safety form.
2. Fill a **400-mL beaker** until it is one-third full of **water.** Place it on a **hot plate** and heat it to boiling.
3. Use a **graduated cylinder** to measure 5.0 mL **10% glucose solution.** Pour it into a **test tube.**
4. Add 3.0 mL **Benedict's solution** to the test tube. Mix the two solutions using a **stirring rod.** Add a **boiling chip** to the test tube. **WARNING:** ***Benedict's solution is an eye and skin irritant.***
5. Using **tongs,** place the test tube in the boiling water bath for 5 min.
6. A color change to yellow or orange indicates the presence of a simple sugar. Record your observations.
7. Repeat the procedure using a **10% starch solution,** a **10% gelatin suspension,** and a few drops of **honey suspended in water.**

Analysis

1. **Describe** the color changes you observed. A color change from blue to yellow was observed in the test tubes containing glucose and honey. The glucose solution served as a positive control. There was no color change in the test tubes containing starch and gelatin.
2. **Classify** Which foods contained a simple sugar? glucose and honey

Inquiry Think about what you had for dinner last night. Which foods likely contained simple sugars? How could you test the food to find out? Answers will vary.

Fats provide more than twice the energy per gram as carbohydrates and proteins. Special lipids, called phospholipids, make up the cellular membranes of living cells.

Introduce the BIGIDEA

Biological Molecules To introduce this chapter's Big Idea, have students brainstorm about proteins, carbohydrates, lipids, and nucleic acids. Ask students what they know about proteins and which foods are good sources of proteins. Accept all reasonable responses about protein knowledge. Possible food sources of proteins are: meats, eggs, cheeses, nuts, and beans. Ask students what they know about carbohydrates. Accept all reasonable responses. Possible food sources are: pastas and potatoes. Continue the discussion with similar questions about lipids or fats and nucleic acids. Accept all reasonable responses.

Tie to Previous Knowledge

Have students review the following concepts before studying this chapter.

- hydrogen bonds
- isomers
- functional groups, polymers

Use the Photo

The Human Body Ask students to look closely at the chapter opener photos. Ask students how the photo showing connective tissue is related to the topics in this chapter. Connective tissue is made of proteins. Ask students how the photo of fat cells is related to the topics in this chapter. Fat cells are examples of cells that contain lipids. Ask students how the photo of skin cells is related to the topics in this chapter. Skin cells are examples of cells that contain nucleic acids.

1 Focus

MAIN IDEA

The Function of Proteins Ask students where proteins are found in their bodies. Accept all reasonable responses. Ask a student volunteer to read the caption to Figure 1. After the volunteer has read the question, ask the same question again. possible responses: hair, nails, muscles, and organs **OL**

2 Teach

Content Background

Protein Proteins were first discovered in the early nineteenth century. The name protein is from the Greek word *proteios* which means *primary*.

There are many different proteins found in living organisms. In fact, proteins are species specific and organ specific. The proteins of one species differ from the proteins found in another species and the organs within a single organism are made up of different types of proteins.

SECTION 1

Proteins

MAIN IDEA Proteins perform essential functions, including structural support, transport of materials, muscle contractions, and regulation of chemical reactions.

CHEM 4 YOU Some cleaning products, such as contact lens cleaning solution, contain enzymes. Did you ever wonder what an enzyme was?

Essential Questions

- How can the structures of amino acids and proteins be described?
- What are the roles of proteins in cells?

Review Vocabulary

polymer: large molecules composed of many repeating units called monomers

New Vocabulary

protein
amino acid
peptide bond
peptide
denaturation
enzyme
substrate
active site

Protein Structure

Enzymes form a class of proteins. **Proteins** are organic polymers made of amino acids linked together in a specific order. Proteins are not just large, randomly arranged chains of amino acids. To function properly, each protein must be folded into a specific three-dimensional structure. All living organisms, including the mountain goat and the plants shown in **Figure 1,** are composed of proteins. In this section, you will read about how proteins are made from their amino-acid building blocks and how different types of proteins function.

Amino acids As you read previously, many different functional groups are found in organic compounds. **Amino acids,** as their name implies, are organic molecules that have both an amino group and an acidic carboxyl group. The general structure of an amino acid is shown below.

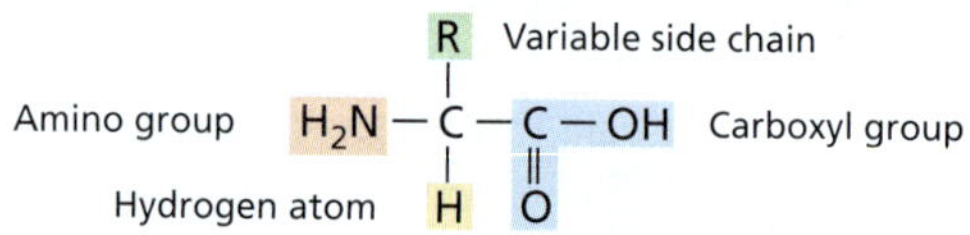

Each amino acid has a central carbon atom around which four groups are arranged: an amino group ($—NH_2$), a carboxyl group ($—COOH$), a hydrogen atom, and a variable side chain, R. The side chains range from a single hydrogen atom to a complex double-ring structure.

Figure 1 All living organisms contain proteins. A goat's hair, hooves, and muscles are made up of structural proteins, as are the roots and leaves of plants.

Demonstration

What's in a protein?

Purpose
to perform a destructive distillation of a protein

Materials
A few peanuts; ovenproof glass test tube and holder; red litmus paper; lead acetate test paper; cobalt chloride test paper; Bunsen burner; tongs

Safety Precautions

CAUTION: ***Inquire whether any student has peanut allergies.***

Disposal All products can be disposed of in the trash.

Procedure
Place a few peanuts in the test tube and heat. **WARNING:** ***Do this in a hood or in a room with good ventilation.*** Place a piece of moist,

Explore **amino acids** with an interactive table. **Concepts In Motion**

Table 1 Amino Acid Examples

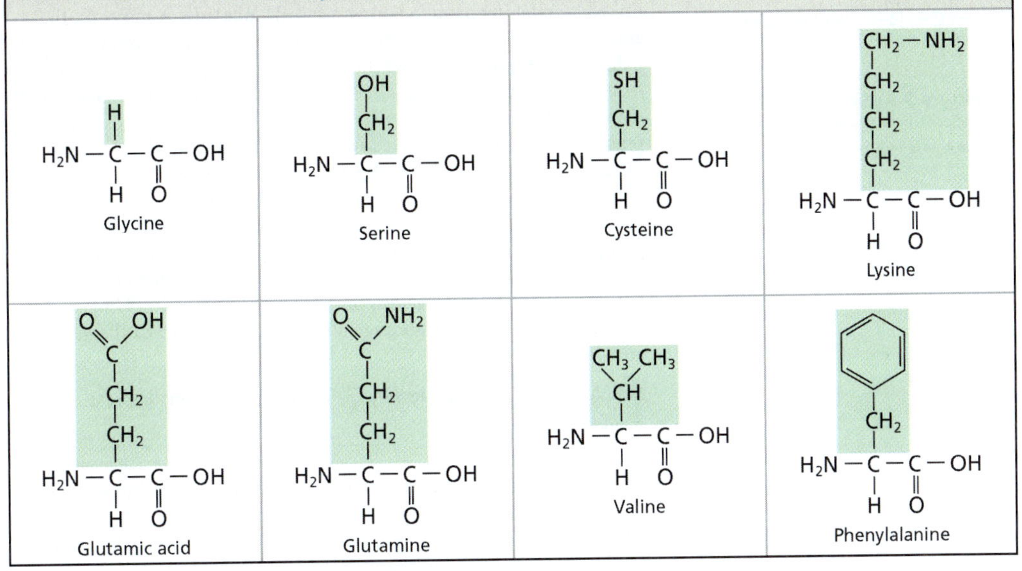

Examine the different side chains of the amino acids shown in **Table 1.** Identify the nonpolar alkanes, polar hydroxyl groups, acidic and basic groups such as carboxyl and amino groups, aromatic rings, and sulfur-containing groups. This wide range of side chains gives the different amino acids a large variety of chemical and physical properties and is an important reason why proteins can perform so many different functions.

The peptide bond The amino and carboxyl groups provide convenient bonding sites for linking amino acids together. Because an amino acid is both an amine and a carboxylic acid, two amino acids can combine to form an amide, releasing water in the process. This reaction is a condensation reaction. As **Figure 2** shows, the carboxyl group of one amino acid reacts with the amino group of another amino acid to form an amide functional group.

☑ READING CHECK **Explain** how an amide functional group forms.

Figure 2 The amino group of one amino acid bonds to the carboxyl group of another amino acid to form a dipeptide and water. The organic functional group formed is an amide linkage called a peptide bond.

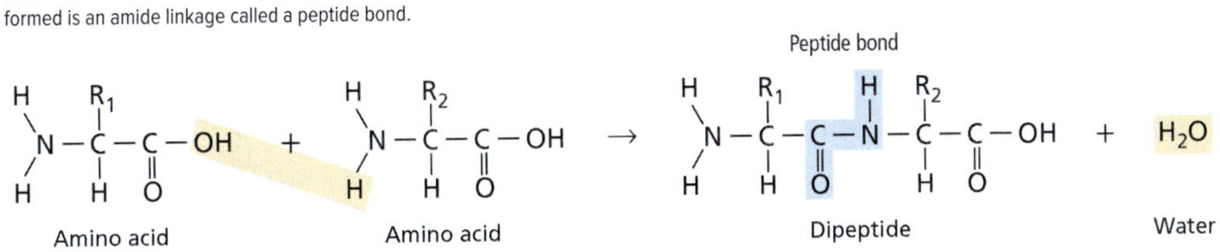

Quick Demo

Model Amino Acids Make ball-and-stick models of the amino acids glycine and alanine. Remind students what a chiral carbon is, and ask them to determine how many chiral carbons there are in each structure. There is one in alanine and none in glycine. Make a model of the mirror image of each structure for which a chiral carbon was found. Ask students what type of isomers those structures represent. They are mirror-image stereoisomers. **OL**

Visual Learning

Table 1 Help students identify the properties of the various side chains in the amino acids shown in the table. **BL** **OL**

☑ READING CHECK An amide functional group forms in a condensation reaction between the carboxyl group of one amino acid and the amino group of a second amino acid.

red litmus paper over the mouth of the test tube while heating to test for ammonia. (Be sure to slant the tube into the flame and point the opening away from people.) In a similar way, use lead acetate paper to test for hydrogen sulfide. Use cobalt chloride test paper to indicate the presence of water.

Results

The protein will blacken when heated, indicating carbon. The ammonia will turn the red litmus to blue. The lead acetate paper turns black for a positive test for H_2S. The cobalt chloride test paper will indicate that water is present by changing color.

Analysis

1. What does the darkening of the protein indicate about what is in a protein? carbon
2. What other elements are shown to be present by the test papers? sulfur, hydrogen, oxygen, and nitrogen

ASSESSMENT

Performance Repeat the demo, but use other types of nuts in place of the peanuts to see if they act in the same way when they are burned.

Math in Chemistry

Peptides The number of possible peptides formed from a fixed number of amino acids can be found by using the formula 20^n, where n represents the number of amino acids in the peptide. Students might understand this formula better if it is explained that for each position in a random peptide, there is a 1/20 chance of any given amino acid being present. For two positions, the chance of any two amino acids being present is $1/20 \times 1/20 = 1/400$. This indicates that there will be one out of 400 possible dipeptides with the given sequence of the two amino acids. The ratio 1/400 is the inverse of 20^2, which also means that there are 400 possible dipeptides.

Enrichment

Protein Assign students into groups of three or four and ask each group to research the name, function, and sequence of a known human protein. Student groups can prepare classroom presentations about their findings or prepare internet pages for use by other classes. OL COOP LEARN

☑ READING CHECK For the dipeptide Glycine-Phenylalanine, glycine has a free amino group and phenylalanine has a free carboxyl group. For the dipeptide Phenylalanine-Glycine, glycine has a free carboxyl group and phenylalanine has a free amino group.

☑ READING CHECK 20^4 or 1.6×10^5

Caption Question Fig. 4 In **a,** the peptide bond is formed between the carboxyl group of glycine and the amino group of phenylalanine. In **b,** the peptide bond is formed between the carboxyl group of phenylalanine and the amino group of glycine.

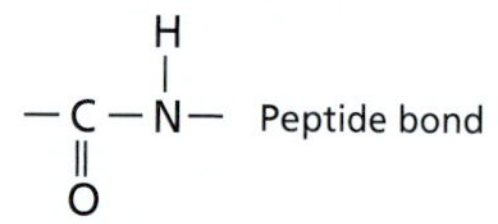

Figure 3 A peptide bond joins two amino acids to form a dipeptide.

The amide bond that joins two amino acids, shown in **Figure 3,** is referred to as a **peptide bond.** A chain of two or more amino acids linked together by peptide bonds is called a **peptide.** A molecule that consists of two amino acids bound together by a peptide bond is called a dipeptide. **Figure 4a** shows the structure of a dipeptide that is formed from the amino acids glycine (Gly) and phenylalanine (Phe). **Figure 4b** shows a different dipeptide, also formed by linking together glycine and phenylalanine. Is Gly-Phe the same compound as Phe-Gly? No, they're different. Examine these two dipeptides to see that the order in which amino acids are linked in a dipeptide is important.

Each end of the two-amino-acid unit in a dipeptide still has a free group—one end has a free amino group and the other end has a free carboxyl group. Each of those groups can be linked to the opposite end of yet another amino acid, forming more peptide bonds. Living cells always build peptides by adding amino acids to the carboxyl end of a growing chain.

☑ READING CHECK **Explain** why Gly-Phe and Phe-Gly are different dipeptides.

Polypeptides As peptide chains increase in length, other ways of referring to them become necessary. A chain of ten or more amino acids joined by peptide bonds is referred to as a polypeptide. When a chain reaches a length of about 50 amino acids, it is called a protein.

Because there are only 20 different amino acids that form proteins, it might seem reasonable to think that only a limited number of different protein structures are possible. However, a protein can have as few as 50 or more than a 1000 amino acids, arranged in any possible sequence. To calculate the number of possible sequences these amino acids can have, consider that each position on the chain can have any of 20 possible amino acids. For a peptide that contains n amino acids, there are 20^n possible sequences of the amino acids. So a dipeptide, with only two amino acids, can have 20^2, or 400, different possible amino acid sequences. Even the smallest protein, containing only 50 amino acids, has 20^{50}, or more than 1×10^{65}, possible arrangements of amino acids! It is estimated that human cells make between 80,000 and 100,000 different proteins. You can see that this is only a small fraction of the total number of proteins possible.

☑ READING CHECK **Calculate** the possible number of sequences for a peptide chain comprised of four amino acids.

Figure 4 Glycine (Gly) and phenylalanine (Phe) can combine in two configurations.
Describe *the difference between the configuration of the peptide bonds in these two dipeptides.*

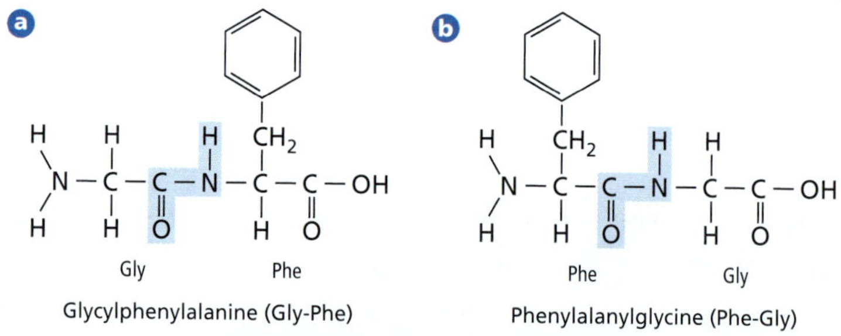

DIFFERENTIATED INSTRUCTION

Struggling Learners Write the two peptides shown in Figure 4 in large letters across two separate pieces of paper. Cut the pieces of paper on both sides of the peptide bond in a zigzag fashion creating puzzlelike pieces. Tell students that the peptide bond holds two different amino acids together to form a complete dipeptide in the same way that two seemingly unrelated puzzle pieces fit together to create a complete picture. BL

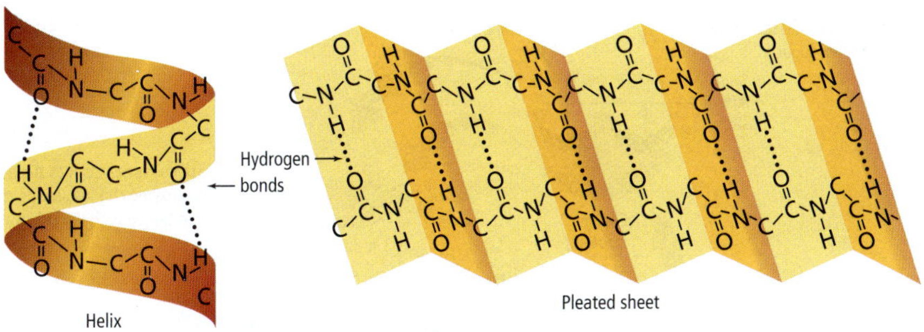

■ **Figure 5** The folding of polypeptide chains into both helices and sheets involves amino acids in the chain held in position by hydrogen bonds. Other interactions among the various side chains are not shown here but play an important role in determining the three-dimensional shape of a polypeptide.

Three-dimensional protein structure Long chains of amino acids start to fold into unique three-dimensional shapes before they are fully synthesized. The three-dimensional shape is determined by the interactions among the amino acids. Some areas of a polypeptide might twirl into helices, which are similar to the coils on a telephone cord. Other areas might bend back and forth repeatedly into a pleated sheet structure, like the folds of an accordion. A polypeptide chain might also fold back on itself and change direction. A given protein might have several helices, sheets, and turns, or none at all. **Figure 5** shows the folding patterns of a typical helix and a sheet. The overall three-dimensional shape of many proteins is globular—shaped like an irregular sphere. Other proteins have a long, fibrous shape. The shape is important to the function of the protein. If the shape of the protein changes, it might not be able to carry out its function in the cell.

Denaturation Changes in temperature, ionic strength, pH, and other factors result in the unfolding and uncoiling of a protein. **Denaturation** is the process in which a protein's natural three-dimensional structure is disrupted. Cooking often denatures the proteins in foods. When an egg is hard-boiled, the protein-rich egg white solidifies due to the denaturation of its protein. Because proteins function properly only when folded, denatured proteins are generally inactive.

The Many Functions of Proteins

Proteins play many roles in living cells. They are involved in speeding up chemical reactions, transport of substances, regulation of cellular processes, structural support of cells, communication within cells and among cells, cellular motion, and even serving as an energy source when other sources are scarce.

Speeding up reactions In most organisms, the largest number of proteins function as enzymes, catalyzing the many reactions that occur in living cells. An **enzyme** is a biological catalyst. You read previously that a catalyst speeds up a chemical reaction without being consumed in the reaction. A catalyst usually lowers the activation energy of a reaction by stabilizing the transition state.

RealWorld CHEMISTRY

Enzymes

PAPAIN An example of an enzyme you might have used is papain, found in papayas, pineapples, and other plant sources. This enzyme catalyzes a reaction that breaks down protein molecules into free amino acids. Papain is the active ingredient in many meat tenderizers. When you sprinkle the dried form of papain onto moist meat, the papain forms a solution that breaks down the tough protein fibers in the meat, making the meat more tender.

Watch a **video about surfactants and enzymes.**

Video

Identify Misconceptions

Students might be familiar with the role of carbohydrates, proteins, and lipids only as components of the foods they eat.

Uncover the Misconception Ask students to name some important functions of proteins, carbohydrates, and lipids.

Demonstrate the Concept Divide the class into three groups. Have each group list examples or bring in samples of other functions of one of these molecules. When students describe the numerous other roles these molecules play in living things, point out that they eat these molecules so that body cells can use the building blocks to construct other needed substances.

Assess New Knowledge Ask students to find additional examples of substances made of proteins, carbohydrates, or lipids. OL COOP LEARN

Use Science Terms

Enzymes Have students write statements explaining the meanings of the terms *enzyme* and *active site.* BL EL

CHEMISTRY PROJECT

Linus Pauling Ask students to research the work of the American scientist Linus Pauling and make class reports. If students do not mention it, point out that Pauling made discoveries in many different areas of chemistry and was awarded two Nobel prizes, one in chemistry and the other for peace. OL

What's CHEMISTRY Got To Do With It?

Video

Squeaky Clean

Build a Model

Hemoglobin Ask students to use pipe cleaners or other materials to build a model of hemoglobin using the structure shown in Figure 7. Tell students to indicate clearly the location of the heme groups. Heme contains iron, to which oxygen binds. OL

Quick Demo

Enzymes Place 5 mL of a freshly made 2% lactose solution in each of two test tubes labeled *lactase* and *control.* Add a ground-up tablet of lactase enzyme to the test tube labeled *lactase.* Mix each test tube gently and let them sit in a beaker of warm water for 2 to 4 minutes. Have a pair of students test each tube for the presence of glucose using Benedict's reagent or glucose test strips. Ask students to explain what happened in each tube. The lactase enzyme broke down the lactose into glucose and galactose, which tested positive for simple sugars. The control tube gave a negative result because no enzyme was present. OL

READING CHECK Sample answer: Enzymes act on a specific substrate. The substrate fits into the active site of the enzyme. The enzyme changes shape to fit with the substrate. Bonds are broken and substrates are converted to products. The enzyme is not altered and can undergo the process multiple times.

ChemLAB

The ChemLab located at the end of the chapter can be used at this point in the lesson.

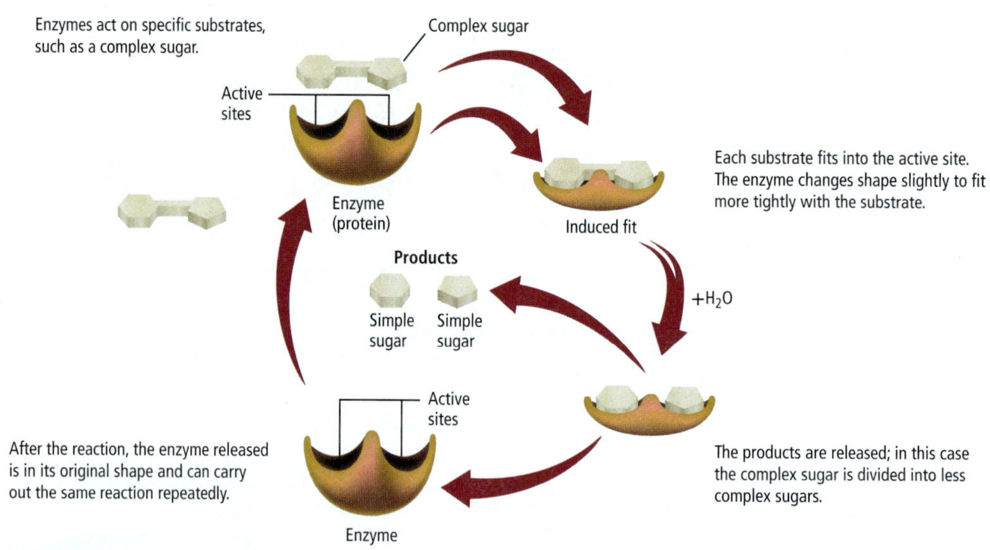

Figure 6 Enzymes lower the activation energy needed for a reaction to occur. Enzymes change the speed at which chemical reactions occur without being altered themselves in the reaction.

How do enzymes function? The term **substrate** refers to a reactant in an enzyme-catalyzed reaction, as shown in **Figure 6.** Substrates bind to specific sites on enzyme molecules, usually pockets or crevices. The spot to which the substrates bind is called the **active site** of the enzyme. After the substrates bind to the active site, the active site changes shape slightly to fit more tightly around the substrates. This recognition process is called induced fit. The shapes of the substrates must fit the shape of the active site, in the same way that puzzle pieces or a lock and key fit together. A molecule that is only slightly different in shape from an enzyme's normal substrate will not bind as well to the active site and might not undergo the catalyzed reaction.

The structure that forms when substrates are bound to an enzyme is called an enzyme-substrate complex. The large size of enzyme molecules allows them to form multiple bonds with their substrates, and the large variety of amino acid side chains in the enzyme allows a number of different intermolecular forces to form. These intermolecular forces lower the activation energy needed for the reaction in which bonds are broken and the substrates are converted to product.

READING CHECK **Describe** in your own words how an enzyme works.

Figure 7 Hemoglobin is a globular protein with four polypeptide chains, each containing an iron group (called a heme) to which oxygen binds.

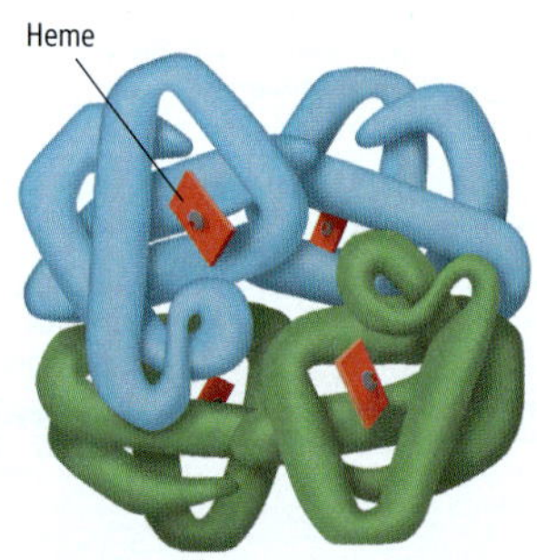

Transport proteins Some proteins are involved in transporting smaller particles throughout the body. **Figure 7** shows the protein hemoglobin, which carries oxygen in the blood from the lungs to the rest of the body. Other proteins combine with biological molecules called lipids to transport them from one part of the body to another through the bloodstream. You will learn about lipids later in this chapter.

DIFFERENTIATED INSTRUCTION

Struggling Learners To better understand the concept of polymer formation, have students use paper clips to represent how amino acid monomers can form into protein polymers. BL EL

CHEMISTRY JOURNAL

Enzymes Have students research an enzyme and create a poster that describes the enzyme's substrates, function, and uses. If possible, display the posters in the classroom. OL

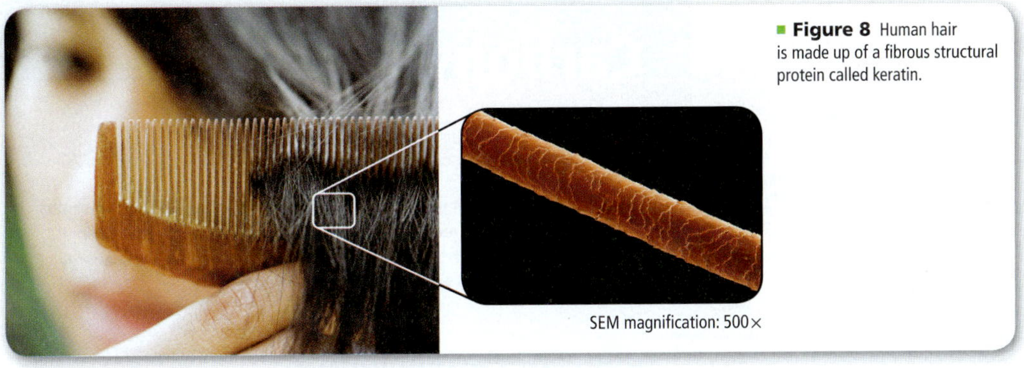

■ **Figure 8** Human hair is made up of a fibrous structural protein called keratin.

Structural support The sole function of certain proteins is to form structures vital to organisms. These molecules are known as structural proteins. The most abundant structural protein in most animals is collagen, which is part of skin, ligaments, tendons, and bones. Other structural proteins make up feathers, fur, wool, hooves, fingernails, cocoons, and hair, as shown in **Figure 8.**

Communication Hormones are messenger molecules that carry signals from one part of the body to another. Some hormones are proteins. Insulin, a familiar example, is a small (51 amino acids) protein hormone made by pancreas cells. When insulin is released into the bloodstream, it signals body cells that blood sugar is abundant and should be stored. A lack of insulin often results in diabetes, a disease that results when there is too much sugar in the bloodstream.

Because modern technology has made possible the laboratory synthesis of proteins, some protein hormones are being synthetically produced for use as medicines. Insulin, thyroid hormones, and growth hormones are some examples. Both natural and synthetic proteins are used in a variety of products—from meat tenderizer to cleaning solutions to health and beauty aids.

FOLDABLES® Incorporate information from this section into your Foldable.

SECTION 1 REVIEW

Section Self-Check

Section Summary

- Proteins are biological polymers made of amino acids that are linked by peptide bonds.
- Protein chains fold into intricate three-dimensional structures.
- Proteins have many functions in the human body, including functions within cells, functions between cells, and functions of structural support.

1. MAINIDEA **Describe** three proteins and identify their functions.
2. **Compare** the structures of amino acids, dipeptides, polypeptides, and proteins. Which has the largest molecular mass? The smallest?
3. **Draw** the structure of the dipeptide Gly-Ser, circling the peptide bond.
4. **Evaluate** How do the properties of proteins make them such useful catalysts? How do they differ from other catalysts you have studied?
5. **Explain** how a change in temperature might affect a protein's function.
6. **Categorize** Identify an amino acid from **Table 1** that can be classified into each of the categories in the following pairs.
   a. nonpolar side chain v. polar side chain
   b. aromatic v. aliphatic
   c. acidic v. basic

FOLDABLES® Rubric

ASSESSMENT

Knowledge Ask students why the hormone insulin must be injected rather than swallowed. Insulin is a protein whose three-dimensional structure must remain intact to function properly. If insulin were swallowed, enzymes in the digestive system would break down the molecule into free amino acids. OL

3 Assess

Check for Understanding

Have students write a short paragraph describing the roles of protein in living things. OL

Reteach

Use a dropper to place a few drops of 3% hydrogen peroxide on some of the following items: cut pieces of fruits or vegetables, uncut pieces of fruits or vegetables, wood, rock, and paper. Ask students to explain the results. Bubbles should form on all sites where hydrogen peroxide was placed on cut living material; few or no bubbles should form on uncut living material or nonliving material. Damaged cells release an enzyme that catalyzes the breakdown of hydrogen peroxide. The bubbles are oxygen gas produced in this reaction. $H_2O_2 \rightarrow H_2O + O_2$
Disposal: Wash hydrogen peroxide off materials with tap water. OL EL

Extension

Ask students to find out which household products contain enzymes. Ask students to share their findings with the class. OL

SECTION 1 REVIEW

1. papain: enzyme that breaks down protein into amino acids; hemoglobin: transports oxygen in the body; collagen: structural protein found in skin, ligaments, tendons, and bones
2. Amino acids are single-unit organic molecules that bond to each other. Two amino acids form a dipeptide, more than ten form a polypeptide, and more than 50 form a protein. From smallest to largest: amino acid, dipeptide, polypeptide, protein.
3. The structure should show the COOH of glycine and the NH_2 of serine contributing to the peptide bond. Refer to the Solutions Manual for structure.
4. Proteins are useful catalysts because of their large size and large number and variety of functional groups on the amino acid side chains. Most inorganic catalysts are much smaller molecules.
5. Changes in temperature can affect the shape of the protein, causing the protein to denature.
6. Student answer may vary. **a.** nonpolar: glycine, polar: serine; **b.** aromatic: phenylalanine, aliphatic: all others; **c.** acidic: glutamic acid, basic: lysine

SECTION 2

1 Focus

MAINIDEA

Carbohydrates To focus students' thoughts on carbohydrates, ask students to name different foods that contain carbohydrates. Write the names of the foods on the board. Sample foods: breads, pastas, and anything that contains sugar. Ask students if they know the primary function of carbohydrates in living organisms. an energy source OL

2 Teach

Content Background

Chitin Chitin is one of the most abundant organic compounds on Earth. It is a polysaccharide composed of the amino sugar glucosamine. Chitin is found in the exoskeleton of insects and crustaceans and in fungal cell walls.

ASSESSMENT

Performance Have students use a modeling kit or gumdrops and toothpicks to make models of glucose in both cyclic and open-chain structures. OL EL

SECTION 2

Essential Questions

- How can the structures of monosaccharides, disaccharides, and polysaccharides be described?
- What are the functions of carbohydrates in living things?

Review Vocabulary

stereoisomers: a class of isomers whose atoms are bonded in the same order but are arranged differently in space

New Vocabulary

carbohydrate
monosaccharide
disaccharide
polysaccharide

Carbohydrates

MAINIDEA Carbohydrates provide energy and structural material for living things.

CHEM 4 YOU A lot of media attention has been focused on carbohydrates. Some recommend low-carb diets as a way of controlling weight. However, carbohydrates are an important energy source for the body.

Kinds of Carbohydrates

Analyzing the term *carbohydrate* offers a hint about the structure of this group of molecules. Early observations that these compounds have the general chemical formula $C_n(H_2O)_n$ and appear to be hydrates of carbon led to their being called carbohydrates. Although scientists now know that there are no full water molecules attached to carbohydrates, the name has stayed.

The main function of carbohydrates in living organisms is as a source of energy, both immediate and stored. Foods rich in carbohydrates include pasta, milk, fruit, bread, and potatoes. **Carbohydrates** are compounds that contain multiple hydroxyl groups (—OH) as well as a carbonyl functional group (C=O). These molecules range in size from single monomers to polymers made of hundreds or even thousands of monomer units.

Monosaccharides The simplest carbohydrates, often called simple sugars, are **monosaccharides.** The most common monosaccharides have either five or six carbon atoms. Examples of monosaccharides are shown in **Figure 9.** Notice that they have a carbonyl group on one carbon and hydroxyl groups on most of the other carbons. The presence of a carbonyl group makes these compounds either aldehydes or ketones, depending on the location of the carbonyl group. Multiple polar groups make monosaccharides water-soluble and give them high melting points.

Figure 9 Glucose, galactose, and fructose are monosaccharides. In aqueous solutions, they exist in an equilibrium between their open-chain and cyclic forms.

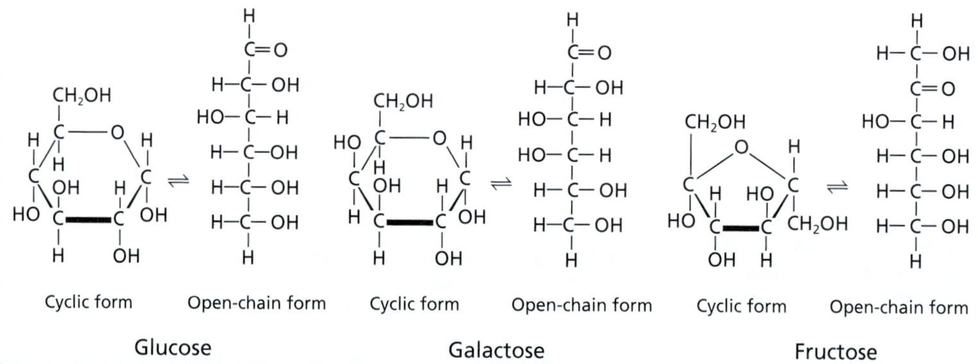

DIFFERENTIATED INSTRUCTION

Advanced Learners Ask students to draw the structural formula for glucose, make a ball-and-stick model of glucose, and determine its IUPAC name. 2,3,4,5,6-pentahydroxyhexanal AL

CHEMISTRY JOURNAL

Low-Carbohydrate Diets Ask students to find information about low-carbohydrate diets in newspapers, magazines, and television. Ask them to research and then write a short report about the safety of that type of diet. OL

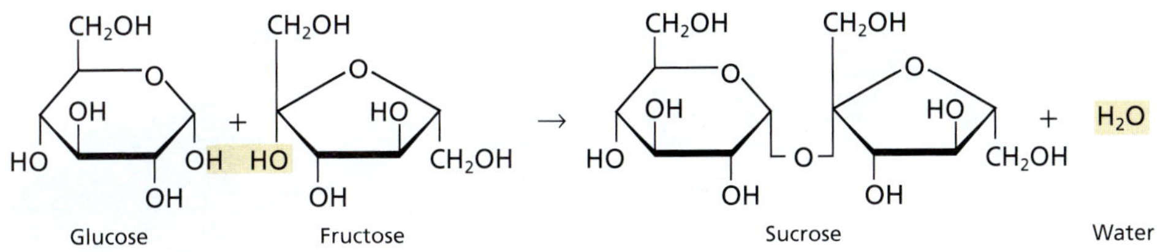

■ **Figure 10** When glucose and fructose bond, the disaccharide sucrose forms. Note that water is also a product of this condensation reaction. Remember that each ring structure is made of carbon atoms, which are not shown for simplicity.

Glucose is a six-carbon sugar that has an aldehyde structure. Glucose is present in high concentration in blood because it serves as the major source of immediate energy for the body. For this reason, glucose is often called blood sugar. Closely related to glucose is galactose, which differs only in how a hydrogen and a hydroxyl group are oriented in space around one of the six carbon atoms. Recall that this relationship makes glucose and galactose stereoisomers. Fructose, also known as fruit sugar because it is the major carbohydrate in most fruits, is a six-carbon monosaccharide that has a ketone structure. Fructose is a structural isomer of glucose.

When monosaccharides are in aqueous solution, they exist in both open-chain and cyclic structures, but they rapidly interconvert forms. The cyclic structures are more stable and are the predominant form of monosaccharides at equilibrium. Note in **Figure 9** that the carbonyl groups are present only in the open-chain structures. In the cyclic structures, they are converted to hydroxyl groups.

Disaccharides Like amino acids, monosaccharides can be linked together by a condensation reaction in which water is released. When two monosaccharides bond together, a **disaccharide** is formed, as shown in **Figure 10.** The new bond formed is an ether functional group (C–O–C).

One common disaccharide is sucrose, also known as table sugar because sucrose is used mainly as a sweetener. Sucrose is formed by the linking of glucose and fructose. Another common disaccharide is lactose, the most important carbohydrate in milk. It is often called milk sugar. Lactose is formed when glucose and galactose bond.

Polysaccharides *Complex carbohydrate* is a term used in some nutrition books and journal articles. Another name for a complex carbohydrate is **polysaccharide,** which is a polymer of simple sugars that contains 12 or more monomers, or subunits. The same type of bond that joins two monosaccharides in a disaccharide also links the monomers in a polysaccharide. Glycogen, shown in **Figure 11,** is a polysaccharide. It is composed of glucose subunits. It stores energy and is found mostly in the liver and muscles of humans and other animals. It is also found in some species of microorganisms including bacteria and fungi.

VOCABULARY

WORD ORIGIN

Polysaccharide
comes from the Greek word *polys,* which means *many* and the ancient Sanskrit word *śarkarā,* which means *sugar*

■ **Figure 11** The glycogen found in the muscle and liver of animals is a polysaccharide made of glucose.

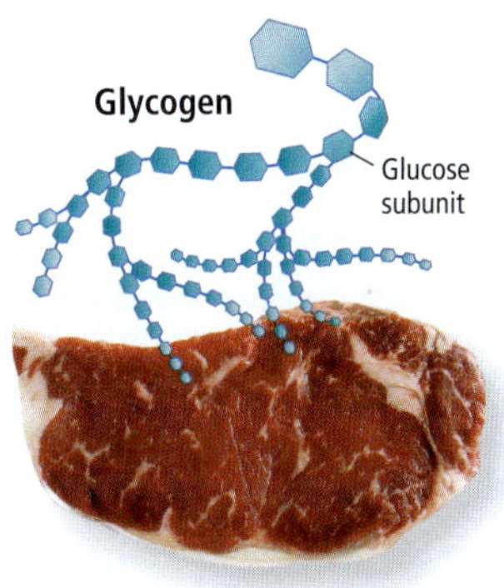

☑ READING CHECK **Explain** the differences among a monosaccharide, a disaccharide, and a polysaccharide.

☑ READING CHECK A monosaccharide is a simple sugar. A disaccharide contains two monosaccharides. A polysaccharide contains 12 or more glucose monomers.

Quick Demo

Testing for Starch Show students a number of materials, including cotton balls, paper, potato pieces, wood, apple pieces, laundry spray starch, bread, crackers, and metal. Ask students to predict which items will give a positive result to a test for starch. The potato, laundry spray starch, bread, and crackers all contain starch. Ask a student volunteer to use a dropper to place 1 or 2 drops of iodine solution on each material. Ask the class to interpret the results. Materials that contain starch will turn a blue-purple color due to the formation of a starch-iodine complex. Materials that contain cellulose (cotton, paper, and wood) turn brown when a similar complex forms. OL AL

ASSESSMENT

Skill Have students determine how many moles of glucose are produced by the hydrolysis of 1.03×10^3 g of sucrose. 3.03 mol glucose

CULTURAL DIVERSITY

What's the source of your energy? Complex carbohydrates are the main dietary source of energy for most people. In the United States, most Americans get their polysaccharides from wheat grain because it is commonly found in breads, pasta, and cereal. Around the world, different cultures use different grains for reasons that include taste, economics, and environmental conditions, such as amount of rainfall and soil type. In Asia, rice provides the most complex carbohydrate calories. In Mexico and South America, corn is the complex carbohydrate most often eaten.

FOLDABLES® Rubric

3 Assess

Check for Understanding

Draw the structures of several different carbohydrates on the board. Ask the students to label each structure as a mono-, di-, or polysaccharide. Have a student who has labeled each correctly explain to the others what information he or she used when answering. Students will have used molecular size, formula, or similarity to a more familiar example of each type to come up with their answers. OL

Reteach

Have students work in groups to make ball-and-stick models of two cyclic monosaccharides. Ask them to join the two monosaccharides to make a disaccharide. Ask them what molecule is released when the structures are joined. water Now, ask the groups to break down their disaccharides into the monosaccharides. Ask them what molecule is needed in addition to the disaccharide to complete the monosaccharides. They will need to add water to hydrolyze the disaccharide into two monosaccharides. OL EL COOP LEARN

Extension

Ask a student volunteer to explain why cellulose is important in the diet. Cellulose is also known as dietary fiber. It provides bulk in the intestines which helps keep the digestive system functioning properly and aids in the elimination of waste products. AL

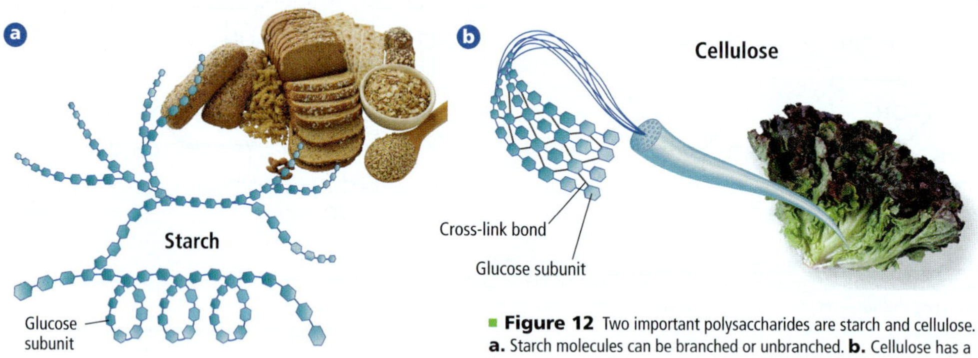

Figure 12 Two important polysaccharides are starch and cellulose. **a.** Starch molecules can be branched or unbranched. **b.** Cellulose has a linear, unbranched structure that resembles a chain-link fence.

Two other important polysaccharides are starch and cellulose, shown in **Figure 12.** Starch and cellulose are also composed solely of glucose subunits. However, that is the only similarity among the three polysaccharides, as all three have different properties and functions. Plants make both starch and cellulose. Starch is a soft, water-insoluble molecule used to store energy, whereas cellulose is a water-insoluble polymer that forms rigid plant-cell walls, such as those found in wood.

FOLDABLES® Incorporate information from this section into your Foldable.

Glycogen, starch, and cellulose are composed of glucose subunits, but they have different properties. The bonds that link the subunits together are oriented differently in space. Because of this difference in bond shape, humans can digest glycogen and starch but not cellulose. Digestive enzymes cannot fit cellulose into their active sites. The cellulose in the fruits, vegetables, and grains that we eat is called *dietary fiber* because it passes through the digestive system largely unchanged.

SECTION 2 REVIEW

Section Self-Check

Section Summary

- Carbohydrates are compounds that contain multiple hydroxyl groups (–OH) and a carbonyl functional group (C=O).
- Carbohydrates range in size from single monomers to polymers composed of hundreds or thousands of monomers.
- Monosaccharides in aqueous solution exist in both open-chain and cyclic structures.

7. MAINIDEA **Explain** the functions of carbohydrates in living things.
8. **Describe** the structures of monosaccharides, disaccharides, and polysaccharides. Which has the largest molecular mass? The smallest?
9. **Compare and contrast** the structures of starch and cellulose. How do the structural differences affect our ability to digest these two polysaccharides?
10. **Calculate** If a carbohydrate has 2^n possible isomers, where n is equal to the number of chiral carbon atoms in the structure, calculate the number of possible isomers for the following monosaccharides: galactose, glucose, and fructose.
11. **Interpret Scientific Illustrations** Copy the illustration of sucrose on a separate sheet of paper, and circle the ether functional group that bonds the monomer sugars together.

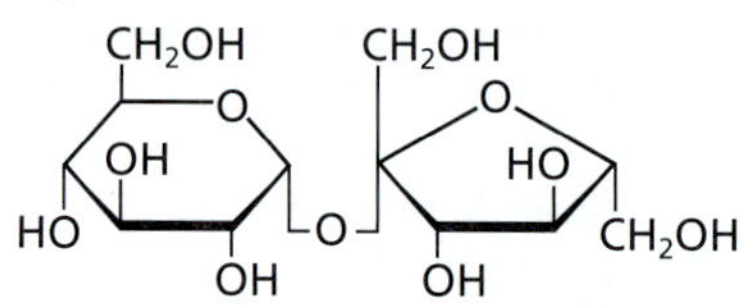

SECTION 2 REVIEW

7. Carbohydrates are the major immediate energy source in living organisms and serve as a form of energy storage.
8. Monosaccharides are molecules that contain multiple hydroxyl groups and an aldehyde or ketone group. Disaccharides are two monosaccharides joined by an ether bond. Polysaccharides are many monosaccharides joined by ether bonds. The order from smallest to largest is monosaccharide, disaccharide, polysaccharide.
9. Both starch and cellulose contain only glucose monomers. They differ in how the bonds that hold glucose together are oriented in space. Because of this shape difference, our digestive enzymes cannot break down cellulose.
10. Galactose: $2^4 = 16$ isomers; glucose: $2^4 = 16$ isomers; fructose: $2^3 = 8$ isomers
11. Refer to the Solutions Manual.

SECTION 3

Essential Questions

- How can the structures of fatty acids, triglycerides, phospholipids, and steroids be described?
- What are the functions of lipids in living organisms?
- What are some reactions that fatty acids undergo?
- How are the structure and function of cell membranes related?

Review Vocabulary

nonpolar: without separate positive and negative areas or dipoles

New Vocabulary

lipid
fatty acid
triglyceride
saponification
phospholipid
wax
steroid

Lipids

MAINIDEA Lipids make cell membranes, store energy, and regulate cellular processes.

CHEM 4 YOU The wax used to polish cars, the fat that drips out of hamburgers, and the vitamin D that fortifies the milk people drink—what do these things have in common? They are all lipids.

What is a lipid?

A **lipid** is a large, nonpolar biological molecule. Because lipids are nonpolar, they are insoluble in water. Lipids have two major functions in living organisms. They store energy efficiently, and they make up most of the structure of cell membranes. Unlike proteins and carbohydrates, lipids are not polymers with repeated monomer subunits.

Fatty acids Although lipids are not polymers, many lipids have a major building block in common. This building block is the **fatty acid,** a long-chain carboxylic acid. Most naturally occurring fatty acids contain between 12 and 24 carbon atoms. Their structure can be represented by the following formula.

$$CH_3(CH_2)_nCOOH$$

Most fatty acids have an even number of carbon atoms, which is a result of being constructed two carbons at a time in enzymatic reactions.

Fatty acids can be grouped into two main categories, depending on the presence or absence of double bonds between carbon atoms. Fatty acids that contain no double bonds are referred to as saturated. Those that have one or more double bonds are called unsaturated. The structures of two common fatty acids are shown in **Figure 13.**

☑ READING CHECK **Explain** why oleic acid is described as *unsaturated.*

■ **Figure 13** Two fatty acids, which are found in many foods, including butter, are the 18-carbon unsaturated oleic acid and the 18-carbon saturated stearic acid.
Explain *how the structure of the molecule is affected by the presence of a double bond.*

Oleic acid

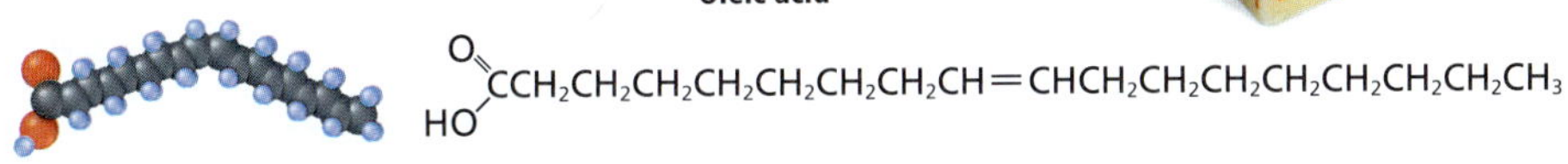

Stearic acid

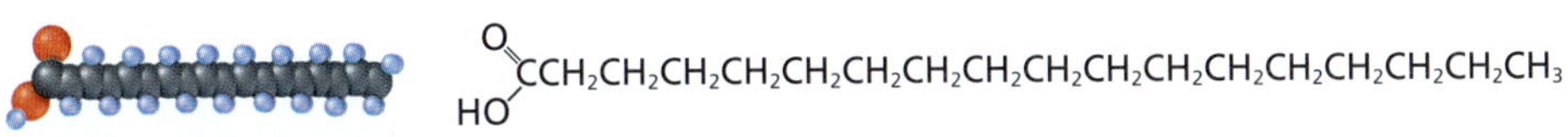

D. Hurst/Alamy

SECTION 3

1 Focus

MAINIDEA

Lipids Ask students to give examples of lipids. fats, oils, waxes, vitamin D, and steroids Ask students if they know what functions lipids serve in living organisms. store energy and make up the composition of cellular membranes OL

2 Teach

Quick Demo

Relative Density Ask students to predict the relative densities of corn syrup, 70% rubbing alcohol, vegetable oil, and water. Density order, from highest to lowest, is corn syrup, water, vegetable oil, rubbing alcohol. Ask a volunteer to test the predictions using a graduated cylinder and the four substances (three with different food colorings added). Instruct the volunteer to pour one substance into the cylinder and then add each of the others VERY slowly. Ask students to determine the actual order of densities by observing the positions of each substance in the density gradient.

Disposal Dilute with soap and water, and flush down a drain. OL

■ **Caption Question Fig. 13** The molecule is bent at the double bond.

☑ **READING CHECK** because it has a double bond

CHEMISTRY JOURNAL

Dietary Lipids Ask students to keep a one-week record in their journals of how many grams of fat they consume each day. They should research to find the recommended amount of fat for someone their age and size to determine if they are reaching or exceeding that recommended amount. Ask students to rank their favorite foods according to fat content and to make suggestions about how to modify their diets if they are eating too much or too little fat. OL

Apply Chemistry

Soap and Detergent While people tend to associate foaming in soaps and detergents with cleaning power, foaming cleansers are not necessarily more effective in removing dirt. Chemists in the detergent industry formulate foaming soaps and detergents mainly to increase product marketability because that is what consumers expect. Foaming can be undesirable. Too much foam formed in washing machines can cause the machine to overflow. Foaming can also denature enzymes that are added to some cleaning agents to promote dirt removal.

☑ READING CHECK Fatty acids are long-chain carboxylic acids; triglycerides have three fatty acids bonded to a glycerol backbone.

■ **Figure 14** Ester bonds in a triglyceride are formed when the hydroxyl groups of glycerol combine with the carboxyl groups of the fatty acids.

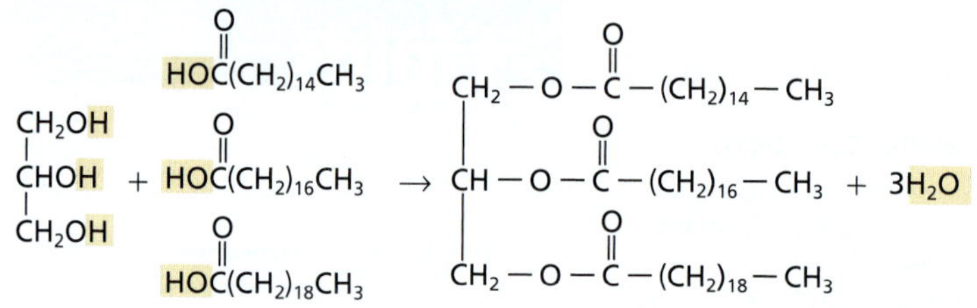

An unsaturated fatty acid can become saturated if it reacts with hydrogen. As you read previously, hydrogenation is an addition reaction in which hydrogen gas reacts with carbon atoms that are linked by multiple bonds. Each unsaturated carbon atom can pick up one hydrogen atom to become saturated. For example, oleic acid, shown in **Figure 13,** can be hydrogenated to form stearic acid.

The double bonds in naturally occurring fatty acids are almost all in the *cis* geometric isomer form. Recall that the *cis* isomer has identical groups oriented on the same side of the molecule around a double bond. Because of the *cis* orientation, unsaturated fatty acids have a kink, or bend, in their structure that prevents them from packing together. They do not form as many intermolecular attractions as saturated fatty acid molecules. As a result, unsaturated fatty acids have lower melting points.

VOCABULARY

SCIENCE USAGE V. COMMON USAGE

Saturate

Science usage: to add something to the point that no more can be absorbed, dissolved, or retained

The salt water solution was saturated with salt.

Common usage: to furnish a market with a product or products to its full purchasing capacity

The shops in the coastal town are saturated with sea shell craft items.

Triglycerides Although fatty acids are abundant in living organisms, they are rarely found alone. They are most often found bonded to glycerol, a molecule with three carbons, each containing a hydroxyl group. When three fatty acids are bonded to a glycerol backbone through ester bonds, a **triglyceride** is formed. The formation of a triglyceride is shown in **Figure 14.** Triglycerides can be either solids or liquids at room temperature, as shown in **Figure 15.** If liquid, they are usually called oils. If solid at room temperature, they are called fats.

☑ READING CHECK **Describe** the difference between fatty acids and triglycerides.

■ **Figure 15** Most mixtures of triglycerides from plant sources are liquids because the triglycerides contain unsaturated fatty acids. Animal fats contain a larger proportion of saturated fatty acids. They are usually solids at room temperature.

Maximilian Stock Ltd./Photolibrary/Getty Images

DIFFERENTIATED INSTRUCTION

Struggling Learners Use a simple analogy to help students understand the concept that a small number of building blocks can be used to make very large and different structures. Show students pictures of two very different buildings, both made of bricks. Ask students to identify the building blocks of the two buildings. Then, ask students why the structures differ so much even though the building blocks are the same. BL EL

CHEMISTRY PROJECT

Dietary Fats and Oils Have students research dietary fats and oils to find out the types of fats and oils that are the most healthful. Students should also find out the quantity of fats and oils that are recommended for a balanced diet. Have students prepare a poster that displays their information. BL OL AL EL

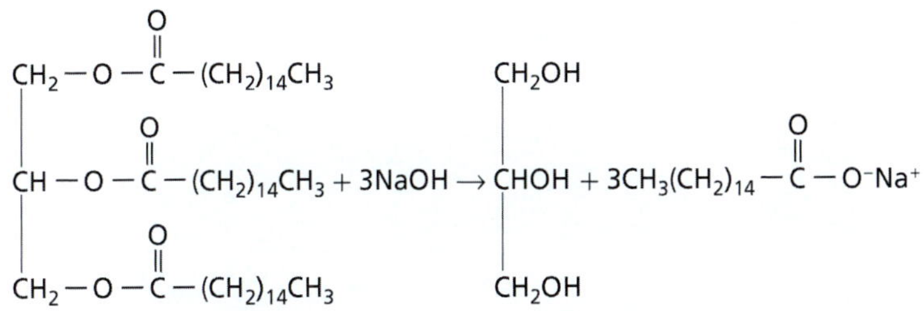

Figure 16 Soap forms by the reaction of a triglyceride and a strong base.

Fatty acids are stored in the fat cells of your body as triglycerides. When energy is abundant, fat cells store the excess energy in the fatty acids of triglycerides. When energy is scarce, the cells break down the triglycerides, releasing the energy used to form them.

Although enzymes break down triglycerides in living cells, the reaction can be duplicated outside of cells by using a strong base, such as sodium hydroxide. This reaction–the hydrolysis of a triglyceride using an aqueous solution of a strong base to form carboxylate salts and glycerol–is **saponification,** as shown in **Figure 16.** Saponification is used to make soaps, which are usually the sodium salts of fatty acids. A soap molecule has both a polar end and a nonpolar end. Soaps are used with water to clean nonpolar dirt and oil because the nonpolar dirt and oil bond to the nonpolar end of the soap molecules, and the polar end of the soap molecules is soluble in water. Thus, the dirt-laden soap molecules can be rinsed away with the water.

MiniLAB

Observe a Saponification Reaction

How is soap made? The reaction between a triglyceride and a strong base is called saponification. A sample chemical reaction is shown in **Figure 16.**

Procedure

1. Read and complete the lab safety form.
2. Place a **250-mL beaker** on a **hot plate.** Add 25 g **solid vegetable shortening** to the beaker. Turn on the hot plate to a medium setting.
3. As the vegetable shortening melts, use a **25-mL graduated cylinder** to slowly add 12 mL **ethanol** and then 5 mL **6.0*M* NaOH** to the beaker.

   WARNING: ***Ethanol is flammable. NaOH causes skin burns. Wear gloves.***
4. Heat the mixture for about 15 min. Use a **stirring rod** to occasionally stir the mixture. Do not allow it to boil.
5. When the mixture begins to thicken, use **tongs** to remove the beaker from the heat. Allow the beaker to cool for 5 min, then place it in a **cold water bath** in a **600-mL beaker.**
6. Add 25 mL **saturated NaCl solution** to the mixture in the beaker. The soap is not very soluble and will appear as small clumps.
7. Collect the solid soap clumps by filtering them through a **cheesecloth-lined funnel.**
8. Using gloved hands, press the soap into an **evaporating dish.** Remove your gloves and wash your hands.

Analysis

1. **Explain** What type of bonds present in the triglycerides are broken during the saponification reaction?
2. **Identify** the type of salt formed in this chemical reaction.
3. **Determine** which is the polar end and which is the nonpolar end of the soap molecule.

MiniLAB

Rubric

Purpose Students will produce a small bar of soap by performing a saponification reaction.

Process Skills classify, compare and contrast

Safety Precautions Approve lab safety forms before work begins. Remind students that they are not to use the lye soap on their skin.

Disposal The contents of the beaker can be flushed down a drain after removing the solid soap. Soap can be discarded in the trash after being rinsed with water.

Teaching Strategies

- Students can prepare their own saturated salt solution by adding 50 g of rock salt to 25 mL of water. This should be done during Step 3 while the soap is being heated.
- A pea-sized piece of crayon can be added to the soap while heating to give it color.

Expected Results The small bar of soap can be molded into an evaporating dish.

Analysis

1. ester bonds
2. a soap
3. The end of the molecule that contains the sodium ion is polar. The other end of the molecule that contains the hydrogen atoms is nonpolar.

ASSESSMENT

Performance Have students do research and write a short paragraph explaining the similarities and differences between soaps and detergents. BL OL

Build a Model

Fatty Acids Students are often confused when they read that fatty acids are nonpolar because the condensed structure is usually shown as $CH_3(CH_2)_nCOOH$. In that structure, the carboxyl group is prominent, making the molecule appear polar. Ask students to build a ball-and-stick model of stearic acid to see for themselves that most of the molecule's volume is occupied by nonpolar CH_2 groups. OL EL

Reinforcement

Waxes Have students bring leaves from various plants to class. Make sure that heavily waxed leaves as well as less-waxed leaves are brought to the classroom. Have students observe the surface of the various types of leaves and note the differences in the amount of lipids on the surface of the leaves. Provide hand lenses, if they are available, for students to use. BL OL EL

Figure 17 A phospholipid has a polar head and two nonpolar tails. The membranes of living cells are formed by a double layer of lipids, called a bilayer. The polar heads are on the outer and inner perimeter of the membrane and the tails are on the inside of the bilayer.

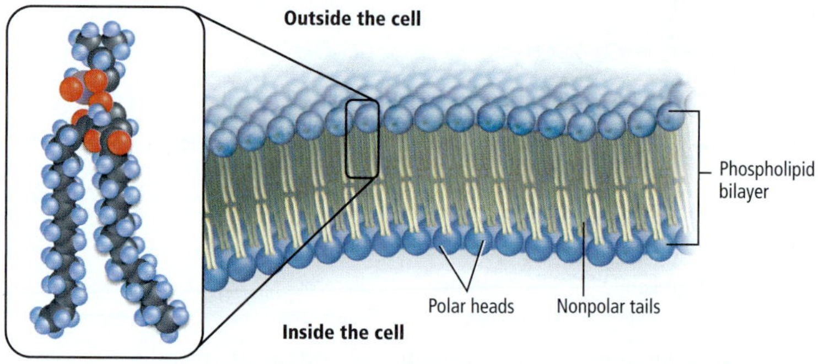

Phospholipids Another important type of triglyceride, a phospholipid, is found in greatest abundance in cellular membranes. A **phospholipid** is a triglyceride in which one of the fatty acids is replaced by a polar phosphate group. As shown in **Figure 17,** the polar part of the molecule forms a head and the nonpolar fatty acids look like tails. A typical cell membrane has two layers of phospholipids, which are arranged with their nonpolar tails pointing inward and their polar heads pointing outward. This arrangement is called a lipid bilayer. Because the lipid bilayer structure acts as a barrier, the cell is able to regulate the materials that enter and leave through the membrane.

Connection to Biology The venom of poisonous snakes contains a class of enzymes known as phospholipases. These enzymes catalyze the breakdown of phospholipids. The venom of the eastern diamond-back rattlesnake contains a phospholipase that hydrolyzes the ester bond at the middle carbon of phospholipids. If the larger of the two breakdown products of this reaction gets into the bloodstream, it dissolves the membranes of red blood cells, causing them to rupture. Because the venom destroys the blood cells, it is referred to as a hemotoxic venom. (The prefix *hemo-* indicates blood.) A bite from the eastern diamond-back can lead to death if it is not treated immediately.

Waxes Another type of lipid, wax, also contains fatty acids. A **wax** is a lipid that is formed by combining a fatty acid with a long-chain alcohol. The general structure of these soft, solid fats with low melting points is shown below, with *x* and *y* representing variable numbers of CH_2 groups.

$$CH_3(CH_2)_x-\overset{\overset{\displaystyle O}{\|}}{C}-O-(CH_2)_yCH_3$$

Figure 18 Plants produce a wax that coats their leaves. The wax protects the leaves from drying out.

Both plants and animals make waxes. Plant leaves are often coated with wax, which prevents water loss. Notice in **Figure 18** how raindrops bead up on the leaves of a plant, indicating the presence of the waxy layer. The honeycombs that bees make are also made of a wax, commonly called beeswax. Combining the 16-carbon fatty acid palmitic acid and a 30-carbon alcohol chain makes a common form of beeswax. Candles are sometimes made of beeswax because it tends to burn slowly and evenly.

DIFFERENTIATED INSTRUCTION

Advanced Learners Have students research how antivenom medications are made and administered, and describe the correct first aid treatment for a snake bite. Have students present an oral report to the class. Encourage students to make visual aids for their report. AL

■ **Figure 19** This Giant Marine toad uses a steroid toxin called bufotoxin as a defense mechanism. The toxin is fatal to some animals, including dogs and cats.

Steroids Not all lipids contain fatty acid chains. **Steroids** are lipids that have multiple cyclic rings in their structures. All steroids are built from the basic four-ring steroid structure shown below.

FOLDABLES® Incorporate information from this section into your Foldable.

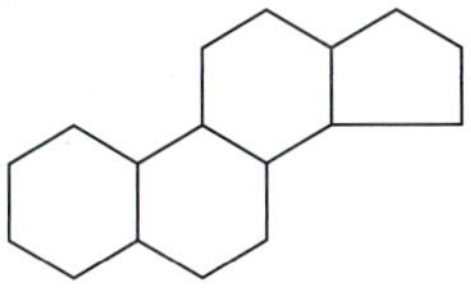

Some hormones, such as many sex hormones, are steroids that function to regulate metabolic processes. Cholesterol, another steroid, is an important structural component of cell membranes. Vitamin D also contains the four-ring steroid structure and plays a role in the formation of bones. The Giant Marine toad, *Bufo marinus,* shown in **Figure 19** uses a steroid called bufotoxin as a defense mechanism. The toad secretes the toxin from warts on its back and from glands just behind the eye. The toxin is only an irritant for humans, but in small animals the toxin causes drooling, loss of coordination, convulsions, and death.

SECTION 3 REVIEW

Section Self-Check

Section Summary

- Fatty acids are long-chain carboxylic acids that usually have between 12 and 24 carbon atoms.
- Saturated fatty acids have no double bonds; unsaturated fatty acids have one or more double bonds.
- Fatty acids can be linked to glycerol backbones to form triglycerides.
- Steroids are lipids that have multiple-ring structures.

12. MAINIDEA **Describe** the function of lipids.

13. Describe the structures of fatty acids, triglycerides, phospholipids, and steroids.

14. List an important function of each of these types of lipids.

a. triglycerides **c.** waxes
b. phospholipids **d.** steroids

15. Identify two reactions that fatty acids undergo.

16. Describe the structure and function of cell membranes.

17. Compare and contrast the structures of a steroid, a phospholipid, and a wax.

18. Write the equation for the complete hydrogenation of the polyunsaturated fatty acid linoleic acid, $CH_3(CH_2)_4CH{=}CHCH_2CH{=}CH(CH_2)_7COOH$.

19. Interpret Scientific Illustrations Draw the general structure of a phospholipid. Label the polar and nonpolar portions of the structure.

FOLDABLES® Rubric

3 Assess

Check for Understanding

Ask groups of students to build ball-and-stick models of three fatty acids and glycerol. Ask them to form a triglyceride and write a word equation for the reaction using only the models as a reference. Students will find that they are forming three ester bonds in this reaction and that each is accompanied by the formation of a molecule of water. The reaction is glycerol + 3 fatty acids ⟶ triglyceride + 3 water. OL COOP LEARN

Reteach

Ask groups of students to use ball-and-stick models to examine the difference in structure between natural saturated and unsaturated fatty acids. Their models should contain only *cis* double bonds in the unsaturated fatty acids. Ask the groups to pool their models to allow them to determine which type of fatty acid packs together better. Students will find that the presence of the cis double bonds in unsaturated fatty acids prevents them from packing together as well as saturated fatty acids. OL COOP LEARN

Extension

Ask students whether saturated or unsaturated fatty acids of the same size will have a greater density. Saturated fatty acids have a higher density because they pack together better and can fit more molecules per unit volume. OL

SECTION 3 REVIEW

12. Lipids store energy efficiently, and they make up most of the structure of living cells.

13. fatty acids: a long-chain carboxylic acid with the formula $CH_3(CH_2)_nCOOH$; triglycerides: three fatty acids ester-bonded to glycerol; phospholipids: two fatty acids and a phosphate group ester-bonded to glycerol; steroid: contains no fatty acids but has a four-ring structure

14. **a.** triglycerides: major fatty acid storage form of lipids
b. phospholipids: make up membrane structure
c. waxes: form protective coatings
d. steroids: hormones, vitamins, and in biological membranes

15. saponification and hydrogenation

16. It has two layers of phospholipids, which are arranged with their nonpolar tails pointing inward and their polar heads pointing out. It acts as a barrier that allows substances into and out of a cell.

17. phospholipid: two fatty acids and a phosphate group ester-bonded to glycerol; wax: a long-chain alcohol and a fatty acid ester-bonded together; steroid: contains no fatty acids but has a four-ring structure

18. $CH_3(CH_2)_4CH = CHCH_2CH = CH(CH_2)_7COOH + 2H_2 \longrightarrow CH_3(CH_2)_{16}COOH$

19. Refer to Solutions Manual. The phosphate group is polar; the fatty acid groups are nonpolar.

SECTION 4

1 Focus

MAINIDEA

Nucleic Acids Ask students what they know about RNA and DNA. Write the correct responses on the board. Many student responses might include things they have seen on crime shows on television. Encourage these responses because students are focusing on the topic and things that are familiar to them. However, correct any misconceptions that students might have. Some students might recall from biology that DNA is responsible for the transmission of inherited traits and that RNA is used to make proteins. BL OL EL

2 Teach

Use Science Terms

Nucleic Acids Have students write statements explaining the meanings of the terms *nucleic acid* and *nucleotide.* BL EL

ASSESSMENT

Performance Have students read James Watson's book, *The Double Helix,* and write reports. The book was published in 1968. It offers insight into the scientific process and how scientists work. It also illustrates the human side of scientists and scientific discovery. OL

SECTION 4

Essential Questions

- What are the structural components of nucleic acids?
- How is the function of DNA related to its structure?
- What are the structure and function of RNA?

Review Vocabulary

genetic information: an inherited sequence of RNA or DNA that causes traits or characteristics to pass from one generation to the next

New Vocabulary

nucleic acid
nucleotide

Nucleic Acids

MAINIDEA Nucleic acids store and transmit genetic information.

CHEM 4 YOU **DNA testing is becoming more routine in medicine, forensic science, genealogy, and identification of victims in disasters. Modern techniques have made it possible to get a useful DNA sample from surprising sources, such as a strand of hair or dried saliva on a postage stamp.**

Structure of Nucleic Acids

Nucleic acids comprise a fourth class of biological molecules. They are the information-storage molecules of the cell. This group of molecules got its name from the cellular location in which the molecules are primarily found—the nucleus. It is from this control center of cells that nucleic acids carry out their major functions. A **nucleic acid** is a nitrogen-containing biological polymer that is involved in the storage and transmission of genetic information. The monomer that makes up a nucleic acid is called a **nucleotide.** Each nucleotide has three parts: an inorganic phosphate group, a five-carbon monosaccharide sugar, and a nitrogen-containing structure called a nitrogenous base. Examine each part of **Figure 20a.** Although the phosphate group is the same in all nucleotides, the sugar and the nitrogen base vary.

In a nucleic acid, the sugar of one nucleotide is bonded to the phosphate of another nucleotide, as shown in **Figure 20b.** Thus, the nucleotides are strung together in a chain, or strand, containing alternating sugar and phosphate groups. Each sugar is also bonded to a nitrogen base that sticks out from the chain. The nitrogen bases on adjoining nucleotide units are stacked one above the other in a slightly askew position, much like the steps in a staircase. This orientation is shown in **Figure 20b.** Intermolecular forces hold each nitrogen base close to the nitrogen bases above and below it.

Figure 20 Nucleotides are the monomers from which nucleic acid polymers are formed.

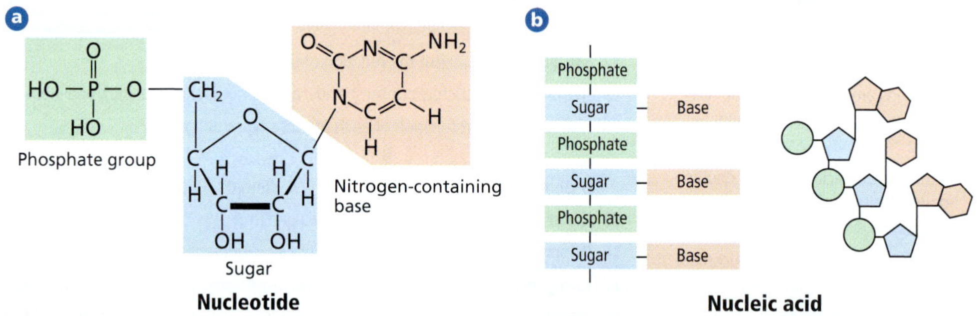

Each nucleotide contains a nitrogen-containing base, a five-carbon sugar, and a phosphate group.

Nucleic acids are linear chains of alternating sugars and phosphates. Attached to every sugar is a nitrogen base. Because the nucleotides are offset, the chains resemble steps in a staircase.

CHEMISTRY PROJECT

Three-Dimensional DNA Ask students to make a 3-D model or a drawing of the DNA double helix. Display the models and drawings in the classroom or in locations around the school. OL EL

DIFFERENTIATED INSTRUCTION

Advanced Learners Have students research the Human Genome Project. The Human Genome Project was an international effort to sequence and map human genes. Have students prepare a short report on this international project. AL

DNA: The Double Helix

You might have heard of DNA (deoxyribonucleic acid), one of the two types of nucleic acids found in living cells. DNA contains the master plans for building all the proteins in an organism's body.

The structure of DNA DNA consists of two long chains of nucleotides wound together to form a spiral structure, as shown in **Figure 21.** Each nucleotide in DNA contains a phosphate group, the five-carbon sugar deoxyribose, and a nitrogenous base. The alternating sugar and phosphate groups in each chain make up the outside, or backbone, of the spiral structure. The nitrogen bases are on the inside of the structure. Because the spiral structure is composed of two chains, it is known as a double helix.

DNA contains four different nitrogenous bases: adenine (A), thymine (T), cytosine (C), and guanine (G). As **Figure 22** shows, both adenine and guanine contain a double ring. Thymine and cytosine are single-ring structures. Looking again at **Figure 22,** notice that each nitrogen base on one strand of the helix is oriented next to a nitrogen base on the opposite strand, in the same way that the teeth of a zipper are oriented. The side-by-side base pairs are close enough so that hydrogen bonds form between them. Because each nitrogen base has a unique arrangement of organic functional groups that can form hydrogen bonds, the nitrogen bases always pair in a specific way so that the optimum number of hydrogen bonds form. As **Figure 22** shows, guanine always binds to cytosine, and adenine always binds to thymine. The G–C and A–T pairs are called complementary base pairs.

☑ READING CHECK **Describe** what forms the teeth of the DNA zipper.

Because of complementary base pairing, the amount of adenine in a molecule of DNA always equals the amount of thymine, and the amount of cytosine always equals the amount of guanine. In 1953, James Watson and Francis Crick used this observation to make one of the greatest scientific discoveries of the twentieth century when they determined the double-helix structure of DNA. They accomplished this feat without performing many laboratory experiments themselves. Instead, they analyzed and synthesized the work of numerous scientists who had carefully carried out studies on DNA.

View an **animation about the structure of DNA.**

Concepts In Motion

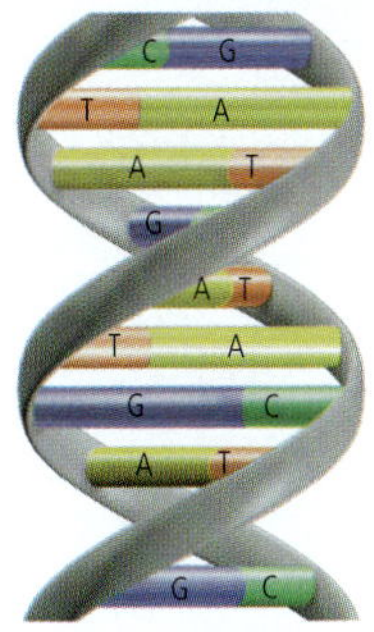

■ **Figure 21** The structure of DNA is a double helix that resembles a twisted zipper. The two sugar-phosphate backbones form the outsides of the zipper.

■ **Figure 22** In DNA, base pairing exists between a double-ringed base and a single-ringed base. Adenine and thymine always pair, forming two hydrogen bonds between them. Guanine and cytosine always form three hydrogen bonds when they pair.

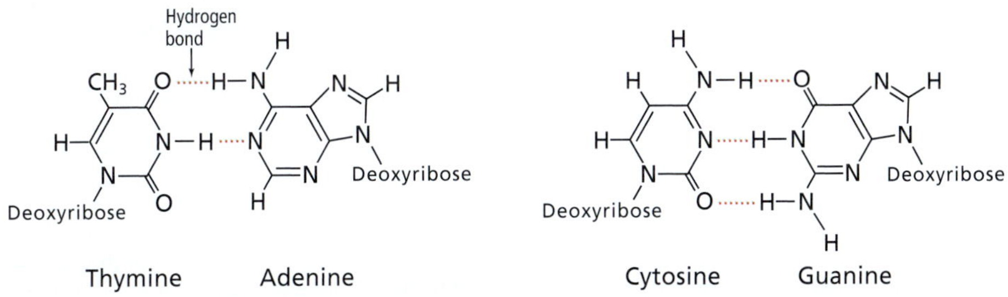

Reinforcement

Nucleic Acid Structure Explain to students that nucleic acid polymers have structures that are more complex than those of protein or carbohydrate polymers. Each nucleotide monomer has three parts: a sugar, a phosphate, and a nitrogen base. Although nucleic acids are linear, the nitrogen bases stick out to the sides like branches. Ask students to practice drawing the structure of DNA or RNA. Student structures should clearly show the three parts of the monomers attached in the correct positions. OL EL

☑ READING CHECK nitrogen bases

Quick Demo

View DNA Ask students what they think DNA would look like if they could see it. Then, let them see if they are right as you extract DNA from wheat germ. Place a small amount of raw wheat germ in a mortar and add about 5 mL of a solution containing 10% dish detergent (to break open cells) and 0.2M citric acid (to protect the DNA by chelating Mg^{2+} and Mn^{2+} ions that DNase enzymes need to break down DNA). Gently grind the wheat germ for 1 minute and strain the mixture through a kitchen strainer or cheesecloth. Slowly add 10 mL of 90–100% alcohol (either ethanol or isopropanol) to the filtrate. Spool strands of DNA onto a rod and let students examine their appearance. OL

CHEMISTRY JOURNAL

Genes and Disease Scientists are finding inherited genes that are responsible for some forms of disease, such as Alzheimer's disease and breast cancer. If you could be tested for the presence of a group of genes that might cause a disease later in your life, would you want the tests performed? Write a summary of your thoughts and feelings about knowing this information in your chemistry journal. OL

ASSESSMENT

Skill Ask students to make a graphic organizer illustrating the relationships among the functions of DNA, RNA, and proteins. Student answers should show that DNA stores genetic information, RNA uses that information to make proteins, and proteins control processes in the cells. OL

Problem-Solving LAB

Purpose Students will model the replication of a small portion of a DNA molecule.

Process Skills sequence, observe and infer, interpret scientific illustrations, apply concepts, hypothesize

Background When DNA replicates, more than a dozen enzymes complete the process of unzipping the strands, base-pairing a new set of nucleotides, and linking them all together. In bacteria, replication takes place at a rate of about 500 nucleotides per second. In higher organisms, the rate is one-tenth as fast.

Teaching Strategies

- Explain or have students research why the complementary base pairs of the DNA molecule are thymine-adenine and cytosine-guanine. These combinations of a purine and a pyrimidine produce a uniform helical diameter. T and A have two hydrogen bonding sites; C and G have three.
- Have students draw a two-dimensional sequence of the hydrogen bonding that occurs between T-A and C-G and the covalent bonding that connects adjacent sugars to phosphates.

Think Critically

1. The base sequence of the new strand is complementary to the sequence of the original strand to which it is bonded.
2. All new DNA molecules will have one red strand and one blue strand. This shows that replication is semiconservative. Each molecule has one original strand and one new strand.
3. The error would be passed on to the RNA, which would then be used to direct the production of a defective protein with an incorrect amino acid. If this error occurred in a sex cell and if the defective protein were vital to life, the new individual might not live. Yes, the effects are permanent because the error would be replicated.

The function of DNA Watson and Crick used their model to predict how DNA's chemical structure enables it to function. DNA stores the genetic information of a cell in the cell's nucleus. Before the cell divides, the DNA is copied so that the new generation of cells gets the same genetic information. Having determined that the two chains of the DNA helix are complementary, Watson and Crick realized that complementary base pairing provides a mechanism by which the genetic material of a cell is copied.

The four nitrogenous bases of DNA serve as the letters of the alphabet in the information-storage language of living cells. The specific sequence of these letters represents an organism's master instructions, just as the sequence of letters in the words of this sentence convey special meaning. The sequence of bases is different in every species of organism, allowing for an enormous diversity of life-forms—all from a language that uses only four letters. It is estimated that the DNA in a human cell has about three billion complementary base pairs, arranged in a sequence unique to humans.

Problem-Solving LAB

Formulate a Model

How does DNA replicate? DNA replicates before a cell divides so that each of the two newly formed cells has a complete set of genetic instructions. When DNA begins to replicate, the two nucleotide strands start to unzip. An enzyme breaks the hydrogen bonds between the nitrogenous bases, and the strands separate. Other enzymes deliver free nucleotides from the surrounding medium to the exposed nitrogenous bases, adenine hydrogen-bonding with thymine, and cytosine bonding with guanine. Thus, each strand builds a complementary strand by base-pairing with free nucleotides. This process is shown in the top diagram at the right. When the free nucleotides have been hydrogen-bonded into place, their sugars and phosphates bond covalently to those on adjacent nucleotides to form the new backbone. Each strand of the original DNA molecule is now bonded to a new strand.

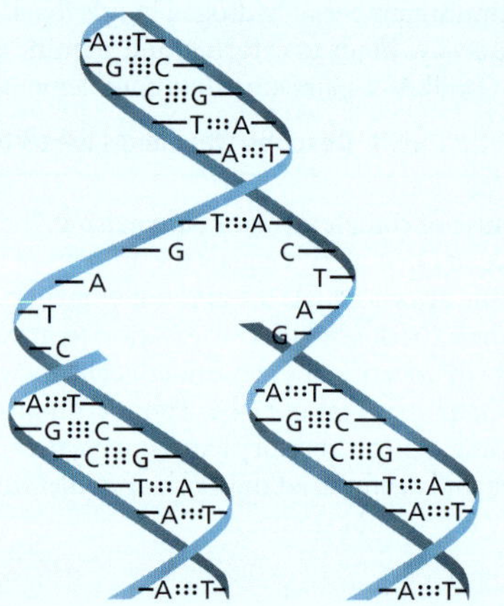

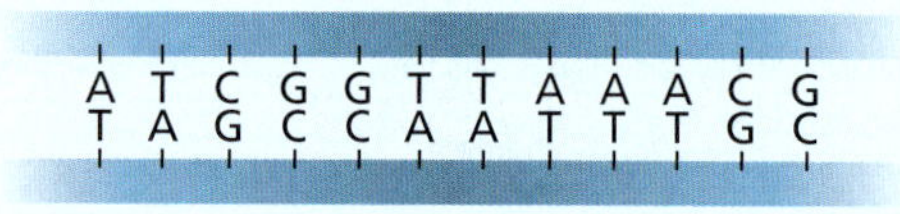

Analysis

The bottom diagram shows a small segment of a DNA molecule. Copy the base sequence onto a clean sheet of paper, being careful not to make copying errors. Show the steps of replication to produce two segments of the DNA.

Think Critically

1. **Describe** how the base sequence of a newly synthesized strand compares with the original strand to which it is bonded.
2. **Explain** If the original DNA segment is colored red and the free nucleotides are colored blue, what pattern of colors will the newly replicated DNA segments have? Will all new segments have the same color pattern?
3. **Explain** how an organism might be affected if an error occurs during replication of its DNA. Are the effects permanent? Explain.

Caption Question Fig. 23 Answers may include the following: DNA contains the sugar deoxyribose; RNA's sugar is ribose. DNA is arranged in a double helix, with hydrogen bonding between the nitrogenous bases; RNA is arranged as a single strand. DNA contains thymine; RNA contains uracil.

ASSESSMENT

Performance If DNA molecular models are available, have students construct a T-A, G-C chain, describe its geometry and bonding, and describe how replication would proceed. This could also be accomplished using three-dimensional computer modeling software or the Internet. OL

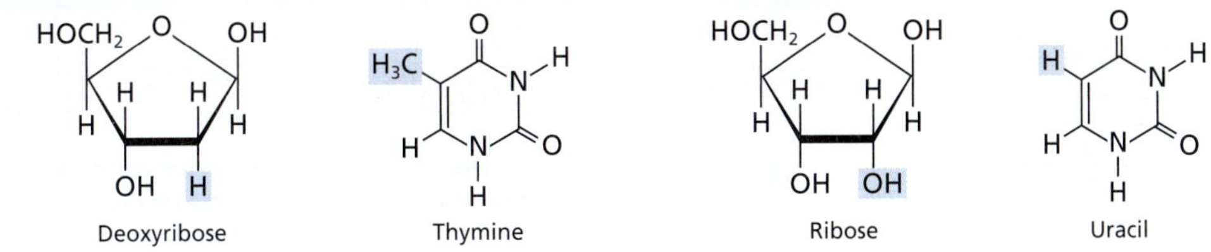

■ **Figure 23** DNA and RNA differ in their components. The two structures on the left are found in DNA. The two structures on the right are found in RNA.
Identify *two differences in the structures of RNA and DNA.*

RNA

RNA (ribonucleic acid) is also a nucleic acid. Its general structure differs from that of DNA in three important ways, as shown in **Figure 23.** First, as you have read, DNA contains the nitrogen bases adenine, cytosine, guanine, and thymine. RNA contains adenine, cytosine, guanine, and uracil. Thymine is never found in RNA. Second, RNA contains the sugar ribose. DNA contains the sugar deoxyribose, which has a hydrogen atom in place of a hydroxyl group at one position.

The third difference between DNA and RNA is a result of these structural differences. DNA is normally arranged in a double helix in which hydrogen bonding links the two chains together through their bases. RNA is usually single-stranded, with no such hydrogen bonds forming among the bases.

Whereas DNA functions to store genetic information, RNA allows cells to use the information found in DNA. You have read that the genetic information of a cell is contained in the sequence of nitrogen bases in the DNA molecule. Cells use this base sequence to make RNA with a corresponding sequence. The RNA is then used to make proteins, each with an amino-acid sequence that is determined by the order of nitrogen bases in RNA. The sequences of bases are referred to as the genetic code. Because proteins are the molecular tools that carry out most activities in a cell, the DNA double helix is ultimately responsible for controlling the thousands of chemical reactions that take place in cells.

FOLDABLES® Incorporate information from this section into your Foldable.

SECTION 4 REVIEW

Section Self-Check

Section Summary

- Nucleic acids are polymers of nucleotides, which consist of a nitrogen base, a phosphate group, and a sugar.
- DNA and RNA are the information-storage molecules of a cell.
- DNA is double stranded, and RNA is single stranded.

20. MAINIDEA **Explain** the primary function of RNA and DNA.

21. **Identify** the specific structural components of both RNA and DNA.

22. **Relate** the function of DNA to its structure.

23. **Relate** the function of RNA to its structure.

24. **Analyze** the structure of nucleic acids to determine what structural feature makes them acidic.

25. **Predict** what might happen if the DNA that coded for a protein contained the wrong base sequence.

FOLDABLES® Rubric

3 Assess

Check for Understanding

Show pictures of DNA and RNA models. Ask students to identify each and explain how they arrived at their answers. Students should look for uracil and ribose in RNA, which is mostly single-stranded. They should look for thymine and deoxyribose in DNA, which is usually double-stranded. OL

Reteach

Ask students to explain why, even though the DNA polymer is so stable that scientists have been able to isolate intact DNA from fossils that are millions of years old, our cells can unzip the strands of the double helix in seconds. Remind students that they should focus on the nature of interparticle forces. Each strand of the DNA polymer is held together by strong covalent ester bonds, making it very stable. The two strands of the helix are held together by the much weaker hydrogen bonds between complementary bases. OL

Extension

Have small groups of students research how the structure of DNA was discovered. Have students prepare a short oral to present to the class. The report should contain information about Rosalind Franklin, Maurice Wilkins, Francis Crick, and James Watson. BL OL AL EL COOP LEARN

SECTION 4 REVIEW

20. The primary function of RNA is protein synthesis. The primary function of DNA is storage of genetic information.

21. RNA contains ribose, phosphate groups, and A, C, G, and U bases. DNA contains deoxyribose, phosphate groups, and A, C, G, and T bases.

22. DNA is a double-stranded structure that unzips and undergoes complementary base pairing. This process ensures that the DNA sequence is copied exactly, passing the genetic information to the new cells.

23. RNA is single-stranded and is used to make proteins with an amino acid sequence determined by the order of nitrogen bases in the RNA.

24. The phosphate groups make nucleic acids acidic.

25. A protein made from DNA with the wrong base sequence might have the wrong amino acid sequence.

SECTION 5

1 Focus

MAINIDEA

Metabolism Ask students if they have ever heard of a person having a high or low metabolism and what that means. Possible answer: A person with high metabolism has a hard time gaining weight, and a person with low metabolism has a hard time losing weight. Ask students what high and low metabolism have to do with this section. This section explains what metabolism is. OL

2 Teach

Content Background

Organic Compounds In the 1800s, scientists placed all compounds into one of two categories. If they were produced by a living organism, they were called *organic;* if they were found in minerals, they were called *inorganic*. It was believed that organic substances could not be synthesized because they possessed a vital force that came from living things. That thinking changed after Friedrich Wohler synthesized urea from two inorganic salts. Today, complex organic molecules have been made in the laboratory.

Caption Question Fig. 24
Answers will vary. Sample answer: french fries: The carbohydrates in the potatoes were broken down into simple sugars and then used to form polysaccharides. The fats from the oil were broken down into fatty acids and then used to form triglycerides.

SECTION 5

Essential Questions

- How do anabolism and catabolism compare?
- What is the role of ATP in metabolism?
- How can the processes of photosynthesis, cellular respiration, and fermentation be compared and contrasted?

Review Vocabulary

redox process: a chemical reaction in which electrons are transferred from one atom to another

New Vocabulary

metabolism
catabolism
anabolism
ATP
photosynthesis
cellular respiration
fermentation

Metabolism

MAINIDEA Metabolism involves many thousands of reactions in living cells.

CHEM 4 YOU You have studied the four major kinds of biological molecules and learned that they are all present in the food you eat. What happens to these molecules after they enter your body?

Anabolism and Catabolism

Many thousands of chemical reactions take place in the cells of a living organism. The set of chemical reactions that occur within an organism is its **metabolism.** Why are so many reactions involved in metabolism? Living organisms must accomplish two major functions in order to survive. They have to extract energy from nutrients in forms that they can use immediately as well as store for future use. In addition, they have to use nutrients to make building blocks for synthesizing all of the molecules needed to perform their life functions. These processes are summarized in **Figure 24.**

The term **catabolism** refers to the metabolic reactions that break down complex biological molecules such as proteins, polysaccharides, triglycerides, and nucleic acids for the purposes of forming smaller building blocks and extracting energy. After you eat a meal of spaghetti and meatballs, your body immediately begins to break down the starch polymer in the pasta into glucose. The glucose is then broken down into smaller molecules in a series of energy-releasing catabolic reactions. Meanwhile, the protein polymers in the meatballs are catabolized into amino acids.

The term **anabolism** refers to the metabolic reactions that use energy and small building blocks to synthesize the complex molecules needed by an organism. After your body has extracted the energy from the starch in the pasta, it uses that energy and the amino-acid building blocks produced from the meat proteins to synthesize the specific proteins that allow your muscles to contract, catalyze metabolic reactions, and perform many other functions in your body.

Figure 24 A large number of different metabolic reactions take place in living cells. Some involve breaking down nutrients to extract energy; these are catabolic processes. Others involve using energy to build large biological molecules; these reactions are anabolic processes.

Describe *Choose one food that you ate recently, and describe how it was metabolized.*

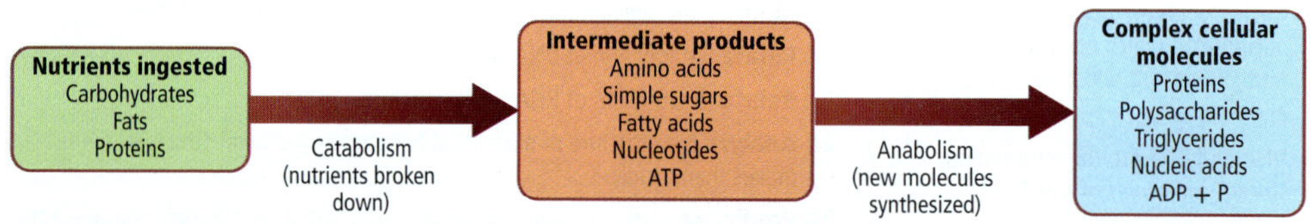

DIFFERENTIATED INSTRUCTION

English Learners Have students write statements explaining the meanings of the terms *anabolism* and *catabolism.* Then, have students write a short paragraph in which they incorporate both terms correctly. EL BL

CHEMISTRY PROJECT

Metabolism Have students work in small groups to create a slide show or animated computer-generated movie showing how anabolism and catabolism work. Students should use Figure 25 as a reference. Have students present their creation to the class. OL AL
COOP LEARN

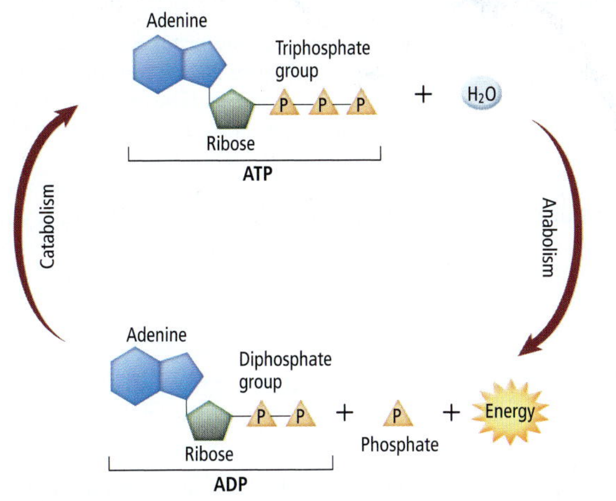

■ **Figure 25** The breakdown of ATP provides energy for cellular processes in living organisms.
Explain *where the energy is stored in ATP.*

Figure 24 shows the relationship between catabolism and anabolism. The nutrients listed on the left side of the diagram are broken down into intermediate products. These intermediate products are used as building blocks for the products listed on the right side of the diagram. Another way of conceptualizing this process is to view the nutrients ingested as the raw materials for the complex cellular molecules formed in a living organism.

Vocabulary
Academic Vocabulary
Conceptualize
visualizing or conceiving an abstract idea in the mind
The atomic cloud model is hard to conceptualize.

☑ READING CHECK **Explain** how the terms *metabolism, catabolism,* and *anabolism* are related.

ATP Catabolism and anabolism are linked by common building blocks that catabolic reactions produce and anabolic reactions use. A common form of potential chemical energy also links the two processes, as shown in **Figure 25.** **ATP** (adenosine triphosphate) is a nucleotide that functions as the universal energy-storage molecule in living cells. During catabolic reactions, cells harness the chemical energy of foods and store it in the bonds of ATP. When these bonds are broken, the chemical energy is released and used by cells to drive anabolic reactions that might not otherwise occur. Most cellular reactions have an efficiency of only about 40% at best; the remaining 60% of the energy in food is lost as heat, which your body uses to keep warm.

During catabolic reactions, cells produce ATP by adding an inorganic phosphate group to the nucleotide adenosine diphosphate (ADP) in an endothermic reaction. One mole of ATP stores approximately 30.5 kJ of energy under normal cellular conditions. During anabolism, the reverse reaction occurs. ATP is broken down to form ADP and inorganic phosphate in an exothermic reaction. Approximately 30.5 kJ of energy is released from each mole of ATP.

☑ READING CHECK **Describe** what occurs when ATP becomes ADP.

Visual Learning

Figure 25 Ask students to examine the structure of ATP and compare and contrast its structure with that of the nucleotide in Figure 20a. Make sure students understand that the adenine rings in ATP contain four nitrogen atoms and the ribose ring contains an oxygen atom. These atoms have been deleted in this shorthand representation of ATP. Both ATP and the nucleotide contain a sugar, phosphate, and nitrogen base. ATP has three phosphates, its sugar is always ribose, and its nitrogen base is always adenine. Nucleotides can have ribose or deoxyribose, any of the five bases, and only one phosphate. **OL** **AL**

■ **Caption Question Fig. 25**
In the bonds of the ATP molecule.

☑ READING CHECK Metabolism is the set of chemical reactions carried out within an organism. Catabolism refers to the metabolic reactions that break down complex biological molecules in an organism. Anabolism refers to the metabolic reactions that synthesize complex biological molecules needed by an organism.

☑ READING CHECK A phosphate group is released from ATP in an exothermic reaction.

DIFFERENTIATED INSTRUCTION

Visually Impaired Have a student or students explain Figure 25 in detail to visually impaired students. Encourage discussion between the students about the figure to ensure that all students understand the processes. **COOP LEARN**

Quick Demo

Form Carbonic Acid Add $3M$ NaOH by the drop to a solution of the acid-base indicator thymol blue just until a blue color appears. Ask each student to pour 5 mL of this solution into a test tube. Ask students to predict what will happen if they blow air into the solution using a straw. Have students test their predictions by using a clean long straw to blow into the solution until a color change takes place.

Disposal Flush the solution down a drain with plenty of water. Students should explain that the carbon dioxide they exhale forms carbonic acid in an aqueous solution. When they have acidified the solution enough, it turns from blue to clear. OL

ASSESSMENT

Knowledge Give students the equation for photosynthesis and ask them to convert it into the equation for cellular respiration. The answer is simply the reverse of the equation for photosynthesis. OL

■ **Figure 26** Grass and other green plants use certain wavelengths of sunlight as an energy source. Other living organisms, such as cows, obtain energy by eating plants or eating other organisms that eat plants.

Get help with **photosynthesis and respiration.**

Personal Tutor

Photosynthesis

What is the source of energy that fuels metabolism? For most living things, including the grass shown in **Figure 26,** certain wavelengths of sunlight provide this energy. Some bacteria and the cells of all plants and algae are able to capture light energy and convert some of it into chemical energy. Animals cannot capture light energy, so they get energy by eating plants or by eating other animals that eat plants. The process that converts energy from sunlight to chemical energy in the bonds of carbohydrates is called **photosynthesis.** During the complex process of photosynthesis, carbon dioxide and water yield a carbohydrate (glucose) and oxygen gas. The following net reaction takes place during photosynthesis.

$$\underset{\text{Carbon dioxide}}{6CO_2} + \underset{\text{Water}}{6H_2O} + \text{light energy} \rightarrow \underset{\text{Glucose}}{C_6H_{12}O_6} + \underset{\text{Oxygen}}{6O_2}$$

Photosynthesis results in the reduction of the carbon atoms in carbon dioxide as glucose is formed. During this redox process, oxygen atoms in water are oxidized to oxygen gas.

Cellular Respiration

Most organisms need oxygen to live. Oxygen that is produced during photosynthesis is used by living things during **cellular respiration,** the process in which glucose is broken down to form carbon dioxide, water, and large amounts of energy. Cellular respiration is the major energy-producing process in living organisms. **Figure 27** shows one use of energy in the body. This energy is stored in the bonds of ATP. Cellular respiration is a redox process; the carbon atoms in glucose are oxidized while oxygen atoms in oxygen gas are reduced to the oxygen in water. The net reaction that takes place during cellular respiration is as follows.

$$\underset{\text{Glucose}}{C_6H_{12}O_6} + \underset{\text{Oxygen}}{6O_2} \rightarrow \underset{\text{Carbon dioxide}}{6CO_2} + \underset{\text{Water}}{6H_2O} + \text{energy}$$

■ **Figure 27** Swimmers need large amounts of energy when they compete in a race. This energy is stored in the bonds of ATP in their cells.

CHEMISTRY JOURNAL

Sources of Energy Ask students to draw a simple food chain in their chemistry journal that starts with a green plant. The food chain should include at least three organisms. BL EL

DIFFERENTIATED INSTRUCTION

Struggling Learners The equation for cellular respiration can be stated as: *When you eat sugar and breathe in air, your body converts sugar and oxygen molecules into the carbon dioxide you breathe out, water that is used by your cells, and energy that allows you to move and talk and work.* BL

■ **Figure 28** Carbon dioxide formed during fermentation, leaves holes in the bread. These holes give bread a light, less-dense texture.

Fermentation

During cellular respiration, glucose is completely oxidized, and oxygen gas is required to act as the oxidizing agent. Cells can extract energy from glucose in the absence of oxygen, but not nearly as efficiently. Without oxygen, only a fraction of the chemical energy of glucose can be released. Whereas cellular respiration produces 38 mol of ATP for every 1 mol of glucose catabolized in the presence of oxygen, only 2 mol of ATP are produced per mole of glucose that is catabolized in the absence of oxygen. This provides enough energy for oxygen-deprived cells so that they do not die. The process by which glucose is broken down in the absence of oxygen is known as **fermentation.** There are two common kinds of fermentation. In one, ethanol and carbon dioxide are produced. In the other, lactic acid is produced.

Alcoholic fermentation Yeast and some bacteria can ferment glucose to produce the alcohol ethanol.

$$\underset{\text{Glucose}}{C_6H_{12}O_6} \rightarrow \underset{\text{Ethanol}}{2CH_3CH_2OH} + \underset{\text{Carbon dioxide}}{2CO_2} + \text{energy}$$

This reaction, called alcoholic fermentation, is important in producing some foods, as shown in **Figure 28.** Alcoholic fermentation is needed to make bread dough rise, form tofu from soybeans, and produce the ethanol in alcoholic beverages. Another use of the ethanol is as an additive to gasoline, as shown in **Figure 29.**

CAREERS IN CHEMISTRY

Baker Using a variety of chemical processes to create tasty and often beautiful creations is the job of a baker. Not only do breads and doughnuts undergo fermentation processes, but cakes and other pastries are often acid-base reactions.

WebQuest

■ **Figure 29** Ethanol is often added to gasoline and used as a fuel in some cars and trucks. Ethanol is made from grain.
Explain *how the use of ethanol can reduce the dependence on fossil fuels.*

CHEMISTRY JOURNAL

Ethanol Using grains to produce ethanol to power cars is one form of renewable energy. Ask students to write a paragraph explaining why finding renewable energy sources is an important scientific endeavor. OL

Quick Demo

Fermentation Place approximately 10 g of sucrose, 100 mL of water, and a packet of yeast in a 250 mL flask. Stretch a balloon over the mouth of the flask. Ask students to observe the setup periodically throughout the class period. After the last observation, ask students to explain what kind of process is taking place. The yeast is carrying out fermentation and the balloon expands as it fills with carbon dioxide gas. OL

CAREERS IN CHEMISTRY

WebQuest

■ **Caption Question Fig. 29** Using a percentage of ethanol in gasoline, reduces the amount of fossil fuel that is used to power our cars, trucks, and buses.

Enrichment

Life Processes Divide the class into small groups. Assign each group the task of giving an oral presentation explaining anabolism and catabolism, photosynthesis, cellular respiration, or fermentation. Students should have at least one visual aid for their presentation. OL COOP LEARN

ASSESSMENT

Performance As students give their oral presentations, have the remaining students write down facts about each topic on a sheet of paper. After the presentations, have each student write one question using their list of facts. Have students exchange questions and try to answer each other's questions. OL COOP LEARN

3 Assess

Check for Understanding

Have students write explanations that compare and contrast catabolism and anabolism. BL OL

Reteach

Ask students why cells use a universal energy currency such as ATP rather than just using chemical energy directly from food molecules. Remind students that most cellular processes depend on the catalytic activity of enzymes. Students should explain that having a universal energy currency allows many different enzymes to use the energy stored in ATP without requiring them to have active sites designed to bind to all the different types of fuel molecules (such as proteins, carbohydrates, and lipids). OL

Extension

Have students research the role of chlorophyll in photosynthesis and summarize their findings in a short report. AL

ASSESSMENT

Skill Have students make a table summarizing cellular respiration, alcoholic fermentation, and lactic acid fermentation. The table should include reactants, products, and moles of ATP produced. OL

Figure 30 During strenuous activity, oxygen can be depleted in cells. Then, energy is produced without oxygen and lactic acid is produced. Soreness in muscles a day or two after the activity is a sign of lactic acid formation.

Lactic acid fermentation Have you ever experienced muscle fatigue while running a race, like the person shown in **Figure 30?** During strenuous activity, muscle cells often use oxygen faster than it can be supplied by the blood. When the supply of oxygen is depleted, cellular respiration stops. Although animal cells cannot undergo alcoholic fermentation, they can produce lactic acid and a small amount of energy from glucose through lactic acid fermentation.

$$\underset{\text{Glucose}}{C_6H_{12}O_6} \rightarrow \underset{\text{Lactic acid}}{2CH_3CH(OH)COOH} + \text{energy}$$

The lactic acid that is produced is moved from the muscles through the blood to the liver. There, it is converted back into glucose that can be used in catabolic processes to yield more energy once oxygen becomes available. However, if lactic acid builds up in muscle cells at a faster rate than the blood can remove it, muscle fatigue results. An immediate burning sensation and soreness a few days later is an indication that lactic acid was produced in the muscles during exercise.

SECTION 5 REVIEW

Section Self-Check

Section Summary

- Living organisms undergo catabolism and anabolism.
- Photosynthesis directly or indirectly provides almost all living things with energy.
- The net equation for cellular respiration is the reverse of the net equation for photosynthesis.

26. MAIN IDEA **Explain** why metabolism is important to living cells.
27. **Compare and contrast** the processes of anabolism and catabolism.
28. **Explain** the role of ATP in the metabolism of living organisms.
29. **Compare and contrast** the processes of photosynthesis, cellular respiration, and fermentation.
30. **Determine** whether each process is anabolic or catabolic.
    a. photosynthesis
    b. cellular respiration
    c. fermentation
31. **Evaluate** Why is it necessary to use sealed casks when making wine?
32. **Calculate** How many moles of ATP would a yeast cell produce if 6 mol of glucose were oxidized completely in the presence of oxygen? How many moles of ATP would the yeast cell produce from 6 mol of glucose if the cell were deprived of oxygen?

SECTION 5 REVIEW

26. Metabolism is important because it is the collection of reactions which organisms use to obtain energy and nutrients to carry out life processes.
27. Anabolism includes the metabolic reactions that build up large molecules from smaller nutrient building blocks, consuming energy in the process. Catabolism includes the metabolic reactions that break down nutrients, extracting energy in the process.
28. ATP functions as an energy currency. The potential chemical energy in food molecules is converted during metabolism to the potential energy of ATP.
29. All are processes of living organisms. Photosynthesis is the process in which energy from the Sun is converted into chemical energy in glucose. Cellular respiration is the process in which living organisms break down glucose to form carbon dioxide, water, and energy. Fermentation is the process in which glucose is broken down by organisms in the absence of oxygen to form ethanol, carbon dioxide, and energy.
30. **a.** photosynthesis, anabolic
    b. cellular respiration, catabolic
    c. fermentation, catabolic
31. to keep oxygen out so cellular respiration does not occur
32. a maximum of 228 moles of ATP in the presence of oxygen; 12 moles of ATP in the absence of oxygen.

CHEMISTRY & careers

Career: Molecular Paleontologist

Acid Test Reveals Surprise

"No right-thinking paleontologist would do what Mary did. We don't go to all this effort to dig this stuff out of the ground and then destroy it in acid." So says a colleague about Mary Schweitzer, the scientist who used the techniques of molecular biology to discover soft tissue where none should be—in the thighbone of a 68-million-year-old *Tyrannosaurus rex.*

Mother Bob When the fossilized *T. rex,* nicknamed Bob, was recovered by paleontologists in 2003 from a remote section of Montana, the bones were encased in plaster for protection during transport. However, the bones and plaster weighed more than the helicopter could lift. So the paleontologists were forced to break the intact thighbone to move the dinosaur out of the remote area. Schweitzer took small fragments from the broken thighbone for further study.

The first surprise came quickly. "Bob" was a female, and she had been producing eggs at the time of her death. The bone Schweitzer studied is called medullary bone. Previously, this bone tissue was known only in birds, as shown in **Figure 1.** Ovulating hens produce medullary bone, then later use the calcium stored in the bone to make eggshells. After egg production, the bone disappears. **Figure 1** shows the medullary bone found in the *T. rex* thighbone.

Figure 1 The hen bone and *T. rex* bone both have a hard outer bone called cortical bone (CB) and softer medullary bone (MB).

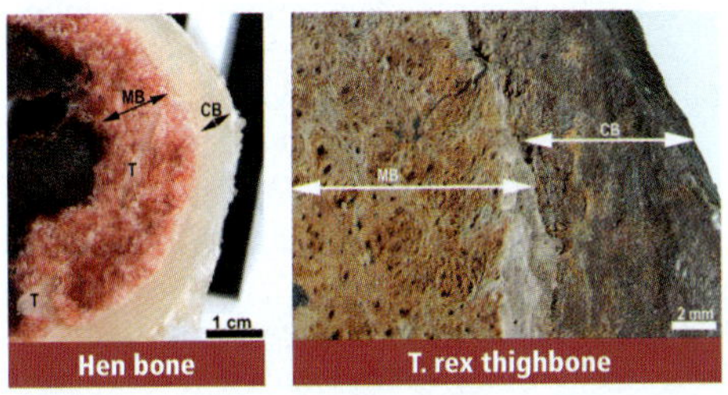

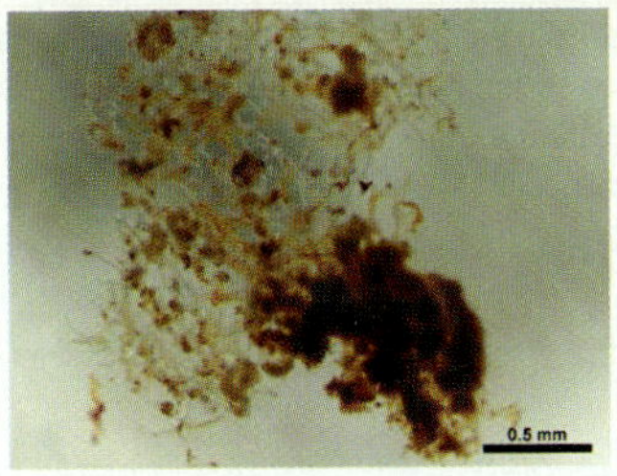

Figure 2 Scientists also found blood vessels and individual cells in the soft tissue of the *T. rex.*

The acid test To study the medullary bone more closely, Schweitzer dissolved fragments of the bone in dilute acid to remove calcium phosphate—a technique normally used to examine fresh tissue. Because a fossilized bone has usually mineralized, it was assumed that the bone would completely dissolve in dilute acid. Yet this step yielded astonishing results—within the bone was soft tissue. Under the microscope, the tissue showed what looked like preserved blood vessels and even individual cells, as shown in **Figure 2.** But how could soft tissue have survived 68 million years in the ground?

More work Schweitzer has since subjected other bones to the same acid test, and found similar soft tissue and fine structures. No one knows yet just what these fine structures are showing, but, says a colleague, "there may be a lot of things out there that we've missed because of our assumption of how preservation works." Clearly, more research is needed.

WRITING IN ▶ Chemistry

Persuasive Writing It is unlikely that dinosaur DNA will be found in these soft tissues. Even so, the discovery brings up the question: Should extinct animals be cloned from recovered DNA? Write a persuasive essay expressing your opinion.

WebQuest

CHEMISTRY & careers

Purpose

Students will investigate the discovery of soft tissue in the bone of a 68-million-year-old fossilized *Tyrannosaurus rex.*

Background

Fossils form when minerals seep into bony structures, such as bones and teeth. Minerals replace organic compounds, leaving behind rock in the shape of the dead organism. Soft parts, such as tissues and blood vessels, normally decay too quickly to be fossilized.

Teaching Strategies

- When Dr. Schweitzer first discovered what looked like blood cells in fossilized dinosaur bone, her mentor told her, "Now see if you can find some evidence to show that that's not what they are." Discuss this comment in class. Why would scientists try to disprove their own hypotheses?
- One critic said of Dr. Schweitzer's work that she should have conducted more tests before publishing. He said, "I'm more in favor of longer work with slam-dunk authenticity." Dr. Schweitzer countered that she had to publish quickly to ensure her funding. Have students discuss the two sides of this issue.

WRITING IN ▶ Chemistry

Rubric WebQuest

Persuasive Writing After performing further research, students will learn that dinosaur DNA is difficult (if not impossible) to find. However, there are other more recently extinct animals whose DNA can be collected. Should we try to revive animals such as the Tasmanian wolf or the woolly mammoth? Ideas such as the causes of extinction and our responsibilities to our fellow species might be discussed.

ChemLAB

Rubric

Preparation

Time Allotment one class period

Process Skills hypothesize, collect and organize data, measure, compare and contrast, recognize cause and effect, make and use graphs

Safety Precautions Approve lab safety forms before work begins. Students should use only GFCI protected circuits for the hot plate. Students should use only 3% hydrogen peroxide, as stated in the lab.

Disposal All solutions can be poured down the drain with water.

Procedure

- Some test tubes might foam over. This is not a problem but will require students to be more creative with their measurements.
- It is important that students take the liquid from the blended potatoes if they are using a disposable pipette to prevent clogging.

Analyze and Conclude

1. The line graph should show that the liver produces more foam than the potato until the enzymes become denatured.
2. The enzymes function more as temperature increases until the enzymes are denatured and lose function.
3. the hot water bath, because the enzymes were denatured
4. Answers will vary.
5. potato: $2H_2O_2(aq) \longrightarrow 2H_2O(g) + O_2(g)$
liver: $2H_2O_2(aq) \longrightarrow 2H_2O(g) + O_2(g)$
The reactions are identical because the enzymes do not participate in the reaction; they only catalyze it.
6. Possible answers: The contents foamed out of the test tube and the height of the foam was an estimate. A possible method to correct this is to use larger test tubes.

INQUIRY EXTENSION

Check student designs.

ChemLAB

Observe Temperature and Enzyme Action

Background: **Enzymes are natural catalysts used by living things to speed reactions. These proteins have specialized structures that enable them to interact with specific substances.**

Question: *How does temperature affect the action of enzymes?*

Materials

red-skin potato pulp
hydrogen peroxide ($3\%\ H_2O_2$)
water
250-mL beaker (4)
test tubes (4)
test-tube rack
test-tube clamp
25-mL graduated cylinder
thermometer
ice
ruler
clock
hot plate
raw fresh liver pulp

Safety Precautions

Procedure

1. Read and complete the lab safety form.
2. Write a hypothesis that identifies the temperature at which the enzymes are the most active.
3. Copy the data table on a separate sheet of paper.
4. Place the four test tubes in the test-tube rack.
5. Measure and place 2.0 mL of red-skin potato pulp into each test tube.
6. Using the hot plate and ice, prepare water baths in the beakers at four different temperatures: ice water, room-temperature water, body-temperature water, and gently boiling water at or near 100°C.
7. Place one test tube in each water bath using a test-tube clamp.
8. Measure and record the temperature of each water bath.
9. After 5 min in the water baths, measure and place 5.0 mL of 3% H_2O_2 in each test tube.
10. Allow the reaction to proceed for 5 min.
11. Measure the height of the foam produced in each test tube.
12. Dispose of the contents of the test tubes as directed by your teacher and wash the test tubes.
13. Repeat Steps 4–12 using 2.0 mL of beef liver pulp instead of potato pulp.
14. **Cleanup and Disposal** Dispose of the remaining solutions as directed by your teacher. Wash and return all lab equipment to its designated location.

Data Table

Water Bath	Temperature (°C)	Height of Foam (cm)
Potato		
Ice water		
Room-temperature water		
Body-temperature water		
Boiling water (near 100°C)		
Liver		
Ice water		
Room-temperature water		
Body-temperature water		
Boiling water (near 100°C)		

Analyze and Conclude

1. **Make and Use Graphs** Make a line graph with temperature on the *x*-axis and height of foam on the *y*-axis. Use a different color for the potato and liver data points and lines.
2. **Summarize** How does temperature affect the action of enzymes? Infer why the maximum reaction occurred at the temperature in which it did for the potato and liver.
3. **Recognize Cause and Effect** Which water bath produced the least amount of foam for each material? Propose explanations for why this happened.
4. **Compare and Contrast** Did the experimental data support your hypothesis in Step 2? Explain.
5. **Model** Write a balanced reaction for the decomposition of hydrogen peroxide for each reaction. How are the reactions similar? Infer why they are similar.
6. **Error Analysis** Identify potential sources of errors for this investigation and suggest methods to correct them.

INQUIRY EXTENSION

Design an Experiment Would a change in pH affect the results? Design an experiment to find out.

STUDY GUIDE

Vocabulary Practice

BIGIDEA Biological molecules—proteins, carbohydrates, lipids, and nucleic acids—interact to carry out activities necessary to living cells.

SECTION 1 Proteins

MAINIDEA Proteins perform essential functions, including regulation of chemical reactions, structural support, transport of materials, and muscle contractions.

- Proteins are biological polymers made of amino acids that are linked by peptide bonds.
- Protein chains fold into intricate three-dimensional structures.
- Proteins have many functions in the human body, including functions within cells, functions between cells, and functions of structural support.

VOCABULARY
- protein
- amino acid
- peptide bond
- peptide
- denaturation
- enzyme
- substrate
- active site

SECTION 2 Carbohydrates

MAINIDEA Carbohydrates provide energy and structural material for living things.

- Carbohydrates are compounds that contain multiple hydroxyl groups (–OH) and a carbonyl functional group (C=O).
- Carbohydrates range in size from single monomers to polymers composed of hundreds or thousands of monomers.
- Monosaccharides in aqueous solution exist in both open-chain and cyclic structures.

VOCABULARY
- carbohydrate
- monosaccharide
- disaccharide
- polysaccharide

SECTION 3 Lipids

MAINIDEA Lipids make cell membranes, store energy, and regulate cellular processes.

- Fatty acids are long-chain carboxylic acids that usually have between 12 and 24 carbon atoms.
- Saturated fatty acids have no double bonds; unsaturated fatty acids have one or more double bonds.
- Fatty acids can be linked to glycerol backbones to form triglycerides.
- Steroids are lipids that have multiple-ring structures.

VOCABULARY
- lipid
- fatty acid
- triglyceride
- saponification
- phospholipid
- wax
- steroid

SECTION 4 Nucleic Acids

MAINIDEA Nucleic acids store and transmit genetic information.

- Nucleic acids are polymers of nucleotides, which consist of a nitrogen base, a phosphate group, and a sugar.
- DNA and RNA are the information-storage molecules of a cell.
- DNA is double stranded, and RNA is single stranded.

VOCABULARY
- nucleic acid
- nucleotide

SECTION 5 Metabolism

MAINIDEA Metabolism involves many thousands of reactions in living cells.

- Living organisms undergo catabolism and anabolism.
- Photosynthesis directly or indirectly provides almost all living things with energy.
- The net equation for cellular respiration is the reverse of the net equation for photosynthesis.

VOCABULARY
- metabolism
- catabolism
- anabolism
- ATP
- photosynthesis
- cellular respiration
- fermentation

Use the Vocabulary

To reinforce chapter vocabulary, have students write a sentence using each term. OL EL

Review Strategies

- Have students describe the differences in structure and function of proteins, carbohydrates, lipids, and nucleic acids. OL
- Have students compare and contrast the metabolic processes of photosynthesis, cellular respiration, and fermentation. OL
- Problems from the Supplemental Problems booklet can be used for review. OL

Students can review with these online resources.

Vocabulary Practice

Vocabulary eGames
Vocabulary eFlashcards
Vocabulary PuzzleMaker

Section Self-Check

Chapter Self-Check

Online Test Practice

Use *eAssessment* to:

- create multiple versions of tests
- edit existing questions and add your own questions
- build tests aligned with select state curriculum standards using built-in state curriculum tags
- track students' progress

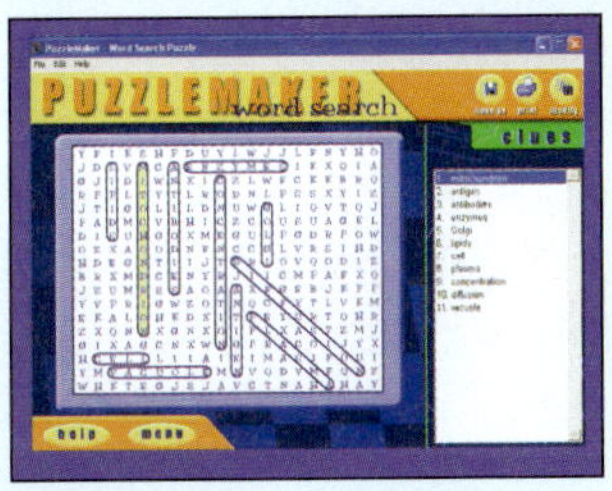

Vocabulary Practice

For additional practice with vocabulary, have students access the Vocabulary PuzzleMaker.

SECTION 1

Mastering Concepts

33. peptide; protein

34. amino and carboxyl groups; dipeptide

35. Sample answers: ■▲◆●; ▲◆●■; ◆●■▲; ●■▲◆

36. Sample answers: skin, ligaments, tendons, bones, and hair

37. Sample answers: enzymes, papain; transport proteins, hemoglobin; structural support, collagen; communication, thyroid hormones

38. An alpha helix is a coiled stretch of a protein chain. A beta sheet is a flat area where a chain folds back and forth repeatedly.

39. **a.** amide group
b. hydroxyl group
c. carboxyl group
d. amino group

40. The active site and the substances bind together. A reaction occurs between the substrates because they are held closely together and the activation energy decreases.

41. phenylalanine

42. nonpolar: glycine, valine, phenylalanine; polar: serine, cysteine, glutamine, lysine, glutamic acid

43. Tryptophan is a large, nonpolar, aromatic amino acid that is somewhat soluble in water and has relatively high melting and boiling points. It is a building block for proteins.

44. No. Each amino acid has a different group attached to the peptide bond.

45. Enzymes form many bonds with the substrates, lowering the activation energy.

46. Yes. The environment of a cell is aqueous, so it makes sense for cell proteins to have more amino acids with polar side chains on the outside of the molecule and fewer on the inside.

Mastering Problems

47. $20^3 = 8.0 \times 10^3$; $20^4 = 1.6 \times 10^5$; $20^5 = 3.2 \times 10^6$

48. 4

49. **a.** 5600 amu
b. 190,000 amu

50. 4 amino acids; 3 peptide bonds

51. about 332

CHAPTER 23 ASSESSMENT

SECTION 1

Mastering Concepts

33. What should you call a chain of eight amino acids? A chain of 200 amino acids?

34. Name the two types of functional groups that react together to form a peptide bond, and name the functional group in the peptide bond itself.

35. Using the symbols below to represent four amino acids, draw peptide structures for four-member chains that link them together in different orders.

Amino acid 1: ■	Amino acid 3: ◆
Amino acid 2: ▲	Amino acid 4: ●

36. Human Anatomy Name five parts of the body that contain structural proteins.

37. List four major functions of proteins, and give one example of a protein that carries out each function.

38. Describe two common shapes found in the three-dimensional folding of proteins.

39. Name the organic functional groups in the side chains of the following amino acids.
a. glutamine
b. serine
c. glutamic acid
d. lysine

40. Explain how the active site of an enzyme functions.

41. Name an example of an amino acid that has an aromatic ring in its side chain.

42. Name two amino acids with nonpolar side chains and two amino acids with polar side chains.

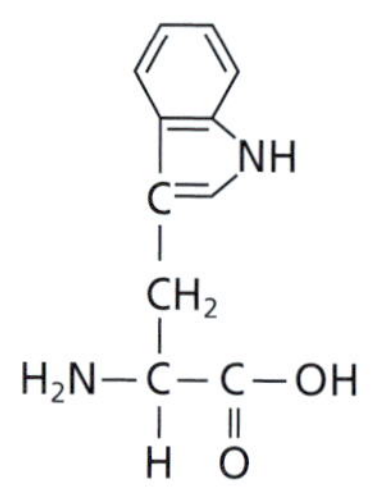

■ **Figure 31**

43. The structure shown in **Figure 31** is tryptophan. Describe some of the properties you would expect tryptophan to have, based on its structure. In what class of large molecules is tryptophan a member? Explain.

44. Is the dipeptide lysine-valine the same compound as the dipeptide valine-lysine? Explain.

45. Enzymes How do enzymes lower the activation energy for a reaction?

46. Cellular Chemistry Most proteins with a globular shape are oriented so that they have mostly amino acids with nonpolar side chains located on the inside and amino acids with polar side chains located on the outer surface. Does this make sense in terms of the nature of the cellular environment? Explain.

Mastering Problems

47. How many different ways can you arrange three different amino acids in a peptide? Four amino acids? Five amino acids?

48. How many peptide bonds are present in a peptide that has five amino acids?

49. Proteins The average molecular weight of an amino acid residue in a polypeptide is 110 amu. What is the approximate molecular weight of the following proteins?
a. Insulin (51 amino acids)
b. Myosin (1750 amino acids)

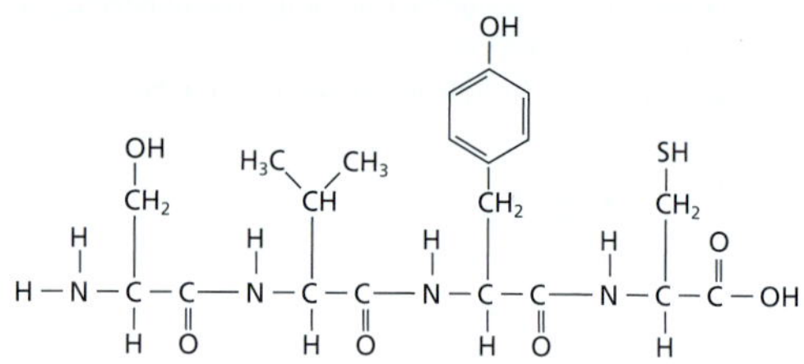

■ **Figure 32**

50. Determine how many amino acids and peptide bonds are in the peptide shown in **Figure 32.**

51. The average molar mass of an amino acid is 110 g/mol. Calculate the approximate number of amino acids in a protein that has a molar mass of 36,500 g/mol.

SECTION 2

Mastering Concepts

52. Carbohydrates Classify the following carbohydrates as monosaccharides, disaccharides, or polysaccharides.
a. starch
b. glucose
c. sucrose
d. ribose
e. cellulose
f. glycogen
g. fructose
h. lactose

53. Name two isomers of glucose.

54. What kind of bond is formed when two monosaccharides combine to form a disaccharide?

55. Sugars Give a scientific term for each of the following.
a. blood sugar
b. fruit sugar
c. table sugar
d. milk sugar

SECTION 2

Mastering Concepts

52. **a.** polysaccharide
b. monosaccharide
c. disaccharide
d. monosaccharide
e. polysaccharide
f. polysaccharide
g. monosaccharide
h. disaccharide

53. fructose and galactose

54. ether bond

55. **a.** glucose
b. fructose
c. sucrose
d. lactose

Chapter Self-Check

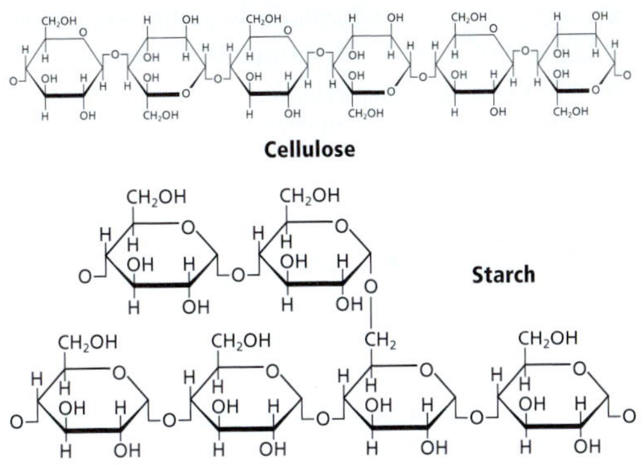

■ **Figure 33**

56. Cellulose and Starch The molecular structures of cellulose and starch are shown in **Figure 33.** Compare and contrast their molecular structures.

57. Chemistry in Plants Compare and contrast the functions of starch and cellulose in plants. Explain why their molecular structures are important to their functions.

58. Infer how the different bonding arrangements in cellulose and starch give them such different properties.

59. The disaccharide maltose is formed from two glucose monomers. Draw its structure.

60. The hydrolysis of cellulose, glycogen, and starch produces only one monosaccharide. Why is this so? What monosaccharide is produced?

61. Digestion Disaccharides and polysaccharides cannot be broken down in the absence of water. Why do you think this is so? Include an equation in your answer.

62. Draw the structure of the open-chain form of fructose. Circle all chiral carbons, and then calculate the number of stereoisomers with the same formula as fructose.

63. Sugars Compare and contrast the molecular formula, molecular weight, and functional groups found in glucose and fructose.

64. Historical Perspective Carbohydrates are not hydrates of carbon as the name suggests. Explain how this misconception occurred.

Mastering Problems

65. Complex Carbohydrates Stachyose is a tetrasaccharide that contains two D-galactose units, one D-glucose unit, and one D-fructose unit. Each sugar unit has a molecular weight of 180 g/mol before it is linked together in this tetrasaccharide, and one water molecule is released for each two sugar units that come together. What is the molecular weight of stachyose?

SECTION 3

Mastering Concepts

66. Compare and contrast the structures of a triglyceride and a phospholipid.

67. Predict whether a triglyceride from beef fat or a triglyceride from olive oil will have a higher melting point. Explain your reasoning.

68. Soaps and Detergents Explain how the structure of soaps makes them effective cleaning agents.

69. Draw a portion of a lipid bilayer membrane, labeling the polar and nonpolar parts of the membrane.

70. Where and in what form are fatty acids stored in the human body?

71. What type of lipid does not contain fatty acid chains? Why are these molecules classified as lipids?

72. Soap Draw the structure of the soap sodium palmitate (palmitate is the conjugate base of the 16-carbon saturated fatty acid, palmitic acid). Label its polar and nonpolar ends.

73. Determine whether each structure is a fatty acid, triglyceride, phospholipid, steroid, or wax. Explain your reasoning.

a.

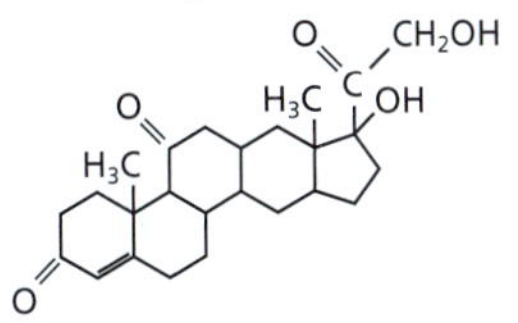

b.

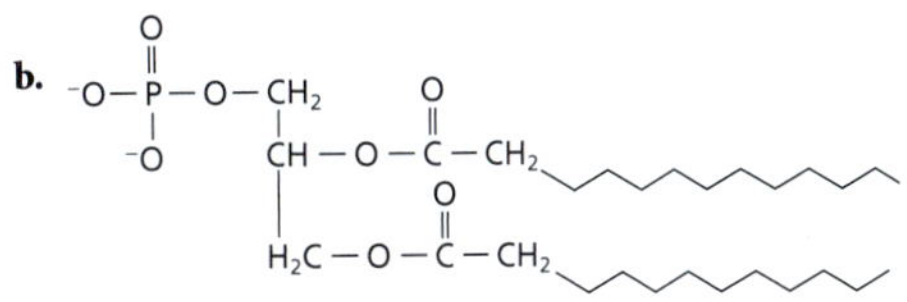

Mastering Problems

74. The fatty acid palmitic acid has a density of 0.853 g/mL at 62°C. What will be the mass of a 0.886-L sample of palmitic acid at that temperature?

75. Polyunsaturated Fats How many moles of hydrogen gas are required for complete hydrogenation of 1 mol of linolenic acid, whose structure is shown below? Write a balanced equation for the hydrogenation reaction.

$$CH_3CH_2CH = (CHCH_2CH)_2 = CH(CH_2)_7COOH$$

56. Both structures contain similar ring structures, but cellulose is a linear structure and starch is a branched structure.

57. Both substances are polysaccharides found in plants. However, starch is used to store energy and cellulose forms the rigid plant-cell walls. The long, linear structure of cellulose allows the chains to stick together tightly forming a strong, rigid structure. Starch is composed of glucose subunits and is insoluble in water, making it a good storage reservoir for energy.

58. The monomers are linked together in different ways. Cellulose is a linear polymer of parallel chains held tightly together in bundles. Starch is a branched polymer; branching keeps the structure from forming tight bundles.

59. The structure should show two glucose units linked with an ether bond. Refer to the Solutions Manual.

60. All three polymers are made only of glucose. Upon hydrolysis, only glucose is released.

61. The ether bonds (COC) holding the sugars together must be broken into two COH bonds by incorporating water. This is a hydrolysis reaction. The equation is the reverse of that in Figure 10.

62. 8 isomers; refer to the Solutions Manual.

63. Glucose and fructose are structural isomers, so both have the same molecular formula ($C_6H_{12}O_6$) and molecular weight (180 g/mol). Both glucose and fructose contain 5 alcohol or hydroxyl groups, but fructose also has a ketone group, while glucose has an aldehyde group.

64. Carbohydrates have the general structural formula $C_n(H_2O)_n$. Early scientists initially thought that these compounds were hydrates of carbon. However, it is now known that no water molecules are attached to the molecules, but the name of the compounds remained the same.

Mastering Problems

65. $(4 \times 180 \text{ g/mol}) - (3 \times 18 \text{ g/mol}) = 666 \text{ g/mol}$

SECTION 3

Mastering Concepts

66. A triglyceride is a glycerol molecule to which three fatty acids are bonded with ester bonds. A phospholipid is a glycerol molecule to which two fatty acids and a phosphate group are bonded with ester bonds.

67. Beef fat contains more saturated fats than olive oil. Saturated fatty acids pack together better than unsaturated fatty acids, so the beef lipid will have a higher melting point than the olive oil.

68. Soaps have both a nonpolar end, which dissolves nonpolar dirt and grease, and a water-soluble polar end, which allows water to wash away the soap and dirt.

69. Diagram should resemble Figure 17. Refer to the Solutions Manual.

70. in fat cells as triglycerides

71. Steroids; They are large, nonpolar biomolecules.

72. $CH_3(CH_2)_{14}COO^-Na^+$. The left end is nonpolar and the charged end is polar.

73. a. steroid

b. phospholipid

Mastering Problems

74. 756 g palmitic acid

75. 3 moles of H_2 are required for complete hydrogenation of linolenic acid. $C_{18}H_{30}O_2 + 3H_2 \longrightarrow C_{18}H_{36}O_2$

SECTION 4

Mastering Concepts

76. What three structures make up a nucleotide?

77. Name two nucleic acids found in organisms.

78. Explain the roles of DNA and RNA in the production of proteins.

79. Where in living cells is DNA found?

80. Describe the types of bonds and attractions that link the monomers together in a DNA molecule.

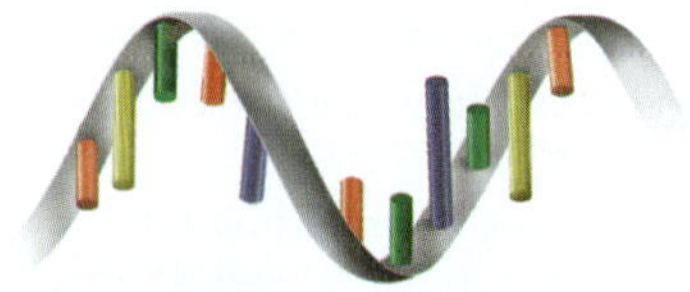

■ **Figure 34**

81. Classify the nucleic acid structure shown in **Figure 34** as DNA or RNA and explain your reasoning.

82. In the double-helical structure of DNA, the base guanine is always bonded to cytosine, and adenine is always bonded to thymine. What do you expect to be the relative proportional amounts of A, T, C, and G in a given length of DNA?

83. DNA Replication One strand in a DNA molecule has the following base sequence. What is the base sequence of the other strand in the DNA molecule?

C-C-G-T-G-G-A-C-A-T-T-A

Mastering Problems

84. A triplet code is a sequence of three bases in RNA that codes for one amino acid in a peptide chain or protein. How many RNA bases are required to code for a protein that contains 577 amino acids?

85. DNA Comparisons A cell of the bacterium *Escherichia coli* has about 4.2×10^6 base pairs of DNA, whereas each human cell has about 3×10^9 base pairs of DNA. What percentage of the size of the human genome does the *E. coli* DNA represent?

SECTION 5

Mastering Problems

86. Energy Calculate and compare the total energy in kJ that is converted to ATP during the processes of cellular respiration and fermentation.

87. How many grams of glucose can be oxidized completely by 2.0 L of O_2 gas at STP during cellular respiration?

88. Life Processes Compare the net reactions for photosynthesis and cellular respiration with respect to reactants, products, and energy.

MIXED REVIEW

89. Draw the carbonyl functional groups present in glucose and fructose. How are the groups similar? How are the groups different?

90. List the names of the monomers that make up proteins, complex carbohydrates, and nucleic acids.

91. Describe the functions of proteins, carbohydrates, lipids, and nucleic acids in living cells.

92. Write balanced equations for photosynthesis, cellular respiration, and the hydrolysis of lactose.

93. Write a balanced equation for the synthesis of sucrose from glucose and fructose.

THINK CRITICALLY

94. Make and Use Graphs A number of saturated fatty acids and values for some of their physical properties are listed in **Table 2**.

a. Make a graph plotting melting point versus number of carbon atoms.

b. Graph density versus the number of carbon atoms.

c. Draw conclusions about the relationships between the number of carbon atoms in a saturated fatty acid and its density and melting point values.

d. Predict the approximate melting point of a saturated fatty acid that has 24 carbon atoms.

Table 2 Physical Properties of Saturated Fatty Acids

Name	Number of Carbon Atoms	Melting Point (°C)	Density (g/mL) (values at 60–80°C)
Palmitic acid	16	63	0.853
Myristic acid	14	58	0.862
Arachidic acid	20	77	0.824
Caprylic acid	8	16	0.910
Docosanoic acid	22	80	0.822
Stearic acid	18	70	0.847
Lauric acid	12	44	0.868

SECTION 4

Mastering Concepts

76. sugar, phosphate, and nitrogen base

77. RNA and DNA

78. DNA holds the instructions for making proteins. These are passed to RNA, which translates the base sequence into an amino acid sequence during protein synthesis.

79. in the nucleus

80. Covalent bonds link sugars and phosphates. Hydrogen bonds hold bases together in the center of the helix.

81. The structure is RNA because uracil is present instead of thymine and it is single-stranded.

82. A and T amounts are equal. C and G amounts are equal.

83. G-G-C-A-C-C-T-G-T-A-A-T

Mastering Problems

84. 1731 RNA bases

85. 0.14 %

SECTION 5

Mastering Problems

86. During fermentation, each mole of glucose produces two moles of ATP. 2 mol ATP $\times$ 30.5 kJ/mol $=$ 61.0 kJ. During cellular respiration, each mole of glucose produces 38 moles of ATP. 38 mol ATP $\times$ 30.5 kJ/mol $=$ 1160 kJ.

87. 2.7 g glucose

88. Photosynthesis: $6CO_2 + 6H_2O +$ energy $\longrightarrow C_6H_{12}O_6 + 6O_2$
Cellular respiration: $C_6H_{12}O_6 + 6O_2 \longrightarrow 6CO_2 + 6H_2O +$ energy

MIXED REVIEW

89. Refer to the Solutions Manual. In glucose, the C=O has an H attached and is an aldehyde. In fructose, the C=O has other C atoms attached and is a ketone.

90. protein monomers: amino acids; complex carbohydrate monomers: monosaccharides; nucleic acid monomers: nucleotides

91. proteins: enzymes, structural, transport, communication and signaling; carbohydrates: energy source, structural in plants; lipids: stored form of energy, make up cell membranes, protection, some hormones and vitamins; nucleic acids: store and transmit genetic information, protein synthesis

92. photosynthesis: $6CO_2 + 6H_2O +$ energy $\longrightarrow C_6H_{12}O_6 + 6O_2$
cellular respiration: $C_6H_{12}O_6 + 6O_2 \longrightarrow 6CO_2 + 6H_2O +$ energy
hydrolysis of lactose: $C_{12}H_{22}O_{11} + H_2O \longrightarrow C_6H_{12}O_6 + C_6H_{12}O_6$

93. $C_6H_{12}O_6 + C_6H_{12}O_6 \longrightarrow C_{12}H_{22}O_{11} + H_2O$

THINK CRITICALLY

94. a. Refer to the Solutions Manual. The graph should plot number of carbons on the *x*-axis and melting point on the *y*-axis. It should show a fairly linear relationship, with melting point increasing as a function of the number of carbons.

b. Refer to Solutions Manual. The graph should show a fairly linear relationship in which density decreases as the number of carbons increases.

c. The more carbon atoms present, the higher the melting point and the lower the density.

d. 83–86°C

95. **Calculate** Approximately 38 mol of ATP are formed when glucose is completely oxidized during cellular respiration. If the heat of combustion for 1 mol of glucose is 2.82×10^3 kJ/mol and each mole of ATP stores 30.5 kJ of energy, what is the efficiency of cellular respiration in terms of the percentage of available energy that is stored in the chemical bonds of ATP?

96. **Recognize Cause and Effect** Some diets suggest severely restricting the intake of lipids. Why is it not a good idea to eliminate all lipids from the diet?

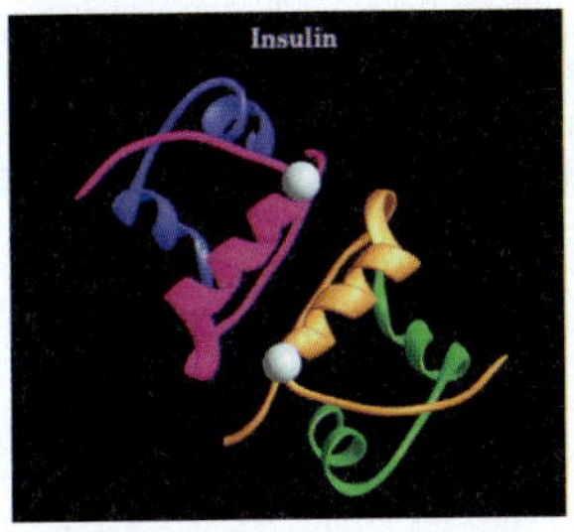

■ **Figure 35**

97. **Analyze** Insulin is a protein that functions as an important hormone in the human body. People who are diabetic often do not produce enough insulin, and must inject themselves with an insulin solution to maintain their health. Use **Figure 35** to infer how a person should care for a bottle of insulin.

98. **Calculate** If a double-stranded section of DNA has adenine as 20% of its bases, what percent of the other three bases are present in the DNA strand?

CHALLENGE PROBLEM

99. Calculate how many moles of ATP a human body can produce from the sugar in a bushel of medium-sized Red Delicious apples. Use the Internet to find the information you need to solve this problem.

CUMULATIVE REVIEW

100. a. Write the balanced equation for the synthesis of ethanol from ethene and water.
    b. If 448 L of ethene gas reacts with excess water at STP, how many grams of ethanol will be produced?

101. Identify whether each of the reactants in these reactions is acting as an acid or a base.
    a. $HBr + H_2O \rightarrow H_3O^+ + Br^-$
    b. $NH_3 + HCOOH \rightarrow NH_4^+ + HCOO^-$
    c. $HCO_3^- + H_2O \rightarrow CO_3^{2-} + H_3O^+$

102. What is a voltaic cell?

Chapter Self-Check

WRITING IN ▶ Chemistry

103. **Cholesterol** Use the library or the Internet to research cholesterol. Write a newspaper article about cholesterol that is written for a teenage audience. Make sure the following questions are answered in the article. Where is this molecule used in your body? What is its function? Why is too much dietary cholesterol considered to be bad for you? Is genetics a factor in high cholesterol?

DBQ Document-Based Questions

Fatty Acids *Omega-3 and omega-6 fatty acids are fatty acids that get their names from their structures. They contain a double bond either three or six carbon atoms from the end of the fatty acid chain. These fatty acids have a beneficial effect on health because they lower bad cholesterol levels and raise good cholesterol levels in the blood. Levels of omega-3 and omega-6 fatty acids were studied in salmon from three different sources as well as in the feed used in salmon farming.*

Figure 36 *shows the percent of omega-3 and omega-6 fatty acids compared to the total amount of lipids in the samples.*

Data obtained from: Hamilton, M.C. et al. 2005. *Environmental Science Technology* 39: 8622–8629.

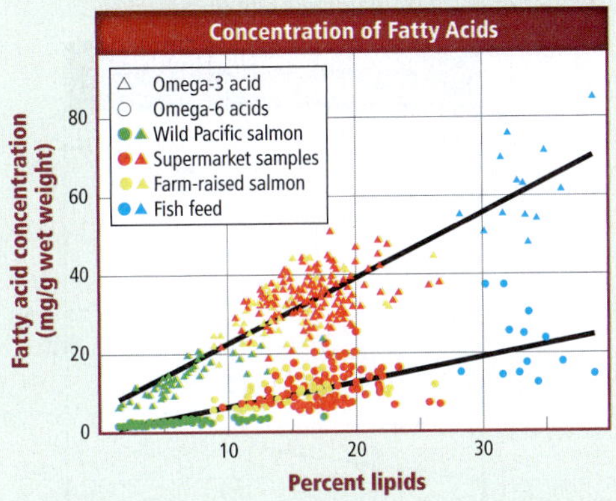

■ **Figure 36**

104. Which type of fish had the most omega fatty acids?

105. Based on this study, which type of salmon would you recommend to people who want to maximize the amounts of omega-3 and omega-6 fatty acids in their diets?

106. Infer from the graph why the farm-raised and supermarket salmon contains more omega-3 and omega-6 fatty acids than wild salmon.

95. 41%

96. The body needs lipids for a number of functions. If the amount of lipids in the diet is severely limited, there may not be enough lipids for the body to carry out those functions.

97. Patients should keep the vials at the recommended temperature (refrigerated for most forms of insulin) and not let them get too hot or cold, they should handle them carefully without shaking to avoid mechanical denaturation, and they should be careful not to contaminate them with anything that might alter the pH.

98. Since A = 20%, T must be 20% also because A and T bind together. The remaining 60% of bases must contain equal amounts of C and G since they bind together, so C = 30% and G = 30%.

CHALLENGE PROBLEM

99. 380 mol ATP/ bushel

CUMULATIVE REVIEW

100. a. $C_2H_4(g) + H_2O(l) \longrightarrow CH_3CH_2OH$
    b. 921 g C_2H_5OH

101. a. HBr is acid, H_2O is base.
    b. HCOOH is acid, NH_3 is base.
    c. HCO_3^- is acid, H_2O is base.

102. A voltaic cell is a chemical system that proceeds spontaneously while producing an electric current.

WRITING IN ▶ Chemistry

Rubric

103. Student answers should include the role of cholesterol in membranes, by the liver to form bile salts, by skin cells to make vitamin D, and by a number of glands to make steroid hormones. Too much dietary cholesterol has been linked to a higher risk for heart problems and strokes.

DBQ Document-Based Questions

Data obtained from: Hamilton, M.C. et al. 2005. *Environmental Science Technology* 39: 8622–8629.

104. farm-raised salmon

105. farm-raised salmon

106. The feed they are given is very rich in omega-3 and omega-6 fatty acids, while the wild salmon do not have supplemented feed.

MULTIPLE CHOICE

1. D
2. D
3. D
4. A
5. A
6. D
7. C
8. B

MULTIPLE CHOICE

1. Which is NOT true of carbohydrates?
   A. Monosaccharides in aqueous solutions convert continuously between an open-chain structure and a cyclic structure.
   B. The monosaccharides in starch are linked together by the same kind of bond that links the monosaccharides in lactose.
   C. All carbohydrates have the general chemical formula $C_n(H_2O)_n$.
   D. Cellulose, made only by plants, is easily digestible by humans.

2. Which is NOT a difference between RNA and DNA?
   A. DNA contains the sugar deoxyribose, while RNA contains the sugar ribose.
   B. RNA contains the nitrogen base uracil, while DNA does not.
   C. RNA is usually single-stranded, while DNA is usually double-stranded.
   D. DNA contains the nitrogen base adenine, while RNA does not.

Use the graph below to answer Question 3.

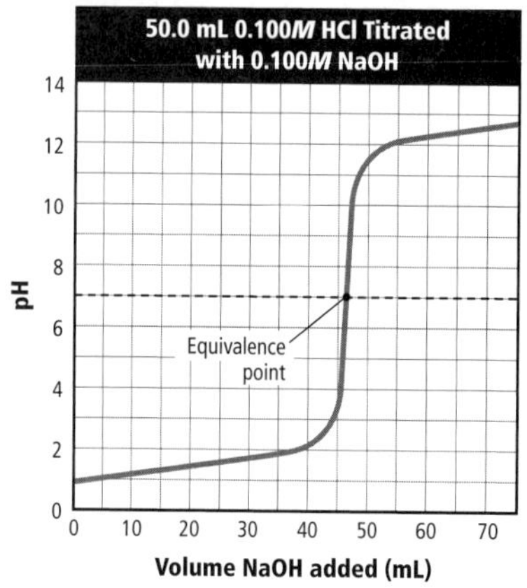

3. Which chemical indicator would be most effective in identifying the equivalence point of this titration?
   A. methyl orange, pH range of 3.2–4.4
   B. phenolphthalein, pH range of 8.2–10
   C. thymol blue, pH range of 8.0–9.6
   D. bromothymol blue, pH range of 6.0–7.6

4. How much NaOH, in grams, is required to completely neutralize 50.0 mL of 0.100*M* HCl?
   A. 0.200 g
   B. 5.00 g
   C. 0.125 g
   D. 200 g

Use the table below to answer Questions 5 to 7.

Nucleotide Data for Samples of Double-Stranded DNA

Sample	Content of Each Nucleotide	A	G	C	T
I	number	165	?	231	?
	percent	20.8	?	29.2	?
II	number	?	402	?	?
	percent	?	32.5	?	?
III	number	?	?	194	234
	percent	?	?	22.7	27.3
IV	number	266	203	?	?
	percent	28.4	21.6	?	?

5. What is the % T of Sample IV?
   A. 28.4%
   B. 78.4%
   C. 71.6%
   D. 21.6%

6. Every nitrogen base found in a DNA molecule is part of a nucleotide of that molecule. The A nucleotide, C nucleotide, G nucleotide, and T nucleotide have molar masses of 347.22 g/mol, 323.20 g/mol, 363.23 g/mol, and 338.21 g/mol respectively. What is the mass of 1 mol of Sample I?
   A. 2.79×10^5 g
   B. 2.7001×10^5 g
   C. 2.6390×10^5 g
   D. 2.72×10^5 g

7. How many molecules of adenine are in one molecule of Sample II?
   A. 402
   B. 434
   C. 216
   D. 175

8. Which is not a structural isomer of $CH_2{=}CHCH_2CH{=}CHCH_3$?
   A. $CH_2{=}CHCH_2CH_2CH{=}CH_2$
   B. $CH_3CH{=}CHCH_2CH{=}CH_2$
   C. $CH_3CH{=}CHCH{=}CHCH_3$
   D. $CH_2{=}C{=}CHCH_2CH_2CH_3$

Online Test Practice

SHORT ANSWER

9. The sequence of bases in RNA determines the sequence of amino acids in a protein. Three bases code for a single amino acid; for example, CAG is the code for glutamine. How many amino acids are coded for in a strand of RNA 2.73×10^4 bases long?

Use the diagram below to answer Question 10.

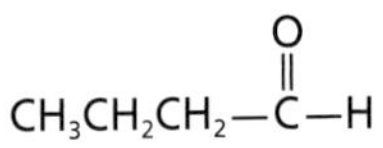

10. Which type of functional group is in this compound?

Use the diagram below to answer Question 11.

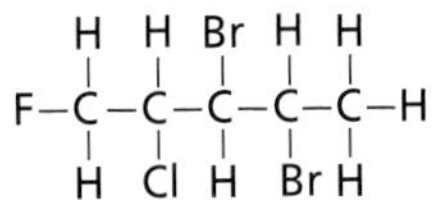

11. Give the IUPAC name for this organic compound.

12. What is the condensed structural formula of heptane?

EXTENDED RESPONSE

Use the diagram below to answer Question 13.

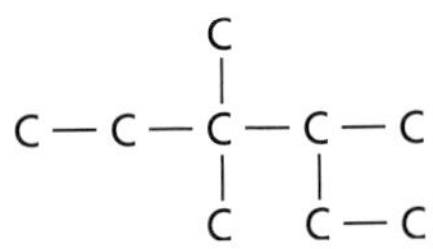

13. A student records the name of the alkane represented by this carbon skeleton as 2-ethyl 3,3-dimethyl pentane. Evaluate whether this is the correct name for the compound.

14. Compare and contrast aliphatic and aromatic compounds.

SAT SUBJECT TEST: CHEMISTRY

Use the table below to answer Questions 15–17.

Data for Elements in the Redox Reaction $Zn + HNO_3 \rightarrow Zn(NO_3)_2 + NO_2 + H_2O$		
Element	Oxidation Number	Complex Ion of which Element is a Part
Zn	0	none
Zn in $Zn(NO_3)_2$	+2	none
H in HNO_3	+1	none
H in H_2O	?	none
N in HNO_3	?	NO_3^-
N in NO_2	+4	none
N in $Zn(NO_3)_2$	?	NO_3^-
O in HNO_3	−2	NO_3^-
O in NO_2	?	none
O in $Zn(NO_3)_2$	?	NO_3^-
O in H_2O	−2	none

15. Which element forms a monatomic ion that is a spectator in the redox reaction?
    A. Zn
    B. O
    C. N
    D. H
    E. O_2

16. What is the oxidation number of N in $Zn(NO_3)_2$?
    A. +1
    B. +2
    C. +3
    D. +5
    E. +6

17. What is the element that is oxidized in this reaction?
    A. Zn
    B. O
    C. N
    D. H
    E. O_3

NEED EXTRA HELP?																	
If You Missed Question . . .	1	2	3	4	5	6	7	8	9	10	11	12	13	14	15	16	17
Review Section . . .	23.2	23.4	19.4	19.4	23.4	23.4	23.4	21.4	23.4	22.1	23.1	21.2	21.2	21.5	9.3	19.1	19.1

SHORT ANSWER

9. 2.73×10^4 bases/3 = 9100 amino acids = 9.10×10^3 amino acids
10. aldehyde
11. 3, 4-dibromo-2-chloro-1-fluoropentane
12. $CH_3(CH_2)_5CH_3$

EXTENDED RESPONSE

13. No, this is not the correct name. Naming rules for branched alkanes require the longest chain to be identified first (six carbons), then the functional groups are identified by their point of attachment so that the number is the lowest possible. The correct name would be 3, 3, 4-trimethylhexane.
14. Both of these are organic compounds, having a hydrocarbon base. Aliphatic compounds have a linear or branched structure, such as alkanes, alkenes, and alkynes. Aromatic compounds, however, have a ring structure whose root is the compound benzene. Members of this family frequently have strong odors associated with them.

SAT SUBJECT TEST: CHEMISTRY

15. D
16. D
17. A

CHAPTER 24 Nuclear Chemistry

BIGIDEA **Nuclear chemistry has a vast range of applications, from the production of electricity to the diagnosis and treatment of diseases.**

ESSENTIAL QUESTIONS	RESOURCES TO ASSESS MASTERY
SECTION 1 Nuclear Radiation **1.** How was radioactivity discovered and studied? **2.** What are the key properties of alpha, beta, and gamma radiations? 1 session 0.5 block	**Progress Monitoring** Formative Assessment, pp. 862, 863 Reading Check, pp. 861, 862, 864 Section Review, p. 864
SECTION 2 Radioactive Decay **1.** Why are certain nuclei radioactive? **2.** How are nuclear equations balanced? **3.** How can you use radioactive decay rates to analyze samples of radioisotopes? 2 sessions 1 block	**Progress Monitoring** Formative Assessment, pp. 866, 867, 868, 870, 871 Reading Check, pp. 866, 867, 868, 870 Graph Check, pp. 866, 870 Section Review, p. 874
SECTION 3 Nuclear Reactions **1.** How are mass and energy related? **2.** How do nuclear fission and nuclear fusion compare and contrast? **3.** What is the process by which nuclear reactors generate electricity? 2 sessions 1 block	**Progress Monitoring** Formative Assessment, p. 881 Reading Check, pp. 880, 882 Section Review, p. 884
SECTION 4 Applications and Effects of Nuclear Reactions **1.** What are several methods used to detect and measure radiation? **2.** How is radiation used in the treatment of disease? **3.** What are some of the damaging effects of radiation on biological systems? 1 session 0.5 block	**Progress Monitoring** Formative Assessment, pp. 886, 888, 889 Reading Check, pp. 886, 887 Section Review, p. 890 **Summative Assessment** Chapter Assessment, p. 894 *eAssessment* Chapter Test (Scaffolded)

Go online!

LEVELED RESOURCES	LAB MATERIALS
Science Notebook 24.1 OL ***FAST FILE Chapter Resources:*** Study Guide, p. 100 BL	LaunchLAB p. 858 / **45 min** set of domino tiles, stopwatch
Science Notebook 24.2 OL ***FAST FILE Chapter Resources:*** MiniLab Worksheet, p. 90 OL Study Guide, p. 101 BL	MiniLAB p. 873 / **25 min** 100 pennies, plastic cup, shoebox
Science Notebook 24.3 OL ***FAST FILE Chapter Resources:*** Study Guide, p. 102 BL	
Science Notebook 24.4 OL ***FAST FILE Chapter Resources:*** ChemLab Worksheet, p. 91 OL Study Guide, p. 105 BL	ChemLAB p. 892 / **60 min** alpha source, beta source, gamma source, Geiger counter, piece of cardboard, piece of plastic, meterstick, clock

ADDITIONAL RESOURCES

eAssessment
connectED Student center
connectED Teacher center
Lesson Presentations
eSolutions
Videos
Animations
Online Quizzes
Virtual Investigations
Multilingual eGlossary
Personal Tutors
WebQuests
Applying Practices Worksheets
PBLs

Practice:

Challenge Problems AL
Supplemental Problems BL OL
Chapter Test (Scaffolded)
Standardized Test Practice

Additional Resources:

Blueprints for Success
Solving Problems: A Chemistry Handbook

Lab Resources:

Laboratory Manual OL
CBL Laboratory Manual OL
Small-Scale Laboratory Manual OL
Forensics Laboratory Manual OL

BL Below Level OL On Level AL Advanced Learners EL English Learners COOP LEARN Cooperative Learning

CHAPTER 24

LaunchLAB

Rubric

How do chain reactions occur?

Purpose Students will model a nuclear chain reaction using dominoes. They will manipulate the arrangement of the dominoes in order to maximize the rate at which the chain reaction occurs.

Safety Precautions Approve lab safety forms before work begins.

Teaching Strategies

- Suggest to students that they start with a single line of dominoes spaced relatively far apart. Have them time how long this arrangement takes to fall over.
- Encourage students to explore different factors that might affect the rate at which the dominoes fall. Factors can include the type of domino used (if more than one type is available), spacing between dominoes, arrangement of dominoes into a single line or a fan-shaped pattern.

Expected Results Dominoes that are tightly spaced and arranged in a fan-shape (where one domino knocks two or more other dominoes over) fall at the fastest rate.

CHAPTER 24

Nuclear Chemistry

BIG IDEA Nuclear chemistry has a vast range of applications, from the production of electricity to the diagnosis and treatment of diseases.

SECTIONS

1 Nuclear Radiation

2 Radioactive Decay

3 Nuclear Reactions

4 Applications and Effects of Nuclear Reactions

LaunchLAB

How do chain reactions occur?

When the products of one nuclear reaction cause additional nuclear reactions to occur, the resulting chain reaction can release large amounts of energy in a short period of time. In this lab, you will explore chain reactions by modeling them with dominoes.

FOLDABLES® Study Organizer

Types of Radiation

Make a layered-look book. Label it as shown. Use it to help you organize information about the different types of radiation.

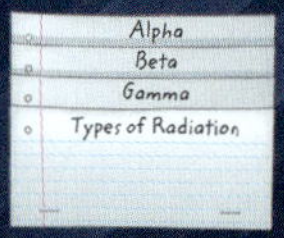

There's a good chance that you or someone you know has been helped by nuclear chemistry. From the magnetic properties of protons that enable MRIs, to the radiation used in X-rays, PET scans, and hundreds of other tests and treatments, research in nuclear chemistry has resulted in some of the most powerful tools available to modern medicine.

Procedure

1. Read and complete the lab safety form.
2. Obtain a **set of domino tiles.**
3. Stand the individual dominoes on end, and arrange them so that when the first domino falls, it causes the other dominoes to fall in series.
4. Practice using different arrangements. Using a **stopwatch,** determine how to cause the most dominoes to fall in the shortest amount of time.
5. Compare the time of your most efficient arrangement with those of your classmates.

Analysis

1. **Discuss** which arrangement caused the most dominoes to fall in the shortest amount of time. Results will vary as to what exact pattern falls the fastest, but in general dominoes that are tightly spaced and arranged in a fan-shape (where one domino knocks two or more other dominoes over) fall at the fastest rate.
2. **Discuss** whether the dominoes fell at a steady rate or an escalating rate. Equally spaced dominoes in a single line fall at a constant rate. Dominoes arranged in a fan shape fall at an escalating rate.
3. **Explain** what happens to the domino chain reaction if a tile does not contact the next tile in the sequence. If the chain of dominoes is broken, the chain reaction stops.

Inquiry What happens if you place an obstacle between two of the dominoes? How can you relate your observation to nuclear chain reactions? If an obstacle is placed between two dominoes, the chain reaction might stop. If the particles that cause the nuclear reaction to be sustained are absorbed, the chain reaction will stop.

Introduce the BIGIDEA

Nuclear Reactions To introduce this chapter's big idea, ask students to describe ways that nuclear chemistry affects their everyday lives. Point out that about one-fifth of the electricity in the United States is generated by nuclear power plants. Students might have benefited from nuclear chemistry by dental or medical X-rays. Bandages used for a cut might have been sterilized using nuclear reactions. Nuclear reactions on the Sun provide light and heat to Earth.

Tie to Previous Knowledge

Have students review the following concepts before studying this chapter.

- nuclear model of the atom, subatomic particles, isotopes, nuclear reactions, stability

Use the Photo

Nuclear Imaging Ask students if they have ever experienced any type of medical imaging. Point out that in the past, X-rays, like the one of the knee shown on this page, were the only type of images available to doctors. Today there are different types of medical imaging. The brain scan was made by PET (positron emission tomography). The image of the spine was made by MRI (magnetic resonance imaging). Explain that each of the images was made using nuclear chemistry.

SECTION 1

1 Focus

MAINIDEA

Changes in the Nucleus Ask students to describe how atoms can change during a chemical reaction. They can gain or lose electrons, but the nucleus is unchanged. Point out that under certain conditions, the nuclei of atoms can change, resulting in a nuclear reaction. Instead of atoms changing by gaining or losing electrons, particles in the nuclei of atoms change. **OL**

2 Teach

Extension

Alchemists Ask students to write essays about the following topic: In ancient times, alchemists experimented to devise a method to change base metals such as lead into gold. Is such a thing possible today? Is it probable? Why or why not? Some elements can be changed into new elements through nuclear reactions. **OL**

SECTION 1

Nuclear Radiation

MAINIDEA Under certain conditions, some nuclei can emit alpha, beta, or gamma radiation.

Essential Questions

- How was radioactivity discovered and studied?
- What are the key properties of alpha, beta, and gamma radiations?

Review Vocabulary

nucleus: the extremely small, positively charged, dense center of an atom that contains positively charged protons, neutral neutrons, and is surrounded by empty space through which one or more negatively charged electrons move

New Vocabulary

radioisotope
X-ray
penetrating power

CHEM 4 YOU If you wake up while it is still dark, the glowing numbers on your clock let you know what time it is. Many clocks use a type of radiation to make the numbers glow. The word *radiation* might cause you to think about nuclear power plants or dangerous, highly radioactive substances. However, less dangerous forms of radiation are often used in everyday objects, such as clocks.

The Discovery of Radioactivity

You have studied various forms of chemical reactions. Atoms can gain, lose, or share valence electrons, but the identity of the atoms does not change. Nuclear reactions, which you will study in this chapter, are different. Nuclear chemistry is concerned with the structure of atomic nuclei and the changes they undergo. Whereas chemical reactions involve only small energy changes, nuclear reactions involve much larger energy changes. **Table 1** offers a comparison of chemical reactions and nuclear reactions.

In 1895, German physicist Wilhelm Roentgen (1845–1923) found that invisible rays were emitted when electrons bombarded the surface of certain materials. These invisible rays caused photographic plates to darken, and Roentgen named these high-energy emissions *X-rays.* At that time, French physicist Henri Becquerel (1852–1908) was studying minerals that emit light after being exposed to sunlight, a phenomenon called phosphorescence. Building on Roentgen's work, Becquerel wanted to determine whether phosphorescent minerals also emitted X-rays.

Table 1 Comparison of Chemical and Nuclear Reactions

Chemical Reactions		Nuclear Reactions	
	• Occur when bonds are broken and formed • Involve only valence electrons • Associated with small energy changes • Atoms keep the same identity although they might gain, lose, or share electrons, and form new substances • Temperature, pressure, concentration, and catalysts affect reaction rates		• Occur when nuclei combine, split, and emit radiation • Can involve protons, neutrons, and electrons • Associated with large energy changes • Atoms of one element are often converted into atoms of another element • Temperature, pressure, and catalysts do not normally affect reaction rates

CHEMISTRY PROJECT

Discovering Radiation Have each student research the contributions of a scientist who contributed to the early knowledge of radiation, such as Wilhelm Roentgen, Henri Becquerel, or Marie Curie. Encourage students to focus on how the scientists' work was one step on the path to an overall understanding of the subject. Have each student prepare a presentation about the scientist. **OL**

Becquerel discovered by chance that phosphorescent uranium salts produced spontaneous emissions that darkened photographic plates. He observed this phenomenon even when the uranium salts were not exposed to light. Chemist Marie Curie (1867–1934) and her husband Pierre Curie (1859–1906) took Becquerel's mineral sample, called pitchblende, and isolated the components emitting the rays. They concluded that the darkening of the photographic plates was due to rays emitted from the uranium atoms present in the mineral sample. Marie Curie named the process by which materials give off such rays *radioactivity;* the rays and particles emitted by a radioactive source are called radiation. **Figure 1** shows the darkening of photographic film that is exposed to radiation emitted by radium salts.

The work of Marie and Pierre Curie was extremely important in establishing the origin of radioactivity and developing the field of nuclear chemistry. In 1898, the Curies identified two new elements, polonium and radium, on the basis of their radioactivity. Henri Becquerel and the Curies shared the 1903 Nobel Prize in Physics for their work. Marie Curie also received the 1911 Nobel Prize in Chemistry for her work with polonium and radium.

Figure 1 Radium salts are placed on a special emulsion on a photographic plate. After the plate is developed, the emulsion shows the dark tracks left by radiation emitted by the radium salts.

☑ READING CHECK **Explain** what Marie and Pierre Curie concluded about the darkening of the photographic plates.

Types of Radiation

After reading about the discovery of radioactivity, you might wonder what types of radiation are emitted by radioactive nuclei or which nuclei are radioactive.

Recall that isotopes are atoms of the same element that have different numbers of neutrons. Isotopes of atoms with unstable nuclei are called **radioisotopes.** These unstable nuclei emit radiation to attain more stable atomic configurations in a process called radioactive decay. During radioactive decay, unstable atoms lose energy by emitting radiation. The three most common types of radiation are alpha (α), beta (β), and gamma (γ). **Table 2** summarizes some of their important properties. Later in this chapter, you will learn about other types of radiation that can be emitted in a nuclear reaction.

Table 2 Properties of Alpha, Beta, and Gamma Radiation

Property	Alpha Radiation	Beta Radiation	Gamma Radiation
Symbol	α	β	γ
Composition	alpha particles	beta particles	high-energy electromagnetic radiation
Description of radiation	helium nuclei, ${}^{4}_{2}He$	electrons	photons
Charge	2+	1−	0
Mass	6.64×10^{-27} kg	9.11×10^{-31} kg	0
Approximate Energy	5 MeV	0.05 to 1 MeV	1 MeV
Relative penetrating power	blocked by paper	blocked by metal foil	not completely blocked by lead or concrete

Identify Misconceptions

Students might think that both chemical change and nuclear change result in the formation of new kinds of atoms.

Uncover the Misconception

Write the following equations on the board:

$H_2O(l) \rightarrow H_2O(s)$

$2H_2(g) + O_2(g) \rightarrow 2H_2O(l)$

${}^{2}_{1}H(g) + {}^{2}_{1}H(g) \rightarrow {}^{4}_{2}He(g)$

$He(g) \rightarrow He(l)$

Ask students which reactions represent physical change (first and fourth), chemical change (second), and nuclear change (third). They should justify their choices.

Demonstrate the Concept

Emphasize the following points.

- Physical change: Atoms that comprise substances undergoing physical change retain their identities and chemical compositions.
- Chemical change: Atoms that comprise substances undergoing chemical change are rearranged to form new substances. The new substances have new properties, but the atoms retain their identities. Chemical change involves changes in chemical bonds.
- Nuclear change: Atoms undergoing nuclear change form new atoms with different identities and properties. Nuclear change involves changes in an atom's nucleus.

Assess New Knowledge

Have students write out other examples of physical, chemical, and nuclear change on the board. At this point, students should be comfortable writing balanced chemical equations, but might have some difficulties writing balanced nuclear equations. OL

CHEMISTRY JOURNAL

Serendipitous Moments Have students look up the meaning of the word *serendipity* in the dictionary and relate it to the works of such scientists as Henri Becquerel, Alexander Fleming, Charles Goodyear, Louis Pasteur, and Roy Plunkett. OL EL

☑ READING CHECK They concluded that the darkening of the film was due to rays emitted from the uranium atoms.

ASSESSMENT

Knowledge Ask students what is the difference between the following symbols: $^{12}_{6}C$, $^{14}_{6}C$, and $^{18}_{6}C$? These are three different isotopes of carbon: the first has six neutrons, the second has eight neutrons, and the third has 12 neutrons. OL

Quick Demo

Becquerel's Experiment

Students can simulate the effect of Becquerel's experiment by arranging small items such as jewelry, hardware, or leaves on top of blueprint paper inside a notebook in semidarkness, then taking their arrangements outside for exposure to the Sun for several minutes. After exposure, students should bring the blueprint paper back inside to semidarkness and place it inside a dome or large inverted beaker filled with ammonia vapor. The vapor will "develop" the blueprint paper and show the rough shapes of the objects placed on it. Blueprint paper can be obtained from technical drawing supply stores or from architect offices. OL EL

FOLDABLES® Rubric

☑ READING CHECK By studying the effect of an electric field on the path of radiation, Rutherford determined that gamma rays are neutral, alpha particles are positively charged, and beta particles are negatively charged.

■ **Caption Question Fig. 3** Radium-226: 88 protons, 138 neutrons; Radon-222: 86 protons, 136 neutrons

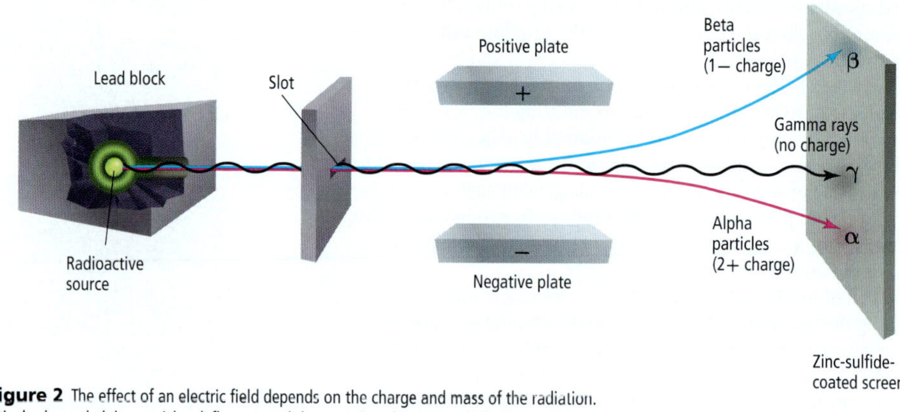

■ **Figure 2** The effect of an electric field depends on the charge and mass of the radiation. Positively charged alpha particles deflect toward the negative plate. Negatively charged beta particles deflect toward the positive plate. The lighter beta particles undergo the larger deflection. Gamma rays have no charge and are not affected by an electric field.

Ernest Rutherford (1871–1937), who performed the famous gold foil experiment that helped define modern atomic structure, identified alpha, beta, and gamma radiation when studying the effects of an electric field on the emissions from a radioactive source. As you can see in **Figure 2,** gamma rays carry no charge and are not affected by the electric field. Alpha particles carry a 2+ charge and are deflected toward the negatively charged plate. Beta particles carry a 1− charge and are deflected toward the positively charged plate. Because beta particles are less massive than alpha particles, they undergo a larger deflection.

☑ READING CHECK **Explain** how Rutherford determined whether each of the three types of radiation had a positive or negative charge or was unchanged.

FOLDABLES® Incorporate information from this section into your Foldable.

Alpha particles An alpha particle (α) has the same composition as a helium nucleus—two protons and two neutrons—and is therefore given the symbol $^{4}_{2}He$. The charge of an alpha particle is 2+ due to the presence of the two protons. Alpha radiation consists of a stream of alpha particles. Because of their mass and charge, alpha particles are relatively slow-moving compared with other types of radiation. Thus, alpha particles are not very penetrating—a single sheet of paper stops alpha particles. As you can see in **Figure 3,** radium-226, an atom whose nucleus contains 88 protons and 138 neutrons, undergoes alpha decay by emitting an alpha particle.

■ **Figure 3** A radium-226 nucleus undergoes alpha decay to form radon-222 and an alpha particle.

Evaluate *What is the number of protons and neutrons in radium-226 and radon-222?*

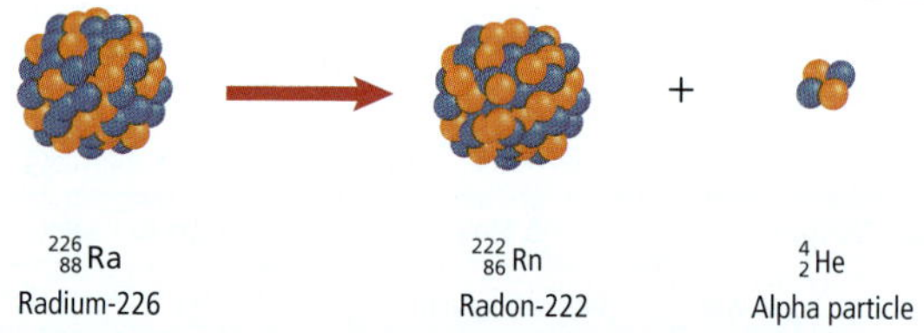

DIFFERENTIATED INSTRUCTION

Struggling Learners Use two different colors of magnetic marbles or gumdrops (with small toothpick pieces) to assemble a nucleus of any given atom. Have students demonstrate an alpha decay by removing two protons of one color and two neutrons of another color. A beta decay can be demonstrated by removing a neutron and substituting a proton in its place. BL EL

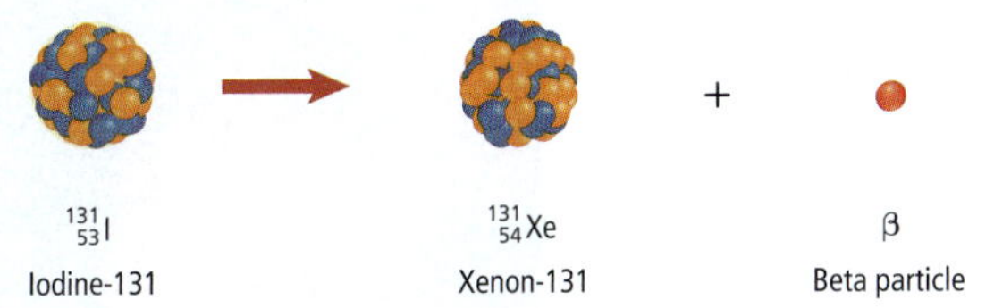

■ **Figure 4** An iodine-131 nucleus undergoes beta decay to form xenon-131 and a beta particle.
Explain ***How does beta decay affect the mass number of the decaying nucleus?***

In examining **Figure 3,** note that the reaction is balanced. That is, the sum of the mass numbers (superscripts) and the sum of the atomic numbers (subscripts) on each side of the arrow are equal. Also note that when a radioactive nucleus emits an alpha particle, the product nucleus has an atomic number that is lower by 2 and a mass number that is lower by 4.

Beta particles A beta particle is a very fast-moving electron that is emitted when a neutron in an unstable nucleus converts into a proton. Beta particles are represented by the symbol β or e^-. They have a 1− charge. Their mass is so small compared with the mass of nuclei involved in nuclear reactions that it can be approximated to zero. Beta radiation consists of a stream of fast-moving electrons. An example of the beta decay process is the decay of iodine-131 into xenon-131 by beta-particle emission, as shown in **Figure 4.** Note that the mass number of the product nucleus is the same as that of the original nucleus (they are both 131), but its atomic number has increased by 1 (54 instead of 53). This change in atomic number occurs because a neutron is converted into a proton, as shown by the following equation.

$$n \rightarrow p + \beta$$

As you might recall, the number of protons in an atom determines its identity. Thus, the formation of an additional proton results in the transformation from iodine-131 to xenon-131. Also, note that the electric charge in the equation above is conserved. The neutron is neutral. The proton has a 1+ charge and the beta particle has a 1− charge. Because beta particles are both lightweight and fast-moving, they have greater penetrating power than alpha particles. A thin sheet of metal foil is required to stop beta particles.

Gamma rays Gamma rays are photons, which are high-energy (short wavelength) electromagnetic radiation. They are denoted by the symbol γ. Because photons have no mass and no charge, the emission of gamma rays does not change the atomic number or mass number of a nucleus. Gamma rays almost always accompany alpha and beta radiation, as they account for most of the energy loss that occurs as a nucleus decays. For example, gamma rays accompany the alpha-decay reaction of uranium-238.

$$^{238}_{92}U \rightarrow {}^{234}_{90}Th + {}^{4}_{2}He + 2\gamma$$

The 2 in front of the γ symbol indicates that two gamma rays of different frequencies are emitted. Because gamma rays have no effect on mass number or atomic number, it is customary to omit them from nuclear equations.

VOCABULARY
WORD ORIGIN
Radiation
comes from the Latin word *radiare* which means *to radiate*

■ **Caption Question Fig. 4** The mass number does not change.

Reinforcement

Alpha and Beta Particles Ask students to describe the similarities and differences between an alpha particle and a helium atom; between a beta particle and an electron. The alpha particle and the helium atom both contain two protons and two neutrons. A helium atom is neutral, whereas an alpha particle has had two electrons removed and, therefore, has an overall charge of +2. A beta particle is exactly like an electron except for the fact that its point of origin comes from the nucleus. Beta particles are ejected from the nucleus during the decay of a neutron into a proton. **OL**

Content Background

Roentgen's Glow Roentgen was working with a cathode ray-tube shielded with black paper when he noticed a glow from across the room. He expected the paper to glow because it was covered with a fluorescent coating, but it was not in the pathway of the electron beam. Roentgen concluded that there was another source of radiation being emitted, which he named X-rays.

ASSESSMENT

Knowledge Have students make lists ranking particles by increasing mass, by increasing energy, by increasing penetrating power, and by increasing charge. **OL**

DIFFERENTIATED INSTRUCTION

Struggling Learners Students might be confused about the use of the notations e^- and β. Point out that both symbols represent the same particle. If the particle is on the reactant side of a nuclear equation, it represents electron capture, and the first symbol should be used. If the particle is on the product side of a nuclear equation, it represents beta decay, and usually, the second symbol should be used. **BL** **EL**

☑ READING CHECK X-rays and gamma rays are a form of high-energy electromagnetic radiation. X-rays are emitted from materials that are in an excited state. Gamma rays are produced by radioactive sources.

3 Assess

Check for Understanding

Divide the class into three groups. Members of the first group will hold a piece of green construction paper in front of them and represent the protons. Members of the second group will represent neutrons with a piece of red paper, and members of the third group will represent electrons with a piece of blue paper. Call for a particular isotope. Students who represent neutrons and protons should cluster together in the center, and the electrons should space themselves outside the nucleus. Then, call for a particular type of decay and have the students carry it out. **OL**

Reteach

Set up two bar magnets with a 3-cm gap between them. Roll a magnetic marble or small ball bearing between the magnets. Repeat the demonstration with a glass marble. Ask students to determine which type of radiation each roll could model. The magnetic marble could be a model for alpha or beta particles. The nonmagnetic glass marble could model gamma rays because it is not affected by the magnets. **OL**

Extension

Ask students to research the effects and penetrating ability of the three types of radiation on various organ systems of the human body. **OL**

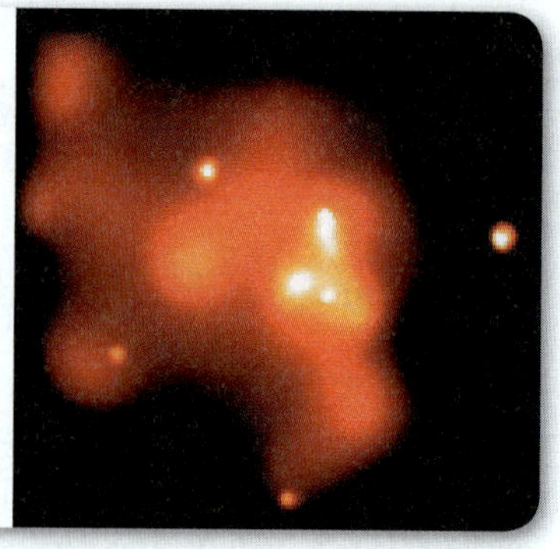

■ **Figure 5** The *Chandra Observatory,* launched in July 1999, photographed X-rays emitted from a cool gas cloud surrounding the black hole at the center of a neighboring galaxy.

As you have learned, the discovery of X-rays helped set the stage for the discovery of radioactivity. **X-rays,** like gamma rays, are a form of high-energy electromagnetic radiation. However, X-rays are not produced by radioactive sources and their energy is lower than that of gamma rays. They are emitted when inner electrons are knocked out and electrons from higher energy levels drop down to fill the vacancy. **Figure 5** shows an X-ray image taken in space. It allows astronomers to observe objects not visible in optical images. The presence of X-rays indicates phenomena such as exploding stars or black holes. Hospitals and dentists have machines that produce X-rays when a beam of electrons strikes a metal target. The familiar X-ray images are produced as the beam of X-rays passes easily through soft tissue but is partly blocked by hard tissue, such as bone.

☑ READING CHECK **Compare and contrast** X-rays and gamma rays.

Penetrating power The ability of radiation to pass through matter is called **penetrating power.** Alpha particles have a low penetrating power because they move slowly due to their large mass, and their 2+ charge causes them to lose energy quickly through interactions with other particles. The penetrating power of beta particles is higher because they are smaller and faster than alpha particles. However, they can still interact with particles and can be stopped by thin shielding. Gamma rays are highly penetrating. Because they have no charge and no mass, the probability of matter stopping them is low.

SECTION 1 REVIEW

Section Self-Check

Section Summary

- Wilhelm Roentgen discovered X-rays in 1895.
- Henri Becquerel, Marie Curie, and Pierre Curie pioneered the fields of radioactivity and nuclear chemistry.
- Radioisotopes emit radiation to attain more stable atomic configurations.

1. **MAIN**IDEA **List** the different types of radiation and their charges.
2. **Compare** the subatomic particles involved in nuclear and chemical reactions.
3. **Explain** how you know whether the reaction is chemical or nuclear when an atom undergoes a reaction and attains a more-stable form.
4. **Calculate** **Table 2** gives approximate energy values in units of MeV. Convert each value into joules using the following conversion factor: $1 \text{MeV} = 1.6 \times 10^{-13}$J.
5. **Summarize** Make a time line that summarizes the major events that led to the understanding of alpha, beta, and gamma radiation.

SECTION 1 REVIEW

1. alpha (2+), beta (1−), and gamma (0)
2. Protons, electrons, and neutrons can be involved in nuclear change; only electrons are involved in chemical change.
3. If the new, more stable form of the atom has an identity different from the original atom or gamma radiation has been released, a nuclear process has occurred.
4. alpha $= 8 \times 10^{-13}$ J; beta $= 8 \times 10^{-15}$ to 1.6×10^{-13} J; gamma $= 1.6 \times 10^{-13}$ J
5. Time lines should include Roentgen's 1895 discovery of X-rays, Becquerel's discovery that uranium exposes photographic film, the Curie's conclusion that uranium atoms emit radiation, and Rutherford's identification of alpha, beta, and gamma radiation.

SECTION 2

Radioactive Decay

MAINIDEA Unstable nuclei can break apart spontaneously, changing the identity of atoms.

Essential Questions

- Why are certain nuclei radioactive?
- How are nuclear equations balanced?
- How can you use radioactive decay rates to analyze samples of radioisotopes?

Review Vocabulary

radioactivity: the process by which some substances spontaneously emit radiation

New Vocabulary

transmutation
nucleon
strong nuclear force
band of stability
positron emission
positron
electron capture
radioactive decay series
half-life
radiochemical dating

CHEM 4 YOU To make sure that containers have the correct amount of fluid, some manufacturing processes use radioactivity. Detectors measure the number of particles produced by radioactive decay after they pass through the containers. For instance, a half-full bottle of juice would allow too much radiation to pass through and would not pass inspection.

Nuclear Stability

Except for the emission of gamma radiation, radioactive decay involves the conversion of an element into another element. Such a reaction, in which an atom's atomic number is altered, is called **transmutation.** Whether an atom spontaneously decays and what type of radiation it emits depends on its neutron-to-proton ratio.

An atom's nucleus contains positively charged protons and neutral neutrons. Protons and neutrons are referred to as **nucleons.** Despite the strong electrostatic repulsion forces among protons, all nucleons remain bound in the dense nucleus because of the strong nuclear force. The **strong nuclear force** acts on subatomic particles that are extremely close together and overcomes the electrostatic repulsion among protons.

The fact that the strong nuclear force acts on both protons and neutrons is important. Two protons repel each other, but because neutrons are neutral, a neutron that is adjacent to a positively charged proton creates no repulsive electrostatic force. Yet these two adjacent particles are held together by the strong nuclear force. Likewise, two adjacent neutrons create no electrostatic force, but they, too, are held together by the strong nuclear force. Thus, the presence of neutrons adds an attractive force within the nucleus, as illustrated in **Figure 6.** The number of neutrons in a nucleus is important because nuclear stability is related to the balance between electrostatic and strong nuclear forces.

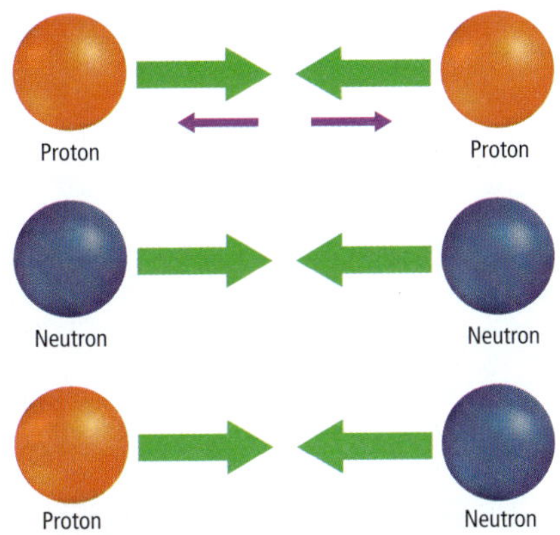

Figure 6 The electrostatic force, represented by the purple arrows, acts between two charged particles. It is repulsive between two protons. The strong nuclear force, represented by the green arrows, acts between any two or more nucleons and is always attractive.

Infer *What is the effect of the electrostatic force between two neutrons? Between a proton and an electron?*

SECTION 2

1 Focus

MAINIDEA

Spontaneous Changes Have students name some things that change spontaneously, or a change that occurs without a person causing it. Possible answers include changes in the weather and fruit that turns brown and breaks down when it is old. Lead students in a discussion of why things break apart. Use a sheet of paper that gets wet as one example. The water causes the fibers of the paper to break apart. OL

2 Teach

Concept Development

Radioactive Decay Remind students that not all radioactive decay is dangerous. We are surrounded by natural radioactivity even though we cannot see the radiation.

Caption Question Fig. 6 The electrostatic force does not affect neutrons. It is attractive between a proton and an electron.

DIFFERENTIATED INSTRUCTION

Hearing Impaired Use two different colors of poker chips to represent neutrons and protons. Assemble different ratios of neutrons to protons and use this as a springboard for discussing stability ratios. OL

ASSESSMENT

Skill Ask students to calculate the neutron-to-proton ratio for $^{104}_{40}Zr$. Ask them to locate its position on the graph in Figure 7 and determine its stability. The n/p ratio is 1.6 : 1, which is slightly above the band of stability. This isotope undergoes beta decay. **OL**

READING CHECK Neutrons are not affected by the electrostatic force. However, the strong nuclear force is attractive among neutrons. Increasing the number of neutrons makes the nucleus more stable.

GRAPH CHECK roughly 20

READING CHECK The band of stability is the area on a graph showing the number of neutrons versus the number of protons that contains all the stable nuclei. It corresponds to a neutron-to-proton ratio of 1:1 to 1.5:1 depending on the atomic number.

Content Background

Conditions for Decay In order for electron capture to occur, the mass of the original nucleus plus the mass of the electron must be greater than the mass of the nucleus that is produced. During positron emission, an electron must also be emitted from the atom in order to balance the charge. Positron emission can therefore only occur if the mass of the original nucleus is greater than the sum of the mass of the nucleus that is produced, the mass of a positron, and the mass of an electron. For this reason, nuclei are much less likely to undergo positron emission.

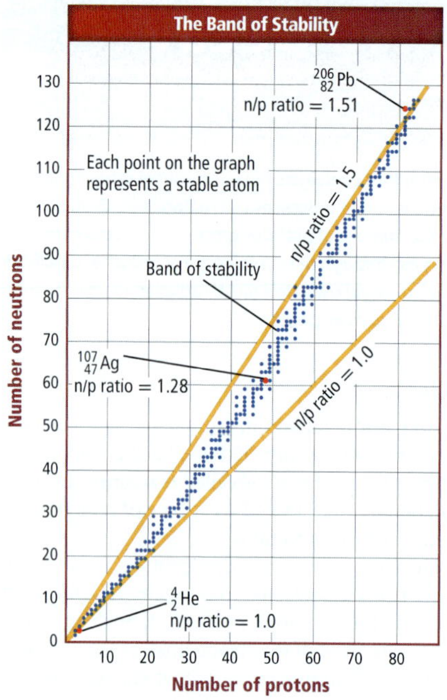

Figure 7 The band of stability is the region where all stable nuclei fall when plotting the number of neutrons versus the number of protons. As the atomic number increases, the neutron-to-proton ratio (n/p) increases from 1:1 to 1.5:1.

GRAPH CHECK
Find the number of protons above which the neutron-to-proton ratio starts to differ from 1:1.

Neutron-to-proton ratio To a certain degree, the stability of a nucleus can be correlated to its neutron-to-proton (n/p) ratio. For atoms with low atomic numbers (<20), the most stable nuclei are those with neutron-to-proton ratios of 1:1. For example, helium ($^{4}_{2}He$) has two neutrons and two protons, and a neutron-to-proton ratio of 1:1. As atomic number increases, more and more neutrons are needed to produce a strong nuclear force that is sufficient to balance the electrostatic repulsion force between protons. Therefore, the neutron-to-proton ratio for stable atoms gradually increases, reaching a maximum of approximately 1.5:1 for the largest atoms. An example of this is lead ($^{206}_{82}Pb$). With 124 neutrons and 82 protons, lead has a neutron-to-proton ratio of 1.51:1.

READING CHECK **Explain** why the neutron-to-proton ratio of stable nuclei increases as the atomic number increases.

The band of stability Examine the plot of the number of neutrons versus the number of protons for all known stable nuclei shown in **Figure 7.** Notice that the slope of the plot indicates that the number of neutrons required for a nucleus to be stable increases as the number of protons increases. This correlates with the increase in the neutron-to-proton ratio of stable nuclei with increasing atomic number. The area on the graph within which all stable nuclei are found is known as the **band of stability.** As shown in **Figure 7,** $^{4}_{2}He$ and $^{206}_{82}Pb$ are both positioned within the band of stability although they have a different neutron-to-proton ratio. All nuclei outside the band of stability—either above or below—are radioactive and undergo decay in order to gain stability. After decay, the new atom is positioned more closely to, if not within, the band of stability. The band of stability ends at lead-208; all elements with atomic numbers greater than 82 are radioactive.

READING CHECK **Define** the band of stability and relate it to the value of the neutron-to-proton ratio.

Types of Radioactive Decay

The type of radioactive decay a particular radioisotope undergoes depends to a large degree on the underlying causes for its instability. Atoms lying above the band of stability generally have too many neutrons to be stable, whereas atoms lying below the band of stability tend to have too many protons to be stable. Depending on the relative number of neutrons and protons, atoms can undergo different types of decay—beta decay, alpha decay, positron emission, or electron capture—to gain stability.

Demonstration

A Cloud Chamber

Purpose
to observe evidence of nuclear decay

Materials
Clear plastic sandwich box; 95% ethanol (50 mL); dry ice (500 g); tongs or thermal gloves; radioactive alpha; high-intensity light source

Safety Precautions

Disposal Save the sources for future use.

Procedure
Construct a cloud chamber. Use black paint to cover the outside of the bottom of a clear plastic box. Wet two pieces of blotter paper with ethanol, and place them on the opposite sides of the box. Place the box on dry ice. WARNING: *Ethanol is flammable. Dry ice can damage the skin, use tongs or thermal gloves.* Place a packaged radioactive source in

Beta decay A radioisotope that lies above the band of stability is unstable because it has too many neutrons relative to its number of protons. For example, unstable $^{14}_{6}C$ has a neutron-to-proton ratio of 1.33:1, whereas stable elements of similar mass, such as $^{12}_{6}C$ and $^{14}_{7}N$, have neutron-to-proton ratios of approximately 1:1. It is not surprising, then, that $^{14}_{6}C$ undergoes beta decay, as this type of decay decreases the number of neutrons in the nucleus.

$$^{14}_{6}C \rightarrow {}^{14}_{7}N + \beta$$

Figure 8a shows the beta decay of carbon-14 into nitrogen-14. Note that the atomic number of the product nucleus, $^{14}_{7}N$, has increased by one. The nitrogen-14 atom now has a stable neutron-to-proton ratio of 1:1. Thus, beta emission has the effect of increasing the stability of a neutron-rich atom by increasing its atomic number, that is by lowering its neutron-to-proton ratio. The resulting atom is closer to, if not within, the band of stability.

☑ READING CHECK **Explain** why radioisotopes above the band of stability are unstable.

Alpha decay All nuclei with more than 82 protons are radioactive and decay spontaneously. Both the number of neutrons and the number of protons must be reduced in order to make these radioisotopes stable. These very heavy nuclei often decay by emitting alpha particles. For example, polonium-210 spontaneously decays into lead-206 by emitting an alpha particle.

$$^{210}_{84}Po \rightarrow {}^{206}_{82}Pb + {}^{4}_{2}He$$

Figure 8b shows the alpha decay of polonium-210 into lead-206. The atomic number of $^{210}_{84}Po$ decreases by 2 and the mass number decreases by 4 as the nucleus decays into $^{206}_{82}Pb$.

☑ READING CHECK **Calculate** how the neutron-to-proton ratio changes when polonium-210 decays into lead-206.

■ **Figure 8** Depending on where nuclei lie on the band of stability, they can emit a beta particle or an alpha particle.
Compare and contrast *beta decay and alpha decay in terms of the atomic number of the nuclei involved in the reaction.*

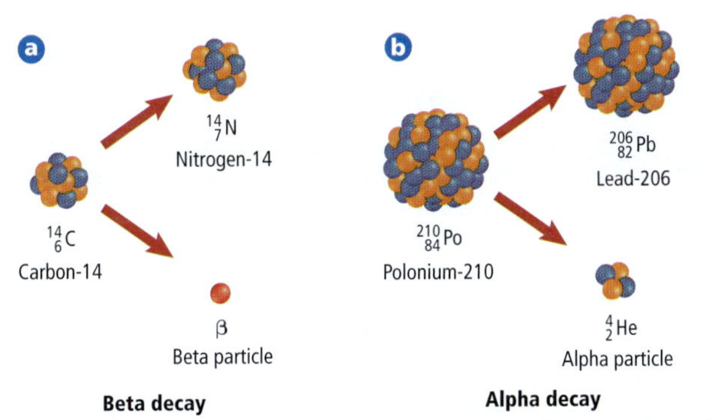

VOCABULARY
SCIENCE USAGE V. COMMON USAGE
Unstable
Science usage: spontaneously radioactive
Unstable atoms decay to reach a more stable state.

Common usage: not firm or fixed in one place
The chair is unstable because one of its legs is shorter than the other.

ASSESSMENT

Knowledge Instead of being powered by electricity, some exit signs are illuminated by tritium, a radioisotope of hydrogen. Ask students to list advantages and disadvantages of using tritium to light the sign. Ask students to write the correct nuclear equation for the decay. An advantage would be that the sign would continue to glow even if electricity were disrupted. A disadvantage is that over time the sign would become dimmer. $^{3}_{1}H \rightarrow {}^{0}_{-1}\beta + {}^{3}_{2}He$ **OL**

☑ READING CHECK They are unstable because they have too many neutrons relative to their number of protons.

☑ READING CHECK The n/p ratio goes from 1.500 : 1 to 1.512 : 1.

■ **Caption Question Fig. 8** The atomic number decreases by 2 in alpha decays. It increases by 1 in beta decays.

the middle of the box, and cover it with the plastic lid. After a minute, darken the room, and shine a high-intensity light beam through the box. The ends of the box might need to be warmed with a lightbulb to increase the saturation of the air with alcohol vapor.

Results

Look straight down into the box to see alpha particles shooting outward from the source, leaving momentary trails of condensed alcohol vapor.

Analysis

Ask these questions.

1. How big are the particles that caused the alcohol vapor to condense in a vapor trail? Submicroscopic, about 0.0012 picometers.
2. The alpha particles that produced the tracks are positively charged. How might you show that they have an electric charge? Put a strong magnet or charged object near the cloud chamber and observe that the tracks curve.

ASSESSMENT

Knowledge Ask students whether the rate of emission of the particles producing the vapor trails is constant. The rate appears constant, but it decreases slowly. **OL**

ASSESSMENT

Performance Have students team up in pairs and quiz each other using Table 3. Sample questions might be "Which kind of decay results in no change in mass number?" Or, "Which kind of decay increases the atomic number by one?" OL

READING CHECK Electron capture, positron emission, and alpha decay result in an increased n/p ratio. Beta decay results in a decreased n/p ratio.

Caption Question Fig. 9 During positron emission and electron capture, the number of protons decreases by 1 and the number of neutrons increases by 1.

Quick Demo

Charge of Alpha Particles

Rub a piece of fleece with a plastic or rubber wand and touch it to an electroscope. The electrons transfer from the fur to the wand and in turn to the electroscope. Because the leaves are now similarly charged, they will repel each other. Touch the top of the electroscope to bleed off the negative charge. Then, hold a radioactive beta source with tongs near an electroscope. The negative beta particles will transfer to the foil leaves and mimic the behavior of the previous step. Ask students what this shows about the nature of the radioactive particles. Students should note that the radiation is negative. Finally, hold a radioactive alpha emitter with tongs to the electroscope. Ask students what they can infer about the nature of the emitter. The positive alpha particles remove electrons. OL

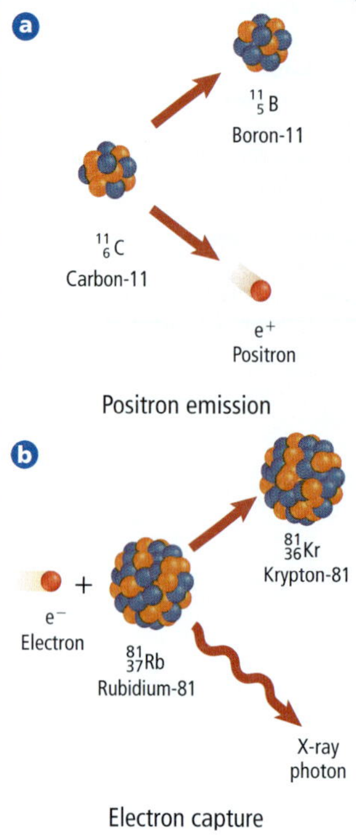

Figure 9 When a nucleus undergoes positron emission or captures an electron, the number of protons decreases by one.

Compare and contrast *how the number of protons and neutrons change during positron emission and electron capture.*

Positron emission and electron capture For nuclei with low neutron-to-proton ratios, two common radioactive decay processes occur: positron emission and electron capture. These two processes tend to increase the neutron-to-proton ratio of the neutron-poor atom, bringing the atom closer to, if not within, the band of stability.

Positron emission is a radioactive decay process that involves the emission of a positron from a nucleus. A **positron** is a particle with the same mass as an electron but opposite charge; thus, it is represented by the symbol β^+ or e^+. During positron emission, a proton in the nucleus is converted into a neutron and a positron, and then the positron is emitted.

$$p \rightarrow n + e^+$$

Figure 9 shows the positron emission of a carbon-11 nucleus. Carbon-11 lies below the band of stability and has a low neutron-to-proton ratio of approximately 0.8:1. Carbon-11 undergoes positron emission to form boron-11. Positron emission decreases the number of protons from six to five, and increases the number of neutrons from five to six. The resulting atom, $^{11}_{5}B$, has a neutron-to-proton ratio of 1.2:1, which is within the band of stability.

Electron capture is the other common radioactive-decay process that decreases the number of protons in unstable nuclei lying below the band of stability. **Electron capture** occurs when the nucleus of an atom draws in a surrounding electron, usually one from the lowest energy level. This captured electron combines with a proton to form a neutron.

$$p + e^- \rightarrow n$$

The atomic number of the nucleus decreases by 1 as a consequence of electron capture. The formation of the neutron also results in an X-ray photon being emitted. These two characteristics of electron capture are shown in the electron capture of rubidium-81 in **Figure 9.** The balanced nuclear equation for the reaction is shown below.

$$e^- + {}^{81}_{37}Rb \rightarrow {}^{81}_{36}Kr + \text{X-ray photon}$$

The five types of radioactive decay you have read about in this chapter are summarized in **Table 3.**

READING CHECK **List** the decay processes that result in an increased neutron-to-proton ratio and a decreased neutron-to-proton ratio.

Explore radioactive decay with an interactive table. Concepts In Motion

Table 3 Summary of Radioactive Decay Processes

Type of Radioactive Decay	Particle Emitted	Change in Mass Number	Change in Atomic Number
Alpha decay	$^{4}_{2}He$	decreases by 4	decreases by 2
Beta decay	β or e^-	no change	increases by 1
Positron emission	β^+ or e^+	no change	decreases by 1
Electron capture	X-ray photon	no change	decreases by 1
Gamma emission	γ	no change	no change

CULTURAL DIVERSITY

Honoring Marie Curie The town of Misasa, Japan, holds a festival honoring radiation pioneer Marie Curie. The city also offers radon "spas" where people bask in the alpha, beta, and gamma radiation emitted from the radon. Participants consider the radon particles to be promoters of health. The atmosphere is comparable to a sauna at a health club.

Writing and Balancing Nuclear Equations

The radioactive decay processes you have just read about are all examples of nuclear reactions. Nuclear reactions are expressed by balanced nuclear equations just as chemical reactions are expressed by balanced chemical equations. However, in balanced chemical equations, numbers and types of atoms are conserved; in balanced nuclear equations, mass numbers and charges are conserved.

EXAMPLE Problem 1

Find help with **algebraic equations.** Math Handbook

EXAMPLE PROBLEM

BALANCING A NUCLEAR EQUATION NASA uses the alpha decay of plutonium-238 $\left(^{238}_{94}\text{Pu}\right)$ as a heat source on spacecraft. Write a balanced equation for this decay.

1 ANALYZE THE PROBLEM

You are given that a plutonium atom undergoes alpha decay and forms an unknown product. Plutonium-238 is the initial reactant, while the alpha particle is one of the products of the reaction. The reaction is summarized below.

$$^{238}_{94}\text{Pu} \rightarrow {}^{A}_{Z}X + {}^{4}_{2}\text{He}$$

You must determine the unknown product of the reaction, *X*.

Known

reactant: plutonium-238 ($^{238}_{94}$Pu)

decay type: alpha particle emission ($^{4}_{2}$He)

Unknown

mass number of the product $A = ?$

atomic number of the product $Z = ?$

reaction product $X = ?$

2 SOLVE FOR THE UNKNOWN

$238 = A + 4$ — **Apply the conservation of mass number.**

$A = 238 - 4 = 234$ — **Solve for A.**

Thus, the mass number of *X* is **234.**

$94 = Z + 2$ — **Apply the conservation of charges.**

$Z = 94 - 2 = 92$ — **Solve for Z.**

Thus, the atomic number of *X* is **92.**

The periodic table identifies the element as uranium (U).

$^{238}_{94}\text{Pu} \rightarrow {}^{234}_{92}\text{U} + {}^{4}_{2}\text{He}$ — **Write the balanced nuclear equation.**

3 EVALUATE THE ANSWER

The correct formula for an alpha particle is used. The sums of the superscripts and subscripts on each side of the equation are equal. Therefore, the charge and the mass number are conserved. The nuclear equation is balanced.

PRACTICE Problems

Do additional problems. Online Practice

6. Write a balanced nuclear equation for the reaction in which oxygen-15 undergoes positron emission.
7. Thorium-229 is used to increase the lifetime of fluorescent bulbs. What type of decay occurs when thorium-229 decays to form radium-225?
8. **Challenge** The figure at right shows one way that bismuth-212 can decay, producing isotopes **A** and **B.**
   a. Write a balanced nuclear equation for this decay.
   b. Identify the isotopes **A** and **B** that are produced.

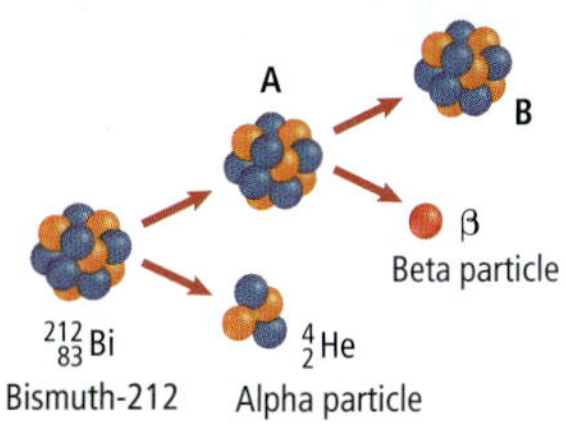

IN-CLASS Example

Question Write a balanced equation for the alpha decay of thorium-232.

Answer $^{232}_{90}\text{Th} \rightarrow X + {}^{4}_{2}\text{He}$

$^{232}_{90}\text{Th} \rightarrow {}^{228}_{88}\text{Ra} + {}^{4}_{2}\text{He}$

PRACTICE Problems

Have students refer to the Selected Solutions appendix for complete solutions to odd-numbered problems. The complete solutions for all problems can be found in the Solutions Manual.

6. $^{15}_{8}\text{O} \rightarrow {}^{0}_{1}\beta + {}^{15}_{7}\text{N}$
7. alpha decay
8. a. $^{212}_{83}\text{Bi} + {}^{208}_{81}\text{Tl} + {}^{4}_{2}\text{He}$
   $^{208}_{81}\text{Tl} + {}^{208}_{82}\text{Pb} + \rightarrow {}^{0}_{-1}\beta$
   b. Isotope A is thallium-208 and isotope B is lead-208.

Reinforcement

Reaction Students who have trouble with math might find it helpful to set up an inventory table where atomic masses and atomic numbers of reactants are tallied and compared to those of products.

CHEMISTRY JOURNAL

Decay Process Have students make an entry in their journals describing the implications of having only one possible type of decay process—only alpha decay or beta decay, for instance. How would this affect the types of isotopes that could be formed? OL

Quick Demo

Atomic Decay Videotape students as they enter the classroom and seat themselves. It would be helpful to have a timer on the video camera or arrange for a clock to be seen in the footage. Students who are seated represent decayed atoms, while those who are standing represent atoms that have not decayed. Replay the videotape and emphasize the randomness of the decay process (some students enter in groups, others individually; there are irregular time intervals between students sitting down; and so on). The half-life of the class can also be determined. OL

GRAPH CHECK U-238
alpha decay to Th-234
beta decay to Pa-234
beta decay to U-234
alpha decay to Th-230
alpha decay to Ra-226
alpha decay to Rn-222
alpha decay to Po-218
alpha decay to Pb-214
beta decay to Bi-214
beta decay to Po-214
alpha decay to Pb-210
beta decay to Bi-210
beta decay to Po-210
alpha decay to Pb-206

READING CHECK A half-life is the time required for one-half of a radioisotope's nuclei to decay into its products.

ASSESSMENT

Skill Have students draw the decay series for americium-241, which is found in smoke detectors. The radioisotope undergoes the following succession of decays: α, α, β, α, α, β, α, α, α, β, α, β. The following nuclei are formed in the decay of americium-241: Np-237, Pa-233, U-233, Th-229, Ra-225, Ac-225, Fr-221, At-217, Bi-213, Po-213, Pb-209, and Bi-209. OL EL

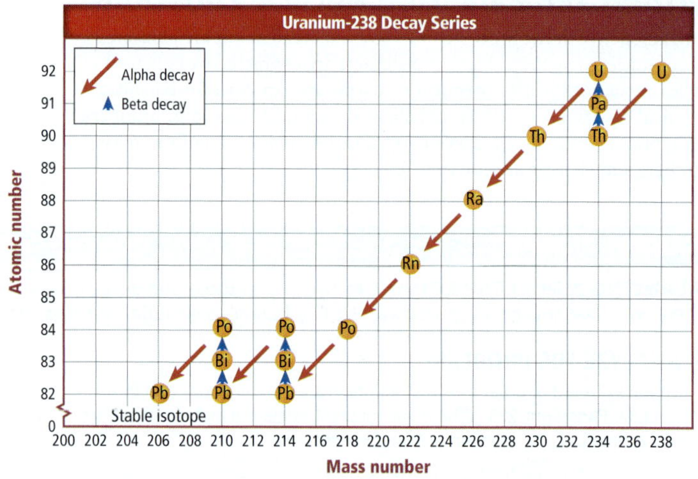

Figure 10 Uranium-238 undergoes 14 different radioactive decay steps before forming stable lead-206.

Radioactive Series

A **radioactive decay series** is a series of nuclear reactions that begins with an unstable nucleus and results in the formation of a stable nucleus. As **Figure 10** shows, uranium-238 first decays to thorium-234, which in turn decays to protactinium-234. Decay reactions continue until a stable nucleus, lead-206, is formed.

GRAPH CHECK **List** each step in the decay of uranium-238. Include the type of decay and the resulting product.

Radioactive Decay Rates

You might wonder how there could be any naturally occurring radioisotopes found on Earth. After all, if radioisotopes undergo continuous radioactive decay, won't they eventually disappear? Furthermore, radioisotopes have been decaying for about 4.6 billion years—the span of Earth's existence. Yet, naturally occurring radioisotopes are not uncommon on Earth. Some radioisotopes, such as carbon-14, are continuously formed in the upper atmosphere of Earth. Others are formed in the universe, during stellar nucleosynthesis for instance. Radioisotopes can also be synthesized in laboratories. The differing decay rates of isotopes also contribute to their presence on Earth.

Radioactive decay rates are measured in half-lives. A **half-life** is the time required for one-half of a radioisotope's nuclei to decay into its products. For example, the half-life of the radioisotope strontium-90 is 29 years. If you had 10.0 g of strontium-90 today, 29 years from now you would have 5.0 g left. **Table 4** shows how this decay continues through four half-lives of strontium-90. **Figure 11** presents the data from the table in terms of the percent of strontium-90 remaining after each half-life. The decay continues until a negligible amount of strontium-90 remains.

Explore **half-life.**

Virtual Investigations

READING CHECK **Define** the term *half-life.*

CHEMISTRY PROJECT

Decay Series Have students work in small groups to research decay series similar to the one for uranium-238 shown in Figure 10. Then, have them use grid paper to model the series. Students should be able to explain whether each type of change on the grid describes alpha decay or beta decay. OL

Table 4 The Decay of Strontium-90		
Number of Half-Lives	Elapsed Time	Amount of Strontium-90 Present
0	0 y	10.0 g
1	29 y	$10.0\text{ g} \times \left(\frac{1}{2}\right) = 5.00\text{ g}$
2	58 y	$10.0\text{ g} \times \left(\frac{1}{2}\right)\left(\frac{1}{2}\right) = 2.50\text{ g}$
3	87 y	$10.0\text{ g} \times \left(\frac{1}{2}\right)\left(\frac{1}{2}\right)\left(\frac{1}{2}\right) = 1.25\text{ g}$
4	116 y	$10.0\text{ g} \times \left(\frac{1}{2}\right)\left(\frac{1}{2}\right)\left(\frac{1}{2}\right)\left(\frac{1}{2}\right) = 0.625\text{ g}$

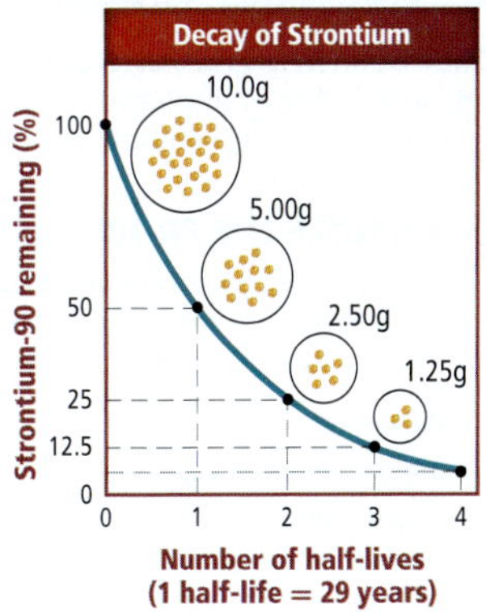

■ **Figure 11** The graph shows how the amount of strontium in a sample changes as a function of the number of half-lives.

Get help with **exponential graphing.**

Personal Tutor

The data in **Table 4** can be summarized in a simple equation representing the decay of any radioactive element.

Remaining Amount of Radioactive Element

$$N = N_0\left(\frac{1}{2}\right)^n$$

N is the remaining amount.
N_0 is the inital amount.
n is the number of half-lives that have passed.

The amount remaining is equal to the initial amount times one-half raised the number of half-lives that have passed.

The exponent n can also be replaced with the equivalent quantity t/T, where t is the elapsed time and T is the duration of the half-life. Note that t and T must have the same units of time.

$$N = N_0\left(\frac{1}{2}\right)^{t/T}$$

This type of expression is known as an exponential decay function. **Figure 11** shows the graph of a typical exponential decay function—in this case, the decay curve for strontium-90.

☑ GRAPH CHECK **Infer** how much strontium remains after 1.5 half-lives.

Each radioisotope has its own characteristic half-life. Half-lives for several radioisotopes are given in **Table 5.** Notice the large range of values for half-lives, from millionths of a second to billions of years!

Table 5 Half-Lives of Several Radioisotopes		
Radioisotope	Symbol	Half-Life
Polonium-214	$^{214}_{84}Po$	163.7 μs
Cobalt-60	$^{60}_{27}Co$	5.272 y
Radon-222	$^{222}_{86}Ra$	3.8 d
Phosphorus-32	$^{32}_{15}P$	14.28 d
Carbon-14	$^{14}_{6}C$	5730 y
Uranium-238	$^{238}_{92}U$	4.46×10^9 y

Extension

Isotopes Have students become familiar with the *CRC Handbook of Chemistry and Physics* by looking up the Table of Isotopes and researching such questions as:

- Which element has the largest number of radioisotopes?
- Which isotope has the longest half-life? The shortest?
- Which decay has the highest associated energy? OL

Visual Learning

Table 5 Ask students to select the best radioisotope from the table for determining the age of an artifact suspected of being 2000 years old. C-14 would be the best choice. The first five radioisotopes have half-lives that are too short, and the half-life of U-238 is too long. OL

ASSESSMENT

Skill Strontium-90 (half-life = 29 years) is produced during nuclear weapon testing. When ingested, it is assimilated into the bones of the body where it damages healthy tissue. In 1963, several countries signed a treaty banning above-ground nuclear testing. Using 1963 as time zero, when the maximum amount of Sr-90 existed in the environment (100%), ask students to calculate the amount of the radioisotope remaining after each of the first 10 half-lives. Sr-90 levels are considered safe after 10 half-lives. Ask students to explain why strontium is easily absorbed into bones. 1963 (100%); 1992 (50%); 2021 (25%); 2050 (12.5%); 2079 (6.25%); 2108 (3.13%); 2137 (1.56%); 2166 (0.781%); 2195 (0.391%); 2224 (0.195%) and 2253 (0.098%). Strontium is in the same family as calcium and is more reactive. OL

☑ GRAPH CHECK 3.54 g remain

DIFFERENTIATED INSTRUCTION

Advanced Learners Have students research information about the particles that result from collisions in particle accelerators. Give them some key terms such as *leptons, quarks,* and *bosons,* to get them started. Have the students report to the class on their findings. AL

IN-CLASS Example

Question The half-life of cobalt-57 is 270 days. How much of a 5.000-mg sample will remain after 810 days?

Answer
Number of half-lives (n) = 810 days/270 days = 3 half-lives

Amount remaining = $(5.000\text{ mg})\left(\frac{1}{2}\right)^3 =$ 0.6250 mg

PRACTICE Problems

Have students refer to the Selected Solutions appendix for complete solutions to odd-numbered problems. The complete solutions for all problems can be found in the Solutions Manual.

9. 5.00 mg; 2.50 mg; 1.25 mg
10. 8.00×10^2 mg
11. Sample B: ~32.3 g
Sample C: ~29.2 g

Apply Chemistry

Radiochemical Dating People who study fine wines might need to determine the age of a particular wine. One method of doing so is to use tritium, a naturally occurring radioisotope of hydrogen. Tritium, which has two neutrons and one proton in its nucleus, constitutes a very small percentage of the hydrogen content of water. It has a half-life of 12.3 years, so it can be used for dating "young" water-containing systems. Once a wine is bottled, it no longer exchanges substances with the outer environment, and its tritium content decays at a rate dictated by its half-life. The tritium level in the environment remains relatively constant.

EXAMPLE Problem 2

Find help with **fractions.** Math Handbook

EXAMPLE PROBLEM

CALCULATING THE AMOUNT OF REMAINING ISOTOPE Krypton-85 is used in indicator lights of appliances. The half-life of krypton-85 is 11 y. How much of a 2.000-mg sample remains after 33 y?

1 ANALYZE THE PROBLEM

You are given a known mass of a radioisotope with a known half-life. You must first determine the number of half-lives that passed during the 33-year period. Then, use the exponential decay equation to calculate the amount of the sample remaining.

Known
Initial amount = 2.000 mg
Elapsed time (t) = 33 y
Half-life (T) = 11 y

Unknown
Amount remaining = ? mg

2 SOLVE FOR THE UNKNOWN

Number of half-lives (n) = $\frac{\text{elapsed time}(t)}{\text{half-life}(T)}$ — **Determine the number of half-lives passed during the 33 y.**

$n = \frac{33\text{ y}}{11\text{ y}} = 3.0$ half-lives — **Substitute t = 33 y and T = 11 y.**

Amount remaining = (initial amount) $\left(\frac{1}{2}\right)^n$ — **Write the exponential decay equation.**

Amount remaining = (2.000 mg) $\left(\frac{1}{2}\right)^{3.0}$ — **Substitute initial amount = 2.000 mg and n = 3.**

Amount remaining = (2.000 mg) $\left(\frac{1}{8}\right)$ = **0.2500 mg**

3 EVALUATE THE ANSWER

Three half-lives are equivalent to $\left(\frac{1}{2}\right)\left(\frac{1}{2}\right)\left(\frac{1}{2}\right)$, or $\left(\frac{1}{8}\right)$. The answer (0.25 mg) is equal to $\left(\frac{1}{8}\right)$ of the initial amount. The answer has two significant figures because the number of years has two significant figures. n does not affect the number of significant figures.

PRACTICE Problems

Do additional problems. Online Practice

PRACTICE PROBLEMS

9. Bandages can be sterilized by exposure to gamma radiation from cobalt-60, which has a half-life of 5.27 y. How much of a 10.0-mg sample of cobalt-60 is left after one half-life? Two half-lives? Three half-lives?

10. If the passing of five half-lives leaves 25.0 mg of a strontium-90 sample, how much was present in the beginning?

11. Challenge The table shows the amounts of radioisotopes in three different samples. To the nearest gram, how much will be in Sample B and Sample C when Sample A has 16.2 g remaining?

Sample	Radioisotope	Half-life	Amount (g)
A	cobalt-60	5.27 y	64.8
B	tritium	12.32 y	58.4
C	strontium-90	28.79 y	37.6

CHEMISTRY PROJECT

Radioactive Decay Have students work in small groups to create a computer presentation about a topic related to radioactive decay. Topics may include the different types of decay, radiochemical dating, a time line of advances in scientific understanding of radioactive decay, or a scientist who made a significant discovery about radioactive decay. OL COOP LEARN

MiniLAB

Model Radioactive Decay

How do radioactive isotopes decay?

Procedure

1. Read and complete the lab safety form.
2. Place **100 pennies** in a **plastic cup.**
3. Place your hand over the top of the cup and shake the cup several times.
4. Pour the pennies into a **shoebox.** Remove all the pennies that land heads-up. These pennies represent atoms of the radioisotope that have undergone radioactive decay.
5. Prepare a data table to record the number of remaining pennies (tails-up pennies).
6. Count the number of pennies that remain, and record this number in your data table.
7. Place all of the tails-up pennies back in the plastic cup.
8. Repeat Steps 3 through 7 as many times as needed until no pennies remain.

Analysis

1. **Construct** a graph of *Trial Number* v. *Number of Pennies Remaining* from your data table. Draw a curve through the plotted points.
2. **Calculate** how many trials it took for 50%, 75%, and 90% of the sample to decay.
3. **Evaluate** the half-life of the radioisotope if the time between each trial is 1 min.
4. **Determine** how the results would change if you used 100 dice instead of pennies. In this case, you would assume that any dice that lands with the six side facing up represents a decayed atom and is removed.

Radiochemical dating Chemical reaction rates are greatly affected by changes in temperature, pressure, and concentration, and by the presence of a catalyst. In contrast, nuclear reaction rates remain constant regardless of such changes. In fact, the half-life of any particular radioisotope is constant. Because of this, radioisotopes can be used to determine the age of an object. The process of determining the age of an object by measuring the amount of a certain radioisotope remaining in that object is called **radiochemical dating.**

Connection to Biology A type of radiochemical dating known as carbon dating is used to measure the age of artifacts that were once part of a living organism. Carbon dating makes use of the radioactive decay of carbon-14, which is formed by cosmic rays in the upper atmosphere at a fairly constant rate. These carbon-14 atoms become evenly spread throughout Earth's biosphere, where they mix with stable carbon-12 and carbon-13 atoms. Plants use carbon dioxide from the environment, which contains all carbon isotopes, to build more complex molecules through the process of photosynthesis. When animals eat plants, the carbon-14 atoms that were part of the plant become part of the animal. Because organisms are constantly taking in carbon compounds, they contain the same ratio of carbon-14 to carbon-12 and carbon-13 found in the atmosphere. However, after they die, organisms no longer ingest new carbon compounds, and the carbon-14 they already contain continues to decay. The carbon-14 undergoes beta decay to form nitrogen-14.

$$^{14}_{6}\text{C} \rightarrow \, ^{14}_{7}\text{N} + \beta$$

Carbon-14 has a half-life of 5730 years. Because the amount of stable carbon in the dead organism remains constant while the carbon-14 continues to decay, the ratio of unstable carbon-14 to stable carbon-12 and carbon-13 decreases.

MiniLAB

Rubric

Purpose Students will use a model to examine the concept of half-life of a radioactive isotope.

Process Skills compare and contrast, make and use graphs, collect data, apply concepts, use numbers, predict, measure, and formulate models

Safety Precautions Approve lab safety forms before work begins.

Teaching Strategy Any object that has two or more distinct states, such as heads and tails, can be used in place of the pennies.

Expected Result The Decay Results table should show that approximately half of the pennies decay during each trial. There will probably be no pennies remaining after 7 to 9 trials.

Analysis

1. The graph should show an exponential decay, and should resemble the shape of the curve seen in Figure 11.
2. It took approximately one trial for 50% of the sample to decay. It took approximately two trials for 75% of the sample to decay. It took approximately four trials for 90% of the sample to decay.
3. The half-life is 1 minute.
4. Approximately one-sixth of the dice will decay with each toss. It would take approximately three tosses to get one-half of the sample to decay.

3 Assess

Check for Understanding

Ask students which member of each of the following pairs would be more stable? Why? OL

- ${}^{12}_{6}C$ or ${}^{13}_{6}C$ — ${}^{12}_{6}C$; n/p = 1
- ${}^{16}_{7}N$ or ${}^{14}_{7}N$ — ${}^{14}_{7}N$; n/p = 1
- Fe or Np — Fe; All elements with atomic numbers greater than 83 are radioactive.

Reteach

Americium-241 is the radioisotope in smoke detectors. It emits alpha particles that ionize the air, thereby producing a current. If smoke interferes with this ionization, the microchip senses a cessation of current and triggers the alarm. Have students write the balanced nuclear reaction for the decay of americium-241 by alpha particle emission. Ask students if they should be concerned about radioactive americium-241 decaying in their homes.

$${}^{241}_{95}Am \longrightarrow {}^{4}_{2}He + {}^{237}_{93}Np$$

The decay poses no risk because alpha particles are easily blocked. Even if the alpha particles escaped their sealed chamber, the skin would easily stop them. OL

Extension

Ask students to use the *CRC Handbook of Chemistry and Physics* to locate an element that has at least two radioisotopes that undergo two different types of decay. Have students write correct nuclear equations for each decay process. OL

Figure 12 Using the radiocarbon dating method on organic materials, such as ash and charcoal found at the Great Pyramid of Giza, scientists estimate the pyramid to be more than 4000 years old.

By measuring this ratio and comparing it to the nearly constant ratio present in the atmosphere, the age of an object can be estimated. For example, if an object's C-14 to (C-12 + C-13) ratio is one-quarter of the ratio measured in the atmosphere, the object is approximately two half-lives, or 11,460 years old. Carbon-14 dating is limited to accurately dating objects up to approximately 45,000 years of age. This method was used to date the Great Pyramid of Giza, shown in **Figure 12.**

Connection to Earth Science The decay process of a different radioisotope, uranium-238 to lead-206, is commonly used to date objects such as rocks. Because the half-life of uranium-238 is 4.5×10^9 years, it can be used to estimate the age of objects that are too old to be dated using carbon-14. By radiochemical dating of meteorites, the age of the solar system has been estimated at 4.6×10^9 years.

SECTION 2 REVIEW

Section Self-Check

Section Summary

- The conversion of an atom of one element to an atom of another by radioactive decay processes is called transmutation.
- Atomic number and mass number are conserved in nuclear reactions.
- A half-life is the time required for half of the atoms in a radioactive sample to decay.
- Radiochemical dating is a technique for determining the age of an object by measuring the amount of certain radioisotopes remaining in the object.

12. MAINIDEA **Describe** what happens to unstable nuclei.
13. **Explain** how you can predict whether or not an isotope is likely to be stable if you know its number of neutrons and protons.
14. **Describe** the forces acting on the particles within a nucleus and explain why neutrons are the glue holding the nucleus together.
15. **Predict** the nuclear equation for the alpha decay of radium-226 used on the tips of older lightning rods.
16. **Calculate** how much of a 10.0-g sample of americium-241 remains after four half-lives. Americium-241 is a radioisotope commonly used in smoke detectors and has a half-life of 430 y.
17. **Calculate** After 2.00 y, 1.986 g of a radioisotope remains from a sample that had an original mass of 2.000 g.
    a. Calculate the half-life.
    b. How much of the radioisotope remains after 10.00 y?
18. **Graph** A sample of polonium-214 originally has a mass of 1.0 g. Express the mass remaining as a percent of the original sample after a period of one, two, and three half-lives. Graph the percent remaining versus the number of half-lives. Approximately how much time has elapsed when 20% of the original sample remains?

Pixtal/SuperStock

SECTION 2 REVIEW

12. Unstable nuclei decay by emitting alpha or beta radiation until they form a stable element.
13. Isotopes are likely to be stable if their n/p ratio is within the range of 1 : 1 to 1.5 : 1. The lightest isotopes have n/p ratios close to 1 : 1, while the heaviest isotopes have ratios close to 1.5 : 1.
14. Protons are positively charged and repel each other. Neutrons do not repel each other or protons. The strong nuclear force is an attractive force that acts on both protons and neutrons in the nucleus. The greater the number of neutrons in the nucleus, the greater the strong nuclear force.
15. ${}^{226}_{88}Ra \longrightarrow {}^{4}_{2}He + {}^{222}_{86}Rn$
16. 0.625 g
17. **a.** 197 years **b.** 1.931 g
18. The graph should resemble the shape seen in Figure 11. Each half-life corresponds to a 50% reduction in the amount remaining. According to table 5, polonium has a half-life of 163.7 μs. From the graph, at 2.3 half-lives, approximately 20% remains. $t = 2.3 \times 163.7 = 380$ μs.

SECTION 3

Nuclear Reactions

MAINIDEA Fission, the splitting of nuclei, and fusion, the combining of nuclei, release tremendous amounts of energy.

Essential Questions

- How are mass and energy related?
- How do nuclear fission and nuclear fusion compare and contrast?
- What is the process by which nuclear reactors generate electricity?

Review Vocabulary

mass number: the number after an element's name, representing the sum of its protons and neutrons

New Vocabulary

induced transmutation
transuranium element
mass defect
nuclear fission
critical mass
breeder reactor
nuclear fusion
thermonuclear reaction

CHEM 4 YOU On a hot summer day, you step outside and feel the intense heat of the Sun. Nuclear reactions within the Sun release enough energy to warm Earth and other planets in the solar system for billions of years. It is no surprise, then, that scientists are trying to use this same type of nuclear reaction to produce electricity.

Induced Transmutation

All nuclear reactions, or transmutations, that have been described thus far are examples of radioactive decay, where one element is converted into another element by the spontaneous emission of radiation. However, transmutations can also be forced, or induced, by bombarding a stable nucleus with a neutron or with high-energy alpha, beta, or gamma radiation. In 1919, Ernest Rutherford performed the first laboratory conversion of one element into another element. By bombarding nitrogen-14 with high-speed alpha particles, oxygen-17 and hydrogen-1 were formed. This transmutation reaction is illustrated in **Figure 13** and the reaction is shown below.

$$^{14}_{7}\mathrm{N} + {}^{4}_{2}\mathrm{He} \rightarrow {}^{17}_{8}\mathrm{O} + {}^{1}_{1}\mathrm{H}$$

As Rutherford demonstrated, nuclear reactions can be induced, in other words, produced artificially. The process, which involves striking nuclei with high-velocity particles, is called **induced transmutation.** In the case of charged particles, such as the alpha particles used by Rutherford, the incident particles must be moving at extremely high speeds to overcome the electrostatic repulsion between themselves and the target nucleus. Because of this, scientists have developed methods to accelerate charged particles to extreme speeds by using very strong electrostatic fields and magnetic fields. Particle accelerators are machines built to produce the high-speed particles needed to induce transmutation. Since Rutherford's first experiments involving induced transmutation, scientists have used the technique to synthesize hundreds of new isotopes in the laboratory.

Figure 13 When an alpha particle bombards a nitrogen-14 atom, an atom of oxygen-17 and an atom of hydrogen-1 are produced.

$^{4}_{2}$He	$^{14}_{7}$N	$^{17}_{8}$O	$^{1}_{1}$H
Bombarding alpha particle	Target nitrogen atom	Oxygen atom	Hydrogen atom

SECTION 3

1 Focus

MAINIDEA

Fission and Fusion Place small balls of modeling clay on a desk. Demonstrate that larger balls can be made by combining some of the smaller balls. Break apart larger balls into several smaller balls. Ask students to name other things that can combine and break apart. Possible answer: If you drip water on a desk, the drops can combine to make a large drop, or a drop can break apart into smaller drops. OL

2 Teach

Concept Development

Transmutation Students might confuse transmutation and induced transmutation. Remind students that transmutation occurs spontaneously as unstable nuclei decay, whereas induced transmutation occurs when nuclei are bombarded with other particles. It is also important to remind students that not all radioactive decay processes result in transmutation. Gamma-ray emission is an exception. OL

CHEMISTRY JOURNAL

Space Invaders Ask students to relate the studies done at particle accelerators to the following story: A visitor from another planet came to Earth and was fascinated by the abundance of different automobiles and their mechanics of operation. To research the automobile, the alien pushed them off a nearby cliff and then climbed to the bottom to study the pieces of wreckage. From this information, the alien was able to return to his home planet and begin the mass assembly of automobiles. OL

Content Background

Dr. Seaborg Glenn Seaborg (1913–1999), for whom element 106 is named, worked on the Manhattan Project during World War II. The Manhattan Project was responsible for creating the atomic bomb. He was a co-discoverer of plutonium, which was formed by bombarding uranium with deuterons. His research on transuranium elements led him to group the lanthanides and actinides in the lower part of the periodic table. In 1951, Seaborg received the Nobel Prize in Chemistry for his work. He served as the chairman of the Atomic Energy Commission (the forerunner of the Nuclear Regulatory Commission) from 1961 to 1971.

Seaborg spent many years campaigning for peaceful uses of atomic energy and against the testing of nuclear weapons. When seaborgium was originally accepted as the official name for element 106, the name was controversial, as it was the first element named in honor of a living scientist.

IN-CLASS Example

Question Write the balanced nuclear equation for the alpha-particle bombardment of aluminum-27, in which one neutron is produced.

Answer $^{27}_{13}\text{Al} + ^{4}_{2}\text{He} \rightarrow ^{30}_{15}\text{P} + \text{n}$

PRACTICE Problems

Have students refer to the Selected Solutions appendix for complete solutions to odd-numbered problems. The complete solutions for all problems can be found in the Solutions Manual.

19. $^{27}_{13}\text{Al} + \text{n} \rightarrow ^{24}_{11}\text{Na} + ^{4}_{2}\text{He}$

20. $^{239}_{94}\text{Pu} + ^{4}_{2}\text{He} \rightarrow ^{1}_{0}\text{n} + ^{242}_{96}\text{Cm}$

21. target: silver-109
unstable isotope: silver-110
$^{1}_{0}\text{n} + ^{109}_{47}\text{Ag} \rightarrow ^{110}_{47}\text{Ag}$
$^{110}_{47}\text{Ag} \rightarrow \text{e}^- + ^{110}_{48}\text{Cd}$

Transuranium elements The elements immediately following uranium in the periodic table—elements with atomic numbers 93 and greater—are known as the **transuranium elements.** All transuranium elements have been produced in the laboratory by induced transmutation and are radioactive. Many transuranium elements have been named in honor of their discoverers or the laboratories at which they were created. Scientists continue their ongoing efforts to synthesize new transuranium elements and study their properties.

EXAMPLE Problem 3

EXAMPLE PROBLEM

INDUCED TRANSMUTATION REACTION EQUATIONS Write a balanced nuclear equation for the induced transmutation of oxygen-16 into nitrogen-13 by proton bombardment. An alpha particle is emitted from the nitrogen atom in the reaction.

1 ANALYZE THE PROBLEM

You are given all of the particles involved in an induced transmutation reaction. Because the proton bombards the oxygen atom, they are reactants and must appear on the reactant side of the reaction arrow.

Known
reactants: oxygen-16 and a proton
products: nitrogen-13 and an α-particle

Unknown
nuclear equation for the reactant = ?

2 SOLVE FOR THE UNKNOWN

Nuclear formula for oxygen-16: $^{16}_{8}\text{O}$ — **Use the periodic table to obtain the atomic number of oxygen.**

Nuclear formula for nitrogen-13: $^{13}_{7}\text{N}$ — **Use the periodic table to obtain the atomic number of nitrogen.**

Nuclear formula for proton: p

Nuclear formula for alpha particle: $^{4}_{2}\text{He}$

$^{16}_{8}\text{O} + \text{p} \rightarrow ^{13}_{7}\text{N} + ^{4}_{2}\text{He}$ — **Write the balanced nuclear equation.**

3 EVALUATE THE ANSWER

A proton has a charge of 1+ and a mass number of 1. Therefore, both charge and mass number are conserved. The formula for each participant in the reaction is also correct. The nuclear equation is written correctly.

PRACTICE Problems

PRACTICE PROBLEMS

19. Write the balanced nuclear equation for the induced transmutation of aluminum-27 into sodium-24 by neutron bombardment. An alpha particle is released in the reaction.

20. Write the balanced nuclear equation for the alpha-particle bombardment of $^{239}_{94}\text{Pu}$. One of the reaction products is a neutron.

21. Challenge Archeologists sometimes use a procedure called neutron activation analysis to identify elements in artifacts. The figure at right shows one type of reaction that can occur when an artifact is bombarded with neutrons. If the product of the process is cadmium-110, what was the target and unstable isotope? Write balanced nuclear equations for the process to support your answer.

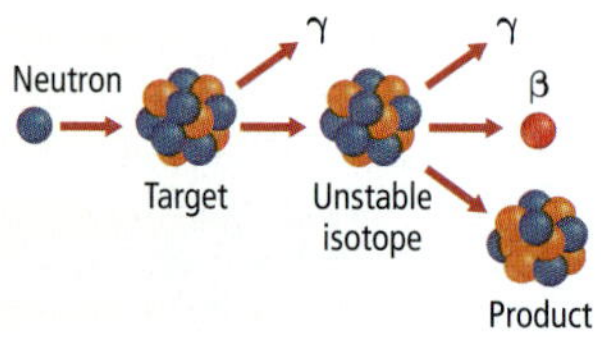

CHEMISTRY JOURNAL

The Name Game Have students research the politics and controversy involved in the IUPAC decision on official names for elements 104 to 110. **OL**

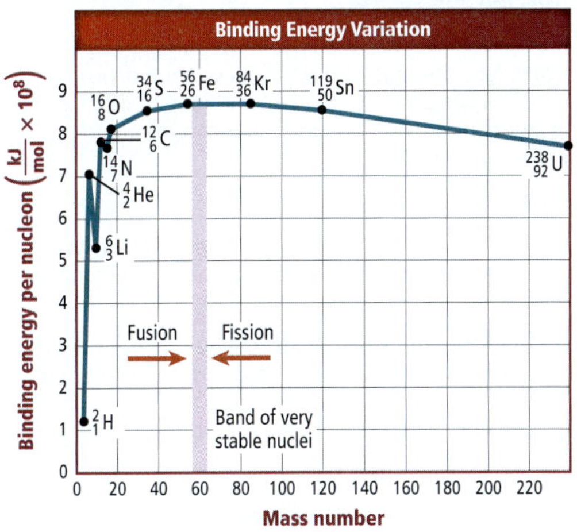

■ **Figure 14** The binding energy per nucleon is a function of the mass number. Light nuclei gain stability by undergoing nuclear fusion. Heavy nuclei gain stability by undergoing nuclear fission.

☑ GRAPH CHECK
Describe how the binding energy varies as a function of the mass number.

Nuclear Reactions and Energy

In your study of chemical reactions, you read that mass is conserved. For most practical situations this is true—but, it is not accurate.

Einstein's equation Albert Einstein's equation relates mass and energy. It states that any reaction produces or consumes energy due to a loss or gain in mass. Energy and mass are equivalent. Note that because c^2 is large, a small change in mass results in a large change in energy.

Energy Equivalent of Mass

$$\Delta E = \Delta mc^2$$

ΔE is the change in energy, in Joules. Δm is the change in mass, in kg. c is the speed of light.

The change in energy is equal to the change in mass times the square of the speed of light.

Mass defect and binding energy Scientists have determined that the mass of the nucleus is always less than the sum of the masses of the individual protons and neutrons that comprise it. This difference in mass between a nucleus and its component nucleons is called the **mass defect.**

When nucleons combine together to form an atom, the energy corresponding to the mass defect is released. Conversely, energy is needed to break apart a nucleus into its nucleons. The nuclear binding energy can be defined as the amount of energy needed to break one mole of nuclei into individual nucleons. The larger the binding energy per nucleon, the more strongly the nucleons are held together, and the more stable the nucleus is. Less-stable atoms have lower binding energies per nucleon. In other words, it is harder to break apart a nucleus with a high binding energy than a nucleus with a low binding energy.

Figure 14 shows the average binding energy per nucleon versus the mass number. Note that the binding energy per nucleon reaches a maximum around a mass number of 60. Elements with a mass number near 60 are the most stable.

Visual Learning

Figure 14 Have students predict whether each of the following isotopes is a candidate for nuclear fusion or fission: lithium-6 (fusion), technetium-97 (fission), and seaborgium-260 (fission). OL

Math in Chemistry

Einstein's Equation Making calculations with Einstein's equation, $\Delta E = \Delta mc^2$, can help students realize the enormous amount of energy stored in atomic nuclei. Have students solve the following problems. For reference, explain that an energy of 1 joule is approximately the energy needed to move an apple a distance of 1 meter.

1. What is the energy equivalent, in joules, of a proton (mass = 1.6726×10^{-27} kg)?
2. What is the energy equivalent, in joules, of a neutron (mass = 1.67492×10^{-27} kg)?
3. What energy is released during a nuclear reaction in which an electron and a positron (both with mass = 9.11×10^{-31} kg) annihilate each other?

1. 1.503×10^{-10} J
2. 1.505×10^{-10} J
3. 1.64×10^{-13} J OL

☑ GRAPH CHECK The binding energy per nucleon increases until the mass number is equal to 60, then it decreases slightly.

DIFFERENTIATED INSTRUCTION

Advanced Learners Present students with the following data regarding the mass defect of a helium atom.

2p: (2)(1.007276 amu)	= 2.014552 amu
2n: (2)(1.008665 amu)	= 2.017330 amu
2e: (2)(0.000548 6 amu)	= 0.001097 amu
Total	= 4.032979 amu

The measured mass of helium is 4.00260 amu, so the mass defect is 0.03038 amu. To calculate the binding energy, the mass is converted to kg, using the relation 1 amu = 1.6605×10^{-27} kg. Then, $E = mc^2$ is used to calculate the binding energy. $E = 4.54 \times 10^{-12}$ J. Have students calculate the binding energy for lithium-6, which has a measured mass of 6.015 amu. 5.2×10^{-12} J AL

Quick Demo

Nuclear Energy Hold up a simulated uranium dioxide pellet made by cutting a pencil into 3-cm sections and painting them dark gray. Explain to the students that as many as 10 million pellets may be used in one nuclear power plant per year. The pellets are loaded into long narrow steel cylinders known as fuel rods. Point out that the volume of coal needed to produce the amount of power produced by the pellets would be huge in comparison.

PROBLEM-SOLVING STRATEGIES

Apply the Strategy

mass defect = (mass of isotope) − (number of protons)m_H − (number of neutrons)m_n
= 7.016003 amu − 3(1.007825 amu)
− 4(1.008665 amu)
= −0.042132 amu

binding energy = (−0.042132 amu)(931.49 MeV/amu) = −39.246 MeV

☑ READING CHECK because they are unstable; The binding energy per nucleon reaches a maximum around an atomic number of 60. Heavy atoms tend to fragment into smaller atoms to increase their stability.

PROBLEM-SOLVING STRATEGIES

Calculating Mass Defect

You can calculate the mass defect of an isotope if you know the mass of the isotope and the number and masses of its components. Applying the equation $\Delta E = \Delta mc^2$, you can then derive the equivalent binding energy.

Mass defect = $m_{\text{nucleus}} - [N_p m_p + N_n m_n]$

where m_{nucleus} is the mass of the nucleus, m_p is the mass of a proton, m_n is the mass of a neutron, N_p is the number of protons, and N_n is the number of neutrons.

If you start with the mass of the atom, you have to take into account the mass of the electrons. To do so, the mass of a hydrogen atom, which is composed of a proton and an electron, is used instead of the mass of a proton. The equation is then:

Mass defect = $m_{\text{isotope}} - [N_p m_H - N_n m_n]$

Use the following values for the calculations: m_H = 1.007825 amu and m_n = 1.008665 amu. The accepted value for c is 3.00×10^8 m/s.

To calculate the energy in Joules, you can convert the masses into kilograms using 1 amu = 1.660540×10^{-27} kg.

Apply the Strategy

Calculate the mass defect and binding energy of lithium-7. The mass of lithium-7 is 7.016003 amu.

In typical chemical reactions, the energy produced or consumed is so small that the accompanying changes in mass are negligible. In contrast, the mass changes and associated energy changes in nuclear reactions are significant. For example, the energy released from the nuclear reaction of 1 kg of uranium is equivalent to the energy released during the chemical combustion of about four billion kilograms of coal.

Nuclear Fission

Binding energies in **Figure 14** indicate that heavy nuclei tend to be unstable. To gain stability, they can fragment into several smaller nuclei. Because atoms with mass numbers around 60 are the most stable, heavy atoms (those with mass numbers greater than 60) tend to fragment into smaller atoms in order to increase their stability. The splitting of a nucleus into fragments is known as **nuclear fission.** The fission of a nucleus is accompanied by a very large release of energy.

Nuclear power plants use nuclear fission to generate power. The first nuclear fission reaction discovered involved uranium-235. As you can see in **Figure 15,** when a neutron strikes a uranium-235 nucleus, it undergoes fission. Barium-141 and krypton-92 are just two of the many possible products of this fission reaction. In fact, scientists have identified more than 200 different product isotopes from the fission of a uranium-235 nucleus.

VOCABULARY
ACADEMIC VOCABULARY
Generate
to bring into existence, to originate by a physical or chemical process
Fire generates a lot of heat.

☑ READING CHECK **Explain** why heavy atoms undergo nuclear fission.

DIFFERENTIATED INSTRUCTION

Struggling Learners Have students locate an electric bill from home and note the kWh (kilowatt-hours) used. Ask them to use dimensional analysis to convert kWh to kilograms of coal consumed and its equivalent in uranium-235 fuel. *The CRC Handbook of Chemistry and Physics* has conversions needed or the conversions can be given to students. BL

- 1 kWh = 3.60×10^6 J
- U-235 undergoes beta decay giving total disintegration energy of 4.68 MeV
- 1 MeV = 1.61×10^{-19} J
- Enthalpy of combustion for carbon is 393.5 kJ/mol

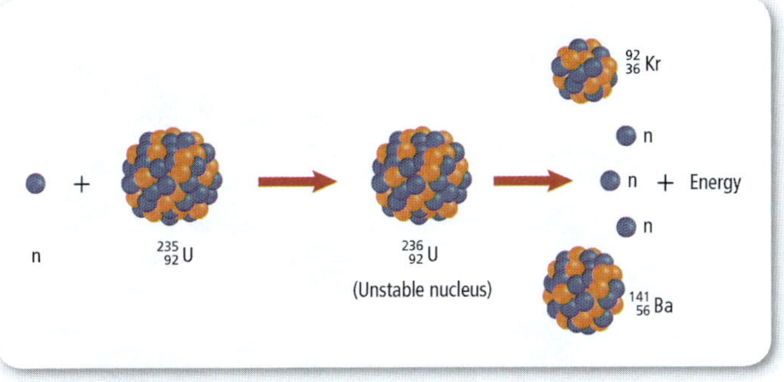

■ **Figure 15** When bombarded with a neutron, uranium-235 forms unstable uranium-236, which then splits into two lighter nuclei and additional neutrons. The fission of uranium-235 is accompanied by a large release of energy.

Chain reactions Each fission of uranium-235 releases additional neutrons, as shown in **Figure 15.** If one fission reaction produces two neutrons, these two neutrons can cause two additional fissions. If those two fissions release four neutrons, those four neutrons could then produce four more fissions, and so on, as shown in **Figure 16.** This self-sustaining process in which one reaction initiates the next is called a chain reaction. As you might imagine, the number of fissions and the amount of energy released can increase rapidly. The explosion from an atomic bomb is an example of an uncontrolled chain reaction.

APPLYING PRACTICES

Develop and Use Models Go to the resources tab in ConnectED to find the Applying Practices worksheet *Modeling Fission, Fusion, and Radioactive Decay.*

■ **Figure 16** When uranium nuclei undergo fission, they release neutrons, which trigger more fission reactions. The ongoing reactions are characteristics of a nuclear chain reaction.

View an **animation about chain reactions.**

Concepts In Motion

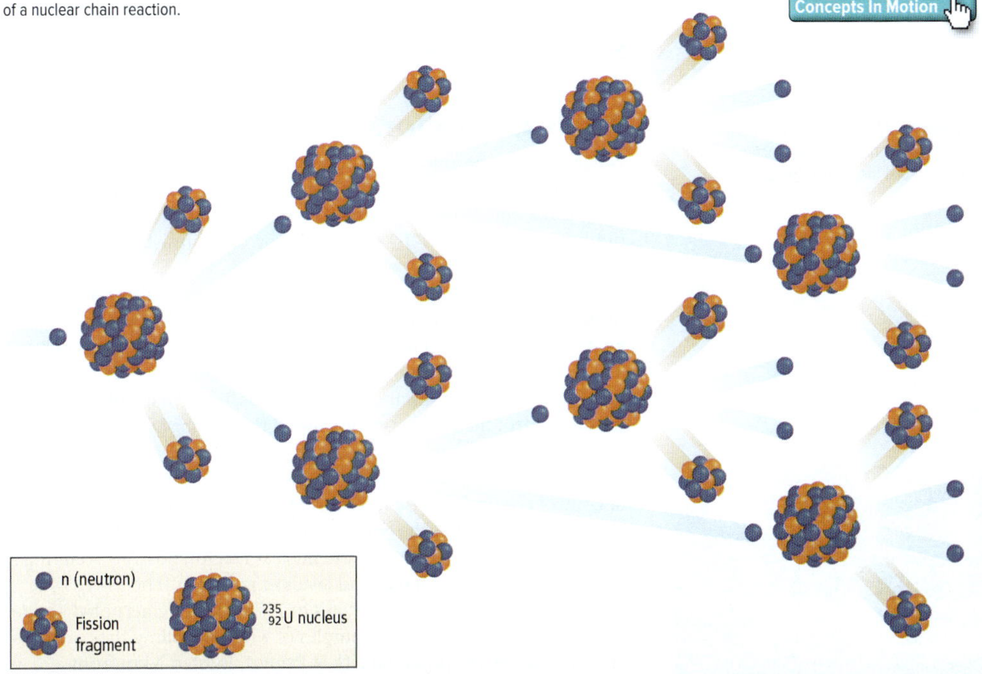

Quick Demo

Chain Reaction Divide a set of dominoes into two groups. Arrange one group in a single row, so that when the lead domino falls, it causes each of the other dominoes in the line to fall sequentially. This demonstrates a chain reaction. Arrange the second group into a bowling pin pattern, where the lead domino causes two dominoes to fall, which causes three more dominoes to fall, and so on. This demonstrates an escalating chain reaction.

Reinforcement

Chain Reaction Ask students to calculate how many cycles of a "chain letter" are needed to reach every student in their high school if they start with just one letter. Ask them to compare this process with that of an escalating nuclear chain reaction. BL EL

CHEMISTRY PROJECT

Nuclear Accidents Have students research one of the major nuclear accidents, such as Three Mile Island or Chernobyl, and prepare a report or poster describing the events. Students should include a description relating what they learn in this section about chain reactions and critical mass to the accident. Suggest that students also include information about how, in hind sight, the accident could have been avoided, long-term effects of the accident, and current safety procedures used in nuclear facilities designed to prevent a similar accident from occurring again. OL AL

Content Background

The Bomb Many students might be confused about the difference between the "A-bomb" (atomic bomb) and the "H-bomb" (hydrogen bomb). The atomic bomb utilizes the principles of fission and was originally developed as part of the Manhattan Project in the early 1940s. The first experimental atomic bomb detonation occurred on July 16, 1945, near Alamogordo, New Mexico. A few weeks later, atomic bombs were dropped on Hiroshima (August 6, 1945) and Nagasaki (August 9, 1945), Japan. Each bomb released energy equal to that of 20,000 tons of TNT. The dropping of the bombs hastened the end of World War II.

The hydrogen bomb has an explosive force 1000 times greater than that of the atomic bomb. In a hydrogen bomb, a fission reaction triggers a fusion reaction of hydrogen isotopes (deuterium and tritium). The United States detonated the first megaton-class hydrogen bomb (equal to 15 million tons of TNT) at Bikini Atoll in the Pacific Ocean on March 1, 1954. The hydrogen bomb has never been used in warfare.

Extension

Atomic Weapons Have students read *Hiroshima* by John Hersey. Hold a debate between teams of students about whether the United States should have used the atomic bomb during World War II. **OL**

READING CHECK If a sample's mass is subcritical, the chain reaction never begins. At critical mass, enough neutrons are released to initiate more reactions and the process is self-sustaining.

Figure 17 Whether a nuclear reaction can be sustained depends on the amount of matter present. In a subcritical mass, the chain reaction does not start because neutrons escape before causing enough fission to sustain the chain reaction. In a supercritical mass, neutrons cause more and more fissions and the chain reaction accelerates.

View an **animation about critical mass.**

Concepts In Motion

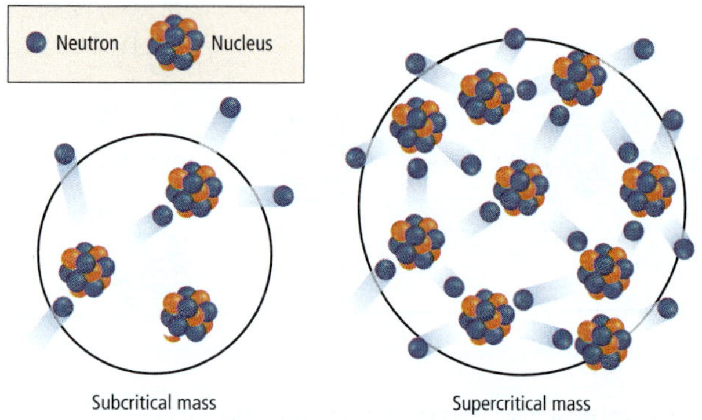

A sample of fissionable material must have sufficient mass in order for a chain reaction to occur. If it does not, neutrons escape from the sample before they can start the chain reaction by striking other nuclei. A sample that is not massive enough to sustain a chain reaction is said to have subcritical mass. A sample that is massive enough to sustain a chain reaction has **critical mass.** When a critical mass is present, the neutrons released in one fission cause other fissions to occur. If much more mass than the critical mass is present, the chain reaction rapidly escalates. This can lead to a violent nuclear explosion. A sample of fissionable material with a mass greater than the critical mass is said to have supercritical mass. **Figure 17** shows the effect of mass on the initiation and progression of a fission reaction.

READING CHECK Compare subcritical mass and critical mass.

Nuclear Reactors

Nuclear fission produces the energy generated by nuclear reactors. This energy is primarily used to generate electricity at nuclear power plants, such as the one shown in **Figure 18**. A common fuel is fissionable uranium (IV) oxide (UO_2) encased in corrosion-resistant rods. U-238 is the most abundant isotope (99%) of uranium. U-235, which makes up 0.7% of the natural uranium, has the rare property of being able to undergo induced fission; U-235 atoms undergo fission when hit by a neutron. The fuel used in nuclear power plants is enriched to contain 3% uranium-235, the amount required to sustain a chain reaction, and is called enriched uranium. Additional rods, often made of cadmium or boron, control the fission process inside the reactor by absorbing neutrons released during the reaction.

Keeping the chain reaction going while preventing it from racing out of control requires precise monitoring and continual adjusting of the control rods. Much of the concern about nuclear power plants focuses on the risk of losing control of the nuclear reactor, possibly resulting in the accidental release of harmful levels of radiation. The Three Mile Island accident in the United States in 1979 and the Chernobyl accident in Ukraine in 1986 provide examples of why controlling the reactor is critical. **Figure 19** shows the city of Pripyat, located 3 km from Chernobyl. The city was completely abandoned after the accident.

Figure 18 The main parts of a nuclear power plant are the reactor under the dome and the cooling tower.

DIFFERENTIATED INSTRUCTION

Advanced Learners Have students research the status of nuclear power generation in the United States. What percentage of United States power needs is provided by nuclear reactors? What government agency regulates nuclear power plants? What are the costs of building and maintaining a plant? What are the advantages and disadvantages of nuclear power for consumers? **AL**

■ **Figure 19** The city of Pripyat was deserted after the accident at the Chernobyl power plant.

The fission within a nuclear reactor is started by a neutron-emitting source and is stopped by positioning the control rods to absorb all of the neutrons produced in the reaction. The reactor core contains a reflector that acts to reflect neutrons back into the core, where they will react with the fuel elements, also called fuel rods. A coolant, usually water, circulates through the reactor core, to carry off the heat generated by the nuclear fission reactions. The hot coolant heats water that is used to power steam-driven turbines, which produce electric power.

Nuclear power plants and fossil-fuel burning power plants are similar; heat from a reaction—nuclear fission or chemical combustion of coal—is used to generate steam. The steam then drives turbines that produce electricity, as shown in the nuclear power plant illustrated in **Figure 20.** The other major components of a nuclear power plant are also illustrated in **Figure 20.**

View an **animation about nuclear power plants.**

Concepts In Motion

■ **Figure 20** A nuclear reactor produces heat that drives the formation of steam. The energy from the steam spins a turbine which produces electricity. The steam is eventually cooled and recycled. The water used to cool the steam enters the cooling tower where steam is released to the atmosphere.

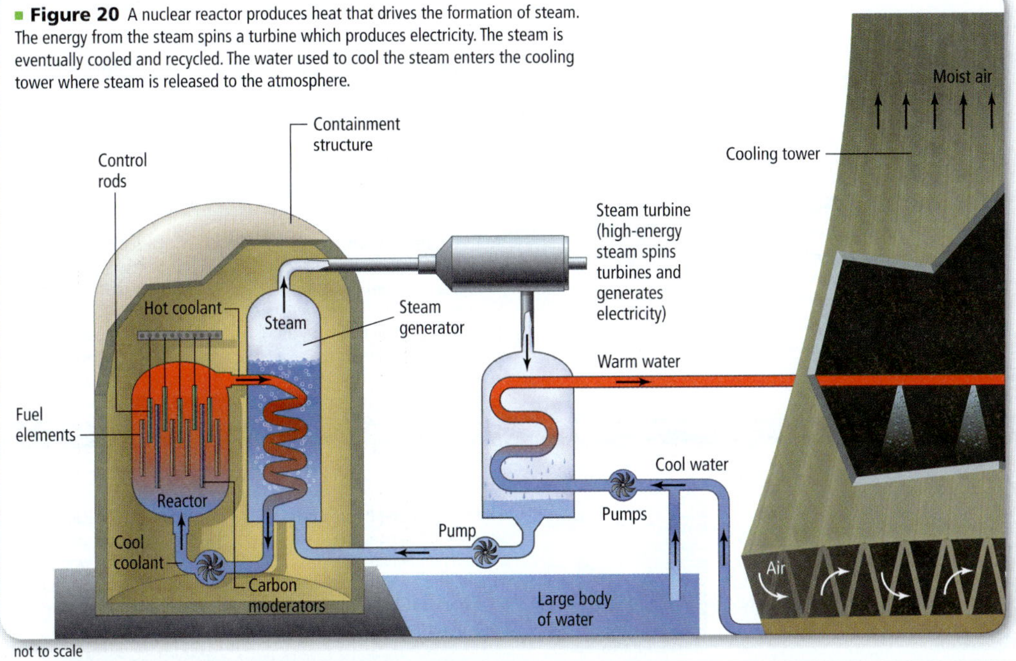

Enrichment

Nuclear Events Have students choose a clip from any current or past popular movie that portrays a nuclear event (*China Syndrome,* for instance). Have them analyze the clip for accuracy and write a report on the clip. Make sure they discuss any pro- or antinuclear sentiments expressed in the clip. **OL**

ASSESSMENT

Performance Have each student make a poster that shows and explains the sequence of events leading to the generation of steam at a nuclear power plant. **OL** **EL**

Extension

Nuclear Power Have students research reasons why some countries rely more heavily on nuclear power than the United States. Lack of geographical access to oil and coal, political agreements with oil-producing countries, and economic factors are a few reasons. **AL**

CHEMISTRY PROJECT

Uranium Enrichment Explain to students that the uranium isotope usable for fission in nuclear reactors makes up only a tiny percent of the uranium ore mined from the ground. Have students research and prepare a report explaining either the gaseous diffusion or gas centrifuge method used to enrich the ore. **OL** **AL**

Build a Model

Nuclear Reactors Have teams of students research the various designs of commercial nuclear reactors and build a scale model of one of them. Each team should present its model to the class and explain its operation. **OL** **EL** **COOP LEARN**

☑ READING CHECK Some nuclear wastes remain radioactive for thousands of years. They need to be stored without damaging the environment.

Content Background

Radiometric Dating Archaeologists rely on radiometric dating to establish the age of some artifacts. The technique relies on the fact that plants absorb a radioactive isotope of carbon, C-14, during photosynthesis. When a plant dies, the C-14, with a half-life of about 5700 years, begins to decay at a predictable rate. By finding the concentration of C-14 left in a plant-based artifact, like wood carvings, parchment, or textiles, scientists can estimate the artifact's age up to 45,000 years.

■ **Figure 21** The interior of a reactor is filled with water. A crane is used to extract and replace fuel rods.

Because of the hazardous radioactive fuels and fission products present at nuclear power plants, a dense concrete structure is usually built to enclose the reactor. The main purpose of the containment structure is to shield personnel and nearby residents from harmful radiation.

As the reactor operates, the fuel rods are gradually depleted and products from the fission reactions accumulate. Because of this, the reactor must be serviced periodically. Spent fuel rods are extracted from the reactor, as shown in **Figure 21,** and can be reprocessed and repackaged to make new fuel rods. Some fission products, however, are extremely radioactive and cannot be used again. These products must be stored as nuclear waste.

Risks of accidents, such as the ones mentioned in **Figure 22,** have to be taken into account when operating nuclear power plants. However, the storage of highly radioactive nuclear waste is still one of the major issues surrounding the debate over the use of nuclear power. Approximately 20 half-lives are required for the radioactivity of nuclear waste materials to reach levels acceptable for biological exposure. For some types of nuclear fuels, the wastes remain substantially radioactive for thousands of years. A considerable amount of scientific research is devoted to the disposal of radioactive wastes. Highly radioactive materials from the reactor core are first treated with advanced technologies that ensure the materials will not deteriorate over a very long period of time. Treated wastes are then stored in sealed containers that are buried deep underground.

Another issue is the limited supply of the uranium-235 used in the fuel rods. One option is to build reactors that produce new quantities of fissionable fuels. Reactors able to produce more fuel than they use are called **breeder reactors.** Although the design of breeder reactors poses many difficult technical problems, they are currently in operation in several countries.

☑ READING CHECK **Infer** how the storage of nuclear wastes affects the environment.

■ **Figure 22**

The Nuclear Age

The discovery of X-rays in 1895 initiated a series of breakthroughs in understanding atomic nuclei. Today, nuclear chemistry applications involving medicine, weaponry, and energy affect the lives of people worldwide.

1890 1900 1920 1940

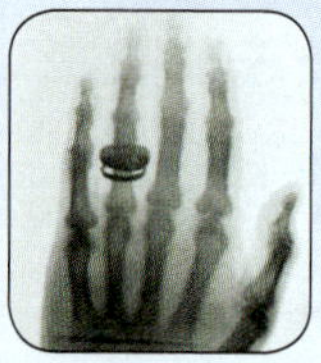

1895 The first X-ray photographs fuel intense interest among the scientific community.

1898 Marie and Pierre Curie discover the radioactive elements polonium and radium. Their work establishes the early framework for the study of nuclear chemistry.

1919 The first artificially induced nuclear reaction causes the transmutation of nitrogen into an isotope of oxygen by bombarding nitrogen gas with alpha particles.

1934 Enrico Fermi's experiments result in the world's first nuclear fission reaction. Fermi's subsequent research will pioneer nuclear power generation.

1941–45 Manhattan Project scientists develop uranium and plutonium bombs, which were dropped on Hiroshima and Nagasaki, Japan, in 1945 and ended World War II.

CHEMISTRY JOURNAL

The NIMBY Syndrome Many strong antinuclear sentiments arise when it comes to the topic of nuclear power and storage of nuclear waste products. While many people might feel comfortable with the idea of using the energy derived from nuclear fission, they are not comfortable with the storage of radioactive waste products in their state, thus leading to the "Not In My Back Yard" syndrome. Ask students to research current radioactive waste disposal options in their state and in the nation. How would they feel living near a nuclear waste disposal site? **OL**

Nuclear Fusion

Recall from the binding energy diagram in **Figure 14** that a mass number of about 60 has the most stable atomic configuration. Thus, it is possible to bind together two or more light (mass number less than 60) and less-stable nuclei to form a single more-stable nucleus. The combining of atomic nuclei is called **nuclear fusion.** Nuclear fusion reactions, which are responsible for the production of the heaviest elements, are capable of releasing very large amounts of energy. You already have some everyday knowledge of this fact—the Sun is powered by a series of fusion reactions as hydrogen atoms fuse to form helium atoms.

$$4\,^{1}_{1}\text{H} \rightarrow 2\beta + \,^{4}_{2}\text{He} + \text{energy}$$

Scientists have spent several decades researching nuclear fusion. It is a promising source of energy and has several advantages compared to nuclear fission. Lightweight isotopes used to fuel the reactions, such as hydrogen, are abundant. Fusion reaction products are not generally radioactive. Nuclear fusion produces large amounts of energy. Fusion reactions produce more energy per unit of mass of fuel than fission reactions. This could solve the problem of the increasing needs for electricity in the world's societies.

Unfortunately, there are major problems that must be overcome on a commercially viable scale. One such problem is that fusion requires extremely high energies to initiate and sustain a reaction. The required energy, which is achieved only at extremely high temperatures, is needed to overcome the electrostatic repulsion between the nuclei in the reaction. Because of the energy requirements, fusion reactions are also known as **thermonuclear reactions.** A temperature of 5,000,000 K is required to fuse hydrogen atoms. This temperature—and even higher temperatures—have been achieved using an atomic explosion to initiate the fusion process, but this approach is not practical for controlled electric power generation.

RealWorld CHEMISTRY
Nuclear Fusion

SOLAR FUSION Nuclear fusion reactions are responsible for the glow and heat from stars such as the Sun. The temperature of the Sun's core is about 15,000,000 K. It is so hot and dense that hydrogen nuclei fuse to produce helium. After billions of years, the Sun's hydrogen will be mostly depleted. Its temperature will rise to about 100,000,000 K, and the fusion process will then change helium into carbon.

1949 Radiocarbon dating allows scientists to determine the age of artifacts made from plant-based materials as old as 45,000 years.

2006 The *Cassini* spacecraft explores the Saturn system. *Cassini* is powered by technology that converts heat from the radioactive decay of plutonium into electricity.

1960 — 1980 — 2000

1960s Scientists research using high-energy radiation to treat cancer. Clinical trials bring dramatic improvement in the treatment and cure of malignant tumors.

1979, 1986 Nuclear power plant accidents at Three Mile Island, Pennsylvania, and Chernobyl, Ukraine, focus world attention on the dangers associated with nuclear power.

2010 Scientists using NASA's Fermi Gamma-ray Space Telescope discover two massive bubbles, one above and one below the core of the Milky Way. They are thought to be the result of activity from either a black hole or star formation.

Enrichment
Stellar Nucleosynthesis Have students research stellar nucleosynthesis, and develop a board game based on their findings. Perhaps dice or a spinner can dictate the fate of a star (for example: supernova, red giant, white dwarf, or neutron star). AL

Enrichment
Nuclear v. Coal Tell students that since the accident at the Three Mile Island nuclear power plant in 1979, no new nuclear power plants have been built in the United States. Ask students to compare the benefits and drawbacks of nuclear and coal-burning power plants. Have students work in groups to develop answers. Burning coal is relatively cheap, but produces CO_2, a greenhouse gas. Nuclear power produces no greenhouse gases, but nuclear waste is deadly and must be controlled to limit environmental impact. AL OL BL COOP LEARN

CHEMISTRY PROJECT
Fission and Fusion Have students work in small groups to prepare a computer graphics demonstration that creatively explains either nuclear fission or nuclear fusion. They might want to present an overview or concentrate on one application, such as fission at a nuclear power plant or solar fusion. OL COOP LEARN

3 Assess

Check for Understanding

Have students respond to the following true-false statements. They should support their choices. This could be done individually or as a group.

1. Great amounts of energy can be liberated from small amounts of matter in nuclear reactions. true
2. The amount of U-235 in a nuclear reactor should always be kept subcritical. false
3. Nuclear power plants do not contribute to pollution. True in terms of air pollution; false if nuclear waste is considered.
4. Inserting boron or cadmium rods into the reactor core can control a fission reaction. true
5. Nuclear power plants are dangerous because they commonly release radioactive particles into the environment. false OL COOP LEARN

Reteach

Place an unlabeled diagram of a nuclear reactor on the board. Have students come up and explain the function of a particular part of the reactor beginning with the fuel rods. If they get stuck, they might "tag" another student to continue the process. OL EL

Extension

Ask students to research information on nuclear fusion as a possible source of energy. Have them write a report on their findings. OL

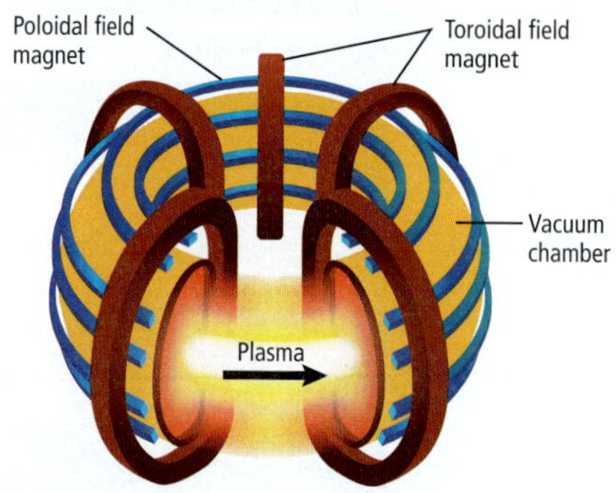

■ **Figure 23** A tokamak reactor, a ring-shaped reactor, uses strong magnetic fields to contain the intensely hot fusion reaction and keep it from direct contact with the reactor interior walls. The poloidal magnets follow the shape of the reactor and the toroidal magnets wrap around the reactor.

Obviously, many problems must be resolved before fusion becomes a practical energy source. Another significant problem is confinement of the reaction. There are currently no materials capable of withstanding the tremendous temperatures that are required by a fusion reaction. Much of the current research centers around an apparatus called a tokamak reactor. The name *tokamak* comes from Russian and means *toroidal chamber with an axial magnetic field.* A tokamak reactor, shown in **Figure 23,** is a donut-shaped device that uses strong magnetic fields to contain the fusion reaction. While significant progress has been made in the field of fusion, temperatures high enough for continuous fusion have not yet been sustained for long periods of time.

SECTION 3 REVIEW

Section Self-Check

Section Summary

- Induced transmutation is the bombardment of nuclei with particles in order to create new elements.
- In a chain reaction, one reaction induces others to occur. A sufficient mass of fissionable material is necessary to initiate the chain reaction.
- Fission and fusion reactions release large amounts of energy.

22. MAINIDEA **Compare and contrast** nuclear fission and nuclear fusion reactions. Describe the particles that are involved in each type of reaction and the changes they undergo.
23. **Describe** the process that occurs during a nuclear chain reaction and explain how to monitor a chain reaction in a nuclear reactor.
24. **Explain** how nuclear fission can be used to generate electric power.
25. **Formulate** an argument supporting or opposing nuclear power as your state's primary power source. Assume the primary source of power currently is the burning of fossil fuels.
26. **Calculate** What is the energy change (ΔE) associated with a change in mass (Δm) of 1.00 mg?
27. **Interpret Graphs** Use the graph in **Figure 14** to answer the following questions.
    - **a.** Why is the isotope $^{56}_{26}Fe$ highest on the curve?
    - **b.** Are more stable isotopes located higher or lower on the curve?
    - **c.** Compare the stability of Li-6 and He-4.

EFDA-JET/Photo Researchers

SECTION 3 REVIEW

22. In fission, heavy nuclei break down into smaller nuclei, releasing large amounts of energy. In fusion, two or more light nuclei join together to form a more stable nucleus and release large amounts of energy.
23. A neutron strikes a target nucleus such as U-235, which then splits into multiple fission fragments and releases several neutrons. These neutrons collide with other nuclei, causing additional fissions and sustaining the reaction. To control the chain reaction in a nuclear reactor, control rods made of cadmium and boron are used to absorb some of the neutrons released during the reaction.
24. Fission reactions heat water and form steam. The steam spins turbines that produce electric power.
25. Answers will vary. See Solutions Manual for example statements.
26. 9.00×10^{10} kg•m^2/s^2 or J
27. **a.** it has the greatest binding energy.
    b. higher
    c. The stability of He-4 is greater.

SECTION 4

Essential Questions

- What are several methods used to detect and measure radiation?
- How is radiation used in the treatment of disease?
- What are some of the damaging effects of radiation on biological systems?

Review Vocabulary

isotope: an atom of the same element with the same number of protons but different number of neutrons

New Vocabulary

ionizing radiation
radiotracer

Applications and Effects of Nuclear Reactions

MAINIDEA Nuclear reactions have many useful applications, but they also have harmful biological effects.

CHEM 4 YOU Almost everyone gets cuts or scrapes from time to time. Usually, the first thing you do is clean the injury and cover it with a bandage to keep out germs. One of the many uses of radiation is to sterilize medical bandages.

Detecting Radioactivity

You read earlier that Becquerel discovered radioactivity because of the effect of radiation on photographic plates. Since this discovery, several other methods have been devised to detect radiation. People who work near radioactive sources, for example, might be required to wear a thermoluminescent dosimeter (TLD) badge, which contains a tiny crystal. Radiation excites electrons within the crystal. To determine the radiation dose, the crystal is heated, and the electrons return to their ground states, emitting light. Radioactivity readers detect this light as a measure of the radiation dose to which a worker has been exposed. Monitoring the radiation dose received by people who work near radioactive sources is important to ensure their safety.

Radiation energetic enough to ionize matter with which it collides is called **ionizing radiation.** The Geiger counter is an ionizing radiation detection device. As shown in **Figure 24,** a Geiger counter consists of a metal tube filled with a gas. In the center of the tube is a wire that is connected to a power supply. When ionizing radiation penetrates the end of the tube, the gas inside the tube absorbs the radiation and forms ions and free electrons. The free electrons are attracted to the wire, causing an electric current. A meter built into the Geiger counter measures the current flow through the ionized gas. This current measurement is used to determine the amount of ionizing radiation present.

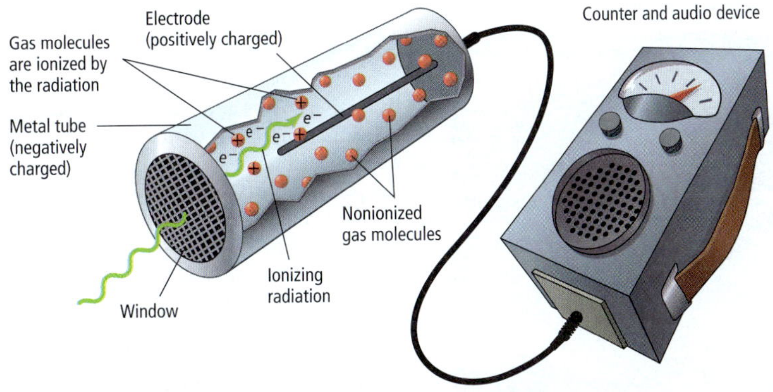

Figure 24 A Geiger counter is used to detect and measure radiation levels. Ionizing radiation produces an electric current in the counter. The current is displayed on a scaled meter, whereas a speaker produces audible sounds.

SECTION 4

1 Focus

MAINIDEA

Useful or Harmful Point out to students that sometimes useful things can also be harmful if used in a different way. For example, jogging can improve health, but jogging on a hard surface with poor shoes can harm your knees and feet. Have students name other things that can be both useful and harmful. Possible examples: Trace amounts of iron are needed in one's diet, but too much can cause illness. Sunlight provides Earth with warmth and light, but too much sunlight can damage skin. OL

2 Teach

Quick Demo

Conservation by Irradiation Check at a grocery store and obtain irradiated and nonirradiated samples of the same food. (Some produce and meats might be irradiated.) Ask employees at the store for help if needed. Place the samples in two different sealed containers in the classroom and observe them for the next several days. The irradiated sample should not spoil as quickly as the nonirradiated sample.

DIFFERENTIATED INSTRUCTION

English Learners Have students look up the definition of ionizing radiation and use it in a sentence. Have them write an explanation of how ionizing radiation is used by a Geiger counter to detect radiation. EL BL

ASSESSMENT

Performance Assign a task that can be achieved or monitored through the use of nuclear technology to teams of students. Each team will represent a company that specializes in this technology. Specific roles for each group member should be outlined (chairman, graphic designer, presenter, and so on). The students will research the use of the particular technology and then sell it to the class using a visually oriented presentation. Possible topics include the use of radioisotopes in gemology and medicine, the environmental cleansing of sulfur dioxides and nitrogen oxides, airport baggage inspection, food preservation (including shrink wrapping), computer disks, and the plastics industry. OL COOP LEARN

☑ READING CHECK Scintillator detectors have a base material containing phosphor. When ionizing radiation strikes the counter, energy is transferred and excites electrons in the phosphor. As the electrons de-excite, they emit energy as light, which is transmitted to a photodetector. The photodetector produces an electrical signal which is related to the amount of ionizing radiation.

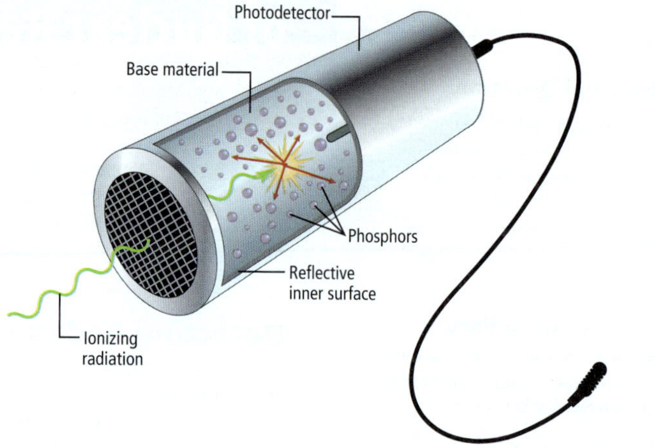

■ **Figure 25** Scintillation counters are used to detect the presence of ionizing radiation. An ionizing radiation excites the electrons in the phosphors. As the electrons return to their ground states, they emit photons, which are then detected by the photodetector.

Another detection device is a scintillation counter. Scintillations are brief flashes of light produced when ionizing radiation excites the electrons in certain types of atoms or molecules called phosphors. A scintillation counter contains a base material—often a plastic, a crystal, or a liquid—containing phosphors, as shown in **Figure 25.** Ionizing radiation that strikes the scintillation counter can transfer energy either directly to the phosphors or to the base material, which then transfers the energy to the phosphors. This energy excites electrons in the phosphors. As these electrons return to their ground states, they release energy in the form of light. This light is transmitted through the base material to a photodetector that convert the light to an electrical signal. The number and brightness of the scintillations give a measure of the amount of ionizing radiation.

☑ READING CHECK **Summarize** how a scintillation detector works.

■ **Figure 26** Gauges such as the one pictured use beta emission from krypton, promethium, or strontium. The radioactive source is placed on one side of the paper, and a detector is on the other side. Most beta particles are absorbed by the paper, but the percentage that are able to travel through to the detector indicates the thickness of the paper.

Uses of Radiation

With proper safety procedures, radiation can be useful in many scientific experiments and industrial applications. For instance, neutron activation analysis is used to detect trace amounts of elements present in a sample. Computer-chip manufacturers use this technique to analyze the composition of highly purified silicon wafers. In the process, the sample is bombarded with a beam of neutrons from a radioactive source, causing some of the atoms in the sample to become radioactive. The type and amount of radiation emitted by the sample is used to determine the types and quantities of elements present. Neutron activation analysis is a highly sensitive measurement technique capable of detecting quantities of less than 1×10^{-9} atoms in a sample. Beta emission is another application of radiation. It is used to measure paper thickness, as shown in **Figure 26.**

CHEMISTRY PROJECT

Radiation in Manufacturing Have students research and prepare a report describing one way that radioactive sources are used in manufacturing. Examples include monitoring the thickness of materials and checking materials for defects. Suggest that students include illustrations with their reports. OL

Using radioisotopes Radioisotopes can also be used to follow the course of an element through a chemical reaction. For example, CO_2 gas containing radioactive carbon-14 isotopes has been used to study glucose formation in photosynthesis.

$$6CO_2 + 6H_2O \xrightarrow{\text{sunlight}} C_6H_{12}O_6 + 6O_2$$

Because the CO_2 containing carbon-14 is used to trace the progress of carbon through the reaction, it is referred to as a radiotracer. A **radiotracer** is a radioisotope that emits non-ionizing radiation and is used to signal the presence of an element or specific substance. The fact that all of an element's isotopes have the same chemical properties makes the use of radioisotopes possible. Thus, replacing a stable atom of an element in a reaction with one of its isotopes does not alter the reaction. Radiotracers are important in a number of areas of chemical research, particularly in analyzing the reaction mechanisms of complex, multistep reactions.

Radiotracers also have important uses in medicine. Iodine-131, for example, is commonly used to detect diseases associated with the thyroid gland. If a problem is suspected, the patient will drink a solution containing a small amount of iodine-131. After the iodine is absorbed, the amount of iodine taken up by the thyroid is measured and used to monitor the functioning of the thyroid gland.

☑ READING CHECK **Define** *radiotracer.*

Treating cancer Radiation can pose serious health problems for humans because it can damage or destroy healthy cells. However, radiation can also destroy unhealthy cells, such as cancer cells. All cancers are characterized by the rapid growth of abnormal cells. This growth can produce masses of abnormal tissue, called malignant tumors. Radiation therapy is used to treat cancer by destroying the cancer cells. In fact, cancer cells are more susceptible to destruction by radiation than healthy ones. **Figure 27** shows a brain scan with a malignant tumor. After radiation treatment, the baseline returns to normal. Unfortunately, in the process of destroying unhealthy cells, radiation also destroys some healthy cells. Despite this major drawback, radiation therapy has become one of the most effective treatment options in the fight against cancer.

CAREERS IN CHEMISTRY

Radiation Therapist Under the supervision of a physician, a radiation therapist administers radiation treatment to patients. Radiation therapists work closely with patients and must be compassionate and supportive. Training programs prepare radiation therapists to use particle accelerators and other forms of technology. Knowledge of radiation hazards is an important part of this job.

WebQuest

Figure 27 Radiation can be used to treat cancer. MRI images taken before treatment and after 4 and 10 months of treatment show the decrease in the size of the tumor.

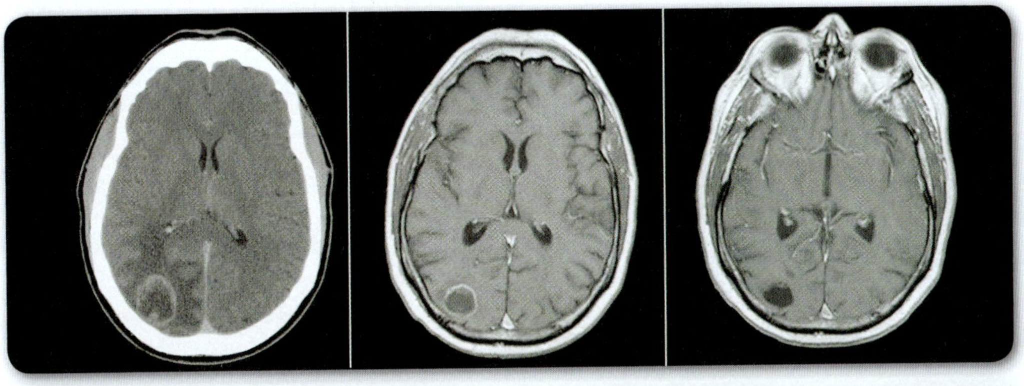

Custom Medical Stock Photo

Extension

Nuclear Medicine Invite a radiologist or radiology technician from the community hospital to discuss and show slides of the various types of nuclear technologies available for diagnosis and treatments of different medical conditions. OL

CAREERS IN CHEMISTRY

WebQuest

☑ READING CHECK A radiotracer is a radioisotope that emits non ionizing radiation.

Content Background

Intensity Units The unit rem is a measure of the radiation dose that will cause the amount of biological damage as one rad of X-rays or gamma rays. The SI unit sievert is equal to 100 rems. Often dose is expressed in millirems (mrem), which is thousandths of a rem. Dose rate is expressed in millirem per second (mrem/s). Radiation intensity is the dose rate per unit area, which is expressed as millirem per second per square meter (mrem/s•m²). Because the unit rem takes into account the effect on biological tissue, the use of the unit mrem/s•m² to describe intensity applies to all types of radiation.

DIFFERENTIATED INSTRUCTION

Struggling Learners Students can better understand the concept of radiotracers if they make a one-page poster describing them. Have students work in pairs to design the posters. Posters should include a definition of a radiotracer in their own words. They should also include a brief description of how radiotracers work and some examples of how they are used in medicine. Encourage students to include hand-drawn figures to explain the concept. BL

Concept Development

PET Scan Radiation Students might wonder why physicians would be willing to inject radioactive substances into a patient, especially radioactive substances that emit high-energy gamma radiation. Remind students about what they learned in Section 2 about half-lives. Point out that the half-lives of the substances used for PET scans is often only minutes or hours. Have students look back at Figure 11. Draw students' attention to how quickly the radioactivity of a substance drops. Although the substances used for PET scans contain radioactive materials, the short half lives means the patients receive only an overall low dose of radiation. **OL**

ASSESSMENT

Performance Set up a quiz-show style game with categories such as nuclear medicine, detecting radiation, nuclear power, fission or fusion. Have teams of students compete for points. **OL** **COOP LEARN**

■ **Caption Question Fig. 29**
somatic

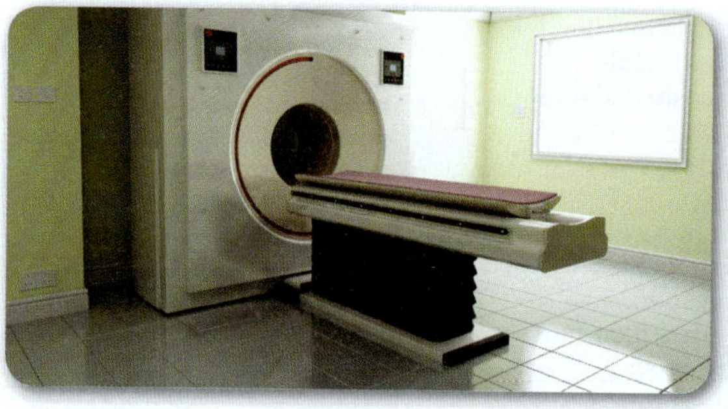

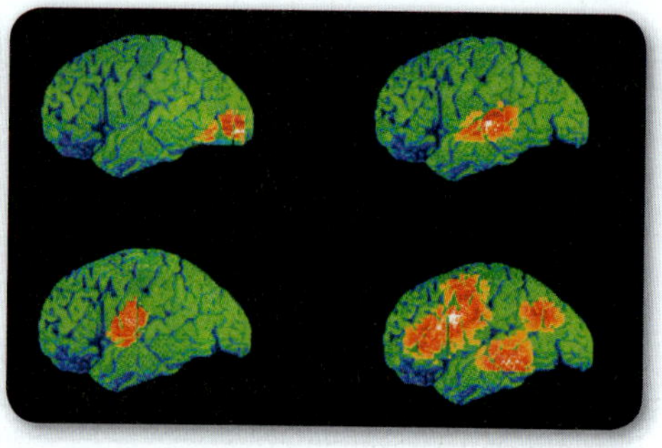

■ **Figure 28** Gamma rays emitted by the radiotracers absorbed by the patients are measured with this detector. The image on the right shows different areas of the brain emitting gamma rays. These images might help doctors locate a tumor or observe a brain function.

Using positron emission Another radiation-based medical diagnostic tool is called positron emission transaxial tomography (PET). In this procedure, a radiotracer that decays by positron emission is injected into the patient's bloodstream. Positrons emitted by the radiotracer cause gamma-ray emissions that are then detected by an array of sensors surrounding the patient, as shown in **Figure 28.** PET scans can be used to diagnose diseases or study the parts of the brain that are activated under given circumstances, also shown in **Figure 28.**

Biological Effects of Radiation

Although radiation has a number of medical and scientific applications, it can be very harmful. The damage produced from ionizing radiation absorbed by the body depends on several factors, such as the type of radiation, its energy, the type of tissue absorbing the radiation, the penetrating power, and the distance from the source. **Figure 29** shows an example of such damage.

Connection to Biology High-energy ionizing radiation is dangerous because it can fragment and ionize molecules within biological tissue. A free radical is an atom or molecule that contains one or more unpaired electrons and is one example of the highly reactive products of ionizing radiation. In a biological system, free radicals can affect a large number of other molecules and ultimately disrupt the operation of normal cells. Ionizing radiation damage to living systems can be classified as either somatic or genetic. Somatic damage affects only nonreproductive body tissue. It includes burns and cancer caused by damage to the cell's growth mechanism. Genetic damage can affect offspring by damaging reproductive tissue. Such damage is difficult to study because it might not become apparent for several generations.

■ **Figure 29** Radiation can disrupt cell processes and damage skin.
Infer *Is the lesion pictured here somatic or genetic?*

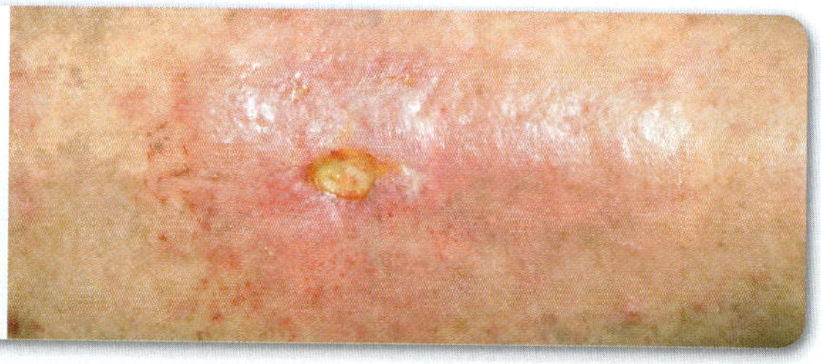

CHEMISTRY JOURNAL

Putting Risk into Perspective On the board, list factors that increase the chance of death by one in a million: smoking 1.4 cigarettes; spending one hour in a coal mine; riding 15 km on a bicycle; traveling 250 km in a car; flying 1500 km by jet; one chest X-ray; living within 8 km of a nuclear reactor for five years; eating 100 barbecued steaks. Ask students to comment on the following. **OL**

- Why do some risks receive more publicity than others?
- Is the concept of "zero risk" possible?
- The difference between voluntary and involuntary risk.
- Some common risks faced daily.
- Is life riskier today than it was 100 years ago?
- How does modern technology affect our perception of risk?

Dose of radiation A dose of radiation refers to the amount of radiation a body absorbs from a radioactive source. Two units, the rad and the rem, are commonly used to measure doses. The rad, which stands for radiation-absorbed dose, is a measure of the amount of radiation that results in the absorption of 0.01 J of energy per kilogram of tissue. The dose in rads, however, does not account for the energy of the radiation, the type of living tissue absorbing the radiation, or the time of the exposure. To account for these factors, the dose in rads is multiplied by a numerical factor that is related to the radiation's effect on the tissue involved. The result of this multiplication is a unit called the rem. The rem, which stands for roentgen equivalent for man, is named after Wilhelm Roentgen, who discovered X-rays in 1895. **Table 6** summarizes the short-term effects of radiation on humans, depending on the dose.

A variety of sources constantly bombard your body with radiation. Your exposure to these sources results in an average annual radiation exposure of 100–300 millirems of high-energy radiation or 0.1–0.3 rems. **Table 7** shows your annual exposure to common radiation sources.

Intensity and distance The intensity of radiation depends on the distance from the source as shown by the equation below. The farther away the source, the lower the intensity. The intensity of radiation is measured in amount of radiation per unit of time and/or surface, such as mrem/s•m^2.

Radiation Intensity and Distance

$$I_1 d_1^2 = I_2 d_2^2$$

d_1 and d_2 are two distances from the source.
I_1 is the intensity at d_1, and I_2 is the intensity at d_2.

The intensity of a radiation at a distance d_1 from the source multiplied by the square of the distance equals the intensity of the radiation at a distance d_2 multiplied by the square of the distance.

Table 6 Effects of Short-term Radiation Exposure

Dose (rem)	Effects on Humans
0–25	no detectable effects
25–50	temporary decrease in white-blood-cell population
100–200	nausea, substantial decrease in white-blood-cell population
500	50% chance of death within 30 days of exposure

Watch a **video about radioisotopes.**

Video

APPLYING PRACTICES

Evaluate Validity and Reliability of Claims
Go to the resources tab in ConnectED to find the Applying Practices worksheet *Human Health and Radiation Frequency.*

Table 7 Average Annual Radiation Exposure

Source	Average Exposure (mrem/y)
Cosmic radiation	20–50
Radiation from the ground	25–175
Radiation from buildings	10–160
Radiation from air	20–260
Human body (internal)	~20
Medical and dental X-rays	50–75
Nuclear weapon testing	<1
Air travel	5
Total average	100–300

CULTURAL DIVERSITY

Music The Polish composer Krysztof Penderecki attempted to commemorate the agony of nuclear war in his 1960 composition *Threnody in Memory of the Victims of Hiroshima.* The horrors of one of the bombings that marked the end of World War II are expressed in the composition by violent torrents of dissonant, percussive sound, some produced by the beating of the bodies of the 52 stringed instruments for which the piece was scored. A 10-minute song of lamentation for the dead begins with a long, screaming tone produced by playing the highest pitches possible on violins.

ChemLAB

The ChemLab located at the end of the chapter can be used at this point in the lesson.

3 Assess

Check for Understanding

Ask students to compare the functioning of a Geiger counter and a smoke detector. Both systems depend on the ionization of a gas (air in a smoke detector and argon [usually] in a Geiger counter). OL

Reteach

Ask students what factors determine the extent of radiation damage to an individual. Dosage, exposure time, nature of radiation, area exposed, and type of tissue are all factors to be considered. BL EL

Extension

Radiation is more destructive to rapidly dividing cells. Ask students how this can be an advantage. Cancer cells divide rapidly and are more easily destroyed by radiation than noncancerous healthy cells. How can this be a disadvantage? Healthy, noncancerous cells could also be damaged. OL

ASSESSMENT

Performance Tell students that light intensity obeys the relationship used in the Problem-Solving Lab on the next page. Have students research the SI unit of light intensity and develop a set of sample problems with answers that illustrate that light obeys the inverse square law. OL

What's CHEMISTRY Got To Do With It?

Video

Isotope Tracers

Problem-Solving LAB

Purpose Students will study how distance affects exposure to sources of radiation.

Process Skills recognize cause and effect, interpret and analyze data from a graph, develop predictions

Teaching Strategies

- Discuss other point like behavior, such as the decrease in light intensity as the distance from the light is squared. Gravitational and electrostatic forces also behave similarly.
- Relate the intensity unit mrem/s•m^2 to sound or light intensity. It's the rate of a radiation falling on a unit area.
- If the distance to a radiation source is much greater than the size of the source, the radiation level (intensity) decreases as the square of the distance (inverse square law). This is a typical behavior of point-like sources.

Think Critically

1. The level at 0.2 m is one-quarter the level at 0.1 m. The level at 0.4 m is one-sixteenth the level at 0.1 m.
2. The radiation decreases with the square of the distance from the source.
3. Set up an inverse square relationship, $\frac{I_1}{I_2} = \frac{d_2^2}{d_1^2}$. Substitute values for I_1 and d_1 from the data given in the graph. Then substitute 0.69 mrem/m^2 for I_2 and solve for d_2. The result is $d_2 = 12$ m.

Problem-Solving LAB

Interpret Graphs

How does distance affect radiation exposure? When one of the reactors at the Chernobyl nuclear power plant exploded, the immediate vicinity of the power plant was highly contaminated and declared a dead zone. The radiation spread over thousands of kilometers. However, the intensity of the radiation decreased with the distance from the reactor.

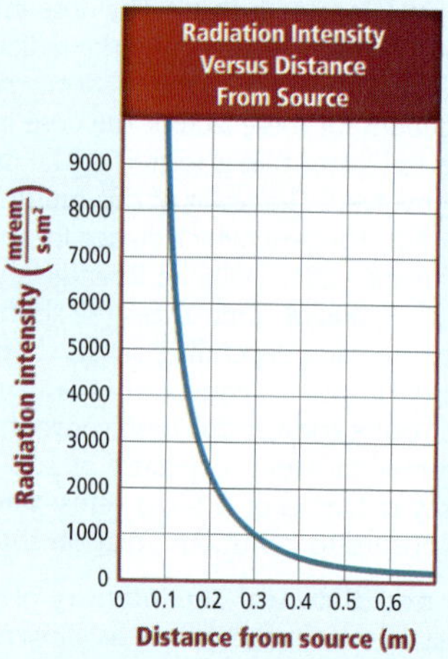

Analysis

The graph to the right shows the intensity of a radioactive source versus the distance from the source. Note how the intensity of the radiation varies with the distance from the source. The unit of radiation intensity is millirems per second per square meter. This is the amount of radiation striking a square meter of area each second.

Think Critically

1. **Evaluate** How does the radiation exposure change as the distance doubles from 0.1 m to 0.2 m? How does it change as the distance quadruples from 0.1 m to 0.4 m?
2. **Formulate** in words the mathematical relationship described in your answer to Question 1.
3. **Interpret Graphs** Determine the distance from the source at which the radiation decreased to 0.69 mrem/s•m^2. This intensity is the maximum radiation exposure intensity considered safe. (*Hint: Use the equation* $I_1/I_2 = d_2^2/d_1^2$.)

SECTION 4 REVIEW

Section Self-Check

Section Summary

- Different types of counters are used to detect and measure radiation.
- Radiotracers are used to diagnose disease and to analyze chemical reactions.
- Short-term and long-term radiation exposure can cause damage to living cells.

28. MAINIDEA **Explain** one way in which nuclear chemistry is used to diagnose or treat disease.
29. **Describe** several methods used to detect and measure radiation.
30. **Compare and contrast** somatic and genetic biological damage.
31. **Explain** why it is safe to use radioisotopes to diagnose medical problems.
32. **Calculate** A lab worker receives an average radiation dose of 21 mrem each month. Her allowed dose is 5,000 mrem/y. On average, what fraction of her yearly dose does she receive?
33. **Interpret Data** Look at the data in **Table 7.** Suppose someone is exposed to the maximum values listed for average annual radiation from the ground, from buildings, and from the air. What fraction would the person receive of the minimum short-term dose (25 rem) that causes a temporary decrease in white blood cell population?

SECTION 4 REVIEW

28. Explanations will vary, but might include radiotracers, PET scans, or radiation treatment to kill cancer cells.
29. Geiger counters, scintillation counters, and film badges can be described.
30. Somatic damage affects the body cells of the organism and will have effects only on that organism in its lifetime. Genetic damage affects the DNA of the organism, and could be transferred to later generations.
31. Radioisotopes used in medical diagnosis have short half-lives, thereby minimizing the patient's exposure.
32. Average yearly dose = 12 × (average monthly dose) = 12 × (21 mrem) = 252 mrem. The fraction of her yearly dose is (252 mrem/5,000 mrem) × 100% = 5.0%.
33. The total annual dose the person receives from the ground, buildings, and air would be 170 mrem + 160 mrem + 260 mrem = 590 mrem. This is a fraction of (590 mrem ÷ 25,000 mrem) × 100% = 2.4%.

CHEMISTRY & careers

Career: Archaeologist

Neutron Activation Analysis

In the Andes mountain range, more than 500 years ago, a young girl was sacrificed to appease the gods. As was the custom of the ancient Incas, pottery and other artifacts were buried with her. Neutron activation analysis performed on pottery such as the vessels in **Figure 1** allowed archaeologists to determine the origin of the soil from which the pottery was made.

Figure 1 Neutron activation analysis lets scientists compare soil and pottery to determine where the pottery was made.

Detecting elements Neutron activation analysis is a method of detecting elements in a material. A small sample of the material is first exposed to a strong neutron source. Neutron bombardment produces radioisotopes in about three-fourths of the elements. When the radioisotopes decay, they emit gamma rays with energies that are characteristic of the element.

A gamma detector is used to measure the sample's radiation output. Gamma rays of different energies produce peaks at different places on graphs, such as the one in **Figure 2.** Each peak corresponds to a specific element. Some elements have more than one peak because they emit gamma rays of different energies. The height of the peak, or the area under the peak, indicates the concentration of the element in the sample.

This method can be used to search for just one element or many elements in a sample. The process can detect extremely low concentrations of elements, as low as parts per billion.

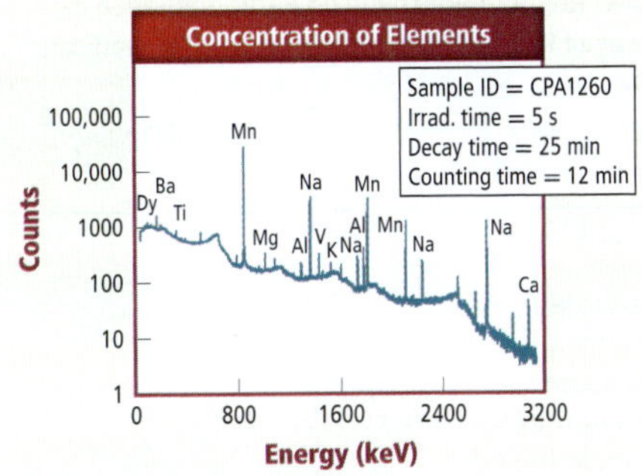

Figure 2 A gamma-ray spectrum indicates the concentration of different elements in a sample.

Advantages Most forms of chemical analysis require vaporization, dissolution, or alteration of the analyzed sample in some way. Neutron activation analysis is a nondestructive process that can be used to study liquid, solid, or gaseous samples. Sensitive items, such as forensic evidence, meteorites, or artifacts, can be analyzed without harm.

Uses Analyzing the composition of artifacts such as pottery allowed scientists to establish the origin of the clay used to make the objects that were buried with the young, sacrificed girl. Astonishingly, the clay did not come from local soil but from the Incan capital and other religious centers. Representatives from the Incan Empire traveled to remote places, bringing pottery and other artifacts with them, to perform rituals.

WRITING IN ▶ Chemistry

Analyze Look at the graph in **Figure 2.** Write an explanation about how a technician could use the graph to determine the elements present in the irradiated sample. Is the height of the peaks important? Which element is found in the greatest concentration in the sample? What are the approximate energies of gamma rays emitted by this element?

WebQuest

CHEMISTRY & careers

Purpose

Students will learn how neutron activation analysis can be used to identify the composition of most elements in a material.

Background

For an element to be detected by neutron activation analysis, the atoms in the element must change to radioisotopes when the nuclei absorb neutrons. Before irradiation, a sample is sealed inside a small vial and its mass is determined. The vial is placed near a neutron source, such as a reactor. Irradiation times will vary from seconds to days, depending on radioisotope decay time. Higher concentrations of an element emit gamma rays at higher rates.

Teaching Strategies

- Point out in Figure 2 that elements have multiple peaks, corresponding to gamma rays of different energies. Peaks only appear for radioisotopes that haven't sufficiently decayed before analysis. Also, have students notice that the graph shows why neutron activation analysis can be used to detect one element or multiple elements.
- Have students imagine a situation in which the soil in an area is suspected of having high levels of arsenic. Neutron activation analysis can determine the arsenic concentration. Ask students to write the chemical equation for the neutron bombardment of arsenic-71, showing the radioisotope that is produced. Answer: ${}^{1}_{0}n + {}^{71}_{33}As = {}^{72}_{33}As$

WRITING IN ▶ Chemistry

Rubric WebQuest

Analyze Students should explain that a technician could determine the elements present in a sample by comparing the energy that corresponds to each peak to the known energies of gamma rays emitted by elements. In the graph, the elements are already listed for each peak. The technician could determine whether elements were present in high or low concentrations by looking at the height of the peaks. The exact concentration of each element could be found by comparing the peak's height to that of a known standard. In this graph, manganese is present in the highest concentration. Although some radioisotopes decayed before the measurements were taken, the gamma ray energies for radioisotopes of manganese shown on the graph are 846 keV, 1810 keV, and 2113 keV. Students can give approximate values.

ChemLAB

Rubric

Preparation

Time Allotment one class period

Process Skills hypothesize; collect and organize data; measure; compare and contrast; formulate models; recognize cause and effect; research

Safety Precautions Approve lab safety forms before work begins. Students should wear safety goggles and lab aprons, and wash their hands after the lab. Remind students that radioactive sources can be harmful if not handled properly. Protective clothing should be worn as needed. Cover working surfaces with absorbent paper. Remind students to handle radioactive material with care and to not physically abuse the containers. They should report any damaged radioactive material. Account for all of the radioactive sources before and after the lab. Arrange for proper storage and disposal of all radioactive materials and equipment.

Preparation of Materials

- The radiation sources are available through science suppliers.
- The plastic shields can be made of waste items such as milk containers.

Procedure

- This lab can be extended by having students test common items, such as lite-salt (KCl), smoke detectors, marble, and dinnerware.
- Students can also test the amount of radiation over time. The determination of total radiation requires a device such as a CBL, Labpro or nuclear rate meter.
- Troubleshooting: Remember that the sources will get weaker over time. Test your sources prior to having the students use them.

Analyze and Conclude

1. As the distance from a radiation source increases, the values obtained decrease dramatically.
2. Answers will vary. Students might be surprised to find that shielding does not stop gamma rays.
3. The goggles and apron provide shielding from some of the radiation.
4. The gamma rays were not affected by the shields. The gamma rays are pure energy. The alpha and beta sources are particles.
5. The particle radiation (alpha and beta) was shielded more as the distance from the source to the shield increased. This is due to the energy of the particles decreasing as they travel from the source.
6. X-ray technicians use distance and shielding to reduce their exposure. The lead shield provides you protection from unnecessary exposure to X-rays. From this you can infer that X-rays cannot penetrate lead or the walls of the room.

INQUIRY EXTENSION

Answers will vary. The students should find a source such as the EPA online calculator and calculate their annual radiation dose. The American Nuclear Society is also a good source of information.

ChemLAB

Investigate Radiation Dosage

Background: Radiation is a term that causes fearful responses in people. However, not all radiation is dangerous. We are surrounded by radiation from space and from natural radioactivity on Earth. Radiation can also be used in a safe and controlled way for medical purposes.

Question: *What methods are effective in minimizing exposure to radiation?*

Materials

alpha source
beta source
gamma source
Geiger counter
piece of cardboard
piece of plastic
meterstick
clock

Safety Precautions

WARNING: ***Radioactive sources can be harmful. Wash hands and arms thoroughly before handling objects which go to the mouth, nose, or eyes. Do not eat or drink in laboratories where radioactive sources are used. Do not handle radioactive sources if you have a break in the skin below the wrist. Do not use–and immediately report to your teacher–any sealed disc containing a radioactive source which is damaged.***

Procedure

1. Read and complete the lab safety form.
2. Using what you know about types of radiation, write a hypothesis about how pieces of cardboard and plastic will affect the radiation dose.
3. Create a table to record your data.
4. Place the meterstick on the lab station with the Geiger counter at the zero-end.
5. Place the alpha source at the 10-cm mark, and record the highest reading on the Geiger counter.
6. Repeat the measurement with the source at 20 cm and 30 cm.
7. Repeat Steps 5 and 6 with the beta source and gamma source.
8. Place the alpha source on the 10-cm mark, and place a heavy piece of cardboard between the source and the Geiger counter.
9. Measure and record the highest reading.
10. Place the source on the 30-cm mark, and place the piece of cardboard on the 10-cm mark. Measure and record the radiation.
11. Place the piece of cardboard on the 20-cm mark, and repeat the measurement.
12. Place the piece of plastic between the source, and counter and record the highest reading.
13. Repeat Steps 8-12 with the beta source and the gamma source.
14. **Cleanup and Disposal** Return all lab equipment and radiation sources to the designated location. Remember to wash your hands with soap and water after completing the lab.

Analyze and Conclude

1. **Summarize** How does distance affect the amount of radiation from a source?
2. **Compare and Contrast** Does the experimental data support your hypothesis?
3. **Explain** Based on the data, explain why you were required to wear goggles and a lab apron in this lab.
4. **Recognize Cause and Effect** Which radiation source was least affected by the cardboard and plastic shields? Explain why this source is different from the other two sources.
5. **Infer** Did the position of the piece of cardboard influence the results? Explain why or why not.
6. **Observe and Infer** What can you say about the penetrating power of X-rays based on the fact that you have to wear a lead shield at the dentist to protect your body from the radiation?

INQUIRY EXTENSION

Research Find references that list and quantify the exposure to radiation that we receive in everyday life. Calculate your average annual exposure, and describe methods that could reduce this dosage.

CHAPTER 24 **STUDY GUIDE**

Vocabulary Practice

BIGIDEA Nuclear chemistry has a vast range of applications, from the production of electricity to the diagnosis and treatment of diseases.

SECTION 1 Nuclear Radiation

MAINIDEA Under certain conditions, some nuclei can emit alpha, beta, or gamma radiation.

- Wilhelm Roentgen discovered X-rays in 1895.
- Henri Becquerel, Marie Curie, and Pierre Curie pioneered the fields of radioactivity and nuclear chemistry.
- Radioisotopes emit radiation to attain more-stable atomic configurations.

VOCABULARY
- radioisotope
- X-ray
- penetrating power

SECTION 2 Radioactive Decay

MAINIDEA Unstable nuclei can break apart spontaneously, changing the identity of atoms.

- The conversion of an atom of one element to an atom of another by radioactive decay processes is called transmutation.
- Atomic number and mass number are conserved in nuclear reactions.
- A half-life is the time required for half of the atoms in a radioactive sample to decay.

$$N = N_0\left(\frac{1}{2}\right)^n \text{ or } N = N_0\left(\frac{1}{2}\right)^{t/T}$$

- Radiochemical dating is a technique for determining the age of an object by measuring the amount of certain radioisotopes remaining in the object.

VOCABULARY
- transmutation
- nucleon
- strong nuclear force
- band of stability
- positron emission
- positron
- electron capture
- radioactive decay series
- half-life
- radiochemical dating

SECTION 3 Nuclear Reactions

MAINIDEA Fission, the splitting of nuclei, and fusion, the combining of nuclei, release tremendous amounts of energy.

- Induced transmutation is the bombardment of nuclei with particles in order to create new elements.
- In a chain reaction, one reaction induces others to occur. A sufficient mass of fissionable material is necessary to initiate the chain reaction.
- Fission and fusion reactions release large amounts of energy.

$$E = mc^2$$

VOCABULARY
- induced transmutation
- transuranium element
- mass defect
- nuclear fission
- critical mass
- breeder reactor
- nuclear fusion
- thermonuclear reaction

SECTION 4 Applications and Effects of Nuclear Reactions

MAINIDEA Nuclear reactions have many useful applications, but they also have harmful biological effects.

- Different types of counters are used to detect and measure radiation.
- Radiotracers are used to diagnose disease and to analyze chemical reactions.
- Short-term and long-term radiation exposure can cause damage to living cells.

$$I_1 d_1^2 = I_2 d_2^2$$

VOCABULARY
- ionizing radiation
- radiotracer

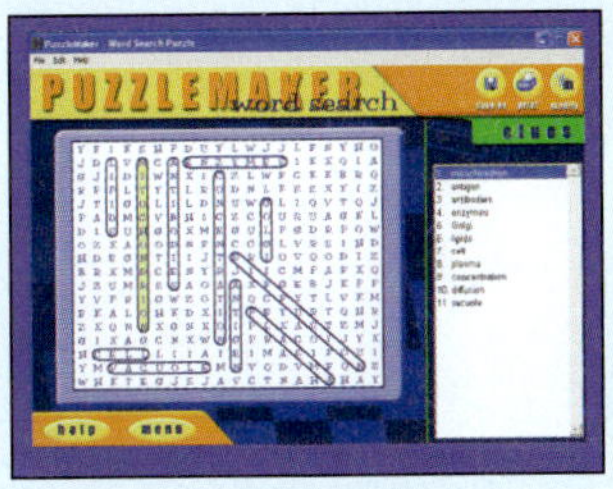

Vocabulary Practice

For additional practice with vocabulary, have students access the Vocabulary PuzzleMaker.

CHAPTER 24 STUDY GUIDE

Use the Vocabulary

To reinforce chapter vocabulary, have students write a sentence using each term. OL EL

Review Strategies

- Have students summarize the five major types of radioactive decay. OL
- Have students write balanced nuclear equations that show each type of nuclear decay. OL
- Have students describe the variable changes that can be manipulated in a radioactive decay problem. OL
- Problems from the Supplemental Practice Problems appendix or the Supplemental Problems booklet can be used for review. OL

Students can review with these online resources.

Vocabulary Practice

Vocabulary eGames
Vocabulary eFlashcards
Vocabulary PuzzleMaker

Section Self-Check

Chapter Self-Check

Online Test Practice

Use *eAssessment* to:
- create multiple versions of tests
- edit existing questions and add your own questions
- build tests aligned with select state curriculum standards using built-in state curriculum tags
- track students' progress

SECTION 1

Mastering Concepts

34. Nuclear reactions release more energy per mole. Nuclear reactions involve neutrons and protons, whereas chemical reactions involve electrons.

35. a. 2
b. 1
c. 3.

36. Radiation **a** must be beta particles because their negative charge would cause them to be deflected away from the negatively-charged plate; **b** must be gamma rays because they have no charge and would not be deflected by the charged plates; **c** must be alpha particles because their positive charge would cause them to be deflected away from the positively charged plate.

37. X-rays are produced by materials that are in an excited electron state. Gamma rays are produced by radioactive sources.

Mastering Problems

38. thorium-230

39. The atomic number and mass number do not change.

40. Beta decay; the mass number does not change.

41. samarium

SECTION 2

Mastering Concepts

42. The strong nuclear force binds nucleons together; protons and neutrons.

43. In positron emission, a proton turns into a neutron: ${}^{1}_{1}p \rightarrow {}^{1}_{0}n + {}^{0}_{1}\beta$. In electron capture, a proton merges with an inner-shell electron: ${}^{1}_{1}p + {}^{0}_{-1}e \rightarrow {}^{1}_{0}n$.

44. a. gamma emission
b. The decay is either positron emission or electron capture.

45. Isotopes lying outside the band will spontaneously decay.

46. a family of decay reactions that continue until a stable, non-radioactive isotope is formed

47. its half-life and how much of the isotope is produced

CHAPTER 24 ASSESSMENT

SECTION 1

Mastering Concepts

34. Compare and contrast chemical reactions and nuclear reactions in terms of energy changes and the particles involved.

35. Match each numbered choice on the right with the correct radiation type on the left.

a. alpha	**1.** high-speed electrons
b. beta	**2.** 2+ charge, blocked easily
c. gamma	**3.** no charge, electromagnetic radiation

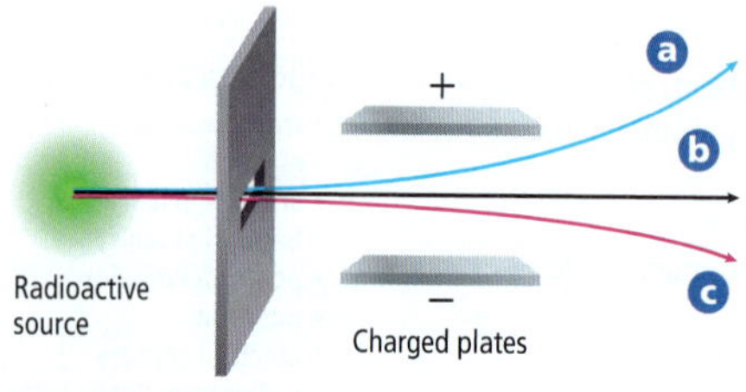

■ **Figure 30**

36. **Figure 30** shows alpha particles, beta particles, and gamma rays passing through a screen and between two charged plates. What can you infer about the identity of **a, b,** and **c?** Explain your answer.

37. What is the difference between X-rays and gamma rays?

Mastering Problems

38. Dental crown Uranium-234 is used to make dental crowns appear brighter. The alpha decay of uranium-234 produces what isotope?

39. Detecting Material Flaws Flaws in welded metal parts of airplanes can be identified by placing the isotope iridium-192 on one side of the weld and photographic film on the other side to detect gamma rays that pass through. How does the gamma ray emission affect the atomic number and mass number of the iridium?

40. Colored Glass Thorium-230 can be used to provide coloring in glass objects. One method of producing thorium-230 is through the radioactive decay of actinium-230. Is this an example of alpha decay or beta decay? How do you know?

41. Plastic Bags Thin sheets of plastic are used to make items such as grocery bags. The sheets move under a source of promethium-147, emitting beta particles. The radiation intensity, measured under the plastic sheets, is used to monitor the thickness of the plastic. During this process, promethium changes into which element?

SECTION 2

Mastering Concepts

42. What is the strong nuclear force? On which particles does it act?

43. Explain the difference between positron emission and electron capture.

44. Categorize each type of radioactive decay.
a. Mass number and atomic number are unchanged.
b. Mass number remains the same and atomic number decreases.

45. What is the significance of the band of stability?

46. What is a radioactive decay series? When does it end?

47. Radioisotopes What are the factors that determine the amount of a given radioisotope in nature?

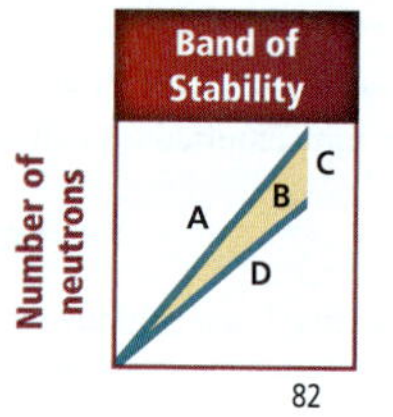

■ **Figure 31**

48. In which region(s) in **Figure 31** are you likely to find
a. stable nuclei?
b. nuclei that undergo alpha decay?
c. nuclei that undergo beta decay?
d. nuclei that undergo positron emission?

49. Carbon-14 Dating Carbon-14 dating makes use of a specific ratio of two different radioisotopes. Define the ratio used in carbon-14 dating. Why is this ratio constant in living organisms?

Mastering Problems

50. Calculate the neutron-to-proton ratio for each atom.

a. tin-134	**c.** carbon-12
b. silver-107	**d.** carbon-14

51. Complete the following equations.
a. ${}^{214}_{83}Bi \rightarrow {}^{4}_{2}He + ?$ **b.** ${}^{239}_{93}Np \rightarrow {}^{239}_{94}Pu + ?$

52. Write a balanced nuclear equation for the alpha decay of americium-241.

53. Write a balanced nuclear equation for the beta decay of cesium-137.

54. Bone Formation The electron capture of strontium-85 can be used by physicians to study bone formation. Write a balanced nuclear equation for this reaction.

48. a. region B
b. region C
c. region A
d. region D

49. C-14 : C-12; Ratio is constant in living organisms because they assimilate new C-14 into their cells, thus replenishing C-14 lost to decay

Mastering Problems

50. a. 1.68
b. 1.28
c. 1.0
d. 1.33

51. a. ${}^{210}_{81}Tl$
b. β

52. ${}^{241}_{95}Am \rightarrow {}^{4}_{2}He + {}^{237}_{93}Np$

53. ${}^{137}_{55}Cs \rightarrow {}^{137}_{56}Ba + \beta$

54. ${}^{0}_{-1}e + {}^{85}_{38}Sr \rightarrow {}^{85}_{37}Rb$

Chapter Self-Check

55. **Nuclear Safety** The half-life of tritium ($^{3}_{1}H$) is 12.3 y. If 48.0 mg of tritium is released from a nuclear power plant during the course of a mishap, what mass of the nuclide will remain after 49.2 y? After 98.4 y?

56. **Static Charge** Static charge can interfere with the production of plastic products by attracting dust and dirt. To reduce it, manufacturers expose the area to polonium-210, which has a half-life of 138 days. How much of a 25.0-g sample will remain after one year (365 days)?

57. The half-life of polonium-218 is 3.0 min. If you start with 20.0 g, how long will it be before only 1.0 g remains?

58. An unknown radioisotope exhibits 8540 decays per second. After 350.0 min, the number of decays has decreased to 1250 per second. What is the half-life?

SECTION 3

Mastering Concepts

59. Define *transmutation.* Are all nuclear reactions also transmutation reactions? Explain.

60. Relate binding energy per nucleon to mass number.

61. Referring to **Figure 7,** would you expect $^{39}_{20}Ca$ to be radioactive? Explain.

62. What is a chain reaction? Give an example of a nuclear chain reaction.

63. Explain the purpose of control rods in a nuclear reactor.

64. Why is the fuel of a nuclear reactor enriched?

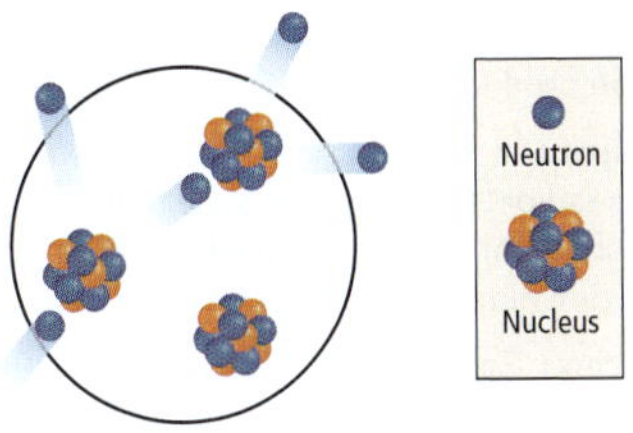

■ **Figure 32**

65. Describe what is meant by the terms *critical mass, subcritical mass,* and *supercritical mass.* Which is shown in **Figure 32?** How can you tell?

66. Explain how it is possible that fission (the splitting of nuclei) and fusion (the combining of nuclei) both release tremendous amounts of energy.

67. Describe the current limitations of fusion as a power source.

68. Why does nuclear fusion require so much heat? How is heat contained within a tokamak reactor?

Mastering Problems

69. **Smoke Detectors** Americium-241, a radioisotope used in smoke detectors, is produced by bombarding plutonium-238 with neutrons to produce plutonium-240, which is bombarded with neutrons to produce plutonium-241. The plutonium-241 decays to americium-241. Write the balanced nuclear equations for each reaction.

70. **Exit Signs** Exit signs are coated with a paint containing phosphors. These phosphors are activated by the radioisotope tritium ($^{3}_{1}H$), produced by bombarding lithium-6 with neutrons to produce lithium-7. The lithium-7 then undergoes alpha decay to produce the tritium. Write balanced nuclear equations for both steps.

71. **Control Rods** Bombarding uranium-235 with neutrons produces samarium-149, which is used in nuclear reactor control rods. What other element is produced?

72. **The Sun** $^{1}_{1}H + ^{2}_{1}H \rightarrow ^{3}_{2}He + \gamma$ is one of the fusion reactions in the Sun. The mass of $^{1}_{1}H$ is 1.007825 amu, the mass of $^{2}_{1}H$ is 2.014102 amu, and the mass of $^{3}_{2}He$ is 3.016029 amu.
   a. What is the mass defect of $^{3}_{2}He$?
   b. What energy is released by the process?

SECTION 4

Mastering Concepts

73. What property of isotopes allows radiotracers to be useful in studying chemical reactions?

74. Which unit of radiation dose, rem or rad, is most useful for describing the effect of radiation on living tissue?

75. **PET Scans** In PET scans, the radiotracer emits positrons, which travel a few millimeters before interacting with electrons. How can the original radiotracer be detected?

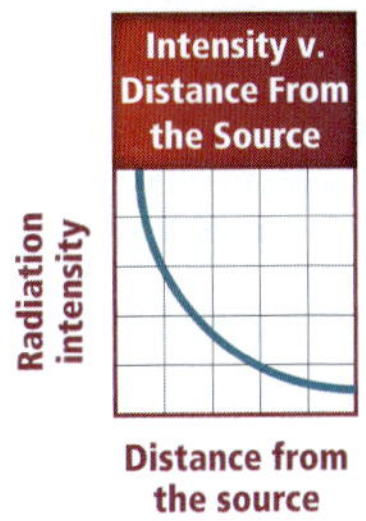

■ **Figure 33**

76. **Figure 33** shows a simplified graph of radiation intensity versus distance from the source. Explain this graph and what it implies about a method of reducing the effects of radiation exposure.

55. After 49.2 years: 3.00 mg; After 98.4 years: 0.188 mg
56. 4.00 g
57. 13 min
58. 126 min

SECTION 3

Mastering Concepts

59. Transmutation is the changing of an atom's nucleus such that a new element is formed.
60. The binding energy per nucleon reaches a maximum value for mass numbers around 60 amu.
61. Ca-39 is likely to be radioactive, because it lies just below the band of stability.
62. A chain reaction occurs when a reaction produces one or more of the particles needed as a reactant. An example is U-235.
63. They absorb neutrons from the fission reaction to control the rate of reaction.
64. Enriching the fuel means increasing the concentration of the fissionable isotope so that the chain reaction is sustained.
65. Critical mass is the minimum mass of a sample of fissionable material needed to sustain a nuclear chain reaction. The mass is subcritical if most of the neutrons escape the material instead of hitting other nuclei. The mass is supercritcal if most neutrons hit other material. The figure shows subcritical mass because most neutrons are escaping the material.
66. With fission, a large nucleus splits into smaller nuclei. The binding energy of the large nucleus is greater than the sum of the binding energies of the smaller nuclei. The difference in energy is released. With fusion, two smaller nuclei combine to form a larger nucleus. The sum of the binding energies of the smaller nuclei is greater than the binding energy of the larger nucleus. Again, the difference in energy is released.
67. A lot of energy is required to initiate the reaction. Currently, no materials are capable of withstanding the very high temperature reached during such reactions.
68. Extremely high temperatures are needed to fuse the positively charged nuclei together. Magnetic fields contain the hot fusion reaction.

Mastering Problems

69. $^{238}_{94}Pu + 2\,^{1}_{0}n \rightarrow ^{240}_{94}Pu$
   $^{240}_{94}Pu + ^{1}_{0}n \rightarrow ^{241}_{94}Pu$
   $^{241}_{94}Pu \rightarrow ^{241}_{95}Am + ^{0}_{-1}\beta$
70. $^{6}_{3}Li + ^{1}_{0}n \rightarrow ^{7}_{3}Li$
   $^{7}_{3}Li + ^{3}_{1}H \rightarrow ^{4}_{2}He$
71. zinc
72. **a.** 0.005898 amu
   b. 5.4939 MeV

SECTION 4

Mastering Concepts

73. Radiotracers have the same chemical properties as the stable isotopes of that element.
74. The rem is most effective because it accounts for the energy of the radiation, the type of living tissue, and the time of the exposure.
75. The positron/electron interaction forms gamma-ray emissions that can be detected.
76. The radiation intensity decreases rapidly with increasing distance from the source of the radiation. The effects of radiation can be decreased by increasing your distance from the source.

Mastering Problems

77. 3.5 m
78. 0.28 mrem/s•m^2
79. 0.66 m

MIXED REVIEW

80. 5.16 g
81. $^{80}_{35}Br$, $^{80}_{34}Se$, $^{80}_{34}Se$
82. 97%
83. 2.2 m
84. Mass number 60; light nuclei can become more stable by undergoing fusion, and heavier nuclei can become more stable by undergoing fission.
85. Place each source behind the paper and use the Geiger counter to identify the source emitting alpha particles, which are blocked by paper. Then, place the remaining two sources behind the foil. The beta particles would be blocked by the foil. Then the remaining source is gamma.
86. 3.80 days
87. Possible answer: The company could use a Geiger counter to determine the intensity of the source at a certain distance. They could move the counter until they reach a safe distance based on the maximum recommended exposure intensity for the source.
88. A and D: alpha decay, atomic number decreases by 2, mass number decreases by 4; B and C: beta decay, atomic number increases by 1, mass number does not change

THINK CRITICALLY

89. The graph should resemble the shape seen in Figure 10, graphing the atomic number on the *x*-axis, and the mass number, plotted on the *y*-axis. Refer to Solutions Manual for detailed graph.
90. Chemical reactions do not affect nuclei; thus, the radioisotopes would continue to emit radiation.
91. Outside the body, alpha radiation is not harmful because it cannot penetrate the skin. Inside the body, alpha radiation is far more damaging because it can directly affect tissue more fragile than the outer layer of the skin.
92. An alpha source would have an open window to prevent absorption of the alpha particles.

Mastering Problems

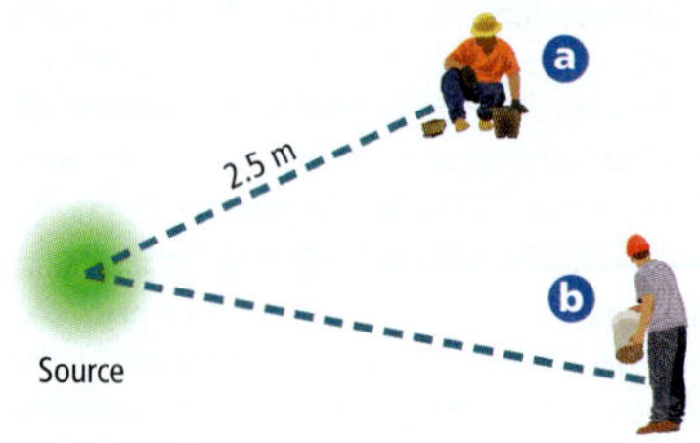

■ **Figure 34**

77. **Figure 34** shows the position of two workers near a radioactive gamma source. The worker at Position A is 2.5 m from the source and receives an exposure of 0.98 mrem/s•m^2. The worker at Position B receives an exposure of 0.50 mrem/s•m^2. What is the distance of the worker at Position B from the source?

78. A worker stands near a machine that uses a cobalt-60 gamma source to sterilize medical equipment. The worker's dose 2.0 m from the source is 0.85 mrem/s•m^2. What is the worker's dose at a distance of 3.5 m?

79. **Safe Exposure** The intensity of a radioactive source is 1.15 mrem/s•m^2 at a distance of 0.50 m. What is the minimum distance a person could be from the source to have a maximum exposure of 0.65 mrem/s•m^2?

MIXED REVIEW

80. Technetium-104 has a half-life of 18.0 min. How much of a 165.0 g sample remains after 90.0 minutes have passed?

81. A bromine-80 nucleus can decay by gamma emission, positron emission, or electron capture. What is the product nucleus in each case?

82. The half-life of plutonium-239 is 24,000 y. How much nuclear waste generated today will remain in 1000 years?

83. **Red Blood Cells** A medical researcher is using a chromium-51 source to study red blood cells. The gamma-emission intensity at a distance of 1.0 m is 0.75 mrem/s•m^2. At what distance would the intensity drop to 0.15 mrem/s•m^2?

84. The binding energy per nucleon reaches a maximum around what mass number? Explain how this number is related to the fission and fusion processes.

85. You have an alpha source, a beta source, and a gamma source. Design a plan to use a Geiger counter, paper, and foil to determine the identity of each source.

86. What is the half-life of radon-222 if a sample initially contains 150 mg and only 18.7 mg after 11.4 days?

87. **Sheet Metal** A company plans to monitor the thickness of sheet metal during production. What would you recommend the company do to determine a safe distance for workers from the gamma source?

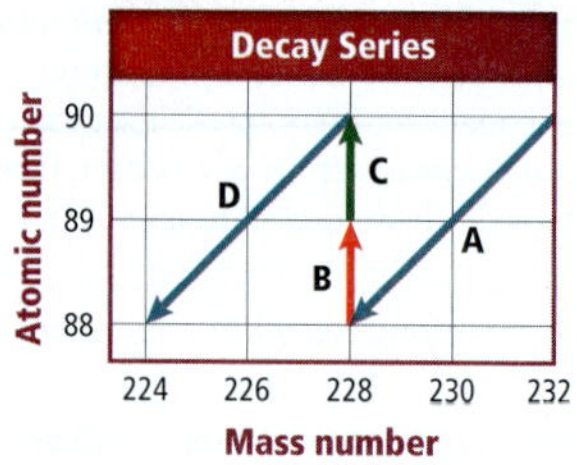

■ **Figure 35**

88. **Figure 35** shows part of the decay series of a radioisotope. For each segment on the graph, tell whether alpha decay or beta decay occurs, and identify the change in atomic number and mass number.

THINK CRITICALLY

89. **Make and Use Graphs** Thorium-231 decays to lead-207 by emitting the following particles in successive steps: β, α, α, β, α, α, α, β, β, α. Plot each step of the decay series on a graph of mass number versus atomic number. Label each plotted point with the symbol of the radioisotope.

90. **Apply** Chemical treatment is often used to destroy harmful chemicals. For example, bases neutralize acids. Why can't chemical treatment be applied to destroy the fission products produced in a nuclear reactor?

91. **Compare** A biological concern about working around some radioactive materials is the radioactive dust a person might inhale. Compare the effect of alpha radiation outside the body and inside the body.

92. **Interpret** Small radioactive sources are often used for laboratory experiments. The radioactive substance is enclosed in a metal container with a small window. A gamma source might be covered with a stainless steel window. What would you expect the window of an alpha source to be like? Why?

93. **Analyze** Some radioisotopes used for medical imaging have half-lives as short as several hours. Why is a short half-life beneficial? Why is it a problem?

94. **Infer** The production of electricity at nuclear fission reactor facilities is controversial. Think about the benefits and dangers of this technology. Explain your opinion about whether nuclear reactors should be used.

93. A short half-life is beneficial because it reduces the long-range exposure to the patient. A short half-life is a problem because it means the radioisotope must be frequently replaced and shipped to the medical facility soon after production or produced at the facility.
94. Those in favor of nuclear facilities might argue that it is a clean source of energy. They might point out that safeguards during the Three Mile Island accident show dangers can be contained and minimized. Opponents might argue that nuclear reactors produce nuclear waste that can pollute the environment and cause storage problems or that safeguards don't always work.

Chapter Self-Check

CHALLENGE PROBLEM

95. Use the information in **Table 8** to calculate the mass defect and binding energy of deuterium ($^{2}_{1}H$), a hydrogen isotope involved in fusion reactions in the Sun.

Table 8 Mass of Particles

Particle	Mass (amu)
Hydrogen	1.007941
Deuterium	2.014102
Neutron	1.008665

a. Find the mass of the nucleons.
b. Find the mass defect by subtracting the mass of the nucleons from the mass of the deuterium.
c. Find the binding energy using the conversion 1 amu = 931.49 MeV.

CUMULATIVE REVIEW

96. Identify each property as chemical or physical.
a. The element mercury has a high density.
b. Solid carbon dioxide sublimes at room temperature.
c. Zinc oxidizes when exposed to air.
d. Sucrose is a white crystalline solid.

97. Why does the second period of the periodic table contain eight elements?

98. Draw each molecule and show the locations of hydrogen bonds between the molecules.
a. two water molecules
b. two ammonia molecules
c. one water molecule and one ammonia molecule

99. What process takes place in each situation?
a. a solid air-freshener cube getting smaller and smaller
b. dewdrops forming on leaves in the morning
c. steam rising from a hot spring
d. a crust of ice forming on top of a pond

100. If the volume of a sample of chlorine gas is 4.5 L at 0.65 atm and 321 K, what volume will the gas occupy at STP?

101. The temperature of 756 g of water in a calorimeter increases from 23.2°C to 37.6°C. How much heat was given off by the reaction in the calorimeter?

102. Explain what a buffer is and why buffers are found in body fluids.

103. Explain how the structure of benzene can be used to explain its unusually high stability compared to other unsaturated cyclic hydrocarbons.

WRITING IN ▸ Chemistry

104. Marie Curie and Irene Curie Joliot Research and report on the lives of Marie Curie and her daughter, Irene Curie Joliot. What kind of scientific training did each receive? What was it like to be a female chemist in their time? What discoveries did each make?

105. Nuclear Waste Evaluate environmental issues associated with nuclear wastes. Research the Yucca Mountain nuclear waste disposal plan, the Hanford nuclear site, or a local nuclear facility. Prepare a poster or multimedia presentation on your findings.

106. Radioactive Sources Students in your school might not realize how beneficial radioactive sources can be. Create a poster showing some common, beneficial uses of radioactive sources. Be sure to point out safeguards that are taken to ensure the sources are safe.

DBQ Document-Based Questions

Half-Lives *The National Institute of Standards and Technology (NIST) maintains a database of radionuclide half-lives. In 1992, researchers at NIST measured the half-lives shown in* **Table 9.**

Data obtained from: Unterweger, M.P., Hoppes, D.D., and Schima, F.J. 1992. New and revised half-life measurements results, *Nucl. Instrum. Meth. Phys. Res.* A312:349-352.

Table 9 Half-Lives

Radionuclide	Half-life
Fluorine-18	1.82951 h
Molybdenum-99	65.9239 h
Samarium-153	46.2853 h

107. Fluorine-18 is used in medical imaging. If a lab has a sample containing 15 g of fluorine-18, how much fluorine-18 will remain in the sample after 8.0 h?

108. Technetium-99 can be used for diagnostic tests of the heart and lungs. Because of technetium-99's very short half-life, medical facilities produce it from molybdenum-99. If the facility has a 25-g sample of molybdenum-99, how much will it have one week (168 h) later?

109. Samarium-153 is used in the production of a drug to treat pain from bone tumors. Radiation released by the samarium hinders the tumor growth, thereby reducing pain. How much of a 1.0 g sample of samarium-153 is left after 4 days (96 h)?

CHALLENGE PROBLEM

95. a. 2.016606 amu
b. −0.002504 amu
c. −2.332 MeV

CUMULATIVE REVIEW

96. a. physical
b. physical
c. chemical
d. physical

97. Because there are 8 electrons in the 2s and 2p orbitals in the second energy level, the period contains eight elements.

98.

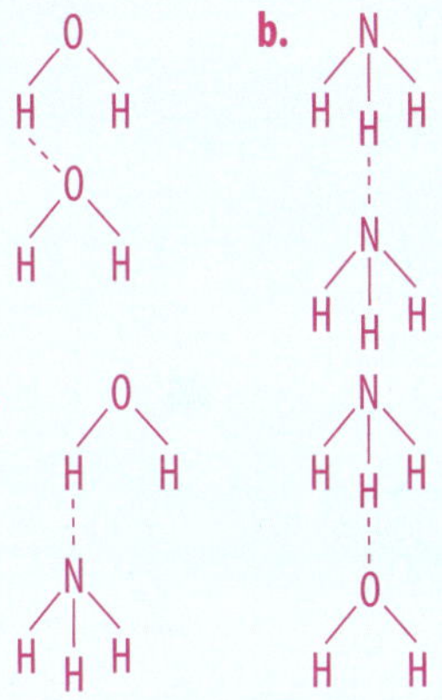

99. a. sublimation
b. condensation
c. evaporation
d. solidification

100. 2.5 L

101. 45.5 kJ

102. A buffer is a solution containing a mixture of an acid and its conjugate base that is able to resist changes in pH. Because many biochemical reactions are sensitive to changes in pH, buffers are needed in the body fluids.

103. The electron pairs in benzene's double bonds are shared among all six carbon atoms in the ring. Thus, the electrons cannot be pulled away as easily as electrons that are held by only two nuclei.

WRITING IN ▸ Chemistry

Rubric

104. Students' answers will vary and might be extensive. For example, Marie Curie and her husband, Pierre, made possible many medical diagnostic techniques that we take for granted today. Marie and her oldest daughter, Irene Curie Joliot, developed and pioneered the use of x-radiography on the battlefields of World War I. Irene Curie Joliot won a Nobel Prize in Chemistry in 1935.

105. Answers will vary according to chosen nuclear site or facility. Projects should include the location and function of the facility, the nature of the issues, and what has been done to remedy the issue.

106. Answers might include the use of radiation to treat cancer, X-ray radiography, radiocarbon dating, and others.

DBQ Document-Based Questions

Data obtained from: Unterweger, M.P., Hoppes, D.D., and Schima, F.J. 1992 *Nuclear Instruments and Methods in Physics Research.* A312:349–352.

107. 0.72 g
108. 4.27 g
109. 0.24 g

MULTIPLE CHOICE

1. C
2. A
3. B
4. C
5. B
6. A
7. B

MULTIPLE CHOICE

1. Geologists use the decay of potassium-40 in volcanic rocks to determine their ages. Potassium-40 has a half-life of 1.26×10^9 years, so it can be used to date very old rocks. If a sample of rock 3.15×10^8 years old contains 2.73×10^{-7} g of potassium-40 today, how much potassium-40 was originally present in the rock?
   A. 1.71×10^{-8} g
   B. 2.30×10^{-7} g
   C. 3.25×10^{-7} g
   D. 4.37×10^{-6} g

2. In the early 1930s, van de Graaf generators were used to generate neutrons by bombarding stable beryllium atoms with deuterons (${}^2_1H$), the nuclei of deuterium atoms. A neutron is released in the reaction. Which is the balanced nuclear equation describing this induced transmutation?
   A. ${}^9_4Be + {}^2_1H \rightarrow {}^{10}_5B + n$
   B. ${}^6_4Be + {}^2_1H \rightarrow {}^8_5B + n$
   C. ${}^9_4Be \rightarrow {}^{10}_5B + {}^2_1H + n$
   D. ${}^9_4Be + {}^2_1H \rightarrow {}^{11}_5 + n$

Use the figure below to answer question 3.

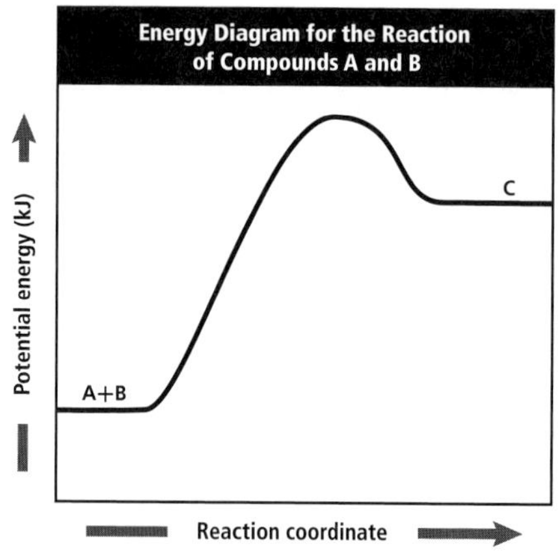

3. Which is NOT a correct description of this reaction?
   A. This is a synthesis reaction.
   B. This reaction releases energy.
   C. This reaction is endothermic.
   D. This reaction will occur spontaneously.

4. Which statement is NOT true of alpha particles?
   A. They carry a charge of 2+.
   B. They are represented by the symbol ${}^4_2He$.
   C. They are more penetrating than β particles.
   D. They have the same composition as helium nuclei.

Use the graph below to answer questions 5 and 6.

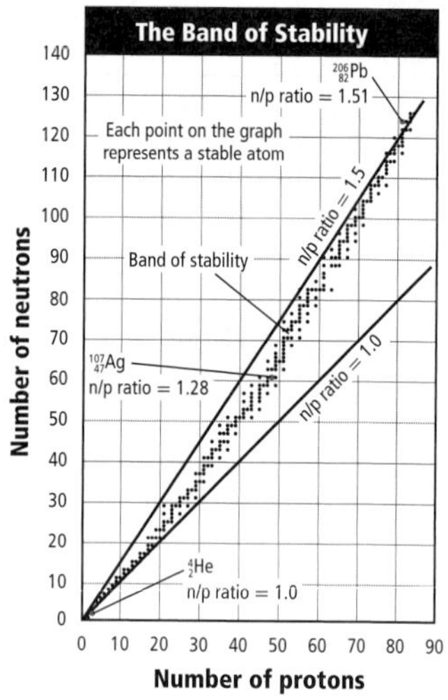

5. Why will calcium-35 undergo positron emission?
   A. It lies above the line of stability.
   B. It lies below the line of stability.
   C. It has a high neutron-to-proton ratio.
   D. It has an overabundance of neutrons.

6. Based on its position relative to the band of stability, which process will ${}^{70}_{30}Zn$ undergo?
   A. beta decay
   B. electron capture
   C. nuclear fusion
   D. positron emission nuclear fusion

7. A solution of 0.600*M* HCl is used to titrate 15.00 mL of KOH solution. The end point of the titration is reached after the addition of 27.13 mL of HCl. What is the concentration of the KOH solution?
   A. 9.00*M*
   B. 1.09*M*
   C. 0.332*M*
   D. 0.0163*M*

SHORT ANSWER

Use the figure below to answer questions 8 to 10.

8. Identify the anode and cathode of this apparatus.
9. Write the oxidation half-reaction.
10. Explain the function of the salt bridge in this apparatus.
11. Predict the products of this reaction.
    $Al(NO_3)_3(aq) + CaSO_4(aq) \rightarrow$

EXTENDED RESPONSE

Use the figure below to answer Questions 12 and 13.

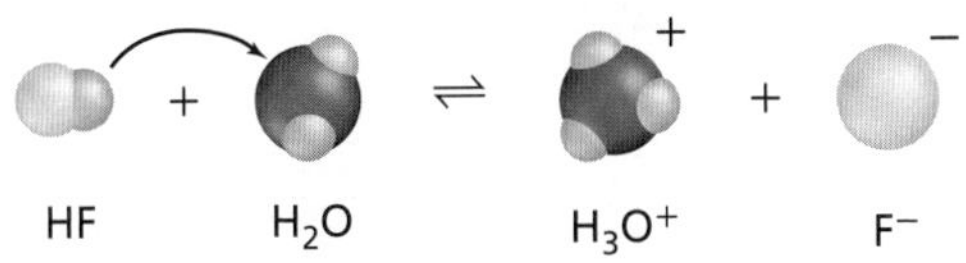

12. Identify the acid and the base for the forward reaction. Explain how you can tell.
13. Explain how you can identify the conjugate acid and conjugate base for the forward reaction. What are they?

SAT SUBJECT TEST: CHEMISTRY

Use the figure below to answer Questions 14 and 15.

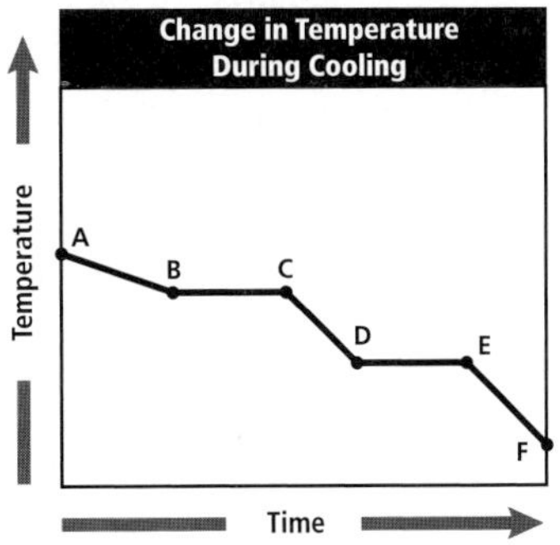

14. During which segments are particles changing states of matter?
    A. AB, CD, EF
    B. AB, EF
    C. BC, CD, DE
    D. BC, EF
    E. BC, DE

15. During which segments are particles losing kinetic energy?
    A. BC, DE
    B. AB, DE
    C. AB, CD, EF
    D. BC, DE, EF
    E. AB, CD, DE

16. In the first steps of its radioactive decay series, thorium-232 decays to radium-228, which then decays to actinium-228. What are the balanced nuclear equations describing these first two decay steps?
    A. ${}^{232}_{90}Th \rightarrow {}^{228}_{88}Ra + e^-$, ${}^{228}_{88}Ra \rightarrow {}^{228}_{89}Ac + e^+$
    B. ${}^{232}_{90}Th \rightarrow {}^{228}_{88}Ra + {}^{4}_{2}He$, ${}^{228}_{88}Ra \rightarrow {}^{228}_{89}Ac + e^-$
    C. ${}^{232}_{90}Th \rightarrow {}^{228}_{88}Ra + e^+$, ${}^{228}_{88}Ra \rightarrow {}^{228}_{89}Ac + e^-$
    D. ${}^{232}_{90}Th \rightarrow {}^{228}_{88}Ra + {}^{4}_{2}He$, ${}^{228}_{88}Ra + e^- \rightarrow {}^{228}_{89}Ac$
    E. ${}^{232}_{90}Th + e^- \rightarrow {}^{228}_{88}Ra$, ${}^{228}_{88}Ra \rightarrow {}^{228}_{89}Ac + e^-$

NEED EXTRA HELP?																
If You Missed Question . . .	1	2	3	4	5	6	7	8	9	10	11	12	13	14	15	16
Review Section . . .	24.2	24.3	15.5	24.1	24.2	24.2	18.4	20.1	20.1	20.1	9.2	18.1	18.1	12.4	12.4	24.2

SHORT ANSWER

8. The anode is where oxidation takes place and electrons are lost, which is the zinc electrode. The cathode is the copper electrode.
9. $Zn \rightarrow Zn^{2+} + 2e^-$
10. It allows ions to pass from one solution to the other, completing the electrical circuit and preventing the buildup of positive or negative ions.
11. $Al_2(SO_4)_3 + Ca(NO_3)_2$

EXTENDED RESPONSE

12. The acid is HF because it is donating its proton. The base is H_2O because it is accepting the proton.
13. The conjugate acid is what remains after the base accepts the proton, and will act as the acid in the reverse reaction by giving up its proton. In this reaction, it is H_3O^+. The conjugate base is what remains after the acid gives up its proton and will serve as the proton acceptor in the reverse reaction. It is the F^- ion in this reaction.

SAT SUBJECT TEST: CHEMISTRY

14. E
15. C
16. B

STUDENT RESOURCES

Rubberball/Getty Images

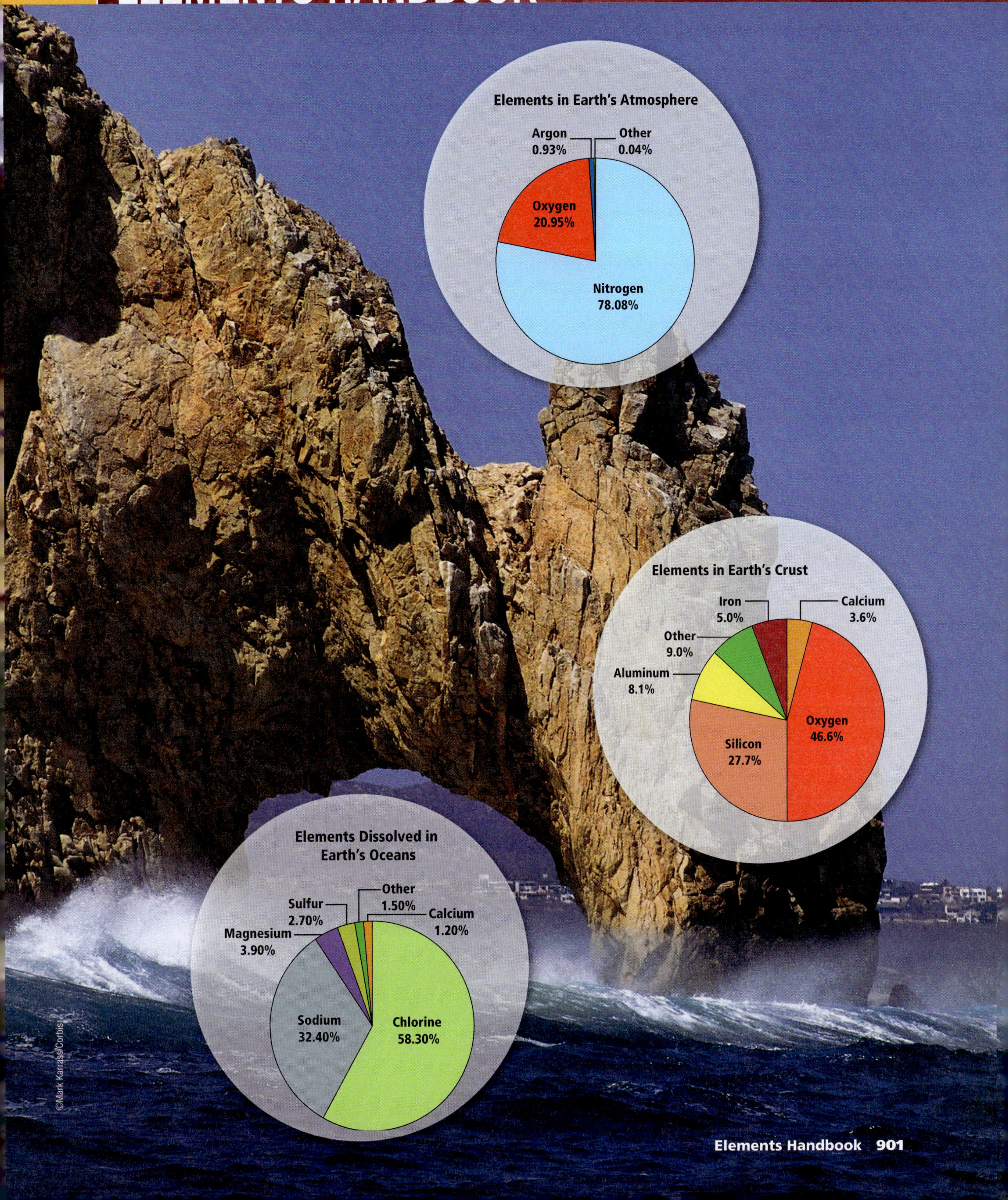
Elements in Earth's Atmosphere
Argon 0.93%
Other 0.04%
Oxygen 20.95%
Nitrogen 78.08%
Elements in Earth's Crust
Iron 5.0%
Calcium 3.6%
Other 9.0%
Aluminum 8.1%
Oxygen 46.6%
Silicon 27.7%
Elements Dissolved in Earth's Oceans
Other 1.50%
Sulfur 2.70%
Calcium 1.20%
Magnesium 3.90%
Sodium 32.40%
Chlorine 58.30%

Table of Contents

How This Handbook Is Organized *The Elements Handbook is divided into 10 sections: hydrogen and groups 1, 2, 3–12, 13, 14, 15, 16, 17, and 18. You will discover physical and atomic properties, common reactions, analytical tests, and real-world applications of the elements in each section. Questions at the end of each section will assess your understanding of the elements.*

How to Use Element Boxes

Each element box on the periodic table contains useful information. In the Elements Handbook, each element box has an element name, symbol, atomic number, and electron configuration. At the beginning of each section, each element box also identifies the state of matter at 25°C and 1 atm. A typical box from the handbook is shown below.

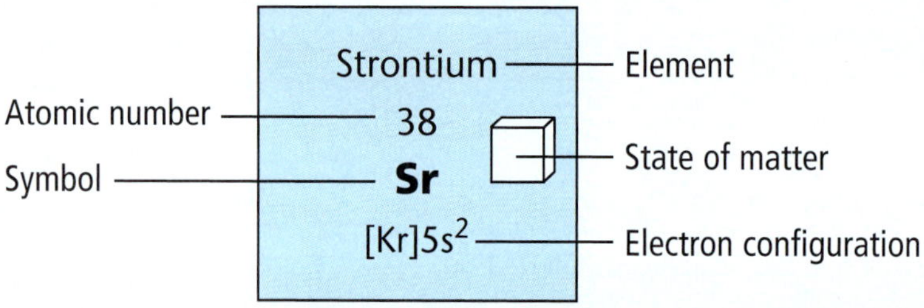

Color Key

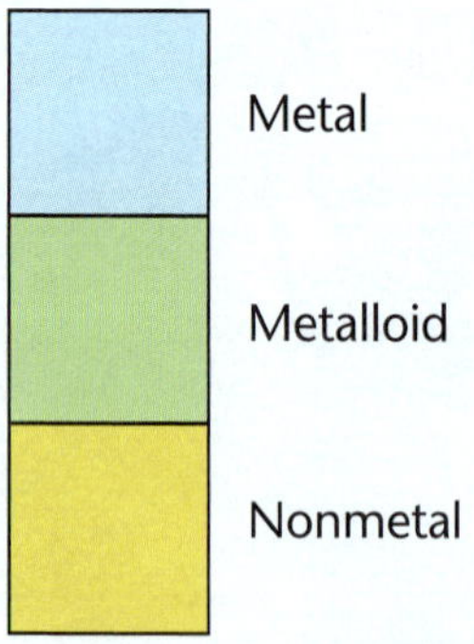

States of Matter Key

View **animations about the elements.**

Concepts In Motion

When you read the Elements Handbook, you need to read for information. Here are some tools that the Elements Handbook has to help you find that information.

See how a group fits in the **Periodic Table.**

Discover the **Physical Properties** and **Atomic Properties** of the elements in a group.

Summarize **Common Reactions** for the elements within a group.

Identify elements by **Analytical Tests.**

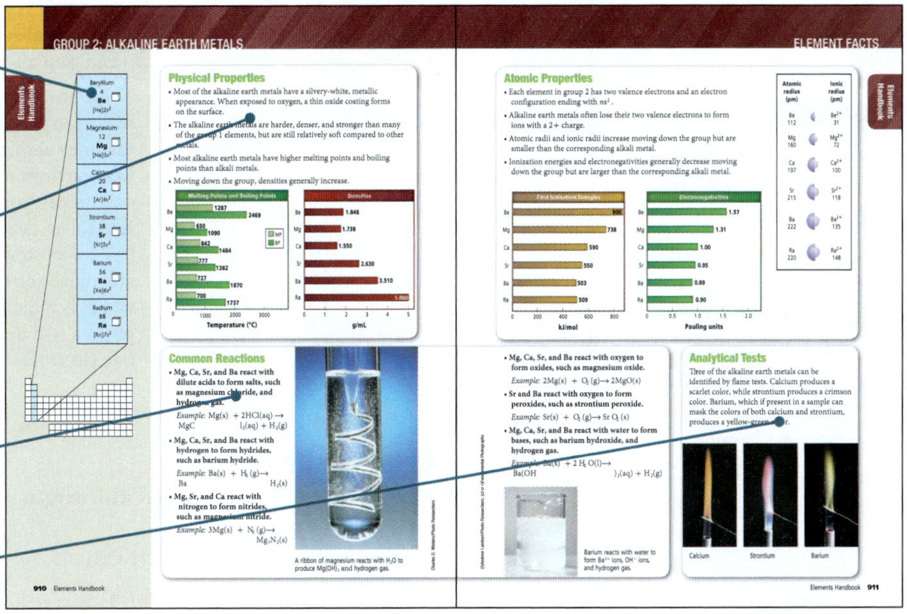

Source: Elements Handbook, p. 910–911

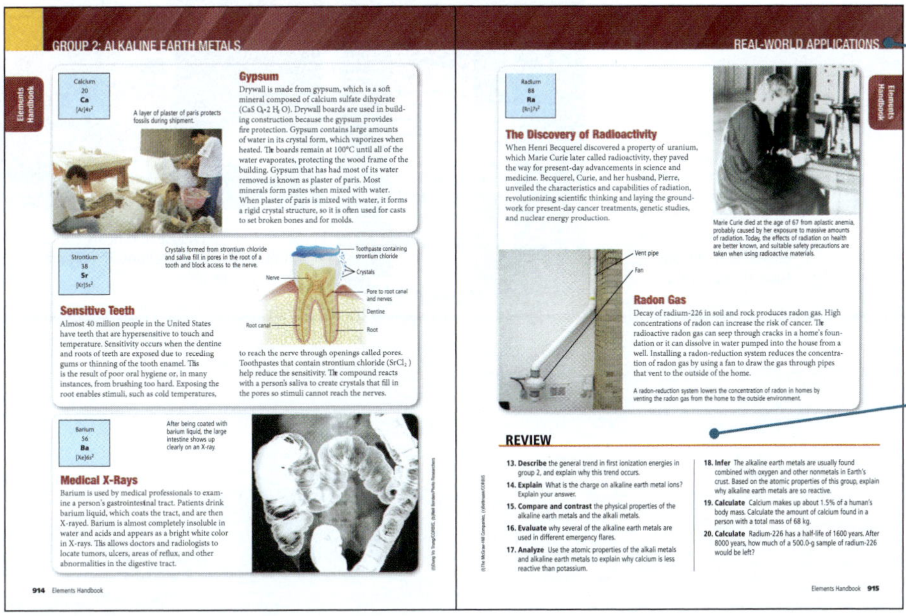

Learn how elements are used every day in **Real-World Applications.**

Test your knowledge of the elements by answering **Assessment** questions.

Source: Elements Handbook, p. 914–915

Physical and Atomic Properties

Activity Have students draw Lewis dot diagrams to illustrate how hydrogen can lose an electron to form an H^+ ion or gain an electron to form an H^- ion. **OL**

Common Reactions

Background The name *hydrogen* comes from Greek in reference to hydrogen's ability to form water when combined with oxygen. *Hydro* means *water* and *genes* means *forming.* Ask students to identify other element names that have the same suffix and research their meaning. Oxygen and nitrogen: the prefix *oxy-* means acid. Thus, *oxygen* means *acid*-forming. The prefix *nitron-* means *native soda.* **OL**

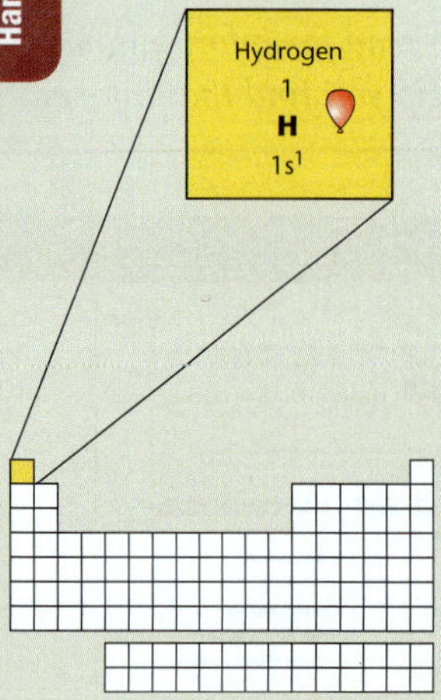

Physical and Atomic Properties

- At constant temperature and pressure, hydrogen gas (H_2) has the lowest density of any gas.
- At very high pressures, such as the interior of planet Jupiter, hydrogen might exist as a solid metal.
- Hydrogen is placed in group 1 because it has one valence electron.
- Hydrogen shares some properties with the group 1 metals. It can lose an electron to form a hydrogen ion (H^+).
- Hydrogen also shares some properties with the group 17 nonmetals. It can gain an electron to form a hydride ion (H^-).
- There are three common hydrogen isotopes. Protium, the most common isotope, has one proton, one electron, and no neutrons. Deuterium, also called heavy hydrogen, has one proton, one neutron, and one electron. Tritium, which is radioactive, has one proton, two neutrons, and one electron.

Physical and Atomic Properties of Hydrogen

Property	Value
Melting point	−259°C
Boiling point	−253°C
Density	8.24×10^{-5} g/mL
Atomic radius	78 pm
First ionization energy	1312 kJ/mol
Electronegativity	2.20 Pauling units

Common Reactions

- **When ignited, hydrogen reacts with oxygen to form water.**

  Example: $2H_2(g) + O_2(g) \rightarrow 2H_2O(l)$

- **Hydrogen reacts with sulfur to form hydrogen sulfide.**

  Example: $H_2(g) + S(g) \rightarrow H_2S(g)$

- **Hydrogen reacts with nitrogen at high temperatures and pressures to form ammonia.**

  Example: $3H_2(g) + N_2(g) \xrightarrow{\text{catalyst}} 2NH_3(g)$

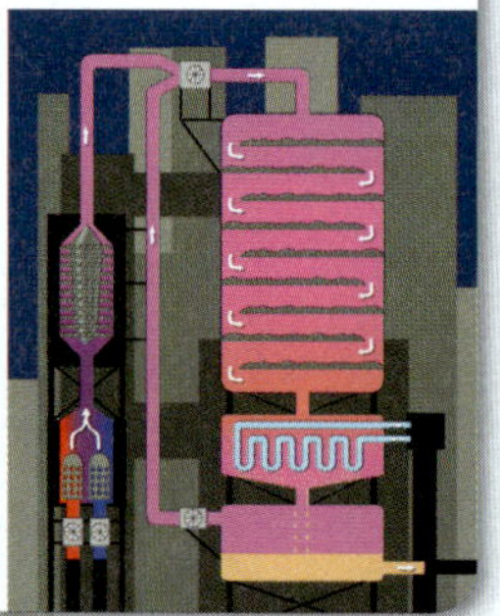

Hydrogen gas in the red tube and nitrogen gas in the blue tube are mixed, then compressed under high pressure and temperature to form liquid ammonia in the orange tube at bottom right.

Analytical Tests

pH is a measure of the hydrogen ion (H^+) concentration of aqueous solutions. When the hydrogen ion concentration is expressed in moles per liter, pH is the negative logarithm of the hydrogen ion concentration, $-\log[H^+]$. For example, if the hydrogen ion concentration is 1×10^{-2} mol/L, the pH is 2.

Some common household items are bases or acids, depending on their H^+ concentrations. The greater the H^+ concentration, the lower the pH.

Hydrogen
1
H
$1s^1$

Identifying Hydrogen in Stars

Spectroscopy is the study of the spectral lines present in an electromagnetic spectrum. The colored lines in an emission spectrum represent the emission of energy. How do scientists know that more than 90% of the atoms in the universe are hydrogen atoms? By analyzing the emission spectra of light from stars or galaxies, astronomers can identify hydrogen. The spectrum of hydrogen consists of four distinct color lines at different wavelengths. The lines are produced when electrons in a gas move to different energy levels in an atom by absorbing and then emitting energy. Each element can be identified by characteristic patterns of spectral lines.

The colorful cloud that makes up this nebula is composed of hydrogen gas.

Hydrogen fuel cells provide the energy to power this electric guitar.

Hydrogen Fuel Cells

Hydrogen fuel cells produce electricity by combining hydrogen (H_2) and oxygen (O_2) without burning. Water and heat are the only by-products of this process. Current demonstration projects that use hydrogen fuel cells as their energy sources include laptop computers, cars, buses, classrooms, and musical instruments. In the future, it might be possible to use a pen-sized container filled with hydrogen gas to power a laptop computer. Or, you might drive a fuel cell car to a filling station and fill a high-pressure gas cylinder with hydrogen gas.

REVIEW

1. **Compare and contrast** hydrogen isotopes.
2. **Write** the balanced equation for the reaction between hydrogen gas and oxygen gas in a fuel cell.
3. **Explain** what happens when hydrogen reacts with a nonmetal element.
4. **Evaluate** at least one advantage and one possible disadvantage of hydrogen fuel cells compared to conventional petroleum engines.
5. **Infer** Hydrogen can gain one electron to reach a stable electron configuration. Why isn't hydrogen placed with the group 17 elements that share this behavior?
6. **Apply** A solution's hydrogen ion concentration is 3.2×10^{-4} mol/L. Determine if this solution is an acid or a base. What is the pH of this solution?

Identifying Hydrogen in Stars

Spectroscopy is a powerful tool in astronomy. Astronomers can identify the elements that comprise stars using their emission spectra. This also enables astronomers to determine the age of stars. Younger stars contain mostly hydrogen and helium. Older stars contain a percentage of other elements. Astronomers count the number of lines in its emission spectrum. A large number of lines indicates a greater number of elements. The intensity of the lines can reveal information about the star's temperature. Intense lines indicate a hot star, with a surface temperature of approximately 220,000°C. A cool star, with a surface temperature of around 55,000°C, is indicated by lines that appear dull or subdued.

Hydrogen Fuel Cells

Fuel cells and batteries operate using the same basic mechanisms: through chemical reactions, substances combine and produce electricity. Disposable and rechargeable batteries have limited life spans. Once the chemicals inside these batteries are depleted, no more electricity can be produced. Fuel cells have a virtually unlimited life span. They receive a constant flow of fresh chemicals from fuels, such as natural gas, that are converted into hydrogen, so they don't run down and they don't need to be recharged.

REVIEW

1. There are three common hydrogen isotopes. Protium is the most common isotope with one proton, one electron, and no neutrons. Deuterium, also called heavy hydrogen, has one proton, one neutron, and one electron. Tritium, which is radioactive, has one proton, two neutrons, and one electron.
2. $2H_2(g) + O_2(g) \longrightarrow 2H_2O(l)$
3. In some cases, hydrogen loses an electron and forms an ion with a 1+ charge. In others it forms a polar covalent bond, with hydrogen at the positive end of the bond.
4. One possible advantage is that the by-products of the reaction are water and heat. No CO_2 is produced. One possible disadvantage is the production and distribution system for hydrogen gas.
5. Elements are placed in groups based on electron configuration. Hydrogen has an electron configuration of $1s^1$, so it is placed in group 1 with the other elements with electron configurations ending in s^1.
6. $pH = -\log [H^+] = -\log (3.2 \times 10^{-4}) = 3.49$
The substance is an acid because its pH is 3.49.

Physical Properties

Background Four of the alkali metals—sodium, potassium, rubidium, and cesium—have melting points below the boiling point of water (100°C). The melting point of rubidium, 39°C, is just slightly higher than normal human body temperature (37°C).

Common Reactions

Activity Have students cut a piece of fruit into eight slices. They should place each slice in a petri dish, and cover all sides of the fruit with different mixtures of the four salts that make up natron: sodium carbonate, sodium chloride, sodium sulfate, and sodium hydrogen carbonate. Have students write brief reports in which they record their results and determine which mixture of salts preserved the fruit the best. Have students share their results with the class. **OL**

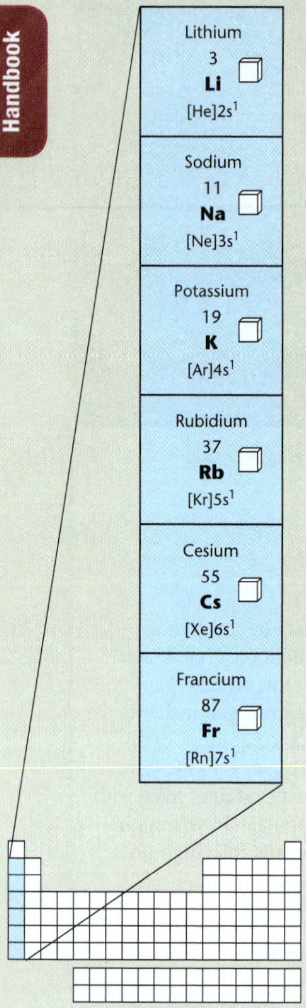

Physical Properties

- Pure alkali metals have a silvery, metallic appearance.
- Solid alkali metals are soft enough to cut with a knife.
- Most of the alkali metals have low densities compared to the solid form of elements from other groups. Lithium, sodium, and potassium metals are less dense than water.
- Compared to other metals, such as silver or gold, alkali metals have low melting points.

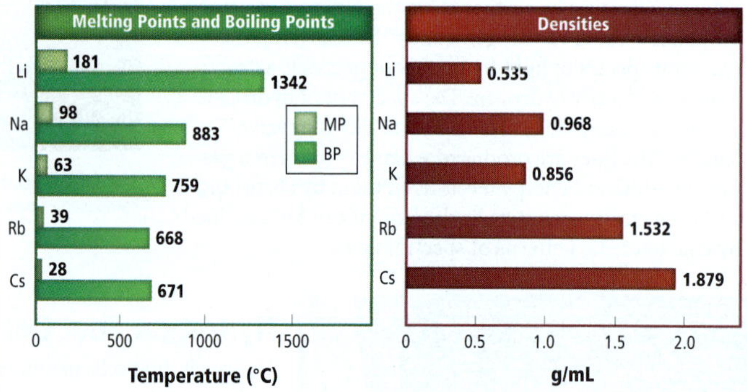

Common Reactions

- **Li, Na, K, Rb, and Cs react vigorously with halogens to form salts, such as lithium chloride.**

  Example: $2Li(s) + Cl_2(g) \rightarrow 2LiCl(s)$

- **Li, Na, K, Rb, and Cs react with oxygen to form oxides (X_2O), peroxides (X_2O_2), or superoxides (XO_2). For example, sodium reacts with oxygen to form sodium peroxide.**

  Example: $2Na(s) + O_2(g) \rightarrow Na_2O_2(s)$

- **Li, Na, K, Rb, and Cs react vigorously with water to form metal hydroxides, such as potassium hydroxide, and hydrogen gas.**

  Example: $2K(s) + 2H_2O(l) \rightarrow 2KOH(aq) + H_2(g)$

Potassium reacts violently with water, producing enough heat to ignite the hydrogen gas produced.

Atomic Properties

- Each element in group 1 has one valence electron and an electron configuration ending with ns^1.
- Group 1 elements lose their valence electrons to form ions with a 1+ charge.
- Going down the elements in group 1, the atomic radii and ionic radii increase.
- Electronegativity decreases going down the elements in group 1.
- The alkali metals are so reactive that they are not found in nature as free metals.
- All the alkali metals have at least one radioactive isotope.
- Because francium is rare and decays rapidly, its properties are not well known.

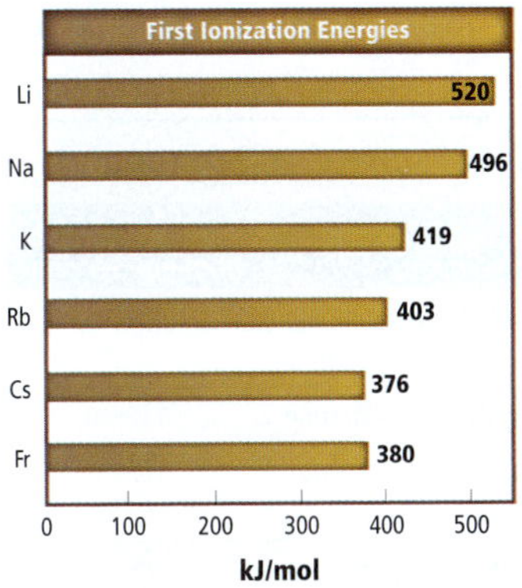

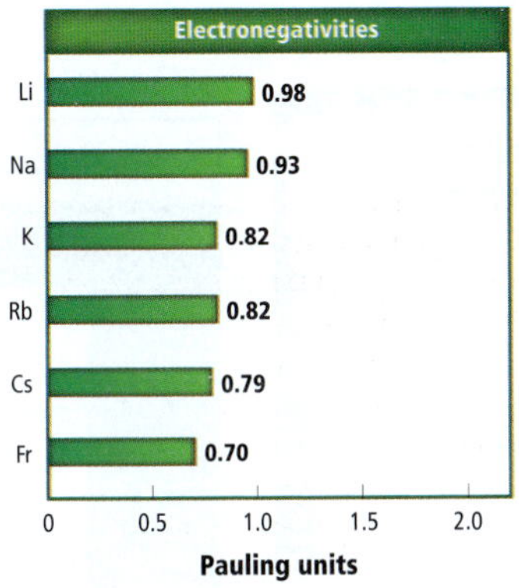

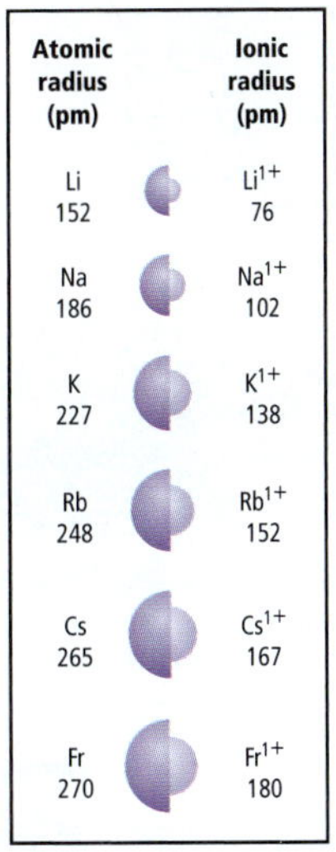

Analytical Tests

Alkali metals can be qualitatively identified by flame tests. Lithium produces a red flame. Sodium produces an orange flame. Potassium, rubidium, and cesium produce violet flames.

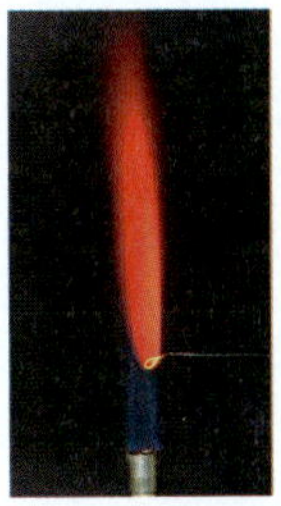
Lithium

Sodium

Potassium

Rubidium

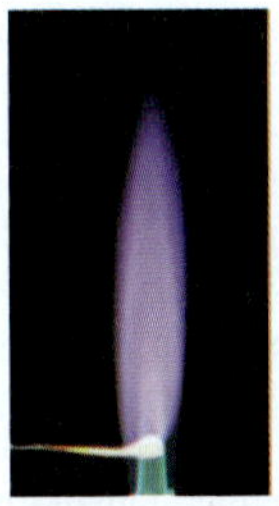
Cesium

Atomic Properties

Activity WARNING: ***To prevent burns, wear gloves when handling sodium, as well as an apron and goggles.*** Fill a 1-L graduated cylinder with 700 mL water, then add 3 drops phenolphathalein. Pour 50 mL mineral oil on the surface of the water. Show students a bottle of sodium, and point out that it is submerged in mineral oil. Remove a piece of the sodium, place it on a glass plate, and cut off a pea-size piece. Place it in the graduated cylinder. (Mineral oil keeps the sodium from "popping" out of the container. The sodium will drop through the mineral oil and react on the surface of the water.) Ask students why sodium is stored in mineral oil. To prevent it from reacting with water. Repeat this demonstration with potassium and lithium. Disposal: Dilute the solutions and flush them down the drain. Have students compare the reactivity of the three metals, and ask them to predict the trend in reactivity of the alkali metals. K is the most reactive; Li is the least reactive. Reactivity increases from top to bottom down the group. **OL**

Analytical Tests

Activity Have students write three or four clues that can be used to identify an individual element or family of elements. The first clue should be general, and each additional clue should be more specific, so choices are narrowed down to one element. Use students' clues to review the properties of alkali metals. **OL**

Environmentally Friendly Batteries

Background In the early 1970s, the first nonrechargeable lithium batteries became commercially available. During the 1980s, research was conducted on rechargeable lithium batteries. However, lithium metal was found to be unstable in rechargeable batteries, and the research shifted to a nonmetallic lithium battery using lithium ions. The first lithium-ion batteries became commercially available in 1991.

Dietary Salt

Background Throughout history, salt has been one of the most important raw materials and has influenced the world's economy. The infrastructure of ancient civilizations was centered around sources of salt. The civilization that controlled the salt supply had the power to control other civilizations. The first tax, levied by the Chinese, was on salt, and Roman soldiers were paid with "salt money." (The word *salary* refers to the money given to Roman soldiers to buy salt.) Salt was a valuable exchange substance, and the "worth of salt" was used to purchase other essential commodities, including food, fuel, and even slaves. Salt is essential for all living things, and it is used as a preservative for fish and meat. It is collected by boiling brine, evaporating seawater, or by mining.

Lithium
3
Li
$[He]2s^1$

The Mars rovers, Spirit and Opportunity, use solar energy to recharge lithium-ion batteries.

Environmentally Friendly Batteries

Someday, electric cars might be powered by lightweight lithium-ion batteries. Lithium batteries have several advantages compared to lead-acid batteries. Unlike lead-acid batteries, lithium batteries do not contain toxic metals or corrosive acids, making them safer for the environment. Lithium's light weight is also an advantage for electric vehicles. However, lithium batteries do have some disadvantages. Researchers are trying to find ways to make lithium batteries that recharge more rapidly. Cost is also a drawback. Lithium batteries are currently used for small applications such as laptop computers, but they will need to be less expensive before they can be routinely used in larger, more energy-demanding applications such as electric or hybrid vehicles.

Sodium
11
Na
$[Ne]3s^1$

Dietary Salt

In 2006, the American Medical Association recommended that the amount of sodium in processed and restaurant foods be reduced by one-half over the next decade. Sodium is essential for humans, but too much might contribute to high blood pressure and heart failure. Current guidelines advise consuming less than 2300 mg of sodium per day, which is about one teaspoon. However, Americans typically consume 4000 to 6000 mg of sodium per day. Foods that contain more than 480 mg of sodium per serving are considered high-sodium foods. To be labeled as low sodium, foods must contain 140 mg or less per serving. The table lists some common foods that are either high or low in sodium.

Sodium Content of Some Common Foods

	Food	Sodium Content (mg) per Serving
High sodium	fast-food submarine sandwich with cold cuts	1310
	canned chicken noodle soup	1106
	fast-food biscuit with egg and sausage	1080
	cottage cheese	851
	dill pickle	833
	fast-food cheeseburger	740
	canned corn	571
	beef hotdog	513
	fried fish fillet	484
Low sodium	wheat bread	133
	low-fat fruit yogurt	132
	fast-food salad with cheese and egg, no dressing	119
	pound cake	111
	oatmeal cookie	96
	raw carrots	76
	canned peaches	16
	frozen corn	2

Sodium
11
Na
$[Ne]3s^1$

Potassium
19
K
$[Ar]4s^1$

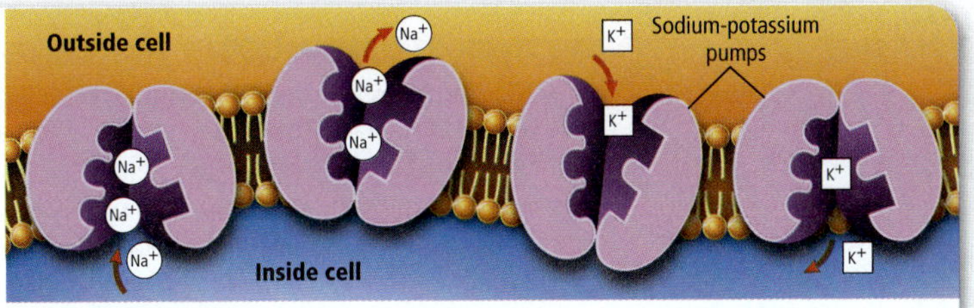

The sodium-potassium pump brings two K^+ ions into a cell for every three Na^+ ions it moves out of a cell.

The Sodium-Potassium Pump

Humans and other vertebrates need to maintain a negative potential charge inside their cells in order to survive. This process requires sodium ions, potassium ions, and a membrane-bound enzyme called sodium/potassium ATPase. Sodium/potassium ATPase uses energy from the hydrolysis of ATP to pump sodium ions out of cells and pump potassium ions into cells. Because of the action of this pump, the sodium ion concentration is low inside cells and high outside cells. The potassium ion concentration is high inside cells and low outside cells. In fact, potassium ions are the most common ions inside living cells. For every three sodium ions pumped out of a cell, sodium/potassium ATPase pumps two potassium ions into the cell. The net result is a negative charge inside the cell and concentration gradients across the cell membrane for both potassium and sodium ions.

Cesium
55
Cs
$[Xe]6s^1$

The cesium fountain atomic clock at NIST is accurate to about 1 second over a period of 100 million years.

Cesium Atomic Clocks

One of the most accurate clocks in the world is located at the United States National Institute of Standards and Technology (NIST) in Boulder, Colorado. This cesium fountain atomic clock provides the official time for the United States. The clock is based on the natural resonance frequency of the cesium atom (9,192,631,770 Hz), which defines the second.

Geoffrey Wheeler

REVIEW

7. **Describe** the trend in density of the alkali metals as atomic number increases.
8. **Compare** lithium-ion batteries and lead-acid batteries.
9. **Write** a balanced equation for the reaction between lithium and water.
10. **Predict** the reactivity of lithium metal with water.
11. **Analyze** Are lithium's atomic and ionic radii closer to sodium or closer to magnesium, which is in group 2? Explain.
12. **Organize** Make a table to summarize the data for physical and atomic properties of the group 1 elements based on the trend of increasing atomic number.

The Sodium-Potassium Pump

Activity Have students form a single line. Each student represents one of the nodes of Ranvier along the axon of a motor neuron. The action of Na^+/K^+ ATPase generates a negative potential inside the axon. Students can show that by putting their hands on their knees. This is the resting potential.

As a nerve impulse travels down an axon, sodium channels open. Sodium ions flow out of the cell. Ask the first student in line to demonstrate this by raising his or her hands over his or her head. When the potential reaches its highest point, sodium channels close and potassium channels open. Potassium flows into the cell. The potential decreases, overshoots the resting potential, and returns to the resting potential. Ask the first student in line to demonstrate this by lowering his or her hands back to his or her knees. Sodium and potassium channels are sensitive to voltage. When the potential at the first "node" reaches the first student's mid thigh, the sodium channels at the next "node" open. Have students continue this process down the line as a wave. **OL**

Cesium Atomic Clocks

Activity Have interested students research the use of cesium ions as a possible fuel for ion engines. They should share their research in a class presentation, and describe the design and function of an ion engine. **AL**

REVIEW

7. In general, density increases as atomic number increases.
8. Lead batteries contain poisonous metals and corrosive acids, while lithium batteries do not. Lithium batteries are lighter weight, more expensive and take a longer time to charge than lead batteries.
9. $2Li(s) + 2H_2O(l) \longrightarrow 2LiOH(aq) + H_2(g)$
10. Following periodic trends, you would expect lithium to react less violently than either sodium or potassium.
11. Atomic radii tend to increase as you move down a group and decrease moving across a period. Thus, elements that are connected diagonally, such as Li and Mg, are often similar in radius and exhibit similar properties.
12.

Properties That Generally Increase With Increasing Atomic Number	Properties That Generally Decrease With Increasing Atomic Number
Atomic radius	first ionization energy
Ionic radius	electronegativity
Density	melting point
Reactivity	boiling point

Physical Properties

Background The alkaline earth metals contain two electrons in their outer energy levels. This allows them to participate in stronger metallic bonding than the group 1 elements. Alkaline earth metals give up two electrons during the metallic bonding process (instead of the alkali metals that give up only one electron) and in doing so create a greater attraction between atoms. This is what makes them harder and stronger than many of the group 1 elements.

Common Reactions

Activity Have students pour 6 mL of 1*M* HCl into a test tube and set it aside. Students should then take a strip of magnesium and shine it by using sandpaper. Instruct students to place the magnesium into the HCl, observe, and explain what happens and why. Have students predict the products of the reaction. Students should see bubbles forming. They should be able to explain that a chemical reaction is taking place between the magnesium and the acid and should be able to predict that the bubbles are hydrogen gas. Ask students why it is important to shine the magnesium first. Students should recall that magnesium forms an oxide coating when exposed to air. Shining the magnesium exposes the magnesium and readies it for reaction. OL

GROUP 2: ALKALINE EARTH METALS

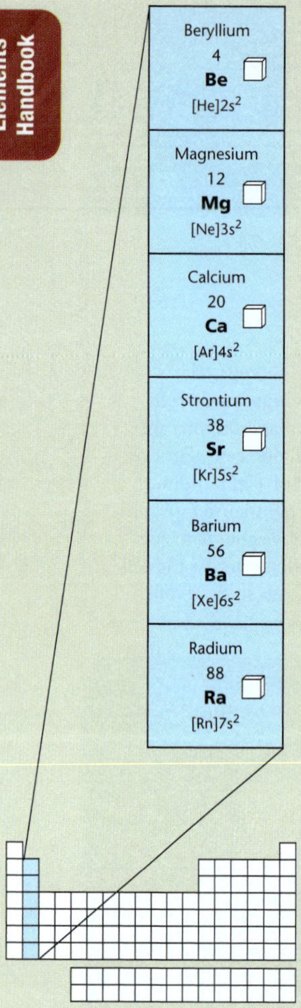

Physical Properties

- Most of the alkaline earth metals have a silvery-white, metallic appearance. When exposed to oxygen, a thin oxide coating forms on the surface.
- The alkaline earth metals are harder, denser, and stronger than many of the group 1 elements, but are still relatively soft compared to other metals.
- Most alkaline earth metals have higher melting points and boiling points than alkali metals.
- Moving down the group, densities generally increase.

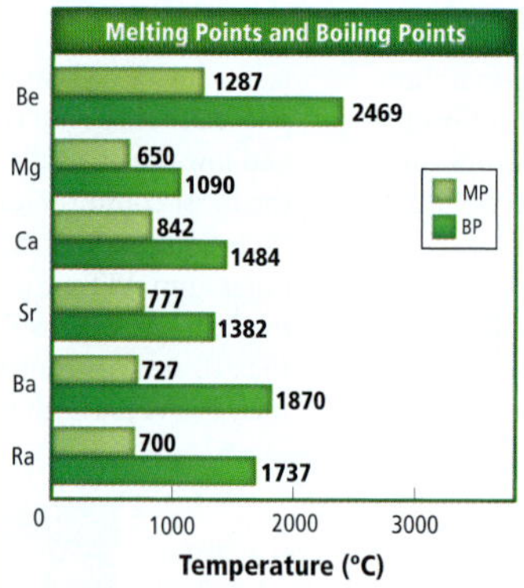

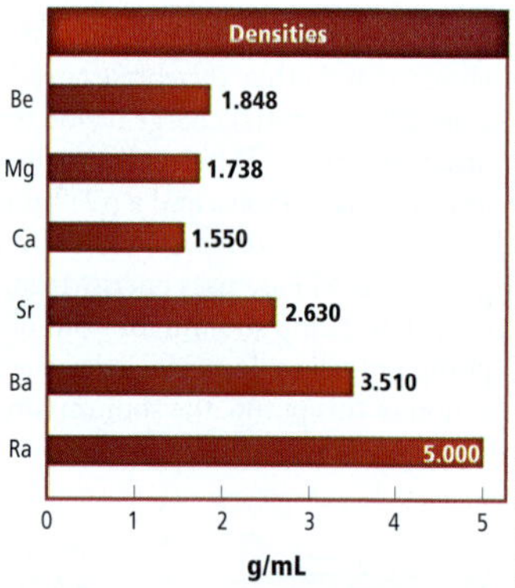

Common Reactions

- **Mg, Ca, Sr, and Ba react with dilute acids to form salts, such as magnesium chloride, and hydrogen gas.**

  Example: $Mg(s) + 2HCl(aq) \rightarrow MgCl_2(aq) + H_2(g)$

- **Mg, Ca, Sr, and Ba react with hydrogen to form hydrides, such as barium hydride.**

  Example: $Ba(s) + H_2(g) \rightarrow BaH_2(s)$

- **Mg, Sr, and Ca react with nitrogen to form nitrides, such as magnesium nitride.**

  Example: $3Mg(s) + N_2(g) \rightarrow Mg_3N_2(s)$

A ribbon of magnesium reacts with H_2O to produce $Mg(OH)_2$ and hydrogen gas.

Atomic Properties

- Each element in group 2 has two valence electrons and an electron configuration ending with ns^2.
- Alkaline earth metals often lose their two valence electrons to form ions with a 2+ charge.
- Atomic radii and ionic radii increase moving down the group but are smaller than the corresponding alkali metal.
- Ionization energies and electronegativities generally decrease moving down the group but are larger than the corresponding alkali metal.

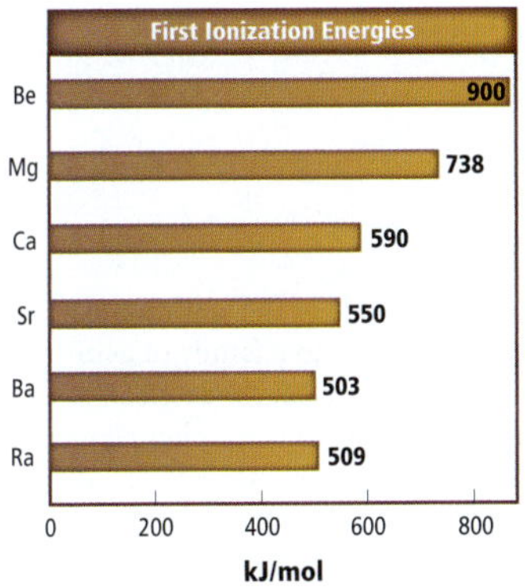

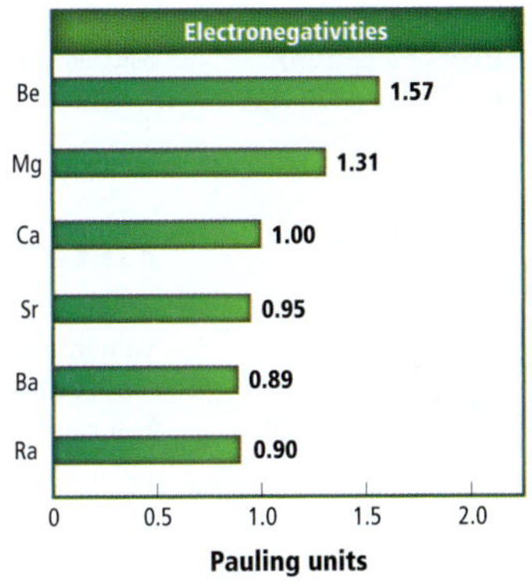

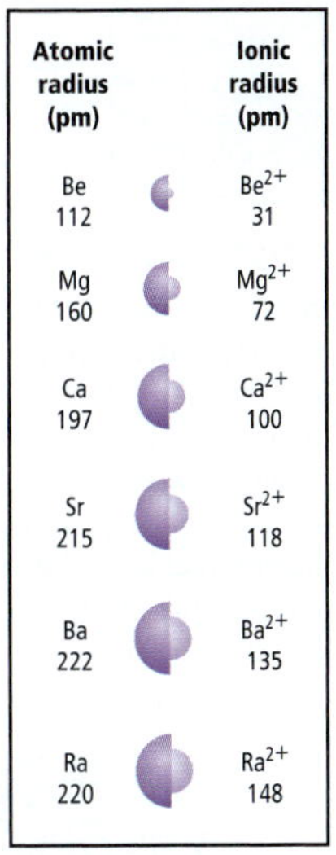

- **Mg, Ca, Sr, and Ba react with oxygen to form oxides, such as magnesium oxide.**

  Example: $2Mg(s) + O_2(g) \rightarrow 2MgO(s)$

- **Sr and Ba react with oxygen to form peroxides, such as strontium peroxide.**

  Example: $Sr(s) + O_2(g) \rightarrow SrO_2(s)$

- **Mg, Ca, Sr, and Ba react with water to form bases, such as barium hydroxide, and hydrogen gas.**

  Example: $Ba(s) + 2H_2O(l) \rightarrow Ba(OH)_2(aq) + H_2(g)$

Barium reacts with water to form Ba^{2+} ions, OH^- ions, and hydrogen gas.

Analytical Tests

Three of the alkaline earth metals can be identified by flame tests. Calcium produces a scarlet color, while strontium produces a crimson color. Barium, which if present in a sample can mask the colors of both calcium and strontium, produces a yellow-green color.

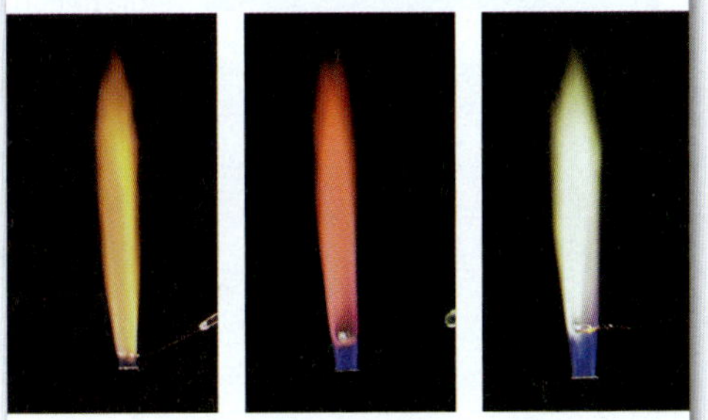

Calcium Strontium Barium

Atomic Properties

Activity Place two petri dishes on an overhead projector. Pour water in each and add 3 drops of phenolphthalein. Show students samples of magnesium and calcium. Ask students to predict which one will be more reactive with water. Students should be able to predict that the calcium will be more reactive because the trend of reactivity increases going down the family and ionization energies decrease. Shine each sample with sandpaper and place magnesium in one petri dish and calcium in the other. Have students observe and explain their observations. Ask students why phenolphthalein is used in the demonstration. Students should observe that calcium is much more reactive than magnesium, which hardly reacts at all. This is because calcium is farther down in the family. Students should also be able to predict that the products will be magnesium hydroxide and calcium hydroxide, bases that turn phenolphthalein pink. Disposal: Dilute the solutions with water and flush down the drain.

Analytical Tests

Background Atoms can exist in either a ground state or an excited state. During a flame test, the Bunsen burner flame excites the valence electrons in the atoms and causes them to "jump" to higher energy states. When the electrons return to lower energy levels, photons are given off that produce visible light of a characteristic wavelength and color.

Space Telescopes

Background The *James Webb Space Telescope (JWST)* is slated to launch in 2014. It will be 2½ times the diameter of the *Hubble Space Telescope,* but will weigh almost two-thirds less. A key feature of the telescope is its primary hexagonal mirror made of beryllium. The mirror will have a diameter of about 6.5 m. The goal of this new technology is to provide NASA scientists with the ability to see into the far reaches of the universe. It is hoped that this knowledge will reveal how the universe began and evolved over billions of years.

Precious Gems

Background Emeralds begin as beryl crystals formed in Earth's interior. Over billions of years, if temperature and pressure conditions are right, these crystals grow molecule-by-molecule into gemstone quality crystals. Hydrothermal veins are environments well-suited for such growth. Almost all of the emeralds mined in Colombia come from these types of veins.

Chlorophyll and Crop Yields

Activity Give students a few spinach leaves and have them grind the leaves with a mortar and pestle. After grinding, have students add a small amount of ethanol (95%) to the mixture, let it sit for a while, and pour the resulting liquid into a beaker. Then, have students place a strip of chromatography paper into the beaker and watch as the colors of the mixture are separated. Students should see red, orange, yellow, blue, and indigo, but not green. Ask students to notice if any color is missing and to explain why. Green is missing because green light is reflected, not absorbed, a consequence of the chlorophyll. OL

Elements Handbook

Beryllium
4
Be
[He]$2s^2$

Space Telescopes

Beryllium and beryllium alloys have properties that make them useful for applications in space: they are hard, they are lighter than aluminum, and they are stable over a wide temperature range. The *Hubble Space Telescope's* reaction plate is made of lightweight beryllium. The reaction plate carries heaters that keep the main mirror at a constant temperature. Beryllium is also being used in the *Hubble's* replacement—the *James Webb Space Telescope (JWST).*

The JWST's large mirror is composed of 18 hexagonal beryllium plates.

◀ Emerald beryl

Precious Gems

Emerald ($Be_3Al_2Si_6O_{18}$), one of the world's most valuable gemstones, belongs to a family of gemstones known as beryls. Pure beryls are clear, colorless crystals. Beryls tinted with other elements form gems such as aquamarine, morganite, and emerald. Trace amounts of chromium or vanadium give emeralds their unique green color.

Magnesium
12
Mg
[Ne]$3s^2$

Amount of Magnesium Removed by Crops from One Hectare of Soil

Crop	Magnesium Removed from Soil (kg)
Alfalfa	44
Corn	58
Cotton	25
Oranges	25
Peanuts	27
Rice	15
Soybeans	27
Tomatoes	40
Wheat	20

Chlorophyll and Crop Yields

In the early 1900s, German chemist Richard Willstätter discovered that a molecule of chlorophyll has a magnesium ion at its center. Chlorophyll, the green pigment in plants, is responsible for photosynthetic processes, which convert sunlight to chemical energy. It is this chemical energy that supports life on Earth. Notice in the table that an average yield of common crops removes large amounts of magnesium from just one hectare of soil. Once the importance of magnesium was revealed, soils deficient in magnesium were fertilized, greatly increasing crop yields. Willstätter's work won him the Nobel Prize in Chemistry in 1915.

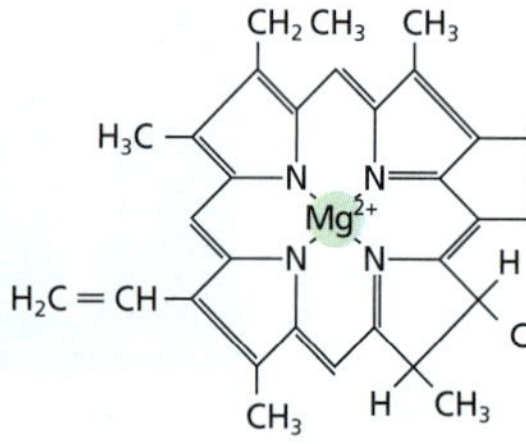

◀ Chlorophyll molecule

(l)Mark A. Schneider/Photo Researchers, (r)courtesy of Northrop Grumman Space Technology

Magnesium 12 **Mg** $[Ne]3s^2$	Calcium 20 **Ca** $[Ar]4s^2$	Strontium 38 **Sr** $[Kr]5s^2$	Barium 56 **Ba** $[Xe]6s^2$

Fireworks

The four main components of fireworks are a container, a fuse, a bursting charge, and stars. Stars contain the chemical compounds needed to produce light of brilliant colors. Many of these compounds contain alkaline earth metals, such as barium chloride ($BaCl_2$), strontium carbonate ($SrCO_3$), and calcium chloride ($CaCl_2$). The table identifies which metals are needed to make the colors seen during a fireworks display.

Metals Used in Fireworks

Color	Metal
Red	strontium, lithium
Orange	calcium
Gold	iron (with carbon)
Yellow	sodium
White	white-hot magnesium or aluminum, barium
Green	barium
Blue	copper
Purple	mixture of strontium (red) and copper (blue)
Silver	aluminum, titanium, or magnesium powder or flakes

New Engineering Alloys

Magnesium alloys are used when strong, but lightweight, materials are needed, such as in backpack frames and aircraft. These alloys also enable automotive engineers to design lighter, more fuel-efficient cars. A new magnesium alloy, introduced in the engine cradle of some 2006 automotive models, replaces traditional aluminum. This alloy reduces the engine cradle's mass by approximately one-third, creating a vehicle that is both agile and controllable. Considered a breakthrough in engineering technology, the new alloy is currently being evaluated for use in other applications.

The magnesium-alloy engine cradle is lighter than the aluminum model, yet it can still withstand the high temperatures produced by the car's engine.

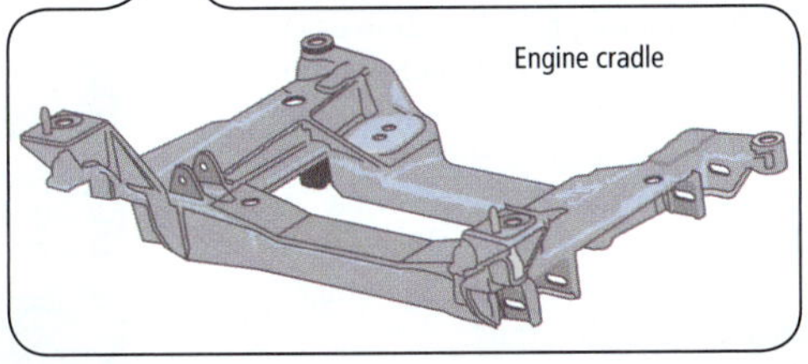

Fireworks

Background Fireworks are launched from a device called a mortar. The mortar is typically a steel tube, which contains black powder that explodes and lights the fuse of the firework. This explosion propels the firework into the sky. The fuse is long enough that the firework does not ignite until an appropriate height is obtained. Once the firework ignites, it explodes, forcing the burning metals to move in all directions from the center, resulting in the typical circular shape of sparkling light.

New Engineering Alloys

Background Alloys are made when pure metals do not provide the physical properties needed for a certain purpose. For example, metals are excellent conductors of electricity. However, nichrome wire, an alloy made from nickel, iron, and chromium, does not conduct electricity as well as a pure metal. This makes the wire suitable for use in heating elements like those found in hair dryers and toasters. As an electric current travels down the heating element, it encounters resistance, which heats up the element.

Gypsum

Background When water changes from a liquid to a gas, the thermal energy absorbed by the water disrupts intermolecular forces instead of increasing molecular kinetic energy. This is why liquid water will remain at 100°C until all of the liquid is vaporized.

Sensitive Teeth

Activity Have student groups grow crystals to observe how they form. Have each group tie a paper clip to 15 cm of dental floss, then tie the floss around a pencil. Place the pencil across the mouth of a glass jar, suspending the paper clip in the jar. Have each group prepare a supersaturated solution of sugar, table salt, borax, or alum crystals by pouring 250 mL water into a beaker, heating it to boiling, slowly adding the crystals, then stirring the solution until no more dissolves. Pour the hot solution into the jar until it covers the paper clip. Have students observe the jars daily until crystals form, then compare results with other groups.
OL COOP LEARN

Medical X-Rays

Background Barium shows up as a bright white outline of soft tissue, such as the colon, on an X-ray film because it absorbs X-rays easily. Air is blown into the colon and shows up as black on the X-ray film. The color contrast allows technologists to see and analyze the colon tissue.

Calcium
20
Ca
$[Ar]4s^2$

A layer of plaster of paris protects fossils during shipment.

Gypsum

Drywall is made from gypsum, which is a soft mineral composed of calcium sulfate dihydrate ($CaSO_4{\cdot}2H_2O$). Drywall boards are used in building construction because the gypsum provides fire protection. Gypsum contains large amounts of water in its crystal form, which vaporizes when heated. The boards remain at 100°C until all of the water evaporates, protecting the wood frame of the building. Gypsum that has had most of its water removed is known as plaster of paris. Most minerals form pastes when mixed with water. When plaster of paris is mixed with water, it forms a rigid crystal structure, so it is often used for casts to set broken bones and for molds.

Strontium
38
Sr
$[Kr]5s^2$

Crystals formed from strontium chloride and saliva fill in pores in the root of a tooth and block access to the nerve.

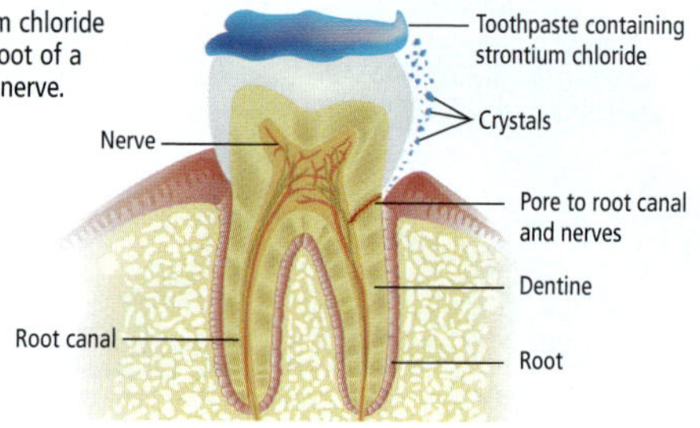

Sensitive Teeth

Almost 40 million people in the United States have teeth that are hypersensitive to touch and temperature. Sensitivity occurs when the dentine and roots of teeth are exposed due to receding gums or thinning of the tooth enamel. This is the result of poor oral hygiene or, in many instances, from brushing too hard. Exposing the root enables stimuli, such as cold temperatures, to reach the nerve through openings called pores. Toothpastes that contain strontium chloride ($SrCl_2$) help reduce the sensitivity. The compound reacts with a person's saliva to create crystals that fill in the pores so stimuli cannot reach the nerves.

Barium
56
Ba
$[Xe]6s^2$

After being coated with barium liquid, the large intestine shows up clearly on an X-ray.

Medical X-Rays

Barium is used by medical professionals to examine a person's gastrointestinal tract. Patients drink barium liquid, which coats the tract, and are then X-rayed. Barium is almost completely insoluble in water and acids and appears as a bright white color in X-rays. This allows doctors and radiologists to locate tumors, ulcers, areas of reflux, and other abnormalities in the digestive tract.

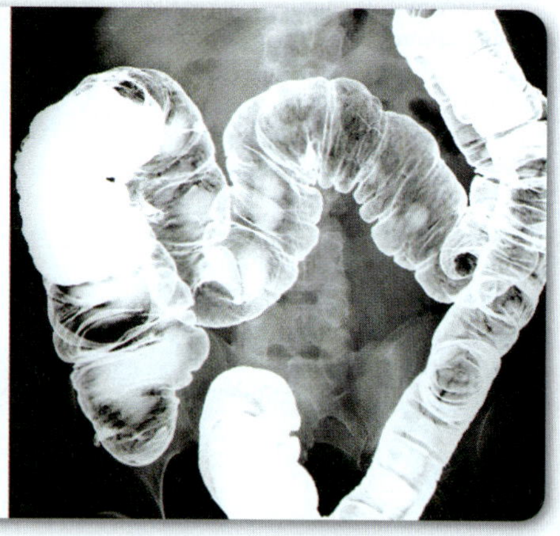

Radium
88
Ra
$[Rn]7s^2$

The Discovery of Radioactivity

When Henri Becquerel discovered a property of uranium, which Marie Curie later called radioactivity, they paved the way for present-day advancements in science and medicine. Becquerel, Curie, and her husband, Pierre, unveiled the characteristics and capabilities of radiation, revolutionizing scientific thinking and laying the groundwork for present-day cancer treatments, genetic studies, and nuclear energy production.

Marie Curie died at the age of 67 from aplastic anemia, probably caused by her exposure to massive amounts of radiation. Today, the effects of radiation on health are better known, and suitable safety precautions are taken when using radioactive materials.

Radon Gas

Decay of radium-226 in soil and rock produces radon gas. High concentrations of radon can increase the risk of cancer. The radioactive radon gas can seep through cracks in a home's foundation or it can dissolve in water pumped into the house from a well. Installing a radon-reduction system reduces the concentration of radon gas by using a fan to draw the gas through pipes that vent to the outside of the home.

A radon-reduction system lowers the concentration of radon in homes by venting the radon gas from the home to the outside environment.

REVIEW

13. **Describe** the general trend in first ionization energies in group 2, and explain why this trend occurs.
14. **Explain** What is the charge on alkaline earth metal ions? Explain your answer.
15. **Compare and contrast** the physical properties of the alkaline earth metals and the alkali metals.
16. **Evaluate** why several of the alkaline earth metals are used in different emergency flares.
17. **Analyze** Use the atomic properties of the alkali metals and alkaline earth metals to explain why calcium is less reactive than potassium.
18. **Infer** The alkaline earth metals are usually found combined with oxygen and other nonmetals in Earth's crust. Based on the atomic properties of this group, explain why alkaline earth metals are so reactive.
19. **Calculate** Calcium makes up about 1.5% of a human's body mass. Calculate the amount of calcium found in a person with a total mass of 68 kg.
20. **Calculate** Radium-226 has a half-life of 1600 years. After 8000 years, how much of a 500.0-g sample of radium-226 would be left?

The Discovery of Radioactivity

Background In the late 1800s, Henri Becquerel discovered that uranium gave off some type of rays. Although, in general, the scientific community ignored his findings, Marie Curie decided to study these mysterious rays. Curie used an instrument developed by her husband, Pierre, to measure electrical changes. These measurements helped Curie conclude that the amount of uranium determined the strength of the rays emitted. This study led to the discovery of other substances that gave off rays, such as thorium and polonium. Marie called these rays radioactive for the way they behaved.

Radon Gas

Background According to EPA estimates, radon is the leading cause of lung cancer among people who do not smoke. In 2005, the World Health Organization (WHO) launched an international effort to raise awareness of the link between radon and lung cancer. Approximately half of a person's radon exposure comes from his or her home. Although new construction procedures greatly diminish the amount of radon exposure, simple measures such as increasing floor ventilation and sealing cracks in basement floors can be taken in older homes. Rooms that are below grade and that have contact with the ground pose the greatest risk for exposure.

REVIEW

13. In general, first ionization energies decrease going down the group because the electrons are farther away from the nucleus and easier to remove.
14. 2+; when alkaline earth metals form ions, they lose their two valence electrons
15. The alkaline earth metals are generally harder, denser, and stronger than the alkali metals. Most also have higher melting points and boiling points than the alkali metals.
16. They give off bright light when they burn.
17. Calcium has two valence electrons, and it takes more energy to remove two valence electrons than one. Potassium has only one valence electron and is larger than calcium; therefore, it takes less energy to remove its valence electrons.
18. Alkaline earth metals lose two valence electrons to form ions with a 2+ charge, so they react easily with nonmetals.
19. 1.0 kg
20. 15.62 g

Physical Properties

Background Transition metals share similar properties such as electrical conductivity, luster, and malleability. However, there are differences. For example, silver is the best conductor of electricity. Iron and titanium are used as structural materials because of their relative strength. Vanadium, manganese, and titanium are used to make different kinds of steel alloys.

Common Reactions

Activity Have students make disposable hand warmers. Place 25 g iron powder and 1 g sodium chloride in a small, resealable plastic bag. Shake the bag to mix the chemicals. Add 5 mL water and seal the bag. Knead the bag to thoroughly mix the contents. Heat is produced by the reaction. BL OL

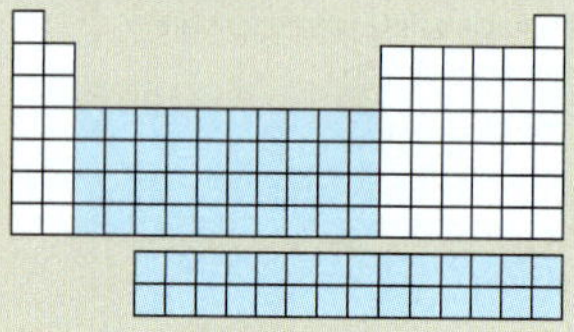

Physical Properties

- The main transition elements include four series of d-block elements with atomic numbers between 21–30, 39–48, 72–80, and 104–112. The inner transition elements include the f-block (rare earth) elements in the lanthanide series (atomic numbers 57–71) and actinide series (atomic numbers 89–103.) All are metals.
- Like many other metals, transition elements are generally good conductors of electricity and heat. They are ductile, which means they can be pulled into wires. Transition metals are also malleable, which means they can be hammered into thin sheets.
- In general, the transition elements have high densities, high melting points, and low vapor pressure. All transition elements are solids at room temperature except for mercury, which is a liquid.
- Strength and availability make transition elements, such as iron, good structural materials.
- Many transition elements reflect visible light at specific wavelengths making some compounds appear brightly colored.
- Transition elements are often paramagnetic, which means they are attracted to an applied magnetic field. Three transition elements—iron, cobalt, and nickel—are ferromagnetic. These elements can form their own magnetic fields.

When exposed to a magnet, iron filings become magnetic and are attracted to the magnet and to each other.

Common Reactions

- **Most transition elements can form stable complex ions and coordinate covalent compounds.** A complex ion is an ion in which a central metal ion is surrounded by weakly bound molecules or ions called ligands.

  Example: Prussian blue, an intense blue pigment used in paints, is a coordinate compound made of iron(III) and an iron(II) cyanide complex: $Fe_4[Fe(CN)_6]_3$.
- **Some transition elements form solutions called alloys.**

  Examples:
  - Brass is an alloy of copper and zinc.
  - Bronze is an alloy of copper and tin.
- **Transition elements and their compounds are often useful as catalysts.**

  Example: Nickel is used as a catalyst in converting unsaturated fats to saturated fats.
- **Transition elements can react with oxygen to form oxides.**

  Example: In the presence of water, iron reacts with oxygen to form rust. The overall reaction is: $4Fe + 3O_2 \rightarrow 2Fe_2O_3$.
- **Some transition elements are important in biochemical reactions.**

  Example: In the protein hemoglobin, iron binds to O_2 to transport oxygen from the lungs to the rest of the body.

Atomic Properties

- The more unpaired electrons in the d sublevel, the greater the hardness and the higher the melting and boiling points. Unpaired d and f electrons produce paramagnetism.
- For transition elements, there is little variation in atomic size, electronegativity, and ionization energy across a period due to shielding. The d electrons shield the outer electrons from an increase in nuclear charge as atomic number increases. Although the number of protons in the nucleus increases across a period, the outer electrons in each atom experience a similar nuclear charge.
- Transition elements can lose s electrons and form ions with a 1+ and/or 2+ charge. Because the s and d electrons are close in energy, many transition elements can also form ions with a charge of 3+ or higher.
- Compounds containing transition elements with partially filled d sublevels absorb specific wavelengths of visible light making compounds appear brightly colored.

Oxidation Numbers of the First Row of Transition Elements

Sc	0			+3				
Ti	0		+2	+3	+4			
V	0	+1	+2	+3	+4	+5		
Cr	0	+1	+2	+3	+4	+5	+6	
Mn	0	+1	+2	+3	+4	+5	+6	+7
Fe	0	+1	+2	+3	+4		+6	
Co	0	+1	+2	+3	+4	+5		
Ni	0	+1	+2	+3	+4			
Cu	0	+1	+2	+3				
Zn	0		+2					

Analytical Tests

Notice the colorful compounds of transition metals in the photo. These compounds absorb different wavelengths of light. The color you see depends on the wavelengths of light that are reflected—the wavelengths least absorbed. Visible spectroscopy uses light absorption at specific wavelengths to measure the concentration of colored compounds in solution. The amount of light absorbed at specific wavelengths relates to the concentration of compound in the sample. Because many transition element compounds are colored, this technique can be used in transition element analysis.

The compounds of transition metals appear colored because of the partially filled d sublevels. The electrons in these sublevels can absorb visible light of specific wavelengths. Compounds with empty or filled d sublevels do not produce brilliant colors.

Martyn F. Chillmaid/Photo Researchers

Atomic Properties

Background The physical properties of transition metals are determined by their electron configurations. Most transition metals are hard solids with relatively high melting and boiling points. Differences in properties among transition metals are based on the ability of unpaired d electrons to move into the valence level. The more unpaired electrons in the d sublevel, the greater the hardness and the higher the melting and boiling points.

Analytical Tests

Activity In addition to identifying different transition elements, color can be used to identify the oxidation state of some transition elements. Label four beakers *2+*, *5+*, *6+*, and *7+*, representing four of the oxidation states of manganese. Place 10 mL of manganese sulfate solution (1 g $MnSO_4$ in 50 mL water) in the 2+ beaker. The solution will be pale pink. Place 10 mL of $MnSO_4$ solution in the 5+ beaker. Add 10 mL of sodium hydroxide solution (25 g NaOH in 25 mL water), and swirl the beaker until the solution turns blue. Place 10 mL of potassium permanganate solution (0.1 g $KMnO_4$ in 100 mL water) in the 6+ beaker. Add 4 mL sodium hydroxide solution (12 g NaOH in 50 mL water), then add 4 mL of sodium sulfite solution (1.0 g Na_2SO_3 in 50 mL water). The solution will turn from purple to green. Place 10 mL of $KMnO_4$ in the 7+ beaker. The solution will be purple.

Lighter but Stronger than Steel

Background Titanium's resistance to corrosion makes it the metal of choice for many applications. Because of it's resistance to salt water, titanium is used to make equipment for desalination plants, saltwater aquariums, fishing, and diving. Titanium and titanium alloys are also resistant to human body fluids. This property makes it ideal for prosthetic devices, such as hip and knee replacements, pacemaker cases, and heart-valve parts.

Strategic and Critical Metals

Background Copper is used in electrical wiring, zinc as a protective coating for other metals, and iron in making steel. Many transition metals are found in alloys used to make items such as jet engines, drill bits, surgical instruments, and armor. The plastics, petroleum, and food industries use transition metals such as platinum, palladium, and nickel to control the conditions at which a reaction will occur.

Titanium
22
Ti
[Ar]$3d^24s^2$

Lighter but Stronger than Steel

The curved surfaces of the Guggenheim Museum in Bilbao, Spain, are covered with 32,000 m^2 of 0.4 mm-thick titanium panels. Titanium's reflective properties give the building a warm look that is ever changing. Titanium is also three times stronger than steel, more resistant to weathering, and weighs less than steel.

The titanium panels that cover the outside of the Guggenheim Museum in Bilbao, Spain, were chosen for the metal's physical properties.

Chromium	Manganese	Cobalt	Tungsten	Platinum
24	25	27	74	78
Cr	**Mn**	**Co**	**W**	**Pt**
[Ar]$3d^54s^1$	[Ar]$3d^54s^2$	[Ar]$3d^74s^2$	[Xe]$4f^{14}5d^46s^2$	[Xe]$4f^{14}5d^96s^1$

Strategic and Critical Materials

Transition metals, such as chromium, manganese, cobalt, and platinum, play a vital role in the economy of many countries because they have a wide variety of uses. As the uses of transition metals increase, so does the demand for these valuable materials. Ores that contain transition metals are located throughout the world.

The United States now imports more than 60 materials that are classified as "strategic and critical" because industry and the military are dependent on these materials.

Iron 26 **Fe** $[Ar]3d^6 4s^2$

Nickel 28 **Ni** $[Ar]3d^8 4s^2$

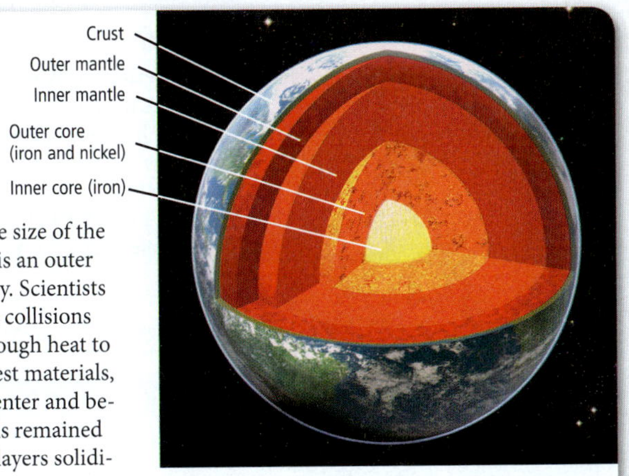

Earth's crust and mantle insulate the hot iron core.

Earth's Iron Core

Earth's core is a solid iron sphere about the size of the Moon. Surrounding the inner core, there is an outer liquid core that contains a nickel-iron alloy. Scientists think the iron core formed when multiple collisions during Earth's early history resulted in enough heat to melt metals. In the molten state, the densest materials, including iron and nickel, settled to the center and became Earth's core. The less-dense materials remained at the surface. As Earth cooled, the outer layers solidified, creating Earth's mantle and crust.

Copper 29 **Cu** $[Ar]3d^{10} 4s^1$

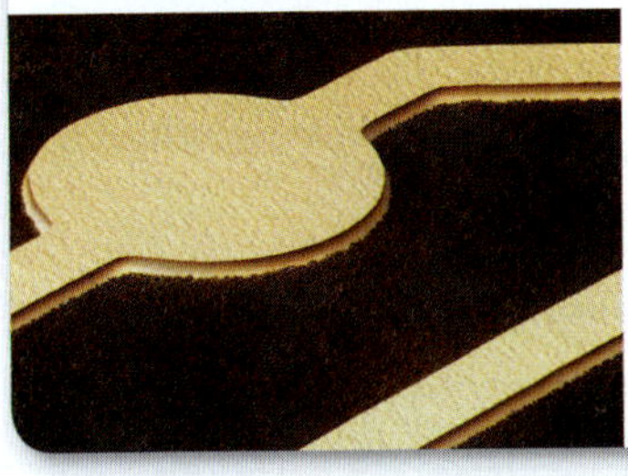

Copper Microchips

For many years, aluminum was used to make computer microchips. Although copper is a better electrical conductor than aluminum, it was not until the late 1990s that the technology existed to use copper in microchips. Combined with the extremely small size of copper wires, this allows copper microchips to be smaller and to operate 25 to 30 times faster than other kinds of microchips. To make wires this small, the copper must be between 99.999% and 99.9999% pure. Copper microchips are part of circuit boards that are used in handheld games, computers, and other electronic devices.

To create a copper microchip, first a layer of tantalum coats a silicon substrate. Then, copper is deposited using a vacuum process.

Titanium 22 **Ti** $[Ar]3d^2 4s^2$

Chromium 24 **Cr** $[Ar]3d^5 4s^1$

Iron 26 **Fe** $[Ar]3d^6 4s^2$

Cobalt 27 **Co** $[Ar]3d^7 4s^2$

Copper 29 **Cu** $[Ar]3d^{10} 4s^1$

Paint Pigments

Paints are a mixture of particles of pigment in a liquid base. Once the liquid evaporates, the pigment particles coat a painted surface. Transition elements and their compounds are often used as paint pigments. Iron oxides are used as red, yellow, and brown pigments. Chromium, copper, and cobalt compounds produce green and blue pigments. Titanium dioxide is often used for white paint.

Artists can create their own paints by mixing dry pigments in a liquid base such as oil, latex, or even egg yolk.

Earth's Iron Core

Activity Have students test aluminum foil, paper clips, coins, hair pins, soda cans, and empty soup cans with a bar magnet and record their observations. Ask students to test as many other items in the classroom as time allows and predict the results before testing each item. Ask students to look at the group of items that were attracted to the magnet. What do these items have in common? Substances that are attracted to the bar magnet are all metals. Plastics, wood and other nonmetal items are all not attracted to a bar magnet. However, some metals are not attracted. Students might wonder why a nickel, which has magnetic character, is not attracted to a bar magnet. A nickel coin contains only 25% nickel. **OL**

Copper Microchips

Background The copper wiring used in computer microchips is less than 1 micron wide. This small size makes the copper interconnects in microchips an example of nanotechnology—technology using structures that are less than 100 nm.

Paint Pigments

Background Ruby red glass has a rich, red color and was first produced by the Egyptians by adding copper to glass. In A.D. 4, the Romans added both gold and silver to glass to produce ruby red glass. Today, most red and pink glass is made from selenium. Incorporating cadmium selenide and zinc sulfide in the mixture makes the glass red. When cadmium sulfide is present, the glass is yellow, but when selenium is added, the glass changes from yellow to orange to a bright ruby red.

Gold Leaf Gilding

Background Copper, silver, gold, platinum, and palladium are the only transition metals that are unreactive enough to be found in nature uncombined with other elements. All other transition metals are found in nature combined with nonmetals in minerals such as oxides and sulfides. Recall that minerals are mixed with other materials in ores.

Touch Sensors for Robot Fingers

Activity Have students work in groups of three. Students will wear work gloves, thin plastic gloves, or no gloves. Tell students that the work gloves simulate the touch capabilities of current robot sensors and that the plastic gloves simulate the nanoparticle sensors. Have students perform an everyday task requiring dexterity, such as threading a nut on a bolt. Which student found it easiest to complete the task? Explain. The student without gloves will be able to perform the task easily because he or she is most able to feel the nut and bolt. OL COOP LEARN

Biotreatment of Acid Mine Wastes

Background Metallurgy, extracting metals from their ores, can be traced back to 6000 B.C. The first tools and weapons, dating to 5000 B.C., were produced from copper. Gold and silver were used for jewelry. Pliny, the Roman scientist and chronicler, described how to purify mercury in 1600 B.C. Around 1200 B.C., wrought iron implements were made. Because of its low melting point, lead was used to make containers and conduits. The predominant metal alloy was bronze, a mixture of tin and copper.

Gold
79
Au
$[Xe]4f^{14}5d^{10}6s^1$

Gilding

Covering an ordinary object with gold foil or gold leaf can make the object look like it is made of solid gold. The process, which is called gilding, has been used for more than 5000 years. To create gold foil, gold is hammered until it is very thin. The thinnest sheets are called gold leaf. They can be as thin as 0.1 mm thick. It takes skill and a special gilder's brush to handle sheets this thin, but the results can be spectacular.

Egyptian King Tutankhamun's coffin was made of wood covered with gold foil. It has lasted more than 3000 years.

Cadmium
48
Cd
$[Kr]4d^{10}5s^2$

Gold
79
Au
$[Xe]4f^{14}5d^{10}6s^1$

Touch Sensors for Robot Fingers

Imagine a surgeon using a robot for microsurgery. In the future, it might be possible for the surgeon to feel what is happening as the robot makes a microsuture. Future robots might use thin, film sensors to mimic the human sense of touch. These sensors are built on a glass base from alternating layers of nanoparticles of gold and cadmium sulfide separated by layers of plastic. The entire sensor is only 100 nm thick and works by transmitting an electroluminescent signal and electric current when regions of the sensor are touched.

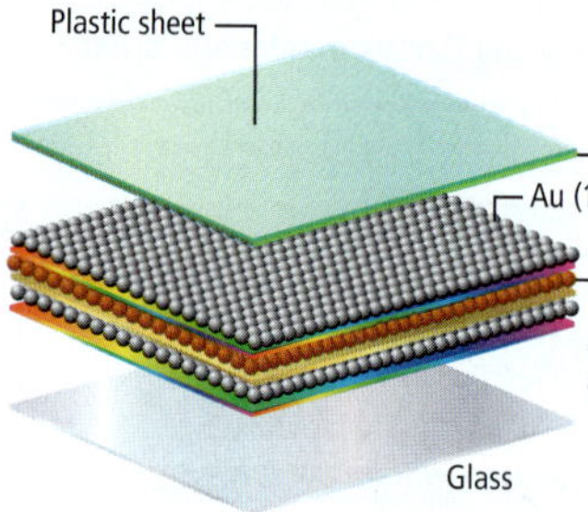

This touch sensor is made from nanoparticles of gold and cadmium sulfide.

Manganese 25 **Mn** $[Ar]3d^54s^2$	Iron 26 **Fe** $[Ar]3d^64s^2$	Copper 29 **Cu** $[Ar]3d^{10}4s^1$	Zinc 30 **Zn** $[Ar]3d^{10}4s^2$	Silver 47 **Ag** $[Kr]4d^{10}5s^1$	Cadmium 48 **Cd** $[Kr]4d^{10}5s^2$

Biotreatment of Acid Mine Wastes

Mining operations can generate acidic wastewater that contain harmful levels of dissolved transition metals, including manganese, iron, copper, zinc, silver, and cadmium. One treatment method uses naturally occurring anaerobic bacteria to remove all of the oxygen. Then sulfate-reducing bacteria convert sulfuric acid in the mine waste to sulfide. Sulfide reacts with metals in the wastewater to form metal sulfide precipitates, which can be recovered and processed for commercial use.

Untreated acid mine drainage can contaminate streams with harmful concentrations of transition metals. The red-orange color of the water comes from iron compounds.

Gadolinium
64
Gd
[Xe]$4f^7 5d^1 6s^2$

Magnetic Resonance Imaging

Gadolinium contrast agents are compounds that enhance differences between normal tissue and abnormal tissue, such as tumors, in magnetic resonance imaging (MRI) scans. The gadolinium compounds are injected directly into the bloodstream prior to an MRI scan. Tumors accumulate more of the gadolinium compounds than normal tissue. Gadolinium enhances MRI images because it is paramagnetic. Magnetic resonance imaging uses a strong magnetic field and radio waves to stimulate water molecules to an excited state. The MRI image is formed as water molecules relax back to their normal state. Gadolinium speeds up the relaxation rate, which improves the contrast between normal and abnormal tissue.

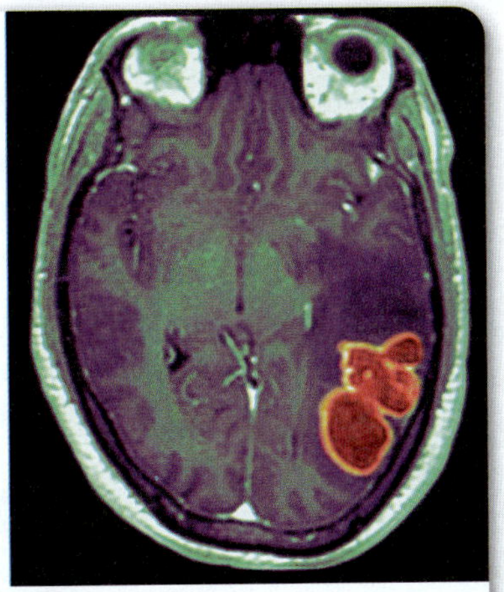

This gadolinium-enhanced MRI scan shows an abcess (red patch) that causes seizures in the patient.

Thorium
90
Th
[Rn]$6d^2 7s^2$

Lawrencium
103
Lr
[Rn]$5f^{14} 6d^1 7s^2$

Reorganizing the Periodic Table

The actinides are a row of radioactive elements from thorium to lawrencium. They were not always separated into their own row in the periodic table. Originally, the actinides were located within the d-block following actinium. In 1944, Glenn Seaborg proposed a reorganization of the periodic chart to reflect what he knew about the chemistry of the actinide elements. He placed the actinide series elements in their own row directly below the lanthanide series. Seaborg had played a major role in the discovery of plutonium in 1941. His reorganization of the periodic table made it possible for him and his coworkers to predict the properties of possible new elements and facilitated the synthesis of nine additional transuranium elements.

Seaborg won the Nobel Prize in Chemistry in 1951 for his work. Element 106, seaborgium, was named in his honor.

REVIEW

21. Compare the electron configurations of the main transition elements and the inner transition elements.

22. Explain how some transition metals can form ions with more than one charge.

23. Identify countries that export only one "strategic and critical" transition metal to the United States.

24. Predict Which elements would you expect to have properties most closely related to gold?

25. Calculate A particular copper-chip manufacturing process specifies that the copper must be 99.999% to 99.9999% pure. Calculate the maximum limit for impurities in the copper in parts per million (ppm).

26. Hypothesize Silver is the best conductor of electricity. Hypothesize why silver is not used for electric wires if it is such a good conductor of electricity.

Magnetic Resonance Imaging

Background Lanthanides are silvery metals with relatively high melting points. Because there is so little variation in properties among inner transition metals, they are found mixed together in nature and are extremely hard to separate. The name of one lanthanide, dysprosium, comes from a Greek word meaning *hard to get at.* The glass in welder's goggles contains neodymium and praseodymium, which absorb high-energy radiation that can damage the eyes. Oxides of yttrium and europium are found in television screens and color computer monitors. An alloy called misch metal, which is 50% cerium, is used by the steel industry to remove carbon from iron and steel. Compounds of lanthanides are used in movie projectors, high-intensity searchlights, lasers, and tinted sunglasses.

Reorganizing the Periodic Table

Background Actinides are radioactive elements. Only three actinides exist in nature. The rest are synthetic elements called transuranium elements. A transuranium element is an element whose atomic number is greater than 92, the atomic number of uranium. Transuranium elements are created in particle accelerators or nuclear reactors. Most transuranium elements decay quickly. One notable exception is plutonium-239. A sample of this isotope can remain radioactive for thousands of years. Plutonium is used as a fuel in nuclear power plants.

REVIEW

21. The final electron of a transition metal enters the d sublevel; the final electron of an inner transition element enters the f sublevel.

22. They can lose two s electrons and form ions with a 2+ charge. When d electrons are lost, ions with a charge of 3+ or higher can form.

23. Venezuela (aluminum), Bolivia (tin), Ireland (zinc), United Kingdom (platinum), Belgium (antimony), Austria (vanadium), Republic of Korea (vanadium), Japan (tantalum), Ukraine (gallium), Kazakhstan (chromium), Gabon (manganese), and Indonesia (tin)

24. copper and silver, because both are in the same group as gold

25. 10 ppm

26. Silver is used in some electric applications, such as certain switches or contacts. However, it is very expensive. Copper conducts electricity almost as well and is less expensive.

Physical Properties

Background One characteristic of metals is that they exhibit luster or shine. Aluminum, a group 13 element, is considered one of the most lustrous metals on the periodic table. Metals are shiny because their electrons move more freely around the nucleus compared to nonmetals. These free electrons readily reflect light which our eyes interpret as shine.

Common Reactions

Background One characteristic that makes aluminum so valuable is that it does not rust. Pure aluminum reacts readily with water; however, aluminum oxide does not. When pure aluminum is exposed to air, it develops a thin layer of aluminum oxide. This layer of aluminum oxide serves as a physical barrier, preventing water from coming in contact with the pure aluminum beneath it.

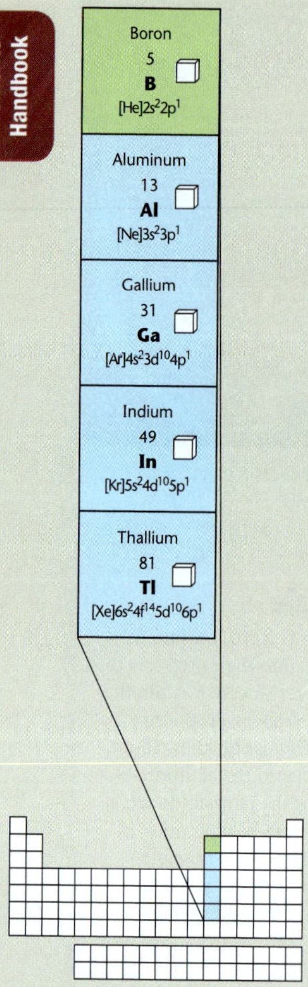

Physical Properties

- Most of the elements in group 13 are metals that have a silvery-white appearance. The exception is boron, which is pure black. Thallium is initially silvery, but oxidizes quickly.
- Boron is a metalloid. The remaining group 13 elements are metals.
- Elements in this group are relatively lightweight and soft, except for boron. Boron is extremely hard—almost as hard as diamond.
- The group 13 elements are solids at room temperature. Gallium melts slightly above room temperature.
- They have higher boiling points than the alkaline earth metals and lower boiling and melting points than the carbon group elements.

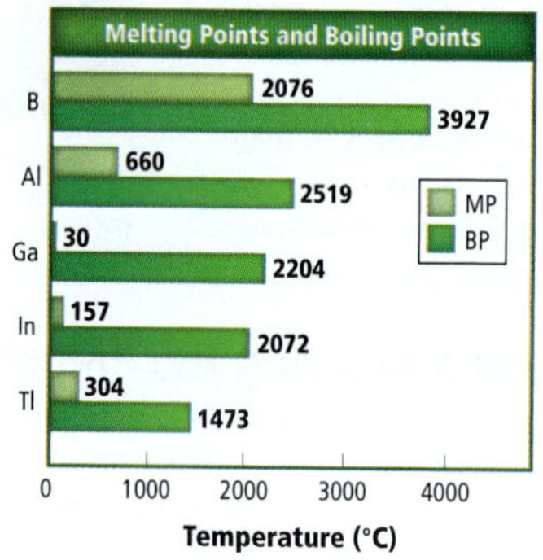

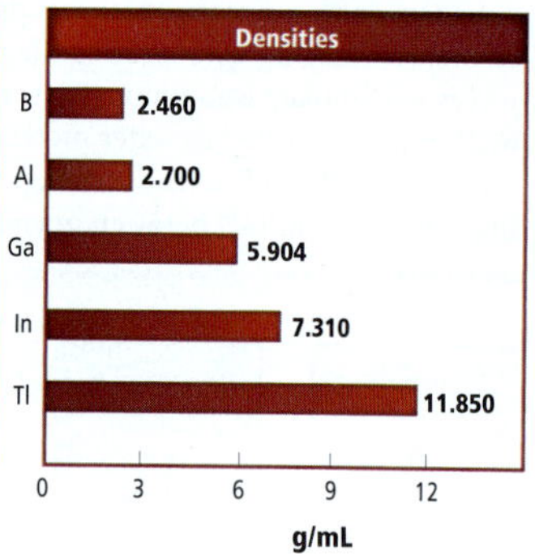

Common Reactions

- **B, Al, Ga, In, and Tl react with oxygen when heated to form metal(III) oxides, such as aluminum(III) oxide.**

  Example: $4Al(s) + 3O_2(g) \rightarrow 2Al_2O_3(s)$
- **B and Al react with nitrogen to form nitrides, such as boron nitride.**

  Example: $2B(s) + N_2(g) \rightarrow 2BN(s)$
- **B, Al, Ga, and In react with halogens to form metal(III) halides, such as gallium(III) fluoride.**

  Example: $2Ga(s) + 3F_2(g) \rightarrow 2GaF_3(g)$
- **Tl reacts with halogens to form metal(I) halides, such as thallium(I) fluoride.**

  Example: $2Tl(s) + F_2(g) \rightarrow 2TlF(s)$
- **Aluminum metal is produced when aluminum(III) oxide reacts with carbon in a redox reaction.**

  Example: $2Al_2O_3(s) + 3C(s) \rightarrow 4Al(s) + 3CO_2(g)$
- **Tl reacts with water to form thallium hydroxide and hydrogen gas.**

  Example: $2Tl(s) + 2H_2O(l) \rightarrow 2TlOH(aq) + H_2(g)$

Atomic Properties

- Each element in group 13 has three valence electrons and an electron configuration ending with ns^2np^1.
- Except for boron and thallium, the group 13 elements lose their three valence electrons to form ions with a 3+ charge. Some of the elements (Ga, In, and Tl) also have the ability to lose just one of their valence electrons to form ions with a 1+ charge.
- Boron participates only in covalent bonding.
- Atomic radii and ionic radii generally increase going down the group and are similar in size to the group 14 elements.
- Except for boron, the group 13 elements have similar first ionization energies.

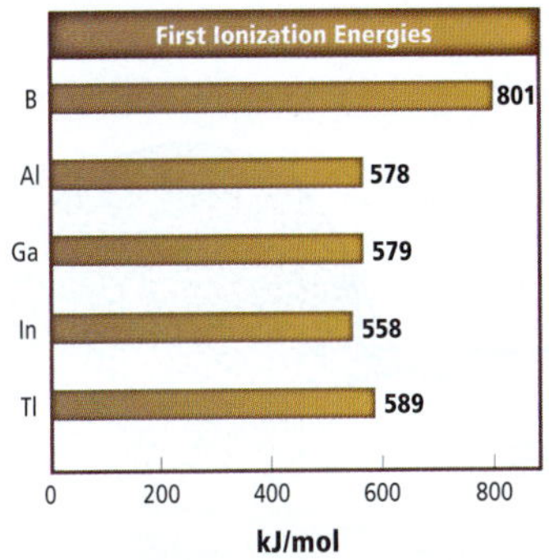

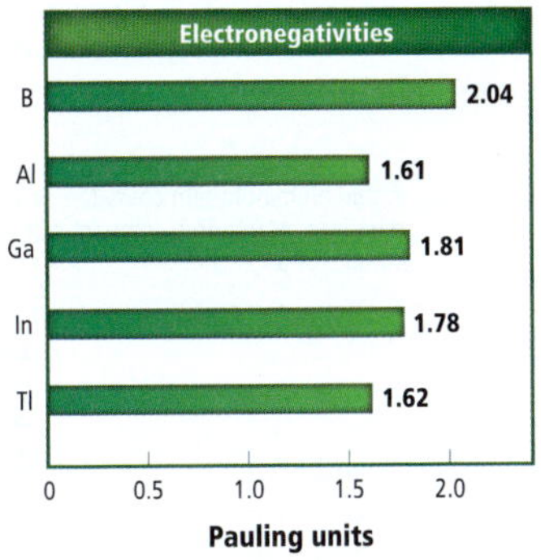

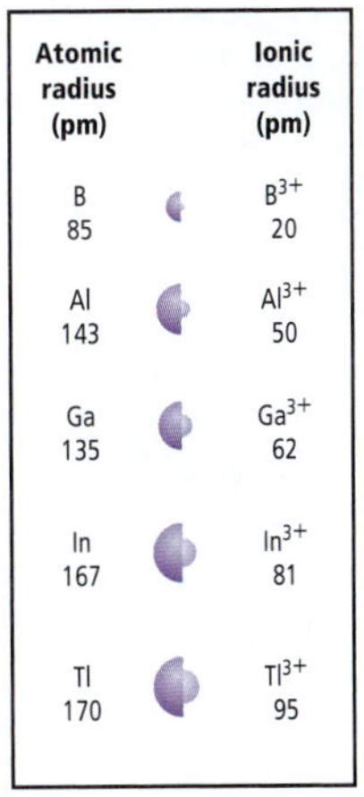

Analytical Tests

With the exception of aluminum, which is one of the most abundant elements in Earth's crust, most of the boron group elements are rare. None of the elements are found free in nature. Three can be identified by flame tests, as shown in the table. Boron produces a bright green color, while indium produces an indigo blue color. Thallium produces a green color. More precise identification methods involve advanced spectral and imaging techniques.

Flame Test Results

Element	Color of Flame
Boron	initial bright green flash
Indium	indigo blue
Thallium	green

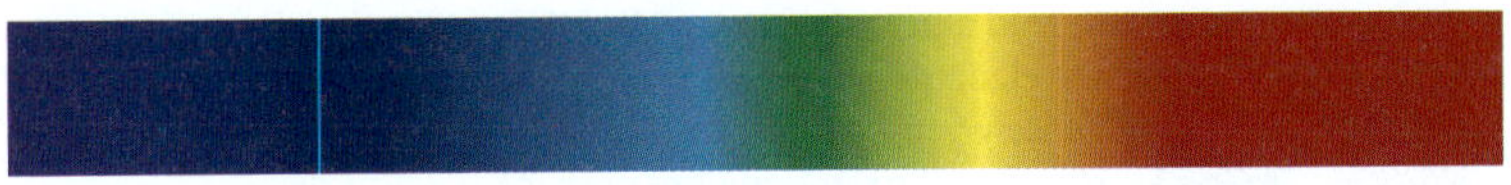

Indium was named after its distinct indigo blue spectral line.

Atomic Properties

Background Boron participates in covalent bonding and can assume chemical structures of varying geometry. One of these structures is the icosahedron. An icosahedron is a geometric solid that is made up of 20 equilateral triangles. This results in 12 vertices. In the boron icosahedron, the vertices represent the position of each boron atom, which corresponds to $B_{12}H_{12}$. Other complex icosahedrons can be formed with boron where not all of the vertices are occupied.

Analytical Tests

Background Colors produced by flame tests are easily masked, and the results are often inconclusive. Scientists use line spectra to more accurately identify elements. Line spectra fall into two categories: emission and absorption. An emission spectrum is the set of frequencies of the electromagnetic waves emitted by atoms of an element. Each element has a unique emission spectrum, which consists of one or more individual lines of color corresponding to the frequencies of the radiation emitted by its atoms. Elements absorb the same specific frequencies of light as the frequencies they emit, which creates an absorption spectrum. In an absorption spectrum, the absorbed frequencies appear as black lines.

Detergent

Background The pH of a wash load of clothes determines how well dirt and stains are removed. An alkaline pH allows the surfaces of dirt particles and clothes to attain negative electrostatic charges that repel one another. Stains that are composed of natural fats and oils also use the alkaline pH to undergo a saponification, or soap-making, process that deposits its product in the wastewater. Borax maintains a proper wash pH by providing a buffering balance of both boric acid H_3BO_3 and its conjugate base $B(OH)_4^-$.

CDs and DVDs

Activity Have students explore how light reflects off a CD. Divide students into groups of two. Give each group a flashlight, white paper, and three CDs. Darken the room and have students shine the flashlight onto the CDs to observe how light reflects off the surface onto the white paper. Have students explore similarities and differences with the CDs. Students should describe their results in writing. Students should observe that light reflects the spectrum of colors because the light is bouncing off an uneven surface (the grooves in the CD). Students might also notice that scratched CDs do not produce as clear of a spectrum. **OL** **COOP LEARN**

HD DVDs

Background Japanese researcher Shuji Nakamura was almost solely responsible for developing the technology that lead to blue lasers. Working alone in the research and development department of a small Japanese company, Nakamura developed blue light-emitting diodes (LEDs). This lead to the development of green LEDs, white LEDs, and blue lasers.

Boron
5
B
$[He]2s^2 2p^1$

Detergent

Sodium perborate ($NaBO_3{\cdot}H_2O$ or $NaBO_3{\cdot}4H_2O$) is one of the key ingredients in powdered laundry detergent. The hydrate, formed by combining borax pentahydrate ($Na_2B_4O_7{\cdot}5H_2O$) with hydrogen peroxide and sodium hydroxide, releases oxygen during the laundering process to help make clothes whiter and brighter. Sodium perborate is the chemical of choice because it remains stable over long periods of time, helps maintain wash water pH, and increases the solubility of detergent ingredients.

Many powder laundry detergents contain boron compounds that help make clothes cleaner.

Aluminum
13
Al
$[Ne]3s^2 3p^1$

A thin aluminum film coats the depressions embedding information in a compact disc and makes the surface of a CD shiny.

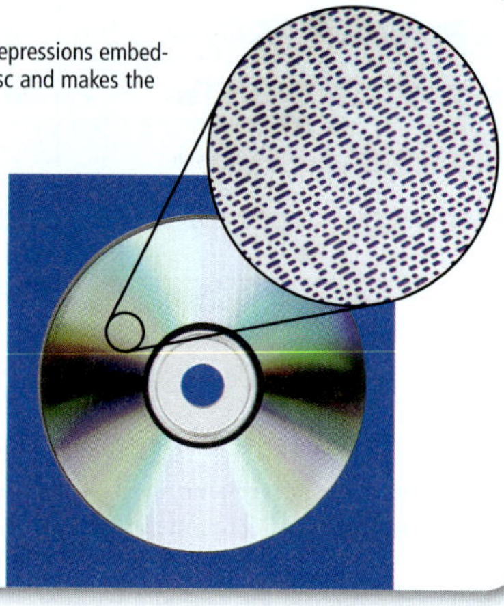

CDs and DVDs

Have you ever wondered what your CDs and DVDs are made of? The inside is made of plastic, about 1 mm thick. A machine embeds digital information, such as sound recordings, into the plastic as a series of bumps and then coats the plastic with aluminum. That is what makes CDs and DVDs so shiny. A thin layer of acrylic protects the aluminum. The shiny surface allows the laser from the CD or DVD player to read the information reflected off the disc's surface.

Gallium
31
Ga
$[Ar]4s^2 3d^{10} 4p^1$

HD DVDs

Videos in high-definition (HD) have higher quality sound and pictures than regular DVDs. However, HD technology requires more information than can be stored on regular DVDs. A red laser is used to read and write data on a regular DVD. Blue lasers made from gallium nitride (GaN) are used to read and write data on HD DVDs. Blue light has a shorter wavelength than red light, so a blue laser can read more densely packed information, allowing more information to be stored in the same amount of space.

HD DVDs store up to 50 gigabytes (GB) of information, compared to 4.7 GB on a regular DVD.

Indium
49
In
$[Kr]5s^24d^{10}5p^1$

Flat-Screen Televisions

Known as ITO in the electronics industry, indium-tin oxide has proven to be the cornerstone of liquid crystal display (LCD) technology. During production, a thin layer of indium-tin oxide (a mixture of In_2O_3 and SnO_2) is used to coat the glass contained within an LCD flat-screen panel. This allows the glass to be both conductive and transparent. About half of the world's indium is used to make LCDs.

Indium-tin oxide is one of the main components in LCD flat-panel televisions.

Thallium
81
Tl
$[Xe]6s^24f^{14}5d^{10}6p^1$

Cardiac Scans

Thallium-201 is a radioisotope used by medical professionals to determine the health of a person's heart. During a thallium-201 scan, also called a heart stress test, a patient performs physical activity and is injected with thallium-201 one to two minutes before stopping the activity. The isotope emits gamma rays that are recorded by a detector to display a two-dimensional image of the heart and its blood supply. If gamma rays are not detected in certain areas in and around the heart, the areas are considered "cold." This means that the blood supply has been impeded or blocked, a condition that often leads to heart attack or stroke.

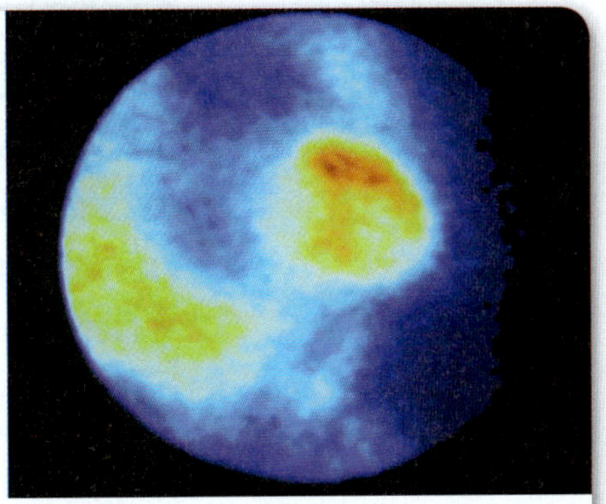

The blue areas in this thallium-201 scan of a heart are areas with low blood supply.

REVIEW

27. **Describe** how the properties of boron are different from the other group 13 elements.
28. **Identify** what an unknown element would be if it produced a green flash of color at the beginning of a flame test.
29. **Describe** any trends in the first ionization energies of the group 13 elements.
30. **Explain** why HD DVDs can store more information than regular DVDs.
31. **Summarize** how "cold" areas in thallium-201 scans could correspond to artery blockages.
32. **Calculate** It is estimated that 123,000 aluminum cans are recycled each minute. Assume that each can has a mass of 14 g. Determine how much aluminum (kg) is recycled during the month of September.

Flat Screen Televisions

Background The gallium nitride in LEDs and indium-tin oxide in flat screen televisions are both semiconductors. A semiconductor is a solid or liquid that can conduct an electrical current at room temperature better than an insulator but not as readily as a metal. Impurities can be added to a semiconductor to increase its ability to carry an electric charge—a process known as *doping*.

Cardiac Scans

Background Nuclear medicine involves the use of radioactive isotopes to study specific regions of the body. During a scan, a device known as a gamma camera uses crystals to capture the scintillations given off by the isotope. The crystals display the nuclear energy in the form of light. A machine then converts this light energy into signals that can be interpreted by a computer. The result is an image that can be electronically colored to study functionality.

REVIEW

27. Boron is a black metalloid that is hard and forms covalent bonds. The rest of the elements are silvery, soft metals that form ionic bonds.
28. By reading the Analytical Tests table, students should see that the element would be boron.
29. Except for boron, group 13 elements have similar first ionization energies.
30. Blue lasers are used to write information on HD optical discs, while red lasers are used to write information on regular DVDs. Because blue light has a shorter wavelength than red light, a blue laser is more precise, and more information can fit in the same amount of space.
31. Thallium-201 is a radioactive tracer that is injected into a person's blood stream. While it is traveling through the blood stream, it emits gamma rays that are detected by a machine. A "cold" spot occurs where no gamma rays are detected. That means no blood is flowing to that area. An arterial blockage is then suspected.
32. 7.43×10^7 kg

Physical Properties

Background Tin has two allotropes, white tin and gray tin. White tin has metallic properties and is used in bakeware as a thin interior coating on steel. Gray tin has no metallic properties and is a dull gray, powdery material with no known uses. When white tin is at temperatures below 13°C, it slowly changes into gray tin. This change from white tin to gray tin is known as tin plague or tin pest.

Common Reactions

Activity Many polymers are made of hydrocarbons. Have students make a polymer out of sodium polyacrylate. Students should use a wooden craft stick or metal scoop to measure out 5 g of sodium polyacrylate. Have students pour the chemical into the bottom of a clear plastic cup. Have students add 10 mL of water and record their observations. (Encourage students to touch the resulting polymer and record their observations.) Students should notice that the product is a gel-like substance that absorbs water so that when the cup is inverted the water does not spill. Have students increase the amount of water in small increments until the polymer can hold no more liquid (or until the cup is full). Then have students add small pinches of salt until they observe a change. Have students record and explain the results. Students should see water seeping back out of the polymer; the polymer will look as if it is deflating. Salt causes the water to leave by changing the osmotic pressure of the system. BL OL

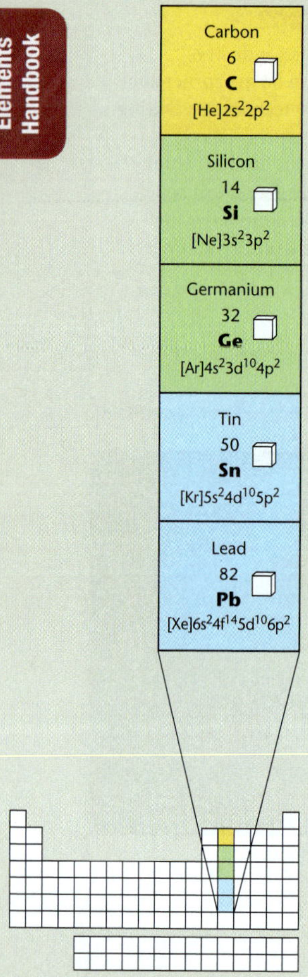

Physical Properties

- Elements in the carbon group increase in metallic character going down the group. Carbon is a nonmetal. Silicon and germanium are metalloids. Tin and lead are metals.
- Carbon can be a black powder; a soft, slippery gray solid; a hard, transparent solid; or an orange-red solid.
- Silicon can be a brown powder or a shiny-gray solid.
- Germanium is a shiny, gray-white solid that breaks easily.
- Tin also occurs in two forms. One form is a silvery-white solid, while the other is a shiny-gray solid. Both forms are ductile and malleable.
- Lead is a shiny-gray solid. It is soft, malleable, and ductile.
- Moving down the group, melting and boiling points decrease and densities increase.

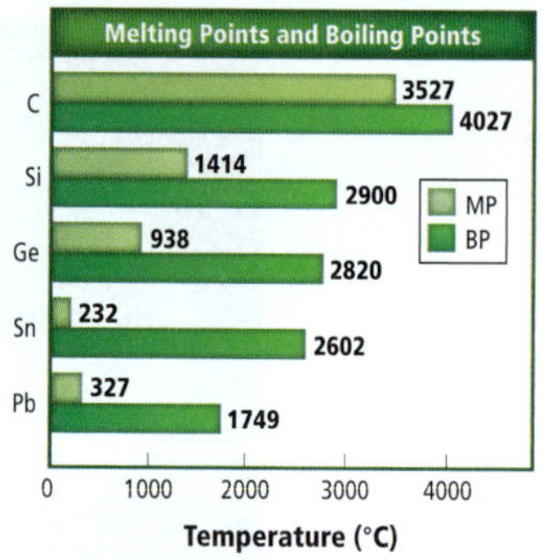

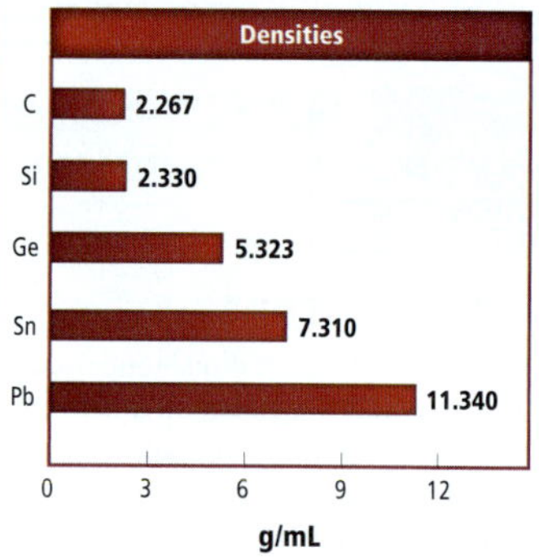

Common Reactions

At room temperature, carbon group elements are generally unreactive. Reactions do occur under elevated temperature conditions.

- **C, Si, Ge, and Sn react with oxygen to form oxides, such as carbon dioxide.**

  Example: $C(s) + O_2(g) \longrightarrow CO_2(g)$

- **C, Si, Ge, and Sn react with halogens to form halides, such as silicon chloride.**

  Example: $Si(s) + 2Cl_2(l) \longrightarrow SiCl_4(g)$

- **Sn and Pb react with bases to form hydroxo ions and hydrogen gas.**

  Example:

  $$Sn(s) + KOH(aq) + 2H_2O(l) \longrightarrow K^+(aq) + Sn(OH)_3^-(aq) + H_2(g)$$

Silicon chloride ($SiCl_4$) reacts with water to form silicon dioxide and hydrochloric acid, which turns litmus paper pink.

Andrew Lambert Photography/Science Photo Library/Photo Researchers

Atomic Properties

- Each element in group 14 has four valence electrons and an electron configuration ending with ns^2np^2.
- Carbon group elements participate in covalent bonding with an oxidation number of +4. Carbon, germanium, tin, and lead can also have an oxidation number of +2. Carbon and silicon have an oxidation number of −4 in some compounds.
- Carbon, silicon, and tin occur as allotropes.
- Atomic and ionic radii increase moving down the group and are similar to their corresponding group 13 elements.
- Except for carbon, the group 14 elements have similar ionization energies and no distinct pattern of electronegativities.

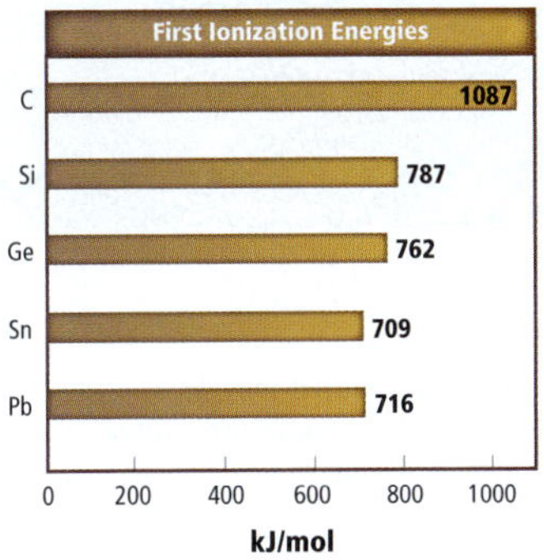

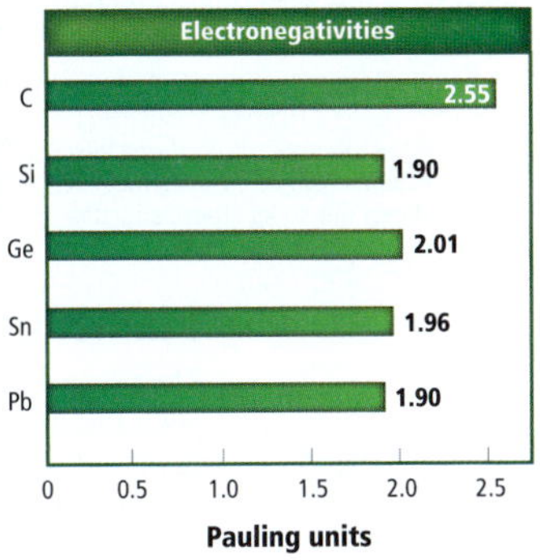

Atomic radius (pm)	Ionic radius (pm)
C 77	C^{4+} 15
Si 118	Si^{4+} 41
Ge 122	Ge^{4+} 53
Sn 140	Sn^{4+} 71
Pb 146	Pb^{4+} 84

- **C reacts with water to form carbon monoxide and hydrogen gas.**

  Example: $C(s) + H_2O(g) \rightarrow CO(g) + H_2(g)$
- **Si reacts with water to form silicon dioxide and hydrogen gas.**

  Example: $Si(s) + 2H_2O(l) \rightarrow SiO_2(s) + 2H_2(g)$
- **Sn and Pb react with acids to form hydrogen gas.**

  Example: $Pb(s) + 2HBr(aq) \rightarrow PbBr_2(aq) + H_2(g)$
- **C reacts with hydrogen to form hydrocarbons, such as propane.**

  Example: $3C(s) + 4H_2(g) \rightarrow C_3H_8(g)$

David Taylor/Photo Researchers

Analytical Tests

Because the group 14 elements bond covalently, they do not lend themselves to identification through flame tests. The exception is lead, which produces a light-blue color. The carbon group elements can be identified through analysis of their physical properties (melting point, boiling point, density), emission spectra, or reactions with other chemicals. For example, tin and lead form precipitates when added to specific solutions.

If lead nitrate is added to potassium iodide, a yellow precipitate of lead iodide forms.

Atomic Properties

Background Students might have difficulty understanding why some groups include metals, metalloids, and nonmetals. Review ionization energy and atomic radii. Point out the correlation of these periodic trends with increasing metallic character. As the ionization energy decreases and the atomic radius increases, atoms are more likely to lose electrons and are considered metals. Also, review electron affinity, and remind students that as the electron affinity increases, atoms are more likely to gain electrons and are considered nonmetals.

Analytical Tests

Activity Perform the precipitate reaction featured in the photo. Place 10 mL of a 1*M* solution of potassium iodide in a 100 mL beaker. Add 10 mL of 1*M* lead nitrate to the beaker. A yellow precipitate will form.

ELEMENTS HANDBOOK

Graphite Golf Shafts

Background Graphite is one of the softest minerals, yet it is a component of many high-strength alloys. The key to graphite's properties lies in its crystal structure. The carbon atoms bond covalently into hexagons that form thin sheets. Layers of sheets are weakly held together by van der Waals forces. These weak bonds enable the layers to slide over each other, resulting in graphite's softness. However, when the sheet layers are rolled into fibers and bonded together with another material, a high-strength composite results.

Diamond Cutting

Activity In a darkened room, have students shine the light from a flashlight or window through the side of a prism and onto a sheet of white paper. What happens to the light as it travels through the prism? The light breaks apart into a rainbow of color. Tell students that the facets on a diamond act like a prism and disperse white light into a spectrum of color. **OL**

Nanotubes

Activity Divide students into groups of three. Have each group make a model of buckminsterfullerene. Students can make ball-and-stick models, gumdrop and toothpick models, or paper, straw, and tape models. Upon construction completion, have students use their models to explain to you why the structure is well suited for nanotubes. **COOP LEARN**

Elements Handbook

Carbon
6
C
$[He]2s^22p^2$

Graphite Golf Shafts

Some golf shafts are created by fusing sheets of graphite together with a binding material. The use of graphite instead of traditional steel allows greater versatility in club design and construction. Graphite sheets can be layered to vary the weight and stiffness of the club, which for many golfers translates into greater shot distance and overall performance. Graphite also offers greater durability than steel for golfers with powerful swings.

Graphite can be easily formed into sheets due to its atomic structure.

Diamond Cutting

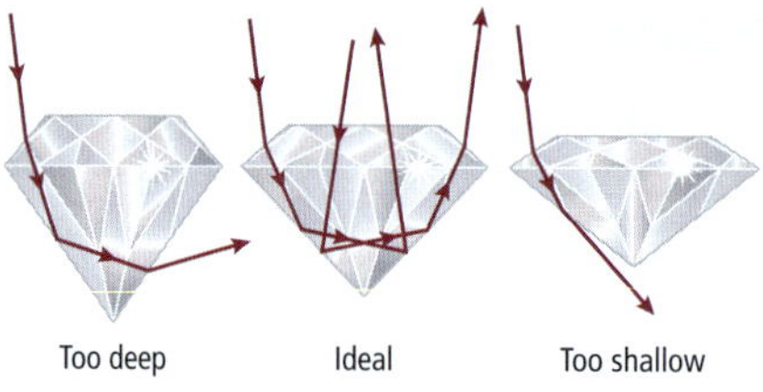

The way a diamond is cut determines how well light is reflected and refracted within the gemstone.

The way a diamond is cut is one of the "4 Cs" that gemologists use to determine a diamond's value. If diamond is the hardest mineral on Earth, then how is it possible to cut a diamond? Diamond cutters use other diamonds and lasers to create facets that reflect and refract light. The more precisely the cuts are made, the greater the gem's brilliance. If a diamond cut is too shallow or too deep, light escapes from the diamond without traveling back to the eye, resulting in a lackluster appearance.

Nanotubes

Fullernes form a group of carbon allotropes. There are spherical fullerenes nicknamed buckyballs and cylindrical fullerenes known as buckytubes or nanotubes. Fullerenes have yet to display all of their capabilities to scientists. One of the most promising areas of fullerene research involves the creation of nanotubes. Nanotubes are sheets of carbon that are rolled up into cylinders. These cylinders are strong—due to the hexagonal structure of the carbon atoms—and have unique conducting properties. Fullerene nano-technology on the horizon includes the development of faster computer chips, smaller electronic components, and more advanced space-exploration vehicles.

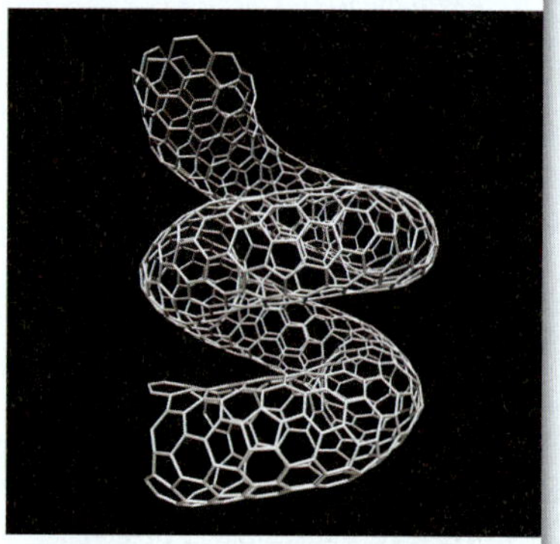

The hexagonal structure of carbon atoms gives extraordinary strength to carbon nanotubes.

(tl)Chemical Design/Science Photo Library/Photo Researchers, (tr)Adrian Muttitt/Alamy, (b)Dr Tim Evans/Science Photo Library/Photo Researchers

Silicon
14
Si
$[Ne]3s^23p^2$

Computer Chips

Computer chips are everywhere. From pet-identification systems to laptop computers—any device that can be programmed contains a computer chip. Silicon's abundance and ability as a semiconductor make it an ideal material for the production of computer chips. The first step in making a computer chip involves cutting pure silicon into wafer-like pieces. Silicon dioxide (SiO_2) is then cultivated on each wafer. Layers upon layers of silicon dioxide and other chemicals are used to create chips for specific functions.

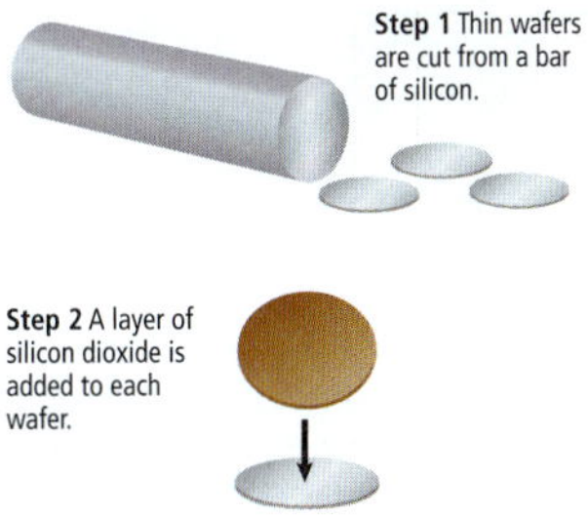

More than 250 steps are needed to create one computer chip.

Glass

Almost 40% of the sand produced in the United States is used for glass production. Glass is created by first melting silicon dioxide (SiO_2) obtained from sand with sodium carbonate and then supercooling the mixture. This results in a solid whose structure resembles a liquid and whose physical properties make it ideal for glassmaking. For manufacturing purposes, sand that yields at least 95% SiO_2 with no impurities is required for making glass products, such as exterior panels on buildings, automotive windshields, and commercial beverage containers. Manufacturers of high precision optical instruments, such as telescopes and microscopes, require sand that contains more than 99.5% SiO_2.

Sand dunes in Michigan provide millions of metric tons of sand each year.

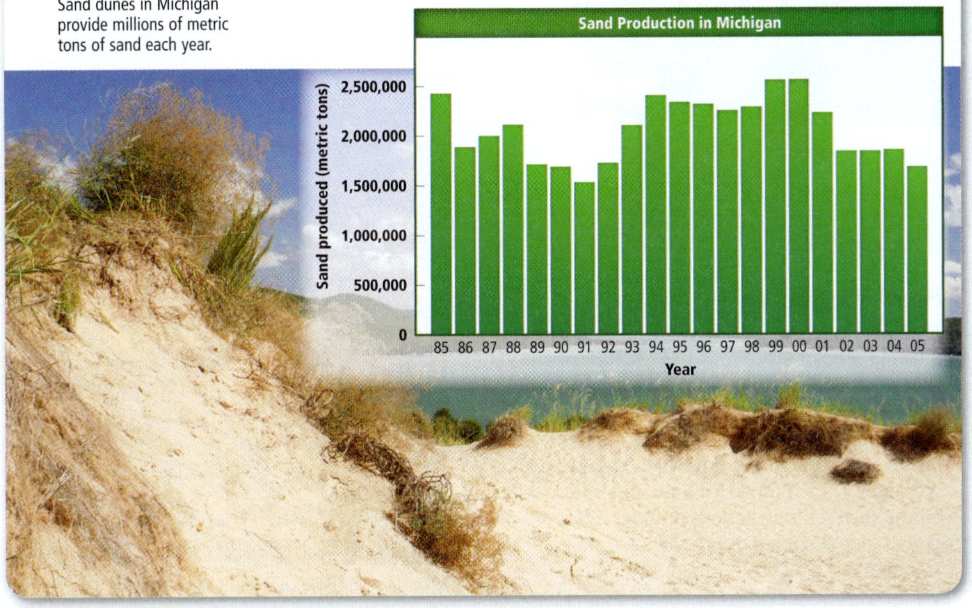

gabes1976/iStock/Getty Images

Computer Chips

Activity In 2004, the Food and Drug Administration approved the implantation of computer chips in humans. Have students research background information on the topic and have a debate on the issues surrounding the procedure. Have students reflect upon the technological, medical, societal, and personal implications as they prepare their arguments. **OL**

Glass

Background Glass making dates back to 2500 B.C. in Mesopotamia, where glass was used to make beads, architectural decorations, and seals. Glass is formed from sand, which is composed of sodium carbonate, limestone, and silicon dioxide. To make glass, sand is melted and then held in the molten state for more than 24 hours to allow the air bubbles to escape. The molten mixture is poured into molds or shaped and allowed to cool. As it cools, the glass solidifies into the desired shapes.

Night Vision

Background The equipment used for night vision captures radiation emitted in the infrared region of the electromagnetic spectrum. The infrared region consists of wavelengths between 750 nm and 10^6 nm. Night vision equipment that accesses the infrared (IR) wavelengths near that of the visible spectrum, such as near infrared (NIR), amplifies the radiation that is reflected off objects. It does so in such a way that the resulting radiation becomes visible to humans. Other equipment accesses the longer IR wavelengths that produce heat. These thermal imaging mechanisms collect the radiation emitted by objects and assemble it into a thermogram that displays hot and cold areas.

Fiber Optic Cables

Activity Darken the classroom and shine a laser on one end of a fiber-optic cable. Have students notice that no light is refracted out of the cable. Then show the students the other end of the cable. Students should see it glowing red from the transmitted laser light. **OL**

Food Packaging

Activity Have students keep a log of how many canned food items they use in a week. Total the class data and have students make a pie chart depicting the percentage of each category (fruit, vegetable, and so on). Students can add drawings, can labels, or magazine images to the pie chart. **BL** **OL** **EL**

Germanium
32
Ge
$[Ar]4s^23d^{10}4p^2$

Night Vision

Lenses that contain germanium are found in an array of night vision equipment including goggles, binoculars, and cameras. Unlike ordinary glass lenses, germanium-containing lenses are transparent to infrared radiation. Infrared radiation is emitted by objects that radiate heat. Infrared radiation is part of the electromagnetic spectrum, a region distinct from the visible spectrum, so special equipment is needed to detect it. Night vision is used for military and security applications, to monitor wildlife, to navigate roads, and to locate objects that have been hidden by criminals.

A night scope is used to observe a military transport plane that takes off with lights out in a conflict zone.

Fiber Optic Cables

Fiber optic cables are responsible for the transmission of information both across the street and across the globe. These cables are made of extremely pure glass that allows light signals to travel the span of the cable without losing a significant amount of energy. Each fiber optic cable consists of three main parts: a core, cladding, and a buffer coating. The core is made by exposing gaseous germanium tetrachloride ($GeCl_4$) to oxygen, resulting in germanium dioxide (GeO_2). The germanium dioxide helps the light signal move effectively along the cable.

Germanium is added to the core of a fiber optic cable to improve the efficiency of the light signal.

Tin
50
Sn
$[Kr]5s^24d^{10}5p^2$

Food Packaging

A quick trip to the grocery store reveals that many different foods are stored in cans. Soft drinks, fruits, vegetables, and even meats can be stored in cans. Cans are made from sheets of steel that are coated on both sides with pure tin. Known as tinplate, the metal is both durable and resistant to rusting and corrosion. These properties allow foods to stay fresh on the shelf for long periods of time, and to be transported long distances. More than 200 million cans are used per day in the United States alone.

More than 2500 different products are packaged in cans.

(t)Everett Collection Inc./age fotostock, (c)JUPITERIMAGES/GOODSHOOT/Alamy, (b)Allan H Shoemake/Taxi/Getty Images

Lead
82
Pb
$[Xe]6s^24f^{14}5d^{10}6p^2$

Leaded or Unleaded?

In the early 1900s, the automotive industry needed to solve a problem that people complained about when they drove their cars—knocking in the engine. At the time, little was known about the chemistry of fuels and fuel additives. Researchers spent seven years searching for a gasoline additive that effectively reduced knocking before discovering tetraethyl lead ($Pb(C_2H_5)_4$). Further research revealed the health and environmental risks posed by lead, leading to the development of unleaded fuels that reduce knocking.

Unleaded fuels reduce knocking in car engines and do not have the health and environmental concerns posed by leaded fuels.

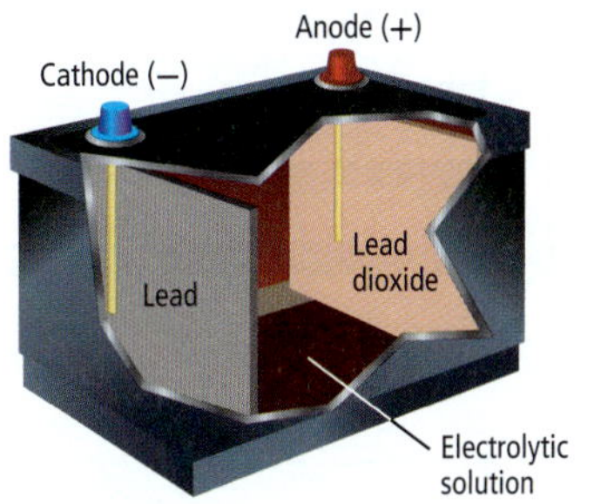

Eighty-five percent of the lead used in the United States goes into making lead-acid batteries.

Batteries

A car battery is composed of three main parts: one electrode made of lead, one electrode made of lead dioxide (PbO_2), and an electrolytic solution made with sulfuric acid (H_2SO_4). That is why car batteries are also called lead-acid batteries. The battery's energy comes from the chemical reactions occurring between the electrodes and the electrolyte. During the chemical reaction, electrons are produced that accumulate on the lead electrode. When a wire connects the electrodes, electrons flow freely from the lead electrode to the lead-dioxide electrode, and the battery discharges. Applying a current reverses the reaction, recharging the battery.

REVIEW

33. **Write** the electron configuration of tin.
34. **Summarize** the physical properties of the elements in group 14.
35. **Compare and contrast** the atomic properties of the group 13 and group 14 elements.
36. **Predict** what product or products will be formed if bromine gas reacts with solid carbon under elevated temperature conditions.
37. **Consider** why graphite is the most suitable carbon allotrope for golf clubs.
38. **Calculate** Pure diamond has a density of 3.52 g/cm^3, while graphite has a density of 2.20 g/cm^3. Recall that density = mass/volume. Samples of diamond and graphite each displace 4.60 mL of water. What is the mass of each sample?

Leaded or Unleaded?

Background A car engine works by compressing gasoline in its cylinders and igniting it with spark plugs. An octane rating describes how well a given gasoline can withstand compression before it will prematurely ignite. Higher octane gasoline can withstand greater compression than lower octane gasoline. Many myths surround the concept of octane ratings. Higher octane gases do not clean car engines better, do not provide more power to the vehicle, and do not offer better gas mileage. What is true, is that it is important to match the fuel type to the manufacturer's engine specifications.

Batteries

Background During the mid-1700s, scientists Luigi Galvani and Alessandro Volta had differing theories about the chemical origins of electricity. Galvani, who performed experiments on the muscles of frogs, thought that something within the frog itself produced the electricity. Volta believed that it had to do with the metals that were used in the frog experiments. This led Volta to create the first electrochemical cell. Known as a voltaic pile, it had an alternating stack of zinc and copper disks separated by cloth that had been immersed in brine. A wire connected to both ends produced a continuous current of electricity. Although Volta's findings temporarily dimmed the work of Galvani, research done years later revealed the electrical nature of nerve impulses.

REVIEW

33. $[Kr]5s^24d^{10}5p^2$
34. Student answers will vary. Carbon is the only nonmetal and has many allotropes, including graphite, diamond, amorphous carbon, such as coal, and fullerene. Silicon and tin have two allotropes. Silicon, germanium, tin, and lead all exhibit some degree of luster and have colors that range from silver-white to gray. Silicon and germanium are metalloids. Tin and lead are metals. Melting and boiling points decrease and densities increase going down the group.
35. The group 13 elements have 3 electrons in their outermost energy level, while the group 14 elements have 4 electrons in their outermost energy level. Atomic and ionic radii in group 14 are similar to the corresponding group 13 elements. First ionization energies of the group 13 elements are less than the corresponding group 14 elements.
36. carbon tetrabromide (CBr_4)
37. The molecular structure of graphite creates sheets that can be folded and shaped. The diamond and fullerene allotropes, while durable, could not provide the flexibility needed for the motion of golf swings.
38. diamond: 16.2 g; graphite: 10.1 g

Physical Properties

Activity Have students measure the mass of a small beaker, then place a mothball in the beaker and measure its mass. Students should measure the mass of the mothball at the end of class and at the beginning and the end of class for three or four days. Students should record their data and observations. Students should notice that the mothball loses mass but does not turn into a liquid. Ask students for other examples of sublimation. Students might suggest dry ice or ice cubes in the freezer. OL

Common Reactions

Background Nitroglycerin ($C_3H_5N_3O_9$), a very unstable oil, was accidentally discovered by an Italian chemist named Ascano Sobrero. Sobrero mixed nitric acid with the organic compound glycerol. The compound was so unstable that any movement or jarring could cause it to instantaneously explode. One such explosion in the laboratory scarred Sobrero, who then deemed it as too dangerous to work with. Undaunted, Alfred Nobel continued the research to later discover that additives could be mixed with nitroglycerin to make it more stable. This new form was called dynamite.

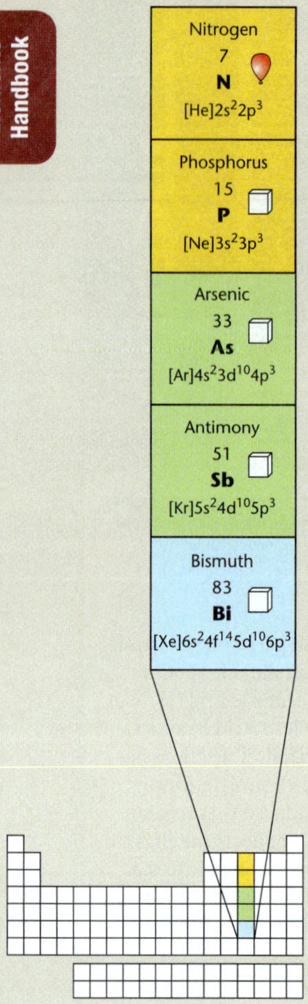

Physical Properties

- Like the elements in group 14, the group 15 elements increase in metallic character going down the group. Nitrogen and phosphorus are nonmetals. Arsenic and antimony are metalloids. Bismuth is a metal.
- Also like group 14, the nitrogen group elements vary in appearance.
- Nitrogen is a colorless, odorless gas (N_2).
- Phosphorus exists in three allotropic forms, which are all solids. The forms are white, red, and black in color.
- Arsenic is a shiny, gray solid that is brittle. Under certain conditions, it can become a dull, yellow solid. Arsenic sublimates when heated.
- Antimony is a shiny, silver-gray solid that is very brittle.
- Bismuth is a shiny, gray solid that has a pink cast to it. It is one of the least conductive metals on the periodic table and is also brittle.
- Boiling points and densities of the group 15 elements generally increase going down the group.

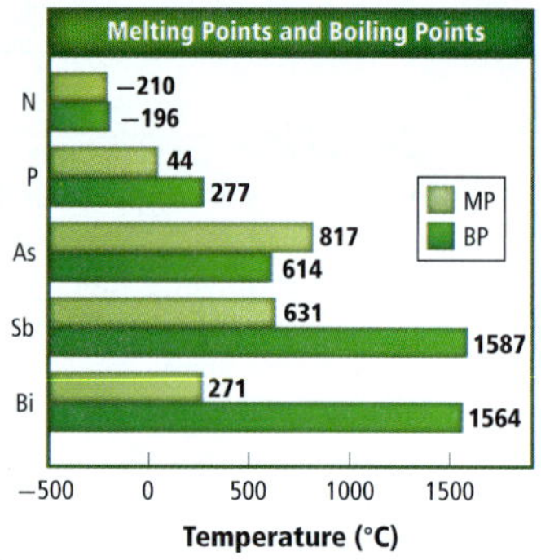

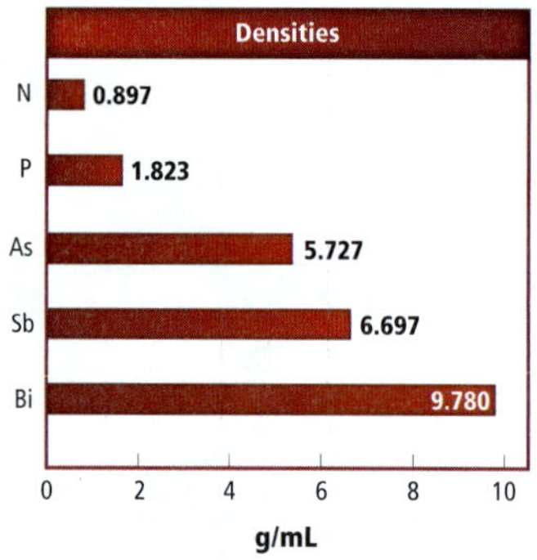

Common Reactions

- **As high temperatures are increased, nitrogen reacts with oxygen to form nitric oxide.**

  Example: $N_2(g) + O_2(g) \rightarrow 2NO(g)$
- **At high temperature and pressure, nitrogen reacts with hydrogen to form ammonia.**

  Example: $N_2(g) + 3H_2(g) \rightarrow 2NH_3(g)$
- **P reacts with an excess of oxygen to form phosphorus(V) oxide.**

  Example: $P_4(s) + 5O_2(g) \rightarrow P_4O_{10}(s)$
- **P, As, Sb, and Bi react with oxygen to form element(III) oxides.**

  Example: $P_4(s) + 3O_2(g) \rightarrow P_4O_6(s)$
- **P, As, Sb, and Bi react with halogens to form trihalides.**

  Example: $2Sb(s) + 3Cl_2(g) \rightarrow 2SbCl_3(s)$

Atomic Properties

- Each element in group 15 has five valence electrons and an electron configuration ending with ns^2p^3.
- Nitrogen is diamagnetic, meaning it is repelled by magnetic fields. This indicates that all of nitrogen's electrons are paired.
- Nitrogen can have oxidation numbers ranging from −3 to +5.
- Phosphorus, arsenic, and antimony can have oxidation numbers of −3, +3, and +5.
- Bismuth commonly has an oxidation number of +3.
- Going down the group, first ionization energies and electronegativities decrease and atomic radii increase.

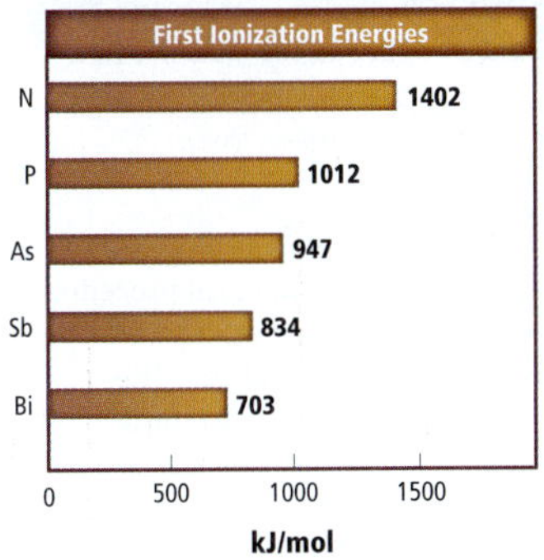

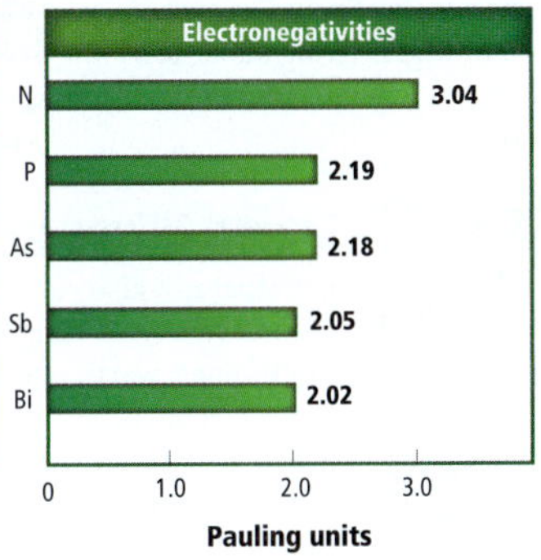

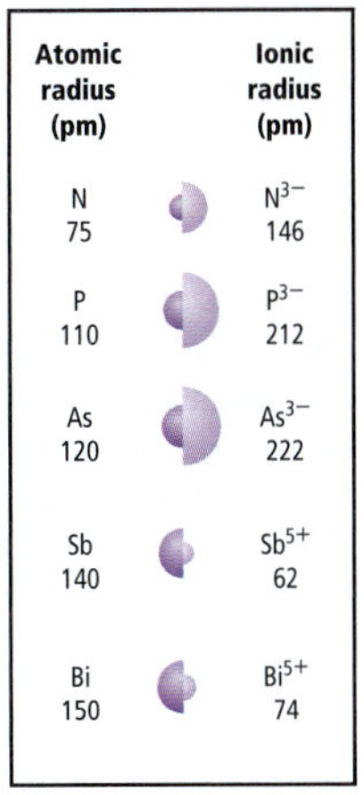

Analytical Tests

Because group 15 elements bond covalently and most are nonmetallic in nature, they do not lend themselves to identification through flame tests. The exceptions are antimony and bismuth. Antimony produces a faint green or blue color when placed in a flame, while bismuth produces a light purple-blue color.

The nitrogen group elements can be identified through analysis of their physical properties (melting point, boiling point, density), emission spectra, or reactions with other chemicals. For example, bismuth ions precipitate when added to tin(II) hydroxide and sodium hydroxide. Another example is the test for ammonium compounds. These compounds, which contain nitrogen, can be identified by their distinct smell when added to sodium hydroxide and by the color change observed when red litmus paper is placed at the opening of the test tube.

The ammonia vapor produced by mixing ammonium compounds (NH_4^+) with sodium hydroxide changes red litmus paper to blue.

Atomic Properties

Activity Have students explore diamagnetism by tying a thin string around a toothpick or straw. Students should then attach the string in such a way that the toothpick or straw is suspended in the air away from metallic objects. (One way is to bridge a meter stick between two desks and attach the string to the middle of it.) Once the string is in place, give students a collection of fruit and vegetable pieces. **WARNING:** ***Remind students to not eat anything in the lab.*** Have students place the foods one at a time on the toothpick or straw and then bring a Neodymium magnet near to it, but not touching it. Have students observe, record, and explain their results. Have students recall if there were any patterns. Students should see that foods with the highest water content move away from the magnet, regardless of which pole is used. This is because water is diamagnetic. **OL**

Analytical Tests

Background During a chemical reaction, elements are rearranged to create products whose properties differ from the original reactants. Because the changes occur at the atomic level, it is sometimes difficult to identify whether or not a chemical reaction occurred. There are at least five indicators that help scientists determine this including: the production of a gas (many times this is observed as bubbling), the release of heat, the absorption of heat, a color change, or the formation of a precipitate. Another indicator is the production of a smell.

Nitrogen-Fixing Bacteria

Background During the fixation process, nitrogen is converted to ammonia, which is used to formulate plant proteins. Some nitrogen-fixing bacteria participate in a symbiotic relationship with the plants they invade, such as legumes. As bacteria invade the legume, it forces the plant to create tiny nodules on its root system that regulate how much oxygen the bacteria come in contact with. This action protects the bacteria from exposure to free oxygen that inhibits the fixation process.

Liquid Nitrogen Cryotherapy

Background The science of cryogenics begins where common refrigeration processes end—at about −157°C. At this temperature, some gases, such as nitrogen, are liquids. The characteristic bubbling produced by liquid nitrogen at room temperature is the vaporization of the liquid into a gas. The white "smoke" forms as moisture in the air comes in contact with the cold nitrogen gas and condenses.

Safety Matches

Activity Obtain packages of safety matches and strike-anywhere matches. Show students how the packages are different. Demonstrate that safety matches need the red phosphorus strip to ignite, while strike-anywhere matches can ignite in a variety of places. The strike-anywhere matches need a frictional surface. Try striking these matches on a smooth lab bench to show how ignition is made more difficult because of the surface.

Nitrogen
7
N
$[He]2s^22p^3$

Nitrogen-Fixing Bacteria

Although nitrogen makes up about 78% of Earth's atmosphere, it occurs in a form that plants cannot use. Some bacteria in the soil convert nitrogen gas (N_2) from the air into a usable form by breaking the molecule's triple bond. This creates a form of nitrogen that plants uptake into their root systems. Plants need nitrogen to build cellular components, to participate in photosynthesis, and to transfer energy effectively. Commercial fertilizers mimic the action of nitrogen-fixing bacteria by providing nitrogen and other nutrients in forms that are easily incorporated into the plant system.

Nitrogen-fixing bacteria are found in protective nodules along plant roots.

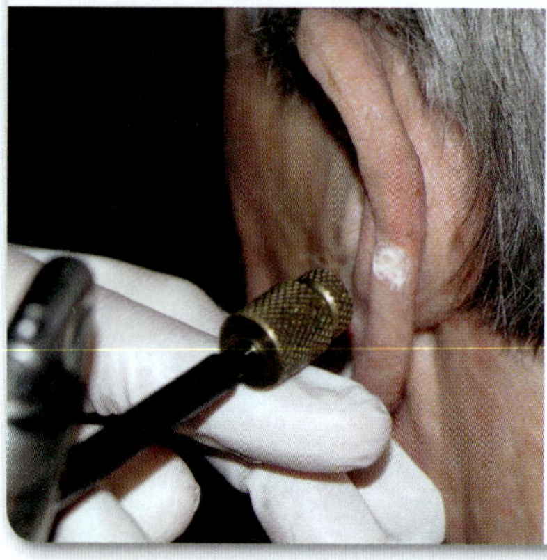

Liquid Nitrogen Cryotherapy

Cryotherapy, also called cryosurgery, is a medical procedure used to remove a variety of skin lesions, including carcinomas, warts, and other tissue abnormalities. The procedure involves dabbing liquid nitrogen onto the affected area to freeze and kill the cells. This is then repeated over time until all of the affected tissue is gone. Research has shown that patients who undergo cryotherapy treatment for certain types of lesions experience a lower recurrence rate than patients who receive radiation or surgical removal.

Doctors use liquid nitrogen as one of the treatment options to remove certain types of skin cancer. More than 1.3 million new cases of skin cancer are recorded each year in the United States.

Phosphorus
15
P
$[Ne]3s^23p^3$

Safety Matches

Safety matches consist of two main parts: the tip and the textured strip on the side of the box. The tip contains potassium chlorate, and the textured strip contains red phosphorus. When these two chemicals come in contact, a chemical reaction occurs, and fire is produced. In safety matches, the chemicals needed for reaction are separate from each other. In strike-anywhere matches, both chemicals are contained in the matchstick so that ignition can occur using almost any surface.

The strike of a match initiates a chemical reaction that produces a flame.

(t)Inga Spence/Photo Researchers; (c)Dr P. Marazzi/Photo Researchers; (b)Joe Belanger/iStock/Getty Images

Antimony
51
Sb
$[Kr]5s^2 4d^{10} 5p^3$

Flame Retardants

Antimony trioxide (Sb_2O_3) is used along with brominated or chlorinated compounds in the making of flame retardants that protect plastics, paints, and some textile products. Antimony trioxide increases the effectiveness of the halogen compounds in preventing the spread of a fire. Research shows that approximately 5000 deaths in the United States are caused by fire each year. The use of flame retardants improves escape time, releases less toxic gases and heat, and decreases fire damage.

Antimony trioxide fire retardants coat electrical wires and components found in a variety of everyday appliances.

Bismuth
83
Bi
$[Xe]6s^2 4f^{14} 5d^{10} 6p^3$

Soothing Upset Stomachs

Originally named *Mixture Cholera Infantum,* the popular pink medicine now used for upset stomachs was created to combat cholera. This mixture, whose active ingredient was bismuth subsalicylate ($C_7H_5BiO_4$), proved effective in treating the nausea and vomiting associated with infant cholera. However, it could not cure the disease itself. Nonetheless, the product became a wide success. As science advanced and doctors realized that cholera was contracted from bacteria (which could be treated with antibiotics), bismuth subsalicylate found its way into medical treatments for a variety of other stomach problems, including heartburn, indigestion, and ulcers.

Bismuth subsalicylate ($C_7H_5BiO_4$) is the active ingredient in some medicines used to treat stomach problems.

REVIEW

39. Identify which elements in the nitrogen group are metals, nonmetals, or metalloids.

40. Infer why nitrogen does not react with other elements under normal temperature conditions.

41. Explain why a compound of antimony is used in flame retardants that protect plastic products.

42. Describe how fertilizers mimic the action of nitrogen-fixing bacteria.

43. Write a balanced chemical equation for the reaction between potassium chlorate ($KClO_3$) and red phosphorus (P_4). The reaction produces potassium chloride (KCl) and phosphorus pentoxide (P_4O_{10}).

44. Predict what product will be formed when bismuth is combined with chlorine.

45. Calculate A 35-kg bag of fertilizer contains 5.25 kg of nitrogen. What percentage of the fertilizer is nitrogen?

Flame Retardants

Activity Demonstrate how flame retardants work. Cut four test strips of coffee filter paper (1-cm wide and 10-cm long). Soak the strips in the following solutions and let dry overnight: Strip 1—water; Strip 2—baking soda solution (1 tsp. baking soda +25 mL of warm water); Strip 3—table salt solution (1 tsp. of salt +25 mL of warm water); Strip 4—nothing. The next day, have students predict how each strip will react when placed in a Bunsen burner flame. Using tongs, test each strip and have students record their observations. (For Strip 4, soak the paper in water and place it in the flame while it is still wet.). Students should observe that Strips 2 and 4 retarded the flames, while the others did not. Draw students' attention to the fact that the flame was not extinguished and that the paper would eventually burn. By looking at the composition of baking soda—$NaHCO_3$—students might guess that one of the products that forms during heating is CO_2. Water vapor is another. Both do not support combustion. The table salt, on the other hand, does not produce retarding chemicals. **OL**

Soothing Upset Stomachs

Background Many ulcers are formed by an imbalance of stomach acid, but some are caused by the presence of *Helicobacter pylori (H. pylori)* bacteria. These bacteria live in the mucous lining of the stomach, which protects them from gastric juice and immune responses that have difficulty penetrating the lining. The resulting inflamed tissue provides an opening for ulcers to form. *H. pylori* spreads through oral contact with infected food and water. Treatments include the use of antibiotics and/or endoscopic treatment.

REVIEW

39. Nitrogen and phosphorus are nonmetals. Arsenic and antimony are metalloids. Bismuth is a metal.

40. Nitrogen exists as a diatomic molecule (N_2). The molecule is bonded with a triple bond, meaning that each nitrogen atom in the molecule is surrounded by eight electrons—an octet that provides stability. In order to force a reaction with nitrogen, heat must be added to break the triple bond and free the molecule for reaction.

41. The antimony compound increases the effectiveness of halogen compounds, which prevent the spread of fire.

42. They provide nitrogen in a form that can be easily taken up by plants.

43. $10KClO_3 + 3P_4 \longrightarrow 10KCl + 3P_4O_{10}$

44. The product is $BiCl_3(s)$.

45. 15%

Physical Properties

Activity Students will observe orthorhombic sulfur and its conversion to the amorphous form, called plastic sulfur. Use a piece of roll sulfur (orthorhombic) to visually reinforce the properties of nonmetals. Cover it with a cloth, and use a hammer to break off a piece. Orthorhombic sulfur is brittle, powdery, and unmalleable. Use a conductivity tester to demonstrate that sulfur does not conduct an electric current. Half-fill a 250-mL beaker with distilled water. Half-fill a test tube with sulfur. Heat the sulfur until it begins to boil and turns dark in color. Pour the hot sulfur into the beaker of water. **WARNING: *The sulfur vapor can ignite.*** After the amorphous sulfur cools, remove it from the water with a spatula. The physical properties of orthorhombic sulfur include yellow color, hardness, brittleness, low melting point, and nonconductivity. Amorphous sulfur is black or dark red in color and elastic, but slowly changes back to the orthorhombic form. Ask students, what are allotropes? They are two or more forms of an element in the same physical state that differ in molecular structure. OL

Common Reactions

Background Many household chemicals contain group 16 elements in the form of sulfate and carbonate groups. Ask students to match each of the following with their chemical formula: gypsum $CaSO_4$, baking soda $NaHCO_3$, Epsom salts $MgSO_4$, washing soda Na_2CO_3, and chalk $CaCO_3$. OL

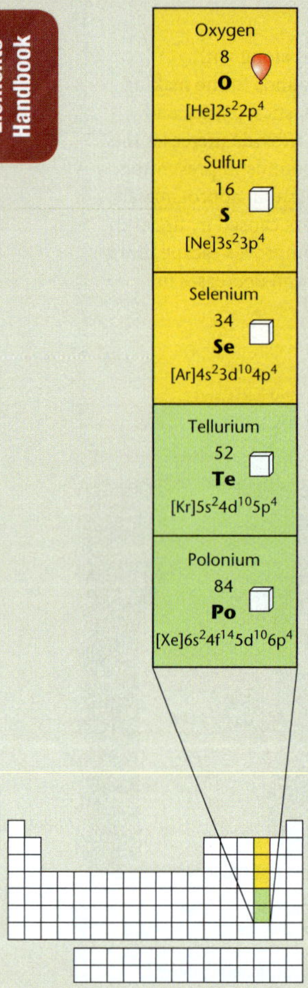

Physical Properties

- At room temperature, oxygen is a clear, odorless gas, while the other group 16 elements are solids.
- Some of the group 16 elements have several common allotropic forms. Oxygen can exist as either O_2 or O_3 (ozone). Sulfur has many allotropes. Selenium has three common allotropes: amorphous gray, red crystalline, and red/black powder.
- Oxygen, sulfur, and selenium are nonmetals. Tellurium and polonium are metalloids.
- O_2 is paramagnetic, which means that a strong magnet will attract oxygen molecules.
- Except for polonium, boiling points and melting points of the group 16 elements increase with increasing atomic number. Density increases with increasing atomic number for all group 16 elements.

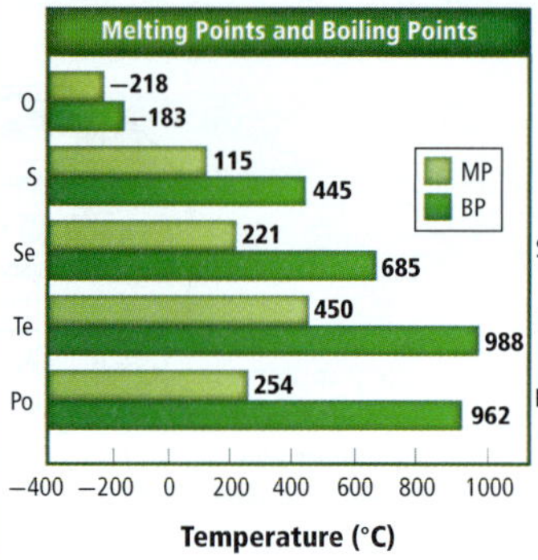

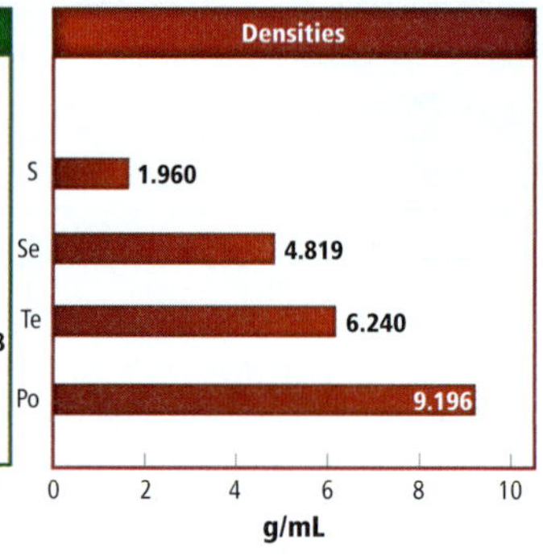

Common Reactions

- **S, Se, Te, and Po react with oxygen to form oxides, such as selenium oxide.**

  Example: $Se(s) + O_2(g) \rightarrow SeO_2(s)$
- **Oxygen also reacts with hydrogen and most of the elements in groups 1, 2, 13, 14, 15, and 17 to form oxides, such as silicon dioxide and magnesium oxide.**

  Examples: $Si + O_2 \rightarrow SiO_2$
  $2Mg + O_2 \rightarrow 2MgO$
- **O, S, Se, Te, and Po react with halogens to form halides, such as sulfur(VI) fluoride.**

  Example: $S(s) + 3F_2(g) \rightarrow SF_6(l)$

Oxides of Main Group Elements

Group	Oxides
H	H_2O, H_2O_2
1	Li_2O, Na_2O, K_2O, Rb_2O, Cs_2O, Fr_2O
2	BeO, MgO, CaO, SrO, BaO, RaO
13	B_2O_3, Al_2O_3, Ga_2O_3, In_2O_3, In_2O, Ti_2O
14	CO_2, SiO_2, GeO_2, SnO_2, SnO, PbO_2, PbO
15	N_2O_5, N_2O_3, N_2O, NO, NO_2, P_4O_{10}, P_4O_6, As_2O_5, As_4O_6, Sb_2O_5, Sb_4O_6, Bi_2O_3
17	Cl_2O_7, Cl_2O, Br_2O, I_2O_5

Atomic Properties

- Each element in group 16 has six valence electrons and an electron configuration ending with ns^2np^4.
- Group 16 elements can have many different oxidation numbers. For example, oxygen can have oxidation numbers of −2 and −1, and sulfur can have oxidation numbers of +6, +4, and −2.
- Going down the elements in group 16, the atomic radii and ionic radii increase.
- Electronegativity and first ionization energy decrease going down the elements in group 16.
- Polonium has 27 known isotopes. All are radioactive.

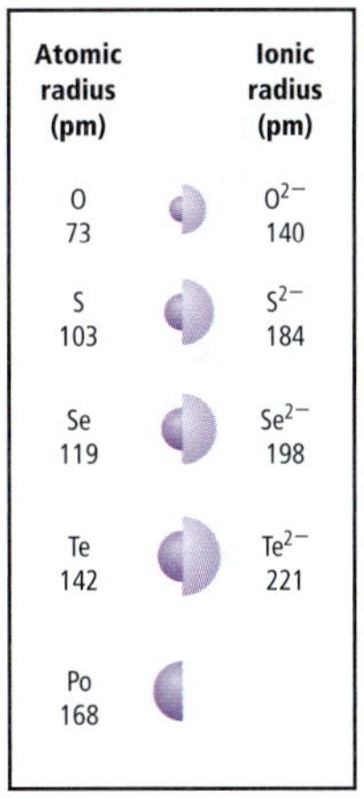

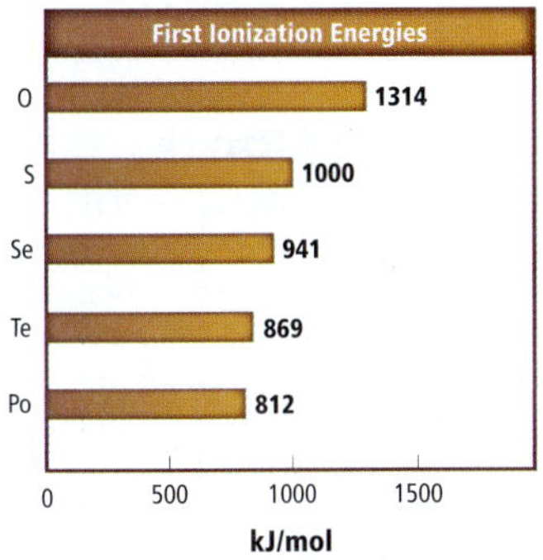

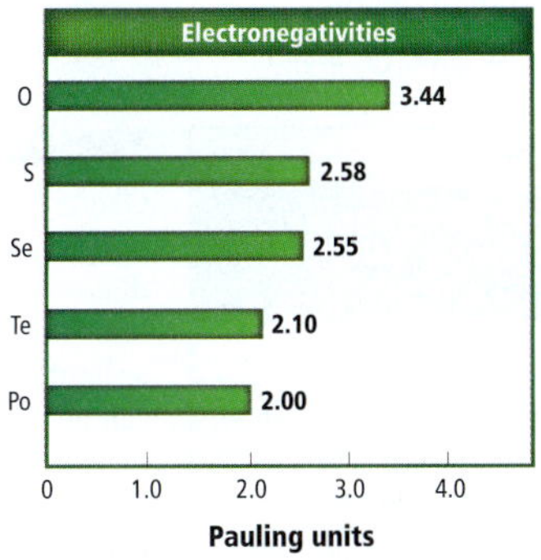

- **Group 16 elements are involved in many important industrial reactions, such as the formation of sulfuric acid.**

  Example: Sulfuric-acid production is a three-step process.

  1) $S(s) + O_2(g) \longrightarrow SO_2(g)$

  2) $2SO_2(g) + O_2(g) \longrightarrow 2SO_3(g)$

  3) $SO_3(g) + H_2O(l) \longrightarrow H_2SO_4(l)$

Analytical Tests

Oxygen can be measured in many different ways and in many different environments. For example, dissolved-oxygen meters measure oxygen in water samples. Dissolved-oxygen meters use an electrochemical reaction that reduces oxygen molecules to hydroxide ions. The meter measures the electric current produced during this reaction. The higher the oxygen concentration, the larger the current.

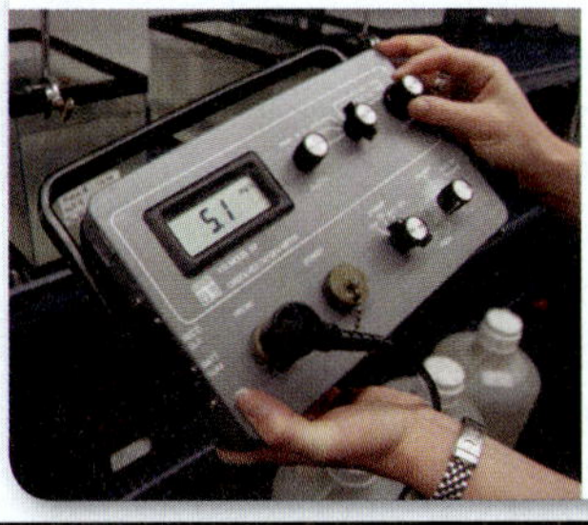

Dissolved-oxygen tests help scientists determine biological activity in water samples.

Atomic Properties

Activity Have students research the allotropes of selenium and tellurium and summarize their findings in one or two paragraphs. **OL**

Analytical Tests

Activity Students will generate oxygen gas and then test for the presence of oxygen using a glowing wooden splint. Students should wear safety glasses while handling hydrogen peroxide. Add 200 mL of 3% hydrogen peroxide to a 250-mL Erlenmeyer flask. Then add 1 mL of dry yeast. The mixture will bubble. Let the reaction go a few minutes to displace the air in the Erlenmeyer flask. Then, place a glowing wooden splint just above the liquid in the flask. In the presence of oxygen, the splint will burst into flame. Flush materials down the sink with water or dispose in the trash. Have students write a balanced chemical equation for the breakdown of hydrogen peroxide. $2H_2O_2 \longrightarrow 2H_2O + O_2$ Ask students: what is the role of the yeast in this reaction? The yeast enzyme, catalase, accelerates the rate of the reaction. **OL** **COOP LEARN**

Photosynthesis Produces O_2 from H_2O

Activity WARNING: ***Remind students to not eat anything in the lab.*** Groups of students can use gumdrop and toothpick models to simulate the process of photosynthesis. Use red gumdrops for oxygen, white gumdrops for hydrogen, and black gumdrops for carbon. Ask each group of students to build 12 water molecules. Simulate the light reactions of photosynthesis by using the atoms in the water molecules to create dioxygen molecules and hydrogen ions. To simulate the dark reactions of photosynthesis, ask each group to build 12 carbon-dioxide models. Use the hydrogen ions and the atoms in the carbon dioxide molecules to build glucose models. How many glucose molecules will this make? two

OL COOP LEARN

The Dual Nature of Ozone

Background Ground-level ozone is produced when oxides of nitrogen react with substances called *volatile organic compounds* in the presence of sunlight. Oxides of nitrogen are produced when fossil fuels are burned, such as in motor vehicle engines. Gasoline vapors, chemical solvents, and paints are a few of the thousands of sources of volatile organic compounds.

Oxygen
8
O
$[He]2s^22p^4$

Photosynthesis Produces O_2 from H_2O

Earth's atmosphere is 21% oxygen by volume. Most of the oxygen in the atmosphere comes from photosynthesis. Photosynthetic organisms, including plants and cyanobacteria, use energy from sunlight to oxidize water. The result is hydrogen ions (H^+) and oxygen (O_2). The reactions involved in this part of photosynthesis are called light reactions because they depend on light energy to proceed. During the dark reactions of photosynthesis, the hydrogen ions derived during the light reactions are combined with carbon dioxide (CO_2) to form glucose ($C_6H_{12}O_6$). The overall reaction for photosynthesis follows:

$$6H_2O + 6CO_2 \rightarrow C_6H_{12}O_6 + 6O_2$$

Photosynthesis captures energy from sunlight and provides hydrogen ions to synthesize glucose from carbon dioxide.

Air Quality Index for Ozone

Index Values	Levels of Health Concern	Cautionary Statements
0–50	good	none
51–100	moderate	Unusually sensitive people should consider reducing prolonged or heavy exertion outdoors.
101–150	unhealthy for sensitive groups	Active children and adults, and people with lung disease, such as asthma, should reduce prolonged or heavy exertion outdoors.
151–200	unhealthy	Active children and adults, and people with lung disease should avoid prolonged or heavy exertion outdoors. Everyone else should reduce prolonged or heavy exertion outdoors.
201–300	very unhealthy	Active children and adults, and people with lung disease, such as asthma, should avoid all outdoor exertion. Everyone else should avoid prolonged or heavy exertion outdoors.
301–500	hazardous	Everyone should avoid all physical activity outdoors.

Data obtained from: Patient Exposure and the Air Quality Index. *U.S. E.P.A.* March 2006

The Dual Nature of Ozone

Ozone (O_3), an allotrope of oxygen, has three oxygen atoms per molecule instead of two. Like diatomic oxygen (O_2), ozone is a gas at room temperature. However, unlike O_2, ozone gas has a slight blue color and a distinctive odor that can be detected during a thunderstorm or near a high-voltage electric motor. Ozone is also more reactive than diatomic oxygen. At ground level, ozone can be a serious potential health hazard, irritating eyes and lungs. High ground-level ozone concentrations are a particular threat on hot sunny days. The table illustrates how ozone affects air quality and health. On the other hand, stratospheric ozone protects Earth from harmful UV radiation by absorbing UV rays from sunlight.

Many cities issue air-quality alerts when ground-level ozone levels are high.

Sulfur
16
S
[Ne]$3s^23p^4$

An Economic Indicator

Sulfuric acid is one of the world's most important industrial raw materials. In the United States, more sulfuric acid is produced than any other industrial chemical. Most sulfuric acid is used in the production of phosphate fertilizers. Sulfuric acid is also important in extracting metals from ore, oil refining, waste treatment, chemical synthesis, and as a component in lead-acid batteries. Sulfuric acid is so important that economists use its production as a measure of a nation's industrial development.

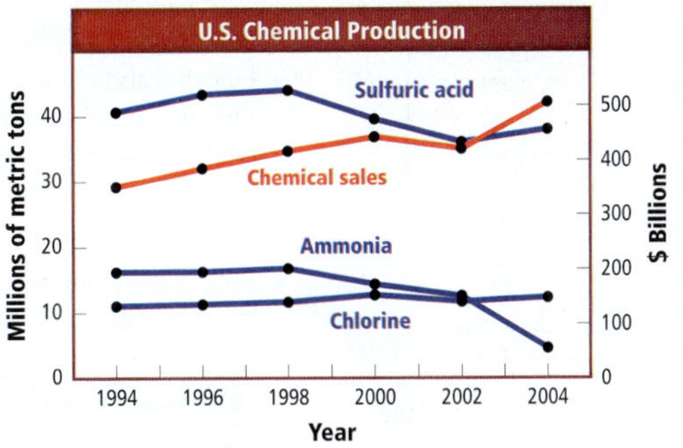

Data obtained from: *Chemical & Engineering News* 83 (2005) and 84 (2006).

Sulfuric acid production in the United States is used to track chemical economic trends.

Selenium
34
Se
[Ar]$4s^23d^{10}4p^4$

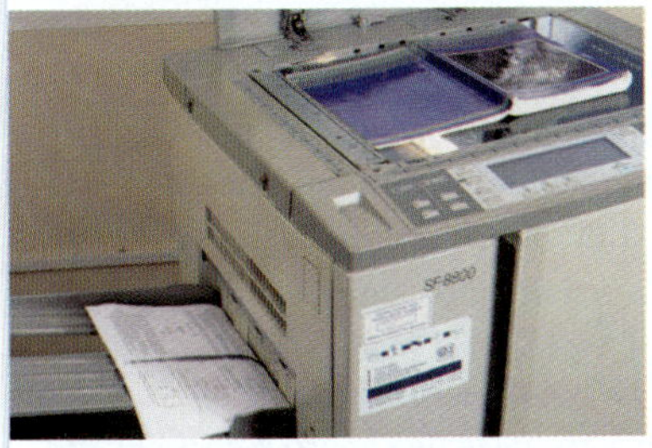

Gray selenium is a key component in many photocopiers.

Photocopies

Gray selenium is a photoconductor, which means it conducts electricity more efficiently in the presence of light than in the dark. Some photocopiers use this property to copy images. In a photocopier, a bright light shines on the original. Mirrors reflect the dark and light areas onto a drum coated with a thin layer of selenium. Because selenium is a photoconductor, the light areas conduct electricity, while the dark areas do not. As current flows through the drum, the light areas develop a negative charge and the dark areas develop a positive charge. Negatively charged toner particles are attracted to the positively charged dark areas to create a copy of the original image. Some of this same technology has been applied in developing new high-resolution digital detectors that use selenium as a photoconductor.

REVIEW

46. Identify the molecule that is the source of oxygen atoms for O_2 production during photosynthesis.

47. Explain why high ozone concentrations are harmful at ground level but beneficial in the upper atmosphere.

48. Calculate Approximately 90% of the sulfur used in the United States is used to make sulfuric acid. In 2004, 38.0 million metric tons of sulfuric acid were produced. How much sulfur did the United States use in 2004?

49. Apply Coal and petroleum products are sometimes contaminated with sulfur. When coal or petroleum containing sulfur is burned, sulfur dioxide (SO_2) can be released into the atmosphere. Use the information about the reactions involved in industrial sulfuric acid production to infer how atmospheric sulfur dioxide contributes to acid precipitation.

An Economic Indicator

Activity Place 13 g of table sugar in a clear glass beaker. Carefully add 15 mL of concentrated sulfuric acid to the sugar. Have baking soda available to neutralize any sulfuric acid spills. Stir with a glass rod and then stand back. The sugar will react with the sulfuric acid to produce black carbon. Ask students to identify the smell. It smells like caramel. The sulfuric acid acts as a catalyst in this reaction. Ask students to write a balanced chemical equation for the production of carbon and water from sugar. $C_{12}H_{22}O_{11} \rightarrow 12C + 11H_2O$ Ask students: where did the water from this reaction go? It vaporized as a result of the heat generated from this reaction. **OL**

Photocopies

Background The process of making photocopies is called *xerography,* from the Greek words *xeros,* meaning *dry,* and *graphein,* meaning *to write.*

REVIEW

46. H_2O

47. Ozone is highly reactive. At ground level, ozone can react with biological molecules and other materials, which is what makes ground level ozone hazardous to living organisms and destructive to materials like rubber and plastic. In the stratosphere, ozone reactions absorb UV light, which help protect Earth's surface from harmful UV radiation.

48. 14 million metric tons

49. The second two steps of industrial sulfuric acid production show how SO_2 can combine with O_2 to form SO_3. SO_3 can combine with water vapor in the air to form H_2SO_4.

Physical Properties

Background Fluorine comes from the Latin word *fluere,* which means *to flow.* The mineral fluorite, which contains fluorine and calcium, is used to lower the melting points of other minerals to make it easier to separate them from their ores.

Common Reactions

Activity Mix 1.0 mL chlorine water (Cl_2 (aq)) with 1.0 mL hexane. Ask students to describe the hexane layer. colorless Repeat, adding 1.0 mL bromine water (Br_2 (aq)) to a fresh sample of hexane, yellow/orange and using 1.0 mL iodine water (I_2 (aq)). pink/purple Observe the hexane layer after combining 1.0 mL of hexane with the following combinations: 1) 1.0 mL Cl_2 (aq) with 1.0 mL of 1.0*M* NaBr; 2) 1.0 mL Cl_2 (aq) with 1.0 mL of 1.0*M* NaI, 3) 1.0 mL Br_2 (aq) with 1.0 mL of 1.0*M* NaCl, 4) 1.0 mL Br_2 (aq) with 1.0 mL of 1.0*M* NaI, 5) 1.0 mL I_2 (aq) with 1.0 mL of 1.0*M* NaCl, 6) 1.0 mL I_2 (aq) with 1.0 mL of 1.0*M* NaBr. Based on their observations, have students draw conclusions about the trend of reactivity among the halogens. Reactivity: Cl > Br > I Disposal: Evaporate or burn off small quantities of the hexane mixtures under the hood. OL

Fluorine
9
F
$[He]2s^22p^5$

Chlorine
17
Cl
$[Ne]3s^23p^5$

Bromine
35
Br
$[Ar]4s^23d^{10}4p^5$

Iodine
53
I
$[Kr]5s^24d^{10}5p^5$

Astatine
85
At
$[Xe]6s^24f^{14}5d^{10}6p^5$

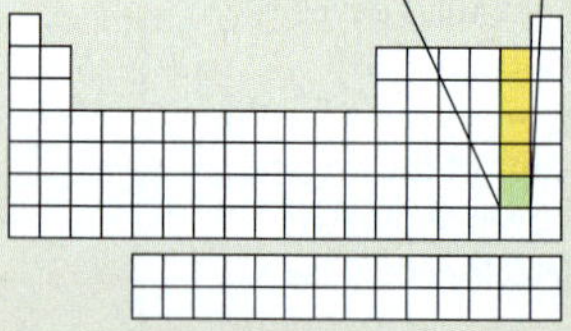

Physical Properties

- Fluorine and chlorine are gases at room temperature. Along with mercury, bromine is one of only two elements that are liquid at room temperature. Iodine is a solid that easily sublimes at room temperature.
- Fluorine gas is pale yellow. Chlorine gas is yellow-green. Bromine is a red-brown liquid. Iodine is a blue-black solid.
- Both boiling points and melting points of the group 17 elements increase with increasing atomic number.

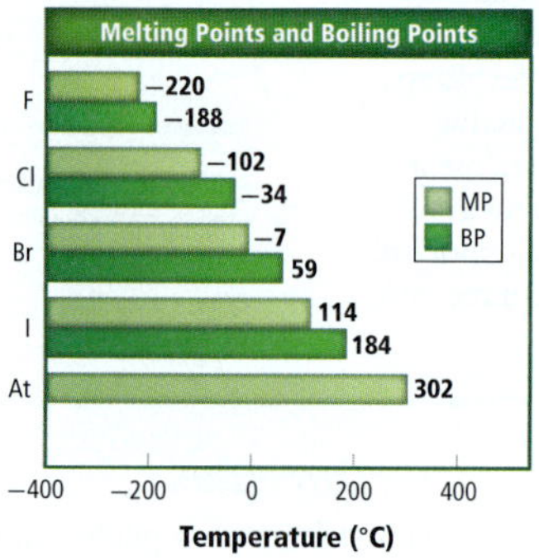

Iodine crystals are a blue-black color. They produce a violet vapor when they sublime at room temperature.

Common Reactions

- **The halogens react with alkali metals and alkaline earth metals to form salts, such as potassium bromide and calcium chloride.**

  Examples: $2K(s) + Br_2(g) \rightarrow 2KBr(s)$ and $Ca(s) + Cl_2(g) \rightarrow CaCl_2(s)$
- **The halogens can form acids, such as hydrochloric acid, by hydrolysis in water.**

  Example: $Cl_2(g) + H_2O(l) \rightarrow HClO(aq) + HCl(aq)$
- **Several important plastic polymers, including nonstick coatings and polyvinyl chloride, contain group 17 elements.**

  Example: Polyvinyl chloride (vinyl) is made by a three-step process.

  1) Ethene reacts with chlorine to form dichloroethane.
     $C_2H_4(g) + Cl_2(g) \rightarrow C_2H_4Cl_2(l)$
  2) At high temperature and pressure, dichloroethane is converted to vinyl chloride and HCl gas.
     $C_2H_4Cl_2(l) \rightarrow C_2H_3Cl(l) + HCl(g)$
  3) Vinyl chloride polymerizes to form polyvinyl chloride.
     $2n(C_2H_3Cl)(l) \rightarrow (\text{—}CH_2\text{–}CHCl\text{–}CH_2\text{–}CHCl\text{—})_n(l)$
- **Fluorine is the most active of all the elements and reacts with every element except helium, neon, and argon.**

  Example: $2Al(s) + 3F_2(g) \rightarrow 2AlF_3(s)$

Matt Meadows/Peter Arnold/Getty Images

Atomic Properties

- Each element in group 17 has seven valence electrons and an electron configuration ending with ns^2np^5.
- Electronegativities and first ionization energies decrease going down the elements in group 17.
- Fluorine is the most electronegative element on the periodic table. Therefore, it has the greatest tendency to attract electrons.
- Astatine is a radioactive element with no known uses.
- The atomic radii and ionic radii of the group 17 elements increase going down the group.

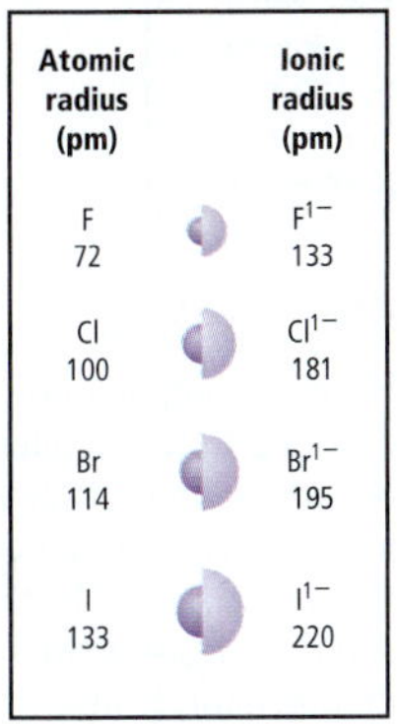

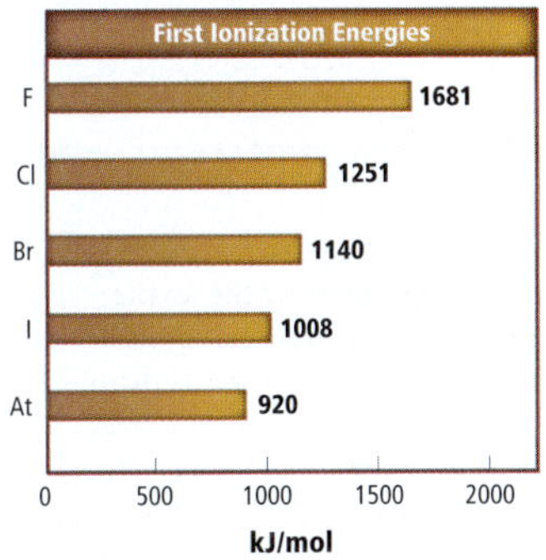

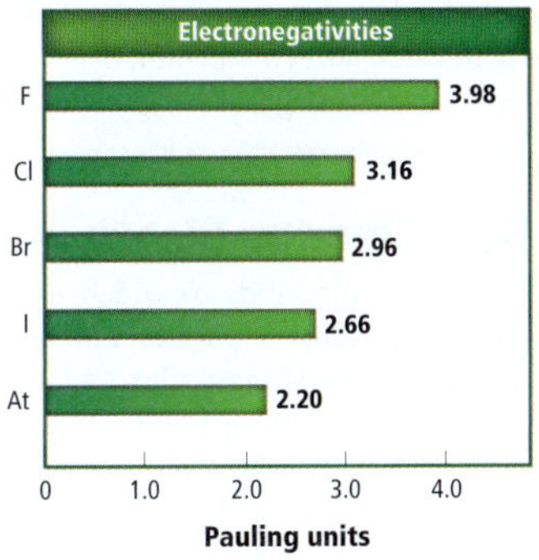

Analytical Tests

Three of the halogens can be identified through precipitation reactions. Chlorine, bromine, and iodine react with silver nitrate, forming distinctive precipitates. Silver chloride is a white precipitate, silver bromide is a cream-colored precipitate, and silver iodide is a yellow precipitate.

Chlorine, bromine, and iodine can also be identified when they dissolve in cyclohexane. As shown in the photo, when these halogens are dissolved in cyclohexane, the solution turns yellow for chlorine, orange for bromine, and violet for iodine.

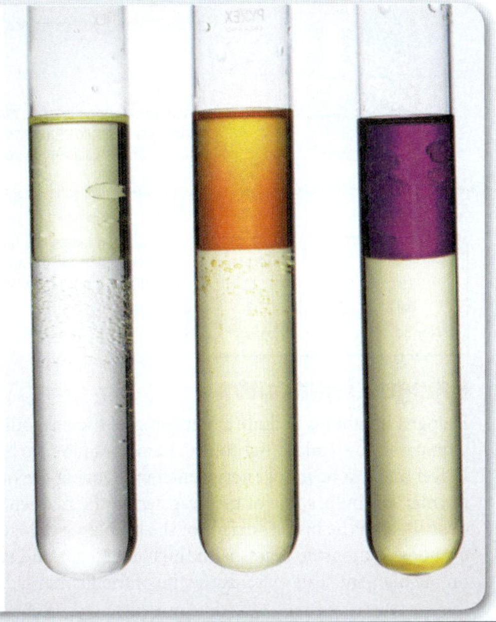

The halogens are only slightly soluble in water (bottom layer). However, in cyclohexane (top layer), chlorine (yellow), bromine (orange), and iodine (violet) readily dissolve.

Andrew Lambert Photography/Science Photo Library/Photo Researchers

Atomic Properties

Background The elements in group 17 are named for their ability to form compounds with almost all metals. Because these compounds are called salts, the group 17 elements are called *salt formers,* or halogens. The halogens are reactive nonmetals that are always found combined with other elements in nature. Fluorine is the halogen with the lowest atomic number. Its atomic size provides little shielding of its valence electrons from the nucleus. Fluorine is the most electronegative element on the periodic table; that is, it has the greatest tendency to attract electrons. Thus, it is logical that fluorine is also the most active of all elements.

Analytical Tests

Activity Students can demonstrate the presence of chlorine in PVC plastics. Students should work in a fume hood. Cut a variety of recycled plastics including PVC into 1-cm to 2-cm pieces. Ask students to record the type of recycled plastic and then test each sample by burning it in a flame. Samples containing chlorine will have a distinct green flame. Ask students how burning plastics might release potentially harmful chemicals to the atmosphere. Burning of PVC plastics produces Cl_2 gas. **OL** **COOP LEARN**

Fluoridation

Activity Have students research the pros and cons of fluoridated drinking water. Divide students into groups, for and against fluoridation, for a debate. OL

How Chlorine Bleach Is Made

Activity Students will observe the oxidizing action of bleach on food colors. Remind students to never mix chlorine bleach with acid or ammonia. This could form highly toxic Cl_2 gas or phosgene gas. Fill three clear glasses with 250 mL of water each. Add one drop of yellow food color to the first glass and one drop of blue food color to the second glass. Stir each with a glass rod. Using a plastic pipette, add 1 mL of chlorine bleach to each glass. The yellow coloring bleaches to clear. The blue color persists. Before adding bleach to the glass containing green food color, ask students to predict what will happen. Green food coloring is a mixture of yellow and blue. The yellow will bleach and the blue will persist. Explain to students that the bleach oxidizes both the yellow and the blue food coloring. However, the yellow food coloring oxidizes faster than the blue. The blue color will also bleach to clear but at a much slower rate. OL

Halogen Lightbulbs

Background The name *bromine* comes from the Greek word *bromos* which means *stench.* Bromine compounds are used as pesticides and in the production of photographic film. Bromine compounds are also sometimes added to flammable materials as fire retardants.

Fluorine
9
F
$[He]2s^22p^5$

Fluoridation

Fluorine compounds added to toothpaste and public drinking-water supplies have greatly reduced the incidence of cavities. Fluoride protects teeth in two ways. As teeth form, fluoride from food and drink is incorporated into the enamel layer. The fluoride makes the enamel stronger and more resistant to decay. Once teeth are present in the mouth, fluoride in saliva bonds to teeth and strengthens the surface enamel. This surface fluoride attracts calcium, which helps to fill in areas where decay has begun.

Many brands of toothpaste contain either stannous fluoride or sodium fluoride, which, like fluoridated water, strengthen teeth and provide protection from cavities.

Chlorine
17
Cl
$[Ne]3s^23p^5$

How Chlorine Bleach Is Made

Chlorine compounds are widely used as bleaching agents by the textile and paper industries. Some chlorine compounds can bleach materials by oxidizing colored molecules. Chlorine compounds are also used as disinfectants. Household bleach is a 5.25% solution of sodium hypochlorite (NaOCl) in water. Chlorine bleach is prepared commercially by passing an electric current through a solution of sodium chloride in water. As the sodium chloride breaks down, sodium hydroxide collects at the cathode and chlorine gas is generated at the anode. Sodium hydroxide and chlorine can then be combined to form sodium hypochlorite.

Household chlorine bleach is made by reacting chlorine gas or liquid chlorine with sodium hydroxide to form sodium hypochlorite.

Bromine	Iodine
35	53
Br	**I**
$[Ar]4s^23d^{10}4p^5$	$[Kr]5s^24d^{10}5p^5$

Halogen Lightbulbs

Halogen lightbulbs include a halogen gas, such as iodine or bromine. Compared to standard lightbulbs, halogen bulbs are brighter and last longer and can be more energy efficient. During the operation of a normal lightbulb, some of the tungsten in the filament evaporates and is deposited on the inside surface of the bulb. In a halogen lamp, the evaporated tungsten reacts with the halogen gas and is redeposited back on the filament. This extends the life of the filament.

Halogen lamps use bromine or other halogen molecules to capture tungsten vapor and return tungsten atoms to the filament.

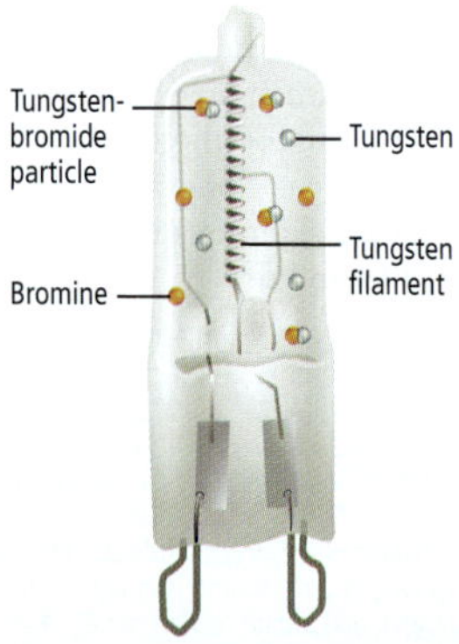

Isolated Products(John Boud)/Alamy

Iodine
53
I
$[Kr]5s^2 4d^{10} 5p^5$

Combating Iodine Deficiency with Salt

The thyroid gland is the only part of the body that absorbs iodine. Thyroid cells use iodine to produce thyroid hormones, which regulate metabolism. Low levels of iodine in the diet can lead to thyroid-hormone deficiencies and goiters, which are enlarged thyroid glands. In serious cases, low levels of thyroid hormones can cause birth defects and brain damage. In the United States, potassium iodide is added to most table salt to protect against dietary iodine deficiency. Even small amounts of added iodine can prevent iodine-deficiency disorders. However, there are parts of the world in which iodine deficiency is still prevalent.

Iodine Deficiency Around the World

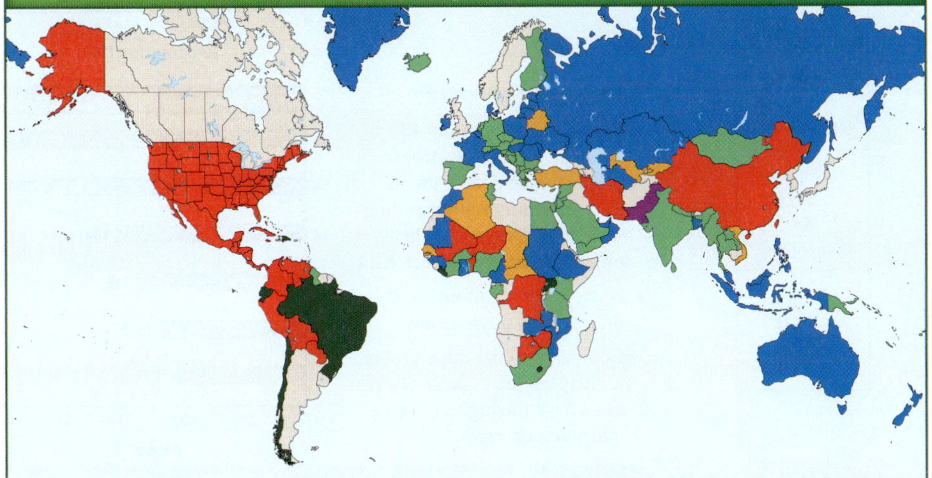

- Severe deficiency (<20 μg/L)
- Moderate deficiency (20–49 μg/L)
- Mild deficiency (50–99 μg/L)
- Optimal (100–199 μg/L)
- Risk of iodine-induced hyperthyroidism (200–299 μg/L)
- Risk of adverse health consequences (>300 μg/L)
- No data

A significant percentage of the world's population was at risk for iodine deficiency in 2004. In 2005, the World Health Organization launched a program to eliminate iodine deficiency worldwide.

REVIEW

50. Compare the risks for iodine deficiency in Europe, Africa, and the United States.

51. Explain why fluorine is the most reactive of all the elements.

52. Evaluate Why does a tungsten filament last longer in a halogen lightbulb than in a normal lightbulb?

53. Calculate Household bleach is typically a 5.25% solution of sodium hypochlorite in water. How many grams of sodium hypochlorite would there be in 300 mL of bleach?

54. Hypothesize In 1962, Neil Bartlett synthesized the first noble gas compound using PtF_6. Hypothesize why Bartlett used a fluorine compound for this synthesis.

Combating Iodine Deficiency with Salt

Activity The World Health Organization has recommended that potassium iodine be added to salt to protect against iodine deficiency disorders. Yet, some individuals are at risk of having too much iodine in their diets. Ask groups of students to research dietary iodine and debate the question: should potassium iodide be added to salt in all countries? **COOP LEARN**

REVIEW

50. The dark purple areas on the WHO map of worldwide iodine deficiency show the most at risk populations. Many European and African countries are at risk for mild to moderate iodine deficiency. Conversely, the United States population is at risk for possible iodine-induced hyperthyroidism.

51. Fluorine's electronegativity makes it the most reactive of all the elements. Fluorine is a small atom where its nucleus is closer to the valence electrons than a larger atom. It will attract bonding electrons more strongly than less reactive elements.

52. In a normal lightbulb, the tungsten evaporates and collects on the glass bulb. In a halogen lightbulb, the tungsten evaporates, reacts with the halogen gas, and redeposits on the filament, which makes the filament last longer.

53. 15.8 g

54. Fluorine is the most reactive of all the elements.

Physical Properties/Atomic Properties

Activity Divide the class into groups, each group representing one of the six noble gases. Have each group complete a chart that summarizes the properties, characteristics, and uses of a gas. Have groups share their charts with the class. **COOP LEARN**

Common Reactions

Background For years after the discovery of the noble gases, scientists tried to react the gases with other chemicals, but to no avail. This, along with the theory of electron configuration, led to the assumption that these gases were inert. In 1962, Neil Bartlett recognized that the first ionization energy of xenon was similar to that of oxygen. Bartlett knew that oxygen would react with platinum hexafluoride and predicted that xenon also might. His predictions proved correct and the first noble gas compound was made—$XePtF_6$. Other compounds also were made resulting in a change in designation from noble gases to rare gases.

Analytical Tests

Background A spectroscope utilizes the principle of diffraction to reveal the unique spectral lines that each noble gas emits. Diffraction occurs when light waves encounter an obstacle and are forced to bend around it. This bending separates the wavelengths of light much like a prism. If the incoming light is of a single wavelength, a single spectral line will be observed on both the left and right sides at equal distances from each slit on the grating.

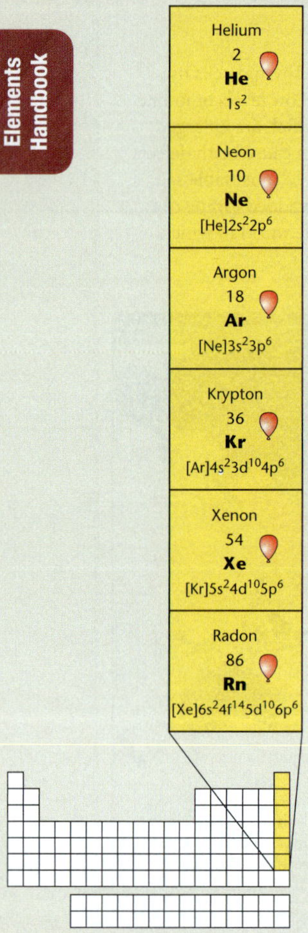

Physical Properties

- The group 18 elements are colorless, odorless gases.
- They are all nonmetals.
- Their melting points and boiling points increase going down the group, but are much lower than those of the other groups in the periodic table.

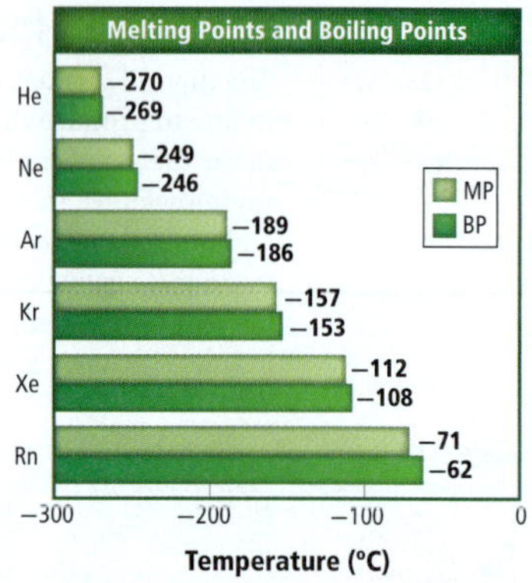

Atomic Properties

- Each element in group 18 has eight valence electrons, producing an octet with an electron configuration ending with ns^2np^6, except for helium, which has two electrons.
- Noble gases are monatomic—they exist as single atoms.
- Compared to the other groups in the periodic table, the noble gases have the highest first ionization energies.

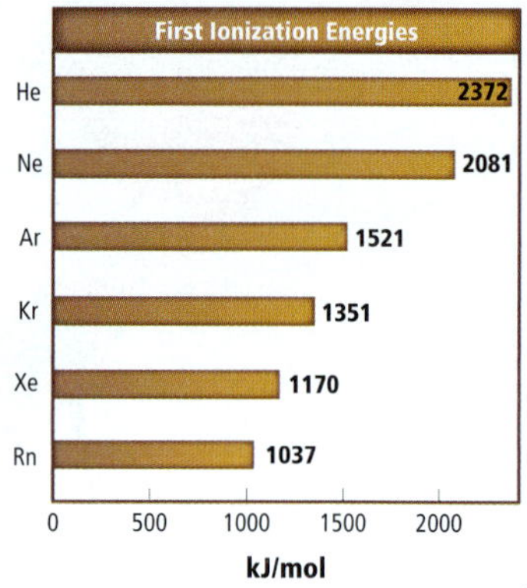

Common Reactions

Although the noble gases are also known as inert gases, a few compounds can be formed if conditions are favorable. Generally, however, noble gases are nonreactive.

Analytical Tests

Because the noble gases are odorless, colorless and generally unreactive, many of the common analytical tests used for identifying elements are not useful. However, the noble gases do emit light of certain colors when exposed to an electric current and have characteristic emission line spectra.

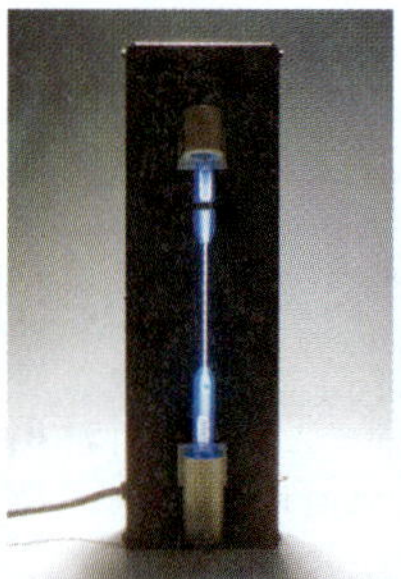

When an electric current passes through xenon, it exhibits a characteristic color (blue) and line spectrum.

Helium
2
He
$1s^2$

The Sun

Only 150 million km away (considered close in astronomical terms), the Sun provides the energy needed to support life on Earth. The Sun makes its energy through the fusion of hydrogen to make helium. Scientists have determined that the core of the Sun is composed of approximately 50% helium, leaving enough hydrogen for the Sun to burn for another 5 billion years.

The Sun's energy comes from a nuclear reaction that produces helium.

Neon	Argon	Krypton	Xenon
10	18	36	54
Ne	**Ar**	**Kr**	**Xe**
$[He]2s^22p^6$	$[Ne]3s^23p^6$	$[Ar]4s^23d^{10}4p^6$	$[Kr]5s^24d^{10}5p^6$

The noble gases are found in many different light sources.

Lighting

Neon, argon, krypton, and xenon are all used in different lighting applications. Neon signs are found in many businesses to advertise products or display the name of the business. Although true neon signs glow with a red-orange color, the term *neon sign* has also come to represent the collection of gas tubes that contain gases that display other colors. Argon is found in everyday lightbulbs such as those in lamps. Because argon is inert, it provides an ideal atmosphere for the filament. Krypton and xenon bulbs produce whiter, sharper light and last longer than traditional argon bulbs. These bulbs are commonly found in chandeliers, flashlights, and luxury car headlights.

REVIEW

55. **Describe** three physical properties of the noble gases.
56. **Write** the reaction for the production of xenon tetroxide.
57. **Analyze** why the noble gases have the highest first ionization energies compared to the rest of the elements on the periodic table.
58. **Hypothesize** why argon is used in everyday lighting even though krypton and xenon produce whiter light and last longer.
59. **Calculate** If the Sun is 150 million km away and light travels at 3.00×10^5 km/s, how long does it take for sunlight to reach Earth?

The Sun

Background The Sun's energy is produced through the fusion of hydrogen. This fusion is known as the proton-proton cycle. During this five-step cycle, two fusion processes occur. Helium nuclei are formed during the first process, which fuse and release two protons. The energy emitted during one cycle is equal to approximately 25 MeV (1 eV = 1.6×10^{-19} J). In any given second, about 635,000,000 metric tons of hydrogen are converted to helium, an action that releases over 350 billion megawatts of energy.

Lighting

Activity Have students make a visual presentation/collage that represents the different types of lighting sources that utilize the noble gases. Students can include pictures from magazines, product labels, and so on, and should also be encouraged to make their own drawings. As an extension, have students research neon lights and the different gases used to produce different colors. Have them design neon signs and identify the gases used to produce the colors. Students should display their work in class. **OL**

REVIEW

55. The noble gas elements are colorless, odorless, nonmetallic gases with low melting and boiling points.
56. $2Xe(g) + 2O_2(g) \longrightarrow XeO_4(g)$
57. The ionization energy is the amount of energy needed to remove one electron from the outermost energy level. Because the noble gases have a full outer energy level—an octet of stability—they will resist the removal of electrons. Thus, a high amount of energy is required.
58. Answers will vary. Students might hypothesize that argon is more common than krypton and xenon, making it a less expensive choice for everyday lighting.
59. 500 seconds, or 8.33 min

MATH HANDBOOK

Mathematics is a language used in science to express and solve problems. Calculations you perform during your study of chemistry require arithmetic operations, such as addition, subtraction, multiplication, and division. Use this handbook to review basic math skills and to reinforce some math skills presented in more depth in the chapters.

Scientific Notation

Scientists must use extremely small and extremely large numbers to describe the objects in **Figure 1.** The mass of the proton at the center of a hydrogen atom is 0.000000000000000000000000001673 kg. HIV, the virus that causes AIDS, is about 0.00000011 m. The temperature at the center of the Sun reaches 15,000,000 K. Such small and large numbers are difficult to read and hard to work with in calculations. Scientists have adopted a method of writing exponential numbers called scientific notation. It is easier than writing numerous zeros when numbers are very large or very small. It is also easier to compare the relative size of numbers when they are written in scientific notation.

A number written in scientific notation has two parts.

$$N \times 10^n$$

The first part (N) is a number in which only one digit is placed to the left of the decimal point and all remaining digits are placed to the right of the decimal point. The second part is an exponent of ten (10^n) by which the decimal portion is multiplied. For example, the number 2.53×10^6 is written in scientific notation.

$$\underbrace{2.53}_{\text{Number between one and ten}} \times \underbrace{10^6}_{\text{Exponent of ten}}$$

The decimal portion is 2.53 and the exponent is 10^6.

Positive exponents are used to express large numbers, and negative exponents are used to express small numbers.

Figure 1 Scientific notation provides a convenient way to express data with extremely large or small numbers. Scientists can express the mass of a proton, the length of HIV, and the temperature of the Sun in scientific notation.

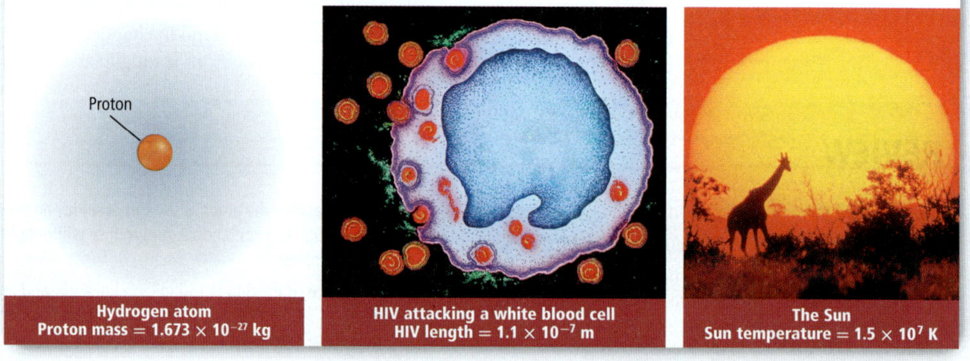

(l)Chris Bjornberg/Photo Researchers, (r)Daniele Pellegrini/Photo Researchers

Positive exponents

When scientists discuss the physical properties of the Moon, shown in **Figure 2,** the numbers are enormously large. A positive exponent of 10, (n) tells how many times a number must be multiplied by 10 to give the long form of the number.

$$2.53 \times 10^6$$
$$= 2.53 \times 10 \times 10 \times 10 \times 10 \times 10 \times 10$$
$$= 2{,}530{,}000$$

You can also think of the positive exponent of 10 as the number of places you move the decimal to the left until only one nonzero digit is to the left of the decimal point.

2,530,000. The decimal point moves six places to the left.

To convert the number 567.98 to scientific notation, first write the number as an exponential number by multiplying by 10^0.

$$567.98 \times 10^0$$

(Remember that multiplying any number by 10^0 is the same as multiplying the number by 1.) Move the decimal point to the left until there is only one digit to the left of the decimal. At the same time, increase the exponent by the same number as the number of places the decimal is moved.

$567.98 \times 10^{0+2}$ The decimal point moves two places to the left.

Thus, 567.98 written in scientific notation is 5.6798×10^2.

Negative exponents

Measurements can also have negative exponents, such as shown by the X-rays in **Figure 3.** Negative exponents are used for numbers that are very small. A negative exponent of 10 tells how many times a number must be divided by 10 to give the long form of the number.

$$6.43 \times 10^{-4} = \frac{6.43}{10 \times 10 \times 10 \times 10} = 0.000643$$

A negative exponent of 10 is the number of places you move the decimal to the right until it is just past the first nonzero digit.

When converting a number that requires the decimal to be moved to the right, the exponent is decreased by the appropriate number. For example, the expression of 0.0098 in scientific notation is as follows:

0.0098×10^0

$0.0098 \times 10^{0-3}$ The decimal point moves three places to the right.

9.8×10^{-3}

Thus, 0.0098 written in scientific notation is 9.8×10^{-3}.

■ **Figure 2** The mass of the Moon is 7.349×10^{22} kg.

■ **Figure 3** Because of their short wavelengths (10^{-8} m to 10^{-13} m), X-rays can pass through some objects.

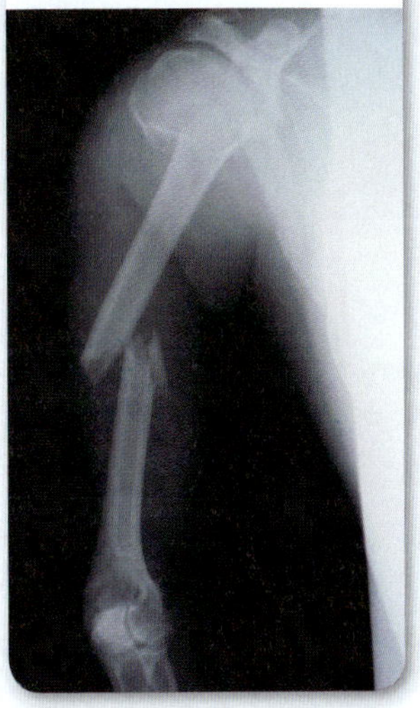

Complete solutions to the Math Handbook Practice Problems can be found in the Solutions Manual.

1. a. 5.8×10^3
b. 4.53×10^5
c. 5.877×10^{-4}
d. 3.6×10^{-3}
2. a. 3.5×10^7
b. 2.0×10^9
c. 3.7×10^{12}
d. 5.6×10^{-8}
3. a. 2.4×10^{-9}
b. 2.7
c. 3.0×10^{-4}
d. 6.0×10^{14}
e. 1.0×10^{16}
f. 3.1

Operations with Scientific Notation

The arithmetic operations performed with ordinary numbers can be done with numbers written in scientific notation. However, the exponential portion of the numbers must also be considered.

1. Addition and subtraction

Before numbers in scientific notation can be added or subtracted, the exponents must be equal. Remember that the decimal is moved to the left to increase the exponent and to the right to decrease the exponent.

$$\begin{aligned}(3.4 \times 10^2) + (4.57 \times 10^3) &= (0.34 \times 10^3) + (4.57 \times 10^3)\\ &= (0.34 + 4.57) \times 10^3\\ &= 4.91 \times 10^3\end{aligned}$$

$$\begin{aligned}(7.52 \times 10^{-4}) - (9.7 \times 10^{-5}) &= (7.52 \times 10^{-4}) - (0.97 \times 10^{-4})\\ &= (7.52 - 0.97) \times 10^{-4}\\ &= 6.55 \times 10^{-4}\end{aligned}$$

2. Multiplication

When numbers in scientific notation are multiplied, only the decimal portion is multiplied. The exponents are added.

$$\begin{aligned}(2.00 \times 10^3)(4.00 \times 10^4) &= (2.00)(4.00) \times 10^{3+4}\\ &= 8.00 \times 10^7\end{aligned}$$

3. Division

When numbers in scientific notation are divided, only the decimal portion is divided, while the exponents are subtracted as follows:

$$\begin{aligned}\frac{9.60 \times 10^7}{1.60 \times 10^4} &= \frac{9.60}{1.60} \times 10^{7-4}\\ &= 6.00 \times 10^3\end{aligned}$$

PRACTICE Problems

1. Express the following numbers in scientific notation.

a. 5800
b. 453,000
c. 0.0005877
d. 0.0036

2. Perform the following operations.

a. $(5.0 \times 10^6) + (3.0 \times 10^7)$
b. $(1.8 \times 10^9) + (2.0 \times 10^8)$
c. $(3.89 \times 10^{12}) - (1.9 \times 10^{11})$
d. $(6.0 \times 10^{-8}) - (4.0 \times 10^{-9})$

3. Perform the following operations.

a. $(6.0 \times 10^{-4}) \times (4.0 \times 10^{-6})$
b. $(4.5 \times 10^9) \times (6.0 \times 10^{-10})$
c. $\frac{4.5 \times 10^{-8}}{1.5 \times 10^{-4}}$
d. $\frac{9.6 \times 10^8}{1.6 \times 10^{-6}}$
e. $\frac{(2.5 \times 10^6)(7.2 \times 10^4)}{1.8 \times 10^{-5}}$
f. $\frac{(6.2 \times 10^{12})(6.0 \times 10^{-7})}{1.2 \times 10^6}$

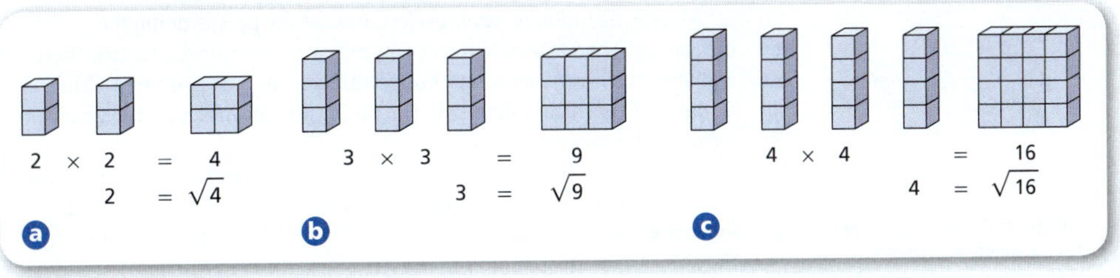

Math Handbook

■ **Figure 4 a.** The number 4 can be expressed as two groups of 2. The identical factors are 2. **b.** The number 9 can be expressed as three groups of 3. Thus, 3 is the square root of 9. **c.** 4 is the square root of 16.

Determine *the cube root of 16 using your calculator.*

Square and Cube Roots

A square root is one of two identical factors of a number. As shown in **Figure 4a,** the number 4 is the product of two identical factors—2. Thus, the square root of 4 is 2. The symbol $\sqrt{\ }$, called a radical sign, is used to indicate a square root. Most scientific calculators have a square root key labeled $\sqrt{\ }$.

$$\sqrt{4} = \sqrt{2 \times 2} = 2$$

This equation is read "the square root of 4 equals 2." What is the square root of 9, shown in **Figure 4b?**

There can be more than two identical factors of a number. You know that $2 \times 4 = 8$. Are there any other factors of the number 8? It is the product of $2 \times 2 \times 2$. A cube root is one of three identical factors of a number. Thus, what is the cube root of 8? It is 2. A cube root is also indicated by a radical.

$$\sqrt[3]{8} = \sqrt[3]{2 \times 2 \times 2} = 2$$

Check your calculator handbook for more information on finding roots.

Significant Figures

Accuracy reflects how close the measurements you make in the laboratory come to the real value. Precision describes the degree of exactness of your measurements. Which ruler in **Figure 5** would give you the most precise length? The top ruler, with the millimeter markings, would allow your measurements to come closer to the actual length of the pencil. The measurement would be more precise.

■ **Figure 5** The estimated digit must be read between the millimeter markings on the top ruler.

Evaluate *Why is the bottom ruler less precise?*

■ **Caption Question Fig. 4**
2.52

■ **In-Text Question** 3

■ **Caption Question Fig. 5**
There is greater distance between the markings on the bottom ruler that will cause greater variation in the estimated digit.

MATH HANDBOOK

■ **Figure 6** If you determine that the length of this pencil is 27.65 cm, that measurement has four significant figures.

Measuring tools are never perfect, nor are the people doing the measuring. Therefore, whenever you measure a physical quantity, there will always be some amount of uncertainty in the measurement. The number of significant figures in the measurement indicates the uncertainty of the measuring tool.

The number of significant figures in a measured quantity is all of the certain digits plus the first uncertain digit. For example, the pencil in **Figure 6** has a length that is between 27.6 and 27.7 cm. You can read the ruler to the nearest millimeter (27.6 cm), but after that you must estimate the next digit in the measurement. If you estimate that the next digit is 5, you would report the measured length of the pencil as 27.65 cm. Your measurement has four significant figures. The first three are certain, and the last is uncertain. The ruler used to measure the pencil has precision to the nearest tenth of a millimeter.

How many significant figures?

When a measurement is provided, the following series of rules will help you to determine how many significant figures there are in that measurement.

1. *All nonzero figures are significant.*
2. *When a zero falls after the decimal point and after a significant figure, that zero is significant.*
3. *When a zero falls between significant figures, the zero is also significant.*
4. *When a zero is used merely to indicate the position of the decimal, it is not significant.*
5. *All counting numbers and exact numbers are treated as if they have an infinite number of significant figures.*

Examine each of the following measurements. Use the rules above to check that all of them have three significant figures.

245 K	**Rule 1**
18.0 L	**Rule 2**
308 km	**Rule 3**
0.00623 g	**Rule 4**
186,000 m	**Rule 4**

Suppose you must do a calculation using the measurement 200 L. You cannot be certain which zero was estimated. To indicate the significance of digits, especially zeros, write measurements in scientific notation. In scientific notation, all digits in the decimal portion are significant. Which measurement is most precise?

200 L has one significant figure.
2×10^2 L has one significant figure.
2.0×10^2 L has two significant figures.
2.00×10^2 L has three significant figures.

The greater the number of digits in a measurement expressed in scientific notation, the more precise the measurement is. In this example, 2.00×10^2 L is the most precise data.

EXAMPLE Problem 1

SIGNIFICANT FIGURES How many significant figures are in the measurement 0.00302 g? 60 min? 5.620 m? 9.80×10^2 m/s^2?

1 ANALYZE THE PROBLEM

To determine the number of significant digits in a series of numbers, review the rules for significant figures.

2 SOLVE FOR THE UNKNOWN

0.00302 g

Not significant (Rule 4) — Significant (Rules 1 and 3)

The measurement 0.00302 g has three significant figures.

60 min
Unlimited significant figures
(Rule 5)

5.620 m
Significant
(Rules 1 and 2)

The measurement 5.620 m has four significant figures.

9.80×10^2 m/s^2
Significant
(Rules 1 and 2)

3 EVALUATE THE ANSWER

The measurements 0.00302 g and 9.80×10^2 m/s^2 have three significant figures. The measurement 60 min has unlimited significant figures. The measurement 5.620 m has four significant figures.

PRACTICE Problems

4. Determine the number of significant figures in each measurement:

a. 35 g	**m.** 0.157 kg
b. 3.57 m	**n.** 28.0 mL
c. 3.507 km	**o.** 2500 m
d. 0.035 kg	**p.** 0.070 mol
e. 0.246 L	**q.** 30.07 nm
f. 0.004 m^3	**r.** 0.106 cm
g. 24.068 kPa	**s.** 0.0076 g
h. 268 K	**t.** 0.0230 cm^3
i. 20.04080 g	**u.** 26.509 cm
j. 20 dozen	**v.** 54.52 cm^3
k. 730,000 kg	**w.** 2.40×10^6 kg
l. 6.751 g	**x.** 4.07×10^{16} m

Math Handbook

Complete solutions to the Math Handbook Practice Problems can be found in the Solutions Manual.

4. a. 2
b. 3
c. 4
d. 2
e. 3
f. 1
g. 5
h. 3
i. 7
j. unlimited
k. 2
l. 4
m. 3
n. 3
o. 2
p. 2
q. 4
r. 3
s. 2
t. 3
u. 5
v. 4
w. 3
x. 3

■ **Caption Question Fig. 7**
Measurements made with the graduated cylinder will be more precise.

MATH HANDBOOK

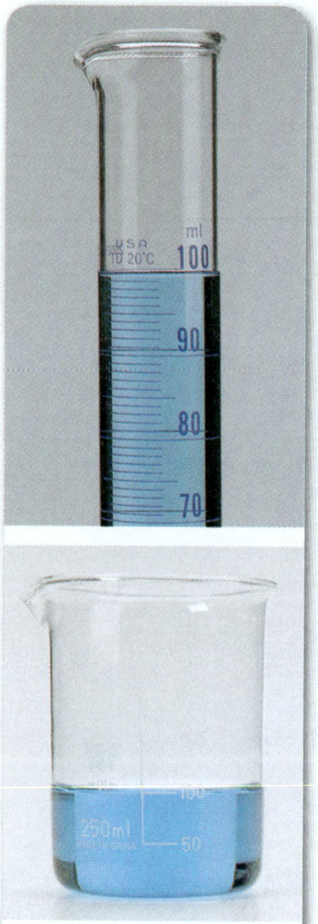

■ **Figure 7** Compare the markings on the graduated cylinder at the top with the markings on the beaker at the bottom.
Analyze ***Which piece of glassware will yield more precise measurements?***

Rounding

Arithmetic operations that involve measurements are done the same way as operations involving any other numbers. However, the results must correctly indicate the uncertainty in the calculated quantities. Perform all of the calculations, and then round the result to the least number of significant figures in any of the measurements used in the calculations. To round a number, use the following rules.

1. *When the left most digit to be dropped is less than 5, that digit and any digits that follow are dropped. Then, the last digit in the rounded number remains unchanged.* For example, when rounding the number 8.7645 to three significant figures, the left most digit to be dropped is 4. Therefore, the rounded number is 8.76.
2. *When the left most digit to be dropped is greater than 5, that digit and any digits that follow are dropped, and the last digit in the rounded number is increased by one.* For example, when rounding the number 8.7676 to three significant figures, the left most digit to be dropped is 7. Therefore, the rounded number is 8.77.
3. *When the left most digit to be dropped is 5 followed by a nonzero number, that digit and any digits that follow are dropped. The last digit in the rounded number increases by one.* For example, 8.7519 rounded to two significant figures equals 8.8.
4. *If the digit to the right of the last significant figure is equal to 5 and is not followed by a nonzero digit, look at the last significant figure. If it is odd, increase it by one; if even, do not round up.* For example, 92.350 rounded to three significant figures equals 92.4, and 92.25 equals 92.2.

Calculations with significant figures

Look at the glassware in **Figure 7.** Would you expect to measure a more precise volume with the beaker or the graduated cylinder? When you perform any calculation using measured quantities such as volume or mass, it is important to remember that the result can never be more precise than the least-precise measurement. That is, your answer cannot have more significant figures than the least precise measurement. Note that it is important to perform all calculations before dropping any insignificant digits.

The following rules determine how to use significant figures in calculations that involve measurements.

1. *To add or subtract measurements, first perform the mathematical operation, then round off the result to the least-precise value.* There should be the same number of digits to the right of the decimal as the measurement with the least number of decimal digits.
2. *To multiply or divide measurements, first perform the calculation, then round the answer to the same number of significant figures as the measurement with the least number of significant figures.* The answer should contain no more significant figures than the fewest number of significant figures in any of the measurements in the calculation.

Matt Meadows

EXAMPLE Problem 2

CALCULATING WITH SIGNIFICANT FIGURES Air contains oxygen (O_2), nitrogen (N_2), carbon dioxide (CO_2), and trace amounts of other gases. Use the known pressures in **Table 1** to calculate the partial pressure of oxygen.

1 ANALYZE THE PROBLEM

The data in **Table 1** contains the gas pressure for nitrogen gas, carbon dioxide gas, and trace gases. To add or subtract measurements, first perform the operation, then round off the result to correspond to the least-precise value involved.

2 SOLVE FOR THE UNKNOWN

$$P_{O_2} = P_{total} - (P_{N_2} + P_{CO_2} + P_{trace})$$

$$P_{O_2} = 101.3 \text{ kPa} - (79.10 \text{ kPa} + 0.040 \text{ kPa} + 0.94 \text{ kPa})$$

$$P_{O_2} = 101.3 \text{ kPa} - 80.080 \text{ kPa}$$

$$P_{O_2} = 21.220 \text{ kPa}$$

The total pressure (P_{total}) was measured to the tenths place. It is the least precise measurement. Therefore, the result should be rounded to the nearest tenth of a kilopascal. The pressure of oxygen is $P_{O_2} = 21.2$ kPa.

3 EVALUATE THE ANSWER

By adding the gas pressure of all the gases, including oxygen, the total gas pressure is 101.3 kPa.

Table 1 Pressures of Gases in Air

	Pressure (kPa)
Nitrogen gas	79.10
Carbon dioxide gas	0.040
Trace gases	0.94
Total gases	101.3

PRACTICE Problems

5. Round off the following measurements to the number of significant figures indicated in parentheses.

- **a.** 2.7518 g (3)
- **b.** 8.6439 m (2)
- **c.** 13.841 g (2)
- **d.** 186.499 m (5)
- **e.** 634,892.34 (4)
- **f.** 355,500 g (2)

6. Perform the following operations.

- **a.** (2.475 m) + (3.5 m) + (4.65 m)
- **b.** (3.45 m) + (3.658 m) + (47 m)
- **c.** $(5.36 \times 10^{-4} \text{ g}) - (6.381 \times 10^{-5} \text{ g})$
- **d.** $(6.46 \times 10^{12} \text{ m}) - (6.32 \times 10^{11} \text{ m})$
- **e.** $(6.6 \times 10^{12} \text{ m}) \times (5.34 \times 10^{18} \text{ m})$
- **f.** $\dfrac{5.634 \times 10^{11} \text{ m}}{3.0 \times 10^{12} \text{ m}}$
- **g.** $\dfrac{(4.765 \times 10^{11} \text{ m})(5.3 \times 10^{-4} \text{ m})}{7.0 \times 10^{-5} \text{ m}}$

Math Handbook

Complete solutions to the Math Handbook Practice Problems can be found in the Solutions Manual.

5. a. 2.75 g
b. 8.6 m
c. 14 g
d. 186.50 m
e. 6.349×10^5 g
f. 3.6×10^5 g

6. a. 10.6 m
b. 54 m
c. 4.72×10^{-4} g
d. 5.83×10^{12} m
e. 3.5×10^{31} m^2
f. 1.9×10^{-1}
g. 3.6×10^{12} m

MATH HANDBOOK

Solving Algebraic Equations

When you are given a problem to solve, it often can be written as an algebraic equation. You can use letters to represent measurements or unspecified numbers in the problem. The laws of chemistry are often written in the form of algebraic equations. For example, the ideal gas law relates pressure, volume, moles, and temperature of the gases. The ideal gas law is written as follows.

$$PV = n\mathrm{R}T$$

The variables are pressure (P), volume (V), number of moles (n), and temperature (T). R is a constant. This is a typical algebraic equation that can be manipulated to solve for any of the individual variables.

When you solve algebraic equations, any operation that you perform on one side of the equal sign must be performed on the other side of the equation. Suppose you are asked to use the ideal gas law to find the pressure of a gas (P). To solve for, or isolate, P requires you to divide the left-hand side of the equation by V. This operation must be performed on the right-hand side of the equation as well, as shown in the second equation below.

$$PV = n\mathrm{R}T$$

$$\frac{PV}{V} = \frac{n\mathrm{R}T}{V}$$

The Vs on the left-hand side of the equation cancel each other out.

$$\frac{PV}{V} = \frac{n\mathrm{R}T}{V}$$

$$P \times \frac{\cancel{V}}{\cancel{V}} = \frac{n\mathrm{R}T}{V}$$

$$P = \frac{n\mathrm{R}T}{V}$$

The ideal gas law equation is now written in terms of pressure. That is, P has been isolated.

■ **Figure 8** When faced with an equation that contains more than one operation, use this flowchart to determine the order in which to perform your calculations.

Order of Operations

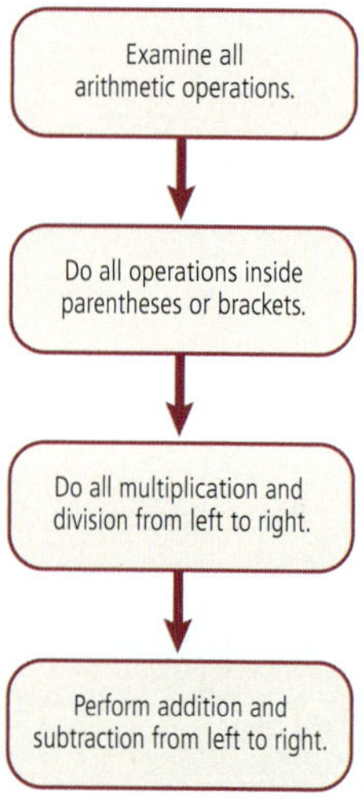

Order of operations

When isolating a variable in an equation, it is important to remember that arithmetic operations have an order of operations, as shown in **Figure 8,** that must be followed. Operations in parentheses (or brackets) take precedence over multiplication and division, which in turn take precedence over addition and subtraction. For example, in the following equation

$$a + b \times c$$

variable b must be multiplied first by variable c. Then, the resulting product is added to variable a. If the equation is written

$$(a + b) \times c$$

the operation in parentheses or brackets must be done first. In the equation above, variable a is added to variable b before the sum is multiplied by variable c.

To see the difference order of operations makes, try replacing a with 2, b with 3, and c with 4.

$$a + (b \times c) = 2 + (3 \times 4) = 14$$
$$(a + b) \times c = (2 + 3) \times 4 = 20$$

To solve algebraic equations, you also must remember the distributive property. To remove parentheses to solve a problem, any number outside the parentheses is distributed across the parentheses as follows.

$$6(x + 2y) = 6(x) + 6(2y) = 6x + 12y$$

Math Handbook

EXAMPLE Problem 3

ORDER OF OPERATIONS The temperature on a cold day was 25°F. What was the temperature on the Celsius scale?

1 ANALYZE THE PROBLEM

The temperature in Celsius can be calculated by using the equation for converting from the Celsius temperature to Fahrenheit temperature. The Celsius temperature is the unknown variable. The known variable is 25°F.

2 SOLVE FOR THE UNKNOWN

Determine the equation for calculating the temperature in Celsius.

$$°F = \frac{9}{5}°C + 32$$

$$°F - 32 = \frac{9}{5}°C + 32 - 32$$ **Rearrange the equation to isolate °C. Begin by subtracting 32 from both sides.**

$$°F - 32 = \frac{9}{5}°C$$

$$5 \times (°F - 32) = 5 \times \frac{9}{5}°C$$ **Then, multiply both sides by 5.**

$$5 \times (°F - 32) = 9°C$$

$$\frac{5 \times (°F - 32)}{9} = \frac{9°C}{9}$$ **Finally, divide both sides by 9.**

$$°C = \frac{5}{9}(°F - 32)$$

$$= \frac{5}{9}(25 - 32)$$ **Substitute the known Fahrenheit temperature.**

$$= -3.9°C$$

The Celsius temperature is −3.9°C.

3 EVALUATE THE ANSWER

To determine if the answer is correct, place the answer, −3.9°C, into the original equation. If the Fahrenheit temperature is 25°, the calculation was done correctly.

Complete solutions to the Math Handbook Practice Problems can be found in the Solutions Manual.

7. $R = \frac{PV}{nT}$

8. $y = \frac{3}{4} - x$

9. $y = \frac{1}{2}\left(\frac{z}{x} - 4\right)$

10. $x = \frac{2}{(3 + y)}$

11. $x = 8.5$

PRACTICE Problems

Isolate the indicated variable in each equation.

7. $PV = nRT$ for R
8. $3 = 4(x + y)$ for y
9. $z = x(4 + 2y)$ for y
10. $\frac{2}{x} = 3 + y$ for x
11. $\frac{2x + 1}{3} = 6$ for x

Dimensional Analysis

The dimensions of a measurement refer to the type of units attached to a quantity. For example, length is a dimensional quantity that can be measured in meters, centimeters, and kilometers. Dimensional analysis is the process of solving algebraic equations for units as well as numbers. It is a way of checking to ensure that you have used the correct equation, and that you have correctly applied the rules of algebra when solving the equation. It can also help you to choose and set up the correct equation, as shown on the next page, when you learn how to do unit conversions. It is good practice to make dimensional analysis a habit by always stating the units as well as the numerical values whenever substituting values into an equation.

■ **Figure 9** Aluminum is a metal that is useful from the kitchen to the sculpture garden.

ABN Images/Alamy

EXAMPLE Problem 4

DIMENSIONAL ANALYSIS The sculpture in **Figure 9** is made from aluminum. The density (D) of aluminum is 2700 kg/m^3. Determine the mass (m) of a piece of aluminum of volume (V) 0.20 m^3.

1 ANALYZE THE PROBLEM

The facts of the problem are density (2700 kg/m^3), volume (0.20 m^3), and the density equation, $D = m/V$.

2 SOLVE FOR THE UNKNOWN

Determine the equation for mass by rearranging the density equation.

The equation for density is

$$D = \frac{m}{V}$$

$$DV = \frac{mV}{V}$$

$$DV = \frac{\cancel{V}}{\cancel{V}} \times m$$

Multiply both sides of the equation by V, and isolate m.

$$m = DV$$

$$m = (2700\ \text{kg}/\cancel{\text{m}^3})(0.20\ \cancel{\text{m}^3}) = 540\ \text{kg}$$

Substitute the known values for D and V.

3 EVALUATE THE ANSWER

Notice that the unit m^3 cancels out, leaving mass in kg, a unit of mass.

PRACTICE Problems

Determine whether the following equations are dimensionally correct. Explain.

12. $v = s \times t$ where v = 24 m/s, s = 12 m, and t = 2 s.

13. $R = \frac{nT}{PV}$ where R is in L•atm/mol•K, n is in mol, T is in K, P is in atm, and V is in L.

14. $t = \frac{v}{s}$ where t is in seconds, v is in m/s, and s is in m.

15. $s = \frac{at^2}{2}$ where s is in m, a is in m/s^2, and t is in s.

Math Handbook

Complete solutions to the Math Handbook Practice Problems can be found in the Solutions Manual.

12. No; units are not the same on both sides of the equation.

13. No; units are not the same on both sides of the equation.

14. No; units are not the same on both sides of the equation.

15. Yes; units are the same on both sides of the equation.

Unit Conversion

Recall that the universal unit system used by scientists is called Le Système Internationale d'Unités, or SI. It is a metric system based on seven base units—meter, second, kilogram, kelvin, mole, ampere, and candela—from which all other units are derived. The size of a unit in the metric system is indicated by a prefix related to the difference between that unit and the base unit. For example, the base unit for length in the metric system is the meter. One-tenth of a meter is a decimeter, where the prefix *deci-* means *one-tenth.* One thousand meters is a kilometer, where the prefix *kilo-* means *one thousand.*

You can use the information in **Table 2** to express a measured quantity in different units. For example, how is 65 m expressed in centimeters? **Table 2** indicates one centimeter and one-hundredth meter are equivalent, that is, 1 cm = 10^{-2} m. This information can be used to form a conversion factor. A conversion factor is a ratio equal to one that relates two units. You can make the following conversion factors from the relationship between meters and centimeters. Be sure when you set up a conversion factor that the measurement in the numerator (the top of the ratio) is equivalent to the measurement in the denominator (the bottom of the ratio).

$$1 = \frac{1\text{ cm}}{10^{-2}\text{ m}} \text{ and } 1 = \frac{10^{-2}\text{ m}}{1\text{ cm}}$$

Table 2 Common SI Prefixes

Prefix	Symbol	Exponential Notation	Prefix	Symbol	Exponential Notation
Peta	P	10^{15}	Deci	d	10^{-1}
Tera	T	10^{12}	Centi	c	10^{-2}
Giga	G	10^{9}	Milli	m	10^{-3}
Mega	M	10^{6}	Micro	μ	10^{-6}
Kilo	k	10^{3}	Nano	n	10^{-9}
Hecto	h	10^{2}	Pico	p	10^{-12}
Deka	da	10^{1}	Femto	f	10^{-15}

Caption Question Fig. 10
There are four significant figures in 54.94 g.

MATH HANDBOOK

Recall that the value of a quantity does not change when it is multiplied by 1. To convert 65 m to centimeters, multiply 65 m by the conversion factor for centimeters.

$$65\ \cancel{\text{m}} \times \frac{1\ \text{cm}}{10^{-2}\ \cancel{\text{m}}}$$
$$= 65 \times 10^{2}\ \text{cm}$$
$$= 6.5 \times 10^{3}\ \text{cm}$$

Note the conversion factor is set up so that the unit meters cancels and the answer is in centimeters as required. When setting up a unit conversion, use dimensional analysis to check that the units cancel to give an answer in the desired units. Always check your answer to be certain the units make sense.

You make unit conversions every day when you determine how many quarters are needed to make a dollar or how many feet are in a yard. One unit that is often used in calculations in chemistry is the mole. Equivalent relationships among moles, grams, and the number of representative particles (atoms, molecules, formula units, or ions) must be used. For example, 1 mol of a substance contains 6.02×10^{23} representative particles. Try the next Example Problem to see how this information can be used in a conversion factor to determine the number of atoms in a sample of manganese.

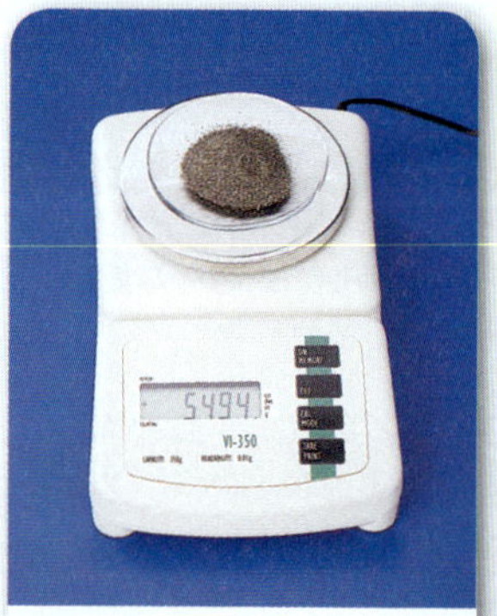

Figure 10 The mass of one mole of manganese equals 54.94 g.
Determine *How many significant figures are in this measurement?*

EXAMPLE Problem 5

UNIT CONVERSIONS One mole of manganese (Mn), shown in **Figure 10,** has a mass of 54.94 g. How many atoms are in 2.0 mol of manganese?

1 ANALYZE THE PROBLEM

You are given the mass of 1 mol of manganese. In order to convert to the number of atoms, you must set up a conversion factor relating the number of moles and the number of atoms.

2 SOLVE FOR THE UNKNOWN

The conversion factors for moles and atoms are shown below.

$$\frac{1\ \text{mol}}{6.02 \times 10^{23}\ \text{atoms}} \text{ and } \frac{6.02 \times 10^{23}\ \text{atoms}}{1\ \text{mol}}$$

Choose the conversion factor that cancels units of moles and gives an answer in number of atoms.

$$2.0\ \cancel{\text{mol}} \times \frac{6.02 \times 10^{23}\ \text{atoms}}{1\ \cancel{\text{mol}}} = 12.04 \times 10^{23}\ \text{atoms}$$
$$= 1.2 \times 10^{24}\ \text{atoms}$$

3 EVALUATE THE ANSWER

The answer is expressed in the desired units (number of atoms). It is expressed in two significant figures because the number of moles (2.0) has two significant figures.

Matt Meadows

PRACTICE Problems

16. Convert the following measurements as indicated.

a. 4 m = ____cm
b. 50.0 cm = ____m
c. 15 cm = ____mm
d. 567 mg = ____g
e. 324 mL = ____L
f. 28 L = ____mL
g. 4.6×10^3 m = ____mm
h. 8.3×10^4 g = ____kg
i. 2.7×10^2 L = ____mL
j. 7.3×10^5 mL = ____L
k. 8.4×10^{10} m = ____km
l. 3.8×10^4 m^2 = ____mm^2
m. 6.9×10^{12} cm^2 = ____m^2
n. 6.3×10^{21} mm^3 = ____cm^3
o. 9.4×10^{12} cm^3 = ____m^3
p. 5.7×10^{20} cm^3 = ____km^3

Drawing Line Graphs

Scientists, such as the one shown in **Figure 11,** as well as you and your classmates, use graphing to analyze data gathered in experiments. Graphs provide a way to visualize data in order to determine the mathematical relationship between the variables in your experiment. Line graphs are used most often.

Figure 11 also shows a line graph. Line graphs are drawn by plotting variables along two axes. Plot the independent variable on the *x*-axis (horizontal axis), also called the abscissa. The independent variable is the quantity controlled by the person doing the experiment. Plot the dependent variable on the *y*-axis (vertical axis), also called the ordinate. The dependent variable is the variable that depends on the independent variable. Label the axes with the variables being plotted and the units attached to those variables.

■ **Figure 11** Once experimental data have been collected, they must be analyzed to determine the relationships between the measured variables.

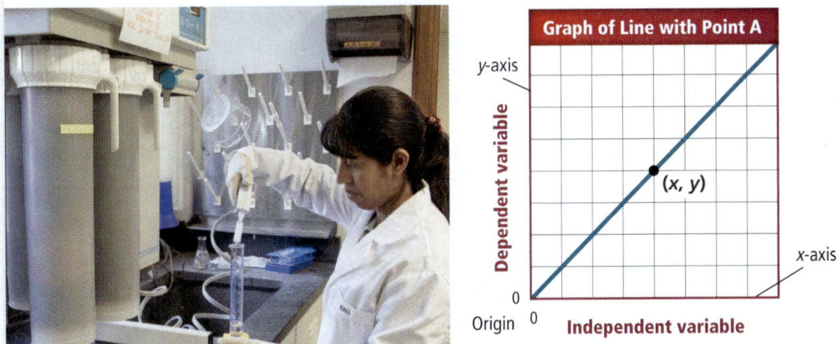

This research scientist might use graphs to analyze the data she collects on ultrapure water.

Any graph of your data should include labeled *x*- and *y*-axes, a suitable scale, and a title.

Math Handbook

Complete solutions to the Math Handbook Practice Problems can be found in the Solutions Manual.

16. a. 4×10^2 cm
b. 5.00×10^{-1} m
c. 150 mm
d. 0.567 g
e. 0.324 L
f. 2.8×10^4 mL
g. 4.6×10^6 mm
h. 83 kg
i. 2.7×10^5 mL
j. 7.3×10^2 L
k. 8.4×10^7 km
l. 3.8×10^{10} mm^2
m. 6.9×10^8 m^2
n. 6.3×10^{12} cm^3
o. 9.4×10^6 m^3
p. 5.7×10^5 km^3

■ **In-Text Question** The positive direction is up. The negative direction is down.

MATH HANDBOOK

■ **Figure 12** To plot a point on a graph, place a dot at the location for each ordered pair (*x*, *y*) determined by your data. In the *Density of Water* graph, the dot marks the ordered pair (40 mL, 40 g). Generally, the line or curve that you draw will not include all of your experimental data points, as shown in the *Experimental Data* graph.

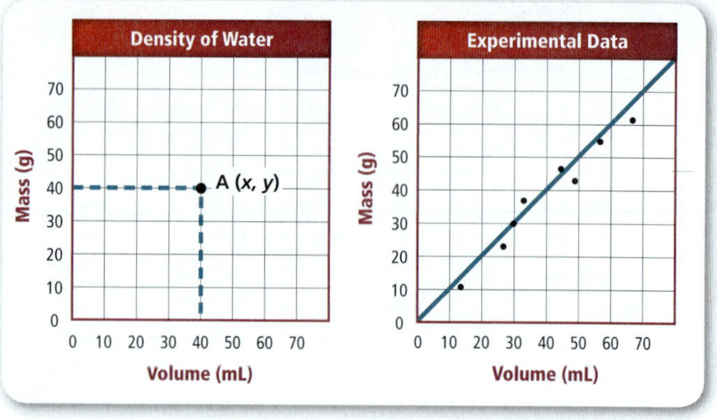

Determining a scale

An important part of graphing is the selection of a scale. Scales should be easy to plot and easy to read. First, examine the data to determine the highest and lowest values. Assign each division on the axis (the square on the graph paper) with an equal value so that all data can be plotted along the axis. Scales divided into multiples of 1, 2, 5, or 10, or decimal values, are often the most convenient. It is not necessary to start at zero, nor is it necessary to plot both variables to the same scale. Scales must, however, be labeled clearly with the appropriate numbers and units.

Plotting data

The values of the independent and dependent variables form ordered pairs of numbers, called the *x*-coordinate and the *y*-coordinate (*x*, *y*), that correspond to points on the graph. The first number in an ordered pair always corresponds to the *x*-axis; the second number always corresponds to the *y*-axis. The ordered pair (0,0) is always the origin. Sometimes, the points are named by using a letter. In **Figure 12,** Point A on the *Density of Water* graph corresponds to Point (*x*, *y*).

After the scales are chosen, plot the data. To graph or plot an ordered pair means to place a dot at the point that corresponds to the values in the ordered pair. The *x*-coordinate indicates how many units to move right (if the number is positive) or left (if the number is negative). The *y*-coordinate indicates how many units to move up or down. Which direction is positive on the *y*-axis? Negative? Locate each pair of *x*- and *y*-coordinates by placing a dot, as shown in **Figure 12** in the *Density of Water* graph. Sometimes, a pair of rulers, one extending from the *x*-axis and the other from the *y*-axis, can ensure that data are plotted correctly.

Drawing a curve

Once the data is plotted, a straight line or a curve is drawn. It is not necessary to make it go through every point plotted, or even any of the points, as shown in the *Experimental Data* graph in **Figure 12.** Graphing data is an averaging process. If the points do not fall along a line, the best-fit line or most-probable smooth curve through the points is drawn. Note that curves do not always go through the origin (0, 0).

Naming a graph

Last but not least, give each graph a title that describes what is being graphed. The title should be placed at the top of the page, or in a box on a clear area of the graph. It should not cross the data curve.

Using Line Graphs

Once the data from an experiment has been collected and plotted, the graph must be interpreted. Much can be learned about the relationship between the independent and dependent variables by examining the shape and slope of the curve. Four common types of curves are shown in **Figure 13.** Each type of curve corresponds to a mathematical relationship between the independent and dependent variables.

Direct and inverse relationships

In your study of chemistry, the most common curves are the linear, representing the direct relationship ($y \propto x$), and the inverse, representing the inverse relationship ($y \propto 1/x$), where x represents the independent variable and y represents the dependent variable. In a direct relationship, y increases in value as x increases in value, or y decreases when x decreases. In an inverse relationship, y decreases in value as x increases.

An example of a typical direct relationship is the increase in volume of a gas with increasing temperature. When the gases inside a hot-air balloon are heated, the balloon gets larger. As the balloon cools, its size decreases. However, a plot of the decrease in pressure as the volume of a gas increases yields a typical inverse curve.

You might also encounter exponential and root curves in your study of chemistry. See **Figure 13.** An exponential curve describes a relationship in which one variable is expressed by an exponent. A root curve describes a relationship in which one variable is expressed by a root.

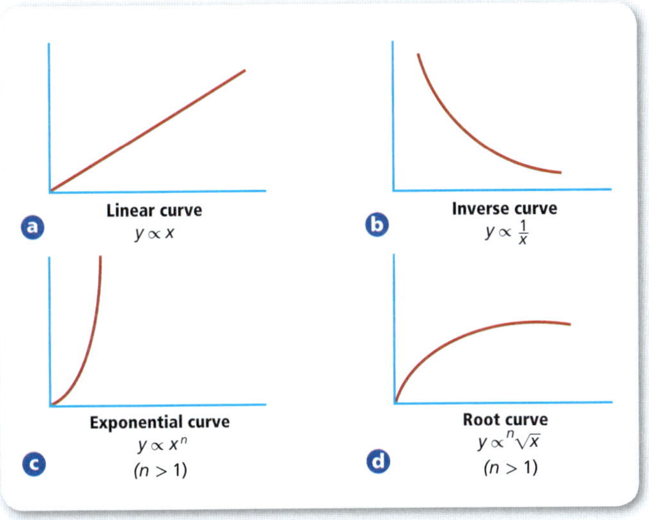

■ **Figure 13** The shape of the curve formed by a plot of experimental data indicates how the variables are related.

Math Handbook

MATH HANDBOOK

■ **Caption Question Fig. 14**
An almost flat line indicates little change in the dependent variable when the independent variable changes.

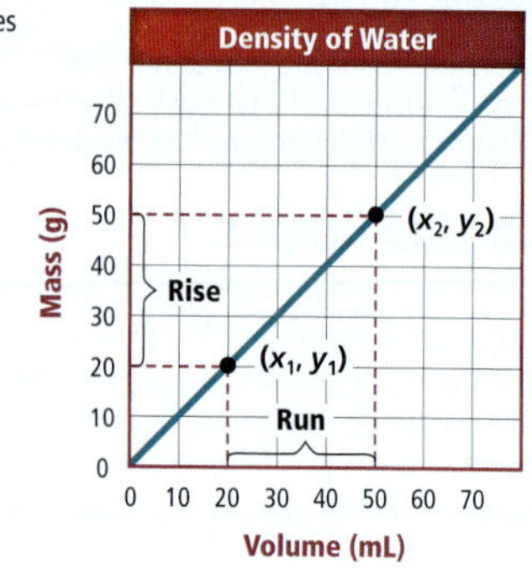

■ **Figure 14** A steep slope indicates that the dependent variable changes rapidly with a change in the independent variable.

Infer ***What would an almost flat line indicate?***

The linear graph

The linear graph is useful in analyzing data because a linear relationship can be translated easily into equation form using the equation for a straight line.

$$y = mx + b$$

In the equation, y stands for the dependent variable, m is the slope of the line, x stands for the independent variable, and b is the y-intercept, the point where the curve crosses the y-axis.

The slope of a linear graph is the steepness of the line. Slope is defined as the ratio of the vertical change (the rise) to the horizontal change (the run) as you move from one point to the next along the line. Use the graph in **Figure 14** to calculate slope. Choose any two points on the line, (x_1, y_1) and (x_2, y_2). The two points need not be actual data points, but both must fall somewhere on the straight line. After selecting two points, calculate slope, m, using the following equation.

$$m = \frac{\text{rise}}{\text{run}} = \frac{\Delta y}{\Delta x} = \frac{y_2 - y_1}{x_2 - x_1}, \text{ where } x_1 \neq x_2$$

The symbol Δ stands for change, x_1 and y_1 are the coordinates or values of the first point, and x_2 and y_2 are the coordinates of the second point.

Choose any two points along the graph of mass v. volume in **Figure 15,** and calculate its slope.

$$m = \frac{135\text{ g} - 54\text{ g}}{50.0\text{ cm}^3 - 20.0\text{ cm}^3} = 2.7\text{ g/cm}^3$$

Note that the units for the slope are the units for density. Plotting a graph of mass versus volume is one way of determining the density of a substance.

Apply the general equation for a straight line to the graph in **Figure 15.**

$$y = mx + b$$
$$mass = (2.7\text{ g/cm}^3)(volume) + 0$$
$$mass = (2.7\text{ g/cm}^3)(volume)$$

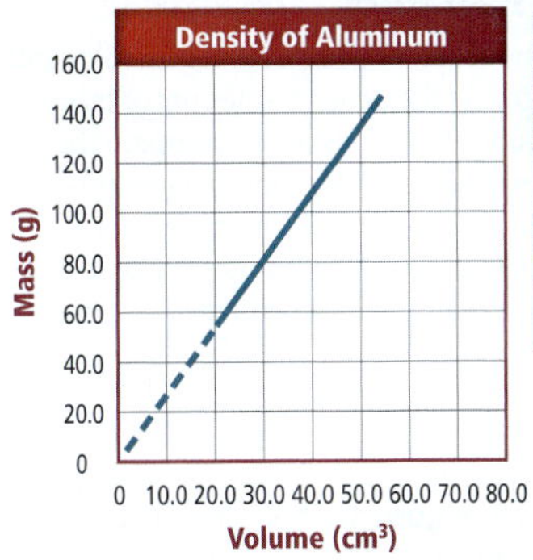

Data	
Volume (cm^3)	Mass (g)
20.0	54.0
30.0	81.0
50.0	135.0

■ **Figure 15** Interpolation and extrapolation will help you determine the values of points you did not plot.

Once the data from the graph in **Figure 15** has been placed in the general equation for a straight line, this equation verifies the direct relationship between mass and volume. For any increase in volume, the mass also increases.

Interpolation and extrapolation

Graphs also serve functions other than determining the relationship between variables. They permit interpolation, the prediction of values of the independent and dependent variables. For example, you can see in the table in **Figure 15** that the mass of 40.0 cm^3 of aluminum was not measured. However, you can interpolate from the graph that the mass would be 108 g.

Graphs also permit extrapolation, which is the determination of points beyond the measured points. To extrapolate, draw a broken line to extend the curve to the desired point. In **Figure 15,** you can determine that the mass at 10.0 cm^3 equals 27 g. One caution regarding extrapolation—some straight-line curves do not remain straight indefinitely. So, extrapolation should only be done where there is a reasonable likelihood that the curve does not change.

PRACTICE Problems

17. Plot the data in each table. Explain whether the graphs represent direct or inverse relationships.

Table 3 Gas Pressure and Volume

Pressure (mm Hg)	Volume (mL)
3040	5.0
1520	10.0
1013	15.0
760	20.0

Table 4 Gas Pressure and Temperature

Pressure (mm Hg)	Temperature (K)
3040	1092
1520	546
1013	410
760	273

Math Handbook

Complete solutions and graphs to the Math Handbook Practice Problems can be found in the Solutions Manual.

17. **Table 3** Graph is a downward curve with a negative slope that indicates an inverse relationship between pressure and volume.
Table 4 Graph is a straight line with positive slope that indicates a direct relationship between pressure and temperature.

MATH HANDBOOK

Ratios, Fractions, and Percents

When you analyze data, you may be asked to compare measured quantities. Or, you may be asked to determine the relative amounts of elements in a compound. Suppose, for example, you are asked to compare the molar masses of the diatomic gases, hydrogen (H_2) and oxygen (O_2). The molar mass of hydrogen gas equals 2.00 g/mol; the molar mass of oxygen equals 32.00 g/mol. The relationship between molar masses can be expressed in three ways: a ratio, a fraction, or a percent.

■ **Figure 16** The mass of one lime would be one-twelfth the mass of one dozen limes.

Ratios

You make comparisons by using ratios in your daily life. For example, if the mass of a dozen limes is shown in **Figure 16,** how does it compare to the mass of one lime? The mass of one dozen limes is 12 times larger than the mass of one lime. In chemistry, the chemical formula for a compound compares the elements that make up that compound, as shown in **Figure 17.** A ratio is a comparison of two numbers by division. One way it can be expressed is with a colon (:). The comparison between the molar masses of oxygen and hydrogen can be expressed as follows.

$$\text{molar mass } (H_2)\text{:molar mass } (O_2)$$
$$2.00 \text{ g/mol:}32.00 \text{ g/mol}$$
$$2.00:32.00$$
$$1:16$$

Notice that the ratio 1:16 is the smallest integer (whole number) ratio. It is obtained by dividing both numbers in the ratio by the smaller number, and then rounding the larger number to remove the digits after the decimal. The ratio of the molar masses is 1 to 16. In other words, the ratio indicates that the molar mass of diatomic hydrogen gas is 16 times smaller than the molar mass of diatomic oxygen gas.

■ **Figure 17** In a crystal of table salt (sodium chloride), each sodium ion is surrounded by chloride ions, yet the ratio of sodium ions to chloride ions is 1:1. The formula for sodium chloride is NaCl.

Fractions

Ratios are often expressed as fractions in simplest form. A fraction is a quotient of two numbers. To express the comparison of the molar masses as a fraction, place the molar mass of hydrogen over the molar mass of oxygen as follows.

$$\frac{\text{molar mass } H_2}{\text{molar mass } O_2}$$
$$= \frac{2.0 \cancel{\text{g/mol}}}{32.00 \cancel{\text{g/mol}}}$$
$$= \frac{2.00}{32.00}$$
$$= \frac{1}{16}$$

In this case, the simplified fraction is calculated by dividing both the numerator (top of the fraction) and the denominator (bottom of the fraction) by 2.00. This fraction yields the same information as the ratio. That is, diatomic hydrogen gas has one-sixteenth the mass of diatomic oxygen gas.

Matt Meadows

Percents

A percent is a ratio that compares a number to 100. The symbol for percent is %. You also are used to working with percents in your daily life. The number of correct answers on an exam can be expressed as a percent. If you answered 90 out of 100 questions correctly, you would receive a grade of 90%. Signs like the one in **Figure 18** indicate a reduction in price. If the item's regular price is \$100, how many dollars would you save? Sixty percent means 60 of every 100, so you would save \$60. How much would you save if the sign said 75% off?

The comparison between molar mass of hydrogen gas and the molar mass of oxygen gas described on the previous page can also be expressed as a percent by taking the fraction, converting it to decimal form, and multiplying by 100 as follows.

$$\frac{\text{molar mass } H_2}{\text{molar mass } O_2} \times 100 = \frac{2.00\ \cancel{g/mol}}{32.00\ \cancel{g/mol}} \times 100 = 0.0625 \times 100 = 6.25\%$$

Diatomic hydrogen gas has 6.25% of the mass of diatomic oxygen gas.

■ **Figure 18** Stores often use percentages when advertising sales.
Analyze *Would the savings be large at this sale? How would you determine the sale price?*

■ **Caption Question Fig. 18** Yes. Multiply the regular price by 0.60 and subtract from the regular price. Or, multiply the regular price by 0.40.

■ **In-Text Question** \$75

Math Handbook

Operations Involving Fractions

Fractions are subject to the same type of operations as other numbers. Remember that the number on the top of a fraction is the numerator and the number on the bottom is the denominator. **Figure 19** shows an example of a fraction.

■ **Figure 19** When two numbers are divided, the one on top is the numerator and the one on the bottom is the denominator. The result is called the quotient. When you perform calculations with fractions, the quotient can be expressed as a fraction or a decimal.

Dividend (numerator)

$$\textbf{Quotient} = \frac{9 \times 10^{8}}{3 \times 10^{-4}}$$

Divisor (denominator)

1. Addition and subtraction

Before two fractions can be added or subtracted, they must have a common denominator. Common denominators are found by finding the least common multiple of the two denominators. Finding the least common multiple is often as easy as multiplying the two denominators together. For example, the least common multiple of the denominators of the fractions $\frac{1}{2}$ and $\frac{1}{3}$ is 2×3 or 6.

$$\frac{1}{2} + \frac{1}{3} = \left(\frac{3}{3} \times \frac{1}{2}\right) + \left(\frac{2}{2} \times \frac{1}{3}\right) = \frac{3}{6} + \frac{2}{6} = \frac{5}{6}$$

Sometimes, one of the denominators will divide into the other, which makes the larger of the two denominators the least common multiple. For example, the fractions $\frac{1}{2}$ and $\frac{1}{6}$ have 6 as the least common multiple denominator.

$$\frac{1}{2} + \frac{1}{6} = \left(\frac{3}{3} \times \frac{1}{2}\right) + \frac{1}{6} = \frac{3}{6} + \frac{1}{6} = \frac{4}{6}$$

In other situations, both denominators will divide into a number that is not the product of the two. For example, the fractions $\frac{1}{4}$ and $\frac{1}{6}$ have the number 12 as their least common multiple denominator, rather than 24, the product of the two denominators.

The least common denominator can be deduced as follows:

$$\frac{1}{6} + \frac{1}{4} = \left(\frac{4}{4} \times \frac{1}{6}\right) + \left(\frac{6}{6} \times \frac{1}{4}\right) = \frac{4}{24} + \frac{6}{24} = \frac{2}{12} + \frac{3}{12} = \frac{5}{12}$$

Because both fractions can be simplified by dividing numerator and denominator by 2, the least common multiple must be 12.

Complete solutions to the Math Handbook Practice Problems can be found in the Solutions Manual.

18. a. $1\frac{5}{12}$
b. $1\frac{1}{10}$
c. $\frac{1}{12}$
d. $\frac{1}{24}$
e. $\frac{1}{4}$
f. $\frac{6}{35}$
g. $2\frac{1}{2}$
h. $1\frac{5}{27}$

2. Multiplication and division

When multiplying fractions, the numerators and denominators are multiplied together as follows:

$$\frac{1}{2} \times \frac{2}{3} = \frac{1 \times 2}{2 \times 3} = \frac{2}{6} = \frac{1}{3}$$

Note the final answer is simplified by dividing the numerator and denominator by 2.

When dividing fractions, the divisor is inverted and multiplied by the dividend as follows:

$$\frac{2}{3} \div \frac{1}{2} = \frac{2}{3} \times \frac{2}{1} = \frac{2 \times 2}{3 \times 1} = \frac{4}{3}$$

PRACTICE Problems

18. Perform the indicated operation:

a. $\frac{2}{3} + \frac{3}{4}$
b. $\frac{4}{5} + \frac{3}{10}$
c. $\frac{1}{4} - \frac{1}{6}$
d. $\frac{7}{8} - \frac{5}{6}$
e. $\frac{1}{3} \times \frac{3}{4}$
f. $\frac{3}{5} \times \frac{2}{7}$
g. $\frac{5}{8} \div \frac{1}{4}$
h. $\frac{4}{9} \div \frac{3}{8}$

Logarithms and Antilogarithms

When you perform calculations, such as the pH of the products in **Figure 20,** you might need to use the log or antilog function on your calculator. A logarithm (log) is the power or exponent to which a number, called a base, must be raised in order to obtain a given positive number.

This textbook uses common logarithms based on a base of 10. Therefore, the common log of any number is the power to which 10 is raised to equal that number. Examine **Table 5** to compare logs and exponents. Note the log of each number is the power of 10 for the exponent of that number. For example, the common log of 100 is 2, and the common log of 0.01 is −2.

Table 5 Comparison Between Exponents and Logs

Exponent	Logarithm
$10^0 = 1$	$\log 1 = 0$
$10^1 = 10$	$\log 10 = 1$
$10^2 = 100$	$\log 100 = 2$
$10^{-1} = 0.1$	$\log 0.1 = -1$
$10^{-2} = 0.01$	$\log 0.01 = -2$

$$\log 10^2 = 2$$
$$\log 10^{-2} = -2$$

A common log can be written in the following general form.

$$\text{If } 10^n = y, \text{ then } \log y = n.$$

In each example in **Table 5,** the log can be determined by inspection. How do you express the common log of 5.34×10^5? Because logarithms are exponents, they have the same properties as exponents, as shown in **Table 6** on the next page.

$$\log 5.34 \times 10^5 = \log 5.34 + \log 10^5$$

Table 6 Properties of Exponents

Exponential Notation	Logarithm
$10^A \times 10^B = 10^{A+B}$	$\log (A \times B) = \log A + \log B$
$10^A \div 10^B = 10^{A-B}$	$\log (A \div B) = \log A - \log B$
A^B	$(\log A) \times B$

Significant figures and logarithms

Most scientific calculators have a button labeled *log* and, in most cases, you enter the number and push the log button to display the log of the number. Note that there is the same number of digits after the decimal in the log as there are significant figures in the original number entered.

$$\log 5.34 \times 10^5 = \log 5.34 + \log 10^5 = 0.728 + 5 = 5.728$$

Antilogarithms

Suppose the pH of the aqueous ammonia in **Figure 20** is 9.54 and you are asked to find the concentration of the hydrogen ions in that solution. By definition, $pH = -\log [H^+]$. Compare this to the general equation for the common log.

Equation for pH: $pH = -\log [H^+]$

General equation: $y = \log 10^n$

To solve the equation for $[H^+]$, you must follow the reverse process and calculate the antilogarithm (antilog) of −9.54 to find $[H^+]$.

Antilogs are the reverse of logs. To find the antilog, use a scientific calculator to input the value of the log. Then, use the inverse function and press the log button. The number of digits after the decimal in the log equals the number of significant figures in the antilog. An antilog can be written in the following general form.

$$\text{If } n = \text{antilog } y, \text{ then } y = 10^n.$$

$$\text{Thus, } [H^+] = \text{antilog}(-9.54) = 10^{-9.54} = 10^{(0.46-10)} = 10^{0.46} \times 10^{-10} = 2.9 \times 10^{-10} M$$

Check the instruction manual for your calculator. The exact procedure to calculate logs and antilogs might vary.

PRACTICE Problems

19. Find the log of each of the following numbers.

a. 367 **b.** 4078 **c.** X^n

20. Find the antilog of each of the following logs.

a. 4.663 **b.** 2.367 **c.** 0.371 **d.** −1.588

Figure 20 Ammonia is a base. That means its hydrogen ion concentration is less than $10^{-7}M$.

Geoff Butler

Math Handbook

Complete solutions to the Math Handbook Practice Problems can be found in the Solutions Manual.

19. a. 2.565
b. 3.6104
c. $(\log x) \times n$

20. a. 4.60×10^4
b. 2.33×10^2
c. 2.35
d. 2.58×10^{-2}

REFERENCE TABLES

Table R-1 Color Key

Carbon	Bromine	Sodium/ Other metals
Hydrogen	Iodine	Gold
Oxygen	Sulfur	Copper
Nitrogen	Phosphorus	Electron
Chlorine	Silicon	Proton
Fluorine	Helium	Neutron

Table R-2 Symbols and Abbreviations

Symbol		Meaning
α	=	rays from radioactive materials, helium nuclei
β	=	rays from radioactive materials, electrons
γ	=	rays from radioactive materials, high-energy quanta
Δ	=	change in
λ	=	wavelength
ν	=	frequency
A	=	ampere *(electric current)*
amu	=	atomic mass unit
Bq	=	becquerel *(nuclear disintegration)*
°C	=	Celsius degree *(temperature)*
C	=	coulomb *(quantity of electricity)*
c	=	speed of light
cd	=	candela *(luminous intensity)*
c	=	specific heat
D	=	density
E	=	energy, electromotive force
F	=	force
G	=	free energy
g	=	gram *(mass)*
Gy	=	gray *(radiation)*
H	=	enthalpy
Hz	=	hertz *(frequency)*
h	=	Planck's constant
h	=	hour *(time)*
J	=	joule *(energy)*
K	=	kelvin *(temperature)*
K_a	=	ionization constant *(acid)*
K_b	=	ionization constant *(base)*
K_{eq}	=	equilibrium constant
K_{sp}	=	solubility product constant
kg	=	kilogram *(mass)*
M	=	molarity
m	=	mass, molality
m	=	meter *(length)*
mol	=	mole *(amount)*
min	=	minute *(time)*
N	=	newton *(force)*
N_A	=	Avogadro's number
n	=	number of moles
P	=	pressure, power
Pa	=	pascal *(pressure)*
q	=	heat
Q_{sp}	=	ion product
R	=	ideal gas constant
S	=	entropy
s	=	second *(time)*
Sv	=	sievert *(absorbed radiation)*
T	=	temperature
V	=	volume
V	=	volt *(electric potential)*
v	=	velocity
W	=	watt *(power)*
w	=	work
X	=	mole fraction

Table R-3 Solubility Product Constants at 298 K

Compound	K_{sp}	Compound	K_{sp}	Compound	K_{sp}
Carbonates		**Halides**		**Hydroxides**	
$BaCO_3$	2.6×10^{-9}	CaF_2	3.5×10^{-11}	$Al(OH)_3$	4.6×10^{-33}
$CaCO_3$	3.4×10^{-9}	$PbBr_2$	6.6×10^{-6}	$Ca(OH)_2$	5.0×10^{-6}
$CuCO_3$	2.5×10^{-10}	$PbCl_2$	1.7×10^{-5}	$Cu(OH)_2$	2.2×10^{-20}
$PbCO_3$	7.4×10^{-14}	PbF_2	3.3×10^{-8}	$Fe(OH)_2$	4.9×10^{-17}
$MgCO_3$	6.8×10^{-6}	PbI_2	9.8×10^{-9}	$Fe(OH)_3$	2.8×10^{-39}
Ag_2CO_3	8.5×10^{-12}	AgCl	1.8×10^{-10}	$Mg(OH)_2$	5.6×10^{-12}
$ZnCO_3$	1.5×10^{-10}	AgBr	5.4×10^{-13}	$Zn(OH)_2$	3×10^{-17}
Hg_2CO_3	3.6×10^{-17}	AgI	8.5×10^{-17}	**Sulfates**	
Chromates		**Phosphates**		$BaSO_4$	1.1×10^{-10}
$BaCrO_4$	1.2×10^{-10}	$AlPO_4$	9.8×10^{-21}	$CaSO_4$	4.9×10^{-5}
$PbCrO_4$	2.3×10^{-13}	$Ca_3(PO_4)_2$	2.1×10^{-33}	$PbSO_4$	2.5×10^{-8}
Ag_2CrO_4	1.1×10^{-12}	$Mg_3(PO_4)_2$	1.0×10^{-24}	Ag_2SO_4	1.2×10^{-5}
Iodates		$FePO_4 \cdot 2H_2O$	9.91×10^{-16}	**Arsenates**	
$Cd(IO_3)_2$	2.3×10^{-8}	$Ni_3(PO_4)_2$	4.7×10^{-32}	$Pb_3(AsO_4)_2$	4.0×10^{-36}

Table R-4 Physical Constants

Quantity	Symbol	Value
Atomic mass unit	amu	1.6605×10^{-27} kg
Avogadro's number	N_A	6.022×10^{23} particles/mole
Ideal gas constant	R	8.31 L•kPa/mol•K 0.0821 L•atm/mol•K 62.4 mm Hg•L/mol•K 62.4 torr•L/mol•K
Mass of an electron	m_e	9.109×10^{-31} kg 5.485799×10^{-4} amu
Mass of a neutron	m_n	1.6749×10^{-27} kg 1.008665 amu
Mass of a proton	m_p	1.6726×10^{-27} kg 1.007276 amu
Molar volume of ideal gas at STP	V	22.414 L/mol
Normal boiling point of water	T_b	373.15 K 100.0°C
Normal freezing point of water	T_f	273.15 K 0.00°C
Planck's constant	h	$6.6260693 \times 10^{-34}$ J•s
Speed of light in a vacuum	c	2.997925×10^{8} m/s

Table R-5 Names and Charges of Polyatomic Ions

1−	2−	3−	4−
Acetate, CH_3COO^-	Carbonate, CO_3^{2-}	Arsenate, AsO_4^{3-}	Hexacyanoferrate(II), $Fe(CN)_6^{4-}$
Amide, NH_2^-	Chromate, CrO_4^{2-}	Arsenite, AsO_3^{3-}	Orthosilicate, SiO_4^{4-}
Astatate, AtO_3^-	Dichromate, $Cr_2O_7^{2-}$	Borate, BO_3^{3-}	Diphosphate, $P_2O_7^{4-}$
Azide, N_3^-	Hexachloroplatinate, $PtCl_6^{2-}$	Citrate, $C_6H_5O_7^{3-}$	
Benzoate, $C_6H_5COO^-$	Hexafluorosilicate, SiF_6^{2-}	Hexacyanoferrate(III), $Fe(CN)_6^{3-}$	
Bismuthate, BiO_3^-	Molybdate, MoO_4^{2-}	Phosphate, PO_4^{3-}	
Bromate, BrO_3^-	Oxalate, $C_2O_4^{2-}$	Phosphite, PO_3^{3-}	
Chlorate, ClO_3^-	Peroxide, O_2^{2-}		
Chlorite, ClO_2^-	Peroxydisulfate, $S_2O_8^{2-}$	**1+**	**2+**
Cyanide, CN^-	Ruthenate, RuO_4^{2-}	Ammonium, NH_4^+	Mercury(I), Hg_2^{2+}
Formate, $HCOO^-$	Selenate, SeO_4^{2-}	Neptunyl(V), NpO_2^+	Neptunyl(VI), NpO_2^{2+}
Hydroxide, OH^-	Selenite, SeO_3^{2-}	Plutonyl(V), PuO_2^+	Plutonyl(VI), PuO_2^{2+}
Hypobromite, BrO^-	Silicate, SiO_3^{2-}	Uranyl(V), UO_2^+	Uranyl(VI), UO_2^{2+}
Hypochlorite, ClO^-	Sulfate, SO_4^{2-}	Vanadyl(V), VO_2^+	Vanadyl(IV), VO^{2+}
Hypophosphite, $H_2PO_2^-$	Sulfite, SO_3^{2-}		
Iodate, IO_3^-	Tartrate, $C_4H_4O_6^{2-}$		
Nitrate, NO_3^-	Tellurate, TeO_4^{2-}		
Nitrite, NO_2^-	Tellurite, TeO_3^{2-}		
Perbromate, BrO_4^-	Tetraborate, $B_4O_7^{2-}$		
Perchlorate, ClO_4^-	Thiosulfate, $S_2O_3^{2-}$		
Periodate, IO_4^-	Tungstate, WO_4^{2-}		
Permanganate, MnO_4^-			
Perrhenate, ReO_4^-			
Thiocyanate, SCN^-			
Vanadate, VO_3^-			

Table R-6 Ionization Constants

Substance	Ionization Constant	Substance	Ionization Constant	Substance	Ionization Constant
$HCOOH$	1.77×10^{-4}	HBO_3^{2-}	1.58×10^{-14}	HS^-	1.00×10^{-19}
CH_3COOH	1.75×10^{-5}	H_2CO_3	4.5×10^{-7}	HSO_4^-	1.02×10^{-2}
$CH_2ClCOOH$	1.36×10^{-3}	HCO_3^-	4.68×10^{-11}	H_2SO_3	1.29×10^{-2}
$CHCl_2COOH$	4.47×10^{-2}	HCN	6.17×10^{-10}	HSO_3^-	6.17×10^{-8}
CCl_3COOH	3.02×10^{-1}	HF	6.3×10^{-4}	$HSeO_4^-$	2.19×10^{-2}
$HOOCCOOH$	5.36×10^{-2}	HNO_2	5.62×10^{-4}	H_2SeO_3	2.29×10^{-3}
$HOOCCOO^-$	1.55×10^{-4}	H_3PO_4	7.08×10^{-3}	$HSeO_3^-$	4.79×10^{-9}
CH_3CH_2COOH	1.34×10^{-5}	$H_2PO_4^-$	6.31×10^{-8}	$HBrO$	2.51×10^{-9}
C_6H_5COOH	6.25×10^{-5}	HPO_4^{2-}	4.17×10^{-13}	$HClO$	2.9×10^{-8}
H_3AsO_4	6.03×10^{-3}	H_3PO_3	5.01×10^{-2}	HIO	3.16×10^{-11}
$H_2AsO_4^-$	1.05×10^{-7}	$H_2PO_3^-$	2.00×10^{-7}	NH_3	5.62×10^{-10}
H_3BO_3	5.75×10^{-10}	H_3PO_2	5.89×10^{-2}	H_2NNH_2	7.94×10^{-9}
$H_2BO_3^-$	1.82×10^{-13}	H_2S	9.1×10^{-8}	H_2NOH	1.15×10^{-6}

Table R-7 Properties of Elements

Element	Symbol	Atomic Number	Atomic Mass* (amu)	Melting Point (°C)	Boiling Point (°C)	Density (g/cm³) (gases measured at STP)	Atomic Radius (pm)	First Ionization Energy (kJ/mol)	Standard Reduction Potential (V) (for elements from or to oxidation state indicated)	Enthalpy of Fusion	Specific Heat	Enthalpy of Vaporization	Abundance in Earth's Crust	Major Oxidation States
Actinium	Ac	89	[227]	1050	3300	10.07	---	499	(3+)−2.13	14	0.120	400	---	3+
Aluminum	Al	13	26.981539	660.32	2519	2.7	143	577.5	(3+)−1.68	10.789	0.897	294	8.2	3+
Americium	Am	95	[243]	1176	2607	13.67	---	578	(3+)−2.07	14.39	0.110	---	---	2+, 3+, 4+
Antimony	Sb	51	121.760	630.6	1587	6.697	140	834	(3+)+0.15	19.79	0.207	68	2×10^{-5}	3+, 5+
Argon	Ar	18	39.948	−189.3	−185.8	0.001784	98	1521	---	1.18	0.520	6.43	1.5×10^{-4}	---
Arsenic	As	33	74.92160	817	614	5.727	120	947	(3+)+0.24	24.44	0.329	32.4	2.1×10^{-4}	3+, 5+
Astatine	At	85	[210]	302	---	---	140	920	(1−)+0.2	6	---	40	---	1−, 5+
Barium	Ba	56	137.327	727	1870	3.51	222	502.9	(2+)−2.92	7.12	0.204	140	0.034	2+
Berkelium	Bk	97	[247]	986	---	14.78	---	601	(3+)−2.01	---	---	---	---	3+, 4+
Beryllium	Be	4	9.012182	1287	2469	1.848	112	899.5	(2+)−1.97	7.895	1.825	297	2×10^{-4}	2+
Bismuth	Bi	83	208.98040	271.3	1564	9.78	150	703	(3+)+0.317	11.145	0.122	151	3×10^{-7}	3+, 5+
Bohrium	Bh	107	[264]	---	---	---	---	---	---	---	---	---	---	---
Boron	B	5	10.811	2076	3927	2.46	85	800.6	(3+)−0.89	50.2	1.026	480	9×10^{-4}	3+
Bromine	Br	35	79.904	−7.3	59	3.119	114	1139.9	(1−)+1.065	10.57	0.474	29.96	3×10^{-4}	1−, 1+, 3+, 5+
Cadmium	Cd	48	112.411	321.07	767	8.65	151	867.8	(2+)−0.4025	6.21	0.232	99.87	1.5×10^{-5}	2+
Calcium	Ca	20	40.078	842	1484	1.55	197	589.8	(2+)−2.84	8.54	0.647	155	5.00	2+
Californium	Cf	98	[251]	900	---	15.1	---	608	(3+)−1.93	---	---	---	---	3+, 4+
Carbon	C	6	12.0107	3527	4027	2.267	77	1086.5	(4−)+0.132	117	0.709	715	0.018	4−, 2+, 4+
Cerium	Ce	58	140.116	795	3360	6.689	---	534.4	(3+)−2.34	5.46	0.192	350	0.006	3+, 4+
Cesium	Cs	55	132.905451	28.4	671	1.879	265	375.7	(1+)−2.923	2.09	0.242	65	1.9×10^{-4}	1+
Chlorine	Cl	17	35.453	−101.5	−34	0.003	100	1251.2	(1−)+1.358	6.40	0.479	20.41	0.017	1−, 1+, 3+, 5+
Chromium	Cr	24	51.9961	1907	2671	7.14	128	652.9	(3+)−0.74	21.0	0.449	339	0.014	2+, 3+, 6+
Cobalt	Co	27	58.9332	1495	2927	8.9	125	760.4	(2+)−0.28	16.06	0.421	375	0.003	2+, 3+
Copper	Cu	29	63.546	1084.62	2570	8.92	128	745.5	(2+)+0.34	12.93	0.385	300	0.0068	1+, 2+
Curium	Cm	96	[247]	1340	3110	13.51	---	581	(3+)−2.06	---	---	---	---	3+, 4+
Darmstadtium	Ds	110	[281]	---	---	---	---	---	---	---	---	---	---	---
Dubnium	Db	105	[262]	---	---	---	---	---	---	---	---	---	---	---
Dysprosium	Dy	66	162.5	1407	2567	8.551	---	573	(3+)−2.29	11.06	0.173	280	6×10^{-4}	2+, 3+
Einsteinium	Es	99	[252]	860	---	---	---	619	(3+)−2	---	---	---	---	3+
Erbium	Er	68	167.259	1497	2868	9.066	---	589.3	(3+)−2.32	19.9	0.168	285	3×10^{-4}	3+
Europium	Eu	63	151.964	826	1527	5.244	---	547.1	(3+)−1.99	9.21	0.182	175	1.8×10^{-4}	2+, 3+
Fermium	Fm	100	[257]	1527	---	---	---	627	(3+)−1.96	---	---	---	---	2+, 3+
Fluorine	F	9	18.9984032	−219.62	−188.12	0.001696	71	1681	(1−)+2.87	0.51	0.824	6.62	0.054	1−
Francium	Fr	87	[223]	---	---	---	270	380	(1+)−2.92	2	---	65	---	1+
Gadolinium	Gd	64	157.25	1312	3250	7.901	---	593.4	(3+)−2.28	10.0	0.236	305	5.2×10^{-4}	3+
Gallium	Ga	31	69.723	29.76	2204	5.904	135	578.8	(3+)−0.53	5.576	0.373	254	0.0019	1+, 3+
Germanium	Ge	32	72.64	938.3	2820	5.323	122	762	(4+)+0.124	36.94	0.320	334	1.4×10^{-4}	2+, 4+
Gold	Au	79	196.966569	1064	2856	19.3	144	890.1	(3+)+1.52	12.72	0.129	324	3×10^{-7}	1+, 3+

*[ ] indicates mass of longest-lived isotope

Reference Tables

Table R-7 Properties of Elements (continued)

Element	Symbol	Atomic Number	Atomic Mass* (amu)	Melting Point (°C)	Boiling Point (°C)	Density (g/cm³) (gases measured at STP)	Atomic Radius (pm)	First Ionization Energy (kJ/mol)	Standard Reduction Potential (V) (for elements from or to oxidation state indicated)	Enthalpy of Fusion	Specific Heat	Enthalpy of Vaporization	Abundance in Earth's Crust	Major Oxidation States
Hafnium	Hf	72	178.49	2233	4603	13.31	159	658.5	(4+)−1.70	27.2	0.144	630	3×10^{-4}	4+
Hassium	Hs	108	[277]	−272.2	---	0.0001785	---	2372.3	---	---	---	0.083	5.5×10^{-4}	---
Helium	He	2	4.002602	−269.7 (2536 kPa)	−268.93	0.00017847	31	2372	---	0.021	5.193	0.08	---	---
Holmium	Ho	67	164.93032	1461	2720	8.795	---	581	(3+)−2.33	17.0	0.165	265	1.2×10^{-4}	3+
Hydrogen	H	1	1.00794	−259.14	−252.87	0.0000899	37	1312	(1+)0.000	0.12	14.304	0.90	0.15	1−, 1+
Indium	In	49	114.818	156.6	2072	7.31	167	558.3	(3+)−0.3382	3.281	0.233	230	1.6×10^{-5}	1+, 3+
Iodine	I	53	126.90447	113.7	184.3	4.94	133	1008.4	(1−)+0.535	15.52	0.214	41.57	4.9×10^{-5}	1−, 1+, 5+, 7+
Iridium	Ir	77	192.217	2466	4428	22.65	136	880	(4+)+0.926	41.12	0.131	560	4×10^{-7}	3+, 4+, 5+
Iron	Fe	26	55.845	1538	2861	7.874	126	762.5	(3+)−0.04	13.81	0.449	347	6.3	2+, 3+
Krypton	Kr	36	83.798	−157.36	−153.22	0.0037493	112	1350.8	---	1.64	0.248	9.08	1.5×10^{-7}	---
Lanthanum	La	57	138.9055	920	3470	6.146	187	538.1	(3+)−2.38	6.20	0.195	400	0.0034	3+
Lawrencium	Lr	103	[262]	1627	---	---	---	---	(3+)−2	---	---	---	---	3+
Lead	Pb	82	207.2	327.46	1749	11.34	146	715.6	(2+)−0.1251	4.782	0.130	179.5	0.001	2+, 4+
Lithium	Li	3	6.941	180.54	1342	0.535	152	520.2	(1+)−3.040	3.00	3.582	147	0.0017	1+
Lutetium	Lu	71	174.967	1652	3402	9.841	160	523.5	(3+)−2.3	22	0.154	415	5.6×10^{-5}	3+
Magnesium	Mg	12	24.305	650	1090	1.738	160	737.7	(2+)−2.356	8.48	1.023	128	2.9	2+
Manganese	Mn	25	54.938045	1246	2061	7.47	127	717.3	(2+)−1.18	12.91	0.479	220	0.11	2+, 3+, 4+, 6+, 7+
Meitnerium	Mt	109	[268]	---	---	---	---	---	---	---	---	---	---	---
Mendelevium	Md	101	[258]	827	---	---	---	635	(3+)−1.7	---	---	---	---	2+, 3+
Mercury	Hg	80	200.59	−38.83	356.73	13.6	151	1007.1	(2+)+0.8535	2.29	0.140	59.11	6.7×10^{-6}	1+, 2+
Molybdenum	Mo	42	95.94	2623	4639	10.28	139	684.3	(6+)+0.114	37.48	0.251	600	1.1×10^{-4}	4+, 5+, 6+
Neodymium	Nd	60	144.24	1024	3100	6.8	---	533.1	(3+)−2.32	7.14	0.190	285	0.0033	2+,3+
Neon	Ne	10	20.1797	−248.59	−246.08	0.0008999	71	2080.7	---	0.328	1.030	1.71	---	---
Neptunium	Np	93	[237]	637	4000	20.45	---	604.5	(4+)−1.30	3.20	0.120	335	---	2+, 3+, 4+, 5+, 6+
Nickel	Ni	28	58.6934	1455	2913	8.908	124	737.1	(2+)−0.257	17.04	0.444	378	0.009	2+, 3+, 4+
Niobium	Nb	41	92.90638	2477	4744	8.57	146	652.1	(5+)−0.65	30	0.265	690	0.0017	4+, 5+
Nitrogen	N	7	14.0067	−210.1	−195.79	0.0012506	75	1402.3	(2−)−0.23	0.71	1.040	5.57	0.002	3−, 2−, 1−, 1+, 2+, 3+, 4+, 5+
Nobelium	No	102	[259]	827	---	---	---	642	(2+)−2.5	---	---	---	---	2+, 3+
Osmium	Os	76	190.23	3033	5012	22.61	135	840	(4+)+0.687	57.85	0.130	630	1.8×10^{-7}	4+, 6+, 8+
Oxygen	O	8	15.9994	−218.79	−182.9	0.001308	73	1313.9	(2−)+1.23	0.44	0.918	6.82	46.0	2−, 1−
Palladium	Pd	46	106.42	1554.9	2963	12.023	137	804.4	(2+)+0.915	16.74	0.246	380	6.3×10^{-7}	2+, 4+
Phosphorus	P	15	30.973462	44.2	277	1.823	110	1011.8	(3−)−0.063	0.66	0.769	12.4	0.10	3−, 3+, 5+
Platinum	Pt	78	195.078	1768.3	3825	21.09	138	870	(4+)+1.15	22.17	0.133	490	3.7×10^{-7}	2+, 4+
Plutonium	Pu	94	[244]	639.4	3230	19.816	---	584.7	(4+)−1.25	2.82	0.130	325	---	3+, 4+, 5+, 6+
Polonium	Po	84	[209]	254	962	9.196	168	812.1	(4+)+0.73	13	---	100	---	2−, 2+, 4+, 6+

*[ ] indicates mass of longest-lived isotope

Table R-7 Properties of Elements (continued)

Potassium	K	19	39.0983	63.38	759	0.856	227	418.8	(1+)−2.925	2.33	0.757	76.9	1.50	1+
Praseodymium	Pr	59	140.90765	935	3290	6.64	---	527	(3+)−2.35	6.89	0.193	330	8.7×10^{-4}	3+, 4+
Promethium	Pm	61	[145]	1100	3000	7.264	---	540	(3+)−2.29	7.7	---	290	---	3+
Protactinium	Pa	91	231.03588	1568	---	15.37	---	568	(5+)−1.19	12.34	---	470	trace	3+, 4+, 5+
Radium	Ra	88	[226]	700	1737	5	220	509.3	(2+)−2.916	8	0.095	125	trace	2+
Radon	Rn	86	[222]	−71	−61.7	0.00973	140	1037	---	3	0.094	17	---	3+
Rhenium	Re	75	186.207	3186	5596	21.02	137	760	(7+)+0.415	60.43	0.137	705	2.6×10^{-7}	3+, 4+, 6+, 7+
Rhodium	Rh	45	102.9055	1964	3695	12.45	134	719.7	(3+)+0.76	26.59	0.243	495	7×10^{-8}	3+, 4+, 5+
Roentgenium	Rg	111	[272]	---	---	---	---	---	---	---	---	---	---	---
Rubidium	Rb	37	85.4678	39.31	688	1.532	248	403	(1+)−2.924	2.19	0.363	72	0.006	1+
Ruthenium	Ru	44	101.07	2334	4150	12.37	134	710.2	(4+)+0.68	38.59	0.238	580	1×10^{-7}	2+, 3+, 4+, 5+
Rutherfordium	Rf	104	[261]	---	---	---	---	---	---	---	---	---	---	---
Samarium	Sm	62	150.36	1072	1803	7.353	---	544.5	(3+)−2.3	8.62	0.197	175	6×10^{-4}	2+, 3+
Scandium	Sc	21	44.95591	1541	2830	2.985	162	633.1	(3+)−2.03	14.1	0.568	318	0.0026	3+
Seaborgium	Sg	106	[266]	---	---	---	---	---	---	---	---	---	---	---
Selenium	Se	34	78.96	221	685	4.819	119	941	(1−)−0.11	6.69	0.321	95.48	5×10^{-6}	2−, 2+, 4+, 6+
Silicon	Si	14	28.0588	1414	2900	2.33	118	786.5	(4−)−0.143	50.21	0.712	359	27.0	2+, 4+
Silver	Ag	47	107.8682	961.78	2162	10.49	144	731	(1+)+0.7991	11.28	0.235	255	8×10^{-6}	1+
Sodium	Na	11	22.989769	97.72	883	0.968	186	495.8	(1+)−2.713	2.60	1.228	97.7	2.3	1+
Strontium	Sr	38	87.62	777	1382	2.63	215	549.5	(2+)−2.89	7.43	0.306	137	0.036	2+
Sulfur	S	16	32.065	115.2	444.7	1.96	103	999.6	(2−)−0.14	1.72	0.708	45	0.042	2−, 4+, 6+
Tantalum	Ta	73	180.9479	3017	5458	16.65	146	761	(5+)−0.81	36.57	0.140	735	1.7×10^{-4}	4+, 5+
Technetium	Tc	43	[98]	2157	4265	11.5	136	702	(6+)+0.83	33.29	0.240	550	---	2+, 4+, 6+, 7+
Tellurium	Te	52	127.60	449.51	988	6.24	142	869.3	(2−)−1.14	17.49	0.202	114.1	1×10^{-7}	2−, 2+, 4+, 6+
Terbium	Tb	65	158.92534	1356	3230	8.219	---	565.8	(3+)−2.31	10.15	0.182	295	1×10^{-4}	3+, 4+
Thallium	Tl	81	204.3822	304	1473	11.85	170	589.4	(1+)−0.3363	4.14	0.129	165	5.3×10^{-5}	1+, 3+
Thorium	Th	90	232.0381	1842	4820	11.72	---	587	(4+)−1.83	13.81	0.118	530	6×10^{-4}	4+
Thulium	Tm	69	168.93421	1545	1950	9.321	---	596.7	(3+)−2.32	16.84	0.160	250	5×10^{-5}	---
Tin	Sn	50	118.710	231.93	2602	7.31	140	708.6	(4+)+0.15	7.173	0.227	290	2.2×10^{-4}	2+, 4+
Titanium	Ti	22	47.867	1668	3287	4.507	147	658.8	(4+)−0.86	14.15	0.523	425	0.66	2+, 3+, 4+
Tungsten	W	74	183.84	3422	5555	19.25	139	770	(6+)−0.09	52.31	0.132	800	1.1×10^{-4}	4+, 5+, 6+
Ununbium	Uub	112	[285]	---	---	---	---	---	---	---	---	---	---	---
Livermorium	Lv	116	[291]	---	---	---	---	---	---	---	---	---	---	---
Ununoctium	Uuo	118	[294]	---	---	---	---	---	---	---	---	---	---	---
Ununpentium	Uup	115	[288]	---	---	---	---	---	---	---	---	---	---	---
Flerovium	Fl	114	[289]	---	---	---	---	---	---	---	---	---	---	---
Ununtrium	Uut	113	[284]	---	---	---	---	---	---	---	---	---	---	---
Uranium	U	92	238.02891	1132.2	3927	19.05	---	597.6	(4+)−1.38	9.14	0.116	420	1.8×10^{-4}	3+, 4+, 5+, 6+
Vanadium	V	23	50.9415	1910	3407	6.11	134	650.9	(5+)−0.236	21.5	0.489	453	0.019	2+, 3+, 4+, 5+
Xenon	Xe	54	131.293	−111.7	−108	0.0058971	131	1170.4	(6+)+2.12	2.27	0.158	12.57	trace	---
Ytterbium	Yb	70	173.04	824	1196	6.57	---	603.4	(3+)−2.22	7.66	0.155	160	2.8×10^{-4}	2+, 3+
Yttrium	Y	39	88.90585	1526	3336	4.472	180	600	(3+)−2.37	11.4	0.298	380	0.0029	3+
Zinc	Zn	30	65.409	419.53	907	7.14	134	906.4	(2+)−0.7926	7.068	0.388	119	0.0079	2+
Zirconium	Zr	40	91.224	1855	4409	6.511	160	640.1	(4+)−1.55	21.00	0.278	580	0.013	4+

*[ ] indicates mass of longest-lived isotope

Table R-8 Solubility Guidelines

A substance is considered soluble if more than three grams of the substance dissolves in 100 mL of water. The more common rules are listed below.

1. All common salts of the group 1 elements and ammonium ions are soluble.
2. All common acetates and nitrates are soluble.
3. All binary compounds of group 17 elements (other than F) with metals are soluble except those of silver, mercury(I), and lead.
4. All sulfates are soluble except those of barium, strontium, lead, calcium, silver, and mercury(I).
5. Except for those in Rule 1, carbonates, hydroxides, oxides, sulfides, and phoshates are insoluble.

Solubility of Compounds in Water

	Acetate	Bromide	Carbonate	Chlorate	Chloride	Chromate	Hydroxide	Iodide	Nitrate	Oxide	Perchlorate	Phosphate	Sulfate	Sufide
Aluminum	S	S	—	S	S	—	I	S	S	I	S	I	S	D
Ammonium	S	S	S	S	S	S	S	S	S	—	S	S	S	S
Barium	S	S	P	S	S	I	S	S	S	S	S	I	I	D
Calcium	S	S	P	S	S	S	S	S	S	P	S	P	I	P
Copper(II)	S	S	—	S	S	—	I	—	S	I	S	I	S	I
Hydrogen	S	S	—	S	S	—	—	S	S	S	S	S	S	S
Iron(II)	—	S	P	S	S	—	I	S	S	I	S	I	S	I
Iron(III)	—	S	—	S	S	I	I	S	S	I	S	P	P	D
Lead(II)	S	I	—	S	I	I	P	P	S	P	S	I	I	I
Lithium	S	S	S	S	S	?	S	S	S	S	S	P	S	S
Magnesium	S	S	P	S	S	S	I	S	S	I	S	P	S	D
Manganese(II)	S	S	P	S	S	—	I	S	S	I	S	P	S	I
Mercury(I)	P	I	I	S	I	P	—	I	S	I	S	I	I	I
Mercury(II)	S	S	—	S	S	P	I	P	S	P	S	I	D	I
Potassium	S	S	S	S	S	S	S	S	S	S	S	S	S	S
Silver	P	I	I	S	I	P	—	I	S	P	S	I	I	I
Sodium	S	S	S	S	S	S	S	S	S	D	S	S	S	S
Strontium	S	S	P	S	S	P	S	S	S	S	S	I	I	S
Tin(II)	D	S	—	S	S	O		S	D	I	S	I	S	I
Tin(IV)	S	S	—	—	S	S	I	D	—	I	S	—	S	I
Zinc	S	S	P	S	S	P	P	S	S	P	S	I	S	I

S – soluble P – partially soluble I – insoluble D – decomposes

Table R-9 Specific Heat Values (J/g·K)

Substance	*c*	Substance	*c*	Substance	*c*
AlF_3	0.8948	Fe_3C	0.5898	$NaVO_3$	1.540
$BaTiO_3$	0.79418	$FeWO_4$	0.37735	$Ni(CO)_4$	1.198
BeO	1.020	HI	0.22795	PbI_2	0.1678
CaC_2	0.9785	K_2CO_3	0.82797	SF_6	0.6660
$CaSO_4$	0.7320	$MgCO_3$	0.8957	SiC	0.6699
CCl_4	0.85651	$Mg(OH)_2$	1.321	SiO_2	0.7395
CH_3OH	2.55	$MgSO_4$	0.8015	$SrCl_2$	0.4769
CH_2OHCH_2OH	2.413	MnS	0.5742	Tb_2O_3	0.3168
CH_3CH_2OH	2.4194	Na_2CO_3	1.0595	$TiCl_4$	0.76535
CdO	0.3382	NaF	1.116	Y_2O_3	0.45397
$CuSO_4 \cdot 5H_2O$	1.12				

Reference Tables

Table R-10 Molal Freezing Point Depression and Boiling Point Elevation Constants

Substance	K_{fp} (°C/*m*)	Freezing Point (°C)	K_{bp} (°C/*m*)	Boiling Point (°C)
Acetic acid	3.90	16.66	3.22	117.90
Benzene	5.12	5.533	2.53	80.100
Camphor	37.7	178.75	5.611	207.42
Cyclohexane	20.0	6.54	2.75	80.725
Cyclohexanol	39.3	25.15	---	---
Nitrobenzene	6.852	5.76	5.24	210.8
Phenol	7.40	40.90	3.60	181.839
Water	1.86	0.000	0.512	100.000

Table R-11 Heat of Formation Values

ΔH_f° (kJ/mol) (concentration of aqueous solutions is 1*M*)

Substance	ΔH_f°	Substance	ΔH_f°	Substance	ΔH_f°	Substance	ΔH_f°
Ag(s)	0	$CsCl$(s)	−443.0	H_3PO_4(aq)	−1271.7	$NaBr$(s)	−361.1
$AgCl$(s)	−127.0	Cs_2SO_4(s)	−1443.0	H_2S(g)	−20.6	$NaCl$(s)	−411.2
$AgCN$(s)	146.0	CuI(s)	−67.8	H_2SO_3(aq)	−608.8	$NaHCO_3$(s)	−950.8
Al_2O_3	−1675.7	CuS(s)	−53.1	H_2SO_4(aq)	−814.0	$NaNO_3$(s)	−467.9
$BaCl_2$(aq)	−855.0	Cu_2S(s)	−79.5	$HgCl_2$(s)	−224.3	$NaOH$(s)	−425.8
$BaSO_4$	−1473.2	$CuSO_4$(s)	−771.4	Hg_2Cl_2(s)	−265.4	Na_2CO_3(s)	−1130.7
BeO(s)	−609.4	F_2(g)	0	Hg_2SO_4(s)	−743.1	Na_2S(s)	−364.8
$BiCl_3$(s)	−379.1	$FeCl_3$(s)	−399.49	I_2(s)	0	Na_2SO_4(s)	−1387.1
Bi_2S_3(s)	−143.1	FeO(s)	−272.0	K(s)	0	NH_4Cl(s)	−314.4
Br_2	0	FeS(s)	−100.0	KBr(s)	−393.8	O_2(g)	0
CCl_4(l)	−128.2	Fe_2O_3(s)	−824.2	$KMnO_4$(s)	−837.2	P_4O_6(s)	−1640.1
CH_4(g)	−74.6	Fe_3O_4(s)	−1118.4	KOH	−424.6	P_4O_{10}(s)	−2984.0
C_2H_2(g)	227.4	H(g)	218.0	$LiBr$(s)	−351.2	$PbBr_2$(s)	−278.7
C_2H_4(g)	52.4	H_2(g)	0	$LiOH$(s)	−487.5	$PbCl_2$(s)	−359.4
C_2H_6(g)	−84.0	HBr(g)	−36.3	Mn(s)	0	SF_6(g)	−1220.5
CO(g)	−110.5	HCl(g)	−92.3	$MnCl_2$(aq)	−555.0	SO_2(g)	−296.8
CO_2(g)	−393.5	HCl(aq)	−167.159	$Mn(NO_3)_2$(aq)	−635.5	SO_3(g)	−395.7
CS_2(l)	89.0	HCN(aq)	108.9	MnO_2(s)	−520.0	SrO(s)	−592.0
Ca(s)	0	$HCHO$	−108.6	MnS(s)	−214.2	TiO_2(s)	−944.0
$CaCO_3$(s)	−1206.9	$HCOOH$	−425.0	N_2(g)	0	TlI(s)	−123.8
CaO(s)	−634.9	HF(g)	−273.3	NH_3(g)	−45.9	UCl_4(s)	−1019.2
$Ca(OH)_2$(s)	−985.2	HI(g)	26.5	NH_4Br(s)	−270.8	UCl_6(s)	−1092.0
Cl_2(g)	0	H_2O(l)	−285.8	NO(g)	91.3	Zn(s)	0
Co_3O_4(s)	−891.0	H_2O(g)	−241.8	NO_2(g)	33.2	$ZnCl_2$(aq)	−415.1
CoO(s)	−237.9	H_2O_2(l)	−187.8	N_2O(g)	81.6	ZnO(s)	−350.5
Cr_2O_3(s)	−1139.7	H_3PO_2(l)	−595.4	Na(s)	0	$ZnSO_4$(s)	−982.8

Safety Symbols

Safety symbols in the following table are used in the lab activities to indicate possible hazards. Learn the meaning of each symbol. **It is recommended that you wear safety goggles and apron at all times in the lab. This might be required in your school district.**

Safety Symbols	Hazard	Examples	Precaution	Remedy
Disposal	Special disposal procedures need to be followed.	certain chemicals, living organisms	Do not dispose of these materials in the sink or trash can.	Dispose of wastes as directed by your teacher.
Biological	Organisms or other biological materials that might be harmful to humans	bacteria, fungi, blood, unpreserved tissues, plant materials	Avoid skin contact with these materials. Wear mask or gloves.	Notify your teacher if you suspect contact with material. Wash hands thoroughly.
Extreme Temperature	Objects that can burn skin by being too cold or too hot	boiling liquids, hot plates, dry ice, liquid nitrogen	Use proper protection when handling.	Go to your teacher for first aid.
Sharp Object	Use of tools or glassware that can easily puncture or slice skin	razor blades, pins, scalpels, pointed tools, dissecting probes, broken glass	Practice common-sense behavior and follow guidelines for use of the tool.	Go to your teacher for first aid.
Fume	Possible danger to respiratory tract from fumes	ammonia, acetone, nail polish remover, heated sulfur, moth balls	Be sure there is good ventilation. Never smell fumes directly. Wear a mask.	Leave foul area and notify your teacher immediately.
Electrical	Possible danger from electrical shock or burn	improper grounding, liquid spills, short circuits, exposed wires	Double-check setup with teacher. Check condition of wires and apparatus. Use GFI-protected outlets.	Do not attempt to fix electrical problems. Notify your teacher immediately.
Irritant	Substances that can irritate the skin or mucous membranes of the respiratory tract	pollen, moth balls, steel wool, fiberglass, potassium permanganate	Wear dust mask and gloves. Practice extra care when handling these materials.	Go to your teacher for first aid.
Chemical	Chemicals that can react with and destroy tissue and other materials	bleaches such as hydrogen peroxide; acids such as sulfuric acid, hydrochloric acid; bases such as ammonia, sodium hydroxide	Wear goggles, gloves, and an apron.	Immediately flush the affected area with water and notify your teacher.
Toxic	Substance may be poisonous if touched, inhaled, or swallowed.	mercury, many metal compounds, iodine, poinsettia plant parts	Follow your teacher's instructions.	Always wash hands thoroughly after use. Go to your teacher for first aid.
Flammable	Flammable chemicals may be ignited by open flame, spark, or exposed heat.	alcohol, kerosene, potassium permanganate	Avoid open flames and heat when using flammable chemicals.	Notify your teacher immediately. Use fire safety equipment if applicable.
Open Flame	Open flame in use, may cause fire.	hair, clothing, paper, synthetic materials	Tie back hair and loose clothing. Follow teacher's instruction on lighting and extinguishing flames.	Notify your teacher immediately. Use fire safety equipment if applicable.

Eye Safety
Proper eye protection should be worn at all times by anyone performing or observing science activities.

Clothing Protection
This symbol appears when substances could stain or burn clothing.

Animal Safety
This symbol appears when safety of animals and students must be ensured.

Radioactivity
This symbol appears when radioactive materials are used.

Handwashing
After the lab, wash hands with soap and water before removing goggles.

SUPPLEMENTAL PRACTICE PROBLEMS

CHAPTER 2

Section 1

1. The density of a substance is 4.8 g/mL. What is the volume of a sample that is 19.2 g?

2. A 2.00-mL sample of Substance A has a density of 18.4 g/mL, and a 5.00-mL sample of Substance B has a density of 35.5 g/mL. Do you have an equal mass of Substances A and B?

Section 2

3. Express the following quantities in scientific notation.
   - **a.** 5,453,000 m
   - **b.** 300.8 kg
   - **c.** 0.00536 ng
   - **d.** 0.0120325 km
   - **e.** 34,800 s
   - **f.** 332,080,000 cm
   - **g.** 0.0002383 ms
   - **h.** 0.3048 mL

4. Solve the following problems. Express your answers in scientific notation.
   - **a.** 3×10^2 m + 5×10^2 m
   - **b.** 8×10^{-5} m + 4×10^{-5} m
   - **c.** 6.0×10^5 m + 2.38×10^6 m
   - **d.** 2.3×10^{-3} L + 5.78×10^{-2} L
   - **e.** 2.56×10^2 g − 1.48×10^2 g
   - **f.** 5.34×10^{-3} L − 3.98×10^{-3} L
   - **g.** 7.623×10^5 nm − 8.32×10^4 nm
   - **h.** 9.052×10^{-2} s − 3.61×10^{-3} s

5. Solve the following problems. Express your answers in scientific notation.
   - **a.** $(8 \times 10^3$ m$) \times (1 \times 10^5$ m$)$
   - **b.** $(4 \times 10^2$ m$) \times (2 \times 10^4$ m$)$
   - **c.** $(5 \times 10^{-3}$ m$) \times (3 \times 10^4$ m$)$
   - **d.** $(3 \times 10^{-4}$ m$) \times (3 \times 10^{-2}$ m$)$
   - **e.** $(8 \times 10^4$ g$) \div (4 \times 10^3$ mL$)$
   - **f.** $(6 \times 10^{-3}$ g$) \div (2 \times 10^{-1}$ mL$)$
   - **g.** $(1.8 \times 10^{-2}$ g$) \div (9 \times 10^{-5}$ mL$)$
   - **h.** $(4 \times 10^{-4}$ g$) \div (1 \times 10^3$ mL$)$

6. Perform the following conversions.
   - **a.** 96 kg to g
   - **b.** 155 mg to g
   - **c.** 15 cg to kg
   - **d.** 584 µs to s
   - **e.** 188 dL to L
   - **f.** 3600 m to km
   - **g.** 24 g to pg
   - **h.** 85 cm to nm

7. How many minutes are there in 5 days?

8. A car is traveling at 118 km/h. What is its speed in Mm/h?

Section 3

9. Three measurements of 34.5 m, 38.4 m, and 35.3 m are taken. If the accepted value of the measurement is 36.7 m, what is the percent error for each measurement?

10. Three measurements of 12.3 mL, 12.5 mL, and 13.1 mL are taken. The accepted value for each measurement is 12.8 mL. Calculate the percent error for each measurement.

Supplemental Practice Problems

SUPPLEMENTAL PRACTICE PROBLEMS

Complete solutions to Supplemental Practice Problems can be found in the Solutions Manual.

CHAPTER 2

Section 1

1. 4.0 mL
2. No, mass A does not equal mass B.

Section 2

3. **a.** 5.453×10^6 m
   b. 3.008×10^2 kg
   c. 5.36×10^{-3} ng
   d. 1.20325×10^{-2} km
   e. 3.48×10^4 s
   f. 3.3208×10^8 cm
   g. 2.383×10^{-4} ms
   h. 3.048×10^{-1} mL
4. **a.** 8×10^2 m
   b. 1.2×10^{-4} m
   c. 2.98×10^6 m
   d. 6.01×10^{-2} L
   e. 1.08×10^2 g
   f. 1.36×10^{-3} L
   g. 6.791×10^5 nm
   h. 8.691×10^{-2} s
5. **a.** 8×10^8 m^2
   b. 8×10^6 m^2
   c. 1.5×10^2 m^2
   d. 9×10^{-6} m^2
   e. 2×10^1 g/mL
   f. 3×10^{-2} g/mL
   g. 2×10^2 g/mL
   h. 4×10^{-7} g/mL
6. **a.** 9.6×10^4 g
   b. 1.55×10^{-1} g
   c. 1.5×10^{-4} kg
   d. 5.84×10^{-4} s
   e. 1.88×10^1 L
   f. 3.6 km
   g. 2.4×10^{13} pg
   h. 8.5×10^8 nm
7. 7200 minutes
8. 0.118 Mm/h

Section 3

9. 5.99%, 4.63%, 3.81%
10. 3.91%, 2.34%, 2.34%

3
d. 8
e. 4
f. 3
g. 4
h. 3

12. a. 0.00309 km
b. 3.08 g
c. 5810 mL
d. 34.7 mg

13. a. 5.832×10^{-3} g
b. 3.86808×10^{5} ns
c. 5.800×10^{-4} km
d. 2.086×10^{3} L

14. a. 103.8 m
b. 0.167 kg
c. 705.2 mL
d. 8.00 mg
e. 2.223 km
f. 21.8 s
g. 462.9 dm
h. 4.81×10^{2} m

15. a. 410 cm^2
b. 0.021 mm^2
c. 604 km^2
d. 2.94 g/mL
e. 1.426 g/mL
f. 15.3 g/mL

CHAPTER 3

Section 2

1. 0.2 kg oxygen
2. 3.1 g hydrochloric acid

Section 4

3. 6.25 g sulfur
4. No. Compound I contains 62.6% tin and 37.4% chlorine, and Compound II contains 45.6% tin and 54.4% chlorine.

CHAPTER 4

Section 3

1. a. 31; 31
b. 14; 14
c. 55; 55
d. 20; 20
e. 42; 42
f. 22; 22

11. Determine the number of significant figures in each measurement.
a. 340,438 g
b. 87,000 ms
c. 4080 kg
d. 961,083,110 m
e. 1.040 s
f. 0.0483 m
g. 0.2080 mL
h. 0.0000481 g

12. Write the following in three significant figures.
a. 0.0030850 km
b. 3.0823 g
c. 5808 mL
d. 34.654 mg

13. Write the answers in scientific notation.
a. 0.005832 g
b. 386,808 ns
c. 0.0005800 km
d. 2086 L

14. Use rounding rules when you complete the following.
a. 34.3 m + 35.8 m + 33.7 m
b. 0.056 kg + 0.0783 kg + 0.0323 kg
c. 309.1 mL + 158.02 mL + 238.1 mL
d. 1.03 mg + 2.58 mg + 4.385 mg
e. 8.376 km − 6.153 km
f. 34.24 s − 12.4 s
g. 804.9 dm − 342.0 dm
h. 6.38×10^{2} m − 1.57×10^{2} m

15. Complete the following calculations. Round off the answers to the correct number of significant figures.
a. 34.3 cm × 12 cm
b. 0.054 mm × 0.3804 mm
c. 45.1 km × 13.4 km
d. 45.5 g ÷ 15.5 mL
e. 35.43 g ÷ 24.84 mL
f. 0.0482 g ÷ 0.003146 mL

CHAPTER 3

Section 2

1. A 3.5-kg iron shovel is left outside through the winter. The shovel, now orange with rust, is rediscovered in the spring. Its mass is 3.7 kg. How much oxygen combined with the iron?
2. When 5.0 g of tin reacts with hydrochloric acid, the mass of the products, tin chloride and hydrogen, totals 8.1 g. How many grams of hydrochloric acid were used?

Section 4

3. A compound is analyzed and found to be 50.0% sulfur and 50.0% oxygen. If the total amount of the sulfur oxide compound is 12.5 g, how many grams of sulfur are there?
4. Two unknown compounds are analyzed. Compound I contains 5.63 g of tin and 3.37 g of chlorine, while Compound II contains 2.5 g of tin and 2.98 g of chlorine. Are the compounds the same?

CHAPTER 4

Section 3

1. How many protons and electrons are in each of the following atoms?
a. gallium
b. silicon
c. cesium
d. calcium
e. molybdenum
f. titanium

2. What is the atomic number of each of the following elements?
   a. an atom that contains 37 electrons
   b. an atom that contains 72 protons
   c. an atom that contains 1 electron
   d. an atom that contains 85 protons

3. Use the periodic table to write the name and the symbol for each element identified in Question 2.

4. An isotope of copper contains 29 electrons, 29 protons, and 36 neutrons. What is the mass number of this isotope?

5. An isotope of uranium contains 92 electrons and 144 neutrons. What is the mass number of this isotope?

6. Use the periodic table to write the symbols for each of the following elements. Then, determine the number of electrons, protons, and neutrons each contains.
   a. yttrium-88
   b. arsenic-75
   c. xenon-129
   d. bromine-79
   e. gold-197
   f. helium-4

7. An element has two naturally occurring isotopes: ^{14}X and ^{15}X. ^{14}X has a mass of 14.00307 amu and a relative abundance of 99.63%. ^{15}X has a mass of 15.00011 amu and a relative abundance of 0.37%. Identify the unknown element.

8. Silver has two naturally occurring isotopes. Ag-107 has an abundance of 51.82% and a mass of 106.9 amu. Ag-109 has a relative abundance of 48.18% and a mass of 108.9 amu. Calculate the atomic mass of silver.

CHAPTER 5

Section 1

1. What is the frequency of an electromagnetic wave that has a wavelength of 4.55×10^{-3} m? 1.00×10^{-12} m?

2. Calculate the wavelength of an electromagnetic wave with a frequency of 8.68×10^{16} Hz; 5.0×10^{14} Hz; and 1.00×10^{6} Hz.

3. What is the energy of a quantum of visible light having a frequency of 5.45×10^{14} s^{-1}?

4. An X-ray has a frequency of 1.28×10^{18} s^{-1}. What is the energy of a quantum of the X-ray?

Section 3

5. Write the ground-state electron configuration for the following.
   a. nickel
   b. cesium
   c. boron
   d. krypton

6. What element has the following ground-state electron configuration $[He]2s^2$? $[Xe]6s^24f^{14}5d^{10}6p^1$?

7. Which element in period 4 has four electrons in its electron-dot structure?

8. Which element in period 2 has six electrons in its electron-dot structure?

9. Draw the electron-dot structure for each element in Question 5.

Supplemental Practice Problems

2. a. 37
   b. 72
   c. 1
   d. 85
3. a. rubidium; Rb
   b. hafnium; Hf
   c. hydrogen; H
   d. astatine; At
4. 65
5. 236
6. a. Y; 39 electrons, 39 protons, 49 neutrons
   b. As; 33 electrons, 33 protons, 42 neutrons
   c. Xe; 54 electrons, 54 protons, 75 neutrons
   d. Br; 35 electrons, 35 protons, 44 neutrons
   e. Au; 79 electrons, 79 protons, 118 neutrons
   f. He; 2 electrons, 2 protons, 2 neutrons
7. nitrogen
8. 107.87 amu

CHAPTER 5

Section 1

1. 6.59×10^{10} Hz; 3.00×10^{20} Hz
2. 3.46×10^{-9} m; 6.0×10^{-7} m; 3.00×10^{2} m
3. 3.61×10^{-19} J
4. 8.48×10^{-16} J

Section 3

5. a. $1s^22s^22p^63s^23p^64s^23d^8$ or $[Ar]4s^23d^8$
   b. $1s^22s^22p^63s^23p^64s^23d^{10}4p^65s^24d^{10}5p^66s^1$ or $[Xe]6s^1$
   c. $1s^22s^22p^1$ or $[He]2s^22p^1$
   d. $1s^22s^22p^63s^23p^64s^23d^{10}4p^6$ or $[Kr]$
6. beryllium; thallium
7. germanium
8. oxygen
9. a. $\dot{\mathrm{Ni}}\cdot$
   b. $\mathrm{Cs}\cdot$
   c. $\cdot\dot{\mathrm{B}}\cdot$
   d. $:\ddot{\underset{\cdot\cdot}{\mathrm{Kr}}}:$

a. 13, 2, p
b. 7, 5, d
c. 7, 6, d

2. a. helium: $1s^2$ or [He]
b. zirconium: $1s^22s^22p^63s^23p^64s^23d^{10}4p^65s^24d^2$ or $[Kr]5s^24d^2$
c. lead: $1s^22s^22p^63s^23p^64s^23d^{10}4p^6$ $5s^24d^{10}5p^66s^24f^{14}5d^{10}6p^2$ or $[Xe]6s^24f^{14}5d^{10}6p^2$
d. francium: $1s^22s^22p^63s^23p^64s^23d^{10}4p^6$ $5s^24d^{10}5p^66s^24f^{14}5d^{10}6p^67s^1$ or $[Rn]7s^1$

Section 3

3. a. magnesium, calcium, and strontium
b. fluorine, oxygen, and lithium
c. fluorine, calcium, and cesium
d. chlorine, selenium, and tellurium
e. krypton, iodine, and magnesium

CHAPTER 7

Section 2

1. Zinc has two valence electrons. Chlorine has seven valence electrons. One zinc atom will lose two electrons forming a 2+ ion. Two chlorine atoms will each gain one electron forming two 1− ions. The oppositely charged ions will attract and form the ionic compound $ZnCl_2$. The two positive charges balance the two negative charges.
2. Barium has two valence electrons. Nitrogen has five valence electrons. Three barium atoms will each lose two electrons forming 2+ ions. Two nitrogen atoms will each gain three electrons forming 3− ions. The oppositely charged ions will attract and form the ionic compound Ba_3N_2. The six positive charges balance the six negative charges.

Supplemental Practice Problems

CHAPTER 6

Section 2

1. Identify the group, period, and block of an atom with the following electron configurations.
   a. $[He]2s^22p^1$ b. $[Kr]5s^24d^5$ c. $[Xe]6s^25f^{14}6d^5$
2. Write the electron configuration for the element fitting each of the following descriptions.
   a. a noble gas in the first period
   b. a group 4 element in the fifth period
   c. a group 14 element in the sixth period
   d. a group 1 element in the seventh period

Section 3

3. Using the periodic table, rank each group of elements in order of increasing size.
   a. calcium, magnesium, and strontium
   b. oxygen, lithium, and fluorine
   c. fluorine, cesium, and calcium
   d. selenium, chlorine, and tellurium
   e. iodine, krypton, and beryllium

CHAPTER 7

Section 2

1. Explain the formation of an ionic compound from zinc and chlorine.
2. Explain the formation of an ionic compound from barium and nitrogen.

Section 3

3. Write the chemical formula of an ionic compound composed of the following pairs of ions.
   a. calcium and arsenide
   b. iron(III) and chloride
   c. magnesium and sulfide
   d. barium and iodide
   e. gallium and phosphide
4. Determine the formula for ionic compounds composed of the following ions.
   a. copper(II) and acetate   c. calcium and hydroxide
   b. ammonium and phosphate   d. gold(III) and cyanide
5. Name the following compounds.
   a. $Co(OH)_2$ c. Na_3PO_4 e. SrI_2
   b. $Ca(ClO_3)_2$ d. $K_2Cr_2O_7$ f. HgF_2

CHAPTER 8

Section 1

1. Draw the Lewis structure for each of the following molecules.
   a. CCl_2H_2 b. HF   c. PCl_3 d. CH_4

Section 2

2. Name the following binary compounds.
   a. S_4N_2 c. SF_6 e. SiO_2
   b. Cl_2O d. NO   f. IF_7
3. Name the following acids: H_3PO_4, HBr, HNO_3.

Section 3

3. a. Ca_3As_2
b. $FeCl_3$
c. MgS
d. BaI_2
e. GaP

4. a. $Cu(C_2H_3O_2)_2$
b. $(NH_4)_3PO_4$
c. $Ca(OH)_2$
d. $Au(CN)_3$

5. a. cobalt(II) hydroxide
b. calcium chlorate
c. sodium phosphate
d. potassium dichromate
e. strontium iodide
f. mercury(II) fluoride

CHAPTER 8

Section 1

1. a. 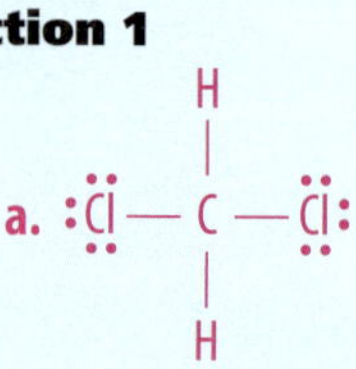

b. 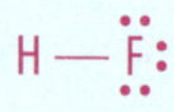

Section 3

4. Draw the Lewis structure for each of the following.
   a. CO c. N_2O e. SiO_2
   a. CH_2O d. Cl_2O f. $AlBr_3$

5. Draw the Lewis resonance structure for CO_3^{2-}.

6. Draw the Lewis resonance structure for $CH_3CO_2^-$.

7. Draw the Lewis structure for NO and IF_4^-.

Section 4

8. Determine the molecular geometry, bond angles, and hybrid of each molecule in Question 4.

Section 5

9. Determine whether each of the following molecules is polar or nonpolar.
   a. CH_2O b. BF_3 c. SiH_4 d. H_2S

CHAPTER 9

Section 1

Write skeleton equations for the following reactions.

1. Solid barium and oxygen gas react to produce solid barium oxide.
2. Solid iron and aqueous hydrogen sulfate react to produce aqueous iron(III) sulfate and gaseous hydrogen.

Write balanced chemical equations for the following reactions.

3. Liquid bromine reacts with solid phosphorus (P_4) to produce solid diphosphorus pentabromide.
4. Aqueous lead(II) nitrate reacts with aqueous potassium iodide to produce solid lead(II) iodide and aqueous potassium nitrate.
5. Solid carbon reacts with gaseous fluorine to produce gaseous carbon tetrafluoride.
6. Aqueous carbonic acid reacts to produce liquid water and gaseous carbon dioxide.
7. Gaseous hydrogen chloride reacts with gaseous ammonia to produce solid ammonium chloride.
8. Solid copper(II) sulfide reacts with aqueous nitric acid to produce aqueous copper(II) sulfate, liquid water, and nitrogen dioxide gas.

Section 2

Classify each of the following reactions into as many types as possible.

9. $2Mo(s) + 3O_2(g) \rightarrow 2MoO_3(s)$
10. $N_2H_4(l) + 3O_2(g) \rightarrow 2NO_2(g) + 2H_2O(l)$

Write balanced chemical equations for the following decomposition reactions.

11. Aqueous hydrogen chlorite decomposes to produce water and gaseous chlorine(III) oxide.
12. Calcium carbonate(s) decomposes to produce calcium oxide(s) and carbon dioxide(g).

Use the activity series to predict whether each of the following single-replacement reactions will occur.

13. $Al(s) + FeCl_3(aq) \rightarrow AlCl_3(aq) + Fe(s)$

c. 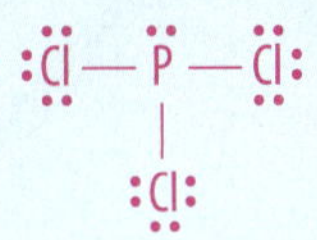

d. 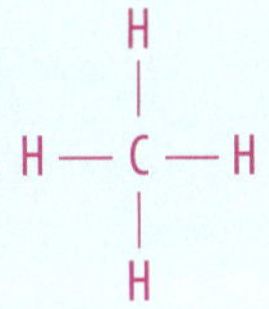

Section 2

2. a. tetrasulfur dinitride
   b. dichlorine monoxide
   c. sulfur hexafluoride
   d. nitrogen monoxide
   e. silicon dioxide
   f. iodine heptafluoride
3. phosphoric acid; hydrobromic acid; nitric acid

Section 3

4. a. 
   b. 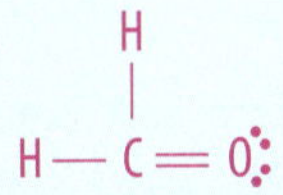
   c. 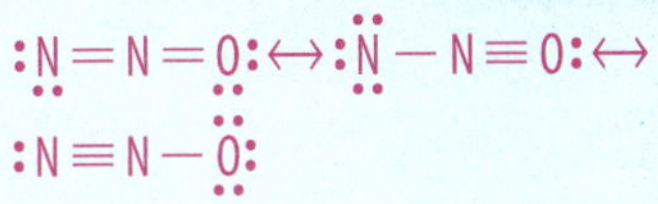
   d. 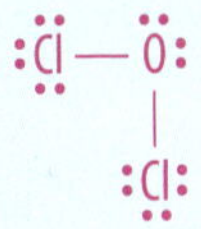
   e. 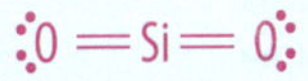
   f. 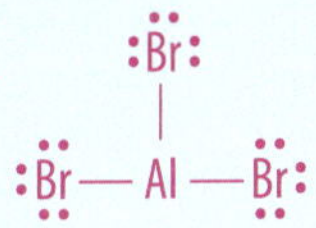
5. 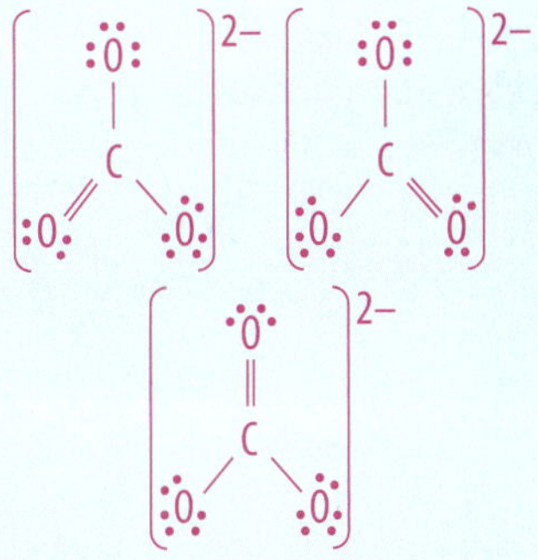
6. 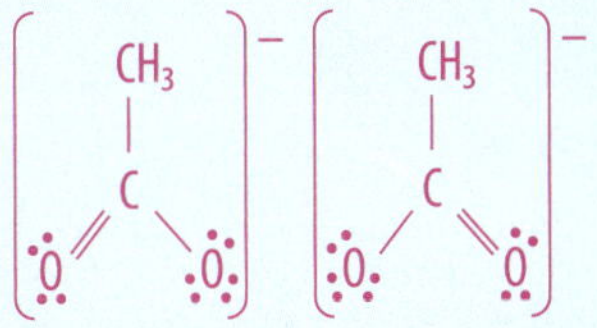
7. 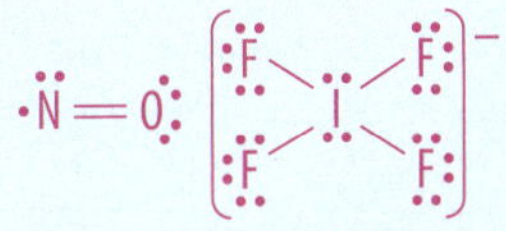

Section 4

8. a. CO: linear, 180°, sp
   b. CH_2O: trigonal planar, 120°, sp^2
   c. N_2O: linear, 180°, sp
   d. Cl_2O: bent, 104.5°, sp^3
   e. SiO_2: linear, 180°, sp
   f. $AlBr_3$: trigonal planar, 120°, sp^2

Section 5

9. a. polar
   b. nonpolar
   c. nonpolar
   d. polar

CHAPTER 9

Section 1

1. $Ba(s) + O_2(g) \rightarrow BaO(s)$
2. $Fe(s) + H_2SO_4(aq) \rightarrow Fe_2(SO_4)_3(aq) + H_2(g)$
3. $5Br_2(l) + P_4(s) \rightarrow 2P_2Br_5(s)$
4. $Pb(NO_3)_2(aq) + 2KI(aq) \rightarrow PbI_2(s) + 2KNO_3(aq)$
5. $C(s) + 2F_2(g) \rightarrow CF_4(g)$
6. $H_2CO_3(aq) \rightarrow H_2O(l) + CO_2(g)$
7. $HCl(g) + NH_3(g) \rightarrow NH_4Cl(s)$
8. $CuS(s) + 8HNO_3(aq) \rightarrow$
   $CuSO_4(aq) + 4H_2O(l) + 8NO_2(g)$

Section 2

9. synthesis; combustion
10. combustion
11. $2HClO_2(aq) \longrightarrow H_2O(l) + Cl_2O_3(g)$
12. $CaCO_3(s) \longrightarrow CaO(s) + CO_2(g)$
13. yes
14. yes
15. no
16. $2Bi(NO_3)_3(aq) + 3Na_2S(aq) \longrightarrow Bi_2S_3(s) + 6NaNO_3(aq)$
17. $MgCl_2(aq) + K_2CO_3(aq) \longrightarrow MgCO_3(s) + 2KCl(aq)$

Section 3

18. $Ba^{2+}(aq) + 2F^-(aq) \longrightarrow BaF_2(s)$
19. $2Cu^+(aq) + S^{2-}(aq) \longrightarrow Cu_2S(s)$
20. $H^+(aq) + OH^-(aq) \longrightarrow H_2O(l)$
21. $H^+(aq) + OH^-(aq) \longrightarrow H_2O(l)$
22. $2H^+(aq) + CO_3^{2-}(aq) \longrightarrow CO_2(g) + H_2O(l)$
23. $H^+(aq) + CN^-(aq) \longrightarrow HCN(g)$

CHAPTER 10

Section 1

1. 2.26×10^{24} atoms Fe
2. 7.53×10^{24} formula units $CaCO_3$
3. 2.09 mol $CaCl_2$
4. 76.2 mol Ag

Section 2

5. 1.47 g S
6. 0.143 g Fe
7. 2.52×10^3 g Ne
8. 1.49 mol Ti
9. 45.9 mol Pb
10. 0.0254 mol As
11. 1.19×10^{21} atoms Na
12. 2.46×10^{25} atoms Ga
13. 741 g Se
14. 4.32×10^{-2} g Li

14. $Br_2(l) + 2LiI(aq) \longrightarrow 2LiBr(aq) + I_2(aq)$
15. $Cu(s) + MgSO_4(aq) \longrightarrow Mg(s) + CuSO_4(aq)$

Write chemical equations for the following chemical reactions.

16. Bismuth(III) nitrate(aq) reacts with sodium sulfide(aq), yielding bismuth(III) sulfide(s) plus sodium nitrate(aq).
17. Magnesium chloride(aq) reacts with potassium carbonate(aq), yielding magnesium carbonate(s) plus potassium chloride(aq).

Section 3 **Write net ionic equations for the following reactions.**

18. Aqueous solutions of barium chloride and sodium fluoride are mixed to form a precipitate of barium fluoride.
19. Aqueous solutions of copper(I) nitrate and potassium sulfide are mixed to form insoluble copper(I) sulfide.
20. Hydrobromic acid reacts with aqueous lithium hydroxide.
21. Perchloric acid reacts with aqueous rubidium hydroxide.
22. Nitric acid reacts with aqueous sodium carbonate.
23. Hydrochloric acid reacts with aqueous lithium cyanide.

CHAPTER 10

Section 1

1. Determine the number of atoms in 3.75 mol of Fe.
2. Calculate the number of formula units in 12.5 mol of $CaCO_3$.
3. How many moles of $CaCl_2$ contain 1.26×10^{24} formula units of $CaCl_2$?
4. How many moles of Ag contain 4.59×10^{25} atoms of Ag?

Section 2

5. Determine the mass in grams of 0.0458 mol of sulfur.
6. Calculate the mass in grams of 2.56×10^{-3} mol of iron.
7. Determine the mass in grams of 125 mol of neon.
8. How many moles of titanium are contained in 71.4 g?
9. How many moles of lead are equivalent to 9.51×10^3 g of Pb?
10. Determine the number of moles of arsenic in 1.90 g of As.
11. Determine the number of atoms in 4.56×10^{-2} g of sodium.
12. How many atoms of gallium are in 2.85×10^3 g of gallium?
13. Determine the mass in grams of 5.65×10^{24} atoms of Se.
14. What is the mass in grams of 3.75×10^{21} atoms of Li?

Section 3

15. How many moles of each element are in 0.0250 mol of K_2CrO_4?
16. How many moles of ammonium ions are in 4.50 mol of $(NH_4)_2CO_3$?
17. Determine the molar mass of silver nitrate.
18. Calculate the molar mass of acetic acid (CH_3COOH).

Section 3

15. 0.0500 mol K; 0.0250 mol Cr; 0.100 mol O
16. 9.00 mol NH_4^+
17. 169.88 g/mol $AgNO_3$
18. 60.05 g/mol CH_3COOH

19. Determine the mass of 8.57 mol of sodium dichromate ($Na_2Cr_2O_7$).

20. Calculate the mass of 42.5 mol of potassium cyanide.

21. Determine the number of moles present in 456 g of $Cu(NO_3)_2$.

22. Calculate the number of moles in 5.67 g of potassium hydroxide.

23. Calculate the number of each atom in 40.0 g of methanol (CH_3OH).

24. What mass of sodium hydroxide contains 4.58×10^{23} formula units?

Section 4

25. What is the percent by mass of each element in sucrose ($C_{12}H_{22}O_{11}$)?

26. Which compound has a greater percent by mass of chromium, K_2CrO_4 or $K_2Cr_2O_7$?

27. Analysis of a compound indicates the percent composition 42.07% Na, 18.89% P, and 39.04% O. Determine its empirical formula.

28. A colorless liquid was found to contain 39.12% C, 8.76% H, and 52.12% O. Determine the empirical formula of the substance.

29. Analysis of a compound used in cosmetics reveals the compound contains 26.76% C, 2.21% H, 71.17% O and has a molar mass of 90.04 g/mol. Determine the molecular formula for this substance.

30. Eucalyptus leaves are the food source for panda bears. Eucalyptol is an oil found in these leaves. Analysis of eucalyptol indicates it has a molar mass of 154 g/mol and contains 77.87% C, 11.76% H, and 10.37% O. Determine the molecular formula of eucalyptol.

31. Beryl is a hard mineral that occurs in a variety of colors. A 50.0-g sample of beryl contains 2.52 g Be, 5.01 g Al, 15.68 g Si, and 26.79 g O. Determine its empirical formula.

32. Analysis of a 15.0-g sample of a compound used to leach gold from low-grade ores is 7.03 g Na, 3.68 g C, and 4.29 g N. Determine the empirical formula for this substance.

Section 5

33. Analysis of a hydrate of iron(III) chloride revealed that in a 10.00-g sample of the hydrate, 6.00 g is anhydrous iron(III) chloride and 4.00 g is water. Determine the formula and name of the hydrate.

34. When 25.00 g of a hydrate of nickel(II) chloride was heated, 11.37 g of water was released. Determine the name and formula of the hydrate.

CHAPTER 11

Section 1

Interpret the following balanced chemical equations in terms of particles, moles, and mass.

1. $Mg + 2HCl \rightarrow MgCl_2 + H_2$
2. $2Al + 3CuSO_4 \rightarrow Al_2(SO_4)_3 + 3Cu$
3. $Cu(NO_3)_2 + 2KOH \rightarrow Cu(OH)_2 + 2KNO_3$
4. Write and balance the equation for the decomposition of aluminum carbonate. Determine the possible mole ratios.

Supplemental Practice Problems

19. 2250 g $Na_2Cr_2O_7$
20. 2770 g KCN
21. 2.43 mol $Cu(NO_3)_2$
22. 0.101 mol KOH
23. 7.52×10^{23} atoms C, 3.01×10^{24} atoms H, 7.52×10^{23} atoms O
24. 32.7 g NaOH

Section 4

25. 42.103% C, 6.480% H, 51.417% O
26. $K_2Cr_2O_7$
27. Na_3PO_4
28. $C_3H_8O_3$
29. $C_2H_2O_4$
30. $C_{10}H_{18}O$
31. $Be_3Al_2Si_6O_{18}$
32. NaCN

Section 5

33. $FeCl_3 \cdot 6H_2O$, iron(III) chloride hexahydrate
34. $NiCl_2 \cdot 6H_2O$, nickel(II) chloride hexahydrate

CHAPTER 11

Section 1

1. 1 atom Mg + 2 molecules HCl $\rightarrow$ 1 formula unit $MgCl_2$ + 1 molecule H_2
   1 mole Mg + 2 moles HCl $\rightarrow$ 1 mole $MgCl_2$ + 1 mole H_2
   24.31 g Mg + 72.92 g HCl $\rightarrow$ 95.21 g $MgCl_2$ + 2.02 g H_2
2. 2 atoms Al + 3 formula units $CuSO_4$ $\rightarrow$ 1 formula unit $Al_2(SO_4)_3$ + 3 atoms Cu
   2 moles Al + 3 moles $CuSO_4$ $\rightarrow$ 1 mole $Al_2(SO_4)_3$ + 3 moles Cu
   53.96 g Al + 478.86 g $CuSO_4$ $\rightarrow$ 342.17 g $Al_2(SO_4)_3$ + 190.65 g Cu
3. 1 formula unit $Cu(NO_3)_2$ + 2 formula units KOH $\rightarrow$ 1 formula unit $Cu(OH)_2$ + 2 formula units KNO_3
   1 mole $Cu(NO_3)_2$ + 2 moles KOH $\rightarrow$ 1 mole $Cu(OH)_2$ + 2 moles KNO_3
   187.57 g $Cu(NO_3)_2$ + 112.22 g KOH $\rightarrow$ 97.57 g $Cu(OH)_2$ + 202.22 g KNO_3
4. $Al_2(CO_3)_3(s) \rightarrow Al_2O_3(s) + 3CO_2(g)$

$$\frac{1 \text{ mol } Al_2(CO_3)_3}{1 \text{ mol } Al_2O_3} \quad \frac{1 \text{ mol } Al_2(CO_3)_3}{3 \text{ mol } CO_2}$$

$$\frac{1 \text{ mol } Al_2O_3}{1 \text{ mol } Al_2(CO_3)_3} \quad \frac{1 \text{ mol } Al_2O_3}{3 \text{ mol } CO_2}$$

$$\frac{3 \text{ mol } CO_2}{1 \text{ mol } Al_2(CO_3)_3} \quad \frac{3 \text{ mol } CO_2}{1 \text{ mol } Al_2O_3}$$

5. $Mg(s) + 2H_2O(l) \longrightarrow Mg(OH)_2(s) + H_2(g)$

$\frac{1\ mol\ Mg}{2\ mol\ H_2O}$ $\frac{1\ mol\ Mg}{1\ mol\ Mg(OH)_2}$ $\frac{1\ mol\ Mg}{1\ mol\ H_2}$

$\frac{2\ mol\ H_2O}{1\ mol\ Mg}$ $\frac{2\ mol\ H_2O}{1\ mol\ Mg(OH)_2}$ $\frac{2\ mol\ H_2O}{1\ mol\ H_2}$

$\frac{1\ mol\ Mg(OH)_2}{1\ mol\ Mg}$ $\frac{1\ mol\ Mg(OH)_2}{2\ mol\ H_2O}$

$\frac{1\ mol\ Mg(OH)_2}{1\ mol\ H_2}$

$\frac{1\ mol\ H_2}{1\ mol\ Mg}$ $\frac{1\ mol\ H_2}{2\ mol\ H_2O}$ $\frac{1\ mol\ H_2}{1\ mol\ Mg(OH)_2}$

Section 2

6. 0.600 mol HCl
7. 2.29 mol Cr_2O_3
8. 924 g CO_2
9. 314 g H_2SO_4
10. 1.31 g CO_2
11. 195.6 g aspirin

Section 3

12. Benzene is the limiting reactant; 4.2 g Cl_2 is in excess; 64.8 g chlorobenzene produced
13. HCl is the limiting reactant; 2.99 g Ni is in excess; 4.44 g $NiCl_2$ produced

Section 4

14. $Sn + 2I_2 \longrightarrow SnI_4$: theoretical yield is 26.4 g SnI_4; 94.7% yield
15. theoretical yield is 41.4 g $NaAu(CN)_2$; percent yield is 93.7%

SUPPLEMENTAL PRACTICE PROBLEMS

Supplemental Practice Problems

5. Write and balance the equation for the formation of magnesium hydroxide and hydrogen from magnesium and water. Determine the possible mole ratios.

Section 2

6. Some antacid tablets contain aluminum hydroxide. The aluminum hydroxide reacts with stomach acid according to the equation: $Al(OH)_3 + 3HCl \longrightarrow AlCl_3 + 3H_2O$. Determine the moles of acid neutralized if a tablet contains 0.200 mol of $Al(OH)_3$.

7. Chromium reacts with oxygen according to the equation: $4Cr + 3O_2 \longrightarrow 2Cr_2O_3$. Determine the moles of chromium(III) oxide produced when 4.58 mol of chromium is allowed to react.

8. Space vehicles use solid lithium hydroxide to remove exhaled carbon dioxide according to the equation: $2LiOH + CO_2 \longrightarrow Li_2CO_3 + H_2O$. Determine the mass of carbon dioxide removed if the space vehicle carries 42.0 mol of LiOH.

9. Some of the sulfur dioxide released into the atmosphere is converted to sulfuric acid according to the equation: $2SO_2 + 2H_2O + O_2 \longrightarrow 2H_2SO_4$. Determine the mass of sulfuric acid formed from 3.20 mol of sulfur dioxide.

10. How many grams of carbon dioxide are produced when 2.50 g of sodium hydrogen carbonate reacts with excess citric acid according to the equation: $3NaHCO_3 + H_3C_6H_5O_7 \longrightarrow Na_3C_6H_5O_7 + 3CO_2 + 3H_2O$?

11. Aspirin ($C_9H_8O_4$) is produced when salicylic acid ($C_7H_6O_3$) reacts with acetic anhydride ($C_4H_6O_3$) according to the equation: $C_7H_6O_3 + C_4H_6O_3 \longrightarrow C_9H_8O_4 + HC_2H_3O_2$. Determine the mass of aspirin produced when 150.0 g of salicylic acid reacts with an excess of acetic anhydride.

Section 3

12. Chlorine reacts with benzene to produce chlorobenzene and hydrogen chloride, $Cl_2 + C_6H_6 \longrightarrow C_6H_5Cl + HCl$. Determine the limiting reactant if 45.0 g of benzene reacts with 45.0 g of chlorine, the mass of the excess reactant after the reaction is complete, and the mass of chlorobenzene produced.

13. Nickel reacts with hydrochloric acid to produce nickel(II) chloride and hydrogen according to the equation: $Ni + 2HCl \longrightarrow NiCl_2 + H_2$. If 5.00 g of Ni and 2.50 g of HCl react, determine the limiting reactant, the mass of the excess reactant after the reaction is complete, and the mass of nickel(II) chloride produced.

Section 4

14. Tin(IV) iodide is prepared by reacting tin with iodine. Write the balanced chemical equation for the reaction. Determine the theoretical yield if a 5.00-g sample of tin reacts in an excess of iodine. Determine the percent yield if 25.0 g of SnI_4 was recovered.

15. Gold is extracted from gold-bearing rock by adding sodium cyanide in the presence of oxygen and water, according to the reaction: $4Au(s) + 8NaCN(aq) + O_2(g) + 2H_2O(l) \longrightarrow 4NaAu(CN)_2(aq) + NaOH(aq)$. Determine the theoretical yield of $NaAu(CN)_2$ if 1000.0 g of gold-bearing rock is used, which contains 3.00% gold by mass. Determine the percent yield of $NaAu(CN)_2$ if 38.790 g of $NaAu(CN)_2$ is recovered.

CHAPTER 12

Section 1

1. Calculate the ratio of effusion rates for methane (CH_4) and nitrogen.
2. Calculate the molar mass of butane. Butane's rate of diffusion is 3.8 times slower than that of helium.
3. What is the total pressure in a canister that contains oxygen gas at a partial pressure of 804 mm Hg, nitrogen at a partial pressure of 220 mm Hg, and hydrogen at a partial pressure of 445 mm Hg?
4. Calculate the partial pressure of neon in a flask that has a total pressure of 1.87 atm. The flask contains krypton at a partial pressure of 0.77 atm and helium at a partial pressure of 0.62 atm.

CHAPTER 13

Section 1

1. The pressure of air in a 2.25-L container is 1.20 atm. What is the new pressure if the sample is transferred to a 6.50-L container? Temperature is constant.
2. The volume of a sample of hydrogen gas at 0.997 atm is 5.00 L. What will be the new volume if the pressure is decreased to 0.977 atm? Temperature is constant.
3. A gas at 55.0°C occupies a volume of 3.60 L. What volume will it occupy at 30.0°C? Pressure is constant.
4. The volume of a gas is 0.668 L at 66.8°C. At what Celsius temperature will the gas have a volume of 0.942 L, assuming pressure remains constant?
5. The pressure in a bicycle tire is 1.34 atm at 33.0°C. At what temperature will the pressure inside the tire be 1.60 atm? Volume is constant.
6. If a sample of oxygen gas has a pressure of 810 torr at 298 K, what will be its pressure if its temperature is raised to 330 K?
7. Air in a tightly sealed bottle has a pressure of 0.978 atm at 25.5°C. What will be its pressure if the temperature is raised to 46.0°C?
8. Hydrogen gas at a temperature of 22.0°C that is confined in a 5.00-L cylinder exerts a pressure of 4.20 atm. If the gas is released into a 10.0-L reaction vessel at a temperature of 33.6°C, what will be the pressure inside the reaction vessel?
9. A sample of neon gas at a pressure of 1.08 atm fills a flask with a volume of 250 mL at a temperature of 24.0°C. If the gas is transferred to another flask at 37.2°C and a pressure of 2.25 atm, what is the volume of the new flask?

Section 2

10. What volume of beaker contains exactly 2.23×10^{-2} mol of nitrogen gas at STP?
11. How many moles of air are in a 6.06-L tire at STP?
12. How many moles of oxygen are in a 5.5-L canister at STP?
13. What mass of helium is in a 2.00-L balloon at STP?
14. What volume will 2.3 kg of nitrogen gas occupy at STP?

Supplemental Practice Problems

CHAPTER 12

Section 1

1. 1.32:1
2. 58 g/mol
3. 1469 mm Hg
4. 0.48 atm

CHAPTER 13

Section 1

1. 0.415 atm
2. 5.10 L
3. 3.33 L
4. 206°C
5. 92°C
6. 9.0×10^2 torr
7. 1.04 atm
8. 2.19 atm
9. 130 mL

Section 2

10. 0.500 L
11. 0.271 mol air
12. 0.25 mol O_2
13. 0.357 g He
14. 1.8×10^3 L N_2

SUPPLEMENTAL PRACTICE PROBLEMS

15. 0.20 mol
16. 4.0×10^1 torr
17. 26 g/mol
18. 5.73 g O_2

Section 3

19. 1.31 L O_2
20. 1.32 L CH_4
21. 1.40 g H_2O_2
22. 343 L O_2

CHAPTER 14

Section 2

1. 0.136%
2. 0.30%
3. 2 mL H_2O_2
4. 10%
5. 1.5*M* K_3PO_4
6. 0.747*M* NH_4Cl
7. 10% solution of NaCl
8. 34 g NaOH
9. 501 mL
10. 194 mL of stock solution and 306 mL of water
11. 0.5*M*

15. Calculate the number of moles of gas that occupy a 3.45-L container at a pressure of 150 kPa and a temperature of 45.6°C.

16. What is the pressure in torr that a 0.44-g sample of carbon dioxide gas will exert at a temperature of 46.2°C when it occupies a volume of 5.00 L?

17. What is the molar mass of a gas that has a density of 1.02 g/L at 0.990 atm pressure and 37°C?

18. Calculate the grams of oxygen gas present in a 2.50-L sample kept at 1.66 atm pressure and a temperature of 10.0°C.

Section 3

19. What volume of oxygen gas is needed to completely combust 0.202 L of butane gas (C_4H_{10})?

20. Determine the volume of methane gas (CH_4) needed to react completely with 0.660 L of O_2 gas to form methanol (CH_3OH).

21. Calculate the mass of hydrogen peroxide needed to obtain 0.460 L of oxygen gas at STP. $2H_2O_2(aq) \rightarrow 2H_2O(l) + O_2(g)$

22. When potassium chlorate is heated in the presence of a catalyst such as manganese dioxide, it decomposes to form solid potassium chloride and oxygen gas: $2KClO_3(s) \rightarrow 2KCl(s) + 3O_2(g)$. How many liters of oxygen will be produced at STP if 1.25 kg of potassium chlorate decomposes completely?

CHAPTER 14

Section 2

1. What is the percent by mass of a sample of ocean water that is found to contain 1.36 g of magnesium ions per 1000 g?

2. What is the percent by mass of iced tea containing 0.75 g of aspartame in 250 g of water?

3. A bottle of hydrogen peroxide is labeled 3%. If you pour out 50 mL of hydrogen peroxide solution, what volume is actually hydrogen peroxide?

4. If 50 mL of pure acetone is mixed with 450 mL of water, what is the percent by volume of acetone?

5. Calculate the molarity of 1270 g of K_3PO_4 in 4.0 L aqueous solution.

6. What is the molarity of 90.0 g of NH_4Cl in 2.25 L aqueous solution?

7. Which is more concentrated, 25 g of NaCl dissolved in 500 mL of water or a 10% solution of NaCl (percent by mass)?

8. Calculate the mass of NaOH required to prepare a 0.343*M* solution dissolved in 2500 mL of water.

9. Calculate the volume required to dissolve 11.2 g of $CuSO_4$ to prepare a 0.140*M* solution.

10. How would you prepare 500 mL of a solution that has a new concentration of 4.5*M* if the stock solution is 11.6*M*?

11. Caustic soda is 19.1*M* NaOH and is diluted for household use. What is the household concentration if 10 mL of the concentrated solution is diluted to 400 mL?

12. What is the molality of a solution containing 63.0 g of HNO_3 in 0.500 kg of water?

13. What is the molality of an acetic acid solution containing 0.500 mol of $HC_2H_3O_2$ in 0.800 kg of water?

14. What mass of ethanol (C_2H_5OH) will be required to prepare a 2.00*m* solution in 8.00 kg of water?

15. A mixture of gases contains 0.215 mol N_2, 0.345 mol O_2, 0.023 mol CO_2, and 0.014 mol SO_2. What is the mole fraction of N_2?

16. A necklace contains 4.85 g of gold, 1.25 g of silver, and 2.40 g of copper. What is the mole fraction of each metal?

Section 3

17. Calculate the mass of gas dissolved at 150.0 kPa, if 0.35 g of the gas dissolves in 2.0 L of water at 30.0 kPa.

18. At which depth, 10 m or 40 m, will a scuba diver have more nitrogen dissolved in the bloodstream?

Section 4

19. Calculate the freezing point and boiling point of a solution containing 6.42 g of sucrose ($C_{12}H_{22}O_{11}$) in 100.0 g of water.

20. Calculate the freezing point and boiling point of a solution containing 23.7 g of copper(II) sulfate in 250.0 g of water.

21. Calculate the freezing point and boiling point of a solution containing 0.15 mol of the molecular compound naphthalene in 175 g of benzene (C_6H_6).

CHAPTER 15

Section 1

1. What is the equivalent in joules of 126 Calories?

2. Convert 455 kilojoules to kilocalories.

3. How much heat is required to warm 122 g of water by 23.0°C?

4. The temperature of 55.6 g of a material decreases by 14.8°C when it loses 3080 J of heat. What is its specific heat?

5. What is the specific heat of a metal if the temperature of a 12.5-g sample increases from 19.5°C to 33.6°C when it absorbs 37.7 J of heat?

Section 2

6. A 75.0-g sample of a metal is placed in boiling water until its temperature is 100.0°C. A calorimeter contains 100.00 g of water at a temperature of 24.4°C. The metal sample is removed from the boiling water and immediately placed in water in the calorimeter. The final temperature of the metal and water in the calorimeter is 34.9°C. Assuming that the calorimeter provides perfect insulation, what is the specific heat of the metal?

Section 3

7. Use **Table 4** to determine how much heat is released when 1.00 mol of gaseous methanol condenses to a liquid.

8. Use **Table 4** to determine how much heat must be supplied to melt 4.60 g of ethanol.

Supplemental Practice Problems

12. 2.00*m*
13. 0.625*m*
14. 737 g ethanol
15. 0.360
16. 0.332 Au, 0.157 Ag, 0.511 Cu

Section 3

17. 1.8 g
18. 40 m

Section 4

19. $T_f = -0.350°C$
 $T_b = 100.096°C$
20. $T_f = -2.20°C$
 $T_b = 100.606°C$
21. $T_f = 1.1°C$
 $T_b = 82.3°C$

CHAPTER 15

Section 1

1. 5.27×10^5 J
2. 109 kcal
3. 1.17×10^4 J
4. 3.74 J/(g•°C)
5. 0.214 J/(g•°C)

Section 2

6. 0.900 J/(g•°C)

Section 3

7. 35.2 kJ
8. 0.494 kJ

Section 4

9. $\Delta H = 52$ kJ
10. $\Delta H = -176$ kJ
11. $\Delta H°_{rxn} = 546$ kJ
12. $\Delta H°_{rxn} = -1124$ kJ

Section 5

13. ΔS_{system} is positive.
14. ΔS_{system} is negative.
15. spontaneous
16. nonspontaneous

CHAPTER 16

Section 1

1. a. 0.20 mol/(L•min)
   b. 0.40 mol/(L•min)

Section 3

2. third order; third order; first order; second order
3. Rate = $k[A]^2$
4. Rate = $k[NO]^2[O_2]$

SUPPLEMENTAL PRACTICE PROBLEMS

Supplemental Practice Problems

Section 4

9. Calculate ΔH_{rxn} for the reaction $2C(s) + 2H_2(g) \rightarrow C_2H_4(g)$, given the following thermochemical equations:
   $2CO_2(g) + 2H_2O(l) \rightarrow C_2H_4(g) + 3O_2(g)$ $\Delta H = 1411$ kJ
   $C(s) + O_2(g) \rightarrow CO_2(g)$ $\Delta H = -393.5$ kJ
   $2H_2(g) + O_2(g) \rightarrow 2H_2O(l)$ $\Delta H = -572$ kJ

10. Calculate ΔH_{rxn} for the reaction $HCl(g) + NH_3(g) \rightarrow NH_4Cl(s)$, given the following thermochemical equations:
   $H_2(g) + Cl_2(g) \rightarrow 2HCl(g)$ $\Delta H = -184$ kJ
   $N_2(g) + 3H_2(g) \rightarrow 2NH_3(g)$ $\Delta H = -92$ kJ
   $N_2(g) + 4H_2(g) + Cl_2(g) \rightarrow 2NH_4Cl(s)$ $\Delta H = -628$ kJ

Use standard enthalpies of formation from Table 5 and Table R-11 to calculate $\Delta H°_{rxn}$ for each of the following reactions.

11. $2HF(g) \rightarrow H_2(g) + F_2(g)$

12. $2H_2S(g) + 3O_2(g) \rightarrow 2H_2O(l) + 2SO_2(g)$

Section 5

Predict the sign of ΔS_{system} for each reaction or process.

13. $FeS(s) \rightarrow Fe^{2+}(aq) + S^{2-}(aq)$

14. $SO_2(g) + H_2O(l) \rightarrow H_2SO_3(aq)$

Determine if each of the following processes or reactions is spontaneous or nonspontaneous.

15. $\Delta H_{system} = 15.6$ kJ, $T = 415$ K, $\Delta S_{system} = 45$ J/K

16. $\Delta H_{system} = 35.6$ kJ, $T = 415$ K, $\Delta S_{system} = 45$ J/K

CHAPTER 16

Section 1

1. In the reaction $A \rightarrow 2B$, suppose that [A] changes from 1.20 mol/L at time = 0 to 0.60 mol/L at time = 3.00 min and that [B] = 0.00 mol/L at time = 0.
   a. What is the average rate at which A is consumed in mol/(L•min)?
   b. What is the average rate at which B is produced in mol/(L•min)?

Section 3

2. What are the overall reaction orders in Practice Problems 19 to 22 in this section on reaction rate laws?

3. If halving [A] in the reaction $A \rightarrow B$ causes the initial rate to decrease to one-fourth its original value, what is the probable rate law for the reaction?

4. Use the data below and the method of initial rates to determine the rate law for the reaction $2NO(g) + O_2(g) \rightarrow 2NO_2(g)$.

Formation of NO_2 Data

Trial	Initial [NO] (*M*)	Initial [O_2] (*M*)	Initial Rate (mol/(L•s))
1	0.030	0.020	0.0041
2	0.060	0.020	0.0164
3	0.060	0.040	0.0328

Section 4

5. When 1 mol of cyclobutane (C_4H_8) decomposes to 2 mol of ethylene (C_2H_4) at 1273 K the rate law for the reaction is Rate = $(87\ s^{-1})[C_4H_8]$. What is the instantaneous rate of this reaction when
   a. $[C_4H_8] = 0.0100$ mol/L?
   b. $[C_4H_8] = 0.200$ mol/L?

CHAPTER 17

Section 1

Write equilibrium constant expressions for the following equilibria.

1. $N_2(g) + O_2(g) \rightleftharpoons 2NO(g)$
2. $3O_2(g) \rightleftharpoons 2O_3(g)$
3. $P_4(g) + 6H_2(g) \rightleftharpoons 4PH_3(g)$
4. $CCl_4(g) + HF(g) \rightleftharpoons CFCl_3(g) + HCl(g)$
5. $4NH_3(g) + 5O_2(g) \rightleftharpoons 4NO(g) + 6H_2O(g)$

Write equilibrium constant expressions for the following equilibria.

6. $NH_4Cl(s) \rightleftharpoons NH_3(g) + HCl(g)$
7. $SO_3(g) + H_2O(l) \rightleftharpoons H_2SO_4(l)$
8. $2Na_2O_2(s) + 2CO_2(g) \rightleftharpoons 2Na_2CO_3(s) + O_2(g)$

Calculate K_{eq} for the following equilibria.

9. $H_2(g) + I_2(g) \rightleftharpoons 2HI(g)$
   $[H_2] = 0.0109$, $[I_2] = 0.00290$, $[HI] = 0.0460$
10. $I_2(s) \rightleftharpoons I_2(g)$
    $[I_2(g)] = 0.0665$

Section 3

11. At a certain temperature, $K_{eq} = 0.0211$ for the equilibrium $PCl_5(g) \rightleftharpoons PCl_3(g) + Cl_2(g)$.
    a. What is $[Cl_2]$ in an equilibrium mixture containing 0.865 mol/L PCl_5 and 0.135 mol/L PCl_3?
    b. What is $[PCl_5]$ in an equilibrium mixture containing 0.100 mol/L PCl_3 and 0.200 mol/L Cl_2?
12. Use the K_{sp} value for zinc carbonate given in **Table 3** to calculate its molar solubility at 298 K.
13. Use the K_{sp} value for iron(II) hydroxide given in **Table 3** to calculate its molar solubility at 298 K.
14. Use the K_{sp} value for silver carbonate given in **Table 3** to calculate $[Ag^+]$ in a saturated solution at 298 K.
15. Use the K_{sp} value for calcium phosphate given in **Table 3** to calculate $[Ca^{2+}]$ in a saturated solution at 298 K.
16. Does a precipitate form when equal volumes of 0.0040M $MgCl_2$ and 0.0020M K_2CO_3 are mixed? If so, identify the precipitate.
17. Does a precipitate form when equal volumes of $1.2 \times 10^{-4}M$ $AlCl_3$ and $2.0 \times 10^{-3}M$ NaOH are mixed? If so, identify the precipitate.

Section 4

5. a. Rate = 0.87 mol/(L•s)
   b. Rate = 17 mol/(L•s)

CHAPTER 17

Section 1

1. $K_{eq} = \frac{[NO]^2}{[N_2][O_2]}$
2. $K_{eq} = \frac{[O_3]^2}{[O_2]^3}$
3. $K_{eq} = \frac{[PH_3]^4}{[P_4][H_2]^6}$
4. $K_{eq} = \frac{[CFCl_3][HCl]}{[CCl_4][HF]}$
5. $K_{eq} = \frac{[NO]^4[H_2O]^6}{[NH_3]^4[O_2]^5}$
6. $K_{eq} = [NH_3][HCl]$
7. $K_{eq} = \frac{1}{[SO_3]}$
8. $K_{eq} = \frac{[O_2]}{[CO_2]^2}$
9. $K_{eq} = 66.9$
10. $K_{eq} = 0.0665$

Section 3

11. a. 0.135M
    b. 0.948M
12. $1.2 \times 10^{-5}M$
13. $2.3 \times 10^{-6}M$
14. $2.6 \times 10^{-4}M$
15. $3.4 \times 10^{-7}M$
16. No precipitate forms.
17. Yes; $Al(OH)_3$

CHAPTER 18

Section 1

1. $Zn(s) + 2HNO_3(aq) \longrightarrow Zn(NO_3)_2(aq) + H_2(g)$
2. $MgCO_3(s) + H_2SO_4(aq) \longrightarrow MgSO_4(aq) + H_2O(l) + CO_2(g)$
3. The base is CH_3NH_2.
4. **a.** acid: H_2SO_4; conjugate base: $Al_2(SO_4)_3$
   b. acid: HBr; conjugate base: $CaBr_2$
5. $H_2S(aq) + H_2O(l) \rightleftharpoons H_3O^+(aq) + HS^-(aq)$
   $HS^-(aq) + H_2O(l) \rightleftharpoons H_3O^+(aq) + S^{2-}(aq)$
6. $H_2CO_3(aq) + H_2O(l) \rightleftharpoons H_3O^+(aq) + HCO_3^-(aq)$
   $HCO_3^-(aq) + H_2O(l) \rightleftharpoons H_3O^+(aq) + CO_3^{2-}(aq)$

Section 2

7. $K_a = \dfrac{[H^+][COOH^-]}{[HCOOH]}$
8. $K_a = \dfrac{[H^+][CO_3^{2-}]}{[HCO_3^-]}$
9. $K_b = \dfrac{[NH_4^+][OH^-]}{[NH_3]}$
10. $K_b = \dfrac{[C_6H_5NH_3^+][OH^-]}{[C_6H_5NH_2]}$

Section 3

11. acidic
12. acidic
13. 3.35
14. pH = 11.94; pOH = 2.06
15. pH = 5.57; pOH = 8.43
16. $1.2 \times 10^{-3}M$
17. $3.6 \times 10^{-1}M$
18. 3.17
19. 9.53
20. $K_a = 2.2 \times 10^{-9}$
21. $K_a = 1.4 \times 10^{-5}$

SUPPLEMENTAL PRACTICE PROBLEMS

CHAPTER 18

Section 1

1. Write the balanced formula equation for the reaction between zinc and nitric acid.
2. Write the balanced formula equation for the reaction between magnesium carbonate and sulfuric acid.
3. Identify the base in the reaction $H_2O(l) + CH_3NH_2(aq) \rightarrow OH^-(aq) + CH_3NH_3^+(aq)$.
4. Identify the conjugate acid-base pair described in the reactions in Practice Problem 1.
5. Write the steps in the complete ionization of hydrosulfuric acid.
6. Write the steps in the complete ionization of carbonic acid.

Section 2

7. Write the acid ionization equation and ionization constant expression for formic acid (HCOOH).
8. Write the acid ionization equation and ionization constant expression for the hydrogen carbonate ion (HCO_3^-).
9. Write the base ionization constant expression for ammonia.
10. Write the base ionization expression for aniline ($C_6H_5NH_2$).

Section 3

11. Is a solution in which $[H^+] = 1.0 \times 10^{-5}M$ acidic, basic, or neutral?
12. Is a solution in which $[OH^-] = 1.0 \times 10^{-11}M$ acidic, basic, or neutral?
13. What is the pH of a solution in which $[H^+] = 4.5 \times 10^{-4}M$?
14. Calculate the pH and pOH of a solution in which $[OH^-] = 8.8 \times 10^{-3}M$.
15. Calculate the pH and pOH of a solution in which $[H^+] = 2.7 \times 10^{-6}M$.
16. What is $[H^+]$ in a solution having a pH of 2.92?
17. What is $[OH^-]$ in a solution having a pH of 13.56?
18. What is the pH of a $0.00067M$ H_2SO_4 solution?
19. What is the pH of a $0.000034M$ NaOH solution?
20. The pH of a $0.200M$ HBrO solution is 4.67. What is the acid's K_a?
21. The pH of a $0.030M$ C_2H_5COOH solution is 3.20. What is the acid's K_a?

Section 4

22. Write the formula equation for the reaction between hydroiodic acid and beryllium hydroxide.
23. Write the formula equation for the reaction between perchloric acid and lithium hydroxide.
24. In a titration, 15.73 mL of $0.2346M$ HI solution neutralizes 20.00 mL of a LiOH solution. What is the molarity of the LiOH?
25. What is the molarity of a caustic soda (NaOH) solution if 35.00 mL of solution is neutralized by 68.30 mL of $1.250M$ HCl?

Section 4

22. $2HI(aq) + Be(OH)_2(aq) \longrightarrow BeI_2(aq) + 2H_2O(l)$
23. $HClO_4(aq) + LiOH(aq) \longrightarrow LiClO_4(aq) + H_2O(l)$
24. $0.1845M$
25. $2.439M$

26. Write the chemical equation for the hydrolysis reaction that occurs when sodium hydrogen carbonate is dissolved in water. Is the resulting solution acidic, basic, or neutral?

27. Write the chemical equation for any hydrolysis reaction that occurs when cesium chloride is dissolved in water. Is the resulting solution acidic, basic, or neutral?

CHAPTER 19

Section 1

Identify the following information for each problem. What element is oxidized? Reduced? What is the oxidizing agent? Reducing agent?

1. $2P + 3Cl_2 \rightarrow 2PCl_3$
2. $C + H_2O \rightarrow CO + H_2$
3. $2ClO_3^- + 3AsO_2^- \rightarrow 3AsO_4^{3-} + 2Cl^-$
4. Determine the oxidation number for each element in the following compounds.
   a. Na_2SeO_3
   b. $HAuCl_4$
   c. H_3BO_3
5. Determine the oxidation number for the following compounds or ions.
   a. P_4O_8
   b. Na_2O_2 *(Hint: This is like H_2O_2.)*
   c. AsO_4^{-3}

Section 2

6. How many electrons will be lost or gained in each of the following half-reactions? Identify whether each is an oxidation or reduction.
   a. $Cr \rightarrow Cr^{3+}$
   b. $O_2 \rightarrow O^{2-}$
   c. $Fe^{+2} \rightarrow Fe^{3+}$
7. Balance the following reaction by the oxidation number method: $MnO_4^- + CH_3OH \rightarrow MnO_2 + HCHO$ (acidic). *(Hint: Assign the oxidation of hydrogen and oxygen as usual, and solve for the oxidation number of carbon.)*
8. Balance the following reaction by the oxidation number method: $Zn + HNO_3 \rightarrow ZnO + NO_2 + NH_3$
9. Use the oxidation number method to balance these net ionic equations.
   a. $SeO_3^{2-} + I^- \rightarrow Se + I_2$ (acidic solution)
   b. $NiO_2 + S_2O_3^{2-} \rightarrow Ni(OH)_2 + SO_3^{2-}$ (acidic solution)

Use the half-reaction method to balance the following redox equations.

10. $Zn(s) + HCl(aq) \rightarrow ZnCl_2(aq) \rightarrow H_2(g)$
11. $MnO_4^-(aq) + H_2SO_3(aq) \rightarrow Mn^{2+}(aq) + HSO_4^-(aq) + H_2O(l)$ (acidic solution)
12. $NO_2(aq) + OH^-(aq) \rightarrow NO_2^-(aq) + NO_3^-(aq) + H_2O(l)$ (basic solution)
13. $HS^-(aq) + IO_3^-(aq) \rightarrow I^-(aq) + S(s) + H_2O(l)$ (acidic solution)

Supplemental Practice Problems

26. $HCO_3^-(aq) + H_2O(l) \rightarrow CO_3^{2-}(aq) + OH^-(aq)$; basic
27. No reaction; neutral

CHAPTER 19

Section 1

1. P is oxidized, Cl is reduced
   Cl_2 is the oxidizing agent
   P is the reducing agent
2. C is oxidized, H is reduced
   H_2O is the oxidizing agent
   C is the reducing agent
3. As is oxidized, Cl is reduced
   Cl is the oxidizing agent
   As is the reducing agent
4. **a.** +1, +4, −2
   b. +1, +3, −1
   c. +1, +3, −2
5. **a.** +4, −2
   b. +1, −1
   c. +5, −2

Section 2

6. **a.** loss of 3 electrons, oxidation
   b. gain of 2 electrons, reduction
   c. loss of 1 electron, oxidation
7. $2MnO_4^- + 3CH_3OH + 2H^+ \rightarrow 2MnO_2 + 3HCHO + 4H_2O$
8. $5Zn + 3HNO_3 \rightarrow 5ZnO + 2NO_2 + NH_3$
9. **a.** $SeO_3^{2-} + 4I^- + 6H^+ \rightarrow Se + 2I_2 + 3H_2O$
   b. $2NiO_2 + S_2O_3^{2-} + 3H_2O \rightarrow 2Ni(OH)_2 + 2SO_3^{2-} + 2H^+$
10. $Zn(s) + 2HCl(aq) \rightarrow ZnCl_2(aq) + H_2(g)$
11. $2MnO_4^-(aq) + H^+(aq) + 5H_2SO_3(aq) \rightarrow 2Mn^{2+}(aq) + 5HSO_4^-(aq) + 3H_2O(l)$
12. $2NO_2(aq) + 2OH^-(aq) \rightarrow NO_2^-(aq) + NO_3^-(aq) + H_2O(l)$
13. $3HS^-(aq) + IO_3^-(aq) + 3H^+ \rightarrow I^-(aq) + 3S(s) + 3H_2O(l)$

SUPPLEMENTAL PRACTICE PROBLEMS

CHAPTER 20

Section 1

1. a. $+1.38\ V$
   b. $+0.509\ V$
   c. $+1.82\ V$
2. a. $+1.41\ V$; spontaneous
   b. $+0.7996\ V$; spontaneous
   c. $-0.789\ V$; not spontaneous; $Cu^{2+}(aq) + Fe(s) \longrightarrow Cu(s) + Fe^{2+}(aq)$

CHAPTER 21

Section 2

1. a. 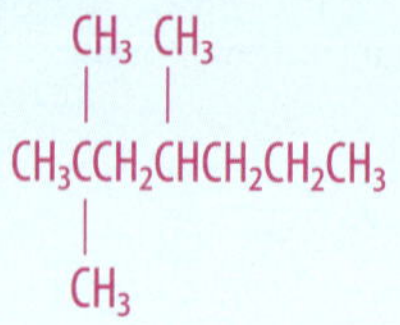

   b. 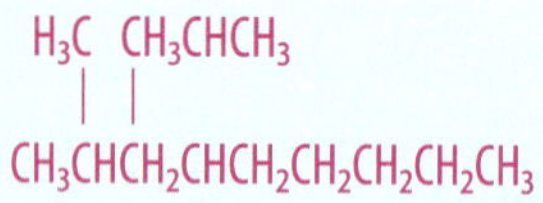

2. a. 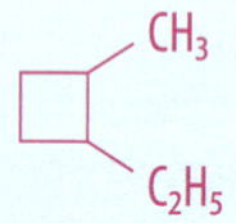

   b. 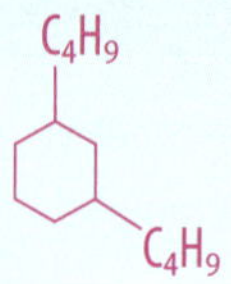

Section 3

3. a. $H_2C{=}CHCH_2CH{=}CHCH_3$

   b. 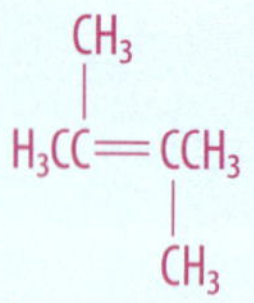

   c. 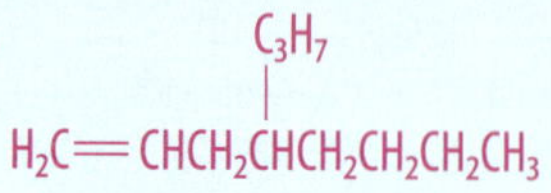

   d. 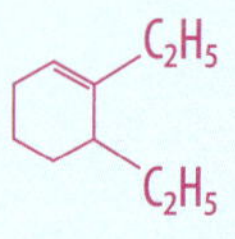

SUPPLEMENTAL PRACTICE PROBLEMS

CHAPTER 20

Section 1

1. Calculate the cell potential for each of the following.
   a. $Co^{2+}(aq) + Al(s) \longrightarrow Co(s) + Al^{3+}(aq)$
   b. $Hg^{2+}(aq) + Cu(s) \longrightarrow Cu^{2+}(aq) + Hg(s)$
   c. $Zn(s) + Br_2(l) \longrightarrow 2Br^-(aq) + Zn^{2+}(aq)$
2. Calculate the cell potential to determine whether the reaction will occur spontaneously or not spontaneously. For each reaction that is not spontaneous, correct the reactants or products so that a reaction would occur spontaneously.
   a. $Ni^{2+}(aq) + Al(s) \longrightarrow Ni(s) + Al^{3+}(aq)$
   b. $Ag^+(aq) + H_2(g) \longrightarrow Ag(s) + H^+(aq)$
   c. $Fe^{2+}(aq) + Cu(s) \longrightarrow Fe(s) + Cu^{2+}(aq)$

CHAPTER 21

Supplemental Practice Problems

Section 2

1. Draw the structure of the following branched alkanes.
   a. 2,2,4-trimethylheptane
   b. 4-isopropyl-2-methylnonane
2. Draw the structure of each of the following cycloalkanes.
   a. 1-ethyl-2-methylcyclobutane
   b. 1,3-dibutylcyclohexane

Section 3

3. Draw the structure of each of the following alkenes.
   a. 1,4-hexadiene
   b. 2,3-dimethyl-2-butene
   c. 4-propyl-1-octene
   d. 2,3-diethylcyclohexene

CHAPTER 22

Section 1

1. Draw the structures of the following alkyl halides.
   a. chloroethane
   b. chloromethane
   c. 1-fluoropentane
   d. 1,3-dibromocyclohexane
   e. 1,2-dibromo-3-chloropropane

CHAPTER 24

Section 2

1. Write balanced equations for each of the following decay processes.
   a. alpha emission of ${}^{244}_{96}Cm$
   b. positron emission of ${}^{70}_{33}As$
   c. beta emission of ${}^{210}_{83}Bi$
   d. electron capture by ${}^{116}_{51}Sb$
2. ${}^{47}_{20}Ca \rightarrow \beta + ?$
3. ${}^{240}_{95}Am + ? \longrightarrow {}^{243}_{97}Bk + n$
4. How much time has passed if 1/8 of an original sample of radon-222 is left? Use **Table 5** for half-life information.
5. If a basement air sample contains 3.64 μg of radon-222, how much radon will remain after 19 days?
6. Cobalt-60, with a half-life of 5 years, is used in cancer radiation treatments. If a hospital purchases 30.0 g, how much would be left after 15 years?

CHAPTER 22

Section 1

1. a. CH_3CH_2Cl
   b. CH_3Cl
   c. $CH_3CH_2CH_2CH_2CH_2F$
   d. 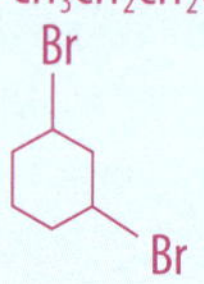
   e. 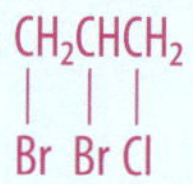

CHAPTER 24

Section 2

1. a. ${}^{244}_{96}Cm \longrightarrow {}^{240}_{94}Pu + {}^{4}_{2}He$
   b. ${}^{70}_{33}As \longrightarrow {}^{70}_{32}Ge + \beta^+$
   c. ${}^{210}_{83}Bi \longrightarrow {}^{210}_{84}Po + \beta^+$
   d. ${}^{116}_{51}Sb + e^- \longrightarrow {}^{116}_{50}Sn$
2. ${}^{47}_{20}Ca \rightarrow \beta + {}^{47}_{21}Sc$
3. ${}^{240}_{95}Am + {}^{4}_{2}He \longrightarrow {}^{243}_{96}Bk + {}^{1}_{1}H$
4. 11.4 days
5. 0.114 μg
6. 3.75 g

SELECTED SOLUTIONS

Solutions to Selected Practice Problems

CHAPTER 1

No practice problems

CHAPTER 2

1. No; the density of aluminum is 2.7 g/cm^3; the density of the cube is $\frac{20\text{ g}}{5\text{ cm}^3} = 4\text{ g/cm}^3$.

3. $\text{volume} = \frac{\text{mass}}{\text{density}} = \frac{147\cancel{\text{g}}}{7.00\cancel{\text{g}}/\text{mL}} = 21.0\text{ mL}$

$\text{volume} = 20.0\text{ mL} + 21.0\text{ mL} = 41.0\text{ mL}$

11. **a.** 7×10^2
b. 3.8×10^4
c. 4.5×10^6
d. 6.85×10^{11}
e. 5.4×10^{-3}
f. 6.87×10^{-6}
g. 7.6×10^{-8}
h. 8×10^{-10}

13. **a.** 7×10^{-5}
b. 3×10^8
c. 2×10^2
d. 5×10^{-12}

15. **a.** $(4 \times 1) \times 10^{2+8} = 4 \times 10^{10}$
b. $(2 \times 3) \times 10^{-4+2} = 6 \times 10^{-2}$
c. $(6 \div 2) \times 10^{2-1} = 3 \times 10^1$
d. $(8 \div 4) \times 10^{4-1} = 2 \times 10^3$

17. **a.** $\frac{16\text{ g salt}}{100\text{ g solution}};\ \frac{100\text{ g solution}}{16\text{ g salt}}$
b. $\frac{1.25\text{ g}}{1\text{ mL}};\ \frac{1\text{ mL}}{1.25\text{ g}}$
c. $\frac{25\text{ m}}{1\text{ s}};\ \frac{1\text{s}}{25\text{ m}}$

19. **a.** $360\cancel{\text{s}} \times \frac{1000\text{ ms}}{1\cancel{\text{s}}} = 360{,}000\text{ ms}$
b. $4800\cancel{\text{g}} \times \frac{1\text{ kg}}{1000\cancel{\text{g}}} = 4.8\text{ kg}$
c. $5600\cancel{\text{dm}} \times \frac{1\text{ m}}{10\cancel{\text{dm}}} = 560\text{ m}$
d. $72\cancel{\text{g}} \times \frac{1000\text{ mg}}{1\cancel{\text{g}}} = 72{,}000\text{ mg}$
e. $2.45 \times 10^2\cancel{\text{ms}} \times \frac{1\text{ s}}{1000\cancel{\text{ms}}} = 0.245\text{ s}$
f. $5\cancel{\mu\text{m}} \times \frac{1\cancel{\text{mm}}}{1000\cancel{\mu\text{m}}} \times \frac{1\cancel{\text{m}}}{1000\cancel{\text{mm}}} \times \frac{1\text{ km}}{1000\cancel{\text{m}}}$
$= 5 \times 10^{-9}\text{ km}$
g. $6.800 \times 10^3\cancel{\text{cm}} \times \frac{1\cancel{\text{m}}}{100\cancel{\text{cm}}} \times \frac{1\text{ km}}{1000\cancel{\text{m}}}$
$= 6.800 \times 10^{-2}\text{ km}$
h. $2.5 \times 10^1\cancel{\text{kg}} \times \frac{1\text{ Mg}}{1000\cancel{\text{kg}}} = 0.025\text{ Mg}$

21. $\frac{65\cancel{\text{mi}}}{1\text{ h}} \times \frac{1\text{ km}}{0.62\cancel{\text{mi}}} = 1.0 \times 10^2\text{ km/h}$

23. $\text{mass} = (\text{volume})(\text{density}) = (185\cancel{\text{mL}})(1.02\text{ g}/\cancel{\text{mL}})$
mass = 189 g vinegar
$(189\cancel{\text{ g vinegar}})\left(\frac{5.00\text{ g acetic acid}}{100\cancel{\text{ g vinegar}}}\right) = 9.45\text{ g acetic acid}$

33. $\frac{0.11}{1.59} \times 100 = 6.92\%$
$\frac{0.10}{1.59} \times 100 = 6.29\%$
$\frac{0.12}{1.59} \times 100 = 7.55\%$
Note: The answers are reported in three significant figures because student error is the difference between the actual value (1.59 g/cm^3) and the measured value.

35. **a.** 4 **b.** 7 **c.** 5 **d.** 3

37. two significant figures: 1.0×10^1, 1.0×10^2, 1.0×10^3
three significant figures: 1.00×10^1, 1.00×10^2, 1.00×10^3
four significant figures: 1.000×10^1, 1.000×10^2, 1.000×10^3

39. **a.** 5.482×10^{-4} g
b. 1.368×10^5 kg
c. 3.087×10^8 mm
d. 2.014 mL

41. **a.** $4.32 \times 10^3\text{ cm} - 1.6 \times 10^6\text{ mm}$
$= 4.32 \times 10^3\text{ cm} - 16 \times 10^6\text{ cm}$
$= 4.32 \times 10^3\text{ cm} - 16{,}000 \times 10^3\text{ cm}$
$= -15{,}995.68 \times 10^3\text{ cm} = -16.0 \times 10^6\text{ cm}$
b. $2.12 \times 10^7\text{ mm} + 1.8 \times 10^3\text{ cm}$
$= 2.12 \times 10^7\text{ mm} + 1.8 \times 10^4\text{ mm}$
$= 2120 \times 10^4\text{ mm} + 1.8 \times 10^4\text{ mm}$
$= 2121.8 \times 10^4\text{ mm} = 2.12 \times 10^7\text{ mm}$

43. **a.** 2.0 m/s
b. 3.00 m/s
c. 2.00 m/s
d. 2.9 m/s

CHAPTER 3

5. amount of bromine that reacted = 100.0 g − 8.5 g = 91.5 g; amount of compound formed = 100.0 g + 10.3 g − 8.5 g = 101.8 g

7. $\text{mass}_{\text{reactants}} = \text{mass}_{\text{products}}$
$\text{mass}_{\text{sodium}} + \text{mass}_{\text{chlorine}} = \text{mass}_{\text{sodium chloride}}$
$\text{mass}_{\text{sodium}} = 15.6\text{ g}$
$\text{mass}_{\text{sodium chloride}} = 39.7\text{ g}$
Substituting and solving for $\text{mass}_{\text{chlorine}}$ yields
$15.6\text{ g} + \text{mass}_{\text{chlorine}} = 39.7\text{ g}$
$\text{mass}_{\text{chlorine}} = 39.7\text{ g} - 15.6\text{ g} = 24.1\text{ g}$ used in the reaction.
Because the sodium reacts with excess chlorine, all of the sodium is used in the reaction; that is, 15.6 g of sodium are used in the reaction.

9. 156.3 g − 106.5 g = 49.8 g
Yes. Mass of reactants equals mass of products.

19. $\text{percent by mass}_{\text{hydrogen}} = \frac{\text{mass}_{\text{hydrogen}}}{\text{mass}_{\text{compound}}} \times 100$
$\text{percent by mass}_{\text{hydrogen}} = \frac{12.4\cancel{\text{g}}}{78.0\cancel{\text{g}}} \times 100 = 15.9\%$

Selected Solutions

21. $mass_{XY} = 3.50\ g + 10.5\ g = 14.0\ g$

$$\text{percent by mass}_X = \frac{mass_X}{mass_{XY}} \times 100$$

$$\text{percent by mass}_X = \frac{3.50\ \cancel{g}}{14.0\ \cancel{g}} \times 100 = 25.0\%$$

$$\text{percent by mass}_Y = \frac{mass_Y}{mass_{XY}} \times 100$$

$$\text{percent by mass}_Y = \frac{10.5\ \cancel{g}}{14.0\ \cancel{g}} \times 100 = 75.0\%$$

23. No, you cannot be sure. Having the same mass percentage of a single element does not guarantee that the composition of each compound is the same.

CHAPTER 4

13. dysprosium

15. Yes. 9

17. 25 protons, 25 electrons, 30 neutrons, manganese

19. N-14 is more abundant because the atomic mass is closer to 14 than 15.

CHAPTER 5

1. $c = \lambda\nu$

$\nu = c / \lambda$

$$\nu = \frac{3.00 \times 10^8\ m/s}{4.90 \times 10^{-7}\ m} = 6.12 \times 10^{14}\ Hz$$

3. 3.00×10^8 m/s

5. **a.** $E_{photon} = \lambda\nu = (6.626 \times 10^{-34}\ J\cdot\cancel{s})(6.32 \times 10^{20}\ \cancel{s^{-1}})$ $= 4.19 \times 10^{-13}\ J$

b. $E_{photon} = \lambda\nu = (6.626 \times 10^{-34}\ J\cdot\cancel{s})(9.50 \times 10^{13}\ \cancel{s^{-1}})$ $= 6.29 \times 10^{-20}\ J$

c. $E_{photon} = \lambda\nu = (6.626 \times 10^{-34}\ J\cdot\cancel{s})(1.05 \times 10^{16}\ \cancel{s^{-1}})$ $= 6.96 \times 10^{-18}\ J$

7. $E_{photon} = hc / \lambda$

$$E_{photon} = \frac{(6.626 \times 10^{-34}\ J\cdot\cancel{s})(3.00 \times 10^8\ \cancel{m}/\cancel{s})}{1.25 \times 10^{-1}\ \cancel{m}}$$

$= 1.59 \times 10^{-24}\ J$

21. **a.** bromine (35 electrons): $[Ar]4s^23d^{10}4p^5$

b. strontium (38 electrons): $[Kr]5s^2$

c. antimony (51 electrons): $[Kr]5s^24d^{10}5p^3$

d. rhenium (75 electrons): $[Xe]6s^24f^{14}5d^5$

e. terbium (65 electrons): $[Xe]6s^24f^9$

f. titanium (22 electrons): $[Ar]4s^23d^2$

23. Sulfur (15 electrons) has the electron configuration $[Ne]3s^23p^4$. Therefore, 6 electrons are in orbitals related to the third energy level of the sulfur atom.

25. $[Xe]6s^2$; barium

27. aluminum; 3 electrons

CHAPTER 6

9. **a.** Sc, Y, La, Ac

b. N, P, As, Sb, Bi

c. Ne, Ar, Kr, Xe, Rn

17. B. The atomic radius increases when going down a group so helium is the smallest and radon is the biggest.

19. **a.** the element in period 2, group 1

b. the element in period 5, group 2

c. the element in period 6, group 15

d. the element in period 4, group 18

CHAPTER 7

7. Three Na atoms each lose 1 e^-, forming 1+ ions. One N atom gains 3 e^-, forming a 3− ion. The ions attract, forming Na_3N.

$$3\ \cancel{\text{Na ions}}\left(\frac{1+}{\cancel{\text{Na ion}}}\right) + 1\ \cancel{\text{N ion}}\left(\frac{3-}{\cancel{\text{N ion}}}\right)$$

$= 3(1+) + 1(3-) = 0$

The overall charge on one formula unit of Na_3N is zero.

9. One Sr atom loses 2 e^-, forming a 2+ ion. Two F atoms each gain 1 e^-, forming 1− ions. The ions attract, forming SrF_2.

$$1\ \cancel{\text{Sr ion}}\left(\frac{2+}{\cancel{\text{Sr ion}}}\right) + 2\ \cancel{\text{F ions}}\left(\frac{1-}{\cancel{\text{F ion}}}\right)$$

$= 1(2+) + 2(1-) = 0$

The overall charge on one formula unit of SrF_2 is zero.

11. Three group 1 atoms lose 1 e^-, forming 1+ ions. One group 15 atom gains 3 e^-, forming a 3− ion. The ions attract, forming X_3Y, where X represents a group 1 atom and Y represents a group 15 atom.

19. KI

21. $AlBr_3$

23. The general formula is XY_2, where X represents the group 2 element and Y represents the group 17 element.

25. $Ca(ClO_3)_2$

27. $MgCO_3$; answers will vary

29. calcium chloride

31. copper(II) nitrate

33. ammonium perchlorate

CHAPTER 8

1.

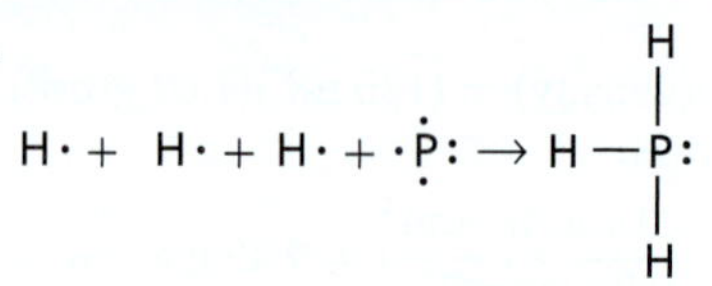

3. H· + ·C̈l: → H—C̈l:

5.

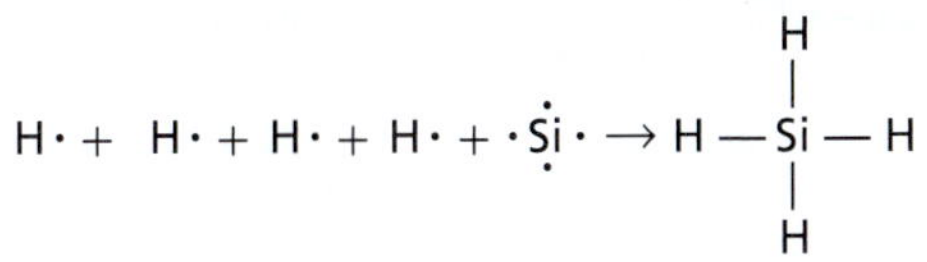

15. sulfur dioxide

17. carbon tetrachloride

19. hydroiodic acid

21. chlorous acid

23. hydrosulfuric acid

25. AgCl

27. ClF_3

29. S_2F_{10}

37.

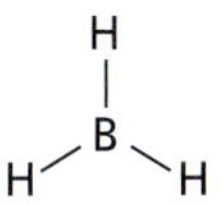

39.

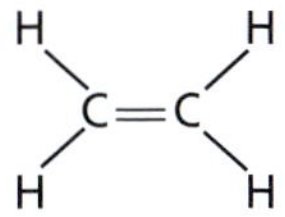

41.

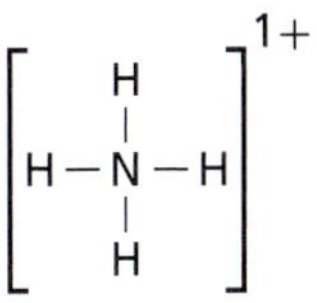

43.

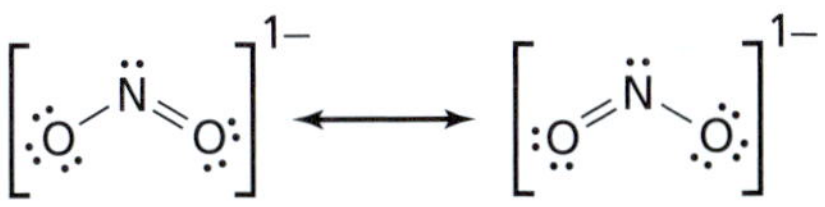

45.

47.

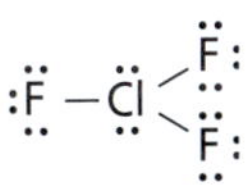

49.

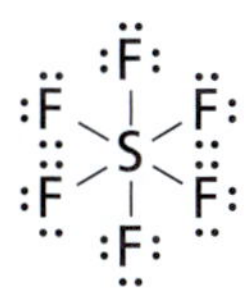

57. bent, 104.5°, sp^3

59. tetrahedral, 109°, sp^3

CHAPTER 9

1. $H_2(g) + Br_2(g) \rightarrow HBr(g)$

3. potassium chlorate(s) → potassium chloride(s) and oxygen(g)
$KClO_3(s) \rightarrow KCl(s) + O_2(g)$

5. $CS_2(l) + 3O_2(g) \rightarrow CO_2(g) + 2SO_2(g)$

15. $H_2O(l) + N_2O_5(g) \rightarrow 2HNO_3(aq)$; synthesis

17. $H_2SO_4(aq) + 2NaOH(aq) \rightarrow Na_2SO_4(aq) + 2H_2O(l)$

19. $Ni(OH)_2(s) \rightarrow NiO(s) + H_2O(l)$

21. Yes. K is above Zn in the metal activity series.
$2K(s) + ZnCl_2(aq) \rightarrow Zn(s) + 2KCl(aq)$

23. No. Fe is below Na in the metal activity series.

25. $LiI(aq) + AgNO_3(aq) \rightarrow AgI(s) + LiNO_3(aq)$

27. $Na_2C_2O_4(aq) + Pb(NO_3)_2(aq) \rightarrow PbC_2O_4(s) + 2NaNO_3(aq)$

35. chemical equation: $KI(aq) + AgNO_3(aq) \rightarrow KNO_3(aq) + AgI(s)$
complete ionic equation:
$K^+(aq) + I^-(aq) + Ag^+(aq) + NO_3^-(aq) \rightarrow K^+(aq) + NO_3^-(aq) + AgI(s)$
net ionic equation: $I^-(aq) + Ag^+(aq) \rightarrow AgI(s)$

37. chemical equation: $AlCl_3(aq) + 3NaOH(aq) \rightarrow Al(OH)_3(s) + 3NaCl(aq)$
complete ionic equation:
$Al^{3+}(aq) + 3Cl^-(aq) + 3Na^+(aq) + 3OH^2(aq) \rightarrow Al(OH)_3(s) + 3Na^+(aq) + 3Cl^-(aq)$
net ionic equation: $Al^{3+}(aq) + 3OH^-(aq) \rightarrow Al(OH)_3(s)$

39. chemical equation: $5Na_2CO_3(aq) + 2MnCl_5(aq) \rightarrow 10NaCl(aq) + Mn_2(CO_3)_5(s)$
complete ionic equation:
$10Na^+(aq) + 5CO_3^{2-}(aq) + 2Mn^{5+}(aq) + 10Cl^-(aq) \rightarrow 10Na^+(aq) + 10Cl^-(aq) + Mn_2(CO_3)_5(s)$
net ionic equation: $5CO_3^{2-}(aq) + 2Mn^{5+}(aq) \rightarrow Mn_2(CO_3)_5(s)$

41. chemical equation: $2HCl(aq) + Ca(OH)_2(aq) \rightarrow 2H_2O(l) + CaCl_2(aq)$
complete ionic equation:
$2H^+(aq) + 2Cl^-(aq) + Ca^{2+}(aq) + 2OH^-(aq) \rightarrow 2H_2O(l) + Ca^{2+}(aq) + 2Cl^-(aq)$
net ionic equation: $H^+(aq) + OH^-(aq) \rightarrow H_2O(l)$

43. chemical equation: $H_2S(aq) + 1\ Ca(OH)_2(aq) \rightarrow 2H_2O(l) + CaS(aq)$
complete ionic equation:
$2H^+(aq) + S^{2-}(aq) + Ca^{2+}(aq) + 2OH^-(aq) \rightarrow 2H_2O(l) + Ca^{2+}(aq) + S^{2-}(aq)$
net ionic equation: $H^+(aq) + OH^-(aq) \rightarrow H_2O(l)$

45. chemical equation: $2HClO_4(aq) + K_2CO_3(aq) \rightarrow H_2O(l) + CO_2(g) + 2KClO_4(aq)$
complete ionic equation:
$2H^+(aq) + 2ClO_4^-(aq) + 2K^+(aq) + CO_3^{2-}(aq) \rightarrow H_2O(l) + CO_2(g) + 2K^+(aq) + 2ClO_4^-(aq)$
net ionic equation: $2H^+(aq) + CO_3^{2-}(aq) \rightarrow H_2O(l) + CO_2(g)$

47. chemical equation: $2HBr(aq) + (NH_4)_2CO_3(aq) \rightarrow H_2O(l) + CO_2(g) + 2NH_4Br(aq)$

complete ionic equation:

$$2H^+(aq) + 2Br^-(aq) + 2NH_4{}^+(aq) + CO_3{}^{2-}(aq) \rightarrow H_2O(l) + CO_2(g) + 2NH_4{}^+(aq) + 2Br^-(aq)$$

net ionic equation: $2H^+(aq) + CO_3{}^{2-}(aq) \rightarrow H_2O(l) + CO_2(g)$

49. chemical equation: $2KI(aq) + Pb(NO_3)_2(aq) \rightarrow 2KNO_3(aq) + PbI_2(s)$

complete ionic equation:

$$2K^+(aq) + 2I^-(aq) + Pb^{2+}(aq) + 2NO_3{}^-(aq) \rightarrow 2K^+(aq) + 2NO_3{}^-(aq) + PbI_2(s)$$

net ionic equation: $Pb^{2+}(aq) + 2I^-(aq) \rightarrow PbI_2(s)$

CHAPTER 10

1. $2.50 \text{ mol Zn} \times \frac{6.02 \times 10^{23} \text{ atoms}}{1 \text{ mol}} = 1.51 \times 10^{24} \text{ atoms of Zn}$

3. $3.25 \text{ mol AgNO}_3 \times \frac{6.02 \times 10^{23} \text{ formula units}}{1 \text{ mol}} = 1.96 \times 10^{24} \text{ formula units of AgNO}_3$

5. a. $5.75 \times 10^{24} \text{ atoms Al} \times \frac{1 \text{ mol}}{6.02 \times 10^{23} \text{ atoms}} = 9.55 \text{ mol Al}$

b. $2.50 \times 10^{20} \text{ atoms Fe} \times \frac{1 \text{ mol}}{6.02 \times 10^{23} \text{ atoms}} = 4.15 \times 10^{-4} \text{ mol Fe}$

15. a. $3.57 \text{ mol Al} \times \frac{26.98 \text{ g Al}}{1 \text{ mol Al}} = 96.3 \text{ g Al}$

b. $42.6 \text{ mol Si} \times \frac{28.09 \text{ g Si}}{1 \text{ mol Si}} = 1.20 \times 10^3 \text{ g Si}$

17. a. $25.5 \text{ g Ag} \times \frac{1 \text{ mol Ag}}{107.9 \text{ g Ag}} = 0.236 \text{ mol Ag}$

b. $300.0 \text{ g S} \times \frac{1 \text{ mol S}}{32.07 \text{ g S}} = 9.355 \text{ mol S}$

19. a. $55.2 \text{ g Li} \times \frac{1 \text{ mol Li}}{6.94 \text{ g Li}} \times \frac{6.02 \times 10^{23} \text{ atoms}}{1 \text{ mol}} = 4.79 \times 10^{24} \text{ atoms Li}$

b. $0.230 \text{ g Pb} \times \frac{1 \text{ mol Pb}}{207.2 \text{ g Pb}} \times \frac{6.02 \times 10^{23} \text{ atoms}}{1 \text{ mol}} = 6.68 \times 10^{20} \text{ atoms Pb}$

c. $11.5 \text{ g Hg} \times \frac{1 \text{ mol Hg}}{200.6 \text{ g Hg}} \times \frac{6.02 \times 10^{23} \text{ atoms}}{1 \text{ mol}} = 3.45 \times 10^{22} \text{ atoms Hg}$

21. a. $4.56 \times 10^3 \text{ g Si} \times \frac{1 \text{ mol Si}}{28.09 \text{ g Si}} \times \frac{6.02 \times 10^{23} \text{ atoms}}{1 \text{ mol}} = 9.77 \times 10^{25} \text{ atoms Si}$

b. $0.120 \text{ kg Ti} \times \frac{1000 \text{ g Ti}}{1 \text{ kg Ti}} \times \frac{1 \text{ mol Ti}}{47.87 \text{ g Ti}} \times \frac{6.02 \times 10^{23} \text{ atoms}}{1 \text{ mol}} = 1.51 \times 10^{24} \text{ atoms Ti}$

29. $2.50 \text{ mol ZnCl}_2 \times \frac{2 \text{ mol Cl}^-}{1 \text{ mol ZnCl}_2} = 5.00 \text{ mol Cl}^-$

31. $3.00 \text{ mol Fe}_2(\text{SO}_4)_3 \times \frac{3 \text{ mol SO}_4{}^{2-}}{1 \text{ mol Fe}_2(\text{SO}_4)_3} = 9.00 \text{ mol SO}_4{}^{2-}$

33. $1.15 \times 10^1 \text{ mol H}_2\text{O} \times \frac{2 \text{ mol H}}{1 \text{ mol H}_2\text{O}} = 23.0 \text{ mol H} = 2.30 \times 10^1 \text{ mol H}$

35. a.

$2 \text{ mol C} \times \frac{12.01 \text{ g C}}{1 \text{ mol C}} = 24.02 \text{ g C}$

$6 \text{ mol H} \times \frac{1.008 \text{ g H}}{1 \text{ mol H}} = 6.048 \text{ g H}$

$1 \text{ mol O} \times \frac{16.00 \text{ g O}}{1 \text{ mol O}} = \underline{16.00 \text{ g O}}$

molar mass C_2H_5OH = 46.07 g/mol

b.

$1 \text{ mol H} \times \frac{1.008 \text{ g H}}{1 \text{ mol H}} = 1.008 \text{ g H}$

$1 \text{ mol C} \times \frac{12.01 \text{ g C}}{1 \text{ mol C}} = 12.01 \text{ g C}$

$1 \text{ mol N} \times \frac{14.01 \text{ g N}}{1 \text{ mol N}} = \underline{14.01 \text{ g N}}$

molar mass HCN = 27.03 g/mol

c.

$1 \text{ mol C} \times \frac{12.01 \text{ g C}}{1 \text{ mol C}} = 12.01 \text{ g C}$

$4 \text{ mol Cl} \times \frac{35.45 \text{ g Cl}}{1 \text{ mol Cl}} = \underline{141.80 \text{ g Cl}}$

molar mass CCl_4 = 153.81 g/mol

37. Step 1: Find the molar mass of H_2SO_4.

$2 \text{ mol H} \times \frac{1.008 \text{ g H}}{1 \text{ mol H}} = 2.016 \text{ g H}$

$1 \text{ mol S} \times \frac{32.07 \text{ g S}}{1 \text{ mol S}} = 32.07 \text{ g S}$

$4 \text{ mol O} \times \frac{16.00 \text{ g O}}{1 \text{ mol O}} = \underline{64.00 \text{ g O}}$

molar mass H_2SO_4 = 98.09 g/mol

Step 2: Make mole → mass conversion.

$3.25 \text{ mol H}_2\text{SO}_4 \times \frac{98.09 \text{ g H}_2\text{SO}_4}{1 \text{ mol H}_2\text{SO}_4} = 319 \text{ g H}_2\text{SO}_4$

39. Potassium permanganate has a formula of $KMnO_4$.

Step 1: Find the molar mass of $KMnO_4$.

$1 \text{ mol K} \times \frac{39.10 \text{ g K}}{1 \text{ mol K}} = 39.10 \text{ g K}$

$1 \text{ mol Mn} \times \frac{54.94 \text{ g Mn}}{1 \text{ mol Mn}} = 54.94 \text{ g Mn}$

$4 \text{ mol O} \times \frac{16.00 \text{ g O}}{1 \text{ mol O}} = \underline{64.00 \text{ g O}}$

molar mass $KMnO_4$ = 158.04 g/mol

Step 2: Make mole → mass conversion.

$2.55 \text{ mol KMnO}_4 \times \frac{158.04 \text{ g KMnO}_4}{1 \text{ mol KMnO}_4} = 403 \text{ g KMnO}_4$

41. a. ionic compound

Step 1: Find the molar mass of Fe_2O_3.

$$2\ \cancel{\text{mol Fe}} \times \frac{55.85\ \text{g Fe}}{1\ \cancel{\text{mol Fe}}} = 111.70\ \text{g Fe}$$

$$3\ \cancel{\text{mol O}} \times \frac{16.00\ \text{g O}}{1\ \cancel{\text{mol O}}} = \underline{48.00\ \text{g O}}$$

$$\text{molar mass } Fe_2O_3 = 159.70\ \text{g/mol}$$

Step 2: Make mass $\rightarrow$ mole conversion.

$$2500\ \cancel{\text{g } Fe_2O_3} \times \frac{1\ \text{mol } Fe_2O_3}{159.70\ \cancel{\text{g } Fe_2O_3}} = 1.57 \times 10^1\ \text{mol } Fe_2O_3$$

b. ionic compound

Step 1: Find the molar mass of $PbCl_4$.

$$1\ \cancel{\text{mol Pb}} \times \frac{207.2\ \text{g Pb}}{1\ \cancel{\text{mol Pb}}} = 207.2\ \text{g Pb}$$

$$4\ \cancel{\text{mol Cl}} \times \frac{35.45\ \text{g Cl}}{1\ \cancel{\text{mol Cl}}} = \underline{141.80\ \text{g CI}}$$

$$\text{molar mass } PbCl_4 = 349.0\ \text{g/mol}$$

Step 2: Make mass $\rightarrow$ mole conversion.

$$25.4\ \cancel{\text{mg } PbCl_4} \times \frac{1\ \cancel{\text{g } PbCl_4}}{1000\ \cancel{\text{mg } PbCl_4}} \times \frac{1\ \text{mol } PbCI_4}{349.0\ \cancel{\text{g } PbCl_4}} =$$

7.28×10^{-5} mol $PbCl_4$

43. a. Step 1: Find the molar mass of Na_2SO_3

$$2\ \cancel{\text{mol Na}} \times \frac{22.99\ \text{g Na}}{1\ \cancel{\text{mol Na}}} = 45.98\ \text{g Na}$$

$$1\ \cancel{\text{mol S}} \times \frac{32.07\ \text{g S}}{1\ \cancel{\text{mol S}}} = 32.07\ \text{g S}$$

$$3\ \cancel{\text{mol O}} \times \frac{16.00\ \text{g O}}{1\ \cancel{\text{mol O}}} = \underline{48.00\ \text{g O}}$$

$$\text{molar mass } Na_2SO_3 = 126.05\ \text{g/mol}$$

Step 2: Make mass $\rightarrow$ mole conversion.

$$2.25\ \cancel{\text{g } Na_2SO_3} \times \frac{1\ \text{mol } Na_2SO_3}{126.05\ \cancel{\text{g } Na_2SO_3}}$$

$= 0.0179$ mol Na_2SO_3

Step 3: Make mole $\rightarrow$ formula unit conversion.

$$0.0179\ \cancel{\text{mol } Na_2SO_3} \times \frac{6.02 \times 10^{23}\ \text{formula units}}{1\ \cancel{\text{mol } Na_2SO_3}}$$

$= 1.08 \times 10^{22}$ formula units Na_2SO_3

Step 4: Determine the number of Na^+ ions.

1.08×10^{22} $\cancel{\text{formula units } Na_2SO_3}$ $\times$

$$\frac{2\ Na^+\ \text{ions}}{1\ \cancel{\text{formula unit } Na_2SO_3}} = 2.16 \times 10^{22}\ Na^+\ \text{ions}$$

b. 1.08×10^{22} $\cancel{\text{formula units } Na_2SO_3}$ $\times$

$$\frac{1\ SO_3^{2-}\ \text{ion}}{1\ \cancel{\text{formula unit } Na_2SO_3}} = 1.08 \times 10^{22}\ SO_3^{2-}\ \text{ions}$$

c. $$\frac{126.05\ \text{g } Na_2SO_3}{1\ \cancel{\text{mol } Na_2SO_3}} \times \frac{1\ \cancel{\text{mol } Na_2SO_3}}{6.02 \times 10^{23}\ \text{formula unit } Na_2SO_3}$$

$= 2.09 \times 10^{-22}$ g Na_2SO_3/formula unit

45. Step 1: Find the number of moles of NaCl.

4.59×10^{24} $\cancel{\text{formula units NaCl}}$ $\times$

$$\frac{1\ \text{mol NaCl}}{6.02 \times 10^{23}\ \cancel{\text{formula unit NaCl}}}$$

$= 7.62$ mol NaCl

Step 2: Find the molar mass of NaCl.

$$1\ \cancel{\text{mol Na}} \times \frac{22.99\ \text{g Na}}{1\ \cancel{\text{mol Na}}} = 22.99\ \text{g Na}$$

$$1\ \cancel{\text{mol Cl}} \times \frac{35.45\ \text{g Cl}}{1\ \cancel{\text{mol Cl}}} = \underline{35.45\ \text{g CI}}$$

$$\text{molar mass NaCl} = 58.44\ \text{g/mol}$$

Step 3: Make mole $\rightarrow$ mass conversion.

$$7.62\ \cancel{\text{mol NaCl}} \times \frac{58.44\ \text{g NaCl}}{1\ \cancel{\text{mol NaCl}}} = 445\ \text{g NaCl}$$

55. Steps 1 and 2: Assume 1 mole; calculate molar mass of H_2SO_3.

$$2\ \cancel{\text{mol H}} \times \frac{1.008\ \text{g H}}{1\ \cancel{\text{mol H}}} = 2.016\ \text{g H}$$

$$1\ \cancel{\text{mol S}} \times \frac{32.06\ \text{g S}}{1\ \cancel{\text{mol S}}} = 32.06\ \text{g S}$$

$$3\ \cancel{\text{mol O}} \times \frac{16.00\ \text{g O}}{1\ \cancel{\text{mol O}}} = \underline{48.00\ \text{g O}}$$

$$\text{molar mass } H_2SO_3 = 82.08\ \text{g/mol}$$

Step 3: Determine percent by mass of S.

$$\text{percent S} = \frac{32.06\ \text{g S}}{82.08\ \text{g } H_2SO_3} \times 100 = 39.06\%\ \text{S}$$

Repeat steps 1 and 2 for $H_2S_2O_8$. Assume 1 mole; calculate molar mass of $H_2S_2O_8$.

$$2\ \cancel{\text{mol H}} \times \frac{1.008\ \text{g H}}{1\ \cancel{\text{mol H}}} = 2.016\ \text{g H}$$

$$2\ \cancel{\text{mol S}} \times \frac{32.06\ \text{g S}}{1\ \cancel{\text{mol S}}} = 64.12\ \text{g S}$$

$$8\ \cancel{\text{mol O}} \times \frac{16.00\ \text{g O}}{1\ \cancel{\text{mol O}}} = \underline{128.00\ \text{g O}}$$

$$\text{molar mass } H_2S_2O_8 = 194.14\ \text{g/mol}$$

Step 3: Determine percent by mass of S.

$$\text{percent S} = \frac{64.12\ \text{g S}}{194.14\ \text{g } H_2S_2O_8} \times 100 = 33.02\%\ \text{S}$$

H_2SO_3 has a larger percent by mass of S.

57. a. sodium, sulfur, and oxygen; Na_2SO_4

b. ionic

c. Steps 1 and 2: Assume 1 mole; calculate molar mass of Na_2SO_4.

$$2\ \cancel{\text{mol Na}} \times \frac{22.99\ \text{g Na}}{1\ \cancel{\text{mol Na}}} = 45.98\ \text{g Na}$$

$$1\ \cancel{\text{mol S}} \times \frac{32.07\ \text{g S}}{1\ \cancel{\text{mol S}}} = 32.07\ \text{g S}$$

$$4\ \cancel{\text{mol O}} \times \frac{16.00\ \text{g O}}{1\ \cancel{\text{mol O}}} = \underline{64.00\ \text{g O}}$$

$$\text{molar mass } Na_2SO_4 = 142.05\ \text{g/mol}$$

Step 3: Determine percent by mass of each element.

$$\text{percent Na} = \frac{45.98 \text{ g Na}}{142.05 \text{ g Na}_2\text{SO}_4} \times 100 = 32.37\% \text{ Na}$$

$$\text{percent S} = \frac{32.07 \text{ g S}}{142.05 \text{ g Na}_2\text{SO}_4} \times 100 = 22.58\% \text{ S}$$

$$\text{percent O} = \frac{64.00 \text{ g O}}{142.05 \text{ g Na}_2\text{SO}_4} \times 100 = 45.05\% \text{ O}$$

59. Step 1: Assume 100 g sample; calculate moles of each element.

$$35.98 \cancel{\text{g Al}} \times \frac{1 \text{ mol Al}}{26.98 \cancel{\text{g Al}}} = 1.333 \text{ mol Al}$$

$$64.02 \cancel{\text{g S}} \times \frac{1 \text{ mol S}}{32.07 \cancel{\text{g S}}} = 1.996 \text{ mol S}$$

Step 2: Calculate mole ratios.

$$\frac{1.333 \text{ mol Al}}{1.333 \text{ mol Al}} = \frac{1.000 \text{ mol Al}}{1.000 \text{ mol Al}} = \frac{1 \text{ mol Al}}{1 \text{ mol Al}}$$

$$\frac{1.996 \text{ mol S}}{1.333 \text{ mol Al}} = \frac{1.497 \text{ mol S}}{1.000 \text{ mol Al}} = \frac{1.5 \text{ mol S}}{1 \text{ mol Al}}$$

The simplest ratio is 1 mol Al: 1.5 mol S.

Step 3: Convert decimal fraction to whole number.

In this case, multiply by 2 because $1.5 \times 2 = 3$. Therefore, the empirical formula is Al_2S_3.

61. Step 1: Assume 100 g sample; calculate moles of each element.

$$60.00 \cancel{\text{g C}} \times \frac{1 \text{ mol C}}{12.01 \cancel{\text{g C}}} = 5.000 \text{ mol C}$$

$$4.44 \cancel{\text{g H}} \times \frac{1 \text{ mol H}}{1.008 \cancel{\text{g H}}} = 4.40 \text{ mol H}$$

$$35.56 \cancel{\text{g O}} \times \frac{1 \text{ mol O}}{16.00 \cancel{\text{g O}}} = 2.22 \text{ mol O}$$

Step 2: Calculate mole ratios.

$$\frac{5.00 \text{ mol C}}{2.22 \text{ mol O}} = \frac{2.25 \text{ mol C}}{1.00 \text{ mol O}} = \frac{2.25 \text{ mol C}}{1 \text{ mol O}}$$

$$\frac{4.40 \text{ mol H}}{2.22 \text{ mol O}} = \frac{1.98 \text{ mol H}}{1.00 \text{ mol O}} = \frac{2 \text{ mol H}}{1 \text{ mol O}}$$

$$\frac{2.22 \text{ mol O}}{2.22 \text{ mol O}} = \frac{1.00 \text{ mol O}}{1.00 \text{ mol O}} = \frac{1 \text{ mol O}}{1 \text{ mol O}}$$

The simplest ratio is 2.25 mol C: 2 mol H: 1 mol O.

Step 3: Convert decimal fraction to whole number.

In this case, multiply by 4 because $2.25 \times 4 = 9$. Therefore, the empirical formula is $C_9H_8O_4$.

63. Step 1: Assume 100 g sample; calculate moles of each element.

$$46.68 \cancel{\text{g N}} \times \frac{1 \text{ mol N}}{14.01 \cancel{\text{g N}}} = 3.332 \text{ mol N}$$

$$53.32 \cancel{\text{g O}} \times \frac{1 \text{ mol O}}{16.00 \cancel{\text{g O}}} = 3.332 \text{ mol O}$$

Step 2: Calculate mole ratios.

$$\frac{3.332 \text{ mol N}}{3.332 \text{ mol N}} = \frac{1.000 \text{ mol N}}{1.000 \text{ mol N}} = \frac{1 \text{ mol N}}{1 \text{ mol N}}$$

$$\frac{3.332 \text{ mol O}}{3.332 \text{ mol N}} = \frac{1.000 \text{ mol O}}{1.000 \text{ mol N}} = \frac{1 \text{ mol O}}{1 \text{ mol N}}$$

The simplest ratio is 1 mol N: 1 mol O.

The empirical formula is NO.

Step 3: Calculate the molar mass of the empirical formula.

$$1 \cancel{\text{mol N}} \times \frac{14.01 \text{ g N}}{1 \cancel{\text{mol N}}} = 14.01 \text{ g N}$$

$$1 \cancel{\text{mol O}} \times \frac{16.00 \text{ g O}}{1 \cancel{\text{mol O}}} = \underline{16.00 \text{ g O}}$$

$$\text{molar mass NO} = 30.01 \text{ g/mol}$$

Step 4: Determine whole number multiplier.

$$\frac{60.01 \cancel{\text{g/mol}}}{30.01 \cancel{\text{g/mol}}} = 2.000$$

The molecular formula is N_2O_2.

65. Step 1: Assume 100 g sample; calculate moles of each element.

$$65.45 \cancel{\text{g C}} \times \frac{1 \text{ mol C}}{12.01 \cancel{\text{g C}}} = 5.450 \text{ mol C}$$

$$5.45 \cancel{\text{g H}} \times \frac{1 \text{ mol H}}{1.008 \cancel{\text{g H}}} = 5.41 \text{ mol H}$$

$$29.09 \cancel{\text{g O}} \times \frac{1 \text{ mol O}}{16.00 \cancel{\text{g O}}} = 1.818 \text{ mol O}$$

Step 2: Calculate mole ratios.

$$\frac{5.450 \text{ mol C}}{1.818 \text{ mol O}} = \frac{3.000 \text{ mol C}}{1.000 \text{ mol O}} = \frac{3 \text{ mol C}}{1 \text{ mol O}}$$

$$\frac{5.41 \text{ mol H}}{1.818 \text{ mol O}} = \frac{2.98 \text{ mol H}}{1.00 \text{ mol O}} = \frac{3 \text{ mol H}}{1 \text{ mol O}}$$

$$\frac{1.818 \text{ mol O}}{1.818 \text{ mol O}} = \frac{1.000 \text{ mol O}}{1.000 \text{ mol O}} = \frac{1 \text{ mol O}}{1 \text{ mol O}}$$

The simplest ratio is 3 mol C: 3 mol H: 1 mol O.

Therefore, the empirical formula is C_3H_3O.

Step 3: Calculate the molar mass of the empirical formula.

$$3 \cancel{\text{mol C}} \times \frac{12.01 \text{ g C}}{1 \cancel{\text{mol C}}} = 36.03 \text{ g C}$$

$$3 \cancel{\text{mol H}} \times \frac{1.008 \text{ g H}}{1 \cancel{\text{mol H}}} = 3.024 \text{ g H}$$

$$1 \cancel{\text{mol O}} \times \frac{16.00 \text{ g O}}{1 \cancel{\text{mol O}}} = \underline{16.00 \text{ g O}}$$

$$\text{molar mass C}_3\text{H}_3\text{O} = 55.05 \text{ g/mol}$$

Step 4: Determine whole number multiplier.

$$\frac{110.00 \cancel{\text{g/mol}}}{55.05 \cancel{\text{g/mol}}} = 1.998, \text{ or } 2$$

The molecular formula is $C_6H_6O_2$.

75. Step 1: Calculate the mass of $CoCl_2$ remaining.

$$0.0712 \cancel{\text{mol CoCl}_2} \times \frac{129.83 \text{ g CoCl}_2}{1 \cancel{\text{mol CoCl}_2}} = 9.24 \text{ g CoCl}_2$$

Step 2: Calculate the mass of water driven off.

mass of hydrated compound − mass of anhydrous compound remaining

$= 11.75 \text{ g CoCl}_2 \cdot x\text{H}_2\text{O} - 9.24 \text{ g CoCl}_2 = 2.51 \text{ g H}_2\text{O}$

Step 3: Calculate moles of each component.

$$9.24\ \cancel{g\ CoCl_2} \times \frac{1\ mol\ CoCl_2}{129.83\ \cancel{g\ CoCl_2}}$$
$$= 0.0712\ mol\ CoCl_2$$

$$2.51\ \cancel{g\ H_2O} \times \frac{1\ mol\ H_2O}{18.02\ \cancel{g\ H_2O}} = 0.139\ mol\ H_2O$$

Step 4: Calculate mole ratios.

$$\frac{0.139\ mol\ H_2O}{0.0712\ mol\ CoCl_2} = \frac{1.95\ mol\ H_2O}{1.00\ mol\ CoCl_2} = \frac{2\ mol\ H_2O}{1\ mol\ CoCl_2}$$

The formula of the hydrate is $CoCl_2{\cdot}2H_2O$. Its name is cobalt(II) chloride dehydrate.

CHAPTER 11

1. a. 1 molecule N_2 + 3 molecules H_2 $\rightarrow$ 2 molecules NH_3

1 mole N_2 + 3 moles H_2 $\rightarrow$ 2 moles NH_3

28.014 g N_2 + 6.048 g H_2 $\rightarrow$ 34.062 g NH_3

b. 1 molecule HCl + 1 formula unit KOH $\rightarrow$ 1 formula unit KCl + 1 molecule H_2O

1 mole HCl + 1 mole KOH $\rightarrow$ 1 mole KCl + 1 mole H_2O

36.461 g HCl + 56.105 g KOH $\rightarrow$ 74.551 g KCl + 18.015 g H_2O

c. 2 atoms Mg + 1 molecule O_2 $\rightarrow$ 2 formula units MgO

2 moles Mg + 1 mole O_2 $\rightarrow$ 2 moles MgO

48.610 g Mg + 31.998 g O_2 $\rightarrow$ 80.608 g MgO

3. a. $\frac{4\ mol\ Al}{3\ mol\ O_2}$ $\frac{3\ mol\ O_2}{2\ mol\ Al_2O_3}$ $\frac{2\ mol\ Al_2O_3}{4\ mol\ Al}$

$\frac{3\ mol\ O_2}{4\ mol\ Al}$ $\frac{2\ mol\ Al_2O_3}{3\ mol\ O_2}$ $\frac{4\ mol\ Al}{2\ mol\ Al_2O_3}$

b. $\frac{3\ mol\ Fe}{4\ mol\ H_2O}$ $\frac{3\ mol\ Fe}{4\ mol\ H_2}$ $\frac{3\ mol\ Fe}{1\ mol\ Fe_3O_4}$

$\frac{4\ mol\ H_2O}{3\ mol\ Fe}$ $\frac{4\ mol\ H_2}{3\ mol\ Fe}$ $\frac{1\ mol\ Fe_3O_4}{3\ mol\ Fe}$

$\frac{1\ mol\ Fe_3O_4}{4\ mol\ H_2}$ $\frac{1\ mol\ Fe_3O_4}{4\ mol\ H_2O}$ $\frac{4\ mol\ H_2O}{4\ mol\ H_2}$

$\frac{4\ mol\ H_2}{1\ mol\ Fe_3O_4}$ $\frac{4\ mol\ H_2O}{1\ mol\ Fe_3O_4}$ $\frac{4\ mol\ H_2}{4\ mol\ H_2O}$

c. $\frac{2\ mol\ HgO}{2\ mol\ Hg}$ $\frac{1\ mol\ O_2}{2\ mol\ Hg}$ $\frac{1\ mol\ O_2}{2\ mol\ HgO}$

$\frac{2\ mol\ Hg}{2\ mol\ HgO}$ $\frac{2\ mol\ Hg}{1\ mol\ O_2}$ $\frac{2\ mol\ HgO}{1\ mol\ O_2}$

11. a. $2CH_4(g) + S_8(s) \rightarrow 2CS_2(l) + 4H_2S(g)$

b. $1.50\ \cancel{mol\ S_8} \times \frac{2\ mol\ CS_2}{1\ \cancel{mol\ S_8}} = 3.00\ mol\ CS_2$

c. $1.50\ \cancel{mol\ S_8} \times \frac{4\ mol\ H_2S}{1\ \cancel{mol\ S_8}} = 6.00\ mol\ H_2S$

13. Step 1: Balance the chemical equation.

$2NaCl(s) \rightarrow 2Na(s) + Cl_2(g)$

Step 2: Make mole $\rightarrow$ mole conversion.

$$2.50\ \cancel{mol\ NaCl} \times \frac{1\ mol\ Cl_2}{2\ \cancel{mol\ NaCl}} = 1.25\ mol\ Cl_2$$

Step 3: Make mole $\rightarrow$ mass conversion.

$$1.25\ \cancel{mol\ Cl_2} \times \frac{70.9\ g\ Cl_2}{1\ \cancel{mol\ Cl_2}} = 88.6\ g\ Cl_2$$

15. $2NaN_3(s) \rightarrow 2Na(s) + 3N_2(g)$

Step 1: Make mass $\rightarrow$ mole conversion.

$$100.0\ \cancel{g\ NaN_3} \times \frac{1\ mol\ NaN_3}{65.02\ \cancel{g\ NaN_3}} = 1.538\ mol\ NaN_3$$

Step 2: Make mole $\rightarrow$ mole conversion.

$$1.538\ \cancel{mol\ NaN_3} \times \frac{3\ mol\ N_2}{2\ \cancel{mol\ NaN_3}} = 2.307\ mol\ N_2$$

Step 3: Make mole $\rightarrow$ mass conversion.

$$2.307\ \cancel{mol\ N_2} \times \frac{28.02\ g\ N_2}{1\ \cancel{mol\ N_2}} = 64.64\ g\ N_2$$

23. Step 1: Make mass $\rightarrow$ mole conversion.

$$100.0\ \cancel{g\ Na} \times \frac{1\ mol\ Na}{22.99\ \cancel{g\ Na}} = 4.350\ mol\ Na$$

$$100.0\ \cancel{g\ Fe_2O_3} \times \frac{1\ mol\ Fe_2O_3}{159.7\ \cancel{g\ Fe_2O_3}} = 0.6261\ mol\ Fe_2O_3$$

Step 2: Make mole ratio comparison.

$\frac{0.6261\ mol\ Fe_2O_3}{4.350\ mol\ Na}$ compared to $\frac{1\ mol\ Fe_2O_3}{6\ mol\ Na}$

0.1439 compared to 0.1667

a. The actual ratio is less than the needed ratio, so iron(III) oxide is the limiting reactant.

b. Sodium is the excess reactant.

c. Step 1: Make mole $\rightarrow$ mole conversion.

$$0.6261\ \cancel{mol\ Fe_2O_3} \times \frac{2\ mol\ Fe}{1\ \cancel{mol\ Fe_2O_3}} = 1.252\ mol\ Fe$$

Step 2: Make mole $\rightarrow$ mass conversion.

$$1.252\ \cancel{mol\ Fe} \times \frac{55.85\ g\ Fe}{1\ \cancel{mol\ Fe}} = 69.92\ g\ Fe$$

d. Step 1: Make mole $\rightarrow$ mole conversion.

$$0.6261\ \cancel{mol\ Fe_2O_3} \times \frac{6\ mol\ Na}{1\ \cancel{mol\ Fe_2O_3}}$$
$$= 3.757\ mol\ Na\ needed$$

Step 2: Make mole $\rightarrow$ mass conversion.

$$3.757\ \cancel{mol\ Na} \times \frac{22.9\ g\ Na}{1\ \cancel{mol\ Na}} = 86.37\ g\ Na\ needed$$

100.0 g Na given − 86.37 g Na needed
= 13.6 g Na in excess

29. a. Step 1: Write the balanced chemical equation.

$Zn(s) + I_2(s) \rightarrow ZnI_2(s)$

Step 2: Make mole $\rightarrow$ mole conversion.

$$1.912\ \cancel{mol\ Zn} \times \frac{1\ mol\ ZnI_2}{1\ \cancel{mol\ Zn}} = 1.912\ mol\ ZnI_2$$

Step 3: Make mole → mass conversion.

$$1.912\ \text{mol ZnI}_2 \times \frac{319.2\ \text{g ZnI}_2}{1\ \text{mol ZnI}_2} = 610.3\ \text{g ZnI}_2$$

610.3 g of ZnI_2 is the theoretical yield.

b. $\%\ \text{yield} = \frac{515.6\ \text{g ZnI}_2}{610.3\ \text{g ZnI}_2} \times 100$

$= 84.48\%$ yield of ZnI_2

CHAPTER 12

1. $\frac{\text{Rate}_{\text{nitrogen}}}{\text{Rate}_{\text{neon}}} = \sqrt{\frac{20.2\ \text{g/mol}}{28.0\ \text{g/mol}}} = 0.849$

3. Rearrange Graham's law to solve for Rate_A.

$\text{Rate}_A = \text{Rate}_B \times \sqrt{\frac{\text{molar mass}_B}{\text{molar mass}_A}}$

$\text{Rate}_B = 3.6\ \text{mol/min}$

$\frac{\text{molar mass}_B}{\text{molar mass}_A} = \frac{1}{2}$

$\text{Rate}_A = 3.6\ \text{mol/min} \times \sqrt{\frac{1}{2}}$

$= 2.5\ \text{mol/min}$

5. $P_{\text{total}} = 5.00\ \text{kPa} + 4.56\ \text{kPa} + 3.02\ \text{kPa} + 1.20\ \text{kPa}$
$= 13.78\ \text{kPa}$

7. $N_2 = 760\ \text{mm Hg} \times 0.78 = 590\ \text{mm Hg}$;
$O_2 = 760\ \text{mm Hg} \times 0.21 = 160\ \text{mm Hg}$;
$Ar = 760\ \text{mm Hg} \times 0.01 = 8\ \text{mm Hg}$

CHAPTER 13

1. $V_2 = \frac{V_1P_1}{P_2} = \frac{(300.0\ \text{mL})(99.0\ \text{kPa})}{188\ \text{kPa}} = 158\ \text{mL}$

3. $P_2 = 1.08\ \text{atm} + (1.08\ \text{atm} \times 0.25) = 1.35\ \text{atm}$

$V_2 = \frac{V_1P_1}{P_2} = \frac{(145.7\ \text{mL})(1.08\ \text{atm})}{1.35\ \text{atm}} = 117\ \text{mL}$

5. $T_1 = 89°\text{C} + 273 = 362\ \text{K}$

$T_2 = \frac{T_1V_2}{V_1} = \frac{(362\ \text{K})(1.12\ \text{L})}{0.67\ \text{L}} = 605\ \text{K}$

$605 - 273 = 332°\text{C} = 330°\text{C}$

7. $V_2 = 0.67\ \text{L} - (0.67\ \text{L} \times 0.45) = 0.37\ \text{L}$

$T_2 = \frac{T_1V_2}{V_1} = \frac{(350\ \text{K})(0.37\ \text{L})}{0.67\ \text{L}} = 190\ \text{K}$

9. $T_2 = 36.5°\text{C} + 273 = 309.5\ \text{K}$

$T_1 = \frac{T_2P_1}{P_2} = \frac{(309.5\ \text{K})(1.12\ \text{atm})}{2.56\ \text{atm}} = 135\ \text{K}$

$135\ \text{K} - 273 = -138°\text{C}$

11. $T_1 = 22.0°\text{C} + 273 = 295\ \text{K}$

$T_2 = 100.0°\text{C} + 273 = 373\ \text{K}$

$V_1 = \frac{V_2T_1P_2}{T_2P_1} = \frac{(0.224\ \text{mL})(295\ \text{K})(1.23\ \text{atm})}{(373\ \text{K})(1.02\ \text{atm})} = 0.214\ \text{mL}$

13. $T_1 = 0.00°\text{C} + 273 = 273\ \text{K}$

$T_2 = 30.0°\text{C} + 273 = 303\ \text{K}$

$\frac{V_2}{V_1} = \frac{P_1T_2}{P_2T_1} = \frac{(1.00\ \text{atm})(303\ \text{K})}{(1.20\ \text{atm})(273\ \text{K})} = 0.92$

This is a ratio, so there are no units. The final volume is less than the original volume, so the piston will move down.

21. $1.0\ \text{L} \times \frac{1\ \text{mol}}{22.4\ \text{L}} = 0.045\ \text{mol}$

$0.045\ \text{mol} \times \frac{44.0\ \text{g}}{1\ \text{mol}} = 2.0\ \text{g}$

23. $0.416\ \text{g} \times \frac{1\ \text{mol}}{83.80\ \text{g}} = 0.00496\ \text{mol}$

$0.00496\ \text{mol} \times \frac{22.4\ \text{L}}{1\ \text{mol}} = 0.111\ \text{L}$

25. $0.860\ \text{g} - 0.205\ \text{g} = 0.655\ \text{g}$ He remaining

Set up the problem as a ratio.

$\frac{V}{0.655\ \text{g}} = \frac{19.2\ \text{L}}{0.860\ \text{g}}$

Solve for V.

$V = \frac{(19.2\ \text{L})(0.655\ \text{g})}{0.860\ \text{g}} = 14.6\ \text{L}$

27. $V = \frac{nRT}{P} = \frac{(0.323\ \text{mol})\left(0.0821\frac{\text{L·atm}}{\text{mol·K}}\right)(265\ \text{K})}{0.900\ \text{atm}} = 7.81\ \text{L}$

29. $n = \frac{PV}{RT} = \frac{(3.81\ \text{atm})(0.44\ \text{L})}{\left(0.0821\frac{\text{L·atm}}{\text{mol·K}}\right)(298\ \text{K})} = 6.9 \times 10^{-3}\ \text{mol}$

39. $2H_2(g) + O_2(g) \rightarrow 2H_2O(g)$

$5.00\ \text{L O}_2 \times \frac{2\ \text{volumes H}_2}{1\ \text{volume O}_2} = 10.0\ \text{L H}_2$

41. $N_2 + O_2 = N_2O$

$2N_2 + O_2 = 2N_2O$

$34\ \text{L N}_2\text{O} \times \frac{1\ \text{volume O}_2}{2\ \text{volumes N}_2\text{O}} = 17\ \text{L O}_2$

43. $2.38\ \text{kg} \times \frac{1000\ \text{g}}{1\ \text{kg}} \times \frac{1\ \text{mol CaCO}_3}{100.09\ \text{g}} \times \frac{1\ \text{mol CO}_2}{1\ \text{mol CaCO}_3}$
$\times \frac{22.4\ \text{L}}{1\ \text{mol}} = 533\ \text{L CO}_2$

45. Molecular mass of sodium bicarbonate = 83.9 g/mol

$28\ \text{g NaHCO}_3 \times \frac{1\ \text{mol NaHCO}_3}{83.9\ \text{g}} = 0.33\ \text{mol NaHCO}_3$

For each mole of sodium bicarbonate, one mole of CO_2 is produced, so 0.33 mol $NaHCO_3$ will produce 0.33 mol CO_2.

For an ideal gas, molar volume is 22.4 L at 273 K and 1 atm.

$T = 20°\text{C} + 273 = 293\ \text{K}$

$0.33\ \text{mol CO}_2 \times \frac{22.4\ \text{L}}{1\ \text{mol}} \times \frac{293\ \text{K}}{273\ \text{K}} = 7.9\ \text{L of CO}_2$

CHAPTER 14

9. $600.0\ \text{mL H}_2\text{O} \times 1.0\ \text{g/mL} = 600.0\ \text{g H}_2\text{O}$

$$\frac{20.0\ \text{g NaHCO}_3}{600.0\ \text{g H}_2\text{O} + 20.0\ \text{g NaHCO}_3} \times 100 = 3\%$$

11. $1500.0\ \text{g} - 54.3\ \text{g} = 1445.7\ \text{g solvent}$

13. $$\frac{35\ \text{mL}}{155\ \text{mL} + 35\ \text{mL}} \times 100 = 18\%$$

15. $$15\% = \frac{18\ \text{mL}}{x\ \text{mL solution}} \times 100 = 120\ \text{mL}$$

17. $$\text{mol KBr} = 1.55\ \text{g} \times \frac{1\ \text{mol}}{119.0\ \text{g}} = 0.0130\ \text{mol KBr}$$

$$\text{molarity} = \frac{\text{mol KBr}}{1.60\ \text{L solution}} = \frac{0.0130\ \text{mol}}{1.60\ \text{L}}$$

$= 8.13 \times 10^{-3}M$

19. $$0.25M = \frac{x\ \text{mol Ca(OH)}_2}{1.5\ \text{L solution}}$$

$x = 0.38\ \text{mol Ca(OH)}_2$

$$0.38\ \text{mol Ca(OH)}_2 \times \frac{74.08\ \text{g}}{1\ \text{mol}}$$

$= 28\ \text{g Ca(OH)}_2$

21. $$\text{mol CaCl}_2 = 500.0\ \text{mL} \times \frac{1\ \text{L}}{1000\ \text{mL}} \times 0.20M$$

$$= 500.0\ \text{mL} \times \frac{1\ \text{L}}{1000\ \text{mL}} \times \frac{0.20\ \text{mol}}{1\ \text{L}} = 0.10\ \text{mol}$$

$$\text{mass CaCl}_2 = 0.10\ \text{mol CaCl}_2 \times \frac{110.98\ \text{g}}{1\ \text{mol}}$$

$= 11\ \text{g}$

23. $$100\ \text{mL} \times \frac{1\ \text{L}}{1000\ \text{mL}} \times \frac{0.15\ \text{mol ethanol}}{1\ \text{L solution}} \times \frac{46\ \text{g ethanol}}{1\ \text{mol ethanol}}$$

$$\times \frac{1\ \text{mL ethanol}}{0.7893\ \text{g ethanol}} = 0.87\ \text{mL ethanol}$$

25. $(5.0M)V_1 = (0.25M)(100.0\ \text{mL})$

$$V_1 = \frac{(0.25M)(100.0\ \text{mL})}{5.0M} = 5.0\ \text{mL}$$

27. $$\text{mol Na}_2\text{SO}_4 = 10.0\ \text{g Na}_2\text{SO}_4 \times \frac{1\ \text{mol}}{142.04\ \text{g Na}_2\text{SO}_4}$$

$= 0.0704\ \text{mol Na}_2\text{SO}_4$

$$\text{molality} = \frac{0.0704\ \text{mol Na}_2\text{SO}_4}{1.0000\ \text{kg H}_2\text{O}} = 0.0704m$$

29. $$22.8\% = \frac{\text{mass NaOH}}{\text{mass NaOH} + \text{mass H}_2\text{O}} \times 100$$

Assume 100.0 g sample.

Then, mass NaOH = 22.8 g

mass H_2O = 100.0 g − (mass NaOH) = 77.2 g

$$\text{mol NaOH} = 22.8\ \text{g} \times \frac{1\ \text{mol}}{40.00\ \text{g}} = 0.570\ \text{mol NaOH}$$

$$\text{mol H}_2\text{O} = 77.2\ \text{g} \times \frac{1\ \text{mol}}{18.02\ \text{g}} = 4.28\ \text{mol H}_2\text{O}$$

$$\text{mol fraction NaOH} = \frac{\text{mol NaOH}}{\text{mol NaOH} + \text{mol H}_2\text{O}}$$

$$= \frac{0.570\ \text{mol NaOH}}{0.570\ \text{mol NaOH} + 4.28\ \text{mol H}_2\text{O}} = \frac{0.570}{4.85}$$

$= 0.118$

The mole fraction of NaOH is 0.118.

37. $$S_2 = \frac{1.5\ \text{g}}{1.0\ \text{L}} = 1.5\ \text{g/L}$$

$$P_2 = P_1 \times \frac{S_2}{S_1} = 10.0\ \text{atm} \times \frac{1.5\ \text{g/L}}{0.66\ \text{g/L}} = 23\ \text{atm}$$

45. $\Delta T_b = 0.512°\text{C}/m \times 0.625m = 0.320°\text{C}$

$T_b = 100°\text{C} + 0.320°\text{C} = 100.320°\text{C}$

$\Delta T_f = 1.86°\text{C}/m \times 0.625m = 1.16°\text{C}$

$T_f = 0.0°\text{C} - 1.16°\text{C} = -1.16°\text{C}$

47. $$K_f = \frac{\Delta T_f}{m}$$

$$= \frac{0.080°\text{C}}{0.045\ m}$$

$= 1.8°\text{C}/m$

It is most likely water because the calculated value is closest to 1.86°C/m.

CHAPTER 15

1. 142 Calories = 142 kcal

$$142\ \text{kcal} \times \frac{1000\ \text{cal}}{1\ \text{kcal}} = 142{,}000\ \text{cal}$$

3. Unit X = 0.1 cal

1 cal = 4.184 J

X = (0.1 cal)(4.184 J/cal) = 0.4184 J

1 cal = 0.001 Calorie

X = (0.1 cal)(1 Cal/1000 cal) = 0.0001 Calorie

5. $q = c \times m \times \Delta T$

$5696\ \text{J} = c \times 155\ \text{g} \times 15.0°\text{C}$

$c = 2.45\ \text{J/(g·°C)}$

The specific heat is very close to the value for ethanol.

13. $q = c \times m \times \Delta T$

$5650\ \text{J} = 4.184\ \text{J/(g·°C)} \times m \times 26.6°\text{C}$

$m = 50.8\ \text{g}$

15. $q = c \times m \times \Delta T$

$9750\ \text{J} = 4.184\ \text{J/(g·°C)} \times 335\ \text{g} \times \Delta T$

$\Delta T = 6.96°\text{C}$

Because the water lost heat, let $\Delta T = -6.96°\text{C}$.

$\Delta T = -6.96°\text{C} = T_f - 65.5°\text{C}$

$T_f = 58.5°\text{C}$

23. $$25.7\ \text{g CH}_3\text{OH} \times \frac{1\ \text{mol CH}_3\text{OH}}{32.04\ \text{g CH}_3\text{OH}} \times \frac{3.22\ \text{kJ}}{1\ \text{mol CH}_3\text{OH}}$$

$= 2.58\ \text{kJ}$

25. $$12{,}880\ \text{kJ} = m \times \frac{1\ \text{mol CH}_4}{16.04\ \text{g CH}_4} \times \frac{891\ \text{kJ}}{1\ \text{mol CH}_4}$$

$$m = 12{,}880\ \text{kJ} \times \frac{16.04\ \text{g CH}_4}{1\ \text{mol CH}_4} \times \frac{1\ \text{mol CH}_4}{891\ \text{kJ}}$$

$m = 232\ \text{g CH}_4$

33. a. $4Al(s) + 3O_2(g) \rightarrow 2Al_2O_3(s) \qquad \Delta H = -3352\ kJ$

b. ΔH for Equation **b** $= -x$ kJ

Add Equation **a** to Equation **b** reversed and tripled.

$4Al(s) + 3O_2(g) \rightarrow 2Al_2O_3(s) \qquad \Delta H = -3352\ kJ$

$3MnO_2(s) \rightarrow 3Mn(s) + 3O_2(g) \qquad \Delta H = 3x\ kJ$

$4Al(s) + 3MnO_2(s) \rightarrow 2Al_2O_3(s) + 3Mn(s)$

$-1789\ kJ = 3x\ kJ + (-3352\ kJ)$

$3x\ kJ = -1789\ kJ + 3352\ kJ = +1563\ kJ$

$x = \frac{1563\ kJ}{3} = +521\ kJ$

Because the direction of Equation **b** was changed, ΔH for Equation **b** $= -521$ kJ.

35. $\Delta H^\circ_{rxn} = [4(33.18\ kJ) + 6(-285.83\ kJ)] - 4(-46.11)\ kJ = -1398\ kJ$

37. Reverse Equation **a** and change the sign of ΔH°_f to obtain Equation **c**.

Add equation **b**.

c. $NO(g) \rightarrow \frac{1}{2}N_2(g) + \frac{1}{2}O_2(g)\ \Delta H^\circ_f = -91.3\ kJ$

b. $\frac{1}{2}N_2(g) + O_2(g) \rightarrow NO_2(g)\ \Delta H^\circ_f = ?$

Add the equations.

$NO(g) + \frac{1}{2}O_2(g) \rightarrow NO_2(g)$

$\Delta H^\circ_{rxn} = -58.1\ kJ = \Delta H^\circ_f(\mathbf{c}) + \Delta H^\circ_f(\mathbf{b})$

$-58.1\ kJ = -91.3\ kJ + \Delta H^\circ_f(\mathbf{b})$

$\Delta H^\circ_f(\mathbf{b}) = -58.1\ kJ + 91.3\ kJ = 33.2\ kJ$

45. The states of the two reactants are the same on both sides of the equation, so it is impossible from the equation alone to predict the sign of ΔS_{system}.

47. For a spontaneous reaction, $\Delta G_{system} < 0$.

$\Delta H_{system} - T\Delta S_{system} < 0$

$T > \frac{\Delta H_{system}}{\Delta S_{system}}$

$T > \frac{-144\ KJ}{(36.8\ J/K)(1\ KJ/1000\ J)}$

$T > 3910\ K$

At any temperature above 3910 K, the reaction is spontaneous.

CHAPTER 16

1. H_2 is consumed. Average reaction rate expression should be negative.

Average reaction rate =

$$-\frac{[H_2] \text{ at time } t_2 - [H_2] \text{ at time } t_1}{t_2 - t_1} = -\frac{\Delta[H_2]}{\Delta t}$$

Average reaction rate $= -\frac{0.020M - 0.030M}{4.00\ s - 0.00\ s}$

$= -\frac{-0.010M}{4.00\ s} = 0.0025\ mol/(L{\cdot}s)$

3. HCl is formed so the average rate expression should be positive.

Average reaction rate =

$$\frac{[HCl] \text{ at time } t_2 - [HCl] \text{ at time } t_1}{t_2 - t_1} = 0.0050\ mol/(L{\cdot}s)$$

$[HCl]_{\text{at time } t_2} =$

$(0.0050\ mol/(L{\cdot}s))(t_2 - t_1) + [HCl]_{\text{at time } t_1}$

$= (0.0050\ mol/L{\cdot}s)(4.00\ s - 0.00\ s) + 0.00\ s$

$= 0.020M$

19. Rate $= k[A]^3$

21. Examining trials 1 and 2, doubling [A] has no effect on the rate; therefore, the reaction is zero order in A. Examining trials 2 and 3, doubling [B] doubles the rate; therefore, the reaction is first order in B.

Rate $= k[A]^0[B] = k[B]$

31. $[NO] = 0.00500M$

$[H_2] = 0.00200M$

$k = 2.90 \times 10^2\ L^2/(mol^2{\cdot}s)$

Rate $= k\ [NO]^2[H_2]$

$= [2.90 \times 10^2 L^2/(mol^2{\cdot}s)](0.00500\ mol/L)^2(0.00200\ mol/L)$

$= [2.90 \times 10^2\ \cancel{L^2}/(\cancel{mol^2}{\cdot}s)](0.00500\ \cancel{mol/L})^2$

$(0.00200\ mol/L)$

$= 1.45 \times 10^{-5}\ mol/(L{\cdot}s)$

33. Rate $= k\ [NO]^2[H_2]$

$[NO] = \sqrt{\frac{Rate}{k[H_2]}}$

$= \sqrt{\frac{9.00 \times 10^{-5}\ mol/(L{\cdot}s)}{(2.90 \times 10^2\ L^2/mol^2{\cdot}s)(0.00300mol/L)}}$

$= 1.02 \times 10^{-2}M$

CHAPTER 17

1. a. $K_{eq} = \frac{[NO_2]^2}{[N_2O_4]}$

b. $K_{eq} = \frac{[H_2]^2[S_2]}{[H_2S]^2}$

c. $K_{eq} = \frac{[CH_4][H_2O]}{[CO][H_2]^3}$

d. $K_{eq} = \frac{[NO]^4[H_2O]^6}{[NH_3]^4[O_2]^5}$

e. $K_{eq} = \frac{[CS_2][H_2]^4}{[CH_4][H_2S]^2}$

3. a. $K_{eq} = [C_{10}H_8(g)]$

b. $K_{eq} = [H_2O(g)]$

c. $K_{eq} = [CO_2(g)]$

d. $K_{eq} = \frac{[CO(g)][H_2(g)]}{[H_2O(g)]}$

e. $K_{eq} = \frac{[CO_2(g)]}{[CO(g)]}$

5. $K_{eq} = \frac{[NO_2]^2}{[N_2O_4]} = \frac{0.0627^2}{0.0185} = 0.213$

7. $\frac{[CO][Cl_2]}{[COCl_2]} = 8.2 \times 10^{-2}$

$$\frac{(0.150)(0.150)}{[COCl_2]} = 8.2 \times 10^{-2}$$

$$[COCl_2] = \frac{(0.150)(0.150)}{8.2 \times 10^{-2}} = 0.27M$$

19. According to the stoichiometry of the equation, the concentration of B is 0.450M; C and D are $1.00 - 0.450 = 0.550M$.

$$K_{eq} = \frac{(0.550)(0.550)}{(0.450)(0.450)} = 1.49$$

21. $K_{sp} = [Pb^{2+}][CO_3^{2-}] = 7.40 \times 10^{-14}$

$(s)(s) = 7.40 \times 10^{-14}$

$s = \sqrt{7.40 \times 10^{-14}} = 2.72 \times 10^{-7}M$

$s = 2.72 \times 10^{-7}$ ~~mol~~/L $\times$ 267.2 g/~~mol~~

$= 7.27 \times 10^{-5}$ g/L

23. $K_{sp} = [Ag^+]^3[PO_4^{3-}] = 2.6 \times 10^{-18}$

$[PO_4^{3-}] = s$, $[Ag^+] = 3s$

$(3s)^3(s) = (27s^3)(s) = 27s^4 = 2.6 \times 10^{-18}$

$$s = \sqrt[4]{\frac{2.6 \times 10^{-18}}{27}} = 1.8 \times 10^{-5} \text{ mol/L}$$

25. a. $PbF_2(s) \rightleftharpoons Pb^{2+}(aq) + 2F^-(aq)$

$Q_{sp} = [Pb^{2+}][F^-]^2 = (0.050M)(0.015M)^2$

$= 1.12 \times 10^{-5}$

$K_{sp} = 3.3 \times 10^{-8}$

$Q_{sp} > K_{sp}$, so a precipitate of PbF_2 will form.

b. $Ag_2SO_4(s) \rightleftharpoons 2Ag^+(aq) + SO_4^{2-}(aq)$

$Q_{sp} = [Ag^+]^2[SO_4^{2-}] = (0.0050M)^2(0.125M)$

$= 3.1 \times 10^{-6}$

$K_{sp} = 1.2 \times 10^{-5}$

$Q_{sp} < K_{sp}$, so a precipitate will not form.

CHAPTER 18

1. a. $2Al(s) + 3H_2SO_4(aq) \rightarrow Al_2(SO_4)_3(aq) + 3H_2(g)$

b. $CaCO_3(s) + 2HBr(aq) \rightarrow CaBr_2(aq) + H_2O(l) + CO_2(g)$

3.

Acid	Conjugate base	Base	Conjugate acid
a. NH_4^+	NH_3	OH^-	H_2O
b. HBr	Br^-	H_2O	H_3O^+
c. H_2O	OH^-	CO_3^{2-}	HCO_3^-

13. $H_2SeO_3(aq) + H_2O(l) \rightleftharpoons HSeO_3^-(aq) + H_3O^+(aq)$

$HSeO_3^-(aq) + H_2O(l) \rightleftharpoons SeO_3^{2-}(aq) + H_3O^+(aq)$

15. a. $C_6H_{13}NH_2(aq) + H_2O(l) \rightleftharpoons C_6H_{13}NH_3^-(aq) + OH^-(aq)$

$$K_b = \frac{[C_6H_{13}NH_3^+][OH^-]}{[C_6H_{13}NH_2]}$$

b. $C_3H_7NH_2(aq) + H_2O(l) \rightleftharpoons C_3H_7NH_3^-(aq) + OH^-(aq)$

$$K_b = \frac{[C_3H_7NH_3^+][OH^-]}{[C_3H_7NH_2]}$$

c. $CO_3^{2-}(aq) + H_2O(l) \rightleftharpoons HCO_3^-(aq) + OH^-(aq)$

$$K_b = \frac{[HCO_3^-][OH^-]}{[CO_3^{2-}]}$$

d. $HSO_3^-(aq) + H_2O(l) \rightleftharpoons H_2SO_3(aq) + OH^-(aq)$

$$K_b = \frac{[H_2SO_3^-][OH^-]}{[HSO_3^-]}$$

23. At 298 K, $[H^+] = [OH^-] = 1.0 \times 10^{-7}M$

$$\text{mol } H^+ = \frac{1.0 \times 10^{-7} \text{ mol}}{1 \cancel{L}} \times \frac{1 \cancel{L}}{1000 \cancel{mL}} \times 300 \cancel{mL} = 3.0 \times 10^{-8} \text{ mol}$$

$$3.0 \times 10^{-8} \cancel{\text{mol } H^+ \text{ ions}} \times \frac{6.02 \times 10^{23} \text{ } H^+ \text{ ions}}{1 \cancel{\text{mol } H^+ \text{ ions}}} = 1.8 \times 10^{16} \text{ } H^+ \text{ ions}$$

Number of H^+ = number of OH^- = 1.8×10^{16} ions

25. a. $[H^+] = 0.0055M$

$pH = -\log[H^+]$

$pH = -\log 0.0055$

$pH = 2.26$

b. $[H^+] = 0.000084M$

$pH = -\log[H^+]$

$pH = -\log 0.000084$

$pH = 4.08$

27. a. $[OH^-] = 1.0 \times 10^{-6}M$

$pOH = -\log[OH^-]$

$pOH = -\log(1.0 \times 10^{-6})$

$pOH = 6.00$

$pH = 14.00 - pOH = 14.00 - 6.00 = 8.00$

b. $[OH^-] = 6.5 \times 10^{-4}M$

$pOH = -\log[OH^-]$

$pOH = -\log(6.5 \times 10^{-4})$

$pOH = 3.19$

$pH = 14.00 - pOH = 14.00 - 3.19 = 10.81$

c. $[H^+] = 3.6 \times 10^{-9}M$

$pH = -\log[H^+]$

$pH = -\log(3.6 \times 10^{-9})$

$pH = 8.44$

$pOH = 14.00 - pH = 14.00 - 8.44 = 5.56$

d. $[H^+] = 2.5 \times 10^{-2}M$

$pH = -\log(-2.5 \times 10^{-2})$

$pH = 1.60$

$pOH = 14.00 - pH = 14.00 - 1.60 = 12.40$

29. $$[HCl] = [H^+] = \frac{1.0 \times 10^{-3} \text{ mol}}{5.0 \text{ L}} = 0.00020M = 2.0 \times 10^{-4}M$$

$pH = -\log(2.0 \times 10^{-4}) = -(-3.70) = 3.70$

$pOH = 14.00 - 3.70 = 10.30$

31. $[OH^-] = \text{antilog}(-pOH)$

$[OH^-] = \text{antilog}(-5.60) = 2.5 \times 10^{-6}M$

$pH = 14.00 - 5.60 = 8.40$

$[H^+] = \text{antilog}(-8.40) = 4.0 \times 10^{-9}M$

33. a. $pH = 14.00 - pOH$

$pH = 14.00 - 10.70 = 3.30$

$[H^+] = \text{antilog}(-pH)$

$[H^+] = \text{antilog}(-3.30) = 5.0 \times 10^{-4}M$

$[C_6H_5COO^-] = [H^+] = 5.0 \times 10^{-4}M$

$[C_6H_5COOH] = 0.00330M - 5.0 \times 10^{-4}M = 0.0028M$

$$K_a = \frac{[H^+][C_6H_5COO^-]}{[C_6H_5COOH]} = \frac{(5.0 \times 10^{-4})(5.0 \times 10^{-4})}{(22.8 \times 10^{-3})}$$

$K_a = 8.9 \times 10^{-5}$

b. $pH = 14.00 - pOH$

$pH = 14.00 - 11.00 = 3.00$

$[H^+] = \text{antilog}(-pH)$

$[H^+] = \text{antilog}(-3.00) = 1.0 \times 10^{-3}M$

$[CNO^-] = [H^+] = 1.0 \times 10^{-3}M$

$[HCNO] = 0.100 - 1.0 \times 10^{-3}M = 0.099M$

$$K_a = \frac{[H^+][CNO^-]}{[HCNO]} = \frac{(1.0 \times 10^{-3})(1.0 \times 10^{-3})}{(0.099)}$$

$K_a = 1.0 \times 10^{-5}$

c. $pH = 14.00 - pOH$

$pH = 14.00 - 11.18 = 2.82$

$[H^+] = \text{antilog}(-pH)$

$[H^+] = \text{antilog}(-2.82) = 1.5 \times 10^{-3}M$

$[C_3H_7COO^-] = [H^+] = 1.5 \times 10^{-3}M$

$[C_3H_7COOH] = 0.150M - 1.5 \times 10^{-3}M = 0.149M$

$$K_a = \frac{[H^+][C_3H_7COO^-]}{[C_3H_7COOH]} = \frac{(1.5 \times 10^{-3})(1.5 \times 10^{-3})}{(0.149)}$$

$K_a = 1.5 \times 10^{-5}$

45. $$49.90\ \text{mL HCl} \times \frac{1\ \text{L}}{1000\ \text{mL}} \times \frac{0.5900\ \text{mol HCl}}{1\ \text{L HCl}} = 2.944 \times 10^{-2}\ \text{mol HCl}$$

$$2.944 \times 10^{-2}\ \text{mol HCl} \times \frac{1\ \text{mol NH}_3}{1\ \text{mol HCl}} = 2.944 \times 10^{-2}\ \text{mol NH}_3$$

$$M_{NH_3} = \frac{2.944 \times 10^{-2}\ \text{mol NH}_3}{0.02500\ \text{L NH}_3} = 1.178M$$

47. a. $NH_4^+(aq) + H_2O(l) \rightleftharpoons NH_3(aq) + H_3O^+(aq)$

The solution is acidic.

b. $SO_4^{2-}(aq) + H_2O(l) \rightleftharpoons HSO_4^-(aq) + OH^-(aq)$

The solution is neutral.

c. $CH_3COO^-(aq) + H_2O(l) \rightleftharpoons CH_3COOH(aq) + OH^-(aq)$

The solution is basic.

d. $CO_3^{2-}(aq) + H_2O(l) \rightleftharpoons HCO_3^-(aq) + OH^-(aq)$

The solution is basic.

CHAPTER 19

1. a. reduction **b.** oxidation **c.** oxidation **d.** reduction

3. Ag^+ is the oxidizing agent, Fe is the reducing agent; Ag^+ is reduced, Fe is oxidized

5. a. +7 **b.** +5 **c.** +3

7. a. −3 **b.** −3 **c.** −2

15.

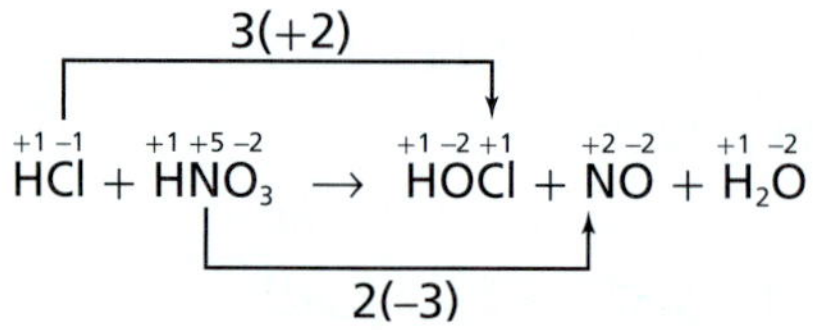

$3HCl + 2HNO_3 \rightarrow 3HOCl + 2NO + H_2O$

17.

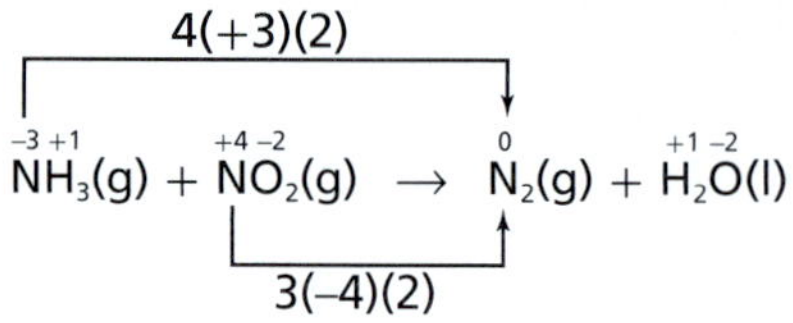

$8NH_3(g) + 6NO_2(g) \rightarrow 7N_2(g) + 12H_2O(l)$

19.

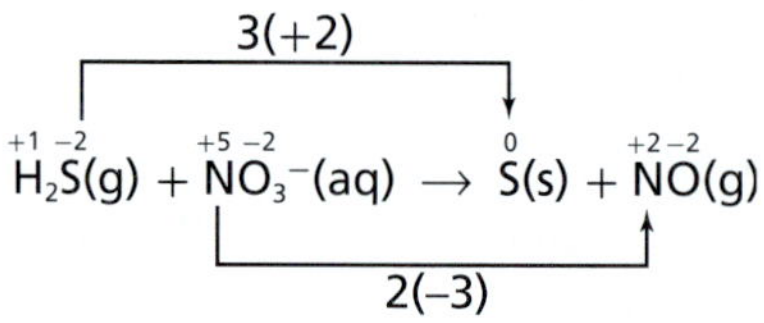

$2H^+(aq) + 3H_2S(g) + 2NO_3^-(aq) \rightarrow 3S(s) + 2NO(g) + 4H_2O(l)$

21.

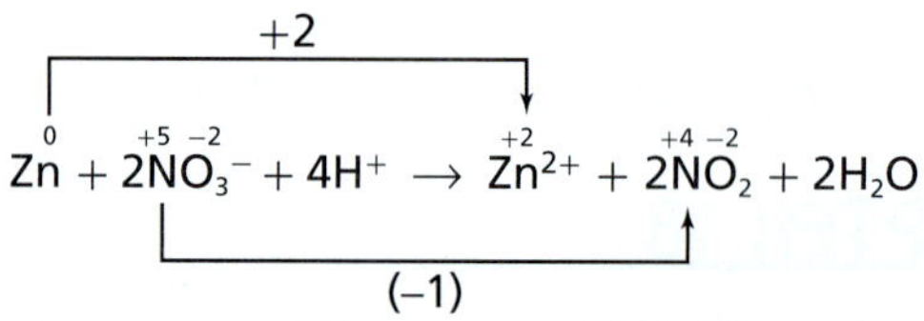

$Zn + 2NO_3^- + 4H^+ \rightarrow Zn^{2+} + 2NO_2 + 2H_2O$

23. $2I^-(aq) \rightarrow I_2(s) + 2e^-$ (oxidation)

$14H^+(aq) + 6e^- + Cr_2O_7^{2-}(aq) \rightarrow 2Cr^{3+}(aq) + 7H_2O(l)$ (reduction)

Multiply oxidation half-reaction by 3 and add to reduction half-reaction.

$14H^+(aq) + 6e^- + CrO_7^{2-}(aq) + 6I^-(aq) \rightarrow 3I_2(s) + 2Cr^{3+}(aq) + 7H_2O(l) + 6e^-$

$14H^+(aq) + CrO_7^{2-}(aq) + 6I^-(aq) \rightarrow 3I_2(s) + 2Cr^{3+}(aq) + 7H_2O(l)$

25. $6OH^-(aq) + N_2O(g) \rightarrow 2NO_2^-(aq) + 4e^- + 3H_2O(l)$ (oxidation)

$ClO^-(aq) + 2e^- + H_2O(l) \rightarrow Cl^-(aq) + 2OH^-(aq)$ (reduction)

Multiply reduction half-reaction by 2 and add to oxidation half-reaction.

$6OH^-(aq) + N_2O(g) + 2ClO^-(aq) + 4e^- + 2H_2O(l) \rightarrow$
$2NO_2^-(aq) + 4e^- + 3H_2O(l) + 2Cl^-(aq) + 4OH^-(aq)$

$N_2O(g) + 2ClO^-(aq) + 2OH^-(aq) \rightarrow$
$2NO_2^-(aq) + 2Cl^-(aq) + H_2O(l)$

CHAPTER 20

1. $Pt^{2+}(aq) + Sn(s) \rightarrow Pt(s) + Sn^{2+}(aq)$

$E^\circ_{cell} = +1.18\ V - (-0.1375\ V)$
$E^\circ_{cell} = +1.32\ V$
$Sn|Sn^{2+}||Pt^{2+}|Pt$

3. $Hg^{2+}(aq) + Cr(s) \rightarrow Hg(l) + Cr^{2+}(aq)$
$E^\circ_{cell} = +0.851\ V - (-0.913\ V)$
$E^\circ_{cell} = +1.764\ V$
$Cr|Cr^{2+}||Hg^{2+}|Hg$

5. $E^\circ_{cell} = +0.3419\ V - (-0.1375\ V)$
$E^\circ_{cell} = +0.4794\ V$
$E^\circ_{cell} > 0$; spontaneous

7. $E^\circ_{cell} = 0.920\ V - (+1.507\ V)$
$E^\circ_{cell} = -0.587\ V$
$E^\circ_{cell} < 0$; not spontaneous

9. $Al|Al^{3+}||Hg^{2+}|Hg_2^{2+}$
$2Al(s) + 6Hg^{2+}(aq) \rightarrow 2Al^{3+}(aq) + 3Hg_2^{2+}(aq)$
$E^\circ_{cell} = 0.920\ V - (-1.662\ V) = +2.582\ V$
The reaction is spontaneous.

CHAPTER 21

1. a.
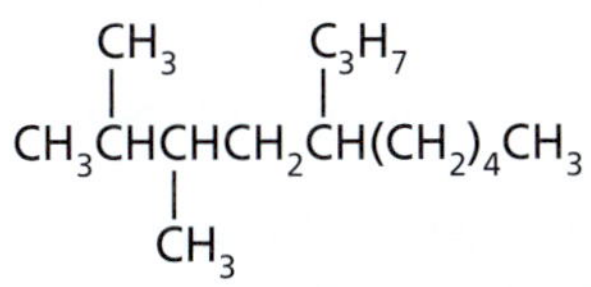

b.
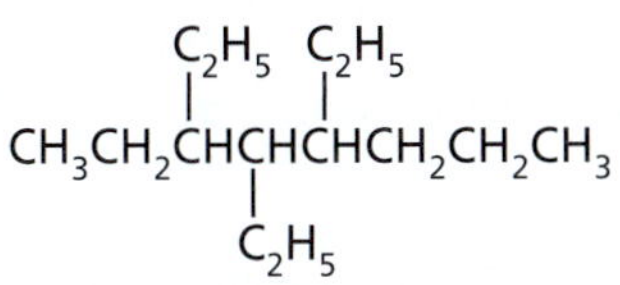

11. a.
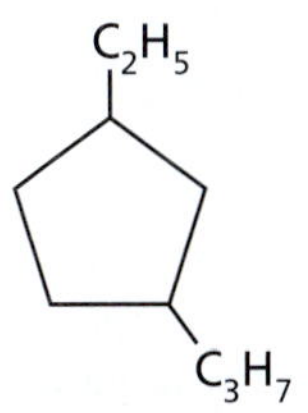

b.
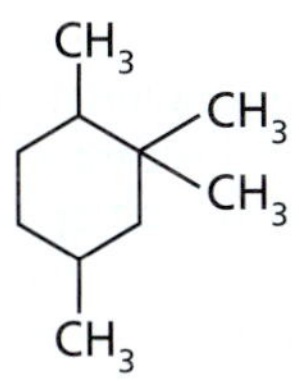

17. a. 4-methyl-2-pentene **b.** 2,2,6-trimethyl-3-octene

31. a. propylbenzene
b. 1-ethyl-2-methylbenzene
c. 1-ethyl-2,3-dimethylbenzene

CHAPTER 22

1. 2,3-difluorobutane

3. 1,3-dibromo-2-chlorobenzene

CHAPTER 23

No practice problems

CHAPTER 24

7. ${}^{229}_{90}Th \rightarrow {}^{4}_{2}He + {}^{225}_{88}Ra$

Alpha decay

9. For one half-life, amount remaining = (initial amount) $\left(\frac{1}{2}\right)^n = (10.0\ mg)\left(\frac{1}{2}\right)^1 = 5.00\ mg$.

For two half-lives, amount remaining = (initial amount) $\left(\frac{1}{2}\right)^n = (10.0\ mg)\left(\frac{1}{2}\right)^2 = 2.50\ mg$.

For three half-lives, amount remaining = (initial amount) $\left(\frac{1}{2}\right)^n = (10.0\ mg)\left(\frac{1}{2}\right)^3 = 1.25\ mg$.

11. For Sample A: $16.2\ g = 64.8\ g\left(\frac{1}{2}\right)^n$; $n = 2$

$t = 2 \times 5.7$ years $= 10.54$ years

For Sample B, amount remaining =
(initial amount)$\left(\frac{1}{2}\right)^{\frac{t}{T}} = (58.4\ g)\left(\frac{1}{2}\right)^{\frac{10.54y}{12.32y}} \approx 32.3\ g$

For Sample C, amount remaining =
(initial amount)$\left(\frac{1}{2}\right)^{\frac{t}{T}} = (37.6\ g)\left(\frac{1}{2}\right)^{\frac{10.54y}{28.79y}} \approx 29.2\ g$

19. ${}^{27}_{13}Al + {}^{1}_{0}n \rightarrow {}^{24}_{11}Na + {}^{4}_{2}He$

21. Let T = target and I = unstable isotope. Then,
${}^{1}_{0}n + T = I$ and $I = {}^{0}_{-1}\beta + {}^{110}_{48}Cd$
Balancing the second equation gives:
${}^{110}_{47}Ag = {}^{0}_{-1}\beta + {}^{110}_{48}Cd$
The first equation must then be: ${}^{1}_{0}n + T = {}^{110}_{47}Ag$
Balancing this equation gives: ${}^{1}_{0}n + {}^{109}_{47}Ag = {}^{110}_{47}Ag$
The target, then, was silver-109, and the unstable isotope was silver-110.

GLOSSARY • GLOSARIO

Multilingual eGlossary

The multilingual science glossary includes Arabic, Bengali, Chinese, English, Haitian, Creole, Hmong, Korean, Portuguese, Russian, Tagalog, Urdu, and Vietnamese.

Como usar el glosario en espanol:
1. Busca el termino en ingles que desees encontrar.
2. El termino en espanol, junto con la definicion, se encuentran en la columna de la derecha.

Pronunciation Key

Use the following key to help you sound out words in the glossary.

a	back (BAK)	**ew**	food (FEWD)
ay	day (DAY)	**yoo**	pure (PYOOR)
ah	father (FAH thur)	**yew**	few (FYEW)
ow	flower (FLOW ur)	**uh**	comma (CAHM uh)
ar	car (CAR)	**u** (+con)	rub (RUB)
e	less (LES)	**sh**	shelf (SHELF)
ee	leaf (LEEF)	**ch**	nature (NAY chur)
ih	trip (TRIHP)	**g**	gift (GIHFT)
i (i+con+e)	idea, life (i DEE uh, life)	**j**	gem (JEM)
oh	go (GOH)	**ing**	sing (SING)
aw	soft (SAWFT)	**zh**	vision (VIHZH un)
or	orbit (OR but)	**k**	cake (KAYK)
oy	coin (COYN)	**s**	seed, cent (SEED, SENT)
oo	foot (FOOT)	**z**	zone, raise (ZOHN, RAYZ)

ENGLISH — A — ESPAÑOL

ENGLISH	ESPAÑOL
absolute zero (p. 445) Zero on the Kelvin scale, which represents the lowest possible theoretical temperature; atoms are all in the lowest possible energy state.	**cero absoluto (pág. 445)** Equivale a cero grados en la escala de Kelvin y representa la temperatura teórica más fría posible; a esta temperatura todos los átomos se encuentran en el menor estado energético posible.
accuracy (p. 47) Refers to how close a measured value is to an accepted value.	**exactitud (pág. 47)** Se refiere a la cercanía entre un valor medido y el valor aceptado.
acid-base indicator (p. 662) A chemical dye whose color is affected by acidic and basic solutions.	**indicador ácido-base (pág. 662)** tinción química cuyo color cambia al entrar en contacto con soluciones ácidas y básicas.
acidic solution (p. 636) Contains more hydrogen ions than hydroxide ions.	**solución ácida (pág. 636)** Solución que contiene más iones hidrógeno que iones hidróxido.
acid ionization constant (p. 647) The value of the equilibrium constant expression for the ionization of a weak acid.	**constante ácida de ionización (pág. 647)** Valor de la expresión de la constante de equilibrio para la ionización de un ácido débil.
actinide series (p. 180) In the periodic table, the f-block elements from period 7 that follow the element actinium.	**serie de actínidos (pág. 180)** Elementos del bloque F del período 7 de la tabla periódica que aparecen después del elemento actinio.
activated complex (p. 564) A short-lived, unstable arrangement of atoms that can break apart and re-form the reactants or can form products; also sometimes referred to as the transition state.	**complejo activado (pág. 564)** Complejo efímero e inestable de átomos que se puede romper para volver a formar los reactivos o para formar los productos; a veces también se le llama estado de transición.
activation energy (p. 564) The minimum amount of energy required by reacting particles in order to form the activated complex and lead to a reaction.	**energía de activación (pág. 564)** La cantidad mínima de energía que requieren las partículas de una reacción para formar el complejo activado y producir la reacción.
active site (p. 830) The pocket or crevice to which a substrate binds in an enzyme-catalyzed reaction.	**sitio activo (pág. 830)** Saliente o hendidura a la que se enlaza un sustrato durante una reacción catalizada por enzimas.

actual yield **(p. 385)** The amount of product produced when a chemical reaction is carried out.

addition polymerization **(p. 811)** Occurs when all the atoms present in the monomers are retained in the polymer product.

addition reaction **(p. 804)** A reaction that occurs when other atoms bond to each of two atoms bonded by double or triple covalent bonds.

alcohol **(p. 792)** An organic compound in which a hydroxyl group replaces a hydrogen atom of a hydrocarbon.

aldehyde **(p. 796)** An organic compound containing the structure in which a carbonyl group at the end of a carbon chain is bonded to a carbon atom on one side and a hydrogen atom on the other side.

aliphatic compounds **(a luh FA tihk • KAHM pownd)** **(p. 771)** Nonaromatic hydrocarbons, such as the alkanes, alkenes, and alkynes.

alkali metals **(p. 177)** Group 1 elements, except for hydrogen, they are reactive and usually exist as compounds with other elements.

alkaline earth metals **(p. 177)** Group 2 elements in the modern periodic table and are highly reactive.

alkane **(p. 750)** Hydrocarbon that contains only single bonds between atoms.

alkene **(p. 759)** An unsaturated hydrocarbon, such as ethene (C_2H_4), with one or more double covalent bonds between carbon atoms in a chain.

alkyl halide **(p. 787)** An organic compound containing a halogen atom covalently bonded to an aliphatic carbon atom.

alkyne **(p. 762)** An unsaturated hydrocarbon, such as ethyne (C_2H_2), with one or more triple bonds between carbon atoms in a chain.

allotrope **(p. 423)** One of two or more forms of an element with different structures and properties when they are in the same state—solid, liquid, or gas.

alloy **(p. 227)** A mixture of elements that has metallic properties; most commonly forms when the elements are either similar in size (substitutional alloy) or the atoms of one element are much smaller than the atoms of the other (interstitial alloy).

alpha particle **(p. 123)** A particle with two protons and two neutrons, with a 2+ charge; is equivalent to a helium-4 nucleus, can be represented by α; and is emitted during radioactive decay.

alpha radiation **(p. 123)** Radiation that is made up of alpha particles; is deflected toward a negatively charged plate when radiation from a radioactive source is directed between two electrically charged plates.

amide **(AM ide)** **(p. 800)** An organic compound in which the $-H$ group of a carboxylic acid is replaced by a nitrogen atom bonded to other atoms.

amines **(A meen)** **(p. 795)** Organic compounds that contain nitrogen atoms bonded to carbon atoms in aliphatic chains or aromatic rings and have the general formula RNH_2.

amino acid **(p. 826)** An organic molecule that has both an amino group ($-NH_2$) and a carboxyl group ($-COOH$).

rendimiento real **(pág. 385)** Cantidad de producto que se obtiene al realizar una reacción química.

polimerización de adición **(pág. 811)** Ocurre cuando todos los átomos presentes en los monómeros forman parte del producto polimérico.

reacción de adición **(pág. 804)** Reacción que ocurre cuando dos átomos unidos entre sí por enlaces covalentes dobles o triples se unen con otros átomos.

alcohol **(pág. 792)** Compuesto orgánico en el que un grupo hidroxilo reemplaza a un átomo de hidrógeno de un hidrocarburo.

aldehído **(pág. 796)** Compuesto orgánico que contiene una estructura en la que un grupo carbonilo, situado al final de una cadena de carbonos, se une a un átomo de carbono por un lado y a un átomo de hidrógeno por el lado opuesto.

compuestos alifáticos **(pág. 771)** Hidrocarburos no aromáticos como los alcanos, los alquenos y los alquinos.

metales alcalinos **(pág. 177)** Incluyen los elementos del grupo 1, a excepción del hidrógeno. Son reactivos y generalmente existen como compuestos con otros elementos.

metales alcalinotérreos **(pág. 177)** Elementos altamente reactivos del grupo 2 de la tabla periódica moderna.

alcano **(pág. 750)** Hidrocarburo que sólo contiene enlaces sencillos entre sus átomos.

alqueno **(pág. 759)** Hidrocarburo no saturado, como el eteno (C_2H_4), que tiene uno o más enlaces covalentes dobles entre los átomos de carbono en una cadena.

haluro de alquilo **(pág. 787)** Compuesto orgánico que contiene un átomo de halógeno enlazado covalentemente a un átomo de carbono alifático.

alquino **(pág. 762)** Hidrocarburo no saturado, como el acetileno (C_2H_2), que tiene uno o más enlaces triples entre los átomos de carbono en una cadena.

alótropos **(pág. 423)** Formas de un elemento que tienen estructura y propiedades distintas cuando están en el mismo estado: sólido, líquido o gaseoso.

aleación **(pág. 227)** Mezcla de elementos que posee propiedades metálicas; en general se forman cuando los elementos tienen un tamaño similar (aleación de sustitución) o cuando los átomos de un elemento son mucho más pequeños que los átomos del otro (aleación intersticial).

partícula alfa **(pág. 123)** Partícula con dos protones y dos neutrones que tiene una carga 2+; equivale a un núcleo de helio 4, se puede representar como α y es emitida durante la desintegración radiactiva.

radiación alfa **(pág. 123)** Radiación compuesta de partículas alfa; si la radiación proveniente de una fuente radiactiva es dirigida hacia dos placas cargadas eléctricamente, este tipo de radiación se desvía hacia la placa con carga negativa.

amida **(pág. 800)** Compuesto orgánico en el que el grupo $-H$ de un ácido carboxílico es sustituido por un átomo de nitrógeno unido a otros átomos.

aminas **(pág. 795)** Compuestos orgánicos que contienen átomos de nitrógeno unidos a átomos de carbono en cadenas alifáticas o anillos aromáticos; su fórmula general es RNH_2.

amino ácido **(pág. 826)** Molécula orgánica que posee un grupo amino ($-NH_2$) y un grupo carboxilo ($-COOH$).

amorphous solid (p. 424) A solid in which particles are not arranged in a regular, repeating pattern that often is formed when molten material cools too quickly to form crystals.

amphoteric (AM foh TAR ihk) (p. 639) Describes water and other substances that can act as both acids and bases.

amplitude (p. 137) The height of a wave from the origin to a crest, or from the origin to a trough.

anabolism (ah NAB oh lih zum) (p. 844) Refers to the metabolic reactions through which cells use energy and small building blocks to build large, complex molecules needed to carry out cell functions and for cell structures.

anion (AN i ahn) (p. 209) An ion that has a negative charge.

anode (p. 710) In an electrochemical cell, the electrode where oxidation takes place.

applied research (p. 17) A type of scientific investigation that is undertaken to solve a specific problem.

aqueous solution (p. 299) A solution in which the solvent is water.

aromatic compounds (p. 771) Organic compounds that contain one or more benzene rings as part of their molecular structure.

Arrhenius model (ah REE nee us • MAH dul) (p. 637) A model of acids and bases; states that an acid is a substance that contains hydrogen and ionizes to produce hydrogen ions in aqueous solution and a base is a substance that contains a hydroxide group and dissociates to produce a hydroxide ion in aqueous solution.

aryl halide (p. 788) An organic compound that contains a halogen atom bonded to a benzene ring or another aromatic group

asymmetric carbon (p. 768) A carbon atom that has four different atoms or groups of atoms attached to it; occurs in chiral compounds.

atmosphere (p. 407) The unit that is often used to report air pressure.

atom (p. 106) The smallest particle of an element that retains all the properties of that element; is electrically neutral, spherically shaped, and composed of electrons, protons, and neutrons.

atomic emission spectrum (p. 144) A set of frequencies of electromagnetic waves given off by atoms of an element; consists of a series of fine lines of individual colors.

atomic mass (p. 119) The weighted average mass of the isotopes of that element.

atomic mass unit (amu) (p. 119) One-twelfth the mass of a carbon-12 atom.

atomic number (p. 115) The number of protons in an atom.

atomic orbital (p. 152) A three-dimensional region around the nucleus of an atom that describes an electron's probable location.

ATP (p. 845) Adenosine triphosphate—a nucleotide that functions as the universal energy-storage molecule in living cells.

sólido amorfo (pág. 424) Sólido cuyas partículas no están ordenadas de modo que formen un patrón regular repetitivo; a menudo se forma cuando el material fundido se enfría demasiado rápido como para formar cristales.

anfotérico (pág. 639) Término que describe al agua y otras sustancias que pueden actuar como ácidos y bases.

amplitud (pág. 137) Altura de una onda desde el origen hasta una cresta o desde el origen hasta un valle.

anabolismo (pág. 844) Reacciones metabólicas en las que las células usan energía y pequeñas unidades básicas para formar las moléculas grandes y complejas que requieren para realizar sus funciones celulares y para construir sus estructuras.

anión (pág. 209) Ion con carga negativa.

ánodo (pág. 710) Electrodo donde sucede la oxidación en una celda electroquímica.

investigación aplicada (pág. 17) Tipo de investigación científica que se realiza para resolver un problema concreto.

solución acuosa (pág. 299) Solución en la que el agua funciona como disolvente.

compuestos aromáticos (pág. 771) Compuestos orgánicos que contienen uno o más anillos de benceno como parte de su estructura molecular.

modelo de Arrhenius (pág. 637) Modelo de ácidos y bases; establece que un ácido es una sustancia que contiene hidrógeno y se ioniza para producir iones hidrógeno en solución acuosa, y que una base es una sustancia que contiene un grupo hidróxido y se disocia para producir un ion hidróxido en solución acuosa.

haluro de arilo (pág. 788) Compuesto orgánico que contiene un átomo de halógeno unido a un anillo de benceno u otro grupo aromático.

carbono asimétrico (pág. 768) Átomo de carbono que está unido a cuatro átomos o grupos de átomos diferentes; se hallan en compuestos quirales.

atmósfera (pág. 407) Unidad que a menudo se usa para reportar la presión atmosférica.

átomo (pág. 106) La partícula más pequeña de un elemento que retiene todas las propiedades de ese elemento; es eléctricamente neutro, de forma esférica y está compuesto de electrones, protones y neutrones.

espectro de emisión atómica (pág. 144) Conjunto de frecuencias de ondas electromagnéticas que emiten los átomos de un elemento; consta de una serie de líneas finas de distintos colores.

masa atómica (pág. 119) La masa promedio ponderada de los isótopos de un elemento.

unidad de masa atómica (uma) (pág. 119) La doceava parte de la masa de un átomo de carbono 12.

número atómico (pág. 115) El número de protones en un átomo.

orbital atómico (pág. 152) Región tridimensional alrededor del núcleo de un átomo que describe la ubicación probable de un electrón.

ATP (pág. 845) Trifosfato de adenosina; nucleótido que sirve como la molécula universal de almacenamiento de energía en las células vivas.

aufbau principle (p. 156) States that each electron occupies the lowest energy orbital available.

Avogadro's number (p. 321) The number 6.0221367×10^{23}, which is the number of representative particles in a mole, and can be rounded to three significant digits 6.02×10^{23}.

Avogadro's principle (p. 452) States that equal volumes of gases at the same temperature and pressure contain equal numbers of particles.

principio de aufbau (pág. 156) Establece que cada electrón ocupa el orbital de energía más bajo disponible.

número de Avogadro (pág. 321) Equivale al número 6.0221367×10^{23}; es el número de partículas representativas en un mol; se puede redondear a tres dígitos significativos: 6.02×10^{23}.

principio de Avogadro (pág. 452) Establece que los volúmenes iguales de gases, a la misma temperatura y presión, contienen igual número de partículas.

B

band of stability (p. 866) The region on a graph within which all stable nuclei are found when plotting the number of neutrons versus the number of protons.

barometer (p. 407) An instrument that is used to measure atmospheric pressure.

base ionization constant (p. 649) The value of the equilibrium constant expression for the ionization of a base.

base unit (p. 33) A defined unit in a system of measurement that is based on an object or event in the physical world and is independent of other units.

basic solution (p. 636) Contains more hydroxide ions than hydrogen ions.

battery (p. 718) One or more electrochemical cells in a single package that generates electrical current.

beta particle (p. 123) A high-speed electron with a 1− charge that is emitted during radioactive decay.

beta radiation (p. 123) Radiation that is made up of beta particles; is deflected toward a positively charged plate when radiation from a radioactive source is directed between two electrically charged plates.

boiling point (p. 427) The temperature at which a liquid's vapor pressure is equal to the external or atmospheric pressure.

boiling-point elevation (p. 500) The temperature difference between a solution's boiling point and a pure solvent's boiling point.

Boyle's law (p. 442) States that the volume of a fixed amount of gas held at a constant temperature varies inversely with the pressure.

breeder reactor (p. 882) A nuclear reactor that is able to produce more fuel than it uses.

Brønsted-Lowry model (p. 638) A model of acids and bases in which an acid is a hydrogen-ion donor and a base is a hydrogen-ion acceptor.

Brownian motion (p. 477) The erratic, random, movements of colloid particles that results from collisions of particles of the dispersion medium with the dispersed particles.

buffer (p. 666) A solution that resists changes in pH when limited amounts of acid or base are added.

buffer capacity (p. 667) The amount of acid or base a buffer solution can absorb without a significant change in pH.

banda de estabilidad (pág. 866) Región de una gráfica en la que se hallan todos los núcleos estables cuando se grafica el número de neutrones contra el número de protones.

barómetro (pág. 407) Instrumento que se utiliza para medir la presión atmosférica.

constante de ionización básica (pág. 649) El valor de la expresión de la constante de equilibrio para la ionización de una base.

unidad básica (pág. 33) Unidad definida en un sistema de medidas; está basada en un objeto o evento del mundo físico y es independiente de otras unidades.

solución básica (pág. 636) Solución que contiene más iones hidróxido que iones hidrógeno.

batería (pág. 718) Una o más celdas electroquímicas contenidas en una sola unidad que genera corriente eléctrica.

partícula beta (pág. 123) Electrón de alta velocidad con una carga 1− que es emitido durante la desintegración radiactiva.

radiación beta (pág. 123) Radiación compuesta de partículas beta; si la radiación proveniente de una fuente radiactiva es dirigida hacia dos placas cargadas eléctricamente, este tipo de radiación se desvía hacia la placa con carga positiva.

punto de ebullición (pág. 427) Temperatura a la cual la presión de vapor de un líquido es igual a la presión externa o atmosférica.

elevación del punto de ebullición (pág. 500) Diferencia de temperatura entre el punto de ebullición de una solución y el punto de ebullición de un disolvente puro.

ley de Boyle (pág. 442) Establece que el volumen de una cantidad dada de gas a temperatura constante varía inversamente según la presión.

reactor generador (pág. 882) Reactor nuclear capaz de producir más combustible del que utiliza.

modelo de Brønsted-Lowry (pág. 638) Modelo de ácidos y bases en el que un ácido es un donante de iones hidrógeno y una base es un receptor de iones hidrógeno.

movimiento browniano (pág. 477) Movimientos erráticos, aleatorios de las partículas coloidales, producidos por el choque entre las partículas del medio de dispersión con las partículas dispersas.

amortiguador (pág. 666) Solución que resiste los cambios de pH cuando se agregan cantidades moderadas del ácido o la base.

capacidad amortiguadora (pág. 667) Cantidad de ácido o base que una solución amortiguadora puede absorber sin sufrir un cambio significativo en el pH.

C

calorie (p. 518) The amount of heat required to raise the temperature of one gram of pure water by one degree Celsius.

calorimeter (p. 523) An insulated device that is used to measure the amount of heat released or absorbed during a physical or chemical process.

carbohydrates (p. 832) Compounds that contain multiple hydroxyl groups, plus an aldehyde or a ketone functional group, and function in living things to provide immediate and stored energy.

carbonyl group (p. 796) Arrangement in which an oxygen atom is double-bonded to a carbon atom.

carboxyl group (p. 798) Consists of a carbonyl group bonded to a hydroxyl group.

carboxylic acid (p. 798) An organic compound that contains a carboxyl group and is polar and reactive.

catabolism (kuh TAB oh lih zum) (p. 844) Refers to metabolic reactions that break down complex biological molecules for the purpose of forming smaller building blocks and extracting energy.

catalyst (p. 571) A substance that increases the rate of a chemical reaction by lowering activation energies but is not itself consumed in the reaction.

cathode (p. 710) In an electrochemical cell, the electrode where reduction takes place.

cathode ray (p. 108) Radiation that originates from the cathode and travels to the anode of a cathode-ray tube.

cation (KAT i ahn) (p. 207) An ion that has a positive charge.

cellular respiration (p. 846) The process in which glucose is broken down in the presence of oxygen gas to produce carbon dioxide, water, and energy.

Charles's law (p. 445) States that the volume of a given mass of gas is directly proportional to its kelvin temperature at constant pressure.

chemical bond (p. 206) The force that holds two atoms together; may form by the attraction of a positive ion for a negative ion or by sharing electrons.

chemical change (p. 77) A process involving one or more substances changing into new substances; also called a chemical reaction.

chemical equation (p. 285) A statement using chemical formulas to describe the identities and relative amounts of the reactants and products involved in the chemical reaction.

chemical equilibrium (p. 596) The state in which forward and reverse reactions balance each other because they occur at equal rates.

chemical potential energy (p. 517) The energy stored in a substance because of its composition; most is released or absorbed as heat during chemical reactions or processes.

chemical property (p. 74) The ability or inability of a substance to combine with or change into one or more new substances.

caloría (pág. 518) Cantidad de calor que se requiere para elevar un grado centígrado la temperatura de un gramo de agua pura.

calorímetro (pág. 523) Dispositivo aislado que sirve para medir la cantidad de calor liberada o absorbida durante un proceso físico o químico.

carbohidratos (pág. 832) Compuestos que contienen múltiples grupos hidroxilo, además de un grupo funcional aldehído o cetona, cuya función en los seres vivos es proporcionar energía inmediata o almacenada.

grupo carbonilo (pág. 796) Grupo formado por un átomo de oxígeno unido por un enlace doble a un átomo de carbono.

grupo carboxilo (pág. 798) Consiste en un grupo carbonilo unido a un grupo hidroxilo.

ácido carboxílico (pág. 798) Compuesto orgánico que contiene un grupo carboxilo; es polar y reactivo.

catabolismo (pág. 844) Reacciones metabólicas en las que se desdoblan moléculas biológicas complejas para obtener unidades básicas más pequeñas y energía.

catalizador (pág. 571) Sustancia que aumenta la velocidad de una reacción química al reducir su energía de activación; el catalizador no es consumido durante la reacción.

cátodo (pág. 710) Electrodo donde sucede la reducción en una celda electroquímica.

rayo catódico (pág. 108) Radiación que se origina en el cátodo y viaja hacia el ánodo de un tubo de rayos catódicos.

catión (pág. 207) Ion con carga positiva.

respiración celular (pág. 846) Proceso en el cual la glucosa es desdoblada en presencia del gas oxígeno para producir dióxido de carbono, agua y energía.

Ley de Charles (pág. 445) Establece que el volumen de una masa dada de gas es directamente proporcional a su temperatura Kelvin a presión constante.

enlace químico (pág. 206) La fuerza que mantiene a dos átomos unidos; puede formarse por la atracción de un ion positivo por un ion negativo compartiendo electrones.

cambio químico (pág. 77) Proceso que involucra una o más sustancias que se transforman en sustancias nuevas; también se conoce como reacción química.

ecuación química (pág. 285) Expresión que utiliza fórmulas químicas para describir las identidades y cantidades relativas de los reactivos y productos presentes en una reacción química.

equilibrio químico (pág. 596) Estado en el que se equilibran mutuamente las reacciones en sentido directo e inverso de una reacción química debido a que suceden a tasas iguales.

energía potencial química (pág. 517) La energía almacenada en una sustancia debido a su composición; la mayoría es liberada o absorbida como calor durante reacciones o procesos químicos.

propiedad química (pág. 74) La capacidad de una sustancia de combinarse con una o más sustancias nuevas o de transformarse en una o más sustancias nuevas.

chemical reaction (p. 282) The process by which the atoms of one or more substances are rearranged to form different substances; occurrence can be indicated by changes in temperature, color, odor, and physical state.

chemistry (p. 4) The study of matter and the changes that it undergoes.

chirality (p. 767) A property of a compound to exist in both left (l–) and right (d–) forms; occurs whenever a compound contains an asymmetric carbon.

chromatography (p. 83) A technique that is used to separate the components of a mixture based on the tendency of each component to travel or be drawn across the surface of a fixed substrate.

coefficient (p. 285) In a chemical equation, the number written in front of a reactant or product; in a balanced equation describes the lowest whole-number ratio of the amounts of all reactants and products.

colligative property (kol LIHG uh tihv • PRAH pur tee) (p. 498) A physical property of a solution that depends on the number, but not the identity, of the dissolved solute particles.

collision theory (p. 563) States that atoms, ions, and molecules must collide in order to react.

colloids (p. 477) A heterogeneous mixture of intermediate-sized particles (between atomic-size of solution particles and the size of suspension particles).

combined gas law (p. 449) A single law combining Boyle's, Charles's, and Gay-Lussac's laws that states the relationship among pressure, volume, and temperature of a fixed amount of gas.

combustion reaction (p. 290) A chemical reaction that occurs when a substance reacts with oxygen, releasing energy in the form of heat and light.

common ion (p. 620) An ion that is common to two or more ionic compounds.

common ion effect (p. 620) The lowering of the solubility of a substance by the presence of a common ion.

complete ionic equation (p. 301) An ionic equation that shows all the particles in a solution as they realistically exist.

complex reaction (p. 580) A chemical reaction that consists of two or more elementary steps.

compound (p. 85) A chemical combination of two or more different elements; can be broken down into simpler substances by chemical means and has properties different from those of its component elements.

concentration (p. 480) A measure of how much solute is dissolved in a specific amount of solvent or solution.

conclusion (p. 15) A judgment based on the information obtained.

condensation (p. 428) The energy-releasing process by which a gas or vapor becomes a liquid.

condensation polymerization (p. 811) Occurs when monomers containing at least two functional groups combine with the loss of a small by-product, usually water.

reacción química (pág. 282) Proceso por el cual los átomos de una o más sustancias se reordenan para formar sustancias diferentes; su pueden identificar cuando suceden cambios en temperatura, color, olor o estado físico.

química (pág. 4) El estudio de la materia y los cambios que ésta experimenta.

quiralidad (pág. 767) Propiedad de un compuesto para existir en forma levógira (i–) o dextrógira (d–); ocurre cuando un compuesto contiene un carbono asimétrico.

cromatografía (pág. 83) Técnica que sirve para separar los componentes de una mezcla según la tendencia de cada componente a desplazarse o ser atraído a lo largo de la superficie de otro material.

coeficiente (pág. 285) Número que precede a un reactivo o un producto en una ecuación química; en una ecuación equilibrada, indica la razón más pequeña expresada en números enteros de las cantidades de reactivos y productos en dicha reacción.

propiedad coligativa (pág. 498) Propiedad física de una solución que depende del número, pero no de la identidad, de las partículas de soluto disueltas.

teoría de colisión (pág. 563) Establece que los átomos, iones y moléculas deben chocar para reaccionar.

coloides (pág. 477) Mezcla heterogénea de partículas de tamaño intermedio (entre el tamaño atómico de partículas en solución y el de partículas en suspensión).

ley combinada de los gases (pág. 449) Ley que combina las leyes de Boyle, Charles y de Gay-Lussac; indica la relación entre la presión, el volumen y la temperatura de una cantidad constante de gas.

reacción de combustión (pág. 290) Reacción química que ocurre al reaccionar una sustancia con el oxígeno, liberando energía en forma de calor y luz.

ion común (pág. 620) Ion común a dos o más compuestos iónicos.

efecto del ion común (pág. 620) Disminución de la solubilidad de una sustancia debida a la presencia de un ion común.

ecuación iónica total (pág. 301) Ecuación iónica que muestra cómo existen realmente todas las partículas en una solución.

reacción compleja (pág. 580) Reacción química que consiste en dos o más pasos elementales.

compuesto (pág. 85) Combinación química de dos o más elementos diferentes; puede ser separado en sustancias más sencillas por medios químicos y exhibe propiedades que difieren de los elementos que lo componen.

concentración (pág. 480) Medida de la cantidad de soluto que se disuelve en una cantidad dada de disolvente o solución.

conclusión (pág. 15) Juicio basado en la información obtenida.

condensación (pág. 428) El proceso de liberación de energía mediante el cual un gas o vapor se convierte en líquido.

polimerización por condensación (pág. 811) Ocurre cuando monómeros que contienen al menos dos grupos funcionales se combinan y pierden un producto secundario pequeño, generalmente agua.

condensation reaction (p. 801) Occurs when two smaller organic molecules combine to form a more complex molecule, accompanied by the loss of a small molecule such as water.

conjugate acid (p. 638) The species produced when a base accepts a hydrogen ion from an acid.

conjugate acid-base pair (p. 638) Consists of two substances related to each other by the donating and accepting of a single hydrogen ion.

conjugate base (p. 638) The species produced when an acid donates a hydrogen ion to a base.

control (p. 14) In an experiment, the standard that is used for comparison.

conversion factor (p. 44) A ratio of equivalent values used to express the same quantity in different units; is always equal to 1 and changes the units of a quantity without changing its value.

coordinate covalent bond (p. 259) Forms when one atom donates a pair of electrons to be shared with an atom or ion that needs two electrons to become stable.

corrosion (p. 724) The loss of metal that results from an oxidation-reduction reaction of the metal with substances in the environment.

covalent bond (p. 241) A chemical bond that results from the sharing of valence electrons.

cracking (p. 748) The process by which heavier fractions of petroleum are converted to gasoline by breaking their large molecules into smaller molecules.

critical mass (p. 880) The minimum mass of a sample of fissionable material necessary to sustain a nuclear chain reaction.

crystal lattice (p. 214) A three-dimensional geometric arrangement of particles in which each positive ion is surrounded by negative ions and each negative ion is surrounded by positive ions; vary in shape due to sizes and relative numbers of the ions bonded.

crystalline solid (p. 420) A solid whose atoms, ions, or molecules are arranged in an orderly, geometric, three-dimensional structure.

crystallization (p. 83) A separation technique that produces pure solid particles of a substance from a solution that contains the dissolved substance.

cyclic hydrocarbon (p. 755) An organic compound that contains a hydrocarbon ring.

cycloalkane (p. 755) Cyclic hydrocarbons that contain single bonds only and can have rings with three, four, five, six, or more carbon atoms.

reacción de condensación (pág. 801) Ocurre cuando dos moléculas orgánicas pequeñas se combinan para formar una molécula más compleja; esta reacción es acompañada de la pérdida de una molécula pequeña como el agua.

ácido conjugado (pág. 638) Especie que se produce cuando una base acepta un ion hidrógeno de un ácido.

par ácido-base conjugado (pág. 638) Consiste en dos sustancias que se relacionan entre sí mediante la donación y aceptación de un solo ion hidrógeno.

base conjugada (pág. 638) Especie que se produce cuando un ácido dona un ion hidrógeno a una base.

control (pág. 14) Estándar de comparación en un experimento.

factor de conversión (pág. 44) Razón de valores equivalentes que sirve para expresar una misma cantidad en unidades diferentes; siempre es igual a 1 y cambia las unidades de una cantidad sin cambiar su valor.

enlace covalente coordinado (pág. 259) Se forma cuando un átomo dona un par de electrones para compartirlos con un átomo o un ion que requieren dos electrones para adquirir estabilidad.

corrosión (pág. 724) Pérdida de metal producida por una reacción de óxido-reducción del metal con sustancias en el ambiente.

enlace covalente (pág. 241) Enlace químico que se produce al compartir electrones de valencia.

cracking (pág. 748) Proceso por el cual las fracciones más pesadas de petróleo son convertidas en gasolina al romper las moléculas grandes en moléculas más pequeñas.

masa crítica (pág. 880) La masa mínima de una muestra de material fisionable que se necesita para sostener una reacción nuclear en cadena.

red cristalina (pág. 214) Ordenamiento geométrico tridimensional de partículas en el que cada ion positivo queda rodeado de iones negativos y cada ion negativo queda rodeado de iones positivos; su forma varía según el tamaño y número de iones enlazados.

sólido cristalino (pág. 420) Sólido cuyos átomos, iones o moléculas forman una estructura tridimensional, ordenada y geométrica.

cristalización (pág. 83) Técnica de separación que produce partículas sólidas puras de una sustancia a partir de una solución que contiene dicha sustancia en solución.

hidrocarburo cíclico (pág. 755) Compuesto orgánico que contiene un anillo de hidrocarburos.

cicloalcano (pág. 755) Hidrocarburos cíclicos que sólo contienen enlaces simples; pueden formar anillos con tres, cuatro, cinco, seis o más átomos de carbono.

D

Dalton's atomic theory (p. 104) States that matter is composed of extremely small particles called atoms; atoms are invisible and indestructable; atoms of a given element are identical in size, mass, and chemical properties; atoms of a specific element are different from those of another element; different atoms combine in simple whole-number ratios to form compounds; in a chemical reaction, atoms are separated, combined, or rearranged.

teoría atómica de Dalton (pág. 104) Establece que la materia se compone de partículas extremadamente pequeñas denominadas átomos; los átomos son invisibles e indestructibles; los átomos de un elemento dado son idénticos en tamaño, masa y propiedades químicas; los átomos de un elemento específico difieren de los de otros elementos; átomos diferentes se combinan en razones simples de números enteros para formar compuestos; los átomos se separan, se combinan o se reordenan durante una reacción química.

Dalton's law of partial pressures (p. 408) States that the total pressure of a mixture of gases is equal to the sum of the pressures of all the gases in the mixture.

de Broglie equation (p. 150) Predicts that all moving particles have wave characteristics and relates each particle's wavelength to its frequency, its mass, and Planck's constant.

decomposition reaction (p. 292) A chemical reaction that occurs when a single compound breaks down into two or more elements or new compounds.

dehydration reaction (p. 803) An elimination reaction in which the atoms removed form water.

dehydrogenation reaction (p. 803) A reaction that eliminates two hydrogen atoms, which form a hydrogen molecule of gas.

delocalized electrons (p. 225) The electrons involved in metallic bonding that are free to move easily from one atom to the next throughout the metal and are not attached to a particular atom.

denaturation (p. 829) The process in which a protein's natural, intricate three-dimensional structure is disrupted.

denatured alcohol (p. 793) Ethanol to which noxious substances have been added in order to make it unfit to drink.

density (p. 36) The amount of mass per unit volume; a physical property.

dependent variable (p. 14) In an experiment, the variable whose value depends on the independent variable.

deposition (p. 429) The energy-releasing process by which a substance changes from a gas or vapor to a solid without first becoming a liquid.

derived unit (p. 35) A unit defined by a combination of base units.

diffusion (p. 404) The movement of one material through another from an area of higher concentration to an area of lower concentration.

dimensional analysis (p. 44) A systematic approach to problem solving that uses conversion factors to move from one unit to another.

dipole-dipole forces (p. 412) The attractions between oppositely charged regions of polar molecules.

disaccharide (p. 833) Forms when two monosaccharides bond together.

dispersion forces (p. 412) The weak forces resulting from temporary shifts in the density of electrons in electron clouds.

distillation (p. 82) A technique that can be used to physically separate most homogeneous mixtures based on the differences in the boiling points of the substances.

double-replacement reaction (p. 296) A chemical reaction that involves the exchange of ions between two compounds and produces either a precipitate, a gas, or water.

dry cell (p. 718) An electrochemical cell that contains a moist electrolytic paste inside a zinc shell.

ley de Dalton de las presiones parciales (pág. 408) Establece que la presión total de una mezcla de gases es igual a la suma de las presiones de todos los gases en la mezcla.

ecuación de deBroglie (pág. 150) Predice que todas las partículas móviles tienen características ondulatorias y relaciona la longitud de onda de cada partícula con su frecuencia, su masa y la constante de Planck.

reacción de descomposición (pág. 292) Reacción química que ocurre cuando un solo compuesto se divide en dos o más elementos o nuevos compuestos.

reacción de deshidratación (pág. 803) Una reacción de eliminación en la que los átomos que se pierden forman agua.

reacción de deshidrogenación (pág. 803) Reacción orgánica en la que se pierden dos átomos de hidrógeno, los cuales se unen y forman una molécula de hidrógeno.

electrones deslocalizados (pág. 225) Los electrones que forman un enlace metálico; estos electrones pasan fácilmente de un átomo a otro a través del metal y no están unidos a ningún átomo en particular.

desnaturalización (pág. 829) Proceso que afecta la estructura tridimensional, compleja y natural de una proteína.

alcohol desnaturalizado (pág. 793) Etanol al cual se añaden sustancias nocivas para evitar que se pueda beber.

densidad (pág. 36) La cantidad de masa por unidad de volumen; una propiedad física.

variable dependiente (pág. 14) Es la variable de un experimento cuyo valor depende de la variable independiente.

depositación (pág. 429) Proceso de liberación de energía por el cual una sustancia cambia de gas o vapor a sólido sin antes convertirse en un líquido.

unidad derivada (pág. 35) Unidad definida por una combinación de unidades básicas.

difusión (pág. 404) El movimiento de un material a través de otro en dirección al área de menor concentración.

análisis dimensional (pág. 44) Un enfoque sistemático para resolver un problema en el que se usan factores de conversión para pasar de una unidad a otra.

fuerzas dipolo-dipolo (pág. 412) La atracción entre regiones con cargas opuestas de moléculas polares.

disacárido (pág. 833) Se forma a partir de la unión de dos monosacáridos.

fuerzas de dispersión (pág. 412) Fuerzas débiles causadas por los cambios temporales en la densidad de electrones en las nubes electrónicas.

destilación (pág. 82) Técnica que se usa para separar físicamente la mayoría de las mezclas homogéneas según las diferencias en los puntos de ebullición de las sustancias.

reacción de sustitución doble (pág. 296) Reacción química en la que dos compuestos intercambian iones positivos, produciendo un precipitado, un gas o agua.

pila seca (pág. 718) Celda electroquímica que contiene una pasta electrolítica húmeda dentro de un armazón de zinc.

E

elastic collision (p. 403) Collision in which no kinetic energy is lost; kinetic energy can be transferred between the colliding particles, but the total kinetic energy of the two particles remains the same.

choque elástico (pág. 403) Colisión en que no se pierde energía cinética; la energía cinética es transferida entre las partículas en choque, pero la energía cinética total de las dos partículas permanece igual.

electrochemical cell (p. 709) An apparatus that uses a redox reaction to produce electrical energy or uses electrical energy to cause a chemical reaction.

electrolysis (p. 728) The process that uses electrical energy to bring about a chemical reaction.

electrolyte (p. 215) An ionic compound whose aqueous solution conducts an electric current.

electrolytic cell (p. 728) An electrochemical cell in which electrolysis occurs.

electromagnetic radiation (p. 137) A form of energy exhibiting wavelike behavior as it travels through space; can be described by wavelength, frequency, amplitude, and speed.

electromagnetic spectrum (p. 139) Includes all forms of electromagnetic radiation; the types of radiation differ in their frequencies and wavelengths.

electron (p. 108) A negatively charged, fast-moving particle with an extremely small mass that is found in all forms of matter and moves through the empty space surrounding an atom's nucleus.

electron capture (p. 868) A radioactive decay process that occurs when an atom's nucleus draws in a surrounding electron, which combines with a proton to form a neutron, resulting in an X-ray photon being emitted.

electron configuration (p. 156) The arrangement of electrons in an atom, which is prescribed by three rules—the aufbau principle, the Pauli exclusion principle, and Hund's rule.

electron-dot structure (p. 161) Consists of an element's symbol, representing the atomic nucleus and inner-level electrons, that is surrounded by dots, representing the atom's valence electrons.

electron sea model (p. 225) Proposes that all metal atoms in a metallic solid contribute their valence electrons to form a "sea" of electrons, and can explain properties of metallic solids such as malleability, conduction, and ductility.

electronegativity (p. 194) Indicates the relative ability of an element's atoms to attract electrons in a chemical bond.

element (p. 84) A pure substance that cannot be broken down into simpler substances by physical or chemical means.

elimination reaction (p. 802) A reaction of organic compounds that occurs when a combination of atoms is removed from two adjacent carbon atoms forming an additional bond between the atoms.

empirical formula (p. 344) A formula that shows the smallest whole-number mole ratio of the elements of a compound, and might or might not be the same as the actual molecular formula.

endothermic (p. 247) A chemical reaction or process in which a greater amount of energy is required to break the existing bonds in the reactants than is released when the new bonds form in the product molecules.

end point (p. 663) The point at which the indicator that is used in a titration changes color.

celda electroquímica (pág. 709) Aparato que usa una reacción redox para producir energía eléctrica o que utiliza energía eléctrica para causar una reacción química.

electrólisis (pág. 728) Proceso que emplea energía eléctrica para producir una reacción química.

electrolito (pág. 215) Compuesto iónico cuya solución acuosa conduce una corriente eléctrica.

celda electrolítica (pág. 728) Celda electroquímica en donde ocurre la electrólisis.

radiación electromagnética (pág. 137) Forma de energía que exhibe un comportamiento ondulatorio al viajar por el espacio; se puede describir por su longitud de onda, su frecuencia, su amplitud y su rapidez.

espectro electromagnético (pág. 139) Incluye toda forma de radiación electromagnética; los distintos tipos de radiación difirien en sus frecuencias y sus longitudes de onda.

electrón (pág. 108) Partícula móvil rápida, de carga negativa y con una masa extremadamente pequeña que se encuentra en todas las formas de materia y que se mueve a través del espacio vacío que rodea el núcleo de un átomo.

captura electrónica (pág. 868) Proceso de desintegración radiactiva que ocurre cuando el núcleo de un átomo atrae un electrón circundante, que luego se combina con un protón para formar un neutrón, provocando la emisión de un fotón de rayos X.

configuración electrónica (pág. 156) El ordenamiento de los electrones en un átomo; está determinado por tres reglas: el principio de aufbau, el principio de exclusión de Pauli y la regla de Hund.

estructura de puntos de electrones (pág. 161) Consiste en el símbolo del elemento, que representa al núcleo atómico y los electrones de los niveles internos, rodeado por puntos que representan los electrones de valencia del átomo.

modelo del mar de electrones (pág. 225) Propone que todos los átomos de metal en un sólido metálico contribuyen con sus electrones de valencia para formar un "mar" de electrones.

electronegatividad (pág. 194) Indica la capacidad relativa de los átomos de un elemento para atraer electrones en un enlace químico.

elemento (pág. 84) Sustancia pura que no puede separarse en sustancias más sencillas por medios físicos ni químicos.

reacción de eliminación (pág. 802) Reacción de compuestos orgánicos que ocurre cuando se pierden un conjunto de átomos en dos átomos adyacentes de carbono, al formarse un enlace entre dichos átomos de carbono.

fórmula empírica (pág. 344) Fórmula que muestra la proporción molar más pequeña expresada en números enteros de los elementos de un compuesto; puede ser distinta de la fórmula molecular real.

endotérmica (pág. 247) Reacción o proceso químico que requiere una mayor cantidad de energía para romper los enlaces existentes en los reactivos, que la que se se libera al formarse los enlaces nuevos en las moléculas del producto.

punto final (pág. 663) Punto en el que el indicador que se utiliza en una titulación cambia de color.

energy (p. 516) The capacity to do work or produce heat; exists as potential energy, which is stored in an object due to its composition or position, and kinetic energy, which is the energy of motion.

energy sublevels (p. 153) The energy levels contained within a principal energy level.

enthalpy (p. 527) The heat content of a system at constant pressure.

enthalpy (heat) of combustion (p. 529) The enthalpy change for the complete burning of one mole of a given substance.

enthalpy (heat) of reaction (p. 527) The change in enthalpy for a reaction—the difference between the enthalpy of the substances that exist at the end of the reaction and the enthalpy of the substances present at the start .

entropy (p. 543) A measure of the number of possible ways that the energy of a system can be distributed; related to the freedom of the system's particles to move and the number of ways they can be arranged.

enzyme (p. 829) A biological catalyst.

equilibrium constant (p. 599) K_{eq} is the numerical value that describes the ratio of product concentrations to reactant concentrations, with each raised to the power corresponding to its coefficient in the balanced equation.

equivalence point (p. 661) The point at which the moles of H^+ ions from the acid equals moles of OH^- ions from the base.

error (p. 48) The difference between an experimental value and an accepted value

ester (p. 799) An organic compound with a carboxyl group in which the hydrogen of the hydroxyl group is replaced by an alkyl group; may be volatile and sweet-smelling and is polar.

ether (p. 794) An organic compound that contains an oxygen atom bonded to two carbon atoms.

evaporation (p. 426) The process in which vaporization occurs only at the surface of a liquid.

excess reactant (p. 379) A reactant that remains after a chemical reaction stops.

exothermic (p. 247) A chemical reaction or process in which more energy is released than is required to break bonds in the initial reactants.

experiment (p. 14) A set of controlled observations that test a hypothesis.

extensive property (p. 73) A physical property, such as mass, length, and volume, that is dependent upon the amount of substance present.

energía (pág. 516) Capacidad de realizar trabajo o producir calor; existe como energía potencial (almacenada en un objeto debido a su composición o posición) o como energía cinética (energía del movimiento).

subniveles de energía (pág. 153) Los niveles de energía dentro de un nivel principal de energía.

entalpía (pág. 527) El contenido de calor en un sistema a presión constante.

entalpía (calor) de combustión (pág. 529) El cambio de entalpía causado por la combustión completa de un mol de una sustancia dada.

entalpía (calor) de reacción (pág. 527) El cambio en la entalpía que ocurre en una reacción; es decir, la diferencia entre la entalpía de las sustancias que existen al final de la reacción y la entalpía de las sustancias presentes al comienzo de la misma.

entropía (pág. 543) Una medida de las formas posibles en que se puede distribuir la energía de un sistema; está relacionada con la libertad de movimiento de las partículas del sistema y el número de maneras en que éstas se pueden ordenar.

enzima (pág. 829) Catalizador biológico.

constante de equilibrio (pág. 599) K_{eq} es el valor numérico que describe la razón de las concentraciones de los productos con respecto a las concentraciones de los reactivos, cada una de ellas elevada a la potencia correspondiente a su coeficiente en la ecuación equilibrada.

punto de equivalencia (pág. 661) Punto en el cual los moles de iones H^+ del ácido equivalen a los moles de iones OH^- de la base.

error (pág. 48) La diferencia entre el valor experimental y el valor aceptado.

éster (pág. 799) Compuesto orgánico con un grupo carboxilo en el que el hidrógeno del grupo de hidroxilo es reemplazado por un grupo alquilo; es polar y puede ser volátil y de olor dulce.

éter (pág. 794) Compuesto orgánico que contiene un átomo de oxígeno unido a dos átomos de carbono.

evaporación (pág. 426) Proceso en el cual la vaporización ocurre sólo en la superficie de un líquido.

reactivo en exceso (pág. 379) Reactivo que sobra luego de finalizar una reacción química.

exotérmica (pág. 247) Reacción o proceso químico en el que se libera más energía que la requerida para romper los enlaces en los reactivos iniciales.

experimento (pág. 14) Conjunto de observaciones controladas que se realizan para probar una hipótesis.

propiedad extensiva (pág. 73) Propiedades físicas, como la masa, la longitud y el volumen, que dependen de la cantidad de sustancia presente.

F

fatty acid (p. 835) A long-chain carboxylic acid that usually has between 12 and 24 carbon atoms and can be saturated (no double bonds), or unsaturated (one or more double bonds).

ácido graso (pág. 835) Ácido carboxílico de cadena larga que tiene generalmente entre 12 y 24 átomos de carbono; puede ser saturado (sin enlaces dobles) o insaturado o no saturado (con uno o más enlaces dobles).

fermentation **(p. 847)** The process in which glucose is broken down in the absence of oxygen, producing either ethanol, carbon dioxide, and energy (alcoholic fermentation) or lactic acid and energy (lactic acid fermentation).

filtration **(p. 82)** A technique that uses a porous barrier to separate a solid from a liquid.

formula unit **(p. 218)** The simplest ratio of ions represented in an ionic compound.

fractional distillation **(p. 747)** The process by which petroleum can be separated into simpler components, called fractions, as they condense at different temperatures.

free energy **(p. 546)** The energy available to do work—the difference between the change in enthalpy and the product of the entropy change and the kelvin temperature.

freezing point **(p. 428)** The temperature at which a liquid is converted into a crystalline solid.

freezing-point depression **(p. 502)** The difference in temperature between a solution's freezing point and the freezing point of its pure solvent.

frequency **(p. 137)** The number of waves that pass a given point per second.

fuel cell **(p. 722)** A voltaic cell in which the oxidation of a fuel, such as hydrogen gas, is used to produce electric energy.

functional group **(p. 786)** An atom or group of atoms that always reacts in a certain way in an organic molecule.

fermentación **(pág. 847)** Proceso en el cual la glucosa es desdoblada en ausencia de oxígeno produciendo etanol, dióxido de carbono y energía (fermentación alcohólica) o ácido láctico y energía (fermentación del ácido láctico).

filtración **(pág. 82)** Técnica que utiliza una barrera porosa para separar un sólido de un líquido.

fórmula unitaria **(pág. 218)** La razón más simple de iones representados en un compuesto iónico.

destilación fraccionaria **(pág. 747)** Proceso mediante el cual se separa el petróleo en componentes más simples llamados fracciones, las cuales se condensan a temperaturas diferentes.

energía libre **(pág. 546)** Energía disponible para hacer trabajo: la diferencia entre el cambio en la entalpía y el producto del cambio de entropía por la temperatura kelvin.

punto de congelación **(pág. 428)** La temperatura a la cual un líquido se convierte en un sólido cristalino.

depresión del punto de congelación **(pág. 502)** Diferencia de temperatura entre el punto de congelación de una solución y el punto de congelación de su disolvente puro.

frecuencia **(pág. 137)** Número de ondas que pasan por un punto dado en un segundo.

celda de combustible **(pág. 722)** Celda voltaica en la cual la oxidación de un combustible, como el gas hidrógeno, se utiliza para producir energía eléctrica.

grupo funcional **(pág. 786)** Átomo o grupo de átomos que siempre reaccionan de cierta manera en una molécula orgánica.

G

galvanization **(p. 727)** The process in which an iron object is dipped into molten zinc or electroplated with zinc to make the iron more resistant to corrosion.

gamma rays **(p. 124)** High-energy radiation that has no electrical charge and no mass, is not deflected by electric or magnetic fields, usually accompanies alpha and beta radiation, and accounts for most of the energy lost during radioactive decay.

gas **(p. 72)** A form of matter that flows to conform to the shape of its container, fills the container's entire volume, and is easily compressed.

Gay-Lussac's law **(p. 447)** States that the pressure of a fixed mass of gas varies directly with the kelvin temperature when the volume remains constant.

geometric isomers **(p. 766)** A category of stereoisomers that results from different arrangements of groups around a double bond.

Graham's law of effusion **(p. 404)** States that the rate of effusion for a gas is inversely proportional to the square root of its molar mass.

graph **(p. 55)** A visual display of data.

ground state **(p. 146)** The lowest allowable energy state of an atom.

group **(p. 177)** A vertical column of elements in the periodic table arranged in order of increasing atomic number; also called a family.

galvanizado **(pág. 727)** Proceso en el cual un objeto de hierro en sumergido o galvanizado en zinc para aumentar la resistencia del hierro a la corrosión.

rayos gamma **(pág. 124)** Radiación de alta energía sin carga eléctrica ni masa; no es desviada por campos eléctricos ni magnéticos; acompaña generalmente a la radiación alfa y beta; representa la mayor parte de la energía perdida durante la desintegración radiactiva.

gas **(pág. 72)** Forma de la materia que fluye para adaptarse a la forma de su contenedor, llena el volumen entero del recipiente y se comprime fácilmente.

ley de Gay-Lussac **(pág. 447)** Establece que la presión de una masa dada de gas varía directamente con la temperatura en grados Kelvin cuando el volumen permanece constante.

isómeros geométricos **(pág. 766)** Categoría de estereoisómeros originada por los diversos ordenamientos posibles de grupos alrededor de un enlace doble.

ley de efusión de Graham **(pág. 404)** Establece que la tasa de efusión de un gas es inversamente proporcional a la raíz cuadrada de su masa molar.

gráfica **(pág. 55)** Representación visual de datos.

estado base **(pág. 146)** Estado de energía más bajo posible de un átomo.

grupo **(pág. 177)** Columna vertical de los elementos en la tabla periódica ordenados en sentido creciente según su número atómico; llamado también familia.

H

half-cells (p. 710) The two parts of an electrochemical cell in which the separate oxidation and reduction reactions occur.

half-life (p. 870) The time required for one-half of a radioisotope's nuclei to decay into its products.

half-reaction (p. 693) One of two parts of a redox reaction—the oxidation half, which shows the number of electrons lost when a species is oxidized, or the reduction half, which shows the number of electrons gained when a species is reduced.

halocarbon (p. 787) Any organic compound containing a halogen substituent.

halogen (p. 180) A highly reactive group 17 element.

halogenation (p. 790) A process by which hydrogen atoms are replaced by halogen atoms.

heat (p. 518) A form of energy that flows from a warmer object to a cooler object.

heat of solution (p. 492) The overall energy change that occurs during the solution formation process.

Heisenberg uncertainty principle (p. 151) States that it is not possible to know precisely both the velocity and the position of a particle at the same time.

Henry's law (p. 496) States that at a given temperature, the solubility of a gas in a liquid is directly proportional to the pressure of the gas above the liquid.

Hess's law (p. 534) States that if two or more thermochemical equations can be added to produce a final equation for a reaction, then the sum of the enthalpy changes for the individual reactions is the enthalpy change for the final reaction.

heterogeneous catalyst (p. 573) A catalyst that exists in a different physical state than the reaction it catalyzes.

heterogeneous equilibrium (p. 602) A state of equilibrium that occurs when the reactants and products of a reaction are present in more than one physical state.

heterogeneous mixture (p. 81) One that does not have a uniform composition and in which the individual substances remain distinct.

homogeneous catalyst (p. 573) A catalyst that exists in the same physical state as the reaction it catalyzes.

homogeneous equilibrium (p. 600) A state of equilibrium that occurs when all the reactants and products of a reaction are in the same physical state.

homogeneous mixture (p. 81) One that has a uniform composition throughout and always has a single phase; also called a solution.

homologous series (p. 751) Describes a series of compounds that differ from one another by a repeating unit.

Hund's rule (p. 157) States that single electrons with the same spin must occupy each equal-energy orbital before additional electrons with opposite spins can occupy the same orbitals.

semiceldas (pág. 710) Las dos partes de una celda electroquímica en las que ocurren las reacciones separadas de oxidación y reducción.

vida media (pág. 870) Tiempo requerido para que la mitad de los núcleos de un radioisótopo se desintegren en sus productos.

semirreacción (pág. 693) Una de dos partes de una reacción redox: la correspondiente a la oxidación muestra el número de electrones que se pierden al oxidarse una especie y la correspondiente a la reducción muestra el número de electrones que se ganan al reducirse una especie.

halocarbono (pág. 787) Cualquier compuesto orgánico que contiene un sustituyente halógeno.

halógeno (pág. 180) Elemento sumamente reactivo del grupo 17.

halogenación (pág. 790) Proceso mediante el cual se reemplazan átomos de hidrógeno por átomos de halógeno.

calor (pág. 518) Forma de energía que fluye hacia cuerpos más fríos.

calor de solución (pág. 492) El cambio global de energía que ocurre durante el proceso de formación de una solución.

principio de incertidumbre de Heisenberg (pág. 151) Establece que no es posible saber con precisión y al mismo tiempo la velocidad y la posición de una partícula.

ley de Henry (pág. 496) Establece que a una temperatura dada, la solubilidad de un gas en un líquido es directamente proporcional a la presión del gas sobre el líquido.

ley de Hess (pág. 534) Establece que si para producir la ecuación final para una reacción se pueden sumar dos o más ecuaciones termoquímicas, entonces la suma de los cambios de entalpía para las reacciones individuales equivale al cambio de entalpía de la reacción final.

catalizador heterogéneo (pág. 573) Catalizador que existe en un estado físico diferente al de la reacción que cataliza.

equilibrio heterogéneo (pág. 602) Estado de equilibrio que ocurre cuando los reactivos y los productos de una reacción están presentes en más de un estado físico.

mezcla heterogénea (pág. 81) Aquella que no tiene una composición uniforme y en la que las sustancias individuales permanecen separadas.

catalizador homogéneo (pág. 573) Catalizador que existe en el mismo estado físico de la reacción que cataliza.

equilibrio homogéneo (pág. 600) Estado de equilibrio que ocurre cuando todos los reactivos y productos de una reacción están en el mismo estado físico.

mezcla homogénea (pág. 81) Aquella que tiene una composición uniforme y siempre tiene una sola fase; también llamada solución.

serie homóloga (pág. 751) Describe una serie de compuestos que difieren entre sí por una unidad repetitiva.

regla de Hund (pág. 157) Establece que los electrones individuales con igual rotación deben ocupar cada uno orbitales distintos con la misma energía, antes de que electrones adicionales con rotación opuesta puedan ocupar los mismos orbitales.

hybridization (p. 262) A process in which atomic orbitals are mixed to form new, identical hybrid orbitals.

hibridación (pág. 262) Proceso mediante el cual se mezclan los orbitales atómicos para formar orbitales híbridos nuevos e idénticos.

hydrate (p. 351) A compound that has a specific number of water molecules bound to its atoms.

hidrato (pág. 351) Compuesto que tiene un número específico de moléculas de agua unidas a sus átomos.

hydration reaction (p. 804) An addition reaction in which a hydrogen atom and a hydroxyl group from a water molecule add to a double or triple bond.

reacción de hidratación (pág. 804) Reacción de adición en la que se añaden el átomo de hidrógeno y el grupo hidroxilo de una molécula de agua a un enlace doble o triple.

hydrocarbon (p. 745) Simplest organic compound composed only of the elements carbon and hydrogen.

hidrocarburo (pág. 745) El compuesto orgánico más simple; está formado sólo por los elementos carbono e hidrógeno.

hydrogenation reaction (p. 804) An addition reaction in which hydrogen is added to atoms in a double or triple bond; usually requires a catalyst.

reacción de hidrogenación (pág. 804) Reacción de adición en la que se agrega hidrógeno a los átomos que forman un enlace doble o triple; requiere generalmente de un catalizador.

hydrogen bond (p. 413) A strong dipole-dipole attraction between molecules that contain a hydrogen atom bonded to a small, highly electronegative atom.

enlace de hidrógeno (pág. 413) Fuerte atracción dipolo-dipolo entre moléculas que contienen un átomo de hidrógeno unido a un átomo pequeño, sumamente electronegativo.

hydroxyl group (p. 792) An oxygen-hydrogen group covalently bonded to a carbon atom.

grupo hidroxilo (pág. 792) Un grupo hidrógeno-oxígeno unido covalentemente a un átomo de carbono.

hypothesis (p. 13) A tentative, testable statement or prediction about what has been observed.

hipótesis (pág. 13) Enunciado tentativo y comprobable o predicción acerca de lo que ha sido observado.

I

ideal gas constant (R) (p. 454) An experimentally determined constant whose value in the ideal gas equation depends on the units that are used for pressure.

constante de los gases ideales (R) (pág. 454) Constante determinada experimentalmente cuyo valor en la ecuación de los gases ideales depende de las unidades en las que se expresa la presión.

ideal gas law (p. 454) Describes the physical behavior of an ideal gas in terms of pressure, volume, temperature, and number of moles of gas.

ley de los gases ideales (pág. 454) Describe el comportamiento físico de un gas ideal en términos de la presión, el volumen, la temperatura y el número de moles del gas.

immiscible (ih MIHS ih bul) (p. 479) Describes two liquids that can be mixed together but separate shortly after you cease mixing them.

inmiscible (pág. 479) Describe dos líquidos que se pueden mezclar entre sí, pero que se separan poco después de que se cesa de mezclarlos.

independent variable (p. 14) In an experiment, the variable that the experimenter plans to change.

variable independiente (pág. 14) La variable de un experimento que el experimentador piensa cambiar.

induced transmutation (p. 875) The process in which nuclei are bombarded with high-velocity charged particles in order to create new elements.

transmutación inducida (pág. 875) Proceso en cual se bombardean núcleos con partículas cargadas de alta velocidad para crear elementos nuevos.

inhibitor (p. 571) A substance that slows down the reaction rate of a chemical reaction or prevents a reaction from happening.

inhibidor (pág. 571) Sustancia que reduce la tasa de reacción de una reacción química o evita que ésta suceda.

inner transition metal (p. 180) A type of group B element that is contained in the f-block of the periodic table and is characterized by a filled outermost orbital, and filled or partially filled 4f and 5f orbitals.

metal de transición interna (pág. 180) Tipo de elemento del grupo B contenido dentro del bloque F de la tabla periódica; se caracteriza por tener el orbital más externo lleno y los orbitales 4f y 5f parcialmente llenos.

insoluble (p. 479) Describes a substance that cannot be dissolved in a given solvent.

insoluble (pág. 479) Describe una sustancia que no se puede disolver en un disolvente dado.

instantaneous rate (p. 578) The rate of decomposition at a specific time, calculated from the rate law, the specific rate constant, and the concentrations of all the reactants.

velocidad instantánea (pág. 578) La tasa de descomposición en un tiempo dado, se calcula a partir de la ley de velocidad de la reacción, la constante de velocidad de la reacción y las concentraciones de los reactivos.

intensive property (p. 73) A physical property that remains the same no matter how much of a substance is present.

propiedad intensiva (pág. 73) Propiedad física que permanece igual sea cual sea la cantidad de sustancia presente.

intermediate (p. 580) A substance produced in one elementary step of a complex reaction and consumed in a subsequent elementary step.

intermediario (pág. 580) Sustancia producida en un paso elemental de una reacción compleja y que es consumida en un paso elemental subsecuente.

ion **(p. 189)** An atom or bonded group of atoms with a positive or negative charge.

ionic bond **(p. 210)** The electrostatic force that holds oppositely charged particles together in an ionic compound.

ionic compounds **(p. 210)** Compounds that contain ionic bonds.

ionization energy **(p. 191)** The energy required to remove an electron from a gaseous atom; generally increases in moving from left-to-right across a period and decreases in moving down a group.

ionizing radiation **(p. 885)** Radiation that is energetic enough to ionize matter it collides with.

ion product constant for water **(p. 650)** The value of the equilibrium constant expression for the self-ionization of water.

isomers **(p. 765)** Two or more compounds that have the same molecular formula but have different molecular structures.

isotopes **(p. 117)** Atoms of the same element with different numbers of neutrons.

ion **(pág. 189)** Átomo o grupo de átomos unidos que tienen carga positiva o negativa.

enlace iónico **(pág. 210)** Fuerza electrostática que mantiene unidas las partículas con carga opuesta en un compuesto iónico.

compuestos iónicos **(pág. 210)** Compuestos que contienen enlaces iónicos.

energía de ionización **(pág. 191)** Energía que se requiere para separar un electrón de un átomo en estado gaseoso; generalmente aumenta al moverse de izquierda a derecha a lo largo de un período de la tabla periódica y disminuye al moverse hacia abajo a lo largo de un grupo.

radiación ionizante **(pág. 885)** Radiación que posee suficiente energía como para ionizar la materia con la que choca.

constante del producto iónico del agua **(pág. 650)** Valor de la expresión de la constante de equilibrio de la ionización del agua.

isómeros **(pág. 765)** Dos o más compuestos que tienen la misma fórmula molecular pero poseen estructuras moleculares diferentes.

isótopos (pág. 117) Átomos del mismo elemento con diferente número de neutrones.

J

joule **(p. 518)** The SI unit of heat and energy.

julio **(pág. 518)** La unidad SI de medida del calor y la energía.

K

kelvin **(p. 35)** The SI base unit of temperature.

ketone **(p. 797)** An organic compound in which the carbon of the carbonyl group is bonded to two other carbon atoms.

kilogram **(p. 34)** The SI base unit for mass.

kinetic-molecular theory **(p. 402)** Describes the behavior of gases in terms of particles in motion; makes several assumptions about size, motion, and energy of gas particles.

kelvin **(pág. 35)** Unidad básica de temperatura del SI.

cetona **(pág. 797)** Compuesto orgánico en el que el carbono del grupo carbonilo está unido a otros dos átomos de carbono.

kilogramo **(pág. 34)** Unidad básica de masa del SI.

teoría cinético-molecular **(pág. 402)** Explica el comportamiento de los gases en términos de partículas en movimiento; hace varias suposiciones acerca del tamaño, movimiento y energía de las partículas de gas.

L

lanthanide series **(p. 180)** In the periodic table, the f-block elements from period 6 that follow the element lanthanum.

lattice energy **(p. 216)** The energy required to separate one mole of the ions of an ionic compound, which is directly related to the size of the ions bonded and is also affected by the charge of the ions.

law of chemical equilibrium **(p. 599)** States that at a given temperature, a chemical system may reach a state in which a particular ratio of reactant and product concentrations has a constant value.

law of conservation of energy **(p. 517)** States that in any chemical reaction or physical process, energy may change from one form to another, but it is neither created nor destroyed.

law of conservation of mass **(p. 77)** States that mass is neither created nor destroyed during a chemical reaction but is conserved.

serie de los lantánidos **(pág. 180)** Los elementos del bloque F del período 6 de la tabla periódica que siguen al elemento lantano.

energía reticular **(pág. 216)** Energía que se requiere para separar un mol de los iones de un compuesto iónico; está directamente relacionada con el tamaño de los iones enlazados y es afectada también por la carga de los iones.

ley del equilibrio químico **(pág. 599)** Establece que a una temperatura dada, un sistema químico puede alcanzar un estado en el que la razón particular de las concentraciones del reactivo y el producto tiene un valor constante.

ley de conservación de la energía **(pág. 517)** Establece que en toda reacción química y en todo proceso físico la energía puede cambiar de una forma a otra, pero no puede ser creada ni destruida.

ley de conservación de la masa **(pág. 77)** Establece que durante una reacción química la masa no se crea ni se destruye, sino que se conserva.

law of definite proportions **(p. 87)** States that, regardless of the amount, a compound is always composed of the same elements in the same proportion by mass.

law of multiple proportions **(p. 89)** States that when different compounds are formed by the combination of the same elements, different masses of one element combine with the same fixed mass of the other element in a ratio of small whole numbers.

Le Châtelier's principle **(luh SHAHT uh lee yays • PRIHN sih puhl) (p. 607)** States that if a stress is applied to a system at equilibrium, the system shifts in the direction that relieves the stress.

Lewis model **(p. 641)** A Lewis acid is an electron-pair acceptor and a Lewis base is an electron-pair donor.

Lewis structure **(p. 242)** A model that uses electron-dot structures to show how electrons are arranged in molecules. Pairs of dots or lines represent bonding pairs.

limiting reactant **(p. 379)** A reactant that is totally consumed during a chemical reaction, limits the extent of the reaction, and determines the amount of product.

lipids **(p. 835)** Large, nonpolar biological molecules that vary in structure, store energy in living organisms, and make up most of the structure of cell membranes.

liquid **(p. 71)** A form of matter that flows, has constant volume, and takes the shape of its container.

liter **(p. 35)** The metric unit for volume equal to one cubic decimeter.

ley de las proporciones definidas **(pág. 87)** Establece que, independientemente de la cantidad, un compuesto siempre se compone de los mismos elementos en la misma proporción por masa.

ley de las proporciones múltiples **(pág. 89)** Establece que cuando la combinación de los mismos elementos forma compuestos diferentes, una masa dada de uno de los elementos se combina con masas diferentes del otro elemento de acuerdo con una razón que se expresa en números enteros pequeños.

Principio de Le Châtelier **(pág. 607)** Establece que si se aplica una perturbación a un sistema en equilibrio, el sistema cambia en la dirección que reduce la perturbación.

modelo de Lewis **(pág. 641)** Un ácido es un receptor de pares de electrones y una base es un donante de pares de electrones.

estructura de Lewis **(pág. 242)** Modelo que utiliza diagramas de puntos de electrones para mostrar la disposición de los electrones en las moléculas. Los pares de puntos o líneas representan pares de electrones enlazados.

reactivo limitante **(pág. 379)** Reactivo que se consume completamente durante una reacción química, limita la duración de la reacción y determina la cantidad del producto.

lípidos **(pág. 835)** Moléculas biológicas no polares de gran tamaño que varían en estructura, almacenan energía en los seres vivos y conforman la mayor parte de la estructura de las membranas celulares.

líquido **(pág. 71)** Forma de materia que fluye, tiene volumen constante y toma la forma de su envase.

litro **(pág. 35)** Unidad de volumen del sistema métrico; equivale a un decímetro cúbico.

M

mass **(p. 9)** A measure that reflects the amount of matter.

mass defect **(p. 877)** The difference in mass between a nucleus and its component nucleons.

mass number **(p. 117)** The number after an element's name, representing the sum of its protons and neutrons.

matter **(p. 4)** Anything that has mass and takes up space.

melting point **(p. 426)** For a crystalline solid, the temperature at which the forces holding a crystal lattice together are broken and it becomes a liquid.

metabolism **(p. 844)** The sum of the many chemical reactions that occur in living cells.

metal **(p. 177)** An element that is solid at room temperature, a good conductor of heat and electricity, and generally is shiny; most metals are ductile and malleable.

metallic bond **(p. 225)** The attraction of a metallic cation for delocalized electrons.

metalloid **(p. 181)** An element that has physical and chemical properties of both metals and nonmetals.

meter **(p. 33)** The SI base unit for length.

masa **(pág. 9)** Medida que refleja la cantidad de materia.

defecto másico **(pág. 877)** La diferencia de masa entre un núcleo y los nucleones que lo componen.

número de masa **(pág. 117)** El número que va después del nombre de un elemento; representa la suma de sus protones y neutrones.

materia **(pág. 4)** Cualquier cosa que tiene masa y ocupa espacio.

punto de fusión **(pág. 426)** Para un sólido cristalino, es la temperatura a la que se rompen las fuerzas que mantienen unida la red cristalina y el sólido se convierte en líquido.

metabolismo **(pág. 844)** El conjunto de las numerosas reacciones químicas que ocurren en las células vivas.

metal **(pág. 177)** Elemento sólido a temperatura ambiente, es buen conductor de calor y electricidad y generalmente es brillante; la mayoría de los metales son dúctiles y maleables.

enlace metálico **(pág. 225)** Atracción de un catión metálico por los electrones deslocalizados.

metaloide **(pág. 181)** Elementos que tienen las propiedades físicas y químicas de metales y de no metales.

metro **(pág. 33)** Unidad básica de longitud del SI.

method of initial rates (p. 576) Determines the reaction order by comparing the initial rates of a reaction carried out with varying reactant concentrations.

miscible (p. 479) Describes two liquids that are soluble in each other.

mixture (p. 80) A physical blend of two or more pure substances in any proportion in which each substance retains its individual properties; can be separated by physical means.

model (p. 10) A visual, verbal, and/or mathematical explanation of data collected from many experiments.

molality (p. 487) The ratio of the number of moles of solute dissolved in one kilogram of solvent; also known as molal concentration.

molar enthalpy (heat) of fusion (p. 530) The amount of heat required to melt one mole of a solid substance.

molar enthalpy (heat) of vaporization (p. 530) The amount of heat required to vaporize one mole of a liquid.

molarity (p. 482) The number of moles of solute dissolved per liter of solution; also known as molar concentration.

molar mass (p. 326) The mass in grams of one mole of any pure substance.

molar volume (p. 452) For a gas, the volume that one mole occupies at 0.00°C and 1.00 atm pressure.

mole (p. 321) The SI base unit used to measure the amount of a substance, abbreviated mol; the number of carbon atoms in exactly 12 g of pure carbon; one mole is the amount of a pure substance that contains 6.02×10^{23} representative particles.

molecular formula (p. 346) A formula that specifies the actual number of atoms of each element in one molecule of a substance.

molecule (p. 241) Forms when two or more atoms covalently bond and is lower in potential energy than its constituent atoms.

mole fraction (p. 488) The ratio of the number of moles of solute in solution to the total number of moles of solute and solvent.

mole ratio (p. 371) In a balanced equation, the ratio between the numbers of moles of any two substances.

monatomic ion (p. 218) An ion formed from only one atom.

monomer (p. 810) A molecule from which a polymer is made.

monosaccharides (p. 832) The simplest carbohydrates, also called simple sugars.

N

net ionic equation (p. 301) An ionic equation that includes only the particles that participate in the reaction.

neutralization reaction (p. 659) A reaction in which an acid and a base react in aqueous solution to produce a salt and water.

método de las velocidades iniciales (pág. 576) Determina el orden de la reacción al comparar las velocidades iniciales de una reacción realizada con diversas concentraciones de reactivo.

miscible (pág. 479) Describe dos líquidos que son solubles entre sí.

mezcla (pág. 80) Combinación física de dos o más sustancias puras en cualquier proporción en la que cada sustancia retiene sus propiedades individuales; las sustancias se pueden separar por medios físicos.

modelo (pág. 10) Explicación matemática, verbal o visual de datos recolectados en muchos experimentos.

molalidad (pág. 487) La razón del número de moles de soluto disueltos en un kilogramo de disolvente; también se conoce como concentración molal.

entalpía (calor) molar de fusión (pág. 530) Cantidad requerida de calor para fundir un mol de una sustancia sólida.

entalpía (calor) molar de vaporización (pág. 530) Cantidad requerida de calor para vaporizar un mol de un líquido.

molaridad (pág. 482) Número de moles de soluto disueltos por litro de solución; también se conoce como concentración molar.

masa molar (pág. 326) Masa en gramos de un mol de cualquier sustancia pura.

volumen molar (pág. 452) Para un gas, es el volumen que ocupa un mol a 0.00°C y una presión de 1.00 atm.

mol (pág. 321) Unidad básica del SI para medir la cantidad de una sustancia, se abrevia mol; el número de átomos de carbono en 12 g exactos de carbono puro; un mol es la cantidad de sustancia pura que contiene 6.02×10^{23} partículas representativas.

fórmula molecular (pág. 346) Fórmula que especifica el número real de átomos de cada elemento en una molécula de la sustancia.

molécula (pág. 241) Se forma cuando dos o más átomos se unen covalentemente y posee menor energía potencial que los átomos que la conforman.

fracción molar (pág. 488) La razón del número de moles de soluto en solución al número total de moles de soluto y disolvente.

razón molar (pág. 371) En una ecuación equilibrada, se refiere a la razón entre el número de moles de dos sustancias cualesquiera.

ion monoatómico (pág. 218) Ion formado de un sólo átomo.

monómero (pág. 810) Molécula a partir de la cual se forma un polímero.

monosacáridos (pág. 832) Los carbohidratos más simples; se llaman también azúcares simples.

ecuación iónica neta (pág. 301) Ecuación iónica que incluye sólo las partículas que participan en la reacción.

reacción de neutralización (pág. 659) Reacción en la que un ácido y una base reaccionan en una solución acuosa para producir sal y agua.

neutron (p. 113) A neutral, subatomic particle in an atom's nucleus that has a mass nearly equal to that of a proton.

noble gas (p. 180) An extremely unreactive group 18 element.

nonmetals (p. 180) Elements that are generally gases or dull, brittle solids that are poor conductors of heat and electricity.

nuclear equation (p. 123) A type of equation that shows the atomic number and mass number of the particles involved.

nuclear fission (p. 878) The splitting of a nucleus into smaller, more stable fragments, accompanied by a large release of energy.

nuclear fusion (p. 883) The process of binding smaller atomic nuclei into a single, larger, and more stable nucleus.

nuclear reaction (p. 122) A reaction that involves a change in the nucleus of an atom.

nucleic acid (p. 840) A nitrogen-containing biological polymer that is involved in the storage and transmission of genetic information.

nucleons (p. 865) The positively charged protons and neutral neutrons contained in an atom's nucleus.

nucleotide (p. 840) The monomer that makes up a nucleic acid; consists of a nitrogen base, an inorganic phosphate group, and a five-carbon monosaccharide sugar.

nucleus (p. 112) The extremely small, positively charged, dense center of an atom that contains positively charged protons and neutral neutrons.

neutrón (pág. 113) Partícula subatómica neutral en el núcleo de un átomo que tiene una masa casi igual a la de un protón.

gas noble (pág. 180) Elemento extremadamente no reactivo del grupo 18.

no metales (pág. 180) Elementos que generalmente son gases o sólidos quebradizos, sin brillo y malos conductores de calor y electricidad.

ecuación nuclear (pág. 123) Tipo de ecuación que muestra el número atómico y el número de masa de las partículas involucradas.

fisión nuclear (pág. 878) Ruptura de un núcleo en fragmentos más pequeños y más estables; se acompaña de una gran liberación de energía.

fusión nuclear (pág. 883) Proceso de unión de núcleos atómicos pequeños en un solo núcleo más grande y más estable.

reacción nuclear (pág. 122) Reacción que implica un cambio en el núcleo de un átomo.

ácido nucleico (pág. 840) Polímero biológico que contiene nitrógeno y que participa en el almacenamiento y transmisión de información genética.

nucleones (pág. 865) Los protones de carga positiva y los neutrones sin carga que contiene el núcleo de un átomo.

nucleótido (pág. 840) Monómeros que forman los ácidos nucleicos; consisten de una base nitrogenada, un grupo fosfato inorgánico y un azúcar monosacárido de cinco carbonos.

núcleo (pág. 112) El diminuto y denso centro con carga positiva de un átomo; contiene protones con su carga positiva y neutrones sin carga.

O

octet rule (p. 193) States that atoms lose, gain, or share electrons in order to acquire the stable electron configuration of a noble gas.

optical isomers (p. 768) Result from different arrangements of four different groups around the same carbon atom and have the same physical and chemical properties except in chemical reactions where chirality is important.

optical rotation (p. 769) An effect that occurs when polarized light passes through a solution containing an optical isomer and the plane of polarization is rotated to the right by a d-isomer or to the left by an l-isomer.

organic compounds (p. 745) All compounds that contain carbon with the primary exceptions of carbon oxides, carbides, and carbonates, all of which are considered inorganic.

osmosis (p. 504) The diffusion of solvent particles across a semipermeable membrane from an area of higher solvent concentration to an area of lower solvent concentration.

osmotic pressure (p. 504) The pressure caused when water molecules move into or out of a solution.

regla del octeto (pág. 193) Establece que los átomos pierden, ganan o comparten electrones para adquirir la configuración electrónica estable de un gas noble.

isómeros ópticos (pág. 768) Son resultado de los distintos ordenamientos que adquieren los cuatro grupos diferentes que rodean a un mismo átomo de carbono; todos poseen las mismas propiedades químicas y físicas, excepto en las reacciones químicas donde la quiralidad es importante.

rotación óptica (pág. 769) Efecto que ocurre cuando la luz polarizada atraviesa una solución que contiene un isómero óptico y el plano de polarización rota a la derecha en los isómeros dextrógiros (−d) y a la izquierda en los isómeros levógiros (−l).

compuestos orgánicos (pág. 745) Todo compuesto que contiene carbono; las excepciones más importantes son los óxidos de carbono, los carburos y los carbonatos, todos los cuales se consideran inorgánicos.

osmosis (pág. 504) Difusión de partículas de disolvente a través de una membrana semipermeable hacia el área donde la concentración del disolvente es menor.

presión osmótica (pág. 504) La presión que causan las moléculas de agua al entrar o salir de una solución.

oxidation (p. 681) The complete or partial loss of electrons from a reacting substance; increases an atom's oxidation number.

oxidation number (p. 219) The positive or negative charge of a monatomic ion.

oxidation-number method (p. 689) The technique that can be used to balance more difficult redox reactions, based on the fact that the number of electrons or ions transferred from atoms must equal the number of electrons or ions accepted by other atoms.

oxidation-reduction reaction (p. 680) Any chemical reaction in which electrons are transferred from one substance to another; also called a redox reaction.

oxidizing agent (p. 683) The substance that oxidizes another substance by accepting its electrons.

oxyacid (p. 250) Any acid that contains hydrogen and an oxyanion.

oxyanion (ahk see AN i ahn) (p. 222) A polyatomic ion composed of an element, usually a nonmetal, bonded to one or more oxygen atoms.

oxidación (pág. 681) Pérdida de electrones de los átomos de una sustancia; aumenta el número de oxidación de un átomo.

número de oxidación (pág. 219) La carga positiva o negativa de un ion monoatómico.

método del número de oxidación (pág. 689) Técnica que sirve para equilibrar las reacciones redox más difíciles; se basa en el hecho de que el número de electrones transferidos por los átomos debe ser igual al número de electrones aceptados por otros átomos.

reacción de oxidación-reducción (pág. 680) Toda reacción química en la que sucede transferencia de electrones de una sustancia a otro; también se llama reacción redox.

agente oxidante (pág. 683) Sustancia que oxida otra sustancia al aceptar sus electrones.

oxiácido (pág. 250) Todo ácido que contiene hidrógeno y un oxianión.

oxianión (pág. 222) Ion poliatómico compuesto de un elemento, generalmente un no metal, unido a uno o a más átomos de oxígeno.

P

parent chain (p. 753) The longest continuous chain of carbon atoms in a branched-chain alkane, alkene, or alkyne.

pascal (p. 407) The SI unit of pressure; one pascal (Pa) is equal to a force of one newton per square meter.

Pauli exclusion principle (p. 157) States that a maximum of two electrons can occupy a single atomic orbital but only if the electrons have opposite spins.

penetrating power (p. 864) The ability of radiation to pass through matter.

peptide (p. 828) A chain of two or more amino acids linked by peptide bonds.

peptide bond (p. 828) The amide bond that joins two amino acids.

percent by mass (p. 87) A percentage determined by the ratio of the mass of each element to the total mass of the compound.

percent composition (p. 342) The percent by mass of each element in a compound.

percent error (p. 48) The ratio of an error to an accepted value.

percent yield (p. 386) The ratio of actual yield (from an experiment) to theoretical yield (from stoichiometric calculations) expressed as a percent.

period (p. 177) A horizontal row of elements in the modern periodic table.

periodic law (p. 176) States that when the elements are arranged by increasing atomic number, there is a periodic repetition of their properties.

periodic table (p. 85) A chart that organizes all known elements into a grid of horizontal rows (periods) and vertical columns (groups or families) arranged by increasing atomic number.

cadena principal (pág. 753) La cadena continua más larga de átomos de carbono en un alcano, un alqueno o un alquino ramificados.

pascal (pág. 407) La unidad SI de presión; un pascal (Pa) es igual a una fuerza de un newton por metro cuadrado.

principio de exclusión de Pauli (pág. 157) Establece que cada orbital atómico sólo puede ser ocupado por un máximo de dos electrones, pero sólo si los electrones tienen giros opuestos.

poder de penetración (pág. 864) La capacidad de la radiación de atravesar la materia.

péptido (pág. 828) Cadena de dos o más aminoácidos unidos por enlaces peptídicos.

enlace peptídico (pág. 828) Enlace amida que une dos aminoácidos.

porcentaje en masa (pág. 87) Porcentaje determinado por la razón de la masa de cada elemento respecto a la masa total del compuesto.

composición porcentual (pág. 342) Porcentaje en masa de cada elemento en un compuesto.

porcentaje de error (pág. 48) La razón del error al valor aceptado.

porcentaje de rendimiento (pág. 386) Razón del rendimiento real (de un experimento) al rendimiento teórico (de cálculos estequiométricos) expresada como porcentaje.

período (pág. 177) Fila horizontal de elementos en la tabla periódica moderna.

ley periódica (pág. 176) Establece que al ordenar los elementos por número atómico en sentido ascendente, existe una repetición periódica de sus propiedades.

tabla periódica (pág. 85) Tabla en la que se organizan todos los elementos conocidos en una cuadrícula de filas horizontales (períodos) y columnas verticales (grupos o familias), ordenados según su número atómico en sentido ascendente.

pH **(p. 652)** The negative logarithm of the hydrogen ion concentration of a solution; acidic solutions have pH values between 0 and 7, basic solutions have values between 7 and 14, and a solution with a pH of 7.0 is neutral.

phase change **(p. 76)** A transition of matter from one state to another.

phase diagram **(p. 429)** A graph of pressure versus temperature that shows which phase a substance exists in under different conditions of temperature and pressure.

phospholipid **(p. 838)** A triglyceride in which one of the fatty acids is replaced by a polar phosphate group.

photoelectric effect **(p. 142)** A phenomenon in which photoelectrons are emitted from a metal's surface when light of a certain frequency shines on the surface.

photon **(p. 143)** A particle of electromagnetic radiation with no mass that carries a quantum of energy.

photosynthesis **(p. 846)** The complex process that converts energy from sunlight to chemical energy in the bonds of carbohydrates.

physical change **(p. 76)** A type of change that alters the physical properties of a substance but does not change its composition.

physical property **(p. 73)** A characteristic of matter that can be observed or measured without changing the sample's composition—for example, density, color, taste, hardness, and melting point.

pi bond **(p. 245)** A bond that is formed when parallel orbitals overlap to share electrons.

Planck's constant (*h*) **(p. 142)** 6.626×10^{-34} J•s, where J is the symbol for the joule.

plastic **(p. 789)** A polymer that can be heated and molded while relatively soft.

pOH **(p. 652)** The negative logarithm of the hydroxide ion concentration of a solution; a solution with a pOH above 7.0 is acidic, a solution with a pOH below 7.0 is basic, and a solution with a pOH of 7.0 is neutral.

polar covalent bond **(p. 266)** A type of bond that forms when electrons are not shared equally.

polyatomic ion **(p. 221)** An ion made up of two or more atoms bonded together that acts as a single unit with a net charge.

polymerization reaction **(p. 810)** A reaction in which monomer units are bonded together to form a polymer.

polymers **(p. 809)** Large molecules formed by combining many repeating structural units (monomers); are synthesized through addition or condensation reactions.

polysaccharide **(p. 833)** A complex carbohydrate, which is a polymer of simple sugars that contains 12 or more monomer units.

positron **(p. 868)** A particle that has the same mass as an electron but an opposite charge.

positron emission **(p. 868)** A radioactive decay process in which a proton in the nucleus is converted into a neutron and a positron, and then the positron is emitted from the nucleus.

pH **(pág. 652)** El logaritmo negativo de la concentración de iones hidrógeno de una solución; las soluciones ácidas poseen valores de pH entre 0 y 7, las soluciones básicas tienen valores entre 7 y 14 y una solución con un pH de 7.0 es neutra.

cambio de fase **(pág. 76)** La transición de la materia de un estado a otro.

diagrama de fase **(pág. 429)** Gráfica de presión contra temperatura que muestra la fase en la que se encuentra una sustancia bajo distintas condiciones de temperatura y presión.

fosfolípido **(pág. 838)** Triglicérido en el que uno de los ácidos grasos es sustituido por un grupo fosfato polar.

efecto fotoeléctrico **(pág. 142)** Fenómeno en el cual la superficie de un metal emiten fotoelectrones cuando una luz de cierta frecuencia ilumina su superficie.

fotón **(pág. 143)** Partícula de radiación electromagnética sin masa que transporta un cuanto de energía.

fotosíntesis **(pág. 846)** Proceso complejo que convierte la energía de la luz solar en la energía química de los enlaces en carbohidratos.

cambio físico **(pág. 76)** Tipo de cambio que altera las propiedades físicas de una sustancia pero no cambia su composición.

propiedad física **(pág. 73)** Característica de la materia que se puede observar o medir sin cambiar la composición de una muestra de la materia; por ejemplo, la densidad, el color, el sabor, la dureza y el punto de fusión.

enlace pi **(pág. 245)** Enlace que se forma cuando orbitales paralelos se superponen para compartir electrones.

constante de Planck (*h*) **(pág. 142)** 6.626×10^{-34} J•s, donde J es el símbolo de julios.

plástico **(pág. 789)** Polímero que se puede calentar y moldear mientras esté relativamente suave.

pOH **(pág. 652)** El logaritmo negativo de la concentración de iones hidróxido de una solución; una solución con un pOH mayor que 7.0 es ácida, una solución con un pOH menor que 7.0 es básica y una solución con un pOH de 7.0 es neutra.

enlace covalente polar **(pág. 266)** Tipo de enlace que se forma cuando los electrones no se comparten de manera equitativa.

ion poliatómico **(pág. 221)** Ion compuesto de dos o más átomos unidos entre sí que actúan como una unidad con carga neta.

reacción de polimerización **(pág. 810)** Reacción en la cual los monómeros se unen para formar un polímero.

polímeros **(pág. 809)** Moléculas grandes formadas por la unión de muchas unidades estructurales repetidas (monómeros); se sintetizan a través de reacciones de adición o de condensación.

polisacárido **(pág. 833)** Carbohidrato complejo; es un polímero de azúcares simples que contiene 12 ó más monómeros.

positrón **(pág. 868)** Partícula que tiene la misma masa que un electrón pero carga opuesta.

emisión de positrones **(pág. 868)** Proceso de desintegración radiactiva en el que un protón del núcleo se convierte en un neutrón y un positrón y luego el positrón es emitido del núcleo.

precipitate (p. 296) A solid produced during a chemical reaction in a solution.

precision (p. 47) Refers to how close a series of measurements are to one another; precise measurements show little variation over a series of trials but might not be accurate.

pressure (p. 406) Force applied per unit area.

primary battery (p. 720) A type of battery that produces electric energy by redox reactions that are not easily reversed, delivers current until the reactants are gone, and then is discarded.

principal energy levels (p. 153) The major energy levels of an atom.

principal quantum number (*n*) (p. 153) Assigned by the quantum mechanical model to indicate the relative sizes and energies of atomic orbitals.

product (p. 283) A substance formed during a chemical reaction.

protein (p. 826) An organic polymer made up of animo acids linked together by peptide bonds that can function as an enzyme, transport important chemical substances, or provide structure in organisms.

proton (p. 113) A subatomic particle in an atom's nucleus that has a positive charge of 1+.

pure research (p. 17) A type of scientific investigation that seeks to gain knowledge for the sake of knowledge itself.

precipitado (pág. 296) Sólido que se produce durante una reacción química en una solución.

precisión (pág. 47) Se refiere a la cercanía de una serie de medidas entre sí; las medidas precisas muestran poca variación durante una serie de pruebas, incluso si no son exactas.

presión (pág. 406) Fuerza aplicada por unidad de área.

batería primaria (pág. 720) Tipo de batería que produce energía eléctrica por reacciones redox que no son fácilmente reversibles, produce corriente hasta que se agotan los reactivos y luego se desecha.

niveles energéticos principales (pág. 153) Los niveles energéticos más importantes de un átomo.

número cuántico principal (*n*) (pág. 153) Asignado por el modelo mecánico cuántico para indicar el tamaño y la energía relativas de los orbitales atómicos.

producto (pág. 283) Sustancia que se forma durante una reacción química.

proteína (pág. 826) Polímero orgánico compuesto de aminoácidos unidos por enlaces peptídicos; puede funcionar como enzima, transportar sustancias químicas importantes o ser parte de la estructura en los organismos.

protón (pág. 113) Partícula subatómica en el núcleo de un átomo con carga positiva 1+.

investigación pura (pág. 17) Tipo de investigación científica que busca obtener conocimiento sin otro interés que satisfacer el interés científico.

Q

qualitative data (p. 13) Information describing color, odor, shape, or some other physical characteristic.

quantitative data (p. 13) Numerical information describing how much, how little, how big, how tall, or how fast.

quantum (p. 141) The minimum amount of energy that can be gained or lost by an atom.

quantum mechanical model of the atom (p. 152) An atomic model in which electrons are treated as waves; also called the wave mechanical model of the atom.

quantum number (p. 147) The number assigned to each orbit of an electron.

datos cualitativos (pág. 13) Información que describe el color, el olor, la forma o alguna otra característica física.

datos cuantitativos (pág. 13) Información numérica que describe cantidad, tamaño o rapidez.

cuanto (pág. 141) La cantidad mínima de energía que puede ganar o perder un átomo.

modelo mecánico cuántico del átomo (pág. 152) Modelo atómico en el cual los electrones se estudian como si fueran ondas; también se denomina modelo mecánico ondulatorio del átomo.

número cuántico (pág. 147) Número que se asigna a cada órbita de un electrón.

R

radiation (p. 122) The rays and particles—alpha and beta particles and gamma rays—that are emitted by radioactive materials.

radioactive decay (p. 122) A spontaneous process in which unstable nuclei lose energy by emitting radiation.

radioactive decay series (p. 870) A series of nuclear reactions that starts with an unstable nucleus and results in the formation of a stable nucleus.

radioactivity (p. 122) The process in which some substances spontaneously emit radiation.

radiochemical dating (p. 873) The process that is used to determine the age of an object by measuring the amount of a certain radioisotope remaining in that object.

radiación (pág. 122) Los rayos y partículas que emiten los materiales radiactivos (partículas alfa y beta y rayos gamma).

desintegración radiactiva (pág. 122) Proceso espontáneo en el que los núcleos inestables pierden energía al emitir radiación.

serie de desintegración radiactiva (pág. 870) Serie de reacciones nucleares que empieza con un núcleo inestable y produce la formación de un núcleo estable.

radiactividad (pág. 122) Proceso en el que algunas sustancias emiten radiación espontáneamente.

datación radioquímica (pág. 873) Proceso que sirve para determinar la edad de un objeto al medir la cantidad restante de cierto radioisótopo en dicho objeto.

radioisotopes **(p. 861)** Isotopes of atoms that have unstable nuclei and emit radiation to attain more stable atomic configurations.

radiotracer **(p. 887)** An isotope that emits non-ionizing radiation and is used to signal the presence of an element or specific substance; can be used to analyze complex chemical reactions mechanisms and to diagnose disease.

rate-determining step **(p. 581)** The slowest elementary step in a complex reaction; limits the instantaneous rate of the overall reaction.

rate law **(p. 574)** The mathematical relationship between the rate of a chemical reaction at a given temperature and the concentrations of reactants.

reactant **(p. 283)** The starting substance in a chemical reaction.

reaction mechanism **(p. 580)** The complete sequence of elementary steps that make up a complex reaction.

reaction order **(p. 575)** For a reactant, describes how the rate is affected by the concentration of that reactant.

reaction rate **(p. 561)** The change in concentration of a reactant or product per unit time, generally calculated and expressed in moles per liter per second.

redox reaction **(p. 680)** An oxidation-reduction reaction.

reducing agent **(p. 683)** The substance that reduces another substance by losing electrons.

reduction **(p. 681)** The complete or partial gain of electrons by a reacting substance; decreases an atom's oxidation number.

reduction potential **(p. 711)** The tendency of a substance to gain electrons.

representative elements **(p. 177)** Elements from groups 1, 2, and 13–18 in the modern periodic table, possessing a wide range of chemical and physical properties.

resonance **(p. 258)** Condition that occurs when more than one valid Lewis structure exists for the same molecule.

reversible reaction **(p. 595)** A reaction that can take place in both the forward and reverse directions; leads to an equilibrium state where the forward and reverse reactions occur at equal rates and the concentrations of reactants and products remain constant.

radioisótopos **(pág. 861)** Isótopos de átomos que poseen núcleos inestables y emiten radiación para obtener una configuración atómica más estable.

radiolocalizador **(pág. 887)** Isótopo que emite radiación no ionizante y se utiliza para señalar la presencia de un elemento o sustancia específica; se usan para analizar los mecanismos de reacciones químicas complejas y para diagnosticar enfermedades.

paso determinante de la velocidad de reacción **(pág. 581)** El paso elemental más lento en una reacción compleja; limita la velocidad instantánea de la reacción general.

ley de velocidad de la reacción **(pág. 574)** Relación matemática entre la velocidad de una reacción química a una temperatura dada y las concentraciones de los reactivos.

reactivo **(pág. 283)** Sustancia inicial en una reacción química.

mecanismo de reacción **(pág. 580)** Sucesión completa de pasos elementales que componen una reacción compleja.

orden de la reacción **(pág. 575)** Describe cómo la concentración de un reactivo afecta la velocidad de la reacción para dicho reactivo.

tasa de reacción **(pág. 561)** Cambio en la concentración de un reactivo o producto por unidad de tiempo, generalmente se calcula y expresa en moles por litro por segundo.

reacción redox **(pág. 680)** Una reacción de oxidorreducción.

agente reductor **(pág. 683)** Sustancia que reduce otra sustancia al perder electrones.

reducción **(pág. 681)** Ganancia de electrones por los átomos de una sustancia; reduce el número de oxidación de los átomos.

potencial de reducción **(pág. 711)** Tendencia de una sustancia a ganar electrones.

elementos representativos **(pág. 177)** Elementos de los grupos 1, 2 y 13 a 18 de la tabla periódica moderna; poseen una gran variedad de propiedades químicas y físicas.

resonancia **(pág. 258)** Condición que ocurre cuando existe más de una estructura válida de Lewis para una misma molécula.

reacción reversible **(pág. 595)** Reacción que puede ocurrir en direcciones normal e inversa; produce un estado de equilibrio donde las reacciones en sentido normal e inverso ocurren a tasas iguales, ocasionando que la concentración de reactivos y productos permanezcan constantes.

S

salt **(p. 659)** An ionic compound made up of a cation from a base and an anion from an acid.

salt bridge **(p. 709)** A pathway constructed to maintain solution neutrality by allowing positive and negative ions to move from one solution to another.

salt hydrolysis **(p. 665)** The process in which anions of the dissociated salt accept hydrogen ions from water, or the cations of the dissociated salt donate hydrogen ions to water.

sal **(pág. 659)** Compuesto iónico formado por un catión proveniente de una base y un anión proveniente de un ácido.

puente salino **(pág. 709)** Medio que permite el movimiento de iones positivos y negativos de una solución a otra.

hidrólisis de sales **(pág. 665)** Proceso en el que los aniones de una sal disociada aceptan iones hidrógeno del agua o en el que los cationes de la sal disociada donan iones hidrógeno al agua.

saponification (suh pahn ih fih KAY shuhn) (p. 837) The hydrolysis of the ester bonds of a triglyceride using an aqueous solution of a strong base to form carboxylate salts and glycerol.

saturated hydrocarbon (p. 746) A hydrocarbon that contains only single bonds.

saturated solution (p. 493) Contains the maximum amount of dissolved solute for a given amount of solvent at a specific temperature and pressure.

scientific law (p. 16) Describes a relationship in nature that is supported by many experiments.

scientific methods (p. 12) A systematic approach used in scientific study; an organized process used by scientists to do research and to verify the work of others.

scientific notation (p. 40) Expresses any number as a number between 1 and 10 (known as a coefficient) multiplied by 10 raised to a power (known as an exponent).

second (p. 33) The SI base unit for time.

second law of thermodynamics (p. 543) The spontaneous processes always proceed in such a way that the entropy of the universe increases.

secondary battery (p. 720) A rechargeable battery that depends on reversible redox reactions.

sigma bond (p. 244) A single covalent bond that is formed when an electron pair is shared by the direct overlap of bonding orbitals.

significant figures (p. 50) The number of all known digits reported in measurements plus one estimated digit.

single-replacement reaction (p. 293) A chemical reaction that occurs when the atoms of one element replace the atoms of another element in a compound.

solid (p. 71) A form of matter that has its own definite shape and volume, is incompressible, and expands only slightly when heated.

solubility (p. 614) The maximum amount of solute that will dissolve in a given amount of solvent at a specific temperature and pressure.

solubility product constant (p. 614) K_{sp}, which is an equilibrium constant for the dissolving of a sparingly soluble ionic compound in water.

soluble (p. 479) Describes a substance that can be dissolved in a given solvent.

solute (p. 299) One or more substances dissolved in a solution.

solution (p. 81) A uniform mixture that can contain solids, liquids, or gases; also called a homogeneous mixture.

solvation (p. 489) The process of surrounding solute particles with solvent particles to form a solution; occurs only where and when the solute and solvent particles come in contact with each other.

solvent (p. 299) The substance that dissolves a solute to form a solution; the most plentiful substance in the solution.

species (p. 693) Any kind of chemical unit involved in a process.

saponificación (pág. 837) La hidrólisis de los enlaces éster de un triglicérido, usando una solución acuosa de una base fuerte, para formar sales de carboxilato y glicerol.

hidrocarburo saturado (pág. 746) Hidrocarburo que sólo contiene enlaces sencillos.

solución saturada (pág. 493) Solución que contiene la cantidad máxima de soluto disuelto para una cantidad dada de disolvente a una temperatura y presión específicas.

ley científica (pág. 16) Describe una relación natural demostrada en muchos experimentos.

métodos científicos (pág. 12) Enfoque sistemático que se usa en los estudios científicos; proceso organizado que siguen los científicos para realizar sus investigaciones y verificar el trabajo realizado por otros científicos.

notación científica (pág. 40) Expresa cualquier número como un número entre 1 y 10 (conocido como coeficiente) multiplicado por 10 elevado a alguna potencia (conocida como exponente).

segundo (pág. 33) Unidad básica de tiempo del SI.

segunda ley de la termodinámica (pág. 543) Los procesos espontáneos siempre proceden de una forma que aumenta la entropía del universo.

batería secundaria (pág. 720) Batería recargable que depende de reacciones redox reversibles.

enlace sigma (pág. 244) Enlace covalente simple que se forma cuando se comparte un par de electrones mediante la superposición directa de los orbitales del enlace.

cifras significativas (pág. 50) El número de dígitos conocidos que se reportan en medidas, más un dígito estimado.

reacción de sustitución simple (pág. 293) Reacción química que ocurre cuando los átomos de un elemento reemplazan a los átomos de otro elemento en un compuesto.

sólido (pág. 71) Forma de la materia que tiene su propia forma y volumen, es incompresible y sólo se expande levemente cuando se calienta.

solubilidad (pág. 614) Cantidad máxima de soluto que se disolverá en una cantidad dada de disolvente a una temperatura y presión específicas.

constante de producto de solubilidad (pág. 614) Se representa como K_{sp}; es la constante de equilibrio para la disolución de un compuesto iónico moderadamente soluble en agua.

soluble (pág. 479) Describe una sustancia que se puede disolver en un disolvente dado.

soluto (pág. 299) Una o más sustancias disueltas en una solución.

solución (pág. 81) Mezcla uniforme que puede contener sólidos, líquidos o gases; llamada también mezcla homogénea.

solvatación (pág. 489) Proceso de rodear las partículas de soluto con partículas del disolvente para formar una solución; ocurre sólo en los lugares y en el momento en que las partículas de soluto y disolvente entran en contacto.

disolvente (pág. 299) Sustancia que disuelve un soluto para formar una solución; la sustancia más abundante en la solución.

especie (pág. 693) Cualquier clase de unidad química que participa en un proceso.

specific heat **(p. 519)** The amount of heat required to raise the temperature of one gram of a given substance by one degree Celsius.

specific rate constant **(p. 574)** A numerical value that relates reaction rate and concentration of reactant at a specific temperature.

spectator ion **(p. 301)** Ion that does not participate in a reaction.

spontaneous process **(p. 542)** A physical or chemical change that occurs without outside intervention and may require energy to be supplied to begin the process.

standard enthalpy (heat) of formation **(p. 537)** The change in enthalpy that accompanies the formation of one mole of a compound in its standard state from its constituent elements in their standard states.

standard hydrogen electrode **(p. 711)** The standard electrode against which the reduction potential of all electrodes can be measured.

states of matter **(p. 71)** The physical forms in which all matter naturally exists on Earth—most commonly as a solid, a liquid, or a gas.

stereoisomers **(p. 766)** A class of isomers whose atoms are bonded in the same order but are arranged differently in space.

steroids **(p. 839)** Lipids that have multiple cyclic rings in their structures.

stoichiometry **(p. 368)** The study of quantitative relationships between the amounts of reactants used and products formed by a chemical reaction; is based on the law of conservation of mass.

strong acid **(p. 644)** An acid that ionizes completely in aqueous solution.

strong base **(p. 648)** A base that dissociates entirely into metal ions and hydroxide ions in aqueous solution.

strong nuclear force **(p. 865)** A force that acts on subatomic particles that are extremely close together.

structural formula **(p. 253)** A molecular model that uses symbols and bonds to show relative positions of atoms; can be predicted for many molecules by drawing the Lewis structure.

structural isomers **(p. 765)** A class of isomers whose atoms are bonded in different orders with the result that they have different chemical and physical properties despite having the same formula.

sublimation **(p. 83)** The energy-requiring process by which a solid changes directly to a gas without first becoming a liquid.

substance **(p. 5)** Matter that has a definite composition; also known as a chemical.

substituent groups **(p. 753)** The side branches that extend from the parent chain; they appear to substitute for a hydrogen atom in the straight chain.

substitution reaction **(p. 790)** A reaction of organic compounds in which one atom or group of atoms in a molecule is replaced by another atom or group of atoms.

calor específico **(pág. 519)** Cantidad de calor requerida para elevar la temperatura de un gramo de una sustancia dada en un grado centígrado (Celsius).

constante de velocidad de la reacción **(pág. 574)** Valor numérico que relaciona la velocidad de la reacción y la concentración de reactivos a una temperatura específica.

ion espectador **(pág. 301)** Ion que no participa en una reacción.

proceso espontáneo **(pág. 542)** Cambio físico o químico que ocurre sin intervención externa; la iniciación del proceso puede requerir un suministro de energía.

entalpía (calor) estándar de formación **(pág. 537)** Cambio en la entalpía que acompaña la formación de un mol de un compuesto en su estado normal, a partir de sus elementos constituyentes en su estado normal.

electrodo normal de hidrógeno **(pág. 711)** Electrodo estándar que sirve de referencia para medir el potencial de reducción de todos los electrodos.

estados de la materia **(pág. 71)** Las formas físicas en las que la materia existe naturalmente en la Tierra, más comúnmente como sólido, líquido o gas.

estereoisómeros **(pág. 766)** Clase de isómeros cuyos átomos se unen en el mismo orden, pero con distinta disposición espacial.

esteroides **(pág. 839)** Lípidos con múltiples anillos en sus estructuras.

estequiometría **(pág. 368)** El estudio de las relaciones cuantitativas entre las cantidades de reactivos utilizados y los productos formados durante una reacción química; se basa en la ley de la conservación de la masa.

ácido fuerte **(pág. 644)** Ácido que se ioniza completamente en solución acuosa.

base fuerte **(pág. 648)** Base que se disocia enteramente en iones metálicos e iones hidróxido en solución acuosa.

fuerza nuclear fuerte **(pág. 865)** Fuerza que actúa sólo en las partículas subatómicas que se encuentran extremadamente cercanas.

fórmula estructural **(pág. 253)** Modelo molecular que usa símbolos y enlaces para mostrar las posiciones relativas de los átomos; esta fórmula se puede predecir para muchas moléculas al trazar su estructura de Lewis.

isómeros estructurales **(pág. 765)** Clase de isómeros cuyos átomos están unidos en distinto orden, por lo que tienen propiedades químicas y físicas diferentes a pesar de tener la misma fórmula.

sublimación **(pág. 83)** Proceso que requiere de energía en el que un sólido se convierte directamente en gas, sin convertirse primero en un líquido.

sustancia **(pág. 5)** Materia con una composición definida; también se conoce como sustancia química.

grupos sustituyentes **(pág. 753)** Las ramas laterales que se extienden desde la cadena principal y parecen sustituir un átomo de hidrógeno de la cadena recta.

reacción de sustitución **(pág. 790)** Reacción de compuestos orgánicos en la cual un átomo o un grupo de átomos en una molécula son sustituidos por otro átomo o grupo de átomos.

substrate (p. 830) A reactant in an enzyme-catalyzed reaction that binds to specific sites on enzyme molecules.

sustrato (pág. 830) Reactivo en una reacción catalizada por enzimas que se enlaza a sitios específicos en las moléculas de la enzima.

supersaturated solution (p. 494) Contains more dissolved solute than a saturated solution at the same temperature.

solución sobresaturada (pág. 494) Aquella que contiene más soluto disuelto que una solución saturada a la misma temperatura.

surface tension (p. 418) The energy required to increase the surface area of a liquid by a given amount; results from an uneven distribution of attractive forces.

tensión superficial (pág. 418) Energía requerida para aumentar el área superficial de un líquido en una cantidad dada; es producida por una distribución desigual de las fuerzas de atracción.

surfactant (p. 419) A compound, such as soap, that lowers the surface tension of water by disrupting hydrogen bonds between water molecules; also called a surface active agent.

surfactante (pág. 419) Compuesto, como el jabón, que reduce la tensión superficial del agua al romper los enlaces de hidrógeno entre las moléculas de agua; llamado también agente tensioactivo.

surroundings (p. 526) In thermochemistry, includes everything in the universe except the system.

alrededores (pág. 526) En termoquímica, incluye todo el universo a excepción del sistema.

suspension (p. 476) A type of heterogeneous mixture whose particles settle out over time and can be separated from the mixture by filtration.

suspensión (pág. 476) Tipo de mezcla heterogénea cuyas partículas se asientan con el tiempo y pueden separarse de la mezcla por filtración.

synthesis reaction (p. 289) A chemical reaction in which two or more substances react to yield a single product.

reacción de síntesis (pág. 289) Reacción química en la que dos o más sustancias reaccionan para generar un solo producto.

system (p. 526) In thermochemistry, the specific part of the universe containing the reaction or process being studied.

sistema (pág. 526) En termoquímica, se refiere a la parte específica del universo que contiene la reacción o el proceso en estudio.

T

technology (p. 9) The practical use of scientific information.

tecnología (pág. 9) Uso práctico de la información científica.

temperature (p. 403) A measure of the average kinetic energy of the particles in a sample of matter.

temperatura (pág. 403) Medida de la energía cinética promedio de las partículas en una muestra de materia.

theoretical yield (p. 385) In a chemical reaction, the maximum amount of product that can be produced from a given amount of reactant.

rendimiento teórico (pág. 385) La cantidad máxima de producto que se puede producir a partir de una cantidad dada de reactivo, durante una reacción química.

theory (p. 16) An explanation supported by many experiments; is still subject to new experimental data, can be modified, and is considered valid if it can be used to make predictions that are proven true.

teoría (pág. 16) Explicación respaldada por muchos experimentos; está sujeta a los resultados obtenidos en nuevos experimentos, se puede modificar y se considera válida si permite hacer predicciones verdaderas.

thermochemical equation (p. 529) A balanced chemical equation that includes the physical states of all the reactants and the energy change, usually expressed as the change in enthalpy.

ecuación termoquímica (pág. 529) Ecuación química equilibrada que incluye el estado físico de todos los reactivos y el cambio de energía, este último usualmente expresado como el cambio en entalpía.

thermochemistry (p. 525) The study of heat changes that accompany chemical reactions and phase changes.

termoquímica (pág. 525) El estudio de los cambios de calor que acompañan a las reacciones químicas y a los cambios de fase.

thermonuclear reaction (p. 883) A nuclear fusion reaction.

reacción termonuclear (pág. 883) Reacción de fusión nuclear.

thermoplastic (p. 813) A type of polymer that can be melted and molded repeatedly into shapes that are retained when it is cooled.

termoplástico (pág. 813) Tipo de polímero que se puede fundir y moldear repetidas veces en formas que el plástico mantiene al enfriarse.

thermosetting (p. 813) A type of polymer that can be molded when it is first prepared but when cool cannot be remelted.

fraguado (pág. 813) Tipo de polímero que se puede moldear la primera vez que es producido, pero que no puede fundirse de nuevo una vez que se ha enfriado.

titrant (p. 661) A solution of known concentration used to titrate a solution of unknown concentration; also called the standard solution.

solución tituladora (pág. 661) Solución de concentración conocida que se usa para titular una solución de concentración desconocida; también conocida como solución estándar.

titration (p. 660) The process in which an acid-base neutralization reaction is used to determine the concentration of a solution of unknown concentration.

titulación (pág. 660) Proceso en el que se usa una reacción de neutralización ácido-base para determinar la concentración de una solución de concentración desconocida.

transition elements (p. 177) Elements in groups 3–12 of the modern periodic table and are further divided into transition metals and inner transition metals.

transition metal (p. 180) An element in groups 3–12 that is contained in the d-block of the periodic table and, with some exceptions, is characterized by a filled outermost s orbital of energy level *n*, and filled or partially filled d orbitals of energy level $n - 1$.

transition state (p. 564) Term used to describe an activated complex because the activated complex is as likely to form reactants as it is to form products.

transmutation (p. 865) The conversion of an atom of one element to an atom of another element.

transuranium element (p. 876) An element with an atomic number of 93 or greater in the periodic table.

triglyceride (p. 836) Forms when three fatty acids are bonded to a glycerol backbone through ester bonds; can be either solid or liquid at room temperature.

triple point (p. 429) The point on a phase diagram representing the temperature and pressure at which the three phases of a substance (solid, liquid, and gas) can coexist.

Tyndall effect (TIHN duhl • ee FEKT) (p. 478) The scattering of light by colloidal particles.

elementos de transición (pág. 177) Elementos de los grupos 3 al 12 de la tabla periódica moderna; se subdividen en metales de transición y metales de transición interna.

metal de transición (pág. 180) Elementos de los grupos 3 al 12 del bloque d de la tabla periódica; con algunas excepciones, se caracterizan por tener lleno el orbital externo s del nivel de energía *n* y por tener orbitales d llenos o parcialmente llenos en el nivel de energía $n - 1$.

estado de transición (pág. 564) Término que se usa para describir un complejo activado por su probabilidad de formar tanto reactivos como productos.

transmutación (pág. 865) Conversión de un átomo de un elemento a un átomo de otro elemento.

elemento transuránico (pág. 876) Elementos de la tabla periódica con un número atómico igual o mayor que 93.

triglicérido (pág. 836) Se forma cuando tres ácidos grasos se enlazan a un cadena principal de glicerol por enlaces éster; puede ser sólido o líquido a temperatura ambiente.

punto triple (pág. 429) El punto en un diagrama de fase que representa la temperatura y la presión en la que coexisten las tres fases de una sustancia (sólido, líquido y gas).

efecto Tyndall (pág. 478) Dispersión de la luz causada por las partículas coloidales.

U

unit cell (p. 421) The smallest arrangement of atoms in a crystal lattice that has the same symmetry as the whole crystal; a small representative part of a larger whole.

universe (p. 526) In thermochemistry, is the system plus the surroundings.

unsaturated hydrocarbon (p. 746) A hydrocarbon that contains at least one double or triple bond between carbon atoms.

unsaturated solution (p. 493) Contains less dissolved solute for a given temperature and pressure than a saturated solution; has further capacity to hold more solute.

celda unitaria (pág. 421) El conjunto más pequeño de átomos en una red cristalina que posee la simetría de todo el cristal; pequeña parte representativa de un entero mayor.

universo (pág. 526) En termoquímica, se refiere el sistema más los alrededores.

hidrocarburo no saturado (pág. 746) Hidrocarburo que contiene por lo menos un enlace doble o triple entre sus átomos de carbono.

solución no saturada (pág. 493) Aquella que contiene menos soluto disuelto a una temperatura y presión dadas que una solución saturada; puede contener cantidades adicionales del soluto.

V

valence electrons (p. 161) The electrons in an atom's outermost orbitals; determine the chemical properties of an element.

vapor (p. 72) Gaseous state of a substance that is a liquid or a solid at room temperature.

vaporization (p. 426) The energy-requiring process by which a liquid changes to a gas or vapor.

vapor pressure (p. 427) The pressure exerted by a vapor over a liquid.

vapor pressure lowering (p. 499) The lowering of vapor pressure of a solvent by the addition of a nonvolatile solute to the solvent.

viscosity (p. 417) A measure of the resistance of a liquid to flow, which is affected by the size and shape of particles, and generally increases as the temperature decreases and as intermolecular forces increase.

electrones de valencia (pág. 161) Los electrones en el orbital más externo de un átomo; determinan las propiedades químicas de un elemento.

vapor (pág. 72) Estado gaseoso de una sustancia que es líquida o sólida a temperatura ambiente.

vaporización (pág. 426) Proceso que requiere energía en el que un líquido se convierte en gas o vapor.

presión de vapor (pág. 427) Presión que ejerce un vapor sobre un líquido.

disminución de la presión de vapor (pág. 499) Reducción de la presión de vapor de un disolvente por la adición de un soluto no volátil al disolvente.

viscosidad (pág. 417) Medida de la resistencia de un líquido a fluir; es afectada por el tamaño y la forma de las partículas y en general aumenta cuando disminuye temperatura y cuando aumentan las fuerzas intermoleculares.

voltaic cell (p. 709) A type of electrochemical cell that converts chemical energy into electrical energy by a spontaneous redox reaction.

VSEPR model (p. 261) Valence Shell Electron Pair Repulsion model, which is based on an arrangement that minimizes the repulsion of shared and unshared pairs of electrons around the central atom.

pila voltaica (pág. 709) Tipo de celda electroquímica que convierte la energía química en energía eléctrica mediante una reacción redox espontánea.

modelo RPCEV (pág. 261) Modelo de Repulsión de los Pares Electrónicos de la Capa de Valencia; se basa en un ordenamiento que minimiza la repulsión de los pares de electrones compartidos y no compartidos alrededor del átomo central.

W

wavelength (p. 137) The shortest distance between equivalent points on a continuous wave; is usually expressed in meters, centimeters, or nanometers.

wax (p. 838) A type of lipid that is formed by combining a fatty acid with a long-chain alcohol; is made by both plants and animals.

weak acid (p. 645) An acid that ionizes only partially in dilute aqueous solution.

weak base (p. 648) A base that ionizes only partially in dilute aqueous solution to form the conjugate acid of the base and hydroxide ion.

weight (p. 9) A measure of an amount of matter and also the effect of Earth's gravitational pull on that matter.

longitud de onda (pág. 137) La distancia más corta entre puntos equivalentes en una onda continua; se expresa generalmente en metros, centímetros o nanómetros.

cera (pág. 838) Tipo de lípido que se forma al combinarse un ácido graso con un alcohol de cadena larga; son elaborados por plantas y animales.

ácido débil (pág. 645) Ácido que se ioniza parcialmente en una solución acuosa diluida.

base débil (pág. 648) Base que se ioniza parcialmente en una solución acuosa diluida para formar el ácido conjugado de la base y el ion hidróxido.

peso (pág. 9) Medida de la cantidad de materia y también del efecto de la fuerza gravitatoria de la Tierra sobre esa materia.

X

X-ray (p. 864) A form of high-energy, penetrating electromagnetic radiation emitted from some materials that are in an excited electron state.

rayos X (pág. 864) Forma de radiación electromagnética penetrante de alta energía que emiten algunos materiales que se encuentran en un estado electrónico excitado.

INDEX

Italic numbers = *illustration/photo* **Bold numbers = vocabulary term**
act. = activity *prob.* = problem

B

C

D

E

Index

F

G

Index

H

Index

I

N

Index

Q

R

Index

S

Index

Index

Index

reviewed – 25 pages
reviewed – 43 pages
reviewed – 62 pages till problem – 77
revised – 96 pages
reviewed – 126 pages
revised – 135 pages
revised – 165 pages
revised – 167 pages

U

V

W

X

Z

PERIODIC TABLE OF THE ELEMENTS

Key	Hydrogen
Element	Hydrogen
Atomic number	1
Symbol	H
Atomic mass	1.008
State of matter	Gas

- Gas
- Liquid
- Solid
- Synthetic

	1	2	3	4	5	6	7	8	9
1	Hydrogen 1 **H** 1.008 (Gas)								
2	Lithium 3 **Li** 6.941 (Solid)	Beryllium 4 **Be** 9.012 (Solid)							
3	Sodium 11 **Na** 22.990 (Solid)	Magnesium 12 **Mg** 24.305 (Solid)							
4	Potassium 19 **K** 39.098 (Solid)	Calcium 20 **Ca** 40.078 (Solid)	Scandium 21 **Sc** 44.956 (Solid)	Titanium 22 **Ti** 47.867 (Solid)	Vanadium 23 **V** 50.942 (Solid)	Chromium 24 **Cr** 51.996 (Solid)	Manganese 25 **Mn** 54.938 (Solid)	Iron 26 **Fe** 55.847 (Solid)	Cobalt 27 **Co** 58.933 (Solid)
5	Rubidium 37 **Rb** 85.468 (Solid)	Strontium 38 **Sr** 87.62 (Solid)	Yttrium 39 **Y** 88.906 (Solid)	Zirconium 40 **Zr** 91.224 (Solid)	Niobium 41 **Nb** 92.906 (Solid)	Molybdenum 42 **Mo** 95.94 (Solid)	Technetium 43 **Tc** (98) (Solid)	Ruthenium 44 **Ru** 101.07 (Solid)	Rhodium 45 **Rh** 102.906 (Solid)
6	Cesium 55 **Cs** 132.905 (Solid)	Barium 56 **Ba** 137.327 (Solid)	Lanthanum 57 **La** 138.905 (Solid)	Hafnium 72 **Hf** 178.49 (Solid)	Tantalum 73 **Ta** 180.948 (Solid)	Tungsten 74 **W** 183.84 (Solid)	Rhenium 75 **Re** 186.207 (Solid)	Osmium 76 **Os** 190.23 (Solid)	Iridium 77 **Ir** 192.217 (Solid)
7	Francium 87 **Fr** (223) (Solid)	Radium 88 **Ra** (226) (Solid)	Actinium 89 **Ac** (227) (Solid)	Rutherfordium 104 **Rf** (261) (Synthetic)	Dubnium 105 **Db** (262) (Synthetic)	Seaborgium 106 **Sg** (266) (Synthetic)	Bohrium 107 **Bh** (264) (Synthetic)	Hassium 108 **Hs** (277) (Synthetic)	Meitnerium 109 **Mt** (268) (Synthetic)

The number in parentheses is the mass number of the longest lived isotope for that element.

Lanthanide series	Cerium 58 **Ce** 140.115 (Solid)	Praseodymium 59 **Pr** 140.908 (Solid)	Neodymium 60 **Nd** 144.242 (Solid)	Promethium 61 **Pm** (145) (Solid)	Samarium 62 **Sm** 150.36 (Solid)	Europium 63 **Eu** 151.965 (Solid)
Actinide series	Thorium 90 **Th** 232.038 (Solid)	Protactinium 91 **Pa** 231.036 (Solid)	Uranium 92 **U** 238.029 (Solid)	Neptunium 93 **Np** (237) (Synthetic)	Plutonium 94 **Pu** (244) (Synthetic)	Americium 95 **Am** (243) (Synthetic)